M·A·N·U·F·A·C·T·U·R·I·N·G
WORLDWIDE

Industry Analyses, Statistics,
Products, and Leading Companies and Countries

ISSN 1084-8738

M·A·N·U·F·A·C·T·U·R·I·N·G
WORLDWIDE

Industry Analyses, Statistics, Products, and Leading Companies and Countries

Third Edition

- Provides detailed comprehensive coverage of 500 manufactured product categories.

- Outlines more than 4,000 companies in 119 manufacturing countries.

- Features data from the United Nations' *General Industrial Statistics* series and *Industrial Commodity Production Statistics* series.

Arsen J. Darnay, Editor

The Gale Group

DETROIT • SAN FRANCISCO • LONDON • BOSTON • WOODBRIDGE, CT

Arsen J. Darnay, *Editor*

Editorial Code & Data Inc. Staff

Monique S. Darnay, *Associate Editor*
Joyce Piwowarski, *Computer Support*

The Gale Group Staff

Jacqueline K. Mueckenheim, *Coordinating Editor*

Evi Seoud, *Assistant Production Manager*
Mary Beth Trimper, *Production Director*

Barbara J. Yarrow, *Graphic Services Manager*
Gary Leach, *Desktop Publisher*

Publication of the statistics contained in *Manufacturing Worldwide* has been made possible with the permission of the United Nations. Definitions and limitations of all statistics and data provided by the United Nations Statistical Division and used in this publication are documented in the United Nations sources cited. The United Nations accepts no responsibility for the accuracy or suitability of the presentation of these statistics in this publication, nor for additional statistics and data provided by the publisher.

While every effort has been made to ensure the reliability of the information presented in this publication, Gale Research Inc. does not guarantee the accuracy of the data contained herein. Gale accepts no payment for listing; and inclusion in the publication of any organization, agency, institution, publication, service, or individual does not imply endorsement of the editors or publisher. Errors brought to the attention of the publisher and verified to the satisfaction of the publisher will be corrected in future editions.

The paper used in this publication meets the minimum requirements of American National Standard for Information Sciences—Permanence Paper for Printed Library Materials, ANSI Z39.48-1984.

This book is printed on recycled paper that meets Environmental Protection Agency Standards.

Copyright © 1999
The Gale Group
27500 Drake Road
Farmington Hills, MI 48331-3535

ISBN 0-7876-1654-0
ISSN 1084-8738

Printed in the United States of America

Table of Contents

Introduction xxv

Part I - Data by ISIC 1

3110 - Food Products 3
  Summary Statistics 3
  Statistics and Ratios 4
  Representative Companies in Sector 18
  Product Tables 21
    Beef and Veal, Fresh (Industrial Production) 21
    Mutton and Lamb (Including Goats), Fresh (Industrial Production) 21
    Pork, Fresh (Industrial Production) 21
    Poultry, Dressed, Fresh (Industrial Production) 22
    Other Meat, Fresh (Industrial Production) 22
    Bacon, Ham and Other Dried, Salted or Smoked Pig Meat 22
    Other Meat and Edible Offals, Dried, Salted or Smoked 23
    Sausages 23
    Meals, Frozen Prepared 23
    Meat, Tinned 24
    Lard (Industrial Production) 24
    Hides, Cattle and Horse, Undressed (Total Production) 24
    Skins, Calf, Goat and Sheep, Undressed (Total Production) 25
    Milk and Cream, Condensed (Total Production) 25
    Milk and Cream, Condensed (Industrial Production) 25
    Milk and Cream, Dried (Total Production) 26
    Milk and Cream, Dried (Industrial Production) 26
    Butter (Industrial Production) 26
    Cheese (Industrial Production) 27
    Ice-Cream 27
    Fruits, Dried 27
    Jams, Marmalades and Fruit Jellies 28
    Fruit & Vegetable Juices, Concentrated 28

Fruit & Vegetable Juices, Unconcentrated 28
Fruits, Frozen 29
Fruits, Tinned or Bottled 29
Vegetables, Frozen 29
Vegetables, Tinned or Bottled 30
Fish, Frozen 30
Fish, Salted, Dried or Smoked 30
Fish, Tinned 31
Margarine, Imitation Lard and Other Prepared Fats 31
Margarine 31
Oils and Fats of Aquatic Animal Origin 32
Oils and Fats of Animals, Unprocessed 32
Oil, Soya Bean, Crude 32
Oil, Soya Bean, Refined 33
Oil, Cotton-Seed, Crude 33
Oil, Cotton-Seed, Refined 33
Oil, Groundnut, Crude 34
Oil, Groundnut, Refined 34
Oil, Olive, Crude 34
Oil, Olive, Refined 35
Oils, Other, of Vegetable Origin, Crude 35
Oils, Other, of Vegetable Origin, Refined 35
Flour, Wheat 36
Meal and Groats of All Cereals 36
Flour, Cereal, Other Than Wheat 36
Cereal Breakfast Food 37
Macaroni and Noodle Products, Uncooked 37
Bread, Ships' Biscuits and Other Ordinary Baker's Wares 37
Biscuits 38
Pastry, Cakes, and Other Fine Bakers' Wares 38
Farinaceous Preparations (e.g. Ravioli, Tortellini etc.) 38
Raw Sugar 39
Refined Sugar 39
Fruit, Glace or Crystallized 39
Sugar Confectionery 40
Cocoa Powder 40
Cocoa Butter 40
Chocolate and Chocolate Products 41
Coffee Extracts, Essences and Concentrates (Including Instant Coffee) . . . 41
Vinegar 41
Prepared Animal Feeds 42

3130 - Beverages 43
Summary Statistics 43

Statistics and Ratios 44
Representative Companies in Sector 58
Product Tables 61
   Distilled Alcoholic Beverages, Excluding Ethyl Alcohol 61
   Ethyl Alcohol for All Purposes 61
   Wine 61
   Malt 62
   Beer 62
   Mineral Waters 62
   Soft Drinks 63

3140 - Tobacco **65**
Summary Statistics 65
Statistics and Ratios 66
Representative Companies in Sector 79
Product Tables 81
   Tobacco, Prepared Leaf 81
   Cigars, Cigarillos and Cheroots 81
   Cigarettes 81
   Tobacco, Manufactured (e.g. Smoking Tobacco, Chewing Tobacco, Snuff) . . 82

3210 - Textiles **83**
Summary Statistics 83
Statistics and Ratios 84
Representative Companies in Sector 97
Product Tables 100
   Cotton Linters 100
   Wool Yarn, Pure and Mixed (Total) 100
   Wool Yarn, Mixed 100
   Wool Yarn, Pure 101
   Cotton Yarn, Pure and Mixed (Total) 101
   Cotton Yarn, Mixed 101
   Cotton Yarn, Pure 102
   Flax, Ramie and True Hemp Yarn 102
   Yarn and Sewing Thread of Man-Made Fibers 102
   Jute Yarn 103
   Yarn of Other Vegetable Textile Fibers 103
   Cotton Woven Fabrics 103
   Silk Fabrics 104
   Woollen Woven Fabrics 104
   Linen Fabrics 104
   Jute Fabrics 105
   Woven Fabrics of Cellulosic Fibers 105
   Woven Fabrics of Non-Cellulosic Fibers 105
   Blankets 106

Bed Linen, Articles 106
Towelling 106
Knitted Fabrics 107
Socks and Other Stockings, Except Women's Stockings 107
Women's Stockings 107
Knitted Undergarments, Excluding Elastic 108
Knitted Sports Shirts 108
Knitted Sweaters 108
Knitted Outer Garments, Other 109
Carpets and Rugs of Wool, Knotted 109
Carpets and Rugs, Other 109
Cordage, Rope and Twine 110
Floor Covering 110

3211 - Spinning, Weaving, etc. **111**
Summary Statistics 111
Statistics and Ratios 112

3220 - Wearing Apparel **121**
Summary Statistics 121
Statistics and Ratios 122
Representative Companies in Sector 135
Product Tables 138
Jackets, Men's and Boys' 138
Overcoats, Men's and Boys' 138
Raincoats, Men's and Boys' 138
Suits, Men's and Boys' 139
Trousers, Men's and Boys' 139
Blouses, Women's and Girls' 139
Coats, Women's and Girls' 140
Dresses, Women's and Girls' 140
Raincoats, Women's and Girls' 140
Skirts, Slacks and Shorts, Women's and Girls' 141
Suits, Women's and Girls' 141
Shirts, Men's and Boys' 141
Underwear, Men's and Boys' 142
Underwear, Women's and Girls' 142

3230 - Leather and Products **143**
Summary Statistics 143
Statistics and Ratios 144
Representative Companies in Sector 156
Product Tables 158
Heavy Leather 158
Light Leather 158

3240 - Footwear **159**
   Summary Statistics 159
   Statistics and Ratios 160
   Representative Companies in Sector 172
   Product Tables 175
      Footwear, Total Production, Excluding Rubber Footwear 175
      Footwear, Leather, Children's 175
      Footwear, Leather, Men's 175
      Footwear, Leather, Women's 176
      Footwear, Other (Sports, Orthopedic etc.) 176
      Footwear, House 176
      Sandals and Similar Light Footwear 177

3310 - Wood Products **179**
   Summary Statistics 179
   Statistics and Ratios 180
   Representative Companies in Sector 193
   Product Tables 196
      Sawnwood, Coniferous 196
      Sawnwood, Broadleaved 196
      Veneer Sheets 196
      Blockboard 197
      Plywood 197
      Particle Board 197

3320 - Furniture, Fixtures **199**
   Summary Statistics 199
   Statistics and Ratios 200
   Representative Companies in Sector 213
   Product Tables 216
      Mattresses 216
      Mattress Supports 216

3410 - Paper and Products **217**
   Summary Statistics 217
   Statistics and Ratios 218
   Representative Companies in Sector 231
   Product Tables 234
      Wood Pulp, Mechanical 234
      Pulp of Fibers Other Than Wood 234
      Wood Pulp, Dissolving Grades 234
      Wood Pulp, Soda and Sulfate 235
      Wood Pulp, Sulfite 235
      Wood Pulp, Semi-Chemical 235
      Newsprint 236

Other Printing and Writing Paper 236
Household and Sanitary Paper 236
Wrapping and Packaging Paper and Paperboard 237
Cigarette Paper 237
Other Machine-Made Paper and Paperboard, Simply Finished 237
Fiberboard, Compressed 238
Packing Containers of Paper or Paperboard 238

3411 - Pulp, Paper, etc. **239**
Summary Statistics 239
Statistics and Ratios 240

3420 - Printing, Publishing **247**
Summary Statistics 247
Statistics and Ratios 248
Representative Companies in Sector 262

3510 - Industrial Chemicals **265**
Summary Statistics 265
Statistics and Ratios 266
Representative Companies in Sector 278
Product Tables 281
Sulfur, Recovered as By-Product 281
Sulfur, Recovered From Pyrites etc. 281
Styrene 281
Acetylene 282
Benzene (Benzol) 282
Butylenes, Butadiene 282
Ethylene 283
Naphthalene 283
Propylene 283
Toluene 284
Xylenes (Orthoxylene, Metaxylene and Paraxylene) 284
Carbon Tetrachloride 284
Dichloromethane 285
Trichloroethylene 285
Methanol (Methyl Alcohol) 285
Butyl Alcohol (Butanols) 286
Ethanediol (Ethylene Glycol) 286
Vinyl Chloride (Chloroethylene) 286
Glycerine (Glycerol) 287
Phenol 287
Diethyl Ether 287
Propylene Glycol 288
Ethylene Oxide 288

Acetaldehyde (Ethanol) 288
Methanol (Formaldehyde) 289
Acetone 289
Acetates (Methyl, Ethyl, Butyl) 289
Acetic Acid 290
Formic Acid 290
Phthalic Anhydride 290
Acetylsalicylic Acid (Aspirin) 291
Maleic Anhydride 291
Tartaric Acid 291
Aniline 292
Acrylonitrile 292
Chlorine 292
Hydrochloric Acid 293
Sulfuric Acid 293
Nitric Acid 293
Phosphoric Acid 294
Carbon Bisulfide 294
Zinc Oxide 294
Titanium Oxides 295
Lead Oxides 295
Ammonia 295
Caustic Soda 296
Aluminum Oxide 296
Hydrated Alumina, in Terms of Al_2O_3 296
Aluminum Sulfate 297
Copper Sulfate 297
Soda Ash 297
Sodium Sulfates, Acid ($NaHSO_4$) or Neutral (Na_2SO_4) 298
Sodium Silicates 298
Hydrogen Peroxide 298
Calcium Carbide 299
Dyestuffs, Synthetic (In Terms of 60 Per Cent Concentration Basis) 299
Activated Carbon 299
Nitrogenous Fertilizers (Total Production) 300
Nitrogenous Fertilizers, N Content 300
Phosphatic Fertilizers (Total Production) 300
Superphosphates, P_2O_5 Content 301
Phosphatic Fertilizers, Other, P_2O_5 Content 301
Potassic Fertilizers (Total Production) 301
Potassic Fertilizers, K_2O Content 302
Multinutrient Fertilizers, N Content 302
Multinutrient Fertilizers, P_2O_5 Content 302
Multinutrient Fertilizers, K_2O Content 303

Insecticides, Fungicides, Disinfectants etc. 303
Rubber, Synthetic 303
Non-Cellulosic Staple and Tow 304
Cellulosic Staple and Tow 304
Artificial Resins and Plastic Materials (Total) 304
Alkyd Resins 305
Amino Plastics 305
Phenolic and Cresylic Plastics 305
Polyethylene 306
Ethylene-Vinyl Acetate Copolymers 306
Polypropylene 306
Acrylic Polymers 307
Polyamides 307
Polystyrene 307
Polyacetals 308
Polyvinyl Chloride 308
Non-Cellulosic Continuous Fibers 308
Cellulosic Continuous Filaments 309

3511 - Basic Chemicals, excl Fertilizers **311**
Summary Statistics 311
Statistics and Ratios 312

3513 - Synthetic Resins, etc. **319**
Summary Statistics 319
Statistics and Ratios 320

3520 - Chemical Products nec **327**
Summary Statistics 327
Statistics and Ratios 328
Representative Companies in Sector 341
Product Tables 344
Paints, Cellulose 344
Paints, Water 344
Paints, Other 344
Mastics 345
Soap 345
Washing Powder and Detergents 345
Carbon Black 346
Printers' Ink 346
Polishes and Creams for Footwear, Furniture, Glass, Metal 346
Explosives 347

3522 - Drugs and Medicines **349**
Summary Statistics 349

Statistics and Ratios 350

3530 - Petroleum Refineries 359
Summary Statistics 359
Statistics and Ratios 360
Representative Companies in Sector 370
Product Tables 373
  Aviation Gasoline 373
  Jet Fuel 373
  Motor Gasoline 373
  Naphthas 374
  Kerosene 374
  White Spirit 374
  Distillate Fuel Oils 375
  Residual Fuel Oils 375
  Lubricating Oils 375
  Paraffin Wax 376
  Petroleum Coke 376
  Bitumen (Asphalt) 376
  Liquefied Petroleum Gas From Natural Gas Plants 377
  Liquefied Petroleum Gas From Petroleum Refineries 377

3540 - Petroleum, Coal Products 379
Summary Statistics 379
Statistics and Ratios 380
Representative Companies in Sector 389
Product Tables 391
  Hard-Coal Briquettes 391
  Brown-Coal Briquettes 391
  Coke 391
  Gas Produced By Cokeries 392
  Tars 392

3550 - Rubber Products 393
Summary Statistics 393
Statistics and Ratios 394
Representative Companies in Sector 406
Product Tables 409
  Inner Tubes, Rubber, for Motor Vehicles 409
  Inner Tubes, Rubber, for Bicycles and Motorcycles 409
  Tires for Agricultural and Other Off-The-Road Vehicles 409
  Tires for Bicycles and Motorcycles 410
  Tires for Road Motor Vehicles, Excluding Bicycles and Motorcycles . . . 410
  Rubber, Reclaimed 410
  Rubber, Unhardened Vulcanized Plates, Sheets etc. 411

Rubber, Unhardened Vulcanized Piping and Tubing 411
Rubber, Hardened 411
Rubber, Transmission, Conveyor, Elevator Belts etc. 412
Rubber Footwear 412
Plastic Footwear 412

3560 - Plastic Products nec **413**
Summary Statistics 413
Statistics and Ratios 414
Representative Companies in Sector 427

3610 - Pottery, China, etc. **431**
Summary Statistics 431
Statistics and Ratios 432
Representative Companies in Sector 443
Product Tables 445
Household Ware of Porcelain or China 445
Household Ware of Other Ceramic Materials 445
Sanitary Ceramic Fittings (e.g. Sinks, Wash Basins etc.) 445

3620 - Glass and Products **447**
Summary Statistics 447
Statistics and Ratios 448
Representative Companies in Sector 459
Product Tables 462
Glass, Drawn or Blown, in Rectangles, Unworked 462
Glass, Cast, Rolled, Drawn or Blown 462
Glass Fibers (Including Glass Wool) 462
Glass, Safety, of Toughened or Laminated Glass 463
Glass Bottles and Other Containers of Common Glass 463

3690 - Nonmetal Products nec **465**
Summary Statistics 465
Statistics and Ratios 466
Representative Companies in Sector 479
Product Tables 482
Building Bricks, Made of Clay 482
Tiles, Roofing, Made of Clay 482
Tiles, Floor and Wall 482
Quicklime 483
Cement 483
Asbestos-Cement Articles 483
Abrasives 484
Concrete Blocks and Bricks - Weight 484

Concrete Blocks and Bricks - Cubage 484
Concrete Pipes - Weight 485
Concrete Pipes - Cubage 485
Concrete, Other Products - Weight 485
Concrete, Other Products - Cubage 486

3710 - Iron and Steel **487**
Summary Statistics 487
Statistics and Ratios 488
Representative Companies in Sector 499
Product Tables 502
    Thomas (Basic) Slag 502
    Spiegeleisen and Ferro-Manganese 502
    Pig Iron, Foundry 502
    Pig Iron, Steel-Making 503
    Other Ferro-Alloys 503
    Ferro-Chromium 503
    Ferro-Nickel 504
    Ferro-Silicon 504
    Crude Steel for Castings 504
    Crude Steel, Ingots 505
    Ingots for Tubes 505
    Semis for Tubes 505
    Wire Rods 506
    Angles, Shapes and Sections (Total Production) 506
    Angles, Shapes and Sections, 80 mm or More (Heavy Sections) . . . 506
    Angles, Shapes and Sections, Less Than 80 mm (Light Sections) . . . 507
    Plates (Heavy), Over 4.75 mm 507
    Plates (Medium), 3 to 4.75 mm 507
    Sheets, Electrical 508
    Sheets Under 3 mm, Cold-Rolled, Uncoated 508
    Sheets Under 3 mm, Hot-Rolled 508
    Tinplate 509
    Sheets, Galvanized 509
    Hoop and Strip, Cold-Reduced 509
    Hoop and Strip, Hot-Rolled 510
    Railway Track Material 510
    Wire, Plain 510
    Tubes, Seamless 511
    Tubes, Welded 511
    Steel Castings in the Rough State 511
    Steel Forgings 512
    Wheels, Wheel Centers, Tires and Axles 512

3720 - Nonferrous metals 513
  Summary Statistics 513
  Statistics and Ratios 514
  Representative Companies in Sector 525
  Product Tables 528
    Copper, Blister and Other Unrefined 528
    Copper, Refined, Unwrought (Total Production) 528
    Copper, Primary, Refined 528
    Copper, Secondary, Refined 529
    Copper-Base Alloys 529
    Copper Bars, Rods, Angles, Shapes, Sections 529
    Copper Wire 530
    Copper Plates, Sheets, Strip, Foil 530
    Copper Tubes and Pipes 530
    Nickel, Unwrought 531
    Alumina, Calcined Equivalent 531
    Aluminum, Unwrought (Total Production) 531
    Aluminum, Unwrought, Primary 532
    Aluminum, Unwrought, Secondary 532
    Aluminum Bars, Rods, Angles, Shapes, Sections 532
    Aluminum Wire 533
    Aluminum Plates, Sheets, Strip, Foil 533
    Aluminum Tubes and Pipes 533
    Lead, Refined, Unwrought (Total Production) 534
    Lead, Secondary, Refined, Soft 534
    Lead-Base Alloys 534
    Zinc, Unwrought (Total Production) 535
    Zinc, Unwrought, Primary 535
    Zinc, Unwrought, Secondary 535
    Zinc-Base Alloys 536
    Zinc Plates, Sheets, Strip, Foil 536
    Tin, Unwrought (Total Production) 536
    Tin, Unwrought, Primary 537
    Tin, Unwrought, Secondary 537
    Cadmium, Unwrought 537
    Magnesium, Unwrought (Total Production) 538
    Magnesium, Unwrought, Primary 538

3810 - Metal Products 539
  Summary Statistics 539
  Statistics and Ratios 540
  Representative Companies in Sector 553
  Product Tables 556
    Boilers, Steam Generating 556

Structures (Except Prefabs) and Parts of Structures, of Iron and Steel 556
Cans, Metal (Capacity Not Exceeding 300 Liters) 556
Casks, Drums etc. (Capacity Over 300 Liters) 557
Compressed Gas Cylinders, Made of Metal 557
Cables 557
Nails, Screws, Nuts, Bolts, Rivets etc. 558
Containers, One Cubic Meter and Over 558
Central-Heating Apparatus, Non-Electric (Boilers, Radiators etc.) 558

3820 - Machinery nec 559
Summary Statistics 559
Statistics and Ratios 560
Representative Companies in Sector 572
Product Tables 575
Steam Turbines 575
Engines, Diesel (Excluding Engines for Vehicles) 575
Engines, Internal Combustion (Excluding Engines for Vehicles) 575
Gas Turbines 576
Hydraulic Turbines 576
Cultivators, Scarifiers, Weeders, Hoes etc. 576
Harrows, Rotary, Animal or Tractor-Operated 577
Ploughs, Animal or Tractor-Operated 577
Seeders, Planters and Transplanters 577
Combine Harvester-Threshers 578
Mowers, Animal or Tractor-Operated and Self-Propelled 578
Rakes, Animal or Tractor-Operated and Self-Propelled 578
Threshing Machines 579
Milking Machines 579
Garden Tractors 579
Tractors of 10 HP and Over, Except Industrial and Road Tractors 580
Fertilizer Distributors, Animal, Hand or Tractor-Operated 580
Drilling and Boring Machines 580
Forging, Stamping and Die-Stamping Machines 581
Grinding and Sharpening Machines 581
Machining Centers 581
Lathes 582
Milling Machines (Cutters) 582
Other Metal-Cutting Machine-Tools 582
Planing, Shaping and Slotting Machines 583
Metal-Working Presses 583
Other Metal-Forming Machine-Tools 583
Rolling Mills for Rolling Metals 584
Machine-Tools for Working Wood 584
Electro-Mechanical Hand Tools 584

Spinning Machines 585
Looms 585
Printing Presses 585
Bulldozers 586
Excavating Machines 586
Graders and Levelers 586
Scrapers 587
Concrete Mixers for Use at Construction Sites 587

3825 - Office, Computing Machinery 589
Summary Statistics 589
Statistics and Ratios 590
Representative Companies in Sector 595
Product Tables 598
Typewriters 598
Calculating Machines 598
Input or Output Units 598
Central and Peripheral Storage Units 599
Scales, Industrial 599
Scales, Other Than Industrial 599

3830 - Electrical Machinery 601
Summary Statistics 601
Statistics and Ratios 602
Representative Companies in Sector 615
Product Tables 618
Ovens for Household Use 618
Stoves, Ranges, Cookers 618
Drying Machines for Household Use 618
Sewing Machines 619
Air-Conditioning Machines 619
Industrial Refrigerators and Freezers 619
Pumps for Liquids, Except Liquid Elevators 620
Compressors 620
Cranes 620
Elevators, for Lifting Goods and Persons 621
Fork-Lift Trucks 621
Refrigerators for Household Use 621
Ball, Roller or Needle Bearings 622
Washing Machines for Household Use 622
Generators for Hydraulic Turbines 622
Generators for Hydraulic Turbines - Kilowatts 623
Generators for Steam Turbines 623
Generators for Steam Turbines - Kilowatts 623
Motors, Electric, Fractional Horsepower 624

Motors, Electric, One Horsepower and Over 624
Transformers, Less Than 5 KVA 624
Transformers, 5 KVA and Over 625
Meters, Electricity-Supply 625
Electric Furnaces 625

3832 - Radio, Television, etc. **627**
Summary Statistics 627
Statistics and Ratios 628
Television Receivers (Total) 635
Radio Receivers (Total) 635
Telephones 635
Electronic Tubes 636
Sound Recorders 636
Sound Reproducers 636
Vacuum Cleaners 637
Shavers and Hair Clippers, Electric 637
Heaters, Electric Space 637
Irons, Electric Smoothing 638
Fuses, Electrical 638
Switches, Electric 638
Wire and Cable, Insulated 639
Batteries and Cells, Primary 639
Accumulators, Electric, for Motor Vehicles 639
Lamps, Electric (Excluding Fluorescent Tubes) 640
Tubes, Fluorescent 640

3840 - Transportation Equipment **641**
Summary Statistics 641
Statistics and Ratios 642
Representative Companies in Sector 655
Product Tables 658
Sea-Going Bulk Carriers, Launched 658
Tankers, Launched 658
Other Sea-Going Merchant Vessels, Launched 658
Locomotives, Electric 659
Locomotives, Diesel 659
Rail Motor Passenger Vehicles 659
Goods Wagons and Vans 660
Rail Passenger Carriages 660
Engines, Compression-Ignition, for Vehicles (Diesel Engines) 660
Engines, Spark-Ignition, for Vehicles (Gasoline Fueled) 661
Bodies for Motor Vehicles 661
Passenger Cars, Assembled From Imported Parts 661
Passenger Cars, Produced 662

Buses and Motor Coaches, Produced 662
Trucks, Including Articulated Vehicles, Assembled From Imported Parts . . 662
Trucks, Including Articulated Vehicles, Produced 663
Road Tractors for Tractor-Trailer Combinations, Produced 663
Trailers and Semi-Trailers 663
Motorcycles, Scooters etc. 664
Bicycles 664
Commercial Passenger and Cargo Planes 664
Perambulators and Push-Chairs for Babies 665

3841 - Shipbuilding, Repair **667**
Summary Statistics 667
Statistics and Ratios 668

3843 - Motor Vehicles **675**
Summary Statistics 675
Statistics and Ratios 676

3850 - Professional Goods **685**
Summary Statistics 685
Statistics and Ratios 686
Representative Companies in Sector 697
Product Tables 700
Gas Meters 700
Liquid Meters 700
Thermostats 700
Binoculars and Refracting Telescopes 701
Optical and Analytical Instruments 701
Cameras, Photographic 701
Frames and Mountings for Spectacles, Goggles or The Like 702
Watches 702
Clocks, With Watch Movements 702
Clocks, Instrument-Panel & Similar Type 703
Clocks, Other, Electric and Non-Electric 703

3900 - Industries nec **705**
Summary Statistics 705
Statistics and Ratios 706
Representative Companies in Sector 719
Product Tables 722
Pianos 722
Musical Instruments, String 722
Organs 722
Musical Instruments, Wind 723
Dolls 723

Fountain Pens, Ball-Point Pens, Propelling Pencils etc. 723
Pencils, Crayons etc. 724
Slide Fasteners (Zippers) 724

Part II - Country Profiles **725**
Worldwide Averages 727
Albania 727
Algeria 728
Angola 728
Argentina 729
Armenia 729
Australia 730
Austria 730
Azerbaijan 731
Bahamas 731
Bahrain 732
Bangladesh 732
Barbados 733
Belgium 733
Belize 734
Benin 734
Bermuda 735
Bolivia 735
Bosnia & Herzegovina 736
Botswana 736
Brazil 737
Bulgaria 737
Burundi 738
Cambodia 738
Cameroon 739
Canada 739
Cape Verde 740
Central African Republic 740
Chile 741
China 741
China (Hong Kong SAR) 742
China (Taiwan Province) 742
Colombia 743
Costa Rica 743
Cote d'Ivoire 744
Croatia 744
Cyprus 745
Czech Republic 745

Denmark 746
Ecuador 746
Egypt 747
El Salvador 747
Equatorial Guinea 748
Ethiopia 748
Fiji 749
Finland 749
Former Yugoslav Republic of Macedonia 750
France 750
Gabon 751
Gambia 751
Germany 752
Ghana 752
Greece 753
Grenada 753
Guatemala 754
Honduras 754
Hungary 755
Iceland 755
India 756
Indonesia 756
Iran (Islamic Republic of) 757
Iraq 757
Ireland 758
Israel 758
Italy 759
Jamaica 759
Japan 760
Jordan 760
Kenya 761
Korea, Republic of 761
Kuwait 762
Kyrgyzstan 762
Latvia 763
Lesotho 763
Liechtenstein 764
Lithuania 764
Luxembourg 765
Macau 765
Malawi 766
Malaysia 766
Malta 767
Mauritius 767

Mexico 768
Mongolia 768
Morocco 769
Mozambique 769
Myanmar 770
Namibia 770
Nepal 771
Netherlands 771
Netherlands Antilles 772
New Zealand 772
Norway 773
Oman 773
Pakistan 774
Panama 774
Paraguay 775
Peru 775
Philippines 776
Poland 776
Portugal 777
Puerto Rico 777
Qatar 778
Republic of Moldova 778
Romania 779
Russian Federation 779
Saint Vincent & the Grenadines 780
Senegal 780
Sierra Leone 781
Singapore 781
Slovakia 782
Slovenia 782
South Africa 783
Spain 783
Sri Lanka 784
Suriname 784
Swaziland 785
Sweden 785
Switzerland 786
Syrian Arab Republic 786
Thailand 787
Tonga 787
Trinidad and Tobago 788
Tunisia 788
Turkey 789
Ukraine 789

United Kingdom 790
United Republic of Tanzania 790
United States of America 791
Uruguay 791
Venezuela 792
Yugoslavia 792
Zambia 793
Zimbabwe 793

Appendix I, Source Notes 795
Alphabetical Index 801

Introduction

Manufacturing Worldwide (*MW*), now in its third edition, presents detailed statistical and company information on manufacturing activities the world over. This edition is a major update, featuring more recent information, more commodities, and dollar-denominated values.

Information is presented, in Part I, on 37 manufacturing sectors (of which 28 are primary sectors, the rest important subsectors), on nearly 500 product categories, and more than 4,000 companies. In Part II, the statistical profiles of 133 manufacturing countries are shown, including a listing of leading industries and summary data. Finally, a comprehensive Alphabetical Index provides access to the data by country, industry, product, and company name.

General Approach and Emphasis

Manufacturing Worldwide (*MW*) represents both a pioneering effort in presenting international statistics as well as a continuation of efforts to bring uniformly formatted and preanalyzed statistics to business and academic users.

Gale's industrial series (*Manufacturing USA*; *Service Industries USA*; *Finance, Insurance, and Real Estate USA*; *Wholesale/Retail USA*, *Transportation and Public Utilities USA*, and *Agriculture and Construction USA*) together present a quite comprehensive view of industrial activity in the United States. These titles combine government statistics from a variety of sources with company information. Data are arranged by industrial code classifications, provide two or more levels of geographical data (country, state, city), and generally provide analysis in the form of precalculated ratios and rankings.

MW presents manufacturing data worldwide in much the same way as the USA- series presents data for the United States.

The level of what might be called the "statistical arts" is nowhere as highly developed as in the United States. Despite the grumblings of professionals who use U.S. government statistics, the fact remains that the U.S. has the best statistical system in the world. This is especially evident when one examines worldwide data. Despite valiant efforts by the United Nations and regional bodies, worldwide statistics invariably reflect the state of the world: there are serious gaps in reporting (about a third of the countries do not report) and a lack of uniformity in definitions; furthermore, worldwide data tend, on the whole, to be much older than U.S. data. It is not an exaggeration to say that we do not really know accurately what the true state of manufacturing is, worldwide. But because the most active producers are also the most efficient in reporting statistical data, it is reasonable to assume that we have a fairly good idea of magnitudes.

For the reasons just given, the emphasis in *MW* is on providing as much comparative information as possible. For every datum provided, one or more ratios are also provided in order to normalize reporting. This enables the user to compare data from country to country. Within the country profiles, at the same time, each datum is shown twice: as measured and as a percentage of the calculated world total or world average to enable the user to compare the value to the world average. Population data are also given so that comparisons to "world share" based on population can be easily visualized.

The emphasis on comparability imposes certain limitations as well. Financial data are reported in U.S. dollars so that all values may be seen in a single currency. This causes the elimination of data for which no reliable conversion information was reported. Dollar-denominated values, also, fail to communicate accurately in all cases in the absence of cost of living data. Comparisons between the U.S., Canada, Germany, France, and Japan will tend to be reasonable because all of these are technologically advanced societies with comparable standards of living. Comparisons between the U.S., Hungary, Romania, and Zambia may be inconclusive without fully understanding the local cost structures.

Sources of Data

Data presented in *MW* are drawn from three sources. The bulk of the statistics were provided by the United Nations

Industrial Development Organization (UNIDO), Industrial Statistics Branch. Commodity data were drawn, as in the first edition, from the United Nations Statistical Division's Industrial Commodity Production Statistics obtained by Gale from the U.N. under a special arrangement. Finally, data on companies participating in each of the major industry sectors were obtained from the *Gale World Business Directory* database as it stood in March 1999.

Contents and Scope

Manufacturing Worldwide provides data for the period 1990 through 1995 (and 1996 for commodities). The database used holds data on 175 countries and 498 products. Only 133 countries had data sufficiently recent and complete to be included in the general statistical presentation and in the profiles. Product Tables, shown as part of the general statistics, include a larger number of countries.

Categories of information covered include:

    Establishments or Enterprises
    Employment
    Female Employment
    Wages and Salaries of Employees
    Output
    Value Added
    Capital Investment
    Indexes of Industrial Production

MW also provides precalculated ratios as follows:

    Employees per Establishment
    Females as Percent of Total Employment
    Wage/Salary per Employee
    Output per Employee
    Output per Establishment (Country Profiles)
    Value Added per Employee
    Capital Investment per Establishment
    Capital Investment per Employee (Profiles)
    Index of Employment

Product tables provide production statistics in actual values (number of items, volumes of product, square meters of textiles, etc.) for the following:

    Total Production
    Regions
    Leading Producers (up to 5)

Organization of Data

MW is presented in two parts. Part I holds general statistics arranged by International Standard Industrial Code (ISIC), each ISIC being a chapter. Within each chapter, information is presented in the same order, as follows:

Each chapter begins with a brief text block describing the products covered by the ISIC. Where possible, the correspondence between the ISIC industries and the SIC codes used in the United States is also described.

A Table of Summary Statistics follows. This table provides, for the years 1989 through 1994, data and ratios for the ISIC industry as a whole.

Next are presented seven tables featuring country-by-country data. These are:

    Establishments and Number Engaged
    Employment and Compensation of Employees
    Female Workers: Total and Percent of Employment
    Output and Output per Employee
    Value Added and Value Added per Employee
    Capital Investment
    Indexes of Industrial Production

Each of these tables shows two sets of values for 1990 through 1995. The first set shows the value (e.g., Value Added) and the second set shows a ratio (e.g., Value Added per Employee).

A table showing Representative Companies follows. This table shows data on companies active in the ISIC industry. Companies are shown in order of employment, with the largest employers shown first. Some chapters—those featuring subsets of major industries—may not have company tables. Company data for these industries are shown in the "parent" ISIC chapter.

The final segment of each chapter features Product Tables. These are tabulations of production (in physical quantities) associated with one or more products that belong under the ISIC industry of the chapter.

Product Tables present information on total production, worldwide; production by major world region (Africa, Asia, etc.); and production by the top five producer nations. Data in these tables are frequently estimated. The methods used in estimating production figures are fully explained in Appendix I, Source Notes.

Part II of *MW* presents 133 country profiles and the World Average. The World Average table begins the presentation.

Each profile shows data, ratios, and percentages of the world total or world average for the years 1990 through 1996. Each table shows the top ranked ISIC industry sectors in the country, in two formats: (1) measured on the basis of employment (highest employing sector is shown first) and (2) measured on the basis of output (ISIC with the highest output, measured in dollars, is shown first). If no output data were available, top ranked industries by output are not shown, of course.

Next, data on the country's population are shown. Population for the year 1998 is used. The country's population, as a percent of world population, is shown to serve as an index for other values.

The body of each profile provides 7 elements of data and 9 ratios—where necessary statistics are available. Data cells for each year (1990 through 1995) show the actual value or ratio and a percentage; the percentage is the value or ratio as a percent of the world value or ratio.

Examples:

In 1992, Australia reported employment in manufacturing of 1.014 million or 0.56 percent of total worldwide employment, as reported in the database (182.5 million). Note that Australia's population ratio in 1998 was 0.31, suggesting that Australia's participation in world manufacturing is nearly twice its share in population.

Australia had an average of 23 employees per establishment in 1992. This was 36.7 percent of the world average (62), indicating that, overall, Australian enterprises were smaller than enterprises worldwide.

In 1994, Austria reported payroll per employee of $32,225. This was 183.5 percent of the world average of $17,560 in that year, meaning that Austrian

manufacturing employees received more than 1.8 times the pay of manufacturing employees worldwide.

The first table in Part II, World Averages, represents the summation of all the data in the database for the years 1990 through 1995. The user should note that the values are derived from a pool of 133 countries. Most of the missing countries are relatively small island nations; some, however, have meaningful manufacturing activity not included in the "world average," e.g., Belarus, Estonia.

ISIC Classification

MW uses the second revision of the International Standard Industrial Classification of All Economic Activities (ISIC). A third revision has been approved (February 1989), but the third revision codes are not yet used in the U.N.'s own database or statistical publication(s).

Important: Differences between second and third revision codes are significant. A user who wishes to translate second to third revision ISIC codes is advised to consult *International Standard Industrial Classification of All Economic Activities*, Statistical Papers, Series M, No. 4, Rev. 3, United Nations, New York, 1990.

Calculations, Conversions, and Estimates

Throughout *MW*, ratios have been calculated, data based on local currencies are reported in U.S. dollars, and missing data are estimated using regression or averaging methods. For a full discussion of these methods the user is referred to Appendix I, Source Notes.

Appendix and Index

All discussion of sources and of methods of calculation and conversion are drawn together for the user's convenience in Appendix I, Source Notes. The Appendix also presents a full listing of countries with and without general data in the database.

Access to the contents of *MW* is provided in a single Alphabetical Index. It lists all countries, all products, and all company names. References are provided to page numbers. Country references are subdivided so that a country's appearance within an ISIC group is easily found.

Acknowledgments

This book would not have been possible without the participation of the United Nations Industrial Development Organization (UNIDO) and, specifically, its provision of statistical data on manufacturing contained in their Industrial Statistics databases. Special thanks to Jill Fraser of UNIDO for her much appreciated assistance in obtaining this data. We would also like to acknowledge the United Nations, Statistical Division, for its provision of commodity production statistics. Please note, however, that neither UNIDO nor the United Nations accepts responsibility for the accuracy or suitability of the presentation of these statistics in this publication nor for additional statistics and data provided by the publisher.

The editor wishes to thank Jacqueline Mueckenheim of The Gale Group for her continuing labors involving data acquisition. She did her job for this edition despite devastating floods, lost computers, and the usual vicissitudes of dealing with many bureaucracies. The editor also appreciates help and advice provided by a select few to members of the ECDI staff.

Comments and Suggestions

Comments on *MW* or suggestions for improvement of its usefulness, format, and coverage are always welcome. Although every effort has been made to maintain accuracy, errors may have occurred; the editor will be grateful if these are called to his attention. Please contact:

Editor, *Manufacturing Worldwide*
The Gale Group
27500 Drake Road
Farmington Hills, MI 48331-3535
Phone: (248)699-GALE

Part I

DATA BY ISIC

ISIC 3110
FOOD PRODUCTS

ISIC 3110—Food Products—includes all food products except beverages. Major subcategories are fresh meat, meat products, meat by-products (including hides), and prepared meats; fats, oils, margarine, and the like, including vegetable and seed oils; dairy products including milk, cream, cheese, butter, ice cream, etc.; fruits and vegetables in fresh, frozen, and canned form; fish and fish products; poultry and related products, including eggs; flours, meals, and cereals; macaroni, noodles, and related products; bread, baked goods, pastries, and the like; raw and refined sugar and related products; cocoa and coffee extracts; condiments, vinegar, and similar products; and prepared animal feeds.

This group corresponds to the 2-digit U.S. SIC series, SIC 20—Food and Kindred Products. Not included are beverages, corresponding to the U.S. SIC categories 2082-2087 (malt beverages; malt; wines, brandy, and brandy spirits; distilled & blended liquors; bottled & canned soft drinks; and flavoring extracts). These will be found under ISIC 3130—Beverages.

Summary Statistics

		1990 Value	N	1991 Value	N	1992 Value	N	1993 Value	N	1994 Value	N	1995 Value	N
Establishments or enterprises	(number)	333,523	91	327,345	101	316,133	95	263,837	94	230,414	79	168,897	40
Number employed	(000)	17,923	102	15,311	104	15,561	98	16,256	100	15,182	92	14,146	69
Total Wages	($ mil.)	138,461	97	148,126	96	162,716	90	137,391	92	135,019	81	124,448	61
Wage/salary per employee	($)	7,888	95	8,700	94	9,752	89	8,540	91	8,643	80	10,006	61
Female workers	(000)	1,171	35	743	33	1,825	28	1,851	34	1,704	28	1,519	16
as % of total employment*	(%)	31	34	29	32	31	28	32	34	35	28	38	16
Output	($ bil.)	1,469	101	1,545	101	1,619	96	1,476	92	1,423	83	1,381	64
per employee	($ 000)	77	96	81	96	87	92	85	89	84	81	103	62
per establishment	($ 000)	6,178	87	6,038	91	5,931	85	5,881	80	5,659	65	4,309	32
Value added	($ bil.)	413	91	417	91	470	85	489	85	435	77	463	63
per employee	($ 000)	23	87	24	86	27	81	23	79	24	72	30	58
per establishment	($ 000)	1,653	80	1,518	84	1,710	75	1,341	72	1,486	58	1,087	29
Capital investment	($ mil.)	40,704	74	38,307	67	36,578	61	37,735	57	29,683	46	16,828	24
per employee	($ 000)	4	71	4	64	4	59	3	53	3	43	3	22
per establishment	($ 000)	366	71	293	64	247	59	200	53	181	44	131	21

Data presented above are drawn from the detailed tables that follow. Columns headed 'N' show the number of countries that provided valid data for inclusion. Values are not strictly comparable one year to the next or one row to the next because the number of countries that report varies from period to period and row to row. However, a general indication of magnitudes can be gleaned from reviewing this summary—especially in earlier years in which more reports are available. For detailed explanations, see Appendix I. *The average for those countries reporting both total and female employment.

Establishments and Number Engaged

Country	Establishments or Enterprises (number)						Employees per Establishment					
	1990	1991	1992	1993	1994	1995	1990	1991	1992	1993	1994	1995
Albania	-	-	-	182	136	91	-	-	-	79	50	45
Argentina	-	-	-	18,130	20,553	-	-	-	-	10	9.666	-
Armenia	86	86	104	114	120	-	256	244	255	195	186	-
Australia	3,322	3,590	3,837	-	-	-	45	41	38	-	-	-
Austria	863	813	808	796	761	-	61	63	63	63	64	-
Azerbaijan	1,129	1,203	1,668	1,384	1,066	-	33	32	23	27	35	-
Bahamas	27	31	-	-	-	-	9.333	7.129	-	-	-	-
Bahrain	-	-	317	-	-	-	-	-	8.927	-	-	-
Bangladesh	6,062	5,736	5,744	-	-	-	18	18	18	-	-	-
Barbados	27	20	23	25	45	-	59	60	49	57	42	-
Belgium	6,997	6,961	6,860	6,813	6,790	-	11	11	11	-	-	-
Belize	77	78	82	-	-	-	-	32	33	-	-	-
Benin	-	-	95	95	-	-	-	-	-	-	-	-
Bolivia	351	367	381	376	381	387	22	22	22	22	23	23
Bosnia & Herzegovina	138	139	-	-	29	-	169	156	-	-	69	-
Botswana	51	57	79	88	90	134	102	91	62	58	58	41
Brazil	5,530	-	5,244	5,029	-	-	100	-	104	103	-	-
Bulgaria	*221	*286	*333	*3,280	*4,703	*5,107	478	305	221	17	13	10
Burundi	*24	*37	-	-	-	-	59	44	-	-	-	-
Cameroon	*38	*89	*70	*53	*60	*40	385	187	230	302	258	397
Canada	3,397	3,176	3,068	3,008	2,957	-	58	60	64	63	64	-
Cape Verde	-	-	257	288	296	-	-	-	-	-	-	-
Central African Republic	-	*7.000	*7.000	*8.000	*8.000	*9.000	-	-	-	-	-	-
Chad	-	10	12	16	19	19	-	-	-	-	-	-
Chile	390	418	424	426	419	-	193	193	200	196	206	-
China	*58,056	*56,522	*53,430	*42,457	*43,130	*46,841	53	57	61	74	74	69
China (Hong Kong SAR)	946	838	790	734	692	-	22	25	26	28	27	-
Colombia	1,357	1,324	1,415	1,345	1,347	-	58	61	73	76	78	-
Costa Rica	1,087	1,079	1,117	1,165	-	-	29	28	30	30	-	-
Croatia	592	711	871	1,102	1,367	1,515	95	74	55	41	33	30
Cyprus	926	923	924	922	993	1,054	6.350	6.320	6.481	6.448	6.437	6.731
Czechoslovakia (Former)	*184	*333	-	-	-	-	913	429	-	-	-	-
Czech Republic	-	*173	*277	*379	-	-	-	572	339	243	-	-
Denmark	2,906	2,851	2,772	-	-	-	30	30	30	-	-	-
Ecuador	316	391	377	386	389	393	86	84	97	94	95	88
Egypt	4,272	3,864	4,165	4,057	-	-	48	63	46	47	-	-
El Salvador	-	68	78	100	107	122	-	-	90	74	87	47
Equatorial Guinea	17	-	-	-	-	-	5.000	-	-	-	-	-
Ethiopia	113	103	100	103	146	130	137	153	147	146	113	119
Fiji	142	124	112	-	-	-	45	39	59	-	-	-
Finland	785	799	738	693	665	741	59	56	56	56	54	52
FYR Macedonia	*53	*137	*255	*287	*388	*663	237	107	58	50	34	21
Gabon	-	18	17	19	15	17	-	147	155	138	185	152
Gambia	-	-	-	37	-	-	-	-	-	32	-	-
Germany	-	4,547	4,408	4,283	4,254	-	-	112	107	105	104	-
Germany (Western Part)	3,648	3,633	3,632	-	-	-	103	111	111	-	-	-
Ghana	-	-	-	43	-	-	-	-	-	173	-	-
Greece	1,144	1,140	1,146	-	-	-	43	42	43	-	-	-
Grenada	-	-	29	24	20	20	-	-	9.103	9.875	13	13
Guatemala	-	95	93	94	93	94	-	206	232	237	240	255
Honduras	-	-	149	166	166	166	-	-	166	163	179	197
Hungary	*493	*1,006	*1,408	*1,896	-	-	343	163	113	71	-	-
Iceland	795	703	676	715	-	-	14	15	15	14	-	-
India	19,760	19,721	21,397	21,491	-	-	57	57	58	57	-	-
Indonesia	3,512	3,372	3,610	3,739	3,863	4,272	112	130	129	137	129	122
Iran (Islamic Republic of)	1,250	854	897	893	-	-	54	83	81	80	-	-
Iraq	-	120	124	-	-	-	-	107	97	-	-	-
Ireland	765	704	-	-	-	-	48	53	-	-	-	-
Israel	1,020	1,139	1,045	1,150	1,064	-	41	38	41	39	42	-
Italy	*1,707	*1,727	*2,308	*2,348	*2,322	-	95	95	83	79	79	-

Continued.

Depending on the table, * means *Enterprises, Engaged,* or *Factor Values;* ± means *Basis Unspecified;* § means *shown in millions.* For additional notes and sources, see Appendix I.

Establishments and Number Engaged

- Continued -

Country	Establishments or Enterprises (number)						Employees per Establishment					
	1990	1991	1992	1993	1994	1995	1990	1991	1992	1993	1994	1995
Jamaica	190	190	-	-	-	-	132	131	-	-	-	-
Japan	46,994	45,941	44,947	45,711	42,838	43,835	24	25	26	26	27	27
Jordan	1,559	1,616	2,041	2,172	1,750	1,913	5.015	5.642	5.045	4.885	7.546	7.413
Kenya	455	635	643	647	746	748	113	84	84	86	77	81
Korea, Republic of	4,064	4,242	4,609	5,410	5,498	5,882	44	42	38	32	32	31
Kuwait	309	296	300	305	296	-	25	16	20	22	23	-
Kyrgyzstan	1,134	1,116	783	855	875	-	24	24	33	29	24	-
Latvia	1,478	1,445	566	761	460	537	25	24	89	48	68	57
Lesotho	-	-	-	10	10	-	-	-	-	226	178	-
Liechtenstein	-	6.000	-	-	-	-	-	91	-	-	-	-
Lithuania	-	-	-	*179	*225	-	-	-	-	307	236	-
Luxembourg	*232	*227	*213	*209	-	-	11	12	13	14	-	-
Macau	131	131	124	126	129	119	7.313	6.702	6.129	6.571	7.426	7.193
Malawi	*12	*12	*12	*12	*12	-	924	492	525	542	658	-
Malaysia	1,314	1,324	1,297	1,400	1,406	-	56	59	60	58	59	-
Malta	374	366	395	389	388	-	6.222	6.366	5.937	5.877	5.899	-
Mauritius	223	110	113	111	118	114	51	116	113	112	106	109
Mexico	417	400	393	388	384	380	233	241	248	244	230	226
Mongolia	66	65	74	79	144	116	181	183	125	123	95	89
Morocco	954	1,495	1,504	1,506	1,515	1,571	22	36	56	62	60	58
Mozambique	69	69	50	50	53	-	445	397	430	416	351	-
Namibia	-	-	-	-	110	-	-	-	-	-	123	-
Nepal	383	609	-	770	798	-	44	35	-	34	31	-
Netherlands	821	854	850	861	934	-	131	129	129	125	113	-
Netherlands Antilles	-	-	-	66	-	-	-	-	-	12	-	-
New Zealand	*1,565	*1,658	*1,668	*1,781	-	-	36	34	35	33	-	-
Norway	1,372	1,351	885	859	886	-	32	32	45	46	46	-
Pakistan	-	858	-	-	-	-	-	98	-	?	-	-
Panama	304	297	-	-	-	-	43	46	-	-	-	-
Paraguay	-	39	-	-	-	-	-	81	-	-	-	-
Peru	2,709	2,853	3,020	-	-	-	16	15	14	-	-	-
Philippines	36,707	39,000	2,824	2,689	-	-	7.260	6.577	56	61	-	-
Poland	*910	*1,086	*1,149	*1,220	-	-	402	331	304	266	-	-
Portugal	*7,343	*7,139	*6,830	*6,773	*7,475	-	14	14	14	14	14	-
Puerto Rico	212	207	201	204	237	242	77	64	72	69	59	59
Qatar	174	177	248	246	247	-	9.448	9.537	8.089	7.919	8.109	-
Republic of Moldova	124	109	105	85	83	85	476	534	638	614	586	560
Romania	-	1,384	2,741	3,609	6,304	-	-	-	-	-	-	-
Russian Federation	-	-	-	*8,473	*9,396	*13,200	-	-	-	181	157	111
Saint Lucia	-	15	16	18	18	21	-	-	-	-	-	-
Saint Vincent & the Grenadines	-	-	-	17	19	21	-	-	-	15	13	11
Senegal	60	50	48	44	53	-	305	381	293	308	293	-
Sierra Leone	-	-	-	47	-	-	-	-	-	128	-	-
Singapore	265	271	293	295	303	-	40	41	39	40	39	-
Slovakia	-	200	200	220	227	-	-	232	210	186	177	-
Slovenia	365	*814	*941	*853	*1,083	*837	-	-	-	-	-	-
South Africa	-	1,881	-		-	-	-	103	-	-	-	-
Spain	35,628	34,612	33,686	-	-	-	7.258	7.629	7.838	-	-	-
Sri Lanka	622	522	541	533	-	-	75	77	87	77	-	-
Suriname	-	-	113	-	-	-	-	-	24	-	-	-
Swaziland	9.000	-	-	-	-	-	1,068	-	-	-	-	-
Sweden	736	831	819	807	795	-	84	76	72	70	69	-
Thailand	4,202	2,726	-	-	-	-	49	80	-	-	-	-
Trinidad and Tobago	96	388	388	406	480	-	133	37	38	41	35	-
Tunisia	-	-	-	3,121	2,628	2,541	-	-	-	13	13	13
Turkey	855	822	2,020	1,824	1,728	-	151	151	69	73	75	-
Ukraine	22,126	22,510	22,878	23,426	26,178	29,802	31	29	27	26	22	19
United Kingdom	8,728	8,487	8,341	8,740	9,114	-	58	58	58	56	53	-
United Republic of Tanzania	147	142	-	-	-	-	243	252	-	-	-	-
USSR (Former)	*9,306	-	-	-	-	-	296	-	-	-	-	-

Continued.

Establishments and Number Engaged

- Continued -

Country	Establishments or Enterprises (number)						Employees per Establishment					
	1990	1991	1992	1993	1994	1995	1990	1991	1992	1993	1994	1995
United States of America . . .	-	-	24,962	-	-	-	-	-	55	-	-	-
Venezuela	2,410	2,573	2,538	2,111	2,139	2,223	34	34	35	39	41	40
Yugoslavia	589	966	1,891	2,585	2,629	2,895	-	101	52	37	35	32
Zambia	105	-	-	-	149	-	177	-	-	-	85	-
Zimbabwe	99	105	100	95	100	-	256	258	266	260	241	-

Employment and Compensation of Employees

Country	Number Employed/Engaged (000)						Wage/Salary per Employee ($)					
	1990	1991	1992	1993	1994	1995	1990	1991	1992	1993	1994	1995
Albania	-	-	-	14	6.766	4.094	-	-	-	439	1,181	861
Argentina	197	-	-	190	199	-	5,138	-	-	12,072	-	-
Armenia.	*22	*21	*26	*22	*22	*21	-	-	407	20,771	-	-
Australia	150	*148	*146	-	-	-	18,896	20,383	19,798	-	-	-
Austria	53	51	51	50	49	49	22,874	24,537	27,091	26,704	27,856	32,557
Azerbaijan	37	39	39	38	37	-	1,384	1,412	-	-	-	-
Bahamas	0.252	0.221	-	-	-	-	13,321	14,575	-	-	-	-
Bahrain	-	-	2.830	-	-	-	-	-	5,806	-	-	-
Bangladesh	111	104	106	-	-	-	641	585	617	-	-	-
Barbados	1.596	1.193	1.120	1.419	1.884	1.727	11,279	12,686	11,635	11,003	12,448	12,300
Belgium	76	76	75	-	-	-	19,170	20,076	22,471	-	-	-
Belize	-	2.486	2.700	-	-	-	-	4,542	4,650	-	-	-
Bermuda	0.217	0.205	0.186	0.169	0.176	0.189	-	-	-	-	-	-
Bolivia	7.625	8.095	8.567	8.441	8.661	8.854	2,023	2,017	1,958	2,274	2,504	2,916
Bosnia & Herzegovina . . .	23	22	-	-	2.013	-	2,903	3,440	-	-	-	-
Botswana	5.200	5.200	4.900	5.100	5.200	5.473	-	-	5,620	5,414	-	-
Brazil	551	-	548	519	-	-	4,704	-	4,247	5,045	-	-
Bulgaria.	106	87	74	56	60	52	1,798	638	1,032	1,404	1,041	1,339
Burundi	1.411	1.620	-	-	-	-	3,720	3,241	-	-	-	-
Cambodia	*8.810	*11	-	-	-	-	-	-	-	-	-	-
Cameroon	15	17	16	16	15	16	4,132	3,862	3,820	3,430	1,708	2,288
Canada	198	192	195	189	189	189	24,433	25,694	25,053	24,095	22,956	23,561
Central African Republic . . .	0.558	-	-	-	-	-	6,872	-	-	-	-	-
Chile	75	81	85	83	86	91	4,028	4,292	5,148	5,450	6,196	6,974
China	*3,050	*3,210	*3,270	*3,130	*3,180	*3,220	-	-	-	-	-	-
China (Hong Kong SAR) . . .	*21	*21	*21	*20	*19	-	9,023	10,790	11,447	13,111	13,729	-
China (Taiwan Province) . . .	106	105	108	108	109	107	10,028	10,836	12,442	12,761	13,556	14,277
Colombia	79	81	103	102	105	107	2,284	2,338	2,526	2,886	3,633	4,066
Costa Rica	31	31	34	35	-	-	2,770	2,405	2,848	3,121	-	-
Croatia	56	53	48	45	45	45	5,798	6,433	1,674	1,983	2,888	4,864
Cyprus	5.880	5.833	5.988	5.945	6.392	7.095	8,517	9,056	10,384	10,509	11,077	14,562
Czechoslovakia (Former) . . .	168	143	-	-	-	-	2,086	1,523	-	-	-	-
Czech Republic	116	99	94	92	-	-	2,104	1,538	1,924	2,275	-	-
Denmark	87	86	83	83	85	83	26,149	26,631	29,547	27,720	30,373	34,287
Ecuador	27	33	37	36	37	34	3,451	3,183	3,139	3,548	2,119	2,708
Egypt	*204	*243	*193	*191	*197	*221	1,542	1,023	1,051	1,163	1,221	1,436
El Salvador.	-	-	*6.995	*7.400	*9.347	*5.778	-	-	-	3,886	5,417	3,996
Equatorial Guinea . . .	0.085	-	-	-	-	-	677	-	-	-	-	-
Ethiopia.	15	16	15	15	17	16	1,537	1,604	1,248	827	818	801
Fiji	6.454	4.817	6.656	6.623	6.948	-	4,218	5,938	3,914	3,949	4,343	-
Finland	47	45	41	38	36	39	26,624	26,208	25,062	19,936	22,745	28,536
FYR Macedonia	13	15	15	14	13	14	5,287	7,447	2,482	2,877	3,445	4,269
France	463	460	457	450	446	443	31,589	30,996	34,988	-	-	-
Gabon	-	2.647	2.627	2.614	2.771	2.588	-	10,787	11,694	11,177	5,839	7,017
Gambia	-	-	-	1.186	-	-	-	-	-	-	-	-
Germany	-	*511	*471	*449	*442	-	-	21,719	25,960	26,151	27,290	-
Germany (Western Part) . . .	*376	*403	*401	-	-	-	24,102	24,762	27,838	-	-	-
Ghana	-	-	-	*7.418	*7.627	*7.842	-	-	-	1,447	1,080	1,103

Continued.

Depending on the table, * means *Enterprises, Engaged,* or *Factor Values;* ± means *Basis Unspecified;* § means *shown in millions.* For additional notes and sources, see Appendix I.

Employment and Compensation of Employees

- Continued -

Country	Number Employed/Engaged (000)						Wage/Salary per Employee ($)					
	1990	1991	1992	1993	1994	1995	1990	1991	1992	1993	1994	1995
Greece	*50	*48	*49	*49	*49	*49	10,569	11,037	11,984	11,774	11,784	13,627
Grenada	-	-	*0.264	*0.237	0.264	0.261	-	-	-	-	-	-
Guatemala	-	*20	*22	*22	*22	*24	-	-	-	-	-	-
Honduras	*19	*23	*25	*27	*30	*33	2,541	1,981	2,128	2,193	1,839	2,187
Hungary	169	164	159	134	135	136	2,605	2,714	3,142	3,451	4,210	5,066
Iceland	11	11	9.982	9.982	10	-	22,779	26,053	28,470	25,333		
India	*1,122	*1,124	*1,240	*1,216	*1,219	*1,248	809	715	681	636	691	825
Indonesia	393	439	467	513	496	520	567	592	644	794	752	1,504
Iran (Islamic Republic of)	68	71	73	72	-	-	3,852	4,136	4,597	4,374	-	-
Iraq	-	13	12	-	-	-	-	8,961	13,780	-	-	-
Ireland	37	38	36	36	35	33	20,059	20,386	24,885	23,151	26,530	33,302
Israel	42	44	43	45	45	47	15,789	15,649	16,510	15,741	17,568	19,066
Italy	163	164	191	186	184	-	36,290	39,895	-	33,944	34,328	-
Jamaica	25	25	26	25	23	-	3,435	3,177	2,989	-	-	-
Japan	*1,138	*1,153	*1,165	*1,185	*1,137	*1,162	18,790	21,008	23,165	26,720	29,282	31,977
Jordan	7.819	9.118	10	11	13	14	2,028	2,043	2,187	2,257	2,173	2,270
Kenya	52	54	54	55	58	61	1,114	1,024	1,019	670	821	1,139
Korea, Republic of	178	177	174	175	177	180	7,790	9,018	9,552	10,737	11,651	13,581
Kuwait	7.727	4.791	5.953	6.744	6.689	6.609	4,201	3,582	7,454	6,956	7,057	9,664
Kyrgyzstan	*28	*27	*26	*25	*21	-	-	-	-	-	-	-
Latvia	36	35	50	36	32	30	-	-	-	-	-	-
Lesotho	-	-	-	*2.260	*1.782	-	-	-	-	2,288	4,723	-
Liechtenstein	-	*0.546	-	-	-	-	-	-	-	-	-	-
Lithuania	-	-	59	55	53	-	-	-	569	644	1,292	-
Luxembourg	2.472	2.722	2.799	2.974	2.860	-	18,945	20,581	22,325	21,372	22,174	-
Macau	0.958	0.878	0.760	0.828	0.958	0.856	3,877	4,332	4,836	5,317	6,046	6,940
Malawi	11	5.900	6.300	6.500	7.900	-	1,071	2,165	2,454	2,023	1,136	-
Malaysia	73	78	78	81	83	86	2,939	3,025	3,594	3,874	4,149	4,573
Malta	2.327	2.330	2.345	2.286	2.289	-	8,402	9,295	10,249	9,059	9,856	-
Mauritius	*11	13	13	12	12	12	2,984	3,015	3,634	3,435	3,861	4,378
Mexico	97	96	97	95	88	86	4,036	5,023	6,065	6,717	7,094	4,301
Mongolia	*12	*12	*9.249	*9.701	*14	*10	1,276	1,169	630	-	-	597
Morocco	21	54	85	93	91	91	4,879	5,183	3,440	2,853	3,435	3,687
Mozambique	31	27	21	21	19	-	1,098	627	-	-	-	-
Namibia	-	-	-	-	14	-	-	-	-	-	6,046	-
Nepal	17	21	-	26	25	-	-	464	-	-	-	-
Netherlands	108	110	110	108	106	106	26,924	27,332	30,545	30,970	33,336	37,969
Netherlands Antilles	-	-	-	0.802	-	-	-	-	-	8,539	-	-
New Zealand	56	56	58	59	-	-	-	-	-	17,090	-	-
Norway	44	44	40	40	41	41	25,473	26,171	34,386	25,336	26,603	30,693
Pakistan	78	84	88	-	-	-	1,688	1,663	1,711	-	-	-
Panama	*13	*14	*15	*16	*17	*19	5,333	5,320	5,577	5,760	6,050	6,337
Paraguay	-	3.158	-	-	-	-	-	-	-	-	-	-
Peru	*43	*43	*43	-	-	-	3,272	3,920	3,650	-	-	-
Philippines	267	256	159	164	-	-	1,476	1,630	2,484	2,458	-	-
Poland	366	360	349	325	338	348	1,265	1,947	2,475	2,421	2,865	3,676
Portugal	*100	*103	*97	*98	*103	*105	-	-	-	-	-	-
Puerto Rico	16	13	14	14	14	14	18,514	19,738	19,105	21,083	22,284	22,362
Qatar	1.644	1.688	2.006	1.948	2.003	-	5,648	5,589	5,333	5,173	5,625	-
Republic of Moldova	59	58	67	52	49	48	-	2,798	-	-	565	681
Russian Federation	-	-	-	1,533	1,475	1,468	-	-	-	963	1,461	1,462
Saint Vincent & the Grenadines	-	-	-	0.251	0.256	0.233	-	-	-	6,653	6,742	-
Senegal	*18	*19	*14	*14	*16	-	4,497	3,990	5,461	5,037	2,403	-
Sierra Leone	-	-	-	*5.998	-	-	-	-	-	561	-	-
Singapore	*11	*11	*12	*12	*12	*12	10,750	12,279	13,625	14,606	16,447	18,628
Slovakia	-	*46	*42	*41	*40	-	-	1,514	1,836	1,989	2,268	-
South Africa	199	194	194	195	176	176	5,893	6,621	7,660	7,404	7,881	8,554
Spain	259	264	264	293	294	289	14,197	15,013	16,490	15,951	15,824	17,453
Sri Lanka	*47	*40	*47	*41	-	-	543	622	628	692	-	-
Suriname	*2.740	*2.782	*2.689	*2.451	-	-	10,305	10,471	15,980	18,880	-	-

Continued.

Employment and Compensation of Employees

- Continued -

Country	Number Employed/Engaged (000)						Wage/Salary per Employee ($)					
	1990	1991	1992	1993	1994	1995	1990	1991	1992	1993	1994	1995
Swaziland	9.616	-	-	-	-	-	1,949	-	-	-	-	-
Sweden	62	*63	*59	56	54	54	22,546	23,644	26,770	21,110	22,768	26,119
Thailand	204	219	-	-	-	-	1,634	2,033	-	-	-	-
Tonga	0.259	0.287	-	-	-	-	1,977	2,051		-	-	-
Trinidad and Tobago	13	15	15	17	17	-	4,603	4,825	5,125	3,615	3,310	-
Tunisia	-	-	-	42	35	34	-	-	3,481	4,378	5,695
Turkey	129	124	139	133	130	130	5,630	7,351	7,594	8,007	4,929	5,633
Ukraine	683	653	618	606	586	556	2,663	3,007	2,538	439	535	654
United Kingdom	503	490	482	490	483	481	18,383	19,917	20,941	18,248	19,337	20,939
United Republic of Tanzania	36	36	-	-	-	-	-	-	-	-	-	-
USSR (Former)	2,753	-	-	-	-	-	2,025	-	-	-	-	-
United States of America	1,333	1,339	1,369	1,385	1,383	1,396	21,928	22,367	23,629	23,951	24,578	25,171
Uruguay	*48	*46	*43	*39	*40	*40	2,841	3,543	4,181	4,914	5,470	6,033
Venezuela	82	88	89	83	87	90	3,498	3,914	4,227	3,730	3,388	4,084
Yugoslavia	-	98	98	95	93	93	-	-	-	-	2,425	1,095
Zambia	19	-	-	-	13	-	1,429	-	-	-	2,114	-
Zimbabwe	25	27	27	25	24	25	3,789	3,635	3,170	2,519	2,590	3,127

Female Workers: Total and Percent of Employment

Country	Female Workers (000)						Female Workers as % of Total Employment					
	1990	1991	1992	1993	1994	1995	1990	1991	1992	1993	1994	1995
Afghanistan	0.530	0.549	-	-	-	-	-	-	-	-	-	-
Albania	-	-	-	6.968	3.369	0.966	-	-	-	48.42	49.79	23.60
Argentina	-	-	-	-	38	-	-	-	-	-	18.98	-
Australia	49	-	-	-	-	-	32.95	-	-	-	-	-
Austria	21	21	20	20	19	-	39.77	41.18	40.20	39.24	39.56	-
Bahrain	-	-	0.098	-	-	-	-	-	3.46	-	-	-
Bangladesh	6.581	5.815	5.637	-	-	-	5.95	5.58	5.31	-	-	-
Bermuda	0.080	0.082	0.076	0.062	0.065	0.063	36.87	40.00	40.86	36.69	36.93	33.33
Bosnia & Herzegovina	9.296	9.036	-	-	-	-	39.87	41.67	-	-	-	-
Botswana	-	-	-	-	-	1.575	-	-	-	-	-	28.78
Bulgaria	-	-	-	30	33	27	-	-	-	53.78	55.95	52.22
Canada	63	62	-	-	-	-	31.85	32.29	-	-	-	-
Chile	19	23	24	22	23	-	25.44	28.79	28.11	26.25	27.09	-
China (Taiwan Province)	48	47	49	49	50	49	44.76	45.31	45.40	45.87	45.63	45.68
Colombia	-	21	-	33	34	-	-	26.36	-	32.33	31.89	-
Croatia	25	23	21	20	20	19	43.77	43.81	43.08	43.13	43.13	42.78
Cyprus	2.608	2.563	2.616	2.591	2.821	-	44.35	43.94	43.69	43.58	44.13	-
Czechoslovakia (Former)	60	54	-	-	-	-	35.89	37.76	-	-	-	-
Czech Republic	58	52	50	49	-	-	50.00	52.53	53.19	53.26	-	-
Denmark	39	38	36	-	-	-	44.55	44.13	43.84	-	-	-
Egypt	-	22	21	22	-	-	-	9.03	11.02	11.62	-	-
El Salvador	-	-	-	2.045	3.464	1.939	-	-	-	27.64	37.06	33.56
Ethiopia	-	2.851	2.737	2.743	3.240	2.888	-	18.13	18.65	18.24	19.62	18.63
FYR Macedonia	4.521	4.685	4.402	4.202	3.925	3.857	35.95	31.94	29.82	29.47	29.41	28.21
Gambia	-	-	-	0.520	-	-	-	-	-	43.84	-	-
Germany (Western Part)	182	-	-	-	-	-	48.37	-	-	-	-	-
Ghana	-	-	-	0.643	-	-	-	-	-	8.67	-	-
Hungary	70	-	65	55	-	-	41.42	-	40.88	41.04	-	-
Indonesia	-	-	-	202	200	210	-	-	-	39.35	40.28	40.29
Iran (Islamic Republic of)	-	4.519	4.715	4.673	-	-	-	6.36	6.49	6.51	-	-
Iraq	-	2.550	-	-	-	-	-	19.91	-	-	-	-
Ireland	8.900	9.300	-	-	-	-	24.25	24.80	-	-	-	-
Italy	-	48	50	47	47	-	-	29.03	26.40	25.20	25.55	-
Jordan	0.728	0.655	0.905	0.648	0.945	0.937	9.31	7.18	8.79	6.11	7.16	6.61
Kenya	9.600	10	11	11	-	-	18.64	19.45	19.54	19.57	-	-
Korea, Republic of	87	85	83	84	87	89	48.79	47.89	47.65	47.97	49.20	49.61

Continued.

Depending on the table, * means *Enterprises, Engaged,* or *Factor Values*; ± means *Basis Unspecified*; § means *shown in millions.* For additional notes and sources, see Appendix I.

Female Workers: Total and Percent of Employment

- Continued -

Country	Female Workers (000)						Female Workers as % of Total Employment					
	1990	1991	1992	1993	1994	1995	1990	1991	1992	1993	1994	1995
Lesotho	-	-	-	0.986	0.844	-	-	-	-	43.63	47.36	-
Macau	0.249	0.235	0.204	0.247	0.375	0.316	25.99	26.77	26.84	29.83	39.14	36.92
Malaysia	25	28	28	28	28	-	34.65	35.29	35.95	34.77	34.14	-
Malta	0.420	0.422	0.418	0.399	0.410	-	18.05	18.11	17.83	17.45	17.91	-
Mongolia	-	-	-	-	7.658	-	-	-	-	-	56.23	-
Morocco	-	-	-	-	-	27	-	-	-	-	-	29.41
Nepal	2.150	2.727	-	2.101	2.745	-	12.62	12.83	-	8.03	11.09	-
New Zealand	15	15	15	17	-	-	27.07	26.81	26.66	29.62	-	-
Panama	2.506	-	-	-	-	-	19.31	-	-	-	-	-
Philippines	-	-	50	51	-	-	-	-	31.52	31.05	-	-
Portugal	19	20	19	19	18	-	19.34	19.37	19.86	19.41	17.54	-
Republic of Moldova	-	-	-	-	28	27	-	-	-	-	57.74	56.76
Saint Vincent & the Grenadines	-	-	-	0.038	0.039	-	-	-	-	15.14	15.23	-
Senegal	2.339	-	-	-	-	-	12.77	-	-	-	-	-
Sri Lanka	16	13	16	14	-	-	33.26	33.08	34.42	34.62	-	-
Sweden	23	-	-	-	-	-	38.05	-	-	-	-	-
Thailand	63	106	-	-	-	-	30.85	48.32	-	-	-	-
Turkey	23	-	-	-	-	-	18.17	-	-	-	-	-
United Kingdom	206	-	211	-	-	-	40.93	-	43.78	-	-	-
United Republic of Tanzania	8.215	8.911	-	-	-	-	23.03	24.95	-	-	-	-
United States of America	-	-	1,033	1,049	1,047	1,058	-	-	75.46	75.74	75.70	75.79
Zambia	-	-	-	-	1.019	-	-	-	-	-	8.04	-

Output and Output per Employee

Country	Output ($ bil.)						Output per Employee ($)					
	1990	1991	1992	1993	1994	1995	1990	1991	1992	1993	1994	1995
Albania	-	-	-	*0.094	*0.030	-	-	-	-	6,512	4,494	-
Argentina	±19	-	-	19	-	-	98,172	-	-	101,678	-	-
Armenia	0.019	0.050	0.162	7.427	0.035	-	842	2,366	6,134	333,611	1,571	-
Australia	*21	*22	*22	-	-	-	142,396	150,712	147,638	-	-	-
Austria	10	10	11	11	11	13	193,503	204,136	221,282	217,443	219,835	256,941
Azerbaijan	±1.089	±1.175	±0.039	±0.178	±0.114	-	29,171	30,443	1,016	4,726	3,068	-
Bahamas	0.013	0.012	-	-	-	-	51,587	54,299	-	-	-	-
Bahrain	-	-	±0.118	-	-	-	-	-	41,750	-	-	-
Bangladesh	1.281	1.233	1.168	-	-	-	11,571	11,839	10,998	-	-	-
Barbados	0.126	0.076	0.065	0.107	0.140	0.127	79,099	63,543	58,416	75,118	74,475	73,265
Belize	0.107	0.116	0.126	-	-	-	-	46,599	46,511	-	-	-
Benin	±0.256	±0.263	±0.288	±0.257	-	-	-	-	-	-	-	-
Bolivia	0.463	0.544	0.557	0.571	0.680	0.693	60,682	67,198	65,032	67,643	78,491	78,253
Bosnia & Herzegovina	±0.812	±1.031	-	-	-	-	34,845	47,568	-	-	-	-
Botswana	-	-	±0.320	±0.297	-	-	-	-	65,344	58,271	-	-
Brazil	*36	-	*36	*40	-	-	65,939	-	66,154	77,452	-	-
Bulgaria	±4.517	±1.313	±1.345	±1.322	±1.055	±1.302	42,737	15,069	18,276	23,783	17,678	25,193
Burundi	±0.089	±0.084	-	-	-	-	63,249	51,604	-	-	-	-
Cameroon	0.301	0.400	0.351	0.288	0.230	0.268	20,607	23,998	21,838	17,982	14,863	16,854
Canada	39	40	39	38	38	39	199,399	207,888	198,857	199,197	198,645	203,695
Central African Republic	*0.029	-	-	-	-	-	52,178	-	-	-	-	-
Chile	4.074	4.862	5.733	5.889	6.669	7.955	54,122	60,299	67,767	70,636	77,318	86,959
China	±29	±31	±34	±41	±39	±48	9,524	9,520	10,379	13,067	12,169	15,024
China (Hong Kong SAR)	1.236	1.308	1.456	1.455	1.535	-	58,862	61,398	70,017	71,328	80,768	-
China (Taiwan Province)	±14	±15	±16	±16	±16	±17	132,175	138,859	150,937	144,800	149,346	162,532
Colombia	4.882	4.645	5.272	5.392	7.458	8.592	61,563	57,207	51,181	53,073	70,826	80,209
Costa Rica	1.246	1.281	1.346	1.551	1.638	2.040	40,148	41,718	39,817	44,472	-	-
Croatia	±3.068	±2.990	±1.724	-	-	-	54,410	56,886	35,686	-	-	-
Cyprus	0.338	0.357	0.402	0.378	0.426	0.538	57,523	61,209	67,129	63,665	66,636	75,817
Czechoslovakia (Former)	±6.633	±3.818	-	-	-	-	39,481	26,701	-	-	-	-
Denmark	*16	*15	*16	*15	*18	*19	179,874	175,412	198,311	186,801	205,737	234,341

Continued.

Depending on the table, * means *Enterprises*, *Engaged*, or *Factor Values*; ± means *Basis Unspecified*; § means *shown in millions*. For additional notes and sources, see Appendix I.

Output and Output per Employee
- Continued -

Country	Output ($ bil.)						Output per Employee ($)					
	1990	1991	1992	1993	1994	1995	1990	1991	1992	1993	1994	1995
Ecuador	1.235	1.579	1.619	1.676	2.103	2.354	45,517	48,208	44,193	46,099	57,083	68,323
Egypt	*4.671	*2.715	*3.173	*3.636	*4.031	*5.598	22,873	11,171	16,440	19,038	20,501	25,321
El Salvador	-	±0.258	±0.304	±0.230	±0.402	±0.335	-	-	43,496	31,078	42,962	57,919
Equatorial Guinea	±0.001	-	-	-	-	-	6,050	-	-	-	-	-
Ethiopia	0.321	0.292	0.148	0.104	0.131	0.154	20,772	18,554	10,087	6,938	7,953	9,932
Fiji	0.332	0.338	0.350	0.355	0.412	-	51,490	70,180	52,543	53,642	59,278	-
Finland	±12	±11	±9.647	±7.739	±8.295	±9.528	251,130	244,033	233,015	201,020	229,144	247,102
FYR Macedonia	±0.357	±0.616	±0.381	±0.335	±0.322	±0.371	28,372	42,013	25,833	23,488	24,125	27,120
France	±96	±94	±103	±94	±97	±111	206,821	205,355	224,833	208,892	218,352	249,818
Gabon	-	*0.119	*0.117	*0.113	*0.068	*0.079	-	44,999	44,604	43,096	24,708	30,515
Germany	-	104	113	105	108	-	-	203,751	238,965	232,916	244,312	-
Germany (Western Part)	91	96	104	-	-	-	241,831	237,536	258,261	-	-	-
Ghana	-	-	-	0.146	0.119	0.125	-	-	-	19,658	15,574	15,894
Greece	*5.579	*5.587	*6.162	*6.067	*6.072	*7.046	112,447	116,725	125,956	124,210	124,507	144,020
Guatemala	-	1.216	1.249	1.411	1.626	1.953	-	62,021	58,021	63,479	72,747	81,423
Honduras	0.585	0.609	0.721	0.718	0.736	0.795	30,322	26,992	29,188	26,514	24,800	24,345
Hungary	±5.215	±5.016	±4.867	±4.712	±4.930	±5.198	30,856	30,588	30,612	35,167	36,508	38,211
Iceland	1.390	1.510	±1.632	±1.382	-	-	124,064	138,612	163,510	138,410	-	-
India	*21	*19	*19	*18	*20	*25	18,670	16,975	15,515	15,207	16,588	19,923
Indonesia	5.638	6.204	8.088	13	9.458	12	14,352	14,146	17,329	24,531	19,050	22,396
Iran (Islamic Republic of)	2.574	2.920	2.975	3.248	-	-	38,074	41,072	40,945	45,256	-	-
Iraq	-	*1.229	*0.799	-	-	-	-	95,955	66,586	-	-	-
Ireland	*11	*11	*13	*12	*13	*16	290,631	287,009	354,667	331,844	382,921	484,451
Israel	4.232	4.189	4.592	4.518	4.771	5.660	100,755	96,076	106,058	99,724	105,561	119,214
Italy	±50	±53	±65	±52	±53	-	305,921	321,845	342,144	278,976	286,036	-
Jamaica	0.755	0.678	0.647	-	-	-	30,180	27,137	25,286	-	-	-
Japan	±180	±203	±223	±251	±248	±271	158,305	176,493	191,098	211,760	218,444	233,289
Jordan	0.219	0.236	0.298	0.315	0.469	0.603	28,035	25,927	28,903	29,715	35,536	42,539
Kenya	*3.169	*3.153	*3.618	*2.521	*3.020	*3.873	61,520	58,872	67,317	45,476	52,371	63,933
Korea, Republic of	17	19	20	22	24	29	96,121	109,505	116,670	125,328	135,844	160,171
Kuwait	0.173	0.130	0.238	0.268	0.298	0.368	22,451	27,098	39,984	39,770	44,545	55,667
Kyrgyzstan	0.997	1.037	0.499	0.103	0.098	-	36,061	38,048	19,429	4,206	4,567	-
Latvia	0.017	0.042	0.335	0.413	0.519	0.605	472	1,207	6,643	11,364	16,468	19,880
Lithuania	-	-	0.450	0.555	0.668	-	-	-	7,611	10,086	12,565	-
Luxembourg	*0.322	*0.372	*0.399	*0.373	*0.374	-	130,082	136,588	142,630	125,268	130,824	-
Macau	0.017	0.017	0.017	0.019	0.024	0.023	17,570	19,628	22,727	23,450	25,381	26,736
Malawi	*0.185	*0.194	*0.218	*0.171	*0.102	-	16,679	32,812	34,668	26,333	12,972	-
Malaysia	*5.920	*6.462	*7.594	*7.863	*9.777	*11	80,760	82,321	97,866	96,954	118,370	131,027
Malta	0.183	0.187	0.200	0.173	0.187	-	78,543	80,455	85,225	75,827	81,614	-
Mauritius	0.518	0.515	0.607	0.552	0.547	0.678	45,398	40,217	47,578	44,214	43,952	54,460
Mexico	7.249	8.304	9.462	9.806	10	7.518	74,749	86,126	97,053	103,451	115,864	87,690
Mongolia	0.289	0.309	0.150	0.057	0.069	0.094	24,160	26,029	16,207	5,902	5,079	9,125
Morocco	±1.452	±3.356	±3.540	±3.813	±4.056	±4.258	70,443	61,765	41,781	40,988	44,782	46,793
Mozambique	-	±0.093	-	±0.045	±0.044	-	-	3,390	-	2,185	2,378	-
Namibia	-	-	-	-	0.691	-	-	-	-	-	50,949	-
Nepal	0.142	0.196	-	0.201	0.209	-	8,320	9,213	-	7,679	8,447	-
Netherlands	32	33	36	34	36	42	297,419	296,113	327,077	319,243	344,967	393,295
New Zealand	7.516	7.246	7.116	7.693	-	-	133,119	129,283	123,530	130,461	-	-
Norway	±10	±11	±11	±8.965	±9.420	±11	234,047	245,150	279,856	226,222	231,123	267,334
Pakistan	2.561	2.615	2.838	-	-	-	32,712	31,171	32,346	-	-	-
Panama	0.630	0.652	0.702	0.737	0.804	0.883	48,522	48,043	47,858	47,277	46,508	45,444
Peru	2.565	2.914	2.863	-	-	-	59,281	68,476	66,246	-	-	-
Philippines	6.005	5.676	5.807	5.940	-	-	22,532	22,128	36,613	36,261	-	-
Poland	±7.087	±8.589	±8.610	±8.416	±10	±13	19,363	23,857	24,670	25,896	29,826	37,741
Portugal	7.309	7.977	8.857	7.409	8.080	9.529	73,264	77,672	91,141	75,654	78,168	90,870
Puerto Rico	±2.509	±2.434	±2.494	±2.627	±2.618	±2.671	152,820	182,292	172,961	185,945	188,617	186,660
Qatar	0.057	0.055	0.059	0.056	0.075	-	34,424	32,388	29,582	28,629	37,444	-
Republic of Moldova	0.000	3.806	0.368	0.344	0.357	0.383	0.098	65,401	5,489	6,602	7,331	8,033
Romania	±7.650	±5.431	±3.807	±4.431	-	-	-	-	-	-	-	-
Russian Federation	-	70	39	17	22	28	-	-	-	10,812	14,710	18,762

Continued.

     Depending on the table, * means *Enterprises, Engaged,* or *Factor Values*; ± means *Basis Unspecified*; § means *shown in millions*. For additional notes and sources, see Appendix I.

Output and Output per Employee
- Continued -

Country	Output ($ bil.)						Output per Employee ($)					
	1990	1991	1992	1993	1994	1995	1990	1991	1992	1993	1994	1995
Senegal	0.815	-	-	-	-	-	44,482	-	-	-	-	-
Sierra Leone	-	-	-	0.088	-	-	-	-	-	14,745	-	-
Singapore	*1.127	*1.217	*1.315	*1.399	*1.623	*1.828	106,215	109,619	113,703	117,795	137,525	154,216
Slovakia	-	±1.463	±1.467	±1.261	±1.310	-		31,491	34,962	30,815	32,545	
Slovenia				1.178	1.189	1.349				49,020	51,690	55,838
South Africa	±9.407	±9.625	±10	±9.559	±9.097	±9.828	47,273	49,614	52,802	49,020	51,690	55,838
Spain	*44	*47	*49	±45	±47	±55	171,855	178,130	184,826	153,219	160,345	189,433
Sri Lanka	0.626	0.530	0.582	0.627	-	-	13,422	13,154	12,314	15,271	-	-
Suriname	*0.174	*0.197	*0.234	0.556	-	-	63,383	70,884	86,877	226,741	-	-
Swaziland	*0.252	-	-	-	-	-	26,167	-	-	-	-	-
Sweden	*14	*16	*17	*11	*11	*13	223,121	252,265	281,186	187,636	201,815	233,068
Thailand	9.817	9.942	-	-	-	-	48,030	45,342	-	-	-	-
Tonga	0.005	0.005	-	-	-	-	17,640	18,243	-	-	-	-
Trinidad and Tobago	±0.493	±0.570	±0.563	±0.500	±0.483	-	38,593	39,277	38,488	30,074	28,494	
Tunisia	1.895	2.074	2.609	2.212	2.471	2.655				52,788	71,515	78,713
Turkey	8.889	10	12	13	10	12	68,783	80,720	86,314	98,346	78,587	90,624
Ukraine	*31	*38	*24	*7.869	*6.947	*6.205	46,117	57,580	39,122	12,985	11,855	11,160
United Kingdom	*72	*74	77	71	75	80	143,681	150,052	159,146	144,411	154,281	166,684
United Republic of Tanzania	*0.185	*0.162	-	-	-	-	5,193	4,530	-	-	-	-
USSR (Former)	±96	-	-	-	-	-	34,871	-	-	-	-	-
United States of America	*337	*339	*356	*371	*377	*392	252,941	252,905	259,971	267,819	272,659	280,663
Uruguay	1.551	1.797	1.821	1.921	2.146	2.367	32,390	38,680	42,469	48,771	54,240	59,780
Venezuela	3.946	4.636	5.090	4.815	4.544	6.968	47,950	52,497	57,189	58,193	52,369	77,515
Yugoslavia	±0.537	±0.612	±1.360	-	±2.692	±1.588	-	6,267	13,804	-	28,849	17,109
Zambia	0.295	-	-	-	0.198	-	15,888	-	-	-	15,603	-
Zimbabwe	*0.865	*0.898	*1.002	*0.823	*0.828	*1.051	34,205	33,091	37,738	33,336	34,346	41,465

Value Added and Value Added per Employee

Country	Value Added ($ bil.)						Value Added per Employee ($)					
	1990	1991	1992	1993	1994	1995	1990	1991	1992	1993	1994	1995
Argentina	±4.695	-	-	4.731	-	-	23,807	-	-	24,874	-	-
Armenia	0.001	0.002	-	-	-	-	55	98	-	-	-	-
Australia	*7.647	*8.028	*7.793				50,979	54,244	53,374			
Austria	2.302	2.417	2.714	2.669	2.565	2.999	43,604	47,394	53,210	53,432	52,329	61,166
Bahamas	0.008	0.008	-	-	-	-	32,429	34,706	-	-	-	-
Bangladesh	0.265	0.233	0.238	-	-	-	2,391	2,232	2,241	-	-	-
Barbados	0.030	0.026	0.025	0.034	0.048	0.043	18,756	21,741	22,025	23,642	25,297	24,934
Belgium	*6.044	*6.156	*6.633	*6.318	*6.887	*8.265	79,739	80,900	88,086	-	-	-
Belize	0.027	0.023	0.027	-	-	-		9,436	10,098	-	-	-
Bolivia	0.115	0.153	0.155	0.137	0.186	0.194	15,027	18,872	18,058	16,285	21,490	21,883
Bosnia & Herzegovina	±0.287	±0.365	-	-	-	-	12,320	16,813	-	-	-	-
Botswana	-	-	±0.069	±0.061	±0.061	-	-	-	14,114	11,863	11,769	-
Brazil	*15	-	*17	*19	-	-	26,817	-	30,365	36,964	-	-
Bulgaria	-	±0.342	±0.348	±0.335	-	-	-	3,926	4,728	6,026	-	-
Burundi	0.059	0.064	-	-	-	-	41,783	39,589	-	-	-	-
Cameroon	0.185	0.123	0.113	0.051	0.076	0.103	12,669	7,396	7,051	3,173	4,949	6,478
Canada	13	14	13	12	12	13	64,213	70,372	68,732	65,210	65,169	67,110
Central African Republic	*0.015	-	-	-	-	-	27,429	-	-	-	-	-
Chile	1.543	1.742	2.173	2.321	2.601	3.103	20,501	21,601	25,686	27,839	30,154	33,918
China	±4.489	±5.037	±5.366	±12	±9.677	±8.476	1,472	1,569	1,641	3,787	3,043	2,632
China (Hong Kong SAR)	0.398	0.447	0.491	0.517	0.521	-	18,944	20,982	23,608	25,341	27,424	-
China (Taiwan Province)	2.935	3.052	3.490	3.351	3.438	3.491	27,625	29,160	32,452	31,137	31,417	32,610
Colombia	1.306	1.263	1.663	1.602	2.542	2.936	16,470	15,557	16,144	15,772	24,140	27,404
Costa Rica	0.292	0.304	0.331	0.365	0.379	0.466	9,407	9,906	9,787	10,466	-	-
Croatia	±1.153	±1.280	±0.709	-	-	-	20,455	24,358	14,675	-	-	-
Cyprus	0.101	0.110	0.125	0.126	0.136	0.177	17,178	18,832	20,851	21,156	21,265	24,980
Czechoslovakia (Former)	±0.916	-	-	-	-	-	5,455	-	-	-	-	-

Continued.

Value Added and Value Added per Employee

- Continued -

Country	Value Added ($ bil.)						Value Added per Employee ($)					
	1990	1991	1992	1993	1994	1995	1990	1991	1992	1993	1994	1995
Denmark	*4.072	*4.405	*4.850	*4.542	*5.178	*5.718	46,607	51,481	58,303	54,934	60,788	69,162
Ecuador	0.228	0.268	0.265	0.329	0.366	0.486	8,407	8,174	7,236	9,044	9,933	14,112
Egypt	*0.949	*0.469	*0.588	*0.908	*1.006	*1.394	4,648	1,931	3,049	4,755	5,117	6,307
El Salvador	-	0.102	-	0.081	0.032	0.108	-	-	-	11,002	3,385	18,744
Ethiopia	0.136	0.135	0.065	0.051	0.067	0.071	8,792	8,569	4,453	3,368	4,067	4,611
Fiji	0.060	0.061	0.060	0.061	0.070		9,335	12,764	9,032	9,187	10,071	
Finland	±2.576	±2.525	±2.270	±1.869	±1.920	±2.326	55,270	56,487	54,819	48,558	53,038	60,335
FYR Macedonia	±0.160	±0.284	±0.158	±0.146	±0.017	±0.144	12,752	19,387	10,696	10,269	1,273	10,515
France	±26	±25	±28	±28	±28	±31	55,255	55,340	61,934	61,485	62,094	70,735
Gabon	-	*0.043	*0.043	*0.042	*0.025	*0.025	-	16,292	16,398	16,073	9,149	9,626
Germany (Western Part)	29	32	35	35	-	-	75,981	78,324	87,294	-	-	-
Ghana	-	-	-	0.051	0.044	0.046	-	-	-	6,889	5,806	5,925
Greece	*1.349	*1.501	*1.813	*1.793	*1.797	*2.090	27,197	31,368	37,058	36,708	36,845	42,715
Haiti	-	±0.026	±0.012	±0.009	±0.008	±0.007	-	-	-	-	-	-
Honduras	*0.123	*0.126	*0.141	*0.149	*0.143	*0.164	6,398	5,566	5,722	5,509	4,808	5,009
Hungary	±0.740	±0.890	±1.123	±1.077	±1.161	±1.261	4,377	5,425	7,061	8,041	8,597	9,273
Iceland	0.285	0.337	0.451	0.397			25,483	30,977	45,162	39,751		
India	*2.212	*1.965	*1.870	*2.241	*2.452	*3.015	1,971	1,747	1,508	1,843	2,011	2,417
Indonesia	*1.910	*2.214	*2.849	*3.124	*2.450	*2.814	4,861	5,048	6,104	6,093	4,934	5,409
Iran (Islamic Republic of)	0.777	0.981	0.931	1.120	-	-	11,491	13,801	12,815	15,607	-	-
Iraq	-	*0.223	*0.490	-	-	-	-	17,419	40,836	-	-	-
Ireland	*3.068	*3.399	*4.107	*3.791	*4.291	*5.271	83,588	90,628	113,362	106,690	124,079	158,451
Israel	1.221	1.241	1.140	1.089	1.167	1.380	29,074	28,470	26,324	24,040	25,819	29,074
Italy	±9.599	±11	±13	±11	±10	-	59,012	66,100	70,470	58,404	56,724	-
Jamaica	0.182	0.164	0.162				7,259	6,556	6,341			
Japan	±67	±77	±86	±99	±98	±110	58,590	66,572	74,112	83,288	86,230	94,512
Jordan	0.058	0.057	0.068	0.074	0.112	0.113	7,437	6,276	6,633	6,996	8,458	7,946
Kenya	*0.252	*0.258	*0.245	*0.166	*0.228	*0.329	4,897	4,819	4,562	2,986	3,954	5,437
Korea, Republic of	6.047	7.333	8.038	8.773	9.578	11	34,027	41,539	46,267	50,069	53,975	63,005
Kuwait	0.069	0.069	0.115	0.118	0.125	0.165	8,949	14,309	19,280	17,484	18,704	24,973
Latvia	-	-	-	*0.139	*0.193	*0.214	-	-	-	3,827	6,112	7,040
Luxembourg	*0.067	*0.087	*0.107	*0.101	*0.101	-	27,007	32,028	38,361	34,026	35,323	-
Macau	0.006	0.006	0.006	0.008	0.010	0.010	5,943	6,841	7,704	9,133	10,225	11,369
Malawi	*0.011	*0.048	*0.077	*0.004	*0.036		989	8,162	12,233	542	4,532	-
Malaysia	*0.865	*0.914	*1.176	*1.306	*1.412	*1.617	11,803	11,640	15,156	16,107	17,099	18,819
Malta	0.056	0.059	0.061	0.051	0.054	-	23,946	25,286	26,139	22,410	23,584	-
Mauritius	0.080	0.081	0.118	0.104	0.088	0.124	7,024	6,317	9,245	8,354	7,067	9,916
Mexico	2.177	2.478	2.817	2.913	3.077	2.261	22,453	25,698	28,896	30,735	34,829	26,370
Mongolia	0.071	0.099	0.050	0.014	0.020	0.019	5,935	8,318	5,456	1,489	1,433	1,846
Morocco	±0.144	±0.747	±0.708	±0.700	±0.771	±0.838	6,991	13,743	8,353	7,522	8,508	9,205
Namibia	-	-	-	-	0.260	-	-	-	-	-	19,159	-
Nepal	0.038	0.052	-	0.061	0.040	-	2,218	2,455	-	2,328	1,610	-
Netherlands	6.037	6.284	6.859	6.575	6.976	7.987	56,019	57,093	62,504	61,051	65,932	75,158
New Zealand	1.676	1.765	1.591	1.656	-	-	29,690	31,492	27,622	28,084	-	-
Norway	±1.307	±1.462	±1.794	±1.378	±1.406	±1.628	29,770	33,544	44,890	34,780	34,488	39,790
Pakistan	0.665	0.655	0.732	-	-	-	8,493	7,814	8,338	-	-	-
Panama	0.205	0.214	0.229	0.240	0.261	0.285	15,809	15,779	15,649	15,415	15,093	14,679
Peru	1.077	1.010	0.916	-	-	-	24,896	23,743	21,198	-	-	-
Philippines	2.206	1.666	1.809	2.074	-	-	8,277	6,495	11,405	12,662	-	-
Poland	±2.595	±2.565	±2.682	±2.389	±2.852	±3.712	7,091	7,126	7,685	7,352	8,440	10,660
Portugal	1.305	1.306	1.690	1.372	1.399	1.650	13,083	12,711	17,395	14,008	13,531	15,735
Puerto Rico	±0.673	±0.396	±0.433	±0.465	±0.455	±0.505	41,005	29,633	30,007	32,944	32,788	35,311
Qatar	0.018	0.018	0.017	0.017	0.027	-	11,196	10,416	8,354	8,603	13,579	-
Romania	±1.649	±1.211	±0.928	±1.303			-	-	-	-	-	-
Russian Federation	-	-	-	±5.648	±8.352	±8.772	-	-	-	3,684	5,662	5,977
Saint Lucia	-	±0.003	±0.004	±0.004	±0.003	±0.004	-	-	-	-	-	-
Senegal	0.221	0.176	0.162	0.166	0.129	-	12,054	9,263	11,533	12,253	8,318	-
Sierra Leone	-	-	-	0.034	-	-	-	-	-	5,680	-	-
Singapore	*0.322	*0.362	*0.401	*0.445	*0.512	*0.583	30,315	32,604	34,697	37,460	43,391	49,207
Slovakia	-	-	-	±0.268	±0.285	-	-	-	-	6,545	7,081	-

Continued.

Depending on the table, * means *Enterprises, Engaged,* or *Factor Values*; ± means *Basis Unspecified*; § means *shown in millions.* For additional notes and sources, see Appendix I.

Value Added and Value Added per Employee
- Continued -

Country	Value Added ($ bil.)						Value Added per Employee ($)					
	1990	1991	1992	1993	1994	1995	1990	1991	1992	1993	1994	1995
Slovenia	-	-	-	0.481	0.405	0.492	-	-	-	-	-	-
South Africa	*2.219	*2.567	*2.864	*2.679	*2.553	*2.768	11,148	13,231	14,761	13,740	14,506	15,725
Spain	*11	*12	*12	±10	±10	±11	41,660	43,868	47,187	34,699	35,364	38,658
Sri Lanka	0.240	0.180	0.163	0.187	-	-	5,151	4,474	3,452	4,563	-	-
Swaziland	*0.074						7,667					
Sweden	*4.249	*3.102	*3.385	*2.527	*2.541	*2.878	69,090	49,344	57,460	44,954	46,635	53,507
Thailand	2.201	2.705	-	-	-	-	10,768	12,337	-	-	-	-
Trinidad and Tobago	±0.114	±0.142	±0.145	±0.102	±0.095	-	8,890	9,757	9,926	6,154	5,580	-
Tunisia	0.315	0.353	0.409	0.361	0.417	0.455	-	-	-	8,620	12,081	13,503
Turkey	2.545	3.288	3.891	4.337	3.096	3.570	19,693	26,449	28,013	32,501	23,850	27,497
United Kingdom	*25	*26	*27	*25	*27	*29	49,986	53,422	56,006	50,803	55,739	60,442
United Republic of Tanzania	*0.027	*0.032	-	-	-	-	766	889	-	-	-	-
United States of America	*120	*123	*133	*142	*146	*154	89,895	91,591	97,388	102,324	105,792	110,648
Uruguay	0.473	0.648	0.622	0.711	0.796	0.879	9,883	13,951	14,518	18,056	20,113	22,190
Venezuela	1.210	1.353	1.569	1.335	1.410	4.243	14,699	15,320	17,634	16,135	16,253	47,201
Yugoslavia	±0.198	±0.246	±0.653	-	±1.001	±0.542		2,518	6,634	-	10,733	5,842
Zambia	0.151	-	-	-	0.059	-	8,138	-	-	-	4,690	-
Zimbabwe	*0.237	*0.314	*0.360	*0.367	*0.241	*0.308	9,372	11,555	13,546	14,859	9,989	12,146

Capital Investment

Country	Gross Fixed Capital Formation ($ mil.)						Per Establishment ($ mil)					
	1990	1991	1992	1993	1994	1995	1990	1991	1992	1993	1994	1995
Argentina	-	-	-	721	-	-	-	-	-	0.040	-	-
Australia	680	-	-	-	-	-	0.205	-	-	-	-	-
Austria	441	455	480	528	544	-	0.511	0.560	0.594	0.663	0.715	-
Bahrain	-	-	16	-	-	-	-	-	0.051	-	-	-
Bangladesh	30	27	24	-	-	-	0.005	0.005	0.004	-	-	-
Barbados	4.120	6.020	1.246	3.309	3.979	-	0.153	0.301	0.054	0.132	0.088	-
Belgium	1,011	1,010	1,029	869	888	989	0.145	0.145	0.150	0.127	0.131	-
Bolivia	12	15	21	24	20	-	0.035	0.042	0.056	0.063	0.052	-
Bosnia & Herzegovina	31	-	-	-	-	-	0.225	-	-	-	-	-
Botswana	-	-	16	11	-	-	-	-	0.206	0.126	-	-
Brazil	1,698	-	996	1,817	-	-	0.307	-	0.190	0.361	-	-
Bulgaria	88	32	447	110	50	69	0.398	0.111	1.342	0.033	0.011	0.014
Cameroon	33	22	39	18	20	23	0.860	0.243	0.557	0.342	0.329	0.586
Canada	884	-	-	-	-	-	0.260	-	-	-	-	-
Chile	231	228	218	307	309	-	0.593	0.546	0.514	0.722	0.738	-
China (Hong Kong SAR)	67	56	54	67	45	-	0.071	0.067	0.069	0.091	0.065	-
Colombia	127	125	203	121	136	-	0.093	0.095	0.143	0.090	0.101	-
Croatia	80	86	54	32	59	55	0.134	0.120	0.062	0.029	0.043	0.036
Cyprus	17	18	19	18	22	35	0.019	0.019	0.020	0.020	0.022	0.033
Czechoslovakia (Former)	285	-	-	-	-	-	1.547	-	-	-	-	-
Denmark	434	482	-	-	-	-	0.149	0.169	-	-	-	-
Ecuador	117	155	162	90	135	25	0.369	0.395	0.429	0.233	0.346	0.062
Egypt	128	207	111	60	-	-	0.030	0.053	0.027	0.015	-	-
El Salvador	-	13	-	23	51	24	-	0.191	-	0.229	0.475	0.196
Ethiopia	9.922	11	3.454	7.032	4.745	3.557	0.088	0.107	0.035	0.068	0.033	0.027
Fiji	22	11	16	-	-	-	0.156	0.090	0.145	-	-	-
Finland	481	355	399	184	217	317	0.613	0.445	0.541	0.266	0.326	0.427
FYR Macedonia	11	21	11	14	6.273	11	0.199	0.151	0.041	0.047	0.016	0.016
Gabon	-	9.829	3.060	1.847	0.973	6.900	-	0.546	0.180	0.097	0.065	0.406
Germany (Western Part)	3,117	3,829	-	4,777	-	-	0.854	1.054	-	-	-	-
Greece	245	244	301	-	-	-	0.214	0.214	0.262	-	-	-
Hungary	243	164	177	200	-	-	0.492	0.163	0.125	0.106	-	-
India	716	661	755	569	-	-	0.036	0.034	0.035	0.026	-	-
Indonesia	393	992	515	256	515	496	0.112	0.294	0.143	0.069	0.133	0.116
Iran (Islamic Republic of)	72	118	157	128	-	-	0.058	0.138	0.175	0.143	-	-

Continued.

Depending on the table, * means *Enterprises, Engaged*, or *Factor Values*; ± means *Basis Unspecified*; § means *shown in millions*. For additional notes and sources, see Appendix I.

13

Capital Investment

- Continued -

Country	Gross Fixed Capital Formation ($ mil.)						Per Establishment ($ mil)					
	1990	1991	1992	1993	1994	1995	1990	1991	1992	1993	1994	1995
Ireland	304	277	-	-	-	-	0.398	0.393	-	-	-	-
Israel	190	173	256	204	230	-	0.187	0.152	0.245	0.177	0.217	-
Italy	2,014	1,982	2,679	1,997	2,020	-	1.180	1.148	1.161	0.850	0.870	-
Japan	5,470	7,082	8,156	8,624	7,925	-	0.116	0.154	0.181	0.189	0.185	-
Jordan	1.184	1.903	5.126	5.670	4.950	9.600	0.001	0.001	0.003	0.003	0.003	0.005
Korea, Republic of	1,320	1,695	1,639	1,851	1,871	2,470	0.325	0.399	0.356	0.342	0.340	0.420
Kuwait	24	24	38	11	13	-	0.076	0.082	0.126	0.037	0.043	-
Latvia	0.425	0.070	7.611	28	27	49	0.000	0.000	0.013	0.037	0.059	0.092
Luxembourg	23	27	22	-	-	-	0.101	0.119	0.105	-	-	-
Macau	0.258	0.723	0.363	1.996	1.953	0.641	0.002	0.006	0.003	0.016	0.015	0.005
Malawi	4.434	6.421	7.632	4.066	2.221	-	0.370	0.535	0.636	0.339	0.185	-
Malaysia	190	213	247	300	251	-	0.145	0.161	0.190	0.214	0.179	-
Malta	3.303	10	12	7.937	17	-	0.009	0.028	0.029	0.020	0.043	-
Mexico	222	305	-	-	-	-	0.532	0.763	-	-	-	-
Mongolia	253	140	38	7.599	14	-	3.840	2.155	0.511	0.096	0.097	-
Morocco	54	168	182	258	192	171	0.057	0.113	0.121	0.171	0.127	0.109
Nepal	46	-	-	-	-	-	0.119	-	-	-	-	-
Netherlands	1,158	1,181	1,336	1,373	-	-	1.410	1.383	1.572	1.595	-	-
New Zealand	219	-	-	-	-	-	0.140	-	-	-	-	-
Norway	336	313	304	237	274	-	0.245	0.232	0.343	0.275	0.309	-
Pakistan	-	81	-	-	-	-	-	0.095	-	-	-	-
Panama	15	22	-	-	-	-	0.050	0.074	-	-	-	-
Peru	94	52	52	-	-	-	0.035	0.018	0.017	-	-	-
Philippines	235	259	286	288	-	-	0.006	0.007	0.101	0.107	-	-
Poland	588	738	706	459	-	-	0.646	0.679	0.614	0.377	-	-
Portugal	367	375	422	262	484	-	0.050	0.052	0.062	0.039	0.065	-
Romania	-	-	-	117	102	-	-	-	-	0.032	0.016	-
Senegal	179	-	-	-	-	-	2.987	-	-	-	-	-
Singapore	63	100	117	93	114	-	0.238	0.370	0.400	0.316	0.375	-
Slovenia	59	54	35	34	45	65	0.160	0.066	0.037	0.039	0.042	0.077
Spain	1,054	1,193	1,278	1,565	1,300	1,618	0.030	0.034	0.038	-	-	-
Sri Lanka	20	14	23	12	-	-	0.033	0.027	0.042	0.022	-	-
Swaziland	9.933	-	-	-	-	-	1.104	-	-	-	-	-
Thailand	3,098	636	-	-	-	-	0.737	0.233	-	-	-	-
Trinidad and Tobago	12	15	15	11	7.038	-	0.123	0.037	0.039	0.027	0.015	-
Tunisia	69	78	86	82	97	118	-	-	-	0.026	0.037	0.047
Turkey	248	252	295	410	231	-	0.290	0.307	0.146	0.225	0.133	-
Ukraine	382	376	494	166	168	56	0.017	0.017	0.022	0.007	0.006	0.002
United Kingdom	2,668	2,761	2,823	-	2,490	-	0.306	0.325	0.338	-	0.273	-
United Republic of Tanzania	56	83	-	-	-	-	0.379	0.582	-	-	-	-
United States of America	7,690	7,900	8,438	8,066	8,493	10,058	-	-	0.338	-	-	-
Uruguay	38	35	53	54	-	-	-	-	-	-	-	-
Venezuela	152	149	212	175	164	125	0.063	0.058	0.083	0.083	0.077	0.056
Yugoslavia	14	13	33	-	63	33	0.023	0.014	0.018	-	0.024	0.011
Zambia	19	-	-	-	-	-	0.181	-	-	-	-	-
Zimbabwe	36	169	53	75	55	-	0.368	1.609	0.530	0.791	0.553	-

Indexes of Industrial Production

Country	Index of Industrial Production (1990=100)						Index of Employment (1990=100)					
	1990	1991	1992	1993	1994	1995	1990	1991	1992	1993	1994	1995
Albania	100	-	-	-	-	-	-	-	-	-	-	-
Algeria	100	99	94	99	95	88	-	-	-	-	-	-
Angola	100	119	100	80	-	-	-	-	-	-	-	-
Argentina	100	109	-	-	-	-	100	-	-	96	101	-
Armenia	100	-	-	-	-	-	100	95	120	101	101	95
Australia	100	103	111	119	-	-	100	99	97	-	-	-
Austria	100	106	107	108	112	113	100	97	97	95	93	93

Continued.

Depending on the table, * means *Enterprises, Engaged,* or *Factor Values*; ± means *Basis Unspecified*; § means *shown in millions*. For additional notes and sources, see Appendix I.

Indexes of Industrial Production

- Continued -

Country	Index of Industrial Production (1990=100)						Index of Employment (1990=100)					
	1990	1991	1992	1993	1994	1995	1990	1991	1992	1993	1994	1995
Azerbaijan	-	-	-	-	-	-	100	103	104	101	99	-
Bahamas	100	-	-	-	-	-	100	88	-	-	-	-
Bahrain	100	-	-	-	-	-	-	-	-	-	-	-
Bangladesh	100	115	108	123	135	136	100	94	96	-	-	-
Barbados	100	108	103	104	107	111	100	75	70	89	118	108
Belgium	100	103	105	106	105	110	100	100	99	-	-	-
Belize	100	102	100	100	105	105	-	-	-	-	-	-
Benin	100	97	-	-	-	-	-	-	-	-	-	-
Bermuda	-	-	-	-	-	-	100	94	86	78	81	87
Bhutan	100	-	-	-	-	-	-	-	-	-	-	-
Bolivia	100	111	110	117	121	128	100	106	112	111	114	116
Bosnia & Herzegovina	-	-	-	-	-	-	100	93	-	-	9	-
Botswana	100	94	93	-	-	-	100	100	94	98	100	105
Brazil	100	104	104	104	106	115	100	-	99	94	-	-
Bulgaria	100	79	68	52	51	52	100	82	70	53	56	49
Burkina-Faso	100	116	101	-	-	-	-	-	-	-	-	-
Burundi	100	96	101	101	-	-	100	115	-	-	-	-
Cambodia	100	102	107	-	-	-	100	127	-	-	-	-
Cameroon	100	98	90	-	-	-	100	114	110	110	106	109
Canada	100	102	104	104	108	109	100	97	99	96	96	96
Central African Republic	100	102	103	-	-	-	100	-	-	-	-	-
Chad	100	82	88	-	-	-	-	-	-	-	-	-
Chile	100	101	119	123	137	143	100	107	112	111	115	122
China	100	-	-	-	-	-	100	105	107	103	104	106
China (Hong Kong SAR)	100	104	109	112	114	113	100	101	99	97	90	-
China (Taiwan Province)	100	104	109	111	117	118	100	99	101	101	103	101
Colombia	100	93	101	105	103	106	100	102	130	128	133	135
Congo	100	59	82	85	-	-	-	-	-	-	-	-
Costa Rica	100	102	113	123	127	133	100	99	109	112	-	-
Cote d'Ivoire	100	106	138	167	144	111	-	-	-	-	-	-
Croatia	100	-	-	-	-	-	100	93	86	81	81	81
Cuba	100	85	85	50	-	-	-	-	-	-	-	-
Cyprus	100	98	104	110	117	121	100	99	102	101	109	121
Czechoslovakia (Former)	100	79	69	-	-	-	100	85	-	-	-	-
Czech Republic	100	-	-	-	-	-	100	85	81	79	-	-
Dem. Rep. of the Congo	100	105	87	-	-	-	-	-	-	-	-	-
Denmark	100	105	105	104	109	104	100	98	95	95	97	95
Dominican Republic	100	99	104	-	-	-	-	-	-	-	-	-
Ecuador	100	109	109	108	-	-	100	121	135	134	136	127
Egypt	100	101	103	124	132	170	100	119	95	94	96	108
El Salvador	100	106	-	-	-	-	-	-	-	-	-	-
Equatorial Guinea	-	-	-	-	-	-	100	-	-	-	-	-
Ethiopia and Eritrea	100	92	96	-	-	-	-	-	-	-	-	-
Ethiopia	-	-	-	-	-	-	100	102	95	97	107	100
Fiji	100	103	107	106	116	112	100	75	103	103	108	-
Finland	100	99	100	103	106	109	100	96	89	83	78	83
FYR Macedonia	100	-	-	-	-	-	100	117	117	113	106	109
France	100	102	104	105	106	110	100	99	99	97	96	96
Gabon	100	101	87	-	-	-	-	-	-	-	-	-
Gambia	100	-	-	-	-	-	-	-	-	-	-	-
Germany (Eastern Part)	100	-	-	-	-	-	-	-	-	-	-	-
Germany (Western Part)	100	107	106	107	109	112	100	107	107	-	-	-
Ghana	100	-	-	-	-	-	-	-	-	-	-	-
Greece	100	110	118	117	116	118	100	96	99	98	98	99
Guatemala	100	98	99	-	-	-	-	-	-	-	-	-
Guyana	100	123	187	187	195	192	-	-	-	-	-	-
Haiti	100	65	46	39	39	39	-	-	-	-	-	-
Honduras	100	125	134	147	143	258	100	117	128	140	154	169
Hungary	100	97	85	82	85	88	100	97	94	79	80	80
Iceland	100	82	-	-	-	-	100	97	89	89	93	-

Continued.

Indexes of Industrial Production

- Continued -

Country	Index of Industrial Production (1990=100)						Index of Employment (1990=100)					
	1990	1991	1992	1993	1994	1995	1990	1991	1992	1993	1994	1995
India	100	104	106	100	102	120	100	100	111	108	109	111
Indonesia	100	108	113	152	201	237	100	112	119	131	126	132
Iran (Islamic Republic of)	100	109	124	131	136	151	100	105	107	106	-	-
Iraq	100	77	91	-	-	-	-	-	-	-	-	-
Ireland	100	104	114	120	129	143	100	102	99	97	94	91
Israel	100	102	105	113	120	130	100	104	103	108	108	113
Italy	100	101	103	103	-	-	100	101	117	115	113	-
Jamaica	100	108	102	98	100		100	100	102	101	93	-
Japan	100	103	103	103	100	101	100	101	102	104	100	102
Jordan	100	84	137	122	114	129	100	117	132	136	169	181
Kenya	100	102	98	98	98	114	100	104	104	108	112	118
Korea, Republic of	100	109	111	113	122	123	100	99	98	99	100	101
Kuwait	100	84	71	-	-	-	100	62	77	87	87	86
Kyrgyzstan	100	-	-	-	-	-	100	99	93	89	77	-
Lao P.D.R.	100	107	117	-	-	-	-	-	-	-	-	-
Latvia	100	-	-	-	-	-	100	96	138	100	86	83
Lebanon	100	125	124	-	-	-	-	-	-	-	-	-
Lesotho	100	104	100	-	-	-	-	-	-	-	-	-
Liberia	100	104	-	-	-	-	-	-	-	-	-	-
Libyan Arab Jamahiriya	100	118	121	-	-	-	-	-	-	-	-	-
Lithuania	100	-	-	-	-	-	-	-	-	-	-	-
Luxembourg	100	100	97	101	96	98	100	110	113	120	116	-
Macau	100	100	100	-	-	-	100	92	79	86	100	89
Madagascar	100	82	83	89	-	-	-	-	-	-	-	-
Malawi	100	92	98	100	101	96	100	53	57	59	71	-
Malaysia	100	96	102	112	120	128	100	107	106	111	113	117
Mali	100	87	105	93	124	93	-	-	-	-	-	-
Malta	100	104	112	116	-	-	100	100	101	98	98	-
Mauritius	100	98	103	91	80	94	100	112	112	109	109	109
Mexico	100	104	105	106	104	105	100	99	101	98	91	88
Mongolia	-	-	-	-	-	-	100	99	77	81	114	86
Morocco	100	104	104	115	120	119	100	264	411	451	439	441
Mozambique	100	96	77	-	-	-	100	89	70	68	61	-
Myanmar	100	93	130	-	-	-	-	-	-	-	-	-
Namibia	100	109	102	-	-	-	-	-	-	-	-	-
Nepal	100	109	117	-	-	-	100	125	-	153	145	-
Netherlands	100	102	105	105	109	108	100	102	102	100	98	99
New Zealand	100	111	115	-	-	-	100	99	102	104	-	-
Nicaragua	100	103	104	-	-	-	-	-	-	-	-	-
Niger	100	100	103	-	-	-	-	-	-	-	-	-
Nigeria	100	106	132	-	-	-	-	-	-	-	-	-
Norway	100	104	105	106	116	118	100	99	91	90	93	93
Pakistan	100	114	116	134	131	139	100	107	112	-	-	-
Panama	100	105	111	116	125	136	100	105	113	120	133	150
Papua New Guinea	100	100	105	-	-	-	-	-	-	-	-	-
Paraguay	100	97	107	101	107	110	-	-	-	-	-	-
Peru	100	97	101	101	121	128	100	98	100	-	-	-
Philippines	100	114	123	128	147	158	100	96	60	61	-	-
Poland	100	97	109	118	133	146	100	98	95	89	92	95
Portugal	100	100	93	96	94	97	100	103	97	98	104	105
Puerto Rico	100	108	97	95	74	-	100	81	88	86	85	87
Qatar	100	-	-	-	-	-	100	103	122	118	122	-
Republic of Moldova	100	-	-	-	-	-	100	99	114	88	82	81
Romania	100	83	69	59	67	69	-	-	-	-	-	-
Russian Federation	100	-	-	-	-	-	-	-	-	-	-	-
Saudi Arabia	100	100	102	-	-	-	-	-	-	-	-	-
Senegal	100	58	82	81	92	106	100	104	77	74	85	-
Sierra Leone	100	103	-	-	-	-	-	-	-	-	-	-
Singapore	100	105	108	112	112	113	100	105	109	112	111	112
Slovakia	100	-	-	-	-	-	-	-	-	-	-	-

Continued.

Depending on the table, * means *Enterprises, Engaged,* or *Factor Values;* ± means *Basis Unspecified;* § means *shown in millions.* For additional notes and sources, see Appendix I.

Indexes of Industrial Production

- Continued -

Country	Index of Industrial Production (1990=100)						Index of Employment (1990=100)					
	1990	1991	1992	1993	1994	1995	1990	1991	1992	1993	1994	1995
Siovenia	100	96	78	76	79	80	-	-	-	-	-	-
Somalia	100	69	31	-	-	-	-	-	-	-	-	-
South Africa	100	99	100	98	94	94	100	97	97	98	88	88
Spain	100	103	99	100	104	102	100	102	102	113	114	112
Sri Lanka	100	102	98	117	125	311	100	86	101	88	-	-
Sudan	100	106	116	-	-	-	-	-	-	-	-	-
Suriname	100	67	69	-	-	-	100	102	98	89	-	-
Swaziland	100	93	98	-	-	-	100	-	-	-	-	-
Sweden	100	97	98	102	107	110	100	102	96	91	89	87
Switzerland	100	100	99	101	100	102	-	-	-	-	-	-
Syrian Arab Republic	100	109	111	105	111	118	-	-	-	-	-	-
Thailand	100	84	99	97	90	103	100	107	-	-	-	-
Tonga	100	-	-	-	-	-	100	111	-	-	-	-
Trinidad and Tobago	100	103	107	99	106	108	100	114	115	130	133	-
Tunisia	100	102	107	107	112	113	-	-	-	-	-	-
Turkey	100	107	105	113	115	115	100	96	107	103	100	100
Uganda	100	130	140	141	177	207	-	-	-	-	-	-
Ukraine	100	-	-	-	-	-	100	96	90	89	86	81
United Arab Emirates	100	102	104	-	-	-	-	-	-	-	-	-
United Kingdom	100	100	101	102	103	104	100	97	96	97	96	96
United Republic of Tanzania	100	120	111	94	81	-	100	100	-	-	-	-
USSR (Former)	100	-	-	-	-	-	100	-	-	-	-	-
United States of America	100	102	103	105	107	109	100	100	103	104	104	105
Uruguay	100	101	104	100	108	110	100	97	90	82	83	83
Venezuela	100	-	-	-	-	-	100	107	108	101	105	109
Yemen	100	101	103	-	-	-	-	-	-	-	-	-
Yugoslavia	100	-	-	-	-	-	-	-	-	-	-	-
Yugoslavia (Former)	100	80	-	-	-	-	-	-	-	-	-	-
Zambia	100	103	129	125	122	121	100	-	-	-	68	-
Zimbabwe	100	102	104	85	90	99	100	107	105	98	95	100

Depending on the table, * means *Enterprises, Engaged,* or *Factor Values*; ± means *Basis Unspecified*; § means *shown in millions*. For additional notes and sources, see Appendix I.

17

Representative Companies in Sector

Name	Address	Tele-phone	Fax	Employ-ment	Y
Canada Safeway Ltd.	1020 64 Ave. NE, Calgary, Alberta T2E 7V8, Canada	403-730-3500	403-730-3888	30,000	98
Maple Leaf Foods Inc.	1500-30 St. Clair Ave. W, Toronto, Ontario M4V 3A2, Canada	416-926-2000	416-926-2018	12,000	98
Highlands & Lowlands BHD	Damansara Heights, Kuala Lumpur 50490, Malaysia	3 2541644	3 2557934	9,556	92
Laemthong Corp. Ltd.	Phyathai, Bangkok 10400, Thailand	2523777	2359922	9,100	94
Gerber Products Co.	445 State St., Fremont, Michigan 49413, U.S.A.	616-928-2000	616-928-2963	9,000	97
Hindustan Lever Ltd.	165/166 Backbay Reclamation, Bombay 400 020, India	22 2870622		9,000	94
Hunt-Wesson Inc.	1645 W. Valencia Dr., Fullerton, California 92833, U.S.A.	714-680-1000	714-449-5119	9,000	97
British Bakeries Ltd.	27-30 King Edward Ct., Windsor SL4 1TJ, United Kingdom	753 857123	753 846537	8,860	93
WLR Foods Inc.	PO Box 7000, Broadway, Virginia 22815-7000, U.S.A.	540-896-7001	540-896-0498	8,500	97
Wm. Wrigley Jr. Co.	410 N. Michigan Ave., Chicago, Illinois 60611, U.S.A.	312-644-2121		8,200	97
National Starch and Chemical Co.	10 Finderne Ave., Bridgewater, New Jersey 08807, U.S.A.	908-685-5000	908-685-5300	8,000	97
J. Bibby & Sons PLC	16 Stratford Pl., London W1N 9AF, United Kingdom	71 6296243	71 4090556	7,976	93
Unicord Public Co. Ltd.	Pathumuran, Bangkok 10330, Thailand	2160200	2161468	7,822	94
McCormick and Company Inc. (Sparks, Maryland)	PO Box 6000, Sparks, Maryland 21152-6000, U.S.A.	410-771-7301	410-771-7462	7,500	97
Swift Armour SA Industria e Comercio	Rua Irineu Jose Bordon 215, Sao Paulo 05120-060, Brazil	11 8358511	11 8354326	7,500	97
Flowers Industries Inc.	PO Box 1338, Thomasville, Georgia 31792-1338, U.S.A.	912-226-9110	912-225-3808	7,300	97
International MultiFoods Corp.	PO Box 2942, Minneapolis, Minnesota 55402-0942, U.S.A.	612-340-3300		7,176	96
Irvin & Johnson Ltd.	70 Prestwich St., Cape Town 8001, Republic of South Africa	21 4029200	21 4029282	7,000	97
Lotte Confectionery Co., Ltd.	Yongdungpo-Gu, Seoul, Republic of Korea	2 6706331	2 6706600	7,000	97
State Meat and Fish Institution	Ankara, Turkey	4 3411100	4 3412212	6,780	89
Ross Young's Holdings Ltd.	Ross House, Grimsby DN31 3SW, United Kingdom	472 35911	472 240640	6,700	87
Friesland Frico Domo Cooperatie BA	Pieter Stuyvesantweg 1, Leeuwarden NL-8937 AC, Netherlands	58 993200	58 993299	6,500	94
Frigobras Cia. Brasileira de Frigorificos	Av. Senador Attillio Fontana 86, Concordia 89700-000, Brazil	11 79264432	11 79264510	6,500	96
Lancaster Colony Corp.	37 W. Broad St., Columbus, Ohio 43215, U.S.A.	614-224-7141	614-469-8219	6,400	97
Convenience Foods Ltd.	Lenton Ln., Nottingham NG7 2NS, United Kingdom	603 868231	602 866247	6,170	93
Sanderson Farms Inc.	PO Box 988, Laurel, Mississippi 39441-0988, U.S.A.	601-649-4030	601-426-1461	6,155	97
Ozeta Trencin	9 Maja 9, Trencin 911 34, Slovakia	831 262800	831 23240	6,045	92
Greggs PLC	Jesmond, Newcastle upon Tyne NE2 1TL, United Kingdom	91 2817721	91 2811444	6,008	94
Cadbury Schweppes Australia Ltd.	Goring-by-Sea, Worthing BN12 6DA, United Kingdom			6,000	90
Langeberg Foods International	PO Box 1055, Bellville 7535, Republic of South Africa	21 9463400	21 9499302	6,000	94
Nabisco Biscuit Co.	PO Box 1911, East Hanover, New Jersey 07936, U.S.A.	201-503-2000		6,000	
Nong Shim Co., Ltd.	Yongsan-gu, Seoul, Republic of Korea	2 32701110	2 7165901	6,000	97
Express Foods Group International Ltd.	Fallside, Glasgow G71 8BG, United Kingdom	698 817326		5,931	
Moran Holdings PLC	Arklow Rd., London SE14 6EB, United Kingdom	81 6948664	81 6948594	5,791	90
Hoops Ltd.	3d Dundee Rd., Slough SL1 4LG, United Kingdom	753 693000	753 533172	5,758	94
MD Foods Amba	Skanderborgvej 277, Viby DK-8260, Denmark	86281000	86281691	5,700	93
Crompton and Knowles Corp.	1 Station Pl., Metro Ctr., Stamford, Connecticut 06902, U.S.A.	203-353-5400	203-353-5424	5,665	96
Golden Poultry Company Inc.	PO Box 2210, Atlanta, Georgia 30301, U.S.A.	404-393-5000	404-393-5347	5,658	96
Crown Corp. Ltd.	PO Box 3148, Wellington, New Zealand	4 724099		5,600	92
Cadbury Ltd.	PO Box 12, Birmingham B30 2NA, United Kingdom	21 4582000		5,590	94
Danish Crown Amba	Marsvej 43, Randers DK-8900, Denmark	89191919	86448066	5,500	93
Riser Foods Inc.	5300 Richmond Rd., Bedford Heights, Ohio 44146, U.S.A.	216-292-7000	216-591-2640	5,500	96
Sasko Pty. Ltd.	PO Box 20, Huguenot 7645, Republic of South Africa	2211 23061	2211 25361	5,500	93
TVK Rt.	PO Box 20, Tiszaujvaros H-3581, Hungary	49 22222	49 21322	5,500	93
F.W. Farnsworth Ltd.	Lenton Lane, Nottingham NG7 2NS, United Kingdom	159 868231	159 866247	5,378	
Clorox Co.	1221 Broadway, Oakland, California 94612, U.S.A.	510-271-7291	510-832-1463	5,300	97
Bangkok Produce Merchandising Co., Ltd.	C.P. Tower, 18th Fl., Bangkok 10500, Thailand	2310221	2310283	5,220	94
Siam Agro Industry (Pineapple and Others) Co., Ltd.	Phrakhanong, Bangkok 10260, Thailand	3961692	3983217	5,217	94
Japfa Comfeed Indonesia PT	Jl. Let Jen Mt Haryono Kav 16, Jakarta 12810, Indonesia	218310310	21 8310309	5,132	97
Rich Products Corp.	1150 Niagara St., Buffalo, New York 14213, U.S.A.	716-878-8000	716-878-8130	5,100	97
Babolna Mezogazdasagi Termelo, Fejleszto Es Kereskedelmi Rt.	Meszaros u. 1, Babolna H-2943, Hungary	34 69111	34 69002	5,000	93
Bajaj Hindustan Ltd.	226 Nariman Pt., Bombay 400 021, India	22 2023626	2021977	5,000	

Continued

Representative Companies in Sector
- Continued -

Name	Address	Tele-phone	Fax	Employ-ment	Y
Lamb-Weston Inc.	PO Box 1900, Pasco, Washington 99302-1900, U.S.A.	509-735-4651	509-736-0399	5,000	97
Parmalat Dairy & Bakery Inc.	405 The West Mall Floor 10, Toronto, Ontario M9C 5J1, Canada	416-626-1973	416-620-3123	5,000	98
Schwan's Sales Enterprises Inc.	115 W. College Dr., Marshall, Minnesota 56258, U.S.A.	507-532-3274	507-537-8450	5,000	95
Sucocitrico Cutrale SA	Av. Da Marginal 3800, 1st Fl., Araquara 14807, Brazil	162 221311	162 226099	5,000	96
Premier Brands Ltd.	11 Walker St., Edinburgh EH3 7NE, United Kingdom			4,911	91
Budgens PLC	Booth Dr., Wellingborough NN8 6GR, United Kingdom	933 400700		4,752	94
Best Foods Div.	PO Box 8000, Englewood Cliffs, New Jersey 07632, U.S.A.	201-894-4000	201-894-2186	4,712	97
Clorox Co.	1221 Broadway Ave., Oakland, California 94612, U.S.A.	510-271-7000	510-832-1463	4,700	93
Pepsico Holdings Ltd.	63 Kew Rd., Richmond SW3 3AA, United Kingdom	81 3320332		4,613	93
Molinos Rio De La Plata SA	Av. Paseo Colon 746, 3rd Fl., Buenos Aires 1063, Argentina	1 3310032	1 3315948	4,600	96
Dalgety Spillers Foods Ltd.	Station Rd., Cambridge CB1 2JN, United Kingdom	223 301132	71 4930892	4,541	93
Lance Inc.	PO Box 32368, Charlotte, North Carolina 28232-2368, U.S.A.	704-554-1421	704-554-5562	4,541	97
Betagro Co. Ltd.	Suan Mali, Bangkok 10100, Thailand	2231371	2257971	4,500	94
Josip Kras	Ravnice bb, Zagreb CT-41000, Croatia	41 215145	41 229899	4,500	90
Kraft Canada Inc.	95 Moatfield Dr., Don Mills, Ontario M3B 3L6, Canada	416-441-5000	416-441-5315	4,500	98
Paterson Zochonis PLC	60 Whitworth St., Manchester M1 6LU, United Kingdom	61 2367111	61 2286719	4,377	94
Agrovale Agro Industrias Do Vale Do Rio Sao Francisco SA	Fazenda Massayao s/n, Juazeiro 48900, Brazil	75 8116008	75 8116008	4,300	97
Asea Brown Boveri AS	Petersmindevej 1, Odense DK-5000, Denmark	66147080	66142580	4,300	93
Corn Products International Inc.	PO Box 345, Argo, Illinois 60501, U.S.A.	708-563-2400		4,300	96
Tropicana Dole Beverages North America	PO Box 338, Bradenton, Florida 34206-0338, U.S.A.	941-747-4461	941-742-3517	4,200	97
Universal Foods Corp.	433 E. Michigan St., Milwaukee, Wisconsin 53202-5106, U.S.A.	414-271-6755	414-347-3785	4,127	97
Marshall Food Group Ltd.	Newbridge, Midlothian EH28 8SW, United Kingdom	31 3333341	31 3333341	4,098	93
Indocement Tunggal Prakarsabogasari Flour Mills Div. PT	Jl. Jendral Sudirman Kav 70-71, Jakarta 12910, Indonesia	212512087	21 5712620	4,085	97
Cargill Agricola SA	Sao Paulo, Sao Paulo 04671-050, Brazil	11 56943388	11 56943380	4,019	97
Foodarama Supermarkets Inc.	922 Hwy. 33, Bldg. 6, Freehold, New Jersey 07728, U.S.A.	732-462-4700		4,000	97
GFPT Co., Ltd.	Ratburana, Bangkok 10140, Thailand	4630040	4635751	4,000	94
National Agricultural and Food Corp.	PO Box 903, Dar es Salaam, United Republic of Tanzania	25961		4,000	91
Nestle Canada Inc.	25 Sheppard Ave. W, North York, Ontario M2M 6S8, Canada	416-512-9000	416-218-2657	4,000	98
Thorn Apple Valley Inc.	26999 Central Park Blvd., Southfield, Michigan 48076, U.S.A.	810-213-1000		4,000	96
Wampler-Longacre Chicken Inc.	PO Box 7000, Broadway, Virginia 22815, U.S.A.	540-896-7000	540-896-0498	4,000	93
Northern Foods Grocery Group Ltd.	Lenton Ln., Nottingham NG7 2NS, United Kingdom	159 868231	159 866247	3,970	93
Tong Yang Confectionery Corp.	Yongsan-Gu, Seoul, Republic of Korea	2 7106000	2 7152582	3,850	97
Unilever Canada Ltd.	160 Bloor St. E Ste. 1500, Toronto, Ontario M4W 3R2, Canada	416-964-1857	416-964-8831	3,816	98
Elite Industries Ltd.	84 Arlozorov St., Ramat Gan 52505, Israel	3 6752111	3 6731110	3,800	97
National Beef Packing Company L.P.	10100 N. Executive Hills Blvd.,, Kansas City, Missouri 64153, U.S.A.	816-891-5900	816-891-5916	3,750	97
Arisco Produtos Alimenticios Ltda.	Parque Ipe, Goiania 74055-020, Brazil	62 2051800	62 2051342	3,700	96
Express Dairy Ltd.	430 Victoria Rd., South Ruislip, United Kingdom	81 8426000	81 8420842	3,681	93
Dongwon Industries Co., Ltd.	Socho-Gu, Seoul, Republic of Korea	2 5893000	2 5893291	3,600	97
Cargill Ltd.	300-240 Graham Ave., Winnipeg, Manitoba R3C 0J7, Canada	204-947-0141	204-947-6222	3,500	98
Dreyer's Grand Ice Cream Inc.	5929 College Ave., Oakland, California 94618, U.S.A.	510-652-8187	510-601-4905	3,500	97
Industrias Alimenticias Noel SA	Av. Guayabal, Medellin, Colombia	4 2851111	4 2853431	3,500	96
Royal Greenland Export	PO Box 8220, Alborg DK-9220, Denmark	98154400	98154420	3,500	93
Takovo	Bul. Jojuod Misica 14, Gornji Milanovac YU-32300, Serbia	32 710380	32 710391	3,500	95
Tulip International AS	Gunnar Clausens Vej 13, Viby DK-8250, Denmark	86282200	86287210	3,500	93
Pauls PLC	47 Key St., Ipswich IP4 1BX, United Kingdom	473 232222	473 230509	3,475	93
United Distillers PLC	33 Ellersly Rd., Edinburgh EH1R GJW, United Kingdom	31 3377373	31 3370163	3,406	93
Bonnita Pty. Ltd.	PO Box 809, Stellenbosch 7599, Republic of South Africa	21 8866640	21 8866905	3,400	96
G.W. Padley Holdings Ltd.	Anwick, Sleaford NG34 9SL, United Kingdom	526 8322661		3,394	93
Curtice-Burns Foods Inc.	PO Box 682, Rochester, New York 14603-0681, U.S.A.	716-383-1850	716-383-1281	3,368	97
General Foods Ltd.	Ruscote Ave., Banbury OX16 7QU, United Kingdom	295 264433	295 259018	3,347	93
MD Foods PLC	Kirkstall Rd., Leeds LS3 1JE, United Kingdom	13 2440141		3,313	93
Saskatchewan Wheat Pool	2625 Victoria Ave., Regina, Saskatchewan S4T 7T9, Canada	306-569-4411	306-569-4708	3,300	98
UniMark Group Inc.	PO Box 229, Argyle, Texas 76226, U.S.A.	817-491-2992	817-491-1272	3,300	95

Continued

Representative Companies in Sector
- Continued -

Name	Address	Tele-phone	Fax	Employ-ment	Y
Interbake Foods Inc.	2821 Emerywood Pkwy., 210, Richmond, Virginia 23230, U.S.A.	804-755-7107	804-359-8825	3,250	97
Mars GB Ltd.	3D Dundэe Rd. Trading Estate, Slough SL1 4LG, United Kingdom	753 693000	753 35396	3,196	94
Foodbrands America Inc.	1601 Northwest Expwy., Ste. 1700, Oklahoma City, Oklahoma 73118, U.S.A.	405-879-4100		3,193	95
Agrifoods International Cooperative Ltd.	3920 Norland Ave. Ste. 300, Burnaby, British Columbia V5G 4K7, Canada	604-298-9600		3,100	98
Cagle's Inc.	2000 Hills Ave. N.W., Atlanta, Georgia 30318, U.S.A.	404-355-2820	404-355-9326	3,100	96
Quintie Confectionery Ltd.	Budafoki ut 64, Budapest H-1117, Hungary	1 1613215	1 1612065	3,100	93
Thai Pineapple Co., Ltd.	Phayathai, Bangkok 10400, Thailand	2710041	2714304	3,047	94
Alpina Productos Alimenticios SA	Carrera 63 No. 15-61, Bogota, Colombia	1 4140011	1 4141480	3,000	98
Beacon Sweets & Chocolates Pty. Ltd.	Jacobs, Durban 4052, Republic of South Africa	31 4607170	31 487583	3,000	96
Chocolates Garoto SA	Praca Meyerfreund 01, Villa Bella 29122-900, Brazil	27 3393999	27 3395397	3,000	96
Culinar Inc., Culinar Canada Division	2700 2 Complexe Desjardins, Montreal, Quebec H5B 1B2, Canada	514-288-3101	514-288-3353	3,000	98
Dole Thailand Ltd.	1126/1 New Petchburi Rd., Bangkok 10400, Thailand	2 2500813	662 2534232	3,000	93
Knott's Berry Farm Foods Inc.	8039 Beach Blvd., Buena Park, California 90620, U.S.A.	714-827-1776		3,000	97
Lykes Meat Group Inc.	PO. Box 1690, Tampa, Florida 33601, U.S.A.	813-223-3981	813-273-5421	3,000	95
Oscar Mayer Foods Corp.	910 Mayer Ave., Madison, Wisconsin 53704, U.S.A.	608-241-3311		3,000	97
President Baking Company Inc.	41 Perimeter Ctr. E., 400, Atlanta, Georgia 30346, U.S.A.	770-673-8600		3,000	95
Procter & Gamble Inc.	4711 Yonge St., North York, Ontario M2N 6K8, Canada	416-730-4711	416-730-4415	3,000	98
Pulau Sambu PT	Pusat, Jakarta, Indonesia	216690208	21 6690412	3,000	97
Stroehmann Bakeries L.C.	PO Box 976, Horsham, Pennsylvania 19044, U.S.A.	215-672-8010	215-672-6988	3,000	95
Alexander and Baldwin Inc.	PO Box 3440, Honolulu, Hawaii 96801-3440, U.S.A.	808-525-6611	808-525-6652	2,960	97
Skogaholms Brod, AB	Box 5025, Eskilstuna S-630 05, Sweden	16 161300	16 135753	2,900	93
Thorntons PLC	Somercotes, Derby DE55 4XJ, United Kingdom	773 540550	773 540757	2,832	94
Kellogg Co. of Great Britain, Ltd.	Talbot Rd., Manchester M16 0PU, United Kingdom	61 8692000	61 8692100	2,831	94
Jacob's Bakery Ltd.	Earley, Reading RG6 1AZ, United Kingdom	734 492000	734 492400	2,811	92
Brake Bros. Foodservice Ltd.	Godinton Rd., Ashford TN23 1EU, United Kingdom	233 637370	233 648200	2,808	93
Citrosuco Paulista SA	Rua Joao Pessoa 305, Centro, Matoa 15990, Brazil	162 821711	162 824378	2,800	96
Osem Investments Ltd.	61 Jabotinsky St., Petah Tikva 49517, Israel	3 9265515	3 9265216	2,800	97
Wampler Foods Inc.	PO Box 7275, Broadway, Virginia 22815-7275, U.S.A.	540-896-7000	540-896-5127	2,800	96
Pataling Rubber Estates Ltd.	1 Great Tower St., London EC3R 5AH, United Kingdom			2,757	94
Riviana Foods Inc.	PO Box 2636, Houston, Texas 77252-2141, U.S.A.	713-529-3251	713-529-1866	2,751	97
London Asiatic Rubber & Produce Co. Ltd.	1 Great Tower St., London EC3R 5AH, United Kingdom	71 6264333	71 7820112	2,748	94
Burtons Gold Medal Biscuits Ltd.	Market St., Bracknell RG12 1LF, United Kingdom	344 412121	344 861973	2,742	93
Conaprole Cooperativa Nacional de Productores de Leche SA	Magallanes 1871, Montevideo, Uruguay	2 946733	2 946672	2,700	96
Michael Foods Inc.	5353 Wayzata Blvd., Ste. 324, Minneapolis, Minnesota 55416, U.S.A.	612-546-1500	612-540-9100	2,700	96
Sam Yang Foods Co., Ltd.	Chongno-Gu, Seoul, Republic of Korea	2 7358951	2 7336180	2,700	97
Stokely USA Inc.	PO Box 248, Oconomowoc, Wisconsin 53066-0248, U.S.A.	414-569-1800		2,700	96
Mastellone Hnos. SA	Av. Leandro N Alem 720, Buenos Aires 1001, Argentina	1 3185000	1 3121560	2,600	96
Valhi Inc.	5430 Lyndon B. Johnson Fwy., 17, Dallas, Texas 75240-2697, U.S.A.	972-233-1700	972-385-0586	2,600	97
Seneca Foods Corp.	1162 Pittsford-Victor Rd., Pittsford, New York 14534, U.S.A.	716-385-9500	716-385-4249	2,572	97
Brooke Bond Foods Ltd.	High St., Croydon CR9 1JQ, United Kingdom	81 6868899	81 7609210	2,537	93
Manor Bakeries Ltd.	PO Box 527, Windsor SL4 1TJ, United Kingdom	753 840401		2,521	93
Bakers	PO Box 724, Westville 3630, Republic of South Africa	31 2669323	31 862404	2,500	94
Bil Mar Foods Inc.	8300 96th Ave., Zeeland, Michigan 49464, U.S.A.	616-875-7711	616-875-7591	2,500	97
Chiquita Processed Food, LLC	PO Box 129, New Richmond, Wisconsin 54017, U.S.A.	715-246-2241	715-243-7330	2,500	94
Good Humor Ice Cream	PO Box 19007, Green Bay, Wisconsin 54307-9007, U.S.A.	920-499-5151		2,500	97
Hill's Pet Nutrition Inc.	PO Box 148, Topeka, Kansas 66601, U.S.A.	913-354-8523	913-231-5770	2,500	97
Icicle Seafoods Inc.	PO Box 79003, Seattle, Washington 98119, U.S.A.	206-282-0988	206-282-7222	2,500	95
Leo Pharmaceutical Products	Industriparken 55, Ballerup DK-2750, Denmark	44923800	44941516	2,500	93
Leprino Foods Co.	PO Box 173400, Denver, Colorado 80217, U.S.A.	303-480-2600	303-480-0260	2,500	95
McCain Foods Ltd.	Main Rd., Florenceville, New Brunswick E0J 1K0, Canada	506-392-5541	506-392-8156	2,500	98

Product Tables
Beef and Veal, Fresh (Industrial Production)

Unit of Measure: Metric tons.

	1990 Value	%	1991 Value	%	1992 Value	%	1993 Value	%	1994 Value	%	1995 Value	%	1996 Value	%
Total Production	35,099,979	100.0	36,918,572	100.0	36,080,604	100.0	35,105,503	100.0	35,163,080	100.0	35,715,652	100.0	36,054,393	100.0
Regions														
Africa	456,000	1.3	486,000	1.3	503,000	1.4	483,000	1.4	384,000	1.1	366,000	1.0	362,000	1.0
America, North	10,581,999	30.1	10,648,051	28.8	10,725,100	29.7	10,721,153	30.5	11,329,198	32.2	12,987,250	36.4	13,092,300	36.3
America, South	3,000,037	8.5	3,082,247	8.3	3,222,075	8.9	3,273,259	9.3	3,510,245	10.0	3,880,253	10.9	4,222,980	11.7
Asia	10,872,606	31.0	10,668,579	28.9	10,298,551	28.5	10,217,800	29.1	9,968,350	28.3	9,645,500	27.0	9,527,887	26.4
Europe	10,189,338	29.0	12,033,695	32.6	11,331,878	31.4	10,410,292	29.7	9,971,287	28.4	8,836,649	24.7	8,849,226	24.5
Leading Producers														
United States of America	10,442,000	29.7	10,509,000	28.5	10,586,000	29.3	10,531,000	30.0	11,144,000	31.7	12,791,000	35.8	12,895,000	35.8
Brazil	2,836,000	8.1	2,921,000	7.9	3,062,000	8.5	3,124,000	8.9	3,333,000	9.5	3,707,000	10.4	4,053,000	11.2
Russian Federation	2,933,000	8.4	2,503,000	6.8	2,123,000	5.9	1,864,000	5.3	1,507,000	4.3	1,106,000	3.1	901,000	2.5
France	1,710,000	4.9	1,819,000	4.9	1,832,000	5.1	1,662,000	4.7	1,587,000	4.5	1,647,000	4.6	1,694,000	4.7
Germany	-		2,133,000	5.8	1,784,000	4.9	1,560,000	4.4	1,406,000	4.0	566,340	1.6	674,504	1.9

Mutton and Lamb (Including Goats), Fresh (Industrial Production)

Unit of Measure: Metric tons.

	1990 Value	%	1991 Value	%	1992 Value	%	1993 Value	%	1994 Value	%	1995 Value	%	1996 Value	%
Total Production	2,063,690	100.0	1,973,983	100.0	1,943,907	100.0	1,852,389	100.0	1,775,292	100.0	1,759,816	100.0	1,743,659	100.0
Regions														
Africa	146,000	7.1	153,000	7.8	141,000	7.3	125,000	6.7	83,000	4.7	79,000	4.5	84,000	4.8
America, North	165,002	8.0	165,002	8.4	158,002	8.1	152,002	8.2	141,002	7.9	146,002	8.3	135,002	7.7
America, South	23,923	1.2	27,462	1.4	16,000	0.8	29,000	1.6	24,000	1.4	23,000	1.3	24,538	1.4
Asia	893,369	43.3	833,991	42.2	835,966	43.0	798,002	43.1	734,182	41.4	679,061	38.6	659,101	37.8
Europe	835,395	40.5	794,529	40.3	792,939	40.8	748,385	40.4	793,108	44.7	832,753	47.3	841,018	48.2
Leading Producers														
United Kingdom	*355,429	17.2	*368,429	18.7	*381,429	19.6	*394,429	21.3	*407,429	22.9	*420,429	23.9	*433,429	24.9
France	156,000	7.6	149,000	7.5	138,000	7.1	131,000	7.1	124,000	7.0	119,000	6.8	123,000	7.1
United States of America	163,000	7.9	163,000	8.3	156,000	8.0	150,000	8.1	139,000	7.8	144,000	8.2	133,000	7.6
South Africa	146,000	7.1	153,000	7.8	141,000	7.3	125,000	6.7	83,000	4.7	79,000	4.5	84,000	4.8
Kazakhstan	157,000	7.6	140,000	7.1	133,000	6.8	123,000	6.6	78,000	4.4	43,000	2.4	28,000	1.6

Pork, Fresh (Industrial Production)

Unit of Measure: Metric tons.

	1990 Value	%	1991 Value	%	1992 Value	%	1993 Value	%	1994 Value	%	1995 Value	%	1996 Value	%
Total Production	31,249,758	100.0	35,018,259	100.0	34,528,936	100.0	33,686,718	100.0	33,932,357	100.0	34,109,021	100.0	34,121,763	100.0
Regions														
Africa	119,000	0.4	114,000	0.3	112,000	0.3	112,000	0.3	97,000	0.3	108,000	0.3	113,000	0.3
America, North	7,064,936	22.6	7,359,614	21.0	7,928,293	23.0	7,909,827	23.5	8,207,362	24.2	9,109,894	26.7	8,732,428	25.6
America, South	749,145	2.4	833,741	2.4	915,783	2.7	908,472	2.7	995,255	2.9	1,187,835	3.5	1,266,751	3.7
Asia	7,627,066	24.4	7,541,494	21.5	7,113,689	20.6	6,847,667	20.3	6,658,703	19.6	6,547,914	19.2	6,527,529	19.1
Europe	15,689,611	50.2	19,169,411	54.7	18,459,171	53.5	17,908,752	53.2	17,974,038	53.0	17,155,378	50.3	17,482,056	51.2
Leading Producers														
United States of America	6,962,000	22.3	7,256,000	20.7	7,819,000	22.6	7,726,000	22.9	8,012,000	23.6	8,925,000	26.2	8,542,000	25.0
Germany	-		3,770,000	10.8	3,559,000	10.3	3,620,000	10.7	3,540,000	10.4	1,944,564	5.7	*2,185,751	6.4
Spain	*1,696,750	5.4	*1,696,750	4.8	*1,696,750	4.9	1,074,000	3.2	1,389,000	4.1	2,208,000	6.5	2,116,000	6.2
France	1,622,000	5.2	1,665,000	4.8	1,734,000	5.0	1,772,000	5.3	1,854,000	5.5	1,872,000	5.5	1,910,000	5.6
Japan	1,555,000	5.0	1,483,000	4.2	1,434,000	4.2	1,440,000	4.3	1,390,000	4.1	1,322,000	3.9	1,266,000	3.7

Commodity data are provided by the United Nations Statistical Division. The symbol * means that data are estimated. For additional notes, see Appendix I.

21

Product Tables

Poultry, Dressed, Fresh (Industrial Production)

Unit of Measure: Metric tons.

	1990 Value	%	1991 Value	%	1992 Value	%	1993 Value	%	1994 Value	%	1995 Value	%	1996 Value	%
Total Production	25,078,667	100.0	26,287,486	100.0	27,177,047	100.0	28,148,263	100.0	29,279,692	100.0	37,577,435	100.0	39,141,398	100.0
Regions														
Africa	163,361	0.7	174,411	0.7	185,461	0.7	196,511	0.7	207,561	0.7	219,611	0.6	230,661	0.6
America, North	11,693,027	46.6	12,286,945	46.7	12,989,872	47.8	13,861,795	49.2	14,444,714	49.3	22,029,638	58.6	23,031,560	58.8
America, South	1,775,372	7.1	1,975,748	7.5	2,072,714	7.6	2,256,714	8.0	2,658,192	9.1	3,013,876	8.0	3,213,414	8.2
Asia	5,031,925	20.1	5,083,292	19.3	4,999,159	18.4	4,930,566	17.5	4,816,892	16.5	4,681,618	12.5	4,736,625	12.1
Europe	6,408,983	25.6	6,761,090	25.7	6,923,932	25.5	6,896,495	24.5	7,145,878	24.4	7,625,965	20.3	7,922,137	20.2
Oceania	6,000	0.0	6,000	0.0	5,909	0.0	6,182	0.0	6,455	0.0	6,727	0.0	7,000	0.0
Leading Producers														
United States of America	11,014,000	43.9	11,596,000	44.1	12,275,000	45.2	12,819,000	45.5	13,356,000	45.6	20,905,000	55.6	21,932,000	56.0
Brazil	1,605,000	6.4	1,801,000	6.9	1,912,000	7.0	2,074,000	7.4	2,459,000	8.4	2,793,000	7.4	3,011,000	7.7
France	1,455,000	5.8	1,569,000	6.0	1,696,000	6.2	1,765,000	6.3	1,822,000	6.2	1,916,000	5.1	2,015,000	5.1
United Kingdom	*1,085,429	4.3	*1,125,536	4.3	*1,165,643	4.3	*1,205,750	4.3	*1,245,857	4.3	*1,285,964	3.4	*1,326,071	3.4
Spain	*916,750	3.7	*916,750	3.5	*916,750	3.4	767,000	2.7	907,000	3.1	990,000	2.6	1,003,000	2.6

Other Meat, Fresh (Industrial Production)

Unit of Measure: Metric tons.

	1990 Value	%	1991 Value	%	1992 Value	%	1993 Value	%	1994 Value	%	1995 Value	%	1996 Value	%
Total Production	1,952,542	100.0	1,898,720	100.0	1,805,635	100.0	1,765,259	100.0	1,669,323	100.0	1,680,772	100.0	1,704,222	100.0
Regions														
America, North	39,067	2.0	42,167	2.2	41,267	2.3	39,367	2.2	39,467	2.4	41,567	2.5	42,667	2.5
America, South	12,000	0.6	19,000	1.0	25,000	1.4	21,000	1.2	26,000	1.6	27,000	1.6	25,000	1.5
Asia	1,554,142	79.6	1,503,220	79.2	1,480,035	82.0	1,502,959	85.1	1,402,894	84.0	1,397,866	83.2	1,421,583	83.4
Europe	347,333	17.8	334,333	17.6	259,333	14.4	201,933	11.4	200,963	12.0	214,340	12.8	214,972	12.6
Leading Producers														
Russian Federation	446,000	22.8	373,000	19.6	325,000	18.0	327,000	18.5	196,000	11.7	164,000	9.8	153,000	9.0
Ukraine	166,000	8.5	186,000	9.8	111,000	6.1	80,000	4.5	65,000	3.9	45,000	2.7	39,000	2.3
Croatia	40,000	2.0	8,000	0.4	2,000	0.1	-		-		-		-	
Spain	*48,000	2.5	*48,000	2.5	*48,000	2.7	30,000	1.7	51,000	3.1	47,000	2.8	64,000	3.8
Belarus	54,000	2.8	45,000	2.4	38,000	2.1	32,000	1.8	27,000	1.6	22,000	1.3	22,000	1.3

Bacon, Ham and Other Dried, Salted or Smoked Pig Meat

Unit of Measure: Metric tons.

	1990 Value	%	1991 Value	%	1992 Value	%	1993 Value	%	1994 Value	%	1995 Value	%	1996 Value	%
Total Production	8,840,505	100.0	9,344,275	100.0	9,413,493	100.0	9,892,814	100.0	10,580,891	100.0	11,033,982	100.0	11,500,399	100.0
Regions														
Africa	16,000	0.2	16,000	0.2	16,000	0.2	15,000	0.2	16,000	0.2	16,000	0.1	17,062	0.1
America, North	6,839,010	77.4	7,386,524	79.0	7,636,038	81.1	8,145,552	82.3	8,742,067	82.6	9,205,234	83.4	9,693,281	84.3
America, South	76,947	0.9	80,493	0.9	84,039	0.9	89,970	0.9	94,049	0.9	97,130	0.9	100,180	0.9
Asia	272,992	3.1	290,935	3.1	289,878	3.1	279,326	2.8	277,264	2.6	277,201	2.5	273,816	2.4
Europe	1,536,356	17.4	1,467,523	15.7	1,282,138	13.6	1,253,966	12.7	1,341,912	12.7	1,327,217	12.0	1,304,260	11.3
Oceania	99,200	1.1	102,800	1.1	105,400	1.1	109,000	1.1	109,600	1.0	111,200	1.0	111,800	1.0
Leading Producers														
United States of America	*6,440,914	72.9	*6,886,286	73.7	*7,331,657	77.9	*7,777,029	78.6	*8,222,400	77.7	*8,667,771	78.6	*9,113,143	79.2
Germany	*260,624	2.9	*260,624	2.8	*260,624	2.8	*260,624	2.6	*260,624	2.5	256,895	2.3	264,352	2.3
Romania	365,000	4.1	260,000	2.8	184,000	2.0	168,000	1.7	175,000	1.7	179,000	1.6	150,000	1.3
Japan	249,000	2.8	248,000	2.7	247,000	2.6	243,000	2.5	245,000	2.3	244,000	2.2	236,000	2.1
France	95,000	1.1	104,000	1.1	104,000	1.1	115,000	1.2	115,000	1.1	122,000	1.1	122,000	1.1

Commodity data are provided by the United Nations Statistical Division. The symbol * means that data are estimated. For additional notes, see Appendix I.

Product Tables
Other Meat and Edible Offals, Dried, Salted or Smoked

Unit of Measure: Metric tons.

	1990 Value	%	1991 Value	%	1992 Value	%	1993 Value	%	1994 Value	%	1995 Value	%	1996 Value	%
Total Production	1,539,141	100.0	1,655,029	100.0	1,765,783	100.0	1,906,233	100.0	1,939,394	100.0	2,057,811	100.0	2,168,422	100.0
Regions														
America, North	1,271,688	82.6	1,384,576	83.7	1,498,464	84.9	1,628,352	85.4	1,738,240	89.6	1,852,129	90.0	1,967,017	90.7
America, South	4,777	0.3	4,777	0.3	4,777	0.3	4,777	0.3	4,777	0.2	4,777	0.2	4,777	0.2
Asia	18,007	1.2	13,007	0.8	11,007	0.6	11,012	0.6	11,507	0.6	12,502	0.6	11,707	0.5
Europe	228,669	14.9	234,669	14.2	233,535	13.2	245,092	12.9	157,870	8.1	167,690	8.1	163,636	7.5
Oceania	16,000	1.0	18,000	1.1	18,000	1.0	17,000	0.9	27,000	1.4	20,714	1.0	21,286	1.0
Leading Producers														
United States of America	*1,217,648	79.1	*1,330,619	80.4	*1,443,590	81.8	*1,556,562	81.7	*1,669,533	86.1	*1,782,505	86.6	*1,895,476	87.4
France	5,000	0.3	4,000	0.2	5,000	0.3	2,000	0.1	2,000	0.1	3,000	0.1	7,000	0.3
Belgium	78,000	5.1	91,000	5.5	91,000	5.2	96,000	5.0	5,000	0.3	7,000	0.3	6,000	0.3
Netherlands	4,000	0.3	4,000	0.2	5,000	0.3	2,000	0.1	2,000	0.1	13,000	0.6	-	
Slovakia	*57,667	3.7	*57,667	3.5	*57,667	3.3	*57,667	3.0	63,000	3.2	61,000	3.0	49,000	2.3

Sausages

Unit of Measure: Metric tons.

	1990 Value	%	1991 Value	%	1992 Value	%	1993 Value	%	1994 Value	%	1995 Value	%	1996 Value	%
Total Production	11,082,797	100.0	12,005,639	100.0	11,345,048	100.0	10,770,300	100.0	10,778,300	100.0	10,735,032	100.0	10,721,349	100.0
Regions														
Africa	145,833	1.3	142,833	1.2	141,833	1.3	140,833	1.3	142,833	1.3	142,833	1.3	143,395	1.3
America, North	2,611,349	23.6	2,619,035	21.8	2,620,146	23.1	2,658,482	24.7	2,653,933	24.6	2,657,977	24.8	2,676,992	25.0
America, South	128,877	1.2	139,873	1.2	149,869	1.3	161,720	1.5	176,859	1.6	185,217	1.7	203,136	1.9
Asia	3,441,404	31.1	3,237,164	27.0	2,651,795	23.4	2,299,947	21.4	2,311,465	21.4	2,402,154	22.4	2,235,957	20.9
Europe	4,747,333	42.8	5,858,733	48.8	5,774,405	50.9	5,500,318	51.1	5,483,210	50.9	5,335,851	49.7	5,448,868	50.8
Oceania	8,000	0.1	8,000	0.1	7,000	0.1	9,000	0.1	10,000	0.1	11,000	0.1	13,000	0.1
Leading Producers														
United States of America	*2,515,533	22.7	*2,513,333	20.9	*2,511,133	22.1	*2,508,933	23.3	*2,506,733	23.3	*2,504,533	23.3	*2,502,333	23.3
Russian Federation	2,283,000	20.6	2,077,000	17.3	1,547,000	13.6	1,493,000	13.9	1,545,000	14.3	1,293,000	12.0	1,296,000	12.1
Germany	-		1,256,000	10.5	1,297,000	11.4	1,321,000	12.3	1,315,000	12.2	1,222,205	11.4	1,218,632	11.4
Poland	714,000	6.4	676,000	5.6	669,000	5.9	676,000	6.3	703,000	6.5	756,000	7.0	845,000	7.9
Ukraine	900,000	8.1	852,000	7.1	758,000	6.7	500,000	4.6	437,000	4.1	276,000	2.6	213,000	2.0

Meals, Frozen Prepared

Unit of Measure: Metric tons.

	1990 Value	%	1991 Value	%	1992 Value	%	1993 Value	%	1994 Value	%	1995 Value	%	1996 Value	%
Total Production	950,230	100.0	948,600	100.0	948,538	100.0	955,982	100.0	965,754	100.0	980,822	100.0	980,932	100.0
Regions														
America, North	721,116	75.9	721,116	76.0	721,116	76.0	720,661	75.4	720,768	74.6	721,461	73.6	721,576	73.6
America, South	7,628	0.8	7,628	0.8	4,898	0.5	6,042	0.6	7,539	0.8	12,031	1.2	7,628	0.8
Asia	1,461	0.2	2,621	0.3	3,778	0.4	3,216	0.3	4,189	0.4	4,247	0.4	3,058	0.3
Europe	220,026	23.2	217,235	22.9	218,746	23.1	226,064	23.6	233,258	24.2	243,083	24.8	248,671	25.4
Leading Producers														
United States of America	*717,880	75.5	*717,880	75.7	*717,880	75.7	*717,880	75.1	*717,880	74.3	*717,880	73.2	*717,880	73.2
France	*90,730	9.5	*98,151	10.3	*105,572	11.1	*112,993	11.8	*120,414	12.5	*127,835	13.0	*135,256	13.8
Sweden	*48,299	5.1	*48,641	5.1	*48,983	5.2	*49,326	5.2	*49,668	5.1	*50,010	5.1	*50,353	5.1
Denmark	*15,739	1.7	*15,739	1.7	*15,739	1.7	*15,739	1.6	*15,739	1.6	*15,739	1.6	*15,739	1.6
Estonia	27,372	2.9	20,362	2.1	13,667	1.4	12,265	1.3	8,638	0.9	9,010	0.9	5,558	0.6

Commodity data are provided by the United Nations Statistical Division. The symbol * means that data are estimated. For additional notes, see Appendix I.

23

Product Tables

Meat, Tinned

Unit of Measure: Metric tons.

	1990		1991		1992		1993		1994		1995		1996	
	Value	%	Value	%	Value	%	Value	%	Value	%	Value	%	Value	%
Total Production	3,550,014	100.0	3,569,210	100.0	3,621,821	100.0	3,557,523	100.0	3,593,448	100.0	3,524,402	100.0	3,568,174	100.0
Regions														
Africa	10,200	0.3	13,843	0.4	12,924	0.4	10,802	0.3	15,478	0.4	14,884	0.4	13,164	0.4
America, North	1,296,133	36.5	1,299,012	36.4	1,301,846	35.9	1,304,584	36.7	1,307,503	36.4	1,310,143	37.2	1,313,015	36.8
America, South	65,108	1.8	117,173	3.3	148,460	4.1	111,771	3.1	108,755	3.0	96,378	2.7	78,689	2.2
Asia	1,027,838	29.0	1,066,747	29.9	1,106,355	30.5	1,084,782	30.5	1,049,788	29.2	1,051,373	29.8	1,083,251	30.4
Europe	1,118,497	31.5	1,055,054	29.6	1,036,122	28.6	1,029,459	28.9	1,098,968	30.6	1,040,666	29.5	1,071,094	30.0
Oceania	32,237	0.9	17,380	0.5	16,113	0.4	16,124	0.5	12,956	0.4	10,958	0.3	8,960	0.3
Leading Producers														
United States of America	*1,222,162	34.4	*1,222,162	34.2	*1,222,162	33.7	*1,222,162	34.4	*1,222,162	34.0	*1,222,162	34.7	*1,222,162	34.3
China	234,500	6.6	300,300	8.4	301,600	8.3	*291,820	8.2	*290,411	8.1	*289,003	8.2	*287,594	8.1
Russian Federation	194,890	5.5	164,225	4.6	194,885	5.4	183,737	5.2	139,179	3.9	137,945	3.9	152,996	4.3
Spain	141,962	4.0	147,386	4.1	150,279	4.1	196,974	5.5	252,888	7.0	227,610	6.5	*223,762	6.3
Brazil	58,695	1.7	110,662	3.1	141,610	3.9	106,044	3.0	101,309	2.8	89,388	2.5	71,807	2.0

Lard (Industrial Production)

Unit of Measure: Metric tons.

	1990		1991		1992		1993		1994		1995		1996	
	Value	%	Value	%	Value	%	Value	%	Value	%	Value	%	Value	%
Total Production	1,708,134	100.0	1,639,307	100.0	1,589,011	100.0	1,506,256	100.0	1,384,927	100.0	1,457,527	100.0	1,429,210	100.0
Regions														
America, North	478,434	28.0	472,579	28.8	466,725	29.4	459,870	30.5	456,015	32.9	448,982	30.8	443,032	31.0
America, South	101,411	5.9	101,353	6.2	99,296	6.2	107,239	7.1	122,182	8.8	112,125	7.7	113,601	7.9
Asia	263,189	15.4	249,134	15.2	252,073	15.9	218,445	14.5	208,345	15.0	200,725	13.8	193,656	13.5
Europe	863,100	50.5	814,340	49.7	769,117	48.4	719,002	47.7	596,785	43.1	694,196	47.6	677,521	47.4
Oceania	2,000	0.1	1,900	0.1	1,800	0.1	1,700	0.1	1,600	0.1	1,500	0.1	1,400	0.1
Leading Producers														
United States of America	*387,629	22.7	*381,657	23.3	*375,686	23.6	*369,714	24.5	*363,743	26.3	*357,771	24.5	*351,800	24.6
Japan	219,000	12.8	178,000	10.9	171,000	10.8	172,000	11.4	168,000	12.1	158,000	10.8	152,000	10.6
Germany	*126,633	7.4	*126,633	7.7	*126,633	8.0	*126,633	8.4	*126,633	9.1	129,260	8.9	124,006	8.7
Spain	99,000	5.8	104,000	6.3	108,000	6.8	85,000	5.6	82,000	5.9	149,000	10.2	144,000	10.1
Netherlands	*86,000	5.0	*86,000	5.2	*86,000	5.4	114,000	7.6	50,000	3.6	94,000	6.4	*86,000	6.0

Hides, Cattle and Horse, Undressed (Total Production)

Unit of Measure: Metric tons.

	1990		1991		1992		1993		1994		1995		1996	
	Value	%	Value	%	Value	%	Value	%	Value	%	Value	%	Value	%
Total Production	6,881,700	100.0	6,948,708	100.0	7,765,408	100.0	7,864,084	100.0	7,487,681	100.0	7,703,158	100.0	7,523,104	100.0
Regions														
Africa	557,668	8.1	574,920	8.3	590,647	7.6	593,400	7.5	541,630	7.2	552,367	7.2	558,245	7.4
America, North	1,276,786	18.6	1,234,840	17.8	1,239,024	16.0	1,231,189	15.7	1,282,992	17.1	1,320,189	17.1	1,320,068	17.5
America, South	1,086,226	15.8	1,107,654	15.9	1,070,795	13.8	1,096,243	13.9	1,103,790	14.7	1,125,048	14.6	1,159,898	15.4
Asia	2,475,594	36.0	2,512,957	36.2	3,181,679	41.0	3,312,785	42.1	2,984,216	39.9	3,136,615	40.7	2,933,571	39.0
Europe	1,234,623	17.9	1,256,307	18.1	1,401,081	18.0	1,353,688	17.2	1,297,794	17.3	1,280,407	16.6	1,270,707	16.9
Oceania	250,804	3.6	262,030	3.8	282,182	3.6	276,778	3.5	277,259	3.7	288,531	3.7	280,616	3.7
Leading Producers														
United States of America	931,313	13.5	924,510	13.3	924,305	11.9	920,769	11.7	974,364	13.0	1,006,900	13.1	1,006,000	13.4
India	830,000	12.1	806,000	11.6	810,000	10.4	826,000	10.5	440,360	5.9	441,840	5.7	443,300	5.9
Brazil	411,500	6.0	448,000	6.4	442,000	5.7	454,500	5.8	448,000	6.0	475,000	6.2	496,000	6.6
Argentina	370,462	5.4	370,347	5.3	351,375	4.5	356,848	4.5	356,410	4.8	347,296	4.5	344,308	4.6
China	263,340	3.8	306,960	4.4	348,940	4.5	440,700	5.6	637,670	8.5	792,200	10.3	914,600	12.2

Commodity data are provided by the United Nations Statistical Division. The symbol * means that data are estimated. For additional notes, see Appendix I.

Product Tables
Skins, Calf, Goat and Sheep, Undressed (Total Production)

Unit of Measure: Metric tons.

	1990 Value	%	1991 Value	%	1992 Value	%	1993 Value	%	1994 Value	%	1995 Value	%	1996 Value	%
Total Production	2,434,443	100.0	2,504,629	100.0	2,543,461	100.0	2,579,890	100.0	2,692,959	100.0	2,816,345	100.0	2,828,058	100.0
Regions														
Africa	298,652	12.3	300,216	12.0	303,009	11.9	305,331	11.8	317,372	11.8	313,350	11.1	316,478	11.2
America, North	34,787	1.4	35,496	1.4	35,162	1.4	35,058	1.4	34,739	1.3	32,653	1.2	31,349	1.1
America, South	97,252	4.0	95,207	3.8	90,496	3.6	90,966	3.5	92,633	3.4	94,462	3.4	87,033	3.1
Asia	1,474,209	60.6	1,533,178	61.2	1,572,548	61.8	1,635,241	63.4	1,739,544	64.6	1,874,716	66.6	1,914,603	67.7
Europe	266,896	11.0	268,513	10.7	270,726	10.6	260,282	10.1	255,702	9.5	253,202	9.0	248,913	8.8
Oceania	262,647	10.8	272,019	10.9	271,520	10.7	253,013	9.8	252,970	9.4	247,962	8.8	229,682	8.1
Leading Producers														
Bangladesh	401,556	16.5	417,034	16.7	434,258	17.1	457,484	17.7	493,400	18.3	534,000	19.0	534,000	18.9
China	246,668	10.1	270,618	10.8	286,668	11.3	308,944	12.0	362,000	13.4	445,000	15.8	488,000	17.3
India	145,080	6.0	155,700	6.2	155,520	6.1	158,760	6.2	160,020	5.9	162,180	5.8	162,000	5.7
Australia	151,736	6.2	160,398	6.4	160,048	6.3	152,025	5.9	152,000	5.6	146,877	5.2	129,000	4.6
Pakistan	124,050	5.1	135,161	5.4	143,548	5.6	152,763	5.9	165,350	6.1	176,770	6.3	177,000	6.3

Milk and Cream, Condensed (Total Production)

Unit of Measure: Metric tons.

	1990 Value	%	1991 Value	%	1992 Value	%	1993 Value	%	1994 Value	%	1995 Value	%	1996 Value	%
Total Production	5,817,495	100.0	5,906,895	100.0	5,818,096	100.0	5,526,373	100.0	5,486,767	100.0	5,359,799	100.0	5,325,112	100.0
Regions														
Africa	51,476	0.9	48,988	0.8	43,442	0.7	43,147	0.8	43,026	0.8	45,289	0.8	42,328	0.8
America, North	1,255,775	21.6	1,291,521	21.9	1,338,695	23.0	1,271,915	23.0	1,270,665	23.2	1,201,655	22.4	1,179,010	22.1
America, South	186,762	3.2	192,152	3.3	184,241	3.2	188,318	3.4	207,419	3.8	228,008	4.3	227,744	4.3
Asia	2,210,090	38.0	2,217,669	37.5	2,227,653	38.3	2,095,803	37.9	2,056,573	37.5	1,960,274	36.6	1,954,086	36.7
Europe	2,025,178	34.8	2,059,830	34.9	1,933,304	33.2	1,821,682	33.0	1,801,228	32.8	1,824,671	34.0	1,826,569	34.3
Oceania	88,214	1.5	96,734	1.6	90,760	1.6	105,507	1.9	107,856	2.0	99,902	1.9	95,375	1.8
Leading Producers														
United States of America	969,540	16.7	981,072	16.6	1,043,775	17.9	980,313	17.7	992,615	18.1	914,956	17.1	881,801	16.6
Germany	543,530	9.3	570,403	9.7	522,783	9.0	536,553	9.7	546,757	10.0	566,724	10.6	538,098	10.1
Netherlands	402,350	6.9	407,142	6.9	386,951	6.7	374,468	6.8	339,781	6.2	359,557	6.7	333,600	6.3
Russian Federation	620,000	10.7	600,000	10.2	582,741	10.0	433,856	7.9	299,040	5.5	225,300	4.2	205,625	3.9
India	390,000	6.7	395,000	6.7	392,000	6.7	407,000	7.4	413,000	7.5	426,000	7.9	*427,668	8.0

Milk and Cream, Condensed (Industrial Production)

Unit of Measure: Metric tons.

	1990 Value	%	1991 Value	%	1992 Value	%	1993 Value	%	1994 Value	%	1995 Value	%	1996 Value	%
Total Production	4,818,155	100.0	4,485,862	100.0	4,101,613	100.0	4,199,748	100.0	4,021,360	100.0	4,068,467	100.0	3,889,935	100.0
Regions														
America, North	1,169,446	24.3	1,198,147	26.7	1,234,849	30.1	1,133,550	27.0	1,100,197	27.4	1,111,687	27.3	1,059,336	27.2
America, South	17,319	0.4	18,706	0.4	20,620	0.5	11,767	0.3	12,291	0.3	16,632	0.4	19,826	0.5
Asia	1,111,211	23.1	892,263	19.9	673,468	16.4	622,386	14.8	606,097	15.1	624,956	15.4	542,973	14.0
Europe	2,433,179	50.5	2,281,746	50.9	2,083,676	50.8	2,328,045	55.4	2,193,642	54.5	2,199,983	54.1	2,146,517	55.2
Oceania	87,000	1.8	95,000	2.1	89,000	2.2	104,000	2.5	109,133	2.7	115,208	2.8	121,283	3.1
Leading Producers														
United States of America	951,000	19.7	968,053	21.6	985,106	24.0	1,002,159	23.9	954,158	23.7	956,000	23.5	897,000	23.1
Germany	*504,750	10.5	474,000	10.6	503,000	12.3	514,000	12.2	528,000	13.1	*504,750	12.4	*504,750	13.0
Netherlands	413,000	8.6	445,000	9.9	428,000	10.4	697,000	16.6	568,000	14.1	525,000	12.9	*508,907	13.1
Estonia	412,000	8.6	296,000	6.6	124,000	3.0	131,000	3.1	133,000	3.3	163,000	4.0	127,000	3.3
Russian Federation	184,000	3.8	133,000	3.0	134,000	3.3	151,000	3.6	136,000	3.4	148,000	3.6	128,000	3.3

Commodity data are provided by the United Nations Statistical Division. The symbol * means that data are estimated. For additional notes, see Appendix I.

25

Product Tables
Milk and Cream, Dried (Total Production)

Unit of Measure: Metric tons.

	1990 Value	%	1991 Value	%	1992 Value	%	1993 Value	%	1994 Value	%	1995 Value	%	1996 Value	%
Total Production	9,104,688	100.0	8,900,490	100.0	8,694,502	100.0	8,904,258	100.0	9,073,718	100.0	9,239,875	100.0	9,247,566	100.0
Regions														
Africa	47,443	0.5	53,701	0.6	39,189	0.5	41,230	0.5	44,321	0.5	50,290	0.5	42,000	0.5
America, North	1,263,388	13.9	1,257,668	14.1	1,285,826	14.8	1,302,638	14.6	1,449,348	16.0	1,452,396	15.7	1,323,016	14.3
America, South	442,640	4.9	402,207	4.5	423,657	4.9	465,574	5.2	494,859	5.5	528,688	5.7	535,000	5.8
Asia	2,318,605	25.5	2,341,896	26.3	2,289,333	26.3	2,340,235	26.3	2,304,501	25.4	2,364,608	25.6	2,431,711	26.3
Europe	4,416,783	48.5	4,173,294	46.9	3,953,015	45.5	3,986,880	44.8	3,913,574	43.1	3,950,992	42.8	3,963,839	42.9
Oceania	615,829	6.8	671,723	7.5	703,482	8.1	767,702	8.6	867,117	9.6	892,900	9.7	952,000	10.3
Leading Producers														
France	1,212,800	13.3	1,176,750	13.2	1,153,710	13.3	1,097,707	12.3	1,156,000	12.7	1,128,000	12.2	1,152,000	12.5
United States of America	1,032,701	11.3	1,006,424	11.3	1,065,376	12.3	1,068,571	12.0	1,194,762	13.2	1,180,126	12.8	1,052,000	11.4
Germany	881,758	9.7	860,741	9.7	736,827	8.5	782,734	8.8	746,723	8.2	790,200	8.6	790,000	8.5
Netherlands	506,319	5.6	471,497	5.3	456,772	5.3	489,509	5.5	423,889	4.7	443,755	4.8	426,000	4.6
Russian Federation	453,000	5.0	432,000	4.9	474,982	5.5	457,870	5.1	420,250	4.6	416,000	4.5	416,000	4.5

Milk and Cream, Dried (Industrial Production)

Unit of Measure: Metric tons.

	1990 Value	%	1991 Value	%	1992 Value	%	1993 Value	%	1994 Value	%	1995 Value	%	1996 Value	%
Total Production	4,897,740	100.0	5,576,500	100.0	5,301,363	100.0	5,472,629	100.0	5,430,848	100.0	5,766,930	100.0	5,500,285	100.0
Regions														
America, North	691,579	14.1	684,221	12.3	676,863	12.8	669,506	12.2	808,883	14.9	864,046	15.0	746,571	13.6
America, South	66,240	1.4	59,502	1.1	58,667	1.1	50,804	0.9	54,584	1.0	57,130	1.0	49,285	0.9
Asia	750,016	15.3	830,958	14.9	878,787	16.6	934,819	17.1	901,601	16.6	995,379	17.3	928,675	16.9
Europe	3,189,701	65.1	3,786,365	67.9	3,462,179	65.3	3,558,872	65.0	3,418,655	62.9	3,594,775	62.3	3,511,676	63.8
Oceania	200,205	4.1	215,454	3.9	224,867	4.2	258,628	4.7	247,125	4.6	255,601	4.4	264,076	4.8
Leading Producers														
France	796,000	16.3	739,000	13.3	653,000	12.3	637,000	11.6	661,000	12.2	651,000	11.3	627,000	11.4
Germany	-		764,696	13.7	627,712	11.8	644,102	11.8	588,594	10.8	753,827	13.1	807,122	14.7
United States of America	591,192	12.1	570,944	10.2	550,695	10.4	530,447	9.7	656,015	12.1	734,314	12.7	625,589	11.4
China	241,525	4.9	293,856	5.3	336,545	6.3	417,368	7.6	424,600	7.8	525,665	9.1	504,100	9.2
Netherlands	220,290	4.5	201,522	3.6	182,812	3.4	252,400	4.6	239,096	4.4	261,099	4.5	*176,105	3.2

Butter (Industrial Production)

Unit of Measure: Metric tons.

	1990 Value	%	1991 Value	%	1992 Value	%	1993 Value	%	1994 Value	%	1995 Value	%	1996 Value	%
Total Production	7,343,765	100.0	7,452,551	100.0	7,117,871	100.0	7,060,539	100.0	6,813,723	100.0	6,814,029	100.0	6,674,054	100.0
Regions														
Africa	5,090	0.1	4,921	0.1	4,135	0.1	3,700	0.1	3,453	0.1	6,119	0.1	6,938	0.1
America, North	605,572	8.2	620,988	8.3	634,029	8.9	613,402	8.7	607,112	8.9	653,510	9.6	605,930	9.1
America, South	78,221	1.1	72,935	1.0	83,112	1.2	72,325	1.0	77,228	1.1	87,827	1.3	90,243	1.4
Asia	3,095,581	42.2	3,010,726	40.4	2,956,828	41.5	2,939,472	41.6	2,853,681	41.9	2,821,506	41.4	2,775,053	41.6
Europe	3,173,115	43.2	3,359,535	45.1	3,058,815	43.0	3,040,104	43.1	2,843,897	41.7	2,831,432	41.6	2,729,284	40.9
Oceania	386,186	5.3	383,447	5.1	380,952	5.4	391,536	5.5	428,353	6.3	413,634	6.1	466,605	7.0
Leading Producers														
United States of America	590,654	8.0	605,897	8.1	619,225	8.7	596,442	8.4	587,995	8.6	632,237	9.3	582,500	8.7
Russian Federation	832,505	11.3	728,997	9.8	648,590	9.1	568,184	8.0	487,777	7.2	421,284	6.2	323,320	4.8
Germany	-		552,640	7.4	473,716	6.7	482,165	6.8	461,271	6.8	479,670	7.0	478,265	7.2
France	451,000	6.1	414,000	5.6	392,000	5.5	377,000	5.3	376,000	5.5	381,000	5.6	404,000	6.1
Ukraine	444,110	6.0	376,405	5.1	302,888	4.3	311,793	4.4	253,659	3.7	221,874	3.3	162,392	2.4

Product Tables
Cheese (Industrial Production)

Unit of Measure: Metric tons.

	1990 Value	%	1991 Value	%	1992 Value	%	1993 Value	%	1994 Value	%	1995 Value	%	1996 Value	%
Total Production	11,151,406	100.0	12,246,899	100.0	12,463,119	100.0	12,650,423	100.0	13,028,025	100.0	13,717,674	100.0	14,173,127	100.0
Regions														
Africa	210,637	1.9	221,223	1.8	234,344	1.9	243,247	1.9	246,207	1.9	247,374	1.8	284,384	2.0
America, North	2,773,407	24.9	2,772,530	22.6	2,971,163	23.8	3,012,055	23.8	3,128,489	24.0	3,556,758	25.9	3,704,700	26.1
America, South	371,139	3.3	399,837	3.3	459,179	3.7	473,729	3.7	502,072	3.9	487,396	3.6	525,567	3.7
Asia	1,745,187	15.6	1,704,067	13.9	1,601,913	12.9	1,643,929	13.0	1,644,280	12.6	1,605,177	11.7	1,626,922	11.5
Europe	5,741,680	51.5	6,834,099	55.8	6,861,088	55.1	6,921,045	54.7	7,078,053	54.3	7,389,132	53.9	7,534,691	53.2
Oceania	309,356	2.8	315,144	2.6	335,432	2.7	356,417	2.8	428,925	3.3	431,837	3.1	496,863	3.5
Leading Producers														
United States of America	2,748,500	24.6	2,746,422	22.4	2,943,024	23.6	2,960,526	23.4	3,053,569	23.4	3,458,448	25.2	3,583,000	25.3
Germany	-		1,247,916	10.2	1,293,411	10.4	1,336,645	10.6	1,399,139	10.7	1,691,289	12.3	1,744,477	12.3
France	1,538,000	13.8	1,571,000	12.8	1,590,000	12.9	1,635,000	12.9	1,674,000	12.8	1,711,000	12.5	1,732,000	12.2
Netherlands	646,476	5.8	663,310	5.4	702,444	5.6	654,500	5.2	669,700	5.1	660,600	4.8	*723,389	5.1
Denmark	331,601	3.0	315,073	2.6	321,344	2.6	340,729	2.7	309,156	2.4	321,386	2.3	*318,271	2.2

Ice-Cream

Unit of Measure: 1,000 Liters.

	1990 Value	%	1991 Value	%	1992 Value	%	1993 Value	%	1994 Value	%	1995 Value	%	1996 Value	%
Total Production	5,928,305	100.0	6,040,225	100.0	6,329,271	100.0	6,302,742	100.0	6,156,879	100.0	7,788,349	100.0	6,789,362	100.0
Regions														
Africa	61,639	1.0	70,483	1.2	70,163	1.1	73,534	1.2	75,655	1.2	88,255	1.1	86,453	1.3
America, North	3,532,533	59.6	3,599,624	59.6	3,619,870	57.2	3,446,929	54.7	3,256,412	52.9	4,727,226	60.7	3,712,818	54.7
America, South	61,876	1.0	66,961	1.1	70,295	1.1	78,554	1.2	91,342	1.5	86,055	1.1	92,869	1.4
Asia	456,790	7.7	483,903	8.0	466,338	7.4	458,778	7.3	465,799	7.6	462,789	5.9	478,192	7.0
Europe	1,553,191	26.2	1,561,126	25.8	1,840,756	29.1	1,983,308	31.5	2,004,591	32.6	2,161,417	27.8	2,156,749	31.8
Oceania	262,276	4.4	258,128	4.3	261,850	4.1	261,639	4.2	263,081	4.3	262,605	3.4	262,280	3.9
Leading Producers														
United States of America	3,412,255	57.6	3,496,092	57.9	3,514,106	55.5	3,338,934	53.0	3,146,186	51.1	4,614,769	59.3	3,598,130	53.0
Germany	-				311,656	4.9	280,617	4.5	316,316	5.1	435,336	5.6	430,143	6.3
United Kingdom	338,712	5.7	343,654	5.7	*311,934	4.9	*324,688	5.2	*337,442	5.5	*350,196	4.5	*362,950	5.3
France	291,544	4.9	286,947	4.8	301,998	4.8	323,285	5.1	355,567	5.8	338,000	4.3	309,000	4.6
Australia	197,121	3.3	192,354	3.2	194,850	3.1	193,149	3.1	*193,235	3.1	*191,737	2.5	*190,238	2.8

Fruits, Dried

Unit of Measure: Metric tons.

	1990 Value	%	1991 Value	%	1992 Value	%	1993 Value	%	1994 Value	%	1995 Value	%	1996 Value	%
Total Production	999,536	100.0	961,089	100.0	990,595	100.0	1,024,477	100.0	1,088,787	100.0	1,133,709	100.0	1,137,774	100.0
Regions														
Africa	760	0.1	731	0.1	703	0.1	674	0.1	645	0.1	616	0.1	775	0.1
America, North	527,896	52.8	545,482	56.8	563,068	56.8	580,654	56.7	598,240	54.9	615,826	54.3	633,412	55.7
America, South	11,554	1.2	7,659	0.8	6,805	0.7	7,523	0.7	7,147	0.7	6,406	0.6	8,063	0.7
Asia	284,443	28.5	271,238	28.2	243,620	24.6	264,113	25.8	305,722	28.1	308,748	27.2	298,386	26.2
Europe	174,883	17.5	135,978	14.1	176,399	17.8	171,513	16.7	177,033	16.3	202,114	17.8	197,139	17.3
Leading Producers														
United States of America	*527,896	52.8	*545,482	56.8	*563,068	56.8	*580,654	56.7	*598,240	54.9	*615,826	54.3	*633,412	55.7
Turkey	157,246	15.7	140,525	14.6	123,726	12.5	143,596	14.0	187,347	17.2	189,615	16.7	180,640	15.9
Greece	98,294	9.8	68,163	7.1	68,036	6.9	*89,835	8.8	*87,807	8.1	*85,779	7.6	*83,752	7.4
France	36,745	3.7	24,800	2.6	50,970	5.1	37,150	3.6	43,100	4.0	61,300	5.4	66,696	5.9
Uzbekistan	26,538	2.7	19,932	2.1	12,162	1.2	17,107	1.7	16,443	1.5	13,138	1.2	*10,301	0.9

Commodity data are provided by the United Nations Statistical Division. The symbol * means that data are estimated. For additional notes, see Appendix I.

27

Product Tables
Jams, Marmalades and Fruit Jellies

Unit of Measure: Metric tons.

	1990		1991		1992		1993		1994		1995		1996	
	Value	%	Value	%	Value	%	Value	%	Value	%	Value	%	Value	%
Total Production	2,011,853	100.0	2,238,910	100.0	2,156,109	100.0	2,114,684	100.0	2,002,757	100.0	2,256,013	100.0	2,268,918	100.0
Regions														
Africa	77,288	3.8	69,413	3.1	68,736	3.2	72,024	3.4	52,386	2.6	59,688	2.6	65,441	2.9
America, North	30,606	1.5	29,217	1.3	30,684	1.4	40,499	1.9	50,315	2.5	60,131	2.7	69,946	3.1
America, South	32,461	1.6	34,373	1.5	40,657	1.9	37,026	1.8	31,546	1.6	37,718	1.7	40,273	1.8
Asia	617,760	30.7	652,575	29.1	577,915	26.8	581,066	27.5	503,010	25.1	491,727	21.8	492,968	21.7
Europe	1,215,862	60.4	1,416,559	63.3	1,399,458	64.9	1,344,879	63.6	1,325,368	66.2	1,566,261	69.4	1,559,445	68.7
Oceania	37,877	1.9	36,773	1.6	38,658	1.8	39,191	1.9	40,131	2.0	40,488	1.8	40,844	1.8
Leading Producers														
Germany	-		218,952	9.8	217,174	10.1	216,524	10.2	232,232	11.6	472,507	20.9	498,829	22.0
France	151,222	7.5	161,822	7.2	154,763	7.2	153,009	7.2	152,159	7.6	149,710	6.6	147,084	6.5
United Kingdom	138,313	6.9	138,179	6.2	*147,278	6.8	*143,244	6.8	*139,210	7.0	*135,177	6.0	*131,143	5.8
Spain	126,805	6.3	149,724	6.7	152,141	7.1	92,605	4.4	74,644	3.7	90,928	4.0	90,704	4.0
Poland	52,058	2.6	42,980	1.9	42,954	2.0	46,787	2.2	51,794	2.6	57,184	2.5	65,102	2.9

Fruit & Vegetable Juices, Concentrated

Unit of Measure: Metric tons.

	1990		1991		1992		1993		1994		1995		1996	
	Value	%	Value	%	Value	%	Value	%	Value	%	Value	%	Value	%
Total Production	2,043,586	100.0	1,914,727	100.0	1,939,520	100.0	1,776,841	100.0	1,589,530	100.0	1,548,473	100.0	1,508,408	100.0
Regions														
Africa	46,623	2.3	48,525	2.5	30,215	1.6	20,471	1.2	5,216	0.3	32,392	2.1	33,113	2.2
America, North	388,649	19.0	322,300	16.8	255,951	13.2	187,304	10.5	123,218	7.8	66,723	4.3	67,883	4.5
America, South	374,267	18.3	374,267	19.5	371,381	19.1	371,458	20.9	383,080	24.1	371,149	24.0	374,267	24.8
Asia	385,307	18.9	393,556	20.6	479,528	24.7	380,794	21.4	311,839	19.6	302,959	19.6	299,611	19.9
Europe	848,740	41.5	776,079	40.5	802,445	41.4	816,814	46.0	766,176	48.2	775,250	50.1	733,535	48.6
Leading Producers														
United States of America	*330,061	16.2	*262,340	13.7	*194,618	10.0	*126,897	7.1	*59,175	3.7	-		-	
Brazil	*370,413	18.1	*370,413	19.3	*370,413	19.1	*370,413	20.8	*370,413	23.3	*370,413	23.9	*370,413	24.6
Germany	*197,315	9.7	*197,315	10.3	*197,315	10.2	*197,315	11.1	*197,315	12.4	*197,315	12.7	197,315	13.1
Philippines	*106,399	5.2	*107,202	5.6	*108,004	5.6	*108,807	6.1	*109,610	6.9	*110,413	7.1	*111,216	7.4
Spain	*81,634	4.0	*81,634	4.3	*81,634	4.2	66,820	3.8	84,459	5.3	93,623	6.0	*81,634	5.4

Fruit & Vegetable Juices, Unconcentrated

Unit of Measure: Metric tons.

	1990		1991		1992		1993		1994		1995		1996	
	Value	%	Value	%	Value	%	Value	%	Value	%	Value	%	Value	%
Total Production	7,924,669	100.0	7,806,865	100.0	7,405,240	100.0	8,092,470	100.0	8,216,357	100.0	8,264,133	100.0	8,432,459	100.0
Regions														
Africa	47,944	0.6	56,674	0.7	46,552	0.6	51,015	0.6	54,248	0.7	51,471	0.6	52,531	0.6
America, North	509,940	6.4	403,142	5.2	291,205	3.9	407,996	5.0	431,872	5.3	423,772	5.1	430,231	5.1
America, South	10,404	0.1	9,759	0.1	11,026	0.1	12,847	0.2	12,587	0.2	14,018	0.2	17,464	0.2
Asia	3,524,328	44.5	3,526,694	45.2	3,254,881	44.0	3,325,829	41.1	3,353,511	40.8	3,491,154	42.2	3,638,847	43.2
Europe	3,832,052	48.4	3,810,596	48.8	3,801,576	51.3	4,294,783	53.1	4,364,139	53.1	4,283,718	51.8	4,293,386	50.9
Leading Producers														
Germany	*2,353,263	29.7	*2,353,263	30.1	*2,353,263	31.8	*2,353,263	29.1	*2,353,263	28.6	2,317,129	28.0	2,389,396	28.3
United States of America	*302,061	3.8	*203,073	2.6	*104,086	1.4	*5,098	0.1	-		-		-	
Spain	312,015	3.9	387,853	5.0	360,975	4.9	775,845	9.6	730,453	8.9	692,127	8.4	647,431	7.7
Ukraine	428,470	5.4	333,588	4.3	286,255	3.9	303,462	3.7	148,533	1.8	97,944	1.2	70,416	0.8
Russian Federation	462,893	5.8	392,956	5.0	264,483	3.6	165,650	2.0	86,838	1.1	79,700	1.0	95,592	1.1

Product Tables
Fruits, Frozen

Unit of Measure: Metric tons.

	1990 Value	%	1991 Value	%	1992 Value	%	1993 Value	%	1994 Value	%	1995 Value	%	1996 Value	%
Total Production	732,638	100.0	851,105	100.0	866,779	100.0	845,203	100.0	846,450	100.0	886,175	100.0	925,218	100.0
Regions														
America, North	407,566	55.6	421,455	49.5	435,344	50.2	449,234	53.2	448,934	53.0	480,200	54.2	469,829	50.8
America, South	5,734	0.8	5,734	0.7	5,702	0.7	7,623	0.9	1,986	0.2	7,623	0.9	5,734	0.6
Asia	14,789	2.0	44,827	5.3	34,647	4.0	39,418	4.7	33,583	4.0	30,986	3.5	39,206	4.2
Europe	304,550	41.6	379,090	44.5	391,085	45.1	348,928	41.3	361,947	42.8	367,366	41.5	410,449	44.4
Leading Producers														
United States of America	*386,801	52.8	*395,959	46.5	*405,117	46.7	*414,275	49.0	*423,433	50.0	*432,591	48.8	*441,749	47.7
Poland	145,551	19.9	212,776	25.0	232,225	26.8	188,827	22.3	187,601	22.2	179,451	20.3	217,823	23.5
France	44,235	6.0	50,860	6.0	30,928	3.6	*47,932	5.7	*50,193	5.9	*52,453	5.9	*54,714	5.9
Yugoslavia	-		33,972	4.0	25,043	2.9	26,071	3.1	24,909	2.9	20,996	2.4	25,449	2.8
Hungary	15,137	2.1	16,446	1.9	26,385	3.0	20,621	2.4	20,579	2.4	15,474	1.7	23,128	2.5

Fruits, Tinned or Bottled

Unit of Measure: Metric tons.

	1990 Value	%	1991 Value	%	1992 Value	%	1993 Value	%	1994 Value	%	1995 Value	%	1996 Value	%
Total Production	6,423,181	100.0	6,526,489	100.0	6,758,308	100.0	6,751,044	100.0	6,527,562	100.0	6,722,742	100.0	6,792,740	100.0
Regions														
Africa	458,013	7.1	400,604	6.1	424,641	6.3	415,466	6.2	409,913	6.3	499,994	7.4	484,980	7.1
America, North	878,576	13.7	847,505	13.0	833,186	12.3	803,220	11.9	776,441	11.9	763,207	11.4	741,172	10.9
America, South	111,408	1.7	104,833	1.6	105,789	1.6	109,750	1.6	122,930	1.9	120,872	1.8	131,338	1.9
Asia	2,771,413	43.1	3,032,312	46.5	3,340,973	49.4	3,325,826	49.3	3,198,592	49.0	3,268,359	48.6	3,369,895	49.6
Europe	2,029,314	31.6	1,972,063	30.2	1,889,832	28.0	1,938,178	28.7	1,866,367	28.6	1,922,276	28.6	1,922,606	28.3
Oceania	174,457	2.7	169,172	2.6	163,888	2.4	158,603	2.3	153,318	2.3	148,034	2.2	142,749	2.1
Leading Producers														
China	581,000	9.0	754,300	11.6	942,100	13.9	*844,040	12.5	*856,637	13.1	*869,234	12.9	*881,831	13.0
United States of America	*744,417	11.6	*718,502	11.0	*692,587	10.2	*666,672	9.9	*640,758	9.8	*614,843	9.1	*588,928	8.7
Philippines	*727,438	11.3	*806,236	12.4	*885,035	13.1	*963,833	14.3	*1,042,631	16.0	*1,121,430	16.7	*1,200,228	17.7
Greece	*346,499	5.4	*372,211	5.7	*397,923	5.9	*423,635	6.3	*449,347	6.9	*475,059	7.1	*500,771	7.4
Spain	219,621	3.4	249,638	3.8	255,440	3.8	384,822	5.7	366,950	5.6	390,031	5.8	*370,838	5.5

Vegetables, Frozen

Unit of Measure: Metric tons.

	1990 Value	%	1991 Value	%	1992 Value	%	1993 Value	%	1994 Value	%	1995 Value	%	1996 Value	%
Total Production	6,968,329	100.0	7,535,624	100.0	7,707,773	100.0	7,855,074	100.0	7,970,403	100.0	8,381,596	100.0	8,601,046	100.0
Regions														
Africa	67,770	1.0	62,177	0.8	66,018	0.9	56,645	0.7	67,752	0.9	79,690	1.0	73,543	0.9
America, North	4,268,190	61.3	4,390,802	58.3	4,513,413	58.6	4,636,024	59.0	4,758,636	59.7	4,881,247	58.2	5,003,859	58.2
Asia	120,282	1.7	150,815	2.0	137,675	1.8	145,403	1.9	144,494	1.8	147,212	1.8	150,262	1.7
Europe	2,183,313	31.3	2,591,217	34.4	2,638,212	34.2	2,652,706	33.8	2,603,912	32.7	2,846,759	34.0	2,927,153	34.0
Oceania	328,774	4.7	340,614	4.5	352,455	4.6	364,296	4.6	395,610	5.0	426,689	5.1	446,230	5.2
Leading Producers														
United States of America	*4,268,190	61.3	*4,390,802	58.3	*4,513,413	58.6	*4,636,024	59.0	*4,758,636	59.7	*4,881,247	58.2	*5,003,859	58.2
United Kingdom	563,515	8.1	585,470	7.8	*595,303	7.7	*607,030	7.7	*618,758	7.8	*630,485	7.5	*642,212	7.5
Belgium	487,328	7.0	555,174	7.4	655,438	8.5	654,226	8.3	440,525	5.5	496,927	5.9	528,813	6.1
France	345,349	5.0	376,674	5.0	405,868	5.3	*414,936	5.3	*439,120	5.5	*463,304	5.5	*487,488	5.7
Australia	*235,419	3.4	*241,814	3.2	*248,208	3.2	*254,603	3.2	*260,998	3.3	*267,393	3.2	*273,787	3.2

Commodity data are provided by the United Nations Statistical Division. The symbol * means that data are estimated. For additional notes, see Appendix I.

29

Product Tables

Vegetables, Tinned or Bottled

Unit of Measure: Metric tons.

	1990 Value	%	1991 Value	%	1992 Value	%	1993 Value	%	1994 Value	%	1995 Value	%	1996 Value	%
Total Production	10,854,257	100.0	10,897,651	100.0	10,570,888	100.0	10,545,467	100.0	10,371,430	100.0	11,537,323	100.0	10,988,501	100.0
Regions														
Africa	142,818	1.3	148,451	1.4	144,175	1.4	126,540	1.2	133,622	1.3	137,082	1.2	146,123	1.3
America, North	1,391,026	12.8	1,242,164	11.4	1,128,446	10.7	960,789	9.1	793,132	7.6	625,475	5.4	457,818	4.2
America, South	27,699	0.3	24,268	0.2	28,881	0.3	37,548	0.4	42,049	0.4	36,479	0.3	29,404	0.3
Asia	3,811,189	35.1	3,838,254	35.2	3,865,682	36.6	3,869,830	36.7	3,734,126	36.0	3,746,130	32.5	3,719,434	33.8
Europe	5,264,309	48.5	5,425,591	49.8	5,208,474	49.3	5,353,025	50.8	5,357,659	51.7	6,672,608	57.8	6,319,265	57.5
Oceania	217,217	2.0	218,923	2.0	195,230	1.8	197,736	1.9	310,843	3.0	319,549	2.8	316,456	2.9
Leading Producers														
United States of America	*1,045,044	9.6	*876,856	8.0	*708,668	6.7	*540,480	5.1	*372,292	3.6	*204,105	1.8	*35,917	0.3
France	1,270,984	11.7	1,384,864	12.7	1,449,452	13.7	1,315,001	12.5	1,323,654	12.8	1,387,054	12.0	1,444,517	13.1
United Kingdom	774,512	7.1	758,038	7.0	*757,559	7.2	*754,814	7.2	*752,069	7.3	*749,325	6.5	*746,580	6.8
China	546,000	5.0	648,000	5.9	651,500	6.2	*689,547	6.5	*711,570	6.9	*733,592	6.4	*755,615	6.9
Spain	464,534	4.3	507,781	4.7	518,252	4.9	984,354	9.3	999,457	9.6	1,221,612	10.6	*929,042	8.5

Fish, Frozen

Unit of Measure: Metric tons.

	1990 Value	%	1991 Value	%	1992 Value	%	1993 Value	%	1994 Value	%	1995 Value	%	1996 Value	%
Total Production	17,327,754	100.0	17,891,444	100.0	17,519,440	100.0	17,390,266	100.0	17,186,751	100.0	17,637,734	100.0	17,600,864	100.0
Regions														
Africa	290,552	1.7	339,333	1.9	327,396	1.9	348,106	2.0	357,440	2.1	337,632	1.9	363,695	2.1
America, North	1,057,231	6.1	1,533,106	8.6	1,408,994	8.0	1,504,302	8.7	1,406,236	8.2	1,267,471	7.2	1,382,196	7.9
America, South	735,709	4.2	757,010	4.2	695,327	4.0	751,155	4.3	863,307	5.0	812,834	4.6	763,877	4.3
Asia	12,117,858	69.9	12,242,443	68.4	12,320,326	70.3	12,469,418	71.7	11,962,296	69.6	12,430,546	70.5	12,340,741	70.1
Europe	2,960,476	17.1	2,779,572	15.5	2,534,145	14.5	2,072,109	11.9	2,385,343	13.9	2,558,160	14.5	2,496,120	14.2
Oceania	165,928	1.0	239,979	1.3	233,251	1.3	245,175	1.4	212,129	1.2	231,092	1.3	254,235	1.4
Leading Producers														
Japan	3,907,245	22.5	3,998,958	22.4	3,925,415	22.4	3,878,035	22.3	3,608,460	21.0	3,279,186	18.6	3,232,305	18.4
Russian Federation	2,028,666	11.7	1,874,898	10.5	1,850,905	10.6	1,693,787	9.7	1,350,623	7.9	1,362,072	7.7	1,555,022	8.8
China	1,298,103	7.5	1,326,085	7.4	1,347,600	7.7	1,446,100	8.3	1,494,400	8.7	2,672,592	15.2	*1,964,803	11.2
United States of America	717,764	4.1	1,238,241	6.9	1,195,632	6.8	1,287,311	7.4	1,243,026	7.2	1,066,370	6.0	1,183,683	6.7
Korea (Dem. Peop. Rep.)	631,700	3.6	630,700	3.5	630,200	3.6	630,312	3.6	630,300	3.7	630,281	3.6	*643,445	3.7

Fish, Salted, Dried or Smoked

Unit of Measure: Metric tons.

	1990 Value	%	1991 Value	%	1992 Value	%	1993 Value	%	1994 Value	%	1995 Value	%	1996 Value	%
Total Production	4,727,602	100.0	4,659,252	100.0	4,329,250	100.0	4,339,763	100.0	4,151,836	100.0	5,206,527	100.0	4,426,519	100.0
Regions														
Africa	436,102	9.2	429,449	9.2	403,251	9.3	388,653	9.0	417,524	10.1	439,174	8.4	454,855	10.3
America, North	96,937	2.1	110,721	2.4	102,327	2.4	93,942	2.2	53,535	1.3	67,016	1.3	103,715	2.3
America, South	74,331	1.6	83,863	1.8	60,923	1.4	88,440	2.0	77,098	1.9	85,165	1.6	85,633	1.9
Asia	3,481,584	73.6	3,412,224	73.2	3,255,253	75.2	3,270,556	75.4	3,072,863	74.0	4,074,993	78.3	3,268,605	73.8
Europe	620,034	13.1	606,206	13.0	494,756	11.4	485,606	11.2	517,947	12.5	527,120	10.1	497,939	11.2
Oceania	18,613	0.4	16,790	0.4	12,740	0.3	12,566	0.3	12,869	0.3	13,059	0.3	15,773	0.4
Leading Producers														
Japan	916,668	19.4	908,745	19.5	839,690	19.4	843,668	19.4	805,677	19.4	831,120	16.0	831,240	18.8
Indonesia	55,945	1.2	20,291	0.4	70,673	1.6	45,571	1.1	53,475	1.3	892,000	17.1	*131,296	3.0
Russian Federation	503,067	10.6	409,225	8.8	227,776	5.3	196,776	4.5	147,279	3.5	110,122	2.1	102,700	2.3
Philippines	244,238	5.2	259,700	5.6	248,900	5.7	234,100	5.4	239,300	5.8	234,056	4.5	*254,218	5.7
China	180,655	3.8	201,000	4.3	202,300	4.7	202,100	4.7	222,000	5.3	538,592	10.3	*290,549	6.6

Product Tables
Fish, Tinned

Unit of Measure: Metric tons.

	1990 Value	%	1991 Value	%	1992 Value	%	1993 Value	%	1994 Value	%	1995 Value	%	1996 Value	%
Total Production	5,231,694	100.0	5,459,411	100.0	5,002,158	100.0	5,541,974	100.0	5,502,975	100.0	5,616,304	100.0	5,547,337	100.0
Regions														
Africa	269,048	5.1	252,965	4.6	282,079	5.6	309,340	5.6	283,882	5.2	300,303	5.3	310,606	5.6
America, North	541,320	10.3	764,353	14.0	692,694	13.8	884,218	16.0	975,995	17.7	1,179,697	21.0	958,769	17.3
America, South	240,496	4.6	252,247	4.6	309,642	6.2	347,122	6.3	300,284	5.5	419,838	7.5	364,857	6.6
Asia	2,932,686	56.1	2,989,693	54.8	2,709,335	54.2	2,965,999	53.5	2,856,038	51.9	2,495,242	44.4	2,760,855	49.8
Europe	1,226,165	23.4	1,169,936	21.4	975,430	19.5	1,005,435	18.1	1,043,116	19.0	1,173,545	20.9	1,110,992	20.0
Oceania	21,980	0.4	30,217	0.6	32,978	0.7	29,860	0.5	43,660	0.8	47,679	0.8	41,257	0.7
Leading Producers														
United States of America	434,227	8.3	659,510	12.1	609,212	12.2	775,218	14.0	862,026	15.7	1,035,175	18.4	*808,169	14.6
Russian Federation	825,014	15.8	700,362	12.8	470,673	9.4	324,251	5.9	136,758	2.5	202,746	3.6	166,522	3.0
Thailand	277,200	5.3	375,000	6.9	327,400	6.5	713,600	12.9	766,000	13.9	373,431	6.6	*607,521	11.0
Japan	194,169	3.7	193,719	3.5	198,129	4.0	189,296	3.4	176,341	3.2	158,371	2.8	147,415	2.7
Spain	122,752	2.3	137,555	2.5	140,563	2.8	207,232	3.7	233,102	4.2	235,631	4.2	*197,250	3.6

Margarine, Imitation Lard and Other Prepared Fats

Unit of Measure: Metric tons.

	1990 Value	%	1991 Value	%	1992 Value	%	1993 Value	%	1994 Value	%	1995 Value	%	1996 Value	%
Total Production	13,145,340	100.0	13,967,044	100.0	14,126,118	100.0	15,432,099	100.0	13,606,565	100.0	13,789,841	100.0	14,523,836	100.0
Regions														
Africa	474,027	3.6	502,417	3.6	478,965	3.4	516,710	3.3	517,526	3.8	573,278	4.2	587,158	4.0
America, North	4,323,868	32.9	4,468,102	32.0	4,648,971	32.9	4,834,219	31.3	4,970,383	36.5	5,053,077	36.6	5,178,888	35.7
America, South	792,431	6.0	845,073	6.1	897,715	6.4	396,817	2.6	398,062	2.9	401,080	2.9	654,295	4.5
Asia	3,715,063	28.3	3,546,136	25.4	3,456,219	24.5	3,359,691	21.8	3,201,272	23.5	3,178,029	23.0	3,200,678	22.0
Europe	3,693,493	28.1	4,458,858	31.9	4,497,789	31.8	6,178,203	40.0	4,372,864	32.1	4,437,919	32.2	4,756,359	32.7
Oceania	146,458	1.1	146,458	1.0	146,458	1.0	146,458	0.9	146,458	1.1	146,458	1.1	146,458	1.0
Leading Producers														
United States of America	*4,009,883	30.5	*4,136,640	29.6	*4,263,396	30.2	*4,390,152	28.4	*4,516,908	33.2	*4,643,664	33.7	*4,770,421	32.8
Germany	-		834,745	6.0	850,574	6.0	827,680	5.4	744,377	5.5	708,896	5.1	710,815	4.9
Japan	696,726	5.3	683,989	4.9	683,091	4.8	700,927	4.5	700,690	5.1	707,203	5.1	710,708	4.9
United Kingdom	598,900	4.6	577,600	4.1	592,100	4.2	616,300	4.0	592,500	4.4	*614,655	4.5	*621,128	4.3
Russian Federation	808,267	6.1	627,300	4.5	560,464	4.0	438,363	2.8	277,986	2.0	197,643	1.4	199,876	1.4

Margarine

Unit of Measure: Metric tons.

	1990 Value	%	1991 Value	%	1992 Value	%	1993 Value	%	1994 Value	%	1995 Value	%	1996 Value	%
Total Production	8,049,478	100.0	8,641,648	100.0	8,775,221	100.0	8,663,547	100.0	8,439,506	100.0	8,817,261	100.0	8,749,378	100.0
Regions														
Africa	288,032	3.6	268,766	3.1	249,736	2.8	288,016	3.3	276,869	3.3	282,299	3.2	306,810	3.5
America, North	1,305,602	16.2	1,268,381	14.7	1,330,235	15.2	1,364,111	15.7	1,246,801	14.8	1,315,107	14.9	1,325,174	15.1
America, South	507,699	6.3	495,249	5.7	499,741	5.7	485,054	5.6	515,381	6.1	486,951	5.5	512,545	5.9
Asia	2,843,130	35.3	2,828,695	32.7	2,858,206	32.6	2,743,897	31.7	2,640,542	31.3	2,844,333	32.3	2,722,373	31.1
Europe	2,947,554	36.6	3,618,388	41.9	3,675,316	41.9	3,622,802	41.8	3,594,765	42.6	3,721,850	42.2	3,714,180	42.5
Oceania	157,461	2.0	162,170	1.9	161,988	1.8	159,668	1.8	165,148	2.0	166,722	1.9	168,296	1.9
Leading Producers														
United States of America	1,255,635	15.6	1,223,813	14.2	1,277,927	14.6	1,311,803	15.1	1,189,657	14.1	*1,259,251	14.3	*1,267,359	14.5
Germany	-		720,557	8.3	745,533	8.5	722,002	8.3	647,816	7.7	656,067	7.4	656,863	7.5
Turkey	485,803	6.0	553,913	6.4	530,899	6.0	492,822	5.7	481,035	5.7	535,244	6.1	535,101	6.1
United Kingdom	347,600	4.3	344,200	4.0	341,500	3.9	338,900	3.9	309,600	3.7	*322,871	3.7	*316,934	3.6
Brazil	368,270	4.6	349,465	4.0	352,995	4.0	348,315	4.0	383,019	4.5	350,178	4.0	365,331	4.2

Commodity data are provided by the United Nations Statistical Division. The symbol * means that data are estimated. For additional notes, see Appendix I.

31

Product Tables

Oils and Fats of Aquatic Animal Origin

Unit of Measure: Metric tons.

	1990		1991		1992		1993		1994		1995		1996	
	Value	%	Value	%	Value	%	Value	%	Value	%	Value	%	Value	%
Total Production	1,814,820	100.0	1,788,201	100.0	1,548,731	100.0	1,595,548	100.0	1,892,105	100.0	1,742,646	100.0	1,566,221	100.0
Regions														
Africa	15,491	0.9	20,084	1.1	23,389	1.5	23,332	1.5	33,040	1.7	40,815	2.3	41,090	2.6
America, North	155,700	8.6	149,100	8.3	101,800	6.6	153,424	9.6	149,400	7.9	129,600	7.4	129,600	8.3
America, South	395,753	21.8	436,275	24.4	326,097	21.1	436,577	27.4	780,137	41.2	576,756	33.1	576,727	36.8
Asia	673,802	37.1	579,041	32.4	423,981	27.4	341,512	21.4	277,339	14.7	301,089	17.3	306,939	19.6
Europe	572,694	31.6	602,513	33.7	672,430	43.4	640,064	40.1	651,089	34.4	692,586	39.7	510,065	32.6
Oceania	1,379	0.1	1,189	0.1	1,034	0.1	639	0.0	1,100	0.1	1,800	0.1	1,800	0.1
Leading Producers														
Japan	370,000	20.4	280,000	15.7	145,000	9.4	98,000	6.1	45,000	2.4	59,000	3.4	67,000	4.3
Chile	188,270	10.4	241,211	13.5	152,990	9.9	191,177	12.0	289,500	15.3	326,100	18.7	326,100	20.8
Peru	192,000	10.6	182,000	10.2	161,000	10.4	235,000	14.7	475,000	25.1	235,000	13.5	235,000	15.0
United Kingdom	124,100	6.8	129,000	7.2	109,000	7.0	105,000	6.6	115,000	6.1	109,800	6.3	109,800	7.0
United States of America	137,000	7.5	124,000	6.9	84,000	5.4	137,944	8.6	132,500	7.0	109,800	6.3	109,800	7.0

Oils and Fats of Animals, Unprocessed

Unit of Measure: Metric tons.

	1990		1991		1992		1993		1994		1995		1996	
	Value	%	Value	%	Value	%	Value	%	Value	%	Value	%	Value	%
Total Production	18,381,827	100.0	19,104,763	100.0	21,515,282	100.0	21,502,912	100.0	22,039,440	100.0	22,555,607	100.0	23,436,056	100.0
Regions														
Africa	216,703	1.2	227,566	1.2	233,146	1.1	239,385	1.1	234,668	1.1	239,848	1.1	240,548	1.0
America, North	5,969,506	32.5	6,137,555	32.1	6,458,533	30.0	6,420,044	29.9	6,614,063	30.0	6,718,153	29.8	6,609,424	28.2
America, South	1,076,465	5.9	1,100,687	5.8	1,124,109	5.2	1,175,807	5.5	1,187,013	5.4	1,203,313	5.3	1,230,041	5.2
Asia	5,277,662	28.7	5,774,450	30.2	7,543,382	35.1	7,929,690	36.9	8,348,180	37.9	8,711,212	38.6	9,046,202	38.6
Europe	5,205,623	28.3	5,191,498	27.2	5,626,809	26.2	5,201,337	24.2	5,113,214	23.2	5,148,925	22.8	5,797,224	24.7
Oceania	635,868	3.5	673,006	3.5	529,302	2.5	536,649	2.5	542,302	2.5	534,155	2.4	512,616	2.2
Leading Producers														
United States of America	5,218,918	28.4	5,378,038	28.2	5,648,210	26.3	5,615,066	26.1	5,788,192	26.3	5,853,149	25.9	5,729,228	24.4
Germany	1,239,628	6.7	1,133,702	5.9	1,080,171	5.0	1,074,815	5.0	1,064,869	4.8	1,041,895	4.6	1,046,519	4.5
China	1,490,036	8.1	1,597,181	8.4	1,719,232	8.0	1,866,227	8.7	2,035,473	9.2	2,139,061	9.5	2,204,576	9.4
Russian Federation	-		-		1,009,020	4.7	908,239	4.2	825,427	3.7	774,567	3.4	702,319	3.0
Brazil	599,025	3.3	628,398	3.3	651,140	3.0	694,220	3.2	693,905	3.1	718,201	3.2	732,743	3.1

Oil, Soya Bean, Crude

Unit of Measure: Metric tons.

	1990		1991		1992		1993		1994		1995		1996	
	Value	%	Value	%	Value	%	Value	%	Value	%	Value	%	Value	%
Total Production	11,875,932	100.0	13,145,176	100.0	13,151,835	100.0	13,117,205	100.0	13,458,885	100.0	14,696,905	100.0	14,748,833	100.0
Regions														
America, North	6,236,802	52.5	6,732,650	51.2	6,997,499	53.2	6,796,347	51.8	7,054,195	52.4	7,652,044	52.1	7,574,892	51.4
America, South	2,436,580	20.5	3,359,580	25.6	3,327,159	25.3	3,539,824	27.0	3,694,326	27.4	4,105,009	27.9	4,319,722	29.3
Asia	1,502,052	12.6	1,440,525	11.0	1,352,488	10.3	1,396,804	10.6	1,404,470	10.4	1,486,808	10.1	1,486,682	10.1
Europe	1,700,499	14.3	1,612,420	12.3	1,474,689	11.2	1,384,231	10.6	1,305,894	9.7	1,453,044	9.9	1,367,537	9.3
Leading Producers														
United States of America	5,721,000	48.2	6,208,000	47.2	6,464,000	49.1	6,254,000	47.7	6,503,000	48.3	7,092,000	48.3	7,006,000	47.5
Brazil	2,073,000	17.5	2,101,000	16.0	1,918,000	14.6	2,071,000	15.8	2,144,000	15.9	2,493,000	17.0	2,473,000	16.8
Argentina	352,000	3.0	1,255,000	9.5	1,402,000	10.7	1,460,000	11.1	1,541,000	11.4	1,599,000	10.9	1,838,000	12.5
Japan	665,000	5.6	629,000	4.8	671,000	5.1	683,000	5.2	664,000	4.9	680,000	4.6	673,000	4.6
China	665,000	5.6	708,000	5.4	555,000	4.2	*686,909	5.2	*714,510	5.3	*742,112	5.0	*769,713	5.2

Product Tables
Oil, Soya Bean, Refined

Unit of Measure: Metric tons.

	1990 Value	%	1991 Value	%	1992 Value	%	1993 Value	%	1994 Value	%	1995 Value	%	1996 Value	%
Total Production	8,801,498	100.0	9,180,202	100.0	9,300,570	100.0	9,315,091	100.0	9,916,760	100.0	10,675,424	100.0	9,373,383	100.0
Regions														
Africa	8,000	0.1	11,000	0.1	18,000	0.2	25,000	0.3	25,000	0.3	13,000	0.1	23,000	0.2
America, North	5,352,929	60.8	5,593,607	60.9	5,818,286	62.6	6,065,964	65.1	6,128,643	61.8	6,182,321	57.9	6,170,000	65.8
America, South	1,200,365	13.6	1,263,775	13.8	1,287,184	13.8	1,249,223	13.4	1,279,300	12.9	1,334,948	12.5	1,320,445	14.1
Asia	1,078,486	12.3	1,125,010	12.3	941,505	10.1	787,676	8.5	1,232,495	12.4	1,819,215	17.0	595,807	6.4
Europe	1,161,719	13.2	1,186,810	12.9	1,235,595	13.3	1,187,228	12.7	1,251,322	12.6	1,325,939	12.4	1,264,131	13.5
Leading Producers														
United States of America	4,985,000	56.6	5,196,000	56.6	5,380,000	57.8	5,586,000	60.0	5,621,000	56.7	5,624,000	52.7	5,567,000	59.4
Brazil	1,083,000	12.3	1,126,000	12.3	1,102,000	11.8	1,055,000	11.3	1,088,000	11.0	1,116,000	10.5	1,125,000	12.0
Belarus	*664,000	7.5	*664,000	7.2	*664,000	7.1	383,000	4.1	816,000	8.2	1,354,000	12.7	103,000	1.1
Germany	*336,235	3.8	*336,235	3.7	*336,235	3.6	*336,235	3.6	*336,235	3.4	328,330	3.1	344,140	3.7
Netherlands	263,000	3.0	276,000	3.0	290,000	3.1	322,000	3.5	329,000	3.3	381,000	3.6	*325,333	3.5

Oil, Cotton-Seed, Crude

Unit of Measure: Metric tons.

	1990 Value	%	1991 Value	%	1992 Value	%	1993 Value	%	1994 Value	%	1995 Value	%	1996 Value	%
Total Production	2,574,671	100.0	2,500,038	100.0	2,534,880	100.0	2,359,196	100.0	2,391,334	100.0	2,487,455	100.0	2,416,396	100.0
Regions														
Africa	224,625	8.7	224,250	9.0	226,250	8.9	225,250	9.5	224,577	9.4	224,387	9.0	224,198	9.3
America, North	634,471	24.6	650,571	26.0	625,171	24.7	591,771	25.1	589,371	24.6	652,971	26.3	616,571	25.5
America, South	287,119	11.2	218,119	8.7	183,119	7.2	102,119	4.3	124,119	5.2	150,119	6.0	111,119	4.6
Asia	1,392,456	54.1	1,384,098	55.4	1,468,340	57.9	1,428,056	60.5	1,438,267	60.1	1,444,978	58.1	1,444,689	59.8
Europe	36,000	1.4	23,000	0.9	32,000	1.3	12,000	0.5	15,000	0.6	15,000	0.6	19,820	0.8
Leading Producers														
United States of America	550,500	21.4	570,000	22.8	548,000	21.6	518,000	22.0	519,000	21.7	586,000	23.6	553,000	22.9
Pakistan	*220,667	8.6	*220,667	8.8	*220,667	8.7	*220,667	9.4	*220,667	9.2	*220,667	8.9	*220,667	9.1
Egypt	*204,500	7.9	*204,500	8.2	*204,500	8.1	*204,500	8.7	*204,500	8.6	*204,500	8.2	*204,500	8.5
China	119,600	4.6	126,000	5.0	193,000	7.6	*161,473	6.8	*164,942	6.9	*168,410	6.8	*171,879	7.1
Brazil	126,000	4.9	156,000	6.2	127,000	5.0	71,000	3.0	76,000	3.2	85,000	3.4	44,000	1.8

Oil, Cotton-Seed, Refined

Unit of Measure: Metric tons.

	1990 Value	%	1991 Value	%	1992 Value	%	1993 Value	%	1994 Value	%	1995 Value	%	1996 Value	%
Total Production	1,620,876	100.0	1,751,496	100.0	1,644,472	100.0	1,588,538	100.0	1,666,428	100.0	1,707,547	100.0	1,615,389	100.0
Regions														
Africa	343,460	21.2	395,534	22.6	344,650	21.0	353,747	22.3	351,783	21.1	340,429	19.9	325,075	20.1
America, North	459,000	28.3	450,000	25.7	462,000	28.1	449,000	28.3	468,000	28.1	444,000	26.0	397,000	24.6
America, South	53,175	3.3	105,410	6.0	90,645	5.5	56,345	3.5	53,987	3.2	56,665	3.3	51,004	3.2
Asia	726,289	44.8	774,885	44.2	725,796	44.1	699,351	44.0	755,848	45.4	823,611	48.2	805,833	49.9
Europe	38,952	2.4	25,667	1.5	21,381	1.3	30,095	1.9	36,810	2.2	42,842	2.5	36,477	2.3
Leading Producers														
Iran, Islamic Republic of	559,000	34.5	620,000	35.4	633,000	38.5	613,000	38.6	691,000	41.5	766,000	44.9	772,000	47.8
United States of America	440,000	27.1	442,000	25.2	453,000	27.5	438,000	27.6	462,000	27.7	444,000	26.0	397,000	24.6
Egypt	317,000	19.6	357,000	20.4	312,000	19.0	329,000	20.7	318,000	19.1	305,000	17.9	288,000	17.8
Brazil	27,000	1.7	65,000	3.7	62,000	3.8	34,000	2.1	40,000	2.4	41,000	2.4	25,000	1.5
Tajikistan	80,000	4.9	74,000	4.2	43,000	2.6	34,000	2.1	26,000	1.6	21,000	1.2	*1,533	0.1

Commodity data are provided by the United Nations Statistical Division. The symbol * means that data are estimated. For additional notes, see Appendix I.

33

Product Tables
Oil, Groundnut, Crude

Unit of Measure: Metric tons.

	1990 Value	%	1991 Value	%	1992 Value	%	1993 Value	%	1994 Value	%	1995 Value	%	1996 Value	%
Total Production	4,015,193	100.0	4,056,276	100.0	4,168,376	100.0	4,152,293	100.0	4,278,797	100.0	4,441,641	100.0	4,568,419	100.0
Regions														
Africa	389,150	9.7	265,123	6.5	297,112	7.1	241,254	5.8	274,677	6.4	286,100	6.4	284,047	6.2
America, North	110,000	2.7	119,000	2.9	173,000	4.2	106,000	2.6	104,000	2.4	151,000	3.4	159,000	3.5
America, South	43,000	1.1	78,000	1.9	96,000	2.3	44,000	1.1	40,000	0.9	49,000	1.1	61,000	1.3
Asia	3,452,200	86.0	3,572,600	88.1	3,580,000	85.9	3,738,067	90.0	3,830,610	89.5	3,924,152	88.3	4,030,267	88.2
Europe	20,843	0.5	21,553	0.5	22,264	0.5	22,973	0.6	29,510	0.7	31,389	0.7	34,106	0.7
Leading Producers														
India	*1,850,000	46.1	*1,850,000	45.6	*1,850,000	44.4	*1,850,000	44.6	*1,850,000	43.2	*1,850,000	41.7	*1,850,000	40.5
Philippines	*1,289,200	32.1	*1,394,600	34.4	*1,500,000	36.0	*1,605,400	38.7	*1,710,800	40.0	*1,816,200	40.9	*1,921,600	42.1
China	275,000	6.8	292,000	7.2	193,000	4.6	*236,667	5.7	*230,810	5.4	*224,952	5.1	*219,095	4.8
Sudan	*148,500	3.7	*148,500	3.7	*148,500	3.6	*148,500	3.6	*148,500	3.5	*148,500	3.3	*148,500	3.3
Senegal	200,000	5.0	76,000	1.9	114,000	2.7	57,000	1.4	80,000	1.9	99,000	2.2	*97,524	2.1

Oil, Groundnut, Refined

Unit of Measure: Metric tons.

	1990 Value	%	1991 Value	%	1992 Value	%	1993 Value	%	1994 Value	%	1995 Value	%	1996 Value	%
Total Production	258,865	100.0	203,546	100.0	247,216	100.0	262,728	100.0	247,015	100.0	247,639	100.0	238,526	100.0
Regions														
Africa	55,499	21.4	28,974	14.2	59,449	24.0	79,924	30.4	88,399	35.8	104,874	42.3	89,034	37.3
America, North	80,714	31.2	59,214	29.1	82,000	33.2	73,000	27.8	68,000	27.5	52,000	21.0	60,000	25.2
America, South	9,000	3.5	9,644	4.7	3,634	1.5	4,624	1.8	3,614	1.5	1,604	0.6	1,594	0.7
Asia	12,006	4.6	12,068	5.9	12,068	4.9	12,068	4.6	12,068	4.9	12,068	4.9	12,068	5.1
Europe	101,646	39.3	93,646	46.0	90,065	36.4	93,113	35.4	74,935	30.3	77,093	31.1	75,830	31.8
Leading Producers														
France	75,000	29.0	70,000	34.4	66,000	26.7	69,000	26.3	55,000	22.3	54,000	21.8	54,000	22.6
United States of America	75,000	29.0	57,000	28.0	82,000	33.2	73,000	27.8	68,000	27.5	52,000	21.0	60,000	25.2
Senegal	48,000	18.5	21,000	10.3	51,000	20.6	71,000	27.0	79,000	32.0	95,000	38.4	*78,686	33.0
Dominican Republic	*5,714	2.2	*2,214	1.1	-		-		-		-		-	
Thailand	*12,000	4.6	*12,000	5.9	*12,000	4.9	*12,000	4.6	*12,000	4.9	*12,000	4.8	*12,000	5.0

Oil, Olive, Crude

Unit of Measure: Metric tons.

	1990 Value	%	1991 Value	%	1992 Value	%	1993 Value	%	1994 Value	%	1995 Value	%	1996 Value	%
Total Production	1,626,540	100.0	2,476,377	100.0	1,986,142	100.0	1,990,418	100.0	2,029,692	100.0	1,761,863	100.0	2,685,008	100.0
Regions														
Africa	287,250	17.7	342,703	13.8	247,007	12.4	303,956	15.3	157,503	7.8	124,908	7.1	398,872	14.9
America, North	2,600	0.2	1,123	0.0	1,891	0.1	1,823	0.1	1,642	0.1	1,784	0.1	2,354	0.1
America, South	9,353	0.6	23,844	1.0	10,994	0.6	9,947	0.5	10,994	0.5	9,138	0.5	5,560	0.2
Asia	204,809	12.6	127,750	5.2	205,376	10.3	137,876	6.9	318,812	15.7	165,646	9.4	370,888	13.8
Europe	1,122,465	69.0	1,980,874	80.0	1,520,762	76.6	1,536,719	77.2	1,540,628	75.9	1,460,267	82.9	1,907,213	71.0
Oceania	64	0.0	82	0.0	112	0.0	96	0.0	112	0.0	120	0.0	120	0.0
Leading Producers														
Spain	702,854	43.2	647,152	26.1	679,399	34.2	612,808	30.8	585,330	28.8	360,000	20.4	1,002,800	37.3
Italy	176,350	10.8	835,800	33.8	469,240	23.6	608,010	30.5	523,197	25.8	677,903	38.5	419,541	15.6
Greece	214,000	13.2	425,500	17.2	339,000	17.1	275,000	13.8	389,800	19.2	362,000	20.5	429,705	16.0
Tunisia	179,000	11.0	281,000	11.3	133,000	6.7	226,000	11.4	81,500	4.0	64,500	3.7	270,000	10.1
Turkey	92,000	5.7	72,000	2.9	65,000	3.3	57,500	2.9	181,500	8.9	50,500	2.9	197,000	7.3

Commodity data are provided by the United Nations Statistical Division. The symbol * means that data are estimated. For additional notes, see Appendix I.

Product Tables
Oil, Olive, Refined

Unit of Measure: Metric tons.

	1990 Value	%	1991 Value	%	1992 Value	%	1993 Value	%	1994 Value	%	1995 Value	%	1996 Value	%
Total Production	460,457	100.0	383,319	100.0	326,338	100.0	322,984	100.0	436,907	100.0	524,072	100.0	400,049	100.0
Regions														
America, South	5,813	1.3	5,813	1.5	5,813	1.8	5,813	1.8	10,756	2.5	870	0.2	5,813	1.5
Asia	10,311	2.2	13,173	3.4	24,192	7.4	34,171	10.6	30,151	6.9	31,202	6.0	24,681	6.2
Europe	444,333	96.5	364,333	95.0	296,333	90.8	283,000	87.6	396,000	90.6	492,000	93.9	369,555	92.4
Leading Producers														
Spain	415,000	90.1	339,000	88.4	261,000	80.0	216,000	66.9	328,000	75.1	424,000	80.9	310,000	77.5
Turkey	8,000	1.7	12,000	3.1	21,000	6.4	32,000	9.9	28,000	6.4	30,000	5.7	23,000	5.7
Greece	16,000	3.5	16,000	4.2	16,000	4.9	39,000	12.1	36,000	8.2	33,000	6.3	*30,984	7.7
Portugal	10,000	2.2	6,000	1.6	16,000	4.9	25,000	7.7	28,000	6.4	32,000	6.1	*25,238	6.3
Colombia	*5,813	1.3	*5,813	1.5	*5,813	1.8	*5,813	1.8	10,756	2.5	870	0.2	*5,813	1.5

Oils, Other, of Vegetable Origin, Crude

Unit of Measure: Metric tons.

	1990 Value	%	1991 Value	%	1992 Value	%	1993 Value	%	1994 Value	%	1995 Value	%	1996 Value	%
Total Production	27,864,929	100.0	28,949,378	100.0	29,455,286	100.0	30,532,050	100.0	31,300,188	100.0	33,936,324	100.0	34,944,561	100.0
Regions														
Africa	1,924,675	6.9	2,014,540	7.0	2,150,196	7.3	2,017,368	6.6	2,106,727	6.7	2,007,455	5.9	2,221,314	6.4
America, North	506,746	1.8	586,899	2.0	577,656	2.0	681,571	2.2	682,518	2.2	611,685	1.8	604,078	1.7
America, South	2,061,000	7.4	2,187,000	7.6	1,992,911	6.8	2,091,431	6.8	2,237,849	7.1	2,342,253	6.9	2,498,120	7.1
Asia	16,013,686	57.5	16,735,820	57.8	17,380,502	59.0	19,205,265	62.9	19,794,281	63.2	22,078,981	65.1	22,302,434	63.8
Europe	6,998,279	25.1	7,028,772	24.3	6,924,572	23.5	6,092,501	20.0	6,014,209	19.2	6,415,893	18.9	6,830,626	19.5
Oceania	360,543	1.3	396,346	1.4	429,448	1.5	443,914	1.5	464,605	1.5	480,058	1.4	487,989	1.4
Leading Producers														
Malaysia	6,095,000	21.9	6,141,000	21.2	6,373,000	21.6	7,403,000	24.2	7,221,000	23.1	7,811,000	23.0	*8,086,229	23.1
China	4,382,000	15.7	5,317,000	18.4	5,609,000	19.0	*6,143,015	20.1	*6,730,889	21.5	*7,318,763	21.6	*7,906,638	22.6
Indonesia	2,544,000	9.1	2,019,000	7.0	2,280,000	7.7	2,539,000	8.3	2,766,000	8.8	3,701,000	10.9	*2,996,619	8.6
Argentina	1,524,000	5.5	1,600,000	5.5	*1,510,364	5.1	*1,596,818	5.2	*1,683,273	5.4	*1,769,727	5.2	*1,856,182	5.3
Japan	1,049,000	3.8	1,083,000	3.7	1,056,000	3.6	1,055,000	3.5	1,052,000	3.4	1,070,000	3.2	1,092,000	3.1

Oils, Other, of Vegetable Origin, Refined

Unit of Measure: Metric tons.

	1990 Value	%	1991 Value	%	1992 Value	%	1993 Value	%	1994 Value	%	1995 Value	%	1996 Value	%
Total Production	9,037,247	100.0	8,517,955	100.0	9,328,805	100.0	9,820,444	100.0	10,096,293	100.0	12,552,762	100.0	11,700,521	100.0
Regions														
Africa	390,869	4.3	348,954	4.1	321,289	3.4	336,408	3.4	341,017	3.4	362,410	2.9	360,330	3.1
America, North	1,138,008	12.6	1,144,589	13.4	1,113,456	11.9	1,220,322	12.4	1,324,189	13.1	1,428,056	11.4	1,534,189	13.1
America, South	370,050	4.1	395,287	4.6	413,524	4.4	482,094	4.9	477,286	4.7	510,293	4.1	520,239	4.4
Asia	1,928,544	21.3	1,688,265	19.8	2,276,795	24.4	2,352,345	24.0	2,567,380	25.4	4,461,021	35.5	3,211,726	27.4
Europe	5,044,074	55.8	4,767,621	56.0	5,023,359	53.8	5,240,878	53.4	5,190,498	51.4	5,587,645	44.5	5,863,261	50.1
Oceania	165,701	1.8	173,238	2.0	180,383	1.9	188,398	1.9	195,923	1.9	203,338	1.6	210,775	1.8
Leading Producers														
Germany	*1,052,497	11.6	*1,052,497	12.4	*1,052,497	11.3	*1,052,497	10.7	*1,052,497	10.4	1,013,515	8.1	1,091,478	9.3
Indonesia	627,000	6.9	280,000	3.3	971,000	10.4	993,000	10.1	1,324,000	13.1	3,164,000	25.2	*1,878,029	16.1
United Kingdom	944,000	10.4	969,000	11.4	1,004,000	10.8	1,017,000	10.4	1,033,000	10.2	*1,175,286	9.4	*1,230,743	10.5
United States of America	*657,500	7.3	*657,500	7.7	*657,500	7.0	*657,500	6.7	*657,500	6.5	*657,500	5.2	*657,500	5.6
Netherlands	881,000	9.7	580,000	6.8	616,000	6.6	765,000	7.8	700,000	6.9	677,000	5.4	*789,676	6.7

Product Tables
Flour, Wheat

Unit of Measure: 1,000 Metric tons.

	1990		1991		1992		1993		1994		1995		1996	
	Value	%	Value	%	Value	%	Value	%	Value	%	Value	%	Value	%
Total Production	274,421	100.0	275,220	100.0	291,002	100.0	282,245	100.0	277,357	100.0	270,722	100.0	281,873	100.0
Regions														
Africa	16,423	6.0	16,342	5.9	16,933	5.8	17,253	6.1	17,972	6.5	18,774	6.9	21,573	7.7
America, North	21,766	7.9	22,151	8.0	22,407	7.7	23,499	8.3	24,156	8.7	23,596	8.7	23,826	8.5
America, South	11,015	4.0	12,057	4.4	12,316	4.2	12,146	4.3	12,558	4.5	12,752	4.7	13,393	4.8
Asia	172,311	62.8	173,556	63.1	188,099	64.6	180,865	64.1	175,015	63.1	166,478	61.5	172,891	61.3
Europe	51,372	18.7	49,590	18.0	49,673	17.1	46,819	16.6	45,893	16.5	47,373	17.5	48,299	17.1
Oceania	1,532	0.6	1,523	0.6	1,574	0.5	1,662	0.6	1,764	0.6	1,749	0.6	1,890	0.7
Leading Producers														
China	70,500	25.7	71,850	26.1	72,900	25.1	74,625	26.4	73,500	26.5	73,500	27.1	78,750	27.9
United States of America	16,073	5.9	16,434	6.0	16,820	5.8	17,571	6.2	17,805	6.4	17,631	6.5	18,043	6.4
Pakistan	14,712	5.4	14,801	5.4	16,481	5.7	16,005	5.7	15,860	5.7	14,841	5.5	15,326	5.4
Russian Federation	15,770	5.7	15,458	5.6	14,803	5.1	12,439	4.4	9,346	3.4	7,498	2.8	9,508	3.4
Italy	7,838	2.9	8,549	3.1	8,666	3.0	7,960	2.8	7,716	2.8	7,880	2.9	8,132	2.9

Meal and Groats of All Cereals

Unit of Measure: Metric tons.

	1990		1991		1992		1993		1994		1995		1996	
	Value	%	Value	%	Value	%	Value	%	Value	%	Value	%	Value	%
Total Production	19,979,802	100.0	20,396,236	100.0	19,647,163	100.0	20,319,609	100.0	20,269,767	100.0	19,318,709	100.0	19,172,788	100.0
Regions														
Africa	4,173,595	20.9	4,196,438	20.6	4,192,281	21.3	4,851,290	23.9	4,997,710	24.7	5,165,350	26.7	5,277,266	27.5
America, North	1,132,567	5.7	1,100,274	5.4	1,076,981	5.5	1,055,688	5.2	1,037,395	5.1	1,000,102	5.2	975,810	5.1
America, South	249,413	1.2	250,219	1.2	255,939	1.3	254,273	1.3	261,658	1.3	271,719	1.4	276,769	1.4
Asia	11,104,459	55.6	11,511,189	56.4	10,931,918	55.6	11,158,793	54.9	11,068,304	54.6	9,897,816	51.2	10,266,194	53.5
Europe	3,202,491	16.0	3,220,491	15.8	3,072,069	15.6	2,881,241	14.2	2,786,027	13.7	2,864,701	14.8	2,257,379	11.8
Oceania	117,277	0.6	117,626	0.6	117,975	0.6	118,324	0.6	118,673	0.6	119,022	0.6	119,370	0.6
Leading Producers														
Uzbekistan	2,892,000	14.5	3,203,000	15.7	3,097,000	15.8	3,248,000	16.0	3,215,000	15.9	1,983,000	10.3	*2,503,867	13.1
Algeria	2,588,000	13.0	2,634,000	12.9	2,539,000	12.9	*3,151,167	15.5	*3,310,590	16.3	*3,470,013	18.0	*3,629,436	18.9
Russian Federation	2,854,000	14.3	2,679,000	13.1	2,011,000	10.2	1,877,000	9.2	1,597,000	7.9	1,418,000	7.3	988,000	5.2
United States of America	*1,097,067	5.5	*1,071,667	5.3	*1,046,267	5.3	*1,020,867	5.0	*995,467	4.9	*970,067	5.0	*944,667	4.9
Ukraine	962,000	4.8	944,000	4.6	804,000	4.1	695,000	3.4	605,000	3.0	532,000	2.8	456,000	2.4

Flour, Cereal, Other Than Wheat

Unit of Measure: Metric tons.

	1990		1991		1992		1993		1994		1995		1996	
	Value	%	Value	%	Value	%	Value	%	Value	%	Value	%	Value	%
Total Production	9,030,865	100.0	9,450,036	100.0	9,012,881	100.0	7,931,062	100.0	7,825,025	100.0	8,007,668	100.0	9,232,233	100.0
Regions														
Africa	532,138	5.9	516,908	5.5	364,171	4.0	367,595	4.6	452,154	5.8	522,686	6.5	467,620	5.1
America, North	1,071,588	11.9	1,058,533	11.2	1,108,478	12.3	1,518,539	19.1	1,929,112	24.7	2,339,685	29.2	2,750,258	29.8
America, South	130,208	1.4	139,599	1.5	112,991	1.3	192,338	2.4	264,085	3.4	279,009	3.5	264,059	2.9
Asia	2,847,670	31.5	2,989,270	31.6	2,682,870	29.8	2,071,636	26.1	1,761,353	22.5	1,552,020	19.4	2,502,220	27.1
Europe	4,370,833	48.4	4,678,333	49.5	4,688,014	52.0	3,735,632	47.1	3,384,036	43.2	3,291,018	41.1	3,235,861	35.0
Oceania	78,429	0.9	67,393	0.7	56,357	0.6	45,321	0.6	34,286	0.4	23,250	0.3	12,214	0.1
Leading Producers														
Russian Federation	2,602,000	28.8	2,747,000	29.1	2,450,000	27.2	1,856,000	23.4	1,563,000	20.0	1,347,000	16.8	2,277,000	24.7
Mexico	932,000	10.3	914,000	9.7	959,000	10.6	1,364,500	17.2	1,770,000	22.6	2,175,500	27.2	2,581,000	28.0
Poland	576,000	6.4	492,000	5.2	454,000	5.0	362,000	4.6	319,000	4.1	280,000	3.5	321,000	3.5
Germany	-		640,000	6.8	631,000	7.0	595,000	7.5	538,000	6.9	500,024	6.2	458,727	5.0
Ukraine	680,000	7.5	588,000	6.2	559,000	6.2	460,000	5.8	309,000	3.9	275,000	3.4	275,000	3.0

Commodity data are provided by the United Nations Statistical Division. The symbol * means that data are estimated. For additional notes, see Appendix I.

Product Tables
Cereal Breakfast Food

Unit of Measure: Metric tons.

	1990 Value	%	1991 Value	%	1992 Value	%	1993 Value	%	1994 Value	%	1995 Value	%	1996 Value	%
Total Production	1,889,953	100.0	1,833,890	100.0	1,759,471	100.0	1,683,195	100.0	1,598,428	100.0	1,558,447	100.0	1,504,979	100.0
Regions														
Africa	24,578	1.3	30,022	1.6	26,971	1.5	32,584	1.9	36,565	2.3	36,905	2.4	34,181	2.3
America, North	1,085,030	57.4	998,755	54.5	912,480	51.9	826,205	49.1	739,930	46.3	653,655	41.9	567,380	37.7
America, South	8,813	0.5	9,995	0.5	11,177	0.6	14,258	0.8	2,041	0.1	20,140	1.3	14,582	1.0
Asia	64,816	3.4	61,721	3.4	53,647	3.0	49,576	2.9	40,117	2.5	36,445	2.3	36,064	2.4
Europe	570,504	30.2	587,669	32.0	608,008	34.6	618,485	36.7	653,601	40.9	679,812	43.6	721,570	47.9
Oceania	136,211	7.2	145,728	7.9	147,187	8.4	142,086	8.4	126,174	7.9	131,489	8.4	131,202	8.7
Leading Producers														
United States of America	*1,083,375	57.3	*997,065	54.4	*910,756	51.8	*824,446	49.0	*738,137	46.2	*651,827	41.8	*565,517	37.6
United Kingdom	278,700	14.7	290,300	15.8	301,500	17.1	307,700	18.3	330,200	20.7	*326,324	20.9	*333,525	22.2
Germany	*151,121	8.0	*151,121	8.2	*151,121	8.6	*151,121	9.0	*151,121	9.5	144,307	9.3	157,935	10.5
Australia	109,696	5.8	118,459	6.5	119,165	6.8	113,310	6.7	96,644	6.0	101,206	6.5	100,165	6.7
Russian Federation	62,300	3.3	59,205	3.2	51,131	2.9	45,054	2.7	38,312	2.4	34,609	2.2	34,162	2.3

Macaroni and Noodle Products, Uncooked

Unit of Measure: Metric tons.

	1990 Value	%	1991 Value	%	1992 Value	%	1993 Value	%	1994 Value	%	1995 Value	%	1996 Value	%
Total Production	16,705,506	100.0	10,182,769	100.0	10,331,167	100.0	10,072,568	100.0	11,432,000	100.0	11,531,987	100.0	11,553,439	100.0
Regions														
Africa	7,303,627	43.7	360,595	3.5	317,232	3.1	301,835	3.0	1,804,116	15.8	1,874,811	16.3	1,894,325	16.4
America, North	1,294,139	7.7	1,321,876	13.0	1,317,517	12.8	1,363,205	13.5	1,403,807	12.3	1,437,606	12.5	1,484,405	12.8
America, South	1,026,334	6.1	1,044,140	10.3	1,046,946	10.1	1,060,292	10.5	1,119,275	9.8	1,107,968	9.6	1,139,354	9.9
Asia	5,296,357	31.7	5,397,698	53.0	5,580,126	54.0	5,342,384	53.0	5,100,994	44.6	5,037,689	43.7	4,960,885	42.9
Europe	1,734,484	10.4	2,005,946	19.7	2,014,883	19.5	1,948,438	19.3	1,945,445	17.0	2,013,601	17.5	2,012,208	17.4
Oceania	50,565	0.3	52,514	0.5	54,463	0.5	56,413	0.6	58,362	0.5	60,312	0.5	62,261	0.5
Leading Producers														
Japan	1,407,000	8.4	1,441,000	14.2	1,458,000	14.1	1,445,000	14.3	1,451,000	12.7	1,456,000	12.6	1,464,000	12.7
Mauritania	6,939,000	41.5	7,988	0.1	8,904	0.1	10,395	0.1	*1,496,109	13.1	*1,538,204	13.3	*1,580,298	13.7
United States of America	*1,020,829	6.1	*1,038,571	10.2	*1,056,314	10.2	*1,074,057	10.7	*1,091,800	9.6	*1,109,543	9.6	*1,127,286	9.8
Russian Federation	1,038,000	6.2	1,115,000	10.9	1,102,000	10.7	836,000	8.3	680,000	5.9	603,000	5.2	444,000	3.8
Brazil	546,000	3.3	545,000	5.4	524,000	5.1	528,000	5.2	589,000	5.2	561,000	4.9	565,000	4.9

Bread, Ships' Biscuits and Other Ordinary Baker's Wares

Unit of Measure: Metric tons.

	1990 Value	%	1991 Value	%	1992 Value	%	1993 Value	%	1994 Value	%	1995 Value	%	1996 Value	%
Total Production	93,862,675	100.0	95,432,248	100.0	92,262,060	100.0	87,158,241	100.0	82,198,977	100.0	78,164,561	100.0	76,363,833	100.0
Regions														
Africa	1,669,364	1.8	1,548,553	1.6	1,582,518	1.7	1,545,423	1.8	1,546,262	1.9	1,481,715	1.9	1,463,274	1.9
America, North	7,895,412	8.4	7,835,229	8.2	7,775,046	8.4	7,683,612	8.8	7,631,429	9.3	7,620,246	9.7	7,563,062	9.9
America, South	105,862	0.1	117,669	0.1	130,476	0.1	143,954	0.2	140,427	0.2	169,778	0.2	150,959	0.2
Asia	57,685,511	61.5	58,682,320	61.5	55,528,812	60.2	53,229,815	61.1	49,427,977	60.1	45,703,889	58.5	43,627,526	57.1
Europe	25,693,455	27.4	26,428,013	27.7	26,417,351	28.6	23,720,186	27.2	22,610,240	27.5	22,338,898	28.6	22,701,583	29.7
Oceania	813,071	0.9	820,464	0.9	827,857	0.9	835,250	1.0	842,643	1.0	850,036	1.1	857,429	1.1
Leading Producers														
Russian Federation	18,242,000	19.4	18,845,000	19.7	16,834,000	18.2	15,030,000	17.2	12,417,000	15.1	11,170,000	14.3	9,851,000	12.9
United States of America	*6,725,295	7.2	*6,656,238	7.0	*6,587,181	7.1	*6,518,124	7.5	*6,449,067	7.8	*6,380,010	8.2	*6,310,952	8.3
Ukraine	6,701,000	7.1	6,685,000	7.0	6,441,000	7.0	5,444,000	6.2	4,816,000	5.9	4,114,000	5.3	3,452,000	4.5
Poland	3,129,000	3.3	2,775,000	2.9	2,634,000	2.9	2,707,000	3.1	2,750,000	3.3	2,750,000	3.5	2,915,000	3.8
Romania	2,988,000	3.2	2,558,000	2.7	2,504,000	2.7	2,145,000	2.5	2,129,000	2.6	2,271,000	2.9	2,425,000	3.2

Commodity data are provided by the United Nations Statistical Division. The symbol * means that data are estimated. For additional notes, see Appendix I.

Product Tables
Biscuits

Unit of Measure: Metric tons.

	1990 Value	%	1991 Value	%	1992 Value	%	1993 Value	%	1994 Value	%	1995 Value	%	1996 Value	%
Total Production	10,029,237	100.0	10,112,367	100.0	10,396,470	100.0	10,026,706	100.0	10,029,162	100.0	10,242,927	100.0	10,179,133	100.0
Regions														
Africa	169,760	1.7	166,976	1.7	173,128	1.7	175,164	1.7	136,270	1.4	181,197	1.8	171,866	1.7
America, North	1,805,875	18.0	1,835,937	18.2	1,874,465	18.0	1,960,410	19.6	2,045,686	20.4	2,132,331	20.8	2,155,531	21.2
America, South	823,544	8.2	850,096	8.4	843,579	8.1	842,373	8.4	859,662	8.6	871,710	8.5	865,166	8.5
Asia	3,477,663	34.7	3,557,129	35.2	3,793,604	36.5	3,297,965	32.9	3,270,850	32.6	3,308,191	32.3	3,187,116	31.3
Europe	3,577,264	35.7	3,522,468	34.8	3,525,018	33.9	3,568,156	35.6	3,534,126	35.2	3,559,637	34.8	3,611,473	35.5
Oceania	175,132	1.7	179,761	1.8	186,676	1.8	182,638	1.8	182,568	1.8	189,862	1.9	187,982	1.8
Leading Producers														
United States of America	*1,306,188	13.0	*1,345,227	13.3	*1,384,266	13.3	*1,423,305	14.2	*1,462,345	14.6	*1,501,384	14.7	*1,540,423	15.1
Italy	*692,720	6.9	*713,700	7.1	*734,680	7.1	*755,660	7.5	*776,640	7.7	*797,620	7.8	*818,600	8.0
United Kingdom	649,900	6.5	670,700	6.6	674,900	6.5	691,500	6.9	684,300	6.8	*679,130	6.6	*682,306	6.7
Brazil	*636,845	6.3	*636,845	6.3	*636,845	6.1	*636,845	6.4	*636,845	6.3	*636,845	6.2	*636,845	6.3
India	664,181	6.6	635,513	6.3	619,105	6.0	630,214	6.3	652,223	6.5	672,710	6.6	621,902	6.1

Pastry, Cakes, and Other Fine Bakers' Wares

Unit of Measure: Metric tons.

	1990 Value	%	1991 Value	%	1992 Value	%	1993 Value	%	1994 Value	%	1995 Value	%	1996 Value	%
Total Production	6,054,746	100.0	6,012,531	100.0	5,237,189	100.0	4,987,924	100.0	4,948,345	100.0	4,787,228	100.0	4,855,006	100.0
Regions														
Africa	92	0.0	102	0.0	90	0.0	175	0.0	198	0.0	244	0.0	-	
America, North	1,555,918	25.7	1,560,326	26.0	1,564,734	29.9	1,563,919	31.4	1,587,037	32.1	1,570,326	32.8	1,581,731	32.6
America, South	29,130	0.5	30,940	0.5	32,750	0.6	33,289	0.7	37,786	0.8	39,246	0.8	39,211	0.8
Asia	2,528,893	41.8	2,506,069	41.7	1,769,209	33.8	1,660,578	33.3	1,535,784	31.0	1,361,790	28.4	1,340,026	27.6
Europe	1,731,145	28.6	1,705,526	28.4	1,660,838	31.7	1,520,395	30.5	1,577,971	31.9	1,606,054	33.5	1,684,470	34.7
Oceania	209,568	3.5	209,568	3.5	209,568	4.0	209,568	4.2	209,568	4.2	209,568	4.4	209,568	4.3
Leading Producers														
United States of America	*1,401,772	23.2	*1,406,180	23.4	*1,410,588	26.9	*1,414,996	28.4	*1,419,403	28.7	*1,423,811	29.7	*1,428,219	29.4
Russian Federation	1,476,352	24.4	1,446,745	24.1	998,513	19.1	906,284	18.2	815,073	16.5	681,111	14.2	603,959	12.4
Germany	*568,366	9.4	*568,366	9.5	*568,366	10.9	*568,366	11.4	*568,366	11.5	557,866	11.7	578,866	11.9
Spain	289,952	4.8	314,023	5.2	307,941	5.9	237,978	4.8	331,636	6.7	372,938	7.8	405,897	8.4
Kazakhstan	258,852	4.3	*258,852	4.3	*258,852	4.9	*258,852	5.2	*258,852	5.2	*258,852	5.4	*258,852	5.3

Farinaceous Preparations (e.g. Ravioli, Tortellini Etc.)

Unit of Measure: Metric tons.

	1990 Value	%	1991 Value	%	1992 Value	%	1993 Value	%	1994 Value	%	1995 Value	%	1996 Value	%
Total Production	1,605,767	100.0	1,622,011	100.0	1,616,889	100.0	1,615,469	100.0	1,623,498	100.0	1,633,720	100.0	1,681,679	100.0
Regions														
Africa	1,080	0.1	1,094	0.1	1,159	0.1	1,175	0.1	1,437	0.1	1,473	0.1	1,510	0.1
America, North	1,052,507	65.5	1,066,902	65.8	1,073,425	66.4	1,086,899	67.3	1,098,395	67.7	1,117,150	68.4	1,131,495	67.3
America, South	103,192	6.4	114,061	7.0	124,930	7.7	120,497	7.5	115,700	7.1	121,014	7.4	144,034	8.6
Asia	108,551	6.8	99,350	6.1	76,645	4.7	73,187	4.5	65,061	4.0	51,557	3.2	51,469	3.1
Europe	320,975	20.0	321,142	19.8	321,269	19.9	314,250	19.5	323,442	19.9	323,064	19.8	333,709	19.8
Oceania	19,462	1.2	19,462	1.2	19,462	1.2	19,462	1.2	19,462	1.2	19,462	1.2	19,462	1.2
Leading Producers														
United States of America	*1,020,509	63.6	*1,033,594	63.7	*1,046,680	64.7	*1,059,765	65.6	*1,072,850	66.1	*1,085,935	66.5	*1,099,020	65.4
United Kingdom	141,648	8.8	*141,197	8.7	*141,197	8.7	*141,197	8.7	*141,197	8.7	*141,197	8.6	*141,197	8.4
France	93,651	5.8	94,509	5.8	96,912	6.0	94,212	5.8	95,700	5.9	95,841	5.9	101,487	6.0
Colombia	76,251	4.7	87,120	5.4	97,989	6.1	93,556	5.8	88,759	5.5	94,073	5.8	*117,093	7.0
Belarus	75,178	4.7	76,114	4.7	58,083	3.6	51,432	3.2	44,055	2.7	31,982	2.0	38,201	2.3

Product Tables
Raw Sugar

Unit of Measure: 1,000 Metric tons.

	1990		1991		1992		1993		1994		1995		1996	
	Value	%	Value	%	Value	%	Value	%	Value	%	Value	%	Value	%
Total Production	122,020	100.0	122,872	100.0	127,348	100.0	121,311	100.0	119,114	100.0	122,222	100.0	132,693	100.0
Regions														
Africa	6,492	5.3	6,572	5.3	5,673	4.5	5,271	4.3	6,081	5.1	6,229	5.1	7,079	5.3
America, North	20,413	16.7	20,219	16.5	20,698	16.3	17,234	14.2	16,874	14.2	15,374	12.6	16,849	12.7
America, South	13,119	10.8	14,379	11.7	12,770	10.0	13,403	11.0	15,226	12.8	17,019	13.9	18,774	14.1
Asia	45,167	37.0	48,467	39.4	54,164	42.5	50,822	41.9	48,245	40.5	53,484	43.8	58,475	44.1
Europe	32,799	26.9	29,592	24.1	29,305	23.0	29,586	24.4	26,897	22.6	24,467	20.0	25,345	19.1
Oceania	4,029	3.3	3,642	3.0	4,737	3.7	4,994	4.1	5,791	4.9	5,648	4.6	6,171	4.7
Leading Producers														
India	12,068	9.9	13,113	10.7	13,873	10.9	11,750	9.7	11,745	9.9	15,337	12.5	18,225	13.7
Brazil	7,900	6.5	8,755	7.1	8,655	6.8	8,744	7.2	10,952	9.2	11,986	9.8	12,231	9.2
Cuba	8,050	6.6	7,233	5.9	7,104	5.6	4,365	3.6	4,024	3.4	3,300	2.7	4,400	3.3
United States of America	6,263	5.1	6,558	5.3	7,039	5.5	6,944	5.7	7,191	6.0	6,804	5.6	6,556	4.9
China	5,820	4.8	6,950	5.7	8,309	6.5	7,713	6.4	5,921	5.0	5,586	4.6	6,402	4.8

Refined Sugar

Unit of Measure: 1,000 Metric tons.

	1990		1991		1992		1993		1994		1995		1996	
	Value	%	Value	%	Value	%	Value	%	Value	%	Value	%	Value	%
Total Production	95,714	100.0	98,039	100.0	103,319	100.0	93,879	100.0	97,743	100.0	102,876	100.0	109,475	100.0
Regions														
Africa	5,410	5.7	5,610	5.7	5,464	5.3	5,185	5.5	5,471	5.6	5,513	5.4	6,216	5.7
America, North	14,920	15.6	14,003	14.3	14,885	14.4	12,540	13.4	13,153	13.5	12,882	12.5	14,135	12.9
America, South	5,482	5.7	5,725	5.8	7,306	7.1	6,111	6.5	6,369	6.5	7,046	6.8	7,404	6.8
Asia	46,445	48.5	51,611	52.6	53,963	52.2	49,062	52.3	52,457	53.7	57,357	55.8	59,547	54.4
Europe	22,605	23.6	20,376	20.8	19,548	18.9	19,523	20.8	18,477	18.9	18,785	18.3	20,169	18.4
Oceania	852	0.9	714	0.7	2,153	2.1	1,458	1.6	1,815	1.9	1,292	1.3	2,004	1.8
Leading Producers														
India	11,241	11.7	12,989	13.2	12,531	12.1	9,973	10.6	12,612	12.9	14,787	14.4	15,307	14.0
United States of America	7,580	7.9	7,784	7.9	8,330	8.1	8,001	8.5	8,218	8.4	7,720	7.5	8,656	7.9
China	7,208	7.5	8,777	9.0	8,650	8.4	6,354	6.8	6,769	6.9	8,573	8.3	7,839	7.2
Germany	4,639	4.8	3,747	3.8	3,970	3.8	3,964	4.2	3,382	3.5	3,497	3.4	3,852	3.5
Japan	2,563	2.7	2,575	2.6	2,523	2.4	2,422	2.6	2,407	2.5	2,368	2.3	2,359	2.2

Fruit, Glace or Crystallized

Unit of Measure: Metric tons.

	1990		1991		1992		1993		1994		1995		1996	
	Value	%	Value	%	Value	%	Value	%	Value	%	Value	%	Value	%
Total Production	23,984	100.0	23,523	100.0	27,059	100.0	24,547	100.0	28,063	100.0	26,842	100.0	30,011	100.0
Regions														
America, South	367	1.5	367	1.6	255	0.9	356	1.5	438	1.6	418	1.6	367	1.2
Asia	344	1.4	525	2.2	1,138	4.2	987	4.0	1,155	4.1	1,284	4.8	1,413	4.7
Europe	22,559	94.1	22,192	94.3	25,539	94.4	23,204	94.5	26,471	94.3	25,140	93.7	28,231	94.1
Oceania	715	3.0	439	1.9	128	0.5	-						-	
Leading Producers														
Germany	*9,609	40.1	*9,609	40.8	*9,609	35.5	*9,609	39.1	*9,609	34.2	8,865	33.0	10,353	34.5
Spain	*9,296	38.8	*9,296	39.5	*9,296	34.4	5,899	24.0	9,846	35.1	9,833	36.6	11,607	38.7
Greece	1,815	7.6	2,091	8.9	2,029	7.5	2,535	10.3	2,519	9.0	1,664	6.2	*2,187	7.3
Australia	715	3.0	*439	1.9	*128	0.5	-		-				-	
Portugal	561	2.3	139	0.6	3,404	12.6	4,022	16.4	3,128	11.1	3,428	12.8	*2,953	9.8

Commodity data are provided by the United Nations Statistical Division. The symbol * means that data are estimated. For additional notes, see Appendix I.

39

Product Tables
Sugar Confectionery

Unit of Measure: Metric tons.

	1990		1991		1992		1993		1994		1995		1996	
	Value	%	Value	%	Value	%	Value	%	Value	%	Value	%	Value	%
Total Production	9,170,429	100.0	9,026,413	100.0	8,655,536	100.0	8,601,259	100.0	8,365,229	100.0	8,507,752	100.0	8,691,609	100.0
Regions														
Africa	260,280	2.8	271,592	3.0	275,124	3.2	281,605	3.3	242,866	2.9	253,867	3.0	276,799	3.2
America, North	956,636	10.4	1,055,529	11.7	1,173,538	13.6	1,216,678	14.1	1,255,798	15.0	1,331,979	15.7	1,401,488	16.1
America, South	3,873	0.0	4,796	0.1	4,817	0.1	4,911	0.1	3,580	0.0	4,035	0.0	4,841	0.1
Asia	4,245,941	46.3	4,087,823	45.3	3,677,960	42.5	3,664,776	42.6	3,503,113	41.9	3,491,067	41.0	3,509,343	40.4
Europe	3,624,129	39.5	3,526,597	39.1	3,440,121	39.7	3,349,535	38.9	3,271,494	39.1	3,337,411	39.2	3,409,224	39.2
Oceania	79,569	0.9	80,076	0.9	83,976	1.0	83,753	1.0	88,379	1.1	89,394	1.1	89,913	1.0
Leading Producers														
United States of America	898,655	9.8	998,022	11.1	1,116,505	12.9	1,160,186	13.5	1,195,674	14.3	1,280,451	15.1	1,346,237	15.5
Russian Federation	1,391,878	15.2	1,193,912	13.2	850,456	9.8	840,040	9.8	715,183	8.5	691,092	8.1	656,075	7.5
United Kingdom	926,990	10.1	928,960	10.3	935,245	10.8	965,725	11.2	987,985	11.8	*1,007,243	11.8	*1,026,603	11.8
Germany	*607,104	6.6	604,776	6.7	609,431	7.0	*607,104	7.1	*607,104	7.3	*607,104	7.1	*607,104	7.0
Ukraine	541,837	5.9	611,594	6.8	490,104	5.7	331,066	3.8	193,519	2.3	153,264	1.8	141,148	1.6

Cocoa Powder

Unit of Measure: Metric tons.

	1990		1991		1992		1993		1994		1995		1996	
	Value	%	Value	%	Value	%	Value	%	Value	%	Value	%	Value	%
Total Production	546,094	100.0	564,147	100.0	638,443	100.0	646,835	100.0	657,917	100.0	678,895	100.0	675,469	100.0
Regions														
Africa	5,022	0.9	5,735	1.0	5,516	0.9	5,395	0.8	5,220	0.8	5,441	0.8	5,821	0.9
America, North	224,929	41.2	225,374	39.9	225,819	35.4	226,264	35.0	226,709	34.5	227,154	33.5	227,599	33.7
America, South	26,347	4.8	28,010	5.0	29,969	4.7	28,773	4.4	32,244	4.9	35,166	5.2	31,471	4.7
Asia	16,352	3.0	14,406	2.6	19,017	3.0	37,649	5.8	35,098	5.3	31,224	4.6	45,520	6.7
Europe	269,624	49.4	286,801	50.8	354,301	55.5	344,933	53.3	354,826	53.9	376,091	55.4	361,238	53.5
Oceania	3,820	0.7	3,820	0.7	3,820	0.6	3,820	0.6	3,820	0.6	3,820	0.6	3,820	0.6
Leading Producers														
United States of America	*197,944	36.2	*197,944	35.1	*197,944	31.0	*197,944	30.6	*197,944	30.1	*197,944	29.2	*197,944	23.3
Netherlands	111,139	20.4	120,652	21.4	116,894	18.3	126,358	19.5	134,800	20.5	161,900	23.8	*151,339	22.4
Germany	-		-		73,768	11.6	46,917	7.3	50,161	7.6	44,547	6.6	34,348	5.1
United Kingdom	*23,734	4.3	*24,267	4.3	*24,800	3.9	*25,334	3.9	*25,867	3.9	*26,401	3.9	*26,934	4.0
Cuba	*21,377	3.9	*21,377	3.8	*21,377	3.3	*21,377	3.3	*21,377	3.2	*21,377	3.1	*21,377	3.2

Cocoa Butter

Unit of Measure: Metric tons.

	1990		1991		1992		1993		1994		1995		1996	
	Value	%	Value	%	Value	%	Value	%	Value	%	Value	%	Value	%
Total Production	337,126	100.0	362,446	100.0	426,802	100.0	427,976	100.0	419,335	100.0	411,789	100.0	427,328	100.0
Regions														
Africa	4,988	1.5	4,550	1.3	4,112	1.0	3,674	0.9	3,235	0.8	2,797	0.7	2,359	0.6
America, North	14,357	4.3	15,196	4.2	16,109	3.8	15,327	3.6	13,082	3.1	13,599	3.3	12,878	3.0
America, South	61,279	18.2	74,158	20.5	60,675	14.2	63,771	14.9	50,534	12.1	44,095	10.7	55,375	13.0
Asia	61,377	18.2	57,014	15.7	62,697	14.7	68,096	15.9	74,360	17.7	75,034	18.2	72,041	16.9
Europe	195,124	57.9	211,528	58.4	283,209	66.4	277,109	64.7	278,124	66.3	276,264	67.1	284,675	66.6
Leading Producers														
Netherlands	118,706	35.2	130,661	36.0	142,284	33.3	151,470	35.4	153,108	36.5	156,354	38.0	*166,850	39.0
Russian Federation	*50,436	15.0	*50,436	13.9	*50,436	11.8	*50,436	11.8	*50,436	12.0	*50,436	12.2	50,436	11.8
Brazil	51,693	15.3	63,053	17.4	48,038	11.3	46,097	10.8	38,726	9.2	31,242	7.6	40,907	9.6
Germany	-		-		61,862	14.5	39,456	9.2	36,589	8.7	29,240	7.1	26,482	6.2
United States of America	12,035	3.6	12,853	3.5	13,671	3.2	12,840	3.0	10,427	2.5	10,968	2.7	10,141	2.4

Commodity data are provided by the United Nations Statistical Division. The symbol * means that data are estimated. For additional notes, see Appendix I.

Product Tables
Chocolate and Chocolate Products

Unit of Measure: Metric tons.

	1990 Value	%	1991 Value	%	1992 Value	%	1993 Value	%	1994 Value	%	1995 Value	%	1996 Value	%
Total Production	5,851,849	100.0	5,832,989	100.0	5,910,298	100.0	6,473,615	100.0	6,283,059	100.0	6,509,870	100.0	6,578,178	100.0
Regions														
Africa	68,733	1.2	68,467	1.2	65,616	1.1	63,942	1.0	69,409	1.1	66,726	1.0	71,512	1.1
America, North	1,294,138	22.1	1,244,358	21.3	1,306,433	22.1	1,370,505	21.2	1,376,402	21.9	1,439,253	22.1	1,479,650	22.5
America, South	172,328	2.9	201,887	3.5	187,923	3.2	174,762	2.7	173,413	2.8	191,040	2.9	205,721	3.1
Asia	1,093,504	18.7	1,125,193	19.3	1,100,141	18.6	1,175,482	18.2	1,206,250	19.2	1,250,231	19.2	1,339,416	20.4
Europe	3,114,506	53.2	3,082,235	52.8	3,134,565	53.0	3,565,460	55.1	3,331,967	53.0	3,435,023	52.8	3,356,255	51.0
Oceania	108,640	1.9	110,849	1.9	115,620	2.0	123,464	1.9	125,618	2.0	127,597	2.0	125,624	1.9
Leading Producers														
United States of America	1,240,797	21.2	1,186,014	20.3	1,243,360	21.0	1,299,397	20.1	1,297,207	20.6	*1,351,502	20.8	*1,383,343	21.0
Germany	*644,608	11.0	650,150	11.1	639,065	10.8	*644,608	10.0	*644,608	10.3	*644,608	9.9	*644,608	9.8
United Kingdom	*559,000	9.6	*575,657	9.9	*592,314	10.0	*608,971	9.4	*625,629	10.0	*642,286	9.9	*658,943	10.0
France	443,526	7.6	490,125	8.4	496,033	8.4	544,310	8.4	553,966	8.8	569,880	8.8	576,785	8.8
Netherlands	218,402	3.7	224,480	3.8	243,704	4.1	351,500	5.4	259,900	4.1	287,400	4.4	*298,708	4.5

Coffee Extracts, Essences and Concentrates (Including Instant Coffee)

Unit of Measure: Metric tons.

	1990 Value	%	1991 Value	%	1992 Value	%	1993 Value	%	1994 Value	%	1995 Value	%	1996 Value	%
Total Production	688,310	100.0	718,346	100.0	781,814	100.0	847,062	100.0	906,275	100.0	862,811	100.0	934,958	100.0
Regions														
Africa	125,606	18.2	123,625	17.2	109,662	14.0	134,997	15.9	129,005	14.2	117,238	13.6	127,720	13.7
America, North	115,861	16.8	120,505	16.8	126,895	16.2	251,069	29.6	288,268	31.8	272,536	31.6	287,784	30.8
America, South	115,718	16.8	114,273	15.9	139,216	17.8	143,270	16.9	157,392	17.4	135,668	15.7	160,063	17.1
Asia	102,077	14.8	109,796	15.3	116,600	14.9	133,969	15.8	137,965	15.2	147,582	17.1	148,433	15.9
Europe	210,757	30.6	231,204	32.2	269,845	34.5	163,509	19.3	172,743	19.1	168,234	19.5	188,752	20.2
Oceania	18,292	2.7	18,944	2.6	19,596	2.5	20,249	2.4	20,901	2.3	21,553	2.5	22,205	2.4
Leading Producers														
Kenya	86,302	12.5	85,805	11.9	69,942	8.9	94,487	11.2	86,367	9.5	80,170	9.3	88,824	9.5
Spain	115,686	16.8	110,770	15.4	118,505	15.2	11,352	1.3	12,636	1.4	12,723	1.5	13,707	1.5
United States of America	*67,536	9.8	*68,666	9.6	*69,797	8.9	*70,927	8.4	*72,057	8.0	*73,188	8.5	*74,318	7.9
Mexico	19,105	2.8	20,193	2.8	24,344	3.1	146,309	17.3	181,300	20.0	163,359	18.9	176,398	18.9
Brazil	53,702	7.8	44,695	6.2	61,661	7.9	64,587	7.6	66,420	7.3	62,878	7.3	62,562	6.7

Vinegar

Unit of Measure: 1,000 Liters.

	1990 Value	%	1991 Value	%	1992 Value	%	1993 Value	%	1994 Value	%	1995 Value	%	1996 Value	%
Total Production	2,407,057	100.0	2,631,765	100.0	2,658,083	100.0	2,623,326	100.0	2,705,405	100.0	2,785,944	100.0	2,893,709	100.0
Regions														
Africa	28,333	1.2	28,333	1.1	28,333	1.1	28,333	1.1	28,349	1.0	28,352	1.0	28,355	1.0
America, North	1,246,724	51.8	1,313,104	49.9	1,379,485	51.9	1,445,865	55.1	1,512,245	55.9	1,578,625	56.7	1,645,005	56.8
America, South	5,572	0.2	5,883	0.2	5,893	0.2	13,183	0.5	7,379	0.3	8,278	0.3	8,928	0.3
Asia	437,652	18.2	471,493	17.9	449,963	16.9	447,883	17.1	443,509	16.4	455,136	16.3	529,597	18.3
Europe	665,874	27.7	789,452	30.0	770,309	29.0	663,363	25.3	688,623	25.5	689,654	24.8	655,323	22.6
Oceania	22,900	1.0	23,500	0.9	24,100	0.9	24,700	0.9	25,300	0.9	25,900	0.9	26,500	0.9
Leading Producers														
United States of America	*1,232,432	51.2	*1,298,981	49.4	*1,365,530	51.4	*1,432,078	54.6	*1,498,627	55.4	*1,565,175	56.2	*1,631,724	56.4
Japan	381,800	15.9	386,100	14.7	391,200	14.7	391,700	14.9	395,300	14.6	402,500	14.4	409,500	14.2
Germany	-		162,800	6.2	167,850	6.3	172,900	6.6	182,600	6.7	179,641	6.4	169,590	5.9
Ukraine	70,200	2.9	68,400	2.6	69,200	2.6	48,700	1.9	47,300	1.7	56,000	2.0	46,000	1.6
France	*15,771	0.7	*6,429	0.2	-		-		-		-		-	

Product Tables
Prepared Animal Feeds

Unit of Measure: 1,000 Metric tons.

	1990 Value	%	1991 Value	%	1992 Value	%	1993 Value	%	1994 Value	%	1995 Value	%	1996 Value	%
Total Production	243,393	100.0	241,995	100.0	243,399	100.0	241,762	100.0	244,197	100.0	250,670	100.0	252,972	100.0
Regions														
Africa	7,227	3.0	6,944	2.9	6,626	2.7	6,595	2.7	6,338	2.6	7,167	2.9	7,283	2.9
America, North	53,023	21.8	54,056	22.3	54,993	22.6	59,705	24.7	60,860	24.9	61,033	24.3	61,511	24.3
America, South	3,805	1.6	3,569	1.5	3,464	1.4	3,493	1.4	5,908	2.4	6,500	2.6	5,198	2.1
Asia	50,657	20.8	51,114	21.1	51,868	21.3	50,613	20.9	49,121	20.1	49,807	19.9	51,134	20.2
Europe	125,202	51.4	122,669	50.7	122,646	50.4	117,396	48.6	117,842	48.3	121,875	48.6	123,399	48.8
Oceania	3,479	1.4	3,643	1.5	3,802	1.6	3,961	1.6	4,128	1.7	4,288	1.7	4,447	1.8
Leading Producers														
United States of America	*47,791	19.6	*48,621	20.1	*49,452	20.3	*50,282	20.8	*51,112	20.9	*51,942	20.7	*52,772	20.9
Japan	25,862	10.6	26,018	10.8	26,024	10.7	26,136	10.8	25,256	10.3	24,866	9.9	24,702	9.8
France	18,178	7.5	18,983	7.8	19,383	8.0	20,269	8.4	20,796	8.5	21,230	8.5	22,003	8.7
Germany	*15,541	6.4	*15,541	6.4	*15,541	6.4	14,991	6.2	14,649	6.0	16,029	6.4	16,497	6.5
Netherlands	13,211	5.4	13,469	5.6	13,894	5.7	15,374	6.4	15,723	6.4	15,622	6.2	*15,072	6.0

ISIC 3130
BEVERAGES

ISIC 3130—Beverages—includes distilled alcoholic beverages, ethyl alcohol, wine, malt, beer, mineral waters, and soft drinks.

This group corresponds to the 3-digit U.S. SIC series 208—Beverages. That classification includes malt beverages; malt; wines, brandy, and brandy spirits; distilled & blended liquors; bottled & canned soft drinks. The U.S. SIC also includes flavoring extracts. Ethyl alcohol, in the U.S. system of classification, falls under SIC 2869, Industrial Organic Chemicals.

Summary Statistics

		1990		1991		1992		1993		1994		1995	
		Value	N	Value	N	Value	N	Value	N	Value	N	Value	N
Establishments or enterprises	(number)	38,175	86	37,635	97	39,434	89	33,051	89	32,133	74	25,054	38
Number employed	(000)	2,974	97	2,677	100	2,695	93	2,673	94	2,550	84	2,384	63
Total Wages	($ mil.)	22,467	92	25,485	91	27,572	83	22,082	84	21,961	73	18,215	55
Wage/salary per employee	($)	10,719	90	12,408	89	13,628	82	11,786	83	12,039	72	14,856	55
Female workers	(000)	108	34	71	32	143	28	142	33	126	28	104	16
as % of total employment*	(%)	20	33	21	31	22	28	23	33	27	28	31	15
Output	($ bil.)	241	98	273	98	289	93	262	89	277	78	262	61
per employee	($ 000)	118	91	129	93	139	87	133	83	145	73	185	57
per establishment	($ 000)	14,294	82	14,531	86	15,849	79	14,726	75	15,296	61	15,230	30
Value added	($ bil.)	101	87	104	88	112	81	113	81	105	71	111	59
per employee	($ 000)	59	82	63	82	69	76	64	74	73	65	93	53
per establishment	($ 000)	8,100	74	7,591	79	8,022	68	7,051	66	7,623	53	7,647	27
Capital investment	($ mil.)	11,151	68	11,006	62	10,692	56	9,880	54	7,493	42	4,368	23
per employee	($ 000)	7	64	8	59	9	54	7	50	6	39	8	21
per establishment	($ 000)	1,132	65	1,021	59	1,275	54	1,123	50	310	40	902	20

Data presented above are drawn from the detailed tables that follow. Columns headed 'N' show the number of countries that provided valid data for inclusion. Values are not strictly comparable one year to the next or one row to the next because the number of countries that report varies from period to period and row to row. However, a general indication of magnitudes can be gleaned from reviewing this summary—especially in earlier years in which more reports are available. For detailed explanations, see Appendix I. *The average for those countries reporting both total and female employment.

Establishments and Number Engaged

Country	Establishments or Enterprises (number)						Employees per Establishment					
	1990	1991	1992	1993	1994	1995	1990	1991	1992	1993	1994	1995
Albania	-	-	-	17	27	10	-	-	-	42	17	42
Argentina	-	-	-	3,328	3,612	-	-	-	-	12	11	-
Armenia	44	43	42	46	44	-	130	117	-	-	-	-
Australia	312	377	386	-	-	-	58	50	44	-	-	-
Austria	242	234	226	222	225	-	52	56	58	56	53	-
Azerbaijan	101	113	143	144	142	-	121	111	88	74	68	-
Bahamas	9.000	4.000	-	-	-	-	54	135	-	-	-	-
Bahrain	-	-	6.000	-	-	-	-	-	99	-	-	-
Bangladesh	11	9.000	12	-	-	-	93	94	143	-	-	-
Belgium	212	217	197	198	200	-	60	58	62	-	-	-
Belize	17	17	17	-	-	-	-	25	26	-	-	-
Benin	-	-	2.000	2.000	-	-	-	-	-	-	-	-
Bolivia	57	57	58	55	53	58	70	70	77	80	84	84
Bosnia & Herzegovina	16	17	-	-	4.000	-	231	219	-	-	182	-
Botswana	7.000	6.000	6.000	5.000	5.000	7.000	186	217	133	180	160	188
Brazil	564	-	508	530	-	-	149	-	167	156	-	-
Bulgaria	*74	*67	*96	*1,855	*2,258	*2,023	292	246	143	6.523	6.643	7.168
Burundi	*3.000	*2.000	-	-	-	-	241	327	-	-	-	-
Cameroon	*7.000	*13	*11	*11	*11	*9.000	1,018	504	543	407	293	171
Canada	260	234	214	194	189	-	92	94	126	139	138	-
Cape Verde	155	151	141	152	161	-	-	-	-	-	-	-
Central African Republic	-	*4.000	*4.000	*4.000	*2.000	*2.000	-	-	-	-	-	-
Chad	-	1.000	1.000	1.000	1.000	1.000	-	-	-	-	-	-
Chile	53	55	51	50	48	-	193	202	228	235	236	-
China	*16,856	*16,881	*16,495	*12,705	*13,161	*14,719	65	69	72	89	89	82
China (Hong Kong SAR)	21	12	11	16	18	-	238	375	373	256	239	-
Colombia	133	131	144	142	141	-	175	177	164	169	167	-
Costa Rica	44	45	44	49	-	-	74	83	81	85	-	-
Croatia	107	117	175	240	307	332	85	69	44	32	24	24
Cyprus	57	57	57	57	62	64	33	34	35	36	33	29
Czechoslovakia (Former)	*39	*73	-	-	-	-	872	438	-	-	-	-
Czech Republic	-	*41	*63	*82	-	-	-	537	349	280	-	-
Denmark	67	61	62	-	-	-	105	114	110	-	-	-
Ecuador	68	70	70	68	68	68	108	113	114	94	103	98
Egypt	44	27	53	38	-	-	416	448	402	400	-	-
El Salvador	-	16	14	10	16	14	-	-	78	97	150	243
Equatorial Guinea	1.000	-	-	-	-	-	3.000	-	-	-	-	-
Ethiopia	21	21	21	21	22	23	373	366	368	353	335	350
Finland	24	22	23	21	21	21	229	245	213	200	200	194
FYR Macedonia	*11	*15	*48	*67	*157	*164	427	148	41	28	18	17
Gabon	-	7.000	7.000	3.000	2.000	2.000	-	115	113	242	331	335
Gambia	-	-	-	2.000	-	-	-	-	-	120	-	-
Germany	-	1,059	1,007	970	945	-	-	106	102	100	95	-
Germany (Western Part)	877	861	851	-	-	-	99	105	105	-	-	-
Ghana	-	-	-	27	-	-	-	-	-	127	-	-
Greece	189	187	186	-	-	-	55	54	52	-	-	-
Grenada	-	-	6.000	5.000	6.000	6.000	-	-	71	91	89	94
Guatemala	-	17	17	17	17	17	-	298	313	308	303	307
Honduras	-	-	16	23	23	23	-	-	271	207	226	239
Hungary	*87	*201	*309	*467	-	-	287	114	65	39	-	-
Iceland	5.000	7.000	9.000	7.000	-	-	67	46	37	42	-	-
India	570	532	564	606	-	-	98	107	105	94	-	-
Indonesia	143	144	180	204	215	249	88	115	93	104	103	99
Iran (Islamic Republic of)	24	25	25	26	-	-	354	341	346	387	-	-
Iraq	-	16	16	-	-	-	-	242	250	-	-	-
Ireland	62	59	-	-	-	-	76	90	-	-	-	-
Israel	36	41	-	-	-	-	92	102	-	-	-	-
Italy	*279	*271	*388	*386	*376	-	90	89	71	76	76	-
Jamaica	19	19	-	-	-	-	173	182	-	-	-	-
Japan	3,746	3,516	3,422	3,637	3,387	3,410	20	20	21	20	26	26

Continued.

Establishments and Number Engaged

- Continued -

Country	Establishments or Enterprises (number)						Employees per Establishment					
	1990	1991	1992	1993	1994	1995	1990	1991	1992	1993	1994	1995
Jordan	15	17	17	23	28	26	55	49	59	56	72	76
Korea, Republic of	590	529	454	411	388	402	41	35	39	42	45	50
Kuwait	7.000	3.000	5.000	6.000	5.000	-	259	97	352	367	410	-
Kyrgyzstan	31	34	41	37	29	-	93	86	70	70	79	-
Latvia	17	17	28	37	36	36	165	160	113	85	82	85
Lesotho	-	-	-	1.000	1.000	-	-	-	-	508	519	-
Lithuania	-	-	-	*24	*32	-	-	-	-	217	173	-
Macau	11	9.000	8.000	6.000	5.000	3.000	20	21	23	32	39	60
Malawi	*4.000	*4.000	*4.000	*4.000	*4.000	-	475	450	450	475	500	-
Malaysia	59	62	56	53	53	-	76	76	79	81	79	-
Malta	30	25	27	27	26	-	41	49	47	47	49	-
Mauritius	13	12	12	13	13	14	174	192	197	198	199	184
Mexico	153	146	138	133	129	128	555	595	656	662	687	631
Mongolia	2.000	2.000	2.000	2.000	20	2.000	209	212	244	503	83	474
Morocco	-	37	39	37	-	-	-	232	211	164	-	-
Mozambique	17	17	10	10	12	-	161	155	84	110	117	-
Namibia	-	-	-	-	6.000	-	-	-	-	-	180	-
Nepal	37	47	-	52	49	-	56	57	-	62	49	-
Netherlands	38	37	34	34	36	-	296	314	335	330	300	-
Netherlands Antilles	-	-	-	8.000	-	-	-	-	-	43	-	-
New Zealand	*128	*128	*145	*151	-	-	23	15	21	-	-	-
Norway	45	43	-	-	-	-	97	102	-	-	-	-
Pakistan	-	47	-	-	-	-	-	117	-	-	-	-
Panama	18	15	-	-	-	-	89	110	-	-	-	-
Paraguay	-	11	-	-	-	-	-	169	-	-	-	-
Peru	570	571	579	-	-	-	22	21	20	-	-	-
Philippines	823	719	97	86	-	-	40	46	290	307	-	-
Poland	*84	*123	*115	*118	-	-	345	252	261	263	-	-
Portugal	*737	*822	*653	*770	*652	-	24	23	27	23	27	-
Puerto Rico	34	33	30	30	32	28	45	45	52	53	52	58
Qatar	5.000	6.000	5.000	5.000	5.000	-	68	70	74	74	85	-
Republic of Moldova	89	89	87	86	100	105	136	130	152	148	109	111
Romania	-	378	636	852	965	-	-	-	-	-	-	-
Russian Federation	-	-	-	*744	*946	*1,255	-	-	-	190	152	122
Saint Lucia	-	6.000	6.000	8.000	8.000	8.000	-	-	-	-	-	-
Saint Vincent & the Grenadines	-	-	-	4.000	4.000	5.000	-	-	-	26	27	22
Senegal	4.000	4.000	5.000	3.000	4.000	-	131	123	84	127	90	-
Sierra Leone	-	-	-	14	-	-	-	-	-	98	-	-
Singapore	14	14	14	13	11	-	174	174	165	168	196	-
Slovakia	-	84	34	35	38	-	-	111	268	253	240	-
Slovenia	62	*23	*29	*40	*45	*69	-	-	-	-	-	-
South Africa	-	258	-	-	-	-	-	140	-	-	-	-
Spain	6,004	5,434	5,432	-	-	-	8.738	8.682	8.378			
Sri Lanka	18	16	17	14	-	-	279	300	304	338	-	-
Suriname	-	-	5.000	-	-	-	-	-	122	-	-	-
Swaziland	3.000	-	-	-	-	-	178	-	-	-	-	-
Sweden	30	33	32	33	34	-	160	146	143	132	122	-
Thailand	61	79	-	-	-	-	1,563	264	-	-	-	-
Trinidad and Tobago	12	19	19	19	21	-	135	55	55	79	105	-
Tunisia	-	-	-	51	53	53	-	-	-	60	62	69
Turkey	75	75	96	96	91	-	185	197	144	142	126	-
Ukraine	573	673	876	836	821	857	117	98	73	75	74	71
United Kingdom	511	478	490	655	683	-	135	144	133	89	82	-
United Republic of Tanzania	20	18	-	-	-	-	314	354	-	-	-	-
USSR (Former)	*1,341	-	-	-	-	-	224	-	-	-	-	-
United States of America	-	-	2,237	-	-	-	-	-	60	-	-	-
Venezuela	113	112	116	100	85	93	142	143	147	169	203	161
Yugoslavia	88	139	383	641	709	748	-	134	47	27	26	26
Zambia	-	-	-	-	9.000	-	-	-	-	-	317	-
Zimbabwe	15	15	17	19	19	-	447	450	353	363	368	-

Depending on the table, * means *Enterprises*, *Engaged*, or *Factor Values*; ± means *Basis Unspecified*; § means *shown in millions*. For additional notes and sources, see Appendix I.

Employment and Compensation of Employees

Country	Number Employed/Engaged (000)						Wage/Salary per Employee ($)					
	1990	1991	1992	1993	1994	1995	1990	1991	1992	1993	1994	1995
Albania	-	-	-	0.708	0.454	0.421	-	-	-	-	1,265	1,090
Argentina	32	-	-	41	41	-	5,881	-	-	16,720	-	-
Armenia	*5.728	*5.038	-	-	-	-	-	-	-	-	-	-
Australia	18	*19	*17	-	-	-	22,873	24,153	24,438	-	-	-
Austria	13	13	13	12	12	12	28,816	29,515	32,501	33,886	36,114	-
Azerbaijan	12	13	13	11	9.647	-	1,368	1,398	-	-	-	-
Bahamas	0.490	0.540	-	-	-	-	19,645	22,757	-	-	-	-
Bahrain	-	-	0.592	-	-	-	-	-	5,580	-	-	-
Bangladesh	1.024	0.845	1.715	-	-	-	661	550	823	-	-	-
Belgium	13	13	12	-	-	-	24,634	25,814	28,636	-	-	-
Belize	-	0.428	0.435	-	-	-	-	7,157	7,131	-	-	-
Bermuda	0.104	0.108	0.108	0.105	0.129	0.131	-	-	-	-	-	-
Bolivia	4.010	3.981	4.450	4.376	4.450	4.894	2,758	3,244	3,153	3,276	3,565	3,625
Bosnia & Herzegovina	3.692	3.728	-	-	0.727	-	3,219	4,382	-	-	-	-
Botswana	1.300	1.300	0.800	0.900	0.800	1.315	-	-	10,902	9,400	-	-
Brazil	84	-	85	83	-	-	5,577	-	6,219	7,110	-	-
Bulgaria	22	16	14	12	15	15	2,021	861	1,451	1,932	1,427	1,833
Burundi	0.722	0.654	-	-	-	-	6,877	7,593	-	-	-	-
Cambodia	*2.246	*2.402	-	-	-	-	-	-	-	-	-	-
Cameroon	7.127	6.556	5.968	4.476	3.224	1.542	11,887	12,209	11,181	16,331	9,188	11,015
Canada	24	22	27	27	26	26	33,818	39,476	35,361	33,532	32,699	33,306
Central African Republic	0.278	-	-	-	-	-	6,883	-	-	-	-	-
Chile	10	11	12	12	11	12	5,874	7,031	7,282	7,587	8,041	9,714
China	*1,100	*1,160	*1,180	*1,130	*1,170	*1,210	-	-	-	-	-	-
China (Hong Kong SAR)	*5.000	*4.500	*4.100	*4.100	*4.300	-	12,991	15,299	17,204	18,823	20,582	-
China (Taiwan Province)	18	18	17	18	19	18	12,731	14,952	18,345	17,292	18,044	18,587
Colombia	23	23	24	24	24	24	3,021	3,022	3,539	3,612	4,824	5,753
Costa Rica	3.277	3.756	3.572	4.183	-	-	3,685	3,305	3,940	4,301	-	-
Croatia	9.140	8.130	7.770	7.630	7.490	7.990	6,076	6,748	1,773	2,171	3,352	5,387
Cyprus	1.905	1.938	1.989	2.039	2.027	1.866	12,531	13,287	15,037	14,585	15,436	20,810
Czechoslovakia (Former)	34	32	-	-	-	-	2,228	1,685	-	-	-	-
Czech Republic	24	22	22	23	-	-	2,205	1,773	2,107	2,561	-	-
Denmark	7.015	6.932	6.832	7.048	5.653	5.355	36,231	35,839	39,866	36,513	-	-
Ecuador	7.350	7.887	7.946	6.402	7.008	6.659	1,974	2,184	2,673	3,690	1,679	3,540
Egypt	*18	*12	*21	*15	*15	*15	2,055	2,006	1,266	1,647	1,693	1,807
El Salvador	-	-	*1.099	*0.971	*2.407	*3.405	-	-	-	3,736	6,540	5,957
Equatorial Guinea	0.003	-	-	-	-	-	1,983	-	-	-	-	-
Ethiopia	7.837	7.694	7.733	7.413	7.381	8.051	1,720	1,854	1,482	970	986	923
Finland	5.500	5.400	4.900	4.200	4.200	4.079	31,851	31,066	30,138	24,804	27,417	36,151
FYR Macedonia	4.692	2.218	1.978	1.894	2.895	2.771	4,683	8,846	2,656	3,449	4,566	5,556
France	44	43	41	41	40	40	-	-	-	-	-	-
Gabon	-	0.806	0.789	0.726	0.661	0.671	-	22,596	25,800	26,608	16,772	17,302
Gambia	-	-	-	0.240	-	-	-	-	-	-	-	-
Germany	-	*112	*103	*97	*89	-	-	28,297	34,190	34,099	36,330	-
Germany (Western Part)	*87	*90	*89	-	-	-	31,583	32,300	36,501	-	-	-
Ghana	-	-	-	*3.418	*3.514	*3.614	-	-	-	1,663	1,242	1,268
Greece	*10	*10	*9.690	*9.731	*9.837	*9.899	14,138	14,434	17,818	18,157	19,520	23,111
Grenada	-	-	*0.424	*0.454	0.532	0.563	-	-	-	-	-	-
Guatemala	-	*5.073	*5.314	*5.242	*5.155	*5.214	-	-	-	-	-	-
Honduras	*4.265	*4.181	*4.336	*4.750	*5.205	*5.502	4,468	3,909	4,473	4,441	3,568	4,133
Hungary	25	23	20	18	18	18	2,432	2,794	3,587	3,797	4,154	4,569
Iceland	0.333	0.325	0.334	0.294	0.306	-	32,511	-	38,423	35,698	-	-
India	*56	*57	*59	*57	*66	*70	1,244	1,079	1,076	989	1,062	1,105
Indonesia	13	16	17	21	22	25	889	970	896	1,167	1,276	1,885
Iran (Islamic Republic of)	8.500	8.532	8.656	10	-	-	4,376	5,079	5,802	5,262	-	-
Iraq	-	3.880	4.000	-	-	-	-	10,410	21,473	-	-	-
Ireland	4.700	5.300	5.348	5.300	4.942	4.743	33,309	32,676	34,939	30,956	36,833	-
Israel	3.300	4.200	4.000	4.500	-	-	17,735	16,193	18,401	17,353	-	-
Italy	25	24	28	29	29	-	-	-	-	38,866	-	-
Jamaica	3.295	3.460	3.536	3.738	3.757	-	7,764	5,628	5,081	-	-	-

Continued.

Depending on the table, * means *Enterprises, Engaged,* or *Factor Values*; ± means *Basis Unspecified*; § means *shown in millions*. For additional notes and sources, see Appendix I.

Employment and Compensation of Employees

- Continued -

Country	Number Employed/Engaged (000)						Wage/Salary per Employee ($)					
	1990	1991	1992	1993	1994	1995	1990	1991	1992	1993	1994	1995
Japan	*74	*72	*71	*73	*89	*89	25,106	28,559	31,694	36,094	-	-
Jordan	0.819	0.829	1.000	1.290	2.020	1.988	2,933	3,389	3,390	3,430	2,358	3,130
Kenya	6.733	6.956	7.029	7.232	-	-	3,044	-	-	-	-	-
Korea, Republic of	24	19	18	17	17	20	10,891	12,910	13,224	14,155	15,952	17,466
Kuwait	1.812	0.290	1.761	2.204	2.050	2.048	4,555	5,634	6,506	6,483	7,523	9,233
Kyrgyzstan	*2.896	*2.941	*2.865	*2.582	*2.284				-	-		
Latvia	2.798	2.713	3.168	3.142	2.947	3.073						
Lesotho	-	-		*0.508	*0.519					6,446	-	
Lithuania	-	-	5.166	5.208	5.523	-	-	-	609	845	1,775	
Macau	0.223	0.187	0.186	0.190	0.193	0.179	5,572	6,415	7,135	7,941	8,101	9,788
Malawi	1.900	1.800	1.800	1.900	2.000	-	1,427	2,358	1,758	2,080	1,156	-
Malaysia	4.500	4.700	4.400	4.300	4.200	4.257	4,929	5,184	6,236	6,469	6,823	7,942
Malta	1.219	1.229	1.271	1.261	1.270	-	11,578	11,837	13,120	12,296	13,011	
Mauritius	*2.264	2.305	2.364	2.577	2.588	2.573	3,480	4,137	4,966	4,094	4,811	5,358
Mexico	85	87	91	88	89	81	4,178	5,078	6,038	7,567	7,352	4,360
Mongolia	*0.418	*0.424	*0.487	*1.006	*1.651	*0.948	2,563	3,840	1,095			737
Morocco	-	8.596	8.210	6.060			-	-	-	-	-	
Mozambique	2.734	2.627	0.839	1.101	1.398	-	1,685	707	683	447	636	-
Namibia	-	-	-	-	1.078		-	-	-	-	9,380	
Nepal	2.081	2.690	-	3.221	2.379	-	605	656	-	437	584	
Netherlands	11	12	11	11	11	11	37,903	37,726	-	39,312	-	-
Netherlands Antilles	-	-	-	0.344	-	-	-	-	-	19,810	-	
New Zealand	2.960	1.897	3.116	-	-	-	-	-	-	-	-	
Norway	4.367	4.389	-	-	-	-	31,643	32,193	-	-	-	
Pakistan	5.636	5.501	5.754	-	-	-	1,629	1,664	1,711			
Panama	*1.599	*1.651	*1.650	*1.646	*1.643	*1.643	7,473	7,697	7,933	8,528	9,063	9,260
Paraguay	-	1.860	-	-	-	-	-	-	-	-	-	
Peru	*13	*12	*12	-	-	-	4,274	6,815	7,899	-	-	
Philippines	33	33	28	26	-	-	3,032	3,282	4,165	4,003	-	
Poland	29	31	30	31	32	33	1,402	2,513	3,507	3,510	4,201	5,419
Portugal	*18	*19	*18	*17	*17	*17	-	-	-	-	-	-
Puerto Rico	1.520	1.500	1.550	1.580	1.650	1.610	33,224	-	-	-	-	
Qatar	0.338	0.419	0.370	0.368	0.425	-	5,734	5,900	5,298	5,805	6,065	
Republic of Moldova	12	12	13	13	11	12		3,588	-	670	609	734
Russian Federation	-	-	-	141	143	153				946	1,374	1,352
Saint Vincent & the Grenadines	-	-	-	0.105	0.109	0.108				937	902	
Senegal	*0.524	*0.493	*0.418	*0.381	*0.359		8,236	8,161	9,779	10,196	6,522	
Sierra Leone	-	-	-	*1.369	-	-				1,206	-	
Singapore	*2.439	*2.440	*2.306	*2.190	*2.156	*2.156	14,404	17,270	18,840	19,152	20,775	23,267
Slovakia	-	*9.284	*9.107	*8.843	*9.106	-		1,571	1,978	2,117	2,454	
South Africa	36	36	37	35	33	34	8,235	9,164	10,102	9,740	10,147	11,411
Spain	52	47	46	52	50	46	22,153	24,416	26,757	25,409	24,701	28,037
Sri Lanka	*5.026	*4.793	*5.170	*4.733	-	-	1,117	1,242	1,171	1,589	-	-
Suriname	*0.589	*0.433	*0.609	*0.652	-	-	15,218	16,302	23,182	37,721	-	-
Swaziland	0.535	-	-	-	-	-	7,384	-	-	-	-	
Sweden	4.800	*4.827	*4.575	4.348	4.141	3.954	27,454	28,778	33,022	25,656	26,320	31,000
Thailand	95	21	-	-	-	-	2,283	4,416	-	-	-	
Tonga	0.038	0.043	-	-	-	-	1,315	2,028	-	-	-	
Trinidad and Tobago	1.623	1.052	1.041	1.501	2.202	-	8,524	13,197	13,698	8,765	7,849	-
Tunisia	-	-	-	3.076	3.296	3.636	-	-	-	8,129	7,606	8,759
Turkey	14	15	14	14	11	12	7,314	9,750	10,160	11,188	7,545	9,932
Ukraine	67	66	64	63	61	61	2,537	3,141	2,724	434	583	840
United Kingdom	69	69	65	58	56	49	24,586	25,830	27,942	24,542	28,604	37,259
United Republic of Tanzania	6.275	6.372	-	-	-	-		430	-	-	-	
USSR (Former)	300	-	-	-	-	-	1,815	-	-	-	-	
United States of America	137	137	134	136	128	130	30,949	31,533	33,164	33,456	35,156	35,200
Uruguay	*5.609	*5.242	*4.998	*5.071	*5.071	*4.444	4,984	6,085	7,333	9,196	9,501	9,132
Venezuela	16	16	17	17	17	15	4,809	6,480	6,478	7,121	5,203	6,623

Continued.

Depending on the table, * means *Enterprises*, *Engaged*, or *Factor Values*; ± means *Basis Unspecified*; § means *shown in millions*. For additional notes and sources, see Appendix I.

47

Employment and Compensation of Employees
- Continued -

Country	Number Employed/Engaged (000)						Wage/Salary per Employee ($)					
	1990	1991	1992	1993	1994	1995	1990	1991	1992	1993	1994	1995
Yugoslavia	-	19	18	18	18	19	-	-	-	-	2,702	1,213
Zambia	-	-	-	-	2.849	-	-	-	-	-	3,013	-
Zimbabwe	6.702	6.747	6.000	6.900	7.000	6.995	4,721	4,565	3,936	3,336	3,315	3,566

Female Workers: Total and Percent of Employment

Country	Female Workers (000)						Female Workers as % of Total Employment					
	1990	1991	1992	1993	1994	1995	1990	1991	1992	1993	1994	1995
Afghanistan	0.035	0.041	-	-	-	-	-	-	-	-	-	-
Albania	-	-	-	0.321	0.308	0.285	-	-	-	45.34	67.84	67.70
Argentina	-	-	-	-	4.706	-	-	-	-	-	11.58	-
Australia	3.651	-	-	-	-	-	20.28	-	-	-	-	-
Austria	2.600	3.000	3.000	2.525	2.488	-	20.47	23.08	23.08	20.45	20.83	-
Bahrain	-	-	0.016	-	-	-	-	-	2.70	-	-	-
Bangladesh	-	-	0.146	-	-	-	-	-	8.51	-	-	-
Bermuda	0.014	0.013	0.015	0.013	0.016	0.017	13.46	12.04	13.89	12.38	12.40	12.98
Bosnia & Herzegovina	0.835	0.788	-	-	-	-	22.62	21.14	-	-	-	-
Botswana	-	-	-	-	-	0.311	-	-	-	-	-	23.65
Bulgaria	-	-	-	5.700	7.100	6.800	-	-	-	47.11	47.33	46.90
Canada	3.000	3.000	-	-	-	-	12.50	13.64	-	-	-	-
Chile	0.846	1.072	0.952	1.013	1.123	-	8.26	9.64	8.18	8.62	9.93	-
China (Taiwan Province)	6.132	6.206	6.458	6.658	6.708	6.365	34.89	35.08	36.97	36.18	35.85	35.11
Colombia	-	3.282	-	3.803	3.582	-	-	14.14	-	15.82	15.21	-
Croatia	3.060	2.730	2.680	2.580	2.580	2.710	33.48	33.58	34.49	33.81	34.45	33.92
Cyprus	0.555	0.573	0.582	0.585	0.557	-	29.13	29.57	29.26	28.69	27.48	-
Czechoslovakia (Former)	9.600	9.600	-	-	-	-	28.24	30.00	-	-	-	-
Czech Republic	10	9.000	10	10	-	-	41.67	40.91	45.45	43.48	-	-
Denmark	1.781	1.776	1.703	-	-	-	25.39	25.62	24.93	-	-	-
Egypt	-	1.314	1.506	1.153	-	-	-	10.86	7.07	7.59	-	-
El Salvador	-	-	-	0.141	0.390	0.296	-	-	-	14.52	16.20	8.69
Ethiopia	-	1.934	1.919	1.960	1.967	2.038	-	25.14	24.82	26.44	26.65	25.31
FYR Macedonia	0.774	0.817	0.698	0.730	0.679	0.610	16.50	36.83	35.29	38.54	23.45	22.01
Gambia	-	-	-	0.009	-	-	-	-	-	3.75	-	-
Germany (Western Part)	15	-	-	-	-	-	17.34	-	-	-	-	-
Ghana	-	-	-	0.372	-	-	-	-	-	10.88	-	-
Hungary	10	-	8.000	7.000	-	-	40.00	-	40.00	38.89	-	-
Indonesia	-	-	-	8.019	7.796	7.674	-	-	-	37.96	35.17	31.19
Iran (Islamic Republic of)	-	0.196	0.226	0.248	-	-	-	2.30	2.61	2.47	-	-
Iraq	-	0.631	-	-	-	-	-	16.26	-	-	-	-
Ireland	0.700	0.800	-	-	-	-	14.89	15.09	-	-	-	-
Italy	-	4.624	5.644	5.912	5.842	-	-	19.19	20.41	20.22	20.34	-
Jordan	0.092	0.107	0.105	0.113	0.137	0.134	11.23	12.91	10.50	8.76	6.78	6.74
Kenya	0.381	0.384	0.398	0.410	-	-	5.66	5.52	5.66	5.67	-	-
Korea, Republic of	6.100	4.281	4.167	4.562	4.381	5.639	25.42	22.90	23.64	26.38	25.21	27.95
Lesotho	-	-	-	0.109	0.109	-	-	-	-	21.46	21.00	-
Macau	0.050	0.040	0.043	0.041	0.042	0.040	22.42	21.39	23.12	21.58	21.76	22.35
Malaysia	1.500	1.600	1.500	1.300	1.400	-	33.33	34.04	34.09	30.23	33.33	-
Malta	0.140	0.133	0.132	0.136	0.144	-	11.48	10.82	10.39	10.79	11.34	-
Mongolia	-	-	-	-	0.924	-	-	-	-	-	55.97	-
Morocco	-	-	-	-	-	0.680	-	-	-	-	-	-
Nepal	0.252	0.293	-	0.423	0.228	-	12.11	10.89	-	13.13	9.58	-
New Zealand	0.871	0.615	0.965	-	-	-	29.43	32.42	30.97	-	-	-
Panama	0.270	-	-	-	-	-	16.89	-	-	-	-	-
Philippines	-	-	2.767	2.622	-	-	-	-	9.85	9.93	-	-
Portugal	2.805	3.009	2.768	2.852	2.871	-	15.57	16.12	15.71	16.35	16.46	-
Republic of Moldova	-	-	-	-	5.148	5.614	-	-	-	-	47.39	48.05
Saint Vincent & the Grenadines	-	-	-	0.038	0.038	-	-	-	-	36.19	34.86	-
Senegal	0.011	-	-	-	-	-	2.10	-	-	-	-	-

Continued.

Depending on the table, * means *Enterprises, Engaged,* or *Factor Values;* ± means *Basis Unspecified;* § means *shown in millions.* For additional notes and sources, see Appendix I.

Female Workers: Total and Percent of Employment

- Continued -

Country	Female Workers (000)						Female Workers as % of Total Employment					
	1990	1991	1992	1993	1994	1995	1990	1991	1992	1993	1994	1995
Sri Lanka	1.184	0.674	0.723	0.813	-	-	23.56	14.06	13.98	17.18	-	-
Sweden	1.300	-	-	-	-	-	27.08	-	-	-	-	-
Thailand	4.583	6.981	-	-	-	-	4.81	33.45	-	-	-	-
Turkey	1.245	-	-	-	-	-	8.96	-	-	-	-	-
United Kingdom	17	-	18	-	-	-	24.93	-	27.69	-	-	-
United Republic of Tanzania	1.034	1.047	-	-	-	-	16.48	16.43	-	-	-	-
United States of America	-	-	68	70	65	65	-	-	50.75	51.47	50.78	50.00
Zambia	-	-	-	-	0.165	-	-	-	-	-	5.79	-

Output and Output per Employee

Country	Output ($ bil.)						Output per Employee ($)					
	1990	1991	1992	1993	1994	1995	1990	1991	1992	1993	1994	1995
Albania	-	-	-	*0.002	*0.006	-	-	-	-	2,720	13,997	-
Argentina	±2.543	-	-	4.847	-	-	79,613	-	-	118,650	-	-
Armenia	0.007	0.028	0.073	1.989	0.014	-	1,144	5,656	-	-	-	-
Australia	*3.741	*4.310	*4.036	-	-	-	207,813	226,847	237,413	-	-	-
Austria	2.287	2.410	2.893	2.749	2.950	3.449	180,050	185,390	222,565	222,651	246,932	289,768
Azerbaijan	±0.902	±1.219	±0.021	±0.020	±0.008	-	73,922	96,967	1,642	1,841	881	-
Bahamas	0.083	0.102	-	-	-	-	169,388	188,889	-	-	-	-
Bahrain	-	-	±0.021	-	-	-	-	-	35,500	-	-	-
Bangladesh	0.008	0.007	0.018	-	-	-	7,966	8,375	10,673	-	-	-
Belgium	2.826	2.870	3.211	2.951	3.102	3.691	220,816	227,774	261,083	-	-	-
Belize	0.019	0.025	0.026	-	-	-	-	58,499	58,718	-	-	-
Benin	±0.047	±0.050	±0.062	±0.048	-	-	-	-	-	-	-	-
Bolivia	0.112	0.156	0.162	0.157	0.182	0.205	27,903	39,211	36,440	35,777	40,805	41,869
Bosnia & Herzegovina	±0.124	±0.160	-	-	-	-	33,456	42,849	-	-	-	-
Botswana	-	-	±0.132	±0.102	-	-	-	-	164,834	112,803	-	-
Brazil	*3.602	-	*3.767	*4.513	-	-	42,872	-	44,344	54,567	-	-
Bulgaria	±0.727	±0.270	±0.310	±0.320	±0.358	±0.497	33,655	16,390	22,617	26,477	23,836	34,298
Burundi	±0.034	±0.032	-	-	-	-	47,218	48,830	-	-	-	-
Cameroon	0.531	0.480	0.555	0.422	0.190	0.203	74,513	73,286	92,983	94,290	58,903	131,331
Canada	5.108	5.394	5.593	5.403	5.185	5.272	212,833	245,186	207,140	200,099	199,405	202,770
Cape Verde	±0.001	-	-	-	-	-	-	-	-	-	-	-
Central African Republic	*0.015	-	-	-	-	-	52,241	-	-	-	-	-
Chile	0.768	0.924	1.166	1.175	1.262	1.599	75,025	83,117	100,221	99,968	111,564	135,126
China	±8.049	±8.687	±10	±13	±12	±14	7,317	7,489	8,735	11,785	10,017	11,436
China (Hong Kong SAR)	0.484	0.482	0.487	0.525	0.612	-	96,896	107,091	118,822	127,948	142,302	-
China (Taiwan Province)	±2.763	±2.896	±3.340	±3.356	±3.515	±3.716	157,218	163,728	191,211	182,400	187,871	204,963
Colombia	1.577	1.629	1.475	1.442	1.937	2.348	67,677	70,188	62,633	59,981	82,279	97,982
Costa Rica	0.237	0.261	0.307	0.319	0.364	0.382	72,461	69,562	85,991	76,247	-	-
Cote d'Ivoire	±0.073	±0.071	±0.076	±0.074	-	-	-	-	-	-	-	-
Croatia	±0.416	±0.541	±0.322	-	-	-	45,558	66,534	41,451	-	-	-
Cyprus	0.156	0.152	0.177	0.158	0.183	0.197	81,635	78,467	89,136	77,505	90,528	105,437
Czechoslovakia (Former)	±0.783	±0.628	-	-	-	-	23,038	19,611	-	-	-	-
Denmark	*1.259	*1.394	*1.591	*1.476	*1.824	*2.183	179,451	201,028	232,805	209,364	322,624	407,644
Ecuador	0.137	0.221	0.209	0.283	0.259	0.325	18,700	27,992	26,289	44,141	36,909	48,779
Egypt	*0.236	*0.212	*0.234	*0.283	*0.291	*0.314	12,891	17,537	10,995	18,645	19,319	20,834
El Salvador	-	-	±0.023	±0.035	±0.121	±0.174	-	-	21,235	36,243	50,403	51,042
Equatorial Guinea	±0.000	-	-	-	-	-	25,711	-	-	-	-	-
Ethiopia	0.210	0.185	0.127	0.095	0.110	0.129	26,770	24,104	16,479	12,813	14,911	15,983
Finland	±1.252	±1.271	±1.072	±0.753	±0.812	±1.065	227,653	235,374	218,688	179,354	193,403	261,104
FYR Macedonia	±0.124	±0.134	±0.077	±0.062	±0.132	±0.163	26,352	60,612	39,044	32,977	45,481	58,776
France	±13	±14	±14	±13	±14	±15	301,395	318,019	333,797	331,994	335,830	377,637
Gabon	-	*0.105	*0.123	*0.129	*0.071	*0.091	-	130,262	155,741	177,119	107,475	134,930
Germany	-	25	27	26	26	-	-	223,462	265,376	268,556	291,131	-
Germany (Western Part)	22	23	25	-	-	-	255,501	259,279	283,705	-	-	-
Ghana	-	-	-	0.097	0.079	0.083	-	-	-	28,512	22,591	23,049

Continued.

Output and Output per Employee
- Continued -

Country	Output ($ bil.)						Output per Employee ($)					
	1990	1991	1992	1993	1994	1995	1990	1991	1992	1993	1994	1995
Greece	*1.265	*1.287	*1.507	*1.539	*1.668	*1.985	122,601	128,037	155,498	158,145	169,609	200,566
Guatemala	-	0.337	0.350	0.473	0.504	0.516	-	66,477	65,869	90,242	97,779	98,922
Honduras	0.135	0.152	0.190	0.192	0.201	0.243	31,565	36,313	43,713	40,504	38,676	44,199
Hungary	±0.666	±0.712	±0.668	±0.788	±0.824	±0.869	26,643	30,950	33,395	43,786	45,326	47,302
Iceland	0.063	0.072	±0.066	±0.054	-	-	189,834	221,924	198,550	182,352	-	-
India	*1.084	*0.984	*0.994	*0.972	*1.203	*1.346	19,365	17,335	16,814	16,991	18,348	19,235
Indonesia	0.247	0.286	0.392	0.441	0.605	0.711	19,650	17,371	23,392	20,893	27,303	28,910
Iran (Islamic Republic of)	0.269	0.215	0.272	0.312	-	-	31,665	25,196	31,390	31,050	-	-
Iraq	-	*0.335	*0.366	-	-	-	-	86,468	91,399	-	-	-
Ireland	*1.273	*1.425	*1.537	*1.353	*1.522	*1.746	270,950	268,906	287,438	255,331	308,051	368,126
Israel	0.371	0.434	0.436	0.518			112,573	103,424	109,085	115,033		
Italy	±7.822	±7.646	±9.875	±8.565	±9.233		309,912	317,366	357,080	292,901	321,515	-
Jamaica	0.242	0.184	0.171				73,564	53,243	48,457			
Japan	±32	±35	±37	±43	±66	±71	435,673	488,601	527,682	583,177	739,060	800,469
Jordan	0.049	0.054	0.055	0.076	0.111	0.116	59,897	65,412	54,696	58,820	55,026	58,589
Kenya	*0.200	*0.193	*0.237	*0.147	*0.163	*0.219	29,684	27,699	33,648	20,360	-	-
Korea, Republic of	3.888	4.137	4.129	4.145	5.275	5.523	162,013	221,299	234,256	239,700	303,592	273,709
Kuwait	0.055	0.011	0.066	0.088	0.094	0.113	30,393	39,388	37,554	40,135	45,978	55,134
Kyrgyzstan	0.044	0.063	0.047	0.005	0.007	-	15,127	21,447	16,278	2,065	3,188	-
Latvia	0.001	0.002	0.025	0.043	0.078	0.113	221	880	7,841	13,735	26,579	36,881
Lithuania	-	-	0.031	0.095	0.159	-	-	-	5,995	18,258	28,712	-
Macau	0.011	0.010	0.009	0.009	0.010	0.010	50,693	51,322	49,517	47,553	50,239	56,664
Malawi	*0.089	*0.099	*0.116	*0.118	*0.070		46,751	54,757	64,339	62,353	34,929	-
Malaysia	*0.374	*0.366	*0.387	*0.370	*0.421	*0.505	83,059	77,908	88,058	86,144	100,317	118,641
Malta	0.080	0.082	0.087	0.080	0.081		65,708	66,728	68,586	63,593	63,950	
Mauritius	0.087	0.093	0.107	0.104	0.109	0.127	38,550	40,440	45,397	40,421	42,026	49,472
Mexico	4.203	5.358	6.056	6.448	6.924	4.596	49,464	61,715	66,875	73,198	78,153	56,942
Mongolia	0.064	0.078	0.039	0.011	0.006	0.008	152,469	185,062	79,453	11,233	3,460	8,781
Morocco		±0.341	±0.415	±0.403	-	-		39,655	50,587	66,511	-	-
Mozambique	-	±0.025	-	±0.020	±0.026	-	-	9,424	-	18,562	18,431	-
Namibia	-	-	-	-	0.111	-	-	-	-	-	103,378	-
Nepal	0.032	0.031		0.040	0.036		15,266	11,485		12,319	15,031	
Netherlands	3.114	3.241	3.620	3.441	3.812	4.621	277,353	279,100	317,472	306,768	353,501	438,747
Norway	±0.983	±1.039	-	-	-	-	225,050	236,738	-	-	-	-
Pakistan	0.131	0.130	0.141				23,259	23,617	24,506			
Panama	0.137	0.144	0.150	0.163	0.175	0.179	85,967	87,498	90,772	98,984	106,365	109,097
Peru	0.818	1.083	1.218	-			63,949	90,976	104,169	-		
Philippines	1.371	1.500	1.667	2.033			41,298	45,171	59,330	77,013		
Poland	±2.224	±3.402	±3.545	±3.450	±4.131	±5.387	76,675	109,737	118,159	111,285	128,846	163,703
Portugal	1.325	1.438	1.660	1.495	1.624	1.796	73,514	77,073	94,182	85,730	93,111	105,796
Puerto Rico	±3.030	±3.243	±3.408	±3.457	±3.855	±3.829	§1,993	§2,162	§2,198	§2,187	§2,336	§2,378
Qatar	0.016	0.016	0.015	0.020	0.023	-	47,955	38,684	41,580	53,751	53,006	-
Republic of Moldova	0.000	1.247	0.069	0.064	0.074	0.118	0.176	107,710	5,239	5,024	6,801	10,084
Romania	±1.712	±1.405	±1.022	±1.142			-	-	-	-		
Russian Federation	-	5.315	2.868	0.981	1.268	1.844	-	-	-	6,942	8,836	12,037
Senegal	0.034	-	-	-	-	-	64,746	-	-	-	-	-
Sierra Leone	-	-	-	0.035	-	-	-	-	-	25,241	-	-
Singapore	*0.282	*0.304	*0.341	*0.345	*0.364	*0.406	115,480	124,577	147,987	157,727	168,744	188,423
Slovakia	-	±0.214	±0.234	±0.227	±0.257		-	23,011	25,656	25,722	28,276	
Slovenia	-	-	-	0.243	0.264	0.345						
South Africa	±3.335	±3.425	±3.921	±3.277	±3.327	±3.798	92,653	95,134	105,985	93,622	100,814	113,308
Spain	*10	*10	*11	±10	±10	±12	192,252	215,178	238,492	194,235	205,822	254,128
Sri Lanka	0.131	0.135	0.140	0.183	-		26,009	28,153	27,113	38,685		
Suriname	*0.086	*0.104	*0.156	0.313	-	-	145,525	239,357	255,734	479,456		
Swaziland	*0.142						265,846					
Sweden	*1.443	*1.807	*1.970	*1.184	*1.435	*1.635	300,586	374,427	430,574	272,335	346,478	413,598
Thailand	3.712	1.921	-				38,931	92,068	-			
Tonga	0.000	0.000	-				8,773	8,882	-			
Trinidad and Tobago	±0.117	±0.117	±0.132	±0.124	±0.121	-	72,244	111,672	126,444	82,433	54,849	-
Tunisia	0.196	0.222	0.249	0.254	0.277	0.305	-	-	-	82,465	84,122	83,876

Continued.

Depending on the table, * means *Enterprises, Engaged,* or *Factor Values*; ± means *Basis Unspecified*; § means *shown in millions.* For additional notes and sources, see Appendix I.

Output and Output per Employee

- Continued -

Country	Output ($ bil.)						Output per Employee ($)					
	1990	1991	1992	1993	1994	1995	1990	1991	1992	1993	1994	1995
Turkey	1.469	1.844	1.826	2.099	1.435	1.931	105,786	124,768	131,739	153,895	124,795	163,834
Ukraine	*2.117	*3.604	*1.836	*0.545	*0.504	*0.590	31,598	54,601	28,692	8,651	8,258	9,676
United Kingdom	*15	*15	20	16	17	19	224,146	224,163	300,870	274,309	301,329	396,484
United Republic of Tanzania	*0.073	*0.087	-	-	-	-	11,631	13,591	-	-	-	-
USSR (Former)	±7.813	-	-	-	-	-	26,042	-	-	-	-	-
United States of America	*47	*49	*51	*52	*54	*57	342,117	357,518	380,851	384,765	421,148	435,385
Uruguay	0.335	0.425	0.515	0.579	0.598	0.504	59,660	81,120	103,085	114,132	118,014	113,450
Venezuela	0.951	1.161	1.372	1.531	1.151	1.588	59,422	72,575	80,706	90,489	66,574	105,863
Yugoslavia	±0.075	±0.089	±0.186	-	±0.499	±0.297	-	4,795	10,283	-	27,434	15,451
Zambia	-	-	-	-	0.102	-	-	-	-	-	35,819	-
Zimbabwe	*0.390	*0.385	*0.298	*0.290	*0.274	*0.297	58,130	57,127	49,599	41,984	39,211	42,516

Value Added and Value Added per Employee

Country	Value Added ($ bil.)						Value Added per Employee ($)					
	1990	1991	1992	1993	1994	1995	1990	1991	1992	1993	1994	1995
Argentina	±0.932	-	-	1.777	-	-	29,165	-	-	43,489	-	-
Armenia	0.000	0.006	-	-	-	-	79	1,273	-	-	-	-
Australia	*1.723	*1.946	*1.838	-	-	-	95,747	102,434	108,143	-	-	-
Austria	0.841	0.851	1.138	1.097	1.190	1.389	66,192	65,486	87,535	88,867	99,576	116,665
Bahamas	0.028	0.047	-	-	-	-	56,898	87,680	-	-	-	-
Bangladesh	0.004	0.005	0.012	-	-	-	3,672	5,497	7,081	-	-	-
Belgium	*0.846	*0.872	*0.985	*0.899	*0.924	*1.078	66,092	69,206	80,062	-	-	-
Belize	0.002	0.008	0.004	-	-	-	-	18,828	10,241	-	-	-
Bolivia	0.052	0.089	0.091	0.084	0.099	0.111	13,010	22,378	20,412	19,094	22,142	22,667
Bosnia & Herzegovina	±0.071	±0.104	-	-	-	-	19,121	28,029	-	-	-	-
Botswana	-	-	±0.031	±0.024	±0.024	-	-	-	38,335	26,504	29,986	-
Brazil	*2.225	-	*2.450	*3.047	-	-	26,490	-	28,839	36,846		
Bulgaria	-	±0.070	±0.080	±0.081	-	-	-	4,272	5,842	6,698	-	-
Burundi	0.023	0.025	-	-	-	-	31,226	37,500	-	-	-	-
Cameroon	0.294	0.216	0.206	0.103	0.089	0.056	41,272	32,975	34,546	23,008	27,550	36,102
Canada	2.948	3.160	3.425	3.263	3.083	3.137	122,843	143,620	126,858	120,863	118,573	120,647
Central African Republic	*0.008	-	-	-	-	-	27,455	-	-	-	-	-
Chile	0.374	0.482	0.652	0.604	0.683	0.866	36,528	43,328	56,016	51,414	60,432	73,203
China	±2.414	±2.863	±3.509	±4.820	±3.834	±4.234	2,195	2,468	2,974	4,266	3,277	3,499
China (Hong Kong SAR)	0.200	0.198	0.210	0.236	0.266	-	40,078	44,009	51,297	57,574	61,778	-
China (Taiwan Province)	1.636	1.561	1.872	1.768	1.791	1.843	93,068	88,261	107,200	96,085	95,735	101,635
Colombia	0.928	1.036	0.928	0.783	1.174	1.422	39,820	44,624	39,419	32,562	49,883	59,336
Costa Rica	0.128	0.143	0.169	0.168	0.185	0.194	39,037	37,948	47,364	40,119		
Cote d'Ivoire	0.066	0.064	0.068	0.067	-	-	-	-	-	-	-	-
Croatia	±0.222	±0.242	±0.150	-	-	-	24,276	29,767	19,309	-	-	-
Cyprus	0.073	0.068	0.086	0.071	0.084	0.082	38,432	35,060	43,188	34,591	41,425	43,809
Czechoslovakia (Former)	±0.258	-	-	-	-	-	7,586	-	-	-	-	-
Denmark	*0.757	*0.832	*0.947	*0.878	*1.079	*1.289	107,910	120,012	138,580	124,643	190,799	240,691
Ecuador	0.033	0.105	0.089	0.141	0.086	0.146	4,538	13,343	11,139	22,000	12,220	21,926
Egypt	*0.046	*0.058	*0.084	*0.125	*0.129	*0.140	2,516	4,802	3,936	8,253	8,560	9,251
El Salvador	-	0.051	-	0.021	0.104	0.116	-	-	-	21,266	43,057	34,059
Ethiopia	0.154	0.137	0.098	0.073	0.082	0.096	19,606	17,760	12,632	9,795	11,126	11,889
Finland	±0.666	±0.731	±0.528	±0.389	±0.366	±0.392	121,054	135,317	107,658	92,574	87,106	96,187
FYR Macedonia	±0.049	±0.069	±0.035	±0.029	±0.067	±0.085	10,414	31,262	17,604	15,056	23,308	30,618
France	±5.382	±5.246	±5.368	±6.021	±6.207	±6.711	122,885	122,276	129,666	148,296	153,637	166,119
Gabon	-	*0.047	*0.054	*0.054	*0.024	*0.035	-	58,224	68,387	74,129	36,579	52,557
Germany (Western Part)	12	13	14	13	-	-	137,686	140,690	157,282	-	-	-
Ghana	-	-	-	0.056	0.048	0.051	-	-	-	16,303	13,741	14,020
Greece	*0.474	*0.512	*0.630	*0.645	*0.701	*0.836	45,941	50,908	64,993	66,327	71,272	84,412
Haiti	-	±0.003	±0.002	±0.001	±0.001	±0.001	-	-	-	-	-	-
Honduras	*0.034	*0.039	*0.060	*0.056	*0.053	*0.058	7,903	9,346	13,741	11,708	10,194	10,597
Hungary	±0.166	±0.181	±0.206	±0.251	±0.264	±0.279	6,627	7,848	10,317	13,948	14,502	15,203

Continued.

Value Added and Value Added per Employee

- Continued -

Country	Value Added ($ bil.)						Value Added per Employee ($)					
	1990	1991	1992	1993	1994	1995	1990	1991	1992	1993	1994	1995
Iceland	0.021	0.023	±0.023	±0.016	-	-	62,462	70,315	68,464	54,958	-	-
India	*0.246	*0.237	*0.222	*0.218	*0.270	*0.301	4,399	4,180	3,753	3,813	4,110	4,304
Indonesia	*0.112	*0.134	*0.195	*0.168	*0.288	*0.261	8,915	8,148	11,636	7,965	12,994	10,618
Iran (Islamic Republic of)	0.160	0.131	0.162	0.183	-	-	18,802	15,346	18,693	18,205	-	-
Iraq	-	*0.205	*0.219	-	-	-	-	52,726	54,839	-	-	-
Ireland	*0.792	*0.882	*0.952	*0.838	*0.944	*1.083	168,413	166,458	177,973	158,136	190,986	228,387
Israel	0.146	0.168	0.125	0.141	-	-	44,338	40,012	31,312	31,251	-	-
Italy	±2.015	±1.932	±2.404	±2.267	±2.194	-	79,834	80,195	86,932	77,511	76,410	-
Jamaica	0.103	0.079	0.074	-	-	-	31,199	22,923	20,949	-	-	-
Japan	±10	±11	±12	±14	±24	±27	139,251	156,715	166,479	185,400	272,517	301,983
Jordan	0.028	0.030	0.032	0.050	0.053	0.052	34,670	35,770	31,776	38,540	26,032	25,916
Kenya	*0.090	*0.092	*0.082	*0.051	*0.068	*0.084	13,351	13,170	11,658	7,104	-	-
Korea, Republic of	1.889	2.145	2.252	2.162	2.851	2.833	78,711	114,754	127,751	125,037	164,096	140,423
Kuwait	0.021	0.006	0.031	0.042	0.041	0.051	11,429	19,937	17,328	18,959	19,797	25,129
Latvia	-	-	-	*0.039	*0.076	*0.109	-	-	-	12,475	25,645	35,320
Macau	0.004	0.003	0.003	0.003	0.003	0.004	18,245	14,570	14,806	14,921	14,674	19,939
Malawi	*0.017	*0.015	*0.027	*0.020	*0.013	-	8,988	8,244	15,264	10,304	6,536	-
Malaysia	*0.201	*0.158	*0.150	*0.148	*0.165	*0.196	44,693	33,654	34,081	34,313	39,185	45,998
Malta	0.042	0.043	0.046	0.046	0.045	-	34,206	35,182	36,415	36,408	35,229	-
Mauritius	0.057	0.062	0.068	0.067	0.072	0.082	25,343	26,958	28,773	25,964	27,870	31,977
Mexico	2.210	2.790	3.287	3.440	3.623	2.405	26,001	32,130	36,301	39,052	40,892	29,793
Mongolia	0.051	0.068	0.032	0.008	0.002	0.004	122,309	160,561	66,195	8,422	1,386	3,862
Morocco	-	±0.207	±0.242	±0.216	-	-	-	24,116	29,473	35,686	-	-
Namibia	-	-	-	-	0.041	-	-	-	-	-	38,427	-
Nepal	0.023	0.022	-	0.024	0.023	-	10,935	8,302	-	7,554	9,754	-
Netherlands	1.500	1.556	1.798	1.689	1.877	2.281	133,614	134,045	157,639	150,576	174,084	216,601
New Zealand	0.216	-	-	-	-	-	72,812	-	-	-	-	-
Norway	±0.660	±0.701	-	-	-	-	151,192	159,805	-	-	-	-
Pakistan	0.069	0.065	0.073	-	-	-	12,258	11,899	12,698	-	-	-
Panama	0.069	0.069	0.071	0.078	0.083	0.085	42,974	41,595	43,164	47,100	50,638	51,948
Peru	0.545	0.610	0.699	-	-	-	42,578	51,252	59,801	-	-	-
Philippines	0.815	0.914	0.964	1.208	-	-	24,559	27,541	34,295	45,775	-	-
Poland	±1.838	±3.031	±2.975	±2.399	±2.842	±3.684	63,394	97,769	99,173	77,382	88,642	111,946
Portugal	0.344	0.376	0.457	0.407	0.423	0.467	19,082	20,131	25,927	23,358	24,228	27,528
Puerto Rico	±1.117	±1.357	±1.691	±1.758	±1.882	±1.990	735,066	904,533	§1,090	§1,112	§1,140	§1,236
Qatar	0.004	0.004	0.003	0.005	0.004	-	11,379	8,524	8,168	12,691	9,050	-
Romania	±0.642	±0.431	±0.304	±0.452	-	-	-	-	-	-	-	-
Russian Federation	-	-	-	±0.521	±0.686	±0.878	-	-	-	3,688	4,784	5,730
Saint Lucia	-	±0.011	±0.009	±0.009	±0.009	±0.011	-	-	-	-	-	-
Senegal	0.017	0.016	0.017	0.017	0.009	-	32,524	32,046	40,880	43,500	25,071	-
Sierra Leone	-	-	-	0.020	-	-	-	-	-	14,537	-	-
Singapore	*0.139	*0.153	*0.170	*0.180	*0.200	*0.223	56,932	62,876	73,892	81,967	92,605	103,624
Slovakia	-	-	-	±0.076	±0.081	-	-	-	-	8,637	8,890	-
Slovenia	-	-	-	0.100	0.102	0.135	-	-	-	-	-	-
South Africa	*1.054	*1.165	*1.419	*1.183	*1.207	*1.387	29,277	32,372	38,342	33,796	36,565	41,365
Spain	*4.014	*3.930	*4.156	±3.575	±3.307	±3.735	76,509	83,304	91,335	68,645	66,081	80,631
Sri Lanka	0.118	0.107	0.109	0.157	-	-	23,496	22,233	21,114	33,081	-	-
Swaziland	*0.101	-	-	-	-	-	188,288	-	-	-	-	-
Sweden	*0.743	*0.508	*0.538	*0.363	*0.424	*0.479	154,869	105,152	117,555	83,548	102,326	121,051
Thailand	2.766	1.474	-	-	-	-	29,012	70,631	-	-	-	-
Trinidad and Tobago	±0.058	±0.057	±0.062	±0.057	±0.051	-	35,848	54,225	59,757	37,911	23,102	-
Tunisia	0.092	0.102	0.122	0.123	0.135	0.149	-	-	-	40,099	40,886	40,858
Turkey	0.893	1.126	1.083	1.216	0.806	1.083	64,305	76,214	78,097	89,194	70,071	91,881
United Kingdom	*6.643	*6.784	*7.193	*5.821	*5.825	*6.638	96,273	98,320	110,655	100,368	104,025	135,279
United Republic of Tanzania	*0.016	*0.019	-	-	-	-	2,593	2,973	-	-	-	-
United States of America	*21	*23	*24	*25	*26	*27	154,307	165,912	178,993	185,206	201,352	204,554
Uruguay	0.181	0.266	0.327	0.373	0.386	0.325	32,282	50,652	65,380	73,547	76,114	73,181
Venezuela	0.583	0.658	0.808	0.805	0.555	1.491	36,447	41,151	47,508	47,575	32,099	99,372

Continued.

Depending on the table, * means *Enterprises, Engaged,* or *Factor Values;* ± means *Basis Unspecified;* § means *shown in millions.* For additional notes and sources, see Appendix I.

Value Added and Value Added per Employee
- Continued -

Country	Value Added ($ bil.)						Value Added per Employee ($)					
	1990	1991	1992	1993	1994	1995	1990	1991	1992	1993	1994	1995
Yugoslavia	±0.037	±0.043	±0.102	-	±0.260	±0.151	-	2,298	5,639	-	14,293	7,861
Zambia	-	-	-	-	0.055	-	-	-	-	-	19,190	-
Zimbabwe	*0.302	*0.295	*0.215	*0.205	*0.191	*0.207	45,044	43,735	35,913	29,715	27,255	29,547

Capital Investment

Country	Gross Fixed Capital Formation ($ mil.)						Per Establishment ($ mil)					
	1990	1991	1992	1993	1994	1995	1990	1991	1992	1993	1994	1995
Argentina	-	-	-	383	-	-	-	-	-	0.115	-	-
Australia	174	-	-	-	-	-	0.558	-	-	-	-	-
Austria	221	266	304	288	289	-	0.913	1.136	1.346	1.296	1.285	-
Bahrain	-	-	1.657	-	-	-	-	-	0.276	-	-	-
Bangladesh	0.315	0.328	0.051	-	-	-	0.029	0.036	0.004	-	-	-
Belgium	304	435	548	385	265	306	1.432	2.005	2.783	1.944	1.326	-
Bolivia	17	13	11	9.717	18	-	0.296	0.226	0.183	0.177	0.345	-
Bosnia & Herzegovina . . .	11	-	-	-	-	-	0.716	-	-	-	-	-
Botswana	-	-	8.959	12	-	-	-	-	1.493	2.311	-	-
Brazil	322	-	318	272	-	-	0.571	-	0.627	0.513	-	-
Bulgaria.	14	9.354	187	42	13	43	0.184	0.140	1.951	0.023	0.006	0.021
Cameroon	31	25	16	74	7.140	7.827	4.383	1.952	1.444	6.692	0.649	0.870
Canada	315	-	-	-	-	-	1.210	-	-	-	-	-
Cape Verde	0.012	-	-	-	-	-	0.000	-	-	-	-	-
Chile.	16	56	89	90	103	-	0.307	1.012	1.743	1.796	2.151	-
China (Hong Kong SAR) . . .	51	-6.305	40	14	-347	-	2.451	-0.525	3.606	0.856	-19	-
Colombia	75	45	40	27	93	-	0.564	0.343	0.280	0.191	0.662	-
Croatia	6.250	15	7.004	28	34	25	0.058	0.124	0.040	0.119	0.111	0.074
Cyprus	9.105	12	19	14	15	15	0.160	0.209	0.327	0.249	0.242	0.241
Czechoslovakia (Former) . . .	72	-	-	-	-	-	1.857	-	-	-	-	-
Denmark	62	89	-	-	-	-	0.928	1.464	-	-	-	-
Ecuador	17	34	42	63	21	35	0.247	0.484	0.597	0.928	0.304	0.507
Egypt	16	8.583	13	10	-	-	0.355	0.318	0.245	0.272	-	-
El Salvador.	-	4.827	-	2.014	3.659	29	-	0.302	-	0.201	0.229	2.059
Ethiopia.	3.190	5.448	5.156	7.473	5.589	3.550	0.152	0.259	0.246	0.356	0.254	0.154
Finland	130	114	197	48	50	111	5.434	5.164	8.551	2.302	2.363	5.285
Former Yugoslav Republic of Macedonia	187	2.832	2.001	3.373	6.806	0.563	0.108	0.189	0.042	0.050	0.043	0.003
Gabon	-	8.862	8.213	4.312	1.981	1.691	-	1.266	1.173	1.437	0.991	0.845
Germany (Western Part) . . .	1,908	2,215	2,452	2,306	-	-	2.175	2.573	2.882	-	-	-
Greece	93	87	90	-	-	-	0.492	0.465	0.484	-	-	-
Hungary	34	58	52	85	-	-	0.396	0.289	0.167	0.181	-	-
India	103	76	127	141	-	-	0.180	0.143	0.225	0.233	-	-
Indonesia	12	44	66	31	49	86	0.087	0.306	0.364	0.153	0.228	0.347
Iran (Islamic Republic of) . . .	7.628	9.334	22	17	-	-	0.318	0.373	0.894	0.659	-	-
Ireland	84	113	-	-	-	-	1.359	1.914	-	-	-	-
Israel.	26	30	-	-	-	-	0.730	0.738	-	-	-	-
Italy	402	432	619	435	420	-	1.442	1.594	1.596	1.127	1.118	-
Japan	1,706	2,041	1,082	1,745	2,240	-	0.455	0.581	0.316	0.480	0.661	-
Jordan	0.620	0.352	0.126	3.307	38	5.458	0.041	0.021	0.007	0.144	1.353	0.210
Korea, Republic of	430	448	479	444	990	975	0.728	0.848	1.055	1.081	2.551	2.425
Kuwait	1.253	0.408	5.167	4.983	9.013	-	0.179	0.136	1.033	0.831	1.803	-
Latvia	0.010	0.025	1.040	2.612	2.691	11	0.001	0.001	0.037	0.071	0.075	0.299
Macau	0.290	2.115	0.204	0.146	0.022	0.118	0.026	0.235	0.025	0.024	0.004	0.039
Malawi	9.015	20	46	47	18	-	2.254	4.941	12	12	4.573	-
Malaysia	27	43	29	25	16	-	0.451	0.686	0.526	0.476	0.302	-
Malta	5.066	5.080	18	8.649	12	-	0.169	0.203	0.666	0.320	0.461	-
Mexico	132	193	-	-	-	-	0.860	1.325	-	-	-	-
Mongolia	10	6.313	1.706	1.919	2.206	-	5.179	3.157	0.853	0.959	0.110	-
Morocco	-	-	-	26	-	-	-	-	-	0.695	-	-
Nepal	16	-	-	-	-	-	0.440	-	-	-	-	-

Continued.

Depending on the table, * means *Enterprises, Engaged,* or *Factor Values;* ± means *Basis Unspecified;* § means *shown in millions.* For additional notes and sources, see Appendix I.

53

Capital Investment
- Continued -

Country	Gross Fixed Capital Formation ($ mil.)						Per Establishment ($ mil)					
	1990	1991	1992	1993	1994	1995	1990	1991	1992	1993	1994	1995
Netherlands	258	271	318	206	-	-	6.792	7.314	9.350	6.065	-	-
Pakistan	-	11	-	-	-	-	-	0.243	-	-	-	-
Panama	4.387	13	-	-	-	-	0.244	0.863	-	-	-	-
Peru	20	49	89	-	-	-	0.036	0.085	0.154	-	-	-
Philippines	211	57	58	86	-	-	0.257	0.079	0.594	0.996	-	-
Poland	57	123	177	196	-	-	0.677	0.998	1.541	1.660	-	-
Portugal	191	169	139	149	165	-	0.260	0.206	0.213	0.194	0.254	-
Romania	-	-	-	74	55	-	-	-	-	0.087	0.057	-
Senegal	2.189	-	-	-	-	-	0.547	-	-	-	-	-
Singapore	66	26	15	15	30	-	4.692	1.891	1.086	1.187	2.707	-
Slovenia	16	30	25	27	32	73	0.265	1.323	0.856	0.681	0.702	1.059
Spain	509	549	386	393	282	431	0.085	0.101	0.071	-	-	-
Sri Lanka	2.098	1.758	5.244	3.343	-	-	0.117	0.110	0.308	0.239	-	-
Swaziland	4.977	-	-	-	-	-	1.659	-	-	-	-	-
Thailand	568	229	-	-	-	-	9.305	2.897	-	-	-	-
Trinidad and Tobago	0.819	0.915	5.835	7.438	13	-	0.068	0.048	0.307	0.391	0.600	-
Tunisia	10	19	21	19	14	17	-	-	-	0.371	0.261	0.319
Turkey	36	37	64	161	45	-	0.486	0.495	0.670	1.675	0.494	-
Ukraine	44	43	77	13	24	7.801	0.077	0.064	0.088	0.015	0.029	0.009
United Kingdom	984	878	797	-	776	-	1.925	1.837	1.626	-	1.137	-
United Republic of Tanzania	27	29	-	-	-	-	1.332	1.601	-	-	-	-
United States of America	1,170	1,370	1,462	1,323	1,602	1,872	-	-	0.654	-	-	-
Uruguay	13	31	18	14	-	-	-	-	-	-	-	-
Venezuela	24	61	61	78	58	302	0.216	0.547	0.528	0.775	0.684	3.250
Yugoslavia	2.714	3.888	7.873	-	9.908	11	0.031	0.028	0.021	-	0.014	0.014
Zimbabwe	54	43	19	15	10	-	3.606	2.878	1.096	0.767	0.527	-

Indexes of Industrial Production

Country	Index of Industrial Production (1990=100)						Index of Employment (1990=100)					
	1990	1991	1992	1993	1994	1995	1990	1991	1992	1993	1994	1995
Albania	100	-	-	-	-	-	-	-	-	-	-	-
Algeria	100	99	94	99	95	88	-	-	-	-	-	-
Angola	100	120	100	80	-	-	-	-	-	-	-	-
Argentina	100	134	-	-	-	-	100	-	-	128	127	-
Armenia	100	-	-	-	-	-	100	88	-	-	-	-
Australia	100	103	111	119	-	-	100	106	94	-	-	-
Austria	100	106	107	108	112	113	100	102	102	97	94	94
Azerbaijan	-	-	-	-	-	-	100	103	103	87	79	-
Bahamas	100	-	-	-	-	-	100	110	-	-	-	-
Bahrain	100	-	-	-	-	-	-	-	-	-	-	-
Bangladesh	100	95	66	67	89	97	100	83	167	-	-	-
Barbados	100	102	98	108	116	118	-	-	-	-	-	-
Belgium	100	95	97	90	83	85	100	98	96	-	-	-
Belize	100	108	112	143	152	152	-	-	-	-	-	-
Benin	100	-	-	-	-	-	-	-	-	-	-	-
Bermuda	-	-	-	-	-	-	100	104	104	101	124	126
Bhutan	100	-	-	-	-	-	-	-	-	-	-	-
Bolivia	100	111	110	117	121	128	100	99	111	109	111	122
Bosnia & Herzegovina	-	-	-	-	-	-	100	101	-	-	20	-
Botswana	100	101	109	-	-	-	100	100	62	69	62	101
Brazil	100	118	99	107	118	138	100	-	101	98	-	-
Bulgaria	100	79	68	52	51	52	100	76	63	56	69	67
Burkina-Faso	100	102	-	-	-	-	-	-	-	-	-	-
Burundi	100	96	101	109	-	-	100	91	-	-	-	-
Cambodia	100	-	-	-	-	-	100	107	-	-	-	-
Cameroon	100	103	82	80	74	54	100	92	84	63	45	22
Canada	100	90	100	104	106	106	100	92	113	113	108	108

Continued.

Depending on the table, * means *Enterprises, Engaged,* or *Factor Values*; ± means *Basis Unspecified*; § means *shown in millions.* For additional notes and sources, see Appendix I.

Indexes of Industrial Production

- Continued -

Country	Index of Industrial Production (1990=100)						Index of Employment (1990=100)					
	1990	1991	1992	1993	1994	1995	1990	1991	1992	1993	1994	1995
Central African Republic . . .	100	87	87	62	77	-	100	-	-	-	-	-
Chad	100	121	116	-	-	-	-	-	-	-	-	-
Chile	100	98	108	115	110	122	100	109	114	115	110	116
China	100	-	-	-	-	-	100	105	107	103	106	110
China (Hong Kong SAR) . . .	100	104	109	112	114	113	100	90	82	82	86	-
China (Taiwan Province) . . .	100	104	109	111	117	118	100	101	99	105	106	103
Colombia	100	99	95	105	109	118	100	100	101	103	101	103
Congo	100	120	129	-	-	-	-	-	-	-	-	-
Costa Rica	100	109	112	115	119	119	100	115	109	128	-	-
Cote d'Ivoire	100	89	86	81	79	99	-	-	-	-	-	-
Croatia	100	-	-	-	-	-	100	89	85	83	82	87
Cuba	100	-	-	-	-	-	-	-	-	-	-	-
Cyprus	100	95	98	84	101	97	100	102	104	107	106	98
Czechoslovakia (Former) . . .	100	114	108	-	-	-	100	94	-	-	-	-
Czech Republic	100	-	-	-	-	-	100	92	92	96	-	-
Dem. Rep. of the Congo . . .	100	-	-	-	-	-	-	-	-	-	-	-
Denmark	100	104	108	106	121	125	100	99	97	100	81	76
Dominican Republic	100	91	95	-	-	-	-	-	-	-	-	-
Ecuador	100	113	114	104	-	-	100	107	108	87	95	91
Egypt	100	86	79	79	78	78	100	66	116	83	82	82
El Salvador	100	104	-	-	-	-	-	-	-	-	-	-
Equatorial Guinea	-	-	-	-	-	-	100	-	-	-	-	-
Ethiopia and Eritrea	100	-	-	-	-	-	-	-	-	-	-	-
Ethiopia	-	-	-	-	-	-	100	98	99	95	94	103
Fiji	100	103	107	106	116	112	-	-	-	-	-	-
Finland	100	101	102	107	97	96	100	98	89	76	76	74
FYR Macedonia	100	-	-	-	-	-	100	47	42	40	62	59
France	100	95	93	95	98	98	100	98	95	93	92	92
Gabon	100	103	98	-	-	-	-	-	-	-	-	-
Gambia	100	-	-	-	-	-	-	-	-	-	-	-
Germany (Eastern Part) . . .	100	-	-	-	-	-	-	-	-	-	-	-
Germany (Western Part) . . .	100	107	110	109	115	105	100	104	103	-	-	-
Ghana	100	-	-	-	-	-	-	-	-	-	-	-
Greece	100	101	105	108	116	121	100	97	94	94	95	96
Guatemala	100	99	103	-	-	-	-	-	-	-	-	-
Guyana	100	95	123	138	131	-	-	-	-	-	-	-
Haiti	100	67	73	60	52	52	-	-	-	-	-	-
Honduras	100	137	197	233	342	297	100	98	102	111	122	129
Hungary	100	97	85	82	85	88	100	92	80	72	73	74
Iceland	100	-	-	-	-	-	100	98	100	88	92	-
India	100	109	107	123	142	152	100	101	106	102	117	125
Indonesia	100	119	105	129	149	177	100	131	133	168	177	196
Iran (Islamic Republic of) . . .	100	107	108	109	122	146	100	100	102	118	-	-
Iraq	100	77	91	-	-	-	-	-	-	-	-	-
Ireland	100	104	103	104	112	117	100	113	114	113	105	101
Israel	100	102	105	113	120	130	100	127	121	136	-	-
Italy	100	99	98	97	-	-	100	95	110	116	114	-
Jamaica	100	87	92	95	93	-	100	105	107	113	114	-
Japan	100	100	100	98	100	93	100	97	96	99	120	120
Jordan	100	116	133	171	178	245	100	101	122	158	247	243
Kenya	100	100	117	116	107	112	100	103	104	107	-	-
Korea, Republic of	100	109	111	113	122	123	100	78	73	72	72	84
Kuwait	100	-	-	-	-	-	100	16	97	122	113	113
Kyrgyzstan	100	-	-	-	-	-	100	102	99	89	79	-
Lao P.D.R.	100	-	-	-	-	-	-	-	-	-	-	-
Latvia	100	-	-	-	-	-	100	97	113	112	105	110
Lebanon	100	110	-	-	-	-	-	-	-	-	-	-
Lesotho	100	-	-	-	-	-	-	-	-	-	-	-
Liberia	100	-	-	-	-	-	-	-	-	-	-	-
Libyan Arab Jamahiriya . . .	100	118	120	-	-	-	-	-	-	-	-	-

Continued.

Depending on the table, * means *Enterprises*, *Engaged*, or *Factor Values*; ± means *Basis Unspecified*; § means *shown in millions*. For additional notes and sources, see Appendix I.

55

Indexes of Industrial Production

- Continued -

Country	Index of Industrial Production (1990=100)						Index of Employment (1990=100)					
	1990	1991	1992	1993	1994	1995	1990	1991	1992	1993	1994	1995
Lithuania	100	-	-	-	-	-	-	-	-	-	-	-
Macau	100	-	-	-	-	-	100	84	83	85	87	80
Madagascar	100	79	76	-	-	-	100	-	-	-	-	-
Malawi	100	92	98	100	101	96	100	95	95	100	105	-
Malaysia	100	102	107	95	109	121	100	104	98	96	93	95
Mali	100	87	105	93	124	93	-	-	-	-	-	-
Malta	100	104	109	119	-	-	100	101	104	103	104	-
Mauritius	100	106	107	106	112	110	100	102	104	114	114	114
Mexico	100	104	105	106	104	105	100	102	107	104	104	95
Mongolia	-	-	-	-	-	-	100	101	117	241	395	227
Morocco	100	118	116	100	107	117	-	-	-	-	-	-
Mozambique	100	91	-	-	-	-	100	96	31	40	51	-
Myanmar	100	125	79	-	-	-	-	-	-	-	-	-
Namibia	100	-	-	-	-	-	-	-	-	-	-	-
Nepal	100	-	-	-	-	-	100	129	-	155	114	-
Netherlands	100	111	122	132	143	150	100	103	102	100	96	94
New Zealand	100	102	101	-	-	-	100	64	105	-	-	-
Nicaragua	100	-	-	-	-	-	-	-	-	-	-	-
Niger	100	-	-	-	-	-	-	-	-	-	-	-
Nigeria	100	103	145	214	305	-	-	-	-	-	-	-
Norway	100	101	100	96	100	102	100	101	-	-	-	-
Pakistan	100	89	100	97	103	123	100	98	102	-	-	-
Panama	100	107	109	118	125	127	100	103	103	103	103	103
Papua New Guinea	100	-	-	-	-	-	-	-	-	-	-	-
Paraguay	100	108	110	126	140	153	-	-	-	-	-	-
Peru	100	129	123	123	138	143	100	93	91	-	-	-
Philippines	100	119	127	124	137	159	100	100	85	80	-	-
Poland	100	97	109	118	133	146	100	107	103	107	111	113
Portugal	100	102	98	99	92	89	100	104	98	97	97	94
Puerto Rico	100	92	77	54	54	-	100	99	102	104	109	106
Qatar	100	-	-	-	-	-	100	124	109	109	126	-
Republic of Moldova	100	-	-	-	-	-	100	96	109	105	90	97
Romania	100	83	69	59	67	69	-	-	-	-	-	-
Russian Federation	100	-	-	-	-	-	-	-	-	-	-	-
Saudi Arabia	100	-	-	-	-	-	-	-	-	-	-	-
Senegal	100	91	96	91	85	103	100	94	80	73	69	-
Sierra Leone	100	-	-	-	-	-	-	-	-	-	-	-
Singapore	100	101	113	112	114	114	100	100	95	90	88	88
Slovakia	100	-	-	-	-	-	-	-	-	-	-	-
Slovenia	100	96	86	82	81	78	-	-	-	-	-	-
Somalia	100	-	-	-	-	-	-	-	-	-	-	-
South Africa	100	96	98	86	88	93	100	100	103	97	92	93
Spain	100	103	99	100	104	102	100	90	87	99	95	88
Sri Lanka	100	102	98	117	125	311	100	95	103	94	-	-
Sudan	100	106	116	-	-	-	-	-	-	-	-	-
Suriname	100	100	-	-	-	-	100	74	103	111	-	-
Swaziland	100	93	98	-	-	-	100	-	-	-	-	-
Sweden	100	97	98	102	107	110	100	101	95	91	86	82
Switzerland	100	100	99	101	100	102	-	-	-	-	-	-
Syrian Arab Republic	100	109	111	105	111	118	-	-	-	-	-	-
Thailand	100	108	124	-	-	-	100	22	-	-	-	-
Tonga	100	-	-	-	-	-	100	113	-	-	-	-
Trinidad and Tobago	100	96	89	84	89	88	100	65	64	92	136	-
Tunisia	100	102	107	107	112	113	-	-	-	-	-	-
Turkey	100	105	114	130	126	147	100	106	100	98	83	85
Uganda	100	114	100	110	147	199	-	-	-	-	-	-
Ukraine	100	-	-	-	-	-	100	99	96	94	91	91
United Arab Emirates	100	-	-	-	-	-	-	-	-	-	-	-
United Kingdom	100	95	93	95	97	105	100	100	94	84	81	71
United Republic of Tanzania	100	100	107	117	125	-	100	102	-	-	-	-

Continued.

Depending on the table, * means *Enterprises, Engaged,* or *Factor Values;* ± means *Basis Unspecified;* § means *shown in millions.* For additional notes and sources, see Appendix I.

Indexes of Industrial Production

- Continued -

Country	Index of Industrial Production (1990=100)						Index of Employment (1990=100)					
	1990	1991	1992	1993	1994	1995	1990	1991	1992	1993	1994	1995
USSR (Former)	100	-	-	-	-	-	100	-	-	-	-	-
United States of America . . .	100	101	105	105	109	111	100	100	98	99	93	95
Uruguay	100	99	106	108	108	84	100	93	89	90	90	79
Venezuela	100	-	-	-	-	-	100	100	106	106	108	94
Yemen	100	-	-	-	-	-	-	-	-	-	-	-
Yugoslavia	100	-	-	-	-	-	-	-	-	-	-	-
Yugoslavia (Former)	100	92	-	-	-	-	-	-	-	-	-	-
Zambia	100	103	129	125	122	121	-	-	-	-	-	-
Zimbabwe	100	103	103	97	98	92	100	101	90	103	104	104

Representative Companies in Sector

Name	Address	Telephone	Fax	Employment	Y
South African Breweries	65 Park Ln., Sandton 2146, Republic of South Africa	11 8818111	11 8818634	10,000	95
Thai Pure Drinks Ltd.	Hua Mark, Bangkok 10240, Thailand	3740251	3749316	9,728	94
CFC Holdings Corp.	1000 Corporate Dr., Fort Lauderdale, Florida 33334, U.S.A.	305-351-5600		9,000	94
Serm Suk Co. Ltd.	Huay Kwang, Bangkok 10310, Thailand	2761041	2762403	8,582	92
Brown-Forman Corp.	850 Dixie Hwy., Louisville, Kentucky 40210, U.S.A.	502-585-1100		7,500	97
McCormick and Company Inc. (Sparks, Maryland)	PO Box 6000, Sparks, Maryland 21152-6000, U.S.A.	410-771-7301	410-771-7462	7,500	97
Irvin & Johnson Ltd.	70 Prestwich St., Cape Town 8001, Republic of South Africa	21 4029200	21 4029282	7,000	97
Whitman Corp.	3501 Algonquin Rd., Rolling Meadows, Illinois 60008, U.S.A.	847-818-5000		6,381	97
Courage Ltd.	1 Bridge St., Staines TW18 4TP, United Kingdom	784 466199	784 468131	5,843	93
Adolph Coors Co.	12th & Ford Sts., Golden, Colorado 80401, U.S.A.	303-279-6565	303-277-6490	5,800	96
MD Foods Amba	Skanderborgvej 277, Viby DK-8260, Denmark	86281000	86281691	5,700	93
Crown Corp. Ltd.	PO Box 3148, Wellington, New Zealand	4 724099		5,600	92
President Enterprises Corp.	Yeongkang, Tainan, Taiwan	6 2532121	6 2536290	5,522	92
E and J Gallo Winery	PO Box 1130, Modesto, California 95353, U.S.A.	209-579-3111	209-579-4361	5,000	97
Lotte Chilsung Beverage Co. Ltd.	Socho-gu, Seoul, Republic of Korea	2 5360222	2 5358619	5,000	97
Premier Brands Ltd.	11 Walker St., Edinburgh EH3 7NE, United Kingdom			4,911	91
Coca-Cola Bottling Company Consolidated	PO Box 31487, Charlotte, North Carolina 28231, U.S.A.	704-551-4400		4,900	97
Suntory International Corp.	1211 Ave. of the Amer., New York, New York 10036, U.S.A.	212-921-9595	212-398-0268	4,800	96
Clorox Co.	1221 Broadway Ave., Oakland, California 94612, U.S.A.	510-271-7000	510-832-1463	4,700	93
CCU Compania Cervecerias Unidas SA	Bandera 84, 6th Fl., Santiago, Chile	2 6703000	2 6703222	4,500	97
Perrier Group of America Inc.	777 W. Putnam Ave., Greenwich, Connecticut 06830, U.S.A.	203-531-4100	203-863-0297	4,500	97
Grupo Embotelladoras Unidas SA de CV	Colonia Ladron de Guevara, Guadalajara 44680, Mexico	36 301655	36 301652	4,120	92
Molson Breweries	1600-175 Bloor St. E, Toronto, Ontario M4W 3R8, Canada	416-975-1786	416-975-4088	4,100	98
Marston, Thompson and Evershed PLC	PO Box 26, Burton on Trent DE14 2BW, United Kingdom	283 531131	283 510378	4,014	94
Coca-Cola Beverages Ltd.	42 Overlea Blvd, Toronto, Ontario M4H 1B8, Canada	416-424-6000	416-424-6079	4,000	98
Holstein & Kappert AG	Juchostrasse 20, Postfach 105026, Dortmund 44143, Germany	231 51850	23 585377	4,000	91
Unilever Canada Ltd.	160 Bloor St. E Ste. 1500, Toronto, Ontario M4W 3R2, Canada	416-964-1857	416-964-8831	3,816	98
Haitai Beverage Co., Ltd.	Pupyong-gu, Inchon, Republic of Korea	32 5402114	32 5425157	3,800	97
Societe Anonyme des Brasseries du Cameroun	BP 4036, Douala, Cameroon	429133	427945	3,679	91
Mansfield Brewery PLC	Littleworth, Mansfield NG18 1AB, United Kingdom	623 25691	623 658620	3,569	94
Companhia Antarctica Paulista Industria y Comercio Brasilerira de Bebidas	Mooca, Sao Paulo 03107-900, Brazil	11 2431000	11 2431900	3,500	96
UST Inc.	100 W. Putnam Ave., Greenwich, Connecticut 06830, U.S.A.	203-661-1100		3,500	96
United Distillers PLC	33 Ellersly Rd., Edinburgh EH1R GJW, United Kingdom	31 3377373	31 3370163	3,406	93
Bonnita Pty. Ltd.	PO Box 809, Stellenbosch 7599, Republic of South Africa	21 8866640	21 8866905	3,400	96
Pepsi-Cola Canada Beverages	1255 Bay St., Toronto, Ontario M5R 2A9, Canada	416-964-1313	416-926-4250	3,310	98
Seagram Distillers PLC	5-7 Mandeville Pl., London W1M 5LB, United Kingdom	71 4084477	71 4084703	3,118	93
Coca-Cola & Schweppes Beverages Ltd.	Charter Pl., Uxbridge UB8 1EZ, United Kingdom	895 31313		3,057	94
Coca Cola de Argentina SA	Paraguay 733, Buenos Aires 1057, Argentina	1 3192000	1 3192010	3,000	97
Osem Investments Ltd.	61 Jabotinsky St., Petah Tikva 49517, Israel	3 9265515	3 9265216	2,800	97
Stroh Brewery Co.	100 River Pl., Detroit, Michigan 48207, U.S.A.	313-446-2000	313-446-2333	2,682	95
Gatorade	5625 E. 14th St., Oakland, California 94621, U.S.A.	510-261-5800		2,600	95
Canandaigua Wine Company Inc.	116 Buffalo St., Canandaigua, New York 14424, U.S.A.	716-394-7900		2,500	96
Coca Cola-Femsa	Av. Dr. Honorio Pueyrredon 1651, Buenos Aires 1414, Argentina	1 8540023	1 8540023	2,500	96
Cerveceria Y Malteria Quilmes SA ic Y G	Teniente General Juan Domingo Peron 667, Buenos Aires 1038, Argentina	1 3266860	1 3262231	2,485	96
Greenalls Midlands Ltd.	New Basford, Nottingham NG7 7FN, United Kingdom			2,300	90
Cerveceria Hondurena SA	PO Box 86, Tegucigalpa, Honduras	530011	531970	2,200	96
Oriental Brewery Co., Ltd.	Chung-Gu, Seoul, Republic of Korea	2 7272114	2 7548216	2,200	97
Oy Hartwall Ab	Ristipellontie 4, Helsinki SF-00391, Finland	54021	5402523	2,200	94

Continued

Representative Companies in Sector
- Continued -

Name	Address	Tele-phone	Fax	Employ-ment	Y
United Distillers Blending and Bottling Ltd.	33 Ellersly Rd., Edinburgh EH12 6JW, United Kingdom			2,178	91
United Malt and Grain Distillers Ltd.	33 Ellersley Rd., Edinburgh EH4 6EA, United Kingdom	31 3377373	31 3379335	2,155	93
Bush Boake Allen Inc.	7 Mercedes Dr., Montvale, New Jersey 07645, U.S.A.	201-391-9870	201-391-0860	2,103	96
Triarc Companies Inc.	280 Park Ave., New York, New York 10017, U.S.A.	212-451-3000		2,000	97
J.M. Smucker Co.	1 Strawberry Ln., Orrville, Ohio 44667-0280, U.S.A.	216-682-3000		1,950	97
Industrial de Gaseosas SA	Carrera 94 24-94, Bogota, Colombia	1 2679300	1 3600987	1,900	97
Union Camp Corp.	1600 Valley Rd., Wayne, New Jersey 07470, U.S.A.	201-628-2000		1,900	97
Ionics Inc.	PO Box 9131, Watertown, Massachusetts 02272-9131, U.S.A.	617-926-2500	617-926-4304	1,875	97
Young & Co.'s Brewery PLC	Wandsworth, London SW18 4JD, United Kingdom	81 8700141	81 8709444	1,832	92
G. Heileman Brewing Company Inc.	100 Harborview Plz., La Crosse, Wisconsin 54601, U.S.A.	608-785-1000	608-785-3468	1,800	95
Guinness Brewing Worldwide Ltd.	Park Royal Brewery, London NW10 7RR, United Kingdom	81 9657700	81 4530222	1,723	91
Central Bottling Co. Ltd.	129 Khanman St., Bnei Brak 51641, Israel	3 6712222	3 6712302	1,700	97
J and J Snack Foods Corp.	6000 Central Hwy., Pennsauken, New Jersey 08109, U.S.A.	609-665-9533	609-665-6359	1,700	97
Schultz Sav-O Stores Inc.	PO Box 419, Sheboygan, Wisconsin 53082-0419, U.S.A.	920-457-4433	920-457-6295	1,680	97
Boon Rawd Brewery Co., Ltd.	999 Samsen Rd., Bangkok 10300, Thailand	2411360	2431740	1,650	94
Joseph Holt PLC	Cheetham, Manchester M3 1JD, United Kingdom	61 8343285		1,612	93
Cold Storage (Malaysia) BHD	10th Fl., Jaya Shopping Centre, Petaling Jaya 46100, Malaysia	3 7588888	3 7581289	1,600	93
Societe de Limonaderies et Brasseries d'Afrique	BP 1304, Abidjan 01, Cote d'Ivoire	249133	359791	1,600	92
National Brands Ltd.--Becketts Division	PO Box 680, Johannesburg 2000, Republic of South Africa	11 8832714	11 8845486	1,550	93
CBR Brewing Company Inc.	433 N. Camden, Ste. 1200, Beverly Hills, California 90210, U.S.A.	310-274-5172		1,537	96
Greene King PLC	Westgate Brewery, Bury St. Edmunds IP33 1QT, United Kingdom	284 763222	284 706502	1,508	93
IND Coope Burton Brewery Ltd.	107 Station St., Burton-on-Trent DE14 1BZ, United Kingdom	283 31111	283 516225	1,500	90
Jim Beam Brands Co.	510 Lake Cook Rd., Deerfield, Illinois 60015, U.S.A.	847-948-8888	847-255-3313	1,500	94
Pepsi-Cola Northwest	PO Box C-14117, Seattle, Washington 98144, U.S.A.	206-323-2932	206-326-7496	1,500	97
Sociedad Anonima el Aguila Fabricas de Cerveza	Edificio Eje Sur, Madrid E-28023, Spain	91 5069700	91 4686234	1,499	97
Fuller, Smith & Turner PLC	Chiswick, London W4 2QB, United Kingdom	81 9940122	81 9950230	1,443	94
Beverage America Inc.	545 E. 32nd St., Holland, Michigan 49423, U.S.A.	616-396-1281	616-392-3291	1,400	97
Cott Corp.	207 Queens Quay W Ste. 800, Toronto, Ontario M5J 1A7, Canada	416-203-3898	416-203-5609	1,350	98
Antarctica da Amazonia SA Industria de Bebidas	Flores, Manaus 69050-002, Brazil	92 6651000	92 2365095	1,300	97
Campbell Soup Co. Ltd.	60 Birmingham St., Toronto, Ontario M8V 2B8, Canada	416-251-1131	416-253-8611	1,300	98
Office National de Commercialisation des Produits Viti-Vinicoles	12 Quai Sud, Algiers, Algeria	79 737275	79 737297	1,300	90
Pabst Brewing Co.	PO Box 1661, San Antonio, Texas 78296, U.S.A.	210-226-0231	210-226-2512	1,300	96
Charles Wells Ltd.	Havelock St., Bedford MK40 4LU, United Kingdom	234 272766	234 279000	1,278	94
Eldridge Pope & Co. PLC	Weymouth Ave., Dorchester DT1 1QT, United Kingdom	305 251251	305 258300	1,271	94
Morland & Co. PLC	Ock St., Abingdon OX14 5DD, United Kingdom	235 553377	235 555225	1,207	94
Guinness Anchor BHD	46710 Petaling Jaya, Selangor, Malaysia	3 7763022	3 7740986	1,200	
Molson Breweries, Ontario Division	4100 Yonge St. Ste. 200, North York, Ontario M2P 2E6, Canada	416-226-1786	416-512-3800	1,200	98
National Beverage Corp.	1 N. University Dr., 400A, Plantation, Florida 33324, U.S.A.	954-581-0922	954-473-4710	1,200	97
Royal Beech-Nut Pty. Ltd.	PO Box 12029, Chloorkop 1624, Republic of South Africa	11 9218388	11 3935121	1,200	95
Schweppes, SA	Sor Angela de la Cruz, 3, Madrid E-28020, Spain	91 5823600	91 5823666	1,200	97
Societe Nationale de Boissons	BP 135, Cotonou, Benin	331124	331048	1,200	91
Tempo Beer Industries Ltd.	Industrial Zone, Netanya 42101, Israel	9 8630630	9 8653679	1,200	97
Vincor International Inc.	6611 Edwards Blvd, Mississauga, Ontario L5T 2H8, Canada	905-564-6900	905-564-6909	1,181	98
Farmer Brothers Co.	20333 S. Normandie Ave., Torrance, California 90502-1254, U.S.A.	310-787-5200		1,168	97
A.G. Barr PLC	1306 Gallowgate, Glasgow G31 4DS, United Kingdom	41 5541899	41 5545768	1,161	93
H.P. Bulmer Ltd.	Plough Ln., Hereford HR4 0LE, United Kingdom	432 352000	432 352084	1,161	94
Spolem PSS	ul. Cyniarska 11, Bielsko-Biala PL-43-300, Poland	30 24576	30 24886	1,100	93
Swire Pacific Holdings Inc.	875 S. West Temple St., Salt Lake City, Utah 84101, U.S.A.	801-530-5300	801-530-5390	1,100	95

Continued

Representative Companies in Sector
- Continued -

Name	Address	Tele-phone	Fax	Employ-ment	Y
William Grant & Sons Ltd.	84 Lower Mortlake Rd., Richmond TW9 2LR, United Kingdom	81 3321188		1,061	92
Conchatoro Vina Concha y Toro SA	San Miguel, Santiago, Chile	2 5567882	2 5567882	1,000	96
Gaseosas Lux SA	Carrera 38 15-40, Bogota, Colombia	1 2512121	1 2353318	1,000	97
Induroman Industrias Roman SA	Bosque Diagonal 21 42-80, Cartagena, Colombia	5 690899	5 681471	985	97
K.W.V.	57 Main St., Suider Paarl 7624, Republic of South Africa	11 73911	11 73000	973	95
McCormick International Div.	PO Box 6000, Sparks, Maryland 21152-6000, U.S.A.	410-771-7336	410-527-8195	972	97
Burtonwood Brewery PLC	Burtonwood, Warrington WA5 4PJ, United Kingdom	925 225131	925 229033	956	94
Quaker Oats Co. of Canada Ltd.	PO Box 4100 Sta. Main Quaker Pk, Peterborough, Ontario K9J 7B2, Canada	705-743-6330	705-876-4192	950	98
Sociedad Anonima Damm	Rosellon, 515, Barcelona E-08025, Spain	93 2909200	93 4565198	944	97
Embotelladora Tica, SA	Apartado 2025-1000, San Jose, Costa Rica	2339333	2216531	900	
Robert Mondavi Corp.	7801 St. Helena Hwy., Oakville, California 94562, U.S.A.	707-259-9463	707-963-1077	890	97
Coca-Cola Bottlers Ulster Ltd.	Lambeg, Lisburn BT27 5SS, United Kingdom	846 674231	846 671049	861	93
Andres Wines Ltd.	697 South Service Rd., Grimsby, Ontario L3M 4E8, Canada	905-643-4131	905-643-4944	850	98
Yeo Hiap Seng Ltd.	950 Dunearn Rd., Singapore 2158, Singapore	4662266	4664641	840	92
C.F. Sauer Company Inc.	PO Box 27366, Richmond, Virginia 23261, U.S.A.	804-359-5786	804-358-4396	800	97
Shepherd Neame Ltd.	17 Court St., Faversham ME13 7AX, United Kingdom	795 532206	795 538907	786	94
Whyte & MacKay Distillers Ltd.	310 St. Vincent St., Glasgow G2 5RG, United Kingdom	41 2485771	41 221993	786	91
Heublein do Brasil, Comercio e Industria Ltda.	Rua Arapore 655, Morumbi 05602-001, Brazil	11 8137300	11 8130306	764	96
Societe Congolaise des Brasseries Kronenbourg	BP 1147, Pointe Noire, Congo	940245	943740	750	91
International Distillers & Vintners Ltd.	Regent's Pk., London NW1 4PU, United Kingdom	71 9354446		729	94
St. Austell Brewery Co. Ltd.	63 Treviathian Rd., St. Austell PL25 4BY, United Kingdom	726 74444	726 68965	725	94
Carlsberg-Tetley UK Ltd.	2410 The Crescent, Birmingham B37 7YE, United Kingdom			709	92
Bavaria BV	Postbus 1, Lieshout NL-5737 ZG, Netherlands	4992 8111	4992 8269	700	
Gray Beverage Inc.	747 Chester Rd., Delta, British Columbia V3M 6E7, Canada	604-520-8000	604-520-8010	700	98
Hiram Walker & Sons Ltd.	2072 Riverside Dr. E, Windsor, Ontario N8Y 4S5, Canada	519-254-5171	519-971-5717	700	98
Southeast Atlantic Corp.	PO Box 17999, Jacksonville, Florida 32245, U.S.A.	904-739-1000		700	93
Pepsico - IVI SA	22nd Km Lavriou Ave., Koropi GR-194 00, Greece	6028541	6646345	675	
San Miguel Brewery Ltd.	Yuen Long, New Territories, Hong Kong	4910411	4917237	663	92
Allied Domecq Spirits & Wine	2072 Riverside Dr. E, Windsor, Ontario N8Y 4S5, Canada	519-254-5171	519-971-5710	650	98
Asia Pacific Breweries Ltd.	438 Alexandra Rd., Singapore 0511, Singapore	2763488	2764287	650	93
Jafora-Tabori Ltd.	Industrial Zone, Rehovot 76360, Israel	8 9344644	8 9467163	650	97
Nutrifood Indonesia, PT	PO Box 4060, Jakarta Pusat 10320, Indonesia	213100083	21 3100144	650	97
Shasta Beverages Inc.	26901 Industrial Blvd., Hayward, California 94545, U.S.A.	510-783-3200	510-785-3228	650	97
Cerveceria y Malteria Paysandu SA	Rambla Baltasar Brum 2933, Montevideo, Uruguay	2 208521	2 237006	640	98
Genesee Corp.	PO Box 762, Rochester, New York 14603, U.S.A.	716-546-1030		622	97
Fab. De Licores y Alcoholes De Antioquia	Carrera 50 12 Sur - 149, Medellin, Colombia	4 3658520	4 2859813	615	97
Alimentos Kern de Guatemala SA	Carretera al Atlantico, Guatemala City, Guatemala	2 560537	2 567978	600	
Carters Drinks Group Ltd.	Side Ley, Kegworth DE7 2FJ, United Kingdom	509 674915	509 673461	583	92
Glenmore Distilleries Co.	1700 Citizens Plz., Louisville, Kentucky 40202, U.S.A.	502-589-0130		581	
Kuwait National Bottling Co.	PO Box 224, Safat 13003, Kuwait	4717155	4714324	565	88
Weider Nutrition International Inc.	1960 South 4250 West, Salt Lake City, Utah 84104-4836, U.S.A.	801-975-5000	801-972-2223	560	97
Taunton Cider PLC	Norton Fitzwarren, Taunton TA2 6RD, United Kingdom	823 332211	823 256433	555	94
Beringer Wine Estates	PO Box 111, St. Helena, California 94574, U.S.A.	707-963-7115		550	95
Premium Beverage Packers Inc.	1090 Spring St., Wyomissing, Pennsylvania 19610, U.S.A.	610-376-6131	610-376-4163	550	96
Saskatchewan Wheat Pool, CSP Foods Division	1-2175 Airport Dr., Saskatoon, Saskatchewan S7L 7E1, Canada	306-978-3400	306-978-3409	550	98
Systems Bio-Industries Inc.	2021 Cabot Blvd., W., Langhorne, Pennsylvania 19047, U.S.A.	215-702-1000	215-638-8168	550	97
Efes Pilsen Beer and Malt Group	Bahcelievler, Istanbul, Turkey	212 5846060	212 5846457	527	93
Taunton Cider Trading Ltd.	Norton Fitzwarren, Taunton TA2 6RD, United Kingdom	823 332211	823 256433	506	92
Barton Inc.	55 E. Monroe St., 1700, Chicago, Illinois 60603, U.S.A.	312-346-9200		500	95
Embotelladora Central SA	24 Calle 6-01, Guatemala City, Guatemala	2 762228		500	
Jel Sert Co.	Hwy. 59 & Conde St., Chicago, Illinois 60607, U.S.A.	630-231-7590	630-231-2149	500	97
William Muir Bond 9 Ltd.	Leith, Edinburgh EH6 7AW, United Kingdom	31 5544355		479	94
Invergordon Distillers Holdings PLC	165 Queen Victoria St., London EC4V 4DD, United Kingdom			471	90
Carters Gold Medal Soft Drinks, Ltd.	Side Ley, Kegworth DE7 2FJ, United Kingdom	509 674915	509 673461	459	94

Product Tables
Distilled Alcoholic Beverages, Excluding Ethyl Alcohol

Unit of Measure: 1,000 Liters.

	1990 Value	%	1991 Value	%	1992 Value	%	1993 Value	%	1994 Value	%	1995 Value	%	1996 Value	%
Total Production	7,741,970	100.0	7,901,122	100.0	8,820,219	100.0	8,407,813	100.0	7,816,939	100.0	7,890,618	100.0	7,302,901	100.0
Regions														
Africa	114,505	1.5	116,260	1.5	117,515	1.3	119,070	1.4	123,072	1.6	124,163	1.6	129,253	1.8
America, North	1,073,798	13.9	927,927	11.7	785,252	8.9	757,966	9.0	686,424	8.8	522,573	6.6	486,456	6.7
America, South	282,106	3.6	315,601	4.0	345,097	3.9	357,838	4.3	356,730	4.6	369,328	4.7	399,532	5.5
Asia	3,634,262	46.9	3,724,683	47.1	4,806,819	54.5	4,490,420	53.4	3,980,028	50.9	3,812,236	48.3	3,342,395	45.8
Europe	2,555,648	33.0	2,744,010	34.7	2,701,905	30.6	2,627,898	31.3	2,625,075	33.6	3,025,718	38.3	2,917,674	40.0
Oceania	81,650	1.1	72,640	0.9	63,630	0.7	54,620	0.6	45,610	0.6	36,600	0.5	27,590	0.4
Leading Producers														
Russian Federation	1,380,000	17.8	1,540,000	19.5	1,520,000	17.2	1,570,000	18.7	1,250,000	16.0	1,230,000	15.6	710,000	9.7
United States of America	601,229	7.8	443,614	5.6	286,000	3.2	268,700	3.2	187,026	2.4	*42,395	0.5	-	
Japan	836,000	10.8	709,200	9.0	792,700	9.0	851,900	10.1	842,200	10.8	833,600	10.6	853,000	11.7
Philippines	*960,024	12.4	*1,046,160	13.2	*1,132,296	12.8	*1,218,432	14.5	*1,304,568	16.7	*1,390,704	17.6	*1,476,840	20.2
Azerbaijan	213,500	2.8	205,000	2.6	1,194,500	13.5	656,400	7.8	372,900	4.8	141,500	1.8	91,800	1.3

Ethyl Alcohol for All Purposes

Unit of Measure: 1,000 Liters.

	1990 Value	%	1991 Value	%	1992 Value	%	1993 Value	%	1994 Value	%	1995 Value	%	1996 Value	%
Total Production	28,423,023	100.0	29,629,618	100.0	30,911,722	100.0	31,726,926	100.0	31,892,498	100.0	34,202,798	100.0	35,282,339	100.0
Regions														
Africa	34,099	0.1	34,436	0.1	34,863	0.1	35,298	0.1	35,746	0.1	36,185	0.1	36,624	0.1
America, North	3,602,784	12.7	3,931,176	13.3	4,259,568	13.8	3,850,473	12.1	3,071,697	9.6	4,229,112	12.4	4,423,059	12.5
America, South	16,854,663	59.3	18,191,955	61.4	19,484,048	63.0	20,760,115	65.4	22,116,965	69.3	23,361,891	68.3	24,659,896	69.9
Asia	4,631,373	16.3	4,296,371	14.5	3,860,204	12.5	3,785,007	11.9	3,363,694	10.5	3,112,090	9.1	2,451,753	6.9
Europe	3,300,104	11.6	3,175,681	10.7	3,273,038	10.6	3,296,032	10.4	3,304,396	10.4	3,463,521	10.1	3,711,007	10.5
Leading Producers														
Brazil	*16,650,914	58.6	*17,971,486	60.7	*19,292,057	62.4	*20,612,629	65.0	*21,933,200	68.8	*23,253,771	68.0	*24,574,343	69.7
United States of America	3,142,725	11.1	3,475,213	11.7	3,807,700	12.3	3,402,700	10.7	2,628,020	8.2	*3,789,530	11.1	*3,987,573	11.3
Russian Federation	1,560,000	5.5	1,500,000	5.1	1,450,000	4.7	1,500,000	4.7	1,280,000	4.0	1,240,000	3.6	830,000	2.4
Croatia	25,900	0.1	24,800	0.1	21,900	0.1	20,100	0.1	12,900	0.1	13,100	0.0	12,900	0.0
France	*1,040,100	3.7	*1,040,100	3.5	*1,040,100	3.4	*1,040,100	3.3	*1,040,100	3.3	*1,040,100	3.0	*1,040,100	2.9

Wine

Unit of Measure: 1,000 Liters.

	1990 Value	%	1991 Value	%	1992 Value	%	1993 Value	%	1994 Value	%	1995 Value	%	1996 Value	%
Total Production	29,421,763	100.0	29,506,899	100.0	28,023,270	100.0	24,636,586	100.0	24,456,396	100.0	24,616,150	100.0	24,545,541	100.0
Regions														
Africa	1,350,403	4.6	1,392,729	4.7	1,214,930	4.3	1,179,213	4.8	1,222,549	5.0	1,296,056	5.3	1,303,475	5.3
America, North	1,906,284	6.5	2,037,854	6.9	2,066,105	7.4	2,113,830	8.6	2,174,554	8.9	2,233,879	9.1	2,307,103	9.4
America, South	2,554,728	8.7	2,539,053	8.6	2,376,918	8.5	2,197,137	8.9	2,142,786	8.8	2,064,028	8.4	2,046,379	8.3
Asia	3,070,095	10.4	3,373,087	11.4	2,519,086	9.0	1,765,352	7.2	1,303,355	5.3	1,203,736	4.9	1,259,506	5.1
Europe	20,046,254	68.1	19,711,076	66.8	19,366,061	69.1	16,890,679	68.6	17,117,673	70.0	17,312,766	70.3	17,116,588	69.7
Oceania	494,000	1.7	453,100	1.5	480,169	1.7	490,375	2.0	495,480	2.0	505,685	2.1	512,491	2.1
Leading Producers														
France	6,553,000	22.3	6,200,000	21.0	*6,199,331	22.1	*6,128,907	24.9	*6,058,484	24.8	*5,988,060	24.3	*5,917,636	24.1
Italy	5,487,000	18.6	5,915,000	20.0	*5,695,735	20.3	*5,511,769	22.4	*5,327,804	21.8	*5,143,838	20.9	*4,959,873	20.2
Spain	3,295,800	11.2	2,877,300	9.8	3,168,700	11.3	1,383,200	5.6	1,626,700	6.7	2,189,100	8.9	1,803,300	7.3
United States of America	1,795,500	6.1	1,860,417	6.3	1,925,333	6.9	1,990,250	8.1	2,055,167	8.4	2,120,083	8.6	2,185,000	8.9
Argentina	1,713,100	5.8	1,711,100	5.8	1,619,300	5.8	1,455,800	5.9	1,417,900	5.8	1,349,100	5.5	1,354,200	5.5

Commodity data are provided by the United Nations Statistical Division. The symbol * means that data are estimated. For additional notes, see Appendix I.

61

Product Tables
Malt

Unit of Measure: Metric tons.

	1990 Value	%	1991 Value	%	1992 Value	%	1993 Value	%	1994 Value	%	1995 Value	%	1996 Value	%
Total Production	15,922,446	100.0	16,501,736	100.0	16,273,112	100.0	15,463,493	100.0	15,385,828	100.0	15,551,954	100.0	15,231,979	100.0
Regions														
Africa	1,404,552	8.8	1,510,381	9.2	1,593,210	9.8	1,656,993	10.7	1,754,545	11.4	1,841,174	11.8	1,932,345	12.7
America, North	2,414,667	15.2	2,418,889	14.7	2,403,111	14.8	2,473,333	16.0	2,449,556	15.9	2,460,778	15.8	2,500,000	16.4
America, South	473,428	3.0	505,334	3.1	431,597	2.7	428,438	2.8	507,047	3.3	590,533	3.8	475,586	3.1
Asia	1,777,750	11.2	2,199,483	13.3	2,097,344	12.9	1,538,355	9.9	1,220,315	7.9	994,876	6.4	680,036	4.5
Europe	9,249,049	58.1	9,263,649	56.1	9,186,851	56.5	8,848,375	57.2	8,897,365	57.8	9,110,594	58.6	9,058,012	59.5
Oceania	603,000	3.8	604,000	3.7	561,000	3.4	518,000	3.3	557,000	3.6	554,000	3.6	586,000	3.8
Leading Producers														
United States of America	2,171,667	13.6	2,176,889	13.2	2,182,111	13.4	2,187,333	14.1	2,192,556	14.3	2,197,778	14.1	2,203,000	14.5
Germany	*1,710,494	10.7	*1,710,494	10.4	*1,710,494	10.5	1,601,000	10.4	1,786,000	11.6	1,761,723	11.3	1,693,254	11.1
France	1,287,000	8.1	1,335,000	8.1	1,318,000	8.1	1,265,000	8.2	1,216,000	7.9	1,269,000	8.2	1,274,000	8.4
United Kingdom	1,214,000	7.6	1,178,000	7.1	1,142,982	7.0	1,107,964	7.2	1,072,946	7.0	*1,083,809	7.0	*1,068,426	7.0
Indonesia	863,000	5.4	1,172,000	7.1	1,188,000	7.3	*753,900	4.9	*516,600	3.4	*279,300	1.8	*42,000	0.3

Beer

Unit of Measure: Millions of Liters.

	1990 Value	%	1991 Value	%	1992 Value	%	1993 Value	%	1994 Value	%	1995 Value	%	1996 Value	%
Total Production	117,466	100.0	133,638	100.0	133,319	100.0	133,982	100.0	135,084	100.0	138,354	100.0	138,109	100.0
Regions														
Africa	5,324	4.5	5,322	4.0	5,728	4.3	6,639	5.0	4,997	3.7	5,212	3.8	5,930	4.3
America, North	30,903	26.3	31,118	23.3	31,270	23.5	31,385	23.4	31,472	23.3	31,411	22.7	31,803	23.0
America, South	7,826	6.7	9,263	6.9	8,399	6.3	8,744	6.5	9,474	7.0	11,691	8.4	10,836	7.8
Asia	27,880	23.7	32,620	24.4	32,519	24.4	32,965	24.6	33,717	25.0	35,746	25.8	34,959	25.3
Europe	43,094	36.7	52,957	39.6	53,136	39.9	52,021	38.8	53,182	39.4	52,073	37.6	52,421	38.0
Oceania	2,439	2.1	2,357	1.8	2,266	1.7	2,228	1.7	2,242	1.7	2,221	1.6	2,160	1.6
Leading Producers														
United States of America	23,667	20.1	23,614	17.7	23,561	17.7	23,508	17.5	23,455	17.4	23,402	16.9	23,348	16.9
Germany	-		11,207	8.4	11,409	8.6	11,107	8.3	11,343	8.4	11,188	8.1	10,894	7.9
China	6,922	5.9	8,380	6.3	10,206	7.7	11,921	8.9	14,142	10.5	15,688	11.3	*14,790	10.7
United Kingdom	7,080	6.0	*6,222	4.7	*6,250	4.7	*6,278	4.7	*6,306	4.7	*6,334	4.6	*6,362	4.6
Japan	6,564	5.6	6,916	5.2	7,011	5.3	6,964	5.2	7,101	5.3	6,797	4.9	6,908	5.0

Mineral Waters

Unit of Measure: 1,000 Liters.

	1990 Value	%	1991 Value	%	1992 Value	%	1993 Value	%	1994 Value	%	1995 Value	%	1996 Value	%
Total Production	36,751,445	100.0	31,653,873	100.0	32,979,931	100.0	31,581,680	100.0	32,713,755	100.0	35,739,562	100.0	35,704,574	100.0
Regions														
Africa	372,865	1.0	337,172	1.1	374,479	1.1	380,986	1.2	402,442	1.2	383,885	1.1	396,948	1.1
America, North	320,060	0.9	301,048	1.0	286,736	0.9	481,817	1.5	377,039	1.2	339,379	0.9	323,467	0.9
America, South	1,874,381	5.1	2,016,296	6.4	2,162,866	6.6	2,328,721	7.4	2,471,596	7.6	2,537,634	7.1	2,739,537	7.7
Asia	8,143,739	22.2	3,600,491	11.4	4,064,468	12.3	2,916,357	9.2	3,575,421	10.9	5,191,936	14.5	4,812,589	13.5
Europe	26,040,400	70.9	25,398,866	80.2	26,091,383	79.1	25,473,800	80.7	25,887,257	79.1	27,286,729	76.3	27,432,034	76.8
Leading Producers														
Germany	*6,909,700	18.8	6,594,500	20.8	7,224,900	21.9	*6,909,700	21.9	*6,909,700	21.1	*6,909,700	19.3	*6,909,700	19.4
France	5,217,000	14.2	5,192,000	16.4	5,300,000	16.1	5,100,000	16.1	5,300,000	16.2	6,154,000	17.2	6,123,000	17.1
Kazakhstan	6,532,100	17.8	1,866,500	5.9	1,127,500	3.4	1,184,800	3.8	1,163,100	3.6	1,392,500	3.9	1,781,200	5.0
Poland	1,735,900	4.7	1,197,800	3.8	832,400	2.5	908,000	2.9	1,076,700	3.3	1,137,400	3.2	1,218,200	3.4
Spain	1,425,000	3.9	1,697,600	5.4	1,747,000	5.3	1,990,000	6.3	2,053,700	6.3	2,370,300	6.6	2,086,600	5.8

Product Tables
Soft Drinks

Unit of Measure: 1,000 Liters.

	1990		1991		1992		1993		1994		1995		1996	
	Value	%	Value	%	Value	%	Value	%	Value	%	Value	%	Value	%
Total Production	74,646,335	100.0	83,151,683	100.0	80,530,497	100.0	83,427,292	100.0	86,110,110	100.0	91,720,768	100.0	95,745,077	100.0
Regions														
Africa	4,696,883	6.3	4,013,995	4.8	4,130,156	5.1	4,022,578	4.8	3,075,283	3.6	3,281,001	3.6	4,028,516	4.2
America, North	9,185,542	12.3	9,383,834	11.3	9,521,272	11.8	12,029,509	14.4	12,568,046	14.6	12,240,730	13.3	12,590,945	13.2
America, South	8,674,465	11.6	9,974,803	12.0	9,745,141	12.1	10,044,523	12.0	10,586,410	12.3	11,922,488	13.0	12,000,560	12.5
Asia	22,066,726	29.6	23,914,245	28.8	21,282,075	26.4	22,138,652	26.5	22,889,915	26.6	23,647,688	25.8	26,102,676	27.3
Europe	28,379,077	38.0	34,187,819	41.1	34,162,381	42.4	33,482,537	40.1	35,147,779	40.8	38,726,645	42.2	39,058,792	40.8
Oceania	1,643,642	2.2	1,676,986	2.0	1,689,473	2.1	1,709,492	2.0	1,842,677	2.1	1,902,215	2.1	1,963,589	2.1
Leading Producers														
Germany	-		6,172,500	7.4	6,471,900	8.0	6,414,400	7.7	6,826,400	7.9	8,781,671	9.6	9,055,981	9.5
Mexico	6,764,500	9.1	6,956,200	8.4	7,083,100	8.8	9,578,600	11.5	10,099,200	11.7	9,754,500	10.6	10,092,300	10.5
United Kingdom	*5,719,281	7.7	*5,847,712	7.0	*5,976,144	7.4	*6,104,576	7.3	*6,233,007	7.2	*6,361,439	6.9	*6,489,871	6.8
Brazil	3,770,400	5.1	4,212,800	5.1	3,409,300	4.2	3,490,400	4.2	3,897,300	4.5	5,098,500	5.6	5,028,900	5.3
Japan	3,145,000	4.2	3,284,000	3.9	3,275,000	4.1	3,196,400	3.8	3,574,300	4.2	3,392,200	3.7	3,383,900	3.5

Commodity data are provided by the United Nations Statistical Division. The symbol * means that data are estimated. For additional notes, see Appendix I.

63

ISIC 3140
TOBACCO

ISIC 3140—Tobacco—includes tobacco, prepared leaf; cigars; cigarettes, manufactured tobacco, including smoking tobacco, chewing tobacco, and snuff.

There is a very close correspondence between ISIC 3140 and the 2-digit U.S. SIC series 21—Tobacco Products. The U.S. structure is divided into essentially the same subcategories: Cigarettes (2111), Cigars (2121), Chewing and Smoking Tobacco (2131) , and Tobacco Stemming and Redrying (2141).

Summary Statistics

		1990		1991		1992		1993		1994		1995	
		Value	N	Value	N	Value	N	Value	N	Value	N	Value	N
Establishments or enterprises	(number)	11,422	79	11,579	92	11,024	85	7,617	83	1,848	68	1,542	34
Number employed	(000)	1,417	91	1,404	96	1,405	90	1,319	89	1,384	78	1,458	58
Total Wages	($ mil.)	6,652	87	7,129	87	7,742	80	6,517	81	5,895	70	5,102	52
Wage/salary per employee	($)	12,011	85	13,073	85	23,492	79	16,538	80	13,633	69	16,961	52
Female workers	(000)	73	36	47	34	68	30	217	34	227	28	241	14
as % of total employment*	(%)	38	36	36	34	37	30	39	34	41	28	48	13
Output	($ bil.)	131	90	153	93	232	88	211	85	142	75	128	57
per employee	($ 000)	205	86	228	89	1,724	84	1,901	81	289	70	351	53
per establishment	($ 000)	56,573	75	68,905	82	489,306	74	517,540	71	86,186	57	76,411	26
Value added	($ bil.)	73	81	77	83	82	76	76	75	61	66	66	56
per employee	($ 000)	126	78	134	79	149	73	143	70	175	60	214	50
per establishment	($ 000)	35,396	68	41,165	74	47,648	63	41,502	60	48,028	47	36,349	24
Capital investment	($ mil.)	1,869	61	2,331	59	2,445	52	2,219	50	1,347	41	747	21
per employee	($ 000)	5	58	6	56	7	50	7	48	6	39	5	19
per establishment	($ 000)	1,383	57	1,759	55	2,726	49	3,194	46	1,707	38	1,185	17

Data presented above are drawn from the detailed tables that follow. Columns headed 'N' show the number of countries that provided valid data for inclusion. Values are not strictly comparable one year to the next or one row to the next because the number of countries that report varies from period to period and row to row. However, a general indication of magnitudes can be gleaned from reviewing this summary—especially in earlier years in which more reports are available. For detailed explanations, see Appendix I. *The average for those countries reporting both total and female employment.

Establishments and Number Engaged

Country	Establishments or Enterprises (number)						Employees per Establishment					
	1990	1991	1992	1993	1994	1995	1990	1991	1992	1993	1994	1995
Albania	-	-	-	3.000	6.000	6.000	-	-	-	751	219	163
Angola	-	-	2.000	2.000	-	-	-	-	247	239	-	-
Argentina	-	-	-	25	27	-	-	-	-	234	184	-
Armenia	5.000	5.000	5.000	4.000	4.000	-	225	210	-	-	-	-
Australia	4.000	5.000	5.000	-	-	-	750	600	600	-	-	-
Austria	15	14	14	14	13	-	93	100	93	89	92	-
Azerbaijan	7.000	7.000	7.000	7.000	8.000	-	214	218	239	262	219	-
Bangladesh	404	418	466	-	-	-	67	93	70	-	-	-
Belgium	55	55	55	43	39	-	89	82	80	-	-	-
Belize	2.000	2.000	2.000	-	-	-	-	-	62	58	-	-
Benin	-	-	1.000	1.000	-	-	-	-	-	-	-	-
Bolivia	1.000	1.000	1.000	1.000	1.000	1.000	158	162	169	173	195	205
Bosnia & Herzegovina	14	14	-	-	2.000	-	156	128	-	-	220	-
Brazil	229	-	194	158	-	-	113	-	121	128	-	-
Bulgaria	*24	*24	*26	*27	*27	*27	638	633	612	537	541	485
Burundi	*1.000	*1.000	-	-	-	-	181	180	-	-	-	-
Cameroon	*3.000	*3.000	*3.000	*3.000	*3.000	*2.000	348	347	320	285	257	283
Canada	18	16	17	17	17	-	278	313	294	294	294	-
Cape Verde	1.000	1.000	1.000	1.000	1.000	-	40	45	45	45	47	-
Central African Republic	-	*2.000	*2.000	*1.000	*1.000	*1.000	-	-	-	-	-	-
Chad	-	1.000	1.000	1.000	1.000	1.000	-	-	-	-	-	-
Chile	2.000	3.000	3.000	3.000	2.000	-	333	234	226	240	361	-
China	*450	*458	*479	*391	*382	*423	622	655	626	742	890	780
China (Hong Kong SAR)	4.000	4.000	4.000	4.000	4.000	-	325	350	325	325	300	-
Colombia	12	12	12	10	10	-	167	195	161	159	150	-
Costa Rica	5.000	4.000	4.000	4.000	-	-	131	159	161	156	-	-
Croatia	12	12	15	17	16	16	294	278	221	186	179	167
Cyprus	7.000	7.000	7.000	7.000	4.000	4.000	63	60	59	51	73	64
Czechoslovakia (Former)	*2.000	*2.000	-	-	-	-	2,000	2,000	-	-	-	-
Czech Republic	-	*1.000	*1.000	*1.000	-	-	-	2,000	2,000	2,000	-	-
Denmark	17	18	17	-	-	-	120	98	99	-	-	-
Egypt	39	27	32	30	-	-	449	615	531	557	-	-
El Salvador	-	2.000	1.000	1.000	1.000	1.000	-	-	273	228	237	250
Ethiopia	1.000	1.000	1.000	1.000	1.000	1.000	868	886	928	984	993	987
Finland	4.000	4.000	4.000	4.000	4.000	6.000	275	275	250	225	225	153
FYR Macedonia	*25	*24	*25	*26	*25	*29	325	297	295	274	230	184
Gabon	-	1.000	1.000	1.000	1.000	1.000	-	77	74	66	47	50
Germany	-	57	54	47	44	-	-	338	325	338	351	-
Germany (Western Part)	50	48	46	-	-	-	311	340	338	-	-	-
Ghana	-	-	-	3.000	-	-	-	-	-	353	-	-
Greece	80	67	64	-	-	-	116	142	149	-	-	-
Grenada	-	-	1.000	1.000	1.000	1.000	-	-	19	19	17	17
Guatemala	-	2.000	2.000	2.000	2.000	2.000	-	479	467	467	426	385
Honduras	-	-	13	13	13	13	-	-	231	253	256	262
Hungary	*7.000	*15	*7.000	*7.000	-	-	714	333	571	429	-	-
Iceland	1.000	1.000	1.000	-	-	-	1.000	10	10	-	-	-
India	7,878	8,299	7,786	5,158	-	-	55	56	63	90	-	-
Indonesia	961	943	902	880	748	815	213	194	203	209	287	423
Iran (Islamic Republic of)	1.000	1.000	1.000	1.000	-	-	10,500	7,919	7,197	7,197	-	-
Iraq	-	1.000	1.000	-	-	-	-	1,293	1,000	-	-	-
Ireland	8.000	6.000	-	-	-	-	163	217	-	-	-	-
Israel	13	13	-	-	-	-	46	46	-	-	-	-
Italy	*22	*18	*42	*40	*32	-	720	837	403	405	441	-
Jamaica	6.000	6.000	-	-	-	-	133	127	-	-	-	-
Japan	36	34	34	31	26	26	250	265	265	290	308	308
Jordan	4.000	4.000	4.000	4.000	6.000	6.000	203	237	239	251	203	211
Korea, Republic of	20	20	20	16	16	16	360	343	324	364	341	315
Kyrgyzstan	6.000	6.000	6.000	6.000	6.000	-	245	254	237	241	216	-
Latvia	1.000	2.000	1.000	1.000	1.000	1.000	358	178	362	343	306	284
Lithuania	-	-	-	*2.000	*2.000	-	-	-	-	-	284	285

Continued.

Depending on the table, * means *Enterprises, Engaged,* or *Factor Values*; ± means *Basis Unspecified*; § means *shown in millions.* For additional notes and sources, see Appendix I.

Establishments and Number Engaged

- Continued -

Country	Establishments or Enterprises (number)						Employees per Establishment					
	1990	1991	1992	1993	1994	1995	1990	1991	1992	1993	1994	1995
Macau	1.000	1.000	1.000	2.000	2.000	-	39	37	33	38	41	-
Malawi	*4.000	*4.000	*4.000	*4.000	*4.000	-	1,500	975	1,025	950	900	-
Malaysia	29	28	33	29	26	-	148	154	142	141	177	-
Malta	6.000	6.000	6.000	5.000	4.000	-	71	61	43	33	40	-
Mauritius	1.000	1.000	1.000	1.000	1.000	1.000	336	350	302	275	211	189
Mexico	7.000	7.000	7.000	7.000	7.000	7.000	749	779	812	828	738	717
Morocco	-	5.000	5.000	6.000			-	604	922	479	-	
Mozambique	4.000	4.000	2.000	2.000	2.000		140	135	141	137	138	
Myanmar	3.000	3.000	3.000	3.000	3.000		545	644	614	558	514	
Nepal	68	85	-	115	82		69	66	-	59	83	
Netherlands	16	15	15	13	13	-	404	425	395	442	427	-
Netherlands Antilles	-	-	-	1.000			-	-	-	16	-	
New Zealand	*4.000	*4.000	*3.000	*3.000	-		161	139	160	-	-	
Norway	2.000	2.000	-	-	-		298	328	-	-	-	
Pakistan	-	19	-	-	-		-	314	-	-	-	
Panama	3.000	3.000	-	-	-		151	157	-	-	-	
Paraguay		2.000	-	-	-		-	125	-	-	-	
Peru	11	11	11	-	-		74	75	74	-	-	
Philippines	31	31	25	22	-		419	435	564	577	-	
Poland	*8.000	*7.000	*8.000	*11	-		1,250	1,571	1,375	1,091	-	
Portugal	*5.000	*4.000	*4.000	*5.000	*5.000	-	361	400	381	307	282	-
Puerto Rico	6.000	5.000	4.000	4.000	4.000	5.000	157	154	203	177	163	138
Republic of Moldova	9.000	9.000	9.000	8.000	8.000	8.000	231	232	246	273	293	287
Romania	-	1.000	1.000	1.000	14	-	-	6,100	5,700	5,800	536	-
Russian Federation	-	-	-	*35	*36	*52	-	-	-	454	419	248
Saint Lucia	-	1.000	1.000	1.000	1.000	1.000	-	-	-	-	-	-
Saint Vincent & the Grenadines	-	-	-	1.000	1.000	1.000	-	-	-	24	24	24
Senegal	2.000	2.000	2.000	2.000	1.000	-	152	215	191	160	321	-
Sierra Leone	-	-	-	1.000			-	-	-	194	-	
Singapore	5.000	4.000	3.000	3.000	3.000	-	146	204	251	254	281	-
Slovakia	-	1.000	1.000	1.000	1.000		-	-	-	-	-	
Slovenia	1.000	*1.000	*1.000	*1.000	*1.000	*1.000	-	-	-	-	-	-
South Africa	-	11	-	-	-		-	364	-	-	-	
Spain	29	28	26	-	-	-	297	304	331	-	-	-
Sri Lanka	281	252	184	143	-		29	28	34	42	-	
Suriname	-	-	2.000	-	-		-	-	43	-	-	
Sweden	4.000	9.000	8.000	8.000	8.000		275	125	124	111	99	-
Thailand	175	164	-	-	-		181	137	-	-	-	
Trinidad and Tobago	1.000	1.000	1.000	1.000	1.000		272	267	267	233	189	-
Tunisia	-	-	-	2.000	2.000	10	-	-	-	1,355	1,250	267
Turkey	50	48	47	46	41	-	643	672	645	636	645	-
Ukraine	17	22	21	17	14	14	353	273	286	294	214	214
United Kingdom	28	26	25	41	23	-	464	462	400	220	391	-
United Republic of Tanzania	3.000	3.000	-	-	-		1,669	1,517	-	-	-	
USSR (Former)	*88	-	-	-	-		375	-	-	-	-	
United States of America	-	-	115	-	-		-	-	330	-	-	
Venezuela	33	30	22	22	15	21	100	110	150	168	202	149
Yugoslavia	22	19	22	23	21	22	-	242	223	209	238	218
Zambia	-	-	-	-	2.000		-	-	-	-	251	
Zimbabwe	8.000	8.000	8.000	7.000	5.000	-	677	744	800	829	560	-

Depending on the table, * means Enterprises, Engaged, or Factor Values; ± means Basis Unspecified; § means shown in millions. For additional notes and sources, see Appendix I.

67

Employment and Compensation of Employees

Country	Number Employed/Engaged (000)						Wage/Salary per Employee ($)					
	1990	1991	1992	1993	1994	1995	1990	1991	1992	1993	1994	1995
Albania	-	-	-	2.253	1.311	0.976	-	-	-	-	993	769
Angola	-	*0.491	*0.494	*0.478	-	-	-	-	-	-	-	-
Argentina	3.340	-	-	5.846	4.962	-	12,770	-	-	25,372	-	-
Armenia	*1.125	*1.050	-	-	-	-	-	-	-	-	-	-
Australia	3.000	*3.000	*3.000	-	-	-	26,094	25,971	27,451	-	-	-
Austria	1.400	1.400	1.300	1.248	1.195	1.195	34,364	35,788	-	-	-	-
Azerbaijan	1.497	1.526	1.672	1.833	1.751	-	1,589	1,888	-	454	-	-
Bangladesh	27	39	33	-	-	-	-	-	-	-	-	-
Belgium	4.900	4.500	4.400	-	-	-	23,414	24,521	27,555	-	-	-
Belize	-	0.123	0.116	-	-	-	-	4,724	6,784	-	-	-
Bolivia	0.158	0.162	0.169	0.173	0.195	0.205	2,127	2,533	3,532	4,093	3,486	3,520
Bosnia & Herzegovina	2.186	1.798	-	-	0.440	-	3,628	5,410	-	-	-	-
Brazil	26	-	23	20	-	-	7,942	-	9,284	12,892	-	-
Bulgaria	15	15	16	15	15	13	1,825	775	1,468	1,878	1,709	2,526
Burundi	0.181	0.180	-	-	-	-	7,988	8,325	-	-	-	-
Cambodia	*1.952	*2.126	-	-	-	-	-	-	-	-	-	-
Cameroon	1.044	1.042	0.959	0.854	0.770	0.567	10,470	10,250	11,082	12,551	5,291	8,688
Canada	5.000	5.000	5.000	5.000	5.000	4.530	-	-	-	39,997	39,690	-
Cape Verde	0.040	0.045	0.045	0.045	0.047	-	5,087	4,445	4,924	4,498	-	-
Central African Republic	0.465	-	-	-	-	-	6,864	-	-	-	-	-
Chile	0.665	0.703	0.679	0.719	0.722	0.723	8,582	10,094	12,275	14,828	15,209	17,511
China	*280	*300	*300	*290	*340	*330	-	-	-	-	-	-
China (Hong Kong SAR)	*1.300	*1.400	*1.300	*1.300	*1.200	-	16,293	18,567	21,763	22,971	24,261	-
China (Taiwan Province)	4.437	5.105	5.105	4.835	5.136	4.141	16,443	18,818	24,917	22,887	24,742	24,806
Colombia	2.000	2.345	1.937	1.591	1.496	1.250	2,699	3,016	2,646	3,468	3,979	4,486
Costa Rica	0.655	0.637	0.643	0.624	-	-	6,638	6,863	7,507	8,324	-	-
Croatia	3.530	3.330	3.320	3.170	2.860	2.680	6,779	8,470	2,246	2,117	3,342	6,223
Cyprus	0.442	0.423	0.411	0.355	0.294	0.255	14,540	15,006	16,664	15,462	17,861	24,386
Czechoslovakia (Former)	4.000	4.000	-	-	-	-	2,228	1,866	-	-	-	-
Czech Republic	2.000	2.000	2.000	2.000	-	-	2,228	2,035	2,831	4,168	-	-
Denmark	2.047	1.763	1.691	1.660	1.600	1.600	30,468	29,797	32,331	30,427	31,764	36,624
Ecuador	0.750	0.599	0.764	0.517	0.429	0.376	4,086	6,729	6,211	8,506	3,710	3,717
Egypt	*18	*17	*17	*17	*17	*17	2,606	1,513	1,700	1,920	2,167	2,477
El Salvador	-	-	*0.273	*0.228	*0.237	*0.250	-	-	-	8,695	10,475	9,274
Ethiopia	0.868	0.886	0.928	0.984	0.993	0.987	2,219	2,327	1,419	943	1,168	1,413
Finland	1.100	1.100	1.000	0.900	0.900	0.919	35,303	32,731	30,629	24,859	26,717	33,810
FYR Macedonia	8.131	7.121	7.370	7.130	5.758	5.325	8,403	14,202	2,780	3,125	3,823	4,863
France	5.100	4.900	4.900	4.800	4.500	4.500	-	39,861	-	-	-	-
Gabon	-	0.077	0.074	0.066	0.047	0.050	-	20,992	24,812	20,226	16,249	18,391
Germany	-	*19	*18	*16	*15	-	-	32,753	37,936	-	-	-
Germany (Western Part)	*16	*16	*16	-	-	-	34,976	36,290	-	-	-	-
Ghana	-	-	-	*1.060	*1.090	*1.121	-	-	-	1,806	1,349	1,376
Greece	*9.301	*9.481	*9.567	*9.548	*9.814	*10	11,526	11,297	12,321	12,123	13,801	17,134
Grenada	-	-	*0.019	*0.019	0.017	0.017	-	-	-	-	-	-
Guatemala	-	*0.958	*0.934	*0.934	*0.853	*0.771	-	791	1,353	1,204	938	1,755
Honduras	*2.848	*2.896	*3.002	*3.289	*3.329	*3.409	1,380	1,263	1,217	1,010	960	1,195
Hungary	5.000	5.000	4.000	3.000	2.970	2.940	2,794	3,415	5,067	5,627	6,191	6,872
Iceland	0.001	0.010	0.010	-	-	-	-	-	-	-	-	-
India	*434	*461	*494	*463	*519	*548	-	-	-	-	-	-
Indonesia	205	183	183	184	215	345	450	537	557	656	629	625
Iran (Islamic Republic of)	10	7.919	7.197	7.197	-	-	3,408	5,221	8,108	6,955	-	-
Iraq	-	1.293	1.000	-	-	-	-	7,351	19,003	-	-	-
Ireland	1.300	1.300	1.419	1.223	1.242	1.339	31,637	31,067	37,854	32,990	34,372	37,850
Israel	0.600	0.600	0.600	0.600	-	-	23,972	25,595	27,788	29,445	-	-
Italy	16	15	17	16	14	-	28,551	33,245	31,324	23,889	24,187	-
Jamaica	0.799	0.764	0.784	0.795	0.765	-	10,302	7,563	4,969	-	-	-
Japan	*9.000	*9.000	*9.000	*9.000	*8.000	*8.000	-	-	-	-	-	-
Jordan	0.810	0.947	0.955	1.003	1.219	1.264	4,795	4,655	4,632	4,974	4,565	4,659
Kenya	0.916	0.939	0.959	0.993	-	-	3,042	-	-	-	-	-
Korea, Republic of	7.200	6.855	6.473	5.823	5.463	5.043	14,796	17,984	19,157	20,983	21,606	26,861

Continued.

Depending on the table, * means *Enterprises, Engaged,* or *Factor Values*; ± means *Basis Unspecified*; § means *shown in millions*. For additional notes and sources, see Appendix I.

Employment and Compensation of Employees
- Continued -

Country	Number Employed/Engaged (000)						Wage/Salary per Employee ($)					
	1990	1991	1992	1993	1994	1995	1990	1991	1992	1993	1994	1995
Kyrgyzstan	*1.471	*1.527	*1.421	*1.444	*1.294	-	453	411	-	-	-	-
Latvia	0.358	0.357	0.362	0.343	0.306	0.284	-	-	-	-	-	-
Lithuania	-	-	0.571	0.568	0.570	-	-	-	548	975	2,190	-
Macau	0.039	0.037	0.033	0.075	0.082	-	2,769	3,057	3,402	4,042	5,683	-
Malawi	6.000	3.900	4.100	3.800	3.600	-	525	695	440	-	-	-
Malaysia	4.300	4.300	4.700	4.100	4.600	4.582	5,159	5,387	5,304	5,609	5,244	5,743
Malta	0.426	0.367	0.257	0.166	0.161	-	10,308	11,287	12,774	11,985	12,899	-
Mauritius	*0.336	0.350	0.302	0.275	0.211	0.189	5,827	10,063	12,416	10,241	14,197	20,081
Mexico	5.240	5.452	5.685	5.798	5.164	5.017	5,655	7,233	9,913	12,002	13,328	8,538
Morocco	-	3.019	4.611	2.872	-	-	-	-	-	-	-	-
Mozambique	0.560	0.542	0.282	0.274	0.275	-	755	535	-	-	491	-
Myanmar	1.635	1.931	1.841	1.675	1.541	-	1,582	1,626	1,644	2,403	1,445	-
Nepal	4.660	5.622	-	6.831	6.823	-	570	585	-	615	408	-
Netherlands	6.461	6.375	5.923	5.740	5.553	5.415	32,300	33,727	37,444	-	-	-
Netherlands Antilles	-	-	-	0.016	-	-				10,894	-	-
New Zealand	0.646	0.557	0.481	-	-	-	-	-	-	-	-	-
Norway	0.596	0.656	-	-	-	-	-	38,798	-	-	-	-
Pakistan	6.804	5.974	6.248	-	-	-	1,547	1,585	1,630	-	-	-
Panama	*0.453	*0.470	*0.466	*0.443	*0.420	*0.400	11,038	12,364	13,116	17,561	24,093	24,280
Paraguay	-	0.250	-	-	-	-	-	-	-	-	-	-
Peru	*0.817	*0.820	*0.814	-	-	-	5,515	8,641	10,096	-	-	-
Philippines	13	14	14	13	-	-	2,069	2,219	2,649	2,399	-	-
Poland	10	11	11	12	12	12	1,507	2,256	2,904	3,285	3,760	4,416
Portugal	*1.804	*1.599	*1.524	*1.535	*1.409	*1.400	-	-	-	-	-	-
Puerto Rico	0.940	0.770	0.810	0.710	0.650	0.690	25,638	31,688	30,370	36,620	-	-
Republic of Moldova	2.079	2.089	2.212	2.182	2.341	2.293	-	3,906	-	860	938	1,083
Romania	6.200	6.100	5.700	5.800	7.500	-	1,964	1,333	935	1,175	1,434	-
Russian Federation	-	-	-	16	15	13	-	-	-	1,344	2,025	2,030
Saint Vincent & the Grenadines	-	-	-	0.024	0.024	0.024	-	-	-	4,224	4,224	-
Senegal	*0.305	*0.430	*0.381	*0.321	*0.321	-	14,198	9,835	12,008	11,728	7,390	-
Sierra Leone	-	-	-	*0.194	-	-	-	-	-	5,392	-	-
Singapore	*0.731	*0.816	*0.753	*0.762	*0.843	*0.845	19,172	20,518	24,620	27,299	31,104	34,291
South Africa	4.000	4.000	4.000	3.000	3.000	3.000	9,952	10,774	10,782	14,587	12,861	13,521
Spain	8.607	8.519	8.594	11	9.703	9.293	24,460	25,112	27,003	27,840	27,884	30,184
Sri Lanka	*8.254	*6.947	*6.280	*5.993	-	-	777	908	1,245	1,329	-	-
Suriname	*0.088	*0.085	*0.085	*0.080	-	-	25,465	38,227	33,613	-	-	-
Sweden	1.100	*1.127	*0.988	0.886	0.796	0.740	23,345	26,855	31,407	28,614	28,460	32,226
Thailand	32	22	-	-	-	-	4,853	6,631	-	-	-	-
Trinidad and Tobago	0.272	0.267	0.267	0.233	0.189	-	17,824	17,297	17,842	14,517	19,200	-
Tunisia	-	-	-	2.710	2.500	2.669	-	-	-	7,051	7,585	9,025
Turkey	32	32	30	29	26	25	5,713	8,185	9,874	10,602	7,752	11,072
Ukraine	6.000	6.000	6.000	5.000	3.000	3.000	2,833	3,526	3,705	527	634	775
United Kingdom	13	12	10	9.000	9.000	8.450	32,280	37,316	-	38,705	-	-
United Republic of Tanzania	5.008	4.551	-	-	-	-	-	-	-	-	-	-
USSR (Former)	33	-	-	-	-	-	2,045	-	-	-	-	-
United States of America	41	40	38	37	35	31	37,073	38,000	-	38,459	-	-
Uruguay	*0.549	*0.533	*0.520	*0.514	*0.514	*0.585	7,886	9,871	11,830	13,696	14,148	11,416
Venezuela	3.300	3.300	3.300	3.702	3.034	3.119	6,978	7,851	11,793	10,195	7,912	5,965
Yugoslavia	-	4.600	4.900	4.800	5.000	4.800	-	-	-	-	2,501	2,190
Zambia	-	-	-	-	0.503	-	-	-	-	-	4,029	-
Zimbabwe	5.414	5.951	6.400	5.800	2.800	2.825	3,408	3,371	2,598	2,483	4,202	4,472

Female Workers: Total and Percent of Employment

Country	Female Workers (000)						Female Workers as % of Total Employment					
	1990	1991	1992	1993	1994	1995	1990	1991	1992	1993	1994	1995
Albania	-	-	-	1.670	0.913	0.685	-	-	-	74.12	69.64	70.18
Angola	-	-	0.126	0.118	-	-	-	-	25.51	24.69	-	-
Argentina	-	-	-	-	0.722	-	-	-	-	-	14.55	-
Australia	1.149	-	-	-	-	-	38.30	-	-	-	-	-
Austria	0.600	0.600	0.500	0.460	0.426	-	42.86	42.86	38.46	36.86	35.65	-
Bangladesh	2.797	2.368	2.597	-	-	-	10.30	6.06	7.91	-	-	-
Bosnia & Herzegovina	1.178	1.067	-	-	-	-	53.89	59.34	-	-	-	-
Bulgaria	-	-	-	9.000	9.000	6.400	-	-	-	62.07	61.64	48.85
Canada	2.000	2.000	-	-	-	-	40.00	40.00	-	-	-	-
Cape Verde	0.022	0.020	0.019	0.020	0.020	-	55.00	44.44	42.22	44.44	42.55	-
Chile	0.059	0.067	0.065	0.057	0.067	-	8.87	9.53	9.57	7.93	9.28	-
China (Taiwan Province)	3.010	3.813	3.952	3.832	4.248	3.653	67.84	74.69	77.41	79.26	82.71	88.22
Colombia	-	0.570	-	0.461	0.411	-	-	24.31	-	28.98	27.47	-
Croatia	1.980	1.820	1.810	1.710	1.530	1.450	56.09	54.65	54.52	53.94	53.50	54.10
Cyprus	0.202	0.222	0.231	0.129	0.166	-	45.70	52.48	56.20	36.34	56.46	-
Czechoslovakia (Former)	1.800	2.000	-	-	-	-	45.00	50.00	-	-	-	-
Czech Republic	1.000	1.000	1.000	1.000	-	-	50.00	50.00	50.00	50.00	-	-
Denmark	1.167	0.980	0.927	-	-	-	57.01	55.59	54.82	-	-	-
Egypt	-	2.672	2.956	2.125	-	-	-	16.10	17.39	12.72	-	-
El Salvador	-	-	-	0.045	0.039	0.037	-	-	-	19.74	16.46	14.80
Ethiopia	-	0.297	0.317	0.362	0.373	0.371	-	33.52	34.16	36.79	37.56	37.59
FYR Macedonia	2.907	2.351	2.406	2.429	2.340	2.159	35.75	33.02	32.65	34.07	40.64	40.54
Germany (Western Part)	5.000	-	-	-	-	-	32.19	-	-	-	-	-
Ghana	-	-	-	0.071	-	-	-	-	-	6.70	-	-
Hungary	2.000	-	2.000	2.000	-	-	40.00	-	50.00	66.67	-	-
Indonesia	-	-	-	147	175	201	-	-	-	79.87	81.32	58.30
Iran (Islamic Republic of)	-	1.237	0.479	0.479	-	-	-	15.62	6.66	6.66	-	-
Iraq	-	0.387	-	-	-	-	-	29.93	-	-	-	-
Ireland	0.500	0.500	-	-	-	-	38.46	38.46	-	-	-	-
Italy	-	1.523	1.947	2.041	0.918	-	-	10.11	11.51	12.60	6.51	-
Jordan	0.036	0.039	0.047	0.037	0.062	0.039	4.44	4.12	4.92	3.69	5.09	3.09
Kenya	0.127	0.128	0.132	0.136	-	-	13.86	13.63	13.76	13.70	-	-
Korea, Republic of	2.200	1.637	1.553	1.281	1.208	1.068	30.56	23.88	23.99	22.00	22.11	21.18
Macau	0.019	0.018	0.015	0.032	0.035	-	48.72	48.65	45.45	42.67	42.68	-
Malaysia	2.000	2.000	2.400	2.000	2.400	-	46.51	46.51	51.06	48.78	52.17	-
Malta	0.283	0.234	0.126	0.049	0.047	-	66.43	63.76	49.03	29.52	29.19	-
Morocco	-	-	-	-	-	0.167	-	-	-	-	-	-
Myanmar	0.547	0.812	0.784	0.732	0.632	-	33.46	42.05	42.59	43.70	41.01	-
Nepal	0.270	0.499	-	0.506	0.516	-	5.79	8.88	-	7.41	7.56	-
New Zealand	0.272	0.239	0.203	-	-	-	42.11	42.91	42.20	-	-	-
Panama	0.090	-	-	-	-	-	19.87	-	-	-	-	-
Philippines	-	-	7.044	6.266	-	-	-	-	49.96	49.34	-	-
Portugal	0.684	0.677	0.596	0.600	0.560	-	37.92	42.34	39.11	39.09	39.74	-
Puerto Rico	0.360	0.290	0.305	0.270	0.230	0.300	38.30	37.66	37.65	38.03	35.38	43.48
Republic of Moldova	-	-	-	-	1.738	1.680	-	-	-	-	74.24	73.27
Saint Vincent & the Grenadines	-	-	-	0.024	0.024	-	-	-	-	100.00	100.00	-
Senegal	0.018	-	-	-	-	-	5.90	-	-	-	-	-
Sri Lanka	4.553	3.298	3.316	3.441	-	-	55.16	47.47	52.80	57.42	-	-
Sweden	0.600	-	-	-	-	-	54.55	-	-	-	-	-
Thailand	16	11	-	-	-	-	49.86	48.35	-	-	-	-
Turkey	12	-	-	-	-	-	37.77	-	-	-	-	-
United Kingdom	4.966	-	3.000	-	-	-	38.20	-	30.00	-	-	-
United Republic of Tanzania	0.801	0.728	-	-	-	-	15.99	16.00	-	-	-	-
United States of America	-	-	27	26	24	22	-	-	71.05	70.27	68.57	70.97
Zambia	-	-	-	-	0.026	-	-	-	-	-	5.17	-

Depending on the table, * means *Enterprises, Engaged*, or *Factor Values*; ± means *Basis Unspecified*; § means *shown in millions*. For additional notes and sources, see Appendix I.

Output and Output per Employee

Country	Output ($ bil.)						Output per Employee ($)					
	1990	1991	1992	1993	1994	1995	1990	1991	1992	1993	1994	1995
Albania	-	-	-	*0.008	*0.008	-	-	-	-	3,655	5,728	-
Angola	-	-	±61	±63	-	-	-	-	§122,6	§131,9	-	-
Argentina	±1.150	-	-	2.344	-	-	344,206	-	-	400,972	-	-
Armenia	0.002	0.004	0.018	0.745	0.007	-	2,027	3,671	-	-	-	-
Australia	*0.630	*0.634	*0.710	-	-	-	209,896	211,401	236,520	-	-	-
Austria	1.671	1.694	1.828	1.713	1.825	2.115	§1,193	§1,210	§1,406	§1,372	§1,527	§1,769
Azerbaijan	±0.262	±0.261	±0.013	±0.027	±0.007	-	175,301	171,240	7,522	14,765	3,973	-
Bangladesh	0.234	0.308	0.285	-	-	-	8,632	7,875	8,688	-	-	-
Belgium	1.685	1.759	1.918	1.695	1.906	2.170	343,784	390,984	436,017	-	-	-
Belize	0.005	0.005	0.005	-	-	-	-	43,199	46,168	-	-	-
Benin	±0.000	±0.001	±0.001	±0.001	-	-	-	-	-	-	-	-
Bolivia	0.032	0.034	0.037	0.033	0.035	0.043	200,245	210,386	219,127	190,460	179,814	211,392
Bosnia & Herzegovina	±0.163	±0.162	-	-	-	-	74,410	89,977	-	-	-	-
Brazil	*2.562	-	*2.249	*2.362	-	-	99,280	-	96,154	116,635	-	-
Bulgaria	±0.824	±0.452	±0.465	±0.297	±0.336	±0.432	53,869	29,762	29,218	20,467	23,037	32,971
Burundi	±0.008	±0.008	-	-	-	-	45,816	43,156	-	-	-	-
Cameroon	0.079	0.076	0.093	0.080	0.046	0.056	76,056	72,755	96,834	93,343	59,267	98,280
Canada	2.425	2.522	2.507	2.256	2.834	2.658	485,087	504,495	501,365	451,128	566,784	586,736
Cape Verde	±0.005	±0.005	±0.006	±0.006	±0.005	-	126,420	117,627	141,083	133,534	105,797	-
Central African Republic	*0.024	-	-	-	-	-	52,109	-	-	-	-	-
Chile	0.348	0.497	0.455	0.511	0.545	0.628	524,057	706,378	669,665	710,954	754,406	869,115
China	±11	±10	±12	±13	±11	±12	38,230	34,279	39,079	46,446	33,063	36,438
China (Hong Kong SAR)	0.684	0.962	1.021	0.809	1.033	-	526,132	687,336	785,369	622,099	861,110	-
China (Taiwan Province)	±1.594	±1.776	±1.939	±1.387	±1.417	±1.382	359,290	347,847	379,777	286,767	275,840	333,620
Colombia	0.245	0.244	0.121	0.117	0.120	0.114	122,447	104,036	62,369	73,360	80,458	90,856
Costa Rica	0.050	0.054	0.064	0.038	0.043	0.059	77,016	84,478	99,325	61,588	-	-
Croatia	±0.203	±0.351	±0.210	-	-	-	57,457	105,291	63,396	-	-	-
Cyprus	0.075	0.078	0.107	0.086	0.087	0.096	168,802	184,224	260,497	242,645	297,084	374,552
Czechoslovakia (Former)	±0.129	±0.088	-	-	-	-	32,173	21,964	-	-	-	-
Denmark	*0.361	*0.434	*0.472	*0.433	*0.430	*0.503	176,180	246,183	279,278	260,768	268,523	314,651
Ecuador	0.023	0.026	0.025	0.029	0.026	0.019	30,128	43,215	33,335	56,978	59,624	49,786
Egypt	*0.513	*0.354	*0.412	*0.481	*0.546	*0.628	29,311	21,316	24,211	28,812	32,375	36,942
El Salvador	-	±0.039	±0.025	±0.031	±0.035	±0.039	-	-	90,277	133,807	148,022	156,028
Ethiopia	0.088	0.100	0.052	0.038	0.035	0.032	101,314	112,547	56,267	38,321	35,238	32,894
Finland	±0.293	±0.271	±0.229	±0.173	±0.186	±0.205	266,023	246,381	229,495	192,372	207,183	223,454
FYR Macedonia	±0.379	±0.517	±0.311	±0.231	±0.198	±0.200	46,602	72,663	42,168	32,395	34,397	37,554
France	±2.448	±2.445	±2.873	±3.069	±3.607	±4.137	479,987	498,918	586,246	639,399	801,473	919,385
Gabon	-	*0.016	*0.018	*0.014	*0.011	*0.012	-	210,657	236,636	218,850	232,847	244,536
Germany	-	18	18	18	18	-	-	915,973	§1,026	§1,118	§1,169	-
Germany (Western Part)	15	16	17	-	-	-	944,746	§1,010	§1,086	-	-	-
Ghana	-	-	-	0.124	0.101	0.106	-	-	-	116,957	92,648	94,527
Greece	*1.095	*1.179	*1.017	*0.999	*1.168	*1.478	117,697	124,325	106,293	104,653	119,001	147,692
Guatemala	-	0.088	0.104	0.118	0.072	0.056	-	91,417	110,818	126,825	83,972	73,207
Honduras	0.055	0.049	0.056	0.051	0.051	0.067	19,346	17,004	18,488	15,429	15,469	19,612
Hungary	±0.182	±0.194	±0.183	±0.200	±0.209	±0.221	36,389	38,804	45,766	66,726	70,508	75,109
India	*1.978	*1.753	*1.822	*1.755	*2.172	*2.430	4,553	3,802	3,690	3,792	4,184	4,432
Indonesia	3.165	2.959	3.609	4.253	4.794	5.505	15,447	16,148	19,743	23,076	22,295	15,970
Iran (Islamic Republic of)	0.114	0.184	0.122	0.179	-	-	10,896	23,198	16,968	24,821	-	-
Iraq	-	*0.047	*0.188	-	-	-	-	36,044	188,424	-	-	-
Ireland	*0.237	*0.231	*0.265	*0.211	*0.222	*0.256	182,421	177,457	186,892	172,474	179,120	191,094
Israel	0.072	0.067	0.068	0.080	-	-	119,862	111,886	112,507	133,682	-	-
Italy	±4.203	±4.266	±5.051	±4.082	±4.186	-	265,277	283,356	298,621	252,070	296,664	-
Jamaica	0.081	0.067	0.051	-	-	-	101,647	87,638	65,588	-	-	-
Japan	±16	±18	±21	±23	±22	±24	§1,805	§2,042	§2,306	§2,579	§2,800	§2,992
Jordan	0.092	0.097	0.101	0.121	0.154	0.168	114,127	102,877	105,855	121,015	126,044	132,667
Kenya	*0.159	*0.151	*0.151	*0.098	*0.117	*0.170	173,407	160,279	157,297	98,274	-	-
Korea, Republic of	3.646	4.057	3.891	4.017	4.050	4.382	506,430	591,787	601,059	689,776	741,384	868,837
Kyrgyzstan	0.292	0.236	0.129	0.027	0.018	-	198,783	154,316	91,127	18,820	13,526	-
Latvia	0.000	0.001	0.005	0.008	0.012	0.010	863	1,838	13,999	23,309	38,756	35,159
Lithuania	-	-	0.013	0.011	0.033	-	-	-	23,179	19,487	58,727	-

Continued.

Depending on the table, * means *Enterprises, Engaged,* or *Factor Values;* ± means *Basis Unspecified;* § means *shown in millions.* For additional notes and sources, see Appendix I.

71

Output and Output per Employee

- Continued -

Country	Output ($ bil.)						Output per Employee ($)					
	1990	1991	1992	1993	1994	1995	1990	1991	1992	1993	1994	1995
Macau	0.000	0.000	0.000	0.006	0.007	-	8,035	10,139	10,288	80,210	82,395	-
Malawi	*0.028	*0.026	*0.020	*0.013	*0.008	-	4,739	6,705	4,968	3,377	2,102	-
Malaysia	*0.323	*0.358	*0.394	*0.433	*0.479	*0.523	75,109	83,177	83,773	105,706	104,152	114,177
Malta	0.046	0.053	0.051	0.041	0.042	-	107,574	145,072	198,529	249,369	262,611	-
Mauritius	0.031	0.035	0.045	0.043	0.049	0.054	91,931	99,266	147,382	157,463	231,109	287,040
Mexico	0.868	1.211	1.574	1.866	1.869	1.315	165,715	222,125	276,821	321,899	361,973	262,017
Morocco	-	±0.383	±0.443	±0.398	-	-	-	126,872	96,015	138,579	-	-
Mozambique	-	±0.010	-	±0.003	±0.002	-	-	18,806	-	10,954	8,455	-
Myanmar	0.050	0.024	0.026	0.029	0.067	-	30,478	12,649	14,365	17,034	43,661	-
Nepal	0.058	0.052	-	0.043	0.049	-	12,502	9,325	-	6,356	7,188	-
Netherlands	3.095	3.125	3.542	3.568	3.835	4.604	478,970	490,212	598,045	621,617	690,621	850,267
Norway	±0.553	±0.590	-	-	-	-	927,420	899,411	-	-	-	-
Pakistan	0.410	0.368	0.400				60,300	61,659	63,987			
Panama	0.036	0.038	0.040	0.053	0.071	0.068	80,185	80,683	86,309	119,797	168,294	169,004
Peru	0.078	0.084	0.085				95,609	102,335	104,081			
Philippines	0.753	0.690	0.868	0.841	-	-	57,954	51,142	61,584	66,209	-	-
Poland	±0.556	±0.940	±1.174	±1.343	±1.573	±1.868	55,558	85,451	106,734	111,887	132,377	157,276
Portugal	0.696	0.799	0.983	0.927	0.874	0.942	385,954	499,971	644,692	603,594	620,468	673,138
Puerto Rico	±0.239	±0.255	±0.317	±0.325	±0.386	±0.353	254,728	330,649	391,852	458,310	593,231	511,739
Republic of Moldova	0.000	0.620	0.017	0.046	0.037	0.036	0.501	296,708	7,850	20,921	15,608	15,810
Romania	±0.620	±0.254	±0.127	±0.114	-	-	99,944	41,633	22,218	19,690	-	-
Russian Federation	-	1.000	0.875	0.374	0.307	0.405	-	-	-	23,522	20,342	31,406
Senegal	0.050	-	-	-	-	-	164,284	-	-	-	-	-
Sierra Leone				0.012						60,131		
Singapore	*0.159	*0.191	*0.227	*0.238	*0.303	*0.334	217,039	234,465	301,634	312,328	359,224	395,770
Slovenia				0.137	0.062	0.073						
South Africa	±0.661	±0.724	±0.791	±0.737	±0.707	±0.747	165,326	180,892	197,755	245,669	235,722	248,854
Spain	*2.326	*2.672	*2.877	±2.048	±2.180	±2.270	270,268	313,632	334,814	194,126	224,680	244,322
Sri Lanka	0.174	0.219	0.196	0.191	-	-	21,129	31,561	31,227	31,906	-	-
Suriname	*0.040	*0.048	*0.086	0.207	-	-	451,999	566,815	§1,008	§2,584	-	-
Sweden	*0.356	*0.446	*0.452	*0.296	*0.321	*0.340	324,072	395,978	457,949	334,449	403,533	460,112
Thailand	2.055	2.249	-	-	-	-	64,815	100,005	-	-	-	-
Trinidad and Tobago	±0.045	±0.046	±0.056	±0.046	±0.041	-	165,495	173,364	209,608	196,261	216,645	-
Tunisia	0.298	0.326	0.354	0.329	0.363	0.429	-	-	-	121,469	145,259	160,565
Turkey	2.115	2.619	2.491	2.812	1.653	2.194	65,788	81,167	82,223	96,092	62,483	86,961
Ukraine	*0.794	*0.740	*0.655	*0.071	*0.026	*0.028	132,358	123,413	109,149	14,127	8,704	9,271
United Kingdom	*3.643	*3.903	13	11	12	12	280,220	325,221	§1,296	§1,268	§1,283	§1,400
United Republic of Tanzania	*0.053	*0.029	-	-	-	-	10,583	6,397	-	-	-	-
USSR (Former)	±3.125	-	-	-	-	-	94,697	-	-	-	-	-
United States of America	*30	*32	*35	*28	*30	*33	729,756	800,750	926,263	767,135	857,743	§1,064
Uruguay	0.114	0.139	0.181	0.197	0.204	0.187	207,081	260,482	348,304	382,998	396,025	319,691
Venezuela	0.348	0.414	0.502	0.571	0.415	-	105,563	125,477	152,264	154,367	136,753	-
Yugoslavia	±0.023	±0.027	±0.087	-	±0.187	±0.154	-	5,915	17,694	-	37,343	32,142
Zambia					0.024						46,789	-
Zimbabwe	*0.099	*0.153	*0.119	*0.100	*0.119	*0.129	18,335	25,637	18,649	17,288	42,594	45,743

Value Added and Value Added per Employee

Country	Value Added ($ bil.)						Value Added per Employee ($)					
	1990	1991	1992	1993	1994	1995	1990	1991	1992	1993	1994	1995
Argentina	±0.480	-	-	1.835	-	-	143,655	-	-	313,873	-	-
Armenia	0.001	0.001	-	-	-	-	478	710	-	-	-	-
Australia	*0.365	*0.355	*0.359	-	-	-	121,615	118,426	119,745	-	-	-
Austria	1.417	1.425	1.566	1.449	1.571	1.820	§1,012	§1,017	§1,204	§1,160	§1,314	§1,523
Bangladesh	0.153	0.228	0.229				5,620	5,826	6,969			
Belgium	*0.310	*0.293	*0.319	*0.282	*0.340	*0.387	63,268	65,096	72,593	-	-	-
Belize	0.002	0.001	0.002	-	-	-	-	9,398	20,116	-	-	-
Bolivia	0.025	0.027	0.029	0.024	0.023	0.028	158,089	168,575	173,205	139,640	116,932	137,467

Continued.

     Depending on the table, * means *Enterprises, Engaged,* or *Factor Values;* ± means *Basis Unspecified;* § means *shown in millions.* For additional notes and sources, see Appendix I.

Value Added and Value Added per Employee
- Continued -

Country	Value Added ($ bil.)						Value Added per Employee ($)					
	1990	1991	1992	1993	1994	1995	1990	1991	1992	1993	1994	1995
Bosnia & Herzegovina	±0.069	±0.069	-	-	-	-	31,607	38,234	-	-	-	-
Brazil	*1.918	-	*1.543	*1.684	-	-	74,318	-	65,982	83,136	-	-
Bulgaria.	-	±0.118	±0.120	±0.075	-	-	-	7,744	7,548	5,184	-	-
Burundi	0.005	0.006	-	-	-	-	30,157	32,986	-	-	-	-
Cameroon	0.023	0.012	0.023	0.025	0.005	0.013	22,136	11,798	23,704	29,076	6,187	22,154
Canada	0.977	1.047	1.051	0.946	1.194	1.120	195,406	209,479	210,143	189,133	238,723	247,267
Central African Republic . . .	*0.013	-	-	-	-	-	27,385	-	-	-	-	-
Chile.	0.303	0.441	0.389	0.440	0.490	0.565	455,685	626,713	573,394	612,282	678,686	782,036
China	±6.220	±5.941	±6.519	±7.322	±6.410	±7.335	22,214	19,804	21,728	25,250	18,854	22,228
China (Hong Kong SAR) . . .	0.394	0.606	0.609	0.495	0.704	-	303,454	432,641	468,459	381,055	586,689	-
China (Taiwan Province) . . .	0.892	0.953	1.112	1.086	1.096	1.072	201,066	186,777	217,831	224,712	213,330	258,893
Colombia	0.173	0.170	0.062	0.052	0.072	0.068	86,609	72,418	32,181	32,893	47,941	54,132
Costa Rica	0.031	0.034	0.041	0.024	0.026	0.035	47,972	53,516	63,123	38,145	-	-
Croatia	±0.124	±0.235	±0.139	-	-	-	34,995	70,446	41,732	-	-	-
Cyprus	0.041	0.043	0.060	0.058	0.063	0.072	92,834	101,430	147,067	162,445	213,403	281,413
Czechoslovakia (Former) . . .	±0.024	-	-	-	-	-	5,989	-	-	-	-	-
Denmark	*0.203	*0.280	*0.306	*0.280	*0.277	*0.326	99,219	159,097	180,869	168,673	173,311	203,605
Ecuador	0.001	0.008	0.006	0.008	0.007	0.005	1,810	14,087	8,224	15,995	16,534	12,328
Egypt	*0.133	*0.084	*0.111	*0.098	*0.113	*0.131	7,626	5,035	6,556	5,885	6,689	7,712
El Salvador.	-	0.031	-	0.025	0.030	0.032	-	-	-	110,158	125,900	128,509
Ethiopia.	0.070	0.082	0.044	0.030	0.028	0.025	81,154	92,672	47,804	30,048	28,385	25,627
Finland	±0.177	±0.166	±0.125	±0.093	±0.100	±0.063	160,470	151,066	124,793	103,869	111,249	68,336
FYR Macedonia	±0.165	±0.209	±0.118	±0.118	±0.098	±0.094	20,237	29,284	15,963	16,521	17,005	17,618
France	±1.919	±1.943	±2.326	±2.593	±3.131	±3.602	376,312	396,552	474,756	540,110	695,765	800,338
Gabon	-	*0.007	*0.008	*0.006	*0.005	*0.004	-	94,925	103,180	94,924	102,972	72,603
Germany (Western Part) . . .	13	13	14	13	-	-	813,379	814,783	905,038	-	-	-
Ghana	-	-	-	0.110	0.095	0.100	-	-	-	103,887	87,547	89,323
Greece	*0.280	*0.266	*0.379	*0.375	*0.440	*0.559	30,104	28,016	39,652	39,237	44,852	55,879
Haiti	-	±0.004	±0.003	±0.003	±0.002	±0.002	-	-	-	-	-	-
Honduras	*0.020	*0.018	*0.016	*0.014	*0.014	*0.017	6,910	6,093	5,332	4,306	4,080	5,034
Hungary	±0.042	±0.041	±0.042	±0.042	±0.043	±0.046	8,489	8,251	10,517	13,836	14,635	15,606
India	*0.489	*0.476	*0.483	*0.471	*0.589	*0.661	1,127	1,032	978	1,019	1,134	1,206
Indonesia	*1.732	*1.637	*2.077	*2.034	*2.867	*2.592	8,451	8,935	11,362	11,037	13,334	7,520
Iran (Islamic Republic of) . . .	0.086	0.152	0.072	0.080	-	-	8,214	19,249	9,969	11,083	-	-
Iraq	-	*0.030	*0.090	-	-	-	-	22,921	90,354	-	-	-
Ireland	*0.166	*0.160	*0.184	*0.147	*0.154	*0.177	127,312	123,400	129,434	119,894	124,317	132,186
Israel.	0.033	0.028	0.020	0.023	-	-	54,558	46,071	33,888	38,279	-	-
Italy	±0.556	±0.594	±0.673	±0.409	±0.712	-	35,082	39,455	39,778	25,241	50,422	-
Jamaica	0.056	0.048	0.038	-	-	-	70,519	62,673	48,171	-	-	-
Japan	±2.003	±2.643	±3.008	±3.786	±3.688	±4.295	222,545	293,635	334,255	420,663	461,061	536,891
Jordan	0.075	0.078	0.085	0.101	0.123	0.135	93,113	82,471	88,824	100,335	100,959	106,541
Kenya	*0.012	*0.012	*0.012	*0.008	*0.011	*0.013	13,339	13,163	12,299	7,640	-	-
Korea, Republic of	2.794	3.014	2.734	2.712	2.769	3.277	388,059	439,645	422,365	465,714	506,928	649,775
Latvia	-	-	-	*0.002	*0.004	*0.006	-	-	-	6,991	12,179	21,687
Macau	0.000	0.000	0.000	0.001	0.001	-	3,946	4,588	5,034	15,482	17,541	-
Malawi	*0.011	*0.013	*0.007	*0.008	*0.005	-	1,753	3,211	1,753	2,206	1,421	-
Malaysia	*0.127	*0.129	*0.149	*0.169	*0.186	*0.204	29,619	29,952	31,739	41,274	40,400	44,580
Malta	0.009	0.010	0.008	0.005	0.005	-	22,104	26,075	32,156	31,146	29,725	-
Mauritius	0.022	0.024	0.034	0.034	0.039	0.042	66,781	68,453	112,574	122,126	184,824	223,921
Mexico	0.680	0.947	1.198	1.443	1.444	1.015	129,729	173,713	210,764	248,837	279,594	202,363
Morocco	-	±0.482	±0.556	±0.538	-	-	-	159,626	120,527	187,406	-	-
Myanmar	±0.073	±0.037	±0.038	±0.039	±0.092	-	44,424	19,298	20,455	23,422	59,735	-
Nepal	0.040	0.037	-	0.032	0.038	-	8,498	6,646	-	4,653	5,543	-
Netherlands	1.848	1.847	2.152	2.081	2.249	2.716	286,022	289,781	363,301	362,540	405,059	501,635
New Zealand	0.045	-	-	-	-	-	69,313	-	-	-	-	-
Norway	±0.478	±0.517	-	-	-	-	802,245	788,190	-	-	-	-
Pakistan	0.336	0.296	0.331	-	-	-	49,375	49,581	52,915	-	-	-
Panama.	0.027	0.027	0.028	0.038	0.050	0.048	59,810	57,077	61,130	85,258	120,221	120,667
Peru	0.059	0.051	0.052	-	-	-	72,821	62,154	63,955	-	-	-
Philippines	0.420	0.404	0.532	0.546	-	-	32,325	29,889	37,727	42,959	-	-

Continued.

Value Added and Value Added per Employee
- Continued -

Country	Value Added ($ bil.)						Value Added per Employee ($)					
	1990	1991	1992	1993	1994	1995	1990	1991	1992	1993	1994	1995
Poland	±0.379	±0.646	±0.874	±0.850	±0.994	±1.180	37,926	58,735	79,414	70,867	83,659	99,295
Portugal	0.592	0.690	0.872	0.842	0.799	0.861	328,188	431,705	572,169	548,216	567,017	615,080
Puerto Rico	±0.171	±0.159	±0.189	±0.176	±0.172	±0.138	181,596	206,753	232,840	247,746	263,846	199,565
Romania	±0.410	±0.127	±0.067	±0.062	-	-	66,150	20,816	11,736	10,616	-	-
Russian Federation	-	-	-	±0.161	±0.144	±0.185	-	-	-	10,144	9,521	14,368
Saint Lucia	-	±0.001	±0.001	±0.001	±0.000	±0.001	-	-	-	-	-	-
Senegal	0.016	0.013	0.016	0.015	0.012	-	52,758	29,603	40,953	46,174	36,657	-
Sierra Leone	-	-	-	0.010	-	-	-	-	-	49,880	-	-
Singapore	*0.064	*0.074	*0.097	*0.112	*0.138	*0.153	87,571	90,336	128,888	147,468	163,431	180,691
Slovenia	-	-	-	0.104	0.025	0.033	-	-	-	-	-	-
South Africa	*0.083	*0.092	*0.122	*0.113	*0.108	*0.114	20,678	23,087	30,505	37,764	36,159	38,036
Spain	*0.912	*1.083	*1.159	±0.598	±0.737	±0.694	105,927	127,116	134,827	56,630	75,935	74,676
Sri Lanka	0.156	0.195	0.136	0.149			18,867	28,043	21,683	24,902		-
Sweden	*0.257	*0.200	*0.239	*0.166	*0.187	*0.198	233,455	177,403	241,885	187,107	234,821	268,025
Thailand	1.623	1.877	-	-	-	-	51,180	83,456	-	-	-	-
Trinidad and Tobago	±0.036	±0.037	±0.045	±0.037	±0.033	-	134,021	137,656	170,034	157,923	175,834	-
Tunisia	0.207	0.218	0.237	0.240	0.273	0.323	-	-	-	88,492	109,072	121,058
Turkey	1.168	1.517	1.327	1.419	0.698	0.921	36,353	46,993	43,794	48,485	26,384	36,518
United Kingdom	*2.375	*2.665	*2.696	*2.432	*2.542	*2.610	182,692	222,124	269,611	270,270	282,457	308,840
United Republic of Tanzania	*0.020	*0.013	-	-	-	-	3,914	2,892	-	-	-	-
United States of America	*23	*24	*27	*21	*22	*25	550,244	612,000	715,974	563,027	632,314	797,290
Uruguay	0.090	0.111	0.148	0.166	0.171	0.157	163,569	208,117	285,322	322,025	333,119	268,957
Venezuela	0.273	0.344	0.423	0.384	0.265	0.738	82,871	104,212	128,106	103,712	87,432	236,498
Yugoslavia	±0.011	±0.018	±0.045	-	±0.090	±0.113	-	3,940	9,120	-	18,042	23,573
Zambia	-	-	-	-	0.021	-	-	-	-	-	41,760	-
Zimbabwe	*0.076	*0.117	*0.094	*0.074	*0.061	*0.066	14,011	19,652	14,720	12,714	21,740	23,276

Capital Investment

Country	Gross Fixed Capital Formation ($ mil.)						Per Establishment ($ mil)					
	1990	1991	1992	1993	1994	1995	1990	1991	1992	1993	1994	1995
Argentina	-	-	-	47	-	-	-	-	-	1.893	-	-
Australia	20	-	-	-	-	-	4.883	-	-	-	-	-
Austria	26	22	31	33	25	-	1.718	1.542	2.197	2.352	1.953	-
Bangladesh	1.863	1.093	0.616	-	-	-	0.005	0.003	0.001	-	-	-
Belgium	43	29	44	23	33	22	0.791	0.526	0.805	0.533	0.857	-
Bolivia	-0.081	0.619	0.305	0.752	1.553	-	-0.081	0.619	0.305	0.752	1.553	-
Bosnia & Herzegovina	2.156	-	-	-	-	-	0.154	-	-	-	-	-
Brazil	88	-	78	299	-	-	0.384	-	0.403	1.892	-	-
Bulgaria	4.840	1.928	207	33	5.526	7.192	0.202	0.080	7.957	1.238	0.205	0.266
Cameroon	14	1.748	4.273	4.231	3.759	15	4.808	0.583	1.424	1.410	1.253	7.337
Cape Verde	0.001	0.001	-	-	-	-	0.001	0.001	-	-	-	-
Chile	10	3.001	11	11	7.628	-	5.010	1.000	3.607	3.621	3.814	-
China (Hong Kong SAR)	8.344	32	31	17	4.529	-	2.086	7.914	7.816	4.201	1.132	-
Colombia	0.589	-0.618	7.476	1.056	-0.713	-	0.049	-0.051	0.623	0.106	-0.071	-
Croatia	3.128	4.953	2.580	7.941	14	28	0.261	0.413	0.172	0.467	0.883	1.752
Cyprus	0.525	0.436	0.271	0.221	0.226	0.272	0.075	0.062	0.039	0.032	0.056	0.068
Czechoslovakia (Former)	3.900	-	-	-	-	-	1.950	-	-	-	-	-
Denmark	22	17	-	-	-	-	1.312	0.929	-	-	-	-
Ecuador	6.425	2.732	11	3.966	2.637	3.465	-	-	-	-	-	-
Egypt	8.900	3.541	15	110	-	-	0.228	0.131	0.464	3.675	-	-
El Salvador	-	0.619	-	0.282	0.316	1.488	-	0.309	-	0.282	0.316	1.488
Ethiopia	0.372	0.056	0.014	0.499	0.271	0.066	0.372	0.056	0.014	0.499	0.271	0.066
Finland	19	12	16	6.162	4.461	-17.170	4.838	2.924	4.057	1.541	1.115	-2.862
FYR Macedonia	6.354	3.692	7.102	16	9.942	3.605	0.254	0.154	0.284	0.625	0.398	0.124
Gabon	-	0.415	0.374	0.198	0.013	0.299	-	0.415	0.374	0.198	0.013	0.299
Germany (Western Part)	211	314	277	-	-	-	4.221	6.541	6.014	-	-	-
Greece	25	33	51	-	-	-	0.313	0.493	0.801	-	-	-

Continued.

Capital Investment
- Continued -

Country	Gross Fixed Capital Formation ($ mil.)						Per Establishment ($ mil)					
	1990	1991	1992	1993	1994	1995	1990	1991	1992	1993	1994	1995
Hungary	3.512	4.349	24	42	-	-	0.502	0.290	3.454	6.065	-	-
India	32	31	49	37	-	-	0.004	0.004	0.006	0.007	-	-
Indonesia	50	64	89	47	99	100	0.052	0.068	0.099	0.054	0.132	0.123
Iran (Islamic Republic of)	-	1.583	24	44	-	-	-	1.583	24	44	-	-
Ireland	4.809	3.069	-	-	-	-	0.601	0.512	-	-	-	-
Israel	0.992	15	-	-	-	-	0.076	1.148	-	-	-	-
Italy	85	107	180	53	42	-	3.870	5.956	4.282	1.335	1.319	-
Japan	276	379	347	450	294	-	7.674	11	10	15	11	-
Jordan	0.002	0.341	0.029	1.338	13	4.848	0.000	0.085	0.007	0.334	2.243	0.808
Korea, Republic of	20	68	58	41	40	49	0.989	3.401	2.901	2.544	2.526	3.068
Latvia	0.002	0.003	0.000	-	4.709	-	0.002	0.001	0.000	-	4.709	-
Macau	-	-	-	1.485	1.035	-	-	-	-	0.742	0.517	-
Malawi	1.136	4.102	0.444	0.750	0.469	-	0.284	1.026	0.111	0.187	0.117	-
Malaysia	5.176	12	11	9.712	7.240	-	0.178	0.416	0.345	0.335	0.278	-
Malta	3.375	1.495	0.396	0.178	0.098	-	0.563	0.249	0.066	0.036	0.024	-
Mexico	14	8.917	-	-	-	-	1.976	1.274	-	-	-	-
Morocco	-	-	-	54	-	-	-	-	-	9.015	-	-
Myanmar	-	-	-	5.613	4.017	-	-	-	-	1.871	1.339	-
Nepal	12	-	-	-	-	-	0.176	-	-	-	-	-
Netherlands	71	116	177	134	-	-	4.428	7.737	12	10	-	-
Pakistan	-	3.603	-	-	-	-	-	0.190	-	-	-	-
Panama	1.393	1.525	-	-	-	-	0.464	0.508	-	-	-	-
Peru	0.312	0.227	0.197	-	-	-	0.028	0.021	0.018	-	-	-
Philippines	11	13	4.037	15	-	-	0.341	0.407	0.161	0.690	-	-
Poland	17	52	30	60	-	-	2.084	7.483	3.770	5.491	-	-
Portugal	10	7.648	8.119	13	4.476	-	2.006	1.912	2.030	2.613	0.895	-
Romania	-	-	3.199	2.948	2.374	-	-	-	3.199	2.948	0.170	-
Senegal	2.696	-	-	-	-	-	1.348	-	-	-	-	-
Singapore	6.345	16	22	17	21	-	1.269	4.101	7.449	5.770	7.036	-
Slovenia	2.739	4.425	8.082	5.060	5.333	4.843	2.739	4.425	8.082	5.060	5.333	4.843
Spain	54	47	44	83	55	58	1.869	1.692	1.676	-	-	-
Sri Lanka	5.541	9.233	5.661	1.762	-	-	0.020	0.037	0.031	0.012	-	-
Thailand	252	255	-	-	-	-	1.440	1.557	-	-	-	-
Trinidad and Tobago	-	-	1.271	1.925	0.827	-	-	-	1.271	1.925	0.827	-
Tunisia	4.328	5.408	5.993	5.579	2.966	2.115	-	-	-	2.790	1.483	0.211
Turkey	20	15	18	70	70	-	0.399	0.305	0.381	1.518	1.712	-
Ukraine	7.554	3.291	6.968	0.513	0.375	3.015	0.444	0.150	0.332	0.030	0.027	0.215
United Kingdom	57	135	92	-	130	-	2.041	5.174	3.675	-	5.659	-
United Republic of Tanzania	18	19	-	-	-	-	5.843	6.431	-	-	-	-
United States of America	280	410	389	388	387	411	-	-	3.383	-	-	-
Uruguay	0.612	1.667	2.684	1.659	-	-	-	-	-	-	-	-
Venezuela	4.478	14	24	8.258	8.007	38	0.136	0.462	1.078	0.375	0.534	1.793
Yugoslavia	1.390	2.391	7.717	-	34	12	0.063	0.126	0.351	-	1.616	0.546
Zimbabwe	8.783	18	16	8.976	6.736	-	1.098	2.261	1.973	1.282	1.347	-

Indexes of Industrial Production

Country	Index of Industrial Production (1990=100)						Index of Employment (1990=100)					
	1990	1991	1992	1993	1994	1995	1990	1991	1992	1993	1994	1995
Albania	100	-	-	-	-	-	-	-	-	-	-	-
Algeria	100	99	94	99	95	88	-	-	-	-	-	-
Angola	100	100	100	-	-	-	-	-	-	-	-	-
Argentina	100	106	-	-	-	-	100	-	-	175	149	-
Armenia	100	-	-	-	-	-	100	93	-	-	-	-
Australia	100	103	111	119	-	-	100	100	100	-	-	-
Austria	100	109	105	108	108	108	100	100	93	89	85	85
Azerbaijan	-	-	-	-	-	-	100	102	112	122	117	-
Bahrain	100	100	100	100	100	100	-	-	-	-	-	-

Continued.

Depending on the table, * means *Enterprises*, *Engaged*, or *Factor Values*; ± means *Basis Unspecified*; § means *shown in millions*. For additional notes and sources, see Appendix I.

75

Indexes of Industrial Production

- Continued -

Country	Index of Industrial Production (1990=100)						Index of Employment (1990=100)					
	1990	1991	1992	1993	1994	1995	1990	1991	1992	1993	1994	1995
Bangladesh	100	132	137	153	171	205	100	144	121	-	-	-
Barbados	100	102	98	108	116	118	-	-	-	-	-	-
Belgium	100	106	98	93	97	98	100	92	90	-	-	-
Belize	100	102	103	104	99	93	-	-	-	-	-	-
Benin	100	-	-	-	-	-	-	-	-	-	-	-
Bhutan	100	-	-	-	-	-	-	-	-	-	-	-
Bolivia	100	111	110	117	121	128	100	103	107	109	123	130
Bosnia & Herzegovina	-	-	-	-	-	-	100	82	-	-	20	-
Botswana	100	100	100	-	-	-	-	-	-	-	-	-
Brazil	100	107	126	132	112	107	100	-	91	78	-	-
Bulgaria	100	79	68	52	51	52	100	99	104	95	95	86
Burkina-Faso	100	120	119	-	-	-	-	-	-	-	-	-
Burundi	100	117	118	135	-	-	100	99	-	-	-	-
Cambodia	100	100	100	-	-	-	100	109	-	-	-	-
Cameroon	100	102	102	-	-	-	100	100	92	82	74	54
Canada	100	97	89	89	103	95	100	100	100	100	100	91
Cape Verde	-	-	-	-	-	-	100	113	113	113	117	-
Central African Republic	100	-	-	-	-	-	100	-	-	-	-	-
Chad	100	182	174	-	-	-	-	-	-	-	-	-
Chile	100	100	108	105	105	106	100	106	102	108	109	109
China	100	-	-	-	-	-	100	107	107	104	121	118
China (Hong Kong SAR)	100	104	109	112	114	113	100	108	100	100	92	-
China (Taiwan Province)	100	105	103	98	99	95	100	115	115	109	116	93
Colombia	100	116	97	85	70	59	100	117	97	80	75	63
Congo	100	58	43	-	-	-	-	-	-	-	-	-
Costa Rica	100	105	104	119	122	131	100	97	98	95	-	-
Cote d'Ivoire	100	103	96	88	89	101	-	-	-	-	-	-
Croatia	100	-	-	-	-	-	100	94	94	90	81	76
Cuba	100	100	-	-	-	-	-	-	-	-	-	-
Cyprus	100	119	134	76	54	54	100	96	93	80	67	58
Czechoslovakia (Former)	100	102	145	-	-	-	100	100	-	-	-	-
Czech Republic	100	-	-	-	-	-	100	100	100	100	-	-
Dem. Rep. of the Congo	100	100	100	-	-	-	-	-	-	-	-	-
Denmark	100	100	99	96	88	89	100	86	83	81	78	78
Dominican Republic	100	90	95	-	-	-	-	-	-	-	-	-
Ecuador	100	97	100	107	-	-	100	80	102	69	57	50
Egypt	100	105	106	111	121	129	100	95	97	95	96	97
El Salvador	100	105	-	-	-	-	-	-	-	-	-	-
Ethiopia and Eritrea	100	102	102	-	-	-	-	-	-	-	-	-
Ethiopia	-	-	-	-	-	-	100	102	107	113	114	114
Fiji	100	103	107	106	116	112	-	-	-	-	-	-
Finland	100	89	89	83	82	78	100	100	91	82	82	84
FYR Macedonia	100	-	-	-	-	-	100	88	91	88	71	65
France	100	92	97	89	88	85	100	96	96	94	88	88
Gabon	100	143	159	-	-	-	-	-	-	-	-	-
Gambia	100	-	-	-	-	-	-	-	-	-	-	-
Germany (Eastern Part)	100	-	-	-	-	-	-	-	-	-	-	-
Germany (Western Part)	100	113	105	99	103	95	100	105	100	-	-	-
Ghana	100	-	-	-	-	-	-	-	-	-	-	-
Greece	100	101	96	95	110	122	100	102	103	103	106	108
Guatemala	100	99	103	-	-	-	-	-	-	-	-	-
Guyana	100	124	129	122	127	129	-	-	-	-	-	-
Haiti	100	99	113	73	77	-	-	-	-	-	-	-
Honduras	100	123	137	138	158	159	100	102	105	115	117	120
Hungary	100	97	85	82	85	88	100	100	80	60	59	59
Iceland	-	-	-	-	-	-	100	1,000	1,000	-	-	-
India	100	109	107	123	142	152	100	106	114	107	120	126
Indonesia	100	75	76	79	93	95	100	89	89	90	105	168
Iran (Islamic Republic of)	100	122	122	-	-	-	100	75	69	69	-	-
Iraq	100	50	22	-	-	-	-	-	-	-	-	-

Continued.

Depending on the table, * means *Enterprises*, *Engaged*, or *Factor Values*; ± means *Basis Unspecified*; § means *shown in millions*. For additional notes and sources, see Appendix I.

Indexes of Industrial Production

- Continued -

Country	Index of Industrial Production (1990 = 100)						Index of Employment (1990 = 100)					
	1990	1991	1992	1993	1994	1995	1990	1991	1992	1993	1994	1995
Ireland	100	108	114	104	105	110	100	100	109	94	96	103
Israel	100	102	105	113	120	130	100	100	100	100	-	-
Italy	100	94	88	89	-	-	100	95	107	102	89	-
Jamaica	100	89	93	91	95	-	100	96	98	99	96	-
Japan	100	101	105	104	103	102	100	100	100	100	89	89
Jordan	100	93	76	-	-	-	100	117	118	124	150	156
Kenya	100	97	108	109	110	119	100	103	105	108	-	-
Korea, Republic of	100	101	105	107	102	99	100	95	90	81	76	70
Kuwait	100	100	100	100	-	-	-	-	-	-	-	-
Kyrgyzstan	100	-	-	-	-	-	100	104	97	98	88	-
Lao P.D.R.	100	100	100	-	-	-	-	-	-	-	-	-
Latvia	100	-	-	-	-	-	100	100	101	96	85	79
Lebanon	100	-	-	-	-	-	-	-	-	-	-	-
Liberia	100	-	-	-	-	-	-	-	-	-	-	-
Libyan Arab Jamahiriya	100	100	100	-	-	-	-	-	-	-	-	-
Lithuania	100	-	-	-	-	-	-	-	-	-	-	-
Macau	100	-	-	-	-	-	100	95	85	192	210	-
Madagascar	100	132	168	-	-	-	-	-	-	-	-	-
Malawi	100	92	98	100	101	96	100	65	68	63	60	-
Malaysia	100	101	96	90	89	90	100	100	109	95	107	107
Mali	100	87	105	93	124	93	-	-	-	-	-	-
Malta	100	118	110	107	-	-	100	86	60	39	38	-
Mauritius	100	113	106	109	106	-	100	104	90	82	63	56
Mexico	100	104	105	106	104	105	100	104	108	111	99	96
Morocco	100	95	91	90	102	104	-	-	-	-	-	-
Mozambique	100	45	-	-	-	-	100	97	50	49	49	-
Myanmar	100	70	41	44	45	77	100	118	113	102	94	-
Namibia	100	-	-	-	-	-	-	-	-	-	-	-
Nepal	100	-	-	-	-	-	100	121	-	147	146	-
Netherlands	100	114	99	98	103	107	100	99	92	89	86	84
New Zealand	100	100	100	-	-	-	100	86	74	-	-	-
Nicaragua	100	100	100	-	-	-	-	-	-	-	-	-
Nigeria	100	100	100	100	100	-	-	-	-	-	-	-
Norway	100	99	96	95	88	82	100	110	-	-	-	-
Pakistan	100	89	100	97	103	123	100	88	92	-	-	-
Panama	100	95	99	130	171	162	100	104	103	98	93	88
Papua New Guinea	100	-	-	-	-	-	-	-	-	-	-	-
Paraguay	100	87	83	95	90	95	-	-	-	-	-	-
Peru	100	104	95	95	105	113	100	100	100	-	-	-
Philippines	100	110	114	111	117	129	100	104	108	98	-	-
Poland	100	95	101	117	129	129	100	110	110	120	119	119
Portugal	100	105	100	97	88	83	100	89	84	85	78	78
Puerto Rico	100	-	-	-	-	-	100	82	86	76	69	73
Republic of Moldova	100	-	-	-	-	-	100	100	106	105	113	110
Romania	100	102	102	89	84	89	100	98	92	94	121	-
Russian Federation	100	-	-	-	-	-	-	-	-	-	-	-
Saudi Arabia	100	-	-	-	-	-	-	-	-	-	-	-
Senegal	100	89	78	76	72	94	100	141	125	105	105	-
Sierra Leone	100	100	100	-	-	-	-	-	-	-	-	-
Singapore	100	111	126	130	182	180	100	112	103	104	115	116
Slovenia	100	93	102	94	91	83	-	-	-	-	-	-
Somalia	100	-	-	-	-	-	-	-	-	-	-	-
South Africa	100	97	93	91	88	86	100	100	100	75	75	75
Spain	100	103	99	100	104	102	100	99	100	123	113	108
Sri Lanka	100	102	98	117	125	311	100	84	76	73	-	-
Sudan	100	100	100	-	-	-	-	-	-	-	-	-
Suriname	100	70	87	-	-	-	100	97	97	91	-	-
Swaziland	100	100	100	-	-	-	-	-	-	-	-	-
Sweden	100	96	106	88	91	87	100	102	90	81	72	67
Switzerland	100	100	99	101	100	102	-	-	-	-	-	-

Continued.

Depending on the table, * means *Enterprises, Engaged,* or *Factor Values;* ± means *Basis Unspecified;* § means *shown in millions.* For additional notes and sources, see Appendix I.

77

Indexes of Industrial Production

- Continued -

Country	Index of Industrial Production (1990=100)						Index of Employment (1990=100)					
	1990	1991	1992	1993	1994	1995	1990	1991	1992	1993	1994	1995
Syrian Arab Republic	100	109	111	105	111	118	-	-	-	-	-	-
Thailand	100	107	108	-	-	-	100	71	-	-	-	-
Tonga	100	-	-	-	-	-	-	-	-	-	-	-
Trinidad and Tobago	100	101	95	92	88	96	100	98	98	86	69	-
Tunisia	100	102	107	107	112	113	-	-	-	-	-	-
Turkey	100	119	107	117	126	145	100	100	94	91	82	79
Uganda	100	114	100	110	147	199	-	-	-	-	-	-
Ukraine	100	-	-	-	-	-	100	100	100	83	50	50
United Arab Emirates	100	100	100	-	-	-	-	-	-	-	-	-
United Kingdom	100	101	106	99	105	101	100	92	77	69	69	65
United Republic of Tanzania . .	100	106	97	114	105	-	100	91	-	-	-	-
USSR (Former)	100	-	-	-	-	-	100	-	-	-	-	-
United States of America . . .	100	94	95	80	99	101	100	98	93	90	85	76
Uruguay	100	103	123	112	112	95	100	97	95	94	94	107
Venezuela	100	-	-	-	-	-	100	100	100	112	92	95
Yemen	100	-	-	-	-	-	-	-	-	-	-	-
Yugoslavia	100	-	-	-	-	-	-	-	-	-	-	-
Yugoslavia (Former)	100	81	-	-	-	-	-	-	-	-	-	-
Zambia	100	103	129	125	122	121	-	-	-	-	-	-
Zimbabwe	100	103	103	97	98	92	100	110	118	107	52	52

Depending on the table, * means *Enterprises, Engaged,* or *Factor Values*; ± means *Basis Unspecified*; § means *shown in millions*. For additional notes and sources, see Appendix I.

Representative Companies in Sector

Name	Address	Tele-phone	Fax	Employ-ment	Y
Tabacalera, SA	Calle Alcala 4, Madrid E-28014, Spain	91 5228487	91 5227586	6,737	97
DIMON Inc.	PO Box 681, Danville, Virginia 24543-0681, U.S.A.	804-792-7511		6,700	97
General Cigar Holdings Inc.	387 Park Ave., S., New York, New York 10016-8899, U.S.A.	212-448-3808		6,000	97
NV Sumatra Tobacco Trading Co.	Pematang, Medan 20232, Indonesia	61517139	622 23410	5,000	97
Mafco Consolidated Group Inc.	35 East 62nd St., New York, New York 10021, U.S.A.	212-572-8600		4,800	96
Consolidated Cigar Corp.	5900 N. Andrews Ave., 700, Fort Lauderdale, Florida 33309-2369, U.S.A.	954-772-9000		4,000	95
DIMON International Inc.	PO Box 166, Farmville, North Carolina 27828-0166, U.S.A.	919-753-8000	919-753-8200	3,700	96
UST Inc.	100 W. Putnam Ave., Greenwich, Connecticut 06830, U.S.A.	203-661-1100		3,500	96
Rothmans International PLC	15 Hill St., London W1X 7FB, United Kingdom	71 4914366	71 4938404	3,011	94
Zaklady Przemyslu Tytoniowego	98 Radom Tytoniowa 2/6, Radom PL-26-600, Poland	22141	21766	2,500	93
Massalin Particulares SA	Av. Leandro N. Alem 466, 9th Fl., Buenos Aires 1003, Argentina	1 3196124	1 3194150	1,800	96
Rothmans of Pall Mall Malaysia BHD	VirginiaPark, Petaling Jaya 46200, Malaysia	3 7566899	3 7558416	1,639	
Brooke Group Ltd.	100 Southeast 2nd St., Miami, Florida 33131, U.S.A.	305-579-8000		1,545	96
BAT UK & Export Ltd.	187 Milton Park, Abingdon OX14 4SJ, United Kingdom	235 834633		1,298	92
Swisher International Group Inc.	20 Thorndal Circle, Darien, Connecticut 06820, U.S.A.	203-656-8000		1,210	97
Philip Morris Ltd.	252 Chesterville Rd., Moorabbin 3189, Australia	3 5560100	3 5533201	1,100	90
Rothmans, Benson & Hedges Inc.	800-1500 Don Mills Rd., Don Mills, Ontario M3B 3L1, Canada	416-449-5525	416-449-4486	997	98
Tobacna Tovarna d.o.o.	Tobacna 5, Ljubljana 1001, Slovenia	61 1255217	61 1259415	962	95
OroAmerica Inc.	443 N Varney St., Burbank, California 91502, U.S.A.	818-848-5555	818-841-4342	851	96
Standard Commercial Tobacco Co.	PO Box 450, Wilson, North Carolina 27894-0450, U.S.A.	919-237-1106		850	95
Pinkerton Group Inc.	6600 W. Broad St., Richmond, Virginia 23230, U.S.A.	804-287-3220	804-287-3208	800	97
Malaysian Tobacco Co. Ltd.	178-3 Jalan Sungei Besi, Kuala Lumpur 57100, Malaysia	3 2213066	3 2213130	725	
K.R. Edwards Leaf Tobacco Company Inc.	PO Box 1337, Smithfield, North Carolina 27577, U.S.A.	919-934-7101		600	95
RJR-Macdonald Inc.	6000-1 First Cdn Pl 60th Floor, Toronto, Ontario M5X 1A4, Canada	416-601-7000	416-601-7001	600	98
Nobel Cigars AS	Tobaksvejen 4, Soeborg DK-2860, Denmark	39696282	39667560	599	93
Tabacalera Andina SA	PO Box 17-01-2155, Quito, Ecuador	460107	674012	560	92
Rothmans of Pall Mall Indonesia PT	Jl. Peltu Sujono S, No. 8, Malang 65148, Indonesia	34161144	341 65995	550	97
Manufacture de Tabacs Heintz van Landewijck SARL	Rue de Hollerich 31, Luxembourg L-1741, Luxembourg	4939391	400344	547	93
House of Blend AB	Rosenlundsg. 36, Stockholm S-118 84, Sweden	8 6580140	8 6687217	530	93
BAT Benelux SA NV	Rue de Koninck 38, Brussels B-1080, Belgium	2 4220211	2 4220422	500	95
Tabacalera Centroamericana SA	Aldea Boca del Monte, Guatemala City, Guatemala	2 330651	2 330154	500	
Carolina Leaf Tobacco Company Inc.	PO Box 5067, Greenville, North Carolina 27835, U.S.A.	919-752-2144	919-752-6988	450	96
Dubek Ltd.	33 Lillenbloom St., Tel Aviv 65133, Israel	3 5140174	3 5140183	450	97
B.A.T. Espana, SA	Orense 4 - 20C, Madrid E-28006, Spain	91 5551904	91 5552185	424	97
Tabacalera Nacional SA	Zona 1, Guatemala City 01001, Guatemala	2 538353	2 538391	400	
Republic Tobacco Co.	Apartado 896-1000, San Jose, Costa Rica	2252333	2250614	380	
Manufacture de Tabacs de l'Ouest Africain	BP 76, Dakar, Senegal	231013	238919	375	
Societe Industrielle des Tabacs du Cameroun	Boulevard de la Republique, Douala, Cameroon	424919	425949	360	91
Manufacture de Tabacs Heintz van Landewijck SARL	31, rue de Hollerich, Luxembourg L-1741, Luxembourg	4939391	400344	350	95
Tabacalera Costarricense SA	Apdo 623-1000, San Jose, Costa Rica	2392211	2242681	350	
Oy P.C. Rettig Ab	Nosturinkatu 5, Turku SF-20101, Finland	21 618811	21 309100	340	94
Tobacco Processors Inc.	PO Box 1989, Wilson, North Carolina 27893, U.S.A.	919-237-5131		300	94
Lancaster Leaf Tobacco Company of Pennsylvania Inc.	PO Box 897, Lancaster, Pennsylvania 17603, U.S.A.	717-394-2676		255	97
Philip Morris Espana, SA	Edificio Centro Colon, Madrid E-28004, Spain	91 3490139	91 3490105	243	97
Agio Tobacco Co. Ltd.	9 Bulebel Industrial Estate, Zejtun ZTN 08, Malta	693510	693548	235	91
Orlik Tobacco Co. AS	Postboks 50, Assen DK-5610, Denmark	64711032	64711160	200	93
Havatampa Inc.	PO Box 1261, Tampa, Florida 33601, U.S.A.	813-621-3535		190	94
Manchester Tobacco Co. Ltd.	Ludgate Hill, Manchester M4 4DA, United Kingdom	61 8317891	61 8346693	170	93
House of Borkum Riff AB	Stockholms Kommun, Arvika S-671 23, Sweden	570 84200	8 6582059	160	93
Gawih Jaya PT	Jl. Raya Darmo 42-44, Surabaya, Indonesia	31 838331	31 581831	150	93

Continued

Representative Companies in Sector
- Continued -

Name	Address	Tele-phone	Fax	Employ-ment	Y
Harald Halberg Tobaksfabrikker AS	Moellergade 56-58, Svendborg DK-5700, Denmark	62213117	62216694	125	93
Hae Yang Knitting Factory Ltd.	Chung-Gu, Seoul, Republic of Korea	2 2523101	2 2352599	120	97
R.J. Reynolds Tobacco (UK) Ltd.	62 London Rd., Staines TW18 4JD, United Kingdom	784 464111	784 465397	114	93
John Middleton Inc.	PO Box 1525, King of Prussia, Pennsylvania 19406, U.S.A.	610-265-1400	610-265-1482	100	95
W.A. Adams Company Inc.	PO Box 450, Wilson, North Carolina 27893, U.S.A.	919-237-9218		100	95
Productora Tabacalera de Colombia SA	Av. 22 37-51, Bogota, Colombia	1 2688200	1 2681434	95	97
Imasco Ltd.	1900-600 Blvd. de Maisonneuve O, Montreal, Quebec H3A 3K7, Canada	514-982-9111	514-982-9369	75	98
Flue-Cured Tobacco Cooperative Stabilization Corp.	PO Box 12300, Raleigh, North Carolina 27605, U.S.A.	919-821-4560	919-821-4564	56	96
Truval Manufacturers CC	Bramley, Johannesburg 2018, Republic of South Africa	11 3331880	11 3332047	40	95
Bastos du Canada Ltee.	371 Rue St.-Marc, Louiseville, Quebec J5V 2L9, Canada	819-228-5531	819-228-2437	30	98
Andrikian Trading Co. Ltd.	13, Archimandritou Kyprianou Str., Nicosia, Cyprus	2 463999	2 452043	20	92
Superior Tobacco Co. NV	Franklinstraat, Oranjestad, Aruba	8 23220	8 22892	17	93
Standard Commercial Tobacco Co. of Canada Ltd.	Hwy 3 W, Tillsonburg, Ontario N4G 4H5, Canada	519-688-1086	519-688-1098	2	98

Product Tables
Tobacco, Prepared Leaf

Unit of Measure: Metric tons.

	1990 Value	%	1991 Value	%	1992 Value	%	1993 Value	%	1994 Value	%	1995 Value	%	1996 Value	%
Total Production	6,256,869	100.0	6,636,206	100.0	7,055,774	100.0	6,970,021	100.0	7,131,449	100.0	7,378,943	100.0	7,627,013	100.0
Regions														
Africa	255,000	4.1	315,200	4.7	356,800	5.1	374,693	5.4	375,693	5.3	397,511	5.4	419,329	5.5
America, North	906,213	14.5	957,012	14.4	969,811	13.7	981,610	14.1	1,023,447	14.4	1,055,572	14.3	1,087,697	14.3
America, South	447,000	7.1	458,000	6.9	577,816	8.2	535,091	7.7	547,457	7.7	566,064	7.7	581,778	7.6
Asia	3,655,646	58.4	4,006,724	60.4	4,400,149	62.4	4,506,200	64.7	4,623,351	64.8	4,819,846	65.3	5,010,280	65.7
Europe	980,009	15.7	886,270	13.4	738,198	10.5	560,427	8.0	549,302	7.7	527,951	7.2	516,129	6.8
Oceania	13,000	0.2	13,000	0.2	13,000	0.2	12,000	0.2	12,200	0.2	12,000	0.2	11,800	0.2
Leading Producers														
China	2,393,000	38.2	2,742,000	41.3	3,084,000	43.7	3,214,000	46.1	*3,390,300	47.5	*3,586,200	48.6	*3,782,100	49.6
United States of America	663,000	10.6	679,000	10.2	703,000	10.0	660,000	9.5	*725,800	10.2	*750,200	10.2	*774,600	10.2
India	508,000	8.1	503,000	7.6	489,000	6.9	472,000	6.8	*463,500	6.5	*463,945	6.3	*464,390	6.1
Brazil	351,000	5.6	337,000	5.1	474,000	6.7	429,000	6.2	*430,308	6.0	*440,143	6.0	*449,978	5.9
United Kingdom	417,257	6.7	280,513	4.2	143,770	2.0	7,026	0.1	6,529	0.1	-		-	

Cigars, Cigarillos and Cheroots

Unit of Measure: Thousands of units.

	1990 Value	%	1991 Value	%	1992 Value	%	1993 Value	%	1994 Value	%	1995 Value	%	1996 Value	%
Total Production	15,291,044	100.0	16,514,449	100.0	14,735,605	100.0	14,100,886	100.0	14,337,470	100.0	14,774,424	100.0	14,589,645	100.0
Regions														
Africa	115,342	0.8	107,330	0.6	101,350	0.7	95,870	0.7	93,577	0.7	91,483	0.6	89,089	0.6
America, North	3,798,690	24.8	3,880,625	23.5	3,962,561	26.9	4,047,497	28.7	4,126,356	28.8	4,207,731	28.5	4,289,106	29.4
America, South	65,740	0.4	59,601	0.4	52,462	0.4	45,762	0.3	38,665	0.3	33,382	0.2	30,188	0.2
Asia	5,215,720	34.1	4,683,170	28.4	2,775,690	18.8	2,061,807	14.6	1,803,423	12.6	1,139,747	7.7	174,584	1.2
Europe	6,095,551	39.9	7,783,723	47.1	7,843,542	53.2	7,849,950	55.7	8,275,451	57.7	9,302,082	63.0	10,006,678	68.6
Leading Producers														
United States of America	3,194,000	20.9	3,301,000	20.0	3,408,000	23.1	3,515,000	24.9	3,622,000	25.3	3,729,000	25.2	3,836,000	26.3
Tajikistan	5,022,000	32.8	4,467,000	27.0	2,607,000	17.7	1,901,000	13.5	1,644,000	11.5	964,000	6.5	-	
Netherlands	1,436,000	9.4	1,426,000	8.6	1,391,000	9.4	1,372,000	9.7	1,321,000	9.2	1,316,000	8.9	*1,239,105	8.5
Germany	-		1,356,000	8.2	1,190,000	8.1	1,024,000	7.3	1,187,000	8.3	1,168,447	7.9	1,343,127	9.2
United Kingdom	518,500	3.4	776,000	4.7	1,033,500	7.0	1,291,000	9.2	*942,846	6.6	*982,214	6.6	*1,021,582	7.0

Cigarettes

Unit of Measure: Millions of units.

	1990 Value	%	1991 Value	%	1992 Value	%	1993 Value	%	1994 Value	%	1995 Value	%	1996 Value	%
Total Production	5,685,994	100.0	5,670,977	100.0	5,635,034	100.0	5,608,939	100.0	5,862,003	100.0	6,036,927	100.0	6,180,298	100.0
Regions														
Africa	170,403	3.0	170,674	3.0	165,931	2.9	163,223	2.9	156,655	2.7	163,040	2.7	178,251	2.9
America, North	846,804	14.9	829,487	14.6	839,226	14.9	821,065	14.6	821,868	14.0	862,882	14.3	890,442	14.4
America, South	236,800	4.2	237,972	4.2	232,320	4.1	242,744	4.3	245,570	4.2	248,724	4.1	258,543	4.2
Asia	3,093,052	54.4	3,081,478	54.3	3,085,824	54.8	3,107,587	55.4	3,328,688	56.8	3,420,396	56.7	3,541,443	57.3
Europe	1,297,651	22.8	1,311,862	23.1	1,273,783	22.6	1,235,918	22.0	1,270,792	21.7	1,303,523	21.6	1,272,855	20.6
Oceania	41,283	0.7	39,505	0.7	37,950	0.7	38,402	0.7	38,430	0.7	38,362	0.6	38,764	0.6
Leading Producers														
China	1,648,765	29.0	1,613,245	28.4	1,642,340	29.1	1,655,630	29.5	*1,868,221	31.9	*1,943,216	32.2	*2,018,211	32.7
United States of America	709,700	12.5	694,500	12.2	703,134	12.5	687,317	12.3	687,223	11.7	721,275	11.9	755,328	12.2
Japan	268,100	4.7	275,000	4.8	279,000	5.0	*261,770	4.7	*257,526	4.4	*253,282	4.2	*249,037	4.0
Germany	*213,760	3.8	*213,760	3.8	*213,760	3.8	204,730	3.7	222,791	3.8	*213,760	3.5	*213,760	3.5
Brazil	173,987	3.1	175,396	3.1	169,000	3.0	*181,796	3.2	*186,150	3.2	*190,503	3.2	*194,857	3.2

Commodity data are provided by the United Nations Statistical Division. The symbol * means that data are estimated. For additional notes, see Appendix I.

81

Product Tables
Tobacco, Manufactured (e.g. Smoking Tobacco, Chewing Tobacco, Snuff)

Unit of Measure: Metric tons.

	1990		1991		1992		1993		1994		1995		1996	
	Value	%	Value	%	Value	%	Value	%	Value	%	Value	%	Value	%
Total Production	418,539	100.0	434,655	100.0	438,043	100.0	438,696	100.0	455,388	100.0	456,053	100.0	452,802	100.0
Regions														
Africa	53,676	12.8	53,153	12.2	53,895	12.3	58,325	13.3	77,202	17.0	61,298	13.4	69,856	15.4
America, North	66,169	15.8	66,776	15.4	67,384	15.4	67,992	15.5	68,599	15.1	69,207	15.2	69,814	15.4
America, South	14,534	3.5	14,534	3.3	14,502	3.3	14,624	3.3	14,504	3.2	14,506	3.2	14,534	3.2
Asia	105,978	25.3	119,931	27.6	128,760	29.4	132,206	30.1	129,576	28.5	140,036	30.7	119,968	26.5
Europe	150,261	35.9	152,775	35.1	148,196	33.8	140,739	32.1	141,540	31.1	147,116	32.3	149,538	33.0
Oceania	27,921	6.7	27,486	6.3	25,306	5.8	24,811	5.7	23,967	5.3	23,891	5.2	29,091	6.4
Leading Producers														
United States of America	57,013	13.6	57,215	13.2	57,417	13.1	57,620	13.1	57,822	12.7	58,024	12.7	58,226	12.9
Indonesia	38,872	9.3	53,587	12.3	68,303	15.6	83,018	18.9	97,733	21.5	112,448	24.7	*98,024	21.6
Romania	44,000	10.5	42,000	9.7	41,000	9.4	23,273	5.3	21,920	4.8	22,538	4.9	25,440	5.6
Netherlands	31,511	7.5	27,477	6.3	30,282	6.9	42,222	9.6	41,651	9.1	41,750	9.2	*38,356	8.5
Azerbaijan	48,707	11.6	46,999	10.8	46,794	10.7	38,137	8.7	20,369	4.5	13,585	3.0	8,539	1.9

Commodity data are provided by the United Nations Statistical Division. The symbol * means that data are estimated. For additional notes, see Appendix I.

ISIC 3210
TEXTILES

ISIC 3210—Textiles—includes all textile producing industries, ranging from yarn mills to fabric/carpet/knitting mills. Yarn products include wool, cotton, synthetic fiber yarn, jute yarn, and yarns made of other fibers. Fabrics include cotton, wool, silk, linen, jute, cellulosic fibers, and non-cellulosic fibers. Also included are products such as blankets, bed linen, towelling, knitted fabrics, socks and stockings, undergarments, knitted sports shirts and sweaters, other knitted outer garments, carpets, and cordage and twine.

In the U.S. SIC system, the corresponding group is the 2-digit SIC 22—Textile Mill Products, ranging from SIC 2211 through 2299.

Summary Statistics

		1990 Value	N	1991 Value	N	1992 Value	N	1993 Value	N	1994 Value	N	1995 Value	N
Establishments or enterprises	(number)	154,224	84	155,750	95	171,746	89	136,283	89	109,356	74	79,492	35
Number employed	(000)	19,186	94	17,458	98	16,838	94	15,770	95	14,985	86	13,585	65
Total Wages	($ mil.)	76,643	89	79,018	91	82,831	86	69,451	86	67,556	76	57,052	59
Wage/salary per employee	($)	7,176	88	7,451	89	10,675	85	11,677	85	7,304	75	8,564	59
Female workers	(000)	1,473	35	1,198	33	1,584	28	1,854	33	1,685	26	1,422	15
as % of total employment*	(%)	48	34	46	32	47	28	45	33	52	26	54	15
Output	($ bil.)	556	94	552	96	545	92	480	90	466	81	425	65
per employee	($ 000)	41	90	41	91	99	88	202	86	42	78	51	62
per establishment	($ 000)	5,230	81	6,251	86	23,960	80	16,161	78	3,430	64	3,093	32
Value added	($ bil.)	180	86	166	84	180	78	180	81	164	72	157	62
per employee	($ 000)	16	83	16	80	19	74	16	75	18	68	21	58
per establishment	($ 000)	1,708	74	1,449	76	1,547	68	1,326	68	1,333	55	1,451	29
Capital investment	($ mil.)	22,092	70	19,835	64	18,917	57	15,089	52	13,410	44	8,612	22
per employee	($ 000)	3	67	5	61	3	55	2	51	2	42	3	21
per establishment	($ 000)	577	66	473	60	323	54	250	49	210	41	310	19

Data presented above are drawn from the detailed tables that follow. Columns headed 'N' show the number of countries that provided valid data for inclusion. Values are not strictly comparable one year to the next or one row to the next because the number of countries that report varies from period to period and row to row. However, a general indication of magnitudes can be gleaned from reviewing this summary—especially in earlier years in which more reports are available. For detailed explanations, see Appendix I. *The average for those countries reporting both total and female employment.

Establishments and Number Engaged

Country	Establishments or Enterprises (number)						Employees per Establishment					
	1990	1991	1992	1993	1994	1995	1990	1991	1992	1993	1994	1995
Albania	-	-	-	16	14	6.000	-	-	-	497	238	29
Angola	-	-	8.000	8.000	8.000	-	-	-	326	68	-	-
Argentina	-	-	-	2,233	2,449	-	-	-	-	21	19	-
Armenia	46	46	49	53	54	-	975	852	687	541	545	-
Australia	916	1,223	1,183	-	-	-	46	31	28	-	-	-
Austria	448	432	410	400	387	-	83	81	80	71	73	-
Azerbaijan	220	223	333	1,072	525	-	395	382	226	58	122	-
Bahamas	8.000	5.000	-	-	-	-	5.875	5.400	-	-	-	-
Bahrain	-	-	5.000	-	-	-	-	-	-	-	-	-
Bangladesh	11,041	12,591	12,811	-	-	-	46	43	46	-	-	-
Belgium	1,211	1,172	1,141	1,537	1,476	-	45	45	43	-	-	-
Benin	-	-	43	43	-	-	-	-	-	-	-	-
Bolivia	118	117	120	117	111	107	30	30	31	37	38	39
Bosnia & Herzegovina	69	75	-	-	4.000	-	595	472	-	-	117	-
Brazil	1,707	-	1,443	1,386	-	-	173	-	168	187	-	-
Bulgaria	*150	*153	*160	*643	*729	*818	747	575	461	100	87	75
Burundi	*3.000	*3.000		-	-	-	690	690	-	-	-	-
Cameroon	-	*6.000	*7.000	*5.000	*7.000	*6.000	-	624	524	566	435	498
Canada	1,337	1,206	1,117	1,057	983	-	54	57	56	57	64	-
Cape Verde	-	-	-	-	1.000	-	-	-	-	-	-	-
Central African Republic	-	*3.000	*3.000	*3.000	*1.000	-	-	-	-	-	-	-
Chad	-	1.000	1.000	1.000	1.000	1.000	-	-	-	-	-	-
Chile	135	131	135	137	131	-	194	198	192	179	171	-
China	*27,684	*27,575	*27,217	*24,613	*24,774	*25,686	269	274	273	278	279	262
China (Hong Kong SAR)	5,556	4,684	4,606	3,619	3,466	-	20	21	19	23	20	-
Colombia	498	486	502	479	468	-	106	111	122	125	127	-
Costa Rica	188	191	195	198	-	-	58	52	57	54	-	-
Croatia	227	278	342	422	530	570	217	140	98	78	59	52
Cyprus	225	230	230	227	251	260	9.693	10	10	10	7.873	8.408
Czechoslovakia (Former)	*89	*113	-	-	-	-	2,247	1,398	-	-	-	-
Czech Republic	-	*89	*144	*180	-	-	-	1,348	750	556	-	-
Denmark	1,030	1,103	1,052	-	-	-	16	15	15	-	-	-
Ecuador	149	152	163	148	143	138	97	90	88	89	85	79
Egypt	873	839	938	940	-	-	309	295	287	297	-	-
El Salvador	-	44	29	51	40	52	-	-	189	146	196	164
Ethiopia	26	25	23	23	29	29	1,096	1,216	1,312	1,289	1,048	1,067
Finland	204	206	186	171	158	148	53	45	44	43	45	40
FYR Macedonia	*29	*48	*76	*73	*85	*104	938	519	306	298	190	120
Gambia	-	-	-	1.000	-	-	-	-	-	34	-	-
Germany	-	2,150	1,949	1,827	1,711	-	-	147	123	115	110	-
Germany (Western Part)	1,789	1,733	1,671	-	-	-	128	129	125	-	-	-
Ghana	-	-	-	29	-	-	-	-	-	328	-	-
Greece	845	828	835	-	-	-	59	54	50	-	-	-
Grenada	-	-	1.000	1.000	1.000	-	-	-	2.000	2.000	16	-
Guatemala	-	29	29	29	29	29	-	245	234	234	238	256
Honduras	-	-	20	25	25	25	-	-	473	414	454	447
Hungary	*236	*367	*574	*726	-	-	322	166	84	67	-	-
Iceland	70	67	65	64	-	-	11	8.642	6.769	6.813	-	-
India	12,837	12,860	14,789	15,168	-	-	109	105	94	94	-	-
Indonesia	1,828	1,794	1,801	1,953	2,017	2,242	227	266	302	297	302	278
Iran (Islamic Republic of)	1,500	627	718	691	-	-	88	212	190	191	-	-
Iraq	-	56	55	-	-	-	-	331	309	-	-	-
Ireland	228	222	-	-	-	-	52	51	-	-	-	-
Israel	456	413	406	393	460	-	31	34	35	38	33	-
Italy	*3,182	*3,137	*3,465	*3,514	*3,511	-	70	74	62	60	60	-
Jamaica	25	25	-	-	-	-	34	37	-	-	-	-
Japan	39,885	38,717	36,807	35,370	31,279	30,450	16	16	16	16	17	16
Jordan	147	148	202	214	308	348	9.299	11	9.599	10	9.438	9.402
Kenya	92	91	92	92	101	101	273	269	267	273	245	247
Korea, Republic of	7,701	8,037	8,136	9,650	9,872	9,772	45	41	39	32	30	28

Continued.

Depending on the table, * means *Enterprises, Engaged,* or *Factor Values;* ± means *Basis Unspecified;* § means *shown in millions.* For additional notes and sources, see Appendix I.

Establishments and Number Engaged

- Continued -

Country	Establishments or Enterprises (number)						Employees per Establishment					
	1990	1991	1992	1993	1994	1995	1990	1991	1992	1993	1994	1995
Kuwait	218	217	260	264	263	-	5.624	4.438	6.031	5.080	5.464	-
Kyrgyzstan	22	22	35	31	29	-	1,349	1,335	942	1,013	863	-
Latvia	53	90	114	153	85	139	696	373	258	118	174	96
Lithuania	-	-	-	*103	*111	-	-	-	-	466	413	-
Macau	290	266	237	227	179	165	35	39	40	39	46	41
Malawi	*6.000	*6.000	*6.000	*6.000	*6.000	-	896	1,000	983	850	950	-
Malaysia	244	271	269	327	318	-	155	165	159	136	137	-
Malta	25	25	23	23	22	-	38	30	31	32	29	-
Mauritius	42	46	49	53	47	46	118	124	121	91	110	107
Mexico	243	241	236	229	224	219	283	264	255	214	196	197
Mongolia	13	13	13	17	29	26	565	524	473	326	219	189
Morocco	662	739	759	774	760	738	83	95	96	93	94	95
Mozambique	21	21	15	15	13	-	511	492	495	479	457	-
Nepal	425	1,427	-	975	1,087	-	98	54	-	67	72	-
Netherlands	217	223	220	223	224	-	98	95	93	88	84	-
Netherlands Antilles	-	-	-	10	-	-	-	-	-	3.300	-	-
New Zealand	*734	*760	*774	*802	-	-	14	12	12	12	-	-
Norway	203	191	129	115	117	-	28	27	39	40	40	-
Pakistan	-	1,478	-	-	-	-	-	168	-	-	-	-
Panama	12	6.000	-	-	-	-	54	90	-	-	-	-
Paraguay	-	10	-	-	-	-	-	94	-	-	-	-
Peru	1,195	1,262	1,283	-	-	-	33	28	25	-	-	-
Philippines	1,850	1,781	592	575	-	-	53	48	134	116	-	-
Poland	*414	*417	*393	*359	-	-	691	528	443	446	-	-
Portugal	*4,300	*4,010	*4,056	*4,241	*3,945	-	37	38	35	32	32	-
Puerto Rico	18	15	12	13	14	13	183	215	269	270	226	235
Qatar	12	11	34	36	35	-	3.333	3.545	5.618	6.028	6.571	-
Republic of Moldova	15	14	16	12	12	12	2,483	2,386	1,946	1,144	1,328	1,051
Romania	*221	937	1,715	2,159	2,568	-	1,875	420	191	127	86	-
Russian Federation	-	-	-	*2,777	*2,690	*3,965	-	-	-	296	260	153
Saint Lucia	-	4.000	4.000	4.000	4.000	4.000	-	-	-	-	-	-
Senegal	12	10	12	7.000	10	-	350	181	153	254	234	-
Sierra Leone	-	-	-	3.000	-	-	-	-	-	11	-	-
Singapore	65	64	64	65	59	-	52	52	51	48	44	-
Slovakia	-	45	45	52	63	-	-	964	846	628	479	-
Slovenia	197	*32	*38	*181	*186	*649	-	-	-	-	-	-
South Africa	-	911	-	-	-	-	-	99	-	-	-	-
Spain	5,060	4,786	4,455	-	-	-	22	22	20	-	-	-
Sri Lanka	502	406	379	404	-	-	92	108	111	99	-	-
Sweden	247	234	209	195	176	-	52	50	50	47	50	-
Thailand	1,002	1,464	-	-	-	-	410	189	-	-	-	-
Tonga	-	6.000	7.000	5.000	5.000	-	-	20	17	4.200	2.200	-
Trinidad and Tobago	22	18	15	16	18	-	19	19	22	12	12	-
Tunisia	-	-	-	525	448	562	-	-	-	38	34	32
Turkey	744	734	1,433	1,344	1,290	-	266	234	131	141	144	-
Ukraine	719	953	1,019	653	637	563	352	227	221	308	275	268
United Kingdom	6,285	5,987	5,815	7,256	6,010	-	36	35	34	27	32	-
United Republic of Tanzania	87	88	-	-	-	-	364	346	-	-	-	-
USSR (Former)	*2,212	-	-	-	-	-	646	-	-	-	-	-
United States of America	-	-	19,485	-	-	-	-	-	43	-	-	-
Venezuela	219	215	227	175	190	198	115	112	106	117	103	99
Yugoslavia	307	492	962	1,227	1,275	1,301	-	190	86	64	59	57
Zambia	31	-	-	-	28	-	267	-	-	-	206	-
Zimbabwe	50	51	50	50	48	-	465	496	474	416	363	-

Employment and Compensation of Employees

Country	Number Employed/Engaged (000)						Wage/Salary per Employee ($)					
	1990	1991	1992	1993	1994	1995	1990	1991	1992	1993	1994	1995
Albania	-		-	7.946	3.338	0.172	-	-	-	-	-	-
Angola	-	*2.248	*2.609	*0.540			-	-	-	-	-	-
Argentina	73			48	47	-	5,839		-	10,879	-	-
Armenia	*45	*39	*34	*29	*29	*25	-	-	-	14,137	-	-
Australia	42	*38	*33	-	-	-	18,595	19,232	20,655	-	-	-
Austria	37	35	33	29	28	28	20,181	21,597	24,534	24,964	25,182	29,409
Azerbaijan	87	85	75	62	64	-	1,304	1,371	-	-	-	-
Bahamas	0.047	0.027	-	-	-	-	8,511	7,000	-	-	-	-
Bahrain	-	-	0.002			-	-	-	2,660	-	-	-
Bangladesh	512	542	583	-	-	-	731	629	572	-	-	-
Belgium	54	52	49	-	-	-	16,042	16,480	18,200	-	-	-
Bolivia	3.554	3.490	3.756	4.330	4.271	4.224	1,254	1,254	1,412	1,416	1,521	1,716
Bosnia & Herzegovina	41	35	-	-	0.469	-	2,166	1,900	-	-	-	-
Brazil	296	-	243	259	-	-	4,499		4,299	5,247	-	-
Bulgaria	112	88	74	65	63	61	1,711	522	786	962	821	1,024
Burundi	2.070	2.070	-	-	-	-	2,290	2,289	-	-	-	-
Cambodia	*18	*15	-	-	-	-	-	-	-	-	-	-
Cameroon	4.456	3.743	3.668	2.832	3.047	2.986	7,562	8,000	7,825	7,760	4,649	5,216
Canada	72	69	62	60	63	63	21,414	22,327	22,698	22,427	21,434	21,462
Chile	26	26	26	25	22	23	3,454	4,051	4,693	5,072	5,891	6,746
China	*7,450	*7,560	*7,430	*6,840	*6,910	*6,730	-	-	-	-	-	-
China (Hong Kong SAR)	*113	*99	*89	*82	*68	-	9,514	10,576	11,672	12,741	13,823	-
China (Taiwan Province)	222	210	206	201	205	196	9,154	10,339	12,031	12,056	12,270	12,933
Colombia	53	54	61	60	59	57	2,138	2,186	2,277	2,706	3,328	4,086
Costa Rica	11	9.916	11	11	-	-	2,417	2,310	2,534	2,795	-	-
Croatia	49	39	34	33	31	30	3,092	3,487	1,378	1,449	1,967	3,049
Cyprus	2.181	2.311	2.401	2.307	1.976	2.186	7,344	7,722	8,798	8,322	9,089	11,788
Czechoslovakia (Former)	200	158	-	-	-	-	1,877	1,286	-	-	-	-
Czech Republic	142	120	108	100	-	-	1,918	1,303	1,556	1,836	-	-
Denmark	16	16	15	14	14	14	24,863	24,292	27,545	25,866	27,007	31,173
Ecuador	14	14	14	13	12	11	2,359	2,633	2,491	2,935	1,489	1,759
Egypt	*269	*247	*269	*279	*279	*279	1,699	1,092	1,133	1,205	1,247	1,385
El Salvador	-	-	*5.470	*7.423	*7.838	*8.549	-	-	-	2,092	3,264	2,301
Ethiopia	28	30	30	30	30	31	1,425	1,187	961	679	702	605
Finland	11	9.200	8.200	7.300	7.100	5.900	21,551	21,228	20,146	17,204	19,684	24,958
FYR Macedonia	27	25	23	22	16	12	3,820	3,677	1,474	1,460	1,853	2,478
France	210	196	182	168	160	154	23,816	24,740	27,896	-	-	-
Gambia		-	-	0.034	-	-	-	-	-	-	-	-
Germany	-	*315	*240	*209	*189	-	-	19,398	25,449	25,615	27,063	-
Germany (Western Part)	*229	*224	*210	-	-	-	23,849	24,493	27,233	-	-	-
Ghana	-	-	-	*9.499	*9.767	*10	-	-	-	1,339	1,000	1,021
Greece	*50	*45	*41	*41	*41	*40	10,541	10,743	11,360	10,829	10,935	12,045
Grenada		-	*0.002	*0.002	0.016	0.016	-	-	-	-	-	-
Guatemala	-	*7.113	*6.795	*6.789	*6.905	*7.418	-	-	-	-	-	-
Honduras	*6.333	*7.512	*9.454	*10	*11	*11	1,440	1,131	1,252	942	649	766
Hungary	76	61	48	49	48	43	2,040	2,140	2,562	2,477	2,545	2,613
Iceland	0.796	0.579	0.440	0.436	0.468	-	23,201	23,247	27,828	21,486	-	-
India	*1,403	*1,351	*1,397	*1,429	*1,442	*1,439	1,196	1,005	979	896	913	954
Indonesia	415	478	543	581	610	624	513	661	776	793	826	1,144
Iran (Islamic Republic of)	133	133	136	132	-	-	3,817	4,433	4,580	4,229	-	-
Iraq	-	19	17	-	-	-	-	7,738	13,661	-	-	-
Ireland	12	11	12	12	12	12	16,331	16,339	18,087	15,914	16,778	18,482
Israel	14	14	14	15	15	16	16,392	15,839	16,266	15,113	16,756	18,430
Italy	224	232	216	210	210	-	28,954	28,995	32,109	25,905	26,809	-
Jamaica	0.859	0.931	0.927	0.764	0.620	-	-	-	-	-	-	-
Japan	*634	*624	*597	*561	*517	*491	18,639	20,973	23,198	26,898	29,635	32,696
Jordan	1.367	1.654	1.939	2.189	2.907	3.272	2,398	2,250	2,408	2,443	2,377	2,332
Kenya	25	24	25	25	25	25	1,163	1,003	935	610	754	1,038
Korea, Republic of	349	333	316	308	296	272	7,825	8,960	9,940	10,631	11,738	14,187
Kuwait	1.226	0.963	1.568	1.341	1.437	1.488	3,858	3,620	7,555	7,188	6,602	9,653

Continued.

Depending on the table, * means *Enterprises, Engaged,* or *Factor Values;* ± means *Basis Unspecified;* § means *shown in millions.* For additional notes and sources, see Appendix I.

Employment and Compensation of Employees

- Continued -

Country	Number Employed/Engaged (000)						Wage/Salary per Employee ($)					
	1990	1991	1992	1993	1994	1995	1990	1991	1992	1993	1994	1995
Kyrgyzstan	*30	*29	*33	*31	*25	-	-	-	-	-	-	-
Latvia	37	34	29	18	15	13	-	-	-	-	-	-
Lithuania	-	-	52	48	46	-	-	-	426	486	971	-
Macau	10	10	9.542	8.884	8.247	6.835	4,355	4,426	4,782	5,245	5,749	6,089
Malawi	5.379	6.000	5.900	5.100	5.700	-	1,179	1,623	1,529	1,461	890	-
Malaysia	38	45	43	44	44	45	2,662	2,752	3,429	3,720	3,997	4,567
Malta	0.947	0.757	0.710	0.730	0.645	-	7,825	8,887	9,988	8,334	9,048	-
Mauritius	*4.963	5.712	5.911	4.802	5.151	4.920	2,135	1,989	2,392	2,139	2,857	3,146
Mexico	69	64	60	49	44	43	3,643	4,270	4,946	5,692	5,756	3,148
Mongolia	*7.349	*6.811	*6.151	*5.534	*6.343	*4.926	1,232	1,474	620	-	-	825
Morocco	55	70	73	72	71	70	2,946	2,596	2,798	2,594	3,232	3,763
Mozambique	11	10	7.425	7.191	5.947	-	1,808	581	411	-	-	-
Nepal	41	78	-	66	78	-	513	494	-	-	404	-
Netherlands	21	21	20	20	19	19	24,576	24,832	27,804	27,863	29,110	33,630
Netherlands Antilles	-	-	-	0.033	-	-	-	-	-	7,686	-	-
New Zealand	10	9.461	9.558	9.843	-	-	13,861	14,750	13,061	-	-	-
Norway	5.747	5.249	5.063	4.604	4.688	4.764	23,183	23,568	30,090	20,728	22,154	25,283
Pakistan	225	248	260	-	-	-	1,226	1,257	1,293	-	-	-
Panama	*0.644	*0.540	*0.482	*0.540	*0.535	*0.497	3,888	4,202	4,297	4,271	4,314	4,365
Paraguay	-	0.942	-	-	-	-	-	-	-	-	-	-
Peru	*39	*35	*32	-	-	-	3,389	4,105	3,818	-	-	-
Philippines	99	86	79	67	-	-	1,453	1,567	1,928	1,849	-	-
Poland	286	220	174	160	186	184	1,103	1,624	2,012	2,042	2,145	2,547
Portugal	*158	*151	*140	*134	*128	*128	-	-	-	-	-	-
Puerto Rico	3.290	3.220	3.230	3.510	3.170	3.060	16,535	7,391	7,554	7,066	7,729	8,039
Qatar	0.040	0.039	0.191	0.217	0.230	-	1,271	1,564	1,464	2,724	3,685	-
Republic of Moldova	37	33	31	14	16	13	-	3,951	-	-	-	-
Romania	414	394	328	273	221	-	1,584	981	562	684	736	-
Russian Federation	-	-	-	821	700	606	-	-	-	532	674	715
Senegal	*4.196	*1.810	*1.835	*1.775	*2.340	-	4,324	4,953	4,297	4,773	4,085	-
Sierra Leone	-	-	-	*0.033	-	-	-	-	-	-	-	-
Singapore	*3.406	*3.343	*3.286	*3.101	*2.613	*1.803	8,933	10,464	11,544	12,490	14,227	18,415
Slovakia	-	*43	*38	*33	*30	-	-	1,249	1,445	1,543	1,696	-
South Africa	97	90	83	66	68	69	5,268	5,710	6,223	6,287	6,490	7,048
Spain	110	104	91	92	95	92	14,286	15,251	17,150	14,728	14,432	16,262
Sri Lanka	*46	*44	*42	*40	-	-	454	695	781	817	-	-
Sweden	13	*12	*10	9.140	8.880	9.100	20,736	22,335	25,293	20,408	21,029	23,370
Switzerland	*31	*29	*27	*24	*24	*22	-	-	-	-	-	-
Thailand	411	276	-	-	-	-	1,887	1,894	-	-	-	-
Tonga	-	0.119	0.121	0.021	0.011	-	-	1,550	1,055	963	1,171	-
Trinidad and Tobago	0.421	0.336	0.323	0.198	0.214	-	4,208	4,784	5,671	4,247	4,101	-
Tunisia	-	-	-	20	15	18	-	-	-	6,970	8,430	9,408
Turkey	198	172	188	189	186	190	4,490	6,819	5,699	6,185	3,574	4,334
Ukraine	253	216	225	201	175	151	2,710	2,980	2,542	429	429	-
United Kingdom	224	208	198	199	192	189	16,159	17,282	18,676	15,996	17,124	18,360
United Republic of Tanzania	32	30	-	-	-	-	-	-	-	-	-	-
USSR (Former)	1,430	-	-	-	-	-	1,920	-	-	-	-	-
United States of America	829	791	831	829	852	842	18,251	18,925	19,869	20,619	21,289	21,645
Uruguay	*20	*18	*16	*14	*14	*14	3,421	4,148	4,698	5,602	5,799	5,044
Venezuela	25	24	24	21	19	20	3,936	4,113	4,285	4,551	3,729	4,020
Yugoslavia	-	94	83	78	75	74	-	-	-	-	1,166	491
Zambia	8.286	-	-	-	5.777	-	1,601	-	-	-	1,834	-
Zimbabwe	23	25	24	21	17	15	3,233	2,673	1,961	2,100	1,929	2,225

Female Workers: Total and Percent of Employment

Country	Female Workers (000)						Female Workers as % of Total Employment					
	1990	1991	1992	1993	1994	1995	1990	1991	1992	1993	1994	1995
Afghanistan	1.825	0.674	-	-	-	-	-	-	-	-	-	-
Albania	-	-	-	4.961	2.079	0.012	-	-	-	62.43	62.28	6.98
Angola	-	-	0.447	0.064	-	-	-	-	17.13	11.85	-	-
Argentina	-	-	-	-	11	-	-	-	-	-	22.62	-
Australia	19	-	-	-	-	-	46.45	-	-	-	-	-
Austria	21	20	18	15	15	-	55.80	57.14	54.55	54.03	53.51	-
Bangladesh	12	11	11	-	-	-	2.27	2.02	1.80	-	-	-
Bosnia & Herzegovina	32	28	-	-	-	-	77.97	79.45	-	-	-	-
Bulgaria	-	-	-	47	47	42	-	-	-	72.56	73.70	69.02
Canada	31	30	-	-	-	-	43.06	43.48	-	-	-	-
Chile	7.978	8.029	8.163	8.153	7.765	-	30.44	31.02	31.43	33.22	34.69	-
China (Taiwan Province)	128	119	115	111	112	107	57.41	56.78	55.92	55.25	54.91	54.27
Colombia	-	19	-	23	24	-	-	35.55	-	38.80	39.92	-
Croatia	34	28	24	24	22	21	69.88	70.93	71.41	71.57	71.25	71.37
Cyprus	1.692	1.783	1.857	1.746	1.457	-	77.58	77.15	77.34	75.68	73.73	-
Czechoslovakia (Former)	105	88	-	-	-	-	52.40	55.38	-	-	-	-
Czech Republic	97	82	74	69	-	-	68.31	68.33	68.52	69.00	-	-
Denmark	9.430	9.487	9.147	-	-	-	58.02	59.08	59.83	-	-	-
Egypt	-	31	38	42	-	-	-	12.57	14.16	15.03	-	-
El Salvador	-	-	-	3.267	3.550	3.238	-	-	-	44.01	45.29	37.88
Ethiopia	-	14	13	13	13	14	-	44.78	44.63	45.21	44.16	44.20
FYR Macedonia	14	11	10	9.350	8.432	6.985	51.47	45.90	44.93	43.02	52.27	56.02
Gambia	-	-	-	0.002	-	-	-	-	-	5.88	-	-
Germany (Western Part)	107	-	-	-	-	-	46.90	-	-	-	-	-
Ghana	-	-	-	0.199	-	-	-	-	-	2.09	-	-
Hungary	45	-	35	32	-	-	59.21	-	72.92	65.31	-	-
Indonesia	-	-	-	309	325	330	-	-	-	53.25	53.33	52.89
Iran (Islamic Republic of)	-	7.393	7.632	6.766	-	-	-	5.57	5.60	5.12	-	-
Iraq	-	4.575	-	-	-	-	-	24.71	-	-	-	-
Ireland	5.500	5.400	-	-	-	-	46.61	47.37	-	-	-	-
Italy	-	133	109	104	104	-	-	57.22	50.32	49.52	49.61	-
Jordan	0.439	0.509	0.635	0.498	0.803	1.076	32.11	30.77	32.75	22.75	27.62	32.89
Kenya	2.154	2.086	2.113	2.121	-	-	8.58	8.53	8.59	8.44	-	-
Korea, Republic of	201	192	172	165	155	140	57.45	57.63	54.35	53.53	52.25	51.43
Macau	6.142	6.871	6.592	6.015	5.904	4.888	60.64	65.50	69.08	67.71	71.59	71.51
Malaysia	22	26	23	23	21	-	58.20	57.81	54.15	52.58	49.08	-
Malta	0.598	0.406	0.388	0.392	0.332	-	63.15	53.63	54.65	53.70	51.47	-
Mongolia	-	-	-	-	4.472	-	-	-	-	-	70.50	-
Morocco	-	-	-	-	-	36	-	-	-	-	-	51.52
Nepal	20	32	-	26	30	-	47.90	41.48	-	39.88	38.82	-
New Zealand	4.590	4.158	4.238	4.444	-	-	44.22	43.95	44.34	45.15	-	-
Panama	0.194	-	-	-	-	-	30.12	-	-	-	-	-
Philippines	-	-	37	35	-	-	-	-	47.12	52.26	-	-
Portugal	55	49	44	41	38	-	34.51	32.48	31.32	30.69	29.72	-
Puerto Rico	1.790	1.730	1.720	1.860	1.680	1.640	54.41	53.73	53.25	52.99	53.00	53.59
Republic of Moldova	-	-	-	-	12	9.497	-	-	-	-	77.95	75.28
Senegal	0.189	-	-	-	-	-	4.50	-	-	-	-	-
Sri Lanka	29	27	25	23	-	-	62.11	62.11	59.82	57.95	-	-
Sweden	6.700	-	-	-	-	-	52.34	-	-	-	-	-
Thailand	274	196	-	-	-	-	66.60	70.82	-	-	-	-
Turkey	63	-	-	-	-	-	31.82	-	-	-	-	-
United Kingdom	107	-	92	-	-	-	47.85	-	46.46	-	-	-
United Republic of Tanzania	9.619	9.081	-	-	-	-	30.36	29.85	-	-	-	-
United States of America	-	-	700	701	717	705	-	-	84.24	84.56	84.15	83.73
Zambia	-	-	-	-	0.632	-	-	-	-	-	10.94	-

Output and Output per Employee

Country	Output ($ bil.)						Output per Employee ($)					
	1990	1991	1992	1993	1994	1995	1990	1991	1992	1993	1994	1995
Albania	-	-	-	*0.010	*0.001	*0.000	-	-	-	1,291	415	1,555
Angola			±13	±7.546	-	-			§4,815	§13,97	-	-
Argentina	±6.124	-	-	3.261	-	-	84,353	-	-	68,285	-	-
Armenia	0.017	0.027	0.093	2.671	0.009		375	699	2,767	93,180	291	-
Australia	*3.948	*3.892	*3.665	-	-	-	94,235	102,413	111,074	-	-	-
Austria	3.597	3.495	3.736	3.285	3.274	3.749	96,952	99,868	113,207	114,903	116,399	136,046
Azerbaijan	±1.439	±1.759	±0.065	±0.168	±0.078		16,563	20,637	859	2,723	1,215	-
Bahamas	0.002	0.001	-	-	-		42,553	37,037	-	-	-	
Bahrain	-	-	±0.000	-	-		-	-	15,958	-	-	
Bangladesh	1.217	1.257	1.361	-	-		2,375	2,320	2,334	-	-	
Belgium	7.251	6.746	7.382	6.584	7.356	7.955	133,775	128,985	150,039	-	-	-
Benin	±0.052	±0.071	±0.085	±0.064	-	-	-	-	-			
Bolivia	0.057	0.058	0.061	0.068	0.074	0.081	16,035	16,656	16,350	15,615	17,270	19,256
Bosnia & Herzegovina	±0.490	±0.395	-	-	-	-	11,930	11,147	-	-	-	-
Brazil	*14	-	*9.674	*11	-	-	47,460	-	39,827	43,351	-	-
Bulgaria	±1.514	±0.375	±0.396	±0.359	±0.355	±0.419	13,519	4,269	5,374	5,559	5,588	6,871
Burundi	±0.018	±0.018					8,505	8,686	-			
Cameroon	0.233	0.163	0.160	0.146	0.112	0.174	52,204	43,629	43,618	51,396	36,803	58,244
Canada	7.182	6.878	6.503	6.364	6.876	6.929	99,751	99,680	104,885	106,064	109,144	109,413
Chile	0.686	0.801	0.867	0.828	0.802	0.955	26,168	30,962	33,372	33,755	35,816	40,991
China	±48	±48	±53	±61	±57	±55	6,430	6,295	7,076	8,933	8,311	8,191
China (Hong Kong SAR)	6.591	6.915	6.745	6.366	5.933		58,531	69,708	75,530	77,634	87,125	-
China (Taiwan Province)	±11	±12	±12	±12	±13	±14	49,302	58,403	60,407	57,790	63,932	69,168
Colombia	1.627	1.632	1.577	1.602	1.820	2.151	30,808	30,194	25,820	26,706	30,611	37,552
Costa Rica	0.093	0.084	0.088	0.097	0.089	0.079	8,558	8,434	7,904	9,016		
Cote d'Ivoire	±0.264	±0.227	±0.257	±0.226	-	-	-	-	-	-		
Croatia	±0.827	±0.801	±0.432	-	-	-	16,814	20,617	12,854	-	-	-
Cyprus	0.079	0.085	0.093	0.078	0.070	0.095	36,140	36,839	38,620	33,597	35,659	43,513
Czechoslovakia (Former)	±2.283	±0.967					11,415	6,123				
Denmark	*1.316	*1.276	*1.388	*1.140	*1.263	*1.413	80,972	79,449	90,797	84,436	88,315	101,680
Ecuador	0.249	0.278	0.263	0.264	0.276	0.323	17,220	20,372	18,364	20,115	22,741	29,703
Egypt	*3.040	*1.923	*2.060	*2.240	*2.332	*2.625	11,284	7,776	7,654	8,022	8,353	9,414
El Salvador	-	±0.174	±0.089	±0.104	±0.127	±0.157	-	-	16,328	14,070	16,216	18,377
Ethiopia	0.163	0.127	0.080	0.078	0.118	0.095	5,731	4,186	2,647	2,618	3,887	3,079
Finland	±0.912	±0.728	±0.672	±0.565	±0.698	±0.725	83,706	79,130	81,920	77,434	98,256	122,883
FYR Macedonia	±0.322	±0.370	±0.223	±0.173	±0.108	±0.103	11,849	14,857	9,564	7,965	6,723	8,250
France	±21	±20	±21	±18	±19	±22	101,249	100,796	115,751	105,558	121,584	141,356
Germany	-	28	29	24	24	-	-	89,842	119,581	116,931	125,047	-
Germany (Western Part)	28	27	28	-	-	-	120,590	121,305	131,903	-	-	-
Ghana	-	-	-	0.089	0.073	0.076	-	-	-	9,413	7,457	7,611
Greece	*3.173	*2.979	*2.790	*2.603	*2.628	*2.844	63,260	66,377	67,399	64,197	64,801	71,364
Guatemala	-	0.220	0.267	0.344	0.232	0.129	-	30,939	39,319	50,665	33,594	17,337
Honduras	0.048	0.043	0.046	0.041	0.038	0.045	7,509	5,726	4,873	3,965	3,306	4,016
Hungary	±1.111	±0.852	±0.633	±0.632	±0.624	±0.568	14,614	13,973	13,197	12,888	13,001	13,203
Iceland	0.060	0.057	±0.055	±0.045	-	-	75,376	97,952	124,635	102,233	-	-
India	*16	*14	*14	*14	*15	*16	11,485	10,140	10,084	10,124	10,415	10,998
Indonesia	4.276	5.369	6.575	7.098	9.665	10	10,313	11,243	12,101	12,227	15,853	16,775
Iran (Islamic Republic of)	2.625	2.799	2.537	2.257	-	-	19,778	21,102	18,622	17,082	-	-
Iraq	-	±0.295	*0.912	-	-	-	-	15,940	53,660	-	-	-
Ireland	*0.899	*0.832	*0.952	*0.854	*0.917	*1.015	76,145	73,024	80,674	70,916	74,680	82,207
Israel	1.290	1.306	1.348	1.245	1.341	1.605	91,458	92,640	95,607	84,134	87,059	102,166
Italy	±33	±32	±33	±27	±31		146,202	136,883	154,156	129,758	147,955	-
Japan	±64	±70	±72	±72	±72	±74	100,886	112,718	120,487	127,935	138,354	150,530
Jordan	0.046	0.054	0.061	0.056	0.073	0.079	33,506	32,898	31,544	25,476	25,114	24,023
Kenya	*0.194	*0.119	*0.102	*0.076	*0.101	*0.150	7,718	4,879	4,163	3,032	4,088	6,028
Korea, Republic of	18	20	20	22	23	26	50,755	58,838	64,589	70,112	76,764	94,696
Kuwait	0.032	0.019	0.045	0.043	0.039	0.060	25,998	19,679	28,745	32,140	26,881	40,015
Kyrgyzstan	0.914	1.462	0.819	0.135	0.103	-	30,802	49,781	24,846	4,309	4,131	-
Latvia	0.009	0.026	0.092	0.082	0.095	0.111	256	763	3,121	4,516	6,428	8,361
Lithuania	-	-	0.212	0.189	0.214	-	-	-	4,080	3,949	4,664	

Continued.

Output and Output per Employee

- Continued -

Country	Output ($ bil.)						Output per Employee ($)					
	1990	1991	1992	1993	1994	1995	1990	1991	1992	1993	1994	1995
Macau	0.327	0.344	0.366	0.323	0.309	0.303	32,275	32,756	38,310	36,376	37,410	44,335
Malawi	*0.057	*0.071	*0.079	*0.093	*0.038	-	10,573	11,825	13,443	18,331	6,613	-
Malaysia	*1.054	*1.271	*1.404	*1.511	*1.807	*2.134	27,895	28,363	32,768	33,963	41,437	47,651
Malta	0.036	0.033	0.035	0.029	0.035		38,541	43,552	49,991	40,407	53,948	
Mauritius	0.076	0.085	0.110	0.086	0.127	0.128	15,407	14,836	18,611	17,921	24,731	26,055
Mexico	1.631	1.674	1.679	1.558	1.451	1.003	23,693	26,309	27,898	31,824	33,092	23,257
Mongolia	0.148	0.118	0.074	0.025	0.036	0.074	20,107	17,304	12,074	4,556	5,754	14,930
Morocco	±1.153	±1.231	±1.323	±1.206	±1.299	±1.424	20,915	17,614	18,076	16,718	18,224	20,229
Mozambique	-	±0.037	-	±0.017	±0.012	-		3,621	-	2,397	2,072	
Nepal	0.136	0.254	-	0.163	0.297	-	3,271	3,272	-	2,479	3,789	
Netherlands	2.847	2.747	2.873	2.679	2.675	3.129	134,106	130,164	140,774	137,140	142,738	169,141
New Zealand	0.733	-	-	-	-	-	70,630	-	-	-	-	-
Norway	±0.507	±0.499	±0.533	±0.383	±0.420	±0.484	88,174	95,125	105,200	83,126	89,524	101,531
Pakistan	4.014	4.688	5.088				17,850	18,865	19,576			
Panama	0.017	0.015	0.014	0.015	0.016	0.014	26,179	27,994	28,165	28,630	28,973	28,992
Peru	1.337	1.242	1.240	-			34,183	35,440	38,760			
Philippines	1.138	1.048	1.146	0.961	-	-	11,503	12,159	14,489	14,434	-	-
Poland	±2.172	±1.994	±1.736	±1.787	±2.185	±2.570	7,593	9,062	9,980	11,169	11,769	13,992
Portugal	5.097	4.998	5.444	4.633	4.963	5.794	32,201	33,059	38,792	34,579	38,889	45,332
Puerto Rico	±0.107	±0.105	±0.107	±0.108	±0.110	±0.102	32,488	32,609	32,972	30,769	34,858	33,366
Qatar	0.000	0.000	0.001	0.003	0.004	-	6,868	7,044	7,192	15,192	17,917	-
Republic of Moldova	0.000	1.695	0.140	0.032	0.030	0.026	0.060	50,761	4,490	2,297	1,857	2,089
Romania	±3.491	±2.846	±1.306	±1.293	-	-	8,425	7,229	3,988	4,734	-	-
Russian Federation	-	55	18	3.640	2.697	3.050	-	-	-	4,435	3,855	5,036
Senegal	0.083	-	-	-	-	-	19,815	-	-	-	-	-
Sierra Leone	-	-	-	0.000	-	-	-	-	-	2,515	-	-
Singapore	*0.218	*0.233	*0.238	*0.228	*0.210	*0.170	63,898	69,551	72,541	73,448	80,375	94,300
Slovakia		±0.405	±0.381	±0.306	±0.312	-		9,341	9,996	9,376	10,348	-
Slovenia	-	-	-	0.538	0.517	0.642	-	-	-	-	-	-
South Africa	±2.452	±2.410	±2.342	±2.302	±2.405	±2.624	25,278	26,779	28,211	34,883	35,362	38,278
Spain	*8.409	*8.529	*8.359	±7.153	±7.791	±9.309	76,688	82,288	91,974	77,378	81,721	101,231
Sri Lanka	0.188	0.191	0.248	0.254	-	-	4,079	4,371	5,888	6,341	-	-
Sweden	*1.250	*1.368	*1.372	*0.887	*0.921	*1.086	97,673	117,639	131,192	97,039	103,667	119,349
Switzerland	3.409	3.103	3.007	2.508	3.144	3.644	109,981	106,986	113,052	103,195	132,106	162,685
Thailand	25	5.286					60,145	19,120				
Tonga		0.000	0.000	0.000	0.000	-		3,326	2,963	6,055	13,497	-
Trinidad and Tobago	±0.012	±0.013	±0.011	±0.005	±0.006	-	28,094	37,362	35,413	25,861	25,790	-
Tunisia	0.688	0.720	0.862	0.853	0.944	1.125	-	-	-	43,178	62,133	62,852
Turkey	7.895	7.773	9.758	10	10	13	39,874	45,240	52,041	55,016	54,463	67,012
Ukraine	*8.461	*11	*6.588	*1.209	*0.904	*0.567	33,442	49,143	29,281	6,013	5,168	3,756
United Kingdom	*16	*15	15	14	15	16	70,081	72,056	77,301	69,529	77,822	83,391
United Republic of Tanzania	*0.100	*0.126	-	-	-	-	3,147	4,149	-	-	-	-
USSR (Former)	±51						35,533					
United States of America	*84	*84	*91	*96	*102	*104	101,846	106,220	109,965	115,586	119,279	123,308
Uruguay	0.572	0.606	0.603	0.530	0.548	0.504	28,568	33,146	36,812	38,933	40,257	34,999
Venezuela	0.711	0.757	0.709	0.563	0.571	0.826	28,335	31,554	29,536	27,436	29,289	42,220
Yugoslavia	±0.155	±0.151	±0.323	-	±0.264	±0.262	-	1,611	3,900	-	3,543	3,561
Zambia	0.114	-	-	-	0.066	-	13,787	-	-	-	11,416	-
Zimbabwe	*0.528	*0.513	*0.348	*0.399	*0.320	*0.327	22,708	20,262	14,677	19,194	18,384	21,791

　Depending on the table, * means *Enterprises, Engaged,* or *Factor Values*; ± means *Basis Unspecified*; § means *shown in millions.* For additional notes and sources, see Appendix I.

Value Added and Value Added per Employee

Country	Value Added ($ bil.)						Value Added per Employee ($)					
	1990	1991	1992	1993	1994	1995	1990	1991	1992	1993	1994	1995
Argentina	±2.209	-	-	1.144	-	-	30,428	-	-	23,944	-	-
Armenia	0.002	0.005	-	-	-	-	53	128	-	-	-	-
Australia	*1.673	*1.648	*1.749	-	-	-	39,939	43,364	52,993			
Austria	1.291	1.267	1.392	1.252	1.238	1.419	34,801	36,187	42,172	43,783	44,004	51,476
Bahamas	0.001	0.001	-	-	-	-	25,213	20,741	-	-	-	-
Bangladesh	0.439	0.432	0.441	-	-	-	856	797	757	-	-	-
Belgium	*2.132	*2.000	*2.185	*1.933	*2.135	*2.317	39,327	38,240	44,401			
Bolivia	0.019	0.019	0.024	0.026	0.027	0.029	5,394	5,554	6,266	5,903	6,290	6,969
Bosnia & Herzegovina	±0.226	±0.237	-	-	-	-	5,498	6,695	-	-	-	-
Brazil	*7.701	-	*5.601	*6.936	-	-	26,004	-	23,059	26,801	-	-
Bulgaria	-	±0.113	±0.124	±0.122	±0.100	±0.129	-	1,290	1,679	1,895	1,575	2,114
Burundi	0.009	0.009	-	-	-	-	4,115	4,348	-	-	-	-
Cameroon	-0.104	0.053	0.040	0.058	0.036	0.090	§-23,4	14,268	10,799	20,338	11,889	30,194
Canada	2.974	2.915	2.829	2.790	2.922	2.944	41,305	42,250	45,637	46,508	46,378	46,494
Central African Republic	-	-	*-0.001	*-0.000	-	-	-	-	-	-	-	-
Chile	0.333	0.363	0.412	0.377	0.361	0.429	12,702	14,043	15,863	15,342	16,108	18,415
China	±10	±9.445	±11	±17	±13	±11	1,382	1,249	1,449	2,413	1,876	1,599
China (Hong Kong SAR)	1.804	1.803	1.912	1.704	1.500	-	16,019	18,172	21,411	20,778	22,031	
China (Taiwan Province)	3.573	4.085	4.774	4.497	4.409	4.223	16,060	19,461	23,118	22,414	21,539	21,506
Colombia	0.816	0.696	0.683	0.693	0.843	0.996	15,460	12,875	11,194	11,558	14,186	17,378
Costa Rica	0.032	0.029	0.030	0.032	0.029	0.026	2,954	2,919	2,737	2,996	-	-
Cote d'Ivoire	0.239	0.206	0.230	0.205	-	-	-	-	-	-	-	-
Croatia	±0.417	±0.442	±0.221	-	-	-	8,486	11,387	6,586	-	-	-
Cyprus	0.032	0.035	0.039	0.033	0.031	0.039	14,761	15,350	16,100	14,431	15,825	17,883
Czechoslovakia (Former)	±0.790	-	-	-	-	-	3,950	-	-	-	-	-
Denmark	*0.610	*0.597	*0.650	*0.533	*0.591	*0.661	37,549	37,174	42,499	39,448	41,313	47,554
Ecuador	0.095	0.103	0.102	0.095	0.103	0.134	6,525	7,548	7,131	7,258	8,508	12,279
Egypt	*0.847	*0.581	*0.531	*0.567	*0.590	*0.661	3,144	2,351	1,974	2,032	2,113	2,371
El Salvador	-	0.068	-	0.044	0.058	0.073	-	-	-	5,935	7,405	8,495
Ethiopia	0.065	0.056	0.036	0.039	0.052	0.042	2,274	1,837	1,209	1,323	1,694	1,355
Finland	±0.386	±0.296	±0.292	±0.242	±0.320	±0.303	35,411	32,146	35,556	33,166	45,083	51,346
FYR Macedonia	±0.164	±0.209	±0.134	±0.095	±0.053	±0.045	6,019	8,405	5,760	4,379	3,262	3,624
France	±7.666	±6.907	±7.266	±5.993	±6.466	±7.263	36,524	35,205	40,013	35,569	40,490	47,073
Germany (Western Part)	12	12	12	11	-	-	51,792	52,979	59,004	-	-	-
Ghana	-	-	-	0.028	0.024	0.026	-	-	-	2,965	2,499	2,550
Greece	*1.109	*1.030	*1.085	*1.011	*1.021	*1.105	22,100	22,959	26,199	24,942	25,170	27,715
Honduras	*0.016	*0.013	*0.015	*0.012	*0.011	*0.014	2,507	1,777	1,631	1,147	955	1,257
Hungary	±0.299	±0.234	±0.192	±0.206	±0.198	±0.177	3,940	3,840	4,006	4,199	4,126	4,122
Iceland	0.021	0.017	±0.020	±0.015	-	-	26,268	29,220	44,451	34,423	-	-
India	*3.264	*2.338	*2.282	*2.808	*2.906	*3.050	2,326	1,731	1,633	1,965	2,015	2,119
Indonesia	*1.306	*1.497	*2.016	*2.033	*3.728	*3.474	3,150	3,135	3,710	3,502	6,115	5,569
Iran (Islamic Republic of)	1.428	1.276	1.085	0.932	-	-	10,763	9,623	7,966	7,057	-	-
Iraq	-	*0.148	*0.249	-	-	-	-	7,986	14,640	-	-	-
Ireland	*0.349	*0.347	*0.397	*0.356	*0.382	*0.423	29,542	30,397	33,607	29,553	31,136	34,283
Israel	0.404	0.431	0.404	0.361	0.393	0.471	28,633	30,558	28,668	24,400	25,512	29,977
Italy	±10	±10	±11	±8.833	±9.617	-	46,088	45,135	49,372	42,126	45,715	-
Japan	±27	±30	±31	±32	±32	±34	42,659	48,109	52,414	56,329	62,223	68,401
Jordan	0.020	0.021	0.017	0.017	0.027	0.029	14,278	12,719	8,584	7,862	9,415	8,996
Kenya	*0.055	*0.052	*0.045	*0.030	*0.040	*0.037	2,190	2,142	1,842	1,193	1,612	1,464
Korea, Republic of	6.833	8.406	9.267	9.638	10	11	19,578	25,238	29,340	31,330	34,674	42,259
Kuwait	0.016	0.010	0.025	0.024	0.021	0.033	12,760	10,680	15,715	18,181	14,590	22,046
Latvia	-	-	-	*0.033	*0.041	*0.040	-	-	-	1,796	2,777	3,004
Macau	0.092	0.097	0.106	0.087	0.084	0.073	9,081	9,202	11,159	9,769	10,211	10,735
Malawi	*0.007	*0.013	*0.050	*0.041	*0.022	-	1,369	2,230	8,523	7,976	3,854	-
Malaysia	*0.297	*0.392	*0.430	*0.499	*0.608	*0.719	7,866	8,748	10,043	11,203	13,943	16,061
Malta	0.017	0.015	0.016	0.013	0.013	-	17,695	20,232	21,884	17,593	19,999	-
Mauritius	0.028	0.032	0.019	0.023	0.038	0.032	5,732	5,574	3,294	4,716	7,328	6,421
Mexico	0.766	0.792	0.762	0.711	0.644	0.445	11,124	12,437	12,657	14,525	14,685	10,319
Mongolia	0.066	0.027	0.039	0.012	0.022	0.055	8,922	3,911	6,351	2,221	3,400	11,078
Morocco	±0.315	±0.350	±0.397	±0.379	±0.446	±0.488	5,725	5,006	5,425	5,257	6,263	6,929

Continued.

Value Added and Value Added per Employee

- Continued -

Country	Value Added ($ bil.)						Value Added per Employee ($)					
	1990	1991	1992	1993	1994	1995	1990	1991	1992	1993	1994	1995
Nepal	0.054	0.109	-	0.059	0.136	-	1,307	1,403	-	898	1,742	-
Netherlands	1.002	0.974	1.014	0.951	0.958	1.106	47,211	46,141	49,673	48,671	51,105	59,796
New Zealand	0.232	-	-	-	-	-	22,374	-	-	-	-	-
Norway	±0.191	±0.193	±0.211	±0.150	±0.161	±0.185	33,274	36,792	41,764	32,516	34,274	38,816
Pakistan	1.100	1.283	1.432	-	-	-	4,890	5,164	5,511	-	-	-
Panama	0.006	0.006	0.006	0.006	0.006	0.006	9,733	11,504	11,670	11,746	11,884	11,956
Peru	0.647	0.474	0.453	-	-	-	16,531	13,523	14,143	-	-	-
Philippines	0.393	0.344	0.371	0.374	-	-	3,974	3,991	4,690	5,618	-	-
Poland	±1.222	±0.878	±0.798	±0.707	±0.848	±0.989	4,272	3,993	4,588	4,417	4,564	5,385
Portugal	1.654	1.578	1.663	1.500	1.649	1.925	10,451	10,436	11,850	11,197	12,922	15,062
Puerto Rico	±0.042	±0.042	±0.045	±0.046	±0.045	±0.050	12,705	13,137	14,056	13,105	14,259	16,340
Qatar	0.000	-	-	0.002	0.002	-	6,868	-	-	7,596	10,750	-
Romania	±1.449	±0.827	±0.368	±0.395	-	-	3,497	2,101	1,124	1,446	-	-
Russian Federation	-	-	-	±1.561	±1.289	±1.215	-	-	-	1,902	1,842	2,007
Saint Lucia	-	±0.001	±0.001	±0.001	±0.001	±0.002	-	-	-	-	-	-
Senegal	0.013	0.008	0.009	0.004	0.016	-	2,984	4,350	4,672	2,203	6,722	-
Sierra Leone	-	-	-	0.000	-	-	-	-	-	1,089	-	-
Singapore	*0.072	*0.073	*0.074	*0.072	*0.072	*0.059	21,024	21,753	22,570	23,283	27,532	32,739
Slovakia	-	-	-	±0.112	±0.107	-	-	-	-	3,435	3,537	-
Slovenia	-	-	-	0.271	0.176	0.200	-	-	-	-	-	-
South Africa	*0.850	*0.831	*0.855	*0.843	*0.882	*0.963	8,766	9,235	10,303	12,766	12,965	14,054
Spain	*3.314	*3.233	*3.104	±2.457	±2.613	±2.944	30,222	31,196	34,152	26,583	27,413	32,019
Sri Lanka	0.082	0.088	0.109	0.105	-	-	1,786	2,020	2,573	2,615	-	-
Sweden	*0.620	*0.472	*0.506	*0.338	*0.367	*0.426	48,440	40,569	48,388	36,950	41,338	46,866
Switzerland	1.408	1.262	1.249	1.068	1.317	1.518	45,422	43,511	46,973	43,965	55,353	67,758
Thailand	7.065	2.013	-	-	-	-	17,181	7,282	-	-	-	-
Trinidad and Tobago	±0.005	±0.006	±0.005	±0.002	±0.002	-	12,388	18,835	16,044	10,005	8,439	-
Tunisia	0.196	0.202	0.264	0.261	0.289	0.334	-	-	-	13,199	19,041	18,672
Turkey	3.222	3.310	4.012	4.241	3.892	4.889	16,274	19,265	21,396	22,432	20,911	25,711
United Kingdom	*7.036	*6.830	*7.145	*6.503	*6.871	*7.256	31,409	32,837	36,085	32,678	35,788	38,453
United Republic of Tanzania	*0.002	*-0.035	-	-	-	-	66	-1,157	-	-	-	-
United States of America	*35	*35	*40	*41	*44	*44	42,171	44,703	47,585	49,630	51,583	52,214
Uruguay	0.235	0.277	0.272	0.245	0.253	0.233	11,749	15,161	16,634	17,981	18,613	16,189
Venezuela	0.291	0.275	0.286	0.192	0.208	0.622	11,602	11,471	11,935	9,378	10,653	31,808
Yugoslavia	±0.077	±0.084	±0.214	-	±0.271	±0.143	-	895	2,593	-	3,637	1,943
Zambia	0.062	-	-	-	0.023	-	7,432	-	-	-	3,972	-
Zimbabwe	*0.255	*0.220	*0.158	*0.132	*0.116	*0.119	10,958	8,698	6,657	6,366	6,655	7,931

Capital Investment

Country	Gross Fixed Capital Formation ($ mil.)						Per Establishment ($ mil)					
	1990	1991	1992	1993	1994	1995	1990	1991	1992	1993	1994	1995
Argentina	-	-	-	133	-	-	-	-	-	0.060	-	-
Australia	109	-	-	-	-	-	0.119	-	-	-	-	-
Austria	270	282	200	142	153	-	0.603	0.654	0.487	0.354	0.394	-
Bangladesh	44	32	50	-	-	-	0.004	0.003	0.004	-	-	-
Belgium	567	494	490	429	540	494	0.468	0.422	0.430	0.279	0.366	-
Bolivia	2.468	3.501	4.337	5.022	6.464	-	0.021	0.030	0.036	0.043	0.058	-
Bosnia & Herzegovina	7.732	-	-	-	-	-	0.112	-	-	-	-	-
Brazil	630	-	333	957	-	-	0.369	-	0.230	0.690	-	-
Bulgaria	84	32	366	82	25	12	0.561	0.207	2.286	0.128	0.035	0.015
Cameroon	-	8.174	10	14	14	17	-	1.362	1.498	2.744	1.957	2.766
Canada	297	-	-	-	-	-	0.222	-	-	-	-	-
Central African Republic	-2.211	-	-	-	-	-	-	-	-	-	-	-
Chile	28	32	34	31	29	-	0.205	0.246	0.253	0.227	0.224	-
China (Hong Kong SAR)	212	233	227	139	113	-	0.038	0.050	0.049	0.039	0.033	-
Colombia	94	85	170	114	-31	-	0.189	0.175	0.340	0.237	-0.066	-
Croatia	25	22	5.500	14	1.483	1.545	0.112	0.078	0.016	0.033	0.003	0.003

Continued.

Depending on the table, * means *Enterprises, Engaged,* or *Factor Values;* ± means *Basis Unspecified;* § means *shown in millions.* For additional notes and sources, see Appendix I.

Capital Investment

- Continued -

Country	Gross Fixed Capital Formation ($ mil.)						Per Establishment ($ mil)					
	1990	1991	1992	1993	1994	1995	1990	1991	1992	1993	1994	1995
Cyprus	7.611	8.477	6.149	3.254	2.741	-0.889	0.034	0.037	0.027	0.014	0.011	-0.003
Czechoslovakia (Former)	202	-	-	-	-	-	2.266	-	-	-	-	-
Denmark	60	52	-	-	-	-	0.058	0.047	-	-	-	-
Ecuador	55	65	56	18	32	35	0.367	0.426	0.347	0.120	0.225	0.253
Egypt	92	84	360	-8.008	-	-	0.105	0.100	0.384	-0.009	-	-
El Salvador	-	14	-	7.415	14	16	-	0.326	-	0.145	0.354	0.301
Ethiopia	37	2.551	8.879	19	6.983	1.099	1.407	0.102	0.386	0.829	0.241	0.038
Finland	75	29	18	31	27	55	0.368	0.140	0.096	0.183	0.170	0.369
FYR Macedonia	6.679	8.085	0.922	1.513	0.551	2.045	0.230	0.168	0.012	0.021	0.006	0.020
Germany (Western Part)	1,406	1,306	1,150	-	-	-	0.786	0.753	0.688	-	-	-
Greece	183	117	156	-	-	-	0.217	0.141	0.187	-	-	-
Hungary	27	29	26	37	-	-	0.113	0.079	0.045	0.051	-	-
India	1,043	926	1,112	1,311	-	-	0.081	0.072	0.075	0.086	-	-
Indonesia	1,010	1,136	825	700	784	746	0.553	0.633	0.458	0.359	0.388	0.333
Iran (Islamic Republic of)	145	193	238	192	-	-	0.097	0.308	0.331	0.278	-	-
Ireland	59	24	-	-	-	-	0.257	0.109	-	-	-	-
Israel	72	63	94	74	83	-	0.158	0.153	0.232	0.189	0.181	-
Italy	2,400	1,752	1,408	1,095	1,282	-	0.754	0.558	0.406	0.312	0.365	-
Japan	2,321	2,977	2,527	2,203	1,722	-	0.058	0.077	0.069	0.062	0.055	-
Jordan	2.071	0.706	1.471	2.909	3.476	2.151	0.014	0.005	0.007	0.014	0.011	0.006
Korea, Republic of	1,896	3,104	2,983	2,402	2,583	2,880	0.246	0.386	0.367	0.249	0.262	0.295
Kuwait	0.366	0.352	0.522	1.013	0.339	-	0.002	0.002	0.002	0.004	0.001	-
Latvia	0.293	0.185	7.686	0.314	1.088	2.937	0.006	0.002	0.067	0.002	0.013	0.021
Macau	7.537	9.287	2.530	32	5.277	4.870	0.026	0.035	0.011	0.141	0.029	0.030
Malawi	1.649	2.105	1.859	0.522	0.927	-	0.275	0.351	0.310	0.087	0.155	-
Malaysia	128	173	368	259	332	-	0.524	0.639	1.369	0.792	1.045	-
Malta	2.817	12	1.912	0.594	0.540	-	0.113	0.466	0.083	0.026	0.025	-
Mexico	65	73	-	-	-	-	0.269	0.301	-	-	-	-
Mongolia	193	112	30	5.701	6.171	-	15	8.625	2.329	0.335	0.213	-
Morocco	115	159	131	84	77	121	0.173	0.216	0.173	0.109	0.101	0.164
Nepal	39	-	-	-	-	-	0.091	-	-	-	-	-
Netherlands	159	154	176	135	-	-	0.734	0.691	0.799	0.606		
New Zealand	13	-	-	-	-	-	0.018	-	-	-	-	-
Niger	-	-0.450	-0.004	-	-0.004	-	-	-	-	-	-	-
Norway	1.118	10	13	11	16	-	0.006	0.053	0.101	0.097	0.137	-
Pakistan	-	291	-	-	-	-	-	0.197	-	-	-	-
Panama	0.925	0.623	-	-	-	-	0.077	0.104	-	-	-	-
Peru	87	60	47	-	-	-	0.073	0.047	0.037	-	-	-
Philippines	65	47	70	45	-	-	0.035	0.026	0.119	0.078	-	-
Poland	124	100	79	80	-	-	0.300	0.240	0.200	0.223	-	-
Portugal	373	408	368	193	351	-	0.087	0.102	0.091	0.045	0.089	-
Romania	97	86	56	58	46	-	0.438	0.092	0.032	0.027	0.018	-
Senegal	6.369	-	-	-	-	-	0.531	-	-	-	-	-
Singapore	14	12	8.359	5.336	7.404	-	0.217	0.184	0.131	0.082	0.125	-
Spain	277	322	354	262	410	392	0.055	0.067	0.079	-	-	-
Sri Lanka	24	24	28	25	-	-	0.047	0.059	0.073	0.061	-	-
Thailand	2,351	613	-	-	-	-	2.347	0.418	-	-	-	-
Trinidad and Tobago	4.852	33	0.518	0.150	0.219	-	0.221	1.856	0.035	0.009	0.012	-
Tunisia	82	92	75	48	40	44	-	-	-	0.091	0.088	0.079
Turkey	715	460	905	606	553	-	0.961	0.626	0.632	0.451	0.429	-
Ukraine	168	92	67	11	19	1.328	0.234	0.096	0.066	0.016	0.029	0.002
United Kingdom	500	473	541	-	547	-	0.080	0.079	0.093	-	0.091	-
United Republic of Tanzania	162	230	-	-	-	-	1.863	2.617	-	-	-	-
United States of America	2,710	2,490	2,657	2,950	3,544	3,544	-	-	0.136	-	-	-
Uruguay	18	19	29	22	-	-	-	-	-	-	-	-
Venezuela	21	54	-1.784	62	17	239	0.095	0.253	-0.008	0.352	0.092	1.205
Yugoslavia	4.000	2.914	3.366	-	17	2.797	0.013	0.006	0.003	-	0.013	0.002
Zambia	13	-	-	-	-	-	0.410	-	-	-	-	-
Zimbabwe	85	108	37	45	29	-	1.709	2.109	0.736	0.890	0.612	-

Indexes of Industrial Production

Country	Index of Industrial Production (1990=100)						Index of Employment (1990=100)					
	1990	1991	1992	1993	1994	1995	1990	1991	1992	1993	1994	1995
Albania	100	-	-	-	-	-	-	-	-	-	-	-
Algeria	100	97	101	95	80	69	-	-	-	-	-	-
Angola	100	-	-	-	-	-	-	-	-	-	-	-
Argentina	100	117	-	-	-	-	100	-	-	66	65	-
Armenia.	100	-	-	-	-	-	100	87	75	64	66	57
Australia	100	104	93	78	-	-	100	91	79	-	-	-
Austria	100	100	98	86	86	85	100	94	89	77	76	74
Azerbaijan	-	-	-	-	-	-	100	98	87	71	74	-
Bahamas	100	-	-	-	-	-	100	57	-	-	-	-
Bahrain	100	-	-	-	-	-	-	-	-	-	-	-
Bangladesh	100	93	93	97	94	87	100	106	114	-	-	-
Barbados	100	-	-	-	-	-	-	-	-	-	-	-
Belgium	100	88	94	85	85	85	100	96	91	-	-	-
Benin	100	-	-	-	-	-	-	-	-	-	-	-
Bolivia	100	107	120	127	135	145	100	98	106	122	120	119
Bosnia & Herzegovina	-	-	-	-	-	-	100	86	-	-	1	-
Brazil	100	103	98	98	101	96	100	-	82	87	-	-
Bulgaria.	100	69	60	50	52	48	100	78	66	58	57	54
Burkina-Faso	100	-	-	-	-	-	-	-	-	-	-	-
Burundi	100	85	78	-	-	-	100	100	-	-	-	-
Cambodia	100	-	-	-	-	-	100	81	-	-	-	-
Cameroon	100	-	-	-	-	-	100	84	82	64	68	67
Canada.	100	96	95	99	111	110	100	96	86	83	88	88
Chile.	100	110	105	101	93	97	100	99	99	94	85	89
China	100	-	-	-	-	-	100	101	100	92	93	90
China (Hong Kong SAR) . . .	100	104	109	101	100	96	100	88	79	73	60	-
China (Taiwan Province) . . .	100	108	108	102	108	102	100	94	93	90	92	88
Colombia	100	101	107	106	107	113	100	102	116	114	113	109
Costa Rica	100	83	75	66	58	50	100	91	102	99	-	-
Cote d'Ivoire	100	95	92	96	98	129	-	-	-	-	-	-
Croatia	100	-	-	-	-	-	100	79	68	67	64	61
Cuba	100	-	-	-	-	-	-	-	-	-	-	-
Cyprus	100	94	108	99	95	82	100	106	110	106	91	100
Czechoslovakia (Former) . . .	100	65	56	-	-	-	100	79	-	-	-	-
Czech Republic	100	-	-	-	-	-	100	85	76	70	-	-
Dem. Rep. of the Congo . . .	100	-	-	-	-	-	-	-	-	-	-	-
Denmark	100	100	99	86	88	85	100	99	94	83	88	86
Dominican Republic.	100	-	-	-	-	-	-	-	-	-	-	-
Ecuador	100	100	89	86	-	-	100	94	99	91	84	75
Egypt	100	108	98	92	92	96	100	92	100	104	104	103
El Salvador.	100	105	-	-	-	-	-	-	-	-	-	-
Ethiopia and Eritrea	100	-	-	-	-	-	-	-	-	-	-	-
Ethiopia.	-	-	-	-	-	-	100	107	106	104	107	109
Finland	100	82	84	86	92	91	100	84	75	67	65	54
FYR Macedonia	100	-	-	-	-	-	100	92	86	80	59	46
France	100	95	93	86	90	89	100	93	87	80	76	74
Gabon	100	-	-	-	-	-	-	-	-	-	-	-
Gambia	100	-	-	-	-	-	-	-	-	-	-	-
Germany (Eastern Part) . . .	100	-	-	-	-	-	-	-	-	-	-	-
Germany (Western Part)	100	99	90	81	79	75	100	98	92	-	-	-
Ghana	100	-	-	-	-	-	-	-	-	-	-	-
Greece	100	91	83	78	78	74	100	89	83	81	81	79
Guatemala	100	-	-	-	-	-	-	-	-	-	-	-
Guyana	100	100	100	100	100	100	-	-	-	-	-	-
Haiti	100	78	-	-	-	-	-	-	-	-	-	-
Honduras	100	141	168	195	243	330	100	119	149	164	179	177
Hungary	100	69	51	48	47	42	100	80	63	64	63	57
Iceland	100	78	-	-	-	-	100	73	55	55	59	-
India.	100	108	113	126	122	123	100	96	100	102	103	103
Indonesia	100	106	115	114	121	129	100	115	131	140	147	150

Continued.

Depending on the table, * means *Enterprises, Engaged,* or *Factor Values*; ± means *Basis Unspecified*; § means *shown in millions*. For additional notes and sources, see Appendix I.

Indexes of Industrial Production
- Continued -

Country	Index of Industrial Production (1990=100)						Index of Employment (1990=100)					
	1990	1991	1992	1993	1994	1995	1990	1991	1992	1993	1994	1995
Iran (Islamic Republic of) . . .	100	114	120	123	134	133	100	100	103	100	-	-
Iraq	100	96	93	-	-	-	-	-	-	-	-	-
Ireland	100	100	105	108	111	112	100	97	100	102	104	105
Israel	100	108	111	111	120	128	100	100	100	105	109	111
Italy	100	101	100	95	-	-	100	104	96	94	94	-
Jamaica	100	-	-	-	-	-	100	108	108	89	72	-
Japan	100	98	95	85	82	77	100	98	94	88	82	77
Jordan	100	87	75	82	81	68	100	121	142	160	213	239
Kenya	100	108	108	125	92	63	100	97	98	100	99	99
Korea, Republic of	100	98	95	87	86	84	100	95	91	88	85	78
Kuwait	100	-	-	-	-	-	100	79	128	109	117	121
Kyrgyzstan	100	-	-	-	-	-	100	99	111	106	84	-
Lao P.D.R.	100	-	-	-	-	-	-	-	-	-	-	-
Latvia	100	-	-	-	-	-	100	91	80	49	40	36
Lebanon	100	-	-	-	-	-	-	-	-	-	-	-
Libyan Arab Jamahiriya . . .	100	-	-	-	-	-	-	-	-	-	-	-
Lithuania	100	-	-	-	-	-	-	-	-	-	-	-
Macau	100	-	-	-	-	-	100	104	94	88	81	67
Madagascar	100	86	55	-	-	-	-	-	-	-	-	-
Malawi	100	149	129	103	93	74	100	112	110	95	106	-
Malaysia	100	106	123	161	192	210	100	119	113	118	115	118
Malta	100	95	92	93	-	-	100	80	75	77	68	-
Mauritius	100	-	-	-	-	-	100	115	119	97	104	99
Mexico	100	96	95	92	90	85	100	92	87	71	64	63
Mongolia	100	84	76	54	38	51	100	93	84	75	86	67
Morocco	100	104	106	103	97	99	100	127	133	131	129	128
Mozambique	100	127	-	-	-	-	100	96	69	67	55	-
Myanmar	100	94	58	66	71	85	-	-	-	-	-	-
Nepal	100	-	-	-	-	-	100	187	-	158	189	-
Netherlands	100	95	88	85	83	84	100	99	96	92	88	87
New Zealand	100	101	99	-	-	-	100	91	92	95	-	-
Nicaragua	100	-	-	-	-	-	-	-	-	-	-	-
Niger	100	-	-	-	-	-	-	-	-	-	-	-
Nigeria	100	-	-	-	-	-	-	-	-	-	-	-
Norway	100	100	96	93	103	104	100	91	88	80	82	83
Pakistan	100	105	107	109	113	-	100	111	116	-	-	-
Panama	100	102	90	102	101	93	100	84	75	84	83	77
Papua New Guinea	100	-	-	-	-	-	-	-	-	-	-	-
Paraguay	100	103	76	76	69	77	-	-	-	-	-	-
Peru	100	97	90	86	120	132	100	90	82	-	-	-
Philippines	100	112	110	117	112	124	100	87	80	67	-	-
Poland	100	81	79	86	99	98	100	77	61	56	65	64
Portugal	100	108	105	97	93	95	100	96	89	85	81	81
Puerto Rico	100	-	-	-	-	-	100	98	98	107	96	93
Qatar	100	-	-	-	-	-	100	98	478	542	575	-
Republic of Moldova	100	-	-	-	-	-	100	90	84	37	43	34
Romania	100	87	61	59	59	59	100	95	79	66	53	-
Russian Federation	100	-	-	-	-	-	-	-	-	-	-	-
Saudi Arabia	100	-	-	-	-	-	-	-	-	-	-	-
Senegal	100	91	92	81	84	70	100	43	44	42	56	-
Sierra Leone	100	-	-	-	-	-	-	-	-	-	-	-
Singapore	100	108	94	83	80	58	100	98	96	91	77	53
Slovakia	100	-	-	-	-	-	-	-	-	-	-	-
Slovenia	100	81	70	70	70	80	-	-	-	-	-	-
Somalia	100	-	-	-	-	-	-	-	-	-	-	-
South Africa	100	97	92	95	100	101	100	93	86	68	70	71
Spain	100	95	89	81	91	93	100	95	83	84	87	84
Sri Lanka	100	109	116	303	331	241	100	95	92	87	-	-
Sudan	100	-	-	-	-	-	-	-	-	-	-	-
Sweden	100	84	78	73	78	83	100	91	82	71	69	71

Continued.

Indexes of Industrial Production
- Continued -

Country	Index of Industrial Production (1990=100)						Index of Employment (1990=100)					
	1990	1991	1992	1993	1994	1995	1990	1991	1992	1993	1994	1995
Switzerland	100	97	94	90	92	96	100	94	86	78	77	72
Syrian Arab Republic	100	93	107	108	116	113	-	-	-	-	-	-
Thailand	100	92	89	-	-	-	100	67	-	-	-	-
Togo.	100	-	-	-	-	-	-	-	-	-	-	-
Tonga	100	-	-	-	-	-	-	-	-	-	-	-
Trinidad and Tobago	100	91	93	59	52	51	100	80	77	47	51	-
Tunisia	100	87	90	95	97	115	-	-	-	-	-	-
Turkey	100	91	96	98	100	109	100	87	95	95	94	96
Uganda	100	92	93	79	62	57	-	-	-	-	-	-
Ukraine	100	-	-	-	-	-	100	85	89	79	69	60
United Arab Emirates	100	-	-	-	-	-	-	-	-	-	-	-
United Kingdom	100	90	89	89	89	88	100	93	88	89	86	84
United Republic of Tanzania . .	100	89	93	93	79	-	100	96	-	-	-	-
USSR (Former)	100	-	-	-	-	-	100	-	-	-	-	-
United States of America . . .	100	99	107	113	119	118	100	95	100	100	103	102
Uruguay	100	108	109	99	99	84	100	91	82	68	68	72
Venezuela	100	-	-	-	-	-	100	96	96	82	78	78
Yemen	100	-	-	-	-	-	-	-	-	-	-	-
Yugoslavia	100	-	-	-	-	-	-	-	-	-	-	-
Yugoslavia (Former)	100	71	-	-	-	-	-	-	-	-	-	-
Zambia	100	84	81	57	54	48	100	-	-	-	70	-
Zimbabwe	100	104	82	89	95	37	100	109	102	89	75	64

Depending on the table, * means *Enterprises, Engaged,* or *Factor Values;* ± means *Basis Unspecified;* § means *shown in millions.* For additional notes and sources, see Appendix I.

Representative Companies in Sector

Name	Address	Tele-phone	Fax	Employ-ment	Y
Cydsa SA y Subsidiarias	Ruiz Cortines, Monterrey 2333, Mexico	83 359090	83 355541	9,210	91
J.P. Stevens and Company Inc.	1185 Ave. of the Amer., New York, New York 10020, U.S.A.	212-930-2000	212-930-3837	8,000	
Avondale Inc.	506 S. Broad St., Monroe, Georgia 30655, U.S.A.	770-267-2226		7,500	97
Delta Woodside Industries Inc.	233 N. Main St., 200, Greenville, South Carolina 29601, U.S.A.	803-232-8301	803-232-6164	7,500	95
Interface Inc. (Atlanta, Georgia)	2859 Paces Ferry Rd., 2000, Atlanta, Georgia 30339, U.S.A.	770-437-6800	770-437-6809	7,300	97
Amoco Fabrics and Fibers Co.	900 Circle 75 Pkwy., 550, Atlanta, Georgia 30339, U.S.A.	770-956-9025	770-618-4023	7,000	97
Hadtex Indosyntec PT	Jalan Garuda No. 153/74, Bandung, Indonesia	22634123	22 631643	7,000	97
Unifi Inc.	7201 W. Friendly Ave., Greensboro, North Carolina 27410, U.S.A.	910-294-4410	910-316-5422	7,000	97
Tultex Corp.	PO Box 5191, Martinsville, Virginia 24115, U.S.A.	540-632-2961	540-632-8000	6,618	96
Guilford Mills Inc.	PO Box 26969, Greensboro, North Carolina 27419-6969, U.S.A.	910-316-4000		6,571	97
Scapa Group PLC	93 Preston New Rd., Blackburn BB2 6AY, United Kingdom	254 580123		6,536	94
Stoddard Sekers International PLC	Glenpatrick Rd., Elderslie PA5 9UJ, United Kingdom	505 27100	505 331330	6,513	94
Associated Furniture Co. Ltd.	PO Box 98800, Sloane Park 2152, Republic of South Africa	11 7066004	11 4635097	6,500	96
Teka Tecelagem Kuehnrich SA	Rua Paulo Kuehnrich 68, Blumenau 89052-900, Brazil	47 3215000	47 3215050	6,500	96
Cone Mills Corp.	PO26540, Greensboro, North Carolina 27415-6540, U.S.A.	910-379-6220	910-379-6287	6,100	97
Ozeta Trencin	9 Maja 9, Trencin 911 34, Slovakia	831 262800	831 23240	6,045	92
Ithaca Industries Inc.	PO Box 620, Wilkesboro, North Carolina 28697, U.S.A.	910-667-5231	910-667-4052	6,000	97
Albany International Corp.	PO Box 1907, Albany, New York 12201-1907, U.S.A.	518-445-2200	518-445-2265	5,881	97
Standard Industries Ltd.	Nariman Point, Bombay 400 021, India	22 258544		5,574	94
Frame Textile Corp. Ltd.	PO Box 32002, Duran 4060, Republic of South Africa	31 425140	31 424822	5,500	97
Delta Galil Industries Ltd.	2 Kaufman St., Tel Aviv 68012, Israel	3 5193636	3 5193705	5,460	97
Galey and Lord Inc.	980 Ave. of the Amer., New York, New York 10018, U.S.A.	212-465-3000		5,451	97
Bibb Co.	PO Box 4207, Macon, Georgia 31208, U.S.A.	912-752-6700	912-752-7332	5,100	95
Union Textiles Industries Corp. Ltd.	Phtakhanong, Bangkok 10250, Thailand	3231085	3115668	5,100	94
Hexcel Corp.	5794 W. Las Positas Blvd., Pleasanton, California 94588-8781, U.S.A.	510-847-9500	510-734-8276	5,013	96
Raymond Woollen Mills Ltd.	Bellard Estate, Bombay 400 038, India	22 2618321	22 2622010	5,010	94
Dan River Inc.	PO Box 261, Danville, Virginia 24543, U.S.A.	804-799-7000	804-799-7276	5,000	97
Gul Ahmed Textile Mills Ltd.	29 W. Wharf Rd., Karachi 74000, Pakistan	21 202704	21 2415936	5,000	
Mount Vernon Mills Inc.	PO Box 3478, Greenville, South Carolina 29602, U.S.A.	864-233-4151	864-370-2315	5,000	97
Textile Manufacturing Co. Djaya PT	Mulia Ctr. 10th Flr. Suite 1008, Jakarta 12940, Indonesia	215229390	21 358936	4,902	97
Jiyajeerao Cotton Mills Ltd.	Birla Nagar, Gwalior 474 004, India			4,850	97
Hindustan Spinning & Weaving Mills Ltd.	Fort, Bombay 400 001, India	255446		4,810	97
Dexter Corp.	1 Elm St., Windsor Locks, Connecticut 06096, U.S.A.	860-292-7675	860-292-7673	4,800	97
Sam Yang Co., Ltd.	Chongno-Gu, Seoul, Republic of Korea	2 7407114	2 7437720	4,800	97
Dainong Corp.	Mapo-Gu, Seoul, Republic of Korea	2 32703500	2 7180033	4,730	97
Laura Ashley Holdings PLC	150 Bath Rd., Maidenhead SL6 4YS, United Kingdom	628 39151	628 71122	4,697	94
David Whitehead Textiles Ltd.	Union Ave., Harare, Zimbabwe	14 730744	14 728120	4,644	92
Formosa Taffeta Co., Ltd.	317 Shu Liou Rd., Tou-Liu 64029, Taiwan	5 5573966	5 5573969	4,626	91
Consoltex Inc.	8555 Rte Transcanadienne, Saint-Laurent, Quebec H4S 1Z6, Canada	514-333-8800	514-335-7020	4,600	98
Dixie Group Inc.	PO Box 751, Chattanooga, Tennessee 37401, U.S.A.	423-698-2501	423-493-7442	4,600	97
Kolon Industries Inc.	Chung-Gu, Seoul, Republic of Korea	2 3117114	2 3118912	4,600	97
Queen Carpet Corp.	PO Box 1527, Dalton, Georgia 30722-1527, U.S.A.	706-277-1900	706-277-3143	4,500	96
Societe Generale des Industries Textiles SA	Ben Arous, Tunis, Tunisia	1 381133		4,500	93
Tong Yang Nylon Co., Ltd.	Mapo-gu, Seoul, Republic of Korea	2 7077114	2 7070130	4,480	97
Indorama Synthetics, PT	PO Box 3375, Jakarta 12930, Indonesia	215261555	21 5731122	4,320	97
Indocement Tunggal Prakarsabogasari Flour Mills Div. PT	Jl. Jendral Sudirman Kav 70-71, Jakarta 12910, Indonesia	212512087	21 5712620	4,085	97
Garware Nylons Ltd.	Nariman Point, Bombay 400 021, India	22 2022312	22 2029377	4,000	94
Romatex Ltd.	PO Box 3739, Tygerpark 7536, Republic of South Africa	21 9488870	21 9488993	4,000	95
Dyersburg Corp.	PO Box 767, Dyersburg, Tennessee 38025-0767, U.S.A.	901-285-2323		3,850	97
Hanes Hosiery Inc.	PO Box 1413, Winston-Salem, North Carolina 27102, U.S.A.	910-768-9540	919-744-8594	3,800	97
Readicut International PLC	Clifton Mills, Brighouse HD6 4ET, United Kingdom	484 721223	484 716135	3,725	94
Sunkyong Industries Ltd.	Chung-gu, Seoul, Republic of Korea	2 2733131	2 2739221	3,679	97

Continued

Representative Companies in Sector
- Continued -

Name	Address	Tele-phone	Fax	Employ-ment	Y
Knoll Inc.	1235 Water St., East Greenville, Pennsylvania 18041, U.S.A.	215-679-7991		3,550	96
Thai Durable Textile Co. Ltd.	Wangburapha, Bangkok 10200, Thailand	2330161	2249770	3,503	94
Ade Textile Industries PT	D/H Jin Tanimbar No. 10, Surakarta 40132, Indonesia	222502742	22 2501134	3,500	97
Aladdin Mills Inc.	PO Box 2208, Dalton, Georgia 30722, U.S.A.	706-277-1100		3,500	93
SBW Ltd.	Kangnam-Gu, Seoul, Republic of Korea	2 5429113	2 5466861	3,500	97
Eclipse Blinds PLC	Inchinnan, Paisley PA4 9RE, United Kingdom	41 8123322	41 8125253	3,434	93
Winner Co. Ltd.	Tsimshatsui East, Kowloon, Hong Kong	7231111	7217537	3,420	
Sherwood Group PLC	Long Eaton, Nottingham NG10 2BQ, United Kingdom	602 461070	602 462720	3,385	93
Maneklal Harilal Mills Ltd.	Saraspur, Ahmedabad 380 018, India	374273		3,300	94
Masland Corp.	PO Box 40, Carlisle, Pennsylvania 17013, U.S.A.	717-249-1866		3,250	95
Karim Jute Mills Ltd.	99 Motijheel C/A, Dhaka 1000, Bangladesh			3,200	
Wellman Inc.	1040 Broad St., Ste. 302, Shrewsbury, New Jersey 07702, U.S.A.	908-542-7300	908-542-9344	3,200	96
Culp Inc.	PO Box 2686, High Point, North Carolina 27261-2686, U.S.A.	910-889-5161	910-887-7089	3,146	97
Fasty	ul. Przedzalniana 8, Bialystok PL-15-679, Poland	85 511070	85 511295	3,100	93
Kitan Consolidated Ltd.	57 Pinhas Rosen St., Tel Aviv 61020, Israel	3 6451515	3 6498605	3,100	97
Zaklady Przemyslu Bawelnianego Bielbawa SA	ul. Piastowska 19, Bielbawa PL-58-260, Poland	74 334341	74 334673	3,100	93
Luckytex Thailand Co., Ltd.	5th Fl., Bubhajit Bldg., Bangkok 10500, Thailand	2383953	2383953	3,028	94
General Directorate of Fine Textiles State Co.	PO Box 5692, Hillah, Iraq			3,000	90
Guilford Mills Inc. Apparel Home Fashions	PO Box 1645, Lumberton, North Carolina 28358, U.S.A.	919-739-8689	919-738-8928	3,000	94
Johnston Industries Inc.	105 13th St., Columbus, Georgia 31901, U.S.A.	706-641-3140		3,000	96
Mannington Mills Inc.	PO Box 30, Salem, New Jersey 08079, U.S.A.	609-935-3000	609-339-5813	3,000	95
Pang Rim Co., Ltd.	Yongdungpo-gu, Seoul, Republic of Korea	2 6302114	2 6755854	3,000	97
Polgat Ltd.	7 Hasivim St., Petah Tikva 49170, Israel	3 9251515	3 9213566	3,000	97
World Carpets Inc.	PO Box 1448, Dalton, Georgia 30722, U.S.A.	706-278-8000	706-278-9454	3,000	96
R.G. Barry Corp.	13405 Yarmouth Rd. NW, Pickerington, Ohio 43147, U.S.A.	614-864-6400		2,900	96
Pharr Yarns Inc.	PO Box 1939, McAdenville, North Carolina 28101-1939, U.S.A.	704-824-3551	704-824-0072	2,800	97
Du Pont Engineering Products SA	Contern, Luxembourg L-2984, Luxembourg	47901	367372	2,754	
Readson Ltd.	96/98 Regent St., Leicester LE1 7DF, United Kingdom	533 549905	533 551382	2,744	87
Melton Medes Ltd.	1 St. Mark's St., Nottingham NG3 1DE, United Kingdom	602 582277	602 585122	2,698	93
Glen Raven Mills Inc.	1831 N. Park Ave., Burlington, North Carolina 27217-1100, U.S.A.	910-227-6211	910-226-8133	2,600	97
Hampshire Group Ltd.	PO Box 2667, Anderson, South Carolina 29622, U.S.A.	803-225-6232	864-225-4421	2,550	95
Synthetic Industries Inc. (Chickamauga, Georgia)	309 Lafayette Rd., Chickamauga, Georgia 30707, U.S.A.	706-375-3121	706-375-6953	2,502	97
Carolina Mills Inc.	PO Box 157, Maiden, North Carolina 28650, U.S.A.	704-428-9911	704-428-2335	2,500	93
Diamond Rug and Carpet Mills Inc.	PO Box 46, Eton, Georgia 30724, U.S.A.	706-695-9446	706-695-3973	2,500	93
Hanil Synthetic Fiber Co., Ltd.	Masan-shi, Kyongnam, Republic of Korea	551 903114	551 903115	2,500	97
Mastercraft Fabrics L.L.C.	PO Box 125, Spindale, North Carolina 28160-0125, U.S.A.	704-286-4811	704-287-8224	2,500	95
Renfro Corp.	PO Box 908, Mount Airy, North Carolina 27030, U.S.A.	910-719-8000		2,500	95
Tong Kook Spinning Co., Ltd.	Chung-Gu, Seoul, Republic of Korea	2 2739850	2 2775574	2,500	97
Union Spinning Mills Pty. Ltd.	Korsten, Port Elizabeth 6014, Republic of South Africa	41 431888	41 435977	2,493	94
Forstmann and Company Inc.	1155 Ave. of the Amer., New York, New York 10036, U.S.A.	212-642-6900	212-642-6992	2,486	97
Sans Fibres Pty. Ltd.	PO Box 2088, Clareinch 7740, Republic of South Africa	21 613130	21 641930	2,470	96
Oneita Industries Inc.	Conifer St. Drawer 24, Andrews, South Carolina 29510, U.S.A.	803-529-5225	803-264-4262	2,450	97
Thomaston Mills Inc.	115 E. Main St., Thomaston, Georgia 30286, U.S.A.	706-647-7131		2,418	97
Budapesti Harisnyagyar	Szentendrei u. 39-53, Budapest, Hungary	1 1889550	1 1689868	2,400	93
Chonbang Co., Ltd.	Chongno-Gu, Seoul, Republic of Korea	2 7226969	2 7201790	2,400	97
Len	Swierczewskiego St. 10, Kamienna Gora PL-58-400, Poland	3011		2,400	93
Ruentex Industrial Co. Ltd.	71 Min Chuan E Rd., Taipei, Taiwan			2,400	90
Shinkong Synthetic Fibers Corp.	8f, 123 Nanking E. Rd., Sec. 2, Taipei 104, Taiwan	02 5071251	02 5072264	2,400	94
ZPL Zyrardow	Marchlewskiego ul. 44, Zyrardow PL-96-300, Poland	300119	4464	2,386	93
Scottish Heritable Trust PLC	18-20 Skeldergate, York YO1 1DH, United Kingdom	904 620021	904 639480	2,332	93
Brasperola Industria e Comercio SA	Rua Guilhermina Guinle, 272, Botafogo 22270-060, Brazil	21 2864422	21 2662688	2,325	96
Allied Textile Cos. PLC	Highburton, Huddersfield HD8 0QJ, United Kingdom	484 604301	484 605740	2,306	93
Horizon Industries Inc. (Calhoun,	PO Box 12069, Calhoun, Georgia 30703, U.S.A.	706-629-7721		2,300	95

Continued

Representative Companies in Sector
- Continued -

Name	Address	Tele-phone	Fax	Employ-ment	Y
Georgia)					
Polymer Group Inc.	4838 Jenkins Ave., North Charleston, South Carolina 29405, U.S.A.	803-566-7293	803-308-0104	2,300	96
Taihan Textile Co., Ltd.	Yongdungpo-Gu, Seoul, Republic of Korea	2 3680114	2 3680461	2,300	97
Hart Holding Company Inc.	1120 Post Rd., Darien, Connecticut 06820, U.S.A.	203-655-6855		2,291	93
Polysindo Eka Perkasa PT	Jl. Hr Rasuna Said Kav X6, No. 8, Jakarta 12940, Indonesia	215229390	21 5229411	2,280	97
Crown Crafts Inc.	1600 RiverEdge Pkwy., 200, Atlanta, Georgia 30328, U.S.A.	770-644-6400		2,200	96
Shuford Industries Inc.	PO Box 2228, Hickory, North Carolina 28603, U.S.A.	704-328-2131	704-328-5792	2,200	95
Standard Commercial Corp.	PO Box 450, Wilson, North Carolina 27894-0450, U.S.A.	919-291-5507	919-237-1109	2,200	96
Swan Mills Ltd.	Nariman Pt., Bombay 400 021, India	22 2027197	22 2027241	2,200	97
Alpargatas SAIC	Avda Regimento de Los Patricios 1142, Buenos Aires 1266, Argentina	1 3030041	1 3031625	2,166	96
Shree Ram Mills Ltd.	Worli, Bombay 400 013, India	22 4932911		2,140	97
Namdinh Silk Weaving Co.	2 Ha Huy Tap St., Nam Dinh City, Vietnam	35 49622	35 49652	2,107	
Magyar Viscosa Rt.	Nyergesujfalu H-2537, Hungary	33 55244	33 55207	2,100	93
Coats Crafts North America	PO Box 24998, Greenville, South Carolina 29616-2498, U.S.A.	803-234-0331		2,044	94
ACS Industries Inc.	160 Hamlet Ave., Woonsocket, Rhode Island 02895, U.S.A.	401-769-4700	401-762-9135	2,000	97
Lancer Industries Inc.	126 E. 56th St., New York, New York 10022, U.S.A.	212-644-8666	212-644-2707	2,000	
Martin Mills Inc.	PO Box 129, St. Martinville, Louisiana 70582, U.S.A.	318-394-6041	318-394-7926	2,000	97
National Spinning Company Inc.	183 Madison Ave., New York, New York 10016, U.S.A.	212-889-3800	212-951-3550	2,000	95
Teijin Indonesia Fiber Corp. PT	5th Fl., Mid Plaza, Jakarta 10220, Indonesia	215706208	21 5706214	2,000	97
Todd & Duncan Ltd.	Lochleven Mills, Kinross KY13 7DH, United Kingdom	577 63521	577 64533	1,992	93
Woo Jung Corp.	Talso-gu, Taegu, Republic of Korea	53 6520855	53 6520856	1,987	97
PAXAR Corp.	105 Corporate Park Dr., White Plains, New York 10604, U.S.A.	914-697-6800	914-697-6893	1,923	95
Brintons Ltd.	Exchange St., Kidderminster DY10 1AG, United Kingdom	562 820000	562 515597	1,908	94
Taris Pamuk Tarim Satis Koop. Birligi	Alsancak, Izmir TR-35249, Turkey	51 217065	51 636555	1,863	92
Chesterfield Manufacturing Corp.	505 Cuthbertson St., Monroe, North Carolina 28110, U.S.A.	704-283-7469	704-283-1212	1,800	95
Industex Eastern Cape Pty. Ltd.	Neave Township, Port Elizabeth 6014, Republic of South Africa	41 431363	41 411558	1,800	97
Intertape Polymer Group Inc.	110 Montee de Liesse, Montreal, Quebec H4T 1N4, Canada	514-731-7591	514-731-5039	1,800	98
Iris Manufacturier de Bas Inc.	6767 Rue Leger, Montreal, Quebec H1G 1L6, Canada	514-328-9334	514-328-9287	1,800	98
Joan Fabrics Corp.	100 Vesper Executive Park, Tyngsboro, Massachusetts 01879, U.S.A.	508-649-5626	508-649-4301	1,800	95
Kyungbang Ltd.	Yongdongpo-Gu, Seoul, Republic of Korea	2 6396000	2 6396249	1,800	97
Markische Faser Aktiengesellschaft Premnitz	Friedrich-Engles-Strasse 1, Premnitz 14727, Germany	3386 40	3386 46001	1,800	92
Unifi Inc. Yadkinville Div.	PO Box 698, Yadkinville, North Carolina 27055, U.S.A.	910-679-8891	910-679-8894	1,800	96
Kabool Ltd.	So-gu, Taegu, Republic of Korea	53 3503114	53 3521989	1,780	93
Ilshin Spinning Co., Ltd.	Yongdungpo-Gu, Seoul, Republic of Korea	2 7865111	2 7865891	1,740	97
Richards PLC	Maberly St., Aberdeen AB9 8DT, United Kingdom	224 630243	224 630260	1,739	94
Complexe Textile de Bujumbura	BP 2899, Bujumbura, Burundi	231900	231750	1,731	91
Badan Tekstil Nasional PT	Jalan Otto Iskandardinata No. 89, Bandung 40111, Indonesia		2256243	1,700	97
Bekaert Textiles NV	Deerlijkseweg 22, Waregem B-8790, Belgium	56 624111	56 624502	1,700	95
Cotton Textile Mills, Ltd.	PO Box 203, Khartoum, Sudan	31414		1,700	91
Ergonbedrijven	Postbus 601, Eindhoven NL-5600 AP, Netherlands	40 595250	40 510660	1,700	91
Fabryka Kabli, Ozarow	ul. Poznanska 55, Ozarow PL-05-850, Poland	2 6286431	2 6286433	1,700	93
Protela Ltda.	Transversal 93 No. 66-18, Bogota, Colombia	1 2761111	1 2760523	1,700	96
Rustom Mills & Industries Ltd.	PO Box 131, Ahmedabad 380 001, India		385851	1,700	91
Tong Yang Polyester Co., Ltd.	Mapo-Gu, Seoul, Republic of Korea	2 7077116	2 7149573	1,680	97
Quaker Fabric Corp.	941 Grinnell St., Fall River, Massachusetts 02721, U.S.A.	508-678-1951	508-678-5979	1,647	97
Bangkok Weaving Mills Ltd.	Bangsoon, Bangkok 10800, Thailand	5855915	5872338	1,638	94
Fabryka Dywanow Kowary SA	Zamkowa ul. 9, Kowary PL-58-530, Poland	75 182032	75 182912	1,628	93
Cotonifico Guilherme Giorgi SA	Av. Guilherme Giorgi 1245, Sao Paulo 03422-001, Brazil	11 9429411	11 2933509	1,620	96
Coronet Industries Inc.	PO Box 1248, Dalton, Georgia 30722-1248, U.S.A.	706-259-4511	706-259-4511	1,600	94
Fab. Industries Inc.	200 Madison Ave., New York, New York 10016, U.S.A.	212-592-2700	212-689-6929	1,600	97
Kyung Nam Wool Textile Ind. Co., Ltd.	Masan-Shi, Kyongnam, Republic of Korea	551 963111	551 963121	1,582	97
Guilford of Maine Inc.	PO Box 179, Guilford, Maine 04443, U.S.A.	207-876-3331		1,550	97
Tah Hsin Industrial Corp.	Taichung Industrial Park, Taichung, Taiwan	4 2522111	4 2525307	1,550	90

Product Tables
Cotton Linters

Unit of Measure: Metric tons.

	1990 Value	%	1991 Value	%	1992 Value	%	1993 Value	%	1994 Value	%	1995 Value	%	1996 Value	%
Total Production	18,977,268	100.0	19,185,671	100.0	19,145,035	100.0	19,107,045	100.0	18,999,983	100.0	19,024,605	100.0	18,957,301	100.0
Regions														
Africa	44,956	0.2	34,111	0.2	21,176	0.1	43,998	0.2	45,879	0.2	47,760	0.3	49,641	0.3
America, North	216,419	1.1	225,733	1.2	216,318	1.1	206,902	1.1	197,487	1.0	189,547	1.0	181,771	1.0
America, South	87,376	0.5	92,296	0.5	89,987	0.5	42,984	0.2	50,196	0.3	58,389	0.3	35,736	0.2
Asia	18,626,735	98.2	18,832,286	98.2	18,816,493	98.3	18,810,460	98.4	18,703,639	98.4	18,726,638	98.4	18,687,790	98.6
Europe	1,782	0.0	1,244	0.0	1,062	0.0	2,701	0.0	2,782	0.0	2,271	0.0	2,362	0.0
Leading Producers														
Indonesia	*18,527,000	97.6	*18,527,000	96.6	*18,527,000	96.8	*18,527,000	97.0	*18,527,000	97.5	*18,527,000	97.4	*18,527,000	97.7
United States of America	209,696	1.1	*220,650	1.2	*212,874	1.1	*205,099	1.1	*197,323	1.0	*189,547	1.0	*181,771	1.0
Azerbaijan	-		171,397	0.9	147,923	0.8	110,825	0.6	73,826	0.4	98,505	0.5	73,857	0.4
Turkey	60,642	0.3	95,735	0.5	101,013	0.5	134,214	0.7	69,962	0.4	70,211	0.4	60,018	0.3
Brazil	73,952	0.4	77,346	0.4	73,510	0.4	39,303	0.2	42,064	0.2	48,844	0.3	22,517	0.1

Wool Yarn, Pure and Mixed (Total)

Unit of Measure: Metric tons.

	1990 Value	%	1991 Value	%	1992 Value	%	1993 Value	%	1994 Value	%	1995 Value	%	1996 Value	%
Total Production	2,314,736	100.0	2,302,088	100.0	2,277,938	100.0	2,125,782	100.0	1,979,350	100.0	1,933,084	100.0	1,915,476	100.0
Regions														
Africa	30,734	1.3	33,280	1.4	32,479	1.4	26,105	1.2	24,782	1.3	24,806	1.3	24,611	1.3
America, North	70,413	3.0	80,847	3.5	80,143	3.5	76,088	3.6	79,739	4.0	74,924	3.9	74,160	3.9
America, South	5,849	0.3	5,612	0.2	5,030	0.2	4,982	0.2	4,724	0.2	4,641	0.2	4,483	0.2
Asia	1,311,226	56.6	1,296,367	56.3	1,286,556	56.5	1,203,598	56.6	1,045,800	52.8	1,068,615	55.3	1,097,899	57.3
Europe	859,039	37.1	849,396	36.9	837,420	36.8	777,426	36.6	779,738	39.4	718,180	37.2	674,105	35.2
Oceania	37,475	1.6	36,587	1.6	36,310	1.6	37,583	1.8	44,568	2.3	41,917	2.2	40,219	2.1
Leading Producers														
Italy	284,706	12.3	306,856	13.3	323,630	14.2	318,925	15.0	335,373	16.9	320,626	16.6	299,920	15.7
China	238,000	10.3	282,800	12.3	350,600	15.4	343,460	16.2	250,600	12.7	325,200	16.8	*367,074	19.2
Russian Federation	239,157	10.3	206,782	9.0	152,924	6.7	116,883	5.5	56,284	2.8	43,989	2.3	29,366	1.5
Japan	105,084	4.5	106,905	4.6	105,512	4.6	84,060	4.0	89,543	4.5	71,654	3.7	64,729	3.4
Belgium	88,236	3.8	87,768	3.8	80,597	3.5	70,542	3.3	*85,044	4.3	*85,019	4.4	*84,994	4.4

Wool Yarn, Mixed

Unit of Measure: Metric tons.

	1990 Value	%	1991 Value	%	1992 Value	%	1993 Value	%	1994 Value	%	1995 Value	%	1996 Value	%
Total Production	682,496	100.0	695,770	100.0	725,640	100.0	689,832	100.0	748,781	100.0	753,673	100.0	748,786	100.0
Regions														
Africa	7,962	1.2	5,775	0.8	4,600	0.6	5,556	0.8	5,149	0.7	4,742	0.6	4,336	0.6
America, North	18,011	2.6	17,814	2.6	17,617	2.4	17,420	2.5	17,223	2.3	17,026	2.3	16,829	2.2
America, South	4,593	0.7	4,297	0.6	3,538	0.5	3,300	0.5	3,022	0.4	2,757	0.4	2,665	0.4
Asia	282,799	41.4	302,644	43.5	331,584	45.7	292,269	42.4	343,340	45.9	358,417	47.6	376,120	50.2
Europe	354,306	51.9	352,945	50.7	357,478	49.3	355,519	51.5	359,493	48.0	344,540	45.7	328,347	43.9
Oceania	14,824	2.2	12,295	1.8	10,823	1.5	15,770	2.3	20,553	2.7	26,190	3.5	20,489	2.7
Leading Producers														
Italy	215,026	31.5	228,663	32.9	236,663	32.6	237,008	34.4	242,648	32.4	235,279	31.2	218,412	29.2
China	208,000	30.5	234,800	33.7	265,400	36.6	231,229	33.5	*279,813	37.4	*297,648	39.5	*315,484	42.1
Poland	52,923	7.8	35,307	5.1	31,976	4.4	32,666	4.7	32,748	4.4	29,308	3.9	33,853	4.5
United Kingdom	28,232	4.1	*31,846	4.6	*31,846	4.4	*31,846	4.6	*31,846	4.3	*31,846	4.2	*31,846	4.3
Japan	26,642	3.9	24,982	3.6	25,444	3.5	22,126	3.2	21,853	2.9	19,757	2.6	16,957	2.3

Commodity data are provided by the United Nations Statistical Division. The symbol * means that data are estimated. For additional notes, see Appendix I.

Product Tables
Wool Yarn, Pure

Unit of Measure: Metric tons.

	1990 Value	%	1991 Value	%	1992 Value	%	1993 Value	%	1994 Value	%	1995 Value	%	1996 Value	%
Total Production	348,929	100.0	383,908	100.0	426,433	100.0	421,675	100.0	414,357	100.0	393,319	100.0	383,546	100.0
Regions														
Africa	2,170	0.6	2,432	0.6	2,253	0.5	1,932	0.5	1,786	0.4	1,640	0.4	1,493	0.4
America, North	2,635	0.8	1,923	0.5	1,210	0.3	498	0.1	31	0.0	28	0.0	26	0.0
America, South	1,477	0.4	1,400	0.4	1,580	0.4	926	0.2	1,034	0.2	910	0.2	1,166	0.3
Asia	160,686	46.1	186,821	48.7	219,884	51.6	223,542	53.0	203,529	49.1	190,266	48.4	193,670	50.5
Europe	157,658	45.2	167,196	43.6	176,077	41.3	168,005	39.8	179,186	43.2	171,957	43.7	159,137	41.5
Oceania	24,303	7.0	24,136	6.3	25,429	6.0	26,772	6.3	28,791	6.9	28,518	7.3	28,055	7.3
Leading Producers														
Japan	78,442	22.5	81,923	21.3	80,068	18.8	61,934	14.7	67,690	16.3	51,897	13.2	47,772	12.5
Italy	69,680	20.0	78,193	20.4	86,967	20.4	81,917	19.4	92,725	22.4	85,347	21.7	81,508	21.3
China	30,000	8.6	48,000	12.5	85,200	20.0	112,231	26.6	*82,667	20.0	*88,585	22.5	*94,503	24.6
United Kingdom	28,959	8.3	*32,210	8.4	*32,210	7.6	*32,210	7.6	*32,210	7.8	*32,210	8.2	*32,210	8.4
Turkey	36,066	10.3	33,582	8.7	33,977	8.0	31,701	7.5	34,257	8.3	32,335	8.2	34,234	8.9

Cotton Yarn, Pure and Mixed (Total)

Unit of Measure: Metric tons.

	1990 Value	%	1991 Value	%	1992 Value	%	1993 Value	%	1994 Value	%	1995 Value	%	1996 Value	%
Total Production	17,585,837	100.0	17,003,338	100.0	17,261,889	100.0	17,029,758	100.0	17,688,318	100.0	18,102,889	100.0	18,105,962	100.0
Regions														
Africa	523,960	3.0	477,582	2.8	507,282	2.9	502,857	3.0	508,755	2.9	539,912	3.0	544,602	3.0
America, North	1,435,830	8.2	1,627,465	9.6	1,770,561	10.3	1,858,005	10.9	2,083,941	11.8	1,961,155	10.8	2,036,987	11.3
America, South	901,678	5.1	890,936	5.2	812,292	4.7	837,708	4.9	914,052	5.2	878,106	4.9	891,320	4.9
Asia	12,657,938	72.0	12,021,787	70.7	12,307,711	71.3	12,057,313	70.8	12,362,981	69.9	12,983,975	71.7	12,936,303	71.4
Europe	2,044,613	11.6	1,961,339	11.5	1,838,578	10.7	1,743,428	10.2	1,782,494	10.1	1,701,831	9.4	1,658,103	9.2
Oceania	21,817	0.1	24,231	0.1	25,465	0.1	30,448	0.2	36,095	0.2	37,910	0.2	38,648	0.2
Leading Producers														
China	4,625,900	26.3	4,104,900	24.1	4,458,800	25.8	4,447,748	26.1	4,894,700	27.7	5,421,800	29.9	5,122,100	28.3
India	1,690,710	9.6	1,687,400	9.9	1,734,490	10.0	1,919,640	11.3	*1,891,562	10.7	*1,956,416	10.8	*2,021,270	11.2
United States of America	1,341,648	7.6	1,533,753	9.0	1,678,019	9.7	1,764,634	10.4	1,989,140	11.2	*1,866,025	10.3	*1,941,376	10.7
Pakistan	911,588	5.2	1,041,248	6.1	1,170,736	6.8	1,218,975	7.2	1,309,622	7.4	1,369,715	7.6	1,464,895	8.1
Russian Federation	1,145,418	6.5	1,118,390	6.6	709,306	4.1	513,904	3.0	265,113	1.5	198,684	1.1	146,887	0.8

Cotton Yarn, Mixed

Unit of Measure: Metric tons.

	1990 Value	%	1991 Value	%	1992 Value	%	1993 Value	%	1994 Value	%	1995 Value	%	1996 Value	%
Total Production	2,997,241	100.0	2,460,205	100.0	2,534,663	100.0	2,766,999	100.0	2,876,651	100.0	3,070,398	100.0	3,233,060	100.0
Regions														
Africa	88,002	2.9	98,827	4.0	107,365	4.2	116,105	4.2	126,539	4.4	132,788	4.3	138,706	4.3
America, North	190,735	6.4	195,383	7.9	173,504	6.8	177,614	6.4	200,465	7.0	205,498	6.7	210,531	6.5
America, South	289,461	9.7	291,751	11.9	231,851	9.1	237,617	8.6	298,363	10.4	304,610	9.9	311,508	9.6
Asia	2,213,508	73.9	1,688,902	68.6	1,832,191	72.3	2,043,915	73.9	2,038,202	70.9	2,224,186	72.4	2,375,771	73.5
Europe	211,510	7.1	181,317	7.4	185,726	7.3	187,722	6.8	209,057	7.3	199,290	6.5	192,518	6.0
Oceania	4,026	0.1	4,026	0.2	4,026	0.2	4,026	0.1	4,026	0.1	4,026	0.1	4,026	0.1
Leading Producers														
China	1,542,400	51.5	983,300	40.0	1,081,100	42.7	1,134,983	41.0	1,306,222	45.4	1,447,700	47.2	1,536,500	47.5
United States of America	189,517	6.3	194,075	7.9	172,107	6.8	176,127	6.4	*198,888	6.9	*203,831	6.6	*208,775	6.5
India	202,010	6.7	225,670	9.2	241,100	9.5	291,460	10.5	*253,342	8.8	*268,900	8.8	*284,457	8.8
Paraguay	218,398	7.3	208,470	8.5	143,381	5.7	151,478	5.5	*202,509	7.0	*211,217	6.9	*219,925	6.8
Indonesia	115,791	3.9	109,378	4.4	117,372	4.6	170,669	6.2	16,678	0.6	*82,890	2.7	*76,051	2.4

Commodity data are provided by the United Nations Statistical Division. The symbol * means that data are estimated. For additional notes, see Appendix I.

101

Product Tables
Cotton Yarn, Pure

Unit of Measure: Metric tons.

	1990 Value	%	1991 Value	%	1992 Value	%	1993 Value	%	1994 Value	%	1995 Value	%	1996 Value	%
Total Production	10,272,000	100.0	10,233,731	100.0	10,876,166	100.0	10,992,932	100.0	10,689,331	100.0	11,198,596	100.0	11,362,464	100.0
Regions														
Africa	413,835	4.0	418,751	4.1	428,848	3.9	409,025	3.7	404,921	3.8	412,040	3.7	374,101	3.3
America, North	1,202,942	11.7	1,389,169	13.6	1,553,338	14.3	1,648,466	15.0	1,682,238	15.7	1,757,935	15.7	1,837,439	16.2
America, South	564,873	5.5	567,867	5.5	576,729	5.3	594,146	5.4	609,288	5.7	565,312	5.0	565,784	5.0
Asia	7,033,230	68.5	6,895,381	67.4	7,401,882	68.1	7,408,332	67.4	7,021,977	65.7	7,521,039	67.2	7,646,116	67.3
Europe	1,039,297	10.1	944,739	9.2	897,545	8.3	915,138	8.3	953,083	8.9	924,446	8.3	921,200	8.1
Oceania	17,824	0.2	17,824	0.2	17,824	0.2	17,824	0.2	17,824	0.2	17,824	0.2	17,824	0.2
Leading Producers														
China	3,083,500	30.0	3,121,600	30.5	3,377,700	31.1	3,312,765	30.1	2,907,509	27.2	3,158,500	28.2	2,763,100	24.3
India	1,488,700	14.5	1,461,730	14.3	1,493,390	13.7	1,622,330	14.8	1,580,200	14.8	1,788,200	16.0	2,088,090	18.4
United States of America	1,152,131	11.2	1,339,678	13.1	1,505,912	13.8	1,588,507	14.5	*1,609,747	15.1	*1,692,759	15.1	*1,775,771	15.6
Pakistan	788,851	7.7	889,123	8.7	994,498	9.1	1,027,458	9.3	1,072,062	10.0	1,155,837	10.3	1,231,817	10.8
Brazil	432,161	4.2	429,925	4.2	434,375	4.0	453,478	4.1	461,777	4.3	409,915	3.7	402,139	3.5

Flax, Ramie and True Hemp Yarn

Unit of Measure: Metric tons.

	1990 Value	%	1991 Value	%	1992 Value	%	1993 Value	%	1994 Value	%	1995 Value	%	1996 Value	%
Total Production	528,476	100.0	471,194	100.0	453,307	100.0	422,822	100.0	378,012	100.0	366,051	100.0	353,133	100.0
Regions														
Africa	565	0.1	516	0.1	479	0.1	354	0.1	446	0.1	532	0.1	425	0.1
America, South	3	0.0	3	0.0	3	0.0	3	0.0	3	0.0	3	0.0	3	0.0
Asia	394,372	74.6	364,535	77.4	357,887	79.0	336,423	79.6	293,599	77.7	291,946	79.8	294,199	83.3
Europe	133,537	25.3	106,140	22.5	94,938	20.9	86,042	20.3	83,964	22.2	73,570	20.1	58,507	16.6
Leading Producers														
Russian Federation	137,121	25.9	107,844	22.9	96,991	21.4	78,614	18.6	40,938	10.8	36,299	9.9	36,632	10.4
Romania	24,435	4.6	21,983	4.7	15,400	3.4	9,332	2.2	7,703	2.0	8,405	2.3	7,030	2.0
Belarus	29,997	5.7	24,370	5.2	27,226	6.0	20,814	4.9	15,060	4.0	16,056	4.4	16,090	4.6
Poland	14,930	2.8	9,549	2.0	9,626	2.1	9,438	2.2	12,844	3.4	8,314	2.3	5,050	1.4
Ukraine	24,659	4.7	19,510	4.1	18,666	4.1	17,955	4.2	13,729	3.6	8,792	2.4	7,954	2.3

Yarn and Sewing Thread of Man-Made Fibers

Unit of Measure: Metric tons.

	1990 Value	%	1991 Value	%	1992 Value	%	1993 Value	%	1994 Value	%	1995 Value	%	1996 Value	%
Total Production	4,337,681	100.0	4,540,343	100.0	4,716,206	100.0	4,557,773	100.0	4,733,920	100.0	4,775,452	100.0	4,762,152	100.0
Regions														
Africa	22,786	0.5	22,786	0.5	22,786	0.5	22,786	0.5	22,786	0.5	22,786	0.5	22,786	0.5
America, North	1,492,087	34.4	1,459,920	32.2	1,537,402	32.6	1,464,714	32.1	1,622,075	34.3	1,573,197	32.9	1,583,081	33.2
America, South	18,375	0.4	27,646	0.6	35,591	0.8	38,010	0.8	38,944	0.8	33,738	0.7	33,404	0.7
Asia	1,729,080	39.9	1,673,116	36.9	1,844,190	39.1	1,826,614	40.1	1,800,291	38.0	1,835,814	38.4	1,857,451	39.0
Europe	1,065,429	24.6	1,345,701	29.6	1,266,190	26.8	1,197,253	26.3	1,240,985	26.2	1,301,457	27.3	1,257,352	26.4
Oceania	9,924	0.2	11,174	0.2	10,048	0.2	8,396	0.2	8,840	0.2	8,459	0.2	8,079	0.2
Leading Producers														
United States of America	1,454,072	33.5	1,420,472	31.3	1,496,034	31.7	1,410,445	30.9	1,554,905	32.8	*1,493,126	31.3	*1,487,505	31.2
Korea, Republic of	805,367	18.6	769,824	17.0	*947,957	20.1	*998,808	21.9	*1,049,659	22.2	*1,100,510	23.0	*1,151,361	24.2
Japan	470,222	10.8	454,198	10.0	431,769	9.2	356,041	7.8	328,654	6.9	296,855	6.2	265,040	5.6
Germany	-		391,190	8.6	368,858	7.8	322,742	7.1	333,705	7.0	401,441	8.4	365,752	7.7
Romania	209,195	4.8	142,101	3.1	112,563	2.4	100,428	2.2	83,306	1.8	92,045	1.9	78,519	1.6

Commodity data are provided by the United Nations Statistical Division. The symbol * means that data are estimated. For additional notes, see Appendix I.

Product Tables
Jute Yarn

Unit of Measure: Metric tons.

	1990 Value	%	1991 Value	%	1992 Value	%	1993 Value	%	1994 Value	%	1995 Value	%	1996 Value	%
Total Production	130,477	100.0	128,153	100.0	165,737	100.0	173,170	100.0	148,354	100.0	151,608	100.0	146,981	100.0
Regions														
Africa	28,398	21.8	27,558	21.5	28,232	17.0	26,885	15.5	21,173	14.3	22,761	15.0	17,910	12.2
America, South	26,157	20.0	21,508	16.8	36,836	22.2	34,200	19.7	31,565	21.3	28,930	19.1	26,295	17.9
Asia	32,359	24.8	29,647	23.1	43,956	26.5	46,958	27.1	48,296	32.6	52,763	34.8	56,647	38.5
Europe	43,564	33.4	49,440	38.6	56,714	34.2	65,126	37.6	47,320	31.9	47,153	31.1	46,130	31.4
Leading Producers														
Brazil	25,859	19.8	21,210	16.6	*36,538	22.0	*33,902	19.6	*31,267	21.1	*28,632	18.9	*25,997	17.7
India	32,083	24.6	27,983	21.8	42,926	25.9	46,702	27.0	*48,169	32.5	*52,072	34.3	*55,975	38.1
Egypt	25,137	19.3	24,139	18.8	25,000	15.1	21,552	12.4	20,777	14.0	19,821	13.1	15,302	10.4
United Kingdom	14,334	11.0	25,556	19.9	36,778	22.2	48,000	27.7	*31,037	20.9	*32,266	21.3	*33,496	22.8
Poland	10,260	7.9	8,577	6.7	4,883	2.9	3,294	1.9	3,989	2.7	3,956	2.6	3,027	2.1

Yarn of Other Vegetable Textile Fibers

Unit of Measure: Metric tons.

	1990 Value	%	1991 Value	%	1992 Value	%	1993 Value	%	1994 Value	%	1995 Value	%	1996 Value	%
Total Production	35,314	100.0	22,535	100.0	29,524	100.0	22,792	100.0	18,872	100.0	15,740	100.0	13,972	100.0
Regions														
America, North	20,404	57.8	6,783	30.1	13,805	46.8	8,123	35.6	8,113	43.0	5,422	34.4	7,200	51.5
America, South	2,660	7.5	2,660	11.8	2,708	9.2	2,576	11.3	2,708	14.3	2,647	16.8	2,660	19.0
Asia	459	1.3	1,670	7.4	890	3.0	232	1.0	198	1.0	330	2.1	64	0.5
Europe	11,792	33.4	11,422	50.7	12,121	41.1	11,861	52.0	7,853	41.6	7,341	46.6	4,049	29.0
Leading Producers														
Mexico	18,491	52.4	5,103	22.6	12,358	41.9	6,909	30.3	7,132	37.8	4,674	29.7	6,685	47.8
Belgium	7,322	20.7	8,129	36.1	8,522	28.9	8,913	39.1	*4,601	24.4	*3,529	22.4	*2,457	17.6
Spain	1,237	3.5	702	3.1	946	3.2	1,006	4.4	1,425	7.6	2,062	13.1	-	
Ecuador	*2,569	7.3	*2,569	11.4	*2,569	8.7	*2,569	11.3	*2,569	13.6	*2,569	16.3	*2,569	18.4

Cotton Woven Fabrics

Unit of Measure: Millions of Square meters.

	1990 Value	%	1991 Value	%	1992 Value	%	1993 Value	%	1994 Value	%	1995 Value	%	1996 Value	%
Total Production	94,256	100.0	94,875	100.0	94,982	100.0	96,773	100.0	96,214	100.0	102,996	100.0	100,937	100.0
Regions														
Africa	1,771	1.9	1,779	1.9	1,801	1.9	1,479	1.5	1,435	1.5	1,340	1.3	1,260	1.2
America, North	4,395	4.7	4,314	4.5	4,432	4.7	4,417	4.6	4,351	4.5	4,545	4.4	4,548	4.5
America, South	1,954	2.1	1,731	1.8	1,667	1.8	1,627	1.7	1,621	1.7	1,411	1.4	1,361	1.3
Asia	76,495	81.2	77,431	81.6	78,213	82.3	81,162	83.9	80,683	83.9	87,849	85.3	86,233	85.4
Europe	9,602	10.2	9,584	10.1	8,830	9.3	8,042	8.3	8,074	8.4	7,792	7.6	7,486	7.4
Oceania	38	0.0	36	0.0	40	0.0	46	0.0	50	0.1	59	0.1	48	0.0
Leading Producers														
China	22,557	23.9	21,719	22.9	22,783	24.0	24,263	25.1	25,243	26.2	31,091	30.2	*28,018	27.8
India	15,177	16.1	16,478	17.4	17,582	18.5	19,648	20.3	*18,729	19.5	*19,506	18.9	*20,284	20.1
Nepal	*15,477	16.4	*16,260	17.1	*17,043	17.9	*17,827	18.4	*18,610	19.3	*19,393	18.8	*20,177	20.0
United States of America	3,732	4.0	3,682	3.9	3,846	4.0	3,682	3.8	3,740	3.9	*3,941	3.8	*3,985	3.9
Russian Federation	6,201	6.6	5,949	6.3	3,799	4.0	2,822	2.9	1,631	1.7	1,506	1.5	1,120	1.1

Commodity data are provided by the United Nations Statistical Division. The symbol * means that data are estimated. For additional notes, see Appendix I.

103

Product Tables
Silk Fabrics

Unit of Measure: 1,000 Square meters.

	1990 Value	%	1991 Value	%	1992 Value	%	1993 Value	%	1994 Value	%	1995 Value	%	1996 Value	%
Total Production	4,377,737	100.0	5,097,270	100.0	4,979,681	100.0	5,100,164	100.0	5,055,710	100.0	4,722,553	100.0	4,876,272	100.0
Regions														
Africa	22,964	0.5	23,126	0.5	23,809	0.5	17,355	0.3	22,880	0.5	23,026	0.5	22,382	0.5
America, North	56,148	1.3	61,002	1.2	65,930	1.3	70,519	1.4	75,216	1.5	80,217	1.7	84,958	1.7
America, South	6,000	0.1	6,000	0.1	6,000	0.1	6,000	0.1	6,000	0.1	6,000	0.1	6,000	0.1
Asia	4,037,820	92.2	4,806,870	94.3	4,729,927	95.0	4,873,666	95.6	4,833,220	95.6	4,479,411	94.9	4,615,962	94.7
Europe	254,805	5.8	200,273	3.9	154,015	3.1	132,624	2.6	118,394	2.3	133,898	2.8	146,969	3.0
Leading Producers														
China	2,345,549	53.6	3,296,015	64.7	3,481,293	69.9	3,866,140	75.8	4,285,360	84.8	*4,045,476	85.7	*4,276,493	87.7
Russian Federation	1,131,802	25.9	1,019,268	20.0	822,521	16.5	656,750	12.9	263,729	5.2	222,538	4.7	148,981	3.1
Romania	107,453	2.5	98,399	1.9	80,958	1.6	73,624	1.4	59,645	1.2	53,790	1.1	56,741	1.2
Japan	83,664	1.9	80,669	1.6	76,971	1.5	71,364	1.4	65,444	1.3	59,577	1.3	58,371	1.2
Uzbekistan	144,096	3.3	125,872	2.5	99,517	2.0	90,397	1.8	83,404	1.6	43,980	0.9	*34,167	0.7

Woollen Woven Fabrics

Unit of Measure: 1,000 Square meters.

	1990 Value	%	1991 Value	%	1992 Value	%	1993 Value	%	1994 Value	%	1995 Value	%	1996 Value	%
Total Production	4,080,636	100.0	4,073,201	100.0	3,860,500	100.0	3,463,760	100.0	3,262,960	100.0	3,784,825	100.0	3,309,341	100.0
Regions														
Africa	58,449	1.4	58,042	1.4	50,385	1.3	37,524	1.1	52,139	1.6	43,908	1.2	37,755	1.1
America, North	126,656	3.1	150,409	3.7	158,063	4.1	166,495	4.8	158,247	4.8	166,768	4.4	173,288	5.2
America, South	5,215	0.1	5,186	0.1	5,203	0.1	5,058	0.1	5,096	0.2	5,100	0.1	5,105	0.2
Asia	2,647,395	64.9	2,580,757	63.4	2,426,879	62.9	2,117,326	61.1	1,928,829	59.1	2,522,565	66.6	2,094,579	63.3
Europe	1,233,730	30.2	1,269,979	31.2	1,210,758	31.4	1,128,530	32.6	1,109,849	34.0	1,038,805	27.4	992,062	30.0
Oceania	9,191	0.2	8,827	0.2	9,212	0.2	8,828	0.3	8,801	0.3	7,679	0.2	6,552	0.2
Leading Producers														
China	486,833	11.9	513,827	12.6	557,568	14.4	387,692	11.2	413,485	12.7	1,078,968	28.5	*717,200	21.7
Italy	417,783	10.2	428,331	10.5	445,145	11.5	426,024	12.3	443,425	13.6	432,640	11.4	430,775	13.0
Japan	334,935	8.2	344,873	8.5	325,744	8.4	286,811	8.3	285,546	8.8	249,262	6.6	246,870	7.5
Russian Federation	582,838	14.3	492,286	12.1	351,382	9.1	269,505	7.8	113,961	3.5	107,110	2.8	66,942	2.0
India	*181,759	4.5	*181,759	4.5	*181,759	4.7	*181,759	5.2	*181,759	5.6	*181,759	4.8	*181,759	5.5

Linen Fabrics

Unit of Measure: 1,000 Square meters.

	1990 Value	%	1991 Value	%	1992 Value	%	1993 Value	%	1994 Value	%	1995 Value	%	1996 Value	%
Total Production	2,213,669	100.0	2,012,943	100.0	1,912,778	100.0	1,767,597	100.0	1,630,242	100.0	1,585,144	100.0	1,562,927	100.0
Regions														
Africa	13,877	0.6	13,197	0.7	14,336	0.7	14,728	0.8	15,120	0.9	15,513	1.0	15,905	1.0
America, South	3,539	0.2	3,539	0.2	3,539	0.2	3,539	0.2	3,539	0.2	3,539	0.2	3,539	0.2
Asia	1,671,703	75.5	1,581,634	78.6	1,527,461	79.9	1,434,527	81.2	1,286,471	78.9	1,283,795	81.0	1,298,768	83.1
Europe	524,550	23.7	414,573	20.6	367,442	19.2	314,803	17.8	325,112	19.9	282,298	17.8	244,715	15.7
Leading Producers														
Russian Federation	588,317	26.6	486,684	24.2	403,365	21.1	301,177	17.0	155,595	9.5	124,054	7.8	111,694	7.1
Romania	95,059	4.3	76,081	3.8	41,486	2.2	25,214	1.4	26,924	1.7	27,951	1.8	22,782	1.5
Belarus	97,552	4.4	84,413	4.2	87,009	4.5	68,361	3.9	41,913	2.6	43,617	2.8	44,068	2.8
Ukraine	102,445	4.6	84,331	4.2	71,619	3.7	61,665	3.5	43,116	2.6	21,803	1.4	21,780	1.4
Poland	36,721	1.7	28,865	1.4	23,537	1.2	21,945	1.2	28,719	1.8	22,412	1.4	13,211	0.8

Product Tables
Jute Fabrics

Unit of Measure: 1,000 Square meters.

	1990 Value	%	1991 Value	%	1992 Value	%	1993 Value	%	1994 Value	%	1995 Value	%	1996 Value	%
Total Production	5,086,088	100.0	4,898,923	100.0	4,254,250	100.0	4,837,083	100.0	4,529,234	100.0	4,452,246	100.0	4,365,842	100.0
Regions														
Africa	50,279	1.0	47,600	1.0	52,450	1.2	48,290	1.0	44,585	1.0	49,929	1.1	40,955	0.9
Asia	4,835,122	95.1	4,657,697	95.1	4,008,335	94.2	4,596,921	95.0	4,300,430	94.9	4,193,537	94.2	4,124,286	94.5
Europe	200,687	3.9	193,626	4.0	193,466	4.5	191,873	4.0	184,219	4.1	208,779	4.7	200,601	4.6
Leading Producers														
India	3,013,175	59.2	3,047,957	62.2	2,458,982	57.8	3,033,834	62.7	*2,851,394	63.0	*2,853,685	64.1	*2,855,976	65.4
Bangladesh	1,113,762	21.9	914,710	18.7	877,695	20.6	940,313	19.4	889,081	19.6	896,693	20.1	852,784	19.5
Thailand	320,970	6.3	287,790	5.9	278,532	6.5	250,669	5.2	247,233	5.5	159,829	3.6	129,627	3.0
Pakistan	201,892	4.0	204,206	4.2	212,716	5.0	205,433	4.2	161,083	3.6	141,771	3.2	148,783	3.4
Egypt	42,333	0.8	40,455	0.8	42,293	1.0	38,406	0.8	34,974	0.8	40,592	0.9	*31,890	0.7

Woven Fabrics of Cellulosic Fibers

Unit of Measure: 1,000 Square meters.

	1990 Value	%	1991 Value	%	1992 Value	%	1993 Value	%	1994 Value	%	1995 Value	%	1996 Value	%
Total Production	9,006,053	100.0	9,346,144	100.0	9,347,043	100.0	9,162,862	100.0	9,045,660	100.0	9,483,016	100.0	9,607,578	100.0
Regions														
Africa	26,553	0.3	26,553	0.3	26,553	0.3	26,553	0.3	26,553	0.3	26,553	0.3	26,553	0.3
America, North	1,344,187	14.9	1,341,071	14.3	1,337,956	14.3	1,334,840	14.6	1,331,725	14.7	1,328,609	14.0	1,325,494	13.8
Asia	4,044,267	44.9	3,923,089	42.0	3,741,772	40.0	3,601,777	39.3	3,445,636	38.1	3,339,795	35.2	3,315,154	34.5
Europe	3,579,962	39.8	4,046,358	43.3	4,232,311	45.3	4,189,563	45.7	4,233,090	46.8	4,779,946	50.4	4,932,807	51.3
Oceania	11,083	0.1	9,072	0.1	8,451	0.1	10,128	0.1	8,656	0.1	8,113	0.1	7,569	0.1
Leading Producers														
France	2,339,046	26.0	2,345,755	25.1	2,596,680	27.8	2,715,720	29.6	2,758,120	30.5	*3,247,886	34.2	*3,467,603	36.1
Thailand	*936,880	10.4	*961,187	10.3	*985,494	10.5	*1,009,801	11.0	*1,034,108	11.4	*1,058,415	11.2	*1,082,722	11.3
India	*808,672	9.0	*751,280	8.0	*693,888	7.4	*636,495	6.9	*579,103	6.4	*521,711	5.5	*464,319	4.8
United States of America	*773,502	8.6	*773,502	8.3	*773,502	8.3	*773,502	8.4	*773,502	8.6	*773,502	8.2	*773,502	8.1
Japan	707,906	7.9	671,043	7.2	585,741	6.3	493,068	5.4	435,139	4.8	408,792	4.3	440,835	4.6

Woven Fabrics of Non-Cellulosic Fibers

Unit of Measure: 1,000 Square meters.

	1990 Value	%	1991 Value	%	1992 Value	%	1993 Value	%	1994 Value	%	1995 Value	%	1996 Value	%
Total Production	16,833,859	100.0	18,041,566	100.0	18,835,252	100.0	17,869,678	100.0	17,919,744	100.0	18,638,337	100.0	19,702,356	100.0
Regions														
Africa	27,428	0.2	24,804	0.1	23,326	0.1	24,038	0.1	23,208	0.1	23,624	0.1	21,386	0.1
America, North	364,000	2.2	364,000	2.0	364,000	1.9	364,000	2.0	364,000	2.0	364,000	2.0	364,000	1.8
America, South	19,120	0.1	19,120	0.1	19,120	0.1	19,120	0.1	19,120	0.1	19,120	0.1	19,120	0.1
Asia	10,321,767	61.3	10,607,827	58.8	10,531,744	55.9	9,884,418	55.3	10,154,470	56.7	10,426,135	55.9	11,348,691	57.6
Europe	5,933,576	35.2	6,850,385	38.0	7,719,985	41.0	7,403,170	41.4	7,168,830	40.0	7,609,806	40.8	7,747,969	39.3
Oceania	167,968	1.0	175,430	1.0	177,077	0.9	174,932	1.0	190,116	1.1	195,652	1.0	201,189	1.0
Leading Producers														
Belgium	3,601,198	21.4	3,544,886	19.6	3,924,278	20.8	4,466,897	25.0	*4,273,111	23.8	*4,468,796	24.0	*4,664,481	23.7
Japan	2,667,884	15.8	2,591,756	14.4	2,589,382	13.7	2,265,105	12.7	2,142,775	12.0	2,049,591	11.0	1,996,686	10.1
Korea, Republic of	3,428,312	20.4	3,478,649	19.3	3,093,687	16.4	2,459,299	13.8	2,540,315	14.2	2,593,760	13.9	*3,257,953	16.5
India	*2,870,711	17.1	*3,130,189	17.3	*3,389,668	18.0	*3,649,147	20.4	*3,908,625	21.8	*4,168,104	22.4	*4,427,582	22.5
Germany	-		955,463	5.3	897,156	4.8	813,018	4.5	791,768	4.4	941,599	5.1	850,675	4.3

Commodity data are provided by the United Nations Statistical Division. The symbol * means that data are estimated. For additional notes, see Appendix I.

105

Product Tables
Blankets

Unit of Measure: Thousands of units.

	1990 Value	%	1991 Value	%	1992 Value	%	1993 Value	%	1994 Value	%	1995 Value	%	1996 Value	%
Total Production	128,350	100.0	123,128	100.0	120,700	100.0	115,509	100.0	118,058	100.0	118,450	100.0	119,304	100.0
Regions														
Africa	19,382	15.1	18,060	14.7	15,892	13.2	13,891	12.0	13,085	11.1	12,551	10.6	11,851	9.9
America, North	14,220	11.1	14,185	11.5	14,150	11.7	14,115	12.2	14,080	11.9	13,805	11.7	14,250	11.9
America, South	13,301	10.4	13,188	10.7	12,118	10.0	9,947	8.6	11,160	9.5	11,425	9.6	11,164	9.4
Asia	41,809	32.6	40,212	32.7	40,580	33.6	47,513	41.1	48,798	41.3	48,411	40.9	51,823	43.4
Europe	39,495	30.8	37,437	30.4	37,960	31.4	30,043	26.0	30,935	26.2	32,258	27.2	30,216	25.3
Oceania	143	0.1	46	0.0	-		-		-		-			
Leading Producers														
China	22,956	17.9	24,000	19.5	24,571	20.4	*31,829	27.6	*33,261	28.2	*34,694	29.3	*36,127	30.3
Spain	11,884	9.3	11,008	8.9	11,483	9.5	5,433	4.7	6,395	5.4	6,524	5.5	7,051	5.9
Brazil	10,791	8.4	10,454	8.5	9,586	7.9	7,938	6.9	8,010	6.8	8,645	7.3	7,932	6.6
Mexico	*9,104	7.1	*9,104	7.4	*9,104	7.5	*9,104	7.9	*9,104	7.7	*9,104	7.7	9,104	7.6
Germany	*9,059	7.1	*9,059	7.4	*9,059	7.5	8,392	7.3	8,963	7.6	10,926	9.2	7,957	6.7

Bed Linen, Articles (e.g. Bed Sheets, Pillow Cases, Mattress Covers Etc.)

Unit of Measure: Thousands of units.

	1990 Value	%	1991 Value	%	1992 Value	%	1993 Value	%	1994 Value	%	1995 Value	%	1996 Value	%
Total Production	866,284	100.0	818,222	100.0	735,218	100.0	723,704	100.0	710,095	100.0	691,493	100.0	709,207	100.0
Regions														
Africa	12,789	1.5	13,560	1.7	12,304	1.7	11,175	1.5	9,973	1.4	13,470	1.9	12,344	1.7
America, North	360,585	41.6	329,691	40.3	358,726	48.8	375,152	51.8	403,142	56.8	419,914	60.7	439,955	62.0
America, South	49,446	5.7	48,479	5.9	54,203	7.4	54,062	7.5	55,853	7.9	55,540	8.0	57,731	8.1
Asia	274,821	31.7	223,196	27.3	113,378	15.4	110,818	15.3	72,665	10.2	48,736	7.0	47,707	6.7
Europe	168,643	19.5	203,296	24.8	196,608	26.7	172,497	23.8	168,462	23.7	153,832	22.2	151,471	21.4
Leading Producers														
United States of America	341,220	39.4	309,312	37.8	337,332	45.9	352,716	48.7	379,368	53.4	395,628	57.2	414,732	58.5
Russian Federation	203,548	23.5	122,532	15.0	59,908	8.1	71,901	9.9	35,171	5.0	15,136	2.2	14,871	2.1
Brazil	45,482	5.3	44,147	5.4	*49,481	6.7	*50,346	7.0	*51,210	7.2	*52,075	7.5	*52,939	7.5
Germany	-		55,904	6.8	47,994	6.5	45,118	6.2	43,972	6.2	36,757	5.3	33,963	4.8
Ukraine	51,622	6.0	34,688	4.2	36,427	5.0	21,535	3.0	8,946	1.3	3,916	0.6	4,134	0.6

Towelling

Unit of Measure: Thousands of units.

	1990 Value	%	1991 Value	%	1992 Value	%	1993 Value	%	1994 Value	%	1995 Value	%	1996 Value	%
Total Production	2,795,910	100.0	3,072,321	100.0	3,502,101	100.0	3,508,091	100.0	3,640,902	100.0	3,750,807	100.0	3,829,516	100.0
Regions														
Africa	8,868	0.3	7,605	0.2	8,398	0.2	11,334	0.3	10,261	0.3	10,381	0.3	11,292	0.3
America, North	538,334	19.3	548,432	17.9	616,718	17.6	613,298	17.5	619,876	17.0	623,586	16.6	580,469	15.2
America, South	18,118	0.6	19,581	0.6	21,086	0.6	25,340	0.7	17,986	0.5	12,106	0.3	22,061	0.6
Asia	2,064,448	73.8	2,338,238	76.1	2,657,553	75.9	2,674,840	76.2	2,795,627	76.8	2,902,829	77.4	3,019,481	78.8
Europe	166,141	5.9	158,465	5.2	198,346	5.7	183,279	5.2	197,152	5.4	201,905	5.4	196,212	5.1
Leading Producers														
China	1,830,000	65.5	2,109,000	68.6	2,352,000	67.2	*2,411,520	68.7	*2,516,398	69.1	*2,621,277	69.9	*2,726,156	71.2
United States of America	508,512	18.2	517,896	16.9	585,468	16.7	582,204	16.6	587,688	16.1	591,756	15.8	543,444	14.2
Vietnam	108,544	3.9	109,224	3.6	209,689	6.0	*178,589	5.1	*196,009	5.4	*213,429	5.7	*230,850	6.0
Germany	-		-		42,527	1.2	40,423	1.2	47,974	1.3	38,627	1.0	32,222	0.8
Russian Federation	82,576	3.0	73,740	2.4	42,224	1.2	22,316	0.6	11,535	0.3	5,279	0.1	3,736	0.1

Product Tables
Knitted Fabrics

Unit of Measure: Metric tons.

	1990 Value	%	1991 Value	%	1992 Value	%	1993 Value	%	1994 Value	%	1995 Value	%	1996 Value	%
Total Production	2,156,053	100.0	2,156,448	100.0	2,161,620	100.0	2,130,405	100.0	2,130,831	100.0	2,079,041	100.0	2,206,790	100.0
Regions														
Africa	45,363	2.1	64,706	3.0	52,164	2.4	52,052	2.4	33,430	1.6	33,230	1.6	48,724	2.2
America, North	862,834	40.0	891,118	41.3	944,100	43.7	992,986	46.6	1,011,891	47.5	1,037,655	49.9	1,068,752	48.4
America, South	308,642	14.3	309,018	14.3	309,394	14.3	307,594	14.4	346,623	16.3	295,207	14.2	282,276	12.8
Asia	357,977	16.6	353,647	16.4	333,599	15.4	317,145	14.9	270,072	12.7	206,057	9.9	297,940	13.5
Europe	556,758	25.8	510,894	23.7	496,321	23.0	436,746	20.5	443,515	20.8	482,351	23.2	484,386	21.9
Oceania	24,479	1.1	27,065	1.3	26,042	1.2	23,882	1.1	25,301	1.2	24,541	1.2	24,713	1.1
Leading Producers														
United States of America	862,048	40.0	890,469	41.3	943,730	43.7	992,426	46.6	1,011,330	47.5	*1,037,093	49.9	*1,068,189	48.4
Brazil	*305,585	14.2	*305,585	14.2	*305,585	14.1	*305,585	14.3	344,930	16.2	292,893	14.1	278,933	12.6
Japan	168,378	7.8	170,001	7.9	159,586	7.4	149,336	7.0	151,934	7.1	150,217	7.2	149,570	6.8
Germany	*71,580	3.3	94,613	4.4	75,475	3.5	60,247	2.8	55,985	2.6	*71,580	3.4	*71,580	3.2
Portugal	84,963	3.9	58,098	2.7	54,872	2.5	71,090	3.3	71,843	3.4	*76,011	3.7	*78,518	3.6

Socks and Other Stockings, Except Women's Stockings

Unit of Measure: 1,000 Pairs.

	1990 Value	%	1991 Value	%	1992 Value	%	1993 Value	%	1994 Value	%	1995 Value	%	1996 Value	%
Total Production	9,634,949	100.0	9,558,184	100.0	9,175,860	100.0	9,041,152	100.0	8,801,891	100.0	8,644,368	100.0	8,737,470	100.0
Regions														
Africa	41,302	0.4	60,564	0.6	60,538	0.7	59,593	0.7	53,107	0.6	56,641	0.7	58,323	0.7
America, North	2,030,721	21.1	2,040,753	21.4	2,058,880	22.4	2,086,339	23.1	2,113,798	24.0	2,141,257	24.8	2,178,215	24.9
America, South	143,830	1.5	155,209	1.6	157,624	1.7	163,087	1.8	156,516	1.8	149,798	1.7	153,977	1.8
Asia	4,723,497	49.0	4,718,241	49.4	4,419,437	48.2	4,496,234	49.7	4,306,312	48.9	4,153,762	48.1	4,257,538	48.7
Europe	2,617,910	27.2	2,504,638	26.2	2,404,497	26.2	2,157,012	23.9	2,092,642	23.8	2,063,682	23.9	2,003,818	22.9
Oceania	77,690	0.8	78,779	0.8	74,885	0.8	78,887	0.9	79,516	0.9	79,228	0.9	85,599	1.0
Leading Producers														
United States of America	*1,894,829	19.7	*1,913,243	20.0	*1,931,657	21.1	*1,950,071	21.6	*1,968,486	22.4	*1,986,900	23.0	*2,005,314	23.0
China	1,120,020	11.6	1,268,440	13.3	1,112,670	12.1	*1,297,050	14.3	*1,341,890	15.2	*1,386,729	16.0	*1,431,568	16.4
Russian Federation	872,442	9.1	742,804	7.8	626,014	6.8	552,406	6.1	352,667	4.0	287,536	3.3	209,291	2.4
Japan	440,808	4.6	455,386	4.8	450,009	4.9	434,875	4.8	421,033	4.8	413,172	4.8	411,769	4.7
Ukraine	443,113	4.6	392,606	4.1	380,924	4.2	297,121	3.3	161,531	1.8	118,595	1.4	66,823	0.8

Women's Stockings

Unit of Measure: 1,000 Pairs.

	1990 Value	%	1991 Value	%	1992 Value	%	1993 Value	%	1994 Value	%	1995 Value	%	1996 Value	%
Total Production	6,483,959	100.0	6,570,343	100.0	6,512,221	100.0	6,679,810	100.0	6,246,010	100.0	6,236,236	100.0	6,456,733	100.0
Regions														
Africa	112,801	1.7	114,976	1.7	90,851	1.4	111,230	1.7	105,497	1.7	107,441	1.7	111,166	1.7
America, North	2,288,481	35.3	2,340,464	35.6	2,376,099	36.5	2,420,256	36.2	2,464,412	39.5	2,508,569	40.2	2,552,726	39.5
America, South	158,294	2.4	164,099	2.5	183,154	2.8	200,063	3.0	195,469	3.1	168,425	2.7	183,384	2.8
Asia	2,030,513	31.3	1,901,656	28.9	1,898,587	29.2	1,972,792	29.5	1,495,926	24.0	1,313,685	21.1	1,529,516	23.7
Europe	1,801,476	27.8	1,968,415	30.0	1,880,272	28.9	1,890,049	28.3	1,896,159	30.4	2,049,900	32.9	1,992,057	30.9
Oceania	92,395	1.4	80,734	1.2	83,258	1.3	85,421	1.3	88,546	1.4	88,215	1.4	87,885	1.4
Leading Producers														
United States of America	*2,117,171	32.7	*2,151,057	32.7	*2,184,943	33.6	*2,218,829	33.2	*2,252,714	36.1	*2,286,600	36.7	*2,320,486	35.9
Japan	1,089,586	16.8	978,204	14.9	999,005	15.3	937,169	14.0	842,576	13.5	734,362	11.8	714,439	11.1
United Kingdom	*653,230	10.1	*678,400	10.3	*703,570	10.8	*728,740	10.9	*753,909	12.1	*779,079	12.5	*804,249	12.5
France	325,037	5.0	341,476	5.2	328,137	5.0	304,824	4.6	296,184	4.7	*333,619	5.3	*334,999	5.2
Korea, Republic of	391,227	6.0	400,429	6.1	421,484	6.5	344,305	5.2	314,785	5.0	257,001	4.1	*424,605	6.6

Commodity data are provided by the United Nations Statistical Division. The symbol * means that data are estimated. For additional notes, see Appendix I.

107

Product Tables

Knitted Undergarments, Excluding Elastic

Unit of Measure: Thousands of units.

	1990		1991		1992		1993		1994		1995		1996	
	Value	%	Value	%	Value	%	Value	%	Value	%	Value	%	Value	%
Total Production	3,253,905	100.0	3,123,915	100.0	2,901,537	100.0	2,766,158	100.0	2,460,699	100.0	2,226,460	100.0	2,267,681	100.0
Regions														
Africa	7,595	0.2	7,771	0.2	7,145	0.2	5,087	0.2	5,107	0.2	5,432	0.2	5,152	0.2
America, North	54,069	1.7	55,225	1.8	56,381	1.9	52,389	1.9	66,475	2.7	57,706	2.6	60,512	2.7
America, South	9,116	0.3	23,718	0.8	38,134	1.3	36,746	1.3	31,003	1.3	35,370	1.6	35,310	1.6
Asia	1,445,334	44.4	1,444,004	46.2	1,203,103	41.5	986,833	35.7	807,283	32.8	716,740	32.2	762,246	33.6
Europe	1,645,524	50.6	1,509,601	48.3	1,521,848	52.4	1,598,072	57.8	1,464,071	59.5	1,324,724	59.5	1,318,242	58.1
Oceania	92,267	2.8	83,597	2.7	74,926	2.6	87,030	3.1	86,759	3.5	86,488	3.9	86,218	3.8
Leading Producers														
Japan	341,021	10.5	350,491	11.2	336,785	11.6	329,766	11.9	306,911	12.5	321,115	14.4	299,679	13.2
Korea, Republic of	343,768	10.6	365,393	11.7	384,550	13.3	315,447	11.4	284,915	11.6	263,802	11.8	*372,258	16.4
Russian Federation	509,186	15.6	446,318	14.3	281,555	9.7	201,239	7.3	124,123	5.0	73,261	3.3	41,455	1.8
Germany	*192,517	5.9	220,792	7.1	192,555	6.6	201,297	7.3	155,425	6.3	*192,517	8.6	*192,517	8.5
France	137,694	4.2	127,542	4.1	119,512	4.1	124,497	4.5	76,984	3.1	*91,077	4.1	*80,804	3.6

Knitted Sports Shirts

Unit of Measure: Thousands of units.

	1990		1991		1992		1993		1994		1995		1996	
	Value	%	Value	%	Value	%	Value	%	Value	%	Value	%	Value	%
Total Production	1,574,104	100.0	1,424,832	100.0	1,621,344	100.0	1,398,555	100.0	1,476,282	100.0	1,632,295	100.0	1,673,532	100.0
Regions														
Africa	6,500	0.4	6,410	0.4	4,452	0.3	5,196	0.4	4,258	0.3	4,394	0.3	2,386	0.1
America, North	1,358,114	86.3	1,221,374	85.7	1,412,378	87.1	1,194,596	85.4	1,283,081	86.9	1,434,804	87.9	1,482,605	88.6
America, South	322	0.0	322	0.0	291	0.0	299	0.0	544	0.0	154	0.0	322	0.0
Asia	62,526	4.0	59,329	4.2	65,405	4.0	65,643	4.7	66,146	4.5	67,386	4.1	66,049	3.9
Europe	130,898	8.3	125,313	8.8	128,288	7.9	120,209	8.6	110,790	7.5	115,244	7.1	113,007	6.8
Oceania	15,745	1.0	12,084	0.8	10,531	0.6	12,612	0.9	11,463	0.8	10,314	0.6	9,164	0.5
Leading Producers														
United States of America	1,357,584	86.2	1,220,844	85.7	1,411,848	87.1	1,194,120	85.4	1,282,500	86.9	*1,434,190	87.9	*1,482,157	88.6
Portugal	*39,057	2.5	*39,057	2.7	39,334	2.4	41,713	3.0	36,125	2.4	*39,057	2.4	*39,057	2.3
Japan	31,130	2.0	28,593	2.0	31,406	1.9	30,835	2.2	31,903	2.2	34,526	2.1	33,222	2.0
Spain	*19,858	1.3	*19,858	1.4	*19,858	1.2	18,716	1.3	19,959	1.4	20,898	1.3	*19,858	1.2
Australia	15,745	1.0	12,084	0.8	10,531	0.6	*12,612	0.9	*11,463	0.8	*10,314	0.6	*9,164	0.5

Knitted Sweaters

Unit of Measure: Thousands of units.

	1990		1991		1992		1993		1994		1995		1996	
	Value	%	Value	%	Value	%	Value	%	Value	%	Value	%	Value	%
Total Production	1,057,970	100.0	998,554	100.0	893,450	100.0	902,648	100.0	893,468	100.0	890,363	100.0	917,572	100.0
Regions														
Africa	19,174	1.8	19,530	2.0	19,855	2.2	20,179	2.2	20,497	2.3	20,835	2.3	21,152	2.3
America, North	77,666	7.3	60,706	6.1	63,794	7.1	75,763	8.4	74,676	8.4	59,435	6.7	53,981	5.9
America, South	1,315	0.1	1,315	0.1	1,315	0.1	1,315	0.1	1,315	0.1	1,315	0.1	1,315	0.1
Asia	325,141	30.7	340,920	34.1	313,230	35.1	320,623	35.5	311,257	34.8	319,770	35.9	355,738	38.8
Europe	620,673	58.7	564,225	56.5	483,270	54.1	474,176	52.5	476,096	53.3	480,957	54.0	478,845	52.2
Oceania	14,001	1.3	11,858	1.2	11,986	1.3	10,592	1.2	9,628	1.1	8,051	0.9	6,540	0.7
Leading Producers														
Romania	202,537	19.1	173,900	17.4	103,000	11.5	101,159	11.2	113,530	12.7	124,046	13.9	111,054	12.1
Hong Kong	170,445	16.1	182,154	18.2	159,606	17.9	185,702	20.6	166,935	18.7	166,467	18.7	*191,967	20.9
United States of America	75,762	7.2	58,860	5.9	61,920	6.9	73,188	8.1	71,400	8.0	*55,458	6.2	*49,288	5.4
United Kingdom	*94,213	8.9	*94,146	9.4	*94,078	10.5	*94,011	10.4	*93,943	10.5	*93,876	10.5	*93,809	10.2
Korea, Republic of	64,411	6.1	52,860	5.3	41,090	4.6	40,268	4.5	40,933	4.6	39,098	4.4	*36,647	4.0

Commodity data are provided by the United Nations Statistical Division. The symbol * means that data are estimated. For additional notes, see Appendix I.

Product Tables
Knitted Outer Garments, Other

Unit of Measure: Thousands of units.

	1990 Value	%	1991 Value	%	1992 Value	%	1993 Value	%	1994 Value	%	1995 Value	%	1996 Value	%
Total Production	602,849	100.0	533,766	100.0	479,247	100.0	398,183	100.0	373,453	100.0	351,718	100.0	380,075	100.0
Regions														
Africa	11,635	1.9	10,012	1.9	9,500	2.0	8,645	2.2	9,776	2.6	10,359	2.9	9,030	2.4
America, North	8,266	1.4	8,275	1.6	8,498	1.8	20,836	5.2	23,867	6.4	27,635	7.9	34,687	9.1
America, South	2,449	0.4	2,449	0.5	2,449	0.5	2,449	0.6	2,449	0.7	2,449	0.7	2,449	0.6
Asia	46,298	7.7	57,490	10.8	23,101	4.8	45,860	11.5	52,482	14.1	45,643	13.0	51,547	13.6
Europe	534,202	88.6	455,540	85.3	435,700	90.9	320,394	80.5	284,879	76.3	265,632	75.5	282,362	74.3
Leading Producers														
Poland	71,630	11.9	49,853	9.3	36,561	7.6	39,804	10.0	41,594	11.1	51,308	14.6	54,782	14.4
Ukraine	95,911	15.9	82,188	15.4	71,760	15.0	58,239	14.6	24,578	6.6	8,203	2.3	5,153	1.4
Spain	50,478	8.4	56,113	10.5	57,584	12.0	6,862	1.7	8,416	2.3	10,093	2.9	*18,846	5.0
Greece	56,741	9.4	54,445	10.2	55,102	11.5	8,575	2.2	11,655	3.1	14,157	4.0	*28,434	7.5
France	23,856	4.0	22,297	4.2	18,335	3.8	26,408	6.6	25,488	6.8	*17,056	4.8	*15,473	4.1

Carpets and Rugs of Wool, Knotted

Unit of Measure: 1,000 Square meters.

	1990 Value	%	1991 Value	%	1992 Value	%	1993 Value	%	1994 Value	%	1995 Value	%	1996 Value	%
Total Production	270,738	100.0	259,332	100.0	253,244	100.0	262,591	100.0	259,932	100.0	257,936	100.0	257,739	100.0
Regions														
Africa	1,110	0.4	925	0.4	653	0.3	980	0.4	855	0.3	839	0.3	680	0.3
America, South	869	0.3	982	0.4	1,095	0.4	739	0.3	671	0.3	661	0.3	1,000	0.4
Asia	184,007	68.0	182,626	70.4	173,136	68.4	177,811	67.7	176,103	67.7	170,227	66.0	169,386	65.7
Europe	31,987	11.8	24,257	9.4	28,387	11.2	31,682	12.1	26,416	10.2	28,431	11.0	28,207	10.9
Oceania	52,766	19.5	50,542	19.5	49,973	19.7	51,379	19.6	55,886	21.5	57,778	22.4	58,467	22.7
Leading Producers														
Australia	45,101	16.7	42,854	16.5	42,262	16.7	42,618	16.2	*46,621	17.9	*47,465	18.4	*48,309	18.7
Poland	11,059	4.1	6,096	2.4	7,066	2.8	8,514	3.2	9,275	3.6	9,521	3.7	8,431	3.3
New Zealand	7,665	2.8	7,688	3.0	7,711	3.0	8,761	3.3	9,265	3.6	10,313	4.0	10,158	3.9
Japan	8,908	3.3	8,179	3.2	7,195	2.8	6,247	2.4	5,754	2.2	*6,097	2.4	*5,712	2.2
Korea, Republic of	6,132	2.3	6,649	2.6	6,995	2.8	7,875	3.0	9,931	3.8	9,323	3.6	*10,314	4.0

Carpets and Rugs, Other

Unit of Measure: 1,000 Square meters.

	1990 Value	%	1991 Value	%	1992 Value	%	1993 Value	%	1994 Value	%	1995 Value	%	1996 Value	%
Total Production	2,709,968	100.0	2,799,619	100.0	2,957,731	100.0	3,088,199	100.0	3,283,432	100.0	3,284,706	100.0	3,067,752	100.0
Regions														
Africa	21,723	0.8	18,506	0.7	17,553	0.6	18,265	0.6	18,211	0.6	16,612	0.5	16,525	0.5
America, North	1,147,432	42.3	1,076,703	38.5	1,194,581	40.4	1,261,868	40.9	1,385,111	42.2	1,346,660	41.0	1,386,017	45.2
America, South	2,230	0.1	2,516	0.1	2,813	0.1	3,475	0.1	4,948	0.2	5,889	0.2	4,891	0.2
Asia	201,975	7.5	187,953	6.7	183,135	6.2	179,056	5.8	166,531	5.1	166,987	5.1	168,669	5.5
Europe	1,336,609	49.3	1,513,941	54.1	1,559,649	52.7	1,625,536	52.6	1,708,631	52.0	1,748,559	53.2	1,491,649	48.6
Leading Producers														
United States of America	1,137,132	42.0	1,068,550	38.2	1,186,565	40.1	1,247,003	40.4	1,370,268	41.7	*1,333,801	40.6	*1,371,288	44.7
Belgium	*692,676	25.6	*692,676	24.7	*692,676	23.4	*692,676	22.4	715,998	21.8	818,202	24.9	543,829	17.7
Germany	-		185,157	6.6	186,611	6.3	172,889	5.6	168,652	5.1	171,225	5.2	173,797	5.7
United Kingdom	131,123	4.8	140,355	5.0	189,333	6.4	238,310	7.7	287,288	8.7	*220,883	6.7	*229,249	7.5
Japan	100,971	3.7	98,511	3.5	92,893	3.1	87,370	2.8	88,458	2.7	*97,238	3.0	*98,507	3.2

Commodity data are provided by the United Nations Statistical Division. The symbol * means that data are estimated. For additional notes, see Appendix I.

109

Product Tables
Cordage, Rope and Twine

Unit of Measure: Metric tons.

	1990 Value	%	1991 Value	%	1992 Value	%	1993 Value	%	1994 Value	%	1995 Value	%	1996 Value	%
Total Production	873,820	100.0	838,641	100.0	838,046	100.0	846,131	100.0	839,650	100.0	899,983	100.0	920,405	100.0
Regions														
Africa	37,473	4.3	36,490	4.4	35,799	4.3	39,405	4.7	36,753	4.4	37,156	4.1	36,956	4.0
America, North	14,119	1.6	14,424	1.7	14,731	1.8	14,141	1.7	15,090	1.8	15,493	1.7	17,302	1.9
America, South	856	0.1	632	0.1	407	0.0	5,345	0.6	3,830	0.5	5,342	0.6	3,225	0.4
Asia	545,827	62.5	545,441	65.0	555,605	66.3	567,969	67.1	579,562	69.0	617,365	68.6	636,558	69.2
Europe	268,764	30.8	234,862	28.0	224,697	26.8	212,452	25.1	197,582	23.5	217,781	24.2	219,506	23.8
Oceania	6,780	0.8	6,793	0.8	6,806	0.8	6,820	0.8	6,833	0.8	6,846	0.8	6,859	0.7
Leading Producers														
Philippines	*117,424	13.4	*133,952	16.0	*150,480	18.0	*167,007	19.7	*183,535	21.9	*200,063	22.2	*216,591	23.5
India	*93,906	10.7	*100,156	11.9	*106,407	12.7	*112,658	13.3	*118,908	14.2	*125,159	13.9	*131,410	14.3
Portugal	65,978	7.6	58,709	7.0	62,909	7.5	62,962	7.4	62,547	7.4	*69,765	7.8	*70,814	7.7
Korea, Republic of	62,666	7.2	57,537	6.9	58,501	7.0	61,685	7.3	67,320	8.0	70,683	7.9	*75,673	8.2
Japan	39,675	4.5	36,258	4.3	33,267	4.0	31,544	3.7	31,482	3.7	29,948	3.3	29,545	3.2

Floor Covering

Unit of Measure: 1,000 Square meters.

	1990 Value	%	1991 Value	%	1992 Value	%	1993 Value	%	1994 Value	%	1995 Value	%	1996 Value	%
Total Production	435,441	100.0	435,433	100.0	432,859	100.0	441,266	100.0	416,452	100.0	410,001	100.0	404,323	100.0
Regions														
America, South	25,314	5.8	32,791	7.5	28,794	6.7	34,679	7.9	36,219	8.7	37,269	9.1	37,478	9.3
Asia	276,412	63.5	279,697	64.2	289,999	67.0	294,690	66.8	273,986	65.8	274,573	67.0	270,993	67.0
Europe	90,778	20.8	81,457	18.7	72,048	16.6	68,781	15.6	57,219	13.7	54,620	13.3	51,928	12.8
Oceania	42,937	9.9	41,487	9.5	42,018	9.7	43,116	9.8	49,028	11.8	43,538	10.6	43,924	10.9
Leading Producers														
Russian Federation	88,035	20.2	89,501	20.6	93,107	21.5	91,877	20.8	64,094	15.4	57,061	13.9	41,099	10.2
Australia	42,937	9.9	41,487	9.5	42,018	9.7	43,116	9.8	49,028	11.8	43,538	10.6	43,924	10.9
Brazil	21,229	4.9	28,706	6.6	26,175	6.0	31,394	7.1	31,468	7.6	31,582	7.7	33,393	8.3
Ukraine	32,156	7.4	29,784	6.8	23,608	5.5	19,632	4.4	9,037	2.2	7,572	1.8	6,012	1.5
Bulgaria	9,487	2.2	3,837	0.9	3,770	0.9	4,435	1.0	4,422	1.1	3,902	1.0	3,715	0.9

ISIC 3211
SPINNING, WEAVING, ETC.

ISIC 3211—Spinning and Weaving, etc. This industry is a subset of ISIC 3210, shown in the last chapter, and breaks out those products which relate specifically to fabric weaving. Mills producing fabrics—including cotton, wool, silk, linen, jute, cellulosic, and non-cellulosic—are included. Also included are producers of objects such as blankets, bed linen, towelling, knitted fabrics, socks and stockings, undergarments, knitted sports shirts and sweaters, other knitted outer garments, carpets, and cordage and twine.

Please note: All commodity tables for this category are shown under the more inclusive ISIC 3210 (previous chapter).

Summary Statistics

		1990		1991		1992		1993		1994		1995	
		Value	N	Value	N	Value	N	Value	N	Value	N	Value	N
Establishments or enterprises	(number)	58,270	48	59,302	52	62,619	55	43,406	54	14,007	41	8,929	18
Number employed	(000)	4,661	52	4,914	53	4,656	53	4,239	53	2,001	38	1,576	21
Total Wages	($ mil.)	28,118	47	30,758	50	32,470	49	25,882	46	13,370	36	12,185	20
Wage/salary per employee	($)	8,907	47	8,305	50	8,847	49	6,801	46	5,678	35	5,476	20
Female workers	(000)	579	24	443	20	575	18	749	22	688	15	625	7
as % of total employment*	(%)	36	22	34	20	36	18	35	22	43	15	43	7
Output	($ bil.)	179	47	183	51	186	48	148	45	89	35	80	20
per employee	($ 000)	48	45	48	49	49	46	40	44	39	32	38	19
per establishment	($ 000)	7,003	43	8,158	48	6,363	45	4,520	42	4,627	32	2,786	15
Value added	($ bil.)	68	45	64	45	66	39	58	39	36	30	33	17
per employee	($ 000)	19	44	18	44	20	38	16	37	18	28	19	16
per establishment	($ 000)	2,114	42	1,740	42	1,977	38	1,642	36	2,076	28	1,284	12
Capital investment	($ mil.)	7,011	20	6,262	24	10,099	27	8,338	27	5,651	18	5,149	9
per employee	($ 000)	5	20	4	24	3	26	3	25	3	18	3	9
per establishment	($ 000)	2,774	20	1,555	24	696	27	454	25	709	17	220	8

Data presented above are drawn from the detailed tables that follow. Columns headed 'N' show the number of countries that provided valid data for inclusion. Values are not strictly comparable one year to the next or one row to the next because the number of countries that report varies from period to period and row to row. However, a general indication of magnitudes can be gleaned from reviewing this summary—especially in earlier years in which more reports are available. For detailed explanations, see Appendix I. *The average for those countries reporting both total and female employment.

Establishments and Number Engaged

Country	Establishments or Enterprises (number)						Employees per Establishment					
	1990	1991	1992	1993	1994	1995	1990	1991	1992	1993	1994	1995
Albania	-	-	-	7.000	10	5.000	-	-	-	890	290	31
Australia	235	287	-	-	-	-	72	56	-	-	-	-
Azerbaijan	56	52	56	155	154	-	460	477	442	167	179	-
Bahrain	-	-	5.000	-	-	-	-	-	-	-	-	-
Bangladesh	9,274	10,887	11,071	-	-	-	52	48	49	-	-	-
Benin	-	-	1.000	1.000	-	-	-	-	-	-	-	-
Bolivia	51	51	52	49	45	42	45	43	45	47	48	50
Bosnia & Herzegovina	7.000	8.000	-	-	3.000	-	827	524	-	-	150	-
Canada	384	338	326	303	283	-	68	71	67	69	78	-
Cape Verde	-	-	-	-	1.000	-	-	-	-	-	-	-
Chile	65	63	64	59	61	-	288	291	279	264	234	-
China (Hong Kong SAR)	2,387	1,861	1,838	1,519	1,573	-	23	26	25	23	19	-
Colombia	227	225	235	208	198	-	155	164	169	178	181	-
Costa Rica	48	48	46	43	-	-	78	68	75	73	-	-
Cyprus	18	18	18	17	15	14	21	20	20	19	22	22
Denmark	179	200	183	-	-	-	25	22	21	-	-	-
Ecuador	69	67	71	71	69	66	147	136	134	122	115	103
Egypt	615	605	593	612	-	-	397	355	391	394	-	-
El Salvador	-	-	-	14	14	17	-	-	-	369	302	274
Finland	44	51	49	48	46	-	75	55	47	46	46	-
Gambia	-	-	-	1.000	-	-	-	-	-	34	-	-
Germany	-	805	711	638	596	-	-	202	166	154	-	-
Germany (Western Part)	642	617	593	-	-	-	184	182	173	-	-	-
Ghana	-	-	-	26	-	-	-	-	-	341	-	-
Greece	439	428	428	-	-	-	71	62	54	-	-	-
Guatemala	-	18	18	18	18	18	-	323	328	327	334	359
Honduras	-	-	10	12	12	12	-	-	616	562	616	618
Hungary	-	-	*130	*159	-	-	-	-	-	-	-	-
Iceland	3.000	6.000	3.000	3.000	-	-	47	33	22	21	-	-
India	10,911	10,840	12,716	12,681	-	-	122	118	104	105	-	-
Indonesia	1,183	1,187	1,270	1,319	1,318	1,502	272	312	327	326	346	318
Iran (Islamic Republic of)	-	459	523	490	-	-	-	252	227	234	-	-
Japan	18,182	17,300	16,153	14,975	-	-	17	17	17	17	-	-
Jordan	20	22	26	25	20	21	37	35	30	38	51	64
Kenya	34	28	29	29	31	31	391	447	421	437	393	388
Korea, Republic of	4,472	4,651	4,694	5,370	5,576	5,458	60	55	51	41	39	37
Kyrgyzstan	16	11	17	17	17	-	1,474	2,121	1,512	1,460	1,126	-
Macau	78	80	69	66	54	50	32	29	28	21	20	16
Malawi	*2.000	*2.000	*2.000	*2.000	*2.000	-	2,182	-	-	-	-	-
Malaysia	104	117	113	147	150	-	196	209	223	182	169	-
Malta	-	-	-	3.000	-	-	-	-	-	71	-	-
Mauritius	-	-	29	30	27	28	-	-	-	-	-	-
Mongolia	1.000	1.000	1.000	1.000	5.000	3.000	1,083	973	1,017	755	318	386
Mozambique	-	-	3.000	3.000	3.000	-	-	-	151	193	133	-
Myanmar	25	24	23	23	17	-	714	594	604	601	772	-
Nepal	216	192	-	163	242	-	38	43	-	48	50	-
Netherlands	78	77	75	75	68	-	118	115	115	106	-	-
New Zealand	*116	*116	*130	*128	-	-	20	19	33	34	-	-
Norway	36	33	27	-	-	-	45	47	55	-	-	-
Pakistan	-	1,236	-	-	-	-	-	187	-	-	-	-
Peru	555	578	604	-	-	-	52	42	38	-	-	-
Philippines	571	560	218	194	-	-	104	89	219	197	-	-
Portugal	*1,169	*1,075	*1,162	*1,098	*1,105	-	78	79	67	66	59	-
Russian Federation	-	-	-	*636	*667	*970	-	-	-	749	606	358
Senegal	8.000	-	8.000	5.000	9.000	-	504	-	219	338	250	-
Sierra Leone	-	-	-	3.000	-	-	-	-	-	11	-	-
Singapore	27	25	23	23	20	-	74	79	83	74	70	-
Slovakia	-	15	15	17	17	-	-	1,366	1,210	863	759	-
South Africa	-	186	-	-	-	-	-	-	-	-	-	-
Spain	2,092	1,852	1,675	-	-	-	30	31	30	-	-	-

Continued.

Depending on the table, * means *Enterprises, Engaged,* or *Factor Values*; ± means *Basis Unspecified*; § means *shown in millions*. For additional notes and sources, see Appendix I.

Establishments and Number Engaged

- Continued -

Country	Establishments or Enterprises (number)						Employees per Establishment					
	1990	1991	1992	1993	1994	1995	1990	1991	1992	1993	1994	1995
Sri Lanka	353	273	268	318	-	-	88	90	88	83	-	-
Swaziland	-	3.000	3.000	3.000	-	-	-	67	72	64	-	-
Sweden	72	74	64	61	56	-	78	72	75	72	73	-
Thailand	682	1,006	-	-	-	-	436	232	-	-	-	-
Tunisia	-	-	-	190	192	154	-	-	-	49	47	52
Turkey	462	444	814	740	711	-	338	299	172	181	185	-
Ukraine	-	-	544	485	452	393	-	-	215	214	212	211
United Kingdom	1,855	-	1,701	-	-	-	52	-	50	-	-	-
United Republic of Tanzania . .	56	60	-	-	-	-	420	387	-	-	-	-
United States of America . . .	-	-	2,972	-	-	-	-	-	112	-	-	-
Venezuela	144	140	147	123	143	145	151	144	139	142	116	114
Zambia	7.000	-	-	-	7.000	-	508	-	-	-	504	-

Employment and Compensation of Employees

Country	Number Employed/Engaged (000)						Wage/Salary per Employee ($)					
	1990	1991	1992	1993	1994	1995	1990	1991	1992	1993	1994	1995
Albania	-	-	-	6.233	2.900	0.154	-	-	-	-	-	-
Australia	17	*16	-	-	-	-	20,280	20,671	-	-	-	-
Azerbaijan	26	25	25	26	28	-	1,479	1,656	-	-	-	-
Bahrain	-	-	0.002	-	-	-	-	-	2,660	-	-	-
Bangladesh	484	518	544	-	-	-	748	639	590	-	-	-
Bolivia	2.273	2.171	2.334	2.323	2.154	2.082	1,343	1,399	1,486	1,535	1,601	2,025
Bosnia & Herzegovina	5.786	4.195	-	-	0.451	-	-	-	-	-	-	-
Canada	26	24	22	21	22	22	23,371	24,294	24,369	24,029	22,235	22,575
Chile	19	18	18	16	14	13	3,639	4,241	4,979	5,448	6,281	7,402
China (Hong Kong SAR) . . .	*55	*48	*45	*36	*30	-	10,272	11,549	12,613	14,409	15,681	-
Colombia	35	37	40	37	36	36	2,340	2,388	2,489	3,017	3,722	4,527
Costa Rica	3.726	3.283	3.452	3.118	-	-	2,443	2,309	2,492	2,793	-	-
Cyprus	0.371	0.369	0.365	0.331	0.337	0.306	8,965	8,809	10,161	9,884	9,720	13,175
Denmark	4.453	4.363	3.797	-	-	-	27,867	26,661	31,895	-	-	-
Ecuador	10	9.119	9.524	8.631	7.909	6.780	2,572	3,087	2,805	3,361	1,654	2,105
Egypt	*244	*215	*232	*241	-	-	1,631	1,099	1,137	1,198	-	-
El Salvador	-	-	-	*5.164	*4.222	*4.650	-	-	-	2,269	3,177	2,152
Finland	3.300	2.800	2.300	2.200	2.100	-	22,957	21,266	22,024	17,753	21,770	-
Gambia	-	-	-	0.034	-	-	-	-	-	-	-	-
Germany	-	*162	*118	*98	-	-	-	19,992	26,646	26,823	-	-
Germany (Western Part) . . .	*118	*113	*103	-	-	-	25,179	25,811	28,743	-	-	-
Ghana	-	-	-	*8.865	-	-	-	-	-	-	-	-
Greece	*31	*27	*23	-	-	-	11,537	11,937	12,512	-	-	-
Guatemala	-	*5.807	*5.913	*5.889	*6.019	*6.458	-	-	-	-	-	-
Honduras	*4.287	*4.893	*6.158	*6.747	*7.392	*7.413	1,206	895	1,004	1,002	758	910
Iceland	0.141	0.201	0.065	0.064	0.070	-	-	16,124	28,499	12,319	-	-
India	*1,331	*1,284	*1,327	*1,337	-	-	1,207	1,016	989	910	-	-
Indonesia	322	370	415	431	456	477	518	566	810	791	823	1,184
Iran (Islamic Republic of) . . .	-	116	119	115	-	-	-	4,466	4,605	4,298	-	-
Japan	*304	*293	*273	*249	-	-	20,992	23,714	26,435	30,951	-	-
Jordan	0.731	0.778	0.793	0.941	1.020	1.354	3,107	3,186	3,277	3,007	3,189	2,938
Kenya	13	13	12	13	12	12	1,154	927	865	553	686	960
Korea, Republic of	266	254	238	221	218	202	7,943	9,138	10,138	10,841	11,863	14,464
Kyrgyzstan	*24	*23	*26	*25	*19	-	-	-	-	-	-	-
Lithuania	-	-	27	24	-	-	-	-	453	511	-	-
Macau	2.468	2.281	1.951	1.404	1.054	0.792	4,615	4,692	4,939	5,453	6,475	5,919
Malawi	4.365	-	-	-	-	-	-	-	-	-	-	-
Malaysia	20	25	25	27	25	-	2,930	3,084	3,639	3,822	4,130	-
Malta	0.186	-	-	0.212	-	-	-	-	-	10,397	-	-
Mongolia	*1.083	*0.973	*1.017	*0.755	*1.591	*1.157	1,220	1,436	760	-	-	651
Mozambique	-	-	0.452	0.580	0.398	-	-	-	-	-	-	-

Continued.

Employment and Compensation of Employees
- Continued -

Country	Number Employed/Engaged (000)						Wage/Salary per Employee ($)					
	1990	1991	1992	1993	1994	1995	1990	1991	1992	1993	1994	1995
Myanmar	18	14	14	14	13	-	1,472	1,737	1,668	2,065	2,422	-
Nepal	8.246	8.217	-	7.873	12	-	512	535	-	-	443	-
Netherlands	9.231	8.868	8.588	7.979	-	-	24,987	25,512	28,937	29,691	-	-
New Zealand	2.337	2.254	4.296	4.405	-	-	-	-	-	-	-	-
Norway	1.609	1.536	1.498	-	-	-	24,127	24,704	-	-	-	-
Pakistan	-	232	-	-	-	-	-	1,247	-	-	-	-
Peru	*29	*24	*23	-	-	-	3,604	4,573	4,181	-	-	-
Philippines	59	50	48	38	-	-	1,608	1,655	2,037	2,072	-	-
Portugal	*91	*85	*78	*72	*65	-	4,569	4,938	5,929	5,325	5,191	-
Russian Federation	-	-	-	476	404	347	-	-	-	503	642	702
Senegal	*4.034	-	*1.752	*1.690	*2.250	-	4,336	-	-	-	4,152	-
Sierra Leone	-	-	-	*0.033	-	-	-	-	-	-	-	-
Singapore	*2.003	*1.971	*1.908	*1.706	*1.396	-	8,577	9,931	11,331	12,496	14,318	-
Slovakia	-	*20	*18	*15	*13	-	-	1,336	1,505	1,643	1,816	-
Spain	62	58	50	-	-	-	15,358	16,489	18,275	-	-	-
Sri Lanka	*31	*25	*24	*26	-	-	405	727	811	824	-	-
Swaziland	-	0.202	0.216	0.192	-	-	-	7,169	3,088	3,535	-	-
Sweden	5.600	*5.343	*4.818	4.391	4.088	-	22,325	23,866	26,577	21,563	22,121	-
Thailand	297	233	-	-	-	-	1,898	1,909	-	-	-	-
Tunisia	-	-	-	9.285	9.074	7.951	-	-	-	-	-	-
Turkey	156	133	140	134	132	-	4,735	7,324	6,295	6,906	3,982	-
Ukraine	-	-	117	104	96	83	-	-	2,583	433	456	413
United Kingdom	97	-	85	-	-	-	17,176	-	20,328	-	-	-
United Republic of Tanzania	24	23	-	-	-	-	-	-	-	-	-	-
United States of America	347	335	334	328	333	328	18,991	19,582	20,868	21,838	22,814	22,982
Uruguay	*13	*13	*11	*8.880	-	-	-	-	-	-	-	-
Venezuela	22	20	20	17	17	16	4,072	4,346	4,480	4,863	3,800	4,036
Zambia	3.559	-	-	-	3.527	-	1,753	-	-	-	1,988	-

Female Workers: Total and Percent of Employment

Country	Female Workers (000)						Female Workers as % of Total Employment					
	1990	1991	1992	1993	1994	1995	1990	1991	1992	1993	1994	1995
Afghanistan	0.021	-	-	-	-	-	-	-	-	-	-	-
Albania	-	-	-	3.758	1.915	0.012	-	-	-	60.29	66.03	7.79
Australia	6.210	-	-	-	-	-	36.53	-	-	-	-	-
Bangladesh	9.385	9.858	9.173	-	-	-	1.94	1.90	1.69	-	-	-
Canada	8.000	8.000	-	-	-	-	30.77	33.33	-	-	-	-
Chile	3.339	3.343	3.310	2.640	2.680	-	17.85	18.25	18.51	16.93	18.79	-
Colombia	-	9.934	-	9.117	9.486	-	-	26.90	-	24.68	26.47	-
Cyprus	0.224	0.220	0.219	0.197	0.214	-	60.38	59.62	60.00	59.52	63.50	-
Denmark	1.942	1.978	1.701	-	-	-	43.61	45.34	44.80	-	-	-
Egypt	-	23	28	31	-	-	-	10.65	12.01	12.64	-	-
El Salvador	-	-	-	1.690	1.660	1.177	-	-	-	32.73	39.32	25.31
Gambia	-	-	-	0.002	-	-	-	-	-	5.88	-	-
Germany (Western Part)	42	-	-	-	-	-	35.62	-	-	-	-	-
Indonesia	-	-	-	215	229	237	-	-	-	49.92	50.37	49.77
Iran (Islamic Republic of)	-	5.978	6.100	5.582	-	-	-	5.16	5.15	4.87	-	-
Jordan	0.133	0.174	0.173	0.109	0.168	0.338	18.19	22.37	21.82	11.58	16.47	24.96
Kenya	0.690	0.596	0.552	0.556	-	-	5.19	4.77	4.52	4.39	-	-
Korea, Republic of	153	145	127	116	111	102	57.63	57.11	53.50	52.71	51.14	50.47
Macau	1.176	1.120	1.036	0.706	0.641	0.461	47.65	49.10	53.10	50.28	60.82	58.21
Malaysia	11	13	13	13	12	-	55.88	51.84	50.00	47.57	45.85	-
Malta	-	-	-	0.022	-	-	-	-	-	10.38	-	-
Myanmar	8.928	7.112	7.160	7.175	7.183	-	49.98	49.92	51.53	51.94	54.73	-
Nepal	2.239	2.300	-	2.100	2.823	-	27.15	27.99	-	26.67	23.32	-
New Zealand	0.752	0.681	1.286	1.301	-	-	32.18	30.21	29.93	29.53	-	-
Panama	0.025	-	-	-	-	-	-	-	-	-	-	-

Continued.

Depending on the table, * means *Enterprises, Engaged,* or *Factor Values*; ± means *Basis Unspecified*; § means *shown in millions.* For additional notes and sources, see Appendix I.

Female Workers: Total and Percent of Employment

- Continued -

Country	Female Workers (000)						Female Workers as % of Total Employment					
	1990	1991	1992	1993	1994	1995	1990	1991	1992	1993	1994	1995
Philippines	-	-	20	19	-	-	-	-	42.18	49.88	-	-
Portugal	31	29	25	23	20	-	34.05	33.73	31.92	31.43	30.35	-
Senegal	0.153	-	-	-	-	-	3.79	-	-	-	-	-
Sri Lanka	16	11	9.551	11	-	-	51.02	45.91	40.37	43.09	-	-
Sweden	2.400	-	-	-	-	-	42.86	-	-	-	-	-
Thailand	228	165	-	-	-	-	76.72	70.57	-	-	-	-
Turkey	45	-	-	-	-	-	28.58	-	-	-	-	-
United Kingdom	-	-	31	-	-	-	-	-	36.47	-	-	-
United Republic of Tanzania	6.587	6.509	-	-	-	-	28.00	28.00	-	-	-	-
United States of America	-	-	291	286	289	284	-	-	87.13	87.20	86.79	86.59
Zambia	-	-	-	-	0.450	-	-	-	-	-	12.76	-

Output and Output per Employee

Country	Output ($ bil.)						Output per Employee ($)					
	1990	1991	1992	1993	1994	1995	1990	1991	1992	1993	1994	1995
Albania	-	-	-	*0.007	*0.001	*0.000	-	-	-	1,198	313	1,421
Australia	*1.760	*1.707	-	-	-	-	103,539	106,691	-	-	-	-
Azerbaijan	±0.816	±0.972	±0.048	±0.123	±0.062	-	31,682	39,217	1,954	4,753	2,232	-
Bahrain	-	-	±0.000	-	-	-	-	-	15,957	-	-	-
Bangladesh	1.102	1.154	1.251	-	-	-	2,278	2,229	2,298	-	-	-
Bolivia	0.037	0.037	0.040	0.037	0.039	0.038	16,323	17,140	17,302	16,007	17,874	18,351
Canada	2.751	2.592	2.333	2.232	2.365	2.441	105,812	108,013	106,049	106,304	107,512	109,714
Chile	0.492	0.562	0.574	0.522	0.516	0.530	26,286	30,702	32,100	33,467	36,170	42,100
China (Hong Kong SAR)	3.423	3.340	3.406	2.871	2.577	-	62,242	69,298	75,350	80,654	87,356	-
Colombia	1.209	1.219	1.164	1.153	1.283	1.581	34,286	33,000	29,245	31,211	35,821	43,699
Costa Rica	0.043	0.044	0.047	0.056	0.052	0.045	11,515	13,367	13,661	17,812	-	-
Cote d'Ivoire	±0.264	±0.227	±0.257	±0.226	-	-	-	-	-	-	-	-
Cyprus	0.020	0.018	0.020	0.018	0.018	0.016	53,537	48,845	53,528	54,138	51,964	51,195
Denmark	*0.356	*0.328	-	-	-	-	79,936	75,253	-	-	-	-
Ecuador	0.194	0.207	0.197	0.193	0.204	0.253	19,112	22,653	20,725	22,363	25,797	37,253
Egypt	*2.616	*1.695	*1.744	*1.896	-	-	10,715	7,883	7,522	7,855	-	-
El Salvador	-	-	-	±0.083	±0.074	±0.089	-	-	-	16,154	17,451	19,052
Finland	±0.304	±0.226	±0.216	±0.201	±0.252	-	92,003	80,719	93,860	91,350	119,879	-
Germany	-	14	14	12	-	-	-	89,141	122,023	117,689	-	-
Germany (Western Part)	15	14	14	-	-	-	125,707	122,866	135,569	-	-	-
Greece	*2.206	*2.030	*1.705	-	-	-	70,707	76,308	73,312	-	-	-
Guatemala	-	0.200	0.245	0.319	0.201	0.098	-	34,455	41,355	54,102	33,353	15,223
Honduras	0.034	0.032	0.034	0.032	0.029	0.035	7,890	6,495	5,515	4,757	3,962	4,757
Iceland	0.017	0.016	±0.010	±0.004	-	-	120,272	77,491	152,467	69,038	-	-
India	*15	*13	*13	*13	-	-	11,234	9,884	9,859	9,973	-	-
Indonesia	3.684	4.442	5.260	5.577	8.261	9.187	11,432	12,002	12,672	12,955	18,136	19,254
Iran (Islamic Republic of)	-	2.296	2.108	1.847	-	-	-	19,834	17,784	16,117	-	-
Japan	±33	±35	±35	±34	-	-	107,006	120,116	127,576	134,639	-	-
Jordan	0.035	0.032	0.030	0.024	0.032	0.039	48,164	41,720	37,913	25,241	31,739	28,694
Korea, Republic of	13	15	15	15	16	19	49,512	59,741	65,131	67,675	75,102	95,686
Kyrgyzstan	0.843	1.380	0.761	0.121	0.093	-	35,763	59,176	29,601	4,875	4,871	-
Lithuania	-	-	0.152	0.122	-	-	-	-	5,721	5,101	-	-
Macau	0.086	0.096	0.073	0.048	0.045	0.032	34,714	42,275	37,421	34,291	42,812	40,779
Malaysia	*0.659	*0.830	*0.953	*1.056	*1.250	-	32,320	33,873	37,819	39,569	49,409	-
Malta	-	-	-	0.014	0.014	-	-	-	-	63,864	-	-
Mauritius	-	-	-	-	0.113	-	-	-	-	-	-	-
Mongolia	0.018	0.016	0.009	0.002	0.005	0.007	16,472	16,928	8,495	2,753	3,241	5,998
Myanmar	0.109	0.152	0.176	0.176	0.237	-	6,096	10,700	12,668	12,743	18,069	-
Nepal	0.043	0.044	-	0.045	0.046	-	5,273	5,318	-	5,705	3,831	-
Netherlands	1.079	1.055	1.064	0.912	-	-	116,904	118,995	123,891	114,242	-	-
Norway	±0.159	±0.158	-	-	-	-	98,790	102,734	-	-	-	-
Pakistan	-	4.359	-	-	-	-	-	18,812	-	-	-	-

Continued.

Output and Output per Employee
- Continued -

Country	Output ($ bil.)						Output per Employee ($)					
	1990	1991	1992	1993	1994	1995	1990	1991	1992	1993	1994	1995
Peru	0.968	0.886	0.878	-	-	-	33,559	36,184	37,793	-	-	-
Philippines	0.815	0.666	0.784	0.668	-	-	13,782	13,290	16,398	17,493	-	-
Portugal	3.052	2.850	3.039	2.476	2.496	-	33,400	33,596	38,814	34,208	38,440	-
Russian Federation	-	-	-	2.241	1.497	1.762	-	-	-	4,703	3,705	5,069
Senegal	0.082	-	-	-	-	-	20,331	-	-	-	-	-
Singapore	*0.116	*0.123	*0.115	*0.103	*0.104	-	58,020	62,619	60,071	60,321	74,674	-
Slovakia	-	±0.234	±0.219	±0.174	±0.165	-	-	11,427	12,073	11,835	12,748	-
South Africa	±1.463	±1.471	±1.428	-	-	-	-	-	-	-	-	-
Spain	*5.117	*5.034	*4.794	-	-	-	82,221	87,320	95,232	-	-	-
Sri Lanka	0.114	0.098	0.138	0.161	-	-	3,676	4,008	5,821	6,107	-	-
Swaziland	-	*0.014	*0.006	*0.005	-	-	-	71,453	28,070	27,480	-	-
Sweden	*0.507	*0.569	*0.568	*0.403	*0.410	-	90,508	106,547	117,940	91,816	100,254	-
Thailand	24	4.864	-	-	-	-	79,567	20,845	-	-	-	-
Turkey	6.298	5.978	7.327	7.402	7.118	-	40,368	44,989	52,190	55,389	54,061	-
Ukraine	-	-	*4.735	*0.817	*0.649	*0.406	-	-	40,466	7,856	6,756	4,888
United Kingdom	-	-	7.157	-	-	-	-	-	84,203	-	-	-
United Republic of Tanzania	*0.060	*0.089	-	-	-	-	2,533	3,816	-	-	-	-
United States of America	*37	*37	*39	*40	*42	*44	105,735	110,090	116,608	122,750	127,156	133,314
Venezuela	0.636	0.677	0.632	0.509	0.499	0.721	29,291	33,493	30,840	29,215	30,060	43,728
Zambia	0.065	-	-	-	0.048	-	18,271	-	-	-	13,560	-

Value Added and Value Added per Employee

Country	Value Added ($ bil.)						Value Added per Employee ($)					
	1990	1991	1992	1993	1994	1995	1990	1991	1992	1993	1994	1995
Australia	*0.722	*0.696	-	-	-	-	42,463	43,485	-	-	-	-
Bangladesh	0.410	0.402	0.404	-	-	-	848	776	742	-	-	-
Bolivia	0.012	0.013	0.016	0.014	0.014	0.014	5,409	6,158	6,806	5,982	6,395	6,566
Canada	1.251	1.187	1.117	1.093	1.098	1.142	48,126	49,460	50,768	52,045	49,928	51,351
Chile	0.241	0.262	0.281	0.247	0.227	0.241	12,867	14,299	15,731	15,842	15,923	19,162
China (Hong Kong SAR)	0.925	0.942	0.968	0.846	0.712	-	16,812	19,537	21,419	23,777	24,150	-
Colombia	0.661	0.542	0.535	0.523	0.622	0.767	18,753	14,665	13,434	14,168	17,354	21,213
Costa Rica	0.014	0.014	0.015	0.018	0.016	0.014	3,709	4,379	4,489	5,704	-	-
Cote d'Ivoire	0.239	0.206	0.230	0.205	-	-	-	-	-	-	-	-
Cyprus	0.008	0.007	0.008	0.007	0.007	0.007	20,431	19,567	22,460	21,640	21,910	22,006
Denmark	*0.174	*0.169	-	-	-	-	39,115	38,738	-	-	-	-
Ecuador	0.073	0.072	0.078	0.067	0.075	0.107	7,206	7,884	8,191	7,802	9,505	15,737
Egypt	*0.746	*0.507	*0.443	*0.451	-	-	3,055	2,360	1,910	1,868	-	-
El Salvador	-	-	-	0.035	0.034	0.037	-	-	-	6,857	7,989	7,851
Finland	±0.110	±0.077	±0.090	±0.080	±0.108	-	33,203	27,642	39,310	36,285	51,507	-
Germany (Western Part)	6.079	5.723	5.958	4.879	-	-	51,556	50,827	58,125	-	-	-
Greece	*0.727	*0.652	*0.658	-	-	-	23,286	24,498	28,306	-	-	-
Honduras	*0.011	*0.009	*0.010	*0.009	*0.008	*0.011	2,502	1,767	1,634	1,351	1,125	1,505
Iceland	0.004	0.002	±0.002	±0.001	-	-	25,286	11,241	37,428	15,855	-	-
India	*3.053	*2.165	*2.103	*2.600	-	-	2,294	1,686	1,585	1,944	-	-
Indonesia	*1.114	*1.241	*1.601	*1.590	*3.280	*3.086	3,456	3,354	3,859	3,692	7,200	6,468
Iran (Islamic Republic of)	-	1.092	0.913	0.775	-	-	-	9,436	7,701	6,764	-	-
Japan	±14	±16	±16	±16	-	-	46,710	53,256	58,221	62,263	-	-
Jordan	0.016	0.013	0.006	0.007	0.015	0.015	21,451	16,870	8,000	7,290	14,519	11,432
Korea, Republic of	4.993	6.572	7.020	6.734	7.418	8.748	18,755	25,917	29,538	30,518	34,080	43,370
Macau	0.023	0.028	0.019	0.016	0.013	0.011	9,407	12,268	9,888	11,160	12,231	13,309
Malaysia	*0.188	*0.258	*0.288	*0.349	*0.407	-	9,210	10,527	11,446	13,056	16,084	-
Malta	-	-	-	0.006	-	-	-	-	-	26,141	-	-
Mauritius	-	-	-	-	0.033	-	-	-	-	-	-	-
Mongolia	0.004	0.002	0.003	0.000	0.001	0.002	3,611	2,202	2,622	573	735	2,127
Myanmar	±0.126	±0.155	±0.153	±0.168	±0.191	-	7,034	10,911	11,009	12,144	14,537	-
Nepal	0.020	0.018	-	0.020	0.011	-	2,424	2,215	-	2,590	904	-
Netherlands	0.429	0.423	0.446	0.388	-	-	46,523	47,707	51,980	48,652	-	-

Continued.

Depending on the table, * means *Enterprises, Engaged,* or *Factor Values*; ± means *Basis Unspecified*; § means *shown in millions*. For additional notes and sources, see Appendix I.

Value Added and Value Added per Employee
- Continued -

Country	Value Added ($ bil.)						Value Added per Employee ($)					
	1990	1991	1992	1993	1994	1995	1990	1991	1992	1993	1994	1995
Norway	±0.058	±0.063	-	-	-	-	36,339	40,772	-	-	-	-
Pakistan	-	1.211	-	-	-	-	-	5,229	-	-	-	-
Peru	0.503	0.357	0.326	-	-	-	17,428	14,589	14,045	-	-	-
Philippines	0.271	0.208	0.252	0.264	-	-	4,589	4,144	5,269	6,904	-	-
Portugal	1.070	0.953	0.986	0.869	0.903	-	11,708	11,238	12,597	12,004	13,906	-
Russian Federation	-	-	-	±0.802	±0.676	±0.707	-	-	-	1,684	1,673	2,036
Senegal	0.012	-	-	-	0.015	-	2,974	-	-	-	6,819	-
Singapore	*0.042	*0.040	*0.039	*0.039	*0.039	-	20,833	20,245	20,650	22,822	27,846	-
Slovakia	-	-	-	±0.057	±0.052	-	-	-	-	3,917	4,047	-
Spain	*2.052	*1.914	*1.795	-	-	-	32,979	33,203	35,663	-	-	-
Sri Lanka	0.052	0.052	0.070	0.071	-	-	1,681	2,104	2,966	2,688	-	-
Swaziland	-	*0.003	*0.001	*-0.000	-	-	-	15,426	2,867	-994	-	-
Sweden	*0.297	*0.223	*0.228	*0.181	*0.182	-	53,098	41,715	47,282	41,246	44,636	-
Thailand	6.698	1.828	-	-	-	-	22,537	7,832	-	-	-	-
Turkey	2.577	2.571	3.101	3.021	2.843	-	16,518	19,351	22,092	22,608	21,592	-
United Kingdom	*3.089	-	*3.260	-	-	-	31,848	-	38,350	-	-	-
United Republic of Tanzania	*-0.005	*-0.041	-	-	-	-	-222	-1,746	-	-	-	-
United States of America	*15	*15	*16	*16	*17	*17	42,795	44,836	48,527	49,942	51,411	52,454
Venezuela	0.266	0.249	0.258	0.173	0.180	0.544	12,263	12,304	12,570	9,942	10,836	32,964
Zambia	0.034	-	-	-	0.016	-	9,516	-	-	-	4,522	-

Capital Investment

Country	Gross Fixed Capital Formation ($ mil.)						Per Establishment ($ mil)					
	1990	1991	1992	1993	1994	1995	1990	1991	1992	1993	1994	1995
Bangladesh	-	30	48	-	-	-	-	0.003	0.004	-	-	-
Bolivia	2.255	2.366	2.844	3.361	2.852	-	0.044	0.046	0.055	0.069	0.063	-
China (Hong Kong SAR)	-	-	189	76	48	-	-	0.103	0.050	0.031	-	-
Colombia	-	73	152	106	-31	-	-	0.324	0.647	0.508	-0.155	-
Cyprus	2.230	2.695	1.162	0.374	0.627	0.759	0.124	0.150	0.065	0.022	0.042	0.054
Ecuador	-	50	39	11	20	24	-	0.749	0.553	0.161	0.290	0.367
El Salvador	-	-	-	6.106	8.797	5.426	-	-	-	0.436	0.628	0.319
Finland	30	8.952	1.362	3.151	7.964	-	0.679	0.176	0.028	0.066	0.173	-
Germany (Western Part)	-	-	615	480	-	-	-	-	1.038	-	-	-
Greece	127	76	108	-	-	-	0.289	0.178	0.252	-	-	-
Hungary	-	-	9.065	17	-	-	-	-	0.070	0.107	-	-
India	1,003	890	1,071	1,217	-	-	0.092	0.082	0.084	0.096	-	-
Indonesia	964	1,024	697	598	713	705	0.815	0.863	0.548	0.454	0.541	0.469
Iran (Islamic Republic of)	-	143	174	136	-	-	-	0.311	0.332	0.278	-	-
Japan	1,437	1,930	1,595	1,376	-	-	0.079	0.112	0.099	0.092	-	-
Jordan	-	-	-	-	-	2.007	-	-	-	-	-	0.096
Korea, Republic of	-	-	2,605	1,899	2,015	2,381	-	-	0.555	0.354	0.361	0.436
Macau	-	-	-	19	1.352	0.595	-	-	-	0.285	0.025	0.012
Malaysia	-	-	-	206	259	-	-	-	-	1.398	1.725	-
Malta	-	-	-	0.005	-	-	-	-	-	0.002	-	-
Mongolia	39	23	6.149	0.876	0.955	-	39	23	6.149	0.876	0.191	-
Myanmar	150	142	141	117	122	-	6.016	5.918	6.150	5.086	7.167	-
Pakistan	-	279	-	-	-	-	-	0.225	-	-	-	-
Peru	71	42	41	-	-	-	0.127	0.072	0.067	-	-	-
Philippines	52	39	39	31	-	-	0.092	0.070	0.179	0.159	-	-
Portugal	230	266	234	78	191	-	0.197	0.247	0.201	0.071	0.173	-
Senegal	6.362	-	-	-	-	-	0.795	-	-	-	-	-
Singapore	10	6.904	4.807	3.160	5.327	-	0.385	0.276	0.209	0.137	0.266	-
Spain	198	228	196	-	-	-	0.095	0.123	0.117	-	-	-
Sri Lanka	21	21	19	16	-	-	0.060	0.077	0.069	0.051	-	-
Swaziland	-	0.048	-	0.001	-	-	-	0.016	-	0.000	-	-
Thailand	1,943	444	-	-	-	-	2.849	0.442	-	-	-	-
Turkey	598	343	389	433	348	-	1.294	0.771	0.477	0.586	0.490	-

Continued.

Depending on the table, * means *Enterprises*, *Engaged*, or *Factor Values*; ± means *Basis Unspecified*; § means *shown in millions*. For additional notes and sources, see Appendix I.

Capital Investment

- Continued -

Country	Gross Fixed Capital Formation ($ mil.)						Per Establishment ($ mil)					
	1990	1991	1992	1993	1994	1995	1990	1991	1992	1993	1994	1995
Ukraine	-	-	50	6.143	16	1.224	-	-	0.091	0.013	0.036	0.003
United Kingdom	-	-	309				-	-	0.182	-	-	-
United Republic of Tanzania . .	127	199	-	-	-	-	2.273	3.310	-	-	-	-
United States of America . . .	-	-	1,364	1,497	1,921	2,029	-	-	0.459	-	-	-
Zambia	0.252	-	-	-	-	-	0.036	-	-	-	-	-

Indexes of Industrial Production

Country	Index of Industrial Production (1990=100)						Index of Employment (1990=100)					
	1990	1991	1992	1993	1994	1995	1990	1991	1992	1993	1994	1995
Australia	-	-	-	-	-	-	100	94	-	-	-	-
Azerbaijan	-	-	-	-	-	-	100	96	96	101	107	-
Bangladesh	-	-	-	-	-	-	100	107	113	-	-	-
Bolivia	-	-	-	-	-	-	100	96	103	102	95	92
Bosnia & Herzegovina . . .	-	-	-	-	-	-	100	73	-	-	8	-
Canada	-	-	-	-	-	-	100	92	85	81	85	86
Chile	-	-	-	-	-	-	100	98	96	83	76	67
China (Hong Kong SAR) . . .	-	-	-	-	-	-	100	88	82	65	54	-
Colombia	-	-	-	-	-	-	100	105	113	105	102	103
Costa Rica	-	-	-	-	-	-	100	88	93	84	-	-
Cyprus	-	-	-	-	-	-	100	99	98	89	91	82
Denmark	-	-	-	-	-	-	100	98	85	-	-	-
Ecuador	-	-	-	-	-	-	100	90	94	85	78	67
Egypt	-	-	-	-	-	-	100	88	95	99	-	-
Finland	-	-	-	-	-	-	100	85	70	67	64	-
Germany (Western Part) . . .	-	-	-	-	-	-	100	95	87	-	-	-
Greece	-	-	-	-	-	-	100	85	75	-	-	-
Honduras	-	-	-	-	-	-	100	114	144	157	172	173
Iceland	-	-	-	-	-	-	100	143	46	45	50	-
India	-	-	-	-	-	-	100	96	100	100	-	-
Indonesia	-	-	-	-	-	-	100	115	129	134	141	148
Japan	-	-	-	-	-	-	100	96	90	82	-	-
Jordan	-	-	-	-	-	-	100	106	108	129	140	185
Kenya	-	-	-	-	-	-	100	94	92	95	92	90
Korea, Republic of	-	-	-	-	-	-	100	95	89	83	82	76
Kyrgyzstan	-	-	-	-	-	-	100	99	109	105	81	-
Macau	-	-	-	-	-	-	100	92	79	57	43	32
Malawi	-	-	-	-	-	-	100	-	-	-	-	-
Malaysia	-	-	-	-	-	-	100	120	124	131	124	-
Malta	-	-	-	-	-	-	100	-	-	114	-	-
Mongolia	-	-	-	-	-	-	100	90	94	70	147	107
Myanmar	-	-	-	-	-	-	100	80	78	77	73	-
Nepal	-	-	-	-	-	-	100	100	-	95	147	-
Netherlands	-	-	-	-	-	-	100	96	93	86	-	-
New Zealand	-	-	-	-	-	-	100	96	184	188	-	-
Norway	-	-	-	-	-	-	100	95	93	-	-	-
Peru	-	-	-	-	-	-	100	85	81	-	-	-
Philippines	-	-	-	-	-	-	100	85	81	65	-	-
Portugal	-	-	-	-	-	-	100	93	86	79	71	-
Senegal	-	-	-	-	-	-	100	-	43	42	56	-
Singapore	-	-	-	-	-	-	100	98	95	85	70	-
Spain	-	-	-	-	-	-	100	93	81	-	-	-
Sri Lanka	-	-	-	-	-	-	100	79	76	85	-	-
Sweden	-	-	-	-	-	-	100	95	86	78	73	-
Thailand	-	-	-	-	-	-	100	79	-	-	-	-
Turkey	-	-	-	-	-	-	100	85	90	86	84	-
United Kingdom	-	-	-	-	-	-	100	-	88	-	-	-
United Republic of Tanzania . .	-	-	-	-	-	-	100	99	-	-	-	-

Continued.

Depending on the table, * means *Enterprises, Engaged,* or *Factor Values;* ± means *Basis Unspecified;* § means *shown in millions.* For additional notes and sources, see Appendix I.

Indexes of Industrial Production
- Continued -

Country	Index of Industrial Production (1990=100)						Index of Employment (1990=100)					
	1990	1991	1992	1993	1994	1995	1990	1991	1992	1993	1994	1995
United States of America . . .	-	-	-	-	-	-	100	97	96	95	96	95
Uruguay	-	-	-	-	-	-	100	96	84	67	-	-
Venezuela	-	-	-	-	-	-	100	93	94	80	77	76
Zambia	-	-	-	-	-	-	100	-	-	-	99	-

ISIC 3220
WEARING APPAREL

ISIC 3220—Wearing Apparel—includes men's and boys' jackets, overcoats, raincoats, suits, trousers, shirts, and underwear; and women's and girls' blouses, coats, dresses, skirts, slacks, shorts, suits, and underwear.

The industry corresponds, in the U.S. system, to U.S. SICs 2311 through 2369 (Men's and Women's Apparel) and 2385 (Waterproof Outerwear).

Summary Statistics

		1990 Value	N	1991 Value	N	1992 Value	N	1993 Value	N	1994 Value	N	1995 Value	N
Establishments or enterprises	(number)	124,230	82	125,005	94	130,713	87	113,613	87	103,401	71	59,902	37
Number employed	(000)	8,415	94	6,420	98	5,938	93	6,128	95	5,475	83	4,798	67
Total Wages	($ mil.)	49,178	88	49,056	91	50,743	85	42,114	86	40,519	74	34,412	60
Wage/salary per employee	($)	5,750	87	6,159	89	11,441	84	10,245	85	6,014	73	6,717	60
Female workers	(000)	1,573	33	1,295	33	1,917	30	1,984	34	1,620	28	1,325	16
as % of total employment*	(%)	73	33	72	33	71	30	69	34	69	28	74	16
Output	($ bil.)	232	91	248	94	269	89	215	86	207	76	181	64
per employee	($ 000)	26	87	28	90	480	86	607	83	30	72	33	61
per establishment	($ 000)	1,924	77	2,077	85	23,508	78	4,080	74	1,570	61	1,090	32
Value added	($ bil.)	89	81	93	82	96	75	94	76	87	68	85	61
per employee	($ 000)	11	79	12	80	13	73	12	72	13	64	15	57
per establishment	($ 000)	751	70	681	76	626	67	586	65	621	53	489	29
Capital investment	($ mil.)	4,640	66	4,019	64	3,912	56	3,084	53	2,816	44	1,533	23
per employee	($ 000)	0.791	63	0.723	61	0.774	55	0.585	51	0.670	43	0.609	22
per establishment	($ 000)	109	62	68	61	58	54	40	49	38	42	30	20

Data presented above are drawn from the detailed tables that follow. Columns headed 'N' show the number of countries that provided valid data for inclusion. Values are not strictly comparable one year to the next or one row to the next because the number of countries that report varies from period to period and row to row. However, a general indication of magnitudes can be gleaned from reviewing this summary—especially in earlier years in which more reports are available. For detailed explanations, see Appendix I. *The average for those countries reporting both total and female employment.

Establishments and Number Engaged

Country	Establishments or Enterprises (number)						Employees per Establishment					
	1990	1991	1992	1993	1994	1995	1990	1991	1992	1993	1994	1995
Albania	-	-	-	30	17	9.000	-	-	-	213	210	169
Angola	-	-	18	18	-	-	-	-	44	3.833	-	-
Argentina	-	-	-	5,727	6,646	-	-	-	-	7.097	6.396	-
Armenia	75	91	55	61	62	-	501	399	478	411	-	-
Australia	1,703	2,091	2,125	-	-	-	27	18	17	-	-	-
Austria	403	369	342	308	280	-	60	60	56	51	48	-
Bahamas	52	48	-	-	-	-	4.038	3.979	-	-	-	-
Bahrain	-	-	1,231	-	-	-	-	-	6.026	-	-	-
Bangladesh	666	727	685	-	-	-	245	288	315	-	-	-
Barbados	14	9.000	8.000	9.000	30	-	79	197	163	61	27	-
Belgium	2,032	1,991	1,915	1,314	1,172	-	18	17	17	-	-	-
Bolivia	82	91	94	96	93	90	11	13	13	16	17	20
Bosnia & Herzegovina	66	67	-	-	8.000	-	500	446	-	-	180	-
Bulgaria	*98	*99	*138	*2,175	*2,950	*3,097	654	534	325	18	16	15
Burundi	*6.000	*8.000	-	-	-	-	37	28	-	-	-	-
Cameroon	-	*11	*7.000	*11	*6.000	*6.000	-	39	50	75	44	26
Canada	2,656	2,274	2,033	1,828	1,674	-	35	37	37	41	44	-
Cape Verde	17	17	20	24	27	-	-	-	-	-	-	-
Central African Republic	-	*1.000	*1.000	*1.000	*1.000	*1.000	-	-	-	-	-	-
Chile	106	115	120	121	118	-	163	157	161	156	162	-
China (Hong Kong SAR)	9,217	7,652	6,711	4,803	3,842	-	24	25	25	27	27	-
Colombia	1,049	972	983	940	901	-	45	49	67	69	74	-
Costa Rica	612	573	554	559	-	-	48	52	60	62	-	-
Croatia	238	312	458	683	879	986	167	112	68	46	35	31
Cyprus	1,430	1,420	1,410	1,350	1,266	1,206	7.883	7.753	7.644	6.430	5.451	5.531
Czechoslovakia (Former)	*26	*29	-	-	-	-	3,577	1,690	-	-	-	-
Czech Republic	-	*20	*83	*133	-	-	-	1,500	386	248	-	-
Denmark	1,354	1,491	1,440	-	-	-	8.549	6.957	6.713	-	-	-
Ecuador	94	118	112	105	111	117	40	40	44	39	37	36
Egypt	333	371	349	344	-	-	75	82	83	86	-	-
El Salvador	-	42	52	61	102	78	-	-	128	171	113	155
Equatorial Guinea	10	-	-	-	-	-	-	-	-	-	-	-
Ethiopia	9.000	10	8.000	8.000	17	13	509	430	485	480	238	305
Finland	265	238	186	172	147	174	55	48	44	42	45	44
FYR Macedonia	*97	*188	*301	*307	*367	*474	307	154	90	84	73	51
Gabon	-	3.000	3.000	3.000	1.000	2.000	-	67	72	85	116	98
Gambia	-	-	-	57	-	-	-	-	-	2.351	-	-
Germany	-	1,941	1,713	1,463	1,244	-	-	96	83	80	82	-
Germany (Western Part)	1,755	1,688	1,546	-	-	-	82	83	81	-	-	-
Ghana	-	-	-	9.000	-	-	-	-	-	100	-	-
Greece	1,180	1,155	1,168	-	-	-	34	32	31	-	-	-
Grenada	-	-	13	16	17	17	-	-	15	12	8.176	15
Guatemala	-	14	14	14	14	14	-	141	154	152	152	142
Honduras	-	-	104	123	153	166	-	-	344	334	333	391
Hungary	*446	*915	*800	*883	-	-	139	62	73	67	-	-
Iceland	95	102	98	85	-	-	4.705	4.833	5.122	5.459	-	-
India	1,740	1,931	2,279	3,119	-	-	64	66	66	67	-	-
Indonesia	1,766	1,699	1,870	1,798	1,862	2,110	137	162	170	195	191	176
Iran (Islamic Republic of)	481	138	133	123	-	-	21	51	43	42	-	-
Iraq	-	24	42	-	-	-	-	317	190	-	-	-
Ireland	251	213	-	-	-	-	43	43	-	-	-	-
Israel	1,082	1,189	1,147	1,373	1,421	-	29	27	28	26	25	-
Italy	*2,739	*2,842	*3,899	*4,020	*3,852	-	56	54	47	44	43	-
Jamaica	123	123	-	-	-	-	107	103	-	-	-	-
Japan	24,611	24,427	23,411	22,703	20,166	19,205	20	20	20	20	20	19
Jordan	1,268	1,321	1,535	1,536	1,306	1,485	1.774	2.178	2.562	2.329	4.312	3.696
Kenya	232	426	518	533	557	557	30	16	13	13	13	13
Korea, Republic of	6,561	6,507	6,468	7,987	8,293	8,545	35	30	29	25	23	22
Kuwait	1,944	1,925	1,912	1,905	1,899	-	4.320	3.686	5.363	5.697	5.190	-
Kyrgyzstan	127	116	10	92	74	-	254	271	1,797	145	126	-

Continued.

Depending on the table, * means *Enterprises, Engaged,* or *Factor Values*; ± means *Basis Unspecified*; § means *shown in millions.* For additional notes and sources, see Appendix I.

Establishments and Number Engaged
- Continued -

Country	Establishments or Enterprises (number)						Employees per Establishment					
	1990	1991	1992	1993	1994	1995	1990	1991	1992	1993	1994	1995
Latvia	82	210	170	169	97	202	237	112	74	56	86	44
Lithuania	-	-	-	*115	*135	-	-	-	-	157	165	-
Macau	745	687	644	588	512	422	48	50	48	48	51	59
Malawi	*8.000	-	-	-	-	-	138	-	-	-	-	-
Malaysia	323	354	363	377	352	-	200	197	197	188	186	-
Malta	142	154	140	135	125	-	42	35	36	36	36	-
Mauritius	401	466	457	433	392	367	190	165	171	174	183	188
Mexico	180	175	171	156	153	151	165	165	166	162	158	173
Mongolia	49	40	43	50	103	59	240	334	172	87	51	118
Morocco	792	824	824	796	781	742	101	109	113	119	129	137
Mozambique	23	23	17	17	10	-	192	188	166	163	182	-
Nepal	95	234	-	325	298	-	122	74	-	84	84	-
Netherlands	135	126	122	104	128	-	59	59	56	59	48	-
Netherlands Antilles	-	-	-	29	-	-	-	-	-	5.345	-	-
New Zealand	*1,087	*1,127	*1,143	*1,212	-	-	11	9.939	9.890	9.408	-	-
Norway	86	90	67	53	51	-	24	23	28	29	29	-
Pakistan	-	153	-	-	-	-	-	128	-	-	-	-
Panama	108	91	-	-	-	-	68	76	-	-	-	-
Paraguay	-	4.000	-	-	-	-	-	54	-	-	-	-
Peru	1,884	1,968	2,066	-	-	-	13	7.936	6.906	-	-	-
Philippines	14,123	14,398	1,861	1,722	-	-	13	13	95	93	-	-
Poland	*524	*728	*681	*691	-	-	305	221	209	210	-	-
Portugal	*8,273	*8,805	*8,405	*8,471	*8,481	-	18	18	18	17	18	-
Puerto Rico	255	246	253	243	219	207	115	114	114	112	107	117
Republic of Moldova	72	32	32	23	24	25	718	1,014	857	641	621	511
Romania	-	881	1,690	2,245	3,963	-	-	278	122	89	53	-
Russian Federation	-	-	-	*9,051	*7,921	*12,673	-	-	-	70	69	47
Saint Lucia	-	27	22	22	22	23	-	-	-	-	-	-
Saint Vincent & the Grenadines	-	-	-	3.000	3.000	4.000	-	-	-	99	96	75
Senegal	1.000	2.000	2.000	2.000	-	-	68	290	204	310	-	-
Sierra Leone	-	-	-	28	-	-	-	-	-	22	-	-
Singapore	370	351	347	311	275	-	75	74	68	67	66	-
Slovakia	-	47	47	69	83	-	-	510	489	340	315	-
Slovenia	178	*399	*2,548	*2,560	*2,927	*1,826	-	-	-	-	-	-
South Africa	-	1,631	-	-	-	-	-	74	-	-	-	-
Spain	5,799	5,430	4,847	-	-	-	15	16	17	-	-	-
Sri Lanka	297	282	246	292	-	-	308	374	428	464	-	-
Suriname	-	-	23	-	-	-	-	-	17	-	-	-
Swaziland	-	14	14	14	-	-	-	110	167	147	-	-
Sweden	183	143	106	85	78	-	37	33	36	31	32	-
Thailand	922	1,327	-	-	-	-	326	171	-	-	-	-
Tonga	-	11	14	14	8.000	-	-	8.091	7.429	5.929	3.375	-
Trinidad and Tobago	130	134	131	130	189	-	21	19	20	19	12	-
Tunisia	-	-	-	1,609	1,716	1,687	-	-	-	48	52	56
Turkey	704	699	1,591	1,533	1,446	-	106	99	54	58	62	-
Ukraine	1,965	3,103	3,250	1,291	1,265	1,162	186	109	80	130	113	103
United Kingdom	8,390	7,841	7,283	7,444	7,855	-	25	23	23	24	21	-
United Republic of Tanzania	53	48	-	-	-	-	36	25	-	-	-	-
USSR (Former)	*5,390	-	-	-	-	-	375	-	-	-	-	-
United States of America	-	-	19,032	-	-	-	-	-	41	-	-	-
Venezuela	873	927	884	871	840	771	33	31	32	32	30	29
Yugoslavia	332	523	949	1,204	1,232	1,229	-	140	79	60	56	55
Zambia	34	-	-	-	46	-	108	-	-	-	50	-
Zimbabwe	111	106	106	96	96	-	187	187	535	179	185	-

Depending on the table, * means *Enterprises, Engaged,* or *Factor Values;* ± means *Basis Unspecified;* § means *shown in millions.* For additional notes and sources, see Appendix I.

Employment and Compensation of Employees

Country	Number Employed/Engaged (000)						Wage/Salary per Employee ($)					
	1990	1991	1992	1993	1994	1995	1990	1991	1992	1993	1994	1995
Albania	-	-	-	6.376	3.565	1.522	-	-	-	-	-	818
Angola	-	*3.113	*0.788	*0.069	-	-	-	-	-	-	-	-
Argentina	33	-	-	41	43	-	4,217	-	-	7,966	-	-
Armenia	*38	*36	*26	*25	-	*20	-	-	-	9,691	-	-
Australia	46	*38	*36	-	-	-	14,023	16,751	15,911		-	-
Austria	24	22	19	16	13	12	14,364	15,070	17,194	17,905	19,022	22,564
Bahamas	0.210	0.191		-	-	-	9,871	6,173	-	-	-	-
Bahrain	-	-	7.418	-	-	-	-	-	2,376	-	-	-
Bangladesh	163	210	216	-	-	-	-	-	-	-	-	-
Barbados	1.103	1.774	1.302	0.546	0.800	1.074	4,801	2,569	2,924	4,931	4,758	4,068
Belgium	36	34	33	-	-	-	12,032	12,310	14,026	-	-	-
Bermuda	0.027	0.030	0.028	0.022	0.023	0.020	-	-	-	-	-	-
Bolivia	0.906	1.144	1.246	1.494	1.584	1.785	1,098	908	865	1,109	1,415	1,549
Bosnia & Herzegovina	33	30	-	-	1.443	-	2,189	1,987	-	-	-	-
Bulgaria	64	53	45	39	48	47	1,521	458	694	862	728	918
Burundi	0.220	0.220	-	-	-	-	1,837	1,833	-	-	-	-
Cambodia	*11	*8.276		-	-	-	-	-	-	-	-	-
Cameroon	0.573	0.427	0.349	0.825	0.266	0.157	5,884	6,276	6,311	5,338	2,912	2,386
Canada	94	84	76	75	73	74	16,402	17,176	16,993	15,906	15,589	15,895
Chile	17	18	19	19	19	18	2,876	3,558	3,991	4,283	5,464	6,351
China (Hong Kong SAR)	*224	*190	*171	*131	*105	-	7,700	8,676	9,641	10,176	11,429	-
China (Taiwan Province)	109	105	97	92	89	82	6,941	7,542	8,402	8,764	9,505	9,933
Colombia	47	47	66	65	66	65	1,363	1,395	1,546	1,797	2,274	2,495
Costa Rica	29	30	33	35	-	-	2,145	2,085	2,427	2,677	-	-
Croatia	40	35	31	31	31	30	3,349	3,647	1,558	1,751	2,195	3,309
Cyprus	11	11	11	8.681	6.901	6.670	6,222	6,683	7,363	7,006	7,656	9,925
Czechoslovakia (Former)	93	49	-	-	-	-	1,743	1,246	-	-	-	-
Czech Republic	35	30	32	33	-	-	1,846	1,221	1,504	1,790	-	-
Denmark	12	10	9.667	7.997	8.525	8.789	19,234	19,790	22,245	21,429	22,837	26,305
Ecuador	3.800	4.755	4.906	4.136	4.136	4.255	1,306	1,225	1,453	1,961	904	1,041
Egypt	*25	*30	*29	*30	*30	*30	1,151	601	769	793	808	850
El Salvador	-	-	*6.652	*10	*12	*12	-	-	-	1,547	2,299	1,906
Equatorial Guinea	0.007	-	-	-	-	-	1,889	-	-	-	-	-
Ethiopia	4.584	4.302	3.882	3.839	4.043	3.962	1,406	1,197	848	600	550	469
Finland	15	11	8.200	7.200	6.600	7.703	19,033	19,038	17,658	13,929	16,539	21,163
FYR Macedonia	30	29	27	26	27	24	3,602	3,459	1,814	1,762	1,924	2,656
France	145	136	132	122	116	112	28,404	28,659	31,285	-	-	-
Gabon	-	0.200	0.217	0.255	0.116	0.196	-	11,609	9,140	8,130	8,167	3,414
Gambia	-	-	-	0.134	-	-	-	-	-	-	-	-
Germany	-	*186	*143	*117	*102	-	-	16,238	20,627	21,252	22,524	-
Germany (Western Part)	*143	*139	*126	-	-	-	18,683	19,452	21,917	-	-	-
Ghana	-	-	-	*0.900	*0.925	*0.951	-	-	-	-	-	-
Greece	*40	*37	*36	*35	*32	*29	7,342	7,357	7,935	7,851	7,781	8,677
Grenada	-	-	*0.191	*0.187	0.139	0.248	-	-	-	-	-	-
Guatemala	-	*1.967	*2.150	*2.135	*2.122	*1.991	-	-	-	-	-	-
Honduras	*12	*34	*36	*41	*51	*65	1,237	904	1,415	1,162	898	1,158
Hungary	62	57	58	59	62	61	1,725	1,884	2,157	2,090	2,267	2,383
Iceland	0.447	0.493	0.502	0.464	0.495	-	18,366	22,582	26,658	20,394	-	-
India	*112	*128	*150	*209	*209	*229	740	652	658	570	575	578
Indonesia	241	276	317	350	356	371	526	543	769	836	827	1,054
Iran (Islamic Republic of)	10	7.071	5.704	5.217	-	-	2,738	3,453	2,913	2,996	-	-
Iraq	-	7.602	8.000	-	-	-	-	4,805	8,322	-	-	-
Ireland	11	9.100	8.811	7.962	7.409	7.001	11,669	12,249	13,442	11,931	12,635	14,030
Israel	31	32	33	35	36	37	9,438	9,229	9,972	9,787	10,772	11,696
Italy	153	154	184	177	167	-	24,964	23,992	24,592	19,513	20,543	-
Jamaica	13	13	14	14	14	-	-	-	-	-	-	-
Japan	*488	*491	*477	*450	*406	*373	12,921	14,726	16,255	18,645	20,580	22,574
Jordan	2.249	2.877	3.933	3.577	5.632	5.489	1,675	1,834	1,672	1,534	1,562	1,795
Kenya	6.868	6.931	6.733	6.820	6.976	7.114	968	872	847	522	703	936
Korea, Republic of	231	198	187	197	190	186	6,466	7,505	8,144	9,150	10,370	12,523

Continued.

Depending on the table, * means *Enterprises, Engaged,* or *Factor Values*; ± means *Basis Unspecified*; § means *shown in millions*. For additional notes and sources, see Appendix I.

Employment and Compensation of Employees

- Continued -

Country	Number Employed/Engaged (000)						Wage/Salary per Employee ($)					
	1990	1991	1992	1993	1994	1995	1990	1991	1992	1993	1994	1995
Kuwait	8.398	7.095	10	11	9.856	11	1,900	2,351	4,909	4,988	5,027	6,724
Kyrgyzstan	*32	*31	*18	*13	*9.324	-	-	-	-	-	-	-
Latvia	19	24	13	9.537	8.370	8.822	-	-	-	-	-	-
Lithuania	-	-	19	18	22	-	-	-	-	539	860	-
Macau	36	34	31	28	26	25	3,830	4,336	4,744	5,120	5,336	5,641
Malawi	1.100	-	-	-	-	-	-	-	-	-	-	-
Malaysia	65	70	72	71	65	65	2,193	2,304	2,825	2,944	3,234	3,499
Malta	5.901	5.446	4.973	4.799	4.503	-	8,544	8,871	9,889	8,185	8,946	-
Mauritius	*76	77	78	75	72	69	1,505	1,604	2,185	2,117	2,338	2,867
Mexico	30	29	28	25	24	26	3,012	3,622	4,039	4,650	4,623	2,633
Mongolia	*12	*13	*7.407	*4.359	*5.278	*6.966	1,589	956	478	-	491	-
Morocco	80	90	93	95	101	102	1,838	1,952	2,104	2,074	2,106	2,428
Mozambique	4.414	4.332	2.817	2.769	1.818	-	1,291	707	-	-	-	-
Nepal	12	17	-	27	25	-	770	636	-	416	453	-
Netherlands	7.913	7.493	6.831	6.126	6.100	5.446	18,461	19,058	21,561	21,709	22,465	26,967
Netherlands Antilles	-	-	-	0.155	-	-	-	-	-	5,140	-	-
New Zealand	12	11	11	11	-	-	-	-	-	-	-	-
Norway	2.081	2.083	1.905	1.549	1.503	1.503	20,113	20,809	25,145	19,747	20,928	23,856
Pakistan	17	20	21	-	-	-	1,513	1,554	1,598	-	-	-
Panama	*7.338	*6.937	*7.323	*7.208	*6.500	*6.200	2,973	3,241	3,314	3,309	3,343	3,388
Paraguay	-	0.217	-	-	-	-	-	-	-	-	-	-
Peru	*24	*16	*14	-	-	-	1,437	1,683	1,580	-	-	-
Philippines	184	183	176	160	-	-	1,428	1,488	1,878	1,730	-	-
Poland	160	161	142	145	148	148	987	1,528	1,960	2,030	2,499	3,084
Portugal	*145	*158	*151	*146	*149	*168	-	-	-	-	-	-
Puerto Rico	29	28	29	27	23	24	14,772	9,936	10,570	11,051	11,726	11,617
Republic of Moldova	52	32	27	15	15	13	-	2,996	-	-	-	-
Romania	258	245	206	200	208	-	1,499	872	558	710	762	-
Russian Federation	-	-	-	630	547	602	-	-	-	549	589	454
Saint Vincent & the Grenadines	-	-	-	0.296	0.289	0.301	-	-	-	2,056	1,927	-
Senegal	*0.068	*0.581	*0.409	*0.621	-	-	1,134	1,617	2,291	2,644	-	-
Sierra Leone	-	-	-	*0.607	-	-	-	-	-	-	-	-
Singapore	*28	*26	*23	*21	*18	*17	6,442	7,192	8,241	8,392	9,523	10,461
Slovakia	-	*24	*23	*23	*26	-	-	1,231	1,477	1,615	1,715	-
South Africa	127	121	114	125	125	133	3,889	4,253	4,853	4,921	4,668	5,460
Spain	87	89	82	119	110	105	11,990	12,831	14,414	10,890	11,183	12,627
Sri Lanka	*91	*105	*105	*136	-	-	499	584	657	653	-	-
Suriname	*0.293	*0.400	*0.390	*0.203	-	-	5,354	6,022	6,608	13,523	-	-
Swaziland	-	1.547	2.343	2.056	-	-	-	4,755	3,783	2,872	-	-
Sweden	6.700	*4.777	*3.823	2.597	2.525	3.329	16,794	18,525	20,568	17,295	18,083	20,090
Thailand	300	227	-	-	-	-	1,982	3,020	-	-	-	-
Tonga	0.290	0.089	0.104	0.083	0.027	-	931	2,011	1,556	1,367	1,515	-
Trinidad and Tobago	2.743	2.571	2.596	2.418	2.270	-	2,790	2,956	2,604	2,064	2,082	-
Tunisia	-	-	-	78	88	94	-	-	-	3,202	2,846	3,512
Turkey	74	69	87	89	90	102	3,061	3,614	3,220	3,438	2,169	2,830
Ukraine	366	338	261	168	143	120	2,255	2,206	1,713	-	-	-
United Kingdom	210	177	169	181	168	168	11,327	13,019	14,092	11,506	12,780	13,603
United Republic of Tanzania	1.910	1.183	-	-	-	-	-	-	-	-	-	-
USSR (Former)	2,023	-	-	-	-	-	1,620	-	-	-	-	-
United States of America	807	776	777	771	734	724	13,408	13,918	14,700	15,109	15,440	15,588
Uruguay	*16	*17	*15	*13	*13	*10	2,377	2,556	2,948	2,971	3,074	3,401
Venezuela	28	29	28	28	25	22	2,663	2,846	3,009	2,800	2,423	3,564
Yugoslavia	-	73	75	73	69	67	-	-	-	-	931	-
Zambia	3.672	-	-	-	2.280	-	997	-	-	-	1,327	-
Zimbabwe	21	20	57	17	18	15	2,309	2,004	492	1,397	1,459	1,658

Depending on the table, * means *Enterprises*, *Engaged*, or *Factor Values*; ± means *Basis Unspecified*; § means *shown in millions*. For additional notes and sources, see Appendix I.

125

Female Workers: Total and Percent of Employment

Country	Female Workers (000)						Female Workers as % of Total Employment					
	1990	1991	1992	1993	1994	1995	1990	1991	1992	1993	1994	1995
Albania	-	-	-	5.587	1.980	0.520	-	-	-	87.63	55.54	34.17
Angola	-	-	0.310	0.041	-	-	-	-	39.34	59.42	-	-
Argentina	-	-	-	-	28	-	-	-	-	-	65.95	-
Australia	35	-	-	-	-	-	77.09	-	-	-	-	-
Austria	21	19	17	14	12	-	88.80	86.36	89.47	86.91	85.67	-
Bahrain	-	-	2.591	-	-	-	-	-	34.93	-	-	-
Bangladesh	116	148	150	-	-	-	71.23	70.47	69.39	-	-	-
Bermuda	0.020	0.022	0.021	0.016	0.014	0.014	74.07	73.33	75.00	72.73	60.87	70.00
Bosnia & Herzegovina	28	26	-	-	-	-	84.28	88.01	-	-	-	-
Bulgaria	-	-	-	35	42	40	-	-	-	89.41	87.42	85.74
Canada	72	64	-	-	-	-	76.60	76.19	-	-	-	-
Chile	11	12	12	12	12	-	65.86	65.73	62.73	64.17	63.36	-
China (Taiwan Province)	86	84	78	73	70	64	79.13	79.53	79.71	79.53	79.08	78.89
Colombia	-	38	-	53	53	-	-	80.16	-	81.20	79.46	-
Croatia	34	30	27	27	27	26	86.20	86.24	85.70	86.48	87.14	86.65
Cyprus	11	10	10	7.991	6.645	-	94.67	93.16	94.03	92.05	96.29	-
Czechoslovakia (Former)	63	34	-	-	-	-	67.85	68.98	-	-	-	-
Czech Republic	29	25	28	29	-	-	82.86	83.33	87.50	87.88	-	-
Denmark	11	9.716	9.074	-	-	-	91.95	93.67	93.87	-	-	-
Egypt	-	16	15	18	-	-	-	53.56	53.24	59.24	-	-
El Salvador	-	-	-	7.965	9.015	10	-	-	-	76.44	77.93	84.30
Ethiopia	-	3.019	2.795	2.771	2.921	2.808	-	70.18	72.00	72.18	72.25	70.87
FYR Macedonia	26	25	23	23	21	18	88.30	87.06	84.99	88.11	78.65	75.47
Gambia	-	-	-	0.009	-	-	-	-	-	6.72	-	-
Germany (Western Part)	121	-	-	-	-	-	84.49	-	-	-	-	-
Hungary	44	-	49	47	-	-	70.97	-	84.48	79.66	-	-
Indonesia	-	-	-	269	272	283	-	-	-	76.75	76.26	76.10
Iran (Islamic Republic of)	-	2.846	2.217	1.881	-	-	-	40.25	38.87	36.06	-	-
Iraq	-	6.582	-	-	-	-	-	86.58	-	-	-	-
Ireland	9.000	7.400	-	-	-	-	82.57	81.32	-	-	-	-
Italy	-	121	146	140	132	-	-	78.71	79.16	78.93	78.64	-
Jordan	0.657	0.980	1.404	1.210	2.345	2.281	29.21	34.06	35.70	33.83	41.64	41.56
Kenya	1.675	1.727	1.354	1.372	-	-	24.39	24.92	20.11	20.12	-	-
Korea, Republic of	168	144	139	139	134	130	72.70	72.67	74.17	70.72	70.63	70.00
Macau	26	26	24	22	21	20	74.52	76.37	76.62	77.28	78.27	78.95
Malaysia	55	60	62	60	53	-	85.27	86.27	86.01	84.30	81.22	-
Malta	4.825	4.457	4.067	3.947	3.618	-	81.77	81.84	81.78	82.25	80.35	-
Mongolia	-	-	-	-	4.040	-	-	-	-	-	76.54	-
Morocco	-	-	-	-	-	82	-	-	-	-	-	80.93
Nepal	1.783	1.991	-	4.796	4.281	-	15.35	11.54	-	17.52	17.12	-
New Zealand	9.971	9.182	9.092	9.269	-	-	81.41	81.97	80.43	81.29	-	-
Panama	4.858	-	-	-	-	-	66.20	-	-	-	-	-
Philippines	-	-	143	129	-	-	-	-	81.36	80.24	-	-
Portugal	48	54	53	49	51	-	33.18	34.17	34.83	33.68	34.09	-
Puerto Rico	26	25	25	24	20	21	88.92	87.95	87.20	86.95	86.57	85.60
Republic of Moldova	-	-	-	-	13	11	-	-	-	-	88.76	85.05
Saint Vincent & the Grenadines	-	-	-	0.134	0.134	-	-	-	-	45.27	46.37	-
Sri Lanka	82	95	93	121	-	-	90.17	90.60	87.98	89.05	-	-
Sweden	5.400	-	-	-	-	-	80.60	-	-	-	-	-
Thailand	252	189	-	-	-	-	84.05	83.13	-	-	-	-
United Kingdom	165	-	134	-	-	-	78.69	-	79.29	-	-	-
United Republic of Tanzania	1.108	0.686	-	-	-	-	58.01	57.99	-	-	-	-
United States of America	-	-	658	656	623	614	-	-	84.68	85.08	84.88	84.81
Zambia	-	-	-	-	0.311	-	-	-	-	-	13.64	-

    Depending on the table, * means *Enterprises, Engaged,* or *Factor Values;* ± means *Basis Unspecified;* § means *shown in millions.* For additional notes and sources, see Appendix I.

Output and Output per Employee

Country	Output ($ bil.)						Output per Employee ($)					
	1990	1991	1992	1993	1994	1995	1990	1991	1992	1993	1994	1995
Albania	-	-	-	*0.004	*0.003	*0.003	-	-	-	693	934	1,671
Angola	-	-	±30	±3.315	-	-	-	-	§38,65	§48,04	-	-
Argentina	±1.421	-	2.405	-	-	-	43,688	-	-	59,170	-	-
Armenia	0.020	0.033	0.055	0.970	0.007	-	528	918	2,090	38,664	-	-
Australia	*2.609	*2.605	*2.491	-	-	-	56,709	68,563	69,199	-	-	-
Austria	1.388	1.314	1.336	1.157	1.067	1.143	57,606	59,730	70,319	72,983	79,371	94,802
Bahamas	0.004	0.005					19,048	26,178	-	-	-	-
Bahrain	-	-	±0.083				-	-	11,146	-	-	-
Bangladesh	0.563	0.750	0.776	-			3,443	3,580	3,597	-	-	-
Barbados	0.011	0.011	0.008	0.005	0.009	0.009	9,609	6,089	6,163	8,534	10,654	8,340
Belgium	2.704	2.578	2.828	2.901	3.134	3.595	75,123	74,940	86,758	-	-	-
Benin	±0.036	±0.032	±0.041	±0.035	-	-						
Bolivia	0.009	0.010	0.013	0.019	0.016	0.018	10,339	8,901	10,394	12,814	10,383	10,155
Bosnia & Herzegovina	±0.273	±0.220	-	-	-	-	8,283	7,364	-	-	-	-
Bulgaria	±0.680	±0.143	±0.154	±0.152	±0.153	±0.188	10,607	2,706	3,424	3,917	3,145	3,993
Burundi	±0.002	±0.002	-	-	-	-	9,510	10,681	-	-	-	-
Cameroon	0.003	0.012	0.007	0.019	0.004	0.004	4,647	27,909	20,860	23,030	14,843	27,154
Canada	5.965	5.481	5.055	4.821	4.665	4.797	63,458	65,254	66,513	64,284	63,899	65,198
Cape Verde	±0.001	±0.001	-	-	-	-						
Chile	0.339	0.445	0.548	0.594	0.644	0.702	19,591	24,684	28,347	31,440	33,602	39,068
China (Hong Kong SAR)	8.247	8.266	8.495	7.423	6.264	-	36,753	43,622	49,652	56,749	59,432	-
China (Taiwan Province)	±6.870	±6.884	±6.426	±5.570	±4.690	±4.648	62,998	65,270	65,922	60,715	52,707	56,907
Colombia	0.569	0.569	0.695	0.749	0.949	1.036	12,167	12,018	10,491	11,518	14,292	15,824
Costa Rica	0.084	0.086	0.119	0.127	0.126	0.133	2,848	2,903	3,560	3,650	-	-
Croatia	±0.580	±0.599	±0.335	-	-	-	14,620	17,106	10,798	-	-	-
Cyprus	0.311	0.326	0.341	0.254	0.212	0.243	27,632	29,617	31,665	29,249	30,785	36,472
Czechoslovakia (Former)	±0.575	±0.199	-	-	-	-	6,188	4,071	-	-	-	-
Denmark	*0.531	*0.522	*0.601	*0.467	*0.541	*0.646	45,866	50,282	62,208	58,459	63,425	73,553
Ecuador	0.033	0.037	0.036	0.039	0.043	0.049	8,804	7,885	7,302	9,456	10,317	11,442
Egypt	*0.254	*0.206	*0.202	*0.214	*0.224	*0.251	10,203	6,782	6,988	7,189	7,550	8,421
El Salvador	-	-	±0.003	±0.048	-	±0.065	-	-	426	4,635	-	5,365
Equatorial Guinea	±0.000	-	-	-	-	-	6,821	-	-	-	-	-
Ethiopia	0.055	0.041	0.016	0.013	0.014	0.010	12,007	9,612	4,074	3,432	3,451	2,642
Finland	±0.899	±0.644	±0.447	±0.340	±0.440	±0.618	61,143	56,014	54,477	47,291	66,715	80,222
FYR Macedonia	±0.384	±0.375	±0.250	±0.144	±0.157	±0.176	12,913	12,919	9,192	5,623	5,845	7,301
France	±14	±13	±14	±12	±12	±13	94,163	95,191	103,381	97,977	105,533	114,726
Gabon	-	*0.009	*0.009	*0.008	*0.003	*0.004	-	42,802	39,591	31,632	27,188	21,802
Germany	-	15	15	14	13	-	-	82,475	107,613	118,601	131,106	-
Germany (Western Part)	14	15	15	-	-	-	100,843	107,140	119,527	-	-	-
Greece	*1.203	*1.123	*1.252	*1.219	*1.106	*1.106	30,289	30,002	34,968	34,615	34,371	38,421
Guatemala	-	0.029	0.031	0.031	0.034	0.041	-	14,662	14,465	14,589	16,041	20,504
Honduras	0.042	0.072	0.092	0.124	0.131	0.133	3,523	2,135	2,575	3,027	2,561	2,048
Hungary	±0.443	±0.452	±0.453	±0.432	±0.468	±0.467	7,145	7,934	7,817	7,319	7,505	7,611
Iceland	0.030	0.031	±0.034	±0.029	-	-	66,307	63,163	67,938	61,860	-	-
India	*1.415	*1.528	*1.681	*2.254	*2.417	*2.759	12,682	11,932	11,195	10,793	11,574	12,027
Indonesia	1.282	1.450	2.285	3.302	2.568	2.879	5,312	5,259	7,204	9,434	7,205	7,753
Iran (Islamic Republic of)	0.267	0.117	0.090	0.063	-	-	25,934	16,583	15,789	12,060	-	-
Iraq	-	*0.170	*0.345	-	-	-	-	22,392	43,127	-	-	-
Ireland	*0.448	*0.350	*0.372	*0.302	*0.299	*0.315	41,109	38,435	42,237	37,875	40,366	45,024
Israel	1.011	1.112	1.219	1.305	1.441	1.613	32,712	34,854	37,512	37,081	40,373	43,997
Italy	±17	±18	±21	±17	±18	-	112,093	113,579	115,718	95,868	109,226	-
Japan	±23	±26	±27	±28	±26	±27	46,223	53,158	56,777	61,851	65,161	73,195
Jordan	0.026	0.033	0.037	0.035	0.060	0.058	11,737	11,417	9,384	9,902	10,678	10,485
Kenya	*0.152	*0.165	*0.203	*0.126	*0.155	*0.059	22,111	23,812	30,149	18,455	22,198	8,254
Korea, Republic of	7.837	7.641	7.378	9.317	11	13	33,855	38,571	39,465	47,366	55,237	72,580
Kuwait	0.105	0.069	0.131	0.165	0.154	0.207	12,486	9,676	12,778	15,226	15,576	19,559
Kyrgyzstan	0.247	0.270	0.107	0.021	0.012	-	7,638	8,587	5,954	1,597	1,296	-
Latvia	0.003	0.008	0.024	0.019	0.029	0.037	139	342	1,859	2,042	3,434	4,177
Lithuania	-	-	0.034	0.034	0.079	-	-	-	1,809	1,897	3,529	-
Macau	0.854	0.880	0.915	0.900	0.902	0.956	24,025	25,619	29,516	31,921	34,357	38,144

Continued.

Output and Output per Employee

- Continued -

Country	Output ($ bil.)						Output per Employee ($)					
	1990	1991	1992	1993	1994	1995	1990	1991	1992	1993	1994	1995
Malaysia	*0.880	*1.014	*1.170	*1.160	*1.206	*1.289	13,649	14,503	16,367	16,409	18,410	19,919
Malta	0.186	0.172	0.180	0.167	0.168	-	31,556	31,615	36,262	34,705	37,364	-
Mauritius	0.645	0.669	0.731	0.783	0.818	0.927	8,465	8,702	9,374	10,400	11,394	13,436
Mexico	0.416	0.457	0.480	0.465	0.438	0.231	14,017	15,824	16,971	18,413	18,103	8,827
Mongolia	0.102	0.093	0.027	0.007	0.009	0.007	8,641	6,966	3,610	1,668	1,700	1,019
Morocco	±0.703	±0.778	±0.827	±0.791	±0.836	±0.949	8,778	8,684	8,892	8,321	8,304	9,324
Mozambique	-	±0.014	-	±0.004	±0.002	-	-	3,263	-	1,286	826	-
Nepal	0.055	0.068	-	0.113	0.099	-	4,730	3,931	-	4,118	3,970	-
Netherlands	0.722	0.705	0.719	0.657	0.680	0.714	91,194	94,149	105,225	107,314	111,493	131,053
Norway	±0.153	±0.152	±0.160	±0.139	±0.159	±0.181	73,466	72,942	83,912	89,455	105,773	120,732
Pakistan	0.273	0.295	0.320	-	-	-	16,291	15,019	15,585	-	-	-
Panama	0.118	0.117	0.129	0.127	0.114	0.110	16,111	16,862	17,600	17,557	17,557	17,767
Peru	0.433	0.296	0.251	-	-	-	18,014	18,921	17,599	-	-	-
Philippines	1.056	1.270	1.275	1.359	-	-	5,753	6,944	7,243	8,476	-	-
Poland	±0.733	±0.954	±0.881	±0.981	±1.159	±1.400	4,579	5,925	6,208	6,765	7,847	9,452
Portugal	3.103	3.447	3.919	3.307	3.382	4.212	21,349	21,798	25,886	22,625	22,692	25,089
Puerto Rico	±1.078	±1.075	±1.142	±1.214	±1.126	±1.259	36,727	38,310	39,718	44,760	47,975	51,794
Republic of Moldova	0.000	0.809	0.029	0.018	0.013	0.011	0.029	24,938	1,069	1,187	903	837
Romania	±1.623	±0.867	±0.575	±0.515	-	-	6,289	3,543	2,789	2,574	-	-
Russian Federation	-	15	3.807	1.350	1.382	1.353	-	-	-	2,141	2,527	2,247
Senegal	0.000	-	-	-	-	-	6,860	-	-	-	-	-
Sierra Leone	-	-	-	0.001	-	-	-	-	-	2,153	-	-
Singapore	*0.954	*1.007	*1.002	*0.823	*0.794	*0.790	34,452	38,845	42,739	39,450	43,516	47,435
Slovakia	-	±0.137	±0.162	±0.160	±0.179	-	-	5,727	7,068	6,816	6,852	-
Slovenia	-	-	-	0.371	0.341	0.433	-	-	-	-	-	-
South Africa	±1.728	±1.756	±1.722	±1.712	±1.752	±2.163	13,607	14,513	15,108	13,696	14,019	16,298
Spain	*5.484	*6.080	*6.179	±6.085	±6.152	±6.607	63,248	68,479	75,318	51,226	55,846	62,757
Sri Lanka	0.343	0.435	0.493	0.570	-	-	3,755	4,127	4,689	4,202	-	-
Suriname	*0.007	*0.010	*0.011	0.012	-	-	22,944	25,210	27,293	60,714	-	-
Swaziland	-	*0.044	*0.039	*0.027	-	-	-	28,625	16,589	12,966	-	-
Sweden	*0.382	*0.392	*0.374	*0.183	*0.217	*0.309	56,988	82,151	97,864	70,311	85,904	92,743
Thailand	4.257	6.470	-	-	-	-	14,185	28,458	-	-	-	-
Tonga	0.002	0.001	0.001	0.001	0.000	-	5,403	8,418	13,184	14,459	7,294	-
Trinidad and Tobago	±0.029	±0.030	±0.029	±0.025	±0.024	-	10,596	11,705	11,357	10,318	10,595	-
Tunisia	1.217	1.331	1.604	1.607	1.927	2.387	-	-	-	20,623	21,786	25,315
Turkey	3.012	3.169	4.391	4.706	4.245	6.434	40,546	45,939	50,683	52,874	47,392	63,080
Ukraine	*5.925	*5.909	*2.740	*0.460	*0.265	*0.195	16,190	17,482	10,496	2,738	1,852	1,621
United Kingdom	*9.045	*8.437	8.894	7.908	8.821	9.395	43,070	47,668	52,627	43,693	52,505	55,924
United Republic of Tanzania	*0.005	*0.003					2,808	2,808	-	-	-	-
USSR (Former)	±20						10,041	-	-	-	-	-
United States of America	*47	*49	*53	*54	*55	*56	58,835	62,616	67,813	69,852	75,471	77,135
Uruguay	0.311	0.351	0.409	0.321	0.325	0.291	19,219	21,157	27,461	24,495	25,390	28,106
Venezuela	0.429	0.467	0.524	0.446	0.422	0.502	15,101	16,103	18,568	16,112	16,570	22,665
Yugoslavia	±0.097	±0.100	±0.210	-	±0.272	±0.138	-	1,363	2,791	-	3,948	2,066
Zambia	0.027	-	-	-	0.014	-	7,353	-	-	-	6,040	-
Zimbabwe	*0.223	*0.215	*0.136	*0.132	*0.133	*0.123	10,764	10,832	2,403	7,662	7,479	8,491

Value Added and Value Added per Employee

Country	Value Added ($ bil.)						Value Added per Employee ($)					
	1990	1991	1992	1993	1994	1995	1990	1991	1992	1993	1994	1995
Argentina	±0.492	-	-	0.877	-	-	15,135	-	-	21,571	-	-
Armenia	0.003	0.007	-	-	-	-	81	192	-	-	-	-
Australia	*1.223	*1.188	*1.269	-	-	-	26,596	31,267	35,249	-	-	-
Austria	0.547	0.519	0.516	0.438	0.397	0.425	22,699	23,588	27,142	27,634	29,546	35,297
Bahamas	0.002	0.002	-	-	-	-	9,005	13,005	-	-	-	-
Bangladesh	0.158	0.199	0.192	-	-	-	965	951	890	-	-	-
Barbados	0.005	0.005	0.005	0.003	0.005	0.006	4,879	3,052	3,570	6,114	6,433	5,151

Continued.

Depending on the table, * means *Enterprises, Engaged,* or *Factor Values;* ± means *Basis Unspecified;* § means *shown in millions.* For additional notes and sources, see Appendix I.

Value Added and Value Added per Employee

- Continued -

Country	Value Added ($ bil.)						Value Added per Employee ($)					
	1990	1991	1992	1993	1994	1995	1990	1991	1992	1993	1994	1995
Belgium	*0.918	*0.885	*0.971	*0.996	*1.073	*1.231	25,494	25,728	29,785	-	-	-
Bolivia	0.004	0.004	0.005	0.006	0.006	0.006	3,935	3,079	4,079	4,312	3,623	3,543
Bosnia & Herzegovina	±0.172	±0.181	-	-			5,213	6,041	-	-		
Bulgaria	-	±0.066	±0.074	±0.086	±0.073	±0.093	-	1,254	1,640	2,218	1,507	1,976
Burundi	0.001	0.001	-	-			2,426	2,604	-	-		
Cameroon	-0.009	0.004	-0.000	0.005	0.001	0.002	§-15,6	9,663	-216.504	6,507	4,706	9,953
Canada	2.828	2.627	2.374	2.217	2.160	2.221	30,088	31,276	31,243	29,558	29,592	30,190
Chile	0.163	0.196	0.239	0.289	0.312	0.341	9,421	10,860	12,352	15,285	16,307	18,961
China (Hong Kong SAR)	2.458	2.394	2.546	2.080	1.763		10,955	12,631	14,878	15,904	16,723	
China (Taiwan Province)	2.214	2.054	2.085	2.202	1.705	1.635	20,302	19,476	21,394	23,999	19,160	20,018
Colombia	0.221	0.236	0.318	0.329	0.441	0.482	4,722	4,989	4,806	5,061	6,642	7,360
Costa Rica	0.032	0.034	0.047	0.049	0.048	0.050	1,099	1,140	1,402	1,401	-	-
Croatia	±0.339	±0.379	±0.209				8,542	10,807	6,736	-		
Cyprus	0.118	0.118	0.126	0.094	0.090	0.093	10,457	10,713	11,659	10,869	13,084	13,956
Czechoslovakia (Former)	±0.223	-	-	-			2,402	-	-	-		
Denmark	*0.259	*0.261	*0.299	*0.235	*0.270	*0.322	22,375	25,126	30,965	29,353	31,648	36,615
Ecuador	0.010	0.011	0.012	0.014	0.013	0.020	2,729	2,343	2,426	3,376	3,258	4,631
Egypt	*0.053	*0.033	*0.056	*0.068	*0.071	*0.080	2,124	1,092	1,953	2,284	2,399	2,676
El Salvador				0.028	0.042	0.053				2,690	3,656	4,416
Ethiopia	0.017	0.013	0.005	0.006	0.006	0.004	3,729	3,100	1,281	1,441	1,392	1,021
Finland	±0.428	±0.332	±0.229	±0.180	±0.209	±0.286	29,139	28,878	27,960	24,946	31,733	37,159
FYR Macedonia	±0.211	±0.213	±0.137	±0.084	±0.090	±0.101	7,104	7,337	5,045	3,253	3,324	4,175
France	±5.807	±5.589	±5.999	±5.511	±5.341	±5.491	40,132	41,004	45,479	45,059	46,087	49,070
Gabon	-	*0.004	*0.004	*0.004	*0.002	*0.002	-	22,349	18,455	16,508	15,387	8,177
Germany (Western Part)	5.887	5.911	6.126	5.426	-	-	41,104	42,401	48,636	-	-	-
Greece	*0.552	*0.523	*0.566	*0.551	*0.500	*0.500	13,900	13,973	15,794	15,645	15,532	17,359
Honduras	*0.019	*0.037	*0.055	*0.070	*0.068	*0.105	1,613	1,098	1,526	1,694	1,335	1,622
Hungary	±0.220	±0.221	±0.232	±0.228	±0.257	±0.263	3,550	3,872	4,003	3,866	4,122	4,288
Iceland	0.010	0.013	±0.014	±0.012	-		22,201	26,766	28,853	26,757	-	
India	*0.316	*0.364	*0.366	*0.704	*0.757	*0.868	2,834	2,842	2,438	3,369	3,625	3,786
Indonesia	*0.458	*0.480	*0.753	*1.644	*1.055	*1.108	1,896	1,742	2,373	4,696	2,960	2,985
Iran (Islamic Republic of)	0.089	0.051	0.038	0.022	-	-	8,669	7,186	6,690	4,307		
Iraq	-	*0.087	*0.079				-	11,490	9,847			
Ireland	*0.207	*0.182	*0.194	*0.157	*0.156	*0.165	18,972	20,043	22,023	19,768	21,079	23,517
Israel	0.427	0.467	0.462	0.478	0.502	0.586	13,820	14,649	14,227	13,572	14,075	15,990
Italy	±4.876	±5.592	±6.853	±5.344	±5.622	-	31,899	36,252	37,219	30,133	33,569	-
Japan	±12	±14	±15	±15	±14	±15	24,428	28,544	30,606	33,573	35,448	39,875
Jordan	0.013	0.015	0.016	0.015	0.024	0.025	5,739	5,083	4,002	4,229	4,336	4,568
Kenya	*0.016	*0.015	*0.014	*0.008	*0.012	*0.014	2,287	2,203	2,028	1,213	1,790	1,968
Korea, Republic of	3.401	3.666	3.602	4.682	5.174	6.618	14,690	18,509	19,266	23,804	27,206	35,651
Kuwait	0.054	0.050	0.102	0.118	0.110	0.153	6,431	7,067	9,990	10,899	11,210	14,463
Latvia	-	-		*0.013	*0.023	*0.026	-	-		1,348	2,710	2,941
Macau	0.231	0.253	0.271	0.241	0.232	0.244	6,515	7,360	8,749	8,567	8,845	9,737
Malaysia	*0.280	*0.309	*0.368	*0.376	*0.373	*0.399	4,346	4,425	5,150	5,321	5,700	6,166
Malta	0.080	0.075	0.076	0.067	0.066	-	13,511	13,840	15,229	14,054	14,761	-
Mauritius	0.204	0.213	0.278	0.269	0.283	0.330	2,675	2,777	3,564	3,576	3,936	4,780
Mexico	0.217	0.233	0.264	0.251	0.236	0.125	7,316	8,074	9,320	9,943	9,768	4,763
Mongolia	0.030	0.046	0.012	0.001	0.003	0.003	2,590	3,432	1,663	293	578	407
Morocco	±0.228	±0.267	±0.300	±0.299	±0.322	±0.385	2,851	2,985	3,221	3,149	3,199	3,779
Nepal	0.024	0.030	-	0.042	0.040	-	2,038	1,765	-	1,539	1,602	-
Netherlands	0.234	0.225	0.223	0.204	0.210	0.224	29,565	29,979	32,633	33,223	34,419	41,125
New Zealand	0.202	-	-	-			16,524	-	-	-		
Norway	±0.058	±0.058	±0.063	±0.055	±0.057	±0.065	27,943	28,066	32,823	35,400	37,992	43,191
Pakistan	0.065	0.064	0.071	-	-		3,872	3,236	3,453	-	-	
Panama	0.034	0.038	0.040	0.040	0.037	0.036	4,576	5,411	5,526	5,534	5,667	5,784
Peru	0.133	0.089	0.076	-	-		5,532	5,709	5,338	-	-	
Philippines	0.509	0.587	0.610	0.642			2,773	3,210	3,465	4,005	-	-
Poland	±0.432	±0.521	±0.523	±0.569	±0.672	±0.812	2,701	3,235	3,686	3,924	4,550	5,480
Portugal	0.985	1.112	1.200	1.110	1.188	1.479	6,775	7,030	7,927	7,593	7,971	8,811
Puerto Rico	±0.486	±0.491	±0.536	±0.548	±0.522	±0.603	16,558	17,512	18,640	20,218	22,259	24,813

Continued.

Depending on the table, * means *Enterprises, Engaged,* or *Factor Values*; ± means *Basis Unspecified*; § means *shown in millions*. For additional notes and sources, see Appendix I.

Value Added and Value Added per Employee
- Continued -

Country	Value Added ($ bil.)						Value Added per Employee ($)					
	1990	1991	1992	1993	1994	1995	1990	1991	1992	1993	1994	1995
Romania	±0.691	±0.302	±0.219	±0.225	-	-	2,678	1,236	1,062	1,126	-	-
Russian Federation	-	?	-	±0.933	±0.819	±0.693	-	-	-	1,481	1,497	1,151
Saint Lucia	-	±0.003	±0.002	±0.002	±0.002	±0.002	-	-	-	-	-	-
Senegal	0.000	-0.001	-0.003	0.005	-	-	2,539	§-1,81	§-7,73	7,859	-	-
Sierra Leone	-	-	-	0.001	-	-	-	-	-	1,458	-	-
Singapore	*0.294	*0.302	*0.303	*0.266	*0.241	*0.242	10,612	11,660	12,938	12,766	13,237	14,527
Slovakia	-	-	-	±0.081	±0.094	-	-	-	-	3,457	3,612	-
Slovenia	-	-	-	0.213	0.194	0.240	-	-	-	-	-	-
South Africa	*0.701	*0.741	*0.775	*0.772	*0.792	*0.980	5,518	6,121	6,800	6,177	6,332	7,381
Spain	*2.242	*2.472	*2.512	±2.216	±2.028	±2.240	25,853	27,843	30,619	18,650	18,412	21,276
Sri Lanka	0.142	0.180	0.239	0.254	-	-	1,555	1,706	2,272	1,876	-	-
Swaziland	-	*0.005	*0.008	*0.006	-	-	-	3,149	3,218	2,928	-	-
Sweden	*0.199	*0.140	*0.130	*0.080	*0.083	*0.134	29,755	29,336	33,937	30,758	33,063	40,161
Thailand	1.050	3.297	-	-	-	-	3,498	14,504	-	-	-	-
Trinidad and Tobago	±0.010	±0.011	±0.009	±0.009	±0.008	-	3,797	4,420	3,480	3,656	3,428	-
Tunisia	0.380	0.411	0.484	0.484	0.584	0.720	-	-	-	6,205	6,599	7,642
Turkey	0.947	1.034	1.416	1.584	1.329	2.014	12,753	14,989	16,349	17,792	14,839	19,743
United Kingdom	*4.679	*4.288	*4.585	*3.883	*4.539	*4.841	22,279	24,229	27,129	21,452	27,018	28,814
United Republic of Tanzania	*0.000	*0.001	-	-	-	-	156	571	-	-	-	-
United States of America	*25	*26	*28	*28	*29	*29	31,574	33,325	35,888	36,446	39,342	40,610
Uruguay	0.111	0.144	0.121	0.111	0.112	0.100	6,860	8,678	8,157	8,461	8,763	9,696
Venezuela	0.160	0.195	0.229	-0.147	0.150	0.371	5,621	6,708	8,119	§-5,30	5,902	16,778
Yugoslavia	±0.051	±0.060	±0.148	-	±0.124	±0.069	-	823	1,972	-	1,807	1,031
Zambia	0.009	-	-	-	0.004	-	2,478	-	-	-	1,668	-
Zimbabwe	*0.102	*0.095	*0.059	*0.049	*0.053	*0.049	4,919	4,766	1,041	2,872	2,954	3,355

Capital Investment

Country	Gross Fixed Capital Formation ($ mil.)						Per Establishment ($ mil)					
	1990	1991	1992	1993	1994	1995	1990	1991	1992	1993	1994	1995
Argentina	-	-	-	45	-	-	-	-	-	0.008	-	-
Australia	44	-	-	-	-	-	0.026	-	-	-	-	-
Austria	40	32	37	37	25	-	0.100	0.087	0.108	0.121	0.090	-
Bahrain	-	-	2.979	-	-	-	-	-	0.002	-	-	-
Bangladesh	3.740	8.198	5.905	-	-	-	0.006	0.011	0.009	-	-	-
Barbados	0.141	0.146	0.042	-0.149	0.145	-	0.010	0.016	0.005	-0.017	0.005	-
Belgium	75	73	88	76	77	75	0.037	0.037	0.046	0.058	0.065	-
Bolivia	0.552	0.210	0.628	0.043	-0.130	-	0.007	0.002	0.007	0.000	-0.001	-
Bosnia & Herzegovina	4.008	-	-	-	-	-	0.061	-	-	-	-	-
Bulgaria	15	16	66	19	8.779	8.694	0.151	0.166	0.477	0.009	0.003	0.003
Cameroon	-	0.121	0.363	0.328	0.411	0.104	-	0.011	0.052	0.030	0.068	0.017
Canada	43	-	-	-	-	-	0.016	-	-	-	-	-
Cape Verde	0.002	0.002	-	-	-	-	0.000	0.000	-	-	-	-
Chile	8.075	10	12	16	27	-	0.076	0.089	0.099	0.130	0.225	-
China (Hong Kong SAR)	184	174	142	82	33	-	0.020	0.023	0.021	0.017	0.009	-
Colombia	18	15	27	11	10	-	0.017	0.015	0.027	0.012	0.011	-
Croatia	14	15	5.744	13	4.090	5.977	0.059	0.049	0.013	0.020	0.005	0.006
Cyprus	9.070	6.568	5.204	2.847	1.316	3.481	0.006	0.005	0.004	0.002	0.001	0.003
Czechoslovakia (Former)	48	-	-	-	-	-	1.843	-	-	-	-	-
Denmark	21	8.755	-	-	-	-	0.016	0.006	-	-	-	-
Ecuador	6.836	6.410	5.796	5.489	-1.990	2.040	0.073	0.054	0.052	0.052	-0.018	0.017
Egypt	0.550	3.631	48	7.681	-	-	0.002	0.010	0.138	0.022	-	-
El Salvador	-	2.351	-	7.124	11	8.518	-	0.056	-	0.117	0.105	0.109
Ethiopia	0.241	0.843	0.707	0.777	0.557	0.129	0.027	0.084	0.088	0.097	0.033	0.010
Finland	15	4.130	5.112	2.591	3.771	9.730	0.055	0.017	0.027	0.015	0.026	0.056
FYR Macedonia	13	7.036	1.625	2.064	1.468	0.768	0.131	0.037	0.005	0.007	0.004	0.002
Gabon	-	0.145	0.344	0.088	0.032	0.072	-	0.048	0.115	0.029	0.032	0.036
Germany (Western Part)	247	232	237	187	-	-	0.141	0.137	0.153	-	-	-

Continued.

Depending on the table, * means *Enterprises*, *Engaged*, or *Factor Values*; ± means *Basis Unspecified*; § means *shown in millions*. For additional notes and sources, see Appendix I.

Capital Investment

- Continued -

Country	Gross Fixed Capital Formation ($ mil.)						Per Establishment ($ mil)					
	1990	1991	1992	1993	1994	1995	1990	1991	1992	1993	1994	1995
Greece	46	28	39	-	-	-	0.039	0.024	0.033	-	-	-
Hungary	16	20	22	21	-	-	0.037	0.022	0.028	0.024	-	-
India	47	55	67	109	-	-	0.027	0.028	0.029	0.035	-	-
Indonesia	49	208	306	180	53	93	0.028	0.123	0.164	0.100	0.029	0.044
Iran (Islamic Republic of)	7.894	5.120	5.005	4.605	-	-	0.016	0.037	0.038	0.037	-	-
Ireland	9.784	9.855	-	-	-	-	0.039	0.046	-	-	-	-
Israel	32	44	59	72	60	-	0.030	0.037	0.051	0.052	0.042	-
Italy	442	378	435	317	374	-	0.162	0.133	0.112	0.079	0.097	-
Japan	414	549	537	405	323	-	0.017	0.022	0.023	0.018	0.016	-
Jordan	0.197	0.562	0.644	0.276	0.149	0.505	0.000	0.000	0.000	0.000	0.000	0.000
Korea, Republic of	348	298	219	264	309	375	0.053	0.046	0.034	0.033	0.037	0.044
Kuwait	0.682	2.673	0.471	1.626	1.101	-	0.000	0.001	0.000	0.001	0.001	-
Latvia	0.017	-0.006	0.356	0.422	2.481	1.561	0.000	-0.000	0.002	0.002	0.026	0.008
Macau	19	21	15	16	19	16	0.025	0.031	0.023	0.027	0.036	0.038
Malaysia	50	49	48	37	40	-	0.156	0.140	0.132	0.098	0.113	-
Malta	2.155	4.755	2.997	3.408	4.214	-	0.015	0.031	0.021	0.025	0.034	-
Mexico	8.986	9.206	-	-	-	-	0.050	0.053	-	-	-	-
Mongolia	73	52	14	1.976	2.777	-	1.500	1.302	0.327	0.040	0.027	-
Morocco	72	55	62	48	70	69	0.091	0.066	0.075	0.060	0.090	0.092
Nepal	3.745	-	-	-	-	-	0.039	-	-	-	-	-
Norway	1.917	2.468	1.626	1.551	3.542	-	0.022	0.027	0.024	0.029	0.069	-
Pakistan	-	9.542	-	-	-	-	-	0.062	-	-	-	-
Panama	0.561	0.572	-	-	-	-	0.005	0.006	-	-	-	-
Peru	8.730	6.444	10	-	-	-	0.005	0.003	0.005	-	-	-
Philippines	19	20	26	21	-	-	0.001	0.001	0.014	0.012	-	-
Poland	37	35	77	70	-	-	0.070	0.048	0.113	0.101	-	-
Portugal	187	161	182	124	183	-	0.023	0.018	0.022	0.015	0.022	-
Romania	29	35	-	37	51	-	-	0.040	-	0.016	0.013	-
Senegal	0.015	-	-	-	-	-	0.015	-	-	-	-	-
Singapore	33	36	20	18	19	-	0.090	0.102	0.057	0.057	0.067	-
Spain	89	115	88	78	147	125	0.015	0.021	0.018	-	-	-
Sri Lanka	17	21	26	31	-	-	0.056	0.076	0.104	0.107	-	-
Swaziland	-	0.431	0.239	1.015	-	-	-	0.031	0.017	0.072	-	-
Thailand	786	396	-	-	-	-	0.852	0.298	-	-	-	-
Trinidad and Tobago	-	-	1.059	0.990	1.131	-	-	-	0.008	0.008	0.006	-
Tunisia	68	69	64	63	62	76	-	-	-	0.039	0.036	0.045
Turkey	66	81	101	96	141	-	0.094	0.116	0.064	0.062	0.097	-
Ukraine	122	32	47	6.932	5.352	1.072	0.062	0.010	0.014	0.005	0.004	0.001
United Kingdom	218	149	134	-	159	-	0.026	0.019	0.018	-	0.020	-
United Republic of Tanzania	1.251	0.589	-	-	-	-	0.024	0.012	-	-	-	-
United States of America	470	410	583	527	575	607	-	-	0.031	-	-	-
Uruguay	3.044	7.612	5.281	5.594	-	-	-	-	-	-	-	-
Venezuela	17	13	13	2.984	3.259	48	0.019	0.014	0.015	0.003	0.004	0.062
Yugoslavia	1.897	1.893	0.583	-	4.196	5.470	0.006	0.004	0.001	-	0.003	0.004
Zambia	2.303	-	-	-	-	-	0.068	-	-	-	-	-
Zimbabwe	8.946	11	8.677	5.160	2.147	-	0.081	0.104	0.082	0.054	0.022	-

Indexes of Industrial Production

Country	Index of Industrial Production (1990=100)						Index of Employment (1990=100)					
	1990	1991	1992	1993	1994	1995	1990	1991	1992	1993	1994	1995
Albania	100	-	-	-	-	-	-	-	-	-	-	-
Algeria	100	100	85	78	75	72	-	-	-	-	-	-
Angola	100	-	-	-	-	-	-	-	-	-	-	-
Argentina	100	115	-	-	-	-	100	-	-	125	131	-
Armenia	100	-	-	-	-	-	100	97	70	67	-	53
Australia	100	93	96	100	-	-	100	83	78	-	-	-
Austria	100	97	87	73	66	61	100	91	79	66	56	50

Continued.

Depending on the table, * means *Enterprises, Engaged,* or *Factor Values*; ± means *Basis Unspecified*; § means *shown in millions*. For additional notes and sources, see Appendix I.

131

Indexes of Industrial Production
- Continued -

Country	Index of Industrial Production (1990=100)						Index of Employment (1990=100)					
	1990	1991	1992	1993	1994	1995	1990	1991	1992	1993	1994	1995
Bahamas	100	-	-	-	-	-	100	91	-	-	-	-
Bahrain	100	-	-	-	-	-	-	-	-	-	-	-
Bangladesh	100	-	-	-	-	-	100	128	132	-	-	-
Barbados	100	80	52	44	27	25	100	161	118	50	73	97
Belgium	100	107	108	117	114	113	100	96	91	-	-	-
Benin	100	-	-	-	-	-	-	-	-	-	-	-
Bermuda	-	-	-	-	-	-	100	111	104	81	85	74
Bolivia	100	110	113	166	187	214	100	126	138	165	175	197
Bosnia & Herzegovina	-	-	-	-	-	-	100	91	-	-	4	-
Brazil	100	87	80	89	87	81	-	-	-	-	-	-
Bulgaria	100	87	74	62	75	63	100	83	70	60	76	73
Burkina-Faso	100	-	-	-	-	-	-	-	-	-	-	-
Burundi	100	-	-	-	-	-	100	100	-	-	-	-
Cambodia	100	-	-	-	-	-	100	75	-	-	-	-
Cameroon	100	-	-	-	-	-	100	75	61	144	46	27
Canada	100	91	87	87	89	90	100	89	81	80	78	78
Chile	100	113	108	97	88	84	100	104	112	109	111	104
China (Hong Kong SAR)	100	97	98	100	100	101	100	84	76	58	47	-
China (Taiwan Province)	100	98	83	72	61	57	100	97	89	84	82	75
Colombia	100	96	101	95	79	77	100	101	142	139	142	140
Costa Rica	100	105	133	135	128	132	100	102	114	119	-	-
Cote d'Ivoire	100	95	92	96	98	129	-	-	-	-	-	-
Croatia	100	-	-	-	-	-	100	88	78	79	79	76
Cuba	100	-	-	-	-	-	-	-	-	-	-	-
Cyprus	100	96	84	71	71	72	100	98	96	77	61	59
Czechoslovakia (Former)	100	59	54	-	-	-	100	53	-	-	-	-
Czech Republic	100	-	-	-	-	-	100	86	91	94	-	-
Dem. Rep. of the Congo	100	-	-	-	-	-	-	-	-	-	-	-
Denmark	100	102	107	88	94	97	100	90	84	69	74	76
Dominican Republic	100	-	-	-	-	-	-	-	-	-	-	-
Ecuador	100	100	89	86	-	-	100	125	129	109	109	112
Egypt	100	115	110	108	109	113	100	122	116	119	119	120
El Salvador	100	105	-	-	-	-	-	-	-	-	-	-
Equatorial Guinea	-	-	-	-	-	-	100	-	-	-	-	-
Ethiopia and Eritrea	100	-	-	-	-	-	-	-	-	-	-	-
Ethiopia	-	-	-	-	-	-	100	94	85	84	88	86
Finland	100	77	64	60	64	52	100	78	56	49	45	52
FYR Macedonia	100	-	-	-	-	-	100	98	91	86	90	81
France	100	94	87	80	82	76	100	94	91	85	80	77
Gabon	100	-	-	-	-	-	-	-	-	-	-	-
Gambia	100	-	-	-	-	-	-	-	-	-	-	-
Germany (Eastern Part)	100	-	-	-	-	-	-	-	-	-	-	-
Germany (Western Part)	100	99	86	76	65	61	100	97	88	-	-	-
Ghana	100	-	-	-	-	-	-	-	-	-	-	-
Greece	100	105	101	99	89	78	100	94	90	89	81	72
Guatemala	100	-	-	-	-	-	-	-	-	-	-	-
Guyana	100	-	-	-	-	-	-	-	-	-	-	-
Haiti	100	78	-	-	-	-	-	-	-	-	-	-
Honduras	100	141	168	195	243	330	100	280	297	341	424	539
Hungary	100	99	90	94	101	99	100	92	94	95	101	99
Iceland	100	-	-	-	-	-	100	110	112	104	111	-
India	100	92	80	76	76	83	100	115	135	187	187	206
Indonesia	100	106	115	114	121	129	100	114	131	145	148	154
Iran (Islamic Republic of)	100	102	100	99	78	82	100	69	55	51	-	-
Iraq	100	96	93	-	-	-	-	-	-	-	-	-
Ireland	100	87	85	79	75	72	100	83	81	73	68	64
Israel	100	104	114	123	137	138	100	103	105	114	116	119
Italy	100	99	103	88	-	-	100	101	120	116	110	-
Jamaica	100	-	-	-	-	-	100	96	105	108	106	-
Japan	100	99	95	88	84	80	100	101	98	92	83	76

Continued.

　Depending on the table, * means *Enterprises, Engaged,* or *Factor Values;* ± means *Basis Unspecified;* § means *shown in millions.* For additional notes and sources, see Appendix I.

Indexes of Industrial Production

- Continued -

Country	Index of Industrial Production (1990=100)						Index of Employment (1990=100)					
	1990	1991	1992	1993	1994	1995	1990	1991	1992	1993	1994	1995
Jordan	100	87	75	82	81	68	100	128	175	159	250	244
Kenya	100	86	85	77	49	40	100	101	98	99	102	104
Korea, Republic of	100	95	87	73	77	79	100	86	81	85	82	80
Kuwait	100	-	-	-	-	-	100	84	122	129	117	126
Kyrgyzstan	100	-	-	-	-	-	100	97	56	41	29	-
Lao P.D.R.	100	-	-	-	-	-	-	-	-	-	-	-
Latvia	100	-	-	-	-	-	100	121	65	49	43	45
Libyan Arab Jamahiriya	100	-	-	-	-	-	-	-	-	-	-	-
Lithuania	100	-	-	-	-	-	-	-	-	-	-	-
Macau	100	-	-	-	-	-	100	97	87	79	74	71
Madagascar	100	86	55	-	-	-	-	-	-	-	-	-
Malawi	100	149	129	103	93	74	100	-	-	-	-	-
Malaysia	100	100	105	106	102	101	100	108	111	110	102	100
Malta	100	86	81	98	-	-	100	92	84	81	76	-
Mauritius	100	-	-	-	-	-	100	101	102	99	94	91
Mexico	100	96	95	92	90	85	100	97	95	85	82	88
Mongolia	100	89	39	19	17	21	100	114	63	37	45	59
Morocco	100	104	106	103	97	99	100	112	116	119	126	127
Mozambique	100	97	-	-	-	-	100	98	64	63	41	-
Myanmar	100	-	-	-	-	-	-	-	-	-	-	-
Nepal	100	-	-	-	-	-	100	149	-	236	215	-
Netherlands	100	100	86	86	87	79	100	95	86	77	77	69
New Zealand	100	85	68	-	-	-	100	91	92	93	-	-
Nicaragua	100	-	-	-	-	-	-	-	-	-	-	-
Niger	100	-	-	-	-	-	-	-	-	-	-	-
Nigeria	100	-	-	-	-	-	-	-	-	-	-	-
Norway	100	98	95	91	95	95	100	100	92	74	72	72
Pakistan	100	-	-	-	-	-	100	117	123	-	-	-
Panama	100	121	131	128	114	109	100	95	100	98	89	84
Paraguay	100	76	68	54	33	34	-	-	-	-	-	-
Peru	100	112	104	105	147	151	100	65	59	-	-	-
Philippines	100	120	130	146	152	165	100	100	96	87	-	-
Poland	100	89	99	107	119	121	100	101	89	91	92	93
Portugal	100	102	100	89	78	85	100	109	104	101	103	116
Puerto Rico	100	-	-	-	-	-	100	96	98	92	80	83
Republic of Moldova	100	-	-	-	-	-	100	63	53	29	29	25
Romania	100	93	65	62	88	110	100	95	80	78	81	-
Russian Federation	100	-	-	-	-	-	-	-	-	-	-	-
Saudi Arabia	100	-	-	-	-	-	-	-	-	-	-	-
Senegal	100	91	92	81	84	70	100	854	601	913	-	-
Sierra Leone	100	-	-	-	-	-	-	-	-	-	-	-
Singapore	100	100	90	71	62	51	100	94	85	75	66	60
Slovakia	100	-	-	-	-	-	-	-	-	-	-	-
Slovenia	100	91	79	80	74	68	-	-	-	-	-	-
Somalia	100	-	-	-	-	-	-	-	-	-	-	-
South Africa	100	98	91	95	98	112	100	95	90	98	98	105
Spain	100	94	86	75	82	80	100	102	95	137	127	121
Sri Lanka	100	109	116	303	331	241	100	115	115	148	-	-
Sudan	100	-	-	-	-	-	-	-	-	-	-	-
Suriname	100	-	-	-	-	-	100	137	133	69	-	-
Sweden	100	86	79	62	74	95	100	71	57	39	38	50
Switzerland	100	103	91	86	82	82	-	-	-	-	-	-
Syrian Arab Republic	100	107	116	-	-	-	-	-	-	-	-	-
Thailand	100	92	89	-	-	-	100	76	-	-	-	-
Togo	100	-	-	-	-	-	-	-	-	-	-	-
Tonga	100	-	-	-	-	-	100	31	36	29	9	-
Trinidad and Tobago	100	91	93	59	52	51	100	94	95	88	83	-
Tunisia	100	87	90	95	97	115	-	-	-	-	-	-
Turkey	100	96	90	81	70	92	100	93	117	120	121	137
Uganda	100	116	104	81	50	46	-	-	-	-	-	-

Continued.

Depending on the table, * means *Enterprises, Engaged,* or *Factor Values*; ± means *Basis Unspecified*; § means *shown in millions.* For additional notes and sources, see Appendix I.

Indexes of Industrial Production

- Continued -

Country	Index of Industrial Production (1990=100)						Index of Employment (1990=100)					
	1990	1991	1992	1993	1994	1995	1990	1991	1992	1993	1994	1995
Ukraine	100	-	-	-	-	-	100	92	71	46	39	33
United Arab Emirates	100	-	-	-	-	-	-	-	-	-	-	-
United Kingdom	100	90	92	92	95	95	100	84	80	86	80	80
United Republic of Tanzania	100	89	93	93	79	-	100	62	-	-	-	-
USSR (Former)	100	-	-	-	-	-	100	-	-	-	-	-
United States of America	100	101	103	106	110	106	100	96	96	96	91	90
Uruguay	100	101	96	95	93	77	100	103	92	81	79	64
Venezuela	100	-	-	-	-	-	100	102	99	97	90	78
Yemen	100	-	-	-	-	-	-	-	-	-	-	-
Yugoslavia	100	-	-	-	-	-	-	-	-	-	-	-
Yugoslavia (Former)	100	77	-	-	-	-	-	-	-	-	-	-
Zambia	100	84	81	57	54	48	100	-	-	-	62	-
Zimbabwe	100	103	86	88	86	69	100	96	273	83	86	70

Depending on the table, * means *Enterprises, Engaged,* or *Factor Values;* ± means *Basis Unspecified;* § means *shown in millions.* For additional notes and sources, see Appendix I.

Representative Companies in Sector

Name	Address	Tele-phone	Fax	Employ-ment	Y
Gerber Products Co.	445 State St., Fremont, Michigan 49413, U.S.A.	616-928-2000	616-928-2963	9,000	97
Hindustan Lever Ltd.	165/166 Backbay Reclamation, Bombay 400 020, India	22 2870622		9,000	94
Lee Apparel Company Inc.	PO Box 2940, Shawnee Mission, Kansas 66201, U.S.A.	913-384-4000	913-384-0190	9,000	97
Maidenform Inc.	154 Ave E., Bayonne, New Jersey 07002, U.S.A.	201-436-9200		8,900	95
Vanity Fair Mills Inc.	624 S. Alabama Ave., Monroeville, Alabama 36462, U.S.A.	334-575-3231	334-743-7582	8,500	95
Oxford Industries Inc.	222 Piedmont Ave., NE, Atlanta, Georgia 30308, U.S.A.	404-659-2424	404-653-1545	8,413	97
Hartmarx Corp.	101 N. Wacker Dr., Chicago, Illinois 60606, U.S.A.	312-372-6300		8,100	97
Cluett, Peabody and Company Inc.	575 5th Ave., New York, New York 10017, U.S.A.	212-930-3000	212-930-3086	8,000	94
Delta Woodside Industries Inc.	233 N. Main St., 200, Greenville, South Carolina 29601, U.S.A.	803-232-8301	803-232-6164	7,500	95
Liz Claiborne Inc.	1441 Broadway, New York, New York 10018, U.S.A.	212-354-4900		7,100	97
Unifirst Corp.	68 Jonspin Rd., Wilmington, Massachusetts 01887-1086, U.S.A.	978-658-8888	978-657-5663	7,000	97
Tultex Corp.	PO Box 5191, Martinsville, Virginia 24115, U.S.A.	540-632-2961	540-632-8000	6,618	96
Haggar Clothing Co.	6113 Lemmon Ave., Dallas, Texas 75209, U.S.A.	214-956-4511	214-956-4446	6,500	96
Haggar Corp.	6113 Lemmon Ave., Dallas, Texas 75209, U.S.A.	214-352-8481	214-956-4446	6,500	97
Jostens Inc.	5501 Norman Center Dr., Minneapolis, Minnesota 55437, U.S.A.	612-830-3300		6,300	96
Ithaca Industries Inc.	PO Box 620, Wilkesboro, North Carolina 28697, U.S.A.	910-667-5231	910-667-4052	6,000	97
New South Africa Garment Manufacturers	27 Leopold St., Durban 4000, Republic of South Africa	31 3008911	31 3055873	6,000	92
Takata Inc.	2500 Takata Dr., Auburn Hills, Michigan 48326, U.S.A.	248-377-6130	248-373-5186	6,000	95
Ozeta Odevne Zavody, Akciova Spolocnost	Velkomoravska 9, Trencin 911 34, Slovakia	831 262111	831 23240	5,800	93
Timberland Co.	200 Domain Dr., Stratham, New Hampshire 03885, U.S.A.	603-772-9500		5,700	97
Frame Textile Corp. Ltd.	PO Box 32002, Duran 4060, Republic of South Africa	31 425140	31 424822	5,500	97
Delta Galil Industries Ltd.	2 Kaufman St., Tel Aviv 68012, Israel	3 5193636	3 5193705	5,460	97
River Island Clothing Co. Ltd.	West Gate, London W5 1DR, United Kingdom	81 9988822	81 9973953	5,309	93
Bibb Co.	PO Box 4207, Macon, Georgia 31208, U.S.A.	912-752-6700	912-752-7332	5,100	95
Cygne Designs Inc.	1372 Broadway, New York, New York 10018, U.S.A.	212-354-6474		5,000	95
G and K Services Inc.	5995 Opus Pkwy., Ste. 500, Minnetonka, Minnesota 55343, U.S.A.	612-912-5500		4,873	97
SR Kent UK Ltd.	Dodworth Rd., Barnsley LN6 8BQ, United Kingdom	226 241434		4,798	91
OshKosh B'Gosh Inc.	112 Otter Ave., Oshkosh, Wisconsin 54901, U.S.A.	414-231-8800	414-231-8621	4,700	96
David Whitehead Textiles Ltd.	Union Ave., Harare, Zimbabwe	14 730744	14 728120	4,644	92
Champion Products Inc. (Winston-Salem, North Carolina)	475 Corporate Square Dr., Winston-Salem, North Carolina 27105, U.S.A.	910-519-6500		4,500	96
Martin International Holdings PLC	182 Kirkby Rd., Sutton in Ashfield NG17 1GP, United Kingdom	623 512808	623 440146	4,395	93
Genesco Inc.	PO Box 731, Nashville, Tennessee 37202-0731, U.S.A.	615-367-7000	615-367-8278	4,300	97
Leslie Fay Companies Inc.	1400 Broadway, New York, New York 10018, U.S.A.	212-221-4000	212-221-4045	4,300	95
Coleman Company Inc.	1526 Cole Blvd., 300, Golden, Colorado 80401, U.S.A.	303-202-2400		4,200	96
Supertiendas y Droguerias Olimpica SA	Cra. 36 No. 38-03, Barranquilla, Colombia	5 3516549	5 3510263	4,200	96
Unitog Co.	1300 Washington, Kansas City, Missouri 64105, U.S.A.	816-474-7000	816-474-0699	4,105	95
Prince Corp. (Holland, Michigan)	1 Prince Ctr., Holland, Michigan 49423, U.S.A.	616-392-5151	616-394-8000	4,000	95
Thai Garment Export Co. Ltd.	95 Rajadamri Rd., Bangkok 10310, Thailand	2520712	2518334	4,000	90
Williamson-Dickie Manufacturing Co.	PO Box 1779, Fort Worth, Texas 76104, U.S.A.	817-336-7201	817-877-5027	4,000	95
Maxxim Medical Inc.	10300 49th St., N., Clearwater, Florida 33762-5000, U.S.A.	813-561-2100	813-561-2180	3,958	97
Farah Inc.	4171 N. Mesa St., Ste. D500, El Paso, Texas 79902-1433, U.S.A.	915-593-4444		3,950	97
K2 Inc.	PO Box 22252, Los Angeles, California 90040, U.S.A.	213-724-2800		3,800	97
Salant Corp.	1058 Claussen Rd., Ste. 101, Augusta, Georgia 30907, U.S.A.	212-221-7500	212-221-5363	3,800	96
Aris Industries Inc.	475 5th Ave., New York, New York 10017, U.S.A.	212-686-5050		3,737	95
Chic By H.I.S. Inc.	1372 Broadway, New York, New York 10018, U.S.A.	212-302-6400		3,700	97
Delami PT	Jl. Soekarno Hatta No. 569, Bandung 40275, Indonesia	22300333	22 300049	3,540	97
Form-O-Uth	PO Box 1600, McAllen, Texas 78501, U.S.A.	210-687-6281	210-687-4896	3,500	94
Vanities Unlimited	Stiles Ln., Pine Brook, New Jersey 07058, U.S.A.	201-575-7290		3,500	94
Sherwood Group PLC	Long Eaton, Nottingham NG10 2BQ, United Kingdom	602 461070	602 462720	3,385	93
Findlay Industries Inc.	4000 Fostoria Rd., Findlay, Ohio 45840, U.S.A.	419-422-1302	419-422-0385	3,300	97

Continued

Representative Companies in Sector
- Continued -

Name	Address	Tele-phone	Fax	Employ-ment	Y
St. John Knits Inc.	17422 Derian Ave., Irvine, California 92713, U.S.A.	714-223-3301		3,255	97
Stirling Group PLC	3 Wyndham Pl., London W1H 1AP, United Kingdom	71 7230821	71 7231411	3,252	94
Masland Corp.	PO Box 40, Carlisle, Pennsylvania 17013, U.S.A.	717-249-1866		3,250	95
Thai Wacoal Co. Ltd.	Sathupradit Rd., Bangkok 10120, Thailand	2893100	2915788	3,250	94
Angelica Uniform Group	700 Rosedale Ave., St. Louis, Missouri 63112, U.S.A.	314-889-1111	314-889-1143	3,000	97
Capital Mercury Shirt Corp.	1372 Broadway, New York, New York 10018-6195, U.S.A.	212-704-4800	212-704-4996	3,000	96
Polgat Ltd.	7 Hasivim St., Petah Tikva 49170, Israel	3 9251515	3 9213566	3,000	97
Jones Apparel Group Inc.	250 Rittenhouse Cir., Bristol, Pennsylvania 19007, U.S.A.	215-785-4000	215-785-1795	2,945	97
Dewhirst Ltd.	Westgate, Driffield YO25 7TH, United Kingdom	377 42561	377 43814	2,923	94
Laura Ashley Ltd.	150 Bath Rd., Maidenhead SL6 4YS, United Kingdom	628 39151	628 782467	2,834	94
Guess ? Inc.	1444 S. Alameda St., Los Angeles, California 90021, U.S.A.	213-765-3100	213-744-7338	2,800	97
Coordinated Apparel Inc.	350 5th Ave., New York, New York 10118, U.S.A.	212-613-9200	212-613-9282	2,750	93
American Marketing Industries Inc.	10450 Holmes Rd., 501, Kansas City, Missouri 64131, U.S.A.	816-943-5180	816-943-5199	2,665	97
Gokaldas Exports	Karnataka, Bangalore 560 027, India	812 2223600	812 2214869	2,650	94
J & J Fashions Ltd.	260-266 York Way, London N7 9PQ, United Kingdom	71 6096261	71 6099845	2,594	91
Hampshire Group Ltd.	PO Box 2667, Anderson, South Carolina 29622, U.S.A.	803-225-6232	864-225-4421	2,550	95
Garan Inc.	350 5th Ave., New York, New York 10118, U.S.A.	212-563-2000	212-695-2488	2,500	97
Vetements Peerless Inc.	8888 Blvd. Pie-Ix, Montreal, Quebec H1Z 4J5, Canada	514-593-9300	514-593-9640	2,500	98
Oneita Industries Inc.	Conifer St. Drawer 24, Andrews, South Carolina 29510, U.S.A.	803-529-5225	803-264-4262	2,450	97
Carhartt Inc.	PO Box 600, Dearborn, Michigan 48121, U.S.A.	313-271-8460	313-271-3455	2,400	95
Dewhirst Ladieswear Ltd.	Fforestfach Industrial Estate, Swansea SA5 4HE, United Kingdom	792 588161	792 580257	2,319	94
James Seddon UK Ltd.	Denton, Manchester M34 3WA, United Kingdom			2,275	93
Pluma Inc.	PO Box 487, Eden, North Carolina 27289, U.S.A.	910-635-4000		2,275	97
Converse Inc.	1 Fordham Rd., North Reading, Massachusetts 01864-2680, U.S.A.	508-664-1100	508-664-8727	2,249	96
Bestform Foundations Inc.	38-01 47th Ave., Long Island City, New York 11101, U.S.A.	718-392-2200	718-392-2498	2,200	95
Crown Crafts Inc.	1600 RiverEdge Pkwy., 200, Atlanta, Georgia 30328, U.S.A.	770-644-6400		2,200	96
Prochnik Co.	Al. Marszalka Smiglego-Rydza 20, Lodz PL-93-115, Poland	42 815920	42 843954	2,200	93
Pentland Group PLC	Finchley, London N3 2QL, United Kingdom	81 3462600	81 3462700	2,070	91
Winner Garment Manufacturing Corp. PT	Jl. Cideng Barat, No. 79, Jakarta 10150, Indonesia	213800402	21 377982	2,050	97
Levi Strauss & Co. (Canada) Inc.	80 Allstate Pky Pkwy, Markham, Ontario L3R 8X6, Canada	905-470-2777	905-470-4514	2,000	98
Rex Trueform Clothing Co. Ltd.	263 Victoria Rd., Salt River 8001, Republic of South Africa	21 4484660	21 4481553	2,000	93
Tiz, Zemun	Georgi Dimitrova 56, Zemun YU-11080, Serbia	11 601333	11 611434	2,000	
Za-Ko RT	Platan sor 8, Zalaegerszeg H-8900, Hungary	92 14000	92 11007	2,000	93
Daks Simpson Group PLC	34 Jermyn St., London SW1Y 6HS, United Kingdom	71 4398781	71 4373633	1,996	94
San East UK PLC	34 Jermyn St., London SW1Y 6HS, United Kingdom			1,996	94
Sukwang Corp.	Kumchon-Gu, Seoul, Republic of Korea	2 8198114	2 8562345	1,995	97
Burberrys Ltd.	18-22 Haymarket, London SW1Y 4DQ, United Kingdom	71 8392434	71 8396691	1,965	93
Shin Woo Co., Ltd.	Kangnam-gu, Seoul, Republic of Korea	2 5165452	2 5464174	1,950	97
Mura European Fashion Design d.d.	Plese 2, Murska Sobota SLO-69000, Slovenia	69 31535	69 32513	1,925	95
CentroTextil	Knez Mihailova 1-3, Belgrade YU-11000, Serbia	11 185333	11 636879	1,908	95
Busana Rama Textile & Garment PT	Jl. Kesehatan III/23, Jakarta 10160, Indonesia	21361815	21 361815	1,900	97
Labod d.d.	Seidlova 35, Novo Mesto SLO-8001, Slovenia	68 321791	68 323093	1,834	95
Pringle of Scotland Ltd.	Victoria Mill, Hawick TD9 7AL, United Kingdom	450 360260		1,825	94
Fenix SA	Av. Boggiani 5086, Asuncion, Paraguay	21 660517	21 663375	1,800	96
Industex Eastern Cape Pty. Ltd.	Neave Township, Port Elizabeth 6014, Republic of South Africa	41 431363	41 411558	1,800	97
Spencers Inc.	PO Box 988, Mount Airy, North Carolina 27030, U.S.A.	919-789-9111	919-789-6824	1,800	
Tah Hsin Industrial Corp.	201-24, Tun Hwa North Rd., Taipei, Taiwan	2 7128311	2 7151446	1,800	94
Varsity Spirit Corp.	PO Box 341609, Memphis, Tennessee 38184-1609, U.S.A.	901-387-4300		1,789	95
Daks-Simpson Ltd.	34 Jermyn St., London SW1Y 6HS, United Kingdom	71 4398781	71 4373633	1,760	94
Richards PLC	Maberly St., Aberdeen AB9 8DT, United Kingdom	224 630243	224 630260	1,739	94
Badan Tekstil Nasional PT	Jalan Otto Iskandarinata No. 89, Bandung 40111, Indonesia		2256243	1,700	97
Intermoda	Ul Mikolaja 8/10, Wroclaw PL-50-951, Poland	71 442886	71 443466	1,700	93
Superior Surgical Manufacturing Company Inc.	PO Box 4002, Seminole, Florida 33775-0002, U.S.A.	813-397-9611	813-391-5401	1,700	97
Bairdwear Racke Ltd.	Springburn, Glasgow G22 5DT, United Kingdom	41 5572711		1,670	90

Continued

For sources and notes, see Appendix I.

Representative Companies in Sector
- Continued -

Name	Address	Tele-phone	Fax	Employ-ment	Y
Corah PLC	Barlestone, Nuncaton CV13 0EL, United Kingdom	455 290513	455 292319	1,656	93
Donnkenny Inc.	1411 Broadway, New York, New York 10018, U.S.A.	212-730-7770	212-228-6036	1,603	96
Dannimac Ltd.	Hollinwood, Oldham OL8 3QN, United Kingdom	61 6812060		1,600	93
K-Products Inc.	PO Box 147, Orange City, Iowa 51041-0147, U.S.A.	712-737-4925	712-737-3818	1,600	94
Kleinert's Inc.	120 W. Germantown Pike, 100, Plymouth Meeting, Pennsylvania 19462-1420, U.S.A.	215-828-7261	215-828-4589	1,585	96
Donna Karan International Inc.	550 7th Ave., New York, New York 10018, U.S.A.	212-789-1500		1,580	96
Tommy Hilfiger Corp.	25 W. 39th St., New York, New York 10018, U.S.A.	212-840-8888		1,570	97
SLM International Inc.	30 Rockfeller Plz., 4314, New York, New York 10112-4399, U.S.A.	212-332-1610		1,550	95
Levi Strauss UK Ltd.	Moulton Park, Northampton NN3 1QG, United Kingdom	604 790436	604 790400	1,549	93
Wardle Storeys PLC	Brantham, Manningtree CO11 1NJ, United Kingdom	206 392401	206 395288	1,522	94
Beales Hunter PLC	Radford Blvd., Nottingham NG7 3AE, United Kingdom	15 9782221	15 9785034	1,518	94
Arden Corp.	26899 Northwestern Hwy., Southfield, Michigan 48034-8420, U.S.A.	248-355-1101	248-355-1230	1,500	97
Catalina	6040 Bandini Blvd., Los Angeles, California 90040, U.S.A.	213-726-1262		1,500	
Hanjoo Corp.	Socho-gu, Seoul, Republic of Korea	2 5809541	2 5486781	1,500	97
I.M. Lockhat	PO Box 2886, Durban 4000, Republic of South Africa	31 3094281	31 3094299	1,500	95
Jantzen Inc.	PO Box 3001, Portland, Oregon 97208, U.S.A.	503-238-5000	503-238-5087	1,500	97
Modus	ul. Torunska 8, Bydgoszcz PL-85-133, Poland	52 395415	52 38036	1,500	93
Riverside Manufacturing Co.	PO Box 460, Moultrie, Georgia 31776, U.S.A.	912-985-5210	912-890-2952	1,500	97
Shin Won Co.	Mapo-Gu, Seoul, Republic of Korea	2 7167610	2 7068191	1,500	97
Wearwel International, PT	Tanjung Priok, Jakarta Utara, Indonesia	214301348	21 496917	1,500	97
Yulinda Duta Fashion, PT	Desa Karang Asem Barat, Jakarta 12220, Indonesia	217394894	21 772375	1,500	97
Morris Cohen Underwear Ltd.	Pengam, Blackwood NP2 1SW, United Kingdom	443 875055	443 833406	1,496	93
Lentex	Powstancow ul. 54, Lubliniec PL-42-701, Poland	334 62641	334 63320	1,490	93
Jos. A. Bank Clothiers Inc.	500 Hanover Pike, Hampstead, Maryland 21074, U.S.A.	410-239-2700	410-239-5716	1,420	96
Rawlings Sporting Goods Company Inc.	1859 Intertech Dr., Fenton, Missouri 63026, U.S.A.	314-349-3500	314-349-3588	1,420	97
Alhos Export-Import	Tesanjska 24a, Sarajevo, Bosnia-Hercegovina	71 39481	71 215999	1,400	93
Best Manufacturing Co.	Edison St., Menlo, Georgia 30731, U.S.A.	706-862-2302	706-862-6000	1,400	96
Horace Small Apparel Co.	350 28th Ave. N, Nashville, Tennessee 37209, U.S.A.	615-320-1000	615-327-1912	1,400	95
Playthe	PO Box 12359, Jacobs 4026, Republic of South Africa	31 4608601	31 4608602	1,400	96
Robinson Manufacturing Company Inc.	520 S. Market St., Dayton, Tennessee 37321, U.S.A.	615-775-2212		1,400	94
Warner's	325 Lafayette St., Bridgeport, Connecticut 06601, U.S.A.	203-579-8100		1,400	
Johnson Worldwide Associates Inc.	PO Box 901, Sturtevant, Wisconsin 53177, U.S.A.	414-884-1500	414-631-4426	1,366	97
Austin Reed Group PLC	103 Regent St., London W1A 2AJ, United Kingdom	71 7346789	71 4942900	1,360	94
Confecciones Colombia SA	Calle 71 65-74, Medellin, Colombia	4 2578563	4 2570636	1,360	97
Hampton Industries Inc.	PO Box 614, Kinston, North Carolina 28502-0614, U.S.A.	919-527-8011	919-527-3538	1,355	97
Alexandra Workwear PLC	Patchway, Bristol BS12 5TP, United Kingdom	272 690805	272 799442	1,354	94
Huber Tricot GmbH	Hauptstrasse 17, Goetzis A-6840, Austria	5523 45550	5523 4555226	1,350	92
Lisca d.d.	Presernova 4, Sevnica 8290, Slovenia	60841041	60841513	1,338	95
Jaeger Tailoring Ltd.	16 Victoria Way, Burgess Hill RH15 9NF, United Kingdom	444 245411	444 247153	1,322	92
United Uniform Services PLC	40 George St., London W1H 5RE, United Kingdom	71 9354650	71 9352730	1,319	92
VF Corp. United Kingdom Ltd.	Donaghadee Rd., New Townards, United Kingdom	247 800200	247 819845	1,306	92
Aranda Tertice Mills Pty. Ltd.	Industrial Sites, Randfontein 1760, Republic of South Africa	11 6933721	11 6935178	1,300	94
Saez Merino, SA	Angel Guimera, 70, Valencia E-46008, Spain	96 3847500	96 3842207	1,300	97
Standard Textile Company Inc.	PO Box 371805, Cincinnati, Ohio 45222-1805, U.S.A.	513-761-9255	513-761-0467	1,300	97
Tom James Co.	PO Box 1469, Brentwood, Tennessee 37024-1469, U.S.A.	615-771-1122	615-771-8971	1,300	97
Fitch Holdings PLC	24-30 Great Titchfield St., London W1TP 7AD, United Kingdom	71 5802266	71 5800428	1,295	90
Macpell Industries Ltd.	40 Hanamal St., Tel Aviv 63506, Israel	3 5460404	3 5460324	1,240	97
Apparel Group	1370 Ave. of the Amer., 5th Fl, New York, New York 10019, U.S.A.	212-399-3500		1,230	94
Authentic Fitness Corp.	6040 Bandini Blvd., Commerce, California 90040, U.S.A.	213-726-1262		1,211	97
Aquascutum Group PLC	100 Regent St., London W1A 2AQ, United Kingdom	71 7346090	71 7340726	1,202	93
Bayly Corp.	5500 S. Valentia Way, Englewood, Colorado 80111, U.S.A.	303-000-0000		1,200	
Boohung Co., Ltd.	Kumchon-gu, Seoul, Republic of Korea	2 8697111	2 8665872	1,200	97
Donnkenny Apparel Inc.	1411 Broadway, New York, New York 10018, U.S.A.	212-730-7770	212-228-6036	1,200	97

Product Tables
Jackets, Men's and Boys'

Unit of Measure: Thousands of units.

	1990 Value	%	1991 Value	%	1992 Value	%	1993 Value	%	1994 Value	%	1995 Value	%	1996 Value	%
Total Production	111,963	100.0	130,886	100.0	134,490	100.0	143,050	100.0	128,795	100.0	119,476	100.0	128,754	100.0
Regions														
Africa	2,580	2.3	2,405	1.8	2,492	1.9	2,307	1.6	2,190	1.7	2,447	2.0	2,139	1.7
America, North	15,344	13.7	15,082	11.5	14,821	11.0	14,310	10.0	15,490	12.0	13,450	11.3	12,871	10.0
America, South	1,116	1.0	1,151	0.9	1,236	0.9	1,366	1.0	1,495	1.2	1,347	1.1	1,478	1.1
Asia	48,592	43.4	66,624	50.9	73,631	54.7	84,162	58.8	69,017	53.6	62,209	52.1	72,717	56.5
Europe	42,855	38.3	44,449	34.0	41,350	30.7	40,055	28.0	39,799	30.9	39,323	32.9	38,910	30.2
Oceania	1,476	1.3	1,175	0.9	960	0.7	850	0.6	803	0.6	700	0.6	640	0.5
Leading Producers														
Hong Kong	16,444	14.7	17,556	13.4	21,130	15.7	14,930	10.4	13,708	10.6	13,552	11.3	*16,726	13.0
United States of America	13,945	12.5	13,683	10.5	13,422	10.0	13,136	9.2	13,864	10.8	*12,127	10.2	*11,397	8.9
Indonesia	10,122	9.0	27,142	20.7	30,972	23.0	49,712	34.8	37,294	29.0	32,333	27.1	*39,655	30.8
Japan	11,275	10.1	10,767	8.2	10,436	7.8	8,979	6.3	7,175	5.6	5,903	4.9	5,202	4.0
Macau	*5,477	4.9	*5,477	4.2	*5,477	4.1	5,717	4.0	4,717	3.7	4,941	4.1	6,534	5.1

Overcoats, Men's and Boys'

Unit of Measure: Number.

	1990 Value	%	1991 Value	%	1992 Value	%	1993 Value	%	1994 Value	%	1995 Value	%	1996 Value	%
Total Production	29,885,910	100.0	27,091,666	100.0	23,798,305	100.0	21,567,164	100.0	19,544,433	100.0	22,220,608	100.0	19,418,366	100.0
Regions														
Africa	201,967	0.7	122,535	0.5	129,158	0.5	69,273	0.3	92,125	0.5	117,444	0.5	97,125	0.5
America, North	4,866,600	16.3	4,930,400	18.2	5,168,200	21.7	5,707,000	26.5	6,156,800	31.5	5,714,943	25.7	5,845,961	30.1
America, South	10,599	0.0	8,143	0.0	3,229	0.0	6,314	0.0	11,400	0.1	12,486	0.1	8,571	0.0
Asia	10,174,168	34.0	9,719,123	35.9	9,116,079	38.3	5,396,346	25.0	3,804,573	19.5	6,109,800	27.5	4,316,243	22.2
Europe	14,129,477	47.3	11,814,265	43.6	8,789,340	36.9	9,782,830	45.4	8,868,035	45.4	9,648,334	43.4	8,509,947	43.8
Oceania	503,100	1.7	497,200	1.8	592,300	2.5	605,400	2.8	611,500	3.1	617,600	2.8	640,519	3.3
Leading Producers														
Dominican Republic	*3,821,600	12.8	*4,136,400	15.3	*4,451,200	18.7	*4,766,000	22.1	*5,080,800	26.0	*5,395,600	24.3	*5,710,400	29.4
Spain	1,632,000	5.5	2,112,000	7.8	1,790,000	7.5	3,465,000	16.1	3,718,000	19.0	3,830,000	17.2	*3,264,219	16.8
Ukraine	6,424,000	21.5	4,155,000	15.3	1,826,000	7.7	1,133,000	5.3	354,000	1.8	186,000	0.8	155,000	0.8
Russian Federation	4,236,000	14.2	4,287,000	15.8	1,807,000	7.6	1,340,000	6.2	554,000	2.8	61,000	0.3	226,000	1.2
Japan	2,209,000	7.4	1,438,000	5.3	1,703,000	7.2	1,261,000	5.8	1,283,000	6.6	1,110,000	5.0	1,008,000	5.2

Raincoats, Men's and Boys'

Unit of Measure: Number.

	1990 Value	%	1991 Value	%	1992 Value	%	1993 Value	%	1994 Value	%	1995 Value	%	1996 Value	%
Total Production	10,932,572	100.0	10,391,310	100.0	8,419,807	100.0	7,450,399	100.0	8,413,544	100.0	6,692,061	100.0	6,691,682	100.0
Regions														
Africa	568,000	5.2	471,000	4.5	319,000	3.8	154,000	2.1	116,000	1.4	112,000	1.7	69,000	1.0
America, North	2,923,000	26.7	2,725,000	26.2	2,535,000	30.1	2,698,803	36.2	2,651,709	31.5	2,606,900	39.0	2,561,258	38.3
America, South	129,778	1.2	130,889	1.3	132,000	1.6	125,000	1.7	93,000	1.1	105,000	1.6	106,048	1.6
Asia	2,144,267	19.6	2,486,337	23.9	1,696,667	20.2	1,478,444	19.8	3,039,222	36.1	1,170,583	17.5	1,644,057	24.6
Europe	5,096,194	46.6	4,501,417	43.3	3,655,141	43.4	2,906,819	39.0	2,420,947	28.8	2,599,579	38.8	2,225,709	33.3
Oceania	71,333	0.7	76,667	0.7	82,000	1.0	87,333	1.2	92,667	1.1	98,000	1.5	85,610	1.3
Leading Producers														
United States of America	2,919,000	26.7	2,722,000	26.2	2,525,000	30.0	*2,689,803	36.1	*2,642,709	31.4	*2,595,614	38.8	*2,548,520	38.1
Belgium	1,036,000	9.5	1,046,000	10.1	778,000	9.2	626,000	8.4	482,500	5.7	339,000	5.1	*91,776	1.4
Hong Kong	1,648,000	15.1	2,039,000	19.6	1,462,667	17.4	886,333	11.9	310,000	3.7	223,000	3.3	*658,733	9.8
Spain	1,508,000	13.8	1,114,000	10.7	759,000	9.0	452,000	6.1	350,000	4.2	269,000	4.0	*609,933	9.1
South Africa	553,000	5.1	449,000	4.3	271,000	3.2	150,000	2.0	111,000	1.3	112,000	1.7	69,000	1.0

Commodity data are provided by the United Nations Statistical Division. The symbol * means that data are estimated. For additional notes, see Appendix I.

Product Tables
Suits, Men's and Boys'

Unit of Measure: Number.

	1990 Value	%	1991 Value	%	1992 Value	%	1993 Value	%	1994 Value	%	1995 Value	%	1996 Value	%
Total Production	79,566,693	100.0	73,616,956	100.0	68,489,960	100.0	63,885,256	100.0	59,259,998	100.0	55,147,063	100.0	54,373,376	100.0
Regions														
Africa	2,135,333	2.7	1,504,333	2.0	1,360,333	2.0	1,300,333	2.0	1,126,333	1.9	1,068,333	1.9	894,333	1.6
America, North	12,091,761	15.2	10,256,144	13.9	10,780,528	15.7	11,066,911	17.3	12,844,294	21.7	10,206,601	18.5	9,904,307	18.2
America, South	2,089,206	2.6	2,242,175	3.0	2,277,143	3.3	2,084,000	3.3	2,452,000	4.1	2,806,000	5.1	2,857,181	5.3
Asia	41,044,571	51.6	40,607,631	55.2	36,703,691	53.6	33,089,360	51.8	28,760,190	48.5	26,864,250	48.7	26,728,390	49.2
Europe	21,784,821	27.4	18,684,673	25.4	17,083,265	24.9	16,055,651	25.1	13,802,124	23.3	13,952,457	25.3	13,771,023	25.3
Oceania	421,000	0.5	322,000	0.4	285,000	0.4	289,000	0.5	275,056	0.5	249,422	0.5	218,141	0.4
Leading Producers														
Japan	14,458,000	18.2	13,621,000	18.5	13,466,000	19.7	11,984,000	18.8	12,381,000	20.9	11,994,000	21.7	11,752,000	21.6
United States of America	11,302,000	14.2	9,506,000	12.9	10,032,000	14.6	10,208,000	16.0	11,948,000	20.2	*9,285,923	16.8	*8,775,246	16.1
Russian Federation	7,713,000	9.7	8,169,000	11.1	7,345,000	10.7	6,978,000	10.9	4,307,000	7.3	3,835,000	7.0	4,077,000	7.5
Korea, Republic of	5,885,000	7.4	5,356,000	7.3	5,323,000	7.8	5,830,000	9.1	6,024,000	10.2	6,134,000	11.1	*6,273,855	11.5
Hong Kong	1,201,000	1.5	2,475,000	3.4	1,598,000	2.3	433,000	0.7	250,000	0.4	219,000	0.4	*313,819	0.6

Trousers, Men's and Boys'

Unit of Measure: Thousands of units.

	1990 Value	%	1991 Value	%	1992 Value	%	1993 Value	%	1994 Value	%	1995 Value	%	1996 Value	%
Total Production	671,904	100.0	693,053	100.0	702,422	100.0	683,861	100.0	594,851	100.0	548,896	100.0	596,417	100.0
Regions														
Africa	29,256	4.4	28,186	4.1	26,171	3.7	27,505	4.0	29,233	4.9	30,615	5.6	30,390	5.1
America, North	133,612	19.9	139,323	20.1	131,003	18.7	139,334	20.4	144,595	24.3	129,655	23.6	129,176	21.7
America, South	38,943	5.8	45,018	6.5	35,226	5.0	37,676	5.5	37,192	6.3	35,087	6.4	34,056	5.7
Asia	220,578	32.8	224,685	32.4	264,784	37.7	250,641	36.7	168,097	28.3	132,261	24.1	178,805	30.0
Europe	239,013	35.6	248,538	35.9	238,448	33.9	222,914	32.6	210,928	35.5	216,668	39.5	219,538	36.8
Oceania	10,501	1.6	7,302	1.1	6,791	1.0	5,791	0.8	4,806	0.8	4,611	0.8	4,452	0.7
Leading Producers														
United States of America	108,007	16.1	114,839	16.6	108,525	15.5	107,228	15.7	112,381	18.9	*101,668	18.5	*98,544	16.5
Hong Kong	93,653	13.9	94,004	13.6	120,099	17.1	94,388	13.8	61,788	10.4	49,499	9.0	*71,517	12.0
Spain	46,771	7.0	47,370	6.8	43,078	6.1	37,835	5.5	40,504	6.8	43,516	7.9	46,091	7.7
Brazil	34,515	5.1	40,152	5.8	29,921	4.3	32,511	4.8	31,422	5.3	30,001	5.5	27,714	4.6
France	*36,974	5.5	*36,153	5.2	*35,332	5.0	*34,512	5.0	*33,691	5.7	*32,870	6.0	*32,050	5.4

Blouses, Women's and Girls'

Unit of Measure: Thousands of units.

	1990 Value	%	1991 Value	%	1992 Value	%	1993 Value	%	1994 Value	%	1995 Value	%	1996 Value	%
Total Production	774,491	100.0	783,687	100.0	756,591	100.0	702,215	100.0	685,453	100.0	562,889	100.0	614,843	100.0
Regions														
Africa	10,085	1.3	10,580	1.4	10,270	1.4	7,333	1.0	7,703	1.1	7,787	1.4	7,814	1.3
America, North	211,993	27.4	182,870	23.3	195,757	25.9	211,721	30.2	247,102	36.0	163,646	29.1	149,517	24.3
America, South	33,624	4.3	35,917	4.6	37,224	4.9	41,264	5.9	33,264	4.9	31,507	5.6	23,507	3.8
Asia	347,272	44.8	349,781	44.6	318,471	42.1	275,770	39.3	243,914	35.6	209,557	37.2	275,349	44.8
Europe	165,633	21.4	199,238	25.4	190,800	25.2	163,191	23.2	151,376	22.1	148,292	26.3	154,579	25.1
Oceania	5,884	0.8	5,300	0.7	4,069	0.5	2,936	0.4	2,094	0.3	2,100	0.4	4,078	0.7
Leading Producers														
United States of America	207,996	26.9	178,980	22.8	191,400	25.3	206,976	29.5	241,968	35.3	*158,122	28.1	*143,605	23.4
Hong Kong	246,406	31.8	247,855	31.6	220,638	29.2	168,806	24.0	152,823	22.3	119,258	21.2	*186,330	30.3
Brazil	31,201	4.0	33,456	4.3	34,713	4.6	39,134	5.6	30,502	4.4	28,445	5.1	20,821	3.4
Germany	-		36,530	4.7	32,450	4.3	31,787	4.5	26,831	3.9	24,650	4.4	20,830	3.4
Japan	21,286	2.7	19,817	2.5	18,318	2.4	18,341	2.6	20,851	3.0	20,584	3.7	20,329	3.3

Commodity data are provided by the United Nations Statistical Division. The symbol * means that data are estimated. For additional notes, see Appendix I.

139

Product Tables

Coats, Women's and Girls'

Unit of Measure: Number.

	1990 Value	%	1991 Value	%	1992 Value	%	1993 Value	%	1994 Value	%	1995 Value	%	1996 Value	%
Total Production	70,004,301	100.0	73,472,783	100.0	64,823,616	100.0	54,758,401	100.0	47,176,839	100.0	48,148,259	100.0	49,238,182	100.0
Regions														
America, North	18,245,250	26.1	15,183,500	20.7	12,120,750	18.7	9,049,000	16.5	8,091,000	17.2	12,761,508	26.5	12,240,978	24.9
America, South	108,411	0.2	121,756	0.2	135,100	0.2	15,000	0.0	55,900	0.1	14,800	0.0	86,814	0.2
Asia	21,704,861	31.0	23,090,972	31.4	20,086,083	31.0	19,539,111	35.7	12,936,889	27.4	9,848,667	20.5	10,842,124	22.0
Europe	29,377,779	42.0	34,419,555	46.8	31,659,683	48.8	25,379,290	46.3	25,507,184	54.1	24,956,553	51.8	25,498,488	51.8
Oceania	568,000	0.8	657,000	0.9	822,000	1.3	776,000	1.4	585,865	1.2	566,731	1.2	569,777	1.2
Leading Producers														
United States of America	18,240,250	26.1	15,172,500	20.7	12,104,750	18.7	9,037,000	16.5	8,088,000	17.1	*12,750,615	26.5	*12,229,359	24.8
Russian Federation	11,880,000	17.0	11,963,000	16.3	9,034,000	13.9	7,268,000	13.3	3,035,000	6.4	2,041,000	4.2	1,402,000	2.8
Germany	-		7,024,000	9.6	6,012,000	9.3	5,000,000	9.1	3,947,000	8.4	3,830,856	8.0	5,064,154	10.3
Japan	3,188,000	4.6	2,914,000	4.0	3,300,000	5.1	3,103,000	5.7	3,401,000	7.2	3,028,000	6.3	2,856,000	5.8
Spain	4,345,000	6.2	4,625,000	6.3	4,694,000	7.2	2,020,000	3.7	2,703,000	5.7	2,854,000	5.9	*3,552,638	7.2

Dresses, Women's and Girls'

Unit of Measure: Thousands of units.

	1990 Value	%	1991 Value	%	1992 Value	%	1993 Value	%	1994 Value	%	1995 Value	%	1996 Value	%
Total Production	592,698	100.0	562,373	100.0	434,024	100.0	397,486	100.0	398,008	100.0	353,535	100.0	332,302	100.0
Regions														
Africa	24,377	4.1	22,889	4.1	18,478	4.3	15,977	4.0	15,101	3.8	14,733	4.2	12,870	3.9
America, North	175,944	29.7	164,798	29.3	161,455	37.2	179,673	45.2	201,852	50.7	178,413	50.5	176,594	53.1
America, South	5,929	1.0	7,727	1.4	7,465	1.7	9,694	2.4	12,035	3.0	12,093	3.4	12,910	3.9
Asia	239,542	40.4	225,866	40.2	131,165	30.2	103,212	26.0	95,530	24.0	78,070	22.1	68,632	20.7
Europe	143,090	24.1	139,131	24.7	115,036	26.5	88,547	22.3	73,018	18.3	69,733	19.7	61,176	18.4
Oceania	3,816	0.6	1,962	0.3	425	0.1	382	0.1	472	0.1	493	0.1	120	0.0
Leading Producers														
United States of America	172,317	29.1	161,320	28.7	157,898	36.4	175,517	44.2	197,096	49.5	*173,058	49.0	*170,640	51.4
Russian Federation	136,482	23.0	121,064	21.5	57,138	13.2	30,781	7.7	22,473	5.6	10,733	3.0	8,587	2.6
Ukraine	60,399	10.2	49,557	8.8	36,090	8.3	15,606	3.9	4,932	1.2	3,218	0.9	2,737	0.8
United Kingdom	20,426	3.4	17,836	3.2	*16,789	3.9	*14,803	3.7	*12,818	3.2	*10,832	3.1	*8,846	2.7
South Africa	20,854	3.5	19,950	3.5	15,523	3.6	13,263	3.3	13,172	3.3	12,893	3.6	11,205	3.4

Raincoats, Women's and Girls'

Unit of Measure: Number.

	1990 Value	%	1991 Value	%	1992 Value	%	1993 Value	%	1994 Value	%	1995 Value	%	1996 Value	%
Total Production	8,636,299	100.0	7,473,054	100.0	6,063,464	100.0	4,582,150	100.0	2,997,685	100.0	2,642,328	100.0	2,118,094	100.0
Regions														
Africa	202,000	2.3	166,600	2.2	153,782	2.5	140,964	3.1	128,145	4.3	115,327	4.4	102,509	4.8
America, North	1,395,000	16.2	1,582,000	21.2	1,131,000	18.7	19,409	0.4	-		-		-	
America, South	77	0.0	77	0.0	143	0.0	77	0.0	10	0.0	77	0.0	77	0.0
Asia	4,554,000	52.7	3,260,000	43.6	2,519,000	41.5	2,493,000	54.4	1,145,000	38.2	746,900	28.3	562,800	26.6
Europe	2,401,333	27.8	2,381,267	31.9	2,177,206	35.9	1,847,145	40.3	1,643,752	54.8	1,700,024	64.3	1,390,994	65.7
Oceania	83,889	1.0	83,111	1.1	82,333	1.4	81,556	1.8	80,778	2.7	80,000	3.0	61,714	2.9
Leading Producers														
United States of America	1,395,000	16.2	1,582,000	21.2	1,131,000	18.7	*19,409	0.4	-		-		-	
Russian Federation	3,524,000	40.8	2,548,000	34.1	1,918,000	31.6	1,856,000	40.5	636,000	21.2	351,000	13.3	193,000	9.1
United Kingdom	1,028,000	11.9	*1,044,933	14.0	*935,539	15.4	*826,145	18.0	*716,752	23.9	*607,358	23.0	*497,964	23.5
Ukraine	602,000	7.0	566,000	7.6	407,000	6.7	390,000	8.5	278,000	9.3	228,000	8.6	167,000	7.9

Product Tables
Skirts, Slacks and Shorts, Women's and Girls'

Unit of Measure: Thousands of units.

	1990		1991		1992		1993		1994		1995		1996	
	Value	%	Value	%	Value	%	Value	%	Value	%	Value	%	Value	%
Total Production	919,061	100.0	974,219	100.0	989,977	100.0	971,368	100.0	916,058	100.0	912,951	100.0	961,965	100.0
Regions														
Africa	16,696	1.8	14,995	1.5	15,621	1.6	17,365	1.8	20,907	2.3	21,330	2.3	22,312	2.3
America, North	455,737	49.6	440,686	45.2	495,576	50.1	530,933	54.7	510,620	55.7	516,852	56.6	523,084	54.4
America, South	1,866	0.2	1,964	0.2	2,063	0.2	5,175	0.5	3,416	0.4	2,562	0.3	3,283	0.3
Asia	234,598	25.5	247,017	25.4	222,545	22.5	177,036	18.2	152,286	16.6	147,818	16.2	188,441	19.6
Europe	193,933	21.1	256,051	26.3	241,239	24.4	227,630	23.4	217,624	23.8	212,708	23.3	213,458	22.2
Oceania	16,232	1.8	13,506	1.4	12,932	1.3	13,230	1.4	11,206	1.2	11,681	1.3	11,386	1.2
Leading Producers														
United States of America	451,428	49.1	436,327	44.8	490,720	49.6	525,289	54.1	*504,188	55.0	*509,633	55.8	*515,077	53.5
Hong Kong	143,134	15.6	152,739	15.7	116,824	11.8	94,072	9.7	74,286	8.1	70,013	7.7	*104,447	10.9
Germany	-		68,170	7.0	60,780	6.1	51,903	5.3	41,806	4.6	38,077	4.2	34,778	3.6
Japan	45,628	5.0	48,300	5.0	46,588	4.7	45,552	4.7	42,514	4.6	38,268	4.2	35,897	3.7
United Kingdom	28,787	3.1	25,783	2.6	22,778	2.3	19,774	2.0	23,297	2.5	*21,870	2.4	*20,676	2.1

Suits, Women's and Girls'

Unit of Measure: Number.

	1990		1991		1992		1993		1994		1995		1996	
	Value	%	Value	%	Value	%	Value	%	Value	%	Value	%	Value	%
Total Production	48,586,736	100.0	56,251,507	100.0	55,095,440	100.0	46,605,742	100.0	46,242,289	100.0	42,115,840	100.0	40,816,847	100.0
Regions														
Africa	1,758,000	3.6	1,792,000	3.2	1,719,000	3.1	1,809,000	3.9	2,132,000	4.6	1,829,000	4.3	1,477,000	3.6
America, North	8,451,000	17.4	10,768,000	19.1	9,843,700	17.9	6,583,109	14.1	5,860,491	12.7	5,137,873	12.2	4,415,255	10.8
America, South	166,611	0.3	189,833	0.3	213,055	0.4	291,055	0.6	372,055	0.8	333,055	0.8	356,495	0.9
Asia	25,802,583	53.1	29,024,917	51.6	28,615,250	51.9	26,679,333	57.2	28,031,667	60.6	21,946,000	52.1	22,684,119	55.6
Europe	12,049,828	24.8	14,286,305	25.4	14,115,435	25.6	10,612,244	22.8	9,782,076	21.2	12,355,911	29.3	11,511,436	28.2
Oceania	358,714	0.7	190,452	0.3	589,000	1.1	631,000	1.4	64,000	0.1	514,000	1.2	372,543	0.9
Leading Producers														
United States of America	7,728,000	15.9	10,404,000	18.5	9,432,000	17.1	*6,226,109	13.4	*5,503,491	11.9	*4,780,873	11.4	*4,058,255	9.9
Indonesia	*8,622,000	17.7	*8,622,000	15.3	8,622,000	15.6	*8,622,000	18.5	*8,622,000	18.6	*8,622,000	20.5	*8,622,000	21.1
Turkey	12,342,000	25.4	14,651,000	26.0	13,439,000	24.4	11,907,000	25.5	12,321,000	26.6	10,716,000	25.4	9,674,000	23.7
Hong Kong	3,975,000	8.2	2,095,000	3.7	2,973,000	5.4	3,348,000	7.2	5,168,000	11.2	880,000	2.1	*2,677,790	6.6
Germany	-		2,506,000	4.5	2,060,000	3.7	1,596,000	3.4	1,765,000	3.8	3,539,780	8.4	2,655,121	6.5

Shirts, Men's and Boys'

Unit of Measure: Thousands of units.

	1990		1991		1992		1993		1994		1995		1996	
	Value	%	Value	%	Value	%	Value	%	Value	%	Value	%	Value	%
Total Production	983,111	100.0	1,013,872	100.0	908,019	100.0	790,753	100.0	758,345	100.0	700,409	100.0	713,200	100.0
Regions														
Africa	38,056	3.9	42,870	4.2	30,150	3.3	26,949	3.4	25,156	3.3	26,431	3.8	20,942	2.9
America, North	131,677	13.4	119,672	11.8	133,646	14.7	113,579	14.4	105,769	13.9	97,959	14.0	90,149	12.6
America, South	91,756	9.3	114,350	11.3	119,906	13.2	117,383	14.8	129,773	17.1	112,365	16.0	106,992	15.0
Asia	454,755	46.3	487,984	48.1	388,007	42.7	343,887	43.5	318,165	42.0	281,692	40.2	319,787	44.8
Europe	250,111	25.4	224,864	22.2	216,112	23.8	174,245	22.0	165,370	21.8	170,201	24.3	164,093	23.0
Oceania	16,755	1.7	24,132	2.4	20,199	2.2	14,710	1.9	14,112	1.9	11,760	1.7	11,236	1.6
Leading Producers														
Hong Kong	173,282	17.6	171,615	16.9	148,344	16.3	132,108	16.7	119,456	15.8	96,323	13.8	*133,456	18.7
United States of America	110,448	11.2	98,952	9.8	113,400	12.5	*92,022	11.6	*82,901	10.9	*73,780	10.5	*64,659	9.1
Brazil	68,610	7.0	88,728	8.8	91,707	10.1	91,950	11.6	99,347	13.1	89,400	12.8	74,981	10.5
Russian Federation	115,025	11.7	101,570	10.0	57,253	6.3	40,512	5.1	25,405	3.4	13,550	1.9	7,159	1.0
Indonesia	31,234	3.2	72,586	7.2	56,104	6.2	60,226	7.6	64,348	8.5	68,052	9.7	*73,860	10.4

Product Tables
Underwear, Men's and Boys'

Unit of Measure: Thousands of units.

	1990 Value	%	1991 Value	%	1992 Value	%	1993 Value	%	1994 Value	%	1995 Value	%	1996 Value	%
Total Production	1,648,039	100.0	1,555,796	100.0	1,688,129	100.0	1,724,851	100.0	1,626,577	100.0	1,703,210	100.0	1,815,458	100.0
Regions														
Africa	16,707	1.0	15,528	1.0	11,472	0.7	12,710	0.7	14,058	0.9	15,443	0.9	13,325	0.7
America, North	1,198,837	72.7	1,196,008	76.9	1,330,995	78.8	1,332,018	77.2	1,348,089	82.9	1,401,960	82.3	1,457,903	80.3
America, South	6,474	0.4	6,969	0.4	7,823	0.5	7,481	0.4	12,480	0.8	11,815	0.7	10,997	0.6
Asia	241,224	14.6	165,094	10.6	180,502	10.7	281,620	16.3	151,121	9.3	143,727	8.4	206,556	11.4
Europe	149,254	9.1	134,301	8.6	123,129	7.3	67,488	3.9	66,010	4.1	94,937	5.6	92,301	5.1
Oceania	35,544	2.2	37,897	2.4	34,208	2.0	23,535	1.4	34,820	2.1	35,329	2.1	34,377	1.9
Leading Producers														
United States of America	1,167,432	70.8	1,161,552	74.7	1,299,780	77.0	1,294,044	75.0	1,303,356	80.1	*1,350,468	79.3	*1,399,652	77.1
Hong Kong	122,773	7.4	77,932	5.0	89,492	5.3	181,873	10.5	45,138	2.8	64,775	3.8	*102,317	5.6
Australia	31,773	1.9	34,763	2.2	31,711	1.9	21,595	1.3	*31,938	2.0	*31,843	1.9	*31,748	1.7
Spain	70,143	4.3	58,680	3.8	52,324	3.1	3,408	0.2	3,642	0.2	4,242	0.2	5,166	0.3
Syrian Arab Republic	18,624	1.1	19,824	1.3	21,600	1.3	24,540	1.4	24,216	1.5	*24,833	1.5	*24,963	1.4

Underwear, Women's and Girls'

Unit of Measure: Thousands of units.

	1990 Value	%	1991 Value	%	1992 Value	%	1993 Value	%	1994 Value	%	1995 Value	%	1996 Value	%
Total Production	2,202,514	100.0	1,905,165	100.0	2,086,056	100.0	1,749,932	100.0	1,850,841	100.0	1,929,472	100.0	2,143,217	100.0
Regions														
Africa	36,823	1.7	39,939	2.1	34,987	1.7	37,379	2.1	44,785	2.4	47,314	2.5	42,089	2.0
America, North	951,390	43.2	998,966	52.4	1,064,254	51.0	1,032,118	59.0	1,041,354	56.3	1,059,402	54.9	1,083,474	50.6
America, South	69,206	3.1	76,815	4.0	62,995	3.0	69,506	4.0	64,626	3.5	67,767	3.5	67,769	3.2
Asia	705,263	32.0	368,714	19.4	484,752	23.2	272,556	15.6	327,232	17.7	319,768	16.6	500,184	23.3
Europe	383,272	17.4	367,183	19.3	389,724	18.7	302,344	17.3	325,586	17.6	383,601	19.9	402,956	18.8
Oceania	56,561	2.6	53,548	2.8	49,343	2.4	36,029	2.1	47,258	2.6	51,620	2.7	46,745	2.2
Leading Producers														
United States of America	899,532	40.8	946,740	49.7	1,011,660	48.5	980,556	56.0	*983,682	53.1	*1,007,244	52.2	*1,030,806	48.1
Hong Kong	558,214	25.3	235,285	12.3	355,038	17.0	130,499	7.5	176,748	9.5	120,307	6.2	*323,491	15.1
United Kingdom	76,001	3.5	71,649	3.8	85,744	4.1	99,839	5.7	107,367	5.8	*103,015	5.3	*106,707	5.0
Germany	*58,272	2.6	*58,272	3.1	*58,272	2.8	26,118	1.5	22,331	1.2	96,901	5.0	87,739	4.1
Brazil	53,024	2.4	60,174	3.2	45,896	2.2	47,531	2.7	46,329	2.5	49,321	2.6	47,105	2.2

ISIC 3230
LEATHER AND PRODUCTS

ISIC 3230—Leather and Products—includes leather in the form of hides and skins in tanned and dressed forms, excluding leather prepared with the hair left on. The industry is divided into heavy leather (e.g., shoe soles) and light leather (e.g., leather for purses).

The industry corresponds to the U.S. SICs 3111—Leather Tanning and Finishing, 3131—Footwear Cut Stock, and (in part) 3199—Leather Goods, nec.

Summary Statistics

		1990 Value	N	1991 Value	N	1992 Value	N	1993 Value	N	1994 Value	N	1995 Value	N
Establishments or enterprises	(number)	23,893	77	25,451	86	26,585	80	25,285	78	23,397	63	18,678	30
Number employed	(000)	1,583	83	1,494	87	1,455	82	1,473	82	1,496	74	1,475	59
Total Wages	($ mil.)	6,124	80	6,326	82	6,783	74	5,335	71	5,223	66	4,083	52
Wage/salary per employee	($)	6,924	78	7,273	80	8,516	73	7,120	70	7,447	65	8,718	52
Female workers	(000)	104	33	84	31	113	27	122	31	100	26	70	13
as % of total employment*	(%)	41	33	38	31	40	27	41	31	44	26	42	12
Output	($ bil.)	47	82	48	87	47	81	47	76	48	71	42	57
per employee	($ 000)	45	77	43	82	49	76	44	72	49	68	55	55
per establishment	($ 000)	3,373	70	3,072	78	2,211	70	2,794	65	2,039	56	1,905	27
Value added	($ bil.)	14	74	15	74	15	68	16	67	15	59	13	51
per employee	($ 000)	16	69	16	70	18	65	16	63	18	57	21	49
per establishment	($ 000)	852	63	732	67	772	59	622	55	761	45	511	22
Capital investment	($ mil.)	876	57	456	55	834	50	558	47	678	37	299	19
per employee	($ 000)	3	54	2	54	2	49	2	44	1	34	2	17
per establishment	($ 000)	388	55	229	53	270	49	144	44	57	35	163	16

Data presented above are drawn from the detailed tables that follow. Columns headed 'N' show the number of countries that provided valid data for inclusion. Values are not strictly comparable one year to the next or one row to the next because the number of countries that report varies from period to period and row to row. However, a general indication of magnitudes can be gleaned from reviewing this summary—especially in earlier years in which more reports are available. For detailed explanations, see Appendix I. *The average for those countries reporting both total and female employment.

Establishments and Number Engaged

Country	Establishments or Enterprises (number)						Employees per Establishment					
	1990	1991	1992	1993	1994	1995	1990	1991	1992	1993	1994	1995
Albania	-	-	-	9.000	1.000	1.000	-	-	-	72	146	10
Argentina	-	-	-	930	1,030	-	-	-	-	15	14	-
Armenia	3.000	4.000	6.000	6.000	6.000	-	618	517	-	-	-	-
Australia	152	229	232	-	-	-	23	13	13	-	-	-
Austria	43	40	41	41	35	-	65	75	73	55	68	-
Azerbaijan	4.000	4.000	4.000	4.000	4.000	-	693	619	516	461	407	-
Bahrain	-	-	1.000	-	-	-	-	-	15	-	-	-
Bangladesh	356	311	309				29	31	35	-	-	-
Belgium	139	128	126	108	95	-	17	16	22	-	-	-
Bolivia	33	35	35	34	33	37	25	25	26	27	29	27
Bosnia & Herzegovina	17	19	-	-	-	-	272	246	-	-	-	-
Bulgaria	*25	*25	*28	*250	*306	*280	480	372	275	27	21	21
Burundi	*1.000	*1.000	-	-	-	-	17	14	-	-	-	-
Canada	204	165	150	135	124		25	30	27	22	24	-
Cape Verde	1.000	1.000	1.000	1.000	1.000		-	-	-	-	-	
Chile	19	18	18	21	20		102	122	112	122	128	
China	*9,053	*9,344	*9,240	*9,370	*9,773	*10,468	87	90	94	91	94	95
China (Hong Kong SAR)	282	206	87	114	78	-	9.574	9.709	13	13	12	-
Colombia	117	109	133	125	111	-	69	73	67	72	70	-
Costa Rica	80	76	81	71	-	-	19	19	18	20	-	-
Croatia	73	97	115	159	192	203	108	62	44	28	24	19
Cyprus	97	95	93	86	88	90	9.691	9.884	9.731	8.686	5.284	3.844
Czechoslovakia (Former)	*8.000	*16	-	-	-	-	3,125	1,063	-	-	-	-
Czech Republic	-	*13	*34	*39	-	-	-	1,000	412	308	-	-
Denmark	160	166	160	-	-	-	6.238	5.855	5.813	-	-	-
Ecuador	24	33	31	30	27	23	41	37	35	36	34	36
Egypt	60	69	60	50	-	-	53	54	52	62	-	-
El Salvador	-	8.000	6.000	10	6.000	1.000	-	-	50	70	61	140
Ethiopia	7.000	7.000	7.000	7.000	7.000	7.000	522	489	446	454	478	481
Finland	54	46	39	37	34	40	26	26	26	24	26	28
FYR Macedonia	*7.000	*15	*23	*26	*41	*51	592	271	193	166	95	65
Gambia	-	-	-	2.000	-	-	-	-	-	10	-	-
Germany	-	395	349	309	278	-	-	78	65	62	61	-
Germany (Western Part)	316	298	287	-	-	-	66	68	67	-	-	-
Ghana	-	-	-	4.000	-	-	-	-	-	37	-	-
Greece	284	287	288	-	-	-	12	13	11	-	-	-
Grenada	-	-	3.000	2.000	-	-	-	-	13	20	-	-
Guatemala	-	8.000	8.000	8.000	8.000	8.000	-	48	51	48	49	49
Honduras	-	-	12	13	13	13	-	-	41	42	45	70
Hungary	*122	*194	*109	*133	-	-	82	67	92	68	-	-
Iceland	21	21	19	20	-	-	13	13	14	11	-	-
India	925	994	969	1,182	-	-	51	51	47	39	-	-
Indonesia	130	161	167	180	199	217	99	129	132	129	103	101
Iran (Islamic Republic of)	125	65	57	56	-	-	34	52	72	58	-	-
Iraq	-	1.000	2.000	-	-	-	-	37	50	-	-	-
Ireland	36	34	-	-	-	-	17	18	-	-	-	-
Israel	56	63	99	82	86	-	16	17	12	15	15	-
Italy	*568	*551	*882	*827	*814	-	48	49	42	40	40	-
Japan	3,977	4,036	3,865	3,636	3,061	2,929	11	11	11	11	12	12
Jordan	59	59	67	68	33	34	6.983	7.407	4.299	3.897	13	14
Kenya	21	23	23	23	26	26	69	63	66	68	62	64
Korea, Republic of	1,257	1,273	1,190	1,382	1,381	1,390	32	30	28	23	22	20
Kuwait	-	-	1.000	2.000	2.000	-	-	-	112	66	63	-
Kyrgyzstan	2.000	3.000	3.000	3.000	2.000	-	573	417	430	421	411	-
Latvia	2.000	10	19	35	22	34	955	191	95	43	56	27
Lithuania	-	-	-	*37	*26	-	-	-	-	149	204	-
Macau	16	12	10	5.000	2.000	1.000	67	73	67	66	92	3.000
Malawi	*1.000	-	-	-	-	-	102	-	-	-	-	-
Malaysia	27	33	38	47	49	-	67	100	79	70	71	-
Malta	10	11	10	11	12	-	12	19	22	22	22	-

Continued.

   Depending on the table, * means *Enterprises, Engaged,* or *Factor Values;* ± means *Basis Unspecified;* § means *shown in millions.* For additional notes and sources, see Appendix I.

Establishments and Number Engaged

- Continued -

Country	Establishments or Enterprises (number)						Employees per Establishment					
	1990	1991	1992	1993	1994	1995	1990	1991	1992	1993	1994	1995
Mauritius	10	10	10	9.000	9.000	11	119	127	114	143	131	140
Mongolia	9.000	12	15	11	18	16	361	268	203	310	130	147
Mozambique	3.000	3.000	1.000	1.000	1.000	-	38	49	48	46	59	-
Myanmar	3.000	3.000	3.000	3.000	2.000	-	198	214	195	119	173	-
Nepal	13	19	-	38	20	-	47	31	-	26	24	-
Netherlands	40	47	41	39	28	-	45	40	43	41	54	-
New Zealand	*191	*186	*187	*198	-	-	11	12	12	12	-	-
Norway	23	20	13	14	14	-	20	23	34	33	34	-
Pakistan	-	80	-	-	-	-	-	117	-	-	-	-
Panama	10	9.000	-	-	-	-	27	28	-	-	-	-
Paraguay	-	3.000	-	-	-	-	-	226	-	-	-	-
Peru	285	288	291	-	-	-	13	13	12	-	-	-
Philippines	339	328	99	83	-	-	24	26	78	65	-	-
Poland	*104	*94	*77	*67	-	-	260	202	195	194	-	-
Portugal	*819	*834	*746	*786	*802	-	12	12	12	12	10	-
Puerto Rico	16	13	9.000	14	12	11	117	135	199	181	221	225
Qatar	3.000	3.000	-	-	-	-	30	17	-	-	-	-
Republic of Moldova	4.000	4.000	4.000	4.000	4.000	4.000	994	943	818	776	778	650
Romania	-	169	314	510	876	-	-	-	-	-	-	-
Russian Federation	-	-	-	*979	*1,080	*1,717	-	-	-	98	83	50
Sierra Leone	-	-	-	1.000	-	-	-	-	-	62	-	-
Singapore	17	19	20	20	20	-	40	41	42	41	38	-
Slovakia	-	11	11	17	17	-	-	677	582	352	323	-
Slovenia	80	*520	*440	*409	*572	*298	-	-	-	-	-	-
South Africa	-	180	-	-	-	-	-	61	-	-	-	-
Spain	1,025	999	993	-	-	-	17	17	16	-	-	-
Sri Lanka	19	14	19	22	-	-	62	83	74	97	-	-
Suriname	-	-	1.000	-	-	-	-	-	-	-	-	-
Swaziland	-	1.000	1.000	1.000	-	-	-	-	-	-	-	-
Sweden	30	23	21	18	19	-	37	37	40	38	39	-
Thailand	66	206	-	-	-	-	184	65	-	-	-	-
Trinidad and Tobago	2.000	5.000	6.000	6.000	7.000	-	-	22	18	16	14	-
Tunisia	-	-	-	89	88	93	-	-	-	35	35	38
Turkey	72	57	151	136	128	-	70	75	39	41	52	-
Ukraine	44	129	147	77	76	75	750	240	150	169	237	213
United Kingdom	1,185	1,081	989	1,591	1,163	-	14	15	14	11	15	-
United Republic of Tanzania	8.000	8.000	-	-	-	-	109	105	-	-	-	-
USSR (Former)	*283	-	-	-	-	-	576	-	-	-	-	-
United States of America	-	-	2,073	-	-	-	-	-	22	-	-	-
Venezuela	108	113	114	93	89	77	38	42	46	48	37	46
Yugoslavia	95	135	237	305	308	523	-	107	58	42	40	22
Zambia	-	-	-	-	3.000	-	-	-	-	-	41	-
Zimbabwe	13	13	15	14	15	-	92	77	153	186	180	-

Employment and Compensation of Employees

Country	Number Employed/Engaged (000)						Wage/Salary per Employee ($)					
	1990	1991	1992	1993	1994	1995	1990	1991	1992	1993	1994	1995
Albania	-	-	-	0.648	0.146	0.010	-	-	-	443	-	2,962
Argentina	15	-	-	14	14	-	5,252	-	-	10,355	-	-
Armenia	*1.853	*2.067	-	-	-	-	-	-	-	-	-	-
Australia	3.500	*3.000	*3.000	-	-	-	16,027	20,777	21,078	-	-	-
Austria	2.800	3.000	3.000	2.242	2.383	2.437	17,465	15,359	16,926	20,285	19,509	22,147
Azerbaijan	2.772	2.475	2.064	1.846	1.630	-	1,654	1,720	-	-	-	-
Bahrain	-	-	0.015	-	-	-	-	-	2,482	-	-	-
Bangladesh	10	9.707	11	-	-	-	557	591	582	-	-	-
Belgium	2.300	2.100	2.800	-	-	-	15,222	16,204	16,685	-	-	-
Bolivia	0.824	0.888	0.906	0.925	0.955	1.008	1,472	1,461	1,545	1,457	1,403	1,393

Continued.

Depending on the table, * means *Enterprises, Engaged,* or *Factor Values*; ± means *Basis Unspecified*; § means *shown in millions*. For additional notes and sources, see Appendix I.

Employment and Compensation of Employees
- Continued -

Country	Number Employed/Engaged (000)						Wage/Salary per Employee ($)					
	1990	1991	1992	1993	1994	1995	1990	1991	1992	1993	1994	1995
Bosnia & Herzegovina	4.619	4.668	-	-	-	-	2,476	1,993	-	-	-	-
Bulgaria.	12	9.300	7.700	6.800	6.500	6.000	1,787	558	781	932	840	1,035
Burundi.	0.017	0.014	-	-	-	-	1,855	1,850	-	-	-	-
Canada	5.000	5.000	4.000	3.000	3.000	3.220	18,684	16,409	16,547	19,378	18,795	18,730
Chile.	1.938	2.194	2.017	2.568	2.554	2.602	3,551	4,609	5,197	5,703	6,050	7,071
China	*790	*840	*870	*850	*920	*990	-	-	-	-	-	-
China (Hong Kong SAR) . . .	*2.700	*2.000	*1.100	*1.500	*0.900	-	7,655	8,429	12,214	11,635	14,952	-
China (Taiwan Province) . .	19	19	20	20	21	20	8,516	9,627	11,088	11,433	12,702	13,250
Colombia	8.100	7.918	8.960	9.040	7.735	7.206	1,782	1,814	1,917	2,258	2,877	2,926
Costa Rica	1.521	1.415	1.463	1.454	-	-	2,445	2,291	2,588	2,601	-	-
Croatia	7.910	6.050	5.080	4.530	4.620	3.930	3,330	3,064	1,295	1,429	1,811	2,827
Cyprus	0.940	0.939	0.905	0.747	0.465	0.346	7,147	7,176	7,786	7,852	9,451	12,144
Czechoslovakia (Former) . . .	25	17	-	-	-	-	1,961	1,477	-	-	-	-
Czech Republic	13	13	14	12	-	-	1,886	1,383	1,668	2,012	-	-
Denmark	0.998	0.972	0.930	0.900	0.896	0.916	20,238	18,176	22,268	20,992	21,827	25,276
Ecuador	0.986	1.224	1.077	1.072	0.930	0.825	1,317	1,460	1,501	1,954	1,122	1,004
Egypt	*3.200	*3.700	*3.100	*3.100	*3.073	*3.054	1,969	1,119	1,121	1,301	1,401	1,623
El Salvador.	-	-	*0.299	*0.698	*0.366	*0.140	-	-	2,121	3,396	3,956
Ethiopia.	3.655	3.424	3.119	3.181	3.349	3.369	1,501	1,767	1,708	1,122	1,235	1,272
Finland	1.400	1.200	1.000	0.900	0.900	1.103	20,304	19,762	18,373	15,950	18,336	22,536
FYR Macedonia	4.143	4.066	4.441	4.324	3.893	3.303	5,093	11,716	2,375	2,457	2,491	4,159
France	20	19	17	15	15	14	37,534	36,636	-	-	-	-
Gambia	-	-	-	0.021	-	-	-	-	-	-	-	-
Germany	-	*31	*23	*19	*17	-	-	16,065	21,721	21,871	23,204	-
Germany (Western Part) . . .	*21	*20	*19	-	-	-	20,239	20,741	23,284	-	-	-
Ghana	-	-	-	*0.149	*0.153	*0.158	-	-	-	-	-	-
Greece	*3.379	*3.622	*3.138	*3.111	*3.089	*3.041	6,800	6,572	7,633	7,292	7,084	7,419
Grenada	-	-	*0.039	*0.039	-	-	-	-	-	-	-	-
Guatemala	-	*0.385	*0.404	*0.385	*0.392	*0.390	-	-	-	-	-	-
Honduras	*0.230	*0.324	*0.493	*0.540	*0.591	*0.908	4,539	3,469	2,466	2,231	1,857	1,696
Hungary	10	13	10	9.000	9.440	9.000	2,395	2,049	2,350	2,180	2,465	2,741
Iceland	0.269	0.272	0.275	0.227	0.184	-	6,927	25,332	26,237	21,028	-	-
India	*47	*50	*45	*46	*49	*52	868	770	783	721	807	887
Indonesia	13	21	22	23	20	22	523	563	696	657	776	1,114
Iran (Islamic Republic of) . . .	4.300	3.368	4.123	3.251	-	-	3,836	3,729	3,685	3,908	-	-
Iraq	-	0.037	0.100	-	-	-	-	6,865	7,910	-	-	-
Ireland	0.600	0.600	0.496	0.459	0.460	0.450	14,649	15,078	18,075	16,401	17,374	18,991
Israel.	0.900	1.100	1.200	1.200	1.300	1.312	11,573	11,568	12,877	12,073	14,051	14,936
Italy	27	27	37	33	32	-	27,749	29,416	30,523	24,497	25,582	-
Japan	*44	*45	*43	*40	*37	*35	18,208	20,291	22,769	26,079	31,467	34,325
Jordan	0.412	0.437	0.288	0.265	0.442	0.477	1,758	2,463	4,182	4,835	2,997	3,077
Kenya	1.442	1.451	1.512	1.564	1.613	1.673	1,431	1,268	1,198	748	799	1,030
Korea, Republic of	41	39	34	32	30	28	8,512	9,935	10,983	11,565	12,781	15,067
Kuwait	-	-	0.112	0.132	0.125	0.126	-	-	6,338	7,701	6,201	9,138
Kyrgyzstan.	*1.146	*1.251	*1.289	*1.263	*0.821	-	511	461	453	-	-	-
Latvia	1.910	1.905	1.797	1.512	1.232	0.921	-	-	-	-	-	-
Lithuania	-	-	5.739	5.524	5.309	-	-	-	467	514	1,014	-
Macau	1.079	0.871	0.668	0.328	0.184	0.003	3,951	4,850	5,142	5,974	7,563	5,750
Malawi	0.102	-	-	-	-	-	-	-	-	-	-	-
Malaysia	1.800	3.300	3.000	3.300	3.500	3.746	1,766	1,565	2,591	2,814	3,179	3,586
Malta	0.121	0.207	0.222	0.239	0.269	-	7,978	7,433	7,734	7,930	8,851	-
Mauritius	*1.187	1.271	1.136	1.283	1.183	1.537	1,808	1,793	2,744	2,252	2,711	2,381
Mongolia	*3.250	*3.215	*3.044	*3.415	*2.336	*2.347	1,676	2,552	966	-	-	827
Mozambique	0.115	0.147	0.048	0.046	0.059	-	-	-	-	-	407	-
Myanmar	0.594	0.643	0.585	0.357	0.346	-	815	747	809	2,089	2,010	-
Nepal	0.615	0.588	-	0.999	0.471	-	562	499	-	453	503	-
Netherlands	1.814	1.879	1.773	1.608	1.523	1.417	22,706	23,056	25,017	24,108	25,654	30,783
New Zealand	2.061	2.214	2.275	2.422	-	-	-	-	-	-	-	-
Norway	0.458	0.459	0.441	0.461	0.477	0.451	23,370	23,188	28,244	23,545	24,358	28,276
Pakistan	8.340	9.359	9.789	-	-	-	1,392	1,382	1,421	-	-	-

Continued.

Depending on the table, * means *Enterprises, Engaged,* or *Factor Values*; ± means *Basis Unspecified*; § means *shown in millions*. For additional notes and sources, see Appendix I.

Employment and Compensation of Employees
- Continued -

Country	Number Employed/Engaged (000)						Wage/Salary per Employee ($)					
	1990	1991	1992	1993	1994	1995	1990	1991	1992	1993	1994	1995
Panama.	*0.265	*0.254	*0.255	*0.256	*0.259	*0.262	4,502	4,890	5,069	5,368	6,286	7,345
Paraguay	-	0.678	-	-	-	-	-	-	-	-	-	-
Peru	*3.677	*3.816	*3.480	-	-	-	1,867	1,996	2,300	-	-	-
Philippines.	8.300	8.600	7.700	5.400	-	-	1,437	1,439	1,858	1,818	-	-
Poland	27	19	15	13	15	16	1,051	1,685	2,031	2,165	2,304	2,743
Portugal	*9.808	*10	*8.860	*9.318	*8.329	*8.238	-	-	-	-	-	-
Puerto Rico	1.880	1.750	1.790	2.540	2.650	2.480	16,330	-	-	-	-	-
Qatar	0.089	0.051	-	-	-	-	4,022	5,548	-	-	-	-
Republic of Moldova	3.975	3.772	3.272	3.103	3.110	2.601	-	3,924	-	-	-	-
Russian Federation	-	-	-	96	89	86	-	-	-	633	837	1,165
Sierra Leone	-	-	-	*0.062	-	-	-	-	-	-	-	-
Singapore	*0.677	*0.788	*0.836	*0.826	*0.768	*0.720	8,440	9,690	11,329	12,745	14,768	16,133
Slovakia	-	*7.445	*6.403	*5.987	*5.494	-	-	1,444	1,752	1,721	1,829	-
South Africa	12	11	11	9.000	9.000	9.140	3,929	4,247	4,909	5,508	5,320	6,128
Spain	18	17	16	20	20	19	16,010	16,621	18,390	14,407	14,835	16,095
Sri Lanka	*1.177	*1.159	*1.397	*2.126	-	-	509	635	535	684	-	-
Sweden.	1.100	*0.856	*0.850	0.681	0.747	0.800	18,891	21,670	22,872	18,770	19,637	21,321
Thailand	12	13	-	-	-	-	3,741	2,250	-	-	-	-
Tonga	0.093	0.102	-	-	-	-	3,417	1,664	-	-	-	-
Trinidad and Tobago	-	0.111	0.109	0.096	0.095	-	-	2,332	2,590	1,752	1,954	-
Tunisia	-	-	-	3.080	3.096	3.496	-	-	-	-	-	-
Turkey	5.063	4.288	5.940	5.588	6.683	6.805	3,407	2,963	2,989	3,340	3,361	3,995
Ukraine	33	31	22	13	18	16	2,776	2,882	2,812	476	472	439
United Kingdom	17	16	14	18	17	17	15,966	16,261	19,813	20,103	16,395	17,468
United Republic of Tanzania	0.876	0.842	-	-	-	-	-	894	-	-	-	-
USSR (Former)	163	-	-	-	-	-	1,979	-	-	-	-	-
United States of America	48	46	46	44	39	36	17,917	18,913	20,348	20,909	21,949	22,056
Uruguay	*6.708	*5.633	*4.138	*3.761	*4.445	*4.445	3,278	4,063	5,742	5,309	5,922	6,418
Venezuela	4.100	4.800	5.200	4.418	3.267	3.516	2,824	2,618	2,891	2,983	2,632	3,643
Yugoslavia.	-	14	14	13	12	11	-	-	-	-	1,183	601
Zambia.	-	-	-	-	0.122	-	-	-	-	-	-	-
Zimbabwe	1.200	1.000	2.300	2.600	2.700	2.169	1,872	1,634	1,391	1,254	995	973

Female Workers: Total and Percent of Employment

Country	Female Workers (000)						Female Workers as % of Total Employment					
	1990	1991	1992	1993	1994	1995	1990	1991	1992	1993	1994	1995
Albania.	-	-	-	0.311	0.120	0.002	-	-	-	47.99	82.19	20.00
Argentina	-	-	-	-	1.889	-	-	-	-	-	13.34	-
Australia	1.094	-	-	-	-	-	31.26	-	-	-	-	-
Austria	1.700	2.000	2.000	1.332	1.405	-	60.71	66.67	66.67	59.41	58.96	-
Bangladesh	0.018	0.017	0.016	-	-	-	0.17	0.18	0.15	-	-	-
Bosnia & Herzegovina	2.584	2.294	-	-	-	-	55.94	49.14	-	-	-	-
Bulgaria.	-	-	-	5.200	4.700	4.000	-	-	-	76.47	72.31	66.67
Canada.	2.000	2.000	-	-	-	-	40.00	40.00	-	-	-	-
Chile.	0.228	0.219	0.164	0.318	0.379	-	11.76	9.98	8.13	12.38	14.84	-
China (Taiwan Province)	8.751	8.522	9.138	8.913	8.332	7.296	46.49	46.02	46.20	44.19	39.62	35.72
Colombia	-	3.839	-	4.177	3.531	-	-	48.48	-	46.21	45.65	-
Croatia	5.080	4.010	3.340	2.970	2.970	2.380	64.22	66.28	65.75	65.56	64.29	60.56
Cyprus	0.673	0.667	0.649	0.530	0.312	-	71.60	71.03	71.71	70.95	67.10	-
Czechoslovakia (Former)	14	9.300	-	-	-	-	54.00	54.71	-	-	-	-
Czech Republic	11	9.000	9.000	9.000	-	-	84.62	69.23	64.29	75.00	-	-
Denmark	0.595	0.558	0.504	-	-	-	59.62	57.41	54.19	-	-	-
Egypt	-	0.349	0.286	0.317	-	-	-	9.43	9.23	10.23	-	-
El Salvador.	-	-	-	0.204	0.120	0.026	-	-	-	29.23	32.79	18.57
Ethiopia.	-	0.545	0.562	0.570	0.613	0.615	-	15.92	18.02	17.92	18.30	18.25
FYR Macedonia	0.729	0.697	0.826	1.076	1.098	0.934	17.60	17.14	18.60	24.88	28.20	28.28
Gambia.	-	-	-	0.002	-	-	-	-	-	9.52	-	-

Continued.

Depending on the table, * means *Enterprises, Engaged,* or *Factor Values;* ± means *Basis Unspecified;* § means *shown in millions.* For additional notes and sources, see Appendix I.

147

Female Workers: Total and Percent of Employment

- Continued -

Country	Female Workers (000)						Female Workers as % of Total Employment					
	1990	1991	1992	1993	1994	1995	1990	1991	1992	1993	1994	1995
Germany (Western Part)	12	-	-	-	-	-	57.39	-	-	-	-	-
Hungary	9.000	-	5.000	5.000	-	-	90.00	-	50.00	55.56	-	-
Indonesia	-	-	-	8.868	8.482	9.295	-	-	-	38.07	41.53	42.36
Iran (Islamic Republic of)	-	0.106	0.553	0.282	-	-	-	3.15	13.41	8.67	-	-
Ireland	0.200	0.200	-	-	-	-	33.33	33.33	-	-	-	-
Italy	-	12	17	16	15	-	-	45.16	44.80	48.95	45.13	-
Jordan	0.062	0.128	0.061	0.027	0.089	0.087	15.05	29.29	21.18	10.19	20.14	18.24
Kenya	0.271	0.275	0.306	0.314	-	-	18.79	18.95	20.24	20.08	-	-
Korea, Republic of	15	15	14	13	12	12	37.28	39.50	40.37	40.91	40.15	41.96
Macau	0.520	0.430	0.287	0.108	0.025	-	48.19	49.37	42.96	32.93	13.59	-
Malaysia	1.100	2.200	2.000	2.100	2.100	-	61.11	66.67	66.67	63.64	60.00	-
Malta	0.082	0.149	0.158	0.171	0.190	-	67.77	71.98	71.17	71.55	70.63	-
Mongolia	-	-	-	-	1.588	-	-	-	-	-	67.98	-
Morocco	-	-	-	-	-	1.674	-	-	-	-	-	-
Myanmar	0.143	0.117	0.133	0.112	0.109	-	24.07	18.20	22.74	31.37	31.50	-
Nepal	0.017	0.032	-	0.091	0.039	-	2.76	5.44	-	9.11	8.28	-
New Zealand	0.692	0.697	0.734	0.749	-	-	33.58	31.48	32.26	30.92	-	-
Panama	0.022	-	-	-	-	-	8.30	-	-	-	-	-
Philippines	-	-	5.146	3.428	-	-	-	-	66.83	63.48	-	-
Portugal	0.512	0.372	0.411	0.503	0.390	-	5.22	3.63	4.64	5.40	4.68	-
Republic of Moldova	-	-	-	-	2.297	1.824	-	-	-	-	73.86	70.13
Sri Lanka	0.561	0.543	0.820	1.341	-	-	47.66	46.85	58.70	63.08	-	-
Sweden	0.500	-	-	-	-	-	45.45	-	-	-	0.152	-
Thailand	6.675	7.055	-	-	-	-	55.11	52.65	-	-	-	-
Turkey	1.151	-	-	-	-	-	22.73	-	-	-	-	-
United Kingdom	7.124	-	5.000	-	-	-	41.91	-	35.71	-	-	-
United Republic of Tanzania	0.317	0.256	-	-	-	-	36.19	30.40	-	-	-	-
United States of America	-	-	36	35	33	30	-	-	78.26	79.55	84.62	83.33
Zambia	-	-	-	-	0.052	-	-	-	-	-	42.62	-

Output and Output per Employee

Country	Output ($ bil.)						Output per Employee ($)					
	1990	1991	1992	1993	1994	1995	1990	1991	1992	1993	1994	1995
Albania	-	-	-	*0.001	*0.000	*0.000	-	-	-	1,255	235	23,153
Argentina	±1.059	-	-	1.159	-	-	70,508	-	-	83,660	-	-
Armenia	0.001	0.003	0.010	0.419	0.003	-	610	1,239	-	-	-	-
Australia	*0.429	*0.425	*0.399	-	-	-	122,545	141,540	132,843	-	-	-
Austria	0.312	0.281	0.304	0.268	0.263	0.311	111,383	93,725	101,495	119,330	110,255	127,730
Azerbaijan	±0.041	±0.047	±0.001	±0.003	±0.001	-	14,822	19,044	390	1,646	871	-
Bahrain	-	-	±0.001	-	-	-	-	-	42,553	-	-	-
Bangladesh	0.267	0.234	0.240	-	-	-	25,661	24,097	22,182	-	-	-
Benin	±0.002	±0.002	±0.002	±0.002	-	-	-	-	-	-	-	-
Bolivia	0.022	0.022	0.021	0.019	0.020	0.017	26,719	24,840	23,224	20,454	20,919	16,759
Bosnia & Herzegovina	±0.057	±0.040	-	-	-	-	12,434	8,520	-	-	-	-
Bulgaria	±0.187	±0.052	±0.057	±0.051	±0.050	±0.058	15,601	5,561	7,406	7,457	7,759	9,585
Burundi	±0.000	±0.000	-	-	-	-	7,831	8,264	-	-	-	-
Canada	0.403	0.314	0.290	0.256	0.256	0.280	80,562	62,844	72,392	85,265	85,433	87,071
Chile	0.085	0.102	0.100	0.124	0.128	0.152	43,630	46,665	49,620	48,267	50,026	58,427
China	±4.163	±4.759	±5.897	±9.905	±9.786	±12	5,269	5,666	6,778	11,653	10,637	11,785
China (Hong Kong SAR)	0.174	0.134	0.100	0.131	0.132	-	64,424	67,107	90,785	87,474	146,214	-
China (Taiwan Province)	±0.658	±0.848	±0.962	±0.907	±0.915	±0.899	34,938	45,806	48,617	44,948	43,509	44,023
Colombia	0.223	0.235	0.239	0.258	0.230	0.218	27,530	29,674	26,633	28,526	29,688	30,185
Costa Rica	0.013	0.011	0.012	0.013	0.012	0.013	8,528	7,602	8,379	9,174	-	-
Croatia	±0.161	±0.116	±0.064	-	-	-	20,315	19,161	12,560	-	-	-
Cyprus	0.030	0.026	0.027	0.019	0.018	0.017	32,103	27,167	29,365	25,939	38,885	47,865
Czechoslovakia (Former)	±0.320	±0.162	-	-	-	-	12,791	9,518	-	-	-	-
Denmark	*0.058	*0.040	*0.046	*0.043	*0.053	*0.057	57,637	40,695	48,996	47,418	59,688	62,234

Continued.

Depending on the table, * means *Enterprises, Engaged,* or *Factor Values*; ± means *Basis Unspecified*; § means *shown in millions*. For additional notes and sources, see Appendix I.

Output and Output per Employee

- Continued -

Country	Output ($ bil.)						Output per Employee ($)					
	1990	1991	1992	1993	1994	1995	1990	1991	1992	1993	1994	1995
Ecuador	0.012	0.018	0.015	0.015	0.016	0.014	12,676	14,858	14,030	14,246	17,370	16,863
Egypt	*0.046	*0.033	*0.027	*0.021	*0.024	*0.027	14,484	9,011	8,822	6,764	7,652	8,760
El Salvador	-	±0.008	±0.002	±0.010	±0.003	±0.003	-	-	6,200	14,626	9,558	18,868
Ethiopia	0.076	0.069	0.045	0.043	0.053	0.078	20,878	20,242	14,552	13,635	15,758	23,234
Finland	±0.121	±0.097	±0.085	±0.066	±0.073	±0.109	86,297	80,778	84,833	73,526	81,469	98,508
FYR Macedonia	±0.101	±0.129	±0.076	±0.074	±0.055	±0.068	24,361	31,616	17,037	17,041	14,134	20,571
France	±2.519	±2.238	±2.340	±1.907	±2.154	±2.613	126,583	117,175	140,137	123,013	148,527	182,758
Germany	-	2.369	2.348	1.953	1.896	-	-	76,645	103,979	102,397	111,002	-
Germany (Western Part)	2.310	2.194	2.220	-	-	-	110,479	108,211	115,891	-	-	-
Greece	*0.207	*0.201	*0.215	*0.202	*0.195	*0.202	61,263	55,365	68,463	64,999	63,173	66,474
Guatemala		0.011	0.010	0.010	0.012	0.017		27,970	23,848	24,873	29,581	42,799
Honduras	0.009	0.010	0.011	0.010	0.009	0.010	41,262	31,751	23,216	19,229	15,595	11,024
Hungary	±0.157	±0.163	±0.119	±0.119	±0.138	±0.146	15,663	12,557	11,941	13,180	14,583	16,267
Iceland	0.019	0.018	±0.020	±0.014	-	-	68,942	64,960	71,323	60,322	-	-
India	*0.909	*0.735	*0.634	*0.711	*0.852	*1.006	19,403	14,604	13,949	15,310	17,343	19,243
Indonesia	0.131	0.235	0.243	0.252	0.247	0.294	10,221	11,364	11,050	10,826	12,116	13,409
Iran (Islamic Republic of)	0.214	0.160	0.131	0.092			49,674	47,580	31,840	28,338	-	-
Iraq	-	*0.002	*0.014				-	61,528	141,479			
Ireland	*0.099	*0.083	*0.077	*0.063	*0.070	*0.075	165,008	138,665	154,575	136,175	151,800	165,938
Israel	0.053	0.057	0.070	0.067	0.086	0.091	58,967	52,253	58,626	56,241	66,165	69,261
Italy	±5.262	±5.161	±7.068	±6.090	±7.342	-	194,181	191,411	190,441	185,344	227,687	-
Jamaica	0.006	0.003	0.003				-	-	-			
Japan	±5.063	±5.605	±5.598	±5.629	±6.154	±6.528	115,057	124,547	130,188	140,737	166,324	186,507
Jordan	0.008	0.014	0.017	0.014	0.016	0.025	19,264	32,877	60,636	53,511	37,067	52,302
Kenya	*0.019	*0.017	*0.015	*0.009	*0.009	*0.010	13,315	12,026	9,854	5,732	5,751	6,276
Korea, Republic of	3.900	3.787	3.684	3.647	4.024	4.209	96,291	98,127	109,433	115,726	133,833	149,646
Kuwait	-	-	0.006	0.007	0.007	0.009	-	-	56,832	50,923	55,919	73,562
Kyrgyzstan	0.019	0.018	0.013	0.003	0.003	-	16,499	14,706	9,799	2,426	3,047	-
Latvia	0.000	0.002	0.006	0.008	0.007	0.008	221	978	3,451	5,167	5,827	8,274
Lithuania			0.026	0.026	0.030				4,556	4,770	5,739	
Macau	0.026	0.026	0.019	0.010	0.004	0.000	23,904	29,996	28,912	30,053	23,517	77,083
Malaysia	*0.025	*0.039	*0.050	*0.054	*0.083	*0.089	14,090	11,834	16,749	16,481	23,778	23,634
Malta	0.004	0.007	0.009	0.010	0.011	-	34,674	32,126	38,812	43,265	39,889	-
Mauritius	0.014	0.013	0.018	0.014	0.012	0.018	11,909	10,285	16,047	10,825	10,425	11,510
Mongolia	0.100	0.117	0.067	0.015	0.012	0.012	30,819	36,462	21,895	4,486	4,949	5,078
Mozambique	-	±0.001	-	±0.000	±0.000	-	-	3,965	-	1,941	3,029	-
Myanmar	0.003	0.003	0.004	0.005	0.008	-	5,312	5,197	7,001	12,739	21,767	-
Nepal	0.015	0.010		0.013	0.011		24,804	17,027		12,687	23,931	
Netherlands	0.267	0.257	0.244	0.226	0.220	0.239	147,134	136,629	137,596	140,631	144,780	168,867
Norway	±0.054	±0.052	±0.058	±0.055	±0.060	±0.066	116,849	114,261	131,311	119,558	126,542	145,374
Pakistan	0.356	0.365	0.396	-	-	-	42,644	38,973	40,442	-	-	-
Panama	0.009	0.008	0.009	0.010	0.013	0.017	32,868	32,185	34,548	38,422	50,310	63,959
Peru	0.064	0.069	0.067	-	-	-	17,324	17,981	19,387	-	-	-
Philippines	0.045	0.059	0.059	0.042	-	-	5,422	6,838	7,636	7,795	-	-
Poland	±0.275	±0.218	±0.173	±0.178	±0.214	±0.273	10,203	11,456	11,537	13,686	14,675	17,536
Portugal	0.560	0.539	0.523	0.485	0.530	0.598	57,121	52,577	59,014	52,061	63,612	72,563
Qatar	0.003	0.002	-	-	-	-	37,042	43,094	-	-	-	-
Republic of Moldova	0.000	0.183	0.007	0.010	0.008	0.051	0.087	48,609	2,286	3,369	2,674	19,776
Romania	±0.165	±0.076	±0.048	±0.046								
Russian Federation		3.287	1.148	0.443	0.440	0.443	-	-	-	4,615	4,922	5,159
Singapore	*0.030	*0.036	*0.046	*0.048	*0.054	*0.056	44,659	46,046	55,514	58,499	70,059	77,231
Slovakia	-	±0.108	±0.083	±0.060	±0.058	-	-	14,452	12,987	9,972	10,627	-
Slovenia	-	-	-	0.279	0.327	0.327	-	-	-	-	-	-
South Africa	±0.323	±0.306	±0.316	±0.288	±0.317	±0.369	26,926	27,786	28,688	31,996	35,197	40,377
Spain	*2.035	*1.845	*1.729	±1.929	±2.257	±2.131	113,611	107,285	109,412	94,334	113,923	112,620
Sri Lanka	0.007	0.008	0.012	0.021	-	-	5,811	6,799	8,390	10,102	-	-
Sweden	*0.112	*0.094	*0.105	*0.075	*0.097	*0.121	101,368	109,237	123,926	109,503	129,271	151,675
Thailand	0.222	1.177	-	-	-	-	18,326	87,834	-	-	-	-
Tonga	0.001	0.001	-	-	-	-	10,804	6,460	-	-	-	-
Trinidad and Tobago	±0.001	±0.001	±0.001	±0.001	±0.001	-	-	11,871	12,736	10,512	13,147	-

Continued.

Depending on the table, * means *Enterprises*, *Engaged*, or *Factor Values*; ± means *Basis Unspecified*; § means *shown in millions*. For additional notes and sources, see Appendix I.

149

Output and Output per Employee
- Continued -

Country	Output ($ bil.)						Output per Employee ($)					
	1990	1991	1992	1993	1994	1995	1990	1991	1992	1993	1994	1995
Turkey	0.217	0.170	0.264	0.268	0.317	0.383	42,931	39,690	44,388	47,913	47,424	56,349
Ukraine	*0.582	*0.583	*0.551	*0.162	*0.146	*0.085	17,627	18,821	25,051	12,482	8,089	5,314
United Kingdom	*1.600	*1.303	1.318	1.489	1.481	1.577	94,118	81,416	94,144	82,749	87,109	92,781
United Republic of Tanzania	*0.003	*0.003	-	-	-	-	3,406	4,010	-	-	-	-
USSR (Former)	±4.125	-	-	-	-	-	25,307	-	-	-	-	-
United States of America	*5.090	*4.840	*5.324	*5.548	*5.160	*4.971	106,042	105,217	115,739	126,091	132,308	138,083
Uruguay	0.216	0.223	0.217	0.192	0.252	0.273	32,197	39,527	52,522	51,007	56,797	61,521
Venezuela	0.127	0.159	0.150	0.119	0.102	0.144	31,062	33,115	28,837	26,855	31,190	41,080
Yugoslavia	±0.030	±0.030	±0.069	-	±0.111	±0.059	-	2,101	5,006	-	8,967	5,174
Zambia	-	-	-	-	0.000	-	-	-	-	-	3,049	-
Zimbabwe	*0.022	*0.017	*0.024	*0.021	*0.019	*0.018	18,723	16,628	10,617	7,927	7,044	8,120

Value Added and Value Added per Employee

Country	Value Added ($ bil.)						Value Added per Employee ($)					
	1990	1991	1992	1993	1994	1995	1990	1991	1992	1993	1994	1995
Argentina	±0.336	-	-	0.300	-	-	22,394	-	-	21,637	-	-
Australia	*0.105	*0.111	*0.117	-	-	-	30,134	36,878	39,159	-	-	-
Austria	0.082	0.084	0.086	0.078	0.064	0.075	29,338	27,949	28,726	34,779	26,930	30,695
Bangladesh	0.042	0.033	0.047	-	-	-	4,059	3,386	4,342	-	-	-
Bolivia	0.006	0.006	0.006	0.005	0.005	0.005	7,184	6,432	6,490	5,247	5,652	4,609
Bosnia & Herzegovina	±0.017	±0.012	-	-	-	-	3,615	2,476	-	-	-	-
Bulgaria	-	±0.017	±0.020	±0.019	-	-	-	1,813	2,626	2,772	-	-
Burundi	0.000	0.000	-	-	-	-	4,465	4,722	-	-	-	-
Canada	0.163	0.131	0.116	0.109	0.103	0.111	32,568	26,185	28,957	36,173	34,173	34,443
Chile	0.037	0.034	0.034	0.042	0.054	0.064	19,037	15,315	16,616	16,452	20,953	24,474
China	±0.944	±1.044	±1.292	±2.635	±2.347	±2.412	1,195	1,242	1,486	3,100	2,551	2,437
China (Hong Kong SAR)	0.038	0.031	0.026	0.032	0.030	-	13,978	15,442	23,254	21,545	33,355	-
China (Taiwan Province)	0.233	0.295	0.309	0.329	0.331	0.307	12,403	15,916	15,602	16,295	15,756	15,044
Colombia	0.066	0.065	0.076	0.073	0.068	0.065	8,111	8,161	8,527	8,027	8,812	8,959
Costa Rica	0.005	0.004	0.005	0.005	0.005	0.005	3,223	2,949	3,245	3,443		
Cote d'Ivoire	0.922	1.028	1.020	0.971	-	-	-	-	-	-	-	-
Croatia	±0.058	±0.054	±0.028	-	-	-	7,290	8,934	5,549	-	-	-
Cyprus	0.011	0.011	0.011	0.008	0.007	0.006	11,567	11,245	12,422	11,189	15,318	18,561
Czechoslovakia (Former)	±0.066	-	-	-	-	-	2,630	-	-	-	-	-
Denmark	*0.025	*0.018	*0.021	*0.019	*0.024	*0.025	24,771	18,498	22,624	21,338	26,322	27,604
Ecuador	0.004	0.006	0.004	0.004	0.005	0.004	3,634	4,728	4,178	4,011	5,869	5,395
Egypt	*0.007	*0.006	*0.009	*0.000	-	-	2,078	1,728	3,015	38	-	-
El Salvador	-	0.004	-	0.005	0.002	0.000	-	-	-	6,689	5,303	3,452
Ethiopia	0.027	0.022	0.012	0.024	0.025	0.022	7,273	6,361	3,964	7,468	7,527	6,563
Finland	±0.048	±0.042	±0.035	±0.029	±0.030	±0.039	33,996	35,237	34,603	31,900	33,396	35,653
FYR Macedonia	±0.040	±0.070	±0.040	±0.036	±0.019	±0.022	9,731	17,320	9,049	8,379	4,894	6,621
France	±1.130	±1.053	±1.123	±0.944	±1.059	±1.341	56,767	55,140	67,235	60,891	73,065	93,754
Germany (Western Part)	0.944	0.952	0.927	0.826	-	-	45,167	46,951	48,378	-	-	-
Greece	*0.068	*0.069	*0.076	*0.072	*0.069	*0.071	20,035	19,104	24,204	22,987	22,342	23,509
Honduras	*0.003	*0.003	*0.003	*0.003	*0.003	*0.003	11,170	8,343	6,470	5,418	4,806	3,360
Hungary	±0.037	±0.030	±0.025	±0.025	±0.028	±0.030	3,713	2,283	2,489	2,820	2,976	3,359
Iceland	0.005	0.009	±0.008	±0.006	-	-	18,631	31,875	30,433	26,216	-	-
India	*0.123	*0.114	*0.095	*0.109	*0.133	*0.158	2,625	2,271	2,085	2,354	2,700	3,024
Indonesia	*0.043	*0.089	*0.087	*0.087	*0.081	*0.096	3,338	4,302	3,947	3,747	3,978	4,354
Iran (Islamic Republic of)	0.072	0.049	0.038	0.030	-	-	16,814	14,506	9,256	9,331	-	-
Iraq	-	*0.001	*0.005	-	-	-	-	21,031	48,103	-	-	-
Ireland	*0.021	*0.017	*0.017	*0.015	*0.016	*0.017	35,102	29,079	35,144	31,989	33,752	36,924
Israel	0.019	0.021	0.022	0.020	0.024	0.027	20,941	19,545	18,638	16,489	18,393	20,444
Italy	±1.234	±1.324	±1.862	±1.489	±1.616	-	45,550	49,116	50,168	45,312	50,126	-
Jamaica	0.002	0.001	0.001	-	-	-	-	-	-	-	-	-
Japan	±1.865	±2.175	±2.235	±2.212	±2.417	±2.637	42,381	48,334	51,965	55,306	65,313	75,332
Jordan	0.004	0.004	0.004	0.004	0.004	0.004	8,663	8,135	15,130	14,033	8,936	8,482

Continued.

Depending on the table, * means *Enterprises*, *Engaged*, or *Factor Values*; ± means *Basis Unspecified*; § means *shown in millions*. For additional notes and sources, see Appendix I.

Value Added and Value Added per Employee

- Continued -

Country	Value Added ($ bil.)						Value Added per Employee ($)					
	1990	1991	1992	1993	1994	1995	1990	1991	1992	1993	1994	1995
Kenya	*0.004	*0.004	*0.004	*0.002	*0.003	*0.003	3,026	3,006	2,463	1,543	1,770	1,860
Korea, Republic of	1.144	1.218	1.283	1.240	1.472	1.406	28,258	31,554	38,111	39,335	48,941	49,974
Kuwait	-	-	0.004	0.004	0.003	0.005	-	-	38,609	29,325	26,523	43,107
Latvia	-	-	-	*0.003	*0.003	*0.003	-	-	-	1,962	2,201	2,856
Macau	0.008	0.008	0.005	0.003	0.002	0.000	7,873	9,052	8,186	9,529	10,557	22,208
Malaysia	*0.006	*0.011	*0.014	*0.018	*0.025	*0.026	3,574	3,262	4,580	5,321	7,109	7,059
Malta	0.002	0.003	0.003	0.004	0.005	-	16,477	12,743	13,386	16,649	18,037	-
Mauritius	0.005	0.005	0.008	0.005	0.004	0.007	4,484	3,911	6,923	3,750	3,784	4,438
Mongolia	0.022	0.033	0.027	0.004	0.003	0.003	6,736	10,344	8,881	1,302	1,427	1,192
Myanmar	±0.003	±0.003	±0.004	±0.002	±0.004	-	4,249	4,455	6,441	5,914	12,577	-
Nepal	0.004	0.003	-	0.003	0.004	-	7,253	4,930	-	3,426	7,938	-
Netherlands	0.070	0.078	0.072	0.061	0.062	0.070	38,449	41,558	40,734	37,836	40,568	49,203
New Zealand	0.054	-	-	-	-	-	26,360	-	-	-	-	-
Norway	±0.016	±0.017	±0.016	±0.016	±0.017	±0.019	34,532	37,975	36,972	35,470	36,537	42,304
Pakistan	0.040	0.051	0.057	-	-	-	4,853	5,421	5,786	-	-	-
Panama	0.003	0.002	0.003	0.003	0.004	0.005	10,570	9,799	10,471	11,575	14,961	18,850
Peru	0.032	0.026	0.020	-	-	-	8,677	6,916	5,866	-	-	-
Philippines	0.025	0.024	0.026	0.018	-	-	2,964	2,835	3,324	3,325	-	-
Poland	±0.120	±0.086	±0.065	±0.059	±0.070	±0.088	4,448	4,539	4,335	4,518	4,781	5,672
Portugal	0.126	0.121	0.123	0.115	0.128	0.145	12,849	11,770	13,937	12,292	15,420	17,582
Qatar	0.001	0.001	-	-	-	-	15,434	10,774	-	-	-	-
Romania	±0.067	±0.031	±0.022	±0.024	-	-	-	-	-	-	-	-
Russian Federation	-	-	-	±0.258	±0.226	±0.211	-	-	-	2,691	2,530	2,464
Singapore	*0.011	*0.013	*0.016	*0.020	*0.022	*0.022	16,157	15,927	19,585	24,553	28,059	30,655
Slovakia	-	-	-	±0.027	±0.018	-	-	-	-	4,549	3,272	-
Slovenia	-	-	-	0.104	0.121	0.118	-	-	-	-	-	-
South Africa	*0.075	*0.116	*0.100	*0.091	*0.101	*0.118	6,248	10,502	9,053	10,114	11,194	12,905
Spain	*0.614	*0.623	*0.562	±0.559	±0.548	±0.549	34,278	36,250	35,540	27,323	27,644	28,994
Sri Lanka	0.003	0.003	0.005	0.005	-	-	2,290	2,619	3,536	2,215	-	-
Sweden	*0.052	*0.035	*0.038	*0.028	*0.033	*0.041	47,612	40,842	44,830	41,685	44,518	50,893
Thailand	0.061	0.604	-	-	-	-	5,065	45,112	-	-	-	-
Trinidad and Tobago	±0.000	±0.000	±0.000	±0.000	±0.000	-	-	3,604	4,317	2,531	4,086	-
Turkey	0.060	0.053	0.075	0.090	0.095	0.115	11,812	12,410	12,567	16,145	14,206	16,877
United Kingdom	*0.536	*0.543	*0.544	*0.646	*0.590	*0.628	31,513	33,960	38,869	35,869	34,682	36,923
United Republic of Tanzania	*0.000	*0.001	-	-	-	-	392	840	-	-	-	-
United States of America	*2.210	*2.110	*2.241	*2.319	*2.201	*2.061	46,042	45,870	48,717	52,705	56,436	57,250
Uruguay	0.067	0.088	0.083	0.064	0.084	0.091	10,061	15,618	20,153	16,904	18,809	20,367
Venezuela	0.040	0.049	0.045	0.029	0.037	0.095	9,636	10,146	8,561	6,612	11,440	27,075
Yugoslavia	±0.012	±0.016	±0.039	-	±0.052	±0.026	-	1,100	2,872	-	4,230	2,287
Zambia	-	-	-	-	0.000	-	-	-	-	-	1,979	-
Zimbabwe	*0.007	*0.006	*0.010	*0.006	*0.006	*0.005	6,025	5,922	4,302	2,490	2,150	2,478

Capital Investment

Country	Gross Fixed Capital Formation ($ mil.)						Per Establishment ($ mil)					
	1990	1991	1992	1993	1994	1995	1990	1991	1992	1993	1994	1995
Argentina	-	-	-	23	-	-	-	-	-	0.025	-	-
Australia	7.813	-	-	-	-	-	0.051	-	-	-	-	-
Austria	11	14	11	14	6.566	-	0.245	0.358	0.269	0.352	0.188	-
Bangladesh	10	2.951	4.056	-	-	-	0.029	0.009	0.013	-	-	-
Belgium	1.736	0.820	0.249	0.231	0.418	0.305	0.012	0.006	0.002	0.002	0.004	-
Bolivia	0.797	1.716	0.685	0.630	0.994	-	0.024	0.049	0.020	0.019	0.030	-
Bosnia & Herzegovina	1.456	-	-	-	-	-	0.086	-	-	-	-	-
Bulgaria	8.767	2.704	25	7.738	0.340	2.276	0.351	0.108	0.888	0.031	0.001	0.008
Canada	3.428	-	-	-	-	-	0.017	-	-	-	-	-
Chile	5.795	1.406	2.069	1.846	2.063	-	0.305	0.078	0.115	0.088	0.103	-
China (Hong Kong SAR)	4.236	4.761	3.230	2.456	1.811	-	0.015	0.023	0.037	0.022	0.023	-
Colombia	4.991	7.064	6.018	2.287	1.515	-	0.043	0.065	0.045	0.018	0.014	-

Continued.

Depending on the table, * means *Enterprises, Engaged,* or *Factor Values;* ± means *Basis Unspecified;* § means *shown in millions.* For additional notes and sources, see Appendix I.

151

Capital Investment

- Continued -

Country	Gross Fixed Capital Formation ($ mil.)						Per Establishment ($ mil)					
	1990	1991	1992	1993	1994	1995	1990	1991	1992	1993	1994	1995
Croatia	0.929	0.325	0.763	0.275	-	0.189	0.013	0.003	0.007	0.002	-	0.001
Cyprus	1.160	0.790	0.942	0.457	0.344	0.281	0.012	0.008	0.010	0.005	0.004	0.003
Czechoslovakia (Former)	18	-	-	-	-	-	2.228	-	-	-	-	-
Denmark	-0.162	1.407	-	-	-	-	-0.001	0.008	-	-	-	-
Ecuador	2.813	3.998	3.374	2.558	-0.188	0.929	0.117	0.121	0.109	0.085	-0.007	0.040
Egypt	3.200	4.291	2.246	-3.381	-	-	0.053	0.062	0.037	-0.068	-	-
El Salvador				0.384	0.583	1.480	-	-		0.038	0.097	1.480
Ethiopia	4.959	3.451	1.161	1.112	2.539	2.728	0.708	0.493	0.166	0.159	0.363	0.390
Finland	6.668	2.992	2.433	1.628	2.393	3.007	0.123	0.065	0.062	0.044	0.070	0.075
FYR Macedonia	1.996	5.362	19	4.975	0.102	0.779	0.285	0.357	0.805	0.191	0.002	0.015
Germany (Western Part)	50	54	53	-	-	-	0.159	0.180	0.185	-	-	-
Greece	3.508	15	2.628	-	-	-	0.012	0.051	0.009	-	-	-
Hungary	3.038	2.061	4.507	2.012	-	-	0.025	0.011	0.041	0.015	-	-
India	30	28	32	21	-	-	0.033	0.028	0.033	0.018	-	-
Indonesia	17	20	11	6.229	7.867	11	0.129	0.124	0.065	0.035	0.040	0.051
Iran (Islamic Republic of)	7.539	4.454	4.590	2.358	-	-	0.060	0.069	0.081	0.042	-	-
Ireland	-	1.131	-	-	-	-	-	0.033	-	-	-	-
Israel	3.472	1.316	1.627	2.120	3.653	-	0.062	0.021	0.016	0.026	0.042	-
Italy	115	107	167	101	139	-	0.203	0.195	0.189	0.122	0.171	-
Japan	55	82	103	45	137	-	0.014	0.020	0.027	0.012	0.045	-
Jordan	-	0.066	0.344	0.278	0.132	0.451	-	0.001	0.005	0.004	0.004	0.013
Korea, Republic of	155	160	132	137	186	134	0.124	0.125	0.111	0.099	0.135	0.096
Kuwait	-	-	1.276	0.020	0.094	-	-	-	1.276	0.010	0.047	-
Latvia	-0.010	-0.039	0.115	0.052	0.974	0.200	-0.005	-0.004	0.006	0.001	0.044	0.006
Macau	0.570	0.370	0.099	-0.004	-	-	0.036	0.031	0.010	-0.001	-	-
Malaysia	4.436	2.909	2.355	1.942	4.954	-	0.164	0.088	0.062	0.041	0.101	-
Malta	0.442	0.282	0.236	0.094	0.339	-	0.044	0.026	0.024	0.009	0.028	-
Mongolia	53	28	7.446	2.368	2.156	-	5.845	2.299	0.496	0.215	0.120	-
Myanmar	22	21	21	13	-	-	7.362	7.108	7.044	4.223	-	-
Nepal	2.520	-	-	-	-	-	0.194	-	-	-	-	-
Norway	1.118	0.771	2.089	0.564	1.134	-	0.049	0.039	0.161	0.040	0.081	-
Pakistan	-	12	-	-	-	-	-	0.152	-	-	-	-
Panama	0.261	0.012	-	-	-	-	0.026	0.001	-	-	-	-
Peru	1.447	1.830	1.020	-	-	-	0.005	0.006	0.004	-	-	-
Philippines	1.193	2.183	0.549	1.069	-	-	0.004	0.007	0.006	0.013	-	-
Poland	12	4.822	15	7.585	-	-	0.118	0.051	0.197	0.113	-	-
Portugal	18	25	21	15	23	-	0.022	0.030	0.029	0.019	0.028	-
Romania	-	-	-	6.064	2.204	-	-	-	-	0.012	0.003	-
Singapore	1.263	2.503	5.017	2.859	0.613	-	0.074	0.132	0.251	0.143	0.031	-
Slovenia	7.068	2.974	4.589	1.130	4.177	3.164	0.088	0.006	0.010	0.003	0.007	0.011
Spain	32	43	23	25	33	33	0.031	0.043	0.024	-	-	-
Sri Lanka	0.150	0.348	0.803	0.411	-	-	0.008	0.025	0.042	0.019	-	-
Thailand	39	-351.738	-	-	-	-	0.597	-1.707	-	-	-	-
Trinidad and Tobago	0.071	0.094	0.047	0.037	0.034	-	0.035	0.019	0.008	0.006	0.005	-
Turkey	12	4.075	6.548	6.464	8.443	-	0.160	0.071	0.043	0.048	0.066	-
Ukraine	6.610	13	7.757	1.178	1.877	8.617	0.150	0.102	0.053	0.015	0.025	0.115
United Kingdom	30	21	28	-	32	-	0.026	0.020	0.029	-	0.028	-
United Republic of Tanzania	5.839	6.808	-	-	-	-	0.730	0.851	-	-	-	-
United States of America	70	70	78	88	64	72	-	-	0.038	-	-	-
Uruguay	9.036	6.564	7.394	7.556								
Venezuela	4.733	5.527	5.806	0.749	4.357	23	0.044	0.049	0.051	0.008	0.049	0.302
Yugoslavia	0.892	0.838	0.925	-	1.176	1.733	0.009	0.006	0.004	-	0.004	0.003

Depending on the table, * means *Enterprises, Engaged,* or *Factor Values;* ± means *Basis Unspecified;* § means *shown in millions.* For additional notes and sources, see Appendix I.

Indexes of Industrial Production

Country	Index of Industrial Production (1990=100)						Index of Employment (1990=100)					
	1990	1991	1992	1993	1994	1995	1990	1991	1992	1993	1994	1995
Albania	100	-	-	-	-	-	-	-	-	-	-	-
Algeria	100	80	57	52	65	59	-	-	-	-	-	-
Angola	100	-	-	-	-	-	-	-	-	-	-	-
Argentina	100	103	-	-	-	-	100	-	-	92	94	-
Armenia	100	-	-	-	-	-	100	112	-	-	-	-
Australia	100	94	96	101	-	-	100	86	86	-	-	-
Austria	100	98	95	88	89	91	100	107	107	80	85	87
Azerbaijan	-	-	-	-	-	-	100	89	74	67	59	-
Bahrain	100	-	-	-	-	-	-	-	-	-	-	-
Bangladesh	100	142	208	242	246	319	100	93	104	-	-	-
Barbados	100	100	100	100	100	100	-	-	-	-	-	-
Belgium	100	55	53	46	42	45	100	91	122	-	-	-
Benin	100	-	-	-	-	-	-	-	-	-	-	-
Bolivia	100	107	120	127	135	145	100	108	110	112	116	122
Bosnia & Herzegovina	-	-	-	-	-	-	100	101	-	-	-	-
Brazil	100	87	80	89	87	81	-	-	-	-	-	-
Bulgaria	100	88	78	65	67	58	100	78	64	57	54	50
Burkina-Faso	100	-	-	-	-	-	-	-	-	-	-	-
Burundi	100	-	-	-	-	-	100	82	-	-	-	-
Cambodia	100	100	100	-	-	-	-	-	-	-	-	-
Cameroon	100	-	-	-	-	-	-	-	-	-	-	-
Canada	100	74	69	64	66	71	100	100	80	60	60	64
Chile	100	115	111	107	97	101	100	113	104	133	132	134
China	100	-	-	-	-	-	100	106	110	108	116	125
China (Hong Kong SAR)	100	95	93	85	81	78	100	74	41	56	33	-
China (Taiwan Province)	100	97	77	67	64	54	100	98	105	107	112	109
Colombia	100	109	105	97	91	77	100	98	111	112	95	89
Costa Rica	100	84	93	104	98	99	100	93	96	96	-	-
Croatia	100	-	-	-	-	-	100	76	64	57	58	50
Cuba	100	-	-	-	-	-	-	-	-	-	-	-
Cyprus	100	84	96	63	57	55	100	100	96	79	49	37
Czechoslovakia (Former)	100	59	45	-	-	-	100	68	-	-	-	-
Czech Republic	100	-	-	-	-	-	100	100	108	92	-	-
Dem. Rep. of the Congo	100	-	-	-	-	-	-	-	-	-	-	-
Denmark	100	104	109	108	125	115	100	97	93	90	90	92
Dominican Republic	100	-	-	-	-	-	-	-	-	-	-	-
Ecuador	100	100	89	86	-	-	100	124	109	109	94	84
Egypt	100	187	171	169	182	192	100	116	97	97	96	95
El Salvador	100	105	-	-	-	-	-	-	-	-	-	-
Ethiopia and Eritrea	100	-	-	-	-	-	-	-	-	-	-	-
Ethiopia	-	-	-	-	-	-	100	94	85	87	92	92
Finland	100	84	84	79	72	60	100	86	71	64	64	79
FYR Macedonia	100	-	-	-	-	-	100	98	107	104	94	80
France	100	94	90	87	86	82	100	96	84	78	73	72
Gabon	100	-	-	-	-	-	-	-	-	-	-	-
Gambia	100	-	-	-	-	-	-	-	-	-	-	-
Germany (Eastern Part)	100	-	-	-	-	-	-	-	-	-	-	-
Germany (Western Part)	100	97	85	74	70	67	100	97	92	-	-	-
Ghana	100	-	-	-	-	-	-	-	-	-	-	-
Greece	100	93	96	91	87	79	100	107	93	92	91	90
Guatemala	100	-	-	-	-	-	-	-	-	-	-	-
Guyana	100	100	100	100	100	100	-	-	-	-	-	-
Haiti	100	-	-	-	-	-	-	-	-	-	-	-
Honduras	100	133	149	163	188	258	100	141	214	235	257	395
Hungary	100	83	62	60	69	46	100	130	100	90	94	90
Iceland	100	-	-	-	-	-	100	101	102	84	68	-
India	100	90	88	110	123	139	100	107	97	99	105	112
Indonesia	100	111	156	171	163	150	100	161	171	181	159	171
Iran (Islamic Republic of)	100	104	92	90	82	77	100	78	96	76	-	-
Iraq	100	-	-	-	-	-	-	-	-	-	-	-

Continued.

Depending on the table, * means *Enterprises, Engaged,* or *Factor Values*; ± means *Basis Unspecified*; § means *shown in millions*. For additional notes and sources, see Appendix I.

153

Indexes of Industrial Production

- Continued -

Country	Index of Industrial Production (1990=100)						Index of Employment (1990=100)					
	1990	1991	1992	1993	1994	1995	1990	1991	1992	1993	1994	1995
Ireland	100	91	77	72	77	75	100	100	83	77	77	75
Israel	100	118	135	145	159	155	100	122	133	133	144	146
Italy	100	96	90	94	~	-	100	99	137	121	119	-
Jamaica	100	-	-	-	-	-	-	-	-	-	-	-
Japan	100	99	96	89	85	79	-	102	98	91	84	80
Jordan	100	111	131	127	113	114	100	106	70	64	107	116
Kenya	100	107	103	93	102	69	100	101	105	108	112	116
Korea, Republic of	100	93	87	66	56	47	100	95	83	78	74	69
Kuwait	100	-	-	-	-	-	-	-	-	-	-	-
Kyrgyzstan	100	-	-	-	-	-	100	109	112	110	72	-
Lao P.D.R.	100	-	-	-	-	-	-	-	-	-	-	-
Latvia	100	-	-	-	-	-	100	100	94	79	65	48
Lebanon	100	-	-	-	-	-	-	-	-	-	-	-
Libyan Arab Jamahiriya	100	-	-	-	-	-	-	-	-	-	-	-
Lithuania	100	-	-	-	-	-	-	-	-	-	-	-
Macau	100	-	-	-	-	-	100	81	62	30	17	0
Madagascar	100	72	70				-	-	-	-	-	-
Malawi	100	149	129	103	93	74	100					
Malaysia	100	-	-	-	-	-	100	183	167	183	194	208
Malta	100	165	217	317	-	-	100	171	183	198	222	-
Mauritius	100	-	-	-	-	-	100	107	96	108	100	129
Mexico	100	96	95	92	90	85	-	-	-	-	-	-
Mongolia	100	66	48	31	27	13	100	99	94	105	72	72
Morocco	100	112	112	111	124	120	-	-	-	-	-	-
Mozambique	100	-	-	-	-	-	100	128	42	40	51	-
Myanmar	100	-	-	-	-	-	100	108	98	60	58	-
Nepal	100	-	-	-	-	-	100	96	-	162	77	-
Netherlands	100	97	91	86	82	77	100	104	98	89	84	78
New Zealand	100	85	68	-	-	-	100	107	110	118	-	-
Nicaragua	100	-	-	-	-	-	-	-	-	-	-	-
Niger	100	-	-	-	-	-	-	-	-	-	-	-
Nigeria	100	-	-	-	-	-	-	-	-	-	-	-
Norway	100	101	107	112	124	118	100	100	96	101	104	98
Pakistan	100	-	-	-	-	-	100	112	117	-	-	-
Panama	100	102	108	120	157	200	100	96	96	97	98	99
Paraguay	100	196	189	236	224	246	-	-	-	-	-	-
Peru	100	85	90	87	83	85	100	104	95	-	-	-
Philippines	100	122	128	137	141	162	100	104	93	65	-	-
Poland	100	82	70	69	78	84	100	70	56	48	54	58
Portugal	100	88	81	74	78	77	100	104	90	95	85	84
Puerto Rico	100	-	-	-	-	-	100	93	95	135	141	132
Qatar	100	-	-	-	-	-	100	57	-	-	-	-
Republic of Moldova	100	-	-	-	-	-	100	95	82	78	78	65
Romania	100	88	63	61	68	74	-	-	-	-	-	-
Russian Federation	100	-	-	-	-	-	-	-	-	-	-	-
Saudi Arabia	100	-	-	-	-	-	-	-	-	-	-	-
Singapore	100	82	95	100	94	87	100	116	123	122	113	106
Slovakia	100	-	-	-	-	-	-	-	-	-	-	-
Slovenia	100	86	87	79	77	72	-	-	-	-	-	-
Somalia	100	-	-	-	-	-	-	-	-	-	-	-
South Africa	100	94	96	92	102	110	100	92	92	75	75	76
Spain	100	100	90	82	86	71	100	96	88	114	111	106
Sri Lanka	100	109	116	303	331	241	100	98	119	181	-	-
Sudan	100	-	-	-	-	-	-	-	-	-	-	-
Sweden	100	83	74	73	75	85	100	78	77	62	68	73
Switzerland	100	91	86	85	80	84	-	-	-	-	-	-
Syrian Arab Republic	100	236	-	-	-	-	-	-	-	-	-	-
Thailand	100	-	-	-	-	-	100	111	-	-	-	-
Togo	100	-	-	-	-	-	-	-	-	-	-	-
Tonga	100	-	-	-	-	-	100	110	-	-	-	-

Continued.

Depending on the table, * means *Enterprises, Engaged,* or *Factor Values*; ± means *Basis Unspecified*; § means *shown in millions.* For additional notes and sources, see Appendix I.

Indexes of Industrial Production

- Continued -

Country	Index of Industrial Production (1990=100)						Index of Employment (1990=100)					
	1990	1991	1992	1993	1994	1995	1990	1991	1992	1993	1994	1995
Trinidad and Tobago	100	111	114	131	148	148	-	-	-	-	-	-
Tunisia	100	-	-	-	-	-	-	-	-	-	-	-
Turkey	100	79	82	83	61	64	100	85	117	110	132	134
Uganda	100	80	106	91	129	218	-	-	-	-	-	-
Ukraine	100	-	-	-	-	-	100	94	67	39	55	48
United Arab Emirates . . .	100	-	-	-	-	-	-	-	-	-	-	-
United Kingdom	100	88	88	95	99	99	100	94	82	106	100	100
United Republic of Tanzania . .	100	69	54	15	2	-	100	96	-	-	-	-
USSR (Former)	100	-	-	-	-	-	100	-	-	-	-	-
United States of America . . .	100	91	93	94	87	80	100	96	96	92	81	75
Uruguay	100	100	84	99	126	126	100	84	62	56	66	66
Venezuela	100	-	-	-	-	-	100	117	127	108	80	86
Yemen	100	-	-	-	-	-	-	-	-	-	-	-
Yugoslavia	100	-	-	-	-	-	-	-	-	-	-	-
Yugoslavia (Former)	100	70	-	-	-	-	-	-	-	-	-	-
Zambia	100	-	-	-	-	-	-	-	-	-	-	-
Zimbabwe	100	103	86	88	86	69	100	83	192	217	225	181

Depending on the table, * means *Enterprises*, *Engaged*, or *Factor Values*; ± means *Basis Unspecified*; § means *shown in millions*. For additional notes and sources, see Appendix I.

155

Representative Companies in Sector

Name	Address	Tele-phone	Fax	Employ-ment	Y
Industrija usnja d.d.	Trzaska 31, Vrhnika 1360, Slovenia	61 754211	61 756128	2,346	95
Feuer Leather Corp.	8 Skyline Dr., Hawthorne, New York 10532, U.S.A.	914-592-0073		1,700	
Ecco Indonesia PT	Jl. Raya Bligo No. 17 Candi, Sidoardjo, Indonesia	3198964555	319 62011	1,600	97
Garden State Tanning Inc.	630 Freedom Business Ctr., King of Prussia, Pennsylvania 19406, U.S.A.	610-265-3400	610-265-6366	1,600	95
Federico Meiners Limitada SA	Moreno 2843, Esperanza 3080, Argentina	496 22301	496 22301	1,200	96
Grupo Canguro SA	Calle 73 Via 40-190, Barranquilla, Colombia	5 3457488	5 3567828	1,070	97
Allana Cold Storage Ltd.	Colaba, Bombay 400 039, India	2874455	2044821	1,000	97
Garnar Booth PLC	Sherbourne Rd., Yeovil BA21 5BA, United Kingdom	935 74321	935 27145	985	91
Taejon Leather Industrial Co., Ltd.	Chung-Gu, Taejon, Republic of Korea	42 5251771	42 5251778	900	97
Agro Fellesslakteri	PO Box 40, Forus N-4033, Norway	4 574500	4 575192	890	91
Cho Kwang Leather Co., Ltd.	Chongju-shi, Chungbuk, Republic of Korea	431 662101	431 622443	890	97
Eagle Ottawa Leather Co.	200 N. Beechtree St., Grand Haven, Michigan 49417, U.S.A.	616-842-4000	616-842-3038	850	97
Prime Tanning Company Inc.	216 Airport Dr., Rochester, New Hampshire 03867, U.S.A.	603-330-3100	603-330-2095	780	97
Bender SA, Curtume	Rua Pres Lucena 4320, Estancia Velha 93600-000, Brazil	51 5612066	51 5611066	555	97
A.L. Gebhardt	PO Box 1164, Milwaukee, Wisconsin 53201-1164, U.S.A.	414-383-6030	414-383-4409	550	97
Irving Tanning Co.	PO Box 400, Hartland, Maine 04943, U.S.A.	207-938-4491	207-938-5100	550	97
Lackawanna Leather	PO Box 939, Conover, North Carolina 28613, U.S.A.	704-322-2015	704-322-1429	550	97
Dong Sung Co., Ltd.	Kangnam-Gu, Seoul, Republic of Korea	2 5169494	2 5153414	540	97
Zakaria Shahid Industries	E-37 Kalindi Colony, New Delhi 110 065, India	11 6830227	11 6845861	500	94
Etienne Aigner Inc.	712 5th Ave., New York, New York 10019, U.S.A.	212-246-8660		400	93
Scottish Tanning Industries Ltd.	Clydesdale Works, Bridge of Weir PA11 3LF, United Kingdom	505 612953	505 615122	399	93
Curtitagui Curtiembres de Itagui	Carrera 53A No. 50-89, Itagui, Colombia	4 3720666	4 2815931	393	97
Titan SA	Yumbo, Cali, Colombia	2 6694475	2 6694707	379	97
Tayun Products Ind. Taiwan Inc.	18 Hsing-ho Rd., Tainan, Taiwan	6 2645526	6 2640186	350	94
Curtume Alianca SA	Av. Lions Clube 188, Jequie 45200-000, Brazil	73 5252821	73 5254605	300	96
La Bilbaina, SA	San Pedro, Costa Rica	2248522	2251774	300	
Curtume Aimore SA	Rua Getulio Vargas 505, Arroio do Meio 95940-000, Brazil	51 7161313	51 7161353	280	96
Mossop Leather	PO Box 12107, Parow 7503, Republic of South Africa	21 9314151	21 9314565	280	94
Pieles Costarricenses SA	Apartado 693-1000, San Jose, Costa Rica	4428585	4428613	275	90
Elmo-Calf AB	Kyrkog 18, Svenljunga S-512 81, Sweden	325 10050	325 11004	254	93
Gibaut Hnos Manufacturas De Cueros SA	Centenario Uruguayo 48, Villa Dominico 1874, Argentina	1 2077532	1 2075122	250	96
Grandoe Corp.	PO Box 713, Gloversville, New York 12078, U.S.A.	518-725-8641	518-725-9088	250	97
Baik San Co., Ltd.	Shihung-shi, Kyonggi, Republic of Korea	345 4990044	345 4990050	248	97
Agroexport d.d.	Marsala Tita 25, Belgrade YU-11000, Serbia	11 341421	11 331974	238	92
Induscuer S.C.A.	Villa de Lujan 1331, Sarandi 1872, Argentina	1 2467660	1 2205557	232	96
G.R. Holdings PLC	54 Jermyn St., London SW1Y 6LX, United Kingdom	71 4081747	71 4951581	214	94
Jia Hsing Enterprise Co. Ltd.	299 Chan Lu Rd., Sho Sew Village, Changhua 500, Taiwan	4 7691949	4 7691383	200	91
Nam Chung Co., Ltd.	Saha-Gu, Pusan, Republic of Korea	51 2650140	51 2650150	200	97
Tanneries Modernes de la Manouba	Rue de la Tannerie, La Manouba, Tunisia	220004		200	93
Teneria Moderna Franco Espanola, SAL	Mollet Del Valles, Barcelona E-08100, Spain	93 5706250	93 5705838	190	97
Lo Chin Seng Co. Ltd.	Phra Khanong, Bangkok 10110, Thailand	2580140	2589959	185	90
Frank Industrial Corp. Ltd.	67 Sung-Chiang Rd., 6-3 Fl., Taipei, Taiwan	2 5062855	2 5072916	180	93
Barbour Corporation Inc.	PO Box 2158, Brockton, Massachusetts 02405, U.S.A.	508-583-8200	508-583-4113	174	97
Cudahy Tanning Company Inc.	5043 S. Packard Ave., Cudahy, Wisconsin 53110, U.S.A.	414-483-8100		155	94
Manufacturas Quintero	Calle 23 69B-42, Bogota, Colombia	1 4114640	1 4110884	150	97
Teneria Primenca, SA	Apartado 72-4050, Alajuela, Costa Rica	4431000	4432000	146	
Envases Industriales Hondurenos	PO Box 688Sula, San Pedro Sula, Honduras	522828	521034	137	90
Sam Woo Co., Ltd.	Kimchon-Shi, Kyongbuk, Republic of Korea	547 327781	547 390748	135	97
Time/System International AS	Gydevang 25, Allerod DK-3450, Denmark	42276611	42276665	135	93
Bridge of Weir Leather Co. Ltd.	Clydesdale Works, Bridge of Weir PA11 3LF, United Kingdom	505 612132	505 614964	132	94
Swewi Svendborg AS	Dronningholmsvej 48, Svendborg DK-5700, Denmark	62211628	62212488	120	93
Curtiembre Becas Sca	Madariaga 796, Avellaneda 1870, Argentina	1 2051041	1 2051594	100	96
NCT Leather Ltd.	Clydedale Works, Bridge of Weir PA11 3RL, United Kingdom	505 612182	505 612123	99	92
Yu Jin Ind. Co., Ltd.	Saha-Gu, Pusan, Republic of Korea	51 2615551	51 2619005	85	97
Buil Leather Co., Ltd.	Saha-gu, Pusan, Republic of Korea	51 2660080	51 2660088	80	97
Shin Kwang Co., Ltd.	Yangsan-Gun, Kyongnam, Republic of Korea	523 830111	523 830113	80	97
Yung Ha Industrial Co., Ltd.	Chungnang-gu, Seoul 131-201, Republic of Korea	2 4911455	2 4937181	78	97

Continued

Representative Companies in Sector
- Continued -

Name	Address	Tele-phone	Fax	Employ-ment	Y
Sung San Corp.	Saha-Gu, Pusan, Republic of Korea	51 2622756	51 2622759	70	97
W.J. & W. Lang Ltd.	1 Seedhill, Paisley PA1 1JL, United Kingdom	41 8893134	41 8893182	64	94
Thomas Legget & Sons Ltd.	24 Blythswood Sq., Glasgow G2 4QJ, United Kingdom			63	89
Cape Cobra Pty. Ltd.	PO Box 3383, Cape Town 8000, Republic of South Africa	21 244334	21 230637	60	94
Fuh Ching Leather Co., Ltd.	Niaosung, Kaohsiung, Taiwan	7 7314375	7 7314365	60	93
Inini d.d.	Bravnicarjeva 11, Ljubljana 1001, Slovenia	61 1590799	61 575992	59	95
Border Sheepskins Ltd.	Bankhead South Industrial Estate, Jedburgh TD8 6ED, United Kingdom	835 862311	835 862142	55	94
Teneria el Progreso SA	Juan Diaz Calle Final E Cludad R, Panama City 7, Panama	660597		50	
Bank Bros & Son Ltd.	350 Weston Rd. Ste. 201, Toronto, Ontario M6N 3P9, Canada	416-762-7771	416-762-0183	40	98
Bum Jin Co., Ltd.	Pochon-Gun, Kyonggi, Republic of Korea	357 334000	357 310592	40	97
Teneria Pirro Antonia Gomez, Ltda.	Apartado 321-3000, Heredia, Costa Rica	2373570	2379648	40	
Yoo Yang Moolsan Co., Ltd.	Chongno-gu, Seoul, Republic of Korea	2 7370714	2 7384570	40	97
European Touch	Amaliegade 8E, Copenhagen DK-1256, Denmark	33919299	33150258	38	93
Dooyang Corp.	Kangnam-Gu, Seoul, Republic of Korea	2 5538198	2 5529124	35	97
Viva Corp.	Songpa-gu, Seoul, Republic of Korea	2 4249930	2 4213673	35	97
Da-E Trading Co., Ltd.	Kangdong-Gu, Seoul, Republic of Korea	2 4758091	2 4758095	30	97
Dae Woo Leather Industrial Co., Ltd.	Yangju-Gun, Kyonggi, Republic of Korea	351 402457	351 402454	30	97
Delta Tanning Corp.	1615 51st St., North Bergen, New Jersey 07047, U.S.A.	201-865-3700	201-865-3142	30	95
Sheffren's Hides & Skins Ltd.	3697 Ch de la Baronnie, Varennes, Quebec J3X 1P7, Canada	450-652-2965	450-652-0991	30	98
Halford Hide & Leather Co. Ltd.	8629 126 Ave. NW, Edmonton, Alberta T5B 1G8, Canada	403-474-4989	403-477-3489	25	98
Seil Leather Co.	Sasang-gu, Pusan, Republic of Korea	51 3021480	51 3020344	24	97
Tudor Corp. Ltd.	2929 15 St. NE, Calgary, Alberta T2E 7L8, Canada	403-250-7225	403-291-5146	17	98
Leather Jacket Land	Shop 32, Sancam Ctr., Alberton 1450, Republic of South Africa	11 9078650	11 9078664	11	95
Vanguard Travellers Bag Inc.	170 Min Chuan E. Rd., 7th Fl., Sec 3, Taipei, Taiwan	2 7123958	2 7127550	10	93
Pellimport, SA	Vic, Barcelona 08500, Spain	3 8851700	3 8862926	5	93
Jinyork Enterprise Co., Ltd.	3 Fl., No. 208, Sec. Roosevelt Rd., Taipei, Taiwan	2 3560292	2 3219110	4	93

Product Tables
Heavy Leather

Unit of Measure: Metric tons.

	1990 Value	%	1991 Value	%	1992 Value	%	1993 Value	%	1994 Value	%	1995 Value	%	1996 Value	%
Total Production	510,313	100.0	497,398	100.0	480,563	100.0	471,027	100.0	464,823	100.0	564,723	100.0	572,250	100.0
Regions														
Africa	4,640	0.9	4,704	0.9	4,586	1.0	4,810	1.0	5,132	1.1	5,014	0.9	5,155	0.9
America, North	22,439	4.4	21,171	4.3	20,562	4.3	19,952	4.2	19,343	4.2	18,733	3.3	18,124	3.2
America, South	50,140	9.8	41,963	8.4	41,867	8.7	54,748	11.6	56,551	12.2	55,900	9.9	59,458	10.4
Asia	324,451	63.6	327,332	65.8	316,773	65.9	306,642	65.1	305,134	65.6	410,306	72.7	416,514	72.8
Europe	108,643	21.3	102,227	20.6	96,776	20.1	84,874	18.0	78,664	16.9	74,770	13.2	73,000	12.8
Leading Producers														
Korea, Republic of	78,942	15.5	*91,677	18.4	*99,091	20.6	*106,505	22.6	*113,918	24.5	*121,332	21.5	*128,746	22.5
Indonesia	69,695	13.7	72,856	14.6	76,016	15.8	79,177	16.8	80,037	17.2	*96,235	17.0	*102,380	17.9
Russian Federation	46,339	9.1	39,939	8.0	28,542	5.9	15,287	3.2	5,545	1.2	84,823	15.0	86,693	15.1
Italy	39,600	7.8	*38,213	7.7	*38,532	8.0	*38,851	8.2	*39,170	8.4	*39,488	7.0	*39,807	7.0
Brazil	21,000	4.1	*22,560	4.5	*23,042	4.8	*23,524	5.0	*24,005	5.2	*24,487	4.3	*24,969	4.4

Light Leather

Unit of Measure: 1,000 Square meters.

	1990 Value	%	1991 Value	%	1992 Value	%	1993 Value	%	1994 Value	%	1995 Value	%	1996 Value	%
Total Production	927,015	100.0	917,587	100.0	904,366	100.0	885,409	100.0	861,473	100.0	862,664	100.0	868,408	100.0
Regions														
Africa	12,612	1.4	12,552	1.4	11,107	1.2	12,590	1.4	12,424	1.4	13,261	1.5	13,669	1.6
America, North	66,355	7.2	65,360	7.1	64,576	7.1	63,792	7.2	63,008	7.3	62,225	7.2	61,441	7.1
America, South	58,470	6.3	57,822	6.3	59,680	6.6	61,539	7.0	63,397	7.4	65,255	7.6	67,113	7.7
Asia	323,026	34.8	321,528	35.0	316,506	35.0	306,631	34.6	291,965	33.9	289,693	33.6	292,308	33.7
Europe	466,551	50.3	460,325	50.2	452,496	50.0	440,858	49.8	430,679	50.0	432,231	50.1	433,877	50.0
Leading Producers														
Italy	149,944	16.2	*151,467	16.5	*154,339	17.1	*157,212	17.8	*160,084	18.6	*162,956	18.9	*165,829	19.1
Korea, Republic of	86,260	9.3	85,544	9.3	*91,023	10.1	*94,284	10.6	*97,544	11.3	*100,805	11.7	*104,065	12.0
United States of America	*62,695	6.8	*61,954	6.8	*61,213	6.8	*60,471	6.8	*59,730	6.9	*58,989	6.8	*58,247	6.7
Spain	52,574	5.7	*61,732	6.7	*63,743	7.0	*65,754	7.4	*67,765	7.9	*69,776	8.1	*71,787	8.3
Brazil	58,470	6.3	*57,822	6.3	*59,680	6.6	*61,539	7.0	*63,397	7.4	*65,255	7.6	*67,113	7.7

Commodity data are provided by the United Nations Statistical Division. The symbol * means that data are estimated. For additional notes, see Appendix I.

ISIC 3240
FOOTWEAR

ISIC 3240—Footwear—includes all shoes made entirely or in part from leather for men, women, and children. All types of footwear—including sandals, slippers, sports shoes, orthopedic shoes, etc.—are included. Rubber shoes and similar items are excluded.

The ISIC corresponds to the U.S. SICs 314—Footwear, Except Rubber.

Summary Statistics

		1990		1991		1992		1993		1994		1995	
		Value	N	Value	N	Value	N	Value	N	Value	N	Value	N
Establishments or enterprises	(number)	17,153	76	18,972	85	18,002	78	17,427	76	16,606	64	9,504	30
Number employed	(000)	1,748	87	1,480	91	1,352	83	1,456	83	1,381	76	1,197	58
Total Wages	($ mil.)	9,835	81	10,223	84	10,231	76	7,805	74	8,110	67	6,152	51
Wage/salary per employee	($)	6,462	80	6,626	82	7,609	75	6,478	73	6,728	66	8,098	51
Female workers	(000)	275	33	307	32	299	29	441	30	419	27	342	13
as % of total employment*	(%)	49	33	48	32	47	29	52	30	54	27	63	12
Output	($ bil.)	51	84	56	89	55	83	49	78	50	71	39	57
per employee	($ 000)	30	79	33	83	38	77	33	73	34	68	40	54
per establishment	($ 000)	5,646	70	4,544	78	3,956	71	3,647	65	2,348	56	1,380	27
Value added	($ bil.)	17	73	18	74	19	68	18	67	18	59	15	51
per employee	($ 000)	13	69	14	70	16	65	13	63	15	57	17	49
per establishment	($ 000)	2,249	62	1,424	67	1,768	59	902	55	942	45	575	22
Capital investment	($ mil.)	1,008	62	1,757	59	1,298	53	1,004	50	932	39	471	20
per employee	($ 000)	1	60	2	57	1	52	1	46	1	36	0.748	18
per establishment	($ 000)	634	60	334	57	247	52	190	46	69	37	63	17

Data presented above are drawn from the detailed tables that follow. Columns headed 'N' show the number of countries that provided valid data for inclusion. Values are not strictly comparable one year to the next or one row to the next because the number of countries that report varies from period to period and row to row. However, a general indication of magnitudes can be gleaned from reviewing this summary—especially in earlier years in which more reports are available. For detailed explanations, see Appendix I. *The average for those countries reporting both total and female employment.

Establishments and Number Engaged

Country	Establishments or Enterprises (number)						Employees per Establishment					
	1990	1991	1992	1993	1994	1995	1990	1991	1992	1993	1994	1995
Albania	-	-	-	4.000	5.000	4.000	-	-	-	562	167	-
Argentina	-	-	-	1,128	1,257	-	-	-	-	14	12	-
Armenia	19	26	23	23	28	-	951	641	-	-	-	-
Australia	170	203	202	-	-	-	59	39	35	-	-	-
Austria	63	57	51	49	47	-	133	123	118	114	110	-
Azerbaijan	12	12	214	333	210	-	650	542	27	11	20	-
Bahrain	-	-	4.000	-	-	-	-	-	5.750	-	-	-
Bangladesh	105	74	77	-	-	-	36	54	69	-	-	-
Belgium	98	93	91	80	81	-	16	16	15	-	-	-
Bolivia	21	24	24	25	26	29	39	34	36	37	35	31
Bosnia & Herzegovina	26	28	-	-	7.000	-	894	733	-	-	358	-
Bulgaria	*25	*24	*36	*246	*389	*459	788	658	422	59	42	34
Burundi	*1.000	*1.000	-	-	-	-	27	47	-	-	-	-
Canada	134	107	100	95	92	-	82	84	90	105	109	-
Cape Verde	1.000	1.000	1.000	2.000	7.000	-	67	83	-	-	-	-
Chile	61	64	67	57	59	-	186	186	187	205	206	-
China (Hong Kong SAR)	371	234	172	121	99	-	11	11	9.884	6.612	6.061	-
Colombia	286	279	297	262	250	-	53	53	57	57	56	-
Costa Rica	131	119	111	102	-	-	24	20	22	17	-	-
Croatia	78	92	102	144	184	195	390	231	184	138	113	102
Cyprus	169	160	155	145	151	156	13	13	13	12	10	8.141
Czechoslovakia (Former)	*4.000	*6.000	-	-	-	-	16,750	7,833	-	-	-	-
Czech Republic	-	*4.000	*32	*41	-	-	-	7,000	906	659	-	-
Denmark	122	120	110	-	-	-	14	13	14	-	-	-
Ecuador	20	20	31	24	27	29	85	104	63	66	56	51
Egypt	94	94	97	91	-	-	102	104	94	76	-	-
El Salvador	-	5.000	2.000	8.000	4.000	5.000	-	-	153	125	102	259
Ethiopia	15	14	14	14	45	43	196	222	226	220	85	85
Finland	56	55	51	53	44	49	57	47	41	36	45	51
FYR Macedonia	*14	*26	*36	*43	*28	*91	603	256	167	65	103	24
Germany	-	302	281	246	228	-	-	157	115	106	101	-
Germany (Western Part)	233	217	206	-	-	-	131	135	128	-	-	-
Ghana	-	-	-	1.000	-	-	-	-	-	106	-	-
Greece	299	292	299	-	-	-	22	20	20	-	-	-
Guatemala	-	13	13	13	13	13	-	130	138	130	125	114
Honduras	-	-	17	17	17	17	-	-	72	79	87	69
Hungary	*122	*209	*224	*252	-	-	230	124	107	87	-	-
Iceland	3.000	3.000	2.000	2.000	-	-	13	13	8.000	6.000	-	-
India	344	372	364	495	-	-	146	128	143	108	-	-
Indonesia	234	281	314	327	345	389	254	462	615	706	769	749
Iran (Islamic Republic of)	166	54	49	50	-	-	68	198	223	207	-	-
Iraq	-	6.000	6.000	-	-	-	-	516	500	-	-	-
Ireland	23	15	-	-	-	-	30	47	-	-	-	-
Israel	131	157	176	169	164	-	19	18	18	20	20	-
Italy	*1,519	*1,544	*1,886	*1,963	*1,962	-	49	48	46	43	44	-
Japan	1,818	1,861	1,836	1,743	2,427	2,949	19	19	19	18	19	18
Jordan	165	162	288	291	184	215	3.703	4.759	3.729	3.505	4.707	4.237
Kenya	5.000	14	15	15	15	15	476	164	151	153	157	161
Korea, Republic of	742	1,649	1,513	1,794	1,746	1,684	39	64	70	43	36	30
Kuwait	-	-	-	-	2.000	-	-	-	-	-	30	-
Kyrgyzstan	10	27	40	20	17	-	704	231	118	167	128	-
Latvia	6.000	20	37	40	24	32	753	228	259	167	186	102
Lithuania	-	-	-	*20	*17	-	-	-	-	417	374	-
Macau	26	25	18	19	19	17	26	31	30	27	44	61
Malawi	*1.000	-	-	-	-	-	273	-	-	-	-	-
Malaysia	6.000	14	16	26	20	-	167	100	119	81	85	-
Malta	18	20	19	19	17	-	55	50	55	56	55	-
Mauritius	16	15	14	17	16	15	49	53	54	48	48	49
Mexico	79	57	54	52	51	51	298	286	282	268	263	220
Mongolia	2.000	5.000	6.000	4.000	39	31	2,061	624	423	404	68	81

Continued.

Depending on the table, * means *Enterprises, Engaged,* or *Factor Values*; ± means *Basis Unspecified*; § means *shown in millions.* For additional notes and sources, see Appendix I.

Establishments and Number Engaged

- Continued -

Country	Establishments or Enterprises (number)						Employees per Establishment					
	1990	1991	1992	1993	1994	1995	1990	1991	1992	1993	1994	1995
Mozambique	8.000	8.000	5.000	5.000	6.000	-	133	118	101	100	65	-
Myanmar	2.000	2.000	2.000	2.000	2.000	-	668	371	810	505	522	-
Nepal	8.000	22	-	35	60	-	76	36	-	78	26	-
Netherlands	41	37	36	40	37	-	57	61	59	50	50	-
Netherlands Antilles . . .	-	-	-	16		-	-	-	-	1.375	-	-
New Zealand	*87	*75	*72	*72		-	21	18	20	25	-	-
Norway	15	14	9.000	9.000	9.000	-	27	28	41	40	43	-
Pakistan	-	24				-	-	218	-	-	-	-
Panama.	19	22	-			-	38	38	-	-	-	-
Paraguay	-	1.000				-	-	37	-	-	-	-
Peru	623	650	675	-	-	-	7.498	5.608	5.227	-	-	-
Philippines.	2,061	2,010	348	363	-	-	6.550	10	44	42	-	-
Poland	*122	*169	*170	*158	-	-	697	426	353	316	-	-
Portugal	*1,742	*1,929	*2,102	*2,134	*2,163	-	34	33	30	31	30	-
Puerto Rico	14	12	12	10	11	12	288	311	308	420	376	360
Republic of Moldova . . .	5.000	6.000	7.000	3.000	3.000	3.000	1,865	1,387	1,152	1,542	1,603	1,215
Romania	-	173	287	376	665	-	-	-	-	-	-	-
Russian Federation . . .	-	-	-	*1,548	*1,341	*1,676	-	-	-	109	106	71
Sierra Leone	-	-	-	1.000		-	-	-	-	13	-	-
Singapore	32	29	27	24	19	-	24	22	20	19	22	-
Slovakia	-	6.000	6.000	21	37	-	-	3,303	3,147	828	416	-
Slovenia	43	*24	*31	*46	*38	*271	-	-	-	-	-	-
South Africa	-	264	-	-	-	-	-	114	-	-	-	-
Spain	2,199	2,032	1,800	-	-	-	14	14	13	-	-	-
Sri Lanka	11	11	10	11	-	-	463	505	483	462	-	-
Suriname	-	-	5.000	-	-	-	-	-	25	-	-	-
Swaziland	-	1.000	1.000	1.000		-	-	460	431	422	-	-
Sweden.	26	20	17	20	16	-	27	30	34	25	30	-
Thailand	64	257	-	-	-	-	278	246	-	-	-	-
Trinidad and Tobago . . .	15	20	14	17	21	-	28	23	29	20	18	-
Tunisia	-	-	-	312	318	294	-	-	-	21	26	27
Turkey	52	46	141	120	112	-	118	129	52	62	53	-
Ukraine	94	425	471	75	104	95	1,149	214	189	1,093	673	611
United Kingdom	746	704	665	620	582	-	60	63	60	65	70	-
United Republic of Tanzania .	9.000	11	-	-	-	-	305	29	-	-	-	-
USSR (Former)	*435	-	-	-	-	-	991	-	-	-	-	-
United States of America . .	-	-	636	-	-	-	-	-	83	-	-	-
Venezuela	526	572	495	497	471	450	31	31	33	32	31	32
Yugoslavia.	76	106	150	192	204	215	-	306	203	149	133	122
Zambia	-	-	-	-	9.000	-	-	-	-	-	108	-
Zimbabwe	14	14	15	14	15	-	432	468	393	443	427	-

Employment and Compensation of Employees

Country	Number Employed/Engaged (000)						Wage/Salary per Employee ($)					
	1990	1991	1992	1993	1994	1995	1990	1991	1992	1993	1994	1995
Albania	-	-	-	2.248	0.836	-	-	-	-	466	-	-
Argentina	19		-	15	16	-	4,540	-	-	9,575	-	-
Armenia.	*18	*17				-	15,891	17,628	17,227	-	-	-
Australia	10	*8.000	*7.000	-	-	-	15,891	17,628	17,227	-	-	-
Austria	8.400	7.000	6.000	5.595	5.152	5.321	15,726	18,206	20,536	19,499	19,559	22,376
Azerbaijan	7.805	6.504	5.689	3.778	4.169	-	1,678	1,688	-	-	-	-
Bahrain	-	-	0.023	-	-	-	-	-	2,313	-	-	-
Bangladesh	3.772	3.961	5.290	-	-	-	1,546	1,697	1,364	-	-	-
Belgium	1.600	1.500	1.400	-	-	-	13,859	13,959	15,463	-	-	-
Bermuda	0.013	0.014	0.015	0.014	0.014	-	-	-	-	-	-	-
Bolivia	0.813	0.820	0.858	0.929	0.902	0.898	2,331	2,479	1,941	1,968	2,168	2,105
Bosnia & Herzegovina . . .	23	21	-	-	2.507	-	2,315	1,905	-	-	-	-

Continued.

Employment and Compensation of Employees
- Continued -

Country	Number Employed/Engaged (000)						Wage/Salary per Employee ($)					
	1990	1991	1992	1993	1994	1995	1990	1991	1992	1993	1994	1995
Bulgaria	20	16	15	15	16	15	1,623	557	818	925	794	1,018
Burundi	0.027	0.047	-	-	-	-	1,838	1,829	-	-	-	-
Cambodia	*3.193	*2.389	-	-	-	-	-	-	-	-	-	-
Canada	11	9.000	9.000	10	10	9.450	18,232	18,814	17,926	16,278	16,183	16,647
Cape Verde	0.067	0.083	-	-	-	-	-	-	-	-	-	-
Chile	11	12	13	12	12	11	2,925	3,477	4,017	4,550	5,170	5,553
China (Hong Kong SAR)	*3.900	*2.600	*1.700	*0.800	*0.600	-	6,715	7,770	7,903	6,464	7,332	-
China (Taiwan Province)	37	34	30	26	23	20	7,489	7,685	8,421	9,214	9,682	10,131
Colombia	15	15	17	15	14	13	1,556	1,625	1,709	1,969	2,485	2,682
Costa Rica	3.140	2.329	2.467	1.750	-	-	2,132	1,934	2,171	2,561	-	-
Croatia	30	21	19	20	21	20	3,172	2,992	1,423	1,580	1,879	2,898
Cyprus	2.139	2.034	2.000	1.723	1.580	1.270	7,722	8,079	9,149	9,011	9,636	12,196
Czechoslovakia (Former)	67	47	-	-	-	-	2,162	1,443	-	-	-	-
Czech Republic	36	28	29	27	-	-	2,228	1,430	1,720	1,930	-	-
Denmark	1.662	1.597	1.533	1.534	1.530	1.440	25,277	26,237	29,071	27,176	32,739	37,539
Ecuador	1.705	2.072	1.938	1.588	1.509	1.473	1,609	1,779	1,913	2,512	1,259	1,401
Egypt	*9.600	*9.800	*9.100	*6.900	*6.909	*6.946	1,557	854	994	1,113	1,199	1,384
El Salvador	-	-	*0.306	*1.003	*0.409	*1.294	-	-	-	2,138	2,465	2,019
Ethiopia	2.943	3.102	3.159	3.086	3.840	3.666	1,791	1,768	1,268	861	732	678
Finland	3.200	2.600	2.100	1.900	2.000	2.482	19,703	19,326	17,285	14,438	15,315	20,481
FYR Macedonia	8.435	6.664	6.028	2.803	2.890	2.198	3,555	4,016	1,642	2,010	3,348	3,005
France	53	48	45	42	39	38	19,786	20,829	22,269	-	-	-
Germany	-	*47	*32	*26	*23	-	-	16,604	23,410	24,454	26,343	-
Germany (Western Part)	*30	*29	*26	-	-	-	21,928	22,826	25,947	-	-	-
Ghana	-	-	-	*0.106	*0.109	*0.112	-	-	-	-	-	-
Greece	*6.526	*5.823	*6.002	*6.524	*6.210	*6.159	7,535	7,594	8,120	9,618	9,293	9,903
Guatemala	-	*1.685	*1.796	*1.685	*1.626	*1.485	-	-	-	-	-	-
Honduras	*0.798	*0.806	*1.225	*1.343	*1.471	*1.177	1,945	1,906	898	1,681	2,018	-
Hungary	28	26	24	22	21	22	1,812	1,915	2,077	2,063	2,042	2,080
Iceland	0.040	0.040	0.016	0.012	0.011	-	14,241	12,670	9,123	7,396	-	-
India	*50	*48	*52	*53	*53	*58	978	822	850	723	726	739
Indonesia	59	130	193	231	265	291	796	532	706	633	781	1,140
Iran (Islamic Republic of)	11	11	11	10	-	-	3,763	3,687	4,092	3,759	-	-
Iraq	-	3.097	3.000	-	-	-	-	7,814	15,827	-	-	-
Ireland	0.700	0.700	0.552	0.501	0.540	0.520	14,688	13,386	15,694	14,116	14,597	16,211
Israel	2.500	2.800	3.100	3.400	3.300	3.477	13,292	12,693	12,987	12,055	13,385	14,704
Italy	74	74	86	85	85	-	21,881	22,738	23,410	19,670	20,571	-
Japan	*34	*35	*34	*32	*47	*52	19,704	21,422	23,920	26,978	29,560	31,690
Jordan	0.611	0.771	1.074	1.020	0.866	0.911	2,443	2,381	1,854	2,281	2,326	2,208
Kenya	2.378	2.290	2.261	2.296	2.350	2.419	1,602	1,400	1,267	897	1,182	1,642
Korea, Republic of	29	106	107	77	63	51	7,055	8,498	8,411	9,430	10,237	12,066
Kuwait	-	-	-	-	0.061	0.061	-	-	-	-	4,511	5,987
Kyrgyzstan	*7.038	*6.242	*4.738	*3.332	*2.175	-	-	-	-	-	-	-
Latvia	4.518	4.557	9.576	6.671	4.469	3.262	-	-	-	-	-	-
Lithuania	-	-	9.843	8.330	6.352	-	-	-	460	-	857	-
Macau	0.681	0.780	0.547	0.504	0.833	1.033	4,127	4,075	4,540	5,134	4,111	3,448
Malawi	0.273	-	-	-	-	-	-	-	-	-	-	-
Malaysia	1.000	1.400	1.900	2.100	1.700	1.702	2,033	2,493	2,500	3,570	3,542	3,777
Malta	0.984	0.992	1.054	1.062	0.930	-	8,303	8,158	8,545	7,454	8,437	-
Mauritius	*0.780	0.801	0.755	0.812	0.774	0.738	1,777	1,864	2,524	1,124	1,539	2,147
Mexico	24	16	15	14	13	11	3,586	4,342	5,534	5,799	5,736	2,728
Mongolia	*4.123	*3.119	*2.538	*1.615	*2.653	*2.506	1,369	2,169	858	-	-	-
Mozambique	1.065	0.947	0.506	0.502	0.391	-	484	-	417	-	-	-
Myanmar	1.336	0.741	1.620	1.010	1.045	-	641	1,441	871	2,195	1,764	-
Nepal	0.606	0.793	-	2.714	1.564	-	761	675	-	606	515	-
Netherlands	2.351	2.267	2.128	1.989	1.856	1.695	20,556	21,233	22,982	22,197	22,914	26,895
Netherlands Antilles	-	-	-	0.022	-	-	-	-	-	7,694	-	-
New Zealand	1.858	1.366	1.418	1.802	-	-	-	-	-	-	-	-
Norway	0.412	0.396	0.371	0.360	0.383	0.334	19,775	21,034	26,812	21,536	21,827	25,790
Pakistan	5.235	5.241	5.482	-	-	-	2,353	2,440	2,510	-	-	-

Continued.

Depending on the table, * means *Enterprises, Engaged,* or *Factor Values;* ± means *Basis Unspecified;* § means *shown in millions.* For additional notes and sources, see Appendix I.

Employment and Compensation of Employees
- Continued -

Country	Number Employed/Engaged (000)						Wage/Salary per Employee ($)					
	1990	1991	1992	1993	1994	1995	1990	1991	1992	1993	1994	1995
Panama.	*0.719	*0.847	*0.851	*0.858	*0.851	*0.843	3,371	4,145	4,214	4,256	4,272	4,282
Paraguay	-	0.037	-	-	-	-	-	-	-	-	-	-
Peru	*4.671	*3.645	*3.528	-	-	-	1,915	2,307	1,890	-	-	-
Philippines	14	20	15	15	-	-	734	791	1,112	1,146	-	-
Poland	85	72	60	50	54	56	990	1,517	1,813	1,804	2,025	2,486
Portugal	*59	*64	*62	*66	*65	*70	-	-	-	-	-	-
Puerto Rico	4.030	3.730	3.700	4.200	4.140	4.320	16,824	-	-	-	-	-
Republic of Moldova	9.327	8.322	8.063	4.626	4.808	3.644	-	3,437	-	-	-	-
Russian Federation	-	-	-	168	142	118	-	-	-	554	681	675
Sierra Leone	-	-	-	*0.013	-	-	-	-	-	-	-	-
Singapore	*0.781	*0.638	*0.536	*0.459	*0.419	*0.386	6,445	7,581	10,187	10,775	13,513	15,161
Slovakia	-	*20	*19	*17	*15	-	-	1,429	1,677	1,559	1,728	-
South Africa	32	30	27	24	26	27	4,409	5,311	5,415	5,623	5,167	5,646
Spain	30	28	24	38	41	42	10,937	11,855	13,109	10,562	10,067	12,245
Sri Lanka	*5.098	*5.553	*4.835	*5.079	-	-	842	1,081	815	1,175	-	-
Suriname	*0.116	*0.131	*0.125	*0.108	-	-	8,693	6,842	8,964	16,599	-	-
Swaziland	-	0.460	0.431	0.422	-	-	-	2,303	2,606	2,759	-	-
Sweden.	0.700	*0.603	*0.570	0.507	0.480	0.553	18,584	20,737	22,492	17,716	17,820	20,537
Thailand	18	63	-	-	-	-	1,780	1,747	-	-	-	-
Tonga	0.004	0.004	-	-	-	-	1,171	579	-	-	-	-
Trinidad and Tobago	0.427	0.455	0.405	0.334	0.373	-	4,258	4,677	3,370	2,518	2,217	-
Tunisia	-	-	-	6.652	8.213	7.976	-	-	-	-	-	-
Turkey	6.147	5.944	7.340	7.460	5.955	5.923	5,239	6,573	5,452	6,041	3,097	3,470
Ukraine	108	91	89	82	70	58	2,719	3,436	2,326	420	431	-
United Kingdom	45	44	40	40	41	39	15,000	15,044	16,166	14,114	15,127	16,150
United Republic of Tanzania	2.741	0.314	-	-	-	-	-	-	-	-	-	-
USSR (Former)	431	-	-	-	-	-	1,934	-	-	-	-	-
United States of America	67	58	53	52	50	48	14,776	14,655	15,566	16,538	16,420	16,604
Uruguay	*3.989	*3.560	*4.307	*3.156	*3.055	*2.287	1,712	1,810	2,207	2,819	2,952	3,537
Venezuela	16	18	16	16	14	15	2,388	2,690	3,067	3,007	2,350	3,262
Yugoslavia.	-	32	30	29	27	26	-	-	-	-	1,057	461
Zambia	-	-	-	-	0.975	-	-	-	-	-	1,329	-
Zimbabwe	6.051	6.558	5.900	6.200	6.400	5.299	3,220	2,862	2,259	1,916	1,825	1,985

Female Workers: Total and Percent of Employment

Country	Female Workers (000)						Female Workers as % of Total Employment					
	1990	1991	1992	1993	1994	1995	1990	1991	1992	1993	1994	1995
Albania	-	-	-	1.670	0.375	-	-	-	-	74.29	44.86	-
Argentina	-	-	-	-	4.394	-	-	-	-	-	28.06	-
Australia	5.439	-	-	-	-	-	54.39	-	-	-	-	-
Austria	5.700	5.000	4.000	3.701	3.396	-	67.86	71.43	66.67	66.15	65.92	-
Bahrain	-	-	0.001	-	-	-	-	-	4.35	-	-	-
Bangladesh	-	-	0.135	-	-	-	-	-	2.55	-	-	-
Bermuda	0.004	0.005	0.006	0.005	0.005	-	30.77	35.71	40.00	35.71	35.71	-
Bosnia & Herzegovina	18	15	-	-	-	-	79.38	73.97	-	-	-	-
Bulgaria.	-	-	-	12	13	12	-	-	-	83.45	81.60	78.71
Canada	7.000	6.000	-	-	-	-	63.64	66.67	-	-	-	-
Chile.	4.281	3.964	4.561	4.148	4.356	-	37.76	33.36	36.32	35.53	35.83	-
China (Taiwan Province)	24	23	20	18	15	13	65.81	66.70	66.50	66.85	66.56	66.30
Colombia	-	7.543	-	7.825	7.099	-	-	50.57	-	52.17	50.45	-
Croatia	23	15	13	14	14	14	74.51	70.01	71.54	71.47	68.93	69.92
Cyprus	1.366	1.260	1.332	1.068	1.024	-	63.86	61.95	66.60	61.98	64.81	-
Czechoslovakia (Former)	47	23	-	-	-	-	70.15	49.15	-	-	-	-
Czech Republic	19	17	21	19	-	-	52.78	60.71	72.41	70.37	-	-
Denmark	0.842	0.801	0.750	-	-	-	50.66	50.16	48.92	-	-	-
Egypt	-	2.153	1.812	1.542	-	-	-	21.97	19.91	22.35	-	-
El Salvador.	-	-	-	-	0.183	0.499	-	-	-	-	44.74	38.56

Continued.

Depending on the table, * means *Enterprises, Engaged,* or *Factor Values*; ± means *Basis Unspecified*; § means *shown in millions*. For additional notes and sources, see Appendix I.

163

Female Workers: Total and Percent of Employment

- Continued -

Country	Female Workers (000)						Female Workers as % of Total Employment					
	1990	1991	1992	1993	1994	1995	1990	1991	1992	1993	1994	1995
Ethiopia	-	1.207	1.256	1.248	1.446	1.403	-	38.91	39.76	40.44	37.66	38.27
FYR Macedonia	5.835	5.094	4.627	2.147	2.866	2.131	69.18	76.44	76.76	76.60	99.17	96.95
Germany (Western Part)	16	-	-	-	-	-	52.53	-	-	-	-	-
Hungary	19	-	17	15	-	-	67.86	-	70.83	68.18	-	-
Indonesia	-	-	-	171	202	224	-	-	-	74.02	76.23	76.82
Iran (Islamic Republic of)	-	0.629	0.772	0.735	-	-	-	5.88	7.07	7.09	-	-
Iraq	-	0.952	-	-	-	-	-	30.74	-	-	-	-
Ireland	0.300	0.300	-	-	-	-	42.86	42.86	-	-	-	-
Italy	-	39	45	43	43	-	-	53.32	51.78	50.40	50.04	-
Jordan	0.076	0.144	0.111	0.097	0.124	0.083	12.44	18.68	10.34	9.51	14.32	9.11
Kenya	0.033	0.027	0.023	0.023	-	-	1.39	1.18	1.02	1.00	-	-
Korea, Republic of	14	67	66	45	35	27	49.30	63.08	61.73	58.26	56.16	52.98
Macau	0.359	0.451	0.321	0.397	0.554	0.818	52.72	57.82	58.68	78.77	66.51	79.19
Malaysia	0.500	0.800	1.100	1.100	0.900	-	50.00	57.14	57.89	52.38	52.94	-
Malta	0.574	0.568	0.643	0.663	0.568	-	58.33	57.26	61.01	62.43	61.08	-
Mongolia	-	-	-	-	1.830	-	-	-	-	-	68.98	-
Morocco	-	-	-	-	-	3.515	-	-	-	-	-	-
Myanmar	0.232	0.566	0.509	0.549	0.616	-	17.37	76.38	31.42	54.36	58.95	-
Nepal	0.106	0.133	-	0.109	0.233	-	17.49	16.77	-	4.02	14.90	-
New Zealand	1.092	0.784	0.827	1.064			58.77	57.39	58.32	59.05		
Panama	0.256	-	-	-	-	-	35.61	-	-	-	-	-
Philippines	-	-	8.373	8.050	-	-	-	-	54.37	53.31	-	-
Portugal	17	19	18	21	19	-	29.29	30.09	28.92	31.94	29.65	-
Republic of Moldova	-	-	-	-	3.645	2.461	-	-	-	-	75.81	67.54
Sri Lanka	2.429	3.102	2.631	2.745	-	-	47.65	55.86	54.42	54.05	-	-
Sweden	0.400	-	-	-	-	-	57.14	-	-	-	-	-
Thailand	14	47	-	-	-	-	76.24	74.80	-	-	-	-
Turkey	1.038	-	-	-	-	-	16.89	-	-	-	-	-
United Kingdom	25	-	21	-	-	-	55.17	-	52.50	-	-	-
United Republic of Tanzania	0.576	0.066	-	-	-	-	21.01	21.02	-	-	-	-
United States of America	-	-	45	44	43	41	-	-	84.91	84.62	86.00	85.42
Zambia	-	-	-	-	0.123	-	-	-	-	-	12.62	-

Output and Output per Employee

Country	Output ($ bil.)						Output per Employee ($)					
	1990	1991	1992	1993	1994	1995	1990	1991	1992	1993	1994	1995
Albania	-	-	-	*0.004	*0.002	*0.001	-	-	-	1,879	1,948	-
Argentina	±0.516	-	-	0.695	-	-	27,216	-	-	45,105	-	-
Armenia	0.007	0.010	0.036	1.957	0.005	-	366	600	-	-	-	-
Australia	*0.551	*0.529	*0.438	-	-	-	55,078	66,128	62,500	-	-	-
Austria	0.708	0.658	0.621	0.541	0.527	0.625	84,276	94,051	103,437	96,710	102,352	117,409
Azerbaijan	±0.092	±0.127	±0.004	±0.011	±0.003	-	11,802	19,596	650	2,807	685	-
Bahrain	-	-	±0.002	-	-	-	-	-	88,344	-	-	-
Bangladesh	0.045	0.049	0.053	-	-	-	11,912	12,480	10,061	-	-	-
Benin	±0.001	±0.001	±0.001	±0.001	-	-	-	-	-	-	-	-
Bolivia	0.018	0.016	0.020	0.019	0.020	0.024	21,962	19,858	22,784	20,114	22,424	26,342
Bosnia & Herzegovina	±0.297	±0.221	-	-	-	-	12,777	10,775	-	-	-	-
Bulgaria	±0.198	±0.060	±0.081	±0.082	±0.090	±0.101	10,036	3,810	5,308	5,659	5,514	6,496
Burundi	±0.000	±0.000	-	-	-	-	6,380	6,541	-	-	-	-
Canada	0.720	0.620	0.587	0.597	0.622	0.605	65,447	68,856	65,268	59,685	62,244	64,001
Cape Verde	±0.001	-	-	-	-	-	8,637	-	-	-	-	-
Chile	0.254	0.330	0.393	0.363	0.402	0.407	22,382	27,774	31,299	31,053	33,043	35,507
China (Hong Kong SAR)	0.123	0.106	0.094	0.017	0.013	-	31,435	40,831	55,171	21,007	21,781	-
China (Taiwan Province)	±2.133	±2.129	±1.590	±1.296	±1.288	±1.072	57,851	62,423	53,295	48,942	55,919	54,790
Colombia	0.263	0.271	0.270	0.239	0.299	0.302	17,290	18,139	15,881	15,928	21,225	22,911
Costa Rica	0.020	0.018	0.026	0.027	0.025	0.026	6,237	7,696	10,590	15,674	-	-
Cote d'Ivoire	±0.011	±0.011	±0.015	±0.014	-	-	-	-	-	-	-	-

Continued.

Depending on the table, * means *Enterprises, Engaged,* or *Factor Values*; ± means *Basis Unspecified*; § means *shown in millions*. For additional notes and sources, see Appendix I.

Output and Output per Employee

- Continued -

Country	Output ($ bil.)						Output per Employee ($)					
	1990	1991	1992	1993	1994	1995	1990	1991	1992	1993	1994	1995
Croatia	±0.401	±0.236	±0.128	-	-	-	13,197	11,114	6,801	-	-	-
Cyprus	0.078	0.079	0.072	0.055	0.060	0.050	36,413	38,910	35,796	32,037	37,762	39,522
Czechoslovakia (Former)	±0.640	±0.249	-	-	-	-	9,554	5,297	-	-	-	-
Denmark	*0.195	*0.200	*0.231	*0.216	*0.271	*0.289	117,148	125,411	150,496	140,859	176,984	200,443
Ecuador	0.017	0.020	0.022	0.021	0.022	0.026	10,163	9,643	11,108	13,264	14,718	17,515
Egypt	*0.085	*0.049	*0.105	*0.036	*0.038	*0.042	8,807	5,050	11,487	5,227	5,486	6,101
El Salvador	-	±0.004	±0.001	±0.010	±0.003	±0.015			1,782	9,884	7,789	11,739
Ethiopia	0.041	0.035	0.019	0.014	0.016	0.019	13,856	11,323	6,144	4,612	4,220	5,285
Finland	±0.226	±0.168	±0.134	±0.106	±0.130	±0.215	70,688	64,673	63,571	55,651	64,995	86,780
FYR Macedonia	±0.152	±0.120	±0.077	±0.030	±0.019	±0.018	18,028	17,977	12,771	10,673	6,672	8,046
France	±2.890	±2.878	±2.953	±2.624	±2.539	±2.694	55,047	60,594	65,611	63,224	65,281	70,156
Germany	-	3.801	3.820	3.599	3.611	-		80,076	118,258	138,403	157,040	-
Germany (Western Part)	3.627	3.642	3.642	-	-	-	119,081	124,350	138,240	-	-	-
Greece	*0.268	*0.229	*0.273	*0.342	*0.317	*0.336	41,097	39,291	45,508	52,485	51,110	54,511
Guatemala	-	0.040	0.039	0.036	0.040	0.031		23,586	21,932	21,258	24,468	21,050
Honduras	0.009	0.009	0.007	0.009	0.008	0.008	11,138	11,383	5,833	6,519	5,388	6,394
Hungary	±0.274	±0.241	±0.184	±0.176	±0.170	±0.176	9,775	9,264	7,659	8,003	7,941	8,133
Iceland	0.002	0.002	±0.001	±0.000			47,011	44,664	47,028	15,039	-	-
India	*0.612	*0.476	*0.498	*0.619	*0.664	*0.757	12,158	9,989	9,565	11,578	12,417	13,112
Indonesia	0.380	0.660	1.337	2.214	2.333	2.447	6,395	5,086	6,917	9,587	8,793	8,397
Iran (Islamic Republic of)	0.210	0.180	0.163	0.138	-	-	18,588	16,803	14,956	13,335	-	-
Iraq		*0.186	*0.423	-	-	-		59,925	140,836	-	-	-
Ireland	*0.038	*0.029	*0.027	*0.022	*0.025	*0.026	54,253	42,003	49,085	44,090	45,698	50,748
Israel	0.123	0.141	0.170	0.178	0.190	0.219	49,003	50,459	54,701	52,274	57,565	62,927
Italy	±8.422	±8.190	±9.735	±8.631	±10	-	113,396	111,130	113,066	101,103	117,952	-
Jamaica	0.025	0.019	0.018	-	-	-				-	-	-
Japan	±3.730	±4.135	±4.280	±4.388	±6.242	±7.538	109,692	118,137	125,868	137,140	132,810	144,957
Jordan	0.008	0.009	0.014	0.017	0.016	0.016	12,955	11,506	12,972	16,638	17,934	17,417
Kenya	*0.039	*0.035	*0.031	*0.022	*0.031	*0.040	16,516	15,240	13,728	9,762	13,058	16,398
Korea, Republic of	1.337	4.526	4.717	3.535	3.159	3.174	46,759	42,636	44,228	45,630	50,234	62,171
Kuwait	-	-	-	-	0.002	0.002					29,926	33,042
Kyrgyzstan	0.055	0.056	0.024	0.004	0.003	-	7,776	8,924	5,122	1,176	1,573	-
Latvia	0.001	0.004	0.028	0.031	0.022	0.014	137	774	2,960	4,655	4,923	4,434
Lithuania	-	-	0.033	0.023	0.029	-			3,315	2,721	4,546	-
Macau	0.015	0.019	0.014	0.015	0.032	0.029	22,570	24,075	25,950	30,621	38,832	27,960
Malaysia	*0.012	*0.018	*0.027	*0.035	*0.028	*0.030	11,867	12,805	14,256	16,520	16,273	17,380
Malta	0.034	0.035	0.038	0.036	0.033	-	34,107	35,270	35,618	33,913	35,313	-
Mauritius	0.007	0.007	0.008	0.006	0.008	0.009	9,230	8,758	10,511	7,746	10,798	12,139
Mexico	0.375	0.352	0.374	0.360	0.305	0.135	15,951	21,604	24,508	25,846	22,739	12,009
Mongolia	0.047	0.060	0.026	0.006	0.004	0.004	11,473	19,156	10,168	3,518	1,615	1,758
Mozambique	-	±0.001	-	±0.001	±0.000	-		1,355	-	1,121	1,198	-
Myanmar	0.008	0.007	0.009	0.009	0.012	-	5,786	9,449	5,562	9,166	11,051	-
Nepal	0.002	0.004	-	0.029	0.006	-	3,203	5,450	-	10,608	3,783	-
Netherlands	0.173	0.168	0.165	0.146	0.142	0.154	73,582	74,317	77,497	73,359	76,657	91,091
Norway	±0.026	±0.029	±0.027	±0.025	±0.025	±0.025	63,978	73,231	73,267	68,915	65,481	74,529
Pakistan	0.087	0.088	0.096	-	-	-	16,681	16,835	17,469	-	-	-
Panama	0.016	0.021	0.021	0.022	0.021	0.021	22,192	24,283	24,816	25,167	25,250	25,281
Peru	0.060	0.072	0.090	-	0.143	-	12,879	19,719	25,484	-	-	-
Philippines	0.047	0.101	0.103	0.143	-	-	3,507	4,973	6,694	9,467	-	-
Poland	±0.528	±0.556	±0.428	±0.374	±0.449	±0.574	6,209	7,717	7,138	7,478	8,364	10,252
Portugal	1.653	1.841	1.968	2.053	2.361	2.823	27,823	28,702	31,694	31,265	36,262	40,615
Republic of Moldova	0.000	0.212	0.019	0.007	0.005	0.005	0.033	25,510	2,389	1,607	1,075	1,267
Romania	±0.892	±0.630	±0.369	±0.384	-	-						
Russian Federation	-	3.325	1.628	0.671	0.534	0.464	-	-	-	3,992	3,760	3,913
Singapore	*0.028	*0.030	*0.029	*0.028	*0.026	*0.026	35,579	47,682	55,026	61,358	61,357	67,608
Slovakia	-	±0.171	±0.155	±0.122	±0.104	-	-	8,607	8,194	7,041	6,731	-
Slovenia	-	-	-	0.172	0.172	0.181	-	-	-	-	-	-
South Africa	±0.659	±0.686	±0.617	±0.567	±0.555	±0.622	20,605	22,863	22,856	23,609	21,357	23,256
Spain	*2.368	*2.351	*2.306	±3.113	±3.543	±4.727	78,405	84,810	95,211	80,995	86,276	113,579
Sri Lanka	0.036	0.042	0.026	0.040	-	-	7,026	7,651	5,437	7,778	-	-

Continued.

Output and Output per Employee
- Continued -

Country	Output ($ bil.)						Output per Employee ($)					
	1990	1991	1992	1993	1994	1995	1990	1991	1992	1993	1994	1995
Suriname	*0.004	*0.009	*0.011	0.015	-	-	38,636	68,424	89,636	140,056	-	-
Swaziland	-	*0.009	*0.008	*0.008	-	-	-	19,411	18,212	18,534	-	-
Sweden	*0.057	*0.063	*0.060	*0.044	*0.039	*0.049	82,060	104,538	105,245	86,841	80,434	87,728
Thailand	0.173	0.626	-	-	-	-	9,722	9,895	-	-	-	-
Tonga	0.000	0.000	-	-	-	-	11,125	7,330	-	-	-	-
Trinidad and Tobago	±0.009	±0.011	±0.008	±0.006	±0.005	-	20,414	23,974	20,566	16,506	14,299	-
Turkey	0.215	0.242	0.275	0.303	0.203	0.232	35,048	40,771	37,408	40,661	34,040	39,087
Ukraine	*1.513	*1.703	*1.334	*0.317	*0.259	*0.131	14,007	18,718	14,985	3,867	3,701	2,253
United Kingdom	*2.514	*2.435	2.479	2.336	2.665	2.726	55,873	55,350	61,970	58,408	64,991	69,348
United Republic of Tanzania	*0.004	*0.001	-	-	-	-	1,616	1,962	-	-	-	-
USSR (Former)	±4.500	-	-	-	-	-	10,441	-	-	-	-	-
United States of America	*4.650	*4.160	*4.215	*4.277	*4.242	*3.980	69,403	71,724	79,528	82,250	84,840	82,917
Uruguay	0.069	0.068	0.066	0.061	0.062	0.056	17,359	18,964	15,264	19,409	20,297	24,315
Venezuela	0.272	0.314	0.324	0.499	0.226	0.341	16,881	17,852	19,666	31,624	15,662	23,367
Yugoslavia	±0.051	±0.041	±0.099	-	±0.170	±0.065	-	1,267	3,251	-	6,247	2,470
Zambia	-	-	-	-	0.008	-	-	-	-	-	8,198	-
Zimbabwe	*0.118	*0.118	*0.095	*0.084	*0.079	*0.074	19,565	17,989	16,037	13,481	12,404	13,873

Value Added and Value Added per Employee

Country	Value Added ($ bil.)						Value Added per Employee ($)					
	1990	1991	1992	1993	1994	1995	1990	1991	1992	1993	1994	1995
Argentina	±0.190	-	-	0.247	-	-	10,033	-	-	16,051	-	-
Australia	*0.279	*0.257	*0.294	-	-	-	27,891	32,139	41,946	-	-	-
Austria	0.213	0.211	0.211	0.186	0.162	0.190	25,370	30,208	35,141	33,282	31,370	35,749
Bangladesh	0.021	0.025	0.026	-	-	-	5,697	6,340	4,911	-	-	-
Bolivia	0.008	0.006	0.007	0.006	0.007	0.008	9,897	7,837	7,608	6,182	7,409	8,704
Bosnia & Herzegovina	±0.115	±0.121	-	-	-	-	4,940	5,871	-	-	-	-
Bulgaria	-	±0.020	±0.029	±0.031	-	-	-	1,245	1,886	2,105	-	-
Burundi	0.000	0.000	-	-	-	-	3,222	3,517	-	-	-	-
Canada	0.334	0.288	0.281	0.279	0.293	0.285	30,386	32,004	31,255	27,905	29,291	30,111
Chile	0.121	0.139	0.173	0.170	0.194	0.196	10,674	11,663	13,749	14,559	15,927	17,118
China (Hong Kong SAR)	0.035	0.028	0.020	0.006	0.006	-	8,920	10,888	11,627	7,756	9,489	-
China (Taiwan Province)	0.453	0.485	0.405	0.374	0.314	0.221	12,277	14,228	13,578	14,133	13,624	11,276
Colombia	0.100	0.102	0.105	0.098	0.117	0.118	6,549	6,856	6,198	6,513	8,313	8,971
Costa Rica	0.008	0.007	0.010	0.011	0.010	0.010	2,452	3,076	4,247	6,126		
Cote d'Ivoire	0.011	0.011	0.015	0.014	-	-						
Croatia	±0.181	±0.124	±0.067	-	-	-	5,961	5,842	3,554	-	-	-
Cyprus	0.030	0.028	0.029	0.024	0.025	0.021	13,901	13,842	14,689	13,929	15,825	16,152
Czechoslovakia (Former)	±0.256	-	-	-	-	-	3,817	-	-	-	-	-
Denmark	*0.065	*0.090	*0.102	*0.096	*0.118	*0.126	39,082	56,195	66,834	62,515	77,041	87,738
Ecuador	0.006	0.006	0.008	0.007	0.007	0.009	3,253	2,955	4,000	4,287	4,885	6,098
Egypt	*0.022	*0.017	*0.075	*0.006	-	-	2,307	1,693	8,235	894	-	-
El Salvador	-	0.002	-	0.005	0.002	0.007	-	-	-	4,784	3,958	5,660
Ethiopia	0.014	0.013	0.008	0.006	0.007	0.008	4,891	4,133	2,494	1,866	1,840	2,084
Finland	±0.093	±0.068	±0.063	±0.049	±0.059	±0.089	28,929	26,250	30,085	25,614	29,482	35,863
FYR Macedonia	±0.075	±0.063	±0.043	±0.014	±0.008	±0.007	8,834	9,434	7,070	4,943	2,835	3,161
France	±1.420	±1.471	±1.523	±1.357	±1.234	±1.258	27,044	30,967	33,838	32,695	31,717	32,754
Germany (Western Part)	1.152	1.183	1.294	1.202	-	-	37,835	40,404	49,108	-	-	-
Greece	*0.100	*0.089	*0.105	*0.131	*0.120	*0.128	15,286	15,290	17,418	20,029	19,249	20,804
Honduras	*0.002	*0.002	*0.002	*0.003	*0.002	*0.002	3,001	2,799	1,653	2,004	1,613	2,057
Hungary	±0.084	±0.079	±0.078	±0.081	±0.076	±0.076	3,015	3,042	3,239	3,663	3,574	3,544
Iceland	0.001	0.000	±0.000	±0.000	-	-	16,471	8,687	1,738	3,575	-	-
India	*0.104	*0.096	*0.098	*0.157	*0.168	*0.192	2,074	2,022	1,890	2,930	3,144	3,323
Indonesia	*0.189	*0.232	*0.531	*0.848	*1.047	*0.941	3,181	1,788	2,748	3,674	3,945	3,228
Iran (Islamic Republic of)	0.090	0.074	0.072	0.053	-	-	7,925	6,940	6,569	5,095	-	-
Iraq	-	*0.123	*0.237	-	-	-	-	39,579	79,100	-	-	-
Ireland	*0.019	*0.013	*0.012	*0.010	*0.011	*0.012	26,534	19,155	21,959	19,560	20,556	22,823

Continued.

Depending on the table, * means *Enterprises, Engaged,* or *Factor Values*; ± means *Basis Unspecified*; § means *shown in millions.* For additional notes and sources, see Appendix I.

Value Added and Value Added per Employee

- Continued -

Country	Value Added ($ bil.)						Value Added per Employee ($)					
	1990	1991	1992	1993	1994	1995	1990	1991	1992	1993	1994	1995
Israel.	0.055	0.060	0.063	0.055	0.057	0.071	21,823	21,468	20,201	16,212	17,209	20,427
Italy .	±2.231	±2.346	±2.842	±2.516	±2.687	-	30,040	31,837	33,007	29,474	31,455	-
Jamaica	0.007	0.006	0.005									
Japan	±1.478	±1.655	±1.698	±1.754	±2.808	±3.370	43,471	47,297	49,929	54,800	59,743	64,811
Jordan	0.003	0.003	0.005	0.007	0.007	0.007	5,423	4,200	5,111	7,006	8,227	7,373
Kenya	*0.013	*0.008	*0.007	*0.005	*0.007	*0.009	5,505	3,492	3,020	2,103	3,037	3,537
Korea, Republic of	0.594	1.957	2.183	1.676	1.529	1.482	20,764	18,435	20,471	21,640	24,303	29,036
Kuwait	-	-	-	-	0.001	0.001	-	-	-	-	15,403	17,204
Latvia				*0.012	*0.009	*0.007				1,778	2,049	2,110
Macau	0.005	0.005	0.004	0.005	0.009	0.006	6,736	6,173	7,113	9,653	10,233	6,090
Malaysia	*0.004	*0.007	*0.012	*0.015	*0.010	*0.011	3,993	5,065	6,198	7,178	6,142	6,554
Malta	0.012	0.013	0.014	0.014	0.013	-	12,298	13,560	13,360	13,158	13,538	-
Mauritius	0.003	0.003	0.003	0.003	0.005	0.005	4,140	3,964	4,281	3,901	6,525	6,141
Mexico	0.142	0.160	0.163	0.158	0.134	0.059	6,055	9,833	10,661	11,353	9,984	5,226
Mongolia	0.020	0.027	0.010	0.001	0.003	0.002	4,955	8,793	3,770	346	949	795
Myanmar	±0.003	±0.004	±0.005	±0.003	±0.005	-	2,480	5,154	2,831	3,377	4,965	-
Nepal	0.001	0.002	-	0.008	0.002	-	1,967	2,268	-	2,935	1,446	-
Netherlands	0.072	0.071	0.071	0.065	0.063	0.068	30,834	31,378	33,404	32,484	33,837	40,086
New Zealand	0.041	-	-	-	-	-	21,850	-	-	-	-	-
Norway .	±0.011	±0.012	±0.012	±0.011	±0.011	±0.011	27,142	31,552	32,525	29,759	29,966	34,193
Pakistan	0.024	0.021	0.024	-	-	-	4,650	4,086	4,360	-	-	-
Panama.	0.005	0.006	0.007	0.007	0.007	0.007	7,106	7,555	7,683	7,761	7,793	7,814
Peru	0.026	0.032	0.027	-	-	-	5,602	8,842	7,781	-	-	-
Philippines.	0.018	0.042	0.040	0.089	-	-	1,353	2,058	2,594	5,927	-	-
Poland	±0.263	±0.230	±0.198	±0.142	±0.168	±0.214	3,089	3,200	3,294	2,845	3,137	3,814
Portugal	0.452	0.463	0.529	0.588	0.631	0.754	7,608	7,218	8,512	8,959	9,693	10,853
Romania	±0.366	±0.219	±0.141	±0.149	-	-	-	-	-	-	-	-
Russian Federation	-	-	-	±0.458	±0.295	±0.203	-	-	-	2,721	2,078	1,710
Singapore	*0.009	*0.009	*0.009	*0.010	*0.010	*0.010	11,601	14,155	17,629	21,178	22,854	25,370
Slovakia	-	-	-	±0.045	±0.041					2,572	2,635	
Slovenia	-	-	-	0.072	0.065	0.075						
South Africa	*0.315	*0.280	*0.244	*0.225	*0.221	*0.249	9,856	9,343	9,038	9,361	8,495	9,307
Spain	*0.781	*0.763	*0.697	±0.777	±0.845	±0.969	25,840	27,530	28,777	20,219	20,568	23,279
Sri Lanka	0.020	0.020	0.018	0.019	-	-	3,853	3,591	3,684	3,810	-	-
Swaziland	-	*0.003	*0.003	*0.003	-	-	-	7,459	6,819	6,565	-	-
Sweden.	*0.027	*0.024	*0.023	*0.017	*0.016	*0.021	38,617	39,307	40,218	33,282	32,863	37,241
Thailand	0.063	0.287	-	-	-	-	3,547	4,533	-	-	-	-
Trinidad and Tobago	±0.003	±0.003	±0.003	±0.002	±0.002	-	6,026	6,800	6,319	5,371	4,344	-
Turkey	0.069	0.095	0.111	0.118	0.076	0.087	11,225	16,010	15,185	15,754	12,789	14,687
United Kingdom .	*1.268	*1.234	*1.247	*1.144	*1.270	*1.299	28,175	28,037	31,184	28,604	30,964	33,044
United Republic of Tanzania	*0.001	*0.000	-	-	-	-	271	392	-	-	-	-
United States of America	*2.320	*2.120	*2.209	*2.253	*2.323	*1.969	34,627	36,552	41,679	43,327	46,460	41,021
Uruguay	0.019	0.019	0.023	0.015	0.015	0.013	4,813	5,270	5,430	4,613	4,816	5,766
Venezuela	0.090	0.111	0.118	0.205	0.074	0.263	5,575	6,306	7,161	13,021	5,155	18,032
Yugoslavia .	±0.021	±0.019	±0.055	-	±0.070	±0.030	-	591	1,816	-	2,571	1,131
Zambia .	-	-	-	-	0.002	-	-	-	-	-	2,266	-
Zimbabwe	*0.066	*0.061	*0.043	*0.038	*0.029	*0.027	10,984	9,346	7,320	6,185	4,561	5,105

Capital Investment

Country	Gross Fixed Capital Formation ($ mil.)						Per Establishment ($ mil)					
	1990	1991	1992	1993	1994	1995	1990	1991	1992	1993	1994	1995
Argentina	-	-	-	27	-	-	-	-	-	0.024	-	-
Australia	15	-	-	-	-	-	0.087	-	-	-	-	-
Austria	20	17	20	20	15	-	0.322	0.305	0.385	0.416	0.326	-
Bahrain.	-	-	0.003	-	-	-	-	-	0.001	-	-	-
Bangladesh	0.691	0.437	1.258	-	-	-	0.007	0.006	0.016	-	-	-
Belgium .	2.424	3.104	1.711	1.503	1.405	1.832	0.025	0.033	0.019	0.019	0.017	-

Continued.

Depending on the table, * means Enterprises, Engaged, or Factor Values; ± means Basis Unspecified; § means shown in millions. For additional notes and sources, see Appendix I.

Capital Investment

- Continued -

Country	Gross Fixed Capital Formation ($ mil.)						Per Establishment ($ mil)					
	1990	1991	1992	1993	1994	1995	1990	1991	1992	1993	1994	1995
Bolivia	0.040	-0.009	-0.022	0.021	0.024	-	0.002	-0.000	-0.001	0.001	0.001	-
Bosnia & Herzegovina	17	-	-	-	-	-	0.661	-	-	-	-	-
Bulgaria	13	2.856	26	7.434	3.008	3.543	0.535	0.119	0.731	0.030	0.008	0.008
Canada	8.570	-	-	-	-	-	0.064	-	-	-	-	-
Cape Verde	0.001	-	-	-	-	-	0.001	-	-	-	-	-
Chile	5.064	8.983	8.699	10	14	-	0.083	0.140	0.130	0.178	0.231	-
China (Hong Kong SAR)	1.540	1.287	0.258	0.017	-	-	0.004	0.005	0.002	0.000	-	-
Colombia	9.067	6.933	13	0.318	-2.401	-	0.032	0.025	0.044	0.001	-0.010	-
Croatia	4.139	0.999	2.073	2.512	3.147	1.372	0.053	0.011	0.020	0.017	0.017	0.007
Cyprus	1.761	1.611	2.193	1.606	1.148	1.176	0.010	0.010	0.014	0.011	0.008	0.008
Czechoslovakia (Former)	26	-	-	-	-	-	6.546	-	-	-	-	-
Denmark	4.524	3.909	-	-	-	-	0.037	0.033	-	-	-	-
Ecuador	1.847	3.699	1.798	1.141	1.370	1.414	0.092	0.185	0.058	0.048	0.051	0.049
Egypt	3.250	1.501	2.157	-7.355	-	-	0.035	0.016	0.022	-0.081	-	-
El Salvador	-	0.371	-	0.497	0.123	1.253	-	0.074	-	0.062	0.031	0.251
Ethiopia	1.050	1.699	0.238	0.216	2.242	0.955	0.070	0.121	0.017	0.015	0.050	0.022
Finland	-0.758	1.039	0.915	1.435	2.929	4.846	-0.014	0.019	0.018	0.027	0.067	0.099
FYR Macedonia	0.531	0.203	0.227	0.085	-	0.100	0.038	0.008	0.006	0.002	-	0.001
Germany (Western Part)	68	72	70	77	-	-	0.292	0.333	0.339	-	-	-
Greece	4.372	4.137	5.519	-	-	-	0.015	0.014	0.018	-	-	-
Hungary	6.613	6.075	4.342	5.112	-	-	0.054	0.029	0.019	0.020	-	-
India	39	16	24	34	-	-	0.114	0.042	0.066	0.069	-	-
Indonesia	43	386	177	171	94	165	0.183	1.374	0.563	0.523	0.272	0.423
Iran (Islamic Republic of)	3.303	4.768	8.873	3.613	-	-	0.020	0.088	0.181	0.072	-	-
Ireland	0.663	0.485	-	-	-	-	0.029	0.032	-	-	-	-
Israel	2.480	4.388	6.913	10	5.646	-	0.019	0.028	0.039	0.061	0.034	-
Italy	180	186	228	199	244	-	0.119	0.121	0.121	0.102	0.124	-
Japan	55	59	79	36	68	-	0.030	0.032	0.043	0.021	0.028	-
Jordan	0.081	0.070	0.285	0.508	0.116	0.060	0.000	0.000	0.001	0.002	0.001	0.000
Korea, Republic of	52	408	266	164	153	157	0.070	0.248	0.176	0.091	0.087	0.093
Kuwait	-	-	-	-	0.094	-	-	-	-	-	0.047	-
Latvia	-0.063	0.013	0.514	0.605	0.436	0.589	-0.010	0.001	0.014	0.015	0.018	0.018
Macau	0.678	1.138	0.281	3.457	1.921	-0.033	0.026	0.046	0.016	0.182	0.101	-0.002
Malaysia	1.848	2.182	1.963	2.331	2.667	-	0.308	0.156	0.123	0.090	0.133	-
Malta	0.035	0.458	0.670	0.314	0.563	-	0.002	0.023	0.035	0.017	0.033	-
Mexico	5.745	9.476	-	-	-	-	0.073	0.166	-	-	-	-
Mongolia	40	32	8.567	0.128	0.142	-	20	6.347	1.428	0.032	0.004	-
Myanmar	11	13	14	11	0.335	-	5.521	6.684	6.798	5.603	0.167	-
Nepal	0.443	-	-	-	-	-	0.055	-	-	-	-	-
Netherlands	12	13	8.530	2.692	-	-	0.281	0.361	0.237	0.067	-	-
Norway	0.320	0.771	0.333	0.282	1.134	-	0.021	0.055	0.037	0.031	0.126	-
Pakistan	-	3.423	-	-	-	-	-	0.143	-	-	-	-
Panama	0.264	-0.005	-	-	-	-	0.014	-0.000	-	-	-	-
Peru	0.483	0.702	0.299	-	-	-	0.001	0.001	0.000	-	-	-
Philippines	1.193	6.259	4.978	1.088	-	-	0.001	0.003	0.014	0.003	-	-
Poland	14	9.644	13	10	-	-	0.119	0.057	0.076	0.065	-	-
Portugal	80	90	111	73	108	-	0.046	0.047	0.053	0.034	0.050	-
Romania	-	-	-	5.663	6.294	-	-	-	-	0.015	0.009	-
Singapore	3.323	1.330	0.438	0.870	0.699	-	0.104	0.046	0.016	0.036	0.037	-
Slovenia	5.036	4.715	2.965	4.212	4.433	5.307	0.117	0.196	0.096	0.092	0.117	0.020
Spain	39	186	38	53	54	62	0.018	0.091	0.021	-	-	-
Sri Lanka	1.498	1.969	1.377	0.930	-	-	0.136	0.179	0.138	0.085	-	-
Swaziland	-	0.101	0.051	0.194	-	-	-	0.101	0.051	0.194	-	-
Thailand	39	51	-	-	-	-	0.617	0.198	-	-	-	-
Trinidad and Tobago	-	-	0.047	0.187	0.169	-	-	-	0.003	0.011	0.008	-
Turkey	4.217	3.835	5.966	5.462	6.957	-	0.081	0.083	0.042	0.046	0.062	-
Ukraine	50	29	15	3.114	3.329	1.058	0.532	0.069	0.032	0.042	0.032	0.011
United Kingdom	41	44	46	-	67	-	0.055	0.063	0.069	-	0.116	-
United Republic of Tanzania	0.774	0.593	-	-	-	-	0.086	0.054	-	-	-	-
United States of America	50	30	55	42	59	35	-	-	0.086	-	-	-

Continued.

Depending on the table, * means *Enterprises, Engaged*, or *Factor Values*; ± means *Basis Unspecified*; § means *shown in millions*. For additional notes and sources, see Appendix I.

Capital Investment
- Continued -

Country	Gross Fixed Capital Formation ($ mil.)						Per Establishment ($ mil)					
	1990	1991	1992	1993	1994	1995	1990	1991	1992	1993	1994	1995
Uruguay	2.677	1.040	0.923	1.259	-	-	-	-	-	-	-	-
Venezuela	7.186	12	12	8.654	5.488	28	0.014	0.021	0.025	0.017	0.012	0.063
Yugoslavia	0.479	0.239	0.341	-	0.578	0.313	0.006	0.002	0.002	-	0.003	0.001
Zimbabwe	3.595	5.193	5.850	4.712	2.196	-	0.257	0.371	0.390	0.337	0.146	-

Indexes of Industrial Production

Country	Index of Industrial Production (1990=100)						Index of Employment (1990=100)					
	1990	1991	1992	1993	1994	1995	1990	1991	1992	1993	1994	1995
Albania	100	-	-	-	-	-	-	-	-	-	-	-
Algeria	100	77	58	48	45	33	-	-	-	-	-	-
Angola	100	-	-	-	-	-	-	-	-	-	-	-
Argentina	100	161	-	-	-	-	100	-	-	81	83	-
Armenia	100	-	-	-	-	-	100	92	-	-	-	-
Australia	100	93	96	100	-	-	100	80	70	-	-	-
Austria	100	98	95	88	89	91	100	83	71	67	61	63
Azerbaijan	-	-	-	-	-	-	100	83	73	48	53	-
Bahrain	100	-	-	-	-	-	-	-	-	-	-	-
Bangladesh	100	76	76	84	106	123	100	105	140	-	-	-
Barbados	100	100	100	100	100	100	-	-	-	-	-	-
Belgium	100	55	53	46	42	45	100	94	88	-	-	-
Benin	100	-	-	-	-	-	-	-	-	-	-	-
Bermuda	-	-	-	-	-	-	100	108	115	108	108	-
Bolivia	100	107	120	127	135	145	100	101	106	114	111	110
Bosnia & Herzegovina	-	-	-	-	-	-	100	88	-	-	11	-
Brazil	100	87	80	89	87	81	-	-	-	-	-	-
Bulgaria	100	88	78	65	67	58	100	80	77	74	83	79
Burkina-Faso	100	-	-	-	-	-	-	-	-	-	-	-
Burundi	100	-	-	-	-	-	100	174	-	-	-	-
Cambodia	100	-	-	-	-	-	100	75	-	-	-	-
Cameroon	100	-	-	-	-	-	-	-	-	-	-	-
Canada	100	81	81	85	90	86	100	82	82	91	91	86
Cape Verde	-	-	-	-	-	-	100	124	-	-	-	-
Chile	100	119	123	118	105	93	100	105	111	103	107	101
China (Hong Kong SAR)	100	95	93	85	81	78	100	67	44	21	15	-
China (Taiwan Province)	100	97	77	67	64	54	100	93	81	72	62	53
Colombia	100	123	124	119	115	104	100	98	112	99	93	87
Congo	100	-	-	-	-	-	-	-	-	-	-	-
Costa Rica	100	84	93	104	98	99	100	74	79	56	-	-
Croatia	100	-	-	-	-	-	100	70	62	65	68	65
Cuba	100	-	-	-	-	-	-	-	-	-	-	-
Cyprus	100	100	85	73	74	66	100	95	94	81	74	59
Czechoslovakia (Former)	100	68	62	-	-	-	100	70	-	-	-	-
Czech Republic	100	-	-	-	-	-	100	78	81	75	-	-
Dem. Rep. of the Congo	100	-	-	-	-	-	-	-	-	-	-	-
Denmark	100	104	109	108	125	115	100	96	92	92	92	87
Dominican Republic	100	-	-	-	-	-	-	-	-	-	-	-
Ecuador	100	111	97	105	-	-	100	122	114	93	89	86
Egypt	100	115	110	108	109	113	100	102	95	72	72	72
El Salvador	100	105	-	-	-	-	-	-	-	-	-	-
Ethiopia and Eritrea	100	-	-	-	-	-	-	-	-	-	-	-
Ethiopia	-	-	-	-	-	-	100	105	107	105	130	125
Finland	100	83	70	70	78	83	100	81	66	59	63	78
FYR Macedonia	100	-	-	-	-	-	100	79	71	33	34	26
France	100	94	90	87	86	82	100	90	86	79	74	73
Gabon	100	-	-	-	-	-	-	-	-	-	-	-
Gambia	100	-	-	-	-	-	-	-	-	-	-	-
Germany (Eastern Part)	100	-	-	-	-	-	-	-	-	-	-	-

Continued.

Depending on the table, * means *Enterprises, Engaged,* or *Factor Values*; ± means *Basis Unspecified*; § means *shown in millions.* For additional notes and sources, see Appendix I.

169

Indexes of Industrial Production
- Continued -

Country	Index of Industrial Production (1990=100)						Index of Employment (1990=100)					
	1990	1991	1992	1993	1994	1995	1990	1991	1992	1993	1994	1995
Germany (Western Part)	100	96	86	77	61	60	100	96	86	-	-	-
Ghana	100	-	-	-	-	-	-	-	-	-	-	-
Greece	100	90	80	101	88	86	100	89	92	100	95	94
Guatemala	100	-	-	-	-	-	-	-	-	-	-	-
Guyana	100	-	-	-	-	-	-	-	-	-	-	-
Haiti	100	-	-	-	-	-	-	-	-	-	-	-
Honduras	100	133	149	163	188	258	100	101	154	168	184	147
Hungary	100	83	63	66	63	64	100	93	86	79	76	77
Iceland	100	-	-	-	-	-	100	100	40	30	28	-
India	100	92	80	76	76	83	100	95	103	106	106	115
Indonesia	100	111	156	171	163	150	100	218	326	389	447	491
Iran (Islamic Republic of)	100	102	100	99	78	82	100	95	97	92	-	-
Iraq	100	96	93	-	-	-	-	-	-	-	-	-
Ireland	100	91	77	72	77	75	100	100	79	72	77	74
Israel	100	104	114	123	137	138	100	112	124	136	132	139
Italy	100	92	95	95	-	-	100	99	116	115	115	-
Jamaica	100	-	-	-	-	-	-	-	-	-	-	-
Japan	100	98	97	88	95	-	100	103	100	94	138	153
Jordan	100	111	131	127	113	114	100	126	176	167	142	149
Kenya	100	107	103	93	102	69	100	96	95	97	99	102
Korea, Republic of	100	93	87	66	56	47	100	371	373	271	220	179
Kuwait	100	100	100	100	-	-	-	-	-	-	-	-
Kyrgyzstan	100	-	-	-	-	-	100	89	67	47	31	-
Lao P.D.R.	100	-	-	-	-	-	-	-	-	-	-	-
Latvia	100	-	-	-	-	-	100	101	212	148	99	72
Liberia	100	-	-	-	-	-	-	-	-	-	-	-
Libyan Arab Jamahiriya	100	-	-	-	-	-	-	-	-	-	-	-
Lithuania	100	-	-	-	-	-	-	-	-	-	-	-
Macau	100	-	-	-	-	-	100	115	80	74	122	152
Madagascar	100	72	70	-	-	-	-	-	-	-	-	-
Malawi	100	149	129	103	93	74	100	-	-	-	-	-
Malaysia	100	100	105	106	102	101	100	140	190	210	170	170
Malta	100	92	97	111	-	-	100	101	107	108	95	-
Mauritius	100	-	-	-	-	-	100	103	97	104	99	95
Mexico	100	96	95	92	90	85	100	69	65	59	57	48
Mongolia	100	66	48	31	27	13	100	76	62	39	64	61
Morocco	100	112	112	111	124	120	-	-	-	-	-	-
Mozambique	100	-	-	-	-	-	100	89	48	47	37	-
Myanmar	100	-	-	-	-	-	100	55	121	76	78	-
Nepal	100	-	-	-	-	-	100	131	-	448	258	-
Netherlands	100	97	91	86	82	77	100	96	91	85	79	72
New Zealand	100	81	-	-	-	-	100	74	76	97	-	-
Nicaragua	100	-	-	-	-	-	-	-	-	-	-	-
Nigeria	100	-	-	-	-	-	-	-	-	-	-	-
Norway	100	96	91	90	92	80	100	96	90	87	93	81
Pakistan	100	-	-	-	-	-	100	100	105	-	-	-
Panama	100	113	114	116	114	112	100	118	118	119	118	117
Papua New Guinea	100	100	100	-	-	-	-	-	-	-	-	-
Paraguay	100	106	108	87	66	69	-	-	-	-	-	-
Peru	100	132	112	121	113	100	100	78	76	-	-	-
Philippines	100	120	130	146	152	165	100	150	114	112	-	-
Poland	100	82	70	69	78	84	100	85	71	59	63	66
Portugal	100	106	103	94	87	91	100	108	105	110	110	117
Puerto Rico	100	-	-	-	-	-	100	93	92	104	103	107
Republic of Moldova	100	-	-	-	-	-	100	89	86	50	52	39
Romania	100	88	63	61	68	74	-	-	-	-	-	-
Russian Federation	100	-	-	-	-	-	-	-	-	-	-	-
Saudi Arabia	100	-	-	-	-	-	-	-	-	-	-	-
Singapore	100	143	120	107	99	90	100	82	69	59	54	49
Slovakia	100	-	-	-	-	-	-	-	-	-	-	-

Continued.

Indexes of Industrial Production

- Continued -

Country	Index of Industrial Production (1990=100)						Index of Employment (1990=100)					
	1990	1991	1992	1993	1994	1995	1990	1991	1992	1993	1994	1995
Slovenia	100	75	75	74	71	60	-	-	-	-	-	-
Somalia	100	-	-	-	-	-	-	-	-	-	-	-
South Africa	100	100	84	81	80	83	100	94	84	75	81	84
Spain	100	94	86	75	82	80	100	92	80	127	136	138
Sri Lanka	100	109	116	303	331	241	100	109	95	100	-	-
Sudan	100	-	-	-	-	-	-	-	-	-	-	-
Suriname	100	-	-	-	-	-	100	113	108	93	-	-
Swaziland	100	100	100	-	-	-	-	-	-	-	-	-
Sweden	100	83	74	73	75	85	100	86	81	72	69	79
Switzerland	100	82	75	-	-	-	-	-	-	-	-	-
Syrian Arab Republic	100	236	-	-	-	-	-	-	-	-	-	-
Thailand	100	-	-	-	-	-	100	356	-	-	-	-
Togo	100	-	-	-	-	-	-	-	-	-	-	-
Tonga	100	-	-	-	-	-	100	100	-	-	-	-
Trinidad and Tobago	100	97	67	51	41	37	100	107	95	78	87	-
Tunisia	100	-	-	-	-	-	-	-	-	-	-	-
Turkey	100	104	101	111	104	103	100	97	119	121	97	96
Uganda	100	80	106	91	129	218	-	-	-	-	-	-
Ukraine	100	-	-	-	-	-	100	84	82	76	65	54
United Arab Emirates	100	-	-	-	-	-	-	-	-	-	-	-
United Kingdom	100	85	80	80	76	73	100	98	89	89	91	87
United Republic of Tanzania	100	79	44	21	21	-	100	11	-	-	-	-
USSR (Former)	100	-	-	-	-	-	100	-	-	-	-	-
United States of America	100	86	89	89	87	81	100	87	79	78	75	72
Uruguay	100	101	96	95	93	77	100	89	108	79	77	57
Venezuela	100	-	-	-	-	-	100	109	102	98	90	91
Yemen	100	-	-	-	-	-	-	-	-	-	-	-
Yugoslavia	100	-	-	-	-	-	-	-	-	-	-	-
Yugoslavia (Former)	100	58	-	-	-	-	-	-	-	-	-	-
Zambia	100	-	-	-	-	-	-	-	-	-	-	-
Zimbabwe	100	103	86	88	86	69	100	108	98	102	106	88

Depending on the table, * means *Enterprises*, *Engaged*, or *Factor Values*; ± means *Basis Unspecified*; § means *shown in millions*. For additional notes and sources, see Appendix I.

171

Representative Companies in Sector

Name	Address	Telephone	Fax	Employment	Y
Calzados Azaleia	Centro, Parobe 95630-000, Brazil	51 5431000	51 5435179	10,000	97
Brown-Forman Corp.	850 Dixie Hwy., Louisville, Kentucky 40210, U.S.A.	502-585-1100		7,500	97
Samsonite Corp.	40301 Fisher Island Dr., Fisher Island, Florida 33109, U.S.A.	305-532-2426		7,100	95
Wolverine World Wide Inc.	9341 Courtland Dr., Rockford, Michigan 49351, U.S.A.	616-866-5500	616-866-0257	6,775	97
TAE HWA Indonesia PT	Desa Cengkudu Balaraja, Tangerang, Indonesia	215406614	21 5406618	6,500	97
Timberland Co.	200 Domain Dr., Stratham, New Hampshire 03885, U.S.A.	603-772-9500		5,700	97
G.H. Bass and Co.	600 Sable Oaks Dr., South Portland, Maine 04116, U.S.A.	207-791-4000	207-791-4900	5,000	95
Acushnet Co.	PO Box 965, Fairhaven, Massachusetts 02719-0965, U.S.A.	508-979-2000		4,780	97
OshKosh B'Gosh Inc.	112 Otter Ave., Oshkosh, Wisconsin 54901, U.S.A.	414-231-8800	414-231-8621	4,700	96
Genesco Inc.	PO Box 731, Nashville, Tennessee 37202-0731, U.S.A.	615-367-7000	615-367-8278	4,300	97
Justin Industries Inc.	PO Box 425, Fort Worth, Texas 76101, U.S.A.	817-336-5125		4,222	97
Stylo PLC	Apperley Bridge, Bradford BD10 0NW, United Kingdom	274 617761	274 616111	3,608	94
Eccolet Sko AS	Industrivej 5, Bredebro DK-6261, Denmark	74711625	74710360	3,500	93
Bangkok Rubber Co., Ltd.	Yannawa, Bangkok 10120, Thailand	2912577	2911353	3,200	94
Flower Indonesia PT	Jl Dinoyo No. 37, Surabaya, Indonesia	021576170		3,000	97
Piyavat Rubber Industry Co., Ltd.	Pathumwan, Bangkok, Thailand	2140780	2155821	3,000	94
Planika d.d.	Savska loka 21, Kranj 4001, Slovenia	64 222721	64 222431	2,989	95
R.G. Barry Corp.	13405 Yarmouth Rd. NW, Pickerington, Ohio 43147, U.S.A.	614-864-6400		2,900	96
Stride Rite Corp.	PO Box 9191, Lexington, Massachusetts 02173-9191, U.S.A.	617-824-6000	617-864-1372	2,900	97
Pirelli UK Tyres Ltd.	40 Chancery Ln., London WC2A 1JH, United Kingdom	71 2428881	71 4301096	2,863	93
Florsheim Shoe Co.	130 S. Canal St., Chicago, Illinois 60606, U.S.A.	312-559-2500		2,810	96
Bata Malaysia BHD	3-1/4 Mile Kapar Rd. Klang, Kapar 42200, Malaysia	3 3425411	3 3426034	2,700	93
C and J Clark America Inc.	156 Oak St., Newton, Massachusetts 02464-1440, U.S.A.	617-243-4100	617-243-4199	2,500	
Munro and Co.	PO Box 1157, Hot Springs, Arkansas 71902, U.S.A.	501-262-1440	501-262-6084	2,500	93
Peter Black Holdings PLC	Lawkholme Ln., Keighley BD21 3JQ, United Kingdom	535 661177	535 609973	2,460	94
Poludniowe Zaklady Przemyslu Skorzanego, Chelmek SA	Plac Kilinskiego 1, Chelm PL-32-580, Poland	61300	61147	2,400	93
Kunnan Enterprise, Ltd.	Tan Tzu Hsiang, Taichung, Taiwan	45 320183	45 333370	2,360	92
Converse Inc.	1 Fordham Rd., North Reading, Massachusetts 01864-2680, U.S.A.	508-664-1100	508-664-8727	2,249	96
Zenco Sales Inc.	2129 Pasong Tamo, Makati 1231, Philippines	2 980388		2,228	93
Peko Trzic d.d.	C.St. Marie aux Mines 5, Trzic 4290, Slovenia	64 53260	64 53367	2,120	95
FII Group PLC	48 George St., London W1H 5PG, United Kingdom	71 9358463	71 4860705	2,107	93
Pentland Group PLC	Finchley, London N3 2QL, United Kingdom	81 3462600	81 3462700	2,070	91
Bata Industries Ltd.	59 Wynford Dr., Don Mills, Ontario M3C 1K3, Canada	416-446-2020	416-443-8861	2,000	98
H.H. Brown Shoe Company Inc.	124 W. Putnam Ave., Greenwich, Connecticut 06830, U.S.A.	203-661-2424	203-661-1818	2,000	95
Red Wing Shoe Company Inc.	314 Main St., River Front Ctr., Red Wing, Minnesota 55066, U.S.A.	612-388-8211	612-388-7415	2,000	97
CentroTextil	Knez Mihailova 1-3, Belgrade YU-11000, Serbia	11 185333	11 636879	1,908	95
Sepatu Bata PT	PO Box 69, Jakarta 100002, Indonesia	21 7992008	21 7995679	1,865	91
Church & Co. PLC	St. James Rd., Northampton NN5 5JB, United Kingdom	604 751251	604 754405	1,841	93
Esquire Co., Ltd.	Songdong-Gu, Seoul, Republic of Korea	2 4699292	2 4699293	1,800	97
Scholl PLC	2-4 Sheet St., Windsor SL4 1BG, United Kingdom	753 833444		1,786	93
Justin Boot Co.	610 W. Daggett Ave., Fort Worth, Texas 76104, U.S.A.	817-332-4385	817-390-2588	1,771	97
Kumkang Shoe Manufacturing Co., Ltd.	Songdong-Gu, Seoul, Republic of Korea	2 2905114	2 2905570	1,700	97
Kukje Corp.	Yongsan-gu, Seoul, Republic of Korea	2 7997114	2 7975444	1,658	97
Calcados Daiby Ltda.	Rua Borges de Medeiros 966, Sapiranga 93800-000, Brazil	51 5992611	51 5991888	1,600	96
Ecco Indonesia PT	Jl. Raya Bligo No. 17 Candi, Sidoardjo, Indonesia	3198964555	319 62011	1,600	97
Donna Karan International Inc.	550 7th Ave., New York, New York 10018, U.S.A.	212-789-1500		1,580	96
Torrebiarte Sucs Miguel	24 Calle 24-75 Zona 12, Guatemala City, Guatemala	760364		1,550	
LaCrosse Footwear Inc.	1319 Saint Andrew St., La Crosse, Wisconsin 54602, U.S.A.	608-782-3020	608-782-8190	1,520	97
Aguila Enterprises Co., Ltd.	Tali, Taichung, Taiwan	4 3303009	4 3302976	1,500	93
Amity Leather Products Co.	820 E Washington St., West Bend, Wisconsin 53095, U.S.A.	414-335-1000	414-335-1010	1,500	94
Se Won Co., Ltd.	Sasang-gu, Pusan, Republic of Korea	51 3285111	51 3230060	1,500	97
SRF Ltd.	9-10 Bahadur Shah Zafar Marg, New Delhi 110 002, India	11 3318155	11 3324052	1,500	94
William H. Kaufman Inc.	410 King St. W, Kitchener, Ontario N2G 4J8, Canada	519-576-1500	519-576-9794	1,500	98
Alpina d.d.	Strojarska 2, Ziri 4226, Slovenia	64 691461	64 692163	1,499	95
Neptun	Pomorska St. 5, Starogard Gdansk PL-83-200, Poland	69 23451	69 22938	1,450	93
Lambert Howarth Group PLC	Healey Rd., Burnley BB11 2HL, United Kingdom	282 25641	282 34542	1,446	93

Continued

Representative Companies in Sector
- Continued -

Name	Address	Tele-phone	Fax	Employ-ment	Y
Texas Boot Co.	127 E. Forest Ave., Lebanon, Tennessee 37087, U.S.A.	615-444-5440		1,400	94
Town and Country Shoes Inc.	9341 Courtland Dr., Rockford, Michigan 49351, U.S.A.	616-866-5500		1,400	
Elcanto Shoe Co., Ltd.	Chung-gu, Seoul, Republic of Korea	2 7783371	2 7520261	1,370	97
Lotus Ltd.	Freeman St., Stafford ST16 3JA, United Kingdom	785 223200	785 223110	1,313	94
Swank Inc.	6 Hazel St., Attleboro, Massachusetts 02703, U.S.A.	508-222-3400	508-226-9598	1,250	97
Gates UK Ltd.	Edinburgh Rd., Dumfries DG1 1QA, United Kingdom	387 253111	387 268937	1,244	93
Acme Boot Company Inc.	1002 Stafford St., Clarksville, Tennessee 37041-0749, U.S.A.	615-552-2000	615-647-3566	1,200	
Bennett Industries Inc.	301 Puritan Rd., Swampscott, Massachusetts 01907, U.S.A.	617-598-5300		1,200	
Norcross Footwear Inc.	PO Box 23569, Louisville, Kentucky 40223, U.S.A.	502-327-6100	502-327-6001	1,200	94
Peter Black Keighley Ltd.	Lawkholme Ln., Keighley BD21 3JQ, United Kingdom	535 661177	535 602177	1,101	93
Hanbee Industries Inc.	Nam-Gu, Inchon, Republic of Korea	32 8659111	32 8659110	1,100	97
Aurora Bahana PT	Jl. Kapuk Kamal Km. 8, 3 No. 388, Jakarta 14470, Indonesia	21 6191665	21 6191547	1,000	92
Bical Ltda.	Tv. Mal Deodoro 56, Birigui 16200-000, Brazil	186 424122	186 424122	1,000	98
Contessa-Skogar	Traktorowa 128, Lodz PL-91-204, Poland	42 529897	42 523910	1,000	93
Klein Tools Inc.	PO Box 599033, Chicago, Illinois 60659-9033, U.S.A.	847-677-9500	847-677-4476	1,000	97
Rocky Shoes and Boots Inc.	39 Canal St., Nelsonville, Ohio 45764, U.S.A.	614-753-1951	614-753-4024	988	95
Tony Lama Co.	PO Box 9518, El Paso, Texas 79985, U.S.A.	915-778-8311		900	97
Crown Footwear Pty. Ltd.	PO Box 2084, Pinetown 3600, Republic of South Africa	31 7001601	31 7001600	850	95
McRae Industries Inc.	402 N. Main St., Mount Gilead, North Carolina 27306, U.S.A.	910-439-6147	919-439-9596	850	96
Sebago Inc.	Bridge St., Westbrook, Maine 04092, U.S.A.	207-854-8474		850	93
Georgia Boot Inc.	1810 Columbia Ave., Franklin, Tennessee 37064, U.S.A.	615-794-1556		800	96
Starsax Ltda., Calcados	Rua Vera Cruz 604, Parobe 95630, Brazil	51 5431100	51 5431100	800	96
Union Industries Corp, Ltd.	Bangkapi, Bangkok 10310, Thailand	5300511	5384247	800	94
Accessories Associates Inc.	PO Box 17417, Smithfield, Rhode Island 02917-0704, U.S.A.	401-231-3800	401-231-4120	750	94
BSC Footwear Supplies Ltd.	Sunningdale Rd., Braunstone LE3 1VR, United Kingdom	16 2320202	16 2320096	750	94
Liberty Footwear Co.	4/42 Punjabi Bagh, New Delhi 132 001, India	11 5417062	11 5455567	750	94
Bata SA Marocaine	228 Blvd. Ibn Tachfine, Casablanca, Morocco	2 401307	2 404553	740	92
Dae Shin Trading Co., Ltd.	Saha-Gu, Pusan, Republic of Korea	51 2619221	51 2619229	710	97
Mason Shoe Manufacturing Inc.	1251 1st Ave., Chippewa Falls, Wisconsin 54729, U.S.A.	715-723-1871		700	94
British Bata Shoe Co. Ltd.	E. Tilbury, Grays RM18 8RL, United Kingdom	375 843400		691	93
Gator Industries Inc.	1000 S.E. 8th St., Hialeah, Florida 33010, U.S.A.	305-888-5000	305-885-3869	675	94
Tandy Brands Accessories Inc.	690 E. Lamar Blvd., Ste. 200, Arlington, Texas 76011, U.S.A.	817-548-0090		669	97
Carolina Shoe Co.	PO Box 1079, Morganton, North Carolina 28680, U.S.A.	704-437-7755	704-438-9020	650	97
Forus SA	Av. Departamental 01053, Santiago, Chile	2 2210422	2 2215895	624	96
HS Corp.	Yonje-Gu, Pusan, Republic of Korea	51 8680711	51 8680710	622	97
Kuan Show International Corp.	3F, No. 38, Shuang Cheng St., Taipei, Taiwan	2 5954260	2 5954276	611	93
Ballet Makers Inc.	1 Campus Rd., Totowa, New Jersey 07512, U.S.A.	201-595-9000		600	93
Calzado Olympic SA	Apartado 4513-1000, San Jose, Costa Rica	2320350	2200731	600	92
Koryo Co., Ltd.	Yongin-Gun, Seoul, Republic of Korea	2 8506114	2 8528011	600	97
Leegin Creative Leather Products Inc.	PO Box 406, City of Industry, California 91746, U.S.A.	818-961-9381	818-961-9380	600	93
G.H. Bass Caribbean, Inc.	Rd. 2, Km. 49.2, Manati 00674, Puerto Rico	787854-2054		550	92
Shaw UK Holdings Ltd.	Newbridge NP1 4XG, United Kingdom	495 243000	495 248612	549	92
Fiona Footwear Ltd.	Waterton Industrial Estate, Bridgend CF31 5XH, United Kingdom	656 646234	656 767032	538	93
Nunn Bush Shoe Co.	PO Box 2047, Milwaukee, Wisconsin 53201, U.S.A.	414-263-8800	414-263-8808	525	97
WEYCO Group Inc.	PO Box 1188, Milwaukee, Wisconsin 53201, U.S.A.	414-263-8800	414-263-8808	515	97
Allen-Edmonds Shoe Corp.	PO Box 998, Port Washington, Wisconsin 53074-0998, U.S.A.	414-284-3461	414-284-1265	500	97
American Trading and Production Corp. Hazel Div.	1200 S. Stafford St., Washington, Missouri 63090, U.S.A.	314-239-2781	314-239-7223	500	97
L.A. Gear Inc.	2850 Ocean Park Blvd., Santa Monica, California 90405, U.S.A.	310-452-4327		500	95
Liu Chiao Industrial Co. Ltd.	30 Ming Sheng E. Rd., 5th Fl., Sec. 5, Taipei 503, Taiwan	2 7568099	2 7616299	500	91
Pegeg & Co.	Jl. Gula 12, Surabaya 60161, Indonesia	31 333182	31 333183	500	93
Pyramid Handbags Inc.	100 W. 33rd St., New York, New York 10001, U.S.A.	212-714-2211		500	94
Tender Tootsies Group Ltd.	1806 Wharncliffe Rd. S, London, Ontario N6L 1K1, Canada	519-287-2518	519-652-3394	488	98
McGuire-Nicholas Co., Inc.	2331 Tubeway Ave., City of Commerce, California 90040, U.S.A.	213-722-6961		480	94
Gali Industries Ltd.	New Industrial Zone, Rishon Lezion 75143, Israel	3 9631050	3 9612897	470	97
Brown Shoe Co. Of Canada, Ltd.	1857 Rogers Rd., Perth, Ontario K7H 1P7, Canada	613-267-2000	613-267-7113	460	98

Continued

Representative Companies in Sector
- Continued -

Name	Address	Telephone	Fax	Employment	Y
Supreme Slipper Manufacturing Corp.	PO Box 1376, Lewiston, Maine 04240, U.S.A.	207-784-2921	207-784-2162	460	93
Weinbrenner Shoe Company Inc.	108 S. Polk St., Merrill, Wisconsin 54452, U.S.A.	715-536-5521	715-536-1172	450	94
Hyde Athletic Industries Inc.	PO Box 6046, Peabody, Massachusetts 01961, U.S.A.	508-532-9000	508-532-6105	442	97
Cincinnati Milacron UK Ltd.	Kingsbury Rd., Birmingham B24 0QU, United Kingdom	21 3514791	21 3517891	438	94
HI-TEC Sports PLC	Aviation Way, Southend-on-Sea SS2 6GH, United Kingdom	702 541741	702 547947	432	93
Curtume Scuck SA	Rua Presidente Lucane 4300, Estancia Velha 93600-000, Brazil	51 5611300	51 5611166	430	96
Fossil Inc.	2280 N. Greenville Ave., Richardson, Texas 75082, U.S.A.	214-234-2525	214-348-1366	430	95
B.B. Walker Co.	P.O. Drawer 1167, Asheboro, North Carolina 27204, U.S.A.	910-625-1380		423	97
Hulera Centroamericana SA	Zona 12, Guatemala City 01012, Guatemala	2 760085	2 766335	420	
Thaibinh Export Leather Products and Shoes Factory	Phu Khanh Sub-dist., Thai Binh Town, Vietnam	36 596		415	
Aneka Regalindo PT	Jl. Pemuda 27-3l, Surabaya 60275, Indonesia	31 519174		400	93
Buxton Co.	PO Box 1650, Springfield, Massachusetts 01102, U.S.A.	413-734-5900	413-785-1367	400	96
Etienne Aigner Inc.	712 5th Ave., New York, New York 10019, U.S.A.	212-246-8660		400	93
Fonicia Kereskedelmi Kft.	Hungaria krt. 140-144, Budapest H-1146, Hungary	1 2515005	1 2517052	400	93
Ilshin Industrial Co., Ltd.	Tongnae-Gu, Pusan, Republic of Korea	51 5025015	51 5042605	400	97
Jumping Jacks Div.	PO Box 6048, Hot Springs, Arkansas 71901, U.S.A.	501-262-6000	501-262-5160	400	94
Red Comercial del Calzado, SA	Inca, Baleares E-07300, Spain	971 500400	971 502977	400	97
MEM Company Inc.	231 Union St., Northvale, New Jersey 07647-2210, U.S.A.	201-767-0100	201-767-0698	391	95
Uniflex, Inc.	PO Box 96, Hicksville, New York 11802, U.S.A.	516-932-2000	516-932-3129	390	96
R. Griggs & Co. Ltd.	Wollaston, Wellingborough NN9 7SW, United Kingdom	933 665381	933 664088	387	94
Headlam Group PLC	Bedford Rd., Northampton NN1 5NH, United Kingdom	604 234121	604 271052	374	91
Steven Madden Ltd.	52-16 Barnett Ave., Long Island City, New York 11104, U.S.A.	718-446-1800	718-446-5599	360	97
Cowtown Boot Company Inc.	PO Box 26428, El Paso, Texas 79926, U.S.A.	915-593-2565		350	94
Sam Yung Chemical Co., Ltd.	Kimhae-Shi, Kyongnam, Republic of Korea	525 361001	525 343535	320	97
Sweet Seventeen Shoes CV	Jl. Kundi No. 17-19 Waru-Sidoarjo, Jatim, Indonesia	31 831917	31 838558	320	93
Terra Footwear Ltd.	5409 Eglinton Ave. W Ste. 103, Toronto, Ontario M9C 5K6, Canada	416-695-3464	416-620-4541	320	98
Saigon Leather Products and Footwear Import-Export Corp.	District 1, Ho Chi Minh City, Vietnam	8 231334	8 299217	302	
American Tourister	91 Main St., Warren, Rhode Island 02885, U.S.A.	401-245-2100	401-247-0988	300	95
Rider Enterprise Co. Ltd.	Wu Ku, Taipei 248, Taiwan	2 9861747	2 9861537	300	93
Zero Enclosures Div.	500 W. 200 N., North Salt Lake, Utah 84054, U.S.A.	801-298-5900	801-292-9450	280	95
Jaclyn Inc.	635 59th St., West New York, New Jersey 07093, U.S.A.	201-868-9400	201-868-3047	262	97
Tolo d.d.	Cesta Leona Dobrotinskega 8, Sentjur pri Celju 3230, Slovenia	63 743691	63 743138	255	95
Boowoon Mulsan Co., Ltd.	Yusong-Gu, Taejon, Republic of Korea	42 8222131	42 8233708	250	97
Fabrica de Calzado Ecco, SA	Apartado 354-7050, Cartago, Costa Rica	5737091	5737139	250	
Greb International, Kodiak Division	6700 Century Ave., Mississauga, Ontario L5N 2V8, Canada	905-567-0030	905-567-0800	250	98
Sang-A Tech Corp.	Songpa-Gu, Seoul, Republic of Korea	2 4177091	2 4177094	250	97
Trina Inc.	PO Box 1431, Fall River, Massachusetts 02722, U.S.A.	508-678-7601		250	93
Wave-Cover International Co., Ltd.	2F, No. 8, Lane 71, Fu Yang St., Taipei, Taiwan	2 7355583	2 7350475	250	94
Hajdusagi Borgyar	Vagohid u. 3, Debrecen H-4030, Hungary	52 68055	52 11793	240	93
Konus Konum d.o.o.	Mestni trg 17, Slovenske Konjice 3210, Slovenia	63 754212	63 754926	232	95
Joongwon Co., Ltd.	Kumjong-Gu, Pusan, Republic of Korea	51 5264001	51 5264047	210	97
Aris Isotoner Inc.	417 5th Ave., New York, New York 10016, U.S.A.	212-532-8627		200	95
Barry Manufacturing Company Inc.	Bubier St., Lynn, Massachusetts 01901, U.S.A.	617-598-1055		200	93
Dae Sung Corp.	Taedok-Gu, Taejon, Republic of Korea	42 6218601	42 6222337	200	97
Fownes Brothers and Company Inc.	411 5th Ave., New York, New York 10016-2203, U.S.A.	212-683-0150	212-683-2832	200	93
G. A. Boulet Inc.	501 St.-Gabriel Rue, St.-Tite, Quebec G0X 3H0, Canada	418-365-5174	418-365-3330	200	98
Goodfit Manufacturing Corp.	Bagumbayan, Taguig M M 1604, Philippines	8238059	28238060	200	97
Lehigh Safety Shoe Co.	1100 E. Main St., Endicott, New York 13760, U.S.A.	607-754-7980	607-757-4354	200	94
Mas Mail	PO Box 3891, Cape Town 8000, Republic of South Africa	21 545130	21 545140	200	93
Ghana Rubber Products Ltd.	PO Box 1069, Accra, Ghana	21 221771	21 220185	190	92
Taiyang Trading Co.	Songpa-Gu, Seoul, Republic of Korea	2 4251293	2 4251292	190	97
Teneria Moderna Franco Espanola, SAL	Mollet Del Valles, Barcelona E-08100, Spain	93 5706250	93 5705838	190	97
Curtidos San Luis SA	Av. Del Maiz 961, Naschel 5359, Argentina	656 91062	656 91068	180	96

Product Tables

Footwear, Total Production, Excluding Rubber Footwear

Unit of Measure: 1,000 Pairs.

	1990 Value	%	1991 Value	%	1992 Value	%	1993 Value	%	1994 Value	%	1995 Value	%	1996 Value	%
Total Production	5,364,981	100.0	5,295,997	100.0	5,168,957	100.0	4,999,230	100.0	4,810,323	100.0	4,879,417	100.0	4,864,997	100.0
Regions														
Africa	149,352	2.8	144,087	2.7	132,791	2.6	129,529	2.6	120,837	2.5	122,296	2.5	116,543	2.4
America, North	237,330	4.4	220,785	4.2	217,577	4.2	230,449	4.6	209,859	4.4	192,868	4.0	171,997	3.5
America, South	203,110	3.8	219,118	4.1	215,190	4.2	252,644	5.1	211,646	4.4	187,654	3.8	235,256	4.8
Asia	3,116,771	58.1	3,260,142	61.6	3,242,847	62.7	3,089,345	61.8	3,059,494	63.6	3,114,416	63.8	3,189,744	65.6
Europe	1,651,566	30.8	1,445,908	27.3	1,356,404	26.2	1,293,456	25.9	1,204,581	25.0	1,258,796	25.8	1,148,801	23.6
Oceania	6,852	0.1	5,957	0.1	4,147	0.1	3,808	0.1	3,905	0.1	3,388	0.1	2,656	0.1
Leading Producers														
China	1,202,565	22.4	1,328,950	25.1	1,613,647	31.2	*1,494,693	29.9	*1,586,862	33.0	*1,679,031	34.4	*1,771,200	36.4
Italy	320,200	6.0	310,200	5.9	295,000	5.7	*310,383	6.2	*306,926	6.4	*303,468	6.2	*300,010	6.2
United States of America	184,568	3.4	168,992	3.2	164,904	3.2	171,733	3.4	156,712	3.3	146,979	3.0	127,315	2.6
Russian Federation	385,262	7.2	336,411	6.4	220,415	4.3	145,889	2.9	76,531	1.6	51,618	1.1	36,764	0.8
France	194,700	3.6	168,084	3.2	160,320	3.1	151,124	3.0	131,051	2.7	*144,741	3.0	*139,990	2.9

Footwear, Leather, Children's

Unit of Measure: 1,000 Pairs.

	1990 Value	%	1991 Value	%	1992 Value	%	1993 Value	%	1994 Value	%	1995 Value	%	1996 Value	%
Total Production	841,836	100.0	810,998	100.0	695,794	100.0	639,148	100.0	595,391	100.0	578,827	100.0	579,145	100.0
Regions														
Africa	7,004	0.8	7,466	0.9	6,359	0.9	5,589	0.9	5,377	0.9	6,379	1.1	6,188	1.1
America, North	32,110	3.8	30,674	3.8	28,872	4.1	31,441	4.9	29,363	4.9	17,334	3.0	13,753	2.4
America, South	10,051	1.2	11,236	1.4	11,563	1.7	11,167	1.7	10,663	1.8	9,051	1.6	9,918	1.7
Asia	634,788	75.4	611,673	75.4	537,320	77.2	505,960	79.2	492,220	82.7	491,654	84.9	498,705	86.1
Europe	155,382	18.5	148,142	18.3	109,992	15.8	83,647	13.1	57,514	9.7	54,214	9.4	50,474	8.7
Oceania	2,501	0.3	1,808	0.2	1,688	0.2	1,344	0.2	254	0.0	195	0.0	107	0.0
Leading Producers														
Russian Federation	170,883	20.3	146,405	18.1	77,319	11.1	39,692	6.2	20,030	3.4	15,014	2.6	10,513	1.8
Ukraine	77,244	9.2	71,003	8.8	51,302	7.4	28,395	4.4	8,796	1.5	3,851	0.7	1,995	0.3
United States of America	17,953	2.1	17,232	2.1	15,363	2.2	13,998	2.2	12,087	2.0	*2,891	0.5	-	
France	24,548	2.9	24,644	3.0	11,090	1.6	12,590	2.0	9,880	1.7	*14,536	2.5	*13,886	2.4
United Kingdom	12,433	1.5	*12,673	1.6	*12,091	1.7	*11,508	1.8	*10,925	1.8	*10,343	1.8	*9,760	1.7

Footwear, Leather, Men's

Unit of Measure: 1,000 Pairs.

	1990 Value	%	1991 Value	%	1992 Value	%	1993 Value	%	1994 Value	%	1995 Value	%	1996 Value	%
Total Production	799,259	100.0	772,255	100.0	679,991	100.0	654,639	100.0	622,289	100.0	574,934	100.0	560,328	100.0
Regions														
Africa	15,266	1.9	15,357	2.0	13,805	2.0	13,546	2.1	13,406	2.2	14,647	2.5	13,378	2.4
America, North	60,757	7.6	57,752	7.5	55,065	8.1	65,151	10.0	62,953	10.1	48,495	8.4	45,204	8.1
America, South	29,711	3.7	30,414	3.9	30,038	4.4	30,503	4.7	29,462	4.7	25,931	4.5	25,928	4.6
Asia	424,483	53.1	421,702	54.6	390,060	57.4	361,780	55.3	356,037	57.2	321,466	55.9	325,619	58.1
Europe	262,355	32.8	241,421	31.3	185,465	27.3	177,706	27.1	156,709	25.2	161,292	28.1	147,874	26.4
Oceania	6,686	0.8	5,609	0.7	5,558	0.8	5,953	0.9	3,722	0.6	3,102	0.5	2,325	0.4
Leading Producers														
United States of America	44,621	5.6	42,945	5.6	41,270	6.1	45,357	6.9	45,363	7.3	*32,848	5.7	*29,408	5.2
Russian Federation	98,795	12.4	87,566	11.3	65,465	9.6	49,772	7.6	30,070	4.8	20,197	3.5	14,021	2.5
France	58,325	7.3	56,119	7.3	11,707	1.7	10,757	1.6	10,028	1.6	*16,171	2.8	*10,972	2.0
Romania	36,434	4.6	35,381	4.6	29,465	4.3	30,660	4.7	37,300	6.0	40,055	7.0	36,335	6.5
Ukraine	53,603	6.7	46,267	6.0	41,151	6.1	33,850	5.2	13,700	2.2	8,550	1.5	6,429	1.1

Commodity data are provided by the United Nations Statistical Division. The symbol * means that data are estimated. For additional notes, see Appendix I.

175

Product Tables
Footwear, Leather, Women's

Unit of Measure: 1,000 Pairs.

	1990 Value	%	1991 Value	%	1992 Value	%	1993 Value	%	1994 Value	%	1995 Value	%	1996 Value	%
Total Production	1,016,311	100.0	958,643	100.0	864,280	100.0	846,361	100.0	787,754	100.0	762,450	100.0	728,127	100.0
Regions														
Africa	18,408	1.8	17,786	1.9	13,904	1.6	16,185	1.9	16,421	2.1	18,460	2.4	17,370	2.4
America, North	87,474	8.6	79,562	8.3	77,987	9.0	78,021	9.2	76,459	9.7	55,028	7.2	49,066	6.7
America, South	70,804	7.0	73,155	7.6	74,317	8.6	81,083	9.6	70,179	8.9	65,688	8.6	73,687	10.1
Asia	385,886	38.0	384,482	40.1	364,311	42.2	345,037	40.8	318,912	40.5	295,812	38.8	295,422	40.6
Europe	441,702	43.5	393,118	41.0	324,892	37.6	318,036	37.6	299,017	38.0	321,288	42.1	287,574	39.5
Oceania	12,037	1.2	10,539	1.1	8,868	1.0	8,000	0.9	6,766	0.9	6,174	0.8	5,008	0.7
Leading Producers														
France	94,618	9.3	88,458	9.2	32,350	3.7	31,844	3.8	31,462	4.0	*41,085	5.4	*35,308	4.8
United States of America	63,082	6.2	55,455	5.8	53,816	6.2	51,473	6.1	50,676	6.4	*29,294	3.8	*21,767	3.0
Brazil	64,969	6.4	66,873	7.0	68,159	7.9	74,967	8.9	64,010	8.1	59,696	7.8	68,193	9.4
Russian Federation	100,277	9.9	100,451	10.5	76,947	8.9	55,144	6.5	26,377	3.3	16,407	2.2	11,973	1.6
Poland	43,273	4.3	28,538	3.0	22,817	2.6	20,935	2.5	22,639	2.9	24,630	3.2	26,884	3.7

Footwear, Other (Sports, Orthopedic Etc.)

Unit of Measure: 1,000 Pairs.

	1990 Value	%	1991 Value	%	1992 Value	%	1993 Value	%	1994 Value	%	1995 Value	%	1996 Value	%
Total Production	1,138,936	100.0	1,110,187	100.0	1,038,570	100.0	1,127,777	100.0	1,138,759	100.0	1,174,570	100.0	1,183,624	100.0
Regions														
Africa	19,009	1.7	16,410	1.5	16,082	1.5	17,165	1.5	13,999	1.2	12,934	1.1	12,008	1.0
America, North	52,016	4.6	49,607	4.5	42,796	4.1	60,095	5.3	56,473	5.0	45,482	3.9	48,568	4.1
America, South	10,899	1.0	7,879	0.7	7,654	0.7	9,957	0.9	9,852	0.9	16,158	1.4	6,630	0.6
Asia	734,259	64.5	780,180	70.3	739,173	71.2	815,901	72.3	837,866	73.6	864,703	73.6	889,607	75.2
Europe	322,041	28.3	255,565	23.0	232,099	22.3	224,033	19.9	220,009	19.3	234,751	20.0	226,290	19.1
Oceania	712	0.1	546	0.0	766	0.1	627	0.1	561	0.0	542	0.0	522	0.0
Leading Producers														
China	*430,265	37.8	*430,265	38.8	*430,265	41.4	*430,265	38.2	*430,265	37.8	*430,265	36.6	*430,265	36.4
India	189,557	16.6	189,641	17.1	198,163	19.1	*263,037	23.3	*283,430	24.9	*303,823	25.9	*324,216	27.4
Poland	54,927	4.8	38,319	3.5	32,364	3.1	26,970	2.4	30,597	2.7	35,153	3.0	39,736	3.4
Indonesia	58,191	5.1	86,072	7.8	49,854	4.8	*66,356	5.9	*72,975	6.4	*79,594	6.8	*86,213	7.3
Romania	39,759	3.5	27,815	2.5	11,772	1.1	11,233	1.0	8,366	0.7	8,184	0.7	8,503	0.7

Footwear, House

Unit of Measure: 1,000 Pairs.

	1990 Value	%	1991 Value	%	1992 Value	%	1993 Value	%	1994 Value	%	1995 Value	%	1996 Value	%
Total Production	487,143	100.0	423,498	100.0	335,415	100.0	331,685	100.0	268,109	100.0	264,006	100.0	297,350	100.0
Regions														
Africa	592	0.1	592	0.1	592	0.2	592	0.2	592	0.2	592	0.2	592	0.2
America, North	53,627	11.0	52,286	12.3	55,429	16.5	61,837	18.6	54,191	20.2	57,712	21.9	47,795	16.1
America, South	53,846	11.1	67,345	15.9	71,244	21.2	94,351	28.4	63,916	23.8	50,992	19.3	94,407	31.7
Asia	179,575	36.9	119,246	28.2	56,807	16.9	44,116	13.3	31,508	11.8	28,011	10.6	35,378	11.9
Europe	199,135	40.9	183,746	43.4	151,146	45.1	130,678	39.4	117,877	44.0	126,700	48.0	119,178	40.1
Oceania	368	0.1	282	0.1	197	0.1	111	0.0	25	0.0	-		-	
Leading Producers														
United States of America	44,718	9.2	42,963	10.1	45,692	13.6	51,686	15.6	43,625	16.3	46,732	17.7	36,401	12.2
Brazil	46,528	9.6	59,108	14.0	62,089	18.5	86,831	26.2	56,616	21.1	46,095	17.5	85,190	28.6
France	61,704	12.7	54,795	12.9	41,938	12.5	41,347	12.5	40,850	15.2	*47,391	18.0	*47,459	16.0
Russian Federation	76,704	15.7	69,954	16.5	34,084	10.2	21,232	6.4	10,001	3.7	6,811	2.6	4,243	1.4
Hong Kong	61,206	12.6	17,243	4.1	3,183	0.9	1,653	0.5	1,116	0.4	1,260	0.5	*12,156	4.1

Commodity data are provided by the United Nations Statistical Division. The symbol * means that data are estimated. For additional notes, see Appendix I.

Product Tables
Sandals and Similar Light Footwear

Unit of Measure: 1,000 Pairs.

	1990 Value	%	1991 Value	%	1992 Value	%	1993 Value	%	1994 Value	%	1995 Value	%	1996 Value	%
Total Production	268,947	100.0	286,726	100.0	226,290	100.0	242,407	100.0	244,151	100.0	232,654	100.0	203,854	100.0
Regions														
Africa	5,557	2.1	5,557	1.9	5,557	2.5	5,557	2.3	5,557	2.3	5,557	2.4	5,557	2.7
America, North	309	0.1	309	0.1	309	0.1	309	0.1	309	0.1	309	0.1	309	0.2
America, South	152,021	56.5	169,114	59.0	119,934	53.0	126,016	52.0	126,875	52.0	129,069	55.5	121,007	59.4
Asia	54,951	20.4	43,321	15.1	45,732	20.2	61,091	25.2	61,497	25.2	37,873	16.3	34,061	16.7
Europe	54,989	20.4	67,304	23.5	53,638	23.7	48,314	19.9	48,793	20.0	58,726	25.2	41,800	20.5
Oceania	1,120	0.4	1,120	0.4	1,120	0.5	1,120	0.5	1,120	0.5	1,120	0.5	1,120	0.5
Leading Producers														
Brazil	150,993	56.1	168,040	58.6	118,626	52.4	124,896	51.5	125,736	51.5	127,999	55.0	119,747	58.7
Indonesia	11,032	4.1	6,592	2.3	28,592	12.6	50,592	20.9	51,828	21.2	29,354	12.6	*26,545	13.0
Spain	22,723	8.4	20,166	7.0	14,749	6.5	11,406	4.7	20,572	8.4	26,715	11.5	*16,501	8.1
Germany	-		19,476	6.8	13,740	6.1	14,690	6.1	9,091	3.7	13,398	5.8	*9,037	4.4
Russian Federation	34,780	12.9	25,197	8.8	9,082	4.0	3,383	1.4	1,706	0.7	1,473	0.6	1,134	0.6

Commodity data are provided by the United Nations Statistical Division. The symbol * means that data are estimated. For additional notes, see Appendix I.

177

ISIC 3310
WOOD PRODUCTS

ISIC 3310—Wood Products—includes wooden railway sleepers, coniferous sawn wood, broadleaved sawn wood, veneer sheets, plywood, and particle board. Sawn wood includes planks, beams, joists, boards, rafters, etc. It may also be sliced, peeled, or rotary cut.

These categories correspond to the outputs of Sawmills and Planing Mills in the U.S. SIC system, SIC 242.

Summary Statistics

		1990		1991		1992		1993		1994		1995	
		Value	N	Value	N	Value	N	Value	N	Value	N	Value	N
Establishments or enterprises	(number)	129,246	82	133,432	90	154,572	87	115,448	85	110,874	71	78,428	34
Number employed	(000)	4,473	97	3,798	98	3,711	96	4,086	96	3,976	87	3,602	64
Total Wages	($ mil.)	36,661	90	37,701	91	40,231	86	34,938	84	36,594	77	34,486	58
Wage/salary per employee	($)	6,864	89	7,851	90	8,890	86	7,423	84	7,623	77	9,365	58
Female workers	(000)	189	35	147	32	521	29	675	33	676	27	619	15
as % of total employment*	(%)	17	34	17	31	19	29	18	33	21	27	27	14
Output	($ bil.)	210	94	219	97	199	91	208	87	224	79	231	62
per employee	($ 000)	39	90	40	92	43	87	40	83	44	76	55	60
per establishment	($ 000)	1,860	78	1,739	85	1,545	77	1,561	72	1,848	61	1,739	30
Value added	($ bil.)	77	84	74	83	82	78	88	77	85	68	87	56
per employee	($ 000)	16	81	15	81	18	77	16	74	18	66	22	54
per establishment	($ 000)	725	71	607	74	655	66	652	62	809	51	651	25
Capital investment	($ mil.)	6,168	68	5,960	64	6,049	57	4,563	52	5,185	42	3,407	22
per employee	($ 000)	3	65	2	61	2	55	2	50	2	41	2	21
per establishment	($ 000)	188	63	134	59	121	53	97	47	148	39	156	19

Data presented above are drawn from the detailed tables that follow. Columns headed 'N' show the number of countries that provided valid data for inclusion. Values are not strictly comparable one year to the next or one row to the next because the number of countries that report varies from period to period and row to row. However, a general indication of magnitudes can be gleaned from reviewing this summary—especially in earlier years in which more reports are available. For detailed explanations, see Appendix I. *The average for those countries reporting both total and female employment.

Establishments and Number Engaged

Country	Establishments or Enterprises (number)						Employees per Establishment					
	1990	1991	1992	1993	1994	1995	1990	1991	1992	1993	1994	1995
Albania	-	-	-	31	16	13	-	-	-	52	111	58
Argentina	-	-	-	5,273	6,120	-	-	-	-	3.486	3.443	-
Armenia	13	12	12	12	10	-	222	163	-	-	-	-
Australia	2,544	3,457	3,345	-	-	-	17	12	12	-	-	-
Austria	1,994	1,955	1,883	1,808	1,771	-	9.679	8.696	9.028	8.683	8.767	-
Azerbaijan	540	533	597	518	289	-	10	11	11	9.616	13	-
Bahrain	-	-	1.000	-	-	-	-	-	15	-	-	-
Bangladesh	1,317	962	1,137	-	-	-	9.825	13	12	-	-	-
Belize	59	61	61	-	-	-	-	11	11	-	-	-
Benin	-	-	42	42	-	-	-	-	-	-	-	-
Bolivia	141	150	159	160	161	159	18	18	17	19	18	17
Bosnia & Herzegovina	106	106	-	-	8.000	-	317	295	-	-	78	-
Brazil	1,048	-	858	826	-	-	76	-	92	101	-	-
Bulgaria	*61	*60	*74	*1,344	*1,961	*2,324	398	305	216	10	7.445	5.852
Cameroon	-	-	-	-	*20	*19	-	-	-	-	276	320
Canada	2,672	2,426	2,288	2,201	2,205	-	39	38	41	45	49	-
Cape Verde	1.000	1.000	1.000	1.000	2.000	-	-	-	-	-	-	-
Central African Republic	-	*8.000	*6.000	*6.000	*6.000	*6.000	-	-	-	-	-	-
Chile	126	120	126	149	154	-	168	165	164	165	146	-
China	*13,562	*13,621	*13,363	*12,410	*13,486	*15,480	52	56	55	59	57	47
China (Hong Kong SAR)	723	583	646	478	483	-	4.841	4.288	4.180	4.603	3.727	-
Colombia	184	183	195	181	186	-	34	36	41	43	41	-
Costa Rica	249	249	252	267	-	-	18	16	17	17	-	-
Croatia	366	466	653	881	1,078	1,196	61	36	25	18	15	12
Cyprus	965	961	962	962	1,089	1,179	1.999	1.896	2.240	2.338	2.127	2.146
Czechoslovakia (Former)	*41	*51	-	-	-	-	1,366	863	-	-	-	-
Czech Republic	-	*32	*103	*146	-	-	-	844	291	205	-	-
Denmark	1,035	1,049	1,019	-	-	-	14	13	14	-	-	-
Ecuador	53	55	55	49	47	68	62	60	60	65	70	55
Egypt	121	98	109	108	-	-	70	64	61	56	-	-
El Salvador	-	3.000	4.000	3.000	7.000	6.000	-	-	32	33	44	28
Ethiopia	11	8.000	10	10	24	22	226	284	318	284	113	116
Fiji	44	43	35	-	-	-	33	29	31	-	-	-
Finland	560	561	507	466	455	495	54	49	48	49	53	50
FYR Macedonia	*13	*19	*31	*35	*56	*79	169	77	60	40	26	19
Gabon	-	12	12	12	12	15	-	151	120	113	120	121
Gambia	-	-	-	2.000	-	-	-	-	-	53	-	-
Germany	-	3,138	3,117	2,173	2,103	-	-	41	39	51	53	-
Germany (Western Part)	2,721	2,771	2,795	-	-	-	38	39	39	-	-	-
Ghana	-	-	-	81	-	-	-	-	-	271	-	-
Greece	248	247	247	-	-	-	30	29	28	-	-	-
Grenada	-	-	4.000	4.000	-	-	-	-	5.500	6.250	-	-
Guatemala	-	24	24	24	24	24	-	64	64	64	63	93
Honduras	-	-	36	41	41	41	-	-	252	242	265	297
Hungary	*296	*673	*753	*945	-	-	51	27	25	19	-	-
Iceland	40	36	37	38	-	-	-	-	-	-	-	-
India	3,117	3,103	3,277	3,237	-	-	20	19	20	20	-	-
Indonesia	1,340	1,253	1,405	1,474	1,589	1,754	243	273	263	257	248	224
Iran (Islamic Republic of)	285	39	35	32	-	-	40	212	183	192	-	-
Ireland	223	223	-	-	-	-	20	21	-	-	-	-
Israel	332	396	364	389	367	-	16	16	16	16	17	-
Italy	*846	*847	*979	*983	*1,018	-	44	44	42	41	41	-
Jamaica	22	22	-	-	-	-	26	29	-	-	-	-
Japan	26,753	25,783	24,793	24,738	23,970	23,844	11	11	11	11	11	11
Jordan	711	727	782	793	1,041	1,240	-	1.282	1.604	1.267	1.463	1.565
Kenya	143	176	177	177	223	223	60	49	48	49	40	41
Korea, Republic of	2,010	2,194	2,150	2,556	2,505	2,490	20	18	17	16	16	15
Kuwait	78	67	109	103	103	-	17	10	11	11	11	-
Kyrgyzstan	536	554	363	316	199	-	7.685	6.551	6.738	8.149	7.432	-
Latvia	1,302	1,258	219	425	265	609	11	12	70	30	53	26

Continued.

   Depending on the table, * means *Enterprises*, *Engaged*, or *Factor Values*; ± means *Basis Unspecified*; § means *shown in millions*. For additional notes and sources, see Appendix I.

Establishments and Number Engaged

- Continued -

Country	Establishments or Enterprises (number)						Employees per Establishment					
	1990	1991	1992	1993	1994	1995	1990	1991	1992	1993	1994	1995
Lithuania	-	-	-	*162	*241	-	-	-	-	45	77	-
Macau	53	45	36	29	27	18	4.038	2.933	2.861	2.793	4.407	3.000
Malaysia	723	768	806	893	917	-	123	125	136	143	150	-
Malta	100	90	81	74	70	-	1.600	1.644	1.481	1.635	1.729	-
Mauritius	17	19	18	21	17	13	54	38	41	38	40	42
Mexico	16	14	14	14	14	14	305	332	301	233	220	264
Mongolia	42	31	47	59	78	47	162	225	99	65	84	119
Mozambique	8.000	8.000	7.000	7.000	7.000	-	384	322	331	270	315	-
Myanmar	100	100	97	97	100	-	161	154	155	148	147	-
Namibia	-	-	-	-	6.000	-	-	-	-	-	43	-
Nepal	112	143	-	168	170	-	20	23	-	12	14	-
Netherlands	206	228	227	218	198	-	61	57	58	55	60	-
Netherlands Antilles	-	-	-	-	34	-	-	-	-	3.647	-	-
New Zealand	*1,551	*1,562	*1,622	*1,763	-	-	8.033	7.883	8.258	8.457	-	-
Norway	607	588	322	286	297	-	25	24	34	36	37	-
Pakistan	-	39	-	-	-	-	-	74	-	-	-	-
Panama	17	13	-	-	-	-	38	44	-	-	-	-
Peru	718	723	734	-	-	-	7.529	6.284	7.480	-	-	-
Philippines	2,745	2,838	539	454	-	-	18	17	65	65	-	-
Poland	*258	*299	*238	*233	-	-	252	194	239	236	-	-
Portugal	*7,783	*7,426	*7,345	*7,654	*7,477	-	7.053	7.825	7.042	6.706	6.837	-
Puerto Rico	93	99	88	85	82	64	9.140	7.273	8.182	7.647	6.463	9.063
Qatar	174	167	88	83	83	-	10	8.557	7.943	7.277	8.687	-
Republic of Moldova	10	14	14	9.000	9.000	13	1,168	409	435	424	167	115
Romania	-	687	1,238	1,358	3,405	-	-	127	72	59	24	-
Russian Federation	-	-	-	*5,144	*5,002	*7,396	-	-	-	100	94	61
Senegal	2.000	2.000	2.000	3.000	2.000	-	35	38	33	57	34	-
Sierra Leone	-	-	-	5.000	-	-	-	-	-	93	-	-
Singapore	82	76	74	72	65	-	31	29	28	26	26	-
Slovakia	-	46	46	63	76	-	-	442	401	230	196	-
Slovenia	301	*2,263	*2,933	*2,932	*3,668	*1,536	65	6.914	5.059	4.742	3.434	7.760
South Africa	-	845	-	-	-	-	-	70	-	-	-	-
Spain	17,437	17,326	16,549	-	-	-	4.230	4.136	4.105	-	-	-
Sri Lanka	130	99	98	98	-	-	30	37	43	44	-	-
Suriname	-	-	35	-	-	-	-	-	36	-	-	-
Sweden	988	875	827	744	758	-	45	46	45	40	41	-
Thailand	521	858	-	-	-	-	51	49	-	-	-	-
Tonga	-	2.000	4.000	4.000	3.000	-	-	11	16	16	19	-
Trinidad and Tobago	96	70	73	74	65	-	10	10	9.397	9.932	11	-
Tunisia	-	-	-	168	178	158	-	-	-	13	13	13
Turkey	120	115	285	262	246	-	114	106	53	54	54	-
Ukraine	15,614	16,076	15,389	14,378	14,114	13,753	5.316	4.976	4.744	4.729	4.322	3.781
United Kingdom	6,917	6,341	6,614	7,767	7,178	-	11	11	9.979	9.012	11	-
United Republic of Tanzania	80	67	-	-	-	-	60	64	-	-	-	-
USSR (Former)	*1,381	-	-	-	-	-	492	-	-	-	-	-
United States of America	-	-	25,254	-	-	-	-	-	19	-	-	-
Venezuela	265	278	304	248	198	249	26	26	25	26	33	28
Yugoslavia	386	787	2,286	2,851	2,982	3,881	-	28	8.793	6.910	6.472	4.999
Zambia	12	-	-	-	20	-	361	-	-	-	156	-
Zimbabwe	29	29	26	24	27	-	192	192	215	242	252	-

Depending on the table, * means *Enterprises, Engaged, or Factor Values*; ± means *Basis Unspecified*; § means *shown in millions*. For additional notes and sources, see Appendix I.

181

Employment and Compensation of Employees

Country	Number Employed/Engaged (000)						Wage/Salary per Employee ($)					
	1990	1991	1992	1993	1994	1995	1990	1991	1992	1993	1994	1995
Albania	-	-	-	1.624	1.784	0.756	-	-	-	-	-	1,255
Argentina	25	-	-	18	21	-	3,289	-	-	7,529	-	-
Armenia	*2.886	*1.958	-	-	-	-	-	-	-	-	-	-
Australia	44	*42	*40	-	-	-	18,093	19,014	18,162	-	-	-
Austria	19	17	17	16	16	16	20,097	21,457	23,285	23,264	24,225	27,974
Azerbaijan	5.663	5.814	6.660	4.981	3.833	-	1,638	1,367	-	409	-	-
Bahrain	-	-	0.015	-	-	-	-	-	4,255	-	-	-
Bangladesh	13	12	13	-	-	-	463	446	415	-	-	-
Belgium	13	13	13	-	-	-	-	-	-	-	-	-
Belize	-	0.676	0.668	-	-	-	-	2,746	3,076	-	-	-
Bermuda	0.070	0.060	0.065	0.067	0.067	0.056	-	-	-	-	-	-
Bolivia	2.477	2.699	2.713	2.987	2.910	2.635	1,311	1,272	1,206	1,250	1,356	1,341
Bosnia & Herzegovina	34	31	-	-	0.623	-	2,484	2,093	-	-	-	-
Brazil	80	-	79	84	-	-	3,855	-	3,049	3,713	-	-
Bulgaria	24	18	16	14	15	14	1,778	584	910	1,117	897	1,144
Cambodia	*7.803	*8.212	-	-	-	-	-	-	-	-	-	-
Cameroon	6.609	5.201	5.271	5.189	5.530	6.087	6,700	6,533	6,851	6,232	3,093	3,793
Canada	103	91	94	100	107	107	27,111	28,257	27,601	26,649	26,184	26,126
Central African Republic	1.795	-	-	-	-	-	2,570	-	-	-	-	-
Chile	21	20	21	25	23	23	3,136	3,521	4,136	4,337	5,111	5,618
China	*700	*760	*730	*730	*770	*730	-	-	-	-	-	-
China (Hong Kong SAR)	*3.500	*2.500	*2.700	*2.200	*1.800	-	6,419	6,897	7,560	9,225	9,633	-
China (Taiwan Province)	62	55	46	38	36	32	7,475	8,204	9,463	9,864	10,620	11,146
Colombia	6.300	6.636	8.019	7.732	7.545	7.506	1,809	1,837	2,060	2,365	3,056	2,996
Costa Rica	4.527	3.977	4.185	4.639	-	-	2,126	2,003	2,301	2,569	-	-
Croatia	22	17	16	16	16	15	3,283	3,548	1,470	1,614	2,065	3,203
Cyprus	1.929	1.822	2.155	2.249	2.316	2.530	8,564	9,298	9,943	9,611	10,787	14,191
Czechoslovakia (Former)	56	44	-	-	-	-	2,049	1,442	-	-	-	-
Czech Republic	30	27	30	30	-	-	2,117	1,457	1,734	2,079	-	-
Denmark	14	14	14	14	16	17	25,497	25,873	28,982	27,593	31,827	38,273
Ecuador	3.282	3.306	3.303	3.193	3.299	3.732	2,141	2,210	2,649	3,348	1,672	1,870
Egypt	*8.500	*6.300	*6.700	*6.000	*6.000	*6.000	1,529	1,129	1,185	1,260	1,357	1,475
El Salvador	-	-	*0.129	*0.099	*0.311	*0.171	-	-	-	1,890	3,151	2,808
Ethiopia	2.488	2.271	3.183	2.838	2.717	2.554	1,525	1,527	764	550	760	702
Fiji	1.474	1.252	1.096	1.113	1.121	-	4,484	4,810	4,565	5,113	5,686	-
Finland	30	27	24	23	24	25	24,604	22,730	21,338	17,905	20,957	27,164
FYR Macedonia	2.191	1.456	1.858	1.392	1.456	1.476	3,404	3,634	1,840	1,849	2,258	2,514
France	93	95	88	82	80	80	25,528	24,021	28,757	-	-	-
Gabon	-	1.814	1.441	1.352	1.441	1.814	-	11,625	12,385	11,700	6,458	6,442
Gambia	-	-	-	0.105	-	-	-	-	-	-	-	-
Germany	-	*129	*122	*112	*112	-	-	23,688	28,269	28,512	29,841	-
Germany (Western Part)	*104	*108	*110	-	-	-	25,332	26,443	29,789	-	-	-
Ghana	-	-	-	*22	*23	*23	-	-	-	966	722	737
Greece	*7.413	*7.054	*6.920	*6.571	*6.210	*6.834	9,961	10,571	11,381	10,934	10,653	12,893
Grenada	-	-	*0.022	*0.025	-	-	-	-	-	-	-	-
Guatemala	-	*1.533	*1.537	*1.544	*1.519	*2.221	-	-	-	-	-	-
Honduras	*8.083	*6.866	*9.062	*9.929	*11	*12	1,912	2,363	2,057	1,925	1,565	1,434
Hungary	15	18	19	18	19	19	2,325	2,316	2,718	2,679	2,936	2,947
Iceland	0.035	0.024	0.017	0.028	0.021	-	16,226	25,708	-	24,460	-	-
India	*63	*59	*66	*66	*66	*66	597	478	497	497	535	635
Indonesia	326	342	370	378	393	393	697	704	711	831	811	1,254
Iran (Islamic Republic of)	11	8.251	6.398	6.131	-	-	2,990	4,896	3,795	3,442	-	-
Iraq	-	0.079	0.078	-	-	-	-	4,925	6,142	-	-	-
Ireland	4.500	4.600	4.720	4.760	5.194	5.429	16,694	17,244	18,828	16,424	17,400	19,218
Israel	5.300	6.200	6.000	6.100	6.300	6.529	15,254	15,003	16,334	15,814	16,922	18,792
Italy	37	37	41	40	41	-	27,449	29,399	31,283	25,543	25,756	-
Jamaica	0.571	0.633	0.712	0.716	0.868	-	4,548	2,722	1,322	-	-	-
Japan	*303	*295	*285	*282	*275	*271	20,287	22,949	25,571	29,657	32,447	35,621
Jordan	0.613	0.932	1.254	1.005	1.523	1.940	1,673	1,687	1,896	1,757	1,839	1,918
Kenya	8.648	8.626	8.547	8.672	8.840	9.160	884	812	748	498	618	835

Continued.

Depending on the table, * means *Enterprises, Engaged,* or *Factor Values*; ± means *Basis Unspecified*; § means *shown in millions*. For additional notes and sources, see Appendix I.

Employment and Compensation of Employees

- Continued -

Country	Number Employed/Engaged (000)						Wage/Salary per Employee ($)					
	1990	1991	1992	1993	1994	1995	1990	1991	1992	1993	1994	1995
Korea, Republic of	41	41	37	40	40	38	8,502	10,463	11,112	11,636	13,151	15,244
Kuwait	1.310	0.695	1.193	1.163	1.095	1.180	3,759	3,926	6,408	5,709	6,160	8,250
Kyrgyzstan	*4.119	*3.629	*2.446	*2.575	*1.479	-	-	-	-	-	-	-
Latvia	15	16	15	13	14	16	-	-	-	-	-	-
Lithuania	-	-	8.264	7.263	19	-	-	-	-	442	876	-
Macau	0.214	0.132	0.103	0.081	0.119	0.054	4,488	5,178	5,993	5,494	4,221	5,683
Malawi	1.800	-	-	-	-	-	-	-	-	-	-	-
Malaysia	89	96	109	128	137	145	2,632	2,690	2,921	3,064	3,084	3,308
Malta	0.160	0.148	0.120	0.121	0.121	-	6,979	7,196	7,600	6,555	7,499	-
Mauritius	*0.911	0.722	0.742	0.797	0.677	0.547	1,617	2,178	2,304	2,282	2,336	3,725
Mexico	4.880	4.649	4.212	3.263	3.078	3.700	3,173	3,771	4,283	4,765	4,516	2,395
Mongolia	*6.823	*6.985	*4.664	*3.822	*6.555	*5.615	1,895	1,653	646	-	-	-
Mozambique	3.074	2.578	2.318	1.890	2.208	-	801	435	414	-	-	-
Myanmar	16	15	15	14	15	-	1,254	1,473	1,610	1,832	1,700	-
Namibia	-	-	-	-	0.255	-	-	-	-	-	5,826	-
Nepal	2.199	3.283	-	2.021	2.350	-	405	416	-	-	400	-
Netherlands	13	13	13	12	12	11	23,551	24,358	27,329	26,214	26,761	31,893
Netherlands Antilles	-	-	-	0.124	-	-	-	-	-	6,672	-	-
New Zealand	12	12	13	15	-	-	-	-	-	-	-	-
Norway	15	14	11	10	11	11	25,338	25,174	32,204	26,222	27,781	31,663
Pakistan	3.014	2.893	3.026	-	-	-	1,280	1,251	1,286	-	-	-
Panama	*0.641	*0.576	*0.629	*0.622	*0.667	*0.610	4,003	4,438	4,518	4,539	4,603	4,640
Peru	*5.406	*4.543	*5.490	-	-	-	1,512	1,579	1,732	-	-	-
Philippines	50	47	35	30	-	-	1,373	1,258	1,573	1,443	-	-
Poland	65	58	57	55	53	52	1,202	1,752	2,213	2,294	3,020	4,286
Portugal	*55	*58	*52	*51	*51	*52	-	-	-	-	-	-
Puerto Rico	0.850	0.720	0.720	0.650	0.530	0.580	15,765	-	-	16,769	24,151	23,621
Qatar	1.800	1.429	0.699	0.604	0.721	-	3,705	4,540	4,251	3,228	4,908	-
Republic of Moldova	12	5.729	6.090	3.815	1.504	1.501	-	2,851	-	-	434	413
Romania	94	87	89	80	83	-	1,455	1,019	690	801	798	-
Russian Federation	-	-	-	515	470	451	-	-	-	620	989	1,069
Senegal	*0.069	*0.076	*0.066	*0.171	*0.068	-	5,057	4,431	5,724	2,354	2,914	-
Sierra Leone	-	-	-	*0.464	-	-	-	-	-	-	-	-
Singapore	*2.574	*2.204	*2.107	*1.864	*1.694	*1.650	9,345	11,016	12,423	12,826	15,237	16,929
Slovakia	-	*20	*18	*15	*15	-	-	1,424	1,607	1,804	2,018	-
Slovenia	20	16	15	14	13	12	7,736	6,173	5,495	5,496	6,528	8,313
South Africa	61	59	57	59	66	67	3,669	3,916	4,220	4,212	4,118	4,582
Spain	74	72	68	78	74	77	11,727	12,689	14,079	12,541	12,480	14,624
Sri Lanka	*3.930	*3.633	*4.172	*4.347	-	-	584	831	569	630	-	-
Suriname	*1.139	*1.237	*1.267	*1.179	-	-	9,001	12,998	14,017	22,048	-	-
Sweden	44	*41	*37	30	31	33	23,077	25,038	26,771	21,629	22,794	25,280
Thailand	27	42	-	-	-	-	1,738	1,872	-	-	-	-
Tonga	0.049	0.022	0.064	0.063	0.056	-	1,418	1,648	1,241	1,674	1,772	-
Trinidad and Tobago	0.968	0.713	0.686	0.735	0.745	-	4,045	3,807	4,084	3,305	3,308	-
Tunisia	-	-	-	2.184	2.288	2.089	-	-	-	-	-	-
Turkey	14	12	15	14	13	14	4,556	6,601	6,235	7,066	4,181	5,521
Ukraine	83	80	73	68	61	52	2,765	2,304	1,786	-	419	412
United Kingdom	78	68	66	70	78	75	19,093	20,432	21,014	18,318	19,476	20,466
United Republic of Tanzania	4.777	4.287	-	-	-	-	-	-	-	-	-	-
USSR (Former)	680	-	-	-	-	-	2,122	-	-	-	-	-
United States of America	508	470	481	500	522	535	19,862	20,319	21,401	22,156	22,301	22,550
Uruguay	*2.529	*2.729	*2.536	*2.249	*2.248	*2.121	2,253	2,645	2,755	3,423	3,508	3,912
Venezuela	6.800	7.200	7.700	6.428	6.545	6.920	2,311	2,599	3,048	2,687	2,522	2,939
Yugoslavia	-	22	20	20	19	19	-	-	-	-	1,351	669
Zambia	4.333	-	-	-	3.126	-	580	-	-	-	1,052	-
Zimbabwe	5.565	5.567	5.600	5.800	6.800	7.350	2,833	2,500	2,307	1,883	3,576	4,147

Depending on the table, * means *Enterprises, Engaged,* or *Factor Values*; ± means *Basis Unspecified*; § means *shown in millions*. For additional notes and sources, see Appendix I.

183

Female Workers: Total and Percent of Employment

Country	Female Workers (000)						Female Workers as % of Total Employment					
	1990	1991	1992	1993	1994	1995	1990	1991	1992	1993	1994	1995
Afghanistan	0.142	0.134	-	-	-	-	-	-	-	-	-	-
Albania	-	-	-	0.379	0.572	0.162	-	-	-	23.34	32.06	21.43
Argentina	-	-	-	-	1.008	-	-	-	-	-	4.78	-
Australia	5.729	-	-	-	-	-	13.02	-	-	-	-	-
Austria	3.200	3.000	3.000	2.596	2.445	-	16.58	17.65	17.65	16.54	15.75	-
Bahrain	-	-	0.004	-	-	-	-	-	26.67	-	-	-
Bangladesh	2.039	2.532	3.248	-	-	-	15.76	20.77	24.51	-	-	-
Bermuda	0.009	0.011	0.013	0.012	0.009	0.008	12.86	18.33	20.00	17.91	13.43	14.29
Bosnia & Herzegovina	7.463	7.459	-	-	-	-	22.24	23.85	-	-	-	-
Bulgaria	-	-	-	6.400	6.400	5.500	-	-	-	46.04	43.84	40.44
Canada	9.000	7.000	-	-	-	-	8.75	7.69	-	-	-	-
Chile	1.138	1.007	0.985	1.294	1.501	-	5.36	5.08	4.76	5.28	6.66	-
China (Taiwan Province)	26	23	20	16	14	13	42.37	42.32	42.81	41.60	39.83	39.47
Colombia	-	1.131	-	1.212	1.471	-	-	17.04	-	15.68	19.50	-
Croatia	6.760	5.060	4.680	4.850	4.840	4.490	30.46	30.34	28.75	29.94	30.63	30.24
Cyprus	0.168	0.173	0.221	0.165	0.241	-	8.71	9.50	10.26	7.34	10.41	-
Czechoslovakia (Former)	15	13	-	-	-	-	27.14	29.09	-	-	-	-
Czech Republic	12	11	10	10	-	-	40.00	40.74	33.33	33.33	-	-
Denmark	2.868	-	2.729	2.706	-	-	20.02	19.35	19.35	-	-	-
Egypt	-	0.171	0.298	0.164	-	-	-	2.71	4.45	2.73	-	-
El Salvador	-	-	-	0.023	0.092	0.033	-	-	-	23.23	29.58	19.30
Ethiopia	-	0.267	0.407	0.277	0.313	0.318	-	11.76	12.79	9.76	11.52	12.45
FYR Macedonia	0.189	0.151	0.294	0.320	0.338	0.300	8.63	10.37	15.82	22.99	23.21	20.33
Gambia	-	-	-	0.003	-	-	-	-	-	2.86	-	-
Germany (Western Part)	19	-	-	-	-	-	18.29	-	-	-	-	-
Ghana	-	-	-	0.751	-	-	-	-	-	3.42	-	-
Hungary	5.000	-	6.000	5.000	-	-	33.33	-	31.58	27.78	-	-
Indonesia	-	-	-	142	143	140	-	-	-	37.44	36.23	35.65
Iran (Islamic Republic of)	-	0.178	0.101	0.106	-	-	-	2.16	1.58	1.73	-	-
Iraq	-	0.008	-	-	-	-	-	10.13	-	-	-	-
Ireland	0.500	0.600	-	-	-	-	11.11	13.04	-	-	-	-
Italy	-	8.236	8.737	8.400	8.727	-	-	22.06	21.32	20.85	21.14	-
Jordan	-	-	0.008	0.005	0.020	0.027	-	-	0.64	0.50	1.31	1.39
Kenya	0.740	0.752	0.735	0.746	-	-	8.56	8.72	8.60	8.60	-	-
Korea, Republic of	11	11	9.910	10	9.915	9.192	26.11	26.46	26.55	25.08	25.04	24.15
Macau	0.027	0.009	0.004	0.004	0.007	0.006	12.62	6.82	3.88	4.94	5.88	11.11
Malaysia	23	27	32	37	41	-	26.04	27.71	28.82	29.09	30.10	-
Malta	0.001	-	0.001	0.002	0.003	-	0.63	-	0.83	1.65	2.48	-
Mongolia	-	-	-	-	1.315	-	-	-	-	-	20.06	-
Morocco	-	-	-	-	-	0.891	-	-	-	-	-	-
Myanmar	0.733	0.733	0.733	0.733	1.089	-	4.56	4.76	4.87	5.09	7.43	-
Nepal	0.119	0.129	-	0.063	0.086	-	5.41	3.93	-	3.12	3.66	-
New Zealand	1.560	1.398	1.477	1.684	-	-	12.52	11.35	11.03	11.29	-	-
Panama	0.036	-	-	-	-	-	5.62	-	-	-	-	-
Philippines	-	-	5.458	4.042	-	-	-	-	15.55	13.66	-	-
Portugal	3.882	3.940	3.539	3.192	3.092	-	7.07	6.78	6.84	6.22	6.05	-
Republic of Moldova	-	-	-	-	0.489	0.438	-	-	-	-	32.51	29.18
Senegal	0.008	-	-	-	-	-	11.59	-	-	-	-	-
Sri Lanka	0.606	0.597	1.065	1.009	-	-	15.42	16.43	25.53	23.21	-	-
Sweden	7.200	-	-	-	-	-	16.36	-	-	-	-	-
Thailand	9.862	15	-	-	-	-	37.09	34.98	-	-	-	-
Turkey	1.096	-	-	-	-	-	8.04	-	-	-	-	-
United Kingdom	11	-	9.000	-	-	-	14.49	-	13.64	-	-	-
United Republic of Tanzania	1.098	0.986	-	-	-	-	22.99	23.00	-	-	-	-
United States of America	-	-	397	417	434	445	-	-	82.54	83.40	83.14	83.18
Zambia	-	-	-	-	0.090	-	-	-	-	-	2.88	-

　　Depending on the table, * means *Enterprises, Engaged,* or *Factor Values*; ± means *Basis Unspecified*; § means *shown in millions.* For additional notes and sources, see Appendix I.

Output and Output per Employee

Country	Output ($ bil.)						Output per Employee ($)					
	1990	1991	1992	1993	1994	1995	1990	1991	1992	1993	1994	1995
Albania	-	-	-	*0.005	*0.003	*0.002	-	-	-	2,842	1,484	3,300
Argentina	±0.552	-	-	0.885	-	-	21,810	-	-	48,128	-	-
Armenia	0.001	0.001	0.003	0.069	0.000		274	431		-	-	
Australia	*3.908	*3.795	*3.530	-	-	-	88,814	90,359	88,254	-	-	-
Austria	3.221	3.001	3.154	2.635	3.021	3.564	166,883	176,501	185,544	167,876	194,606	225,816
Azerbaijan	±0.068	±0.071	±0.004	±0.011	±0.002		11,936	12,193	623	2,253	483	
Bahrain			±0.001					-	57,979			
Bangladesh	0.045	0.033	0.034				3,443	2,730	2,550	-		
Belgium	1.195	1.124	1.129	1.031	1.099	1.280	93,389	87,839	89,580			
Belize	0.007	0.008	0.009						11,666	12,942		
Benin	±0.003	±0.003	±0.004	±0.003			-	-	-	-		
Bolivia	0.047	0.054	0.049	0.057	0.063	0.061	19,035	19,992	17,990	19,073	21,647	23,042
Bosnia & Herzegovina	±0.506	±0.558					15,069	17,846				
Brazil	*1.918	-	*1.589	*2.523	-	-	24,050	-	20,190	30,105	-	-
Bulgaria	±0.289	±0.110	±0.112	±0.112	±0.108	±0.144	11,913	6,030	7,019	8,060	7,372	10,555
Cameroon	0.240	0.178	0.188	0.215	0.180	0.280	36,302	34,135	35,648	41,367	32,489	46,080
Canada	13	11	12	15	17	17	121,602	123,443	130,085	147,198	157,406	157,129
Cape Verde		±0.000	-	-	-	-		-	-	-	-	-
Central African Republic	*0.033						18,477					
Chile	0.680	0.716	0.754	0.997	1.186	1.318	32,014	36,141	36,435	40,645	52,616	57,834
China	±2.158	±2.297	±2.852	±4.835	±4.251	±4.856	3,083	3,022	3,906	6,624	5,521	6,652
China (Hong Kong SAR)	0.177	0.114	0.144	0.135	0.134		50,432	45,553	53,494	61,287	74,545	
China (Taiwan Province)	±1.678	±1.725	±1.613	±1.403	±1.578	±1.382	26,876	31,342	35,125	36,451	44,020	43,486
Colombia	0.111	0.119	0.157	0.191	0.237	0.231	17,540	17,961	19,526	24,748	31,425	30,828
Costa Rica	0.052	0.049	0.053	0.056	0.061	0.054	11,516	12,294	12,693	12,089		
Cote d'Ivoire	±0.283	±0.245	±0.264	±0.261	-	-	-	-	-	-		
Croatia	±0.492	±0.428	±0.314	-	-	-	22,164	25,674	19,259			
Cyprus	0.092	0.100	0.108	0.103	0.112	0.139	47,793	55,042	50,102	45,645	48,375	55,001
Czechoslovakia (Former)	±0.697	±0.381					12,455	8,665				
Denmark	*1.050	*1.054	*1.205	*1.161	*1.470	*1.868	73,312	74,737	86,137	82,014	94,287	113,167
Ecuador	0.047	0.051	0.056	0.061	0.069	0.095	14,219	15,276	16,883	19,014	20,781	25,586
Egypt	*0.078	*0.041	*0.065	*0.037	*0.041	*0.043	9,182	6,545	9,768	6,213	6,892	7,205
El Salvador	-	±0.001	±0.001	±0.001	±0.004	±0.001	-	-	4,227	6,736	11,382	7,301
Ethiopia	0.013	0.011	0.006	0.006	0.011	0.011	5,251	4,846	1,968	2,026	4,039	4,196
Fiji	0.036	0.037	0.039	0.046	0.052	-	24,706	29,369	35,128	40,962	46,192	-
Finland	±4.564	±3.293	±2.975	±2.667	±3.665	±4.427	150,125	120,615	123,451	117,484	152,707	177,793
FYR Macedonia	±0.026	±0.025	±0.020	±0.012	±0.014	±0.015	11,642	17,137	10,656	8,767	9,444	10,145
France	±10	±10	±11	±9.584	±11	±13	112,581	106,545	124,117	116,596	135,750	160,454
Gabon	-	*0.049	*0.042	*0.044	*0.037	*0.041	-	26,812	29,398	32,805	25,447	22,854
Germany		15	-	-	-	-		114,760	-	-	-	-
Germany (Western Part)	14	14	-	-	-	-	132,123	131,787	-	-	-	-
Ghana	-	-	-	0.163	0.133	0.140	-	-	-	7,426	5,883	6,003
Greece	*0.500	*0.510	*0.535	*0.488	*0.449	*0.599	67,426	72,241	77,240	74,243	72,326	87,622
Guatemala	-	0.015	0.015	0.022	0.025	0.044	-	10,058	9,633	14,215	16,614	19,853
Honduras	0.080	0.082	0.107	0.105	0.106	0.109	9,890	11,872	11,830	10,535	9,728	8,908
Hungary	±0.250	±0.305	±0.389	±0.374	±0.425	±0.418	16,665	16,949	20,461	20,781	22,530	22,496
Iceland	0.004	0.004	±0.004	±0.003	-	-	114,758	158,132	222,329	119,764	-	-
India	*0.482	*0.365	*0.385	*0.412	*0.446	*0.537	7,673	6,143	5,859	6,238	6,753	8,104
Indonesia	3.980	4.763	5.117	5.803	6.352	6.511	12,222	13,939	13,822	15,348	16,149	16,566
Iran (Islamic Republic of)	0.237	0.152	0.134	0.104			21,013	18,474	20,996	17,042		
Iraq	-	*0.002	*0.003				-	20,107	36,771	-		
Ireland	*0.436	*0.440	*0.493	*0.434	*0.502	*0.580	96,849	95,631	104,546	91,261	96,609	106,769
Israel	0.303	0.351	0.362	0.375	0.410	0.467	57,085	56,545	60,252	61,459	65,050	71,465
Italy	±5.285	±5.564	±6.578	±5.430	±6.051	-	142,350	149,004	160,539	134,815	146,561	-
Jamaica	0.008	0.007	0.004	-	-	-	14,073	11,085	5,381	-	-	-
Japan	±36	±39	±40	±46	±50	±53	119,075	132,312	140,046	164,230	181,409	195,173
Jordan	0.010	0.014	0.020	0.017	0.026	0.031	16,421	14,653	15,632	16,809	17,045	16,112
Kenya	*0.077	*0.073	*0.047	*0.032	*0.036	*0.075	8,881	8,513	5,520	3,658	4,077	8,236
Korea, Republic of	2.458	2.933	2.781	3.403	3.626	4.047	60,553	72,392	74,502	84,079	91,579	106,307
Kuwait	0.018	0.013	0.039	0.031	0.031	0.048	13,429	18,279	32,450	26,778	28,574	40,437

Continued.

Output and Output per Employee

- Continued -

Country	Output ($ bil.)						Output per Employee ($)					
	1990	1991	1992	1993	1994	1995	1990	1991	1992	1993	1994	1995
Kyrgyzstan	0.033	0.027	0.015	0.003	0.001	-	7,920	7,313	5,977	1,088	961	-
Latvia	0.002	0.004	0.039	0.062	0.119	0.193	104	259	2,511	4,952	8,556	12,313
Lithuania	-	-	0.064	0.028	0.106	-	-	-	7,713	3,849	5,673	-
Macau	0.003	0.002	0.003	0.002	0.001	0.001	15,535	17,688	24,584	19,556	11,766	20,428
Malaysia	*1.951	*2.125	*2.656	*3.792	*4.110	*4.707	21,892	22,138	24,300	29,649	29,959	32,498
Malta	0.005	0.004	0.005	0.004	0.005	-	29,594	30,186	39,413	35,632	37,868	-
Mauritius	0.010	0.007	0.008	0.008	0.006	0.009	10,598	9,017	10,877	9,477	9,203	15,911
Mexico	0.130	0.118	0.114	0.112	0.114	0.098	26,617	25,402	27,032	34,374	37,155	26,451
Mongolia	0.072	0.051	0.022	0.006	0.008	0.006	10,594	7,257	4,708	1,485	1,150	1,123
Mozambique	-	±0.003	-	±0.001	±0.001	-	-	1,246	-	745	428	-
Myanmar	0.021	0.023	0.025	0.029	0.056	-	1,301	1,469	1,668	2,004	3,836	-
Namibia	-	-	-	-	0.010	-	-	-	-	-	37,682	-
Nepal	0.006	0.017	-	0.010	0.010	-	2,772	5,045	-	4,913	4,059	-
Netherlands	1.397	1.392	1.575	1.391	1.422	1.547	110,952	106,920	119,216	116,340	118,973	137,421
Norway	±2.036	±1.730	±1.485	±1.326	±1.679	±1.954	133,586	124,141	134,807	128,865	152,034	173,561
Pakistan	0.035	0.033	0.036	-	-	-	11,447	11,497	11,930	-	-	-
Panama	0.012	0.012	0.015	0.015	0.017	0.015	18,738	20,918	23,475	23,450	25,387	24,021
Peru	0.073	0.079	0.110	-	-	-	13,459	17,420	19,991	-	-	-
Philippines	0.447	0.467	0.390	0.390	-	-	8,996	9,903	11,125	13,168	-	-
Poland	±0.835	±0.944	±0.878	±0.917	±1.075	±1.408	12,839	16,271	15,400	16,665	20,131	27,153
Portugal	2.304	2.465	2.495	2.209	2.422	2.952	41,977	42,425	48,242	43,040	47,371	57,226
Qatar	0.030	0.024	0.010	0.009	0.018	-	16,789	16,726	14,149	14,100	25,529	-
Republic of Moldova	0.000	0.112	0.006	0.006	0.007	0.006	0.019	19,632	1,055	1,448	4,612	4,090
Romania	±0.713	±0.573	±0.398	±0.456	-	-	7,588	6,575	4,497	5,738	-	-
Russian Federation	-	6.510	3.999	1.537	2.182	2.570	-	-	-	2,983	4,644	5,693
Senegal	0.004	-	-	-	-	-	52,859	-	-	-	-	-
Sierra Leone	-	-	-	0.001	-	-	-	-	-	1,774	-	-
Singapore	*0.193	*0.172	*0.180	*0.180	*0.179	*0.195	75,026	78,116	85,374	96,418	105,901	118,079
Slovakia	-	±0.244	±0.212	±0.159	±0.170	-	-	11,990	11,508	10,944	11,444	-
Slovenia	0.730	0.580	0.460	0.460	0.505	0.604	37,231	37,062	31,000	33,077	40,093	50,706
South Africa	±1.170	±1.162	±1.142	±1.099	±1.137	±1.290	19,179	19,703	20,029	18,623	17,221	19,192
Spain	*5.835	*5.713	*5.825	±5.280	±5.523	±7.126	79,110	79,736	85,755	67,690	74,421	92,337
Sri Lanka	0.013	0.011	0.014	0.012	-	-	3,417	2,968	3,286	2,766	-	-
Suriname	*0.022	*0.053	*0.061	0.101	-	-	19,182	43,024	48,196	85,530	-	-
Sweden	*7.697	*7.398	*7.024	*4.407	*5.387	*6.532	174,937	182,428	188,744	146,736	174,305	199,759
Thailand	0.623	0.899	-	-	-	-	23,420	21,424	-	-	-	-
Tonga	0.000	0.000	0.000	0.000	0.000	-	3,617	9,399	3,317	3,670	3,016	-
Trinidad and Tobago	±0.013	±0.008	±0.007	±0.006	±0.007	-	13,408	11,253	10,496	8,441	9,991	-
Turkey	0.599	0.585	0.837	0.958	0.536	0.717	43,928	47,933	55,254	67,796	40,171	53,111
Ukraine	*1.251	*1.124	*0.866	*0.230	*0.210	*0.179	15,074	14,046	11,863	3,389	3,441	3,450
United Kingdom	*8.346	*7.343	7.154	6.640	8.104	8.173	107,005	107,991	108,390	94,852	103,899	109,246
United Republic of Tanzania	*0.013	*0.017	-	-	-	-	2,654	3,866	-	-	-	-
USSR (Former)	±7.188	-	-	-	-	-	10,570	-	-	-	-	-
United States of America	*54	*52	*60	*68	*75	*75	106,969	111,128	123,784	136,294	144,029	139,989
Uruguay	0.038	0.036	0.030	0.032	0.036	0.038	15,103	13,368	11,947	14,404	15,860	17,686
Venezuela	0.090	0.111	0.148	0.103	0.082	0.126	13,279	15,442	19,267	16,061	12,600	18,263
Yugoslavia	±0.044	±0.047	±0.113	-	±0.210	±0.120	-	2,078	5,629	-	10,870	6,188
Zambia	0.015	-	-	-	0.018	-	3,565	-	-	-	5,914	-
Zimbabwe	*0.082	*0.074	*0.066	*0.061	*0.091	*0.113	14,796	13,217	11,743	10,522	13,316	15,425

Depending on the table, * means *Enterprises, Engaged,* or *Factor Values;* ± means *Basis Unspecified;* § means *shown in millions.* For additional notes and sources, see Appendix I.

Value Added and Value Added per Employee

Country	Value Added ($ bil.)						Value Added per Employee ($)					
	1990	1991	1992	1993	1994	1995	1990	1991	1992	1993	1994	1995
Argentina	±0.255	-	-	0.348	-	-	10,100	-	-	18,934	-	-
Armenia	0.000	0.001	-	-	-	-	164	317	-	-	-	-
Australia	*1.728	*1.705	*1.808	-	-	-	39,276	40,588	45,210	-	-	-
Austria	0.879	0.849	0.878	0.736	0.810	0.953	45,548	49,967	51,667	46,912	52,172	60,404
Bangladesh	0.014	0.011	0.014				1,062	888	1,029			
Belgium	*0.321	*0.298	*0.300	*0.274	*0.285	*0.335	25,103	23,260	23,792	-	-	-
Belize	0.002	0.003	0.003				-	4,478	4,870			
Bolivia	0.015	0.016	0.017	0.019	0.022	0.021	5,988	6,097	6,171	6,355	7,501	7,985
Bosnia & Herzegovina	±0.231	±0.255	-	-			6,875	8,142		-		
Brazil	*1.215	-	*1.053	*1.769	-	-	15,238	-	13,375	21,112	-	-
Bulgaria		±0.043	±0.044	±0.054		-	-	2,334	2,721	3,854	-	
Cameroon	0.084	0.082	0.032	0.050	0.077	0.062	12,730	15,847	5,985	9,562	13,926	10,167
Canada	4.465	3.998	4.650	6.108	7.052	7.037	43,394	43,929	49,464	61,081	65,905	65,788
Central African Republic	*0.012						6,859					
Chile	0.270	0.289	0.325	0.421	0.467	0.518	12,719	14,596	15,686	17,151	20,702	22,746
China	±0.502	±0.528	±0.669	±1.655	±1.155	±1.138	718	695	917	2,267	1,500	1,560
China (Hong Kong SAR)	0.038	0.029	0.036	0.033	0.033	-	10,967	11,581	13,493	15,043	18,259	-
China (Taiwan Province)	0.683	0.732	0.674	0.587	0.466	0.469	10,936	13,298	14,684	15,262	12,983	14,765
Colombia	0.054	0.060	0.068	0.087	0.099	0.097	8,612	9,097	8,491	11,240	13,186	12,928
Costa Rica	0.022	0.021	0.023	0.023	0.025·	0.022	4,831	5,232	5,418	5,030	-	-
Cote d'Ivoire	0.257	0.223	0.242	0.237	-	-	-	-	-	-	-	-
Croatia	±0.186	±0.193	±0.148	-	-	-	8,397	11,576	9,095	-	-	-
Cyprus	0.039	0.043	0.047	0.045	0.048	0.059	20,295	23,778	21,920	20,184	20,906	23,500
Czechoslovakia (Former)	±0.289	-	-	-	-	-	5,153	-	-	-	-	-
Denmark	*0.486	*0.489	*0.560	*0.540	*0.686	*0.873	33,887	34,664	40,017	38,128	44,005	52,907
Ecuador	0.016	0.017	0.021	0.020	0.025	0.042	4,857	5,238	6,216	6,336	7,620	11,366
Egypt	*0.017	*0.011	*0.024	*0.005	-	-	1,971	1,758	3,576	914	-	-
El Salvador		0.001	-	0.000	0.003	0.002	-	-	-	4,063	9,389	13,201
Ethiopia	0.008	0.007	0.004	0.004	0.007	0.007	3,243	2,870	1,152	1,351	2,673	2,923
Fiji	0.011	0.012	0.011	0.013	0.015		7,353	9,660	10,162	11,678	13,104	-
Finland	±1.578	±0.951	±1.015	±0.992	±1.381	±1.365	51,923	34,846	42,101	43,696	57,560	54,819
FYR Macedonia	±0.011	±0.011	±0.008	±0.005	±0.006	±0.006	4,972	7,355	4,238	3,744	4,134	4,315
France	±4.183	±4.000	±4.480	±4.043	±4.188	±5.057	44,877	42,198	50,847	49,189	52,285	63,530
Gabon	-	*0.015	*0.009	*0.014	*0.012	*0.014	-	8,334	6,523	10,119	8,261	7,556
Germany (Western Part)	6.179	6.723	7.654	7.452	-	-	59,496	62,275	69,886	-	-	-
Ghana	-	-	-	0.093	0.081	0.085	-	-	-	4,229	3,564	3,637
Greece	*0.176	*0.185	*0.195	*0.178	*0.164	*0.219	23,697	26,163	28,179	27,118	26,409	32,082
Honduras	*0.022	*0.023	*0.027	*0.026	*0.025	*0.028	2,699	3,288	2,981	2,654	2,266	2,312
Hungary	±0.088	±0.093	±0.102	±0.107	±0.120	±0.117	5,872	5,146	5,378	5,954	6,363	6,307
Iceland	0.001	0.001	±0.002	±0.001	-	-	27,942	49,156	90,771	52,988	-	-
India	*0.102	*0.077	*0.075	*0.088	*0.095	*0.114	1,617	1,294	1,146	1,335	1,441	1,721
Indonesia	*1.382	*1.599	*1.905	*1.941	*2.158	*2.323	4,243	4,678	5,146	5,133	5,487	5,910
Iran (Islamic Republic of)	0.114	0.070	0.067	0.058	-	-	10,080	8,527	10,498	9,529	-	-
Iraq	-	*0.001	*0.001	-	-	-	-	11,437	10,059	-	-	-
Ireland	*0.170	*0.174	*0.195	*0.172	*0.198	*0.229	37,848	37,824	41,332	36,074	38,157	42,150
Israel	0.116	0.142	0.127	0.137	0.132	0.161	21,898	22,858	21,146	22,417	20,875	24,621
Italy	±1.616	±1.660	±2.065	±1.699	±1.778	-	43,523	44,466	50,385	42,174	43,059	-
Jamaica	0.005	0.004	0.002	-	-	-	9,208	5,737	2,799	-	-	-
Japan	±14	±15	±16	±18	±20	±21	46,226	52,291	55,658	63,587	71,119	77,598
Jordan	0.004	0.005	0.008	0.006	0.012	0.010	6,405	5,248	6,224	5,794	7,577	5,100
Kenya	*0.017	*0.016	*0.014	*0.009	*0.012	*0.015	1,917	1,854	1,598	1,074	1,413	1,613
Korea, Republic of	0.876	1.229	1.102	1.328	1.464	1.683	21,566	30,344	29,531	32,803	36,978	44,213
Kuwait	0.010	0.005	0.015	0.012	0.014	0.019	7,639	7,762	12,945	10,204	13,006	16,425
Latvia	-	-	-	*0.029	*0.056	*0.079	-	-	-	2,325	4,013	5,006
Macau	0.002	0.001	0.001	0.001	0.001	0.000	7,519	8,433	9,155	6,792	5,556	8,669
Malaysia	*0.584	*0.650	*0.758	*1.232	*1.231	*1.407	6,556	6,771	6,932	9,634	8,969	9,714
Malta	0.002	0.002	0.002	0.002	0.002	-	15,241	15,187	20,100	20,207	19,699	-
Mauritius	0.005	0.003	0.004	0.004	0.003	0.004	5,502	4,823	4,919	4,692	3,989	7,357
Mexico	0.053	0.051	0.047	0.045	0.047	0.040	10,937	11,064	11,272	13,907	15,223	10,827
Mongolia	0.022	0.015	0.006	0.001	0.003	0.002	3,162	2,083	1,282	380	447	392

Continued.

Value Added and Value Added per Employee

- Continued -

Country	Value Added ($ bil.)						Value Added per Employee ($)					
	1990	1991	1992	1993	1994	1995	1990	1991	1992	1993	1994	1995
Myanmar	±0.024	±0.024	±0.028	±0.016	±0.025	-	1,516	1,550	1,831	1,101	1,701	-
Namibia	-	-	-	-	0.005	-	-	-	-	-	19,512	-
Nepal	0.003	0.007	-	0.006	0.004	-	1,146	2,167	-	2,921	1,739	-
Netherlands	0.477	0.487	0.558	0.497	0.508	0.556	37,900	37,420	42,264	41,573	42,501	49,376
New Zealand	0.323	-	-	-	-	-	25,924	-	-	-	-	-
Norway	±0.619	±0.506	±0.442	±0.411	±0.522	±0.604	40,622	36,343	40,127	39,936	47,248	53,638
Pakistan	0.013	0.012	0.014	-	-	-	4,351	4,292	4,580	-	-	-
Panama	0.004	0.004	0.005	0.005	0.005	0.005	6,674	6,594	7,315	7,312	7,858	7,488
Peru	0.039	0.025	0.036	-	-	-	7,282	5,497	6,622	-	-	-
Philippines	0.164	0.182	0.140	0.138	-	-	3,309	3,850	3,984	4,659	-	-
Poland	±0.325	±0.321	±0.324	±0.315	±0.367	±0.478	4,996	5,541	5,677	5,729	6,867	9,219
Portugal	0.532	0.615	0.625	0.559	0.617	0.750	9,694	10,586	12,079	10,896	12,062	14,546
Qatar	0.017	0.011	0.004	0.004	0.007	-	9,310	7,882	5,502	5,913	9,907	-
Romania	±0.312	±0.242	±0.178	±0.219	-	-	3,320	2,777	2,005	2,756	-	-
Russian Federation	-	-	-	±0.850	±0.911	±0.962	-	-	-	1,649	1,939	2,131
Senegal	0.001	0.001	0.001	0.001	0.001	-	8,091	8,302	10,762	4,688	10,595	-
Sierra Leone	-	-	-	0.000	-	-	-	-	-	543	-	-
Singapore	*0.055	*0.046	*0.047	*0.049	*0.052	*0.056	21,182	20,775	22,483	26,098	30,515	34,008
Slovakia	-	-	-	±0.064	±0.062	-	-	-	-	4,429	4,185	-
Slovenia	0.193	0.147	0.128	0.130	0.147	0.161	9,845	9,425	8,622	9,331	11,666	13,497
South Africa	*0.468	*0.361	*0.361	*0.347	*0.359	*0.407	7,679	6,120	6,330	5,883	5,437	6,056
Spain	*2.164	*2.135	*2.177	±1.799	±1.787	±2.207	29,334	29,797	32,053	23,064	24,075	28,596
Sri Lanka	0.009	0.008	0.009	0.008	-	-	2,261	2,145	2,203	1,887	-	-
Sweden	*3.046	*1.949	*1.779	*1.320	*1.723	*2.053	69,230	48,069	47,801	43,957	55,752	62,781
Thailand	0.186	0.394	-	-	-	-	7,003	9,389	-	-	-	-
Trinidad and Tobago	-	±0.004	±0.003	±0.003	±0.004	-	-	5,478	4,836	4,246	5,958	-
Turkey	0.187	0.181	0.291	0.334	0.149	0.199	13,752	14,832	19,196	23,614	11,185	14,755
United Kingdom	*3.214	*2.853	*2.825	*2.794	*3.260	*3.285	41,209	41,957	42,804	39,918	41,799	43,908
United Republic of Tanzania	*0.003	*0.002	-	-	-	-	701	383	-	-	-	-
United States of America	*21	*20	*24	*28	*30	*29	41,004	42,298	50,792	55,428	57,082	54,211
Uruguay	0.019	0.019	0.016	0.017	0.019	0.020	7,693	7,039	6,134	7,730	8,338	9,298
Venezuela	0.036	0.042	0.056	0.036	0.033	0.099	5,324	5,786	7,328	5,675	5,103	14,366
Yugoslavia	±0.017	±0.021	±0.061	-	±0.078	±0.046	-	918	3,016	-	4,022	2,378
Zambia	0.008	-	-	-	0.010	-	1,733	-	-	-	3,287	-
Zimbabwe	*0.043	*0.042	*0.039	*0.037	*0.047	*0.059	7,773	7,601	6,969	6,316	6,931	8,035

Capital Investment

Country	Gross Fixed Capital Formation ($ mil.)						Per Establishment ($ mil)					
	1990	1991	1992	1993	1994	1995	1990	1991	1992	1993	1994	1995
Argentina	-	-	-	38	-	-	-	-	-	0.007	-	-
Australia	179	-	-	-	-	-	0.070	-	-	-	-	-
Austria	262	247	224	175	179	-	0.131	0.126	0.119	0.097	0.101	-
Bangladesh	0.781	0.355	1.001	-	-	-	0.001	0.000	0.001	-	-	-
Belgium	75	34	30	23	25	35	-	-	-	-	-	-
Bolivia	2.392	1.877	1.125	1.989	1.623	-	0.017	0.013	0.007	0.012	0.010	-
Bosnia & Herzegovina	25	-	-	-	-	-	0.239	-	-	-	-	-
Brazil	59	-	62	103	-	-	0.056	-	0.072	0.124	-	-
Bulgaria	15	1.546	82	35	3.571	3.015	0.249	0.026	1.111	0.026	0.002	0.001
Cameroon	-	-	-	-	12	27	-	-	-	-	0.625	1.438
Canada	539	-	-	-	-	-	0.202	-	-	-	-	-
Cape Verde	-	0.000	-	-	-	-	-	0.000	-	-	-	-
Chile	33	19	27	61	58	-	0.263	0.160	0.212	0.407	0.380	-
China (Hong Kong SAR)	4.236	2.445	1.034	1.551	2.458	-	0.006	0.004	0.002	0.003	0.005	-
Colombia	9.161	2.294	5.505	-0.091	6.949	-	0.050	0.013	0.028	-0.001	0.037	-
Cote d'Ivoire	11	11	11	11	-	-	-	-	-	-	-	-
Croatia	9.807	4.024	4.593	7.561	3.820	1.958	0.027	0.009	0.007	0.009	0.004	0.002
Cyprus	2.571	5.557	4.818	4.247	3.062	3.419	0.003	0.006	0.005	0.004	0.003	0.003

Continued.

   Depending on the table, * means *Enterprises, Engaged,* or *Factor Values*; ± means *Basis Unspecified*; § means *shown in millions*. For additional notes and sources, see Appendix I.

Capital Investment

- Continued -

Country	Gross Fixed Capital Formation ($ mil.)						Per Establishment ($ mil)					
	1990	1991	1992	1993	1994	1995	1990	1991	1992	1993	1994	1995
Czechoslovakia (Former) . . .	94	-	-	-	-	-	2.296	-	-	-	-	-
Denmark	88	68	-	-	-	-	0.085	0.065	-	-	-	-
Ecuador	11	7.558	10	7.280	16	16	0.202	0.137	0.189	0.149	0.351	0.237
Egypt	0.750	0.510	4.014	-1.394	-	-	0.006	0.005	0.037	-0.013	-	-
El Salvador.	-	-	-	0.060	0.225	0.117	-	-	-	0.020	0.032	0.019
Ethiopia.	0.411	0.121	0.027	0.010	0.043	0.097	0.037	0.015	0.003	0.001	0.002	0.004
Fiji	2.253	0.822	0.816	-	-	-	0.051	0.019	0.023	-	-	-
Finland	327	247	159	128	189	231	0.583	0.441	0.314	0.274	0.416	0.467
FYR Macedonia	3.415	0.882	0.801	0.403	0.417	0.442	0.263	0.046	0.026	0.012	0.007	0.006
Gabon	-	3.740	2.033	2.440	5.974	5.097	-	0.312	0.169	0.203	0.498	0.340
Germany (Western Part) . . .	-	844	1,007	-	-	-	-	0.304	0.360	-	-	-
Greece	43	44	43	-	-	-	0.174	0.180	0.175	-	-	-
Hungary	23	15	18	18	-	-	0.076	0.022	0.024	0.020	-	-
India	16	10	18	24	-	-	0.005	0.003	0.005	0.007	-	-
Indonesia	299	379	435	318	311	185	0.223	0.303	0.309	0.216	0.196	0.106
Iran (Islamic Republic of) . . .	14	16	5.691	15	-	-	0.050	0.405	0.163	0.472	-	-
Ireland	21	20	-	-	-	-	0.093	0.090	-	-	-	-
Israel.	5.456	17	15	18	16	-	0.016	0.043	0.041	0.045	0.043	-
Italy	366	380	384	239	231	-	0.433	0.448	0.392	0.243	0.226	-
Japan	566	772	624	629	734	-	0.021	0.030	0.025	0.025	0.031	-
Jordan	0.020	0.119	-	0.098	0.053	0.029	0.000	0.000	-	0.000	0.000	0.000
Korea, Republic of	165	236	227	222	195	423	0.082	0.107	0.105	0.087	0.078	0.170
Kuwait	0.969	0.542	-0.290	0.795	0.205	-	0.012	0.008	-0.003	0.008	0.002	-
Latvia	0.028	0.064	1.715	3.831	9.630	20	0.000	0.000	0.008	0.009	0.036	0.033
Macau	0.082	0.232	0.007	0.000	0.017	0.000	0.002	0.005	0.000	0.000	0.001	0.000
Malaysia	180	254	338	431	604	-	0.250	0.330	0.419	0.483	0.658	-
Malta	0.025	0.012	0.003	0.026	0.061	-	0.000	0.000	0.000	0.000	0.001	-
Mexico	1.733	2.255	-	-	-	-	0.108	0.161	-	-	-	-
Mongolia	123	72	19	2.660	2.697	-	2.927	2.308	0.411	0.045	0.035	-
Morocco	20	11	10	-	-	-	-	-	-	-	-	-
Myanmar	1.578	2.567	2.922	3.333	14	-	0.016	0.026	0.030	0.034	0.139	-
Nepal	3.030	-	-	-	-	-	0.027	-	-	-	-	-
Netherlands	85	96	82	71	-	-	0.411	0.420	0.361	0.326	-	-
Norway	118	74	67	41	79	-	0.195	0.126	0.209	0.145	0.266	-
Pakistan	-	2.004	-	-	-	-	-	0.051	-	-	-	-
Panama.	0.343	0.070	-	-	-	-	0.020	0.005	-	-	-	-
Peru	0.329	1.126	1.484	-	-	-	0.000	0.002	0.002	-	-	-
Philippines	17	14	14	14	-	-	0.006	0.005	0.025	0.030	-	-
Poland	66	50	73	87	-	-	0.256	0.168	0.305	0.375	-	-
Portugal	177	133	157	91	126	-	0.023	0.018	0.021	0.012	0.017	-
Romania	-	-	14	23	27	-	-	-	0.011	0.017	0.008	-
Senegal.	0.029	-	-	-	-	-	0.015	-	-	-	-	-
Singapore	3.307	4.911	4.751	6.622	6.752	-	0.040	0.065	0.064	0.092	0.104	-
Slovenia	9.719	2.938	10	2.764	4.433	5.586	0.032	0.001	0.003	0.001	0.001	0.004
Spain	227	271	194	161	177	264	0.013	0.016	0.012	-	-	-
Sri Lanka	0.176	0.825	0.110	0.610	-	-	0.001	0.008	0.001	0.006	-	-
Thailand	168	91	-	-	-	-	0.322	0.106	-	-	-	-
Trinidad and Tobago	1.313	2.631	6.165	2.859	-	-	0.014	0.038	0.084	0.039	-	-
Turkey	32	15	48	47	42	-	0.268	0.129	0.168	0.180	0.172	-
Ukraine	18	23	84	38	6.233	-15.171	0.001	0.001	0.005	0.003	0.000	-0.001
United Kingdom	238	184	224	-	191	-	0.034	0.029	0.034	-	0.027	-
United Republic of Tanzania . .	4.686	6.060	-	-	-	-	0.059	0.090	-	-	-	-
United States of America . . .	1,380	1,230	1,275	1,444	1,860	2,165	-	-	0.050	-	-	-
Uruguay	1.321	0.496	0.820	0.847	-	-	-	-	-	-	-	-
Venezuela	3.412	12	6.201	2.653	2.417	34	0.013	0.042	0.020	0.011	0.012	0.138
Yugoslavia	2.150	1.807	1.653	-	1.850	1.253	0.006	0.002	0.001	-	0.001	0.000
Zambia	6.409	-	-	-	-	-	0.534	-	-	-	-	-
Zimbabwe	6.332	12	6.851	6.891	34	-	0.218	0.417	0.263	0.287	1.251	-

Depending on the table, * means *Enterprises*, *Engaged*, or *Factor Values*; ± means *Basis Unspecified*; § means *shown in millions*. For additional notes and sources, see Appendix I.

189

Indexes of Industrial Production

Country	Index of Industrial Production (1990=100)						Index of Employment (1990=100)					
	1990	1991	1992	1993	1994	1995	1990	1991	1992	1993	1994	1995
Albania	100	-	-	-	-	-	-	-	-	-	-	-
Algeria	100	66	54	67	55	45	-	-	-	-	-	-
Angola	100	-	-	-	-	-	-	-	-	-	-	-
Argentina	100	107	-	-	-	-	100	-	-	73	83	-
Armenia	100	-	-	-	-	-	100	68	-	-	-	-
Australia	100	97	92	96	-	-	100	95	91	-	-	-
Austria	100	109	112	109	111	113	100	88	88	81	80	82
Azerbaijan	-	-	-	-	-	-	100	103	118	88	68	-
Bahrain	100	-	-	-	-	-	-	-	-	-	-	-
Bangladesh	100	100	100	-	-	-	100	94	102	-	-	-
Barbados	100	100	100	100	100	100	-	-	-	-	-	-
Belgium	100	98	92	89	91	96	100	100	98	-	-	-
Belize	100	125	130	128	122	107	-	-	-	-	-	-
Benin	100	-	-	-	-	-	-	-	-	-	-	-
Bermuda	-	-	-	-	-	-	100	86	93	96	96	80
Bhutan	100	-	-	-	-	-	-	-	-	-	-	-
Bolivia	100	104	104	112	114	114	100	109	110	121	117	106
Bosnia & Herzegovina	-	-	-	-	-	-	100	93	-	-	2	-
Brazil	100	95	95	-	-	-	100	-	99	105	-	-
Bulgaria	100	83	74	65	75	74	100	75	66	57	60	56
Burkina-Faso	100	-	-	-	-	-	-	-	-	-	-	-
Cambodia	100	-	-	-	-	-	100	105	-	-	-	-
Cameroon	100	75	89	94	119	131	100	79	80	79	84	92
Canada	100	90	97	102	107	105	100	88	91	97	104	104
Central African Republic	100	127	108	91	142	-	100	-	-	-	-	-
Chile	100	94	82	70	72	70	100	93	97	116	106	107
China	100	-	-	-	-	-	100	109	104	104	110	104
China (Hong Kong SAR)	100	95	93	85	81	78	100	71	77	63	51	-
China (Taiwan Province)	100	104	89	73	58	45	100	88	74	62	57	51
Colombia	100	96	95	107	117	102	100	105	127	123	120	119
Costa Rica	100	92	93	88	97	84	100	88	92	102	-	-
Cote d'Ivoire	100	90	77	74	79	76	-	-	-	-	-	-
Croatia	100	-	-	-	-	-	100	75	73	73	71	67
Cuba	100	100	100	-	-	-	-	-	-	-	-	-
Cyprus	100	117	120	103	105	115	100	94	112	117	120	131
Czechoslovakia (Former)	100	76	60	-	-	-	100	79	-	-	-	-
Czech Republic	100	-	-	-	-	-	100	90	100	100	-	-
Dem. Rep. of the Congo	100	83	83	-	-	-	-	-	-	-	-	-
Denmark	100	100	104	106	124	136	100	98	98	99	109	115
Dominican Republic	100	-	-	-	-	-	-	-	-	-	-	-
Ecuador	100	112	131	141	-	-	100	101	101	97	101	114
Egypt	100	92	89	92	98	95	100	74	79	71	71	71
El Salvador	100	108	-	-	-	-	-	-	-	-	-	-
Ethiopia and Eritrea	100	67	67	-	-	-	-	-	-	-	-	-
Ethiopia	-	-	-	-	-	-	100	91	128	114	109	103
Fiji	100	94	84	97	104	100	100	85	74	76	76	-
Finland	100	78	81	95	108	105	100	90	79	75	79	82
FYR Macedonia	100	-	-	-	-	-	100	66	85	64	66	67
France	100	100	96	89	97	97	100	102	95	88	86	85
Gabon	100	100	100	-	-	-	-	-	-	-	-	-
Gambia	100	-	-	-	-	-	-	-	-	-	-	-
Germany (Eastern Part)	100	-	-	-	-	-	-	-	-	-	-	-
Germany (Western Part)	100	105	108	107	117	115	100	104	105	-	-	-
Ghana	100	-	-	-	-	-	-	-	-	-	-	-
Greece	100	102	99	91	83	97	100	95	93	89	84	92
Guatemala	100	146	146	-	-	-	-	-	-	-	-	-
Guyana	100	104	130	163	323	370	-	-	-	-	-	-
Haiti	100	-	-	-	-	-	-	-	-	-	-	-
Honduras	100	143	185	200	277	370	100	85	112	123	135	151
Hungary	100	116	123	136	153	148	100	120	127	120	126	124

Continued.

   Depending on the table, * means *Enterprises, Engaged,* or *Factor Values;* ± means *Basis Unspecified;* § means *shown in millions.* For additional notes and sources, see Appendix I.

Indexes of Industrial Production

- Continued -

Country	Index of Industrial Production (1990=100)						Index of Employment (1990=100)					
	1990	1991	1992	1993	1994	1995	1990	1991	1992	1993	1994	1995
Iceland	100	-	-	-	-	-	100	69	49	80	60	-
India	100	101	95	104	105	121	100	95	105	105	105	105
Indonesia	100	107	115	104	109	108	100	105	114	116	121	121
Iran (Islamic Republic of)	100	99	110	110	103	109	100	73	57	54	-	-
Iraq	100	-	-	-	-	-	-	-	-	-	-	-
Ireland	100	100	103	104	115	121	100	102	105	106	115	121
Israel	100	111	125	140	158	168	100	117	113	115	119	123
Italy	100	100	110	109	-	-	100	101	110	108	111	-
Jamaica	100	80	70	-	-	-	100	111	125	125	152	-
Japan	100	97	93	89	86	81	100	97	94	93	91	89
Jordan	100	50	75	77	67	72	100	152	205	164	248	316
Kenya	100	107	109	108	112	108	100	100	99	100	102	106
Korea, Republic of	100	108	103	86	88	87	100	100	92	100	98	94
Kuwait	100	-	-	-	-	-	100	53	91	89	84	90
Kyrgyzstan	100	-	-	-	-	-	100	88	59	63	36	-
Lao P.D.R.	100	95	119	-	-	-	-	-	-	-	-	-
Latvia	100	-	-	-	-	-	100	107	106	87	96	108
Lebanon	100	86	69	-	-	-	-	-	-	-	-	-
Liberia	100	100	100	-	-	-	-	-	-	-	-	-
Libyan Arab Jamahiriya	100	-	-	-	-	-	-	-	-	-	-	-
Lithuania	100	-	-	-	-	-	-	-	-	-	-	-
Luxembourg	100	106	90	96	102	94	-	-	-	-	-	-
Macau	100	-	-	-	-	-	100	62	48	38	56	25
Malawi	100	98	96	85	87	104	100	-	-	-	-	-
Malaysia	100	105	117	143	149	158	100	108	123	144	154	163
Malta	100	96	95	111	-	-	100	93	75	76	76	-
Mauritius	100	-	-	-	-	-	100	79	81	87	74	60
Mexico	100	104	106	97	102	81	100	95	86	67	63	76
Mongolia	100	72	46	41	32	39	100	102	68	56	96	82
Morocco	100	108	102	111	118	103	-	-	-	-	-	-
Mozambique	100	70	68	-	-	-	100	84	75	61	72	-
Myanmar	100	81	96	-	-	-	100	96	94	90	91	-
Namibia	100	-	-	-	-	-	-	-	-	-	-	-
Nepal	100	-	-	-	-	-	100	149	-	92	107	-
Netherlands	100	101	103	102	102	96	100	103	105	95	95	89
New Zealand	100	107	117	-	-	-	100	99	108	120	-	-
Nicaragua	100	80	80	-	-	-	-	-	-	-	-	-
Niger	100	-	-	-	-	-	-	-	-	-	-	-
Nigeria	100	100	100	-	-	-	-	-	-	-	-	-
Norway	100	91	91	92	101	103	100	91	72	68	72	74
Pakistan	100	106	102	-	-	-	100	96	100	-	-	-
Panama	100	98	118	116	133	114	100	90	98	97	104	95
Papua New Guinea	100	100	100	-	-	-	-	-	-	-	-	-
Paraguay	100	103	103	109	121	128	-	-	-	-	-	-
Peru	100	125	102	125	150	137	100	84	102	-	-	-
Philippines	100	113	110	127	107	110	100	95	71	60	-	-
Poland	100	164	204	212	234	258	100	89	88	85	82	80
Portugal	100	109	114	124	105	112	100	106	94	94	93	94
Puerto Rico	100	-	-	-	-	-	100	85	85	76	62	68
Qatar	100	-	-	-	-	-	100	79	39	34	40	-
Republic of Moldova	100	-	-	-	-	-	100	49	52	33	13	13
Romania	100	80	65	59	49	49	100	93	94	85	88	-
Russian Federation	100	-	-	-	-	-	-	-	-	-	-	-
Saudi Arabia	100	-	-	-	-	-	-	-	-	-	-	-
Senegal	100	110	108	106	90	103	100	110	96	248	99	-
Sierra Leone	100	83	83	-	-	-	-	-	-	-	-	-
Singapore	100	85	86	79	73	71	100	86	82	72	66	64
Slovakia	100	-	-	-	-	-	-	-	-	-	-	-
Slovenia	100	86	75	74	79	78	100	80	76	71	64	61
South Africa	100	98	93	94	98	103	100	97	93	97	108	110

Continued.

Depending on the table, * means *Enterprises, Engaged,* or *Factor Values*; ± means *Basis Unspecified*; § means *shown in millions*. For additional notes and sources, see Appendix I.

191

Indexes of Industrial Production

- Continued -

Country	Index of Industrial Production (1990=100)						Index of Employment (1990=100)					
	1990	1991	1992	1993	1994	1995	1990	1991	1992	1993	1994	1995
Spain	100	107	90	86	91	95	100	97	92	106	101	105
Sri Lanka	100	115	76	111	83	153	100	92	106	111	-	-
Suriname	100	92	98	-	-	-	100	109	111	104	-	-
Sweden	100	92	82	87	96	105	100	92	85	68	70	74
Switzerland	100	94	92	86	94	91	-	-	-	-	-	-
Syrian Arab Republic	100	40	35	101	73	113	-	-	-	-	-	-
Thailand	100	107	114	-	-	-	100	158	-	-	-	-
Togo	100	-	-	-	-	-	-	-	-	-	-	-
Tonga	100	-	-	-	-	-	100	45	131	129	114	-
Trinidad and Tobago	100	59	69	91	66	67	100	74	71	76	77	-
Tunisia	100	100	111	-	-	-	-	-	-	-	-	-
Turkey	100	96	97	97	94	109	100	90	111	104	98	99
Uganda	100	100	138	176	185	193	-	-	-	-	-	-
Ukraine	100	-	-	-	-	-	100	96	88	82	73	63
United Arab Emirates	100	-	-	-	-	-	-	-	-	-	-	-
United Kingdom	100	88	87	90	94	89	100	87	85	90	100	96
United Republic of Tanzania	100	77	63	50	41	-	100	90	-	-	-	-
USSR (Former)	100	-	-	-	-	-	100	-	-	-	-	-
United States of America	100	93	98	99	104	105	100	93	95	98	103	105
Uruguay	100	86	108	-	-	-	100	108	100	89	89	84
Venezuela	100	-	-	-	-	-	100	106	113	95	96	102
Yugoslavia	100	-	-	-	-	-	-	-	-	-	-	-
Yugoslavia (Former)	100	74	-	-	-	-	-	-	-	-	-	-
Zambia	100	101	106	109	76	71	100	-	-	-	72	-
Zimbabwe	100	113	118	106	118	128	100	100	101	104	122	132

Depending on the table, * means *Enterprises, Engaged,* or *Factor Values;* ± means *Basis Unspecified;* § means *shown in millions.* For additional notes and sources, see Appendix I.

Representative Companies in Sector

Name	Address	Telephone	Fax	Employment	Y
Stone-Consolidated Corp.	800 Blvd. Rene-Levesque O, Montreal, Quebec H3B 1X9, Canada	514-875-2160	514-875-6284	10,000	98
Kimball International Inc.	1600 Royal St., Jasper, Indiana 47549-1001, U.S.A.	812-482-1600		8,949	97
Paper Industries Corp. of the Philippines	Ortigas Complex, Pasig, Philippines	2 8181916	2 8175423	7,600	94
Oakwood Homes Corp.	PO Box 27081, Greensboro, North Carolina 27425-7081, U.S.A.	336-664-2400		7,561	97
SPX Corp.	PO Box 3301, Muskegon, Michigan 49443-3301, U.S.A.	616-724-5000	616-724-5720	7,125	96
Potlatch Corp.	301 W. Riverside Ave., 1100, Spokane, Washington 99201, U.S.A.	509-835-1500	509-835-1555	6,700	97
Associated Furniture Co. Ltd.	PO Box 98800, Sloane Park 2152, Republic of South Africa	11 7066004	11 4635097	6,500	96
Furnel International Ltd.	Al Jerozolimskie 65/79, Warsaw PL-00-697, Poland	22 300467	22 306525	6,300	93
Donohue Inc.	800-500 Rue Sherbrooke O, Montreal, Quebec H3A 3C6, Canada	418-847-7700	418-847-7754	6,000	98
Clayton Homes Inc.	PO Box 15169, Knoxville, Tennessee 37901, U.S.A.	423-970-7200	423-970-1238	5,991	97
Hunter PLC	32 Hampstead High St., London NW3, United Kingdom	71 7940677		5,983	90
Canfor Corp.	1055 Dunsmuir St. Ste. 3000, Vancouver, British Columbia V7X 1B5, Canada	604-661-5241	604-661-5472	5,700	98
West Fraser Timber Co. Ltd.	1100 Melville St. Ste. 1000, Vancouver, British Columbia V6E 4A6, Canada	604-895-2700	604-681-6061	5,500	98
Autoglass International UK Ltd.	Old Deer Park, Richmond TW9 2AZ, United Kingdom	81 9409177	81 9487913	5,269	93
Longaberger Co.	95 N. Chestnut St., Dresden, Ohio 43821-9600, U.S.A.	614-754-5000	614-754-5240	5,200	96
MascoTech Inc.	21001 Van Born Rd., Taylor, Michigan 48180, U.S.A.	313-274-7405	313-274-8959	5,100	96
Hollis PLC	Headington Hill Hall OX3 0BW, United Kingdom	865 64881		5,050	90
Canadian Forest Products Ltd.	2900-1055 Dunsmuir St., Vancouver, British Columbia V7X 1B5, Canada	604-661-5241	604-661-5235	5,000	98
Griffon Corp.	100 Jericho Quadrangle, Jericho, New York 11753, U.S.A.	516-938-5544	516-938-5644	5,000	97
Triangle Pacific Corp.	PO Box 660100, Dallas, Texas 75266-0100, U.S.A.	214-887-2000		4,967	96
Masonite Corp.	1 S. Wacker Dr., Chicago, Illinois 60606, U.S.A.	312-750-0900	312-750-0958	4,900	97
Weyerhaeuser Canada Ltd.	1075 W Georgia St. 25th Floor, Vancouver, British Columbia V6E 3C9, Canada	604-687-0431	604-691-2445	4,800	98
Greif Bros. Corp.	425 Winter Rd., Delaware, Ohio 43015, U.S.A.	740-549-6000		4,500	97
Butler Manufacturing Co.	PO Box 419917, Kansas City, Missouri 64141-0917, U.S.A.	816-968-3000		4,350	96
Nusantara Plywood PT	Jl. Haji Fachrudin, No. 19, Jakarta 10250, Indonesia	21337339	21 330048	4,342	97
Coachmen Industries Inc.	PO Box 3300, Elkhart, Indiana 46515, U.S.A.	219-262-0123		4,274	97
Premdor Inc.	1600 Britannia Rd. E, Mississauga, Ontario L4W 1J2, Canada	905-670-6500	905-670-6520	4,250	98
Palm Harbor Homes Inc.	15303 Dallas Pkwy.,Ste. 800, Dallas, Texas 75248, U.S.A.	214-991-2422	214-991-5949	4,010	96
Ibstock PLC	Lutterworth House, Lutterworth LE17 4PS, United Kingdom	455 553071		3,681	94
Ford France SA	BP 32, Blanquefort F-33290, France	56 954000	56 954378	3,624	91
Andersen Corp.	100 4th Ave. N., Bayport, Minnesota 55003, U.S.A.	612-439-5150	612-430-5107	3,600	97
HL & H Mining Timber	PO Box 5906, Johannesburg 2000, Republic of South Africa	11 4800400	11 4800601	3,500	97
Skyline Corp.	PO Box 743, Elkhart, Indiana 46515-0743, U.S.A.	219-294-6521		3,500	97
Weldwood of Canada Ltd.	1055 Hastings St. W, Vancouver, British Columbia V6B 3V8, Canada	604-687-7366	604-662-2798	3,500	98
Magnet Ltd.	Roydings Ave., Keighley BD21 4BY, United Kingdom	535 661133	535 662896	3,227	92
Atlantic Veneer do Brasil SA	Rodovia Br 101 Norte Km 264, Serra 29160-900, Brazil	27 3280222	27 3284256	3,200	97
Norco Windows Inc.	200 E. Mallard Dr., Boise, Idaho 83706-3980, U.S.A.	208-345-8251		3,150	97
TJ International Inc.	PO Box 65, Boise, Idaho 83707, U.S.A.	208-364-3300	208-364-3370	3,150	97
International Forest Products Ltd.	1055 Dunsmuir St. Ste. 3500, Vancouver, British Columbia V7X 1H7, Canada	604-689-6800	604-688-0313	3,100	98
Bukoza AS	Ul. Hencovska, Vranov nad Toplou 090 02, Slovakia	931 23241	931 23447	3,000	93
Clopay Corp.	312 Walnut St., 1600, Cincinnati, Ohio 45202-4036, U.S.A.	513-381-4800	513-762-3965	3,000	97
Doman Industries Ltd.	300-435 Trunk Rd., Duncan, British Columbia V9L 2P9, Canada	604-748-3711	604-748-6045	3,000	98
E.B. Eddy Forest Products Ltd.	1600 Scott St., Ottawa, Ontario K1Y 4N7, Canada	613-725-6700	613-725-6820	3,000	98
Mannington Mills Inc.	PO Box 30, Salem, New Jersey 08079, U.S.A.	609-935-3000	609-339-5813	3,000	95
Steeledale Reinforcing & Engineering Industries Ltd.	Tulisa Park, Johannesburg 2197, Republic of South Africa	11 6132331	11 6131394	2,950	93
Erna Djuliawati PT	Jl. Abdul Muis No. 50, Jakarta 10160, Indonesia	213800051	21 373887	2,904	97
Rayonier Inc.	1177 Summer St., Stamford, Connecticut 06905, U.S.A.	203-348-7000		2,700	96

Continued

Representative Companies in Sector
- Continued -

Name	Address	Tele-phone	Fax	Employ-ment	Y
Universal Forest Products, Inc.	2801 E Beltline, NE, Grand Rapids, Michigan 49505, U.S.A.	616-364-6161	616-361-7534	2,700	96
Erskine House Group PLC	Oak Hill Rd., Sevenoaks TN13 1NW, United Kingdom	732 460044	732 461108	2,677	93
Intercraft Co.	1 Intercraft Plz., Taylor, Texas 76574, U.S.A.	512-352-8501	512-352-4870	2,670	97
American Homestar Corp.	PO Box 580484, Houston, Texas 77258-0484, U.S.A.	281-334-9700	281-334-9737	2,661	97
Perawang Lumber PT	Jl. Gatot Subroto, No. 5, Pekanbaru, Indonesia	76133322	761 31869	2,625	97
Pacific Coast Building Products Inc.	PO Box 160488, Sacramento, California 95816, U.S.A.	916-444-9304	916-325-3697	2,600	97
Valhi Inc.	5430 Lyndon B. Johnson Fwy., 17, Dallas, Texas 75240-2697, U.S.A.	972-233-1700	972-385-0586	2,600	97
Schult Homes Corp.	PO Box 151, Middlebury, Indiana 46540, U.S.A.	219-825-5881		2,563	97
RA Multiproduct	Sector 2, Bucharest, Romania	1 6873490	1 6873490	2,508	93
Hiag AG	Riehen CH-4125, Switzerland	41 61496799	41 61497887	2,500	91
MacMillan Bloedel Inc.	PO Box 235016, Montgomery, Alabama 36123-5016, U.S.A.	334-213-6100	334-262-0976	2,500	97
Potlatch Corp. Western Wood Products	PO Box 1016, Lewiston, Idaho 83501-1016, U.S.A.	208-799-1850	208-799-1837	2,500	95
John Carr Group PLC	169 Watch House Ln., Doncaster DN5 9LR, United Kingdom	302 783333	302 787383	2,481	93
Federated Co-operatives Ltd.	401 22nd St. E, Saskatoon, Saskatchewan S7K 0H2, Canada	306-244-3311	306-244-3403	2,450	98
Anglian Group PLC	Horsford, Norwich NR10 3AQ, United Kingdom			2,407	94
Plum Creek Timber Company L.P.	999 3rd Ave., 2300, Seattle, Washington 98104-4096, U.S.A.	206-467-3600	206-467-3795	2,400	97
HLTH Timber Processors Pty. Ltd.	PO Box 784460, Sandton 2146, Republic of South Africa	11 8845220	11 8836157	2,355	94
Northwood Inc.	PO Box 9000 Sta. A, Prince George, British Columbia V2L 4W2, Canada	250-962-9611	250-962-3582	2,350	98
Southern Energy Homes Inc.	PO Box 390, Addison, Alabama 35540, U.S.A.	205-747-8589	205-747-2963	2,338	97
Kodeco Batu Licin Plywood PT	Kav 58 Wisma Nusantara, 10th Flr., Jakarta 10045, Indonesia	21336905	21 335298	2,304	97
Pope and Talbot Inc.	1500 S.W. 1st Ave., Portland, Oregon 97201, U.S.A.	503-228-9161		2,300	96
Springs Window Fashions Div.	7549 Graber Rd., Middleton, Wisconsin 53562, U.S.A.	608-836-1011	608-831-2184	2,300	97
Bucina, AS	Areal Buclny 1335/21, Zvolen 960 96, Slovakia	855 27777	855 26951	2,275	93
International Paper Holdings (UK) Ltd.	Rd. 3 Industrial Estate, Winsford CW7 3RJ, United Kingdom			2,270	94
Belmont Homes Inc.	Hwy. 25 S., Industrial Park Dr., Belmont, Mississippi 38827, U.S.A.	601-454-9217		2,251	96
Buchanan Forest Products Ltd.	233 Court St. S, Thunder Bay, Ontario P7B 2X9, Canada	807-345-0571	807-345-4004	2,200	98
Satya Raya Indah Wood-Based Industries PT	Jl. Letjend S Parman Kav 62-63, Jakarta 11410, Indonesia	215707099	21 5482503	2,200	97
Simpson Timber Co.	215 N. 3rd St., Shelton, Washington 98584, U.S.A.	206-426-3381	206-427-8186	2,200	95
Jason Inc.	411 E. Wisconsin Ave., 2500, Milwaukee, Wisconsin 53202, U.S.A.	414-277-9300		2,172	97
American Woodmark Corp.	3102 Shawnee Dr., Winchester, Virginia 22601-4208, U.S.A.	540-665-9100		2,154	97
Cavalier Homes Inc.	PO Box 300, Addison, Alabama 35540, U.S.A.	205-747-1575	205-747-1605	2,000	95
Commonwealth Plywood Co. Ltd.	15 Blvd. Labelle, Ste-Therese-De-Blainville, Quebec J7E 2X1, Canada	450-435-6541	450-435-3814	2,000	98
Coastal Lumber Co.	PO Box 829, Weldon, North Carolina 27890, U.S.A.	919-536-4211	919-536-3102	1,900	97
PRIDE Industries Inc.	1 Sierragate Plz., A 200, Roseville, California 95678, U.S.A.	916-783-5266	916-783-8234	1,900	96
British Tissues Ltd.	43-51 Lowlands Rd., Harrow HA1 3BW, United Kingdom	81 8645411		1,824	87
ABT Building Products Corp.	1 Neenah Center, Ste. 600, Neenah, Wisconsin 54956-3070, U.S.A.	414-751-8611	414-751-0370	1,809	96
Drevoindustria AS	Dolne Rudiny 3, Zilina 010 97, Slovakia	89 30161	89 61086	1,800	93
West Lumber Company Inc.	5775 Glenridge Dr., Atlanta, Georgia 30328, U.S.A.	404-847-7801		1,800	95
Cashway Building Centres Ltd.	205 Peter St., Port Hope, Ontario L1A 3V6, Canada	905-885-3700	905-885-3720	1,700	98
Werner Ladder Co.	93 Werner Rd., Greenville, Pennsylvania 16125, U.S.A.	412-588-8600	412-588-0315	1,600	97
Morgan Products Ltd.	469 McLaws Cir., Williamsburg, Virginia 23185, U.S.A.	804-564-1700		1,593	97
Russ Berrie and Company Inc.	111 Bauer Dr., Oakland, New Jersey 07436, U.S.A.	201-337-9000	201-337-9634	1,580	96
Boulton & Paul Manufacturing Ltd.	Riverside Works, Norwich NR1 1EB, United Kingdom	603 660133	603 626972	1,561	93
Etz Lavud Ltd.	Harakevet Qut, Petah Tikva 49100, Israel	3 9234931	3 9231931	1,500	97
Horton Homes Inc.	PO Drawer 4410, Eatonton, Georgia 31024, U.S.A.	706-485-8506	706-485-4446	1,500	95
Junckers Industrier AS	Vaerftsvej 4A, Koge DK-4600, Denmark	53651895	53659936	1,500	93
Patrick Industries Inc.	PO Box 638, Elkhart, Indiana 46515, U.S.A.	219-294-7511	219-522-5213	1,497	96
African Timber & Plywood Ghana Ltd.	PO Box 1, Samreboi, Ghana	314833	31 4655	1,458	92
Western Veneer & Lumber Co. Ltd.	PO Box 99, Takoradi, Ghana	31 3564	31 4718	1,450	92
Stimson Lumber Co.	308 Pacific Bldg., Portland, Oregon 97204, U.S.A.	503-222-1676	503-222-2682	1,400	95

Continued

Representative Companies in Sector
- Continued -

Name	Address	Tele-phone	Fax	Employ-ment	Y
Wayne-Dalton Corp.	PO Box 67, Mount Hope, Ohio 44660, U.S.A.	216-674-7015	216-674-4983	1,400	95
Johnson Controls UK Ltd.	147 Victoria Rd., Swindon SN1 3BU, United Kingdom			1,360	95
R.G. Carter Holdings Ltd.	Drayton, Norwich NR8 5AH, United Kingdom	603 867355	603 260151	1,328	93
Tolko Industries Ltd.	3203 30 Ave., Vernon, British Columbia V1T 2C6, Canada	250-545-4411	250-545-4783	1,300	98
Seddon Group Ltd.	Crewe, Northwich CW4 7BA, United Kingdom	477 534422		1,290	94
Durospand Co., Ltd.	Bangsue Dusit, Bangkok 10800, Thailand	5854900	5870516	1,286	94
Simpson Manufacturing Company Inc.	4637 Chabot Dr., 200, Pleasanton, California 94588, U.S.A.	510-460-9912		1,272	97
Liberty Homes Inc.	PO Box 35, Goshen, Indiana 46527-0035, U.S.A.	219-533-0431	219-533-0438	1,250	96
Preglejka AS	Bystricka 657/7, Zarnovica 966 81, Slovakia	858 2281	858 2280	1,228	93
Yule Catto Group Ltd.	Temple Fields, Harlow CM20 2AH, United Kingdom	279 442791	279 641360	1,228	94
Ainsworth Lumber Co. Ltd.	Exeter Rd., 100 Mile House, British Columbia V0K 2E0, Canada	250-395-6200	250-395-6201	1,200	98
Chief Industries Inc.	PO Box 2078, Grand Island, Nebraska 68802-2078, U.S.A.	308-382-8820	308-389-7221	1,200	95
Ferrum SA	Espana 496, Avellaneda 1870, Argentina	1 2221500	1 2221516	1,200	97
Industry Modernization & Mining Enterprise, Budex	Sobieskiego 9, Czestochowa PL-42-200, Poland	34 44627	34 44761	1,200	93
Rose Art Industries Inc.	6 Regent St., Livingston, New Jersey 07039, U.S.A.	973-535-1313		1,200	97
Skeena Cellulose Inc.	666 Burrard St. Ste. 2300, Vancouver, British Columbia V6C 2X8, Canada	604-688-2225	604-688-0310	1,200	98
TimberWest Forest Ltd.	2300-1055 Georgia St. W, Vancouver, British Columbia V6E 3P3, Canada	604-654-4600	604-654-4571	1,200	98
Wellborn Cabinet Inc.	PO Box 1210, Ashland, Alabama 36251, U.S.A.	205-354-7151	205-354-7022	1,169	97
Binhdinh Exported Rattan Enterprise	94 Lam Son St., Quy Nhon City, Vietnam	56 22203		1,150	
WTD Industries Inc.	PO Box 5805, Portland, Oregon 97228-5805, U.S.A.	503-246-3440	503-245-4229	1,150	97
Baltek Corp.	PO Box 195, Northvale, New Jersey 07647-0195, U.S.A.	201-767-1400	201-387-6631	1,131	96
Conestoga Wood Specialties Inc.	PO Box 158, East Earl, Pennsylvania 17519, U.S.A.	717-445-6701	717-445-7096	1,100	94
Crestbrook Forest Industries Ltd.	220 Cranbrook St. N, Cranbrook, British Columbia V1C 3R2, Canada	604-426-6241	604-426-3406	1,100	98
Shelman Swiss-Hellenic Wood Products Manufacturers SA	32 Voucourestiou Str., Athens GR-106 71, Greece	3602211	3648017	1,097	
Piloimpregna	Festivalove Nam, Kosice 042 50, Slovakia	95 55824	95 775172	1,026	92
Spruce Falls Inc.	PO Box 100 Sta. Main, Kapuskasing, Ontario P5N 2Y2, Canada	705-337-1311	705-337-9700	1,010	98
Banks Lumber Company Inc.	PO Box 2299, Elkhart, Indiana 46515, U.S.A.	219-294-5671	219-294-1032	1,000	97
Batts Inc.	200 Franklin St., Zeeland, Michigan 49464, U.S.A.	616-772-4635		1,000	95
Panca Wana Indonesia PT	Jl. Jati Kalang Krian, Sidoardjo, Indonesia	318971046	319 333728	1,000	97
Therma-Tru Corp.	1687 Woodlands Dr., Maumee, Ohio 43537, U.S.A.	419-891-7400	419-891-7411	1,000	97
Timber Products Co.	PO Box 269, Springfield, Oregon 97477, U.S.A.	541-747-4577	503-747-4577	1,000	95
Weyerhaeuser Saskatchewan Ltd., Prince Albert Pulp & Paper	PO Box 3001, Prince Albert, Saskatchewan S6V 5T1, Canada	306-764-1521	306-764-8886	1,000	98
York Timbers Ltd.	PO Box 380, Pretoria 0001, Republic of South Africa	12 869730	12 869780	1,000	95
Placas do Parana SA	Rua Roberto Hauer 411, Vila Hauer, Curitiba 81610-180, Brazil	41 3213131	41 3213456	996	96
Harrow Industries Inc.	2627 E. Beltline Ave. S.E., Grand Rapids, Michigan 49546, U.S.A.	616-942-1440	616-942-2170	984	97
Blandin Paper Co.	115 1st St. S.W., Grand Rapids, Minnesota 55744-3699, U.S.A.	218-327-6200	218-327-6212	975	97
Commodore Corp.	PO Box 577, Goshen, Indiana 46526, U.S.A.	219-533-7100	219-534-2716	950	97
Malette Inc.	PO Box 1100, Timmins, Ontario P4N 7H9, Canada	705-268-1462	705-360-1258	950	98
Boskor Timber Processors	PO Box 1, Kleinbos 6310, Republic of South Africa	42 5411611	42 5411632	911	94
Northwest Hardwoods Div.	10220 S.W. Greenburg Rd., Ste. 5, Portland, Oregon 97223, U.S.A.	503-246-5700	503-245-5838	900	97
Eagon Industrial Co., Ltd.	Nam-ku, Incheon City, Republic of Korea	32 8708700	32 8689555	885	97
Timber Products Co. Medford	PO Box 1669, Medford, Oregon 97501, U.S.A.	541-773-6681	541-770-1509	876	95
Ponderosa Industrial SA de CV	Fracc. Hacienda Sta. Fa., Chihuahua 31237, Mexico	14 300011	14 300184	814	92
Besse Forest Products Group	PO Box 352, Gladstone, Michigan 49837, U.S.A.	906-428-3113	906-428-3310	800	97
Decorel Inc.	444 E. Courtland St., Mundelein, Illinois 60060, U.S.A.	708-566-4444		800	95
Indian Head Div.	PO Box 605, Newport, Vermont 05855, U.S.A.	802-334-6711		800	95

Product Tables
Sawnwood, Coniferous

Unit of Measure: 1,000 Cubic meters.

	1990 Value	%	1991 Value	%	1992 Value	%	1993 Value	%	1994 Value	%	1995 Value	%	1996 Value	%
Total Production	377,117	100.0	370,310	100.0	385,671	100.0	388,584	100.0	400,996	100.0	398,777	100.0	406,727	100.0
Regions														
Africa	2,472	0.7	2,403	0.6	2,408	0.6	2,279	0.6	2,438	0.6	2,521	0.6	2,506	0.6
America, North	140,776	37.3	132,171	35.7	139,976	36.3	139,313	35.9	144,316	36.0	138,218	34.7	145,151	35.7
America, South	11,291	3.0	11,819	3.2	12,012	3.1	12,037	3.1	12,460	3.1	13,320	3.3	13,332	3.3
Asia	144,707	38.4	141,028	38.1	145,940	37.8	148,917	38.3	148,393	37.0	148,113	37.1	149,649	36.8
Europe	74,222	19.7	79,221	21.4	81,244	21.1	81,484	21.0	88,552	22.1	91,464	22.9	90,920	22.4
Oceania	3,651	1.0	3,669	1.0	4,091	1.1	4,555	1.2	4,837	1.2	5,142	1.3	5,170	1.3
Leading Producers														
United States of America	84,500	22.4	78,260	21.1	81,481	21.1	77,736	20.0	80,493	20.1	75,982	19.1	80,004	19.7
Canada	53,702	14.2	51,037	13.8	55,512	14.4	58,651	15.1	60,648	15.1	59,343	14.9	61,927	15.2
Japan	26,421	7.0	25,075	6.8	24,200	6.3	23,298	6.0	22,984	5.7	21,730	5.4	21,730	5.3
China	14,778	3.9	11,526	3.1	11,180	2.9	15,294	3.9	15,229	3.8	15,229	3.8	16,341	4.0
Sweden	11,798	3.1	11,250	3.0	11,928	3.1	12,538	3.2	13,616	3.4	14,737	3.7	14,170	3.5

Sawnwood, Broadleaved

Unit of Measure: 1,000 Cubic meters.

	1990 Value	%	1991 Value	%	1992 Value	%	1993 Value	%	1994 Value	%	1995 Value	%	1996 Value	%
Total Production	132,583	100.0	120,787	100.0	133,873	100.0	131,368	100.0	130,495	100.0	127,732	100.0	118,225	100.0
Regions														
Africa	5,635	4.2	5,437	4.5	5,594	4.2	5,405	4.1	6,720	5.1	6,575	5.1	6,561	5.5
America, North	27,490	20.7	26,094	21.6	28,638	21.4	30,451	23.2	31,439	24.1	31,628	24.8	31,758	26.9
America, South	13,992	10.6	5,083	4.2	5,661	4.2	4,405	3.4	5,641	4.3	5,632	4.4	5,558	4.7
Asia	65,775	49.6	65,882	54.5	75,403	56.3	75,027	57.1	70,279	53.9	67,055	52.5	58,547	49.5
Europe	17,662	13.3	16,461	13.6	16,698	12.5	14,302	10.9	14,513	11.1	14,892	11.7	14,121	11.9
Oceania	2,030	1.5	1,828	1.5	1,879	1.4	1,779	1.4	1,902	1.5	1,949	1.5	1,679	1.4
Leading Producers														
United States of America	25,300	19.1	24,103	20.0	26,456	19.8	27,780	21.1	29,054	22.3	29,344	23.0	29,650	25.1
India	14,960	11.3	14,960	12.4	14,960	11.2	14,960	11.4	14,960	11.5	14,960	11.7	14,960	12.7
China	7,965	6.0	8,578	7.1	7,720	5.8	9,557	7.3	9,516	7.3	9,516	7.4	1,021	0.9
Malaysia	8,780	6.6	8,924	7.4	9,300	6.9	9,310	7.1	8,758	6.7	8,232	6.4	8,232	7.0
Indonesia	9,007	6.8	8,500	7.0	8,300	6.2	8,200	6.2	6,700	5.1	6,500	5.1	7,200	6.1

Veneer Sheets

Unit of Measure: Cubic meters.

	1990 Value	%	1991 Value	%	1992 Value	%	1993 Value	%	1994 Value	%	1995 Value	%	1996 Value	%
Total Production	5,553,298	100.0	5,442,848	100.0	6,124,361	100.0	7,097,171	100.0	7,258,738	100.0	6,931,614	100.0	7,026,944	100.0
Regions														
Africa	566,467	10.2	438,900	8.1	393,000	6.4	410,650	5.8	504,200	6.9	477,100	6.9	486,100	6.9
America, North	519,000	9.3	531,000	9.8	636,000	10.4	648,000	9.1	668,400	9.2	669,600	9.7	669,600	9.5
America, South	496,000	8.9	516,900	9.5	577,850	9.4	656,930	9.3	704,200	9.7	719,200	10.4	735,750	10.5
Asia	1,612,532	29.0	1,879,248	34.5	2,395,965	39.1	3,196,882	45.0	3,070,165	42.3	2,794,878	40.3	2,846,394	40.5
Europe	2,281,500	41.1	1,983,500	36.4	2,016,245	32.9	2,098,409	29.6	2,192,473	30.2	2,125,536	30.7	2,131,800	30.3
Oceania	77,800	1.4	93,300	1.7	105,300	1.7	86,300	1.2	119,300	1.6	145,300	2.1	157,300	2.2
Leading Producers														
Malaysia	480,000	8.6	705,000	13.0	1,200,000	19.6	2,122,000	29.9	2,067,000	28.5	1,800,000	26.0	1,800,000	25.6
Canada	501,000	9.0	501,000	9.2	501,000	8.2	501,000	7.1	501,000	6.9	501,000	7.2	501,000	7.1
Italy	622,000	11.2	436,000	8.0	483,000	7.9	500,000	7.0	500,000	6.9	500,000	7.2	500,000	7.1
Germany	494,000	8.9	442,000	8.1	419,000	6.8	380,000	5.4	392,000	5.4	392,000	5.7	392,000	5.6
Japan	307,000	5.5	303,000	5.6	274,000	4.5	268,000	3.8	242,000	3.3	242,000	3.5	242,000	3.4

Product Tables
Blockboard

Unit of Measure: Cubic meters.

	1990 Value	%	1991 Value	%	1992 Value	%	1993 Value	%	1994 Value	%	1995 Value	%	1996 Value	%
Total Production	46,157,850	100.0	34,173,150	100.0	49,390,365	100.0	75,180,727	100.0	54,314,470	100.0	83,606,325	100.0	60,546,323	100.0
Regions														
Africa	37,750	0.1	39,150	0.1	40,550	0.1	41,950	0.1	43,350	0.1	44,750	0.1	46,150	0.1
America, North	135,000	0.3	111,000	0.3	94,000	0.2	157,000	0.2	165,000	0.3	205,000	0.2	205,000	0.3
America, South	100,150	0.2	100,150	0.3	100,150	0.2	100,150	0.1	100,150	0.2	100,150	0.1	100,150	0.2
Asia	39,483,000	85.5	28,247,000	82.7	43,620,000	88.3	69,855,000	92.9	48,777,000	89.8	78,227,220	93.6	55,094,840	91.0
Europe	6,401,950	13.9	5,675,850	16.6	5,535,665	11.2	5,026,627	6.7	5,228,970	9.6	5,029,205	6.0	5,100,183	8.4
Leading Producers														
Malaysia	36,354,000	78.8	25,578,000	74.8	40,762,000	82.5	67,153,000	89.3	46,826,000	86.2	77,028,000	92.1	*53,564,500	88.5
Mexico	135,000	0.3	111,000	0.3	94,000	0.2	157,000	0.2	165,000	0.3	205,000	0.2	205,000	0.3
France	5,113,000	11.1	4,472,000	13.1	4,232,000	8.6	3,746,000	5.0	*3,975,733	7.3	*3,865,194	4.6	*3,754,655	6.2
Belarus	1,669,000	3.6	1,442,000	4.2	1,188,000	2.4	794,000	1.1	361,000	0.7	201,000	0.2	133,000	0.2
Switzerland	722,000	1.6	699,000	2.0	762,000	1.5	742,000	1.0	656,000	1.2	564,000	0.7	*712,295	1.2

Plywood

Unit of Measure: Cubic meters.

	1990 Value	%	1991 Value	%	1992 Value	%	1993 Value	%	1994 Value	%	1995 Value	%	1996 Value	%
Total Production	48,923,101	100.0	47,533,729	100.0	51,102,720	100.0	51,759,660	100.0	53,227,660	100.0	58,049,059	100.0	58,571,559	100.0
Regions														
Africa	494,356	1.0	500,690	1.1	479,823	0.9	458,115	0.9	447,816	0.8	487,817	0.8	486,018	0.8
America, North	20,946,644	42.8	18,441,822	38.8	19,146,900	37.5	19,107,700	36.9	19,400,700	36.4	19,164,900	33.0	18,983,400	32.4
America, South	1,634,850	3.3	1,572,850	3.3	1,766,550	3.5	1,924,200	3.7	2,331,000	4.4	2,428,500	4.2	2,435,000	4.2
Asia	21,872,800	44.7	23,525,917	49.5	26,030,633	50.9	26,358,200	50.9	26,954,517	50.6	31,716,633	54.6	32,417,850	55.3
Europe	3,760,950	7.7	3,316,950	7.0	3,469,814	6.8	3,674,945	7.1	3,801,127	7.1	3,924,709	6.8	3,928,791	6.7
Oceania	213,500	0.4	175,500	0.4	209,000	0.4	236,500	0.5	292,500	0.5	326,500	0.6	320,500	0.5
Leading Producers														
United States of America	18,771,000	38.4	16,508,000	34.7	17,109,000	33.5	17,093,000	33.0	17,380,000	32.7	17,140,000	29.5	16,975,000	29.0
Indonesia	8,250,000	16.9	9,600,000	20.2	10,100,000	19.8	10,050,000	19.4	9,836,000	18.5	9,500,000	16.4	9,575,000	16.3
Japan	6,415,000	13.1	6,174,000	13.0	5,954,000	11.7	5,263,000	10.2	4,865,000	9.1	4,421,000	7.6	4,421,000	7.5
China	759,000	1.6	1,054,000	2.2	1,565,000	3.1	2,125,000	4.1	2,610,000	4.9	7,590,000	13.1	7,950,000	13.6
Canada	1,971,000	4.0	1,705,000	3.6	1,838,000	3.6	1,824,000	3.5	1,834,000	3.4	1,831,000	3.2	1,814,000	3.1

Particle Board

Unit of Measure: Cubic meters.

	1990 Value	%	1991 Value	%	1992 Value	%	1993 Value	%	1994 Value	%	1995 Value	%	1996 Value	%
Total Production	61,228,870	100.0	60,872,368	100.0	68,558,238	100.0	70,471,742	100.0	78,718,784	100.0	83,754,925	100.0	86,586,367	100.0
Regions														
Africa	504,225	0.8	486,625	0.8	594,225	0.9	413,697	0.6	793,799	1.0	861,601	1.0	865,903	1.0
America, North	10,432,895	17.0	9,936,526	16.3	11,276,157	16.4	12,182,400	17.3	19,650,700	25.0	20,232,500	24.2	22,814,500	26.3
America, South	1,165,000	1.9	1,268,300	2.1	1,391,100	2.0	1,587,100	2.3	1,325,600	1.7	1,475,000	1.8	1,569,000	1.8
Asia	12,139,750	19.8	12,899,917	21.2	19,439,983	28.4	20,054,200	28.5	19,250,667	24.5	22,186,133	26.5	22,135,600	25.6
Europe	36,092,000	58.9	35,487,000	58.3	35,052,773	51.1	35,346,345	50.2	36,697,018	46.6	37,968,691	45.3	38,210,364	44.1
Oceania	895,000	1.5	794,000	1.3	804,000	1.2	888,000	1.3	1,001,000	1.3	1,031,000	1.2	991,000	1.1
Leading Producers														
Germany	7,939,000	13.0	7,441,000	12.2	7,451,000	10.9	7,961,000	11.3	8,639,000	11.0	8,902,000	10.6	8,629,000	10.0
United States of America	6,877,000	11.2	6,781,000	11.1	7,207,000	10.5	7,506,000	10.7	14,664,000	18.6	14,429,000	17.2	15,563,000	18.0
Canada	3,112,000	5.1	2,650,000	4.4	3,557,000	5.2	4,173,000	5.9	4,493,000	5.7	5,309,000	6.3	6,757,000	7.8
Russian Federation	-		-		4,522,000	6.6	3,941,000	5.6	2,626,000	3.3	2,206,000	2.6	1,460,000	1.7
France	2,464,000	4.0	2,638,000	4.3	2,668,000	3.9	2,461,000	3.5	2,567,000	3.3	2,733,000	3.3	3,030,000	3.5

Commodity data are provided by the United Nations Statistical Division. The symbol * means that data are estimated. For additional notes, see Appendix I.

197

ISIC 3320
FURNITURE, FIXTURES

ISIC 3320—Furniture, Fixtures—includes wood and metal household furniture, mattresses and bedsprings, office furniture, and related fixtures. The industry corresponds to the U.S. SIC 25 of the same name.

Summary Statistics

		1990 Value	N	1991 Value	N	1992 Value	N	1993 Value	N	1994 Value	N	1995 Value	N
Establishments or enterprises	(number)	82,935	85	85,359	93	100,040	87	69,115	85	66,217	72	38,605	34
Number employed	(000)	3,483	99	2,948	99	2,894	94	2,956	95	2,826	85	2,364	63
Total Wages	($ mil.)	31,516	91	34,729	91	37,895	85	30,463	85	32,218	74	27,007	54
Wage/salary per employee	($)	6,810	90	7,475	90	8,251	85	7,024	85	7,269	74	9,263	54
Female workers	(000)	203	36	131	34	461	30	491	34	491	28	460	15
as % of total employment*	(%)	19	36	19	34	20	30	23	34	23	28	29	14
Output	($ bil.)	148	94	165	95	176	91	154	87	164	77	145	60
per employee	($ 000)	33	91	35	92	38	88	33	83	36	73	43	57
per establishment	($ 000)	1,891	79	1,850	84	1,515	77	1,508	72	1,477	59	833	28
Value added	($ bil.)	63	82	62	83	68	78	69	77	62	69	64	57
per employee	($ 000)	15	80	15	81	17	77	14	73	15	65	20	53
per establishment	($ 000)	716	70	551	75	588	67	462	63	511	51	308	25
Capital investment	($ mil.)	4,834	67	7,313	59	4,128	55	2,846	51	3,298	41	1,834	22
per employee	($ 000)	2	64	3	57	2	53	1	49	2	39	1	20
per establishment	($ 000)	156	63	99	55	87	52	69	47	64	38	25	19

Data presented above are drawn from the detailed tables that follow. Columns headed 'N' show the number of countries that provided valid data for inclusion. Values are not strictly comparable one year to the next or one row to the next because the number of countries that report varies from period to period and row to row. However, a general indication of magnitudes can be gleaned from reviewing this summary—especially in earlier years in which more reports are available. For detailed explanations, see Appendix I. *The average for those countries reporting both total and female employment.

Establishments and Number Engaged

Country	Establishments or Enterprises (number)						Employees per Establishment					
	1990	1991	1992	1993	1994	1995	1990	1991	1992	1993	1994	1995
Albania	-	-	-	13	-	-	-	-	-	108	-	-
Argentina	-	-	-	4,862	5,582	-	-	-	-	3.381	3.439	-
Armenia	22	24	22	22	20	-	265	219	-	-	-	-
Australia	2,251	3,268	3,245	-	-	-	16	10	9.553	-	-	-
Austria	609	582	591	568	544	-	51	57	56	57	60	-
Azerbaijan	20	17	19	20	17	-	246	276	217	161	147	-
Bahamas	35	33	-	-	-	-	4.057	3.848	-	-	-	-
Bahrain	-	-	311	-	-	-	-	-	7.032	-	-	-
Bangladesh	408	257	226	-	-	-	13	11	10	-	-	-
Belize	47	47	48	-	-	-	-	10	11	-	-	-
Bolivia	100	114	126	133	133	150	9.920	9.930	11	9.887	9.549	8.660
Bosnia & Herzegovina	56	60	-	-	9.000	-	461	373	-	-	142	-
Brazil	915	-	843	785	-	-	87	-	85	90	-	-
Bulgaria	*62	*62	*75	*616	*1,052	*1,130	361	305	213	20	15	14
Cameroon	-	-	-	-	*6.000	*8.000	-	-	-	-	73	13
Canada	2,607	2,258	2,074	1,965	1,839	-	25	24	24	25	28	-
Cape Verde	87	87	96	113	120	-	-	-	-	-	-	-
Central African Republic	-	*3.000	*3.000	*3.000	*3.000	*3.000	-	-	-	-	-	-
Chile	34	34	38	48	47	-	148	155	156	130	152	-
China	*11,792	*11,241	*10,762	*8,014	*8,171	*8,760	36	36	37	42	43	40
China (Hong Kong SAR)	986	768	713	475	453	-	5.375	5.078	4.769	4.211	3.974	-
Colombia	238	222	262	239	262	-	37	33	34	39	42	-
Costa Rica	506	466	459	486	-	-	5.350	5.116	5.244	5.218	-	-
Croatia	122	155	195	271	396	465	206	126	82	57	36	28
Cyprus	708	700	702	702	730	742	3.278	3.300	3.406	3.617	3.715	2.968
Czechoslovakia (Former)	*21	*27	-	-	-	-	2,810	1,185	-	-	-	-
Czech Republic	-	*15	*93	*119	-	-	-	1,200	290	218	-	-
Denmark	1,608	1,644	1,606	-	-	-	13	12	12	-	-	-
Ecuador	58	75	78	76	79	88	54	42	30	30	32	33
Egypt	97	128	113	116	-	-	60	64	59	69	-	-
El Salvador	-	11	7.000	10	19	34	-	-	60	82	52	38
Equatorial Guinea	13	-	-	-	-	-	16	-	-	-	-	-
Ethiopia	26	19	19	20	54	63	64	80	77	72	39	36
Fiji	96	93	102	-	-	-	9.958	9.000	9.422	-	-	-
Finland	337	358	319	283	264	305	32	29	28	28	28	29
FYR Macedonia	*48	*104	*209	*236	*292	*363	166	64	32	26	22	13
Gabon	-	3.000	4.000	6.000	5.000	9.000	-	43	23	7.833	6.800	9.222
Gambia	-	-	-	16	-	-	-	-	-	8.000	-	-
Germany	-	1,645	1,614	1,580	1,534	-	-	110	105	101	101	-
Germany (Western Part)	1,334	1,335	1,337	-	-	-	106	111	111	-	-	-
Ghana	-	-	-	29	-	-	-	-	-	98	-	-
Greece	414	410	415	-	-	-	15	15	15	-	-	-
Grenada	-	-	-	-	4.000	6.000	-	-	-	-	14	7.833
Guatemala	-	8.000	8.000	9.000	8.000	8.000	-	78	78	77	94	111
Honduras	-	-	16	16	16	16	-	-	369	404	443	496
Hungary	*183	*277	*377	*459	-	-	142	83	53	44	-	-
Iceland	240	277	335	344	-	-	4.017	4.350	3.131	2.660	-	-
India	305	318	331	311	-	-	21	25	21	22	-	-
Indonesia	606	695	722	782	898	1,159	131	147	143	158	147	124
Iran (Islamic Republic of)	95	54	57	48	-	-	46	70	76	80	-	-
Ireland	212	206	-	-	-	-	15	15	-	-	-	-
Israel	440	485	571	565	594	-	13	12	12	13	14	-
Italy	*1,417	*1,404	*1,924	*1,927	*1,952	-	49	49	48	46	46	-
Jamaica	62	62	-	-	-	-	34	32	-	-	-	-
Japan	10,918	10,772	10,371	10,100	9,537	9,433	15	15	15	15	17	16
Jordan	1,495	1,528	2,002	2,015	1,985	2,195	1.548	1.762	2.176	2.273	2.462	2.469
Kenya	93	211	262	270	292	292	42	18	14	14	13	14
Korea, Republic of	1,881	2,056	2,132	2,684	2,713	2,909	22	22	21	18	18	16
Kuwait	316	303	345	334	326	-	12	7.683	9.643	11	11	-
Kyrgyzstan	9.000	9.000	9.000	8.000	8.000	-	408	377	332	330	188	-

Continued.

    Depending on the table, * means *Enterprises, Engaged,* or *Factor Values*; ± means *Basis Unspecified*; § means *shown in millions.* For additional notes and sources, see Appendix I.

Establishments and Number Engaged

- Continued -

Country	Establishments or Enterprises (number)						Employees per Establishment					
	1990	1991	1992	1993	1994	1995	1990	1991	1992	1993	1994	1995
Latvia	13	49	75	104	67	130	569	172	115	74	100	47
Lesotho	92	-	1.000	1.000	1.000	-	9.739	-	-	218	79	-
Lithuania	-	-	-	*62	*77	-	-	-	-	207	143	-
Macau	194	165	158	145	129	105	3.696	3.388	3.063	2.800	2.310	1.810
Malaysia	342	389	406	507	495	-	43	56	59	60	66	-
Malta	337	347	376	379	385	-	4.231	4.248	4.184	4.388	4.434	-
Mauritius	33	29	26	29	32	33	43	43	49	47	46	46
Mexico	68	68	68	68	64	61	106	107	105	97	107	83
Mongolia	2.000	4.000	3.000	2.000	23	-	1,094	150	179	278	39	-
Mozambique	27	27	14	14	18	-	45	42	44	54	44	-
Myanmar	6.000	6.000	6.000	6.000	7.000	-	271	271	271	271	232	-
Namibia	-	-	-	-	17	-	-	-	-	-	41	-
Nepal	117	255	-	289	320	-	11	12	-	9.796	12	-
Netherlands	224	231	243	247	339	-	45	46	44	43	31	-
Netherlands Antilles	-	-	-	22	-	-	-	-	-	7.909	-	-
New Zealand	*1,317	*1,356	*1,382	*1,514	-	-	4.010	3.701	3.732	3.783	-	-
Norway	247	250	154	169	178	-	28	26	37	38	38	-
Pakistan	-	59	-	-	-	-	-	44	-	-	-	-
Panama	103	90	-	-	-	-	13	16	-	-	-	-
Paraguay	-	5.000	-	-	-	-	-	32	-	-	-	-
Peru	783	788	805	-	-	-	7.656	7.787	5.514	-	-	-
Philippines	5,344	5,714	638	557	-	-	9.150	7.350	47	41	-	-
Poland	*290	*302	*289	*283	-	-	266	272	249	247	-	-
Portugal	*6,584	*6,811	*6,560	*6,450	*6,715	-	5.786	7.238	7.831	6.225	6.039	-
Puerto Rico	126	113	109	113	109	123	15	13	15	14	21	18
Qatar	66	66	365	361	356	-	3.727	6.561	5.775	5.030	5.287	-
Republic of Moldova	13	12	12	9.000	9.000	9.000	677	690	649	637	605	544
Romania	-	741	1,515	2,337	1,731	-	-	248	107	70	87	-
Russian Federation	-	-	-	*2,947	*3,076	*4,816	-	-	-	92	85	49
Saint Lucia	-	11	11	11	11	9.000	-	-	-	-	-	-
Senegal	4.000	3.000	4.000	4.000	5.000	-	18	22	50	16	17	-
Sierra Leone	-	-	-	48	-	-	-	-	-	19	-	-
Singapore	141	139	135	141	137	-	46	46	47	45	44	-
Slovakia	-	50	50	61	87	-	-	379	359	258	-	-
Slovenia	126	*95	*142	*250	*277	*1,754	-	-	-	-	-	-
South Africa	-	1,516	-	-	-	-	-	28	-	-	-	-
Spain	11,462	10,659	10,117	-	-	-	4.970	5.629	6.067	-	-	-
Sri Lanka	67	63	55	64	-	-	26	23	22	18	-	-
Suriname	-	-	28	-	-	-	-	-	6.929	-	-	-
Swaziland	-	7.000	7.000	3.000	-	-	-	49	44	109	-	-
Sweden	279	247	217	209	210	-	40	43	44	40	43	-
Thailand	238	333	-	-	-	-	135	87	-	-	-	-
Trinidad and Tobago	242	261	257	204	206	-	4.087	5.716	5.599	7.672	8.058	-
Tunisia	-	-	-	1,463	1,511	1,732	-	-	-	5.517	5.523	4.990
Turkey	47	48	192	181	172	-	92	83	37	44	43	-
Ukraine	228	613	869	342	342	335	583	207	139	327	295	275
United Kingdom	7,798	7,342	7,289	6,403	7,624	-	15	15	14	19	17	-
United Republic of Tanzania	133	129	-	-	-	-	51	19	-	-	-	-
USSR (Former)	*983	-	-	-	-	-	573	-	-	-	-	-
United States of America	-	-	19,867	-	-	-	-	-	21	-	-	-
Venezuela	779	791	942	732	779	688	16	18	16	20	21	19
Yugoslavia	95	201	420	641	664	672	-	112	51	33	31	29
Zambia	12	-	-	-	33	-	119	-	-	-	55	-
Zimbabwe	44	44	45	45	47	-	112	106	109	113	121	-

Depending on the table, * means *Enterprises*, *Engaged*, or *Factor Values*; ± means *Basis Unspecified*; § means *shown in millions*. For additional notes and sources, see Appendix I.

Employment and Compensation of Employees

Country	Number Employed/Engaged (000)						Wage/Salary per Employee ($)					
	1990	1991	1992	1993	1994	1995	1990	1991	1992	1993	1994	1995
Albania	-	-	-	1.398	-	-	-	-	-	504	-	-
Argentina	14	-	-	16	19	-	3,242	-	-	7,646	-	-
Armenia	*5.827	*5.246	-	-	-	-	-	-	-	-	-	-
Australia	35	*34	*31	-	-	-	17,230	17,003	16,556	-	-	-
Austria	31	33	33	33	33	33	18,701	20,290	23,555	23,377	24,720	28,951
Azerbaijan	4.910	4.690	4.129	3.227	2.506	-	1,845	1,873	-	-	-	-
Bahamas	0.142	0.127	-	-	-	-	8,542	12,622	-	-	-	-
Bahrain	-	-	2.187	-	-	-	-	-	3,591	-	-	-
Bangladesh	5.189	2.811	2.304	-	-	-	-	418	468	-	-	-
Belgium	20	21	21	-	-	-	-	-	-	-	-	-
Belize	-	0.480	0.540	-	-	-	-	1,649	1,463	-	-	-
Bermuda	0.052	0.047	0.033	0.037	0.039	0.036	-	-	-	-	-	-
Bolivia	0.992	1.132	1.327	1.315	1.270	1.299	1,062	1,041	941	976	1,021	969
Bosnia & Herzegovina	26	22	-	-	1.281	-	2,294	2,664	-	-	-	-
Brazil	80	-	72	70	-	-	4,217	-	3,262	3,854	-	-
Bulgaria	22	19	16	12	16	15	1,735	556	743	973	841	975
Cambodia	*1.878	*1.977	-	-	-	-	-	-	-	-	-	-
Cameroon	0.200	0.400	0.090	0.050	0.435	0.104	6,593	6,824	6,758	6,569	2,998	2,735
Canada	66	55	49	50	52	56	20,011	21,186	21,409	20,107	19,265	19,824
Central African Republic	0.100	-	-	-	-	-	3,269	-	-	-	-	-
Chile	5.026	5.264	5.934	6.242	7.155	7.255	3,069	3,449	4,146	4,684	5,044	5,797
China	*420	*400	*400	*340	*350	*350	-	-	-	-	-	-
China (Hong Kong SAR)	*5.300	*3.900	*3.400	*2.000	*1.800	-	8,235	8,810	9,537	9,049	9,058	-
China (Taiwan Province)	33	31	31	29	29	26	8,883	10,048	11,169	11,550	12,077	12,587
Colombia	8.700	7.274	8.926	9.385	11	11	1,400	1,431	1,481	1,798	2,318	2,665
Costa Rica	2.707	2.384	2.407	2.536	-	-	1,746	1,623	1,876	2,179	-	-
Croatia	25	19	16	15	14	13	3,289	3,399	1,408	1,530	1,985	3,146
Cyprus	2.321	2.310	2.391	2.539	2.712	2.202	8,385	9,070	9,618	9,201	9,742	13,551
Czechoslovakia (Former)	59	32	-	-	-	-	1,945	1,389	-	-	-	-
Czech Republic	18	18	27	26	-	-	2,043	1,413	1,651	1,958	-	-
Denmark	20	21	20	19	21	22	24,877	23,410	27,158	24,880	28,240	33,504
Ecuador	3.147	3.153	2.378	2.253	2.510	2.901	2,051	1,790	1,648	2,301	1,267	1,685
Egypt	*5.800	*8.200	*6.700	*8.000	*7.976	*7.928	1,655	1,058	1,131	1,138	1,139	1,151
El Salvador	-	-	*0.417	*0.823	*0.994	*1.279	-	-	-	2,222	3,215	2,541
Equatorial Guinea	0.208	-	-	-	-	-	2,080	-	-	-	-	-
Ethiopia	1.661	1.521	1.455	1.433	2.106	2.241	1,620	1,687	1,309	889	769	723
Fiji	0.956	0.837	0.961	1.088	1.097	-	2,223	2,992	3,050	2,520	2,752	-
Finland	11	10	8.800	7.900	7.400	8.913	23,209	22,124	20,117	15,603	19,258	25,520
FYR Macedonia	7.959	6.648	6.730	6.141	6.360	4.861	4,093	4,158	1,672	1,832	1,645	2,880
France	84	83	82	77	75	74	28,564	28,816	31,172	-	-	-
Gabon	-	0.128	0.093	0.047	0.034	0.083	-	15,065	12,187	15,779	5,827	10,645
Gambia	-	-	-	0.128	-	-	-	-	-	-	-	-
Germany	-	*182	*169	*160	*154	-	-	24,943	29,819	29,717	31,071	-
Germany (Western Part)	*142	*148	*148	-	-	-	27,542	28,399	31,861	-	-	-
Ghana	-	-	-	*2.855	*2.936	*3.018	-	-	-	818	611	623
Greece	*6.331	*6.143	*6.219	*6.219	*6.219	*6.219	7,104	7,371	8,018	7,949	8,482	9,459
Grenada	-	-	-	-	0.055	0.047	-	-	-	-	-	-
Guatemala	-	*0.625	*0.625	*0.694	*0.750	*0.889	-	-	-	-	-	-
Honduras	*3.493	*4.474	*5.905	*6.470	*7.089	*7.939	995	897	647	748	657	-
Hungary	26	23	20	20	20	20	2,083	2,175	2,498	2,608	2,465	2,678
Iceland	0.964	1.205	1.049	0.915	0.904	-	34,904	31,169	35,716	26,304	-	-
India	*6.424	*7.918	*6.819	*6.752	*6.726	*6.359	863	766	758	646	698	881
Indonesia	79	102	103	123	132	144	516	612	615	646	672	911
Iran (Islamic Republic of)	4.400	3.765	4.319	3.843	-	-	3,654	4,172	4,101	3,883	-	-
Iraq	-	0.933	0.922	-	-	-	-	6,548	8,157	-	-	-
Ireland	3.100	3.100	3.094	3.092	3.070	3.059	13,320	13,810	15,465	13,604	15,719	18,155
Israel	5.700	6.000	7.100	7.200	8.100	7.938	13,487	12,797	14,147	14,330	15,457	16,840
Italy	69	69	92	88	89	-	28,022	30,130	31,735	25,950	25,835	-
Jamaica	2.111	1.994	2.090	2.333	2.416	-	4,311	4,001	3,048	-	-	-
Japan	*168	*165	*159	*152	*160	*155	21,583	24,475	27,213	31,120	36,139	39,988

Continued.

Depending on the table, * means *Enterprises, Engaged*, or *Factor Values*; ± means *Basis Unspecified*; § means *shown in millions*. For additional notes and sources, see Appendix I.

Employment and Compensation of Employees

- Continued -

Country	Number Employed/Engaged (000)						Wage/Salary per Employee ($)					
	1990	1991	1992	1993	1994	1995	1990	1991	1992	1993	1994	1995
Jordan	2.315	2.693	4.356	4.581	4.888	5.419	1,915	1,832	1,998	2,028	2,066	2,184
Kenya	3.898	3.745	3.759	3.830	3.907	4.050	1,378	1,178	1,103	669	868	1,172
Korea, Republic of	42	46	45	48	48	47	8,273	10,371	10,837	11,700	12,486	15,019
Kuwait	3.652	2.328	3.327	3.576	3.542	3.564	4,265	3,710	6,405	5,993	5,492	8,067
Kyrgyzstan	*3.668	*3.396	*2.986	*2.638	*1.507	-	-	-	-	-	-	-
Latvia	7.396	8.405	8.654	7.716	6.687	6.080	-	-	-	-	-	-
Lesotho	*0.896	-	-	*0.218	*0.079	-	-	-	-	702	2,870	-
Lithuania	-	-	14	13	11	-				558	1,026	-
Macau	0.717	0.559	0.484	0.406	0.298	0.190	4,521	4,831	5,545	6,341	7,228	7,371
Malawi	0.205	-	-	-	-	-	-	-	-	-	-	-
Malaysia	15	22	24	30	33	35	2,118	2,220	2,756	2,977	3,139	3,360
Malta	1.426	1.474	1.573	1.663	1.707	-	8,172	9,078	9,644	8,389	8,955	-
Mauritius	*1.430	1.238	1.274	1.355	1.472	1.510	2,310	2,932	4,274	4,006	3,760	4,323
Mexico	7.240	7.277	7.131	6.619	6.821	5.080	3,352	4,121	5,114	5,569	5,352	2,925
Mongolia	*2.189	*0.598	*0.537	*0.557	*0.897	*0.124	-	2,758	984	-	-	-
Mozambique	1.224	1.126	0.617	0.750	0.790	-	723	536	-	-	-	-
Myanmar	1.623	1.623	1.623	1.623	1.623	-	1,128	1,233	1,330	2,598	2,056	-
Namibia	-	-	-	-	0.700	-	-	-	-	-	2,883	-
Nepal	1.254	3.167	-	2.831	3.954	-	405	503	-	459	423	-
Netherlands	9.995	11	11	11	11	10	22,473	23,320	26,337	25,297	25,838	30,417
Netherlands Antilles	-	-	-	0.174	-	-	-	-	-	9,908	-	-
New Zealand	5.281	5.019	5.157	5.727	-	-	-	-	-	-	-	-
Norway	6.813	6.598	5.674	6.405	6.833	6.951	23,964	24,057	31,250	25,133	26,771	30,632
Pakistan	2.225	2.576	2.694	-	-	-	955	874	899	-	-	-
Panama	*1.325	*1.401	*1.420	*1.438	*1.477	*1.482	3,881	3,727	3,800	3,853	3,973	4,000
Paraguay	-	0.158	-	-	-	-	-	-	-	-	-	-
Peru	*5.995	*6.136	*4.439	-	-	-	1,885	2,274	2,073	-	-	-
Philippines	49	42	30	23	-	-	1,085	1,162	1,411	1,369	-	-
Poland	77	82	72	70	68	64	1,069	1,658	2,136	2,213	3,112	5,222
Portugal	*38	*49	*51	*40	*41	*43	-	-	-	-	-	-
Puerto Rico	1.870	1.420	1.600	1.570	2.240	2.250	15,455	16,549	16,125	12,994	9,866	10,356
Qatar	0.246	0.433	2.108	1.816	1.882	-	2,292	2,585	4,213	4,282	4,293	-
Republic of Moldova	8.806	8.275	7.785	5.733	5.445	4.898	-	3,908	-	586	440	484
Romania	204	183	162	163	150	-	1,569	1,035	666	824	821	-
Russian Federation	-	-	-	271	261	237	-	-	-	656	864	858
Senegal	*0.071	*0.066	*0.200	*0.063	*0.087	-	1,811	1,826	963	3,027	1,346	-
Sierra Leone	-	-	-	*0.897	-	-	-	-	-	-	-	-
Singapore	*6.521	*6.463	*6.310	*6.344	*5.964	*5.096	8,397	9,241	10,359	11,335	12,924	13,934
Slovakia	-	*19	*18	*16	-	-	-	1,296	1,522	1,737	-	-
South Africa	43	42	41	46	46	49	4,638	4,889	5,678	6,127	6,104	6,906
Spain	57	60	61	101	96	95	12,896	13,672	15,052	13,546	13,424	15,071
Sri Lanka	*1.742	*1.462	*1.214	*1.134	-	-	-	-	-	504	-	-
Suriname	*0.212	*0.208	*0.194	*0.187	-	-	6,606	7,811	7,508	11,684	-	-
Swaziland	-	0.343	0.309	0.326	-	-	-	2,227	2,403	2,466	-	-
Sweden	11	*11	*9.613	8.263	9.061	9.061	20,776	22,337	24,423	19,930	21,228	23,386
Thailand	32	29	-	-	-	-	7,004	1,937	-	-	-	-
Tonga	0.066	-	-	-	-	-	2,153	-	-	-	-	-
Trinidad and Tobago	0.989	1.492	1.439	1.565	1.660	-	2,891	4,274	4,709	2,782	2,979	-
Tunisia	-	-	-	8.071	8.346	8.642	-	-	-	-	-	-
Turkey	4.330	3.968	7.045	8.043	7.318	7.673	3,895	5,195	4,069	4,788	2,681	3,217
Ukraine	133	127	121	112	101	92	2,847	2,647	2,265	409	435	-
United Kingdom	114	107	104	119	128	124	20,457	21,123	21,983	19,116	19,920	20,680
United Republic of Tanzania	6.769	2.415	-	-	-	-	-	-	-	-	-	-
USSR (Former)	563	-	-	-	-	-	2,107	-	-	-	-	-
United States of America	438	406	414	425	439	460	18,630	19,212	20,486	21,031	21,601	21,913
Uruguay	*2.681	*3.041	*2.391	*2.197	*2.274	*2.145	1,542	2,106	2,280	2,481	2,756	3,074
Venezuela	12	14	15	15	16	13	2,295	2,670	2,785	3,078	2,662	3,433
Yugoslavia	-	23	21	21	20	19	-	-	-	-	1,435	721
Zambia	1.432	-	-	-	1.814	-	632	-	-	-	838	-
Zimbabwe	4.934	4.672	4.900	5.100	5.700	5.972	2,211	2,541	1,787	1,366	1,285	1,512

Depending on the table, * means *Enterprises, Engaged,* or *Factor Values*; ± means *Basis Unspecified*; § means *shown in millions*. For additional notes and sources, see Appendix I.

203

Female Workers: Total and Percent of Employment

Country	Female Workers (000)						Female Workers as % of Total Employment					
	1990	1991	1992	1993	1994	1995	1990	1991	1992	1993	1994	1995
Albania	-	-	-	0.581	-	-	-	-	-	41.56	-	-
Argentina	-	-	-	-	1.402	-	-	-	-	-	7.30	-
Australia	7.379	-	-	-	-	-	21.08	-	-	-	-	-
Austria	6.200	7.000	7.000	6.665	6.674	-	20.13	21.21	21.21	20.42	20.51	-
Bahrain	-	-	0.015	-	-	-	-	-	0.69	-	-	-
Bangladesh	0.310	0.306	0.001	-	-	-	5.97	10.89	0.04	-	-	-
Bermuda	0.014	0.013	0.010	0.011	0.011	0.010	26.92	27.66	30.30	29.73	28.21	27.78
Bosnia & Herzegovina	7.461	6.866	-	-	-	-	28.92	30.71	-	-	-	-
Bulgaria	-	-	-	6.400	7.700	7.000	-	-	-	52.46	49.36	45.75
Canada	15	13	-	-	-	-	22.73	23.64	-	-	-	-
Chile	0.473	0.450	0.555	0.715	0.744	-	9.41	8.55	9.35	11.45	10.40	-
China (Taiwan Province)	12	12	12	11	11	10	37.90	37.09	37.41	38.00	37.97	38.23
Colombia	-	1.611	-	2.226	2.769	-	-	22.15	-	23.72	24.89	-
Croatia	8.960	6.680	5.510	4.970	4.980	4.560	35.64	34.27	34.33	32.38	34.51	34.42
Cyprus	0.494	0.504	0.479	0.625	1.035	-	21.28	21.82	20.03	24.62	38.16	-
Czechoslovakia (Former)	23	12	-	-	-	-	38.64	37.19	-	-	-	-
Czech Republic	10	9.000	12	12	-	-	55.56	50.00	44.44	46.15	-	-
Denmark	5.567	5.617	5.398	-	-	-	27.57	27.34	27.04	-	-	-
Egypt	-	0.401	0.250	0.370	-	-	-	4.89	3.73	4.63	-	-
El Salvador	-	-	-	0.112	0.096	0.116	-	-	-	13.61	9.66	9.07
Ethiopia	-	0.194	0.196	0.222	0.281	0.295	-	12.75	13.47	15.49	13.34	13.16
FYR Macedonia	1.767	1.636	1.596	1.545	1.323	1.043	22.20	24.61	23.71	25.16	20.80	21.46
Germany (Western Part)	32	-	-	-	-	-	22.58	-	-	-	-	-
Ghana	-	-	-	0.103	-	-	-	-	-	3.61	-	-
Hungary	8.000	-	7.000	6.000	-	-	30.77	-	35.00	30.00	-	-
Indonesia	-	-	-	40	42	42	-	-	-	32.38	31.61	29.42
Iran (Islamic Republic of)	-	0.074	0.092	0.087	-	-	-	1.97	2.13	2.26	-	-
Iraq	-	0.094	-	-	-	-	-	10.08	-	-	-	-
Ireland	0.200	0.300	-	-	-	-	6.45	9.68	-	-	-	-
Italy	-	17	23	22	22	-	-	23.92	24.88	24.52	24.95	-
Jordan	0.010	0.010	0.110	0.054	0.052	0.075	0.43	0.37	2.53	1.18	1.06	1.38
Kenya	0.106	0.110	0.119	0.121	-	-	2.72	2.94	3.17	3.16	-	-
Korea, Republic of	12	13	13	14	14	13	27.96	28.86	29.17	28.75	28.34	27.23
Lesotho	-	-	-	0.110	0.021	-	-	-	-	50.46	26.58	-
Macau	0.042	0.026	0.021	0.023	0.022	0.021	5.86	4.65	4.34	5.67	7.38	11.05
Malaysia	4.400	6.600	7.200	8.800	9.300	-	29.73	30.56	29.88	28.85	28.35	-
Malta	0.083	0.072	0.082	0.083	0.115	-	5.82	4.88	5.21	4.99	6.74	-
Mongolia	-	-	-	-	0.300	-	-	-	-	-	33.44	-
Morocco	-	-	-	-	-	0.389	-	-	-	-	-	-
Myanmar	0.120	0.120	0.120	0.120	0.120	-	7.39	7.39	7.39	7.39	7.39	-
Nepal	0.015	0.126	-	0.011	0.083	-	1.20	3.98	-	0.39	2.10	-
New Zealand	1.188	1.143	1.151	1.250	-	-	22.50	22.77	22.32	21.83	-	-
Panama	0.089	-	-	-	-	-	6.72	-	-	-	-	-
Philippines	-	-	8.018	5.378	-	-	-	-	26.73	23.28	-	-
Portugal	1.101	1.075	1.107	1.146	1.272	-	2.89	2.18	2.15	2.85	3.14	-
Puerto Rico	0.540	0.420	0.570	0.520	0.530	0.500	28.88	29.58	35.63	33.12	23.66	22.22
Republic of Moldova	-	-	-	-	2.533	2.251	-	-	-	-	46.52	45.96
Senegal	0.001	-	-	-	-	-	1.41	-	-	-	-	-
Sri Lanka	0.089	0.093	0.037	0.130	-	-	5.11	6.36	3.05	11.46	-	-
Sweden	3.500	-	-	-	-	-	31.53	-	-	-	-	-
Thailand	15	14	-	-	-	-	46.02	48.88	-	-	-	-
Turkey	0.294	-	-	-	-	-	6.79	-	-	-	-	-
United Kingdom	25	-	20	-	-	-	21.83	-	19.23	-	-	-
United Republic of Tanzania	0.745	0.266	-	-	-	-	11.01	11.01	-	-	-	-
United States of America	-	-	335	344	361	378	-	-	80.92	80.94	82.23	82.17
Zambia	-	-	-	-	0.050	-	-	-	-	-	2.76	-

Depending on the table, * means *Enterprises, Engaged,* or *Factor Values;* ± means *Basis Unspecified;* § means *shown in millions.* For additional notes and sources, see Appendix I.

Output and Output per Employee

Country	Output ($ bil.)						Output per Employee ($)					
	1990	1991	1992	1993	1994	1995	1990	1991	1992	1993	1994	1995
Albania	-	-	-	*0.003	-	-	-	-	-	1,997	-	-
Argentina	±0.513	-	-	0.929	-	-	36,172	-	-	56,535	-	-
Armenia	0.002	0.003	0.006	0.322	0.001	-	408	655	-	-	-	-
Australia	*2.317	*2.304	*2.087	-	-	-	66,205	67,760	67,315	-	-	-
Austria	2.483	2.813	3.162	3.064	3.284	3.874	80,615	85,243	95,815	93,862	100,900	118,200
Azerbaijan	±0.072	±0.062	±0.001	±0.004	±0.001	-	14,576	13,266	348	1,242	501	-
Bahamas	0.004	0.005	-	-	-	-	28,169	39,370	-	-	-	-
Bahrain	-	-	±0.086	-	-	-	-	-	39,216	-	-	-
Bangladesh	0.011	0.007	0.007	-	-	-	2,152	2,323	3,053	-	-	-
Belgium	4.953	4.998	5.094	4.596	4.810	5.775	242,801	241,464	247,303	-	-	-
Belize	0.006	0.005	0.005	-	-	-	-	10,195	9,693	-	-	-
Benin	±0.008	±0.008	±0.008	±0.007	-	-	-	-	-	-	-	-
Bolivia	0.008	0.009	0.011	0.011	0.011	0.010	8,436	8,286	8,258	8,499	8,688	7,718
Bosnia & Herzegovina	±0.388	±0.372	-	-	-	-	15,021	16,632	-	-	-	-
Brazil	*2.299	-	*1.461	*2.053	-	-	28,787	-	20,296	29,164	-	-
Bulgaria	±0.280	±0.085	±0.088	±0.091	±0.088	±0.110	12,516	4,473	5,484	7,469	5,671	7,202
Cameroon	0.007	0.014	0.003	0.001	0.015	0.001	35,738	35,651	35,303	22,885	35,452	10,267
Canada	4.414	3.788	3.574	3.542	3.727	4.142	66,875	68,874	72,941	70,847	71,679	73,450
Central African Republic	*0.002	-	-	-	-	-	23,433	-	-	-	-	-
Chile	0.106	0.118	0.171	0.190	0.221	0.257	21,151	22,357	28,741	30,462	30,821	35,410
China	±1.701	±1.697	±2.069	±2.666	±2.538	±2.706	4,051	4,243	5,172	7,842	7,253	7,733
China (Hong Kong SAR)	0.171	0.163	0.151	0.081	0.062	-	32,214	41,838	44,494	40,268	34,289	-
China (Taiwan Province)	±1.141	±1.276	±1.376	±1.370	±2.411	±2.245	34,718	40,838	44,362	46,736	83,394	85,445
Colombia	0.084	0.068	0.081	0.091	0.157	0.181	9,612	9,362	9,067	9,694	14,084	16,188
Costa Rica	0.068	0.055	0.063	0.056	0.055	0.058	25,291	23,228	26,161	22,216	-	-
Croatia	±0.371	±0.393	±0.249	-	-	-	14,739	20,143	15,527	-	-	-
Cyprus	0.079	0.081	0.096	0.085	0.093	0.103	33,947	34,875	39,972	33,570	34,325	46,906
Czechoslovakia (Former)	±0.413	±0.222	-	-	-	-	7,006	6,933	-	-	-	-
Denmark	*1.279	*1.315	*1.445	*1.276	*1.590	*1.972	63,325	63,988	72,365	66,580	76,224	90,712
Ecuador	0.028	0.019	0.017	0.020	0.028	0.047	9,031	5,920	7,056	8,663	11,138	16,303
Egypt	*0.064	*0.060	*0.042	*0.072	*0.074	*0.078	11,026	7,342	6,227	8,979	9,288	9,881
El Salvador	-	±0.009	±0.002	±0.012	±0.011	±0.019	-	-	5,753	14,996	11,540	14,932
Equatorial Guinea	±0.001	-	-	-	-	-	4,415	-	-	-	-	-
Ethiopia	0.010	0.010	0.008	0.005	0.008	0.009	6,315	6,367	5,171	3,385	3,886	3,915
Fiji	0.011	0.012	0.013	0.013	0.014	-	11,220	14,581	13,448	11,969	13,068	-
Finland	±1.025	±0.814	±0.664	±0.485	±0.583	±0.922	93,998	78,226	75,421	61,338	78,724	103,388
FYR Macedonia	±0.150	±0.136	±0.072	±0.056	±0.048	±0.048	18,837	20,435	10,739	9,094	7,479	9,956
France	±9.167	±9.008	±9.415	±8.625	±8.832	±10	109,786	109,055	114,672	112,593	118,232	137,995
Gabon	-	*0.003	*0.003	*0.002	*0.001	*0.003	-	25,228	34,286	48,014	23,097	36,448
Germany	-	19	21	20	20	-	-	105,373	125,742	126,548	131,783	-
Germany (Western Part)	16	18	20	-	-	-	116,042	123,434	136,139	-	-	-
Ghana	-	-	-	0.010	0.008	0.009	-	-	-	3,614	2,863	2,922
Greece	*0.210	*0.203	*0.231	*0.229	*0.244	*0.273	33,139	32,992	37,070	36,809	39,315	43,865
Guatemala	-	0.007	0.007	0.008	0.012	0.026	-	11,953	11,299	11,794	16,277	29,483
Honduras	0.022	0.022	0.022	0.023	0.030	0.028	6,336	4,918	3,715	3,480	4,231	3,519
Hungary	±0.348	±0.310	±0.319	±0.308	±0.286	±0.304	13,387	13,497	15,971	15,395	14,435	15,267
Iceland	0.099	0.109	±0.107	±0.079	-	-	102,578	90,246	102,199	86,553	-	-
India	*0.029	*0.031	*0.026	*0.028	*0.030	*0.036	4,571	3,898	3,836	4,138	4,498	5,729
Indonesia	0.314	0.492	0.549	0.710	0.792	0.944	3,955	4,805	5,321	5,754	5,992	6,556
Iran (Islamic Republic of)	0.086	0.086	0.073	0.054	-	-	19,602	22,839	16,965	14,033	-	-
Iraq	-	*0.028	*0.040	-	-	-	-	30,452	43,272	-	-	-
Ireland	*0.192	*0.193	*0.216	*0.190	*0.220	*0.254	62,055	62,223	69,934	61,604	71,670	83,088
Israel	0.302	0.312	0.381	0.419	0.480	0.520	52,905	51,994	53,667	58,253	59,205	65,503
Italy	±9.669	±9.769	±14	±11	±12	-	139,856	141,039	152,160	123,940	138,606	-
Jamaica	0.056	0.049	0.039	-	-	-	26,513	24,692	18,814	-	-	-
Japan	±20	±22	±22	±23	±29	±31	116,754	131,056	137,059	151,990	178,493	199,529
Jordan	0.037	0.038	0.054	0.063	0.072	0.077	16,065	14,094	12,338	13,728	14,748	14,274
Kenya	*0.037	*0.025	*0.039	*0.024	*0.029	*0.015	9,404	6,795	10,404	6,302	7,306	3,649
Korea, Republic of	2.032	2.805	3.200	3.289	3.770	4.281	48,159	60,927	70,925	67,978	79,071	90,849
Kuwait	0.056	0.031	0.094	0.098	0.089	0.131	15,330	13,452	28,163	27,477	25,241	36,800

Continued.

Depending on the table, * means Enterprises, Engaged, or Factor Values; ± means Basis Unspecified; § means shown in millions. For additional notes and sources, see Appendix I.

205

Output and Output per Employee
- Continued -

Country	Output ($ bil.)						Output per Employee ($)					
	1990	1991	1992	1993	1994	1995	1990	1991	1992	1993	1994	1995
Kyrgyzstan	0.037	0.036	0.020	0.004	0.003	-	10,019	10,533	6,655	1,491	1,916	-
Latvia	0.001	0.003	0.019	0.036	0.043	0.046	161	338	2,229	4,667	6,492	7,572
Lithuania	-	-	0.041	0.050	0.053	-	-	-	2,829	3,862	4,837	-
Macau	0.011	0.009	0.009	0.008	0.006	0.005	14,802	16,468	17,665	20,794	21,529	27,409
Malaysia	*0.199	*0.303	*0.404	*0.582	*0.686	*0.785	13,424	14,018	16,745	19,079	20,913	22,500
Malta	0.047	0.048	0.055	0.053	0.058	-	33,223	32,697	34,883	31,739	34,244	-
Mauritius	0.017	0.017	0.023	0.020	0.020	0.025	11,574	13,934	18,046	14,929	13,375	16,569
Mexico	0.175	0.197	0.222	0.236	0.235	0.116	24,240	27,024	31,198	35,689	34,509	22,847
Mongolia	0.003	0.008	0.003	0.000	0.000	-	1,313	13,297	5,561	784	414	-
Mozambique	-	±0.002	-	±0.001	±0.001	-	-	1,872	-	1,673	1,543	-
Myanmar	0.006	0.008	0.010	0.012	0.016	-	3,975	4,812	6,385	7,200	9,797	-
Namibia	-	-	-	-	0.007	-	-	-	-	-	10,351	-
Nepal	0.005	0.012	-	0.008	0.014	-	3,856	3,822	-	2,814	3,537	-
Netherlands	1.016	1.073	1.157	1.109	1.134	1.234	101,649	100,650	107,139	104,223	106,582	122,527
Norway	±0.682	±0.675	±0.644	±0.658	±0.782	±0.912	100,030	102,375	113,497	102,712	114,485	131,250
Pakistan	0.021	0.024	0.026	-	-	-	9,536	9,174	9,521	-	-	-
Panama	0.021	0.022	0.023	0.024	0.026	0.026	15,715	15,371	15,981	16,400	17,370	17,611
Peru	0.077	0.105	0.089	-	-	-	12,778	17,156	20,064	-	-	-
Philippines	0.214	0.213	0.197	0.218	-	-	4,368	5,079	6,562	9,418	-	-
Poland	±0.592	±0.894	±0.917	±0.933	±1.133	±1.674	7,692	10,904	12,733	13,332	16,722	26,053
Portugal	0.688	0.925	1.103	0.881	0.869	1.059	18,064	18,759	21,463	21,954	21,421	24,620
Qatar	0.002	0.004	0.038	0.037	0.032	-	6,701	9,517	17,985	20,120	16,933	-
Republic of Moldova	0.000	0.260	0.018	0.022	0.019	0.020	0.047	31,363	2,328	3,888	3,520	4,093
Romania	±0.887	±0.687	±0.486	±0.636	-	-	4,357	3,745	3,008	3,904	-	-
Russian Federation	-	4.977	2.346	1.104	1.302	1.408	-	-	-	4,072	4,995	5,936
Senegal	0.001	-	-	-	-	-	12,105	-	-	-	-	-
Sierra Leone	-	-	-	0.002	-	-	-	-	-	2,485	-	-
Singapore	*0.285	*0.299	*0.326	*0.382	*0.416	*0.381	43,675	46,323	51,682	60,144	69,834	74,792
Slovakia	-	±0.171	±0.181	±0.165	-	-	-	9,030	10,103	10,530	-	-
Slovenia	-	-	-	0.682	0.643	0.654	-	-	-	-	-	-
South Africa	±0.870	±0.906	±0.929	±0.875	±0.906	±1.082	20,233	21,574	22,654	19,017	19,699	22,298
Spain	*3.745	*4.192	*4.361	±5.808	±5.797	±6.814	65,749	69,858	71,053	57,734	60,531	71,465
Sri Lanka	0.002	0.002	0.001	0.003	-	-	931	1,344	1,180	3,077	-	-
Suriname	*0.006	*0.010	*0.010	0.015	-	-	26,426	48,481	51,980	80,888	-	-
Swaziland	-	*0.006	*0.005	*0.004	-	-	-	16,216	15,977	11,569	-	-
Sweden	*1.215	*1.277	*1.222	*0.830	*1.037	*1.149	109,435	120,196	127,165	100,441	114,396	126,834
Thailand	1.983	0.546	-	-	-	-	61,895	18,823	-	-	-	-
Tonga	0.000	-	-	-	-	-	5,772	-	-	-	-	-
Trinidad and Tobago	±0.015	±0.025	±0.029	±0.027	±0.029	-	15,643	16,811	20,128	16,956	17,590	-
Turkey	0.192	0.181	0.305	0.392	0.257	0.336	44,444	45,670	43,291	48,750	35,094	43,743
Ukraine	*1.930	*1.880	*1.604	*0.419	*0.334	*0.230	14,512	14,800	13,256	3,740	3,305	2,496
United Kingdom	*9.729	*9.094	9.226	9.377	11	11	85,338	84,989	88,713	78,797	83,784	86,924
United Republic of Tanzania	*0.008	*0.009	-	-	-	-	1,235	3,775	-	-	-	-
USSR (Former)	±6.750	-	-	-	-	-	11,989	-	-	-	-	-
United States of America	*33	*32	*35	*39	*41	*45	75,662	78,128	85,655	90,727	93,560	97,276
Uruguay	0.027	0.039	0.034	0.043	0.041	0.044	10,064	12,769	14,196	19,364	18,244	20,350
Venezuela	0.165	0.232	0.276	0.266	0.248	0.249	13,518	16,581	18,429	18,059	15,180	19,530
Yugoslavia	±0.044	±0.049	±0.115	-	±0.217	±0.114	-	2,182	5,396	-	10,575	5,897
Zambia	0.013	-	-	-	0.010	-	8,843	-	-	-	5,347	-
Zimbabwe	*0.064	*0.064	*0.046	*0.046	*0.049	*0.061	12,874	13,613	9,414	8,997	8,567	10,238

Depending on the table, * means *Enterprises, Engaged,* or *Factor Values*; ± means *Basis Unspecified*; § means *shown in millions.* For additional notes and sources, see Appendix I.

Value Added and Value Added per Employee

Country	Value Added ($ bil.)						Value Added per Employee ($)					
	1990	1991	1992	1993	1994	1995	1990	1991	1992	1993	1994	1995
Argentina	±0.246	-	-	0.351	-	-	17,362	-	-	21,327	-	-
Australia	*1.032	*1.007	*1.036	-	-	-	29,487	29,629	33,412			
Austria	0.994	1.195	1.371	1.303	1.387	1.637	32,270	36,202	41,551	39,905	42,629	49,962
Bahamas	0.003	0.003	-	-	-	-	20,479	23,567				
Bangladesh	0.006	0.003	0.002	-	-	-	1,142	933	847			
Belgium	*1.613	*1.703	*1.737	*1.567	*1.632	*1.960	79,084	82,293	84,300			
Belize	0.003	0.003	0.002	-	-	-		5,278	2,950			
Bolivia	0.003	0.003	0.004	0.004	0.004	0.003	2,796	2,992	2,905	2,941	2,920	2,594
Bosnia & Herzegovina	±0.175	±0.168	-	-	-	-	6,774	7,500				
Brazil	*1.508	-	*0.840	*1.320	-	-	18,886	-	11,673	18,753		
Bulgaria	-	±0.033	±0.034	±0.044	-	-	-	1,731	2,129	3,580	-	-
Cameroon	0.003	0.007	0.001	0.000	0.003	0.005	12,525	16,563	5,919	3,673	5,851	47,176
Canada	2.245	1.877	1.820	1.791	1.882	2.090	34,022	34,120	37,146	35,811	36,191	37,072
Central African Republic	*0.001	-	-	-	-	-	8,705	-	-	-	-	-
Chile	0.053	0.059	0.089	0.089	0.115	0.134	10,562	11,244	15,060	14,275	16,021	18,402
China	±0.455	±0.452	±0.547	±0.838	±0.682	±0.676	1,082	1,129	1,368	2,465	1,947	1,931
China (Hong Kong SAR)	0.066	0.055	0.052	0.029	0.022		12,522	14,221	15,275	14,414	12,364	
China (Taiwan Province)	0.743	0.810	0.847	0.755	0.707	0.628	22,601	25,916	27,312	25,752	24,468	23,890
Colombia	0.038	0.033	0.038	0.041	0.076	0.088	4,348	4,559	4,280	4,404	6,804	7,822
Costa Rica	0.021	0.018	0.020	0.018	0.017	0.018	7,926	7,403	8,365	6,923	-	-
Croatia	±0.171	±0.186	±0.112	-	-	-	6,802	9,547	6,988			
Cyprus	0.036	0.037	0.042	0.038	0.042	0.046	15,463	16,058	17,598	14,859	15,572	21,013
Czechoslovakia (Former)	±0.154	-	-	-	-	-	2,616	-	-	-	-	-
Denmark	*0.642	*0.662	*0.727	*0.642	*0.799	*0.989	31,790	32,200	36,399	33,511	38,277	45,513
Ecuador	0.009	0.008	0.007	0.008	0.011	0.014	2,975	2,392	2,806	3,477	4,491	4,945
Egypt	*0.013	*0.018	*0.013	*0.017	*0.018	*0.019	2,233	2,156	1,976	2,180	2,258	2,407
El Salvador	-	0.004	-	0.006	0.004	0.005	-	-	-	6,800	4,188	3,789
Ethiopia	0.006	0.006	0.004	0.003	0.004	0.005	3,546	3,732	2,900	1,854	2,071	2,044
Fiji	0.003	0.004	0.004	0.004	0.005	-	3,221	4,624	4,560	3,772	4,118	-
Finland	±0.515	±0.377	±0.290	±0.213	±0.245	±0.419	47,239	36,260	32,954	26,968	33,063	46,982
FYR Macedonia	±0.062	±0.055	±0.029	±0.026	±0.022	±0.024	7,774	8,319	4,337	4,237	3,480	4,883
France	±3.973	±3.974	±4.273	±4.052	±4.101	±4.739	47,581	48,117	52,052	52,904	54,895	63,869
Gabon	-	*0.001	*0.001	*0.001	*0.000	*0.001	-	10,662	13,365	15,028	7,311	12,576
Germany (Western Part)	7.885	8.735	10	9.773	-	-	55,648	59,174	68,340	-	-	-
Ghana	-	-	-	0.005	0.004	0.005	-	-	-	1,740	1,467	1,497
Greece	*0.093	*0.092	*0.103	*0.102	*0.109	*0.122	14,698	14,950	16,560	16,437	17,552	19,582
Honduras	*0.007	*0.007	*0.007	*0.007	*0.008	*0.009	2,111	1,509	1,183	1,073	1,103	1,126
Hungary	±0.120	±0.109	±0.104	±0.103	±0.095	±0.101	4,613	4,745	5,210	5,146	4,824	5,096
Iceland	0.040	0.045	±0.049	±0.034	-	-	41,986	37,687	46,550	36,780	-	-
India	*0.008	*0.007	*0.006	*0.007	*0.007	*0.009	1,192	944	900	1,030	1,114	1,409
Indonesia	*0.117	*0.188	*0.230	*0.249	*0.271	*0.332	1,476	1,832	2,228	2,019	2,047	2,304
Iran (Islamic Republic of)	0.034	0.037	0.037	0.028	-	-	7,726	9,862	8,482	7,309	-	-
Iraq	-	*0.013	*0.014	-	-	-	-	13,468	15,191	-	-	-
Ireland	*0.086	*0.092	*0.103	*0.090	*0.104	*0.121	27,764	29,548	33,204	29,247	34,015	39,428
Israel	0.131	0.134	0.139	0.158	0.179	0.193	22,972	22,377	19,645	21,986	22,099	24,371
Italy	±2.900	±3.013	±4.147	±3.402	±3.464	-	41,954	43,502	45,319	38,723	38,792	-
Jamaica	0.019	0.016	0.009	-	-	-	8,767	8,173	4,545	-	-	-
Japan	±8.730	±9.739	±9.862	±10	±13	±14	51,964	59,027	62,024	68,274	80,227	89,716
Jordan	0.014	0.015	0.020	0.030	0.027	0.027	5,948	5,440	4,651	6,582	5,497	4,952
Kenya	*0.011	*0.011	*0.007	*0.004	*0.006	*0.007	2,823	2,846	1,817	1,080	1,461	1,632
Korea, Republic of	0.972	1.479	1.611	1.621	1.979	2.138	23,038	32,115	35,704	33,490	41,508	45,358
Kuwait	0.030	0.016	0.047	0.050	0.045	0.066	8,200	6,918	13,979	13,985	12,618	18,429
Latvia	-	-	-	*0.016	*0.021	*0.020	-	-	-	2,081	3,140	3,349
Macau	0.005	0.005	0.004	0.004	0.003	0.002	7,058	8,384	8,757	9,878	11,108	11,703
Malaysia	*0.070	*0.112	*0.139	*0.201	*0.246	*0.282	4,729	5,200	5,782	6,597	7,511	8,076
Malta	0.028	0.026	0.029	0.027	0.029	-	19,403	17,803	18,624	16,179	16,967	-
Mauritius	0.004	0.006	0.010	0.008	0.008	0.010	3,058	4,485	7,712	5,888	5,269	6,731
Mexico	0.068	0.078	0.085	0.091	0.092	0.045	9,392	10,746	11,982	13,775	13,515	8,949
Mongolia	0.000	0.002	0.001	0.000	0.000	-	171	3,267	1,982	240	143	-
Myanmar	±0.004	±0.004	±0.004	±0.006	±0.007	-	2,366	2,171	2,453	3,402	4,125	-

Continued.

Depending on the table, * means *Enterprises, Engaged*, or *Factor Values*; ± means *Basis Unspecified*; § means *shown in millions*. For additional notes and sources, see Appendix I.

207

Value Added and Value Added per Employee

- Continued -

Country	Value Added ($ bil.)						Value Added per Employee ($)					
	1990	1991	1992	1993	1994	1995	1990	1991	1992	1993	1994	1995
Namibia	-	-	-	-	0.005	-	-	-	-	-	6,600	-
Nepal	0.002	0.005	-	0.003	0.007	-	1,602	1,644	-	1,097	1,781	-
Netherlands	0.362	0.395	0.436	0.415	0.424	0.464	36,209	37,010	40,348	39,008	39,881	46,068
New Zealand	0.126	-	-	-	-	-	23,853	-	-	-	-	-
Norway	±0.236	±0.229	±0.220	±0.232	±0.280	±0.325	34,656	34,764	38,812	36,291	40,933	46,755
Pakistan	0.007	0.007	0.008	-	-	-	2,929	2,843	3,034	-	-	-
Panama	0.007	0.008	0.009	0.009	0.010	0.010	5,654	5,869	6,091	6,243	6,596	6,684
Peru	0.030	0.042	0.035	-	-	-	4,972	6,900	7,818	-	-	-
Philippines	0.103	0.104	0.084	0.081	-	-	2,100	2,470	2,784	3,511	-	-
Poland	±0.307	±0.346	±0.394	±0.361	±0.434	±0.636	3,986	4,224	5,467	5,163	6,403	9,896
Portugal	0.233	0.321	0.372	0.291	0.300	0.366	6,107	6,502	7,243	7,255	7,408	8,510
Qatar	0.001	0.002	0.019	0.019	0.015	-	4,467	4,441	9,123	10,438	7,737	-
Romania	±0.321	±0.246	±0.156	±0.209	-	-	1,576	1,341	967	1,283	-	-
Russian Federation	-	-	-	±0.620	±0.560	±0.519	-	-	-	2,287	2,147	2,186
Saint Lucia	-	±0.001	±0.002	±0.001	±0.001	±0.001	-	-	-	-	-	-
Senegal	0.000	0.000	0.000	0.000	0.000	-	2,949	3,706	1,908	3,307	3,126	-
Sierra Leone	-	-	-	0.001	-	-	-	-	-	1,271	-	-
Singapore	*0.089	*0.099	*0.113	*0.124	*0.134	*0.123	13,698	15,355	17,964	19,609	22,414	24,205
Slovenia	-	-	-	0.246	0.120	0.175	-	-	-	-	-	-
South Africa	*0.307	*0.301	*0.278	*0.334	*0.352	*0.419	7,146	7,157	6,790	7,251	7,653	8,637
Spain	*1.534	*1.693	*1.799	±2.264	±2.119	±2.438	26,937	28,212	29,318	22,506	22,127	25,566
Sri Lanka	0.001	0.001	0.001	0.002	-	-	444	692	675	1,681	-	-
Swaziland	-	*0.002	*0.004	*0.001	-	-	-	4,945	14,297	2,715	-	-
Sweden	*0.551	*0.428	*0.414	*0.299	*0.373	*0.412	49,619	40,302	43,032	36,186	41,166	45,455
Thailand	0.643	0.277	-	-	-	-	20,078	9,553	-	-	-	-
Trinidad and Tobago	-	±0.010	±0.008	±0.009	±0.010	-	-	6,671	5,805	5,947	5,816	-
Turkey	0.081	0.089	0.133	0.171	0.119	0.155	18,680	22,533	18,899	21,302	16,208	20,215
United Kingdom	*4.554	*4.331	*4.392	*4.550	*5.164	*5.208	39,944	40,476	42,233	38,232	40,343	41,862
United Republic of Tanzania	*0.002	*0.002	-	-	-	-	342	926	-	-	-	-
United States of America	*17	*16	*18	*19	*20	*22	38,607	39,754	43,995	45,144	46,312	47,213
Uruguay	0.013	0.020	0.019	0.019	0.021	0.022	4,880	6,547	7,836	8,658	9,113	10,165
Venezuela	0.065	0.091	0.108	0.095	0.087	0.166	5,308	6,500	7,194	6,426	5,325	13,037
Yugoslavia	±0.019	±0.023	±0.062	-	±0.087	±0.050	-	1,004	2,903	-	4,264	2,583
Zambia	0.007	-	-	-	0.003	-	5,204	-	-	-	1,923	-
Zimbabwe	*0.032	*0.028	*0.020	*0.016	*0.020	*0.025	6,400	5,902	3,998	3,099	3,541	4,210

Capital Investment

Country	Gross Fixed Capital Formation ($ mil.)						Per Establishment ($ mil)					
	1990	1991	1992	1993	1994	1995	1990	1991	1992	1993	1994	1995
Argentina	-	-	-	25	-	-	-	-	-	0.005	-	-
Australia	67	-	-	-	-	-	0.030	-	-	-	-	-
Austria	140	171	194	180	186	-	0.229	0.293	0.328	0.317	0.341	-
Bahrain	-	-	1.112	-	-	-	-	-	0.004	-	-	-
Bangladesh	0.043	-	-	-	-	-	0.000	-	-	-	-	-
Belgium	302	285	220	188	229	223	-	-	-	-	-	-
Bolivia	0.161	0.592	0.266	0.292	0.021	-	0.002	0.005	0.002	0.002	0.000	-
Bosnia & Herzegovina	8.078	-	-	-	-	-	0.144	-	-	-	-	-
Brazil	176	-	63	118	-	-	0.192	-	0.075	0.151	-	-
Bulgaria	10	1.906	39	27	8.387	1.743	0.163	0.031	0.522	0.044	0.008	0.002
Cameroon	-	-	-	-	0.670	0.048	-	-	-	-	0.112	0.006
Canada	55	-	-	-	-	-	0.021	-	-	-	-	-
Chile	2.847	4.055	8.070	8.806	7.294	-	0.084	0.119	0.212	0.183	0.155	-
China (Hong Kong SAR)	8.601	2.188	1.679	0.259	0.776	-	0.009	0.003	0.002	0.001	0.002	-
Colombia	4.259	2.979	2.949	2.440	3.391	-	0.018	0.013	0.011	0.010	0.013	-
Croatia	18	1.734	7.351	5.005	5.467	3.859	0.151	0.011	0.038	0.018	0.014	0.008
Cyprus	3.947	2.469	3.889	3.922	2.116	3.359	0.006	0.004	0.006	0.006	0.003	0.005
Czechoslovakia (Former)	34	-	-	-	-	-	1.618	-	-	-	-	-

Continued.

Depending on the table, * means *Enterprises, Engaged,* or *Factor Values;* ± means *Basis Unspecified;* § means *shown in millions.* For additional notes and sources, see Appendix I.

Capital Investment

- Continued -

Country	Gross Fixed Capital Formation ($ mil.)						Per Establishment ($ mil)					
	1990	1991	1992	1993	1994	1995	1990	1991	1992	1993	1994	1995
Denmark	62	54	-	-	-	-	0.038	0.033	-	-	-	-
Ecuador	4.012	0.187	2.197	2.459	2.654	3.461	0.069	0.002	0.028	0.032	0.034	0.039
Egypt	2.700	3.781	1.348	4.627	-	-	0.028	0.030	0.012	0.040	-	-
El Salvador	-	-	-	0.512	1.084	2.245	-	-	-	0.051	0.057	0.066
Ethiopia	0.057	0.158	0.266	0.091	0.497	0.916	0.002	0.008	0.014	0.005	0.009	0.015
Fiji	0.417	1.808	1.112	-	-	-	0.004	0.019	0.011	-	-	-
Finland	47	35	38	8.893	16	27	0.140	0.098	0.120	0.031	0.062	0.087
FYR Macedonia	5.034	3.015	3.086	0.691	0.405	0.284	0.105	0.029	0.015	0.003	0.001	0.001
Gabon	-	0.039	0.121	0.085	0.016	0.218	-	0.013	0.030	0.014	0.003	0.024
Germany (Western Part)	566	-	752	-	-	-	0.425	-	0.562	-	-	-
Greece	18	11	11	-	-	-	0.043	0.027	0.026	-	-	-
Hungary	19	8.269	6.811	14	-	-	0.102	0.030	0.018	0.030	-	-
India	1.143	0.572	0.926	1.181	-	-	0.004	0.002	0.003	0.004	-	-
Indonesia	18	62	34	35	37	34	0.030	0.089	0.047	0.045	0.041	0.030
Iran (Islamic Republic of)	1.946	7.543	10	3.324	-	-	0.020	0.140	0.178	0.069	-	-
Ireland	10	8.239	-	-	-	-	0.049	0.040	-	-	-	-
Israel	7.936	13	12	24	38	-	0.018	0.026	0.021	0.043	0.063	-
Italy	411	422	473	362	395	-	0.290	0.300	0.246	0.188	0.203	-
Japan	477	586	632	333	587	-	0.044	0.054	0.061	0.033	0.062	-
Jordan	0.235	0.351	0.669	1.140	1.176	0.217	0.000	0.000	0.000	0.001	0.001	0.000
Korea, Republic of	155	4,353	240	200	231	281	0.083	2.117	0.113	0.074	0.085	0.096
Kuwait	0.240	16	2.119	2.669	2.950	-	0.001	0.052	0.006	0.008	0.009	-
Latvia	0.032	0.036	0.675	2.654	2.613	4.928	0.002	0.001	0.009	0.026	0.039	0.038
Macau	0.407	0.184	0.223	0.555	1.142	0.067	0.002	0.001	0.001	0.004	0.009	0.001
Malaysia	33	46	43	59	72	-	0.097	0.118	0.105	0.116	0.145	-
Malta	1.208	4.604	2.142	2.730	2.283	-	0.004	0.013	0.006	0.007	0.006	-
Mexico	3.400	3.155	-	-	-	-	0.050	0.046	-	-	-	-
Mongolia	0.214	-	-	-	-	-	0.107	-	-	-	-	-
Morocco	2.548	1.493	0.703	-	-	-	0.064	0.167	0.632	0.689	0.383	-
Myanmar	0.383	1.004	3.792	4.134	2.678	-	0.011	0.167	0.632	0.689	0.383	-
Nepal	1.260	-	-	-	-	-	0.011	-	-	-	-	-
Netherlands	91	79	61	61	-	-	0.405	0.343	0.250	0.248	-	-
Norway	25	20	13	18	27	-	0.102	0.081	0.087	0.108	0.150	-
Pakistan	-	0.207	-	-	-	-	-	0.004	-	-	-	-
Panama	0.220	0.927	-	-	-	-	0.002	0.010	-	-	-	-
Peru	0.328	1.494	2.037	-	-	-	0.000	0.002	0.003	-	-	-
Philippines	5.841	4.767	6.507	6.633	-	-	0.001	0.001	0.010	0.012	-	-
Poland	42	38	43	42	-	-	0.143	0.127	0.149	0.147	-	-
Portugal	54	57	67	65	129	-	0.008	0.008	0.010	0.010	0.019	-
Romania	-	-	-	38	31	-	-	-	-	0.016	0.018	-
Senegal	0.018	-	-	-	-	-	0.005	-	-	-	-	-
Singapore	12	9.349	13	23	20	-	0.087	0.067	0.096	0.165	0.143	-
Slovenia	20	22	9.066	13	15	28	0.156	0.234	0.064	0.051	0.053	0.016
Spain	124	123	212	186	182	281	0.011	0.012	0.021	-	-	-
Sri Lanka	0.075	0.054	0.132	0.202	-	-	0.001	0.001	0.002	0.003	-	-
Swaziland	-	0.108	0.786	0.040	-	-	-	0.015	0.112	0.013	-	-
Thailand	889	86	-	-	-	-	3.737	0.259	-	-	-	-
Trinidad and Tobago	-	-	-	0.187	-	-	-	-	-	0.001	-	-
Turkey	5.750	6.951	6.111	14	6.214	-	0.122	0.145	0.032	0.076	0.036	-
Ukraine	39	18	38	3.679	2.944	1.546	0.170	0.030	0.044	0.011	0.009	0.005
United Kingdom	211	205	224	-	254	-	0.027	0.028	0.031	-	0.033	-
United Republic of Tanzania	1.984	2.811	-	-	-	-	0.015	0.022	-	-	-	-
United States of America	620	510	609	745	785	908	-	-	0.031	-	-	-
Uruguay	1.240	1.968	-0.064	0.910	-	-	-	-	-	-	-	-
Venezuela	3.987	3.467	5.645	1.299	0.949	29	0.005	0.004	0.006	0.002	0.001	0.042
Yugoslavia	2.714	1.980	0.477	-	4.315	1.524	0.029	0.010	0.001	-	0.006	0.002
Zambia	2.266	-	-	-	-	-	0.189	-	-	-	-	-
Zimbabwe	4.289	5.397	5.065	6.103	4.748	-	0.097	0.123	0.113	0.136	0.101	-

Depending on the table, * means *Enterprises*, *Engaged*, or *Factor Values*; ± means *Basis Unspecified*; § means *shown in millions*. For additional notes and sources, see Appendix I.

209

Indexes of Industrial Production

Country	Index of Industrial Production (1990=100)						Index of Employment (1990=100)					
	1990	1991	1992	1993	1994	1995	1990	1991	1992	1993	1994	1995
Albania	100	-	-	-	-	-	-	-	-	-	-	-
Algeria	100	85	70	64	56	56	-	-	-	-	-	-
Angola	100	-	-	-	-	-	-	-	-	-	-	-
Argentina	100	191	-	-	-	-	100	-	-	116	135	-
Armenia	100	-	-	-	-	-	100	90	-	-	-	-
Australia	100	97	92	96	-	-	100	97	89	-	-	-
Austria	100	109	112	109	111	113	100	107	107	106	106	106
Azerbaijan	-	-	-	-	-	-	100	96	84	66	51	-
Bahamas	100	-	-	-	-	-	100	89	-	-	-	-
Bahrain	100	-	-	-	-	-	-	-	-	-	-	-
Bangladesh	100	-	-	-	-	-	100	54	44	-	-	-
Barbados	100	64	48	19	23	19	-	-	-	-	-	-
Belgium	100	98	92	89	91	96	100	101	101	-	-	-
Belize	100	125	130	128	122	107	-	-	-	-	-	-
Benin	100	-	-	-	-	-	-	-	-	-	-	-
Bermuda	-	-	-	-	-	-	100	90	63	71	75	69
Bhutan	100	-	-	-	-	-	-	-	-	-	-	-
Bolivia	100	106	107	109	118	112	100	114	134	133	128	131
Bosnia & Herzegovina	-	-	-	-	-	-	100	87	-	-	5	-
Brazil	100	95	95	-	-	-	100	-	90	88	-	-
Bulgaria	100	83	74	65	75	74	100	84	71	54	70	68
Burkina-Faso	100	-	-	-	-	-	-	-	-	-	-	-
Cambodia	100	-	-	-	-	-	100	105	-	-	-	-
Cameroon	100	75	89	94	119	131	100	200	45	25	217	52
Canada	100	83	82	86	97	106	100	83	74	76	79	85
Central African Republic	100	-	-	-	-	-	100	-	-	-	-	-
Chile	100	100	98	98	103	105	100	105	118	124	142	144
China	100	-	-	-	-	-	100	95	95	81	83	83
China (Hong Kong SAR)	100	95	93	85	81	78	100	74	64	38	34	-
China (Taiwan Province)	100	106	106	95	89	78	100	95	94	89	88	80
Colombia	100	94	104	121	149	154	100	84	103	108	128	129
Costa Rica	100	84	80	64	61	63	100	88	89	94	-	-
Cote d'Ivoire	100	60	30	40	40	75	-	-	-	-	-	-
Croatia	100	-	-	-	-	-	100	78	64	61	57	53
Cuba	100	-	-	-	-	-	-	-	-	-	-	-
Cyprus	100	90	91	87	90	76	100	100	103	109	117	95
Czechoslovakia (Former)	100	73	68	-	-	-	100	54	-	-	-	-
Czech Republic	100	-	-	-	-	-	100	100	150	144	-	-
Dem. Rep. of the Congo	100	83	83	-	-	-	-	-	-	-	-	-
Denmark	100	106	106	99	114	122	100	102	99	95	103	108
Dominican Republic	100	-	-	-	-	-	-	-	-	-	-	-
Ecuador	100	112	131	141	-	-	100	100	76	72	80	92
Egypt	100	90	101	104	103	101	100	141	116	138	138	137
El Salvador	100	109	-	-	-	-	-	-	-	-	-	-
Equatorial Guinea	-	-	-	-	-	-	100	-	-	-	-	-
Ethiopia and Eritrea	100	67	67	-	-	-	-	-	-	-	-	-
Ethiopia	-	-	-	-	-	-	100	92	88	86	127	135
Fiji	100	-	-	-	-	-	100	88	101	114	115	-
Finland	100	84	78	76	82	86	100	95	81	72	68	82
FYR Macedonia	100	-	-	-	-	-	100	84	85	77	80	61
France	100	98	93	90	87	85	100	99	98	92	89	89
Gabon	100	-	-	-	-	-	-	-	-	-	-	-
Gambia	100	-	-	-	-	-	-	-	-	-	-	-
Germany (Eastern Part)	100	-	-	-	-	-	-	-	-	-	-	-
Germany (Western Part)	100	109	107	102	99	99	100	104	105	-	-	-
Ghana	100	-	-	-	-	-	-	-	-	-	-	-
Greece	100	95	86	86	91	89	100	97	98	98	98	98
Guatemala	100	146	146	-	-	-	-	-	-	-	-	-
Guyana	100	-	-	-	-	-	-	-	-	-	-	-
Haiti	100	-	-	-	-	-	-	-	-	-	-	-

Continued.

Depending on the table, * means *Enterprises, Engaged,* or *Factor Values;* ± means *Basis Unspecified;* § means *shown in millions.* For additional notes and sources, see Appendix I.

Indexes of Industrial Production

- Continued -

Country	Index of Industrial Production (1990=100)						Index of Employment (1990=100)					
	1990	1991	1992	1993	1994	1995	1990	1991	1992	1993	1994	1995
Honduras	100	143	185	200	277	370	100	128	169	185	203	227
Hungary	100	86	70	74	68	71	100	88	77	77	76	77
Iceland	100	-	-	-	-	-	100	125	109	95	94	-
India	100	101	95	104	105	121	100	123	106	105	105	99
Indonesia	100	107	115	104	109	108	100	129	130	155	166	181
Iran (Islamic Republic of)	100	99	110	110	103	109	100	86	98	87	-	-
Iraq	100	-	-	-	-	-	-	-	-	-	-	-
Ireland	100	100	103	104	115	121	100	100	100	100	99	99
Israel	100	111	125	140	158	168	100	105	125	126	142	139
Italy	100	106	104	103	-	-	100	100	132	127	129	-
Jamaica	100	80	70	-	-	-	100	94	99	111	114	-
Japan	100	97	93	89	86	81	100	98	95	90	95	92
Jordan	100	50	75	77	67	72	100	116	188	198	211	234
Kenya	100	97	64	68	69	72	100	96	96	98	100	104
Korea, Republic of	100	102	95	88	86	86	100	109	107	115	113	112
Kuwait	100	-	-	-	-	-	100	64	91	98	97	98
Kyrgyzstan	100	-	-	-	-	-	100	93	81	72	41	-
Lao P.D.R.	100	-	-	-	-	-	-	-	-	-	-	-
Latvia	100	-	-	-	-	-	100	114	117	104	90	82
Lebanon	100	-	-	-	-	-	-	-	-	-	-	-
Lesotho	100	-	-	-	-	-	100	-	-	24	9	-
Liberia	100	100	100	-	-	-	-	-	-	-	-	-
Libyan Arab Jamahiriya	100	-	-	-	-	-	-	-	-	-	-	-
Lithuania	100	-	-	-	-	-	-	-	-	-	-	-
Luxembourg	100	106	90	96	102	94	-	-	-	-	-	-
Macau	100	-	-	-	-	-	100	78	68	57	42	26
Malawi	100	108	-	-	-	-	100	-	-	-	-	-
Malaysia	100	105	117	143	149	158	100	146	163	206	222	236
Malta	100	114	124	139	-	-	100	103	110	117	120	-
Mauritius	100	-	-	-	-	-	100	87	89	95	103	106
Mexico	100	104	106	97	102	81	100	101	98	91	94	70
Mongolia	100	72	46	41	32	39	100	27	25	25	41	6
Morocco	100	108	102	111	118	103	-	-	-	-	-	-
Mozambique	100	70	68	-	-	-	100	92	50	61	65	-
Myanmar	100	81	96	-	-	-	100	100	100	100	100	-
Namibia	100	-	-	-	-	-	-	-	-	-	-	-
Nepal	100	-	-	-	-	-	100	253	-	226	315	-
Netherlands	100	101	103	102	102	96	100	107	108	106	106	101
New Zealand	100	107	117	-	-	-	100	95	98	108	-	-
Nicaragua	100	80	80	-	-	-	-	-	-	-	-	-
Nigeria	100	100	100	-	-	-	-	-	-	-	-	-
Norway	100	96	92	86	92	94	100	97	83	94	100	102
Pakistan	100	106	102	-	-	-	100	116	121	-	-	-
Panama	100	113	117	121	130	131	100	106	107	109	111	112
Papua New Guinea	100	-	-	-	-	-	-	-	-	-	-	-
Paraguay	100	110	111	82	89	88	-	-	-	-	-	-
Peru	100	125	102	125	150	137	100	102	74	-	-	-
Philippines	100	107	108	116	129	145	100	86	61	47	-	-
Poland	100	110	120	133	152	189	100	106	94	91	88	83
Portugal	100	109	114	124	105	112	100	129	135	105	106	113
Puerto Rico	100	-	-	-	-	-	100	76	86	84	120	120
Qatar	100	-	-	-	-	-	100	176	857	738	765	-
Republic of Moldova	100	-	-	-	-	-	100	94	88	65	62	56
Romania	100	99	81	104	116	152	100	90	79	80	74	-
Russian Federation	100	-	-	-	-	-	-	-	-	-	-	-
Saudi Arabia	100	-	-	-	-	-	-	-	-	-	-	-
Senegal	100	110	108	106	90	103	100	93	282	89	123	-
Sierra Leone	100	-	-	-	-	-	-	-	-	-	-	-
Singapore	100	106	107	118	133	109	100	99	97	97	91	78
Slovakia	100	-	-	-	-	-	-	-	-	-	-	-

Continued.

Depending on the table, * means *Enterprises, Engaged,* or *Factor Values*; ± means *Basis Unspecified*; § means *shown in millions.* For additional notes and sources, see Appendix I.

211

Indexes of Industrial Production

- Continued -

Country	Index of Industrial Production (1990=100)						Index of Employment (1990=100)					
	1990	1991	1992	1993	1994	1995	1990	1991	1992	1993	1994	1995
Slovenia	100	95	88	92	95	104	-	-	-	-	-	-
Somalia	100	-	-	-	-	-	-	-	-	-	-	-
South Africa	100	98	92	91	95	105	100	98	95	107	107	113
Spain	100	107	90	86	91	95	100	105	108	177	168	167
Sri Lanka	100	115	76	111	83	153	100	84	70	65	-	-
Suriname	100	92	98	-	-	-	100	98	92	88	-	-
Sweden	100	94	89	83	92	92	100	96	87	74	82	82
Switzerland	100	94	92	86	94	91	-	-	-	-	-	-
Syrian Arab Republic	100	40	35	101	73	113	-	-	-	-	-	-
Thailand	100	107	114	-	-	-	100	90	-	-	-	-
Togo	100	-	-	-	-	-	-	-	-	-	-	-
Tonga	100	-	-	-	-	-	100	-	-	-	-	-
Trinidad and Tobago	100	97	58	68	32	68	100	151	146	158	168	-
Tunisia	100	100	111	-	-	-	-	-	-	-	-	-
Turkey	100	85	87	108	75	85	100	92	163	186	169	177
Uganda	100	85	92	78	63	111	-	-	-	-	-	-
Ukraine	100	-	-	-	-	-	100	95	91	84	76	69
United Arab Emirates	100	-	-	-	-	-	-	-	-	-	-	-
United Kingdom	100	88	87	90	94	89	100	94	91	104	112	109
United Republic of Tanzania	100	77	63	50	41	-	100	36	-	-	-	-
USSR (Former)	100	-	-	-	-	-	100	-	-	-	-	-
United States of America	100	94	99	104	107	108	100	93	95	97	100	105
Uruguay	100	86	108	-	-	-	100	113	89	82	85	80
Venezuela	100	-	-	-	-	-	100	115	123	121	134	104
Yugoslavia	100	-	-	-	-	-	-	-	-	-	-	-
Yugoslavia (Former)	100	83	-	-	-	-	-	-	-	-	-	-
Zambia	100	101	106	109	76	71	100	-	-	-	127	-
Zimbabwe	100	113	118	106	118	128	100	95	99	103	116	121

Depending on the table, * means *Enterprises, Engaged,* or *Factor Values*; ± means *Basis Unspecified*; § means *shown in millions.* For additional notes and sources, see Appendix I.

Representative Companies in Sector

Name	Address	Tele-phone	Fax	Employ-ment	Y
Kimball International Inc.	1600 Royal St., Jasper, Indiana 47549-1001, U.S.A.	812-482-1600		8,949	97
Godrej & Boyce Manufacturing Co. Ltd.	Godrej Bhavan Homji St., Bombay 400 001, India	22 5171166	22 5172688	8,810	94
Herman Miller Inc.	PO Box 302, Zeeland, Michigan 49464-0302, U.S.A.	616-654-3000		7,425	97
MFI Furniture Group PLC	Colindale, London NW9 6TD, United Kingdom	81 2008000		7,134	94
Broyhill Furniture Industries Inc.	1 Broyhill Park, Lenoir, North Carolina 28633, U.S.A.	704-758-3111	704-758-3720	7,000	97
HON Industries Inc.	PO Box 1109, Muscatine, Iowa 52761-7109, U.S.A.	319-264-7085	319-264-7217	6,900	97
Associated Furniture Companies Ltd.	PO Box 98800, Sloane Park 2152, Republic of South Africa	11 7066004	11 4635097	6,600	94
Thomasville Furniture Industries Inc.	PO Box 339, Thomasville, North Carolina 27361-0339, U.S.A.	336-472-4000	336-472-4071	6,500	97
Ethan Allen Interiors Inc.	Ethan Allen Dr., Danbury, Connecticut 06801, U.S.A.	203-743-8000	203-743-8298	6,417	97
Klaussner Furniture Industries Inc.	PO Box 220, Asheboro, North Carolina 27204, U.S.A.	336-625-6174	336-626-0905	6,000	97
LADD Furniture Inc.	1 Plaza Ctr., Box HP3, High Point, North Carolina 27261-1500, U.S.A.	910-889-0333	910-888-6446	5,900	97
Ethan Allen Inc.	Ethan Allen Dr., Danbury, Connecticut 06811, U.S.A.	203-743-8000	203-743-8298	5,884	97
Bassett Furniture Industries Inc.	PO Box 626, Bassett, Virginia 24055, U.S.A.	540-629-6000	540-629-6400	5,700	97
Chesapeake Corp.	PO Box 2350, Richmond, Virginia 23218-2350, U.S.A.	804-697-1000	804-697-1199	5,305	95
Consolidated Furniture Corp.	4401 Fair Lakes Ct., Fairfax, Virginia 22033, U.S.A.	703-968-8050	703-968-8023	5,000	
Triangle Pacific Corp.	PO Box 660100, Dallas, Texas 75266-0100, U.S.A.	214-887-2000		4,967	96
Invacare Corp.	PO Box 4028, Elyria, Ohio 44036, U.S.A.	216-329-6000	216-366-9008	4,470	97
Drexel Heritage Furnishings Inc.	101 N. Main St., Drexel, North Carolina 28619, U.S.A.	704-433-3000	704-433-3349	4,400	95
Spalding and Evenflo Companies Inc.	PO Box 30101, Tampa, Florida 33630, U.S.A.	813-204-5200	813-204-5219	4,400	97
Sunrise Medical Inc.	2382 Faraday Ave., Ste. 200, Carlsbad, California 92008-7220, U.S.A.	760-930-1500	760-930-1580	4,254	97
ACCO USA Inc.	300 Tower Pkwy., Lincolnshire, Illinois 60069, U.S.A.	847-541-9500		4,000	97
Hill Rom Company Inc.	1069 State Rte. 46, Batesville, Indiana 47006-9167, U.S.A.	812-934-7777	812-934-8189	4,000	97
Universal Furniture Industries Inc.	2622 Uwharrie Rd., High Point, North Carolina 27263, U.S.A.	919-861-7200	919-431-2124	4,000	94
Knoll Inc.	1235 Water St., East Greenville, Pennsylvania 18041, U.S.A.	215-679-7991		3,550	96
ACCO Europe PLC	Gate House Rd., Aylesbury HP19 3DY, United Kingdom	296 397444	296 895399	3,484	91
Silentnight Holdings PLC	Salterforth, Colne BB8 6BL, United Kingdom	282 812711	282 816840	3,380	94
Ashley Furniture Industries Inc.	1 Ashley Way, Arcadia, Wisconsin 54612, U.S.A.	608-323-3377	608-323-3021	3,200	97
Bush Industries Inc.	1 Mason Dr., Jamestown, New York 14702-0460, U.S.A.	716-665-2000	716-665-2618	3,200	97
Sauder Woodworking Co.	502 Middle St., Archbold, Ohio 43502, U.S.A.	419-446-2711		3,200	97
JDI Group Inc.	PO Box 7240, St. Louis, Missouri 63177, U.S.A.	314-291-0400	314-291-6260	3,000	97
Krueger International Inc.	PO Box 8100, Green Bay, Wisconsin 54308-8100, U.S.A.	920-468-8100	920-468-0280	3,000	97
Palliser Furniture Ltd.	55-1155 Gateway Rd., Winnipeg, Manitoba R2G 1B9, Canada	204-988-5600	204-663-5267	2,909	98
Stanley Furniture	PO Box 30, Stanleytown, Virginia 24168, U.S.A.	703-629-7561		2,900	94
Virco Manufacturing Corp.	2027 Harpers Way, Torrance, California 90501, U.S.A.	310-533-0474	310-533-6841	2,800	97
Stanley Furniture Company Inc.	PO Box 30, Stanleytown, Virginia 24168, U.S.A.	540-627-2000		2,700	97
Aaron Rents Inc.	309 E. Paces Ferry Rd., N.E., Atlanta, Georgia 30305-2377, U.S.A.	404-231-0011	404-240-6584	2,550	97
Berkline Corp.	1 Berkline Dr., Morristown, Tennessee 37814, U.S.A.	615-585-1500		2,500	
Carolina Mills Inc.	PO Box 157, Maiden, North Carolina 28650, U.S.A.	704-428-9911	704-428-2335	2,500	93
Kirsch	PO Box 0370, Sturgis, Michigan 49091, U.S.A.	616-659-5100	616-659-5606	2,500	97
Interlake Corp.	550 Warrenville Rd., Lisle, Illinois 60532-4387, U.S.A.	630-852-8800	630-719-7277	2,491	97
Peter Black Holdings PLC	Lawkholme Ln., Keighley BD21 3JQ, United Kingdom	535 661177	535 609973	2,460	94
Home Improvement Holdings Ltd.	Farnham Common, Slough SL2 3PQ, United Kingdom	753 645622	753 646627	2,430	90
Hampson Industries PLC	77 Birmingham Rd., West Bromwich B70 6PY, United Kingdom	21 5534681	21 5253733	2,417	94
Dorel Industries Inc.	4750 Blvd. des Grandes-Prairies, Montreal, Quebec H1R 1A3, Canada	514-323-5701	514-323-9444	2,400	98
Flexsteel Industries Inc.	PO Box 877, Dubuque, Iowa 52004-0877, U.S.A.	319-556-7730		2,400	97
HLTH Timber Processors Pty. Ltd.	PO Box 784460, Sandton 2146, Republic of South Africa	11 8845220	11 8836157	2,355	94
Chromcraft Revington Inc.	1100 N. Washington St., Delphi, Indiana 46923, U.S.A.	765-564-3500	765-564-3722	2,300	97
Springs Window Fashions Div.	7549 Graber Rd., Middleton, Wisconsin 53562, U.S.A.	608-836-1011	608-831-2184	2,300	97
Hoffman Engineering Co.	900 Ehlen Dr., Anoka, Minnesota 55303, U.S.A.	612-421-2240	612-422-2299	2,250	97
Little Tikes Co.	2180 Barlow Rd., Hudson, Ohio 44236, U.S.A.	330-650-3000	330-650-3349	2,200	
Kinetic Concepts Inc.	8023 Vantage Dr., San Antonio, Texas 78230, U.S.A.	210-524-9000		2,066	96
O'Sullivan Industries Inc.	1900 Gulf St., Lamar, Missouri 64759-1899, U.S.A.	417-682-3322	417-682-6010	2,000	97

Continued

Representative Companies in Sector
- Continued -

Name	Address	Tele-phone	Fax	Employ-ment	Y
Singer Furniture Co.	PO Box 11067, High Point, North Carolina 27265, U.S.A.	910-802-4600	910-889-2414	2,000	95
ACCO-Rexel Ltd.	Gatehouse Rd., Aylesbury HP19 3DT, United Kingdom	270 582411		1,872	93
Henredon Furniture Industries Inc.	PO Box 70, Morganton, North Carolina 28680, U.S.A.	704-437-5261	704-437-5264	1,800	97
Pulaski Furniture Corp.	PO Box 1371, Pulaski, Virginia 24301, U.S.A.	703-980-7330	703-980-0617	1,800	97
River Oaks Furniture Inc.	PO Box 277, Fulton, Mississippi 38843, U.S.A.	601-891-4550		1,800	96
OSF Inc.	650 Barmac Dr., Weston, Ontario M9L 2X8, Canada	416-749-7700	416-746-3229	1,700	98
SMED International Inc.	10 Smed Lane SE, Calgary, Alberta T2C 4T5, Canada	403-203-6000	403-203-6001	1,700	98
Shelby Williams Industries Inc.	11-111 Merchandise Mart, Chicago, Illinois 60654, U.S.A.	312-527-3593	312-527-3597	1,667	96
Century Furniture Industries	PO Box 608, Hickory, North Carolina 28603, U.S.A.	704-328-1851	704-328-2176	1,600	95
Madix Inc.	PO Box 729, Terrell, Texas 75160, U.S.A.	972-563-5744	972-563-0792	1,600	97
Tah Hsin Industrial Corp.	Taichung Industrial Park, Taichung, Taiwan	4 2522111	4 2525307	1,550	90
Airsprung Group PLC	Canal Rd., Trowbridge BA14 8RQ, United Kingdom	225 754411	225 763256	1,540	94
Rowe Furniture Corp.	1650 Tysons Blvd., 710, McLean, Virginia 22102, U.S.A.	703-847-8670		1,525	96
Hygena Ltd.	Howden, Goole DN14 7PA, United Kingdom	430 430905	430 431540	1,512	94
Junckers Industrier AS	Vaerftsvej 4A, Koge DK-4600, Denmark	53651895	53659936	1,500	93
William H. Kaufman Inc.	410 King St. W, Kitchener, Ontario N2G 4J8, Canada	519-576-1500	519-576-9794	1,500	98
Baker Furniture Co.	1661 Monroe Ave. N.W., Grand Rapids, Michigan 49505, U.S.A.	616-361-7321		1,400	
Lozier Corp.	6336 Pershing Dr., Omaha, Nebraska 68110, U.S.A.	402-457-8000	402-457-8338	1,400	97
Kingsway Group PLC	289-293 High Holborn, London WC1V 7HU, United Kingdom	71 8315304	71 8310038	1,364	93
Aladdin Industries Inc.	703 Murfreesboro Rd., Nashville, Tennessee 37210, U.S.A.	615-748-3000	615-748-3030	1,350	97
Allsteel Inc.	Rte. 31 Allsteel Dr., Aurora, Illinois 60507-0871, U.S.A.	708-859-2600	708-844-7187	1,317	
Falcon Products Inc.	9387 Dielman Industrial Dr., St. Louis, Missouri 63132-2299, U.S.A.	314-991-9200	314-991-9295	1,309	97
Brodart Co.	500 Arch St., Williamsport, Pennsylvania 17705, U.S.A.	717-326-2461	717-326-6769	1,300	97
Cleveland Chair Co.	PO Box 1359, Cleveland, Tennessee 37364-0159, U.S.A.	615-476-8544		1,300	94
Wagon Storage Products Ltd.	Halesfield, Telford TF7 4LN, United Kingdom	952 680051	952 680182	1,300	
Walker & Homer Group PLC	Saltbrook Rd., Halesowen B63 2QJ, United Kingdom	384 633223	384 294885	1,265	
Hadinata Brothers & Co. PT	Jl. Tapos Km. 1 Cibinong, Jawa Barat, Indonesia	218752291	21 8752261	1,200	97
Hunt Corp. (Philadelphia, Pennsylvania)	2005 Market St., 1 Commerce Sq., Philadelphia, Pennsylvania 19103-7085, U.S.A.	215-732-7700		1,200	97
Unaka Company Inc.	PO Box 877, Greeneville, Tennessee 37744, U.S.A.	423-639-1163	615-639-7270	1,200	96
Hooker Furniture Corp.	PO Box 4708, Martinsville, Virginia 24115, U.S.A.	703-632-2133		1,150	
Companias CIC SA	Av. Esquina Blanca 960, Santiago, Chile	2 5304000	2 5574362	1,100	96
Cosco Inc.	2525 State St., Columbus, Indiana 47201, U.S.A.	812-372-0141	812-372-0911	1,100	97
Interlake Material Handling Div.	1240 E. Diehl Rd., Naperville, Illinois 60563, U.S.A.	630-245-8800	630-245-8908	1,100	97
Knape and Vogt Manufacturing Co.	2700 Oak Industrial Dr., NE, Grand Rapids, Michigan 49505-6083, U.S.A.	616-459-3311	616-459-3290	1,061	97
Pikolin, SA	Carretera N-232 A Logrono KM 245, Zaragoza E-50011, Spain	976 342200	976 343545	1,050	97
Sherrill Furniture Co.	PO Box 189, Hickory, North Carolina 28603, U.S.A.	704-322-2640		1,050	94
Eastern Asia Woods Industrial Corp. Group	5 Shin-Yi Rd., Sec. 5, 4G29, Taipei 100, Taiwan	2 7252758	2 7231721	1,000	91
Franklin Corp. (Houston, Mississippi)	PO Box 569, Houston, Mississippi 38851, U.S.A.	601-456-4286	601-456-3156	1,000	94
Furnsteel Pty. Ltd.	PO Box 1870, Benoni 1500, Republic of South Africa	11 4214401	11 4219254	1,000	95
Gomma Gomma Holdings Pty. Ltd.	PO Box 911006, Rosslyn 0200, Republic of South Africa	1461 3181	1461 31130	1,000	94
Keller Industries Inc.	3499 N.W. 53rd St., Fort Lauderdale, Florida 33309-6323, U.S.A.	954-777-2060	954-651-0053	1,000	95
Pilliod Furniture Inc.	4620 Grand Over Pkwy., Greenboro, North Carolina 27265, U.S.A.	336-884-3929	336-410-6722	1,000	97
RHC/Spacemaster Corp.	1400 N. 25th Ave., Melrose Park, Illinois 60160, U.S.A.	708-345-2500	708-345-3823	1,000	93
York Timbers Ltd.	PO Box 380, Pretoria 0001, Republic of South Africa	12 869730	12 869780	1,000	95
Relyon Group PLC	Station Mills, Wellington TA21 8NN, United Kingdom	823 667501	823 666079	967	93
Krause's Furniture Inc.	200 N. Berry St., Brea, California 92821-3903, U.S.A.	714-990-3100	714-256-4297	960	97
Meridian Inc.	PO Box 530, Spring Lake, Michigan 49456-0530, U.S.A.	616-846-0280	616-846-9236	950	97
Norwalk Furniture Corp.	100 Furniture Pkwy., Norwalk, Ohio 44857, U.S.A.	419-668-4461		950	
Zala Butorgyar RT	Malom Ut 2, Zalaegerszeg H-8900, Hungary	92 14250	92 14039	950	93
Bernstein Group PLC	Middleton, Manchester M24 1AR, United Kingdom	61 6539191	61 6535392	924	93
Diethelm Singapore Pte. Ltd.	34 Boon Leat Terr., Singapore 0511, Singapore	4711466	4799104	920	93
Martin Industries Inc.	PO Box 128, Florence, Alabama 35631-0128, U.S.A.	205-767-0330	205-740-5192	907	97
Fisher-Price Inc.	636 Girard Ave. E, East Aurora, New York 14052-1824, U.S.A.	716-687-3000		900	97

Continued

Representative Companies in Sector
- Continued -

Name	Address	Tele-phone	Fax	Employ-ment	Y
Henredon Furniture Industries Inc. Spruce Pine Div.	Altapass Rd., Spruce Pine, North Carolina 28777, U.S.A.	704-765-9641	704-765-6664	900	93
Super Sagless Inc.	PO Box 197, Tupelo, Mississippi 38802-0197, U.S.A.	601-842-5704	601-844-3536	900	97
Bostinco BV	PO Box 2803, Jakarta 10160, Indonesia	213841213	21 3805932	887	97
Vaughan-Bassett Furniture Co.	PO Box 1549, Galax, Virginia 24333, U.S.A.	703-236-6161		875	93
Hickory Chair Co.	PO Box 2147, Hickory, North Carolina 28603, U.S.A.	704-328-1801	704-328-8954	850	97
Dongsuh Furniture Co., Ltd.	Nam-Gu, Inchon, Republic of Korea	32 7606888	32 7621588	810	97
American Drew	4620 Grandover P, Greensboro, North Carolina 27417, U.S.A.	336-294-5233	336-315-4392	800	97
American Seating Co.	401 American Seating Center, Grand Rapids, Michigan 49504, U.S.A.	616-732-6600	616-732-6401	800	96
Gunlocke Co.	1 Gunlocke Dr., Wayland, New York 14572, U.S.A.	716-728-5111	716-728-8353	800	94
Hickory White Co.	PO Box 1600, High Point, North Carolina 27261, U.S.A.	919-885-1200	919-885-1678	800	93
Knoll North America Corp.	1000 Arrow Rd., Weston, Ontario M9M 2Y7, Canada	416-741-5453	416-741-1494	800	98
Schnadig Corp.	1111 E. Touhy Ave., Des Plaines, Illinois 60018, U.S.A.	847-803-6000		800	97
Tye-Sil Corp. Ltd.	12225 Blvd. Industriel, Montreal, Quebec H1B 5M7, Canada	514-640-9727	514-640-6323	800	98
Whole Space Plastic Manufacturing Co., Ltd.	19 Pa Teh Rd., 3rd Fl., Ln. 768, Sec. 4, Taipei, Taiwan	2 7882808	2 7882431	800	93
Woodbridge Foam Corp.	4240 Sherwoodtowne Blvd, Mississauga, Ontario L4Z 2G6, Canada	905-896-3626	905-896-9262	800	98
WinsLoew Furniture Inc.	201 Cahaba Valley Pkwy., Pelham, Alabama 35124, U.S.A.	205-403-0206	205-403-0403	797	97
Asian Design Manufacturing Corp.	Mandaue City, Cebu 6014, Philippines	461156	32460407	775	97
Clarson Enterprises, Inc.	PO Box 161, Mandaue, Philippines	82060	32460221	750	97
Tapei Hisz	Regesz Ter 8, Tape H-6753, Hungary	62 326288	62 325424	750	93
Hispano-Suiza International PLC	21 Holburn Viaduct, London EC1A 2AY, United Kingdom			741	
Scandinavian Mobility Export	Sdr. Ringvej 39, Broendby DK-2605, Denmark	43437788	43435063	741	93
ERA Group PLC	Maxted Rd., Hemel Hempstead HP2 7BT, United Kingdom	442 249355	442 248744	718	93
Ameriwood Industries International Corp.	168 Louis Campau Promenade NW, Ste. 400, Grand Rapids, Michigan 49503-2638, U.S.A.	616-336-9400		700	97
Bodilsen Holding AS	Glyngoere, Roslev DK-7870, Denmark	97731377	97731885	700	93
Giroflex SA	Rua Dr. Rubens Gomes Bueno 691, Sao Paulo 04730-903, Brazil	11 5240922	11 5240832	700	96
Good Tables Inc.	1118 E. 223rd St., Carson, California 90745, U.S.A.	310-513-6060		700	96
HTH Kokkener AS	Industrievej 6, Olgod DK-6870, Denmark	75246000	75245898	700	93
Muebles y Almacenamiento Tecnico Carvajal Mepal SA	Av. 6 AN No. 29N-73, Cali, Colombia	2 6602100	2 6612110	700	97
Simula Inc.	401 W. Baseline Rd., 204, Tempe, Arizona 85283, U.S.A.	602-752-8918		700	95
Kroehler-Dunmore Furniture Industries	PO Box 3740, Hickory, North Carolina 28603, U.S.A.	704-322-6503	704-322-7187	680	
Gomme Ltd.	Spring Gdns., High Wycombe HP137AD, United Kingdom	494 526250		679	91
Syratech Corp.	175 McClellan Hwy., East Boston, Massachusetts 02128, U.S.A.	617-561-2200	617-561-0275	670	96
Reliable Manufacturing Works, Inc.	Mandaue City, Cebu 6014, Philippines	460315	32460315	652	97
Brookwood Furniture Company Inc.	PO Box 540, Pontotoc, Mississippi 38863, U.S.A.	601-489-1100		650	93
Thomson Crown Wood Products Inc.	PO Box 647, Mocksville, North Carolina 27028, U.S.A.	704-634-6241	704-634-8200	650	
Premier Spring Industries Pty. Ltd.	PO Box 1302, Durban 4000, Republic of South Africa	31 4690461	31 429233	640	93
Best Chairs Inc.	PO Box 158, Ferdinand, Indiana 47532, U.S.A.	812-367-1761		600	
Child Craft Industries Inc.	PO Box 444, Salem, Indiana 47167, U.S.A.	812-883-3111		600	97
F.M. Thorpe Manufacturing Co.	1300 E. 12th St., Lamar, Missouri 64759, U.S.A.	417-682-3375		600	95
Kenney Manufacturing Co.	1000 Jefferson Blvd., Warwick, Rhode Island 02886, U.S.A.	401-739-2200	401-821-4240	600	95
Kessler Industries Inc. (El Paso, Texas)	8600 Gateway Blvd. E., El Paso, Texas 79907, U.S.A.	915-591-8161		600	94
Steelcase Canada Ltd.	1 Steelcase Rd. W, Markham, Ontario L3R 0T3, Canada	905-475-6333	905-475-6947	600	98
U.S. Furniture Industries Inc.	PO Box 2127, High Point, North Carolina 27261, U.S.A.	919-884-7375	919-884-7015	600	93
Webb Furniture Enterprises Inc.	PO Box 1277, Galax, Virginia 24333, U.S.A.	703-236-2984		600	94
Kewaunee Scientific Corp.	PO Box 1842, Statesville, North Carolina 28687-1842, U.S.A.	704-873-7202	704-872-4355	560	97
Bosal Benelux NV	Nijverheidsstraat 12, Oevel-Werlo B-2260, Belgium	14 574711	14 585060	550	95
Ikeda Hoover Ltd.	Cherry Blossom Way, Sunderland SR5 3TW, United Kingdom	91 4156000	91 4153857	534	
Northern Upholstery Group Ltd.	Adwick-le-St., Doncaster DN6 7BD, United Kingdom	302 330365	302 330880	531	92
Confortluxe	Menensteenweg 40, Wervik B-8940, Belgium	56 312073	56 311670	520	95
Northern Upholstery Ltd.	Adwick Le St., Doncaster DN6 7Bd, United Kingdom	302 330305	302 330880	512	93

Product Tables

Mattresses

Unit of Measure: Number.

	1990 Value	%	1991 Value	%	1992 Value	%	1993 Value	%	1994 Value	%	1995 Value	%	1996 Value	%
Total Production	40,436,078	100.0	48,283,692	100.0	49,649,509	100.0	50,416,840	100.0	52,081,175	100.0	52,576,937	100.0	54,896,417	100.0
Regions														
Africa	1,187,516	2.9	1,111,849	2.3	1,101,182	2.2	1,079,516	2.1	1,214,849	2.3	1,082,358	2.1	1,227,248	2.2
America, North	20,030,952	49.5	20,701,229	42.9	21,371,505	43.0	22,041,781	43.7	22,915,457	44.0	23,074,133	43.9	23,893,810	43.5
America, South	895,556	2.2	1,033,778	2.1	1,184,000	2.4	1,269,061	2.5	1,468,288	2.8	1,600,515	3.0	1,541,419	2.8
Asia	384,700	1.0	1,344,800	2.8	677,900	1.4	824,000	1.6	815,100	1.6	1,086,486	2.1	1,074,336	2.0
Europe	16,854,354	41.7	23,099,036	47.8	24,245,922	48.8	24,109,483	47.8	24,687,904	47.4	24,774,813	47.1	26,221,918	47.8
Oceania	1,083,000	2.7	993,000	2.1	1,069,000	2.2	1,093,000	2.2	979,577	1.9	958,632	1.8	937,687	1.7
Leading Producers														
United States of America	*16,939,200	41.9	*17,537,100	36.3	*18,135,000	36.5	*18,732,900	37.2	*19,330,800	37.1	*19,928,700	37.9	*20,526,600	37.4
Germany	-		6,039,000	12.5	6,401,000	12.9	6,315,000	12.5	6,626,000	12.7	6,946,417	13.2	7,787,888	14.2
United Kingdom	3,232,000	8.0	3,226,000	6.7	*3,205,000	6.5	*3,205,000	6.4	*3,205,000	6.2	*3,205,000	6.1	*3,205,000	5.8
Spain	*2,628,750	6.5	*2,628,750	5.4	*2,628,750	5.3	2,537,000	5.0	2,723,000	5.2	2,573,000	4.9	2,682,000	4.9
Mexico	1,703,200	4.2	1,798,800	3.7	1,894,400	3.8	1,990,000	3.9	2,289,000	4.4	1,873,000	3.6	2,118,000	3.9

Mattress Supports

Unit of Measure: Number.

	1990 Value	%	1991 Value	%	1992 Value	%	1993 Value	%	1994 Value	%	1995 Value	%	1996 Value	%
Total Production	21,928,198	100.0	24,239,264	100.0	24,477,698	100.0	25,103,343	100.0	25,599,328	100.0	28,635,294	100.0	28,665,318	100.0
Regions														
Africa	536,667	2.4	620,361	2.6	649,344	2.7	678,328	2.7	707,311	2.8	736,294	2.6	765,278	2.7
America, North	12,780,200	58.3	13,114,600	54.1	13,449,000	54.9	13,783,400	54.9	14,117,800	55.1	14,452,200	50.5	14,786,600	51.6
America, South	6,333	0.0	6,667	0.0	7,000	0.0	7,000	0.0	9,000	0.0	10,000	0.0	9,019	0.0
Asia	157,000	0.7	499,000	2.1	426,000	1.7	365,000	1.5	362,000	1.4	399,000	1.4	331,000	1.2
Europe	8,044,998	36.7	9,668,636	39.9	9,629,353	39.3	9,942,615	39.6	10,103,678	39.5	12,758,733	44.6	12,514,828	43.7
Oceania	403,000	1.8	330,000	1.4	317,000	1.3	327,000	1.3	299,538	1.2	279,066	1.0	258,593	0.9
Leading Producers														
United States of America	*12,291,200	56.1	*12,625,600	52.1	*12,960,000	52.9	*13,294,400	53.0	*13,628,800	53.2	*13,963,200	48.8	*14,297,600	49.9
Germany	-		1,654,000	6.8	1,908,000	7.8	2,149,000	8.6	1,876,000	7.3	4,133,220	14.4	4,100,482	14.3
United Kingdom	2,270,915	10.4	2,245,553	9.3	2,220,191	9.1	2,194,829	8.7	2,556,752	10.0	*2,442,404	8.5	*2,478,561	8.6
Spain	*1,727,750	7.9	*1,727,750	7.1	*1,727,750	7.1	1,464,000	5.8	1,551,000	6.1	1,889,000	6.6	2,007,000	7.0
Belgium	*1,518,667	6.9	*1,518,667	6.3	*1,518,667	6.2	*1,518,667	6.0	1,519,000	5.9	1,512,000	5.3	1,525,000	5.3

Commodity data are provided by the United Nations Statistical Division. The symbol * means that data are estimated. For additional notes, see Appendix I.

ISIC 3410
PAPER AND PRODUCTS

ISIC 3410—Paper and Products—includes the entire category of paper and board raw materials and finished products. Specifically, it includes mechanical wood pulp, pulp of fibers other than wood, dissolving grades of wood pulp, sulphate and soda wood pulp, sulphite pulp, semi-chemical pulp, newsprint, other printing and writing paper, kraft paper and paperboard, other paperboard, compressed and noncompressed fiberboard, and packaging papers and board. Photographic paper is under ISIC 3520.

This industry corresponds, in the U.S. system of classification, to SIC 26—Paper and Allied Products. That group is subdivided further into Pulp Mills, Paper Mills, Paperboard Mills, Paperboard Containers and Boxes, and Miscellaneous Converted Paper Products.

Summary Statistics

		1990 Value	N	1991 Value	N	1992 Value	N	1993 Value	N	1994 Value	N	1995 Value	N
Establishments or enterprises	(number)	41,314	84	42,523	94	50,117	88	41,580	85	38,976	70	30,406	34
Number employed	(000)	4,227	94	4,105	99	4,137	94	4,043	94	3,862	85	3,637	63
Total Wages	($ mil.)	56,044	89	61,925	90	66,266	84	55,378	85	56,564	75	52,313	56
Wage/salary per employee	($)	9,856	88	10,698	88	18,714	84	15,375	84	10,433	74	12,908	56
Female workers	(000)	217	35	122	33	580	29	564	33	547	27	522	15
as % of total employment*	(%)	21	34	21	32	23	29	23	33	25	27	32	14
Output	($ bil.)	376	91	396	95	442	92	392	89	417	79	451	63
per employee	($ 000)	77	88	74	91	254	89	690	85	81	75	106	60
per establishment	($ 000)	9,341	79	8,600	85	17,346	82	16,548	76	8,880	61	7,824	30
Value added	($ bil.)	159	84	152	85	161	78	158	79	154	71	183	60
per employee	($ 000)	29	81	28	82	31	77	27	75	31	67	41	56
per establishment	($ 000)	3,377	73	2,857	78	3,285	70	2,653	66	3,321	53	2,725	27
Capital investment	($ mil.)	34,477	70	29,564	65	24,272	58	20,172	53	17,185	41	12,937	23
per employee	($ 000)	8	67	9	62	8	57	5	50	5	39	5	21
per establishment	($ 000)	1,314	66	1,394	62	1,046	56	565	48	529	38	434	19

Data presented above are drawn from the detailed tables that follow. Columns headed 'N' show the number of countries that provided valid data for inclusion. Values are not strictly comparable one year to the next or one row to the next because the number of countries that report varies from period to period and row to row. However, a general indication of magnitudes can be gleaned from reviewing this summary—especially in earlier years in which more reports are available. For detailed explanations, see Appendix I. *The average for those countries reporting both total and female employment.

Establishments and Number Engaged

Country	Establishments or Enterprises (number)						Employees per Establishment					
	1990	1991	1992	1993	1994	1995	1990	1991	1992	1993	1994	1995
Albania	-	-	-	6.000	2.000	1.000	-	-	-	77	23	10
Angola	-	-	4.000	4.000	-	-	-	-	47	12	-	-
Argentina	-	-	-	874	961	-	-	-	-	27	25	-
Armenia	1.000	1.000	1.000	1.000	2.000	-	589	527	-	-	-	-
Australia	241	289	312	-	-	-	87	69	61	-	-	-
Austria	148	147	141	141	136	-	135	136	142	134	135	-
Azerbaijan	3.000	4.000	7.000	7.000	5.000	-	237	169	84	70	82	-
Bahrain	-	-	4.000	-	-	-	-	-	60	-	-	-
Bangladesh	51	58	91	-	-	-	321	303	184	-	-	-
Belgium	308	307	292	289	291	-	58	59	59	-	-	-
Belize	2.000	2.000	2.000	-	-	-	-	33	33	-	-	-
Benin	-	-	1.000	1.000	-	-	-	-	-	-	-	-
Bolivia	11	13	13	14	13	18	28	29	37	43	49	36
Bosnia & Herzegovina	23	22	-	-	8.000	-	406	432	-	-	43	-
Brazil	902	-	824	788	-	-	139	-	142	149	-	-
Bulgaria	*20	*19	*22	*101	*167	*176	745	705	545	106	82	81
Burundi	*2.000	*2.000	-	-	-	-	42	41	-	-	-	-
Cambodia	3.000	7.000	-	-	-	-	97	36	-	-	-	-
Cameroon	*8.000	*8.000	*10	*11	*12	*10	82	94	80	82	70	79
Canada	715	668	667	651	656	-	159	163	156	155	152	-
Chile	33	40	41	41	46	-	273	262	283	271	251	-
China	*12,163	*12,373	*12,619	*11,940	*12,282	*13,890	99	101	101	105	104	96
China (Hong Kong SAR)	1,764	1,517	1,456	1,235	1,117	-	9.524	9.756	9.341	10	8.863	-
Colombia	150	144	164	158	163	-	79	82	84	90	91	-
Costa Rica	38	41	45	46	-	-	64	66	63	64	-	-
Croatia	79	109	147	196	239	253	124	76	50	34	27	26
Cyprus	60	60	60	60	56	54	13	14	13	12	14	14
Czechoslovakia (Former)	*23	*26	-	-	-	-	1,957	1,500	-	-	-	-
Czech Republic	-	*17	*44	*52	-	-	-	1,353	568	423	-	-
Denmark	228	244	246	-	-	-	48	45	42	-	-	-
Ecuador	27	31	31	37	36	46	148	136	143	134	123	112
Egypt	99	99	98	96	-	-	235	337	196	194	-	-
El Salvador	-	6.000	5.000	9.000	6.000	8.000	-	-	183	204	273	202
Ethiopia	4.000	4.000	3.000	4.000	3.000	4.000	320	363	432	322	452	340
Fiji	9.000	7.000	8.000	-	-	-	25	37	30	-	-	-
Finland	148	162	165	162	159	195	300	258	243	241	243	198
FYR Macedonia	*7.000	*12	*37	*39	*65	*103	258	134	48	40	26	18
Gabon	-	2.000	-	-	-	2.000	-	-	-	-	-	18
Germany	-	1,286	1,253	1,219	1,161	-	-	152	144	137	137	-
Germany (Western Part)	1,072	1,089	1,104	-	-	-	152	157	152	-	-	-
Ghana	-	-	-	13	-	-	-	-	-	112	-	-
Greece	149	146	150	-	-	-	61	56	53	-	-	-
Grenada	-	-	1.000	1.000	1.000	-	-	-	30	30	37	-
Guatemala	-	7.000	7.000	7.000	7.000	7.000	-	215	226	229	225	237
Honduras	-	-	17	18	18	18	-	-	102	106	116	173
Hungary	*62	*135	*115	*150	-	-	210	89	104	73	-	-
Iceland	9.000	10	9.000	10	-	-	26	23	26	23	-	-
India	2,163	2,111	2,278	2,602	-	-	67	69	68	60	-	-
Indonesia	184	217	258	268	305	311	235	272	286	276	258	285
Iran (Islamic Republic of)	130	76	83	75	-	-	75	160	144	158	-	-
Iraq	-	-	15	-	-	-	-	-	267	-	-	-
Ireland	87	86	-	-	-	-	39	42	-	-	-	-
Israel	165	184	178	207	193	-	39	35	38	35	38	-
Italy	*686	*701	*782	*793	*806	-	90	88	81	78	78	-
Jamaica	25	25	-	-	-	-	54	52	-	-	-	-
Japan	9,974	9,760	9,503	9,611	10,392	10,538	26	26	27	26	26	25
Jordan	26	27	27	35	64	66	62	70	61	66	49	43
Kenya	48	48	49	48	60	60	149	150	148	154	126	129
Korea, Republic of	1,901	2,062	2,129	2,307	2,405	2,472	31	29	28	26	26	26
Kuwait	16	13	13	13	12	-	58	41	64	67	76	-

Continued.

Depending on the table, * means *Enterprises, Engaged,* or *Factor Values;* ± means *Basis Unspecified;* § means *shown in millions.* For additional notes and sources, see Appendix I.

Establishments and Number Engaged

- Continued -

Country	Establishments or Enterprises (number)						Employees per Establishment					
	1990	1991	1992	1993	1994	1995	1990	1991	1992	1993	1994	1995
Kyrgyzstan	1.000	1.000	2.000	2.000	1.000	-	142	120	67	51	82	-
Latvia	12	16	26	24	27	27	312	253	153	101	70	61
Liechtenstein	-	2.000	-	-	-	-	-	1.500	-	-	-	-
Lithuania	-	-	-	*16	*13	-	-	-	-	365	385	-
Macau	47	44	41	37	35	28	12	11	9.195	7.270	6.171	6.536
Malaysia	131	143	153	171	175	-	103	113	115	112	111	-
Malta	15	17	18	17	18	-	21	18	19	20	19	-
Mauritius	15	16	13	13	15	15	27	50	57	56	55	55
Mexico	132	132	129	128	123	121	251	244	241	219	196	213
Mongolia	-	-	-	-	1.000	-	-	-	-	-	19	-
Morocco	-	89	82	91	-	-	-	81	94	92	-	-
Mozambique	4.000	4.000	4.000	4.000	4.000	-	162	156	149	146	128	-
Namibia	-	-	-	-	3.000	-	-	-	-	-	23	-
Nepal	22	47	-	46	88	-	68	31	-	41	23	-
Netherlands	146	161	163	153	166	-	167	158	155	157	141	-
Netherlands Antilles	-	-	-	5.000	-	-	-	-	-	36	-	-
New Zealand	*118	*119	*114	*124	-	-	80	75	76	69	-	-
Norway	103	99	84	82	80	-	115	115	127	124	127	-
Pakistan	-	74	-	-	-	-	-	112	-	-	-	-
Panama	22	20	-	-	-	-	48	52	-	-	-	-
Paraguay	-	1.000	-	-	-	-	-	19	-	-	-	-
Peru	203	207	223	-	-	-	31	28	23	-	-	-
Philippines	307	342	197	200	-	-	50	49	92	78	-	-
Poland	*66	*92	*93	*88	-	-	636	424	355	352	-	-
Portugal	*423	*424	*419	*442	*452	-	44	40	40	39	33	-
Puerto Rico	45	47	42	38	38	39	38	34	40	42	43	44
Qatar	3.000	4.000	3.000	3.000	3.000	-	22	19	19	22	20	-
Republic of Moldova	2.000	2.000	2.000	2.000	2.000	2.000	714	680	752	735	674	669
Romania	*17	76	83	103	223	-	2,547	505	431	312	127	-
Russian Federation	-	-	-	*378	*383	*502	-	-	-	489	430	342
Saint Lucia	-	4.000	4.000	4.000	4.000	4.000	-	-	-	-	-	-
Saint Vincent & the Grenadines	-	-	-	1.000	1.000	2.000	-	-	-	87	87	51
Senegal	6.000	5.000	4.000	4.000	3.000	-	39	56	71	73	93	-
Sierra Leone	-	-	-	3.000	-	-	-	-	-	12	-	-
Singapore	92	95	100	102	105	-	50	52	50	51	52	-
Slovakia	-	15	15	17	19	-	-	924	857	859	738	-
Slovenia	67	*230	*328	*325	*323	*264	-	-	-	-	-	-
South Africa	-	386	-	-	-	-	-	130	-	-	-	-
Spain	1,097	1,070	1,046	-	-	-	38	39	38	-	-	-
Sri Lanka	17	19	15	26	-	-	247	214	273	194	-	-
Suriname	-	-	6.000	-	-	-	-	-	4.833	-	-	-
Swaziland	-	5.000	5.000	5.000	-	-	-	711	663	686	-	-
Sweden	214	253	247	235	233	-	241	200	191	178	182	-
Thailand	136	124	-	-	-	-	135	140	-	-	-	-
Trinidad and Tobago	29	27	27	26	29	-	37	41	40	38	38	-
Tunisia	-	-	-	75	79	89	-	-	-	49	57	55
Turkey	97	92	150	140	143	-	222	227	135	140	136	-
Ukraine	75	169	192	92	88	80	387	172	146	283	273	288
United Kingdom	3,192	3,088	3,041	3,094	3,269	-	46	46	45	41	39	-
United Republic of Tanzania	8.000	11	-	-	-	-	652	497	-	-	-	-
USSR (Former)	*249	-	-	-	-	-	1,020	-	-	-	-	-
United States of America	-	-	6,810	-	-	-	-	-	86	-	-	-
Venezuela	109	105	104	83	93	97	122	145	135	142	131	126
Yugoslavia	98	208	589	829	863	904	-	68	24	16	15	14
Zambia	12	-	-	-	13	-	131	-	-	-	72	-
Zimbabwe	15	14	12	12	12	-	313	326	308	333	300	-

Employment and Compensation of Employees

Country	Number Employed/Engaged (000)						Wage/Salary per Employee ($)					
	1990	1991	1992	1993	1994	1995	1990	1991	1992	1993	1994	1995
Albania	2.486	-	-	0.463	0.046	0.010	441	-	-	-	-	1,398
Angola	-	*0.430	*0.188	*0.048	-	-	-	-	-	-	-	-
Argentina	30	-	-	24	24	-	5,917	-	-	14,546	-	-
Armenia	*0.589	*0.527	-	-	-	-	-	-	-	-	-	-
Australia	21	*20	*19	-	-	-	25,383	27,892	28,676	-	-	-
Austria	20	20	20	19	18	18	30,831	32,383	35,886	34,470	36,448	-
Azerbaijan	0.712	0.675	0.590	0.489	0.412	-	1,917	1,966	-	-	-	-
Bahrain	-	-	0.241	-	-	-	-	-	4,547	-	-	-
Bangladesh	16	18	17	-	-	-	1,231	1,180	1,451	-	-	-
Belgium	18	18	17	-	-	-	23,787	24,814	27,640	-	-	-
Belize	-	0.067	0.067	-	-	-	-	4,045	4,940	-	-	-
Bolivia	0.306	0.378	0.480	0.595	0.633	0.641	1,414	1,262	1,293	1,394	1,649	1,590
Bosnia & Herzegovina	9.339	9.498	-	-	0.345	-	3,022	2,832	-	-	-	-
Brazil	125	-	117	118	-	-	7,732	-	6,930	9,220	-	-
Bulgaria	15	13	12	11	14	14	1,669	605	921	1,243	909	1,285
Burundi	0.083	0.081	-	-	-	-	950	1,007				
Cambodia	*0.291	*0.250	-	-	-	-	-	-	-	-	-	-
Cameroon	0.654	0.749	0.803	0.898	0.845	0.794	7,116	7,955	8,878	9,132	4,148	5,109
Canada	114	109	104	101	100	100	34,876	37,219	36,323	35,134	34,629	35,490
Chile	9.011	10	12	11	12	12	6,291	6,718	8,175	8,828	11,080	12,783
China	*1,200	*1,250	*1,270	*1,250	*1,280	*1,330	-	-	-	-	-	-
China (Hong Kong SAR)	*17	*15	*14	*13	*9.900	-	8,948	10,451	12,301	14,040	14,063	-
China (Taiwan Province)	60	60	63	63	64	65	10,512	11,326	13,393	13,329	14,152	15,326
Colombia	12	12	14	14	15	15	3,247	3,515	3,901	4,546	5,655	6,587
Costa Rica	2.424	2.719	2.816	2.932	-	-	4,245	3,901	4,547	5,104	-	-
Croatia	9.790	8.320	7.410	6.740	6.340	6.620	5,125	5,162	1,630	1,771	2,478	4,219
Cyprus	0.779	0.810	0.759	0.748	0.756	0.738	9,668	9,941	12,373	10,970	12,678	15,123
Czechoslovakia (Former)	45	39	-	-	-	-	2,105	1,583	-	-	-	-
Czech Republic	25	23	25	22	-	-	2,184	1,652	1,868	2,304	-	-
Denmark	11	11	10	10	10	10	32,248	33,422	36,660	33,889	39,578	-
Ecuador	3.987	4.204	4.428	4.948	4.436	5.144	3,111	4,162	3,940	4,516	2,792	2,933
Egypt	*23	*33	*19	*19	*19	*18	1,545	1,084	1,345	1,542	1,608	1,632
El Salvador	-	-	*0.916	*1.837	*1.636	*1.614	-	-	-	8,618	5,962	2,178
Ethiopia	1.280	1.450	1.297	1.290	1.357	1.362	1,823	1,707	1,534	1,018	963	997
Fiji	0.229	0.259	0.240	0.267	0.290	-	5,151	5,206	6,371	6,588	7,102	-
Finland	44	42	40	39	39	39	34,974	34,850	32,299	26,113	30,084	39,568
FYR Macedonia	1.809	1.605	1.767	1.541	1.694	1.863	4,796	9,753	2,713	2,703	3,075	4,224
France	107	106	105	102	101	100	35,387	35,691	37,516	-	-	-
Gabon	-	-	-	-	-	0.036	-	-	-	-	-	3,506
Germany	-	*195	*180	*167	*159	-	-	27,401	32,358	31,834	33,398	-
Germany (Western Part)	*163	*171	*168	-	-	-	29,200	29,941	33,633	-	-	-
Ghana	-	-	-	*1.461	*1.502	*1.545	-	-	-	1,381	1,032	1,052
Greece	*9.141	*8.220	*7.889	*7.812	*7.880	*7.917	12,125	13,348	14,473	13,420	14,431	17,040
Grenada	-	-	*0.030	*0.030	0.037	-	-	-	-	-	-	-
Guatemala	-	*1.508	*1.585	*1.602	*1.572	*1.660	-	-	-	-	-	-
Honduras	*1.263	*1.465	*1.734	*1.900	*2.082	*3.114	3,842	2,944	3,612	3,760	3,398	2,479
Hungary	13	12	12	11	11	11	2,976	3,201	3,687	4,090	4,365	5,546
Iceland	0.231	0.231	0.233	0.231	0.224	-	28,625	34,356	32,338	25,992	-	-
India	*144	*147	*156	*157	*164	*172	1,361	1,180	1,147	1,065	1,199	1,329
Indonesia	43	59	74	74	79	89	969	927	1,158	1,121	1,421	2,390
Iran (Islamic Republic of)	9.800	12	12	12	-	-	4,538	5,065	6,171	5,797	-	-
Iraq	-	4.404	4.000	-	-	-	-	11,343	18,975	-	-	-
Ireland	3.400	3.600	3.630	3.630	3.675	3.675	23,510	23,829	26,152	22,780	24,103	26,421
Israel	6.400	6.500	6.700	7.200	7.300	7.518	20,227	20,116	21,000	19,532	21,428	23,722
Italy	62	62	63	62	63	-	36,779	38,666	-	32,695	33,882	-
Jamaica	1.354	1.303	1.232	1.301	1.262	-	-	-	-	-	-	-
Japan	*257	*255	*254	*252	*270	*268	27,169	30,363	33,479	38,969	-	-
Jordan	1.605	1.892	1.637	2.307	3.125	2.861	3,024	2,899	3,082	3,363	2,973	3,387
Kenya	7.142	7.196	7.232	7.378	7.547	7.759	1,806	1,571	1,492	917	1,204	1,556
Korea, Republic of	60	60	59	60	63	64	9,699	11,261	12,321	12,975	14,399	17,432

Continued.

Depending on the table, * means *Enterprises, Engaged,* or *Factor Values;* ± means *Basis Unspecified;* § means *shown in millions.* For additional notes and sources, see Appendix I.

Employment and Compensation of Employees

- Continued -

Country	Number Employed/Engaged (000)						Wage/Salary per Employee ($)					
	1990	1991	1992	1993	1994	1995	1990	1991	1992	1993	1994	1995
Kuwait	0.924	0.534	0.834	0.874	0.912	0.894	4,514	5,025	7,616	8,555	8,676	11,185
Kyrgyzstan	*0.142	*0.120	*0.133	*0.102	*0.082	-	-	-	-	-	-	-
Latvia	3.738	4.051	3.983	2.418	1.887	1.637	-	-	-	-	-	-
Liechtenstein	-	*0.003	-	-	-	-	-	-	-	-	-	-
Lithuania	-	-	6.735	5.837	5.000		-	-	487	566	1,239	
Macau	0.573	0.471	0.377	0.269	0.216	0.183	4,048	4,550	4,771	5,342	5,312	5,251
Malaysia	14	16	18	19	20	22	3,338	3,369	4,117	4,418	4,752	5,560
Malta	0.314	0.300	0.340	0.337	0.334	-	9,635	10,134	10,627	9,516	10,154	-
Mauritius	*0.402	0.795	0.737	0.732	0.832	0.820	2,293	1,223	2,689	2,299	2,095	2,620
Mexico	33	32	31	28	24	26	5,080	5,951	7,031	7,765	7,942	4,602
Mongolia	-	-	-	-	*0.019		-	-	-	-	-	
Morocco	-	7.246	7.716	8.410			-	-	-	-		
Mozambique	0.648	0.623	0.596	0.584	0.514	-	1,510	1,041	611	562	595	-
Namibia	-	-	-	-	0.070	-	-	-	-	-	8,219	-
Nepal	1.504	1.436	-	1.895	1.999	-	482	514	-	435	-	-
Netherlands	24	25	25	24	23	23	28,775	29,900	34,047	33,142	36,126	
Netherlands Antilles	-	-	-	0.179	-	-	-	-	-	12,256	-	-
New Zealand	9.493	8.894	8.635	8.581	-		25,219	25,326	23,368	27,283		
Norway	12	11	11	10	10	9.965	29,752	30,394	39,596	31,303	33,158	39,503
Pakistan	8.631	8.270	8.650	-	-		1,700	1,828	1,880			
Panama	*1.052	*1.046	*1.053	*1.050	*1.116	*1.119	9,146	8,897	9,087	9,060	10,051	10,155
Paraguay	-	0.019	-	-	-	-	-	-	-	-	-	-
Peru	*6.264	*5.815	*5.175	-	-	-	4,207	5,062	3,763			
Philippines	15	17	18	16	-	-	2,248	2,221	2,675	2,466	-	-
Poland	42	39	33	31	29	28	1,254	1,893	2,541	2,700	3,789	5,573
Portugal	*19	*17	*17	*17	*15	*15	-	-	-	-	-	-
Puerto Rico	1.700	1.620	1.700	1.600	1.650	1.700	23,529	24,383	23,765	23,937	23,818	24,176
Qatar	0.066	0.076	0.058	0.065	0.061	-	3,838	2,035	3,486	2,844	4,152	
Republic of Moldova	1.427	1.359	1.504	1.470	1.348	1.338	-	2,788	-	-	486	584
Romania	43	38	36	32	28		1,765	1,261	830	890	925	
Russian Federation	-	-	-	185	165	172	-	-	-	796	1,384	2,025
Saint Vincent & the Grenadines	-	-	-	0.087	0.087	0.102	-	-	-	4,677	5,502	
Senegal	*0.234	*0.282	*0.285	*0.291	*0.278	-	7,189	7,001	7,423	7,282	3,920	-
Sierra Leone	-	-	-	*0.036	-	-	-	-	-	989	-	-
Singapore	*4.604	*4.932	*5.032	*5.184	*5.459	*5.748	11,818	13,499	15,167	16,309	18,264	21,032
Slovakia	-	*14	*13	*15	*14	-	-	1,532	1,927	2,012	2,276	-
South Africa	49	50	51	50	49	51	9,197	9,278	10,629	10,637	10,891	12,249
Spain	41	42	40	47	47	46	21,202	22,214	25,249	21,372	20,948	24,217
Sri Lanka	*4.207	*4.075	*4.100	*5.044	-	-	943	1,134	1,131	1,152	-	-
Suriname	*0.062	*0.066	*0.029	*0.029	-	-	9,036	7,639	9,659	9,659	-	-
Swaziland	-	3.555	3.313	3.432	-	-	-	6,198	7,215	6,123	-	-
Sweden	52	*51	*47	42	42	42	26,858	29,202	32,424	25,285	26,597	29,470
Switzerland	*17	*17	*16	*16	*15	*15	-	-	-	-	-	-
Thailand	18	17	-	-	-	-	2,417	1,999	-	-	-	-
Tonga	0.007	-	-	-	-	-	1,450	-	-	-	-	-
Trinidad and Tobago	1.081	1.099	1.093	0.988	1.106	-	7,929	7,117	7,454	4,842	4,075	-
Tunisia	-	-	-	3.710	4.488	4.907	-	-	-	-	-	-
Turkey	22	21	20	20	20	20	7,825	9,450	9,510	10,412	6,569	7,876
Ukraine	29	29	28	26	24	23	2,800	2,756	2,109	-	484	661
United Kingdom	148	142	136	126	126	127	23,637	25,103	26,866	23,345	24,660	26,585
United Republic of Tanzania	5.215	5.463	-	-	-	-	-	410	-	-	-	-
USSR (Former)	254	-	-	-	-	-	2,325	-	-	-	-	-
United States of America	590	586	587	588	583	589	30,644	31,587	33,211	33,847	35,163	35,885
Uruguay	*3.498	*3.313	*3.169	*3.034	*3.030	*3.034	3,749	5,410	6,086	7,142	8,037	8,037
Venezuela	13	15	14	12	12	12	5,407	5,545	6,040	5,598	5,168	5,775
Yugoslavia	-	14	14	13	13	13	-	-	-	-	2,182	1,000
Zambia	1.575	-	-	-	0.941	-	1,241	-	-	-	1,332	-
Zimbabwe	4.691	4.562	3.700	4.000	3.600	3.326	5,256	4,493	3,528	3,171	3,436	3,858

Depending on the table, * means *Enterprises*, *Engaged*, or *Factor Values*; ± means *Basis Unspecified*; § means *shown in millions*. For additional notes and sources, see Appendix I.

221

Female Workers: Total and Percent of Employment

Country	Female Workers (000)						Female Workers as % of Total Employment					
	1990	1991	1992	1993	1994	1995	1990	1991	1992	1993	1994	1995
Afghanistan	0.017	0.019	-	-	-	-	-	-	-	-	-	-
Albania	-	-	-	0.100	0.015	0.003	-	-	-	21.60	32.61	30.00
Angola	-	-	0.019	0.004	-	-	-	-	10.11	8.33	-	-
Argentina	-	-	-	-	3.253	-	-	-	-	-	13.68	-
Australia	3.702						17.63					
Austria	4.000	4.000	4.000	3.624	3.314	-	20.00	20.00	20.00	19.22	18.10	-
Bahrain	-	-	0.003	-	-	-	-	-	1.24	-	-	-
Bangladesh	0.104	0.169	0.231	-	-	-	0.63	0.96	1.38	-	-	-
Bosnia & Herzegovina	3.127	2.946	-	-	-	-	33.48	31.02	-	-	-	-
Bulgaria	-	-	-	5.300	7.600	7.200	-	-	-	49.53	55.47	50.35
Canada	16	14	-	-	-	-	14.04	12.84	-	-	-	-
Chile	0.509	0.648	0.735	0.707	1.037	-	5.65	6.17	6.35	6.37	8.98	-
China (Taiwan Province)	19	19	21	22	22	22	31.75	32.08	33.04	34.57	34.31	33.24
Colombia	-	2.361	-	2.885	2.945	-	-	20.08	-	20.21	19.80	-
Croatia	4.160	3.580	3.010	2.590	2.560	2.600	42.49	43.03	40.62	38.43	40.38	39.27
Cyprus	0.326	0.362	0.296	0.308	0.290	-	41.85	44.69	39.00	41.18	38.36	-
Czechoslovakia (Former)	12	11	-	-	-	-	26.22	27.44	-	-	-	-
Czech Republic	11	10	11	10	-	-	44.00	43.48	44.00	45.45	-	-
Denmark	2.904	3.007	2.824	-	-	-	26.49	27.31	27.49	-	-	-
Egypt	-	2.529	1.786	1.531	-	-	-	7.57	9.30	8.23	-	-
El Salvador	-	-	-	0.636	0.416	0.455	-	-	-	34.62	25.43	28.19
Ethiopia	-	0.375	0.382	0.384	0.373	0.383	-	25.86	29.45	29.77	27.49	28.12
FYR Macedonia	0.494	0.475	0.461	0.448	0.436	0.412	27.31	29.60	26.09	29.07	25.74	22.11
Germany (Western Part)	47	-	-	-	-	-	28.79	-	-	-	-	-
Ghana	-	-	-	0.124	-	-	-	-	-	8.49	-	-
Hungary	5.000	-	6.000	5.000	-	-	38.46	-	50.00	45.45	-	-
Indonesia	-	-	-	18	17	19	-	-	-	23.99	22.11	21.12
Iran (Islamic Republic of)	-	0.316	0.361	0.365	-	-	-	2.60	3.02	3.08	-	-
Iraq	-	1.110	-	-	-	-	-	25.20	-	-	-	-
Ireland	0.800	0.900	-	-	-	-	23.53	25.00	-	-	-	-
Italy	-	12	12	12	12	-	-	20.17	18.53	19.68	19.44	-
Jordan	0.222	0.264	0.268	0.307	0.524	0.469	13.83	13.95	16.37	13.31	16.77	16.39
Kenya	0.308	0.347	0.354	0.361	-	-	4.31	4.82	4.89	4.89	-	-
Korea, Republic of	14	14	14	14	14	14	23.58	23.93	23.96	23.59	22.55	22.35
Macau	0.060	0.055	0.054	0.036	0.034	0.030	10.47	11.68	14.32	13.38	15.74	16.39
Malaysia	4.500	5.300	5.600	6.100	6.400	-	33.33	32.72	31.82	31.94	32.82	-
Malta	0.056	0.035	0.055	0.045	0.046	-	17.83	11.67	16.18	13.35	13.77	-
Mongolia	-	-	-	-	0.004	-	-	-	-	-	21.05	-
Morocco	-	-	-	-	-	1.422	-	-	-	-	-	-
Nepal	0.080	0.169	-	0.254	0.255	-	5.32	11.77	-	13.40	12.76	-
New Zealand	1.609	1.514	1.559	1.622	-	-	16.95	17.02	18.05	18.90	-	-
Panama	0.127	-	-	-	-	-	12.07	-	-	-	-	-
Philippines	-	-	3.328	3.543	-	-	-	-	18.39	22.57	-	-
Portugal	2.536	2.000	2.351	2.015	1.559	-	13.70	11.73	13.97	11.80	10.56	-
Puerto Rico	0.430	0.400	0.400	0.280	0.320	0.350	25.29	24.69	23.53	17.50	19.39	20.59
Republic of Moldova	-	-	-	-	0.569	0.574	-	-	-	-	42.21	42.90
Saint Vincent & the Grenadines	-	-	-	0.007	0.007	-	-	-	-	8.05	8.05	-
Senegal	0.009	-	-	-	-	-	3.85	-	-	-	-	-
Sri Lanka	0.557	0.568	0.450	0.565	-	-	13.24	13.94	10.98	11.20	-	-
Sweden	11	-	-	-	-	-	21.32	-	-	-	-	-
Thailand	6.556	7.668	-	-	-	-	35.59	44.09	-	-	-	-
Turkey	1.575	-	-	-	-	-	7.31	-	-	-	-	-
United Kingdom	43	-	40	-	-	-	28.74	-	29.41	-	-	-
United Republic of Tanzania	0.363	0.343	-	-	-	-	6.96	6.28	-	-	-	-
United States of America	-	-	448	449	449	453	-	-	76.32	76.36	77.02	76.91
Zambia	-	-	-	-	0.039	-	-	-	-	-	4.14	-

　　Depending on the table, * means *Enterprises*, *Engaged*, or *Factor Values*; ± means *Basis Unspecified*; § means *shown in millions*. For additional notes and sources, see Appendix I.

Output and Output per Employee

Country	Output ($ bil.)						Output per Employee ($)					
	1990	1991	1992	1993	1994	1995	1990	1991	1992	1993	1994	1995
Albania	-	-	-	*0.000	*0.000	*0.000	-	-	-	952	3,368	2,980
Angola	-	-	±2.893	±2.517	-	-	-	-	§15,38	§52,44	-	-
Argentina	±2.133	-	2.284	-	-	-	71,437	-	-	96,991	-	-
Armenia	0.000	0.001	0.001	0.075	0.000	-	661	1,210	-	-	-	-
Australia	*3.268	*3.811	*3.337	-	-	5.045	155,618	190,573	175,619	-	-	-
Austria	4.173	4.103	4.288	3.717	4.244	5.045	208,672	205,160	214,424	197,065	231,851	277,688
Azerbaijan	±0.014	±0.027	±0.001	±0.001	±0.000	-	19,546	39,663	991	2,154	396	-
Bahrain	-	-	±0.009	-	-	-	-	-	39,287	-	-	-
Bangladesh	0.171	0.185	0.182	-	-	-	10,419	10,512	10,879	-	-	-
Belgium	3.975	3.750	3.935	3.218	3.561	4.437	222,043	208,350	227,433	-	-	-
Belize	0.002	0.002	0.002	-	-	-	-	28,224	31,104			
Benin	±0.001	±0.001	±0.001	±0.001	-	-	-	-	-	-	-	-
Bolivia	0.008	0.010	0.010	0.014	0.014	0.014	24,955	26,517	20,669	23,763	21,472	22,319
Bosnia & Herzegovina	±0.291	±0.360	-	-	-	-	31,164	37,915	-	-	-	-
Brazil	*8.507	-	*7.155	*8.453	-	-	68,064	-	61,226	71,890	-	-
Bulgaria	±0.290	±0.175	±0.121	±0.130	±0.144	±0.262	19,491	13,096	10,122	12,110	10,525	18,355
Burundi	±0.000	±0.000	-	-	-	-	1,266	1,428	-	-	-	-
Cameroon	0.041	0.038	0.050	0.054	0.032	0.039	62,878	50,644	61,728	60,312	37,293	48,773
Canada	21	19	17	17	20	20	184,114	172,083	168,092	169,455	197,349	203,178
Chile	1.127	1.249	1.590	1.498	1.945	2.349	125,045	119,045	137,301	134,911	168,513	194,434
China	±8.127	±7.954	±8.939	±11	±8.803	±12	6,772	6,363	7,038	8,486	6,877	9,133
China (Hong Kong SAR)	1.093	1.067	1.177	1.245	0.904	-	65,088	72,078	86,528	96,533	91,281	-
China (Taiwan Province)	±4.733	±5.056	±5.793	±4.859	±5.750	±7.104	78,933	83,645	92,164	76,800	89,893	108,740
Colombia	0.840	0.923	0.927	0.979	1.260	1.537	71,204	78,533	67,424	68,572	84,706	101,491
Costa Rica	0.180	0.173	0.189	0.185	0.170	0.227	74,218	63,628	67,177	63,189	-	-
Croatia	±0.453	±0.461	±0.238	-	-	-	46,296	55,386	32,156	-	-	-
Cyprus	0.064	0.054	0.057	0.049	0.055	0.068	81,631	66,581	74,982	65,387	73,092	91,974
Czechoslovakia (Former)	±1.040	±0.555	-	-	-	-	23,114	14,230	-	-	-	-
Denmark	*1.360	*1.407	*1.547	*1.392	*1.705	*2.010	124,009	127,781	150,561	139,160	163,106	191,564
Ecuador	0.228	0.250	0.264	0.286	0.274	0.442	57,227	59,429	59,694	57,760	61,689	85,864
Egypt	*0.375	*0.368	*0.374	*0.344	*0.368	*0.371	16,099	11,029	19,477	18,507	19,602	20,245
El Salvador	-	±0.032	±0.027	±0.056	±0.059	±0.080	-	-	30,001	30,687	35,796	49,376
Ethiopia	0.018	0.020	0.011	0.013	0.020	0.014	14,412	13,536	8,298	10,149	14,538	10,304
Fiji	0.016	0.017	0.017	0.020	0.024	-	67,802	64,146	71,574	75,483	81,736	-
Finland	±12	±10	±9.752	±8.534	±11	±17	262,131	242,334	243,203	218,808	277,884	448,423
FYR Macedonia	±0.041	±0.061	±0.036	±0.026	±0.030	±0.043	22,692	37,825	20,534	16,553	17,532	23,067
France	±20	±19	±20	±17	±19	±24	186,362	180,169	193,394	167,725	187,652	243,389
Gabon	-	*0.001	-	-	-	*0.001	-	-	-	-	-	26,879
Germany	-	30	31	27	28	-	-	153,928	170,940	158,693	176,138	-
Germany (Western Part)	-	-	30	-	-	-	-	-	179,158	-	-	-
Ghana	-	-	-	0.025	0.020	0.021	-	-	-	17,190	13,620	13,894
Greece	*0.724	*0.704	*0.763	*0.702	*0.757	*0.896	79,171	85,625	96,661	89,877	96,111	113,136
Guatemala	-	0.046	0.050	0.065	0.122	0.139	-	30,436	31,274	40,323	77,678	83,526
Honduras	0.060	0.053	0.061	0.068	0.081	0.093	47,210	36,386	35,462	35,671	38,916	29,725
Hungary	±0.454	±0.396	±0.495	±0.391	±0.406	±0.484	34,929	33,006	41,276	35,570	37,174	45,727
Iceland	0.024	0.027	±0.027	±0.024	-	-	105,581	117,485	114,243	103,501	-	-
India	*2.899	*2.476	*2.413	*2.318	*2.813	*3.350	20,092	16,882	15,501	14,788	17,176	19,525
Indonesia	1.347	1.731	1.996	1.923	2.510	3.217	31,181	29,356	27,066	25,969	31,946	36,251
Iran (Islamic Republic of)	0.314	0.393	0.432	0.389	-	-	31,995	32,333	36,099	32,835	-	-
Iraq	-	*0.122	*0.244	-	-	-	-	27,683	60,932	-	-	-
Ireland	*0.477	*0.472	*0.524	*0.457	*0.491	*0.539	140,328	131,081	144,352	125,855	133,608	146,686
Israel	0.690	0.688	0.723	0.711	0.764	0.896	107,799	105,777	107,915	98,691	104,681	119,150
Italy	±13	±12	±13	±11	±13	-	208,910	199,383	207,296	173,674	206,962	-
Japan	±57	±62	±64	±70	±79	±90	222,246	243,981	253,007	279,348	292,390	335,884
Jordan	0.066	0.071	0.061	0.081	0.116	0.130	41,311	37,266	37,297	35,204	37,100	45,478
Kenya	*0.161	*0.181	*0.131	*0.082	*0.099	*0.178	22,485	25,158	18,112	11,077	13,096	23,005
Korea, Republic of	6.255	6.897	7.245	7.405	9.100	12	104,600	115,021	122,592	123,365	143,936	191,640
Kuwait	0.062	0.023	0.049	0.059	0.068	0.080	67,174	42,708	59,072	67,498	74,900	89,717
Kyrgyzstan	0.002	0.002	0.001	0.000	0.000	-	14,066	12,821	10,753	1,503	1,682	-
Latvia	0.001	0.003	0.016	0.009	0.008	0.012	209	818	4,088	3,607	4,450	7,034

Continued.

Depending on the table, * means *Enterprises, Engaged,* or *Factor Values*; ± means *Basis Unspecified*; § means *shown in millions*. For additional notes and sources, see Appendix I.

223

Output and Output per Employee
- Continued -

Country	Output ($ bil.)						Output per Employee ($)					
	1990	1991	1992	1993	1994	1995	1990	1991	1992	1993	1994	1995
Lithuania	-	-	0.029	0.020	0.036	-			4,354	3,475	7,200	-
Macau	0.013	0.012	0.011	0.009	0.008	0.006	22,612	25,088	29,203	33,399	36,868	35,208
Malaysia	*0.484	*0.558	*0.679	*0.761	*0.936	*1.134	35,842	34,436	38,587	39,827	47,982	52,707
Malta	0.019	0.017	0.020	0.018	0.017		59,284	57,595	58,361	53,700	52,261	
Mauritius	0.009	0.010	0.017	0.013	0.019	0.020	22,695	12,079	22,781	18,160	23,396	23,950
Mexico	2.262	2.345	2.384	2.475	2.534	2.076	68,246	72,770	76,660	88,186	105,252	80,377
Mongolia	-	-	-	-	0.000		-	-	-	-	587	
Morocco		±0.395	±0.433	±0.412	-	-		54,524	56,103	49,012	-	-
Mozambique		±0.005	-	±0.005	±0.003	-		8,779	-	7,778	6,058	-
Namibia	-	-	-	-	0.002					-	25,544	
Nepal	0.009	0.007	-	0.008	0.010	-	5,909	5,140	-	4,289	4,993	-
Netherlands	4.782	4.613	4.805	4.043	4.390	5.123	195,739	181,737	190,276	168,837	187,616	219,748
New Zealand	1.632	1.494	1.341	1.424	-	-	171,941	168,035	155,353	165,905	-	-
Norway	±2.966	±2.731	±2.495	±2.118	±2.387	±2.821	250,763	238,838	234,488	208,707	234,837	283,135
Pakistan	0.209	0.234	0.254	-	-		24,177	28,326	29,393	-	-	
Panama	0.105	0.091	0.094	0.094	0.113	0.115	100,212	86,812	89,471	89,303	101,437	102,953
Peru	0.302	0.243	0.204	-	-		48,289	41,853	39,489	-	-	
Philippines	0.516	0.556	0.743	0.530	-	-	33,316	33,504	41,055	33,761	-	-
Poland	±0.769	±0.826	±0.830	±0.815	±1.080	±1.520	18,308	21,187	25,144	26,282	36,948	54,393
Portugal	1.890	1.852	2.014	1.581	1.945	2.325	102,142	108,621	119,650	92,600	131,734	156,901
Puerto Rico	±0.267	±0.266	±0.261	±0.263	±0.260	±0.294	157,346	164,074	153,412	164,313	157,758	173,000
Qatar	0.001	0.001	0.001	0.001	0.001	-	8,325	7,230	18,947	12,680	18,015	-
Republic of Moldova	0.000	0.064	0.008	0.008	0.012	0.012	0.040	47,137	5,439	5,700	8,665	8,641
Romania	±0.602	±0.517	±0.310	±0.259	-	-	13,899	13,466	8,662	8,066	-	-
Russian Federation	-	7.206	4.649	1.201	2.148	4.590	-	-	-	6,500	13,033	26,747
Senegal	0.017	-	-	-	-	-	71,811	-	-	-	-	-
Sierra Leone	-	-	-	0.000	-	-	-	-	-	8,459	-	-
Singapore	*0.450	*0.503	*0.493	*0.528	*0.621	*0.749	97,791	102,051	97,886	101,915	113,744	130,371
Slovakia		±0.334	±0.373	±0.324	±0.398	-		24,086	29,015	22,207	28,385	-
Slovenia	-	-	-	0.832	0.982	1.246	-	-	-	-	-	-
South Africa	±3.347	±3.489	±3.612	±3.332	±3.590	±4.340	68,300	69,778	70,827	66,640	73,275	84,393
Spain	*6.917	*7.057	*7.122	±6.348	±7.201	±10	167,642	167,322	177,699	134,122	154,362	223,142
Sri Lanka	0.030	0.031	0.034	0.065	-	-	7,143	7,511	8,386	12,833	-	-
Suriname	*0.010	*0.006	*0.005	0.005	-	-	162,646	84,882	173,863	173,863	-	-
Swaziland	-	*0.103	*0.110	*0.099	-	-	-	28,871	33,115	28,844	-	-
Sweden	*12	*12	*12	*8.538	*10	*11	227,522	237,748	252,618	203,725	245,230	272,098
Switzerland	3.099	2.891	2.912	2.594	3.092	3.769	183,344	172,096	180,858	165,195	203,440	244,743
Thailand	1.217	0.572	-	-	-	-	66,050	32,911	-	-	-	-
Tonga	0.000	-	-	-	-	-	40,373	-	-	-	-	-
Trinidad and Tobago	±0.054	±0.054	±0.054	±0.047	±0.050	-	49,511	48,930	49,306	47,646	45,033	-
Turkey	1.350	1.281	1.396	1.630	1.458	1.784	62,635	61,277	69,102	83,246	74,734	90,315
Ukraine	*0.808	*1.055	*0.844	*0.183	*0.179	*0.240	27,873	36,363	30,159	7,049	7,461	10,443
United Kingdom	*19	*18	19	16	18	19	125,977	129,253	136,653	123,147	141,216	152,514
United Republic of Tanzania	*0.031	*0.068	-	-	-	-	5,941	12,445	-	-	-	-
USSR (Former)	±5.250	-	-	-	-	-	20,669	-	-	-	-	-
United States of America	*126	*124	*127	*128	*138	*166	213,475	211,160	217,109	217,148	236,264	281,365
Uruguay	0.126	0.147	0.150	0.141	0.157	0.157	35,940	44,233	47,403	46,337	51,974	51,898
Venezuela	0.738	0.959	0.827	0.741	0.751	1.010	55,466	63,068	59,080	62,869	61,765	82,472
Yugoslavia	±0.052	±0.078	±0.138	-	±0.262	±0.197	-	5,519	9,766	-	20,813	15,076
Zambia	0.015	-	-	-	0.006	-	9,345	-	-	-	6,411	-
Zimbabwe	*0.155	*0.138	*0.112	*0.096	*0.103	*0.109	33,127	30,341	30,347	23,986	28,630	32,882

Depending on the table, * means *Enterprises, Engaged,* or *Factor Values*; ± means *Basis Unspecified*; § means *shown in millions*. For additional notes and sources, see Appendix I.

Value Added and Value Added per Employee

Country	Value Added ($ bil.)						Value Added per Employee ($)					
	1990	1991	1992	1993	1994	1995	1990	1991	1992	1993	1994	1995
Albania	±0.004	-	-	-	-	-	1,448	-	-	-	-	-
Argentina	±0.882	-	-	0.583	-	-	29,529	-	-	24,753	-	-
Armenia	0.000	0.000	-	-	-	-	65	158	-	-	-	-
Australia	*1.302	*1.459	*1.446	-	-	-	62,016	72,926	76,129	-	-	-
Austria	1.333	1.307	1.303	1.126	1.352	1.604	66,662	65,326	65,152	59,714	73,848	88,314
Bangladesh	0.053	0.042	0.055	-	-	-	3,237	2,399	3,306	-	-	-
Belgium	*1.042	*0.983	*1.031	*0.842	*0.929	*1.155	58,223	54,597	59,573	-	-	-
Belize	0.000	0.001	0.001	-	-	-	-	7,724	10,194	-	-	-
Bolivia	0.002	0.004	0.004	0.005	0.004	0.005	5,093	11,460	8,590	7,961	6,933	7,102
Bosnia & Herzegovina	±0.128	±0.158	-	-	-	-	13,709	16,679	-	-	-	-
Brazil	*4.539	-	*4.167	*5.072	-	-	36,316	-	35,653	43,135	-	-
Bulgaria	-	±0.042	±0.021	±0.033	±0.032	±0.055	-	3,150	1,774	3,072	2,368	3,858
Burundi	0.000	0.000	-	-	-	-	1,112	1,265	-	-	-	-
Cameroon	0.011	0.012	0.011	0.013	0.007	0.007	16,669	15,887	13,414	14,622	7,784	9,255
Canada	8.750	6.887	6.279	6.116	7.989	8.199	76,758	63,180	60,380	60,553	79,892	82,216
Chile	0.561	0.652	0.757	0.673	1.014	1.224	62,240	62,102	65,319	60,628	87,871	101,339
China	±1.949	±1.854	±2.129	±2.622	±2.226	±2.782	1,625	1,483	1,676	2,097	1,739	2,092
China (Hong Kong SAR)	0.275	0.310	0.334	0.354	0.255	-	16,398	20,945	24,546	27,408	25,774	-
China (Taiwan Province)	1.383	1.443	1.492	1.350	1.520	1.549	23,058	23,871	23,738	21,343	23,763	23,716
Colombia	0.301	0.339	0.323	0.320	0.508	0.621	25,478	28,815	23,508	22,425	34,188	40,987
Costa Rica	0.045	0.044	0.048	0.046	0.042	0.055	18,551	16,176	17,148	15,751	-	-
Croatia	±0.203	±0.202	±0.092	-	-	-	20,706	24,254	12,367	-	-	-
Cyprus	0.017	0.017	0.019	0.017	0.021	0.022	21,660	20,777	25,557	22,127	27,139	29,902
Czechoslovakia (Former)	±0.255						5,658	-	-	-	-	-
Denmark	*0.628	*0.663	*0.729	*0.656	*0.806	*0.951	57,288	60,213	70,997	65,579	77,113	90,621
Ecuador	0.034	0.036	0.065	0.069	0.027	0.068	8,644	8,638	14,679	13,846	6,161	13,261
Egypt	*0.084	*0.079	*0.130	*0.088	*0.094	*0.095	3,605	2,376	6,747	4,720	5,001	5,165
El Salvador	-	0.012	-	0.021	0.025	0.035	-	-	-	11,539	15,110	21,907
Ethiopia	0.003	0.007	0.005	0.006	0.011	0.005	2,419	5,003	3,647	4,513	7,928	3,804
Fiji	0.005	0.004	0.006	0.007	0.008	-	22,064	17,014	23,861	25,059	27,076	
Finland	±3.603	±2.758	±3.017	±2.909	±3.825	±6.278	81,149	65,985	75,241	74,585	98,838	162,626
FYR Macedonia	±0.013	±0.027	±0.016	±0.010	±0.010	±0.014	7,258	16,660	9,250	6,187	5,944	7,642
France	±6.823	±6.639	±6.819	±5.799	±5.979	±7.730	63,528	62,573	64,882	56,912	59,311	77,070
Gabon	-	*0.000	-	-	-	*0.000	-	-	-	-	-	6,511
Germany (Western Part)	13	15	16	14	-	-	82,631	86,781	93,142	-	-	-
Ghana	-	-	-	0.011	0.010	0.010	-	-	-	7,632	6,433	6,563
Greece	*0.272	*0.280	*0.320	*0.295	*0.320	*0.380	29,745	34,120	40,501	37,797	40,591	47,944
Honduras	*0.011	*0.011	*0.013	*0.014	*0.014	*0.016	9,022	7,352	7,407	7,418	6,768	5,239
Hungary	±0.128	±0.120	±0.117	±0.111	±0.112	±0.129	9,884	9,974	9,746	10,086	10,287	12,159
Iceland	0.009	0.011	±0.010	±0.010	-	-	40,643	46,551	42,183	41,450	-	-
India	*0.574	*0.459	*0.428	*0.433	*0.523	*0.620	3,979	3,129	2,749	2,761	3,191	3,614
Indonesia	*0.477	*0.674	*0.713	*0.662	*0.889	*1.030	11,043	11,425	9,662	8,935	11,310	11,608
Iran (Islamic Republic of)	0.137	0.156	0.137	0.139	-	-	14,004	12,869	11,480	11,767	-	-
Iraq	-	*0.039	*0.228	-	-	-	-	8,763	56,994	-	-	-
Ireland	*0.190	*0.192	*0.213	*0.186	*0.200	*0.219	55,995	53,267	58,690	51,177	54,356	59,691
Israel	0.241	0.242	0.229	0.232	0.253	0.290	37,664	37,262	34,232	32,194	34,712	38,611
Italy	±3.878	±3.894	±4.041	±3.331	±3.822	-	62,684	63,021	64,014	53,581	61,036	-
Japan	±22	±24	±26	±29	±34	±39	86,722	94,786	100,749	116,086	125,558	144,001
Jordan	0.020	0.022	0.013	0.021	0.034	0.029	12,577	11,609	7,839	9,002	10,895	10,107
Kenya	*0.042	*0.039	*0.034	*0.021	*0.031	*0.033	5,866	5,456	4,721	2,898	4,113	4,310
Korea, Republic of	2.123	2.670	2.912	3.051	3.766	4.629	35,495	44,528	49,276	50,840	59,563	72,881
Kuwait	0.031	0.012	0.017	0.024	0.027	0.031	33,105	23,257	20,339	27,755	29,580	34,375
Latvia	-	-	-	*0.003	*0.004	*0.003	-	-	-	1,372	1,915	1,734
Macau	0.004	0.004	0.003	0.002	0.002	0.002	6,907	7,570	8,812	9,021	10,514	10,559
Malaysia	*0.155	*0.181	*0.233	*0.252	*0.315	*0.382	11,504	11,147	13,226	13,209	16,141	17,774
Malta	0.007	0.007	0.009	0.007	0.008	-	21,851	24,345	26,202	21,649	23,168	-
Mauritius	0.004	0.004	0.007	0.006	0.008	0.008	8,737	4,565	9,381	8,500	9,423	10,149
Mexico	0.706	0.736	0.743	0.722	0.723	0.591	21,294	22,841	23,901	25,736	30,042	22,906
Mongolia	-	-	-	-	0.000	-	-	-	-	-	179	-
Morocco	±0.151	±0.084	±0.088	±0.081	-	-	-	11,618	11,415	9,577	-	-

Continued.

Depending on the table, * means *Enterprises*, *Engaged*, or *Factor Values*; ± means *Basis Unspecified*; § means *shown in millions*. For additional notes and sources, see Appendix I.

225

Value Added and Value Added per Employee
- Continued -

Country	Value Added ($ bil.)						Value Added per Employee ($)					
	1990	1991	1992	1993	1994	1995	1990	1991	1992	1993	1994	1995
Namibia	-	-	-	-	0.000	-	-	-	-	-	3,730	-
Nepal	0.003	0.003	-	0.005	0.003	-	2,083	1,851	-	2,526	1,428	-
Netherlands	1.618	1.667	1.688	1.516	1.653	1.932	66,250	65,657	66,856	63,317	70,622	82,855
New Zealand	0.553	0.540	0.468	0.473	-	-	58,236	60,677	54,152	55,134	-	-
Norway	±0.787	±0.672	±0.589	±0.551	±0.641	±0.757	66,553	58,792	55,370	54,246	63,096	75,968
Pakistan	0.064	0.073	0.082	-	-	-	7,363	8,863	9,458	-	-	-
Panama	0.048	0.028	0.029	0.029	0.035	0.035	45,420	26,731	27,550	27,498	31,231	31,697
Peru	0.135	0.067	0.060	-	-	-	21,630	11,565	11,676	-	-	-
Philippines	0.184	0.188	0.257	0.189	-	-	11,862	11,305	14,211	12,012	-	-
Poland	±0.348	±0.261	±0.268	±0.246	±0.321	±0.447	8,293	6,694	8,124	7,940	10,966	15,997
Portugal	0.577	0.513	0.529	0.367	0.613	0.734	31,160	30,078	31,451	21,515	41,536	49,521
Puerto Rico	±0.066	±0.071	±0.074	±0.079	±0.080	±0.088	38,588	43,827	43,529	49,375	48,667	51,529
Qatar	0.000	0.000	0.000	0.000	0.001	-	4,162	3,615	4,737	4,227	9,007	-
Romania	±0.169	±0.158	±0.083	±0.070	-	-	3,912	4,125	2,331	2,181	-	-
Russian Federation	-	-	-	±0.528	±0.854	±1.834	-	-	-	2,860	5,182	10,686
Saint Lucia	-	±0.003	±0.005	±0.004	±0.004	±0.005	-	-	-	-	-	-
Senegal	0.005	0.005	0.006	0.005	0.004	-	19,291	19,245	19,897	18,192	13,399	-
Sierra Leone	-	-	-	0.000	-	-	-	-	-	4,685	-	-
Singapore	*0.189	*0.208	*0.215	*0.248	*0.303	*0.368	41,092	42,121	42,650	47,920	55,571	63,960
Slovakia	-	-	-	±0.073	±0.123	-	-	-	-	4,966	8,756	-
Slovenia	-	-	-	0.272	0.299	0.358	-	-	-	-	-	-
South Africa	*1.208	*1.213	*1.207	*1.113	*1.200	*1.451	24,649	24,257	23,664	22,268	24,495	28,223
Spain	*2.101	*2.119	*2.134	±1.931	±2.191	±3.011	50,924	50,237	53,239	40,800	46,958	66,162
Sri Lanka	0.019	0.013	0.017	0.032	-	-	4,515	3,071	4,233	6,342	-	-
Swaziland	-	*0.042	*0.052	*0.028	-	-	-	11,756	15,563	8,238	-	-
Sweden	*4.524	*3.188	*3.067	*2.685	*3.441	*3.745	87,682	62,971	65,128	64,080	80,983	89,814
Switzerland	1.080	1.092	1.116	0.970	1.144	1.406	63,892	65,001	69,323	61,769	75,273	91,289
Thailand	0.281	0.161	-	-	-	-	15,253	9,235	-	-	-	-
Trinidad and Tobago	±0.017	±0.014	±0.017	±0.013	±0.013	-	15,635	12,963	15,242	13,562	11,506	-
Turkey	0.559	0.557	0.455	0.608	0.600	0.735	25,929	26,630	22,522	31,026	30,728	37,185
United Kingdom	*8.036	*8.241	*8.488	*7.072	*7.988	*8.688	54,295	58,033	62,409	56,128	63,395	68,608
United Republic of Tanzania	*0.006	*0.035	-	-	-	-	1,075	6,443	-	-	-	-
United States of America	*57	*56	*57	*57	*60	*77	96,949	95,427	97,641	96,498	103,736	130,148
Uruguay	0.051	0.071	0.074	0.064	0.071	0.071	14,473	21,347	23,449	20,963	23,517	23,484
Venezuela	0.277	0.368	0.313	0.261	0.261	0.689	20,814	24,180	22,350	22,130	21,452	56,269
Yugoslavia	±0.018	±0.035	±0.073	-	±0.106	±0.081	-	2,456	5,183	-	8,381	6,167
Zambia	0.006	-	-	-	0.003	-	3,931	-	-	-	3,379	-
Zimbabwe	*0.064	*0.062	*0.052	*0.040	*0.034	*0.036	13,707	13,684	14,150	9,977	9,390	10,765

Capital Investment

Country	Gross Fixed Capital Formation ($ mil.)						Per Establishment ($ mil)					
	1990	1991	1992	1993	1994	1995	1990	1991	1992	1993	1994	1995
Argentina	-	-	-	172	-	-	-	-	-	0.197	-	-
Australia	127	-	-	-	-	-	0.525	-	-	-	-	-
Austria	840	522	487	319	266	-	5.673	3.553	3.453	2.261	1.958	-
Bahrain	-	-	1.021	-	-	-	-	-	0.255	-	-	-
Bangladesh	13	1.394	0.976	-	-	-	0.262	0.024	0.011	-	-	-
Belgium	287	629	587	284	467	463	0.931	2.050	2.009	0.984	1.605	-
Bolivia	0.750	0.459	0.961	0.652	4.013	-	0.068	0.035	0.074	0.047	0.309	-
Bosnia & Herzegovina	1.457	-	-	-	-	-	0.063	-	-	-	-	-
Brazil	1,245	-	665	658	-	-	1.380	-	0.806	0.835	-	-
Bulgaria	33	6.728	61	84	5.511	5.529	1.642	0.354	2.757	0.830	0.033	0.031
Cameroon	8.485	0.751	3.831	2.613	1.691	2.374	1.061	0.094	0.383	0.238	0.141	0.237
Canada	3,678	-	-	-	-	-	5.145	-	-	-	-	-
Chile	98	1,021	608	80	82	-	2.983	26	15	1.939	1.789	-
China (Hong Kong SAR)	62	52	53	77	44	-	0.035	0.034	0.036	0.062	0.040	-
Colombia	76	42	61	-3.721	-0.399	-	0.509	0.290	0.370	-0.024	-0.002	-

Continued.

Depending on the table, * means *Enterprises, Engaged,* or *Factor Values*; ± means *Basis Unspecified*; § means *shown in millions*. For additional notes and sources, see Appendix I.

Capital Investment

- Continued -

Country	Gross Fixed Capital Formation ($ mil.)						Per Establishment ($ mil)					
	1990	1991	1992	1993	1994	1995	1990	1991	1992	1993	1994	1995
Croatia	7.219	9.402	7.321	5.555	2.721	6.305	0.091	0.086	0.050	0.028	0.011	0.025
Cyprus	5.120	4.616	2.842	4.179	1.670	2.242	0.085	0.077	0.047	0.070	0.030	0.042
Czechoslovakia (Former)	217	-	-	-	-	-	9.447	-	-	-	-	-
Denmark	80	164	-	-	-	-	0.350	0.671	-	-	-	-
Ecuador	22	50	59	23	33	63	0.822	1.627	1.899	0.623	0.915	1.361
Egypt	30	12	11	-15.778	-	-	0.305	0.118	0.115	-0.164	-	-
El Salvador	-	0.495	-	3.403	7.046	2.230	-	0.083	-	0.378	1.174	0.279
Ethiopia	0.186	4.853	0.413	0.040	0.102	0.360	0.047	1.213	0.138	0.010	0.034	0.090
Fiji	0.384	0.559	1.183	-	-	-	0.043	0.080	0.148	-	-	-
Finland	1,780	1,302	964	699	664	855	12	8.035	5.840	4.317	4.178	4.386
FYR Macedonia	1.666	1.653	0.209	0.165	0.123	1.276	0.238	0.138	0.006	0.004	0.002	0.012
France	1,931	1,850	1,598	806	801	1,203	-	-	-	-	-	-
Gabon	-	0.025	-	-	-	0.012	-	0.012	-	-	-	0.006
Germany (Western Part)	2,325	2,419	2,363	2,047	-	-	2.169	2.221	2.141	-	-	-
Greece	20	27	39	-	-	-	0.134	0.186	0.258	-	-	-
Hungary	15	58	64	27	-	-	0.236	0.433	0.554	0.182	-	-
India	301	127	234	390	-	-	0.139	0.060	0.103	0.150	-	-
Indonesia	774	563	201	184	368	235	4.206	2.594	0.777	0.685	1.206	0.755
Iran (Islamic Republic of)	10	26	21	26	-	-	0.079	0.344	0.251	0.346	-	-
Ireland	31	20	-	-	-	-	0.360	0.237	-	-	-	-
Israel	35	41	36	45	70	-	0.213	0.222	0.203	0.218	0.365	-
Italy	1,315	1,035	890	753	660	-	1.916	1.476	1.139	0.950	0.818	-
Japan	4,254	5,716	3,498	3,849	3,718	-	0.427	0.586	0.368	0.400	0.358	-
Jordan	2.181	7.395	1.768	0.420	2.568	2.859	0.084	0.274	0.065	0.012	0.040	0.043
Korea, Republic of	702	1,230	1,102	960	1,037	1,611	0.369	0.596	0.518	0.416	0.431	0.652
Kuwait	0.558	0.451	3.208	2.490	2.960	-	0.035	0.035	0.247	0.192	0.247	-
Latvia	0.001	0.074	0.228	0.276	0.765	0.373	0.000	0.005	0.009	0.012	0.028	0.014
Macau	0.203	0.463	-0.032	-0.293	0.020	0.001	0.004	0.011	-0.001	-0.008	0.001	0.000
Malaysia	96	120	103	97	92	-	0.731	0.842	0.672	0.566	0.525	-
Malta	1.565	0.282	0.415	0.314	0.767	-	0.104	0.017	0.023	0.018	0.043	-
Mexico	104	198	-	-	-	-	0.787	1.500	-	-	-	-
Morocco	36	45	59	23	-	-	-	0.505	0.716	0.253	-	-
Nepal	6.367	-	-	-	-	-	0.289	-	-	-	-	-
Netherlands	347	490	330	265	-	-	2.377	3.043	2.027	1.735	-	-
New Zealand	79	-	-	-	-	-	0.673	-	-	-	-	-
Norway	184	203	477	132	88	-	1.788	2.054	5.681	1.612	1.103	-
Pakistan	-	36	-	-	-	-	-	0.487	-	-	-	-
Panama	1.625	3.494	-	-	-	-	0.074	0.175	-	-	-	-
Peru	8.772	2.500	2.221	-	-	-	0.043	0.012	0.010	-	-	-
Philippines	19	21	26	32	-	-	0.061	0.060	0.130	0.162	-	-
Poland	89	58	68	132	-	-	1.352	0.629	0.729	1.497	-	-
Portugal	269	496	205	291	52	-	0.636	1.169	0.489	0.657	0.116	-
Romania	22	19	9.122	9.966	17	-	1.267	0.248	0.110	0.097	0.075	-
Senegal	0.393	-	-	-	-	-	0.066	-	-	-	-	-
Singapore	41	49	49	77	54	-	0.442	0.520	0.493	0.760	0.511	-
Slovenia	54	74	23	13	16	11	0.807	0.324	0.071	0.039	0.049	0.040
Spain	468	480	422	373	373	501	0.427	0.449	0.403	-	-	-
Sri Lanka	1.482	0.629	0.277	1.184	-	-	0.087	0.033	0.018	0.046	-	-
Swaziland	-	8.479	6.793	11	-	-	-	1.696	1.359	2.239	-	-
Thailand	441	144	-	-	-	-	3.246	1.160	-	-	-	-
Trinidad and Tobago	5.784	13	1.929	1.887	1.772	-	0.199	0.486	0.071	0.073	0.061	-
Turkey	64	98	6.693	63	49	-	0.664	1.060	0.045	0.447	0.342	-
Ukraine	111	29	12	1.047	2.512	2.266	1.486	0.172	0.061	0.011	0.029	0.028
United Kingdom	886	958	958	-	1,025	-	0.277	0.310	0.315	-	0.313	-
United Republic of Tanzania	100	123	-	-	-	-	12	11	-	-	-	-
United States of America	10,600	8,830	7,766	7,137	7,097	7,942	-	-	1.140	-	-	-
Uruguay	2.742	3.135	4.990	3.927	-	-	-	-	-	-	-	-
Venezuela	85	54	75	20	70	24	0.780	0.514	0.725	0.237	0.750	0.249

Continued.

Depending on the table, * means Enterprises, Engaged, or Factor Values; ± means Basis Unspecified; § means shown in millions. For additional notes and sources, see Appendix I.

227

Capital Investment

- Continued -

Country	Gross Fixed Capital Formation ($ mil.)						Per Establishment ($ mil)					
	1990	1991	1992	1993	1994	1995	1990	1991	1992	1993	1994	1995
Yugoslavia	1.268	1.000	2.112	-	1.772	2.714	0.013	0.005	0.004	-	0.002	0.003
Zambia	0.758	-	-	-	-	-	0.063	-	-	-	-	-
Zimbabwe	21	61	41	5.577	5.865	-	1.370	4.384	3.442	0.465	0.489	-

Indexes of Industrial Production

Country	Index of Industrial Production (1990=100)						Index of Employment (1990=100)					
	1990	1991	1992	1993	1994	1995	1990	1991	1992	1993	1994	1995
Albania	100	-	-	-	-	-	100	-	-	19	2	0
Algeria	100	89	79	89	82	77	-	-	-	-	-	-
Angola	100	-	-	-	-	-	-	-	-	-	-	-
Argentina	100	124	-	-	-	-	100	-	-	79	80	-
Armenia	100	-	-	-	-	-	100	89	-	-	-	-
Australia	100	96	92	105	-	-	100	95	90	-	-	-
Austria	100	104	107	108	116	119	100	100	100	94	92	91
Azerbaijan	-	-	-	-	-	-	100	95	83	69	58	-
Bahrain	100	-	-	-	-	-	-	-	-	-	-	-
Bangladesh	100	97	95	96	97	89	100	107	102	-	-	-
Barbados	100	-	-	-	-	-	-	-	-	-	-	-
Belgium	100	98	102	93	97	95	100	101	97	-	-	-
Belize	100	-	-	-	-	-	-	-	-	-	-	-
Benin	100	-	-	-	-	-	-	-	-	-	-	-
Bhutan	100	-	-	-	-	-	-	-	-	-	-	-
Bolivia	100	103	118	121	111	106	100	124	157	194	207	209
Bosnia & Herzegovina . . .	-	-	-	-	-	-	100	102	-	-	4	-
Brazil	100	107	105	110	113	113	100	-	94	94	-	-
Bulgaria	100	71	65	63	74	87	100	90	81	72	92	96
Burkina-Faso	100	-	-	-	-	-	-	-	-	-	-	-
Burundi	100	-	-	-	-	-	100	98	-	-	-	-
Cambodia	100	-	-	-	-	-	100	86	-	-	-	-
Cameroon	100	-	-	-	-	-	100	115	123	137	129	121
Canada	100	96	96	99	104	105	100	96	91	89	88	87
Central African Republic . . .	100	100	100	100	100	-	-	-	-	-	-	-
Chad	100	-	-	-	-	-	-	-	-	-	-	-
Chile	100	109	131	132	141	149	100	116	129	123	128	134
China	100	-	-	-	-	-	100	104	106	104	107	111
China (Hong Kong SAR) . . .	100	110	129	146	150	150	100	88	81	77	59	-
China (Taiwan Province) . . .	100	107	113	110	113	117	100	101	105	106	107	109
Colombia	100	112	122	113	122	133	100	100	117	121	126	128
Congo	100	-	-	-	-	-	-	-	-	-	-	-
Costa Rica	100	104	89	92	85	92	100	112	116	121	-	-
Croatia	100	-	-	-	-	-	100	85	76	69	65	68
Cuba	100	98	98	-	-	-	-	-	-	-	-	-
Cyprus	100	91	85	89	102	116	100	104	97	96	97	95
Czechoslovakia (Former) . .	100	82	80	-	-	-	100	87	-	-	-	-
Czech Republic	100	-	-	-	-	-	100	92	100	88	-	-
Dem. Rep. of the Congo . .	100	-	-	-	-	-	-	-	-	-	-	-
Denmark	100	104	104	99	112	114	100	100	94	91	95	96
Dominican Republic.	100	-	-	-	-	-	-	-	-	-	-	-
Ecuador	100	115	105	104	-	-	100	105	111	124	111	129
Egypt	100	101	105	75	77	72	100	143	82	80	81	79
El Salvador.	100	100	-	-	-	-	-	-	-	-	-	-
Ethiopia and Eritrea	100	72	29	-	-	-	-	-	-	-	-	-
Ethiopia.	-	-	-	-	-	-	100	113	101	101	106	106
Fiji	100	109	111	127	141	129	100	113	105	117	127	-
Finland	100	97	101	109	121	122	100	94	90	88	87	87
FYR Macedonia	100	-	-	-	-	-	100	89	98	85	94	103
France	100	103	108	106	112	110	100	99	98	95	94	93

Continued.

Depending on the table, * means *Enterprises*, *Engaged*, or *Factor Values*; ± means *Basis Unspecified*; § means *shown in millions*. For additional notes and sources, see Appendix I.

Indexes of Industrial Production

- Continued -

Country	Index of Industrial Production (1990=100)						Index of Employment (1990=100)					
	1990	1991	1992	1993	1994	1995	1990	1991	1992	1993	1994	1995
Gabon	100	-	-	-	-	-	-	-	-	-	-	-
Gambia	100	-	-	-	-	-	-	-	-	-	-	-
Germany (Eastern Part)	100	-	-	-	-	-	-	-	-	-	-	-
Germany (Western Part)	100	103	102	99	104	100	100	105	103	-	-	-
Ghana	100	-	-	-	-	-	-	-	-	-	-	-
Greece	100	108	110	102	109	113	100	90	86	85	86	87
Guatemala	100	100	100	-	-	-	-	-	-	-	-	-
Guyana	100	-	-	-	-	-	-	-	-	-	-	-
Haiti	100	-	-	-	-	-	-	-	-	-	-	-
Honduras	100	128	129	150	155	234	100	116	137	150	165	247
Hungary	100	80	78	69	71	83	100	92	92	85	84	81
Iceland	100	-	-	-	-	-	100	100	101	100	97	-
India	100	106	110	114	129	147	100	102	108	109	114	119
Indonesia	100	104	150	153	175	210	100	137	171	171	182	205
Iran (Islamic Republic of)	100	159	159	-	-	-	100	124	122	121	-	-
Iraq	100	-	-	-	-	-	-	-	-	-	-	-
Ireland	100	102	104	104	107	107	100	106	107	107	108	108
Israel	100	102	109	113	118	121	100	102	105	113	114	117
Italy	100	97	100	100	-	-	100	100	102	100	101	-
Jamaica	100	-	-	-	-	-	100	96	91	96	93	-
Japan	100	103	101	99	101	105	100	99	99	98	105	104
Jordan	100	93	91	110	117	104	100	118	102	144	195	178
Kenya	100	108	130	91	79	77	100	101	101	103	106	109
Korea, Republic of	100	104	111	120	134	143	100	100	99	100	106	106
Kuwait	100	-	-	-	-	-	100	58	90	95	99	97
Kyrgyzstan	100	-	-	-	-	-	100	85	94	72	58	-
Lao P.D.R.	100	-	-	-	-	-	-	-	-	-	-	-
Latvia	100	-	-	-	-	-	100	108	107	65	50	44
Lebanon	100	-	-	-	-	-	-	-	-	-	-	-
Liberia	100	-	-	-	-	-	-	-	-	-	-	-
Libyan Arab Jamahiriya	100	-	-	-	-	-	-	-	-	-	-	-
Lithuania	100	-	-	-	-	-	-	-	-	-	-	-
Macau	100	-	-	-	-	-	100	82	66	47	38	32
Madagascar	100	64	64	-	-	-	-	-	-	-	-	-
Malawi	100	108	-	-	-	-	-	-	-	-	-	-
Malaysia	100	155	275	-	-	-	100	120	130	141	144	159
Malta	100	145	155	185	-	-	100	96	108	107	106	-
Mauritius	100	-	-	-	-	-	100	198	183	182	207	204
Mexico	100	97	98	98	94	96	100	97	94	85	73	78
Morocco	100	92	99	94	104	114	-	-	-	-	-	-
Mozambique	100	-	-	-	-	-	100	96	92	90	79	-
Myanmar	100	-	-	-	-	-	-	-	-	-	-	-
Nepal	100	-	-	-	-	-	100	95	-	126	133	-
Netherlands	100	99	100	97	103	104	100	104	103	98	96	95
New Zealand	100	-	-	-	-	-	100	94	91	90	-	-
Nicaragua	100	-	-	-	-	-	-	-	-	-	-	-
Niger	100	-	-	-	-	-	-	-	-	-	-	-
Nigeria	100	101	101	-	-	-	-	-	-	-	-	-
Norway	100	99	97	103	113	117	100	97	90	86	86	84
Pakistan	100	92	91	92	86	86	100	96	100	-	-	-
Panama	100	103	105	104	124	125	100	99	100	100	106	106
Papua New Guinea	100	-	-	-	-	-	-	-	-	-	-	-
Paraguay	100	103	102	72	50	53	-	-	-	-	-	-
Peru	100	76	46	49	96	132	100	93	83	-	-	-
Philippines	100	112	87	85	91	114	100	107	117	101	-	-
Poland	100	97	106	113	141	167	100	93	79	74	70	67
Portugal	100	111	106	103	109	114	100	92	91	92	80	80
Puerto Rico	100	-	-	-	-	-	100	95	100	94	97	100
Qatar	100	-	-	-	-	-	100	115	88	98	92	-
Republic of Moldova	100	-	-	-	-	-	100	95	105	103	94	94

Continued.

Depending on the table, * means *Enterprises, Engaged,* or *Factor Values;* ± means *Basis Unspecified;* § means *shown in millions.* For additional notes and sources, see Appendix I.

229

Indexes of Industrial Production

- Continued -

Country	Index of Industrial Production (1990=100)						Index of Employment (1990=100)					
	1990	1991	1992	1993	1994	1995	1990	1991	1992	1993	1994	1995
Romania	100	69	52	47	47	62	100	89	83	74	65	-
Russian Federation	100	-	-	-	-	-	-	-	-	-	-	-
Saudi Arabia	100	-	-	-	-	-	-	-	-	-	-	-
Senegal	100	100	82	75	85	77	100	121	122	124	119	-
Sierra Leone	100	-	-	-	-	-	-	-	-	-	-	-
Singapore	100	104	92	93	99	107	100	107	109	113	119	125
Slovakia	100	-	-	-	-	-	-	-	-	-	-	-
Slovenia	100	88	77	72	80	78	-	-	-	-	-	-
Somalia	100	-	-	-	-	-	-	-	-	-	-	-
South Africa	100	96	96	93	101	113	100	102	104	102	100	105
Spain	100	105	106	105	110	109	100	102	97	115	113	110
Sri Lanka	100	104	126	329	312	244	100	97	97	120	-	-
Sudan	100	50	50	-	-	-	-	-	-	-	-	-
Suriname	100	-	-	-	-	-	100	106	47	47	-	-
Swaziland	100	134	138	-	-	-	-	-	-	-	-	-
Sweden	100	99	100	104	112	110	100	98	91	81	82	81
Switzerland	100	100	102	101	104	104	100	99	95	93	90	91
Syrian Arab Republic	100	114	121	169	183	211	-	-	-	-	-	-
Thailand	100	98	61	-	-	-	100	94	-	-	-	-
Togo	100	-	-	-	-	-	-	-	-	-	-	-
Tonga	100	-	-	-	-	-	100	-	-	-	-	-
Trinidad and Tobago	100	93	89	134	118	137	100	102	101	91	102	-
Tunisia	100	101	104	104	102	110	-	-	-	-	-	-
Turkey	100	97	117	130	131	139	100	97	94	91	91	92
Uganda	100	123	133	164	220	271	-	-	-	-	-	-
Ukraine	100	-	-	-	-	-	100	100	97	90	83	79
United Arab Emirates	100	-	-	-	-	-	-	-	-	-	-	-
United Kingdom	100	98	99	102	105	107	100	96	92	85	85	86
United Republic of Tanzania	100	85	67	130	83	-	100	105	-	-	-	-
USSR (Former)	100	-	-	-	-	-	100	-	-	-	-	-
United States of America	100	101	104	108	113	115	100	99	99	100	99	100
Uruguay	100	114	122	120	130	120	100	95	91	87	87	87
Venezuela	100	-	-	-	-	-	100	114	105	89	91	92
Yemen	100	-	-	-	-	-	-	-	-	-	-	-
Yugoslavia	100	-	-	-	-	-	-	-	-	-	-	-
Yugoslavia (Former)	100	79	-	-	-	-	-	-	-	-	-	-
Zambia	100	98	95	88	88	68	100	-	-	-	60	-
Zimbabwe	100	105	105	109	124	114	100	97	79	85	77	71

Depending on the table, * means *Enterprises, Engaged,* or *Factor Values*; ± means *Basis Unspecified*; § means *shown in millions.* For additional notes and sources, see Appendix I.

Representative Companies in Sector

Name	Address	Telephone	Fax	Employment	Y
Pentair Inc.	1500 County Rd. B2 W., Ste. 400, St. Paul, Minnesota 55113-3105, U.S.A.	612-636-7920		10,000	97
Stone-Consolidated Corp.	800 Blvd. Rene-Levesque O, Montreal, Quebec H3B 1X9, Canada	514-875-2160	514-875-6284	10,000	98
Bemis Company Inc.	222 S. 9th St., 2300, Minneapolis, Minnesota 55402-4099, U.S.A.	612-376-3000		9,275	97
New Oji Paper Co., Ltd.	Chuo-ku, Tokyo 104, Japan	3 35631111	3 35631130	9,100	94
Media General Inc.	PO Box 85333, Richmond, Virginia 23293-0001, U.S.A.	804-649-6000	804-649-6865	8,800	97
Riverwood International Corp.	3350 Cumberland Cir., 1600, Atlanta, Georgia 30339, U.S.A.	404-664-3000		8,500	95
Johns Manville Corp.	PO Box 5108, Denver, Colorado 80217-5108, U.S.A.	303-978-2000	303-978-2041	8,300	97
Tunzl PLC	110 Park St., London W1Y 3RB, United Kingdom	71 4954950	71 4954953	8,046	93
Pabrik Kertas Tjiwi Kimia PT	Desa Kramat Tumenggung Kecamatan Tarik, Sidoarjo, Indonesia	32121552	321 21615	8,000	97
J. Bibby & Sons PLC	16 Stratford Pl., London W1N 9AF, United Kingdom	71 6296243	71 4090556	7,976	93
Paper Industries Corp. of the Philippines	Ortigas Complex, Pasig, Philippines	2 8181916	2 8175423	7,600	94
PBH UK Ltd.	1 Redcliffe St., Bristol BS997QY, United Kingdom			7,200	90
Gibson Greetings Inc.	2100 Section Rd., Cincinnati, Ohio 45237, U.S.A.	513-841-6600	513-841-6739	7,100	97
Fort Howard Corp.	1919 S. Broadway, Green Bay, Wisconsin 54304, U.S.A.	414-435-8821	414-251-2844	7,000	96
Kruger Inc.	3285 Ch Bedford, Montreal, Quebec H3S 1G5, Canada	514-737-1131	514-343-3124	7,000	98
Weyerhaeuser Paper Co.	33663 Weyerhaeuser Way S., Federal Way, Washington 98003, U.S.A.	253-924-2345		7,000	97
Banta Corp.	Box 8003, Menasha, Wisconsin 54952-8003, U.S.A.	920-751-7777	920-722-6495	6,900	97
Potlatch Corp.	301 W. Riverside Ave., 1100, Spokane, Washington 99201, U.S.A.	509-835-1500	509-835-1555	6,700	97
Stora Papyrus AB	PO Box 183, Molndal S-431 23, Sweden	31 670700	31 876310	6,700	93
CCL Industries Inc.	800-105 Gordon Baker Rd., Willowdale, Ontario M2H 3P8, Canada	416-756-8500	416-756-8555	6,500	98
Standard Register Co.	600 Albany St., Dayton, Ohio 45401, U.S.A.	937-443-1000		6,400	97
Donohue Inc.	800-500 Rue Sherbrooke O, Montreal, Quebec H3A 3C6, Canada	418-847-7700	418-847-7754	6,000	98
Quebecor Printing Inc.	612 Rue Saint-Jacques, Montreal, Quebec H3C 1C7, Canada	514-954-0101	514-954-9624	6,000	98
Rock-Tenn Co.	PO Box 4098, Norcross, Georgia 30091, U.S.A.	770-448-2193	770-263-4483	6,000	97
Consolidated Papers Inc.	PO Box 8050, Wisconsin Rapids, Wisconsin 54495-8050, U.S.A.	715-422-3111	715-422-3052	5,871	96
Canfor Corp.	1055 Dunsmuir St. Ste. 3000, Vancouver, British Columbia V7X 1B5, Canada	604-661-5241	604-661-5472	5,700	98
ACX Technologies Inc.	16000 Table Mountain Pkwy., Golden, Colorado 80403, U.S.A.	303-271-7000	303-271-7003	5,600	97
Menasha Corp.	PO Box 367, Neenah, Wisconsin 54957-0367, U.S.A.	920-751-1000	920-751-1236	5,500	97
Mobil Chemical Co. Plastics Div.	1159 Pittsford Victor, Pittsford, New York 14534, U.S.A.	716-248-5700	716-248-8920	5,500	93
West Fraser Timber Co. Ltd.	1100 Melville St. Ste. 1000, Vancouver, British Columbia V6E 4A6, Canada	604-895-2700	604-681-6061	5,500	98
Chesapeake Corp.	PO Box 2350, Richmond, Virginia 23218-2350, U.S.A.	804-697-1000	804-697-1199	5,305	95
Simpson Paper Co.	1201 3rd Ave., 900, Seattle, Washington 98101, U.S.A.	206-224-5700		5,300	94
Press Corp. of South Africa Ltd.	Doornfontein, Johannesburg 2001, Republic of South Africa	11 7769111	11 4028036	5,244	93
Bowater Inc.	PO Box 1028, Greenville, South Carolina 29602, U.S.A.	803-271-7733		5,025	96
Canadian Forest Products Ltd.	2900-1055 Dunsmuir St., Vancouver, British Columbia V7X 1B5, Canada	604-661-5241	604-661-5235	5,000	98
Griffon Corp.	100 Jericho Quadrangle, Jericho, New York 11753, U.S.A.	516-938-5544	516-938-5644	5,000	97
Tembec Inc.	2790-800 Blvd. Rene-Levesque O, Montreal, Quebec H3B 1X9, Canada	514-871-0137	514-397-0896	5,000	98
Envirodyne Industries Inc.	701 Harger Rd., 190, Oak Brook, Illinois 60521, U.S.A.	630-571-8400		4,900	96
First Brands Corp.	PO Box 1911, Danbury, Connecticut 06813-1911, U.S.A.	203-731-2300		4,800	97
Weyerhaeuser Canada Ltd.	1075 W Georgia St. 25th Floor, Vancouver, British Columbia V6E 3C9, Canada	604-687-0431	604-691-2445	4,800	98
Bowater Packaging Ltd.	Ditchmore Ln., Stevenage SG1 3LD, United Kingdom	438 313300	438 313300	4,547	93
Greif Board Corp.	425 Winter Rd., Delaware, Ohio 43015, U.S.A.	740-549-6000		4,500	97
Borsodchem Rt.	PO Box 208, Kazincbarcika H-3702, Hungary	48 10211	46 354496	4,200	93
Caraustar Industries Inc.	PO Box 115, Austell, Georgia 30001, U.S.A.	770-948-3101	770-732-3401	4,048	97

Continued

Representative Companies in Sector
- Continued -

Name	Address	Tele-phone	Fax	Employ-ment	Y
Mobil Chemical Company Inc.	3225 Gallows Rd., Fairfax, Virginia 22037-0001, U.S.A.	703-846-3000	703-846-2313	4,000	97
Norampac Inc.	772 Rue Sherbrooke O, Montreal, Quebec H3A 1G1, Canada	514-282-2635	514-282-2677	4,000	98
CSS Industries Inc.	1845 Walnut St., 800, Philadelphia, Pennsylvania 19103, U.S.A.	215-569-9900		3,924	97
Gaylord Container Corp.	500 Lake Cook Rd., 400, Deerfield, Illinois 60015-4965, U.S.A.	847-405-5500	847-405-5628	3,900	97
Longview Fibre Co.	PO Box 639, Longview, Washington 98632, U.S.A.	360-425-1550		3,900	97
Crown Vantage Inc.	300 Lakeside Dr., 1400, Oakland, California 94612-3592, U.S.A.	510-874-3400	510-874-3531	3,850	97
Zaklady Celulozowo-Papiernicze Kwidzyn	Lotnicza 1, Kwidzyn PL-82-500, Poland	50 5558000	50 5558451	3,728	93
Fraser Papers Inc.	27 Rice St., Edmundston, New Brunswick E3V 1S9, Canada	506-735-5551	506-737-2100	3,700	98
Ibstock PLC	Lutterworth House, Lutterworth LE17 4PS, United Kingdom	455 553071		3,681	94
SCA Packaging Ltd.	Cowley, Uxbridge UB8 2JP, United Kingdom	895 445533	895 422306	3,562	93
American Business Products Inc.	PO Box 105684, Atlanta, Georgia 30348, U.S.A.	770-953-8300	770-952-2343	3,520	97
Cubic Corp.	PO Box 85587, San Diego, California 92186-5587, U.S.A.	619-277-6780	619-277-1878	3,500	97
Printpack Inc.	4335 Wendell Dr., SW, Atlanta, Georgia 30336, U.S.A.	404-691-5830	404-699-6116	3,500	97
Weldwood of Canada Ltd.	1055 Hastings St. W, Vancouver, British Columbia V6B 3V8, Canada	604-687-7366	604-662-2798	3,500	98
Continental Can Company Inc. (Syosset, New York)	1 Aerial Way, Syosset, New York 11791, U.S.A.	516-822-4940	516-931-6344	3,463	96
IPC Inc.	100 Tri-State Dr., 200, Lincolnshire, Illinois 60069, U.S.A.	847-945-9100	847-945-9184	3,226	97
Intermec Technologies Corp.	6001 36th Ave., W., Everett, Washington 98203, U.S.A.	425-348-2600	425-355-9551	3,100	97
Kimberly-Clark Ltd.	Larkfield, Maidstone ME20 7PS, United Kingdom	622 717700	622 718361	3,042	93
P.H. Glatfelter Co.	228 S. Main St., Spring Grove, Pennsylvania 17362, U.S.A.	717-225-4711		3,029	96
Doman Industries Ltd.	300-435 Trunk Rd., Duncan, British Columbia V9L 2P9, Canada	604-748-3711	604-748-6045	3,000	98
E.B. Eddy Forest Products Ltd.	1600 Scott St., Ottawa, Ontario K1Y 4N7, Canada	613-725-6700	613-725-6820	3,000	98
Fletcher Challenge Canada Ltd.	700 W Georgia St. 9th Floor, Vancouver, British Columbia V7Y 1J7, Canada	604-654-4000	604-654-4049	3,000	98
Procter & Gamble Inc.	4711 Yonge St., North York, Ontario M2N 6K8, Canada	416-730-4711	416-730-4415	3,000	98
Tambrands Inc.	777 Westchester Ave., White Plains, New York 10604, U.S.A.	914-696-6000	914-696-6758	2,900	96
Clarcor Inc.	PO Box 7007, Rockford, Illinois 61125, U.S.A.	815-962-8867	815-962-0417	2,872	97
American Biltrite Inc.	57 River St., Wellesley Hills, Massachusetts 02181, U.S.A.	617-237-6655		2,835	95
BPB Paper & Packaging Ltd.	Gadbrook Pk., Northwich CW9 7TH, United Kingdom	606 40411	606 42818	2,700	90
Rayonier Inc.	1177 Summer St., Stamford, Connecticut 06905, U.S.A.	203-348-7000		2,700	96
Shorewood Packaging Corp.	277 Park Ave., New York, New York 10172-0124, U.S.A.	212-371-1500		2,700	97
Melton Medes Ltd.	1 St. Mark's St., Nottingham NG3 1DE, United Kingdom	602 582277	602 585122	2,698	93
Inti Indorayon Utama PT	Jalan Let Jend M. T. Haryono A1, Medan 20231, Indonesia	61532532	61 532355	2,650	97
Green Bay Packaging Inc.	PO Box 19017, Green Bay, Wisconsin 54307-9017, U.S.A.	920-433-5111		2,600	96
Lawson Mardon Group UK Ltd.	6 Hill St., London W1X 7FU, United Kingdom	71 4937323		2,575	92
Kimberly-Clark Inc.	90 Burnhamthorpe Rd. W, Mississauga, Ontario L5B 3Y5, Canada	905-277-6500	905-277-6894	2,500	98
MacMillan Bloedel Inc.	PO Box 235016, Montgomery, Alabama 36123-5016, U.S.A.	334-213-6100	334-262-0976	2,500	97
W.H. Brady Co.	PO Box 571, Milwaukee, Wisconsin 53201-0571, U.S.A.	414-358-6600		2,500	97
Schweitzer-Mauduit International Inc.	100 North Point Ctr. E., 600, Alpharetta, Georgia 30022-8246, U.S.A.	770-569-4200	770-569-4275	2,465	97
Kedaung Industrial Ltd.	Jl. Tubagur Angke Kp-Polgar, Jakarta 11710, Indonesia	215402273	21 6190123	2,430	97
Gilman Paper Co.	50 Tice Blvd., Woodcliff Lake, New Jersey 07675, U.S.A.	201-307-0600	201-307-1664	2,400	97
Solo Cup Co.	1505 E. Main St., Urbana, Illinois 61801, U.S.A.	217-384-1800		2,400	94
Nashua Corp.	PO Box 2002, Nashua, New Hampshire 03061-2002, U.S.A.	603-880-2323	603-880-5671	2,398	96
Smurfit Carton de Colombia SA	Yumbo, Cali, Colombia	2 6694015	2 4425822	2,380	97
Smead Manufacturing Co.	600 E. Smead Blvd., Hastings, Minnesota 55033, U.S.A.	612-437-4111	612-437-9134	2,370	96
Northwood Inc.	PO Box 9000 Sta. A, Prince George, British Columbia V2L 4W2, Canada	250-962-9611	250-962-3582	2,350	98
Park Electrochemical Corp.	5 Dakota Dr., New Hyde Park, New York 11042, U.S.A.	516-354-4100	516-354-4128	2,340	97
Cheng Loong Co. Ltd.	Panchiao, Taipei, Taiwan	2 2225131	2 2226110	2,300	93
Paloma Sladkogorska Tovarna Papirja d.d.	Sladki vrh 1, Sladki Vrh SLO-62214, Slovenia	62 644460	62 644460	2,300	96
Pope and Talbot Inc.	1500 S.W. 1st Ave., Portland, Oregon 97201, U.S.A.	503-228-9161		2,300	96

Continued

For sources and notes, see Appendix I.

Representative Companies in Sector
- Continued -

Name	Address	Tele-phone	Fax	Employ-ment	Y
Yuhan-Kimberly Ltd.	Kangnam-Gu, Seoul, Republic of Korea	2 5281001	2 5281086	2,300	97
Shuford Industries Inc.	PO Box 2228, Hickory, North Carolina 28603, U.S.A.	704-328-2131	704-328-5792	2,200	95
Field Group PLC	Rectory Way, Old Amersham HP7 0DD, United Kingdom	494 433711	494 431138	2,114	94
David S. Smith Packaging Ltd.	Burwell, Cambridge CB5 0AJ, United Kingdom	638 743074	638 741755	2,108	
American Israeli Paper Mills Ltd.	Industrial Zone, Hadera 38101, Israel	6 6349349	6 6339740	2,100	97
Athens Papermill Co. SA	1 Hartergaton Str., Athens GR-118 55, Greece	1 3466015	3451970	2,100	97
Mead Fine Paper Div.	PO Box 2500, Chillicothe, Ohio 45601, U.S.A.	614-772-3111	614-772-0024	2,100	97
3M Canada Co.	1840 Oxford St. E, London, Ontario N5V 3R6, Canada	519-451-2500	519-452-6262	2,000	98
Carton de Colombia SA	Pto. Isaacs Carretera Yumbo Km.13, Cali, Colombia	2 4425800	2 6645702	2,000	96
Lancer Industries Inc.	126 E. 56th St., New York, New York 10022, U.S.A.	212-644-8666	212-644-2707	2,000	
Promon Eletronica	Campinas Moji Mirim, Campinas 13088-061, Brazil	19 7893030	19 7894252	2,000	97
St. Joe Paper Co.	1650 Prudential Dr., 400, Jacksonville, Florida 32207, U.S.A.	904-396-6600	904-396-4442	2,000	96
Sango Ceramic Indonesia PT	Semarang Plaza, Semarang 50138, Indonesia	24518391	24 289335	2,000	97
Ja/Mont UK Ltd.	43-51 Lowlands Rd., Harrow HA1 3BW, United Kingdom			1,958	94
Union Camp Corp.	1600 Valley Rd., Wayne, New Jersey 07470, U.S.A.	201-628-2000		1,900	97
Wausau Paper Mills Co.	PO Box 1408, Wausau, Wisconsin 54402-1408, U.S.A.	715-845-5266	715-848-2652	1,890	97
Dixie Toga SA	Av. Guido Caloi 864, Sao Paulo 05802-140, Brazil	11 5151177	11 5150202	1,850	96
Scott Ltd.	Scott House, East Grinstead RH19 1UR, United Kingdom	342 327191		1,840	93
Intertape Polymer Group Inc.	110 Montee de Liesse, Montreal, Quebec H4T 1N4, Canada	514-731-7591	514-731-5039	1,800	98
Spencers Inc.	PO Box 988, Mount Airy, North Carolina 27030, U.S.A.	919-789-9111	919-789-6824	1,800	
Smurfit Espana, SA	Capitan Haya, 38-50, Madrid E-28020, Spain	91 5700505	91 5704523	1,750	97
Celulosa Argentina SA	Aristobulo del Valle 594, Zarate 2800, Argentina	487 25800	487 22732	1,700	96
Cleo Inc.	4025 Viscount Ave.., Memphis, Tennessee 38118, U.S.A.	901-369-6300	901-362-1099	1,700	97
Danapak AMBA	Kongevejen 100, Holte DK-2840, Denmark	45411210	45411810	1,700	93
Scott Paper Ltd.	200-1900 Minnesota Crt, Mississauga, Ontario L5N 3C9, Canada	905-812-6900	905-812-6910	1,700	98
Suparma PT	Jl. Sulung Sekolahan 6A, Surabaya 60174, Indonesia	31333842	31 333827	1,631	97
Hansol Paper Co. Ltd.	Chung-ku, Seoul 100-101, Republic of Korea	2 3994161	2 3994050	1,626	93
Borregaard Industries Ltd.	Birchwood, Warrington WA3 6QQ, United Kingdom	925 838659	925 812186	1,621	93
Sealright Company Inc.	7101 College Blvd., 1400, Overland Park, Kansas 66210-1891, U.S.A.	913-344-9000	913-344-9005	1,603	96
Playtex Products Inc.	300 Nyala Farms Rd., Westport, Connecticut 06880, U.S.A.	203-341-4000		1,600	96
Duropack Holding AG-Gruppe	Brunnerstr. 75, Vienna-Liesing A-1235, Austria	222 43186300	222 8660316	1,560	95
Engraph Inc.	2635 Century Pkwy. N.E., 900, Atlanta, Georgia 30345, U.S.A.	404-329-0332	404-320-7460	1,531	
Arrow Industries Inc.	PO Box 810489, Dallas, Texas 75381, U.S.A.	972-416-6500	972-417-8371	1,500	97
Crown Packaging Ltd.	8255 Wiggins St., Burnaby, British Columbia V3N 2V7, Canada	604-522-6889	604-522-0758	1,500	98
Junckers Industrier AS	Vaerftsvej 4A, Koge DK-4600, Denmark	53651895	53659936	1,500	93
MacFarlane Group Clansman PLC	21 Newton Pl., Glasgow G3 7PY, United Kingdom			1,454	93
Hiang Seng Fibre Container Co. Ltd.	Klongtoey Phrakhanong, Bangkok 10110, Thailand	2490251	2495713	1,450	94
Pall Europe Ltd.	Havant St., Portsmouth PO1 3PD, United Kingdom	705 303303	705 831324	1,434	94
Atlantic Packaging Products Ltd.	111 Progress Ave., Scarborough, Ontario M1P 2Y9, Canada	416-298-8101	416-297-2218	1,400	98
Curtis 1000 Inc.	PO Box 105683, Atlanta, Georgia 30348, U.S.A.	770-951-1000	770-955-0707	1,400	97
Kertas Leces (Persero) PT	Desa Leces, Probolinggo, Indonesia	33521993	335 21628	1,400	97
Westvaco Corp. Kraft Div.	PO Box 118005, Charleston, South Carolina 29423-8005, U.S.A.	803-745-3000		1,400	97
Borden Decorative Products Ltd.	Belgrave Rd., Darwen BB3 2RR, United Kingdom	254 704988	254 873340	1,362	92
Bonar Inc.	2380 McDowell Rd., Burlington, Ontario L7R 4A1, Canada	905-637-5611	905-637-9954	1,350	98
Arjo Wiggings Belgium SA	Place des Deportes 12, Nivelles B-1400, Belgium	67 281211	67 281640	1,339	95
Paragon Trade Brands Inc.	180 Technology Pkwy., Norcross, Georgia 30092, U.S.A.	770-300-4000		1,319	96
Mosinee Paper Corp.	1244 Kronenwetter Dr., Mosinee, Wisconsin 54455-9099, U.S.A.	715-693-4470	715-693-4803	1,303	96
Daishowa Inc.	10 Blvd. des Capucins, Quebec, Quebec G1J 3R4, Canada	418-525-2500	418-525-2832	1,300	98
Fox Valley Corp.	PO Box 727, Appleton, Wisconsin 54912, U.S.A.	414-739-8982	800-635-4481	1,300	95
Kieleckie Zaklady Wyrobow Papierowych SA	Ul. Malikow 150, Kielce PL-25-639, Poland	41 673900	41 56440	1,300	93
Printpak Ltd.	PO Box 784324, Sandton 2146, Republic of South Africa	11 4447418	11 4444735	1,300	96
William E. Coutts Co. Ltd.	2 Hallcrown Pl, North York, Ontario M2J 1P6, Canada	416-492-1300	416-494-0027	1,300	98
Berli Jucker Co. Ltd.	PO Box 173 BMC, Bangkok 10000, Thailand	3671111	3671000	1,264	90

Product Tables

Wood Pulp, Mechanical

Unit of Measure: Metric tons.

	1990 Value	%	1991 Value	%	1992 Value	%	1993 Value	%	1994 Value	%	1995 Value	%	1996 Value	%
Total Production	38,111,278	100.0	37,705,428	100.0	38,264,287	100.0	38,186,096	100.0	39,207,955	100.0	40,738,147	100.0	38,630,835	100.0
Regions														
Africa	403,100	1.1	401,300	1.1	319,300	0.8	321,400	0.8	288,900	0.7	400,800	1.0	400,800	1.0
America, North	16,439,400	43.1	16,550,800	43.9	16,151,800	42.2	16,190,800	42.4	16,901,800	43.1	17,165,400	42.1	16,385,800	42.4
America, South	593,000	1.6	629,300	1.7	625,700	1.6	740,000	1.9	717,000	1.8	751,000	1.8	784,000	2.0
Asia	5,387,278	14.1	5,399,528	14.3	6,663,178	17.4	6,286,178	16.5	6,185,728	15.8	6,696,011	16.4	6,572,390	17.0
Europe	14,282,500	37.5	13,686,500	36.3	13,363,309	34.9	13,538,718	35.5	14,087,527	35.9	14,594,936	35.8	13,420,845	34.7
Oceania	1,006,000	2.6	1,038,000	2.8	1,141,000	3.0	1,109,000	2.9	1,027,000	2.6	1,130,000	2.8	1,067,000	2.8
Leading Producers														
Canada	10,537,000	27.6	10,630,000	28.2	10,212,000	26.7	10,589,000	27.7	11,000,000	28.1	11,550,000	28.4	10,979,000	28.4
United States of America	5,772,000	15.1	5,810,000	15.4	5,898,000	15.4	5,586,000	14.6	5,884,000	15.0	5,593,000	13.7	5,371,000	13.9
Finland	3,293,000	8.6	3,156,000	8.4	3,170,000	8.3	3,401,000	8.9	3,631,000	9.3	3,797,000	9.3	3,489,000	9.0
Sweden	2,953,000	7.7	2,709,000	7.2	2,525,000	6.6	2,722,000	7.1	2,858,000	7.3	2,861,000	7.0	2,753,000	7.1
Japan	2,048,000	5.4	2,072,000	5.5	1,861,000	4.9	1,644,000	4.3	1,636,000	4.2	1,673,000	4.1	1,705,000	4.4

Pulp of Fibers Other Than Wood

Unit of Measure: Metric tons.

	1990 Value	%	1991 Value	%	1992 Value	%	1993 Value	%	1994 Value	%	1995 Value	%	1996 Value	%
Total Production	14,903,775	100.0	15,631,175	100.0	16,926,484	100.0	18,515,518	100.0	20,477,627	100.0	25,278,936	100.0	25,309,345	100.0
Regions														
Africa	227,000	1.5	187,700	1.2	188,000	1.1	188,000	1.0	201,000	1.0	201,000	0.8	201,000	0.8
America, North	706,000	4.7	576,000	3.7	526,000	3.1	390,000	2.1	453,000	2.2	456,000	1.8	504,000	2.0
America, South	508,100	3.4	487,600	3.1	551,600	3.3	414,600	2.2	390,600	1.9	497,600	2.0	492,600	1.9
Asia	12,989,200	87.2	13,934,400	89.1	15,217,200	89.9	17,107,700	92.4	19,123,700	93.4	23,764,000	94.0	23,750,800	93.8
Europe	461,475	3.1	433,475	2.8	434,684	2.6	409,218	2.2	303,327	1.5	353,336	1.4	353,945	1.4
Oceania	12,000	0.1	12,000	0.1	9,000	0.1	6,000	0.0	6,000	0.0	7,000	0.0	7,000	0.0
Leading Producers														
China	11,043,000	74.1	11,786,000	75.4	12,923,000	76.3	14,861,000	80.3	17,105,000	83.5	21,760,000	86.1	21,760,000	86.0
India	921,000	6.2	1,009,000	6.5	1,099,000	6.5	1,096,000	5.9	900,000	4.4	920,000	3.6	920,000	3.6
United States of America	353,000	2.4	240,000	1.5	187,000	1.1	170,000	0.9	237,000	1.2	240,000	0.9	251,000	1.0
Mexico	254,000	1.7	237,000	1.5	240,000	1.4	121,000	0.7	117,000	0.6	117,000	0.5	154,000	0.6
Pakistan	183,000	1.2	159,000	1.0	145,000	0.9	145,000	0.8	150,000	0.7	160,000	0.6	165,000	0.7

Wood Pulp, Dissolving Grades

Unit of Measure: Metric tons.

	1990 Value	%	1991 Value	%	1992 Value	%	1993 Value	%	1994 Value	%	1995 Value	%	1996 Value	%
Total Production	4,819,950	100.0	4,718,950	100.0	5,021,395	100.0	4,892,689	100.0	4,541,407	100.0	4,561,114	100.0	4,479,793	100.0
Regions														
Africa	470,000	9.8	470,000	10.0	568,000	11.3	521,000	10.6	521,000	11.5	490,000	10.7	490,000	10.9
America, North	1,398,000	29.0	1,466,000	31.1	1,500,000	29.9	1,415,000	28.9	1,324,000	29.2	1,347,000	29.5	1,245,000	27.8
America, South	95,000	2.0	76,000	1.6	78,000	1.6	64,000	1.3	75,000	1.7	75,000	1.6	148,000	3.3
Asia	1,305,700	27.1	1,289,200	27.3	1,616,200	32.2	1,647,025	33.7	1,516,025	33.4	1,526,025	33.5	1,472,025	32.9
Europe	1,551,250	32.2	1,417,750	30.0	1,254,195	25.0	1,240,664	25.4	1,100,382	24.2	1,118,089	24.5	1,119,768	25.0
Oceania	-		-		5,000	0.1	5,000	0.1	5,000	0.1	5,000	0.1	5,000	0.1
Leading Producers														
United States of America	1,173,000	24.3	1,243,000	26.3	1,255,000	25.0	1,277,000	26.1	1,293,000	28.5	1,197,000	26.2	1,095,000	24.4
South Africa	400,000	8.3	400,000	8.5	498,000	9.9	451,000	9.2	451,000	9.9	420,000	9.2	420,000	9.4
Russian Federation	-		-		320,000	6.4	293,000	6.0	191,000	4.2	210,000	4.6	180,000	4.0
Sweden	296,000	6.1	319,000	6.8	319,000	6.4	319,000	6.5	319,000	7.0	319,000	7.0	319,000	7.1
China	221,000	4.6	241,000	5.1	250,000	5.0	250,000	5.1	250,000	5.5	250,000	5.5	250,000	5.6

Commodity data are provided by the United Nations Statistical Division. The symbol * means that data are estimated. For additional notes, see Appendix I.

Product Tables
Wood Pulp, Soda and Sulfate

Unit of Measure: 1,000 Metric tons.

	1990		1991		1992		1993		1994		1995		1996	
	Value	%	Value	%	Value	%	Value	%	Value	%	Value	%	Value	%
Total Production	96,779	100.0	100,546	100.0	105,904	100.0	105,798	100.0	115,200	100.0	113,977	100.0	112,651	100.0
Regions														
Africa	1,264	1.3	1,298	1.3	1,397	1.3	1,414	1.3	1,054	0.9	1,292	1.1	1,292	1.1
America, North	55,518	57.4	57,947	57.6	58,250	55.0	57,610	54.5	65,667	57.0	61,970	54.4	60,574	53.8
America, South	5,032	5.2	5,617	5.6	6,721	6.3	7,211	6.8	7,658	6.6	7,857	6.9	8,166	7.2
Asia	16,390	16.9	17,004	16.9	20,546	19.4	20,036	18.9	20,426	17.7	22,035	19.3	22,605	20.1
Europe	17,632	18.2	17,670	17.6	18,016	17.0	18,455	17.4	19,347	16.8	19,767	17.3	19,043	16.9
Oceania	942	1.0	1,010	1.0	974	0.9	1,072	1.0	1,047	0.9	1,057	0.9	971	0.9
Leading Producers														
United States of America	45,028	46.5	46,758	46.5	46,957	44.3	46,368	43.8	53,050	46.1	49,099	43.1	48,013	42.6
Canada	10,097	10.4	10,830	10.8	11,013	10.4	11,033	10.4	12,457	10.8	12,592	11.0	12,237	10.9
Japan	8,731	9.0	9,118	9.1	8,850	8.4	8,534	8.1	8,576	7.4	9,089	8.0	9,155	8.1
Sweden	5,954	6.2	6,035	6.0	6,064	5.7	6,252	5.9	6,338	5.5	6,348	5.6	5,840	5.2
Finland	4,870	5.0	4,763	4.7	4,859	4.6	5,465	5.2	5,844	5.1	5,782	5.1	5,720	5.1

Wood Pulp, Sulfite

Unit of Measure: Metric tons.

	1990		1991		1992		1993		1994		1995		1996	
	Value	%	Value	%	Value	%	Value	%	Value	%	Value	%	Value	%
Total Production	10,936,656	100.0	10,482,222	100.0	11,581,171	100.0	11,111,640	100.0	11,072,846	100.0	11,375,118	100.0	10,963,924	100.0
Regions														
Africa	56,100	0.5	58,100	0.6	75,100	0.6	90,100	0.8	63,100	0.6	66,100	0.6	66,100	0.6
America, North	2,947,500	27.0	2,520,200	24.0	2,320,900	20.0	2,295,600	20.7	2,303,300	20.8	2,292,000	20.1	2,124,700	19.4
America, South	101,000	0.9	109,000	1.0	112,000	1.0	118,000	1.1	110,333	1.0	110,233	1.0	102,833	0.9
Asia	3,386,056	31.0	3,460,522	33.0	4,684,989	40.5	4,362,122	39.3	4,301,858	38.9	4,528,594	39.8	4,413,330	40.3
Europe	4,403,000	40.3	4,291,400	40.9	4,336,182	37.4	4,193,818	37.7	4,245,255	38.3	4,301,191	37.8	4,173,961	38.1
Oceania	43,000	0.4	43,000	0.4	52,000	0.4	52,000	0.5	49,000	0.4	77,000	0.7	83,000	0.8
Leading Producers														
United States of America	1,416,000	12.9	1,371,000	13.1	1,450,000	12.5	1,462,000	13.2	1,462,000	13.2	1,368,000	12.0	1,338,000	12.2
Canada	1,515,000	13.9	1,132,000	10.8	853,000	7.4	815,000	7.3	822,000	7.4	904,000	7.9	766,000	7.0
Sweden	723,000	6.6	733,000	7.0	725,000	6.3	715,000	6.4	722,000	6.5	727,000	6.4	643,000	5.9
Russian Federation	-		-		1,110,000	9.6	756,000	6.8	557,000	5.0	735,000	6.5	528,000	4.8
Germany	661,000	6.0	829,000	7.9	720,000	6.2	682,000	6.1	698,000	6.3	684,000	6.0	683,000	6.2

Wood Pulp, Semi-Chemical

Unit of Measure: Metric tons.

	1990		1991		1992		1993		1994		1995		1996	
	Value	%	Value	%	Value	%	Value	%	Value	%	Value	%	Value	%
Total Production	8,144,206	100.0	8,024,114	100.0	8,110,459	100.0	7,684,477	100.0	8,264,164	100.0	7,682,494	100.0	7,407,510	100.0
Regions														
Africa	165,700	2.0	164,900	2.1	123,900	1.5	111,900	1.5	144,900	1.8	188,900	2.5	188,900	2.6
America, North	4,478,000	55.0	4,344,000	54.1	4,232,000	52.2	4,074,000	53.0	4,472,000	54.1	3,965,000	51.6	3,880,000	52.4
America, South	271,000	3.3	270,100	3.4	272,000	3.4	141,000	1.8	198,000	2.4	205,000	2.7	197,000	2.7
Asia	1,156,631	14.2	1,207,239	15.0	1,598,847	19.7	1,510,689	19.7	1,430,258	17.3	1,442,827	18.8	1,376,396	18.6
Europe	1,794,875	22.0	1,741,875	21.7	1,749,711	21.6	1,715,888	22.3	1,786,006	21.6	1,731,767	22.5	1,563,214	21.1
Oceania	278,000	3.4	296,000	3.7	134,000	1.7	131,000	1.7	233,000	2.8	149,000	1.9	202,000	2.7
Leading Producers														
United States of America	3,828,000	47.0	3,714,000	46.3	3,721,000	45.9	3,640,000	47.4	4,099,000	49.6	3,609,000	47.0	3,499,000	47.2
Japan	324,000	4.0	334,000	4.2	293,000	3.6	243,000	3.2	214,000	2.6	211,000	2.7	198,000	2.7
Canada	650,000	8.0	630,000	7.9	511,000	6.3	434,000	5.6	373,000	4.5	356,000	4.6	381,000	5.1
Finland	434,000	5.3	433,000	5.4	458,000	5.6	472,000	6.1	487,000	5.9	509,000	6.6	468,000	6.3
Russian Federation	-		-		382,000	4.7	300,000	3.9	215,000	2.6	266,000	3.5	200,000	2.7

Commodity data are provided by the United Nations Statistical Division. The symbol * means that data are estimated. For additional notes, see Appendix I.

235

Product Tables
Newsprint

Unit of Measure: Metric tons.

	1990		1991		1992		1993		1994		1995		1996	
	Value	%	Value	%	Value	%	Value	%	Value	%	Value	%	Value	%
Total Production	34,420,043	100.0	35,550,101	100.0	35,200,895	100.0	35,627,006	100.0	37,036,106	100.0	38,058,490	100.0	39,190,593	100.0
Regions														
Africa	384,000	1.1	379,429	1.1	359,357	1.0	169,286	0.5	340,537	0.9	342,321	0.9	344,105	0.9
America, North	15,228,000	44.2	15,341,000	43.2	15,468,000	43.9	15,642,000	43.9	15,770,000	42.6	15,693,000	41.2	16,443,571	42.0
America, South	684,523	2.0	669,812	1.9	634,101	1.8	617,582	1.7	593,214	1.6	677,028	1.8	621,000	1.6
Asia	8,564,020	24.9	8,526,360	24.0	7,835,174	22.3	7,683,259	21.6	8,174,459	22.1	8,896,942	23.4	8,741,530	22.3
Europe	8,878,500	25.8	9,908,500	27.9	10,104,264	28.7	10,727,878	30.1	11,427,319	30.9	11,666,309	30.7	12,263,183	31.3
Oceania	681,000	2.0	725,000	2.0	800,000	2.3	787,000	2.2	730,577	2.0	782,890	2.1	777,203	2.0
Leading Producers														
Canada	9,068,000	26.3	8,976,000	25.2	8,931,000	25.4	9,132,000	25.6	9,322,000	25.2	9,226,000	24.2	9,580,000	24.4
United States of America	6,000,000	17.4	6,206,000	17.5	6,424,000	18.2	6,412,000	18.0	6,340,000	17.1	6,352,000	16.7	*6,719,333	17.1
Japan	3,479,000	10.1	3,516,000	9.9	3,253,000	9.2	2,917,000	8.2	2,972,000	8.0	3,098,000	8.1	3,140,000	8.0
Sweden	2,549,000	7.4	2,307,000	6.5	2,357,000	6.7	2,592,000	7.3	2,706,000	7.3	2,679,000	7.0	*2,932,448	7.5
Germany	-		1,256,000	3.5	1,263,000	3.6	1,270,000	3.6	1,428,000	3.9	1,770,905	4.7	1,710,534	4.4

Other Printing and Writing Paper

Unit of Measure: Metric tons.

	1990		1991		1992		1993		1994		1995		1996	
	Value	%	Value	%	Value	%	Value	%	Value	%	Value	%	Value	%
Total Production	74,292,972	100.0	75,424,156	100.0	78,142,630	100.0	78,206,680	100.0	87,913,855	100.0	87,364,931	100.0	88,509,407	100.0
Regions														
Africa	623,125	0.8	600,425	0.8	544,225	0.7	644,212	0.8	608,050	0.7	517,688	0.6	517,225	0.6
America, North	24,281,000	32.7	24,028,000	31.9	24,431,000	31.3	26,204,000	33.5	30,591,000	34.8	28,462,000	32.6	27,777,000	31.4
America, South	1,868,500	2.5	1,940,300	2.6	2,030,900	2.6	2,309,900	3.0	2,592,100	2.9	2,503,200	2.9	2,560,300	2.9
Asia	18,486,922	24.9	19,381,006	25.7	20,952,889	26.8	18,353,422	23.5	20,362,906	23.2	21,837,389	25.0	23,623,872	26.7
Europe	28,589,425	38.5	29,049,425	38.5	29,862,616	38.2	30,381,145	38.8	33,397,800	38.0	33,666,655	38.5	33,666,009	38.0
Oceania	444,000	0.6	425,000	0.6	321,000	0.4	314,000	0.4	362,000	0.4	378,000	0.4	365,000	0.4
Leading Producers														
United States of America	20,092,000	27.0	19,872,000	26.3	20,281,000	26.0	21,511,000	27.5	25,714,000	29.2	23,042,000	26.4	22,550,000	25.5
Japan	9,250,000	12.5	9,727,000	12.9	9,610,000	12.3	9,543,000	12.2	9,805,000	11.2	10,565,000	12.1	10,812,000	12.2
Germany	4,982,000	6.7	5,114,000	6.8	5,173,000	6.6	4,928,000	6.3	5,865,000	6.7	5,872,000	6.7	5,553,000	6.3
Finland	4,768,000	6.4	4,778,000	6.3	5,045,000	6.5	5,567,000	7.1	6,159,000	7.0	6,457,000	7.4	6,014,000	6.8
China	4,486,000	6.0	4,654,000	6.2	4,986,000	6.4	2,387,000	3.1	3,844,000	4.4	3,986,000	4.6	4,420,000	5.0

Household and Sanitary Paper

Unit of Measure: Metric tons.

	1990		1991		1992		1993		1994		1995		1996	
	Value	%	Value	%	Value	%	Value	%	Value	%	Value	%	Value	%
Total Production	14,840,292	100.0	15,358,992	100.0	15,829,125	100.0	15,517,004	100.0	17,743,165	100.0	18,285,150	100.0	18,952,712	100.0
Regions														
Africa	125,000	0.8	147,000	1.0	144,000	0.9	125,000	0.8	161,000	0.9	178,000	1.0	178,000	0.9
America, North	6,172,967	41.6	6,102,967	39.7	6,222,567	39.3	6,490,967	41.8	7,128,300	40.2	7,137,956	39.0	7,142,689	37.7
America, South	776,575	5.2	829,475	5.4	849,475	5.4	869,475	5.6	930,475	5.2	979,475	5.4	1,016,475	5.4
Asia	2,934,750	19.8	3,305,550	21.5	3,616,350	22.8	2,893,500	18.6	3,956,500	22.3	4,476,500	24.5	5,052,500	26.7
Europe	4,591,000	30.9	4,763,000	31.0	4,811,733	30.4	4,906,062	31.6	5,329,890	30.0	5,253,219	28.7	5,298,048	28.0
Oceania	240,000	1.6	211,000	1.4	185,000	1.2	232,000	1.5	237,000	1.3	260,000	1.4	265,000	1.4
Leading Producers														
United States of America	5,264,000	35.5	5,143,000	33.5	5,247,000	33.1	5,450,000	35.1	6,032,000	34.0	5,988,000	32.7	6,008,000	31.7
Japan	1,366,000	9.2	1,438,000	9.4	1,475,000	9.3	1,523,000	9.8	1,548,000	8.7	1,560,000	8.5	1,649,000	8.7
China	865,000	5.8	998,000	6.5	1,150,000	7.3	360,000	2.3	1,358,000	7.7	1,800,000	9.8	2,300,000	12.1
Germany	829,000	5.6	889,000	5.8	879,000	5.6	847,000	5.5	864,000	4.9	877,000	4.8	886,000	4.7
Canada	467,000	3.1	515,000	3.4	530,000	3.3	542,000	3.5	584,000	3.3	617,000	3.4	605,000	3.2

Commodity data are provided by the United Nations Statistical Division. The symbol * means that data are estimated. For additional notes, see Appendix I.

Product Tables

Wrapping and Packaging Paper and Paperboard

Unit of Measure: 1,000 Metric tons.

	1990 Value	%	1991 Value	%	1992 Value	%	1993 Value	%	1994 Value	%	1995 Value	%	1996 Value	%
Total Production	107,076	100.0	111,208	100.0	117,798	100.0	119,205	100.0	130,969	100.0	138,097	100.0	136,479	100.0
Regions														
Africa	1,474	1.4	1,472	1.3	1,476	1.3	1,296	1.1	1,110	0.8	1,325	1.0	1,341	1.0
America, North	41,802	39.0	43,044	38.7	44,441	37.7	44,899	37.7	51,506	39.3	51,493	37.3	51,774	37.9
America, South	3,710	3.5	4,327	3.9	4,248	3.6	4,141	3.5	4,324	3.3	4,723	3.4	4,783	3.5
Asia	27,572	25.7	28,891	26.0	33,256	28.2	33,766	28.3	36,554	27.9	43,089	31.2	41,102	30.1
Europe	31,112	29.1	32,005	28.8	32,951	28.0	33,589	28.2	35,883	27.4	35,864	26.0	36,244	26.6
Oceania	1,406	1.3	1,469	1.3	1,427	1.2	1,514	1.3	1,592	1.2	1,603	1.2	1,235	0.9
Leading Producers														
United States of America	36,870	34.4	37,929	34.1	39,265	33.3	39,874	33.4	46,019	35.1	45,682	33.1	45,834	33.6
Japan	11,716	10.9	12,023	10.8	11,755	10.0	11,618	9.7	11,986	9.2	12,255	8.9	12,295	9.0
China	5,663	5.3	5,972	5.4	6,590	5.6	7,155	6.0	9,428	7.2	14,660	10.6	12,480	9.1
Germany	4,250	4.0	4,685	4.2	5,019	4.3	5,024	4.2	5,275	4.0	5,348	3.9	5,694	4.2
Sweden	3,974	3.7	3,986	3.6	3,949	3.4	4,067	3.4	4,495	3.4	4,456	3.2	4,287	3.1

Cigarette Paper

Unit of Measure: Metric tons.

	1990 Value	%	1991 Value	%	1992 Value	%	1993 Value	%	1994 Value	%	1995 Value	%	1996 Value	%
Total Production	104,260	100.0	97,540	100.0	125,091	100.0	107,820	100.0	102,738	100.0	114,332	100.0	106,374	100.0
Regions														
America, North	3,258	3.1	2,884	3.0	2,706	2.2	2,812	2.6	2,578	2.5	3,012	2.6	3,015	2.8
America, South	11,134	10.7	11,086	11.4	11,038	8.8	19,135	17.7	17,260	16.8	11,501	10.1	14,692	13.8
Asia	38,682	37.1	36,747	37.7	65,611	52.5	44,972	41.7	42,662	41.5	53,350	46.7	46,770	44.0
Europe	51,187	49.1	46,824	48.0	45,737	36.6	40,901	37.9	40,237	39.2	46,469	40.6	41,897	39.4
Leading Producers														
Japan	18,348	17.6	19,322	19.8	15,967	12.8	13,947	12.9	10,973	10.7	11,130	9.7	11,562	10.9
Spain	18,377	17.6	18,637	19.1	20,025	16.0	15,057	14.0	*18,051	17.6	*18,669	16.3	*19,287	18.1
Indonesia	8,915	8.6	6,272	6.4	38,233	30.6	19,115	17.7	22,894	22.3	32,725	28.6	*26,799	25.2
Brazil	*8,922	8.6	*9,026	9.3	*9,131	7.3	*9,235	8.6	*9,340	9.1	*9,444	8.3	*9,548	9.0
Finland	8,626	8.3	8,506	8.7	9,225	7.4	9,479	8.8	10,754	10.5	13,272	11.6	14,436	13.6

Other Machine-Made Paper and Paperboard, Simply Finished

Unit of Measure: Metric tons.

	1990 Value	%	1991 Value	%	1992 Value	%	1993 Value	%	1994 Value	%	1995 Value	%	1996 Value	%
Total Production	23,391,222	100.0	22,443,292	100.0	23,004,274	100.0	26,654,359	100.0	25,099,906	100.0	21,881,371	100.0	23,920,653	100.0
Regions														
Africa	343,880	1.5	348,960	1.6	383,540	1.7	403,920	1.5	494,800	2.0	477,080	2.2	487,560	2.0
America, North	7,177,055	30.7	7,029,425	31.3	7,398,796	32.2	7,465,501	28.0	7,980,205	31.8	7,961,909	36.4	7,994,280	33.4
America, South	814,986	3.5	377,357	1.7	399,929	1.7	324,500	1.2	380,500	1.5	335,000	1.5	305,500	1.3
Asia	9,435,327	40.3	9,320,508	41.5	10,161,489	44.2	13,835,620	51.9	11,763,301	46.9	8,237,982	37.6	10,552,113	44.1
Europe	5,195,975	22.2	5,148,042	22.9	4,648,520	20.2	4,611,818	17.3	4,409,100	17.6	4,795,400	21.9	4,567,200	19.1
Oceania	424,000	1.8	219,000	1.0	12,000	0.1	13,000	0.0	72,000	0.3	74,000	0.3	14,000	0.1
Leading Producers														
United States of America	3,738,000	16.0	3,574,000	15.9	3,943,000	17.1	3,996,000	15.0	4,496,000	17.9	4,463,000	20.4	4,478,000	18.7
Japan	2,277,000	9.7	2,350,000	10.5	2,229,000	9.7	2,163,000	8.1	2,216,000	8.8	2,186,000	10.0	2,118,000	8.9
China	2,583,000	11.0	2,752,000	12.3	2,950,000	12.8	7,409,000	27.8	6,020,000	24.0	2,778,000	12.7	5,900,000	24.7
Canada	*3,317,200	14.2	*3,334,200	14.9	*3,351,200	14.6	*3,368,200	12.6	*3,385,200	13.5	*3,402,200	15.5	*3,419,200	14.3
Sweden	234,000	1.0	216,000	1.0	204,000	0.9	207,000	0.8	18,000	0.1	17,000	0.1	16,000	0.1

Commodity data are provided by the United Nations Statistical Division. The symbol * means that data are estimated. For additional notes, see Appendix I.

237

Product Tables

Fiberboard, Compressed

Unit of Measure: Cubic meters.

	1990 Value	%	1991 Value	%	1992 Value	%	1993 Value	%	1994 Value	%	1995 Value	%	1996 Value	%
Total Production	14,915,266	100.0	14,889,449	100.0	15,838,301	100.0	17,106,425	100.0	18,076,449	100.0	17,645,045	100.0	18,121,496	100.0
Regions														
Africa	86,033	0.6	85,033	0.6	85,200	0.5	83,200	0.5	61,200	0.3	84,630	0.5	84,084	0.5
America, North	3,693,000	24.8	3,667,000	24.6	3,943,000	24.9	4,037,000	23.6	4,244,000	23.5	4,204,445	23.8	4,291,780	23.7
America, South	838,650	5.6	852,600	5.7	912,700	5.8	1,165,900	6.8	1,217,800	6.7	1,096,104	6.2	1,135,450	6.3
Asia	5,528,816	37.1	5,757,049	38.7	6,436,956	40.6	6,911,289	40.4	7,062,822	39.1	7,074,992	40.1	7,312,555	40.4
Europe	4,330,767	29.0	4,045,767	27.2	3,965,445	25.0	4,398,036	25.7	4,899,627	27.1	4,585,137	26.0	4,664,496	25.7
Oceania	438,000	2.9	482,000	3.2	495,000	3.1	511,000	3.0	591,000	3.3	599,736	3.4	633,130	3.5
Leading Producers														
United States of America	3,281,000	22.0	3,229,000	21.7	3,533,000	22.3	3,603,000	21.1	3,778,000	20.9	*3,720,549	21.1	*3,787,699	20.9
China	1,172,000	7.9	1,174,000	7.9	1,445,000	9.1	1,810,000	10.6	1,930,000	10.7	*1,854,577	10.5	*1,944,659	10.7
Russian Federation	*1,030,333	6.9	*1,030,333	6.9	1,270,000	8.0	1,088,000	6.4	733,000	4.1	*1,030,333	5.8	*1,030,333	5.7
Germany	703,000	4.7	526,000	3.5	538,000	3.4	634,000	3.7	804,000	4.4	*734,000	4.2	*765,000	4.2
Brazil	637,000	4.3	637,000	4.3	637,000	4.0	637,000	3.7	637,000	3.5	*674,516	3.8	*686,633	3.8

Packing Containers of Paper or Paperboard

Unit of Measure: Metric tons.

	1990 Value	%	1991 Value	%	1992 Value	%	1993 Value	%	1994 Value	%	1995 Value	%	1996 Value	%
Total Production	40,802,831	100.0	42,565,765	100.0	44,980,597	100.0	45,625,913	100.0	47,643,921	100.0	49,222,031	100.0	49,556,768	100.0
Regions														
Africa	1,113,769	2.7	1,136,348	2.7	1,034,595	2.3	1,149,841	2.5	1,169,407	2.5	1,280,520	2.6	1,284,915	2.6
America, North	1,593,390	3.9	1,620,119	3.8	1,652,848	3.7	2,529,576	5.5	2,626,305	5.5	2,634,033	5.4	2,792,762	5.6
America, South	480,333	1.2	480,333	1.1	480,333	1.1	480,333	1.1	480,333	1.0	480,333	1.0	480,333	1.0
Asia	23,485,643	57.6	20,582,065	48.4	23,164,170	51.5	22,683,132	49.7	23,984,237	50.3	25,241,615	51.3	25,461,992	51.4
Europe	13,463,695	33.0	18,080,900	42.5	17,982,651	40.0	18,117,030	39.7	18,717,639	39.3	18,919,529	38.4	18,870,765	38.1
Oceania	666,000	1.6	666,000	1.6	666,000	1.5	666,000	1.5	666,000	1.4	666,000	1.4	666,000	1.3
Leading Producers														
Indonesia	11,000,000	27.0	*8,040,667	18.9	*8,040,667	17.9	*8,040,667	17.6	*8,040,667	16.9	*8,040,667	16.3	*8,040,667	16.2
Japan	8,275,000	20.3	8,568,000	20.1	8,426,000	18.7	8,394,000	18.4	8,748,000	18.4	9,119,000	18.5	9,048,000	18.3
Germany	-		4,996,000	11.7	4,930,000	11.0	4,797,000	10.5	4,928,000	10.3	4,760,871	9.7	4,712,917	9.5
India	2,348,000	5.8	2,467,000	5.8	2,563,000	5.7	2,745,000	6.0	*3,215,962	6.7	*3,436,945	7.0	*3,657,929	7.4
France	*1,870,000	4.6	*1,870,000	4.4	*1,870,000	4.2	*1,870,000	4.1	*1,870,000	3.9	*1,870,000	3.8	*1,870,000	3.8

ISIC 3411
PULP, PAPER, ETC.

ISIC 3411—Pulp, Paper, Etc.—is a subset of the more comprehensive category shown in the last chapter. This ISIC excludes packaging containers of paper and paperboard.

Please note that commodities that fall under this ISIC were all shown as part of the last chapter.

Summary Statistics

		1990		1991		1992		1993		1994		1995	
		Value	N	Value	N	Value	N	Value	N	Value	N	Value	N
Establishments or enterprises	(number)	3,852	42	4,135	50	4,571	49	3,413	45	6,323	35	765	16
Number employed	(000)	889	44	986	50	952	49	951	45	696	33	552	20
Total Wages	($ mil.)	21,901	42	24,423	47	24,131	46	19,672	40	16,348	29	12,489	18
Wage/salary per employee	($)	15,144	42	14,826	47	15,453	46	11,518	40	11,882	29	9,818	18
Female workers	(000)	37	19	20	18	166	16	167	17	163	12	160	5
as % of total employment*	(%)	11	19	11	18	14	16	16	17	20	12	29	5
Output	($ bil.)	159	40	164	45	171	43	138	36	122	26	109	17
per employee	($ 000)	125	39	113	44	116	42	97	36	122	26	124	17
per establishment	($ 000)	29,047	38	25,783	44	30,056	41	25,216	34	33,017	25	14,016	12
Value added	($ bil.)	70	41	64	44	62	38	56	34	45	24	52	15
per employee	($ 000)	46	40	42	42	42	36	38	31	48	22	62	13
per establishment	($ 000)	11,154	39	8,994	42	10,285	37	8,789	31	12,484	22	4,713	9
Capital investment	($ mil.)	9,950	22	10,626	28	13,192	29	10,595	24	7,131	15	6,736	7
per employee	($ 000)	18	21	16	27	13	27	10	21	15	14	15	6
per establishment	($ 000)	5,531	21	3,987	27	3,124	28	1,812	21	4,538	13	1,314	5

Data presented above are drawn from the detailed tables that follow. Columns headed 'N' show the number of countries that provided valid data for inclusion. Values are not strictly comparable one year to the next or one row to the next because the number of countries that report varies from period to period and row to row. However, a general indication of magnitudes can be gleaned from reviewing this summary—especially in earlier years in which more reports are available. For detailed explanations, see Appendix I. *The average for those countries reporting both total and female employment.

Establishments and Number Engaged

Country	Establishments or Enterprises (number)						Employees per Establishment					
	1990	1991	1992	1993	1994	1995	1990	1991	1992	1993	1994	1995
Albania	-	-	-	4.000	2.000	1.000	-	-	-	70	23	10
Australia	50	78	93	-	-	-	160	90	75	-	-	-
Austria	44	42	39	38	35	-	286	286	308	298	313	-
Bangladesh	41	48	81	-	-	-	370	337	191	-	-	-
Bolivia	1.000	1.000	1.000	1.000	1.000	2.000	19	18	18	18	18	9.000
Bosnia & Herzegovina	4.000	5.000	-	-	5.000	-	859	1,157	-	-	50	-
Canada	146	155	156	158	162	-	534	484	449	430	414	-
Chile	12	13	13	12	14	-	397	417	452	395	378	-
China	-	-	-	-	*4,756	-	-	-	-	-	-	-
China (Hong Kong SAR)	43	43	25	59	25	-	30	28	48	22	44	-
Colombia	32	28	33	29	32	-	170	199	185	216	199	-
Costa Rica	2.000	5.000	5.000	6.000	-	-	4.500	6.800	6.000	6.167	-	-
Denmark	12	15	14	-	-	-	171	132	105	-	-	-
Ecuador	9.000	9.000	9.000	10	11	7.000	220	211	188	189	158	123
Egypt	31	34	33	27	-	-	219	438	321	378	-	-
El Salvador	-	-	-	2.000	2.000	-	-	-	-	94	133	-
Finland	74	78	72	72	72	-	501	444	446	433	429	-
FYR Macedonia	-	-	-	-	-	*5.000	-	-	-	-	-	119
Germany	-	232	205	197	186	-	-	279	277	262	256	-
Germany (Western Part)	165	163	161	-	-	-	307	315	310	-	-	-
Ghana	-	-	-	1.000	-	-	-	-	-	284	-	-
Greece	36	35	33	-	-	-	124	109	92	-	-	-
Grenada	-	-	1.000	1.000	1.000	-	-	-	30	30	37	-
Guatemala	-	2.000	2.000	2.000	2.000	2.000	-	298	333	333	315	298
Hungary	-	-	*13	*20	-	-	-	-	-	-	-	-
Iceland	9.000	10	9.000	10	-	-	26	23	26	23	-	-
India	895	839	927	1,005	-	-	127	139	131	119	-	-
Indonesia	99	109	124	123	135	134	342	413	438	381	379	472
Iran (Islamic Republic of)	-	21	20	20	-	-	-	301	279	276	-	-
Italy	*143	*142	-	-	-	-	164	164	-	-	-	-
Japan	654	643	643	665	-	-	99	101	101	96	-	-
Jordan	5.000	4.000	3.000	8.000	13	10	56	57	74	79	50	43
Kenya	2.000	1.000	1.000	1.000	1.000	1.000	1,816	3,639	3,646	3,726	3,828	3,945
Korea, Republic of	308	253	264	267	261	275	71	80	76	78	81	78
Macau	2.000	2.000	1.000	-	-	-	9.000	16	27	-	-	-
Malaysia	12	9.000	12	15	16	-	233	311	250	207	181	-
Mexico	67	67	65	65	61	59	318	309	301	273	242	278
Mozambique	-	-	3.000	3.000	3.000	-	-	-	177	173	149	-
Myanmar	5.000	5.000	5.000	5.000	4.000	-	522	487	460	501	633	-
Netherlands	27	28	28	21	26	-	318	317	312	375	-	-
New Zealand	*15	*13	*16	*18	-	-	335	347	251	-	-	-
Norway	45	45	43	-	-	-	182	174	170	-	-	-
Pakistan	-	26	-	-	-	-	-	138	-	-	-	-
Peru	25	22	24	-	-	-	133	126	89	-	-	-
Philippines	41	48	48	46	-	-	166	144	181	154	-	-
Portugal	*91	*97	*100	*101	*84	-	134	110	104	101	93	-
Russian Federation	-	-	-	*126	*130	*144	-	-	-	1,275	1,101	1,040
Saint Lucia	-	2.000	2.000	2.000	2.000	2.000	-	-	-	-	-	-
Saint Vincent & the Grenadines	-	-	-	1.000	1.000	1.000	-	-	-	87	87	94
Sierra Leone	-	-	-	3.000	-	-	-	-	-	12	-	-
Slovakia	-	4.000	4.000	5.000	5.000	-	-	2,236	2,111	2,120	2,018	-
South Africa	-	37	-	-	-	-	-	-	-	-	-	-
Spain	149	149	132	-	-	-	109	105	110	-	-	-
Sri Lanka	6.000	3.000	4.000	3.000	-	-	627	1,068	834	1,053	-	-
Swaziland	-	4.000	4.000	4.000	-	-	-	840	792	823	-	-
Sweden	80	112	110	104	110	-	485	337	316	304	295	-
Thailand	21	19	-	-	-	-	259	290	-	-	-	-
Tunisia	-	-	-	56	61	60	-	-	-	52	58	59
Turkey	22	23	60	56	45	-	606	582	232	240	281	-
Ukraine	-	-	21	22	22	23	-	-	1,000	909	818	783

Continued.

Depending on the table, * means *Enterprises*, *Engaged*, or *Factor Values*; ± means *Basis Unspecified*; § means *shown in millions*. For additional notes and sources, see Appendix I.

Establishments and Number Engaged

- Continued -

Country	Establishments or Enterprises (number)						Employees per Establishment					
	1990	1991	1992	1993	1994	1995	1990	1991	1992	1993	1994	1995
United Kingdom	391	373	354	-	-	-	84	83	82	-	-	-
United Republic of Tanzania . .	4.000	6.000	-	-	-	-	1,097	792	-	-	-	-
United States of America . . .	-	-	529	-	-	-	-	-	374	-	-	-
Venezuela	32	33	26	19	35	39	250	279	308	368	229	197
Zambia	-	-	-	-	2.000	-	-	-	-	-	-	-

Employment and Compensation of Employees

Country	Number Employed/Engaged (000)						Wage/Salary per Employee ($)					
	1990	1991	1992	1993	1994	1995	1990	1991	1992	1993	1994	1995
Albania	-	-	-	0.279	0.046	0.010	-	-	-	407	-	1,398
Australia	8.000	*7.000	*7.000	-	-	-	27,920	32,500	32,668	-	-	-
Austria	13	12	12	11	11	-	34,482	36,920	-	38,037	39,855	-
Bangladesh	15	16	15	-	-	-	1,258	1,209	1,497	-	-	-
Bolivia	0.019	0.018	0.018	0.018	0.018	0.018	4,280	4,732	3,532	4,572	4,557	4,433
Bosnia & Herzegovina . . .	3.437	5.784	-	-	0.252	-	-	-	-	-	-	-
Canada	78	75	70	68	67	62	38,424	-	-	39,076	38,461	-
Chile	4.767	5.422	5.877	4.744	5.299	4.161	7,722	8,726	10,275	11,866	14,860	18,189
China (Hong Kong SAR) . . .	*1.300	*1.200	*1.200	*1.300	*1.100	-	11,159	13,511	16,149	14,518	17,174	-
Colombia	5.445	5.562	6.108	6.253	6.361	6.379	4,073	4,375	5,007	5,842	6,929	8,357
Costa Rica	0.009	0.034	0.030	0.037	-	-	1,885	1,787	1,903	2,247	-	-
Denmark	2.051	1.975	1.467	-	-	-	33,245	34,753	-	-	-	-
Ecuador	1.981	1.898	1.695	1.891	1.736	0.860	2,719	4,162	3,837	4,112	2,525	3,009
Egypt	*6.800	*15	*11	*10	-	-	2,618	1,505	1,269	1,547	-	-
El Salvador	-	-	-	*0.189	*0.266	-	-	-	-	2,130	1,520	-
Finland	37	35	32	31	31	-	35,856	35,845	33,600	27,015	30,977	-
FYR Macedonia	-	-	-	-	-	0.596	-	-	-	-	-	3,488
Germany	-	*65	*57	*52	*48	-	-	29,246	35,607	35,199	37,546	-
Germany (Western Part) . . .	*51	*51	*50	-	-	-	33,430	34,156	38,339	-	-	-
Ghana	-	-	-	*0.284	-	-	-	-	-	-	-	-
Greece	*4.468	*3.824	*3.039	-	-	-	13,137	14,846	16,719	-	-	-
Grenada	-	-	*0.030	*0.030	0.037	-	-	-	-	-	-	-
Guatemala	-	*0.596	*0.666	*0.666	*0.631	*0.596	-	-	-	-	-	-
Iceland	0.231	0.231	0.233	0.231	0.224	-	28,625	34,356	32,338	25,992	-	-
India	*114	*117	*121	*120	-	-	1,497	1,294	1,254	1,189	-	-
Indonesia	34	45	54	47	51	63	1,028	940	1,228	1,245	1,605	2,666
Iran (Islamic Republic of) . . .	-	6.322	5.586	5.520	-	-	-	5,037	6,504	7,384	-	-
Italy	23	23	-	-	-	-	-	-	-	-	-	-
Japan	*65	*65	*65	*64	-	-	35,702	39,172	-	-	-	-
Jordan	0.278	0.229	0.222	0.634	0.651	0.433	2,129	2,379	2,835	3,806	3,521	3,416
Kenya	3.632	3.639	3.646	3.726	3.828	3.945	1,558	1,326	1,214	747	992	1,214
Korea, Republic of	22	20	20	21	21	21	11,470	13,572	15,367	15,786	17,361	21,440
Lithuania	-	-	5.358	4.527	-	-	-	-	474	511	-	-
Macau	0.018	0.032	0.027	-	-	-	3,486	4,574	4,769	-	-	-
Malaysia	2.800	2.800	3.000	3.100	2.900	3.559	5,097	4,961	6,176	6,278	6,806	9,139
Mexico	21	21	20	18	15	16	5,277	6,158	7,218	8,055	8,000	4,586
Mozambique	-	-	0.531	0.520	0.448	-	-	-	623	563	603	-
Myanmar	2.609	2.436	2.299	2.505	2.534	-	1,660	1,887	2,135	2,265	3,033	-
Netherlands	8.588	8.884	8.742	7.865	-	-	30,439	31,787	35,647	34,776	-	-
New Zealand	5.019	4.509	4.016	-	-	-	31,986	-	-	-	-	-
Norway	8.186	7.839	7.313	-	-	-	30,210	30,894	-	-	-	-
Pakistan	-	3.596	-	-	-	-	-	1,638	-	-	-	-
Peru	*3.330	*2.783	*2.137	-	-	-	5,378	5,968	3,091	-	-	-
Philippines	6.800	6.900	8.700	7.100	-	-	2,383	2,500	2,708	3,116	-	-
Portugal	*12	*11	*10	*10	*7.800	-	10,800	13,255	15,809	11,541	13,574	-
Russian Federation	-	-	-	161	143	150	-	-	-	832	1,441	2,133
Saint Vincent & the Grenadines .	-	-	-	0.087	0.087	0.094	-	-	-	4,677	5,502	-
Sierra Leone	-	-	-	*0.036	-	-	-	-	-	-	-	-

Continued.

Depending on the table, * means *Enterprises, Engaged,* or *Factor Values;* ± means *Basis Unspecified;* § means *shown in millions.* For additional notes and sources, see Appendix I.

241

Employment and Compensation of Employees
- Continued -

Country	Number Employed/Engaged (000)						Wage/Salary per Employee ($)					
	1990	1991	1992	1993	1994	1995	1990	1991	1992	1993	1994	1995
Slovakia	-	*8.946	*8.443	*11	*10	-	-	1,574	1,903	1,959	2,211	-
Spain	16	16	14	-	-	-	25,032	27,012	29,994	-	-	-
Sri Lanka	*3.759	*3.205	*3.337	*3.158	-	-	930	1,145	1,148	1,177	-	-
Swaziland	-	3.360	3.166	3.290	-	-	-	6,228	7,170	6,090	-	-
Sweden	39	*38	*35	32	32	-	27,693	30,099	33,370	25,733	27,029	-
Thailand	5.434	5.506	-	-	-	-	1,341	1,762	-	-	-	-
Tunisia	-	-	-	2.890	3.562	3.532	-	-	-	-	-	-
Turkey	13	13	14	13	13	-	8,892	10,529	10,815	11,736	7,621	-
Ukraine	-	-	21	20	18	18	-	-	2,147	-	504	692
United Kingdom	33	31	29	-	-	-	25,054	27,405	29,061	-	-	-
United Republic of Tanzania	4.389	4.750	-	-	-	-	-	422	-	-	-	-
United States of America	199	198	198	194	190	189	39,095	-	-	-	-	-
Uruguay	*2.124	*1.731	*1.614	*1.537	-	-	-	-	-	-	-	-
Venezuela	8.000	9.200	8.000	6.984	8.028	7.672	6,146	6,036	6,706	5,885	5,442	6,045

Female Workers: Total and Percent of Employment

Country	Female Workers (000)						Female Workers as % of Total Employment					
	1990	1991	1992	1993	1994	1995	1990	1991	1992	1993	1994	1995
Albania	-	-	-	0.031	0.015	0.003	-	-	-	11.11	32.61	30.00
Australia	0.740	-	-	-	-	-	9.25	-	-	-	-	-
Austria	1.500	2.000	1.000	1.264	1.139	-	11.90	16.67	8.33	11.16	10.39	-
Bangladesh	0.104	0.169	0.231	-	-	-	0.69	1.05	1.49	-	-	-
Canada	6.000	6.000	-	-	-	-	7.69	8.00	-	-	-	-
Chile	0.170	0.250	0.277	0.222	0.350	-	3.57	4.61	4.71	4.68	6.61	-
Colombia	-	0.804	-	0.893	0.900	-	-	14.46	-	14.28	14.15	-
Denmark	0.349	0.376	0.283	-	-	-	17.02	19.04	19.29	-	-	-
Egypt	-	1.054	0.768	0.562	-	-	-	7.07	7.25	5.51	-	-
El Salvador	-	-	-	0.056	0.015	-	-	-	-	29.63	5.64	-
Germany (Western Part)	7.000	-	-	-	-	-	13.81	-	-	-	-	-
Indonesia	-	-	-	9.266	10	12	-	-	-	19.79	20.31	19.29
Iran (Islamic Republic of)	-	0.103	0.101	0.098	-	-	-	1.63	1.81	1.78	-	-
Italy	-	2.060	-	-	-	-	-	8.83	-	-	-	-
Jordan	0.042	0.029	0.006	0.046	0.132	0.034	15.11	12.66	2.70	7.26	20.28	7.85
Kenya	0.087	0.091	0.096	0.098	-	-	2.40	2.50	2.63	2.63	-	-
Korea, Republic of	3.200	2.827	2.827	2.977	2.838	2.664	14.55	13.91	14.10	14.27	13.45	12.45
Macau	0.001	0.003	0.003	-	-	-	5.56	9.38	11.11	-	-	-
Malaysia	0.500	0.400	0.500	0.500	0.500	-	17.86	14.29	16.67	16.13	17.24	-
Philippines	-	-	1.333	1.390	-	-	-	-	15.32	19.58	-	-
Portugal	1.816	1.413	1.597	1.243	0.914	-	14.94	13.27	15.34	12.17	11.72	-
Saint Vincent & the Grenadines	-	-	-	0.007	0.007	-	-	-	-	8.05	8.05	-
Sri Lanka	0.456	0.451	0.357	0.341	-	-	12.13	14.07	10.70	10.80	-	-
Sweden	7.300	-	-	-	-	-	18.81	-	-	-	-	-
Thailand	1.539	1.725	-	-	-	-	28.32	31.33	-	-	-	-
Turkey	0.647	-	-	-	-	-	4.86	-	-	-	-	-
United Kingdom	5.527	-	5.000	-	-	-	16.75	-	17.24	-	-	-
United Republic of Tanzania	0.132	0.143	-	-	-	-	3.01	3.01	-	-	-	-
United States of America	-	-	152	148	146	145	-	-	76.77	76.29	76.84	76.72

Output and Output per Employee

Country	Output ($ bil.)						Output per Employee ($)					
	1990	1991	1992	1993	1994	1995	1990	1991	1992	1993	1994	1995
Albania	-	-	-	*0.000	*0.000	*0.000	-	-	-	773	3,368	2,980
Australia	*1.525	*1.736	*1.436	-	-	-	190,625	247,983	205,147	-	-	-
Austria	3.154	2.999	3.110	2.633	3.045	-	250,290	249,886	259,191	232,470	277,669	-
Bangladesh	0.163	0.174	0.171	-	-	-	10,722	10,783	11,058	-	-	-
Bolivia	0.001	0.002	0.000	0.002	0.001	0.002	78,167	123,676	27,717	86,672	81,785	122,955
Canada	16	14	13	13	15	15	206,899	186,320	185,441	186,716	225,695	237,299
Chile	0.772	0.865	1.070	0.955	1.365	1.313	161,921	159,603	182,046	201,309	257,581	315,430
China (Hong Kong SAR)	0.148	0.125	0.160	0.157	0.127	-	113,659	103,909	133,172	120,919	115,630	-
Colombia	0.478	0.536	0.525	0.516	0.637	0.806	87,758	96,334	85,882	82,454	100,209	126,282
Denmark	*0.342	*0.287	-	-	-	-	166,855	145,265	-	-	-	-
Ecuador	0.053	0.054	0.053	0.061	0.064	0.071	26,841	28,374	31,381	32,013	36,672	83,058
Egypt	*0.164	*0.203	*0.135	*0.122	-	-	24,088	13,625	12,767	11,985	-	-
El Salvador	-	-	-	±0.002	±0.003	-	-	-	-	8,579	9,445	-
Finland	±10	±8.857	±8.471	±7.411	±9.406	-	277,591	255,985	263,887	237,516	304,412	-
FYR Macedonia	-	-	-	-	-	±0.011	-	-	-	-	-	18,280
Germany	-	12	12	9.958	11	-	-	190,556	212,074	193,032	239,156	-
Germany (Western Part)	-	-	12	-	-	-	-	-	232,152	-	-	-
Greece	*0.318	*0.303	*0.286	-	-	-	71,283	79,241	94,049	-	-	-
Guatemala	-	0.018	0.022	0.028	0.033	0.061	-	30,914	33,532	42,455	53,044	102,540
Iceland	0.024	0.027	±0.027	±0.024	-	-	105,581	117,485	114,243	103,501	-	-
India	*2.403	*2.080	*1.981	*1.874	-	-	21,092	17,855	16,314	15,674	-	-
Indonesia	1.202	1.510	1.672	1.352	1.790	2.503	35,481	33,536	30,783	28,885	34,997	39,600
Iran (Islamic Republic of)	-	0.145	0.137	0.135	-	-	-	22,875	24,594	24,424	-	-
Italy	±5.215	±4.825	-	-	-	-	222,347	206,783	-	-	-	-
Japan	±25	±26	±27	±29	-	-	379,010	404,972	414,103	452,029	-	-
Jordan	0.015	0.009	0.010	0.024	0.040	0.030	54,444	39,474	45,125	38,001	61,011	68,697
Korea, Republic of	3.289	3.403	3.725	3.634	4.467	6.369	149,511	167,417	185,858	174,239	211,710	297,679
Lithuania	-	-	0.023	0.013	-	-	-	-	-	4,318	2,795	-
Macau	0.000	0.001	0.000	-	-	-	18,174	18,695	15,287	-	-	-
Malaysia	*0.117	*0.113	*0.132	*0.117	*0.168	*0.244	41,895	40,466	44,045	37,783	57,946	68,628
Mexico	1.599	1.629	1.610	1.831	1.872	1.439	75,097	78,671	82,385	103,261	126,967	87,857
Myanmar	0.027	0.039	0.071	0.066	0.091	-	10,483	16,114	30,874	26,473	35,996	-
Netherlands	2.079	1.938	1.903	1.460	-	-	242,104	218,116	217,722	185,656	-	-
New Zealand	1.054	-	-	-	-	-	210,067	-	-	-	-	-
Norway	±2.431	±2.194	±1.948	-	-	-	296,925	279,893	266,392	-	-	-
Pakistan	-	0.093	-	-	-	-	-	25,932	-	-	-	-
Peru	0.152	0.090	0.053	-	-	-	45,630	32,293	24,748	-	-	-
Philippines	0.272	0.284	0.465	0.328	-	-	40,051	41,217	53,484	46,138	-	-
Portugal	1.508	1.350	1.487	1.068	1.354	-	124,034	126,794	142,827	104,606	173,622	-
Russian Federation	-	-	-	1.086	1.928	4.204	-	-	-	6,760	13,472	28,061
Slovakia	-	±0.239	±0.264	±0.236	±0.289	-	-	26,694	31,232	22,262	28,617	-
South Africa	±1.747	±1.739	±1.834	-	-	-	-	-	-	-	-	-
Spain	*3.353	*3.206	*2.959	-	-	-	205,660	204,549	204,248	-	-	-
Sri Lanka	0.026	0.022	0.025	0.033	-	-	6,839	6,772	7,408	10,589	-	-
Swaziland	-	*0.095	*0.104	*0.093	-	-	-	28,211	32,843	28,251	-	-
Sweden	*9.635	*9.605	*9.481	*7.036	*8.610	-	248,327	254,442	272,580	222,802	265,374	-
Thailand	0.475	0.238	-	-	-	-	87,334	43,175	-	-	-	-
Turkey	0.738	0.736	0.914	0.981	0.831	-	55,368	55,007	65,578	73,084	65,612	-
Ukraine	-	-	*0.698	*0.157	*0.155	*0.208	-	-	33,222	7,855	8,608	11,551
United Kingdom	*5.787	*5.591	5.313	-	-	-	175,379	180,360	183,197	-	-	-
United Republic of Tanzania	*0.023	*0.061	-	-	-	-	5,255	12,778	-	-	-	-
United States of America	*57	*54	*54	*53	*58	*77	288,844	271,162	274,707	271,490	306,547	405,148
Venezuela	0.513	0.625	0.525	0.493	0.541	0.651	64,163	67,908	65,564	70,605	67,399	84,811

Depending on the table, * means Enterprises, Engaged, or Factor Values; ± means Basis Unspecified; § means shown in millions. For additional notes and sources, see Appendix I.

243

Value Added and Value Added per Employee

Country	Value Added ($ bil.)						Value Added per Employee ($)					
	1990	1991	1992	1993	1994	1995	1990	1991	1992	1993	1994	1995
Australia	*0.631	*0.689	-	-	-	-	78,906	98,392	-	-	-	-
Austria	0.986	0.923	0.871	0.726	0.910	-	78,241	76,896	72,580	64,113	82,944	-
Bangladesh	0.050	0.038	0.053	-	-	-	3,268	2,333	3,423	-	-	-
Belgium	*0.482	*0.439	*0.450	*0.345	*0.399	*0.463	-	-	-	-	-	-
Bolivia	0.000	0.000	0.000	0.000	0.000	0.001	12,757	17,672	7,649	20,958	19,034	28,609
Canada	6.891	5.036	4.575	4.395	6.195	6.209	88,342	67,150	65,359	64,632	92,464	99,887
Chile	0.407	0.472	0.516	0.397	0.710	0.661	85,305	87,067	87,743	83,661	134,050	158,908
China (Hong Kong SAR)	0.024	0.036	0.050	0.042	0.040	-	18,565	30,133	41,986	32,318	36,583	-
Colombia	0.173	0.205	0.174	0.156	0.248	0.310	31,812	36,844	28,479	25,023	38,956	48,650
Denmark	*0.147	*0.131	-	-	-	-	71,532	66,418	-	-	-	-
Ecuador	0.017	0.019	0.016	0.018	0.015	0.017	8,336	10,033	9,546	9,458	8,688	19,616
Egypt	*0.043	*0.045	*0.029	*0.026	-	-	6,257	2,993	2,699	2,573	-	-
El Salvador	-	-	-	0.001	0.001	-	-	-	-	3,438	3,806	-
Finland	±3.113	±2.314	±2.530	±2.497	±3.351	-	83,922	66,880	78,817	80,017	108,441	-
FYR Macedonia	-	-	-	-	-	±0.002	-	-	-	-	-	3,930
Germany (Western Part)	5.599	5.569	5.274	4.288	-	-	110,434	108,476	105,586	-	-	-
Greece	*0.127	*0.123	*0.123	-	-	-	28,313	32,221	40,480	-	-	-
Iceland	0.009	0.011	±0.010	±0.010	-	-	40,643	46,551	42,183	41,450	-	-
India	*0.493	*0.397	*0.371	*0.375	-	-	4,329	3,409	3,054	3,137	-	-
Indonesia	*0.447	*0.616	*0.624	*0.470	*0.655	*0.861	13,183	13,679	11,491	10,040	12,803	13,623
Iran (Islamic Republic of)	-	0.070	0.037	0.060	-	-	-	11,128	6,558	10,787	-	-
Italy	±1.592	±1.618	-	-	-	-	67,864	69,331	-	-	-	-
Japan	±9.365	±10	±11	±12	-	-	144,081	154,292	162,532	191,238	-	-
Jordan	0.004	0.002	0.003	0.007	0.012	0.006	15,234	7,297	12,447	10,470	18,517	13,646
Korea, Republic of	1.081	1.310	1.554	1.517	1.879	2.513	49,144	64,455	77,516	72,720	89,043	117,447
Macau	0.000	0.000	0.000	-	-	-	4,903	7,176	6,995	-	-	-
Malaysia	*0.044	*0.042	*0.049	*0.031	*0.053	*0.083	15,831	14,960	16,474	10,013	18,185	23,334
Mexico	0.485	0.503	0.485	0.505	0.504	-	22,796	24,299	24,809	28,496	34,201	-
Myanmar	±0.027	±0.041	±0.051	±0.049	±0.028	-	10,219	16,808	22,142	19,662	11,162	-
Netherlands	0.766	0.780	0.686	0.557	-	-	89,206	87,836	78,515	70,785	-	-
New Zealand	0.372	-	-	-	-	-	74,106	-	-	-	-	-
Norway	±0.610	±0.490	±0.412	-	-	-	74,490	62,496	56,322	-	-	-
Pakistan	-	0.028	-	-	-	-	-	7,905	-	-	-	-
Peru	0.076	0.018	0.012	-	-	-	22,722	6,369	5,819	-	-	-
Philippines	0.115	0.112	0.185	0.127	0.468	-	16,937	16,207	21,270	17,889	-	-
Portugal	0.504	0.405	0.426	0.271	0.468	-	41,460	37,983	40,918	26,583	60,002	-
Russian Federation	-	-	-	±0.457	±0.742	±1.671	-	-	-	2,843	5,186	11,153
Saint Lucia	-	±0.001	±0.000	±0.001	±0.000	±0.000	-	-	-	-	-	-
Slovakia	-	-	-	±0.051	±0.099	-	-	-	-	4,807	9,829	-
Spain	*0.977	*0.903	*0.732	-	-	-	59,936	57,635	50,554	-	-	-
Sri Lanka	0.017	0.010	0.014	0.017	-	-	4,615	2,969	4,236	5,524	-	-
Swaziland	-	*0.040	*0.051	*0.027	-	-	-	11,837	16,207	8,103	-	-
Sweden	*3.639	*2.446	*2.293	*2.112	*2.801	-	93,792	64,792	65,932	66,879	86,337	-
Thailand	0.078	0.050	-	-	-	-	14,400	9,153	-	-	-	-
Turkey	0.313	0.352	0.281	0.396	0.345	-	23,483	26,268	20,189	29,525	27,275	-
United Kingdom	*2.214	*2.335	*2.182	-	-	-	67,100	75,307	75,241	-	-	-
United Republic of Tanzania	*0.005	*0.034	-	-	-	-	1,214	7,230	-	-	-	-
United States of America	*28	*25	*26	*24	*26	*39	141,407	127,323	129,283	123,675	135,984	206,582
Venezuela	0.202	0.244	0.218	0.182	0.192	0.440	25,211	26,567	27,263	26,005	23,905	57,378

Depending on the table, * means *Enterprises, Engaged,* or *Factor Values;* ± means *Basis Unspecified;* § means *shown in millions.* For additional notes and sources, see Appendix I.

Capital Investment

Country	Gross Fixed Capital Formation ($ mil.)						Per Establishment ($ mil)					
	1990	1991	1992	1993	1994	1995	1990	1991	1992	1993	1994	1995
Austria	777	434	372	229	177	-	18	10	9.534	6.031	5.050	-
Bangladesh	-	1.202	0.693	-	-	-	-	0.025	0.009	-	-	-
Belgium	125	425	454	186	351	326	-	-	-	-	-	-
Bolivia	0.001	0.066	0.000	0.126	-	-	0.001	0.066	0.000	0.126	-	-
China (Hong Kong SAR)	-	-	5.684	19	26	-	-	-	0.227	0.324	1.035	-
Colombia	-	34	50	-15	-14	-	-	1.198	1.526	-0.515	-0.451	-
Denmark	27	106	-	-	-	-	2.276	7.088	-	-	-	-
Ecuador	-	6.902	14	22	9.308	11	-	0.767	1.575	2.204	0.846	1.529
El Salvador	-	-	-	0.012	6.111	-	-	-	-	0.006	3.055	-
Finland	1,701	1,153	918	666	612	-	23	15	13	9.251	8.507	-
Germany (Western Part)	1,292	1,213	1,126	953	-	-	7.828	7.442	6.996	-	-	-
Greece	6.019	6.874	9.522	-	-	-	0.167	0.196	0.289	-	-	-
Hungary	-	-	7.963	11	-	-	-	-	0.613	0.536	-	-
India	277	107	208	275	-	-	0.310	0.128	0.225	0.274	-	-
Indonesia	760	519	183	144	309	210	7.674	4.765	1.478	1.169	2.287	1.564
Iran (Islamic Republic of)	-	17	-1.685	12	-	-	-	0.798	-0.084	0.595	-	-
Italy	869	659	-	-	-	-	6.076	4.643	-	-	-	-
Japan	2,997	4,135	1,958	2,203	-	-	4.583	6.430	3.045	3.313	-	-
Jordan	-	-	-	-	-	1.500	-	-	-	-	-	0.150
Korea, Republic of	-	-	761	573	622	887	-	-	2.884	2.148	2.384	3.225
Malaysia	-	-	-	19	18	-	-	-	-	1.269	1.143	-
Mexico	91	184	-	-	-	-	1.363	2.752	-	-	-	-
Myanmar	-	-	58	-	137	-	-	-	12	-	34	-
Netherlands	151	266	107	114	-	-	5.593	9.513	3.839	5.435	-	-
New Zealand	63	-	-	-	-	-	4.219	-	-	-	-	-
Norway	-	-	461	-	-	-	-	-	11	-	-	-
Pakistan	-	33	-	-	-	-	-	1.288	-	-	-	-
Peru	2.349	1.115	0.424	-	-	-	0.094	0.051	0.018	-	-	-
Philippines	12	5.240	13	20	-	-	0.293	0.109	0.274	0.444	-	-
Portugal	246	476	174	188	22	-	2.704	4.908	1.742	1.863	0.257	-
Spain	323	286	241	-	-	-	2.171	1.918	1.823	-	-	-
Sri Lanka	1.398	0.385	0.000	0.328	-	-	0.233	0.128	0.000	0.109	-	-
Swaziland	-	8.316	6.881	11	-	-	-	2.079	1.720	2.765	-	-
Thailand	108	38	-	-	-	-	5.139	2.005	-	-	-	-
Turkey	26	86	33	37	25	-	1.185	3.731	0.558	0.654	0.554	-
Ukraine	-	-	53	1.006	2.324	2.331	-	-	2.547	0.046	0.106	0.101
United Kingdom	-	306	249	-	-	-	-	0.820	0.704	-	-	-
United Republic of Tanzania	94	118	-	-	-	-	24	20	-	-	-	-
United States of America	-	-	5,725	4,925	4,829	5,299	-	-	11	-	-	-

Indexes of Industrial Production

Country	Index of Industrial Production (1990=100)						Index of Employment (1990=100)					
	1990	1991	1992	1993	1994	1995	1990	1991	1992	1993	1994	1995
Australia	-	-	-	-	-	-	100	88	88	-	-	-
Austria	-	-	-	-	-	-	100	95	95	90	87	-
Bangladesh	-	-	-	-	-	-	100	107	102	-	-	-
Bolivia	-	-	-	-	-	-	100	95	95	95	95	95
Bosnia & Herzegovina	-	-	-	-	-	-	100	168	-	-	7	-
Canada	-	-	-	-	-	-	100	96	90	87	86	80
Chile	-	-	-	-	-	-	100	114	123	100	111	87
China (Hong Kong SAR)	-	-	-	-	-	-	100	92	92	100	85	-
Colombia	-	-	-	-	-	-	100	102	112	115	117	117
Costa Rica	-	-	-	-	-	-	100	378	333	411	-	-
Denmark	-	-	-	-	-	-	100	96	72	-	-	-
Ecuador	-	-	-	-	-	-	100	96	86	95	88	43
Egypt	-	-	-	-	-	-	100	219	156	150	-	-
Finland	-	-	-	-	-	-	100	93	87	84	83	-

Continued.

Depending on the table, * means *Enterprises*, *Engaged*, or *Factor Values*; ± means *Basis Unspecified*; § means *shown in millions*. For additional notes and sources, see Appendix I.

245

Indexes of Industrial Production
- Continued -

Country	Index of Industrial Production (1990=100)						Index of Employment (1990=100)					
	1990	1991	1992	1993	1994	1995	1990	1991	1992	1993	1994	1995
Germany (Western Part) . . .	-	-	-	-	-	-	100	101	99	-	-	-
Greece	-	-	-	-	-	-	100	86	68	-	-	-
Iceland	-	-	-	-	-	-	100	100	101	100	97	-
India	-	-	-	-	-	-	100	102	107	105	-	-
Indonesia	-	-	-	-	-	-	100	133	160	138	151	187
Italy	-	-	-	-	-	-	100	99	-	-	-	-
Japan	-	-	-	-	-	-	100	100	100	98	-	-
Jordan	-	-	-	-	-	-	100	82	80	228	234	156
Kenya	-	-	-	-	-	-	100	100	100	103	105	109
Korea, Republic of	-	-	-	-	-	-	100	92	91	95	96	97
Macau	-	-	-	-	-	-	100	178	150	-	-	-
Malaysia	-	-	-	-	-	-	100	100	107	111	104	127
Mexico	-	-	-	-	-	-	100	97	92	83	69	77
Myanmar	-	-	-	-	-	-	100	93	88	96	97	-
Netherlands	-	-	-	-	-	-	100	103	102	92	-	-
New Zealand	-	-	-	-	-	-	100	90	80	-	-	-
Norway	-	-	-	-	-	-	100	96	89	-	-	-
Peru	-	-	-	-	-	-	100	84	64	-	-	-
Philippines	-	-	-	-	-	-	100	101	128	104	-	-
Portugal	-	-	-	-	-	-	100	88	86	84	64	-
Spain	-	-	-	-	-	-	100	96	89	-	-	-
Sri Lanka	-	-	-	-	-	-	100	85	89	84	-	-
Sweden	-	-	-	-	-	-	100	97	90	81	84	-
Thailand	-	-	-	-	-	-	100	101	-	-	-	-
Turkey	-	-	-	-	-	-	100	100	105	101	95	-
United Kingdom	-	-	-	-	-	-	100	94	88	-	-	-
United Republic of Tanzania . .	-	-	-	-	-	-	100	108	-	-	-	-
United States of America . . .	-	-	-	-	-	-	100	99	99	97	95	95
Uruguay	-	-	-	-	-	-	100	81	76	72	-	-
Venezuela	-	-	-	-	-	-	100	115	100	87	100	96

Depending on the table, * means *Enterprises, Engaged,* or *Factor Values;* ± means *Basis Unspecified;* § means *shown in millions.* For additional notes and sources, see Appendix I.

ISIC 3420
PRINTING, PUBLISHING

ISIC 3420—Printing, Publishing—includes all direct printing activities (letterpress, offset, gravure, etc.), printing support activities, and all publishing activities, including the publishing of newspapers, magazines, books, blankbooks, etc. No details by "subindustry" are available for this ISIC category.

The industry corresponds to SIC 27—Printing and Publishing—in the U.S. classification system.

Summary Statistics

		1990 Value	N	1991 Value	N	1992 Value	N	1993 Value	N	1994 Value	N	1995 Value	N
Establishments or enterprises	(number)	124,515	90	127,693	101	223,680	96	123,497	92	117,060	76	62,252	36
Number employed	(000)	5,940	101	5,837	106	5,864	101	5,631	100	5,462	89	5,064	63
Total Wages	($ mil.)	106,481	93	116,559	94	126,038	89	109,009	88	109,958	77	105,007	56
Wage/salary per employee	($)	9,826	92	10,917	93	14,674	89	15,098	87	10,845	76	12,970	56
Female workers	(000)	412	36	232	35	1,052	31	960	35	930	29	892	15
as % of total employment*	(%)	34	35	32	34	34	31	32	35	35	29	38	14
Output	($ bil.)	451	98	509	101	520	96	496	92	500	81	513	63
per employee	($ 000)	44	94	51	98	104	93	236	88	49	76	60	60
per establishment	($ 000)	2,212	85	2,643	91	7,291	86	4,182	79	2,200	63	1,445	31
Value added	($ bil.)	251	87	279	87	279	80	282	81	276	72	294	60
per employee	($ 000)	23	84	31	85	27	79	24	77	25	68	31	56
per establishment	($ 000)	1,162	75	1,608	80	1,118	72	973	68	1,090	54	621	28
Capital investment	($ mil.)	20,350	71	18,958	67	19,270	60	17,006	55	15,896	44	9,575	23
per employee	($ 000)	4	68	5	65	3	59	3	52	5	42	3	21
per establishment	($ 000)	224	68	216	64	165	58	182	50	235	41	100	19

Data presented above are drawn from the detailed tables that follow. Columns headed 'N' show the number of countries that provided valid data for inclusion. Values are not strictly comparable one year to the next or one row to the next because the number of countries that report varies from period to period and row to row. However, a general indication of magnitudes can be gleaned from reviewing this summary—especially in earlier years in which more reports are available. For detailed explanations, see Appendix I. *The average for those countries reporting both total and female employment.

Establishments and Number Engaged

Country	Establishments or Enterprises (number)						Employees per Establishment					
	1990	1991	1992	1993	1994	1995	1990	1991	1992	1993	1994	1995
Albania	-	-	-	20	17	2.000	-	-	-	53	47	94
Angola	-	-	6.000	6.000	-	-	-	-	88	11	-	-
Argentina	-	-	-	6,266	7,378	-	-	-	-	5.447	4.848	-
Armenia	52	52	55	54	52	-	61	57	-	-	-	-
Australia	3,157	4,405	4,601	-	-	-	27	19	18	-	-	-
Austria	333	324	329	324	313	-	74	74	70	67	67	-
Azerbaijan	48	46	94	43	42	-	76	75	33	62	57	-
Bahamas	49	31	-	-	-	-	8.347	12	-	-	-	-
Bahrain	-	-	15	-	-	-	-	-	95	-	-	-
Bangladesh	680	665	744	-	-	-	22	17	21	-	-	-
Belgium	2,557	2,654	2,632	2,561	2,520	-	14	13	13	-	-	-
Belize	13	14	14	-	-	-	-	8.857	11	-	-	-
Benin	-	-	65	65	-	-	-	-	-	-	-	-
Bolivia	155	160	171	182	185	166	16	15	16	15	16	18
Bosnia & Herzegovina	28	27	-	-	5.000	-	156	170	-	-	234	-
Brazil	988	-	914	852	-	-	118	-	113	115	-	-
Bulgaria	*37	*40	*44	*341	*515	*547	241	183	127	21	13	12
Burundi	*6.000	*9.000	-	-	-	-	40	29	-	-	-	-
Cambodia	8.000	1.000	-	-	-	-	120	800	-	-	-	-
Cameroon	*8.000	*14	*14	*11	*12	*12	161	90	84	103	68	70
Canada	5,522	5,067	4,894	4,655	4,472	-	26	26	26	27	28	-
Cape Verde	11	14	14	15	18	-	14	10	10	11	8.111	-
Central African Republic	-	*3.000	*3.000	*4.000	*5.000	*6.000	-	-	-	-	-	-
Chad	-	2.000	2.000	4.000	4.000	4.000	-	-	-	-	-	-
Chile	49	53	60	63	58	-	174	170	165	179	191	-
China	*14,329	*14,641	*14,895	*13,174	*13,576	*15,436	68	70	69	71	73	63
China (Hong Kong SAR)	4,996	4,978	4,925	4,688	4,651	-	8.407	8.276	7.898	8.660	9.052	-
Colombia	370	373	426	423	413	-	59	56	63	64	69	-
Costa Rica	284	295	309	347	-	-	15	15	15	14	-	-
Croatia	584	863	1,281	1,744	2,289	2,613	31	19	11	6.944	4.871	4.313
Cyprus	330	320	320	320	340	350	5.585	5.319	5.534	5.681	6.018	5.574
Czechoslovakia (Former)	*40	*53	-	-	-	-	700	434	-	-	-	-
Czech Republic	-	*34	*75	*85	-	-	-	471	240	188	-	-
Denmark	3,655	3,689	3,616	-	-	-	15	14	14	-	-	-
Ecuador	83	87	82	81	85	78	50	53	56	54	56	50
Egypt	144	164	163	154	-	-	99	95	125	122	-	-
El Salvador	-	19	23	36	41	79	-	-	61	54	52	38
Equatorial Guinea	10	-	-	-	-	-	-	-	-	-	-	-
Ethiopia	20	19	19	19	25	28	160	165	163	165	130	130
Fiji	41	41	44	-	-	-	21	21	21	-	-	-
Finland	668	722	670	637	608	632	57	50	49	47	47	43
FYR Macedonia	*60	*101	*120	*217	*171	*235	65	46	37	18	23	16
Gabon	-	4.000	6.000	6.000	8.000	6.000	-	74	42	47	30	42
Gambia	-	-	-	10	-	-	-	-	-	32	-	-
Germany	-	2,332	2,391	2,379	2,299	-	-	90	86	82	81	-
Germany (Western Part)	2,126	2,197	2,272	-	-	-	85	86	83	-	-	-
Ghana	-	-	-	48	-	-	-	-	-	91	-	-
Greece	252	251	257	-	-	-	40	40	41	-	-	-
Grenada	-	-	3.000	1.000	3.000	-	-	-	12	100	12	-
Guatemala	-	13	13	13	13	13	-	162	175	187	180	168
Honduras	-	-	24	28	28	28	-	-	113	106	116	130
Hungary	*369	*663	*1,632	*1,975	-	-	57	32	10	12	-	-
Iceland	311	332	344	359	-	-	6.611	6.193	5.599	5.047	-	-
India	3,136	3,111	3,287	3,261	-	-	46	47	46	46	-	-
Indonesia	518	486	500	511	528	594	84	89	90	95	101	101
Iran (Islamic Republic of)	330	113	116	113	-	-	30	70	73	78	-	-
Iraq	-	-	19	-	-	-	-	-	158	-	-	-
Ireland	347	361	-	-	-	-	32	30	-	-	-	-
Israel	833	866	861	897	964	-	19	18	20	21	21	-
Italy	*1,028	*1,048	*1,345	*1,354	*1,316	-	80	80	65	64	63	-

Continued.

Establishments and Number Engaged

- Continued -

Country	Establishments or Enterprises (number)						Employees per Establishment					
	1990	1991	1992	1993	1994	1995	1990	1991	1992	1993	1994	1995
Jamaica	48	48	-	-	-	-	44	44	-	-	-	-
Japan	31,073	30,445	29,486	30,076	26,461	27,145	19	20	20	20	20	20
Jordan	216	228	283	281	202	203	8.352	8.632	9.481	10	14	14
Kenya	147	256	331	338	367	367	41	23	18	18	17	18
Korea, Republic of	3,032	3,276	3,512	4,272	4,454	4,900	24	22	22	20	20	18
Kuwait	61	60	54	55	57	-	45	31	59	62	65	-
Kyrgyzstan	23	23	21	26	21	-	96	88	85	69	-	-
Latvia	64	147	140	181	97	311	83	33	33	24	43	17
Lesotho	6.000	-	2.000	2.000	3.000	-	60	-	77	72	47	-
Liechtenstein	-	29	-	-	-	-	-	9.000	-	-	-	-
Lithuania	-	-	-	*103	*141	-	-	-	-	40	38	-
Macau	94	92	96	97	103	92	14	13	13	13	12	12
Malaysia	241	249	258	290	291	-	92	96	99	96	103	-
Malta	72	74	72	74	75	-	19	19	20	20	20	-
Mauritius	30	32	27	32	40	46	54	40	46	43	40	39
Mexico	73	72	70	70	70	68	153	154	156	135	137	124
Mongolia	29	32	34	34	39	44	50	26	20	23	27	20
Morocco	-	312	317	322	-	-	-	19	19	21	-	-
Mozambique	24	24	20	20	23	-	109	107	107	97	88	-
Myanmar	16	16	15	-	15	-	40	40	42	-	43	-
Namibia	-	-	-	-	17	-	-	-	-	-	51	-
Nepal	88	135	-	152	159	-	26	19	-	16	15	-
Netherlands	658	710	743	736	759	-	92	88	83	83	80	-
Netherlands Antilles	-	-	-	52	-	-	-	-	-	8.731	-	-
New Zealand	*1,266	*1,296	*1,337	*1,392	-	-	12	11	11	11	-	-
Norway	861	836	467	447	479	-	38	38	63	65	63	-
Pakistan	-	110	-	-	-	-	-	70	-	-	-	-
Panama	65	84	-	-	-	-	26	23	-	-	-	-
Paraguay	-	10	-	-	-	-	-	80	-	-	-	-
Peru	894	909	926	-	-	-	15	12	10	-	-	-
Philippines	2,362	2,490	644	630	-	-	12	12	36	42	-	-
Poland	*148	*136	*118	*115	-	-	270	243	237	226	-	-
Portugal	*2,433	*2,554	*2,549	*2,923	*2,960	-	14	14	13	12	12	-
Puerto Rico	118	182	173	169	166	164	16	12	12	13	15	17
Qatar	17	17	23	22	19	-	59	62	73	58	65	-
Republic of Moldova	12	11	12	16	17	17	310	296	243	156	142	123
Romania	*17	127	214	296	1,608	-	1,547	183	111	108	17	-
Russian Federation	-	-	-	*3,058	*4,730	*3,874	-	-	-	43	28	32
Saint Lucia	-	15	15	20	20	20	-	-	-	-	-	-
Senegal	20	17	18	16	18	-	38	43	40	42	48	-
Sierra Leone	-	-	-	16	-	-	-	-	-	61	-	-
Singapore	327	342	361	383	380	-	48	49	48	47	47	-
Slovakia	-	47	47	67	69	-	-	190	175	127	123	-
Slovenia	190	*487	*571	*813	*880	*1,235	-	-	-	-	-	-
South Africa	-	1,627	-	-	-	-	-	31	-	-	-	-
Spain	6,813	6,407	6,598	-	-	-	12	13	12	-	-	-
Sri Lanka	78	59	68	72	-	-	83	127	89	104	-	-
Suriname	-	-	17	-	-	-	-	-	11	-	-	-
Swaziland	-	19	19	19	-	-	-	19	21	21	-	-
Sweden	849	949	922	863	866	-	49	56	52	49	51	-
Thailand	330	420	-	-	-	-	77	77	-	-	-	-
Tonga	-	4.000	4.000	4.000	4.000	-	-	29	28	29	28	-
Trinidad and Tobago	187	181	175	187	247	-	11	11	11	11	8.530	-
Tunisia	-	-	-	254	268	274	-	-	-	16	15	15
Turkey	120	113	245	225	212	-	114	106	60	66	67	-
Ukraine	560	680	844	548	547	554	-	54	40	58	55	54
United Kingdom	18,762	18,148	19,674	23,283	25,054	-	16	16	15	13	13	-
United Republic of Tanzania	53	53	-	-	-	-	40	65	-	-	-	-
USSR (Former)	*2,383	-	-	-	-	-	100	-	-	-	-	-
United States of America	-	-	95,664	-	-	-	-	-	16	-	-	-

Continued.

Depending on the table, * means *Enterprises, Engaged,* or *Factor Values;* ± means *Basis Unspecified;* § means *shown in millions.* For additional notes and sources, see Appendix I.

Establishments and Number Engaged

- Continued -

Country	Establishments or Enterprises (number)						Employees per Establishment					
	1990	1991	1992	1993	1994	1995	1990	1991	1992	1993	1994	1995
Venezuela	633	690	619	529	518	522	29	30	31	33	31	32
Yugoslavia	365	582	1,098	1,503	1,536	1,587	-	54	27	18	16	16
Zambia	23	-	-	-	41	-	111	-	-	-	53	-
Zimbabwe	69	74	68	68	73	-	86	87	74	74	75	-

Employment and Compensation of Employees

Country	Number Employed/Engaged (000)						Wage/Salary per Employee ($)					
	1990	1991	1992	1993	1994	1995	1990	1991	1992	1993	1994	1995
Albania	3.169	-	-	1.067	0.803	0.187	-	-	-	-	-	1,543
Angola	-	*0.608	*0.530	*0.064	-	-	-	-	-	-	-	-
Argentina	28	-	-	34	36	-	7,622	-	-	20,863	-	-
Armenia	*3.148	*2.953	-	-	-	-	-	-	-	-	-	-
Australia	84	*85	*81	-	-	-	21,548	22,989	22,476	-	-	-
Austria	25	24	23	22	21	21	30,926	33,359	38,758	37,481	38,843	-
Azerbaijan	3.633	3.439	3.120	2.653	2.403	-	1,638	1,625	-	-	-	-
Bahamas	0.409	0.357	-	-	-	-	14,961	16,389	-	-	-	-
Bahrain	-	-	1.421	-	-	-	-	-	6,641	-	-	-
Bangladesh	15	11	15	-	-	-	721	560	736	-	-	-
Belgium	35	34	34	-	-	-	24,201	24,996	27,590	-	-	-
Belize	-	0.124	0.154	-	-	-	-	5,581	3,776	-	-	-
Bermuda	0.416	0.413	0.383	0.380	0.369	0.375	-	-	-	-	-	-
Bolivia	2.458	2.454	2.658	2.719	2.869	2.947	1,949	1,875	2,137	2,088	2,275	2,093
Bosnia & Herzegovina	4.377	4.587	-	-	1.168	-	2,805	3,394	-	-	-	-
Brazil	116	-	103	98	-	-	8,067	-	7,874	11,407	-	-
Bulgaria	8.900	7.300	5.600	7.300	6.900	6.300	1,933	781	1,240	1,857	1,573	2,087
Burundi	0.240	0.265	-	-	-	-	2,105	2,017	-	-	-	-
Cambodia	*0.964	*0.800	-	-	-	-	-	-	-	-	-	-
Cameroon	1.289	1.254	1.179	1.131	0.820	0.842	8,400	7,847	7,364	7,363	3,809	3,176
Canada	142	134	129	125	123	124	28,252	29,826	28,982	27,756	26,642	27,107
Cape Verde	0.152	0.144	0.144	0.162	0.146	-	-	-	-	-	-	-
Central African Republic	0.149	-	-	-	-	-	6,606	-	-	-	-	-
Chile	8.533	8.990	9.874	11	11	11	6,678	7,808	9,823	10,067	11,594	14,956
China	*980	*1,020	*1,030	*930	*990	*970	-	-	-	-	-	-
China (Hong Kong SAR)	*42	*41	*39	*41	*42	-	9,762	11,263	12,328	13,994	16,803	-
China (Taiwan Province)	51	51	54	56	59	58	11,083	12,638	14,862	15,403	15,811	16,892
Colombia	22	21	27	27	28	27	2,167	2,144	2,450	2,765	3,521	3,800
Costa Rica	4.378	4.285	4.571	4.846	-	-	3,333	3,195	3,743	4,279	-	-
Croatia	18	16	14	12	11	11	5,531	5,980	1,471	1,848	2,848	4,848
Cyprus	1.843	1.702	1.771	1.818	2.046	1.951	11,031	12,048	13,446	12,958	13,008	17,164
Czechoslovakia (Former)	28	23	-	-	-	-	2,129	1,563	-	-	-	-
Czech Republic	18	16	18	16	-	-	2,105	1,526	2,045	2,738	-	-
Denmark	55	53	52	51	52	54	25,953	26,194	28,469	26,559	28,940	34,213
Ecuador	4.162	4.636	4.616	4.406	4.762	3.871	2,907	3,031	3,648	4,657	2,575	3,029
Egypt	*14	*15	*20	*19	*17	*15	2,636	1,375	2,401	2,969	3,193	3,234
El Salvador	-	-	*1.413	*1.962	*2.139	*3.008	-	-	-	2,753	4,824	3,112
Equatorial Guinea	0.007	-	-	-	-	-	1,403	-	-	-	-	-
Ethiopia	3.193	3.127	3.102	3.139	3.243	3.636	1,837	1,751	1,632	1,099	1,086	1,012
Fiji	0.843	0.860	0.924	1.024	1.108	-	3,930	4,364	5,009	5,212	5,630	-
Finland	38	36	33	30	29	27	31,316	30,478	27,682	22,257	25,054	31,756
FYR Macedonia	3.886	4.659	4.400	3.946	3.856	3.762	5,316	6,694	2,283	3,271	1,699	3,351
France	234	233	229	222	219	219	36,740	39,098	-	-	-	-
Gabon	-	0.298	0.254	0.282	0.238	0.251	-	17,890	24,944	21,766	13,690	14,439
Gambia	-	-	-	0.323	-	-	-	-	-	-	-	-
Germany	-	*210	*206	*195	*186	-	-	31,521	36,150	35,933	37,753	-
Germany (Western Part)	*180	*188	*190	-	-	-	32,268	33,231	37,168	-	-	-
Ghana	-	-	-	*4.391	*4.515	*4.642	-	-	-	1,150	859	877
Greece	*10	*10	*10	*10	*11	*11	12,207	12,695	14,170	13,157	12,918	14,701

Continued.

Depending on the table, * means *Enterprises, Engaged,* or *Factor Values*; ± means *Basis Unspecified*; § means *shown in millions.* For additional notes and sources, see Appendix I.

Employment and Compensation of Employees

- Continued -

Country	Number Employed/Engaged (000)						Wage/Salary per Employee ($)					
	1990	1991	1992	1993	1994	1995	1990	1991	1992	1993	1994	1995
Grenada	-	-	*0.037	*0.100	0.037	-	-	-	-	-	-	-
Guatemala	-	*2.105	*2.278	*2.429	*2.343	*2.183	-	-	-	483	458	-
Honduras	*2.332	*2.288	*2.709	*2.968	*3.252	*3.642	3,126	2,758	3,170	3,558	2,874	2,584
Hungary	21	21	17	24	24	24	3,158	3,258	3,822	4,641	4,486	4,585
Iceland	2.056	2.056	1.926	1.812	1.745	-	27,871	32,555	32,893	30,697	-	-
India	*146	*147	*152	*148	*148	*147	1,548	1,322	1,279	1,237	1,474	1,732
Indonesia	43	43	45	49	54	60	1,010	1,094	1,094	1,343	1,217	1,901
Iran (Islamic Republic of)	9.900	7.872	8.439	8.775	-	-	3,968	5,307	6,027	5,536	-	-
Iraq	-	3.303	3.000	-	-	-	-	5,583	9,341	-	-	-
Ireland	11	11	11	11	11	11	25,207	26,318	31,374	29,530	31,234	38,565
Israel	16	16	17	19	21	20	18,030	19,013	19,461	19,847	20,488	22,945
Italy	83	84	88	87	83	-	-	-	-	-	-	-
Jamaica	2.107	2.102	1.999	2.020	1.908	-	-	-	-	-	-	-
Japan	*581	*595	*594	*590	*535	*542	31,775	35,595	39,027	-	-	-
Jordan	1.804	1.968	2.683	2.815	2.825	2.774	3,166	3,104	3,345	3,471	3,175	4,011
Kenya	5.992	5.870	5.965	6.063	6.284	6.575	2,259	2,100	1,920	1,159	1,333	1,737
Korea, Republic of	71	72	77	87	89	90	11,266	12,011	13,107	14,634	15,959	18,036
Kuwait	2.775	1.868	3.213	3.384	3.691	-	4,966	4,550	9,429	9,239	10,134	-
Kyrgyzstan	*2.212	*2.015	*1.793	*1.787	-	-	433	-	-	-	-	-
Latvia	5.327	4.848	4.571	4.294	4.161	5.378	-	-	-	-	-	-
Lesotho	*0.362	-	*0.154	*0.144	*0.140	-	-	-	2,206	2,338	2,289	-
Liechtenstein	-	*0.261	-	-	-	-	-	-	-	-	-	-
Lithuania	-	-	4.442	4.095	5.365	-	-	-	-	644	1,766	-
Macau	1.283	1.218	1.248	1.241	1.278	1.139	4,961	5,543	6,055	6,640	7,194	8,667
Malaysia	22	24	25	28	30	32	4,400	4,530	5,260	5,483	5,846	6,949
Malta	1.337	1.397	1.434	1.466	1.487	-	11,314	12,637	13,212	11,432	12,012	-
Mauritius	*1.606	1.286	1.238	1.362	1.580	1.781	3,393	4,888	6,211	4,805	5,744	5,182
Mexico	11	11	11	9.438	9.594	8.410	4,878	5,977	7,736	8,585	8,575	5,128
Mongolia	*1.463	*0.833	*0.687	*0.783	*1.038	*0.868	1,233	2,232	735	-	-	528
Morocco	-	5.774	6.027	6.702	-	-	-	-	-	-	-	-
Mozambique	2.621	2.579	2.134	1.945	2.027	-	814	657	573	512	558	-
Myanmar	0.637	0.638	0.634	-	0.639	-	4,606	4,764	4,961	-	6,365	-
Namibia	-	-	-	-	0.867	-	-	-	-	-	7,354	-
Nepal	2.261	2.581	-	2.410	2.445	-	624	-	-	405	438	-
Netherlands	61	63	62	61	61	61	29,756	30,543	34,424	32,814	34,598	-
Netherlands Antilles	-	-	-	0.454	-	-	-	-	-	13,879	-	-
New Zealand	16	15	15	15	-	-	21,380	20,266	19,089	24,343	-	-
Norway	32	32	30	29	30	30	27,747	28,154	35,615	28,215	28,722	33,940
Pakistan	8.902	7.743	8.099	-	-	-	2,070	2,150	2,211	-	-	-
Panama	*1.671	*1.923	*2.056	*2.108	*2.165	*2.200	5,786	6,872	7,428	7,637	7,912	8,092
Paraguay	-	0.799	-	-	-	-	-	-	-	-	-	-
Peru	*13	*11	*9.607	-	-	-	3,289	5,015	4,968	-	-	-
Philippines	28	30	23	26	-	-	1,735	1,778	2,519	2,524	-	-
Poland	40	33	28	26	26	25	1,161	1,966	2,461	2,531	2,907	4,166
Portugal	*34	*35	*34	*36	*35	*35	-	-	-	-	-	-
Puerto Rico	1.890	2.260	2.090	2.200	2.420	2.760	26,085	-	-	-	-	-
Qatar	1.010	1.062	1.675	1.270	1.243	-	8,437	8,231	7,846	8,040	8,289	-
Republic of Moldova	3.720	3.254	2.914	2.498	2.410	2.099	-	2,824	-	-	591	792
Romania	26	23	24	32	28	-	1,755	1,187	782	938	1,097	-
Russian Federation	-	-	-	132	132	123	-	-	-	644	1,150	1,256
Senegal	*0.766	*0.739	*0.711	*0.672	*0.870	-	6,440	5,286	7,896	7,310	2,447	-
Sierra Leone	-	-	-	*0.979	-	-	-	-	-	-	-	-
Singapore	*16	*17	*17	*18	*18	*18	13,222	15,906	17,354	18,333	21,327	24,215
Slovakia	-	*8.943	*8.232	*8.501	*8.473	-	-	1,733	2,209	2,603	2,987	-
South Africa	50	51	54	54	53	52	10,412	11,156	11,902	11,720	11,897	12,572
Spain	84	82	80	107	104	105	21,099	23,381	25,470	21,211	21,355	23,277
Sri Lanka	*6.473	*7.510	*6.032	*7.520	-	-	540	492	532	1,348	-	-
Suriname	*0.187	*0.177	*0.195	*0.135	-	-	12,583	15,193	15,801	29,879	-	-
Swaziland	-	0.368	0.396	0.406	-	-	-	4,195	4,304	9,118	-	-
Sweden	42	*53	*48	42	44	44	28,328	27,806	31,635	25,080	25,048	26,717

Continued.

Depending on the table, * means *Enterprises, Engaged,* or *Factor Values*; ± means *Basis Unspecified*; § means *shown in millions.* For additional notes and sources, see Appendix I.

251

Employment and Compensation of Employees
- Continued -

Country	Number Employed/Engaged (000)						Wage/Salary per Employee ($)					
	1990	1991	1992	1993	1994	1995	1990	1991	1992	1993	1994	1995
Switzerland	*73	*72	*65	*62	*62	*62	-	-	-	-	-	-
Thailand	26	32	-	-	-	-	2,634	4,666	-	-	-	-
Tonga	0.110	0.115	0.110	0.115	0.114	-	2,158	2,657	2,801	2,720	4,784	-
Trinidad and Tobago	2.114	2.056	1.942	2.087	2.107	-	6,466	7,135	7,215	4,316	4,590	-
Tunisia	-	-	-	4.059	4.057	4.238	-	-	-	-	-	-
Turkey	14	12	15	15	14	15	7,315	9,146	8,594	7,955	4,873	6,066
Ukraine	-	37	34	32	30	30	-	2,707	1,824	-	530	682
United Kingdom	304	292	291	304	325	325	27,156	28,519	29,969	24,518	26,778	28,763
United Republic of Tanzania	2.144	3.439	-	-	-	-	-	-	-	-	-	-
USSR (Former)	239	-	-	-	-	-	2,032	-	-	-	-	-
United States of America	1,538	1,488	1,495	1,500	1,502	1,534	25,234	26,035	27,603	28,105	28,621	29,168
Uruguay	*8.060	*6.506	*6.378	*6.769	*6.778	*6.778	2,850	4,263	5,056	6,776	6,759	7,322
Venezuela	19	20	19	18	16	16	4,028	4,002	4,604	4,485	4,384	5,429
Yugoslavia	-	32	29	27	24	26	-	-	-	-	2,048	980
Zambia	2.546	-	-	-	2.183	-	1,167	-	-	-	1,206	-
Zimbabwe	5.927	6.427	5.000	5.000	5.500	5.351	6,716	5,254	4,817	4,039	3,781	4,065

Female Workers: Total and Percent of Employment

Country	Female Workers (000)						Female Workers as % of Total Employment					
	1990	1991	1992	1993	1994	1995	1990	1991	1992	1993	1994	1995
Afghanistan	0.540	0.576	-	-	-	-	-	-	-	-	-	-
Albania	-	-	-	0.643	0.425	-	-	-	-	60.26	52.93	-
Angola	-	-	0.129	0.017	-	-	-	-	24.34	26.56	-	-
Argentina	-	-	-	-	7.561	-	-	-	-	-	21.14	-
Australia	32	-	-	-	-	-	38.14	-	-	-	-	-
Austria	8.500	8.000	8.000	7.550	7.270	-	34.55	33.33	34.78	34.69	34.65	-
Bahrain	-	-	0.082	-	-	-	-	-	5.77	-	-	-
Bangladesh	0.187	0.140	0.079	-	-	-	1.28	1.24	0.51	-	-	-
Bermuda	0.197	0.197	0.175	0.179	0.166	0.164	47.36	47.70	45.69	47.11	44.99	43.73
Bosnia & Herzegovina	2.360	2.311	-	-	-	-	53.92	50.38	-	-	-	-
Bulgaria	-	-	-	4.200	3.600	3.500	-	-	-	57.53	52.17	55.56
Canada	58	55	-	-	-	-	40.85	41.04	-	-	-	-
Chile	1.345	1.385	1.643	1.886	1.807	-	15.76	15.41	16.64	16.69	16.32	-
China (Taiwan Province)	18	18	19	20	21	21	34.90	35.20	34.87	34.95	36.04	35.97
Colombia	-	7.619	-	11	11	-	-	36.37	-	39.17	39.46	-
Croatia	8.920	7.930	6.590	5.840	5.350	5.500	49.92	49.59	48.35	48.22	47.98	48.80
Cyprus	0.706	0.593	0.687	0.741	0.865	-	38.31	34.84	38.79	40.76	42.28	-
Czechoslovakia (Former)	9.800	8.300	-	-	-	-	35.00	36.09	-	-	-	-
Czech Republic	9.000	8.000	10	9.000	-	-	50.00	50.00	55.56	56.25	-	-
Denmark	21	20	20	-	-	-	37.54	37.79	37.97	-	-	-
Egypt	-	1.205	1.606	1.801	-	-	-	7.77	7.87	9.58	-	-
El Salvador	-	-	-	0.567	0.656	0.759	-	-	-	28.90	30.67	25.23
Ethiopia	-	1.180	1.174	1.175	1.229	1.394	-	37.74	37.85	37.43	37.90	38.34
FYR Macedonia	1.187	1.103	1.006	0.781	0.931	0.791	30.55	23.67	22.86	19.79	24.14	21.03
Gambia	-	-	-	0.081	-	-	-	-	-	25.08	-	-
Germany (Western Part)	49	-	-	-	-	-	27.17	-	-	-	-	-
Ghana	-	-	-	1.286	-	-	-	-	-	29.29	-	-
Hungary	10	-	12	12	-	-	47.62	-	70.59	50.00	-	-
Indonesia	-	-	-	15	17	18	-	-	-	31.04	31.86	30.71
Iran (Islamic Republic of)	-	0.512	0.536	0.571	-	-	-	6.50	6.35	6.51	-	-
Iraq	-	0.832	-	-	-	-	-	25.19	-	-	-	-
Ireland	3.200	3.200	-	-	-	-	29.09	29.09	-	-	-	-
Italy	-	24	23	23	21	-	-	28.02	26.67	26.48	24.67	-
Jordan	0.081	0.121	0.161	0.115	0.169	0.189	4.49	6.15	6.00	4.09	5.98	6.81
Kenya	0.839	0.805	0.853	0.867	-	-	14.00	13.71	14.30	14.30	-	-
Korea, Republic of	21	21	23	25	26	27	28.89	29.76	30.11	28.99	29.02	30.44
Lesotho	-	-	-	0.095	0.086	-	-	-	-	65.97	61.43	-

Continued.

Female Workers: Total and Percent of Employment

- Continued -

Country	Female Workers (000)						Female Workers as % of Total Employment					
	1990	1991	1992	1993	1994	1995	1990	1991	1992	1993	1994	1995
Macau	0.295	0.292	0.329	0.330	0.353	0.346	22.99	23.97	26.36	26.59	27.62	30.38
Malaysia	9.100	10	11	11	13	-	41.18	42.50	43.53	41.22	42.52	-
Malta	0.359	0.422	0.397	0.398	0.399	-	26.85	30.21	27.68	27.15	26.83	-
Mongolia	-	-	-	-	0.550	-	-	-	-	-	52.99	-
Morocco	-	-	-	-	-	1.227	-	-	-	-	-	-
Myanmar	0.883	0.895	0.949	-	0.208	-	138.62	140.28	149.68	-	32.55	-
Nepal	0.185	0.319	-	0.118	0.239	-	8.18	12.36	-	4.90	9.78	-
New Zealand	6.362	5.950	5.818	5.977	-	-	40.25	39.97	39.31	39.11	-	-
Panama	0.578	-	-	-	-	-	34.59	-	-	-	-	-
Philippines	-	-	7.219	8.196	-	-	-	-	31.12	31.28	-	-
Portugal	3.323	3.080	3.069	2.932	2.967	-	9.88	8.74	8.98	8.25	8.49	-
Puerto Rico	0.900	1.040	1.040	1.060	1.170	1.300	47.62	46.02	49.76	48.18	48.35	47.10
Republic of Moldova	-	-	-	-	1.546	1.397	-	-	-	-	64.15	66.56
Senegal	0.056	-	-	-	-	-	7.31	-	-	-	-	-
Sri Lanka	0.855	0.914	1.004	0.843	-	-	13.21	12.17	16.64	11.21	-	-
Sweden	15	-	-	-	-	-	35.25	-	-	-	-	-
Thailand	11	16	-	-	-	-	43.60	48.80	-	-	-	-
United Kingdom	108	-	104	-	-	-	35.45	-	35.74	-	-	-
United Republic of Tanzania	0.493	0.791	-	-	-	-	22.99	23.00	-	-	-	-
United States of America	-	-	787	786	784	809	-	-	52.64	52.40	52.20	52.74
Zambia	-	-	-	-	0.373	-	-	-	-	-	17.09	-

Output and Output per Employee

Country	Output ($ bil.)						Output per Employee ($)					
	1990	1991	1992	1993	1994	1995	1990	1991	1992	1993	1994	1995
Albania	-	-	-	*0.003	*0.003	*0.003	-	-	-	2,727	3,843	17,206
Angola	-	-	±2.628	±1.066	-	-	-	-	§4,959	§16,64	-	-
Argentina	±1.770	-	-	3.415	-	-	63,545	-	-	100,058	-	-
Armenia	0.000	0.001	0.004	0.081	0.001	-	144	251	-	-	-	-
Australia	*7.280	*7.714	*7.201	-	-	-	86,672	90,754	88,907	-	-	-
Austria	2.831	2.991	3.181	2.856	2.989	3.518	115,065	124,636	138,296	131,236	142,484	167,869
Azerbaijan	±0.015	±0.015	±0.001	±0.002	±0.001	-	4,195	4,386	373	741	465	-
Bahamas	0.017	0.016	-	-	-	-	41,565	44,818	-	-	-	-
Bahrain	-	-	±0.051	-	-	-	-	-	36,160	-	-	-
Bangladesh	0.054	0.037	0.055	-	-	-	3,680	3,234	3,553	-	-	-
Belgium	4.028	3.982	4.339	4.384	4.734	5.911	115,092	116,104	129,130	-	-	-
Belize	0.002	0.002	0.003	-	-	-	-	18,629	19,227	-	-	-
Benin	±0.005	±0.005	±0.005	±0.005	-	-	-	-	-	-	-	-
Bolivia	0.047	0.051	0.054	0.056	0.059	0.066	18,985	20,588	20,334	20,696	20,683	22,265
Bosnia & Herzegovina	±0.102	±0.108	-	-	-	-	23,416	23,446	-	-	-	-
Brazil	*4.100	-	*3.449	*4.196	-	-	35,295	-	33,418	42,723	-	-
Bulgaria	±0.149	±0.074	±0.040	±0.105	±0.123	±0.159	16,726	10,141	7,207	14,389	17,829	25,314
Burundi	±0.001	±0.001	-	-	-	-	3,795	4,324	-	-	-	-
Cameroon	0.020	0.021	0.017	0.014	0.006	0.021	15,168	17,102	14,391	12,172	7,659	25,164
Canada	12	12	11	11	10	11	86,791	89,237	86,453	84,087	84,778	86,322
Cape Verde	±0.001	±0.002	-	-	-	-	6,886	10,622	-	-	-	-
Central African Republic	*0.007	-	-	-	-	-	47,428	-	-	-	-	-
Chile	0.399	0.484	0.594	0.624	0.701	0.931	46,709	53,836	60,146	55,246	63,318	81,563
China	±3.625	±4.065	±4.758	±5.875	±4.694	±4.928	3,699	3,986	4,619	6,317	4,742	5,081
China (Hong Kong SAR)	2.079	2.204	2.376	2.714	3.161	-	49,497	53,502	61,071	66,859	75,088	-
China (Taiwan Province)	±2.720	±3.169	±3.464	±2.434	±2.413	±2.526	53,634	61,726	64,469	43,541	41,227	43,551
Colombia	0.468	0.430	0.542	0.595	0.805	0.826	21,365	20,518	20,226	21,868	28,275	30,544
Costa Rica	0.098	0.094	0.113	0.145	0.146	0.148	22,462	22,036	24,795	29,860	-	-
Croatia	±0.729	±0.660	±0.270	-	-	-	40,769	41,290	19,785	-	-	-
Cyprus	0.088	0.086	0.098	0.090	0.100	0.119	47,507	50,465	55,591	49,535	48,666	61,034
Czechoslovakia (Former)	±0.339	±0.166	-	-	-	-	12,097	7,197	-	-	-	-
Denmark	*2.448	*2.391	*2.707	*2.509	*2.851	*3.490	44,278	45,158	52,306	48,941	54,338	64,783

Continued.

Depending on the table, * means *Enterprises*, *Engaged*, or *Factor Values*; ± means *Basis Unspecified*; § means *shown in millions*. For additional notes and sources, see Appendix I.

253

Output and Output per Employee

- Continued -

Country	Output ($ bil.)						Output per Employee ($)					
	1990	1991	1992	1993	1994	1995	1990	1991	1992	1993	1994	1995
Ecuador	0.087	0.106	0.110	0.111	0.151	0.152	20,899	22,875	23,915	25,265	31,655	39,291
Egypt	*0.260	*0.180	*0.306	*0.264	*0.250	*0.227	18,178	11,590	15,019	14,029	15,041	15,130
El Salvador	-	±0.022	±0.012	±0.048	-	±0.036	-	-	8,798	24,405	-	11,846
Equatorial Guinea	±0.000	-	-	-	-	-	3,148	-	-	-	-	-
Ethiopia	0.030	0.028	0.022	0.015	0.018	0.019	9,550	8,950	6,968	4,866	5,681	5,108
Fiji	0.018	0.019	0.020	0.024	0.028	-	21,185	22,497	21,872	23,155	25,169	-
Finland	±4.759	±4.274	±3.452	±2.629	±3.034	±4.010	125,910	118,386	104,918	87,043	105,340	146,230
FYR Macedonia	±0.080	±0.103	±0.050	±0.039	±0.024	±0.041	20,524	22,126	11,443	9,793	6,237	10,926
France	±30	±31	±33	±32	±33	±38	128,936	133,288	145,370	142,112	152,394	175,006
Gabon	-	*0.020	*0.022	*0.024	*0.011	*0.015	-	65,518	88,010	83,931	48,071	59,520
Germany	-	23	25	22	23	-	-	109,710	120,246	114,523	121,493	-
Germany (Western Part)	21	22	24	-	-	-	114,853	117,044	124,389	-	-	-
Ghana	-	-	-	0.024	0.020	0.021	-	-	-	5,507	4,363	4,453
Greece	*0.558	*0.530	*0.615	*0.576	*0.568	*0.648	55,176	52,399	59,009	54,954	54,005	61,609
Guatemala	-	0.042	0.064	0.058	0.060	0.068	-	19,908	28,204	23,778	25,576	31,312
Honduras	0.027	0.027	0.040	0.037	0.034	0.039	11,382	11,742	14,586	12,455	10,436	10,796
Hungary	±0.426	±0.488	±0.892	±0.879	±0.862	±0.903	20,266	23,257	52,468	36,605	35,874	37,627
Iceland	0.157	0.180	±0.185	±0.157	-	-	76,247	87,588	96,286	86,645	-	-
India	*1.319	*1.289	*1.363	*1.459	*1.770	*2.108	9,047	8,745	8,952	9,823	11,966	14,306
Indonesia	0.495	0.538	0.609	0.741	0.872	1.145	11,390	12,456	13,598	15,194	16,295	19,098
Iran (Islamic Republic of)	0.196	0.154	0.156	0.152		-	19,756	19,612	18,450	17,284	-	-
Iraq	-	*0.047	*0.173	-	-	-	-	14,268	57,663	-	-	-
Ireland	*0.878	*0.895	*1.081	*1.027	*1.089	*1.364	79,858	81,348	97,602	92,260	97,730	121,363
Israel	0.917	0.936	1.070	1.206	1.316	1.461	57,677	58,889	63,709	64,128	63,597	73,284
Italy	±13	±13	±15	±12	±13	-	163,195	159,571	171,876	138,861	151,046	
Japan	±91	±103	±109	±122	±122	±139	156,854	173,644	182,706	207,155	228,392	256,117
Jordan	0.032	0.036	0.061	0.079	0.053	0.070	17,764	18,510	22,860	28,058	18,819	25,101
Kenya	*0.076	*0.072	*0.151	*0.092	*0.098	*0.050	12,672	12,262	25,289	15,185	15,615	7,630
Korea, Republic of	4.234	4.404	4.973	6.117	7.047	8.883	59,380	61,140	64,469	70,672	79,251	98,809
Kuwait	0.038	0.042	0.070	0.066	0.077	-	13,821	22,475	21,632	19,466	20,904	-
Kyrgyzstan	0.010	0.010	0.004	0.001	0.002	-	4,515	5,058	2,144	647	-	-
Latvia	0.001	0.002	0.007	0.018	0.036	0.072	158	481	1,593	4,279	8,728	13,345
Lithuania	-	-	0.005	0.011	0.036	-	-	-	1,128	2,713	6,785	-
Macau	0.020	0.023	0.027	0.030	0.031	0.033	15,334	18,699	21,918	23,857	24,261	28,660
Malaysia	*0.554	*0.647	*0.765	*0.866	*0.983	*1.253	25,079	26,940	29,988	31,057	32,652	39,279
Malta	0.062	0.075	0.097	0.085	0.087	-	46,498	53,372	67,742	57,909	58,726	-
Mauritius	0.026	0.027	0.033	0.030	0.038	0.040	16,255	20,916	26,449	22,066	23,893	22,405
Mexico	0.386	0.427	0.450	0.419	0.396	0.297	34,462	38,396	41,241	44,441	41,324	35,295
Mongolia	0.014	0.012	0.004	0.001	0.002	0.002	9,606	13,922	5,147	1,331	1,852	2,578
Morocco	-	±0.140	±0.158	±0.149	-	-	-	24,167	26,235	22,191	-	-
Mozambique	-	±0.005	-	±0.005	±0.005	-	-	1,944	-	2,527	2,525	-
Myanmar	0.024	0.026	0.023	-	0.014	-	37,445	40,881	35,915	-	22,263	-
Namibia	-	-	-	-	0.024	-	-	-	-	-	27,621	-
Nepal	0.005	0.009	-	0.007	0.008	-	2,379	3,328	-	2,939	3,224	-
Netherlands	8.193	8.403	9.084	8.104	8.536	9.960	134,892	134,312	146,519	132,161	140,130	163,874
New Zealand	1.192	1.245	1.174	1.297	-	-	75,397	83,631	79,338	84,848	-	-
Norway	±3.188	±3.175	±3.059	±2.708	±2.875	±3.376	98,373	98,746	103,169	93,085	95,767	113,183
Pakistan	0.140	0.158	0.171	-	-	-	15,672	20,362	21,128	-	-	-
Panama	0.055	0.060	0.071	0.075	0.081	0.085	33,127	31,051	34,437	35,719	37,433	38,566
Peru	0.336	0.467	0.395	-	-	-	25,140	41,588	41,148	-	-	-
Philippines	0.301	0.301	0.365	0.545	-	-	10,552	10,104	15,721	20,786	-	-
Poland	±0.282	±0.392	±0.222	±0.335	±0.381	±0.530	7,039	11,877	7,929	12,869	14,876	21,422
Portugal	1.376	1.705	1.934	1.739	1.709	2.043	40,917	48,392	56,569	48,966	48,908	57,840
Puerto Rico	±0.390	±0.427	±0.450	±0.467	±0.483	±0.528	206,152	188,850	215,502	212,182	199,711	191,377
Qatar	0.033	0.024	0.054	0.034	0.038	-	32,369	22,506	31,983	26,607	30,500	-
Republic of Moldova	0.000	0.053	0.003	0.005	0.007	0.009	0.019	16,378	1,155	1,874	2,962	4,521
Romania	±0.303	±0.225	±0.148	±0.173	-	-	11,526	9,705	6,222	5,415	-	-
Russian Federation	-	1.414	0.633	0.335	0.598	0.747	-	-	-	2,535	4,543	6,090
Senegal	0.031	-	-	-	-	-	41,083	-	-	-	-	-
Sierra Leone	-	-	-	0.005	-	-	-	-	-	4,978	-	-

Continued.

Depending on the table, * means *Enterprises, Engaged,* or *Factor Values*; ± means *Basis Unspecified*; § means *shown in millions*. For additional notes and sources, see Appendix I.

Output and Output per Employee

- Continued -

Country	Output ($ bil.)						Output per Employee ($)					
	1990	1991	1992	1993	1994	1995	1990	1991	1992	1993	1994	1995
Singapore	*0.961	*1.117	*1.310	*1.425	*1.562	*1.822	60,825	66,331	75,907	79,260	87,832	100,369
Slovakia	-	±0.124	±0.132	±0.150	±0.169	-	-	13,822	16,047	17,590	19,984	-
Slovenia	-	-	-	0.428	0.444	0.603	-	-	-	-	-	-
South Africa	±1.527	±1.637	±1.744	±1.781	±1.768	±1.835	30,549	32,103	32,290	32,989	33,350	35,293
Spain	*9.684	*11	*11	±10	±11	±13	115,940	129,026	139,714	96,983	104,890	123,732
Sri Lanka	0.039	0.041	0.039	0.043	-	-	6,027	5,457	6,453	5,699	-	-
Suriname	*0.010	*0.007	*0.011	0.019	-	-	50,929	41,146	57,459	141,093	-	-
Swaziland	-	*0.006	*0.007	*0.010	-	-	-	16,835	18,575	24,123	-	-
Sweden	*4.994	*6.843	*7.044	*4.974	*5.508	*5.826	119,762	129,453	147,177	117,695	124,652	131,535
Switzerland	8.012	7.620	7.385	6.624	8.250	10	110,361	106,570	113,794	106,840	132,219	163,395
Thailand	0.391	22	-	-	-	-	15,335	691,720	-	-	-	-
Tonga	0.001	0.001	0.001	0.001	0.001	-	8,687	7,165	6,465	6,521	5,708	-
Trinidad and Tobago	±0.047	±0.050	±0.057	±0.048	±0.049	-	22,326	24,457	29,114	22,834	23,110	-
Turkey	0.856	0.951	1.268	2.031	1.333	1.780	62,799	79,256	86,190	137,439	94,217	122,573
Ukraine	*0.362	*0.442	*0.280	*0.068	*0.081	*0.105	-	11,944	8,229	2,124	2,691	3,499
United Kingdom	*32	*32	35	31	34	37	104,928	109,019	119,291	100,645	105,840	113,947
United Republic of Tanzania	*0.013	*0.024	-	-	-	-	6,088	7,060	-	-	-	-
USSR (Former)	±1.438	-	-	-	-	-	6,015	-	-	-	-	-
United States of America	*157	*157	*166	*173	*176	*188	102,120	105,302	111,300	115,352	117,431	122,842
Uruguay	0.159	0.197	0.188	0.276	0.276	0.299	19,732	30,261	29,547	40,798	40,689	44,073
Venezuela	0.492	0.576	0.578	0.689	0.531	0.680	26,609	28,088	30,275	39,125	33,179	41,338
Yugoslavia	±0.104	±0.119	±0.189	-	±0.441	±0.282	-	3,775	6,474	-	18,309	10,814
Zambia	0.050	-	-	-	0.021	-	19,495	-	-	-	9,636	-
Zimbabwe	*0.145	*0.121	*0.088	*0.079	*0.088	*0.093	24,398	18,836	17,589	15,821	15,929	17,373

Value Added and Value Added per Employee

Country	Value Added ($ bil.)						Value Added per Employee ($)					
	1990	1991	1992	1993	1994	1995	1990	1991	1992	1993	1994	1995
Albania	±0.003	-	-	-	-	-	1,010	-	-	-	-	-
Argentina	±0.695	-	-	1.483	-	-	24,949	-	-	43,452	-	-
Armenia	0.000	0.000	-	-	-	-	30	47	-	-	-	-
Australia	*4.058	*4.349	*4.256	-	-	-	48,307	51,165	52,545	-	-	-
Austria	1.163	1.227	1.260	1.176	1.230	1.440	47,272	51,138	54,770	54,015	58,610	68,727
Bahamas	0.011	0.008	-	-	-	-	26,042	23,286	-	-	-	-
Bangladesh	0.023	0.015	0.023	-	-	-	1,577	1,365	1,516	-	-	-
Belgium	*1.677	*1.658	*1.806	*1.825	*1.968	*2.458	47,913	48,334	53,758	-	-	-
Belize	0.001	0.001	0.001	-	-	-	-	7,702	5,795	-	-	-
Bolivia	0.015	0.017	0.020	0.021	0.020	0.022	6,025	6,875	7,624	7,608	6,981	7,515
Bosnia & Herzegovina	±0.053	±0.055	-	-	-	-	12,011	12,023	-	-	-	-
Brazil	*3.133	-	*2.637	*3.108	-	-	26,976	-	25,549	31,642	-	-
Bulgaria	-	±0.023	±0.033	±0.038	±0.035	±0.045	-	3,196	5,853	5,273	5,047	7,130
Burundi	0.001	0.001	-	-	-	-	2,727	3,106	-	-	-	-
Cameroon	0.006	0.008	0.008	0.007	0.003	0.005	4,798	6,753	6,710	6,520	3,104	5,827
Canada	7.671	7.428	7.132	6.589	6.554	6.725	54,018	55,431	55,284	52,709	53,284	54,237
Central African Republic	*0.003	-	-	-	-	-	17,601	-	-	-	-	-
Chile	0.224	0.272	0.340	0.381	0.427	0.567	26,305	30,202	34,389	33,760	38,527	49,664
China	±1.036	±1.114	±1.329	±2.150	±1.443	±1.475	1,058	1,092	1,290	2,312	1,458	1,520
China (Hong Kong SAR)	0.878	0.907	1.010	1.172	1.287	-	20,913	22,016	25,954	28,863	30,578	-
China (Taiwan Province)	0.751	0.808	0.973	0.967	0.899	0.847	14,815	15,738	18,098	17,304	15,356	14,607
Colombia	0.213	0.196	0.274	0.309	0.436	0.448	9,728	9,351	10,229	11,340	15,325	16,568
Costa Rica	0.034	0.033	0.040	0.049	0.049	0.049	7,715	7,697	8,689	10,198	-	-
Croatia	±0.394	±0.336	±0.133	-	-	-	22,030	20,996	9,790	-	-	-
Cyprus	0.037	0.038	0.046	0.041	0.044	0.052	20,108	22,310	25,766	22,763	21,675	26,726
Czechoslovakia (Former)	±0.127	-	-	-	-	-	4,536	-	-	-	-	-
Denmark	*1.592	*1.551	*1.754	*1.626	*1.844	*2.255	28,791	29,295	33,893	31,715	35,152	41,864
Ecuador	0.027	0.030	0.032	0.040	0.054	0.046	6,434	6,367	6,978	9,134	11,352	11,756
Egypt	*0.076	*0.059	*0.080	*0.107	*0.102	*0.092	5,311	3,830	3,903	5,703	6,117	6,160

Continued.

Depending on the table, * means *Enterprises, Engaged,* or *Factor Values*; ± means *Basis Unspecified*; § means *shown in millions*. For additional notes and sources, see Appendix I.

255

Value Added and Value Added per Employee

- Continued -

Country	Value Added ($ bil.)						Value Added per Employee ($)					
	1990	1991	1992	1993	1994	1995	1990	1991	1992	1993	1994	1995
El Salvador	-	0.013	-	0.027	0.039	0.048	-	-	-	13,740	18,277	15,930
Ethiopia	0.019	0.018	0.014	0.010	0.011	0.011	5,889	5,719	4,559	3,050	3,302	2,985
Fiji	0.006	0.006	0.007	0.008	0.009	-	7,356	6,873	7,422	7,755	8,372	
Finland	±2.113	±1.885	±1.531	±1.166	±1.365	±1.665	55,913	52,223	46,542	38,612	47,382	60,707
FYR Macedonia	±0.043	±0.067	±0.030	±0.024	±0.011	±0.023	11,080	14,463	6,911	5,970	2,906	6,065
France	±13	±13	±15	±14	±15	±17	53,512	57,199	63,550	64,640	68,232	76,004
Gabon	-	*0.008	*0.009	*0.011	*0.005	*0.006	-	27,811	37,185	38,710	19,396	24,839
Germany (Western Part)	10	11	12	11			56,859	58,404	64,331			
Ghana	-	-	-	0.008	0.007	0.007				1,860	1,568	1,600
Greece	*0.289	*0.273	*0.337	*0.316	*0.311	*0.356	28,611	27,028	32,310	30,136	29,631	33,836
Honduras	*0.010	*0.009	*0.015	*0.014	*0.012	*0.013	4,320	4,146	5,382	4,697	3,545	3,499
Hungary	±0.190	±0.217	±0.261	±0.298	±0.292	±0.304	9,034	10,352	15,375	12,430	12,147	12,680
Iceland	0.084	0.096	±0.093	±0.074			40,920	46,596	48,151	40,978		
India	*0.340	*0.357	*0.383	*0.503	*0.609	*0.725	2,331	2,421	2,517	3,384	4,120	4,923
Indonesia	*0.150	*0.172	*0.269	*0.315	*0.396	*0.471	3,448	3,995	6,010	6,459	7,390	7,857
Iran (Islamic Republic of)	0.120	0.100	0.096	0.079	-	-	12,095	12,661	11,396	8,962	-	-
Iraq	-	*0.027	*0.088				-	8,296	29,368			
Ireland	*0.561	*0.571	*0.689	*0.654	*0.694	*0.868	50,972	51,946	62,233	58,769	62,232	77,180
Israel	0.470	0.503	0.502	0.564	0.623	0.686	29,541	31,652	29,894	30,015	30,098	34,412
Italy	±6.171	±6.454	±6.599	±5.233	±4.958		74,619	76,517	75,323	60,366	59,446	
Japan	±48	±54	±58	±65	±67	±76	82,510	91,588	97,674	110,825	125,232	139,779
Jordan	0.012	0.014	0.025	0.029	0.020	0.031	6,653	7,163	9,207	10,411	7,253	11,293
Kenya	*0.027	*0.027	*0.023	*0.014	*0.018	*0.021	4,515	4,583	3,851	2,332	2,896	3,135
Korea, Republic of	2.531	2.625	2.944	3.830	4.333	5.340	35,491	36,440	38,168	44,246	48,735	59,399
Kuwait	0.005	0.018	0.037	0.023	0.026	-	1,963	9,766	11,498	6,812	6,924	
Latvia	-	-	-	*0.012	*0.024	*0.044	-	-	-	2,740	5,868	8,189
Macau	0.009	0.011	0.014	0.015	0.014	0.014	7,083	8,728	10,997	11,870	10,880	12,648
Malaysia	*0.266	*0.306	*0.395	*0.440	*0.505	*0.642	12,029	12,769	15,471	15,762	16,775	20,122
Malta	0.034	0.041	0.055	0.046	0.046		25,196	29,287	38,690	31,590	30,874	
Mauritius	0.012	0.013	0.016	0.017	0.019	0.021	7,541	10,135	13,085	12,169	12,299	11,638
Mexico	0.170	0.190	0.203	0.192	0.183	0.137	15,196	17,086	18,603	20,390	19,087	16,347
Mongolia	0.008	0.007	0.002	0.001	0.001	0.001	5,432	8,676	3,174	696	793	727
Morocco	±0.043	±0.043	±0.049	±0.056	-		-	7,519	8,181	8,328		
Namibia	-	-	-	-	0.011		-	-	-	-	12,524	
Nepal	0.002	0.002	-	0.004	0.004	-	693	915	-	1,516	1,736	-
Netherlands	3.217	3.340	3.770	3.635	3.822	4.453	52,966	53,385	60,803	59,288	62,751	73,266
New Zealand	0.537	0.580	0.547	0.599	-	-	33,996	38,937	36,979	39,204		
Norway	±1.381	±1.390	±1.370	±1.249	±1.316	±1.545	42,604	43,221	46,210	42,945	43,820	51,774
Pakistan	0.077	0.106	0.118				8,640	13,637	14,553			
Panama	0.029	0.030	0.035	0.038	0.040	0.042	17,227	15,828	17,255	17,798	18,536	19,027
Peru	0.151	0.207	0.169	-	-	-	11,264	18,476	17,615	-	-	-
Philippines	0.125	0.126	0.147	0.220			4,372	4,245	6,341	8,381	-	
Poland	±0.166	±0.216	±0.157	±0.151	±0.170	±0.234	4,142	6,533	5,599	5,813	6,630	9,449
Portugal	0.523	0.613	0.720	0.663	0.692	0.826	15,546	17,399	21,051	18,658	19,795	23,396
Puerto Rico	±0.164	±0.172	±0.176	±0.194	±0.199	±0.212	86,614	76,018	84,067	88,273	82,314	76,775
Qatar	0.021	0.013	0.037	0.020	0.026	-	20,400	12,158	22,306	16,008	20,776	-
Romania	±0.143	±0.055	±0.038	±0.048	-		5,424	2,370	1,596	1,497	-	
Russian Federation	-	-	-	±0.256	±0.411	±0.429			-	1,940	3,121	3,500
Saint Lucia	-	±0.002	±0.002	±0.002	±0.002	±0.002						
Senegal	0.010	0.010	0.017	0.008	0.005		13,071	13,085	23,396	12,434	5,445	
Sierra Leone	-	-	-	0.002	-		-	-	-	2,107	-	
Singapore	*0.514	*0.604	*0.746	*0.836	*0.974	*1.138	32,545	35,880	43,243	46,478	54,753	62,689
Slovakia	-	-	-	±0.059	±0.069	-	-	-	-	6,900	8,187	
Slovenia	-	-	-	0.181	0.190	0.250	-	-	-			
South Africa	*0.763	*0.893	*0.878	*0.892	*0.883	*0.915	15,251	17,511	16,265	16,516	16,653	17,590
Spain	*4.403	*4.705	*5.041	±4.385	±4.420	±4.962	52,717	57,350	63,046	41,045	42,401	47,086
Sri Lanka	0.015	0.012	0.017	0.014	-	-	2,275	1,603	2,811	1,839	-	-
Swaziland	-	*0.001	*0.002	*0.005			-	3,067	4,308	12,443		
Sweden	*3.158	*2.487	*2.682	*1.879	*2.033	*2.161	75,722	47,056	56,035	44,467	46,012	48,785
Switzerland	3.888	3.718	3.676	3.338	4.059	5.005	53,548	51,995	56,635	53,836	65,052	80,472

Continued.

Depending on the table, * means *Enterprises, Engaged,* or *Factor Values*; ± means *Basis Unspecified*; § means *shown in millions*. For additional notes and sources, see Appendix I.

Value Added and Value Added per Employee

- Continued -

Country	Value Added ($ bil.)						Value Added per Employee ($)					
	1990	1991	1992	1993	1994	1995	1990	1991	1992	1993	1994	1995
Thailand	0.203	22	-	-	-	-	7,958	684,770	-	-	-	-
Trinidad and Tobago	±0.021	±0.025	±0.026	±0.021	±0.020	-	9,960	12,063	13,471	10,235	9,685	-
Turkey	0.434	0.432	0.619	0.921	0.612	0.818	31,878	36,024	42,101	62,355	43,279	56,294
United Kingdom	*20	*20	*22	*19	*22	*23	64,615	67,851	75,000	62,530	66,293	71,379
United Republic of Tanzania	*0.004	*0.005	-	-	-	-	1,645	1,441	-	-	-	-
United States of America	*103	*104	*113	*117	*121	*126	67,087	69,738	75,383	78,088	80,295	82,146
Uruguay	0.081	0.110	0.096	0.147	0.146	0.159	10,088	16,924	15,008	21,650	21,597	23,399
Venezuela	0.182	0.222	0.258	0.313	0.238	0.658	9,859	10,851	13,499	17,808	14,870	39,994
Yugoslavia	±0.060	±0.065	±0.114	-	±0.220	±0.137		2,056	3,895	-	9,140	5,263
Zambia	0.031	-	-	-	0.007	-	12,145	-	-	-	3,379	-
Zimbabwe	*0.094	*0.080	*0.055	*0.055	*0.035	*0.037	15,776	12,509	10,950	10,948	6,305	6,850

Capital Investment

Country	Gross Fixed Capital Formation ($ mil.)						Per Establishment ($ mil)					
	1990	1991	1992	1993	1994	1995	1990	1991	1992	1993	1994	1995
Argentina	-	-	-	205	-	-	-	-	-	0.033	-	-
Australia	366			-		-	0.116			-		-
Austria	177	301	208	205	169	-	0.531	0.930	0.631	0.634	0.542	-
Bahrain	-		2.638				-		0.176			
Bangladesh	1.157	1.202	1.181	-	-	-	0.002	0.002	0.002	-	-	-
Belgium	528	436	508	408	472	525	0.206	0.164	0.193	0.159	0.187	-
Bolivia	3.116	0.759	2.408	4.580	0.876	-	0.020	0.005	0.014	0.025	0.005	-
Bosnia & Herzegovina	3.641	-	-	-	-	-	0.130	-	-	-	-	-
Brazil	278	-	182	114	-	-	0.282	-	0.199	0.134	-	-
Bulgaria	6.027	4.103	36	10	1.866	4.846	0.163	0.103	0.828	0.030	0.004	0.009
Cameroon	0.951	2.393	0.132	1.533	0.510	2.416	0.119	0.171	0.009	0.139	0.042	0.201
Canada	520	-	-	-	-	-	0.094	-	-	-	-	-
Cape Verde	0.002	0.002	-	-	-	-	0.000	0.000	-	-	-	-
Chile	14	29	43	27	43	-	0.290	0.553	0.724	0.422	0.741	-
China (Hong Kong SAR)	155	159	213	144	233	-	0.031	0.032	0.043	0.031	0.050	-
Colombia	29	19	31	20	41	-	0.079	0.050	0.074	0.048	0.099	-
Croatia	12	20	5.077	2.907	12	10	0.020	0.024	0.004	0.002	0.005	0.004
Cyprus	7.665	8.991	6.509	5.841	7.245	7.138	0.023	0.028	0.020	0.018	0.021	0.020
Czechoslovakia (Former)	36	-	-	-	-	-	0.905	-	-	-	-	-
Denmark	95	84	-	-	-	-	0.026	0.023	-	-	-	-
Ecuador	8.033	23	19	9.400	20	33	0.097	0.264	0.231	0.116	0.241	0.424
Egypt	5.950	2.581	26	269	-	-	0.041	0.016	0.157	1.745	-	-
El Salvador	-	0.990	-	8.637	4.977	4.899	-	0.052	-	0.240	0.121	0.062
Ethiopia	0.522	0.367	0.244	0.300	1.509	2.443	0.026	0.019	0.013	0.016	0.060	0.087
Fiji	1.132	1.358	0.566	-	-	-	0.028	0.033	0.013	-	-	-
Finland	341	239	197	92	144	138	0.510	0.330	0.295	0.144	0.237	0.218
FYR Macedonia	2.921	4.415	1.531	2.254	1.595	3.866	0.049	0.044	0.013	0.010	0.009	0.016
France	1,928	1,758	1,532	1,234	1,332	1,611	-	-	-	-	-	-
Gabon	-	1.205	1.938	0.724	0.629	0.547	-	0.301	0.323	0.121	0.079	0.091
Germany (Western Part)	1,378	1,422	1,418	1,537	-	-	0.648	0.647	0.624	-	-	-
Greece	41	33	30	-	-	-	0.163	0.130	0.118	-	-	-
Hungary	18	21	42	41	-	-	0.050	0.031	0.026	0.021	-	-
India	109	74	97	84	-	-	0.035	0.024	0.030	0.026	-	-
Indonesia	98	56	82	39	63	152	0.189	0.116	0.165	0.077	0.119	0.256
Iran (Islamic Republic of)	16	11	21	14	-	-	0.048	0.101	0.181	0.128	-	-
Ireland	75	51	-	-	-	-	0.216	0.140	-	-	-	-
Israel	66	79	58	91	131	-	0.079	0.091	0.067	0.101	0.136	-
Italy	584	545	557	513	508	-	0.568	0.520	0.414	0.379	0.386	-
Japan	3,667	4,543	4,943	4,658	3,855	-	0.118	0.149	0.168	0.155	0.146	-
Jordan	0.301	0.376	2.494	0.993	9.484	1.198	0.001	0.002	0.009	0.004	0.047	0.006
Korea, Republic of	353	379	458	498	519	683	0.116	0.116	0.131	0.117	0.116	0.139
Kuwait	1.462	23	18	12	7.534	-	0.024	0.385	0.340	0.225	0.132	-

Continued.

Depending on the table, * means *Enterprises*, *Engaged*, or *Factor Values*; ± means *Basis Unspecified*; § means *shown in millions*. For additional notes and sources, see Appendix I.

257

Capital Investment

- Continued -

Country	Gross Fixed Capital Formation ($ mil.)						Per Establishment ($ mil)					
	1990	1991	1992	1993	1994	1995	1990	1991	1992	1993	1994	1995
Latvia	0.015	0.024	2.532	0.417	0.873	9.722	0.000	0.000	0.018	0.002	0.009	0.031
Macau	2.007	2.922	2.229	3.462	2.616	2.034	0.021	0.032	0.023	0.036	0.025	0.022
Malaysia	90	89	98	91	97	-	0.374	0.356	0.379	0.313	0.334	-
Malta	2.719	3.474	0.579	5.309	10	-	0.038	0.047	0.008	0.072	0.135	-
Mexico	25	23	-	-	-	-	0.340	0.318	-	-	-	-
Mongolia	13	6.271	1.692	0.461	0.550		0.465	0.196	0.050	0.014	0.014	
Morocco	-	8.958	14	16	-		-	0.029	0.043	0.050	-	
Myanmar	50	56	-	-	61		3.125	3.505	-	-	4.073	
Nepal	4.018	-	-	-			0.046	-	-	-		
Netherlands	669	578	639	511			1.017	0.814	0.860	0.695		
New Zealand	16	-	-	-	-		0.012	-	-	-	-	
Norway	124	97	108	87	126		0.145	0.116	0.232	0.195	0.262	
Pakistan	-	7.018					-	0.064				
Panama	2.228	2.686	-	-	-		0.034	0.032	-	-	-	
Peru	5.288	7.026	11	-	-		0.006	0.008	0.012	-	-	
Philippines	5.224	20	13	41	-		0.002	0.008	0.019	0.065	-	
Poland	36	32	47	57	-		0.241	0.235	0.394	0.495	-	
Portugal	113	139	91	88	162		0.046	0.054	0.036	0.030	0.055	
Romania	3.968	8.339	6.186	23	21		0.233	0.066	0.029	0.079	0.013	
Senegal	1.789	-	-	-			0.089	-	-	-		
Singapore	100	132	102	119	123	-	0.306	0.387	0.284	0.312	0.325	-
Slovenia	8.482	19	11	13	19	27	0.045	0.040	0.019	0.016	0.021	0.022
Spain	318	398	462	445	317	583	0.047	0.062	0.070	-	-	-
Sri Lanka	1.148	1.868	4.266	10	-		0.015	0.032	0.063	0.144	-	
Swaziland	-	0.108	0.082	0.642	-		-	0.006	0.004	0.034	-	
Thailand	243	39	-	-	-		0.738	0.092	-	-	-	
Trinidad and Tobago	-	-	6.635	4.747	2.464		-	-	0.038	0.025	0.010	
Turkey	67	113	104	195	129	-	0.559	0.999	0.426	0.869	0.608	-
Ukraine	26	25	8.011	1.934	2.233	1.113	0.047	0.037	0.009	0.004	0.004	0.002
United Kingdom	1,609	1,651	1,337	-	1,557		0.086	0.091	0.068	-	0.062	
United Republic of Tanzania	11	15	-	-	-	-	0.207	0.284	-	-	-	-
United States of America	5,810	5,040	5,380	4,870	5,656	5,615	-	-	0.056	-	-	-
Uruguay	3.022	9.182	8.321	25	-		-	-	-	-	-	
Venezuela	25	77	47	136	12	151	0.039	0.112	0.076	0.256	0.024	0.290
Yugoslavia	3.972	5.254	11	-	8.148	6.263	0.011	0.009	0.010	-	0.005	0.004
Zambia	12	-	-	-	-		0.526	-	-	-	-	
Zimbabwe	21	15	9.207	6.551	9.178	-	0.300	0.207	0.135	0.096	0.126	-

Indexes of Industrial Production

Country	Index of Industrial Production (1990=100)						Index of Employment (1990=100)					
	1990	1991	1992	1993	1994	1995	1990	1991	1992	1993	1994	1995
Albania	100	-	-	-	-	-	100	-	-	34	25	6
Algeria	100	89	78	88	81	76	-	-	-	-	-	-
Angola	100	-	-	-	-	-	-	-	-	-	-	-
Argentina	100	99	-	-	-	-	100	-	-	123	128	-
Armenia	100	-	-	-	-	-	100	94	-	-	-	-
Australia	100	96	92	105	-	-	100	101	96	-	-	-
Austria	100	108	114	116	128	130	100	98	93	88	85	85
Azerbaijan	-	-	-	-	-	-	100	95	86	73	66	-
Bahamas	100	-	-	-	-	-	100	87	-	-	-	-
Bahrain	100	-	-	-	-	-	-	-	-	-	-	-
Bangladesh	100	-	-	-	-	-	100	77	105	-	-	-
Barbados	100	-	-	-	-	-	-	-	-	-	-	-
Belgium	100	98	102	93	97	95	100	98	96	-	-	-
Belize	100	-	-	-	-	-	-	-	-	-	-	-
Benin	100	-	-	-	-	-	-	-	-	-	-	-
Bermuda	-	-	-	-	-	-	100	99	92	91	89	90

Continued.

Depending on the table, * means *Enterprises, Engaged,* or *Factor Values*; ± means *Basis Unspecified*; § means *shown in millions*. For additional notes and sources, see Appendix I.

Indexes of Industrial Production

- Continued -

Country	Index of Industrial Production (1990=100)						Index of Employment (1990=100)					
	1990	1991	1992	1993	1994	1995	1990	1991	1992	1993	1994	1995
Bhutan	100	-	-	-	-	-	-	-	-	-	-	-
Bolivia	100	103	118	121	111	106	100	100	108	111	117	120
Bosnia & Herzegovina	-	-	-	-	-	-	100	105	-	-	27	-
Brazil	100	107	105	110	113	113	100	-	89	85	-	-
Bulgaria	100	109	115	154	181	152	100	82	63	82	78	71
Burkina-Faso	100						-	-	-	-	-	-
Burundi	100	-	-	-	-	-	100	110	-	-	-	-
Cambodia	100	-	-	-	-	-	100	83	-	-	-	-
Cameroon	100	-	-	-	-	-	100	97	91	88	64	65
Canada	100	89	83	80	79	92	100	94	91	88	87	87
Cape Verde	-	-	-	-	-	-	100	95	95	107	96	-
Central African Republic	100	-	-	-	-	-	100	-	-	-	-	-
Chad	100	-	-	-	-	-	-	-	-	-	-	-
Chile	100	130	181	184	198	230	100	105	116	132	130	134
China	100	-	-	-	-	-	100	104	105	95	101	99
China (Hong Kong SAR)	100	110	129	146	150	150	100	98	93	97	100	-
China (Taiwan Province)	100	102	114	118	116	112	100	101	106	110	115	114
Colombia	100	104	105	114	132	121	100	96	122	124	130	123
Congo	100	-	-	-	-	-	-	-	-	-	-	-
Costa Rica	100	97	108	128	130	118	100	98	104	111	-	-
Cote d'Ivoire	100	-	-	-	-	-	-	-	-	-	-	-
Croatia	100	-	-	-	-	-	100	89	76	68	62	63
Cuba	100	98	98	-	-	-	-	-	-	-	-	-
Cyprus	100	87	105	96	100	97	100	92	96	99	111	106
Czechoslovakia (Former)	100	73	64				100	82	-	-	-	-
Czech Republic	100	-	-	-	-	-	100	89	100	89	-	-
Dem. Rep. of the Congo	100	-	-	-	-	-	-	-	-	-	-	-
Denmark	100	100	103	101	106	112	100	96	94	93	95	97
Dominican Republic	100	-	-	-	-	-	-	-	-	-	-	-
Ecuador	100	123	137	144	-	-	100	111	111	106	114	93
Egypt	100	98	94	77	70	59	100	108	143	131	116	105
El Salvador	100	100	-	-	-	-	-	-	-	-	-	-
Equatorial Guinea	-	-	-	-	-	-	100	-	-	-	-	-
Ethiopia and Eritrea	100	-	-	-	-	-	-	-	-	-	-	-
Ethiopia	-	-	-	-	-	-	100	98	97	98	102	114
Fiji	100	109	111	127	141	129	100	102	110	121	131	-
Finland	100	90	84	81	85	90	100	96	87	80	76	73
FYR Macedonia	100	-	-	-	-	-	100	120	113	102	99	97
France	100	96	94	92	93	95	100	100	98	95	94	94
Gabon	100	-	-	-	-	-	-	-	-	-	-	-
Gambia	100	-	-	-	-	-	-	-	-	-	-	-
Germany (Eastern Part)	100	-	-	-	-	-	-	-	-	-	-	-
Germany (Western Part)	100	106	106	101	99	103	100	104	105	-	-	-
Ghana	100	-	-	-	-	-	-	-	-	-	-	-
Greece	100	92	89	84	82	82	100	100	103	104	104	104
Guatemala	100	100	100	-	-	-	-	-	-	-	-	-
Guyana	100	-	-	-	-	-	-	-	-	-	-	-
Haiti	100	-	-	-	-	-	-	-	-	-	-	-
Honduras	100	123	147	150	181	236	100	98	116	127	139	156
Hungary	100	100	87	110	107	110	100	100	81	114	114	114
Iceland	100	-	-	-	-	-	100	100	94	88	85	-
India	100	106	110	114	129	147	100	101	104	102	101	101
Indonesia	100	104	150	153	175	210	100	99	103	112	123	138
Iran (Islamic Republic of)	100	159	159	-	-	-	100	80	85	89	-	-
Iraq	100	-	-	-	-	-	-	-	-	-	-	-
Ireland	100	110	122	133	135	154	100	100	101	101	101	102
Israel	100	98	107	126	130	135	100	100	106	118	130	125
Italy	100	97	90	86	-	-	100	102	106	105	101	-
Jamaica	100	-	-	-	-	-	100	100	95	96	91	-
Japan	100	104	106	111	114	117	100	102	102	102	92	93

Continued.

Depending on the table, * means *Enterprises, Engaged,* or *Factor Values*; ± means *Basis Unspecified*; § means *shown in millions*. For additional notes and sources, see Appendix I.

259

Indexes of Industrial Production

- Continued -

Country	Index of Industrial Production (1990=100)						Index of Employment (1990=100)					
	1990	1991	1992	1993	1994	1995	1990	1991	1992	1993	1994	1995
Jordan	100	-	-	-	-	-	100	109	149	156	157	154
Kenya	100	103	105	105	108	115	100	98	100	101	105	110
Korea, Republic of	100	103	115	111	121	140	100	101	108	121	125	126
Kuwait	100	-	-	-	-	-	100	67	116	122	133	-
Kyrgyzstan	100	-	-	-	-	-	100	91	81	81	-	-
Lao P.D.R.	100	-	-	-	-	-	-	-	-	-	-	-
Latvia	100	-	-	-	-	-	100	91	86	81	78	101
Lebanon	100	-	-	-	-	-	-	-	-	-	-	-
Lesotho	100	-	-	-	-	-	100	-	43	40	39	-
Liberia	100	-	-	-	-	-	-	-	-	-	-	-
Libyan Arab Jamahiriya	100	-	-	-	-	-	-	-	-	-	-	-
Lithuania	100	-	-	-	-	-	-	-	-	-	-	-
Macau	100	-	-	-	-	-	100	95	97	97	100	89
Madagascar	100	64	64	-	-	-	-	-	-	-	-	-
Malawi	100	108	-	-	-	-	-	-	-	-	-	-
Malaysia	100	155	275	-	-	-	100	109	115	126	136	144
Malta	100	122	159	167	-	-	100	104	107	110	111	-
Mauritius	100	-	-	-	-	-	100	80	77	85	98	111
Mexico	100	97	98	98	94	96	100	99	98	84	86	75
Mongolia	-	-	-	-	-	-	100	57	47	54	71	59
Morocco	100	92	99	94	104	114	-	-	-	-	-	-
Mozambique	100	-	-	-	-	-	100	98	81	74	77	-
Myanmar	100	100	100	-	-	-	100	100	100	-	100	-
Namibia	100	-	-	-	-	-	-	-	-	-	-	-
Nepal	100		-	-	-	-	100	114	-	107	108	-
Netherlands	100	102	100	100	103	104	100	103	102	101	100	100
New Zealand	100	-	-	-	-	-	100	94	94	97	-	-
Nicaragua	100	-	-	-	-	-	-	-	-	-	-	-
Niger	100	-	-	-	-	-	-	-	-	-	-	-
Nigeria	100	101	101	-	-	-	-	-	-	-	-	-
Norway	100	100	100	100	104	107	100	99	91	90	93	92
Pakistan	100	113	100	113	-	-	100	87	91	-	-	-
Panama	100	103	120	127	135	140	100	115	123	126	130	132
Papua New Guinea	100	-	-	-	-	-	-	-	-	-	-	-
Paraguay	100	105	106	127	138	152	-	-	-	-	-	-
Peru	100	101	111	108	188	269	100	84	72	-	-	-
Philippines	100	120	131	139	155	171	100	105	81	92	-	-
Poland	100	87	91	126	135	158	100	83	70	65	64	62
Portugal	100	111	106	103	109	114	100	105	102	106	104	105
Puerto Rico	100	-	-	-	-	-	100	120	111	116	128	146
Qatar	100	-	-	-	-	-	100	105	166	126	123	-
Republic of Moldova	100	-	-	-	-	-	100	87	78	67	65	56
Romania	100	83	84	122	211	264	100	88	90	122	106	-
Russian Federation	100	-	-	-	-	-	-	-	-	-	-	-
Saudi Arabia	100	-	-	-	-	-	-	-	-	-	-	-
Senegal	100	100	82	75	85	77	100	96	93	88	114	-
Sierra Leone	100	-	-	-	-	-	-	-	-	-	-	-
Singapore	100	109	116	126	135	141	100	107	109	114	113	115
Slovakia	100	-	-	-	-	-	-	-	-	-	-	-
Slovenia	100	104	87	96	99	82	-	-	-	-	-	-
Somalia	100	-	-	-	-	-	-	-	-	-	-	-
South Africa	100	100	96	103	103	99	100	102	108	108	106	104
Spain	100	105	106	105	110	109	100	98	96	128	125	126
Sri Lanka	100	-	-	-	-	-	100	116	93	116	-	-
Sudan	100	50	50	-	-	-	-	-	-	-	-	-
Suriname	100	-	-	-	-	-	100	95	104	72	-	-
Swaziland	100	134	138	-	-	-	-	-	-	-	-	-
Sweden	100	95	85	83	87	83	100	127	115	101	106	106
Switzerland	100	98	95	93	102	105	100	98	89	85	86	86
Syrian Arab Republic	100	114	121	169	183	211	-	-	-	-	-	-

Continued.

Depending on the table, * means *Enterprises, Engaged,* or *Factor Values;* ± means *Basis Unspecified;* § means *shown in millions.* For additional notes and sources, see Appendix I.

Indexes of Industrial Production

- Continued -

Country	Index of Industrial Production (1990=100)						Index of Employment (1990=100)					
	1990	1991	1992	1993	1994	1995	1990	1991	1992	1993	1994	1995
Thailand	100	98	61	-	-	-	100	127	-	-	-	-
Togo	100	-	-	-	-	-	-	-	-	-	-	-
Tonga	100	-	-	-	-	-	100	105	100	105	104	-
Trinidad and Tobago	100	111	120	126	115	125	100	97	92	99	100	-
Tunisia	100	101	104	104	102	110	-	-	-	-	-	-
Turkey	100	91	76	97	76	88	100	88	108	108	104	107
Uganda	100	123	133	164	220	271	-	-	-	-	-	-
Ukraine	100	-	-	-	-	-	-	-	-	-	-	-
United Arab Emirates	100	-	-	-	-	-	-	-	-	-	-	-
United Kingdom	100	94	95	98	100	101	100	96	96	100	107	107
United Republic of Tanzania	100	85	67	130	83	-	100	160	-	-	-	-
USSR (Former)	100	-	-	-	-	-	100	-	-	-	-	-
United States of America	100	96	97	98	98	97	100	97	97	98	98	100
Uruguay	100	119	114	117	113	113	100	81	79	84	84	84
Venezuela	100	-	-	-	-	-	100	111	103	95	86	89
Yemen	100	-	-	-	-	-	-	-	-	-	-	-
Yugoslavia	100	-	-	-	-	-	-	-	-	-	-	-
Yugoslavia (Former)	100	80	-	-	-	-	-	-	-	-	-	-
Zambia	100	98	95	88	88	68	100	-	-	-	86	-
Zimbabwe	100	105	105	109	124	114	100	108	84	84	93	90

Depending on the table, * means *Enterprises*, *Engaged*, or *Factor Values*; ± means *Basis Unspecified*; § means *shown in millions*. For additional notes and sources, see Appendix I.

261

Representative Companies in Sector

Name	Address	Tele-phone	Fax	Employ-ment	Y
Communications Quebecor Inc.	900-612 Rue Saint-Jacques, Montreal, Quebec H3C 4M8, Canada	514-877-9777	514-877-9757	38,000	98
Moore Corp. Ltd.	PO Box 78 Sta. 1st Can Place, Toronto, Ontario M5X 1G5, Canada	416-364-2600	416-364-1667	20,000	98
Southeast Publishing Ventures Inc.	401 N. Wabash Ave., Ste. 740, Chicago, Illinois 60611, U.S.A.	312-321-2299		9,692	95
Reynolds and Reynolds Co.	PO Box 2608, Dayton, Ohio 45401, U.S.A.	937-485-2000		9,138	97
Media General Inc.	PO Box 85333, Richmond, Virginia 23293-0001, U.S.A.	804-649-6000	804-649-6865	8,800	97
De la Rue PLC	6 Agar St., London WC2N 4DE, United Kingdom	71 8368383		8,599	93
McClatchy Newspapers Inc.	2100 Q St., Sacramento, California 95816, U.S.A.	916-321-1846		7,590	96
Washington Post Co.	1150 15th St. N.W., Washington, District of Columbia 20071, U.S.A.	202-334-6000		7,440	97
Gibson Greetings Inc.	2100 Section Rd., Cincinnati, Ohio 45237, U.S.A.	513-841-6600	513-841-6739	7,100	97
Clarin AGEAS.A.	Tacuari 1842, Buenos Aires 1139, Argentina	1 3070330	1 3070311	7,000	96
Banta Corp.	Box 8003, Menasha, Wisconsin 54952-8003, U.S.A.	920-751-7777	920-722-6495	6,900	97
Pittway Corp.	200 S. Wacker Dr., 700, Chicago, Illinois 60606-5802, U.S.A.	312-831-1070	312-831-0808	6,800	96
E.W. Scripps Co.	PO Box 5380, Cincinnati, Ohio 45201, U.S.A.	513-977-3000	513-977-3721	6,700	96
Journal Communications Inc.	PO Box 661, Milwaukee, Wisconsin 53201, U.S.A.	414-224-2000	414-224-2469	6,550	95
G.T.C. Transcontinental Group Ltd.	1 Place Ville-Marie Ste. 3315, Montreal, Quebec H3B 3N2, Canada	514-954-4000	514-954-4016	6,500	98
Moore Business Forms and Systems Inc.	275 N. Field Dr., Lake Forest, Illinois 60045, U.S.A.	847-615-6000	847-615-7300	6,500	97
Standard Register Co.	600 Albany St., Dayton, Ohio 45401, U.S.A.	937-443-1000		6,400	97
PRIMEDIA Inc.	745 5th Ave., New York, New York 10151, U.S.A.	212-745-0100	212-745-0121	6,310	97
Jostens Inc.	5501 Norman Center Dr., Minneapolis, Minnesota 55437, U.S.A.	612-830-3300		6,300	96
Lee Enterprises Inc.	215 N. Main St., Davenport, Iowa 52801-1924, U.S.A.	319-383-2100		6,100	97
Jostens Inc. School Products Group Div.	5501 Norman Center Dr., Minneapolis, Minnesota 55437, U.S.A.	612-830-3300	612-830-3364	6,000	97
Quebecor Printing Inc.	612 Rue Saint-Jacques, Montreal, Quebec H3C 1C7, Canada	514-954-0101	514-954-9624	6,000	98
Reader's Digest Association Inc.	Reader's Digest Rd., Pleasantville, New York 10570-7000, U.S.A.	914-238-1000		6,000	97
Insilco Corp.	425 Metro Pl. N., 5th Fl., Dublin, Ohio 43017, U.S.A.	614-792-0468	614-791-3197	5,764	96
Commerce Clearing House Inc.	2700 Lake Cook Rd., Riverwoods, Illinois 60015, U.S.A.	847-940-4600		5,700	94
John H. Harland Co.	PO Box 105250, Atlanta, Georgia 30348, U.S.A.	404-981-9460		5,599	96
Harte-Hanks Communications Inc.	PO Box 269, San Antonio, Texas 78291-0269, U.S.A.	210-829-9000		5,550	97
Central Newspapers Inc.	135 N. Pennsylvania St., 1200, Indianapolis, Indiana 46204-2400, U.S.A.	317-231-9200		5,341	96
Press Corp. of South Africa Ltd.	Doornfontein, Johannesburg 2001, Republic of South Africa	11 7769111	11 4028036	5,244	93
Scholastic Corp.	555 Broadway, New York, New York 10012, U.S.A.	212-343-6100	212-343-6933	5,025	97
Harcourt Brace and Co.	6277 Sea Harbor Dr., Orlando, Florida 32887, U.S.A.	407-345-2000	407-351-7832	5,000	97
International Thomson Publishing Inc.	10 Davis Dr., Belmont, California 94002, U.S.A.	650-595-2350		5,000	97
Addison Wesley Longman	1 Jacob Way, Reading, Massachusetts 01867, U.S.A.	781-944-3700	781-944-9338	4,800	97
Standex International Corp.	6 Manor Pkwy., Salem, New Hampshire 03079, U.S.A.	603-893-9701	603-893-7324	4,800	97
Franklin Covey Co.	2200 W. Parkway Blvd., Salt Lake City, Utah 84119-2331, U.S.A.	801-975-1776		4,741	97
Berlitz International Inc.	Research Park, 293 Wall St., Princeton, New Jersey 08540, U.S.A.	609-924-8500	609-683-9138	4,611	96
Wallace Computer Services Inc.	2275 E. Cabot Dr., Lisle, Illinois 60532-3630, U.S.A.	630-588-5000	630-588-5105	4,610	97
Imprimeries Transcontinental Inc.	395 Blvd. Lebeau, Saint-Laurent, Quebec H4N 1S2, Canada	514-337-8560	514-339-5230	4,400	98
Justin Industries Inc.	PO Box 425, Fort Worth, Texas 76101, U.S.A.	817-336-5125		4,222	97
Journal Register Co.	50 W. State St., 12th Fl., Trenton, New Jersey 08608-1298, U.S.A.	609-396-2200		4,200	97
Big Flower Holdings Inc.	3 East 54th St., New York, New York 10022, U.S.A.	212-521-1600	212-521-1697	4,100	96
News International PLC	PO Box 495, London E1 9XY, United Kingdom	71 7826000	71 8959017	4,001	94
ACCO USA Inc.	300 Tower Pkwy., Lincolnshire, Illinois 60069, U.S.A.	847-541-9500		4,000	97
Macmillan Publishing Co.	1663 Broadway, New York, New York 10019, U.S.A.	212-654-8500		4,000	95
CSS Industries Inc.	1845 Walnut St., 800, Philadelphia, Pennsylvania 19103, U.S.A.	215-569-9900		3,924	97
Guardian Media Group PLC	64 Deansgate, Manchester M60 2RR, United Kingdom	61 8327200	61 8325351	3,762	94

Continued

Representative Companies in Sector
- Continued -

Name	Address	Tele-phone	Fax	Employ-ment	Y
A.H. Belo Corp.	PO Box 655237, Dallas, Texas 75265, U.S.A.	214-977-6606	214-977-6603	3,760	97
Westminster Press Ltd.	8-16 Great New St., London EC4P 4ER, United Kingdom	71 3531030	71 3537526	3,691	94
Yattendon Investment Trust PLC	Edgbaston, Birmingham B15 3BU, United Kingdom	21 4564004	21 4546937	3,651	94
American Bank Note Co.	200 Park Ave., 49th Fl., New York, New York 10166-4999, U.S.A.	212-557-9100	212-338-0757	3,600	96
American Business Products Inc.	PO Box 105684, Atlanta, Georgia 30348, U.S.A.	770-953-8300	770-952-2343	3,520	97
HarperCollins Publishers Inc.	10 East 53rd St., New York, New York 10022, U.S.A.	212-207-7000	212-207-7065	3,500	97
Bowne and Company Inc.	345 Hudson St., New York, New York 10014, U.S.A.	212-924-5500	212-229-3421	3,400	97
Pulitzer Publishing Co.	900 N. Tucker Blvd., St. Louis, Missouri 63101, U.S.A.	314-622-7000		3,300	97
Saskatchewan Wheat Pool	2625 Victoria Ave., Regina, Saskatchewan S4T 7T9, Canada	306-569-4411	306-569-4708	3,300	98
American Banknote Corp.	200 Park Ave., New York, New York 10166-4999, U.S.A.	212-557-9100		3,260	96
Cadmus Communications Corp.	PO Box 27367, Richmond, Virginia 23261-7367, U.S.A.	804-287-5680	804-287-6267	3,200	97
Her Majesty's Stationery Office	Duke St., Norwich NR3 1PD, United Kingdom	603 622211	603 695582	3,200	92
Mirror Group Newspapers PLC	33 Holborn Circus, London EC1P 1DQ, United Kingdom	71 3530246	71 8223405	3,125	94
Sun Media Corp.	333 King St. E, Toronto, Ontario M5A 3X5, Canada	416-947-2222	416-947-3119	3,102	98
ATC Communications Group Inc.	5950 Berkshire Ln., 1650, Dallas, Texas 75225, U.S.A.	214-361-9870	214-361-9874	3,076	97
Jordan Industries Inc.	1751 Lake Cook Rd., 550, Deerfield, Illinois 60015, U.S.A.	847-945-5591	847-945-9645	2,800	95
Ameriscribe Corp.	75 Varick St., New York, New York 10013, U.S.A.	212-219-0800	212-219-0517	2,700	
National Education Corp.	2601 Main St., Irvine, California 92614, U.S.A.	714-474-9400	714-474-9494	2,700	96
Shorewood Packaging Corp.	277 Park Ave., New York, New York 10172-0124, U.S.A.	212-371-1500		2,700	97
Houghton Mifflin Co.	222 Berkeley St., Boston, Massachusetts 02116-3764, U.S.A.	617-351-5000	617-227-5409	2,550	97
Merrill Corp.	1 Merrill Cir., St. Paul, Minnesota 55108, U.S.A.	612-646-4501	612-649-1348	2,320	96
O Globo Empresa Jornalistica Brasileira	Cidade Nova RJ, Rio de Janeiro 02023-900, Brazil	21 2922000	21 2922000	2,293	97
IPL Ltd.	21 Holborn Viaduct, London EC1A 2DY, United Kingdom	71 6281388	71 3210584	2,252	90
Devon Group Inc.	281 Tresser Blvd., 501, Stamford, Connecticut 06901-3227, U.S.A.	203-964-1444	203-964-1036	2,200	97
John Wiley and Sons Inc.	605 3rd Ave., New York, New York 10158-0012, U.S.A.	212-850-6000	212-850-6088	2,170	97
Associated Newspapers PLC	2 Derry St., London W8 5TT, United Kingdom	71 9386000	71 9386681	2,158	93
D.C. Thomson & Co., Ltd.	2 Albert Sq., Dundee DD1 9QJ, United Kingdom	382 223131	382 222214	2,152	93
Portsmouth & Sunderland Newspapers PLC	37 Abingdon Rd., London W8 6AH, United Kingdom	71 9379741	71 9371479	2,119	94
Meredith Corp.	1716 Locust St., Des Moines, Iowa 50309-3023, U.S.A.	515-284-3000		2,102	97
Miami Herald Publishing Company Inc.	1 Herald Plz., Miami, Florida 33132, U.S.A.	305-350-2111		2,100	97
American Media Inc. (Lantana, Florida)	600 S. East Coast Ave., Lantana, Florida 33462, U.S.A.	561-540-1000	561-547-1017	2,030	96
IPC Magazines Ltd. Overseas	Stamford St., London SE1 9LS, United Kingdom	71 261500		2,014	92
New England Business Service Inc.	500 Main St., Groton, Massachusetts 01471, U.S.A.	508-448-6111	508-448-6305	2,014	97
Carlton Cards Ltd.	1460 The Queensway, Toronto, Ontario M8Z 1S7, Canada	416-255-9131	416-503-6523	2,000	98
Houston Chronicle Publishing Co.	PO Box 4260, Houston, Texas 77210, U.S.A.	281-220-7171	281-220-6611	2,000	94
Waltons Stationery Co.–Export Div.	PO Box 39531, Booysens 2016, Republic of South Africa	11 4937616	11 4932484	2,000	95
William Collins Sons & Co., Ltd.	Bishopbriggs, Glasgow G64 2QT, United Kingdom	41 7723200	41 3063119	1,990	90
MacMillan Ltd.	Houndmills, Basingstoke RG21 6XS, United Kingdom	256 29242	256 479476	1,977	93
Carlton Cards Ltd.	Saville Town, Dewsbury WF12 9AW, United Kingdom	924 465200	924 466641	1,942	93
Harper Collins Publishers Ltd.	Bishopbriggs, Glasgow G64 2QT, United Kingdom	41 7723200		1,940	94
PAXAR Corp.	105 Corporate Park Dr., White Plains, New York 10604, U.S.A.	914-697-6800	914-697-6893	1,923	95
Reed International Books Ltd.	81 Fulham Rd., London SW3 6RB, United Kingdom	71 5819393	71 5894819	1,897	93
Eastern Counties Newspapers Group Ltd.	Rouen Rd., Norwich NR1 1RE, United Kingdom	603 628311	603 628311	1,885	93
ACCO-Rexel Ltd.	Gatehouse Rd., Aylesbury HP19 3DT, United Kingdom	270 582411		1,872	93
Bowes Publishers Ltd.	1147 Gainsborough Rd., London, Ontario N6H 5L5, Canada	519-471-8520	519-471-1892	1,800	98
Penguin Putnam Inc.	200 Madison Ave., New York, New York 10016, U.S.A.	212-951-8400	212-213-6706	1,750	97
Polychrome Corp.	222 Bridge Plaza S., Fort Lee, New Jersey 07024, U.S.A.	201-346-8800	201-346-8849	1,700	95
Marvel Entertainment Group Inc.	387 Park Ave. S., New York, New York 10016, U.S.A.	212-696-0808	212-576-8598	1,600	94
Thomson Corp., The	TD Bank Tower Ste. 2706, Toronto, Ontario M5K 1A1, Canada	416-360-8700	416-360-8812	1,600	98
Express Newspapers PLC	245 Blackfriars Rd., London SE1 9UX, United Kingdom	71 9288000		1,580	93
Russ Berrie and Company Inc.	111 Bauer Dr., Oakland, New Jersey 07436, U.S.A.	201-337-9000	201-337-9634	1,580	96
Argus Press Ltd.	2 Queensway, Redhill RH1 1QS, United Kingdom	737 768611	737 770036	1,573	94
Southern Newspapers PLC	45 Above Bar St., Southampton SO9 7BA, United Kingdom	703 634134	703 630428	1,573	94

Continued

Representative Companies in Sector
- Continued -

Name	Address	Telephone	Fax	Employment	Y
Bemrose Corp. PLC	PO Box 52, Derby DE2 6XP, United Kingdom	332 294242	332 290366	1,556	94
Brooke Group Ltd.	100 Southeast 2nd St., Miami, Florida 33131, U.S.A.	305-579-8000		1,545	96
Bantam Doubleday Dell Publishing Group Inc.	1540 Broadway, New York, New York 10036, U.S.A.	212-354-6500	212-492-8904	1,500	96
Cheler Corp.	PO Box 1750, Seattle, Washington 98111-1750, U.S.A.	206-624-9699	206-762-8014	1,500	94
Matthews International Corp.	2 Northshore Ctr., Pittsburgh, Pennsylvania 15212-5851, U.S.A.	412-442-8200	412-442-8290	1,500	97
Moody's Investors Service	99 Church St., New York, New York 10007, U.S.A.	212-553-0300	212-553-4063	1,500	97
Consolidated Graphics Inc.	2210 W. Dallas St., Houston, Texas 77019, U.S.A.	713-529-4200	713-525-4305	1,417	96
MGN Ltd.	Canary Wharf, London E14 5AP, United Kingdom			1,408	94
Curtis 1000 Inc.	PO Box 105683, Atlanta, Georgia 30348, U.S.A.	770-951-1000	770-955-0707	1,400	97
Mosby Inc.	11830 Westline Indstl., St. Louis, Missouri 63146, U.S.A.	314-872-8370	314-432-1380	1,400	97
Walsworth Publishing Company Inc.	306 N. Kansas Ave., Marceline, Missouri 64658, U.S.A.	660-376-3543	660-258-7898	1,400	97
Midland News Association Ltd.	51-53 Queen St., Wolverhampton WV1 3BU, United Kingdom	902 313131	902 335490	1,361	94
Dispatch Printing Co.	34 S. 3rd St., Columbus, Ohio 43215, U.S.A.	614-461-5000	614-461-5533	1,300	97
Mack Printing Company Inc.	1991 Northampton St., Easton, Pennsylvania 18042-3189, U.S.A.	215-258-9111	215-250-7202	1,300	
St. Joseph Printing Ltd.	50 MacIntosh Blvd, Concord, Ontario L4K 4P3, Canada	905-660-3111	905-660-6820	1,300	98
Thomas Nelson Inc.	501 Nelson Place, Nashville, Tennessee 37214-1000, U.S.A.	615-889-9000		1,300	97
William E. Coutts Co. Ltd.	2 Hallcrown Pl, North York, Ontario M2J 1P6, Canada	416-492-1300	416-494-0027	1,300	98
Britannia Products Ltd.	Dudley Hill, Bradford BD4 6HW, United Kingdom	274 688221		1,299	94
Ennis Business Forms Inc.	107 N. Sherman St., Ennis, Texas 75119, U.S.A.	214-875-6581	214-875-4915	1,270	95
Valassis Communications Co.	19975 Victor Pkwy., Livonia, Michigan 48152-7001, U.S.A.	313-591-3000	313-591-4994	1,270	95
Melham Holdings Ltd.	6 Union Rd., Nottingham NG3 1FH, United Kingdom	15 9582277	15 9585122	1,262	93
Biber Paper Converting Ltd.	Apsley, Hemel Hempstead HP3 9SS, United Kingdom			1,243	93
Bristol Evening Post PLC	Temple Way, Bristol BS99 7HD, United Kingdom	272 260080	272 279568	1,241	93
Reed Business Publishing Ltd.	The Quadrant, Sutton SM2 5AS, United Kingdom	81 6613500	81 6618948	1,237	93
BPP Holdings PLC	142-144 Uxbridge Rd., London W12 8AA, United Kingdom	81 7401111	81 7461060	1,218	93
Courier Corp.	15 Wellman Ave., North Chelmsford, Massachusetts 01863, U.S.A.	978-251-6000	978-251-8228	1,202	97
Crain-Drummond Inc.	2000-1570 Rue Ampere, Boucherville, Quebec J4B 7L4, Canada	450-449-7171	450-449-8700	1,200	98
Data Documents Inc.	4205 South 96th St., Omaha, Nebraska 68127-1290, U.S.A.	402-339-0900	402-339-0485	1,200	96
Golden Books Family Entertainment Inc.	888 7th Ave., New York, New York 10106, U.S.A.	212-547-6700		1,200	97
LSI Industries Inc.	PO Box 42728, Cincinnati, Ohio 45242-0728, U.S.A.	513-793-3200	513-791-0813	1,200	97
Morgan-Grampian PLC	30 Calderwood St., London SE18 6QH, United Kingdom	81 8557777	81 8547476	1,197	93
Longman Group Ltd.	Burnt Mill, Harlow CM20 2JE, United Kingdom	279 623623	279 431059	1,182	93
Data Business Forms Ltd.	2 Shaftsbury Lane, Brampton, Ontario L6T 3X7, Canada	905-791-3151	905-791-3277	1,175	98
Metroland Printing, Publishing & Distribution Ltd.	3125 Wolfedale Rd., Mississauga, Ontario L5C 1W1, Canada	905-279-0440	905-279-5103	1,155	98
Jarrold & Sons Ltd.	Whitefriars, Norwich NR3 1SH, United Kingdom	603 660211	603 630162	1,151	94
Conde Nast Publications Ltd.	Hanover Sq., London W1R 9HG, United Kingdom	71 4999080	71 4931345	1,148	92
Day Runner Inc.	PO Box 57027, Irvine, California 92619-7027, U.S.A.	714-680-3500	714-441-4848	1,146	97
BBC Enterprises Ltd.	80 Wood Ln., London W12 0TT, United Kingdom	81 5762000	81 7490538	1,122	93
Bell Group UK Holdings Ltd.	1 Northumberland Ave., London WC2 5BW, United Kingdom	71 8725864		1,116	89
American Business Information Inc.	PO Box 27347, Omaha, Nebraska 68127-0347, U.S.A.	402-593-4500	402-331-1505	1,100	96
Topps Company Inc.	1 Whitehall St., New York, New York 10004, U.S.A.	212-514-8190		1,100	
Midland Newspapers Ltd.	Queensway, Birmingham B4 6AX, United Kingdom	21 2363366		1,066	93
United Methodist Publishing House	201 8th Ave. S., Nashville, Tennessee 37203, U.S.A.	615-749-6000	615-749-6079	1,064	
Express Gifts Ltd.	Church, Accrington BB5 4EE, United Kingdom	254 382121	254 351012	1,056	94
Telegraph PLC	Canary Wharf, London E14 5DT, United Kingdom			1,054	93
Dai Nippon Printing Indonesia PT	Jl. Pulogadung Kav li Blk H, No. 2-3, Jakarta 13930, Indonesia	214722310	21 4881287	1,050	97
Landesverlag Holding GmbH-Gruppe	Hafenstrasse 1-3, Linz A-4010, Austria	732 76160	732 7616450	1,050	92
Boosey & Hawkes PLC	Deansbrook Rd., Edgware HA8 9BB, United Kingdom	81 9527711	81 9511314	1,043	93
Scientific Games Holdings Corp.	1500 Bluegrass Lakes Pkwy., Alpharetta, Georgia 30201, U.S.A.	770-664-3700	770-343-8798	1,022	96
Gray Communications Systems Inc.	PO Box 48, Albany, Georgia 31702-0048, U.S.A.	912-888-9390		1,020	97

For sources and notes, see Appendix I.

ISIC 3510
INDUSTRIAL CHEMICALS

ISIC 3510—Industrial Chemicals—includes chemicals, fertilizers, and synthetic resins. Chapters on Basic Chemicals (ISIC 3511) and Synthetic Resins (ISIC 3513) further break out details. Significant detail, by commodity, is shown in this chapter.

In the U.S. SIC system, this industry combines elements of SIC 281—Industrial Inorganic Chemicals, 282—Plastics Materials and Synthetics, 286—Industrial Organic Chemicals, and 287—Agricultural Chemicals.

Summary Statistics

		1990		1991		1992		1993		1994		1995	
		Value	N	Value	N	Value	N	Value	N	Value	N	Value	N
Establishments or enterprises	(number)	29,658	75	30,606	87	35,610	81	13,338	77	9,949	63	4,850	32
Number employed	(000)	7,656	84	6,568	89	6,868	85	6,967	83	2,943	74	2,448	56
Total Wages	($ mil.)	65,193	79	64,624	82	82,303	77	59,044	71	58,532	64	45,845	52
Wage/salary per employee	($)	13,432	78	13,806	80	16,398	76	14,898	71	15,380	64	18,189	51
Female workers	(000)	188	33	103	31	338	27	325	29	296	24	280	14
as % of total employment*	(%)	18	32	18	30	19	27	20	29	25	24	25	13
Output	($ bil.)	558	83	573	86	592	82	513	78	507	70	566	59
per employee	($ 000)	126	77	123	79	134	75	128	70	152	64	186	53
per establishment	($ 000)	20,542	68	17,549	73	20,213	69	16,447	64	18,026	52	11,971	27
Value added	($ bil.)	241	80	232	80	242	74	236	72	207	63	236	55
per employee	($ 000)	50	73	50	73	54	68	51	64	64	57	76	48
per establishment	($ 000)	8,417	63	7,301	67	8,260	61	5,604	56	7,552	45	5,063	23
Capital investment	($ mil.)	39,836	61	41,491	59	36,751	52	29,770	47	27,925	41	21,018	21
per employee	($ 000)	10	57	12	55	12	49	11	44	16	37	14	19
per establishment	($ 000)	1,747	56	1,654	54	2,199	48	1,341	41	1,470	36	1,372	17

Data presented above are drawn from the detailed tables that follow. Columns headed 'N' show the number of countries that provided valid data for inclusion. Values are not strictly comparable one year to the next or one row to the next because the number of countries that report varies from period to period and row to row. However, a general indication of magnitudes can be gleaned from reviewing this summary—especially in earlier years in which more reports are available. For detailed explanations, see Appendix I. *The average for those countries reporting both total and female employment.

Establishments and Number Engaged

Country	Establishments or Enterprises (number)						Employees per Establishment					
	1990	1991	1992	1993	1994	1995	1990	1991	1992	1993	1994	1995
Albania	-	-	-	8.000	3.000	1.000	-	-	-	263	58	17
Argentina				598	624					30	28	-
Armenia	7.000	7.000	7.000	7.000	8.000	-	1,291	1,204	-	-	-	-
Australia	269	353	379		-		63	45	40	-	-	
Austria	143	145	143	143	136	-	142	138	140	125	126	-
Azerbaijan	12	10	11	11	11		1,334	1,498	1,407	1,415	1,263	-
Bahrain	-	-	8.000	-		-	-	-	90	-		
Bangladesh	56	57	51	-			194	176	191	-		
Belize	3.000	3.000	4.000				-	-	-	-		
Benin	-	-	38	38			-	-				
Bolivia	24	26	26	25	28	28	15	14	15	15	16	16
Bosnia & Herzegovina	21	22	-	-	3.000		473	390	-	-	308	-
Bulgaria	*18	*16	*18	*79	*146	*90	1,817	1,813	1,339	290	153	271
Burundi	*2.000	*2.000	-	-			95	94	-	-		
Canada	475	446	435	436	444	-	69	72	71	71	63	-
Cape Verde	1.000	1.000	1.000	1.000	1.000		-	-	-	-	-	
Central African Republic	-	*1.000	*1.000	*1.000	*1.000	*1.000	-	-	-	-	-	-
Chile	25	28	30	31	31		145	143	163	165	181	
China	*16,350	*16,940	*17,772	-	-		221	224	220	-	-	
China (Hong Kong SAR)	360	252	248	268	188		8.333	9.127	9.274	8.582	11	
Colombia	146	147	150	159	163		114	103	100	93	91	-
Costa Rica	42	47	45	49	-		59	47	54	53	-	
Croatia	61	61	81	113	128	135	258	208	165	102	89	81
Cyprus	6.000	6.000	6.000	7.000	10	12	16	16	16	22	18	10
Czechoslovakia (Former)	*31	*35	-	-	-		2,871	2,229	-	-	-	
Czech Republic	-	*25	*25	*27	-		-	1,680	1,360	1,111	-	
Denmark	206	219	221	-		-	51	48	60	-		-
Ecuador	43	48	49	49	43	33	40	47	50	46	51	36
Egypt	52	60	46	52	-	-	977	950	1,002	742	-	-
El Salvador	-	11	4.000	4.000	6.000	4.000	-	-	109	141	85	231
Ethiopia	1.000	1.000	1.000	1.000	2.000	1.000	177	178	179	178	95	170
Finland	160	165	157	147	151	115	86	81	80	81	77	-
FYR Macedonia	*11	*21	*43	*43	*67	*82	806	382	169	153	89	68
Gabon	-	6.000	7.000	8.000	8.000	10	-	34	27	27	19	24
Germany	-	390	391	382	390	-	-	-	884	821	737	-
Germany (Western Part)	328	330	332	-	-		939	933	896	-	-	
Ghana	-	-	-	7.000	-		-	-	-	73	-	
Greece	58	58	57	-	-		110	101	87	-	-	
Grenada	-	-	1.000	1.000	-		-	-	5.000	5.000	-	
Guatemala	-	9.000	6.000	6.000	7.000	7.000	-	84	111	125	121	133
Honduras	-	-	10	8.000	8.000	8.000	-	-	43	59	65	66
Hungary	*134	*218	*152	*186	-		276	151	191	124	-	
Iceland	6.000	6.000	6.000	7.000	-		43	41	41	33	-	
India	2,220	2,175	2,345	2,496	-		102	102	107	96	-	
Indonesia	304	285	307	325	343	403	165	177	176	185	184	175
Iran (Islamic Republic of)	217	55	63	71	-		70	195	189	159	-	
Iraq	-	-	10	-			-	-	700	-		
Israel	73	84	74	83	83		108	93	105	98	99	
Italy	*337	*325	*351	*368	*363		257	262	227	210	196	
Jamaica	14	14	-	-			58	56	-	-		
Japan	1,853	1,870	1,875	1,848	1,405	1,426	97	96	98	98	114	107
Jordan	24	20	23	26	51	59	61	86	81	76	51	45
Kenya	42	33	33	31	38	38	85	106	109	117	97	101
Korea, Republic of	777	685	725	761	810	873	67	67	64	72	69	66
Kuwait	7.000	8.000	7.000	7.000	7.000	-	164	138	169	182	184	-
Kyrgyzstan	-	1.000	1.000	1.000	-		-	262	250	-	-	
Latvia	4.000	9.000	14	13	8.000	14	1,725	737	466	385	523	292
Liechtenstein	-	5.000	-	-	-		-	23	-	-	-	
Lithuania	-	-	-	*31	*9.000		-	-	-	188	576	
Malawi	*6.000	*6.000	*6.000	*6.000	*6.000	-	133	133	133	133	183	-

Continued.

Establishments and Number Engaged
- Continued -

Country	Establishments or Enterprises (number)						Employees per Establishment					
	1990	1991	1992	1993	1994	1995	1990	1991	1992	1993	1994	1995
Malaysia	103	111	116	126	120	-	95	105	107	106	102	-
Malta	5.000	5.000	5.000	5.000	4.000	-	19	19	23	21	25	-
Mauritius	11	7.000	7.000	7.000	7.000	8.000	50	71	73	72	77	67
Mexico	177	174	173	169	165	161	324	317	287	240	222	251
Mongolia	-	-	-	-	2.000	4.000	-	-	-	-	4.000	-
Mozambique	6.000	6.000	4.000	4.000	5.000	-	32	30	67	71	56	-
Nepal	1.000	1.000	-	-	-	-	-	-	-	-	-	-
Netherlands	103	108	107	119	149	-	559	527	503	356	283	-
Netherlands Antilles	-	-	-	1.000	-	-	-	-	-	17	-	-
New Zealand	*139	*146	*149	*176	-	-	32	30	29	25	-	-
Norway	47	52	43	50	52	-	176	148	171	165	160	-
Pakistan	-	120	-	-	-	-	-	148	-	-	-	-
Panama	7.000	7.000	-	-	-	-	23	23	-	-	-	-
Paraguay	-	3.000	-	-	-	-	-	50	-	-	-	-
Peru	317	322	325	-	-	-	28	23	19	-	-	-
Philippines	169	203	149	148	-	-	69	64	83	78	-	-
Poland	*65	*67	*77	*70	-	-	1,662	1,313	1,026	1,071	-	-
Portugal	*174	*190	*185	*208	*201	-	71	47	52	43	37	-
Qatar	6.000	6.000	13	10	9.000	-	247	249	121	93	169	-
Republic of Moldova	3.000	3.000	3.000	2.000	2.000	2.000	321	262	210	144	156	161
Romania	-	-	-	-	254	-	-	-	-	-	-	-
Russian Federation	-	-	-	*584	*607	*750	-	-	-	1,192	1,055	772
Saint Lucia	-	2.000	2.000	2.000	2.000	2.000	-	-	-	-	-	-
Saint Vincent & the Grenadines	-	-	-	-	-	1.000	-	-	-	-	-	7.000
Senegal	14	8.000	11	10	10	-	102	140	126	131	136	-
Sierra Leone	-	-	-	2.000	-	-	-	-	-	46	-	-
Singapore	77	77	74	75	82	-	62	63	66	70	69	-
Slovakia	-	13	13	13	13	-	-	2,323	1,982	1,675	1,528	-
Slovenia	50	*25	*38	*27	*27	*73	-	-	-	-	-	-
South Africa	-	267	-	-	-	-	-	146	-	-	-	-
Spain	525	529	575	-	-	-	79	74	64	-	-	-
Sri Lanka	20	17	11	19	-	-	49	60	70	65	-	-
Swaziland	-	1.000	1.000	1.000	-	-	-	4.000	6.000	6.000	-	-
Sweden	175	199	193	187	202	-	102	95	90	82	77	-
Thailand	184	245	-	-	-	-	63	77	-	-	-	-
Trinidad and Tobago	6.000	13	11	13	14	-	166	96	123	88	78	-
Tunisia	-	-	-	53	64	47	-	-	-	93	84	112
Turkey	72	67	92	86	89	-	433	429	305	300	275	-
Ukraine	87	128	120	102	85	88	2,195	1,477	1,550	1,784	2,000	1,864
United Kingdom	1,576	1,515	1,573	1,724	1,671	-	91	92	85	66	64	-
United Republic of Tanzania	11	10	-	-	-	-	154	760	-	-	-	-
USSR (Former)	*413	-	-	-	-	-	2,966	-	-	-	-	-
United States of America	-	-	4,426	-	-	-	-	-	90	-	-	-
Venezuela	112	123	118	113	92	91	110	109	112	114	127	127
Yugoslavia	64	89	200	263	276	283	-	302	126	94	89	86
Zambia	10	-	-	-	15	-	206	-	-	-	118	-

Employment and Compensation of Employees

Country	Number Employed/Engaged (000)						Wage/Salary per Employee ($)					
	1990	1991	1992	1993	1994	1995	1990	1991	1992	1993	1994	1995
Albania	-	-	-	2.102	0.175	0.017	-	-	-	-	-	1,187
Argentina	23	-	-	18	18	-	10,249	-	-	20,006	-	-
Armenia	*9.036	*8.425	-	-	-	-	-	-	-	-	-	-
Australia	17	*16	*15	-	-	-	28,194	33,259	31,569	-	-	-
Austria	20	20	20	18	17	17	31,675	33,586	35,567	36,494	38,342	-
Azerbaijan	16	15	15	16	14	-	1,773	2,000	-	-	-	-
Bahrain	-	-	0.718	-	-	-	-	-	15,083	-	-	-

Continued.

Employment and Compensation of Employees

- Continued -

Country	Number Employed/Engaged (000)						Wage/Salary per Employee ($)					
	1990	1991	1992	1993	1994	1995	1990	1991	1992	1993	1994	1995
Bangladesh	11	10	9.766	-	-	-	1,539	1,475	1,506	-	-	-
Bermuda	0.017	0.016	0.017	0.016	0.017	0.014	-	-	-	-	-	-
Bolivia	0.357	0.370	0.378	0.375	0.453	0.442	1,884	2,020	2,195	2,402	2,367	2,541
Bosnia & Herzegovina	9.924	8.585	-	-	0.924	-	3,483	3,506	-	-	-	-
Bulgaria	33	29	24	23	22	24	2,170	758	1,398	1,793	1,450	2,065
Burundi	0.190	0.189	-	-	-	-	2,157	2,169	-	-	-	-
Cambodia	*0.215	*0.187	-	-	-	-	-	-	-	-	-	-
Cameroon	0.961	1.128	1.181	1.055	-	-	8,454	7,712	7,764	-	-	-
Canada	33	32	31	31	28	28	-	-	39,926	37,381	38,942	-
Central African Republic	0.159	-	-	-	-	-	3,511	-	-	-	-	-
Chile	3.620	4.000	4.898	5.106	5.621	5.878	8,391	9,620	10,065	10,904	12,018	13,663
China	*3,620	*3,790	*3,910	*3,810	-	-	-	-	-	-	-	-
China (Hong Kong SAR)	*3.000	*2.300	*2.300	*2.300	*2.000	-	9,928	14,546	14,716	-	-	-
China (Taiwan Province)	63	63	64	63	64	65	14,688	16,560	18,751	18,283	18,801	21,241
Colombia	17	15	15	15	15	15	4,103	4,149	4,627	5,078	6,768	7,987
Costa Rica	2.484	2.222	2.448	2.574	-	-	4,075	3,836	4,238	4,443	-	-
Croatia	16	13	13	11	11	11	6,837	6,376	1,585	1,761	2,928	4,201
Cyprus	0.094	0.096	0.099	0.153	0.179	0.123	11,127	11,812	14,299	11,454	12,878	18,123
Czechoslovakia (Former)	89	78	-	-	-	-	2,385	1,731	-	-	-	-
Czech Republic	47	42	34	30	-	-	2,465	1,736	2,238	2,688	-	-
Denmark	11	11	13	13	13	14	34,629	34,601	-	37,731	-	-
Ecuador	1.729	2.267	2.445	2.251	2.182	1.193	4,394	3,996	4,120	4,698	2,487	4,063
Egypt	*51	*57	*46	*39	*39	*39	2,891	1,675	1,901	2,293	2,432	2,539
El Salvador	-	-	*0.438	*0.564	*0.508	*0.923	-	-	-	6,341	3,609	6,509
Ethiopia	0.177	0.178	0.179	0.178	0.190	0.170	1,815	2,017	1,541	1,061	1,106	1,042
Finland	14	13	13	12	12	-	32,788	32,143	30,266	24,557	27,985	-
FYR Macedonia	8.867	8.015	7.265	6.559	5.964	5.597	5,778	6,807	2,271	2,817	3,749	5,440
France	118	113	108	104	101	100	-	-	-	-	-	-
Gabon	-	0.204	0.187	0.215	0.155	0.241	-	16,038	19,274	16,623	12,097	11,630
Germany	-	-	*346	*314	*287	-	-	-	-	-	-	-
Germany (Western Part)	*308	*308	*297	-	-	-	-	-	-	-	-	-
Ghana	-	-	-	*0.512	*0.526	*0.541	-	-	-	1,973	1,475	1,505
Greece	*6.399	*5.881	*4.941	*4.915	*4.928	*5.066	20,572	21,283	22,568	22,187	22,743	29,185
Grenada	-	-	*0.005	*0.005	-	-	-	-	-	-	-	-
Guatemala	-	*0.754	*0.668	*0.752	*0.850	*0.932	-	1,089	1,376	1,157	988	1,508
Honduras	*0.452	*0.423	*0.431	*0.472	*0.517	*0.528	1,381	1,809	1,654	1,525	1,299	1,440
Hungary	37	33	29	23	22	20	3,130	3,375	3,873	4,082	5,352	6,455
Iceland	0.259	0.248	0.246	0.234	0.217	-	34,070	39,765	-	35,805	-	-
India	*225	*221	*252	*241	*246	*255	2,536	2,137	2,354	1,962	2,120	2,337
Indonesia	50	50	54	60	63	70	1,700	1,890	1,875	2,077	2,273	3,663
Iran (Islamic Republic of)	15	11	12	11	-	-	5,007	6,513	6,276	6,713	-	-
Iraq	-	7.286	7.000	-	-	-	-	10,021	17,354	-	-	-
Ireland	4.000	4.200	4.321	4.393	4.537	4.655	37,106	37,695	-	-	-	-
Israel	7.900	7.800	7.800	8.100	8.200	8.551	34,091	34,933	-	37,734	-	-
Italy	87	85	80	77	71	-	-	-	-	39,834	-	-
Jamaica	0.816	0.782	0.701	0.643	0.660	-	-	-	-	-	-	-
Japan	*179	*180	*184	*181	*160	*153	38,931	-	-	-	-	-
Jordan	1.474	1.725	1.866	1.985	2.626	2.672	5,187	5,015	5,109	5,048	5,001	5,268
Kenya	3.557	3.485	3.581	3.612	3.671	3.839	1,907	1,739	1,627	1,012	1,377	1,966
Korea, Republic of	52	46	47	55	56	57	13,764	14,740	16,178	17,380	19,008	23,687
Kuwait	1.147	1.106	1.186	1.276	1.287	1.279	29,849	24,224	34,573	36,014	36,141	-
Kyrgyzstan	-	*0.262	*0.250	-	-	-	-	587	-	-	-	-
Latvia	6.898	6.631	6.525	5.004	4.187	4.084	-	-	-	-	-	-
Liechtenstein	-	*0.113	-	-	-	-	-	-	-	-	-	-
Lithuania	-	-	6.523	5.832	5.185	-	-	-	568	615	1,411	-
Malawi	0.800	0.800	0.800	0.800	1.100	-	1,878	2,140	2,220	2,073	1,113	-
Malaysia	9.800	12	12	13	12	14	7,171	7,298	8,120	8,798	9,533	10,304
Malta	0.093	0.096	0.115	0.106	0.099	-	10,685	10,352	9,680	9,557	10,154	-
Mauritius	*0.547	0.500	0.512	0.503	0.541	0.536	3,973	5,505	6,691	6,579	7,101	7,835
Mexico	57	55	50	40	37	40	6,575	7,828	9,451	10,834	11,056	6,783

Continued.

Depending on the table, * means *Enterprises, Engaged,* or *Factor Values;* ± means *Basis Unspecified;* § means *shown in millions.* For additional notes and sources, see Appendix I.

Employment and Compensation of Employees
- Continued -

Country	Number Employed/Engaged (000)						Wage/Salary per Employee ($)					
	1990	1991	1992	1993	1994	1995	1990	1991	1992	1993	1994	1995
Mongolia	-	-	-	-	*0.008	-	-	-	-	-	-	-
Mozambique	0.194	0.178	0.268	0.285	0.281	-	1,060	854	1,036	1,340	893	
Netherlands	58	57	54	42	42	42	37,980	38,252	-	-	-	-
Netherlands Antilles	-	-	-	0.017	-	-	-	-	-	23,825	-	-
New Zealand	4.428	4.428	4.371	4.447	-	-	21,303	21,184	19,820	-	-	-
Norway	8.274	7.721	7.354	8.231	8.324	8.324	36,665	38,578	-	-	-	-
Pakistan	17	18	19	-	-	-	3,340	3,503	3,603	-	-	-
Panama	*0.162	*0.163	*0.172	*0.177	*0.182	*0.186	7,284	9,000	9,447	9,644	9,868	10,052
Paraguay		0.149	-	-	-	-	-	-	-	-	-	-
Peru	*8.734	*7.302	*6.159	-	-	-	5,452	6,630	6,857	-	-	-
Philippines	12	13	12	12	-	-	3,368	3,275	4,277	4,398	-	-
Poland	108	88	79	75	72	70	1,458	1,945	2,536	2,628	3,373	4,645
Portugal	*12	*8.906	*9.700	*8.930	*7.432	*7.460	-	-	-	-	-	-
Qatar	1.482	1.492	1.573	0.930	1.521	-	37,683	22,829	29,011	-	32,223	-
Republic of Moldova	0.964	0.786	0.629	0.288	0.311	0.322	-	2,777	-	-	-	518
Russian Federation	-	-	-	696	641	579	-	-	-	751	1,165	1,406
Saint Vincent & the Grenadines	-	-	-	-	-	0.007	-	-	-	-	-	-
Senegal	*1.428	*1.119	*1.382	*1.311	*1.362	-	7,961	9,348	9,199	8,399	4,799	-
Sierra Leone	-	-	-	*0.091	-	-	-	-	-	-	-	-
Singapore	*4.750	*4.814	*4.862	*5.275	*5.635	*5.685	22,676	25,667	28,690	31,115	33,442	37,496
Slovakia	-	*30	*26	*22	*20	-	-	1,717	2,112	2,349	2,626	-
South Africa	41	39	36	-	-	-	12,905	13,910	14,921	-	-	-
Spain	41	39	37	44	43	41	26,635	28,548	31,528	28,880	29,518	32,406
Sri Lanka	*0.987	*1.021	*0.771	*1.244	-	-	910	1,306	1,404	1,378	-	-
Swaziland		0.004	0.006	0.006	-	-	-	1,996	2,047	2,554	-	-
Sweden	18	*19	*17	15	15	15	28,391	30,384	33,212	26,406	27,966	31,594
Thailand	12	19	-	-	-	-	2,329	8,120	-	-	-	-
Trinidad and Tobago	0.996	1.251	1.354	1.150	1.096	-	22,112	16,683	17,204	15,194	14,830	-
Tunisia	-	-	-	4.949	5.388	5.283	-	-	-	9,274	8,428	12,478
Turkey	31	29	28	26	25	25	10,545	13,097	13,812	14,743	10,013	12,141
Ukraine	191	189	186	182	170	164	3,036	2,963	2,869	-	531	648
United Kingdom	144	140	134	114	107	110	30,060	33,287	35,692	30,992	33,791	38,324
United Republic of Tanzania	1.697	7.595	-	-	-	-	731	-	-	-	-	-
USSR (Former)	1,225	-	-	-	-	-	2,122	-	-	-	-	-
United States of America	402	401	397	382	370	365	38,010	39,426	-	-	-	-
Uruguay	*1.728	*1.567	*1.259	*1.223	*1.246	*1.220	5,837	7,767	9,823	11,743	12,840	13,157
Venezuela	12	13	13	13	12	12	6,024	10,182	10,313	12,026	10,132	12,012
Yugoslavia	-	27	25	25	25	24	-	-	-	-	2,090	917
Zambia	2.056	-	-	-	1.769	-	2,065	-	-	-	3,770	-

Female Workers: Total and Percent of Employment

Country	Female Workers (000)						Female Workers as % of Total Employment					
	1990	1991	1992	1993	1994	1995	1990	1991	1992	1993	1994	1995
Afghanistan	0.026	0.031	-	-	-	-	-	-	-	-	-	-
Albania	-	-	-	0.681	0.147	0.003	-	-	-	32.40	84.00	17.65
Argentina	-	-	-	-	1.612	-	-	-	-	-	9.09	-
Australia	2.410	-	-	-	-	-	14.18	-	-	-	-	-
Austria	3.700	4.000	3.000	3.176	2.866	-	18.23	20.00	15.00	17.72	16.75	-
Bahrain	-	-	0.038	-	-	-	-	-	5.29	-	-	-
Bangladesh	0.300	0.316	0.292	-	-	-	2.77	3.15	2.99	-	-	-
Bermuda	0.004	0.005	0.005	0.005	0.004	0.003	23.53	31.25	29.41	31.25	23.53	21.43
Bosnia & Herzegovina	1.593	1.586	-	-	-	-	16.05	18.47	-	-	-	-
Bulgaria	-	-	-	9.200	8.900	9.600	-	-	-	40.17	39.91	39.34
Canada	5.000	5.000	-	-	-	-	15.15	15.63	-	-	-	-
Chile	0.257	0.318	0.512	0.422	0.487	-	7.10	7.95	10.45	8.26	8.66	-
China (Taiwan Province)	17	17	16	16	16	16	26.43	26.14	25.73	25.34	25.22	24.90
Colombia	-	1.989	-	2.432	2.532	-	-	13.18	-	16.47	17.08	-

Continued.

Depending on the table, * means *Enterprises, Engaged,* or *Factor Values*; ± means *Basis Unspecified*; § means *shown in millions*. For additional notes and sources, see Appendix I.

269

Female Workers: Total and Percent of Employment
- Continued -

Country	Female Workers (000)						Female Workers as % of Total Employment					
	1990	1991	1992	1993	1994	1995	1990	1991	1992	1993	1994	1995
Croatia	3.910	3.080	3.420	2.880	2.850	2.640	24.83	24.23	25.54	25.11	25.00	24.20
Cyprus	0.024	0.025	0.026	0.037	0.044	-	25.53	26.04	26.26	24.18	24.58	-
Czechoslovakia (Former)	19	18	-	-	-	-	20.79	22.69	-	-	-	-
Czech Republic	17	15	11	10	-	-	36.17	35.71	32.35	33.33	-	-
Denmark	2.816	2.691	4.018	-	-	-	26.56	25.37	30.53		-	-
Egypt	-	3.337	2.035	1.873	-	-	-	5.85	4.41	4.85	-	-
El Salvador	-	-	-	0.069	0.070	0.159	-	-	-	12.23	13.78	17.23
Ethiopia	-	0.042	0.042	0.043	0.048	0.041	-	23.60	23.46	24.16	25.26	24.12
FYR Macedonia	1.649	1.666	1.123	1.146	1.355	1.130	18.60	20.79	15.46	17.47	22.72	20.19
Germany (Western Part)	54	-	-	-	-	-	17.53	-	-	-	-	-
Ghana	-	-	-	0.035	-	-	-	-	-	6.84	-	-
Hungary	14	-	9.000	9.000	-	-	37.84	-	31.03	39.13	-	-
Indonesia	-	-	-	12	13	13	-	-	-	19.86	19.98	18.91
Iran (Islamic Republic of)	-	0.266	0.322	0.366	-	-	-	2.48	2.71	3.24	-	-
Iraq	-	0.774	-	-	-	-	-	10.62	-	-	-	-
Ireland	1.200	1.300	-	-	-	-	30.00	30.95	-	-	-	-
Italy	-	9.180	8.121	7.034	6.913	-	-	10.79	10.19	9.08	9.70	-
Jordan	0.088	0.088	0.080	0.091	0.085	0.100	5.97	5.10	4.29	4.58	3.24	3.74
Kenya	0.587	0.456	0.481	0.487	-	-	16.50	13.08	13.43	13.48	-	-
Korea, Republic of	8.200	8.223	8.545	11	11	10	15.68	17.86	18.29	19.54	20.13	17.85
Malaysia	1.600	1.800	2.000	2.100	2.000	-	16.33	15.52	16.13	15.79	16.26	-
Malta	0.024	0.029	0.032	0.028	0.028	-	25.81	30.21	27.83	26.42	28.28	-
Mongolia	-	-	-	-	0.004	-	-	-	-	-	50.00	-
Morocco	-	-	-	-	-	1.173	-	-	-	-	-	-
New Zealand	0.927	0.960	0.926	0.935	-	-	20.93	21.68	21.19	21.03	-	-
Panama	0.042	-	-	-	-	-	25.93	-	-	-	-	-
Philippines	-	-	2.207	2.018	-	-	-	-	17.80	17.40	-	-
Portugal	1.391	0.931	0.961	0.861	0.615	-	11.20	10.45	9.91	9.64	8.28	-
Republic of Moldova	-	-	-	-	0.115	0.119	-	-	-	-	36.98	36.96
Senegal	0.064	-	-	-	-	-	4.48	-	-	-	-	-
Sri Lanka	0.080	0.067	0.051	0.186	-	-	8.11	6.56	6.61	14.95	-	-
Sweden	4.200	-	-	-	-	-	23.46	-	-	-	-	-
Thailand	1.612	4.023	-	-	-	-	13.95	21.25	-	-	-	-
Turkey	1.393	-	-	-	-	-	4.47	-	-	-	-	-
United Kingdom	25	-	25	-	-	-	17.35	-	18.66	-	-	-
United Republic of Tanzania	0.167	1.132	-	-	-	-	9.84	14.90	-	-	-	-
United States of America	-	-	238	231	225	225	-	-	59.95	60.47	60.81	61.64
Zambia	-	-	-	-	0.124	-	-	-	-	-	7.01	-

Output and Output per Employee

Country	Output ($ bil.)						Output per Employee ($)					
	1990	1991	1992	1993	1994	1995	1990	1991	1992	1993	1994	1995
Albania	-	-	-	*0.004	*0.001	*0.000	-	-	-	1,776	5,590	21,987
Argentina	±4.054	-	-	2.730	-	-	175,652	-	-	150,027	-	-
Armenia	0.003	0.007	0.083	0.865	0.013	-	330	880	-	-	-	-
Australia	*4.592	*4.783	*4.208	-	-	-	270,129	298,938	280,539	-	-	-
Austria	4.619	4.446	4.375	3.659	4.119	5.133	227,532	222,315	218,746	204,159	240,728	303,837
Azerbaijan	±0.328	±0.395	±0.020	±0.076	±0.028	-	20,487	26,375	1,324	4,878	2,048	-
Bahrain	-	-	±0.076	-	-	-	-	-	105,568	-	-	-
Bangladesh	0.320	0.309	0.262	-	-	-	29,495	30,847	26,830	-	-	-
Belgium	14	13	15	13	15	20	-	-	-	-	-	-
Benin	±0.002	±0.002	±0.002	±0.002	-	-	-	-	-	-	-	-
Bolivia	0.006	0.007	0.007	0.007	0.009	0.011	16,061	18,663	17,845	19,680	19,753	24,343
Bosnia & Herzegovina	±0.381	±0.364	-	-	-	-	38,373	42,386	-	-	-	-
Bulgaria	±0.648	±0.510	±0.480	±0.341	±0.447	±0.680	19,829	17,575	19,926	14,872	20,056	27,870
Burundi	±0.015	±0.014	-	-	-	-	78,422	73,140	-	-	-	-
Canada	13	11	11	11	12	13	388,008	356,769	343,744	351,811	440,413	470,481

Continued.

Depending on the table, * means *Enterprises, Engaged,* or *Factor Values;* ± means *Basis Unspecified;* § means *shown in millions.* For additional notes and sources, see Appendix I.

Output and Output per Employee
- Continued -

Country	Output ($ bil.)						Output per Employee ($)					
	1990	1991	1992	1993	1994	1995	1990	1991	1992	1993	1994	1995
Central African Republic . . .	*0.004	-	-	-	-	-	23,655	-	-	-	-	-
Chile.	0.463	0.556	0.778	1.090	1.130	1.345	128,020	138,881	158,864	213,421	200,959	228,836
China	±31	±30	±34	±40	-	-	8,463	7,959	8,702	10,552	-	-
China (Hong Kong SAR)	0.459	0.442	0.406	-	-	-	152,849	192,237	176,315	-	-	-
China (Taiwan Province) . . .	±12	±14	±15	±13	±17	±24	197,716	222,394	227,314	205,660	269,394	364,706
Colombia	1.589	1.550	1.292	1.416	1.741	2.117	95,139	102,675	86,107	95,917	117,427	138,567
Costa Rica	0.136	0.136	0.184	0.151	0.190	0.298	54,803	61,384	75,100	58,570	-	-
Croatia	±0.825	±0.993	±0.576	-	-	-	52,368	78,146	43,020	-	-	-
Cyprus	0.009	0.007	0.012	0.011	0.019	0.018	96,722	73,502	119,753	73,303	106,662	147,499
Czechoslovakia (Former) . . .	±3.090	±0.941	-	-	-	-	34,716	12,068	-	-	-	-
Denmark	*2.172	*2.074	*2.459	*2.282	*2.745	*3.525	204,894	195,593	186,839	175,519	205,747	259,940
Ecuador	0.065	0.091	0.086	0.099	0.118	0.095	37,874	40,104	35,282	44,129	54,202	79,574
Egypt	*1.109	*0.831	*0.720	*0.736	*0.800	*0.827	21,836	14,571	15,609	19,076	20,343	21,423
El Salvador.	-	±0.044	±0.027	±0.045	±0.021	±0.031	-	-	61,497	80,103	40,837	33,775
Ethiopia.	0.001	0.001	0.001	0.001	0.001	0.001	8,286	6,690	4,890	4,845	5,824	5,094
Finland	±3.534	±2.936	±2.679	±2.304	±2.827	-	257,937	220,748	212,596	193,597	243,677	-
FYR Macedonia	±0.273	±0.314	±0.202	±0.099	±0.118	±0.166	30,837	39,196	27,871	15,091	19,857	29,706
France	±30	±28	±28	±25	±27	±33	256,840	243,103	258,517	236,290	269,471	323,429
Gabon	-	*0.025	*0.028	*0.020	*0.012	*0.014	-	121,771	147,120	93,792	77,299	57,733
Ghana	-	-	-	0.024	0.020	0.021	-	-	-	47,841	37,933	38,699
Greece	*0.845	*0.701	*0.613	*0.593	*0.607	*0.791	132,119	119,196	124,120	120,721	123,133	156,165
Guatemala.	-	0.035	0.046	0.060	0.069	0.147	-	45,789	69,178	79,485	80,593	157,375
Honduras	0.006	0.007	0.006	0.007	0.008	0.008	13,520	16,168	13,726	15,397	15,040	14,872
Hungary	±1.753	±1.535	±1.095	±0.969	±1.325	±1.471	47,378	46,508	37,747	42,122	60,217	72,629
Iceland	0.041	0.048	±0.042	±0.031	-	-	158,338	193,479	172,079	131,373	-	-
India.	*11	*9.687	*11	*10	*12	*14	48,418	43,866	44,311	41,745	46,906	53,587
Indonesia	2.285	2.601	3.196	2.932	3.501	3.827	45,459	51,561	59,193	48,775	55,609	54,385
Iran (Islamic Republic of) . . .	0.516	0.426	0.368	0.616	-	-	33,929	39,701	30,948	54,480	-	-
Iraq	-	*0.631	*2.399	-	-	-	-	86,580	342,673	-	-	-
Ireland	*1.768	*1.875	*2.394	*2.291	*2.866	*3.632	441,998	446,457	553,957	521,568	631,717	780,244
Israel.	1.513	1.504	1.651	1.619	1.877	2.094	191,487	192,834	211,616	199,836	228,950	244,845
Italy	±22	±21	±20	±17	±19	-	252,591	250,019	250,090	220,611	267,503	-
Japan	±84	±93	±94	±98	±95	±110	467,253	514,191	513,526	543,891	595,404	719,816
Jordan	0.196	0.191	0.402	0.315	0.464	0.525	132,719	110,761	215,539	158,690	176,741	196,468
Kenya	*0.166	*0.144	*0.167	*0.104	*0.128	*0.184	46,619	41,308	46,631	28,736	34,894	47,913
Korea, Republic of	13	10	11	12	13	20	241,864	218,417	230,722	218,036	238,968	345,962
Kuwait	0.093	0.026	0.061	0.097	0.133	0.128	80,863	23,066	51,252	76,011	103,432	100,220
Kyrgyzstan	-	0.003	0.001	0.000	-	-	-	10,765	4,156	-	-	-
Latvia	0.002	0.004	0.058	0.059	0.053	0.074	284	601	8,901	11,761	12,646	18,062
Lithuania	-	-	0.076	0.073	0.111	-	-	-	11,669	12,464	21,501	-
Malawi	*0.027	*0.032	*0.033	*0.037	*0.025	-	33,713	40,443	41,247	46,022	22,362	-
Malaysia	*1.796	*2.314	*2.695	*3.014	*3.159	*3.856	183,307	199,504	217,331	226,627	256,830	274,361
Malta	0.004	0.005	0.006	0.005	0.005	-	45,555	50,890	52,748	46,725	54,754	-
Mauritius	0.034	0.041	0.043	0.036	0.036	0.046	62,669	81,574	83,971	71,691	66,198	85,217
Mexico	5.059	5.387	5.216	4.739	5.012	5.382	88,265	97,551	104,885	117,055	137,058	132,909
Mongolia	-	-	-	-	-	0.001	-	-	-	-	-	-
Mozambique	-	±0.002	-	±0.003	±0.003	-	-	8,875	-	9,367	11,633	-
Netherlands	19	17	16	13	14	17	326,172	291,682	299,013	299,320	335,642	399,316
New Zealand	0.789	-	-	-	-	-	178,242	-	-	-	-	-
Norway	±2.530	±2.421	±2.292	±2.303	±2.681	±3.061	305,776	313,518	311,601	279,835	322,125	367,681
Pakistan	0.774	0.848	0.921	-	-	-	44,813	47,677	49,475	-	-	-
Panama.	0.019	0.019	0.022	0.023	0.024	0.026	114,296	117,896	125,856	129,362	133,756	137,370
Peru	0.489	0.425	0.401	-	-	-	55,984	58,178	65,070	-	-	-
Philippines.	0.758	0.876	0.946	0.805	-	-	64,784	67,422	76,308	69,411	-	-
Poland	±2.561	±2.401	±2.325	±2.212	±2.761	±3.714	23,714	27,287	29,429	29,495	38,132	52,707
Portugal	1.533	1.288	1.427	1.148	1.209	1.362	123,464	144,605	147,082	128,598	162,711	182,578
Puerto Rico	±0.115	±0.111	±0.115	±0.100	±0.100	±0.111	-	-	-	-	-	-
Qatar	0.343	0.424	0.388	0.359	0.366	-	231,718	284,300	246,956	386,092	240,588	-
Republic of Moldova	0.000	0.021	0.001	0.001	0.001	0.001	0.045	27,167	1,486	1,835	1,790	3,771
Romania	±1.271	±1.153	±0.866	±1.027	-	-	-	-	-	-	-	-

Continued.

Depending on the table, * means *Enterprises, Engaged,* or *Factor Values*; ± means *Basis Unspecified*; § means *shown in millions.* For additional notes and sources, see Appendix I.

271

Output and Output per Employee

- Continued -

Country	Output ($ bil.)						Output per Employee ($)					
	1990	1991	1992	1993	1994	1995	1990	1991	1992	1993	1994	1995
Russian Federation	-	50	58	6.010	8.424	12	-	-	-	8,634	13,152	20,121
Senegal	0.129	-	-	-	-	-	90,152	-	-	-	-	-
Singapore	*1.737	*1.790	*1.623	*1.643	*2.211	*2.510	365,777	371,759	333,718	311,407	392,370	441,454
Slovakia	-	±1.123	±0.901	±0.529	±0.618	-	-	37,201	34,960	24,314	31,102	-
Slovenia	-	-	-	0.774	1.085	1.362	-	-	-	-	-	-
South Africa	±3.613	±3.678	±3.646	±3.689	±3.754	±4.232	88,113	94,316	101,274	-	-	-
Spain	*12	*11	*10	±8.833	±11	±14	281,772	280,604	277,049	198,894	260,209	337,358
Sri Lanka	0.020	0.021	0.021	0.032	-	-	20,333	20,895	27,699	25,385	-	-
Swaziland	-	*0.000	*0.000	*0.000	-	-	-	32,021	19,066	19,049	-	-
Sweden	*4.398	*5.100	*4.963	*3.303	*3.950	*4.478	245,682	271,104	286,183	216,506	255,395	290,754
Thailand	0.405	2.423	-	-	-	-	35,068	127,977	-	-	-	-
Trinidad and Tobago	±0.328	±0.384	±0.312	±0.345	±0.577	-	329,246	306,691	230,463	299,719	526,342	-
Tunisia	1.037	1.043	1.045	0.878	1.055	1.292	-	-	-	177,480	195,837	244,527
Turkey	3.649	3.522	3.793	3.811	3.274	4.093	117,114	122,580	135,135	147,526	133,554	162,356
Ukraine	*5.596	*5.159	*7.463	*1.420	*1.796	*1.823	29,298	27,298	40,123	7,802	10,563	11,117
United Kingdom	*36	*35	35	27	29	34	252,183	248,154	260,798	240,675	274,763	312,883
United Republic of Tanzania	*0.026	*0.013	-	-	-	-	15,471	1,769	-	-	-	-
USSR (Former)	±25	-	-	-	-	-	20,663	-	-	-	-	-
United States of America	*159	*156	*159	*159	*171	*189	395,821	388,279	401,481	416,084	461,486	518,608
Uruguay	0.184	0.140	0.117	0.126	0.139	0.140	106,625	89,154	93,297	102,897	111,893	114,351
Venezuela	1.041	1.403	1.377	1.459	1.531	2.280	84,620	104,736	104,314	113,411	131,104	197,098
Yugoslavia	±0.123	±0.135	±0.175	-	±0.383	±0.188	-	5,008	6,907	-	15,642	7,724
Zambia	0.104	-	-	-	.028	-	50,592	-	-	-	16,083	-

Value Added and Value Added per Employee

Country	Value Added ($ bil.)						Value Added per Employee ($)					
	1990	1991	1992	1993	1994	1995	1990	1991	1992	1993	1994	1995
Argentina	±1.844	-	-	0.655	-	-	79,916	-	-	35,983	-	-
Armenia	-0.001	-0.000	-	-	-	-	-153.102	-50.697	-	-	-	-
Australia	*1.660	*1.808	*1.785	-	-	-	97,656	112,972	118,967	-	-	-
Austria	1.277	1.215	1.218	1.071	1.239	1.539	62,904	60,749	60,920	59,757	72,427	91,102
Bangladesh	0.134	0.149	0.106	-	-	-	12,334	14,866	10,805	-	-	-
Belgium	*4.483	*4.278	*4.677	*4.066	*4.771	*6.189	-	-	-	-	-	-
Bolivia	0.002	0.003	0.003	0.003	0.003	0.004	6,876	8,214	8,594	8,499	7,683	9,204
Bosnia & Herzegovina	±0.162	±0.221	-	-	-	-	16,328	25,796	-	-	-	-
Bulgaria	-	±0.065	±0.073	±0.057	-	-	-	2,241	3,010	2,499	-	-
Burundi	0.000	0.000	-	-	-	-	2,406	2,408	-	-	-	-
Cameroon	0.017	0.020	0.020	-	-	-	17,639	17,595	16,664	-	-	-
Canada	4.808	3.823	3.773	3.883	4.855	5.226	145,698	119,468	121,698	125,272	173,393	185,851
Central African Republic	*0.001	-	-	-	-	-	7,762	-	-	-	-	-
Chile	0.247	0.280	0.389	0.410	0.472	0.561	68,223	70,085	79,342	80,383	83,931	95,422
China	±8.459	±8.651	±9.913	±13	-	-	2,337	2,283	2,535	3,446	-	-
China (Hong Kong SAR)	0.064	0.110	0.096	-	-	-	21,267	47,835	41,677	-	-	-
China (Taiwan Province)	3.371	3.703	3.816	3.552	4.720	6.586	53,756	58,456	59,679	56,246	73,233	100,729
Colombia	0.522	0.554	0.536	0.578	0.702	0.855	31,236	36,695	35,723	39,121	47,356	55,991
Costa Rica	0.033	0.034	0.045	0.036	0.045	0.070	13,281	15,132	18,523	14,054	-	-
Cote d'Ivoire	1.370	1.290	1.447	1.261	-	-	-	-	-	-	-	-
Croatia	±0.152	±0.355	±0.146	-	-	-	9,635	27,953	10,907	-	-	-
Cyprus	0.003	0.003	0.004	0.003	0.005	0.006	33,987	26,413	36,453	20,555	29,541	50,215
Czechoslovakia (Former)	±0.698	-	-	-	-	-	7,843	-	-	-	-	-
Denmark	*1.107	*1.118	*1.328	*1.232	*1.487	*1.912	104,390	105,396	100,899	94,780	111,416	141,016
Ecuador	0.017	0.025	0.025	0.034	0.040	0.027	10,104	10,933	10,364	15,025	18,273	22,714
Egypt	*0.338	*0.220	*0.243	*0.232	*0.252	*0.261	6,663	3,865	5,262	6,013	6,416	6,757
El Salvador	-	0.022	-	0.013	0.005	0.018	-	-	-	23,318	10,247	19,855
Ethiopia	0.001	0.001	0.001	0.001	0.001	0.001	5,571	4,125	3,329	3,260	3,564	3,550
Finland	±1.371	±1.002	±1.077	±0.883	±1.054	-	100,098	75,337	85,471	74,173	90,836	-
FYR Macedonia	±0.092	±0.085	±0.080	±0.004	±0.045	±0.064	10,403	10,583	11,015	678	7,569	11,458

Continued.

Depending on the table, * means *Enterprises, Engaged,* or *Factor Values*; ± means *Basis Unspecified*; § means *shown in millions*. For additional notes and sources, see Appendix I.

Value Added and Value Added per Employee

- Continued -

Country	Value Added ($ bil.)						Value Added per Employee ($)					
	1990	1991	1992	1993	1994	1995	1990	1991	1992	1993	1994	1995
France	±11	±9.361	±8.526	±7.070	±7.872	±10	91,908	82,550	78,800	67,782	77,713	101,956
Gabon	-	*0.009	*0.011	*0.008	*0.005	*0.005	-	45,508	56,347	38,683	32,432	21,023
Germany (Western Part)	36	33	34	29	-	-	115,396	106,842	114,385			
Ghana	-	-	-	0.005	0.004	0.005	-	-	-	10,100	8,519	8,691
Greece	*0.290	*0.244	*0.231	*0.225	*0.230	*0.301	45,319	41,554	46,807	45,709	46,710	59,506
Honduras	*0.002	*0.003	*0.002	*0.002	*0.002	*0.002	5,018	6,230	5,297	5,262	4,034	4,475
Hungary	±0.404	±0.270	±0.189	±0.202	±0.216	±0.240	10,928	8,180	6,520	8,772	9,808	11,829
Iceland	0.017	0.017	±0.017	±0.012	-	-	64,880	68,649	70,647	49,289	-	-
India	*1.833	*1.368	*2.348	*2.546	*2.931	*3.496	8,131	6,196	9,316	10,576	11,932	13,689
Indonesia	*0.687	*0.945	*1.283	*1.114	*1.309	*1.430	13,664	18,735	23,754	18,531	20,789	20,325
Iran (Islamic Republic of)	0.239	0.225	0.182	0.413	-	-	15,738	20,943	15,282	36,585	-	-
Iraq	-	*0.230	*0.602	-	-	-	-	31,544	85,944	-	-	-
Ireland	*0.757	*0.840	*1.078	*1.034	*1.299	*1.650	189,345	200,054	249,431	235,403	286,207	354,427
Israel	0.498	0.486	0.553	0.530	0.632	0.696	63,034	62,328	70,904	65,434	77,032	81,390
Italy	±5.906	±5.202	±5.430	±4.705	±5.247	-	68,182	61,129	68,136	60,742	73,624	-
Japan	±38	±42	±46	±48	±48	±56	212,714	234,289	248,245	267,399	301,585	366,197
Jordan	0.044	0.049	0.039	0.025	0.056	0.076	29,956	28,153	20,727	12,405	21,260	28,268
Kenya	*0.017	*0.017	*0.015	*0.009	*0.014	*0.016	4,907	4,798	4,160	2,482	3,693	4,254
Korea, Republic of	4.182	4.116	4.498	5.110	5.631	7.982	79,952	89,379	96,284	93,682	100,504	138,969
Kuwait	0.043	0.002	0.029	0.063	0.098	0.082	37,826	1,675	24,219	49,474	76,026	63,898
Latvia	-	-	-	*0.019	*0.013	*0.018	-	-	-	3,787	3,221	4,499
Malawi	*0.006	*0.008	*0.008	*0.020	*0.010	-	7,421	9,810	10,268	24,927	8,689	-
Malaysia	*0.748	*1.085	*1.172	*1.310	*1.127	*1.377	76,294	93,501	94,499	98,518	91,657	97,976
Malta	0.002	0.002	0.003	0.002	0.003	-	25,576	25,768	24,036	22,523	26,215	-
Mauritius	0.012	0.016	0.014	0.014	0.011	0.016	21,353	31,728	27,321	27,712	20,779	29,116
Mexico	2.090	2.202	2.259	1.970	2.147	2.307	36,459	39,873	45,427	48,667	58,703	56,976
Mongolia	-	-	-	-	-	0.000	-	-	-	-	-	-
Morocco	±0.403	-	-	-	-	-	-	-	-	-	-	-
Netherlands	5.592	4.532	4.280	3.445	3.807	4.501	97,091	79,595	79,461	81,308	90,316	106,948
New Zealand	0.249	-	-	-	-	-	56,223	-	-	-	-	-
Norway	±0.811	±0.694	±0.642	±0.673	±0.825	±0.942	98,064	89,942	87,310	81,724	99,136	113,163
Pakistan	0.342	0.366	0.409	-	-	-	19,823	20,579	21,961	-	-	-
Panama	0.006	0.007	0.007	0.008	0.008	0.009	39,759	40,276	42,650	43,685	45,006	46,097
Peru	0.237	0.179	0.162	-	-	-	27,112	24,463	26,285	-	-	-
Philippines	0.277	0.343	0.384	0.360	-	-	23,710	26,361	30,956	31,041	-	-
Poland	±1.056	±0.627	±0.755	±0.602	±0.742	±0.992	9,777	7,119	9,561	8,026	10,250	14,076
Portugal	0.432	0.289	0.333	0.286	0.294	0.332	34,815	32,491	34,330	31,987	39,621	44,536
Puerto Rico	±0.039	±0.054	±0.044	±0.034	±0.035	±0.040	-	-	-	-	-	-
Qatar	0.264	0.327	0.219	0.190	0.240	-	178,330	219,486	139,371	204,715	157,863	-
Romania	±0.111	±0.234	±0.180	±0.209	-	-	-	-	-	-	-	-
Russian Federation	-	-	-	±2.692	±3.747	±3.602	-	-	-	3,867	5,850	6,223
Saint Lucia	-	±0.000	±0.000	±0.000	±0.000	±0.000	-	-	-	-	-	-
Senegal	0.013	0.036	0.033	0.027	0.052	-	8,897	32,400	23,723	20,863	38,053	-
Singapore	*0.584	*0.556	*0.513	*0.540	*0.754	*0.857	122,878	115,517	105,516	102,439	133,796	150,694
Slovakia	-	-	-	±0.131	±0.167	-	-	-	-	6,012	8,409	-
Slovenia	-	-	-	0.268	0.389	0.458	-	-	-	-	-	-
South Africa	*0.932	*1.191	*1.252	*1.135	*1.155	*1.302	22,728	30,541	34,790	-	-	-
Spain	*3.427	*3.344	*3.133	±2.375	±3.177	±3.964	83,095	85,973	85,107	53,470	73,327	97,462
Sri Lanka	0.010	0.005	0.009	0.011	-	-	9,888	5,277	11,401	8,754	-	-
Swaziland	-	*0.000	*0.000	*0.000	-	-	-	3,628	2,690	3,320	-	-
Sweden	*1.983	*1.386	*1.394	*1.147	*1.369	*1.548	110,807	73,679	80,416	75,180	88,486	100,484
Thailand	0.154	1.013	-	-	-	-	13,309	53,505	-	-	-	-
Trinidad and Tobago	±0.109	±0.163	±0.086	±0.126	±0.270	-	109,261	130,531	63,794	109,981	246,609	-
Tunisia	0.039	0.093	0.094	0.096	0.147	0.199	-	-	-	19,487	27,203	37,618
Turkey	1.421	1.322	1.567	1.483	1.557	1.953	45,613	46,016	55,829	57,395	63,532	77,488
United Kingdom	*14	*14	*14	*11	*12	*14	98,462	98,078	105,888	97,584	112,937	130,206
United Republic of Tanzania	*0.011	*0.005	-	-	-	-	6,235	691	-	-	-	-
United States of America	*73	*70	*73	*73	*79	*90	182,786	174,464	182,801	190,414	213,108	245,970
Uruguay	0.068	0.059	0.050	0.055	0.061	0.061	39,070	37,507	39,974	45,188	49,226	50,352
Venezuela	0.443	0.648	0.646	0.569	0.562	1.924	36,043	48,366	48,952	44,242	48,179	166,260

Continued.

Value Added and Value Added per Employee

- Continued -

Country	Value Added ($ bil.)						Value Added per Employee ($)					
	1990	1991	1992	1993	1994	1995	1990	1991	1992	1993	1994	1995
Yugoslavia	±0.029	±0.040	±0.101	-	±0.171	±0.068	-	1,491	3,986	-	6,964	2,784
Zambia	0.011	-	-	-	0.015	-	5,434	-	-	-	8,524	-

Capital Investment

Country	Gross Fixed Capital Formation ($ mil.)						Per Establishment ($ mil)					
	1990	1991	1992	1993	1994	1995	1990	1991	1992	1993	1994	1995
Argentina	-	-	-	111	-	-	-	-	-	0.186	-	-
Australia	328	-	-	-	-	-	1.220	-	-	-	-	-
Austria	323	363	435	271	204	-	2.257	2.501	3.041	1.898	1.502	-
Bahrain	-	-	6.960	-	-	-	-	-	0.870	-	-	-
Bangladesh	77	62	52	-	-	-	1.373	1.084	1.013	-	-	-
Barbados	-	-	-	-	-0.942	-	-	-	-	-	-	-
Belgium	2,601	2,224	2,057	1,696	1,622	2,689	-	-	-	-	-	-
Bolivia	0.423	0.929	0.541	0.460	2.798	-	0.018	0.036	0.021	0.018	0.100	-
Bosnia & Herzegovina	21	-	-	-	-	-	0.993	-	-	-	-	-
Bulgaria	54	36	589	105	10	25	3.026	2.222	33	1.323	0.072	0.281
Canada	1,196	-	-	-	-	-	2.517	-	-	-	-	-
Chile	23	47	30	60	69	-	0.909	1.690	0.996	1.951	2.231	-
China (Hong Kong SAR)	-	19	-	-	-	-	-	0.075	-	-	-	-
Colombia	115	79	32	8.087	-18.018	-	0.785	0.539	0.211	0.051	-0.111	-
Croatia	22	8.815	6.780	6.077	4.060	6.932	0.361	0.145	0.084	0.054	0.032	0.051
Cyprus	1.470	0.795	1.644	0.833	0.916	0.409	0.245	0.132	0.274	0.119	0.092	0.034
Czechoslovakia (Former)	378	-	-	-	-	-	12	-	-	-	-	-
Denmark	160	186	-	-	-	-	0.778	0.848	-	-	-	-
Ecuador	6.324	15	9.035	3.796	11	19	0.147	0.321	0.184	0.077	0.253	0.570
Egypt	277	307	125	27	-	-	5.334	5.113	2.707	0.510	-	-
El Salvador	-	2.475	-	1.860	0.592	2.144	-	0.225	-	0.465	0.099	0.536
Ethiopia	0.036	0.055	0.015	0.013	0.099	0.061	0.036	0.055	0.015	0.013	0.050	0.061
Finland	333	184	149	182	189	-	2.081	1.112	0.947	1.237	1.255	-
FYR Macedonia	13	2.815	15	1.297	0.847	2.037	1.155	0.134	0.358	0.030	0.013	0.025
France	2,717	2,603	2,217	1,602	1,685	2,106	-	-	-	-	-	-
Gabon	-	0.510	0.763	0.805	0.267	1.793	-	0.085	0.109	0.101	0.033	0.179
Greece	21	18	15	-	-	-	0.359	0.311	0.265	-	-	-
Hungary	131	66	48	80	-	-	0.976	0.301	0.315	0.432	-	-
India	1,137	986	905	1,122	-	-	0.512	0.453	0.386	0.449	-	-
Indonesia	177	384	144	165	701	362	0.582	1.346	0.470	0.509	2.043	0.899
Iran (Islamic Republic of)	42	40	22	29	-	-	0.195	0.725	0.354	0.409	-	-
Israel	99	63	132	279	390	-	1.352	0.747	1.780	3.363	4.701	-
Italy	1,665	1,662	1,606	891	760	-	4.941	5.114	4.575	2.422	2.094	-
Japan	6,430	7,386	8,504	8,354	7,582	-	3.470	3.950	4.535	4.521	5.397	-
Jordan	1.741	8.924	2.437	4.237	46	20	0.073	0.446	0.106	0.163	0.892	0.336
Korea, Republic of	2,596	5,598	2,406	2,172	2,533	3,160	3.340	8.173	3.319	2.854	3.128	3.620
Kuwait	9.082	62	7.898	7.841	13	-	1.297	7.773	1.128	1.120	1.861	-
Latvia	0.064	0.036	0.126	-	0.039	67	0.016	0.004	0.009	-	0.005	4.799
Malawi	3.042	1.427	2.692	3.929	6.788	-	0.507	0.238	0.449	0.655	1.131	-
Malaysia	118	669	1,099	847	196	-	1.145	6.024	9.476	6.724	1.632	-
Malta	0.025	0.015	0.270	0.463	0.153	-	0.005	0.003	0.054	0.093	0.038	-
Mexico	178	267	-	-	-	-	1.005	1.533	-	-	-	-
Netherlands	1,674	1,372	1,355	908	-	-	16	13	13	7.633	-	-
New Zealand	18	-	-	-	-	-	0.129	-	-	-	-	-
Norway	157	215	120	92	107	-	3.345	4.129	2.791	1.849	2.060	-
Pakistan	-	39	-	-	-	-	-	0.329	-	-	-	-
Panama	0.010	0.056	-	-	-	-	0.001	0.008	-	-	-	-
Peru	11	13	7.908	-	-	-	0.034	0.041	0.024	-	-	-
Philippines	35	22	24	29	-	-	0.206	0.111	0.162	0.195	-	-
Poland	178	316	226	156	-	-	2.739	4.721	2.932	2.229	-	-
Portugal	74	-100.353	194	62	50	-	0.427	-0.528	1.049	0.298	0.246	-

Continued.

Depending on the table, * means *Enterprises*, *Engaged*, or *Factor Values*; ± means *Basis Unspecified*; § means *shown in millions*. For additional notes and sources, see Appendix I.

Capital Investment
- Continued -

Country	Gross Fixed Capital Formation ($ mil.)						Per Establishment ($ mil)					
	1990	1991	1992	1993	1994	1995	1990	1991	1992	1993	1994	1995
Romania	-	-	-	61	41	-	-	-	-	-	0.160	-
Senegal	28	-	-	-	-	-	1.992	-	-	-	-	-
Singapore	79	130	72	337	153	-	1.032	1.693	0.975	4.491	1.866	-
Slovenia	15	22	12	13	22	61	0.302	0.884	0.317	0.468	0.829	0.839
Spain	797	729	492	435	403	692	1.519	1.377	0.856	-	-	-
Sri Lanka	0.774	1.327	1.412	2.606	-	-	0.039	0.078	0.128	0.137	-	-
Swaziland	-	0.008	0.002	-	-	-	-	0.008	0.002	-	-	-
Thailand	271	162	-	-	-	-	1.474	0.662	-	-	-	-
Trinidad and Tobago	12	34	46	27	198	-	2.041	2.639	4.186	2.100	14	-
Tunisia	15	14	12	33	46	36	-	-	-	0.630	0.726	0.765
Turkey	204	124	179	175	134	-	2.827	1.850	1.942	2.039	1.510	-
Ukraine	163	153	83	15	42	45	1.878	1.197	0.695	0.144	0.499	0.509
United Kingdom	3,709	3,688	2,894	-	1,551	-	2.353	2.435	1.840	-	0.928	-
United Republic of Tanzania	5.122	7.273	-	-	-	-	0.466	0.727	-	-	-	-
United States of America	11,010	11,100	10,186	9,274	9,026	10,826	-	-	2.301	-	-	-
Uruguay	2.934	0.664	2.232	2.775	-	-	-	-	-	-	-	-
Venezuela	88	89	219	114	129	891	0.785	0.724	1.860	1.006	1.399	9.795
Yugoslavia	8.685	7.066	3.425	-	11	4.697	0.136	0.079	0.017	-	0.042	0.017
Zambia	27	-	-	-	-	-	2.669	-	-	-	-	-

Indexes of Industrial Production

Country	Index of Industrial Production (1990=100)						Index of Employment (1990=100)					
	1990	1991	1992	1993	1994	1995	1990	1991	1992	1993	1994	1995
Albania	100	-	-	-	-	-	-	-	-	-	-	-
Algeria	100	97	80	83	97	89	-	-	-	-	-	-
Angola	100	-	-	-	-	-	-	-	-	-	-	-
Argentina	100	99	-	-	-	-	100	-	-	79	77	-
Armenia	100	-	-	-	-	-	100	93	-	-	-	-
Australia	100	100	101	110	-	-	100	94	88	-	-	-
Austria	100	101	101	98	106	114	100	99	99	88	84	83
Azerbaijan	-	-	-	-	-	-	100	94	97	97	87	-
Bahrain	100	100	-	-	-	-	-	-	-	-	-	-
Bangladesh	100	94	107	126	146	131	100	92	90	-	-	-
Barbados	100	96	76	73	68	94	-	-	-	-	-	-
Belgium	100	100	112	108	107	112	-	-	-	-	-	-
Benin	100	-	-	-	-	-	-	-	-	-	-	-
Bermuda	-	-	-	-	-	-	100	94	100	94	100	82
Bhutan	100	-	-	-	-	-	-	-	-	-	-	-
Bolivia	100	102	101	99	106	109	100	104	106	105	127	124
Bosnia & Herzegovina	-	-	-	-	-	-	100	87	-	-	9	-
Brazil	100	92	92	96	102	102	-	-	-	-	-	-
Bulgaria	100	82	68	60	82	96	100	89	74	70	68	75
Burkina-Faso	100	177	94	-	-	-	-	-	-	-	-	-
Burundi	100	87	64	-	-	-	100	99	-	-	-	-
Cambodia	100	-	-	-	-	-	100	87	-	-	-	-
Cameroon	100	-	-	-	-	-	100	117	123	110	-	-
Canada	100	87	94	99	109	115	100	97	94	94	85	85
Central African Republic	100	-	-	-	-	-	100	-	-	-	-	-
Chile	100	108	116	112	119	124	100	110	135	141	155	162
China	100	-	-	-	-	-	100	105	108	105	-	-
China (Hong Kong SAR)	100	94	90	81	74	70	100	77	77	77	67	-
China (Taiwan Province)	100	113	129	137	164	173	100	101	102	101	103	104
Colombia	100	104	101	102	104	113	100	90	90	88	89	91
Costa Rica	100	99	133	102	129	163	100	89	99	104	-	-
Cote d'Ivoire	100	98	96	101	120	119	-	-	-	-	-	-
Croatia	100	-	-	-	-	-	100	81	85	73	72	69
Cyprus	100	98	135	128	132	141	100	102	105	163	190	131

Continued.

Depending on the table, * means *Enterprises, Engaged,* or *Factor Values*; ± means *Basis Unspecified*; § means *shown in millions*. For additional notes and sources, see Appendix I.

275

Indexes of Industrial Production

- Continued -

Country	Index of Industrial Production (1990=100)						Index of Employment (1990=100)					
	1990	1991	1992	1993	1994	1995	1990	1991	1992	1993	1994	1995
Czechoslovakia (Former) . . .	100	68	63	-	-	-	100	88	-	-	-	-
Czech Republic	100	-	-	-	-	-	100	89	72	64	-	-
Dem. Rep. of the Congo . .	100	80	-	-	-	-	-	-	-	-	-	-
Denmark	100	102	110	108	120	133	100	100	124	123	126	128
Dominican Republic.	100	-	-	-	-	-	-	-	-	-	-	-
Ecuador	100	121	128	132	-	-	100	131	141	130	126	69
Egypt	100	101	84	91	95	91	100	112	91	76	77	76
El Salvador.	100	105	-	-	-	-	-	-	-	-	-	-
Ethiopia and Eritrea . . .	100	-	-	-	-	-	-	-	-	-	-	-
Ethiopia.	-	-	-	-	-	-	100	101	101	101	107	96
Fiji	100	100	100	100	100	100	-	-	-	-	-	-
Finland	100	92	98	100	114	117	100	97	92	87	85	-
FYR Macedonia	100	-	-	-	-	-	100	90	82	74	67	63
France	100	100	104	103	110	109	100	96	91	88	86	85
Gabon	100	-	-	-	-	-	-	-	-	-	-	-
Gambia.	100	-	-	-	-	-	-	-	-	-	-	-
Germany (Eastern Part) . . .	100	-	-	-	-	-	-	-	-	-	-	-
Germany (Western Part) . . .	100	97	97	94	103	103	100	100	97	-	-	-
Ghana	100	-	-	-	-	-	-	-	-	-	-	-
Greece	100	95	78	76	77	88	100	92	77	77	77	79
Guatemala	100	-	-	-	-	-	-	-	-	-	-	-
Guyana	100	100	100	100	100	100	-	-	-	-	-	-
Haiti	100	121	113	110	129	-	-	-	-	-	-	-
Honduras	100	146	173	194	219	310	100	94	95	104	114	117
Hungary	100	74	47	-	-	-	100	89	78	62	59	55
Iceland	100	100	127	-	-	-	100	96	95	90	84	-
India	100	104	120	130	139	158	100	98	112	107	109	113
Indonesia	100	103	93	99	109	100	100	100	107	120	125	140
Iran (Islamic Republic of) . .	100	101	105	107	149	161	100	71	78	74	-	-
Iraq	100	100	102	-	-	-	-	-	-	-	-	-
Ireland	100	122	143	157	188	217	100	105	108	110	113	116
Israel.	100	104	115	127	140	145	100	99	99	103	104	108
Italy	100	94	95	90	-	-	100	98	92	89	82	-
Jamaica	100	96	88	-	-	-	100	96	86	79	81	-
Japan	100	102	102	100	105	112	100	101	103	101	89	85
Jordan	100	90	81	71	66	59	100	117	127	135	178	181
Kenya	100	118	118	124	108	106	100	98	101	102	103	108
Korea, Republic of	100	116	138	152	165	176	100	88	89	104	107	110
Kuwait	100	91	79	-	-	-	100	96	103	111	112	112
Kyrgyzstan	100	-	-	-	-	-	-	-	-	-	-	-
Lao P.D.R.	100	-	-	-	-	-	-	-	-	-	-	-
Latvia	100	-	-	-	-	-	100	96	95	73	61	59
Liberia	100	-	-	-	-	-	-	-	-	-	-	-
Libyan Arab Jamahiriya . .	100	100	-	-	-	-	-	-	-	-	-	-
Lithuania	100	-	-	-	-	-	-	-	-	-	-	-
Macau	100	100	100	-	-	-	-	-	-	-	-	-
Madagascar	100	-	-	-	-	-	-	-	-	-	-	-
Malawi	-	-	-	-	-	-	100	100	100	100	138	-
Malaysia	100	128	119	126	138	156	100	118	127	136	126	143
Malta	100	174	206	267	-	-	100	103	124	114	106	-
Mauritius	100	100	-	-	-	-	100	91	94	92	99	98
Mexico	100	106	109	106	112	109	100	96	87	71	64	71
Morocco	100	91	93	98	87	85	-	-	-	-	-	-
Mozambique	100	-	-	-	-	-	100	92	138	147	145	-
Myanmar	100	68	69	-	-	-	-	-	-	-	-	-
Namibia	100	-	-	-	-	-	-	-	-	-	-	-
Nepal	100	-	-	-	-	-	-	-	-	-	-	-
Netherlands	100	97	98	99	108	111	100	99	94	74	73	73
New Zealand	100	82	82	-	-	-	100	100	99	100	-	-
Nicaragua	100	-	-	-	-	-	-	-	-	-	-	-

Continued.

Depending on the table, * means *Enterprises, Engaged,* or *Factor Values;* ± means *Basis Unspecified;* § means *shown in millions.* For additional notes and sources, see Appendix I.

Indexes of Industrial Production

- Continued -

Country	Index of Industrial Production (1990=100)						Index of Employment (1990=100)					
	1990	1991	1992	1993	1994	1995	1990	1991	1992	1993	1994	1995
Nigeria	100	110	110	-	-	-	-	-	-	-	-	-
Norway	100	95	92	100	102	102	100	93	89	99	101	101
Pakistan	100	102	106	126	135	103	100	103	108	-	-	-
Panama.	100	103	114	120	126	131	100	101	106	109	112	115
Papua New Guinea	100	-	-	-	-	-	-	-	-	-	-	-
Paraguay	100	101	92	256	245	246	-	-	-	-	-	-
Peru	100	101	88	99	107	113	100	84	71	-	-	-
Philippines.	100	109	117	136	147	168	100	111	106	99	-	-
Poland	100	110	97	103	121	137	100	81	73	69	67	65
Portugal	100	83	76	69	69	68	100	72	78	72	60	60
Puerto Rico	100	-	-	-	-	-	-	-	-	-	-	-
Qatar	100	-	-	-	-	-	100	101	106	63	103	-
Republic of Moldova	100	-	-	-	-	-	100	82	65	30	32	33
Romania	100	65	57	59	54	59	-	-	-	-	-	-
Russian Federation	100	-	-	-	-	-	-	-	-	-	-	-
Saudi Arabia	100	107	-	-	-	-	-	-	-	-	-	-
Senegal.	-	-	-	-	-	-	100	78	97	92	95	-
Singapore	100	102	97	102	121	123	100	101	102	111	119	120
Slovakia	100	-	-	-	-	-	-	-	-	-	-	-
Slovenia	100	82	79	73	87	91	-	-	-	-	-	-
Somalia.	100	-	-	-	-	-	-	-	-	-	-	-
South Africa	100	97	91	92	96	103	100	95	88	-	-	-
Spain	100	95	95	94	107	110	100	94	89	108	105	99
Sri Lanka	100	88	98	137	135	163	100	103	78	126	-	-
Sudan	100	100	100	-	-	-	-	-	-	-	-	-
Sweden.	100	114	126	134	133	136	100	105	97	85	86	86
Switzerland	100	101	106	113	130	143	-	-	-	-	-	-
Syrian Arab Republic . . .	100	99	100	106	110	116	-	-	-	-	-	-
Thailand	100	121	146	-	-	-	100	164	-	-	-	-
Tonga	100	-	-	-	-	-	-	-	-	-	-	-
Trinidad and Tobago	100	104	98	101	105	117	100	126	136	115	110	-
Tunisia	100	102	102	103	116	122	-	-	-	-	-	-
Turkey	100	96	97	98	95	103	100	92	90	83	79	81
Uganda.	100	105	137	185	209	279	-	-	-	-	-	-
Ukraine	100	-	-	-	-	-	100	99	97	95	89	86
United Arab Emirates . . .	100	-	-	-	-	-	-	-	-	-	-	-
United Kingdom	100	107	109	107	111	122	100	97	93	79	74	76
United Republic of Tanzania . .	100	97	52	44	38	-	100	448	-	-	-	-
USSR (Former)	100	-	-	-	-	-	100	-	-	-	-	-
United States of America . . .	100	96	99	97	101	102	100	100	99	95	92	91
Uruguay	100	97	97	98	105	97	100	91	73	71	72	71
Venezuela	100	-	-	-	-	-	100	109	107	105	95	94
Yugoslavia.	100	-	-	-	-	-	-	-	-	-	-	-
Yugoslavia (Former). . . .	100	76	-	-	-	-	-	-	-	-	-	-
Zambia	100	101	96	90	68	60	100	-	-	-	86	-
Zimbabwe	100	100	87	82	94	84	-	-	-	-	-	-

Depending on the table, * means *Enterprises, Engaged,* or *Factor Values;* ± means *Basis Unspecified;* § means *shown in millions.* For additional notes and sources, see Appendix I.

277

Representative Companies in Sector

Name	Address	Tele-phone	Fax	Employ-ment	Y
Alcan Aluminium Ltd.	1188 Sherbrooke St. W, Montreal, Quebec H3A 3G1, Canada	514-848-8000	514-848-8115	11,000	98
Elf Aquitaine Inc.	280 Park Ave., 36th Fl., New York, New York 10017, U.S.A.	212-922-3000	212-922-3001	10,000	
Olin Corp.	PO Box 4500, Norwalk, Connecticut 06856-4500, U.S.A.	203-750-3000	203-750-3595	10,000	97
Huntsman Corp.	500 Huntsman Way, Salt Lake City, Utah 84108, U.S.A.	801-532-5200	801-584-5781	9,500	97
Cydsa SA y Subsidiarias	Ruiz Cortines, Monterrey 2333, Mexico	83 359090	83 355541	9,210	91
Dow Corning Corp.	PO Box 994, Midland, Michigan 48686-0994, U.S.A.	517-496-4000	517-496-4586	9,000	97
Novo Nordisk AS	1191 Corniche El Nil, Cairo, Egypt	2 773665	2 773894	9,000	91
IMC Global Inc.	2100 Sanders Rd., Northbrook, Illinois 60062-6146, U.S.A.	847-272-9200		8,950	97
Inco Ltd.	145 King St. W Ste. 1500, Toronto, Ontario M5H 4B7, Canada	416-361-7511	416-361-7781	8,600	98
Eregli Iron and Steel Works Co.	Uzunkum No. 7, Eregli TR-67330, Turkey	388 19500	388 13969	8,349	89
Zaklady Azotowe W Tarnowie-Moscicach SA	ul E. Kwiatkowskiego 8, Tarnow PL-33-101, Poland	14 330781	14 330718	8,205	93
Israel Chemicals Ltd.	21 Shazar Blvd., Beer Sheba 84106, Israel	7 6405700	7 6286563	8,000	97
Mallinckrodt Inc.	PO Box 5840, St. Louis, Missouri 63134, U.S.A.	314-654-2000		8,000	97
National Starch and Chemical Co.	10 Finderne Ave., Bridgewater, New Jersey 08807, U.S.A.	908-685-5000	908-685-5300	8,000	97
Pabrik Kertas Tjiwi Kimia PT	Desa Kramat Tumenggung Kecamatan Tarik, Sidoarjo, Indonesia	32121552	321 21615	8,000	97
Rhone-Poulenc Agrochimie	14/20, rue Pierre Baizet, Lyon F-69009, France	7 2292525	7 2292799	8,000	91
Foamex International Inc.	1000 Columbia Ave., Linwood, Pennsylvania 19061, U.S.A.	610-859-3000		7,700	94
Laporte PLC	3 Bedford Sq., London WC1B 3RA, United Kingdom	71 5800223		7,330	94
Interface Inc. (Atlanta, Georgia)	2859 Paces Ferry Rd., 2000, Atlanta, Georgia 30339, U.S.A.	770-437-6800	770-437-6809	7,300	97
RWE-DEA AG fur Mineraloel und Chemie	Ueberseering 40, Hamburg D-22297, Germany	40 63750	40 63753496	7,243	94
Witco Corp.	1 American Ln., Greenwich, Connecticut 06831-2559, U.S.A.	203-552-2000		7,200	97
Sentrachem Ltd.	PO Box 781811, Sandton 2146, Republic of South Africa	11 7803600	11 7832180	7,000	95
Vulcan Materials Co.	PO Box 530187, Birmingham, Alabama 35209, U.S.A.	205-877-3000	205-877-3094	6,963	97
Sodaso	Bratstva i Jedinstva 17, Tuzla 75001, Bosnia-Hercegovina	75 211111	75 212172	6,927	93
Ferro Corp.	1000 Lakeside Ave., Cleveland, Ohio 44114, U.S.A.	216-641-8580	216-696-4786	6,851	97
Sigma-Aldrich Corp.	3050 Spruce St., St. Louis, Missouri 63103, U.S.A.	314-771-5765	314-534-2674	6,666	97
MAPCO Inc.	PO Box 645, Tulsa, Oklahoma 74101, U.S.A.	918-599-3712	918-581-1893	6,508	97
Nalco Chemical Co.	1 Nalco Ctr., Naperville, Illinois 60563-1198, U.S.A.	630-305-1000	630-305-2900	6,500	97
Occidental Chemical Corp.	PO Box 809050, Dallas, Texas 75380, U.S.A.	972-404-3800	972-404-3669	6,500	97
Ashland Chemical Co.	PO Box 2219, Columbus, Ohio 43216, U.S.A.	614-790-3333	614-790-4119	6,436	97
Engelhard Corp.	101 Wood Ave., Iselin, New Jersey 08830-0770, U.S.A.	908-205-6000	908-906-0337	6,300	96
Hercules Inc.	1313 N. Market St., Hercules Plz, Wilmington, Delaware 19894-0001, U.S.A.	302-594-5000		6,221	97
M.A. Hanna Co.	200 Public Sq., 36-500, Cleveland, Ohio 44114-2304, U.S.A.	216-589-4000	216-589-4109	6,068	97
ZSNP Aluminium Works AS	Ziar nad Hronom 965 63, Slovakia	857 2201	857 2240	5,880	93
Cominco Ltd.	200 Burrard St. Ste. 500, Vancouver, British Columbia V6C 3L7, Canada	604-682-0611	604-685-3091	5,743	98
Crompton and Knowles Corp.	1 Station Pl., Metro Ctr., Stamford, Connecticut 06902, U.S.A.	203-353-5400	203-353-5424	5,665	96
Chemlon AS	Chemslonska 1, Humenne 066 33, Slovakia	933 63741	933 62425	5,661	93
Standard Industries Ltd.	Nariman Point, Bombay 400 021, India	22 258544		5,574	94
Frame Textile Corp. Ltd.	PO Box 32002, Duran 4060, Republic of South Africa	31 425140	31 424822	5,500	97
TVK Rt.	PO Box 20, Tiszaujvaros H-3581, Hungary	49 22222	49 21322	5,500	93
Elana JSC	Sklodowskiej Curie 73, Torun PL-87-100, Poland	56 484125	56 484069	5,300	93
Foamex L.P.	1000 Columbia Ave., Linwood, Pennsylvania 19061, U.S.A.	610-859-3000		5,295	95
Cytec Industries Inc.	5 Garret Mountain Plaza, West Paterson, New Jersey 07424, U.S.A.	973-357-3100	973-357-3065	5,200	97
Hiram Walker Group Ltd.	Kilver St., Shepton Mallet BA4 5ND, United Kingdom	749 343300	749 345653	5,196	
Great Lakes Chemical Corp.	PO Box 2200, West Lafayette, Indiana 47906-0200, U.S.A.	765-497-6100		5,100	97
Hexcel Corp.	5794 W. Las Positas Blvd., Pleasanton, California 94588-8781, U.S.A.	510-847-9500	510-734-8276	5,013	96
Bajaj Hindustan Ltd.	226 Nariman Pt., Bombay 400 021, India	22 2023626	2021977	5,000	
Pioneer Hi-Bred International Inc.	700 Capital Sq., 400 Locust St., Des Moines, Iowa 50309-2340, U.S.A.	515-248-4800		5,000	97
Transelco Div.	1789 Transelco Dr., Penn Yan, New York 14527, U.S.A.	315-536-3357	315-536-8091	5,000	96
BP Chemicals Ltd.	1 Finsbury Circus, London EC2M 7BA, United Kingdom	71 4964000	71 4964630	4,983	92
Jiyajeerao Cotton Mills Ltd.	Birla Nagar, Gwalior 474 004, India			4,850	97
Albright & Wilson Ltd.	Oldbury, Warley B68 0NN, United Kingdom	21 4294942	21 4205151	4,812	92

Continued

Representative Companies in Sector
- Continued -

Name	Address	Telephone	Fax	Employment	Y
Cabot Corp.	75 State St., Boston, Massachusetts 02109-1806, U.S.A.	617-345-0100		4,800	97
Clorox Co.	1221 Broadway Ave., Oakland, California 94612, U.S.A.	510-271-7000	510-832-1463	4,700	93
International Flavors and Fragrances Inc.	521 W. 57th St., New York, New York 10019-2960, U.S.A.	212-765-5500		4,629	96
Agrium Inc.	426-10333 Southport Rd. SW, Calgary, Alberta T2W 3X6, Canada	403-258-4600	403-258-4692	4,500	98
Falconbridge Ltd.	95 Wellington St. W Ste. 1200, Toronto, Ontario M5J 2V4, Canada	416-956-5700	416-956-5757	4,500	98
Zaklady Wlokien Chemicznych Wistom SA	ul. Spalska 103/105, Tomaszow Mazowiecki PL-97-200, Poland	3 9121707	3 5248	4,500	93
Terra Industries Inc.	PO Box 6000, Sioux City, Iowa 51102-6000, U.S.A.	712-277-1340	712-233-3648	4,435	97
Instrochem SP	Nobelova 34, Bratislava 836 05, Slovakia	7 255888	7 258288	4,300	93
Lubrizol Corp.	29400 Lakeland Blvd., Wickliffe, Ohio 44092-2298, U.S.A.	440-943-4200	440-943-5337	4,291	97
Duslo SP	Sala 927 03, Slovakia	706 2561	706 5643	4,290	93
ARCO Chemical Co.	3801 W. Chester Pike, Newtown Square, Pennsylvania 19073-2387, U.S.A.	610-359-2000	610-359-2722	4,200	97
Borsodchem Rt.	PO Box 208, Kazincbarcika H-3702, Hungary	48 10211	46 354496	4,200	93
Chevron Chemical Co.	PO Box 5047, San Ramon, California 94583-0947, U.S.A.	510-842-5500	510-842-5775	4,200	
Elektrosvit AS	Komarnanska cesta 3, Nove Zamky 940 37, Slovakia	817 22886	817 27866	4,200	93
Millennium Chemicals Inc.	99 Wood Ave. S., Iselin, New Jersey 08830, U.S.A.	732-603-6600		4,200	97
Formosa Plastics Corporation U.S.A.	9 Peach Tree Hill Rd., Livingston, New Jersey 07039, U.S.A.	973-992-2090	973-992-9627	4,000	97
Garware Nylons Ltd.	Nariman Point, Bombay 400 021, India	22 2022312	22 2029377	4,000	94
Mobil Chemical Company Inc.	3225 Gallows Rd., Fairfax, Virginia 22037-0001, U.S.A.	703-846-3000	703-846-2313	4,000	97
UOP	25 E. Algonquin Rd., Des Plaines, Illinois 60017-5017, U.S.A.	847-391-2000	708-391-2253	4,000	95
ZA Pulawy SA	Panstwa Polskiego 13, Pulawy PL-24-110, Poland	7 3431	7 5444	4,000	93
Zaklady Azotowe, Wloclawek SA	ul. Torunska 222, Wloclawek PL-87-810, Poland	54 362551	54 361983	3,900	93
Kerr-McGee Corp.	PO Box 25861, Oklahoma City, Oklahoma 73125, U.S.A.	405-270-3949	405-270-3949	3,851	97
Unilever Canada Ltd.	160 Bloor St. E Ste. 1500, Toronto, Ontario M4W 3R2, Canada	416-964-1857	416-964-8831	3,816	98
Readicut International PLC	Clifton Mills, Brighouse HD6 4ET, United Kingdom	484 721223	484 716135	3,725	94
Helmerich and Payne Inc.	Utica at 21st St., Tulsa, Oklahoma 74114, U.S.A.	918-742-5531		3,627	97
Ford France SA	BP 32, Blanquefort F-33290, France	56 954000	56 954378	3,624	91
PMC Inc. (Sun Valley, California)	PO Box 1367, Sun Valley, California 91353-1367, U.S.A.	818-896-1101	818-897-0180	3,600	97
Cargill Ltd.	300-240 Graham Ave., Winnipeg, Manitoba R3C 0J7, Canada	204-947-0141	204-947-6222	3,500	98
Reichhold Chemicals Inc.	PO Box 13582, Research Triangle Park, North Carolina 27709, U.S.A.	919-990-7500	919-990-7711	3,500	97
Pauls PLC	47 Key St., Ipswich IP4 1BX, United Kingdom	473 232222	473 230509	3,475	93
Carter-Wallace Inc.	1345 Ave. of the Amer., New York, New York 10105, U.S.A.	212-339-5000	212-339-5100	3,460	96
Henkel Corp.	2200 Renaissance Blvd. Ste. 200, King of Prussia, Pennsylvania 19406-2755, U.S.A.	610-270-8100	610-270-8102	3,300	94
Monsanto Co. Agricultural Sector	800 N. Lindbergh Blvd., St. Louis, Missouri 63167, U.S.A.	314-694-1321	314-694-5190	3,300	97
Saskatchewan Wheat Pool	2625 Victoria Ave., Regina, Saskatchewan S4T 7T9, Canada	306-569-4411	306-569-4708	3,300	98
Wellman Inc.	1040 Broad St., Ste. 302, Shrewsbury, New Jersey 07702, U.S.A.	908-542-7300	908-542-9344	3,200	96
NL Industries Inc.	16825 Northchase Dr., 1200, Houston, Texas 77060-2544, U.S.A.	281-423-3300		3,100	96
Zaklady Przemyslu Bawelnianego Bielbawa SA	ul. Piastowska 19, Bielbawa PL-58-260, Poland	74 334341	74 334673	3,100	93
Akzo Nobel Chemicals Inc.	300 S. Riverside Plz., 2200, Chicago, Illinois 60606, U.S.A.	312-906-7500	312-906-7680	3,000	97
Bukoza AS	Ul. Hencovska, Vranov nad Toplou 090 02, Slovakia	931 23241	931 23447	3,000	93
Cenex/Land O'Lakes Ag Services	PO Box 64089, St. Paul, Minnesota 55164-0089, U.S.A.	612-451-5151	612-451-5568	3,000	97
Suzano	Av. Brig. Faria Lima 2100, 5th Fl., Sao Paulo 01452-002, Brazil	11 2106564	11 2106564	2,877	96
Hickson International PLC	Chancellor House, Leeds LS2 6HG, United Kingdom	13 2426161	13 2341133	2,840	93
Albemarle Corp.	PO Box 1335, Richmond, Virginia 23219, U.S.A.	804-788-6000		2,800	97
Uniroyal Chemical Company Inc.	Benson Rd., Middlebury, Connecticut 06762, U.S.A.	203-573-2000	203-573-2265	2,800	95
Allied Colloids Group PLC	Low Moor, Bradford BD12 0JZ, United Kingdom	274 671267	274 606499	2,796	93
Dusun Durian Plantations Ltd.	1 Great Tower St., London EC3R 5AH, United Kingdom			2,759	94
Pataling Rubber Estates Ltd.	1 Great Tower St., London EC3R 5AH, United Kingdom			2,757	94
London Asiatic Rubber & Produce Co. Ltd.	1 Great Tower St., London EC3R 5AH, United Kingdom	71 6264333	71 7820112	2,748	94

Continued

For sources and notes, see Appendix I.

Representative Companies in Sector
- Continued -

Name	Address	Tele-phone	Fax	Employ-ment	Y
Uniroyal Chemical Corp.	Benson Rd., Middlebury, Connecticut 06762, U.S.A.	203-573-2000		2,730	94
Nova Chemicals Ltd.	645 7 Ave. SW 23rd Floor, Calgary, Alberta T2P 4G8, Canada	403-750-3600	403-269-7410	2,710	98
Puritan-Bennett Corp.	9401 Indian Creek Pkwy., Bldg. 4, Overland Park, Kansas 66210, U.S.A.	913-661-0444	913-661-0234	2,700	94
Inti Indorayon Utama PT	Jalan Let Jend M. T. Haryono A1, Medan 20231, Indonesia	61532532	61 532355	2,650	97
Inco Europe Ltd.	50 Victoria St., London SW1H 0XB, United Kingdom	71 931773		2,621	94
Millennium Inorganic Chemicals Inc.	200 International Cir., 5000, Hunt Valley, Maryland 21030, U.S.A.	410-229-4400		2,600	96
International Specialty Products Inc.	1361 Alps Rd., Wayne, New Jersey 07470, U.S.A.	201-628-4000		2,500	95
NutraSweet Kelco Co.	200 World Trade Center, Chicago, Illinois 60654, U.S.A.	847-940-9800	847-405-7812	2,500	97
Park-Ohio Industries Inc.	23000 Euclid Ave., Cleveland, Ohio 44117, U.S.A.	216-692-7200		2,500	96
Hoechst UK Ltd.	Salisbury Rd., Hounslow TW4 6JH, United Kingdom	81 5707712	81 5771854	2,479	93
Bayer Inc.	77 Belfield Rd., Toronto, Ontario M9W 1G6, Canada	416-248-0771	416-248-1438	2,450	98
Rhodia SA	Av. Maria Coelho de Aguiar 215, Sao Paulo 05804-902, Brazil	11 37417973	11 37417159	2,439	96
Shinkong Synthetic Fibers Corp.	8f, 123 Nanking E. Rd., Sec. 2, Taipei 104, Taiwan	02 5071251	02 5072264	2,400	94
Dow Quimica SA	Rua Alexandre Dumas 1671, 4th Fl., Sao Paulo 04717-1203, Brazil	11 5469122	11 5469385	2,382	98
Air Products PLC	Molesey Rd., Walton-on-Thames KT12 4RZ, United Kingdom	932 249200	932 249565	2,377	93
KCM SA	Assenovgradsko Shosse, Plovdiv BG-4009, Bulgaria	32 23496	32 269044	2,287	93
Minerals Technologies Inc.	405 Lexington Ave., New York, New York 10174-1901, U.S.A.	212-878-1800	212-878-1801	2,250	97
Atomic Energy Corp. of South Africa Ltd.	PO Box 582, Pretoria 0001, Republic of South Africa	12 3163270	12 3163222	2,200	97
Courtaulds Chemicals Holdings Ltd.	50 George St., London W1A 2BB, United Kingdom			2,174	94
Scotts Co.	14111 Scottslawn Rd., Marysville, Ohio 43041, U.S.A.	513-644-0011	513-644-7072	2,075	97
Achema	Taurastos 26, Jonava 5000, Lithuania	19 56625	19 556619	2,000	95
AECI Industrial Chemicals	PO Box 5, Modderfontein 1645, Republic of South Africa	11 6062516	11 6081861	2,000	95
Champion Papel Celulose Ltda.	Rodovia SP 340 Km. No. 171, Mogi Guacu 13840-970, Brazil	192 611657	192 611098	2,000	96
Destilaria Alta Mogiana Ltda.	Fazenda Santa Ana, Sao Joaquim da Barra 14600-000, Brazil	16 7285199	16 7284333	2,000	98
ECC International	100 Mansell Court E., Ste. 300, Roswell, Georgia 30076, U.S.A.	770-594-0660	770-645-3384	2,000	97
Engelhard Corp. Catalysts and Chemicals	101 Wood Ave., Iselin, New Jersey 08830, U.S.A.	732-205-6000	732-205-5915	2,000	97
Hoescht do Brasil Quimica e Farmaceutica SA	Av. Das Nacoes Unidas 18001, Santo Amaro 04795-900, Brazil	11 5257233	11 2476640	2,000	96
Millennium Petrochemicals Inc.	11500 Northlake Dr., Cincinnati, Ohio 45249-1644, U.S.A.	513-530-6500		2,000	97
Samator Gas Industry Group PT	Jl. Kusuma Bangswa 51 A-B, Surabaya, Indonesia	31 516733	31 516674	2,000	93
Zimco Industries Pty. Ltd.	PO Box 78069, Sandton 2146, Republic of South Africa	11 7837023	11 7830732	2,000	95
Allied Colloids Ltd.	Low Moor, Bradford BD120JZ, United Kingdom	274 671267	274 606499	1,978	94
Schering Holdings Ltd.	Huntingdon Rd., Cambridge CB3 0DA, United Kingdom	223 323222	223 66853	1,975	92
Cipla Ltd.	289 J B B Marg Bombay Central, Bombay 400 008, India	22 3082891	22 3070013	1,900	94
Union Camp Corp.	1600 Valley Rd., Wayne, New Jersey 07470, U.S.A.	201-628-2000		1,900	97
Zaklady Chemiczne, Organika-Sarzyna Przedsiebiorstwo Panst.	ul. Chemikow 1, Nowa Sarzyna PL-37-310, Poland	17 38621	17 628624	1,900	93
Makhteshim Chemical Works Ltd.	Industrial Zone, Beer Sheba 84100, Israel	7 6296624	7 6280364	1,855	97
Rotem Amfert Negev Ltd.	Science Park, Ashdod 77101, Israel	8 8511540	8 8566219	1,850	97
Kiwi Holdings Ltd.	225 Bath Rd., Slough SL1 4AU, United Kingdom			1,832	91
Dorogi Szenbanyak RT	Hantken Miksa u. 8-9, Dorog H-2510, Hungary	33 31614	33 31819	1,820	93
Geon Co.	1 Geon Ctr., Avon Lake, Ohio 44012, U.S.A.	216-930-1000	216-930-1002	1,808	96
Dead Sea Works Ltd.	Potash House, Beer Sheva 84585, Israel	7 6465349	7 6233177	1,803	97
Noranda Mining and Exploration Inc.	PO Box 3000 Sta. Main, Bathurst, New Brunswick E2A 3Z8, Canada	506-546-6671	506-547-6076	1,800	98
Cambrex Corp.	1 Meadowlands Plz., East Rutherford, New Jersey 07073-2100, U.S.A.	201-804-3000	201-804-9852	1,790	97
Plastic Specialties and Technologies, Inc.	65 Railroad Ave., Ridgefield, New Jersey 07657, U.S.A.	201-941-2900	201-941-4670	1,750	97
Schering Agrochemicals Ltd.	Hauxton, Cambridge CB2 5HU, United Kingdom	223 870312	223 872142	1,731	93
Mississippi Chemical Corp.	PO Box 388, Yazoo City, Mississippi 39194-0388, U.S.A.	601-746-4131	601-746-9158	1,700	97
Sterling Chemicals Inc.	1200 Smith St., Ste. 1900, Houston, Texas 77002-4312, U.S.A.	713-650-3700		1,700	97
Crown Berger Ltd.	PO Box 37, Darwen BB3 0BG, United Kingdom	254 704951	254 774414	1,698	93
Soda Sanayii AS	Kazanli Bucagi Yani P.K. 654, Mersin TR-33004, Turkey	741 23550	741 23555	1,661	89

For sources and notes, see Appendix I.

Product Tables
Sulfur, Recovered as By-Product

Unit of Measure: Metric tons.

	1990 Value	%	1991 Value	%	1992 Value	%	1993 Value	%	1994 Value	%	1995 Value	%	1996 Value	%
Total Production	37,960,510	100.0	44,152,335	100.0	41,695,926	100.0	42,341,201	100.0	45,012,043	100.0	46,251,125	100.0	47,448,164	100.0
Regions														
Africa	375,000	1.0	360,000	0.8	355,000	0.9	373,000	0.9	387,000	0.9	404,923	0.9	412,418	0.9
America, North	15,379,006	40.5	15,715,806	35.6	16,149,369	38.7	15,565,933	36.8	17,029,211	37.8	16,688,166	36.1	17,023,389	35.9
America, South	663,000	1.7	784,000	1.8	827,000	2.0	930,000	2.2	912,115	2.0	1,005,725	2.2	1,059,278	2.2
Asia	15,427,278	40.6	21,101,444	47.8	18,155,611	43.5	19,069,778	45.0	20,329,681	45.2	21,900,382	47.4	22,674,922	47.8
Europe	5,820,226	15.3	5,890,085	13.3	5,833,945	14.0	6,011,491	14.2	5,962,036	13.2	5,858,807	12.7	5,866,849	12.4
Oceania	296,000	0.8	301,000	0.7	375,000	0.9	391,000	0.9	392,000	0.9	393,122	0.8	411,309	0.9
Leading Producers														
Canada	7,753,000	20.4	7,821,000	17.7	7,309,000	17.5	6,077,000	14.4	6,731,000	15.0	*7,228,308	15.6	*7,183,958	15.1
United States of America	6,536,000	17.2	6,650,000	15.1	7,050,000	16.9	7,720,000	18.2	7,160,000	15.9	7,250,000	15.7	7,470,000	15.7
Indonesia	1,481,000	3.9	6,836,000	15.5	4,109,000	9.9	4,205,000	9.9	*5,413,267	12.0	*5,910,381	12.8	*6,407,495	13.5
Japan	2,580,000	6.8	2,590,000	5.9	2,710,000	6.5	2,890,000	6.8	2,900,000	6.4	*2,825,747	6.1	*2,873,066	6.1
Russian Federation	2,499,000	6.6	2,695,000	6.1	2,383,000	5.7	2,424,000	5.7	2,334,000	5.2	2,764,000	6.0	2,519,000	5.3

Sulfur, Recovered from Pyrites Etc.

Unit of Measure: Metric tons.

	1990 Value	%	1991 Value	%	1992 Value	%	1993 Value	%	1994 Value	%	1995 Value	%	1996 Value	%
Total Production	16,604,000	100.0	16,699,522	100.0	15,365,644	100.0	15,384,055	100.0	15,243,170	100.0	16,036,708	100.0	15,915,883	100.0
Regions														
Africa	743,000	4.5	569,000	3.4	623,000	4.1	579,000	3.8	508,000	3.3	609,423	3.8	606,435	3.8
America, North	339,944	2.0	344,067	2.1	348,189	2.3	352,311	2.3	356,433	2.3	360,556	2.2	364,678	2.3
America, South	46,000	0.3	66,000	0.4	25,000	0.2	10,000	0.1	10,000	0.1	31,319	0.2	27,352	0.2
Asia	12,554,556	75.6	13,183,156	78.9	12,209,756	79.5	12,477,689	81.1	12,168,289	79.8	12,643,438	78.8	12,674,921	79.6
Europe	2,920,500	17.6	2,537,300	15.2	2,159,700	14.1	1,965,055	12.8	2,200,448	14.4	2,391,972	14.9	2,242,497	14.1
Leading Producers														
Pakistan	5,468,000	32.9	5,757,000	34.5	5,612,000	36.5	*5,612,333	36.5	*5,612,333	36.8	*5,612,333	35.0	*5,612,333	35.3
China	4,400,000	26.5	4,940,000	29.6	4,500,000	29.3	5,000,000	32.5	5,000,000	32.8	*5,650,000	35.2	*5,942,286	37.3
Spain	950,000	5.7	800,000	4.8	406,000	2.6	327,000	2.1	350,000	2.3	*435,857	2.7	*374,657	2.4
Finland	659,000	4.0	568,000	3.4	667,000	4.3	665,000	4.3	840,000	5.5	875,000	5.5	813,000	5.1
South Africa	452,000	2.7	293,000	1.8	296,000	1.9	323,000	2.1	252,000	1.7	*298,330	1.9	*280,059	1.8

Styrene

Unit of Measure: Metric tons.

	1990 Value	%	1991 Value	%	1992 Value	%	1993 Value	%	1994 Value	%	1995 Value	%	1996 Value	%
Total Production	12,708,766	100.0	12,833,646	100.0	13,730,786	100.0	14,082,253	100.0	15,538,190	100.0	15,958,940	100.0	16,755,246	100.0
Regions														
America, North	4,489,864	35.3	4,609,880	35.9	5,081,282	37.0	5,420,742	38.5	6,266,003	40.3	6,377,354	40.0	6,668,697	39.8
America, South	381,958	3.0	401,376	3.1	416,074	3.0	432,961	3.1	450,455	2.9	482,546	3.0	795,990	4.8
Asia	4,036,584	31.8	4,268,494	33.3	4,546,399	33.1	4,485,158	31.8	4,930,904	31.7	5,296,613	33.2	5,324,854	31.8
Europe	3,799,639	29.9	3,553,175	27.7	3,686,309	26.8	3,742,670	26.6	3,890,106	25.0	3,801,705	23.8	3,964,982	23.7
Oceania	722	0.0	722	0.0	722	0.0	722	0.0	722	0.0	722	0.0	722	0.0
Leading Producers														
United States of America	3,636,269	28.6	3,680,516	28.7	4,082,245	29.7	4,352,032	30.9	5,127,619	33.0	5,169,297	32.4	5,390,967	32.2
Japan	2,161,102	17.0	2,226,993	17.4	2,182,290	15.9	2,166,811	15.4	2,621,601	16.9	2,939,111	18.4	3,085,351	18.4
Germany	*1,127,499	8.9	*1,127,499	8.8	*1,127,499	8.2	*1,127,499	8.0	1,150,723	7.4	1,080,531	6.8	1,151,244	6.9
Canada	780,000	6.1	*850,047	6.6	*913,999	6.7	*977,951	6.9	*1,041,904	6.7	*1,105,856	6.9	*1,169,808	7.0
Netherlands	*784,926	6.2	*784,926	6.1	*784,926	5.7	*784,926	5.6	*784,926	5.1	784,926	4.9	*784,926	4.7

Commodity data are provided by the United Nations Statistical Division. The symbol * means that data are estimated. For additional notes, see Appendix I.

281

Product Tables
Acetylene

Unit of Measure: Metric tons.

	1990 Value	%	1991 Value	%	1992 Value	%	1993 Value	%	1994 Value	%	1995 Value	%	1996 Value	%
Total Production	1,119,326	100.0	1,074,467	100.0	1,237,516	100.0	1,068,373	100.0	1,214,008	100.0	1,206,361	100.0	1,084,033	100.0
Regions														
Africa	9,599	0.9	9,606	0.9	8,336	0.7	7,425	0.7	18,364	1.5	19,496	1.6	13,746	1.3
America, North	136,618	12.2	146,073	13.6	144,249	11.7	143,265	13.4	136,544	11.2	127,780	10.6	128,035	11.8
America, South	23,256	2.1	25,441	2.4	16,911	1.4	19,404	1.8	18,754	1.5	15,971	1.3	15,197	1.4
Asia	511,241	45.7	485,194	45.2	548,317	44.3	434,829	40.7	403,875	33.3	371,493	30.8	342,249	31.6
Europe	438,574	39.2	408,116	38.0	519,667	42.0	463,413	43.4	636,434	52.4	671,585	55.7	584,770	53.9
Oceania	37	0.0	37	0.0	37	0.0	37	0.0	37	0.0	37	0.0	37	0.0
Leading Producers														
Germany	-		-		155,930	12.6	102,673	9.6	119,150	9.8	248,901	20.6	164,084	15.1
United States of America	127,568	11.4	136,595	12.7	134,562	10.9	133,446	12.5	126,836	10.4	*117,837	9.8	*117,706	10.9
Russian Federation	141,674	12.7	123,128	11.5	198,872	16.1	100,740	9.4	76,453	6.3	51,265	4.2	27,960	2.6
Italy	49,265	4.4	38,768	3.6	15,926	1.3	17,109	1.6	19,105	1.6	19,943	1.7	18,559	1.7
Romania	37,788	3.4	26,612	2.5	12,821	1.0	8,949	0.8	4,843	0.4	4,416	0.4	3,876	0.4

Benzene (Benzol)

Unit of Measure: Metric tons.

	1990 Value	%	1991 Value	%	1992 Value	%	1993 Value	%	1994 Value	%	1995 Value	%	1996 Value	%
Total Production	24,150,084	100.0	23,627,285	100.0	23,957,804	100.0	23,310,677	100.0	24,963,901	100.0	25,720,073	100.0	25,600,177	100.0
Regions														
America, North	6,412,078	26.6	5,925,242	25.1	5,798,309	24.2	6,233,740	26.7	7,651,861	30.7	7,920,114	30.8	7,716,050	30.1
America, South	751,945	3.1	727,966	3.1	717,523	3.0	687,800	3.0	802,086	3.2	843,356	3.3	814,356	3.2
Asia	8,051,020	33.3	8,259,793	35.0	8,903,987	37.2	8,891,229	38.1	8,936,400	35.8	9,375,670	36.5	9,429,550	36.8
Europe	8,935,041	37.0	8,714,284	36.9	8,537,985	35.6	7,497,908	32.2	7,573,553	30.3	7,580,933	29.5	7,640,220	29.8
Leading Producers														
United States of America	5,644,128	23.4	5,209,209	22.0	5,110,290	21.3	5,547,615	23.8	6,880,988	27.6	7,193,351	28.0	7,021,651	27.4
Japan	3,011,861	12.5	3,284,557	13.9	3,527,076	14.7	3,501,531	15.0	3,620,000	14.5	3,898,318	15.2	4,176,637	16.3
Germany	*1,664,850	6.9	*1,664,850	7.0	*1,664,850	6.9	1,518,515	6.5	1,944,311	7.8	1,598,753	6.2	1,597,822	6.2
Netherlands	1,331,000	5.5	1,201,100	5.1	1,337,700	5.6	905,158	3.9	472,617	1.9	556,224	2.2	590,627	2.3
Russian Federation	1,443,665	6.0	1,173,561	5.0	1,129,642	4.7	941,326	4.0	667,177	2.7	737,817	2.9	492,000	1.9

Butylenes, Butadiene

Unit of Measure: Metric tons.

	1990 Value	%	1991 Value	%	1992 Value	%	1993 Value	%	1994 Value	%	1995 Value	%	1996 Value	%
Total Production	10,870,787	100.0	12,969,711	100.0	13,397,634	100.0	13,885,150	100.0	14,992,351	100.0	15,236,384	100.0	15,828,121	100.0
Regions														
America, North	3,247,518	29.9	3,691,219	28.5	3,926,423	29.3	4,161,627	30.0	4,396,830	29.3	4,657,984	30.6	4,877,973	30.8
America, South	259,028	2.4	220,134	1.7	246,343	1.8	304,080	2.2	333,049	2.2	369,747	2.4	385,309	2.4
Asia	4,099,915	37.7	4,291,633	33.1	4,318,266	32.2	4,341,788	31.3	4,856,044	32.4	5,490,759	36.0	5,514,662	34.8
Europe	3,264,325	30.0	4,766,725	36.8	4,906,602	36.6	5,077,655	36.6	5,406,427	36.1	4,717,894	31.0	5,050,177	31.9
Leading Producers														
Japan	3,070,227	28.2	3,200,905	24.7	3,280,127	24.5	3,111,760	22.4	3,310,026	22.1	3,784,532	24.8	3,867,712	24.4
United States of America	2,893,629	26.6	*3,332,549	25.7	*3,562,971	26.6	*3,793,394	27.3	*4,023,817	26.8	*4,254,240	27.9	*4,484,663	28.3
Germany	-		1,545,500	11.9	1,504,234	11.2	1,670,522	12.0	1,965,163	13.1	1,164,198	7.6	*1,479,421	9.3
Netherlands	554,595	5.1	549,105	4.2	543,615	4.1	538,126	3.9	592,832	4.0	618,207	4.1	665,838	4.2
Russian Federation	811,217	7.5	783,579	6.0	618,929	4.6	342,705	2.5	236,760	1.6	332,072	2.2	301,289	1.9

Commodity data are provided by the United Nations Statistical Division. The symbol * means that data are estimated. For additional notes, see Appendix I.

Product Tables

Ethylene

Unit of Measure: Metric tons.

	1990 Value	%	1991 Value	%	1992 Value	%	1993 Value	%	1994 Value	%	1995 Value	%	1996 Value	%
Total Production	57,923,958	100.0	60,775,223	100.0	63,919,278	100.0	61,675,368	100.0	66,767,479	100.0	69,954,970	100.0	71,245,697	100.0
Regions														
Africa	41,000	0.1	69,612	0.1	53,400	0.1	49,600	0.1	78,700	0.1	61,197	0.1	61,538	0.1
America, North	20,063,214	34.6	21,724,653	35.7	22,339,056	34.9	22,019,316	35.7	24,377,861	36.5	25,676,215	36.7	26,782,002	37.6
America, South	1,799,505	3.1	1,727,367	2.8	1,816,012	2.8	2,007,964	3.3	2,177,787	3.3	2,200,570	3.1	2,213,149	3.1
Asia	15,371,972	26.5	16,421,569	27.0	17,675,329	27.7	17,930,243	29.1	18,754,132	28.1	20,357,080	29.1	20,593,031	28.9
Europe	20,648,267	35.6	20,832,022	34.3	22,035,480	34.5	19,668,245	31.9	21,378,999	32.0	21,659,908	31.0	21,595,977	30.3
Leading Producers														
United States of America	16,541,341	28.6	18,123,454	29.8	18,562,930	29.0	18,149,407	29.4	20,249,185	30.3	21,322,424	30.5	22,290,173	31.3
Japan	5,809,627	10.0	6,141,798	10.1	6,103,155	9.5	5,772,626	9.4	6,125,000	9.2	6,944,470	9.9	7,137,543	10.0
Germany	*4,016,398	6.9	*4,016,398	6.6	*4,016,398	6.3	3,904,814	6.3	4,182,722	6.3	4,163,377	6.0	3,814,680	5.4
France	2,252,460	3.9	2,421,395	4.0	2,654,440	4.2	2,535,779	4.1	2,796,903	4.2	2,667,255	3.8	*2,815,046	4.0
Canada	2,425,000	4.2	2,425,300	4.0	2,521,200	3.9	2,535,956	4.1	2,715,696	4.1	2,861,784	4.1	2,920,795	4.1

Naphthalene

Unit of Measure: Metric tons.

	1990 Value	%	1991 Value	%	1992 Value	%	1993 Value	%	1994 Value	%	1995 Value	%	1996 Value	%
Total Production	825,760	100.0	824,021	100.0	819,018	100.0	777,833	100.0	812,036	100.0	808,589	100.0	813,916	100.0
Regions														
Africa	994	0.1	994	0.1	994	0.1	994	0.1	994	0.1	994	0.1	994	0.1
America, North	165,000	20.0	165,475	20.1	175,184	21.4	184,893	23.8	194,602	24.0	204,311	25.3	214,020	26.3
America, South	15,557	1.9	19,544	2.4	17,367	2.1	15,567	2.0	14,942	1.8	16,623	2.1	16,354	2.0
Asia	333,403	40.4	339,187	41.2	330,964	40.4	316,415	40.7	320,590	39.5	324,896	40.2	329,570	40.5
Europe	310,806	37.6	298,822	36.3	294,509	36.0	259,964	33.4	280,908	34.6	261,765	32.4	252,978	31.1
Leading Producers														
Japan	202,198	24.5	207,663	25.2	199,202	24.3	184,405	23.7	188,027	23.2	192,725	23.8	197,518	24.3
United States of America	145,000	17.6	*145,475	17.7	*155,184	18.9	*164,893	21.2	*174,602	21.5	*184,311	22.8	*194,020	23.8
Germany	*81,177	9.8	*81,177	9.9	80,793	9.9	71,939	9.2	90,800	11.2	*81,177	10.0	*81,177	10.0
Ukraine	59,845	7.2	48,900	5.9	43,345	5.3	19,970	2.6	19,024	2.3	13,273	1.6	6,090	0.7
Czech Republic	*24,713	3.0	*24,713	3.0	*24,713	3.0	*24,713	3.2	29,106	3.6	22,763	2.8	22,271	2.7

Propylene

Unit of Measure: Metric tons.

	1990 Value	%	1991 Value	%	1992 Value	%	1993 Value	%	1994 Value	%	1995 Value	%	1996 Value	%
Total Production	28,799,564	100.0	31,160,853	100.0	33,189,420	100.0	32,630,684	100.0	34,851,799	100.0	37,188,734	100.0	36,380,810	100.0
Regions														
America, North	10,967,090	38.1	10,799,280	34.7	11,701,032	35.3	10,843,941	33.2	11,941,239	34.3	12,790,211	34.4	12,537,521	34.5
America, South	890,540	3.1	887,424	2.8	974,497	2.9	1,151,782	3.5	1,280,330	3.7	1,280,632	3.4	1,309,256	3.6
Asia	7,699,158	26.7	8,237,339	26.4	8,812,493	26.6	8,871,060	27.2	9,201,536	26.4	9,966,414	26.8	9,786,620	26.9
Europe	9,242,776	32.1	11,236,809	36.1	11,701,397	35.3	11,763,901	36.1	12,428,694	35.7	13,151,477	35.4	12,747,413	35.0
Leading Producers														
United States of America	9,909,380	34.4	9,774,421	31.4	10,623,870	32.0	9,738,909	29.8	10,870,226	31.2	11,663,832	31.4	11,400,300	31.3
Japan	4,214,464	14.6	4,431,384	14.2	4,535,939	13.7	4,272,069	13.1	4,435,000	12.7	4,956,100	13.3	5,143,283	14.1
Germany	-		1,837,100	5.9	2,220,307	6.7	2,439,013	7.5	2,668,131	7.7	2,817,361	7.6	2,827,178	7.8
France	1,437,198	5.0	1,691,016	5.4	1,804,687	5.4	1,651,214	5.1	1,866,210	5.4	1,946,852	5.2	*1,953,758	5.4
Netherlands	*944,000	3.3	*944,000	3.0	*944,000	2.8	*944,000	2.9	*944,000	2.7	*944,000	2.5	*944,000	2.6

Commodity data are provided by the United Nations Statistical Division. The symbol * means that data are estimated. For additional notes, see Appendix I.

283

Product Tables
Toluene

Unit of Measure: Metric tons.

	1990 Value	%	1991 Value	%	1992 Value	%	1993 Value	%	1994 Value	%	1995 Value	%	1996 Value	%
Total Production	9,300,088	100.0	9,872,483	100.0	9,366,248	100.0	9,342,406	100.0	10,474,999	100.0	9,842,145	100.0	9,771,280	100.0
Regions														
Africa	461	0.0	461	0.0	461	0.0	461	0.0	461	0.0	461	0.0	461	0.0
America, North	3,573,727	38.4	3,579,577	36.3	3,095,517	33.0	3,016,938	32.3	3,781,374	36.1	3,447,920	35.0	3,497,985	35.8
America, South	237,390	2.6	231,121	2.3	220,324	2.4	296,144	3.2	315,046	3.0	305,346	3.1	317,280	3.2
Asia	3,604,658	38.8	3,641,622	36.9	3,795,272	40.5	3,776,709	40.4	3,734,070	35.6	3,749,967	38.1	3,707,147	37.9
Europe	1,883,852	20.3	2,419,703	24.5	2,254,674	24.1	2,252,155	24.1	2,644,049	25.2	2,338,451	23.8	2,248,407	23.0
Leading Producers														
United States of America	2,816,418	30.3	2,856,521	28.9	2,342,423	25.0	2,276,572	24.4	3,068,182	29.3	*2,740,337	27.8	*2,759,357	28.2
Japan	1,110,517	11.9	1,150,517	11.7	1,181,438	12.6	1,219,515	13.1	1,245,000	11.9	1,307,657	13.3	1,370,315	14.0
Germany	-		595,700	6.0	522,380	5.6	449,060	4.8	596,948	5.7	514,617	5.2	702,986	7.2
Korea, Republic of	407,478	4.4	569,185	5.8	773,760	8.3	859,759	9.2	953,665	9.1	979,095	9.9	*959,771	9.8
Russian Federation	749,856	8.1	616,010	6.2	589,033	6.3	477,054	5.1	344,888	3.3	289,924	2.9	189,000	1.9

Xylenes (Orthoxylene, Metaxylene and Paraxylene)

Unit of Measure: Metric tons.

	1990 Value	%	1991 Value	%	1992 Value	%	1993 Value	%	1994 Value	%	1995 Value	%	1996 Value	%
Total Production	12,169,218	100.0	12,937,572	100.0	13,789,438	100.0	13,847,298	100.0	15,061,177	100.0	16,000,514	100.0	15,444,340	100.0
Regions														
Africa	12	0.0	12	0.0	12	0.0	12	0.0	12	0.0	12	0.0	12	0.0
America, North	3,728,329	30.6	3,693,756	28.6	3,792,018	27.5	3,819,639	27.6	4,134,605	27.5	4,078,607	25.5	4,127,505	26.7
America, South	467,808	3.8	412,021	3.2	400,290	2.9	413,152	3.0	420,026	2.8	430,776	2.7	389,078	2.5
Asia	5,637,552	46.3	6,162,463	47.6	6,986,420	50.7	7,213,164	52.1	7,747,657	51.4	8,654,377	54.1	8,133,117	52.7
Europe	2,335,517	19.2	2,669,320	20.6	2,610,698	18.9	2,401,331	17.3	2,758,877	18.3	2,836,742	17.7	2,794,628	18.1
Leading Producers														
United States of America	2,815,656	23.1	2,866,253	22.2	2,981,649	21.6	3,003,812	21.7	3,239,163	21.5	*3,107,102	19.4	*3,169,601	20.5
Japan	2,651,508	21.8	2,917,705	22.6	3,208,752	23.3	3,462,229	25.0	3,627,189	24.1	4,154,176	26.0	3,991,134	25.8
Korea, Republic of	572,554	4.7	750,223	5.8	1,236,399	9.0	1,296,461	9.4	1,552,916	10.3	1,863,810	11.6	*1,554,247	10.1
Russian Federation	869,675	7.1	835,172	6.5	805,908	5.8	652,137	4.7	650,701	4.3	620,510	3.9	525,497	3.4
Germany	-		544,100	4.2	552,641	4.0	561,182	4.1	669,727	4.4	680,829	4.3	713,885	4.6

Carbon Tetrachloride

Unit of Measure: Metric tons.

	1990 Value	%	1991 Value	%	1992 Value	%	1993 Value	%	1994 Value	%	1995 Value	%	1996 Value	%
Total Production	798,207	100.0	715,349	100.0	732,260	100.0	682,271	100.0	572,285	100.0	508,167	100.0	495,078	100.0
Regions														
America, North	187,470	23.5	142,944	20.0	209,055	28.5	198,389	29.1	187,723	32.8	177,057	34.8	166,391	33.6
America, South	27,266	3.4	31,670	4.4	42,921	5.9	44,342	6.5	31,270	5.5	21,646	4.3	26,023	5.3
Asia	52,039	6.5	51,475	7.2	49,539	6.8	53,171	7.8	36,705	6.4	20,642	4.1	5,293	1.1
Europe	531,432	66.6	489,260	68.4	430,745	58.8	386,369	56.6	316,587	55.3	288,821	56.8	297,370	60.1
Leading Producers														
United States of America	187,470	23.5	142,944	20.0	*209,055	28.5	*198,389	29.1	*187,723	32.8	*177,057	34.8	*166,391	33.6
Ukraine	260,422	32.6	210,940	29.5	157,154	21.5	111,946	16.4	66,612	11.6	51,465	10.1	50,527	10.2
Germany	*106,955	13.4	*106,955	15.0	106,955	14.6	*106,955	15.7	*106,955	18.7	*106,955	21.0	*106,955	21.6
Italy	40,474	5.1	55,595	7.8	57,544	7.9	45,817	6.7	24,433	4.3	14,876	2.9	*27,349	5.5
Japan	52,039	6.5	51,475	7.2	49,539	6.8	53,171	7.8	36,705	6.4	20,642	4.1	5,293	1.1

Commodity data are provided by the United Nations Statistical Division. The symbol * means that data are estimated. For additional notes, see Appendix I.

Product Tables
Dichloromethane

Unit of Measure: Metric tons.

	1990 Value	%	1991 Value	%	1992 Value	%	1993 Value	%	1994 Value	%	1995 Value	%	1996 Value	%
Total Production	855,418	100.0	851,141	100.0	861,716	100.0	901,356	100.0	919,535	100.0	932,552	100.0	953,730	100.0
Regions														
America, North	549,732	64.3	537,197	63.1	544,709	63.2	574,004	63.7	603,300	65.6	603,017	64.7	615,061	64.5
Asia	77,466	9.1	82,259	9.7	83,519	9.7	93,349	10.4	88,877	9.7	96,944	10.4	100,845	10.6
Europe	228,220	26.7	231,685	27.2	233,489	27.1	234,003	26.0	227,358	24.7	232,591	24.9	237,824	24.9
Leading Producers														
Mexico	*340,616	39.8	*360,549	42.4	*380,482	44.2	*400,414	44.4	*420,347	45.7	*440,279	47.2	*460,212	48.3
United States of America	209,116	24.4	176,648	20.8	164,227	19.1	173,590	19.3	182,953	19.9	*162,738	17.5	*154,849	16.2
Germany	*76,716	9.0	*76,716	9.0	*76,716	8.9	77,448	8.6	71,021	7.7	76,472	8.2	81,922	8.6
Japan	77,466	9.1	82,259	9.7	83,519	9.7	93,349	10.4	88,877	9.7	96,944	10.4	100,845	10.6
Spain	14,177	1.7	17,642	2.1	19,446	2.3	19,228	2.1	19,011	2.1	18,793	2.0	18,575	1.9

Trichloroethylene

Unit of Measure: Metric tons.

	1990 Value	%	1991 Value	%	1992 Value	%	1993 Value	%	1994 Value	%	1995 Value	%	1996 Value	%
Total Production	616,008	100.0	611,772	100.0	665,669	100.0	707,621	100.0	739,993	100.0	786,619	100.0	822,642	100.0
Regions														
America, North	479,952	77.9	513,530	83.9	547,108	82.2	580,686	82.1	614,264	83.0	647,841	82.4	681,419	82.8
America, South	3,774	0.6	2,460	0.4	2,364	0.4	2,075	0.3	1,348	0.2	276	0.0	-	
Asia	56,850	9.2	51,679	8.4	61,080	9.2	68,416	9.7	77,159	10.4	83,049	10.6	90,350	11.0
Europe	75,432	12.2	44,102	7.2	55,117	8.3	56,444	8.0	47,222	6.4	55,453	7.0	50,873	6.2
Leading Producers														
United States of America	*479,952	77.9	*513,530	83.9	*547,108	82.2	*580,686	82.1	*614,264	83.0	*647,841	82.4	*681,419	82.8
Japan	56,850	9.2	51,679	8.4	61,080	9.2	68,416	9.7	77,159	10.4	83,049	10.6	90,350	11.0
Romania	17,335	2.8	9,357	1.5	12,629	1.9	13,401	1.9	7,571	1.0	17,216	2.2	14,051	1.7
Spain	8,122	1.3	4,065	0.7	8	0.0	*1,977	0.3	-		-		-	
Slovakia	15,779	2.6	6,157	1.0	*11,632	1.7	*10,831	1.5	*10,030	1.4	*9,229	1.2	*8,427	1.0

Methanol (Methyl Alcohol)

Unit of Measure: Metric tons.

	1990 Value	%	1991 Value	%	1992 Value	%	1993 Value	%	1994 Value	%	1995 Value	%	1996 Value	%
Total Production	14,269,989	100.0	16,283,058	100.0	16,072,780	100.0	17,282,125	100.0	18,160,640	100.0	18,990,604	100.0	19,215,057	100.0
Regions														
Africa	93,000	0.7	81,000	0.5	94,900	0.6	84,000	0.5	80,200	0.4	101,548	0.5	104,922	0.5
America, North	6,297,600	44.1	6,762,466	41.5	6,477,743	40.3	7,691,939	44.5	8,971,043	49.4	7,998,482	42.1	8,555,672	44.5
America, South	242,298	1.7	299,476	1.8	300,767	1.9	334,022	1.9	334,942	1.8	337,584	1.8	325,875	1.7
Asia	5,059,240	35.5	5,460,556	33.5	5,568,551	34.6	5,831,902	33.7	5,081,890	28.0	7,069,775	37.2	6,818,616	35.5
Europe	2,577,850	18.1	3,679,560	22.6	3,630,819	22.6	3,340,262	19.3	3,692,565	20.3	3,483,214	18.3	3,409,971	17.7
Leading Producers														
United States of America	3,784,957	26.5	3,948,035	24.2	3,666,060	22.8	4,765,318	27.6	5,522,196	30.4	*4,540,952	23.9	*4,644,593	24.2
Canada	2,000,000	14.0	2,222,527	13.6	2,153,989	13.4	2,253,195	13.0	*2,307,368	12.7	*2,374,398	12.5	*2,441,429	12.7
Germany	-		1,231,541	7.6	1,290,994	8.0	1,202,189	7.0	1,438,327	7.9	1,425,795	7.5	1,546,958	8.1
Russian Federation	*859,970	6.0	*859,970	5.3	*859,970	5.4	*859,970	5.0	1,909	0.0	1,523,000	8.0	1,055,000	5.5
Trinidad and Tobago	402,536	2.8	452,881	2.8	481,718	3.0	492,757	2.9	1,019,524	5.6	963,025	5.1	1,354,462	7.0

Commodity data are provided by the United Nations Statistical Division. The symbol * means that data are estimated. For additional notes, see Appendix I.

285

Product Tables
Butyl Alcohol (Butanols)

Unit of Measure: Metric tons.

	1990 Value	%	1991 Value	%	1992 Value	%	1993 Value	%	1994 Value	%	1995 Value	%	1996 Value	%
Total Production	3,111,637	100.0	3,699,765	100.0	3,772,903	100.0	3,840,073	100.0	4,023,356	100.0	4,186,606	100.0	4,262,259	100.0
Regions														
America, North	1,402,408	45.1	1,483,824	40.1	1,565,240	41.5	1,646,656	42.9	1,728,072	43.0	1,809,488	43.2	1,890,904	44.4
America, South	52,383	1.7	53,826	1.5	52,305	1.4	50,990	1.3	53,362	1.3	37,715	0.9	51,375	1.2
Asia	917,240	29.5	961,299	26.0	1,005,014	26.6	1,031,286	26.9	1,037,154	25.8	1,126,259	26.9	1,046,002	24.5
Europe	739,606	23.8	1,200,816	32.5	1,150,344	30.5	1,111,141	28.9	1,204,769	29.9	1,213,144	29.0	1,273,978	29.9
Leading Producers														
United States of America	*1,402,408	45.1	*1,483,824	40.1	*1,565,240	41.5	*1,646,656	42.9	*1,728,072	43.0	*1,809,488	43.2	*1,890,904	44.4
Germany	-		470,800	12.7	426,387	11.3	381,974	9.9	465,085	11.6	473,108	11.3	537,419	12.6
Japan	300,096	9.6	334,018	9.0	371,188	9.8	388,099	10.1	380,295	9.5	424,198	10.1	414,978	9.7
Russian Federation	*201,540	6.5	*201,540	5.4	*201,540	5.3	*201,540	5.2	205,597	5.1	239,022	5.7	160,000	3.8
China	*45,832	1.5	*45,832	1.2	*45,832	1.2	*45,832	1.2	*45,832	1.1	*45,832	1.1	*45,832	1.1

Ethanediol (Ethylene Glycol)

Unit of Measure: Metric tons.

	1990 Value	%	1991 Value	%	1992 Value	%	1993 Value	%	1994 Value	%	1995 Value	%	1996 Value	%
Total Production	4,052,044	100.0	4,332,595	100.0	4,744,198	100.0	4,798,449	100.0	5,321,470	100.0	5,505,507	100.0	5,459,754	100.0
Regions														
America, North	2,299,942	56.8	2,181,568	50.4	2,326,153	49.0	2,358,831	49.2	2,761,510	51.9	2,617,949	47.6	2,674,491	49.0
America, South	99,494	2.5	114,770	2.6	99,224	2.1	111,455	2.3	114,084	2.1	123,788	2.2	114,231	2.1
Asia	760,260	18.8	982,068	22.7	1,221,842	25.8	1,187,965	24.8	1,299,358	24.4	1,533,272	27.8	1,486,806	27.2
Europe	892,347	22.0	1,054,188	24.3	1,096,978	23.1	1,140,198	23.8	1,146,518	21.5	1,230,498	22.4	1,184,227	21.7
Leading Producers														
United States of America	2,299,942	56.8	2,181,568	50.4	2,326,153	49.0	2,358,831	49.2	2,761,510	51.9	*2,617,949	47.6	*2,674,491	49.0
Japan	500,604	12.4	608,943	14.1	560,490	11.8	526,812	11.0	567,136	10.7	709,418	12.9	750,754	13.8
Belgium	*385,063	9.5	*385,063	8.9	*385,063	8.1	*385,063	8.0	*385,063	7.2	385,063	7.0	*385,063	7.1
Germany	-		200,900	4.6	216,245	4.6	231,590	4.8	222,326	4.2	277,842	5.0	242,610	4.4
Korea, Republic of	69,895	1.7	151,067	3.5	371,192	7.8	356,366	7.4	400,688	7.5	434,783	7.9	*371,661	6.8

Vinyl Chloride (Chloroethylene)

Unit of Measure: 1,000 Metric tons.

	1990 Value	%	1991 Value	%	1992 Value	%	1993 Value	%	1994 Value	%	1995 Value	%	1996 Value	%
Total Production	4,823,976	100.0	3,621,306	100.0	2,418,709	100.0	1,215,967	100.0	13,170	100.0	874,621	100.0	417,167	100.0
Regions														
America, North	4,818,754	99.9	3,615,750	99.8	2,412,745	99.8	1,209,741	99.5	6,736	51.1	868,029	99.2	410,341	98.4
America, South	161	0.0	129	0.0	96	0.0	90	0.0	109	0.8	114	0.0	141	0.0
Asia	3,201	0.1	3,449	0.1	3,644	0.2	3,729	0.3	3,792	28.8	3,902	0.4	4,034	1.0
Europe	1,861	0.0	1,978	0.1	2,224	0.1	2,407	0.2	2,532	19.2	2,576	0.3	2,651	0.6
Leading Producers														
United States of America	4,818,754	99.9	3,615,750	99.8	2,412,745	99.8	1,209,741	99.5	6,736	51.1	*868,029	99.2	*410,341	98.4
Japan	*1,896	0.0	*1,971	0.1	*2,046	0.1	*2,121	0.2	*2,197	16.7	*2,272	0.3	*2,347	0.6
Belgium	*881	0.0	*881	0.0	*881	0.0	*881	0.1	904	6.9	857	0.1	*881	0.2
Germany	-		-		346	0.0	460	0.0	435	3.3	589	0.1	620	0.1
Korea, Republic of	374	0.0	547	0.0	667	0.0	680	0.1	687	5.2	675	0.1	*756	0.2

Product Tables
Glycerine (Glycerol)

Unit of Measure: Metric tons.

	1990 Value	%	1991 Value	%	1992 Value	%	1993 Value	%	1994 Value	%	1995 Value	%	1996 Value	%
Total Production	578,015	100.0	621,163	100.0	608,496	100.0	596,595	100.0	633,543	100.0	625,472	100.0	623,074	100.0
Regions														
Africa	12,151	2.1	9,974	1.6	10,122	1.7	10,219	1.7	8,348	1.3	6,805	1.1	3,639	0.6
America, North	181,296	31.4	185,066	29.8	188,419	31.0	198,415	33.3	207,468	32.7	213,977	34.2	219,446	35.2
America, South	30,025	5.2	32,435	5.2	27,894	4.6	22,732	3.8	24,682	3.9	21,684	3.5	24,611	3.9
Asia	112,517	19.5	141,961	22.9	142,522	23.4	129,602	21.7	145,873	23.0	147,222	23.5	141,317	22.7
Europe	233,577	40.4	243,279	39.2	231,090	38.0	227,179	38.1	238,724	37.7	227,336	36.3	225,613	36.2
Oceania	8,448	1.5	8,448	1.4	8,448	1.4	8,448	1.4	8,448	1.3	8,448	1.4	8,448	1.4
Leading Producers														
United States of America	*170,667	29.5	*176,790	28.5	*182,913	30.1	*189,036	31.7	*195,159	30.8	*201,283	32.2	*207,406	33.3
Germany	*87,040	15.1	*87,040	14.0	80,226	13.2	84,543	14.2	96,350	15.2	*87,040	13.9	*87,040	14.0
Japan	52,279	9.0	48,959	7.9	47,406	7.8	41,831	7.0	48,247	7.6	48,825	7.8	51,421	8.3
France	18,660	3.2	25,100	4.0	23,028	3.8	20,616	3.5	21,180	3.3	*22,065	3.5	*21,616	3.5
Russian Federation	-		32,548	5.2	30,207	5.0	20,430	3.4	23,027	3.6	19,143	3.1	9,522	1.5

Phenol

Unit of Measure: Metric tons.

	1990 Value	%	1991 Value	%	1992 Value	%	1993 Value	%	1994 Value	%	1995 Value	%	1996 Value	%
Total Production	4,663,384	100.0	5,094,491	100.0	5,222,533	100.0	4,918,417	100.0	5,416,201	100.0	5,714,632	100.0	6,027,087	100.0
Regions														
Africa	143	0.0	143	0.0	143	0.0	143	0.0	143	0.0	143	0.0	143	0.0
America, North	1,697,585	36.4	1,725,680	33.9	1,857,974	35.6	1,641,147	33.4	1,990,057	36.7	2,015,429	35.3	2,074,051	34.4
America, South	97,138	2.1	98,445	1.9	91,354	1.7	105,259	2.1	112,799	2.1	114,819	2.0	112,114	1.9
Asia	1,517,491	32.5	1,554,360	30.5	1,560,637	29.9	1,473,971	30.0	1,557,700	28.8	1,736,477	30.4	1,665,841	27.6
Europe	1,351,027	29.0	1,715,863	33.7	1,712,425	32.8	1,697,897	34.5	1,755,503	32.4	1,847,763	32.3	2,174,938	36.1
Leading Producers														
United States of America	1,604,623	34.4	1,631,620	32.0	1,762,481	33.7	1,544,222	31.4	1,891,700	34.9	*1,915,640	33.5	*1,972,830	32.7
Germany	-		469,700	9.2	478,997	9.2	488,293	9.9	497,590	9.2	586,863	10.3	929,595	15.4
Japan	403,071	8.6	495,041	9.7	538,083	10.3	512,542	10.4	669,837	12.4	771,351	13.5	761,610	12.6
Italy	370,000	7.9	350,915	6.9	337,566	6.5	345,548	7.0	367,997	6.8	351,340	6.1	*348,278	5.8
Russian Federation	450,546	9.7	352,471	6.9	280,698	5.4	202,238	4.1	135,290	2.5	172,806	3.0	99,975	1.7

Diethyl Ether

Unit of Measure: Metric tons.

	1990 Value	%	1991 Value	%	1992 Value	%	1993 Value	%	1994 Value	%	1995 Value	%	1996 Value	%
Total Production	75,560	100.0	75,944	100.0	70,913	100.0	77,536	100.0	89,820	100.0	97,894	100.0	111,962	100.0
Regions														
America, North	35,677	47.2	35,677	47.0	35,677	50.3	28,430	36.7	36,346	40.5	40,416	41.3	37,515	33.5
America, South	1,825	2.4	1,809	2.4	1,786	2.5	1,704	2.2	1,689	1.9	1,584	1.6	1,649	1.5
Asia	48	0.1	44	0.1	31	0.0	26	0.0	13	0.0	30	0.0	31	0.0
Europe	38,010	50.3	38,414	50.6	33,420	47.1	47,376	61.1	51,772	57.6	55,864	57.1	72,768	65.0
Leading Producers														
Bulgaria	33,188	43.9	33,531	44.2	28,464	40.1	42,330	54.6	46,655	51.9	50,327	51.4	67,853	60.6
Mexico	*35,677	47.2	*35,677	47.0	*35,677	50.3	28,430	36.7	36,346	40.5	40,416	41.3	37,515	33.5
Germany	*4,045	5.4	*4,045	5.3	*4,045	5.7	*4,045	5.2	*4,045	4.5	4,391	4.5	3,698	3.3
Colombia	*1,249	1.7	*1,249	1.6	*1,249	1.8	*1,249	1.6	1,249	1.4	*1,249	1.3	*1,249	1.1
Sweden	740	1.0	825	1.1	900	1.3	*990	1.3	*1,063	1.2	*1,136	1.2	*1,209	1.1

Commodity data are provided by the United Nations Statistical Division. The symbol * means that data are estimated. For additional notes, see Appendix I.

287

Product Tables
Propylene Glycol

Unit of Measure: 1,000 Metric tons.

	1990		1991		1992		1993		1994		1995		1996	
	Value	%	Value	%	Value	%	Value	%	Value	%	Value	%	Value	%
Total Production	342,803	100.0	395,361	100.0	414,455	100.0	433,570	100.0	452,669	100.0	471,789	100.0	490,851	100.0
Regions														
America, North	342,388	99.9	394,983	99.9	414,078	99.9	433,173	99.9	452,258	99.9	471,365	99.9	490,457	99.9
America, South	0	0.0	0	0.0	0	0.0	0	0.0	0	0.0	0	0.0	0	0.0
Asia	69	0.0	65	0.0	69	0.0	63	0.0	66	0.0	68	0.0	70	0.0
Europe	346	0.1	313	0.1	308	0.1	335	0.1	345	0.1	356	0.1	324	0.1
Leading Producers														
United States of America	342,204	99.8	*394,799	99.9	*413,893	99.9	*432,988	99.9	*452,082	99.9	*471,176	99.9	*490,270	99.9
Mexico	*184	0.1	*184	0.0	*184	0.0	185	0.0	176	0.0	189	0.0	187	0.0
Bulgaria	112	0.0	78	0.0	76	0.0	98	0.0	107	0.0	117	0.0	*84	0.0
Japan	69	0.0	65	0.0	69	0.0	62	0.0	66	0.0	68	0.0	70	0.0
Portugal	*29	0.0	*29	0.0	*29	0.0	*29	0.0	*29	0.0	*29	0.0	*29	0.0

Ethylene Oxide

Unit of Measure: Metric tons.

	1990		1991		1992		1993		1994		1995		1996	
	Value	%	Value	%	Value	%	Value	%	Value	%	Value	%	Value	%
Total Production	6,724,931	100.0	6,662,449	100.0	6,777,298	100.0	6,652,993	100.0	6,919,369	100.0	6,516,525	100.0	6,479,072	100.0
Regions														
America, North	3,588,966	53.4	2,906,865	43.6	3,178,626	46.9	2,960,662	44.5	3,207,429	46.4	3,197,483	49.1	3,225,117	49.8
America, South	127,243	1.9	150,371	2.3	142,648	2.1	149,487	2.2	163,474	2.4	161,338	2.5	163,711	2.5
Asia	1,180,860	17.6	1,808,714	27.1	1,647,959	24.3	1,511,630	22.7	1,474,427	21.3	1,564,605	24.0	1,520,190	23.5
Europe	1,827,861	27.2	1,796,499	27.0	1,808,066	26.7	2,031,214	30.5	2,074,039	30.0	1,593,100	24.4	1,570,054	24.2
Leading Producers														
United States of America	3,048,110	45.3	2,380,363	35.7	2,643,813	39.0	2,417,537	36.3	2,655,992	38.4	*2,637,734	40.5	*2,657,057	41.0
Japan	673,985	10.0	760,096	11.4	720,873	10.6	680,272	10.2	716,098	10.3	802,506	12.3	839,509	13.0
Germany	*389,875	5.8	*389,875	5.9	*389,875	5.8	623,281	9.4	656,930	9.5	148,302	2.3	130,989	2.0
Canada	400,000	5.9	*377,333	5.7	*377,333	5.6	*377,333	5.7	*377,333	5.5	*377,333	5.8	*377,333	5.8
Russian Federation	-		474,374	7.1	434,604	6.4	350,489	5.3	280,684	4.1	283,984	4.4	199,175	3.1

Acetaldehyde (Ethanol)

Unit of Measure: Metric tons.

	1990		1991		1992		1993		1994		1995		1996	
	Value	%	Value	%	Value	%	Value	%	Value	%	Value	%	Value	%
Total Production	2,391,859	100.0	2,720,972	100.0	2,713,969	100.0	2,675,479	100.0	2,686,495	100.0	2,767,813	100.0	2,738,876	100.0
Regions														
America, North	152,193	6.4	153,748	5.7	155,304	5.7	156,859	5.9	158,415	5.9	159,971	5.8	161,526	5.9
America, South	66,169	2.8	92,723	3.4	88,848	3.3	75,738	2.8	60,341	2.2	61,642	2.2	86,250	3.1
Asia	806,559	33.7	807,529	29.7	814,923	30.0	773,703	28.9	800,165	29.8	841,278	30.4	815,828	29.8
Europe	1,366,938	57.1	1,666,972	61.3	1,654,894	61.0	1,669,179	62.4	1,667,574	62.1	1,704,923	61.6	1,675,272	61.2
Leading Producers														
France	*853,000	35.7	*853,000	31.3	*853,000	31.4	*853,000	31.9	*853,000	31.8	*853,000	30.8	*853,000	31.1
Germany	-		337,400	12.4	327,656	12.1	348,179	13.0	358,949	13.4	385,850	13.9	346,776	12.7
Japan	383,466	16.0	384,127	14.1	391,212	14.4	349,684	13.1	369,364	13.7	395,283	14.3	418,714	15.3
Mexico	*152,193	6.4	*153,748	5.7	*155,304	5.7	*156,859	5.9	*158,415	5.9	*159,971	5.8	*161,526	5.9
Brazil	66,169	2.8	83,323	3.1	80,525	3.0	72,278	2.7	55,056	2.0	54,532	2.0	*76,840	2.8

Commodity data are provided by the United Nations Statistical Division. The symbol * means that data are estimated. For additional notes, see Appendix I.

Product Tables
Methanol (Formaldehyde)

Unit of Measure: Metric tons.

	1990		1991		1992		1993		1994		1995		1996	
	Value	%	Value	%	Value	%	Value	%	Value	%	Value	%	Value	%
Total Production	7,745,077	100.0	7,641,956	100.0	8,358,675	100.0	8,204,450	100.0	8,567,484	100.0	8,512,782	100.0	8,405,522	100.0
Regions														
America, North	3,139,784	40.5	3,236,364	42.3	3,989,844	47.7	4,000,709	48.8	4,008,282	46.8	3,975,079	46.7	4,091,493	48.7
America, South	196,291	2.5	212,325	2.8	221,585	2.7	262,388	3.2	277,982	3.2	292,738	3.4	296,662	3.5
Asia	1,327,951	17.1	1,327,362	17.4	1,373,764	16.4	1,265,823	15.4	1,562,823	18.2	1,460,997	17.2	1,304,448	15.5
Europe	3,081,050	39.8	2,865,906	37.5	2,773,482	33.2	2,675,530	32.6	2,718,397	31.7	2,783,969	32.7	2,712,920	32.3
Leading Producers														
United States of America	2,908,500	37.6	2,999,191	39.2	3,754,955	44.9	3,713,887	45.3	3,703,109	43.2	*3,682,254	43.3	*3,788,707	45.1
Germany	*728,163	9.4	*728,163	9.5	*728,163	8.7	*728,163	8.9	724,970	8.5	678,867	8.0	780,652	9.3
Japan	534,174	6.9	*547,640	7.2	*565,430	6.8	*583,219	7.1	*601,009	7.0	*618,798	7.3	*636,588	7.6
Romania	330,578	4.3	*376,418	4.9	*356,899	4.3	*337,380	4.1	*317,862	3.7	*298,343	3.5	*278,824	3.3
Sweden	244,452	3.2	219,668	2.9	202,860	2.4	237,412	2.9	260,113	3.0	279,231	3.3	*261,503	3.1

Acetone

Unit of Measure: Metric tons.

	1990		1991		1992		1993		1994		1995		1996	
	Value	%	Value	%	Value	%	Value	%	Value	%	Value	%	Value	%
Total Production	2,315,438	100.0	2,255,818	100.0	2,247,948	100.0	2,148,252	100.0	2,219,349	100.0	2,294,015	100.0	2,291,862	100.0
Regions														
America, North	1,105,636	47.8	1,109,136	49.2	1,152,067	51.2	1,146,123	53.4	1,249,237	56.3	1,228,230	53.5	1,254,404	54.7
America, South	76,093	3.3	80,445	3.6	76,851	3.4	82,614	3.8	81,293	3.7	88,108	3.8	85,113	3.7
Asia	704,315	30.4	687,324	30.5	637,990	28.4	547,101	25.5	506,881	22.8	577,426	25.2	552,202	24.1
Europe	429,394	18.5	378,912	16.8	381,040	17.0	372,414	17.3	381,937	17.2	400,251	17.4	400,144	17.5
Leading Producers														
United States of America	1,056,654	45.6	1,064,701	47.2	1,104,319	49.1	1,102,426	51.3	1,207,704	54.4	*1,188,860	51.8	*1,217,198	53.1
Japan	333,708	14.4	369,100	16.4	377,340	16.8	350,312	16.3	352,776	15.9	396,120	17.3	417,119	18.2
Russian Federation	312,405	13.5	255,661	11.3	198,872	8.8	137,435	6.4	90,780	4.1	110,622	4.8	66,732	2.9
United Kingdom	128,462	5.5	125,885	5.6	*128,860	5.7	*128,707	6.0	*128,553	5.8	*128,400	5.6	*128,247	5.6
France	*108,037	4.7	*108,037	4.8	*108,037	4.8	*108,037	5.0	*108,037	4.9	*108,037	4.7	*108,037	4.7

Acetates (Methyl, Ethyl, Butyl)

Unit of Measure: Metric tons.

	1990		1991		1992		1993		1994		1995		1996	
	Value	%	Value	%	Value	%	Value	%	Value	%	Value	%	Value	%
Total Production	1,574,633	100.0	1,684,154	100.0	1,652,082	100.0	1,658,698	100.0	1,735,321	100.0	1,706,716	100.0	1,742,797	100.0
Regions														
America, North	238,052	15.1	285,767	17.0	248,780	15.1	241,015	14.5	288,678	16.6	292,970	17.2	306,101	17.6
America, South	42,628	2.7	54,836	3.3	57,437	3.5	65,710	4.0	59,929	3.5	58,912	3.5	61,589	3.5
Asia	206,503	13.1	207,759	12.3	216,435	13.1	225,111	13.6	233,787	13.5	242,463	14.2	251,139	14.4
Europe	1,087,450	69.1	1,135,792	67.4	1,129,430	68.4	1,126,862	67.9	1,152,927	66.4	1,112,371	65.2	1,123,968	64.5
Leading Producers														
Germany	*404,725	25.7	*404,725	24.0	*404,725	24.5	377,730	22.8	431,721	24.9	*404,725	23.7	*404,725	23.2
United States of America	238,052	15.1	285,767	17.0	248,780	15.1	241,015	14.5	288,678	16.6	*292,970	17.2	*306,101	17.6
Japan	201,225	12.8	*202,481	12.0	*211,157	12.8	*219,833	13.3	*228,509	13.2	*237,185	13.9	*245,861	14.1
Italy	114,100	7.2	145,302	8.6	142,477	8.6	128,110	7.7	126,427	7.3	113,429	6.6	124,678	7.2
Brazil	42,588	2.7	54,796	3.3	57,401	3.5	65,670	4.0	59,885	3.5	58,872	3.4	61,549	3.5

Commodity data are provided by the United Nations Statistical Division. The symbol * means that data are estimated. For additional notes, see Appendix I.

289

Product Tables
Acetic Acid

Unit of Measure: Metric tons.

	1990 Value	%	1991 Value	%	1992 Value	%	1993 Value	%	1994 Value	%	1995 Value	%	1996 Value	%
Total Production	3,463,405	100.0	3,905,222	100.0	3,827,141	100.0	3,647,439	100.0	3,977,327	100.0	4,123,770	100.0	4,125,185	100.0
Regions														
America, North	1,769,813	51.1	1,714,927	43.9	1,673,495	43.7	1,544,255	42.3	1,852,072	46.6	1,773,474	43.0	1,804,006	43.7
America, South	101,984	2.9	123,894	3.2	116,738	3.1	106,731	2.9	79,816	2.0	86,127	2.1	90,916	2.2
Asia	925,861	26.7	1,079,990	27.7	1,053,108	27.5	1,072,798	29.4	1,056,018	26.6	1,171,954	28.4	1,244,822	30.2
Europe	665,748	19.2	986,412	25.3	983,800	25.7	923,655	25.3	989,421	24.9	1,092,215	26.5	985,440	23.9
Leading Producers														
United States of America	1,701,303	49.1	1,639,897	42.0	1,630,099	42.6	1,508,940	41.4	1,807,387	45.4	*1,738,051	42.1	*1,776,069	43.1
Japan	456,940	13.2	453,641	11.6	446,057	11.7	479,275	13.1	514,202	12.9	568,661	13.8	593,138	14.4
Germany	-		321,000	8.2	330,823	8.6	340,646	9.3	405,577	10.2	473,759	11.5	429,053	10.4
Ukraine	156,343	4.5	166,466	4.3	161,943	4.2	93,067	2.6	92,696	2.3	123,892	3.0	71,308	1.7
Brazil	87,904	2.5	109,926	2.8	104,301	2.7	99,517	2.7	71,159	1.8	72,189	1.8	75,574	1.8

Formic Acid

Unit of Measure: Metric tons.

	1990 Value	%	1991 Value	%	1992 Value	%	1993 Value	%	1994 Value	%	1995 Value	%	1996 Value	%
Total Production	170,856	100.0	169,762	100.0	163,870	100.0	178,679	100.0	181,525	100.0	170,360	100.0	179,342	100.0
Regions														
America, South	2,930	1.7	3,106	1.8	2,802	1.7	2,978	1.7	2,995	1.6	3,011	1.8	3,027	1.7
Asia	20,188	11.8	19,684	11.6	20,350	12.4	20,841	11.7	22,807	12.6	17,705	10.4	20,549	11.5
Europe	147,738	86.5	146,972	86.6	140,719	85.9	154,859	86.7	155,723	85.8	149,644	87.8	155,766	86.9
Leading Producers														
Germany	*92,645	54.2	92,316	54.4	87,056	53.1	98,563	55.2	*92,645	51.0	*92,645	54.4	*92,645	51.7
Finland	35,702	20.9	34,554	20.4	33,538	20.5	35,069	19.6	37,356	20.6	32,060	18.8	39,010	21.8
Japan	11,664	6.8	12,177	7.2	12,006	7.3	11,613	6.5	10,845	6.0	10,677	6.3	10,848	6.0
Sweden	8,921	5.2	9,477	5.6	9,617	5.9	10,705	6.0	15,185	8.4	14,388	8.4	*13,545	7.6
Indonesia	8,233	4.8	7,251	4.3	8,166	5.0	9,080	5.1	11,929	6.6	6,946	4.1	*9,663	5.4

Phthalic Anhydride

Unit of Measure: Metric tons.

	1990 Value	%	1991 Value	%	1992 Value	%	1993 Value	%	1994 Value	%	1995 Value	%	1996 Value	%
Total Production	2,535,798	100.0	2,668,265	100.0	2,800,916	100.0	2,706,720	100.0	2,839,877	100.0	2,806,664	100.0	2,829,824	100.0
Regions														
America, North	462,908	18.3	310,763	11.6	450,815	16.1	431,257	15.9	486,115	17.1	457,244	16.3	459,084	16.2
America, South	86,862	3.4	100,474	3.8	119,986	4.3	122,240	4.5	137,187	4.8	119,763	4.3	134,752	4.8
Asia	996,907	39.3	1,039,996	39.0	1,053,271	37.6	1,028,450	38.0	1,088,002	38.3	1,131,381	40.3	1,152,778	40.7
Europe	989,121	39.0	1,217,032	45.6	1,176,844	42.0	1,124,773	41.6	1,128,573	39.7	1,098,276	39.1	1,083,210	38.3
Leading Producers														
United States of America	426,483	16.8	266,277	10.0	407,350	14.5	387,113	14.3	435,130	15.3	*410,421	14.6	*412,722	14.6
Japan	301,228	11.9	308,025	11.5	309,045	11.0	276,605	10.2	307,911	10.8	316,020	11.3	342,201	12.1
Germany	-		198,700	7.4	203,820	7.3	208,940	7.7	206,069	7.3	191,880	6.8	184,412	6.5
Belgium	123,124	4.9	172,278	6.5	185,340	6.6	152,842	5.6	160,971	5.7	149,285	5.3	137,599	4.9
Italy	136,419	5.4	144,924	5.4	119,572	4.3	132,383	4.9	131,787	4.6	117,960	4.2	*126,066	4.5

Commodity data are provided by the United Nations Statistical Division. The symbol * means that data are estimated. For additional notes, see Appendix I.

Product Tables
Acetylsalicylic Acid (Aspirin)

Unit of Measure: Metric tons.

	1990 Value	%	1991 Value	%	1992 Value	%	1993 Value	%	1994 Value	%	1995 Value	%	1996 Value	%
Total Production	15,493	100.0	15,269	100.0	14,768	100.0	15,148	100.0	14,398	100.0	14,272	100.0	14,328	100.0
Regions														
America, North	10,230	66.0	10,250	67.1	9,900	67.0	9,550	63.0	9,200	63.9	8,850	62.0	8,500	59.3
America, South	1,646	10.6	1,621	10.6	1,335	9.0	1,746	11.5	1,553	10.8	1,468	10.3	1,737	12.1
Europe	3,618	23.3	3,397	22.3	3,533	23.9	3,852	25.4	3,644	25.3	3,954	27.7	4,091	28.6
Leading Producers														
United States of America	10,230	66.0	*10,250	67.1	*9,900	67.0	*9,550	63.0	*9,200	63.9	*8,850	62.0	*8,500	59.3
Spain	2,343	15.1	2,123	13.9	2,258	15.3	*2,405	15.9	*2,542	17.7	*2,679	18.8	*2,816	19.7
Germany	*1,274	8.2	*1,274	8.3	*1,274	8.6	1,447	9.6	1,102	7.7	*1,274	8.9	*1,274	8.9
Brazil	1,017	6.6	992	6.5	559	3.8	*960	6.3	*1,009	7.0	*1,059	7.4	*1,109	7.7
Colombia	*629	4.1	*629	4.1	776	5.3	786	5.2	544	3.8	409	2.9	*629	4.4

Maleic Anhydride

Unit of Measure: 1,000 Metric tons.

	1990 Value	%	1991 Value	%	1992 Value	%	1993 Value	%	1994 Value	%	1995 Value	%	1996 Value	%
Total Production	192,751	100.0	144,658	100.0	96,592	100.0	48,525	100.0	451	100.0	28,397	100.0	6,545	100.0
Regions														
America, North	192,529	99.9	144,452	99.9	96,374	99.8	48,297	99.5	220	48.7	28,166	99.2	6,306	96.4
America, South	8	0.0	8	0.0	10	0.0	10	0.0	17	3.8	13	0.0	17	0.3
Asia	115	0.1	115	0.1	119	0.1	130	0.3	126	28.0	133	0.5	133	2.0
Europe	99	0.1	83	0.1	88	0.1	88	0.2	88	19.5	85	0.3	88	1.3
Leading Producers														
United States of America	192,529	99.9	144,452	99.9	96,374	99.8	48,297	99.5	220	48.7	*28,166	99.2	*6,306	96.4
Japan	101	0.1	100	0.1	104	0.1	115	0.2	112	24.7	117	0.4	118	1.8
Germany	*28	0.0	*28	0.0	*28	0.0	*28	0.1	*28	6.3	27	0.1	30	0.5
Korea, Republic of	*12	0.0	*12	0.0	*12	0.0	*12	0.0	*13	2.8	*13	0.0	*13	0.2
Hungary	9	0.0	9	0.0	9	0.0	10	0.0	10	2.3	9	0.0	12	0.2

Tartaric Acid

Unit of Measure: Metric tons.

	1990 Value	%	1991 Value	%	1992 Value	%	1993 Value	%	1994 Value	%	1995 Value	%	1996 Value	%
Total Production	13,108	100.0	12,602	100.0	8,837	100.0	12,086	100.0	12,671	100.0	13,328	100.0	13,432	100.0
Regions														
America, South	4,482	34.2	3,713	29.5	3,249	36.8	2,448	20.3	2,882	22.7	3,307	24.8	3,250	24.2
Asia	914	7.0	914	7.3	914	10.3	914	7.6	914	7.2	914	6.9	914	6.8
Europe	7,712	58.8	7,975	63.3	4,674	52.9	8,725	72.2	8,875	70.0	9,107	68.3	9,268	69.0
Leading Producers														
Spain	7,127	54.4	7,620	60.5	4,439	50.2	*8,502	70.3	*8,708	68.7	*8,913	66.9	*9,118	67.9
Argentina	4,356	33.2	3,587	28.5	3,123	35.3	2,322	19.2	2,756	21.8	*3,181	23.9	*3,124	23.3
Japan	*908	6.9	*908	7.2	*908	10.3	*908	7.5	*908	7.2	*908	6.8	*908	6.8
Ukraine	426	3.2	240	1.9	100	1.1	94	0.8	48	0.4	83	0.6	48	0.4
Colombia	*126	1.0	*126	1.0	*126	1.4	126	1.0	*126	1.0	*126	0.9	*126	0.9

Product Tables

Aniline

Unit of Measure: Metric tons.

	1990 Value	%	1991 Value	%	1992 Value	%	1993 Value	%	1994 Value	%	1995 Value	%	1996 Value	%
Total Production	1,236,862	100.0	1,232,413	100.0	1,330,639	100.0	1,298,145	100.0	1,465,505	100.0	1,486,501	100.0	1,438,826	100.0
Regions														
America, North	450,166	36.4	437,510	35.5	459,300	34.5	450,886	34.7	574,233	39.2	532,588	35.8	490,944	34.1
Asia	160,757	13.0	177,551	14.4	208,269	15.7	197,837	15.2	218,043	14.9	262,194	17.6	247,387	17.2
Europe	625,939	50.6	617,351	50.1	663,070	49.8	649,422	50.0	673,229	45.9	691,719	46.5	700,496	48.7
Leading Producers														
United States of America	448,620	36.3	436,021	35.4	457,867	34.4	449,509	34.6	572,913	39.1	531,324	35.7	489,736	34.0
Belgium	*175,416	14.2	*175,416	14.2	*175,416	13.2	*175,416	13.5	175,416	12.0	*175,416	11.8	*175,416	12.2
Japan	147,730	11.9	163,537	13.3	192,489	14.5	184,001	14.2	204,023	13.9	241,468	16.2	222,374	15.5
Germany	*126,820	10.3	*126,820	10.3	135,500	10.2	113,984	8.8	130,977	8.9	*126,820	8.5	*126,820	8.8
Czech Republic	*71,437	5.8	*71,437	5.8	*71,437	5.4	*71,437	5.5	66,576	4.5	75,135	5.1	72,600	5.0

Acrylonitrile

Unit of Measure: Metric tons.

	1990 Value	%	1991 Value	%	1992 Value	%	1993 Value	%	1994 Value	%	1995 Value	%	1996 Value	%
Total Production	2,896,106	100.0	2,860,966	100.0	3,009,707	100.0	2,816,409	100.0	3,154,214	100.0	3,387,635	100.0	3,421,989	100.0
Regions														
America, North	1,262,443	43.6	1,248,854	43.7	1,330,576	44.2	1,175,937	41.8	1,398,886	44.3	1,501,788	44.3	1,576,329	46.1
America, South	78,310	2.7	63,780	2.2	74,469	2.5	74,568	2.6	76,832	2.4	80,135	2.4	78,750	2.3
Asia	811,318	28.0	802,209	28.0	856,145	28.4	860,325	30.5	906,220	28.7	981,026	29.0	932,799	27.3
Europe	744,035	25.7	746,123	26.1	748,517	24.9	705,579	25.1	772,275	24.5	824,685	24.3	834,111	24.4
Leading Producers														
United States of America	1,213,875	41.9	1,200,857	42.0	1,283,150	42.6	1,129,082	40.1	1,352,602	42.9	1,456,075	43.0	1,531,186	44.7
Japan	592,495	20.5	603,797	21.1	620,979	20.6	593,815	21.1	610,191	19.3	663,243	19.6	674,666	19.7
Germany	*410,732	14.2	*410,732	14.4	*410,732	13.6	358,677	12.7	412,986	13.1	427,650	12.6	443,614	13.0
Italy	*118,508	4.1	*118,508	4.1	*118,508	3.9	*118,508	4.2	*118,508	3.8	*118,508	3.5	*118,508	3.5
Spain	84,106	2.9	96,496	3.4	98,627	3.3	96,600	3.4	100,988	3.2	120,198	3.5	120,493	3.5

Chlorine

Unit of Measure: Metric tons.

	1990 Value	%	1991 Value	%	1992 Value	%	1993 Value	%	1994 Value	%	1995 Value	%	1996 Value	%
Total Production	35,472,130	100.0	37,559,493	100.0	36,386,663	100.0	35,947,400	100.0	36,809,664	100.0	37,515,321	100.0	37,117,744	100.0
Regions														
Africa	33,000	0.1	33,000	0.1	35,000	0.1	37,227	0.1	36,301	0.1	37,374	0.1	35,448	0.1
America, North	13,642,000	38.5	13,232,067	35.2	12,280,133	33.7	12,322,200	34.3	12,369,267	33.6	12,636,333	33.7	12,719,400	34.3
America, South	1,203,671	3.4	1,270,767	3.4	1,264,863	3.5	1,300,959	3.6	1,311,017	3.6	1,318,821	3.5	1,293,457	3.5
Asia	10,518,399	29.7	10,680,637	28.4	10,711,078	29.4	10,597,940	29.5	10,620,680	28.9	10,868,501	29.0	10,834,098	29.2
Europe	10,075,059	28.4	12,343,022	32.9	12,095,589	33.2	11,689,074	32.5	12,472,400	33.9	12,654,292	33.7	12,235,341	33.0
Leading Producers														
United States of America	11,809,000	33.3	11,572,000	30.8	10,605,000	29.1	10,846,000	30.2	10,956,000	29.8	11,244,000	30.0	11,303,000	30.5
Indonesia	*8,757,000	24.7	*8,757,000	23.3	*8,757,000	24.1	8,757,000	24.4	*8,757,000	23.8	*8,757,000	23.3	*8,757,000	23.6
Germany	-		3,033,000	8.1	2,942,500	8.1	2,852,000	7.9	3,135,000	8.5	3,281,236	8.7	3,099,946	8.4
France	1,340,000	3.8	1,268,000	3.4	1,405,000	3.9	1,388,000	3.9	1,476,000	4.0	1,419,527	3.8	*1,265,236	3.4
Canada	1,497,000	4.2	1,402,000	3.7	1,364,000	3.7	1,170,000	3.3	1,074,000	2.9	1,032,000	2.8	1,028,000	2.8

Commodity data are provided by the United Nations Statistical Division. The symbol * means that data are estimated. For additional notes, see Appendix I.

Product Tables
Hydrochloric Acid

Unit of Measure: Metric tons.

	1990		1991		1992		1993		1994		1995		1996	
	Value	%	Value	%	Value	%	Value	%	Value	%	Value	%	Value	%
Total Production	12,360,963	100.0	12,610,319	100.0	13,748,057	100.0	13,555,920	100.0	14,069,276	100.0	14,291,880	100.0	14,552,493	100.0
Regions														
Africa	205,008	1.7	210,121	1.7	212,769	1.5	208,513	1.5	206,020	1.5	213,620	1.5	174,096	1.2
America, North	3,093,445	25.0	3,316,779	26.3	3,440,245	25.0	3,344,921	24.7	3,562,261	25.3	3,683,117	25.8	3,883,727	26.7
America, South	185,237	1.5	147,280	1.2	122,603	0.9	139,742	1.0	163,654	1.2	193,260	1.4	219,493	1.5
Asia	4,538,438	36.7	4,814,867	38.2	4,957,750	36.1	5,087,274	37.5	5,235,885	37.2	5,516,360	38.6	5,540,748	38.1
Europe	4,272,864	34.6	4,056,548	32.2	4,951,213	36.0	4,713,240	34.8	4,834,809	34.4	4,618,146	32.3	4,666,322	32.1
Oceania	65,970	0.5	64,724	0.5	63,477	0.5	62,231	0.5	66,647	0.5	67,377	0.5	68,106	0.5
Leading Producers														
United States of America	2,847,875	23.0	3,067,136	24.3	3,234,803	23.5	3,167,471	23.4	3,387,551	24.1	3,541,479	24.8	3,733,477	25.7
China	2,622,800	21.2	2,851,700	22.6	3,028,000	22.0	*3,174,070	23.4	*3,331,741	23.7	*3,489,412	24.4	*3,647,083	25.1
Japan	798,906	6.5	823,160	6.5	819,674	6.0	824,118	6.1	829,660	5.9	864,056	6.0	845,700	5.8
Italy	739,212	6.0	561,389	4.5	575,028	4.2	553,338	4.1	562,449	4.0	609,296	4.3	562,954	3.9
Germany	-		-		878,205	6.4	826,596	6.1	865,292	6.2	328,948	2.3	334,080	2.3

Sulfuric Acid

Unit of Measure: 1,000 Metric tons.

	1990		1991		1992		1993		1994		1995		1996	
	Value	%	Value	%	Value	%	Value	%	Value	%	Value	%	Value	%
Total Production	166,259	100.0	137,512	100.0	131,482	100.0	124,598	100.0	125,607	100.0	133,636	100.0	133,047	100.0
Regions														
Africa	4,627	2.8	4,628	3.4	4,856	3.7	4,762	3.8	5,340	4.3	5,406	4.0	5,527	4.2
America, North	44,906	27.0	17,281	12.6	16,714	12.7	16,196	13.0	16,141	12.9	15,753	11.8	16,146	12.1
America, South	4,472	2.7	5,253	3.8	4,851	3.7	5,421	4.4	6,089	4.8	6,268	4.7	6,482	4.9
Asia	78,232	47.1	78,673	57.2	76,582	58.2	72,971	58.6	72,341	57.6	77,901	58.3	78,524	59.0
Europe	32,558	19.6	30,692	22.3	27,664	21.0	24,379	19.6	24,863	19.8	27,472	20.6	25,623	19.3
Oceania	1,464	0.9	986	0.7	816	0.6	868	0.7	833	0.7	837	0.6	745	0.6
Leading Producers														
United States of America	40,222	24.2	12,842	9.3	12,340	9.4	11,900	9.6	11,300	9.0	11,500	8.6	11,806	8.9
China	11,969	7.2	13,329	9.7	14,087	10.7	13,365	10.7	15,365	12.2	18,110	13.6	18,836	14.2
Russian Federation	12,767	7.7	11,597	8.4	9,704	7.4	8,243	6.6	6,334	5.0	6,946	5.2	5,764	4.3
Japan	6,887	4.1	7,057	5.1	7,100	5.4	6,937	5.6	6,594	5.2	6,888	5.2	6,851	5.1
Canada	3,830	2.3	3,676	2.7	3,776	2.9	3,713	3.0	4,059	3.2	3,844	2.9	3,929	3.0

Nitric Acid

Unit of Measure: Metric tons.

	1990		1991		1992		1993		1994		1995		1996	
	Value	%	Value	%	Value	%	Value	%	Value	%	Value	%	Value	%
Total Production	26,880,439	100.0	21,457,573	100.0	20,777,248	100.0	19,688,549	100.0	19,999,224	100.0	21,275,273	100.0	20,854,747	100.0
Regions														
Africa	80,000	0.3	78,000	0.4	81,000	0.4	81,000	0.4	78,000	0.4	78,000	0.4	78,000	0.4
America, North	8,434,972	31.4	2,799,189	13.0	2,879,906	13.9	3,040,622	15.4	2,906,339	14.5	2,954,056	13.9	2,970,772	14.2
America, South	507,733	1.9	520,834	2.4	512,535	2.5	532,763	2.7	685,481	3.4	728,607	3.4	719,250	3.4
Asia	3,868,900	14.4	4,178,800	19.5	3,762,700	18.1	3,577,600	18.2	3,799,000	19.0	4,214,900	19.8	3,993,464	19.1
Europe	13,777,667	51.3	13,666,667	63.7	13,324,107	64.1	12,236,647	62.2	12,307,571	61.5	13,073,960	61.5	12,864,594	61.7
Oceania	211,167	0.8	214,083	1.0	217,000	1.0	219,917	1.1	222,833	1.1	225,750	1.1	228,667	1.1
Leading Producers														
United States of America	7,194,000	26.8	1,610,000	7.5	1,680,000	8.1	1,840,000	9.3	1,710,000	8.6	1,770,000	8.3	1,740,000	8.3
United Kingdom	*3,067,333	11.4	*3,067,333	14.3	*3,067,333	14.8	*3,067,333	15.6	*3,067,333	15.3	*3,067,333	14.4	*3,067,333	14.7
Poland	1,577,000	5.9	1,438,000	6.7	1,388,000	6.7	1,608,000	8.2	1,701,000	8.5	1,931,000	9.1	1,929,000	9.2
Belgium	1,453,000	5.4	1,434,000	6.7	1,626,000	7.8	1,320,000	6.7	*1,462,269	7.3	*1,467,681	6.9	*1,473,093	7.1
Bulgaria	1,213,000	4.5	874,000	4.1	712,000	3.4	609,000	3.1	719,000	3.6	1,033,000	4.9	1,055,000	5.1

Commodity data are provided by the United Nations Statistical Division. The symbol * means that data are estimated. For additional notes, see Appendix I.

293

Product Tables
Phosphoric Acid

Unit of Measure: Metric tons.

	1990 Value	%	1991 Value	%	1992 Value	%	1993 Value	%	1994 Value	%	1995 Value	%	1996 Value	%
Total Production	26,321,746	100.0	26,093,405	100.0	26,516,746	100.0	24,993,526	100.0	27,321,377	100.0	28,836,565	100.0	32,573,315	100.0
Regions														
Africa	423,748	1.6	439,994	1.7	468,961	1.8	468,477	1.9	530,039	1.9	554,213	1.9	481,822	1.5
America, North	11,503,296	43.7	11,320,695	43.4	11,837,693	44.6	10,296,127	41.2	12,163,560	44.5	12,350,994	42.8	12,821,219	39.4
America, South	616,553	2.3	671,633	2.6	493,492	1.9	604,998	2.4	696,094	2.5	713,140	2.5	755,244	2.3
Asia	8,350,677	31.7	8,769,342	33.6	9,245,389	34.9	9,756,305	39.0	9,935,384	36.4	10,703,034	37.1	11,344,103	34.8
Europe	5,413,176	20.6	4,887,081	18.7	4,471,211	16.9	3,867,620	15.5	3,996,299	14.6	4,515,183	15.7	7,170,926	22.0
Oceania	14,296	0.1	4,659	0.0	-		-		-		-		-	
Leading Producers														
United States of America	10,917,518	41.5	10,807,000	41.4	11,349,000	42.8	9,720,000	38.9	11,500,000	42.1	11,600,000	40.2	11,982,791	36.8
Belgium	535,857	2.0	647,801	2.5	682,831	2.6	694,855	2.8	687,022	2.5	741,341	2.6	3,234,020	9.9
Brazil	609,241	2.3	664,321	2.5	487,162	1.8	600,639	2.4	688,571	2.5	702,104	2.4	747,932	2.3
Tunisia	419,742	1.6	435,240	1.7	464,994	1.8	463,320	1.9	525,360	1.9	549,441	1.9	*476,973	1.5
Spain	597,500	2.3	463,288	1.8	383,300	1.4	393,700	1.6	448,126	1.6	431,990	1.5	506,540	1.6

Carbon Bisulfide

Unit of Measure: Metric tons.

	1990 Value	%	1991 Value	%	1992 Value	%	1993 Value	%	1994 Value	%	1995 Value	%	1996 Value	%
Total Production	483,281	100.0	455,865	100.0	444,382	100.0	443,499	100.0	446,689	100.0	451,068	100.0	437,720	100.0
Regions														
America, North	175,950	36.4	175,877	38.6	175,877	39.6	175,877	39.7	175,877	39.4	175,877	39.0	175,877	40.2
America, South	28,168	5.8	24,854	5.5	25,532	5.7	26,136	5.9	25,827	5.8	23,681	5.2	22,851	5.2
Asia	72,514	15.0	71,196	15.6	61,749	13.9	58,143	13.1	50,077	11.2	49,058	10.9	45,901	10.5
Europe	206,649	42.8	183,938	40.3	181,224	40.8	183,343	41.3	194,908	43.6	202,452	44.9	193,091	44.1
Leading Producers														
United States of America	*175,877	36.4	*175,877	38.6	*175,877	39.6	*175,877	39.7	*175,877	39.4	*175,877	39.0	*175,877	40.2
Poland	99,701	20.6	70,075	15.4	62,903	14.2	64,319	14.5	73,033	16.3	71,944	15.9	56,449	12.9
Japan	72,514	15.0	68,164	15.0	59,499	13.4	57,623	13.0	49,662	11.1	47,938	10.6	45,251	10.3
Austria	*65,566	13.6	*71,342	15.6	*77,118	17.4	*82,894	18.7	*88,671	19.9	*94,447	20.9	*100,223	22.9
Spain	23,383	4.8	21,397	4.7	19,649	4.4	*21,898	4.9	*21,955	4.9	*22,012	4.9	*22,069	5.0

Zinc Oxide

Unit of Measure: Metric tons.

	1990 Value	%	1991 Value	%	1992 Value	%	1993 Value	%	1994 Value	%	1995 Value	%	1996 Value	%
Total Production	477,586	100.0	482,038	100.0	496,936	100.0	516,652	100.0	542,925	100.0	549,044	100.0	596,402	100.0
Regions														
Africa	326	0.1	325	0.1	324	0.1	322	0.1	321	0.1	320	0.1	388	0.1
America, North	156,360	32.7	156,881	32.5	169,763	34.2	181,952	35.2	200,081	36.9	212,209	38.7	247,337	41.5
America, South	21,353	4.5	23,676	4.9	24,303	4.9	29,040	5.6	28,602	5.3	26,501	4.8	27,121	4.5
Asia	91,948	19.3	95,236	19.8	97,614	19.6	89,186	17.3	92,435	17.0	85,789	15.6	90,953	15.3
Europe	205,252	43.0	204,156	42.4	203,753	41.0	215,556	41.7	221,474	40.8	224,225	40.8	230,602	38.7
Oceania	2,347	0.5	1,763	0.4	1,180	0.2	596	0.1	12	0.0	-		-	
Leading Producers														
United States of America	98,047	20.5	94,564	19.6	103,939	20.9	102,000	19.7	106,000	19.5	104,000	18.9	125,000	21.0
Japan	83,190	17.4	84,932	17.6	82,334	16.6	75,203	14.6	73,888	13.6	75,973	13.8	76,008	12.7
Canada	*55,490	11.6	*59,376	12.3	*63,262	12.7	*67,147	13.0	*71,033	13.1	*74,919	13.6	*78,804	13.2
Germany	*47,977	10.0	*47,977	10.0	*47,977	9.7	49,225	9.5	46,729	8.6	*47,977	8.7	*47,977	8.0
France	44,588	9.3	40,059	8.3	32,512	6.5	*35,995	7.0	*35,091	6.5	*34,187	6.2	*33,284	5.6

Commodity data are provided by the United Nations Statistical Division. The symbol * means that data are estimated. For additional notes, see Appendix I.

Product Tables
Titanium Oxides

Unit of Measure: Metric tons.

	1990 Value	%	1991 Value	%	1992 Value	%	1993 Value	%	1994 Value	%	1995 Value	%	1996 Value	%
Total Production	2,107,286	100.0	2,150,884	100.0	2,640,376	100.0	2,691,552	100.0	2,835,520	100.0	2,981,764	100.0	3,189,908	100.0
Regions														
America, North	1,019,338	48.4	1,032,679	48.0	1,180,679	44.7	1,200,679	44.6	1,290,679	45.5	1,290,679	43.3	1,258,362	39.4
America, South	46,050	2.2	41,740	1.9	42,570	1.6	59,427	2.2	58,918	2.1	58,591	2.0	58,605	1.8
Asia	477,679	22.7	496,878	23.1	490,715	18.6	503,879	18.7	528,415	18.6	547,685	18.4	550,313	17.3
Europe	564,220	26.8	579,587	26.9	926,413	35.1	927,567	34.5	957,507	33.8	1,084,810	36.4	1,322,628	41.5
Leading Producers														
United States of America	978,659	46.4	992,000	46.1	1,140,000	43.2	1,160,000	43.1	1,250,000	44.1	1,250,000	41.9	1,217,683	38.2
Germany	-		-		317,209	12.0	300,074	11.1	332,355	11.7	431,207	14.5	656,135	20.6
United Kingdom	*285,481	13.5	*298,694	13.9	*311,907	11.8	*325,120	12.1	*338,333	11.9	*351,546	11.8	*364,759	11.4
Japan	279,802	13.3	279,054	13.0	252,326	9.6	245,992	9.1	237,956	8.4	249,290	8.4	237,942	7.5
Belgium	57,458	2.7	69,458	3.2	77,340	2.9	76,190	2.8	*71,430	2.5	*70,796	2.4	*70,162	2.2

Lead Oxides

Unit of Measure: Metric tons.

	1990 Value	%	1991 Value	%	1992 Value	%	1993 Value	%	1994 Value	%	1995 Value	%	1996 Value	%
Total Production	461,169	100.0	892,587	100.0	736,375	100.0	882,452	100.0	907,617	100.0	1,021,166	100.0	831,310	100.0
Regions														
Africa	2,579	0.6	2,861	0.3	2,447	0.3	1,609	0.2	1,125	0.1	791	0.1	701	0.1
America, North	29,889	6.5	458,161	51.3	309,638	42.0	310,122	35.1	331,505	36.5	656,889	64.3	676,272	81.4
America, South	3,462	0.8	3,462	0.4	2,485	0.3	2,959	0.3	4,763	0.5	3,639	0.4	3,462	0.4
Asia	74,256	16.1	77,993	8.7	67,600	9.2	68,155	7.7	65,274	7.2	58,541	5.7	53,211	6.4
Europe	350,983	76.1	350,111	39.2	354,205	48.1	499,607	56.6	504,949	55.6	301,307	29.5	97,664	11.7
Leading Producers														
United States of America	-		428,475	48.0	280,899	38.1	276,000	31.3	292,000	32.2	612,000	59.9	626,000	75.3
Germany	*317,818	68.9	*317,818	35.6	*317,818	43.2	468,246	53.1	472,453	52.1	267,675	26.2	62,897	7.6
Japan	63,047	13.7	64,405	7.2	56,840	7.7	57,112	6.5	55,266	6.1	48,053	4.7	43,982	5.3
Mexico	29,889	6.5	29,686	3.3	28,739	3.9	34,122	3.9	39,506	4.4	44,889	4.4	50,272	6.0
Spain	18,111	3.9	15,817	1.8	22,346	3.0	*16,636	1.9	*17,087	1.9	*17,539	1.7	*17,991	2.2

Ammonia

Unit of Measure: 1,000 Metric tons.

	1990 Value	%	1991 Value	%	1992 Value	%	1993 Value	%	1994 Value	%	1995 Value	%	1996 Value	%
Total Production	116,997	100.0	119,221	100.0	119,802	100.0	116,697	100.0	120,398	100.0	127,182	100.0	133,706	100.0
Regions														
Africa	962	0.8	1,572	1.3	1,711	1.4	1,577	1.4	1,275	1.1	1,403	1.1	1,472	1.1
America, North	21,583	18.4	20,425	17.1	21,408	17.9	20,442	17.5	21,883	18.2	21,535	16.9	22,494	16.8
America, South	1,991	1.7	1,737	1.5	1,602	1.3	1,782	1.5	1,751	1.5	1,918	1.5	1,923	1.4
Asia	67,342	57.6	69,569	58.4	69,888	58.3	69,534	59.6	72,325	60.1	77,559	61.0	81,888	61.2
Europe	25,119	21.5	25,435	21.3	24,733	20.6	22,886	19.6	22,672	18.8	24,254	19.1	25,417	19.0
Oceania	-		484	0.4	460	0.4	476	0.4	491	0.4	513	0.4	511	0.4
Leading Producers														
China	21,290	18.2	22,016	18.5	22,981	19.2	21,925	18.8	24,368	20.2	27,659	21.7	30,942	23.1
United States of America	15,424	13.2	12,800	10.7	13,400	11.2	12,600	10.8	13,400	11.1	13,000	10.2	14,021	10.5
Russian Federation	12,592	10.8	11,936	10.0	10,529	8.8	9,900	8.5	8,838	7.3	9,657	7.6	9,650	7.2
Ukraine	4,941	4.2	4,642	3.9	4,821	4.0	3,938	3.4	3,655	3.0	3,782	3.0	4,018	3.0
Canada	3,861	3.3	3,669	3.1	3,770	3.1	4,181	3.6	4,368	3.6	4,246	3.3	4,263	3.2

Commodity data are provided by the United Nations Statistical Division. The symbol * means that data are estimated. For additional notes, see Appendix I.

295

Product Tables
Caustic Soda

Unit of Measure: Metric tons.

	1990 Value	%	1991 Value	%	1992 Value	%	1993 Value	%	1994 Value	%	1995 Value	%	1996 Value	%
Total Production	44,110,680	100.0	43,557,658	100.0	43,689,293	100.0	42,969,278	100.0	53,644,338	100.0	45,601,308	100.0	46,109,174	100.0
Regions														
Africa	70,000	0.2	75,000	0.2	75,000	0.2	75,106	0.2	65,340	0.1	64,575	0.1	54,809	0.1
America, North	13,105,778	29.7	12,927,378	29.7	13,091,478	30.0	12,904,578	30.0	12,974,678	24.2	13,262,899	29.1	13,483,348	29.2
America, South	1,514,855	3.4	1,586,276	3.6	1,579,958	3.6	1,569,130	3.7	1,578,025	2.9	1,635,709	3.6	1,765,833	3.8
Asia	14,945,810	33.9	15,067,350	34.6	15,230,467	34.9	15,025,350	35.0	25,207,086	47.0	16,745,184	36.7	17,164,949	37.2
Europe	14,474,237	32.8	13,901,655	31.9	13,712,390	31.4	13,395,114	31.2	13,819,208	25.8	13,892,942	30.5	13,640,235	29.6
Leading Producers														
United States of America	11,116,000	25.2	11,114,000	25.5	11,292,000	25.8	11,308,000	26.3	11,388,000	21.2	*11,734,121	25.7	*11,932,470	25.9
Japan	3,800,000	8.6	3,788,000	8.7	3,751,000	8.6	3,664,000	8.5	3,672,000	6.8	3,885,000	8.5	3,940,000	8.5
Germany	*3,280,054	7.4	*3,280,054	7.5	3,187,000	7.3	3,121,000	7.3	3,367,000	6.3	3,445,215	7.6	*3,280,054	7.1
China	3,354,000	7.6	3,541,000	8.1	3,795,000	8.7	3,954,000	9.2	4,295,000	8.0	5,318,000	11.7	5,738,000	12.4
India	964,000	2.2	1,027,000	2.4	1,078,000	2.5	1,109,000	2.6	11,171,000	20.8	1,357,000	3.0	1,456,000	3.2

Aluminum Oxide

Unit of Measure: Metric tons.

	1990 Value	%	1991 Value	%	1992 Value	%	1993 Value	%	1994 Value	%	1995 Value	%	1996 Value	%
Total Production	11,346,005	100.0	10,786,080	100.0	10,676,067	100.0	11,008,733	100.0	10,542,885	100.0	10,883,499	100.0	11,224,425	100.0
Regions														
America, North	4,801,320	42.3	4,620,995	42.8	4,459,318	41.8	4,493,472	40.8	4,054,727	38.5	4,321,434	39.7	4,303,313	38.3
America, South	1,871,789	16.5	2,027,181	18.8	2,182,573	20.4	2,337,966	21.2	2,493,358	23.6	2,648,750	24.3	2,804,143	25.0
Asia	626,980	5.5	311,699	2.9	308,348	2.9	459,397	4.2	310,139	2.9	210,333	1.9	501,333	4.5
Europe	4,045,916	35.7	3,826,206	35.5	3,725,828	34.9	3,717,898	33.8	3,684,661	34.9	3,702,981	34.0	3,615,636	32.2
Leading Producers														
United States of America	4,801,320	42.3	4,620,995	42.8	4,459,318	41.8	4,493,472	40.8	4,054,727	38.5	4,321,434	39.7	4,303,313	38.3
Brazil	*1,871,789	16.5	*2,027,181	18.8	*2,182,573	20.4	*2,337,966	21.2	*2,493,358	23.6	*2,648,750	24.3	*2,804,143	25.0
Spain	952,761	8.4	954,503	8.8	891,438	8.3	*970,253	8.8	*996,490	9.5	*1,022,728	9.4	*1,048,966	9.3
Germany	*845,968	7.5	863,222	8.0	856,972	8.0	840,038	7.6	823,638	7.8	*845,968	7.8	*845,968	7.5
France	466,543	4.1	403,765	3.7	365,633	3.4	*269,272	2.4	*204,812	1.9	*140,352	1.3	*75,891	0.7

Hydrated Alumina, in Terms of Al$_2$O$_3$

Unit of Measure: Metric tons.

	1990 Value	%	1991 Value	%	1992 Value	%	1993 Value	%	1994 Value	%	1995 Value	%	1996 Value	%
Total Production	1,831,936	100.0	2,911,635	100.0	2,871,708	100.0	2,856,697	100.0	2,577,727	100.0	1,894,718	100.0	1,850,282	100.0
Regions														
America, South	513	0.0	513	0.0	513	0.0	513	0.0	513	0.0	513	0.0	513	0.0
Asia	2,208	0.1	2,208	0.1	2,208	0.1	2,170	0.1	1,671	0.1	2,520	0.1	2,470	0.1
Europe	1,829,215	99.9	2,908,914	99.9	2,868,987	99.9	2,854,014	99.9	2,575,543	99.9	1,891,685	99.8	1,847,299	99.8
Leading Producers														
Germany	-		1,148,310	39.4	1,119,898	39.0	1,109,978	38.9	950,734	36.9	322,436	17.0	333,610	18.0
France	606,000	33.1	538,000	18.5	508,000	17.7	504,177	17.6	*379,405	14.7	*318,300	16.8	*257,194	13.9
Spain	50,393	2.8	49,782	1.7	68,267	2.4	*67,037	2.3	*72,582	2.8	*78,128	4.1	*83,673	4.5
Slovenia	*2,208	0.1	*2,208	0.1	*2,208	0.1	2,170	0.1	1,671	0.1	2,520	0.1	2,470	0.1
Colombia	*513	0.0	*513	0.0	*513	0.0	*513	0.0	*513	0.0	*513	0.0	*513	0.0

Commodity data are provided by the United Nations Statistical Division. The symbol * means that data are estimated. For additional notes, see Appendix I.

Product Tables
Aluminum Sulfate

Unit of Measure: Metric tons.

	1990 Value	%	1991 Value	%	1992 Value	%	1993 Value	%	1994 Value	%	1995 Value	%	1996 Value	%
Total Production	3,924,505	100.0	3,903,664	100.0	3,791,391	100.0	3,675,755	100.0	3,927,819	100.0	4,013,231	100.0	3,992,741	100.0
Regions														
Africa	13,479	0.3	11,018	0.3	11,400	0.3	7,671	0.2	9,863	0.3	8,523	0.2	7,837	0.2
America, North	1,451,140	37.0	1,431,502	36.7	1,332,715	35.2	1,321,447	36.0	1,357,302	34.6	1,370,717	34.2	1,386,641	34.7
America, South	243,198	6.2	269,976	6.9	211,055	5.6	132,620	3.6	252,463	6.4	226,822	5.7	145,985	3.7
Asia	1,386,126	35.3	1,438,945	36.9	1,454,294	38.4	1,445,216	39.3	1,551,717	39.5	1,617,833	40.3	1,676,465	42.0
Europe	830,562	21.2	752,223	19.3	781,926	20.6	768,800	20.9	756,474	19.3	789,335	19.7	775,812	19.4
Leading Producers														
United States of America	1,241,245	31.6	1,222,079	31.3	1,143,865	30.2	1,109,309	30.2	1,163,657	29.6	1,192,230	29.7	1,206,107	30.2
Japan	959,704	24.5	953,887	24.4	924,246	24.4	881,643	24.0	942,760	24.0	954,036	23.8	984,780	24.7
Korea, Republic of	*334,681	8.5	*365,623	9.4	*396,565	10.5	*427,507	11.6	*458,450	11.7	*489,392	12.2	*520,334	13.0
United Kingdom	*273,367	7.0	*273,367	7.0	*273,367	7.2	293,037	8.0	253,697	6.5	*273,367	6.8	*273,367	6.8
Brazil	194,256	4.9	219,892	5.6	159,828	4.2	129,632	3.5	162,393	4.1	119,436	3.0	76,480	1.9

Copper Sulfate

Unit of Measure: Metric tons.

	1990 Value	%	1991 Value	%	1992 Value	%	1993 Value	%	1994 Value	%	1995 Value	%	1996 Value	%
Total Production	384,960	100.0	395,991	100.0	399,645	100.0	392,792	100.0	393,131	100.0	397,237	100.0	396,374	100.0
Regions														
Africa	11,977	3.1	12,101	3.1	12,226	3.1	12,350	3.1	12,475	3.2	12,599	3.2	12,724	3.2
America, North	34,286	8.9	40,200	10.2	46,800	11.7	46,400	11.8	48,400	12.3	52,000	13.1	43,400	10.9
America, South	3,570	0.9	4,880	1.2	6,193	1.5	6,750	1.7	6,980	1.8	7,896	2.0	6,578	1.7
Asia	20,608	5.4	29,746	7.5	26,142	6.5	24,017	6.1	23,348	5.9	23,419	5.9	24,819	6.3
Europe	314,518	81.7	309,063	78.0	308,285	77.1	303,275	77.2	301,929	76.8	301,322	75.9	308,853	77.9
Leading Producers														
United Kingdom	*268,430	69.7	*268,430	67.8	*268,430	67.2	*268,430	68.3	268,430	68.3	*268,430	67.6	*268,430	67.7
United States of America	34,286	8.9	40,200	10.2	46,800	11.7	46,400	11.8	48,400	12.3	52,000	13.1	43,400	10.9
Japan	*16,089	4.2	*16,089	4.1	*16,089	4.0	*16,089	4.1	*16,089	4.1	*16,089	4.1	*16,089	4.1
South Africa	*11,977	3.1	*12,101	3.1	*12,226	3.1	*12,350	3.1	*12,475	3.2	*12,599	3.2	*12,724	3.2
France	*9,180	2.4	*8,846	2.2	*8,513	2.1	*8,180	2.1	*7,846	2.0	*7,513	1.9	*7,179	1.8

Soda Ash

Unit of Measure: Metric tons.

	1990 Value	%	1991 Value	%	1992 Value	%	1993 Value	%	1994 Value	%	1995 Value	%	1996 Value	%
Total Production	37,756,735	100.0	36,873,585	100.0	37,260,208	100.0	36,115,987	100.0	36,724,207	100.0	39,040,772	100.0	39,522,484	100.0
Regions														
Africa	244,000	0.6	307,000	0.8	305,000	0.8	342,000	0.9	398,000	1.1	426,000	1.1	343,000	0.9
America, North	9,605,000	25.4	9,459,000	25.7	9,820,000	26.4	9,400,000	26.0	9,760,000	26.6	10,550,000	27.0	10,690,162	27.0
America, South	317,000	0.8	329,000	0.9	342,000	0.9	352,000	1.0	340,000	0.9	325,000	0.8	337,410	0.9
Asia	16,090,307	42.6	16,123,640	43.7	16,370,973	43.9	16,687,307	46.2	16,697,640	45.5	17,029,973	43.6	17,357,707	43.9
Europe	11,500,429	30.5	10,654,945	28.9	10,422,235	28.0	9,334,680	25.8	9,528,567	25.9	10,709,799	27.4	10,794,205	27.3
Leading Producers														
United States of America	9,156,000	24.2	9,010,000	24.4	9,380,000	25.2	8,960,000	24.8	9,320,000	25.4	10,100,000	25.9	10,200,000	25.8
China	3,795,000	10.1	3,936,000	10.7	4,549,000	12.2	5,349,000	14.8	5,814,000	15.8	5,977,000	15.3	6,693,000	16.9
Russian Federation	3,240,000	8.6	3,048,000	8.3	2,679,000	7.2	1,992,000	5.5	1,585,000	4.3	1,823,000	4.7	1,449,000	3.7
Germany	*1,888,865	5.0	*1,888,865	5.1	*1,888,865	5.1	1,586,000	4.4	1,380,000	3.8	2,215,558	5.7	2,373,904	6.0
France	1,180,000	3.1	1,140,000	3.1	1,100,000	3.0	1,222,000	3.4	1,200,000	3.3	1,200,000	3.1	*1,073,324	2.7

Commodity data are provided by the United Nations Statistical Division. The symbol * means that data are estimated. For additional notes, see Appendix I.

297

Product Tables

Sodium Sulfates, Acid (NaHSO$_4$) or Neutral (Na$_2$SO$_4$)

Unit of Measure: Metric tons.

	1990 Value	%	1991 Value	%	1992 Value	%	1993 Value	%	1994 Value	%	1995 Value	%	1996 Value	%
Total Production	3,753,753	100.0	3,917,188	100.0	3,376,834	100.0	3,209,751	100.0	3,329,640	100.0	3,450,992	100.0	3,337,138	100.0
Regions														
America, North	1,500,657	40.0	1,537,334	39.2	1,043,010	30.9	1,056,686	32.9	1,137,789	34.2	1,186,899	34.4	1,130,051	33.9
America, South	26,705	0.7	26,705	0.7	23,140	0.7	23,895	0.7	28,368	0.9	31,416	0.9	26,705	0.8
Asia	1,353,451	36.1	1,239,139	31.6	1,217,958	36.1	1,139,423	35.5	1,042,471	31.3	1,121,754	32.5	1,036,003	31.0
Europe	872,940	23.3	1,114,011	28.4	1,092,726	32.4	989,747	30.8	1,121,012	33.7	1,110,922	32.2	1,144,379	34.3
Leading Producers														
United States of America	713,000	19.0	730,000	18.6	216,000	6.4	210,000	6.5	293,000	8.8	318,000	9.2	246,000	7.4
Korea, Republic of	*610,200	16.3	*634,500	16.2	*658,800	19.5	*683,100	21.3	*707,400	21.2	*731,700	21.2	*756,000	22.7
Mexico	465,757	12.4	485,636	12.4	505,514	15.0	525,393	16.4	523,698	15.7	548,011	15.9	563,365	16.9
Russian Federation	707,506	18.8	568,876	14.5	523,440	15.5	420,653	13.1	299,347	9.0	354,330	10.3	244,279	7.3
Canada	*318,186	8.5	*318,186	8.1	*318,186	9.4	318,186	9.9	*318,186	9.6	*318,186	9.2	*318,186	9.5

Sodium Silicates

Unit of Measure: Metric tons.

	1990 Value	%	1991 Value	%	1992 Value	%	1993 Value	%	1994 Value	%	1995 Value	%	1996 Value	%
Total Production	2,533,353	100.0	2,684,889	100.0	2,560,197	100.0	2,707,311	100.0	2,957,734	100.0	2,994,847	100.0	3,002,645	100.0
Regions														
Africa	5,622	0.2	5,442	0.2	7,536	0.3	8,450	0.3	7,624	0.3	13,952	0.5	18,328	0.6
America, North	1,066,320	42.1	1,161,122	43.2	1,185,121	46.3	1,253,362	46.3	1,249,819	42.3	1,299,606	43.4	1,335,981	44.5
America, South	105,262	4.2	144,339	5.4	99,320	3.9	106,477	3.9	211,679	7.2	240,932	8.0	179,114	6.0
Asia	930,249	36.7	931,858	34.7	883,766	34.5	938,844	34.7	1,028,583	34.8	1,001,366	33.4	1,031,730	34.4
Europe	425,900	16.8	442,128	16.5	384,454	15.0	400,178	14.8	460,029	15.6	438,991	14.7	437,493	14.6
Leading Producers														
United States of America	837,099	33.0	904,729	33.7	927,634	36.2	1,001,721	37.0	999,822	33.8	*1,034,797	34.6	*1,065,615	35.5
Japan	788,303	31.1	786,102	29.3	721,473	28.2	728,550	26.9	830,746	28.1	836,169	27.9	799,513	26.6
Mexico	223,813	8.8	251,379	9.4	252,867	9.9	247,416	9.1	246,166	8.3	*261,372	8.7	*267,323	8.9
United Kingdom	*167,497	6.6	*167,497	6.2	*167,497	6.5	158,660	5.9	176,335	6.0	*167,497	5.6	*167,497	5.6
Brazil	74,517	2.9	110,959	4.1	63,306	2.5	84,596	3.1	187,169	6.3	211,043	7.0	145,747	4.9

Hydrogen Peroxide

Unit of Measure: Metric tons.

	1990 Value	%	1991 Value	%	1992 Value	%	1993 Value	%	1994 Value	%	1995 Value	%	1996 Value	%
Total Production	680,160	100.0	797,816	100.0	829,837	100.0	825,366	100.0	869,106	100.0	881,870	100.0	910,729	100.0
Regions														
America, North	224,049	32.9	322,865	40.5	380,219	45.8	409,058	49.6	438,920	50.5	431,050	48.9	452,612	49.7
America, South	31,761	4.7	34,020	4.3	37,458	4.5	40,146	4.9	43,155	5.0	45,553	5.2	47,188	5.2
Asia	278,057	40.9	294,408	36.9	267,422	32.2	223,960	27.1	212,879	24.5	206,419	23.4	206,706	22.7
Europe	146,293	21.5	146,523	18.4	144,738	17.4	152,202	18.4	174,152	20.0	198,848	22.5	204,223	22.4
Leading Producers														
United States of America	216,127	31.8	230,492	28.9	254,022	30.6	273,564	33.1	288,839	33.2	*289,868	32.9	*305,289	33.5
Canada	-		83,111	10.4	118,860	14.3	128,447	15.6	142,553	16.4	132,796	15.1	138,687	15.2
Japan	144,613	21.3	157,831	19.8	159,045	19.2	143,791	17.4	141,610	16.3	142,575	16.2	143,118	15.7
Sweden	*55,805	8.2	*60,298	7.6	*64,792	7.8	*69,285	8.4	*73,778	8.5	*78,271	8.9	*82,764	9.1
Russian Federation	84,640	12.4	87,773	11.0	59,573	7.2	34,583	4.2	21,341	2.5	14,080	1.6	13,650	1.5

Product Tables
Calcium Carbide

Unit of Measure: Metric tons.

	1990 Value	%	1991 Value	%	1992 Value	%	1993 Value	%	1994 Value	%	1995 Value	%	1996 Value	%
Total Production	7,046,026	100.0	6,954,664	100.0	6,854,474	100.0	6,888,936	100.0	7,035,808	100.0	7,613,409	100.0	7,179,515	100.0
Regions														
America, North	221,877	3.1	219,725	3.2	216,572	3.2	224,420	3.3	225,267	3.2	220,294	2.9	219,653	3.1
America, South	143,192	2.0	160,648	2.3	147,104	2.1	160,560	2.3	148,516	2.1	137,001	1.8	123,000	1.7
Asia	3,951,050	56.1	3,891,367	56.0	3,901,067	56.9	3,947,517	57.3	4,112,578	58.5	4,645,440	61.0	4,210,864	58.7
Europe	2,729,907	38.7	2,682,925	38.6	2,589,731	37.8	2,556,439	37.1	2,549,447	36.2	2,610,674	34.3	2,625,999	36.6
Leading Producers														
China	2,280,000	32.4	2,358,000	33.9	2,425,000	35.4	2,642,000	38.4	2,920,000	41.5	3,457,000	45.4	3,092,000	43.1
United Kingdom	*1,494,000	21.2	*1,494,000	21.5	*1,494,000	21.8	1,494,000	21.7	*1,494,000	21.2	*1,494,000	19.6	*1,494,000	20.8
Japan	294,000	4.2	303,000	4.4	270,000	3.9	245,000	3.6	246,000	3.5	273,000	3.6	263,000	3.7
Poland	297,000	4.2	283,000	4.1	221,000	3.2	180,000	2.6	174,000	2.5	178,000	2.3	145,000	2.0
Germany	*292,821	4.2	*292,821	4.2	*292,821	4.3	300,000	4.4	285,000	4.1	293,464	3.9	*292,821	4.1

Dyestuffs, Synthetic (In Terms of 60 Per Cent Concentration Basis)

Unit of Measure: Metric tons.

	1990 Value	%	1991 Value	%	1992 Value	%	1993 Value	%	1994 Value	%	1995 Value	%	1996 Value	%
Total Production	1,339,466	100.0	1,348,560	100.0	1,343,519	100.0	1,239,940	100.0	1,265,827	100.0	1,381,246	100.0	1,398,551	100.0
Regions														
America, North	43,180	3.2	52,655	3.9	62,130	4.6	71,604	5.8	81,079	6.4	90,553	6.6	100,028	7.2
America, South	29,114	2.2	30,584	2.3	33,714	2.5	34,483	2.8	35,251	2.8	36,019	2.6	36,788	2.6
Asia	269,074	20.1	276,742	20.5	249,652	18.6	221,112	17.8	222,128	17.5	216,861	15.7	219,784	15.7
Europe	998,093	74.5	988,574	73.3	998,019	74.3	912,736	73.6	927,365	73.3	1,037,806	75.1	1,041,946	74.5
Oceania	5	0.0	5	0.0	5	0.0	5	0.0	5	0.0	5	0.0	5	0.0
Leading Producers														
Netherlands	*424,650	31.7	*424,650	31.5	*424,650	31.6	*424,650	34.2	*424,650	33.5	*424,650	30.7	*424,650	30.4
Germany	*178,764	13.3	*178,764	13.3	*178,764	13.3	110,580	8.9	119,081	9.4	241,121	17.5	244,275	17.5
Japan	74,880	5.6	77,114	5.7	73,707	5.5	67,395	5.4	71,161	5.6	70,818	5.1	69,081	4.9
United Kingdom	54,399	4.1	61,196	4.5	67,993	5.1	74,790	6.0	81,420	6.4	*76,084	5.5	*78,968	5.6
France	45,967	3.4	47,183	3.5	52,104	3.9	60,155	4.9	57,600	4.6	*57,874	4.2	*59,472	4.3

Activated Carbon

Unit of Measure: Metric tons.

	1990 Value	%	1991 Value	%	1992 Value	%	1993 Value	%	1994 Value	%	1995 Value	%	1996 Value	%
Total Production	801,682	100.0	826,906	100.0	791,268	100.0	861,922	100.0	890,681	100.0	879,188	100.0	887,966	100.0
Regions														
America, North	117,686	14.7	116,850	14.1	119,616	15.1	130,446	15.1	142,869	16.0	134,148	15.3	137,699	15.5
Asia	67,991	8.5	100,892	12.2	100,118	12.7	88,075	10.2	102,693	11.5	98,338	11.2	101,801	11.5
Europe	616,005	76.8	609,164	73.7	571,534	72.2	643,401	74.6	645,119	72.4	646,702	73.6	648,465	73.0
Leading Producers														
Spain	115,702	14.4	108,809	13.2	71,172	9.0	*143,121	16.6	*144,854	16.3	*146,588	16.7	*148,321	16.7
United States of America	117,686	14.7	116,850	14.1	119,616	15.1	130,446	15.1	142,869	16.0	*134,148	15.3	*137,699	15.5
Japan	64,553	8.1	66,824	8.1	68,657	8.7	65,125	7.6	82,758	9.3	80,544	9.2	83,544	9.4
Germany	*14,727	1.8	*14,727	1.8	*14,727	1.9	*14,727	1.7	*14,727	1.7	14,727	1.7	*14,727	1.7
Russian Federation	-		29,609	3.6	20,821	2.6	10,295	1.2	7,241	0.8	7,803	0.9	4,259	0.5

Commodity data are provided by the United Nations Statistical Division. The symbol * means that data are estimated. For additional notes, see Appendix I.

299

Product Tables

Nitrogenous Fertilizers (Total Production)

Unit of Measure: 1,000 Metric tons.

	1990 Value	%	1991 Value	%	1992 Value	%	1993 Value	%	1994 Value	%	1995 Value	%	1996 Value	%
Total Production	110,887	100.0	115,621	100.0	115,846	100.0	113,963	100.0	117,934	100.0	123,369	100.0	123,955	100.0
Regions														
Africa	6,858	6.2	6,785	5.9	7,772	6.7	8,277	7.3	8,753	7.4	8,911	7.2	6,272	5.1
America, North	18,711	16.9	19,193	16.6	19,121	16.5	19,868	17.4	21,097	17.9	21,169	17.2	22,125	17.8
America, South	1,424	1.3	1,408	1.2	1,221	1.1	1,333	1.2	1,448	1.2	1,574	1.3	1,620	1.3
Asia	55,792	50.3	61,944	53.6	60,791	52.5	60,400	53.0	62,640	53.1	65,789	53.3	67,284	54.3
Europe	27,812	25.1	26,013	22.5	26,632	23.0	23,729	20.8	23,690	20.1	25,601	20.8	26,313	21.2
Oceania	290	0.3	278	0.2	309	0.3	356	0.3	306	0.3	325	0.3	340	0.3
Leading Producers														
China	14,636	13.2	15,101	13.1	15,705	13.6	15,256	13.4	17,363	14.7	16,980	13.8	*17,673	14.3
United States of America	12,576	11.3	13,124	11.4	13,557	11.7	14,006	12.3	14,415	12.2	14,017	11.4	*14,624	11.8
Ukraine	8,346	7.5	8,110	7.0	7,723	6.7	6,230	5.5	6,024	5.1	5,813	4.7	6,515	5.3
India	7,063	6.4	7,697	6.7	7,797	6.7	7,786	6.8	8,160	6.9	9,480	7.7	8,764	7.1
Russian Federation	*5,329	4.8	6,680	5.8	5,815	5.0	4,770	4.2	4,050	3.4	*5,329	4.3	*5,329	4.3

Nitrogenous Fertilizers, N Content

Unit of Measure: 1,000 Metric tons.

	1990 Value	%	1991 Value	%	1992 Value	%	1993 Value	%	1994 Value	%	1995 Value	%	1996 Value	%
Total Production	93,554	100.0	92,443	100.0	93,165	100.0	93,326	100.0	98,184	100.0	101,816	100.0	106,646	100.0
Regions														
Africa	5,995	6.4	5,784	6.3	6,694	7.2	7,078	7.6	7,505	7.6	6,904	6.8	7,101	6.7
America, North	16,056	17.2	16,490	17.8	16,270	17.5	16,985	18.2	18,484	18.8	17,801	17.5	18,180	17.0
America, South	1,167	1.2	1,132	1.2	1,111	1.2	1,161	1.2	1,217	1.2	1,311	1.3	1,347	1.3
Asia	52,160	55.8	52,889	57.2	53,106	57.0	53,410	57.2	55,778	56.8	59,216	58.2	63,275	59.3
Europe	18,118	19.4	16,084	17.4	15,928	17.1	14,629	15.7	15,125	15.4	16,507	16.2	16,662	15.6
Oceania	59	0.1	63	0.1	57	0.1	63	0.1	74	0.1	78	0.1	81	0.1
Leading Producers														
China	14,636	15.6	15,101	16.3	15,705	16.9	15,256	16.3	17,363	17.7	18,592	18.3	21,361	20.0
United States of America	10,077	10.8	10,509	11.4	10,820	11.6	11,102	11.9	11,768	12.0	11,247	11.0	*11,181	10.5
Russian Federation	7,186	7.7	6,680	7.2	5,815	6.2	4,770	5.1	4,050	4.1	4,879	4.8	4,807	4.5
India	5,991	6.4	6,148	6.7	6,154	6.6	*6,973	7.5	*7,394	7.5	*7,816	7.7	*8,237	7.7
Egypt	4,600	4.9	4,339	4.7	5,342	5.7	5,588	6.0	5,918	6.0	*5,506	5.4	*5,630	5.3

Phosphatic Fertilizers (Total Production)

Unit of Measure: Metric tons.

	1990 Value	%	1991 Value	%	1992 Value	%	1993 Value	%	1994 Value	%	1995 Value	%	1996 Value	%
Total Production	54,570,111	100.0	56,277,778	100.0	52,390,086	100.0	50,417,441	100.0	51,255,162	100.0	55,680,630	100.0	54,285,225	100.0
Regions														
Africa	3,336,333	6.1	2,987,333	5.3	3,015,279	5.8	2,684,815	5.3	2,919,352	5.7	3,319,597	6.0	3,209,594	5.9
America, North	11,627,000	21.3	11,496,000	20.4	11,552,000	22.0	11,614,000	23.0	11,379,923	22.2	12,270,110	22.0	11,803,821	21.7
America, South	1,183,000	2.2	1,220,000	2.2	1,200,203	2.3	1,344,916	2.7	1,494,302	2.9	1,371,773	2.5	1,387,968	2.6
Asia	23,008,444	42.2	28,243,311	50.2	27,206,178	51.9	25,835,044	51.2	26,494,411	51.7	29,084,606	52.2	27,722,057	51.1
Europe	12,858,333	23.6	10,795,133	19.2	7,071,426	13.5	6,038,666	12.0	6,183,174	12.1	7,075,862	12.7	7,510,079	13.8
Oceania	2,557,000	4.7	1,536,000	2.7	2,345,000	4.5	2,900,000	5.8	2,784,000	5.4	2,558,681	4.6	2,651,705	4.9
Leading Producers														
United States of America	9,910,000	18.2	9,928,000	17.6	10,518,000	20.1	10,684,000	21.2	10,223,000	19.9	11,055,000	19.9	*10,517,000	19.4
China	4,114,000	7.5	4,196,000	7.5	4,622,000	8.8	4,190,000	8.3	5,044,000	9.8	6,626,000	11.9	*5,629,305	10.4
France	3,669,000	6.7	3,420,000	6.1	2,117,000	4.0	1,867,000	3.7	1,831,000	3.6	*2,668,110	4.8	*2,616,363	4.8
Russian Federation	-		4,275,000	7.6	3,015,000	5.8	2,512,000	5.0	1,718,000	3.4	1,929,000	3.5	1,584,000	2.9
India	2,237,000	4.1	2,973,000	5.3	2,804,000	5.4	2,098,000	4.2	3,085,000	6.0	3,769,000	6.8	2,803,000	5.2

Product Tables
Superphosphates, P$_2$O$_5$ Content

Unit of Measure: 1,000 Metric tons.

	1990 Value	%	1991 Value	%	1992 Value	%	1993 Value	%	1994 Value	%	1995 Value	%	1996 Value	%
Total Production	34,047	100.0	32,847	100.0	30,638	100.0	29,341	100.0	30,111	100.0	31,418	100.0	31,461	100.0
Regions														
Africa	2,100	6.2	2,542	7.7	2,752	9.0	2,650	9.0	2,835	9.4	2,955	9.4	3,195	10.2
America, North	2,155	6.3	2,079	6.3	1,475	4.8	1,223	4.2	1,552	5.2	1,443	4.6	1,519	4.8
America, South	617	1.8	637	1.9	653	2.1	723	2.5	825	2.7	775	2.5	788	2.5
Asia	23,054	67.7	23,341	71.1	22,152	72.3	21,719	74.0	21,712	72.1	22,934	73.0	22,565	71.7
Europe	3,564	10.5	2,712	8.3	1,973	6.4	1,432	4.9	1,522	5.1	1,559	5.0	1,626	5.2
Oceania	2,557	7.5	1,536	4.7	1,633	5.3	1,594	5.4	1,664	5.5	1,752	5.6	1,767	5.6
Leading Producers														
China	4,114	12.1	4,597	14.0	4,622	15.1	4,190	14.3	5,044	16.8	6,626	21.1	6,512	20.7
Russian Federation	4,943	14.5	4,275	13.0	3,015	9.8	2,512	8.6	1,718	5.7	*829	2.6	*7	0.0
Australia	2,351	6.9	1,368	4.2	1,422	4.6	1,374	4.7	1,518	5.0	1,622	5.2	1,653	5.3
Egypt	*1,175	3.5	*1,249	3.8	*1,323	4.3	*1,397	4.8	*1,471	4.9	*1,545	4.9	*1,619	5.1
United States of America	887	2.6	867	2.6	930	3.0	694	2.4	774	2.6	*602	1.9	*547	1.7

Phosphatic Fertilizers, Other, P$_2$O$_5$ Content

Unit of Measure: Metric tons.

	1990 Value	%	1991 Value	%	1992 Value	%	1993 Value	%	1994 Value	%	1995 Value	%	1996 Value	%
Total Production	10,578,435	100.0	10,067,404	100.0	10,415,861	100.0	10,260,252	100.0	10,838,241	100.0	11,309,044	100.0	11,564,809	100.0
Regions														
Africa	8,000	0.1	13,000	0.1	10,000	0.1	7,000	0.1	7,000	0.1	6,000	0.1	8,162	0.1
America, North	7,278,250	68.8	7,230,250	71.8	7,638,250	73.3	7,307,250	71.2	8,318,250	76.7	8,800,228	77.8	9,175,406	79.3
America, South	26,000	0.2	24,000	0.2	25,000	0.2	26,000	0.3	29,000	0.3	22,396	0.2	19,905	0.2
Asia	873,707	8.3	635,729	6.3	873,158	8.4	854,836	8.3	705,515	6.5	623,193	5.5	569,872	4.9
Europe	2,392,478	22.6	2,164,425	21.5	1,869,453	17.9	2,065,166	20.1	1,778,476	16.4	1,857,227	16.4	1,791,464	15.5
Leading Producers														
United States of America	7,260,000	68.6	7,212,000	71.6	7,620,000	73.2	7,289,000	71.0	8,300,000	76.6	*8,781,978	77.7	*9,157,156	79.2
France	888,000	8.4	772,000	7.7	605,000	5.8	765,000	7.5	497,000	4.6	*604,956	5.3	*562,241	4.9
Romania	*717,000	6.8	*717,000	7.1	*717,000	6.9	*717,000	7.0	*717,000	6.6	*717,000	6.3	*717,000	6.2
Turkey	357,000	3.4	83,000	0.8	352,000	3.4	356,000	3.5	217,000	2.0	144,000	1.3	110,000	1.0
Belgium	245,586	2.3	224,000	2.2	175,000	1.7	160,000	1.6	*151,854	1.4	*134,274	1.2	*116,695	1.0

Potassic Fertilizers (Total Production)

Unit of Measure: Metric tons.

	1990 Value	%	1991 Value	%	1992 Value	%	1993 Value	%	1994 Value	%	1995 Value	%	1996 Value	%
Total Production	43,462,405	100.0	51,358,098	100.0	39,328,950	100.0	36,642,915	100.0	39,046,107	100.0	43,508,061	100.0	42,888,125	100.0
Regions														
America, North	10,160,250	23.4	10,992,433	21.4	10,381,617	26.4	10,884,391	29.7	11,073,781	28.4	13,070,170	30.0	12,219,731	28.5
America, South	81,840	0.2	114,840	0.2	97,255	0.2	181,806	0.5	249,040	0.6	209,590	0.5	237,252	0.6
Asia	22,486,083	51.7	30,618,900	59.6	21,336,883	54.3	19,455,533	53.1	20,526,183	52.6	21,907,833	50.4	22,172,766	51.7
Europe	10,734,232	24.7	9,631,926	18.8	7,513,195	19.1	6,121,185	16.7	7,197,104	18.4	8,320,467	19.1	8,258,375	19.3
Leading Producers														
Canada	6,774,000	15.6	7,520,000	14.6	7,014,000	17.8	7,289,000	19.9	7,293,000	18.7	9,104,000	20.9	*8,068,171	18.8
Belarus	4,994,000	11.5	12,128,000	23.6	3,311,000	8.4	1,947,000	5.3	2,515,000	6.4	2,796,000	6.4	2,717,000	6.3
France	4,950,000	11.4	4,439,000	8.6	2,374,000	6.0	960,000	2.6	1,806,000	4.6	*3,128,000	7.2	*3,097,514	7.2
Russian Federation	*3,170,500	7.3	4,086,000	8.0	3,470,000	8.8	2,628,000	7.2	2,498,000	6.4	*3,170,500	7.3	*3,170,500	7.4
United States of America	3,360,000	7.7	3,446,000	6.7	3,341,000	8.5	*3,568,591	9.7	*3,753,797	9.6	*3,939,003	9.1	*4,124,210	9.6

Commodity data are provided by the United Nations Statistical Division. The symbol * means that data are estimated. For additional notes, see Appendix I.

301

Product Tables
Potassic Fertilizers, K₂O Content

Unit of Measure: Metric tons.

	1990 Value	%	1991 Value	%	1992 Value	%	1993 Value	%	1994 Value	%	1995 Value	%	1996 Value	%
Total Production	41,584,703	100.0	41,569,022	100.0	39,702,942	100.0	37,394,056	100.0	38,760,140	100.0	41,458,174	100.0	40,470,222	100.0
Regions														
America, North	8,487,000	20.4	9,269,000	22.3	8,719,000	22.0	8,770,227	23.5	8,756,224	22.6	10,549,220	25.4	9,621,178	23.8
America, South	68,000	0.2	101,000	0.2	77,000	0.2	174,000	0.5	229,000	0.6	210,286	0.5	233,786	0.6
Asia	21,743,302	52.3	21,361,285	51.4	20,041,269	50.5	18,160,919	48.6	19,242,310	49.6	20,290,989	48.9	20,454,640	50.5
Europe	11,286,401	27.1	10,837,737	26.1	10,865,673	27.4	10,288,910	27.5	10,532,606	27.2	10,407,679	25.1	10,160,618	25.1
Leading Producers														
Canada	6,774,000	16.3	7,520,000	18.1	7,014,000	17.7	7,289,000	19.5	7,293,000	18.8	9,104,000	22.0	*8,193,962	20.2
Germany	*3,498,736	8.4	*3,498,736	8.4	*3,498,736	8.8	*3,498,736	9.4	3,498,736	9.0	*3,498,736	8.4	*3,498,736	8.6
Belarus	4,994,000	12.0	4,238,000	10.2	3,311,000	8.3	1,947,000	5.2	2,515,000	6.5	2,796,000	6.7	2,717,000	6.7
Russian Federation	3,848,000	9.3	4,086,000	9.8	3,470,000	8.7	2,628,000	7.0	2,498,000	6.4	2,831,000	6.8	2,685,000	6.6
United States of America	1,713,000	4.1	1,749,000	4.2	1,705,000	4.3	*1,481,227	4.0	*1,463,224	3.8	*1,445,220	3.5	*1,427,217	3.5

Multinutrient Fertilizers, N Content

Unit of Measure: Metric tons.

	1990 Value	%	1991 Value	%	1992 Value	%	1993 Value	%	1994 Value	%	1995 Value	%	1996 Value	%
Total Production	16,422,968	100.0	16,204,579	100.0	16,856,764	100.0	15,368,052	100.0	14,674,292	100.0	15,100,084	100.0	16,571,961	100.0
Regions														
Africa	645,000	3.9	589,000	3.6	584,479	3.5	615,730	4.0	684,082	4.7	676,215	4.5	727,944	4.4
America, North	2,751,762	16.8	2,809,526	17.3	2,967,581	17.6	3,138,637	20.4	2,878,692	19.6	3,008,747	19.9	3,655,448	22.1
America, South	205,895	1.3	193,905	1.2	164,942	1.0	176,590	1.1	193,933	1.3	189,021	1.3	186,684	1.1
Asia	4,323,608	26.3	4,440,659	27.4	5,004,710	29.7	4,455,518	29.0	3,993,176	27.2	4,513,882	29.9	4,809,606	29.0
Europe	8,496,703	51.7	8,171,489	50.4	8,135,052	48.3	6,981,577	45.4	6,924,409	47.2	6,712,220	44.5	7,192,279	43.4
Leading Producers														
Ukraine	5,324,000	32.4	5,291,000	32.7	5,179,000	30.7	4,158,000	27.1	4,089,000	27.9	3,942,000	26.1	4,432,000	26.7
United States of America	2,499,000	15.2	2,615,000	16.1	2,737,000	16.2	2,904,000	18.9	2,647,000	18.0	2,770,000	18.3	*3,429,600	20.7
Belarus	1,750,000	10.7	1,662,000	10.3	1,444,000	8.6	983,000	6.4	710,000	4.8	893,000	5.9	1,006,000	6.1
India	756,000	4.6	845,000	5.2	1,148,000	6.8	*1,047,758	6.8	*1,102,925	7.5	*1,158,093	7.7	*1,213,261	7.3
Turkey	645,000	3.9	551,000	3.4	966,000	5.7	992,000	6.5	695,000	4.7	904,000	6.0	914,000	5.5

Multinutrient Fertilizers, P₂O₅ Content

Unit of Measure: Metric tons.

	1990 Value	%	1991 Value	%	1992 Value	%	1993 Value	%	1994 Value	%	1995 Value	%	1996 Value	%
Total Production	16,157,521	100.0	15,816,419	100.0	15,326,383	100.0	14,569,721	100.0	14,654,708	100.0	15,448,357	100.0	15,970,513	100.0
Regions														
Africa	159,467	1.0	166,124	1.1	184,448	1.2	146,150	1.0	128,641	0.9	152,704	1.0	153,123	1.0
America, North	5,583,107	34.6	5,508,164	34.8	5,652,221	36.9	5,581,279	38.3	5,565,336	38.0	5,572,393	36.1	5,528,162	34.6
America, South	526,500	3.3	549,500	3.5	480,500	3.1	565,500	3.9	617,423	4.2	555,456	3.6	551,646	3.5
Asia	2,976,102	18.4	3,327,881	21.0	4,409,860	28.8	3,977,347	27.3	3,843,469	26.2	4,115,197	26.6	4,413,853	27.6
Europe	6,912,345	42.8	6,264,750	39.6	4,599,355	30.0	4,299,445	29.5	4,499,838	30.7	5,052,607	32.7	5,323,728	33.3
Leading Producers														
United States of America	*4,986,929	30.9	*4,990,321	31.6	*4,993,714	32.6	*4,997,107	34.3	*5,000,500	34.1	*5,003,893	32.4	*5,007,286	31.4
France	2,141,000	13.3	2,124,000	13.4	1,500,000	9.8	1,570,000	10.8	1,516,000	10.3	*2,122,264	13.7	*2,187,699	13.7
India	1,293,000	8.0	1,467,000	9.3	2,084,000	13.6	1,947,000	13.4	*2,038,731	13.9	*2,151,582	13.9	*2,264,434	14.2
Ukraine	2,098,000	13.0	1,682,000	10.6	569,000	3.7	389,000	2.7	406,000	2.8	255,000	1.7	419,000	2.6
Japan	436,000	2.7	415,000	2.6	407,000	2.7	403,000	2.8	396,000	2.7	378,000	2.4	363,000	2.3

Commodity data are provided by the United Nations Statistical Division. The symbol * means that data are estimated. For additional notes, see Appendix I.

Product Tables
Multinutrient Fertilizers, K$_2$O Content

Unit of Measure: Metric tons.

	1990 Value	%	1991 Value	%	1992 Value	%	1993 Value	%	1994 Value	%	1995 Value	%	1996 Value	%
Total Production	14,705,780	100.0	13,183,350	100.0	11,205,026	100.0	7,642,308	100.0	8,851,971	100.0	10,368,005	100.0	14,965,332	100.0
Regions														
America, North	26,250	0.2	26,433	0.2	26,617	0.2	26,800	0.4	26,983	0.3	27,167	0.3	27,350	0.2
America, South	7,000	0.0	7,000	0.1	7,000	0.1	7,000	0.1	7,000	0.1	7,000	0.1	7,000	0.0
Asia	9,146,000	62.2	8,310,000	63.0	6,246,736	55.7	3,679,000	48.1	4,548,000	51.4	5,013,000	48.4	9,398,000	62.8
Europe	5,526,530	37.6	4,839,917	36.7	4,924,674	44.0	3,929,508	51.4	4,269,987	48.2	5,320,838	51.3	5,532,982	37.0
Leading Producers														
Belarus	8,819,000	60.0	7,890,000	59.8	5,850,000	52.2	3,341,000	43.7	4,218,000	47.7	4,706,000	45.4	9,081,000	60.7
France	3,654,000	24.8	3,311,000	25.1	3,308,000	29.5	2,507,000	32.8	2,725,000	30.8	*3,869,055	37.3	*4,068,196	27.2
Japan	327,000	2.2	315,000	2.4	313,000	2.8	310,000	4.1	304,000	3.4	292,000	2.8	272,000	1.8
Ukraine	509,000	3.5	324,000	2.5	318,000	2.8	206,000	2.7	186,000	2.1	62,000	0.6	83,000	0.6
Germany	*167,000	1.1	*167,000	1.3	*167,000	1.5	166,000	2.2	168,000	1.9	*167,000	1.6	*167,000	1.1

Insecticides, Fungicides, Disinfectants Etc.

Unit of Measure: Metric tons.

	1990 Value	%	1991 Value	%	1992 Value	%	1993 Value	%	1994 Value	%	1995 Value	%	1996 Value	%
Total Production	2,387,062	100.0	2,589,737	100.0	2,484,324	100.0	2,478,544	100.0	2,485,061	100.0	2,559,517	100.0	2,573,958	100.0
Regions														
Africa	66,773	2.8	67,984	2.6	59,933	2.4	68,099	2.7	78,255	3.1	87,953	3.4	86,436	3.4
America, North	133,130	5.6	145,360	5.6	137,496	5.5	175,471	7.1	208,514	8.4	176,057	6.9	181,374	7.0
America, South	92,309	3.9	90,555	3.5	94,753	3.8	87,833	3.5	90,857	3.7	89,746	3.5	94,544	3.7
Asia	929,913	39.0	947,596	36.6	924,810	37.2	922,374	37.2	826,463	33.3	982,181	38.4	1,007,210	39.1
Europe	1,164,936	48.8	1,338,243	51.7	1,267,331	51.0	1,224,768	49.4	1,280,971	51.5	1,223,580	47.8	1,204,393	46.8
Leading Producers														
France	326,397	13.7	314,669	12.2	303,534	12.2	301,896	12.2	*328,398	13.2	*333,162	13.0	*337,925	13.1
China	227,800	9.5	255,000	9.8	280,800	11.3	257,134	10.4	289,500	11.6	416,549	16.3	447,500	17.4
Germany	-		251,688	9.7	204,829	8.2	191,462	7.7	204,374	8.2	127,491	5.0	105,877	4.1
Korea, Republic of	181,887	7.6	189,993	7.3	171,045	6.9	221,180	8.9	141,372	5.7	156,929	6.1	*178,254	6.9
Mexico	126,668	5.3	138,820	5.4	130,879	5.3	168,776	6.8	201,742	8.1	169,207	6.6	174,447	6.8

Rubber, Synthetic

Unit of Measure: Metric tons.

	1990 Value	%	1991 Value	%	1992 Value	%	1993 Value	%	1994 Value	%	1995 Value	%	1996 Value	%
Total Production	10,870,294	100.0	10,698,946	100.0	11,410,456	100.0	11,432,218	100.0	11,989,087	100.0	12,833,033	100.0	12,831,734	100.0
Regions														
Africa	45,900	0.4	34,900	0.3	33,500	0.3	35,100	0.3	48,600	0.4	56,500	0.4	57,500	0.4
America, North	2,429,000	22.3	2,477,200	23.2	2,482,100	21.8	2,450,500	21.4	2,709,200	22.6	2,819,000	22.0	2,687,623	20.9
America, South	313,400	2.9	293,936	2.7	297,789	2.6	295,215	2.6	322,730	2.7	340,554	2.7	336,845	2.6
Asia	5,383,361	49.5	5,418,510	50.6	5,498,793	48.2	5,518,675	48.3	5,598,465	46.7	6,202,484	48.3	6,240,453	48.6
Europe	2,657,833	24.5	2,438,900	22.8	3,058,374	26.8	3,086,328	27.0	3,265,492	27.2	3,366,295	26.2	3,464,174	27.0
Oceania	40,800	0.4	35,500	0.3	39,900	0.3	46,400	0.4	44,600	0.4	48,200	0.4	45,140	0.4
Leading Producers														
United States of America	2,114,500	19.5	2,190,000	20.5	2,150,000	18.8	2,170,000	19.0	2,390,000	19.9	2,540,000	19.8	*2,392,952	18.6
Japan	1,425,751	13.1	1,377,290	12.9	1,389,895	12.2	1,309,792	11.5	1,350,822	11.3	1,497,575	11.7	1,519,938	11.8
Russian Federation	*754,925	6.9	*754,925	7.1	*754,925	6.6	*754,925	6.6	631,854	5.3	836,887	6.5	796,033	6.2
Germany	-		-		556,829	4.9	583,709	5.1	643,120	5.4	478,531	3.7	563,097	4.4
France	514,644	4.7	470,707	4.4	498,624	4.4	486,401	4.3	518,400	4.3	619,400	4.8	*553,152	4.3

Commodity data are provided by the United Nations Statistical Division. The symbol * means that data are estimated. For additional notes, see Appendix I.

303

Product Tables
Non-Cellulosic Staple and Tow

Unit of Measure: Metric tons.

	1990 Value	%	1991 Value	%	1992 Value	%	1993 Value	%	1994 Value	%	1995 Value	%	1996 Value	%
Total Production	8,712,396	100.0	9,389,160	100.0	10,131,187	100.0	9,775,530	100.0	10,694,544	100.0	10,607,526	100.0	9,553,428	100.0
Regions														
Africa	74,300	0.9	75,700	0.8	60,400	0.6	73,200	0.7	89,900	0.8	98,750	0.9	102,840	1.1
America, North	1,832,044	21.0	1,827,972	19.5	1,890,909	18.7	1,898,605	19.4	2,009,355	18.8	1,962,859	18.5	2,000,891	20.9
America, South	183,200	2.1	207,100	2.2	218,354	2.2	215,004	2.2	223,439	2.1	226,080	2.1	207,500	2.2
Asia	4,257,156	48.9	4,563,958	48.6	5,218,262	51.5	4,995,111	51.1	5,632,544	52.7	5,816,416	54.8	4,733,432	49.5
Europe	2,363,696	27.1	2,712,330	28.9	2,741,022	27.1	2,591,230	26.5	2,736,786	25.6	2,500,762	23.6	2,505,966	26.2
Oceania	2,000	0.0	2,100	0.0	2,240	0.0	2,380	0.0	2,520	0.0	2,660	0.0	2,800	0.0
Leading Producers														
United States of America	1,626,300	18.7	1,599,200	17.0	1,655,700	16.3	1,622,700	16.6	1,698,000	15.9	1,640,100	15.5	1,642,100	17.2
Korea, Republic of	1,391,927	16.0	1,492,949	15.9	1,615,289	15.9	1,672,939	17.1	1,838,686	17.2	1,953,530	18.4	658,100	6.9
China	811,600	9.3	889,900	9.5	911,800	9.0	905,200	9.3	1,164,200	10.9	1,108,000	10.4	1,423,000	14.9
Japan	722,900	8.3	714,700	7.6	729,700	7.2	694,200	7.1	708,200	6.6	731,200	6.9	400,875	4.2
Italy	691,617	7.9	672,790	7.2	693,184	6.8	641,176	6.6	713,961	6.7	649,146	6.1	663,077	6.9

Cellulosic Staple and Tow

Unit of Measure: Metric tons.

	1990 Value	%	1991 Value	%	1992 Value	%	1993 Value	%	1994 Value	%	1995 Value	%	1996 Value	%
Total Production	2,459,867	100.0	2,429,475	100.0	2,696,230	100.0	2,618,859	100.0	2,735,849	100.0	2,949,375	100.0	2,824,694	100.0
Regions														
Africa	5,200	0.2	5,200	0.2	5,200	0.2	5,200	0.2	5,300	0.2	5,200	0.2	5,300	0.2
America, North	186,468	7.6	178,234	7.3	172,814	6.4	177,538	6.8	176,079	6.4	187,681	6.4	168,936	6.0
America, South	79,619	3.2	81,319	3.3	82,043	3.0	79,337	3.0	85,218	3.1	94,060	3.2	75,638	2.7
Asia	1,014,421	41.2	1,094,431	45.0	1,135,348	42.1	1,188,414	45.4	1,264,250	46.2	1,393,978	47.3	1,341,498	47.5
Europe	1,174,159	47.7	1,070,290	44.1	1,300,825	48.2	1,168,371	44.6	1,205,003	44.0	1,268,456	43.0	1,233,322	43.7
Leading Producers														
Japan	176,700	7.2	169,900	7.0	168,000	6.2	162,000	6.2	140,700	5.1	134,900	4.6	86,300	3.1
France	259,404	10.5	260,148	10.7	287,976	10.7	184,512	7.0	171,744	6.3	*293,687	10.0	*311,582	11.0
China	157,600	6.4	185,000	7.6	189,200	7.0	212,100	8.1	270,900	9.9	365,200	12.4	359,000	12.7
Germany	-		-		183,400	6.8	178,376	6.8	196,975	7.2	141,688	4.8	142,564	5.0
United States of America	135,700	5.5	124,000	5.1	124,800	4.6	126,100	4.8	123,800	4.5	131,500	4.5	116,200	4.1

Artificial Resins and Plastic Materials (Total)

Unit of Measure: 1,000 Metric tons.

	1990 Value	%	1991 Value	%	1992 Value	%	1993 Value	%	1994 Value	%	1995 Value	%	1996 Value	%
Total Production	63,420	100.0	404,018	100.0	266,983	100.0	88,505	100.0	91,910	100.0	91,163	100.0	237,744	100.0
Regions														
Africa	12	0.0	10	0.0	8	0.0	9	0.0	9	0.0	-			
America, North	20,992	33.1	21,031	5.2	21,117	7.9	21,246	24.0	21,309	23.2	21,373	23.4	21,451	9.0
America, South	27	0.0	27	0.0	26	0.0	31	0.0	19	0.0	32	0.0	27	0.0
Asia	18,387	29.0	349,636	86.5	211,897	79.4	33,986	38.4	35,789	38.9	35,662	39.1	182,340	76.7
Europe	23,382	36.9	32,703	8.1	33,333	12.5	32,640	36.9	34,200	37.2	33,522	36.8	33,362	14.0
Oceania	620	1.0	611	0.2	602	0.2	592	0.7	583	0.6	574	0.6	564	0.2
Leading Producers														
Yugoslavia	-		331,365	82.0	194,055	72.7	16,425	18.6	18,677	20.3	18,118	19.9	165,027	69.4
United States of America	*19,721	31.1	*19,721	4.9	*19,721	7.4	*19,721	22.3	*19,721	21.5	*19,721	21.6	*19,721	8.3
Germany	-		10,116	2.5	10,081	3.8	9,925	11.2	11,307	12.3	10,313	11.3	10,001	4.2
Japan	*7,700	12.1	*7,700	1.9	*7,700	2.9	*7,700	8.7	*7,700	8.4	*7,700	8.4	*7,700	3.2
Netherlands	3,627	5.7	3,509	0.9	3,915	1.5	*3,825	4.3	*3,954	4.3	*4,083	4.5	*4,212	1.8

Commodity data are provided by the United Nations Statistical Division. The symbol * means that data are estimated. For additional notes, see Appendix I.

Product Tables
Alkyd Resins

Unit of Measure: Metric tons.

	1990 Value	%	1991 Value	%	1992 Value	%	1993 Value	%	1994 Value	%	1995 Value	%	1996 Value	%
Total Production	2,277,984	100.0	2,190,174	100.0	2,127,409	100.0	2,053,109	100.0	2,030,560	100.0	2,075,287	100.0	2,069,052	100.0
Regions														
Africa	41,867	1.8	46,865	2.1	51,863	2.4	56,862	2.8	61,860	3.0	66,858	3.2	71,856	3.5
America, North	333,579	14.6	335,322	15.3	335,842	15.8	341,245	16.6	341,474	16.8	341,008	16.4	344,950	16.7
America, South	18,438	0.8	18,930	0.9	20,687	1.0	22,396	1.1	30,760	1.5	24,580	1.2	28,568	1.4
Asia	983,598	43.2	947,141	43.2	886,606	41.7	835,590	40.7	802,412	39.5	808,050	38.9	815,969	39.4
Europe	898,845	39.5	839,194	38.3	832,410	39.1	797,016	38.8	794,054	39.1	834,791	40.2	807,709	39.0
Oceania	1,657	0.1	2,724	0.1	-		-		-		-		-	
Leading Producers														
Japan	441,550	19.4	426,855	19.5	410,248	19.3	385,950	18.8	383,257	18.9	400,403	19.3	400,875	19.4
United States of America	*323,801	14.2	*323,801	14.8	*323,801	15.2	*323,801	15.8	*323,801	15.9	*323,801	15.6	*323,801	15.6
Germany	*147,457	6.5	*147,457	6.7	*147,457	6.9	154,020	7.5	156,841	7.7	140,226	6.8	138,740	6.7
United Kingdom	122,570	5.4	105,725	4.8	88,880	4.2	72,035	3.5	54,419	2.7	*81,846	3.9	*80,479	3.9
Russian Federation	134,337	5.9	112,564	5.1	87,260	4.1	72,154	3.5	42,775	2.1	47,611	2.3	45,874	2.2

Amino Plastics

Unit of Measure: Metric tons.

	1990 Value	%	1991 Value	%	1992 Value	%	1993 Value	%	1994 Value	%	1995 Value	%	1996 Value	%
Total Production	3,144,165	100.0	4,114,006	100.0	3,949,002	100.0	3,894,194	100.0	4,026,754	100.0	3,708,129	100.0	3,756,835	100.0
Regions														
America, North	575,856	18.3	573,346	13.9	570,835	14.5	568,325	14.6	570,262	14.2	570,367	15.4	575,279	15.3
America, South	10,760	0.3	10,864	0.3	10,968	0.3	11,072	0.3	11,176	0.3	11,280	0.3	11,385	0.3
Asia	659,750	21.0	636,595	15.5	576,327	14.6	557,930	14.3	535,298	13.3	504,718	13.6	515,858	13.7
Europe	1,897,799	60.4	2,893,201	70.3	2,790,871	70.7	2,756,867	70.8	2,910,017	72.3	2,621,764	70.7	2,654,314	70.7
Leading Producers														
Germany	-		1,073,942	26.1	974,330	24.7	874,719	22.5	884,793	22.0	620,446	16.7	627,078	16.7
Japan	625,693	19.9	609,968	14.8	555,036	14.1	542,579	13.9	521,440	12.9	490,523	13.2	502,269	13.4
United States of America	*552,830	17.6	*552,830	13.4	*552,830	14.0	*552,830	14.2	*552,830	13.7	*552,830	14.9	*552,830	14.7
Spain	210,915	6.7	181,647	4.4	213,349	5.4	255,962	6.6	322,735	8.0	325,299	8.8	335,240	8.9
United Kingdom	*176,928	5.6	*180,191	4.4	*183,454	4.6	*186,717	4.8	*189,980	4.7	*193,244	5.2	*196,507	5.2

Phenolic and Cresylic Plastics

Unit of Measure: Metric tons.

	1990 Value	%	1991 Value	%	1992 Value	%	1993 Value	%	1994 Value	%	1995 Value	%	1996 Value	%
Total Production	3,295,596	100.0	3,074,510	100.0	2,865,333	100.0	2,723,712	100.0	2,501,284	100.0	2,610,032	100.0	2,240,760	100.0
Regions														
America, North	545,524	16.6	524,719	17.1	503,914	17.6	483,109	17.7	460,276	18.4	439,126	16.8	424,174	18.9
America, South	4,110	0.1	3,890	0.1	3,669	0.1	3,449	0.1	3,229	0.1	3,009	0.1	2,789	0.1
Asia	2,072,167	62.9	1,920,573	62.5	1,729,332	60.4	1,600,199	58.8	1,371,803	54.8	1,494,409	57.3	1,105,282	49.3
Europe	667,042	20.2	618,769	20.1	622,053	21.7	630,783	23.2	659,999	26.4	667,705	25.6	702,926	31.4
Oceania	6,754	0.2	6,560	0.2	6,365	0.2	6,171	0.2	5,977	0.2	5,783	0.2	5,589	0.2
Leading Producers														
Russian Federation	1,641,323	49.8	1,491,463	48.5	1,336,374	46.6	1,240,169	45.5	1,007,192	40.3	1,130,660	43.3	780,729	34.8
United States of America	*529,124	16.1	*508,002	16.5	*486,880	17.0	*465,758	17.1	*444,636	17.8	*423,514	16.2	*402,392	18.0
Japan	384,879	11.7	382,980	12.5	355,822	12.4	328,234	12.1	330,032	13.2	327,045	12.5	293,774	13.1
Germany	*258,573	7.8	*258,573	8.4	*258,573	9.0	*258,573	9.5	*258,573	10.3	*258,573	9.9	258,573	11.5
Finland	80,206	2.4	62,154	2.0	73,633	2.6	83,122	3.1	104,640	4.2	100,626	3.9	110,023	4.9

Commodity data are provided by the United Nations Statistical Division. The symbol * means that data are estimated. For additional notes, see Appendix I.

305

Product Tables
Polyethylene

Unit of Measure: Metric tons.

	1990 Value	%	1991 Value	%	1992 Value	%	1993 Value	%	1994 Value	%	1995 Value	%	1996 Value	%
Total Production	29,905,502	100.0	31,460,890	100.0	34,079,214	100.0	34,965,330	100.0	36,597,086	100.0	36,907,527	100.0	37,611,749	100.0
Regions														
Africa	38,185	0.1	35,255	0.1	32,518	0.1	26,500	0.1	40,309	0.1	30,690	0.1	30,269	0.1
America, North	9,782,189	32.7	10,610,832	33.7	11,414,390	33.5	12,218,591	34.9	13,100,465	35.8	12,794,952	34.7	13,277,488	35.3
America, South	1,331,868	4.5	1,330,408	4.2	1,314,008	3.9	1,545,782	4.4	1,635,462	4.5	1,700,108	4.6	1,735,373	4.6
Asia	6,802,468	22.7	7,482,924	23.8	8,839,107	25.9	8,870,505	25.4	9,234,807	25.2	9,935,378	26.9	9,964,074	26.5
Europe	11,709,107	39.2	11,754,267	37.4	12,226,467	35.9	12,045,711	34.5	12,322,284	33.7	12,177,119	33.0	12,329,746	32.8
Oceania	241,685	0.8	247,204	0.8	252,723	0.7	258,242	0.7	263,760	0.7	269,279	0.7	274,798	0.7
Leading Producers														
United States of America	8,007,520	26.8	8,695,140	27.6	9,382,760	27.5	10,070,380	28.8	10,758,000	29.4	*10,334,481	28.0	*10,684,895	28.4
Japan	2,887,555	9.7	2,982,445	9.5	2,980,911	8.7	2,761,533	7.9	2,944,112	8.0	3,193,046	8.7	3,313,198	8.8
Netherlands	*1,861,659	6.2	*1,861,659	5.9	*1,861,659	5.5	*1,861,659	5.3	1,771,296	4.8	2,007,331	5.4	1,806,351	4.8
Germany	*1,734,672	5.8	*1,734,672	5.5	*1,734,672	5.1	1,623,391	4.6	1,766,745	4.8	1,790,734	4.9	1,757,819	4.7
Canada	1,454,231	4.9	1,566,689	5.0	1,654,062	4.9	1,742,077	5.0	1,907,765	5.2	*1,997,205	5.4	*2,100,762	5.6

Ethylene-Vinyl Acetate Copolymers

Unit of Measure: 1,000 Metric tons.

	1990 Value	%	1991 Value	%	1992 Value	%	1993 Value	%	1994 Value	%	1995 Value	%	1996 Value	%
Total Production	281,027	100.0	334,548	100.0	359,141	100.0	383,790	100.0	408,383	100.0	432,925	100.0	457,628	100.0
Regions														
America, North	279,950	99.6	333,332	99.6	357,859	99.6	382,386	99.6	406,914	99.6	431,442	99.7	455,969	99.6
America, South	18	0.0	18	0.0	2	0.0	26	0.0	12	0.0	31	0.0	18	0.0
Asia	429	0.2	573	0.2	651	0.2	729	0.2	808	0.2	886	0.2	964	0.2
Europe	630	0.2	625	0.2	629	0.2	649	0.2	649	0.2	567	0.1	677	0.1
Leading Producers														
United States of America	279,944	99.6	*333,326	99.6	*357,854	99.6	*382,381	99.6	*406,909	99.6	*431,436	99.7	*455,964	99.6
Turkey	426	0.2	*570	0.2	*648	0.2	*726	0.2	*803	0.2	*881	0.2	*959	0.2
Italy	*413	0.1	*413	0.1	*413	0.1	*413	0.1	*413	0.1	*413	0.1	*413	0.1
Slovakia	*137	0.0	*137	0.0	*137	0.0	*137	0.0	*137	0.0	83	0.0	190	0.0
Romania	*31	0.0	*31	0.0	*31	0.0	*31	0.0	*31	0.0	*31	0.0	*31	0.0

Polypropylene

Unit of Measure: Metric tons.

	1990 Value	%	1991 Value	%	1992 Value	%	1993 Value	%	1994 Value	%	1995 Value	%	1996 Value	%
Total Production	12,976,336	100.0	13,325,191	100.0	14,294,604	100.0	14,958,949	100.0	16,295,729	100.0	16,501,244	100.0	17,326,942	100.0
Regions														
America, North	3,574,080	27.5	3,769,462	28.3	4,026,479	28.2	4,283,656	28.6	4,542,669	27.9	4,659,519	28.2	4,861,065	28.1
America, South	359,659	2.8	415,962	3.1	476,049	3.3	596,870	4.0	665,923	4.1	723,109	4.4	770,511	4.4
Asia	3,383,291	26.1	3,627,071	27.2	4,184,194	29.3	4,392,568	29.4	4,812,372	29.5	5,182,131	31.4	5,424,697	31.3
Europe	5,489,476	42.3	5,335,450	40.0	5,423,220	37.9	5,493,777	36.7	6,075,270	37.3	5,729,574	34.7	6,056,345	35.0
Oceania	169,831	1.3	177,247	1.3	184,662	1.3	192,078	1.3	199,494	1.2	206,910	1.3	214,325	1.2
Leading Producers														
United States of America	3,296,840	25.4	3,554,380	26.7	3,811,920	26.7	4,069,460	27.2	4,327,000	26.6	*4,444,550	26.9	*4,643,782	26.8
Japan	1,942,054	15.0	1,955,237	14.7	2,038,210	14.3	2,031,122	13.6	2,224,755	13.7	2,501,858	15.2	2,730,162	15.8
Belgium	813,803	6.3	794,854	6.0	825,911	5.8	888,874	5.9	897,141	5.5	905,408	5.5	1,013,468	5.8
Korea, Republic of	571,362	4.4	769,907	5.8	1,222,580	8.6	1,435,840	9.6	1,607,116	9.9	1,619,205	9.8	*1,557,542	9.0
Germany	*691,546	5.3	*691,546	5.2	*691,546	4.8	*691,546	4.6	706,909	4.3	668,482	4.1	699,246	4.0

Commodity data are provided by the United Nations Statistical Division. The symbol * means that data are estimated. For additional notes, see Appendix I.

Product Tables
Acrylic Polymers

Unit of Measure: 1,000 Metric tons.

	1990 Value	%	1991 Value	%	1992 Value	%	1993 Value	%	1994 Value	%	1995 Value	%	1996 Value	%
Total Production	684,940	100.0	742,098	100.0	755,663	100.0	769,226	100.0	782,909	100.0	796,425	100.0	810,070	100.0
Regions														
America, North	683,832	99.8	740,969	99.8	754,539	99.9	768,101	99.9	781,680	99.8	795,249	99.9	808,833	99.8
America, South	12	0.0	12	0.0	14	0.0	11	0.0	10	0.0	14	0.0	12	0.0
Asia	69	0.0	66	0.0	54	0.0	35	0.0	30	0.0	29	0.0	22	0.0
Europe	1,027	0.1	1,051	0.1	1,056	0.1	1,078	0.1	1,188	0.2	1,134	0.1	1,202	0.1
Leading Producers														
United States of America	683,771	99.8	*740,908	99.8	*754,478	99.8	*768,049	99.8	*781,620	99.8	*795,190	99.8	*808,761	99.8
United Kingdom	*204	0.0	*204	0.0	*204	0.0	*204	0.0	204	0.0	*204	0.0	*204	0.0
Germany	*194	0.0	*194	0.0	*194	0.0	191	0.0	197	0.0	*194	0.0	*194	0.0
Italy	166	0.0	174	0.0	170	0.0	163	0.0	177	0.0	175	0.0	203	0.0
France	111	0.0	125	0.0	129	0.0	141	0.0	236	0.0	*175	0.0	*185	0.0

Polyamides

Unit of Measure: Metric tons.

	1990 Value	%	1991 Value	%	1992 Value	%	1993 Value	%	1994 Value	%	1995 Value	%	1996 Value	%
Total Production	9,669,907	100.0	13,517,290	100.0	13,992,396	100.0	13,991,001	100.0	14,150,188	100.0	14,123,320	100.0	14,185,822	100.0
Regions														
Asia	416,441	4.3	429,188	3.2	436,040	3.1	440,893	3.2	447,019	3.2	462,238	3.3	466,100	3.3
Europe	9,253,466	95.7	13,088,101	96.8	13,556,356	96.9	13,550,108	96.8	13,703,170	96.8	13,661,082	96.7	13,719,722	96.7
Leading Producers														
Bulgaria	8,299,000	85.8	*12,140,000	89.8	*12,140,000	86.8	*12,140,000	86.8	*12,140,000	85.8	*12,140,000	86.0	*12,140,000	85.6
Germany	-		-		442,544	3.2	466,105	3.3	589,472	4.2	536,168	3.8	581,707	4.1
Netherlands	*217,817	2.3	*217,817	1.6	*217,817	1.6	*217,817	1.6	218,116	1.5	210,283	1.5	225,051	1.6
Japan	*161,314	1.7	*171,371	1.3	*181,429	1.3	*191,486	1.4	*201,543	1.4	*211,600	1.5	*221,657	1.6
Belgium	*93,430	1.0	*93,430	0.7	*93,430	0.7	*93,430	0.7	96,331	0.7	91,091	0.6	92,867	0.7

Polystyrene

Unit of Measure: Metric tons.

	1990 Value	%	1991 Value	%	1992 Value	%	1993 Value	%	1994 Value	%	1995 Value	%	1996 Value	%
Total Production	9,973,195	100.0	10,054,653	100.0	9,783,170	100.0	10,050,778	100.0	10,244,685	100.0	10,432,793	100.0	10,488,955	100.0
Regions														
America, North	2,412,722	24.2	2,373,433	23.6	2,335,173	23.9	2,463,654	24.5	2,472,204	24.1	2,499,787	24.0	2,502,280	23.9
America, South	231,657	2.3	268,089	2.7	260,257	2.7	277,416	2.8	294,234	2.9	313,804	3.0	295,899	2.8
Asia	3,807,776	38.2	3,960,067	39.4	3,808,322	38.9	3,963,652	39.4	4,169,008	40.7	4,307,248	41.3	4,382,912	41.8
Europe	3,514,970	35.2	3,446,993	34.3	3,373,348	34.5	3,339,985	33.2	3,303,169	32.2	3,305,884	31.7	3,301,793	31.5
Oceania	6,071	0.1	6,071	0.1	6,071	0.1	6,071	0.1	6,071	0.1	6,071	0.1	6,071	0.1
Leading Producers														
United States of America	*2,178,972	21.8	*2,128,378	21.2	*2,077,785	21.2	*2,027,191	20.2	*1,976,597	19.3	*1,926,003	18.5	*1,875,409	17.9
Japan	2,092,186	21.0	2,121,383	21.1	2,005,022	20.5	1,965,623	19.6	2,098,527	20.5	2,149,071	20.6	2,177,728	20.8
Germany	*612,740	6.1	*612,740	6.1	*612,740	6.3	*612,740	6.1	656,156	6.4	627,556	6.0	554,509	5.3
France	541,759	5.4	529,581	5.3	509,054	5.2	481,071	4.8	533,429	5.2	*552,985	5.3	*560,177	5.3
Korea, Republic of	592,023	5.9	758,471	7.5	754,255	7.7	808,972	8.0	869,400	8.5	904,572	8.7	*982,342	9.4

Commodity data are provided by the United Nations Statistical Division. The symbol * means that data are estimated. For additional notes, see Appendix I.

307

Product Tables
Polyacetals

Unit of Measure: 1,000 Metric tons.

	1990		1991		1992		1993		1994		1995		1996	
	Value	%	Value	%	Value	%	Value	%	Value	%	Value	%	Value	%
Total Production	289,577	100.0	337,890	100.0	355,949	100.0	374,017	100.0	392,032	100.0	410,179	100.0	428,190	100.0
Regions														
America, North	288,562	99.6	336,843	99.7	354,896	99.7	372,949	99.7	391,002	99.7	409,055	99.7	427,108	99.7
Asia	142	0.0	153	0.0	164	0.0	174	0.0	185	0.0	196	0.0	207	0.0
Europe	872	0.3	894	0.3	889	0.2	894	0.2	845	0.2	928	0.2	875	0.2
Leading Producers														
United States of America	288,561	99.6	*336,841	99.7	*354,894	99.7	*372,947	99.7	*391,000	99.7	*409,053	99.7	*427,106	99.7
Netherlands	*445	0.2	*445	0.1	*445	0.1	*445	0.1	440	0.1	472	0.1	422	0.1
Belgium	*156	0.1	*156	0.0	*156	0.0	*156	0.0	116	0.0	187	0.0	166	0.0
Japan	*138	0.0	*149	0.0	*160	0.0	*170	0.0	*181	0.0	*191	0.0	*202	0.0
Italy	103	0.0	131	0.0	123	0.0	127	0.0	128	0.0	110	0.0	127	0.0

Polyvinyl Chloride

Unit of Measure: Metric tons.

	1990		1991		1992		1993		1994		1995		1996	
	Value	%	Value	%	Value	%	Value	%	Value	%	Value	%	Value	%
Total Production	16,744,652	100.0	18,398,812	100.0	18,953,783	100.0	19,273,094	100.0	19,986,958	100.0	20,025,047	100.0	20,890,793	100.0
Regions														
Africa	14,207	0.1	12,122	0.1	6,958	0.0	11,953	0.1	17,150	0.1	12,623	0.1	12,953	0.1
America, North	4,821,034	28.8	5,038,218	27.4	5,348,950	28.2	5,594,850	29.0	5,847,950	29.3	6,002,418	30.0	6,225,831	29.8
America, South	726,072	4.3	732,492	4.0	715,763	3.8	730,166	3.8	722,716	3.6	943,736	4.7	891,832	4.3
Asia	4,885,389	29.2	5,041,442	27.4	5,163,400	27.2	5,130,919	26.6	5,345,805	26.7	5,459,942	27.3	5,836,586	27.9
Europe	6,279,165	37.5	7,555,753	41.1	7,699,927	40.6	7,786,420	40.4	8,034,552	40.2	7,587,542	37.9	7,904,806	37.8
Oceania	18,785	0.1	18,785	0.1	18,785	0.1	18,785	0.1	18,785	0.1	18,785	0.1	18,785	0.1
Leading Producers														
United States of America	4,199,280	25.1	4,395,960	23.9	4,592,640	24.2	4,789,320	24.8	4,986,000	24.9	*5,196,593	26.0	*5,399,352	25.8
Japan	2,048,823	12.2	2,055,121	11.2	1,982,917	10.5	1,979,848	10.3	2,110,745	10.6	2,274,235	11.4	2,510,973	12.0
Germany	-		1,328,233	7.2	1,280,295	6.8	1,209,722	6.3	1,263,591	6.3	1,228,393	6.1	1,294,197	6.2
France	1,061,741	6.3	1,050,504	5.7	1,124,060	5.9	1,203,032	6.2	1,177,861	5.9	1,061,658	5.3	*1,230,793	5.9
China	785,300	4.7	813,100	4.4	904,200	4.8	*895,555	4.6	*940,455	4.7	*985,356	4.9	*1,030,257	4.9

Non-Cellulosic Continuous Fibers

Unit of Measure: Metric tons.

	1990		1991		1992		1993		1994		1995		1996	
	Value	%	Value	%	Value	%	Value	%	Value	%	Value	%	Value	%
Total Production	6,216,568	100.0	6,688,238	100.0	7,176,266	100.0	7,362,669	100.0	7,907,702	100.0	8,327,659	100.0	8,644,320	100.0
Regions														
Africa	68,000	1.1	67,200	1.0	89,300	1.2	106,200	1.4	109,700	1.4	120,200	1.4	119,900	1.4
America, North	1,494,564	24.0	1,545,829	23.1	1,566,994	21.8	1,602,286	21.8	1,762,549	22.3	1,873,061	22.5	1,953,743	22.6
America, South	224,500	3.6	238,100	3.6	233,100	3.2	254,100	3.5	256,400	3.2	260,800	3.1	259,000	3.0
Asia	2,983,266	48.0	3,108,935	46.5	3,519,311	49.0	3,709,548	50.4	3,960,737	50.1	4,245,221	51.0	4,541,356	52.5
Europe	1,430,938	23.0	1,714,073	25.6	1,755,762	24.5	1,678,035	22.8	1,809,016	22.9	1,819,277	21.8	1,757,021	20.3
Oceania	15,300	0.2	14,100	0.2	11,800	0.2	12,500	0.2	9,300	0.1	9,100	0.1	13,300	0.2
Leading Producers														
United States of America	1,259,700	20.3	1,303,700	19.5	1,324,900	18.5	1,353,400	18.4	1,495,300	18.9	1,549,500	18.6	1,611,900	18.6
Japan	702,100	11.3	715,100	10.7	719,100	10.0	670,900	9.1	665,800	8.4	672,800	8.1	661,700	7.7
China	531,200	8.5	599,000	9.0	821,900	11.5	966,700	13.1	1,135,000	14.4	1,156,000	13.9	1,204,000	13.9
Germany	-		375,000	5.6	376,600	5.2	368,918	5.0	410,734	5.2	427,578	5.1	407,837	4.7
India	259,300	4.2	263,100	3.9	315,400	4.4	357,100	4.9	378,900	4.8	429,900	5.2	549,500	6.4

Commodity data are provided by the United Nations Statistical Division. The symbol * means that data are estimated. For additional notes, see Appendix I.

Product Tables
Cellulosic Continuous Filaments

Unit of Measure: Metric tons.

	1990		1991		1992		1993		1994		1995		1996	
	Value	%	Value	%	Value	%	Value	%	Value	%	Value	%	Value	%
Total Production	1,199,759	100.0	1,043,846	100.0	964,588	100.0	899,478	100.0	816,020	100.0	851,076	100.0	781,346	100.0
Regions														
Africa	7,900	0.7	7,900	0.8	7,500	0.8	7,600	0.8	7,600	0.9	7,500	0.9	7,500	1.0
America, North	115,487	9.6	118,329	11.3	120,397	12.5	127,663	14.2	127,191	15.6	137,200	16.1	128,800	16.5
America, South	25,300	2.1	22,600	2.2	20,606	2.1	24,114	2.7	25,179	3.1	26,982	3.2	27,745	3.6
Asia	842,767	70.2	714,939	68.5	641,894	66.5	576,215	64.1	491,835	60.3	509,290	59.8	445,872	57.1
Europe	208,305	17.4	180,078	17.3	174,190	18.1	163,886	18.2	164,215	20.1	170,104	20.0	171,428	21.9
Leading Producers														
Russian Federation	356,638	29.7	243,314	23.3	207,201	21.5	167,415	18.6	95,975	11.8	111,327	13.1	58,810	7.5
United States of America	93,500	7.8	96,700	9.3	99,700	10.3	102,700	11.4	102,100	12.5	108,600	12.8	99,700	12.8
Japan	99,100	8.3	96,500	9.2	85,800	8.9	82,400	9.2	78,200	9.6	77,700	9.1	74,200	9.5
China	56,700	4.7	57,000	5.5	59,900	6.2	64,100	7.1	65,200	8.0	70,200	8.2	73,000	9.3
India	58,200	4.9	59,000	5.7	58,800	6.1	60,900	6.8	66,400	8.1	69,300	8.1	67,300	8.6

Commodity data are provided by the United Nations Statistical Division. The symbol * means that data are estimated. For additional notes, see Appendix I.

309

ISIC 3511
BASIC CHEMICALS, EXCL FERTILIZERS

ISIC 3511—Basic Chemicals, excluding Fertilizers—is a subset of ISIC 3510, covered in the previous chapter. Shown are statistics for countries that report separately on chemical production that excludes synthetic resins and fertilizers.

Please note that commodity details are shown under the more inclusive ISIC 3510.

Summary Statistics

		1990		1991		1992		1993		1994		1995	
		Value	N	Value	N	Value	N	Value	N	Value	N	Value	N
Establishments or enterprises	(number)	6,352	45	6,624	54	9,249	51	5,360	47	7,159	37	1,549	19
Number employed	(000)	781	48	797	53	744	49	1,033	46	769	35	745	24
Total Wages	($ mil.)	21,241	46	22,631	52	22,231	49	17,469	42	12,055	34	12,129	22
Wage/salary per employee	($)	13,181	44	14,044	51	15,047	48	11,604	42	12,056	34	13,098	22
Female workers	(000)	31	20	17	20	156	19	135	19	125	13	122	5
as % of total employment*	(%)	11	20	11	20	13	19	15	19	14	13	20	5
Output	($ bil.)	213	47	213	51	205	47	171	41	128	33	141	22
per employee	($ 000)	138	44	137	49	131	45	117	41	140	32	160	21
per establishment	($ 000)	13,441	42	13,463	48	12,595	44	9,897	39	11,738	31	8,397	15
Value added	($ bil.)	92	44	90	48	89	41	79	38	58	33	65	21
per employee	($ 000)	54	43	56	46	56	39	54	37	64	31	74	19
per establishment	($ 000)	5,085	41	4,866	45	4,940	40	4,160	37	5,107	31	3,074	14
Capital investment	($ mil.)	5,997	26	9,473	31	15,384	31	11,171	25	6,837	19	8,916	9
per employee	($ 000)	15	25	10	30	15	29	15	24	22	19	15	9
per establishment	($ 000)	1,480	25	1,064	30	1,164	30	1,234	24	1,687	18	722	8

Data presented above are drawn from the detailed tables that follow. Columns headed 'N' show the number of countries that provided valid data for inclusion. Values are not strictly comparable one year to the next or one row to the next because the number of countries that report varies from period to period and row to row. However, a general indication of magnitudes can be gleaned from reviewing this summary—especially in earlier years in which more reports are available. For detailed explanations, see Appendix I. *The average for those countries reporting both total and female employment.

Establishments and Number Engaged

Country	Establishments or Enterprises (number)						Employees per Establishment					
	1990	1991	1992	1993	1994	1995	1990	1991	1992	1993	1994	1995
Albania	-	-	-	5.000	-	-	-	-	-	106	-	-
Australia	140	182	201	-	-	-	57	38	35	-	-	-
Austria	56	54	54	53	50	-	132	130	130	117	117	-
Azerbaijan	11	9.000	10	10	10	-	81	108	105	101	82	-
Bahrain	-	-	7.000	-	-	-	-	-	94	-	-	-
Bangladesh	32	34	32	-	-	-	47	47	35	-	-	-
Benin	-	-	35	35	-	-	-	-	-	-	-	-
Bolivia	15	16	16	15	16	18	18	17	18	18	20	17
Bosnia & Herzegovina	16	16	-	-	3.000	-	468	374	-	-	308	-
Canada	210	202	198	195	211	-	105	104	101	87	71	-
Central African Republic	-	*1.000	*1.000	*1.000	*1.000	*1.000	-	-	-	-	-	-
Chile	18	22	23	21	23	-	134	126	120	146	180	-
China	-	-	-	-	*4,917	-	-	-	-	-	-	-
China (Hong Kong SAR)	65	51	101	73	92	-	15	18	12	14	9.783	-
Colombia	94	97	98	104	103	-	72	64	59	50	49	-
Costa Rica	10	13	13	17	-	-	8.100	8.077	8.692	9.353	-	-
Cyprus	1.000	1.000	1.000	1.000	1.000	1.000	29	27	28	27	32	35
Denmark	54	52	56	-	-	-	86	88	128	-	-	-
Ecuador	20	21	22	23	21	16	28	28	30	30	32	40
Egypt	21	22	17	24	-	-	624	714	624	463	-	-
El Salvador	-	-	-	2.000	1.000	1.000	-	-	-	33	24	35
Finland	56	59	57	57	57	-	88	83	95	82	81	-
FYR Macedonia	-	-	-	-	-	*42	-	-	-	-	-	18
Gabon	-	1.000	1.000	1.000	1.000	1.000	-	39	39	43	43	73
Ghana	-	-	-	1.000	-	-	-	-	-	91	-	-
Greece	31	30	28	-	-	-	53	37	36	-	-	-
Guatemala	-	3.000	2.000	2.000	2.000	2.000	-	69	91	91	91	91
Honduras	-	-	4.000	4.000	4.000	4.000	-	-	79	86	94	97
Hungary	-	-	*96	*120	-	-	-	-	-	-	-	-
India	1,377	1,405	1,444	1,522	-	-	67	67	67	64	-	-
Indonesia	182	171	196	204	212	240	88	94	98	107	109	109
Iran (Islamic Republic of)	-	39	44	50	-	-	-	74	79	77	-	-
Italy	*120	*118	-	-	-	-	357	321	-	-	-	-
Japan	1,279	1,287	1,299	1,277	-	-	72	71	73	73	-	-
Jordan	11	10	12	11	13	14	40	34	40	43	36	38
Kenya	27	22	22	21	28	28	94	113	115	121	92	97
Korea, Republic of	617	437	456	483	479	506	59	58	56	54	52	53
Kuwait	6.000	7.000	6.000	6.000	6.000	-	59	31	38	38	37	-
Malaysia	49	52	59	65	66	-	108	119	117	114	97	-
Mauritius	9.000	6.000	6.000	6.000	6.000	7.000	19	21	23	22	23	24
Mexico	77	77	76	75	73	72	261	248	243	202	170	198
Myanmar	1.000	2.000	2.000	2.000	2.000	-	170	81	77	74	73	-
Nepal	1.000	1.000	-	-	-	-	-	-	-	-	-	-
New Zealand	*35	*37	*35	*40	-	-	31	33	29	26	-	-
Norway	34	39	31	-	-	-	109	101	111	-	-	-
Pakistan	-	30	-	-	-	-	-	150	-	-	-	-
Panama	4.000	4.000	-	-	-	-	23	25	-	-	-	-
Peru	230	232	231	-	-	-	19	16	14	-	-	-
Philippines	90	100	82	76	-	-	61	57	66	64	-	-
Portugal	*79	*86	*87	*96	*94	-	53	51	49	41	36	-
Russian Federation	-	-	-	*425	*416	*514	-	-	-	1,079	1,029	738
Saint Lucia	-	1.000	1.000	1.000	1.000	1.000	-	-	-	-	-	-
Senegal	7.000	-	-	-	6.000	-	172	-	-	-	187	-
Sierra Leone	-	-	-	2.000	-	-	-	-	-	46	-	-
Slovakia	-	7.000	7.000	8.000	7.000	-	-	1,379	1,086	763	804	-
South Africa	-	166	-	-	-	-	-	-	-	-	-	-
Spain	273	286	317	-	-	-	67	65	56	-	-	-
Sri Lanka	14	10	4.000	10	-	-	44	51	105	61	-	-
Swaziland	-	1.000	1.000	1.000	-	-	-	4.000	6.000	6.000	-	-
Sweden	-	84	82	78	88	-	-	97	93	82	75	-

Continued.

Depending on the table, * means *Enterprises, Engaged*, or *Factor Values*; ± means *Basis Unspecified*; § means *shown in millions*. For additional notes and sources, see Appendix I.

Establishments and Number Engaged

- Continued -

Country	Establishments or Enterprises (number)						Employees per Establishment					
	1990	1991	1992	1993	1994	1995	1990	1991	1992	1993	1994	1995
Thailand	68	118	-	-	-	-	57	53	-	-	-	-
Trinidad and Tobago	-	11	9.000	11	12	-	-	109	145	100	87	-
Tunisia	-	-	-	6.000	22	20	-	-	-	47	40	80
Turkey	37	34	50	42	39	-	203	182	327	122	110	-
United Kingdom	791	777	838	-	-	-	130	129	117	-	-	-
United Republic of Tanzania	5.000	6.000	-	-	-	-	120	1,250	-	-	-	-
United States of America	-	-	2,705	-	-	-	-	-	84	-	-	-
Venezuela	74	75	74	78	68	61	82	88	88	88	96	105
Zambia	5.000	-	-	-	8.000	-	95	-	-	-	59	-

Employment and Compensation of Employees

Country	Number Employed/Engaged (000)						Wage/Salary per Employee ($)					
	1990	1991	1992	1993	1994	1995	1990	1991	1992	1993	1994	1995
Albania	-	-	-	0.530	-	-	-	-	-	-	-	-
Australia	8.000	*7.000	*7.000	-	-	-	27,188	33,280	32,038	-	-	-
Austria	7.400	7.000	7.000	6.226	5.851	5.532	35,073	39,348	39,208	-	-	-
Azerbaijan	0.888	0.975	1.049	1.012	0.818	-	1,470	1,593	-	-	-	-
Bahrain	-	-	0.661	-	-	-	-	-	15,994	-	-	-
Bangladesh	1.503	1.596	1.121	-	-	-	1,761	1,609	847	-	-	-
Bolivia	0.272	0.279	0.284	0.277	0.324	0.312	2,132	2,268	2,585	2,843	2,752	3,036
Bosnia & Herzegovina	7.486	5.979	-	-	0.924	-	-	-	-	-	-	-
Canada	22	21	20	17	15	15	-	-	-	-	-	-
Chile	2.416	2.772	2.754	3.066	4.136	3.433	9,792	11,211	12,671	12,150	12,681	15,877
China (Hong Kong SAR)	*1.000	*0.900	*1.200	*1.000	*0.900	-	12,837	17,586	17,548	21,201	21,565	-
Colombia	6.787	6.229	5.819	5.244	5.020	5.313	3,134	3,416	3,602	3,432	4,673	5,673
Costa Rica	0.081	0.105	0.113	0.159	-	-	2,834	3,010	3,603	4,111	-	-
Cyprus	0.029	0.027	0.028	0.027	0.032	0.035	13,808	16,319	19,048	17,960	19,329	19,460
Denmark	4.646	4.563	7.176	-	-	-	36,030	35,463	-	-	-	-
Ecuador	0.559	0.593	0.666	0.700	0.679	0.639	5,084	4,778	4,477	4,915	2,536	4,417
Egypt	*13	*16	*11	*11	-	-	3,172	1,479	1,890	2,052	-	-
El Salvador	-	-	-	*0.067	*0.024	*0.035	-	-	-	2,928	5,076	7,982
Finland	4.900	4.900	5.400	4.700	4.600	-	33,382	32,308	31,552	25,082	28,908	-
FYR Macedonia	-	-	-	-	-	0.769	-	-	-	-	-	-
Gabon	-	0.039	0.039	0.043	0.043	0.073	-	30,085	29,643	24,228	17,299	17,427
Ghana	-	-	-	*0.091	-	-	-	-	-	-	-	-
Greece	*1.642	*1.121	*1.014	-	-	-	18,992	18,265	19,096	-	-	-
Guatemala	-	*0.207	*0.183	*0.183	*0.183	*0.183	-	-	-	-	-	-
Honduras	*0.306	*0.309	*0.315	*0.345	*0.378	*0.386	986	1,250	1,337	1,143	773	764
India	*93	*93	*97	*97	-	-	1,908	1,744	2,243	1,580	-	-
Indonesia	16	16	19	22	23	26	999	1,235	1,390	1,476	1,733	2,898
Iran (Islamic Republic of)	-	2.869	3.465	3.836	-	-	-	6,389	6,843	6,485	-	-
Italy	43	38	-	-	-	-	-	-	-	-	-	-
Japan	*92	*92	*95	*93	-	-	39,487	-	-	-	-	-
Jordan	0.444	0.338	0.477	0.475	0.472	0.537	3,300	3,615	3,875	4,277	3,325	3,783
Kenya	2.531	2.480	2.540	2.551	2.574	2.708	2,000	1,834	1,733	1,047	1,391	1,930
Korea, Republic of	36	25	26	26	25	27	14,141	15,262	17,242	18,069	20,350	23,925
Kuwait	0.352	0.220	0.225	0.227	0.219	-	5,040	5,762	9,329	9,263	10,757	-
Lithuania	-	-	0.027	0.060	-	-	-	-	-	737	-	-
Malawi	0.194	-	-	-	-	-	-	-	-	-	-	-
Malaysia	5.300	6.200	6.900	7.400	6.400	8.826	7,582	7,519	8,204	8,893	9,538	10,304
Mauritius	*0.169	0.124	0.140	0.131	0.141	0.166	3,868	6,322	10,155	12,544	9,359	-
Mexico	20	19	18	15	12	14	7,631	9,221	11,175	12,690	13,460	7,441
Myanmar	0.170	0.161	0.154	0.149	0.145	-	1,049	1,364	1,213	2,638	3,075	-
New Zealand	1.090	1.236	1.002	1.044	-	-	18,622	17,802	18,259	-	-	-
Norway	3.718	3.923	3.435	-	-	-	34,975	34,798	-	-	-	-
Pakistan	-	4.485	-	-	-	-	-	3,496	-	-	-	-
Panama	*0.090	*0.100	-	-	-	-	9,922	11,670	-	-	-	-

Continued.

Depending on the table, * means *Enterprises, Engaged,* or *Factor Values*; ± means *Basis Unspecified*; § means *shown in millions*. For additional notes and sources, see Appendix I.

313

Employment and Compensation of Employees
- Continued -

Country	Number Employed/Engaged (000)						Wage/Salary per Employee ($)					
	1990	1991	1992	1993	1994	1995	1990	1991	1992	1993	1994	1995
Peru	*4.259	*3.624	*3.345	-	-	-	5,676	6,254	6,580	-	-	-
Philippines	5.500	5.700	5.400	4.900	-	-	3,298	3,397	4,326	4,683	-	-
Portugal	*4.212	*4.343	*4.234	*3.919	*3.377	-	-	9,811	12,910	12,135	11,967	-
Russian Federation	-	-	-	459	428	379	-	-	-	765	1,243	1,470
Senegal	*1.206	-	-	-	*1.120	-	8,610	-	-	-	4,789	-
Sierra Leone	-	-	-	*0.091	-	-	-	-	-	-	-	-
Slovakia	-	*9.650	*7.604	*6.101	*5.625	-	-	1,680	2,029	2,307	2,624	-
Spain	18	19	18	-	-	-	27,893	30,016	32,902	-	-	-
Sri Lanka	*0.618	*0.505	*0.420	*0.611	-	-	1,050	1,597	1,513	1,476	-	-
Swaziland	-	0.004	0.006	0.006	-	-	-	1,996	2,047	2,553	-	-
Sweden	-	*8.172	*7.666	6.416	6.566	-	-	31,664	34,228	28,410	29,381	-
Thailand	3.857	6.256	-	-	-	-	3,151	4,203	-	-	-	-
Trinidad and Tobago	-	1.201	1.302	1.099	1.047	-	-	15,947	16,698	14,556	13,896	-
Tunisia	-	-	-	0.283	0.876	1.609	-	-	-	15,462	4,930	3,897
Turkey	7.514	6.197	16	5.107	4.280	-	8,469	10,599	13,484	11,712	7,575	-
Ukraine	-	-	-	-	-	48	-	-	-	-	-	659
United Kingdom	103	100	98	-	-	-	30,756	34,336	36,922	-	-	-
United Republic of Tanzania	0.602	7.498	-	-	-	-	1,320	-	-	-	-	-
United States of America	227	231	228	218	202	199	39,780	-	-	-	-	-
Uruguay	*0.787	*0.744	*0.654	*0.634	-	-	-	-	-	-	-	-
Venezuela	6.100	6.600	6.500	6.844	6.554	6.391	5,694	9,064	9,079	12,106	9,463	9,932
Zambia	0.476	-	-	-	0.475	-	2,237	-	-	-	4,470	-

Female Workers: Total and Percent of Employment

Country	Female Workers (000)						Female Workers as % of Total Employment					
	1990	1991	1992	1993	1994	1995	1990	1991	1992	1993	1994	1995
Albania	-	-	-	0.166	-	-	-	-	-	31.32	-	-
Australia	1.322	-	-	-	-	-	16.53	-	-	-	-	-
Austria	1.200	1.000	1.000	0.988	0.809	-	16.22	14.29	14.29	15.87	13.83	-
Bahrain	-	-	0.037	-	-	-	-	-	5.60	-	-	-
Bangladesh	0.002	0.018	0.001	-	-	-	0.13	1.13	0.09	-	-	-
Canada	3.000	3.000	-	-	-	-	13.64	14.29	-	-	-	-
Chile	0.128	0.220	0.212	0.203	0.251	-	5.30	7.94	7.70	6.62	6.07	-
Colombia	-	0.791	-	0.909	0.899	-	-	12.70	-	17.33	17.91	-
Cyprus	0.002	0.002	0.003	0.003	0.003	-	6.90	7.41	10.71	11.11	9.38	-
Denmark	1.270	1.273	2.600	-	-	-	27.34	27.90	36.23	-	-	-
Egypt	-	1.414	0.615	0.759	-	-	-	9.01	5.80	6.84	-	-
El Salvador	-	-	-	0.003	0.002	0.004	-	-	-	4.48	8.33	11.43
Indonesia	-	-	-	3.764	4.004	4.283	-	-	-	17.20	17.33	16.30
Iran (Islamic Republic of)	-	0.081	0.106	0.143	-	-	-	2.82	3.06	3.73	-	-
Italy	-	2.805	-	-	-	-	-	7.41	-	-	-	-
Jordan	0.013	0.017	0.021	0.016	0.018	0.026	2.93	5.03	4.40	3.37	3.81	4.84
Kenya	0.244	0.193	0.204	0.205	-	-	9.64	7.78	8.03	8.04	-	-
Korea, Republic of	3.800	2.538	2.621	2.735	2.395	2.647	10.47	10.02	10.21	10.52	9.69	9.94
Malaysia	0.700	0.800	1.000	1.000	0.900	-	13.21	12.90	14.49	13.51	14.06	-
Myanmar	0.011	0.012	0.012	0.013	0.017	-	6.47	7.45	7.79	8.72	11.72	-
New Zealand	0.229	0.271	0.215	0.206	-	-	21.01	21.93	21.46	19.73	-	-
Panama	0.008	-	-	-	-	-	8.89	-	-	-	-	-
Philippines	-	-	0.805	0.726	-	-	-	-	14.91	14.82	-	-
Portugal	-	0.322	0.353	0.340	0.268	-	-	7.41	8.34	8.68	7.94	-
Senegal	0.060	-	-	-	-	-	4.98	-	-	-	-	-
Sri Lanka	0.052	0.034	0.023	0.127	-	-	8.41	6.73	5.48	20.79	-	-
Thailand	0.523	1.284	-	-	-	-	13.56	20.52	-	-	-	-
Turkey	0.314	-	-	-	-	-	4.18	-	-	-	-	-
United Kingdom	18	-	19	-	-	-	17.33	-	19.39	-	-	-

Continued.

Depending on the table, * means *Enterprises, Engaged*, or *Factor Values*; ± means *Basis Unspecified*; § means *shown in millions*. For additional notes and sources, see Appendix I.

Female Workers: Total and Percent of Employment

- Continued -

Country	Female Workers (000)						Female Workers as % of Total Employment					
	1990	1991	1992	1993	1994	1995	1990	1991	1992	1993	1994	1995
United Republic of Tanzania	0.090	1.125	-	-	-	-	14.95	15.00	-	-	-	-
United States of America	-	-	127	123	115	115	-	-	55.70	56.42	56.93	57.79
Zambia	-	-	-	-	0.035	-	-	-	-	-	7.37	-

Output and Output per Employee

Country	Output ($ bil.)						Output per Employee ($)					
	1990	1991	1992	1993	1994	1995	1990	1991	1992	1993	1994	1995
Albania	-	-	-	*0.001	-	-	-	-	-	1,257	-	-
Australia	*1.923	*2.049	*1.945	-	-	-	240,332	292,726	277,836	-	-	-
Austria	2.009	1.814	1.785	1.340	1.568	1.970	271,423	259,176	254,956	215,200	268,068	356,137
Azerbaijan	±0.298	±0.359	±0.017	±0.070	±0.028	-	335,077	368,179	16,035	69,006	33,808	-
Bahrain	-	-	±0.075	-	-	-	-	-	112,740	-	-	-
Bangladesh	0.046	0.038	0.011	-	-	-	30,477	23,850	9,779	-	-	-
Bolivia	0.004	0.005	0.005	0.005	0.006	0.006	15,878	18,782	18,374	18,430	17,942	20,176
Canada	8.236	7.314	6.594	6.007	6.678	7.041	374,373	348,300	329,693	353,370	445,226	467,532
Chile	0.360	0.451	0.465	0.711	0.802	0.791	148,963	162,806	168,952	231,788	194,024	230,417
China (Hong Kong SAR)	0.089	0.093	0.140	0.125	0.113	-	89,091	103,659	116,593	125,394	125,655	-
Colombia	0.331	0.314	0.277	0.336	0.375	0.444	48,697	50,410	47,535	64,128	74,648	83,522
Costa Rica	0.001	0.001	0.001	0.001	0.001	0.001	8,668	5,508	7,230	6,105	-	-
Cyprus	0.002	0.002	0.002	0.002	0.002	0.003	54,327	69,514	81,190	84,880	74,771	75,314
Denmark	*1.064	*1.060	-	-	-	-	228,907	232,312	-	-	-	-
Ecuador	0.028	0.032	0.031	0.034	0.039	0.044	50,451	54,020	46,133	48,621	57,784	68,791
Egypt	*0.313	*0.209	*0.157	*0.159	-	-	23,859	13,336	14,846	14,308	-	-
El Salvador	-	-	-	±0.003	±0.002	±0.002	-	-	-	51,559	73,524	59,277
Finland	±1.428	±1.204	±1.056	±0.986	±1.266	-	291,446	245,766	195,587	209,738	275,136	-
FYR Macedonia	-	-	-	-	-	±0.023	-	-	-	-	-	30,388
Gabon	-	*0.009	*0.009	*0.008	*0.006	*0.005	-	220,136	237,917	197,357	129,474	68,610
Greece	*0.242	*0.108	*0.122	-	-	-	147,549	96,033	120,054	-	-	-
Guatemala	-	0.003	0.003	0.005	0.006	0.015	-	14,218	18,952	25,332	35,365	83,893
Honduras	0.003	0.003	0.004	0.004	0.004	0.004	8,825	10,517	11,519	10,724	11,104	10,203
India	*3.241	*2.932	*2.818	*2.702	-	-	34,882	31,368	28,950	27,912	-	-
Indonesia	0.544	0.646	0.736	1.015	1.473	1.546	34,056	40,172	38,469	46,385	63,737	58,835
Iran (Islamic Republic of)	-	0.172	0.139	0.148	-	-	-	59,943	40,089	38,660	-	-
Italy	±10	±9.818	-	-	-	-	235,659	259,374	-	-	-	-
Japan	±43	±47	±47	±48	-	-	462,814	509,307	494,026	517,038	-	-
Jordan	0.034	0.026	0.031	0.025	0.027	0.032	75,922	77,223	64,179	52,258	57,807	59,011
Korea, Republic of	9.163	5.575	6.396	6.375	6.878	11	252,424	220,032	249,243	245,135	278,411	418,004
Kuwait	0.024	0.010	0.012	0.014	0.013	-	69,340	43,262	51,832	61,061	58,886	-
Lithuania	-	-	0.000	0.000	-	-	-	-	480	3,633	-	-
Malaysia	*1.234	*1.656	*1.885	*2.073	*1.921	*2.861	232,848	267,064	273,146	280,166	300,080	324,101
Mauritius	0.010	0.012	0.014	0.016	0.008	-	59,000	96,453	100,743	122,541	55,600	-
Mexico	2.262	2.310	2.372	2.120	2.141	2.575	112,515	120,742	128,358	139,626	172,492	180,819
Myanmar	0.001	0.001	0.002	0.003	0.003	-	6,839	4,626	14,243	17,822	19,622	-
New Zealand	0.186	-	-	-	-	-	170,889	-	-	-	-	-
Norway	±1.175	±1.113	±1.096	-	-	-	316,067	283,732	319,091	-	-	-
Pakistan	-	0.134	-	-	-	-	-	29,838	-	-	-	-
Panama	0.007	0.008	-	-	-	-	75,356	77,740	-	-	-	-
Peru	0.233	0.217	0.217	-	-	-	54,733	59,923	64,973	-	-	-
Philippines	0.296	0.320	0.420	0.322	-	-	53,908	56,151	77,799	65,639	-	-
Portugal	0.480	0.497	0.556	0.445	0.371	-	113,900	114,487	131,307	113,548	109,816	-
Russian Federation	-	-	-	3.782	5.910	7.665	-	-	-	8,246	13,804	20,219
Senegal	0.117	-	-	-	-	-	97,330	-	-	-	-	-
Slovakia	-	±0.263	±0.243	±0.170	±0.203	-	-	27,281	31,937	27,934	36,116	-
South Africa	±1.191	±1.202	±1.209	-	-	-	-	-	-	-	-	-
Spain	*6.196	*6.044	*5.624	-	-	-	339,429	326,294	316,675	-	-	-
Sri Lanka	0.004	0.007	0.008	0.009	-	-	6,745	12,966	19,790	15,342	-	-
Swaziland	-	*0.000	*0.000	*0.000	-	-	-	32,021	19,066	19,049	-	-

Continued.

Output and Output per Employee

- Continued -

Country	Output ($ bil.)						Output per Employee ($)					
	1990	1991	1992	1993	1994	1995	1990	1991	1992	1993	1994	1995
Sweden	-	*2.297	*2.327	*1.451	*1.765	-	-	281,107	303,543	226,137	268,812	-
Thailand	0.211	0.213				-	54,711	34,078				
Trinidad and Tobago	±0.007	±0.307	±0.248	±0.281	±0.507	-	-	255,963	190,548	255,592	484,350	-
Tunisia	0.073	0.059	0.064	0.048	0.046	0.056			170,746	52,556	34,894	
Turkey	0.444	0.396	1.657	0.502	0.286	-	59,130	63,901	101,243	98,220	66,838	
United Kingdom	*24	*23	25	-	-	-	237,517	234,053	252,272			
United Republic of Tanzania . .	*0.012	*0.012	-	-	-	-	20,413	1,621	-	-	-	
United States of America . . .	*92	*91	*92	*91	*95	*103	407,004	393,680	403,219	415,445	468,000	518,899
Venezuela	0.389	0.600	0.575	0.693	0.851	1.204	63,788	90,902	88,531	101,293	129,878	188,313
Zambia	0.021	-	-	-	0.014	-	43,327	-	-	-	29,088	

Value Added and Value Added per Employee

Country	Value Added ($ bil.)						Value Added per Employee ($)					
	1990	1991	1992	1993	1994	1995	1990	1991	1992	1993	1994	1995
Australia	*0.815	*0.876	-	-	-	-	101,855	125,104	-	-	-	-
Austria	0.528	0.488	0.489	0.412	0.480	0.543	71,371	69,691	69,875	66,238	81,969	98,066
Bangladesh	0.033	0.024	0.004				21,895	15,135	3,138			
Bolivia	0.002	0.003	0.003	0.003	0.003	0.003	8,074	9,706	10,353	9,626	8,578	9,646
Canada	3.342	2.575	2.507	2.488	3.032	3.201	151,931	122,612	125,341	146,363	202,109	212,553
Chile	0.202	0.241	0.251	0.236	0.370	0.337	83,745	87,007	91,231	76,970	89,542	98,251
China (Hong Kong SAR) . .	0.039	0.039	0.057	0.052	0.059	-	39,025	42,893	47,692	51,838	65,703	
Colombia	0.133	0.152	0.153	0.205	0.217	0.259	19,655	24,461	26,257	39,057	43,253	48,671
Costa Rica	0.000	0.000	0.001	0.001	0.001	0.001	5,298	3,423	4,507	3,712		
Cyprus	0.001	0.001	0.001	0.001	0.001	0.001	33,955	42,477	51,349	47,246	39,611	39,679
Denmark	*0.618	*0.652	-	-	-	-	133,094	142,779	-	-	-	-
Ecuador	0.008	0.009	0.010	0.012	0.013	0.016	14,428	15,819	15,343	16,759	19,165	25,763
Egypt	*0.112	*0.067	*0.057	*0.046	-	-	8,546	4,274	5,377	4,171	-	
El Salvador	-	-	-	0.002	0.001	0.001	-	-	-	25,592	44,648	28,516
Finland	±0.653	±0.450	±0.404	±0.397	±0.519		133,315	91,847	74,828	84,402	112,743	
FYR Macedonia	-	-	-	-	-	±0.010	-	-	-	-	-	12,901
Gabon	-	*0.004	*0.005	*0.004	*0.003	*0.003	-	108,614	115,956	102,991	77,533	40,672
Greece	*0.082	*0.049	*0.055	-	-	-	49,728	43,676	54,400	-		
Honduras	*0.001	*0.002	*0.002	*0.002	*0.002	*0.002	4,222	5,184	5,453	5,296	3,971	4,264
India	*0.718	*0.701	*0.621	*0.744	-	-	7,734	7,495	6,376	7,685	-	-
Indonesia	*0.178	*0.182	*0.267	*0.357	*0.561	*0.484	11,158	11,335	13,931	16,299	24,299	18,431
Iran (Islamic Republic of) . .	-	0.064	0.043	0.069	-	-	-	22,479	12,335	17,866		
Italy	±3.031	±2.480	-	-	-	-	70,811	65,525	-	-		
Japan	±19	±21	±22	±23	-	-	204,869	227,058	229,559	245,417	-	
Jordan	0.009	0.009	0.012	0.006	0.006	0.007	19,544	26,927	24,661	13,309	11,687	12,287
Korea, Republic of	2.974	2.114	2.560	2.635	2.889	4.456	81,937	83,425	99,762	101,332	116,950	167,272
Kuwait	0.010	0.007	0.008	0.010	0.009	-	29,265	30,138	36,375	42,550	42,000	
Malaysia	*0.613	*0.905	*0.953	*1.082	*0.819	*1.176	115,681	146,024	138,078	146,280	127,939	133,263
Mauritius	0.003	0.004	0.005	0.005	0.003	-	17,159	32,099	34,606	35,209	20,574	
Mexico	1.078	1.089	1.100	0.924	0.916	-	53,621	56,915	59,546	60,844	73,762	
Myanmar	±0.002	±0.001	±0.002	±0.002	±0.001	-	9,243	9,222	13,541	11,053	4,617	
New Zealand	0.078	-	-	-	-	-	71,204	-	-	-	-	
Norway	±0.431	±0.355	±0.335				115,968	90,514	97,466	-	-	
Pakistan	-	0.062	-	-	-	-	-	13,817	-	-		
Panama	0.005	0.005	-	-	-	-	51,500	46,240	-	-		
Peru	0.113	0.096	0.090	-	-	-	26,524	26,480	27,005	-		
Philippines	0.105	0.134	0.127	0.119	-	-	19,108	23,495	23,591	24,297	-	
Portugal	0.125	0.141	0.171	0.142	0.123	-	29,589	32,578	40,340	36,166	36,295	
Russian Federation	-	-	-	±1.891	±2.848	±2.555	-	-	-	4,123	6,653	6,741
Saint Lucia	-	±0.000	±0.000	±0.000	±0.000	±0.000						
Senegal	0.011	-	-	-	0.047	-	8,896	-	-	-	41,755	
Slovakia	-	-	-	±0.042	±0.058	-	-	-	-	6,904	10,247	
Spain	*1.971	*1.975	*1.791	-	-	-	108,008	106,629	100,855	-	-	

Continued.

Depending on the table, * means *Enterprises*, *Engaged*, or *Factor Values*; ± means *Basis Unspecified*; § means *shown in millions*. For additional notes and sources, see Appendix I.

Value Added and Value Added per Employee

- Continued -

Country	Value Added ($ bil.)						Value Added per Employee ($)					
	1990	1991	1992	1993	1994	1995	1990	1991	1992	1993	1994	1995
Sri Lanka	0.003	0.002	0.006	0.006	-	-	4,443	4,604	15,037	9,799	-	-
Swaziland	-	*0.000	*0.000	*0.000	-	-	-	3,628	2,690	3,319	-	-
Sweden	-	*0.677	*0.676	*0.505	*0.607	-	-	82,797	88,147	78,648	92,464	-
Thailand	0.084	0.090	-	-	-	-	21,838	14,364	-	-	-	-
Trinidad and Tobago	-	±0.135	±0.074	±0.114	±0.254	-	-	112,338	57,071	103,964	242,626	-
Tunisia	0.022	0.017	0.019	0.017	0.016	0.020	-	-	-	59,145	18,193	12,431
Turkey	0.192	0.195	0.631	0.252	0.157	-	25,611	31,409	38,542	49,395	36,575	-
United Kingdom	*9.804	*9.607	*10	-	-	-	95,180	96,071	103,862	-	-	-
United Republic of Tanzania	*0.006	*0.005	-	-	-	-	9,827	607	-	-	-	-
United States of America	*45	*42	*43	*43	*44	*50	197,841	182,554	188,491	195,844	217,876	253,688
Venezuela	0.189	0.319	0.283	0.295	0.315	1.110	30,948	48,266	43,603	43,067	48,081	173,706
Zambia	0.012	-	-	-	0.009	-	24,364	-	-	-	19,976	-

Capital Investment

Country	Gross Fixed Capital Formation ($ mil.)						Per Establishment ($ mil)					
	1990	1991	1992	1993	1994	1995	1990	1991	1992	1993	1994	1995
Austria	145	144	211	141	76	-	2.584	2.661	3.901	2.657	1.518	-
Bahrain	-	-	6.960	-	-	-	-	-	0.994	-	-	-
Bangladesh	-	1.284	0.180	-	-	-	-	0.038	0.006	-	-	-
Bolivia	0.423	0.929	0.516	0.430	2.808	-	0.028	0.058	0.032	0.029	0.175	-
China (Hong Kong SAR)	-	-	6.201	-0.776	11	-	-	-	0.061	-0.011	0.115	-
Colombia	-	12	19	1.470	3.422	-	-	0.123	0.191	0.014	0.033	-
Cyprus	1.173	0.181	0.876	0.169	0.431	0.188	1.173	0.181	0.876	0.169	0.431	0.188
Denmark	116	137	-	-	-	-	2.148	2.628	-	-	-	-
Ecuador	-	4.767	3.928	-0.531	2.069	4.677	-	0.227	0.179	-0.023	0.099	0.292
El Salvador	-	-	-	0.151	0.005	0.078	-	-	-	0.075	0.005	0.078
Finland	195	82	88	84	112	-	3.479	1.385	1.537	1.472	1.963	-
Gabon	-	0.121	0.155	0.141	0.059	0.076	-	0.121	0.155	0.141	0.059	0.076
Greece	9.678	6.787	10	-	-	-	0.312	0.226	0.360	-	-	-
Hungary	-	-	16	58	-	-	-	-	0.164	0.480	-	-
India	353	426	260	239	-	-	0.256	0.303	0.180	0.157	-	-
Indonesia	95	121	71	72	274	187	0.522	0.708	0.362	0.352	1.290	0.779
Iran (Islamic Republic of)	-	8.194	8.501	16	-	-	-	0.210	0.193	0.310	-	-
Italy	898	839	-	-	-	-	7.484	7.111	-	-	-	-
Japan	3,398	3,853	3,901	3,471	-	-	2.657	2.994	3.003	2.718	-	-
Jordan	-	-	-	-	-	0.080	-	-	-	-	-	0.006
Korea, Republic of	-	-	1,368	1,024	1,134	2,071	-	-	2.999	2.120	2.368	4.093
Kuwait	0.658	1.697	3.215	1.209	0.490	-	0.110	0.242	0.536	0.201	0.082	-
Malaysia	-	-	-	710	108	-	-	-	-	11	1.634	-
Mexico	63	67	-	-	-	-	0.824	0.872	-	-	-	-
Myanmar	4.504	4.257	6.985	6.948	7.029	-	4.504	2.128	3.493	3.474	3.515	-
New Zealand	9.552	-	-	-	-	-	0.273	-	-	-	-	-
Norway	-	-	59	-	-	-	-	-	1.896	-	-	-
Pakistan	-	22	-	-	-	-	-	0.748	-	-	-	-
Panama	0.017	0.056	-	-	-	-	0.004	0.014	-	-	-	-
Peru	7.323	5.175	4.196	-	-	-	0.032	0.022	0.018	-	-	-
Philippines	18	11	10	19	-	-	0.201	0.107	0.123	0.247	-	-
Portugal	68	70	52	28	27	-	0.863	0.815	0.592	0.294	0.287	-
Senegal	26	-	-	-	-	-	3.743	-	-	-	-	-
Spain	373	455	303	-	-	-	1.366	1.591	0.957	-	-	-
Sri Lanka	0.300	1.002	0.429	1.335	-	-	0.021	0.100	0.107	0.133	-	-
Swaziland	-	0.008	0.002	-	-	-	-	0.008	0.002	-	-	-
Thailand	180	45	-	-	-	-	2.642	0.379	-	-	-	-
Trinidad and Tobago	-	-	46	27	196	-	-	-	5.090	2.411	16	-
Tunisia	2.392	4.651	4.749	3.587	1.977	5.287	-	-	-	0.598	0.090	0.264
Turkey	27	24	79	28	14	-	0.736	0.719	1.586	0.661	0.358	-
United Kingdom	-	3,120	2,493	-	-	-	-	4.015	2.975	-	-	-

Continued.

Capital Investment
- Continued -

Country	Gross Fixed Capital Formation ($ mil.)						Per Establishment ($ mil)					
	1990	1991	1992	1993	1994	1995	1990	1991	1992	1993	1994	1995
United Republic of Tanzania . .	4.137	7.200	-	-	-	-	0.827	1.200	-	-	-	-
United States of America . . .	-	-	6,354	5,242	4,868	6,647	-	-	2.349	-	-	-
Zambia	1.073	-	-	-	-	-	0.215	-	-	-	-	-

Indexes of Industrial Production

Country	Index of Industrial Production (1990=100)						Index of Employment (1990=100)					
	1990	1991	1992	1993	1994	1995	1990	1991	1992	1993	1994	1995
Australia	-	-	-	-	-	-	100	88	88	-	-	-
Austria	-	-	-	-	-	-	100	95	95	84	79	75
Azerbaijan	-	-	-	-	-	-	100	110	118	114	92	-
Bangladesh	-	-	-	-	-	-	100	106	75	-	-	-
Bolivia	-	-	-	-	-	-	100	103	104	102	119	115
Bosnia & Herzegovina . . .	-	-	-	-	-	-	100	80	-	-	12	-
Canada	-	-	-	-	-	-	100	95	91	77	68	68
Chile	-	-	-	-	-	-	100	115	114	127	171	142
China (Hong Kong SAR) . . .	-	-	-	-	-	-	100	90	120	100	90	-
Colombia	-	-	-	-	-	-	100	92	86	77	74	78
Costa Rica	-	-	-	-	-	-	100	130	140	196	-	-
Cyprus	-	-	-	-	-	-	100	93	97	93	110	121
Denmark	-	-	-	-	-	-	100	98	154	-	-	-
Ecuador	-	-	-	-	-	-	100	106	119	125	121	114
Egypt	-	-	-	-	-	-	100	120	81	85	-	-
Finland	-	-	-	-	-	-	100	100	110	96	94	-
Greece	-	-	-	-	-	-	100	68	62	-	-	-
Honduras	-	-	-	-	-	-	100	101	103	113	124	126
India	-	-	-	-	-	-	100	101	105	104	-	-
Indonesia	-	-	-	-	-	-	100	101	120	137	145	165
Italy	-	-	-	-	-	-	100	88	-	-	-	-
Japan	-	-	-	-	-	-	100	100	103	101	-	-
Jordan	-	-	-	-	-	-	100	76	107	107	106	121
Kenya	-	-	-	-	-	-	100	98	100	101	102	107
Korea, Republic of	-	-	-	-	-	-	100	70	71	72	68	73
Kuwait	-	-	-	-	-	-	100	63	64	64	62	-
Malawi	-	-	-	-	-	-	100	-	-	-	-	-
Malaysia	-	-	-	-	-	-	100	117	130	140	121	167
Mauritius	-	-	-	-	-	-	100	73	83	78	83	98
Mexico	-	-	-	-	-	-	100	95	92	76	62	71
Myanmar	-	-	-	-	-	-	100	95	91	88	85	-
New Zealand	-	-	-	-	-	-	100	113	92	96	-	-
Norway	-	-	-	-	-	-	100	106	92	-	-	-
Panama	-	-	-	-	-	-	100	111	-	-	-	-
Peru	-	-	-	-	-	-	100	85	79	-	-	-
Philippines	-	-	-	-	-	-	100	104	98	89	-	-
Portugal	-	-	-	-	-	-	100	103	101	93	80	-
Senegal	-	-	-	-	-	-	100	-	-	-	93	-
Spain	-	-	-	-	-	-	100	101	97	-	-	-
Sri Lanka	-	-	-	-	-	-	100	82	68	99	-	-
Thailand	-	-	-	-	-	-	100	162	-	-	-	-
Turkey	-	-	-	-	-	-	100	82	218	68	57	-
United Kingdom	-	-	-	-	-	-	100	97	95	-	-	-
United Republic of Tanzania . .	-	-	-	-	-	-	100	1,246	-	-	-	-
United States of America . . .	-	-	-	-	-	-	100	102	100	96	89	88
Uruguay	-	-	-	-	-	-	100	95	83	81	-	-
Venezuela	-	-	-	-	-	-	100	108	107	112	107	105
Zambia	-	-	-	-	-	-	100	-	-	-	100	-

Depending on the table, * means *Enterprises*, *Engaged*, or *Factor Values*; ± means *Basis Unspecified*; § means *shown in millions*. For additional notes and sources, see Appendix I.

ISIC 3513
SYNTHETIC RESINS, ETC.

ISIC 3513—Synthetic Resins, etc.—is a subset of ISIC 3510. Shown here are statistics for countries that report on synthetic resins (plastics) production separately.

Please note that commodity details were shown in the chapter on ISIC 3510.

Summary Statistics

		1990		1991		1992		1993		1994		1995	
		Value	N	Value	N	Value	N	Value	N	Value	N	Value	N
Establishments or enterprises	(number)	2,909	36	3,006	42	3,308	42	1,970	35	1,131	26	578	14
Number employed	(000)	480	39	512	43	482	42	621	36	460	26	393	18
Total Wages	($ mil.)	12,278	37	13,223	41	11,988	38	11,136	30	7,176	24	6,878	17
Wage/salary per employee	($)	14,683	37	14,411	41	13,996	38	11,662	30	12,221	24	11,207	17
Female workers	(000)	16	16	15	16	97	14	100	12	101	8	98	5
as % of total employment*	(%)	16	15	13	15	18	14	21	12	24	8	31	5
Output	($ bil.)	130	37	132	39	124	37	116	30	79	24	83	18
per employee	($ 000)	165	36	156	38	143	36	131	30	173	23	152	17
per establishment	($ 000)	21,336	35	23,403	38	24,339	35	21,731	28	22,776	22	14,422	12
Value added	($ bil.)	53	37	54	38	53	31	52	28	33	24	34	16
per employee	($ 000)	54	36	52	37	54	31	49	28	56	23	64	15
per establishment	($ 000)	7,591	35	7,369	36	8,054	31	8,751	27	8,094	22	5,014	10
Capital investment	($ mil.)	4,656	18	5,462	22	9,217	23	9,572	20	4,825	15	4,356	7
per employee	($ 000)	11	18	10	22	13	22	16	19	11	15	12	7
per establishment	($ 000)	1,667	18	1,624	22	1,675	23	2,353	19	1,246	14	1,335	6

Data presented above are drawn from the detailed tables that follow. Columns headed 'N' show the number of countries that provided valid data for inclusion. Values are not strictly comparable one year to the next or one row to the next because the number of countries that report varies from period to period and row to row. However, a general indication of magnitudes can be gleaned from reviewing this summary—especially in earlier years in which more reports are available. For detailed explanations, see Appendix I. *The average for those countries reporting both total and female employment.

Establishments and Number Engaged

Country	Establishments or Enterprises (number)						Employees per Establishment					
	1990	1991	1992	1993	1994	1995	1990	1991	1992	1993	1994	1995
Australia	77	98	98	-	-	-	65	51	41	-	-	-
Bahrain	-	-	1.000	-	-	-	-	-	57	-	-	-
Bangladesh	3.000	3.000	3.000	-	-	-	13	14	14	-	-	-
Benin	-	-	2.000	2.000	-	-	-	-	-	-	-	-
Bolivia	7.000	7.000	7.000	7.000	9.000	7.000	11	11	12	12	13	16
Bosnia & Herzegovina	5.000	5.000	-	-	-	-	488	463	-	-	-	-
Canada	100	91	85	91	89	-	70	77	82	99	101	-
Chile	6.000	6.000	4.000	5.000	4.000	-	186	205	181	113	104	-
China (Hong Kong SAR)	292	197	145	194	95	·	6.507	7.107	7.586	6.701	12	-
Colombia	26	25	27	27	30	·	216	220	209	214	185	-
Costa Rica	18	17	18	18	-	-	49	37	37	33	-	-
Denmark	133	146	142	-	-	-	37	33	34	-	-	-
Ecuador	18	23	23	23	22	15	42	57	62	60	61	31
Egypt	13	14	7.000	8.000	-	-	831	771	1,486	50	-	-
El Salvador	-	-	-	-	2.000	1.000	-	-	-	-	118	87
Finland	91	94	89	83	85	-	71	66	65	66	66	-
FYR Macedonia	-	-	-	-	-	*33	-	-	-	-	-	32
Gabon	-	5.000	6.000	7.000	7.000	9.000	-	33	25	25	16	19
Ghana	-	-	-	3.000	-	-	-	-	99	-	-	-
Greece	13	13	15	-	-	-	54	79	73	-	-	-
Grenada	-	-	1.000	1.000	-	-	-	-	5.000	5.000	-	-
Guatemala	-	1.000	-	-	-	-	-	21	-	-	-	-
Hungary	-	-	*33	*40	-	-	-	-	-	-	-	-
India	274	261	268	322	-	-	167	168	203	158	-	-
Indonesia	53	54	49	92	89	105	159	165	188	214	237	228
Iran (Islamic Republic of)	-	7.000	10	11	-	-	-	77	150	131	-	-
Italy	*172	*161	-	-	-	-	195	232	-	-	-	-
Japan	280	276	279	281	-	-	261	268	272	267	-	-
Jordan	10	8.000	6.000	10	28	35	26	46	44	39	34	30
Korea, Republic of	53	102	120	90	129	129	187	130	111	228	179	171
Malaysia	23	27	27	31	29	-	91	96	111	119	131	-
Mexico	50	48	48	47	46	43	510	522	478	407	381	456
New Zealand	*64	*63	*68	*83	-	-	34	32	33	29	-	-
Norway	10	11	10	-	-	-	172	166	177	-	-	-
Pakistan	-	61	-	-	-	-	-	64	-	-	-	-
Peru	45	46	47	-	-	-	65	55	38	-	-	-
Philippines	54	72	41	47	-	-	56	57	83	72	-	-
Portugal	*72	*74	*70	*91	*88	-	42	43	43	32	28	-
Russian Federation	-	-	-	*127	*151	*155	-	-	-	1,484	1,100	1,002
Saint Lucia	-	1.000	1.000	1.000	1.000	1.000	-	-	-	-	-	-
Senegal	7.000	-	-	-	2.000	-	32	-	-	-	53	-
Singapore	29	31	29	31	32	-	61	65	68	79	81	-
Slovakia	-	4.000	4.000	4.000	4.000	-	-	4,278	3,748	3,139	2,838	-
South Africa	-	48	-	-	-	-	-	-	-	-	-	-
Spain	116	117	126	-	-	-	127	119	102	-	-	-
Sri Lanka	3.000	3.000	4.000	2.000	-	-	68	76	53	72	-	-
Sweden	105	108	104	102	106	-	87	91	87	81	79	-
Thailand	30	49	-	-	-	-	98	200	-	-	-	-
Tunisia	-	-	-	29	24	13	-	-	-	24	23	33
Turkey	16	15	26	25	31	-	881	900	240	504	411	-
Ukraine	-	-	7.000	11	9.000	10	-	-	4,000	2,455	2,667	2,300
United Kingdom	617	581	570	-	-	-	47	50	46	-	-	-
United States of America	-	-	658	-	-	-	-	-	195	-	-	-
Venezuela	24	33	30	24	16	22	108	94	103	114	125	108
Zambia	-	-	-	-	3.000	-	-	-	-	-	-	-

Depending on the table, * means *Enterprises, Engaged,* or *Factor Values;* ± means *Basis Unspecified;* § means *shown in millions.* For additional notes and sources, see Appendix I.

Employment and Compensation of Employees

Country	Number Employed/Engaged (000)						Wage/Salary per Employee ($)					
	1990	1991	1992	1993	1994	1995	1990	1991	1992	1993	1994	1995
Australia	5.000	*5.000	*4.000	-	-	-	31,719	31,632	35,846	-	-	-
Azerbaijan	14	13	13	13	12	-	1,789	2,041	-	-	-	-
Bahrain	-	-	0.057	-	-	-	-	-	4,526	-	-	-
Bangladesh	0.040	0.043	0.041	-	-	-	-	635	626	-	-	-
Bolivia	0.076	0.080	0.081	0.085	0.114	0.115	1,049	1,194	984	1,037	1,374	1,330
Bosnia & Herzegovina	2.438	2.314	-	-	-	-	-	-	-	-	-	-
Canada	7.000	7.000	7.000	9.000	9.000	7.995	38,934	39,028	39,594	35,828	37,021	39,772
Chile	1.118	1.228	0.725	0.563	0.415	0.404	5,762	6,028	7,970	12,696	17,267	18,907
China (Hong Kong SAR)	*1.900	*1.400	*1.100	*1.300	*1.100	-	8,648	12,317	11,627	15,115	20,115	-
Colombia	5.613	5.489	5.648	5.786	5.555	6.061	4,374	4,497	4,646	5,480	7,419	8,789
Costa Rica	0.888	0.627	0.666	0.601			3,694	3,439	4,092	4,976	-	-
Denmark	4.936	4.878	4.854	-	-	-	32,571	33,398	37,544	-	-	-
Ecuador	0.759	1.300	1.415	1.381	1.344	0.468	3,981	3,581	3,901	4,640	2,521	3,731
Egypt	*11	*11	*10	*0.400	-	-	2,134	1,409	1,503	1,038	-	-
El Salvador	-	-	-	-	*0.237	*0.087	-	-	-	-	2,964	2,789
Finland	6.500	6.200	5.800	5.500	5.600	-	31,662	31,397	29,437	23,417	26,522	-
FYR Macedonia	-	-	-	-	-	1.048	-	-	-	-	-	4,445
Gabon		0.165	0.148	0.172	0.112	0.168		12,718	16,542	14,722	10,099	9,111
Ghana				*0.296								
Greece	*0.704	*1.027	*1.095	-	-	-	17,681	19,296	20,979			
Grenada	-	-	*0.005	*0.005			-	-	-	-	-	-
Guatemala	-	*0.021					-	-	-	-	-	-
India	*46	*44	*54	*51	-	-	3,248	2,428	2,631	2,495		
Indonesia	8.438	8.927	9.235	20	21	24	1,200	1,387	1,577	1,472	1,589	1,956
Iran (Islamic Republic of)	-	0.538	1.498	1.446			-	4,156	7,119	6,444		
Italy	34	37	-	-	-	-						
Japan	*73	*74	*76	*75	-	-	39,453					
Jordan	0.256	0.370	0.266	0.388	0.959	1.064	3,606	2,877	2,178	2,596	3,280	2,553
Korea, Republic of	9.900	13	13	20	23	22	13,687	14,246	15,181	16,819	18,025	24,193
Lithuania	-	-	2.137	1.873			-	-	406	472		
Malaysia	2.100	2.600	3.000	3.700	3.800	3.000	4,788	5,566	7,288	7,738	8,413	8,855
Mexico	26	25	23	19	18	20	5,770	6,913	8,580	9,341	9,265	6,258
New Zealand	2.162	2.040	2.230	2.444	-	-	21,815	22,991	19,787			
Norway	1.719	1.821	1.773	-	-	-	-	36,339	-	-	-	-
Pakistan	-	3.874	-	-	-	-	-	2,723	-	-	-	-
Peru	*2.912	*2.546	*1.773	-	-	-	4,951	6,353	7,016	-	-	-
Philippines	3.000	4.100	3.400	3.400	-	-	3,085	3,080	3,908	3,568	-	-
Portugal	*2.992	*3.176	*2.994	*2.912	*2.502	-	9,721	10,750				
Russian Federation	-	-	-	189	166	155				734	996	1,280
Senegal	*0.222				*0.107	-	4,434				3,030	-
Singapore	*1.777	*2.011	*1.969	*2.438	*2.583	-	20,776	21,658	24,439	29,383	30,869	-
Slovakia	-	*17	*15	*13	*11	-	-	1,701	2,138	2,408	2,708	-
Spain	15	14	13	-	-	-	27,363	28,136	32,341			
Sri Lanka	*0.203	*0.228	*0.210	*0.143	-	-	615	630	830	854	-	-
Sweden	9.100	*9.825	*9.005	8.257	8.325	-	26,883	29,226	32,048	24,776	26,773	-
Thailand	2.926	9.793	-	-	-	-	1,711	10,524	-	-	-	-
Tunisia	-	-	-	0.697	0.543	0.430	-	-	-	-	-	-
Turkey	14	13	6.251	13	13	-	11,584	14,638	13,943	15,847	10,797	-
Ukraine	-	-	28	27	24	23	-	-	3,331		487	587
United Kingdom	29	29	26	-	-	-	27,833	29,234	31,666			
United States of America	132	129	128	126	128	126	36,364	37,907	39,883	-	-	-
Uruguay	*0.351	*0.317	*0.224	*0.210	-	-						
Venezuela	2.600	3.100	3.100	2.735	1.995	2.377	7,044	8,176	9,105	10,257	10,217	12,185

Depending on the table, * means *Enterprises, Engaged,* or *Factor Values;* ± means *Basis Unspecified;* § means *shown in millions.* For additional notes and sources, see Appendix I.

321

Female Workers: Total and Percent of Employment

Country	Female Workers (000)						Female Workers as % of Total Employment					
	1990	1991	1992	1993	1994	1995	1990	1991	1992	1993	1994	1995
Afghanistan	0.026	0.031	-	-	-	-	-	-	-	-	-	-
Australia	0.632	-	-	-	-	-	12.64	-	-	-	-	-
Bahrain	-	-	0.001	-	-	-	-	-	1.75	-	-	-
Bangladesh	-	0.002	0.001	-	-	-	-	4.65	2.44	-	-	-
Canada	1.000	1.000	-	-	-	-	14.29	14.29	-	-	-	-
Chile	0.103	0.098	0.173	0.045	0.038	-	9.21	7.98	23.86	7.99	9.16	-
Colombia	-	0.592	-	0.763	0.738	-	-	10.79	-	13.19	13.29	-
Denmark	1.372	1.177	1.181	-	-	-	27.80	24.13	24.33	-	-	-
Egypt	-	0.553	0.393	0.045	-	-	-	5.12	3.78	11.25	-	-
El Salvador	-	-	-	-	0.040	0.021	-	-	-	-	16.88	24.14
Indonesia	-	-	-	7.043	7.334	7.724	-	-	-	35.81	34.80	32.22
Iran (Islamic Republic of)	-	0.052	0.076	0.079	-	-	-	9.67	5.07	5.46	-	-
Italy	-	5.074	-	-	-	-	-	13.58	-	-	-	-
Jordan	0.010	0.008	0.030	0.048	0.031	0.045	3.91	2.16	11.28	12.37	3.23	4.23
Korea, Republic of	3.000	3.844	3.989	6.050	7.114	5.650	30.30	29.09	29.87	29.51	30.86	25.59
Malaysia	0.500	0.500	0.500	0.700	0.700	-	23.81	19.23	16.67	18.92	18.42	-
New Zealand	0.506	0.497	0.525	0.569	-	-	23.40	24.36	23.54	23.28	-	-
Philippines	-	-	0.583	0.588	-	-	-	-	17.15	17.29	-	-
Portugal	0.262	0.303	-	-	-	-	8.76	9.54	-	-	-	-
Senegal	0.004	-	-	-	-	-	1.80	-	-	-	-	-
Sri Lanka	0.010	0.010	0.011	0.014	-	-	4.93	4.39	5.24	9.79	-	-
Sweden	2.300	-	-	-	-	-	25.27	-	-	-	-	-
Thailand	0.758	1.700	-	-	-	-	25.91	17.36	-	-	-	-
Turkey	0.701	-	-	-	-	-	4.97	-	-	-	-	-
United Kingdom	5.048	-	4.000	-	-	-	17.41	-	15.38	-	-	-
United States of America	-	-	86	84	85	85	-	-	67.19	66.67	66.41	67.46

Output and Output per Employee

Country	Output ($ bil.)						Output per Employee ($)					
	1990	1991	1992	1993	1994	1995	1990	1991	1992	1993	1994	1995
Australia	*1.459	*1.507	*1.183	-	-	-	291,719	301,363	295,772	-	-	-
Bahrain	-	-	±0.001	-	-	-	-	-	22,396	-	-	-
Bangladesh	0.000	0.000	0.000	-	-	-	2,170	1,906	1,879	-	-	-
Bolivia	0.001	0.001	0.001	0.001	0.002	0.003	17,186	18,743	15,053	17,256	20,642	25,863
Canada	3.094	2.819	2.813	3.488	4.123	4.265	441,990	402,748	401,849	387,567	458,081	533,460
Chile	0.100	0.104	0.049	0.093	0.107	0.100	89,258	84,875	66,900	165,697	258,409	246,306
China (Hong Kong SAR)	0.367	0.346	0.266	0.346	0.495	-	193,168	247,250	241,583	266,003	449,817	-
Colombia	0.603	0.627	0.540	0.606	0.759	0.954	107,478	114,233	95,553	104,744	136,578	157,429
Costa Rica	0.007	0.006	0.010	0.011	0.012	0.009	7,968	10,294	15,545	17,764	-	-
Denmark	*0.865	*0.795	-	-	-	-	175,260	162,918	-	-	-	-
Ecuador	0.028	0.051	0.050	0.061	0.070	0.029	37,224	39,506	35,467	44,150	52,160	62,989
Egypt	*0.195	*0.112	*0.112	*0.013	-	-	18,028	10,404	10,769	33,365	-	-
El Salvador	-	-	-	-	±0.004	±0.002	-	-	-	-	18,822	19,683
Finland	±1.452	±1.214	±1.123	±0.999	±1.206	-	223,367	195,750	193,568	181,586	215,270	-
FYR Macedonia	-	-	-	-	-	±0.023	-	-	-	-	-	21,620
Gabon	-	*0.016	*0.018	*0.012	*0.006	*0.009	-	98,522	123,194	67,901	57,267	53,007
Greece	*0.156	*0.216	*0.209	-	-	-	221,012	210,384	190,836	-	-	-
India	*2.566	*1.460	*2.831	*2.660	-	-	56,040	33,238	52,036	52,214	-	-
Indonesia	0.609	0.696	1.172	0.806	0.852	0.992	72,160	77,956	126,876	41,008	40,446	41,401
Iran (Islamic Republic of)	-	0.025	0.055	0.057	-	-	-	47,222	36,685	39,625	-	-
Italy	±9.401	±9.337	-	-	-	-	280,125	249,884	-	-	-	-
Japan	±36	±40	±42	±44	-	-	495,758	542,106	551,562	585,492	-	-
Jordan	0.020	0.018	0.010	0.014	0.043	0.064	78,619	47,565	38,738	36,930	45,085	60,074
Korea, Republic of	2.064	2.873	2.681	3.539	4.532	6.206	208,440	217,415	200,794	172,636	196,572	281,076
Lithuania	-	-	0.008	0.009	-	-	-	-	3,739	4,574	-	-
Malaysia	*0.227	*0.304	*0.410	*0.541	*0.809	*0.546	108,322	117,059	136,584	146,186	212,949	182,134
Mexico	1.912	2.043	2.051	1.733	1.941	2.021	74,917	81,552	89,456	90,616	110,749	103,065

Continued.

Depending on the table, * means *Enterprises, Engaged,* or *Factor Values*; ± means *Basis Unspecified*; § means *shown in millions*. For additional notes and sources, see Appendix I.

Output and Output per Employee

- Continued -

Country	Output ($ bil.)						Output per Employee ($)					
	1990	1991	1992	1993	1994	1995	1990	1991	1992	1993	1994	1995
New Zealand	0.300	-	-	-	-	-	138,622	-	-	-	-	-
Norway	±0.626	±0.601	-	-	-	-	364,205	330,274	-	-	-	-
Pakistan	-	0.169	-	-	-	-	-	43,688	-	-	-	-
Peru	0.143	0.142	0.128	-	-	-	49,133	55,755	72,299	-	-	-
Philippines	0.166	0.209	0.176	0.176	-	-	55,229	51,037	51,844	51,692	-	-
Portugal	0.709	0.648	0.629	0.522	0.639	-	236,858	204,019	210,151	179,377	255,553	-
Russian Federation	-	-	-	1.784	1.948	3.211	-	-	-	9,466	11,725	20,676
Senegal	0.011	-	-	-	-	-	51,157	-	-	-	-	-
Singapore	*0.716	*0.763	*0.673	*0.736	*0.964	-	402,989	379,319	342,014	301,704	373,022	-
Slovakia	-	±0.366	±0.391	±0.295	±0.345	-	-	21,399	26,096	23,472	30,414	-
South Africa	±1.458	±1.421	±1.499	-	-	-	-	-	-	-	-	-
Spain	*3.821	*3.480	*3.368	-	-	-	259,862	249,285	262,636	-	-	-
Sri Lanka	0.005	0.003	0.003	0.002	-	-	24,961	14,783	16,549	15,261	-	-
Sweden	*2.212	*2.529	*2.406	*1.728	*2.053	-	243,024	257,444	267,177	209,321	246,565	-
Thailand	0.077	1.523	-	-	-	-	26,262	155,478	-	-	-	-
Turkey	1.831	1.668	0.897	1.587	1.937	-	129,894	123,645	143,508	125,820	151,919	-
Ukraine	-	-	*0.903	*0.160	*0.151	*0.106	-	-	32,255	5,911	6,312	4,617
United Kingdom	*7.398	*7.136	6.698	-	-	-	255,111	246,079	257,611	-	-	-
United States of America	*48	*46	*49	*49	*55	*64	366,818	358,450	379,727	392,643	432,141	504,016
Venezuela	0.383	0.434	0.415	0.430	0.401	0.650	147,409	140,073	133,743	157,120	200,848	273,663

Value Added and Value Added per Employee

Country	Value Added ($ bil.)						Value Added per Employee ($)					
	1990	1991	1992	1993	1994	1995	1990	1991	1992	1993	1994	1995
Australia	*0.476	*0.488	-	-	-	-	95,156	97,546	-	-	-	-
Bangladesh	0.000	0.000	0.000	-	-	-	579	635	626	-	-	-
Bolivia	0.000	0.000	0.000	0.000	0.000	0.001	3,061	3,351	2,387	2,871	3,805	4,770
Canada	0.968	0.829	0.836	0.930	1.238	1.389	138,352	118,455	119,373	103,351	137,506	173,746
Chile	0.042	0.039	0.022	0.038	0.041	0.037	37,691	31,886	29,863	67,269	98,237	92,018
China (Hong Kong SAR)	0.024	0.071	0.039	0.043	0.049	-	12,770	50,461	35,116	33,014	44,699	-
Colombia	0.203	0.213	0.204	0.215	0.288	0.355	36,181	38,748	36,125	37,097	51,919	58,599
Costa Rica	0.001	0.001	0.002	0.002	0.002	0.001	1,311	1,723	2,610	2,907	-	-
Denmark	*0.385	*0.362	-	-	-	-	78,072	74,264	-	-	-	-
Ecuador	0.007	0.014	0.014	0.021	0.021	0.007	9,143	11,127	10,044	15,211	15,892	14,138
Egypt	*0.047	*0.005	*0.030	*0.002	-	-	4,389	478	2,846	5,413	-	-
El Salvador	-	-	-	-	0.002	0.001	-	-	-	-	9,435	9,019
Finland	±0.449	±0.362	±0.415	±0.365	±0.421	-	69,078	58,310	71,631	66,364	75,175	-
FYR Macedonia	-	-	-	-	-	±0.012	-	-	-	-	-	11,275
Gabon	-	*0.005	*0.006	*0.004	*0.002	*0.002	-	30,592	40,639	22,606	15,117	12,486
Germany (Western Part)	1.354	1.255	-	-	-	-	-	-	-	-	-	-
Greece	*0.044	*0.062	*0.053	-	-	-	61,914	59,896	48,762	-	-	-
India	*0.439	*0.002	*0.736	*0.910	-	-	9,586	44	13,526	17,858	-	-
Indonesia	*0.169	*0.307	*0.491	*0.268	*0.293	*0.310	20,032	34,379	53,138	13,636	13,925	12,938
Iran (Islamic Republic of)	-	0.008	0.029	0.026	-	-	-	14,530	19,361	17,870	-	-
Italy	±2.322	±2.214	-	-	-	-	69,192	59,257	-	-	-	-
Japan	±17	±19	±22	±23	-	-	239,080	258,513	286,014	310,192	-	-
Jordan	0.006	0.004	0.002	0.002	0.015	0.026	22,131	12,077	7,867	6,196	15,632	24,739
Korea, Republic of	0.717	1.357	1.268	1.642	1.943	2.494	72,444	102,681	94,968	80,116	84,287	112,971
Malaysia	*0.047	*0.070	*0.094	*0.128	*0.183	*0.101	22,587	26,964	31,483	34,491	48,143	33,615
Mexico	0.790	0.854	0.886	0.749	0.919	-	30,972	34,078	38,656	39,167	52,422	-
New Zealand	0.097	-	-	-	-	-	44,735	-	-	-	-	-
Norway	±0.134	±0.119	-	-	-	-	77,692	65,225	-	-	-	-
Pakistan	-	0.059	-	-	-	-	-	15,311	-	-	-	-
Peru	0.057	0.049	0.049	-	-	-	19,444	19,420	27,901	-	-	-
Philippines	0.070	0.089	0.064	0.067	-	-	23,432	21,799	18,965	19,818	-	-
Portugal	0.190	0.105	0.094	0.084	0.120	-	63,626	33,005	31,478	28,688	47,871	-
Russian Federation	-	-	-	±0.637	±0.707	±0.890	-	-	-	3,380	4,254	5,732

Continued.

Depending on the table, * means *Enterprises, Engaged,* or *Factor Values*; ± means *Basis Unspecified*; § means *shown in millions*. For additional notes and sources, see Appendix I.

323

Value Added and Value Added per Employee
- Continued -

Country	Value Added ($ bil.)						Value Added per Employee ($)					
	1990	1991	1992	1993	1994	1995	1990	1991	1992	1993	1994	1995
Senegal	0.002	-	-	-	0.001	-	8,901	-	-	-	6,885	-
Singapore	*0.224	*0.231	*0.214	*0.240	*0.326	-	126,198	114,648	108,501	98,477	126,162	-
Slovakia	-	-	-	±0.084	±0.104	-	-	-	-	6,695	9,143	-
Spain	*0.947	*0.919	*0.958	-	-	-	64,409	65,829	74,726	-	-	-
Sri Lanka	0.002	0.001	0.001	0.001	-	-	9,099	2,543	4,384	5,130	-	-
Sweden	*0.873	*0.663	*0.672	*0.598	*0.726	-	95,984	67,466	74,578	72,431	87,172	-
Thailand	0.025	0.702	-	-	-	-	8,522	71,708	-	-	-	-
Turkey	0.816	0.555	0.494	0.646	1.070	-	57,891	41,162	79,028	51,243	83,908	-
United Kingdom	*2.607	*2.501	*2.523	-	-	-	89,901	86,237	97,037	-	-	-
United States of America	*21	*20	*21	*21	*24	*28	155,379	152,481	163,531	166,683	191,359	219,730
Venezuela	0.138	0.188	0.152	0.154	0.119	0.428	53,133	60,773	49,046	56,351	59,589	180,107

Capital Investment

Country	Gross Fixed Capital Formation ($ mil.)						Per Establishment ($ mil)					
	1990	1991	1992	1993	1994	1995	1990	1991	1992	1993	1994	1995
Bolivia	-	-	0.025	0.029	-0.009	-	-	-	0.004	0.004	-0.001	-
China (Hong Kong SAR)	-	-	38	13	8.152	-	-	-	0.263	0.065	0.086	-
Colombia	-	48	1.486	6.936	-2.631	-	-	1.939	0.055	0.257	-0.088	-
Denmark	36	37	-	-	-	-	0.270	0.251	-	-	-	-
Ecuador	-	8.988	5.272	4.032	7.962	2.672	-	0.391	0.229	0.175	0.362	0.178
El Salvador	-	-	-	-	0.085	-	-	-	-	-	0.043	-
Finland	115	86	52	92	68	-	1.260	0.913	0.582	1.105	0.797	-
Gabon	-	0.390	0.608	0.664	0.207	1.717	-	0.078	0.101	0.095	0.030	0.191
Greece	4.290	7.034	3.772	-	-	-	0.330	0.541	0.251	-	-	-
Hungary	-	-	20	15	-	-	-	-	0.611	0.383	-	-
India	468	162	361	269	-	-	1.709	0.620	1.347	0.837	-	-
Indonesia	34	17	18	31	18	117	0.645	0.323	0.362	0.339	0.198	1.110
Iran (Islamic Republic of)	-	1.859	1.812	1.775	-	-	-	0.266	0.181	0.161	-	-
Italy	573	692	-	-	-	-	3.334	4.296	-	-	-	-
Japan	2,832	3,303	4,295	4,550	-	-	10	12	15	16	-	-
Jordan	-	-	-	-	-	0.100	-	-	-	-	-	0.003
Korea, Republic of	-	-	916	786	970	835	-	-	7.635	8.734	7.516	6.469
Malaysia	-	-	-	120	47	-	-	-	-	3.872	1.629	-
Mexico	84	170	-	-	-	-	1.670	3.537	-	-	-	-
New Zealand	2.388	-	-	-	-	-	0.037	-	-	-	-	-
Pakistan	-	6.929	-	-	-	-	-	0.114	-	-	-	-
Peru	2.965	2.852	3.043	-	-	-	0.066	0.062	0.065	-	-	-
Philippines	6.623	5.895	5.723	6.036	-	-	0.123	0.082	0.140	0.128	-	-
Portugal	38	82	96	30	13	-	0.526	1.105	1.367	0.329	0.144	-
Senegal	1.693	-	-	-	-	-	0.242	-	-	-	-	-
Singapore	42	52	24	213	107	-	1.446	1.662	0.829	6.859	3.347	-
Spain	319	235	168	-	-	-	2.747	2.010	1.335	-	-	-
Sri Lanka	0.324	0.226	0.336	0.079	-	-	0.108	0.075	0.084	0.039	-	-
Thailand	26	83	-	-	-	-	0.856	1.687	-	-	-	-
Turkey	72	47	57	126	103	-	4.528	3.100	2.205	5.029	3.314	-
Ukraine	-	-	4.429	1.066	0.673	0.584	-	-	0.633	0.097	0.075	0.058
United Kingdom	-	415	302	-	-	-	-	0.714	0.530	-	-	-
United States of America	-	-	2,843	3,306	3,486	3,400	-	-	4.321	-	-	-

Depending on the table, * means *Enterprises*, *Engaged*, or *Factor Values*; ± means *Basis Unspecified*; § means *shown in millions*. For additional notes and sources, see Appendix I.

Indexes of Industrial Production

Country	Index of Industrial Production (1990=100)						Index of Employment (1990=100)					
	1990	1991	1992	1993	1994	1995	1990	1991	1992	1993	1994	1995
Australia	-	-	-	-	-	-	100	100	80	-	-	-
Azerbaijan	-	-	-	-	-	-	100	93	96	97	87	-
Bangladesh	-	-	-	-	-	-	100	108	102	-	-	-
Bolivia	-	-	-	-	-	-	100	105	107	112	150	151
Bosnia & Herzegovina	-	-	-	-	-	-	100	95	-	-	-	-
Canada	-	-	-	-	-	-	100	100	100	129	129	114
Chile	-	-	-	-	-	-	100	110	65	50	37	36
China (Hong Kong SAR)	-	-	-	-	-	-	100	74	58	68	58	-
Colombia	-	-	-	-	-	-	100	98	101	103	99	108
Costa Rica	-	-	-	-	-	-	100	71	75	68	-	-
Denmark	-	-	-	-	-	-	100	99	98	-	-	-
Ecuador	-	-	-	-	-	-	100	171	186	182	177	62
Egypt	-	-	-	-	-	-	100	100	96	4	-	-
Finland	-	-	-	-	-	-	100	95	89	85	86	-
Greece	-	-	-	-	-	-	100	146	156	-	-	-
India	-	-	-	-	-	-	100	96	119	111	-	-
Indonesia	-	-	-	-	-	-	100	106	109	233	250	284
Italy	-	-	-	-	-	-	100	111	-	-	-	-
Japan	-	-	-	-	-	-	100	101	104	103	-	-
Jordan	-	-	-	-	-	-	100	145	104	152	375	416
Korea, Republic of	-	-	-	-	-	-	100	133	135	207	233	223
Malaysia	-	-	-	-	-	-	100	124	143	176	181	143
Mexico	-	-	-	-	-	-	100	98	90	75	69	77
New Zealand	-	-	-	-	-	-	100	94	103	113	-	-
Norway	-	-	-	-	-	-	100	106	103	-	-	-
Peru	-	-	-	-	-	-	100	87	61	-	-	-
Philippines	-	-	-	-	-	-	100	137	113	113	-	-
Portugal	-	-	-	-	-	-	100	106	100	97	84	-
Senegal	-	-	-	-	-	-	100	-	-	-	48	-
Singapore	-	-	-	-	-	-	100	113	111	137	145	-
Spain	-	-	-	-	-	-	100	95	87	-	-	-
Sri Lanka	-	-	-	-	-	-	100	112	103	70	-	-
Sweden	-	-	-	-	-	-	100	108	99	91	91	-
Thailand	-	-	-	-	-	-	100	335	-	-	-	-
Turkey	-	-	-	-	-	-	100	96	44	89	90	-
United Kingdom	-	-	-	-	-	-	100	100	90	-	-	-
United States of America	-	-	-	-	-	-	100	98	97	95	97	95
Uruguay	-	-	-	-	-	-	100	90	64	60	-	-
Venezuela	-	-	-	-	-	-	100	119	119	105	77	91

Depending on the table, * means *Enterprises, Engaged,* or *Factor Values;* ± means *Basis Unspecified;* § means *shown in millions.* For additional notes and sources, see Appendix I.

325

ISIC 3520
CHEMICAL PRODUCTS NEC

ISIC 3520—Chemical Products nec—includes statistics on paints, drugs and medicines, soaps and detergents, carbon black, printers' ink, explosives, photographic film, and photographic paper. A separate chapter breaks out drugs and medicines for those countries that report on these products separately.

These products correspond, in the U.S. SIC classification, to SICs 283—Drugs, 284—Soaps, Cleaners, and Toilet Goods, 285—Paints and Allied Products, and 289—Miscellaneous Chemical Products.

Summary Statistics

		1990 Value	N	1991 Value	N	1992 Value	N	1993 Value	N	1994 Value	N	1995 Value	N
Establishments or enterprises	(number)	36,435	84	39,639	93	51,188	88	27,778	82	21,962	70	11,661	33
Number employed	(000)	4,775	95	4,869	97	4,961	91	4,730	91	3,204	82	2,747	58
Total Wages	($ mil.)	63,536	89	77,596	91	89,227	84	67,501	80	69,807	73	56,934	55
Wage/salary per employee	($)	10,874	88	11,881	88	13,323	82	11,698	79	11,981	72	14,216	54
Female workers	(000)	381	33	211	31	510	25	522	30	487	26	427	14
as % of total employment*	(%)	37	32	36	30	38	25	39	30	41	25	48	13
Output	($ bil.)	500	96	586	97	646	91	592	87	604	79	565	61
per employee	($ 000)	78	89	84	89	99	83	99	79	103	73	119	56
per establishment	($ 000)	7,472	79	7,799	82	8,398	77	9,521	69	8,186	59	5,080	28
Value added	($ bil.)	247	87	264	86	295	77	308	76	287	68	300	57
per employee	($ 000)	37	79	39	78	46	71	45	68	52	62	60	51
per establishment	($ 000)	3,272	69	3,132	72	3,797	64	3,399	59	3,951	48	2,388	24
Capital investment	($ mil.)	19,846	67	22,744	63	25,104	54	23,379	50	19,577	41	12,592	21
per employee	($ 000)	8	64	6	60	5	53	4	47	5	39	7	20
per establishment	($ 000)	614	62	559	58	536	51	355	44	340	38	281	18

Data presented above are drawn from the detailed tables that follow. Columns headed 'N' show the number of countries that provided valid data for inclusion. Values are not strictly comparable one year to the next or one row to the next because the number of countries that report varies from period to period and row to row. However, a general indication of magnitudes can be gleaned from reviewing this summary—especially in earlier years in which more reports are available. For detailed explanations, see Appendix I. *The average for those countries reporting both total and female employment.

Establishments and Number Engaged

Country	Establishments or Enterprises (number)						Employees per Establishment					
	1990	1991	1992	1993	1994	1995	1990	1991	1992	1993	1994	1995
Albania	-	-	-	4.000	3.000	3.000	-	-	-	169	76	39
Argentina				1,723	1,830					25	24	-
Armenia	7.000	7.000	6.000	7.000	7.000	-	435	387	-			
Australia	581	735	772	-	-	-	52	42	39	-	-	-
Austria	283	281	257	253	242	-	66	64	70	70	71	-
Azerbaijan	11	11	11	12	11	-	594	551	464	396	490	-
Bahamas	7.000	9.000	-	-	-	-	40	31	-	-	-	-
Bahrain	-	-	15	-	-	-	-	-	6.733	-	-	-
Bangladesh	611	488	532	-	-	-	60	69	63	-	-	-
Barbados	5.000	3.000	7.000	5.000	12	-	40	35	27	36	25	-
Belize	3.000	3.000	3.000	-	-	-	-	-	-	-	-	-
Benin	-	-	7.000	7.000	-	-	-	-	-	-	-	-
Bolivia	58	59	59	56	58	62	29	30	31	31	35	34
Bosnia & Herzegovina	19	19	-	-	3.000	-	334	309	-	-	404	-
Bulgaria	*52	*51	*54	*256	*386	*385	608	510	450	89	58	54
Burundi	*8.000	*8.000					41	41	-			
Canada	1,154	1,034	1,003	960	961	-	59	64	68	69	67	-
Cape Verde	1.000	1.000	2.000	3.000	3.000	-	48	49	24	24	25	-
Central African Republic	-	*8.000	*8.000	*8.000	*8.000	-						
Chad	-	2.000	2.000	3.000	3.000	3.000						
Chile	96	95	102	105	105	-	172	176	170	183	176	-
China	*10,146	*10,549	*11,099	-	-	-	133	134	132	-	-	-
China (Hong Kong SAR)	464	434	374	416	400	-	14	12	14	12	13	-
Colombia	331	325	351	356	358	-	79	89	98	108	106	-
Costa Rica	173	172	171	164	-	-	29	30	32	33	-	-
Croatia	97	128	204	273	334	369	159	108	62	45	35	31
Cyprus	103	108	106	107	113	115	12	13	13	14	14	13
Czechoslovakia (Former)	*24	*23	-	-	-	-	1,125	826	-	-	-	-
Czech Republic	-	*16	*37	*47	-	-	-	750	378	298	-	-
Denmark	393	423	445	-	-	-	45	42	36	-	-	-
Ecuador	79	84	83	85	83	86	82	92	75	70	71	72
Egypt	164	187	178	178	-	-	307	277	303	300	-	-
El Salvador	-	69	38	51	50	56	-	-	98	91	85	77
Equatorial Guinea	2.000	-	-	-	-	-	10	-	-	-	-	-
Ethiopia	11	10	9.000	10	22	26	144	156	175	163	86	97
Fiji	26	19	20	-	-	-	19	28	25	-	-	-
Finland	92	110	105	107	103	104	112	93	89	87	90	-
FYR Macedonia	*8.000	*19	*32	*46	*60	*79	425	156	95	55	34	31
Gabon	-	4.000	4.000	4.000	4.000	4.000	-	38	34	33	28	18
Germany		1,457	1,459	1,459	1,444	-		227	213	204	197	-
Germany (Western Part)	1,280	1,289	1,309	-	-	-	224	224	222	-	-	-
Ghana	-	-	-	42	-	-	-	-	-	106	-	-
Greece	276	281	284	-	-	-	58	58	56	-	-	-
Grenada	-	-	5.000	4.000	3.000	-	-	-	24	30	25	-
Guatemala	-	35	35	35	35	35	-	120	121	124	124	127
Honduras	-	-	49	46	46	47	-	-	55	64	67	94
Hungary	*81	*154	*283	*327	-	-	321	162	81	80	-	-
Iceland	25	26	25	22	-	-	12	16	15	17	-	-
India	4,695	5,079	5,541	6,041	-	-	70	71	70	67	-	-
Indonesia	560	529	545	567	579	605	151	164	173	177	183	188
Iran (Islamic Republic of)	352	176	191	196	-	-	72	128	127	122	-	-
Iraq	-	-	15	-	-	-	-	-	200	-	-	-
Italy	*480	*472	*870	*848	*852	-	106	109	144	142	137	-
Jamaica	47	47	-	-	-	-	64	63	-	-	-	-
Japan	3,499	3,521	3,465	3,492	3,834	3,883	63	64	67	66	63	62
Jordan	40	49	87	89	135	147	83	78	59	59	43	40
Kenya	111	132	137	134	148	148	78	66	65	70	65	68
Korea, Republic of	1,062	1,088	1,157	1,302	1,332	1,395	72	69	67	58	57	53
Kuwait	15	12	15	16	15	-	43	26	34	32	37	-
Kyrgyzstan	6.000	6.000	5.000	7.000	5.000	-	133	82	83	77	73	-

Continued.

   Depending on the table, * means *Enterprises, Engaged,* or *Factor Values;* ± means *Basis Unspecified;* § means *shown in millions.* For additional notes and sources, see Appendix I.

Establishments and Number Engaged

- Continued -

Country	Establishments or Enterprises (number)						Employees per Establishment					
	1990	1991	1992	1993	1994	1995	1990	1991	1992	1993	1994	1995
Latvia	11	51	63	69	42	52	774	164	130	87	124	81
Lesotho	14	-	2.000	2.000	2.000	-	41	-	94	90	91	-
Liechtenstein	-	7.000	-	-	-	-	-	10	-	-	-	-
Lithuania	-	-	-	-	*29	-	-	-	-	-	118	-
Luxembourg	*11	*11	*11	*11	-	-	42	80	100	99	-	-
Macau	14	11	11	10	10	10	22	25	23	26	26	25
Malawi	*7.000	*7.000	*7.000	*7.000	*7.000	-	129	157	200	200	200	-
Malaysia	149	152	155	164	164	-	83	85	93	93	97	-
Mauritius	28	23	23	22	23	21	39	48	48	52	51	58
Mexico	239	234	219	211	206	204	277	289	308	298	304	306
Mongolia	7.000	7.000	7.000	7.000	13	7.000	112	99	91	73	40	79
Mozambique	18	18	13	13	14	-	105	108	133	135	124	-
Nepal	85	90	-	112	110	-	61	48	-	70	63	-
Netherlands	160	154	165	165	179	-	200	206	194	198	183	-
Netherlands Antilles	-	-	-	7.000			-	-	-	29		
New Zealand	*234	*241	*243	*261	-	-	23	23	25	23	-	-
Norway	75	80	60	49	50	-	73	66	90	102	99	-
Pakistan	-	280	-	-	-	-	-	96	-	-	-	-
Panama	48	45	-	-	-	-	36	38	-	-	-	-
Paraguay	-	18	-	-	-	-	-	55	-	-	-	-
Peru	759	787	799	-	-	-	20	18	16	-	-	-
Philippines	409	466	328	291	-	-	78	69	95	106	-	-
Poland	*142	*183	*169	*169	-	-	387	306	337	325	-	-
Portugal	*712	*693	*729	*707	*701	-	33	33	32	31	30	-
Qatar	3.000	3.000	5.000	5.000	6.000	-	57	73	53	45	49	-
Republic of Moldova	13	13	13	11	10	8.000	162	159	134	166	165	193
Romania	-	-	-	-	785	-	-	-	-	-	-	-
Russian Federation	-	-	-	*1,305	*1,610	*2,169	-	-	-	152	105	89
Saint Lucia	-	6.000	6.000	7.000	9.000	9.000	-	-	-	-	-	-
Saint Vincent & the Grenadines	-	-	-	-	-	1.000	-	-	-	-	-	10
Senegal	19	14	14	11	15	-	86	94	109	69	61	-
Sierra Leone	-	-	-	20	-	-	-	-	-	31	-	-
Singapore	93	94	97	95	100	-	56	56	57	57	59	-
Slovakia	-	15	15	19	19	-	-	540	505	401	403	-
Slovenia	119	*76	*103	*95	*108	*166	-	92	-	-	-	-
South Africa	-	663	-	-	-	-	-	92	-	-	-	-
Spain	1,902	1,936	1,904	-	-	-	43	44	43	-	-	-
Sri Lanka	68	61	41	73	-	-	60	79	97	64	-	-
Suriname	-	-	12	-	-	-	-	-	28	-	-	-
Sweden	150	172	174	169	174	-	159	121	113	107	117	-
Thailand	280	312	-	-	-	-	79	108	-	-	-	-
Tonga	3.000	3.000	3.000	3.000	2.000	-	4.667	5.000	5.333	4.000	7.000	-
Trinidad and Tobago	72	67	65	68	58	-	20	20	20	21	24	-
Tunisia	-	-	-	193	221	211	-	-	-	20	22	26
Turkey	174	170	265	265	255	-	158	155	102	103	101	-
Ukraine	153	389	471	153	159	154	386	141	113	320	277	279
United Kingdom	1,748	1,666	1,712	2,114	2,195	-	97	100	97	82	78	-
United Republic of Tanzania	39	35	-	-	-	-	72	87	-	-	-	-
USSR (Former)	*465	-	-	-	-	-	667	-	-	-	-	-
United States of America	-	-	10,871	-	-	-	-	-	44	-	-	-
Venezuela	291	294	296	268	253	278	97	105	105	108	106	92
Yugoslavia	111	195	493	705	763	819	-	92	35	24	23	21
Zambia	25	-	-	-	37	-	117	-	-	-	69	-
Zimbabwe	47	50	51	49	46	-	123	119	108	108	115	-

Depending on the table, * means *Enterprises, Engaged,* or *Factor Values*; ± means *Basis Unspecified*; § means *shown in millions.* For additional notes and sources, see Appendix I.

Employment and Compensation of Employees

Country	Number Employed/Engaged (000)						Wage/Salary per Employee ($)					
	1990	1991	1992	1993	1994	1995	1990	1991	1992	1993	1994	1995
Albania	-	-	-	0.675	0.228	0.118	-	-	-	514	-	1,227
Argentina	38	-	-	44	45	-	11,331	-	-	23,839	-	-
Armenia	*3.047	*2.709	-	-	-	-	-	-	-	-	-	-
Australia	30	*31	*30	-	-	-	23,443	25,208	25,515	-	-	-
Austria	19	18	18	18	17	17	29,315	31,594	34,585	34,678	36,641	-
Azerbaijan	6.535	6.063	5.108	4.749	5.390	-	1,646	1,466	-	-	-	-
Bahamas	0.279	0.282	-	-	-	-	35,050	-	-	-	-	-
Bahrain	-	-	0.101	-	-	-	-	-	5,609	-	-	-
Bangladesh	36	34	33	-	-	-	1,053	1,161	1,231	-	-	-
Barbados	0.199	0.105	0.191	0.180	0.296	0.259	13,143	4,343	14,419	6,719	11,772	11,814
Bolivia	1.668	1.757	1.829	1.735	2.038	2.105	2,591	2,924	2,913	3,382	3,408	3,238
Bosnia & Herzegovina	6.341	5.878	-	-	1.213	-	3,125	3,116	-	-	-	-
Bulgaria	32	26	24	23	22	21	1,889	783	1,543	1,829	1,475	1,817
Burundi	0.328	0.331	-	-	-	-	2,978	3,024	-	-	-	-
Cambodia	*1.239	*1.077	-	-	-	-	-	-	-	-	-	-
Cameroon	0.688	0.807	0.845	0.755	-	-	12,188	11,130	11,200	-	-	-
Canada	68	66	68	66	64	64	29,530	31,871	29,967	29,666	30,001	30,368
Cape Verde	0.048	0.049	0.048	0.073	0.076	-	-	-	-	-	-	-
Central African Republic	0.160	-	-	-	-	-	7,667	-	-	-	-	-
Chile	17	17	17	19	19	20	7,737	8,986	9,851	10,162	11,843	13,929
China	*1,350	*1,410	*1,460	*1,420	-	-	-	-	-	-	-	-
China (Hong Kong SAR)	*6.600	*5.400	*5.200	*5.000	*5.000	-	11,262	11,414	9,813	-	-	-
China (Taiwan Province)	56	55	57	60	62	63	10,407	11,538	13,721	13,644	14,757	15,398
Colombia	26	29	35	38	38	38	3,142	3,417	3,821	4,578	6,006	6,749
Costa Rica	5.048	5.167	5.434	5.347	-	-	4,054	3,878	4,352	3,534	-	-
Croatia	15	14	13	12	12	11	5,379	5,604	1,920	2,187	2,928	3,751
Cyprus	1.287	1.368	1.360	1.477	1.537	1.552	9,236	9,929	11,559	11,466	13,277	16,375
Czechoslovakia (Former)	27	19	-	-	-	-	2,084	1,607	-	-	-	-
Czech Republic	13	12	14	14	-	-	2,228	1,555	2,022	2,600	-	-
Denmark	18	18	16	16	17	18	36,019	35,679	39,247	36,612	-	-
Ecuador	6.458	7.703	6.230	5.916	5.926	6.200	3,955	4,110	5,711	6,757	3,900	4,113
Egypt	*50	*52	*54	*53	*54	*53	2,652	1,988	1,839	2,033	2,127	2,173
El Salvador	-	-	*3.733	*4.628	*4.243	*4.339	-	-	-	3,879	5,694	1,273
Equatorial Guinea	0.020	-	-	-	-	-	3,173	-	-	-	-	-
Ethiopia	1.589	1.562	1.575	1.630	1.900	2.511	2,008	1,994	1,469	998	953	813
Fiji	0.493	0.538	0.496	0.524	0.515	-	4,879	6,417	6,098	6,175	6,501	-
Finland	10	10	9.300	9.300	9.300	-	29,538	29,584	26,984	21,548	24,877	-
FYR Macedonia	3.396	2.969	3.031	2.537	2.043	2.451	9,946	11,411	2,864	3,949	4,634	6,431
France	184	185	183	180	177	178	-	-	-	-	-	-
Gabon	-	0.151	0.137	0.134	0.111	0.073	-	16,010	22,971	22,903	14,458	17,070
Germany	-	*330	*311	*297	*284	-	-	33,396	-	-	-	-
Germany (Western Part)	*286	*288	*290	-	-	-	35,836	36,666	-	-	-	-
Ghana	-	-	-	*4.455	*4.581	*4.710	-	-	-	2,087	1,559	1,591
Greece	*16	*16	*16	*16	*16	*17	13,350	14,284	15,873	16,390	16,750	20,434
Grenada	-	-	*0.118	*0.120	0.075	-	-	-	-	-	-	-
Guatemala	-	*4.206	*4.222	*4.343	*4.354	*4.437	-	841	913	913	946	957
Honduras	*3.032	*2.627	*2.676	*2.932	*3.082	*4.425	2,904	3,540	2,983	2,787	3,032	2,984
Hungary	26	25	23	26	26	26	3,250	3,697	4,609	5,067	5,328	5,034
Iceland	0.302	0.412	0.366	0.377	0.353	-	26,776	21,793	24,784	24,625	-	-
India	*328	*361	*388	*404	*428	*420	1,644	1,329	1,371	1,243	1,388	1,400
Indonesia	84	87	94	100	106	114	1,280	1,393	1,537	1,639	1,781	2,577
Iran (Islamic Republic of)	25	22	24	24	-	-	4,466	5,180	5,366	5,187	-	-
Iraq	-	3.122	3.000	-	-	-	-	8,074	13,981	-	-	-
Ireland	9.600	10	10	11	11	11	25,515	25,864	31,527	29,398	35,012	-
Italy	51	51	126	120	117	-	-	-	-	-	-	-
Jamaica	3.006	2.960	2.892	2.688	2.709	-	-	-	-	-	-	-
Japan	*222	*226	*232	*232	*242	*241	34,968	38,595	-	-	-	-
Jordan	3.328	3.823	5.121	5.222	5.773	5.833	3,565	3,319	3,183	3,809	3,858	4,451
Kenya	8.663	8.720	8.962	9.355	9.693	10	2,970	2,637	2,434	1,488	1,703	2,308
Korea, Republic of	76	75	77	76	76	74	10,690	11,904	12,668	14,138	15,763	18,111

Continued.

Depending on the table, * means *Enterprises, Engaged,* or *Factor Values*; ± means *Basis Unspecified*; § means *shown in millions*. For additional notes and sources, see Appendix I.

Employment and Compensation of Employees
- Continued -

Country	Number Employed/Engaged (000)						Wage/Salary per Employee ($)					
	1990	1991	1992	1993	1994	1995	1990	1991	1992	1993	1994	1995
Kuwait	0.652	0.307	0.507	0.517	0.552	0.538	6,944	7,008	9,142	9,460	9,423	12,615
Kyrgyzstan	*0.801	*0.490	*0.413	*0.536	*0.364	-	731	510	537	-	-	-
Latvia	8.514	8.346	8.193	6.024	5.216	4.235	-	-	-	-	-	-
Lesotho	*0.578	-	*0.187	*0.179	*0.181		-	-	3,971	3,248	2,956	-
Liechtenstein	-	*0.073					-	-	-	-	-	-
Lithuania	-	-	5.422	4.248	3.435		-	-	-	501	1,179	-
Luxembourg	0.463	0.885	1.095	1.084	1.245	-	21,199	27,828	29,087	28,851	30,514	-
Macau	0.311	0.276	0.254	0.261	0.262	0.246	5,015	5,613	6,572	8,156	9,495	9,782
Malawi	0.900	1.100	1.400	1.400	1.400	-	4,357	4,378	3,806	3,293	1,782	-
Malaysia	12	13	14	15	16	16	4,662	5,001	5,193	5,870	6,363	6,844
Mauritius	*1.089	1.102	1.115	1.154	1.163	1.210	2,848	2,942	4,154	3,820	4,156	4,556
Mexico	66	68	67	63	63	62	7,233	9,037	11,256	12,859	13,296	9,068
Mongolia	*0.786	*0.693	*0.634	*0.508	*0.519	*0.551	1,113	1,470	493	-	409	1,024
Mozambique	1.886	1.936	1.734	1.751	1.738	-	1,098	1,012	499	445	441	-
Nepal	5.175	4.275	-	7.808	6.878	-	547	497	-	-	445	-
Netherlands	32	32	32	33	33	33	31,793	32,599	36,504	36,456	39,796	-
Netherlands Antilles	-	-	-	0.201	-		-	-	-	8,813		
New Zealand	5.355	5.594	6.175	6.107	-	-	20,960	21,013	16,993	-	-	-
Norway	5.485	5.269	5.376	4.983	4.929	4.827	31,310	33,286	-	37,850	39,038	-
Pakistan	27	27	28	-	-	-	2,957	3,138	3,228	-	-	-
Panama	*1.712	*1.732	*1.896	*1.950	*1.964	*1.150	6,950	7,094	7,305	7,368	7,465	7,563
Paraguay	-	0.983	-	-	-	-	-	-	-	-	-	-
Peru	*16	*14	*13	-	-	-	5,038	6,411	7,479	-	-	-
Philippines	32	32	31	31	-	-	4,998	5,297	6,624	6,364		-
Poland	55	56	57	55	53	51	1,369	2,163	2,761	3,201	4,347	6,217
Portugal	*24	*23	*23	*22	*21	*21						
Qatar	0.170	0.219	0.267	0.225	0.297	-	7,605	6,655	6,155	6,897	6,632	-
Republic of Moldova	2.103	2.070	1.736	1.829	1.654	1.545	-	3,356	-	-	539	620
Russian Federation	-	-	-	198	169	194	-	-	-	695	1,180	1,310
Saint Vincent & the Grenadines	-	-	-	-	-	0.010	-	-	-	-	-	-
Senegal	*1.629	*1.311	*1.530	*0.759	*0.909	-	4,886	5,792	5,576	7,459	6,216	-
Sierra Leone	-	-	-	*0.611	-	-	-	-	-	-	-	-
Singapore	*5.168	*5.282	*5.506	*5.444	*5.923	*5.935	16,596	18,783	21,470	23,534	27,989	30,830
Slovakia	-	*8.098	*7.575	*7.620	*7.655	-	-	1,667	2,172	2,619	2,988	-
South Africa	59	61	61	-	-	-	12,244	13,186	15,186	-	-	-
Spain	83	84	82	92	89	88	25,413	27,362	30,516	27,822	28,148	31,836
Sri Lanka	*4.113	*4.802	*3.974	*4.653	-	-	1,287	1,406	917	1,556	-	-
Suriname	*0.335	*0.312	*0.338	*0.346	-	-	18,730	23,881	21,547	31,088	-	-
Sweden	24	*21	*20	18	20	20	27,940	30,717	34,480	27,465	28,254	31,828
Thailand	22	34	-	-	-	-	2,369	4,065	-	-	-	-
Tonga	0.014	0.015	0.016	0.012	0.014	-	3,011	3,395	3,619	5,238	4,545	-
Trinidad and Tobago	1.435	1.357	1.323	1.425	1.393	-	6,657	6,323	6,782	5,311	4,507	-
Tunisia	-	-	-	3.894	4.822	5.384	-	-	-	10,119	8,364	10,874
Turkey	28	26	27	27	26	27	10,538	13,429	13,654	14,576	9,267	12,818
Ukraine	59	55	53	49	44	43	3,105	2,992	2,340	-	589	665
United Kingdom	170	167	166	173	172	171	26,366	28,350	30,866	27,444	28,607	32,176
United Republic of Tanzania	2.801	3.036	-	-	-	-	-	479	-	-	-	-
USSR (Former)	310	-	-	-	-	-	2,123	-	-	-	-	-
United States of America	485	479	483	489	485	506	32,371	33,800	35,522	36,771	38,581	39,308
Uruguay	*6.878	*6.020	*5.771	*5.681	*5.668	*5.683	6,057	7,459	8,816	10,958	12,211	12,209
Venezuela	28	31	31	29	27	26	5,787	6,796	7,314	7,166	6,167	7,597
Yugoslavia	-	18	17	17	17	17	-	-	-	-	3,155	1,456
Zambia	2.916	-	-	-	2.568	-	1,751	-	-	-	1,943	-
Zimbabwe	5.797	5.933	5.500	5.300	5.300	4.954	5,736	5,017	5,368	2,985	4,524	4,972

Depending on the table, * means *Enterprises, Engaged,* or *Factor Values;* ± means *Basis Unspecified;* § means *shown in millions.* For additional notes and sources, see Appendix I.

Female Workers: Total and Percent of Employment

Country	Female Workers (000)						Female Workers as % of Total Employment					
	1990	1991	1992	1993	1994	1995	1990	1991	1992	1993	1994	1995
Afghanistan	0.439	0.458	-	-	-	-	-	-	-	-	-	-
Albania	-	-	-	0.273	0.081	0.023	-	-	-	40.44	35.53	19.49
Argentina	-	-	-	-	13	-	35.73	-	-	-	28.81	-
Australia	11	-	-	-	-	-	35.73	-	-	-	-	-
Austria	7.500	7.000	7.000	6.957	6.717	-	40.11	38.89	38.89	39.56	39.04	-
Bangladesh	1.502	1.604	1.383				4.12	4.73	4.14			
Bosnia & Herzegovina	3.964	3.992	-	-	-	-	62.51	67.91	-	-	-	-
Bulgaria	-	-	-	13	12	11	-	-	-	55.02	55.41	54.11
Canada	25	25	-	-	-	-	36.76	37.88	-	-	-	-
Chile	5.298	5.323	5.416	6.166	5.774	-	32.04	31.90	31.24	32.00	31.16	-
China (Taiwan Province)	25	24	25	27	28	28	43.85	44.13	44.79	45.08	44.67	44.55
Colombia	-	12	-	18	18	-	-	42.63	-	46.97	46.03	-
Croatia	7.910	7.060	6.390	6.220	6.060	5.830	51.30	50.86	50.63	50.61	51.23	51.32
Cyprus	0.593	0.665	0.658	0.686	0.674	-	46.08	48.61	48.38	46.45	43.85	-
Czechoslovakia (Former)	9.700	5.900	-	-	-	-	35.93	31.05	-	-	-	-
Czech Republic	7.000	7.000	8.000	8.000	-	-	53.85	58.33	57.14	57.14	-	-
Denmark	8.471	8.572	7.831	-	-	-	48.09	48.12	48.53	-	-	-
Egypt	-	13	15	14	-	-	-	24.41	27.21	26.05	-	-
El Salvador	-	-	-	2.366	1.913	2.853	-	-	-	51.12	45.09	65.75
Ethiopia	-	0.516	0.507	0.503	0.659	0.741	-	33.03	32.19	30.86	34.68	29.51
FYR Macedonia	2.577	2.386	2.548	2.400	1.913	1.837	75.88	80.36	84.06	94.60	93.64	74.95
Germany (Western Part)	108	-	-	-	-	-	37.71	-	-	-	-	-
Ghana	-	-	-	0.692	-	-	-	-	-	15.53	-	-
Hungary	13	-	13	12	-	-	50.00	-	56.52	46.15	-	-
Indonesia	-	-	-	51	54	59	-	-	-	51.22	50.59	51.69
Iran (Islamic Republic of)	-	3.648	3.994	3.761	-	-	-	16.24	16.49	15.78	-	-
Iraq	-	0.332	-	-	-	-	-	10.63	-	-	-	-
Ireland	2.800	3.000	-	-	-	-	29.17	29.70	-	-	-	-
Italy	-	14	34	33	31	-	-	27.59	27.16	27.20	26.58	-
Jordan	0.918	1.038	1.244	1.383	1.262	1.321	27.58	27.15	24.29	26.48	21.86	22.65
Kenya	1.498	1.494	1.642	1.715	-	-	17.29	17.13	18.32	18.33	-	-
Korea, Republic of	28	28	27	25	25	25	36.97	37.32	35.05	33.73	33.24	34.34
Lesotho	-	-	-	0.091	0.091	-	-	-	-	50.84	50.28	-
Macau	0.157	0.138	0.121	0.121	0.121	0.115	50.48	50.00	47.64	46.36	46.18	46.75
Malaysia	5.100	5.200	5.800	5.800	6.000	-	41.46	40.31	40.28	37.91	37.74	-
Mongolia	-	-	-	-	0.324	-	-	-	-	-	62.43	-
Morocco	-	-	-	-	-	5.345	-	-	-	-	-	-
Myanmar	-	-	-	-	1.158	-	-	-	-	-	-	-
Nepal	0.879	0.865	-	0.634	0.857	-	16.99	20.23	-	8.12	12.46	-
New Zealand	2.185	2.191	2.424	2.373	-	-	40.80	39.17	39.26	38.86	-	-
Panama	0.425	-	-	-	-	-	24.82	-	-	-	-	-
Philippines	-	-	9.546	9.439	-	-	-	-	30.69	30.55	-	-
Portugal	6.031	5.630	5.769	5.291	4.951	-	25.34	24.47	24.78	23.88	23.51	-
Republic of Moldova	-	-	-	-	0.926	1.021	-	-	-	-	55.99	66.08
Senegal	0.119	-	-	-	-	-	7.31	-	-	-	-	-
Sri Lanka	1.245	1.430	0.953	1.301	-	-	30.27	29.78	23.98	27.96	-	-
Sweden	10	-	-	-	-	-	42.44	-	-	-	-	-
Thailand	12	18	-	-	-	-	55.89	53.72	-	-	-	-
Turkey	6.273	-	-	-	-	-	22.81	-	-	-	-	-
United Kingdom	66	-	61	-	-	-	38.92	-	36.75	-	-	-
United Republic of Tanzania	0.775	0.850	-	-	-	-	27.67	28.00	-	-	-	-
United States of America	-	-	263	263	267	285	-	-	54.45	53.78	55.05	56.32
Zambia	-	-	-	-	0.297	-	-	-	-	-	11.57	-

Depending on the table, * means *Enterprises, Engaged,* or *Factor Values*; ± means *Basis Unspecified*; § means *shown in millions.* For additional notes and sources, see Appendix I.

Output and Output per Employee

Country	Output ($ bil.)						Output per Employee ($)					
	1990	1991	1992	1993	1994	1995	1990	1991	1992	1993	1994	1995
Albania	-	-	*0.002	*0.000	*0.005	-	-	-	-	2,903	1,760	42,015
Argentina	±4.520	-	-	6.582	-	-	117,666	-	-	150,995	-	-
Armenia	0.002	0.005	0.034	0.942	0.003	-	806	1,757	-	-	-	-
Australia	*4.878	*5.582	*5.458	-	-	-	162,604	180,052	181,936	-	-	-
Austria	3.385	3.533	3.767	3.687	3.781	4.712	181,005	196,305	209,260	209,669	219,751	277,211
Azerbaijan	±0.099	±0.092	±0.003	±0.007	±0.003	-	15,211	15,175	560	1,576	474	-
Bahamas	0.015	0.017	-	-	-	-	53,763	60,284	-	-	-	-
Bahrain	-	-	±0.005	-	-	-	-	-	52,507	-	-	-
Bangladesh	0.333	0.356	0.392	-	-	-	9,133	10,509	11,723	-	-	-
Barbados	0.018	0.002	0.022	0.014	0.027	0.026	89,332	19,724	113,911	78,961	89,973	100,980
Belgium	3.319	3.604	4.084	3.898	4.294	5.711	-	-	-	-	-	-
Benin	±0.018	±0.016	±0.018	±0.016	-	-	-	-	-	-	-	-
Bolivia	0.042	0.046	0.046	0.050	0.060	0.071	25,019	25,998	24,919	28,603	29,662	33,695
Bosnia & Herzegovina	±0.212	±0.344	-	-	-	-	33,483	58,476	-	-	-	-
Bulgaria	±0.715	±0.472	±0.365	±0.319	±0.315	±0.404	22,629	18,137	15,027	13,946	14,178	19,531
Burundi	±0.012	±0.011	-	-	-	-	35,958	34,441	-	-	-	-
Canada	12	12	12	12	12	12	171,283	180,252	172,889	176,167	182,383	184,761
Cape Verde	±0.001	±0.000	±0.008	±0.008	±0.009	-	16,616	5,147	164,204	110,420	124,315	-
Central African Republic	*0.008	-	-	-	-	-	51,812	-	-	-	-	-
Chile	1.152	1.310	1.528	1.689	1.774	2.290	69,661	78,496	88,133	87,683	95,764	112,690
China	±14	±15	±18	±21	-	-	10,148	10,633	12,113	14,716	-	-
China (Hong Kong SAR)	0.441	0.397	0.388	-	-	-	66,773	73,610	74,681	-	-	-
China (Taiwan Province)	±3.838	±4.289	±4.945	±5.300	±5.771	±6.238	68,554	77,824	86,939	88,580	92,447	99,363
Colombia	1.386	1.566	1.584	1.947	2.334	2.657	53,093	54,131	45,863	50,775	61,236	69,284
Costa Rica	0.168	0.187	0.234	0.302	0.378	0.346	33,255	36,174	43,048	56,450		
Cote d'Ivoire	±0.147	±0.124	±0.136	±0.117	-	-	-	-	-	-	-	-
Croatia	±0.851	±0.823	±0.575	-	-	-	55,194	59,311	45,576	-	-	-
Cyprus	0.102	0.120	0.124	0.119	0.138	0.154	78,940	87,367	91,428	80,349	89,548	99,177
Czechoslovakia (Former)	±0.693	±0.255	-	-	-	-	25,668	13,408	-	-	-	-
Denmark	*2.610	*2.730	*3.236	*3.003	*3.613	*4.640	148,176	153,276	200,534	187,748	214,781	262,572
Ecuador	0.237	0.277	0.264	0.322	1.239	0.339	36,669	35,978	42,352	54,384	209,038	54,714
Egypt	*1.418	*1.013	*1.004	*1.261	*1.371	*1.416	28,195	19,558	18,593	23,617	25,290	26,522
El Salvador	-	±0.174	0.066	±0.193	±0.157	±0.074	-	-	17,557	41,710	37,061	16,962
Equatorial Guinea	±0.000	-	-	-	-	-	12,855	-	-	-	-	-
Ethiopia	0.036	0.027	0.019	0.029	0.043	0.045	22,620	17,448	11,797	17,704	22,893	17,917
Fiji	0.029	0.038	0.036	0.038	0.040	-	58,655	71,137	71,728	73,386	77,748	-
Finland	±1.553	±1.520	±1.373	±1.170	±1.314	-	150,734	148,974	147,629	125,780	141,338	-
FYR Macedonia	±0.122	±0.137	±0.065	±0.056	±0.055	±0.072	35,797	46,195	21,522	22,247	27,023	29,172
France	±35	±35	±39	±38	±41	±47	188,909	188,377	215,182	212,578	232,290	265,026
Gabon	-	*0.012	*0.014	*0.013	*0.007	*0.008	-	82,491	99,000	97,487	64,436	103,765
Germany	-	58	64	60	64	-	-	177,033	206,985	201,131	223,478	-
Germany (Western Part)	55	57	63	-	-	-	191,342	196,102	216,140	-	-	-
Ghana	-	-	-	0.142	0.116	0.121	-	-	-	31,878	25,252	25,772
Greece	*1.605	*1.713	*1.928	*2.031	*2.089	*2.628	99,680	105,248	121,957	126,085	128,926	157,287
Guatemala	-	0.262	0.277	0.316	0.405	0.460	-	62,309	65,523	72,771	93,129	103,768
Honduras	0.081	0.094	0.119	0.115	0.117	0.141	26,714	35,726	44,632	39,072	38,084	31,973
Hungary	±0.924	±1.060	±1.135	±1.235	±1.267	±1.185	35,537	42,390	49,331	47,499	48,790	45,342
Iceland	0.046	0.049	±0.047	±0.042	-	-	151,116	117,846	128,825	111,970	-	-
India	*8.033	*7.681	*8.197	*8.391	*10	*10	24,498	21,264	21,112	20,768	23,744	24,308
Indonesia	1.672	2.182	2.054	2.740	3.303	3.667	19,804	25,177	21,774	27,350	31,119	32,224
Iran (Islamic Republic of)	0.886	0.903	0.837	1.026	-	-	35,143	40,190	34,544	43,041	-	-
Iraq	-	*0.036	*0.137	-	-	-	-	11,498	45,552	-	-	-
Ireland	*2.306	*2.446	*3.122	*2.988	*3.738	*4.737	240,171	242,134	299,456	281,355	339,541	418,126
Italy	±13	±13	±32	±26	±26	-	250,475	257,365	256,454	213,362	223,545	-
Japan	±79	±89	±97	±110	±125	±140	357,679	391,763	417,795	475,805	515,791	579,131
Jordan	0.161	0.177	0.259	0.264	0.274	0.258	48,319	46,332	50,633	50,533	47,504	44,280
Kenya	*0.904	*0.765	*0.966	*0.593	*0.625	*0.873	104,372	87,714	107,779	63,400	64,457	86,821
Korea, Republic of	9.424	10	10	12	13	14	124,001	134,188	135,081	155,858	171,422	188,343
Kuwait	0.036	0.025	0.040	0.047	0.048	0.062	54,605	82,443	78,121	90,653	87,169	114,720
Kyrgyzstan	0.018	0.025	0.009	0.001	0.001	-	22,608	50,890	20,899	1,552	2,832	-

Continued.

Output and Output per Employee
- Continued -

Country	Output ($ bil.)						Output per Employee ($)					
	1990	1991	1992	1993	1994	1995	1990	1991	1992	1993	1994	1995
Latvia	0.004	0.009	0.043	0.051	0.059	0.068	491	1,117	5,204	8,526	11,330	16,006
Lithuania	-	-	0.019	0.016	0.025	-	-	-	3,530	3,833	7,318	-
Luxembourg	*0.075	*0.153	*0.304	*0.273	*0.352	-	162,805	172,628	277,353	252,165	282,736	-
Macau	0.012	0.011	0.012	0.016	0.016	0.018	37,584	39,488	47,341	60,521	61,021	74,046
Malawi	*0.048	*0.036	*0.035	*0.035	*0.021	-	53,624	32,429	24,660	24,741	15,281	-
Malaysia	*0.663	*0.794	*0.959	*1.011	*1.203	*1.299	53,943	61,531	66,598	66,060	75,684	81,725
Mauritius	0.041	0.036	0.046	0.044	0.046	0.054	37,323	32,751	41,429	38,010	39,588	44,706
Mexico	4.684	5.381	6.051	6.128	6.932	4.853	70,748	79,591	89,685	97,394	110,701	77,660
Mongolia	0.012	0.008	0.004	0.001	0.002	0.003	15,653	11,020	6,185	2,593	4,401	5,671
Mozambique	-	±0.019	-	±0.017	±0.014	-		9,957	-	9,805	7,781	
Nepal	0.040	0.035	-	0.047	0.048	-	7,711	8,144	-	6,004	7,048	-
Netherlands	6.301	6.474	7.295	7.113	7.935	9.425	197,291	203,603	227,538	218,224	242,285	287,344
New Zealand	0.795	-	-	-	-	-	148,390	-	-	-	-	-
Norway	±1.027	±1.001	±1.224	±1.125	±1.139	±1.361	187,159	189,909	227,633	225,715	231,179	282,040
Pakistan	0.877	0.946	1.027	-	-	-	32,917	35,141	36,464	-		
Panama	0.096	0.110	0.124	0.128	0.131	0.077	55,911	63,227	65,139	65,712	66,583	67,263
Peru	0.818	1.013	1.056	-	-	-	52,764	72,284	82,119	-		
Philippines	1.722	1.893	2.244	2.226	-	-	54,157	59,149	72,148	72,045	-	-
Poland	±1.247	±1.766	±1.328	±1.801	±2.248	±3.024	22,672	31,542	23,297	32,751	42,496	58,903
Portugal	1.881	2.061	2.387	1.970	1.927	2.171	79,040	89,573	102,539	88,888	91,507	103,416
Puerto Rico	±9.214	±10	±12	±13	±15	±16	-	-	-	-	-	-
Qatar	0.010	0.015	0.016	0.013	0.017	-	56,561	70,249	60,707	59,829	58,275	-
Republic of Moldova	0.000	0.144	0.007	0.009	0.007	0.007	0.096	69,679	4,165	5,181	4,492	4,408
Romania	±2.011	±1.560	±1.138	±0.894	-	-						
Russian Federation		8.837	4.410	1.433	2.030	3.003			-	7,247	12,034	15,473
Senegal	0.078	-	-	-	-	-	48,008					
Singapore	*0.977	*1.303	*1.407	*1.469	*1.590	*1.789	189,109	246,764	255,596	269,900	268,418	301,453
Slovakia	-	±0.197	±0.255	±0.275	±0.273	-		24,320	33,648	36,052	35,617	
Slovenia	-	-	-	0.381	0.853	1.066						
South Africa	±3.905	±4.157	±4.237	±4.147	±4.221	±4.758	66,183	68,143	69,453		-	-
Spain	*15	*16	*17	±14	±15	±17	178,739	185,964	205,375	156,026	164,725	198,758
Sri Lanka	0.066	0.118	0.130	0.134	-	-	16,054	24,598	32,620	28,737	-	-
Suriname	*0.036	*0.038	*0.058	0.163	-	-	107,028	120,305	170,719	471,171		-
Sweden	*3.982	*4.703	*5.082	*3.905	*4.581	*5.194	167,315	225,088	258,384	216,799	225,471	255,512
Thailand	0.745	2.345	-	-	-	-	33,705	69,754				
Tonga	0.001	0.002	0.003	0.003	0.003	-	75,115	118,149	217,875	281,410	232,324	-
Trinidad and Tobago	±0.058	±0.056	±0.057	±0.055	±0.054	-	40,566	41,435	42,845	38,910	38,760	-
Tunisia	0.305	0.323	0.366	0.366	0.432	0.500	-	-	-	93,900	89,686	92,782
Turkey	3.322	3.511	3.838	4.553	3.755	5.592	120,794	132,918	142,173	166,650	145,183	203,916
Ukraine	*2.191	*2.567	*1.649	*0.377	*0.440	*0.434	37,131	46,664	31,109	7,696	10,004	10,085
United Kingdom	*28	*28	31	31	34	38	162,059	170,272	189,674	178,644	196,152	220,651
United Republic of Tanzania	*0.039	*0.052	-	-	-	-	13,745	17,248	-	-	-	-
USSR (Former)	±8.563	-	-	-	-	-	27,621	-	-	-		
United States of America	*134	*142	*152	*162	*169	*179	276,186	296,848	314,571	330,446	347,812	354,219
Uruguay	0.348	0.348	0.398	0.477	0.529	0.529	50,601	57,734	68,992	84,022	93,299	93,112
Venezuela	1.281	1.575	1.806	1.697	1.317	1.946	45,592	50,959	58,264	58,717	48,996	76,293
Yugoslavia	±0.119	±0.151	±0.426	-	±0.693	±0.566	-	8,367	24,478	-	40,291	32,691
Zambia	0.102	-	-	-	0.067	-	34,815	-	-	-	26,006	
Zimbabwe	*0.268	*0.236	*0.196	*0.126	*0.206	*0.212	46,175	39,848	35,656	23,700	38,801	42,814

Depending on the table, * means *Enterprises, Engaged*, or *Factor Values*; ± means *Basis Unspecified*; § means *shown in millions*. For additional notes and sources, see Appendix I.

Value Added and Value Added per Employee

Country	Value Added ($ bil.)						Value Added per Employee ($)					
	1990	1991	1992	1993	1994	1995	1990	1991	1992	1993	1994	1995
Argentina	±1.791	-	-	2.187	-	-	46,636	-	-	50,163	-	-
Armenia	0.000	0.000	-	-	-	-	69	96	-	-	-	-
Australia	*2.291	*2.563	*2.430	-	-	-	76,354	82,687	81,004	-	-	-
Austria	1.070	1.162	1.245	1.272	1.258	1.572	57,196	64,572	69,170	72,363	73,104	92,496
Bahamas	0.011	0.013	-	-	-	-	38,194	45,163	-	-	-	-
Bangladesh	0.166	0.160	0.195	-	-	-	4,542	4,729	5,822	-	-	-
Barbados	0.004	0.001	0.008	0.005	0.012	0.010	20,786	10,262	43,969	29,256	39,064	40,215
Belgium	*1.199	*1.243	*1.409	*1.352	*1.491	*2.026	-	-	-	-	-	-
Bolivia	0.017	0.017	0.019	0.022	0.024	0.029	9,948	9,768	10,421	12,478	11,994	13,556
Bosnia & Herzegovina	±0.102	±0.164	-	-	-	-	16,024	27,982	-	-	-	-
Bulgaria	-	±0.060	±0.081	±0.059	-	-	-	2,296	3,318	2,559	-	-
Burundi	0.001	0.001	-	-	-	-	3,965	3,930	-	-	-	-
Cameroon	0.021	0.024	0.024	-	-	-	29,987	29,939	28,346	-	-	-
Canada	6.256	6.485	6.288	6.085	5.939	5.981	92,006	98,259	92,467	92,194	92,793	94,038
Central African Republic	*0.003	-	-	-	-	-	16,987	-	-	-	-	-
Chile	0.617	0.743	0.854	0.933	0.991	1.279	37,323	44,502	49,261	48,434	53,463	62,916
China	±3.372	±3.448	±3.951	±5.234	-	-	2,498	2,446	2,706	3,686	-	-
China (Hong Kong SAR)	0.154	0.151	0.144	-	-	-	23,282	27,952	27,701	-	-	-
China (Taiwan Province)	1.097	1.265	1.626	1.745	1.740	1.437	19,595	22,962	28,582	29,158	27,872	22,881
Colombia	0.597	0.724	0.818	0.998	1.276	1.456	22,885	25,007	23,682	26,038	33,473	37,964
Costa Rica	0.050	0.056	0.071	0.091	0.112	0.101	9,882	10,929	13,126	17,052	-	-
Cote d'Ivoire	0.132	0.113	0.121	0.106	-	-	-	-	-	-	-	-
Croatia	±0.380	±0.441	±0.314	-	-	-	24,629	31,769	24,877	-	-	-
Cyprus	0.028	0.033	0.037	0.040	0.046	0.047	21,379	23,851	26,864	27,378	29,834	30,491
Czechoslovakia (Former)	±0.177	-	-	-	-	-	6,541	-	-	-	-	-
Denmark	*1.537	*1.587	*1.893	*1.756	*2.130	*2.749	87,255	89,088	117,286	109,790	126,609	155,574
Ecuador	0.075	0.089	0.074	0.114	0.923	0.075	11,663	11,573	11,924	19,194	155,768	12,032
Egypt	*0.486	*0.254	*0.217	*0.372	*0.405	*0.418	9,670	4,901	4,024	6,973	7,470	7,836
El Salvador	-	0.101	-	0.090	0.085	0.025	-	-	-	19,350	20,075	5,873
Ethiopia	0.013	0.010	0.007	0.012	0.015	0.014	8,038	6,626	4,567	7,063	7,814	5,520
Fiji	0.007	0.010	0.009	0.010	0.010	-	14,764	18,470	18,311	18,746	19,859	-
Finland	±0.707	±0.664	±0.619	±0.544	±0.606	-	68,626	65,093	66,589	58,448	65,193	-
FYR Macedonia	±0.073	±0.073	±0.031	±0.034	±0.029	±0.035	21,498	24,442	10,366	13,428	13,982	14,215
France	±12	±12	±14	±15	±16	±17	67,430	66,992	77,285	81,481	87,575	94,177
Gabon	-	*0.006	*0.006	*0.005	*0.003	*0.002	-	37,231	45,060	40,007	27,228	21,983
Germany (Western Part)	28	31	34	33	-	-	97,563	107,378	117,998	-	-	-
Ghana	-	-	-	0.054	0.047	0.049	-	-	-	12,120	10,214	10,424
Greece	*0.628	*0.683	*0.771	*0.814	*0.838	*1.056	39,023	41,988	48,785	50,559	51,737	63,208
Honduras	*0.019	*0.019	*0.020	*0.020	*0.019	*0.027	6,301	7,233	7,564	6,675	6,187	6,103
Hungary	±0.437	±0.407	±0.397	±0.463	±0.475	±0.443	16,793	16,264	17,272	17,825	18,277	16,967
Iceland	0.015	0.017	±0.021	±0.019	-	-	48,308	41,207	57,326	50,572	-	-
India	*1.647	*1.740	*1.800	*1.926	*2.339	*2.352	5,023	4,817	4,636	4,767	5,466	5,599
Indonesia	*0.535	*0.773	*0.598	*0.902	*1.201	*1.099	6,342	8,919	6,343	9,004	11,313	9,662
Iran (Islamic Republic of)	0.434	0.326	0.432	0.456	-	-	17,234	14,528	17,839	19,140	-	-
Iraq	-	*0.028	*0.074	-	-	-	-	9,023	24,544	-	-	-
Ireland	*1.718	*1.906	*2.447	*2.348	*2.951	*3.751	178,949	188,679	234,680	221,107	268,031	331,081
Israel	0.420	0.425	-	-	-	-	-	-	-	-	-	-
Italy	±3.974	±4.052	±11	±9.173	±9.422	-	78,280	78,869	90,386	76,290	80,687	-
Japan	±47	±52	±57	±66	±75	±84	210,650	230,420	245,756	284,165	310,372	349,605
Jordan	0.042	0.053	0.082	0.082	0.079	0.063	12,658	13,971	16,045	15,747	13,697	10,855
Kenya	*0.067	*0.064	*0.057	*0.034	*0.046	*0.054	7,757	7,337	6,373	3,686	4,785	5,378
Korea, Republic of	4.926	5.238	5.527	6.368	7.079	7.352	64,812	69,823	71,346	84,270	92,707	99,491
Kuwait	0.015	0.012	0.017	0.021	0.021	0.027	22,796	39,661	33,409	41,112	38,627	50,586
Latvia	-	-	-	*0.027	*0.021	*0.031	-	-	-	4,489	4,007	7,339
Luxembourg	*0.028	*0.030	*0.098	*0.087	*0.112	-	60,559	33,983	89,052	80,687	90,304	-
Macau	0.006	0.005	0.004	0.005	0.004	0.005	18,649	18,170	16,490	17,536	15,620	21,087
Malawi	*0.012	*0.006	*0.014	*0.015	*0.012	-	13,396	5,481	10,050	11,048	8,871	-
Malaysia	*0.232	*0.299	*0.360	*0.365	*0.415	*0.447	18,846	23,216	25,026	23,870	26,077	28,138
Mauritius	0.010	0.010	0.014	0.014	0.017	0.018	9,589	8,650	12,505	12,477	14,545	14,850
Mexico	2.025	2.282	2.646	2.799	3.256	2.280	30,586	33,749	39,216	44,486	51,990	36,493

Continued.

Depending on the table, * means *Enterprises*, *Engaged*, or *Factor Values*; ± means *Basis Unspecified*; § means *shown in millions*. For additional notes and sources, see Appendix I.

335

Value Added and Value Added per Employee

- Continued -

Country	Value Added ($ bil.)						Value Added per Employee ($)					
	1990	1991	1992	1993	1994	1995	1990	1991	1992	1993	1994	1995
Mongolia	0.005	0.004	0.002	0.001	0.001	0.001	6,498	5,533	3,209	996	1,930	2,282
Morocco	±0.030	-	-	-	-	-	-	-	-	-	-	-
Nepal	0.013	0.013	-	0.015	0.020	-	2,599	3,070	-	1,875	2,913	-
Netherlands	1.846	1.944	2.204	2.439	2.689	3.176	57,791	61,128	68,751	74,828	82,114	96,825
New Zealand	0.211						39,467					
Norway	±0.393	±0.359	±0.461	±0.432	±0.410	±0.491	71,619	68,065	85,794	86,620	83,192	101,658
Pakistan	0.327	0.335	0.374	-	-	-	12,285	12,445	13,281	-	-	-
Panama	0.042	0.047	0.053	0.055	0.056	0.034	24,754	27,207	27,948	28,170	28,531	29,253
Peru	0.427	0.407	0.408	-	-	-	27,571	29,065	31,720	-	-	-
Philippines	0.767	0.941	1.205	1.184			24,130	29,419	38,756	38,332		
Poland	±0.649	±0.832	±0.665	±0.732	±0.904	±1.209	11,807	14,850	11,659	13,315	17,086	23,541
Portugal	0.481	0.615	0.717	0.574	0.576	0.649	20,207	26,741	30,792	25,918	27,367	30,926
Puerto Rico	±5.334	±6.096	±6.847	±7.867	±8.158	±8.905	-	-	-	-	-	-
Qatar	0.000	0.001	0.001	0.003	0.004	-	1,616	3,763	4,116	14,652	12,025	-
Romania	±0.441	±0.357	±0.319	±0.245								
Russian Federation	-	-	-	±0.693	±1.141	±1.368	-	-	-	3,501	6,766	7,049
Saint Lucia	-	±0.000	±0.000	±0.001	±0.001	±0.001	-	-	-	-	-	-
Senegal	0.024	0.019	0.029	0.019	0.020	-	14,665	14,241	19,075	24,702	22,101	-
Singapore	*0.600	*0.861	*0.915	*0.982	*1.028	*1.159	116,136	162,940	166,116	180,408	173,562	195,335
Slovakia				±0.098	±0.098	-				12,919	12,768	
Slovenia	-	-	-	0.172	0.216	0.376	-	-	-	-	-	-
South Africa	*1.254	*1.295	*1.318	*1.306	*1.329	*1.498	21,258	21,236	21,613	-	-	-
Spain	*5.609	*5.900	*6.454	±5.142	±5.013	±5.739	67,848	69,950	79,083	55,639	56,453	65,479
Sri Lanka	0.033	0.041	0.052	0.061	-	-	7,963	8,443	12,993	13,182	-	-
Sweden	*2.544	*1.998	*2.394	*2.099	*2.525	*2.860	106,905	95,648	121,735	116,566	124,274	140,680
Thailand	0.289	0.802	-	-	-	-	13,064	23,848	-	-	-	-
Trinidad and Tobago	±0.023	±0.017	±0.016	±0.017	±0.013	-	16,288	12,327	12,268	11,947	9,572	-
Tunisia	0.095	0.098	0.111	0.118	0.141	0.162	-	-	-	30,268	29,141	30,028
Turkey	1.449	1.614	1.867	2.403	1.853	2.761	52,675	61,101	69,162	87,974	71,628	100,706
United Kingdom	*15	*16	*17	*16	*17	*20	87,605	93,265	105,219	93,353	101,232	114,103
United Republic of Tanzania	*0.003	*0.007	-	-	-	-	1,060	2,175	-	-	-	-
United States of America	*82	*87	*95	*101	*106	*111	168,598	182,296	195,888	206,971	217,928	218,532
Uruguay	0.162	0.172	0.206	0.249	0.277	0.277	23,533	28,515	35,757	43,903	48,801	48,728
Venezuela	0.662	0.789	0.961	0.673	0.517	1.736	23,557	25,541	31,007	23,293	19,245	68,048
Yugoslavia	±0.051	±0.084	±0.302	-	±0.370	±0.211	-	-	4,684	17,331	21,508	12,176
Zambia	0.043	-	-	-	0.033	-	14,640				12,826	
Zimbabwe	*0.127	*0.125	*0.085	*0.047	*0.087	*0.089	21,889	21,136	15,451	8,952	16,372	18,040

Capital Investment

Country	Gross Fixed Capital Formation ($ mil.)						Per Establishment ($ mil)					
	1990	1991	1992	1993	1994	1995	1990	1991	1992	1993	1994	1995
Argentina	-	-	-	209	-	-	-	-	-	0.122	-	-
Australia	141	-	-	-	-	-	0.243	-	-	-	-	-
Austria	239	223	267	289	268	-	0.846	0.795	1.037	1.142	1.106	-
Bahrain	-	-	0.176	-	-	-	-	-	0.012	-	-	-
Bangladesh	14	17	13	-	-	-	0.023	0.034	0.025	-	-	-
Barbados	0.347	0.046	1.367	0.199	1.242	-	0.069	0.015	0.195	0.040	0.103	-
Bolivia	1.881	1.699	2.388	2.315	-0.923	-	0.032	0.029	0.040	0.041	-0.016	-
Bosnia & Herzegovina	5.376	-	-	-	-	-	0.283	-	-	-	-	-
Bulgaria	38	5.970	221	78	13	11	0.722	0.117	4.095	0.303	0.033	0.028
Canada	388	-	-	-	-	-	0.336	-	-	-	-	-
Cape Verde	0.000	0.000	-	-	-	-	0.000	0.000	-	-	-	-
Chile	30	27	32	65	54	-	0.313	0.280	0.317	0.617	0.514	-
China (Hong Kong SAR)	-	-28.438	-	-	-	-	-	-0.066	-	-	-	-
Colombia	29	37	58	41	51	-	0.087	0.115	0.166	0.114	0.142	-
Croatia	20	23	52	16	18	29	0.204	0.180	0.255	0.059	0.055	0.077
Cyprus	7.512	5.868	6.260	5.718	7.282	6.079	0.073	0.054	0.059	0.053	0.064	0.053

Continued.

Depending on the table, * means *Enterprises, Engaged,* or *Factor Values;* ± means *Basis Unspecified;* § means *shown in millions.* For additional notes and sources, see Appendix I.

Capital Investment

- Continued -

Country	Gross Fixed Capital Formation ($ mil.)						Per Establishment ($ mil)					
	1990	1991	1992	1993	1994	1995	1990	1991	1992	1993	1994	1995
Czechoslovakia (Former)	50	-	-	-	-	-	2.066	-	-	-	-	-
Denmark	171	204	-	-	-	-	0.436	0.482	-	-	-	-
Ecuador	24	54	36	1.678	27	42	0.308	0.647	0.433	0.020	0.319	0.486
Egypt	62	66	15	68	-	-	0.377	0.355	0.086	0.382	-	-
El Salvador	-	6.064	-	6.188	6.145	5.838	-	0.088	-	0.121	0.123	0.104
Ethiopia	1.060	1.082	0.237	0.167	0.778	2.080	0.096	0.108	0.026	0.017	0.035	0.080
Fiji	5.515	0.660	0.470	-	-	-	0.212	0.035	0.023	-	-	-
Finland	132	82	67	40	61		1.432	0.742	0.643	0.375	0.593	-
FYR Macedonia	0.729	3.728	2.497	1.178	2.488	8.479	0.091	0.196	0.078	0.026	0.041	0.107
France	1,522	1,670	2,042	1,673	1,781	2,189	-	-	-	-	-	-
Gabon	-	0.610	0.797	0.858	0.077	2.292	-	0.152	0.199	0.215	0.019	0.573
Germany (Western Part)	2,468	2,573	3,319	2,906			1.928	1.996	2.536	-	-	-
Greece	42	47	56				0.151	0.169	0.196	-	-	-
Hungary	47	50	92	95			0.579	0.328	0.324	0.290	-	-
India	502	389	588	670	-	-	0.107	0.077	0.106	0.111	-	-
Indonesia	124	333	163	216	167	195	0.221	0.629	0.299	0.380	0.288	0.323
Iran (Islamic Republic of)	54	51	65	49	-	-	0.153	0.289	0.342	0.250	-	-
Italy	607	615	1,558	1,165	1,219	-	1.264	1.303	1.791	1.374	1.431	-
Japan	4,075	4,922	5,077	5,782	4,902		1.165	1.398	1.465	1.656	1.278	-
Jordan	2.078	5.859	8.691	4.580	15	6.747	0.052	0.120	0.100	0.051	0.108	0.046
Korea, Republic of	774	1,788	887	1,372	1,276	1,857	0.729	1.644	0.766	1.054	0.958	1.331
Kuwait	0.736	2.394	0.949	1.242	0.983		0.049	0.200	0.063	0.078	0.066	-
Latvia	0.123	0.342	0.533	1.673	4.003	8.343	0.011	0.007	0.008	0.024	0.095	0.160
Luxembourg	102	52	15	-	-	-	9.271	4.736	1.394	-	-	-
Macau	0.229	0.907	0.444	-11.515	0.610	0.323	0.016	0.082	0.040	-1.151	0.061	0.032
Malawi	2.748	1.427	1.166	1.317	0.767	-	0.393	0.204	0.167	0.188	0.110	-
Malaysia	54	55	56	94	84	-	0.360	0.364	0.362	0.576	0.511	-
Mexico	127	132	-	-	-	-	0.533	0.563	-	-	-	-
Mongolia	20	15	3.943	0.483	1.257	-	2.901	2.086	0.563	0.069	0.097	-
Nepal	18	-	-	-	-	-	0.216	-	-	-	-	-
Netherlands	397	640	547	368	-	-	2.482	4.154	3.315	2.232	-	-
New Zealand	27	-	-	-	-	-	0.115	-	-	-	-	-
Norway	82	71	59	52	59	-	1.097	0.889	0.983	1.062	1.179	-
Pakistan	-	36	-	-	-	-	-	0.127	-	-	-	-
Panama	1.541	1.670	-	-	-	-	0.032	0.037	-	-	-	-
Peru	11	18	16	-	-	-	0.014	0.023	0.021	-	-	-
Philippines	69	55	73	82		-	0.168	0.117	0.222	0.282	-	-
Poland	90	109	100	141	-	-	0.636	0.593	0.589	0.834	-	-
Portugal	82	88	130	96	101	-	0.115	0.127	0.178	0.136	0.144	-
Romania	2.184	3.233	-	20	45	-	-	-	-	-	0.057	-
Senegal	4.143	-	-	-	-	-	0.218	-	-	-	-	-
Singapore	57	57	53	43	80	-	0.613	0.610	0.542	0.449	0.800	-
Slovenia	-	-	-	56	45	112	-	-	-	0.593	0.413	0.672
Spain	578	711	660	597	574	611	0.304	0.367	0.346	-	-	-
Sri Lanka	3.245	4.503	4.784	5.605	-	-	0.048	0.074	0.117	0.077	-	-
Thailand	141	302	-	-	-	-	0.503	0.969	-	-	-	-
Trinidad and Tobago	5.272	11	1.506	2.336	2.684	-	0.073	0.171	0.023	0.034	0.046	-
Tunisia	18	18	23	15	12	25	-	-	-	0.079	0.054	0.120
Turkey	151	193	126	196	106	-	0.870	1.135	0.475	0.738	0.415	-
Ukraine	51	35	41	4.402	6.300	4.337	0.333	0.089	0.088	0.029	0.040	0.028
United Kingdom	1,577	1,595	1,837	-	1,692	-	0.902	0.957	1.073	-	0.771	-
United Republic of Tanzania	18	27					0.463	0.769	-	-	-	-
United States of America	4,510	5,240	6,629	6,768	6,789	7,220	-	-	0.610	-	-	-
Uruguay	7.165	12	11	13	-	-	-	-	-	-	-	-
Venezuela	39	58	57	65	24	221	0.134	0.198	0.194	0.242	0.094	0.796
Yugoslavia	1.296	1.203	7.977	-	52	36	0.012	0.006	0.016	-	0.068	0.044
Zambia	8.425	-	-	-	-	-	0.337	-	-	-	-	-
Zimbabwe	12	24	16	12	32	-	0.262	0.481	0.309	0.236	0.688	-

Indexes of Industrial Production

Country	Index of Industrial Production (1990=100)						Index of Employment (1990=100)					
	1990	1991	1992	1993	1994	1995	1990	1991	1992	1993	1994	1995
Albania	100	-	-	-	-	-	-	-	-	-	-	-
Algeria	100	97	80	83	97	89	-	-	-	-	-	-
Angola	100	-	-	-	-	-	-	-	-	-	-	-
Argentina	100	119	-	-	-	-	100	-	-	113	116	-
Armenia	100	-	-	-	-	-	100	89	-	-	-	-
Australia	100	100	101	110	-	-	100	103	100	-	-	-
Austria	100	101	101	98	106	114	100	96	96	94	92	91
Azerbaijan	-	-	-	-	-	-	100	93	78	73	82	-
Bahamas	100	-	-	-	-	-	100	101	-	-	-	-
Bahrain	100	-	-	-	-	-	-	-	-	-	-	-
Bangladesh	100	105	125	152	165	174	100	93	92	-	-	-
Barbados	100	96	76	73	68	94	100	53	96	90	149	130
Belgium	100	100	112	108	107	112	-	-	-	-	-	-
Benin	100	-	-	-	-	-	-	-	-	-	-	-
Bhutan	100	-	-	-	-	-	-	-	-	-	-	-
Bolivia	100	102	101	99	106	109	100	105	110	104	122	126
Bosnia & Herzegovina	-	-	-	-	-	-	100	93	-	-	19	-
Brazil	100	92	92	96	102	102	-	-	-	-	-	-
Bulgaria	100	82	68	60	82	96	100	82	77	72	70	66
Burkina-Faso	100	177	94	-	-	-	-	-	-	-	-	-
Burundi	100	121	115	-	-	-	100	101	-	-	-	-
Cambodia	100	-	-	-	-	-	100	87	-	-	-	-
Cameroon	100	-	-	-	-	-	100	117	123	110	-	-
Canada	100	95	96	99	100	99	100	97	100	97	94	94
Cape Verde	-	-	-	-	-	-	100	102	100	152	158	-
Central African Republic	100	-	-	-	-	-	100	-	-	-	-	-
Chile	100	106	121	130	139	157	100	101	105	117	112	123
China	100	-	-	-	-	-	100	104	108	105	-	-
China (Hong Kong SAR)	100	94	90	81	74	70	100	82	79	76	76	-
China (Taiwan Province)	100	107	116	124	137	145	100	98	102	107	111	112
Colombia	100	101	108	114	119	121	100	111	132	147	146	147
Costa Rica	100	112	126	157	195	171	100	102	108	106	-	-
Cote d'Ivoire	100	98	96	101	120	119	-	-	-	-	-	-
Croatia	100	-	-	-	-	-	100	90	82	80	77	74
Cyprus	100	95	99	97	108	107	100	106	106	115	119	121
Czechoslovakia (Former)	100	69	68	-	-	-	100	70	-	-	-	-
Czech Republic	100	-	-	-	-	-	100	92	108	108	-	-
Dem. Rep. of the Congo	100	80	-	-	-	-	-	-	-	-	-	-
Denmark	100	102	110	108	120	133	100	101	92	91	95	100
Dominican Republic	100	-	-	-	-	-	-	-	-	-	-	-
Ecuador	100	121	128	132	-	-	100	119	96	92	92	96
Egypt	100	101	84	91	95	91	100	103	107	106	108	106
El Salvador	100	105	-	-	-	-	-	-	-	-	-	-
Equatorial Guinea	-	-	-	-	-	-	100	-	-	-	-	-
Ethiopia and Eritrea	100	-	-	-	-	-	-	-	-	-	-	-
Ethiopia	-	-	-	-	-	-	100	98	99	103	120	158
Fiji	100	107	111	117	115	108	100	109	101	106	104	-
Finland	100	98	92	91	102	102	100	99	90	90	90	-
FYR Macedonia	100	-	-	-	-	-	100	87	89	75	60	72
France	100	104	111	114	122	126	100	100	99	97	96	96
Gabon	100	112	111	-	-	-	-	-	-	-	-	-
Gambia	100	-	-	-	-	-	-	-	-	-	-	-
Germany (Eastern Part)	100	-	-	-	-	-	-	-	-	-	-	-
Germany (Western Part)	100	107	110	104	108	106	100	101	101	-	-	-
Ghana	100	-	-	-	-	-	-	-	-	-	-	-
Greece	100	96	99	105	107	118	100	101	98	100	101	104
Guatemala	100	-	-	-	-	-	-	-	-	-	-	-
Guyana	100	-	-	-	-	-	-	-	-	-	-	-
Haiti	100	121	113	110	129	-	-	-	-	-	-	-
Honduras	100	146	173	194	219	310	100	87	88	97	102	146

Continued.

Depending on the table, * means *Enterprises, Engaged,* or *Factor Values*; ± means *Basis Unspecified*; § means *shown in millions.* For additional notes and sources, see Appendix I.

Indexes of Industrial Production

- Continued -

Country	Index of Industrial Production (1990=100)						Index of Employment (1990=100)					
	1990	1991	1992	1993	1994	1995	1990	1991	1992	1993	1994	1995
Hungary	100	77	59	60	61	56	100	96	88	100	100	100
Iceland	100	-	-	-	-	-	100	136	121	125	117	-
India	100	95	89	93	105	101	100	110	118	123	130	128
Indonesia	100	96	133	181	184	170	100	103	112	119	126	135
Iran (Islamic Republic of)	100	101	105	107	149	161	100	89	96	95	-	-
Iraq	100	-	-	-	-	-	-	-	-	-	-	-
Ireland	100	122	143	157	188	217	100	105	109	111	115	118
Israel	100	104	115	127	140	145	-	-	-	-	-	-
Italy	100	101	104	-	-	-	100	101	248	237	230	-
Jamaica	100	73	89	-	-	-	100	98	96	89	90	-
Japan	100	102	103	101	106	112	100	102	105	105	109	109
Jordan	100	90	81	71	66	59	100	115	154	157	173	175
Kenya	100	129	122	119	115	124	100	101	103	108	112	116
Korea, Republic of	100	116	138	152	165	176	100	99	102	99	100	97
Kuwait	100	-	-	-	-	-	100	47	78	79	85	83
Kyrgyzstan	100	-	-	-	-	-	100	61	52	67	45	-
Lao P.D.R.	100	-	-	-	-	-	-	-	-	-	-	-
Latvia	100	-	-	-	-	-	100	98	96	71	61	50
Lesotho	100	-	-	-	-	-	100	-	32	31	31	-
Liberia	100	-	-	-	-	-	-	-	-	-	-	-
Libyan Arab Jamahiriya	100	-	-	-	-	-	-	-	-	-	-	-
Lithuania	100	-	-	-	-	-	-	-	-	-	-	-
Luxembourg	100	105	121	128	156	153	100	191	237	234	269	-
Macau	100	-	-	-	-	-	100	89	82	84	84	79
Madagascar	100	90	75	-	-	-	-	-	-	-	-	-
Malawi	100	-	-	-	-	-	100	122	156	156	156	-
Malaysia	100	111	126	138	159	159	100	105	117	124	129	129
Malta	100	121	139	160	-	-	-	-	-	-	-	-
Mauritius	100	-	-	-	-	-	100	101	102	106	107	111
Mexico	100	106	109	106	112	109	100	102	102	95	95	94
Mongolia	-	-	-	-	-	-	100	88	81	65	66	70
Morocco	100	102	111	109	129	145	-	-	-	-	-	-
Mozambique	100	-	-	-	-	-	100	103	92	93	92	-
Myanmar	100	116	108	-	-	-	-	-	-	-	-	-
Nepal	100	-	-	-	-	-	100	83	-	151	133	-
Netherlands	100	97	98	99	108	111	100	100	100	102	103	103
New Zealand	100	-	-	-	-	-	100	104	115	114	-	-
Nicaragua	100	-	-	-	-	-	-	-	-	-	-	-
Niger	100	-	-	-	-	-	-	-	-	-	-	-
Nigeria	100	-	-	-	-	-	-	-	-	-	-	-
Norway	100	95	100	100	107	112	100	96	98	91	90	88
Pakistan	100	129	163	-	-	-	100	101	106	-	-	-
Panama	100	111	123	127	128	75	100	101	111	114	115	67
Papua New Guinea	100	-	-	-	-	-	-	-	-	-	-	-
Paraguay	100	108	82	107	105	107	-	-	-	-	-	-
Peru	100	109	114	122	143	185	100	90	83	-	-	-
Philippines	100	109	117	136	147	168	100	101	98	97	-	-
Poland	100	110	97	103	121	137	100	102	104	100	96	93
Portugal	100	83	76	69	69	68	100	97	98	93	89	88
Puerto Rico	100	-	-	-	-	-	-	-	-	-	-	-
Qatar	100	-	-	-	-	-	100	129	157	132	175	-
Republic of Moldova	100	-	-	-	-	-	100	98	83	87	79	73
Romania	100	75	43	-	-	-	-	-	-	-	-	-
Russian Federation	100	-	-	-	-	-	-	-	-	-	-	-
Saudi Arabia	100	-	-	-	-	₁	-	-	-	-	-	-
Senegal	-	-	-	-	-	-	100	80	94	47	56	-
Singapore	100	131	120	124	128	129	100	102	107	105	115	115
Slovakia	100	-	-	-	-	-	-	-	-	-	-	-
Slovenia	100	90	61	61	71	71	-	-	-	-	-	-
Somalia	100	-	-	-	-	-	-	-	-	-	-	-

Continued.

Depending on the table, * means *Enterprises*, *Engaged*, or *Factor Values*; ± means *Basis Unspecified*; § means *shown in millions*. For additional notes and sources, see Appendix I.

339

Indexes of Industrial Production

- Continued -

Country	Index of Industrial Production (1990=100)						Index of Employment (1990=100)					
	1990	1991	1992	1993	1994	1995	1990	1991	1992	1993	1994	1995
South Africa	100	100	100	99	101	106	100	103	103	-	-	-
Spain	100	95	95	94	107	110	100	102	99	112	107	106
Sri Lanka	100	88	98	137	135	163	100	117	97	113	-	-
Sudan	100	-	-	-	-	-	-	-	-	-	-	-
Suriname	100	-	-	-	-	-	100	93	101	103	-	-
Sweden	100	114	126	134	133	136	100	88	83	76	85	85
Switzerland	100	101	106	113	130	143	-	-	-	-	-	-
Syrian Arab Republic	100	99	100	106	110	116	-	-	-	-	-	-
Thailand	100	105	-	-	-	-	100	152	-	-	-	-
Tonga	100	-	-	-	-	-	100	107	114	86	100	-
Trinidad and Tobago	100	99	95	82	83	75	100	95	92	99	97	-
Tunisia	100	102	102	103	116	122	-	-	-	-	-	-
Turkey	100	111	110	121	103	133	100	96	98	99	94	100
Uganda	100	105	137	185	209	279	-	-	-	-	-	-
Ukraine	100	-	-	-	-	-	100	93	90	83	75	73
United Arab Emirates	100	-	-	-	-	-	-	-	-	-	-	-
United Kingdom	100	102	107	112	117	123	100	98	98	102	101	101
United Republic of Tanzania	100	106	93	98	97	-	100	108	-	-	-	-
USSR (Former)	100	-	-	-	-	-	100	-	-	-	-	-
United States of America	100	102	106	109	112	115	100	99	100	101	100	104
Uruguay	100	97	97	98	105	97	100	88	84	83	82	83
Venezuela	100	-	-	-	-	-	100	110	110	103	96	91
Yugoslavia	100	-	-	-	-	-	-	-	-	-	-	-
Yugoslavia (Former)	100	73	-	-	-	-	-	-	-	-	-	-
Zambia	100	101	96	90	68	60	100	-	-	-	88	-
Zimbabwe	100	100	87	82	94	84	100	102	95	91	91	85

Depending on the table, * means *Enterprises, Engaged,* or *Factor Values;* ± means *Basis Unspecified;* § means *shown in millions.* For additional notes and sources, see Appendix I.

Representative Companies in Sector

Name	Address	Tele-phone	Fax	Employ-ment	Y
DiverseyLever	1500-1 Robert Speck Pky, Mississauga, Ontario L4Z 3S9, Canada	905-897-5600	905-897-5640	13,000	98
Olin Corp.	PO Box 4500, Norwalk, Connecticut 06856-4500, U.S.A.	203-750-3000	203-750-3595	10,000	97
Estee Lauder Inc.	767 5th Ave., New York, New York 10153, U.S.A.	212-572-4200	212-572-4292	9,900	95
General Nutrition Companies Inc.	921 Penn Ave., Pittsburgh, Pennsylvania 15222, U.S.A.	412-288-4600	412-288-4764	9,610	95
Alcon Laboratories Inc.	6201 South Fwy., Fort Worth, Texas 76134-2099, U.S.A.	817-293-0450	817-551-4629	9,500	97
Akzo Nobel Inc.	300 S. Riverside Plaza, Chicago, Illinois 60606, U.S.A.	312-906-9584	312-906-7811	9,000	96
Novo Nordisk AS	1191 Corniche El Nil, Cairo, Egypt	2 773665	2 773894	9,000	91
G.D. Searle and Co.	5200 Old Orchard Rd., Skokie, Illinois 60077, U.S.A.	847-982-7000	847-470-1480	8,500	97
Glaxo Wellcome Inc.	PO Box 13398, Research Triangle Park, North Carolina 27709, U.S.A.	919-483-2100		8,400	97
Zaklady Azotowe W Tarnowie-Moscicach SA	ul E. Kwiatkowskiego 8, Tarnow PL-33-101, Poland	14 330781	14 330718	8,205	93
IVAX Corp.	4400 Biscayne Blvd., Miami, Florida 33137, U.S.A.	305-590-2200		8,162	96
Mallinckrodt Inc.	PO Box 5840, St. Louis, Missouri 63134, U.S.A.	314-654-2000		8,000	97
Schering-Plough Pharmaceuticals	2000 Galloping Hill Rd., Kenilworth, New Jersey 07033, U.S.A.	908-298-4000		7,325	
Sigma-Aldrich Corp.	3050 Spruce St., St. Louis, Missouri 63103, U.S.A.	314-771-5765	314-534-2674	6,666	97
RPM Inc. (Medina, Ohio)	PO Box 777, Medina, Ohio 44258, U.S.A.	330-273-5090	330-225-8743	6,651	97
CCL Industries Inc.	800-105 Gordon Baker Rd., Willowdale, Ontario M2H 3P8, Canada	416-756-8500	416-756-8555	6,500	98
Baxter Healthcare Corp. of Puerto Rico	PO Box 2025, Carolina 00984, Puerto Rico	787750-7373		6,000	92
Gessy Lever Ltda. Industrial	Av. Maria Coelho Aguiar 215 Bl.C 7th Fl., Sao Paulo 05855-000, Brazil	11 5452956	11 5452783	6,000	97
H.B. Fuller Co.	PO Box 64683, St. Paul, Minnesota 55164-0683, U.S.A.	612-645-3401		6,000	97
Orion Corp.	Orionintie 1, Espoo SF-02101, Finland	4291	4293815	5,900	94
Teva Pharmeceutical Industries Ltd.	Har Hatotzvim Industrial Park, Jerusalem 97774, Israel	3 9267267	3 9234050	5,900	97
Insilco Corp.	425 Metro Pl. N., 5th Fl., Dublin, Ohio 43017, U.S.A.	614-792-0468	614-791-3197	5,764	96
Crompton and Knowles Corp.	1 Station Pl., Metro Ctr., Stamford, Connecticut 06902, U.S.A.	203-353-5400	203-353-5424	5,665	96
Amgen Inc.	1840 Dehavilland Dr., Thousand Oaks, California 91320-1789, U.S.A.	805-447-1000	805-499-3507	5,308	97
Clorox Co.	1221 Broadway, Oakland, California 94612, U.S.A.	510-271-7291	510-832-1463	5,300	97
Ciba-Geigy PLC	Hulley Rd., Macclesfield SK10 2NX, United Kingdom	625 421933	625 619637	5,085	93
Minnesota Mining & Manufacturing Co. UK Holdings PLC	Market Pl., Bracknell RG12 1JU, United Kingdom	344 858220	344 858278	5,012	94
Bayer Corp. Diagnostics Div.	511 Benedict Ave., Tarrytown, New York 10591, U.S.A.	914-631-8000	914-524-2132	4,900	97
Dexter Corp.	1 Elm St., Windsor Locks, Connecticut 06096, U.S.A.	860-292-7675	860-292-7673	4,800	97
Ohmeda Inc.	110 Allen Rd., Liberty Corner, New Jersey 07938, U.S.A.	908-647-9200		4,800	96
Paterson Zochonis PLC	60 Whitworth St., Manchester M1 6LU, United Kingdom	61 2367111	61 2286719	4,377	94
Universal Foods Corp.	433 E. Michigan St., Milwaukee, Wisconsin 53202-5106, U.S.A.	414-271-6755	414-347-3785	4,127	97
Perrigo Co.	117 Water St., Allegan, Michigan 49010, U.S.A.	616-673-8451	616-673-9122	4,122	97
Ares Serono SA	2 chemin des Mines, Geneva CH-1202, Switzerland			3,935	93
Unilever Canada Ltd.	160 Bloor St. E Ste. 1500, Toronto, Ontario M4W 3R2, Canada	416-964-1857	416-964-8831	3,816	98
Block Drug Company Inc.	257 Cornelison Ave., Jersey City, New Jersey 07302-9988, U.S.A.	201-434-3000	201-434-3032	3,600	95
Genzyme Corp.	1 Kendall Sq., Cambridge, Massachusetts 02139, U.S.A.	617-252-7500	617-374-7368	3,516	96
Mary Kay Inc.	16251 N. Dallas Pkwy., Dallas, Texas 75248-2696, U.S.A.	214-630-8787	214-905-5931	3,500	95
Carter-Wallace Inc.	1345 Ave. of the Amer., New York, New York 10105, U.S.A.	212-339-5000	212-339-5100	3,460	96
St. John Knits Inc.	17422 Derian Ave., Irvine, California 92713, U.S.A.	714-223-3301		3,255	97
Genentech Inc.	460 Point San Bruno Blvd., South San Francisco, California 94080-4990, U.S.A.	415-225-1000	415-266-2501	3,242	97
Bayer do Brasil SA	Rua Domingos Jorge 1000, Sao Paulo 04779-900, Brazil	11 5255166	11 5255615	3,200	96
Unilever De Argentina SA	Fraga 1163, Buenos Aires 1427, Argentina	1 5552555	1 3094800	3,200	96
Valspar Corp.	1101 3rd St. S., Minneapolis, Minnesota 55415, U.S.A.	612-332-7371	612-375-7723	3,200	97
Glaxo Operations UK Ltd.	Hamre Rd. St., Barnard Castle DL12 8DT, United Kingdom	833 6906004		3,171	94
AB Wilh. Becker	Lovholmsgrand 12, Stockholm S-117 83, Sweden	8 7756400	8 7442181	3,050	93
Rhone-Poulenc Rorer Holdings Ltd.	St. Leonard's Rd., Eastbourne BN21 3YG, United Kingdom	323 214222	323 32362	3,039	93

Continued

Representative Companies in Sector
- Continued -

Name	Address	Tele-phone	Fax	Employ-ment	Y
Akzo Nobel Chemicals Inc.	300 S. Riverside Plz., 2200, Chicago, Illinois 60606, U.S.A.	312-906-7500	312-906-7680	3,000	97
Marmoran Pty. Ltd.	16 Turfhall Rd., Lansdanne 7780, Republic of South Africa	21 7974250	21 7970165	3,000	94
Osothsapha Teck Heng Yoo Co. Ltd.	Huamark, Bangkok 10240, Thailand	3740121	3747014	3,000	94
Procter & Gamble Inc.	4711 Yonge St., North York, Ontario M2N 6K8, Canada	416-730-4711	416-730-4415	3,000	98
Robert McBride Ltd.	Middleton, Manchester M24 4DP, United Kingdom	61 6539037	61 6552278	2,965	92
Polfa	Starogard, Gdansk PL-83-200, Poland	58 23401	58 22353	2,902	93
Baxter Healthcare Corp. Baxter Diagnostics Inc.	PO Box 520672, Miami, Florida 33152, U.S.A.	305-592-2311		2,840	
Akzo Nobel Coatings Inc.	PO Box 37230, Louisville, Kentucky 40233, U.S.A.	502-367-6111	502-375-5475	2,800	96
Ancock Ingram International	Private Bag 1, Industria 2042, Republic of South Africa	11 4707000	11 4744823	2,800	93
Natura	Rua Amador Bueno 491, Santo Amaro 04752-900, Brazil	11 5477655	11 5243339	2,750	98
Alpha Therapeutic Corp.	5555 Valley Blvd., Los Angeles, California 90032, U.S.A.	213-225-2221	213-227-7027	2,700	97
Plascon Paints Natal Pty. Ltd.	PO Box 1386, Durban 4000, Republic of South Africa	31 4690841	31 420376	2,700	94
Glaxochem Ltd.	Greenford, London UB6 0HE, United Kingdom	81 4223434	81 4234232	2,699	94
Alpharma Inc.	PO Box 1399, Fort Lee, New Jersey 07024, U.S.A.	201-947-7774	201-947-4879	2,600	97
Krka, Tovarna Zdravil d.d.	Smarjeska 6, Novo Mesto SLO-61526, Slovenia	68 312111	68 321537	2,575	95
Dial Corp.	15501 N. Dial Blvd., Scottsdale, Arizona 85260-1619, U.S.A.	602-754-3425		2,533	97
Nobel Industries Sweden (UK) Ltd.	23 Grosvenor Rd., St. Albans AL1 3AW, United Kingdom	727 41421	727 41529	2,502	91
Apotex Inc.	150 Signet Dr., Weston, Ontario M9L 1T9, Canada	416-749-9300	416-749-9578	2,500	98
Bio-Rad Laboratories Inc.	1000 Alfred Nobel Dr., Hercules, California 94547, U.S.A.	510-724-7000	510-741-5817	2,500	96
Leo Pharmaceutical Products	Industriparken 55, Ballerup DK-2750, Denmark	44923800	44941516	2,500	93
Sherwin-Williams Diversified Brands Inc.	31500 Solon Rd., Solon, Ohio 44139-3908, U.S.A.	440-498-2300		2,500	97
Hoechst UK Ltd.	Salisbury Rd., Hounslow TW4 6JH, United Kingdom	81 5707712	81 5771854	2,479	93
Peter Black Holdings PLC	Lawkholme Ln., Keighley BD21 3JQ, United Kingdom	535 661177	535 609973	2,460	94
Bristol-Myers Squibb Holdings Ltd.	Ickenham, Uxbridge UB10 8NS, United Kingdom	895 639911	895 636975	2,459	93
Body Shop International PLC	Watersmead Business Park, Littlehampton BN17 6LF, United Kingdom	903 731500	903 726250	2,456	94
North American Biologicals Inc.	PO Box 310701, Boca Raton, Florida 33431-0701, U.S.A.	407-989-5800	407-989-5801	2,400	96
East Asiatic Co. Malaysia BHD	1 Jalan 205, Petaling Jaya 40650, Malaysia	3 7914233	3 7919416	2,300	
Medeva PLC	10 St. James St., London SW1A 1EF, United Kingdom	71 8393888		2,221	93
Ache Laboratorios Farmaceuticos SA	Rodovia Presidente Dutra Km. No. 227, Guarulhos 07034-090, Brazil	11 9616000	11 9408418	2,200	96
Tejicondor SA Tejidos El Condor	Carrera 65 No. 44-79, Medellin, Colombia	4 2600999	4 2603646	2,200	96
Reckitt & Colman Products Ltd.	1 Burlington Ln., London W4 2RW, United Kingdom	81 9946464	81 9948920	2,132	94
Alkaloida Chemical Co.	Kabay Janos u. 29, Tiszavasvari H-4440, Hungary	42 72511	42 72512	2,130	93
Lilly Industries Ltd.	Kingsclere Rd., Basingstoke RG21 2XA, United Kingdom	256 473241		2,116	93
Bush Boake Allen Inc.	7 Mercedes Dr., Montvale, New Jersey 07645, U.S.A.	201-391-9870	201-391-0860	2,103	96
IDEXX Laboratories Inc.	1 IDEXX Dr., Westbrook, Maine 04092, U.S.A.	207-856-0300	207-856-0346	2,100	97
Instituto Nacional de Seguro	Calles 9-9B, San Jose 1000, Costa Rica	2235800	2553427	2,100	
Lilly Industries Inc.	733 S. West St., Indianapolis, Indiana 46225, U.S.A.	317-687-6700	317-687-6710	2,100	97
Avon Cosmetics Ltd.	Nun Mills Rd., Northampton NN1 5PA, United Kingdom	604 232425	604 232444	2,076	93
Dong-A Pharmaceutical Co., Ltd.	Tongdaemun-Gu, Seoul, Republic of Korea	2 9208114	2 9242662	2,000	97
May & Baker Exports Ltd.	Rainham Rd. S., Dagenham RM10 7XS, United Kingdom	81 5923060		2,000	90
Reckitt and Colman Inc.	1655 Valley Rd., Wayne, New Jersey 07474-0943, U.S.A.	973-633-3600	973-633-3734	2,000	97
Slovakfarma AS	Zelenicna 12, Hlohovec 920 27, Slovakia	804 24752	804 21241	2,000	93
Zimco Industries Pty. Ltd.	PO Box 78069, Sandton 2146, Republic of South Africa	11 7837023	11 7830732	2,000	95
Warner-Lambert UK Ltd.	Chestnut Ave., Eastleigh SO5 3ZQ, United Kingdom	703 620500	703 629811	1,991	93
Benjamin Moore and Co.	51 Chestnut Ridge Rd., Montvale, New Jersey 07645, U.S.A.	201-573-9600	201-573-0048	1,968	95
Glaxo Pharmaceuticals UK Ltd.	Stackley Pk. W., Uxbridge UB11 1BT, United Kingdom	81 9909000		1,935	94
Sandoz Holdings Great Britain Ltd.	27 Austin Friars, London EC2N 2AA, United Kingdom	71 6286839		1,907	93
Cipla Ltd.	289 J B B Marg Bombay Central, Bombay 400 008, India	22 3082891	22 3070013	1,900	94
Schein Pharmaceutical Inc.	100 Campus Dr., Florham Park, New Jersey 07932, U.S.A.	973-593-5500	973-593-5590	1,900	96
Lever Brothers Ltd.	3 St. James' Rd., Kingston upon Thames KT1 2BA, United Kingdom	81 5418200	81 4591422	1,872	93
Taris Pamuk Tarim Satis Koop. Birligi	Alsancak, Izmir TR-35249, Turkey	51 217065	51 636555	1,863	92
Kiwi Holdings Ltd.	225 Bath Rd., Slough SL1 4AU, United Kingdom			1,832	91
Grupo Billi	Av. Vereador Jose Diniz 3465, Sao Paulo 04603-003, Brazil	11 5363688	11 5310277	1,800	96
Turtle Wax Inc.	5655 W. 73rd St., Chicago, Illinois 60638, U.S.A.	708-563-3600	708-563-4302	1,800	97

Continued

For sources and notes, see Appendix I.

Representative Companies in Sector
- Continued -

Name	Address	Tele-phone	Fax	Employ-ment	Y
Wesley-Jessen Co.	400 W. Superior St., Chicago, Illinois 60610, U.S.A.	312-751-6357		1,800	93
Scholl PLC	2-4 Sheet St., Windsor SL4 1BG, United Kingdom	753 833444		1,786	93
Mylan Laboratories Inc.	130 7th St., 1030 Century Bldg., Pittsburgh, Pennsylvania 15222, U.S.A.	412-232-0100		1,750	96
Parke Davis & Co. Ltd.	Chestnut Ave., Eastleigh SO5 3ZQ, United Kingdom	703 620500	703 629818	1,733	92
Alza Corp.	PO Box 10950, Palo Alto, California 94303-0802, U.S.A.	415-494-5000	415-494-5151	1,700	97
Lord Corp.	2000 W. Grandview Blvd., Erie, Pennsylvania 16514, U.S.A.	814-868-3611	814-864-3452	1,700	96
Unijax Div.	7785 Baymeadows way, 200, Jacksonville, Florida 32256, U.S.A.	904-783-0550		1,700	97
Forest Laboratories Inc.	909 3rd Ave., New York, New York 10022, U.S.A.	212-421-7850		1,663	96
Bayer PLC	Strawberry Hill, Newbury RG13 1JA, United Kingdom	635 39000	635 39393	1,634	
Gillette do Brasil Ltda.	Av. Suburbana 577, 1st Fl., Rio de Janeiro 20911, Brazil	21 5742112	21 5742075	1,600	96
Hankook Cosmetics Co., Ltd.	Chongno-Gu, Seoul, Republic of Korea	2 7243450	2 7373322	1,600	97
Novopharm Ltd.	30 Novopharm Crt, Toronto, Ontario M1B 2K9, Canada	416-291-8876	416-291-2162	1,600	98
Stiefel Laboratories Inc.	255 Alhambra Circle, Coral Gables, Florida 33134, U.S.A.	305-443-3807	305-443-3467	1,600	96
Toco Holdings Ltd.	PO Box 79496, Senderwood 2145, Republic of South Africa	11 8231957	11 8232423	1,600	91
Life Technologies Inc.	9800 Medical Center Dr., Rockville, Maryland 20850, U.S.A.	301-840-8000	301-670-1394	1,586	97
L'Oreal UK Ltd.	30 Kensington Church St., London W8 4HA, United Kingdom	71 9375454	71 9385757	1,584	93
Jeyes Group PLC	Brunel Way, Thetford IP24 1HA, United Kingdom	842 754567	842 754439	1,554	94
Old Bond Street Holding Co. Ltd.	Vale Rd., Camberley GU15 3XA, United Kingdom	276 62211	276 23402	1,509	93
Dunn Edwards Corp.	PO Box 30389, Los Angeles, California 90030-0389, U.S.A.	213-771-3330	213-771-4440	1,500	97
Elizabeth Arden Co.	1345 Ave. of the Amer., New York, New York 10105, U.S.A.	212-261-1000	212-261-1303	1,500	97
Junckers Industrier AS	Vaerftsvej 4A, Koge DK-4600, Denmark	53651895	53659936	1,500	93
Merck Sharp & Dohme Ltd.	Hertford Rd., Hoddesdon EN11 9BU, United Kingdom	992 467272	992 468175	1,497	93
Diagnostic Products Corp.	5700 W. 96th St., Los Angeles, California 90045, U.S.A.	213-776-0180		1,467	97
NBTY Inc.	90 Orville Dr., Bohemia, New York 11716, U.S.A.	516-567-9500	516-563-1180	1,460	97
American Home Products Holdings (UK) Ltd.	Taplow, Maidenhead SL6 0PH, United Kingdom			1,449	93
Berlex Laboratories Inc.	300 Fairfield Rd., Wayne, New Jersey 07470-7358, U.S.A.	973-694-4100	973-942-1610	1,435	97
Chong Kun Dang Corp.	Kuro-Gu, Seoul, Republic of Korea	2 6773986	2 6752570	1,400	97
Cooper Companies Inc.	6140 Stoneridge Mall Rd., Ste.59, Pleasanton, California 94588, U.S.A.	510-460-3600	510-460-3648	1,400	97
Pasteur Merieux Connaught Laboratories USA	PO Box 187, Swiftwater, Pennsylvania 18370-0187, U.S.A.	717-839-4267	717-839-0561	1,400	97
Del Laboratories Inc.	565 Broad Hollow Rd., Farmingdale, New York 11735, U.S.A.	516-293-7070		1,329	96
Henkel Corp. Chemicals Group	5051 Estecreek Rd., Cincinnati, Ohio 45232-1446, U.S.A.	513-482-3000		1,300	95
Tambour Paint and Chemicals Ltd.	Industrial Zone, Acre 24121, Israel	4 9853611	4 9853629	1,300	97
Warner-Lambert Canada Inc.	2200 Eglinton Ave. E, Scarborough, Ontario M1L 2N3, Canada	416-288-2200	416-288-2424	1,300	98
Il Yang Pharm Co., Ltd.	Songbuk-Gu, Seoul, Republic of Korea	2 9125291	2 9127875	1,281	97
Roussel Laboratories Ltd.	Denham, Uxbridge UB9 5HP, United Kingdom	895 834343	895 834479	1,281	93
Stepan Co.	22 W. Frontage Rd., Northfield, Illinois 60093, U.S.A.	847-446-7500	847-501-2100	1,270	97
Berli Jucker Co. Ltd.	PO Box 173 BMC, Bangkok 10000, Thailand	3671111	3671000	1,264	90
Harcros Chemicals UK Ltd.	Eccles, Manchester M30 0BH, United Kingdom	61 7897300		1,260	92
Mylan Pharmaceuticals Inc.	PO Box 4310, Morgantown, West Virginia 26504, U.S.A.	304-599-2595		1,250	96
Sociedad Espanola de Carburos Metalicos, SA	Aragon, 300, Barcelona E-08009, Spain	93 2902600	93 2902611	1,250	97
Swank Inc.	6 Hazel St., Attleboro, Massachusetts 02703, U.S.A.	508-222-3400	508-226-9598	1,250	97
Glaxo Wellcome Inc.	7333 Mississauga Rd. N, Mississauga, Ontario L5N 6L4, Canada	905-819-3000	905-819-3099	1,240	98
Baker Petrolite	3920 Essex Lane, Houston, Texas 77027, U.S.A.	713-599-7400		1,200	95
Banner Pharmacaps Inc.	4125 Premier Dr., High Point, North Carolina 27265, U.S.A.	910-812-8700		1,200	95
Dulux Pty. Ltd. Republic of South Africa	PO Box 2262, Bedfordview 2008, Republic of South Africa	11 6202000		1,200	94
Schering Manati Inc.	PO Box 486, Manati 00674, Puerto Rico	787854-2700		1,200	92
Smith & Nephew Ltd.	PO Box 92, Pinetown 3600, Republic of South Africa	31 722411	31 7017297	1,200	98
Societe Nationale Pour l'Industrie des Corps Gras	BP 312, Cotonou, Benin	330701	331520	1,179	91
Choongwae Pharma Corp.	Tongjak-Gu, Seoul, Republic of Korea	2 8406777	2 8314300	1,167	97
Alen Mak Perfumery & Cosmetics Ad	148 Vassil Levski St., Plovdiv BG-4003, Bulgaria	32 55800	32 551687	1,150	93

Product Tables
Paints, Cellulose

Unit of Measure: Metric tons.

	1990 Value	%	1991 Value	%	1992 Value	%	1993 Value	%	1994 Value	%	1995 Value	%	1996 Value	%
Total Production	1,691,959	100.0	1,543,715	100.0	1,324,904	100.0	1,195,360	100.0	1,195,141	100.0	1,288,210	100.0	1,298,867	100.0
Regions														
Africa	4,203	0.2	3,916	0.3	3,587	0.3	4,601	0.4	4,076	0.3	4,140	0.3	4,205	0.3
America, North	1,026	0.1	1,026	0.1	1,026	0.1	1,021	0.1	1,032	0.1	1,026	0.1	1,026	0.1
America, South	138	0.0	64	0.0	54	0.0	1	0.0	2	0.0	8	0.0	-	
Asia	933,910	55.2	874,744	56.7	709,970	53.6	614,836	51.4	631,757	52.9	703,074	54.6	699,250	53.8
Europe	752,683	44.5	663,964	43.0	610,267	46.1	574,901	48.1	558,275	46.7	579,962	45.0	594,386	45.8
Leading Producers														
India	320,252	18.9	301,974	19.6	300,764	22.7	310,893	26.0	407,143	34.1	500,786	38.9	473,764	36.5
Russian Federation	300,631	17.8	284,187	18.4	173,852	13.1	122,431	10.2	88,255	7.4	66,527	5.2	51,835	4.0
Romania	134,604	8.0	94,595	6.1	58,488	4.4	55,827	4.7	45,835	3.8	49,026	3.8	43,628	3.4
Belarus	208,305	12.3	178,082	11.5	125,380	9.5	80,336	6.7	45,866	3.8	47,619	3.7	54,572	4.2
Bulgaria	94,806	5.6	36,350	2.4	34,707	2.6	29,134	2.4	31,566	2.6	36,469	2.8	42,051	3.2

Paints, Water

Unit of Measure: Metric tons.

	1990 Value	%	1991 Value	%	1992 Value	%	1993 Value	%	1994 Value	%	1995 Value	%	1996 Value	%
Total Production	3,750,017	100.0	3,756,296	100.0	3,614,521	100.0	3,569,719	100.0	3,629,990	100.0	3,814,725	100.0	3,819,644	100.0
Regions														
Africa	2,262	0.1	1,871	0.0	1,837	0.1	1,710	0.0	1,742	0.0	1,792	0.0	1,864	0.0
America, North	65,061	1.7	67,184	1.8	72,752	2.0	72,260	2.0	71,337	2.0	67,651	1.8	68,037	1.8
America, South	365,582	9.7	367,887	9.8	371,841	10.3	376,137	10.5	381,196	10.5	381,688	10.0	383,377	10.0
Asia	738,472	19.7	750,077	20.0	756,114	20.9	743,038	20.8	776,519	21.4	802,205	21.0	877,613	23.0
Europe	2,564,597	68.4	2,554,936	68.0	2,397,341	66.3	2,361,642	66.2	2,383,967	65.7	2,545,865	66.7	2,472,932	64.7
Oceania	14,043	0.4	14,339	0.4	14,636	0.4	14,932	0.4	15,229	0.4	15,525	0.4	15,821	0.4
Leading Producers														
Germany	*817,839	21.8	*817,839	21.8	695,598	19.2	812,911	22.8	854,804	23.5	908,044	23.8	*817,839	21.4
United Kingdom	368,951	9.8	363,475	9.7	357,998	9.9	352,521	9.9	366,509	10.1	*399,924	10.5	*408,366	10.7
Japan	398,591	10.6	382,109	10.2	374,150	10.4	363,570	10.2	373,815	10.3	376,998	9.9	411,329	10.8
Brazil	*297,875	7.9	*297,875	7.9	*297,875	8.2	*297,875	8.3	*297,875	8.2	*297,875	7.8	*297,875	7.8
Spain	382,251	10.2	408,234	10.9	371,889	10.3	227,967	6.4	240,364	6.6	249,466	6.5	260,418	6.8

Paints, Other

Unit of Measure: Metric tons.

	1990 Value	%	1991 Value	%	1992 Value	%	1993 Value	%	1994 Value	%	1995 Value	%	1996 Value	%
Total Production	9,620,205	100.0	9,950,182	100.0	9,982,567	100.0	9,822,093	100.0	9,889,506	100.0	10,119,195	100.0	10,105,790	100.0
Regions														
Africa	181,443	1.9	160,033	1.6	166,641	1.7	158,651	1.6	178,424	1.8	170,315	1.7	170,219	1.7
America, North	184,583	1.9	190,073	1.9	195,563	2.0	200,052	2.0	204,542	2.1	207,617	2.1	211,858	2.1
America, South	254,821	2.6	173,215	1.7	293,658	2.9	350,129	3.6	212,552	2.1	259,268	2.6	260,129	2.6
Asia	5,829,123	60.6	5,884,350	59.1	5,956,911	59.7	5,819,899	59.3	6,021,707	60.9	6,110,164	60.4	6,240,686	61.8
Europe	3,153,268	32.8	3,525,857	35.4	3,353,456	33.6	3,277,335	33.4	3,256,568	32.9	3,356,430	33.2	3,207,811	31.7
Oceania	16,965	0.2	16,652	0.2	16,339	0.2	16,026	0.2	15,713	0.2	15,400	0.2	15,087	0.1
Leading Producers														
Japan	1,753,000	18.2	1,688,000	17.0	1,647,000	16.5	1,554,000	15.8	1,597,000	16.1	1,581,000	15.6	1,623,000	16.1
Germany	-		728,000	7.3	722,000	7.2	634,000	6.5	669,000	6.8	635,000	6.3	666,000	6.6
Ukraine	593,000	6.2	*593,000	6.0	*593,000	5.9	*593,000	6.0	*593,000	6.0	*593,000	5.9	*593,000	5.9
United Kingdom	289,370	3.0	273,335	2.7	257,299	2.6	241,263	2.5	239,533	2.4	*212,691	2.1	*196,946	1.9
Netherlands	289,000	3.0	312,000	3.1	307,000	3.1	*339,850	3.5	*358,392	3.6	*376,935	3.7	*395,477	3.9

Commodity data are provided by the United Nations Statistical Division. The symbol * means that data are estimated. For additional notes, see Appendix I.

Product Tables
Mastics

Unit of Measure: Metric tons.

	1990 Value	%	1991 Value	%	1992 Value	%	1993 Value	%	1994 Value	%	1995 Value	%	1996 Value	%
Total Production	688,445	100.0	1,015,910	100.0	1,024,538	100.0	1,043,301	100.0	1,104,605	100.0	1,446,856	100.0	1,488,770	100.0
Regions														
Africa	1,491	0.2	2,465	0.2	2,150	0.2	1,336	0.1	1,331	0.1	2,352	0.2	2,388	0.2
America, North	3	0.0	1	0.0	3	0.0	3	0.0	3	0.0	3	0.0	3	0.0
America, South	773	0.1	776	0.1	773	0.1	849	0.1	859	0.1	940	0.1	910	0.1
Asia	9,942	1.4	13,740	1.4	10,968	1.1	8,045	0.8	10,779	1.0	10,914	0.8	9,468	0.6
Europe	676,236	98.2	998,928	98.3	1,010,643	98.6	1,033,068	99.0	1,091,633	98.8	1,432,648	99.0	1,476,001	99.1
Leading Producers														
Germany	-		340,924	33.6	377,897	36.9	421,259	40.4	476,076	43.1	851,285	58.8	882,803	59.3
Sweden	102,060	14.8	86,844	8.5	71,627	7.0	56,410	5.4	41,194	3.7	25,977	1.8	*49,849	3.3
United Kingdom	69,443	10.1	60,230	5.9	51,018	5.0	41,805	4.0	75,836	6.9	*62,306	4.3	*60,440	4.1
Spain	41,653	6.1	57,354	5.6	48,501	4.7	60,676	5.8	54,022	4.9	49,790	3.4	45,953	3.1
Denmark	40,631	5.9	37,407	3.7	49,461	4.8	37,562	3.6	30,339	2.7	30,067	2.1	*33,157	2.2

Soap

Unit of Measure: Metric tons.

	1990 Value	%	1991 Value	%	1992 Value	%	1993 Value	%	1994 Value	%	1995 Value	%	1996 Value	%
Total Production	10,278,971	100.0	9,796,426	100.0	9,423,343	100.0	9,216,797	100.0	8,950,023	100.0	9,216,810	100.0	8,855,098	100.0
Regions														
Africa	1,225,339	11.9	1,187,905	12.1	1,198,791	12.7	1,302,573	14.1	1,082,766	12.1	1,066,256	11.6	1,160,119	13.1
America, North	520,679	5.1	542,586	5.5	522,045	5.5	495,213	5.4	489,979	5.5	486,679	5.3	496,428	5.6
America, South	829,018	8.1	829,915	8.5	840,937	8.9	841,792	9.1	850,931	9.5	893,091	9.7	878,699	9.9
Asia	6,180,203	60.1	5,792,461	59.1	5,535,980	58.7	5,401,890	58.6	5,405,959	60.4	5,674,429	61.6	5,244,474	59.2
Europe	1,460,515	14.2	1,381,102	14.1	1,264,525	13.4	1,115,367	12.1	1,061,396	11.9	1,038,753	11.3	1,018,732	11.5
Oceania	63,217	0.6	62,457	0.6	61,065	0.6	59,961	0.7	58,993	0.7	57,601	0.6	56,647	0.6
Leading Producers														
India	1,551,060	15.1	1,547,800	15.8	1,469,900	15.6	1,464,811	15.9	1,571,728	17.6	1,798,468	19.5	1,477,913	16.7
China	1,067,100	10.4	904,500	9.2	840,200	8.9	844,747	9.2	798,573	8.9	826,500	9.0	733,300	8.3
Brazil	*525,432	5.1	*525,432	5.4	*525,432	5.6	*525,432	5.7	*525,432	5.9	*525,432	5.7	*525,432	5.9
Egypt	327,865	3.2	289,333	3.0	290,072	3.1	318,190	3.5	256,993	2.9	198,065	2.1	184,864	2.1
Russian Federation	516,599	5.0	410,933	4.2	344,329	3.7	257,736	2.8	173,354	1.9	167,742	1.8	139,512	1.6

Washing Powder and Detergents

Unit of Measure: Metric tons.

	1990 Value	%	1991 Value	%	1992 Value	%	1993 Value	%	1994 Value	%	1995 Value	%	1996 Value	%
Total Production	20,466,762	100.0	19,936,252	100.0	20,108,749	100.0	20,429,152	100.0	21,331,909	100.0	22,122,957	100.0	22,630,063	100.0
Regions														
Africa	611,439	3.0	622,618	3.1	570,578	2.8	597,157	2.9	618,557	2.9	653,614	3.0	670,262	3.0
America, North	903,296	4.4	948,259	4.8	977,157	4.9	884,796	4.3	935,090	4.4	905,632	4.1	906,086	4.0
America, South	922,137	4.5	941,439	4.7	963,067	4.8	985,449	4.8	1,030,738	4.8	1,025,671	4.6	1,058,781	4.7
Asia	7,213,120	35.2	6,965,501	34.9	7,150,991	35.6	7,232,041	35.4	7,758,141	36.4	8,356,918	37.8	8,560,426	37.8
Europe	10,500,365	51.3	10,139,504	50.9	10,125,501	50.4	10,405,730	50.9	10,662,880	50.0	10,852,093	49.1	11,102,956	49.1
Oceania	316,406	1.5	318,931	1.6	321,455	1.6	323,980	1.6	326,504	1.5	329,028	1.5	331,553	1.5
Leading Producers														
Germany	*2,229,569	10.9	*2,229,569	11.2	*2,229,569	11.1	*2,229,569	10.9	2,392,276	11.2	2,130,364	9.6	2,166,068	9.6
Italy	*1,756,191	8.6	*1,756,191	8.8	*1,756,191	8.7	*1,756,191	8.6	1,748,664	8.2	1,712,098	7.7	1,807,812	8.0
China	1,513,900	7.4	1,462,000	7.3	1,660,000	8.3	1,883,291	9.2	2,174,800	10.2	2,997,962	13.6	2,622,000	11.6
Spain	1,326,118	6.5	1,363,236	6.8	1,442,106	7.2	1,853,518	9.1	1,961,783	9.2	2,040,022	9.2	1,763,708	7.8
United Kingdom	1,272,835	6.2	1,130,670	5.7	988,504	4.9	846,339	4.1	892,159	4.2	*1,099,763	5.0	*1,101,713	4.9

Commodity data are provided by the United Nations Statistical Division. The symbol * means that data are estimated. For additional notes, see Appendix I.

345

Product Tables
Carbon Black

Unit of Measure: Metric tons.

	1990		1991		1992		1993		1994		1995		1996	
	Value	%	Value	%	Value	%	Value	%	Value	%	Value	%	Value	%
Total Production	6,885,910	100.0	7,088,922	100.0	7,384,464	100.0	7,581,509	100.0	7,894,332	100.0	8,348,835	100.0	8,630,776	100.0
Regions														
Africa	57,000	0.8	57,000	0.8	57,000	0.8	57,000	0.8	57,000	0.7	57,000	0.7	57,000	0.7
America, North	1,672,701	24.3	1,688,614	23.8	1,733,961	23.5	1,751,538	23.1	1,816,455	23.0	1,826,231	21.9	1,870,454	21.7
America, South	397,248	5.8	426,291	6.0	445,239	6.0	474,650	6.3	503,496	6.4	529,328	6.3	568,535	6.6
Asia	2,966,247	43.1	3,187,030	45.0	3,402,493	46.1	3,572,470	47.1	3,788,395	48.0	4,101,348	49.1	4,327,085	50.1
Europe	1,792,714	26.0	1,729,988	24.4	1,745,771	23.6	1,725,850	22.8	1,728,986	21.9	1,834,928	22.0	1,807,702	20.9
Leading Producers														
United States of America	1,354,081	19.7	1,391,697	19.6	1,429,313	19.4	1,466,929	19.3	1,504,545	19.1	*1,500,450	18.0	*1,528,702	17.7
Japan	788,035	11.4	792,942	11.2	770,711	10.4	701,881	9.3	703,520	8.9	756,596	9.1	756,923	8.8
Germany	*320,039	4.6	*320,039	4.5	*320,039	4.3	334,620	4.4	299,152	3.8	330,799	4.0	315,587	3.7
France	254,296	3.7	226,226	3.2	232,061	3.1	205,002	2.7	235,200	3.0	259,202	3.1	*246,224	2.9
Korea, Republic of	215,525	3.1	231,808	3.3	247,936	3.4	300,133	4.0	310,564	3.9	323,409	3.9	*341,627	4.0

Printers' Ink

Unit of Measure: Metric tons.

	1990		1991		1992		1993		1994		1995		1996	
	Value	%	Value	%	Value	%	Value	%	Value	%	Value	%	Value	%
Total Production	1,191,872	100.0	1,482,453	100.0	1,513,774	100.0	1,583,263	100.0	1,669,156	100.0	1,698,513	100.0	1,736,836	100.0
Regions														
Africa	3,496	0.3	3,495	0.2	3,661	0.2	3,257	0.2	3,269	0.2	2,926	0.2	2,828	0.2
America, North	9,537	0.8	9,548	0.6	9,558	0.6	10,604	0.7	8,814	0.5	8,675	0.5	10,247	0.6
America, South	28,637	2.4	29,269	2.0	29,712	2.0	28,854	1.8	29,434	1.8	29,811	1.8	30,441	1.8
Asia	500,133	42.0	518,137	35.0	527,225	34.8	549,826	34.7	575,011	34.4	585,892	34.5	614,908	35.4
Europe	625,089	52.4	896,821	60.5	918,229	60.7	965,131	61.0	1,026,834	61.5	1,045,209	61.5	1,052,209	60.6
Oceania	24,979	2.1	25,183	1.7	25,387	1.7	25,591	1.6	25,795	1.5	25,999	1.5	26,203	1.5
Leading Producers														
Japan	386,094	32.4	386,932	26.1	379,096	25.0	383,266	24.2	398,502	23.9	409,503	24.1	426,056	24.5
Germany	-		265,093	17.9	270,529	17.9	275,964	17.4	304,908	18.3	316,225	18.6	321,864	18.5
United Kingdom	96,259	8.1	106,122	7.2	115,984	7.7	125,847	7.9	138,577	8.3	*129,736	7.6	*135,763	7.8
Italy	74,398	6.2	74,127	5.0	78,445	5.2	87,875	5.6	89,943	5.4	94,067	5.5	91,675	5.3
France	66,672	5.6	65,136	4.4	69,108	4.6	68,820	4.3	76,836	4.6	*77,025	4.5	*79,224	4.6

Polishes and Creams for Footwear, Furniture, Glass, Metal

Unit of Measure: Metric tons.

	1990		1991		1992		1993		1994		1995		1996	
	Value	%	Value	%	Value	%	Value	%	Value	%	Value	%	Value	%
Total Production	476,322	100.0	482,414	100.0	721,440	100.0	758,130	100.0	685,716	100.0	575,708	100.0	576,387	100.0
Regions														
Africa	1,000	0.2	1,000	0.2	1,000	0.1	1,000	0.1	1,000	0.1	1,000	0.2	1,000	0.2
America, North	2,781	0.6	2,781	0.6	2,781	0.4	3,400	0.4	2,745	0.4	2,312	0.4	2,666	0.5
America, South	21,875	4.6	21,994	4.6	19,381	2.7	23,190	3.1	24,591	3.6	21,502	3.7	22,429	3.9
Asia	2,042	0.4	2,035	0.4	2,029	0.3	2,023	0.3	2,017	0.3	2,011	0.3	2,000	0.3
Europe	448,625	94.2	454,603	94.2	696,249	96.5	728,517	96.1	655,362	95.6	548,883	95.3	548,292	95.1
Leading Producers														
Germany	-		-		248,599	34.5	218,730	28.9	231,588	33.8	83,026	14.4	84,847	14.7
United Kingdom	*156,241	32.8	*156,241	32.4	*156,241	21.7	203,365	26.8	109,117	15.9	*156,241	27.1	*156,241	27.1
Spain	22,672	4.8	20,770	4.3	22,160	3.1	28,104	3.7	36,246	5.3	28,494	4.9	*29,550	5.1
Colombia	*20,874	4.4	*20,874	4.3	17,862	2.5	21,924	2.9	23,391	3.4	20,319	3.5	*20,874	3.6
Sweden	2,807	0.6	1,368	0.3	9,087	1.3	10,181	1.3	10,258	1.5	13,005	2.3	*8,572	1.5

Product Tables
Explosives

Unit of Measure: Metric tons.

	1990		1991		1992		1993		1994		1995		1996	
	Value	%	Value	%	Value	%	Value	%	Value	%	Value	%	Value	%
Total Production	2,878,523	100.0	2,621,254	100.0	2,654,325	100.0	2,621,200	100.0	3,092,817	100.0	3,129,531	100.0	3,107,762	100.0
Regions														
Africa	22,440	0.8	21,048	0.8	22,555	0.8	25,362	1.0	27,670	0.9	27,277	0.9	24,938	0.8
America, North	2,156,000	74.9	1,850,000	70.6	1,893,000	71.3	1,880,000	71.7	2,320,000	75.0	2,280,000	72.9	2,240,000	72.1
America, South	72,026	2.5	71,019	2.7	72,763	2.7	84,311	3.2	102,382	3.3	100,874	3.2	108,762	3.5
Asia	231,834	8.1	285,479	10.9	272,710	10.3	241,837	9.2	265,637	8.6	340,272	10.9	360,824	11.6
Europe	396,223	13.8	393,709	15.0	393,296	14.8	389,690	14.9	377,128	12.2	381,108	12.2	373,239	12.0
Leading Producers														
United States of America	2,156,000	74.9	1,850,000	70.6	1,893,000	71.3	1,880,000	71.7	2,320,000	75.0	2,280,000	72.9	2,240,000	72.1
India	105,193	3.7	138,466	5.3	138,788	5.2	122,857	4.7	152,572	4.9	224,085	7.2	235,998	7.6
Germany	*80,750	2.8	*80,750	3.1	*80,750	3.0	74,627	2.8	80,924	2.6	86,562	2.8	80,889	2.6
Spain	66,307	2.3	67,490	2.6	68,371	2.6	*78,376	3.0	*79,392	2.6	*80,407	2.6	*81,423	2.6
Japan	78,508	2.7	76,882	2.9	71,017	2.7	64,400	2.5	62,395	2.0	62,675	2.0	61,660	2.0

Commodity data are provided by the United Nations Statistical Division. The symbol * means that data are estimated. For additional notes, see Appendix I.

347

ISIC 3522
DRUGS AND MEDICINES

ISCI 3522—Drugs and Medicines—is a subset of Chemical Products nec (see last chapter). This includes all medicinals and botanicals, pharmaceutical preparations, and biological products. No further breakdown of this category is available. It corresponds to SIC 283 in the U.S. classification system.

Summary Statistics

		1990 Value	N	1991 Value	N	1992 Value	N	1993 Value	N	1994 Value	N	1995 Value	N
Establishments or enterprises	(number)	7,315	53	7,529	55	9,837	58	6,999	54	2,964	44	1,888	19
Number employed	(000)	1,001	56	1,120	57	1,201	57	1,042	53	636	41	502	23
Total Wages	($ mil.)	25,195	55	30,526	57	34,600	57	23,511	49	17,213	41	13,045	24
Wage/salary per employee	($)	12,187	52	12,929	54	14,292	54	11,028	48	10,844	40	11,618	22
Female workers	(000)	146	27	68	23	178	20	169	21	160	16	155	6
as % of total employment*	(%)	42	26	39	22	39	20	40	21	40	16	46	6
Output	($ bil.)	183	54	217	56	243	55	208	48	155	40	137	25
per employee	($ 000)	90	50	101	52	110	51	99	46	107	37	96	21
per establishment	($ 000)	15,228	49	17,527	52	19,246	51	17,676	45	18,486	37	17,807	16
Value added	($ bil.)	107	51	116	50	130	44	126	42	77	35	74	22
per employee	($ 000)	51	48	56	47	65	41	60	39	67	32	52	18
per establishment	($ 000)	7,507	46	7,892	45	9,638	41	8,062	38	8,341	31	2,306	12
Capital investment	($ mil.)	4,571	28	6,094	31	10,820	34	9,356	30	5,340	21	5,701	10
per employee	($ 000)	7	26	7	29	7	31	6	27	7	20	9	9
per establishment	($ 000)	1,571	26	1,524	29	1,481	32	1,251	27	1,470	19	475	8

Data presented above are drawn from the detailed tables that follow. Columns headed 'N' show the number of countries that provided valid data for inclusion. Values are not strictly comparable one year to the next or one row to the next because the number of countries that report varies from period to period and row to row. However, a general indication of magnitudes can be gleaned from reviewing this summary—especially in earlier years in which more reports are available. For detailed explanations, see Appendix I. *The average for those countries reporting both total and female employment.

Establishments and Number Engaged

Country	Establishments or Enterprises (number)						Employees per Establishment					
	1990	1991	1992	1993	1994	1995	1990	1991	1992	1993	1994	1995
Australia	110	137	143	-	-	-	82	73	77	-	-	-
Austria	64	65	59	61	58	-	122	123	136	134	138	-
Azerbaijan	4.000	4.000	4.000	4.000	4.000	-	388	397	387	369	350	-
Bahrain	-	-	12	-	-	-	-	-	2.667	-	-	-
Bangladesh	290	196	237	-	-	-	74	93	76	-	-	-
Benin	-	-	2.000	2.000	-	-	-	-	-	-	-	-
Bolivia	25	27	26	23	26	28	41	39	43	46	50	47
Bosnia & Herzegovina	4.000	4.000	-	-	1.000	-	163	274	-	-	751	-
Brazil	467	-	399	407	-	-	106	-	124	126	-	-
Canada	136	122	117	119	112	-	147	172	179	185	179	-
Cape Verde	-	-	1.000	1.000	1.000	-	-	-	26	32	34	-
Chad	-	-	-	1.000	1.000	1.000	-	-	-	-	-	-
Chile	35	35	37	39	39	-	162	169	162	166	165	-
China (Hong Kong SAR)	226	226	228	249	229	-	9.292	10	11	9.639	10	-
Colombia	118	121	132	130	130	-	97	101	119	123	129	-
Costa Rica	45	45	44	40	-	-	28	27	29	31	-	-
Cyprus	4.000	4.000	4.000	4.000	9.000	11	51	62	67	72	39	40
Denmark	83	88	91	-	-	-	135	128	104	-	-	-
Ecuador	30	31	32	31	29	31	94	97	93	91	92	88
Egypt	24	25	26	26	-	-	983	1,040	1,069	992	-	-
El Salvador	-	-	-	20	27	35	-	-	-	109	81	96
Finland	16	19	21	19	19	-	238	200	186	195	195	-
Germany	-	339	348	353	351	-	-	312	307	302	286	-
Germany (Western Part)	312	308	316	-	-	-	293	309	313	-	-	-
Ghana	-	-	-	18	-	-	-	-	-	80	-	-
Greece	94	96	102	-	-	-	74	74	69	-	-	-
Grenada	-	-	2.000	1.000	1.000	-	-	-	42	81	-	-
Guatemala	-	11	11	11	11	11	-	147	155	158	158	161
Honduras	-	-	14	14	14	14	-	-	59	65	71	82
Hungary	-	-	*72	*74	-	-	-	-	-	-	-	-
India	1,794	1,886	2,112	2,352	-	-	75	79	79	75	-	-
Indonesia	215	217	223	223	221	226	170	175	184	183	202	216
Iran (Islamic Republic of)	-	51	54	53	-	-	-	221	222	220	-	-
Japan	1,144	1,125	1,113	1,104	-	-	87	88	90	91	-	-
Jordan	8.000	10	11	12	16	16	242	203	221	239	191	175
Kenya	21	22	23	23	29	29	84	77	78	86	76	83
Korea, Republic of	314	307	319	338	342	348	101	101	99	90	91	85
Kyrgyzstan	2.000	2.000	2.000	3.000	3.000	-	304	167	144	140	91	-
Macau	6.000	7.000	6.000	5.000	5.000	5.000	18	14	16	18	18	20
Malawi	*1.000	-	-	-	-	-	95	-	-	-	-	-
Malaysia	33	38	41	37	33	-	76	71	76	86	106	-
Mauritius	1.000	1.000	1.000	1.000	1.000	1.000	90	88	80	80	84	81
Mexico	72	71	71	71	71	71	364	379	389	391	384	379
Mozambique	-	-	2.000	2.000	2.000	-	-	-	157	152	157	-
Myanmar	2.000	2.000	2.000	2.000	2.000	-	1,067	992	937	864	874	-
Nepal	12	20	-	15	22	-	104	70	-	35	92	-
Netherlands	38	40	38	36	39	-	334	317	333	361	-	-
New Zealand	*40	*50	*54	*57	-	-	29	23	25	24	-	-
Norway	15	16	17	-	-	-	131	125	131	-	-	-
Pakistan	-	146	-	-	-	-	-	123	-	-	-	-
Panama	10	7.000	-	-	-	-	31	39	-	-	-	-
Peru	162	170	170	-	-	-	34	30	29	-	-	-
Philippines	111	133	107	93	-	-	131	110	130	151	-	-
Portugal	*103	*106	*111	*121	*125	-	90	87	87	74	70	-
Puerto Rico	83	77	78	79	81	85	-	-	-	-	-	-
Russian Federation	-	-	-	*438	*635	*852	-	-	-	205	131	97
Senegal	3.000	-	-	-	3.000	-	77	-	-	-	79	-
Sierra Leone	-	-	-	1.000	-	-	-	-	-	33	-	-
Singapore	19	18	19	19	19	-	88	95	92	88	91	-
Slovakia	-	7.000	7.000	7.000	7.000	-	-	773	708	722	710	-

Continued.

Depending on the table, * means *Enterprises, Engaged,* or *Factor Values;* ± means *Basis Unspecified;* § means *shown in millions.* For additional notes and sources, see Appendix I.

Establishments and Number Engaged

- Continued -

Country	Establishments or Enterprises (number)						Employees per Establishment					
	1990	1991	1992	1993	1994	1995	1990	1991	1992	1993	1994	1995
South Africa	-	88	-	-	-	-	-	-	-	-	-	-
Spain	329	327	326	-	-	-	107	111	114	-	-	-
Sri Lanka	18	22	9.000	19	-	-	61	57	114	70	-	-
Sweden	31	39	40	41	43	-	306	246	227	227	218	-
Thailand	129	141	-	-	-	-	99	96	-	-	-	-
Trinidad and Tobago	8.000	8.000	6.000	6.000	2.000	-	15	16	15	16	23	-
Tunisia	-	-	-	19	24	20	-	-	-	29	39	45
Turkey	54	53	65	67	61	-	231	230	188	188	190	-
Ukraine	-	-	104	47	48	49	-	-	173	319	292	286
United Kingdom	362	347	366	-	-	-	210	225	216	-	-	-
United Republic of Tanzania	7.000	7.000	-	-	-	-	67	97	-	-	-	-
United States of America	-	-	1,800	-	-	-	-	-	108	-	-	-
Venezuela	74	65	60	61	59	55	119	142	152	145	158	154
Zambia	7.000	-	-	-	9.000	-	121	-	-	-	81	-

Employment and Compensation of Employees

Country	Number Employed/Engaged (000)						Wage/Salary per Employee ($)					
	1990	1991	1992	1993	1994	1995	1990	1991	1992	1993	1994	1995
Australia	9.000	*10	*11	-	-	-	25,668	26,256	25,401	-	-	-
Austria	7.800	8.000	8.000	8.162	7.977	8.584	30,884	31,946	35,672	35,054	37,975	-
Azerbaijan	1.550	1.587	1.548	1.478	1.401	-	1,496	1,452	-	-	-	-
Bahrain	-	-	0.032	-	-	-	-	-	3,740	-	-	-
Bangladesh	22	18	18	-	-	-	1,084	1,314	1,295	-	-	-
Bolivia	1.019	1.060	1.116	1.056	1.306	1.320	2,940	3,315	3,240	3,676	3,627	3,496
Bosnia & Herzegovina	0.650	1.098	-	-	0.751	-	-	-	-	-	-	-
Brazil	50	-	49	51	-	-	10,627	-	15,118	15,549	-	-
Canada	20	21	21	22	20	22	31,454	33,874	33,527	31,816	34,344	33,585
Cape Verde	0.024	0.025	0.026	0.032	0.034	-	3,219	4,806	4,852	3,357	4,554	-
Chile	5.666	5.919	6.001	6.485	6.443	6.663	8,696	9,949	10,910	11,347	12,543	15,600
China (Hong Kong SAR)	*2.100	*2.300	*2.500	*2.400	*2.300	-	8,497	9,175	10,490	10,288	11,589	-
Colombia	11	12	16	16	17	16	3,804	4,007	4,541	5,428	7,183	8,088
Costa Rica	1.246	1.236	1.271	1.240	-	-	5,157	4,836	5,487	-	-	-
Cyprus	0.204	0.248	0.267	0.287	0.351	0.439	7,916	8,631	10,337	10,250	13,906	15,424
Denmark	11	11	9.465	-	-	-	39,008	38,803	-	-	-	-
Ecuador	2.813	3.021	2.983	2.824	2.681	2.733	4,543	5,478	6,174	7,962	4,739	4,757
Egypt	*24	*26	*28	*26	-	-	3,267	2,465	2,280	2,652	-	-
El Salvador	-	-	-	*2.182	*2.188	*3.364	-	-	-	3,681	5,459	-
Finland	3.800	3.800	3.900	3.700	3.700	-	28,965	29,173	26,612	20,771	25,074	-
Germany	-	*106	*107	*107	*100	-	-	36,142	-	-	-	-
Germany (Western Part)	*91	*95	*99	-	-	-	38,179	38,673	-	-	-	-
Ghana	-	-	-	*1.438	-	-	-	-	-	-	-	-
Greece	*6.951	*7.057	*7.037	-	-	-	13,155	14,098	16,039	-	-	-
Grenada	-	-	0.083	*0.081	-	-	-	-	-	-	-	-
Guatemala	-	*1.618	*1.707	*1.741	*1.739	*1.769	-	1,052	1,218	1,227	1,188	1,232
Honduras	*0.766	*0.811	*0.826	*0.905	*0.992	*1.144	3,572	3,232	1,969	2,420	2,239	2,210
India	*135	*148	*167	*175	-	-	2,096	1,704	1,700	1,542	-	-
Indonesia	37	38	41	41	45	49	1,736	1,885	1,850	2,184	2,315	3,371
Iran (Islamic Republic of)	-	11	12	12	-	-	-	5,587	5,536	5,192	-	-
Japan	*99	*99	*100	*100	-	-	35,091	39,291	-	-	-	-
Jordan	1.938	2.035	2.431	2.873	3.063	2.804	3,889	3,580	3,601	4,525	4,774	5,160
Kenya	1.774	1.691	1.784	1.987	2.218	2.412	3,428	2,970	2,692	1,689	1,976	2,506
Korea, Republic of	32	31	32	30	31	29	10,597	11,753	12,246	13,941	15,230	17,375
Kyrgyzstan	*0.608	*0.334	*0.289	*0.420	*0.272	-	810	557	-	-	-	-
Lithuania	-	-	3.163	2.361	-	-	-	-	-	472	-	-
Macau	0.105	0.099	0.096	0.091	0.091	0.099	7,654	8,398	8,914	11,996	13,206	12,438
Malawi	0.095	-	-	-	-	-	-	-	-	-	-	-
Malaysia	2.500	2.700	3.100	3.200	3.500	3.564	4,155	4,727	4,419	4,783	4,758	5,699

Continued.

Depending on the table, * means *Enterprises*, *Engaged*, or *Factor Values*; ± means *Basis Unspecified*; § means *shown in millions*. For additional notes and sources, see Appendix I.

351

Employment and Compensation of Employees

- Continued -

Country	Number Employed/Engaged (000)						Wage/Salary per Employee ($)					
	1990	1991	1992	1993	1994	1995	1990	1991	1992	1993	1994	1995
Mauritius	*0.090	0.088	0.080	0.080	0.084	0.081	3,538	3,361	6,337	5,241	5,104	-
Mexico	26	27	28	28	27	27	7,510	9,372	11,094	13,693	14,084	9,309
Mozambique	-	-	0.313	0.304	0.313	-	-	-	868	691	722	-
Myanmar	2.133	1.983	1.874	1.728	1.748	-	1,612	1,709	1,767	2,171	2,716	-
Nepal	1.251	1.399	-	0.524	2.025	-	619	550	-	469	465	-
Netherlands	13	13	13	13	-	-	32,717	33,874	37,863	38,633	-	-
New Zealand	1.165	1.153	1.360	1.343	-	-	-	-	-	-	-	-
Norway	1.965	1.994	2.233	-	-	-	33,007	35,353	-	-	-	-
Pakistan	-	18	-	-	-	-	-	3,603	-	-	-	-
Panama	*0.313	*0.273	-	-	-	-	8,827	8,341	-	-	-	-
Peru	*5.463	*5.084	*4.909	-	-	-	5,739	7,716	8,997	-	-	-
Philippines	15	15	14	14	-	-	5,966	5,995	7,577	7,334	-	-
Portugal	*9.236	*9.249	*9.689	*8.987	*8.754	-	8,961	10,510	12,799	12,660	13,352	-
Russian Federation	-	-	-	90	83	83	-	-	-	644	1,157	1,251
Senegal	*0.232	-	-	-	*0.237	-	8,185	-	-	-	9,576	-
Sierra Leone	-	-	-	*0.033	-	-	-	-	-	-	-	-
Singapore	*1.664	*1.705	*1.753	*1.672	*1.723	-	16,204	17,865	19,603	22,768	25,613	-
Slovakia	-	*5.409	*4.955	*5.055	*4.971	-	-	1,700	2,142	2,604	3,051	-
Spain	35	36	37	-	-	-	27,811	30,491	33,702	-	-	-
Sri Lanka	*1.099	*1.259	*1.028	*1.331	-	-	1,067	932	1,118	1,204	-	-
Sweden	9.500	*9.604	*9.076	9.304	9.363	-	29,130	33,536	37,775	29,081	30,136	-
Thailand	13	13	-	-	-	-	2,435	4,378	-	-	-	-
Trinidad and Tobago	0.118	0.127	0.089	0.098	0.046	-	12,123	8,336	9,523	8,009	4,770	-
Tunisia	-	-	-	0.550	0.926	0.903	-	-	-	12,463	7,947	11,829
Turkey	12	12	12	13	12	-	11,594	13,829	15,549	16,595	10,683	-
Ukraine	-	-	18	15	14	14	-	-	2,122	431	742	799
United Kingdom	76	78	79	-	-	-	28,524	31,200	34,150	-	-	-
United Republic of Tanzania	0.470	0.676	-	-	-	-	-	425	-	-	-	-
United States of America	183	184	194	202	206	218	37,432	39,130	-	-	-	-
Uruguay	*2.716	*2.491	*2.131	*2.203	-	-	-	-	-	-	-	-
Venezuela	8.800	9.200	9.100	8.864	9.331	8.461	6,501	6,845	7,417	7,802	6,890	8,865
Zambia	0.846	-	-	-	0.727	-	2,282	-	-	-	2,198	-

Female Workers: Total and Percent of Employment

Country	Female Workers (000)						Female Workers as % of Total Employment					
	1990	1991	1992	1993	1994	1995	1990	1991	1992	1993	1994	1995
Afghanistan	0.196	0.200	-	-	-	-	-	-	-	-	-	-
Australia	4.488	-	-	-	-	-	49.87	-	-	-	-	-
Austria	3.800	4.000	4.000	3.914	3.789	-	48.72	50.00	50.00	47.95	47.50	-
Bangladesh	1.179	1.371	1.024	-	-	-	5.47	7.50	5.66	-	-	-
Canada	9.000	10	-	-	-	-	45.00	47.62	-	-	-	-
Cape Verde	0.014	0.013	0.013	0.013	0.015	-	58.33	52.00	50.00	40.63	44.12	-
Chile	2.207	2.290	2.310	2.558	2.392	-	38.95	38.69	38.49	39.44	37.13	-
Colombia	-	5.443	-	7.309	7.635	-	-	44.41	-	45.83	45.61	-
Cyprus	0.139	0.154	0.179	0.182	0.176	-	68.14	62.10	67.04	63.41	50.14	-
Denmark	5.901	5.929	5.129	-	-	-	52.59	52.57	54.19	-	-	-
Egypt	-	5.925	11	9.509	-	-	-	22.79	40.18	36.86	-	-
El Salvador	-	-	-	1.095	1.065	1.898	-	-	-	50.18	48.67	56.42
Germany (Western Part)	47	-	-	-	-	-	51.44	-	-	-	-	-
Indonesia	-	-	-	21	23	26	-	-	-	51.33	52.17	53.90
Iran (Islamic Republic of)	-	2.762	2.929	2.737	-	-	-	24.54	24.48	23.43	-	-
Jordan	0.675	0.748	0.868	1.054	0.966	0.905	34.83	36.76	35.71	36.69	31.54	32.28
Kenya	0.347	0.276	0.293	0.326	-	-	19.56	16.32	16.42	16.41	-	-
Korea, Republic of	14	13	13	13	13	13	44.62	42.24	41.80	43.13	41.20	44.77
Macau	0.031	0.030	0.031	0.034	0.034	0.038	29.52	30.30	32.29	37.36	37.36	38.38
Malaysia	1.300	1.300	1.500	1.500	1.600	-	52.00	48.15	48.39	46.88	45.71	-
Myanmar	0.806	0.770	0.738	0.706	0.708	-	37.79	38.83	39.38	40.86	40.50	-

Continued.

Depending on the table, * means *Enterprises, Engaged,* or *Factor Values*; ± means *Basis Unspecified*; § means *shown in millions*. For additional notes and sources, see Appendix I.

Female Workers: Total and Percent of Employment

- Continued -

Country	Female Workers (000)						Female Workers as % of Total Employment					
	1990	1991	1992	1993	1994	1995	1990	1991	1992	1993	1994	1995
Nepal	0.477	0.561	-	0.139	0.365	-	38.13	40.10	-	26.53	18.02	-
New Zealand	0.624	0.599	0.706	0.703	-	-	53.56	51.95	51.91	52.35	-	-
Panama	0.142	-	-	-	-	-	45.37	-	-	-	-	-
Philippines	-	-	4.699	4.663	-	-	-	-	33.81	33.31	-	-
Portugal	3.460	3.472	3.622	3.406	3.294	-	37.46	37.54	37.38	37.90	37.63	-
Senegal	0.059	-	-	-	-	-	25.43	-	-	-	-	-
Sri Lanka	0.291	0.377	0.335	0.396	-	-	26.48	29.94	32.59	29.75	-	-
Sweden	5.200	-	-	-	-	-	54.74	-	-	-	-	-
Thailand	8.232	8.118	-	-	-	-	64.71	60.28	-	-	-	-
Turkey	3.740	-	-	-	-	-	30.00	-	-	-	-	-
United Kingdom	32	-	32	-	-	-	42.76	-	40.51	-	-	-
United Republic of Tanzania	0.136	0.196	-	-	-	-	28.94	28.99	-	-	-	-
United States of America	-	-	93	95	102	113	-	-	47.94	47.03	49.51	51.83
Zambia	-	-	-	-	0.141	-	-	-	-	-	19.39	-

Output and Output per Employee

Country	Output ($ bil.)						Output per Employee ($)					
	1990	1991	1992	1993	1994	1995	1990	1991	1992	1993	1994	1995
Australia	*1.438	*1.694	*1.777	-	-	-	159,809	169,381	161,564	-	-	-
Austria	1.494	1.635	1.731	1.770	1.838	2.482	191,563	204,319	216,330	216,862	230,373	289,105
Azerbaijan	±0.036	±0.040	±0.001	±0.001	±0.001	-	23,515	24,948	460	893	485	-
Bahrain	-	-	±0.002	-	-	-	-	-	51,529	-	-	-
Bangladesh	0.188	0.213	0.228	-	-	-	8,697	11,651	12,606	-	-	-
Bolivia	0.021	0.024	0.023	0.026	0.033	0.038	21,057	22,688	20,851	24,180	24,969	28,832
Brazil	*4.085	-	*4.013	*4.834	-	-	82,361	-	81,221	93,946	-	-
Canada	3.848	4.067	4.170	4.255	4.123	4.519	192,407	193,685	198,560	193,431	206,137	208,732
Cape Verde	-	-	±0.005	±0.005	±0.005	-	-	-	203,127	148,803	160,198	-
Chile	0.348	0.414	0.503	0.586	0.608	0.752	61,452	69,924	83,897	90,412	94,299	112,921
China (Hong Kong SAR)	0.084	0.106	0.123	0.155	0.172	-	40,224	46,269	49,299	64,690	74,767	-
Colombia	0.508	0.548	0.674	0.694	0.904	0.963	44,310	44,682	42,773	43,536	53,997	60,787
Costa Rica	0.048	0.058	0.074	0.125	0.175	0.133	38,708	47,318	58,071	100,655	-	-
Cyprus	0.018	0.027	0.030	0.030	0.039	0.049	88,503	109,280	112,784	106,016	112,360	110,832
Denmark	*1.555	*1.619	-	-	-	-	138,610	143,524	-	-	-	-
Ecuador	0.060	0.077	0.083	0.106	0.972	0.107	21,161	25,335	27,989	37,701	362,556	39,064
Egypt	*0.662	*0.464	*0.495	*0.548	-	-	28,030	17,850	17,815	21,224	-	-
El Salvador	-	-	-	±0.072	±0.053	±0.000	-	-	-	33,090	24,375	62
Finland	±0.550	±0.578	±0.540	±0.391	±0.444	-	144,723	152,207	138,526	105,604	120,040	-
FYR Macedonia	-	-	-	-	-	±0.041	-	-	-	-	-	-
France	±15	±15	±17	±17	±18	±21	-	-	-	-	-	-
Germany	-	19	21	20	21	-	-	178,943	197,668	183,048	208,190	-
Germany (Western Part)	17	18	20	-	-	-	186,730	187,864	205,370	-	-	-
Greece	*0.631	*0.673	*0.756	-	-	-	90,831	95,346	107,467	-	-	-
Guatemala	-	0.113	0.119	0.153	0.215	0.253	-	69,551	69,904	88,006	123,735	142,846
Honduras	0.020	0.017	0.023	0.020	0.018	0.021	26,651	21,558	28,072	21,930	18,029	18,685
India	*3.422	*3.313	*3.713	*3.980	-	-	25,438	22,323	22,191	22,680	-	-
Indonesia	0.701	0.992	0.729	1.026	1.405	1.352	19,202	26,175	17,819	25,087	31,429	27,675
Iran (Islamic Republic of)	-	0.397	0.307	0.349	-	-	-	35,242	25,641	29,910	-	-
Japan	±36	±40	±44	±51	-	-	360,885	401,761	437,110	510,881	-	-
Jordan	0.073	0.074	0.118	0.142	0.147	0.109	37,905	36,523	48,651	49,586	47,959	38,845
Korea, Republic of	3.627	3.637	3.894	4.194	4.692	5.164	114,785	117,542	123,086	137,964	150,523	175,474
Kyrgyzstan	0.012	0.019	0.005	0.000	0.000	-	20,149	56,426	15,735	1,015	1,816	-
Lithuania	-	-	0.009	0.008	-	-	-	-	2,953	3,490	-	-
Macau	0.009	0.007	0.007	0.008	0.008	0.011	85,350	73,846	70,271	87,872	86,176	112,910
Malaysia	*0.080	*0.096	*0.117	*0.121	*0.128	*0.160	32,016	35,608	37,850	37,902	36,603	44,948
Mauritius	0.002	0.002	0.003	0.002	0.003	-	24,445	23,450	37,188	29,748	29,894	-
Mexico	1.592	1.784	2.055	2.615	2.868	2.140	60,654	66,260	74,362	94,227	105,180	79,516
Myanmar	0.023	0.025	0.024	0.025	0.031	-	11,002	12,421	12,943	14,486	17,713	-

Continued.

Depending on the table, * means *Enterprises, Engaged,* or *Factor Values*; ± means *Basis Unspecified*; § means *shown in millions*. For additional notes and sources, see Appendix I.

353

Output and Output per Employee

- Continued -

Country	Output ($ bil.)						Output per Employee ($)					
	1990	1991	1992	1993	1994	1995	1990	1991	1992	1993	1994	1995
Nepal	0.004	0.006	-	0.003	0.007	-	3,457	4,106	-	5,380	3,675	-
Netherlands	2.350	2.649	3.021	2.917	-	-	185,223	209,157	238,587	224,302	-	-
Norway	±0.447	±0.456	±0.600	-	-	-	227,555	228,593	268,658	-	-	-
Pakistan	-	0.593	-	-	-	-	-	32,888	-	-	-	-
Panama	0.017	0.018	-	-	-	-	52,958	66,791	-	-	-	-
Peru	0.303	0.405	0.453	-	-	-	55,494	79,612	92,334	-	-	-
Philippines	0.688	0.725	0.902	0.865	-	-	47,437	49,624	64,898	61,765	-	-
Portugal	0.565	0.688	0.853	0.720	0.776	-	61,214	74,371	88,079	80,123	88,598	-
Puerto Rico	±8.307	±9.372	±10	±12	±13	±15	-	-	-	-	-	-
Russian Federation	-	-	-	0.475	0.865	1.015	-	-	-	5,276	10,433	12,263
Senegal	0.012	-	-	-	-	-	50,076	-	-	-	-	-
Singapore	*0.563	*0.833	*0.845	*0.889	*0.874	-	338,380	488,523	481,895	531,572	507,344	-
Slovakia	-	±0.121	±0.128	±0.156	±0.158	-	-	22,432	25,830	30,770	31,702	-
South Africa	±1.037	±1.126	±1.234	-	-	-	-	-	-	-	-	-
Spain	*5.845	*6.712	*7.764	-	-	-	166,412	184,850	209,006	-	-	-
Sri Lanka	0.008	0.009	0.010	0.012	-	-	7,200	7,466	9,333	9,219	-	-
Sweden	*1.914	*2.409	*2.780	*2.544	*2.916	-	201,492	250,875	306,282	273,418	311,492	-
Thailand	0.321	0.740	-	-	-	-	25,210	54,971	-	-	-	-
Trinidad and Tobago	±0.006	±0.004	±0.004	±0.003	±0.001	-	47,577	34,863	43,363	34,324	28,252	-
Tunisia	0.026	0.033	0.040	0.041	0.060	0.067	-	-	-	74,090	64,356	74,610
Turkey	1.564	1.717	1.715	2.094	1.599	-	125,507	140,664	140,501	165,961	138,123	-
Ukraine	-	-	*0.367	*0.103	*0.146	*0.147	-	-	20,395	6,894	10,402	10,519
United Kingdom	*12	*13	15	-	-	-	160,785	170,842	192,691	-	-	-
United Republic of Tanzania	*0.009	*0.011	-	-	-	-	18,969	15,700	-	-	-	-
United States of America	*54	*61	*68	*71	*76	*81	293,552	330,652	348,778	350,609	370,087	371,028
Venezuela	0.377	0.436	0.480	0.544	0.458	0.550	42,806	47,392	52,774	61,407	49,031	64,960
Zambia	0.011	-	-	-	0.010	-	12,565	-	-	-	13,706	-

Value Added and Value Added per Employee

Country	Value Added ($ bil.)						Value Added per Employee ($)					
	1990	1991	1992	1993	1994	1995	1990	1991	1992	1993	1994	1995
Australia	*0.685	*0.792	-	-	-	-	76,128	79,236	-	-	-	-
Austria	0.525	0.609	0.642	0.680	0.676	0.938	67,271	76,107	80,228	83,347	84,785	109,292
Bangladesh	0.105	0.102	0.110	-	-	-	4,877	5,574	6,061	-	-	-
Belgium	*0.841	*0.903	*1.026	*0.997	*1.104	*1.582	-	-	-	-	-	-
Bolivia	0.010	0.010	0.011	0.012	0.017	0.020	9,911	9,629	9,843	11,406	12,829	14,813
Brazil	*2.870	-	*2.964	*3.864	-	-	57,859	-	59,982	75,091	-	-
Canada	2.511	2.618	2.598	2.612	2.380	2.686	125,557	124,690	123,706	118,736	118,995	124,070
Chile	0.192	0.249	0.301	0.353	0.366	0.447	33,901	42,005	50,083	54,474	56,729	67,161
China (Hong Kong SAR)	0.035	0.047	0.051	0.067	0.063	-	16,444	20,365	20,309	28,063	27,229	-
Colombia	0.247	0.275	0.392	0.390	0.543	0.589	21,547	22,437	24,919	24,425	32,456	37,176
Costa Rica	0.016	0.019	0.025	0.041	0.056	0.042	12,683	15,767	19,412	32,792	-	-
Cyprus	0.005	0.006	0.008	0.010	0.012	0.014	22,086	22,339	30,920	35,376	35,168	31,141
Denmark	*1.051	*1.091	-	-	-	-	93,640	96,709	-	-	-	-
Ecuador	0.008	0.012	0.009	0.034	0.900	0.027	2,968	3,969	3,136	12,163	335,777	9,811
Egypt	*0.302	*0.124	*0.133	*0.201	-	-	12,801	4,787	4,799	7,778	-	-
El Salvador	-	-	-	0.040	0.035	0.000	-	-	-	18,494	16,076	13
Finland	±0.287	±0.300	±0.271	±0.193	±0.222	-	75,562	78,934	69,606	52,234	60,020	-
FYR Macedonia	-	-	-	-	-	±0.023	-	-	-	-	-	-
France	±4.509	±4.563	±5.559	±5.816	±5.713	±6.545	-	-	-	-	-	-
Germany (Western Part)	11	12	13	12	-	-	118,952	125,044	135,015	-	-	-
Greece	*0.222	*0.230	*0.251	-	-	-	31,953	32,556	35,622	-	-	-
Honduras	*0.006	*0.005	*0.005	*0.005	*0.005	*0.006	7,714	6,098	6,646	5,992	4,638	5,151
India	*0.756	*0.798	*0.851	*1.009	-	-	5,617	5,378	5,088	5,750	-	-
Indonesia	*0.274	*0.451	*0.251	*0.383	*0.608	*0.437	7,497	11,913	6,123	9,356	13,605	8,946
Iran (Islamic Republic of)	-	0.154	0.188	0.167	-	-	-	13,715	15,691	14,330	-	-
Japan	±25	±27	±30	±35	-	-	248,985	275,264	296,881	351,259	-	-

Continued.

Depending on the table, * means *Enterprises, Engaged,* or *Factor Values*; ± means *Basis Unspecified*; § means *shown in millions*. For additional notes and sources, see Appendix I.

Value Added and Value Added per Employee

- Continued -

Country	Value Added ($ bil.)						Value Added per Employee ($)					
	1990	1991	1992	1993	1994	1995	1990	1991	1992	1993	1994	1995
Jordan	0.026	0.031	0.046	0.053	0.053	0.036	13,647	15,154	19,002	18,613	17,165	13,007
Korea, Republic of	2.216	2.238	2.465	2.664	2.852	3.104	70,131	72,327	77,925	87,624	91,486	105,473
Macau	0.004	0.003	0.002	0.002	0.002	0.003	37,199	31,859	17,254	18,802	19,150	27,841
Malaysia	*0.034	*0.038	*0.048	*0.054	*0.061	*0.066	13,546	14,141	15,398	16,887	17,528	18,615
Mauritius	0.001	0.000	0.001	0.001	0.001	-	5,831	5,663	11,486	8,500	10,075	-
Mexico	0.861	0.925	1.093	1.423	1.624	-	32,816	34,348	39,548	51,275	59,573	-
Myanmar	±0.016	±0.015	±0.013	±0.014	±0.018	-	7,283	7,777	7,157	8,108	10,054	-
Nepal	0.002	0.003	-	0.001	0.003	-	1,497	2,053	-	1,825	1,428	-
Netherlands	0.683	0.756	0.862	1.007	-	-	53,793	59,723	68,046	77,473	-	-
Norway	±0.196	±0.175	±0.262	-	-	-	99,591	87,724	117,488	-	-	-
Pakistan	-	0.216	-	-	-	-	-	11,988	-	-	-	-
Panama	0.009	0.011	-	-	-	-	29,821	41,414	-	-	-	-
Peru	0.171	0.184	0.206	-	-	-	31,389	36,189	42,063	-	-	-
Philippines	0.332	0.384	0.538	0.504	-	-	22,913	26,297	38,729	35,974	-	-
Portugal	0.184	0.258	0.298	0.239	0.249	-	19,940	27,913	30,794	26,615	28,455	-
Russian Federation	-	-	-	±0.238	±0.485	±0.483	-	-	-	2,643	5,847	5,831
Senegal	0.005	-	-	-	0.005	-	20,138	-	-	-	22,815	-
Singapore	*0.446	*0.685	*0.698	*0.750	*0.732	-	268,187	401,591	398,127	448,839	424,629	-
Slovakia	-	-	-	±0.066	±0.065	-	-	-	-	13,058	13,064	-
Spain	*2.346	*2.706	*3.193	-	-	-	66,793	74,517	85,947	-	-	-
Sri Lanka	0.003	0.004	0.004	0.004	-	-	2,294	2,785	3,819	3,299	-	-
Sweden	*1.532	*1.330	*1.675	*1.571	*1.891	-	161,300	138,467	184,534	168,830	201,927	-
Thailand	0.107	0.376	-	-	-	-	8,400	27,887	-	-	-	-
Trinidad and Tobago	±0.003	±0.001	±0.001	±0.001	±0.000	-	26,230	8,584	7,762	8,962	5,871	-
Tunisia	0.010	0.013	0.016	0.017	0.025	0.029	-	-	-	31,520	27,290	31,747
Turkey	0.740	0.901	0.922	1.211	0.909	-	59,386	73,858	75,515	95,950	78,499	-
United Kingdom	*7.982	*8.627	*9.825	-	-	-	105,028	110,597	124,368	-	-	-
United Republic of Tanzania	*0.000	*0.004	-	-	-	-	502	5,427	-	-	-	-
United States of America	*38	*43	*49	*51	*56	*57	209,016	235,054	250,216	253,950	269,558	260,174
Venezuela	0.219	0.234	0.261	0.266	0.190	0.482	24,905	25,429	28,663	30,057	20,380	56,937
Zambia	0.003	-	-	-	0.006	-	3,688	-	-	-	7,717	-

Capital Investment

Country	Gross Fixed Capital Formation ($ mil.)						Per Establishment ($ mil)					
	1990	1991	1992	1993	1994	1995	1990	1991	1992	1993	1994	1995
Austria	152	121	146	195	180	-	2.373	1.860	2.479	3.198	3.110	-
Bahrain	-	-	0.003	-	-	-	-	-	0.000	-	-	-
Bangladesh	-	12	13	-	-	-	-	0.060	0.054	-	-	-
Belgium	199	194	257	217	287	380	-	-	-	-	-	-
Bolivia	0.783	1.052	1.699	2.363	-0.394	-	0.031	0.039	0.065	0.103	-0.015	-
Brazil	220	-	150	-77.593	-	-	0.470	-	0.375	-0.191	-	-
China (Hong Kong SAR)	-	-	2.325	4.395	1.682	-	-	-	0.010	0.018	0.007	-
Colombia	-	16	20	24	28	-	-	0.130	0.148	0.182	0.217	-
Cyprus	0.696	1.305	2.060	1.889	1.915	2.621	0.174	0.326	0.515	0.472	0.213	0.238
Denmark	150	167	-	-	-	-	1.810	1.903	-	-	-	-
Ecuador	-	22	24	0.982	5.887	9.873	-	0.694	0.758	0.032	0.203	0.318
El Salvador	-	-	-	3.925	3.456	-	-	-	-	0.196	0.128	-
Finland	72	43	38	21	33	-	4.473	2.278	1.821	1.104	1.748	-
Germany (Western Part)	839	956	1,176	1,159	-	-	2.688	3.103	3.720	-	-	-
Greece	18	22	21	-	-	-	0.187	0.233	0.208	-	-	-
Hungary	-	-	72	76	-	-	-	-	0.999	1.030	-	-
India	259	211	268	273	-	-	0.144	0.112	0.127	0.116	-	-
Indonesia	69	222	69	93	82	108	0.323	1.021	0.309	0.417	0.371	0.476
Iran (Islamic Republic of)	-	20	32	20	-	-	-	0.387	0.593	0.372	-	-
Japan	1,865	2,138	2,211	2,617	-	-	1.630	1.900	1.986	2.370	-	-
Jordan	-	-	-	-	-	3.483	-	-	-	-	-	0.218
Korea, Republic of	-	-	376	266	411	658	-	-	1.180	0.787	1.201	1.890

Continued.

Depending on the table, * means *Enterprises*, *Engaged*, or *Factor Values*; ± means *Basis Unspecified*; § means *shown in millions*. For additional notes and sources, see Appendix I.

355

Capital Investment
- Continued -

Country	Gross Fixed Capital Formation ($ mil.)						Per Establishment ($ mil)					
	1990	1991	1992	1993	1994	1995	1990	1991	1992	1993	1994	1995
Macau	-	-	-	0.036	0.376	0.190	-	-	-	0.007	0.075	0.038
Malaysia	-	-	-	14	12	-	-	-	-	0.378	0.358	-
Mexico	58	70	-	-	-	-	0.805	0.989	-	-	-	-
Myanmar	32	32	29	31	35	-	16	16	15	15	18	-
Netherlands	153	119	220	153	-	-	4.018	2.982	5.776	4.262	-	-
Norway	-	-	35	-	-	-	-	-	2.030	-	-	-
Pakistan	-	28	-	-	-	-	-	0.189	-	-	-	-
Panama	0.108	0.083	-	-	-	-	0.011	0.012	-	-	-	-
Peru	3.348	9.470	7.407	-	-	-	0.021	0.056	0.044	-	-	-
Philippines	20	18	24	29	-	-	0.182	0.133	0.227	0.310	-	-
Portugal	30	41	53	43	45	-	0.296	0.389	0.482	0.354	0.357	-
Senegal	1.524	-	-	-	-	-	0.508	-	-	-	-	-
Singapore	26	26	24	18	27	-	1.382	1.467	1.278	0.966	1.399	-
Spain	246	347	328	-	-	-	0.749	1.060	1.006	-	-	-
Sri Lanka	0.574	0.781	0.571	1.299	-	-	0.032	0.036	0.063	0.068	-	-
Thailand	49	26	-	-	-	-	0.379	0.185	-	-	-	-
Trinidad and Tobago	-	-	0.188	0.019	0.084	-	-	-	0.031	0.003	0.042	-
Tunisia	8.314	8.652	10	1.793	5.931	12	-	-	-	0.094	0.247	0.582
Turkey	97	113	62	124	40	-	1.796	2.139	0.956	1.849	0.652	-
Ukraine	-	-	5.557	1.047	2.345	1.974	-	-	0.053	0.022	0.049	0.040
United Kingdom	-	1,101	1,260	-	-	-	-	3.174	3.442	-	-	-
United Republic of Tanzania	2.640	8.250	-	-	-	-	0.377	1.179	-	-	-	-
United States of America	-	-	3,882	4,043	4,140	4,526	-	-	2.157		-	-
Zambia	0.448	-	-	-	-	-	0.064	-	-	-	-	-

Indexes of Industrial Production

Country	Index of Industrial Production (1990=100)						Index of Employment (1990=100)					
	1990	1991	1992	1993	1994	1995	1990	1991	1992	1993	1994	1995
Australia	-	-	-	-	-	-	100	111	122	-	-	-
Austria	-	-	-	-	-	-	100	103	103	105	102	110
Azerbaijan	-	-	-	-	-	-	100	102	100	95	90	-
Bangladesh	-	-	-	-	-	-	100	85	84	-	-	-
Bolivia	-	-	-	-	-	-	100	104	110	104	128	130
Bosnia & Herzegovina	-	-	-	-	-	-	100	169	-	-	116	-
Brazil	-	-	-	-	-	-	100	-	100	104	-	-
Canada	-	-	-	-	-	-	100	105	105	110	100	108
Cape Verde	-	-	-	-	-	-	100	104	108	133	142	-
Chile	-	-	-	-	-	-	100	104	106	114	114	118
China (Hong Kong SAR)	-	-	-	-	-	-	100	110	119	114	110	-
Colombia	-	-	-	-	-	-	100	107	137	139	146	138
Costa Rica	-	-	-	-	-	-	100	99	102	100	-	-
Cyprus	-	-	-	-	-	-	100	122	131	141	172	215
Denmark	-	-	-	-	-	-	100	101	84	-	-	-
Ecuador	-	-	-	-	-	-	100	107	106	100	95	97
Egypt	-	-	-	-	-	-	100	110	118	109	-	-
Finland	-	-	-	-	-	-	100	100	103	97	97	-
Germany (Western Part)	-	-	-	-	-	-	100	104	108	-	-	-
Greece	-	-	-	-	-	-	100	102	101	-	-	-
Honduras	-	-	-	-	-	-	100	106	108	118	130	149
India	-	-	-	-	-	-	100	110	124	130	-	-
Indonesia	-	-	-	-	-	-	100	104	112	112	122	134
Japan	-	-	-	-	-	-	100	100	101	101	-	-
Jordan	-	-	-	-	-	-	100	105	125	148	158	145
Kenya	-	-	-	-	-	-	100	95	101	112	125	136
Korea, Republic of	-	-	-	-	-	-	100	98	100	96	99	93
Kyrgyzstan	-	-	-	-	-	-	100	55	48	69	45	-
Macau	-	-	-	-	-	-	100	94	91	87	87	94

Continued.

Depending on the table, * means *Enterprises*, *Engaged*, or *Factor Values*; ± means *Basis Unspecified*; § means *shown in millions*. For additional notes and sources, see Appendix I.

Indexes of Industrial Production

- Continued -

Country	Index of Industrial Production (1990=100)						Index of Employment (1990=100)					
	1990	1991	1992	1993	1994	1995	1990	1991	1992	1993	1994	1995
Malawi	-	-	-	-	-	-	100	-	-	-	-	-
Malaysia	-	-	-	-	-	-	100	108	124	128	140	143
Mauritius	-	-	-	-	-	-	100	98	89	89	93	90
Mexico	-	-	-	-	-	-	100	103	105	106	104	103
Myanmar	-	-	-	-	-	-	100	93	88	81	82	-
Nepal	-	-	-	-	-	-	100	112	-	42	162	-
Netherlands	-	-	-	-	-	-	100	100	100	102	-	-
New Zealand	-	-	-	-	-	-	100	99	117	115	-	-
Norway	-	-	-	-	-	-	100	101	114	-	-	-
Panama	-	-	-	-	-	-	100	87	-	-	-	-
Peru	-	-	-	-	-	-	100	93	90	-	-	-
Philippines	-	-	-	-	-	-	100	101	96	97	-	-
Portugal	-	-	-	-	-	-	100	100	105	97	95	-
Senegal	-	-	-	-	-	-	100	-	-	-	102	-
Singapore	-	-	-	-	-	-	100	102	105	100	104	-
Spain	-	-	-	-	-	-	100	103	106	-	-	-
Sri Lanka	-	-	-	-	-	-	100	115	94	121	-	-
Sweden	-	-	-	-	-	-	100	101	96	98	99	-
Thailand	-	-	-	-	-	-	100	106	-	-	-	-
Trinidad and Tobago	-	-	-	-	-	-	100	108	75	83	39	-
Turkey	-	-	-	-	-	-	100	98	98	101	93	-
United Kingdom	-	-	-	-	-	-	100	103	104	-	-	-
United Republic of Tanzania	-	-	-	-	-	-	100	144	-	-	-	-
United States of America	-	-	-	-	-	-	100	101	106	110	113	119
Uruguay	-	-	-	-	-	-	100	92	78	81	-	-
Venezuela	-	-	-	-	-	-	100	105	103	101	106	96
Zambia	-	-	-	-	-	-	100	-	-	-	86	-

Depending on the table, * means *Enterprises, Engaged,* or *Factor Values*; ± means *Basis Unspecified*; § means *shown in millions*. For additional notes and sources, see Appendix I.

357

ISIC 3530
PETROLEUM REFINERIES

ISIC 3530—Petroleum Refineries—is an industry group producing a wide variety of items normally appearing as primary or secondary outputs of refineries. These include jet (aviation) fuels, motor gasoline, naphta, kerosene, white spirit, distillate fuel oils, heavy distillates and residuals, lubricating oils, petroleum coke, asphalt, liquefied petroleum gas (LPG), and similar products.

The industry group corresponds to the U.S. SIC Group 29—Petroleum and Coal Products, including 2911—Petroleum refining, 2951—Asphalt paving mixtures & blocks, 2952—Asphalt felts & coatings, and 2992—Lubricating oils & greases. The U.S. SIC 2999—Petroleum and coal products, nec, is not included in this chapter but corresponds to ISIC 3540—Petroleum, Coal Products (see next chapter).

Summary Statistics

		1990 Value	N	1991 Value	N	1992 Value	N	1993 Value	N	1994 Value	N	1995 Value	N
Establishments or enterprises	(number)	1,688	53	1,731	61	2,097	58	965	53	880	43	536	22
Number employed	(000)	1,088	60	1,028	66	1,016	61	963	59	458	52	440	44
Total Wages	($ mil.)	10,921	57	11,000	61	11,285	57	9,181	50	9,255	45	9,826	40
Wage/salary per employee	($)	18,882	55	19,491	57	59,989	54	25,334	49	20,329	44	25,715	38
Female workers	(000)	25	27	23	26	66	23	65	22	53	17	50	9
as % of total employment*	(%)	13	25	13	25	14	23	17	22	17	17	24	9
Output	($ bil.)	470	59	450	64	631	59	406	54	361	49	405	45
per employee	($ 000)	883	55	858	59	10,147	55	1,509	51	839	46	1,051	41
per establishment	($ 000)	480,782	48	388,958	54	4,045,942	49	667,949	42	811,506	34	243,026	19
Value added	($ bil.)	94	56	94	58	92	52	96	49	96	44	106	41
per employee	($ 000)	215	52	239	53	210	49	229	46	229	42	295	38
per establishment	($ 000)	141,764	44	108,623	48	138,098	43	190,866	37	250,698	30	84,268	16
Capital investment	($ mil.)	11,887	43	12,266	40	13,046	35	14,999	32	12,705	26	9,434	16
per employee	($ 000)	30	42	29	39	27	35	30	32	41	24	38	16
per establishment	($ 000)	16,148	39	12,836	37	18,492	33	37,910	28	22,263	22	27,116	13

Data presented above are drawn from the detailed tables that follow. Columns headed 'N' show the number of countries that provided valid data for inclusion. Values are not strictly comparable one year to the next or one row to the next because the number of countries that report varies from period to period and row to row. However, a general indication of magnitudes can be gleaned from reviewing this summary—especially in earlier years in which more reports are available. For detailed explanations, see Appendix I. *The average for those countries reporting both total and female employment.

Establishments and Number Engaged

Country	Establishments or Enterprises (number)						Employees per Establishment					
	1990	1991	1992	1993	1994	1995	1990	1991	1992	1993	1994	1995
Albania	-	-	-	5.000	4.000	4.000	-	-	-	517	479	419
Angola	-	-	1.000	1.000	-	-	-	-	346	235	-	-
Argentina	-	-	-	83	85	-	-	-	-	85	71	-
Australia	21	14	30	-	-	-	190	286	133	-	-	-
Austria	30	29	27	27	25	-	123	138	148	123	142	-
Azerbaijan	3.000	3.000	3.000	3.000	3.000	-	1,577	1,555	1,624	1,891	2,067	-
Bahrain	-	-	1.000				-	-	2,729			
Bangladesh	1.000	1.000	1.000				815	934	889			
Belgium	21	21	21	-	-	-	-	-	-	-	-	-
Bolivia	4.000	4.000	5.000	5.000	5.000	4.000	150	153	215	214	217	274
Bosnia & Herzegovina	2.000	2.000	-	-	-	-	1,311	1,199	-	-	-	-
Canada	62	61	60	62	61	-	242	230	200	210	197	-
Chile	2.000	2.000	2.000	2.000	2.000	-	621	642	658	660	662	-
China	*955	*980	*1,076	-	-	-	534	582	548	-	-	-
China (Hong Kong SAR)	-	-	-	1.000	2.000	-	-	-	-	-	-	-
Colombia	6.000	6.000	6.000	6.000	6.000	-	900	798	1,050	1,104	1,050	-
Costa Rica	2.000	2.000	2.000	2.000	-	-	1,058	1,380	1,159	1,194	-	-
Croatia	5.000	5.000	6.000	6.000	6.000	6.000	1,272	1,468	902	858	835	828
Cyprus	1.000	1.000	1.000	1.000	1.000	1.000	143	141	157	152	153	143
Czechoslovakia (Former)	*8.000	*7.000					2,875	3,429				
Czech Republic	-	*5.000	*4.000	*4.000			-	3,200	3,500	3,500	-	-
Denmark	3.000	3.000	4.000				207	210	168			
Egypt	8.000	8.000	8.000	8.000	-	-	2,150	2,300	2,188	2,213	-	-
El Salvador	-	1.000	1.000	1.000	1.000	1.000	-	-	70	64	72	69
Finland	4.000	4.000	4.000	3.000	3.000	14	775	775	775	1,000	900	215
FYR Macedonia	*1.000	*1.000	*1.000	*2.000	*1.000	*2.000	1,419	1,338	1,304	659	552	233
Gabon	-	2.000	2.000	2.000	2.000	2.000	-	251	217	238	228	204
Ghana	-	-	-	1.000	-	-	-	-	-	328	-	-
Greece	17	17	17	-	-	-	273	242	242	-	-	-
Guatemala	-	1.000	1.000	1.000	1.000	1.000	-	94	94	94	94	97
Honduras	-	-	2.000	2.000	2.000	1.000	-	-	110	80	88	-
Hungary	*5.000	*12	*6.000	*13			1,200	500	3,667	1,538	-	-
India	79	82	79	120	-	-	281	263	309	229	-	-
Indonesia	-	-	-	-	1.000	4.000	-	-	-	-	340	145
Iran (Islamic Republic of)	9.000	10	10	10	-	-	233	242	175	174	-	-
Iraq	-	-	3.000				-	-	2,667			
Italy	*27	*25	*115	*117	*111	-	484	615	-	-	-	-
Jamaica	1.000	1.000	-	-	-	-	-	-	-	-	-	-
Japan	181	190	183	186	268	274	116	111	115	118	90	84
Jordan	1.000	1.000	1.000	1.000	1.000	1.000	3,374	3,404	3,544	3,844	3,947	3,863
Kenya	1.000	1.000	1.000	1.000	2.000	2.000	264	264	264	264	132	133
Korea, Republic of	10	7.000	14	11	15	9.000	710	1,039	548	721	550	1,107
Kuwait	1.000	1.000	1.000	1.000	1.000	-	4,856	3,758	4,133	4,753	5,065	-
Latvia	1.000	1.000	2.000	2.000	1.000	2.000	133	123	60	198	86	44
Malaysia	11	10	8.000	8.000	6.000	-	100	120	150	163	450	-
Mozambique	2.000	2.000	2.000	2.000	3.000	-	228	227	228	230	248	-
Netherlands	7.000	7.000	7.000	7.000	1.000	-	915	914	900	735	5,041	-
New Zealand	*7.000	*7.000	*3.000	*2.000	-	-	107	107	222	-	-	-
Norway	3.000	3.000	3.000	59	65	-	397	390	391	27	25	-
Pakistan	-	3.000	-	-	-	-	-	750	-	-	-	-
Paraguay	-	1.000	-	-	-	-	-	27	-	-	-	-
Peru	26	26	28	-	-	-	145	145	107	-	-	-
Philippines	4.000	4.000	4.000	4.000	-	-	625	600	625	650	-	-
Poland	*8.000	*9.000	*8.000	*10	-	-	2,000	2,000	2,000	1,600	-	-
Portugal	*2.000	*2.000	*2.000	*1.000	*1.000	-	-	-	-	3,727	3,553	-
Qatar	1.000	1.000	1.000	1.000	1.000	-	619	599	571	564	558	-
Romania	-	10	11	11	14	-	-	2,550	2,300	2,482	-	-
Russian Federation	-	-	-	*122	*133	*171	-	-	-	884	838	683
Senegal	1.000	1.000	1.000	1.000	1.000	-	326	281	245	299	228	-
Sierra Leone	-	-	-	1.000	-	-	-	-	-	156	-	-

Continued.

Depending on the table, * means *Enterprises, Engaged,* or *Factor Values*; ± means *Basis Unspecified*; § means *shown in millions.* For additional notes and sources, see Appendix I.

Establishments and Number Engaged

- Continued -

Country	Establishments or Enterprises (number)						Employees per Establishment					
	1990	1991	1992	1993	1994	1995	1990	1991	1992	1993	1994	1995
Slovakia	-	3.000	3.000	3.000	3.000	-	-	2,850	2,674	2,503	2,156	-
Slovenia	3.000	*3.000	*3.000	*5.000	*4.000	*11	-	-	-	-	-	-
South Africa	-	32	-	-	-	-	-	563	-	-	-	-
Spain	10	10	10	-	-	-	723	708	705	-	-	-
Sweden	8.000	11	12	8.000	6.000	-	200	130	147	181	225	-
Thailand	2.000	10	-	-	-	-	-	243	-	-	-	-
Tunisia	-	-	-	1.000	1.000	1.000	-	-	-	484	484	484
Turkey	5.000	5.000	5.000	5.000	5.000	-	1,004	984	988	1,016	1,021	-
Ukraine	10	14	11	10	10	10	1,500	1,071	1,364	1,500	1,600	1,700
United Kingdom	25	25	27	-	-	-	360	320	296	-	-	-
United Republic of Tanzania	1.000	1.000	-	-	-	-	359	485	-	-	-	-
USSR (Former)	*60	-	-	-	-	-	2,333	-	-	-	-	-
United States of America	-	-	232	-	-	-	-	-	323	-	-	-
Venezuela	14	13	11	10	10	9.000	507	600	700	659	648	676
Yugoslavia	6.000	7.000	4.000	4.000	5.000	6.000	-	671	1,175	1,150	900	733
Zambia	-	-	-	-	1.000	-	-	-	-	-	-	-

Employment and Compensation of Employees

Country	Number Employed/Engaged (000)						Wage/Salary per Employee ($)					
	1990	1991	1992	1993	1994	1995	1990	1991	1992	1993	1994	1995
Albania	-	-	-	2.586	1.917	1.675	-	-	-	405	673	826
Angola	-	*0.348	*0.346	*0.235	-	-	-	-	-	-	-	-
Argentina	10	-	-	7.071	6.002	-	16,832	-	-	34,046	-	-
Australia	4.000	*4.000	*4.000	-	-	-	-	-	-	-	-	-
Austria	3.700	4.000	4.000	3.308	3.552	3.677	-	-	-	-	-	-
Azerbaijan	4.732	4.666	4.872	5.672	6.202	-	2,235	2,439	-	774	427	-
Bahrain	-	-	2.729	-	-	-	-	-	28,046	-	-	-
Bangladesh	0.815	0.934	0.889	-	-	-	3,858	3,920	4,332	-	-	-
Bermuda	0.019	0.036	0.060	0.065	0.070	-	-	-	-	-	-	-
Bolivia	0.602	0.612	1.075	1.068	1.084	1.097	5,906	6,415	6,587	5,145	7,898	7,620
Bosnia & Herzegovina	2.622	2.397	-	-	-	-	4,618	4,885	-	-	-	-
Cambodia	*0.215	*0.187	-	-	-	-	-	-	-	-	-	-
Canada	15	14	12	13	12	12	-	-	-	-	-	-
Chile	1.242	1.284	1.315	1.319	1.324	1.276	16,148	17,250	18,834	20,060	22,090	28,634
China	*510	*570	*590	*460	-	-	-	-	-	-	-	-
China (Taiwan Province)	14	13	14	14	14	14	23,290	24,080	29,659	28,881	32,150	34,213
Colombia	5.400	4.789	6.303	6.623	6.301	6.301	5,614	5,225	6,596	7,435	9,153	10,214
Costa Rica	2.115	2.760	2.318	2.389	-	-	5,166	6,214	7,035	7,869	-	-
Croatia	6.360	7.340	5.410	5.150	5.010	4.970	9,757	7,867	1,972	2,174	3,446	5,259
Cyprus	0.143	0.141	0.157	0.152	0.153	0.143	26,916	29,472	31,253	34,602	34,043	-
Czechoslovakia (Former)	23	24	-	-	-	-	2,640	2,064	-	-	-	-
Czech Republic	13	16	14	14	-	-	2,786	2,014	2,578	2,869	-	-
Denmark	0.621	0.630	0.674	0.674	0.672	0.671	-	-	-	-	-	-
Ecuador	1.665	1.410	1.410	2.632	2.277	2.539	4,914	9,160	8,946	11,636	8,193	6,105
Egypt	*17	*18	*18	*18	*18	*17	4,927	3,019	3,765	4,253	4,268	4,261
El Salvador	-	-	*0.070	*0.064	*0.072	*0.069	-	-	-	19,154	23,497	25,271
Finland	3.100	3.100	3.100	3.000	2.700	3.014	39,142	39,373	36,497	29,334	33,283	-
FYR Macedonia	1.419	1.338	1.304	1.318	0.552	0.465	6,756	7,406	6,973	4,771	5,158	6,338
France	23	21	20	18	18	17	-	-	36,921	-	-	-
Gabon	-	0.502	0.433	0.476	0.456	0.408	-	-	-	-	28,822	34,510
Ghana	-	-	-	*0.328	*0.337	*0.347	-	-	-	-	-	-
Greece	*4.633	*4.111	*4.108	*3.834	*4.152	*4.217	18,289	19,841	22,062	19,318	21,416	24,545
Guatemala	-	*0.094	*0.094	*0.094	*0.094	*0.097	-	1,144	1,107	1,016	1,008	1,544
Honduras	*0.214	*0.212	*0.220	*0.160	*0.175	-	9,097	6,596	6,378	602	593	-
Hungary	6.000	6.000	22	20	20	20	4,472	5,506	5,833	6,389	6,908	7,399
India	*22	*22	*24	*27	*28	*29	3,231	2,993	2,790	2,384	2,586	2,717
Indonesia	-	-	-	-	0.340	0.581	-	-	-	-	3,206	4,259

Continued.

Depending on the table, * means *Enterprises, Engaged,* or *Factor Values*; ± means *Basis Unspecified*; § means *shown in millions*. For additional notes and sources, see Appendix I.

361

Employment and Compensation of Employees
- Continued -

Country	Number Employed/Engaged (000)						Wage/Salary per Employee ($)					
	1990	1991	1992	1993	1994	1995	1990	1991	1992	1993	1994	1995
Iran (Islamic Republic of)	2.100	2.423	1.751	1.741	-	-	2,795	3,241	3,564	4,547	-	-
Iraq	-	8.328	8.000	-	-	-	-	16,608	28,767	-	-	-
Italy	13	15	-	-	-	-	-	-	-	-	-	-
Japan	*21	*21	*21	*22	*24	*23	-	-	-	-	-	-
Jordan	3.374	3.404	3.544	3.844	3.947	3.863	4,732	4,759	5,108	4,958	5,008	5,563
Kenya	0.264	0.264	0.264	0.264	0.264	0.266	6,953	6,592	6,373	3,774	5,434	8,935
Korea, Republic of	7.100	7.275	7.670	7.926	8.247	9.966	20,955	19,041	19,594	21,356	25,718	28,671
Kuwait	4.856	3.758	4.133	4.753	5.065	4.753	36,013	37,028	-	-	-	-
Latvia	0.133	0.123	0.120	0.396	0.086	0.088	-	-	-	-	-	-
Malaysia	1.100	1.200	1.200	1.300	2.700	3.084	14,049	14,181	16,455	15,988	15,468	16,924
Mozambique	0.457	0.454	0.455	0.460	0.745	-	1,135	1,648	1,435	1,311	1,983	-
Netherlands	6.403	6.398	6.303	5.148	5.041	4.915	-	-	-	-	-	-
New Zealand	0.750	0.750	0.667	-	-	-	33,194	-	-	-	-	-
Norway	1.192	1.169	1.173	1.574	1.639	1.556	-	-	-	-	-	-
Pakistan	2.325	2.250	2.353	-	-	-	4,452	4,677	4,812	-	-	-
Paraguay	-	0.027	-	-	-	-	-	-	-	-	-	-
Peru	*3.782	*3.780	*2.994	-	-	-	24,466	12,348	9,390	-	-	-
Philippines	2.500	2.400	2.500	2.600	-	-	11,534	12,585	14,252	14,249	-	-
Poland	16	18	16	16	16	16	2,513	2,368	4,235	5,232	5,994	7,570
Portugal	-	-	-	*3.727	*3.553	*3.625	-	-	-	-	-	-
Qatar	0.619	0.599	0.571	0.564	0.558	-	27,101	35,123	35,865	35,546	36,704	-
Romania	26	25	25	27	-	-	2,038	1,572	1,175	1,326	-	-
Russian Federation	-	-	-	108	111	117	-	-	-	1,450	2,502	2,853
Senegal	*0.326	*0.281	*0.245	*0.299	*0.228	-	13,160	13,864	16,376	11,906	9,029	-
Sierra Leone	-	-	-	*0.156	-	-	-	-	-	-	-	-
Slovakia	-	*8.549	*8.023	*7.510	*6.468	-	-	2,111	2,682	2,982	3,836	-
South Africa	18	18	18	-	-	-	12,089	13,460	15,506	-	-	-
Spain	7.229	7.082	7.051	7.947	7.395	7.109	34,362	36,196	-	-	39,727	-
Sweden	1.600	*1.429	*1.764	1.450	1.348	1.352	29,777	36,426	38,793	31,954	32,229	35,372
Thailand	-	2.430	-	-	-	-	-	11,313	-	-	-	-
Tunisia	-	-	-	0.484	0.484	0.484	-	-	-	-	-	-
Turkey	5.021	4.921	4.942	5.078	5.107	5.332	14,201	20,361	21,847	22,696	15,369	18,677
Ukraine	15	15	15	15	16	17	3,022	3,416	3,957	495	654	1,425
United Kingdom	9.000	8.000	8.000	-	-	-	36,905	-	-	-	-	-
United Republic of Tanzania	0.359	0.485	-	-	-	-	1,214	1,411	-	-	-	-
USSR (Former)	140	-	-	-	-	-	2,205	-	-	-	-	-
United States of America	72	74	75	73	72	70	-	-	-	-	-	-
Uruguay	*2.061	*1.813	*1.486	*0.615	*1.146	*1.081	-	-	-	-	-	-
Venezuela	7.100	7.800	7.700	6.592	6.476	6.087	8,595	16,608	14,654	16,741	11,615	9,362
Yugoslavia	-	4.700	4.700	4.600	4.500	4.400	-	-	-	-	2,778	1,754

Female Workers: Total and Percent of Employment

Country	Female Workers (000)						Female Workers as % of Total Employment					
	1990	1991	1992	1993	1994	1995	1990	1991	1992	1993	1994	1995
Albania	-	-	-	0.975	0.809	0.650	-	-	-	37.70	42.20	38.81
Angola	-	-	0.016	0.084	-	-	-	-	4.62	35.74	-	-
Argentina	-	-	-	-	0.404	-	-	-	-	-	6.73	-
Australia	0.299	-	-	-	-	-	7.47	-	-	-	-	-
Austria	0.700	1.000	1.000	0.645	0.730	-	18.92	25.00	25.00	19.50	20.55	-
Bahrain	-	-	0.151	-	-	-	-	-	5.53	-	-	-
Bangladesh	0.012	0.015	0.014	-	-	-	1.47	1.61	1.57	-	-	-
Bermuda	0.002	0.004	0.010	0.010	0.011	-	10.53	11.11	16.67	15.38	15.71	-
Bosnia & Herzegovina	0.415	0.395	-	-	-	-	15.83	16.48	-	-	-	-
Canada	3.000	3.000	-	-	-	-	20.00	21.43	-	-	-	-
Chile	0.088	0.094	0.094	0.085	0.095	-	7.09	7.32	7.15	6.44	7.18	-
China (Taiwan Province)	1.256	1.290	1.286	1.334	1.327	1.307	9.02	9.71	9.30	9.20	9.35	9.41
Colombia	-	0.537	-	0.848	0.699	-	-	11.21	-	12.80	11.09	-

Continued.

Depending on the table, * means *Enterprises, Engaged,* or *Factor Values;* ± means *Basis Unspecified;* § means *shown in millions.* For additional notes and sources, see Appendix I.

Female Workers: Total and Percent of Employment
- Continued -

Country	Female Workers (000)						Female Workers as % of Total Employment					
	1990	1991	1992	1993	1994	1995	1990	1991	1992	1993	1994	1995
Croatia	1.360	1.640	1.330	1.220	1.190	1.230	21.38	22.34	24.58	23.69	23.75	24.75
Cyprus	0.010	0.010	0.011	0.010	0.010	-	6.99	7.09	7.01	6.58	6.54	-
Czechoslovakia (Former)	3.800	4.600	-	-	-	-	16.52	19.17	-	-	-	-
Czech Republic	6.000	5.000	5.000	5.000	-	-	46.15	31.25	35.71	35.71	-	-
Denmark	0.064	0.068	0.079	-	-	-	10.31	10.79	11.72	-	-	-
Egypt	-	1.426	1.298	1.305	-	-	-	7.75	7.42	7.37	-	-
El Salvador	-	-	-	0.004	0.002	0.002	-	-	-	6.25	2.78	2.90
FYR Macedonia	0.289	0.280	0.268	0.266	0.263	0.262	20.37	20.93	20.55	20.18	47.64	56.34
Germany (Western Part)	3.000	-	-	-	-	-						
Hungary	2.000	-	5.000	5.000	-	-	33.33	-	22.73	25.00	-	-
Indonesia	-	-	-	-	0.008	0.077	-	-	-	-	2.35	13.25
Iran (Islamic Republic of)	-	0.059	0.070	0.053	-	-	-	2.43	4.00	3.04	-	-
Iraq	-	0.885	-	-	-	-	-	10.63	-	-	-	-
Italy	-	1.163	-	-	-	-	-	7.56	-	-	-	-
Jordan	0.064	0.068	0.067	0.071	0.072	0.070	1.90	2.00	1.89	1.85	1.82	1.81
Kenya	0.019	0.019	0.019	0.019	-	-	7.20	7.20	7.20	7.20	-	-
Korea, Republic of	0.300	0.278	0.360	0.235	0.271	0.300	4.23	3.82	4.69	2.96	3.29	3.01
Malaysia	0.100	0.100	0.100	0.100	0.300	-	9.09	8.33	8.33	7.69	11.11	-
New Zealand	0.079	0.084	0.071	-	-	-	10.53	11.20	10.64	-	-	-
Philippines	-	-	0.324	0.362	-	-	-	-	12.96	13.92	-	-
Portugal	0.861	0.755	-	0.719	0.698	-	-	-	-	19.29	19.65	-
Senegal	0.015	-	-	-	-	-	4.60	-	-	-	-	-
Sweden	0.100	-	-	-	-	-	6.25	-	-	-	-	-
Thailand	-	0.580	-	-	-	-	-	23.87	-	-	-	-
Turkey	0.190	-	-	-	-	-	3.78	-	-	-	-	-
United Kingdom	1.341	-	1.000	-	-	-	14.90	-	12.50	-	-	-
United Republic of Tanzania	0.098	0.132	-	-	-	-	27.30	27.22	-	-	-	-
United States of America	-	-	48	47	46	46	-	-	64.00	64.38	63.89	65.71

Output and Output per Employee

Country	Output ($ bil.)						Output per Employee ($)					
	1990	1991	1992	1993	1994	1995	1990	1991	1992	1993	1994	1995
Albania	-	-	-	*0.048	*0.043	*0.068	-	-	-	18,399	22,267	40,327
Angola	-	-	±177	±8.265	-	-	-	-	§510,6	§35,17	-	-
Argentina	±12	-	-	8.000	-	-	§1,208	-	-	§1,131	-	-
Australia	*5.642	*6.844	*6.633	-	-	-	§1,410	§1,710	§1,658	-	-	-
Austria	3.177	3.043	2.835	2.648	2.694	2.974	858,661	760,813	708,709	800,366	758,375	808,820
Azerbaijan	±0.511	±0.761	±0.060	±0.139	±0.182	-	107,974	163,000	12,412	24,566	29,328	-
Bahrain	-	-	±0.194	-	-	-	-	-	71,236	-	-	-
Bangladesh	0.014	0.018	0.013	-	-	-	17,406	19,660	14,728	-	-	-
Bolivia	0.435	0.540	0.468	0.447	0.477	0.464	723,275	882,836	435,590	418,148	440,357	422,610
Bosnia & Herzegovina	±0.428	±0.345	-	-	-	-	163,365	143,970	-	-	-	-
Canada	17	15	14	13	5.785	15	§1,121	§1,105	§1,197	§1,036	482,084	§1,282
Chile	1.862	1.898	1.948	2.138	2.175	2.715	§1,499	§1,478	§1,481	§1,620	§1,642	§2,127
China	±10	±13	±16	±23	-	-	20,574	23,559	27,669	49,431	-	-
China (Taiwan Province)	±8.147	±9.274	±9.609	±9.487	±9.580	±11	585,262	698,364	695,020	654,485	675,340	805,246
Colombia	0.812	0.721	0.701	0.741	0.926	1.037	150,431	150,544	111,193	111,835	146,947	164,500
Costa Rica	0.132	0.095	0.146	0.137	0.129	0.178	62,328	34,599	63,174	57,430	-	-
Cote d'Ivoire	±0.507	±0.362	±0.404	±0.388	-	-	-	-	-	-	-	-
Croatia	±1.486	±0.815	±0.487	-	-	-	233,579	111,075	90,021	-	-	-
Cyprus	0.117	0.129	0.118	0.109	0.123	0.125	820,829	914,082	748,861	714,445	805,045	870,987
Czechoslovakia (Former)	±2.320	±1.235	-	-	-	-	100,860	51,461	-	-	-	-
Denmark	*1.479	*1.773	*1.968	*1.879	*2.171	*2.651	§2,380	§2,815	§2,920	§2,787	§3,230	§3,951
Ecuador	0.562	0.627	0.817	0.936	1.511	1.325	337,341	444,678	579,594	355,531	663,624	521,684
Egypt	*2.280	*2.091	*2.689	*3.257	*3.304	*3.220	132,555	113,646	153,671	183,987	187,784	189,384
El Salvador	-	±0.162	-	±0.215	±0.144	±0.142	-	-	§3,364	§1,994	§2,058	
Finland	±2.902	±2.518	±2.274	±1.940	±2.098	±2.287	936,277	812,275	733,681	646,733	777,115	758,657

Continued.

Depending on the table, * means *Enterprises*, *Engaged*, or *Factor Values*; ± means *Basis Unspecified*; § means *shown in millions*. For additional notes and sources, see Appendix I.

363

Output and Output per Employee

- Continued -

Country	Output ($ bil.)						Output per Employee ($)					
	1990	1991	1992	1993	1994	1995	1990	1991	1992	1993	1994	1995
FYR Macedonia	±0.321	±0.263	±0.057	±0.099	±0.040	±0.045	225,875	196,791	43,963	75,069	72,506	96,265
France	±36	±35	±35	±34	±35	±39	§1,517	§1,713	§1,767	§1,891	§1,927	§2,281
Gabon	-	*0.141	*0.143	*0.132	*0.098	*0.135	-	280,852	331,339	277,280	215,972	331,986
Greece	*3.234	*2.992	*2.884	*2.520	*2.948	*3.462	698,076	727,790	702,046	657,288	710,121	821,079
Guatemala	-	0.112	0.101	0.105	0.117	0.155	-	§1,195	§1,073	§1,120	§1,248	§1,600
Honduras	0.199	0.166	0.135	0.004	0.004	0.003	931,118	783,581	614,218	25,966	20,988	-
Hungary	±1.685	±1.793	±2.214	±2.079	±2.148	±2.213	280,828	298,834	100,651	103,955	109,209	113,509
India	*11	*5.270	*6.780	*7.122	*8.005	*8.595	486,214	244,677	277,791	259,275	283,971	300,369
Indonesia	-	-	-	-	0.029	0.034	-	-	-	-	84,937	58,750
Iran (Islamic Republic of)	0.029	0.041	0.082	0.070	-	-	14,013	16,938	47,076	40,009	-	-
Iraq	-	*0.655	*2.492	-	-	-	-	78,676	311,495	-	-	-
Italy	±19	±21	-	-	-	-	§1,453	§1,391	-	-	-	-
Jamaica	0.423	0.331	0.298	-	-	-	-	-	-	-	-	-
Japan	±53	±60	±61	±65	±70	±74	§2,500	§2,855	§2,909	§2,952	§2,915	§3,217
Jordan	0.416	0.403	0.520	0.540	0.545	0.569	123,415	118,438	146,798	140,526	138,047	147,284
Kenya	*1.012	*1.100	*0.827	*0.487	*0.626	*0.556	§3,834	§4,166	§3,132	§1,844	§2,371	§2,092
Korea, Republic of	10	12	15	15	16	21	§1,412	§1,711	§1,923	§1,939	§1,916	§2,066
Kuwait	4.285	0.615	2.650	3.160	4.360	4.560	882,334	163,742	641,142	664,866	860,732	959,462
Latvia	0.000	0.000	0.001	0.001	0.001	0.002	405	1,065	10,968	1,679	7,679	21,266
Malaysia	*1.727	*1.740	*1.757	*1.651	*1.909	*2.433	§1,570	§1,450	§1,464	§1,270	707,166	788,779
Mozambique	-	±0.003	-	±0.001	±0.001	-	-	6,936	-	2,271	884	-
Netherlands	11	11	9.585	10	11	13	§1,771	§1,668	§1,520	§1,960	§2,131	§2,649
New Zealand	0.428	-	-	-	-	-	570,746	-	-	-	-	-
Norway	±2.753	±2.311	±2.338	±2.328	±2.224	±2.285	§2,309	§1,977	§1,993	§1,479	§1,357	§1,468
Pakistan	1.219	1.276	1.385	-	-	-	524,442	567,057	588,519	-	-	-
Peru	1.906	2.242	2.067	-	-	-	503,839	593,119	690,534	-	-	-
Philippines	2.751	3.255	3.401	3.376		-	§1,100	§1,356	§1,360	§1,298		-
Poland	±2.716	±3.264	±4.289	±5.015	±5.708	±7.177	169,770	181,312	268,040	313,421	357,638	450,539
Portugal	-	-	-	3.683	4.186	5.109	-	-	-	988,204	§1,178	§1,409
Puerto Rico	±2.732	±2.280	±2.117	±1.394	±1.295	±1.317	-	-	-	-	-	-
Qatar	0.460	0.430	0.451	0.353	0.380	-	743,400	718,689	789,534	626,413	681,890	-
Romania	±3.183	±2.104	±1.471	±1.596	-	-	122,894	82,497	58,156	58,455	-	-
Russian Federation	-	11	25	8.456	8.597	11	-	-	-	78,438	77,103	92,050
Senegal	0.145	-	-	-	-	-	443,808	-	-	-	-	-
Slovakia	-	±1.123	±0.901	±0.799	±0.844	-	-	131,416	112,305	106,365	130,522	-
South Africa	±3.658	±3.847	±3.653	±3.763	±3.830	±4.318	203,235	213,706	202,956	-	-	-
Spain	*9.215	*9.791	*10	±8.628	±8.434	±9.839	§1,274	§1,382	§1,459	§1,085	§1,140	§1,384
Sweden	*4.168	*3.768	*3.678	*3.064	*2.684	*3.008	§2,604	§2,636	§2,084	§2,113	§1,990	§2,224
Thailand	-	5.213	-	-	-	-	-	§2,145	-	-	-	-
Turkey	9.076	8.328	8.635	8.674	6.908	8.748	§1,807	§1,692	§1,747	§1,708	§1,352	§1,640
Ukraine	*2.452	*2.533	*8.780	*1.472	*0.831	*0.677	163,488	168,876	585,360	98,120	51,961	39,806
United Kingdom	*21	*19	32	-	-	-	§2,294	§2,334	§3,940	-	-	-
United Republic of Tanzania	*0.010	*0.016	-	-	-	-	27,061	32,834	-	-	-	-
USSR (Former)	±13	-	-	-	-	-	90,625	-	-	-	-	-
United States of America	*159	*145	*137	*130	*128	*136	§2,214	§1,964	§1,821	§1,778	§1,779	§1,943
Uruguay	-	0.657	0.580	0.165	0.502	0.528	-	362,329	390,534	268,683	437,793	488,325
Venezuela	5.861	4.255	4.306	3.939	2.732	2.156	825,505	545,486	559,280	597,577	421,815	354,155
Yugoslavia	±0.135	±0.142	±0.126	-	±0.251	±0.128	-	30,154	26,871	-	55,789	29,085

Value Added and Value Added per Employee

Country	Value Added ($ bil.)						Value Added per Employee ($)					
	1990	1991	1992	1993	1994	1995	1990	1991	1992	1993	1994	1995
Argentina	±6.069	-	-	2.650	-	-	599,846	-	-	374,838	-	-
Australia	*1.659	*1.896	*1.316	-	-	-	414,648	474,094	328,888	-	-	-
Austria	0.489	0.456	0.439	0.434	0.549	0.615	132,116	114,016	109,860	131,268	154,421	167,338
Bangladesh	0.009	0.015	0.006	-	-	-	11,202	16,267	7,133	-	-	-
Bolivia	0.336	0.426	0.333	0.306	0.336	0.326	557,803	696,400	310,075	286,854	309,558	297,082

Continued.

Depending on the table, * means *Enterprises, Engaged,* or *Factor Values*; ± means *Basis Unspecified*; § means *shown in millions*. For additional notes and sources, see Appendix I.

Value Added and Value Added per Employee

- Continued -

Country	Value Added ($ bil.)						Value Added per Employee ($)					
	1990	1991	1992	1993	1994	1995	1990	1991	1992	1993	1994	1995
Bosnia & Herzegovina	±0.096	±0.077	-	-	-	-	36,562	32,206	-	-	-	-
Bulgaria.	-	±0.241	±0.293	±0.246	-	-	-	-	-	-	-	-
Canada	2.271	2.077	1.688	1.744	1.919	1.953	151,411	148,381	140,647	134,158	159,881	162,868
Chile.	0.481	0.550	0.635	0.878	0.863	1.085	386,909	428,719	482,809	665,839	651,942	850,022
China	±2.714	±3.170	±3.454	±5.702	-	-	5,322	5,562	5,854	12,395	-	-
China (Taiwan Province) . . .	2.353	4.722	4.527	4.819	5.072	5.612	169,038	355,632	327,452	332,481	357,560	403,978
Colombia	0.151	0.139	0.220	0.135	0.307	0.345	28,021	29,024	34,974	20,439	48,778	54,793
Costa Rica	0.035	0.026	0.040	0.036	0.033	0.046	16,539	9,337	17,103	15,153	-	-
Cote d'Ivoire	0.459	0.330	0.363	0.353	-	-	-	-	-	-	-	-
Croatia	±0.123	±-0.755	±0.192	-	-	-	19,368	§-102,	35,418	-	-	-
Cyprus	0.007	0.008	0.009	0.010	0.010	0.014	47,956	57,473	59,122	66,888	62,434	97,441
Czechoslovakia (Former) . . .	±0.316	-	-	-	-	-	13,734	-	-	-	-	-
Denmark	*0.118	*0.174	*0.193	*0.184	*0.212	*0.259	189,417	275,470	285,756	272,723	316,051	386,516
Ecuador	0.374	0.435	0.653	0.640	0.989	0.949	224,623	308,311	462,837	243,177	434,260	373,824
Egypt	*1.156	*1.234	*1.740	*1.601	*1.625	*1.583	67,201	67,089	99,426	90,468	92,334	93,122
El Salvador.	-	0.049	-	0.060	0.007	0.084	-	-	-	937,289	95,133	§1,217
Finland	±0.674	±0.581	±0.370	±0.319	±0.432	±0.188	217,556	187,534	119,400	106,495	159,961	62,464
FYR Macedonia	±0.084	±0.022	±0.011	±0.055	±0.036	±0.036	58,972	16,734	8,619	41,762	64,790	77,759
France	±15	±16	±16	±16	±16	±19	646,559	774,197	793,743	889,623	910,273	§1,097
Gabon	-	*0.034	*0.040	*0.043	*0.022	*0.026	-	68,049	92,365	90,656	48,114	62,636
Greece	*0.295	*0.464	*0.542	*0.476	*0.571	*0.678	63,600	112,953	131,853	124,248	137,505	160,709
Honduras	*0.008	*0.006	*0.006	*0.001	*0.001	*0.001	36,743	29,930	26,128	7,218	6,065	-
Hungary	±0.460	±0.665	±1.009	±0.989	±1.057	±1.126	76,736	110,901	45,860	49,472	53,747	57,768
India	*1.072	*0.593	*1.079	*1.114	*1.271	*1.378	48,248	27,514	44,219	40,551	45,102	48,166
Indonesia	-	-	-	-	*0.028	*0.028	-	-	-	-	83,817	48,108
Iran (Islamic Republic of) . . .	0.016	0.023	0.012	0.025	-	-	7,852	9,679	6,974	14,197	-	-
Iraq	-	*0.562	*1.470	-	-	-	-	67,428	183,722	-	-	-
Israel.	0.115	0.112	-	-	-	-	-	-	-	-	-	-
Italy	±1.718	±2.065	-	-	-	-	131,525	134,309	-	-	-	-
Jamaica	0.151	0.111	0.072	-	-	-	-	-	-	-	-	-
Japan	±4.841	±8.292	±10	±14	±17	±15	230,547	394,852	497,058	614,372	720,331	653,145
Jordan	0.055	0.052	0.041	0.049	0.044	0.054	16,164	15,134	11,473	12,875	11,040	14,038
Kenya	*0.007	*0.007	*0.006	*0.004	*0.006	*0.008	26,447	27,540	23,514	14,368	21,625	29,239
Korea, Republic of	2.865	3.312	3.909	3.895	4.756	6.753	403,535	455,270	509,607	491,479	576,685	677,633
Kuwait	1.652	0.237	1.048	1.270	1.752	1.824	340,128	63,120	253,472	267,189	345,900	383,700
Latvia	-	-	-	*0.000	*0.001	*0.001	-	-	-	485	8,131	6,418
Malaysia	*0.199	*0.212	*0.194	*0.187	*0.466	*0.603	180,917	176,630	161,276	144,188	172,448	195,609
Netherlands	1.095	1.246	0.851	0.887	0.909	1.037	171,023	194,778	134,971	172,255	180,228	210,914
New Zealand	0.137	-	-	-	-	-	183,085	-	-	-	-	-
Norway	±0.195	±0.229	±0.143	±0.255	±0.221	±0.230	163,371	195,685	121,720	161,739	134,948	148,126
Pakistan	0.132	0.096	0.107	-	-	-	56,821	42,450	45,309	-	-	-
Peru	1.409	1.963	1.184	-	-	-	372,513	519,217	395,298	-	-	-
Philippines.	0.489	0.634	0.779	0.491	-	-	195,648	264,293	311,524	188,662	-	-
Poland	±1.419	±1.175	±1.979	±1.964	±2.223	±2.785	88,684	65,258	123,716	122,757	139,258	174,819
Portugal	-	-	-	1.952	2.178	2.647	-	-	-	523,681	612,977	730,228
Qatar	0.437	0.369	0.382	0.293	0.252	-	705,232	615,953	669,733	520,224	452,460	-
Romania	±-0.134	±0.295	±0.237	±0.267	-	-	§-5,16	11,551	9,357	9,779	-	-
Russian Federation	-	-	-	±1.815	±2.099	±3.283	-	-	-	16,832	18,825	28,104
Senegal.	0.027	0.010	0.016	0.013	0.017	-	82,923	34,261	67,141	43,702	73,049	-
Slovakia	-	-	-	±0.189	±0.183	-	-	-	-	25,112	28,326	-
South Africa	*1.243	*1.284	*1.307	*1.294	*1.317	*1.485	69,076	71,343	72,600	-	-	-
Spain	*1.348	*2.068	*2.217	±1.392	±1.073	±1.373	186,407	291,939	314,397	175,191	145,050	193,102
Sweden.	*1.325	*0.849	*0.355	*0.366	*0.289	*0.323	827,843	594,384	201,235	252,431	214,495	238,682
Thailand	-	5.140	-	-	-	-	-	§2,115	-	-	-	-
Turkey	4.525	4.566	5.101	5.026	3.557	4.503	901,298	927,839	§1,032	989,803	696,408	844,586
United Kingdom	*4.429	*2.627	*2.525	-	-	-	492,063	328,319	315,592	-	-	-
United Republic of Tanzania . .	*0.005	*0.006	-	-	-	-	13,038	11,816	-	-	-	-
United States of America . . .	*23	*20	*19	*19	*23	*26	316,944	267,568	254,827	254,575	325,486	375,243

Continued.

Depending on the table, * means *Enterprises, Engaged,* or *Factor Values;* ± means *Basis Unspecified;* § means *shown in millions.* For additional notes and sources, see Appendix I.

Value Added and Value Added per Employee
- Continued -

Country	Value Added ($ bil.)						Value Added per Employee ($)					
	1990	1991	1992	1993	1994	1995	1990	1991	1992	1993	1994	1995
Uruguay	0.239	0.319	0.279	0.086	0.246	0.259	115,741	175,813	187,656	139,787	214,430	239,180
Venezuela	4.734	2.688	3.232	2.957	1.980	2.015	666,813	344,626	419,768	448,535	305,689	330,965
Yugoslavia	±0.026	±0.024	±-0.034	-	±0.150	±0.060	-	5,152	-7,263	-	33,256	13,617

Capital Investment

Country	Gross Fixed Capital Formation ($ mil.)						Per Establishment ($ mil)					
	1990	1991	1992	1993	1994	1995	1990	1991	1992	1993	1994	1995
Argentina	-	-	-	280	-	-	-	-	-	3.371	-	-
Australia	180	-	-	-	-	-	8.594	-	-	-	-	-
Austria	106	100	127	54	54	-	3.541	3.461	4.715	2.003	2.157	-
Bahrain	-	-	35	-	-	-	-	-	35	-	-	-
Bangladesh	2.080	0.246	2.234	-	-	-	2.080	0.246	2.234	-	-	-
Bolivia	8.237	8.769	5.706	16	2.287	-	2.059	2.192	1.141	3.236	0.457	-
Bosnia & Herzegovina	7.411	-	-	-	-	-	3.706	-	-	-	-	-
Canada	858	-	-	-	-	-	14	-	-	-	-	-
Chile	11	16	32	36	51	-	5.625	7.923	16	18	26	-
Colombia	163	49	66	272	71	-	27	8.108	11	45	12	-
Croatia	14	13	12	23	29	33	2.748	2.639	2.083	3.752	4.758	5.449
Cyprus	0.385	0.315	0.431	0.489	3.093	1.170	0.385	0.315	0.431	0.489	3.093	1.170
Czechoslovakia (Former)	62	-	-	-	-	-	7.730	-	-	-	-	-
Denmark	29	47	-	-	-	-	9.748	16	-	-	-	-
Ecuador	97	215	19	-647	108	226	-	-	-	-	-	-
Egypt	303	190	211	237	-	-	38	24	26	30	-	-
El Salvador	-	9.406	-	1.406	3.495	2.149	-	9.406	-	1.406	3.495	2.149
Finland	70	102	56	89	20	48	18	25	14	30	6.598	3.427
Former Yugoslav Republic of Macedonia	0.188	0.424	0.118	1.581	3.634	0.337	0.188	0.424	0.118	0.790	3.634	0.168
Gabon	-	13	5.792	3.546	0.634	1.549	-	6.271	2.896	1.773	0.317	0.774
Greece	94	58	54	-	-	-	5.517	3.417	3.205	-	-	-
Hungary	60	94	293	221	-	-	12	7.857	49	17	-	-
India	377	353	384	405	-	-	4.769	4.302	4.866	3.371	-	-
Indonesia	-	-	-	-	-	0.196	-	-	-	-	-	0.049
Iran (Islamic Republic of)	1.809	0.934	2.564	7.347	-	-	0.201	0.093	0.256	0.735	-	-
Italy	712	853	-	-	-	-	26	34	-	-	-	-
Japan	1,264	1,975	2,337	4,703	2,896	-	6.983	10	13	25	11	-
Jordan	3.267	1.628	15	8.939	4.166	7.054	3.267	1.628	15	8.939	4.166	7.054
Korea, Republic of	1,325	-	1,020	730	1,056	2,430	133	-	73	66	70	270
Kuwait	82	63	84	182	148	-	82	63	84	182	148	-
Latvia	0.001	0.001	0.002	-	-	0.124	0.001	0.001	0.001	-	-	0.062
Malaysia	52	85	48	49	397	-	4.705	8.509	6.036	6.070	66	-
New Zealand	3.582	-	-	-	-	-	0.512	-	-	-	-	-
Norway	221	37	33	49	37	-	74	12	11	0.824	0.569	-
Pakistan	-	3.816	-	-	-	-	-	1.272	-	-	-	-
Peru	25	0.700	6.206	-	-	-	0.966	0.027	0.222	-	-	-
Philippines	67	162	194	386	-	-	17	40	48	96	-	-
Poland	76	125	127	164	-	-	9.496	14	16	16	-	-
Portugal	-	-	-	392	-33	-	-	-	-	392	-33	-
Romania	385	307	-	24	23	-	-	31	-	2.180	1.673	-
Senegal	0.635	-	-	-	-	-	0.635	-	-	-	-	-
Spain	603	539	366	324	248	246	60	54	37	-	-	-
Thailand	-	147	-	-	-	-	-	15	-	-	-	-
Turkey	40	77	194	53	3.884	-	7.974	15	39	11	0.777	-
Ukraine	7.554	14	32	14	110	44	0.755	0.974	2.929	1.439	11	4.436
United Kingdom	573	710	707	-	484	-	23	28	26	-	-	-
United Republic of Tanzania	1.251	1.237	-	-	-	-	1.251	1.237	-	-	-	-
United States of America	3,820	5,600	6,140	5,986	5,518	5,876	-	-	26	-	-	-

Continued.

Capital Investment
- Continued -

Country	Gross Fixed Capital Formation ($ mil.)						Per Establishment ($ mil)					
	1990	1991	1992	1993	1994	1995	1990	1991	1992	1993	1994	1995
Uruguay	1.121	0.412	0.009	6.480	-	-	-	-	-	-	-	-
Venezuela	178	296	434	929	1,464	516	13	23	39	93	146	57
Yugoslavia	1.061	1.173	1.064	-	2.533	3.173	0.177	0.168	0.266	-	0.507	0.529

Indexes of Industrial Production

Country	Index of Industrial Production (1990=100)						Index of Employment (1990=100)					
	1990	1991	1992	1993	1994	1995	1990	1991	1992	1993	1994	1995
Albania	100	-	-	-	-	-	-	-	-	-	-	-
Algeria	100	100	102	99	91	96	-	-	-	-	-	-
Angola	100	100	101	101	-	-	-	-	-	-	-	-
Argentina	100	102	-	-	-	-	100	-	-	70	59	-
Australia	100	100	101	110	-	-	100	100	100	-	-	-
Austria	100	103	105	106	106	101	100	108	108	89	96	99
Azerbaijan	-	-	-	-	-	-	100	99	103	120	131	-
Bahrain	100	103	103	99	98	101	-	-	-	-	-	-
Bangladesh	100	140	131	171	155	177	100	115	109	-	-	-
Barbados	100	95	81	90	92	96	-	-	-	-	-	-
Belgium	100	91	86	74	73	70	-	-	-	-	-	-
Belize	100	100	100	100	100	100	-	-	-	-	-	-
Bermuda	-	-	-	-	-	-	100	189	316	342	368	-
Bhutan	100	-	-	-	-	-	-	-	-	-	-	-
Bolivia	100	102	101	99	106	109	100	102	179	177	180	182
Bosnia & Herzegovina . . .	-	-	-	-	-	-	100	91	-	-	-	-
Botswana	100	100	100	-	-	-	-	-	-	-	-	-
Brazil	100	92	92	96	102	102	-	-	-	-	-	-
Bulgaria.	100	82	68	60	82	96	-	-	-	-	-	-
Burkina-Faso	100	100	100	-	-	-	-	-	-	-	-	-
Burundi	100	100	100	100	-	-	-	-	-	-	-	-
Cambodia	100	-	-	-	-	-	100	87	-	-	-	-
Canada	100	99	96	99	101	102	100	93	80	87	80	80
Central African Republic . .	100	100	100	100	100	-	-	-	-	-	-	-
Chile	100	102	106	112	119	130	100	103	106	106	107	103
China	100	-	-	-	-	-	100	112	116	90	-	-
China (Hong Kong SAR) . .	100	100	100	100	100	100	-	-	-	-	-	-
China (Taiwan Province) . .	100	102	104	122	118	144	100	95	99	104	102	100
Colombia	100	103	90	95	98	98	100	89	117	123	117	117
Costa Rica	100	90	128	131	140	183	100	130	110	113	-	-
Croatia	100	-	-	-	-	-	100	115	85	81	79	78
Cyprus	100	111	105	-	-	-	100	99	110	106	107	100
Czechoslovakia (Former) . .	100	90	84	-	-	-	100	104	-	-	-	-
Czech Republic	100	-	-	-	-	-	100	123	108	108	-	-
Dem. Rep. of the Congo . .	100	105	105	-	-	-	-	-	-	-	-	-
Denmark	100	102	103	104	111	117	100	101	109	109	108	108
Dominican Republic. . . .	100	126	125	-	-	-	-	-	-	-	-	-
Ecuador	100	130	130	156	-	-	100	85	85	158	137	152
Egypt	100	122	121	117	114	103	100	107	102	103	102	99
El Salvador.	100	106	-	-	-	-	-	-	-	-	-	-
Ethiopia and Eritrea	100	105	109	-	-	-	-	-	-	-	-	-
Fiji	100	100	100	100	100	100	-	-	-	-	-	-
Finland	100	105	107	103	124	118	100	100	100	97	87	97
FYR Macedonia	100	-	-	-	-	-	100	94	92	93	39	33
France	100	105	102	107	104	106	100	88	85	77	77	74
Gabon	100	103	101	-	-	-	-	-	-	-	-	-
Gambia	100	-	-	-	-	-	-	-	-	-	-	-
Germany (Eastern Part) . .	100	-	-	-	-	-	-	-	-	-	-	-
Ghana	100	-	-	-	-	-	-	-	-	-	-	-
Greece	100	87	100	88	102	105	100	89	89	83	90	91

Continued.

Depending on the table, * means *Enterprises, Engaged,* or *Factor Values*; ± means *Basis Unspecified*; § means *shown in millions.* For additional notes and sources, see Appendix I.

367

Indexes of Industrial Production
- Continued -

Country	Index of Industrial Production (1990=100)						Index of Employment (1990=100)					
	1990	1991	1992	1993	1994	1995	1990	1991	1992	1993	1994	1995
Guatemala	100	97	120	-	-	-	-	-	-	-	-	-
Guyana	100	100	100	100	100	100	-	-	-	-	-	-
Haiti	100	100	100	100	100	100	-	-	-	-	-	-
Honduras	100	109	140	36	70	231	100	99	103	75	82	-
Hungary	100	86	88	84	86	87	100	100	367	333	328	325
Iceland	100	100	100	-	-	-	-	-	-	-	-	-
India	100	99	106	104	109	112	100	97	110	124	127	129
Indonesia	100	103	109	108	-	-	-	-	-	-	-	-
Iran (Islamic Republic of)	100	107	113	120	120	123	100	115	83	83	-	-
Iraq	100	78	126	132	-	-	-	-	-	-	-	-
Israel	100	104	115	127	140	145	-	-	-	-	-	-
Italy	100	100	103	105	-	-	100	118	-	-	-	-
Jamaica	100	91	123	-	-	-	-	-	-	-	-	-
Japan	100	106	112	115	120	121	100	100	100	105	114	110
Jordan	100	90	109	110	113	119	100	101	105	114	117	114
Kenya	-	-	-	-	-	-	100	100	100	100	100	101
Korea, Republic of	100	129	164	179	184	208	100	102	108	112	116	140
Kuwait	100	18	58	83	-	-	100	77	85	98	104	98
Kyrgyzstan	100	-	-	-	-	-	-	-	-	-	-	-
Latvia	100	-	-	-	-	-	100	92	90	298	65	66
Lebanon	100	123	77	77	-	-	-	-	-	-	-	-
Libyan Arab Jamahiriya	100	101	101	100			-	-	-	-	-	-
Lithuania	100	-	-	-	-	-	-	-	-	-	-	-
Luxembourg	100	100	100	100	100	100	-	-	-	-	-	-
Macau	100	100	100	-	-	-	-	-	-	-	-	-
Madagascar	100	115	118	-	-	-	-	-	-	-	-	-
Malawi	100	100	100	100	100	100	-	-	-	-	-	-
Malaysia	100	106	110	119	139	164	100	109	109	118	245	280
Malta	100	100	100	100	-	-	-	-	-	-	-	-
Mauritius	100	100	100	100	100	100	-	-	-	-	-	-
Mexico	100	106	109	106	112	109	-	-	-	-	-	-
Morocco	100	94	105	107	114	105	-	-	-	-	-	-
Mozambique	100	-	-	-	-	-	100	99	100	101	163	-
Myanmar	100	100	102	104	-	-	-	-	-	-	-	-
Nepal	100	100	100	-	-	-	-	-	-	-	-	-
Netherlands	100	99	101	98	102	107	100	100	98	80	79	77
Netherlands Antilles	100	96	102	-	-	-	-	-	-	-	-	-
New Zealand	100	94	104	-	-	-	100	100	89	-	-	-
Nicaragua	100	105	108	-	-	-	-	-	-	-	-	-
Norway	100	95	106	106	110	99	100	98	98	132	138	131
Pakistan	100	127	130	-	-	-	100	97	101	-	-	-
Panama	100	97	150	136	97	112	-	-	-	-	-	-
Papua New Guinea	100	100	100	-	-	-	-	-	-	-	-	-
Paraguay	100	89	87	70	72	52	-	-	-	-	-	-
Peru	100	102	103	107	111	109	100	100	79	-	-	-
Philippines	100	108	114	-	-	-	100	96	100	104	-	-
Poland	100	86	100	112	120	127	100	113	100	100	100	100
Portugal	100	89	101	97	118	126	-	-	-	-	-	-
Puerto Rico	100	90	79	-	-	-	-	-	-	-	-	-
Qatar	100	-	-	-	-	-	100	97	92	91	90	-
Republic of Moldova	100	-	-	-	-	-	-	-	-	-	-	-
Romania	100	64	57	56	65	66	100	98	98	105	-	-
Russian Federation	100	-	-	-	-	-	-	-	-	-	-	-
Saudi Arabia	100	93	99	101	-	-	-	-	-	-	-	-
Senegal	-	-	-	-	-	-	100	86	75	92	70	-
Slovakia	100	-	-	-	-	-	-	-	-	-	-	-
Slovenia	100	111	111	110	71	108	-	-	-	-	-	-
Somalia	100	-	-	-	-	-	-	-	-	-	-	-
South Africa	100	99	103	118	125	140	100	100	100	-	-	-
Spain	100	102	106	101	103	97	100	98	98	110	102	98

Continued.

Depending on the table, * means *Enterprises, Engaged,* or *Factor Values*; ± means *Basis Unspecified*; § means *shown in millions*. For additional notes and sources, see Appendix I.

Indexes of Industrial Production
- Continued -

Country	Index of Industrial Production (1990=100)						Index of Employment (1990=100)					
	1990	1991	1992	1993	1994	1995	1990	1991	1992	1993	1994	1995
Sudan	100	105	106	-	-	-	-	-	-	-	-	-
Sweden	100	85	97	114	93	94	100	89	110	91	84	84
Switzerland	100	136	108	-	-	-	-	-	-	-	-	-
Syrian Arab Republic	100	99	100	106	110	116	-	-	-	-	-	-
Thailand	100	93	115	-	-	-	-	-	-	-	-	-
Tonga	100	-	-	-	-	-	-	-	-	-	-	-
Trinidad and Tobago	100	92	105	99	97	88	-	-	-	-	-	-
Tunisia	100	103	98	104	99	107	-	-	-	-	-	-
Turkey	100	98	101	111	112	123	100	98	98	101	102	106
Uganda	100	100	100	100	100	100	-	-	-	-	-	-
Ukraine	100	-	-	-	-	-	100	100	100	100	107	113
United Arab Emirates	100	102	104	-	-	-	-	-	-	-	-	-
United Kingdom	100	104	104	108	104	104	100	89	89	-	-	-
United Republic of Tanzania . .	100	108	111	106	105	-	100	135	-	-	-	-
USSR (Former)	100	-	-	-	-	-	100	-	-	-	-	-
United States of America . . .	100	99	100	102	102	104	100	103	104	101	100	97
Uruguay	100	102	91	30	2	90	100	88	72	30	56	52
Venezuela	100	-	-	-	-	-	100	110	108	93	91	86
Yemen	100	107	140	137	-	-	-	-	-	-	-	-
Yugoslavia	100	-	-	-	-	-	-	-	-	-	-	-
Yugoslavia (Former)	100	60	-	-	-	-	-	-	-	-	-	-
Zambia	100	101	96	90	68	60	-	-	-	-	-	-
Zimbabwe	100	100	87	82	94	84	-	-	-	-	-	-

Depending on the table, * means *Enterprises, Engaged,* or *Factor Values;* ± means *Basis Unspecified;* § means *shown in millions.* For additional notes and sources, see Appendix I.

369

Representative Companies in Sector

Name	Address	Telephone	Fax	Employment	Y
Mitie Group PLC	Barley Wood, Wrington BS18 7SA, United Kingdom	934 862006	934 862239	9,750	94
Amerada Hess Corp.	1185 Ave. of the Amer., New York, New York 10036, U.S.A.	212-997-8500		9,085	96
Phillips 66 Div.	974 Adams Bldg., Bartlesville, Oklahoma 74004, U.S.A.	918-661-4400	918-661-7636	9,000	
Chevron U.S.A. Products Company Inc.	575 Market St., San Francisco, California 94105, U.S.A.	415-894-7700	415-894-8552	8,900	96
Conoco UK Ltd.	116 Park St., London W1Y 4NN, United Kingdom	71 4086000	71 4086660	7,286	93
RWE-DEA AG fur Mineraloel und Chemie	Ueberseering 40, Hamburg D-22297, Germany	40 63750	40 63753496	7,243	94
MAPCO Inc.	PO Box 645, Tulsa, Oklahoma 74101, U.S.A.	918-599-3712	918-581-1893	6,508	97
Quaker State Corp.	225 E. John Carpenter Fwy., Irving, Texas 75062, U.S.A.	214-868-0400		6,002	97
CITGO Petroleum Corp.	PO Box 3758, Tulsa, Oklahoma 74102, U.S.A.	918-495-4000		5,200	97
Petroecuador	PO Box 50075008, Quito, Ecuador	521436	569738	5,102	92
Ashland Petroleum Co.	PO Box 391, Ashland, Kentucky 41114, U.S.A.	606-329-3333	606-329-3447	5,079	95
Mobil Holdings Ltd.	54-60 Victoria St., London SW1E 6QB, United Kingdom	71 8289777		4,260	93
Ultramar PLC	141 Moorgate, London EC2M 6TX, United Kingdom	71 2566080	71 2568556	4,227	92
Esso Petroleum Co. Ltd.	Victoria St., London SO4 1TX, United Kingdom	703 892511		4,226	91
Mobil Chemical Company Inc.	3225 Gallows Rd., Fairfax, Virginia 22037-0001, U.S.A.	703-846-3000	703-846-2313	4,000	97
Shell Canada Ltd.	400 4 Ave. SW, Calgary, Alberta T2P 0J4, Canada	403-691-3111	403-269-8031	3,700	98
Przyjazn	Dabrowa Gornicza PL-42-523, Poland	32 641213	32 255410	3,200	93
ARCO Products Co.	333 S. Hope St., Los Angeles, California 90071, U.S.A.	213-486-3511		3,126	97
Crown Central Petroleum Corp.	PO Box 1168, Baltimore, Maryland 21203, U.S.A.	410-539-7400	410-659-4747	2,904	97
Nova Chemicals Ltd.	645 7 Ave. SW 23rd Floor, Calgary, Alberta T2P 4G8, Canada	403-750-3600	403-269-7410	2,710	98
FINA Inc.	PO Box 2159, Dallas, Texas 75221, U.S.A.	214-750-2400	214-750-2355	2,664	97
Federated Co-operatives Ltd.	401 22nd St. E, Saskatoon, Saskatchewan S7K 0H2, Canada	306-244-3311	306-244-3403	2,450	98
Giant Industries Inc.	23733 N. Scottsdale Rd., Scottsdale, Arizona 85255, U.S.A.	602-585-8888	602-585-8893	2,450	97
Champion Laboratories Inc.	200 S. 4th St., Albion, Illinois 62806-1313, U.S.A.	618-445-6011	618-445-3107	2,275	97
Koch Refining Co.	PO Box 2302, Wichita, Kansas 67201, U.S.A.	316-828-5119	316-828-8088	2,000	94
Societe Zaire Shell	BP 2799, Kinshasa, Zaire	21368	21550	2,000	
Shell International Petroleum Co. Ltd.	2 York Rd., London SE1 7NA, United Kingdom	71 9341234	71 9348060	1,898	94
Valero Energy Corp.	PO Box 500, San Antonio, Texas 78292, U.S.A.	210-370-2000	210-246-2103	1,855	97
Mobil Oil Co. Ltd.	54-60 Victoria St., London SW1E 6QB, United Kingdom	71 8289777		1,753	93
United Refining Co.	15 Bradley St., Warren, Pennsylvania 16365, U.S.A.	814-723-1500	814-723-8899	1,706	97
Oil Refineries Ltd.	Hahistadrut Blvd., Haifa Bay 31000, Israel	4 8788111	4 8728319	1,650	97
Badak Natural Gas Liquefaction Co. PT	Jl. M.H. Thamrin No. 59, Jakarta 10350, Indonesia	21330243	21 3322974	1,500	97
Ethyl Corp.	PO Box 2189, Richmond, Virginia 23218, U.S.A.	804-788-5000	804-788-5612	1,500	97
Unocal Corp.	2141 Rosecrans Ave., Ste. 4000, El Segundo, California 90245, U.S.A.	310-726-7600	213-977-7591	1,400	97
Motor Oil Hellas Corinth Refineries SA	2 Karageorgi Servias, Athens GR-105 62, Greece	3246311	3222337	1,325	
Rexene Corp.	5005 LBJ Frwy., Ste. 500, Dallas, Texas 75244, U.S.A.	972-450-9000	972-450-9197	1,320	96
Lyondell Petrochemical Co.	PO Box 3646, Houston, Texas 77253-3646, U.S.A.	713-652-7200	713-652-4598	1,300	97
Canadian Occidental Petroleum Ltd.	635 8 Ave. SW Ste. 1500, Calgary, Alberta T2P 3Z1, Canada	403-234-6700	403-263-8673	1,216	98
Refinadora Costarricense de Petroleo, SA Recope	Apartado 4351-1000, San Jose, Costa Rica	2576544	2554993	1,200	
Coalite Products Ltd.	Bolsover, Chesterfield S44 6AB, United Kingdom	246 822281		1,172	93
BHP Hawaii Inc.	PO Box 3379, Honolulu, Hawaii 96842, U.S.A.	808-547-3111	808-547-3145	1,150	95
Thai Oil Refinery	Yannawa, Bangkok 10120, Thailand	2 2409200		1,018	94
Esso Singapore Private Ltd.	1 Raffles Pl., 28th-38th Fls., Singapore 0104, Singapore	5355533	5352796	1,000	
Tesoro Petroleum Corp.	PO Box 17536, San Antonio, Texas 78217, U.S.A.	210-828-8484	210-828-8600	1,000	97
Petroleos del Norte, SA	Av. Zugazarte, 29, Guecho E-48930, Spain	94 4818700	94 4635966	849	97
Nafta Lendava d.o.o.	Rudarska 1, Lendava SLO-69220, Slovenia	69 75201	69 75621	817	95
Ergon Inc.	PO Drawer 1639, Jackson, Mississippi 39215-1639, U.S.A.	601-948-3472		800	
Louisiana Land and Exploration Co.	PO Box 60350, New Orleans, Louisiana 70160, U.S.A.	504-566-6500		745	95
Kenyon UK	Pontardawe, Swansea SA8 4EW, United Kingdom	792 862424	792 830354	744	91
Elf Oil Co. Ltd.	Wembley, Middlesex HA9 0ND, United Kingdom	81 9028820	81 9020448	689	93
Petroleum Corp. of N.Z.	PO Box 1818, Wellington, New Zealand	04 739812		653	87
Holly Corp.	100 Crescent Court, Ste. 1600, Dallas, Texas 75201-6927, U.S.A.	214-871-3555	214-871-3560	572	97

Continued

Representative Companies in Sector
- Continued -

Name	Address	Telephone	Fax	Employment	Y
Uniwax SA	BP 3994, Abidjan 1, Cote d'Ivoire	454412	454942	524	93
Fuchs (UK) PLC	Belper, Derby DE5 1WF, United Kingdom	773 824151	773 823659	504	93
BP Oil Espana, SA	Maria de Molina, 6, Madrid E-28006, Spain	91 3629292	91 3629608	501	97
Consumers' Co-Operative Refineries Ltd.	550 East 9th Ave. N, Regina, Saskatchewan S4N 7L5, Canada	306-721-5353	306-721-5348	460	98
Aberdeen Service Co. North Sea Ltd.	PO Box 19, Peterhead AB42 6UW, United Kingdom	779 474712	779 470549	452	93
Chevron Canada Ltd.	1050 W Pender St. Ste. 1500, Vancouver, British Columbia V6E 3T4, Canada	604-668-5300	604-668-5304	450	98
Pride Companies L.P.	PO Box 3237, Abilene, Texas 79604, U.S.A.	915-674-8000	915-676-8792	437	96
Repsol (UK) Ltd.	66 Hammersmith Rd., London W14 8YH, United Kingdom	71 3711166		413	93
Navajo Refining Co.	501 E. Main, Artesia, New Mexico 88210, U.S.A.	505-748-3311	505-746-6410	400	95
Fedgas Pty. Ltd.	PO Box 4200, Alrode 1451, Republic of South Africa	11 8642130	11 8644519	380	95
Texas Olefins Co.	8707 Katy Fwy., 300, Houston, Texas 77024, U.S.A.	713-461-2223		360	95
Texas Petrochemicals Corp.	8600 Park Place Blvd., Houston, Texas 77017, U.S.A.	713-477-9211		360	95
Thanomwongse Service Co. Ltd.	Phayathai, Bangkok 10400, Thailand	2710205	2711601	350	90
Coastal Refining and Marketing Inc.	9 Greenway Plz., Houston, Texas 77046, U.S.A.	713-877-1400		329	95
Marathon Oil Co. Texas Refining Div.	PO Box 1191, Texas City, Texas 77592-1191, U.S.A.	409-945-2331		300	94
Wainoco Oil Corp.	10000 Memorial Dr., 600, Houston, Texas 77024-3411, U.S.A.	713-688-9600	713-688-0616	291	97
Gulf Oil Refining Ltd.	Waterston, Milford Haven SA73 1DR, United Kingdom	646 692461	646 25837	282	93
International Group Inc.	PO Box 383, Wayne, Pennsylvania 19087, U.S.A.	610-687-9030		250	94
Carless Refining & Marketing Ltd.	Eastern Rd., Romford RM1 3NH, United Kingdom	708 755557	708 753890	220	93
Suprachem	PO Box 98048, Westpark 0146, Republic of South Africa	12 3861165	12 3861141	212	95
PTT Exploration & Production Co., Ltd.	Chatuchak, Bangkok 10900, Thailand	5373600	5373645	200	94
Boardman Bros. Pty. Ltd.	PO Box 175, Stutterheim 4930, Republic of South Africa	436 31328	436 32835	188	92
Heto Holten AS	Gydevang 17-19, Alleroed DK-3450, Denmark	48142777	42274655	180	93
Tesoro Alaska Petroleum Co.	PO Box 196272, Anchorage, Alaska 99519-0272, U.S.A.	907-561-5521	907-561-5047	180	96
Parkland Industries Ltd.	4919 59 St. Ste. 236, Red Deer, Alberta T4N 6C9, Canada	403-343-1515	403-346-3015	170	98
Se An Industrial Co., Ltd.	So-Gu, Inchon, Republic of Korea	32 5711955	32 5716180	170	97
Mi-Chang Oil Industries Co., Ltd.	Yongdo-Gu, Pusan, Republic of Korea	51 4036442	51 4036440	155	97
Chemlite Holdings Pty. Ltd.	Denver 2027, Johannesburg 25720, Republic of South Africa	11 8739744	11 8252957	150	98
Cie. Franco-Tunisienne des Petroles	10, rue Jacques Cartier, Tunis, Tunisia	1 287114	1 894400	150	93
CITGO Asphalt Refining Co.	PO Box 3000, Blue Bell, Pennsylvania 19422, U.S.A.	215-542-4020	215-542-4040	150	97
International Group Inc.	50 Salome Dr., Scarborough, Ontario M1S 2A8, Canada	416-293-4151	416-293-0344	140	98
Hankook Shell Oil Co., Ltd.	Sodaemun-gu, Seoul, Republic of Korea	2 3643631	2 3933274	132	97
Ziegler Chemical and Mineral Corp.	100 Jericho Quadrangle, Jericho, New York 11753, U.S.A.	516-681-9600	516-681-9604	130	95
Tarconord AS	Avernakke, Nyborg DK-5800, Denmark	65301110	65301120	120	93
Repsol Derivados, SA	Orense 34, Tercero A, Madrid E-28020, Spain	91 3487800	91 5557779	103	97
Chunmi Lubricants Co., Ltd.	Yongdungpo-gu, Seoul, Republic of Korea	2 7844151	2 7862207	100	97
Cross Oil Refining and Marketing Inc. of Arkansas	PO Box 105, Smackover, Arkansas 71762, U.S.A.	501-725-3611		100	95
DeMenno-Kerdoon	2000 N. Alameda St., Compton, California 90222, U.S.A.	310-537-7100		90	94
Huntway Partners L.P.	PO Box 7033, Van Nuys, California 91409, U.S.A.	805-286-1582		81	96
Sleaford Trading Co. Ltd.	East Rd. Industrial Estate, Sleaford NG34 7JX, United Kingdom	529 305000	529 413720	81	93
Quaker State Inc.	1101 Blair Rd., Burlington, Ontario L7M 1T3, Canada	905-335-5577	905-332-6406	70	98
Powerine Oil Co.	PO Box 2108, Santa Fe Springs, California 90670-9883, U.S.A.	310-944-6111		60	95
OEMV - MAPETROL d.o.o.	Linhartova 17A, Maribor 2001, Slovenia	62 108108	62 31990	59	95
Enterprise Oil Norge Ltd.	Trafalgar Sq., London WC2N 5EJ, United Kingdom	71 9254000		56	93
Eastham Refinery Ltd.	Ellesmere Port, South Wirral L65 1AJ, United Kingdom	51 3272222	51 3276893	52	93
Hemisphere Oil Co. Inc.	Rd. 901, Km. 2.8, Barrio Camino Nuevo, Yabucoa 00767, Puerto Rico	787893-2320		50	92
Enerchem International Inc.	1400 8 St., Nisku, Alberta T9E 7M1, Canada	403-955-3388	403-955-2064	40	98
Moore and Munger Marketing Inc.	Two Corporate Dr., Ste. 434, Shelton, Connecticut 06484, U.S.A.	203-925-4300		25	95
Battenfeld Grease (Canada) Ltd.	68 Titan Rd., Toronto, Ontario M8Z 2J8, Canada	416-239-1548	416-239-3449	24	98
Parkland Refining Ltd.	236-4919 59 St., Red Deer, Alberta T4N 6C9, Canada	403-357-6400	403-346-3015	23	98

Continued

Representative Companies in Sector
- Continued -

Name	Address	Tele- phone	Fax	Employ- ment	Y
Britoil PLC	301 St. Vincent St., Glasgow G2 5DD, United Kingdom	41 2042525	41 2255050	18	93
Nelson House Forest Industries	PO Box 457, Nelson House, Manitoba R0B 1A0, Canada	204-484-2084	204-484-2197	17	98
BP Turkey Refining Ltd.	Moor Ln., London EC2Y 9BU, United Kingdom	71 9208000	71 9208263	8	92

For sources and notes, see Appendix I.

Product Tables
Aviation Gasoline

Unit of Measure: Metric tons.

	1990 Value	%	1991 Value	%	1992 Value	%	1993 Value	%	1994 Value	%	1995 Value	%	1996 Value	%
Total Production	2,242,722	100.0	2,008,467	100.0	2,054,665	100.0	2,065,807	100.0	2,006,636	100.0	2,009,949	100.0	1,988,866	100.0
Regions														
Africa	193,694	8.6	201,078	10.0	199,461	9.7	201,844	9.8	205,228	10.2	213,611	10.6	224,936	11.3
America, North	1,148,000	51.2	1,075,000	53.5	1,064,000	51.8	1,037,000	50.2	1,083,000	54.0	1,078,000	53.6	1,003,305	50.4
America, South	138,000	6.2	154,000	7.7	112,982	5.5	132,918	6.4	158,855	7.9	147,791	7.4	132,689	6.7
Asia	460,528	20.5	315,000	15.7	412,000	20.1	340,000	16.5	288,000	14.4	295,000	14.7	347,543	17.5
Europe	143,500	6.4	139,389	6.9	167,222	8.1	229,044	11.1	164,554	8.2	166,547	8.3	154,365	7.8
Oceania	159,000	7.1	124,000	6.2	99,000	4.8	125,000	6.1	107,000	5.3	109,000	5.4	126,029	6.3
Leading Producers														
United States of America	991,000	44.2	932,000	46.4	927,000	45.1	890,000	43.1	916,000	45.6	909,000	45.2	*867,381	43.6
Australia	159,000	7.1	124,000	6.2	99,000	4.8	125,000	6.1	107,000	5.3	109,000	5.4	*126,029	6.3
Canada	93,000	4.1	87,000	4.3	80,000	3.9	88,000	4.3	107,000	5.3	110,000	5.5	*86,867	4.4
Iran, Islamic Republic of	150,000	6.7	106,000	5.3	110,000	5.4	90,000	4.4	100,000	5.0	100,000	5.0	*118,914	6.0

Jet Fuel

Unit of Measure: 1,000 Metric tons.

	1990 Value	%	1991 Value	%	1992 Value	%	1993 Value	%	1994 Value	%	1995 Value	%	1996 Value	%
Total Production	168,504	100.0	164,255	100.0	170,830	100.0	171,939	100.0	180,426	100.0	186,012	100.0	195,840	100.0
Regions														
Africa	8,860	5.3	6,801	4.1	7,956	4.7	9,258	5.4	12,925	7.2	16,256	8.7	12,960	6.6
America, North	78,803	46.8	76,018	46.3	74,334	43.5	74,952	43.6	76,536	42.4	75,074	40.4	83,401	42.6
America, South	7,725	4.6	8,045	4.9	8,106	4.7	8,276	4.8	8,546	4.7	8,594	4.6	9,529	4.9
Asia	37,401	22.2	37,957	23.1	43,713	25.6	41,278	24.0	42,188	23.4	46,195	24.8	47,578	24.3
Europe	30,262	18.0	29,848	18.2	30,886	18.1	31,979	18.6	33,774	18.7	33,420	18.0	35,631	18.2
Oceania	5,453	3.2	5,586	3.4	5,835	3.4	6,196	3.6	6,458	3.6	6,472	3.5	6,740	3.4
Leading Producers														
United States of America	69,957	41.5	67,605	41.2	65,932	38.6	66,840	38.9	68,044	37.7	66,548	35.8	*74,265	37.9
Russian Federation	*10,752	6.4	*10,752	6.5	13,681	8.0	11,126	6.5	9,184	5.1	9,015	4.8	*10,752	5.5
United Kingdom	7,541	4.5	7,037	4.3	7,681	4.5	8,341	4.9	7,697	4.3	7,837	4.2	*8,686	4.4
Singapore	6,370	3.8	6,425	3.9	6,460	3.8	6,600	3.8	6,800	3.8	6,900	3.7	*7,975	4.1
Netherlands	4,940	2.9	4,730	2.9	4,978	2.9	4,938	2.9	5,500	3.0	5,997	3.2	*5,943	3.0

Motor Gasoline

Unit of Measure: 1,000 Metric tons.

	1990 Value	%	1991 Value	%	1992 Value	%	1993 Value	%	1994 Value	%	1995 Value	%	1996 Value	%
Total Production	773,252	100.0	799,992	100.0	857,175	100.0	878,820	100.0	880,315	100.0	901,827	100.0	911,776	100.0
Regions														
Africa	33,418	4.3	32,002	4.0	32,647	3.8	33,245	3.8	33,774	3.8	33,331	3.7	36,657	4.0
America, North	349,351	45.2	350,443	43.8	351,903	41.1	365,758	41.6	369,533	42.0	380,966	42.2	384,754	42.2
America, South	36,328	4.7	36,130	4.5	37,990	4.4	40,618	4.6	42,366	4.8	44,722	5.0	45,115	4.9
Asia	165,183	21.4	168,650	21.1	211,671	24.7	213,591	24.3	209,679	23.8	215,962	23.9	217,103	23.8
Europe	172,790	22.3	196,518	24.6	206,448	24.1	208,679	23.7	208,032	23.6	209,795	23.3	211,120	23.2
Oceania	16,182	2.1	16,249	2.0	16,516	1.9	16,929	1.9	16,930	1.9	17,051	1.9	17,028	1.9
Leading Producers														
United States of America	298,814	38.6	299,534	37.4	303,897	35.5	316,055	36.0	313,977	35.7	325,854	36.1	331,722	36.4
Japan	31,067	4.0	32,678	4.1	34,013	4.0	35,295	4.0	36,671	4.2	37,409	4.1	*36,737	4.0
Russian Federation	-		-		35,289	4.1	30,147	3.4	26,903	3.1	28,140	3.1	26,800	2.9
Germany	-		24,622	3.1	25,330	3.0	26,848	3.1	26,419	3.0	25,875	2.9	26,104	2.9
Canada	26,357	3.4	25,827	3.2	25,507	3.0	26,555	3.0	27,077	3.1	27,567	3.1	*26,867	2.9

Commodity data are provided by the United Nations Statistical Division. The symbol * means that data are estimated. For additional notes, see Appendix I.

373

Product Tables
Naphthas

Unit of Measure: 1,000 Metric tons.

	1990 Value	%	1991 Value	%	1992 Value	%	1993 Value	%	1994 Value	%	1995 Value	%	1996 Value	%
Total Production	115,546	100.0	122,288	100.0	132,970	100.0	131,300	100.0	135,332	100.0	140,347	100.0	140,699	100.0
Regions														
Africa	14,826	12.8	12,144	9.9	13,424	10.1	14,455	11.0	15,275	11.3	15,081	10.7	16,537	11.8
America, North	24,720	21.4	24,435	20.0	27,431	20.6	25,849	19.7	23,905	17.7	24,219	17.3	23,750	16.9
America, South	8,512	7.4	7,641	6.2	7,419	5.6	7,476	5.7	7,608	5.6	7,280	5.2	7,227	5.1
Asia	34,418	29.8	38,616	31.6	43,740	32.9	43,647	33.2	43,986	32.5	47,489	33.8	48,472	34.5
Europe	32,852	28.4	39,153	32.0	40,757	30.7	39,613	30.2	44,371	32.8	46,029	32.8	44,441	31.6
Oceania	217	0.2	298	0.2	199	0.1	260	0.2	187	0.1	249	0.2	272	0.2
Leading Producers														
United States of America	19,241	16.7	19,077	15.6	21,491	16.2	20,306	15.5	19,679	14.5	19,419	13.8	*18,588	13.2
Japan	8,004	6.9	10,386	8.5	11,793	8.9	12,794	9.7	12,869	9.5	13,136	9.4	*11,263	8.0
Netherlands	9,790	8.5	9,601	7.9	9,773	7.3	8,579	6.5	11,227	8.3	11,215	8.0	10,681	7.6
Germany	-		7,179	5.9	7,588	5.7	8,393	6.4	9,531	7.0	9,685	6.9	9,301	6.6
Korea, Republic of	4,060	3.5	6,069	5.0	7,467	5.6	7,492	5.7	7,777	5.7	10,105	7.2	11,800	8.4

Kerosene

Unit of Measure: 1,000 Metric tons.

	1990 Value	%	1991 Value	%	1992 Value	%	1993 Value	%	1994 Value	%	1995 Value	%	1996 Value	%
Total Production	113,945	100.0	114,254	100.0	121,187	100.0	122,498	100.0	123,381	100.0	124,336	100.0	124,778	100.0
Regions														
Africa	11,963	10.5	11,023	9.6	11,768	9.7	11,797	9.6	11,110	9.0	10,851	8.7	12,061	9.7
America, North	5,734	5.0	5,822	5.1	5,941	4.9	6,837	5.6	7,417	6.0	6,831	5.5	5,387	4.3
America, South	2,738	2.4	2,491	2.2	2,511	2.1	2,431	2.0	2,164	1.8	2,169	1.7	1,952	1.6
Asia	85,720	75.2	87,594	76.7	93,435	77.1	94,422	77.1	94,711	76.8	96,463	77.6	97,572	78.2
Europe	7,494	6.6	7,022	6.1	7,271	6.0	6,718	5.5	7,704	6.2	7,727	6.2	7,598	6.1
Oceania	296	0.3	301	0.3	262	0.2	293	0.2	275	0.2	295	0.2	208	0.2
Leading Producers														
Japan	18,819	16.5	19,918	17.4	21,168	17.5	21,945	17.9	22,138	17.9	22,218	17.9	*21,235	17.0
Indonesia	5,956	5.2	6,178	5.4	6,396	5.3	6,197	5.1	6,694	5.4	6,193	5.0	*6,837	5.5
India	5,686	5.0	5,270	4.6	5,397	4.5	5,376	4.4	5,206	4.2	5,325	4.3	5,969	4.8
China	3,925	3.4	4,062	3.6	3,945	3.3	3,729	3.0	4,072	3.3	4,458	3.6	*4,115	3.3
Iran, Islamic Republic of	3,454	3.0	3,800	3.3	4,000	3.3	4,400	3.6	4,500	3.6	4,600	3.7	*4,263	3.4

White Spirit

Unit of Measure: Metric tons.

	1990 Value	%	1991 Value	%	1992 Value	%	1993 Value	%	1994 Value	%	1995 Value	%	1996 Value	%
Total Production	3,988,929	100.0	3,965,060	100.0	4,224,857	100.0	4,004,170	100.0	4,117,739	100.0	4,114,114	100.0	4,264,346	100.0
Regions														
Africa	30,000	0.8	32,000	0.8	26,000	0.6	28,000	0.7	29,000	0.7	30,000	0.7	31,257	0.7
America, North	372,250	9.3	353,250	8.9	357,250	8.5	260,250	6.5	225,250	5.5	195,250	4.7	181,012	4.2
America, South	435,000	10.9	460,000	11.6	527,000	12.5	505,000	12.6	539,346	13.1	488,165	11.9	509,850	12.0
Asia	1,257,679	31.5	1,300,643	32.8	1,343,607	31.8	1,351,571	33.8	1,387,536	33.7	1,588,500	38.6	1,483,798	34.8
Europe	1,854,000	46.5	1,775,167	44.8	1,929,000	45.7	1,808,348	45.2	1,891,607	45.9	1,676,199	40.7	2,004,801	47.0
Oceania	40,000	1.0	44,000	1.1	42,000	1.0	51,000	1.3	45,000	1.1	136,000	3.3	53,629	1.3
Leading Producers														
Belgium	352,000	8.8	361,000	9.1	371,000	8.8	361,000	9.0	391,000	9.5	334,000	8.1	*405,771	9.5
Hungary	539,000	13.5	333,000	8.4	420,000	9.9	323,000	8.1	368,000	8.9	223,000	5.4	*485,438	11.4
Canada	281,000	7.0	262,000	6.6	265,000	6.3	168,000	4.2	133,000	3.2	103,000	2.5	*89,010	2.1
Brazil	215,000	5.4	271,000	6.8	309,000	7.3	286,000	7.1	304,000	7.4	250,000	6.1	269,000	6.3
Netherlands	255,000	6.4	244,000	6.2	211,000	5.0	182,000	4.5	181,000	4.4	220,000	5.3	*236,057	5.5

Commodity data are provided by the United Nations Statistical Division. The symbol * means that data are estimated. For additional notes, see Appendix I.

Product Tables
Distillate Fuel Oils

Unit of Measure: 1,000 Metric tons.

	1990 Value	%	1991 Value	%	1992 Value	%	1993 Value	%	1994 Value	%	1995 Value	%	1996 Value	%
Total Production	822,580	100.0	885,653	100.0	990,008	100.0	1,012,099	100.0	1,030,302	100.0	1,037,374	100.0	1,068,560	100.0
Regions														
Africa	59,596	7.2	51,536	5.8	55,903	5.6	59,775	5.9	65,765	6.4	66,353	6.4	68,227	6.4
America, North	190,121	23.1	192,117	21.7	193,669	19.6	202,030	20.0	206,674	20.1	202,498	19.5	211,778	19.8
America, South	48,855	5.9	52,158	5.9	50,842	5.1	51,437	5.1	53,899	5.2	52,989	5.1	56,683	5.3
Asia	314,452	38.2	334,714	37.8	418,777	42.3	423,427	41.8	421,214	40.9	437,684	42.2	445,636	41.7
Europe	195,710	23.8	241,498	27.3	257,538	26.0	261,486	25.8	268,227	26.0	263,049	25.4	271,400	25.4
Oceania	13,846	1.7	13,629	1.5	13,279	1.3	13,944	1.4	14,523	1.4	14,801	1.4	14,836	1.4
Leading Producers														
United States of America	147,653	17.9	149,525	16.9	150,536	15.2	158,124	15.6	161,790	15.7	159,307	15.4	167,842	15.7
Russian Federation	-		-		65,131	6.6	56,702	5.6	46,721	4.5	47,282	4.6	46,700	4.4
Japan	48,996	6.0	54,510	6.2	56,785	5.7	58,276	5.8	60,841	5.9	62,640	6.0	*63,160	5.9
Germany	-		41,685	4.7	45,527	4.6	47,324	4.7	48,840	4.7	45,288	4.4	47,326	4.4
Italy and San Marino	29,133	3.5	30,517	3.4	31,642	3.2	32,215	3.2	31,817	3.1	31,460	3.0	31,462	2.9

Residual Fuel Oils

Unit of Measure: 1,000 Metric tons.

	1990 Value	%	1991 Value	%	1992 Value	%	1993 Value	%	1994 Value	%	1995 Value	%	1996 Value	%
Total Production	835,773	100.0	843,989	100.0	900,982	100.0	885,606	100.0	861,693	100.0	851,290	100.0	866,750	100.0
Regions														
Africa	69,299	8.3	60,119	7.1	69,760	7.7	69,039	7.8	68,742	8.0	69,543	8.2	73,625	8.5
America, North	99,866	11.9	96,492	11.4	96,772	10.7	93,712	10.6	92,521	10.7	89,526	10.5	90,230	10.4
America, South	42,210	5.1	43,703	5.2	41,013	4.6	39,979	4.5	40,180	4.7	39,674	4.7	38,169	4.4
Asia	481,954	57.7	490,306	58.1	520,884	57.8	520,086	58.7	509,203	59.1	506,653	59.5	521,441	60.2
Europe	134,319	16.1	146,137	17.3	165,212	18.3	155,543	17.6	143,891	16.7	138,713	16.3	138,017	15.9
Oceania	8,125	1.0	7,232	0.9	7,340	0.8	7,246	0.8	7,156	0.8	7,181	0.8	5,269	0.6
Leading Producers														
Russian Federation	*78,809	9.4	*78,809	9.3	89,262	9.9	85,789	9.7	72,294	8.4	67,891	8.0	*78,809	9.1
United States of America	52,346	6.3	51,516	6.1	49,316	5.5	46,033	5.2	45,528	5.3	43,439	5.1	*43,333	5.0
Japan	41,023	4.9	40,956	4.9	43,488	4.8	43,556	4.9	47,357	5.5	44,321	5.2	*36,549	4.2
China	32,679	3.9	32,440	3.8	32,322	3.6	32,300	3.6	30,490	3.5	29,607	3.5	*32,670	3.8
Italy and San Marino	23,653	2.8	24,508	2.9	24,616	2.7	22,767	2.6	20,949	2.4	18,850	2.2	*18,887	2.2

Lubricating Oils

Unit of Measure: Metric tons.

	1990 Value	%	1991 Value	%	1992 Value	%	1993 Value	%	1994 Value	%	1995 Value	%	1996 Value	%
Total Production	41,440,103	100.0	41,275,056	100.0	43,541,177	100.0	42,389,843	100.0	43,076,542	100.0	45,290,369	100.0	44,464,980	100.0
Regions														
Africa	1,308,306	3.2	1,431,056	3.5	1,507,806	3.5	1,544,556	3.6	1,745,306	4.1	1,772,056	3.9	1,754,460	3.9
America, North	11,171,714	27.0	10,257,000	24.9	10,457,000	24.0	10,398,000	24.5	11,185,000	26.0	11,336,000	25.0	11,184,819	25.2
America, South	1,401,000	3.4	1,423,000	3.4	1,537,000	3.5	1,497,000	3.5	1,544,615	3.6	1,468,802	3.2	1,491,332	3.4
Asia	19,495,667	47.0	19,256,167	46.7	20,616,167	47.3	19,695,848	46.5	18,838,530	43.7	20,647,602	45.6	20,300,233	45.7
Europe	7,503,417	18.1	8,285,833	20.1	8,813,205	20.2	8,651,439	20.4	9,137,091	21.2	9,379,909	20.7	9,080,756	20.4
Oceania	560,000	1.4	622,000	1.5	610,000	1.4	603,000	1.4	626,000	1.5	686,000	1.5	653,381	1.5
Leading Producers														
United States of America	8,751,000	21.1	8,156,000	19.8	8,247,000	18.9	8,356,000	19.7	8,884,000	20.6	9,113,000	20.1	*8,898,876	20.0
Russian Federation	*3,033,500	7.3	*3,033,500	7.3	4,069,000	9.3	3,315,000	7.8	2,200,000	5.1	2,550,000	5.6	*3,033,500	6.8
Japan	2,232,000	5.4	2,213,000	5.4	2,209,000	5.1	2,260,000	5.3	2,422,000	5.6	2,437,000	5.4	*2,420,305	5.4
China	1,800,000	4.3	1,850,000	4.5	2,000,000	4.6	2,100,000	5.0	2,150,000	5.0	3,936,000	8.7	*2,630,295	5.9
Germany	-		1,551,000	3.8	1,534,000	3.5	1,421,000	3.4	1,539,000	3.6	1,552,000	3.4	*1,521,500	3.4

Commodity data are provided by the United Nations Statistical Division. The symbol * means that data are estimated. For additional notes, see Appendix I.

375

Product Tables
Paraffin Wax

Unit of Measure: Metric tons.

	1990		1991		1992		1993		1994		1995		1996	
	Value	%	Value	%	Value	%	Value	%	Value	%	Value	%	Value	%
Total Production	4,845,750	100.0	5,072,750	100.0	5,699,205	100.0	5,415,295	100.0	5,265,386	100.0	5,336,477	100.0	5,404,119	100.0
Regions														
Africa	89,000	1.8	93,000	1.8	105,000	1.8	100,000	1.8	102,000	1.9	100,000	1.9	125,883	2.3
America, North	871,000	18.0	935,000	18.4	942,000	16.5	1,014,000	18.7	1,058,000	20.1	1,069,000	20.0	1,028,857	19.0
America, South	50,000	1.0	55,000	1.1	59,000	1.0	61,000	1.1	66,000	1.3	74,000	1.4	67,502	1.2
Asia	3,241,250	66.9	3,182,250	62.7	3,700,750	64.9	3,429,750	63.3	3,199,750	60.8	3,192,750	59.8	3,391,237	62.8
Europe	574,500	11.9	789,500	15.6	877,455	15.4	792,545	14.6	821,636	15.6	877,727	16.4	767,440	14.2
Oceania	20,000	0.4	18,000	0.4	15,000	0.3	18,000	0.3	18,000	0.3	23,000	0.4	23,200	0.4
Leading Producers														
United States of America	788,000	16.3	850,000	16.8	857,000	15.0	928,000	17.1	971,000	18.4	981,000	18.4	*946,648	17.5
China	550,000	11.4	560,000	11.0	610,000	10.7	670,000	12.4	710,000	13.5	750,000	14.1	*705,743	13.1
Russian Federation	*378,750	7.8	*378,750	7.5	700,000	12.3	461,000	8.5	191,000	3.6	163,000	3.1	*378,750	7.0
Germany	-		252,000	5.0	231,000	4.1	205,000	3.8	187,000	3.6	196,000	3.7	*167,400	3.1
Japan	137,000	2.8	129,000	2.5	129,000	2.3	113,000	2.1	111,000	2.1	118,000	2.2	*112,857	2.1

Petroleum Coke

Unit of Measure: Metric tons.

	1990		1991		1992		1993		1994		1995		1996	
	Value	%	Value	%	Value	%	Value	%	Value	%	Value	%	Value	%
Total Production	50,862,583	100.0	53,851,583	100.0	56,628,250	100.0	58,530,250	100.0	58,752,365	100.0	59,706,715	100.0	62,509,822	100.0
Regions														
Africa	128,000	0.3	124,000	0.2	134,000	0.2	137,000	0.2	141,000	0.2	156,000	0.3	158,486	0.3
America, North	37,596,000	73.9	38,776,000	72.0	40,739,000	71.9	42,363,000	72.4	42,492,000	72.3	43,132,000	72.2	45,313,333	72.5
America, South	2,228,000	4.4	2,138,000	4.0	2,497,000	4.4	2,335,000	4.0	2,260,115	3.8	2,369,132	4.0	2,628,215	4.2
Asia	4,709,750	9.3	4,714,750	8.8	4,975,250	8.8	5,244,250	9.0	5,201,250	8.9	5,351,250	9.0	5,312,702	8.5
Europe	5,769,833	11.3	7,638,833	14.2	7,767,000	13.7	7,945,000	13.6	8,051,000	13.7	8,103,333	13.6	8,523,047	13.6
Oceania	431,000	0.8	460,000	0.9	516,000	0.9	506,000	0.9	607,000	1.0	595,000	1.0	574,038	0.9
Leading Producers														
United States of America	36,550,000	71.9	37,602,000	69.8	39,509,000	69.8	40,959,000	70.0	41,153,000	70.0	41,678,000	69.8	*43,957,810	70.3
Germany	-		1,574,000	2.9	1,561,000	2.8	1,626,000	2.8	1,735,000	3.0	1,624,000	2.7	*1,706,200	2.7
United Kingdom	1,562,000	3.1	1,598,000	3.0	1,582,000	2.8	1,691,000	2.9	1,626,000	2.8	1,906,000	3.2	*1,855,724	3.0
Argentina	1,248,000	2.5	1,284,000	2.4	1,675,000	3.0	1,508,000	2.6	1,215,000	2.1	1,213,000	2.0	1,310,000	2.1
China	1,100,000	2.2	1,150,000	2.1	1,220,000	2.2	1,380,000	2.4	1,400,000	2.4	1,435,000	2.4	*1,424,476	2.3

Bitumen (Asphalt)

Unit of Measure: 1,000 Metric tons.

	1990		1991		1992		1993		1994		1995		1996	
	Value	%	Value	%	Value	%	Value	%	Value	%	Value	%	Value	%
Total Production	95,021	100.0	95,770	100.0	98,135	100.0	98,773	100.0	97,941	100.0	98,603	100.0	100,952	100.0
Regions														
Africa	2,415	2.5	2,320	2.4	2,431	2.5	2,567	2.6	2,909	3.0	3,062	3.1	2,411	2.4
America, North	31,676	33.3	30,339	31.7	29,625	30.2	31,931	32.3	32,413	33.1	33,286	33.8	33,709	33.4
America, South	3,700	3.9	3,071	3.2	3,241	3.3	3,602	3.6	3,942	4.0	3,869	3.9	4,079	4.0
Asia	37,372	39.3	37,109	38.7	39,126	39.9	38,126	38.6	35,812	36.6	35,701	36.2	37,169	36.8
Europe	19,148	20.2	22,230	23.2	22,985	23.4	21,763	22.0	22,008	22.5	21,886	22.2	22,762	22.5
Oceania	710	0.7	702	0.7	728	0.7	785	0.8	858	0.9	798	0.8	822	0.8
Leading Producers														
United States of America	27,112	28.5	25,933	27.1	25,330	25.8	27,218	27.6	27,247	27.8	28,174	28.6	*28,982	28.7
Russian Federation	*6,037	6.4	*6,037	6.3	7,394	7.5	6,333	6.4	5,138	5.2	5,281	5.4	*6,037	6.0
Japan	6,185	6.5	5,989	6.3	6,150	6.3	6,086	6.2	6,078	6.2	5,963	6.0	*6,545	6.5
Germany	-		3,859	4.0	4,089	4.2	3,804	3.9	4,200	4.3	3,707	3.8	*3,874	3.8
Canada	2,793	2.9	2,633	2.7	2,517	2.6	2,938	3.0	3,376	3.4	3,305	3.4	*3,017	3.0

Product Tables

Liquefied Petroleum Gas from Natural Gas Plants

Unit of Measure: 1,000 Metric tons.

	1990		1991		1992		1993		1994		1995		1996	
	Value	%	Value	%	Value	%	Value	%	Value	%	Value	%	Value	%
Total Production	91,655	100.0	93,971	100.0	98,540	100.0	105,902	100.0	111,173	100.0	114,596	100.0	113,212	100.0
Regions														
Africa	20,469	22.3	19,081	20.3	21,546	21.9	23,889	22.6	25,763	23.2	26,294	22.9	26,377	23.3
America, North	54,633	59.6	57,873	61.6	59,200	60.1	61,769	58.3	62,918	56.6	63,350	55.3	63,398	56.0
America, South	4,566	5.0	4,627	4.9	4,796	4.9	5,800	5.5	5,673	5.1	6,158	5.4	6,193	5.5
Asia	6,956	7.6	7,220	7.7	7,765	7.9	9,017	8.5	10,621	9.6	12,610	11.0	11,053	9.8
Europe	2,888	3.2	3,139	3.3	3,171	3.2	3,258	3.1	3,841	3.5	3,757	3.3	3,757	3.3
Oceania	2,143	2.3	2,031	2.2	2,062	2.1	2,169	2.0	2,357	2.1	2,427	2.1	2,433	2.1
Leading Producers														
United States of America	39,168	42.7	41,854	44.5	42,905	43.5	43,934	41.5	43,878	39.5	44,738	39.0	*43,665	38.6
S.Arabia	12,699	13.9	13,530	14.4	14,830	15.0	15,550	14.7	17,030	15.3	17,100	14.9	*16,480	14.6
Canada	9,405	10.3	9,687	10.3	9,930	10.1	11,712	11.1	12,240	11.0	12,000	10.5	*12,076	10.7
Mexico	6,060	6.6	6,332	6.7	6,365	6.5	6,123	5.8	6,800	6.1	6,612	5.8	*7,657	6.8
Algeria	4,533	4.9	4,580	4.9	4,530	4.6	4,570	4.3	4,510	4.1	4,384	3.8	*5,745	5.1

Liquefied Petroleum Gas from Petroleum Refineries

Unit of Measure: Metric tons.

	1990		1991		1992		1993		1994		1995		1996	
	Value	%	Value	%	Value	%	Value	%	Value	%	Value	%	Value	%
Total Production	74,916,056	100.0	80,288,539	100.0	86,251,990	100.0	87,631,586	100.0	91,157,630	100.0	94,974,674	100.0	96,756,054	100.0
Regions														
Africa	2,629,000	3.5	2,701,000	3.4	2,856,000	3.3	2,875,000	3.3	3,011,000	3.3	3,024,000	3.2	3,244,896	3.4
America, North	19,345,500	25.8	20,605,500	25.7	22,829,500	26.5	23,053,000	26.3	23,808,115	26.1	25,043,231	26.4	25,510,689	26.4
America, South	4,881,000	6.5	4,976,000	6.2	5,339,000	6.2	5,516,000	6.3	5,908,500	6.5	5,684,000	6.0	5,792,605	6.0
Asia	33,379,222	44.6	34,366,706	42.8	37,611,630	43.6	37,724,804	43.0	38,523,978	42.3	41,174,153	43.4	41,722,976	43.1
Europe	13,906,333	18.6	16,822,333	21.0	16,781,861	19.5	17,486,782	20.0	18,933,036	20.8	18,972,291	20.0	19,453,298	20.1
Oceania	775,000	1.0	817,000	1.0	834,000	1.0	976,000	1.1	973,000	1.1	1,077,000	1.1	1,031,590	1.1
Leading Producers														
United States of America	15,640,000	20.9	16,791,000	20.9	19,070,000	22.1	18,551,000	21.2	19,159,000	21.0	20,500,000	21.6	*21,461,133	22.2
Russian Federation	*5,800,750	7.7	*5,800,750	7.2	7,165,000	8.3	6,113,000	7.0	4,886,000	5.4	5,039,000	5.3	*5,800,750	6.0
Japan	4,450,000	5.9	4,576,000	5.7	4,724,000	5.5	4,601,000	5.3	4,585,000	5.0	4,921,000	5.2	*4,755,733	4.9
Germany	-		2,468,000	3.1	2,655,000	3.1	2,873,000	3.3	3,425,000	3.8	3,281,000	3.5	*3,659,200	3.8
Brazil	3,044,000	4.1	2,946,000	3.7	3,178,000	3.7	3,388,000	3.9	3,511,000	3.9	3,336,000	3.5	3,281,000	3.4

ISIC 3540
PETROLEUM, COAL PRODUCTS

ISIC 3540—Petroleum, Coal Products—includes, broadly, all products resulting from the preparation or coking of coal. Specific products are hard coal briquettes, brown coal briquettes, coke, coking gases, and tar.

This industry corresponds most closely to a single U.S. 4-digit SIC code, 2999—Petroleum and Coal Products, nec. Tar production, in the U.S. system, is classified as SIC 295—Asphalt Paving and Roofing Materials.

Summary Statistics

		1990		1991		1992		1993		1994		1995	
		Value	N	Value	N	Value	N	Value	N	Value	N	Value	N
Establishments or enterprises	(number)	6,902	52	7,742	56	9,693	50	3,430	49	2,650	37	2,266	16
Number employed	(000)	712	54	575	56	546	51	530	49	293	41	284	36
Total Wages	($ mil.)	3,831	53	3,528	53	3,432	47	2,958	43	2,985	36	3,153	33
Wage/salary per employee	($)	11,707	52	11,949	51	13,461	45	11,712	41	12,904	35	15,653	32
Female workers	(000)	11	15	11	17	30	13	35	15	35	13	33	7
as % of total employment*	(%)	13	15	14	17	17	13	18	15	24	13	29	7
Output	($ bil.)	48	55	46	56	48	53	37	47	36	39	38	36
per employee	($ 000)	129	52	134	52	131	49	134	42	156	34	182	33
per establishment	($ 000)	18,417	47	21,506	48	15,124	45	22,680	38	11,873	27	22,502	14
Value added	($ bil.)	12	49	11	48	12	44	12	40	11	33	12	31
per employee	($ 000)	45	46	43	46	48	43	49	37	55	31	66	29
per establishment	($ 000)	5,174	40	4,023	42	4,518	39	6,808	32	3,399	22	5,864	10
Capital investment	($ mil.)	1,234	37	1,042	32	1,599	29	1,068	25	839	19	577	10
per employee	($ 000)	7	36	8	31	10	29	6	25	9	17	9	10
per establishment	($ 000)	863	34	751	30	3,048	28	453	23	733	16	1,569	9

Data presented above are drawn from the detailed tables that follow. Columns headed 'N' show the number of countries that provided valid data for inclusion. Values are not strictly comparable one year to the next or one row to the next because the number of countries that report varies from period to period and row to row. However, a general indication of magnitudes can be gleaned from reviewing this summary—especially in earlier years in which more reports are available. For detailed explanations, see Appendix I. *The average for those countries reporting both total and female employment.

Establishments and Number Engaged

Country	Establishments or Enterprises (number)						Employees per Establishment					
	1990	1991	1992	1993	1994	1995	1990	1991	1992	1993	1994	1995
Argentina	-	-	-	24	26	-	-	-	-	11	12	-
Armenia	2.000	2.000	2.000	2.000	1.000	-	145	80	-	-	-	-
Australia	20	29	32	-	-	-	50	34	31	-	-	-
Austria	32	29	27	29	34	-	34	34	37	31	28	-
Azerbaijan	95	97	71	69	60	-	136	126	159	166	167	-
Bangladesh	5.000	4.000	6.000	-	-	-	36	42	34	-	-	-
Belgium	4.000	4.000	4.000	-	-	-	-	-	-	-	-	-
Bosnia & Herzegovina	3.000	3.000	-	-	-	-	883	818	-	-	-	-
Bulgaria	*6.000	*6.000	*7.000	*15	*12	*14	2,350	2,400	1,986	900	1,133	993
Canada	92	92	102	108	119	-	33	22	29	28	17	-
Chile	9.000	11	8.000	7.000	8.000	-	168	133	134	94	106	-
China	*2,922	*3,850	*3,770	-	-	-	86	73	77	-	-	-
China (Hong Kong SAR)	3.000		-	3.000	3.000	-	67	-	-	-	-	-
Colombia	28	24	32	30	34	-	36	40	47	42	44	-
Costa Rica	2.000	2.000	2.000	2.000	-	-	43	25	15	13	-	-
Croatia	8.000	7.000	10	11	11	10	116	116	73	62	59	28
Czechoslovakia (Former)	*5.000	*6.000	-	-	-	-	4,400	3,000	-	-	-	-
Czech Republic	-	*5.000	-	-	-	-	-	2,800	-	-	-	-
Denmark	91	94	100	-	-	-	31	27	26	-	-	-
Ecuador	5.000	5.000	5.000	5.000	5.000	5.000	45	34	51	39	48	16
Egypt	16	17	16	15	-	-	75	82	56	73	-	-
El Salvador	-	1.000	1.000	2.000	1.000	3.000	-	-	14	19	14	12
Finland	20	20	20	17	20	1.000	35	45	40	41	45	198
Gambia	-	-	-	2.000	-	-	-	-	-	9.000	-	-
Ghana	-	-	-	1.000	-	-	-	-	-	71	-	-
Greece	43	43	44	-	-	-	26	24	23	-	-	-
Guatemala	-	2.000	2.000	2.000	2.000	2.000	-	42	44	45	46	46
Hungary	-	-	-	*1.000	-	-	-	-	-	1,000	-	-
Iceland	2.000	2.000	2.000	2.000	-	-	4.000	5.000	5.000	4.500	-	-
India	515	511	567	566	-	-	70	72	77	75	-	-
Indonesia	5.000	9.000	11	13	11	21	147	171	80	59	110	77
Iran (Islamic Republic of)	83	45	59	60	-	-	33	46	26	26	-	-
Iraq	-	-	42	-	-	-	-	-	48	-	-	-
Italy	*69	*77	*1.000	*1.000	*1.000	-	82	75	-	-	-	-
Jamaica	1.000	1.000	-	-	-	-	-	-	-	-	-	-
Japan	893	903	909	939	801	813	13	14	14	14	12	12
Korea, Republic of	351	342	358	369	359	339	31	30	26	22	20	18
Kuwait	2.000	-	6.000	6.000	6.000	-	12	-	25	30	38	-
Kyrgyzstan	19	16	16	14	15	-	31	15	25	31	26	-
Lithuania	-	-	-	*2.000	-	-	-	-	-	1,559	-	-
Malaysia	27	28	32	30	35	-	37	32	31	33	34	-
Mexico	23	23	22	21	21	20	190	178	167	179	180	185
Mongolia	-	-	-	1.000	1.000	-	-	-	-	-	94	-
Netherlands	17	19	19	19	16	-	107	97	97	81	95	-
New Zealand	*21	*19	*23	*24	-	-	9.381	9.421	9.783	-	-	-
Norway	26	24	12	-	-	-	31	35	58	-	-	-
Pakistan	-	17	-	-	-	-	-	102	-	-	-	-
Paraguay	-	1.000	-	-	-	-	-	22	-	-	-	-
Peru	12	12	12	-	-	-	6.833	5.500	5.167	-	-	-
Philippines	14	17	17	16	-	-	36	35	41	50	-	-
Poland	*4.000	*5.000	*4.000	*4.000	-	-	3,250	2,200	1,750	1,750	-	-
Portugal	*32	*26	*23	*25	*23	-	-	-	-	13	14	-
Qatar	4.000	4.000	4.000	5.000	1.000	-	22	38	71	15	19	-
Republic of Moldova	1.000	1.000	1.000	1.000	1.000	1.000	3,165	1,059	1,064	644	-	530
Romania	-	1.000	1.000	1.000	-	-	-	7,400	7,300	7,900	-	-
Russian Federation	-	-	-	*122	*196	*339	-	-	-	640	409	245
Senegal	-	1.000	-	2.000	2.000	-	-	24	-	16	14	-
Slovakia	-	-	-	1.000	1.000	-	-	-	-	-	-	-
Slovenia	15	-	-	-	-	-	-	-	-	-	-	-
South Africa	-	69	-	-	-	-	-	87	-	-	-	-

Continued.

Depending on the table, * means *Enterprises, Engaged*, or *Factor Values*; ± means *Basis Unspecified*; § means *shown in millions*. For additional notes and sources, see Appendix I.

Establishments and Number Engaged

- Continued -

Country	Establishments or Enterprises (number)						Employees per Establishment					
	1990	1991	1992	1993	1994	1995	1990	1991	1992	1993	1994	1995
Spain	149	159	170	-	-	-	29	31	29	-	-	-
Sweden	48	58	46	53	49	-	38	26	26	26	28	-
Thailand	2.000	5.000	-	-	-	-	-	36	-	-	-	-
Tunisia	-	-	-	2.000	6.000	2.000	-	-	-	24	15	22
Turkey	26	27	32	32	33	-	192	191	134	129	144	-
Ukraine	828	832	803	729	707	670	60	59	61	66	68	70
United Kingdom	129	113	110	-	-	-	62	62	55	-	-	-
USSR (Former)	*146	-	-	-	-	-	1,055	-	-	-	-	-
United States of America	-	-	2,108	-	-	-	-	-	19	-	-	-
Venezuela	26	21	19	23	24	22	42	62	68	59	42	53
Yugoslavia	1.000	1.000	3.000	2.000	2.000	4.000	-	1,000	400	450	450	250
Zambia	-	-	-	-	3.000	-	-	-	-	-	-	-

Employment and Compensation of Employees

Country	Number Employed/Engaged (000)						Wage/Salary per Employee ($)					
	1990	1991	1992	1993	1994	1995	1990	1991	1992	1993	1994	1995
Argentina	1.769	-	-	0.267	0.303	-	7,864	-	-	20,216	-	-
Armenia	*0.290	*0.160	-	-	-	-	-	-	-	-	-	-
Australia	1.000	*1.000	*1.000	-	-	-	9,688	10,908	11,029	-	-	-
Austria	1.100	1.000	1.000	0.885	0.941	0.900	33,741	35,029	32,032	38,274	-	-
Azerbaijan	13	12	11	11	10	-	1,614	1,353	-	-	-	-
Bangladesh	0.180	0.168	0.204	-	-	-	804	651	629	-	-	-
Bosnia & Herzegovina	2.648	2.455	-	-	-	-	3,161	1,947	-	-	-	-
Bulgaria	14	14	14	14	14	14	2,980	1,472	2,567	3,429	2,787	3,525
Canada	3.000	2.000	3.000	3.000	2.000	1.880	25,426	34,040	25,372	23,771	33,319	31,615
Chile	1.510	1.460	1.074	0.661	0.845	0.919	6,964	5,354	7,332	10,241	10,427	11,708
China	*250	*280	*290	*220	-	-	-	-	-	-	-	-
China (Hong Kong SAR)	*0.200	-	-	-	-	-	20,540	-	-	-	-	-
China (Taiwan Province)	3.623	3.100	3.239	3.461	3.884	4.218	12,258	15,623	18,884	18,290	18,178	23,172
Colombia	1.000	0.969	1.517	1.265	1.487	1.604	2,192	2,302	3,540	6,802	5,471	6,029
Costa Rica	0.086	0.049	0.029	0.025	-	-	2,206	1,984	2,295	2,525	-	-
Croatia	0.930	0.810	0.730	0.680	0.650	0.280	4,704	4,616	1,731	1,976	3,160	4,268
Czechoslovakia (Former)	22	18	-	-	-	-	2,558	2,016	-	-	-	-
Czech Republic	17	14	-	-	-	-	2,589	2,108	-	-	-	-
Denmark	2.827	2.491	2.585	2.591	2.630	2.662	35,093	34,584	38,326	36,078	38,979	-
Ecuador	0.223	0.172	0.256	0.195	0.240	0.079	8,627	11,453	7,316	14,125	8,094	5,311
Egypt	*1.200	*1.400	*0.900	*1.100	*1.000	*1.000	2,542	1,758	2,895	2,939	3,321	3,187
El Salvador	-	-	*0.014	*0.038	*0.014	*0.036	-	-	-	5,976	1,388	3,024
Finland	0.700	0.900	0.800	0.700	0.900	0.198	33,174	32,668	32,008	27,034	25,887	39,554
Gambia	-	-	-	0.018	-	-	-	-	-	-	-	-
Ghana	-	-	-	*0.071	*0.073	*0.075	-	-	-	-	-	-
Greece	*1.113	*1.030	*0.992	*0.979	*0.990	*0.968	15,372	16,486	18,451	17,573	18,698	24,450
Guatemala	-	*0.084	*0.088	*0.090	*0.092	*0.092	-	-	-	4,873	6,183	7,437
Hungary	-	-	-	1.000	0.910	0.840	-	-	-	4,873	6,183	7,437
Iceland	0.008	0.010	0.010	0.009	-	-	25,950	22,205	28,673	18,408	-	-
India	*36	*37	*43	*43	*42	*42	1,743	1,471	1,379	1,270	1,693	1,998
Indonesia	0.737	1.541	0.882	0.772	1.205	1.618	980	1,198	1,059	906	1,082	1,973
Iran (Islamic Republic of)	2.700	2.053	1.561	1.552	-	-	3,279	3,545	4,149	3,760	-	-
Iraq	-	2.083	2.000	-	-	-	-	12,141	21,048	-	-	-
Italy	5.679	5.810	-	-	-	-	38,801	-	-	-	-	-
Japan	*12	*13	*13	*13	*10	*10	31,655	33,691	39,479	-	-	-
Korea, Republic of	11	10	9.135	8.124	7.114	6.177	10,712	12,744	13,544	14,372	16,028	18,859
Kuwait	0.024	-	0.148	0.177	0.225	0.187	6,279	-	8,579	8,998	9,620	12,305
Kyrgyzstan	*0.583	*0.233	*0.398	*0.438	*0.392	-	-	-	-	-	-	-
Lithuania	-	-	3.008	3.119	-	-	-	-	731	1,345	-	-
Malaysia	1.000	0.900	1.000	1.000	1.200	1.223	3,586	3,959	5,182	5,594	7,526	7,642
Mexico	4.360	4.105	3.664	3.768	3.774	3.710	5,571	6,794	8,934	9,791	11,201	6,532

Continued.

Employment and Compensation of Employees
- Continued -

Country	Number Employed/Engaged (000)						Wage/Salary per Employee ($)					
	1990	1991	1992	1993	1994	1995	1990	1991	1992	1993	1994	1995
Mongolia	-	-	-	-	*0.094	-	-	-	-	-	-	-
Netherlands	1.819	1.842	1.845	1.538	1.527	1.514	32,607	33,101	36,062	36,058	38,019	-
New Zealand	0.197	0.179	0.225	-	-	-	19,092	-	-	-	-	-
Norway	0.817	0.834	0.691	-	-	-	36,174	35,511	-	-	-	-
Pakistan	1.251	1.739	1.819	-	-	-	1,775	1,525	1,569	-	-	-
Paraguay	-	0.022	-	-	-	-	-	-	-	-	-	-
Peru	*0.082	*0.066	*0.062	-	-	-	3,657	4,102	3,999	-	-	-
Philippines	0.500	0.600	0.700	0.800	-	-	1,810	1,820	2,688	2,116	-	-
Poland	13	11	7.000	7.000	7.480	7.900	1,821	3,058	3,886	3,999	3,917	4,421
Portugal	-	-	-	*0.315	*0.328	*0.314	-	-	-	-	-	-
Qatar	0.088	0.151	0.282	0.077	0.019	-	4,639	4,709	3,151	2,951	7,880	-
Republic of Moldova	3.165	1.059	1.064	0.644	-	0.530	-	4,016	-	-	-	451
Romania	6.600	7.400	7.300	7.900	-	-	2,336	1,591	1,189	1,338	-	-
Russian Federation	-	-	-	78	80	83	-	-	-	1,261	2,178	2,003
Senegal	-	*0.024	-	*0.032	*0.028	-	-	7,680	-	7,946	3,409	-
South Africa	6.000	6.000	6.000	-	-	-	11,982	13,339	15,311	-	-	-
Spain	4.369	4.944	4.878	-	-	-	21,045	23,362	28,536	-	-	-
Sweden	1.800	*1.507	*1.187	1.372	1.383	1.386	27,032	29,124	31,480	24,455	25,786	28,776
Thailand	-	0.180	-	-	-	-	-	-	3,353	-	-	-
Tunisia	-	-	-	0.048	0.091	0.044	-	-	-	-	-	-
Turkey	4.979	5.156	4.296	4.140	4.750	5.073	12,550	14,273	15,648	15,876	9,172	11,460
Ukraine	50	49	49	48	48	47	3,381	3,310	4,086	579	616	916
United Kingdom	8.000	7.000	6.000	-	-	-	27,679	30,341	35,336	-	-	-
USSR (Former)	154	-	-	-	-	-	2,468	-	-	-	-	-
United States of America	40	39	40	41	41	40	32,000	32,821	33,250	34,195	34,829	36,600
Uruguay	*0.148	*0.137	*0.081	*0.078	*0.088	*0.083	-	-	-	-	-	-
Venezuela	1.100	1.300	1.300	1.361	1.007	1.162	3,702	3,683	4,028	3,996	3,972	3,601
Yugoslavia	-	1.000	1.200	0.900	0.900	1.000	-	-	-	-	2,140	1,090

Female Workers: Total and Percent of Employment

Country	Female Workers (000)						Female Workers as % of Total Employment					
	1990	1991	1992	1993	1994	1995	1990	1991	1992	1993	1994	1995
Argentina	-	-	-	-	0.024	-	-	-	-	-	7.92	-
Australia	0.056	-	-	-	-	-	5.60	-	-	-	-	-
Austria	0.100	0.100	0.100	0.130	0.136	-	9.09	10.00	10.00	14.69	14.45	-
Bangladesh	0.003	-	-	-	-	-	1.67	-	-	-	-	-
Bosnia & Herzegovina	0.416	0.357	-	-	-	-	15.71	14.54	-	-	-	-
Bulgaria	-	-	-	5.200	5.200	5.300	-	-	-	38.52	38.24	38.13
Chile	0.062	0.054	0.082	0.062	0.075	-	4.11	3.70	7.64	9.38	8.88	-
China (Taiwan Province)	0.650	0.546	0.478	0.568	0.727	0.552	17.94	17.61	14.76	16.41	18.72	13.09
Colombia	-	0.118	-	0.211	0.256	-	-	12.18	-	16.68	17.22	-
Croatia	0.120	0.120	0.120	0.120	0.120	0.110	12.90	14.81	16.44	17.65	18.46	39.29
Czechoslovakia (Former)	2.600	2.200	-	-	-	-	11.82	12.22	-	-	-	-
Czech Republic	5.000	5.000	-	-	-	-	29.41	35.71	-	-	-	-
Denmark	0.304	0.288	0.304	-	-	-	10.75	11.56	11.76	-	-	-
Egypt	-	0.071	0.083	0.077	-	-	-	5.07	9.22	7.00	-	-
El Salvador	-	-	-	0.005	0.004	0.002	-	-	-	13.16	28.57	5.56
Gambia	-	-	-	0.002	-	-	-	-	-	11.11	-	-
Indonesia	-	-	-	0.083	0.400	0.426	-	-	-	10.75	33.20	26.33
Iran (Islamic Republic of)	-	0.009	0.007	0.020	-	-	-	0.44	0.45	1.29	-	-
Iraq	-	0.221	-	-	-	-	-	10.61	-	-	-	-
Italy	-	0.569	-	-	-	-	-	9.79	-	-	-	-
Korea, Republic of	1.300	1.221	1.154	1.126	0.945	0.823	11.82	11.83	12.63	13.86	13.28	13.32
Malaysia	0.200	0.100	0.100	0.100	0.200	-	20.00	11.11	10.00	10.00	16.67	-
Mongolia	-	-	-	-	0.026	-	-	-	-	-	27.66	-
New Zealand	0.044	0.040	0.052	-	-	-	22.34	22.35	23.11	-	-	-
Philippines	-	-	0.132	0.176	-	-	-	-	18.86	22.00	-	-

Continued.

   Depending on the table, * means *Enterprises*, *Engaged*, or *Factor Values*; ± means *Basis Unspecified*; § means *shown in millions*. For additional notes and sources, see Appendix I.

Female Workers: Total and Percent of Employment

- Continued -

Country	Female Workers (000)						Female Workers as % of Total Employment					
	1990	1991	1992	1993	1994	1995	1990	1991	1992	1993	1994	1995
Sweden	0.200	-	-	-	-	-	11.11	-	-	-	-	-
Thailand	-	0.055	-	-	-	-	-	30.56	-	-	-	-
United Kingdom	0.398	-	1.000	-	-	-	4.97	-	16.67	-	-	-
United States of America	-	-	26	27	27	26	-	-	65.00	65.85	65.85	65.00

Output and Output per Employee

Country	Output ($ bil.)						Output per Employee ($)					
	1990	1991	1992	1993	1994	1995	1990	1991	1992	1993	1994	1995
Argentina	±0.365	-	-	0.080	-	-	206,377	-	-	301,301	-	-
Armenia	0.000	0.000	0.002	0.067	0.001	-	927	1,779	-	-	-	-
Australia	*0.101	*0.135	*0.099	-	-	-	100,781	134,788	99,265	-	-	-
Austria	0.198	0.190	0.160	0.172	0.200	0.220	179,739	189,705	159,978	194,185	212,037	244,764
Azerbaijan	±0.075	±0.082	±0.002	±0.006	±0.004	-	5,783	6,694	151	487	358	-
Bangladesh	0.004	0.004	0.004	-	-	-	24,669	23,096	18,248	-	-	-
Bosnia & Herzegovina	±0.100	±0.073		-	-	-	37,904	29,931	-	-	-	-
Bulgaria	±1.222	±0.769	±0.781	±0.780	±1.132	±1.401	86,693	53,381	56,151	57,791	83,264	100,827
Canada	0.720	0.663	0.645	0.605	0.630	0.563	239,973	331,675	215,107	201,535	314,880	299,247
Chile	0.151	0.162	0.205	0.199	0.249	0.305	100,035	111,244	190,909	301,247	295,067	331,571
China	±1.492	±1.585	±1.685	±2.359	-	-	5,967	5,660	5,810	10,723	-	-
China (Hong Kong SAR)	0.032	-	-	-	-	-	159,824	-	-	-	-	-
China (Taiwan Province)	±0.502	±0.540	±0.676	±0.724	±0.824	±1.169	138,669	174,192	208,634	209,274	212,059	277,055
Colombia	0.100	0.100	0.113	0.141	0.196	0.231	99,550	103,600	74,383	111,404	131,861	144,309
Costa Rica	0.000	0.000	0.000	0.000	0.000	0.000	3,136	3,000	6,486	7,934	-	-
Cote d'Ivoire	±0.055	±0.064	±0.060	±0.053	-	-	-	-	-	-	-	-
Croatia	±0.060	±0.051	±0.034	-	-	-	64,711	62,922	46,168	-	-	-
Czechoslovakia (Former)	±0.365	±0.168	-	-	-	-	16,586	9,347	-	-	-	-
Denmark	*0.352	*0.359	*0.399	*0.380	*0.440	*0.537	124,483	144,171	154,193	146,824	167,158	201,688
Ecuador	0.023	0.035	0.036	0.033	0.047	0.004	102,805	206,239	139,937	169,321	194,492	51,561
Egypt	*0.095	*0.068	*0.068	*0.085	*0.086	*0.084	79,458	48,424	75,214	77,002	85,943	83,739
El Salvador	-	-	±0.011	-	±0.001		-	-	-	302,289	-	29,053
Finland	±0.283	±0.313	±0.273	±0.267	±0.340	±0.144	403,840	347,840	341,564	382,133	377,780	724,793
Greece	*0.112	*0.123	*0.143	*0.135	*0.145	*0.185	100,339	119,667	144,160	137,456	146,064	191,345
Guatemala	-	0.008	0.012	0.014	0.018	0.024	-	92,667	137,572	150,848	196,320	257,777
Hungary	-	-	-	±0.027	±0.031	±0.034	-	-	-	26,650	33,630	40,448
Iceland	0.001	0.001	±0.001	±0.001	-	-	88,575	84,243	115,560	79,385	-	-
India	*1.134	*1.207	*1.226	*1.294	*1.571	*1.863	31,421	33,016	28,214	30,450	37,511	44,269
Indonesia	0.027	0.076	0.044	0.025	0.029	0.036	36,741	49,375	50,368	32,058	24,221	22,242
Iran (Islamic Republic of)	0.064	0.053	0.056	0.045	-	-	23,583	25,702	36,182	28,956	-	-
Iraq	-	*0.104	*0.395	-	-	-	-	49,934	197,749	-	-	-
Italy	±1.548	±3.998	-	-	-	-	272,633	688,135	-	-	-	-
Japan	±5.435	±5.857	±6.040	±6.610	±6.193	±6.719	452,955	450,540	464,636	508,439	619,313	671,912
Korea, Republic of	2.153	2.019	1.848	1.649	1.415	1.346	195,713	195,738	202,326	202,973	198,854	217,878
Kuwait	0.001	-	0.009	0.010	0.016	0.016	37,957	-	63,071	55,655	71,976	84,370
Kyrgyzstan	0.005	0.005	0.005	0.001	0.002	-	9,136	22,009	11,328	2,613	3,836	-
Lithuania	-	-	0.116	0.103	-	-	-	-	38,464	33,098	-	-
Malaysia	*0.082	*0.094	*0.124	*0.129	*0.198	*0.213	82,443	104,885	123,655	129,171	165,346	174,181
Mexico	0.549	0.596	0.577	0.586	0.670	0.531	126,001	145,198	157,436	155,452	177,480	143,023
Mongolia	-	-	-	0.000	-	-	-	-	-	-	-	-
Mozambique	-	±0.010	-	-	-	-	-	-	-	-	-	-
Netherlands	0.442	0.437	0.469	0.372	0.395	0.479	242,738	237,225	254,282	241,552	258,954	316,635
New Zealand	0.037	-	-	-	-	-	187,893	-	-	-	-	-
Norway	±0.240	±0.217	±0.184	-	-	-	294,280	259,861	265,964	-	-	-
Pakistan	0.094	0.102	0.111	-	-	-	75,157	58,745	60,955	-	-	-
Peru	0.006	0.004	0.004	-	-	-	76,503	67,857	64,036	-	-	-
Philippines	0.022	0.042	0.037	0.041	-	-	43,108	70,114	53,252	51,171	-	-
Poland	±0.680	±0.771	±0.419	±0.438	±0.499	±0.627	52,324	70,116	59,927	62,608	66,689	79,396
Portugal	-	-	-	0.036	0.036	0.043	-	-	-	112,967	111,030	138,110

Continued.

Depending on the table, * means *Enterprises*, *Engaged*, or *Factor Values*; ± means *Basis Unspecified*; § means *shown in millions*. For additional notes and sources, see Appendix I.

383

Output and Output per Employee
- Continued -

Country	Output ($ bil.)						Output per Employee ($)					
	1990	1991	1992	1993	1994	1995	1990	1991	1992	1993	1994	1995
Puerto Rico	±0.129	±0.134	±0.150	±0.124	±0.154	±0.180	-	-	-	-	-	-
Qatar	0.005	0.009	0.006	-	0.002	-	56,194	58,220	22,407	-	101,215	-
Republic of Moldova	0.000	0.038	0.005	0.003	0.005	0.005	0.037	35,439	4,460	5,316	-	8,514
Romania	±0.388	±0.295	±0.157	±0.362			58,763	39,803	21,530	45,800	-	-
Russian Federation	-	3.126	3.155	1.284	1.797	2.796				16,444	22,405	33,725
South Africa	±0.640	±0.674	±0.640	±0.659	±0.670	±0.756	106,674	112,265	106,591	-	-	-
Spain	*1.273	*1.408	*1.340	-			291,427	284,855	274,754			
Sweden	*0.620	*0.532	*0.469	*0.384	*0.427	*0.478	344,465	353,045	395,257	280,225	308,461	344,929
Thailand	-	0.033	-	-	-	-	-	182,014	-	-	-	-
Turkey	2.723	2.002	1.771	1.232	0.931	1.234	546,803	388,195	412,209	297,650	196,038	243,300
Ukraine	*3.035	*2.069	*7.481	*0.947	*1.320	*0.928	60,699	42,217	152,682	19,735	27,503	19,741
United Kingdom	*1.855	*1.793	1.737	-			231,920	256,131	289,458	-		
USSR (Former)	±5.125	-	-	-			33,279	-	-	-		
United States of America	*13	*13	*14	*15	*15	*15	329,500	325,385	341,050	363,805	363,805	380,950
Uruguay		0.003	0.003	0.004	0.004	0.004	-	22,902	33,264	52,462	40,614	45,307
Venezuela	0.039	0.053	0.053	0.062	0.037	0.050	35,336	40,955	40,894	45,763	36,565	42,659
Yugoslavia	±0.002	±0.002	±0.003	-	±0.010	±0.006		2,132	2,516	-	10,878	6,054

Value Added and Value Added per Employee

Country	Value Added ($ bil.)						Value Added per Employee ($)					
	1990	1991	1992	1993	1994	1995	1990	1991	1992	1993	1994	1995
Argentina	±0.122	-	-	0.028	-	-	68,688	-	-	106,535	-	-
Australia	*0.034	*0.043	*0.038	-	-	-	34,375	42,852	38,118	-	-	-
Austria	0.065	0.069	0.057	0.061	0.056	0.061	58,767	68,945	56,875	68,387	59,080	68,222
Bangladesh	0.001	0.001	0.001	-	-	-	3,182	3,090	3,272	-	-	-
Bosnia & Herzegovina	±0.020	±0.015	-	-	-	-	7,608	5,994	-	-	-	-
Bulgaria	-	±0.063	±0.077	±0.065	-	-	-	4,345	5,511	4,790	-	-
Canada	0.291	0.236	0.273	0.240	0.242	0.215	97,132	117,832	91,007	80,097	120,826	114,135
Chile	0.069	0.055	0.088	0.094	0.130	0.159	45,836	37,496	81,978	141,479	153,437	172,658
China	±0.208	±0.229	±0.181	±0.300	-	-	833	819	623	1,364	-	-
China (Hong Kong SAR)	0.013	-	-	-	-	-	66,112	-	-	-	-	-
China (Taiwan Province)	0.135	0.149	0.191	0.200	0.247	0.374	37,164	48,071	58,830	57,760	63,615	88,661
Colombia	0.034	0.035	0.043	0.047	0.076	0.090	33,847	36,518	28,061	37,398	51,198	56,138
Costa Rica	0.000	0.000	0.000	0.000	0.000	0.000	1,028	1,000	2,179	2,617	-	-
Cote d'Ivoire	0.051	0.057	0.053	0.049	-	-	-	-	-	-	-	-
Croatia	±-0.002	±0.011	±0.015	-	-	-	§-2,55	13,664	20,878	-	-	-
Czechoslovakia (Former)	±0.209	-	-	-	-	-	9,521	-	-	-	-	-
Denmark	*0.207	*0.182	*0.203	*0.195	*0.231	*0.285	73,215	72,870	78,683	75,293	87,687	107,046
Ecuador	0.004	0.008	0.006	0.007	0.010	0.001	17,464	48,853	23,870	38,186	41,708	18,875
Egypt	*0.036	*0.025	*0.022	*0.016	*0.017	*0.016	29,917	17,556	24,594	14,856	16,569	16,159
El Salvador	-	-	-	0.003	0.000	0.000	-	-	-	65,862	3,102	7,624
Finland	±0.121	±0.111	±0.089	±0.085	±0.099	±0.041	172,968	123,365	111,343	122,042	109,548	204,895
Greece	*0.027	*0.036	*0.048	*0.046	*0.049	*0.063	24,402	34,894	48,796	46,784	49,641	65,198
Hungary	-	-	-	0.007	±0.008	±0.009	-	-	-	7,310	8,648	10,401
Iceland	0.000	0.000	±0.001	±0.000	-	-	29,382	28,307	53,349	34,844	-	-
India	*0.146	*0.171	*0.188	*0.254	*0.257	*0.305	4,057	4,690	4,335	5,970	6,132	7,236
Indonesia	*0.008	*0.017	*0.013	*0.010	*0.010	*0.015	10,564	10,932	14,184	13,423	8,488	9,253
Iran (Islamic Republic of)	0.040	0.027	0.029	0.026	-	-	14,939	13,327	18,580	16,650		
Iraq		*0.080	*0.209	-	-	-		38,343	104,502	-	-	-
Israel	0.115	0.112					-	-	-	-	-	-
Italy	±0.406	±0.495					71,575	85,184	-	-	-	-
Japan	±1.540	±1.707	±1.840	±2.122	±2.045	±2.211	128,347	131,336	141,517	163,254	204,481	221,135
Korea, Republic of	0.517	0.565	0.621	0.608	0.506	0.464	46,998	54,774	67,926	74,838	71,112	75,134
Kuwait	0.000	-	0.004	0.005	0.007	0.007	15,411	-	24,029	27,500	30,261	36,207
Malaysia	*0.032	*0.036	*0.047	*0.056	*0.092	*0.091	32,312	39,595	46,714	56,292	76,751	74,715
Mexico	0.172	0.189	0.191	0.174	0.196	0.155	39,523	46,140	52,005	46,245	51,845	41,786
Mongolia	-	-	-	0.000	-	-	-	-	-	-	-	-

Continued.

Depending on the table, * means *Enterprises, Engaged,* or *Factor Values*; ± means *Basis Unspecified*; § means *shown in millions*. For additional notes and sources, see Appendix I.

Value Added and Value Added per Employee
- Continued -

Country	Value Added ($ bil.)						Value Added per Employee ($)					
	1990	1991	1992	1993	1994	1995	1990	1991	1992	1993	1994	1995
Netherlands	0.131	0.121	0.138	0.111	0.116	0.135	72,157	65,622	74,589	72,466	75,869	89,411
New Zealand	0.009	-	-	-	-	-	45,458	-	-	-	-	-
Norway	±0.063	±0.054	±0.055	-	-	-	77,432	64,179	79,762	-	-	-
Pakistan	0.039	0.043	0.048	-	-	-	31,162	24,838	26,505	-	-	-
Peru	0.005	0.002	0.002	-	-	-	54,915	34,282	30,358	-	-	-
Philippines	0.003	0.005	0.013	0.014	-	-	5,841	9,037	18,759	17,211	-	-
Poland	±0.249	±0.107	±0.093	±0.072	±0.080	±0.100	19,134	9,722	13,315	10,244	10,759	12,700
Portugal	-	-	-	0.009	0.009	0.010	-	-	-	27,166	28,212	33,370
Qatar	0.002	0.002	0.002	-	0.001	-	24,975	12,736	6,819	-	43,378	-
Romania	±0.036	±0.038	±0.047	±0.115	-	-	5,404	5,130	6,495	14,539	-	-
Russian Federation	-	-	-	±0.479	±0.685	±0.969	-	-	-	6,131	8,542	11,692
Senegal	-	0.000	-	0.000	0.000	-	-	13,883	-	7,173	8,813	-
South Africa	*0.217	*0.224	*0.228	*0.226	*0.230	*0.259	36,202	37,361	38,043	-	-	-
Spain	*0.383	*0.482	*0.459	-	-	-	87,703	97,563	94,035	-	-	-
Sweden	*0.218	*0.150	*0.134	*0.123	*0.119	*0.134	121,079	99,337	113,159	89,872	86,117	96,325
Thailand	-	0.013	-	-	-	-	-	70,977	-	-	-	-
Turkey	0.458	0.364	0.424	0.427	0.328	0.439	91,929	70,619	98,734	103,087	68,998	86,590
United Kingdom	*0.750	*0.719	*0.721	-	-	-	93,750	102,655	120,141	-	-	-
United States of America	*4.390	*4.230	*4.797	*5.367	*5.351	*5.333	109,750	108,462	119,925	130,902	130,512	133,325
Uruguay	0.001	0.001	0.001	0.002	0.001	0.001	9,671	9,426	11,518	20,946	15,758	17,579
Venezuela	0.019	0.026	0.022	0.014	0.011	0.034	17,426	19,645	16,605	9,999	11,395	29,038
Yugoslavia	±0.001	±0.001	±0.002	-	±0.006	±0.003	-	812	1,433	-	6,216	2,923

Capital Investment

Country	Gross Fixed Capital Formation ($ mil.)						Per Establishment ($ mil)					
	1990	1991	1992	1993	1994	1995	1990	1991	1992	1993	1994	1995
Argentina	-	-	-	1.092	-	-	-	-	-	0.046	-	-
Australia	0.781	-	-	-	-	-	0.039	-	-	-	-	-
Austria	20	7.965	9.464	20	19	-	0.610	0.275	0.351	0.694	0.548	-
Bangladesh	0.049	0.055	0.026	-	-	-	0.010	0.014	0.004	-	-	-
Bosnia & Herzegovina	3.512	-	-	-	-	-	1.171	-	-	-	-	-
Bulgaria	5.571	6.841	463	19	74	111	0.928	1.140	66	1.289	6.145	7.934
Canada	9.427	-	-	-	-	-	0.102	-	-	-	-	-
Chile	-3.168	1.515	1.600	3.508	2.516	-	-0.352	0.138	0.200	0.501	0.314	-
China (Hong Kong SAR)	7.189	-	-	-	-	-	2.396	-	-	-	-	-
Colombia	1.921	-0.019	2.608	1.010	1.884	-	0.069	-0.001	0.081	0.034	0.055	-
Croatia	-	0.157	0.325	-	-	-	-	0.022	0.033	-	-	-
Czechoslovakia (Former)	57	-	-	-	-	-	11	-	-	-	-	-
Denmark	25	27	-	-	-	-	0.279	0.289	-	-	-	-
Ecuador	5.372	7.355	3.199	0.050	6.627	0.036	1.074	1.471	0.640	0.010	1.325	0.007
Egypt	0.145	-	0.449	0.326	-	-	0.009	-	0.028	0.022	-	-
El Salvador	-	0.124	-	0.161	0.213	0.583	-	0.124	-	0.081	0.213	0.194
Finland	7.218	56	104	10	6.720	4.929	0.361	2.787	5.188	0.617	0.336	4.929
Greece	8.536	8.893	7.796	-	-	-	0.199	0.207	0.177	-	-	-
Hungary	-	-	-	3.296	-	-	-	-	-	3.296	-	-
India	54	58	36	50	-	-	0.104	0.113	0.064	0.089	-	-
Indonesia	0.543	1.538	1.478	-	-	0.408	0.109	0.171	0.134	-	-	0.019
Iran (Islamic Republic of)	7.514	5.142	3.857	3.174	-	-	0.091	0.114	0.065	0.053	-	-
Italy	165	164	-	-	-	-	2.395	2.136	-	-	-	-
Japan	111	141	213	441	137	-	0.124	0.156	0.235	0.469	0.171	-
Korea, Republic of	192	-	102	107	103	101	0.547	-	0.286	0.291	0.288	0.299
Kuwait	-	-	0.741	0.772	1.114	-	-	-	0.123	0.129	0.186	-
Malaysia	2.588	4.363	3.140	3.108	11	-	0.096	0.156	0.098	0.104	0.316	-
Mexico	3.534	15	-	-	-	-	0.154	0.631	-	-	-	-
New Zealand	1.194	-	-	-	-	-	0.057	-	-	-	-	-
Norway	11	6.633	2.306	-	-	-	0.430	0.276	0.192	-	-	-
Pakistan	-	1.505	-	-	-	-	-	0.089	-	-	-	-

Continued.

Capital Investment
- Continued -

Country	Gross Fixed Capital Formation ($ mil.)						Per Establishment ($ mil)					
	1990	1991	1992	1993	1994	1995	1990	1991	1992	1993	1994	1995
Peru	0.021	-	0.011	-	-	-	0.002	-	0.001	-	-	-
Philippines	0.247	0.582	0.118	0.147	-	-	0.018	0.034	0.007	0.009	-	-
Poland	18	33	35	2.931	-	-	4.597	6.547	8.697	0.733	-	-
Portugal	-	-	-	3.706	1.151	-	-	-	-	0.148	0.050	-
Romania	13	1.833	-	0.367	1.505	-	-	1.833	-	0.367	-	-
Spain	26	49	49	-	-	-	0.173	0.307	0.288	-	-	-
Thailand	-	0.333	-	-	-	-	-	0.067	-	-	-	-
Turkey	41	77	35	43	47	-	1.563	2.841	1.105	1.346	1.424	-
Ukraine	32	26	74	37	12	7.318	0.039	0.031	0.092	0.051	0.017	0.011
United Kingdom	66	37	42	-	4.594	-	0.512	0.329	0.385	-	-	-
United States of America	340	300	399	315	409	337	-	-	0.189	-	-	-
Uruguay	0.077	0.092	0.002	0.065								
Venezuela	1.642	4.735	9.228	1.189	0.613	14	0.063	0.225	0.486	0.052	0.026	0.630
Yugoslavia	0.009	0.010	0.106	-	0.617	0.397	0.009	0.010	0.035	-	0.308	0.099

Indexes of Industrial Production

Country	Index of Industrial Production (1990=100)						Index of Employment (1990=100)					
	1990	1991	1992	1993	1994	1995	1990	1991	1992	1993	1994	1995
Albania	100	-	-	-	-	-	-	-	-	-	-	-
Algeria	100	-	-	-	-	-	-	-	-	-	-	-
Angola	100	100	100	100	-	-	-	-	-	-	-	-
Argentina	100	125	-	-	-	-	100	-	-	15	17	-
Armenia	100	-	-	-	-	-	100	55	-	-	-	-
Australia	100	100	101	110	-	-	100	100	100	-	-	-
Austria	100	103	105	106	106	101	100	91	91	80	86	82
Azerbaijan	-	-	-	-	-	-	100	94	87	88	77	-
Bahrain	100	100	100	100	100	100	-	-	-	-	-	-
Bangladesh	100	-	-	-	-	-	100	93	113	-	-	-
Barbados	100	100	100	100	100	100	-	-	-	-	-	-
Belgium	100	91	86	74	73	70	-	-	-	-	-	-
Belize	100	100	100	100	100	100	-	-	-	-	-	-
Bhutan	100	-	-	-	-	-	-	-	-	-	-	-
Bosnia & Herzegovina	-	-	-	-	-	-	100	93	-	-	-	-
Botswana	100	100	100	-	-	-	-	-	-	-	-	-
Brazil	100	92	92	96	102	102	-	-	-	-	-	-
Bulgaria	100	82	68	60	82	96	100	102	99	96	96	99
Burkina-Faso	100	100	100	-	-	-	-	-	-	-	-	-
Burundi	100	100	100	100	-	-	-	-	-	-	-	-
Cambodia	100	100	100	-	-	-	-	-	-	-	-	-
Canada	100	86	105	96	107	94	100	67	100	100	67	63
Central African Republic	100	100	100	100	100	-	-	-	-	-	-	-
Chile	100	131	129	136	144	154	100	97	71	44	56	61
China	100	-	-	-	-	-	100	112	116	88	-	-
China (Hong Kong SAR)	100	-	-	-	-	-	100	-	-	-	-	-
China (Taiwan Province)	100	102	104	122	118	144	100	86	89	96	107	116
Colombia	100	92	96	95	110	116	100	97	152	126	149	160
Costa Rica	100	-	-	-	-	-	100	57	34	29	-	-
Croatia	100	-	-	-	-	-	100	87	78	73	70	30
Cyprus	100	100	100	100	100	100	-	-	-	-	-	-
Czechoslovakia (Former)	100	87	-	-	-	-	100	82	-	-	-	-
Czech Republic	100	-	-	-	-	-	100	82	-	-	-	-
Dem. Rep. of the Congo	100	100	100	-	-	-	-	-	-	-	-	-
Denmark	100	102	103	104	111	117	100	88	91	92	93	94
Dominican Republic	100	-	-	-	-	-	-	-	-	-	-	-
Ecuador	100	130	130	156	-	-	100	77	115	87	108	35
Egypt	100	122	121	117	114	103	100	117	75	92	83	83
El Salvador	100	-	-	-	-	-	-	-	-	-	-	-

Continued.

Depending on the table, * means *Enterprises, Engaged,* or *Factor Values;* ± means *Basis Unspecified;* § means *shown in millions.* For additional notes and sources, see Appendix I.

Indexes of Industrial Production

- Continued -

Country	Index of Industrial Production (1990=100)						Index of Employment (1990=100)					
	1990	1991	1992	1993	1994	1995	1990	1991	1992	1993	1994	1995
Ethiopia and Eritrea	100	100	100	-	-	-	-	-	-	-	-	-
Fiji	100	100	100	100	100	100	-	-	-	-	-	-
Finland	100	86	82	84	91	89	100	129	114	100	129	28
FYR Macedonia	100	-	-	-	-	-	-	-	-	-	-	-
France	100	96	94	85	82	80	-	-	-	-	-	-
Gabon	100	100	100	-	-	-	-	-	-	-	-	-
Gambia	100	-	-	-	-	-	-	-	-	-	-	-
Germany (Eastern Part) . .	100	-	-	-	-	-	-	-	-	-	-	-
Ghana	100	-	-	-	-	-	-	-	-	-	-	-
Greece	100	104	113	121	114	128	100	93	89	88	89	87
Guatemala	100	-	-	-	-	-	-	-	-	-	-	-
Guyana	100	100	100	100	100	100	-	-	-	-	-	-
Haiti	100	100	100	100	100	100	-	-	-	-	-	-
Honduras	100	100	100	100	100	100	-	-	-	-	-	-
Hungary	100	108	56	-	-	-	-	-	-	-	-	-
Iceland	100	-	-	-	-	-	100	125	125	113	-	-
India	100	108	111	-	-	-	100	101	120	118	116	117
Indonesia	100	-	-	-	-	-	100	209	120	105	164	220
Iran (Islamic Republic of) . . .	100	113	115	-	-	-	100	76	58	57	-	-
Iraq	100	-	-	-	-	-	-	-	-	-	-	-
Israel	100	104	115	127	140	145	-	-	-	-	-	-
Italy	100	100	103	105	-	-	100	102	-	-	-	-
Jamaica	100	91	123	-	-	-	-	-	-	-	-	-
Japan	100	98	91	90	88	90	100	108	108	108	83	83
Jordan	100	100	100	100	100	100	-	-	-	-	-	-
Korea, Republic of	100	129	164	179	184	208	100	94	83	74	65	56
Kuwait	100	-	-	-	-	-	100	-	617	738	938	779
Kyrgyzstan	100	-	-	-	-	-	100	40	68	75	67	-
Lebanon	100	100	100	100	-	-	-	-	-	-	-	-
Libyan Arab Jamahiriya . . .	100	100	100	100	-	-	-	-	-	-	-	-
Lithuania	100	-	-	-	-	-	-	-	-	-	-	-
Macau	100	100	100	-	-	-	-	-	-	-	-	-
Madagascar	100	100	100	100	-	-	-	-	-	-	-	-
Malawi	100	100	100	100	100	100	-	-	-	-	-	-
Malaysia	100	-	-	-	-	-	100	90	100	100	120	122
Malta	100	-	-	-	-	-	-	-	-	-	-	-
Mauritius	100	100	100	100	100	100	-	-	-	-	-	-
Mexico	100	106	109	106	112	109	100	94	84	86	87	85
Morocco	100	-	-	-	-	-	-	-	-	-	-	-
Mozambique	100	-	-	-	-	-	-	-	-	-	-	-
Myanmar	100	100	100	100	100	100	-	-	-	-	-	-
Nepal	100	100	100	-	-	-	-	-	-	-	-	-
Netherlands	100	99	101	98	102	107	100	101	101	85	84	83
New Zealand	100	100	100	-	-	-	100	91	114	-	-	-
Nicaragua	100	-	-	-	-	-	-	-	-	-	-	-
Nigeria	100	100	100	100	100	-	-	-	-	-	-	-
Norway	100	97	93	93	98	99	100	102	85	-	-	-
Pakistan	100	108	110	-	-	-	100	139	145	-	-	-
Panama	100	97	150	136	97	112	-	-	-	-	-	-
Papua New Guinea	100	100	100	-	-	-	-	-	-	-	-	-
Paraguay	100	-	-	-	-	-	-	-	-	-	-	-
Peru	100	102	103	107	111	109	100	80	76	-	-	-
Philippines	100	121	114	109	113	120	100	120	140	160	-	-
Poland	100	86	100	112	120	127	100	85	54	54	58	61
Portugal	100	89	101	97	118	126	-	-	-	-	-	-
Puerto Rico	100	-	-	-	-	-	-	-	-	-	-	-
Qatar	100	-	-	-	-	-	100	172	320	88	22	-
Republic of Moldova	100	-	-	-	-	-	100	33	34	20	-	17
Romania	100	63	66	-	-	-	100	112	111	120	-	-
Russian Federation	100	-	-	-	-	-	-	-	-	-	-	-

Continued.

Indexes of Industrial Production

- Continued -

Country	Index of Industrial Production (1990=100)						Index of Employment (1990=100)					
	1990	1991	1992	1993	1994	1995	1990	1991	1992	1993	1994	1995
Saudi Arabia	100	-	-	-	-	-	-	-	-	-	-	-
Slovakia	100	-	-	-	-	-	-	-	-	-	-	-
Slovenia	100	85	87	94	121	110	-	-	-	-	-	-
Somalia.	100	100	100	-	-	-	-	-	-	-	-	-
South Africa	100	99	103	118	125	140	100	100	100	-	-	-
Spain	100	100	97	93	95	89	100	113	112	-	-	-
Sudan	100	100	100	-	-	-	-	-	-	-	-	-
Sweden.	100	85	97	114	93	94	100	84	66	76	77	77
Switzerland	100	136	108	-	-	-	-	-	-	-	-	-
Syrian Arab Republic	100	99	100	106	110	116	-	-	-	-	-	-
Thailand	100	93	115	-	-	-	-	-	-	-	-	-
Tonga	100	-	-	-	-	-	-	-	-	-	-	-
Trinidad and Tobago	100	92	105	99	97	88	-	-	-	-	-	-
Tunisia	100	-	-	-	-	-	-	-	-	-	-	-
Turkey	100	96	101	102	87	100	100	104	86	83	95	102
Uganda.	100	100	100	100	100	100	-	-	-	-	-	-
Ukraine	100	-	-	-	-	-	100	98	98	96	96	94
United Arab Emirates . . .	100	-	-	-	-	-	-	-	-	-	-	-
United Kingdom	100	95	85	79	80	75	100	88	75	-	-	-
United Republic of Tanzania . .	100	100	100	100	100	-	-	-	-	-	-	-
USSR (Former)	100	-	-	-	-	-	100	-	-	-	-	-
United States of America . . .	100	99	99	102	102	104	100	98	100	102	102	100
Uruguay	100	102	91	30	2	90	100	93	55	53	59	56
Venezuela	100	-	-	-	-	-	100	118	118	124	92	106
Yugoslavia	100	-	-	-	-	-	-	-	-	-	-	-
Yugoslavia (Former).	100	60	-	-	-	-	-	-	-	-	-	-
Zambia	100	101	96	90	68	60	-	-	-	-	-	-
Zimbabwe	100	132	121	-	-	-	-	-	-	-	-	-

Depending on the table, * means *Enterprises, Engaged,* or *Factor Values;* ± means *Basis Unspecified;* § means *shown in millions.* For additional notes and sources, see Appendix I.

Representative Companies in Sector

Name	Address	Tele-phone	Fax	Employ-ment	Y
RWE-DEA AG fur Mineraloel und Chemie	Ueberseering 40, Hamburg D-22297, Germany	40 63750	40 63753496	7,243	94
Witco Corp.	1 American Ln., Greenwich, Connecticut 06831-2559, U.S.A.	203-552-2000		7,200	97
Vulcan Materials Co.	PO Box 530187, Birmingham, Alabama 35209, U.S.A.	205-877-3000	205-877-3094	6,963	97
Lafarge Canada Inc.	606 Rue Cathcart Ste. 900, Montreal, Quebec H3B 1L7, Canada	514-861-1411	514-861-1123	5,037	98
First Brands Corp.	PO Box 1911, Danbury, Connecticut 06813-1911, U.S.A.	203-731-2300		4,800	97
Lubrizol Corp.	29400 Lakeland Blvd., Wickliffe, Ohio 44092-2298, U.S.A.	440-943-4200	440-943-5337	4,291	97
Valvoline Co.	PO Box 14000, Lexington, Kentucky 40512, U.S.A.	606-357-7777	606-357-7381	3,500	96
Granite Construction Inc.	PO Box 50085, Watsonville, California 95077-5085, U.S.A.	408-724-1011		3,309	97
Emco Ltd.	620 Richmond St., London, Ontario N6A 5J9, Canada	519-645-3900	519-645-3939	3,300	98
Wynn's International Inc.	PO Box 14143, Orange, California 92613, U.S.A.	714-938-3700		2,073	97
Edward C. Levy Co.	8800 Dix Ave., Detroit, Michigan 48209, U.S.A.	313-843-7200	313-849-9448	2,000	97
CalMat Co.	3200 San Fernando Rd., Los Angeles, California 90065, U.S.A.	213-258-2777	213-258-5920	1,880	97
Dixon Ticonderoga Co.	PO Box 958413, Heathrow, Florida 32795-8413, U.S.A.	407-875-9000		1,770	97
Miller Paving Ltd.	505 Miller Ave., Markham, Ontario L6G 1B2, Canada	905-475-6660	905-475-3852	1,500	98
Repsol Quimica SA	Paseo de la Castellana 278-280, Madrid E-28046, Spain	91 3488000	91 5767935	1,492	97
Gencor Industries Inc.	5201 N. Orange Blossom Trail, Orlando, Florida 32810, U.S.A.	407-290-6000	407-578-0577	1,351	97
Fjeldhammer Brug AS	Postboks 55, Fjellhamer N-1472, Norway	67 979000	67 705877	1,150	93
Acheson Colloids Co.	PO Box 611747, Port Huron, Michigan 48061-0489, U.S.A.	810-984-5581	810-984-1446	1,100	95
Klein Tools Inc.	PO Box 599033, Chicago, Illinois 60659-9033, U.S.A.	847-677-9500	847-677-4476	1,000	97
Reilly Industries Inc.	300 N. Meridian, 1500, Indianapolis, Indiana 46204-1763, U.S.A.	317-638-7531	317-248-6472	950	97
Quaker Chemical Corp.	Elm & Lee Sts., Conshohocken, Pennsylvania 19428, U.S.A.	610-832-4000	610-832-8682	835	97
Elcor Corp.	14643 Dallas Pkwy., Ste. 1000, Dallas, Texas 75240-8871, U.S.A.	972-851-0500	972-851-0550	800	97
Esso Malaysia BHD	Jalan Sultan Ismail, Kuala Lumpur 50250, Malaysia	3 422322	3 2413270	700	93
Petro-Canada, Petro-Canada Lubricants Division	385 Southdown Rd., Mississauga, Ontario L5J 2Y3, Canada	905-822-4222	905-403-6738	592	98
Century Lubricants Co.	2140 S. 88th St., Kansas City, Kansas 66111, U.S.A.	913-422-4022	913-441-2333	500	97
Lord Corp. Chemical Products Div.	2000 W. Grandview Blvd., Erie, Pennsylvania 16514-0038, U.S.A.	814-868-3611	814-864-3452	500	96
Simard-Beaudry Inc.	4230 Blvd. Saint-Elzear E, Laval, Quebec H7E 4P2, Canada	514-329-4747	514-329-4608	500	98
Safety-Kleen Canada Inc.	300 Woolwich St., Breslau, Ontario N0B 1M0, Canada	519-648-2295	519-648-2807	450	98
Shell Espana, SA	Rio Bullaque, 2, Madrid E-28034, Spain	91 5370100	91 5370106	430	97
Redland Stone Products Co.	17910 I-10 W., San Antonio, Texas 78257, U.S.A.	210-696-8500		420	96
Monsey Products Co.	PO Box 368, Kimberton, Pennsylvania 19442, U.S.A.	215-933-8888	215-933-4598	400	94
Norton Chemical Process Products Corp.	3855 Fishcreek Rd., Stow, Ohio 44224, U.S.A.	216-673-5860	216-677-7245	400	97
Houghton International Inc.	PO Box 930, Valley Forge, Pennsylvania 19482, U.S.A.	610-666-4000	610-666-1376	370	97
U.S. Intec Inc.	PO Box 2845, Port Arthur, Texas 77643, U.S.A.	409-724-7024	409-724-2348	356	94
Thanomwongse Service Co. Ltd.	Phayathai, Bangkok 10400, Thailand	2710205	2711601	350	90
E/M Corp.	PO Box 2400, West Lafayette, Indiana 47906-0400, U.S.A.	765-497-6346	765-497-6348	340	96
H.B. Fuller Automotive Products Inc.	31601 Research Park Dr., Madison Heights, Michigan 48071-4626, U.S.A.	248-585-2200	248-776-0960	320	97
D. Crupi & Sons Ltd.	85 Passmore Ave., Scarborough, Ontario M1V 4S9, Canada	416-291-1986	416-291-3252	300	98
Superior Graphite Co.	120 S. Riverside Plz., Chicago, Illinois 60606, U.S.A.	312-559-2999	312-559-9064	300	94
Texas Refinery Corp.	PO Box 711, Fort Worth, Texas 76101, U.S.A.	817-332-1161	817-336-8440	300	97
Garland Company Inc.	3800 E. 91st St., Cleveland, Ohio 44105, U.S.A.	216-641-7500		280	97
Great Lakes Carbon Corp.	110 East 59th St., New York, New York 10022, U.S.A.	212-527-3002		254	97
Amrep Inc.	990 Industrial Park Dr., Marietta, Georgia 30062-2433, U.S.A.	404-422-2071	404-422-1737	250	95
Betterroads Asphalt Corp.	Marginal y Andes St., Rio Piedras 00926, Puerto Rico	787764-1000	787764-6720	250	92
California Products Corp.	PO Box 569, Cambridge, Massachusetts 02139, U.S.A.	617-547-5300		250	94
Societat Catalana de Petrolis, SA	7A Planta (Edificio Master's), Barcelona E-08034, Spain	93 4005070	93 2803320	250	97
Recalit SA	Carretera A Paraiso, Cartago, Costa Rica			230	90
Seyang Ind. Co., Ltd.	Kwanjin-Gu, Seoul 143-200, Republic of Korea	2 4564511	2 4562725	201	97
Consolidated Fiber Glass Products Co.	PO Box 5248, Bakersfield, California 93388-5248, U.S.A.	805-323-6026		200	94
W.R. Meadows Inc.	PO Box 338, Hampshire, Illinois 60140, U.S.A.	708-683-4500	708-683-4544	200	93

Continued

Representative Companies in Sector
- Continued -

Name	Address	Tele-phone	Fax	Employ-ment	Y
Mobil Oil do Brasil Industria e Comercio Ltda.	Av. Paulista 1009, 5th Fl., Sao Paulo 01311-919, Brazil	11 2455300	11 2455331	190	96
Paz Lubricants and Chemicals Ltd.	8 El Gazel Ltd., Haifa 35025, Israel	4 8352056	4 8670006	167	97
WD-40 Company	1061 Cudahy Pl., San Diego, California 92110, U.S.A.	619-275-1400	619-275-5823	166	97
Bird Corp. (Norwood, Massachusetts)	1077 Pleasant St., Norwood, Massachusetts 02062, U.S.A.	617-551-0656	617-255-0114	164	96
Gold Eagle Co.	4400 S. Kildare Ave., Chicago, Illinois 60632, U.S.A.	312-376-4400	312-376-3245	160	94
Battenfeld-American Inc.	1575 Clinton St., Buffalo, New York 14206, U.S.A.	716-822-8410	716-822-8410	150	93
Battenfeld Grease and Oil Corp.	PO Box 728, North Tonawanda, New York 14120-0728, U.S.A.	716-695-2100	716-695-0367	150	97
Carriere Bernier Ltee.	25 Rue Petit, St.-Jean-Sur-Richelieu, Quebec J2X 4J3, Canada	514-875-2841	450-347-4200	150	98
P. Baillargeon Ltee.	800 Rue des Carrieres, St.-jean-sur-Richelieu, Quebec J3B 2P2, Canada	514-866-8333	450-346-6897	150	98
Southwestern Petroleum Corp.	PO Box 961005, Fort Worth, Texas 76161-0005, U.S.A.	817-332-2336	817-877-4047	135	97
Master Chemical Corp.	501 W. Boundary St., Perrysburg, Ohio 43551, U.S.A.	419-874-7902	419-874-0684	122	95
Hard-Rock Paving Co. Ltd.	198 Welland St., Port Colborne, Ontario L3K 1V3, Canada	905-835-8413	905-834-3811	120	98
National Automobiles	Ruwi, Muscat 112, Oman	702081	698942	120	92
Matheson & MacMillan Ltd.	355 Sherwood Rd., Charlottetown, Prince Edward Island C1A 7L1, Canada	902-892-1057	902-892-1059	115	98
H L Blachford Ltd.	2323 Royal Windsor Dr., Mississauga, Ontario L5J 1K5, Canada	905-823-3200	905-823-9290	111	98
Agip Espana, SA	Capitan Haya, 48-20, Madrid E-28020, Spain	91 5962100	91 5715359	104	97
B A Blacktop Ltd.	175 Harbour Ave., North Vancouver, British Columbia V7J 2E7, Canada	604-985-0611	604-985-0485	100	98
Cross Oil Refining and Marketing Inc. of Arkansas	PO Box 105, Smackover, Arkansas 71762, U.S.A.	501-725-3611		100	95
Elf Lubricants North America	5 N. Stiles St., Linden, New Jersey 07036, U.S.A.	908-862-2500	908-862-6885	100	96
Gulf States Asphalt Company Inc.	300 Christy Pl., South Houston, Texas 77587, U.S.A.	713-941-4410	713-947-4906	100	93
LPS Laboratories Inc.	P.O. 105052, Tucker, Georgia 30084, U.S.A.	404-934-7800	404-493-9206	100	94
Lubricating Specialties Co.	8015 Paramount Blvd., Pico Rivera, California 90660-4888, U.S.A.	310-928-3311		100	94
Monsey Bakor Inc.	284 Watline Ave., Mississauga, Ontario L4Z 1P4, Canada	905-890-4800	905-890-4866	100	98
Pavages Maska Inc.	767 Rue Principale, St.-Dominique-de-Bagot, Quebec J0H 1L0, Canada	514-773-2591	514-773-9153	100	98
Cantol Ltd.	199 Steelcase Rd. W, Markham, Ontario L3R 2M4, Canada	905-475-6141	905-475-1584	90	98
Canroof Corp. Inc.	560 Commissioners St., Toronto, Ontario M4M 1A7, Canada	416-461-8122	416-461-2926	80	98
Katepal Oy	Nurmisaarentie 2, Lempaala SF-37501, Finland	31 3759111	31 3759111	80	95
Lunday-Thagard Co.	PO Box 1519, South Gate, California 90280, U.S.A.	310-928-7000		80	95
Spanjaard Ltd.	PO Box 7294, Johannesburg 2000, Republic of South Africa	11 7862210	11 7865685	75	94
Lubrizol Canada Ltd.	5800 Thorold Stone Rd., Niagara Falls, Ontario L2J 1A2, Canada	905-358-5778	905-358-0253	70	98
Quaker State Inc.	1101 Blair Rd., Burlington, Ontario L7M 1T3, Canada	905-335-5577	905-332-6406	70	98
Bertrand Construction L'Orignal Inc.	56 Longueil St., L'Orignal, Ontario K0B 1K0, Canada	613-675-4614	613-675-2610	55	98
Neste Alfa Oy	Kimmeltie 3, Espoo SF-02151, Finland	4501	4504640	55	94
Dimol Manufacturing Pty. Ltd.	PO Box 6163, Ansfrere 1711, Republic of South Africa	11 7621255	11 7625617	54	94
Hemisphere Oil Co. Inc.	Rd. 901, Km. 2.8, Barrio Camino Nuevo, Yabucoa 00767, Puerto Rico	787893-2320		50	92
Arno Hentschel GmbH	Hauptstrasse 71, Oberoderwitz 02744, Germany	522927443	522927450	46	91
Hal Industries Inc.	9681 187 St., Surrey, British Columbia V4N 3N3, Canada	604-888-0777	604-888-1656	45	98
Impex Inc. 2000 C.A.	Apartado Postal 88090, Caracas, Venezuela	2 9796884	2 9799883	40	97
Belgin Madeni Yaglar AS	Goztepe, Istanbul TR-81080, Turkey	1 3561896	1 3686217	35	90
Bitum Petrochemicals Industries Ltd.	41 Hayetzira St., Haifa Bay 26111, Israel	4 8416217	4 8416219	35	97
Municipal Ready Mix Ltd.	19 MacRae Ave., Sydney, Nova Scotia B1S 1M1, Canada	902-564-4541	902-562-6057	35	98
Apco Industries Co. Ltd.	10 Industrial St., Toronto, Ontario M4G 1Z1, Canada	416-421-6161	416-421-1096	30	98
Viscosity Oil Co.	600-H Joliet Rd., Willowbrook, Illinois 60521, U.S.A.	630-850-4000	630-850-4020	30	95
Forsythe Lubrication Association Ltd.	120 Chatham St., Hamilton, Ontario L8P 2B5, Canada	905-525-7192	905-525-7024	27	98
Flor-Quim, Inc.	Rd. 3, Km. 111, Barrio Guardarraya, Patillas 00723, Puerto Rico	787839-2710		25	92

Product Tables
Hard-Coal Briquettes

Unit of Measure: Metric tons.

	1990 Value	%	1991 Value	%	1992 Value	%	1993 Value	%	1994 Value	%	1995 Value	%	1996 Value	%
Total Production	21,910,902	100.0	18,804,645	100.0	23,091,606	100.0	17,150,620	100.0	10,922,583	100.0	9,439,029	100.0	7,801,553	100.0
Regions														
Asia	20,108,102	91.8	16,301,045	86.7	12,294,956	53.2	8,969,254	52.3	5,859,551	53.6	4,598,848	48.7	3,199,567	41.0
Europe	1,798,000	8.2	2,499,000	13.3	10,792,250	46.7	8,177,167	47.7	5,059,032	46.3	4,836,381	51.2	4,598,386	58.9
Oceania	4,800	0.0	4,600	0.0	4,400	0.0	4,200	0.0	4,000	0.0	3,800	0.0	3,600	0.0
Leading Producers														
Korea, Republic of	18,779,000	85.7	14,996,000	79.7	11,069,000	47.9	7,747,000	45.2	4,684,000	42.9	3,005,000	31.8	1,960,000	25.1
Ukraine	-		-		8,571,000	37.1	5,972,000	34.8	3,057,000	28.0	3,057,000	32.4	3,000,000	38.5
Poland	77,000	0.4	1,000	0.0	3,250	0.0	5,500	0.0	7,750	0.1	10,000	0.1	-	
United Kingdom	496,000	2.3	582,000	3.1	555,000	2.4	665,000	3.9	602,000	5.5	510,000	5.4	*502,524	6.4
Germany	-		860,000	4.6	677,000	2.9	585,000	3.4	460,000	4.2	379,000	4.0	*238,500	3.1

Brown-Coal Briquettes

Unit of Measure: 1,000 Metric tons.

	1990 Value	%	1991 Value	%	1992 Value	%	1993 Value	%	1994 Value	%	1995 Value	%	1996 Value	%
Total Production	65,673	100.0	117,285	100.0	82,090	100.0	78,358	100.0	73,880	100.0	70,280	100.0	63,293	100.0
Regions														
Asia	6,736	10.3	6,631	5.7	7,113	8.7	6,929	8.8	6,233	8.4	4,973	7.1	6,075	9.6
Europe	58,230	88.7	109,939	93.7	74,256	90.5	70,913	90.5	67,068	90.8	64,735	92.1	56,691	89.6
Oceania	706	1.1	715	0.6	721	0.9	516	0.7	580	0.8	572	0.8	528	0.8
Leading Producers														
Germany	-		52,011	44.3	16,936	20.6	14,060	17.9	11,424	15.5	8,943	12.7	-	
Hungary	1,791	2.7	1,924	1.6	683	0.8	611	0.8	487	0.7	344	0.5	*860	1.4
Russian Federation	*1,421	2.2	*1,421	1.2	2,078	2.5	1,983	2.5	1,391	1.9	232	0.3	*1,421	2.2
Bulgaria	1,451	2.2	1,317	1.1	1,412	1.7	1,627	2.1	1,168	1.6	1,011	1.4	*1,311	2.1
Ukraine	*1,320	2.0	*1,320	1.1	1,916	2.3	1,244	1.6	800	1.1	*1,320	1.9	*1,320	2.1

Coke

Unit of Measure: 1,000 Metric tons.

	1990 Value	%	1991 Value	%	1992 Value	%	1993 Value	%	1994 Value	%	1995 Value	%	1996 Value	%
Total Production	414,423	100.0	413,932	100.0	433,847	100.0	431,987	100.0	430,576	100.0	473,524	100.0	454,703	100.0
Regions														
Africa	3,647	0.9	3,854	0.9	4,294	1.0	5,219	1.2	4,992	1.2	5,269	1.1	4,999	1.1
America, North	31,173	7.5	27,701	6.7	27,077	6.2	26,644	6.2	26,265	6.1	27,168	5.7	26,307	5.8
America, South	9,355	2.3	9,967	2.4	10,082	2.3	10,231	2.4	10,157	2.4	10,205	2.2	10,396	2.3
Asia	258,461	62.4	261,186	63.1	268,691	61.9	275,390	63.7	278,352	64.6	320,240	67.6	298,940	65.7
Europe	107,292	25.9	106,988	25.8	119,452	27.5	110,402	25.6	106,263	24.7	106,025	22.4	109,789	24.1
Oceania	4,496	1.1	4,236	1.0	4,250	1.0	4,101	0.9	4,547	1.1	4,617	1.0	4,272	0.9
Leading Producers														
China	73,283	17.7	73,516	17.8	79,839	18.4	93,177	21.6	97,807	22.7	135,018	28.5	*111,889	24.6
Japan	45,887	11.1	45,458	11.0	43,403	10.0	41,767	9.7	41,287	9.6	42,010	8.9	*42,365	9.3
Russian Federation	*28,116	6.8	*28,116	6.8	32,000	7.4	27,370	6.3	25,392	5.9	27,702	5.9	*28,116	6.2
United States of America	25,054	6.0	21,814	5.3	21,237	4.9	21,030	4.9	20,581	4.8	21,545	4.5	20,934	4.6
Ukraine	*21,407	5.2	*21,407	5.2	26,667	6.1	22,582	5.2	18,190	4.2	18,190	3.8	*21,407	4.7

Product Tables
Gas Produced by Cokeries

Unit of Measure: Terajoules.

	1990 Value	%	1991 Value	%	1992 Value	%	1993 Value	%	1994 Value	%	1995 Value	%	1996 Value	%
Total Production	2,488,888	100.0	2,601,110	100.0	2,619,877	100.0	2,459,245	100.0	2,427,708	100.0	2,467,557	100.0	2,442,513	100.0
Regions														
Africa	47,000	1.9	45,023	1.7	38,949	1.5	28,554	1.2	31,058	1.3	30,895	1.3	33,698	1.4
America, North	285,414	11.5	251,249	9.7	238,984	9.1	160,120	6.5	145,381	6.0	136,564	5.5	147,405	6.0
America, South	73,345	2.9	77,599	3.0	78,925	3.0	81,275	3.3	79,841	3.3	79,907	3.2	81,637	3.3
Asia	1,413,525	56.8	1,439,599	55.3	1,506,443	57.5	1,453,497	59.1	1,446,441	59.6	1,493,049	60.5	1,483,144	60.7
Europe	634,947	25.5	756,034	29.1	720,874	27.5	700,866	28.5	689,822	28.4	691,932	28.0	663,129	27.1
Oceania	34,658	1.4	31,606	1.2	35,703	1.4	34,932	1.4	35,165	1.4	35,210	1.4	33,501	1.4
Leading Producers														
China	307,001	12.3	334,621	12.9	324,665	12.4	330,000	13.4	375,592	15.5	390,000	15.8	*389,283	15.9
Japan	299,939	12.1	297,592	11.4	280,029	10.7	332,730	13.5	327,130	13.5	337,011	13.7	*295,250	12.1
Russian Federation	*203,799	8.2	*203,799	7.8	300,000	11.5	193,109	7.9	148,546	6.1	173,542	7.0	*203,799	8.3
United States of America	225,622	9.1	192,085	7.4	179,955	6.9	102,512	4.2	94,199	3.9	88,343	3.6	*95,094	3.9
Germany	-		162,000	6.2	126,021	4.8	104,993	4.3	96,000	4.0	100,000	4.1	*71,597	2.9

Tars

Unit of Measure: Metric tons.

	1990 Value	%	1991 Value	%	1992 Value	%	1993 Value	%	1994 Value	%	1995 Value	%	1996 Value	%
Total Production	6,581,643	100.0	6,153,464	100.0	6,612,673	100.0	6,133,956	100.0	6,703,931	100.0	6,606,288	100.0	6,484,195	100.0
Regions														
America, North	56,000	0.9	53,000	0.9	42,000	0.6	59,000	1.0	68,000	1.0	73,000	1.1	69,000	1.1
Asia	2,589,500	39.3	2,585,500	42.0	2,586,500	39.1	2,587,300	42.2	2,587,500	38.6	2,587,700	39.2	2,587,900	39.9
Europe	3,936,143	59.8	3,514,964	57.1	3,984,173	60.3	3,487,656	56.9	4,048,431	60.4	3,945,588	59.7	3,827,295	59.0
Leading Producers														
Japan	*2,579,500	39.2	*2,579,500	41.9	*2,579,500	39.0	*2,579,500	42.1	*2,579,500	38.5	*2,579,500	39.0	*2,579,500	39.8
United Kingdom	*777,500	11.8	*777,500	12.6	*777,500	11.8	500,000	8.2	1,055,000	15.7	*777,500	11.8	*777,500	12.0
Poland	647,000	9.8	550,000	8.9	554,000	8.4	514,000	8.4	578,000	8.6	585,000	8.9	526,000	8.1
Germany	-		-		574,000	8.7	456,000	7.4	391,000	5.8	488,637	7.4	469,334	7.2
France	278,000	4.2	264,000	4.3	256,000	3.9	230,000	3.7	215,000	3.2	*206,527	3.1	*194,369	3.0

ISIC 3550
RUBBER PRODUCTS

ISIC 3550—Rubber Products—includes inner tubes for automobiles and bicycles/motorcycles; tires for agricultural and offroad vehicles, for bicycles and motorcycles, and for road vehicles; reclaimed rubber; sheet rubber; rubber belting; and rubber footwear.

The industry's coverage corresponds to the U.S. SICs 301—Tires and Inner Tubes, 302—Rubber and Plastics Footwear, 305—Hose & Belting & Gaskets & Packing, and 306—Fabricated Rubber Products, nec.

Summary Statistics

		1990 Value	N	1991 Value	N	1992 Value	N	1993 Value	N	1994 Value	N	1995 Value	N
Establishments or enterprises	(number)	19,342	78	19,243	87	21,605	82	18,345	77	14,573	64	11,503	29
Number employed	(000)	2,823	85	2,427	89	2,417	84	2,372	84	2,214	76	2,076	57
Total Wages	($ mil.)	25,592	80	27,041	81	29,187	76	22,661	74	22,274	68	19,635	52
Wage/salary per employee	($)	9,879	79	10,064	79	11,776	75	10,268	73	10,435	67	12,765	52
Female workers	(000)	256	32	128	31	269	24	293	29	276	22	239	11
as % of total employment*	(%)	26	31	23	30	27	24	26	29	27	22	37	10
Output	($ bil.)	135	83	138	86	148	82	129	80	127	72	126	58
per employee	($ 000)	50	78	50	82	60	77	62	75	57	68	69	55
per establishment	($ 000)	7,424	72	6,315	76	7,066	71	6,222	67	6,300	55	4,194	26
Value added	($ bil.)	58	78	56	78	61	73	61	74	54	65	56	55
per employee	($ 000)	22	73	22	74	27	69	23	68	27	61	33	51
per establishment	($ 000)	2,564	67	2,166	70	2,722	64	2,006	61	2,453	48	1,639	23
Capital investment	($ mil.)	7,131	64	6,628	58	6,450	51	5,404	50	4,265	40	2,221	20
per employee	($ 000)	7	60	22	55	9	49	8	46	8	37	4	18
per establishment	($ 000)	2,114	61	2,097	55	2,532	49	2,371	46	3,481	38	363	17

Data presented above are drawn from the detailed tables that follow. Columns headed 'N' show the number of countries that provided valid data for inclusion. Values are not strictly comparable one year to the next or one row to the next because the number of countries that report varies from period to period and row to row. However, a general indication of magnitudes can be gleaned from reviewing this summary—especially in earlier years in which more reports are available. For detailed explanations, see Appendix I. *The average for those countries reporting both total and female employment.

Establishments and Number Engaged

Country	Establishments or Enterprises (number)						Employees per Establishment					
	1990	1991	1992	1993	1994	1995	1990	1991	1992	1993	1994	1995
Albania	-	-	-	-	2.000	2.000	-	-	-	-	172	176
Argentina	-	-	-	547	585	-	-	-	-	19	17	-
Armenia	3.000	5.000	5.000	5.000	5.000	-	1,041	502	-	-	-	-
Australia	214	250	251	-	-	-	42	32	32	-	-	-
Austria	39	37	35	33	33	-	182	162	171	162	156	-
Azerbaijan	3.000	3.000	3.000	3.000	4.000	-	1,383	1,337	1,270	1,194	856	-
Bangladesh	125	130	180	-	-	-	24	25	20	-	-	-
Belgium	98	100	103	98	97	-	47	46	44	-	-	-
Belize	2.000	2.000	2.000	-	-	-	-	-	-	-	-	-
Bolivia	7.000	6.000	7.000	8.000	9.000	10	8.143	9.500	8.571	9.875	9.222	8.400
Bosnia & Herzegovina	6.000	7.000	-	-	1.000	-	292	210	-	-	180	-
Brazil	454	-	439	433	-	-	141	-	124	121	-	-
Bulgaria	*15	*15	*17	*227	*329	*429	1,253	1,013	841	57	36	26
Cambodia	3.000	-	-	-	-	-	360	-	-	-	-	-
Canada	178	172	176	173	175	-	140	134	131	133	137	-
Central African Republic	-	*1.000	*1.000	*1.000	-	-	-	-	-	-	-	-
Chile	17	20	19	24	23	-	178	167	194	196	186	-
China	*4,260	*4,316	*4,354	*4,155	*4,242	*4,663	176	178	186	183	182	165
China (Hong Kong SAR)	180	136	106	96	80	-	6.667	8.088	7.547	5.208	6.250	-
Colombia	78	76	105	100	91	-	82	84	72	75	77	-
Costa Rica	40	46	47	51	-	-	62	69	73	67	-	-
Croatia	31	39	51	80	89	101	257	142	65	40	28	24
Cyprus	15	15	15	15	19	23	9.333	9.067	11	9.867	8.158	6.870
Czechoslovakia (Former)	*14	*13	-	-	-	-	1,929	1,692	-	-	-	-
Czech Republic	-	*10	*14	*21	-	-	-	1,500	857	571	-	-
Denmark	133	140	130	-	-	-	22	20	20	-	-	-
Ecuador	18	23	23	20	23	18	90	80	85	98	91	101
Egypt	22	32	36	24	-	-	127	303	231	304	-	-
El Salvador	-	10	25	11	9.000	11	-	-	8.800	39	41	158
Ethiopia	4.000	4.000	4.000	4.000	5.000	4.000	210	217	212	214	175	235
Fiji	7.000	7.000	6.000	-	-	-	17	15	17	-	-	-
Finland	26	29	29	28	32	34	100	86	79	79	78	73
FYR Macedonia	*1.000	*1.000	*6.000	*8.000	*11	*14	-	1.000	-	1.000	1.182	2.143
Gambia	-	-	-	1.000	-	-	-	-	-	15	-	-
Germany	-	332	330	332	317	-	-	341	311	275	265	-
Germany (Western Part)	274	279	289	-	-	-	359	348	330	-	-	-
Ghana	-	-	-	6.000	-	-	-	-	-	457	-	-
Greece	45	44	40	-	-	-	53	45	38	-	-	-
Grenada	-	-	1.000	1.000	1.000	-	-	-	-	5.000	7.000	-
Guatemala	-	6.000	6.000	6.000	6.000	6.000	-	221	215	208	202	217
Honduras	-	-	19	20	20	20	-	-	98	68	74	123
Hungary	*32	*81	*77	*101	-	-	250	99	78	50	-	-
India	1,997	2,078	2,172	2,259	-	-	56	54	56	56	-	-
Indonesia	511	468	473	448	448	441	324	288	313	271	292	289
Iran (Islamic Republic of)	32	44	49	52	-	-	341	252	231	231	-	-
Iraq	-	-	4.000	-	-	-	-	-	500	-	-	-
Ireland	42	49	-	-	-	-	57	53	-	-	-	-
Israel	50	57	47	48	53	-	44	40	43	44	42	-
Italy	*335	*342	*258	*266	*272	-	140	131	134	120	123	-
Jamaica	11	11	-	-	-	-	68	72	-	-	-	-
Japan	4,260	4,238	4,124	4,149	3,339	3,505	36	36	37	36	38	37
Jordan	16	15	10	10	27	25	9.500	6.800	7.000	10	6.963	6.440
Kenya	25	44	56	56	67	67	88	52	41	42	36	36
Korea, Republic of	1,684	650	686	772	801	858	108	48	46	43	42	41
Kuwait	5.000	3.000	2.000	2.000	3.000	-	46	49	57	53	42	-
Kyrgyzstan	-	-	1.000	1.000	1.000	-	-	-	262	216	144	-
Latvia	1.000	8.000	6.000	8.000	7.000	12	2,680	306	36	20	21	16
Lithuania	-	-	-	*42	*11	-	-	-	-	5.310	30	-
Macau	1.000	1.000	1.000	1.000	-	-	16	14	13	9.000	-	-
Malaysia	368	387	381	379	366	-	161	167	170	182	189	-

Continued.

Depending on the table, * means *Enterprises, Engaged,* or *Factor Values*; ± means *Basis Unspecified*; § means *shown in millions.* For additional notes and sources, see Appendix I.

Establishments and Number Engaged

- Continued -

Country	Establishments or Enterprises (number)						Employees per Establishment					
	1990	1991	1992	1993	1994	1995	1990	1991	1992	1993	1994	1995
Malta	8.000	8.000	8.000	9.000	9.000	-	73	84	81	86	102	-
Mauritius	10	11	11	11	12	11	34	33	33	30	26	29
Mexico	48	48	46	46	45	43	355	346	346	296	282	288
Morocco	-	32	26	28	-	-	-	79	98	94	-	-
Mozambique	5.000	5.000	6.000	6.000	5.000	-	191	191	149	145	161	-
Myanmar	3.000	3.000	3.000	3.000	2.000	-	431	472	404	528	715	-
Nepal	33	51	-	88	25	-	22	21	-	16	31	-
Netherlands	28	31	32	37	34	-	200	172	154	129	132	-
New Zealand	*84	*95	*91	*96	-	-	23	19	20	20	-	-
Norway	27	29	13	16	24	-	44	40	40	39	30	-
Pakistan	-	49	-	-	-	-	-	170	-	-	-	-
Panama	8.000	7.000	-	-	-	-	19	20	-	-	-	-
Paraguay	-	2.000	-	-	-	-	-	32	-	-	-	-
Peru	119	121	120	-	-	-	25	18	14	-	-	-
Philippines	419	410	186	178	-	-	71	66	159	127	-	-
Poland	*45	*48	*50	*44	-	-	756	604	460	523	-	-
Portugal	*267	*250	*252	*243	*243	-	25	22	26	24	24	-
Puerto Rico	18	13	13	13	12	11	67	85	59	37	52	33
Qatar	1.000	1.000	-	-	-	-	30	34	-	-	-	-
Romania	-	237	357	440	572	-	-	-	-	-	-	-
Russian Federation	-	-	-	*488	*444	*642	-	-	-	359	339	221
Saint Lucia	-	1.000	1.000	1.000	1.000	1.000	-	-	-	-	-	-
Senegal	-	1.000	1.000	1.000	1.000	-	-	9.000	9.000	9.000	6.000	-
Singapore	30	31	28	29	31	-	54	57	60	61	60	-
Slovakia	-	6.000	6.000	9.000	9.000	-	-	1,298	1,128	704	700	-
Slovenia	33	*169	*90	*86	*254	*213	-	-	-	-	-	-
South Africa	-	185	-	-	-	-	-	97	-	-	-	-
Spain	962	1,025	855	-	-	-	35	32	36	-	-	-
Sri Lanka	170	146	139	170	-	-	80	91	116	89	-	-
Suriname	-	-	1.000	-	-	-	-	-	20	-	-	-
Sweden	79	75	73	67	68	-	91	86	82	75	72	-
Thailand	213	377	-	-	-	-	178	145	-	-	-	-
Trinidad and Tobago	28	34	30	35	33	-	20	16	17	13	5.152	-
Tunisia	-	-	-	93	97	105	-	-	-	27	28	25
Turkey	96	86	156	154	143	-	134	127	83	87	89	-
Ukraine	35	130	159	38	37	37	1,314	346	283	1,132	1,000	946
United Kingdom	605	606	602	577	649	-	107	96	90	88	76	-
United Republic of Tanzania	7.000	7.000	-	-	-	-	122	123	-	-	-	-
USSR (Former)	*166	-	-	-	-	-	1,681	-	-	-	-	-
United States of America	-	-	2,862	-	-	-	-	-	71	-	-	-
Venezuela	52	62	62	69	65	64	113	119	119	96	102	108
Yugoslavia	41	57	92	132	135	133	-	333	193	133	127	119
Zambia	8.000	-	-	-	6.000	-	190	-	-	-	163	-
Zimbabwe	12	12	13	13	14	-	239	166	238	231	221	-

Employment and Compensation of Employees

Country	Number Employed/Engaged (000)						Wage/Salary per Employee ($)					
	1990	1991	1992	1993	1994	1995	1990	1991	1992	1993	1994	1995
Albania	-	-	-	-	0.344	0.353	-	-	-	-	-	882
Argentina	13	-	-	10	10	-	8,217	-	-	16,872	-	-
Armenia	*3.123	*2.508	-	-	-	-	-	-	-	-	-	-
Australia	9.000	*8.000	*8.000	-	-	-	23,446	29,120	25,000	-	-	-
Austria	7.100	6.000	6.000	5.339	5.142	4.823	31,092	33,188	35,581	34,394	36,913	-
Azerbaijan	4.148	4.010	3.809	3.582	3.423	-	2,035	1,819	-	-	-	-
Bangladesh	3.031	3.199	3.663	-	-	-	447	470	484	-	-	-
Belgium	4.600	4.600	4.500	-	-	-	25,676	25,815	28,892	-	-	-
Bolivia	0.057	0.057	0.060	0.079	0.083	0.084	1,344	1,739	1,637	1,493	1,836	1,771

Continued.

Depending on the table, * means *Enterprises, Engaged,* or *Factor Values*; ± means *Basis Unspecified*; § means *shown in millions*. For additional notes and sources, see Appendix I.

395

Employment and Compensation of Employees

- Continued -

Country	Number Employed/Engaged (000)						Wage/Salary per Employee ($)					
	1990	1991	1992	1993	1994	1995	1990	1991	1992	1993	1994	1995
Bosnia & Herzegovina	1.754	1.471	-	-	0.180	-	2,763	2,156	-	-	-	-
Brazil	64	-	54	52	-	-	6,869	-	6,371	9,255	-	-
Bulgaria	19	15	14	13	12	11	1,904	640	1,042	1,185	945	1,237
Cambodia	*1.079	*1.100	-	-	-	-	-	-	-	-	-	-
Canada	25	23	23	23	24	24	28,762	30,435	30,036	28,410	27,552	28,572
Chile	3.022	3.346	3.686	4.715	4.279	4.274	5,906	6,561	7,155	7,832	9,731	10,978
China	*750	*770	*810	*760	*770	*770	-	-	-	-	-	-
China (Hong Kong SAR)	*1.200	*1.100	*0.800	*0.500	*0.500	-	7,061	8,306	8,397	11,893	11,645	-
China (Taiwan Province)	52	50	49	47	47	46	9,270	10,197	11,873	11,814	12,354	12,911
Colombia	6.400	6.351	7.585	7.513	7.012	6.500	3,366	3,701	3,892	4,152	5,366	6,067
Costa Rica	2.484	3.189	3.442	3.419			3,463	3,092	3,497	3,934		
Croatia	7.970	5.550	3.310	3.230	2.470	2.380	3,462	2,724	1,112	1,302	2,059	3,458
Cyprus	0.140	0.136	0.161	0.148	0.155	0.158	9,222	10,339	10,794	12,113	12,299	16,068
Czechoslovakia (Former)	27	22	-	-	-	-	2,352	1,773	-	-	-	-
Czech Republic	16	15	12	12	-	-	2,437	1,696	2,153	2,830	-	-
Denmark	2.909	2.841	2.654	2.633	2.645	2.624	28,550	28,617	32,523	30,693	31,917	36,940
Ecuador	1.615	1.849	1.953	1.954	2.104	1.812	5,937	5,515	6,490	6,848	4,753	4,735
Egypt	*2.800	*9.700	*8.300	*7.300	*7.216	*7.110	1,804	-	1,563	1,897	1,879	1,897
El Salvador	-	-	*0.220	*0.425	*0.373	*1.743	-	-	-	2,947	3,941	2,730
Ethiopia	0.840	0.869	0.846	0.856	0.875	0.940	2,486	2,655	2,069	1,311	1,305	1,242
Fiji	0.121	0.107	0.103	0.102	0.102	-	3,415	4,059	4,703	4,950	5,327	-
Finland	2.600	2.500	2.300	2.200	2.500	2.497	25,738	24,204	23,722	20,403	23,770	31,549
FYR Macedonia	-	0.001	0.005	0.008	0.013	0.030	-	5,092	12,039	19,502	5,342	7,895
France	94	91	86	83	82	83	26,626	27,005	31,994	-	-	-
Gambia	-	-	-	0.015	-	-	-	-	-	-	-	-
Germany	-	*113	*103	*91	*84		-	28,098	33,902	33,231	35,864	-
Germany (Western Part)	*98	*97	*95				30,592	31,272	35,337			
Ghana	-	-	-	*2.741	*2.818	*2.898	-	-	-	829	619	632
Greece	*2.384	*1.959	*1.524	*1.533	*1.537	*1.540	15,171	14,888	15,466	15,299	15,407	17,608
Grenada	-	-	-	*0.005	0.007	-	-	-	-	-	-	-
Guatemala	-	*1.325	*1.287	*1.247	*1.210	*1.299	-	463	449	496	-	521
Honduras	*1.738	*1.795	*1.862	*1.356	*1.486	*2.451	2,030	2,202	2,128	2,503	2,168	1,357
Hungary	8.000	8.000	6.000	5.000	4.360	4.230	3,415	3,062	3,838	3,955	4,109	4,405
India	*113	*113	*122	*126	*132	*138	1,533	1,211	1,188	1,171	1,249	1,309
Indonesia	166	135	148	122	131	127	560	581	682	836	740	1,199
Iran (Islamic Republic of)	11	11	11	12	-	-	4,578	5,325	5,827	5,086	-	-
Iraq	-	2.085	2.000	-	-	-	-	6,250	10,847	-	-	-
Ireland	2.400	2.600	2.561	2.561	2.561	2.531	23,148	22,990	26,749	23,488	24,349	28,182
Israel	2.200	2.300	2.000	2.100	2.200	2.234	18,938	19,649	22,366	20,864	22,342	25,111
Italy	47	45	35	32	33	-	35,642	37,117	-	32,821	33,816	-
Jamaica	0.746	0.787	0.683	0.683	0.659	-	-	-	-	-	-	-
Japan	*152	*153	*152	*149	*128	*128	27,990	31,246	34,856	38,929	-	-
Jordan	0.152	0.102	0.070	0.102	0.188	0.161	2,388	2,073	2,815	2,730	2,047	2,536
Kenya	2.202	2.306	2.296	2.334	2.398	2.440	2,963	2,584	2,350	1,498	1,629	2,453
Korea, Republic of	182	31	31	33	33	35	7,353	11,195	11,734	13,061	14,184	17,367
Kuwait	0.229	0.147	0.114	0.106	0.127	0.119	2,782	3,282	3,114	4,530	6,024	6,134
Kyrgyzstan	-	-	*0.262	*0.216	*0.144	-	-	-	-	-	-	-
Latvia	2.680	2.451	0.218	0.163	0.149	0.197	-	-	-	-	-	-
Lithuania	-	-	0.280	0.223	0.331	-	-	-	-	-	665	-
Macau	0.016	0.014	0.013	0.009	-	-	4,625	4,821	4,846	4,694		-
Malawi	0.138	-	-	-	-	-	-	-	-	-	-	-
Malaysia	59	65	65	69	69	73	2,636	2,813	3,269	3,549	3,668	4,218
Malta	0.580	0.674	0.648	0.772	0.915	-	16,752	14,423	17,771	12,984	12,996	-
Mauritius	*0.340	0.366	0.359	0.331	0.317	0.316	2,295	2,192	2,968	2,996	3,566	4,005
Mexico	17	17	16	14	13	12	7,427	8,904	10,239	12,504	12,896	7,095
Morocco	-	2.525	2.560	2.646	-	-	-	-	-	-	-	-
Mozambique	0.953	0.953	0.892	0.870	0.805	-	4,527	2,853	1,604	1,159	1,216	-
Myanmar	1.294	1.417	1.213	1.583	1.430	-	1,129	1,031	1,179	1,727	1,891	-
Nepal	0.723	1.092	-	1.410	0.768	-	446	437	-	-	442	-
Netherlands	5.613	5.337	4.934	4.756	4.502	4.502	28,276	28,561	31,926	32,377	38,343	-

Continued.

Depending on the table, * means *Enterprises, Engaged,* or *Factor Values*; ± means *Basis Unspecified*; § means *shown in millions.* For additional notes and sources, see Appendix I.

Employment and Compensation of Employees

- Continued -

Country	Number Employed/Engaged (000)						Wage/Salary per Employee ($)					
	1990	1991	1992	1993	1994	1995	1990	1991	1992	1993	1994	1995
New Zealand	1.936	1.779	1.854	1.876	-	-	21,555					
Norway	1.194	1.168	0.519	0.630	0.720	0.720	28,231	28,262	37,659	28,864	31,290	35,408
Pakistan	8.346	8.348	8.732	-	-	-	1,504	1,572	1,617	-	-	-
Panama	*0.154	*0.140	*0.158	*0.163	*0.164	*0.163	4,338	4,507	4,660	4,715	4,758	4,731
Paraguay	-	0.063	-	-	-	-	-	-	-	-	-	-
Peru	*2.942	*2.222	*1.650	-	-	-	5,923	8,238	8,227			-
Philippines	30	27	30	23	-	-	2,089	2,128	2,681	3,052	-	-
Poland	34	29	23	23	22	21	1,132	1,701	2,395	2,694	3,473	5,047
Portugal	*6.614	*5.594	*6.434	*5.865	*5.834	*5.566	-	-	-	-	-	-
Puerto Rico	1.210	1.100	0.770	0.480	0.620	0.360	19,008	-	-	-	-	-
Qatar	0.030	0.034					4,020	4,573	-			
Russian Federation	-			175	151	142	-	-	-	678	953	1,152
Senegal	-	*0.009	*0.009	*0.009	*0.006	-	-	4,332	3,358	3,139	1,501	-
Singapore	*1.611	*1.760	*1.691	*1.760	*1.867	*1.850	11,016	12,621	14,735	15,014	16,626	18,190
Slovakia	-	*7.790	*6.765	*6.338	*6.300	-	-	1,829	2,302	2,723	3,165	-
South Africa	18	18	17	16	16	16	9,555	9,194	11,179	11,438	11,617	13,687
Spain	33	33	30	26	28	29	22,066	23,153	25,417	24,549	23,123	26,418
Sri Lanka	*14	*13	*16	*15	-	-	515	728	735	942	-	-
Suriname	*0.028	*0.027	*0.020	*0.026	-	-	8,003	10,375	25,210	26,718		
Sweden	7.200	*6.431	*6.005	5.027	4.904	4.800	23,465	24,958	26,389	21,571	23,067	25,431
Thailand	38	55	-	-	-	-	2,659	2,392	-	-	-	-
Trinidad and Tobago	0.570	0.561	0.506	0.451	0.170	-	9,735	10,100	9,657	7,127	4,468	
Tunisia	-		-	2.465	2.670	2.633	-		-	5,214	5,205	7,662
Turkey	13	11	13	13	13	13	9,659	13,433	13,180	13,269	9,875	13,126
Ukraine	46	45	45	43	37	35	3,079	3,061	3,082	413	577	681
United Kingdom	65	58	54	51	49	51	21,868	24,168	26,764	23,229	25,159	27,337
United Republic of Tanzania	0.857	0.859	-	-	-	-	604	1,461	-	-	-	-
USSR (Former)	279	-					2,153	-				
United States of America	205	197	204	208	209	217	26,146	27,005	28,716	29,423	29,708	29,926
Uruguay	*2.728	*2.378	*2.100	*2.051	*2.074	*2.320	5,205	5,727	5,168	6,337	6,565	7,209
Venezuela	5.900	7.400	7.400	6.656	6.633	6.923	5,544	5,233	5,648	4,908	4,820	6,701
Yugoslavia		19	18	18	17	16	-	-	-	-	1,703	749
Zambia	1.517	-	-	-	0.980	-	5,890	-	-	-	2,525	-
Zimbabwe	2.869	1.996	3.100	3.000	3.100	2.766	4,546	3,391	3,191	2,992	3,289	3,586

Female Workers: Total and Percent of Employment

Country	Female Workers (000)						Female Workers as % of Total Employment					
	1990	1991	1992	1993	1994	1995	1990	1991	1992	1993	1994	1995
Afghanistan	0.007	0.005	-	-	-	-	-	-	-	-	-	-
Albania	-	-	-	-	0.238	0.180	-	-	-	-	69.19	50.99
Argentina	-	-	-	-	1.021	-	-	-	-	-	10.09	-
Australia	1.324	-	-	-	-	-	14.71	-	-	-	-	-
Austria	1.700	1.000	1.000	1.074	1.068	-	23.94	16.67	16.67	20.12	20.77	-
Bangladesh	-	-	0.028	-	-	-	-	-	0.76	-	-	-
Bosnia & Herzegovina	0.658	0.654	-	-	-	-	37.51	44.46	-	-	-	-
Bulgaria	-	-	-	6.600	5.900	5.400	-	-	-	50.77	50.43	47.79
Canada	5.000	4.000	-	-	-	-	20.00	17.39	-	-	-	-
Chile	0.317	0.328	0.255	0.681	0.505	-	10.49	9.80	6.92	14.44	11.80	-
China (Taiwan Province)	24	22	22	21	20	20	45.41	44.47	44.69	44.06	43.32	42.90
Colombia	-	0.869	-	1.656	1.654	-	-	13.68	-	22.04	23.59	-
Croatia	3.910	2.590	1.460	1.350	1.010	1.000	49.06	46.67	44.11	41.80	40.89	42.02
Cyprus	0.035	0.036	0.042	0.042	0.053	-	25.00	26.47	26.09	28.38	34.19	-
Czechoslovakia (Former)	6.800	5.900	-	-	-	-	25.19	26.82	-	-	-	-
Czech Republic	7.000	6.000	4.000	4.000	-	-	43.75	40.00	33.33	33.33		-
Denmark	0.755	0.713	0.697	-	-	-	25.95	25.10	26.26	-	-	-
Egypt	-	0.288	0.970	0.511	-	-	-	2.97	11.69	7.00	-	-
El Salvador	-	-	-	0.110	0.079	0.913	-	-	-	25.88	21.18	52.38

Continued.

Depending on the table, * means *Enterprises*, *Engaged*, or *Factor Values*; ± means *Basis Unspecified*; § means *shown in millions*. For additional notes and sources, see Appendix I.

397

Female Workers: Total and Percent of Employment
- Continued -

Country	Female Workers (000)						Female Workers as % of Total Employment					
	1990	1991	1992	1993	1994	1995	1990	1991	1992	1993	1994	1995
Ethiopia	-	0.082	0.087	0.090	0.095	0.096	-	9.44	10.28	10.51	10.86	10.21
Gambia	-	-	-	0.001	-	-	-	-	-	6.67	-	-
Germany (Western Part)	24	-	-	-	-	-	24.38	-	-	-	-	-
Ghana	-	-	-	0.965	-	-	-	-	-	35.21	-	-
Hungary	3.000	-	2.000	2.000	-	-	37.50	-	33.33	40.00	-	-
Indonesia	-	-	-	30	32	30	-	-	-	24.94	24.74	23.42
Iran (Islamic Republic of)	-	0.332	0.338	0.382	-	-	-	2.99	2.99	3.18	-	-
Iraq	-	0.221	-	-	-	-	-	10.60	-	-	-	-
Ireland	0.600	0.700	-	-	-	-	25.00	26.92	-	-	-	-
Italy	-	8.138	4.901	5.077	5.223	-	-	18.22	14.18	15.94	15.61	-
Jordan	0.009	0.008	-	-	0.011	0.002	5.92	7.84	-	-	5.85	1.24
Kenya	0.099	0.139	0.130	0.132	-	-	4.50	6.03	5.66	5.66	-	-
Korea, Republic of	107	7.330	7.091	6.522	6.735	7.275	58.74	23.45	22.55	19.85	20.14	20.84
Macau	0.002	0.002	0.002	0.002	-	-	12.50	14.29	15.38	22.22	-	-
Malaysia	30	33	33	34	33	-	50.25	50.77	50.23	49.71	47.10	-
Malta	0.190	0.246	0.239	0.287	0.333	-	32.76	36.50	36.88	37.18	36.39	-
Morocco	-	-	-	-	-	0.118	-	-	-	-	-	-
Myanmar	0.472	0.447	0.436	0.250	0.331	-	36.48	31.55	35.94	15.79	23.15	-
Nepal	0.036	0.094	-	0.028	0.022	-	4.98	8.61	-	1.99	2.86	-
New Zealand	0.357	0.331	0.357	0.343	-	-	18.44	18.61	19.26	18.28	-	-
Panama	0.032	-	-	-	-	-	20.78	-	-	-	-	-
Philippines	-	-	14	8.994	-	-	-	-	47.52	39.80	-	-
Portugal	0.436	0.391	-	0.426	0.361	-	6.59	6.99	-	7.26	6.19	-
Sri Lanka	4.467	4.104	5.225	3.765	-	-	33.01	30.90	32.51	25.02	-	-
Sweden	2.200	-	-	-	-	-	30.56	-	-	-	-	-
Thailand	15	28	-	-	-	-	40.60	51.27	-	-	-	-
Turkey	0.999	-	-	-	-	-	7.74	-	-	-	-	-
United Kingdom	16	-	13	-	-	-	24.04	-	24.07	-	-	-
United Republic of Tanzania	0.055	0.054	-	-	-	-	6.42	6.29	-	-	-	-
United States of America	-	-	158	162	166	174	-	-	77.45	77.88	79.43	80.18
Zambia	-	-	-	-	0.058	-	-	-	-	-	5.92	-

Output and Output per Employee

Country	Output ($ bil.)						Output per Employee ($)					
	1990	1991	1992	1993	1994	1995	1990	1991	1992	1993	1994	1995
Albania	-	-	-	-	*0.001	*0.001	-	-	-	-	1,842	3,342
Argentina	±1.061	-	-	0.661	-	-	79,134	-	-	64,807	-	-
Armenia	0.001	0.003	0.016	0.328	0.002	-	442	1,092	-	-	-	-
Australia	*1.031	*0.993	*0.869	-	-	-	114,583	124,075	108,640	-	-	-
Austria	0.944	0.934	0.995	0.832	0.898	1.139	132,954	155,618	165,863	155,853	174,658	236,070
Azerbaijan	±0.066	±0.073	±0.004	±0.008	±0.003	-	15,816	18,148	1,171	2,245	776	-
Bangladesh	0.017	0.018	0.022	-	-	-	5,635	5,475	5,993	-	-	-
Belgium	0.752	0.716	0.870	0.621	0.755	0.904	163,496	155,563	193,392	-	-	-
Bolivia	0.001	0.001	0.001	0.001	0.001	0.001	10,241	11,622	10,195	10,592	10,300	10,205
Bosnia & Herzegovina	±0.026	±0.027	-	-	-	-	14,558	18,105	-	-	-	-
Brazil	*3.704	-	*3.000	*3.460	-	-	57,927	-	55,069	66,130	-	-
Bulgaria	±0.327	±0.117	±0.117	±0.088	±0.078	±0.121	17,415	7,674	8,212	6,806	6,646	10,695
Canada	3.257	3.064	3.897	4.736	5.038	5.233	130,271	133,201	169,424	205,916	209,920	219,408
Chile	0.146	0.177	0.203	0.281	0.282	0.318	48,442	53,032	55,176	59,642	65,981	74,418
China	±5.957	±5.960	±6.914	±7.774	±6.383	±7.422	7,942	7,740	8,536	10,229	8,290	9,640
China (Hong Kong SAR)	0.047	0.037	0.034	0.027	0.029	-	38,940	33,457	42,794	54,812	57,192	-
China (Taiwan Province)	±2.071	±2.173	±2.395	±2.381	±2.363	±2.590	39,605	43,322	49,202	50,557	50,541	56,094
Colombia	0.307	0.325	0.297	0.320	0.406	0.418	47,908	51,130	39,164	42,598	57,870	64,381
Costa Rica	0.050	0.043	0.049	0.048	0.057	0.065	20,266	13,387	14,177	14,074	-	-
Cote d'Ivoire	±0.026	±0.025	±0.026	±0.025	-	-	-	-	-	-	-	-
Croatia	±0.175	±0.083	±0.031	-	-	-	22,005	14,951	9,467	-	-	-
Cyprus	0.007	0.006	0.009	0.008	0.009	0.010	51,751	43,927	55,735	51,498	56,719	64,103

Continued.

Depending on the table, * means *Enterprises, Engaged,* or *Factor Values;* ± means *Basis Unspecified;* § means *shown in millions.* For additional notes and sources, see Appendix I.

Output and Output per Employee

- Continued -

Country	Output ($ bil.)						Output per Employee ($)					
	1990	1991	1992	1993	1994	1995	1990	1991	1992	1993	1994	1995
Czechoslovakia (Former) . . .	±0.597	±0.215	-	-	-	-	22,098	9,776	-	-	-	-
Denmark	*0.213	*0.187	*0.217	*0.208	*0.212	*0.249	73,151	65,819	81,906	78,899	80,298	94,764
Ecuador	0.052	0.083	0.083	0.084	0.090	0.099	32,343	44,960	42,535	42,965	42,776	54,674
Egypt	*0.025	*0.042	*0.080	*0.079	*0.078	*0.077	9,071	4,297	9,635	10,868	10,751	10,850
El Salvador.	-	-	-	±0.011	-	±0.012	-	-	-	25,370	-	7,079
Ethiopia.	0.019	0.012	0.011	0.011	0.017	0.012	22,123	14,070	12,464	13,197	19,800	13,187
Fiji	0.004	0.003	0.004	0.004	0.004	-	33,895	29,500	35,263	37,680	40,739	-
Finland	±0.278	±0.220	±0.223	±0.208	±0.280	±0.385	106,916	87,834	96,966	94,374	111,956	154,212
FYR Macedonia	±0.000	±0.000	±0.000	±0.000	±0.001	±0.001	-	106,936	73,373	58,263	112,179	49,123
France	±6.938	±6.764	±7.458	±6.396	±6.803	±7.810	73,804	73,919	87,226	77,147	83,265	94,213
Germany	-	12	13	11	12	-	-	105,261	129,785	121,327	139,912	-
Germany (Western Part) . . .	11	12	13	-	-	-	116,574	119,147	136,234	-	-	-
Ghana	-	-	-	0.006	0.005	0.005	-	-	-	2,248	1,781	1,817
Greece	*0.224	*0.206	*0.178	*0.169	*0.168	*0.189	93,763	105,381	116,528	110,365	109,547	122,946
Guatemala.	-	0.052	0.058	0.057	0.092	0.090	-	38,927	45,275	45,476	75,828	69,433
Honduras	0.031	0.028	0.027	0.027	0.027	0.028	17,623	15,374	14,705	20,265	17,984	11,538
Hungary	±0.245	±0.198	±0.163	±0.087	±0.088	±0.092	30,654	24,754	27,148	17,306	20,284	21,632
India	*3.041	*2.264	*2.541	*2.560	*2.909	*3.238	27,026	20,098	20,847	20,260	22,092	23,509
Indonesia	1.711	1.642	2.221	1.750	2.153	2.900	10,323	12,197	14,985	14,406	16,437	22,752
Iran (Islamic Republic of) . . .	0.240	0.248	0.446	0.422	-	-	21,993	22,366	39,407	35,197	-	-
Iraq	-	*0.026	*0.100	-	-	-	-	12,655	50,161	-	-	-
Ireland	*0.237	*0.250	*0.288	*0.252	*0.262	*0.301	98,673	96,185	112,424	98,535	102,393	118,872
Israel.	0.169	0.182	0.192	0.171	0.193	0.227	76,652	79,169	96,173	81,269	87,857	101,526
Italy	±5.842	±5.522	±4.667	±3.831	±4.625	-	124,856	123,619	135,070	120,238	138,273	-
Japan	±23	±26	±27	±29	±28	±32	153,398	170,931	179,941	192,893	217,154	249,342
Jordan	0.005	0.003	0.001	0.002	0.003	0.004	33,945	32,550	20,588	15,166	14,009	22,992
Kenya	*0.118	*0.113	*0.091	*0.054	*0.059	*0.074	53,507	48,870	39,744	23,343	24,700	30,281
Korea, Republic of	7.271	2.922	3.047	3.417	3.810	4.549	39,953	93,505	96,910	104,009	113,915	130,329
Kuwait	0.005	0.005	0.004	0.005	0.007	0.007	22,627	36,696	37,782	44,171	57,443	61,162
Kyrgyzstan.	-	-	0.003	0.000	0.000	-	-	-	9,936	1,250	1,621	-
Latvia	0.000	0.001	0.001	0.001	0.001	0.002	185	501	3,028	5,182	7,383	9,644
Lithuania	-	-	0.001	0.000	0.001	-	-	-	2,200	2,067	2,798	-
Macau	0.000	0.000	0.000	0.000	-	-	18,305	23,750	18,106	21,292	-	-
Malaysia	*1.953	*2.163	*2.337	*2.414	*2.781	*3.383	33,053	33,484	36,014	35,089	40,311	46,101
Malta	0.024	0.027	0.029	0.026	0.031	-	41,368	39,940	44,874	33,580	33,946	-
Mauritius	0.005	0.006	0.006	0.005	0.006	0.007	15,297	15,466	17,343	16,246	17,810	21,697
Mexico	0.990	1.037	1.004	0.895	0.993	0.722	58,089	62,481	63,121	65,779	78,179	58,350
Morocco	-	±0.121	±0.098	±0.102	-	-	-	47,850	38,157	38,407	-	-
Mozambique	-	±0.005	-	±0.003	±0.003	-	-	4,960	-	3,461	3,211	-
Myanmar	0.026	0.015	0.035	0.010	0.019	-	19,831	10,906	28,544	6,602	13,460	-
Nepal	0.005	0.008	-	0.009	0.003	-	6,970	7,325	-	6,661	3,730	-
Netherlands	0.662	0.636	0.633	0.569	0.629	0.728	117,898	119,255	128,394	119,661	139,827	161,595
New Zealand	0.206	-	-	-	-	-	106,390	-	-	-	-	-
Norway	±0.137	±0.119	±0.064	±0.058	±0.076	±0.086	115,064	101,822	122,400	91,514	105,088	119,949
Pakistan	0.123	0.128	0.138	-	-	-	14,737	15,275	15,850	-	-	-
Panama.	0.005	0.004	0.006	0.007	0.007	0.007	32,052	30,486	38,135	40,600	42,018	41,501
Peru	0.137	0.114	0.115	-	-	-	46,536	51,467	69,504	-	-	-
Philippines.	0.366	0.362	0.544	0.413	-	-	12,312	13,470	18,457	18,289	-	-
Poland	±0.396	±0.412	±0.453	±0.488	±0.601	±0.836	11,650	14,212	19,687	21,208	27,228	39,467
Portugal	0.250	0.208	0.273	0.242	0.296	0.416	37,766	37,174	42,385	41,327	50,765	74,783
Qatar	0.001	0.001	-	-	-	-	45,788	40,401	-	-	-	-
Romania	±0.468	±0.373	±0.271	±0.318	-	-	-	-	-	-	-	-
Russian Federation	-	4.978	4.714	1.270	1.378	2.254	-	-	-	7,250	9,149	15,908
Singapore	*0.081	*0.093	*0.100	*0.104	*0.129	*0.165	50,236	53,058	59,017	58,974	68,926	89,038
Slovakia	-	±0.196	±0.215	±0.189	±0.203	-	-	25,186	31,745	29,787	32,217	-
Slovenia	-	-	-	0.207	0.221	0.285	-	-	-	-	-	-
South Africa	±1.040	±0.939	±0.977	±0.930	±0.923	±1.087	57,760	52,149	57,441	58,114	57,662	68,208
Spain	*2.962	*3.021	*3.208	±2.371	±2.677	±3.470	88,999	92,283	105,544	90,860	94,635	121,587
Sri Lanka	0.064	0.080	0.102	0.131	-	-	4,718	6,046	6,316	8,681	-	-
Suriname	-	*0.001	*0.005	0.023	-	-	-	20,749	252,101	883,430	-	-

Continued.

Output and Output per Employee
- Continued -

Country	Output ($ bil.)						Output per Employee ($)					
	1990	1991	1992	1993	1994	1995	1990	1991	1992	1993	1994	1995
Sweden	*0.735	*0.736	*0.830	*0.499	*0.568	*0.612	102,072	114,442	138,225	99,204	115,802	127,428
Thailand	1.818	2.734	-	-	-	-	48,076	50,075	-	-	-	-
Trinidad and Tobago	±0.020	±0.022	±0.014	±0.011	±0.003	-	34,648	39,233	28,215	23,909	19,956	-
Tunisia	0.080	0.080	0.075	0.077	0.081	0.096	-	-	-	31,324	30,508	36,628
Turkey	0.909	1.049	1.277	1.350	1.115	1.543	70,464	95,934	99,126	100,223	87,353	116,515
Ukraine	*1.533	*1.510	*1.875	*0.390	*0.329	*0.509	33,317	33,558	41,666	9,067	8,892	14,538
United Kingdom	*5.782	*5.625	5.618	4.730	5.158	5.856	88,956	96,979	104,044	92,740	105,260	114,550
United Republic of Tanzania	*0.023	*0.029	-	-	-	-	27,206	33,847	-	-	-	-
USSR (Former)	±5.750	-	-	-	-	-	20,609	-	-	-	-	-
United States of America	*26	*25	*27	*29	*30	*32	125,073	128,629	131,137	138,779	142,885	149,088
Uruguay	0.095	0.074	0.066	0.074	0.077	0.095	34,666	31,096	31,599	35,920	37,177	40,770
Venezuela	0.290	0.344	0.362	0.398	0.424	0.543	49,145	46,473	48,946	59,758	63,910	78,438
Yugoslavia	±0.057	±0.056	±0.086	-	±0.192	±0.085	-	2,944	4,824	-	11,140	5,391
Zambia	0.028	-	-	-	0.016	-	18,221	-	-	-	15,853	-
Zimbabwe	*0.096	*0.053	*0.107	*0.092	*0.107	*0.104	33,315	26,728	34,384	30,591	34,514	37,435

Value Added and Value Added per Employee

Country	Value Added ($ bil.)						Value Added per Employee ($)					
	1990	1991	1992	1993	1994	1995	1990	1991	1992	1993	1994	1995
Argentina	±0.368	-	-	0.273	-	-	27,418	-	-	26,787	-	-
Armenia	-0.000	0.000	-	-	-	-	-92.750	34	-	-	-	-
Australia	*0.546	*0.554	*0.552	-	-	-	60,677	69,244	68,961	-	-	-
Austria	0.311	0.328	0.357	0.305	0.348	0.431	43,802	54,585	59,544	57,050	67,766	89,418
Bangladesh	0.005	0.004	0.009	-	-	-	1,749	1,213	2,383	-	-	-
Belgium	*0.272	*0.271	*0.329	*0.235	*0.285	*0.341	59,132	58,893	73,205	-	-	-
Bolivia	0.000	0.000	0.000	0.000	0.000	0.000	3,030	4,679	4,875	4,122	4,454	4,412
Bosnia & Herzegovina	±0.011	±0.011	-	-	-	-	6,045	7,581	-	-	-	-
Brazil	*1.962	-	*1.916	*2.293	-	-	30,681	-	35,172	43,823	-	-
Bulgaria	-	±0.029	±0.032	±0.021	-	-	-	1,886	2,232	1,645	-	-
Canada	1.397	1.318	1.679	1.659	1.736	1.797	55,879	57,303	73,021	72,121	72,313	75,343
Chile	0.072	0.085	0.091	0.131	0.136	0.153	23,912	25,499	24,773	27,890	31,832	35,907
China	±1.603	±1.650	±2.020	±2.180	±1.605	±1.657	2,138	2,143	2,494	2,869	2,085	2,152
China (Hong Kong SAR)	0.016	0.015	0.012	0.010	0.010	-	13,586	13,804	15,341	20,942	19,668	-
China (Taiwan Province)	0.738	0.791	0.936	0.943	0.863	0.839	14,117	15,773	19,229	20,016	18,458	18,165
Colombia	0.131	0.153	0.147	0.129	0.216	0.222	20,532	24,107	19,387	17,190	30,842	34,213
Costa Rica	0.017	0.014	0.017	0.016	0.019	0.021	6,758	4,540	4,831	4,728	-	-
Cote d'Ivoire	0.022	0.021	0.023	0.021	-	-	-	-	-	-	-	-
Croatia	±0.063	±0.034	±0.016	-	-	-	7,902	6,206	4,846	-	-	-
Cyprus	0.003	0.002	0.003	0.003	0.004	0.004	21,225	17,088	21,629	21,807	24,507	26,663
Czechoslovakia (Former)	±0.131	-	-	-	-	-	4,869	-	-	-	-	-
Denmark	*0.122	*0.111	*0.129	*0.123	*0.126	*0.148	41,936	39,073	48,624	46,859	47,727	56,260
Ecuador	0.017	0.024	0.023	0.025	0.027	0.027	10,616	13,063	11,770	12,689	12,791	14,806
Egypt	*0.008	*0.019	*0.016	*0.019	*0.019	*0.019	2,696	1,958	1,938	2,604	2,580	2,606
El Salvador	-	-	-	0.005	0.004	0.012	-	-	-	11,019	9,587	6,854
Ethiopia	0.008	0.006	0.005	0.006	0.009	0.006	9,251	6,447	5,681	6,618	10,090	6,277
Fiji	0.001	0.001	0.001	0.001	0.001	-	7,148	7,174	7,461	7,888	8,500	-
Finland	±0.133	±0.096	±0.110	±0.096	±0.126	±0.185	51,195	38,279	47,755	43,765	50,541	74,148
FYR Macedonia	±0.000	±0.000	±0.000	±0.000	±0.001	±0.001	-	20,369	39,098	26,483	74,786	32,456
France	±3.341	±3.389	±3.810	±3.252	±3.275	±3.526	35,543	37,035	44,561	39,229	40,088	42,538
Germany (Western Part)	6.414	6.275	6.870	6.055	-	-	65,166	64,554	72,032	-	-	-
Ghana	-	-	-	0.003	0.003	0.003	-	-	-	1,257	1,059	1,081
Greece	*0.084	*0.078	*0.077	*0.073	*0.073	*0.082	35,111	39,620	50,367	47,657	47,288	53,049
Honduras	*0.007	*0.006	*0.006	*0.006	*0.005	*0.006	3,953	3,372	3,200	4,405	3,429	2,476
Hungary	±0.079	±0.053	±0.038	±0.030	±0.029	±0.030	9,904	6,667	6,277	5,963	6,761	7,148
India	*0.566	*0.417	*0.484	*0.459	*0.521	*0.579	5,029	3,702	3,969	3,636	3,955	4,202
Indonesia	*0.494	*0.475	*0.605	*0.427	*0.544	*0.565	2,983	3,530	4,084	3,512	4,150	4,435
Iran (Islamic Republic of)	0.123	0.121	0.200	0.175	-	-	11,264	10,919	17,679	14,603	-	-

Continued.

Depending on the table, * means *Enterprises, Engaged,* or *Factor Values;* ± means *Basis Unspecified;* § means *shown in millions.* For additional notes and sources, see Appendix I.

Value Added and Value Added per Employee

- Continued -

Country	Value Added ($ bil.)						Value Added per Employee ($)					
	1990	1991	1992	1993	1994	1995	1990	1991	1992	1993	1994	1995
Iraq	-	*0.015	*0.039	-	-	-	-	7,187	19,614	-	-	-
Ireland	*0.118	*0.139	*0.161	*0.141	*0.146	*0.169	49,198	53,374	62,681	54,996	57,197	66,626
Israel	0.076	0.086	0.080	0.070	0.076	0.092	34,493	37,391	40,055	33,315	34,720	41,219
Italy	±2.254	±2.313	±2.267	±1.877	±2.373	-	48,183	51,768	65,615	58,912	70,945	-
Japan	±11	±13	±14	±15	±15	±16	75,018	83,646	91,217	98,740	114,577	127,080
Jordan	0.001	0.001	0.001	0.001	0.001	0.002	7,173	6,781	10,399	7,356	5,882	11,589
Kenya	*0.033	*0.032	*0.027	*0.017	*0.021	*0.026	15,061	13,873	11,896	7,387	8,630	10,838
Korea, Republic of	3.063	1.364	1.471	1.558	1.786	2.168	16,828	43,649	46,780	47,423	53,404	62,132
Kuwait	0.002	0.002	0.001	0.001	0.003	0.002	8,285	12,983	7,125	8,497	23,331	16,575
Latvia	-	-	-	*0.000	*0.000	*0.001	-	-	-	2,925	2,916	5,762
Macau	0.000	0.000	0.000	0.000	-	-	12,094	8,348	7,481	8,431	-	-
Malaysia	*0.528	*0.592	*0.671	*0.737	*0.797	*0.974	8,935	9,161	10,337	10,709	11,549	13,267
Malta	0.017	0.019	0.021	0.017	0.021	-	28,478	28,351	32,223	21,800	22,491	-
Mauritius	0.002	0.002	0.002	0.002	0.002	0.003	5,679	5,324	6,479	6,180	6,130	7,949
Mexico	0.464	0.472	0.452	0.395	0.450	0.327	27,236	28,435	28,442	29,019	35,443	26,446
Morocco	-	±0.061	±0.056	±0.054	-	-	-	24,335	22,006	20,484	-	-
Myanmar	±0.015	±0.007	±0.018	±0.004	±0.007	-	11,745	5,054	14,990	2,708	4,565	-
Nepal	0.002	0.003	-	0.004	0.001	-	2,920	2,974	-	2,512	1,336	-
Netherlands	0.284	0.278	0.286	0.246	0.287	0.336	50,681	52,112	57,973	51,623	63,768	74,608
New Zealand	0.062	-	-	-	-	-	32,071	-	-	-	-	-
Norway	±0.058	±0.050	±0.026	±0.025	±0.030	±0.034	48,166	43,185	50,840	40,275	41,130	46,835
Pakistan	0.043	0.045	0.050	-	-	-	5,178	5,372	5,733	-	-	-
Panama	0.002	0.002	0.002	0.002	0.002	0.002	12,584	10,721	12,754	13,427	13,832	13,690
Peru	0.074	0.047	0.043	-	-	-	25,222	20,946	26,004	-	-	-
Philippines	0.158	0.154	0.247	0.200	-	-	5,316	5,736	8,382	8,855	-	-
Poland	±0.209	±0.169	±0.207	±0.200	±0.244	±0.337	6,158	5,823	9,001	8,684	11,044	15,919
Portugal	0.054	0.053	0.085	0.088	0.098	0.138	8,192	9,484	13,266	15,018	16,762	24,767
Qatar	0.000	0.001	-	-	-	-	9,158	24,240	-	-	-	-
Romania	±0.129	±0.119	±0.102	±0.127	-	-	-	-	-	-	-	-
Russian Federation	-	-	-	±0.484	±0.642	±0.806	-	-	-	2,762	4,264	5,689
Saint Lucia	-	±0.000	±0.000	±0.000	±0.000	±0.000	-	-	-	-	-	-
Senegal	-	-0.000	-0.000	-0.000	0.000	-	-	§-1,57	§-4,19	§-1,17	7,505	-
Singapore	*0.035	*0.043	*0.048	*0.054	*0.061	*0.067	21,548	24,319	28,225	30,446	32,753	36,087
Slovakia	-	-	-	±0.060	±0.057	-	-	-	-	9,486	9,065	-
Slovenia	-	-	-	0.079	0.079	0.094	-	-	-	-	-	-
South Africa	*0.400	*0.375	*0.361	*0.344	*0.342	*0.403	22,245	20,823	21,265	21,526	21,368	25,302
Spain	*1.490	*1.549	*1.672	±1.092	±1.212	±1.460	44,772	47,317	55,010	41,829	42,842	51,155
Sri Lanka	0.035	0.039	0.051	0.054	-	-	2,604	2,970	3,168	3,556	-	-
Sweden	*0.387	*0.261	*0.260	*0.198	*0.251	*0.271	53,735	40,538	43,216	39,344	51,171	56,357
Thailand	0.473	1.021	-	-	-	-	12,516	18,695	-	-	-	-
Trinidad and Tobago	±0.008	±0.009	±0.006	±0.003	±0.001	-	14,790	15,267	12,230	7,417	6,751	-
Tunisia	0.028	0.029	0.031	0.034	0.036	0.043	-	-	13,944	13,578	16,468
Turkey	0.452	0.530	0.683	0.717	0.586	0.811	35,039	48,405	52,991	53,224	45,895	61,274
United Kingdom	*3.018	*2.970	*2.977	*2.468	*2.671	*3.049	46,429	51,205	55,130	48,401	54,505	59,651
United Republic of Tanzania . .	*0.001	*0.007	-	-	-	-	1,107	8,260	-	-	-	-
United States of America . . .	*13	*14	*14	*16	*16	*17	65,512	68,528	70,152	75,630	78,033	78,539
Uruguay	0.058	0.041	0.036	0.035	0.037	0.045	21,333	17,133	16,999	17,297	17,859	19,536
Venezuela	0.139	0.143	0.166	0.182	0.198	0.495	23,555	19,268	22,378	27,418	29,909	71,499
Yugoslavia	±0.027	±0.027	±0.053	-	±0.101	±0.043	-	1,437	2,968	-	5,895	2,695
Zambia	0.016	-	-	-	0.006	-	10,300	-	-	-	5,732	-
Zimbabwe	*0.037	*0.023	*0.038	*0.037	*0.042	*0.040	12,773	11,698	12,240	12,236	13,449	14,470

Depending on the table, * means *Enterprises, Engaged,* or *Factor Values;* ± means *Basis Unspecified;* § means *shown in millions.* For additional notes and sources, see Appendix I.

401

Capital Investment

Country	Gross Fixed Capital Formation ($ mil.)						Per Establishment ($ mil)					
	1990	1991	1992	1993	1994	1995	1990	1991	1992	1993	1994	1995
Argentina	-	-	-	23	-	-	-	-	-	0.042	-	-
Australia	20	-	-	-	-	-	0.091	-	-	-	-	-
Austria	48	47	57	33	31	-	1.225	1.266	1.635	0.993	0.939	-
Bangladesh	0.208	0.191	0.257	-	-	-	0.002	0.001	0.001	-	-	-
Belgium	106	65	78	54	55	59	1.079	0.648	0.753	0.553	0.565	-
Bolivia	0.002	0.030	0.018	0.022	0.031	-	0.000	0.005	0.003	0.003	0.003	-
Bosnia & Herzegovina	0.340	-	-	-	-	-	0.057	-	-	-	-	-
Brazil	249	-	71	143	-	-	0.548	-	0.161	0.329	-	-
Bulgaria	19	1.282	164	25	0.225	2.205	1.236	0.085	9.669	0.112	0.001	0.005
Canada	426	-	-	-	-	-	2.393	-	-	-	-	-
Chile	8.291	8.997	8.696	13	5.517	-	0.488	0.450	0.458	0.554	0.240	-
China (Hong Kong SAR)	3.980	2.188	0.904	0.646	0.388	-	0.022	0.016	0.009	0.007	0.005	-
Colombia	21	7.059	34	4.150	-1.704	-	0.270	0.093	0.328	0.041	-0.019	-
Croatia	1.349	0.047	0.221	0.943	0.799	0.374	0.044	0.001	0.004	0.012	0.009	0.004
Cyprus	0.573	0.140	0.880	0.288	0.529	0.272	0.038	0.009	0.059	0.019	0.028	0.012
Czechoslovakia (Former)	60	-	-	-	-	-	4.298	-	-	-	-	-
Denmark	7.109	12	-	-	-	-	0.053	0.084	-	-	-	-
Ecuador	2.955	8.625	13	-0.042	12	5.725	0.164	0.375	0.545	-0.002	0.505	0.318
Egypt	0.700	0.450	13	-1.750	-	-	0.032	0.014	0.360	-0.073	-	-
El Salvador	-	0.619	-	1.181	0.619	3.568	-	0.062	-	0.107	0.069	0.324
Ethiopia	0.085	6.623	1.836	18	7.189	6.879	0.021	1.656	0.459	4.436	1.438	1.720
Fiji	0.128	0.175	0.038	-	-	-	0.018	0.025	0.006	-	-	-
Finland	23	8.432	13	5.129	20	36	0.899	0.291	0.461	0.183	0.631	1.050
FYR Macedonia	0.808	0.826	0.042	0.013	0.035	0.039	0.808	0.826	0.007	0.002	0.003	0.003
Germany (Western Part)	672	657	628	605	-	-	2.451	2.356	2.171	-	-	-
Greece	16	13	5.818	-	-	-	0.345	0.296	0.145	-	-	-
Hungary	28	11	2.481	4.340	-	-	0.873	0.136	0.032	0.043	-	-
India	199	349	246	221	-	-	0.100	0.168	0.113	0.098	-	-
Indonesia	169	197	151	99	134	116	0.330	0.422	0.319	0.221	0.300	0.262
Iran (Islamic Republic of)	6.550	14	30	21	-	-	0.205	0.324	0.613	0.400	-	-
Ireland	28	60	-	-	-	-	0.659	1.220	-	-	-	-
Israel	8.432	7.459	16	14	16	-	0.169	0.131	0.346	0.287	0.301	-
Italy	387	331	234	215	238	-	1.156	0.969	0.907	0.809	0.877	-
Japan	1,423	2,034	2,053	1,475	1,340	-	0.334	0.480	0.498	0.355	0.401	-
Jordan	0.030	-	-	-	0.021	0.007	0.002	-	-	-	0.001	0.000
Korea, Republic of	535	442	527	443	454	619	0.318	0.680	0.768	0.574	0.566	0.722
Kuwait	-	2.148	0.027	-0.083	2.040	-	-	0.716	0.014	-0.041	0.680	-
Latvia	0.008	0.003	-	-	0.003	0.103	0.008	0.000	-	-	0.000	0.009
Macau	-	-	-	0.296	-	-	-	-	-	0.296	-	-
Malaysia	204	124	159	184	181	-	0.555	0.319	0.417	0.485	0.496	-
Malta	0.978	0.303	0.085	0.045	2.780	-	0.122	0.038	0.011	0.005	0.309	-
Mexico	16	33	-	-	-	-	0.332	0.686	-	-	-	-
Morocco	-	-	-	6.130	-	-	-	-	-	0.219	-	-
Myanmar	291	283	292	281	244	-	97	94	97	94	122	-
Nepal	2.996	-	-	-	-	-	0.091	-	-	-	-	-
Netherlands	44	30	35	27	-	-	1.569	0.983	1.084	0.728	-	-
New Zealand	4.776	-	-	-	-	-	0.057	-	-	-	-	-
Norway	8.786	-8.175	1.496	1.692	1.842	-	0.325	-0.282	0.115	0.106	0.077	-
Pakistan	-	1.322	-	-	-	-	-	0.027	-	-	-	-
Panama	0.039	0.082	-	-	-	-	0.005	0.012	-	-	-	-
Peru	2.503	2.265	2.536	-	-	-	0.021	0.019	0.021	-	-	-
Philippines	23	35	29	32	-	-	0.056	0.086	0.156	0.182	-	-
Poland	14	17	30	45	-	-	0.307	0.351	0.596	1.025	-	-
Portugal	-29	-15	55	62	44	-	-0.110	-0.061	0.218	0.254	0.183	-
Romania	-	-	-	11	19	-	-	-	-	0.025	0.033	-
Singapore	7.142	9.626	14	12	8.249	-	0.238	0.311	0.506	0.400	0.266	-
Slovenia	17	14	14	28	17	30	0.506	0.081	0.154	0.326	0.067	0.139
Spain	162	105	125	123	109	167	0.168	0.102	0.146	-	-	-
Sri Lanka	25	15	15	11	-	-	0.146	0.103	0.106	0.065	-	-
Thailand	351	480	-	-	-	-	1.650	1.273	-	-	-	-

Continued.

Capital Investment
- Continued -

Country	Gross Fixed Capital Formation ($ mil.)						Per Establishment ($ mil)					
	1990	1991	1992	1993	1994	1995	1990	1991	1992	1993	1994	1995
Trinidad and Tobago	-	-	0.918	0.280	0.203	-	-	-	0.031	0.008	0.006	-
Tunisia	3.417	6.381	5.654	6.974	3.954	7.401	-	-	0.075	0.041	0.070
Turkey	156	90	62	74	79	-	1.629	1.045	0.396	0.482	0.552	-
Ukraine	29	24	20	1.038	3.335	3.725	0.836	0.181	0.129	0.027	0.090	0.101
United Kingdom	248	219	237	-	173	-	0.410	0.362	0.393	-	0.267	-
United Republic of Tanzania	7.675	7.310	-	-	-	-	1.096	1.044	-	-	-	-
United States of America	1,050	850	952	1,059	1,041	1,067	-	-	0.333	-	-	-
Uruguay	1.320	1.896	2.463	2.182	-	-	-	-	-	-	-	-
Venezuela	10	11	30	14	9.044	86	0.194	0.182	0.476	0.202	0.139	1.341
Yugoslavia	1.493	1.218	13	-	9.178	11	0.036	0.021	0.139	-	0.068	0.086
Zambia	3.952	-	-	-	-	-	0.494	-	-	-	-	-
Zimbabwe	5.106	2.771	6.772	5.330	2.933	-	0.426	0.231	0.521	0.410	0.209	-

Indexes of Industrial Production

Country	Index of Industrial Production (1990=100)						Index of Employment (1990=100)					
	1990	1991	1992	1993	1994	1995	1990	1991	1992	1993	1994	1995
Albania	100	-	-	-	-	-	-	-	-	-	-	-
Algeria	100	97	80	83	97	89	-	-	-	-	-	-
Angola	100	-	-	-	-	-	-	-	-	-	-	-
Argentina	100	89	-	-	-	-	100	-	-	76	75	-
Armenia	100	-	-	-	-	-	100	80	-	-	-	-
Australia	100	87	100	114	109	-	100	89	89	-	-	-
Austria	100	94	94	80	85	93	100	85	85	75	72	68
Azerbaijan	-	-	-	-	-	-	100	97	92	86	83	-
Bahrain	100	100	100	100	100	100	-	-	-	-	-	-
Bangladesh	100	80	59	65	69	61	100	106	121	-	-	-
Belgium	100	95	103	88	97	110	100	100	98	-	-	-
Bhutan	100	-	-	-	-	-	-	-	-	-	-	-
Bolivia	100	102	101	99	106	109	100	100	105	139	146	147
Bosnia & Herzegovina	-	-	-	-	-	-	100	84	-	-	10	-
Brazil	100	99	99	108	112	112	100	-	85	82	-	-
Bulgaria	100	82	68	60	82	96	100	81	76	69	62	60
Burkina-Faso	100	85	70	-	-	-	-	-	-	-	-	-
Burundi	100	100	100	100	-	-	-	-	-	-	-	-
Cambodia	100	-	-	-	-	-	100	102	-	-	-	-
Canada	100	90	112	130	140	143	100	92	92	92	96	95
Central African Republic	100	100	100	100	100	-	-	-	-	-	-	-
Chile	100	142	163	150	141	139	100	111	122	156	142	141
China	100	-	-	-	-	-	100	103	108	101	103	103
China (Hong Kong SAR)	100	94	90	81	74	70	100	92	67	42	42	-
China (Taiwan Province)	100	104	107	97	95	97	100	96	93	90	89	88
Colombia	100	104	112	111	100	85	100	99	119	117	110	102
Costa Rica	100	90	95	96	112	114	100	128	139	138	-	-
Cote d'Ivoire	100	98	100	79	93	86	-	-	-	-	-	-
Croatia	100	-	-	-	-	-	100	70	42	41	31	30
Cyprus	100	94	116	81	77	77	100	97	115	106	111	113
Czechoslovakia (Former)	100	70	65	-	-	-	100	81	-	-	-	-
Czech Republic	100	-	-	-	-	-	100	94	75	75	-	-
Dem. Rep. of the Congo	100	-	-	-	-	-	-	-	-	-	-	-
Denmark	100	104	110	97	104	92	100	98	91	91	91	90
Dominican Republic	100	-	-	-	-	-	-	-	-	-	-	-
Ecuador	100	130	130	157	-	-	100	114	121	121	130	112
Egypt	100	105	98	82	77	71	100	346	296	261	258	254
El Salvador	100	107	-	-	-	-	-	-	-	-	-	-
Ethiopia and Eritrea	100	-	-	-	-	-	-	-	-	-	-	-
Ethiopia	-	-	-	-	-	-	100	103	101	102	104	112
Fiji	100	101	94	97	99	89	100	88	85	84	84	-

Continued.

Depending on the table, * means *Enterprises, Engaged,* or *Factor Values;* ± means *Basis Unspecified;* § means *shown in millions.* For additional notes and sources, see Appendix I.

403

Indexes of Industrial Production

- Continued -

Country	Index of Industrial Production (1990=100)						Index of Employment (1990=100)					
	1990	1991	1992	1993	1994	1995	1990	1991	1992	1993	1994	1995
Finland	100	83	92	102	125	140	100	96	88	85	96	96
FYR Macedonia	100	-	-	-	-	-	-	-	-	-	-	-
France	100	98	101	93	101	105	100	97	91	88	87	88
Gabon	100	-	-	-	-	-	-	-	-	-	-	-
Gambia	100	-	-	-	-	-	-	-	-	-	-	-
Germany (Eastern Part) . . .	100	-	-	-	-	-	-	-	-	-	-	-
Germany (Western Part) . . .	100	91	89	75	80	90	100	99	97			
Ghana	100	-	-	-	-	-	-	-	-	-	-	-
Greece	100	83	74	71	70	69	100	82	64	64	64	65
Guatemala	100	-	-	-	-	-	-	-	-	-	-	-
Guyana	100	100	100	100	100	100	-	-	-	-	-	-
Haiti	100	-	-	-	-	-	-	-	-	-	-	-
Honduras	100	138	140	155	197	260	100	103	107	78	86	141
Hungary	100	94	86	96	123	130	100	100	75	63	54	53
Iceland	100	-	-	-	-	-	-	-	-	-	-	-
India	100	99	97	101	107	114	100	100	108	112	117	122
Indonesia	100	131	142	140	126	129	100	81	89	73	79	77
Iran (Islamic Republic of) . . .	100	122	117	115	96	111	100	102	104	110	-	-
Iraq	100	-	-	-	-	-	-	-	-	-	-	-
Ireland	100	105	111	100	111	116	100	108	107	107	107	105
Israel	100	111	125	141	158	184	100	105	91	95	100	102
Italy	100	94	89	87	-	-	100	95	74	68	71	
Jamaica	100	94	-	-	-	-	100	105	92	92	88	
Japan	100	100	96	89	88	94	100	101	100	98	84	84
Jordan	100	-	-	-	-	-	100	67	46	67	124	106
Kenya	100	104	204	202	199	201	100	105	104	106	109	111
Korea, Republic of	100	108	115	120	130	136	100	17	17	18	18	19
Kuwait	100	-	-	-	-	-	100	64	50	46	55	52
Kyrgyzstan	100	-	-	-	-	-	-	-	-	-	-	-
Lao P.D.R.	100	-	-	-	-	-	-	-	-	-	-	-
Latvia	100	-	-	-	-	-	100	91	8	6	6	7
Lebanon	100	-	-	-	-	-	-	-	-	-	-	-
Libyan Arab Jamahiriya . . .	100	-	-	-	-	-	-	-	-	-	-	-
Lithuania	100	-	-	-	-	-	-	-	-	-	-	-
Luxembourg	100	93	97	94	102	114	-	-	-	-	-	-
Macau	100	-	-	-	-	-	100	88	81	56	-	-
Madagascar	100	-	-	-	-	-	-	-	-	-	-	-
Malawi	100	-	-	-	-	-	100					
Malaysia	100	111	124	146	166	187	100	109	110	116	117	124
Malta	100	114	122	130	-	-	100	116	112	133	158	
Mauritius	100	-	-	-	-	-	100	108	106	97	93	93
Mexico	100	106	109	106	112	109	100	97	93	80	75	73
Mongolia	100	100	100	100	100	100	-	-	-	-	-	-
Morocco	100	104	95	97	104	113	-	-	-	-	-	-
Mozambique	100	-	-	-	-	-	100	100	94	91	84	-
Myanmar	100	101	101	101	-	-	100	110	94	122	111	
Nepal	100	-	-	-	-	-	100	151	-	195	106	
Netherlands	100	101	100	98	106	106	100	95	88	85	80	80
New Zealand	100	-	-	-	-	-	100	92	96	97	-	-
Nicaragua	100	-	-	-	-	-	-	-	-	-	-	-
Nigeria	100	-	-	-	-	-	-	-	-	-	-	-
Norway	100	85	49	47	50	50	100	98	43	53	60	60
Pakistan	100	99	94	-	-	-	100	100	105	-	-	-
Panama	100	93	129	141	145	141	100	91	103	106	106	106
Papua New Guinea	100	-	-	-	-	-	-	-	-	-	-	-
Paraguay	100	109	111	126	106	108	-	-	-	-	-	-
Peru	100	86	79	87	89	95	100	76	56	-	-	-
Philippines	100	119	122	112	110	127	100	91	99	76	-	-
Poland	100	125	140	168	195	228	100	85	68	68	65	62
Portugal	100	87	62	56	74	91	100	85	97	89	88	84

Continued.

Depending on the table, * means *Enterprises, Engaged,* or *Factor Values;* ± means *Basis Unspecified;* § means *shown in millions.* For additional notes and sources, see Appendix I.

Indexes of Industrial Production

- Continued -

Country	Index of Industrial Production (1990=100)						Index of Employment (1990=100)					
	1990	1991	1992	1993	1994	1995	1990	1991	1992	1993	1994	1995
Puerto Rico	100	-	-	-	-	-	100	91	64	40	51	30
Qatar	100	-	-	-	-	-	100	113	-	-	-	-
Republic of Moldova	100	-	-	-	-	-	-	-	-	-	-	-
Romania	100	75	53	56	45	48	-	-	-	-	-	-
Russian Federation	100	-	-	-	-	-	-	-	-	-	-	-
Saudi Arabia	100	-	-	-	-	-	-	-	-	-	-	-
Singapore	100	114	105	110	123	141	100	109	105	109	116	115
Slovakia	100	-	-	-	-	-	-	-	-	-	-	-
Slovenia	100	101	101	100	106	111	-	-	-	-	-	-
Somalia	100	100	100				-	-	-	-	-	-
South Africa	100	88	88	88	88	96	100	100	94	89	89	89
Spain	100	100	102	96	103	114	100	98	91	78	85	86
Sri Lanka	100	88	98	137	135	163	100	98	119	111	-	-
Sudan	100	100	100				-	-	-	-	-	-
Suriname	100	-	-	-	-	-	100	96	71	93	-	-
Swaziland	100	100	100	-			-	-	-	-	-	-
Sweden	100	96	90	90	105	102	100	89	83	70	68	67
Switzerland	100	91	86	85	80	84	-	-	-	-	-	-
Syrian Arab Republic	100	99	100	106	110	116	-	-	-	-	-	-
Thailand	100	106	-	-	-	-	100	144	-	-	-	-
Tonga	100	-	-	-	-	-	-	-	-	-	-	-
Trinidad and Tobago	100	111	114	131	148	148	100	98	89	79	30	-
Tunisia	100	96	92	-	-	-	-	-	-	-	-	-
Turkey	100	126	141	149	135	162	100	85	100	104	99	103
Uganda	100	175	241	405	621	1,069	-	-	-	-	-	-
Ukraine	100	-	-	-	-	-	100	98	98	93	80	76
United Arab Emirates	100	-	-	-	-	-	-	-	-	-	-	-
United Kingdom	100	94	89	87	91	97	100	89	83	78	75	79
United Republic of Tanzania	100	89	77	86	85	-	100	100	-	-	-	-
USSR (Former)	100	-	-	-	-	-	100	-	-	-	-	-
United States of America	100	96	106	112	118	121	100	96	100	101	102	106
Uruguay	100	83	82	82	83	94	100	87	77	75	76	85
Venezuela	100	-	-	-	-	-	100	125	125	113	112	117
Yemen	100	-	-	-	-	-	-	-	-	-	-	-
Yugoslavia	100	-	-	-	-	-	-	-	-	-	-	-
Yugoslavia (Former)	100	80	-	-	-	-	-	-	-	-	-	-
Zambia	100	101	96	90	68	60	100	-	-	-	65	-
Zimbabwe	100	-	-	-	-	-	100	70	108	105	108	96

Depending on the table, * means *Enterprises, Engaged,* or *Factor Values*; ± means *Basis Unspecified*; § means *shown in millions*. For additional notes and sources, see Appendix I.

405

Representative Companies in Sector

Name	Address	Telephone	Fax	Employment	Y
Aeroquip Corp.	3000 Strayer Rd., Maumee, Ohio 43537, U.S.A.	419-867-2200	419-867-2590	10,000	97
Cooper Tire and Rubber Co.	701 Lima Ave., Findlay, Ohio 45840, U.S.A.	419-423-1321		8,284	96
Michelin Tyre PLC	Campbell Rd., Stoke-on-Trent ST4 4EY, United Kingdom	782 48101	782 402011	7,849	93
Witco Corp.	1 American Ln., Greenwich, Connecticut 06831-2559, U.S.A.	203-552-2000		7,200	97
Kelly-Springfield Tire Co.	12501 Willow Brook Rd. SE, Cumberland, Maryland 21502-2599, U.S.A.	301-777-6000	301-777-6008	7,000	96
Carlisle Companies Inc.	250 S. Clinton St., 201, Syracuse, New York 13202-1258, U.S.A.	315-477-9108	315-474-2008	6,900	97
Continental General Tire Inc.	1800 Continental Blvd., Charlotte, North Carolina 28273, U.S.A.	704-583-3900	704-583-8698	6,700	97
Hankook Tire Manufacturing Co., Ltd.	Kangnam-Gu, Seoul, Republic of Korea	2 2221000	2 2221100	6,500	97
Goodyear do Brasil Produtos de Borracha Ltda.	Av. Paulista 854, 8th Fl., Sao Paulo 01310, Brazil	11 3852244	11 2881520	6,300	97
Bakrie Sumatra Plantation PT	Wisma Bakrie, Jakarta 12920, Indonesia	215250192	21 520437	6,200	97
Taurus Hungarian Rubber Works	Kerepesi ut 17, Budapest H-1087, Hungary	1 1130830	1 1135434	6,000	93
Avon Rubber PLC	Bath Rd., Melksham SN12 8AA, United Kingdom	225 703101	225 707880	5,238	93
Gajah Tunggal, Pt	Jl. Hayam Wuruk No. 8, Jakarta 10120, Indonesia	213459302	21 3804908	5,090	97
Pou Chen Corp.	Fu Hsing Hsiang, Changhua, Taiwan	4 7695147	4 7695150	4,809	92
Gatic SACFI Y A	Eva Peron 2535, San Martin 1650, Argentina	1 7539040	1 7525536	4,700	96
Hevea-Cameroun	BP 1298, Douala, Cameroon	427564	428141	4,257	91
Bata Chile SA	Maipu, Santiago, Chile	2 5575609	2 5332906	4,200	97
Hickory Springs Manufacturing Co.	PO Box 128, Hickory, North Carolina 28603, U.S.A.	704-328-2201	704-328-5501	4,000	97
Dunlop Ltd.	Vincent Sq., London SW1P 2PL, United Kingdom	71 8343848	71 8288508	3,965	93
Gumarne Barum, AS	T Vansovej 1054/45, Puchov 020 32, Slovakia	825 2531	825 41507	3,900	93
Continental UK Group Holdings Ltd.	Yiewsley, West Drayton UB7 7DJ, United Kingdom			3,800	93
Bridgestone/Firestone do Brasil Industria e Comercio Ltda.	Casa Branca, Santo Andre 09015, Brazil	11 4111666	11 4126655	3,774	98
Readicut International PLC	Clifton Mills, Brighouse HD6 4ET, United Kingdom	484 721223	484 716135	3,725	94
Goodyear SA	Ave. T. Gordon Smith, Colmar-Berg L-7750, Luxembourg	81991	81992496	3,550	95
Michelin North America (Canada) Inc.	100 Granton Rd. RR 3, New Glasgow, Nova Scotia B2H 5C6, Canada	902-755-6040	902-752-0612	3,500	98
TriMas Corp.	315 E. Eisenhower Pkwy., Ann Arbor, Michigan 48108, U.S.A.	313-747-7025	313-747-6565	3,500	95
Dunlop Tire Corp.	PO Box 1109, Buffalo, New York 14240, U.S.A.	716-639-5200	716-639-5320	3,200	97
Tillotson Corp.	59 Waters Ave., Everett, Massachusetts 02149, U.S.A.	617-387-9400	617-389-9639	3,000	95
Cooper Tire and Rubber Co. Engineered Products Div.	725 W. 11th St., Auburn, Indiana 46706, U.S.A.	219-925-0700	219-925-1473	2,900	97
Pirelli UK Tyres Ltd.	40 Chancery Ln., London WC2A 1JH, United Kingdom	71 2428881	71 4301096	2,863	93
American Biltrite Inc.	57 River St., Wellesley Hills, Massachusetts 02181, U.S.A.	617-237-6655		2,835	95
Melton Medes Ltd.	1 St. Mark's St., Nottingham NG3 1DE, United Kingdom	602 582277	602 585122	2,698	93
Tubos e Conexoes Tigre Ltda.	Rua Xavantes 54 Bairro Atiradores, Joinville 89203-900, Brazil	47 4415000	47 4384517	2,642	96
Bandag Inc.	2905 N. Hwy. 61, Muscatine, Iowa 52761-5886, U.S.A.	319-262-1400	319-262-1284	2,591	97
Aeroquip Industrial Americas Group	1695 Indian Wood Cir., Maumee, Ohio 43537, U.S.A.	419-891-7600		2,500	94
Safeskin Corp.	12671 High Bluff Dr., San Diego, California 92130, U.S.A.	619-794-8111		2,500	95
Hoechst UK Ltd.	Salisbury Rd., Hounslow TW4 6JH, United Kingdom	81 5707712	81 5771854	2,479	93
John Crane Inc.	6400 W. Oakton St., Morton Grove, Illinois 60053, U.S.A.	847-967-2400	847-967-3915	2,100	97
ITT Higbie Baylock Manufacturing Company Inc.	3000 University Dr., Auburn Hills, Michigan 48321-7016, U.S.A.	313-340-3000	313-340-3707	2,000	
Zimco Industries Pty. Ltd.	PO Box 78069, Sandton 2146, Republic of South Africa	11 7837023	11 7830732	2,000	95
Hwa Fong Rubber Industrial Co. Ltd.	Rm. 815, 8th Fl., 152 Sungchiang Rd., Taipei, Taiwan	2 5319833	2 5631853	1,900	92
Bridgestone/Firestone Canada Inc.	5770 Hurontario St. Ste. 400, Mississauga, Ontario L5R 3G5, Canada	905-890-1990	905-890-1991	1,800	98
Chicago Raw Hide Manufacturing Co.	900 N. State St., Elgin, Illinois 60123, U.S.A.	847-742-7840	847-742-7845	1,800	97
Fel-Pro Inc.	7450 N. McCormick Blvd., Skokie, Illinois 60076, U.S.A.	847-674-7700	708-674-5816	1,800	95
Industex Eastern Cape Pty. Ltd.	Neave Township, Port Elizabeth 6014, Republic of South Africa	41 431363	41 411558	1,800	97
Argha Karya Prima Industry PT	Jl. Jend Sudirman Kav. 70-71, Jakarta 12910, Indonesia	215703778	21 5710811	1,792	97
Plastic Specialties and Technologies, Inc.	65 Railroad Ave., Ridgefield, New Jersey 07657, U.S.A.	201-941-2900	201-941-4670	1,750	97

Continued

For sources and notes, see Appendix I.

Representative Companies in Sector
- Continued -

Name	Address	Telephone	Fax	Employment	Y
Meiwa Indonesia, PT	Jl. Karet Tengsin, No. 19, Jakarta 10220, Indonesia	215703335	21 5738189	1,701	97
Catecu SA	Camino a Melipilla 9460, Santiago, Chile	2 5332366	2 5332931	1,700	96
Graphic Controls Corp.	PO Box 1271, Buffalo, New York 14240, U.S.A.	716-853-7500	800-347-2420	1,650	97
Grupo Zillo Lorenzetti	Rua Libero Badaro 377, 14th Fl., Sao Paulo 01074-900, Brazil	11 2323300	11 2323054	1,620	96
Goshen Rubber Company Inc.	PO Box 517, Goshen, Indiana 46527, U.S.A.	219-533-1111	219-533-5332	1,600	96
Nylex (Malaysia) SDN BHD	PO Box 7033, Shah Alam-Selangor 40910, Malaysia	3 5591706	3 5500088	1,600	94
Playtex Products Inc.	300 Nyala Farms Rd., Westport, Connecticut 06880, U.S.A.	203-341-4000		1,600	96
LSB Industries Inc.	16 S. Pennsylvania Ave., Oklahoma City, Oklahoma 73107, U.S.A.	405-235-4546	405-235-5067	1,563	96
LaCrosse Footwear Inc.	1319 Saint Andrew St., La Crosse, Wisconsin 54602, U.S.A.	608-782-3020	608-782-8190	1,520	97
Polmo Praszka SA	ul. Kaliszka 72, Praszka PL-46-320, Poland	83 4444		1,501	93
Arden Corp.	26899 Northwestern Hwy., Southfield, Michigan 48034-8420, U.S.A.	248-355-1101	248-355-1230	1,500	97
A.W. Chesterton Co.	225 Fallon Rd., Stoneham, Massachusetts 02180, U.S.A.	617-438-7000		1,500	97
Dana Corp. Plumley Div.	PO Box 758, Paris, Tennessee 38242, U.S.A.	901-642-5582	901-642-6872	1,500	97
Fayette Tubular Products Inc.	1835 Technology Dr., Troy, Michigan 48083, U.S.A.	810-589-7710		1,500	95
Inoue Rubber Indonesia PT	Jl. Gajah Mada 19-26, Jakarta 10130, Indonesia	212311011	21 3807508	1,500	97
Kingston-Warren Corp.	PO Box 169, Newfields, New Hampshire 03056, U.S.A.	603-772-3771	603-772-4357	1,500	95
Siebe North Inc.	PO Box 70729, Charleston, South Carolina 29415-0729, U.S.A.	803-554-0660	803-744-2857	1,500	97
Standard Products (Canada) Ltd.	1030 Erie St., Stratford, Ontario N5A 6V7, Canada	519-271-3360	519-272-2281	1,500	98
William H. Kaufman Inc.	410 King St. W, Kitchener, Ontario N2G 4J8, Canada	519-576-1500	519-576-9794	1,500	98
Icollantas Industria Colombiana de Llantas	Calle 7 No. 22-01, Cali, Colombia	2 8823236	2 8839747	1,450	96
Tillotson Healthcare Corp.	360 Rte. 101, Bedford, New Hampshire 03110, U.S.A.	603-472-6600		1,440	95
DMIB BHD	4 Jalan Tandang, Petaling Jaya 46700, Malaysia	3 7918833	3 7925414	1,369	
Pirelli Neumaticos, SA	Numancia, 1, Barcelona E-08029, Spain	934 304455	934 395488	1,352	97
Uniroyal Engelbert Tyres Ltd.	Newbridge Industrial Est., Newbridge EH28 8LG, United Kingdom	31 3332700	31 3334668	1,332	93
Dongil Rubber Belt Co., Ltd.	Kumjong-Gu, Pusan, Republic of Korea	51 5209114	51 5230289	1,325	97
Goody Products Inc.	600 W. Park Dr., Peachtree City, Georgia 30269, U.S.A.	770-486-1995		1,300	95
Minnesota Rubber	3630 Wooddale Ave., Minneapolis, Minnesota 55416, U.S.A.	612-927-1400	612-927-2192	1,300	97
Siam Tyre Co. Ltd.	Klang Amphur, Sampurakarn 10130, Thailand	3940121	3941924	1,300	94
Vulkan d.d.	6 POB. 74, Nis YU-18106, Serbia	18 63220	18 65856	1,300	92
Jones Stroud Holdings PLC	Long Eaton, Nottingham NG10 1HF, United Kingdom	602 734421	602 461083	1,291	94
Vesper Corp.	11 Bala Ave., Bala Cynwyd, Pennsylvania 19004, U.S.A.	610-667-7312	215-667-4839	1,282	
Vans Inc.	15700 Shoemaker Ave., Santa Fe Springs, California 90670, U.S.A.	562-565-8267		1,269	97
Woodhead Industries Inc.	2150 E. Lake Cook Rd., 400, Buffalo Grove, Illinois 60089, U.S.A.	847-465-8300	847-272-8126	1,259	97
Kenda Rubber Industrial Co. Ltd.	Yuan Lin Chen, Changhua, Taiwan	4 8945171	4 831865	1,250	93
Gates UK Ltd.	Edinburgh Rd., Dumfries DG1 1QA, United Kingdom	387 253111	387 268937	1,244	93
Alenco Holdings Ltd.	55 Maylands Ave., Hemel Hempstead HP2 4SJ, United Kingdom	442 238100	442 238111	1,225	93
Lexington Precision Corp.	767 3rd Ave., New York, New York 10017, U.S.A.	212-319-4657	212-949-1588	1,200	96
Pharma Plast International AS	Engmosen 1, Lynge DK-3540, Denmark	42188388	42189555	1,200	93
Thai Bridgestone Co. Ltd.	Bangrak, Bangkok 10500, Thailand	2368858	2366780	1,200	94
LRC Products Ltd.	N. Circular Rd., London E4 8QA, United Kingdom	81 5272377	81 5033100	1,136	94
Alltrista Corp.	PO Box 5004, Muncie, Indiana 47307-5004, U.S.A.	765-281-5000	765-281-5400	1,100	97
Goetze Corporation of America	331 W. Laketon Ave., Muskegon, Michigan 49441, U.S.A.	616-726-5226	616-727-9434	1,100	
Pirelli Armstrong Tire Corp.	PO Box 2001, New Haven, Connecticut 06536-0201, U.S.A.	203-784-2200		1,100	97
Stanton y Compania SA	Carrera 42 12-64, Bogota, Colombia	1 7814501	1 7812314	1,100	97
Yale-South Haven Inc.	400 Aylworth Ave., South Haven, Michigan 49090, U.S.A.	616-637-2116	616-637-8315	1,100	96
Goodyear Indonesia PT	Jalan Pemuda No. 27, Bogor 16000, Indonesia	251322071	251 328088	1,095	97
Firestone De La Argentina SA IC	Av. Antartida Argentina 2715, Lavallol 1836, Argentina	1 2980031	1 2981467	1,050	96
HS Chemical Co., Ltd.	Yangsan-Gun, Kyongnam, Republic of Korea	523 828811	523 3878870	1,050	97
American Consumer Products Inc.	31100 Solon Rd., Solon, Ohio 44139, U.S.A.	216-248-7000	216-248-8051	1,024	94
Akro Corp.	1212 7th St. S.W., Canton, Ohio 44707, U.S.A.	330-456-4543		1,000	
ATH Albarus Transmissoes Homocineticas	Rua Joaquina Silveira 557, Sao Sebastiao 91060-320, Brazil	51 3402855	51 3401135	1,000	97

Continued

Representative Companies in Sector
- Continued -

Name	Address	Tele-phone	Fax	Employ-ment	Y
Camoplast Inc.	2144 Rue King O Bureau 110, Sherbrooke, Quebec J1J 2E8, Canada	819-823-1777	819-823-8772	1,000	98
Contessa-Skogar	Traktorowa 128, Lodz PL-91-204, Poland	42 529897	42 523910	1,000	93
Gleason Corp. (Los Angeles, California)	10474 Santa Monica Blvd., 400, Los Angeles, California 90025, U.S.A.	310-470-6001	310-474-2994	1,000	97
Magla Products Inc.	700 Shunpike Rd., Chatham, New Jersey 07928, U.S.A.	201-377-0500	201-377-2790	1,000	
Rahman Hydraulic Tin BHD	38 Jalan Sultan Ismail, Kuala Lumpur 50250, Malaysia	3 2412077	3 2418242	996	93
Rogers Corp.	PO Box 188, Rogers, Connecticut 06263-0188, U.S.A.	860-774-9605	860-774-5852	993	97
Dong Ah Tire Ind. Co., Ltd.	Yangsan-Gun, Kyongnam, Republic of Korea	523 3890011	523 827736	970	97
Rubatex Corp.	5221 Valley Park Dr., Roanoke, Virginia 24019-3074, U.S.A.	703-586-2611	703-586-8779	960	94
BP Belgium NV	Nieuwe Weg 1, Zwijndrecht B-2070, Belgium	3 2502111	3 2195909	955	95
Alliance Tire Co. Ltd.	Industrial Zone, Hadera 38500, Israel	6 6240500	6 6240555	920	97
Jabur Pneus SA	Rua Nassim Jabur 249, Londrina 86079-440, Brazil	43 3290202	43 3292831	900	96
Pretty Products Inc.	PO Box 6002, Coshocton, Ohio 43812-6002, U.S.A.	614-622-3522	614-622-4915	900	97
Cable Design Technologies Inc.	661 Andersen Dr., Foster Plz., B, Pittsburgh, Pennsylvania 15220, U.S.A.	412-937-2300	412-937-9690	867	96
Greene, Tweed and Co.	2075 Detwiler Rd., Kulpsville, Pennsylvania 19443, U.S.A.	215-256-9521	215-256-0189	850	95
Poly Unggul PT	Desa Sukadamai, Tangerang, Indonesia	82131380	82 3810833	850	97
Korea Kumho Petrochemical Co., Ltd.	Chongno-Gu, Seoul, Republic of Korea	2 3997560	2 7209742	840	97
Don & Low Holdings Ltd.	Forfar, Angus DD8 2AL, United Kingdom	307 452200	307 468474	825	93
Condere Corp.	2750 Dixwell Ave., Hamden, Connecticut 06518, U.S.A.	203-287-2200	203-287-2278	800	93
Crotty Corp.	PO Box 37, Quincy, Michigan 49082, U.S.A.	517-639-8787	517-639-4309	800	94
Vernay Laboratories Inc.	PO Box 310, Yellow Springs, Ohio 45387, U.S.A.	513-767-7261	513-767-7913	800	94
Fidelity Tire Manufacturing Co.	PO Box 927, Natchez, Mississippi 39120, U.S.A.	601-446-6381		780	
Lister & Co. PLC	Longwood, Huddersfield HD3 4UT, United Kingdom	484 646647	484 651833	743	94
Chomerics Inc.	77 Dragon Court, Woburn, Massachusetts 01888-4014, U.S.A.	781-935-4850	781-933-4318	740	93
Indian Head Industries Inc.	8530 Cliff Cameron Dr., Charlotte, North Carolina 28269-9786, U.S.A.	704-547-7411	704-547-9367	715	97
Acadia Polymers Inc.	5251 Concourse Dr., Roanoke, Virginia 24012, U.S.A.	540-265-2700	540-265-2764	700	97
Cadillac Rubber and Plastics Inc.	805 W. 13th St., Cadillac, Michigan 49601, U.S.A.	616-775-1345		700	95
Kirkhill Rubber Co.	300 E. Cypress Ave., Brea, California 92621, U.S.A.	714-529-4901	714-529-6775	700	97
Starcan Corp.	BCE Pl Ste. 4550, Toronto, Ontario M5J 2S1, Canada	416-361-0255	416-361-1227	700	98
Stearns Manufacturing Co.	PO Box 1498, St. Cloud, Minnesota 56302, U.S.A.	320-252-1642		700	97
Ajover	Transv 93 65 A 30, Bogota, Colombia	1 2230600	1 2763728	660	96
Jasper Rubber Products Inc.	1010 1st Ave., Jasper, Indiana 47546, U.S.A.	812-482-3242	812-482-0816	640	94
Instrument Specialties Company Inc.	PO Box 650, Delaware Water Gap, Pennsylvania 18327, U.S.A.	717-424-8510	717-424-6213	625	97
Milbank Manufacturing Co.	PO Box 419028, Kansas City, Missouri 64141-0028, U.S.A.	816-483-5314	816-483-6357	625	95
Vincke, SA	Palamos, Gerona E-17230, Spain	972 314450	972 314692	620	97
Ames Industries Inc.	2537 Curtiss St., Downers Grove, Illinois 60515, U.S.A.	708-964-2440	708-964-0497	600	94
Calzado Olympic SA	Apartado 4513-1000, San Jose, Costa Rica	2320350	2200731	600	92
Chardon Rubber Co.	373 Washington St., Chardon, Ohio 44024, U.S.A.	440-285-2161	440-286-8422	600	97
Great American Industries Inc.	300 Plaza Dr., Vestal, New York 13850-3646, U.S.A.	607-729-9331	607-797-7103	600	
Mercur SA	Rua Cristovao Colombo 53, Santa Cruz do Sul 96825-010, Brazil	51 7151611	51 7152834	600	96
NRI Industries Inc.	394 Symington Ave., Toronto, Ontario M6N 2W3, Canada	416-657-1111	416-656-1231	600	98
Oliver Rubber Co.	PO Box 8447, Oakland, California 94662, U.S.A.	510-654-7711	510-655-6319	600	96
Ashworth Inc.	2791 Loker Ave. W., Carlsbad, California 92008, U.S.A.	760-438-6610	760-438-6657	574	97
S. Jerome & Sons Holdings PLC	Victoria Works, Shipley BD17 7EF, United Kingdom	274 587251	274 588654	574	93
Eurofloor SA	BP 10, Wiltz L-9501, Luxembourg	9578111	957807	570	
American Fuel Cell and Coated Fabrics Co.	PO Box 887, Magnolia, Arkansas 71753, U.S.A.	501-234-3381	501-235-7270	560	93
Codan Gummi AS	Kobenhavnsvej 104, Koge DK-4600, Denmark	53654465	53660911	550	93
Goodyear Thailand Ltd.	Klong Toey, Phra Khanong, Bangkok 10110, Thailand	2607171	2607166	550	94
JMK International Inc.	4800 Bryant Irvin Ct., Fort Worth, Texas 76107, U.S.A.	817-737-3703	817-735-1669	550	97
Petindo Jaya Sakti PT	Jl. Gajah Mada, No. 3-5, Block D-11-13, Jakarta 11045, Indonesia	213800044	21 3860120	540	97
Brasway PLC	All Saints Rd., Wednesbury WS10 9TS, United Kingdom	21 5263089	21 5264980	511	93

Product Tables
Inner Tubes, Rubber, for Motor Vehicles

Unit of Measure: Thousands of units.

	1990		1991		1992		1993		1994		1995		1996	
	Value	%	Value	%	Value	%	Value	%	Value	%	Value	%	Value	%
Total Production	258,946	100.0	259,356	100.0	255,710	100.0	236,964	100.0	220,343	100.0	219,437	100.0	238,175	100.0
Regions														
Africa	5,505	2.1	5,303	2.0	4,711	1.8	4,229	1.8	4,329	2.0	4,016	1.8	4,302	1.8
America, North	31,995	12.4	31,787	12.3	31,527	12.3	32,577	13.7	32,728	14.9	32,636	14.9	33,568	14.1
America, South	18,273	7.1	17,672	6.8	17,357	6.8	17,284	7.3	17,123	7.8	16,833	7.7	16,443	6.9
Asia	156,485	60.4	162,405	62.6	160,044	62.6	145,104	61.2	129,628	58.8	132,924	60.6	149,214	62.6
Europe	43,031	16.6	38,588	14.9	38,523	15.1	34,278	14.5	33,098	15.0	29,646	13.5	31,320	13.1
Oceania	3,657	1.4	3,602	1.4	3,547	1.4	3,492	1.5	3,437	1.6	3,382	1.5	3,327	1.4
Leading Producers														
Korea, Republic of	53,772	20.8	55,702	21.5	61,655	24.1	56,104	23.7	56,457	25.6	57,557	26.2	*69,391	29.1
Russian Federation	47,676	18.4	44,788	17.3	41,816	16.4	36,541	15.4	19,770	9.0	18,843	8.6	21,164	8.9
United States of America	*29,000	11.2	*29,000	11.2	*29,000	11.3	*29,000	12.2	*29,000	13.2	*29,000	13.2	*29,000	12.2
Japan	18,363	7.1	18,584	7.2	15,737	6.2	12,320	5.2	10,614	4.8	11,725	5.3	12,168	5.1
Italy	11,045	4.3	13,578	5.2	14,773	5.8	12,134	5.1	12,003	5.4	10,381	4.7	10,899	4.6

Inner Tubes, Rubber, for Bicycles and Motorcycles

Unit of Measure: Thousands of units.

	1990		1991		1992		1993		1994		1995		1996	
	Value	%	Value	%	Value	%	Value	%	Value	%	Value	%	Value	%
Total Production	191,749	100.0	189,858	100.0	200,954	100.0	211,608	100.0	209,084	100.0	209,250	100.0	218,244	100.0
Regions														
Africa	2,080	1.1	2,081	1.1	2,250	1.1	1,831	0.9	1,827	0.9	1,655	0.8	1,944	0.9
America, North	1,728	0.9	2,432	1.3	3,336	1.7	3,388	1.6	2,727	1.3	2,303	1.1	2,076	1.0
America, South	25,880	13.5	19,386	10.2	20,770	10.3	22,155	10.5	23,539	11.3	24,923	11.9	26,307	12.1
Asia	141,315	73.7	145,890	76.8	152,111	75.7	164,423	77.7	162,213	77.6	159,281	76.1	166,513	76.3
Europe	20,745	10.8	20,069	10.6	22,487	11.2	19,812	9.4	18,778	9.0	21,089	10.1	21,404	9.8
Leading Producers														
Japan	26,824	14.0	26,415	13.9	24,653	12.3	24,019	11.4	20,302	9.7	17,724	8.5	16,746	7.7
Korea, Republic of	28,573	14.9	24,720	13.0	22,999	11.4	24,851	11.7	18,109	8.7	12,022	5.7	*19,862	9.1
Indonesia	23,007	12.0	26,667	14.0	30,326	15.1	40,793	19.3	45,958	22.0	49,229	23.5	*45,560	20.9
India	22,806	11.9	22,878	12.1	25,826	12.9	25,788	12.2	24,300	11.6	25,056	12.0	25,394	11.6
Thailand	20,887	10.9	23,548	12.4	*25,995	12.9	*29,154	13.8	*32,313	15.5	*35,472	17.0	*38,631	17.7

Tires for Agricultural and Other Off-The-Road Vehicles

Unit of Measure: Number.

	1990		1991		1992		1993		1994		1995		1996	
	Value	%	Value	%	Value	%	Value	%	Value	%	Value	%	Value	%
Total Production	25,694,392	100.0	24,210,139	100.0	21,807,123	100.0	19,320,104	100.0	18,431,913	100.0	19,653,798	100.0	19,874,605	100.0
Regions														
Africa	328,000	1.3	267,000	1.1	285,000	1.3	349,000	1.8	411,000	2.2	391,000	2.0	395,562	2.0
America, North	4,330,261	16.9	4,314,878	17.8	4,336,081	19.9	4,297,037	22.2	4,293,978	23.3	4,320,390	22.0	4,320,002	21.7
America, South	2,852,818	11.1	2,918,695	12.1	3,183,571	14.6	3,294,357	17.1	3,508,143	19.0	3,670,973	18.7	3,905,631	19.7
Asia	10,082,400	39.2	10,011,200	41.4	7,063,159	32.4	4,717,681	24.4	3,405,204	18.5	3,004,726	15.3	3,562,248	17.9
Europe	8,100,913	31.5	6,698,367	27.7	6,939,311	31.8	6,662,029	34.5	6,813,589	37.0	8,266,709	42.1	7,691,163	38.7
Leading Producers														
United States of America	*4,113,000	16.0	*4,113,000	17.0	*4,113,000	18.9	*4,113,000	21.3	*4,113,000	22.3	*4,113,000	20.9	*4,113,000	20.7
Russian Federation	6,627,400	25.8	6,133,200	25.3	4,028,000	18.5	2,037,000	10.5	782,000	4.2	546,000	2.8	751,000	3.8
Brazil	*2,545,000	9.9	*2,719,786	11.2	*2,894,571	13.3	*3,069,357	15.9	*3,244,143	17.6	*3,418,929	17.4	*3,593,714	18.1
India	1,724,000	6.7	1,761,000	7.3	1,792,000	8.2	1,825,000	9.4	1,874,000	10.2	1,787,000	9.1	1,933,000	9.7
Poland	995,000	3.9	569,000	2.4	795,000	3.6	874,000	4.5	1,148,000	6.2	1,186,000	6.0	1,028,000	5.2

Commodity data are provided by the United Nations Statistical Division. The symbol * means that data are estimated. For additional notes, see Appendix I.

409

Product Tables
Tires for Bicycles and Motorcycles

Unit of Measure: Thousands of units.

	1990 Value	%	1991 Value	%	1992 Value	%	1993 Value	%	1994 Value	%	1995 Value	%	1996 Value	%
Total Production	223,517	100.0	247,239	100.0	214,537	100.0	211,157	100.0	197,863	100.0	179,276	100.0	201,696	100.0
Regions														
Africa	4,865	2.2	31,019	12.5	3,491	1.6	3,239	1.5	9,391	4.7	9,768	5.4	9,635	4.8
America, North	1,697	0.8	2,336	0.9	3,425	1.6	3,100	1.5	2,581	1.3	2,871	1.6	2,848	1.4
America, South	4,992	2.2	4,265	1.7	4,509	2.1	4,456	2.1	4,254	2.2	4,568	2.5	4,942	2.5
Asia	177,858	79.6	177,645	71.9	172,879	80.6	172,247	81.6	154,424	78.0	135,507	75.6	160,372	79.5
Europe	34,105	15.3	31,974	12.9	30,233	14.1	28,115	13.3	27,213	13.8	26,561	14.8	23,899	11.8
Leading Producers														
Japan	29,409	13.2	30,207	12.2	28,353	13.2	26,247	12.4	23,519	11.9	20,638	11.5	19,806	9.8
India	32,403	14.5	28,558	11.6	25,275	11.8	19,963	9.5	18,949	9.6	18,278	10.2	18,246	9.0
Korea, Republic of	24,637	11.0	23,930	9.7	20,385	9.5	19,189	9.1	17,228	8.7	12,625	7.0	*18,493	9.2
Indonesia	26,239	11.7	31,491	12.7	36,744	17.1	43,553	20.6	30,541	15.4	19,448	10.8	*37,163	18.4
Thailand	17,369	7.8	18,373	7.4	*19,441	9.1	*21,334	10.1	*23,227	11.7	*25,120	14.0	*27,013	13.4

Tires for Road Motor Vehicles, Excluding Bicycles and Motorcycles

Unit of Measure: Thousands of units.

	1990 Value	%	1991 Value	%	1992 Value	%	1993 Value	%	1994 Value	%	1995 Value	%	1996 Value	%
Total Production	913,009	100.0	925,699	100.0	1,034,455	100.0	978,075	100.0	1,010,394	100.0	1,032,143	100.0	1,078,065	100.0
Regions														
Africa	14,263	1.6	13,761	1.5	13,502	1.3	14,014	1.4	14,634	1.4	15,786	1.5	15,708	1.5
America, North	244,927	26.8	239,300	25.9	267,602	25.9	272,698	27.9	279,169	27.6	272,887	26.4	278,634	25.8
America, South	41,569	4.6	42,344	4.6	45,014	4.4	47,645	4.9	50,669	5.0	52,494	5.1	54,185	5.0
Asia	372,446	40.8	390,899	42.2	412,228	39.8	361,256	36.9	361,627	35.8	381,749	37.0	414,064	38.4
Europe	230,654	25.3	230,767	24.9	287,391	27.8	273,656	28.0	295,399	29.2	300,242	29.1	306,400	28.4
Oceania	9,150	1.0	8,628	0.9	8,717	0.8	8,806	0.9	8,896	0.9	8,985	0.9	9,074	0.8
Leading Producers														
United States of America	210,660	23.1	202,391	21.9	230,256	22.3	237,444	24.3	243,600	24.1	*238,768	23.1	*243,038	22.5
Japan	153,226	16.8	153,677	16.6	154,900	15.0	142,595	14.6	139,172	13.8	152,040	14.7	157,819	14.6
France	54,536	6.0	57,876	6.3	59,928	5.8	53,390	5.5	66,744	6.6	*64,120	6.2	*65,678	6.1
Germany	-		-		50,993	4.9	45,595	4.7	46,415	4.6	48,534	4.7	48,703	4.5
Italy	30,767	3.4	32,447	3.5	29,978	2.9	29,138	3.0	31,520	3.1	33,551	3.3	33,600	3.1

Rubber, Reclaimed

Unit of Measure: Metric tons.

	1990 Value	%	1991 Value	%	1992 Value	%	1993 Value	%	1994 Value	%	1995 Value	%	1996 Value	%
Total Production	454,527	100.0	420,876	100.0	380,393	100.0	352,605	100.0	338,246	100.0	333,489	100.0	329,851	100.0
Regions														
America, North	19,859	4.4	11,745	2.8	3,632	1.0	3,599	1.0	3,599	1.1	3,599	1.1	3,599	1.1
America, South	25,105	5.5	25,139	6.0	25,172	6.6	26,866	7.6	25,496	7.5	26,449	7.9	26,019	7.9
Asia	306,772	67.5	303,677	72.2	279,480	73.5	250,777	71.1	242,403	71.7	240,640	72.2	237,193	71.9
Europe	102,630	22.6	80,155	19.0	71,948	18.9	71,203	20.2	66,588	19.7	62,640	18.8	62,880	19.1
Oceania	161	0.0	161	0.0	161	0.0	161	0.0	161	0.0	161	0.0	161	0.0
Leading Producers														
Japan	42,091	9.3	41,377	9.8	39,102	10.3	36,149	10.3	31,363	9.3	31,305	9.4	29,941	9.1
United States of America	*16,260	3.6	*8,147	1.9	*33	0.0	-		-		-		-	
Russian Federation	61,707	13.6	57,565	13.7	41,376	10.9	22,201	6.3	20,374	6.0	20,402	6.1	18,208	5.5
Brazil	*24,439	5.4	*24,439	5.8	*24,439	6.4	*24,439	6.9	*24,439	7.2	*24,439	7.3	*24,439	7.4
India	*23,947	5.3	*23,947	5.7	*23,947	6.3	*23,947	6.8	*23,947	7.1	*23,947	7.2	*23,947	7.3

Commodity data are provided by the United Nations Statistical Division. The symbol * means that data are estimated. For additional notes, see Appendix I.

Product Tables

Rubber, Unhardened Vulcanized Plates, Sheets Etc.

Unit of Measure: Metric tons.

	1990 Value	%	1991 Value	%	1992 Value	%	1993 Value	%	1994 Value	%	1995 Value	%	1996 Value	%
Total Production	552,641	100.0	546,132	100.0	530,815	100.0	514,785	100.0	512,994	100.0	517,527	100.0	528,922	100.0
Regions														
America, South	1,271	0.2	1,271	0.2	1,271	0.2	1,271	0.2	1,271	0.2	1,271	0.2	1,271	0.2
Asia	8,723	1.6	13,533	2.5	12,458	2.3	10,808	2.1	10,391	2.0	9,618	1.9	9,681	1.8
Europe	542,646	98.2	531,328	97.3	517,086	97.4	502,706	97.7	501,332	97.7	506,638	97.9	517,970	97.9
Leading Producers														
Germany	*139,465	25.2	*139,465	25.5	*139,465	26.3	137,624	26.7	141,306	27.5	*139,465	26.9	*139,465	26.4
Austria	142,083	25.7	133,554	24.5	131,130	24.7	114,941	22.3	122,807	23.9	134,693	26.0	*130,421	24.7
Spain	46,977	8.5	55,613	10.2	48,428	9.1	58,321	11.3	54,219	10.6	54,453	10.5	56,176	10.6
Belgium	*27,850	5.0	*27,850	5.1	*27,850	5.2	*27,850	5.4	27,850	5.4	*27,850	5.4	*27,850	5.3
United Kingdom	*22,915	4.1	*22,915	4.2	*22,915	4.3	22,755	4.4	23,075	4.5	*22,915	4.4	*22,915	4.3

Rubber, Unhardened Vulcanized Piping and Tubing

Unit of Measure: Metric tons.

	1990 Value	%	1991 Value	%	1992 Value	%	1993 Value	%	1994 Value	%	1995 Value	%	1996 Value	%
Total Production	161,986	100.0	228,571	100.0	232,763	100.0	233,604	100.0	249,906	100.0	275,563	100.0	255,242	100.0
Regions														
Asia	40,648	25.1	43,277	18.9	41,947	18.0	55,156	23.6	51,492	20.6	66,930	24.3	61,757	24.2
Europe	121,338	74.9	185,295	81.1	190,816	82.0	178,447	76.4	198,413	79.4	208,632	75.7	193,485	75.8
Leading Producers														
Germany	-		56,751	24.8	64,889	27.9	52,139	22.3	54,544	21.8	70,491	25.6	58,738	23.0
France	24,741	15.3	31,572	13.8	33,456	14.4	36,209	15.5	51,408	20.6	*39,499	14.3	*41,184	16.1
Japan	31,382	19.4	32,226	14.1	30,290	13.0	28,888	12.4	31,733	12.7	*33,399	12.1	*34,199	13.4
Spain	11,152	6.9	12,755	5.6	13,721	5.9	12,669	5.4	14,632	5.9	20,573	7.5	14,292	5.6
Turkey	7,603	4.7	5,793	2.5	5,195	2.2	23,996	10.3	17,182	6.9	31,025	11.3	*25,632	10.0

Rubber, Hardened (Ebonite and Vulcanite in Blocks, Plates, Sheets Etc.)

Unit of Measure: Metric tons.

	1990 Value	%	1991 Value	%	1992 Value	%	1993 Value	%	1994 Value	%	1995 Value	%	1996 Value	%
Total Production	101,483	100.0	108,112	100.0	104,400	100.0	95,438	100.0	91,183	100.0	101,095	100.0	100,610	100.0
Regions														
Asia	252	0.2	252	0.2	252	0.2	252	0.3	252	0.3	252	0.2	252	0.3
Europe	101,231	99.8	107,860	99.8	104,148	99.8	95,186	99.7	90,931	99.7	100,843	99.8	100,358	99.7
Leading Producers														
Germany	*39,555	39.0	47,621	44.0	45,114	43.2	34,067	35.7	31,416	34.5	*39,555	39.1	*39,555	39.3
United Kingdom	*12,455	12.3	*12,455	11.5	*12,455	11.9	13,738	14.4	11,172	12.3	*12,455	12.3	*12,455	12.4
Romania	4,420	4.4	3,381	3.1	2,452	2.3	2,667	2.8	3,246	3.6	3,007	3.0	2,787	2.8
Hungary	274	0.3	239	0.2	109	0.1	110	0.1	75	0.1	51	0.0	26	0.0
Spain	862	0.8	546	0.5	650	0.6	1,316	1.4	1,821	2.0	2,590	2.6	2,349	2.3

Commodity data are provided by the United Nations Statistical Division. The symbol * means that data are estimated. For additional notes, see Appendix I.

Product Tables
Rubber, Transmission, Conveyor, Elevator Belts Etc.

Unit of Measure: Metric tons.

	1990		1991		1992		1993		1994		1995		1996	
	Value	%	Value	%	Value	%	Value	%	Value	%	Value	%	Value	%
Total Production	252,153	100.0	252,194	100.0	228,098	100.0	214,037	100.0	218,673	100.0	230,672	100.0	237,923	100.0
Regions														
America, South	514	0.2	548	0.2	581	0.3	65	0.0	30	0.0	82	0.0	236	0.1
Asia	55,662	22.1	59,746	23.7	49,775	21.8	44,086	20.6	42,280	19.3	46,090	20.0	51,235	21.5
Europe	195,977	77.7	191,901	76.1	177,742	77.9	169,886	79.4	176,363	80.7	184,500	80.0	186,452	78.4
Leading Producers														
Poland	33,739	13.4	24,675	9.8	21,969	9.6	24,858	11.6	25,841	11.8	30,251	13.1	28,524	12.0
Germany	*40,963	16.2	50,412	20.0	42,463	18.6	31,362	14.7	39,614	18.1	*40,963	17.8	*40,963	17.2
Japan	35,838	14.2	36,506	14.5	31,494	13.8	28,958	13.5	28,195	12.9	29,414	12.8	30,964	13.0
France	13,684	5.4	*11,892	4.7	*11,687	5.1	*11,481	5.4	*11,275	5.2	*11,069	4.8	*10,864	4.6
United Kingdom	12,494	5.0	14,345	5.7	16,196	7.1	18,047	8.4	13,979	6.4	*15,641	6.8	*16,272	6.8

Rubber Footwear

Unit of Measure: 1,000 Pairs.

	1990		1991		1992		1993		1994		1995		1996	
	Value	%	Value	%	Value	%	Value	%	Value	%	Value	%	Value	%
Total Production	1,974,048	100.0	1,973,489	100.0	1,879,472	100.0	1,910,192	100.0	2,228,419	100.0	2,675,720	100.0	2,382,307	100.0
Regions														
Africa	10,092	0.5	7,393	0.4	5,673	0.3	6,532	0.3	6,068	0.3	5,541	0.2	8,229	0.3
America, North	145,815	7.4	156,869	7.9	151,045	8.0	121,106	6.3	118,365	5.3	127,684	4.8	135,172	5.7
America, South	107,596	5.5	116,486	5.9	82,789	4.4	86,769	4.5	88,661	4.0	101,119	3.8	101,434	4.3
Asia	1,598,608	81.0	1,598,537	81.0	1,555,663	82.8	1,634,557	85.6	1,963,640	88.1	2,393,054	89.4	2,094,941	87.9
Europe	110,945	5.6	93,212	4.7	83,311	4.4	60,237	3.2	50,694	2.3	47,330	1.8	41,540	1.7
Oceania	992	0.1	992	0.1	992	0.1	992	0.1	992	0.0	992	0.0	992	0.0
Leading Producers														
China	897,580	45.5	929,960	47.1	904,165	48.1	1,103,438	57.8	1,467,944	65.9	1,832,450	68.5	1,521,910	63.9
Brazil	100,889	5.1	109,925	5.6	76,669	4.1	79,487	4.2	81,511	3.7	95,117	3.6	94,312	4.0
United States of America	105,708	5.4	113,184	5.7	108,810	5.8	80,321	4.2	76,963	3.5	*90,336	3.4	*90,223	3.8
Indonesia	114,636	5.8	127,392	6.5	140,148	7.5	40,048	2.1	31,817	1.4	*113,849	4.3	*121,578	5.1
Russian Federation	123,144	6.2	90,458	4.6	72,372	3.9	52,544	2.8	28,420	1.3	22,667	0.8	20,184	0.8

Plastic Footwear

Unit of Measure: 1,000 Pairs.

	1990		1991		1992		1993		1994		1995		1996	
	Value	%	Value	%	Value	%	Value	%	Value	%	Value	%	Value	%
Total Production	165,010	100.0	197,056	100.0	231,728	100.0	261,080	100.0	168,184	100.0	174,732	100.0	189,486	100.0
Regions														
Africa	7,216	4.4	57,461	29.2	81,691	35.3	114,469	43.8	31,577	18.8	30,945	17.7	32,524	17.2
America, North	17,051	10.3	15,648	7.9	16,541	7.1	14,608	5.6	13,569	8.1	10,610	6.1	13,884	7.3
America, South	6,551	4.0	7,369	3.7	8,960	3.9	7,964	3.1	7,636	4.5	5,780	3.3	7,842	4.1
Asia	93,998	57.0	86,450	43.9	82,800	35.7	77,939	29.9	71,618	42.6	79,997	45.8	82,514	43.5
Europe	40,194	24.4	30,128	15.3	41,736	18.0	46,100	17.7	43,785	26.0	47,401	27.1	52,722	27.8
Leading Producers														
Nigeria	-		48,987	24.9	73,732	31.8	105,825	40.5	22,098	13.1	20,531	11.7	*21,671	11.4
Indonesia	*32,223	19.5	*32,223	16.4	35,420	15.3	32,429	12.4	28,821	17.1	*32,223	18.4	*32,223	17.0
Iran, Islamic Republic of	18,725	11.3	14,213	7.2	9,492	4.1	8,874	3.4	6,700	4.0	5,155	3.0	5,362	2.8
United Kingdom	2,042	1.2	9,537	4.8	17,032	7.4	24,527	9.4	23,380	13.9	*29,014	16.6	*33,355	17.6
Mexico	14,836	9.0	13,579	6.9	14,618	6.3	12,831	4.9	11,938	7.1	9,126	5.2	12,546	6.6

PLASTIC PRODUCTS NEC

ISIC 3560—Plastic Products nec—includes plastics products not included in any of the other major ISIC categories covered thus far. The industry corresponds to the U.S. SIC system's SIC 308—Miscellaneous Plastics Products, nec., which includes film and sheet, plastic plate, plastic pipe, bottles, foam products, etc.

Summary Statistics

		1990		1991		1992		1993		1994		1995	
		Value	N	Value	N	Value	N	Value	N	Value	N	Value	N
Establishments or enterprises	(number)	75,570	82	80,593	92	98,902	86	77,094	83	70,550	67	52,496	32
Number employed	(000)	4,366	90	4,527	93	4,627	88	4,400	89	4,243	78	3,995	59
Total Wages	($ mil.)	56,159	86	67,650	87	75,008	81	62,216	80	64,822	71	57,443	54
Wage/salary per employee	($)	8,635	84	9,059	84	10,436	79	8,516	78	8,906	69	10,557	54
Female workers	(000)	392	33	268	31	822	27	861	31	883	24	842	13
as % of total employment*	(%)	29	32	28	30	29	26	31	30	32	23	43	12
Output	($ bil.)	324	88	375	91	406	86	369	84	389	75	377	59
per employee	($ 000)	50	83	51	85	59	80	54	79	55	71	67	57
per establishment	($ 000)	3,154	76	3,173	81	3,073	75	2,726	70	2,829	58	1,916	28
Value added	($ bil.)	136	79	142	80	160	74	164	75	151	66	162	55
per employee	($ 000)	21	75	21	75	25	70	22	70	23	62	28	52
per establishment	($ 000)	1,245	69	1,051	73	1,186	66	974	63	1,118	49	744	24
Capital investment	($ mil.)	16,504	65	17,115	59	18,265	54	16,389	52	14,560	40	9,692	20
per employee	($ 000)	5	62	5	56	4	52	4	48	4	37	4	19
per establishment	($ 000)	434	62	298	56	326	52	260	48	286	38	149	18

Data presented above are drawn from the detailed tables that follow. Columns headed 'N' show the number of countries that provided valid data for inclusion. Values are not strictly comparable one year to the next or one row to the next because the number of countries that report varies from period to period and row to row. However, a general indication of magnitudes can be gleaned from reviewing this summary—especially in earlier years in which more reports are available. For detailed explanations, see Appendix I. *The average for those countries reporting both total and female employment.

Establishments and Number Engaged

Country	Establishments or Enterprises (number)						Employees per Establishment					
	1990	1991	1992	1993	1994	1995	1990	1991	1992	1993	1994	1995
Albania	-	-	-	4.000	4.000	1.000	-	-	-	351	172	438
Argentina	-	-	-	2,913	3,189	-	-	-	-	13	12	-
Armenia	3.000	5.000	5.000	4.000	6.000	-	676	339	-	-	-	-
Australia	1,062	1,385	1,374	-	-	-	34	25	23	-	-	-
Austria	245	249	256	257	252	-	56	56	55	57	59	-
Azerbaijan	13	14	12	13	13	-	259	217	195	164	225	-
Bahrain	-	-	4.000	-	-	-	-	-	110	-	-	-
Bangladesh	142	200	167	-	-	-	20	18	19	-	-	-
Belgium	567	573	582	585	571	-	33	32	32	-	-	-
Belize	3.000	3.000	3.000	-	-	-	-	-	-	-	-	-
Bolivia	63	66	74	75	82	86	21	22	21	20	18	17
Bosnia & Herzegovina	17	18	-	-	3.000	-	162	176	-	-	35	-
Brazil	915	-	828	833	-	-	129	-	128	130	-	-
Bulgaria	*30	*30	*33	*740	*939	*980	427	347	294	10	10	9.694
Burundi	*3.000	*3.000	-	-	-	-	28	27	-	-	-	-
Canada	1,369	1,276	1,223	1,221	1,198	-	45	47	49	51	54	-
Central African Republic	-	*1.000	*1.000	*1.000	-	-	-	-	-	-	-	-
Chile	87	93	98	101	109	-	124	121	132	140	134	-
China	*16,916	*17,116	*17,553	*16,274	*16,826	*19,255	59	60	59	62	60	57
China (Hong Kong SAR)	4,762	3,795	3,511	2,374	2,041	-	10	10	9.285	9.478	7.594	-
Colombia	372	370	428	416	404	-	50	58	58	61	59	-
Costa Rica	78	89	95	108	-	-	67	64	64	60	-	-
Croatia	217	274	413	575	695	766	46	28	16	12	9.309	8.381
Cyprus	43	43	43	43	55	62	27	28	28	28	21	22
Czechoslovakia (Former)	*15	*17	-	-	-	-	533	412	-	-	-	-
Czech Republic	-	*17	*50	*66	-	-	-	412	300	242	-	-
Denmark	506	517	517	-	-	-	25	27	28	-	-	-
Ecuador	79	98	97	97	107	110	74	64	59	60	64	64
Egypt	218	217	214	225	-	-	84	127	87	80	-	-
El Salvador	-	15	15	21	17	16	-	-	92	107	96	118
Ethiopia	3.000	3.000	5.000	5.000	11	9.000	290	289	183	187	103	118
Fiji	13	12	14	-	-	-	20	21	13	-	-	-
Finland	154	169	164	153	141	227	49	41	40	39	43	45
FYR Macedonia	*30	*54	*107	*107	*145	*186	58	32	19	16	19	13
Gambia	-	-	-	2.000	-	-	-	-	-	44	-	-
Germany	-	2,580	2,696	2,750	2,800	-	-	123	116	108	104	-
Germany (Western Part)	2,366	2,457	2,545	-	-	-	121	123	118	-	-	-
Ghana	-	-	-	31	-	-	-	-	-	96	-	-
Greece	289	290	299	-	-	-	33	32	32	-	-	-
Guatemala	-	22	24	23	23	23	-	92	91	96	97	98
Honduras	-	-	18	27	27	27	-	-	131	64	70	102
Hungary	*373	*652	*574	*694	-	-	48	28	26	22	-	-
Iceland	59	59	58	52	-	-	8.780	8.576	8.362	8.250	-	-
India	2,694	2,953	3,153	3,217	-	-	31	30	30	31	-	-
Indonesia	679	702	760	801	854	938	120	148	123	149	166	168
Iran (Islamic Republic of)	398	163	175	172	-	-	36	82	74	73	-	-
Iraq	-	-	46	-	-	-	-	-	22	-	-	-
Ireland	218	218	-	-	-	-	32	33	-	-	-	-
Israel	406	452	457	482	489	-	30	30	32	33	35	-
Italy	*1,429	*1,411	*1,334	*1,411	*1,470	-	64	64	57	56	55	-
Jamaica	17	17	-	-	-	-	92	95	-	-	-	-
Japan	21,634	21,846	21,137	21,201	18,832	19,404	21	22	22	22	23	23
Jordan	58	58	94	96	174	162	36	43	33	36	23	26
Kenya	42	43	49	48	64	64	81	83	80	84	66	70
Korea, Republic of	3,940	3,889	4,192	5,038	5,209	5,451	25	29	27	24	24	23
Kuwait	23	15	21	22	24	-	48	49	60	68	67	-
Kyrgyzstan	2.000	2.000	2.000	3.000	4.000	-	600	522	407	297	117	-
Latvia	5.000	47	53	53	20	50	575	68	46	28	81	30
Lithuania	-	-	-	-	*49	-	-	-	-	-	48	-
Macau	30	29	26	24	21	18	11	19	19	17	11	7.889

Continued.

Depending on the table, * means *Enterprises*, *Engaged*, or *Factor Values*; ± means *Basis Unspecified*; § means *shown in millions*. For additional notes and sources, see Appendix I.

Establishments and Number Engaged

- Continued -

Country	Establishments or Enterprises (number)						Employees per Establishment					
	1990	1991	1992	1993	1994	1995	1990	1991	1992	1993	1994	1995
Malaysia	314	378	403	476	509	-	114	120	122	114	122	-
Malta	19	18	20	26	24	-	37	40	37	31	32	-
Mauritius	35	22	19	23	26	27	34	46	54	44	44	40
Mexico	165	186	182	177	173	172	163	173	172	141	148	155
Morocco	-	222	236	258	-	-	-	35	36	35	-	-
Mozambique	6.000	6.000	7.000	7.000	6.000	-	163	170	136	123	121	-
Myanmar	3.000	3.000	3.000	3.000	3.000	-	-	-	-	-	-	-
Nepal	55	110	-	133	139	-	24	19	-	21	19	-
Netherlands	278	299	313	319	320	-	92	90	88	83	88	-
Netherlands Antilles	-	-	-	6.000	-	-	-	-	-	9.333	-	-
New Zealand	*405	*426	*384	*395	-	-	16	15	16	17	-	-
Norway	207	197	125	110	120	-	28	29	41	40	40	-
Pakistan	-	67	-	-	-	-	-	76	-	-	-	-
Panama	32	31	-	-	-	-	45	51	-	-	-	-
Paraguay	-	4.000	-	-	-	-	-	66	-	-	-	-
Peru	560	585	598	-	-	-	17	16	16	-	-	-
Philippines	374	405	372	377	-	-	54	49	67	63	-	-
Poland	*135	*147	*147	*148	-	-	304	286	252	243	-	-
Portugal	*715	*699	*672	*795	*874	-	24	24	24	20	18	-
Puerto Rico	62	60	61	66	69	69	45	44	43	36	36	37
Qatar	15	19	19	18	20	-	20	21	24	27	29	-
Republic of Moldova	2.000	2.000	2.000	1.000	1.000	1.000	1,817	1,859	1,415	1,751	1,607	1,217
Romania	-	304	702	771	971	-	-	-	-	-	-	-
Russian Federation	-	-	-	*1,157	*1,239	*1,866	-	-	-	71	59	43
Saint Lucia	-	5.000	5.000	5.000	5.000	3.000	-	-	-	-	-	-
Saint Vincent & the Grenadines	-	-	-	1.000	1.000	2.000	-	-	-	-	-	-
Senegal	19	16	17	14	16	-	70	58	58	61	159	-
Sierra Leone	-	-	-	2.000	-	-	-	-	-	51	-	-
Singapore	293	294	302	313	312	-	51	55	54	56	58	-
Slovakia	-	16	16	24	29	-	-	453	435	295	262	-
Slovenia	195	*1,018	*1,114	*1,076	*1,254	*1,084	-	-	-	-	-	-
South Africa	-	837	-	-	-	-	-	55	-	-	-	-
Spain	2,958	2,727	2,907	-	-	-	22	23	21	-	-	-
Sri Lanka	22	29	24	45	-	-	134	159	120	197	-	-
Suriname	-	-	6.000	-	-	-	-	-	11	-	-	-
Swaziland	-	4.000	4.000	4.000	-	-	-	43	62	63	-	-
Sweden	282	279	262	252	263	-	45	44	45	40	42	-
Thailand	127	490	-	-	-	-	164	54	-	-	-	-
Trinidad and Tobago	2.000	2.000	2.000	2.000	2.000	-	111	124	125	119	151	-
Tunisia	-	-	-	141	172	176	-	-	-	31	34	36
Turkey	175	170	397	376	381	-	86	81	44	47	44	-
Ukraine	89	559	621	77	75	64	236	38	45	221	200	188
United Kingdom	4,596	4,527	4,732	4,526	5,033	-	37	36	35	40	36	-
United Republic of Tanzania	8.000	9.000	-	-	-	-	96	55	-	-	-	-
USSR (Former)	*203	-	-	-	-	-	591	-	-	-	-	-
United States of America	-	-	17,307	-	-	-	-	-	41	-	-	-
Venezuela	424	440	446	420	396	398	50	53	49	52	53	135
Yugoslavia	177	332	854	1,172	1,215	799	-	32	12	8.703	7.819	12
Zambia	9.000	-	-	-	13	-	59	-	-	-	53	-
Zimbabwe	22	22	22	20	21	-	144	235	173	170	167	-

Depending on the table, * means *Enterprises, Engaged,* or *Factor Values*; ± means *Basis Unspecified*; § means *shown in millions*. For additional notes and sources, see Appendix I.

415

Employment and Compensation of Employees

Country	Number Employed/Engaged (000)						Wage/Salary per Employee ($)					
	1990	1991	1992	1993	1994	1995	1990	1991	1992	1993	1994	1995
Albania	-	-	-	1.405	0.686	0.438	-	-	-	-	-	1,010
Argentina	27	-	-	37	38	-	4,852	-	-	12,220	-	-
Armenia	*2.028	*1.693	-	-	-	-	-	-	-	-	-	-
Australia	36	*35	*32	-	-	-	19,336	20,947	21,691	-	-	-
Austria	14	14	14	15	15	15	21,301	22,531	26,592	25,585	26,640	34,346
Azerbaijan	3.365	3.034	2.345	2.133	2.931	-	1,518	1,701	-	-	-	-
Bahrain	-	-	0.441	-	-	-	-	-	5,934	-	-	-
Bangladesh	2.786	3.551	3.214	-	-	-	588	562	727	-	-	-
Belgium	19	19	19	-	-	-	22,779	23,963	26,580	-	-	-
Bolivia	1.319	1.454	1.530	1.518	1.454	1.463	1,524	1,612	1,669	1,879	1,934	2,089
Bosnia & Herzegovina	2.752	3.165	-	-	0.105	-	2,684	3,105	-	-	-	-
Brazil	118	-	106	108	-	-	6,315	-	5,581	6,282	-	-
Bulgaria	13	10	9.700	7.600	9.800	9.500	1,859	677	1,232	1,605	1,234	1,674
Burundi	0.083	0.081	-	-	-	-	2,441	2,503	-	-	-	-
Cambodia	*0.592	*0.515	-	-	-	-	-	-	-	-	-	-
Canada	62	60	60	62	65	65	22,573	23,290	22,614	22,079	21,574	21,883
Chile	11	11	13	14	15	14	3,968	4,460	5,487	6,045	6,847	7,823
China	*1,000	*1,020	*1,030	*1,010	*1,010	*1,090	-	-	-	-	-	-
China (Hong Kong SAR)	*50	*39	*33	*23	*15	-	7,826	9,064	10,458	11,364	11,570	-
China (Taiwan Province)	218	208	201	191	194	188	8,647	10,143	11,843	12,187	12,776	13,565
Colombia	19	22	25	25	24	23	2,110	2,225	2,502	2,891	3,495	3,938
Costa Rica	5.219	5.689	6.035	6.480	-	-	3,132	2,828	3,254	3,636	-	-
Croatia	9.920	7.610	6.590	7.090	6.470	6.420	5,382	5,602	1,915	2,181	2,922	5,027
Cyprus	1.175	1.192	1.194	1.194	1.152	1.334	9,919	10,478	11,915	11,855	12,910	15,546
Czechoslovakia (Former)	8.000	7.000	-	-	-	-	2,019	1,551	-	-	-	-
Czech Republic	7.000	7.000	15	16	-	-	2,149	1,551	1,934	2,382	-	-
Denmark	13	14	14	14	15	15	27,477	28,218	31,092	28,941	32,391	37,430
Ecuador	5.835	6.303	5.734	5.788	6.819	7.032	2,272	2,609	2,836	3,115	2,169	2,383
Egypt	*18	*28	*19	*18	*17	*16	1,587	1,083	1,119	1,246	1,231	1,228
El Salvador	-	-	*1.374	*2.239	*1.640	*1.884	-	-	-	2,298	3,759	2,856
Ethiopia	0.870	0.868	0.914	0.933	1.128	1.064	1,776	1,916	1,513	1,078	988	1,034
Fiji	0.256	0.255	0.177	0.176	0.175	-	3,508	3,842	4,218	4,201	4,466	-
Finland	7.600	7.000	6.500	6.000	6.000	10	26,904	26,201	23,575	19,403	22,348	28,870
FYR Macedonia	1.746	1.710	1.998	1.732	2.683	2.405	5,329	8,234	1,854	2,558	2,606	4,081
France	121	124	124	119	118	119	32,985	34,101	36,502	-	-	-
Gambia	-	-	-	0.089	-	-	-	-	-	-	-	-
Germany	-	*317	*311	*296	*291	-	-	26,846	30,783	30,115	31,752	-
Germany (Western Part)	*286	*301	*300	-	-	-	27,038	27,714	31,327	-	-	-
Ghana	-	-	-	*2.963	*3.047	*3.133	-	-	-	861	643	656
Greece	*9.654	*9.299	*9.660	*9.611	*9.512	*9.500	10,615	11,055	11,884	12,514	14,269	16,358
Guatemala	-	*2.030	*2.182	*2.204	*2.234	*2.263	-	-	-	-	-	-
Honduras	*1.875	*2.276	*2.362	*1.719	*1.884	*2.760	2,874	2,012	3,659	5,069	4,371	3,311
Hungary	18	18	15	15	16	16	2,743	2,739	3,086	3,371	4,243	4,547
Iceland	0.518	0.506	0.485	0.429	0.474	-	29,565	34,383	38,675	36,760	-	-
India	*83	*89	*94	*100	*96	*97	1,027	890	1,017	875	1,145	1,351
Indonesia	82	104	93	120	141	158	522	576	1,805	1,253	823	1,161
Iran (Islamic Republic of)	14	13	13	13	-	-	3,874	4,460	4,590	4,127	-	-
Iraq	-	1.042	1.000	-	-	-	-	10,476	18,129	-	-	-
Ireland	7.000	7.300	7.535	7.720	7.993	8.305	19,095	19,762	21,846	19,291	20,510	22,953
Israel	12	14	15	16	17	17	14,799	14,983	15,906	15,956	18,742	19,410
Italy	92	91	76	79	81	-	32,659	34,098	37,133	30,181	30,649	-
Jamaica	1.562	1.617	1.728	1.819	1.801	-	-	-	-	-	-	-
Japan	*456	*477	*466	*463	*439	*445	23,522	26,705	29,618	34,301	38,756	-
Jordan	2.087	2.503	3.118	3.491	4.024	4.143	2,461	2,227	2,420	2,424	2,432	2,627
Kenya	3.420	3.587	3.897	4.051	4.217	4.465	1,426	1,321	1,328	849	1,010	1,469
Korea, Republic of	100	112	111	120	123	125	8,459	11,140	11,727	12,670	14,253	16,812
Kuwait	1.112	0.741	1.252	1.492	1.615	1.484	3,988	3,079	7,169	6,505	6,649	9,127
Kyrgyzstan	*1.201	*1.045	*0.813	*0.892	*0.468	-	-	-	-	-	-	-
Latvia	2.876	3.188	2.416	1.462	1.622	1.521	-	-	-	-	-	-
Lithuania	-	-	3.280	2.712	2.359	-	-	-	-	544	1,030	-

Continued.

Depending on the table, * means *Enterprises, Engaged,* or *Factor Values;* ± means *Basis Unspecified;* § means *shown in millions.* For additional notes and sources, see Appendix I.

Employment and Compensation of Employees

- Continued -

Country	Number Employed/Engaged (000)						Wage/Salary per Employee ($)					
	1990	1991	1992	1993	1994	1995	1990	1991	1992	1993	1994	1995
Macau	0.335	0.557	0.481	0.405	0.223	0.142	4,423	3,939	5,133	6,143	5,368	5,296
Malawi	0.962	-	-	-	-	-	-	-	-	-	-	-
Malaysia	36	46	49	54	62	71	2,244	2,421	3,061	3,361	3,759	4,154
Malta	0.711	0.727	0.747	0.800	0.768	-	9,158	9,599	10,499	9,116	10,048	-
Mauritius	*1.195	1.001	1.028	1.013	1.144	1.084	2,275	3,213	3,247	3,658	3,339	3,989
Mexico	27	32	31	25	26	27	3,927	4,769	5,450	6,388	6,401	3,946
Morocco	-	7.823	8.593	9.021	-	-	-	-	-	-	-	-
Mozambique	0.980	1.020	0.955	0.860	0.726	-	3,674	2,183	598	501	467	-
Nepal	1.306	2.135	-	2.764	2.669	-	412	467	-	440	525	-
Netherlands	26	27	28	27	28	29	26,450	27,045	30,252	29,606	30,830	35,772
Netherlands Antilles	-	-	-	0.056	-	-	-	-	-	6,584	-	-
New Zealand	6.439	6.402	6.295	6.760	-	-	17,338	17,908	17,182	-	-	-
Norway	5.695	5.648	5.099	4.384	4.750	4.731	27,013	27,529	36,003	29,228	30,307	35,246
Pakistan	4.721	5.069	5.302	-	-	-	1,593	1,619	1,665	-	-	-
Panama	*1.446	*1.589	*1.846	*2.037	*2.108	*2.090	5,988	6,505	6,532	6,502	6,563	6,636
Paraguay	-	0.265	-	-	-	-	-	-	-	-	-	-
Peru	*9.650	*9.520	*9.354	-	-	-	2,678	3,667	4,106	-	-	-
Philippines	20	20	25	24	-	-	1,552	1,874	2,165	1,963	-	-
Poland	41	42	37	36	36	37	1,196	1,929	2,493	2,626	3,337	4,692
Portugal	*17	*17	*16	*16	*16	*16	-	-	-	-	-	-
Puerto Rico	2.760	2.650	2.600	2.400	2.510	2.580	23,080	-	-	-	-	-
Qatar	0.307	0.397	0.462	0.487	0.587	-	4,670	4,551	4,890	5,299	5,719	-
Republic of Moldova	3.633	3.717	2.831	1.751	1.607	1.217	-	2,537	-	-	-	-
Russian Federation	-	-	-	83	73	81	-	-	-	530	882	986
Senegal	*1.329	*0.932	*0.990	*0.855	*2.545	-	3,913	4,054	3,347	3,709	910	-
Sierra Leone	-	-	-	*0.102	-	-	-	-	-	-	-	-
Singapore	*15	*16	*16	*18	*18	*20	9,229	10,162	11,776	13,390	14,936	16,507
Slovakia	-	*7.246	*6.952	*7.078	*7.594	-	-	1,723	1,975	2,172	2,367	-
South Africa	44	46	47	47	43	46	7,431	7,243	8,035	7,696	7,774	8,609
Spain	65	62	60	60	63	64	18,128	19,530	22,012	18,945	18,713	20,618
Sri Lanka	*2.954	*4.598	*2.876	*8.862	-	-	617	779	627	543	-	-
Suriname	*0.018	*0.062	*0.066	*0.032	-	-	6,225	1,717	1,783	1,751	-	-
Swaziland	-	0.171	0.248	0.251	-	-	-	2,990	2,363	3,259	-	-
Sweden	13	*12	*12	10	11	11	22,742	25,025	27,776	21,515	23,265	25,260
Thailand	21	26	-	-	-	-	5,054	1,771	-	-	-	-
Trinidad and Tobago	0.223	0.247	0.250	0.239	0.302	-	7,893	7,554	6,219	8,757	6,036	-
Tunisia	-	-	-	4.400	5.858	6.314	-	-	-	3,088	2,354	2,549
Turkey	15	14	17	18	17	17	5,270	6,531	5,381	5,685	3,449	6,432
Ukraine	21	21	28	17	15	12	2,788	2,575	2,283	-	413	-
United Kingdom	169	164	167	179	183	189	21,101	22,448	23,783	20,199	21,423	23,071
United Republic of Tanzania	0.767	0.491	-	-	-	-	475	966	-	-	-	-
USSR (Former)	120	-	-	-	-	-	2,094	-	-	-	-	-
United States of America	670	646	710	737	766	804	22,269	23,297	24,468	25,094	25,668	25,845
Uruguay	*5.030	*4.731	*4.801	*4.352	*4.616	*4.536	3,138	3,607	4,570	5,714	6,276	6,686
Venezuela	21	23	22	22	21	54	3,357	3,530	3,925	3,917	3,583	1,814
Yugoslavia	-	10	10	10	9.500	9.700	-	-	-	-	2,006	914
Zambia	0.527	-	-	-	0.685	-	1,595	-	-	-	1,395	-
Zimbabwe	3.161	5.177	3.800	3.400	3.500	3.203	4,462	3,605	3,084	2,586	2,300	3,012

Female Workers: Total and Percent of Employment

Country	Female Workers (000)						Female Workers as % of Total Employment					
	1990	1991	1992	1993	1994	1995	1990	1991	1992	1993	1994	1995
Albania	-	-	-	0.867	0.220	0.180	-	-	-	61.71	32.07	41.10
Argentina	-	-	-	-	6.952	-	-	-	-	-	18.08	-
Australia	11	-	-	-	-	-	30.48	-	-	-	-	-
Austria	5.000	5.000	5.000	4.679	4.779	-	36.50	35.71	35.71	31.99	32.09	-
Bahrain	-	-	0.024	-	-	-	-	-	5.44	-	-	-

Continued.

Depending on the table, * means *Enterprises, Engaged,* or *Factor Values*; ± means *Basis Unspecified*; § means *shown in millions*. For additional notes and sources, see Appendix I.

417

Female Workers: Total and Percent of Employment

- Continued -

Country	Female Workers (000)						Female Workers as % of Total Employment					
	1990	1991	1992	1993	1994	1995	1990	1991	1992	1993	1994	1995
Bangladesh	0.026	0.023	0.029	-	-	-	0.93	0.65	0.90	-	-	-
Bosnia & Herzegovina	1.267	1.301	-	-	-	-	46.04	41.11	-	-	-	-
Bulgaria	-	-	-	4.100	5.300	5.100	-	-	-	53.95	54.08	53.68
Canada	22	21					35.48	35.00				
Chile	1.818	1.894	2.266	2.255	2.413	-	16.92	16.85	17.49	15.98	16.53	-
China (Taiwan Province)	116	108	102	93	92	87	53.11	52.18	50.62	48.86	47.70	46.24
Colombia	-	7.430	-	8.796	8.534		-	34.33	-	34.65	35.51	-
Croatia	4.120	3.060	2.710	2.790	2.420	2.320	41.53	40.21	41.12	39.35	37.40	36.14
Cyprus	0.393	0.400	0.412	0.369	0.357		33.45	33.56	34.51	30.90	30.99	
Czechoslovakia (Former)	2.600	2.500	-	-			32.50	35.71	-	-		
Czech Republic	4.000	3.000	7.000	8.000	-	-	57.14	42.86	46.67	50.00	-	-
Denmark	4.898	5.652	5.775	-			38.42	40.05	40.22	-		
Egypt	-	3.267	2.564	2.386			-	11.84	13.78	13.26		
El Salvador				0.517	0.479	0.729				23.09	29.21	38.69
Ethiopia	-	0.276	0.284	0.292	0.380	0.342	-	31.80	31.07	31.30	33.69	32.14
Gambia	-	-	-	0.012			-	-	-	13.48		
Germany (Western Part)	86	-	-	-	-	-	30.05	-	-	-	-	-
Ghana	-	-	-	0.155			-	-	-	5.23		
Hungary	8.000	-	7.000	6.000	-	-	44.44	-	46.67	40.00	-	-
Indonesia	-	-	-	61	73	79	-	-	-	50.85	51.49	50.25
Iran (Islamic Republic of)	-	0.538	0.550	0.817	-	-	-	4.02	4.25	6.52	-	-
Iraq	-	0.111	-	-			-	10.65	-	-		
Ireland	1.900	2.000	-	-			27.14	27.40	-	-		
Italy		26	21	21	22			28.80	27.32	26.29	27.54	
Jordan	0.067	0.086	0.118	0.135	0.243	0.273	3.21	3.44	3.78	3.87	6.04	6.59
Kenya	0.289	0.334	0.347	0.361	-		8.45	9.31	8.90	8.91	-	
Korea, Republic of	28	29	27	30	31	31	28.23	25.72	24.51	24.93	24.96	25.03
Macau	0.047	0.164	0.150	0.146	0.079	0.060	14.03	29.44	31.19	36.05	35.43	42.25
Malaysia	19	24	26	27	29	-	53.35	53.63	52.14	49.45	45.91	-
Malta	0.158	0.184	0.158	0.216	0.202	-	22.22	25.31	21.15	27.00	26.30	-
Morocco	-	-	-	-	-	2.348	-	-	-	-	-	-
Myanmar	0.062	0.050	0.051	0.069	0.109							
Nepal	0.143	0.228	-	0.153	0.104		10.95	10.68	-	5.54	3.90	
New Zealand	1.719	1.703	1.659	1.729	-		26.70	26.60	26.35	25.58	-	
Panama	0.260						17.98	:				
Philippines	-	-	7.658	6.468	-		-	-	30.76	27.41	-	
Portugal	2.237	2.233	2.193	1.926	1.689		13.28	13.22	13.46	12.01	10.60	
Republic of Moldova	-	-	-	-	1.019	0.789	-	-	-	-	63.41	64.83
Senegal	0.038	-	-	-			2.86	-	-	-		
Sri Lanka	1.793	1.899	1.378	3.846	-		60.70	41.30	47.91	43.40	-	
Sweden	5.000	-	-	-	-	-	39.06	-	-	-	-	-
Thailand	6.959	15	-	-	-	-	33.37	58.61	-	-	-	-
Turkey	3.159	-	-	-			20.88	-	-	-		
United Kingdom	54	-	54	-	-	-	32.09	-	32.34	-	-	-
United Republic of Tanzania	0.121	0.078	-	-			15.78	15.89	-	-		
United States of America	-	-	546	572	601	632	-	-	76.90	77.61	78.46	78.61
Zambia	-	-	-	-	0.034		-	-	-	-	4.96	

Output and Output per Employee

Country	Output ($ bil.)						Output per Employee ($)					
	1990	1991	1992	1993	1994	1995	1990	1991	1992	1993	1994	1995
Albania	-	-	-	*0.003	*0.001	*0.001	-	-	-	1,967	1,117	1,237
Argentina	±1.108	-	-	2.653			41,759	-	-	72,307	-	
Armenia	0.000	0.001	0.003	0.058	0.000	-	156	450	-	-	-	-
Australia	*3.855	*4.146	*3.877	-	-		107,096	118,448	121,163	-	-	
Austria	1.496	1.604	1.732	1.686	1.761	2.233	109,181	114,539	123,708	115,298	118,276	153,809
Azerbaijan	±0.030	±0.031	±0.001	±0.005	±0.006	-	8,862	10,315	311	2,140	1,925	-

Continued.

 Depending on the table, * means *Enterprises*, *Engaged*, or *Factor Values*; ± means *Basis Unspecified*; § means *shown in millions*. For additional notes and sources, see Appendix I.

Output and Output per Employee

- Continued -

Country	Output ($ bil.)						Output per Employee ($)					
	1990	1991	1992	1993	1994	1995	1990	1991	1992	1993	1994	1995
Bahrain	-	-	±0.019	-	-	-	-	-	44,007	-	-	-
Bangladesh	0.036	0.032	0.029	-	-	-	12,761	9,065	8,947	-	-	-
Benin	±0.001	±0.001	±0.001	±0.001	-	-	-	-	-	-	-	-
Bolivia	0.024	0.028	0.031	0.032	0.032	0.034	17,919	19,415	20,199	21,045	21,698	23,103
Bosnia & Herzegovina	±0.077	±0.125	-	-	-	-	28,125	39,579	-	-	-	-
Brazil	*5.725	-	*4.263	*5.039	-	-	48,419	-	40,097	46,647	-	-
Bulgaria	±0.301	±0.110	±0.123	±0.108	±0.130	±0.184	23,509	10,604	12,712	14,245	13,301	19,316
Burundi	±0.001	±0.001	-	-	-	-	16,202	15,725	-	-	-	-
Canada	6.616	6.311	6.172	6.581	7.081	7.201	106,716	105,176	102,865	106,143	108,941	110,779
Chile	0.419	0.474	0.578	0.690	0.807	0.909	38,998	42,165	44,588	48,905	55,273	63,203
China	±7.314	±8.251	±10	±12	±11	±14	7,314	8,089	9,965	12,234	10,645	12,388
China (Hong Kong SAR)	2.465	2.100	2.050	1.425	1.095	-	49,603	54,259	62,871	63,349	70,623	-
China (Taiwan Province)	±11	±12	±12	±11	±12	±12	51,068	57,299	58,958	57,824	60,384	63,555
Colombia	0.621	0.703	0.713	0.792	0.964	1.043	33,397	32,463	28,848	31,183	40,115	45,101
Costa Rica	0.103	0.107	0.121	0.137	0.149	0.173	19,659	18,736	20,085	21,161	-	-
Croatia	±0.360	±0.280	±0.162	-	-	-	36,300	36,843	24,554	-	-	-
Cyprus	0.065	0.071	0.071	0.066	0.068	0.087	55,572	59,805	59,801	54,874	58,747	65,092
Czechoslovakia (Former)	±0.124	±0.054	-	-	-	-	15,529	7,705	-	-	-	-
Denmark	*1.160	*1.151	*1.375	*1.253	*1.519	*1.760	90,997	81,564	95,722	89,255	100,855	116,852
Ecuador	0.133	0.179	0.154	0.156	0.186	0.251	22,743	28,363	26,846	26,987	27,303	35,750
Egypt	*0.399	*0.295	*0.268	*0.237	*0.232	*0.231	21,811	10,705	14,392	13,171	13,660	14,655
El Salvador	-	±0.027	-	±0.041	±0.036	±0.027	-	-	-	18,271	21,933	14,193
Ethiopia	0.016	0.011	0.010	0.009	0.015	0.014	18,436	13,058	10,822	9,318	13,698	12,958
Fiji	0.010	0.011	0.008	0.009	0.009	-	39,698	44,019	45,427	48,343	52,567	-
Finland	±0.909	±0.714	±0.642	±0.534	±0.646	±1.581	119,639	101,985	98,846	88,931	107,655	153,331
FYR Macedonia	±0.038	±0.056	±0.030	±0.017	±0.031	±0.037	21,542	32,745	15,186	9,810	11,397	15,527
France	±18	±18	±20	±18	±19	±22	149,818	148,097	159,463	151,311	162,568	187,544
Germany	-	37	41	37	41	-	-	117,696	130,739	125,364	139,596	-
Germany (Western Part)	34	37	40	-	-	-	119,698	122,343	133,189	-	-	-
Ghana	-	-	-	0.043	0.035	0.037	-	-	-	14,657	11,610	11,848
Greece	*0.746	*0.726	*0.804	*0.844	*0.954	*1.088	77,323	78,028	83,228	87,837	100,297	114,505
Guatemala	-	0.096	0.101	0.156	0.152	0.179	-	47,524	46,489	70,745	68,234	79,061
Honduras	0.055	0.052	0.062	0.055	0.059	0.076	29,513	22,818	26,049	32,020	31,484	27,699
Hungary	±0.416	±0.468	±0.449	±0.530	±0.685	±0.738	23,117	26,018	29,950	35,326	44,033	47,009
Iceland	0.066	0.067	±0.069	±0.060	-	-	128,054	132,701	142,448	138,957	-	-
India	*1.872	*1.799	*1.865	*1.910	*2.349	*2.786	22,639	20,316	19,894	19,156	24,340	28,726
Indonesia	0.779	0.991	1.735	1.491	1.700	2.460	9,534	9,558	18,598	12,467	12,026	15,614
Iran (Islamic Republic of)	0.650	0.405	0.359	0.338	-	-	45,449	30,200	27,744	26,972	-	-
Iraq	-	*0.082	*0.311	-	-	-	-	78,473	310,932	-	-	-
Ireland	*0.732	*0.764	*0.875	*0.792	*0.875	*1.020	104,525	104,610	116,063	102,640	109,420	122,843
Israel	1.077	1.206	1.410	1.425	1.596	1.803	87,543	89,347	95,909	89,625	93,358	106,712
Italy	±15	±15	±13	±12	±13	-	162,768	165,127	177,113	148,903	165,569	-
Japan	±74	±89	±91	±99	±100	±111	163,092	185,942	194,666	213,535	228,726	248,467
Jordan	0.064	0.071	0.088	0.061	0.107	0.134	30,776	28,323	28,086	17,526	26,715	32,448
Kenya	*0.086	*0.109	*0.112	*0.074	*0.082	*0.134	25,264	30,404	28,832	18,301	19,461	30,048
Korea, Republic of	7.058	12	13	15	18	24	70,648	107,891	120,457	121,205	148,389	187,491
Kuwait	0.035	0.020	0.058	0.080	0.104	0.108	31,740	27,219	46,566	53,346	64,195	72,686
Kyrgyzstan	0.013	0.008	0.002	0.001	0.001	-	10,754	7,545	2,577	1,126	1,364	-
Latvia	0.000	0.001	0.004	0.005	0.010	0.014	158	371	1,851	3,372	6,060	9,039
Lithuania	-	-	0.005	0.006	0.012	-	-	-	1,648	2,219	5,014	-
Macau	0.010	0.031	0.035	0.039	0.008	0.006	28,358	55,777	72,459	96,062	34,789	40,453
Malaysia	*0.753	*0.996	*1.205	*1.425	*1.905	*2.393	21,047	21,887	24,545	26,292	30,571	33,750
Malta	0.025	0.026	0.029	0.026	0.027	-	35,676	35,384	38,658	32,634	34,891	-
Mauritius	0.019	0.022	0.016	0.019	0.021	0.022	15,652	22,301	15,951	18,593	18,023	20,470
Mexico	0.852	0.975	1.057	1.015	1.104	0.796	31,729	30,237	33,841	40,786	43,134	29,884
Morocco	-	±0.193	±0.224	±0.203	-	-	-	24,708	26,088	22,531	-	-
Mozambique	-	±0.005	-	±0.002	±0.001	-	-	4,634	-	2,036	1,783	-
Myanmar	0.004	0.003	0.003	0.004	0.012	-	-	-	-	-	-	-
Nepal	0.010	0.017	-	0.018	0.027	-	7,378	8,185	-	6,519	10,242	-
Netherlands	3.952	4.039	4.326	3.870	4.317	5.128	154,118	150,711	157,318	145,841	153,278	178,669

Continued.

Depending on the table, * means *Enterprises, Engaged,* or *Factor Values;* ± means *Basis Unspecified;* § means *shown in millions.* For additional notes and sources, see Appendix I.

419

Output and Output per Employee
- Continued -

Country	Output ($ bil.)						Output per Employee ($)					
	1990	1991	1992	1993	1994	1995	1990	1991	1992	1993	1994	1995
New Zealand	0.626	-	-	-	-	-	97,169	-	-	-	-	-
Norway	±0.778	±0.754	±0.714	±0.567	±0.653	±0.760	136,666	133,578	140,016	129,355	137,545	160,574
Pakistan	0.081	0.083	0.091	-	-	-	17,067	16,454	17,073	-	-	-
Panama	0.054	0.060	0.069	0.076	0.079	0.079	37,421	37,673	37,464	37,094	37,354	37,766
Peru	0.288	0.325	0.357	-	-	-	29,892	34,172	38,216	-	-	-
Philippines	0.387	0.463	0.551	0.487	-	-	19,071	23,147	22,141	20,633	-	-
Poland	±0.549	±0.794	±0.768	±0.816	±1.006	±1.398	13,381	18,899	20,751	22,664	27,647	37,978
Portugal	0.806	0.836	0.940	0.788	0.877	1.233	47,853	49,522	57,665	49,164	55,076	75,252
Qatar	0.010	0.013	0.013	0.016	0.018	-	31,320	32,524	29,138	32,155	30,421	-
Republic of Moldova	0.000	0.049	0.003	0.002	0.001	0.001	0.035	13,230	959	907	703	499
Romania	±1.190	±0.784	±0.343	±0.304	-	-	-	-	-	-	-	-
Russian Federation	-	1.117	0.796	0.335	0.448	0.712	-	-	-	4,052	6,148	8,787
Senegal	0.039	-	-	-	-	-	29,530	-	-	-	-	-
Singapore	*0.914	*1.010	*1.031	*1.203	*1.421	*1.722	61,226	62,436	63,057	68,485	78,777	86,673
Slovakia	-	±0.142	±0.150	±0.142	±0.167	-	-	19,559	21,531	20,111	22,055	-
Slovenia	-	-	-	0.206	0.246	0.433	-	-	-	-	-	-
South Africa	±1.761	±1.817	±1.930	±1.800	±1.934	±2.308	40,012	39,497	41,061	38,308	44,973	49,954
Spain	*7.080	*7.211	*7.570	±5.782	±6.746	±8.833	109,379	116,751	126,029	96,841	106,298	137,705
Sri Lanka	0.022	0.035	0.023	0.051	-	-	7,351	7,699	8,109	5,744	-	-
Suriname	*0.001	*0.007	*0.004	0.010	-	-	31,124	108,430	67,906	297,619	-	-
Swaziland	-	*0.004	*0.010	*0.007	-	-	-	24,387	39,734	27,904	-	-
Sweden	*1.458	*1.630	*1.692	*1.053	*1.235	*1.331	113,908	132,498	142,788	103,251	113,012	122,640
Thailand	2.402	0.453	-	-	-	-	115,163	17,201	-	-	-	-
Trinidad and Tobago	±0.008	±0.011	±0.009	±0.007	±0.011	-	35,622	43,037	35,887	30,260	35,432	-
Tunisia	0.142	0.143	0.161	0.155	0.171	0.206	-	-	-	35,165	29,157	32,637
Turkey	0.973	0.973	1.219	1.420	1.246	2.488	64,301	71,048	70,354	80,754	73,543	143,795
Ukraine	*0.354	*0.398	*0.390	*0.054	*0.050	*0.040	16,862	18,940	13,945	3,160	3,321	3,343
United Kingdom	*18	*17	19	17	20	22	104,036	104,770	113,646	96,608	108,227	116,641
United Republic of Tanzania	*0.008	*0.006	-	-	-	-	11,002	11,514	-	-	-	-
USSR (Former)	±1.688	-	-	-	-	-	14,063	-	-	-	-	-
United States of America	*78	*78	*89	*96	*107	*115	116,791	119,969	125,772	130,653	139,107	143,618
Uruguay	0.144	0.147	0.176	0.174	0.202	0.212	28,598	31,000	36,557	39,950	43,814	46,644
Venezuela	0.547	0.708	0.635	0.663	0.599	0.885	25,551	30,515	28,878	30,482	28,726	16,503
Yugoslavia	±0.043	±0.050	±0.089	-	±0.197	±0.131	-	4,743	8,570	-	20,685	13,473
Zambia	0.012	-	-	-	0.008	-	23,540	-	-	-	11,374	-
Zimbabwe	*0.085	*0.122	*0.084	*0.066	*0.065	*0.073	26,880	23,497	22,213	19,494	18,475	22,681

Value Added and Value Added per Employee

Country	Value Added ($ bil.)						Value Added per Employee ($)					
	1990	1991	1992	1993	1994	1995	1990	1991	1992	1993	1994	1995
Argentina	±0.436	-	-	1.032	-	-	16,434	-	-	28,134	-	-
Armenia	0.000	0.000	-	-	-	-	75	267	-	-	-	-
Australia	*1.702	*1.828	*1.818	-	-	-	47,287	52,223	56,817	-	-	-
Austria	0.545	0.616	0.699	0.706	0.721	0.921	39,751	43,967	49,894	48,236	48,411	63,475
Bangladesh	0.014	0.008	0.008	-	-	-	4,874	2,193	2,428	-	-	-
Bolivia	0.008	0.009	0.012	0.012	0.010	0.011	5,858	6,341	7,910	8,094	7,102	7,561
Bosnia & Herzegovina	±0.033	±0.054	-	-	-	-	12,040	16,942	-	-	-	-
Brazil	*3.426	-	*2.639	*3.165	-	-	28,977	-	24,820	29,301	-	-
Bulgaria	-	±0.021	±0.026	±0.017	-	-	-	1,978	2,633	2,265	-	-
Burundi	0.000	0.000	-	-	-	-	5,494	5,768	-	-	-	-
Canada	2.897	2.819	2.854	2.915	3.156	3.209	46,723	46,987	47,572	47,008	48,556	49,376
Chile	0.178	0.184	0.224	0.322	0.372	0.420	16,591	16,323	17,271	22,807	25,513	29,179
China	±1.736	±1.845	±2.377	±3.621	±2.572	±2.695	1,736	1,809	2,308	3,585	2,547	2,472
China (Hong Kong SAR)	0.760	0.674	0.623	0.492	0.323	-	15,299	17,417	19,117	21,856	20,828	-
China (Taiwan Province)	3.482	3.746	4.089	4.080	4.027	3.762	15,976	18,052	20,379	21,348	20,786	20,029
Colombia	0.223	0.273	0.305	0.338	0.419	0.454	11,989	12,634	12,332	13,310	17,427	19,630
Costa Rica	0.036	0.038	0.043	0.048	0.051	0.059	6,896	6,684	7,188	7,381	-	-

Continued.

Depending on the table, * means *Enterprises*, *Engaged*, or *Factor Values*; ± means *Basis Unspecified*; § means *shown in millions*. For additional notes and sources, see Appendix I.

Value Added and Value Added per Employee

- Continued -

Country	Value Added ($ bil.)						Value Added per Employee ($)					
	1990	1991	1992	1993	1994	1995	1990	1991	1992	1993	1994	1995
Croatia	±0.135	±0.134	±0.070	-	-	-	13,592	17,646	10,549	-	-	-
Cyprus	0.025	0.029	0.032	0.028	0.029	0.037	21,413	24,332	26,473	23,633	25,206	27,452
Czechoslovakia (Former)	±0.049	-	-	-	-	-	6,128	-	-	-	-	-
Denmark	*0.635	*0.627	*0.749	*0.683	*0.829	*0.961	49,833	44,405	52,186	48,649	55,070	63,817
Ecuador	0.042	0.057	0.047	0.053	0.057	0.079	7,197	9,037	8,278	9,101	8,366	11,211
Egypt	*0.072	*0.062	*0.082	*0.043	*0.042	*0.041	3,959	2,248	4,388	2,364	2,453	2,633
El Salvador	-	0.013	-	0.019	0.017	0.010	-	-	-	8,555	10,139	5,519
Ethiopia	0.008	0.005	0.005	0.004	0.006	0.006	9,127	5,835	5,327	4,572	5,345	5,812
Fiji	0.003	0.004	0.002	0.003	0.003	-	13,565	13,808	14,066	14,972	16,281	-
Finland	±0.425	±0.333	±0.310	±0.250	±0.303	±0.645	55,949	47,619	47,740	41,694	50,509	62,552
FYR Macedonia	±0.019	±0.031	±0.018	±0.009	±0.014	±0.018	10,748	18,308	8,857	5,456	5,366	7,583
France	±6.663	±6.819	±7.437	±6.795	±6.854	±8.045	54,974	55,077	60,168	56,866	58,279	67,434
Germany (Western Part)	17	19	21	19	-	-	60,501	62,852	68,425	-	-	-
Ghana	-	-	-	0.016	0.014	0.014	-	-	-	5,292	4,459	4,551
Greece	*0.276	*0.281	*0.330	*0.348	*0.393	*0.449	28,571	30,167	34,191	36,164	41,350	47,256
Honduras	*0.015	*0.015	*0.014	*0.014	*0.014	*0.019	8,224	6,698	6,056	8,234	7,562	6,845
Hungary	±0.145	±0.134	±0.131	±0.164	±0.209	±0.224	8,042	7,463	8,718	10,922	13,452	14,281
Iceland	0.027	0.027	±0.030	±0.026	-	-	52,698	54,171	61,509	59,910	-	-
India	*0.297	*0.271	*0.296	*0.318	*0.373	*0.442	3,593	3,061	3,160	3,184	3,862	4,558
Indonesia	*0.228	*0.274	*0.448	*0.521	*0.519	*0.804	2,795	2,638	4,806	4,361	3,673	5,103
Iran (Islamic Republic of)	0.173	0.145	0.114	0.125	-	-	12,116	10,796	8,854	9,946	-	-
Iraq	-	*0.034	*0.089	-	-	-	-	32,772	89,389	-	-	-
Ireland	*0.332	*0.365	*0.419	*0.380	*0.420	*0.491	47,358	50,037	55,617	49,239	52,572	59,114
Israel	0.468	0.547	0.567	0.571	0.601	0.707	38,025	40,497	38,563	35,934	35,153	41,834
Italy	±4.799	±4.917	±4.515	±3.922	±4.285	-	52,236	54,152	59,767	49,954	52,804	-
Japan	±31	±37	±39	±43	±45	±49	67,536	77,222	83,329	92,181	101,470	110,305
Jordan	0.017	0.019	0.024	0.023	0.039	0.030	8,114	7,733	7,606	6,622	9,663	7,332
Kenya	*0.024	*0.025	*0.026	*0.017	*0.023	*0.029	7,145	7,094	6,690	4,256	5,500	6,532
Korea, Republic of	2.734	4.336	5.107	5.664	7.432	9.973	27,367	38,561	45,938	47,290	60,278	79,536
Kuwait	0.016	0.011	0.030	0.038	0.048	0.052	14,481	14,332	23,615	25,578	29,667	35,027
Latvia	-	-	-	*0.003	*0.006	*0.007	-	-	-	2,103	3,743	4,290
Macau	0.003	0.008	0.007	0.007	0.002	0.002	8,504	14,208	13,674	16,698	9,752	11,681
Malaysia	*0.261	*0.339	*0.459	*0.561	*0.719	*0.905	7,293	7,447	9,354	10,351	11,537	12,769
Malta	0.011	0.011	0.014	0.013	0.014	-	15,693	15,654	18,333	16,342	17,809	-
Mauritius	0.006	0.009	0.008	0.008	0.010	0.010	5,405	8,610	8,038	7,423	8,834	9,469
Mexico	0.441	0.471	0.512	0.488	0.540	0.387	16,430	14,622	16,401	19,613	21,087	14,538
Morocco	-	±0.050	±0.059	±0.055	-	-	-	6,401	6,924	6,068	-	-
Myanmar	±0.003	±0.003	±0.002	±0.002	±0.001	-	-	-	-	-	-	-
Nepal	0.003	0.006	-	0.006	0.008	-	2,555	2,753	-	2,170	3,070	-
Netherlands	1.305	1.354	1.486	1.366	1.516	1.796	50,887	50,537	54,052	51,502	53,823	62,563
New Zealand	0.229	-	-	-	-	-	35,511	-	-	-	-	-
Norway	±0.278	±0.263	±0.276	±0.214	±0.236	±0.274	48,837	46,620	54,172	48,906	49,726	57,975
Pakistan	0.023	0.026	0.028	-	-	-	4,969	5,032	5,370	-	-	-
Panama	0.023	0.024	0.028	0.031	0.032	0.032	15,932	15,355	15,327	15,206	15,326	15,495
Peru	0.144	0.124	0.122	-	-	-	14,905	12,992	13,028	-	-	-
Philippines	0.111	0.152	0.189	0.187	-	-	5,463	7,591	7,599	7,945	-	-
Poland	±0.274	±0.306	±0.290	±0.290	±0.354	±0.488	6,678	7,276	7,825	8,063	9,720	13,257
Portugal	0.237	0.254	0.292	0.252	0.259	0.363	14,043	15,037	17,916	15,742	16,261	22,145
Qatar	0.003	0.005	0.007	0.008	0.007	-	10,738	12,456	15,461	16,359	12,636	-
Romania	±0.397	±0.233	±0.112	±0.103	-	-	-	-	-	-	-	-
Russian Federation	-	-	-	±0.182	±0.268	±0.298	-	-	-	2,209	3,681	3,683
Saint Lucia	-	±0.001	±0.001	±0.001	±0.001	±0.001	-	-	-	-	-	-
Senegal	0.011	0.009	0.024	0.008	0.007	-	8,017	9,714	23,821	8,901	2,626	-
Singapore	*0.327	*0.372	*0.410	*0.493	*0.574	*0.698	21,887	22,963	25,061	28,087	31,840	35,104
Slovakia	-	-	-	±0.047	±0.054	-	-	-	-	6,580	7,056	-
Slovenia	-	-	-	0.065	0.080	0.141	-	-	-	-	-	-
South Africa	*0.559	*0.618	*0.633	*0.590	*0.633	*0.755	12,711	13,431	13,473	12,561	14,725	16,332
Spain	*2.452	*2.690	*2.879	±2.151	±2.369	±2.811	37,881	43,562	47,931	36,018	37,332	43,817
Sri Lanka	0.009	0.016	0.010	0.018	-	-	3,076	3,403	3,639	2,070	-	-
Swaziland	-	*0.001	*0.003	*0.001	-	-	-	3,554	10,631	5,695	-	-

Continued.

Depending on the table, * means *Enterprises*, *Engaged*, or *Factor Values*; ± means *Basis Unspecified*; § means *shown in millions*. For additional notes and sources, see Appendix I.

Value Added and Value Added per Employee

- Continued -

Country	Value Added ($ bil.)						Value Added per Employee ($)					
	1990	1991	1992	1993	1994	1995	1990	1991	1992	1993	1994	1995
Sweden	*0.786	*0.588	*0.627	*0.428	*0.504	*0.543	61,375	47,784	52,900	41,926	46,123	50,069
Thailand	0.892	0.213	-	-	-	-	42,779	8,090	-	-	-	-
Trinidad and Tobago	±0.005	±0.004	±0.004	±0.003	±0.004	-	20,434	17,585	15,831	13,918	13,133	-
Tunisia	0.035	0.035	0.047	0.048	0.051	0.062	-	-	-	10,914	8,623	9,796
Turkey	0.328	0.356	0.424	0.543	0.458	0.915	21,687	25,982	24,473	30,858	27,015	52,875
United Kingdom	*8.250	*8.154	*9.339	*8.450	*9.571	*11	48,817	49,719	55,923	47,209	52,302	56,390
United Republic of Tanzania	*0.000	*-0.000	-	-	-	-	548	-585	-	-	-	-
United States of America	*37	*37	*45	*48	*54	*57	55,701	58,034	63,785	65,756	70,000	70,292
Uruguay	0.065	0.065	0.085	0.086	0.100	0.105	12,908	13,806	17,752	19,747	21,679	23,090
Venezuela	0.215	0.257	0.256	0.212	0.232	0.659	10,041	11,091	11,639	9,762	11,143	12,289
Yugoslavia	±0.018	±0.022	±0.050	-	±0.093	±0.059	-	2,142	4,827	-	9,790	6,069
Zambia	0.006	-	-	-	0.004	-	11,580	-	-	-	5,659	-
Zimbabwe	*0.047	*0.053	*0.027	*0.020	*0.030	*0.027	14,880	10,263	7,212	6,016	8,540	8,326

Capital Investment

Country	Gross Fixed Capital Formation ($ mil.)						Per Establishment ($ mil)					
	1990	1991	1992	1993	1994	1995	1990	1991	1992	1993	1994	1995
Argentina	-	-	-	303	-	-	-	-	-	0.104	-	-
Australia	438	-	-	-	-	-	0.413	-	-	-	-	-
Austria	150	144	149	131	152	-	0.612	0.580	0.584	0.510	0.605	-
Bahrain	-	-	2.088	-	-	-	-	-	0.522	-	-	-
Bangladesh	0.628	1.694	0.539	-	-	-	0.004	0.008	0.003	-	-	-
Bolivia	1.599	1.911	1.401	-0.207	1.923	-	0.025	0.029	0.019	-0.003	0.023	-
Bosnia & Herzegovina	0.988	-	-	-	-	-	0.058	-	-	-	-	-
Brazil	322	-	203	238	-	-	0.352	-	0.245	0.286	-	-
Bulgaria	16	2.743	89	35	5.465	3.787	0.545	0.091	2.690	0.047	0.006	0.004
Canada	184	-	-	-	-	-	0.135	-	-	-	-	-
Chile	29	27	33	59	50	-	0.337	0.292	0.334	0.589	0.460	-
China (Hong Kong SAR)	99	114	102	57	34	-	0.021	0.030	0.029	0.024	0.016	-
Colombia	34	41	53	33	25	-	0.093	0.110	0.125	0.080	0.061	-
Croatia	2.828	3.635	3.748	3.586	2.288	0.815	0.013	0.013	0.009	0.006	0.003	0.001
Cyprus	6.079	7.760	6.056	7.030	2.004	4.761	0.141	0.180	0.141	0.163	0.036	0.077
Czechoslovakia (Former)	9.471	-	-	-	-	-	0.631	-	-	-	-	-
Denmark	84	120	-	-	-	-	0.165	0.232	-	-	-	-
Ecuador	25	31	38	8.668	33	38	0.317	0.314	0.389	0.089	0.311	0.342
Egypt	18	9.453	7.099	-21	-	-	0.081	0.044	0.033	-0.094	-	-
El Salvador	-	0.619	-	7.505	2.948	5.842	-	0.041	-	0.357	0.173	0.365
Ethiopia	0.067	0.892	0.011	0.184	0.543	0.423	0.022	0.297	0.002	0.037	0.049	0.047
Fiji	0.198	0.316	0.825	-	-	-	0.015	0.026	0.059	-	-	-
Finland	53	30	20	30	44	106	0.346	0.179	0.124	0.196	0.313	0.469
FYR Macedonia	0.994	0.395	0.434	0.178	0.130	0.461	0.033	0.007	0.004	0.002	0.001	0.002
Germany (Western Part)	2,379	2,370	2,413	2,075	-	-	1.005	0.965	0.948	-	-	-
Greece	59	53	50	-	-	-	0.204	0.183	0.167	-	-	-
Hungary	37	20	30	69	-	-	0.100	0.031	0.052	0.099	-	-
India	135	97	130	319	-	-	0.050	0.033	0.041	0.099	-	-
Indonesia	86	93	109	103	118	191	0.126	0.133	0.144	0.129	0.138	0.204
Iran (Islamic Republic of)	16	28	49	22	-	-	0.040	0.172	0.281	0.125	-	-
Ireland	48	49	-	-	-	-	0.221	0.224	-	-	-	-
Israel	60	115	128	148	219	-	0.148	0.253	0.280	0.307	0.448	-
Italy	952	1,010	869	696	714	-	0.666	0.716	0.652	0.493	0.485	-
Japan	3,488	5,211	4,998	4,442	4,050	-	0.161	0.239	0.236	0.210	0.215	-
Jordan	0.985	2.203	3.332	1.538	4.014	3.182	0.017	0.038	0.035	0.016	0.023	0.020
Korea, Republic of	781	1,145	2,458	2,012	2,094	2,968	0.198	0.294	0.586	0.399	0.402	0.545
Kuwait	0.784	5.398	4.181	5.563	12	-	0.034	0.360	0.199	0.253	0.504	-
Latvia	0.014	0.008	0.349	0.035	2.095	0.500	0.003	0.000	0.007	0.001	0.105	0.010
Macau	0.611	7.384	0.532	0.205	-0.531	0.006	0.020	0.255	0.020	0.009	-0.025	0.000
Malaysia	107	157	212	223	259	-	0.341	0.415	0.525	0.468	0.510	-

Continued.

Depending on the table, * means *Enterprises, Engaged,* or *Factor Values;* ± means *Basis Unspecified;* § means *shown in millions.* For additional notes and sources, see Appendix I.

Capital Investment
- Continued -

Country	Gross Fixed Capital Formation ($ mil.)						Per Establishment ($ mil)					
	1990	1991	1992	1993	1994	1995	1990	1991	1992	1993	1994	1995
Malta	1.798	3.291	1.528	1.895	0.989	-	0.095	0.183	0.076	0.073	0.041	-
Mexico	34	35	-	-	-	-	0.207	0.190	-	-	-	-
Morocco	-	-	-	23	-	-	-	-	-	0.089	-	-
Myanmar	7.593	7.786	9.555	11	11	-	2.531	2.595	3.185	3.555	3.738	-
Nepal	2.792	-	-	-	-	-	0.051	-	-	-	-	-
Netherlands	334	342	342	387	-	-	1.203	1.143	1.094	1.214	-	-
New Zealand	48	-	-	-	-	-	0.118	-	-	-	-	-
Norway	40	56	28	27	36	-	0.193	0.285	0.224	0.246	0.302	-
Pakistan	-	2.733	-	-	-	-	-	0.041	-	-	-	-
Panama	1.901	2.887	-	-	-	-	0.059	0.093	-	-	-	-
Peru	6.391	12	17	-	-	-	0.011	0.020	0.029	-	-	-
Philippines	16	33	49	50	-	-	0.043	0.081	0.130	0.131	-	-
Poland	40	47	74	74	-	-	0.299	0.322	0.506	0.501	-	-
Portugal	68	81	94	75	133	-	0.095	0.115	0.139	0.094	0.153	-
Romania	-	-	-	13	9.087	-	-	-	-	0.017	0.009	-
Senegal	5.741	-	-	-	-	-	0.302	-	-	-	-	-
Singapore	74	87	86	117	124	-	0.254	0.296	0.284	0.373	0.399	-
Slovenia	-	-	-	16	17	26	-	-	-	0.015	0.014	0.024
Spain	388	404	390	325	378	460	0.131	0.148	0.134	-	-	-
Sri Lanka	1.772	1.894	2.821	6.178	-	-	0.081	0.065	0.118	0.137	-	-
Swaziland	-	0.106	0.107	0.522	-	-	-	0.027	0.027	0.130	-	-
Thailand	1,230	523	-	-	-	-	9.681	1.068	-	-	-	-
Trinidad and Tobago	-	-	0.518	0.168	0.017	-	-	-	0.259	0.084	0.008	-
Tunisia	13	16	19	20	19	21	-	-	-	0.141	0.109	0.120
Turkey	53	66	61	110	201	-	0.300	0.388	0.153	0.293	0.527	-
Ukraine	8.499	17	28	1.060	0.988	6.165	0.095	0.031	0.044	0.014	0.013	0.096
United Kingdom	1,021	809	899	-	975	-	0.222	0.179	0.190	-	0.194	-
United Republic of Tanzania	13	13	-	-	-	-	1.627	1.475	-	-	-	-
United States of America	3,410	3,470	3,934	4,085	4,796	5,711	-	-	0.227	-	-	-
Uruguay	4.728	6.186	9.825	6.094	-	-	-	-	-	-	-	-
Venezuela	29	162	44	26	18	141	0.068	0.367	0.098	0.061	0.047	0.354
Yugoslavia	0.901	1.772	0.640	-	4.817	2.192	0.005	0.005	0.001	-	0.004	0.003
Zambia	7.004	-	-	-	-	-	0.778	-	-	-	-	-
Zimbabwe	17	17	12	6.118	9.264	-	0.752	0.774	0.562	0.306	0.441	-

Indexes of Industrial Production

Country	Index of Industrial Production (1990=100)						Index of Employment (1990=100)					
	1990	1991	1992	1993	1994	1995	1990	1991	1992	1993	1994	1995
Albania	100	-	-	-	-	-	-	-	-	-	-	-
Algeria	100	97	80	83	97	89	-	-	-	-	-	-
Angola	100	-	-	-	-	-	-	-	-	-	-	-
Argentina	100	108		-	-	-	100	-	-	138	145	-
Armenia	100	-	-	-	-	-	100	83	-	-	-	-
Australia	100	105	103	-	-	-	100	97	89	-	-	-
Austria	100	94	94	80	85	93	100	102	102	107	109	106
Azerbaijan	-	-	-	-	-	-	100	90	70	63	87	-
Bahrain	100	-	-	-	-	-	-	-	-	-	-	-
Bangladesh	100	-	-	-	-	-	100	127	115	-	-	-
Belgium	100	106	114	105	106	112	100	101	102	-	-	-
Benin	100	-	-	-	-	-	-	-	-	-	-	-
Bhutan	100	-	-	-	-	-	-	-	-	-	-	-
Bolivia	100	102	101	99	106	109	100	110	116	115	110	111
Bosnia & Herzegovina	-	-	-	-	-	-	100	115	-	-	4	-
Brazil	100	100	89	95	99	109	100	-	90	91	-	-
Bulgaria	100	82	68	60	82	96	100	81	76	59	77	74
Burkina-Faso	100	85	70	-	-	-	-	-	-	-	-	-
Burundi	100	-	-	-	-	-	100	98	-	-	-	-

Continued.

Depending on the table, * means *Enterprises*, *Engaged*, or *Factor Values*; ± means *Basis Unspecified*; § means *shown in millions*. For additional notes and sources, see Appendix I.

423

Indexes of Industrial Production

- Continued -

Country	Index of Industrial Production (1990=100)						Index of Employment (1990=100)					
	1990	1991	1992	1993	1994	1995	1990	1991	1992	1993	1994	1995
Cambodia	100	-	-	-	-	-	100	87	-	-	-	-
Canada	100	93	96	108	122	122	100	97	97	100	105	105
Central African Republic	100	100	100	100	100	-	-	-	-	-	-	-
Chile	100	142	163	150	141	139	100	105	121	131	136	134
China	100	-	-	-	-	-	100	102	103	101	101	109
China (Hong Kong SAR)	100	92	86	72	59	50	100	78	66	45	31	-
China (Taiwan Province)	100	102	98	93	94	85	100	95	92	88	89	86
Colombia	100	99	115	128	148	143	100	116	133	136	129	124
Costa Rica	100	104	113	131	142	145	100	109	116	124	-	-
Cote d'Ivoire	100	86	103	69	52	52	-	-	-	-	-	-
Croatia	100	-	-	-	-	-	100	77	66	71	65	65
Cyprus	100	94	116	81	77	77	100	101	102	102	98	114
Czechoslovakia (Former)	100	-	-	-	-	-	100	88	-	-	-	-
Czech Republic	100	-	-	-	-	-	100	100	214	229	-	-
Dem. Rep. of the Congo	100	100	100	-	-	-	-	-	-	-	-	-
Denmark	100	104	113	109	122	122	100	111	113	110	118	118
Dominican Republic	100	-	-	-	-	-	-	-	-	-	-	-
Ecuador	100	130	130	157	-	-	100	108	98	99	117	121
Egypt	100	105	98	82	77	71	100	151	102	98	93	86
El Salvador	100	107	-	-	-	-	-	-	-	-	-	-
Ethiopia and Eritrea	100	-	-	-	-	-	-	-	-	-	-	-
Ethiopia	-	-	-	-	-	-	100	100	105	107	130	122
Fiji	100	101	94	97	99	89	100	100	69	69	68	-
Finland	100	87	87	94	100	102	100	92	86	79	79	136
FYR Macedonia	100	-	-	-	-	-	100	98	114	99	154	138
France	100	99	100	97	100	98	100	102	102	99	97	98
Gabon	100	-	-	-	-	-	-	-	-	-	-	-
Gambia	100	-	-	-	-	-	-	-	-	-	-	-
Germany (Eastern Part)	100	-	-	-	-	-	-	-	-	-	-	-
Germany (Western Part)	100	108	99	93	99	112	100	105	105	-	-	-
Ghana	100	-	-	-	-	-	-	-	-	-	-	-
Greece	100	97	87	92	103	87	100	96	100	100	99	98
Guatemala	100	-	-	-	-	-	-	-	-	-	-	-
Guyana	100	-	-	-	-	-	-	-	-	-	-	-
Haiti	100	-	-	-	-	-	-	-	-	-	-	-
Honduras	100	138	140	157	199	263	100	121	126	92	100	147
Hungary	100	94	86	96	123	130	100	100	83	83	86	87
Iceland	100	-	-	-	-	-	100	98	94	83	92	-
India	100	110	116	-	-	-	100	107	113	121	117	117
Indonesia	100	131	142	140	126	129	100	127	114	146	173	193
Iran (Islamic Republic of)	100	122	117	115	96	111	100	94	90	88	-	-
Iraq	100	-	-	-	-	-	-	-	-	-	-	-
Ireland	100	97	102	106	112	119	100	104	108	110	114	119
Israel	100	111	125	141	158	184	100	110	120	129	139	137
Italy	100	94	89	87	-	-	100	99	82	85	88	-
Jamaica	100	94	-	-	-	-	100	104	111	116	115	-
Japan	100	101	98	95	95	95	100	105	102	102	96	98
Jordan	100	110	129	138	149	122	100	120	149	167	193	199
Kenya	100	125	148	163	166	176	100	105	114	118	123	131
Korea, Republic of	100	108	115	120	130	136	100	113	111	120	123	126
Kuwait	100	-	-	-	-	-	100	67	113	134	145	133
Kyrgyzstan	100	-	-	-	-	-	100	87	68	74	39	-
Lao P.D.R.	100	-	-	-	-	-	-	-	-	-	-	-
Latvia	100	-	-	-	-	-	100	111	84	51	56	53
Lebanon	100	100	100	100	-	-	-	-	-	-	-	-
Liberia	100	-	-	-	-	-	-	-	-	-	-	-
Libyan Arab Jamahiriya	100	-	-	-	-	-	-	-	-	-	-	-
Lithuania	100	-	-	-	-	-	-	-	-	-	-	-
Macau	100	-	-	-	-	-	100	166	144	121	67	42
Madagascar	100	-	-	-	-	-	-	-	-	-	-	-

Continued.

Depending on the table, * means *Enterprises, Engaged,* or *Factor Values*; ± means *Basis Unspecified*; § means *shown in millions.* For additional notes and sources, see Appendix I.

Indexes of Industrial Production

- Continued -

Country	Index of Industrial Production (1990=100)						Index of Employment (1990=100)					
	1990	1991	1992	1993	1994	1995	1990	1991	1992	1993	1994	1995
Malawi	100	-	-	-	-	-	100	-	-	-	-	-
Malaysia	100	112	131	145	171	199	100	127	137	151	174	198
Malta	100	103	113	123	-	-	100	102	105	113	108	-
Mauritius	100	-	-	-	-	-	100	84	86	85	96	91
Mexico	100	106	109	106	112	109	100	120	116	93	95	99
Mongolia	100	100	100	100	100	100	-	-	-	-	-	-
Morocco	100	101	96	98	100	99	-	-	-	-	-	-
Mozambique	100	-	-	-	-	-	100	104	97	88	74	-
Myanmar	100	-	-	-	-	-	-	-	-	-	-	-
Nepal	100	-	-	-	-	-	100	163	-	212	204	-
Netherlands	100	97	98	99	108	111	100	105	107	103	110	112
New Zealand	100	-	-	-	-	-	100	99	98	105	-	-
Nicaragua	100											
Nigeria	100	-	-	-	-	-						
Norway	100	95	94	100	107	109	100	99	90	77	83	83
Pakistan	100	-	-	-	-	-	100	107	112	-	-	-
Panama	100	111	126	137	141	140	100	110	128	141	146	145
Papua New Guinea	100	-	-	-	-	-	-	-	-	-	-	-
Paraguay	100	109	111	126	106	108	-	-	-	-	-	-
Peru	100	118	121	166	216	219	100	99	97	-	-	-
Philippines	100	119	122	112	110	127	100	99	123	116	-	-
Poland	100	125	140	168	195	228	100	102	90	88	89	90
Portugal	100	87	62	56	74	91	100	100	97	95	95	97
Puerto Rico	100	-	-	-	-	-	100	96	94	87	91	93
Qatar	100	-	-	-	-	-	100	129	150	159	191	-
Republic of Moldova	100	-	-	-	-	-	100	102	78	48	44	33
Romania	100	75	53	56	45	48	-	-	-	-	-	-
Russian Federation	100	-	-	-	-	-	-	-	-	-	-	-
Saudi Arabia	100	-	-	-	-	-	-	-	-	-	-	-
Senegal	-	-	-	-	-	-	100	70	74	64	191	-
Singapore	100	110	108	116	129	140	100	108	110	118	121	133
Slovakia	100	-	-	-	-	-	-	-	-	-	-	-
Slovenia	100	89	67	72	77	83	-	-	-	-	-	-
Somalia	100	-	-	-	-	-	-	-	-	-	-	-
South Africa	100	97	99	97	105	116	100	105	107	107	98	105
Spain	100	100	102	96	103	114	100	95	93	92	98	99
Sri Lanka	100	88	98	137	135	163	100	156	97	300	-	-
Sudan	100	100	100	-	-	-	-	-	-	-	-	-
Suriname	100	-	-	-	-	-	100	344	367	178	-	-
Sweden	100	96	90	90	105	102	100	96	93	80	85	85
Switzerland	100	91	86	85	80	84	-	-	-	-	-	-
Syrian Arab Republic	100	99	100	106	110	116	-	-	-	-	-	-
Thailand	100	106	-	-	-	-	100	126	-	-	-	-
Tonga	100	-	-	-	-	-	-	-	-	-	-	-
Trinidad and Tobago	100	111	114	131	148	148	100	111	112	107	135	-
Tunisia	100	96	92	-	-	-	-	-	-	-	-	-
Turkey	100	112	110	121	108	187	100	90	115	116	112	114
Uganda	100	175	241	405	621	1,069	-	-	-	-	-	-
Ukraine	100	-	-	-	-	-	100	100	133	81	71	57
United Arab Emirates	100	-	-	-	-	-	-	-	-	-	-	-
United Kingdom	100	94	98	104	116	121	100	97	99	106	108	112
United Republic of Tanzania	100	98	124	112	204	-	100	64	-	-	-	-
USSR (Former)	100	-	-	-	-	-	100	-	-	-	-	-
United States of America	100	100	110	118	129	133	100	96	106	110	114	120
Uruguay	100	104	106	104	117	113	100	94	95	87	92	90
Venezuela	100	-	-	-	-	-	100	108	103	102	97	251
Yemen	100	-	-	-	-	-	-	-	-	-	-	-
Yugoslavia	100	-	-	-	-	-	-	-	-	-	-	-

Continued.

Depending on the table, * means *Enterprises, Engaged,* or *Factor Values;* ± means *Basis Unspecified;* § means *shown in millions.* For additional notes and sources, see Appendix I.

425

Indexes of Industrial Production

- Continued -

Country	Index of Industrial Production (1990=100)						Index of Employment (1990=100)					
	1990	1991	1992	1993	1994	1995	1990	1991	1992	1993	1994	1995
Yugoslavia (Former).	100	75	-	-	-	-	-	-	-	-	-	-
Zambia	100	101	96	90	68	60	100	-	-	-	130	-
Zimbabwe	100	-	-	-	-	-	100	164	120	108	111	101

Depending on the table, * means *Enterprises, Engaged,* or *Factor Values*; ± means *Basis Unspecified*; § means *shown in millions.* For additional notes and sources, see Appendix I.

Representative Companies in Sector

Name	Address	Telephone	Fax	Employment	Y
Aeroquip Corp.	3000 Strayer Rd., Maumee, Ohio 43537, U.S.A.	419-867-2200	419-867-2590	10,000	97
Budd Co.	PO Box 2601, Troy, Michigan 48007-2601, U.S.A.	313-643-3500		10,000	96
Huntsman Corp.	500 Huntsman Way, Salt Lake City, Utah 84108, U.S.A.	801-532-5200	801-584-5781	9,500	97
GenCorp Inc.	175 Ghent Rd., Fairlawn, Ohio 44333-3300, U.S.A.	330-869-4200	330-869-4211	9,460	97
Bemis Company Inc.	222 S. 9th St., 2300, Minneapolis, Minnesota 55402-4099, U.S.A.	612-376-3000		9,275	97
Gerber Products Co.	445 State St., Fremont, Michigan 49413, U.S.A.	616-928-2000	616-928-2963	9,000	97
Raychem Corp.	300 Constitution Dr., Menlo Park, California 94025-1164, U.S.A.	415-361-3333	415-361-4913	8,650	97
Tunzl PLC	110 Park St., London W1Y 3RB, United Kingdom	71 4954950	71 4954953	8,046	93
Ball Corp.	PO Box 2407, Muncie, Indiana 47307-0407, U.S.A.	765-747-6100		7,900	97
McCormick and Company Inc. (Sparks, Maryland)	PO Box 6000, Sparks, Maryland 21152-6000, U.S.A.	410-771-7301	410-771-7462	7,500	97
Witco Corp.	1 American Ln., Greenwich, Connecticut 06831-2559, U.S.A.	203-552-2000		7,200	97
Eagle Industries Inc.	2 N. Riverside Plz., 1100, Chicago, Illinois 60606, U.S.A.	312-906-8700	312-906-8372	7,000	94
Owens-Brockway Div.	1 Seagate, Toledo, Ohio 43666, U.S.A.	419-247-5000		7,000	94
Carlisle Companies Inc.	250 S. Clinton St., 201, Syracuse, New York 13202-1258, U.S.A.	315-477-9108	315-474-2008	6,900	97
Ferro Corp.	1000 Lakeside Ave., Cleveland, Ohio 44114, U.S.A.	216-641-8580	216-696-4786	6,851	97
Tupperware Corp.	PO Box 2363, Orlando, Florida 32802-2353, U.S.A.	407-826-5050		6,600	95
Carpenter Co.	PO Box 27205, Richmond, Virginia 23261, U.S.A.	804-359-0800	804-353-0694	6,500	97
Baxter Healthcare Corp. of Puerto Rico	PO Box 2025, Carolina 00984, Puerto Rico	787750-7373		6,000	92
Wavin BV	Handellaan 251, Zwolle NL-8031 EM, Netherlands	38 294911	38 294238	5,800	90
Heywood Williams Group PLC	Edgerton Rd., Huddersfield HD3 3AR, United Kingdom	484 435477	484 547511	5,588	93
Lawson Mardon Wheaton Inc.	1101 Wheaton Ave., Millville, New Jersey 08332-2047, U.S.A.	609-825-1400	609-825-0146	5,500	97
Menasha Corp.	PO Box 367, Neenah, Wisconsin 54957-0367, U.S.A.	920-751-1000	920-751-1236	5,500	97
TVK Rt.	PO Box 20, Tiszaujvaros H-3581, Hungary	49 22222	49 21322	5,500	93
Jannock Ltd.	5205-40 King St. W, Toronto, Ontario M5H 3Y2, Canada	416-364-8586	416-364-9342	5,400	98
West Company Inc.	PO Box 645, Exton, Pennsylvania 19341-0645, U.S.A.	610-594-2900	610-648-2424	5,210	97
Hollis PLC	Headington Hill Hall OX3 0BW, United Kingdom	865 64881		5,050	90
Perfekta Enterprises Ltd.	141 Connaught Rd. W., Hong Kong, Hong Kong	5475131	5592450	5,000	
Envirodyne Industries Inc.	701 Harger Rd., 190, Oak Brook, Illinois 60521, U.S.A.	630-571-8800		4,900	96
Pou Chen Corp.	Fu Hsing Hsiang, Changhua, Taiwan	4 7695147	4 7695150	4,809	92
Newman Tonks Group PLC	Birmingham Business Park, Birmingham B37 7YX, United Kingdom	21 7177777	21 7177776	4,749	93
Spalding and Evenflo Companies Inc.	PO Box 30101, Tampa, Florida 33630, U.S.A.	813-204-5200	813-204-5219	4,400	97
Low & Bonar PLC	Faraday St., Dundee DD1 9JA, United Kingdom	382 818171	382 816262	4,236	93
AEP Industries, Inc.	125 Phillips Ave., South Hackensack, New Jersey 07606-1546, U.S.A.	201-641-6600	201-807-2490	4,200	97
Coleman Company Inc.	1526 Cole Blvd., 300, Golden, Colorado 80401, U.S.A.	303-202-2400		4,200	96
Sealed Air Corp.	Park 80 E, Saddle Brook, New Jersey 07663-5291, U.S.A.	201-791-7600	201-703-4205	4,200	97
Royal Group Technologies Ltd.	1 Royal Gate Blvd., Woodbridge, Ontario L4L 8Z7, Canada	905-264-0701	905-264-0702	4,000	98
Dunlop Ltd.	Vincent Sq., London SW1P 2PL, United Kingdom	71 8343848	71 8288508	3,965	93
Andersen Corp.	100 4th Ave. N., Bayport, Minnesota 55003, U.S.A.	612-439-5150	612-430-5107	3,600	97
PMC Inc. (Sun Valley, California)	PO Box 1367, Sun Valley, California 91353-1367, U.S.A.	818-896-1101	818-897-0180	3,600	97
Signal P Co.	ul. Zelinskogo, 11, Chisinau 277032, Moldova	553076	553087	3,500	92
Continental Can Company Inc.	1 Aerial Way, Syosset, New York 11791, U.S.A.	516-822-4940	516-931-6344	3,463	96
Furon Co.	29982 Ivy Glenn Dr., Laguna Niguel, California 92677, U.S.A.	714-831-5350	714-643-1548	3,456	96
Emco Ltd.	620 Richmond St., London, Ontario N6A 5J9, Canada	519-645-3900	519-645-3939	3,300	98
Tredegar Industries Inc.	1100 Boulders Pkwy., Richmond, Virginia 23225, U.S.A.	804-330-1000	804-330-1260	3,300	95
Clopay Corp.	312 Walnut St., 1600, Cincinnati, Ohio 45202-4036, U.S.A.	513-381-4800	513-762-3965	3,000	97
Dart Container Corp.	500 Hogsback Rd., Mason, Michigan 48854-1447, U.S.A.	517-676-3800	517-676-3883	3,000	96
DataCard Corp.	PO Box 9355, Minneapolis, Minnesota 55343, U.S.A.	612-933-1223	612-933-7971	3,000	97
Teepak Inc.	3 Westbrook Corporate Ctr., Westchester, Illinois 60154, U.S.A.	708-409-3000	708-409-3679	3,000	94
Varlen Corp.	PO Box 3089, Naperville, Illinois 60566-7089, U.S.A.	630-420-0400	630-420-7123	2,850	97
Hickson International PLC	Chancellor House, Leeds LS2 6HG, United Kingdom	13 2426161	13 2341133	2,840	93
Indomobil Suzuki International, PT	Wisma Indomobil 8th Flr., Jakarta 13330, Indonesia	218564530	21 4893048	2,700	97

Continued

Representative Companies in Sector
- Continued -

Name	Address	Tele-phone	Fax	Employ-ment	Y
Intercraft Co.	1 Intercraft Plz., Taylor, Texas 76574, U.S.A.	512-352-8501	512-352-4870	2,670	97
Lawson Mardon Group UK Ltd.	6 Hill St., London W1X 7FU, United Kingdom	71 4937323		2,575	92
Borden UK Ltd.	North Baddesley, Southampton S052 9ZB, United Kingdom	703 732131		2,574	93
JSJ Corp.	700 Robbins Rd., Grand Haven, Michigan 49417, U.S.A.	616-842-6350	616-847-3112	2,500	95
Park-Ohio Industries Inc.	23000 Euclid Ave., Cleveland, Ohio 44117, U.S.A.	216-692-7200		2,500	96
Interlake Corp.	550 Warrenville Rd., Lisle, Illinois 60532-4387, U.S.A.	630-852-8800	630-719-7277	2,491	97
Adwest Group PLC	Woodley, Reading RG5 4SN, United Kingdom	734 697171	734 690121	2,483	94
Gamrat Plastic Works	ul. Mickiewicza 108, Jaslo PL-38-200, Poland	136 2021	136 5658	2,450	93
Lin Pac Containers International Ltd.	8 Coney Gene, Olney MK46 4AE, United Kingdom	234 241362		2,440	91
Lin Pac Plastics International Ltd.	1 Charles St., Louth LN11 0HA, United Kingdom	507 600700	507 600701	2,427	92
Fibreboard Corp.	2200 Ross Ave., Ste. 3600, Dallas, Texas 75201-6731, U.S.A.	214-954-9500	214-954-9512	2,400	96
Shinkong Synthetic Fibers Corp.	8f, 123 Nanking E. Rd., Sec. 2, Taipei 104, Taiwan	02 5071251	02 5072264	2,400	94
Mueller Industries Inc.	2959 N. Rock Rd., Wichita, Kansas 67226-1191, U.S.A.	316-636-6300	316-682-9650	2,350	97
A. Schulman Inc.	3550 W. Market St., Akron, Ohio 44333, U.S.A.	330-666-3751	330-668-7204	2,181	97
Elco Textron Inc.	PO Box 7009, Rockford, Illinois 61125-7009, U.S.A.	815-397-5151	815-395-8270	2,180	97
Speciality Packaging (UK) PLC	Rock Valley Rd., Mansfield NG18 2EZ, United Kingdom	623 22651	623 653481	2,134	93
Spartech Corp.	7733 Forsyth Blvd., 1450, Clayton, Missouri 63105-1817, U.S.A.	314-721-4242		2,125	97
ESSEF Corp.	220 Park Dr., Chardon, Ohio 44024, U.S.A.	216-286-2200	216-286-2206	2,100	97
Magyar Viscosa Rt.	Nyergesujfalu H-2537, Hungary	33 55244	33 55207	2,100	93
Wihuri Oy	Wihurinaukio 2, Helsinki SF-00571, Finland	825851	6848918	2,030	94
Contico International Inc.	1101 Warson Rd., St. Louis, Missouri 63132, U.S.A.	314-997-5900	314-997-1270	2,000	97
James Walker Group Ltd.	Oriental Rd., Woking GU22 8AP, United Kingdom	483 757575	483 755711	1,970	94
PureTec Corp.	65 Railroad Ave., Ridgefield, New Jersey 07657, U.S.A.	201-941-6550	201-941-0602	1,900	97
Norsk Hydro UK Ltd.	69 London Rd., Twickenham TW1 1EE, United Kingdom	81 8911366	81 8921686	1,862	93
ABT Building Products Corp.	1 Neenah Center, Ste. 600, Neenah, Wisconsin 54956-3070, U.S.A.	414-751-8611	414-751-0370	1,809	96
Industex Eastern Cape Pty. Ltd.	Neave Township, Port Elizabeth 6014, Republic of South Africa	41 431363	41 411558	1,800	97
Tah Hsin Industrial Corp.	201-24, Tun Hwa North Rd., Taipei, Taiwan	2 7128311	2 7151446	1,800	94
Lin Pac Mouldings Ltd.	Witton, Birmingham B6 7HY, United Kingdom	21 3282606	21 3270899	1,789	92
Tuscarora Inc.	PO Box 448, New Brighton, Pennsylvania 15066, U.S.A.	412-843-8200	412-843-4845	1,716	97
Danapak AMBA	Kongevejen 100, Holte DK-2840, Denmark	45411210	45411810	1,700	93
Bailey Corp.	PO Box 307, Seabrook, New Hampshire 03874, U.S.A.	603-474-3011		1,668	95
Autobar Group Ltd.	41-42 Kew Bridge Rd., Brentford TW8 0EB, United Kingdom	81 5600151		1,649	93
Sealright Company Inc.	7101 College Blvd., 1400, Overland Park, Kansas 66210-1891, U.S.A.	913-344-9000	913-344-9005	1,603	96
Playtex Products Inc.	300 Nyala Farms Rd., Westport, Connecticut 06880, U.S.A.	203-341-4000		1,600	96
Werner Ladder Co.	93 Werner Rd., Greenville, Pennsylvania 16125, U.S.A.	412-588-8600	412-588-0315	1,600	97
Plysu PLC	Woburn Sands, Milton Keynes MK17 8SE, United Kingdom	908 584222	908 585450	1,590	94
Central Sprinkler Corp.	451 N. Cannon Ave., Lansdale, Pennsylvania 19446, U.S.A.	215-362-0700	215-362-5385	1,575	97
Metalrax Group PLC	Kings Norton, Birmingham B38 9PN, United Kingdom	21 4333444	21 4333325	1,575	93
Tah Hsin Industrial Corp.	Taichung Industrial Park, Taichung, Taiwan	4 2522111	4 2525307	1,550	90
Bowater Windows Ltd.	Hockley Heath, Solihull, United Kingdom	564 783113	603 415682	1,527	93
Wardle Storeys PLC	Brantham, Manningtree CO11 1NJ, United Kingdom	206 392401	206 395288	1,522	94
Berlina PT	Pandaan K. 43, East Java 67156, Indonesia	343 31901	343 31902	1,500	93
CamVec Corp.	1190 Meyerside Dr., Mississauga, Ontario L5T 1R7, Canada	905-670-7111	905-795-3455	1,500	98
Igloo Products Corp.	PO Box 19322, Houston, Texas 77224-9322, U.S.A.	713-461-5955	713-935-7702	1,500	95
Norton Performance Plastics Corp.	150 Dey Rd., Wayne, New Jersey 07470-4699, U.S.A.	201-696-4700	201-696-4056	1,500	93
Wheaton Plastic Products Div.	1101 Wheaton Ave., Millville, New Jersey 08332, U.S.A.	609-825-1400	609-825-0146	1,500	97
Lentex	Powstancow ul. 54, Lubliniec PL-42-701, Poland	334 62641	334 63320	1,490	93
Plastika Nitra	Novozamocka 222, Nitra 949 53, Slovakia	87 14311	87 410727	1,440	92
ITW Ltd.	Waybridge Trading Estate, Addlestone KT15 2SF, United Kingdom	932 848666	932 845548	1,422	94
Leslie's Poolmart Inc.	20222 Plummer St., Chatsworth, California 91311, U.S.A.	818-993-4212	818-349-1059	1,416	95
Imacc Corp.	5801 Christie Ave., 255, Emeryville, California 94608, U.S.A.	510-652-6847	510-271-6280	1,400	95
Raven Industries Inc.	PO Box 5107, Sioux Falls, South Dakota 57117-5107, U.S.A.	605-336-2750		1,368	95
IPEX Inc.	50 Valleybrook Dr., Don Mills, Ontario M3B 2S9, Canada	416-445-3400	416-445-4461	1,352	98
Aladdin Industries Inc.	703 Murfreesboro Rd., Nashville, Tennessee 37210, U.S.A.	615-748-3000	615-748-3030	1,350	97
Atlantis Plastics Inc.	1870 The Exchange, 200, Atlanta, Georgia 30339, U.S.A.	770-953-4567		1,350	97

     For sources and notes, see Appendix I.

Representative Companies in Sector
- Continued -

Name	Address	Telephone	Fax	Employ-ment	Y
Bonar Inc.	2380 McDowell Rd., Burlington, Ontario L7R 4A1, Canada	905-637-5611	905-637-9954	1,350	98
Victaulic PLC	2 N. Fourth St., Milton Keynes MK9 1NW, United Kingdom	908 691000	908 690099	1,348	93
Addis Ltd.	Ware Rd., Hertford SG13 7HL, United Kingdom	992 584221	992 553050	1,343	92
BXL Plastics Ltd.	794 Parklane, Croydon CR0 1JO, United Kingdom	81 6887770		1,320	92
Canron Inc.	100 Disco Rd., Etobicoke, Ontario M9W 1M1, Canada	416-675-6400	416-674-7008	1,300	98
Goody Products Inc.	600 W. Park Dr., Peachtree City, Georgia 30269, U.S.A.	770-486-1995		1,300	95
Minnesota Rubber	3630 Wooddale Ave., Minneapolis, Minnesota 55416, U.S.A.	612-927-1400	612-927-2192	1,300	97
NMC Group PLC	40 S. Audley St., London W1Y 5DHQ, United Kingdom	71 4912880	71 7300457	1,300	94
Penn Engineering and Manufacturing Corp.	PO Box 1000, Danboro, Pennsylvania 18916, U.S.A.	215-766-8853	215-766-7366	1,300	96
Ventra Group Inc.	1 Mitten Crt, Cambridge, Ontario N1R 5S9, Canada	519-658-6777	519-658-5422	1,300	98
Western Industries Inc.	1215 N. 62nd St., Milwaukee, Wisconsin 53213, U.S.A.	414-771-7700	414-771-7371	1,300	96
Ekco Group Inc.	98 Spit Brook Rd., Ste. 102, Nashua, New Hampshire 03062-5738, U.S.A.	603-888-1212	603-888-1427	1,250	96
Chris-Craft Industries Inc.	767 5th Ave., New York, New York 10153, U.S.A.	212-421-0200		1,246	95
Congoleum Corp.	PO Box 3127, Mercerville, New Jersey 08619-0127, U.S.A.	609-584-3000	609-584-3523	1,234	97
Alenco Holdings Ltd.	55 Maylands Ave., Hemel Hempstead HP2 4SJ, United Kingdom	442 238100	442 238111	1,225	93
Courtaulds Packaging Ltd.	Severalls Business Park, Colchester CO4 4QR, United Kingdom	206 753400		1,218	93
Raychem Ltd.	Dorcan, Swindon SN3 5HH, United Kingdom	793 528171	793 430114	1,218	93
Blessings Corp.	200 Enterprise Dr., Newport News, Virginia 23603, U.S.A.	804-887-2100		1,200	97
Flambeau Corp.	801 Lynn Ave., Baraboo, Wisconsin 53913, U.S.A.	608-356-5551	608-356-5260	1,200	97
Graboplast Textil-Es Muborgyarto Rt.	Fehervari u. 16/B, Gyor H-9023, Hungary	96 14155	96 14330	1,200	93
Miniature Precision Components Inc.	PO Box 1901, Walworth, Wisconsin 53184-9901, U.S.A.	414-275-5791	414-275-6346	1,200	94
Packaging Resources Inc.	1 Conway Park, Lake Forest, Illinois 60045, U.S.A.	708-295-6100	708-295-3707	1,200	94
Pharma Plast International AS	Engmosen 1, Lynge DK-3540, Denmark	42188388	42189555	1,200	93
RPC Group PLC	Raunds, Wellingborough NN9 6ED, United Kingdom	933 623311	933 622126	1,171	94
Fjeldhammer Brug AS	Postboks 55, Fjellhamer N-1472, Norway	67 979000	67 705877	1,150	93
Newcor Inc.	1825 S. Woodward Ave., Ste. 240, Bloomfield Hills, Michigan 48302-0574, U.S.A.	248-253-2400	248-253-2413	1,150	97
Sheldahl Inc.	1150 Sheldahl Rd., Northfield, Minnesota 55057-9444, U.S.A.	507-663-8000	507-663-8545	1,129	97
Reunion Industries Inc.	62 Southfield Ave., 1 Stamford L, Stamford, Connecticut 06902, U.S.A.	203-324-8858		1,127	96
Automotive Plastic Technologies Inc.	6600 E. 15 Mile Rd., Sterling Heights, Michigan 48312, U.S.A.	313-979-5000	313-979-0499	1,100	93
O'Sullivan Corp.	PO Box 3510, Winchester, Virginia 22604-2547, U.S.A.	540-667-6666	540-667-3351	1,100	97
Uniroyal Technology Corp.	2 N. Tamiami Tr., 900, Sarasota, Florida 34236-5568, U.S.A.	941-361-2100	941-361-2214	1,100	97
Avon Polymer Products Ltd.	Bath Rd., Melksham SN12 8AA, United Kingdom	225 703101	225 707880	1,056	93
Fluoroware Inc.	3500 Lyman Blvd., Chaska, Minnesota 55318, U.S.A.	612-448-3131	612-448-5576	1,050	97
DRG Plastics Ltd.	Yate, Bristol BS17 5AA, United Kingdom	454 313131	454 319697	1,037	93
Elexmetal	Klonowa St. 1, Gdansk PL-80-264, Poland	58 416860	58 416860	1,025	93
Advanced Drainage Systems Inc.	3300 Riverside Dr., Columbus, Ohio 43221, U.S.A.	614-457-3051	614-459-0169	1,000	95
Batts Inc.	200 Franklin St., Zeeland, Michigan 49464, U.S.A.	616-772-4635		1,000	95
Camoplast Inc.	2144 Rue King O Bureau 110, Sherbrooke, Quebec J1J 2E8, Canada	819-823-1777	819-823-8772	1,000	98
Comar Inc.	1 Comar Pl., Buena, New Jersey 08310, U.S.A.	609-692-6100	609-692-9251	1,000	96
DSM Engineering Plastic Products Inc.	PO Box 14235, Reading, Pennsylvania 19612-4235, U.S.A.	610-320-6600	215-320-6845	1,000	95
Huls America Inc.	220 Davidson Ave., Somerset, New Jersey 08873, U.S.A.	732-560-6800	732-980-6970	1,000	97
Ludlow Corp.	2 Tyco Park, Exeter, New Hampshire 03833, U.S.A.	603-778-1500	603-778-0108	1,000	97
Shape Inc.	PO Box 366, Biddeford, Maine 04005, U.S.A.	207-282-6155	207-283-9130	1,000	95
Material Sciences Corp.	2300 E. Pratt Blvd., Elk Grove Village, Illinois 60007-5995, U.S.A.	847-439-8270	847-439-0737	988	96
Bespak PLC	Bergen Way, King's Lynn PE30 2JJ, United Kingdom	553 691000	553 691622	955	94
Armstrong World Industries Ltd.	38 Market Sq., Uxbridge UB8 1LH, United Kingdom	895 251122	895 31571	953	93
Alenco Div.	615 Carson St., Bryan, Texas 77801, U.S.A.	409-779-7770	409-822-3259	950	97
Laitram Corp.	PO Box 50699, New Orleans, Louisiana 70150, U.S.A.	504-733-6000		950	94
Courtaulds Films Holdings Ltd.	PO Box 1, Bridgewater TA6 4PA, United Kingdom	278 424321	278 421999	914	93
Epwin Group PLC	Alders Way, Paignton TQ4 7QE, United Kingdom	803 525354	803 520129	901	93

ISIC 3610
POTTERY, CHINA, ETC.

ISIC 3610—Pottery, China, etc.—includes articles made of porcelain or china, stoneware, earthenware, and common pottery. The category encompasses laboratory, chemical, and industrial goods made of ceramics, ceramic sanitary fixtures, electrical insulators, and similar products.

The industry corresponds to the U.S. 3-digit SIC 326—Pottery and Related Products, which are divided into vitreous plumbing fixtures (3261), vitreous china table & kitchenware (3262), semivitreous table & kitchenware (3263), porcelain electrical supplies (3264), and pottery products, nec (3269).

Summary Statistics

		1990		1991		1992		1993		1994		1995	
		Value	N	Value	N	Value	N	Value	N	Value	N	Value	N
Establishments or enterprises	(number)	15,021	68	15,655	76	15,730	71	8,978	64	7,972	55	5,371	28
Number employed	(000)	1,178	76	1,046	81	985	76	1,001	71	578	61	479	50
Total Wages	($ mil.)	7,959	69	8,761	71	7,962	66	6,694	59	6,912	53	6,341	44
Wage/salary per employee	($)	8,380	68	10,390	70	10,376	65	9,075	59	9,336	53	10,051	44
Female workers	(000)	96	30	79	28	106	26	110	27	91	20	78	11
as % of total employment*	(%)	37	30	35	28	37	26	39	27	39	20	39	10
Output	($ bil.)	30	71	33	77	29	72	28	63	25	58	24	46
per employee	($ 000)	28	68	30	73	32	68	32	59	34	54	39	44
per establishment	($ 000)	4,020	62	3,771	69	3,851	63	4,502	51	4,502	44	3,989	21
Value added	($ bil.)	16	66	16	68	15	63	15	58	13	52	14	45
per employee	($ 000)	18	63	18	64	19	60	19	54	20	48	23	42
per establishment	($ 000)	2,382	56	2,126	60	2,227	55	2,223	47	2,558	39	1,995	20
Capital investment	($ mil.)	2,021	55	1,923	52	1,385	44	1,136	37	1,271	31	594	14
per employee	($ 000)	3	52	2	50	3	42	2	35	3	29	3	14
per establishment	($ 000)	502	52	380	49	383	42	243	32	262	28	379	12

Data presented above are drawn from the detailed tables that follow. Columns headed 'N' show the number of countries that provided valid data for inclusion. Values are not strictly comparable one year to the next or one row to the next because the number of countries that report varies from period to period and row to row. However, a general indication of magnitudes can be gleaned from reviewing this summary—especially in earlier years in which more reports are available. For detailed explanations, see Appendix I. *The average for those countries reporting both total and female employment.

Establishments and Number Engaged

Country	Establishments or Enterprises (number)						Employees per Establishment					
	1990	1991	1992	1993	1994	1995	1990	1991	1992	1993	1994	1995
Albania	-	-	-	-	5.000	1.000	-	-	-	-	-	-
Argentina	-	-	-	221	249	-	-	-	-	14	12	-
Armenia	2.000	2.000	2.000	2.000	2.000	-	518	496	-	-	-	-
Australia	110	188	202	-	-		18	11	9.901	-	-	-
Austria	32	29	29	29	26	-	94	103	103	84	94	-
Azerbaijan	2.000	2.000	2.000	53	42	-	1,421	1,397	1,284	46	62	-
Bahrain	-	-	2.000	-	-	-	-	-	8.500	-	-	-
Bangladesh	429	418	349	-	-	-	8.713	8.938	9.705	-	-	-
Bolivia	1.000	1.000	1.000	-	1.000	1.000	28	28	13	-	7.000	7.000
Bosnia & Herzegovina	4.000	6.000	-	-	-	-	439	249	-	-	-	-
Bulgaria	*8.000	*8.000	*10	*59	*97	*111	725	650	480	64	41	32
Canada	48	39	37	34	31	-	21	26	27	29	32	-
Cape Verde	4.000	4.000	3.000	3.000	13	-	-	10	13	13	-	-
Central African Republic	-	*1.000	*1.000	*1.000	*1.000	*1.000	-	-	-	-	-	-
Chile	8.000	7.000	8.000	8.000	7.000	-	262	373	349	357	292	-
China	*3,418	*3,410	*3,384	-	-	-	102	111	118	-	-	-
China (Hong Kong SAR)	96	47	37	23	23	-	6.250	8.511	5.405	4.348	4.348	-
Colombia	26	24	19	17	15	-	204	233	309	373	424	-
Costa Rica	75	61	52	56	-	-	11	13	16	16	-	-
Croatia	13	14	15	28	39	47	235	185	161	84	56	43
Cyprus	63	63	66	66	62	57	1.571	1.476	2.500	2.727	2.500	1.649
Czechoslovakia (Former)	*8.000	*7.000	-	-	-	-	1,250	1,143	-	-	-	-
Czech Republic	-	*7.000	*19	*24	-	-	-	-	1,143	526	458	-
Denmark	431	497	495	-	-	-	4.399	3.602	3.307	-	-	-
Ecuador	12	14	13	13	11	11	106	92	89	80	70	75
Egypt	23	21	16	15	-	-	365	552	388	373	-	-
El Salvador	-	2.000	2.000	1.000	3.000	5.000	-	-	10	12	14	13
Finland	13	13	10	9.000	9.000	-	100	100	110	100	111	-
FYR Macedonia	*5.000	*13	*25	*23	*26	*33	751	261	124	98	89	59
Germany	-	211	208	199	182	-	-	242	228	195	186	-
Germany (Western Part)	165	160	162	-	-	-	224	226	233	-	-	-
Ghana	-	-	-	4.000	-	-	-	-	-	46	-	-
Greece	69	66	66	-	-	-	45	42	38	-	-	-
Guatemala	-	1.000	1.000	1.000	1.000	1.000	-	276	276	279	276	276
Honduras	-	-	5.000	5.000	5.000	4.000	-	-	28	31	33	53
Hungary	*42	*62	*57	*75	-	-	286	194	175	80	-	-
Iceland	23	24	24	21	-	-	1.217	-	-	1.048	-	-
India	532	557	579	624	-	-	53	55	45	41	-	-
Indonesia	65	76	80	86	95	95	366	407	413	450	436	483
Iran (Islamic Republic of)	24	15	20	25	-	-	142	280	280	266	-	-
Iraq	-	-	1.000	-	-	-	-	-	3.000	-	-	-
Ireland	16	17	-	-	-	-	44	41	-	-	-	-
Israel	24	31	29	29	35	-	42	35	38	41	37	-
Italy	*510	*506	*213	*203	*198	-	102	101	84	78	74	-
Japan	3,750	3,775	3,650	3,483	3,342	3,270	18	18	18	18	18	19
Jordan	4.000	4.000	5.000	9.000	11	13	100	146	162	65	5.727	69
Kenya	3.000	1.000	1.000	1.000	1.000	1.000	70	230	240	221	228	237
Korea, Republic of	555	637	625	666	672	664	29	25	24	23	23	22
Latvia	16	22	8.000	9.000	11	12	197	137	361	232	152	115
Lithuania	-	-	-	*7.000	-	-	-	-	-	-	197	-
Macau	13	8.000	2.000	1.000	-	-	48	38	68	83	-	-
Malaysia	37	42	44	48	48	-	192	186	175	173	169	-
Malta	7.000	7.000	7.000	7.000	7.000	-	8.286	8.857	8.286	8.714	9.143	-
Mauritius	3.000	2.000	2.000	2.000	2.000	2.000	22	14	16	15	15	25
Mexico	5.000	5.000	5.000	5.000	5.000	5.000	1,492	1,407	1,447	1,335	1,311	1,373
Mongolia	1.000	1.000	1.000	1.000	3.000	4.000	316	303	301	250	99	28
Mozambique	1.000	1.000	-	-	-	-	80	85	-	-	-	-
Netherlands	80	81	75	71	69	-	79	77	80	81	87	-
New Zealand	*193	*203	*209	*223	-	-	1.394	1.571	1.694	1.744	-	-
Norway	7.000	7.000	5.000	7.000	6.000	-	117	111	141	104	122	-

Continued.

Depending on the table, * means *Enterprises, Engaged,* or *Factor Values*; ± means *Basis Unspecified*; § means *shown in millions*. For additional notes and sources, see Appendix I.

Establishments and Number Engaged

- Continued -

Country	Establishments or Enterprises (number)						Employees per Establishment					
	1990	1991	1992	1993	1994	1995	1990	1991	1992	1993	1994	1995
Pakistan	-	20	-	-	-	-	-	157	-	-	-	-
Panama	4.000	-	-	-	-	-	6.750	-	-	-	-	-
Paraguay	-	1.000	-	-	-	-	-	10	-	-	-	-
Peru	65	67	69	-		-	20	18	17	-	-	-
Philippines	754	787	73	61	-	-	7.162	8.132	162	154	-	-
Poland	*34	*38	*40	*32	-	-	706	553	425	531	-	-
Portugal	*957	*941	*977	*954	*973	-	27	26	25	26	26	-
Puerto Rico	8.000	7.000	7.000	5.000	5.000	8.000	65	69	67	86	86	53
Romania	-	-	-	-	75	-	-	-	-	-	-	-
Russian Federation	-	-	-	*320	*431	*739	-	-	-	306	209	47
Saint Lucia	-	2.000	2.000	2.000	2.000	2.000	-	-	-	-	-	-
Sierra Leone	-	-	-	1.000	-	-	-	-	-	36	-	-
Slovakia	-	4.000	4.000	4.000	4.000	-	-	182	172	165	144	-
Slovenia	9.000	*60	*28	*34	*82	*60	-	-	-	-	-	-
South Africa	-	123	-	-	-	-	-	49	-	-	-	-
Spain	975	1,007	997	-	-	-	15	15	13	-	-	-
Sri Lanka	37	36	28	28	-	-	164	204	197	276	-	-
Sweden	18	18	18	20	20	-	128	103	95	77	76	-
Thailand	153	198	-	-	-	-	123	93	-	-	-	-
Trinidad and Tobago	39	12	10	11	10	-	-	7.000	7.300	7.182	8.100	-
Tunisia	-	-	-	70	74	87	-	-	-	66	68	59
Turkey	29	25	46	46	55	-	454	435	214	228	205	-
Ukraine	31	65	66	35	32	28	806	385	379	686	750	821
United Kingdom	791	758	730	765	758	-	56	53	51	48	49	-
United Republic of Tanzania	3.000	1.000	-	-	-	-	27	42	-	-	-	-
USSR (Former)	*88	-	-	-	-	-	2,364	-	-	-	-	-
United States of America	-	-	1,674	-	-	-	-	-	22	-	-	-
Venezuela	30	38	29	29	19	44	77	79	107	121	128	123
Yugoslavia	15	22	44	65	64	65	-	318	159	94	98	102
Zambia	-	-	-	-	1.000	-	-	-	-	-	-	-
Zimbabwe	5.000	5.000	5.000	6.000	5.000	-	160	160	180	133	60	-

Employment and Compensation of Employees

Country	Number Employed/Engaged (000)						Wage/Salary per Employee ($)					
	1990	1991	1992	1993	1994	1995	1990	1991	1992	1993	1994	1995
Argentina	8.828	-	-	3.163	3.005	-	6,948	-	-	17,970	-	-
Armenia	*1.036	*0.992	-	-	-	-	-	-	-	-	-	-
Australia	2.000	*2.000	*2.000	-	-	-	20,469	19,868	18,015	-	-	-
Austria	3.000	3.000	3.000	2.440	2.448	2.440	23,366	24,238	26,087	29,103	29,577	33,454
Azerbaijan	2.841	2.793	2.569	2.421	2.603	-	1,439	1,584	-	-	-	-
Bahrain	-	-	0.017	-	-	-	-	-	3,129	-	-	-
Bangladesh	3.738	3.736	3.387	-	-	-	638	644	432	-	-	-
Belgium	3.200	3.300	3.300	-	-	-	-	-	-	-	-	-
Bermuda	0.008	0.010	0.008	0.012	0.010	0.007	-	-	-	-	-	-
Bolivia	0.028	0.028	0.013	-	0.007	0.007	-	-	749	-	680	655
Bosnia & Herzegovina	1.757	1.496	-	-	-	-	2,760	1,816	-	-	-	-
Bulgaria	5.800	5.200	4.800	3.800	4.000	3.600	1,999	599	926	1,167	1,048	1,433
Cambodia	*2.759	*2.222	-	-	-	-	-	-	-	-	-	-
Cameroon	0.181	0.176	0.180	0.167	-	-	17,918	16,576	18,239	19,413	-	-
Canada	1.000	1.000	1.000	1.000	1.000	0.990	29,997	29,676	29,784	21,704	18,307	18,533
Cape Verde	-	0.041	0.040	0.040	-	-	-	-	-	-	-	-
Chile	2.094	2.613	2.794	2.859	2.046	2.064	1,798	2,897	3,766	4,464	5,942	7,004
China	*350	*380	*400	*380	-	-	-	-	-	-	-	-
China (Hong Kong SAR)	*0.600	*0.400	*0.200	*0.100	*0.100	-	6,847	6,434	-	-	10,351	-
China (Taiwan Province)	29	30	30	31	33	,32	8,434	9,413	11,639	12,208	13,157	14,283
Colombia	5.300	5.600	5.878	6.337	6.366	6.359	2,203	2,232	2,598	3,275	4,192	4,667
Costa Rica	0.831	0.788	0.845	0.903	-	-	2,971	2,718	2,903	3,294	-	-

Continued.

Depending on the table, * means *Enterprises, Engaged,* or *Factor Values*; ± means *Basis Unspecified*; § means *shown in millions*. For additional notes and sources, see Appendix I.

Employment and Compensation of Employees

- Continued -

Country	Number Employed/Engaged (000)						Wage/Salary per Employee ($)					
	1990	1991	1992	1993	1994	1995	1990	1991	1992	1993	1994	1995
Croatia	3.050	2.590	2.410	2.360	2.170	2.000	4,450	5,074	1,779	1,791	2,372	3,706
Cyprus	0.099	0.093	0.165	0.180	0.155	0.094	7,117	6,990	7,529	6,383	6,865	9,551
Czechoslovakia (Former)	10	8.000	-	-	-	-	1,950	1,526	-	-	-	-
Czech Republic	9.000	8.000	10	11	-	-	2,105	1,526	1,840	2,064	-	-
Denmark	1.896	1.790	1.637	1.593	1.679	1.719	28,719	28,999	32,487	29,635	34,219	-
Ecuador	1.266	1.293	1.162	1.036	0.770	0.824	2,418	2,919	3,484	2,360	3,113	3,214
Egypt	*8.400	*12	*6.200	*5.600	*5.230	*5.785	1,720	1,532	1,353	1,552	1,470	1,597
El Salvador	-	-	*0.020	*0.012	*0.043	*0.065	-	-	-	1,538	1,884	1,785
Finland	1.300	1.300	1.100	0.900	1.000	-	25,487	25,318	23,522	18,848	22,724	-
FYR Macedonia	3.756	3.392	3.112	2.246	2.323	1.948	3,747	4,502	1,509	1,820	2,132	2,756
Germany	-	*51	*47	*39	*34	-	-	18,946	23,429	23,343	25,602	-
Germany (Western Part)	*37	*36	*38	-	-	-	22,733	22,820	25,645	-	-	-
Ghana	-	-	-	*0.185	*0.190	*0.196	-	-	-	-	-	-
Greece	*3.090	*2.744	*2.539	*2.496	*2.603	*2.686	12,066	12,118	13,401	13,202	13,791	16,151
Guatemala	-	*0.276	*0.276	*0.279	*0.276	*0.276	-	-	-	-	-	-
Honduras	*0.142	*0.131	*0.139	*0.153	*0.167	*0.213	1,397	1,374	1,764	1,949	1,716	1,205
Hungary	12	12	10	6.000	6.150	6.000	2,526	2,500	3,164	3,729	3,513	3,961
Iceland	0.028	0.008	0.023	0.022	0.008	-	33,947	-	-	21,449	-	-
India	*28	*31	*26	*26	*26	*26	1,124	948	967	851	945	1,114
Indonesia	24	31	33	39	41	46	686	683	788	891	1,009	1,629
Iran (Islamic Republic of)	3.400	4.207	5.603	6.652	-	-	3,954	4,606	4,423	4,223	-	-
Iraq	-	0.004	0.003	-	-	-	-	7,235	13,934	-	-	-
Ireland	0.700	0.700	0.684	0.650	0.661	0.628	16,584	16,386	19,067	16,797	19,106	24,242
Israel	1.000	1.100	1.100	1.200	1.300	1.275	18,351	18,747	20,702	19,139	20,948	23,282
Italy	52	51	18	16	15	-	35,528	37,879	36,068	28,789	29,317	-
Japan	*69	*68	*66	*63	*61	*62	21,721	24,235	26,917	30,975	34,644	-
Jordan	0.399	0.583	0.811	0.586	0.063	0.903	2,370	2,282	1,699	2,458	1,907	2,971
Kenya	0.209	0.230	0.240	0.221	0.228	0.237	1,570	1,398	1,290	781	913	1,264
Korea, Republic of	16	16	15	16	16	15	7,232	8,178	8,668	9,624	10,680	12,902
Latvia	3.151	3.023	2.885	2.089	1.668	1.380	-	-	-	-	-	-
Lithuania	-	-	1.254	1.045	1.377	-	-	-	447	536	714	-
Macau	0.620	0.305	0.135	0.083	-	-	4,230	4,244	5,127	5,580	-	-
Malaysia	7.100	7.800	7.700	8.300	8.100	8.850	2,057	2,317	2,722	2,719	3,274	3,614
Malta	0.058	0.062	0.058	0.061	0.064	-	7,614	7,390	8,675	6,094	7,316	-
Mauritius	*0.065	0.028	0.033	0.031	0.031	0.049	1,449	3,514	5,783	4,387	2,335	3,254
Mexico	7.460	7.037	7.235	6.674	6.553	6.866	4,294	5,417	6,575	7,861	6,863	4,308
Mongolia	*0.316	*0.303	*0.301	*0.250	*0.296	*0.110	1,300	1,907	578	-	-	-
Mozambique	0.080	0.085	-	-	-	-	444	541	-	-	-	-
Netherlands	6.314	6.227	6.005	5.745	6.000	6.000	24,963	26,455	28,883	28,491	29,901	34,572
New Zealand	0.269	0.319	0.354	0.389	-	-	-	-	-	-	-	-
Norway	0.816	0.775	0.707	0.731	0.731	0.740	23,493	24,083	32,616	-	-	-
Pakistan	3.474	3.141	3.285	-	-	-	1,276	1,357	1,396	-	-	-
Panama	*0.027	-	-	-	-	-	1,074	-	-	-	-	-
Paraguay	-	0.010	-	-	-	-	-	-	-	-	-	-
Peru	*1.321	*1.175	*1.148	-	-	-	2,442	3,227	3,422	-	-	-
Philippines	5.400	6.400	12	9.400	-	-	1,996	2,064	2,166	2,112	-	-
Poland	24	21	17	17	19	20	1,126	1,626	2,204	2,331	2,576	3,103
Portugal	*26	*24	*25	*25	*26	*26	-	-	-	-	-	-
Puerto Rico	0.520	0.480	0.470	0.430	0.430	0.420	-	-	-	-	-	-
Russian Federation	-	-	-	98	90	35	-	-	-	693	949	870
Sierra Leone	-	-	-	*0.036	-	-	-	-	-	-	-	-
Slovakia	-	*0.728	*0.686	*0.659	*0.578	-	-	1,351	1,805	1,973	2,376	-
South Africa	6.000	6.000	6.000	-	-	-	5,025	4,527	4,675	-	-	-
Spain	15	15	13	15	16	16	16,044	16,950	19,363	17,232	16,443	19,171
Sri Lanka	*6.066	*7.328	*5.528	*7.716	-	-	930	1,021	1,188	1,028	-	-
Sweden	2.300	*1.856	*1.708	1.536	1.519	1.600	21,229	25,180	29,404	22,147	23,086	25,140
Thailand	19	18	-	-	-	-	1,151	1,816	-	-	-	-
Trinidad and Tobago	-	0.084	0.073	0.079	0.081	-	-	4,482	4,512	3,312	3,542	-
Tunisia	-	-	-	4.647	5.023	5.142	-	-	-	9,182	8,866	9,553
Turkey	13	11	9.851	10	11	12	6,549	7,828	8,405	9,207	5,317	6,511

Continued.

Depending on the table, * means *Enterprises, Engaged,* or *Factor Values*; ± means *Basis Unspecified*; § means *shown in millions*. For additional notes and sources, see Appendix I.

Employment and Compensation of Employees

- Continued -

Country	Number Employed/Engaged (000)						Wage/Salary per Employee ($)					
	1990	1991	1992	1993	1994	1995	1990	1991	1992	1993	1994	1995
Ukraine	25	25	25	24	24	· 23	2,606	3,009	2,320	493	655	521
United Kingdom	44	40	37	37	37	38	17,330	18,805	20,724	17,247	18,542	19,704
United Republic of Tanzania . .	0.082	0.042	-	-	-	-	-	-	-	-	-	-
USSR (Former)	208	-	-	-	-	-	1,971	-	-	-	-	-
United States of America . . .	38	37	37	37	40	44	21,842	22,162	22,838	23,730	24,350	24,477
Uruguay	*2.022	*2.112	*2.121	*2.063	*2.076	*2.011	3,131	3,967	5,274	5,947	6,166	6,666
Venezuela	2.300	3.000	3.100	3.498	2.435	5.396	4,227	3,884	4,973	4,935	5,512	4,956
Yugoslavia	-	7.000	7.000	6.100	6.300	6.600	-	-	-	-	1,964	699
Zimbabwe	0.800	0.800	0.900	0.800	0.300	0.281	2,145	1,714	1,287	1,024	1,636	1,842

Female Workers: Total and Percent of Employment

Country	Female Workers (000)						Female Workers as % of Total Employment					
	1990	1991	1992	1993	1994	1995	1990	1991	1992	1993	1994	1995
Argentina	-	-	-	-	0.412	-	-	-	-	-	13.71	-
Australia	0.640	-	-	-	-	-	32.00	-	-	-	-	-
Austria	1.400	1.400	1.000	1.091	1.058	-	46.67	46.67	33.33	44.71	43.22	-
Bangladesh	0.152	0.061	0.051	-	-	-	4.07	1.63	1.51	-	-	-
Bermuda	0.004	0.003	0.004	0.006	0.006	0.004	50.00	30.00	50.00	50.00	60.00	57.14
Bosnia & Herzegovina . . .	0.683	0.491	-	-	-	-	38.87	32.82	-	-	-	-
Bulgaria.	-	-	-	1.900	2.000	1.200	-	-	-	50.00	50.00	33.33
Chile	0.299	0.494	0.441	0.412	0.317	-	14.28	18.91	15.78	14.41	15.49	-
China (Taiwan Province) . .	15	16	16	15	15	14	53.27	52.57	51.92	49.23	45.60	43.65
Colombia	-	1.183	-	1.389	1.249	-	-	21.12	-	21.92	19.62	-
Croatia	1.250	1.100	0.990	0.960	0.910	0.900	40.98	42.47	41.08	40.68	41.94	45.00
Cyprus	0.075	0.068	0.120	0.129	0.096	-	75.76	73.12	72.73	71.67	61.94	-
Czechoslovakia (Former) . .	4.400	4.000	-	-	-	\	44.00	50.00	-	-	-	-
Czech Republic	5.000	5.000	6.000	7.000	-	-	55.56	62.50	60.00	63.64	-	-
Denmark	1.322	1.320	1.242	-	-	-	69.73	73.74	75.87	-	-	-
Egypt	-	1.686	1.382	1.111	-	-	-	14.53	22.29	19.84	-	-
El Salvador.	-	-	-	0.007	0.021	0.017	-	-	-	58.33	48.84	26.15
Germany (Western Part) . .	18	-	-	-	-	-	48.75	-	-	-	-	-
Hungary	6.000	-	4.000	3.000	-	-	50.00	-	40.00	50.00	-	-
Indonesia	-	-	-	14	14	16	-	-	-	37.36	34.93	35.62
Iran (Islamic Republic of) . . .	-	0.700	0.868	0.989	-	-	-	16.64	15.49	14.87	-	-
Iraq	-	0.001	-	-	-	-	-	25.00	-	-	-	-
Ireland	0.300	0.300	-	-	-	-	42.86	42.86	-	-	-	-
Italy	-	16	4.165	4.017	3.590	-	-	30.62	23.29	25.52	24.42	-
Jordan	0.032	0.027	0.021	0.024	0.011	0.059	8.02	4.63	2.59	4.10	17.46	6.53
Kenya	0.008	0.013	0.013	0.012	-	-	3.83	5.65	5.42	5.43	-	-
Korea, Republic of . . .	8.000	8.017	7.666	8.017	8.204	7.693	50.00	51.07	51.17	51.70	52.48	52.73
Macau	0.183	0.091	0.031	0.017	-	-	29.52	29.84	22.96	20.48	-	-
Malaysia	4.200	4.400	4.500	4.800	4.200	-	59.15	56.41	58.44	57.83	51.85	-
Malta	0.007	0.011	0.009	0.009	0.012	-	12.07	17.74	15.52	14.75	18.75	-
Mongolia	-	-	-	-	0.166	-	-	-	-	-	56.08	-
Morocco	-	-	-	-	-	1.304	-	-	-	-	-	-
New Zealand	0.117	0.140	0.159	0.203	-	-	43.49	43.89	44.92	52.19	-	-
Panama.	0.004	-	-	-	-	-	14.81	-	-	-	-	-
Philippines	-	-	5.944	4.023	-	-	-	-	50.37	42.80	-	-
Portugal	5.918	5.931	6.381	6.627	6.835	-	23.14	24.67	25.70	26.91	26.77	-
Puerto Rico	-	-	0.050	0.050	0.050	0.050	-	-	10.64	11.63	11.63	11.90
Sri Lanka	3.296	3.084	2.371	4.618	-	-	54.34	42.09	42.89	59.85	-	-
Sweden.	0.900	-	-	-	-	-	39.13	-	-	-	-	-
Thailand	10	8.125	-	-	-	-	54.16	44.36	-	-	-	-
Turkey	1.851	-	-	-	-	-	14.05	-	-	-	-	-
United Kingdom	6.900	-	14	-	-	-	15.68	-	37.84	-	-	-
United Republic of Tanzania . .	0.010	0.005	-	-	-	-	12.20	11.90	-	-	-	-
United States of America . . .	-	-	29	30	32	36	-	-	78.38	81.08	80.00	81.82

Output and Output per Employee

Country	Output ($ bil.)						Output per Employee ($)					
	1990	1991	1992	1993	1994	1995	1990	1991	1992	1993	1994	1995
Argentina	±0.363	-	-	0.149	-	-	41,064	-	-	46,988	-	-
Armenia	0.000	0.000	0.001	0.053	0.000		202	434	-	-	-	-
Australia	*0.118	*0.134	*0.129	-	-	-	58,984	67,004	64,706	-	-	-
Austria	0.199	0.197	0.217	0.201	0.213	0.240	66,461	65,633	72,406	82,411	87,121	98,517
Azerbaijan	±0.011	±0.012	±0.000	±0.002	±0.000	-	3,965	4,185	120	906	146	-
Bahrain	-	-	±0.000	-	-	-	-	-	7,822	-	-	-
Bangladesh	0.020	0.022	0.024	-	-	-	5,388	5,822	6,943	-	-	-
Bolivia	0.000	0.000	0.000	-	0.000	0.000	2,747	3,700	10,866	-	2,288	1,637
Bosnia & Herzegovina	±0.023	±0.020	-	-	-	-	13,075	13,649	-	-	-	-
Bulgaria	±0.066	±0.026	±0.034	±0.036	±0.039	±0.038	11,337	4,973	7,141	9,347	9,657	10,454
Canada	0.111	0.105	0.108	0.085	0.073	0.074	111,416	104,739	107,554	85,265	73,228	74,300
Cape Verde	-	±0.000	±0.000	-	-	-	-	702	763	-	-	-
Chile	0.017	0.035	0.047	0.050	0.035	0.042	8,033	13,530	16,684	17,409	17,085	20,294
China	±1.676	±1.903	±2.502	±3.927	-	-	4,788	5,009	6,255	10,335	-	-
China (Hong Kong SAR)	0.015	0.008	-	-	0.003	-	25,033	20,589	-	-	34,936	-
China (Taiwan Province)	±0.690	±0.706	±0.824	±0.988	±1.082	±1.101	24,003	23,673	27,108	31,393	32,591	34,370
Colombia	0.108	0.114	0.134	0.173	0.216	0.240	20,286	20,303	22,747	27,348	33,903	37,752
Costa Rica	0.010	0.010	0.012	0.012	0.013	0.012	11,554	12,518	14,000	13,794		
Croatia	±0.075	±0.065	±0.043	-	-	-	24,700	25,090	17,845	-	-	-
Cyprus	0.003	0.003	0.005	0.004	0.004	0.003	31,165	31,329	27,946	24,816	24,783	30,913
Czechoslovakia (Former)	±0.074	±0.027	-	-	-	-	7,354	3,350	-	-	-	-
Denmark	*0.080	*0.077	*0.083	*0.073	*0.090	*0.111	42,354	43,061	50,673	46,080	53,792	64,528
Ecuador	0.018	0.016	0.017	0.019	0.022	0.025	13,962	12,591	14,627	18,803	29,037	30,844
Egypt	*0.149	*0.126	*0.049	*0.044	*0.043	*0.051	17,762	10,863	7,836	7,870	8,221	8,825
El Salvador	-	±0.000	-	±0.000	±0.000	±0.000	-	-	-	9,890	5,475	3,648
Finland	±0.114	±0.096	±0.080	±0.063	±0.084	-	87,504	74,184	72,453	70,024	84,235	-
FYR Macedonia	±0.059	±0.073	±0.033	±0.022	±0.019	±0.011	15,700	21,528	10,604	9,640	7,982	5,579
Germany	-	2.307	2.816	2.272	2.280	-	-	45,247	59,488	58,459	67,480	-
Germany (Western Part)	2.139	2.101	2.584	-	-	-	57,931	58,095	68,494	-	-	-
Greece	*0.126	*0.105	*0.114	*0.110	*0.119	*0.144	40,907	38,393	44,985	44,120	45,892	53,554
Guatemala	-	0.012	0.012	0.011	0.015	0.020	-	44,298	43,082	40,328	54,665	71,813
Honduras	0.001	0.001	0.001	0.001	0.001	0.001	8,722	8,503	10,012	9,530	4,805	4,997
Hungary	±0.136	±0.140	±0.076	±0.089	±0.085	±0.091	11,339	11,708	7,609	14,764	13,772	15,176
Iceland	0.002	0.002	±0.002	±0.001	-	-	69,304	231,160	80,767	58,900	-	-
India	*0.195	*0.191	*0.163	*0.154	*0.173	*0.208	6,926	6,236	6,214	6,005	6,747	8,072
Indonesia	0.183	0.255	0.253	0.423	0.518	0.482	7,701	8,250	7,645	10,942	12,506	10,502
Iran (Islamic Republic of)	0.058	0.049	0.073	0.073	-	-	17,161	11,613	13,077	10,978	-	-
Iraq	-	*0.000	*0.000	-	-	-	-	12,862	50,375	-	-	-
Ireland	*0.048	*0.042	*0.047	*0.039	*0.046	*0.055	68,467	59,312	68,919	60,650	68,930	87,313
Israel	0.053	0.074	0.072	0.069	0.069	0.086	52,574	67,411	65,804	57,124	52,881	67,185
Italy	±6.347	±6.285	±1.585	±1.123	±1.143	-	121,638	122,397	88,648	71,336	77,766	-
Japan	±5.477	±5.983	±6.222	±6.951	±7.915	±9.271	79,375	87,989	94,271	110,340	129,755	149,527
Jordan	0.008	0.010	0.017	0.011	0.001	0.019	19,525	16,624	20,416	19,242	10,355	21,378
Kenya	*0.001	*0.001	*0.001	*0.001	*0.001	*0.001	6,264	6,322	5,173	3,121	3,130	3,282
Korea, Republic of	0.413	0.458	0.483	0.568	0.626	0.701	25,812	29,164	32,259	36,632	40,032	48,061
Latvia	0.000	0.001	0.005	0.005	0.007	0.005	56	182	1,885	2,416	4,124	3,698
Lithuania	-	-	0.002	0.002	0.002	-	-	-	1,480	1,894	1,755	-
Macau	0.006	0.003	0.001	0.001	-	-	10,384	10,414	9,414	8,901	-	-
Malaysia	*0.070	*0.079	*0.083	*0.100	*0.109	*0.131	9,831	10,144	10,808	12,015	13,426	14,781
Malta	0.001	0.001	0.001	0.001	0.002	-	19,526	18,426	20,928	17,895	24,926	-
Mauritius	0.000	0.001	0.001	0.001	0.000	0.001	6,211	19,395	28,817	24,493	7,723	15,929
Mexico	0.226	0.250	0.303	0.299	0.303	0.212	30,287	35,549	41,917	44,778	46,303	30,813
Mongolia	0.002	0.003	0.001	0.000	0.000	0.000	7,007	8,840	3,411	835	568	1,512
Mozambique	-	±0.000	-	±0.000	±0.000	-	-	49	-	-	-	-
Netherlands	0.587	0.592	0.657	0.637	0.725	0.847	92,892	95,082	109,472	110,963	120,861	141,088
Norway	±0.048	±0.045	±0.044	-	-	-	58,928	58,516	62,913	-	-	-
Pakistan	0.026	0.025	0.027	-	-	-	7,368	8,005	8,307	-	-	-
Panama	0.000	-	-	-	-	-	1,667	-	-	-	-	-
Peru	0.016	0.016	0.021	-	-	-	12,248	14,017	18,257	-	-	-
Philippines	0.049	0.058	0.091	0.087	-	-	9,133	8,996	7,703	9,250	-	-

Continued.

Depending on the table, * means *Enterprises, Engaged,* or *Factor Values;* ± means *Basis Unspecified;* § means *shown in millions.* For additional notes and sources, see Appendix I.

Output and Output per Employee
- Continued -

Country	Output ($ bil.)						Output per Employee ($)					
	1990	1991	1992	1993	1994	1995	1990	1991	1992	1993	1994	1995
Poland	±0.161	±0.184	±0.177	±0.186	±0.228	±0.284	6,724	8,753	10,391	10,963	12,054	14,488
Portugal	0.590	0.619	0.737	0.695	0.770	0.913	23,078	25,764	29,670	28,229	30,151	34,744
Romania	±1.707	±1.539	±0.951	±0.852	-	-	-	-	-	-	-	-
Russian Federation	-	0.371	0.246	0.341	0.420	0.585	-	-	-	3,477	4,666	16,798
Slovakia	-	±0.003	±0.006	±0.006	±0.007	-	-	3,867	8,563	9,271	11,500	-
South Africa	±0.071	±0.063	±0.056	±0.064	±0.065	±0.074	11,788	10,563	9,292	-	-	-
Spain	*0.781	*0.783	*0.761	±1.028	±1.075	±1.121	52,173	52,163	59,630	67,546	69,139	70,236
Sri Lanka	0.029	0.039	0.035	0.041	-	-	4,761	5,305	6,412	5,323	-	-
Sweden	*0.174	*0.151	*0.163	*0.119	*0.127	*0.150	75,659	81,440	95,571	77,526	83,490	93,984
Thailand	0.110	0.317	-	-	-	-	5,838	17,289	-	-	-	-
Trinidad and Tobago	-	±0.002	±0.001	±0.001	±0.001	-	-	18,487	19,662	15,613	15,211	-
Tunisia	0.159	0.177	0.218	0.208	0.222	0.257	-	-	-	44,659	44,219	50,032
Turkey	0.703	0.708	0.641	0.782	0.755	0.964	53,408	65,113	65,052	74,568	66,848	82,709
Ukraine	*0.174	*0.233	*0.200	*0.063	*0.065	*0.058	6,950	9,328	7,990	2,639	2,700	2,517
United Kingdom	*2.162	*2.074	2.065	1.688	1.975	2.152	49,148	51,858	55,821	45,613	53,392	56,839
United Republic of Tanzania . .	*0.000	*0.000	-	-	-	-	500	760	-	-	-	-
USSR (Former)	±0.500	-	-	-	-	-	2,404	-	-	-	-	-
United States of America . . .	*2.610	*2.560	*2.817	*2.933	*3.301	*3.532	68,684	69,189	76,135	79,270	82,525	80,273
Uruguay	0.033	0.043	0.048	0.044	0.046	0.048	16,565	20,214	22,481	21,203	21,939	23,680
Venezuela	0.033	0.044	0.058	0.075	0.057	-	14,221	14,609	18,781	21,372	23,520	-
Yugoslavia	±0.014	±0.014	±0.025	-	±0.045	±0.025	-	1,951	3,559	-	7,215	3,796
Zimbabwe	*0.004	*0.005	*0.004	*0.003	*0.002	*0.002	5,106	5,834	4,798	3,283	5,726	6,519

Value Added and Value Added per Employee

Country	Value Added ($ bil.)						Value Added per Employee ($)					
	1990	1991	1992	1993	1994	1995	1990	1991	1992	1993	1994	1995
Argentina	±0.156	-	-	0.090	-	-	17,646	-	-	28,495	-	-
Australia	*0.076	*0.079	*0.079	-	-	-	37,891	39,735	39,460	-	-	-
Austria	0.112	0.111	0.114	0.110	0.114	0.129	37,203	36,942	38,129	44,993	46,636	52,759
Bangladesh	0.010	0.010	0.011	-	-	-	2,781	2,626	3,305	-	-	-
Bolivia	0.000	0.000	0.000	-	0.000	0.000	1,711	1,576	3,905	-	2,010	1,428
Bosnia & Herzegovina	±0.011	±0.010	-	-	-	-	6,085	6,365	-	-	-	-
Bulgaria	-	±0.006	±0.010	±0.011	-	-	-	1,081	2,044	2,947	-	-
Cameroon	0.008	0.008	0.010	-	-	-	42,594	45,578	56,712	-	-	-
Canada	0.069	0.061	0.058	0.047	0.037	0.037	68,564	61,098	57,913	46,508	36,614	37,169
Chile	0.009	0.019	0.028	0.028	0.021	0.025	4,400	7,277	9,957	9,799	10,449	12,275
China	±0.504	±0.560	±0.759	±1.371	-	-	1,441	1,475	1,897	3,608	-	-
China (Hong Kong SAR) . . .	0.006	0.003	-	-	0.002	-	10,484	7,721	-	-	15,527	-
China (Taiwan Province) . . .	0.332	0.379	0.502	0.496	0.523	0.529	11,565	12,699	16,507	15,749	15,768	16,523
Colombia	0.060	0.067	0.080	0.107	0.139	0.154	11,270	11,889	13,591	16,941	21,759	24,260
Costa Rica	0.003	0.004	0.004	0.004	0.004	0.004	4,075	4,489	5,037	4,836	-	-
Croatia	±0.045	±0.032	±0.018	-	-	-	14,698	12,490	7,337	-	-	-
Cyprus	0.002	0.002	0.003	0.003	0.002	0.002	17,682	18,184	17,104	13,906	14,413	20,232
Czechoslovakia (Former) . . .	±0.046	-	-	-	-	-	4,568	-	-	-	-	-
Denmark	*0.071	*0.064	*0.068	*0.061	*0.074	*0.090	37,326	35,549	41,786	38,145	43,934	52,441
Ecuador	0.007	0.006	0.006	0.007	0.007	0.010	5,563	4,627	5,411	6,700	8,940	11,752
Egypt	*0.062	*0.027	*0.009	*0.013	*0.013	*0.015	7,393	2,323	1,449	2,298	2,405	2,547
El Salvador	-	0.000	-	0.000	0.000	0.000	-	-	-	7,516	4,221	4,028
Finland	±0.073	±0.056	±0.045	±0.038	±0.052	-	56,526	43,179	41,199	42,015	51,881	-
FYR Macedonia	±0.033	±0.037	±0.017	±0.012	±0.009	±0.006	8,725	10,918	5,404	5,452	4,036	3,188
Germany (Western Part) . . .	1.555	1.621	1.645	1.472	-	-	42,108	44,820	43,616	-	-	-
Greece	*0.073	*0.063	*0.068	*0.066	*0.072	*0.086	23,479	23,025	26,906	26,400	27,469	32,065
Honduras	*0.001	*0.000	*0.001	*0.000	*0.000	*0.000	3,689	3,624	3,871	3,134	2,321	1,958
Hungary	±0.055	±0.044	±0.037	±0.049	±0.047	±0.050	4,568	3,667	3,707	8,231	7,639	8,318
Iceland	0.001	0.000	±-0.000	±0.001	-	-	34,682	54,029	§-1,51	36,846	-	-
India	*0.053	*0.043	*0.037	*0.042	*0.046	*0.055	1,887	1,395	1,405	1,626	1,807	2,133
Indonesia	*0.077	*0.120	*0.114	*0.204	*0.274	*0.197	3,248	3,874	3,442	5,282	6,603	4,287

Continued.

Depending on the table, * means *Enterprises*, *Engaged*, or *Factor Values*; ± means *Basis Unspecified*; § means *shown in millions*. For additional notes and sources, see Appendix I.

437

Value Added and Value Added per Employee

- Continued -

Country	Value Added ($ bil.)						Value Added per Employee ($)					
	1990	1991	1992	1993	1994	1995	1990	1991	1992	1993	1994	1995
Iran (Islamic Republic of)	0.043	0.031	0.043	0.048	-	-	12,759	7,432	7,628	7,262	-	-
Iraq	-	*0.000	*0.000	-	-	-	-	8,842	31,083	-	-	-
Ireland	*0.028	*0.025	*0.029	*0.024	*0.028	*0.033	40,038	36,234	41,994	36,991	41,867	52,848
Israel	0.030	0.047	0.036	0.035	0.028	0.040	29,759	42,680	32,532	29,151	21,459	31,578
Italy	±2.860	±2.861	±0.872	±0.623	±0.634		54,804	55,712	48,751	39,584	43,127	
Japan	±2.984	±3.370	±3.608	±4.020	±4.510	±5.220	43,241	49,562	54,672	63,806	73,940	84,195
Jordan	0.003	0.005	0.009	0.006	0.000	0.010	8,247	8,584	11,275	10,914	3,951	11,579
Kenya	*0.001	*0.001	*0.001	*0.000	*0.001	*0.001	4,176	3,161	2,587	1,560	3,130	3,282
Korea, Republic of	0.275	0.300	0.304	0.365	0.407	0.458	17,158	19,106	20,293	23,552	26,025	31,380
Latvia	-	-	-	*0.003	*0.005	*0.003	-	-	-	1,322	3,018	2,189
Macau	0.003	0.002	0.001	0.001	-	-	5,250	6,274	7,239	7,890	-	-
Malaysia	*0.036	*0.040	*0.045	*0.054	*0.060	*0.073	5,082	5,156	5,863	6,454	7,466	8,216
Malta	0.001	0.001	0.001	0.001	0.001		12,292	12,684	12,687	11,630	19,015	
Mauritius	0.000	0.000	0.000	0.000	0.000	0.000	2,070	7,758	12,072	6,763	3,233	5,769
Mexico	0.146	0.158	0.191	0.190	0.191	0.133	19,508	22,507	26,417	28,416	29,175	19,408
Mongolia	0.001	0.002	0.001	0.000	0.000	0.000	3,278	6,413	2,599	226	231	501
Netherlands	0.304	0.309	0.348	0.345	0.390	0.455	48,099	49,645	57,956	59,980	65,025	75,817
Norway	±0.027	±0.025	±0.026	-	-	-	33,086	31,846	36,754	-	-	-
Pakistan	0.013	0.012	0.014	-	-	-	3,774	3,892	4,154	-	-	-
Panama	0.000	-	-	-	-	-	1,370	-	-	-	-	-
Peru	0.010	0.010	0.010	-	-	-	7,712	8,108	9,117	-	-	-
Philippines	0.029	0.032	0.050	0.048	-	-	5,294	5,038	4,225	5,129	-	-
Poland	±0.107	±0.101	±0.098	±0.099	±0.120	±0.148	4,478	4,786	5,772	5,803	6,316	7,555
Portugal	0.291	0.302	0.357	0.336	0.383	0.454	11,391	12,580	14,366	13,662	14,988	17,270
Romania	±0.571	±0.499	±0.368	±0.337	-	-	-	-	-	-	-	-
Russian Federation	-	-	-	±0.219	±0.245	±0.276	-	-	-	2,232	2,718	7,936
Saint Lucia	-	±0.000	±0.000	±0.000	±0.000	±0.000	-	-	-	-	-	-
Slovakia	-	-	-	±0.003	±0.002	-	-	-	-	4,044	4,211	-
South Africa	*0.042	*0.032	*0.046	*0.040	*0.041	*0.046	6,957	5,372	7,597	-	-	-
Spain	*0.432	*0.431	*0.437	±0.450	±0.478	±0.557	28,863	28,722	34,290	29,610	30,723	34,911
Sri Lanka	0.017	0.017	0.017	0.020	-	-	2,724	2,336	3,042	2,640	-	-
Sweden	*0.123	*0.070	*0.084	*0.067	*0.072	*0.085	53,622	37,526	49,223	43,529	47,603	53,176
Thailand	0.059	0.192	-	-	-	-	3,154	10,483	-	-	-	-
Trinidad and Tobago	-	±0.000	±0.000	±0.001	±0.000	-	-	12,045	6,769	7,806	5,834	-
Tunisia	0.067	0.077	0.093	0.088	0.096	0.109	-	-	-	19,039	19,019	21,288
Turkey	0.467	0.470	0.441	0.517	0.519	0.663	35,421	43,262	44,756	49,336	45,970	56,906
United Kingdom	*1.464	*1.427	*1.406	*1.111	*1.311	*1.428	33,279	35,664	38,010	30,030	35,429	37,721
United Republic of Tanzania	*0.000	*0.000	-	-	-	-	250	435	-	-	-	-
United States of America	*1.840	*1.760	*1.995	*2.073	*2.310	*2.506	48,421	47,568	53,919	56,027	57,750	56,955
Uruguay	0.020	0.028	0.032	0.025	0.026	0.027	10,086	13,030	14,860	12,246	12,668	13,670
Venezuela	0.018	0.027	0.037	0.039	0.031	0.166	7,824	9,047	11,922	11,196	12,776	30,707
Yugoslavia	±0.008	±0.007	±0.016	-	±0.027	±0.015	-	1,066	2,289	-	4,218	2,277
Zambia	-	-	-	-	-0.000	-	-	-	-	-	-	-
Zimbabwe	*0.003	*0.003	*0.003	*0.001	*0.001	*0.001	3,217	4,084	3,403	1,661	3,354	3,825

Capital Investment

Country	Gross Fixed Capital Formation ($ mil.)						Per Establishment ($ mil)					
	1990	1991	1992	1993	1994	1995	1990	1991	1992	1993	1994	1995
Argentina	-	-	-	3.430	-	-	-	-	-	0.016	-	-
Australia	3.125	-	-	-	-	-	0.028	-	-	-	-	-
Austria	24	9.592	9.282	7.651	13	-	0.759	0.331	0.320	0.264	0.502	-
Bahrain	-	-	0.011	-	-	-	-	-	0.005	-	-	-
Bangladesh	0.726	0.847	7.548	-	-	-	0.002	0.002	0.022	-	-	-
Bolivia	0.004	-	-	-	-	-	0.004	-	-	-	-	-
Bosnia & Herzegovina	0.442	-	-	-	-	-	0.111	-	-	-	-	-
Bulgaria	9.817	4.025	33	2.374	3.859	4.169	1.227	0.503	3.317	0.040	0.040	0.038
Canada	10	-	-	-	-	-	0.214	-	-	-	-	-

Continued.

Depending on the table, * means *Enterprises, Engaged,* or *Factor Values;* ± means *Basis Unspecified;* § means *shown in millions.* For additional notes and sources, see Appendix I.

Capital Investment

- Continued -

Country	Gross Fixed Capital Formation ($ mil.)						Per Establishment ($ mil)					
	1990	1991	1992	1993	1994	1995	1990	1991	1992	1993	1994	1995
Cape Verde	-	0.001	-	-	-	-	-	0.000	-	-	-	-
Chile	0.971	0.633	1.023	1.425	1.316	-	0.121	0.090	0.128	0.178	0.188	-
China (Hong Kong SAR)	0.128	-0.386	-	-	0.129	-	0.001	-0.008	-	-	0.006	-
Colombia	6.915	5.219	4.909	1.466	6.272	-	0.266	0.217	0.258	0.086	0.418	-
Croatia	4.497	13	6.986	0.510	0.725	0.215	0.346	0.953	0.466	0.018	0.019	0.005
Cyprus	0.282	0.028	4.762	0.352	0.287	0.150	0.004	0.000	0.072	0.005	0.005	0.003
Czechoslovakia (Former)	6.128		-	-	-	-	0.766	-	-	-	-	-
Denmark	0.162	0.782	-	-	-	-	0.000	0.002	-	-	-	-
Ecuador	2.351	1.322	4.202	4.402	5.634	1.807	0.196	0.094	0.323	0.339	0.512	0.164
Egypt	5.800	3.481	5.901	2.877	-	-	0.252	0.166	0.369	0.192	-	-
El Salvador	-	-		0.090	0.067	0.077	-	-		0.090	0.022	0.015
Finland	7.374	5.959	33	2.871	6.030	-	0.567	0.458	3.284	0.319	0.670	-
FYR Macedonia	0.467	0.025	-	-	-	-	0.093	0.002	-	-	-	-
Germany (Western Part)	123	128	121	137	-	-	0.743	0.802	0.747		-	-
Greece	5.937	15	3.169	-	-	-	0.086	0.232	0.048		-	-
Hungary	8.243	4.871	4.140	9.529	-	-	0.196	0.079	0.073	0.127	-	-
India	16	7.299	21	18	-	-	0.029	0.013	0.036	0.030	-	-
Indonesia	49	77	83	69	134	165	0.760	1.019	1.035	0.797	1.413	1.742
Iran (Islamic Republic of)	3.077	4.858	9.582	12	-	-	0.128	0.324	0.479	0.500	-	-
Ireland	1.493	1.454		-	-	-	0.093	0.086		-	-	-
Israel	5.456	1.755	5.693	4.240	5.314	-	0.227	0.057	0.196	0.146	0.152	-
Italy	446	423	81	61	60	-	0.874	0.836	0.382	0.301	0.304	-
Japan	345	431	387	270	499	-	0.092	0.114	0.106	0.077	0.149	-
Jordan	0.577	1.244	0.703	-	-	-	0.144	0.311	0.141	-	-	-
Korea, Republic of	55	67	61	106	73	81	0.099	0.104	0.097	0.160	0.109	0.122
Latvia	0.003	0.003	0.023	0.024	0.033	0.054	0.000	0.000	0.003	0.003	0.003	0.005
Macau	0.231	-0.311	-	-	-	-	0.018	-0.039	-	-	-	-
Malaysia	16	8.000	9.029	10	14	-	0.430	0.190	0.205	0.219	0.286	-
Malta	0.066	0.025	0.009	0.003	0.011	-	0.009	0.004	0.001	0.000	0.002	-
Mexico	21	12	-	-	-	-	4.282	2.463	-	-	-	-
Mongolia	3.482	0.756	0.204	0.070	0.058	-	3.482	0.756	0.204	0.070	0.019	-
Netherlands	85	60	44	45	-	-	1.057	0.740	0.584	0.629	-	-
Niger	-	-	-	-	-0.004	-	-	-	-	-	-	-
Norway	1.757	1.851	1.197	-	-	-	0.251	0.264	0.239	-	-	-
Pakistan	-	37	-	-	-	-	-	1.850	-	-	-	-
Peru	0.134	0.042	0.181	-	-	-	0.002	0.001	0.003	-	-	-
Philippines	5.306	1.674	3.645	5.229		-	0.007	0.002	0.050	0.086		-
Poland	12	18	15	16	-	-	0.344	0.470	0.367	0.486	-	-
Portugal	117	61	65	67	74	-	0.122	0.065	0.067	0.070	0.076	-
Romania	-	-	-	6.407	4.891	-	-	-	-		0.065	-
Slovenia	17	17	13	-	-	-	1.914	0.285	0.474	-	-	-
Spain	50	53	45	44	39	60	0.051	0.053	0.045	-	-	-
Sri Lanka	5.017	1.310	2.655	7.579	-	-	0.136	0.036	0.095	0.271	-	-
Thailand	190	99	-	-	-	-	1.239	0.500	-	-	-	-
Trinidad and Tobago	-	-	0.047	-	0.422	-	-	-	0.005	-	0.042	-
Tunisia	26	28	42	30	51	48	-	-		0.434	0.695	0.547
Turkey	98	94	69	65	82	-	3.371	3.778	1.509	1.405	1.492	-
Ukraine	9.443	4.701	3.977	0.295	0.572	0.547	0.305	0.072	0.060	0.008	0.018	0.020
United Kingdom	96	65	92	-	75	-	0.122	0.086	0.126	-	0.099	-
United Republic of Tanzania	0.036	0.032	-	-	-	-	0.012	0.032	-	-	-	-
United States of America	110	140	85	111	120	149	-	-	0.051	-	-	-
Uruguay	1.902	4.344	2.559	1.108			-	-	-	-		
Venezuela	14	5.632	3.203	12	0.189	83	0.466	0.148	0.110	0.424	0.010	1.885
Yugoslavia	0.563	1.442	0.100	-	0.759	0.355	0.038	0.066	0.002		0.012	0.005

Depending on the table, * means *Enterprises, Engaged,* or *Factor Values;* ± means *Basis Unspecified;* § means *shown in millions.* For additional notes and sources, see Appendix I.

439

Indexes of Industrial Production

Country	Index of Industrial Production (1990=100)						Index of Employment (1990=100)					
	1990	1991	1992	1993	1994	1995	1990	1991	1992	1993	1994	1995
Albania	100	-	-	-	-	-	-	-	-	-	-	-
Algeria	100	100	104	98	87	91	-	-	-	-	-	-
Argentina	100	121	-	-	-	-	100	-	-	36	34	-
Armenia	100	-	-	-	-	-	100	96	-	-	-	-
Australia	100	95	104	118	-	-	100	100	100	-	-	-
Austria	100	100	101	102	110	107	100	100	100	81	82	81
Azerbaijan	-	-	-	-	-	-	100	98	90	85	92	-
Bahrain	100	-	-	-	-	-	-	-	-	-	-	-
Bangladesh	100	-	-	-	-	-	100	100	91	-	-	-
Barbados	100	100	100	100	100	100	-	-	-	-	-	-
Belgium	100	94	97	97	103	103	100	103	103	-	-	-
Belize	100	100	100	100	100	100	-	-	-	-	-	-
Bermuda	-	-	-	-	-	-	100	125	100	150	125	88
Bhutan	100	-	-	-	-	-	-	-	-	-	-	-
Bolivia	100	105	111	118	131	146	100	100	46	-	25	25
Bosnia & Herzegovina	-	-	-	-	-	-	100	85	-	-	-	-
Botswana	100	100	100	-	-	-	-	-	-	-	-	-
Brazil	100	101	93	98	100	105	-	-	-	-	-	-
Bulgaria	100	81	67	66	80	84	100	90	83	66	69	62
Burkina-Faso	100	-	-	-	-	-	-	-	-	-	-	-
Burundi	100	100	100	100	-	-	-	-	-	-	-	-
Cambodia	100	-	-	-	-	-	100	81	-	-	-	-
Cameroon	100	-	-	-	-	-	100	97	99	92	-	-
Canada	100	82	79	80	82	81	100	100	100	100	100	99
Central African Republic	100	100	100	100	100	-	-	-	-	-	-	-
Chad	100	100	100	-	-	-	-	-	-	-	-	-
Chile	100	95	93	99	103	108	100	125	133	137	98	99
China	100	-	-	-	-	-	100	109	114	109	-	-
China (Hong Kong SAR)	100	94	90	81	74	70	100	67	33	17	17	-
China (Taiwan Province)	100	108	113	122	130	129	100	104	106	110	115	111
Colombia	100	112	124	141	157	156	100	106	111	120	120	120
Costa Rica	100	104	114	123	126	119	100	95	102	109	-	-
Cote d'Ivoire	100	15	25	53	21	26	-	-	-	-	-	-
Croatia	100	-	-	-	-	-	100	85	79	77	71	66
Cuba	100	70	-	-	-	-	-	-	-	-	-	-
Cyprus	100	72	70	49	99	124	100	94	167	182	157	95
Czechoslovakia (Former)	100	67	70	-	-	-	100	80	-	-	-	-
Czech Republic	100	-	-	-	-	-	100	89	111	122	-	-
Dem. Rep. of the Congo	100	89	68	-	-	-	-	-	-	-	-	-
Denmark	100	96	94	88	100	106	100	94	86	84	89	91
Dominican Republic	100	-	-	-	-	-	-	-	-	-	-	-
Ecuador	100	109	116	100	-	-	100	102	92	82	61	65
Egypt	100	93	93	95	89	98	100	138	74	67	62	69
El Salvador	100	104	-	-	-	-	-	-	-	-	-	-
Ethiopia and Eritrea	100	100	100	-	-	-	-	-	-	-	-	-
Fiji	100	100	100	100	100	100	-	-	-	-	-	-
Finland	100	89	86	77	92	91	100	100	85	69	77	-
FYR Macedonia	100	-	-	-	-	-	100	90	83	60	62	52
France	100	95	86	79	85	86	-	-	-	-	-	-
Gabon	100	100	100	-	-	-	-	-	-	-	-	-
Gambia	100	-	-	-	-	-	-	-	-	-	-	-
Germany (Eastern Part)	100	-	-	-	-	-	-	-	-	-	-	-
Germany (Western Part)	100	97	87	78	73	77	100	98	102	-	-	-
Ghana	100	-	-	-	-	-	-	-	-	-	-	-
Greece	100	68	69	67	72	76	100	89	82	81	84	87
Guatemala	100	102	-	-	-	-	-	-	-	-	-	-
Guyana	100	100	100	100	100	100	-	-	-	-	-	-
Haiti	100	100	100	100	100	100	-	-	-	-	-	-
Honduras	100	121	126	176	165	227	100	92	98	108	118	150
Hungary	100	92	81	76	72	76	100	100	83	50	51	50

Continued.

　Depending on the table, * means *Enterprises, Engaged,* or *Factor Values;* ± means *Basis Unspecified;* § means *shown in millions.* For additional notes and sources, see Appendix I.

Indexes of Industrial Production

- Continued -

Country	Index of Industrial Production (1990=100)						Index of Employment (1990=100)					
	1990	1991	1992	1993	1994	1995	1990	1991	1992	1993	1994	1995
Iceland	100	-	-	-	-	-	100	29	82	79	29	-
India	100	100	100	101	106	122	100	108	93	91	91	91
Indonesia	100	-	-	-	-	-	100	130	139	163	174	193
Iran (Islamic Republic of)	100	111	118	115	109	121	100	124	165	196	-	-
Iraq	100	39	19	-	-	-	-	-	-	-	-	-
Ireland	100	94	98	94	104	114	100	100	98	93	94	90
Israel	100	130	151	145	155	193	100	110	110	120	130	127
Italy	100	99	96	90	-	-	100	98	34	30	28	-
Jamaica	100	89	107	102	100	-	-	-	-	-	-	-
Japan	100	100	94	91	93	93	100	99	96	91	88	90
Jordan	100	-	-	-	-	-	100	146	203	147	16	226
Kenya	100	77	170	355	524	621	100	110	115	106	109	113
Korea, Republic of	100	116	124	125	134	142	100	98	94	97	98	91
Kuwait	100	-	-	-	-	-	-	-	-	-	-	-
Kyrgyzstan	100	-	-	-	-	-	-	-	-	-	-	-
Latvia	100	-	-	-	-	-	100	96	92	66	53	44
Lebanon	100	-	-	-	-	-	-	-	-	-	-	-
Libyan Arab Jamahiriya	100	-	-	-	-	-	-	-	-	-	-	-
Lithuania	100	-	-	-	-	-	-	-	-	-	-	-
Macau	100	-	-	-	-	-	100	49	22	13	-	-
Madagascar	100	100	100	100	-	-	-	-	-	-	-	-
Malawi	100	100	100	100	100	100	-	-	-	-	-	-
Malaysia	100	120	132	140	158	176	100	110	108	117	114	125
Malta	100	132	147	195	-	-	100	107	100	105	110	-
Mauritius	100	100	-	-	-	-	100	43	51	48	48	75
Mexico	100	105	113	114	120	101	100	94	97	89	88	92
Mongolia	-	-	-	-	-	-	100	96	95	79	94	35
Morocco	100	88	114	103	110	108	-	-	-	-	-	-
Mozambique	100	-	-	-	-	-	100	106	-	-	-	-
Myanmar	100	-	-	-	-	-	-	-	-	-	-	-
Nepal	100	100	100	-	-	-	-	-	-	-	-	-
Netherlands	100	96	90	89	99	100	100	99	95	91	95	95
New Zealand	100	85	87	-	-	-	100	119	132	145	-	-
Nicaragua	100	-	-	-	-	-	-	-	-	-	-	-
Nigeria	100	115	114	110	104	-	-	-	-	-	-	-
Norway	100	96	97	101	114	116	100	95	87	90	90	91
Pakistan	100	108	110	113	107	120	100	90	95	-	-	-
Panama	100	160	131	221	229	198	100	-	-	-	-	-
Papua New Guinea	100	100	100	-	-	-	-	-	-	-	-	-
Paraguay	100	102	123	128	120	137	-	-	-	-	-	-
Peru	100	125	142	169	244	315	100	89	87	-	-	-
Philippines	100	132	133	153	185	224	100	119	219	174	-	-
Poland	100	91	102	111	128	134	100	88	71	71	79	82
Portugal	100	99	102	105	106	110	100	94	97	96	100	103
Puerto Rico	100	94	95	96	104	-	100	92	90	83	83	81
Republic of Moldova	100	-	-	-	-	-	-	-	-	-	-	-
Romania	100	81	73	-	-	-	-	-	-	-	-	-
Russian Federation	100	-	-	-	-	-	-	-	-	-	-	-
Saudi Arabia	100	-	-	-	-	-	-	-	-	-	-	-
Senegal	100	100	100	100	100	100	-	-	-	-	-	-
Slovakia	100	-	-	-	-	-	-	-	-	-	-	-
Slovenia	100	106	94	93	92	94	-	-	-	-	-	-
South Africa	100	86	72	72	78	85	100	100	100	-	-	-
Spain	100	99	92	87	96	102	100	100	85	102	104	107
Sri Lanka	100	72	106	105	91	88	100	121	91	127	-	-
Sudan	100	100	100	-	-	-	-	-	-	-	-	-
Swaziland	100	100	100	-	-	-	-	-	-	-	-	-
Sweden	100	90	78	71	72	77	100	81	74	67	66	70
Switzerland	100	87	83	77	85	81	-	-	-	-	-	-
Syrian Arab Republic	100	109	111	-	-	-	-	-	-	-	-	-

Continued.

Depending on the table, * means *Enterprises, Engaged,* or *Factor Values;* ± means *Basis Unspecified;* § means *shown in millions.* For additional notes and sources, see Appendix I.

441

Indexes of Industrial Production
- Continued -

Country	Index of Industrial Production (1990=100)						Index of Employment (1990=100)					
	1990	1991	1992	1993	1994	1995	1990	1991	1992	1993	1994	1995
Thailand	100	106	120	146	166	190	100	97	-	-	-	-
Trinidad and Tobago	100	114	119	116	124	140	-	-	-	-	-	-
Tunisia	100	-	-	-	-	-	-	-	-	-	-	-
Turkey	100	110	128	134	139	154	100	83	75	80	86	88
Uganda	100	105	132	169	161	238	-	-	-	-	-	-
Ukraine	100	-	-	-	-	-	100	100	100	96	96	92
United Arab Emirates	100	-	-	-	-	-	-	-	-	-	-	-
United Kingdom	100	94	89	87	88	90	100	91	84	84	84	86
United Republic of Tanzania	100	100	100	100	100	-	100	51	-	-	-	-
USSR (Former)	100	-	-	-	-	-	100	-	-	-	-	-
United States of America	100	91	93	95	100	102	100	97	97	97	105	116
Uruguay	100	96	139	143	144	139	100	104	105	102	103	99
Venezuela	100	-	-	-	-	-	100	130	135	152	106	235
Yugoslavia	100	-	-	-	-	-	-	-	-	-	-	-
Yugoslavia (Former)	100	78	-	-	-	-	-	-	-	-	-	-
Zambia	100	97	96	87	60	63	-	-	-	-	-	-
Zimbabwe	100	106	98	81	105	97	100	100	113	100	38	35

Depending on the table, * means *Enterprises, Engaged,* or *Factor Values;* ± means *Basis Unspecified;* § means *shown in millions.* For additional notes and sources, see Appendix I.

Representative Companies in Sector

Name	Address	Tele-phone	Fax	Employ-ment	Y
Grupo Industrial Saltillo SA de CV	25280 Saltillo, Coahuila, Mexico	84 155620	84 153313	9,575	92
Siemens Energy and Automation Inc.	3333 Old Milton Pkwy., Alpharetta, Georgia 30202, U.S.A.	770-751-2000	770-751-2001	9,500	97
Brown-Forman Corp.	850 Dixie Hwy., Louisville, Kentucky 40210, U.S.A.	502-585-1100		7,500	97
Eagle Industries Inc.	2 N. Riverside Plz., 1100, Chicago, Illinois 60606, U.S.A.	312-906-8700	312-906-8372	7,000	94
NKT Holding AS	NKT Alle 1, Brondby DK-2605, Denmark	43482000	43961820	6,425	93
Royal Doulton (UK) Ltd.	London Rd., Stoke-on-Trent ST4 7QD, United Kingdom	782 744766	782 416962	6,388	93
ACX Technologies Inc.	16000 Table Mountain Pkwy., Golden, Colorado 80403, U.S.A.	303-271-7000	303-271-7003	5,600	97
Longaberger Co.	95 N. Chestnut St., Dresden, Ohio 43821-9600, U.S.A.	614-754-5000	614-754-5240	5,200	96
Oneida Ltd.	5-9 Pembroke Rd., Ruislip HA4 8NQ, United Kingdom	895 639452		4,982	91
Great American Management Inc.	2 N. Riverside Plz., Chicago, Illinois 60606, U.S.A.	312-648-5656	312-454-9946	4,600	95
Oneida Ltd.	Kenwood Ave., Oneida, New York 13421-2899, U.S.A.	315-361-3000	315-829-3950	4,525	96
Lenox Inc.	100 Lenox Dr., Lawrenceville, New Jersey 08648-2394, U.S.A.	609-896-2800	609-896-3704	4,200	97
Libbey Inc.	PO Box 10060, Toledo, Ohio 43699-0060, U.S.A.	419-727-2100		4,150	97
Ibstock PLC	Lutterworth House, Lutterworth LE17 4PS, United Kingdom	455 553071		3,681	94
Armitage Shanks Ltd.	Showell Rd., Wolverhampton WS15 4BT, United Kingdom	543 490250		3,134	94
Mikasa Inc.	PO Box 6239, Carson, California 90749-6239, U.S.A.	310-886-3700	310-635-1546	2,750	95
Coors Ceramics Co.	600 9th St., Golden, Colorado 80401, U.S.A.	303-278-4000	303-277-4901	2,100	95
Sango Ceramic Indonesia PT	Semarang Plaza, Semarang 50138, Indonesia	24518391	24 289335	2,000	97
Herend Porcelain Factory	Kossuth u. 140, Herend H-8440, Hungary	80 61144	80 61518	1,550	93
Caradon Bathrooms Ltd.	24 Queens Rd., Waybridge KT13 9UX, United Kingdom	932 823327	932 823328	1,421	93
Zortech International Ltd.	Hadzor, Droitwich WR9 7DJ, United Kingdom	905 794211	905 795193	1,358	94
Staffordshire Tableware, Ltd.	Meir Pk., Stoke-on-Trent ST3 7AA, United Kingdom	782 315251	782 311433	1,346	94
Zsolnay Porcelangyar	Zsolnay u. 69, Pecs H-7630, Hungary	72 25266	72 13645	1,300	93
Villeroy & Boch SARL	Rue de Rollingergrund 330, Luxembourg L-2441, Luxembourg	468211	462580	1,283	93
Briggs Industries Inc.	4350 W. Cypress St., Tampa, Florida 33607, U.S.A.	813-878-0178	813-874-1394	1,200	95
Ideal-Standard Ltd.	National Ave., Hull HU5 4JE, United Kingdom	482 46461	482 445886	998	92
Industria de Ceramica Costarricense SA	Apartado 4120-1000, San Jose, Costa Rica	2325266		906	
Gerber Plumbing Fixtures Corp.	4656 W. Touhy Ave., Chicago, Illinois 60646, U.S.A.	312-675-6570		900	
Kokomo Sanitary Pottery Div.	PO Box 829, Kokomo, Indiana 46903-0829, U.S.A.	317-459-5113		900	93
WorldCrisa Corp.	PO Box 5020, Wallingford, Connecticut 06492, U.S.A.	203-265-8000	203-265-8024	850	94
Lapp Insulator Co.	130 Gilbert St., Le Roy, New York 14482, U.S.A.	716-768-6221	716-768-6219	775	93
John Tams Group PLC	Longton, Stoke-on-Trent ST3 2PG, United Kingdom	782 599226	782 599149	767	92
Continental China Pty. Ltd.	Range Rd., Blackheath 7581, Republic of South Africa	21 9051120	21 9051424	740	94
Hackman Tabletop Oy Ab	Hameentie 135, Helsinki SF-00560, Finland	39391	791329	600	94
Intermagnetics General Corp.	PO Box 461, Latham, New York 12110-0461, U.S.A.	518-782-1122	518-783-2601	537	97
Alfold Porcelan Edenygyar Rt	Erzsebeti Ut 7, Hodmezovasarhely H-6800, Hungary	62 345689	62 345005	530	93
Syracuse China Co.	PO Box 4820, Syracuse, New York 13221, U.S.A.	315-455-5671	315-455-4575	500	94
Teka Industrial, SA	Cajo, 17, Santander E-39011, Spain	942 335100	942 347694	458	97
Treasure Craft Co.	2320 N. Alameda St., Compton, California 90222, U.S.A.	213-636-9777		450	93
LK-Products Oy	Takatie 6, Kempele SF-90440, Finland	81 516122	81 515667	420	94
Mansfield Plumbing Products Inc. (Kilgore, Texas)	PO Box 472, Kilgore, Texas 75662, U.S.A.	903-984-3525		400	94
Hall China Co.	PO Box 989, East Liverpool, Ohio 43920, U.S.A.	216-385-2900		380	94
Porcelanas Lladro, SA	(Poligono Lladro) Tavernes Blanques, Valencia E-46016, Spain	96 1850177	96 3523082	370	97
Duncan Enterprises	PO Box 7827, Fresno, California 93727, U.S.A.	209-291-4444	209-291-9444	300	95
Ceradyne Inc.	3169 Redhill Ave., Costa Mesa, California 92626, U.S.A.	714-549-0421	714-549-5787	280	97
Carron Phoenix Ltd.	Carron, Falkirk FK2 8DW, United Kingdom	324 38321	324 20978	276	93
Sam Hyun Co., Ltd.	Chung-gu, Seoul, Republic of Korea	2 2775271	2 2775270	270	97
MAAX Inc.	600 Rte Cameron, Sainte-Marie, Quebec G6E 1B2, Canada	418-387-4155	418-387-3507	250	98
Negev Ceramics Ltd.	154 Abba Hillel St., Ramat Gan 52572, Israel	3 6103333	3 7529343	250	97
Armitage Shanks South Africa Pty. Ltd.	Private Bag X8, Olifantsfontein 1665, Republic of South Africa	11 3162571	11 3164236	245	93
Department 56 Inc.	6436 City West Pkwy., One Villag, Eden Prairie, Minnesota 55344, U.S.A.	612-944-5600	612-943-4500	203	96
Daniel Montgomery & Son Ltd.	Kirkintilloch, Glasgow G66 1ST, United Kingdom	41 7765276	41 7776251	156	93

Continued

Representative Companies in Sector
- Continued -

Name	Address	Tele-phone	Fax	Employ-ment	Y
Precidio Inc.	35 Precidio Crt, Brampton, Ontario L6S 6B7, Canada	905-790-0790	905-790-0791	150	98
Craftco Industries Inc.	410 Wentworth St. N, Hamilton, Ontario L8L 5W3, Canada	905-572-7333	905-572-1164	119	98
Isofoam Insulating Materials Plants	PO Box 23053 Safat, Kuwait City 13091, Kuwait	4763820	4763819	100	
Royal China and Porcelain Companies Inc.	PO Box 1012, Moorestown, New Jersey 08057-0912, U.S.A.	609-866-2900		100	93
V. Schierholzesche Porzellanmanufactur Plaue GmbH	Am Spring 1, Plaue 99338, Germany	36207	244	81	93
Trek Plastics CC	PO Box 2395, Port Elizabeth 6056, Republic of South Africa	41 414073	41 433593	65	95
Elektro Isola AS	Groenlandsvej 197, Vejle DK-7100, Denmark	75827588	75827336	51	93
Hwa Shin Instrument Co., Ltd.	Kangnam-Ku, Seoul 135-080, Republic of Korea	2 5547011	2 5555729	50	97
Melba-Wain England Ltd.	Longton, Stoke-on-Trent ST3 2JY, United Kingdom	782 319501	782 324708	48	
Permafreeze	Tulisa Park, Johannesburg 2136, Republic of South Africa	11 9073444	11 9073039	47	95
CV Samson	Jalan Mojoklanggru Kidul, E-56, Surabaya, Indonesia	519284	5942872	25	94
Taiwan Fairmost International Corp.	No. 247 Pateh Rd., Sec. 3, Taipei, Taiwan	2 7724218	2 7524876	25	93
Yi Wai Industrial Co., Ltd.	No. 44, 17th St., Chung Teo, Tainan, Taiwan	6 2677751	6 2697632	25	93
Cerart International Co., Ltd.	Yongdeungpo-gu, Seoul 150-010, Republic of Korea	2 7849990	2 7841917	20	97
Herbst GmbH	Warmekeramik, Durrohrsdorf 01833, Germany	275		20	91
Londero Mosaik ApS	Teglholmsgade 30 ST, Copenhagen DK-2450, Denmark	31223033	31225733	17	93
Golden Building Material Co., Ltd.	PO Box 67-526, Taipei, Taiwan	2 5455373	2 5455370	15	93
Sherston Earl Vineyards	Near Chippenham, Hullavington SN16 0PY, United Kingdom	666 837979	666 837110	12	91
Kingsley Enamels Ltd.	Stoke Prior, Bromsgrove B60 4EP, United Kingdom	527 32292	527 579049	10	92
Santo Arden Co., Ltd.	11 Fl., No. 311, Nan King E. Rd., Taipei, Taiwan	2 5455366	2 5455758	6	94
Ross Catherall Group PLC	Killamarsh, Sheffield S31 8BA, United Kingdom	742 486404	742 475999	4	93

Product Tables

Household Ware of Porcelain or China

Unit of Measure: Thousands of units.

	1990 Value	%	1991 Value	%	1992 Value	%	1993 Value	%	1994 Value	%	1995 Value	%	1996 Value	%
Total Production	2,927,293	100.0	2,913,413	100.0	2,781,756	100.0	2,646,050	100.0	2,438,125	100.0	2,196,511	100.0	2,248,020	100.0
Regions														
Africa	1,089	0.0	961	0.0	589	0.0	459	0.0	444	0.0	447	0.0	445	0.0
America, North	9,369	0.3	10,771	0.4	12,172	0.4	13,573	0.5	14,974	0.6	16,375	0.7	17,777	0.8
America, South	17,624	0.6	17,624	0.6	17,624	0.6	17,624	0.7	17,624	0.7	17,624	0.8	17,624	0.8
Asia	2,358,303	80.6	2,384,283	81.8	2,249,208	80.9	2,122,190	80.2	1,929,596	79.1	1,720,505	78.3	1,815,319	80.8
Europe	540,908	18.5	499,775	17.2	502,164	18.1	492,204	18.6	475,487	19.5	441,559	20.1	396,856	17.7
Leading Producers														
Russian Federation	551,747	18.8	531,779	18.3	551,898	19.8	541,774	20.5	340,600	14.0	278,961	12.7	200,224	8.9
Ukraine	293,251	10.0	273,229	9.4	291,579	10.5	274,670	10.4	248,648	10.2	204,705	9.3	147,174	6.5
Korea, Republic of	246,171	8.4	194,854	6.7	171,462	6.2	145,704	5.5	126,671	5.2	121,943	5.6	*174,300	7.8
Vietnam	140,300	4.8	162,733	5.6	129,627	4.7	*119,630	4.5	*107,635	4.4	*95,640	4.4	*83,645	3.7
Poland	113,065	3.9	96,421	3.3	87,067	3.1	99,976	3.8	108,314	4.4	119,751	5.5	128,824	5.7

Household Ware of Other Ceramic Materials

Unit of Measure: Thousands of units.

	1990 Value	%	1991 Value	%	1992 Value	%	1993 Value	%	1994 Value	%	1995 Value	%	1996 Value	%
Total Production	205,197	100.0	179,328	100.0	182,848	100.0	161,210	100.0	157,617	100.0	156,039	100.0	159,844	100.0
Regions														
America, North	132,652	64.6	133,414	74.4	134,176	73.4	134,938	83.7	135,700	86.1	136,462	87.5	137,224	85.8
America, South	19,082	9.3	15,274	8.5	11,467	6.3	7,660	4.8	3,852	2.4	45	0.0	-	
Asia	30,430	14.8	7,744	4.3	17,106	9.4	9,627	6.0	4,134	2.6	3,183	2.0	9,269	5.8
Europe	23,034	11.2	22,896	12.8	20,100	11.0	8,986	5.6	13,931	8.8	16,349	10.5	13,351	8.4
Leading Producers														
United States of America	*126,000	61.4	*126,000	70.3	*126,000	68.9	*126,000	78.2	*126,000	79.9	*126,000	80.7	*126,000	78.8
Colombia	19,082	9.3	15,274	8.5	11,467	6.3	7,660	4.8	3,852	2.4	45	0.0	-	
Indonesia	30,386	14.8	7,700	4.3	17,062	9.3	9,583	5.9	4,090	2.6	3,139	2.0	*9,225	5.8
Finland	8,616	4.2	7,414	4.1	8,105	4.4	3,770	2.3	1,530	1.0	3,830	2.5	4,150	2.6
Cuba	*6,651	3.2	*7,414	4.1	*8,176	4.5	*8,938	5.5	*9,700	6.2	*10,462	6.7	*11,224	7.0

Sanitary Ceramic Fittings (e.g. Sinks, Wash Basins Etc.)

Unit of Measure: Number.

	1990 Value	%	1991 Value	%	1992 Value	%	1993 Value	%	1994 Value	%	1995 Value	%	1996 Value	%
Total Production	52,941,192	100.0	54,687,689	100.0	55,123,404	100.0	62,969,059	100.0	74,550,757	100.0	63,713,062	100.0	70,194,166	100.0
Regions														
Africa	1,565,819	3.0	1,671,362	3.1	1,772,905	3.2	1,869,448	3.0	1,934,990	2.6	2,056,533	3.2	2,066,295	2.9
America, North	3,184,139	6.0	2,754,322	5.0	3,156,506	5.7	4,805,189	7.6	6,453,872	8.7	8,102,556	12.7	9,751,239	13.9
America, South	2,884,697	5.4	2,996,848	5.5	3,109,000	5.6	5,627,000	8.9	6,991,000	9.4	3,352,000	5.3	5,007,489	7.1
Asia	14,509,500	27.4	16,660,500	30.5	16,875,718	30.6	19,107,909	30.3	20,572,500	27.6	19,264,091	30.2	20,475,768	29.2
Europe	29,657,394	56.0	29,375,763	53.7	28,891,132	52.4	30,152,121	47.9	37,101,751	49.8	29,351,989	46.1	31,218,232	44.5
Oceania	1,139,643	2.2	1,228,893	2.2	1,318,143	2.4	1,407,393	2.2	1,496,643	2.0	1,585,893	2.5	1,675,143	2.4
Leading Producers														
United Kingdom	*7,182,991	13.6	*7,182,991	13.1	*7,182,991	13.0	7,446,861	11.8	6,919,122	9.3	*7,182,991	11.3	*7,182,991	10.2
Spain	*6,929,250	13.1	*6,929,250	12.7	*6,929,250	12.6	6,479,000	10.3	6,871,000	9.2	7,330,000	11.5	7,037,000	10.0
Mexico	2,757,000	5.2	2,315,000	4.2	2,705,000	4.9	4,341,500	6.9	5,978,000	8.0	7,614,500	12.0	9,251,000	13.2
Germany	*4,135,248	7.8	*4,135,248	7.6	*4,135,248	7.5	*4,135,248	6.6	*4,135,248	5.5	*4,135,248	6.5	4,135,248	5.9
Portugal	*3,317,000	6.3	*3,317,000	6.1	2,974,000	5.4	3,463,000	5.5	3,514,000	4.7	*3,317,000	5.2	*3,317,000	4.7

Commodity data are provided by the United Nations Statistical Division. The symbol * means that data are estimated. For additional notes, see Appendix I.

445

ISIC 3620
GLASS AND PRODUCTS

ISIC 3620—Glass and Porducts—includes glass in various forms (sheets, plates, tubes, rods) and of various chemical compositions, including fused quartz and other fused silica. Virtually all articles of glass are included in this industry except fabrics woven of glass yarn, insulation glass wool, syringes, and optical glass components.

The industry categories corresponding to this grouping in the U.S. SIC system are industries 3210—Flat Glass, 3220—Glass and Glassware, Pressed or Blown, and 3230—Products of Purchased Glass.

Summary Statistics

		1990 Value	N	1991 Value	N	1992 Value	N	1993 Value	N	1994 Value	N	1995 Value	N
Establishments or enterprises	(number)	11,361	78	11,874	87	14,779	78	7,989	73	6,529	61	3,184	28
Number employed	(000)	1,726	81	1,701	86	1,518	81	1,518	81	918	70	788	57
Total Wages	($ mil.)	16,189	77	18,745	81	19,628	76	14,721	72	14,621	62	11,747	51
Wage/salary per employee	($)	10,361	75	10,762	78	12,046	74	10,542	71	10,990	61	13,295	51
Female workers	(000)	118	32	90	31	170	26	163	28	145	23	133	13
as % of total employment*	(%)	20	32	20	31	22	26	23	28	26	23	35	12
Output	($ bil.)	78	78	88	84	90	80	79	74	77	64	69	52
per employee	($ 000)	48	73	49	78	54	74	50	71	56	62	69	51
per establishment	($ 000)	6,726	68	7,488	76	6,249	69	6,800	61	7,704	49	7,691	23
Value added	($ bil.)	40	71	40	73	43	69	42	65	38	53	39	49
per employee	($ 000)	27	66	27	68	29	65	27	62	31	52	37	48
per establishment	($ 000)	3,401	60	3,401	65	3,227	59	3,076	52	4,166	38	3,283	20
Capital investment	($ mil.)	6,532	57	6,512	55	5,532	51	4,580	48	4,099	35	2,939	19
per employee	($ 000)	7	54	7	53	8	49	9	45	8	34	5	19
per establishment	($ 000)	1,914	53	2,091	51	1,965	48	2,163	42	2,562	32	290	16

Data presented above are drawn from the detailed tables that follow. Columns headed 'N' show the number of countries that provided valid data for inclusion. Values are not strictly comparable one year to the next or one row to the next because the number of countries that report varies from period to period and row to row. However, a general indication of magnitudes can be gleaned from reviewing this summary—especially in earlier years in which more reports are available. For detailed explanations, see Appendix I. *The average for those countries reporting both total and female employment.

Establishments and Number Engaged

Country	Establishments or Enterprises (number)						Employees per Establishment					
	1990	1991	1992	1993	1994	1995	1990	1991	1992	1993	1994	1995
Albania	-	-	-	3.000	-	-	-	-	-	83	-	-
Argentina	-	-	-	244	277	-	-	-	-	32	28	-
Armenia	5.000	5.000	5.000	5.000	5.000	-	666	638	-	-	-	-
Australia	88	172	210	-	-	-	68	35	29	-	-	-
Austria	74	73	73	72	75	-	122	123	123	116	115	-
Azerbaijan	7.000	7.000	7.000	7.000	8.000	-	673	606	534	505	413	-
Bangladesh	22	29	25	-	-	-	70	72	82	-	-	-
Belgium	144	144	149	157	165	-	93	88	86	-	-	-
Belize	2.000	2.000	2.000	-	-	-	-	-	-	-	-	-
Bolivia	5.000	5.000	7.000	8.000	8.000	12	75	77	50	46	49	32
Bosnia & Herzegovina	1.000	1.000	-	-	-	-	360	333	-	-	-	-
Bulgaria	*12	*10	*10	*174	*162	*115	1,400	1,490	1,330	72	77	114
Cambodia	561	475	-	-	-	-	3.166	2.265	-	-	-	-
Canada	196	174	163	155	150	-	61	63	61	65	60	-
Cape Verde	1.000	1.000	1.000	1.000	1.000	-	-	-	13	14	-	-
Chile	11	10	11	12	13	-	181	197	201	211	188	-
China	*3,264	*3,229	*3,268	-	-	-	169	173	162	-	-	-
China (Hong Kong SAR)	108	137	68	88	46	-	10	10	13	9.091	13	-
Colombia	74	72	80	80	75	-	95	92	94	94	96	-
Costa Rica	24	27	36	49	-	-	34	30	24	19	-	-
Croatia	17	16	30	42	49	53	313	292	122	81	68	59
Cyprus	33	33	33	33	25	23	2.576	2.788	2.242	2.939	3.960	5.000
Czechoslovakia (Former)	*38	*50	-	-	-	-	1,921	1,240	-	-	-	-
Czech Republic	-	*41	*49	*48	-	-	-	1,293	939	646	-	-
Denmark	117	115	112	-	-	-	23	21	21	-	-	-
Ecuador	11	12	11	14	12	13	54	57	51	41	57	44
Egypt	70	67	70	60	-	-	201	199	213	245	-	-
El Salvador	-	3.000	1.000	1.000	1.000	4.000	-	-	-	24	25	19
Ethiopia	1.000	1.000	1.000	1.000	1.000	1.000	320	448	368	371	434	468
Finland	49	49	46	38	37	43	65	63	63	71	76	75
FYR Macedonia	*3.000	*6.000	*4.000	*4.000	*9.000	*13	101	139	47	46	81	46
Germany	-	437	431	433	437	-	-	212	190	172	161	-
Germany (Western Part)	347	355	364	-	-	-	203	204	193	-	-	-
Ghana	-	-	-	2.000	-	-	-	-	-	142	-	-
Greece	36	36	36	-	-	-	53	46	37	-	-	-
Grenada	-	-	-	-	1.000	1.000	-	-	-	-	-	4.000
Guatemala	-	1.000	1.000	1.000	1.000	1.000	-	567	522	521	672	832
Hungary	*28	*51	*79	*96	-	-	536	275	139	104	-	-
Iceland	17	23	22	24	-	-	5.647	4.913	4.636	4.083	-	-
India	598	628	590	667	-	-	89	105	99	87	-	-
Indonesia	44	52	51	57	56	71	321	368	374	353	354	300
Iran (Islamic Republic of)	64	40	40	38	-	-	130	203	209	216	-	-
Iraq	-	-	7.000	-	-	-	-	-	286	-	-	-
Ireland	47	49	-	-	-	-	83	82	-	-	-	-
Israel	28	39	39	39	42	-	29	23	28	28	31	-
Italy	*276	*276	*320	*318	*307	-	109	111	96	93	90	-
Japan	1,420	1,411	1,368	1,387	1,094	1,104	49	50	50	48	51	50
Jordan	17	18	19	20	42	24	32	31	28	9.900	5.786	8.083
Kenya	3.000	3.000	3.000	3.000	3.000	3.000	479	481	483	496	491	505
Korea, Republic of	356	389	431	534	573	599	72	63	59	47	44	43
Kuwait	16	15	13	16	15	-	29	14	33	29	32	-
Kyrgyzstan	2.000	2.000	2.000	2.000	2.000	-	649	711	1,081	1,013	919	-
Latvia	5.000	17	14	15	17	22	1,092	308	344	186	71	52
Lithuania	-	-	-	-	*5.000	-	-	-	-	-	427	-
Macau	14	13	13	13	-	7.000	2.071	1.846	1.615	1.538	-	1.286
Malaysia	26	26	27	28	29	-	138	138	137	150	179	-
Malta	8.000	9.000	9.000	9.000	10	-	13	11	13	13	11	-
Mauritius	1.000	2.000	2.000	2.000	3.000	3.000	51	38	38	35	29	31
Mexico	22	22	22	21	21	21	1,109	1,079	1,034	972	915	878
Mongolia	1.000	1.000	1.000	-	1.000	-	-	-	-	-	24	-

Continued.

Depending on the table, * means *Enterprises, Engaged,* or *Factor Values;* ± means *Basis Unspecified;* § means *shown in millions.* For additional notes and sources, see Appendix I.

Establishments and Number Engaged

- Continued -

Country	Establishments or Enterprises (number)						Employees per Establishment					
	1990	1991	1992	1993	1994	1995	1990	1991	1992	1993	1994	1995
Mozambique	3.000	3.000	2.000	2.000	1.000	-	341	342	452	427	680	-
Myanmar	2.000	2.000	2.000	2.000	2.000	-	879	824	579	730	733	-
Nepal	1.000	2.000	-	2.000	3.000	-	-	-	-	-	12	-
Netherlands	26	23	25	28	36	-	237	247	228	213	154	-
New Zealand	*156	*162	*182	*209	-	-	7.282	6.333	6.467	5.842	-	-
Norway	32	33	27	30	27	-	46	50	57	51	59	-
Pakistan	-	31					-	157				
Panama	7.000	5.000	-				33	42	-			
Paraguay	-	2.000					-	9.500	-			
Peru	131	136	141	-			26	22	18	-		
Philippines	67	85	49	53		-	127	100	147	109	-	-
Poland	*85	*92	*87	*87		-	529	457	425	402	-	-
Portugal	*484	*430	*484	*504	*461	-	22	24	19	18	19	-
Puerto Rico	34	36	34	31	32	36	44	33	34	38	42	37
Qatar	1.000	1.000	17	18	18	-						
Republic of Moldova	5.000	5.000	3.000	3.000	3.000	3.000	1,096	1,095	985	796	823	780
Romania	-	86	115	149	115	-	-	-	-	-	-	-
Russian Federation	-	-	-	*542	*549	*727	-	-	-	209	184	132
Sierra Leone	-	-	-	1.000	-	-	-	-	-	18	-	-
Slovakia	-	12	12	14	14	-	-	840	773	627	609	-
Slovenia	20	*17	*61	*60	*68	*67	-	-	-	-	-	-
South Africa	-	76	-	-	-	-	-	132	-	-	-	-
Spain	730	675	774	-	-	-	28	30	26	-	-	-
Sri Lanka	8.000	5.000	8.000	2.000	-	-	127	163	116	283	-	-
Swaziland	-	1.000	1.000	1.000	-	-	-	48	48	57	-	-
Sweden	45	47	43	37	38	-	102	97	93	88	94	-
Thailand	17	27	-	-	-	-	1,248	656	-	-	-	-
Trinidad and Tobago	1.000	1.000	1.000	1.000	1.000	-	461	471	516	490	437	-
Tunisia	-	-	-	33	37	41	-	-	-	29	30	27
Turkey	36	39	75	71	71	-	419	340	160	151	147	-
Ukraine	50	104	140	65	65	63	800	442	479	938	831	794
United Kingdom	770	745	885	899	1,083	-	55	55	44	41	37	-
United Republic of Tanzania	1.000	1.000					673	649	-			
USSR (Former)	*235						519					
United States of America	-	-	3,126	-	-	-	-	-	43	-	-	-
Venezuela	78	74	60	62	63	58	79	89	117	100	112	123
Yugoslavia	17	30	44	57	58	56	-	353	220	167	155	146
Zambia	-	-	-	-	1.000		-	-	-	-	-	
Zimbabwe	5.000	5.000	5.000	5.000	5.000	-	180	180	180	180	180	-

Employment and Compensation of Employees

Country	Number Employed/Engaged (000)						Wage/Salary per Employee ($)					
	1990	1991	1992	1993	1994	1995	1990	1991	1992	1993	1994	1995
Albania	-	-	-	0.248	-	-	-	-	-	-	-	-
Argentina	8.368	-	-	7.921	7.797	-	8,104	-	-	17,792	-	-
Armenia	*3.331	*3.190	-	-	-	-	-	-	-	-	-	-
Australia	6.000	*6.000	*6.000	-	-	-	27,617	32,204	27,819	-	-	-
Austria	9.000	9.000	9.000	8.375	8.641	8.600	25,750	26,788	30,637	30,333	30,963	38,192
Azerbaijan	4.710	4.240	3.737	3.535	3.305	-	2,108	1,977	-	-	-	-
Bangladesh	1.530	2.089	2.051	-	-	-	909	733	914	-	-	-
Belgium	13	13	13	-	-	-	22,677	24,124	26,648	-	-	-
Bermuda	0.060	0.049	0.057	0.065	0.063	0.017	-	-	-	-	-	-
Bolivia	0.373	0.385	0.352	0.368	0.392	0.388	1,979	1,992	1,523	1,685	2,526	2,783
Bosnia & Herzegovina	0.360	0.333	-	-	-	-	3,769	4,501	-	-	-	-
Bulgaria	17	15	13	13	13	13	1,994	597	915	1,222	1,024	1,453
Cambodia	*1.776	*1.076	-	-	-	-	-	-	-	-	-	-
Cameroon	0.140	0.136	0.139	0.130	-	-	17,893	16,603	18,265	19,288	-	-

Continued.

Depending on the table, * means *Enterprises, Engaged,* or *Factor Values*; ± means *Basis Unspecified*; § means *shown in millions*. For additional notes and sources, see Appendix I.

449

Employment and Compensation of Employees

- Continued -

Country	Number Employed/Engaged (000)						Wage/Salary per Employee ($)					
	1990	1991	1992	1993	1994	1995	1990	1991	1992	1993	1994	1995
Canada	12	11	10	10	9.000	8.970	27,640	29,041	28,874	27,285	27,420	27,459
Cape Verde	-	-	0.013	0.014	-	-	-	-	3,393	2,769	-	-
Chile	1.988	1.972	2.216	2.527	2.446	2.489	5,423	6,594	7,565	7,022	3,505	4,231
China	*550	*560	*530	*520	-	-	-	-	-	-		
China (Hong Kong SAR)	*1.100	*1.400	*0.900	*0.800	*0.600	-	7,119	7,261	11,196	14,382		
China (Taiwan Province)	23	22	21	21	21	21	10,029	10,870	12,312	12,748	13,139	13,199
Colombia	7.000	6.606	7.554	7.555	7.169	7.198	2,676	2,562	2,953	3,497	4,625	5,349
Costa Rica	0.811	0.797	0.869	0.914	-	-	2,831	2,586	2,987	3,210	-	-
Croatia	5.320	4.670	3.650	3.400	3.350	3.150	4,431	5,079	1,779	1,791	2,372	3,706
Cyprus	0.085	0.092	0.074	0.097	0.099	0.115	8,238	9,672	10,841	11,015	11,139	12,807
Czechoslovakia (Former)	73	62					2,099	1,587	-	-	-	-
Czech Republic	60	53	46	31			2,089	1,568	2,062	2,437		
Denmark	2.705	2.468	2.376	2.284	2.466	2.554	28,254	29,141	32,075	29,546	33,353	39,416
Ecuador	0.594	0.679	0.565	0.576	0.681	0.578	3,568	4,137	5,018	6,297	3,517	3,524
Egypt	*14	*13	*15	*15	*14	*15	1,961	1,083	1,347	1,388	1,364	1,536
El Salvador	-	-	-	*0.024	*0.025	*0.074	-	-	-	1,639	2,112	25,220
Ethiopia	0.320	0.448	0.368	0.371	0.434	0.468	1,350	1,530	1,501	590	855	853
Finland	3.200	3.100	2.900	2.700	2.800	3.232	28,635	27,352	24,719	22,103	25,004	31,980
FYR Macedonia	0.303	0.832	0.188	0.183	0.728	0.601	3,062	3,617	2,151	4,321	2,989	4,773
France	56	56	55	53	52	51	32,221	32,796	34,834	-	-	-
Germany	-	*93	*82	*74	*70	-	-	25,534	30,817	30,862	33,069	-
Germany (Western Part)	*70	*72	*70				28,948	29,608	33,246			
Ghana	-	-	-	*0.285	*0.293	*0.301	-	-	-	-	-	-
Greece	*1.894	*1.648	*1.330	*1.526	*1.504	*1.504	13,603	12,804	11,948	13,286	13,277	15,168
Grenada	-	-	-	-	-	0.004	-	-	-	-	-	-
Guatemala	-	*0.567	*0.522	*0.521	*0.672	*0.832	-	-	-	-	-	484
Hungary	15	14	11	10	10	9.970	2,511	2,649	2,977	3,235	3,282	3,262
Iceland	0.096	0.113	0.102	0.098	0.091	-	27,059	28,186	31,654	24,105	-	-
India	*53	*66	*58	*58	*58	*58	1,032	919	904	792	879	1,036
Indonesia	14	19	19	20	20	21	1,002	1,032	1,019	1,241	1,509	2,298
Iran (Islamic Republic of)	8.300	8.120	8.358	8.199	-	-	3,921	4,715	6,058	5,446	-	-
Iraq	-	1.764	2.000	-	-	-	-	10,434	18,177	-	-	-
Ireland	3.900	4.000	4.008	4.000	4.020	4.038	23,770	22,900	25,941	22,612	24,988	29,925
Israel	0.800	0.900	1.100	1.100	1.300	1.240	24,179	23,401	19,593	18,310	21,715	22,868
Italy	30	31	31	29	28	-	36,288	38,067	39,836	32,938	34,032	-
Japan	*70	*71	*69	*66	*56	*55	30,586	33,980	37,534	-	-	-
Jordan	0.542	0.556	0.531	0.198	0.243	0.194	2,918	2,847	2,853	2,871	2,190	2,237
Kenya	1.436	1.443	1.450	1.487	1.472	1.516	1,565	1,406	1,339	797	884	1,286
Korea, Republic of	25	25	25	25	25	26	12,134	13,644	14,928	15,709	17,014	20,699
Kuwait	0.460	0.212	0.432	0.463	0.480	0.469	5,144	5,896	8,841	8,711	10,431	12,607
Kyrgyzstan	*1.297	*1.421	*2.162	*2.027	*1.838	-	585	456	-	-	-	-
Latvia	5.459	5.240	4.815	2.786	1.210	1.149	-	-	-	-	-	-
Lithuania	-	-	2.049	1.842	2.134	-	-	-	481	613	1,333	-
Macau	0.029	0.024	0.021	0.020	-	0.009	5,302	7,750	7,661	8,106	-	6,861
Malaysia	3.600	3.600	3.700	4.200	5.200	5.552	4,775	5,737	6,058	5,744	6,177	6,777
Malta	0.100	0.103	0.116	0.121	0.115	-	8,580	8,266	9,488	7,572	9,363	-
Mauritius	*0.051	0.077	0.076	0.069	0.087	0.092	3,034	2,142	3,214	3,203	3,712	3,391
Mexico	24	24	23	20	19	18	6,018	7,192	9,323	10,685	10,352	6,604
Mongolia	-	-	-	-	*0.024	-	-	-	-	-	-	-
Mozambique	1.023	1.027	0.904	0.854	0.680	-	3,357	1,914	968	823	651	-
Myanmar	1.758	1.649	1.158	1.459	1.465	-	1,082	937	1,404	1,613	1,829	-
Nepal	-	-	-	-	0.037	-	-	-	-	-	527	-
Netherlands	6.152	5.684	5.697	5.953	5.550	5.514	28,923	31,711	34,038	33,645	-	-
New Zealand	1.136	1.026	1.177	1.221	-	-	-	-	-	-		
Norway	1.485	1.661	1.530	1.522	1.597	1.572	28,723	28,510	37,713	29,637	30,698	36,797
Pakistan	4.916	4.861	5.084	-	-	-	1,570	1,564	1,609	-	-	-
Panama	*0.232	*0.210	*0.176	*0.270	*0.275	*0.240	4,677	5,114	5,223	5,225	5,312	5,436
Paraguay	-	0.019	-	-	-	-	-	-	-	-	-	-
Peru	*3.361	*2.986	*2.522	-	-	-	2,859	3,651	2,733	-	-	-
Philippines	8.500	8.500	7.200	5.800	-	-	4,036	3,750	3,969	3,314	-	-

Continued.

Depending on the table, * means *Enterprises, Engaged,* or *Factor Values;* ± means *Basis Unspecified;* § means *shown in millions.* For additional notes and sources, see Appendix I.

Employment and Compensation of Employees
- Continued -

Country	Number Employed/Engaged (000)						Wage/Salary per Employee ($)					
	1990	1991	1992	1993	1994	1995	1990	1991	1992	1993	1994	1995
Poland	45	42	37	35	38	39	1,195	1,770	2,176	2,267	2,562	3,106
Portugal	*11	*10	*9.191	*8.949	*8.988	*8.912	-	-	-	-	-	-
Puerto Rico	1.490	1.190	1.170	1.170	1.350	1.340	24,765	17,143	17,265	26,239	23,926	23,134
Republic of Moldova	5.481	5.473	2.954	2.389	2.470	2.341	-	4,306		-	509	763
Russian Federation	-	-	-	114	101	96	-	-	-	576	847	1,021
Sierra Leone	-	-	-	*0.018	-	-	-	-	-	-	-	-
Slovakia	-	*10	*9.275	*8.774	*8.524	-	-	1,601	2,018	2,130	2,394	-
South Africa	10	10	9.000	9.000	10	9.940	9,701	10,394	13,168	13,261	12,223	14,472
Spain	20	20	20	23	22	22	20,268	21,659	23,761	19,428	19,187	21,936
Sri Lanka	*1.014	*0.816	*0.924	*0.567	-	-	1,206	1,249	821	1,803	-	-
Swaziland		0.048	0.048	0.057	-	-		2,842	2,895	2,575	-	-
Sweden	4.600	*4.582	*4.001	3.264	3.590	3.738	23,469	25,372	27,131	22,543	23,808	26,480
Thailand	21	18	-	-	-	-	3,546	3,927	-	-	-	-
Trinidad and Tobago	0.461	0.471	0.516	0.490	0.437	-	11,555	13,121	11,548	10,450	9,540	-
Tunisia	-	-	-	0.970	1.107	1.090	-	-	-	6,583	5,999	6,719
Turkey	15	13	12	11	10	11	10,260	12,847	14,722	14,029	8,585	10,919
Ukraine	40	46	67	61	54	50	3,116	2,668	2,186	-	471	508
United Kingdom	42	41	39	37	40	42	22,959	24,865	25,505	21,995	21,746	23,183
United Republic of Tanzania	0.673	0.649	-	-	-	-	-	443	-	-	-	-
USSR (Former)	122	-	-	-	-	-	2,049	-	-	-	-	-
United States of America	141	133	133	131	132	132	27,092	28,120	29,271	30,229	30,977	31,667
Uruguay	*1.284	*1.134	*0.939	*0.938	*0.939	*0.935	3,971	4,804	6,092	7,984	8,319	8,745
Venezuela	6.200	6.600	7.000	6.174	7.040	7.136	4,103	4,309	4,530	4,542	4,433	5,362
Yugoslavia	-	11	9.700	9.500	9.000	8.200	-	-	-	-	1,410	927
Zimbabwe	0.900	0.900	0.900	0.900	0.900	0.847	4,902	4,797	3,730	3,553	3,790	4,261

Female Workers: Total and Percent of Employment

Country	Female Workers (000)						Female Workers as % of Total Employment					
	1990	1991	1992	1993	1994	1995	1990	1991	1992	1993	1994	1995
Albania	-	-	-	0.038	-	-	-	-	-	15.32	-	-
Argentina	-	-	-	-	0.439	-	-	-	-	-	5.63	-
Australia	0.856	-	-	-	-	-	14.27	-	-	-	-	-
Austria	3.100	3.000	3.000	2.765	2.986	-	34.44	33.33	33.33	33.01	34.56	-
Bangladesh	0.070	0.231	0.216	-	-	-	4.58	11.06	10.53	-	-	-
Bermuda	0.021	0.017	0.023	0.028	0.027	0.013	35.00	34.69	40.35	43.08	42.86	76.47
Bosnia & Herzegovina	0.116	0.152	-	-	-	-	32.22	45.65	-	-	-	-
Bulgaria	-	-	-	6.100	6.200	6.300	-	-	-	48.80	49.60	48.09
Canada	3.000	2.000	-	-	-	-	25.00	18.18	-	-	-	-
Chile	0.100	0.100	0.093	0.137	0.125	-	5.03	5.07	4.20	5.42	5.11	-
China (Taiwan Province)	8.284	8.150	7.780	7.747	7.808	7.555	36.30	37.25	36.94	37.69	37.02	36.37
Colombia	-	0.936	-	1.245	1.235	-	-	14.17	-	16.48	17.23	-
Croatia	2.080	1.810	1.320	1.290	1.340	1.260	39.10	38.76	36.16	37.94	40.00	40.00
Cyprus	0.021	0.022	0.018	0.025	0.025	-	24.71	23.91	24.32	25.77	25.25	-
Czechoslovakia (Former)	27	25	-	-	-	-	36.30	39.52	-	-	-	-
Czech Republic	29	27	21	14	-	-	48.33	50.94	45.65	45.16	-	-
Denmark	0.830	0.743	0.705	-	-	-	30.68	30.11	29.67	-	-	-
Egypt	-	1.188	1.444	1.446	-	-	-	8.93	9.69	9.84	-	-
El Salvador	-	-	-	0.008	0.008	0.019	-	-	-	33.33	32.00	25.68
Ethiopia	-	0.069	0.075	0.082	0.081	0.082	-	15.40	20.38	22.10	18.66	17.52
Germany (Western Part)	15	-	-	-	-	-	21.31	-	-	-	-	-
Hungary	6.000	-	4.000	4.000	-	-	40.00	-	36.36	40.00	-	-
Indonesia	-	-	-	4.839	3.532	3.508	-	-	-	24.05	17.81	16.49
Iran (Islamic Republic of)	-	0.193	0.204	0.181	-	-	-	2.38	2.44	2.21	-	-
Iraq	-	0.216	-	-	-	-	-	12.24	-	-	-	-
Ireland	0.700	0.700	-	-	-	-	17.95	17.50	-	-	-	-
Italy	-	4.066	4.493	4.577	4.139	-	-	13.26	14.58	15.55	15.03	-
Jordan	0.007	0.006	0.005	0.005	0.004	0.001	1.29	1.08	0.94	2.53	1.65	0.52

Continued.

Female Workers: Total and Percent of Employment
- Continued -

Country	Female Workers (000)						Female Workers as % of Total Employment					
	1990	1991	1992	1993	1994	1995	1990	1991	1992	1993	1994	1995
Kenya	0.005	0.006	0.007	0.007	-	-	0.35	0.42	0.48	0.47	-	-
Korea, Republic of	5.700	5.382	5.358	5.620	5.288	5.405	22.35	21.90	21.17	22.62	20.86	21.12
Macau	0.001	0.001				0.001	3.45	4.17	-	-	-	11.11
Malaysia	0.500	0.500	0.500	0.500	0.700	-	13.89	13.89	13.51	11.90	13.46	-
Malta	0.014	0.013	0.015	0.014	0.015	-	14.00	12.62	12.93	11.57	13.04	-
Mongolia	-	-	-	-	0.010	-	-	-	-	-	41.67	-
Morocco	-	-	-	-	-	0.110	-	-	-	-	-	-
Myanmar	0.343	0.319	0.294	0.297	0.363	-	19.51	19.35	25.39	20.36	24.78	-
Nepal	-	-	-	-	0.002	-	-	-	-	-	5.41	-
New Zealand	0.179	0.163	0.174	0.182	-	-	15.76	15.89	14.78	14.91	-	-
Panama	0.017	-					7.33	-	-	-	-	-
Philippines			0.627	0.522			-	-	8.71	9.00	-	-
Portugal	1.431	1.278	1.157	1.011	0.913	-	13.51	12.20	12.59	11.30	10.16	-
Republic of Moldova	-	-	-	-	1.090	1.067	-	-	-	-	44.13	45.58
Sri Lanka	0.059	0.081	0.097	0.045			5.82	9.93	10.50	7.94	-	-
Sweden	1.400	-					30.43	-	-	-	-	-
Thailand	7.572	6.864					35.68	38.75	-	-	-	-
Turkey	0.809	-					5.36	-	-	-	-	-
United Kingdom	3.751	-	9.000				8.93	-	23.08	-	-	-
United Republic of Tanzania	0.082	0.079					12.18	12.17	-	-	-	-
United States of America	-	-	108	106	109	108	-	-	81.20	80.92	82.58	81.82

Output and Output per Employee

Country	Output ($ bil.)						Output per Employee ($)					
	1990	1991	1992	1993	1994	1995	1990	1991	1992	1993	1994	1995
Albania	-	-	-	*0.000	-	-	-	-	-	909	-	-
Argentina	±0.543			0.583	-	-	64,859	-	-	73,583	-	-
Armenia	0.001	0.002	0.004	0.301	0.001	-	315	682	-	-	-	-
Australia	*0.869	*0.802	*0.732	-			144,792	133,749	121,936	-	-	-
Austria	0.978	0.993	1.103	1.009	1.063	1.314	108,619	110,321	122,607	120,429	122,992	152,737
Azerbaijan	±0.061	±0.070	±0.003	±0.004	±0.002	-	12,944	16,541	738	1,046	521	-
Bangladesh	0.012	0.012	0.014				7,990	5,939	6,960	-	-	-
Bolivia	0.007	0.008	0.009	0.009	0.011	0.011	18,105	19,906	24,943	23,218	27,449	27,948
Bosnia & Herzegovina	±0.021	±0.019	-				58,658	56,427	-	-	-	-
Bulgaria	±0.189	±0.073	±0.077	±0.083	±0.092	±0.132	11,280	4,878	5,805	6,675	7,339	10,071
Canada	1.380	1.239	1.167	1.031	1.069	1.067	114,987	112,674	116,654	103,093	118,792	118,997
Cape Verde	-	-	±0.001	-	-	-	-	-	115,126	-	-	-
Chile	0.082	0.092	0.128	0.145	0.166	0.205	41,243	46,659	57,964	57,357	67,714	82,295
China	±2.607	±2.756	±3.379	±5.304			4,740	4,921	6,376	10,200	-	-
China (Hong Kong SAR)	0.053	0.073	0.067	0.060			47,965	52,207	74,499	75,463	-	-
China (Taiwan Province)	±0.710	±0.727	±0.850	±0.975	±1.021	±1.008	31,119	33,243	40,330	47,441	48,423	48,549
Colombia	0.205	0.204	0.203	0.230	0.296	0.344	29,296	30,860	26,810	30,378	41,287	47,769
Costa Rica	0.030	0.028	0.029	0.030	0.032	0.036	37,320	35,325	33,412	32,467	-	-
Cote d'Ivoire	±0.037	±0.032	±0.038	±0.035			-	-	-	-	-	-
Croatia	±0.193	±0.208	±0.090	-	-	-	36,367	44,552	24,629	-	-	-
Cyprus	0.004	0.006	0.005	0.005	0.006	0.008	52,619	61,579	64,775	53,247	56,105	66,381
Czechoslovakia (Former)	±0.670	±0.318	-	-	-	-	9,173	5,126	-	-	-	-
Denmark	*0.213	*0.206	*0.221	*0.196	*0.241	*0.296	78,787	83,369	93,195	85,791	97,767	115,934
Ecuador	0.022	0.024	0.032	0.033	0.035	0.043	37,540	35,415	57,061	57,231	50,798	73,983
Egypt	*0.146	*0.075	*0.109	*0.117	*0.114	*0.135	10,323	5,603	7,313	7,953	7,867	9,152
El Salvador	-	±0.000	-	±0.000	±0.000	±0.000	-	-	-	20,737	15,886	6,446
Ethiopia	0.001	0.004	0.002	0.001	0.003	0.003	3,193	9,094	4,608	2,691	7,792	7,251
Finland	±0.347	±0.321	±0.312	±0.273	±0.308	±0.447	108,443	103,698	107,542	101,081	109,874	138,284
FYR Macedonia	±0.004	±0.008	±0.003	±0.002	±0.011	±0.012	12,306	9,156	17,595	10,651	15,072	20,273
France	±5.738	±5.571	±5.957	±5.229	±5.673	±6.860	102,092	100,199	107,913	98,482	110,150	133,721
Germany	-	9.024	9.550	8.608	9.224	-	-	97,384	116,791	115,893	130,899	-
Germany (Western Part)	8.279	8.532	9.047	-	-	-	117,622	118,015	128,489	-	-	-

Continued.

Depending on the table, * means *Enterprises, Engaged,* or *Factor Values;* ± means *Basis Unspecified;* § means *shown in millions.* For additional notes and sources, see Appendix I.

Output and Output per Employee

- Continued -

Country	Output ($ bil.)						Output per Employee ($)					
	1990	1991	1992	1993	1994	1995	1990	1991	1992	1993	1994	1995
Greece	*0.112	*0.099	*0.094	*0.118	*0.116	*0.132	58,931	59,927	70,928	77,193	77,154	88,017
Guatemala	-	0.020	0.020	0.019	0.035	0.065	-	36,125	37,582	35,838	51,373	78,280
Hungary	±0.220	±0.217	±0.197	±0.201	±0.203	±0.200	14,661	15,483	17,952	20,128	20,313	20,089
Iceland	0.010	0.011	±0.011	±0.008	-	-	103,463	95,912	103,021	82,897	-	-
India	*0.500	*0.635	*0.485	*0.481	*0.541	*0.652	9,374	9,641	8,335	8,317	9,348	11,197
Indonesia	0.217	0.274	0.351	0.456	0.379	0.476	15,335	14,315	18,427	22,639	19,125	22,398
Iran (Islamic Republic of)	0.139	0.209	0.238	0.195	-	-	16,749	25,752	28,491	23,843	-	-
Iraq	-	*0.034	*0.060	-	-	-	-	19,422	29,904	-	-	-
Ireland	*0.255	*0.255	*0.289	*0.252	*0.280	*0.337	65,399	63,732	72,217	62,940	69,592	83,377
Israel	0.078	0.093	0.104	0.092	0.116	0.127	97,957	103,842	94,639	83,518	89,157	102,268
Italy	±4.270	±4.251	±4.341	±3.381	±3.671	-	141,934	138,651	140,914	114,876	133,340	-
Japan	±15	±16	±15	±16	±17	±17	209,959	221,132	222,684	241,034	294,736	307,927
Jordan	0.012	0.009	0.007	0.007	0.005	0.003	21,234	16,517	13,210	33,867	20,400	17,866
Kenya	*0.011	*0.009	*0.006	*0.004	*0.004	*0.005	7,901	6,550	4,281	2,551	2,666	3,335
Korea, Republic of	1.707	2.040	2.247	2.428	2.792	3.559	66,922	83,030	88,763	97,739	110,113	139,095
Kuwait	0.030	0.021	0.022	0.024	0.025	0.033	64,488	98,459	50,507	52,515	52,349	69,922
Kyrgyzstan	0.011	0.009	0.030	0.010	0.012	-	8,110	6,316	13,752	4,731	6,380	-
Latvia	0.001	0.002	0.015	0.006	0.007	0.009	103	348	3,205	2,029	5,718	8,249
Lithuania	-	-	0.008	0.008	0.011	-	-	-	3,853	4,250	5,334	-
Macau	0.001	0.001	0.001	0.001	-	0.001	26,004	35,161	46,256	50,963	-	68,875
Malaysia	*0.147	*0.177	*0.177	*0.192	*0.369	*0.438	40,954	49,271	47,743	45,777	70,891	78,922
Malta	0.003	0.003	0.004	0.003	0.003	-	32,429	28,585	33,046	26,394	27,122	-
Mauritius	0.000	0.001	0.001	0.000	0.001	0.001	7,915	7,634	6,848	6,816	8,384	7,406
Mexico	1.057	1.248	1.275	1.352	1.331	0.979	43,356	52,563	56,045	66,253	69,296	53,086
Mongolia	0.002	0.001	0.001	-	-	-	-	-	-	-	-	-
Mozambique	-	±0.005	-	±0.003	±0.003	-	-	5,330	-	3,359	4,740	-
Myanmar	0.027	0.011	0.007	0.026	0.046	-	15,165	6,466	6,366	17,700	31,417	-
Nepal	-	-	-	-	0.000	-	-	-	-	-	3,337	-
Netherlands	0.801	0.776	0.825	0.786	0.894	1.043	130,243	136,440	144,837	131,958	161,008	189,181
Norway	±0.182	±0.207	±0.194	±0.168	±0.194	±0.231	122,423	124,535	126,594	110,121	121,728	146,800
Pakistan	0.064	0.065	0.070	-	-	-	12,995	13,313	13,816	-	-	-
Panama	0.012	0.012	0.010	0.017	0.018	0.016	50,772	58,800	58,463	64,603	66,550	66,561
Peru	0.110	0.152	0.118	-	-	-	32,672	50,901	46,830	-	-	-
Philippines	0.159	0.197	0.187	0.182	-	-	18,665	23,218	25,963	31,361	-	-
Poland	±0.419	±0.515	±0.483	±0.464	±0.568	±0.706	9,301	12,269	13,065	13,244	14,897	18,022
Portugal	0.426	0.480	0.565	0.462	0.500	0.593	40,257	45,823	61,422	51,664	55,632	66,543
Republic of Moldova	0.000	0.326	0.011	0.013	0.019	0.023	0.092	59,479	3,832	5,356	7,870	9,970
Romania	±0.290	±0.207	±0.145	±0.125	-	-	-	-	-	-	-	-
Russian Federation	-	1.699	1.236	0.370	0.507	0.742	-	-	-	3,260	5,006	7,703
Slovakia	-	±0.110	±0.121	±0.123	±0.133	-	-	10,953	13,048	13,975	15,588	-
South Africa	±0.556	±0.550	±0.495	±0.454	±0.514	±0.605	55,579	54,974	54,971	50,405	51,359	60,833
Spain	*2.299	*2.332	*2.365	±1.921	±2.190	±2.744	112,441	117,069	116,637	82,539	98,658	122,164
Sri Lanka	0.007	0.006	0.008	0.004	-	-	6,720	7,674	9,008	7,672	-	-
Swaziland	-	*0.001	*0.001	*0.001	-	-	-	19,783	19,995	14,106	-	-
Sweden	*0.510	*0.483	*0.485	*0.321	*0.406	*0.481	110,918	105,402	121,234	98,430	112,977	128,655
Thailand	0.220	0.697	-	-	-	-	10,353	39,360	-	-	-	-
Trinidad and Tobago	±0.018	±0.021	±0.020	±0.021	±0.020	-	38,592	44,548	38,543	42,105	44,956	-
Tunisia	0.040	0.039	0.049	0.044	0.050	0.048	-	-	45,399	45,587	44,231	
Turkey	0.922	0.792	0.938	0.991	0.755	0.988	61,077	59,786	78,064	92,412	72,401	92,380
Ukraine	*0.754	*0.877	*0.865	*0.193	*0.201	*0.204	18,839	19,072	12,912	3,166	3,714	4,081
United Kingdom	*3.662	*3.651	3.652	3.299	3.775	4.230	87,202	89,057	93,640	89,157	94,372	101,198
United Republic of Tanzania . .	*0.003	*0.005	-	-	-	-	3,969	8,113	-	-	-	-
USSR (Former)	±2.250	-	-	-	-	-	18,443	-	-	-	-	-
United States of America . . .	*17	*17	*18	*19	*20	*21	122,979	127,895	135,301	145,580	152,742	155,364
Uruguay	0.037	0.034	0.028	0.039	0.040	0.042	29,060	29,945	29,374	41,114	42,764	44,903
Venezuela	0.190	0.234	0.272	0.330	0.303	-	30,676	35,441	38,926	53,465	43,066	-
Yugoslavia	±0.020	±0.025	±0.053	-	±0.084	±0.047	-	2,394	5,511	-	9,324	5,677
Zimbabwe	*0.018	*0.020	*0.017	*0.019	*0.019	*0.021	20,425	22,365	18,758	20,772	21,541	24,405

Depending on the table, * means *Enterprises, Engaged,* or *Factor Values*; ± means *Basis Unspecified*; § means *shown in millions.* For additional notes and sources, see Appendix I.

Value Added and Value Added per Employee

Country	Value Added ($ bil.)						Value Added per Employee ($)					
	1990	1991	1992	1993	1994	1995	1990	1991	1992	1993	1994	1995
Argentina	±0.249	-	-	0.261	-	-	29,786	-	-	32,980	-	-
Australia	*0.528	*0.552	*0.540	-	-	-	88,021	91,936	90,021	-	-	-
Austria	0.518	0.520	0.610	0.570	0.608	0.756	57,510	57,801	67,765	68,057	70,326	87,893
Bangladesh	0.007	0.007	0.007	-	-	-	4,447	3,257	3,417	-	-	-
Bolivia	0.003	0.004	0.005	0.005	0.004	0.005	8,793	10,886	14,327	13,215	11,419	11,626
Bosnia & Herzegovina	±0.008	±0.010	-	-	-	-	22,334	29,360	-	-	-	-
Bulgaria.	-	±0.016	±0.022	±0.026	-	-	-	1,056	1,669	2,096	-	-
Cameroon	0.006	0.006	0.008	-	-	-	42,580	45,612	56,779	-	-	-
Canada	0.643	0.576	0.571	0.543	0.571	0.570	53,565	52,370	57,086	54,259	63,464	63,563
Chile.	0.051	0.052	0.064	0.079	0.099	0.122	25,479	26,334	28,927	31,385	40,278	48,914
China	±0.705	±0.783	±1.060	±1.917	-	-	1,283	1,398	2,001	3,686	-	-
China (Hong Kong SAR)	0.019	0.024	0.025	0.030	-	-	17,622	17,372	27,847	37,327	-	-
China (Taiwan Province)	0.422	0.445	0.527	0.524	0.531	0.526	18,507	20,350	25,026	25,498	25,179	25,312
Colombia	0.113	0.110	0.109	0.121	0.174	0.202	16,212	16,604	14,478	16,040	24,270	28,106
Costa Rica	0.011	0.010	0.011	0.011	0.011	0.013	13,519	13,014	12,349	11,695	-	-
Cote d'Ivoire	0.033	0.028	0.034	0.032	-	-	-	-	-	-	-	-
Croatia	±0.105	±0.112	±0.047	-	-	-	19,739	24,057	13,010	-	-	-
Cyprus	0.001	0.002	0.002	0.002	0.002	0.003	17,197	20,377	22,523	19,893	20,716	22,614
Czechoslovakia (Former)	±0.298	-	-	-	-	-	4,083	-	-	-	-	-
Denmark	*0.114	*0.113	*0.121	*0.108	*0.129	*0.158	42,112	45,675	50,951	47,235	52,513	61,702
Ecuador	0.008	0.007	0.010	0.011	0.011	0.018	13,611	10,382	17,000	19,332	16,072	31,254
Egypt	*0.042	*0.026	*0.032	*0.042	*0.041	*0.048	2,965	1,977	2,129	2,839	2,810	3,247
El Salvador.	-	0.000	-	0.000	0.000	-0.000	-	-	-	4,311	3,314	§-1,48
Ethiopia.	0.000	0.003	0.001	0.001	0.002	0.002	1,034	6,065	2,225	1,508	5,250	4,837
Finland	±0.163	±0.153	±0.154	±0.135	±0.146	±0.189	50,830	49,376	53,271	50,054	52,305	58,402
FYR Macedonia	±0.002	±0.003	±0.002	±0.001	±0.004	±0.005	7,028	4,113	11,741	7,409	4,960	7,619
France	±3.090	±2.971	±3.035	±2.643	±2.796	±3.333	54,979	53,440	54,990	49,781	54,290	64,976
Germany (Western Part)	4.791	5.018	5.423	4.827	-	-	68,066	69,417	77,016	-	-	-
Greece	*0.049	*0.044	*0.042	*0.051	*0.051	*0.058	25,835	26,749	31,397	33,676	33,619	38,264
Hungary	±0.082	±0.071	±0.064	±0.071	±0.071	±0.069	5,478	5,052	5,806	7,110	7,087	6,924
Iceland	0.004	0.004	±0.005	±0.004	-	-	37,836	37,201	47,158	35,984	-	-
India.	*0.111	*0.145	*0.103	*0.101	*0.113	*0.136	2,080	2,196	1,767	1,750	1,959	2,334
Indonesia	*0.064	*0.109	*0.173	*0.208	*0.128	*0.205	4,563	5,707	9,069	10,357	6,442	9,651
Iran (Islamic Republic of)	0.077	0.137	0.137	0.099	-	-	9,230	16,847	16,437	12,058	-	-
Iraq	-	*0.028	*0.053	-	-	-	-	15,928	26,527	-	-	-
Ireland	*0.144	*0.151	*0.171	*0.149	*0.165	*0.198	37,037	37,722	42,684	37,229	41,061	49,102
Israel.	0.037	0.039	0.036	0.030	0.040	0.043	46,498	43,877	32,902	27,625	30,400	34,716
Italy	±1.673	±1.712	±1.772	±1.366	±1.604	-	55,625	55,839	57,515	46,421	58,259	-
Japan	±8.467	±8.782	±8.859	±9.344	±10	±9.994	120,963	123,688	128,392	141,569	180,476	181,702
Jordan	0.003	0.002	0.001	0.003	0.002	0.001	5,977	4,247	2,506	15,042	6,282	5,144
Kenya	*0.005	*0.004	*0.004	*0.002	*0.003	*0.004	3,647	3,023	2,997	1,623	1,939	2,565
Korea, Republic of	0.991	1.219	1.316	1.301	1.656	2.143	38,874	49,610	51,975	52,386	65,319	83,769
Kuwait	0.012	0.005	0.012	0.014	0.015	0.018	26,608	21,243	27,549	29,365	30,425	39,144
Latvia	-	-	-	*0.004	*0.004	*0.003	-	-	-	1,271	3,025	2,594
Macau	0.000	0.000	0.000	0.000	-	0.000	11,138	13,870	19,446	20,706	-	25,972
Malaysia	*0.073	*0.085	*0.090	*0.091	*0.141	*0.168	20,190	23,585	24,190	21,690	27,179	30,213
Malta	0.002	0.002	0.002	0.002	0.002	-	15,836	15,540	16,781	14,755	14,792	-
Mauritius	0.000	0.000	0.000	0.000	0.000	0.000	3,958	3,319	3,635	3,531	4,928	4,078
Mexico	0.627	0.716	0.753	0.806	0.792	0.582	25,704	30,153	33,086	39,470	41,204	31,560
Mongolia	0.001	0.000	0.000	-	-	-	-	-	-	-	-	-
Myanmar	±0.020	±0.015	±0.016	±0.021	±0.033	-	11,486	9,361	14,005	14,583	22,506	-
Netherlands	0.358	0.356	0.380	0.368	0.411	0.478	58,203	62,668	66,679	61,774	74,054	86,620
Norway	±0.077	±0.083	±0.076	±0.071	±0.080	±0.095	52,067	50,055	49,487	46,401	50,306	60,594
Pakistan	0.031	0.032	0.036	-	-	-	6,384	6,673	7,122	-	-	-
Panama.	0.007	0.006	0.005	0.009	0.009	0.008	31,862	29,633	29,372	32,706	33,709	33,663
Peru	0.037	0.050	0.047	-	-	-	11,099	16,858	18,531	-	-	-
Philippines	0.086	0.102	0.094	0.090	-	-	10,119	12,005	13,049	15,475	-	-
Poland	±0.227	±0.217	±0.221	±0.196	±0.238	±0.294	5,041	5,158	5,982	5,604	6,238	7,510
Portugal	0.173	0.200	0.225	0.185	0.222	0.264	16,375	19,058	24,527	20,637	24,719	29,579
Romania	±0.120	±0.080	±0.059	±0.053	-	-	-	-	-	-	-	-

Continued.

Depending on the table, * means *Enterprises, Engaged,* or *Factor Values;* ± means *Basis Unspecified;* § means *shown in millions.* For additional notes and sources, see Appendix I.

Value Added and Value Added per Employee

- Continued -

Country	Value Added ($ bil.)						Value Added per Employee ($)					
	1990	1991	1992	1993	1994	1995	1990	1991	1992	1993	1994	1995
Russian Federation	-	-	-	±0.189	±0.234	±0.347	-	-	-	1,661	2,309	3,608
Slovakia				±0.064	±0.064					7,286	7,553	-
South Africa	*0.292	*0.301	*0.338	*0.310	*0.354	*0.419	29,181	30,058	37,517	34,471	35,390	42,150
Spain	*1.128	*1.133	*1.126	±0.894	±1.010	±1.209	55,156	56,912	55,544	38,400	45,478	53,832
Sri Lanka	0.004	0.003	0.004	0.002	-		3,840	3,987	4,188	3,682		
Swaziland	-	*0.000	*0.000	*0.000	-	-		6,161	5,746	1,693	-	
Sweden.	*0.294	*0.192	*0.174	*0.136	*0.160	*0.188	63,906	41,841	43,558	41,697	44,666	50,387
Thailand	0.092	0.428	-	-	-	-	4,321	24,157	-	-	-	-
Trinidad and Tobago	±0.010	±0.012	±0.011	±0.010	±0.009	-	20,700	25,808	21,488	21,205	19,736	
Tunisia	0.016	0.017	0.021	0.019	0.021	0.020	-	-	-	19,105	19,044	18,465
Turkey	0.531	0.427	0.560	0.614	0.452	0.593	35,148	32,209	46,611	57,238	43,399	55,410
United Kingdom	*2.089	*2.050	*2.027	*1.749	*2.011	*2.251	49,745	49,989	51,962	47,277	50,268	53,842
United Republic of Tanzania . .	*-0.001	*0.001	-	-	-	-	§-1.55	1,835				
United States of America . . .	*10	*9.700	*11	*11	*12	*12	71,489	72,932	82,429	84,618	91,818	94,068
Uruguay	0.021	0.019	0.014	0.021	0.021	0.022	16,369	16,545	14,740	21,951	22,825	23,961
Venezuela	0.109	0.130	0.167	0.165	0.196	0.613	17,622	19,723	23,820	26,705	27,873	85,834
Yugoslavia.	±0.008	±0.012	±0.038	-	±0.027	±0.023	-	1,173	3,929	-	2,953	2,775
Zimbabwe	*0.009	*0.011	*0.008	*0.008	*0.005	*0.005	10,122	12,252	8,419	8,995	5,453	6,135

Capital Investment

Country	Gross Fixed Capital Formation ($ mil.)						Per Establishment ($ mil)					
	1990	1991	1992	1993	1994	1995	1990	1991	1992	1993	1994	1995
Argentina	-	-	-	36	-	-	-	-	-	0.147	-	-
Australia	138	-	-	-	-	-	1.571	-	-	-	-	-
Austria	170	133	154	90	105	-	2.301	1.826	2.112	1.249	1.397	-
Bangladesh	4.481	0.656	0.770	-	-	-	0.204	0.023	0.031	-	-	-
Bolivia	0.076	0.084	0.105	14	1.533	-	0.015	0.017	0.015	1.724	0.192	-
Bosnia & Herzegovina	0.603	-	-	-	-	-	0.603	-	-	-	-	-
Bulgaria.	18	0.534	56	45	0.515	3.247	1.526	0.053	5.634	0.256	0.003	0.028
Canada.	164	-	-	-	-	-	0.835	-	-	-	-	-
Chile.	-0.174	11	3.723	24	6.849	-	-0.016	1.143	0.338	2.037	0.527	-
China (Hong Kong SAR) . . .	7.446	0.129	-5.814	5.947	-	-	0.069	0.001	-0.085	0.068	-	-
Colombia	4.052	8.642	17	15	0.318	-	0.055	0.120	0.218	0.184	0.004	-
Croatia	8.922	4.802	36	11	12	1.161	0.525	0.300	1.198	0.257	0.240	0.022
Cyprus	0.742	0.307	0.222	0.654	0.118	0.383	0.022	0.009	0.007	0.020	0.005	0.017
Czechoslovakia (Former) . . .	55	-	-	-	-	-	1.437	-	-	-	-	-
Denmark	28	9.225	-	-	-	-	0.238	0.080	-	-	-	-
Ecuador	3.545	2.779	9.284	-0.203	1.821	9.618	0.322	0.232	0.844	-0.014	0.152	0.740
Egypt	14	11	31	21	-	-	0.194	0.160	0.447	0.346	-	-
El Salvador.	-	-	-	0.103	-	0.009	-	-	-	0.103	-	0.002
Ethiopia.	0.074	0.187	0.168	0.242	0.167	0.333	0.074	0.187	0.168	0.242	0.167	0.333
Finland	23	15	13	11	25	36	0.469	0.305	0.275	0.280	0.671	0.844
FYR Macedonia	-	0.002	0.231	-	-	-	-	0.000	0.058	-	-	-
France	450	204	385	339	425	555	-	-	-	-	-	-
Germany (Western Part) . . .	796	780	692	702	-	-	2.294	2.198	1.902	-	-	-
Greece	4.940	9.217	10	-	-	-	0.137	0.256	0.279	-	-	-
Hungary	75	27	5.355	4.057	-	-	2.670	0.539	0.068	0.042	-	-
India.	31	91	72	68	-	-	0.051	0.146	0.123	0.101	-	-
Indonesia	205	33	45	37	56	53	4.662	0.631	0.889	0.656	1.008	0.740
Iran (Islamic Republic of) . . .	24	15	49	5.669	-	-	0.375	0.382	1.232	0.149	-	-
Ireland	17	16	-	-	-	-	0.367	0.317	-	-	-	-
Israel.	8.432	7.898	24	98	21	-	0.301	0.203	0.626	2.519	0.498	-
Italy	404	430	375	229	254	-	1.464	1.560	1.173	0.721	0.829	-
Japan	1,257	1,603	1,224	962	988	-	0.885	1.136	0.895	0.694	0.903	-
Jordan	0.149	0.094	0.025	0.563	0.014	0.014	0.009	0.005	0.001	0.028	0.000	0.001
Korea, Republic of	349	501	395	373	503	896	0.980	1.287	0.917	0.699	0.877	1.496
Kuwait	-	6.701	4.874	2.907	1.201	-	-	0.447	0.375	0.182	0.080	-

Continued.

Depending on the table, * means *Enterprises, Engaged,* or *Factor Values*; ± means *Basis Unspecified*; § means *shown in millions.* For additional notes and sources, see Appendix I.

455

Capital Investment
- Continued -

Country	Gross Fixed Capital Formation ($ mil.)						Per Establishment ($ mil)					
	1990	1991	1992	1993	1994	1995	1990	1991	1992	1993	1994	1995
Latvia	0.011	0.014	0.110	0.018	0.109	1.170	0.002	0.001	0.008	0.001	0.006	0.053
Macau	0.004	-	0.011	0.000	-	0.001	0.000	-	0.001	0.000	-	0.000
Malaysia	50	32	104	52	151	-	1.934	1.217	3.853	1.873	5.190	-
Malta	0.420	0.065	0.302	0.079	0.029	-	0.052	0.007	0.034	0.009	0.003	-
Mexico	52	230	-	-	-	-	2.370	10	-	-	-	-
Mongolia	1.768	0.987	0.266	0.043	0.050	-	1.768	0.987	0.266	-	0.050	-
Myanmar	111	114	122	126	133	-	56	57	61	63	66	-
Netherlands	95	48	76	47	-	-	3.654	2.070	3.048	1.692	-	-
Norway	7.988	5.090	9.282	3.806	4.959	-	0.250	0.154	0.344	0.127	0.184	-
Pakistan	-	8.573	-	-	-	-	-	0.277	-	-	-	-
Panama	0.294	2.007	-	-	-	-	0.042	0.401	-	-	-	-
Peru	5.983	4.307	3.754	-	-	-	0.046	0.032	0.027	-	-	-
Philippines	32	32	24	29	-	-	0.475	0.378	0.484	0.554	-	-
Poland	37	41	36	30	-	-	0.438	0.441	0.412	0.342	-	-
Portugal	80	114	79	48	27	-	0.165	0.266	0.164	0.095	0.059	-
Romania	-	-	-	8.707	9.409	-	-	-	-	0.058	0.082	-
Spain	206	285	133	112	75	206	0.282	0.422	0.172	-	-	-
Sri Lanka	1.148	0.002	0.440	12	-	-	0.144	0.000	0.055	5.766		
Swaziland	-	-	0.031	0.071	-	-	-	-	0.031	0.071	-	-
Thailand	30	147	-	-	-	-	1.775	5.460	-	-	-	-
Trinidad and Tobago	-	-	2.729	2.467	0.726	-	-	-	2.729	2.467	0.726	-
Tunisia	4.100	3.677	7.010	0.996	1.977	2.115	-	-	-	0.030	0.053	0.052
Turkey	228	222	44	95	55	-	6.347	5.704	0.592	1.340	0.780	-
Ukraine	24	20	26	2.176	3.896	19	0.472	0.194	0.186	0.033	0.060	0.295
United Kingdom	368	234	210	-	204	-	0.478	0.314	0.238	-	0.188	-
United Republic of Tanzania	0.205	6.552	-	-	-	-	0.205	6.552	-	-	-	-
United States of America	920	990	1,000	881	989	1,155	-	-	0.320	-	-	-
Uruguay	0.693	1.989	1.991	1.388	-	-	-	-	-	-	-	-
Venezuela	14	46	50	33	43	0.130	0.175	0.620	0.829	0.529	0.681	0.002
Yugoslavia	1.897	0.249	1.402	-	0.643	0.919	0.112	0.008	0.032	-	0.011	0.016

Indexes of Industrial Production

Country	Index of Industrial Production (1990=100)						Index of Employment (1990=100)					
	1990	1991	1992	1993	1994	1995	1990	1991	1992	1993	1994	1995
Albania	100	-	-	-	-	-	-	-	-	-	-	-
Algeria	100	100	104	98	87	91	-	-	-	-	-	-
Angola	100	-	-	-	-	-	-	-	-	-	-	-
Argentina	100	70	-	-	-	-	100	-	-	95	93	-
Armenia	100	-	-	-	-	-	100	96	-	-	-	-
Australia	100	95	104	118	-	-	100	100	100	-	-	-
Austria	100	105	108	102	105	112	100	100	100	93	96	96
Azerbaijan	-	-	-	-	-	-	100	90	79	75	70	-
Bahrain	100	100	100	100	100	100	-	-	-	-	-	-
Bangladesh	100	106	169	173	168	221	100	137	134	-	-	-
Barbados	100	96	67	68	88	94	-	-	-	-	-	-
Belgium	100	94	97	97	103	103	100	95	96	-	-	-
Bermuda	-	-	-	-	-	-	100	82	95	108	105	28
Bhutan	100	-	-	-	-	-	-	-	-	-	-	-
Bolivia	100	105	111	118	131	146	100	103	94	99	105	104
Bosnia & Herzegovina	-	-	-	-	-	-	100	93	-	-	-	-
Botswana	100	100	100	-	-	-	-	-	-	-	-	-
Brazil	100	101	93	98	100	105	-	-	-	-	-	-
Bulgaria	100	81	67	66	80	84	100	89	79	74	74	78
Burkina-Faso	100	100	100	-	-	-	-	-	-	-	-	-
Burundi	100	100	100	100	-	-	-	-	-	-	-	-
Cambodia	100	-	-	-	-	-	100	61	-	-	-	-
Cameroon	100	-	-	-	-	-	100	97	99	93	-	-

Continued.

Depending on the table, * means *Enterprises, Engaged,* or *Factor Values;* ± means *Basis Unspecified;* § means *shown in millions.* For additional notes and sources, see Appendix I.

Indexes of Industrial Production

- Continued -

Country	Index of Industrial Production (1990=100)						Index of Employment (1990=100)					
	1990	1991	1992	1993	1994	1995	1990	1991	1992	1993	1994	1995
Canada	100	90	96	105	110	108	100	92	83	83	75	75
Central African Republic . . .	100	100	100	100	100	-	-	-	-	-	-	-
Chad	100	100	100	-	-	-	-	-	-	-	-	-
Chile	100	109	125	134	146	158	100	99	111	127	123	125
China	100	-	-	-	-	-	100	102	96	95	-	-
China (Hong Kong SAR) . . .	100	94	90	81	74	70	100	127	82	73	55	-
China (Taiwan Province) . . .	100	108	113	122	130	129	100	96	92	90	92	91
Colombia	100	99	98	101	106	110	100	94	108	108	102	103
Costa Rica	100	104	114	123	126	119	100	98	107	113	-	-
Cote d'Ivoire	100	138	179	236	90	231	-	-	-	-	-	-
Croatia	100	-	-	-	-	-	100	88	69	64	63	59
Cuba	100	70	-	-	-	-	-	-	-	-	-	-
Cyprus	100	99	101	102	99	97	100	108	87	114	116	135
Czechoslovakia (Former) . . .	100	75	75	-	-	-	100	85	-	-	-	-
Czech Republic	100	-	-	-	-	-	100	88	77	52	-	-
Dem. Rep. of the Congo . . .	100	89	68	-	-	-	-	-	-	-	-	-
Denmark	100	96	94	88	100	106	100	91	88	84	91	94
Dominican Republic. . . .	100	85	-	-	-	-	-	-	-	-	-	-
Ecuador	100	109	116	100	-	-	100	114	95	97	115	97
Egypt	100	93	93	95	89	98	100	94	106	104	103	105
El Salvador.	100	104	-	-	-	-	-	-	-	-	-	-
Ethiopia and Eritrea . . .	100	89	-	-	-	-	-	-	-	-	-	-
Ethiopia.	-	-	-	-	-	-	100	140	115	116	136	146
Fiji	100	100	100	100	100	100	-	-	-	-	-	-
Finland	100	96	104	109	115	120	100	97	91	84	88	101
FYR Macedonia	100	-	-	-	-	-	100	275	62	60	240	198
France	100	99	99	94	101	107	100	99	98	94	92	91
Gabon	100	-	-	-	-	-	-	-	-	-	-	-
Gambia	100	-	-	-	-	-	-	-	-	-	-	-
Germany (Eastern Part) . . .	100	-	-	-	-	-	-	-	-	-	-	-
Germany (Western Part) . . .	100	104	108	103	109	101	100	103	100			
Ghana	100	-	-	-	-	-	-	-	-	-	-	-
Greece	100	95	66	83	81	81	100	87	70	81	79	79
Guatemala	100	102	-	-	-	-	-	-	-	-	-	-
Guyana	100	100	100	100	100	100	-	-	-	-	-	-
Haiti	100	100	100	100	100	100	-	-	-	-	-	-
Honduras	100	100	100	100	100	100	-	-	-	-	-	-
Hungary	100	84	82	94	94	91	100	93	73	67	67	66
Iceland	100	-	-	-	-	-	100	118	106	102	95	-
India	100	100	99	100	105	121	100	123	109	108	109	109
Indonesia	100	155	145	177	243	290	100	136	135	142	140	151
Iran (Islamic Republic of) . . .	100	111	118	115	109	121	100	98	101	99	-	-
Iraq	100	39	19	-	-	-	-	-	-	-	-	-
Ireland	100	94	98	94	104	114	100	103	103	103	103	104
Israel.	100	130	151	145	155	193	100	113	138	138	163	155
Italy	100	99	96	90	-	-	100	102	102	98	92	-
Jamaica	100	89	107	102	100	-	-	-	-	-	-	-
Japan	100	100	94	91	93	93	100	101	99	94	80	79
Jordan	100	-	-	-	-	-	100	103	98	37	45	36
Kenya	100	77	170	355	524	621	100	100	101	104	103	106
Korea, Republic of	100	116	124	125	134	142	100	96	99	97	99	100
Kuwait	100	-	-	-	-	-	100	46	94	101	104	102
Kyrgyzstan	100	-	-	-	-	-	100	110	167	156	142	-
Lao P.D.R.	100	-	-	-	-	-	-	-	-	-	-	-
Latvia	100	-	-	-	-	-	100	96	88	51	22	21
Lebanon	100	-	-	-	-	-	-	-	-	-	-	-
Liberia	100	-	-	-	-	-	-	-	-	-	-	-
Libyan Arab Jamahiriya . . .	100	-	-	-	-	-	-	-	-	-	-	-
Lithuania	100	-	-	-	-	-	-	-	-	-	-	-
Macau	100	-	-	-	-	-	100	83	72	69	-	31

Continued.

Depending on the table, * means *Enterprises, Engaged,* or *Factor Values;* ± means *Basis Unspecified;* § means *shown in millions.* For additional notes and sources, see Appendix I.

457

Indexes of Industrial Production

- Continued -

Country	Index of Industrial Production (1990=100)						Index of Employment (1990=100)					
	1990	1991	1992	1993	1994	1995	1990	1991	1992	1993	1994	1995
Madagascar	100	-	-	-	-	-	-	-	-	-	-	-
Malawi	100	100	100	100	100	100	-	-	-	-	-	-
Malaysia	100	113	120	114	129	142	100	100	103	117	144	154
Malta	100	132	147	195	-	-	100	103	116	121	115	-
Mauritius	100	100	-	-	-	-	100	151	149	135	171	180
Mexico	100	105	113	114	120	101	100	97	93	84	79	76
Morocco	100	88	114	103	110	108	-	-	-	-	-	-
Mozambique	100	-	-	-	-	-	100	100	88	83	66	-
Myanmar	100	105	109	97	114	125	100	94	66	83	83	-
Nepal	100	93	-	-	-	-	-	-	-	-	-	-
Netherlands	100	96	90	89	99	100	100	92	93	97	90	90
New Zealand	100	85	87	-	-	-	100	90	104	107	-	-
Nicaragua	100	-	-	-	-	-	-	-	-	-	-	-
Nigeria	100	115	114	110	104	-	-	-	-	-	-	-
Norway	100	97	95	88	100	104	100	112	103	102	108	106
Pakistan	100	108	110	113	107	120	100	99	103	-	-	-
Panama	100	160	131	221	229	198	100	91	76	116	119	103
Papua New Guinea	100	-	-	-	-	-	-	-	-	-	-	-
Paraguay	100	103	105	60	34	35	-	-	-	-	-	-
Peru	100	91	76	71	86	128	100	89	75	-	-	-
Philippines	100	132	133	153	185	224	100	100	85	68	-	-
Poland	100	91	102	111	128	134	100	93	82	78	85	87
Portugal	100	99	102	105	106	110	100	99	87	85	85	84
Puerto Rico	100	94	95	96	104	-	100	80	79	79	91	90
Republic of Moldova	100	-	-	-	-	-	100	100	54	44	45	43
Romania	100	81	73	-	-	-	-	-	-	-	-	-
Russian Federation	100	-	-	-	-	-	-	-	-	-	-	-
Saudi Arabia	100	-	-	-	-	-	-	-	-	-	-	-
Senegal	100	100	100	100	100	100	-	-	-	-	-	-
Slovakia	100	-	-	-	-	-	-	-	-	-	-	-
Slovenia	100	84	80	81	98	104	-	-	-	-	-	-
Somalia	100	100	100	-	-	-	-	-	-	-	-	-
South Africa	100	95	81	78	89	97	100	100	90	90	100	99
Spain	100	99	92	87	96	102	100	97	99	114	109	110
Sri Lanka	100	72	106	105	91	88	100	80	91	56	-	-
Sudan	100	112	-	-	-	-	-	-	-	-	-	-
Sweden	100	90	78	71	72	77	100	100	87	71	78	81
Switzerland	100	87	83	77	85	81	-	-	-	-	-	-
Syrian Arab Republic	100	105	115	127	132	148	-	-	-	-	-	-
Thailand	100	106	120	146	166	190	100	83	-	-	-	-
Trinidad and Tobago	100	114	119	116	124	140	100	102	112	106	95	-
Tunisia	100	95	85	90	94	-	-	-	-	-	-	-
Turkey	100	94	110	111	111	126	100	88	80	71	69	71
Uganda	100	100	100	100	100	100	-	-	-	-	-	-
Ukraine	100	-	-	-	-	-	100	115	167	152	135	125
United Arab Emirates	100	-	-	-	-	-	-	-	-	-	-	-
United Kingdom	100	96	95	102	96	101	100	98	93	88	95	100
United Republic of Tanzania	100	100	92	-	-	-	100	96	-	-	-	-
USSR (Former)	100	-	-	-	-	-	100	-	-	-	-	-
United States of America	100	94	98	96	100	96	100	94	94	93	94	94
Uruguay	100	96	139	143	144	139	100	88	73	73	73	73
Venezuela	100	-	-	-	-	-	100	106	113	100	114	115
Yemen	100	-	-	-	-	-	-	-	-	-	-	-
Yugoslavia	100	-	-	-	-	-	-	-	-	-	-	-
Yugoslavia (Former)	100	78	-	-	-	-	-	-	-	-	-	-
Zambia	100	97	96	87	60	63	-	-	-	-	-	-
Zimbabwe	100	106	98	81	105	97	100	100	100	100	100	94

　　Depending on the table, * means *Enterprises, Engaged,* or *Factor Values*; ± means *Basis Unspecified*; § means *shown in millions.* For additional notes and sources, see Appendix I.

Representative Companies in Sector

Name	Address	Telephone	Fax	Employment	Y
J. Bibby & Sons PLC	16 Stratford Pl., London W1N 9AF, United Kingdom	71 6296243	71 4090556	7,976	93
Safelite Glass Corp.	PO Box 2000, Columbus, Ohio 43216-2000, U.S.A.	614-842-3000	614-842-3037	7,500	97
Rugby Group PLC	Crown House, Rugby CV21 2DT, United Kingdom	788 542666		7,142	93
Apogee Enterprises Inc.	7900 Xerxes Ave. S., Minneapolis, Minnesota 55431-1159, U.S.A.	612-835-1874	612-835-3196	6,553	96
Lancaster Colony Corp.	37 W. Broad St., Columbus, Ohio 43215, U.S.A.	614-224-7141	614-469-8219	6,400	97
Royal Doulton (UK) Ltd.	London Rd., Stoke-on-Trent ST4 7QD, United Kingdom	782 744766	782 416962	6,388	93
Schott Glaswerke/Schott Gruppe	Hattenbergstrassse 10, Mainz 55122, Germany	6131 660	6131 662000	5,899	93
Heywood Williams Group PLC	Edgerton Rd., Huddersfield HD3 3AR, United Kingdom	484 435477	484 547511	5,588	93
Lawson Mardon Wheaton Inc.	1101 Wheaton Ave., Millville, New Jersey 08332-2047, U.S.A.	609-825-1400	609-825-0146	5,500	97
Autoglass International UK Ltd.	Old Deer Park, Richmond TW9 2AZ, United Kingdom	81 9409177	81 9487913	5,269	93
Donnelly Corp.	414 E. 40th St., Holland, Michigan 49423-5368, U.S.A.	616-786-7000	616-786-6034	5,000	97
Oneida Ltd.	5-9 Pembroke Rd., Ruislip HA4 8NQ, United Kingdom	895 639452		4,982	91
Lenox Inc.	100 Lenox Dr., Lawrenceville, New Jersey 08648-2394, U.S.A.	609-896-2800	609-896-3704	4,200	97
Tropicana Dole Beverages North America	PO Box 338, Bradenton, Florida 34206-0338, U.S.A.	941-747-4461	941-742-3517	4,200	97
Libbey Inc.	PO Box 10060, Toledo, Ohio 43699-0060, U.S.A.	419-727-2100		4,150	97
Libbey Glass Inc.	PO Box 10060, Toledo, Ohio 43699-0060, U.S.A.	419-325-2100		4,000	96
Consol Glass	PO Box 562, Germiston 1400, Republic of South Africa	11 8274311	11 8242935	3,616	95
Andersen Corp.	100 4th Ave. N., Bayport, Minnesota 55003, U.S.A.	612-439-5150	612-430-5107	3,600	97
Ford Motor Co. Glass Div.	300 Renaissance Ctr., Detroit, Michigan 48243, U.S.A.	313-446-5945		3,500	93
Samsung Corning Co., Ltd.	Kangnam-gu, Seoul, Republic of Korea	2 34579638	2 34579539	3,500	97
Belopal	Beloslav, Varna BG-9150, Bulgaria	55223340	51523380	3,000	93
Keumkang Ltd.	Socho-Gu, Seoul, Republic of Korea	2 34505000	2 5522139	3,000	97
Consumers Packaging Inc.	777 Kipling Ave., Etobicoke, Ontario M8Z 5Z4, Canada	416-232-3000	416-232-3635	2,900	98
Mikasa Inc.	PO Box 6239, Carson, California 90749-6239, U.S.A.	310-886-3700	310-635-1546	2,750	95
Pilkington Glass Ltd.	Prescot Rd., St. Helens WA10 3TT, United Kingdom	744 28882	744 692660	2,626	94
Taiwan Glass Industry Corp.	Section 3, Taipei 104, Taiwan	2 7130333	2 7150333	2,600	93
Kedaung Industrial Ltd.	Jl. Tubagur Angke Kp-Polgar, Jakarta 11710, Indonesia	215402273	21 6190123	2,430	97
Anglian Group PLC	Horsford, Norwich NR10 3AQ, United Kingdom			2,407	94
Hankuk Glass Industries Inc.	Youngdungpo-Gu, Seoul, Republic of Korea	2 37720311	2 7862170	2,375	97
Pilkington Distribution Services Ltd.	Prescot Rd., St. Helens WA10 3TT, United Kingdom	744 28882		2,364	94
Quartz	Industrialna Zona, Sliven BG-8800, Bulgaria	44 23263	44 80691	2,200	93
Vicasa, SA	Edicio Ederra, Madrid E-28046, Spain	91 3972354	91 3972095	2,171	97
Asahimas Flat Glass Co. Ltd. PT	Jl. Ancol Ix/5 Barat, Jakarta 14430, Indonesia	216904041	21 675802	2,139	97
United Glass Ltd.	Valley Rd., St. Albans AL3 6NY, United Kingdom	727 859261	727 842661	2,025	93
Slovakfarma AS	Zelenicna 12, Hlohovec 920 27, Slovakia	804 24752	804 21241	2,000	93
Weather Shield Manufacturing Inc.	PO Box 309, Medford, Wisconsin 54451, U.S.A.	715-748-2100	715-748-6999	2,000	97
GB Glass Ltd.	Harworth, Doncaster DN11 8NF, United Kingdom	302 742631	302 750102	1,953	
Hankuk Electric Glass Co. Ltd.	Kumi-shi, Kyongbuk, Republic of Korea	546 4631191	546 4618032	1,946	97
Mulia Glass PT	Jl. H.R. Rasuna Said Kav C11-14, Jakarta 12940, Indonesia	215200777	21 5200946	1,852	97
Cristaleria Peldar S.A	Apdo. 215, Medellin, Colombia	4 3330422	4 2704025	1,600	96
Picvue Electronics, Ltd.	Hsin Chu Hsien, Taipei, Taiwan	2 82616	2 7003497	1,600	93
Tecnoquimicas SA	Calle 23 No. 7-39, Cali, Colombia	2 8825555	2 8838859	1,600	96
Owens-Corning Canada Inc.	3450 McNicoll, Scarborough, Ontario M1V 1Z5, Canada	416-292-4000	416-412-7091	1,500	98
St. Glass Oblosuveggyarto Es Forgalmazo RT	Huta ut 1, Salgotarjan H-3100, Hungary	32 10433	32 10808	1,500	93
Sandomierz Glass Works	ul. Portowa 24, Sandomierz PL-27-600, Poland	153 23041	153 23925	1,494	93
Huta Szkla, Jaroslaw SA	Morawska ul. 1, Jaroslaw PL-37-500, Poland	4001	2980	1,460	93
Anchor Hocking Corp.	PO Box 600, Lancaster, Ohio 43130-0600, U.S.A.	614-687-2500	614-687-2365	1,400	97
Staffordshire Tableware, Ltd.	Meir Pk., Stoke-on-Trent ST3 7AA, United Kingdom	782 315251	782 311433	1,346	94
Han Kuk Fiber Glass Co., Ltd.	Miryang-Gun, Kyongnam, Republic of Korea	527 3550081	527 534924	1,300	97
Loceria Colombiana SA	Antioquia, Medellin, Colombia	4 2780400	4 3031133	1,300	97
Thai Glass Industries Ltd.	Rat Burana, Bangkok 10140, Thailand	4270060	4276603	1,250	90
PLM Redfearn Ltd.	Monk Bretton, Barnsley S71 2QG, United Kingdom	226 710211		1,118	93
Alltrista Corp.	PO Box 5004, Muncie, Indiana 47307-5004, U.S.A.	765-281-5000	765-281-5400	1,100	97
Photronics Inc.	1061 E. Indiantown Rd., Jupiter, Florida 33477, U.S.A.	561-745-1222		1,050	97
Kalbe Farma PT	Blok A/9 Pulomas, Jakarta Timur 13210, Indonesia	214892808	21 4893549	1,004	97
Gemtron Corp.	615 Hwy. 68, Sweetwater, Tennessee 37874, U.S.A.	615-337-3522	615-337-7979	1,000	97
Kimble Glass Inc.	537 Crystal Ave., Vineland, New Jersey 08360, U.S.A.	609-692-3600	609-794-9762	1,000	97

Continued

Representative Companies in Sector
- Continued -

Name	Address	Tele-phone	Fax	Employ-ment	Y
Lionheart PLC	Adams Hill, Knutsford WA16 6BA, United Kingdom	565 650003	565 650004	962	93
Ifo Sanitar AB	Ifotorget, Bromolla S-295 22, Sweden	456 48000	456 25698	950	93
Caradon Everest Ltd.	Cuffley, Potters Bar EN6 4SG, United Kingdom	707 875700	707 875621	940	93
BSN Vidrio Espana, SA	Km. 8,200, Madrid E-28041, Spain	91 3170290	91 3173303	916	97
Dott Industries Inc.	395 DeMille Rd., Lapeer, Michigan 48446-3055, U.S.A.	810-667-3460	810-667-3091	900	97
Africa Glass Holdings Pty. Ltd.	Cleveland, Johannesburg 2022, Republic of South Africa	11 6221754	11 6155915	780	94
Cape Building Products Ltd.	Iver Ln., Uxbridge UB8 2JQ, United Kingdom	895 37111	895 59262	768	94
John Tams Group PLC	Longton, Stoke-on-Trent ST3 2PG, United Kingdom	782 599226	782 599149	767	92
Iglas (Persero) PT	Jl. Nagel No. 153, Surabaya 60246, Indonesia	31575597	31 574796	705	97
Anchor Hocking Specialty Glass Co.	400 9th St., Monaca, Pennsylvania 15061, U.S.A.	412-773-3286		700	94
Corning Consumers Products Co.	100 8th St., Charleroi, Pennsylvania 15022, U.S.A.	412-483-6531	412-489-2211	700	97
Industria Centroamericana de Vidrio SA	Apdo 1759, Guatemala City, Guatemala	2 760406		700	
Pacific Industries Inc.	Yongin-shi, Kyonggi, Republic of Korea	331 2842300	331 2842313	700	97
CCX Inc.	1901 Roxborough Rd., 205, Charlotte, North Carolina 28211, U.S.A.	704-365-0560	704-365-8195	675	97
National Electrical Coil	800 King Ave., Columbus, Ohio 43212, U.S.A.	614-488-1151	614-488-8892	650	96
Beatson Clark PLC	Greasbrough Rd., Rotherham S60 1RD, United Kingdom	709 828141	709 828476	632	93
Rauch Industries Inc.	PO Box 609, Gastonia, North Carolina 28053-0609, U.S.A.	704-867-5333	704-864-2081	631	94
Vidrieria y Cristaleria de Lamiaco SA	Columbia, 63, Madrid E-20816, Spain	94 4637400	94 4645242	623	97
Owens Corning Fiberglass Ltd.	Wrexham Industrial Estate, Wrexham LL13 9JU, United Kingdom	978 661151	978 660382	622	93
Creation Windows Inc.	PO Box 1046, Elkhart, Indiana 46515, U.S.A.	219-264-3131	219-264-9725	600	96
Hackman Tabletop Oy Ab	Hameentie 135, Helsinki SF-00560, Finland	39391	791329	600	94
Kangar Consolidated Industries PT	Cakung, Jakarta 13960, Indonesia	214600792	21 4600797	600	97
Litton Poly-Scientific	1213 N. Main St., Blacksburg, Virginia 24060-3100, U.S.A.	540-552-3011	540-953-1841	600	97
Advanced Lighting Technologies, Inc.	3200 Aurora Rd., Cleveland, Ohio 44139, U.S.A.	216-963-6680		564	96
Vidriera Centroamericana, SA	Apartado 355-7050, Cartago, Costa Rica	5512684	5514473	555	
Fenton Art Glass Co.	700 Elizabeth St., Williamstown, West Virginia 26187, U.S.A.	304-375-6122	304-375-6459	545	97
Abbas Ali Hazeem & Sons Co.	Safat, Kuwait City 13003, Kuwait	2448290		500	
Operadora de Factorias SA	Ave. Petapa 48-01 2n 12, Guatemala City, Guatemala	760406		500	
Schott Glass Technologies Inc.	400 York Ave., Duryea, Pennsylvania 18642, U.S.A.	717-457-7485	717-457-6960	500	96
Owens Corning Fiberglas Ltd.	Bowling-Bank, Wrexham LL13 9JU, United Kingdom	978 661151	978 660382	493	93
Ortel Corp.	2015 W. Chestnut St., Alhambra, California 91803, U.S.A.	818-281-3636	818-281-8231	464	96
Ahlstrom Riihimaen Lasi Oy	Ryttyla SF-12310, Finland	14 67311	14 67824	420	94
Action Performance Companies Inc.	2401 West 1st St., 130, Tempe, Arizona 85281, U.S.A.	602-894-0100	602-967-1403	418	97
Vidrieria Leonesa, SA	Trobajo Del Cerecedo, Leon E-24192, Spain	987 204251	987 214200	414	97
Hankook Chinaware Co., Ltd.	Chongju-Shi, Chungbuk, Republic of Korea	431 625251	431 625215	400	97
Hollinee Corp.	25 W Skippeck Pik 203, Broad Axe, Pennsylvania 19002, U.S.A.	215-628-3850		400	
Lampart Vegyipari Gepgyar	PO Box 41, Budapest H-1105, Hungary	1 1570111	1 1572029	400	93
Luxguard I SA	Rte. de Luxembourg, Bascharage L-4940, Luxembourg	50301	503745	400	95
Phonecia America-Israel (Flat Glass) Ltd.	Industrial Zone, Zipporit 17053, Israel	6 6410222	6 6519237	400	97
Vidrieria Rovira, SA	Pgno. Ind. Zona Franca, Barcelona E-08004, Spain	93 3359951	93 3355486	390	97
Vidrala SA	Barrio Areta, Llodio, Alava E-01400, Spain	94 6721550	94 6726039	360	97
AFG Industries Ltd.	350 Danforth Rd., Scarborough, Ontario M1L 3X7, Canada	416-694-3401	416-694-4778	330	98
Crystal Clear Industries Inc.	2 Bergen Tpk., Ridgefield Park, New Jersey 07660, U.S.A.	201-440-4200		325	97
Darby Group PLC	Sunningdale Rd., Scunthorpe DN17 2SS, United Kingdom	724 842030	724 868295	317	94
Gemco-Ware Inc.	PO Box 0813, Freeport, New York 11520-0813, U.S.A.	516-623-9300	516-378-6699	300	97
L.B.L. Inc.	225 Montee de Liesse, Saint-Laurent, Quebec H4T 1P5, Canada	514-737-4533	514-342-7772	300	98
ODL Inc.	215 E. Roosevelt Ave., Zeeland, Michigan 49464, U.S.A.	616-772-9111	616-772-9110	300	94
SpecTran Corp.	50 Hall Rd., Sturbridge, Massachusetts 01566, U.S.A.	508-347-2261	508-347-2747	300	97
Olbernhauer Glas, Pech & Kunte GmbH	Bruchbergweg 6, Olbernhau 09526, Germany	7668 2786	7668 3387	270	91
Corning Ltd.	Wear Glass Works, Sunderland SR4 6EJ, United Kingdom	91 5676222	91 5670116	256	93
Evanite Fiber Corp.	PO Box E, Corvallis, Oregon 97339, U.S.A.	503-753-1211	503-753-0388	241	97
Gulf Glass Manufacturing Co. KSC	PO Box 26996, Safat 13130, Kuwait	3260455	3263136	230	

Continued

For sources and notes, see Appendix I.

Representative Companies in Sector
- Continued -

Name	Address	Tele-phone	Fax	Employ-ment	Y
Tudor PLC	Wordsley, Stourbridge DY8 4AD, United Kingdom	384 480221	384 336580	228	93
Align-Rite International Inc.	2428 Ontario St., Burbank, California 91504, U.S.A.	818-843-7220		220	95
Department 56 Inc.	6436 City West Pkwy., One Villag, Eden Prairie, Minnesota 55344, U.S.A.	612-944-5600	612-943-4500	203	96
Hartung Agalite Glass Co.	17830 W. Valley Hwy., Tukwila, Washington 98188, U.S.A.	206-656-2626	206-656-2601	200	94
Mediakreasi Lokanusa Industry	Jl. Kebon Jeruk Raya No. 6, Jakarta 11530, Indonesia	21 5307590	21 5307589	200	91
Sambo Chinaware Co., Ltd.	Kimpo-Gun, Kyonggi, Republic of Korea	341 9871330	341 9884288	180	97
SWP Group PLC	Spring Ln. S., Malvern Link WR14 1AQ, United Kingdom	684 892602	684 892801	172	94
Cheil Glass Ind. Co., Ltd.	Nam Dong-Gu, Incheon, Republic of Korea	32 8115111	32 8115130	155	97
Daeseong Co., Ltd.	Ansan-shi, Kyonggi, Republic of Korea	345 4912101	345 4930521	150	97
Para Press SA	Rte. de Luxembourg, Bettembourg L-3201, Luxembourg	513513	519696	150	95
Art	Woyska Polskiego 29, Szczecin PL-70-473, Poland	91 47496		140	93
Adel Rootstein Display Mannequins Ltd.	Chelsea, London SW3 4BB, United Kingdom	71 3511247	71 3765084	130	91
Giesserei und Glasformenbau Radeberg GmbH	Strasse des Friedens 8-12, Radeberg 01471, Germany	5191 4241	5191 2192	130	91
Suh Il Ceramics Co., Ltd.	Changpyong-Myon, Chongyang-Gun, Republic of Korea	454 427871	454 427875	120	97
Noma Inc., Danbel Industries Division	375 Kennedy Rd., Scarborough, Ontario M1K 2A3, Canada	416-267-4614	416-267-5211	100	98
Cisper Da Amazonia SA	Av. Santa Cruz Machado 200, Japiu Manaus 69078-000, Brazil	92 2378348	11 2432168	80	98
Fly Bridges Enterprise Co., Ltd.-- David Hughes Trading Co., Ltd.	293-811 Sung-Chiang Rd., Taipei, Taiwan	2 5170157	2 5023937	80	94
National Glass Ltd.	5744 198 St., Langley, British Columbia V3A 7J2, Canada	604-530-2311	604-530-4662	65	98
Anderson Millwork Ltd.	46 Cheryl Cres, Sundridge, Ontario P0A 1Z0, Canada	705-384-5341	705-384-7514	55	98
Optica Industrial, SA	Apartado 757-1000, San Jose, Costa Rica	2230719	2226967	51	
Colony Glass	4460 Lake Forest Dr., 200, Cincinnati, Ohio 45242, U.S.A.	513-563-7767	513-563-9639	50	95
Farbglashutte Lauscha GmbH	Postfach 53, Lauscha 98724, Germany	3 6702417	3 67023340	50	93
Tempa Glass Industries Ltd.	1500 Railway St., North Vancouver, British Columbia V7J 1B5, Canada	604-980-0576	604-980-6839	50	98
United Glass Co. WLL	PO Box 24430, Safat 13105, Kuwait	3984705	3984925	40	
C.L.O. Glass Ltd.	237 Toryork Dr., Weston, Ontario M9L 1Y2, Canada	416-749-6161	416-749-1014	35	98
Mulgrave Machine Works Ltd.	149 Mill St., Mulgrave, Nova Scotia B0E 2G0, Canada	902-747-2157	902-747-2227	35	98
Vivian Regina Sales Ltd.	PO Box 853, Springs 1560, Republic of South Africa	11 8134147	11 8133743	33	96
Jumbo Enterprise Co.Ltd.	Youngdungpo-Gu, Seoul 150-104, Republic of Korea	2 6790123	2 6315200	30	96
Swedal International Inc.	5 Hsin-Yi Rd., Sec. 5, Taipei, Taiwan	2 7252627	2 7252493	15	93

Product Tables

Glass, Drawn or Blown, in Rectangles, Unworked

Unit of Measure: 1,000 Square meters.

	1990		1991		1992		1993		1994		1995		1996	
	Value	%	Value	%	Value	%	Value	%	Value	%	Value	%	Value	%
Total Production	1,211,454	100.0	1,152,344	100.0	1,222,969	100.0	1,210,275	100.0	1,181,123	100.0	1,194,826	100.0	1,181,112	100.0
Regions														
America, North	491,909	40.6	458,326	39.8	508,607	41.6	552,985	45.7	586,522	49.7	595,308	49.8	610,205	51.7
America, South	37,979	3.1	38,021	3.3	38,940	3.2	39,361	3.3	39,442	3.3	39,218	3.3	43,120	3.7
Asia	453,017	37.4	455,903	39.6	454,167	37.1	422,041	34.9	379,235	32.1	379,004	31.7	366,618	31.0
Europe	228,549	18.9	200,094	17.4	221,256	18.1	195,888	16.2	175,924	14.9	181,297	15.2	161,169	13.6
Leading Producers														
United States of America	440,087	36.3	398,676	34.6	433,033	35.4	461,134	38.1	*486,769	41.2	*499,069	41.8	*511,368	43.3
Russian Federation	130,432	10.8	127,462	11.1	117,601	9.6	101,271	8.4	58,576	5.0	58,830	4.9	44,814	3.8
Poland	57,191	4.7	53,412	4.6	50,469	4.1	46,326	3.8	42,923	3.6	37,829	3.2	24,258	2.1
Mexico	51,181	4.2	58,959	5.1	74,833	6.1	91,061	7.5	98,912	8.4	95,349	8.0	97,897	8.3
Romania	44,683	3.7	35,016	3.0	36,038	2.9	37,320	3.1	29,630	2.5	36,200	3.0	36,805	3.1

Glass, Cast, Rolled, Drawn or Blown

Unit of Measure: Metric tons.

	1990		1991		1992		1993		1994		1995		1996	
	Value	%	Value	%	Value	%	Value	%	Value	%	Value	%	Value	%
Total Production	5,598,000	100.0	5,514,000	100.0	5,567,000	100.0	5,664,019	100.0	5,716,256	100.0	5,711,493	100.0	5,628,838	100.0
Regions														
Africa	26,000	0.5	25,000	0.5	24,000	0.4	22,000	0.4	21,000	0.4	23,000	0.4	21,000	0.4
America, North	58,000	1.0	55,000	1.0	26,000	0.5	94,000	1.7	92,000	1.6	85,000	1.5	96,000	1.7
Asia	1,116,000	19.9	1,118,000	20.3	1,156,000	20.8	1,160,429	20.5	1,161,036	20.3	1,178,643	20.6	1,182,250	21.0
Europe	4,398,000	78.6	4,316,000	78.3	4,361,000	78.3	4,387,591	77.5	4,442,220	77.7	4,424,850	77.5	4,329,588	76.9
Leading Producers														
Germany	*1,585,000	28.3	1,524,000	27.6	1,646,000	29.6	*1,585,000	28.0	*1,585,000	27.7	*1,585,000	27.8	*1,585,000	28.2
Italy	816,000	14.6	859,000	15.6	826,000	14.8	863,000	15.2	892,000	15.6	880,000	15.4	796,000	14.1
Indonesia	*637,000	11.4	*637,000	11.6	*637,000	11.4	*637,000	11.2	637,000	11.1	*637,000	11.2	*637,000	11.3
Turkey	440,000	7.9	449,000	8.1	482,000	8.7	493,000	8.7	497,000	8.7	518,000	9.1	525,000	9.3
France	*405,000	7.2	*405,000	7.3	*405,000	7.3	*405,000	7.2	*405,000	7.1	*405,000	7.1	*405,000	7.2

Glass Fibers (Including Glass Wool)

Unit of Measure: Metric tons.

	1990		1991		1992		1993		1994		1995		1996	
	Value	%	Value	%	Value	%	Value	%	Value	%	Value	%	Value	%
Total Production	3,602,806	100.0	3,554,847	100.0	3,641,298	100.0	3,499,389	100.0	3,691,606	100.0	3,969,704	100.0	3,885,861	100.0
Regions														
America, North	2,265,143	62.9	2,259,619	63.6	2,256,095	62.0	2,258,058	64.5	2,253,152	61.0	2,251,924	56.7	2,246,904	57.8
America, South	1,260	0.0	1,260	0.0	703	0.0	1,810	0.1	1,665	0.0	860	0.0	1,260	0.0
Asia	679,818	18.9	748,996	21.1	769,362	21.1	696,628	19.9	859,529	23.3	1,111,184	28.0	1,048,204	27.0
Europe	656,585	18.2	544,972	15.3	615,138	16.9	542,892	15.5	577,260	15.6	605,736	15.3	589,493	15.2
Leading Producers														
United States of America	*2,260,143	62.7	*2,254,619	63.4	*2,249,095	61.8	*2,243,571	64.1	*2,238,048	60.6	*2,232,524	56.2	*2,227,000	57.3
Japan	*534,400	14.8	*560,400	15.8	*586,400	16.1	*612,400	17.5	*638,400	17.3	*664,400	16.7	*690,400	17.8
China	86,900	2.4	120,800	3.4	115,300	3.2	12,100	0.3	124,900	3.4	328,057	8.3	254,800	6.6
Romania	75,684	2.1	3,376	0.1	2,999	0.1	-		-		-		-	
Sweden	113,000	3.1	93,000	2.6	70,101	1.9	74,558	2.1	64,636	1.8	54,713	1.4	*37,231	1.0

Commodity data are provided by the United Nations Statistical Division. The symbol * means that data are estimated. For additional notes, see Appendix I.

Product Tables

Glass, Safety, of Toughened or Laminated Glass

Unit of Measure: Metric tons.

	1990		1991		1992		1993		1994		1995		1996	
	Value	%	Value	%	Value	%	Value	%	Value	%	Value	%	Value	%
Total Production	1,461,271	100.0	1,356,114	100.0	1,413,962	100.0	1,490,601	100.0	1,604,888	100.0	1,536,375	100.0	1,618,216	100.0
Regions														
Africa	2,000	0.1	2,000	0.1	2,000	0.1	2,000	0.1	3,000	0.2	3,000	0.2	3,000	0.2
America, North	571,000	39.1	464,000	34.2	509,000	36.0	648,000	43.5	735,000	45.8	728,700	47.4	783,000	48.4
Asia	26,714	1.8	36,357	2.7	40,000	2.8	40,143	2.7	40,286	2.5	40,929	2.7	48,571	3.0
Europe	861,557	59.0	853,757	63.0	862,962	61.0	800,459	53.7	826,603	51.5	763,747	49.7	783,645	48.4
Leading Producers														
United States of America	526,000	36.0	413,000	30.5	455,000	32.2	592,000	39.7	678,000	42.2	*677,700	44.1	*726,000	44.9
Germany	*272,000	18.6	*272,000	20.1	*272,000	19.2	261,000	17.5	283,000	17.6	*272,000	17.7	*272,000	16.8
Italy	227,000	15.5	223,000	16.4	234,000	16.5	182,000	12.2	172,000	10.7	159,000	10.3	145,000	9.0
Mexico	45,000	3.1	51,000	3.8	54,000	3.8	56,000	3.8	57,000	3.6	51,000	3.3	57,000	3.5
Sweden	43,000	2.9	42,000	3.1	40,000	2.8	39,000	2.6	47,000	2.9	5,000	0.3	*39,495	2.4

Glass Bottles and Other Containers of Common Glass

Unit of Measure: Thousands of units.

	1990		1991		1992		1993		1994		1995		1996	
	Value	%	Value	%	Value	%	Value	%	Value	%	Value	%	Value	%
Total Production	76,652,825	100.0	75,733,871	100.0	76,040,470	100.0	74,753,387	100.0	76,470,810	100.0	76,136,118	100.0	74,332,434	100.0
Regions														
Africa	28,534	0.0	38,962	0.1	32,055	0.0	30,318	0.0	39,178	0.1	46,867	0.1	38,197	0.1
America, North	47,103,671	61.5	46,087,078	60.9	46,611,485	61.3	47,760,892	63.9	47,942,299	62.7	45,928,706	60.3	43,983,113	59.2
America, South	4,375,501	5.7	4,438,113	5.9	4,511,724	5.9	4,513,521	6.0	4,681,035	6.1	4,782,328	6.3	4,787,091	6.4
Asia	4,136,124	5.4	4,960,943	6.6	6,507,175	8.6	5,510,079	7.4	5,114,733	6.7	5,742,505	7.5	6,099,936	8.2
Europe	21,008,995	27.4	20,208,775	26.7	18,378,031	24.2	16,938,577	22.7	18,693,565	24.4	19,635,711	25.8	19,424,095	26.1
Leading Producers														
United States of America	41,717,000	54.4	40,480,000	53.5	41,191,000	54.2	41,994,000	56.2	41,554,000	54.3	38,969,000	51.2	36,557,000	49.2
Spain	*5,666,500	7.4	*5,666,500	7.5	*5,666,500	7.5	5,276,000	7.1	5,904,000	7.7	5,712,000	7.5	5,774,000	7.8
Mexico	4,942,000	6.4	5,141,000	6.8	4,933,000	6.5	5,258,000	7.0	5,858,000	7.7	6,408,000	8.4	6,853,000	9.2
United Kingdom	4,539,780	5.9	3,290,561	4.3	2,041,341	2.7	792,121	1.1	803,844	1.1	*1,580,703	2.1	*1,198,058	1.6
Poland	3,145,000	4.1	3,205,000	4.2	2,914,000	3.8	3,190,000	4.3	3,407,000	4.5	3,611,000	4.7	3,760,000	5.1

Commodity data are provided by the United Nations Statistical Division. The symbol * means that data are estimated. For additional notes, see Appendix I.

463

ISIC 3690
NONMETAL PRODUCTS NEC

ISIC 3690—Nonmetal Products nec—includes refractory ceramic products (e.g., bricks, blocks, tiles, metallurgical furnace liners, refractory cements, retorts, crucibles, mufflers, nozzles, etc.); ceramic building materials such as brick, roofing tiles, pipes, etc.; cement, lime, plaster, and products made of these; cut and shaped stone; and miscellaneous mineral products including asbestos yarns, insulating materials, abrasives, etc.

In the U.S. SIC system, corresponding industries are 3250—Structural Clay Products, 3270—Concrete, Gypsum, and Plaster Products, 3280—Cut Stone and Stone Products, and 3290—Miscellaneous Nonmetallic Mineral Products.

Summary Statistics

		1990 Value	N	1991 Value	N	1992 Value	N	1993 Value	N	1994 Value	N	1995 Value	N
Establishments or enterprises	(number)	127,433	83	129,270	91	142,977	84	67,020	83	52,902	69	36,574	33
Number employed	(000)	8,416	93	6,699	96	6,520	90	7,066	88	3,758	79	3,313	58
Total Wages	($ mil.)	48,620	85	50,576	88	55,240	80	42,642	77	45,063	70	38,216	53
Wage/salary per employee	($)	9,473	83	10,563	86	12,024	79	10,071	77	10,485	70	12,545	53
Female workers	(000)	155	34	118	32	345	26	367	30	360	25	336	13
as % of total employment*	(%)	14	33	14	31	15	26	15	30	17	25	22	12
Output	($ bil.)	291	89	317	93	338	86	295	80	282	73	256	56
per employee	($ 000)	59	83	61	87	70	80	64	75	72	70	84	54
per establishment	($ 000)	3,488	77	3,576	82	3,910	73	4,247	67	3,726	58	3,262	27
Value added	($ bil.)	127	80	128	82	139	75	138	71	118	64	123	53
per employee	($ 000)	28	75	29	77	33	71	29	66	32	61	39	50
per establishment	($ 000)	1,630	67	1,693	71	1,864	63	1,673	56	1,863	48	1,286	23
Capital investment	($ mil.)	18,926	65	17,851	62	15,791	54	15,305	51	12,017	40	7,280	20
per employee	($ 000)	7	61	8	58	5	52	5	49	6	39	4	20
per establishment	($ 000)	536	60	592	57	270	51	314	45	262	37	115	17

Data presented above are drawn from the detailed tables that follow. Columns headed 'N' show the number of countries that provided valid data for inclusion. Values are not strictly comparable one year to the next or one row to the next because the number of countries that report varies from period to period and row to row. However, a general indication of magnitudes can be gleaned from reviewing this summary—especially in earlier years in which more reports are available. For detailed explanations, see Appendix I. *The average for those countries reporting both total and female employment.

Establishments and Number Engaged

Country	Establishments or Enterprises (number)						Employees per Establishment					
	1990	1991	1992	1993	1994	1995	1990	1991	1992	1993	1994	1995
Albania	-	-	-	40	26	14	-	-	-	166	178	272
Argentina	-	-	-	3,542	4,037	-	-	-	-	7.888	7.181	-
Armenia	85	73	48	49	44	-	223	251	-	-	-	-
Australia	1,231	1,327	1,345	-	-	-	27	24	22	-	-	-
Austria	442	433	435	425	422	-	59	58	57	55	55	-
Azerbaijan	344	180	699	869	572	-	33	62	16	12	16	-
Bahamas	16	14	-	-	-	-	9.188	11	-	-	-	-
Bahrain	-	-	40	-	-	-	-	-	40	-	-	-
Bangladesh	522	525	594	-	-	-	37	37	35	-	-	-
Belize	29	30	30	-	-	-	-	11	11	-	-	-
Benin	-	-	3.000	3.000	-	-	-	-	-	-	-	-
Bolivia	117	124	134	137	136	134	23	22	23	24	23	27
Bosnia & Herzegovina	32	32	-	-	2.000	-	177	171	-	-	298	-
Bulgaria	*108	*109	*102	*289	*360	*392	341	273	229	74	60	51
Burundi	*1.000	*1.000	-	-	-	-	190	186	-	-	-	-
Canada	1,447	1,380	1,350	1,330	1,341	-	28	28	25	24	24	-
Cape Verde	5.000	5.000	4.000	4.000	3.000	-	-	-	-	-	-	-
Central African Republic	-	*1.000	*1.000	*1.000	*1.000	*1.000	-	-	-	-	-	-
Chile	41	48	46	52	54	-	144	141	164	165	164	-
China	*55,275	*54,110	*53,854	-	-	-	54	57	58	-	-	-
China (Hong Kong SAR)	206	120	167	112	165	-	14	23	16	23	21	-
Colombia	312	314	327	326	330	-	66	65	65	70	70	-
Costa Rica	209	202	188	192	-	-	19	18	20	21	-	-
Croatia	187	234	281	354	416	457	110	73	53	40	32	28
Cyprus	235	236	236	236	275	307	10	10	11	11	9.160	8.664
Czechoslovakia (Former)	*92	*118	-	-	-	-	815	534	-	-	-	-
Czech Republic	-	*66	*110	*129	-	-	-	561	336	287	-	-
Denmark	818	813	788	-	-	-	18	18	17	-	-	-
Ecuador	81	89	84	75	83	89	58	56	52	56	54	48
Egypt	529	611	571	584	-	-	98	95	105	103	-	-
El Salvador	-	14	1.000	19	30	38	-	-	1,258	71	70	60
Equatorial Guinea	3.000	-	-	-	-	-	17	-	-	-	-	-
Ethiopia	21	18	18	17	41	48	141	163	153	169	102	101
Finland	362	401	350	303	279	-	43	35	32	28	29	-
FYR Macedonia	*3.000	*3.000	*6.000	*7.000	*8.000	*9.000	793	593	233	195	175	137
Gabon	-	2.000	3.000	1.000	3.000	1.000	-	187	105	423	118	305
Gambia	-	-	-	4.000	-	-	-	-	-	27	-	-
Germany	-	2,728	2,758	2,850	2,978	-	-	71	62	61	58	-
Germany (Western Part)	2,361	2,393	2,428	-	-	-	59	61	59	-	-	-
Ghana	-	-	-	24	-	-	-	-	-	118	-	-
Greece	533	537	544	-	-	-	35	34	32	-	-	-
Grenada	-	-	12	7.000	4.000	4.000	-	-	6.583	8.286	8.250	7.750
Guatemala	-	14	14	14	13	13	-	134	156	156	173	174
Honduras	-	-	29	32	32	30	-	-	160	167	181	209
Hungary	*149	*260	*329	*375	-	-	188	92	58	51	-	-
Iceland	65	74	66	71	-	-	12	11	12	9.859	-	-
India	8,311	8,734	9,196	9,535	-	-	43	42	42	38	-	-
Indonesia	1,218	1,265	1,534	1,355	1,452	1,861	62	62	54	66	65	59
Iran (Islamic Republic of)	2,575	1,954	2,049	1,956	-	-	32	29	26	28	-	-
Iraq	-	-	172	-	-	-	-	-	105	-	-	-
Ireland	243	178	-	-	-	-	27	35	-	-	-	-
Israel	302	369	411	385	391	-	22	21	21	22	23	-
Italy	*992	*970	*1,797	*1,791	*1,717	-	65	67	64	62	62	-
Japan	15,516	15,427	15,152	15,243	14,763	14,923	20	21	21	21	21	21
Jordan	1,342	1,368	1,689	1,711	1,895	2,008	4.455	5.135	5.303	6.553	5.891	5.313
Kenya	49	47	48	47	48	48	110	116	117	121	117	126
Korea, Republic of	2,747	3,065	3,204	3,334	3,233	3,269	30	30	28	27	27	26
Kuwait	125	113	85	90	102	-	36	18	40	43	47	-
Kyrgyzstan	264	255	190	162	130	-	73	76	88	81	68	-
Latvia	209	133	114	87	65	92	75	121	111	72	71	44

Continued.

Depending on the table, * means *Enterprises, Engaged,* or *Factor Values*; ± means *Basis Unspecified*; § means *shown in millions.* For additional notes and sources, see Appendix I.

Establishments and Number Engaged

- Continued -

Country	Establishments or Enterprises (number)						Employees per Establishment					
	1990	1991	1992	1993	1994	1995	1990	1991	1992	1993	1994	1995
Lesotho	41	-	-	3.000	3.000	-	27	-	-	60	71	-
Lithuania	-	-	-	*144	*145	-	-	-	-	161	139	-
Macau	10	11	12	14	13	16	3.700	5.000	8.667	18	34	31
Malawi	*3.000	*3.000	*3.000	*3.000	*3.000		300	300	367	333	400	-
Malaysia	345	366	357	373	376	-	72	78	81	80	84	-
Malta	86	97	96	95	91	-	8.140	9.794	11	10	11	-
Mauritius	15	27	27	28	34	33	83	50	53	50	45	48
Mexico	166	161	156	152	144	142	186	188	185	168	166	141
Mongolia	44	41	48	33	22	36	82	112	68	66	114	101
Mozambique	10	10	7.000	7.000	7.000	-	342	285	195	192	173	-
Myanmar	19	19	19	19	-	-	-	-	-	-	-	-
Namibia	-	-	-	-	28	-	-	-	-	-	42	-
Nepal	636	707	-	752	727	-	90	88	-	72	84	-
Netherlands	193	195	204	203	217	-	87	87	83	79	68	-
Netherlands Antilles	-	-	-	25		-	-	-	-	2.800	-	
New Zealand	*313	*287	*297	*310		-	13	14	14	14	-	
Norway	252	228	109	140	140	-	24	23	39	30	31	-
Pakistan	-	110	-	-	-	-	-	185	-	-	-	-
Panama	48	49	-	-	-	-	32	35	-	-	-	-
Paraguay	-	7.000	-	-	-	-	-	168	-	-	-	-
Peru	679	687	683	-	-	-	16	15	13	-	-	-
Philippines	2,561	2,738	417	340		-	11	10	56	58	-	
Poland	*333	*378	*379	*322		-	306	251	219	227	-	
Portugal	*2,811	*2,832	*2,456	*2,605	*2,670	-	13	14	14	14	13	-
Puerto Rico	117	113	-	115	120	118	16	17	-	15	15	16
Republic of Moldova	33	37	40	37	37	35	587	576	526	374	160	164
Romania	-	-	-	-	515	-	-	-	-	-	-	-
Russian Federation	-	-	-	*5,980	*4,964	*6,698	-	-	-	166	182	130
Saint Lucia	-	1.000	1.000	1.000	1.000	1.000	-	-	-	-	-	-
Senegal	4.000	3.000	2.000	4.000	2.000	-	123	156	358	184	363	-
Sierra Leone	-	-	-	15	-	-	-	-	-	27	-	-
Singapore	65	67	71	75	76	-	67	64	66	68	70	-
Slovakia	-	72	72	76	80	-	-	377	321	259	226	-
Slovenia	112	*218	*307	*326	*451	*289	-	-	-	-	-	-
South Africa	-	1,155	-	-	-	-	-	59	-	-	-	-
Spain	6,666	6,693	6,218	-	-	-	14	14	15	-	-	-
Sri Lanka	167	117	118	116	-	-	61	55	51	56	-	-
Swaziland	-	10	10	9.000	-	-	-	33	30	29	-	-
Sweden	319	333	321	307	295	-	49	47	41	34	34	-
Thailand	573	1,049	-	-	-	-	117	60	-	-	-	-
Trinidad and Tobago	63	61	40	43	39	-	17	17	28	28	34	-
Tunisia	-	-	-	340	349	359	-	-	-	46	44	44
Turkey	419	416	739	710	706	-	109	101	65	62	61	-
Ukraine	3,777	5,115	5,506	4,141	4,151	4,108	101	73	60	83	70	65
United Kingdom	2,724	2,566	2,698	-	-	-	45	42	37	-	-	-
United Republic of Tanzania	17	17					195	179	-	-	-	-
USSR (Former)	*3,380	-	-				559	-	-			
United States of America	-	-	17,044				-	-	20	-	-	-
Venezuela	488	506	498	446	468	477	43	41	44	46	46	37
Yugoslavia	144	221	356	511	521	524	-	154	97	65	61	59
Zambia	13	-	-	-	24	-	200	-	-	-	106	-
Zimbabwe	33	30	30	31	32	-	179	200	233	216	222	-

Depending on the table, * means *Enterprises, Engaged,* or *Factor Values;* ± means *Basis Unspecified;* § means *shown in millions.* For additional notes and sources, see Appendix I.

467

Employment and Compensation of Employees

Country	Number Employed/Engaged (000)						Wage/Salary per Employee ($)					
	1990	1991	1992	1993	1994	1995	1990	1991	1992	1993	1994	1995
Albania	25	-	-	6.638	4.640	3.801	462	-	-	475	2,354	2,353
Argentina	51	-	-	28	29	-	5,598	-	-	12,505	-	-
Armenia	*19	*18	-	-	-	-	-	-	-	-	-	-
Australia	33	*32	*30	-	-	-	22,959	24,080	23,235	-	-	-
Austria	26	25	25	23	23	23	29,567	31,017	34,591	34,316	35,751	-
Azerbaijan	11	11	11	11	9.273	-	1,900	1,756	-	-	-	-
Bahamas	0.147	0.151	-	-	-	-	15,912	20,199	-	-	-	-
Bahrain	-	-	1.582	-	-	-	-	-	5,993	-	-	-
Bangladesh	19	19	21	-	-	-	-	-	-	-	-	-
Belgium	20	20	20	-	-	-	-	-	-	-	-	-
Belize	-	0.340	0.340	-	-	-	-	2,749	3,565	-	-	-
Bermuda	0.006	0.005	-	-	-	-	-	-	-	-	-	-
Bolivia	2.652	2.747	3.140	3.265	3.152	3.573	2,132	2,351	2,347	2,786	2,784	2,655
Bosnia & Herzegovina	5.659	5.478	-	-	0.596	-	2,611	2,719	-	-	-	-
Bulgaria	37	30	23	22	22	20	1,962	700	1,083	1,416	1,172	1,531
Burundi	0.190	0.186	-	-	-	-	2,754	2,811	-	-	-	-
Cambodia	*7.236	*5.828	-	-	-	-	-	-	-	-	-	-
Cameroon	0.376	0.365	0.373	0.347	-	-	17,876	16,578	18,242	19,357	-	-
Canada	41	38	34	32	32	32	29,156	28,895	28,738	27,977	26,568	26,860
Chile	5.900	6.750	7.525	8.566	8.865	9.319	6,811	7,550	8,633	9,166	10,464	12,481
China	*3,000	*3,070	*3,150	*3,060	-	-	-	-	-	-	-	-
China (Hong Kong SAR)	*2.900	*2.700	*2.700	*2.600	*3.400	-	13,059	14,822	-	18,446	18,153	-
China (Taiwan Province)	49	47	50	52	54	51	10,522	11,699	14,710	14,777	15,485	15,469
Colombia	20	20	21	23	23	23	2,470	2,505	2,860	3,135	4,039	4,543
Costa Rica	3.931	3.604	3.817	4.046	-	-	2,887	2,792	3,192	3,551	-	-
Croatia	20	17	15	14	13	13	4,116	4,383	1,572	1,637	2,286	3,770
Cyprus	2.399	2.428	2.552	2.486	2.519	2.660	12,776	13,389	14,050	14,288	14,908	20,345
Czechoslovakia (Former)	75	63	-	-	-	-	2,303	1,610	-	-	-	-
Czech Republic	45	37	37	37	-	-	2,303	1,604	2,066	2,498	-	-
Denmark	15	15	14	13	14	15	32,391	32,034	35,866	33,668	36,712	-
Ecuador	4.717	4.948	4.395	4.223	4.479	4.240	4,405	5,057	6,432	6,459	2,911	3,975
Egypt	*52	*58	*60	*60	*58	*61	2,082	1,387	1,489	1,687	1,686	1,875
El Salvador	-	-	*1.258	*1.355	*2.112	*2.284	-	-	-	3,210	4,982	3,841
Equatorial Guinea	0.050	-	-	-	-	-	1,613	-	-	-	-	-
Ethiopia	2.954	2.929	2.753	2.865	4.172	4.864	1,279	1,324	1,090	797	733	673
Finland	16	14	11	8.500	8.100	-	29,014	27,101	24,067	19,730	22,340	-
FYR Macedonia	2.379	1.779	1.400	1.366	1.399	1.236	1,619	5,593	1,954	1,991	2,548	3,683
France	89	88	83	78	75	74	-	-	-	-	-	-
Gabon	-	0.374	0.316	0.423	0.355	0.305	-	13,658	18,077	12,323	9,884	12,776
Gambia	-	-	-	0.109	-	-	-	-	-	-	-	-
Germany	-	*193	*172	*173	*174	-	-	26,261	32,984	32,699	34,912	-
Germany (Western Part)	*139	*145	*143	-	-	-	30,533	31,449	35,775	-	-	-
Ghana	-	-	-	*2.835	*2.915	*2.997	-	-	-	-	-	-
Greece	*19	*18	*18	*17	*18	*18	13,289	14,137	15,303	14,983	15,524	17,876
Grenada	-	-	*0.079	*0.058	0.033	0.031	-	-	-	-	-	-
Guatemala	-	*1.877	*2.190	*2.190	*2.252	*2.265	-	-	-	-	-	-
Honduras	*2.924	*4.324	*4.632	*5.332	*5.791	*6.283	3,431	2,256	2,740	1,723	1,571	1,870
Hungary	28	24	19	19	22	25	2,626	2,891	3,424	3,655	3,653	3,695
Iceland	0.770	0.815	0.767	0.700	0.650	-	30,141	33,025	34,336	25,762	-	-
India	*360	*369	*382	*366	*377	*409	878	761	787	769	822	896
Indonesia	75	78	83	90	94	111	660	755	995	1,137	1,056	1,801
Iran (Islamic Republic of)	83	56	54	54	-	-	3,644	4,769	4,870	4,678	-	-
Iraq	-	16	18	-	-	-	-	8,006	13,942	-	-	-
Ireland	6.500	6.300	6.232	6.180	6.137	5.992	25,003	25,848	29,557	25,877	28,875	35,408
Israel	6.700	7.700	8.700	8.400	9.000	9.263	21,690	22,850	22,343	23,009	24,650	26,790
Italy	64	65	114	111	107	-	34,490	36,915	39,784	32,373	32,883	-
Jamaica	2.173	2.145	2.100	2.264	2.203	-	-	-	-	-	-	-
Japan	*316	*317	*317	*313	*314	*308	26,249	29,834	32,878	37,954	-	-
Jordan	5.978	7.024	8.956	11	11	11	2,625	2,463	2,604	2,308	2,563	3,158
Kenya	5.403	5.467	5.639	5.703	5.640	6.037	2,609	2,303	2,087	1,284	1,477	2,031

Continued.

Depending on the table, * means *Enterprises, Engaged,* or *Factor Values*; ± means *Basis Unspecified*; § means *shown in millions*. For additional notes and sources, see Appendix I.

Employment and Compensation of Employees

- Continued -

Country	Number Employed/Engaged (000)						Wage/Salary per Employee ($)					
	1990	1991	1992	1993	1994	1995	1990	1991	1992	1993	1994	1995
Korea, Republic of	83	91	91	91	88	86	10,047	11,845	13,028	13,702	15,234	18,206
Kuwait	4.549	1.993	3.381	3.912	4.824	4.123	5,091	5,452	8,223	8,017	8,143	10,964
Kyrgyzstan	*19	*19	*17	*13	*8.843	-	432	-	-	-	-	-
Latvia	16	16	13	6.284	4.614	4.087	-	-	-	-	-	-
Lesotho	*1.098	-	*0.240	*0.179	*0.214	-			2,130	2,052	1,791	-
Lithuania	-	-	30	23	20	-	-	-	-	518	1,162	-
Macau	0.037	0.055	0.104	0.245	0.444	0.491	3,686	5,236	9,237	10,222	10,900	9,526
Malawi	0.900	0.900	1.100	1.000	1.200	-	1,099	1,149	933	886	506	-
Malaysia	25	28	29	30	32	34	3,619	3,789	4,449	4,911	5,179	5,719
Malta	0.700	0.950	1.010	0.962	1.010	-	8,738	8,124	8,752	7,927	8,610	-
Mauritius	*1.252	1.356	1.439	1.399	1.539	1.584	2,461	3,821	5,040	5,776	5,351	5,923
Mexico	31	30	29	26	24	20	6,195	7,781	9,519	10,603	11,775	7,544
Mongolia	*3.609	*4.603	*3.241	*2.176	*2.500	*3.631	3,958	2,424	805	-	972	791
Mozambique	3.424	2.853	1.368	1.345	1.208	-	1,969	1,572	769	713	991	-
Namibia	-	-	-	-	1.170	-	-	-	-	-	3,710	-
Nepal	57	62	-	54	61	-	-	-	-	-	-	-
Netherlands	17	17	17	16	15	15	27,465	28,590	31,784	30,887	37,573	-
Netherlands Antilles	-	-	-	0.070	-	-	-	-	-	5,890	-	-
New Zealand	4.219	4.070	4.153	4.220	-	-	-	-	-	-	-	-
Norway	6.093	5.188	4.232	4.262	4.347	4.940	31,594	31,576	-	33,008	34,616	39,489
Pakistan	20	20	21	-	-	-	2,717	2,769	2,848	-	-	-
Panama	*1.553	*1.736	*1.986	*2.204	*2.321	*2.150	5,483	5,729	6,240	6,725	7,061	7,152
Paraguay	-	1.177	-	-	-	-	-	-	-	-	-	-
Peru	*11	*11	*9.191	-	-	-	2,935	3,411	3,686	-	-	-
Philippines	29	28	23	20	-	-	1,708	1,881	2,911	2,868	-	-
Poland	102	95	83	73	84	89	1,180	1,877	2,374	2,498	2,643	3,136
Portugal	*38	*41	*35	*37	*35	*35	-	-	-	-	-	-
Puerto Rico	1.910	1.950	1.820	1.770	1.820	1.860	-	36,974	39,505	35,650	36,264	38,011
Republic of Moldova	19	21	21	14	5.918	5.728	-	3,307	-	-	-	-
Russian Federation	-	-	-	995	901	869	-	-	-	812	1,337	1,359
Senegal	*0.491	*0.469	*0.716	*0.735	*0.725	-	8,625	8,918	5,804	5,785	3,250	-
Sierra Leone	-	-	-	*0.405	-	-	-	-	-	-	-	-
Singapore	*4.356	*4.305	*4.719	*5.071	*5.336	*5.304	11,105	14,197	15,516	16,547	18,663	21,433
Slovakia	-	*27	*23	*20	*18	-	-	1,605	1,997	2,248	2,486	-
South Africa	71	68	66	-	-	-	6,347	6,487	7,342	-	-	-
Spain	91	93	91	110	108	110	17,337	18,732	21,049	18,372	18,091	20,469
Sri Lanka	*10	*6.429	*5.975	*6.526	-	-	980	1,184	1,021	1,012	-	-
Swaziland	-	0.333	0.299	0.258	-	-	-	3,659	5,628	3,344	-	-
Sweden	15	*16	*13	11	9.956	10	25,386	26,680	30,255	23,793	25,175	27,899
Thailand	67	63	-	-	-	-	6,179	4,851	-	-	-	-
Trinidad and Tobago	1.077	1.064	1.107	1.211	1.344	-	8,014	9,021	9,322	7,824	7,321	-
Tunisia	-	-	-	16	15	16	-	-	-	8,112	8,337	9,467
Turkey	46	42	48	44	43	45	7,199	7,624	7,500	7,912	4,718	5,760
Ukraine	382	374	328	345	291	268	2,877	2,950	2,573	-	565	632
United Kingdom	123	108	100	-	-	-	23,926	24,844	26,060	-	-	-
United Republic of Tanzania	3.317	3.048	-	-	-	-	-	496	-	-	-	-
USSR (Former)	1,889	-	-	-	-	-	2,220	-	-	-	-	-
United States of America	364	339	333	334	347	364	25,687	26,165	27,787	28,389	29,124	29,626
Uruguay	*4.251	*4.481	*4.610	*4.377	*4.392	*4.316	3,190	3,866	4,178	4,930	5,116	5,442
Venezuela	21	21	22	20	21	18	4,049	4,348	5,068	5,027	3,896	5,161
Yugoslavia	-	34	35	33	32	31	-	-	-	-	2,182	975
Zambia	2.604	-	-	-	2.554	-	1,899	-	-	-	1,780	-
Zimbabwe	5.900	6.000	7.000	6.700	7.100	6.929	3,746	3,369	2,434	2,421	2,468	2,680

Depending on the table, * means Enterprises, Engaged, or Factor Values; ± means Basis Unspecified; § means shown in millions. For additional notes and sources, see Appendix I.

469

Female Workers: Total and Percent of Employment

Country	Female Workers (000)						Female Workers as % of Total Employment					
	1990	1991	1992	1993	1994	1995	1990	1991	1992	1993	1994	1995
Afghanistan	0.045	0.056	-	-	-	-	-	-	-	-	-	-
Albania	-	-	-	1.043	1.026	0.731	-	-	-	15.71	22.11	19.23
Argentina	-	-	-	-	1.236	-	-	-	-	-	4.26	-
Australia	3.046	-	-	-	-	-	9.23	-	-	-	-	-
Austria	3.100	3.000	3.000	2.908	2.891	-	11.97	12.00	12.00	12.42	12.36	-
Bahrain	-	-	0.009	-	-	-	-	-	0.57	-	-	-
Bangladesh	0.077	0.241	0.317	-	-	-	0.40	1.25	1.54	-	-	-
Bermuda	0.003	0.002	-	-	-	-	50.00	40.00	-	-	-	-
Bosnia & Herzegovina	1.210	1.163	-	-	-	-	21.38	21.23	-	-	-	-
Bulgaria	-	-	-	7.700	7.400	6.400	-	-	-	35.81	34.26	32.00
Canada	3.000	3.000	-	-	-	-	7.35	7.89	-	-	-	-
Chile	0.297	0.300	0.357	0.425	0.473	-	5.03	4.44	4.74	4.96	5.34	-
China (Taiwan Province)	12	12	12	12	12	12	25.36	25.19	23.96	23.97	22.89	22.97
Colombia	-	1.965	-	2.334	2.403	-	-	9.61	-	10.28	10.33	-
Croatia	3.610	3.150	2.840	2.640	2.420	2.240	17.63	18.53	19.16	18.72	18.45	17.45
Cyprus	0.274	0.277	0.324	0.325	0.306	-	11.42	11.41	12.70	13.07	12.15	-
Czechoslovakia (Former)	12	11	-	-	-	-	16.53	17.62	-	-	-	-
Czech Republic	12	11	10	10	-	-	26.67	29.73	27.03	27.03	-	-
Denmark	2.441	2.537	2.300	-		-	16.83	17.05	16.90	-	-	-
Egypt	-	1.854	2.180	2.283	-	-	-	3.21	3.65	3.79	-	-
El Salvador	-	-	-	0.086	0.122	0.130	-	-	-	6.35	5.78	5.69
Ethiopia	-	0.301	0.384	0.433	0.435	0.433	-	10.28	13.95	15.11	10.43	8.90
Gambia	-	-	-	0.006	-	-	-	-	-	5.50	-	-
Germany (Western Part)	21	-	-	-	-	-	15.12	-	-	-	-	-
Hungary	8.000	-	6.000	5.000	-	-	28.57	-	31.58	26.32	-	-
Indonesia	-	-	-	20	21	26	-	-	-	22.41	22.09	23.14
Iran (Islamic Republic of)	-	0.982	0.897	1.133	-	-	-	1.75	1.66	2.09	-	-
Iraq	-	1.948	-	-	-	-	-	12.27	-	-	-	-
Ireland	0.600	0.600	-	-	-	-	9.23	9.52	-	-	-	-
Italy	-	4.617	17	16	17	-	-	7.10	14.82	14.92	15.71	-
Jordan	0.045	0.080	0.063	0.071	0.094	0.073	0.75	1.14	0.70	0.63	0.84	0.68
Kenya	0.148	0.155	0.175	0.174	-	-	2.74	2.84	3.10	3.05	-	-
Korea, Republic of	15	18	17	16	15	14	18.65	19.31	18.28	17.20	16.61	16.15
Lesotho	-	-	-	0.009	0.012	-	-	-	-	5.03	5.61	-
Macau	0.005	0.007	0.015	0.019	0.045	0.074	13.51	12.73	14.42	7.76	10.14	15.07
Malaysia	4.900	5.500	5.800	5.600	5.700	-	19.60	19.37	20.07	18.79	18.10	-
Malta	0.027	0.040	0.036	0.027	0.025	-	3.86	4.21	3.56	2.81	2.48	-
Mongolia	-	-	-	-	1.138	-	-	-	-	-	45.52	-
Morocco	-	-	-	-	-	0.834	-	-	-	-	-	-
Nepal	10	13	-	7.785	9.081	-	17.55	20.25	-	14.37	14.96	-
New Zealand	0.463	0.457	0.476	0.523	-	-	10.97	11.23	11.46	12.39	-	-
Panama	0.131	-	-	-	-	-	8.44	-	-	-	-	-
Philippines	-	-	2.232	2.284	-	-	-	-	9.62	11.54	-	-
Portugal	1.938	1.654	1.680	1.780	1.489	-	5.12	4.08	4.78	4.88	4.25	-
Republic of Moldova	-	-	-	-	1.723	1.618	-	-	-	-	29.11	28.25
Senegal	0.004	-	-	-	-	-	0.81	-	-	-	-	-
Sri Lanka	2.400	1.425	1.166	1.347	-	-	23.73	22.17	19.51	20.64	-	-
Sweden	2.100	-	-	-	-	-	13.55	-	-	-	-	-
Thailand	11	18	-	-	-	-	16.80	28.68	-	-	-	-
Turkey	2.370	-	-	-	-	-	5.19	-	-	-	-	-
United Kingdom	19	-	15	-	-	-	15.51	-	15.00	-	-	-
United Republic of Tanzania	0.586	0.535	-	-	-	-	17.67	17.55	-	-	-	-
United States of America	-	-	244	246	257	272	-	-	73.27	73.65	74.06	74.73
Zambia	-	-	-	-	0.091	-	-	-	-	-	3.56	-

Output and Output per Employee

Country	Output ($ bil.)						Output per Employee ($)					
	1990	1991	1992	1993	1994	1995	1990	1991	1992	1993	1994	1995
Albania	-	-	-	*0.017	*0.026	*0.021	-	-	-	2,614	5,665	5,485
Argentina	±1.343	-	-	1.918	-	-	26,540	-	-	68,651	-	-
Armenia	0.007	0.012	0.061	1.639	0.013	-	357	633	-	-	-	-
Australia	*5.124	*4.919	*4.521	-	-	-	155,279	153,730	150,686	-	-	-
Austria	3.783	3.965	4.240	4.059	4.479	5.034	146,080	158,595	169,588	173,305	191,561	218,907
Azerbaijan	±0.121	±0.120	±0.008	±0.033	±0.012	-	10,580	10,635	689	3,076	1,251	-
Bahamas	0.014	0.015	-	-	-	-	95,238	99,338	-	-	-	-
Bahrain	-	-	±0.099				-	-	62,761			
Bangladesh	0.072	0.052	0.070	-	-	-	3,749	2,709	3,399	-	-	-
Belize	0.008	0.006	0.007	-	-	-	-	18,063	21,044	-	-	-
Benin	±0.067	±0.067	±0.085	±0.067	-	-	-	-	-	-	-	-
Bolivia	0.065	0.074	0.078	0.090	0.109	0.118	24,451	26,862	24,854	27,431	34,665	32,995
Bosnia & Herzegovina	±0.240	±0.292	-	-	-	-	42,437	53,236	-	-	-	-
Bulgaria	±0.711	±0.216	±0.207	±0.205	±0.192	±0.269	19,320	7,249	8,834	9,554	8,891	13,452
Burundi	±0.004	±0.004	-	-	-	-	18,949	19,096	-	-	-	-
Canada	5.819	4.931	4.368	4.193	4.262	4.281	142,631	129,776	128,480	131,046	133,183	134,790
Chile	0.414	0.501	0.715	0.850	0.881	1.106	70,209	74,240	95,061	99,187	99,362	118,675
China	±14	±15	±20	±31	-	-	4,779	4,940	6,321	10,215	-	-
China (Hong Kong SAR)	0.548	0.550	0.507	0.541	0.764	-	188,974	203,887	187,707	208,178	224,725	-
China (Taiwan Province)	±3.673	±3.837	±4.724	±6.529	±6.597	±6.358	74,683	81,117	95,203	125,694	121,213	123,556
Colombia	0.657	0.641	0.761	0.950	1.368	1.553	32,050	31,344	35,971	41,834	58,802	66,591
Costa Rica	0.092	0.080	0.095	0.110	0.119	0.121	23,440	22,207	24,895	27,179	-	-
Cote d'Ivoire	±1.146	±1.106	±1.050	±1.211	-	-	-	-	-	-	-	-
Croatia	±0.509	±0.495	±0.310	-	-	-	24,848	29,132	20,905	-	-	-
Cyprus	0.165	0.168	0.186	0.174	0.184	0.251	68,729	69,003	72,914	69,989	72,976	94,476
Czechoslovakia (Former)	±1.180	±0.502	-	-	-	-	15,733	7,963	-	-	-	-
Denmark	*1.587	*1.510	*1.625	*1.438	*1.769	*2.173	109,441	101,451	119,351	111,817	123,180	143,840
Ecuador	0.156	0.191	0.210	0.220	0.281	0.307	33,139	38,603	47,894	52,096	62,626	72,508
Egypt	*0.933	*0.729	*1.031	*1.087	*1.060	*1.259	18,086	12,609	17,238	18,057	18,134	20,631
El Salvador	-	±0.020	-	±0.031	±0.095	±0.098	-	-	-	22,875	44,917	43,101
Equatorial Guinea	±0.000	-	-	-	-	-	2,938	-	-	-	-	-
Ethiopia	0.034	0.033	0.023	0.025	0.037	0.051	11,544	11,305	8,242	8,576	8,988	10,541
Finland	±2.229	±1.697	±1.183	±0.807	±0.968	-	142,873	122,074	105,602	94,904	119,498	-
FYR Macedonia	±0.038	±0.044	±0.020	±0.015	±0.014	±0.011	15,933	24,857	14,070	10,857	10,192	8,516
France	±17	±17	±18	±16	±17	±19	190,422	193,537	213,421	200,741	225,469	259,077
Gabon	-	*0.029	*0.028	*0.026	*0.015	*0.019	-	76,306	87,026	62,249	43,603	62,342
Germany	-	24	28	28	32	-	-	122,078	160,466	160,129	183,056	-
Germany (Western Part)	20	22	25	-	-	-	146,906	153,057	177,015	-	-	-
Greece	*1.527	*1.515	*1.585	*1.538	*1.607	*1.853	81,541	83,972	90,033	87,929	90,974	104,622
Guatemala	-	0.091	0.102	0.136	0.138	0.170	-	48,540	46,434	62,049	61,471	74,837
Honduras	0.070	0.070	0.068	0.095	0.104	0.122	24,058	16,161	14,709	17,826	17,884	19,345
Hungary	±0.628	±0.519	±0.489	±0.488	±0.566	±0.637	22,432	21,632	25,749	25,684	25,413	25,508
Iceland	0.083	0.092	±0.087	±0.070	-	-	107,766	112,847	113,683	99,879	-	-
India	*4.571	*4.509	*4.397	*3.965	*4.463	*5.367	12,681	12,235	11,518	10,826	11,848	13,133
Indonesia	0.979	1.182	1.231	1.486	1.596	1.934	13,050	15,060	14,806	16,596	16,955	17,470
Iran (Islamic Republic of)	1.225	1.079	1.313	1.123	-	-	14,842	19,219	24,247	20,731	-	-
Iraq	-	*0.236	*2.527	-	-	-	-	14,900	140,389	-	-	-
Ireland	*1.165	*1.018	*1.156	*0.939	*1.117	*1.345	179,156	161,602	185,487	151,936	182,054	224,395
Israel	0.802	1.097	1.226	1.173	1.327	1.518	119,628	142,458	140,973	139,655	147,454	163,834
Italy	±11	±12	±21	±16	±16	-	170,290	177,971	182,265	141,706	148,254	-
Japan	±54	±60	±63	±69	±76	±81	169,385	190,806	200,284	220,999	241,011	264,442
Jordan	0.158	0.191	0.263	0.297	0.333	0.311	26,486	27,135	29,394	26,469	29,875	29,109
Kenya	*0.175	*0.198	*0.123	*0.087	*0.093	*0.120	32,306	36,174	21,797	15,177	16,575	19,904
Korea, Republic of	8.173	11	11	11	12	15	98,356	117,717	119,714	122,915	139,277	174,953
Kuwait	0.260	0.129	0.181	0.276	0.358	0.361	57,064	64,864	53,534	70,553	74,145	87,677
Kyrgyzstan	0.185	0.179	0.140	0.023	0.023	-	9,648	9,189	8,355	1,752	2,621	-
Latvia	0.002	0.005	0.030	0.021	0.034	0.035	122	296	2,364	3,356	7,365	8,522
Lithuania	-	-	0.083	0.085	0.109	-	-	-	2,791	3,655	5,411	-
Macau	0.001	0.004	0.014	0.034	0.087	0.064	16,389	66,718	130,352	139,554	195,070	130,392
Malawi	*0.016	*0.018	*0.015	*0.010	*0.007	-	18,200	19,540	13,296	9,539	5,790	-

Continued.

Depending on the table, * means *Enterprises, Engaged,* or *Factor Values*; ± means *Basis Unspecified*; § means *shown in millions*. For additional notes and sources, see Appendix I.

471

Output and Output per Employee
- Continued -

Country	Output ($ bil.)						Output per Employee ($)					
	1990	1991	1992	1993	1994	1995	1990	1991	1992	1993	1994	1995
Malaysia	*0.942	*1.166	*1.383	*1.489	*1.843	*2.216	37,690	41,068	47,854	49,950	58,492	65,211
Malta	0.046	0.063	0.071	0.049	0.058		66,142	66,045	70,266	51,355	57,122	
Mauritius	0.028	0.032	0.055	0.060	0.063	0.070	22,748	23,728	37,959	42,735	40,611	44,392
Mexico	1.863	2.401	2.876	3.032	3.028	1.762	60,186	79,460	99,461	118,638	126,429	87,849
Mongolia	0.125	0.053	0.019	0.005	0.015	0.017	34,547	11,422	5,780	2,388	5,893	4,574
Mozambique	-	±0.015	-	±0.015	±0.014	-	-	5,164	-	10,842	11,797	
Namibia	-	-	-	-	0.028		-	-	-	-	24,004	
Nepal	0.057	0.069	-	0.049	0.069		992	1,119	-	908	1,138	-
Netherlands	2.564	2.557	2.834	2.591	2.948	3.441	152,812	151,338	166,397	160,886	198,466	233,458
Norway	±0.988	±0.815	±0.758	±0.722	±0.824	±1.076	162,217	157,136	179,188	169,439	189,508	217,797
Pakistan	0.579	0.584	0.634	-	-	-	28,904	28,759	29,842			
Panama	0.043	0.067	0.092	0.117	0.134	0.123	27,508	38,446	46,419	53,292	57,622	57,319
Peru	0.409	0.390	0.431	-	-	-	38,592	36,839	46,929			
Philippines	0.573	0.602	0.736	0.684	-	-	19,897	21,126	31,718	34,566	-	
Poland	±1.191	±1.420	±1.298	±1.290	±1.581	±1.966	11,676	14,942	15,635	17,675	18,714	22,209
Portugal	1.767	2.073	2.291	2.049	2.169	2.573	46,703	51,150	65,227	56,134	61,858	73,190
Puerto Rico	±0.133	±0.123	±0.130	±0.133	±0.147	±0.148	69,672	63,128	71,538	75,085	80,549	79,624
Republic of Moldova	0.000	0.449	0.038	0.027	0.034	0.029	0.030	21,049	1,787	1,977	5,755	5,026
Romania	±0.134	±0.107	±0.064	±0.139			-					
Russian Federation	-	16	11	4.576	6.715	9.046	-			4,599	7,449	10,404
Senegal	0.068	-	-	-	-	-	138,151					
Singapore	*0.476	*0.655	*0.790	*0.943	*1.080	*1.246	109,205	152,250	167,495	186,028	202,384	234,879
Slovakia	-	±0.328	±0.322	±0.297	±0.284	-	-	12,086	13,954	15,056	15,739	-
South Africa	±1.833	±1.772	±1.752	±1.807	±1.839	±2.073	25,819	26,059	26,552	-		
Spain	*11	*11	*12	±11	±12	±14	117,884	121,396	130,968	103,330	108,007	126,722
Sri Lanka	0.084	0.083	0.094	0.083	-	-	8,340	12,938	15,752	12,648	-	
Swaziland	-	*0.007	*0.008	*0.005	-	-	-	21,719	25,362	18,533	-	
Sweden	*2.142	*2.754	*2.486	*1.181	*1.225	*1.453	138,210	176,069	190,085	111,640	123,069	141,052
Thailand	6.227	4.808	-	-	-	-	92,544	76,545	-	-		
Trinidad and Tobago	±0.046	±0.055	±0.059	±0.056	±0.069	-	42,534	51,802	53,487	46,557	51,161	
Tunisia	0.715	0.696	0.730	0.675	0.733	0.875	-	-	-	43,481	47,341	55,181
Turkey	2.464	2.418	2.935	3.465	2.532	3.250	53,937	57,691	61,100	78,128	58,861	72,320
Ukraine	*4.953	*5.433	*6.161	*1.645	*1.453	*1.491	12,965	14,528	18,784	4,768	4,994	5,564
United Kingdom	*17	*14	13	-	-	-	135,990	130,449	134,770	-		
United Republic of Tanzania	*0.027	*0.078	-	-	-	-	7,995	25,746	-	-		
USSR (Former)	±19	-	-	-	-	-	9,992	-	-	-		
United States of America	*47	*43	*45	*47	*52	*56	127,802	126,696	135,354	140,713	149,669	155,135
Uruguay	0.103	0.129	0.132	0.152	0.158	0.165	24,339	28,788	28,731	34,706	36,015	38,318
Venezuela	0.552	0.652	0.767	0.817	0.680	-	26,307	31,214	34,880	40,070	31,772	-
Yugoslavia	±0.074	±0.089	±0.181	-	±0.406	±0.235	-	2,621	5,227	-	12,795	7,570
Zambia	0.062	-	-	-	0.032	-	23,684	-	-	-	12,674	-
Zimbabwe	*0.103	*0.114	*0.104	*0.093	*0.111	*0.118	17,378	18,961	14,919	13,882	15,640	17,087

Value Added and Value Added per Employee

Country	Value Added ($ bil.)						Value Added per Employee ($)					
	1990	1991	1992	1993	1994	1995	1990	1991	1992	1993	1994	1995
Albania	±0.013	-	-	-	-	-	512	-	-	-	-	-
Argentina	±0.932	-	-	0.723	-	-	18,412	-	-	25,885	-	-
Australia	*2.170	*2.136	*2.173	-	-	-	65,743	66,761	72,441	-	-	-
Austria	1.473	1.610	1.671	1.522	1.736	1.940	56,866	64,388	66,834	64,998	74,239	84,366
Bahamas	0.005	0.007	-	-	-	-	30,850	43,464	-	-		
Bangladesh	0.031	0.022	0.031	-	-	-	1,623	1,155	1,513	-	-	-
Belize	0.003	0.003	0.004	-	-	-	-	9,471	10,866	-	-	-
Bolivia	0.031	0.031	0.032	0.041	0.054	0.059	11,508	11,345	10,272	12,474	17,232	16,408
Bosnia & Herzegovina	±0.107	±0.130	-	-	-	-	18,986	23,816	-	-		
Bulgaria	-	±0.070	±0.066	±0.083	±0.073	±0.095	-	2,360	2,809	3,876	3,368	4,734
Burundi	0.002	0.002	-	-	-	-	8,602	9,005	-	-	-	-

Continued.

Depending on the table, * means *Enterprises, Engaged,* or *Factor Values;* ± means *Basis Unspecified;* § means *shown in millions.* For additional notes and sources, see Appendix I.

Value Added and Value Added per Employee

- Continued -

Country	Value Added ($ bil.)						Value Added per Employee ($)					
	1990	1991	1992	1993	1994	1995	1990	1991	1992	1993	1994	1995
Cameroon	0.016	0.017	0.021	-	-	-	42,513	45,567	56,741	-	-	-
Canada	2.803	2.339	2.077	2.046	2.080	2.089	68,690	61,557	61,077	63,949	64,990	65,775
Chile	0.218	0.284	0.399	0.484	0.461	0.579	37,022	42,126	53,058	56,507	52,012	62,105
China	±4.524	±5.023	±6.802	±12	-	-	1,508	1,636	2,159	4,018	-	-
China (Hong Kong SAR)	0.095	0.104	0.082	0.105	0.148	-	32,890	38,699	30,431	40,323	43,461	-
China (Taiwan Province)	1.601	1.711	2.133	2.535	2.480	2.093	32,544	36,169	42,980	48,801	45,567	40,667
Colombia	0.338	0.324	0.414	0.481	0.721	0.820	16,511	15,825	19,593	21,173	31,013	35,134
Costa Rica	0.036	0.031	0.036	0.041	0.043	0.043	9,072	8,733	9,557	10,252	-	-
Cote d'Ivoire	1.043	1.007	0.945	0.908	-	-	-	-	-	-	-	-
Croatia	±0.243	±0.246	±0.159	-	-	-	11,878	14,459	10,707	-	-	-
Cyprus	0.069	0.070	0.077	0.078	0.085	0.103	28,860	28,814	30,226	31,435	33,669	38,820
Czechoslovakia (Former)	±0.411	-	-	-	-	-	5,482	-	-	-	-	-
Denmark	*0.941	*0.898	*0.966	*0.854	*1.055	*1.297	64,894	60,329	70,993	66,448	73,444	85,866
Ecuador	0.060	0.075	0.085	0.079	0.099	0.102	12,654	15,167	19,269	18,816	22,115	23,979
Egypt	*0.331	*0.140	*0.375	*0.401	*0.391	*0.464	6,405	2,424	6,263	6,655	6,685	7,597
El Salvador	-	0.011	-	0.016	0.056	0.052	-	-	-	11,822	26,378	22,927
Ethiopia	0.016	0.014	0.013	0.011	0.019	0.024	5,423	4,842	4,551	3,885	4,489	4,990
Finland	±1.053	±0.773	±0.513	±0.365	±0.446	-	67,505	55,647	45,765	42,962	55,069	-
FYR Macedonia	±0.017	±0.023	±0.011	±0.007	±0.008	±0.005	7,060	13,176	7,724	5,428	5,560	3,896
France	±7.523	±7.524	±7.823	±6.967	±7.427	±8.765	84,436	85,215	94,373	89,547	99,155	117,655
Gabon	-	*0.014	*0.011	*0.011	*0.005	*0.005	-	37,400	35,676	26,933	14,572	15,167
Germany (Western Part)	12	13	15	15	-	-	86,604	89,511	105,829	-	-	-
Greece	*0.641	*0.637	*0.639	*0.622	*0.650	*0.750	34,255	35,316	36,306	35,521	36,788	42,345
Haiti	-	±0.004	±0.002	±0.001	±0.000	±0.001	-	-	-	-	-	-
Honduras	*0.025	*0.022	*0.033	*0.035	*0.034	*0.043	8,668	5,101	7,048	6,565	5,879	6,770
Hungary	±0.199	±0.170	±0.151	±0.170	±0.188	±0.204	7,120	7,065	7,963	8,934	8,446	8,166
Iceland	0.037	0.038	±0.040	±0.030	-	-	48,021	47,024	51,946	42,652	-	-
India	*1.122	*1.254	*0.875	*0.858	*0.965	*1.158	3,112	3,403	2,293	2,344	2,561	2,833
Indonesia	*0.374	*0.469	*0.487	*0.586	*0.677	*0.713	4,987	5,976	5,863	6,549	7,194	6,441
Iran (Islamic Republic of)	0.725	0.586	0.792	0.657	-	-	8,791	10,445	14,633	12,127	-	-
Iraq	-	*0.181	*1.074	-	-	-	-	11,434	59,646	-	-	-
Ireland	*0.560	*0.518	*0.589	*0.478	*0.571	*0.690	86,082	82,160	94,548	77,393	93,092	115,137
Israel	0.306	0.413	0.404	0.394	0.426	0.498	45,675	53,621	46,461	46,944	47,343	53,729
Italy	±4.299	±4.545	±8.164	±6.205	±6.098	-	66,964	69,865	71,495	56,125	56,989	-
Japan	±27	±30	±32	±36	±39	±43	84,343	95,871	101,798	114,206	125,694	138,210
Jordan	0.085	0.086	0.128	0.151	0.167	0.147	14,178	12,256	14,296	13,508	14,968	13,762
Kenya	*0.042	*0.041	*0.035	*0.022	*0.026	*0.032	7,753	7,447	6,165	3,809	4,555	5,282
Korea, Republic of	3.697	5.190	5.064	5.490	5.742	7.150	44,492	57,048	55,942	60,069	65,084	83,367
Kuwait	0.072	0.043	0.084	0.104	0.154	0.152	15,904	21,344	24,829	26,495	31,961	36,903
Latvia	-	-	-	*0.010	*0.017	*0.017	-	-	-	1,634	3,686	4,164
Macau	0.000	0.001	0.003	0.007	0.019	0.014	5,649	20,282	26,847	30,056	43,011	28,172
Malawi	*0.002	*0.004	*0.005	*0.000	*0.001	-	1,792	3,924	4,415	454	792	-
Malaysia	*0.441	*0.512	*0.629	*0.630	*0.883	*1.063	17,623	18,035	21,760	21,126	28,046	31,267
Malta	0.015	0.020	0.022	0.017	0.022	-	21,938	20,730	21,798	17,568	21,580	-
Mauritius	0.011	0.013	0.024	0.023	0.024	0.028	8,480	9,254	16,588	16,270	15,738	17,776
Mexico	0.986	1.297	1.525	1.631	1.646	0.958	31,851	42,910	52,728	63,828	68,727	47,758
Mongolia	0.022	0.009	0.003	0.001	0.005	0.005	6,140	1,944	1,029	454	1,806	1,311
Namibia	-	-	-	-	0.012	-	-	-	-	-	10,448	-
Nepal	0.034	0.043	-	0.027	0.040	-	601	701	-	500	666	-
Netherlands	1.016	1.011	1.118	1.064	1.164	1.344	60,562	59,839	65,637	66,086	78,340	91,205
Norway	±0.361	±0.325	±0.267	±0.288	±0.317	±0.410	59,176	62,706	63,083	67,571	72,981	82,939
Pakistan	0.316	0.309	0.345	-	-	-	15,765	15,218	16,240	-	-	-
Panama	0.019	0.030	0.041	0.051	0.058	0.054	12,235	17,276	20,476	23,259	25,029	24,971
Peru	0.204	0.165	0.218	-	-	-	19,270	15,547	23,679	-	-	-
Philippines	0.240	0.241	0.333	0.317	-	-	8,347	8,471	14,364	16,020	-	-
Poland	±0.602	±0.530	±0.503	±0.443	±0.535	±0.662	5,906	5,575	6,062	6,071	6,340	7,475
Portugal	0.724	0.834	0.909	0.830	0.893	1.060	19,124	20,584	25,877	22,731	25,471	30,153
Romania	±0.040	±0.024	±0.020	±0.046	-	-	-	-	-	-	-	-
Russian Federation	-	-	-	±2.301	±3.350	±3.530	-	-	-	2,313	3,717	4,059
Senegal	0.031	0.028	0.033	0.030	0.020	-	63,473	59,829	45,858	40,409	27,715	-

Continued.

Depending on the table, * means *Enterprises*, *Engaged*, or *Factor Values*; ± means *Basis Unspecified*; § means *shown in millions*. For additional notes and sources, see Appendix I.

473

Value Added and Value Added per Employee
- Continued -

Country	Value Added ($ bil.)						Value Added per Employee ($)					
	1990	1991	1992	1993	1994	1995	1990	1991	1992	1993	1994	1995
Singapore	*0.149	*0.210	*0.258	*0.286	*0.332	*0.383	34,109	48,840	54,761	56,355	62,276	72,147
Slovakia	-	-	-	±0.108	±0.107	-	-	-	-	5,498	5,922	-
South Africa	*0.794	*0.769	*0.901	*0.830	*0.845	*0.953	11,181	11,312	13,659	-	-	-
Spain	*4.797	*4.618	*4.720	±4.216	±4.619	±5.466	52,479	49,599	52,047	38,168	42,655	49,646
Sri Lanka	0.023	0.031	0.035	0.037	-	-	2,281	4,852	5,939	5,627	-	-
Swaziland	-	*0.002	*0.002	*0.001	-	-	-	6,907	7,725	5,114	-	-
Sweden	*1.129	*0.980	*0.895	*0.530	*0.579	*0.670	72,811	62,686	68,394	50,098	58,152	65,078
Thailand	2.769	2.264	-	-	-	-	41,152	36,036	-	-	-	-
Trinidad and Tobago	±0.023	±0.023	±0.028	±0.025	±0.028	-	21,543	21,834	25,096	20,894	20,658	-
Tunisia	0.263	0.256	0.264	0.236	0.264	0.317	-	-	-	15,185	17,050	19,976
Turkey	1.365	1.291	1.593	2.033	1.404	1.803	29,872	30,805	33,160	45,848	32,646	40,135
United Kingdom	*9.036	*7.565	*7.244	-	-	-	73,461	70,043	72,438	-	-	-
United Republic of Tanzania	*0.005	*0.016	-	-	-	-	1,490	5,313	-	-	-	-
United States of America	*24	*22	*24	*25	*28	*30	65,879	65,044	71,039	73,743	79,994	83,058
Uruguay	0.043	0.055	0.064	0.067	0.070	0.073	10,195	12,329	13,889	15,249	15,824	16,837
Venezuela	0.290	0.329	0.386	0.356	0.310	1.150	13,823	15,719	17,527	17,444	14,507	65,571
Yugoslavia	±0.036	±0.042	±0.096	-	±0.173	±0.103	-	1,227	2,781	-	5,471	3,313
Zambia	0.033	-	-	-	0.010	-	12,853	-	-	-	3,916	-
Zimbabwe	*0.054	*0.070	*0.061	*0.039	*0.063	*0.067	9,215	11,601	8,777	5,770	8,900	9,723

Capital Investment

Country	Gross Fixed Capital Formation ($ mil.)						Per Establishment ($ mil)					
	1990	1991	1992	1993	1994	1995	1990	1991	1992	1993	1994	1995
Argentina	-	-	-	139	-	-	-	-	-	0.039	-	-
Australia	318	-	-	-	-	-	0.258	-	-	-	-	-
Austria	295	343	381	302	366	-	0.667	0.792	0.875	0.712	0.866	-
Bahrain	-	-	4.160	-	-	-	-	-	0.104	-	-	-
Bangladesh	1.996	0.929	1.643	-	-	-	0.004	0.002	0.003	-	-	-
Bolivia	4.050	6.464	3.880	26	-0.601	-	0.035	0.052	0.029	0.193	-0.004	-
Bosnia & Herzegovina	5.811	-	-	-	-	-	0.182	-	-	-	-	-
Bulgaria	33	3.665	103	69	6.043	8.397	0.309	0.034	1.010	0.237	0.017	0.021
Canada	290	-	-	-	-	-	0.200	-	-	-	-	-
Chile	42	52	19	59	53	-	1.013	1.092	0.417	1.134	0.987	-
China (Hong Kong SAR)	14	30	-	28	27	-	0.069	0.247	-	0.250	0.165	-
Colombia	82	43	76	20	63	-	0.263	0.138	0.232	0.063	0.191	-
Croatia	10	9.387	12	12	8.237	27	0.054	0.040	0.043	0.033	0.020	0.059
Cyprus	11	11	15	15	22	17	0.048	0.048	0.062	0.065	0.079	0.057
Czechoslovakia (Former)	101	-	-	-	-	-	1.096	-	-	-	-	-
Denmark	127	121	-	-	-	-	0.156	0.148	-	-	-	-
Ecuador	28	49	62	34	38	-93	0.346	0.554	0.733	0.456	0.454	-1.047
Egypt	404	193	30	38	-	-	0.764	0.316	0.053	0.065	-	-
El Salvador	-	0.990	-	2.775	5.101	-33	-	0.071	-	0.146	0.170	-0.862
Ethiopia	0.932	2.207	1.425	4.138	2.746	3.740	0.044	0.123	0.079	0.243	0.067	0.078
Finland	261	87	47	41	35	-	0.721	0.218	0.133	0.137	0.126	-
FYR Macedonia	0.149	0.302	0.349	0.246	0.387	0.045	0.050	0.101	0.058	0.035	0.048	0.005
France	1,582	1,516	857	1,230	1,401	2,019	-	-	-	-	-	-
Gabon	-	2.488	3.548	3.885	1.871	1.679	-	1.244	1.183	3.885	0.624	1.679
Germany (Western Part)	1,621	1,848	2,152	2,971	-	-	0.687	0.772	0.886	-	-	-
Greece	115	111	98	-	-	-	0.216	0.206	0.179	-	-	-
Hungary	62	41	42	40	-	-	0.415	0.160	0.129	0.105	-	-
India	456	268	512	569	-	-	0.055	0.031	0.056	0.060	-	-
Indonesia	54	418	196	198	436	189	0.045	0.330	0.128	0.146	0.301	0.102
Iran (Islamic Republic of)	111	101	147	104	-	-	0.043	0.052	0.072	0.053	-	-
Ireland	61	58	-	-	-	-	0.252	0.327	-	-	-	-
Israel	51	129	90	136	183	-	0.168	0.348	0.220	0.352	0.468	-
Italy	1,023	1,161	1,643	1,171	1,166	-	1.032	1.197	0.915	0.654	0.679	-
Japan	2,252	2,992	2,850	2,995	2,837	-	0.145	0.194	0.188	0.196	0.192	-

Continued.

Depending on the table, * means *Enterprises, Engaged,* or *Factor Values*; ± means *Basis Unspecified*; § means *shown in millions*. For additional notes and sources, see Appendix I.

Capital Investment
- Continued -

Country	Gross Fixed Capital Formation ($ mil.)						Per Establishment ($ mil)					
	1990	1991	1992	1993	1994	1995	1990	1991	1992	1993	1994	1995
Jordan	1.974	1.574	16	4.237	7.305	5.363	0.001	0.001	0.009	0.002	0.004	0.003
Korea, Republic of	1,416	2,074	2,194	1,714	1,732	1,650	0.515	0.677	0.685	0.514	0.536	0.505
Kuwait	5.130	63	23	13	30	-	0.041	0.561	0.273	0.142	0.294	-
Latvia	0.033	0.210	3.255	0.317	0.954	3.914	0.000	0.002	0.029	0.004	0.015	0.043
Macau	0.017	1.201	0.515	1.526	0.246	3.844	0.002	0.109	0.043	0.109	0.019	0.240
Malawi	1.539	0.392	0.333	0.386	0.229		0.513	0.131	0.111	0.129	0.076	
Malaysia	72	199	244	179	381	-	0.208	0.542	0.683	0.480	1.012	-
Malta	2.369	9.827	3.129	3.387	2.595	-	0.028	0.101	0.033	0.036	0.029	-
Mexico	70	576	-	-	-	-	0.420	3.577	-	-	-	-
Mongolia	283	156	42	2.458	2.175	-	6.429	3.813	0.879	0.074	0.099	
Nepal	67	-	-	-	-	-	0.105	-	-			-
Netherlands	284	231	191	218	-		1.474	1.182	0.934	1.074		
Norway	115	97	-5.606	34	60	-	0.457	0.427	-0.051	0.243	0.428	
Pakistan	-	29	-	-	-			0.265	-	-		
Panama	0.677	1.361	-	-	-		0.014	0.028	-			
Peru	16	11	12	-	-		0.024	0.017	0.018			
Philippines	55	39	117	66	-		0.022	0.014	0.280	0.193		
Poland	72	95	105	122	-		0.217	0.251	0.276	0.378		
Portugal	205	390	72	164	180	-	0.073	0.138	0.029	0.063	0.068	
Romania	73	32	-	26	30	-	-	-	-	-	0.059	
Senegal	0.492	-	-	-	-	-	0.123	-	-	-	-	
Singapore	16	26	31	39	33	-	0.242	0.381	0.438	0.518	0.436	-
Slovenia	15	6.420	4.010	-	-		0.132	0.029	0.013	-		
Spain	524	708	676	579	528	759	0.079	0.106	0.109			
Sri Lanka	112	0.547	2.894	1.341	-	-	0.671	0.005	0.025	0.012	-	-
Swaziland	-	0.084	0.456	0.053	-	-		0.008	0.046	0.006	-	-
Thailand	2,551	841	-	-	-	-	4.451	0.802	-	-	-	-
Trinidad and Tobago	-	-	11	6.167	11	-	-	-	0.278	0.143	0.286	-
Tunisia	65	80	84	118	120	77	-	-	-	0.348	0.343	0.215
Turkey	175	213	304	237	273	-	0.417	0.513	0.412	0.334	0.387	-
Ukraine	157	234	364	27	34	17	0.042	0.046	0.066	0.007	0.008	0.004
United Kingdom	1,127	550	415	-	-	-	0.414	0.215	0.154	-	-	-
United Republic of Tanzania	32	184	-	-	-	-	1.907	11	-	-	-	-
United States of America	1,800	1,320	1,460	1,503	1,869	2,212	-	-	0.086	-	-	-
Uruguay	2.006	2.008	2.112	3.938	-	-						
Venezuela	144	77	67	36	55	406	0.295	0.153	0.135	0.080	0.117	0.851
Yugoslavia	2.798	2.315	4.907	-	17	5.950	0.019	0.010	0.014	-	0.033	0.011
Zambia	46	-	-	-	-	-	3.502	-	-	-	-	-

Indexes of Industrial Production

Country	Index of Industrial Production (1990=100)						Index of Employment (1990=100)					
	1990	1991	1992	1993	1994	1995	1990	1991	1992	1993	1994	1995
Albania	100	-	-	-	-	-	100	-	-	27	19	15
Algeria	100	100	104	98	87	91	-	-	-	-	-	-
Angola	100	91	91	-	-	-	-	-	-	-	-	-
Argentina	100	148	-	-	-	-	100	-	-	55	57	-
Armenia	100	-	-	-	-	-	100	97	-	-	-	-
Australia	100	95	104	118	-	-	100	97	91	-	-	-
Austria	100	107	111	-	-	-	100	97	97	90	90	89
Azerbaijan	-	-	-	-	-	-	100	98	96	93	81	-
Bahamas	100	-	-	-	-	-	100	103	-	-	-	-
Bahrain	100	-	-	-	-	-	-	-	-	-	-	-
Bangladesh	100	81	81	62	96	94	100	101	107	-	-	-
Barbados	100	96	67	68	88	94	-	-	-	-	-	-
Belgium	100	94	97	97	103	103	100	100	100	-	-	-
Belize	100	-	-	-	-	-	-	-	-	-	-	-
Benin	100	100	-	-	-	-	-	-	-	-	-	-

Continued.

Indexes of Industrial Production

- Continued -

Country	Index of Industrial Production (1990=100)						Index of Employment (1990=100)					
	1990	1991	1992	1993	1994	1995	1990	1991	1992	1993	1994	1995
Bermuda	-	-	-	-	-	-	100	83	-	-	-	-
Bhutan	100	-	-	-	-	-	-	-	-	-	-	-
Bolivia	100	105	111	118	131	146	100	104	118	123	119	135
Bosnia & Herzegovina	-	-	-	-	-	-	100	97	-	-	11	-
Botswana	100	100	100	-	-	-	-	-	-	-	-	-
Brazil	100	101	93	98	100	105	-	-	-	-	-	-
Bulgaria	100	60	52	48	52	57	100	81	64	58	59	54
Burkina-Faso	100	-	-	-	-	-	-	-	-	-	-	-
Burundi	100	-	-	-	-	-	100	98	-	-	-	-
Cambodia	100	-	-	-	-	-	100	81	-	-	-	-
Cameroon	100	-	-	-	-	-	100	97	99	92	-	-
Canada	100	82	79	80	82	81	100	93	83	78	78	78
Central African Republic	100	100	100	100	100	-	-	-	-	-	-	-
Chad	100	-	-	-	-	-	-	-	-	-	-	-
Chile	100	108	130	152	152	167	100	114	128	145	150	158
China	100	-	-	-	-	-	100	102	105	102	-	-
China (Hong Kong SAR)	100	94	90	81	74	70	100	93	93	90	117	-
China (Taiwan Province)	100	108	113	122	130	129	100	96	101	106	111	105
Colombia	100	111	116	128	136	138	100	100	103	111	113	114
Costa Rica	100	104	114	123	126	119	100	92	97	103	-	-
Cote d'Ivoire	100	86	90	88	95	105	-	-	-	-	-	-
Croatia	100	-	-	-	-	-	100	83	72	69	64	63
Cuba	100	70	-	-	-	-	-	-	-	-	-	-
Cyprus	100	99	101	102	99	97	100	101	106	104	105	111
Czechoslovakia (Former)	100	69	62	-	-	-	100	84	-	-	-	-
Czech Republic	100	-	-	-	-	-	100	82	82	-	-	-
Dem. Rep. of the Congo	100	89	68	-	-	-	-	-	-	-	-	-
Denmark	100	96	94	88	100	106	100	103	94	89	99	104
Dominican Republic	100	85	-	-	-	-	-	-	-	-	-	-
Ecuador	100	109	141	161	-	-	100	105	93	90	95	90
Egypt	100	93	93	95	89	98	100	112	116	117	113	118
El Salvador	100	104	-	-	-	-	-	-	-	-	-	-
Equatorial Guinea	-	-	-	-	-	-	100	-	-	-	-	-
Ethiopia and Eritrea	100	89	-	-	-	-	-	-	-	-	-	-
Ethiopia	-	-	-	-	-	-	100	99	93	97	141	165
Fiji	100	101	109	102	120	117	-	-	-	-	-	-
Finland	100	83	74	69	68	70	100	89	72	54	52	-
FYR Macedonia	100	-	-	-	-	-	100	75	59	57	59	52
France	100	95	86	79	85	86	100	99	93	87	84	84
Gabon	100	98	105	-	-	-	-	-	-	-	-	-
Gambia	100	-	-	-	-	-	-	-	-	-	-	-
Germany (Eastern Part)	100	-	-	-	-	-	-	-	-	-	-	-
Germany (Western Part)	100	105	112	115	128	122	100	104	103	-	-	-
Ghana	100	-	-	-	-	-	-	-	-	-	-	-
Greece	100	90	89	87	90	91	100	96	94	93	94	95
Guatemala	100	102	-	-	-	-	-	-	-	-	-	-
Guyana	100	100	100	100	100	100	-	-	-	-	-	-
Haiti	100	118	-	-	-	-	-	-	-	-	-	-
Honduras	100	121	126	176	165	227	100	148	158	182	198	215
Hungary	100	67	58	74	85	94	100	86	68	68	80	89
Iceland	100	99	-	-	-	-	100	106	100	91	84	-
India	100	100	100	101	106	122	100	102	106	102	104	113
Indonesia	100	105	119	130	172	200	100	105	111	119	125	148
Iran (Islamic Republic of)	100	111	118	115	109	121	100	68	66	66	-	-
Iraq	100	39	19	-	-	-	-	-	-	-	-	-
Ireland	100	94	98	94	104	114	100	97	96	95	94	92
Israel	100	130	151	145	155	193	100	115	130	125	134	138
Italy	100	99	96	90	-	-	100	101	178	172	167	-
Jamaica	100	89	107	102	100	-	100	99	97	104	101	-
Japan	100	100	94	91	93	93	100	100	100	99	99	97

Continued.

Depending on the table, * means *Enterprises, Engaged,* or *Factor Values;* ± means *Basis Unspecified;* § means *shown in millions.* For additional notes and sources, see Appendix I.

Indexes of Industrial Production

- Continued -

Country	Index of Industrial Production (1990=100)						Index of Employment (1990=100)					
	1990	1991	1992	1993	1994	1995	1990	1991	1992	1993	1994	1995
Jordan	100	99	99	111	110	110	100	117	150	188	187	178
Kenya	100	119	140	137	144	143	100	101	104	106	104	112
Korea, Republic of	100	116	124	125	134	142	100	109	109	110	106	103
Kuwait	100	55	40	-	-	-	100	44	74	86	106	91
Kyrgyzstan	100	-	-	-	-	-	100	102	87	68	46	-
Lao P.D.R.	100	-	-	-	-	-	-	-	-	-	-	-
Latvia	100	-	-	-	-	-	100	103	80	40	29	26
Lebanon	100	100	100	-	-	-	-	-	-	-	-	-
Lesotho	100	-	-	-	-	-	100	-	22	16	19	-
Liberia	100	-	-	-	-	-	-	-	-	-	-	-
Libyan Arab Jamahiriya	100	101	101	-	-	-	-	-	-	-	-	-
Lithuania	100	-	-	-	-	-	-	-	-	-	-	-
Macau	100	-	-	-	-	-	100	149	281	662	1,200	1,327
Madagascar	100	161	161	152	-	-	-	-	-	-	-	-
Malawi	100	111	111	117	121	123	100	100	122	111	133	-
Malaysia	100	120	132	140	158	176	100	114	116	119	126	136
Mali	100	110	-	-	-	-	-	-	-	-	-	-
Malta	100	132	147	195	-	-	100	136	144	137	144	-
Mauritius	100	100	-	-	-	-	100	108	115	112	123	127
Mexico	100	105	113	114	120	101	100	98	93	83	77	65
Mongolia	100	100	110	-	-	-	100	128	90	60	69	101
Morocco	100	107	116	105	112	110	-	-	-	-	-	-
Mozambique	100	82	-	-	-	-	100	83	40	39	35	-
Myanmar	100	105	109	97	114	125	-	-	-	-	-	-
Nepal	100	93	-	-	-	-	100	108	-	95	106	-
Netherlands	100	96	90	89	99	100	100	101	102	96	89	88
New Zealand	100	85	87	-	-	-	100	96	98	100	-	-
Nicaragua	100	100	-	-	-	-	-	-	-	-	-	-
Niger	100	103	-	-	-	-	-	-	-	-	-	-
Nigeria	100	115	114	110	104	-	-	-	-	-	-	-
Norway	100	86	90	90	104	119	100	85	69	70	71	81
Pakistan	100	108	110	113	107	120	100	101	106	-	-	-
Panama	100	154	209	265	298	272	100	112	128	142	149	138
Papua New Guinea	100	-	-	-	-	-	-	-	-	-	-	-
Paraguay	100	102	123	128	120	137	-	-	-	-	-	-
Peru	100	103	99	123	154	178	100	100	87	-	-	-
Philippines	100	132	133	153	185	224	100	99	81	69	-	-
Poland	100	91	102	111	128	134	100	93	81	72	83	87
Portugal	100	99	102	105	106	110	100	107	93	96	93	93
Puerto Rico	100	94	95	96	104	-	100	102	95	93	95	97
Republic of Moldova	100	-	-	-	-	-	100	110	109	71	31	30
Romania	100	76	52	52	47	53	-	-	-	-	-	-
Russian Federation	100	-	-	-	-	-	-	-	-	-	-	-
Saudi Arabia	100	110	110	-	-	-	-	-	-	-	-	-
Senegal	100	101	117	116	139	140	100	96	146	150	148	-
Singapore	100	118	130	148	151	156	100	99	108	116	122	122
Slovakia	100	-	-	-	-	-	-	-	-	-	-	-
Slovenia	100	86	79	72	79	79	-	-	-	-	-	-
Somalia	100	-	-	-	-	-	-	-	-	-	-	-
South Africa	100	93	87	86	86	94	100	96	93	-	-	-
Spain	100	99	92	87	96	102	100	102	99	121	118	120
Sri Lanka	100	72	106	105	91	88	100	64	59	65	-	-
Sudan	100	112	-	-	-	-	-	-	-	-	-	-
Sweden	100	90	78	71	72	77	100	101	84	68	64	66
Switzerland	100	87	83	77	85	81	-	-	-	-	-	-
Syrian Arab Republic	100	105	115	127	132	148	-	-	-	-	-	-
Thailand	100	106	120	146	166	190	100	93	-	-	-	-
Trinidad and Tobago	100	114	119	116	124	140	100	99	103	112	125	-
Tunisia	100	99	101	104	102	104	-	-	-	-	-	-
Turkey	100	108	115	124	115	128	100	92	105	97	94	98

Continued.

Depending on the table, * means *Enterprises, Engaged,* or *Factor Values*; ± means *Basis Unspecified*; § means *shown in millions*. For additional notes and sources, see Appendix I.

477

Indexes of Industrial Production

- Continued -

Country	Index of Industrial Production (1990=100)						Index of Employment (1990=100)					
	1990	1991	1992	1993	1994	1995	1990	1991	1992	1993	1994	1995
Uganda	100	105	132	169	161	238	-	-	-	-	-	-
Ukraine	100	-	-	-	-	-	100	98	86	90	76	70
United Arab Emirates	100	99	99	-	-	-	-	-	-	-	-	-
United Kingdom	100	87	82	85	93	90	100	88	81	-	-	-
United Republic of Tanzania . .	100	154	102	112	98	-	100	92	-	-	-	-
USSR (Former)	100	-	-	-	-	-	100	-	-	-	-	-
United States of America . . .	100	91	93	94	100	102	100	93	91	92	95	100
Uruguay	100	96	139	143	144	139	100	105	108	103	103	102
Venezuela	100	-	-	-	-	-	100	100	105	97	102	84
Yemen	100	114	-	-	-	-	-	-	-	-	-	-
Yugoslavia	100	-	-	-	-	-	-	-	-	-	-	-
Yugoslavia (Former)	100	78	-	-	-	-	-	-	-	-	-	-
Zambia	100	97	96	87	60	63	100	-	-	-	98	-
Zimbabwe	100	106	98	81	105	97	100	102	119	114	120	117

Depending on the table, * means *Enterprises, Engaged,* or *Factor Values*; ± means *Basis Unspecified*; § means *shown in millions*. For additional notes and sources, see Appendix I.

Representative Companies in Sector

Name	Address	Tele-phone	Fax	Employment	Y
Marley PLC	Riverhead, Sevenoaks TN13 2DS, United Kingdom	732 455255	732 740694	9,773	93
Grupo Industrial Saltillo SA de CV	25280 Saltillo, Coahuila, Mexico	84 155620	84 153313	9,575	92
Centex Corp.	PO Box 199000, Dallas, Texas 75219-9000, U.S.A.	214-559-6500		8,926	96
Eregli Iron and Steel Works Co.	Uzunkum No. 7, Eregli TR-67330, Turkey	388 19500	388 13969	8,349	89
Johns Manville Corp.	PO Box 5108, Denver, Colorado 80217-5108, U.S.A.	303-978-2000	303-978-2041	8,300	97
Tambang Timah PT	Jl. Jend Sudirman Pangkal Pinang No. 51, Bangka 12730, Indonesia	71722355	717 22323	8,300	97
Dal-Tile International Inc.	7834 Hawn Fwy., 1704, Dallas, Texas 75217, U.S.A.	214-398-1411		7,600	96
Eagle-Picher Industries Inc.	PO Box 779, Cincinnati, Ohio 45201, U.S.A.	513-721-7010	513-721-2341	7,500	95
Lafarge Corp.	PO Box 4600, Reston, Virginia 22090, U.S.A.	703-264-3600	703-264-0634	7,300	97
Rugby Group PLC	Crown House, Rugby CV21 2DT, United Kingdom	788 542666		7,142	93
Indocement Tunggal Prakarsa PT	J. Jend Sudirman Kav 70-71, Jakarta 12910, Indonesia	212512121	21 5710222	7,000	94
HON Industries Inc.	PO Box 1109, Muscatine, Iowa 52761-7109, U.S.A.	319-264-7085	319-264-7217	6,900	97
Ferro Corp.	1000 Lakeside Ave., Cleveland, Ohio 44114, U.S.A.	216-641-8580	216-696-4786	6,851	97
United States Gypsum Co.	125 S. Franklin St., Chicago, Illinois 60606, U.S.A.	312-606-4000	312-606-4093	6,600	97
Siam Cement Co., Ltd.	Bangsue, Bangkok 10800, Thailand	5863333	5872199	6,500	94
Waterford Wedgwood Holdings PLC	Barlaston, Stoke-on-Trent ST12 9ES, United Kingdom	782 204141	782 204402	6,366	
Cookson America Inc.	1 Cookson Place, Providence, Rhode Island 02903, U.S.A.	401-521-1000	401-521-7258	6,000	97
Tolmex, SA de CV	Av. Constitucion, 444 Pte., Monterrey 64900, Mexico	83 452000	83 452025	5,695	92
Jannock Ltd.	5205-40 King St. W, Toronto, Ontario M5H 3Y2, Canada	416-364-8586	416-364-9342	5,400	98
Lafarge Canada Inc.	606 Rue Cathcart Ste. 900, Montreal, Quebec H3B 1L7, Canada	514-861-1411	514-861-1123	5,037	98
Raymond Woollen Mills Ltd.	Bellard Estate, Bombay 400 038, India	22 2618321	22 2622010	5,010	94
Bajaj Hindustan Ltd.	226 Nariman Pt., Bombay 400 021, India	22 2023626	2021977	5,000	
Transelco Div.	1789 Transelco Dr., Penn Yan, New York 14527, U.S.A.	315-536-3357	315-536-8091	5,000	96
First Brands Corp.	PO Box 1911, Danbury, Connecticut 06813-1911, U.S.A.	203-731-2300		4,800	97
GAF Corp.	1361 Alps Rd., Wayne, New Jersey 07470, U.S.A.	973-628-3000	973-628-4081	4,300	97
Justin Industries Inc.	PO Box 425, Fort Worth, Texas 76101, U.S.A.	817-336-5125		4,222	97
Martin Marietta Materials Inc.	PO Box 30013, Raleigh, North Carolina 27622, U.S.A.	919-781-4550	919-783-4552	4,000	96
Vencemos Mara C.A.	Edificio Las Fundaciones, Caracas, Venezuela	2 5077423	2 613061	4,000	96
Precoria Portland Cement Co. Ltd.	PO Box 3811, Johannesburg 2000, Republic of South Africa	11 4881884	11 4881972	3,900	96
Corobrik Pty. Ltd.	PO Box 5198, Durban 4000, Republic of South Africa	31 5603911	31 5603411	3,757	96
Ibstock PLC	Lutterworth House, Lutterworth LE17 4PS, United Kingdom	455 553071		3,681	94
Semen Padang (Persero) PT	PO Box 94, Padang 25237, Indonesia	75132250	751 34590	3,500	97
Tanzania Saruji Corp.	PO Box 4123, Dar es Salaam, United Republic of Tanzania	31243	26429	3,500	91
Texas Industries Inc.	1341 W. Mockingbird Lane, Ste. 7, Dallas, Texas 75247-6913, U.S.A.	214-647-6700	214-647-3878	3,400	97
Granite Construction Inc.	PO Box 50085, Watsonville, California 95077-5085, U.S.A.	408-724-1011		3,309	97
Carborundum Co.	PO Box 337, Niagara Falls, New York 14302-0337, U.S.A.	716-278-2000	716-278-2900	3,200	94
Mannington Mills Inc.	PO Box 30, Salem, New Jersey 08079, U.S.A.	609-935-3000	609-339-5813	3,000	95
Siam Fibre-Cement Co., Ltd.	Bangsue Dusit, Bangkok 10800, Thailand	5863333	5872199	3,000	94
Steeledale Reinforcing & Engineering Industries Ltd.	Tulisa Park, Johannesburg 2197, Republic of South Africa	11 6132331	11 6131394	2,950	93
Clarcor Inc.	PO Box 7007, Rockford, Illinois 61125, U.S.A.	815-962-8867	815-962-0417	2,872	97
American Biltrite Inc.	57 River St., Wellesley Hills, Massachusetts 02181, U.S.A.	617-237-6655		2,835	95
Ameron International Corp.	245 S. Los Robles Ave., Pasadena, California 91101-2894, U.S.A.	626-683-4000	626-683-4060	2,761	97
Pacific Coast Building Products Inc.	PO Box 160488, Sacramento, California 95816, U.S.A.	916-444-9304	916-325-3697	2,600	97
National Gypsum Co.	2001 Rexford Rd., Charlotte, North Carolina 28211, U.S.A.	704-365-7300	704-365-7579	2,581	94
Park-Ohio Industries Inc.	23000 Euclid Ave., Cleveland, Ohio 44117, U.S.A.	216-692-7200		2,500	96
St. Lawrence Cement Inc.	1945 Blvd. Graham, Montreal, Quebec H3R 1H1, Canada	514-340-1881	514-342-8154	2,500	98
Florida Rock Industries Inc.	PO Box 4667, Jacksonville, Florida 32201-4667, U.S.A.	904-355-1781	904-355-0817	2,448	97
American Olean Tile Co.	PO Box 271, Lansdale, Pennsylvania 19446-0271, U.S.A.	215-855-1111	215-393-2784	2,409	95
Fibreboard Corp.	2200 Ross Ave., Ste. 3600, Dallas, Texas 75201-6731, U.S.A.	214-954-9500	214-954-9512	2,400	96
George Koch Sons Inc.	10 S. 11th Ave., Evansville, Indiana 47744, U.S.A.	812-465-9600	812-465-9724	2,400	97
INDRESCO Inc.	PO Box 219022, Dallas, Texas 75221, U.S.A.	214-953-4500	214-953-4596	2,400	95
Southdown Inc.	1200 Smith St., 2400, Houston, Texas 77002-4486, U.S.A.	713-650-6200	713-653-6815	2,400	97
Lehigh Portland Cement Co.	7660 Imperial Way, Allentown, Pennsylvania 18195, U.S.A.	610-366-4600	610-366-4680	2,300	97
Blue Circle Cement (Marietta, Georgia)	1800 Parkway Pl., 1200, Marietta, Georgia 30067, U.S.A.	770-423-4700	770-423-4738	2,200	94

Continued

Representative Companies in Sector
- Continued -

Name	Address	Tele-phone	Fax	Employ-ment	Y
Semen Tonasa (Persero) PT	Tonasa Pangkep, Pangkajene, Indonesia	411320672	411 311973	2,200	97
Tong Yang Cement Corp.	Yongdungpo-gu, Seoul, Republic of Korea	2 37703000	2 37703304	2,170	97
Redland Aggregates Ltd.	21A Buckden Rd., Huntingdon PE18 8PR, United Kingdom	480 434236		2,125	93
Coors Ceramics Co.	600 9th St., Golden, Colorado 80401, U.S.A.	303-278-4000	303-277-4901	2,100	95
Hepworth Building Products Ltd.	Stocksbridge, Sheffield S30 5HG, United Kingdom	226 763561	226 764827	2,067	93
Blue Circle Canada Inc.	55 Industrial St., Toronto, Ontario M4G 3W9, Canada	416-696-4411	416-696-4434	2,000	98
ECC International	100 Mansell Court E., Ste. 300, Roswell, Georgia 30076, U.S.A.	770-594-0660	770-645-3384	2,000	97
Edward C. Levy Co.	8800 Dix Ave., Detroit, Michigan 48209, U.S.A.	313-843-7200	313-849-9448	2,000	97
Grupo Empresarial Maya SA de CV	Avda. Constitucion 444 Pte., Monterrey 64000, Mexico	83 452000	83 452025	2,000	92
Malaysian Mosaics BHD	424 Jalan Tun Razak, Kuala Lumpur 50400, Malaysia	3 9847333	3 9843664	2,000	94
Semen Gresik (Persero) PT	Jl. Veteran, No. 7, Gresik 61122, Indonesia	31981745	319 83209	2,000	97
Serinco Djaja Marmer Industries PT	Jl. Mangga, Jakarta 10730, Indonesia	216252202	21 6011444	2,000	97
Statyba Stock Co.	33 Gedimino St., Kaunas 3000, Lithuania	127 205482	127 203475	2,000	92
A.P. Green Industries Inc.	Green Blvd., Mexico, Missouri 65265, U.S.A.	314-473-3626	314-473-5308	1,966	95
Heracles General Cement Co. SA	PO Box 3500, Athens GR-102 20, Greece	2898111	2819406	1,955	
CalMat Co.	3200 San Fernando Rd., Los Angeles, California 90065, U.S.A.	213-258-2777	213-258-5920	1,880	97
Engelhard Corp. Pigments and Additives	101 Wood Ave., Iselin, New Jersey 08830, U.S.A.	908-205-5000	908-321-1598	1,840	95
Dixon Ticonderoga Co.	PO Box 958413, Heathrow, Florida 32795-8413, U.S.A.	407-875-9000		1,770	97
Metallurg Inc.	6 East 43rd St., 12th Fl., New York, New York 10017-4609, U.S.A.	212-687-9470	212-687-9623	1,700	97
Norton SA Industria e Comercio	Rua Joao Zacharias 119, Bairro Macedo, Guarulhos 07111-150, Brazil	11 9645127	11 2091533	1,700	96
Soda Sanayii AS	Kazanli Bucagi Yani P.K. 654, Mersin TR-33004, Turkey	741 23550	741 23555	1,661	89
Perlmooser Zementwerke AG-Gruppe	Pf. 126, Vienna A-1043, Austria	222 588890	222 58887470	1,510	88
Byucksan Corp.	Chung-Gu, Seoul, Republic of Korea	2 2606114	2 2750050	1,500	97
Owens-Corning Canada Inc.	3450 McNicoll, Scarborough, Ontario M1V 1Z5, Canada	416-292-4000	416-412-7091	1,500	98
Semen Cibinong PT	BRI Bldg. II, 26th Flr., Jakarta 10220, Indonesia	212512377	21 8198362	1,500	97
Titan Cement Co. SA	22 A.Halkidos Str., Athens GR-111 43, Greece	2591111	2183058	1,500	
Vesuvius U.S.A. Corp.	PO Box 4014, Champaign, Illinois 61824, U.S.A.	217-351-5000	217-351-5031	1,500	97
H & R Johnson Tiles Ltd.	Tunstall, Stoke-on-Trent ST6 4JX, United Kingdom	782 575575		1,371	95
Halla Cement Manufacturing Corp.	Kangnung-shi, Kangwon, Republic of Korea	394 341000	394 340716	1,365	97
R.G. Carter Holdings Ltd.	Drayton, Norwich NR8 5AH, United Kingdom	603 867355	603 260151	1,328	93
Morrison Construction Ltd.	12 Atholl Cres., Edinburgh EH3 8HA, United Kingdom	31 2284188	31 3371880	1,311	94
LAB Chrysotile Inc.	835 Rue Mooney, Thetford-Mines, Quebec G6G 5T5, Canada	418-338-7500	418-338-9522	1,300	98
Raytech Corp.	1 Corporate Dr., 512, Shelton, Connecticut 06484, U.S.A.	203-925-8023		1,300	95
Aalborg Portland AS	PO Box 165, Alborg DK-9100, Denmark	98167777	98101186	1,200	93
Cold Spring Granite Co.	202 S. 3rd Ave., Cold Spring, Minnesota 56320, U.S.A.	612-685-3621	612-685-8490	1,200	95
Halkis Cement Co. SA	2-4 Messoghion Ave., Athens GR-115 27, Greece	7706811	7783475	1,200	97
Cementos Progreso SA	Zona 6, Guatemala City, Guatemala	2 566411	2 562217	1,182	92
Castle Cement Ltd.	Keeton, Stamford PE9 3SX, United Kingdom	780 720501	780 721316	1,181	93
Cementos de Honduras SA	3A Avda No. 40, San Pedro Sula, Honduras			1,170	
Cementownia Ozarow SA	Ozarow PL-27-530, Poland	15 221625	16 721714	1,154	93
Shelter Components Corp.	PO Box 4026, Elkhart, Indiana 46514, U.S.A.	219-262-4541	219-262-3936	1,127	96
Medusa Corp.	PO Box 5668, Cleveland, Ohio 44101, U.S.A.	216-371-4000	216-371-2912	1,086	96
Watts, Blake, Bearne & Co. PLC	Courtenay Pk., Newton Abbot TQ12 4PS, United Kingdom	626 332345	626 332344	1,066	93
Lone Star Industries Inc.	PO Box 120014, Stamford, Connecticut 06912-0014, U.S.A.	203-969-8600	203-969-8546	1,060	97
Centex Construction Products Inc.	PO Box 19000, Dallas, Texas 75219, U.S.A.	214-559-6500	214-559-9812	1,053	96
Briggs Amasco Ltd.	West St., Dorking RH4 1UJ, United Kingdom	306 885933	306 887019	1,035	93
Wimpey Hobbs Ltd.	Nailsea, Bristol BS19 1BW, United Kingdom	272 858151	272 857636	1,012	91
Eichsfelder Zementwerke	Werkstrasse 7, Deuna 37355, Germany		82255	1,000	91
Pilkington Insulation Ltd.	PO Box 10, St. Helens WA10 3NS, United Kingdom	744 24022	744 612007	996	92
GE Superabrasives	6325 Huntley Rd., Worthington, Ohio 43085, U.S.A.	614-438-2000	614-438-2829	950	95
Puerto Rican Cement Company Inc.	PO Box 364487, San Juan, Puerto Rico 00936-4487, U.S.A.	809-783-3000		939	95
Associated Pan Malaysia Cement SDN BHD	2 Jalan Kilang, Petaling Jaya 46050, Malaysia	3 7918344	3 7918309	924	93
Oglebay Norton Co.	1100 Superior Ave., Cleveland, Ohio 44114-2598, U.S.A.	216-861-3300	216-861-2863	908	97

Continued

Representative Companies in Sector
- Continued -

Name	Address	Tele-phone	Fax	Employ-ment	Y
Industri Keramik Angsa Daya PT	Pusat, Jakarta 10730, Indonesia	216011606	21 3803475	900	97
Southwestern Portland Cement Co.	2601 Saturn St., 200, Brea, California 92822, U.S.A.	714-985-4000	714-985-4060	900	95
Vesuvius Crucible Company Inc.	661 Willet Rd., Buffalo, New York 14218, U.S.A.	716-825-7900		900	
Siam Iron and Steel Co. Ltd.	Bangsue Dusit, Bangkok 10501, Thailand	5864071	5862986	890	94
Juan Minetti SA	Hipolito Yrigoyen 434, Buenos Aires 1086, Argentina	1 3430116	1 3430116	880	97
Gyproc AB	Box 505, Malmo S-201 25, Sweden	40 105600	40 129991	859	93
Redland Roof Tiles Ltd.	Castlefield Rd., Reigate RH2 0SJ, United Kingdom	737 242488	737 240247	856	93
National Refractories and Minerals Corp.	1852 Rutan Dr., Livermore, California 94550, U.S.A.	510-449-5010	510-455-8363	850	96
Nesher Israel Cement Enterprises Ltd.	99 Ben Yehuda St., Tel Aviv 63401, Israel	3 5209500	3 5278338	850	97
Georgia Marble Co.	1201 Roberts Blvd.B, 100, Kennesaw, Georgia 30144-3619, U.S.A.	404-421-6500	404-421-6507	825	94
John Fyfe Ltd.	Westhill, Aberdeen AB3 3TYL, United Kingdom	224 744144	224 744500	806	93
Scancem International Ans	PO Box 1344 Vika, Oslo N-0113, Norway	22 833800	22 830935	804	93
Sherman International Corp.	1400 Urban Circle Dr., 200, Birmingham, Alabama 35242-2547, U.S.A.	205-970-7500		800	95
Minnesota Mining & Manufacturing Co. South Africa Pty. Ltd.	PO Box 926, Isando 1600, Republic of South Africa	11 9229111	11 9222388	780	95
Graymont Ltd.	215-10451 Shellbridge Way, Richmond, British Columbia V6C 2W8, Canada	604-207-4292	604-207-9014	750	98
Sunnen Products Co.	7910 Manchester Ave., St. Louis, Missouri 63143, U.S.A.	314-781-2100	314-781-2268	750	97
Continental Materials Corp.	225 W. Wacker Dr., 18th Fl., Chicago, Illinois 60606, U.S.A.	312-541-7200		748	97
Dravo Corp.	11 Stanwix St., Pittsburgh, Pennsylvania 15222-2682, U.S.A.	412-566-3000	412-566-3116	738	97
Corhart Refractories Corp.	PO Box 740009, Louisville, Kentucky 40201-7409, U.S.A.	502-778-3311	502-775-7300	725	94
Industrial Acoustics Company Inc.	1160 Commerce Ave., Bronx, New York 10462, U.S.A.	718-931-8000	718-863-1138	710	96
Iglas (Persero) PT	Jl. Nagel No. 153, Surabaya 60246, Indonesia	31575597	31 574796	705	97
Ladrillera Santafe	Cr. 9 No. 74-08, 6th Fl., Bogota, Colombia	1 2178111	1 3458188	700	97
Blue Circle Cement Pty. Ltd.	PO Box 1320, Parklands 2121, Republic of South Africa	11 7882160	11 4471725	694	95
Belapatfalvi Cement Es Meszipari Rt.	Bela ut 1, Belapatfalva H-3346, Hungary	36 354377	36 354405	675	93
Republic Group Inc.	PO Box 1307, Hutchinson, Kansas 67504-1307, U.S.A.	316-727-2700	316-727-2727	675	97
Asland, SA	Orense, 81-4A Planta, Madrid E-28020, Spain	91 5720045	91 5790525	673	97
Cementos del Nare SA	Carrera 46 No. 56-11, 12th Fl., Medellin, Colombia	4 2514646	4 2517167	670	97
Oil-Dri Corporation of America	410 N. Michigan Ave., 400, Chicago, Illinois 60611-4211, U.S.A.	312-321-1515		665	97
Reliable Manufacturing Works, Inc.	Mandaue City, Cebu 6014, Philippines	460315	32460315	652	97
Lamino Oy	PO Box 476, Tampere SF-33101, Finland	31 3499111	31 3499111	650	95
Porcelanosa, SA	Villarreal de Los Infantes, Castellon E-12540, Spain	964 521262	964 534776	640	97
Monarch Tile Inc.	PO Box 999, Florence, Alabama 35631-0999, U.S.A.	205-764-6181	205-760-8686	625	94
Tauell, SA	S/N, Castellon E-12006, Spain	964 250105	964 252771	602	97
Alony Haim Combine Ltd.	Bar Yehuda Rd., Nesher IL-36602, Israel		4 8307528	600	97
Bentonite Co.	97 Belomorskiy Blvd., Kurdzhali BG-6600, Bulgaria	361 24701	361 26572	600	93
Caterpillar Paving Products Inc.	PO Box 1362, Minneapolis, Minnesota 55440-1362, U.S.A.	612-425-4100	612-493-1490	600	95
Goldcorp Inc.	145 King St. W Ste. 2700, Toronto, Ontario M5H 1J8, Canada	416-865-0326	416-361-5741	600	98
Inland Cement Ltd.	12640 156 St. NW, Edmonton, Alberta T5V 1K2, Canada	403-420-2500	403-420-2503	600	98
Puerto Rican Cement Co., Inc.	Amelida Ind. Park, Guaynabo 00970, Puerto Rico			600	92
Westroc Inc.	2424 Lakeshore Rd. W, Mississauga, Ontario L5J 1K4, Canada	905-823-9881	905-823-4860	600	98
Ytong Industries Ltd.	Pardess Hana, Pardess Hana 37100, Israel	6 6377998	6 6377984	600	97
Central Pre-Mix Concrete Co.	PO Box 3366, Spokane, Washington 99220, U.S.A.	509-534-6221	509-534-3839	580	
KPT Industries, Ltd.	Kung Yi Li, Cha-Nana Chen, Miaoli, Taiwan	37 581102	2 7003617	579	92
Sheffield Insulations Group PLC	Langsett Rd., Sheffield S6 2LW, United Kingdom	742 852852	742 337191	577	94
Monarch Cement Co.	PO Box 1000, Humboldt, Kansas 66748-1000, U.S.A.	316-473-2225		575	96
Thomas Roberts Westminster Ltd.	Lancaster PL., London WC2E 7HX, United Kingdom	71 8365801	71 8360407	556	94
Ralston Purina Canada Inc.	2500 Royal Windsor Dr., Mississauga, Ontario L5J 1K8, Canada	905-822-1611	905-855-5700	551	98
J.E. Baker Co.	PO Box 1189, York, Pennsylvania 17405, U.S.A.	717-848-1501	717-845-9555	550	96
Temtex Industries Inc.	5400 Lyndon B. Johnson Fwy., 13, Dallas, Texas 75240, U.S.A.	972-726-7175		540	97
Cementos del Caribe SA	A.A. 2739, Barranquilla, Colombia	5 3681082	5 3536047	530	97

Product Tables
Building Bricks, Made of Clay

Unit of Measure: Millions of units.

	1990 Value	%	1991 Value	%	1992 Value	%	1993 Value	%	1994 Value	%	1995 Value	%	1996 Value	%
Total Production	575,213	100.0	576,276	100.0	699,847	100.0	761,868	100.0	820,994	100.0	869,735	100.0	827,763	100.0
Regions														
Africa	2,156	0.4	2,049	0.4	2,249	0.3	2,160	0.3	1,997	0.2	2,039	0.2	2,203	0.3
America, North	7,340	1.3	6,144	1.1	6,243	0.9	7,057	0.9	7,446	0.9	7,576	0.9	7,798	0.9
America, South	1,038	0.2	1,359	0.2	1,320	0.2	911	0.1	1,334	0.2	1,248	0.1	1,191	0.1
Asia	533,103	92.7	537,892	93.3	662,873	94.7	726,796	95.4	785,382	95.7	835,485	96.1	794,799	96.0
Europe	29,655	5.2	27,179	4.7	25,510	3.6	23,170	3.0	22,966	2.8	21,714	2.5	20,356	2.5
Oceania	1,922	0.3	1,653	0.3	1,652	0.2	1,775	0.2	1,869	0.2	1,673	0.2	1,416	0.2
Leading Producers														
China	448,507	78.0	456,041	79.1	591,193	84.5	657,446	86.3	720,431	87.8	771,043	88.7	731,563	88.4
Russian Federation	24,477	4.3	23,668	4.1	21,697	3.1	18,959	2.5	14,655	1.8	13,880	1.6	10,943	1.3
United States of America	7,116	1.2	5,939	1.0	6,029	0.9	6,804	0.9	7,153	0.9	7,244	0.8	7,426	0.9
Iran, Islamic Republic of	12,850	2.2	10,292	1.8	2,502	0.4	2,208	0.3	2,077	0.3	1,916	0.2	1,842	0.2
Ukraine	7,241	1.3	6,662	1.2	5,785	0.8	5,142	0.7	4,008	0.5	2,735	0.3	1,879	0.2

Tiles, Roofing, Made of Clay

Unit of Measure: 1,000 Square meters.

	1990 Value	%	1991 Value	%	1992 Value	%	1993 Value	%	1994 Value	%	1995 Value	%	1996 Value	%
Total Production	440,670	100.0	448,423	100.0	631,120	100.0	3,594,786	100.0	2,031,953	100.0	1,846,619	100.0	1,351,285	100.0
Regions														
Africa	17,435	4.0	17,435	3.9	19,435	3.1	19,435	0.5	20,435	1.0	21,436	1.2	22,435	1.7
Asia	49,113	11.1	26,580	5.9	55,046	8.7	2,420,513	67.3	774,480	38.1	831,946	45.1	797,413	59.0
Europe	369,801	83.9	400,408	89.3	552,638	87.6	1,150,838	32.0	1,237,038	60.9	993,237	53.8	531,437	39.3
Oceania	4,321	1.0	4,000	0.9	4,000	0.6	4,000	0.1	-		-		-	
Leading Producers														
Ukraine	285,000	64.7	315,000	70.2	467,000	74.0	1,066,000	29.7	1,151,000	56.6	906,000	49.1	444,000	32.9
Russian Federation	23,000	5.2	9,000	2.0	46,000	7.3	2,412,000	67.1	727,000	35.8	772,000	41.8	518,000	38.3
Belarus	3,000	0.7	2,000	0.4	1,000	0.2	2,000	0.1	33,000	1.6	45,000	2.4	264,000	19.5
France	*44,500	10.1	*44,500	9.9	*44,500	7.1	*44,500	1.2	*44,500	2.2	*44,500	2.4	*44,500	3.3
Italy	*26,000	5.9	*26,000	5.8	*26,000	4.1	*26,000	0.7	*26,000	1.3	*26,000	1.4	*26,000	1.9

Tiles, Floor and Wall

Unit of Measure: 1,000 Square meters.

	1990 Value	%	1991 Value	%	1992 Value	%	1993 Value	%	1994 Value	%	1995 Value	%	1996 Value	%
Total Production	896,610	100.0	921,729	100.0	985,883	100.0	1,184,845	100.0	1,238,130	100.0	1,325,459	100.0	1,354,115	100.0
Regions														
Africa	16,241	1.8	15,622	1.7	24,867	2.5	25,072	2.1	23,746	1.9	25,031	1.9	23,339	1.7
America, North	54,403	6.1	52,453	5.7	53,783	5.5	57,830	4.9	61,279	4.9	63,926	4.8	66,026	4.9
America, South	134,176	15.0	134,406	14.6	134,709	13.7	135,014	11.4	136,520	11.0	135,875	10.3	135,991	10.0
Asia	225,090	25.1	249,490	27.1	271,134	27.5	279,117	23.6	292,512	23.6	313,484	23.7	331,092	24.5
Europe	458,259	51.1	461,045	50.0	492,404	49.9	678,554	57.3	714,542	57.7	777,338	58.6	787,588	58.2
Oceania	8,440	0.9	8,713	0.9	8,986	0.9	9,259	0.8	9,532	0.8	9,805	0.7	10,078	0.7
Leading Producers														
Brazil	*129,363	14.4	*129,363	14.0	*129,363	13.1	*129,363	10.9	*129,363	10.4	*129,363	9.8	*129,363	9.6
Italy	*99,082	11.1	*99,082	10.7	*99,082	10.1	*99,082	8.4	*99,082	8.0	*99,082	7.5	*99,082	7.3
Spain	20,503	2.3	22,557	2.4	20,936	2.1	256,971	21.7	304,762	24.6	347,718	26.2	357,057	26.4
Germany	*70,055	7.8	71,933	7.8	71,798	7.3	70,432	5.9	66,056	5.3	*70,055	5.3	*70,055	5.2
China	*62,265	6.9	*70,154	7.6	*78,042	7.9	*85,930	7.3	*93,819	7.6	*101,707	7.7	*109,595	8.1

Product Tables
Quicklime

Unit of Measure: 1,000 Metric tons.

	1990 Value	%	1991 Value	%	1992 Value	%	1993 Value	%	1994 Value	%	1995 Value	%	1996 Value	%
Total Production	117,684	100.0	122,345	100.0	120,545	100.0	111,758	100.0	112,468	100.0	109,400	100.0	109,951	100.0
Regions														
Africa	3,154	2.7	3,694	3.0	3,625	3.0	3,511	3.1	3,432	3.1	3,413	3.1	3,457	3.1
America, North	22,019	18.7	22,454	18.4	23,014	19.1	17,411	15.6	18,090	16.1	18,652	17.0	19,817	18.0
America, South	4,508	3.8	4,259	3.5	4,031	3.3	3,928	3.5	4,037	3.6	3,844	3.5	4,274	3.9
Asia	46,973	39.9	47,114	38.5	46,113	38.3	45,635	40.8	45,169	40.2	45,323	41.4	45,552	41.4
Europe	40,924	34.8	44,729	36.6	43,656	36.2	41,063	36.7	41,575	37.0	38,049	34.8	36,774	33.4
Oceania	106	0.1	96	0.1	105	0.1	211	0.2	164	0.1	118	0.1	78	0.1
Leading Producers														
United States of America	13,392	11.4	13,200	10.8	13,700	11.4	14,200	12.7	14,800	13.2	15,800	14.4	16,500	15.0
Japan	8,983	7.6	9,045	7.4	8,049	6.7	7,958	7.1	7,712	6.9	7,871	7.2	7,744	7.0
Germany	-		7,533	6.2	7,542	6.3	7,483	6.7	8,511	7.6	5,432	5.0	4,958	4.5
Ukraine	8,677	7.4	7,648	6.3	7,484	6.2	5,924	5.3	4,663	4.1	3,902	3.6	3,570	3.2
Mexico	6,000	5.1	6,500	5.3	6,500	5.4	399	0.4	460	0.4	557	0.5	614	0.6

Cement

Unit of Measure: 1,000 Metric tons.

	1990 Value	%	1991 Value	%	1992 Value	%	1993 Value	%	1994 Value	%	1995 Value	%	1996 Value	%
Total Production	1,292,289	100.0	1,312,934	100.0	1,408,962	100.0	1,494,616	100.0	1,574,316	100.0	1,650,392	100.0	1,680,757	100.0
Regions														
Africa	50,815	3.9	53,166	4.0	53,059	3.8	50,608	3.4	49,961	3.2	52,941	3.2	56,725	3.4
America, North	117,285	9.1	109,911	8.4	113,130	8.0	120,009	8.0	128,910	8.2	122,087	7.4	125,845	7.5
America, South	49,593	3.8	52,562	4.0	53,474	3.8	65,091	4.4	56,215	3.6	62,578	3.8	70,298	4.2
Asia	785,790	60.8	827,323	63.0	892,877	63.4	973,858	65.2	1,045,300	66.4	1,134,072	68.7	1,146,012	68.2
Europe	281,448	21.8	263,520	20.1	289,751	20.6	277,559	18.6	286,023	18.2	271,218	16.4	274,542	16.3
Oceania	7,358	0.6	6,453	0.5	6,671	0.5	7,492	0.5	7,908	0.5	7,495	0.5	7,334	0.4
Leading Producers														
China	209,711	16.2	244,656	18.6	308,217	21.9	367,878	24.6	421,180	26.8	475,606	28.8	491,189	29.2
Japan	84,445	6.5	89,564	6.8	88,252	6.3	88,046	5.9	91,624	5.8	90,474	5.5	94,492	5.6
United States of America	70,944	5.5	67,193	5.1	69,585	4.9	73,807	4.9	77,948	5.0	76,906	4.7	79,265	4.7
Russian Federation	83,034	6.4	77,463	5.9	61,699	4.4	49,903	3.3	37,220	2.4	36,466	2.2	27,792	1.7
India	46,170	3.6	52,013	4.0	53,936	3.8	57,326	3.8	63,717	4.0	67,722	4.1	73,496	4.4

Asbestos-Cement Articles

Unit of Measure: Metric tons.

	1990 Value	%	1991 Value	%	1992 Value	%	1993 Value	%	1994 Value	%	1995 Value	%	1996 Value	%
Total Production	7,337,874	100.0	6,946,491	100.0	6,426,952	100.0	6,273,308	100.0	6,860,786	100.0	7,038,846	100.0	6,932,420	100.0
Regions														
Africa	229,809	3.1	216,605	3.1	213,858	3.3	194,903	3.1	182,356	2.7	210,563	3.0	213,160	3.1
America, North	256,000	3.5	258,000	3.7	291,000	4.5	281,000	4.5	259,000	3.8	174,000	2.5	227,000	3.3
America, South	1,678,000	22.9	1,687,000	24.3	1,696,000	26.4	1,714,000	27.3	1,750,000	25.5	2,089,000	29.7	1,840,057	26.5
Asia	2,583,367	35.2	2,590,733	37.3	2,252,850	35.1	2,121,550	33.8	2,370,372	34.5	2,511,213	35.7	2,678,807	38.6
Europe	2,590,698	35.3	2,194,153	31.6	1,973,244	30.7	1,961,855	31.3	2,299,058	33.5	2,054,070	29.2	1,973,396	28.5
Leading Producers														
Brazil	*1,408,000	19.2	*1,408,000	20.3	*1,408,000	21.9	*1,408,000	22.4	*1,408,000	20.5	*1,408,000	20.0	*1,408,000	20.3
India	802,000	10.9	690,000	9.9	495,000	7.7	681,000	10.9	758,000	11.0	980,000	13.9	1,083,000	15.6
Thailand	833,000	11.4	854,000	12.3	*720,750	11.2	*720,750	11.5	*720,750	10.5	*720,750	10.2	*720,750	10.4
France	648,000	8.8	583,000	8.4	457,000	7.1	391,000	6.2	720,000	10.5	*497,143	7.1	*481,600	6.9
Spain	523,000	7.1	492,000	7.1	404,000	6.3	*434,424	6.9	*429,746	6.3	*425,068	6.0	*420,389	6.1

Commodity data are provided by the United Nations Statistical Division. The symbol * means that data are estimated. For additional notes, see Appendix I.

483

Product Tables

Abrasives, Agglomerated or Not (Millstones, Grindstones, Etc.)

Unit of Measure: Metric tons.

	1990 Value	%	1991 Value	%	1992 Value	%	1993 Value	%	1994 Value	%	1995 Value	%	1996 Value	%
Total Production	245,570	100.0	240,846	100.0	229,334	100.0	205,536	100.0	213,217	100.0	223,814	100.0	216,650	100.0
Regions														
America, South	805	0.3	851	0.4	897	0.4	524	0.3	815	0.4	553	0.2	791	0.4
Asia	77,970	31.8	76,395	31.7	68,744	30.0	62,400	30.4	62,571	29.3	65,675	29.3	63,432	29.3
Europe	166,795	67.9	163,601	67.9	159,693	69.6	142,612	69.4	149,831	70.3	157,586	70.4	152,427	70.4
Leading Producers														
Germany	*88,085	35.9	*88,085	36.6	*88,085	38.4	*88,085	42.9	*88,085	41.3	91,604	40.9	84,566	39.0
Japan	70,999	28.9	68,859	28.6	60,954	26.6	55,848	27.2	56,138	26.3	58,169	26.0	55,081	25.4
Czech Republic	*20,658	8.4	*20,658	8.6	*20,658	9.0	*20,658	10.1	21,159	9.9	20,174	9.0	20,641	9.5
Spain	22,432	9.1	23,400	9.7	24,157	10.5	7,973	3.9	14,971	7.0	19,729	8.8	21,724	10.0
United Kingdom	*12,800	5.2	*12,800	5.3	*12,800	5.6	12,310	6.0	13,289	6.2	*12,800	5.7	*12,800	5.9

Concrete Blocks and Bricks - Weight

Unit of Measure: 1,000 Metric tons.

	1990 Value	%	1991 Value	%	1992 Value	%	1993 Value	%	1994 Value	%	1995 Value	%	1996 Value	%
Total Production	77,072	100.0	76,886	100.0	76,496	100.0	70,744	100.0	75,018	100.0	73,973	100.0	73,022	100.0
Regions														
Africa	737	1.0	537	0.7	570	0.7	594	0.8	582	0.8	707	1.0	880	1.2
America, North	999	1.3	1,013	1.3	1,163	1.5	2,817	4.0	2,889	3.9	1,739	2.4	2,029	2.8
Asia	343	0.4	348	0.5	353	0.5	319	0.5	304	0.4	316	0.4	316	0.4
Europe	74,993	97.3	74,988	97.5	74,410	97.3	67,015	94.7	71,243	95.0	71,212	96.3	69,798	95.6
Leading Producers														
Germany	*19,959	25.9	*19,959	26.0	*19,959	26.1	*19,959	28.2	*19,959	26.6	20,810	28.1	19,108	26.2
France	*18,232	23.7	*18,232	23.7	*18,232	23.8	*18,232	25.8	*18,232	24.3	*18,232	24.6	*18,232	25.0
United Kingdom	*15,615	20.3	*15,615	20.3	*15,615	20.4	13,878	19.6	17,352	23.1	*15,615	21.1	*15,615	21.4
Spain	9,772	12.7	9,920	12.9	9,844	12.9	3,247	4.6	3,128	4.2	3,891	5.3	3,960	5.4
Netherlands	*3,953	5.1	*3,953	5.1	*3,953	5.2	3,761	5.3	4,307	5.7	3,790	5.1	*3,953	5.4

Concrete Blocks and Bricks - Cubage

Unit of Measure: 1,000 Cubic meters.

	1990 Value	%	1991 Value	%	1992 Value	%	1993 Value	%	1994 Value	%	1995 Value	%	1996 Value	%
Total Production	319,727	100.0	309,428	100.0	295,312	100.0	286,797	100.0	267,430	100.0	258,941	100.0	252,040	100.0
Regions														
Africa	2	0.0	3	0.0	3	0.0	1	0.0	1	0.0	0	0.0	-	
Asia	249,159	77.9	247,540	80.0	233,029	78.9	227,493	79.3	212,336	79.4	210,776	81.4	206,416	81.9
Europe	70,566	22.1	61,884	20.0	62,280	21.1	59,303	20.7	55,093	20.6	48,164	18.6	45,625	18.1
Leading Producers														
Russian Federation	79,398	24.8	75,100	24.3	58,639	19.9	50,426	17.6	32,952	12.3	28,085	10.8	20,003	7.9
Germany	*17,506	5.5	13,967	4.5	16,108	5.5	18,686	6.5	21,264	8.0	*17,506	6.8	*17,506	6.9
Ukraine	23,310	7.3	22,494	7.3	20,255	6.9	15,281	5.3	8,501	3.2	5,630	2.2	2,981	1.2
Poland	5,164	1.6	4,358	1.4	4,030	1.4	3,698	1.3	4,238	1.6	4,014	1.6	4,136	1.6
Belarus	7,424	2.3	7,134	2.3	6,011	2.0	4,431	1.5	2,680	1.0	1,719	0.7	1,372	0.5

Product Tables
Concrete Pipes - Weight

Unit of Measure: Metric tons.

	1990 Value	%	1991 Value	%	1992 Value	%	1993 Value	%	1994 Value	%	1995 Value	%	1996 Value	%
Total Production	23,085,724	100.0	22,887,039	100.0	22,401,821	100.0	22,592,039	100.0	22,832,720	100.0	20,575,218	100.0	20,062,600	100.0
Regions														
Africa	888,202	3.8	730,812	3.2	722,421	3.2	675,031	3.0	666,948	2.9	637,217	3.1	605,486	3.0
America, North	758,000	3.3	440,000	1.9	344,000	1.5	621,000	2.7	662,000	2.9	396,000	1.9	310,000	1.5
America, South	1,464,500	6.3	1,464,500	6.4	1,460,000	6.5	1,467,000	6.5	1,462,000	6.4	1,469,000	7.1	1,464,500	7.3
Asia	3,262,026	14.1	3,238,474	14.1	3,017,421	13.5	3,105,043	13.7	2,890,057	12.7	2,811,027	13.7	2,732,399	13.6
Europe	16,712,996	72.4	17,013,254	74.3	16,857,978	75.3	16,723,965	74.0	17,151,715	75.1	15,261,973	74.2	14,950,215	74.5
Leading Producers														
Germany	*4,569,404	19.8	*4,569,404	20.0	*4,569,404	20.4	5,510,000	24.4	5,598,000	24.5	3,700,600	18.0	3,469,018	17.3
France	*2,497,000	10.8	*2,497,000	10.9	*2,497,000	11.1	*2,497,000	11.1	*2,497,000	10.9	*2,497,000	12.1	*2,497,000	12.4
Japan	*1,826,300	7.9	*1,687,200	7.4	*1,548,100	6.9	*1,409,000	6.2	*1,269,900	5.6	*1,130,800	5.5	*991,700	4.9
Spain	1,984,000	8.6	2,419,000	10.6	2,344,000	10.5	1,303,000	5.8	1,464,000	6.4	1,535,000	7.5	1,754,000	8.7
Brazil	*1,448,000	6.3	*1,448,000	6.3	*1,448,000	6.5	*1,448,000	6.4	*1,448,000	6.3	*1,448,000	7.0	*1,448,000	7.2

Concrete Pipes - Cubage

Unit of Measure: Cubic meters.

	1990 Value	%	1991 Value	%	1992 Value	%	1993 Value	%	1994 Value	%	1995 Value	%	1996 Value	%
Total Production	2,503,000	100.0	2,269,000	100.0	1,759,000	100.0	1,465,848	100.0	1,333,587	100.0	1,302,058	100.0	1,222,663	100.0
Regions														
America, North	422,000	16.9	422,000	18.6	422,000	24.0	422,000	28.8	422,000	31.6	422,000	32.4	422,000	34.5
America, South	188,000	7.5	187,000	8.2	187,000	10.6	187,000	12.8	185,133	13.9	185,000	14.2	185,000	15.1
Asia	1,035,000	41.4	872,000	38.4	517,000	29.4	331,000	22.6	215,000	16.1	190,000	14.6	147,000	12.0
Europe	858,000	34.3	788,000	34.7	633,000	36.0	525,848	35.9	511,453	38.4	505,058	38.8	468,663	38.3
Leading Producers														
Mexico	*422,000	16.9	*422,000	18.6	*422,000	24.0	*422,000	28.8	*422,000	31.6	*422,000	32.4	*422,000	34.5
Russian Federation	632,000	25.2	529,000	23.3	293,000	16.7	201,000	13.7	128,000	9.6	124,000	9.5	83,000	6.8
Netherlands	268,000	10.7	246,000	10.8	245,000	13.9	*264,182	18.0	*266,787	20.0	*269,392	20.7	*271,997	22.2
Ecuador	*185,000	7.4	*185,000	8.2	*185,000	10.5	*185,000	12.6	*185,000	13.9	*185,000	14.2	*185,000	15.1
Ukraine	404,000	16.1	358,000	15.8	200,000	11.4	58,000	4.0	36,000	2.7	31,000	2.4	9,000	0.7

Concrete, Other Products - Weight

Unit of Measure: 1,000 Metric tons.

	1990 Value	%	1991 Value	%	1992 Value	%	1993 Value	%	1994 Value	%	1995 Value	%	1996 Value	%
Total Production	122,942	100.0	122,559	100.0	116,712	100.0	125,521	100.0	133,524	100.0	140,205	100.0	133,210	100.0
Regions														
Africa	688	0.6	621	0.5	694	0.6	636	0.5	511	0.4	511	0.4	511	0.4
America, North	345	0.3	345	0.3	345	0.3	345	0.3	345	0.3	345	0.2	345	0.3
America, South	2,448	2.0	2,688	2.2	2,912	2.5	13,887	11.1	20,120	15.1	23,806	17.0	15,454	11.6
Asia	13,431	10.9	15,366	12.5	15,349	13.2	15,839	12.6	15,840	11.9	16,079	11.5	16,374	12.3
Europe	106,031	86.2	103,539	84.5	97,412	83.5	94,815	75.5	96,708	72.4	99,464	70.9	100,526	75.5
Leading Producers														
Germany	*46,743	38.0	*46,743	38.1	*46,743	40.0	*46,743	37.2	*46,743	35.0	47,325	33.8	46,160	34.7
Poland	14,948	12.2	13,149	10.7	10,782	9.2	9,908	7.9	10,408	7.8	9,197	6.6	9,737	7.3
Japan	*12,816	10.4	*13,228	10.8	*13,641	11.7	*14,054	11.2	*14,466	10.8	*14,879	10.6	*15,292	11.5
Netherlands	*11,828	9.6	*11,828	9.7	*11,828	10.1	11,012	8.8	12,257	9.2	12,216	8.7	*11,828	8.9
Spain	7,684	6.3	7,808	6.4	6,960	6.0	7,347	5.9	8,669	6.5	10,752	7.7	11,364	8.5

Commodity data are provided by the United Nations Statistical Division. The symbol * means that data are estimated. For additional notes, see Appendix I.

485

Product Tables
Concrete, Other Products - Cubage

Unit of Measure: Cubic meters.

	1990		1991		1992		1993		1994		1995		1996	
	Value	%	Value	%	Value	%	Value	%	Value	%	Value	%	Value	%
Total Production	39,355,834	100.0	38,260,277	100.0	37,307,065	100.0	37,593,263	100.0	38,673,332	100.0	38,600,401	100.0	39,301,696	100.0
Regions														
Africa	795,362	2.0	852,905	2.2	910,448	2.4	967,990	2.6	1,025,533	2.7	1,083,076	2.8	1,140,619	2.9
America, North	989,389	2.5	1,011,622	2.6	1,033,856	2.8	1,056,089	2.8	1,078,322	2.8	1,100,556	2.9	1,122,789	2.9
America, South	6,891,667	17.5	6,919,833	18.1	6,948,000	18.6	7,158,000	19.0	7,217,000	18.7	7,316,000	19.0	7,293,444	18.6
Asia	6,382,000	16.2	6,365,000	16.6	6,348,000	17.0	6,435,000	17.1	6,417,000	16.6	6,371,000	16.5	6,430,000	16.4
Europe	24,297,417	61.7	23,110,917	60.4	22,066,762	59.1	21,976,183	58.5	22,935,476	59.3	22,729,769	58.9	23,314,843	59.3
Leading Producers														
Poland	8,683,000	22.1	7,560,000	19.8	6,913,000	18.5	6,527,000	17.4	7,118,000	18.4	6,433,000	16.7	6,922,000	17.6
Brazil	*6,672,000	17.0	*6,672,000	17.4	*6,672,000	17.9	*6,672,000	17.7	*6,672,000	17.3	*6,672,000	17.3	*6,672,000	17.0
Romania	*6,582,000	16.7	*6,582,000	17.2	*6,582,000	17.6	*6,582,000	17.5	*6,582,000	17.0	*6,582,000	17.1	*6,582,000	16.7
Uzbekistan	6,300,000	16.0	*6,300,000	16.5	*6,300,000	16.9	*6,300,000	16.8	*6,300,000	16.3	*6,300,000	16.3	*6,300,000	16.0
Netherlands	4,177,000	10.6	4,308,000	11.3	4,179,000	11.2	*4,329,970	11.5	*4,434,145	11.5	*4,538,319	11.8	*4,642,494	11.8

ISIC 3710
IRON AND STEEL

ISIC 3710—Iron and Steel—includes primary and secondary products made of iron and steel, including pig iron, rod, blooms, billets, shapes, wire, tubes, pipes, forgings, and castings. Sintered ores are excluded.

This industry is classified under two 3-digit codes in the U.S. system: SIC 331—Blast Furnace and Basic Steel Products and SIC 332—Iron and Steel Foundries.

Summary Statistics

		1990 Value	1990 N	1991 Value	1991 N	1992 Value	1992 N	1993 Value	1993 N	1994 Value	1994 N	1995 Value	1995 N
Establishments or enterprises	(number)	23,533	70	24,359	76	27,349	68	21,783	64	16,937	56	14,878	24
Number employed	(000)	7,858	78	6,780	79	6,933	75	7,177	73	6,901	70	6,527	51
Total Wages	($ mil.)	68,921	74	66,523	74	76,346	68	55,783	66	55,777	62	47,949	46
Wage/salary per employee	($)	11,584	73	12,138	72	14,120	67	11,746	65	12,448	62	13,830	46
Female workers	(000)	159	29	111	27	383	24	371	28	333	24	329	13
as % of total employment*	(%)	10	28	9	27	12	24	13	28	15	24	22	12
Output	($ bil.)	537	76	514	78	563	73	488	69	489	66	478	51
per employee	($ 000)	95	73	92	74	102	69	98	65	109	63	120	49
per establishment	($ 000)	40,777	65	29,800	69	30,953	62	27,956	57	26,093	50	29,542	22
Value added	($ bil.)	179	72	171	71	174	64	177	63	162	59	174	48
per employee	($ 000)	33	69	31	67	36	61	34	58	40	56	44	45
per establishment	($ 000)	11,496	60	8,144	61	8,430	53	8,617	50	8,991	42	8,591	19
Capital investment	($ mil.)	31,688	61	30,942	56	28,601	50	24,224	45	19,627	37	14,231	19
per employee	($ 000)	9	57	9	52	9	48	10	40	14	34	12	17
per establishment	($ 000)	2,838	55	2,232	50	2,630	46	2,196	37	2,540	33	1,820	15

Data presented above are drawn from the detailed tables that follow. Columns headed 'N' show the number of countries that provided valid data for inclusion. Values are not strictly comparable one year to the next or one row to the next because the number of countries that report varies from period to period and row to row. However, a general indication of magnitudes can be gleaned from reviewing this summary—especially in earlier years in which more reports are available. For detailed explanations, see Appendix I. *The average for those countries reporting both total and female employment.

Establishments and Number Engaged

Country	Establishments or Enterprises (number)						Employees per Establishment					
	1990	1991	1992	1993	1994	1995	1990	1991	1992	1993	1994	1995
Albania	-	-	-	1.000	2.000	1.000	-	-	-	1,985	860	1,700
Argentina	-	-	-	521	550	-	-	-	-	50	47	-
Armenia	3.000	3.000	3.000	4.000	4.000	-	718	714	-	-	-	-
Australia	446	548	605	-	-	-	92	71	60	-	-	-
Austria	111	115	109	103	105	-	329	287	284	271	264	-
Azerbaijan	8.000	8.000	8.000	8.000	8.000	-	1,104	1,049	1,040	1,014	953	-
Bahamas	9.000	8.000	-	-	-	-	9.889	13	-	-	-	-
Bahrain	-	-	1.000	-	-	-	-	-	315	-	-	-
Bangladesh	198	201	192	-	-	-	71	78	77	-	-	-
Bolivia	7.000	8.000	9.000	9.000	9.000	12	17	20	32	29	28	21
Bosnia & Herzegovina	18	18	-	-	1.000	-	1,124	1,056	-	-	404	-
Bulgaria	*10	*10	*9.000	*18	*18	*18	2,590	2,180	2,933	1,422	1,411	1,472
Canada	214	195	192	187	185	-	248	262	245	246	249	-
Chile	20	17	20	18	23	-	397	444	375	398	318	-
China	*3,710	*3,731	*4,086	*5,839	*6,400	*7,299	809	823	783	584	541	474
China (Hong Kong SAR)	72	62	74	75	32	-	21	21	15	16	28	-
Colombia	68	71	86	90	81	-	137	133	115	102	112	-
Costa Rica	12	20	20	28	-	-	3.667	23	24	20	-	-
Croatia	24	21	21	22	20	20	445	487	372	332	322	281
Czechoslovakia (Former)	*21	*22	-	-	-	-	8,000	6,909	-	-	-	-
Czech Republic	-	*18	*37	*56	-	-	-	6,611	3,054	1,893	-	-
Denmark	56	69	66	-	-	-	78	61	61	-	-	-
Ecuador	13	16	17	14	19	16	112	97	95	111	99	117
Egypt	86	105	102	103	-	-	597	453	469	472	-	-
El Salvador	-	4.000	7.000	5.000	7.000	7.000	-	-	214	92	97	128
Ethiopia	3.000	3.000	3.000	3.000	3.000	3.000	381	368	375	379	373	394
Finland	51	55	56	54	54	-	251	218	204	207	215	-
FYR Macedonia	*4.000	*4.000	*4.000	*5.000	*4.000	*4.000	4,224	3,286	3,016	2,272	2,427	1,748
Germany	-	584	559	548	509	-	-	-	492	431	394	-
Germany (Western Part)	483	484	474	-	-	-	537	518	502	-	-	-
Ghana	-	-	-	3.000	-	-	-	-	-	454	-	-
Greece	64	63	64	-	-	-	89	86	74	-	-	-
Grenada	-	-	-	-	1.000	1.000	-	-	-	-	20	22
Guatemala	-	6.000	6.000	6.000	6.000	6.000	-	118	120	131	298	471
Honduras	-	-	3.000	5.000	5.000	5.000	-	-	151	99	109	148
Hungary	*91	*137	*146	*177	-	-	484	263	205	181	-	-
Iceland	1.000	1.000	1.000	1.000	-	-	209	188	190	161	-	-
India	3,304	3,108	3,388	3,362	-	-	141	134	145	139	-	-
Indonesia	62	74	81	87	93	103	410	372	365	362	365	315
Iran (Islamic Republic of)	112	52	60	62	-	-	427	1,006	840	791	-	-
Ireland	32	33	-	-	-	-	47	45	-	-	-	-
Israel	87	85	72	44	54	-	31	35	46	64	56	-
Italy	*1,005	*1,003	*677	*655	*645	-	141	135	164	120	142	-
Japan	6,477	6,407	6,205	6,194	4,628	4,616	52	53	53	52	61	58
Jordan	-	11	10	9.000	10	18	-	67	105	103	125	70
Korea, Republic of	950	1,055	1,075	1,176	1,165	1,245	93	83	79	73	74	71
Kuwait	4.000	4.000	6.000	6.000	6.000	-	163	83	90	116	123	-
Latvia	1.000	1.000	2.000	1.000	2.000	2.000	2,557	2,633	1,391	2,491	1,374	1,203
Lesotho	-	-	-	1.000	1.000	-	-	-	-	-	61	61
Lithuania	-	-	-	-	*4.000	-	-	-	-	-	442	-
Luxembourg	*8.000	*9.000	*10	*12	-	-	1,270	1,063	892	663	-	-
Malaysia	127	145	140	157	159	-	108	107	114	110	116	-
Mexico	94	92	92	88	85	79	504	472	411	350	330	371
Mongolia	1.000	2.000	3.000	2.000	2.000	-	198	117	54	50	49	-
Mozambique	10	10	7.000	7.000	6.000	-	171	167	210	193	171	-
Myanmar	1.000	1.000	2.000	2.000	2.000	-	395	398	497	499	190	-
Nepal	24	15	-	15	19	-	65	104	-	101	92	-
New Zealand	*73	*68	*70	*96	-	-	45	42	43	33	-	-
Norway	40	40	38	37	48	-	184	171	165	120	117	-
Pakistan	-	187	-	-	-	-	-	236	-	-	-	-

Continued.

Depending on the table, * means *Enterprises*, *Engaged*, or *Factor Values*; ± means *Basis Unspecified*; § means *shown in millions*. For additional notes and sources, see Appendix I.

Establishments and Number Engaged

- Continued -

Country	Establishments or Enterprises (number)						Employees per Establishment					
	1990	1991	1992	1993	1994	1995	1990	1991	1992	1993	1994	1995
Panama.	3.000	3.000	-	-	-	-	-	-	-	-	-	-
Paraguay	-	1.000	-	-	-	-	-	1.000	-	-	-	-
Peru	82	86	93	-	-	-	94	82	69	-	-	-
Philippines	193	216	180	178		-	99	90	110	106	-	-
Poland	*47	*50	*48	*47	-	-	2,894	2,500	2,458	2,319	-	-
Portugal	*254	*267	*258	*335	*295	-	59	52	47	36	27	-
Qatar	1.000	2.000	3.000	3.000	3.000	-	1,074	549	387	390	387	-
Republic of Moldova	2.000	2.000	2.000	-	-	-	1,627	1,607	1,646	-	-	-
Romania	*36	-	-	-	176	-	-	-	-	-	-	-
Russian Federation	-	-	-	*636	*798	*1,115	-	-	-	1,087	805	577
Singapore	14	13	15	16	16	-	124	137	117	107	109	-
Slovakia	-	10	10	16	18	-	-	2,330	2,144	1,423	1,232	-
Slovenia	16	*8.000	*9.000	*14	*16	*15	-	438	-	-	-	-
South Africa	-	178	-	-	-	-	-	438	-	-	-	-
Spain	1,159	1,175	1,113	-	-	-	66	66	65	-	-	-
Sri Lanka	8.000	7.000	4.000	7.000	-	-	235	274	388	248	-	-
Sweden	92	102	85	90	94	-	332	280	316	274	257	-
Thailand	226	258	-	-	-	-	123	134	-	-	-	-
Trinidad and Tobago	5.000	17	14	12	10	-	340	86	104	122	146	-
Tunisia	-	-	-	4.000	6.000	8.000	-	-	-	701	461	372
Turkey	185	176	297	262	239	-	337	330	189	211	214	-
Ukraine	46	76	90	52	58	61	6,435	3,816	3,356	5,654	4,810	4,443
United Kingdom	2,587	2,429	2,343	-	-	-	58	56	53	-	-	-
United Republic of Tanzania	2.000	34	-	-	-	-	257	108	-	-	-	-
USSR (Former)	*162	-	-	-	-	-	6,037	-	-	-	-	-
United States of America	-	-	3,657	-	-	-	-	-	104	-	-	-
Venezuela	150	198	159	142	171	192	181	139	159	161	132	126
Yugoslavia	21	22	28	30	31	32	-	841	632	607	565	497
Zambia	5.000	-	-	-	10	-	248	-	-	-	103	-
Zimbabwe	22	20	23	23	21	-	727	810	743	635	567	-

Employment and Compensation of Employees

Country	Number Employed/Engaged (000)						Wage/Salary per Employee ($)					
	1990	1991	1992	1993	1994	1995	1990	1991	1992	1993	1994	1995
Albania	-	-	-	1.985	1.720	1.700	-	-	-	405	-	521
Argentina	37	-	-	26	26	-	13,569	-	-	18,064	-	-
Armenia	*2.154	*2.141	-	-	-	-	-	-	-	-	-	-
Australia	41	*39	*36	-	-	-	26,094	28,668	29,085	-	-	-
Austria	37	33	31	28	28	25	29,395	29,813	33,139	32,758	34,273	-
Azerbaijan	8.836	8.395	8.321	8.115	7.624	-	2,521	2,607	-	477	-	-
Bahamas	0.089	0.103	-	-	-	-	4,843	11,019	-	-	-	-
Bahrain	-	-	0.315	-	-	-	-	-	15,164	-	-	-
Bangladesh	14	16	15	-	-	-	1,126	1,123	1,024	-	-	-
Belgium	40	39	37									
Bolivia	0.120	0.160	0.285	0.265	0.253	0.256	1,295	871	1,019	1,206	1,280	1,234
Bosnia & Herzegovina	20	19	-	-	0.404	-	3,478	2,082	-	-	-	-
Bulgaria	26	22	26	26	25	27	2,508	852	1,670	2,204	1,878	2,214
Cameroon	1.017	0.797	1.042	0.961	-	-	16,996	18,262	17,219	-	-	-
Canada	53	51	47	46	46	46	36,562	38,849	38,374	37,021	36,168	37,242
Chile	7.931	7.540	7.500	7.171	7.324	6.859	7,396	8,367	9,581	11,285	13,180	19,148
China	*3,000	*3,070	*3,200	*3,410	*3,460	*3,460	-	-	-	-	-	-
China (Hong Kong SAR)	*1.500	*1.300	*1.100	*1.200	*0.900	-	11,725	14,650	15,150	18,745	21,565	-
China (Taiwan Province)	68	70	76	83	85	84	14,422	15,836	18,000	17,808	18,387	18,940
Colombia	9.300	9.427	9.870	9.153	9.041	8.947	2,849	3,254	4,820	4,057	4,547	5,261
Costa Rica	0.044	0.452	0.476	0.558	-	-	1,941	3,040	3,719	3,911	-	-
Croatia	11	10	7.820	7.300	6.450	5.620	5,112	4,679	1,617	1,510	1,841	2,859
Czechoslovakia (Former)	168	152	-	-	-	-	2,620	1,957	-	-	-	-

Continued.

Employment and Compensation of Employees

- Continued -

Country	Number Employed/Engaged (000)						Wage/Salary per Employee ($)					
	1990	1991	1992	1993	1994	1995	1990	1991	1992	1993	1994	1995
Czech Republic	133	119	113	106	-	-	2,572	1,890	2,414	2,924	-	-
Denmark	4.364	4.231	4.027	3.670	3.800	3.851	30,360	30,264	33,365	31,476	32,767	38,036
Ecuador	1.457	1.548	1.616	1.559	1.872	1.880	4,547	4,411	4,982	5,120	2,637	2,614
Egypt	*51	*48	*48	*49	*50	*51	2,866	2,046	2,137	2,401	2,603	2,885
El Salvador.	-	-	*1.500	*0.462	*0.679	*0.895	-	-	-	2,001	3,837	3,127
Ethiopia.	1.144	1.103	1.125	1.137	1.120	1.183	2,516	2,064	1,727	1,206	1,172	1,019
Finland	13	12	11	11	12	-	32,169	30,718	29,678	23,755	26,736	-
FYR Macedonia	17	13	12	11	9.708	6.992	5,821	9,558	2,601	2,453	2,401	3,387
France	188	184	175	162	156	158	26,101	25,632	25,250	-	-	-
Germany	-	-	*275	*236	*201		-	-	32,994	32,199	35,419	
Germany (Western Part)	*259	*251	*238				31,436	32,178	35,901	-	-	-
Ghana	-	-	-	*1.362	*1.400	*1.440	-	-	-	1,012	756	772
Greece	*5.667	*5.413	*4.750	*4.900	*4.800	*4.811	17,363	17,722	19,986	17,300	18,034	20,637
Grenada	-	-	-	-	0.020	0.022	-	-	-	-	-	-
Guatemala.	-	*0.710	*0.720	*0.787	*1.788	*2.823	-	-	-	-	-	-
Honduras	*0.378	*0.350	*0.452	*0.495	*0.543	*0.742	1,615	1,836	2,141	1,854	1,246	839
Hungary	44	36	30	32	32	31	2,955	3,037	3,477	3,839	4,012	4,235
Iceland	0.209	0.188	0.190	0.161	0.157	-	36,047	37,426	-	34,573	-	-
India.	*467	*416	*490	*467	*468	*469	1,912	1,216	1,660	1,575	1,707	1,818
Indonesia	25	28	30	31	34	32	1,139	1,303	1,667	1,754	1,889	3,950
Iran (Islamic Republic of)	48	52	50	49	-	-	5,597	6,572	8,569	8,059	-	-
Ireland	1.500	1.500	1.379	1.439	1.379	1.470	26,755	25,417	27,812	24,035	25,112	27,187
Israel.	2.700	3.000	3.300	2.800	3.000	3.236	24,799	27,496	25,385	26,122	25,904	29,601
Italy	142	136	111	79	92	-	37,701	39,532	-	33,326	35,833	-
Japan	*338	*340	*331	*322	*281	*269	38,579	-	-	-	-	-
Jordan	0.703	0.740	1.046	0.931	1.248	1.262	4,220	3,619	3,215	3,588	3,775	4,207
Korea, Republic of	88	88	85	86	87	88	13,459	15,750	16,235	16,827	19,083	21,654
Kuwait	0.651	0.332	0.541	0.696	0.740	0.672	4,750	2,758	5,810	6,799	7,491	9,132
Latvia	2.557	2.633	2.782	2.491	2.749	2.407	-	-	-	-	-	-
Lesotho.	-	-	-	*0.061	*0.061	-	-	-	-	5,017	5,235	-
Lithuania	-	-	3.190	2.555	1.768	-	-	-	-	440	666	1,189
Luxembourg	10	9.564	8.919	7.951	7.729	-	35,257	35,234	36,914	36,509	38,800	-
Malaysia	14	15	16	17	18	20	4,380	4,638	5,810	6,087	6,289	6,869
Mexico	47	43	38	31	28	29	6,284	6,626	8,770	9,597	9,797	6,316
Mongolia	*0.198	*0.233	*0.163	*0.100	*0.098	-	1,353	1,398	706	-	-	-
Mozambique	1.713	1.668	1.473	1.350	1.026	-	1,027	786	535	521	624	-
Myanmar	0.395	0.398	0.995	0.998	0.379	-	3,730	3,579	1,967	3,971	11,287	-
Nepal	1.565	1.558	-	1.516	1.753	-	564	603	-	508	603	-
New Zealand	3.296	2.881	3.003	3.161	-	-	-	-	-	-	-	-
Norway	7.365	6.844	6.251	4.444	5.595	5.271	32,493	33,131	-	-	36,214	-
Pakistan	42	44	46	-	-	-	2,848	3,056	3,144	-	-	-
Paraguay	-	0.001	-	-	-	-	-	-	-	-	-	-
Peru.	*7.734	*7.035	*6.400	-	-	-	6,100	5,430	4,504	-	-	-
Philippines.	19	19	20	19	-	-	2,809	2,848	3,542	3,379	-	-
Poland	136	125	118	109	120	132	1,749	2,460	2,937	3,016	3,319	4,117
Portugal	*15	*14	*12	*12	*7.963	*7.940	-	-	-	-	-	-
Qatar	1.074	1.097	1.162	1.169	1.161	-	18,406	19,529	19,572	21,680	21,569	-
Republic of Moldova	3.254	3.214	3.291	-	0.218	0.178	-	4,178	-	-	-	426
Russian Federation	-	-	-	691	643	644	-	-	-	918	1,310	1,577
Singapore	*1.730	*1.787	*1.756	*1.714	*1.739	*1.729	15,548	17,236	20,367	22,641	24,088	26,523
Slovakia	-	*23	*21	*23	*22	-	-	2,203	2,662	2,970	3,503	-
South Africa	80	78	73	67	61	63	11,793	12,540	14,174	14,031	14,885	18,055
Spain	77	77	73	60	53	53	23,871	25,299	28,178	23,947	24,599	27,318
Sri Lanka	*1.877	*1.918	*1.551	*1.739	-	-	1,237	1,383	1,521	2,039	-	-
Sweden.	30	*29	*27	25	24	24	27,087	28,072	31,500	24,662	27,011	29,898
Thailand	28	35	-	-	-	-	5,014	3,686	-	-	-	-
Trinidad and Tobago	1.700	1.467	1.460	1.462	1.457	-	10,661	11,323	13,448	10,430	10,287	-
Tunisia	-	-	-	2.805	2.768	2.979	-	-	-	13,773	14,253	14,917
Turkey	62	58	56	55	51	51	9,499	15,498	15,277	15,125	9,727	11,313
Ukraine	296	290	302	294	279	271	3,477	3,763	4,541	629	617	931

Continued.

Depending on the table, * means *Enterprises, Engaged*, or *Factor Values*; ± means *Basis Unspecified*; § means *shown in millions*. For additional notes and sources, see Appendix I.

Employment and Compensation of Employees
- Continued -

Country	Number Employed/Engaged (000)						Wage/Salary per Employee ($)					
	1990	1991	1992	1993	1994	1995	1990	1991	1992	1993	1994	1995
United Kingdom	151	135	124	-	-	-	24,149	25,618	27,029	-	-	-
United Republic of Tanzania	0.514	3.672	-	-	-	-	838	-	-	-	-	-
USSR (Former)	978	-	-	-	-	-	2,543	-	-	-	-	-
United States of America	413	392	379	367	364	373	33,680	33,852	35,681	37,534	39,791	-
Uruguay	*1.169	*1.056	*0.767	*0.849	*0.905	*0.818	4,051	5,208	4,763	7,740	6,810	9,523
Venezuela	27	28	25	23	23	24	7,636	8,042	4,681	5,822	4,442	13,782
Yugoslavia	-	19	18	18	18	16	-	-	-	-	1,771	954
Zambia	1.238	-	-	-	1.035	-	1,948	-	-	-	2,678	-
Zimbabwe	16	16	17	15	12	12	5,778	5,150	3,443	3,356	3,923	4,383

Female Workers: Total and Percent of Employment

Country	Female Workers (000)						Female Workers as % of Total Employment					
	1990	1991	1992	1993	1994	1995	1990	1991	1992	1993	1994	1995
Albania	-	-	-	0.349	0.335	0.300	-	-	-	17.58	19.48	17.65
Argentina	-	-	-	-	0.820	-	-	-	-	-	3.18	-
Australia	3.115	-	-	-	-	-	7.60	-	-	-	-	-
Austria	3.600	3.000	3.000	2.809	2.723	-	9.86	9.09	9.68	10.06	9.82	-
Bahrain	-	-	0.017	-	-	-	-	-	5.40	-	-	-
Bangladesh	0.004	0.004	0.014	-	-	-	0.03	0.03	0.09	-	-	-
Bosnia & Herzegovina	3.010	2.766	-	-	-	-	14.88	14.55	-	-	-	-
Bulgaria	-	-	-	8.700	8.500	8.700	-	-	-	33.98	33.46	32.83
Canada	4.000	3.000	-	-	-	-	7.55	5.88	-	-	-	-
Chile	0.231	0.229	0.228	0.215	0.232	-	2.91	3.04	3.04	3.00	3.17	-
China (Taiwan Province)	11	11	12	13	14	13	15.55	15.27	15.71	15.76	16.26	15.69
Colombia	-	0.869	-	0.997	0.991	-	-	9.22	-	10.89	10.96	-
Croatia	1.190	1.310	1.250	1.230	1.090	1.030	11.13	12.81	15.98	16.85	16.90	18.33
Czechoslovakia (Former)	28	27	-	-	-	-	16.79	18.03	-	-	-	-
Czech Republic	38	35	29	27	-	-	28.57	29.41	25.66	25.47	-	-
Denmark	0.698	0.701	0.647	-	-	-	15.99	16.57	16.07	-	-	-
Egypt	-	1.172	1.203	1.118	-	-	-	2.46	2.52	2.30	-	-
El Salvador	-	-	-	0.019	0.036	0.381	-	-	-	4.11	5.30	42.57
Ethiopia	-	0.076	0.077	0.088	0.088	0.147	-	6.89	6.84	7.74	7.86	12.43
FYR Macedonia	1.037	0.990	0.956	0.847	0.851	0.607	6.14	7.53	7.93	7.46	8.77	8.68
Germany (Western Part)	22	-	-	-	-	-	8.48	-	-	-	-	-
Ghana	-	-	-	0.384	-	-	-	-	-	28.19	-	-
Hungary	12	-	10	9.000	-	-	27.27	-	33.33	28.13	-	-
Indonesia	-	-	-	1.680	1.798	1.524	-	-	-	5.34	5.30	4.69
Iran (Islamic Republic of)	-	0.402	0.483	0.451	-	-	-	0.77	0.96	0.92	-	-
Ireland	0.100	0.100	-	-	-	-	6.67	6.67	-	-	-	-
Italy	-	9.286	4.455	4.339	4.904	-	-	6.84	4.02	5.51	5.35	-
Jordan	0.010	0.015	0.008	0.008	0.012	0.015	1.42	2.03	0.76	0.86	0.96	1.19
Korea, Republic of	6.600	7.377	6.447	7.025	7.128	7.431	7.47	8.38	7.59	8.15	8.22	8.42
Lesotho	-	-	-	0.006	0.006	-	-	-	-	9.84	9.84	-
Malaysia	1.600	1.900	1.900	2.000	2.100	-	11.68	12.26	11.88	11.63	11.41	-
Mongolia	-	-	-	-	0.020	-	-	-	-	-	20.41	-
Morocco	-	-	-	-	-	0.080	-	-	-	-	-	-
Myanmar	0.027	0.028	0.087	0.087	0.165	-	6.84	7.04	8.74	8.72	43.54	-
Nepal	0.011	0.023	-	0.011	0.010	-	0.70	1.48	-	0.73	0.57	-
New Zealand	0.366	0.317	0.327	0.336	-	-	11.10	11.00	10.89	10.63	-	-
Panama	0.001	-	-	-	-	-	-	-	-	-	-	-
Philippines	-	-	1.823	1.714	-	-	-	-	9.21	9.12	-	-
Portugal	0.975	0.826	0.701	0.635	0.336	-	6.46	6.00	5.82	5.29	4.22	-
Republic of Moldova	-	-	-	-	0.070	0.050	-	-	-	-	32.11	28.09
Sri Lanka	0.128	0.078	0.078	0.089	-	-	6.82	4.07	5.03	5.12	-	-
Sweden	4.600	-	-	-	-	-	15.08	-	-	-	-	-
Thailand	5.023	3.528	-	-	-	-	18.01	10.17	-	-	-	-
Turkey	1.908	-	-	-	-	-	3.06	-	-	-	-	-

Continued.

Female Workers: Total and Percent of Employment

- Continued -

Country	Female Workers (000)						Female Workers as % of Total Employment					
	1990	1991	1992	1993	1994	1995	1990	1991	1992	1993	1994	1995
United Kingdom	10	-	14	-	-	-	6.85	-	11.29	-	-	-
United Republic of Tanzania . .	0.028	0.202	-	-	-	-	5.45	5.50	-	-	-	-
United States of America . . .	-	-	294	287	287	296	-	-	77.57	78.20	78.85	79.36
Zambia	-	-	-	-	0.015	-	-	-	-	-	1.45	-

Output and Output per Employee

Country	Output ($ bil.)						Output per Employee ($)					
	1990	1991	1992	1993	1994	1995	1990	1991	1992	1993	1994	1995
Albania	-	-	-	*0.005	*0.007	*0.008	-	-	-	2,759	4,357	4,834
Argentina	±3.587	-	-	2.951	-	-	96,787	-	-	112,578	-	-
Armenia.	0.001	0.001	0.006	0.107	0.001	-	290	480	-	-	-	-
Australia	*7.373	*7.293	*7.017	-	-	-	179,840	187,009	194,914	-	-	-
Austria	4.841	4.478	4.520	3.904	4.374	5.596	132,620	135,699	145,791	139,769	157,812	222,020
Azerbaijan	±0.144	±0.137	±0.012	±0.032	±0.006	-	16,350	16,305	1,385	3,900	773	-
Bahamas	0.003	0.004	-	-	-	-	33,708	38,835	-	-	-	-
Bahrain	-	-	±0.042	-	-	-	-	-	133,941	-	-	-
Bangladesh	0.284	0.283	0.299	-	-	-	20,221	18,140	20,188	-	-	-
Belgium	7.063	6.072	5.737	4.732	5.594	7.208	177,029	155,690	155,482	-	-	-
Bolivia	0.001	0.001	0.006	0.005	0.006	0.014	12,074	7,281	21,183	19,981	22,377	55,081
Bosnia & Herzegovina	±0.918	±0.621	-	-	-	-	45,368	32,658	-	-	-	-
Bulgaria.	±0.734	±0.522	±0.371	±0.478	±0.548	±0.766	28,332	23,929	14,059	18,682	21,575	28,911
Canada	8.570	8.222	7.810	8.581	9.505	9.838	161,707	161,217	166,171	186,538	206,630	213,683
Chile	0.637	0.620	0.663	0.723	0.787	1.066	80,301	82,196	88,422	100,847	107,412	155,379
China	±27	±29	±38	±68	±48	±44	9,051	9,414	11,791	20,008	13,968	12,667
China (Hong Kong SAR) . . .	0.199	0.190	0.166	0.208	0.226	-	132,909	145,904	151,269	173,225	251,454	-
China (Taiwan Province) . . .	±8.632	±9.551	±10	±13	±13	±15	127,640	137,178	136,764	152,530	151,898	177,575
Colombia	0.637	0.595	0.653	0.572	0.697	0.792	68,508	63,163	66,152	62,461	77,053	88,505
Croatia	±0.711	±0.438	±0.279	-	-	-	66,512	42,813	35,694	-	-	-
Czechoslovakia (Former) . . .	±5.020	±2.149	-	-	-	-	29,881	14,138	-	-	-	-
Denmark	*0.606	*0.520	*0.565	*0.475	*0.524	*0.622	138,770	122,868	140,422	129,516	138,024	161,427
Ecuador	0.107	0.129	0.119	0.121	0.153	0.170	73,415	83,485	73,550	77,505	81,968	90,527
Egypt	*1.343	*0.842	*0.930	*1.068	*1.199	*1.373	26,181	17,688	19,450	21,974	24,049	26,989
El Salvador.	-	±0.007	-	±0.011	±0.024	±0.054	-	-	-	23,913	36,008	60,734
Ethiopia.	0.038	0.023	0.011	0.012	0.040	0.036	33,550	21,154	9,531	10,239	35,765	30,551
Finland	±3.238	±2.831	±3.054	±2.796	±3.371	-	252,966	235,884	267,932	249,634	290,630	-
FYR Macedonia	±0.652	±0.699	±0.343	±0.185	±0.109	±0.141	38,569	53,208	28,474	16,304	11,269	20,204
France	±27	±24	±24	±20	±23	±27	143,342	131,485	137,878	123,744	148,317	172,738
Germany	-	-	39	31	34	-	-	-	140,000	132,470	171,864	-
Germany (Western Part) . . .	40	37	36	-	-	-	154,636	147,941	153,042	-	-	-
Ghana	-	-	-	0.016	0.013	0.014	-	-	-	11,875	9,411	9,600
Greece	*1.249	*1.068	*0.973	*0.870	*0.886	*1.019	220,416	197,363	204,938	177,538	184,632	211,751
Guatemala	-	0.067	0.065	0.071	0.077	0.084	-	94,108	90,378	89,829	42,919	29,739
Honduras	0.013	0.010	0.012	0.011	0.010	0.012	34,186	27,948	25,447	22,849	18,078	15,979
Hungary	±1.821	±1.323	±0.974	±0.885	±1.197	±1.385	41,387	36,760	32,480	27,645	37,410	44,666
Iceland	0.041	0.031	±0.033	±0.040	-	-	194,987	167,195	175,878	251,073	-	-
India	*16	*12	*14	*12	*14	*15	33,551	29,142	28,983	26,685	29,392	31,723
Indonesia	2.395	2.252	2.155	3.096	3.657	4.467	94,146	81,746	72,910	98,400	107,845	137,512
Iran (Islamic Republic of) . . .	2.335	1.997	3.006	2.956	-	-	48,846	38,188	59,622	60,294	-	-
Ireland	*0.267	*0.233	*0.231	*0.211	*0.210	*0.248	178,351	155,089	167,208	146,603	152,288	168,463
Israel	0.422	0.603	0.647	0.518	0.541	0.698	156,326	201,103	195,933	185,001	180,333	215,686
Italy	±29	±27	±23	±16	±22	-	207,122	196,517	210,019	197,504	243,309	-
Japan	±126	±139	±130	±134	±124	±140	374,139	408,546	393,238	415,847	442,081	519,205
Jordan	0.088	0.105	0.139	0.106	0.110	0.114	124,518	141,779	133,333	113,331	88,456	90,531
Korea, Republic of	17	20	20	22	25	31	196,561	224,447	232,033	253,392	287,385	349,210
Kuwait	0.029	0.007	0.026	0.042	0.050	0.052	44,357	22,463	47,611	60,026	67,672	77,831
Latvia	0.001	0.001	0.028	0.050	0.066	0.055	350	469	10,073	20,077	24,167	22,787
Lithuania	-	-	0.006	0.009	0.011	-	-	-	1,952	3,586	6,151	-

Continued.

Depending on the table, * means *Enterprises, Engaged,* or *Factor Values*; ± means *Basis Unspecified*; § means *shown in millions*. For additional notes and sources, see Appendix I.

Output and Output per Employee
- Continued -

Country	Output ($ bil.)						Output per Employee ($)					
	1990	1991	1992	1993	1994	1995	1990	1991	1992	1993	1994	1995
Luxembourg	*2.257	*1.947	*1.894	*1.665	*1.700	-	222,143	203,575	212,359	209,353	219,976	-
Malaysia	*1.466	*1.705	*2.001	*2.303	*2.670	*3.265	107,005	110,007	125,079	133,876	145,132	165,809
Mexico	5.531	5.419	5.300	5.222	5.170	5.235	116,757	124,732	140,280	169,797	184,501	178,718
Mongolia	0.001	0.001	0.001	0.000	0.000	-	7,125	5,906	4,281	532	895	-
Mozambique	-	±0.012	-	±0.007	±0.004	-	-	7,068	-	5,053	4,296	-
Myanmar	0.052	0.054	0.043	0.064	0.098	-	131,435	135,424	43,102	63,681	259,220	-
Nepal	0.039	0.039	-	0.037	0.049	-	24,846	25,274	-	24,730	27,872	-
Norway	±1.444	±1.298	±1.294	-	±1.210	±1.470	196,127	189,615	207,001	-	216,272	278,860
Pakistan	0.914	0.919	0.997	-	-	-	21,549	20,847	21,633		-	-
Peru	0.327	0.308	0.271	-	-	-	42,313	43,835	42,282	-	-	-
Philippines	1.212	1.165	1.330	1.437	-	-	63,104	60,063	67,180	76,417	-	-
Poland	±4.755	±3.821	±3.366	±3.429	±4.226	±5.829	34,964	30,566	28,525	31,463	35,283	44,267
Portugal	0.889	0.753	0.765	0.712	0.648	0.780	58,856	54,715	63,561	59,241	81,430	98,248
Qatar	0.196	0.173	0.190	0.220	0.208	-	182,383	158,023	163,606	188,007	178,891	-
Republic of Moldova . . .	0.000	0.138	0.051	-	-	0.000	0.068	42,827	15,567		-	2,142
Romania	±3.303	±2.650	±1.702	±1.637	-	-	-	-	-	-	-	-
Russian Federation	-	18	23	8.727	12	18	-	-	-	12,627	18,136	27,262
Singapore	*0.321	*0.311	*0.322	*0.349	*0.357	*0.388	185,532	173,761	183,174	203,903	205,175	224,427
Slovakia	-	±1.214	±1.198	±1.173	±1.301	-	-	52,115	55,855	51,496	58,673	-
Slovenia	-	-	-	1.497	1.594	1.986	-	-	-	-	-	-
South Africa	±5.049	±4.720	±4.415	±4.152	±4.326	±5.341	63,111	60,515	60,486	61,977	70,921	84,967
Spain	*12	*11	*10	±8.491	±9.493	±14	157,863	147,294	143,837	140,359	178,654	258,378
Sri Lanka	0.024	0.033	0.031	0.032	-	-	12,966	17,027	20,019	18,393	-	-
Sweden	*6.273	*5.501	*5.368	*4.710	*5.983	*6.686	205,673	192,541	199,562	191,021	247,918	279,202
Thailand	2.104	3.294	-	-	-	-	75,428	94,950	-	-	-	-
Trinidad and Tobago . . .	±0.169	±0.186	±0.217	±0.210	±0.243	-	99,523	126,770	148,389	143,545	166,903	-
Tunisia	0.354	0.340	0.361	0.321	0.335	0.386	-	-	-	114,514	121,169	129,560
Turkey	5.851	5.895	6.145	7.889	7.004	8.218	93,910	101,478	109,676	142,643	137,184	160,494
Ukraine	*12	*11	*25	*4.393	*5.298	*6.107	40,253	38,377	83,113	14,943	18,988	22,533
United Kingdom	*21	*18	18	-	-	-	137,015	136,703	141,713	-	-	-
United Republic of Tanzania . .	*0.032	*0.030	-	-	-	-	62,168	8,122	-	-	-	-
USSR (Former)	±24	-	-	-	-	-	24,987	-	-	-	-	-
United States of America . . .	*76	*69	*71	*76	*85	*92	183,390	175,663	188,116	206,041	232,720	245,349
Uruguay	0.057	0.050	0.038	0.067	0.063	0.079	48,408	47,164	50,141	78,786	69,316	96,910
Venezuela	1.465	1.580	1.550	1.468	1.559	2.103	54,045	57,253	61,264	64,056	68,944	86,628
Yugoslavia	±0.088	±0.065	±0.171	-	±0.250	±0.128	-	3,515	9,657	-	14,266	8,075
Zambia	0.021	-	-	-	0.012	-	16,645	-	-	-	11,223	-
Zimbabwe	*0.406	*0.400	*0.502	*0.293	*0.325	*0.357	25,352	24,670	29,376	20,043	27,334	31,043

Value Added and Value Added per Employee

Country	Value Added ($ bil.)						Value Added per Employee ($)					
	1990	1991	1992	1993	1994	1995	1990	1991	1992	1993	1994	1995
Argentina	±1.651	-	-	0.705	-	-	44,562	-	-	26,896	-	-
Armenia	0.000	0.000	-	-	-	-	16	41	-	-	-	-
Australia	*2.431	*2.455	*2.492	-	-	-	59,299	62,949	69,210	-	-	-
Austria	2.088	1.809	1.615	1.386	1.556	2.032	57,216	54,816	52,108	49,626	56,136	80,638
Bahamas	0.002	0.002	-	-	-	-	22,573	21,282	-	-	-	-
Bangladesh	0.043	0.048	0.068	-	-	-	3,023	3,098	4,569	-	-	-
Belgium	*2.197	*1.674	*2.002	*1.651	*1.955	*2.519	55,062	42,912	54,262	-	-	-
Bolivia	0.000	0.000	0.003	0.002	0.002	0.005	3,583	2,510	11,048	6,982	7,518	18,507
Bosnia & Herzegovina	±0.209	±0.220	-	-	-	-	10,348	11,563	-	-	-	-
Bulgaria	-	±0.035	±0.069	±0.103	±0.086	±0.160	-	1,596	2,623	4,011	3,390	6,044
Cameroon	0.038	0.030	0.024	-	-	-	37,437	37,778	22,686	-	-	-
Canada	3.231	3.046	2.954	3.349	3.603	3.725	60,963	59,729	62,842	72,795	78,322	80,900
Chile	0.284	0.283	0.257	0.284	0.323	0.437	35,766	37,554	34,285	39,634	44,128	63,766
China	±6.571	±7.038	±9.471	±22	±15	±13	2,190	2,292	2,960	6,540	4,327	3,645
China (Hong Kong SAR) . . .	0.044	0.034	0.039	0.044	0.032	-	29,098	26,330	35,233	36,627	35,942	-

Continued.

Depending on the table, * means *Enterprises, Engaged,* or *Factor Values*; ± means *Basis Unspecified*; § means *shown in millions*. For additional notes and sources, see Appendix I.

493

Value Added and Value Added per Employee

- Continued -

Country	Value Added ($ bil.)						Value Added per Employee ($)					
	1990	1991	1992	1993	1994	1995	1990	1991	1992	1993	1994	1995
China (Taiwan Province) . . .	2.989	3.170	3.620	4.103	4.133	4.691	44,195	45,530	47,345	49,715	48,593	55,710
Colombia	0.281	0.266	0.350	0.211	0.279	0.316	30,186	28,207	35,439	23,085	30,815	35,348
Croatia	±0.229	±0.114	±0.110	-	-	-	21,427	11,147	14,005	-	-	-
Czechoslovakia (Former) . . .	±1.271	-	-	-	-	-	7,564	-	-	-	-	-
Denmark	*0.281	*0.261	*0.284	*0.240	*0.263	*0.312	64,386	61,711	70,486	65,262	69,318	80,993
Ecuador	0.019	0.019	0.021	0.018	0.033	0.019	12,738	12,415	12,814	11,680	17,473	9,907
Egypt	*0.314	*0.246	*0.190	*0.204	*0.229	*0.262	6,114	5,159	3,985	4,207	4,598	5,152
El Salvador.	-	0.003	-	0.003	0.009	0.014	-	-	-	5,445	13,961	15,322
Ethiopia.	0.006	0.007	0.004	0.005	0.012	0.015	5,419	6,350	3,497	4,732	11,080	13,087
Finland	±0.850	±0.746	±0.913	±0.866	±1.019	-	66,419	62,150	80,055	77,339	87,849	-
FYR Macedonia	±0.165	±0.187	±0.065	±0.062	±0.026	±0.032	9,757	14,229	5,374	5,479	2,709	4,562
France	±8.434	±7.350	±7.062	±5.932	±6.703	±8.373	44,766	39,903	40,472	36,709	42,912	53,126
Germany (Western Part) . . .	19	18	18	13	-	-	74,018	72,525	76,907	-	-	-
Ghana	-	-	-	0.004	0.004	0.004	-	-	-	3,104	2,617	2,669
Greece	*0.280	*0.217	*0.205	*0.183	*0.185	*0.216	49,389	40,138	43,057	37,284	38,644	44,863
Haiti	-	±0.021	±0.004	±0.002	±0.003	±0.002	-	-	-	-	-	-
Honduras	*0.004	*0.003	*0.003	*0.003	*0.002	*0.003	9,483	7,885	6,634	5,602	4,404	3,422
Hungary	±0.337	±0.213	±0.141	±0.136	±0.144	±0.143	7,670	5,917	4,701	4,258	4,500	4,612
Iceland	0.009	0.006	±0.006	±0.014	-	-	44,043	29,546	33,410	84,729	-	-
India	*2.551	*1.105	*1.857	*2.178	*2.387	*2.569	5,459	2,655	3,788	4,662	5,100	5,474
Indonesia	*1.045	*0.699	*0.929	*1.347	*1.602	*1.868	41,072	25,385	31,444	42,798	47,250	57,509
Iran (Islamic Republic of) . . .	0.942	0.821	1.385	1.283	-	-	19,705	15,694	27,464	26,171	-	-
Ireland	*0.092	*0.084	*0.083	*0.076	*0.076	*0.089	61,139	56,004	60,428	52,941	55,011	60,785
Israel.	0.113	0.150	0.153	0.133	0.131	0.170	41,883	49,873	46,457	47,323	43,506	52,392
Italy	±8.117	±7.140	±5.686	±3.701	±5.506	-	57,110	52,556	51,324	46,973	60,080	-
Japan	±49	±55	±53	±54	±51	±58	143,607	161,021	161,160	167,707	181,365	215,792
Jordan	0.024	0.024	0.042	0.030	0.023	0.030	34,546	31,887	40,291	31,687	18,705	23,698
Korea, Republic of	6.187	7.617	7.667	7.814	9.163	11	70,071	86,565	90,250	90,657	105,696	129,345
Kuwait	0.011	0.003	0.017	0.024	0.029	0.031	16,182	9,598	31,827	34,497	38,994	46,812
Latvia	-	-	-	*0.001	*0.019	*0.012	-	-	-	548	6,925	4,912
Luxembourg	*0.623	*0.473	*0.495	*0.340	*0.349	-	61,318	49,456	55,481	42,726	45,126	-
Malaysia	*0.287	*0.323	*0.381	*0.474	*0.352	*0.426	20,951	20,846	23,799	27,537	19,117	21,619
Mexico	1.799	1.721	1.673	1.679	1.753	1.775	37,978	39,603	44,292	54,580	62,553	60,594
Mongolia	0.001	0.001	0.000	0.000	0.000	-	3,157	3,697	1,600	212	438	-
Myanmar	±0.017	±0.018	±0.028	±0.052	±0.089	-	42,613	45,301	28,219	52,452	233,607	-
Nepal	0.008	0.010	-	0.007	0.010	-	5,265	6,254	-	4,455	5,846	-
New Zealand	0.113	-	-	-	-	-	34,234	-	-	-	-	-
Norway	±0.347	±0.314	±0.320	-	±0.327	±0.390	47,069	45,933	51,115	-	58,399	73,901
Pakistan	0.265	0.258	0.288	-	-	-	6,246	5,859	6,253	-	-	-
Peru	0.177	0.108	0.104	-	-	-	22,875	15,386	16,326	-	-	-
Philippines.	0.236	0.300	0.384	0.349	-	-	12,276	15,444	19,369	18,545	-	-
Poland	±1.887	±1.103	±0.968	±0.905	±1.102	±1.509	13,873	8,826	8,206	8,302	9,197	11,456
Portugal	0.273	0.227	0.185	0.185	0.140	0.169	18,047	16,475	15,371	15,378	17,644	21,335
Qatar	0.106	0.088	0.098	0.123	0.088	-	98,993	80,640	83,931	104,814	75,484	-
Romania	±0.651	±0.554	±0.337	±0.344	-	-	-	-	-	-	-	-
Russian Federation	-	-	-	±4.228	±4.668	±5.162	-	-	-	6,118	7,263	8,021
Singapore	*0.097	*0.102	*0.122	*0.100	*0.099	*0.109	56,006	57,161	69,471	58,545	56,955	62,840
Slovakia	-	-	-	±0.196	±0.259	-	-	-	-	8,614	11,679	-
Slovenia	-	-	-	0.430	0.469	0.597	-	-	-	-	-	-
South Africa	*2.342	*2.123	*2.334	*2.208	*2.321	*2.904	29,273	27,217	31,979	32,958	38,043	46,194
Spain	*3.762	*3.349	*3.133	±2.292	±2.603	±4.012	49,005	43,437	43,087	37,888	48,993	76,243
Sri Lanka	0.008	0.011	0.010	0.013	-	-	4,122	5,851	6,712	7,664	-	-
Sweden.	*2.097	*1.281	*1.285	*1.227	*1.702	*1.892	68,742	44,847	47,784	49,739	70,528	79,004
Thailand	0.430	1.141	-	-	-	-	15,408	32,881	-	-	-	-
Trinidad and Tobago	±0.037	±0.032	±0.039	±0.045	±0.052	-	21,882	21,957	26,692	30,575	35,980	-
Tunisia	0.069	0.065	0.068	0.060	0.064	0.080	-	-	-	21,560	23,086	26,941
Turkey	1.403	1.793	1.829	2.612	2.531	2.977	22,518	30,856	32,640	47,233	49,579	58,139
United Kingdom	*8.089	*7.131	*6.852	-	-	-	53,571	52,822	55,255	-	-	-
United Republic of Tanzania . .	*0.003	*0.001	-	-	-	-	5,356	244	-	-	-	-
United States of America . . .	*32	*27	*30	*33	*37	*40	76,949	69,082	79,425	88,798	101,310	106,070

Continued.

     Depending on the table, * means *Enterprises, Engaged,* or *Factor Values*; ± means *Basis Unspecified*; § means *shown in millions*. For additional notes and sources, see Appendix I.

Value Added and Value Added per Employee

- Continued -

Country	Value Added ($ bil.)						Value Added per Employee ($)					
	1990	1991	1992	1993	1994	1995	1990	1991	1992	1993	1994	1995
Uruguay	0.031	0.019	0.013	0.028	0.026	0.033	26,101	18,351	16,992	32,632	28,696	40,086
Venezuela	0.498	0.726	0.775	0.489	0.676	1.894	18,394	26,320	30,614	21,328	29,891	78,014
Yugoslavia	±0.013	±0.006	±0.068	-	±0.105	±0.037	-	316	3,861	-	5,994	2,337
Zambia	0.008	-	-	-	0.004	-	6,478	-	-	-	3,837	-
Zimbabwe	*0.184	*0.202	*0.314	*0.149	*0.176	*0.194	11,512	12,475	18,382	10,194	14,775	16,834

Capital Investment

Country	Gross Fixed Capital Formation ($ mil.)						Per Establishment ($ mil)					
	1990	1991	1992	1993	1994	1995	1990	1991	1992	1993	1994	1995
Argentina	-	-	-	161	-	-	-	-	-	0.309	-	-
Australia	345	-	-	-	-	-	0.772	-	-	-	-	-
Austria	321	306	349	214	249	-	2.888	2.664	3.201	2.076	2.375	-
Bahrain	-	-	0.146	-	-	-	-	-	0.146	-	-	-
Bangladesh	63	15	16	-	-	-	0.317	0.072	0.081	-	-	-
Belgium	528	677	710	504	590	632	-	-	-	-	-	-
Bolivia	0.228	-	0.260	0.025	0.016	-	0.033	-	0.029	0.003	0.002	-
Bosnia & Herzegovina . . .	24	-	-	-	-	-	1.321	-	-	-	-	-
Bulgaria	37	1.866	180	32	232	65	3.740	0.187	20	1.773	13	3.627
Canada	626	-	-	-	-	-	2.928	-	-	-	-	-
Chile	26	34	68	43	30	-	1.324	1.988	3.375	2.381	1.284	-
China (Hong Kong SAR) . .	20	11	3.746	56	28	-	0.271	0.174	0.051	0.752	0.873	-
Colombia	44	5.891	25	25	11	-	0.652	0.083	0.286	0.279	0.133	-
Croatia	14	7.071	3.839	1.519	0.382	1.282	0.586	0.337	0.183	0.069	0.019	0.064
Czechoslovakia (Former) . .	242	-	-	-	-	-	12	-	-	-	-	-
Denmark	33	34	-	-	-	-	0.589	0.496	-	-	-	-
Ecuador	-3.792	16	11	8.894	-3.069	28	-0.292	1.018	0.666	0.635	-0.162	1.721
Egypt	68	97	46	77	-	-	0.796	0.927	0.450	0.744	-	-
El Salvador	-	0.619	-	0.354	2.071	7.383	-	0.155	-	0.071	0.296	1.055
Ethiopia	2.265	0.878	0.332	0.076	0.314	0.440	0.755	0.293	0.111	0.025	0.105	0.147
Finland	214	177	109	67	102	-	4.198	3.224	1.938	1.241	1.885	-
FYR Macedonia	11	8.148	2.974	1.970	1.889	2.687	2.820	2.037	0.743	0.394	0.472	0.672
France	988	918	788	-196	-43	45	-	-	-	-	-	-
Germany (Western Part) . .	1,959	2,258	2,340	1,863	-	-	4.056	4.665	4.936	-	-	-
Greece	29	50	51	-	-	-	0.448	0.796	0.794	-	-	-
Hungary	60	41	46	37	-	-	0.658	0.298	0.317	0.212	-	-
India	2,182	2,412	2,257	2,033	-	-	0.661	0.776	0.666	0.605	-	-
Indonesia	113	213	74	188	125	168	1.821	2.882	0.918	2.159	1.349	1.630
Iran (Islamic Republic of) . .	36	61	194	240	-	-	0.320	1.168	3.241	3.873	-	-
Ireland	7.131	9.370	-	-	-	-	0.223	0.284	-	-	-	-
Israel	4.960	9.214	9.760	16	23	-	0.057	0.108	0.136	0.369	0.431	-
Italy	2,325	2,600	1,938	645	942	-	2.313	2.592	2.863	0.984	1.461	-
Japan	7,376	9,450	10,257	11,385	9,079	-	1.139	1.475	1.653	1.838	1.962	-
Jordan	0.794	0.354	1.059	1.166	0.534	2.604	-	0.032	0.106	0.130	0.053	0.145
Korea, Republic of	5,200	3,849	2,890	1,584	3,159	6,947	5.473	3.649	2.688	1.347	2.711	5.580
Kuwait	0.555	1.289	4.048	5.649	3.221	-	0.139	0.322	0.675	0.942	0.537	-
Latvia	0.022	0.030	0.011	-	0.000	0.548	0.022	0.030	0.005	-	0.000	0.274
Luxembourg	108	105	66	-	-	-	13	12	6.616	-	-	-
Malaysia	276	330	373	563	288	-	2.175	2.275	2.664	3.585	1.814	-
Mexico	72	79	-	-	-	-	0.769	0.857	-	-	-	-
Myanmar	47	45	83	78	74	-	47	45	42	39	37	-
Nepal	11	-	-	-	-	-	0.463	-	-	-	-	-
Netherlands	-	-	-	122	-	-	-	-	-	-	-	-
Norway	99	70	35	-	47	-	2.472	1.747	0.927	-	0.971	-
Pakistan	-	37	-	-	-	-	-	0.196	-	-	-	-
Peru	11	9.776	8.888	-	-	-	0.128	0.114	0.096	-	-	-
Philippines	68	12	104	106	-	-	0.353	0.057	0.576	0.598	-	-
Poland	265	300	239	161	-	-	5.643	6.010	4.969	3.431	-	-

Continued.

Depending on the table, * means *Enterprises*, *Engaged*, or *Factor Values*; ± means *Basis Unspecified*; § means *shown in millions*. For additional notes and sources, see Appendix I.

495

Capital Investment
- Continued -

Country	Gross Fixed Capital Formation ($ mil.)						Per Establishment ($ mil)					
	1990	1991	1992	1993	1994	1995	1990	1991	1992	1993	1994	1995
Portugal	196	71	122	46	12	-	0.772	0.264	0.475	0.136	0.041	-
Romania	170	100	-	49	47	-	4.734	-	-	-	0.267	-
Singapore	22	21	41	31	16	-	1.594	1.617	2.726	1.909	0.984	-
Slovenia	6.273	7.182	8.820	6.376	9.720	15	0.392	0.898	0.980	0.455	0.607	0.980
Spain	746	719	480	370	261	554	0.644	0.612	0.432	-	-	-
Sri Lanka	0.200	0.313	0.054	1.264	-	-	0.025	0.045	0.014	0.181	-	-
Thailand	303	503	-	-	-	-	1.341	1.950	-	-	-	-
Trinidad and Tobago	20	26	28	11	80	-	4.097	1.533	1.992	0.920	8.022	-
Tunisia	12	16	13	12	12	13	-	-	-	3.014	1.977	1.586
Turkey	1,267	167	357	449	276	-	6.846	0.948	1.201	1.712	1.157	-
Ukraine	185	141	199	26	56	76	4.023	1.856	2.215	0.497	0.972	1.248
United Kingdom	1,077	816	624	-	-	-	0.416	0.336	0.266	-	-	-
United Republic of Tanzania	7.449	3.586	-	-	-	-	3.724	0.105	-	-	-	-
United States of America	3,540	3,900	3,126	2,777	3,740	4,038	-	-	0.855	-	-	-
Uruguay	1.955	4.918	2.546	33	-	-	-	-	-	-	-	-
Venezuela	275	189	315	391	169	1,633	1.836	0.954	1.984	2.751	0.985	8.505
Yugoslavia	4.188	5.934	1.351	-	4.779	1.983	0.199	0.270	0.048	-	0.154	0.062
Zambia	0.890	-	-	-	-	-	0.178	-	-	-	-	-

Indexes of Industrial Production

Country	Index of Industrial Production (1990=100)						Index of Employment (1990=100)					
	1990	1991	1992	1993	1994	1995	1990	1991	1992	1993	1994	1995
Albania	100	-	-	-	-	-	-	-	-	-	-	-
Algeria	100	97	85	93	86	97	-	-	-	-	-	-
Angola	100	-	-	-	-	-	-	-	-	-	-	-
Argentina	100	104	-	-	-	-	100	-	-	71	70	-
Armenia	100	-	-	-	-	-	100	99	-	-	-	-
Australia	100	101	101	107	-	-	100	95	88	-	-	-
Austria	100	94	90	88	96	106	100	90	85	77	76	69
Azerbaijan	-	-	-	-	-	-	100	95	94	92	86	-
Bahamas	100	-	-	-	-	-	100	116	-	-	-	-
Bahrain	100	-	-	-	-	-	-	-	-	-	-	-
Bangladesh	100	80	62	66	118	196	100	111	105	-	-	-
Barbados	100	100	100	100	100	100	-	-	-	-	-	-
Belgium	100	98	89	95	103	106	100	98	92	-	-	-
Belize	100	100	100	100	100	100	-	-	-	-	-	-
Bhutan	100	-	-	-	-	-	-	-	-	-	-	-
Bolivia	100	118	119	130	134	103	100	133	238	221	211	213
Bosnia & Herzegovina	-	-	-	-	-	-	100	94	-	-	2	-
Botswana	100	100	100	-	-	-	-	-	-	-	-	-
Brazil	100	94	94	101	111	109	-	-	-	-	-	-
Bulgaria	100	55	50	63	81	91	100	84	102	99	98	102
Burundi	100	100	100	100	-	-	-	-	-	-	-	-
Cambodia	100	100	100	-	-	-	-	-	-	-	-	-
Cameroon	100	-	-	-	-	-	100	78	102	94	-	-
Canada	100	92	95	108	112	114	100	96	89	87	87	87
Central African Republic	100	100	100	100	100	-	-	-	-	-	-	-
Chile	100	101	120	129	124	147	100	95	95	90	92	86
China	100	-	-	-	-	-	100	102	107	114	115	115
China (Hong Kong SAR)	100	98	94	85	84	83	100	87	73	80	60	-
China (Taiwan Province)	100	109	121	134	143	144	100	103	113	122	126	125
Colombia	100	97	108	113	130	132	100	101	106	98	97	96
Congo	100	100	100	100	-	-	-	-	-	-	-	-
Costa Rica	100	-	-	-	-	-	100	1,027	1,082	1,268	-	-
Croatia	100	-	-	-	-	-	100	96	73	68	60	53
Cuba	100	100	-	-	-	-	-	-	-	-	-	-
Cyprus	100	100	100	100	100	100	-	-	-	-	-	-

Continued.

Depending on the table, * means *Enterprises, Engaged,* or *Factor Values*; ± means *Basis Unspecified*; § means *shown in millions*. For additional notes and sources, see Appendix I.

Indexes of Industrial Production

- Continued -

Country	Index of Industrial Production (1990=100)						Index of Employment (1990=100)					
	1990	1991	1992	1993	1994	1995	1990	1991	1992	1993	1994	1995
Czechoslovakia (Former) . . .	100	79	67	-	-	-	100	90	-	-	-	-
Czech Republic	100	-	-	-	-	-	100	89	85	80	-	-
Dem. Rep. of the Congo . . .	100	69	38	-	-	-	-	-	-	-	-	-
Denmark	100	97	96	82	87	89	100	97	92	84	87	88
Dominican Republic.	100	84	-	-	-	-	-	-	-	-	-	-
Ecuador	100	127	127	140	-	-	100	106	111	107	128	129
Egypt	100	112	91	90	97	103	100	93	93	95	97	99
El Salvador.	100	100	-	-	-	-	-	-	-	-	-	-
Ethiopia and Eritrea	100	-	-	-	-	-	-	-	-	-	-	-
Ethiopia.	-	-	-	-	-	-	100	96	98	99	98	103
Finland	100	103	117	129	134	136	100	94	89	88	91	-
FYR Macedonia	100	-	-	-	-	-	100	78	71	67	57	41
France	100	97	93	84	94	95	100	98	93	86	83	84
Gabon	100	-	-	-	-	-	-	-	-	-	-	-
Gambia	100	-	-	-	-	-	-	-	-	-	-	-
Germany (Eastern Part) . . .	100	-	-	-	-	-	-	-	-	-	-	-
Germany (Western Part) . . .	100	99	93	83	92	93	100	97	92	-	-	-
Ghana	100	-	-	-	-	-	-	-	-	-	-	-
Greece	100	99	90	81	73	86	100	96	84	86	85	85
Guatemala.	100	100	-	-	-	-	-	-	-	-	-	-
Guyana	100	100	100	100	100	100	-	-	-	-	-	-
Haiti	100	-	-	-	-	-	-	-	-	-	-	-
Honduras	100	-	-	-	-	-	100	93	120	131	144	196
Hungary	100	66	46	44	59	67	100	82	68	73	73	70
Iceland	100	80	-	-	-	-	100	90	91	77	75	-
India.	100	109	109	138	142	147	100	89	105	100	100	100
Indonesia	100	158	165	213	236	185	100	108	116	124	133	128
Iran (Islamic Republic of) . . .	100	-	-	-	-	-	100	109	105	103	-	-
Ireland	100	89	81	85	81	87	100	100	92	96	92	98
Israel.	100	111	116	121	139	166	100	111	122	104	111	120
Italy	100	98	97	94	-	-	100	96	78	55	64	-
Japan	100	101	91	89	88	91	100	101	98	95	83	80
Jordan	100	112	133	105	89	97	100	105	149	132	178	180
Korea, Republic of	100	111	116	129	140	154	100	100	96	98	98	100
Kuwait	100	-	-	-	-	-	100	51	83	107	114	103
Kyrgyzstan.	100	-	-	-	-	-	-	-	-	-	-	-
Latvia	100	-	-	-	-	-	100	103	109	97	108	94
Lesotho	100	-	-	-	-	-	-	-	-	-	-	-
Libyan Arab Jamahiriya . . .	100	-	-	-	-	-	-	-	-	-	-	-
Lithuania	100	-	-	-	-	-	-	-	-	-	-	-
Luxembourg	100	97	92	90	87	73	100	94	88	78	76	-
Macau	100	100	100	-	-	-	-	-	-	-	-	-
Madagascar	100	100	100	100	-	-	-	-	-	-	-	-
Malawi	100	100	100	100	100	100	-	-	-	-	-	-
Malaysia	100	113	135	152	174	197	100	113	117	126	134	144
Mali	100	100	100	100	100	100	-	-	-	-	-	-
Malta	100	100	100	100	-	-	-	-	-	-	-	-
Mexico	100	97	97	104	108	117	100	92	80	65	59	62
Mongolia	100	-	-	-	-	-	100	118	82	51	49	-
Morocco	100	107	103	100	104	121	-	-	-	-	-	-
Mozambique	100	-	-	-	-	-	100	97	86	79	60	-
Myanmar	100	-	-	-	-	-	100	101	252	253	96	-
Nepal	100	-	-	-	-	-	100	100	-	97	112	-
New Zealand	100	92	-	-	-	-	100	87	91	96	-	-
Nicaragua	100	-	-	-	-	-	-	-	-	-	-	-
Nigeria	100	-	-	-	-	-	-	-	-	-	-	-
Norway	100	95	98	99	109	116	100	93	85	60	76	72
Pakistan	100	-	-	-	-	-	100	104	109	-	-	-
Panama.	100	146	169	196	170	202	-	-	-	-	-	-
Papua New Guinea	100	-	-	-	-	-	-	-	-	-	-	-

Continued.

Indexes of Industrial Production

- Continued -

Country	Index of Industrial Production (1990=100)						Index of Employment (1990=100)					
	1990	1991	1992	1993	1994	1995	1990	1991	1992	1993	1994	1995
Paraguay	100	100	96	82	92	100	-	-	-	-	-	-
Peru	100	112	113	126	171	179	100	91	83	-	-	-
Philippines	100	97	155	-	-	-	100	101	103	98	-	-
Poland	100	76	74	75	87	101	100	92	87	80	88	97
Portugal	100	89	96	90	95	100	100	91	80	80	53	53
Puerto Rico	100	-	-	-	-	-	-	-	-	-	-	-
Qatar	100	-	-	-	-	-	100	102	108	109	108	-
Republic of Moldova	100	-	-	-	-	-	100	99	101	-	7	5
Romania	100	74	52	55	57	67	-	-	-	-	-	-
Russian Federation	100	-	-	-	-	-	-	-	-	-	-	-
Saudi Arabia	100	-	-	-	-	-	-	-	-	-	-	-
Senegal	100	100	100	100	100	100	-	-	-	-	-	-
Singapore	100	95	98	107	106	97	100	103	102	99	101	100
Slovakia	100	-	-	-	-	-	-	-	-	-	-	-
Slovenia	100	70	68	62	71	72	-	-	-	-	-	-
Somalia	100	100	100	-	-	-	-	-	-	-	-	-
South Africa	100	88	81	80	84	96	100	98	91	84	76	79
Spain	100	101	96	94	103	110	100	100	95	79	69	69
Sri Lanka	100	95	64	72	94	53	100	102	83	93	-	-
Sudan	100	100	100	-	-	-	-	-	-	-	-	-
Swaziland	100	100	100	-	-	-	-	-	-	-	-	-
Sweden	100	92	96	106	125	126	100	94	88	81	79	79
Switzerland	100	90	90	83	83	86	-	-	-	-	-	-
Syrian Arab Republic	100	100	100	100	100	100	-	-	-	-	-	-
Thailand	100	105	-	-	-	-	100	124	-	-	-	-
Tonga	100	-	-	-	-	-	-	-	-	-	-	-
Trinidad and Tobago	100	-	-	-	-	-	100	86	86	86	86	-
Tunisia	100	103	99	97	97	96	-	-	-	-	-	-
Turkey	100	95	102	117	114	116	100	93	90	89	82	82
Uganda	100	139	177	241	362	455	-	-	-	-	-	-
Ukraine	100	-	-	-	-	-	100	98	102	99	94	92
United Arab Emirates	100	-	-	-	-	-	-	-	-	-	-	-
United Kingdom	100	92	88	89	91	94	100	89	82	-	-	-
United Republic of Tanzania	100	105	110	-	-	-	100	714	-	-	-	-
USSR (Former)	100	-	-	-	-	-	100	-	-	-	-	-
United States of America	100	90	94	101	106	109	100	95	92	89	88	90
Uruguay	100	104	98	86	78	91	100	90	66	73	77	70
Venezuela	100	-	-	-	-	-	100	102	93	85	83	90
Yugoslavia	100	-	-	-	-	-	-	-	-	-	-	-
Yugoslavia (Former)	100	65	-	-	-	-	-	-	-	-	-	-
Zambia	100	102	115	120	113	109	100	-	-	-	84	-
Zimbabwe	100	102	90	74	83	79	100	101	107	91	74	72

Representative Companies in Sector

Name	Address	Telephone	Fax	Employment	Y
Cosipa Cia. Siderurgica Paulista	Av. Do Cafe 277 Torre B, 7th /8th Fl., Sao Paulo 04311-000, Brazil	11 55841800	11 55841863	10,000	96
Cawsl Corp.	PO Box 235, Wynnewood, Pennsylvania 19096, U.S.A.	215-649-3210	215-889-9250	9,500	
National Steel Corp.	4100 Edison Lakes Pkwy., Mishawaka, Indiana 46545-3440, U.S.A.	219-273-7000	219-273-7869	9,417	97
Precision Castparts Corp.	4650 S.W. Macadam, 440, Portland, Oregon 97201, U.S.A.	503-417-4855	503-417-4817	9,280	96
Bumar-Labedy	Mechanikow St. 9, Gliwice PL-44-109, Poland	32 345111	32 346966	9,014	93
AMSTED Industries Inc.	205 N. Michigan Ave., Chicago, Illinois 60601, U.S.A.	312-645-1700	312-819-8523	9,000	95
Eregli Iron and Steel Works Co.	Uzunkum No. 7, Eregli TR-67330, Turkey	388 19500	388 13969	8,349	89
Walter Industries Inc.	PO Box 31601, Tampa, Florida 33631-3601, U.S.A.	813-871-4811	813-871-4430	7,584	97
Engineering Equip. Inc.	Ortigas Industrial Estate, Quezon City, Philippines	2 6350851	2 7220644	7,200	92
Rugby Group PLC	Crown House, Rugby CV21 2DT, United Kingdom	788 542666		7,142	93
Dofasco Inc.	1330 Burlington St. E, Hamilton, Ontario L8N 3J5, Canada	905-544-3761	905-545-3236	7,100	98
Hindustan Cables Ltd.	9 Elgin Rd., Calcutta 700020, India	332471808	332471657	6,790	97
Nucor Corp.	2100 Rexford Rd., Charlotte, North Carolina 28211, U.S.A.	704-366-7000	704-362-4208	6,600	97
Scapa Group PLC	93 Preston New Rd., Blackburn BB2 6AY, United Kingdom	254 580123		6,536	94
Intermet Corp.	5445 Corporate Dr., 200, Troy, Michigan 48098-2683, U.S.A.	248-952-2500	248-952-2501	6,500	97
NKT Holding AS	NKT Alle 1, Brondby DK-2605, Denmark	43482000	43961820	6,425	93
Cyclops Industries Inc.	650 Washington Rd., Pittsburgh, Pennsylvania 15228, U.S.A.	412-343-4000	412-571-6356	6,300	
Armco Inc.	1 Oxford Centre, 301 Grant St., Pittsburgh, Pennsylvania 15219-1415, U.S.A.	412-255-9800	800-231-1054	6,000	97
Elkem AS	PO Box 4282 Torshov, Oslo N-0401, Norway	22 450100	22 450155	6,000	93
AK Steel Holding Corp.	703 Curtis St., Middletown, Ohio 45043-0001, U.S.A.	513-425-5000		5,800	97
Zelezorudne Bane SP	Stara Cesta 3, Spisska Nova Ves 052 54, Slovakia	965 23009	965 21370	5,800	93
Citation Corp.	2 Office Park Cir., 204, Birmingham, Alabama 35223, U.S.A.	205-871-5731	205-870-8211	5,778	97
Cominco Ltd.	200 Burrard St. Ste. 500, Vancouver, British Columbia V6C 3L7, Canada	604-682-0611	604-685-3091	5,743	98
WHX Corp.	110 E. 59th St., New York, New York 10022, U.S.A.	212-355-5200		5,706	97
Algoma Steel Inc.	PO Box 1400 Sta. Main, Sault Ste Marie, Ontario P6A 5P2, Canada	705-945-2351	705-945-2203	5,400	98
Weirton Steel Corp.	400 Three Springs Dr., Weirton, West Virginia 26062-4989, U.S.A.	304-797-2000	304-797-2887	5,373	97
Carpenter Technology Corp.	PO Box 14662, Reading, Pennsylvania 19612-4662, U.S.A.	610-208-2000		5,081	97
Grede Foundries Inc.	PO Box 26499, Milwaukee, Wisconsin 53226, U.S.A.	414-257-3600	414-256-9399	5,000	97
Poongsan Corp.	Chung-gu, Seoul, Republic of Korea	2 2733021	2 2733835	4,984	97
Huta Zawiercie SA	ul. Okolna 10, Zawiercie PL-42-400, Poland	21621	22536	4,700	93
Zaklady Starachowickie, Star Spolka Akcyjna	1 Maja ul. 12, Starachowice PL-27-200, Poland	8831	7038	4,430	93
Shaft Sinkers Pty. Ltd.	PO Box 783501, Sandton 2146, Republic of South Africa	11 4445600	11 4444410	4,324	97
Energomontaz-Polnoc SA	Przemyslowa St. 30, Warsaw PL-00-450, Poland	22 211041	22 296324	4,317	93
Republic Engineered Steels Inc.	PO Box 579, Massillon, Ohio 44648-0579, U.S.A.	330-837-6000	330-837-6099	4,094	97
Acindar Industria Arg. de Aceros SA	Edificio Fortabat, Buenos Aires 1106, Argentina	1 3168000	1 3168001	3,964	96
Oriental Precision & Engineering Co., Ltd.	Saha-Gu, Pusan, Republic of Korea	51 2020101	51 2025877	3,790	97
Quanex Corp.	1900 W. Loop S., 1500, Houston, Texas 77027, U.S.A.	713-961-4600	713-877-5333	3,771	97
Hitchiner Manufacturing Company Inc.	PO Box 2001, Milford, New Hampshire 03055, U.S.A.	603-673-1100	603-672-5926	3,605	97
Mostostal, Zabrze	Wolnosci 191, Zabrze PL-41-800, Poland	32 713221	32 713151	3,600	93
Goodyear SA	Ave. T. Gordon Smith, Colmar-Berg L-7750, Luxembourg	81991	81992496	3,550	95
Acominas Aco Minas Gerais SA	Rodovia Minas Gerais 443 Km. No. 0, 5, Ouro Branco 36406-000, Brazil	31 7492749	31 7492233	3,500	96
Columbus McKinnon Corp.	140 Audubon Pkwy., Amherst, New York 14228-1197, U.S.A.	716-689-5400	716-689-5644	3,479	96
Lukens Inc.	50 S. 1st Ave., Coatesville, Pennsylvania 19320-0911, U.S.A.	610-383-2000	610-383-3324	3,450	96
Compania Siderurgica Huachipato SA	Gran Bretana 2910, Talcahuano, Chile	41 502000	41 502699	3,400	96
Texas Industries Inc.	1341 W. Mockingbird Lane, Ste. 7, Dallas, Texas 75247-6913, U.S.A.	214-647-6700	214-647-3878	3,400	97
Intermet Foundries Inc.	PO Box 11589, Lynchburg, Virginia 24506-1589, U.S.A.	804-528-8200	804-528-8243	3,300	93
Acos Villares SA	Av. Interlagos 4455, Sao Paulo 04669-900, Brazil	11 5253322	11 5482212	3,200	96
Przyjazn	Dabrowa Gornicza PL-42-523, Poland	32 641213	32 255410	3,200	93

Continued

Representative Companies in Sector
- Continued -

Name	Address	Tele-phone	Fax	Employ-ment	Y
Rouge Steel Co.	PO Box 1699, Dearborn, Michigan 48121-1699, U.S.A.	313-317-8900	313-845-0199	3,128	97
Aarque Steel Corp.	PO Box 310, Jamestown, New York 14702-0310, U.S.A.	716-664-6014	716-664-5057	3,000	95
American Cast Iron Pipe Co.	PO Box 2727, Birmingham, Alabama 35202, U.S.A.	205-325-7701	205-325-8014	3,000	97
American Steel Foundries	10 S. Riverside Plz., 10th Fl., Chicago, Illinois 60606, U.S.A.	312-258-8000	312-258-5466	3,000	97
Nanjing Ferro Alloy Plant	Zhongxin Cun, Zhongyangmen Wai, Nanjing 210038, People's Republic of China	25 5504328	25 5501327	3,000	96
Sandvik Inc.	1702 Nevins Rd., Fair Lawn, New Jersey 07410, U.S.A.	201-794-5000	201-794-5011	3,000	97
Steeledale Reinforcing & Engineering Industries Ltd.	Tulisa Park, Johannesburg 2197, Republic of South Africa	11 6132331	11 6131394	2,950	93
Atchison Casting Corp.	PO Box 188, Atchison, Kansas 66002-0188, U.S.A.	913-367-2121	913-367-2155	2,800	97
Carpenter Steel Div.	PO Box 14662, Reading, Pennsylvania 19612-4662, U.S.A.	215-208-2000	215-208-2633	2,643	95
Veitscher Magnesitwerke AG	Postfach 143, Vienna A-1011, Austria	222 515130	222 5134315	2,621	91
Geneva Steel Co.	PO Box 2500, Provo, Utah 84603, U.S.A.	801-227-9000	801-227-9431	2,600	97
Inco United States Inc.	1 New York Plz., New York, New York 10004, U.S.A.	212-612-5500	212-612-5603	2,555	96
Drotovna, AS Hlohovec	Hlohovec 920 28, Slovakia	804 22241	804 22742	2,550	93
Mostaren Brezno AS	Mostarenska 9, Brezno 977 56, Slovakia	867 2171	867 2440	2,550	93
RA Multiproduct	Sector 2, Bucharest, Romania	1 6873490	1 6873490	2,508	93
Kia Steel Co., Ltd.	Kunsan-shi, Chonbuk, Republic of Korea	654 608114	654 608560	2,500	97
Mostostal Krakow SA	ul. Ujastek 7, Krakow PL-30 969, Poland	12 442454	12 445658	2,500	93
Acme Metals Inc.	13500 S. Perry Ave., Riverdale, Illinois 60627-1182, U.S.A.	708-849-2500	708-849-4503	2,471	97
Andoria SA	ul. Krakowska 140, Andrychow PL-34-120, Poland	387 53240	30 21606	2,400	93
Bakrie & Brothers, PT	Wisma Bakrie, Jakarta 12920, Indonesia	215250212	21 5200361	2,400	97
Oregon Steel Mills Inc.	1000 S.W. Broadway, Bldg. 1000,, Portland, Oregon 97205, U.S.A.	503-223-9228	503-240-5250	2,380	97
Mueller Industries Inc.	2959 N. Rock Rd., Wichita, Kansas 67226-1191, U.S.A.	316-636-6300	316-682-9650	2,350	97
Societe Tunisienne de Siderurgie Elfouladh	BP 23 & 24, Menzel Bourguiba 7050, Tunisia	2 60522	2 61533	2,350	93
Allegheny Teledyne Inc.	1000 Six PPG Place, Pittsburgh, Pennsylvania 15222-5479, U.S.A.	412-394-2861	412-394-3034	2,300	97
Huta Bobrek	Konstytucji 61, Katowice PL-41-905, Poland	32 814251	32 811688	2,300	93
WCI Steel Inc.	1040 Pine Ave. S.E., Warren, Ohio 44483-6528, U.S.A.	330-841-8314		2,251	97
Lukens Steel Co.	ARC Bldg., Modena Rd., Coatesville, Pennsylvania 19320, U.S.A.	215-383-2000	215-383-2436	2,238	94
Atomic Energy Corp. of South Africa Ltd.	PO Box 582, Pretoria 0001, Republic of South Africa	12 3163270	12 3163222	2,200	97
Prakash Tubes Ltd.	Prakash Nagar Sankhol Bahadurgarh, Haryana 124 507, India	8310540		2,110	97
Northwestern Steel and Wire Co.	PO Box 618, Sterling, Illinois 61081, U.S.A.	815-625-2500	815-625-0440	2,100	97
UNR Industries Inc.	332 S. Michigan Ave., Chicago, Illinois 60604-4385, U.S.A.	312-341-1234		2,100	93
Amalgamated Metal Corp. PLC	London Bridge, London EC4R 9DP, United Kingdom	71 6264521	71 6236015	2,099	93
Dae Won Kang-Up Co., Ltd.	Chung-Gu, Seoul, Republic of Korea	2 7565711	2 7565730	2,070	97
Lone Star Technologies Inc.	PO Box 803546, Dallas, Texas 75380-3546, U.S.A.	972-386-3981	972-656-6303	2,044	97
Kumbong Container PT	Jl. M.T. Haryono Kav 62, Jakarta, Indonesia	217975959	21 7976405	2,037	97
Keystone Consolidated Industries Inc.	5430 Lyndon B. Johnson Fwy., 17, Dallas, Texas 75240-2697, U.S.A.	214-458-0028		2,000	95
Hopkinsons Group PLC	Trafford Park, Manchester M17 1HP, United Kingdom	61 8728291		1,989	94
Sturm, Ruger and Company Inc.	One Lacey Pl., Southport, Connecticut 06490, U.S.A.	203-259-7843	203-254-2195	1,978	97
Huta-Zabrze SA	Bytomska ul. 1, Zabrze PL-41-800, Poland	32 713211	32 716901	1,950	93
NS Group Inc.	PO Box 1670, Newport, Kentucky 41072, U.S.A.	606-292-6809	606-292-0593	1,948	97
Lone Star Steel Co.	PO Box 1000, Lone Star, Texas 75668-1000, U.S.A.	972-386-3981	903-656-6879	1,941	97
William Cook PLC	Parkway Ave., Sheffield S9 4WA, United Kingdom	142 700895		1,919	94
Gulf States Steel Incorporated of Alabama	174 South 26th St., Gadsden, Alabama 35904-1935, U.S.A.	205-543-6100	205-543-6768	1,900	97
Waupaca Foundry Inc.	PO Box 249, Waupaca, Wisconsin 54981, U.S.A.	715-258-6611	715-258-9268	1,900	95
Fabryka Urzadzen Gornictwa Odkrywkowego, Fugo SA	ul. Przemyslowa 85, Konin PL-62-510, Poland	63 421581	63 424499	1,874	93
CF and I Steel Corp.	PO Box 316, Pueblo, Colorado 81002, U.S.A.	719-561-7103		1,850	
QIT-Fer & Titane Inc.	1625 Rte Marie-Victorin, Tracy, Quebec J3R 1M6, Canada	514-746-3000	514-286-9336	1,850	98
Rudne Bane SP	ul. J Krala 1, Banska Bystrica 974 32, Slovakia	88 32841	88 32842	1,837	93
Huta-Buczek SA	ul. Nowopogonska 1, Sosnowiec 41-200, Poland	32 691603	32 662286	1,800	93
Glynwed Steels Ltd.	Sheldon, Birmingham B26 3AZ, United Kingdom	21 7422366		1,790	93

Continued

For sources and notes, see Appendix I.

Representative Companies in Sector
- Continued -

Name	Address	Tele-phone	Fax	Employ-ment	Y
Birmingham Steel Corp.	1000 Urban Center Dr., Ste. 300, Birmingham, Alabama 35242-2516, U.S.A.	205-970-1200		1,789	97
Worthington Steel Co.	1127 Dearborn Dr., Columbus, Ohio 43085, U.S.A.	614-438-3210		1,750	95
Union Steel Manufacturing Co., Ltd.	Chongno-Gu, Seoul, Republic of Korea	2 7324111	2 7323263	1,707	97
Fabryka Kabli, Ozarow	ul. Poznanska 55, Ozarow PL-05-850, Poland	2 6286431	2 6286433	1,700	93
Lubelskie Zaklady Naprawy Samochowdow	Droga Meczennikow Majdanka ul. 12, Lublin PL-20-325, Poland	81 42661	81 42823	1,700	93
Korea Iron & Steel Co., Ltd.	Masan-Shi, Kyongnam, Republic of Korea	551 493000	551 449212	1,653	97
New Zealana Steel Ltd.	Private Bag, Auckland 1, New Zealand	9 758999		1,646	87
Cold Storage (Malaysia) BHD	10th Fl., Jaya Shopping Centre, Petaling Jaya 46100, Malaysia	3 7588888	3 7581289	1,600	93
Laclede Steel Co.	1 Metropolitan Sq., St. Louis, Missouri 63102-2738, U.S.A.	314-425-1400	314-425-1561	1,600	96
Malayawata Steel BHD	Province Wellesley, Pulau Penang 12700, Malaysia	4 307144	4 308863	1,600	94
McLouth Steel	1650 W. Jefferson St., Trenton, Michigan 48183, U.S.A.	313-285-1200	313-246-4059	1,600	
Chun Yuan Steel Industry Co., Ltd.	7F, No. 502 Fu Hsing North Rd., Taipei, Taiwan	2 5018111	2 5055390	1,508	93
Centrilift	200 W. Stuart Roosa Dr., Claremore, Oklahoma 74017-3095, U.S.A.	918-341-9600	918-342-0260	1,500	96
Seshin Commercial Co., Ltd.	Yangsan-Gun, Kyongnam, Republic of Korea	523 3881321	523 3881332	1,500	97
Vesuvius U.S.A. Corp.	PO Box 4014, Champaign, Illinois 61824, U.S.A.	217-351-5000	217-351-5031	1,500	97
Daelim Trading Co., Ltd.	Chongno-Gu, Seoul, Republic of Korea	2 7309811	2 7391341	1,482	97
Frederick Cooper PLC	Whittington, Worcester WR5 2RL, United Kingdom	905 764467	922 743600	1,458	93
Biwater Industries Ltd.	Clay Cross, Chesterfield S45 9NG, United Kingdom	246 250740	246 250741	1,447	93
YIT-Industry	Panuntie 11, Helsinki SF-00621, Finland	15941	15943726	1,430	94
National-Standard Co.	1618 Terminal Rd., Niles, Michigan 49120, U.S.A.	616-683-8100	616-683-6249	1,406	97
Bangkok Steel Industry Co., Ltd.	Samphanthawong, Bangkok 10100, Thailand	2227166	2247698	1,400	94
Bekaert Corp.	3200 W. Market St., Akron, Ohio 44333, U.S.A.	330-867-3325	330-867-3328	1,400	97
CSC Ltd.	4000 Mahoning Ave. N.W., Warren, Ohio 44483, U.S.A.	330-841-6011	330-841-7125	1,400	97
Falk Corp.	PO Box 492, Milwaukee, Wisconsin 53201, U.S.A.	414-937-4140	414-937-4739	1,400	97
Buckeye Steel Castings Co.	2211 Parsons Ave., Columbus, Ohio 43207, U.S.A.	614-444-2121		1,395	97
Avis Industrial Corp.	PO Box 548, Upland, Indiana 46989, U.S.A.	765-998-8100	765-998-8111	1,360	97
Copperweld Corp.	4 Gateway Center, 22nd Fl., Pittsburgh, Pennsylvania 15222-1211, U.S.A.	412-263-3200		1,350	95
Ampco-Pittsburgh Corp.	600 Grant St., 4600, Pittsburgh, Pennsylvania 15219, U.S.A.	412-456-4400		1,324	97
Birmid Components Ltd.	Shirley, Solihull B90 4LE, United Kingdom	21 7114555		1,304	93
J and L Specialty Steel Inc.	PO Box 3373, Pittsburgh, Pennsylvania 15230-3373, U.S.A.	412-338-1600	412-338-1723	1,302	97
Hyundai Pipe Co., Ltd.	Ulsan-Shi, Kyongnam, Republic of Korea	522 800114	522 878916	1,300	97
Ispat Indo PT	Jl Jend Sudirman Kav 44-46, Jakarta 10210, Indonesia	315713500	31 838220	1,300	97
Tong Yang Moolsan Co., Ltd.	Yongsan-gu, Seoul, Republic of Korea	2 7275041	2 7275112	1,300	97
Henry Barrett Group PLC	Dudley Hill, Bradford BD4 9HU, United Kingdom	274 682281	274 651052	1,268	93
Crane Ltd.	Nacton Rd., Ipswich IP3 9QH, United Kingdom	454 414686		1,236	93
Allied Tube and Conduit Co.	16100 S. Lathrop Ave., Harvey, Illinois 60426, U.S.A.	708-339-1610	708-339-9827	1,200	97
ALZ NV	Industrieterrein Genk-Zuid Zone 6A, Genk B-3600, Belgium	89 301900	89 301905	1,200	95
CSC Industries Inc.	4000 Mahoning Ave. N.W., Warren, Ohio 44483, U.S.A.	216-841-6011	216-841-6973	1,200	94
Dongkuk Steel Mill Co., Ltd.	Chung-Gu, Seoul, Republic of Korea	2 3171114	2 3171391	1,200	97
Titan PO Factory of Metal Products and Foundry	Kovinarska 28, Kaminik 61240, Slovakia	61 816221	61 812610	1,187	93
Chaparral Steel Co.	300 Ward Rd., Midlothian, Texas 76065-9651, U.S.A.	214-775-8241	214-775-1930	1,183	96
Schnitzer Steel Industries Inc.	PO Box 10047, Portland, Oregon 97210-0047, U.S.A.	503-224-9900	503-323-2804	1,183	97
Danish Steel Works Ltd.	Staalvaerksvej 5, Frederiksvaerk DK-3300, Denmark	47770333	42124666	1,150	93
Roanoke Electric Steel Corp.	PO Box 13948, Roanoke, Virginia 24038-3948, U.S.A.	540-342-1831	540-342-9437	1,150	97
Voksel Electric PT	Jakarta Barat, Jakarta 11120, Indonesia	216592196	21 6297866	1,134	97
Consani Engineering Pty. Ltd.	PO Box 1, Elsies River 7480, Republic of South Africa	21 5903400	21 5912825	1,120	96
Tyler Corp.	2121 San Jacinto St., Ste.. 3200, Dallas, Texas 75201, U.S.A.	214-754-7800	214-754-7821	1,118	96
Atlantic Steel Industries Inc.	PO Box 1714, Atlanta, Georgia 30301, U.S.A.	404-897-4500	404-897-4623	1,100	95
Dunaferr Fejleszto es Karbantarto Kft.	Vasmu Ter 1-3, Dunaujvaros H-2401, Hungary	25 81412	25 19857	1,100	93
FMC Corp. (UK) Ltd.	Commercial Rd., Bromborough L62 3NL, United Kingdom	51 3348085	51 3348501	1,089	92
Maverick Tube Corp.	400 Chesterfield Ctr., 2nd Fl., Chesterfield, Missouri 63017, U.S.A.	314-537-1314	314-537-1363	1,079	97
Seamless Tubes Ltd.	Wednesfield, Wolverhampton WV11 3SQ, United Kingdom	902 305000	902 307277	1,068	93

Product Tables
Thomas (Basic) Slag

Unit of Measure: Metric tons.

	1990 Value	%	1991 Value	%	1992 Value	%	1993 Value	%	1994 Value	%	1995 Value	%	1996 Value	%
Total Production	1,687,509	100.0	1,586,176	100.0	968,147	100.0	3,477,619	100.0	749,091	100.0	617,691	100.0	1,102,191	100.0
Regions														
America, South	855,500	50.7	855,500	53.9	284,000	29.3	2,765,000	79.5	173,000	23.1	200,000	32.4	855,500	77.6
Asia	77,000	4.6	77,000	4.9	77,000	8.0	77,000	2.2	77,000	10.3	77,000	12.5	77,000	7.0
Europe	755,009	44.7	653,676	41.2	607,147	62.7	635,619	18.3	499,091	66.6	340,691	55.2	169,691	15.4
Leading Producers														
Colombia	*854,500	50.6	*854,500	53.9	283,000	29.2	2,764,000	79.5	172,000	23.0	199,000	32.2	*854,500	77.5
Luxembourg	603,000	35.7	536,000	33.8	519,000	53.6	578,000	16.6	472,000	63.0	328,000	53.1	157,000	14.2
France	*97,667	5.8	*76,600	4.8	*55,533	5.7	*34,467	1.0	*13,400	1.8	-		-	
Indonesia	*77,000	4.6	*77,000	4.9	*77,000	8.0	77,000	2.2	*77,000	10.3	*77,000	12.5	*77,000	7.0
Sweden	38,846	2.3	29,385	1.9	19,923	2.1	10,462	0.3	1,000	0.1	-		-	

Spiegeleisen and Ferro-Manganese

Unit of Measure: Metric tons.

	1990 Value	%	1991 Value	%	1992 Value	%	1993 Value	%	1994 Value	%	1995 Value	%	1996 Value	%
Total Production	4,776,664	100.0	4,722,738	100.0	5,550,048	100.0	5,477,569	100.0	5,910,765	100.0	5,969,542	100.0	6,078,020	100.0
Regions														
Africa	2,000	0.0	2,000	0.0	282,000	5.1	425,000	7.8	628,692	10.6	544,659	9.1	517,626	8.5
America, North	411,214	8.6	224,821	4.8	305,938	5.5	372,918	6.8	354,897	6.0	398,877	6.7	421,857	6.9
America, South	416,200	8.7	475,000	10.1	500,000	9.0	503,652	9.2	465,854	7.9	315,057	5.3	465,260	7.7
Asia	2,108,500	44.1	2,083,167	44.1	2,301,833	41.5	2,242,500	40.9	2,408,167	40.7	2,539,833	42.5	2,547,500	41.9
Europe	1,778,750	37.2	1,892,750	40.1	2,105,277	37.9	1,858,500	33.9	1,953,154	33.0	2,061,115	34.5	2,015,777	33.2
Oceania	60,000	1.3	45,000	1.0	55,000	1.0	75,000	1.4	100,000	1.7	110,000	1.8	110,000	1.8
Leading Producers														
China	370,000	7.7	390,000	8.3	750,000	13.5	740,000	13.5	917,000	15.5	1,010,000	16.9	950,000	15.6
Japan	514,000	10.8	531,000	11.2	434,000	7.8	491,000	9.0	476,000	8.1	496,000	8.3	523,000	8.6
South Africa	-		-		270,000	4.9	393,000	7.2	591,000	10.0	507,000	8.5	480,000	7.9
Brazil	387,000	8.1	441,000	9.3	479,000	8.6	486,000	8.9	448,000	7.6	297,000	5.0	447,000	7.4
Germany	-		381,000	8.1	346,000	6.2	264,000	4.8	291,000	4.9	413,000	6.9	*341,700	5.6

Pig Iron, Foundry

Unit of Measure: Metric tons.

	1990 Value	%	1991 Value	%	1992 Value	%	1993 Value	%	1994 Value	%	1995 Value	%	1996 Value	%
Total Production	20,666,907	100.0	18,502,060	100.0	17,282,794	100.0	13,803,345	100.0	12,997,656	100.0	12,184,220	100.0	12,947,209	100.0
Regions														
Africa	273,000	1.3	289,000	1.6	221,000	1.3	262,036	1.9	270,300	2.1	237,564	1.9	175,827	1.4
America, North	673,500	3.3	675,567	3.7	675,033	3.9	674,500	4.9	673,967	5.2	673,433	5.5	672,900	5.2
America, South	234,667	1.1	225,267	1.2	223,067	1.3	310,200	2.2	308,000	2.4	37,800	0.3	214,267	1.7
Asia	12,763,500	61.8	11,489,500	62.1	10,638,500	61.6	8,468,000	61.3	7,898,222	60.8	7,566,756	62.1	8,246,289	63.7
Europe	6,722,240	32.5	5,822,726	31.5	5,525,194	32.0	4,088,608	29.6	3,847,167	29.6	3,668,667	30.1	3,637,926	28.1
Leading Producers														
Kazakhstan	5,226,000	25.3	4,953,000	26.8	4,666,000	27.0	3,552,000	25.7	2,435,000	18.7	2,530,000	20.8	2,536,000	19.6
India	*2,701,500	13.1	*2,701,500	14.6	*2,701,500	15.6	2,239,000	16.2	2,880,000	22.2	2,431,000	20.0	3,256,000	25.1
Russian Federation	3,266,000	15.8	1,990,000	10.8	1,969,000	11.4	1,405,000	10.2	1,026,000	7.9	1,182,000	9.7	793,000	6.1
Ukraine	2,706,000	13.1	1,797,000	9.7	1,598,000	9.2	1,052,000	7.6	657,000	5.1	316,000	2.6	278,000	2.1
Japan	1,105,000	5.3	1,209,000	6.5	732,000	4.2	697,000	5.0	918,000	7.1	807,000	6.6	900,000	7.0

Product Tables
Pig Iron, Steel-Making

Unit of Measure: 1,000 Metric tons.

	1990		1991		1992		1993		1994		1995		1996	
	Value	%	Value	%	Value	%	Value	%	Value	%	Value	%	Value	%
Total Production	622,378	100.0	635,570	100.0	638,553	100.0	641,354	100.0	652,369	100.0	669,777	100.0	661,839	100.0
Regions														
Africa	7,830	1.3	8,397	1.3	8,804	1.4	8,091	1.3	8,016	1.2	8,301	1.2	7,951	1.2
America, North	59,353	9.5	54,704	8.6	58,218	9.1	59,298	9.2	60,871	9.3	63,020	9.4	62,442	9.4
America, South	24,500	3.9	25,747	4.1	25,854	4.0	26,759	4.2	28,342	4.3	28,379	4.2	26,836	4.1
Asia	348,954	56.1	351,285	55.3	354,015	55.4	362,775	56.6	370,356	56.8	385,718	57.6	388,189	58.7
Europe	175,553	28.2	189,836	29.9	185,269	29.0	177,222	27.6	177,335	27.2	176,805	26.4	168,861	25.5
Oceania	6,188	1.0	5,600	0.9	6,394	1.0	7,209	1.1	7,449	1.1	7,554	1.1	7,560	1.1
Leading Producers														
Japan	79,124	12.7	78,776	12.4	72,412	11.3	73,041	11.4	72,858	11.2	74,098	11.1	73,697	11.1
China	62,380	10.0	67,000	10.5	75,890	11.9	87,389	13.6	97,410	14.9	105,293	15.7	107,225	16.2
United States of America	49,668	8.0	44,123	6.9	47,377	7.4	48,155	7.5	49,400	7.6	50,900	7.6	49,400	7.5
Russian Federation	55,812	9.0	46,638	7.3	44,021	6.9	39,339	6.1	35,454	5.4	38,494	5.7	36,286	5.5
Germany	-		29,878	4.7	27,577	4.3	26,322	4.1	29,202	4.5	29,279	4.4	27,172	4.1

Other Ferro-Alloys

Unit of Measure: Metric tons.

	1990		1991		1992		1993		1994		1995		1996	
	Value	%	Value	%	Value	%	Value	%	Value	%	Value	%	Value	%
Total Production	11,311,690	100.0	12,058,226	100.0	11,783,197	100.0	12,123,807	100.0	12,947,068	100.0	15,127,572	100.0	14,397,785	100.0
Regions														
Africa	2,124,000	18.8	2,118,000	17.6	1,836,000	15.6	1,965,000	16.2	2,429,000	18.8	2,970,000	19.6	2,819,000	19.6
America, North	1,191,000	10.5	1,007,000	8.4	978,000	8.3	911,000	7.5	1,007,000	7.8	1,021,000	6.7	951,000	6.6
America, South	711,000	6.3	551,000	4.6	584,000	5.0	600,000	4.9	540,000	4.2	645,000	4.3	786,000	5.5
Asia	3,876,333	34.3	4,994,333	41.4	5,155,333	43.8	5,393,333	44.5	5,821,000	45.0	7,150,000	47.3	6,630,810	46.1
Europe	3,291,357	29.1	3,264,893	27.1	3,107,864	26.4	3,149,473	26.0	3,020,068	23.3	3,201,572	21.2	3,085,975	21.4
Oceania	118,000	1.0	123,000	1.0	122,000	1.0	105,000	0.9	130,000	1.0	140,000	0.9	125,000	0.9
Leading Producers														
China	2,442,000	21.6	2,160,000	17.9	2,650,000	22.5	2,999,000	24.7	3,361,000	26.0	4,319,000	28.6	4,180,000	29.0
South Africa	1,878,000	16.6	1,861,000	15.4	1,585,000	13.5	1,789,000	14.8	2,170,000	16.8	2,624,000	17.3	2,478,000	17.2
Russian Federation	-		1,377,000	11.4	1,087,000	9.2	978,000	8.1	765,000	5.9	876,000	5.8	697,000	4.8
United States of America	875,000	7.7	777,000	6.4	773,000	6.6	715,000	5.9	784,000	6.1	793,000	5.2	699,000	4.9
Norway	717,000	6.3	773,000	6.4	684,000	5.8	794,000	6.5	871,000	6.7	925,000	6.1	895,000	6.2

Ferro-Chromium (Including Ferro-Silico-Chromium and Charge Chrome)

Unit of Measure: Metric tons.

	1990		1991		1992		1993		1994		1995		1996	
	Value	%	Value	%	Value	%	Value	%	Value	%	Value	%	Value	%
Total Production	4,762,750	100.0	4,966,267	100.0	4,584,290	100.0	4,203,733	100.0	4,717,120	100.0	5,630,706	100.0	5,025,898	100.0
Regions														
Africa	1,289,000	27.1	1,405,000	28.3	1,035,000	22.6	1,014,000	24.1	1,398,000	29.6	1,883,000	33.4	1,721,000	34.2
America, North	115,028	2.4	74,211	1.5	67,294	1.5	69,578	1.7	74,161	1.6	79,444	1.4	43,928	0.9
America, South	91,000	1.9	90,000	1.8	100,000	2.2	89,000	2.1	87,000	1.8	104,000	1.8	80,000	1.6
Asia	2,584,222	54.3	2,635,556	53.1	2,590,889	56.5	2,283,222	54.3	2,367,556	50.2	2,738,889	48.6	2,418,127	48.1
Europe	683,500	14.4	761,500	15.3	791,106	17.3	747,933	17.8	790,403	16.8	825,373	14.7	762,843	15.2
Leading Producers														
South Africa	1,051,000	22.1	1,190,000	24.0	824,000	18.0	878,000	20.9	1,179,000	25.0	1,582,000	28.1	1,426,000	28.4
Kazakhstan	389,000	8.2	372,000	7.5	350,000	7.6	315,000	7.5	312,000	6.6	512,000	9.1	346,000	6.9
Russian Federation	476,000	10.0	505,000	10.2	428,000	9.3	292,000	6.9	343,000	7.3	354,000	6.3	135,000	2.7
Japan	319,000	6.7	297,000	6.0	296,000	6.5	225,000	5.4	216,000	4.6	236,000	4.2	211,000	4.2
China	340,000	7.1	380,000	7.7	410,000	8.9	372,000	8.8	370,000	7.8	400,000	7.1	450,000	9.0

Commodity data are provided by the United Nations Statistical Division. The symbol * means that data are estimated. For additional notes, see Appendix I.

503

Product Tables
Ferro-Nickel

Unit of Measure: Metric tons.

	1990 Value	%	1991 Value	%	1992 Value	%	1993 Value	%	1994 Value	%	1995 Value	%	1996 Value	%
Total Production	521,292	100.0	635,554	100.0	686,445	100.0	641,451	100.0	659,163	100.0	823,511	100.0	806,989	100.0
Regions														
America, North	83,292	16.0	88,554	13.9	71,928	10.5	75,451	11.8	96,163	14.6	97,511	11.8	108,989	13.5
America, South	78,000	15.0	83,000	13.1	80,000	11.7	83,000	12.9	82,000	12.4	84,000	10.2	87,000	10.8
Asia	259,000	49.7	334,000	52.6	317,000	46.2	331,000	51.6	332,000	50.4	484,000	58.8	466,000	57.7
Europe	69,000	13.2	96,000	15.1	185,517	27.0	115,000	17.9	109,000	16.5	116,000	14.1	103,000	12.8
Oceania	32,000	6.1	34,000	5.3	32,000	4.7	37,000	5.8	40,000	6.1	42,000	5.1	42,000	5.2
Leading Producers														
Japan	234,000	44.9	295,000	46.4	237,000	34.5	257,000	40.1	242,000	36.7	351,000	42.6	329,000	40.8
Dominican Republic	71,800	13.8	76,000	12.0	58,313	8.5	60,774	9.5	80,425	12.2	80,711	9.8	78,489	9.7
Russian Federation	-		-		46,000	6.7	47,000	7.3	59,000	9.0	77,000	9.4	75,000	9.3
Greece	61,000	11.7	87,000	13.7	73,000	10.6	52,000	8.1	77,000	11.7	82,000	10.0	86,000	10.7
Colombia	44,000	8.4	49,000	7.7	47,000	6.8	47,000	7.3	48,000	7.3	53,000	6.4	57,000	7.1

Ferro-Silicon

Unit of Measure: Metric tons.

	1990 Value	%	1991 Value	%	1992 Value	%	1993 Value	%	1994 Value	%	1995 Value	%	1996 Value	%
Total Production	5,655,236	100.0	5,306,169	100.0	5,274,880	100.0	5,522,858	100.0	5,502,300	100.0	5,563,378	100.0	5,735,463	100.0
Regions														
Africa	140,000	2.5	146,000	2.8	172,691	3.3	172,882	3.1	173,073	3.1	167,264	3.0	145,455	2.5
America, North	535,000	9.5	419,000	7.9	406,000	7.7	381,485	6.9	415,341	7.5	414,000	7.4	418,000	7.3
America, South	468,800	8.3	292,000	5.5	320,000	6.1	343,000	6.2	300,000	5.5	288,000	5.2	307,000	5.4
Asia	2,852,806	50.4	2,978,139	56.1	3,042,472	57.7	3,334,942	60.4	3,338,931	60.7	3,411,920	61.3	3,677,871	64.1
Europe	1,638,630	29.0	1,452,030	27.4	1,316,717	25.0	1,271,959	23.0	1,256,389	22.8	1,263,652	22.7	1,168,620	20.4
Oceania	20,000	0.4	19,000	0.4	17,000	0.3	18,591	0.3	18,566	0.3	18,542	0.3	18,517	0.3
Leading Producers														
China	727,000	12.9	817,000	15.4	834,000	15.8	1,040,000	18.8	1,100,000	20.0	1,163,000	20.9	1,375,000	24.0
Norway	460,000	8.1	414,000	7.8	377,000	7.1	401,000	7.3	453,000	8.2	386,000	6.9	350,000	6.1
Ukraine	594,000	10.5	502,000	9.5	432,000	8.2	376,000	6.8	273,000	5.0	321,000	5.8	254,000	4.4
United States of America	434,000	7.7	338,000	6.4	346,000	6.6	323,000	5.8	359,000	6.5	358,000	6.4	362,000	6.3
Brazil	389,000	6.9	213,000	4.0	267,000	5.1	267,000	4.8	240,000	4.4	222,000	4.0	240,000	4.2

Crude Steel for Castings

Unit of Measure: Metric tons.

	1990 Value	%	1991 Value	%	1992 Value	%	1993 Value	%	1994 Value	%	1995 Value	%	1996 Value	%
Total Production	39,309,833	100.0	39,514,787	100.0	42,025,567	100.0	37,792,725	100.0	38,278,914	100.0	38,778,801	100.0	39,732,613	100.0
Regions														
Africa	304,571	0.8	324,971	0.8	316,028	0.8	318,086	0.8	321,144	0.8	341,202	0.9	339,650	0.9
America, North	1,251,357	3.2	1,227,357	3.1	1,227,857	2.9	1,201,357	3.2	1,218,750	3.2	1,214,918	3.1	1,211,085	3.0
America, South	95,611	0.2	83,711	0.2	86,811	0.2	80,911	0.2	74,011	0.2	54,111	0.1	49,211	0.1
Asia	28,228,071	71.8	28,816,726	72.9	31,486,381	74.9	28,621,172	75.7	29,302,886	76.6	29,526,524	76.1	30,945,238	77.9
Europe	9,430,222	24.0	9,062,022	22.9	8,908,489	21.2	7,571,198	20.0	7,362,123	19.2	7,642,047	19.7	7,187,429	18.1
Leading Producers														
Austria	4,291,000	10.9	4,082,000	10.3	3,846,000	9.2	4,018,000	10.6	4,400,000	11.5	4,823,000	12.4	4,307,000	10.8
Turkey	3,852,000	9.8	3,751,000	9.5	4,508,000	10.7	4,362,000	11.5	4,604,000	12.0	3,743,000	9.7	4,505,000	11.3
Russian Federation	4,893,000	12.4	4,330,000	11.0	5,062,000	12.0	3,888,000	10.3	2,946,000	7.7	2,834,000	7.3	2,625,000	6.6
China	2,508,000	6.4	2,881,000	7.3	3,312,000	7.9	*3,292,636	8.7	*3,425,696	8.9	*3,558,755	9.2	*3,691,815	9.3
Indonesia	2,890,000	7.4	3,250,000	8.2	3,171,000	7.5	1,948,000	5.2	3,220,000	8.4	3,500,000	9.0	3,400,000	8.6

Product Tables
Crude Steel, Ingots

Unit of Measure: 1,000 Metric tons.

	1990		1991		1992		1993		1994		1995		1996	
	Value	%	Value	%	Value	%	Value	%	Value	%	Value	%	Value	%
Total Production	910,574	100.0	921,299	100.0	912,486	100.0	917,179	100.0	916,757	100.0	938,536	100.0	933,683	100.0
Regions														
Africa	12,595	1.4	13,487	1.5	13,049	1.4	12,710	1.4	11,914	1.3	12,712	1.4	11,500	1.2
America, North	110,498	12.1	100,457	10.9	106,237	11.6	111,371	12.1	113,918	12.4	119,742	12.8	120,083	12.9
America, South	29,182	3.2	30,440	3.3	31,090	3.4	32,814	3.6	34,769	3.8	34,495	3.7	35,294	3.8
Asia	493,605	54.2	491,326	53.3	485,903	53.3	490,016	53.4	485,154	52.9	497,495	53.0	503,279	53.9
Europe	256,400	28.2	277,642	30.1	270,243	29.6	261,787	28.5	262,429	28.6	265,299	28.3	254,683	27.3
Oceania	8,295	0.9	7,947	0.9	5,964	0.7	8,481	0.9	8,573	0.9	8,793	0.9	8,844	0.9
Leading Producers														
Japan	109,548	12.0	108,914	11.8	97,518	10.7	99,063	10.8	97,715	10.7	101,046	10.8	98,207	10.5
United States of America	89,726	9.9	79,738	8.7	84,322	9.2	88,793	9.7	91,200	9.9	95,172	10.1	95,500	10.2
China	66,350	7.3	71,000	7.7	80,940	8.9	89,556	9.8	92,617	10.1	95,360	10.2	100,056	10.7
Russian Federation	84,730	9.3	72,770	7.9	61,966	6.8	54,458	5.9	45,866	5.0	48,756	5.2	46,628	5.0
Germany	-		41,267	4.5	39,337	4.3	37,322	4.1	40,533	4.4	41,732	4.4	39,496	4.2

Ingots for Tubes

Unit of Measure: Metric tons.

	1990		1991		1992		1993		1994		1995		1996	
	Value	%	Value	%	Value	%	Value	%	Value	%	Value	%	Value	%
Total Production	8,221,787	100.0	8,364,134	100.0	10,917,067	100.0	10,555,463	100.0	10,940,859	100.0	11,343,305	100.0	10,587,708	100.0
Regions														
America, North	231,200	2.8	204,600	2.4	196,000	1.8	273,000	2.6	355,000	3.2	288,220	2.5	182,506	1.7
America, South	146,155	1.8	142,774	1.7	139,393	1.3	136,012	1.3	132,631	1.2	129,250	1.1	125,869	1.2
Asia	3,107,139	37.8	3,052,989	36.5	4,219,839	38.7	4,023,689	38.1	3,704,539	33.9	3,976,389	35.1	3,594,239	33.9
Europe	4,737,294	57.6	4,963,771	59.3	6,361,835	58.3	6,122,762	58.0	6,748,690	61.7	6,949,447	61.3	6,685,094	63.1
Leading Producers														
Germany	-		-		1,378,000	12.6	1,308,000	12.4	1,990,000	18.2	2,202,000	19.4	2,139,000	20.2
Turkey	-		-		1,183,000	10.8	1,060,000	10.0	822,000	7.5	1,013,000	8.9	712,000	6.7
Russian Federation	1,059,000	12.9	1,000,000	12.0	968,000	8.9	872,000	8.3	769,000	7.0	834,000	7.4	719,000	6.8
Ukraine	809,000	9.8	736,000	8.8	710,000	6.5	571,000	5.4	354,000	3.2	340,000	3.0	276,000	2.6
France	604,222	7.3	667,000	8.0	628,000	5.8	583,000	5.5	742,000	6.8	648,000	5.7	628,000	5.9

Semis for Tubes

Unit of Measure: Metric tons.

	1990		1991		1992		1993		1994		1995		1996	
	Value	%	Value	%	Value	%	Value	%	Value	%	Value	%	Value	%
Total Production	41,703,305	100.0	41,438,270	100.0	41,610,325	100.0	41,706,380	100.0	41,473,936	100.0	42,701,463	100.0	43,854,564	100.0
Regions														
America, North	1,669,364	4.0	1,772,636	4.3	1,875,909	4.5	1,979,182	4.7	2,082,455	5.0	2,185,727	5.1	2,289,000	5.2
Asia	31,818,817	76.3	32,535,167	78.5	32,381,517	77.8	33,159,867	79.5	33,211,217	80.1	34,103,567	79.9	34,847,917	79.5
Europe	8,215,125	19.7	7,130,467	17.2	7,352,899	17.7	6,567,332	15.7	6,180,264	14.9	6,412,169	15.0	6,717,647	15.3
Leading Producers														
Korea, Republic of	*18,211,800	43.7	*19,149,000	46.2	*20,086,200	48.3	*21,023,400	50.4	*21,960,600	53.0	*22,897,800	53.6	*23,835,000	54.4
Japan	3,220,000	7.7	3,492,000	8.4	2,649,000	6.4	2,804,000	6.7	2,596,000	6.3	2,339,000	5.5	2,189,000	5.0
United States of America	1,669,364	4.0	1,772,636	4.3	1,875,909	4.5	1,979,182	4.7	2,082,455	5.0	2,185,727	5.1	2,289,000	5.2
Russian Federation	2,763,000	6.6	2,213,000	5.3	2,015,000	4.8	1,733,000	4.2	1,105,000	2.7	1,234,000	2.9	1,103,000	2.5
Ukraine	2,118,000	5.1	1,750,000	4.2	1,904,000	4.6	1,143,000	2.7	491,000	1.2	605,000	1.4	915,000	2.1

Commodity data are provided by the United Nations Statistical Division. The symbol * means that data are estimated. For additional notes, see Appendix I.

505

Product Tables
Wire Rods

Unit of Measure: Metric tons.

	1990 Value	%	1991 Value	%	1992 Value	%	1993 Value	%	1994 Value	%	1995 Value	%	1996 Value	%
Total Production	63,330,975	100.0	66,268,630	100.0	69,861,362	100.0	73,267,453	100.0	75,243,584	100.0	75,829,674	100.0	77,817,536	100.0
Regions														
Africa	720,731	1.1	714,702	1.1	708,674	1.0	702,645	1.0	696,617	0.9	690,588	0.9	684,560	0.9
America, North	5,977,000	9.4	5,440,000	8.2	6,343,000	9.1	6,966,000	9.5	7,168,000	9.5	7,520,000	9.9	7,590,000	9.8
America, South	3,193,000	5.0	3,125,000	4.7	3,487,000	5.0	3,761,000	5.1	3,650,000	4.9	3,398,088	4.5	3,509,528	4.5
Asia	30,747,694	48.6	31,941,544	48.2	33,511,661	48.0	35,994,825	49.1	36,540,990	48.6	37,296,451	49.2	39,799,935	51.1
Europe	22,234,750	35.1	24,617,783	37.1	25,409,627	36.4	25,469,783	34.8	26,842,977	35.7	26,607,748	35.1	25,944,913	33.3
Oceania	457,800	0.7	429,600	0.6	401,400	0.6	373,200	0.5	345,000	0.5	316,800	0.4	288,600	0.4
Leading Producers														
China	9,990,000	15.8	10,999,000	16.6	12,571,000	18.0	13,895,000	19.0	15,713,000	20.9	16,872,000	22.2	18,340,000	23.6
Japan	7,335,000	11.6	7,023,000	10.6	6,544,000	9.4	7,123,000	9.7	6,826,000	9.1	6,820,000	9.0	6,950,000	8.9
Germany	-		4,626,000	7.0	4,908,000	7.0	5,073,000	6.9	5,423,000	7.2	5,369,172	7.1	5,233,575	6.7
United States of America	3,873,000	6.1	3,961,000	6.0	4,092,000	5.9	4,423,000	6.0	4,368,000	5.8	4,447,000	5.9	4,510,000	5.8
Italy	2,813,000	4.4	3,020,000	4.6	3,199,000	4.6	3,122,000	4.3	3,585,000	4.8	3,805,000	5.0	3,484,000	4.5

Angles, Shapes and Sections (Total Production)

Unit of Measure: 1,000 Metric tons.

	1990 Value	%	1991 Value	%	1992 Value	%	1993 Value	%	1994 Value	%	1995 Value	%	1996 Value	%
Total Production	145,651	100.0	167,897	100.0	162,059	100.0	162,552	100.0	158,516	100.0	160,899	100.0	158,793	100.0
Regions														
Africa	1,842	1.3	1,920	1.1	1,927	1.2	1,926	1.2	1,918	1.2	1,858	1.2	2,028	1.3
America, North	18,883	13.0	18,757	11.2	18,704	11.5	20,102	12.4	21,909	13.8	23,075	14.3	23,806	15.0
America, South	440	0.3	475	0.3	456	0.3	536	0.3	573	0.4	541	0.3	526	0.3
Asia	70,241	48.2	93,755	55.8	88,535	54.6	89,569	55.1	87,377	55.1	89,492	55.6	88,686	55.8
Europe	54,133	37.2	52,946	31.5	52,436	32.4	50,419	31.0	46,738	29.5	45,933	28.5	43,747	27.5
Oceania	111	0.1	44	0.0	-		-		-		-		-	
Leading Producers														
Japan	21,908	15.0	20,814	12.4	18,776	11.6	17,940	11.0	17,659	11.1	18,440	11.5	19,271	12.1
Russian Federation	-		23,347	13.9	19,330	11.9	19,709	12.1	16,517	10.4	16,292	10.1	*13,962	8.8
United States of America	12,863	8.8	12,252	7.3	11,640	7.2	12,337	7.6	13,432	8.5	14,140	8.8	14,189	8.9
Ukraine	15,212	10.4	10,967	6.5	11,651	7.2	9,462	5.8	6,420	4.1	5,550	3.4	4,450	2.8
Italy	8,791	6.0	9,020	5.4	9,860	6.1	9,169	5.6	8,711	5.5	9,301	5.8	9,789	6.2

Angles, Shapes and Sections, 80 mm or More (Heavy Sections)

Unit of Measure: Metric tons.

	1990 Value	%	1991 Value	%	1992 Value	%	1993 Value	%	1994 Value	%	1995 Value	%	1996 Value	%
Total Production	33,285,781	100.0	47,762,721	100.0	45,743,866	100.0	44,504,038	100.0	46,289,147	100.0	47,574,365	100.0	48,727,855	100.0
Regions														
America, North	6,346,000	19.1	6,561,467	13.7	6,513,352	14.2	6,738,236	15.1	6,976,121	15.1	7,332,006	15.4	7,304,891	15.0
America, South	117,000	0.4	117,000	0.2	143,000	0.3	174,000	0.4	205,000	0.4	169,110	0.4	168,505	0.3
Asia	10,497,750	31.5	20,223,750	42.3	19,444,750	42.5	19,448,000	43.7	21,406,000	46.2	22,591,000	47.5	24,160,800	49.6
Europe	16,325,031	49.0	20,860,505	43.7	19,642,764	42.9	18,143,802	40.8	17,702,026	38.2	17,482,249	36.7	17,093,658	35.1
Leading Producers														
Russian Federation	-		10,543,000	22.1	10,458,000	22.9	11,193,000	25.2	13,360,000	28.9	13,969,000	29.4	*14,830,800	30.4
Japan	10,191,000	30.6	9,362,000	19.6	8,795,000	19.2	8,045,000	18.1	7,925,000	17.1	8,353,000	17.6	9,005,000	18.5
United States of America	5,436,000	16.3	5,311,000	11.1	5,186,000	11.3	5,269,000	11.8	5,390,000	11.6	5,695,000	12.0	5,571,000	11.4
Ukraine	-		4,578,000	9.6	4,219,000	9.2	2,569,000	5.8	2,576,000	5.6	1,839,000	3.9	1,828,000	3.8
Poland	*2,849,650	8.6	*2,957,100	6.2	*3,064,550	6.7	*3,172,000	7.1	*3,279,450	7.1	*3,386,900	7.1	*3,494,350	7.2

Product Tables
Angles, Shapes and Sections, Less Than 80 mm (Light Sections)

Unit of Measure: Metric tons.

	1990 Value	%	1991 Value	%	1992 Value	%	1993 Value	%	1994 Value	%	1995 Value	%	1996 Value	%
Total Production	50,083,557	100.0	67,772,843	100.0	61,845,165	100.0	61,299,795	100.0	55,101,926	100.0	56,263,463	100.0	54,430,294	100.0
Regions														
America, North	11,305,714	22.6	11,100,786	16.4	10,960,857	17.7	11,995,429	19.6	13,428,000	24.4	14,100,571	25.1	14,722,143	27.0
America, South	146,000	0.3	180,000	0.3	148,000	0.2	176,000	0.3	182,000	0.3	172,121	0.3	166,699	0.3
Asia	11,850,500	23.7	24,461,500	36.1	19,018,500	30.8	18,598,536	30.3	13,089,071	23.8	12,599,607	22.4	11,262,143	20.7
Europe	26,781,343	53.5	32,030,557	47.3	31,717,808	51.3	30,529,831	49.8	28,402,855	51.5	29,391,164	52.2	28,279,310	52.0
Leading Producers														
Japan	11,717,000	23.4	11,452,000	16.9	9,981,000	16.1	9,895,000	16.1	9,734,000	17.7	10,087,000	17.9	10,266,000	18.9
United States of America	7,427,000	14.8	6,940,500	10.2	6,454,000	10.4	7,068,000	11.5	8,042,000	14.6	8,445,000	15.0	8,618,000	15.8
Italy	7,865,000	15.7	7,968,000	11.8	8,982,000	14.5	8,559,000	14.0	8,097,000	14.7	8,329,000	14.8	8,302,000	15.3
Russian Federation	-		12,804,000	18.9	8,872,000	14.3	8,516,000	13.9	3,157,000	5.7	2,323,000	4.1	904,000	1.7
Ukraine	-		6,389,000	9.4	5,308,000	8.6	4,387,000	7.2	3,110,000	5.6	3,141,000	5.6	2,568,000	4.7

Plates (Heavy), Over 4.75 mm

Unit of Measure: Metric tons.

	1990 Value	%	1991 Value	%	1992 Value	%	1993 Value	%	1994 Value	%	1995 Value	%	1996 Value	%
Total Production	74,062,495	100.0	69,980,161	100.0	67,409,051	100.0	64,466,017	100.0	65,013,875	100.0	65,881,576	100.0	64,617,211	100.0
Regions														
Africa	1,858,190	2.5	1,856,952	2.7	1,855,714	2.8	1,854,476	2.9	1,853,238	2.9	1,852,000	2.8	1,850,762	2.9
America, North	4,919,786	6.6	4,557,821	6.5	4,641,857	6.9	3,392,893	5.3	3,181,929	4.9	3,005,964	4.6	2,752,714	4.3
America, South	2,415,250	3.3	2,527,875	3.6	2,690,500	4.0	2,629,125	4.1	2,910,750	4.5	2,923,661	4.4	2,991,657	4.6
Asia	33,036,833	44.6	31,759,726	45.4	30,274,619	44.9	29,817,179	46.3	29,556,161	45.5	29,457,375	44.7	28,635,664	44.3
Europe	31,832,436	43.0	29,277,786	41.8	27,946,360	41.5	26,772,344	41.5	27,511,797	42.3	28,642,577	43.5	28,386,413	43.9
Leading Producers														
Japan	*8,361,357	11.3	*7,871,107	11.2	*7,380,857	10.9	*6,890,607	10.7	*6,400,357	9.8	*5,910,107	9.0	*5,419,857	8.4
Russian Federation	7,290,000	9.8	6,071,000	8.7	4,962,000	7.4	4,678,000	7.3	3,654,000	5.6	4,120,000	6.3	4,007,000	6.2
Ukraine	5,754,000	7.8	4,784,000	6.8	3,768,000	5.6	2,930,000	4.5	2,140,000	3.3	2,716,000	4.1	2,794,000	4.3
China	684,000	0.9	719,000	1.0	854,000	1.3	1,003,000	1.6	*1,821,090	2.8	*1,569,410	2.4	*1,317,731	2.0
United States of America	*2,029,214	2.7	*1,683,893	2.4	*1,338,571	2.0	*993,250	1.5	*647,929	1.0	*302,607	0.5	-	

Plates (Medium), 3 to 4.75 mm

Unit of Measure: Metric tons.

	1990 Value	%	1991 Value	%	1992 Value	%	1993 Value	%	1994 Value	%	1995 Value	%	1996 Value	%
Total Production	26,658,661	100.0	25,691,005	100.0	24,430,722	100.0	22,125,798	100.0	21,904,624	100.0	22,671,258	100.0	22,120,995	100.0
Regions														
America, North	281,786	1.1	235,179	0.9	188,571	0.8	141,964	0.6	95,357	0.4	48,750	0.2	2,143	0.0
America, South	50,764	0.2	73,882	0.3	97,000	0.4	123,000	0.6	119,000	0.5	142,000	0.6	166,336	0.8
Asia	14,342,250	53.8	13,554,345	52.8	13,015,440	53.3	11,884,786	53.7	11,248,881	51.4	11,566,976	51.0	11,836,071	53.5
Europe	11,983,861	45.0	11,827,599	46.0	11,129,710	45.6	9,976,048	45.1	10,441,386	47.7	10,913,532	48.1	10,116,444	45.7
Leading Producers														
Russian Federation	5,110,000	19.2	4,357,000	17.0	3,829,000	15.7	2,666,000	12.0	2,069,000	9.4	2,405,000	10.6	2,579,000	11.7
Germany	*3,013,500	11.3	3,301,000	12.8	3,014,000	12.3	2,771,000	12.5	2,968,000	13.5	*3,013,500	13.3	*3,013,500	13.6
Ukraine	2,191,000	8.2	2,087,000	8.1	1,863,000	7.6	1,384,000	6.3	1,184,000	5.4	1,320,000	5.8	1,000,000	4.5
Slovakia	*1,079,000	4.0	*1,079,000	4.2	*1,079,000	4.4	979,000	4.4	1,027,000	4.7	1,122,000	4.9	1,188,000	5.4
United Kingdom	*1,047,571	3.9	*1,108,107	4.3	*1,168,643	4.8	*1,229,179	5.6	*1,289,714	5.9	*1,350,250	6.0	*1,410,786	6.4

Commodity data are provided by the United Nations Statistical Division. The symbol * means that data are estimated. For additional notes, see Appendix I.

507

Product Tables
Sheets, Electrical

Unit of Measure: Metric tons.

	1990 Value	%	1991 Value	%	1992 Value	%	1993 Value	%	1994 Value	%	1995 Value	%	1996 Value	%
Total Production	12,213,829	100.0	12,738,241	100.0	12,695,362	100.0	12,605,915	100.0	12,596,891	100.0	14,820,488	100.0	15,156,088	100.0
Regions														
America, North	2,580,500	21.1	2,467,500	19.4	2,567,500	20.2	2,624,500	20.8	2,721,500	21.6	3,867,500	26.1	4,167,500	27.5
America, South	80,750	0.7	85,500	0.7	90,250	0.7	95,000	0.8	105,000	0.8	102,000	0.7	105,000	0.7
Asia	8,091,222	66.2	8,374,206	65.7	8,177,189	64.4	8,005,422	63.5	7,786,406	61.8	8,948,312	60.4	9,024,848	59.5
Europe	1,461,357	12.0	1,811,036	14.2	1,860,423	14.7	1,880,993	14.9	1,983,985	15.7	1,902,676	12.8	1,858,740	12.3
Leading Producers														
Mexico	2,130,000	17.4	2,043,000	16.0	2,162,000	17.0	2,212,000	17.5	2,313,000	18.4	3,472,000	23.4	3,741,000	24.7
Japan	1,476,000	12.1	1,730,000	13.6	1,574,000	12.4	1,631,000	12.9	1,649,000	13.1	1,981,000	13.4	1,774,000	11.7
Belarus	1,112,000	9.1	1,123,000	8.8	1,105,000	8.7	946,000	7.5	880,000	7.0	744,000	5.0	886,000	5.8
Russian Federation	1,113,000	9.1	931,000	7.3	597,000	4.7	377,000	3.0	391,000	3.1	406,000	2.7	278,000	1.8
Moldavia	712,000	5.8	623,000	4.9	653,000	5.1	497,000	3.9	1,000	0.0	*664,923	4.5	*701,352	4.6

Sheets Under 3 mm, Cold-Rolled, Uncoated

Unit of Measure: 1,000 Metric tons.

	1990 Value	%	1991 Value	%	1992 Value	%	1993 Value	%	1994 Value	%	1995 Value	%	1996 Value	%
Total Production	115,119	100.0	110,635	100.0	112,831	100.0	114,098	100.0	117,186	100.0	119,083	100.0	120,645	100.0
Regions														
Africa	246	0.2	225	0.2	261	0.2	260	0.2	257	0.2	254	0.2	250	0.2
America, North	15,009	13.0	15,240	13.8	15,471	13.7	15,701	13.8	15,932	13.6	16,163	13.6	16,393	13.6
America, South	326	0.3	301	0.3	434	0.4	385	0.3	390	0.3	383	0.3	380	0.3
Asia	50,413	43.8	46,128	41.7	47,324	41.9	49,765	43.6	49,876	42.6	51,956	43.6	53,030	44.0
Europe	49,125	42.7	48,741	44.1	49,341	43.7	47,987	42.1	50,731	43.3	50,327	42.3	50,591	41.9
Leading Producers														
Japan	26,561	23.1	26,977	24.4	*28,297	25.1	*29,030	25.4	*29,763	25.4	*30,496	25.6	*31,229	25.9
United States of America	*13,597	11.8	*13,824	12.5	*14,050	12.5	*14,277	12.5	*14,504	12.4	*14,730	12.4	*14,957	12.4
Germany	*10,663	9.3	*10,663	9.6	10,586	9.4	10,098	8.9	11,305	9.6	*10,663	9.0	*10,663	8.8
France	6,372	5.5	6,420	5.8	6,132	5.4	5,448	4.8	6,408	5.5	*6,421	5.4	*6,478	5.4
Russian Federation	6,643	5.8	6,193	5.6	5,555	4.9	4,679	4.1	3,829	3.3	4,923	4.1	4,844	4.0

Sheets Under 3 mm, Hot-Rolled

Unit of Measure: 1,000 Metric tons.

	1990 Value	%	1991 Value	%	1992 Value	%	1993 Value	%	1994 Value	%	1995 Value	%	1996 Value	%
Total Production	102,930	100.0	103,360	100.0	105,041	100.0	106,313	100.0	107,832	100.0	110,971	100.0	113,910	100.0
Regions														
Africa	2,704	2.6	2,896	2.8	2,835	2.7	3,026	2.8	3,049	2.8	3,225	2.9	3,318	2.9
America, North	17,349	16.9	17,716	17.1	18,083	17.2	18,451	17.4	18,818	17.5	19,185	17.3	19,553	17.2
America, South	208	0.2	214	0.2	203	0.2	286	0.3	262	0.2	265	0.2	269	0.2
Asia	63,231	61.4	63,750	61.7	64,362	61.3	65,082	61.2	66,083	61.3	67,807	61.1	69,925	61.4
Europe	19,439	18.9	18,784	18.2	19,558	18.6	19,468	18.3	19,620	18.2	20,490	18.5	20,845	18.3
Leading Producers														
Japan	*44,772	43.5	*45,460	44.0	*46,147	43.9	*46,834	44.1	*47,521	44.1	*48,208	43.4	*48,895	42.9
United States of America	*12,120	11.8	*12,300	11.9	*12,480	11.9	*12,660	11.9	*12,840	11.9	*13,020	11.7	*13,200	11.6
Belgium	*4,813	4.7	*4,975	4.8	*5,137	4.9	*5,300	5.0	*5,462	5.1	*5,624	5.1	*5,787	5.1
United Kingdom	*4,693	4.6	*4,839	4.7	*4,984	4.7	*5,130	4.8	*5,275	4.9	*5,421	4.9	*5,567	4.9
Canada	*3,542	3.4	*3,715	3.6	*3,889	3.7	*4,063	3.8	*4,236	3.9	*4,410	4.0	*4,584	4.0

Commodity data are provided by the United Nations Statistical Division. The symbol * means that data are estimated. For additional notes, see Appendix I.

Product Tables

Tinplate

Unit of Measure: Metric tons.

	1990 Value	%	1991 Value	%	1992 Value	%	1993 Value	%	1994 Value	%	1995 Value	%	1996 Value	%
Total Production	13,987,667	100.0	15,739,246	100.0	16,306,609	100.0	15,607,349	100.0	16,069,197	100.0	16,360,574	100.0	16,223,912	100.0
Regions														
Africa	298,000	2.1	321,467	2.0	329,297	2.0	337,127	2.2	344,958	2.1	352,788	2.2	360,618	2.2
America, North	2,965,000	21.2	3,277,000	20.8	3,831,000	23.5	3,953,000	25.3	3,924,000	24.4	3,649,121	22.3	3,806,527	23.5
America, South	721,000	5.2	830,767	5.3	927,097	5.7	891,942	5.7	957,993	6.0	1,055,044	6.4	1,078,180	6.6
Asia	3,935,845	28.1	4,078,224	25.9	4,206,875	25.8	3,890,526	24.9	3,907,344	24.3	4,120,422	25.2	3,951,421	24.4
Europe	5,757,822	41.2	6,990,789	44.4	6,906,340	42.4	6,428,753	41.2	6,769,518	42.1	7,035,837	43.0	6,897,825	42.5
Oceania	310,000	2.2	241,000	1.5	106,000	0.7	106,000	0.7	165,385	1.0	147,363	0.9	129,341	0.8
Leading Producers														
United States of America	2,467,000	17.6	2,753,000	17.5	3,354,000	20.6	3,466,000	22.2	3,466,000	21.6	3,266,000	20.0	3,431,000	21.1
Japan	1,711,000	12.2	1,820,000	11.6	1,909,000	11.7	1,650,000	10.6	1,711,000	10.6	1,695,000	10.4	1,583,000	9.8
Germany	-		1,175,000	7.5	1,079,000	6.6	895,000	5.7	974,000	6.1	1,027,000	6.3	960,000	5.9
France	996,000	7.1	984,000	6.3	960,000	5.9	876,000	5.6	984,000	6.1	1,052,000	6.4	1,084,000	6.7
United Kingdom	851,000	6.1	742,000	4.7	794,000	4.9	831,000	5.3	762,000	4.7	783,000	4.8	733,000	4.5

Sheets, Galvanized

Unit of Measure: Metric tons.

	1990 Value	%	1991 Value	%	1992 Value	%	1993 Value	%	1994 Value	%	1995 Value	%	1996 Value	%
Total Production	46,232,964	100.0	48,477,110	100.0	48,281,195	100.0	48,269,613	100.0	50,950,550	100.0	52,964,595	100.0	53,847,760	100.0
Regions														
Africa	589,000	1.3	625,000	1.3	608,000	1.3	610,000	1.3	625,942	1.2	644,812	1.2	653,682	1.2
America, North	11,419,543	24.7	12,595,077	26.0	13,304,228	27.6	14,016,380	29.0	14,729,531	28.9	15,441,647	29.2	16,154,013	30.0
America, South	418,909	0.9	392,455	0.8	455,000	0.9	541,000	1.1	750,000	1.5	653,813	1.2	660,636	1.2
Asia	18,303,607	39.6	19,510,893	40.2	18,743,179	38.8	17,827,464	36.9	18,014,750	35.4	19,055,336	36.0	18,767,912	34.9
Europe	15,501,905	33.5	15,353,686	31.7	15,170,788	31.4	15,274,769	31.6	16,830,327	33.0	17,168,987	32.4	17,611,517	32.7
Leading Producers														
Japan	12,949,000	28.0	13,551,000	28.0	12,711,000	26.3	11,539,000	23.9	11,451,000	22.5	11,704,000	22.1	11,331,000	21.0
United States of America	10,167,000	22.0	*11,171,333	23.0	*11,837,248	24.5	*12,503,164	25.9	*13,169,079	25.8	*13,834,994	26.1	*14,500,909	26.9
Germany	*3,842,333	8.3	*3,842,333	7.9	3,455,000	7.2	3,682,000	7.6	4,390,000	8.6	*3,842,333	7.3	*3,842,333	7.1
France	2,196,000	4.7	2,292,000	4.7	2,304,000	4.8	2,016,000	4.2	2,280,000	4.5	*2,537,264	4.8	*2,658,127	4.9
Korea, Republic of	1,820,000	3.9	2,344,000	4.8	2,559,000	5.3	2,813,000	5.8	2,825,000	5.5	3,132,000	5.9	*3,215,990	6.0

Hoop and Strip, Cold-Reduced

Unit of Measure: Metric tons.

	1990 Value	%	1991 Value	%	1992 Value	%	1993 Value	%	1994 Value	%	1995 Value	%	1996 Value	%
Total Production	30,600,683	100.0	33,007,863	100.0	15,372,413	100.0	15,191,984	100.0	49,159,772	100.0	55,922,652	100.0	35,226,636	100.0
Regions														
Africa	115,000	0.4	147,000	0.4	169,000	1.1	118,000	0.8	131,000	0.3	160,000	0.3	144,000	0.4
America, North	740,000	2.4	685,000	2.1	755,000	4.9	734,545	4.8	723,706	1.5	712,867	1.3	702,028	2.0
America, South	19,227,250	62.8	19,227,250	58.3	1,406,000	9.1	1,182,000	7.8	34,568,000	70.3	39,753,000	71.1	19,227,250	54.6
Asia	5,433,694	17.8	6,023,611	18.2	6,077,055	39.5	6,859,181	45.1	7,183,907	14.6	7,761,432	13.9	8,018,954	22.8
Europe	5,084,738	16.6	6,925,002	21.0	6,965,358	45.3	6,298,258	41.5	6,553,159	13.3	7,535,353	13.5	7,134,404	20.3
Leading Producers														
Colombia	*19,210,250	62.8	*19,210,250	58.2	1,389,000	9.0	1,165,000	7.7	34,551,000	70.3	39,736,000	71.1	*19,210,250	54.5
Korea, Republic of	2,649,000	8.7	3,267,000	8.9	3,395,000	22.1	4,330,000	28.5	4,618,000	9.4	5,239,000	9.4	*5,258,396	14.9
Germany	-		2,231,000	6.8	2,180,000	14.2	1,563,000	10.3	1,796,000	3.7	2,666,590	4.8	2,393,552	6.8
United States of America	740,000	2.4	685,000	2.1	755,000	4.9	*734,545	4.8	*723,706	1.5	*712,867	1.3	*702,028	2.0
Japan	591,000	1.9	588,000	1.8	*614,527	4.0	*618,736	4.1	*622,945	1.3	*627,155	1.1	*631,364	1.8

Commodity data are provided by the United Nations Statistical Division. The symbol * means that data are estimated. For additional notes, see Appendix I.

509

Product Tables

Hoop and Strip, Hot-Rolled

Unit of Measure: Metric tons.

	1990 Value	%	1991 Value	%	1992 Value	%	1993 Value	%	1994 Value	%	1995 Value	%	1996 Value	%
Total Production	44,520,214	100.0	44,215,306	100.0	43,470,437	100.0	41,009,331	100.0	39,892,378	100.0	40,302,343	100.0	41,997,117	100.0
Regions														
Africa	208,000	0.5	242,000	0.5	222,000	0.5	231,000	0.6	259,000	0.6	289,000	0.7	309,000	0.7
America, North	804,786	1.8	309,679	0.7	-		-		-				-	
America, South	226,357	0.5	237,036	0.5	245,464	0.6	259,143	0.6	270,821	0.7	279,500	0.7	290,429	0.7
Asia	29,992,000	67.4	31,768,310	71.8	31,835,619	73.2	31,282,929	76.3	30,580,930	76.7	31,557,800	78.3	33,295,925	79.3
Europe	11,570,171	26.0	10,122,482	22.9	9,814,654	22.6	8,066,659	19.7	7,795,126	19.5	7,372,642	18.3	7,481,463	17.8
Oceania	1,718,900	3.9	1,535,800	3.5	1,352,700	3.1	1,169,600	2.9	986,500	2.5	803,400	2.0	620,300	1.5
Leading Producers														
Korea, Republic of	6,378,000	14.3	7,770,000	17.6	8,718,000	20.1	10,073,000	24.6	10,152,000	25.4	10,256,000	25.4	*11,439,255	27.2
Russian Federation	6,282,000	14.1	5,523,000	12.5	4,244,000	9.8	2,410,000	5.9	1,174,000	2.9	1,626,000	4.0	1,368,000	3.3
United States of America	*804,786	1.8	*309,679	0.7	-		-		-				-	
Ukraine	4,540,000	10.2	3,680,000	8.3	2,965,000	6.8	1,470,000	3.6	906,000	2.3	686,000	1.7	945,000	2.3
Japan	*2,137,000	4.8	*2,227,643	5.0	*2,318,286	5.3	*2,408,929	5.9	*2,499,571	6.3	*2,590,214	6.4	*2,680,857	6.4

Railway Track Material

Unit of Measure: Metric tons.

	1990 Value	%	1991 Value	%	1992 Value	%	1993 Value	%	1994 Value	%	1995 Value	%	1996 Value	%
Total Production	13,710,659	100.0	13,558,363	100.0	12,651,612	100.0	12,713,930	100.0	13,139,497	100.0	12,560,042	100.0	12,228,162	100.0
Regions														
Africa	98,000	0.7	57,000	0.4	77,000	0.6	72,000	0.6	63,000	0.5	61,000	0.5	57,000	0.5
America, North	486,000	3.5	826,500	6.1	844,000	6.7	992,000	7.8	961,000	7.3	1,014,000	8.1	1,094,000	8.9
America, South	22,000	0.2	30,000	0.2	10,000	0.1	40,000	0.3	49,000	0.4	14,000	0.1	6,000	0.0
Asia	9,413,083	68.7	8,930,033	65.9	8,390,983	66.3	8,994,433	70.7	9,466,883	72.0	8,794,833	70.0	8,469,438	69.3
Europe	3,500,004	25.5	3,515,687	25.9	3,122,915	24.7	2,401,211	18.9	2,377,757	18.1	2,446,780	19.5	2,364,724	19.3
Oceania	191,571	1.4	199,143	1.5	206,714	1.6	214,286	1.7	221,857	1.7	229,429	1.8	237,000	1.9
Leading Producers														
Russian Federation	2,646,000	19.3	2,084,000	15.4	1,527,000	12.1	1,575,000	12.4	1,668,000	12.7	1,603,000	12.8	1,424,000	11.6
China	1,661,000	12.1	1,638,000	12.1	1,550,000	12.3	2,013,000	15.8	2,262,000	17.2	1,611,000	12.8	1,455,000	11.9
Ukraine	1,603,000	11.7	1,335,000	9.8	1,060,000	8.4	492,000	3.9	429,000	3.3	353,000	2.8	408,000	3.3
United States of America	470,000	3.4	473,500	3.5	477,000	3.8	586,000	4.6	572,000	4.4	571,000	4.5	606,000	5.0
Japan	389,000	2.8	413,000	3.0	441,000	3.5	451,000	3.5	471,000	3.6	455,000	3.6	412,000	3.4

Wire, Plain

Unit of Measure: Metric tons.

	1990 Value	%	1991 Value	%	1992 Value	%	1993 Value	%	1994 Value	%	1995 Value	%	1996 Value	%
Total Production	24,915,695	100.0	26,418,526	100.0	25,675,380	100.0	24,121,481	100.0	23,990,960	100.0	23,371,377	100.0	22,858,265	100.0
Regions														
Africa	365,000	1.5	364,000	1.4	347,000	1.4	347,000	1.4	372,000	1.6	389,000	1.7	409,400	1.8
America, North	1,236,900	5.0	1,252,650	4.7	1,209,400	4.7	1,253,150	5.2	1,256,900	5.2	1,092,650	4.7	1,115,400	4.9
America, South	1,269,750	5.1	1,224,750	4.6	1,144,000	4.5	1,308,273	5.4	1,311,969	5.5	1,471,664	6.3	1,364,110	6.0
Asia	14,191,295	57.0	13,815,043	52.3	13,034,790	50.8	12,221,288	50.7	11,956,036	49.8	12,284,572	52.6	12,071,557	52.8
Europe	7,412,250	29.7	9,321,583	35.3	9,499,690	37.0	8,551,270	35.5	8,653,555	36.1	7,692,990	32.9	7,457,298	32.6
Oceania	440,500	1.8	440,500	1.7	440,500	1.7	440,500	1.8	440,500	1.8	440,500	1.9	440,500	1.9
Leading Producers														
China	*4,629,000	18.6	*4,629,000	17.5	*4,629,000	18.0	*4,629,000	19.2	*4,629,000	19.3	*4,629,000	19.8	*4,629,000	20.3
Japan	4,387,000	17.6	4,292,000	16.2	4,087,000	15.9	3,894,000	16.1	3,847,000	16.0	3,887,000	16.6	3,900,000	17.1
Germany	-		2,397,000	9.1	2,376,000	9.3	2,301,000	9.5	2,570,000	10.7	1,474,691	6.3	1,337,953	5.9
Russian Federation	2,794,000	11.2	2,485,000	9.4	1,880,000	7.3	1,234,000	5.1	747,000	3.1	821,000	3.5	655,000	2.9
United States of America	835,000	3.4	817,000	3.1	799,000	3.1	719,000	3.0	714,000	3.0	593,000	2.5	592,000	2.6

Product Tables
Tubes, Seamless

Unit of Measure: Metric tons.

	1990		1991		1992		1993		1994		1995		1996	
	Value	%	Value	%	Value	%	Value	%	Value	%	Value	%	Value	%
Total Production	29,824,261	100.0	28,805,933	100.0	28,131,060	100.0	27,675,778	100.0	26,277,165	100.0	27,034,337	100.0	27,382,185	100.0
Regions														
Africa	12,000	0.0	14,000	0.0	21,000	0.1	23,000	0.1	24,000	0.1	13,731	0.1	12,374	0.0
America, North	1,734,000	5.8	1,755,000	6.1	1,507,000	5.4	1,915,000	6.9	2,062,000	7.8	2,429,000	9.0	2,673,000	9.8
America, South	843,000	2.8	991,000	3.4	858,000	3.1	964,152	3.5	946,739	3.6	1,032,326	3.8	1,161,914	4.2
Asia	17,850,978	59.9	17,852,778	62.0	16,792,578	59.7	16,744,378	60.5	15,890,178	60.5	15,498,978	57.3	15,609,778	57.0
Europe	9,384,283	31.5	8,193,156	28.4	8,952,482	31.8	8,029,248	29.0	7,354,248	28.0	8,060,302	29.8	7,925,120	28.9
Leading Producers														
Japan	2,794,000	9.4	3,084,000	10.7	2,327,000	8.3	2,484,000	9.0	2,300,000	8.8	2,061,000	7.6	1,938,000	7.1
Russian Federation	4,049,000	13.6	3,553,000	12.3	2,967,000	10.5	2,594,000	9.4	1,857,000	7.1	1,934,000	7.2	1,669,000	6.1
China	2,111,000	7.1	2,314,000	8.0	2,656,000	9.4	2,831,000	10.2	3,039,000	11.6	2,737,000	10.1	3,162,000	11.5
Ukraine	2,925,000	9.8	2,517,000	8.7	2,402,000	8.5	1,729,000	6.2	836,000	3.2	893,000	3.3	971,000	3.5
United States of America	1,416,000	4.7	1,374,000	4.8	1,240,000	4.4	1,591,000	5.7	1,695,000	6.5	1,960,000	7.3	2,064,000	7.5

Tubes, Welded

Unit of Measure: Metric tons.

	1990		1991		1992		1993		1994		1995		1996	
	Value	%	Value	%	Value	%	Value	%	Value	%	Value	%	Value	%
Total Production	54,488,365	100.0	53,206,258	100.0	51,332,424	100.0	48,072,794	100.0	48,324,783	100.0	49,786,933	100.0	51,603,927	100.0
Regions														
Africa	685,000	1.3	611,000	1.1	640,000	1.2	708,000	1.5	728,000	1.5	709,736	1.4	638,758	1.2
America, North	4,577,083	8.4	4,717,233	8.9	4,244,383	8.3	4,687,533	9.8	5,407,683	11.2	5,277,833	10.6	5,903,983	11.4
America, South	815,667	1.5	893,933	1.7	886,109	1.7	886,618	1.8	896,127	1.9	884,636	1.8	870,139	1.7
Asia	34,799,389	63.9	34,583,772	65.0	32,040,156	62.4	30,327,456	63.1	28,814,839	59.6	30,107,722	60.5	30,791,877	59.7
Europe	13,199,083	24.2	12,002,319	22.6	13,137,918	25.6	11,093,472	23.1	12,122,562	25.1	12,465,577	25.0	13,071,883	25.3
Oceania	412,143	0.8	398,000	0.7	383,857	0.7	369,714	0.8	355,571	0.7	341,429	0.7	327,286	0.6
Leading Producers														
Japan	7,733,000	14.2	7,802,000	14.7	7,177,000	14.0	6,939,000	14.4	6,406,000	13.3	6,617,000	13.3	6,981,000	13.5
Russian Federation	7,870,000	14.4	6,938,000	13.0	5,114,000	10.0	3,210,000	6.7	1,733,000	3.6	1,801,000	3.6	1,851,000	3.6
United States of America	2,804,000	5.1	2,698,000	5.1	2,568,000	5.0	2,443,000	5.1	2,809,000	5.8	2,972,000	6.0	3,284,000	6.4
Italy	1,919,000	3.5	2,062,000	3.9	2,029,000	4.0	1,937,000	4.0	2,808,000	5.8	2,800,000	5.6	2,946,000	5.7
Korea, Republic of	2,942,000	5.4	3,152,000	5.9	2,840,000	5.5	2,923,000	6.1	3,141,000	6.5	3,685,000	7.4	*3,589,305	7.0

Steel Castings in the Rough State

Unit of Measure: Metric tons.

	1990		1991		1992		1993		1994		1995		1996	
	Value	%	Value	%	Value	%	Value	%	Value	%	Value	%	Value	%
Total Production	6,525,267	100.0	6,551,572	100.0	6,222,345	100.0	6,622,788	100.0	6,598,428	100.0	5,631,139	100.0	5,226,141	100.0
Regions														
Africa	115,010	1.8	111,524	1.7	97,038	1.6	80,552	1.2	84,067	1.3	77,581	1.4	79,095	1.5
America, North	1,398,556	21.4	1,243,022	19.0	1,309,489	21.0	1,276,653	19.3	1,271,406	19.3	1,266,159	22.5	1,260,913	24.1
Asia	2,199,893	33.7	2,265,179	34.6	2,045,464	32.9	2,516,000	38.0	1,871,000	28.4	1,935,000	34.4	1,791,152	34.3
Europe	2,811,810	43.1	2,931,848	44.8	2,770,354	44.5	2,749,583	41.5	3,371,955	51.1	2,352,399	41.8	2,094,980	40.1
Leading Producers														
Ukraine	*1,072,250	16.4	*1,072,250	16.4	*1,072,250	17.2	1,187,000	17.9	1,795,000	27.2	748,000	13.3	559,000	10.7
Russian Federation	*997,500	15.3	*997,500	15.2	*997,500	16.0	1,535,000	23.2	862,000	13.1	869,000	15.4	724,000	13.9
United States of America	1,028,000	15.8	868,000	13.2	930,000	14.9	*892,697	13.5	*882,984	13.4	*873,270	15.5	*863,557	16.5
Japan	485,000	7.4	450,000	6.9	377,000	6.1	349,000	5.3	356,000	5.4	370,000	6.6	375,000	7.2
Korea, Republic of	423,000	6.5	515,000	7.9	387,000	6.2	373,000	5.6	458,000	6.9	504,000	9.0	*525,152	10.0

Commodity data are provided by the United Nations Statistical Division. The symbol * means that data are estimated. For additional notes, see Appendix I.

Product Tables
Steel Forgings

Unit of Measure: Metric tons.

	1990		1991		1992		1993		1994		1995		1996	
	Value	%	Value	%	Value	%	Value	%	Value	%	Value	%	Value	%
Total Production	6,952,562	100.0	6,306,412	100.0	6,384,515	100.0	5,631,334	100.0	5,923,953	100.0	6,072,229	100.0	5,956,473	100.0
Regions														
America, North	4,583	0.1	5,767	0.1	4,950	0.1	52,133	0.9	46,317	0.8	65,500	1.1	94,683	1.6
America, South	2,000	0.0	2,000	0.0	2,000	0.0	2,000	0.0	2,000	0.0	2,000	0.0	2,000	0.0
Asia	3,554,722	51.1	3,391,689	53.8	3,227,656	50.6	2,922,122	51.9	3,189,409	53.8	3,267,630	53.8	3,249,622	54.6
Europe	3,335,899	48.0	2,855,135	45.3	3,101,624	48.6	2,610,328	46.4	2,645,013	44.6	2,699,420	44.5	2,576,025	43.2
Oceania	55,357	0.8	51,821	0.8	48,286	0.8	44,750	0.8	41,214	0.7	37,679	0.6	34,143	0.6
Leading Producers														
Russian Federation	945,000	13.6	795,000	12.6	694,000	10.9	390,000	6.9	548,000	9.3	528,000	8.7	495,000	8.3
Japan	671,000	9.7	593,000	9.4	534,000	8.4	506,000	9.0	536,000	9.0	569,000	9.4	583,000	9.8
Romania	423,000	6.1	293,000	4.6	186,000	2.9	188,000	3.3	163,000	2.8	164,000	2.7	192,000	3.2
Poland	336,000	4.8	233,000	3.7	210,000	3.3	185,000	3.3	252,000	4.3	285,000	4.7	271,000	4.5
Italy	333,000	4.8	349,000	5.5	617,000	9.7	288,000	5.1	*340,731	5.8	*335,066	5.5	*329,401	5.5

Wheels, Wheel Centers, Tires and Axles

Unit of Measure: Metric tons.

	1990		1991		1992		1993		1994		1995		1996	
	Value	%	Value	%	Value	%	Value	%	Value	%	Value	%	Value	%
Total Production	3,093,879	100.0	2,785,751	100.0	2,686,658	100.0	2,536,497	100.0	2,454,836	100.0	2,559,175	100.0	2,673,607	100.0
Regions														
America, North	464,000	15.0	371,000	13.3	395,000	14.7	470,182	18.5	490,864	20.0	511,545	20.0	532,227	19.9
America, South	4,750	0.2	4,750	0.2	5,000	0.2	4,000	0.2	4,000	0.2	6,000	0.2	4,750	0.2
Asia	1,928,889	62.3	1,796,489	64.5	1,643,089	61.2	1,516,689	59.8	1,527,289	62.2	1,626,889	63.6	1,658,489	62.0
Europe	696,240	22.5	613,512	22.0	643,569	24.0	545,626	21.5	432,683	17.6	414,740	16.2	478,140	17.9
Leading Producers														
Russian Federation	735,000	23.8	607,000	21.8	439,000	16.3	309,000	12.2	295,000	12.0	379,000	14.8	415,000	15.5
United States of America	464,000	15.0	371,000	13.3	395,000	14.7	*470,182	18.5	*490,864	20.0	*511,545	20.0	*532,227	19.9
Ukraine	379,000	12.2	319,000	11.5	332,000	12.4	222,000	8.8	127,000	5.2	109,000	4.3	165,000	6.2
Poland	64,000	2.1	46,000	1.7	24,000	0.9	33,000	1.3	27,000	1.1	29,000	1.1	31,000	1.2
Japan	51,000	1.6	37,000	1.3	40,000	1.5	34,000	1.3	48,000	2.0	54,000	2.1	40,000	1.5

ISIC 3720
NONFERROUS METALS

ISIC 3720—Nonferrous Metals—includes primary and secondary nonferrous metal products, including copper, aluminum, nickel, tin, chrome, lead, manganese, zinc, etc. and the precious metals of gold, silver, and platinum.

The products of this industry are classified in the U.S. system under SIC 333—Primary Nonferrous Metals, SIC 334—Secondary Nonferrous Metals, SIC 335—Nonferrous Rolling and Drawing, SIC 336—Nonferrous Foundries (Castings), and SIC 339—Miscellaneous Primary Metal Products.

Summary Statistics

		1990 Value	N	1991 Value	N	1992 Value	N	1993 Value	N	1994 Value	N	1995 Value	N
Establishments or enterprises	(number)	14,951	62	15,635	67	19,197	63	16,270	58	12,732	51	10,053	21
Number employed	(000)	2,596	69	2,340	71	2,313	69	2,458	65	2,550	63	2,244	47
Total Wages	($ mil.)	27,317	64	30,945	64	32,090	62	24,619	57	25,207	55	21,846	42
Wage/salary per employee	($)	11,990	64	13,129	64	14,727	62	12,886	57	12,750	55	14,139	42
Female workers	(000)	61	27	35	24	207	22	208	23	209	20	216	10
as % of total employment*	(%)	14	25	13	23	16	22	18	23	19	20	29	9
Output	($ bil.)	261	70	269	72	270	69	213	65	229	62	239	47
per employee	($ 000)	128	66	121	67	133	65	120	59	128	58	148	44
per establishment	($ 000)	29,309	59	23,620	62	26,642	58	22,819	52	19,633	45	12,461	18
Value added	($ bil.)	72	65	67	62	70	58	69	58	69	53	79	45
per employee	($ 000)	40	62	35	60	41	56	36	53	40	50	49	42
per establishment	($ 000)	7,839	53	6,181	53	6,956	48	5,990	45	6,605	36	6,339	16
Capital investment	($ mil.)	12,496	54	10,687	51	11,813	45	10,916	41	8,091	33	6,013	14
per employee	($ 000)	8	50	7	48	8	44	8	36	8	30	17	11
per establishment	($ 000)	1,277	50	1,056	46	1,473	42	931	33	729	29	1,909	9

Data presented above are drawn from the detailed tables that follow. Columns headed 'N' show the number of countries that provided valid data for inclusion. Values are not strictly comparable one year to the next or one row to the next because the number of countries that report varies from period to period and row to row. However, a general indication of magnitudes can be gleaned from reviewing this summary—especially in earlier years in which more reports are available. For detailed explanations, see Appendix I. *The average for those countries reporting both total and female employment.

Establishments and Number Engaged

Country	Establishments or Enterprises (number)						Employees per Establishment					
	1990	1991	1992	1993	1994	1995	1990	1991	1992	1993	1994	1995
Albania	-	-	-	-	4.000	4.000	-	-	-	-	539	546
Argentina	-	-	-	469	501	-	-	-	-	17	16	-
Armenia	3.000	3.000	3.000	3.000	3.000	-	1,507	1,431	-	-	-	-
Australia	228	261	284	-	-	-	123	107	92	-	-	-
Austria	57	57	54	51	49	-	168	158	148	137	138	-
Azerbaijan	6.000	6.000	7.000	7.000	7.000	-	1,393	1,361	1,239	1,128	1,060	-
Bahrain	-	-	4.000	-	-	-	-	-	771	-	-	-
Bangladesh	-	-	11				24	17	42	-	-	-
Bolivia	3.000	3.000										
Bolivia	6.000	6.000	6.000	6.000	5.000	5.000	164	163	145	180	220	228
Bosnia & Herzegovina	7.000	8.000	-	-	-	-	834	739	-	-	-	-
Bulgaria	*9.000	*9.000	*11	*38	*44	*34	1,367	1,244	964	279	241	324
Canada	279	259	246	230	228	-	158	158	163	170	167	-
Cape Verde	-	-	-	-	1.000	-	-	-	-	-	-	-
Chile	23	23	22	25	21	-	370	374	358	329	367	-
China	*2,742	*2,795	*2,883	*3,460	*3,609	*4,621	314	315	319	266	280	219
China (Hong Kong SAR)	157	98	90	101	89	-	11	16	16	16	15	-
Colombia	34	34	36	38	41	-	65	67	69	65	60	-
Costa Rica	6.000	7.000	5.000	6.000	-	-	53	44	63	57	-	-
Croatia	15	21	34	53	63	67	372	237	119	50	40	38
Czechoslovakia (Former)	*22	*21	-	-	-	-	1,091	1,000	-	-	-	-
Czech Republic	-	*18	*17	*18	-	-	-	667	471	500	-	-
Denmark	78	79	79	-	-	-	24	23	22	-	-	-
Ecuador	4.000	4.000	5.000	5.000	5.000	6.000	76	90	78	80	85	71
Egypt	39	25	30	22	-	-	528	788	683	859	-	-
El Salvador	-	-	2.000	-	2.000	-	-	-	81	-	10	-
Finland	22	23	21	19	19	-	195	183	200	221	221	-
FYR Macedonia	*3.000	*6.000	*13	*17	*24	*24	672	369	140	107	45	35
Germany	-	536	536	522	500	-	-	248	216	198	191	-
Germany (Western Part)	463	484	488	-	-	-	227	229	216	-	-	-
Ghana	-	-	-	5.000	-	-	-	-	-	497	-	-
Greece	43	43	45	-	-	-	133	127	122	-	-	-
Guatemala	-	1.000	1.000	1.000	1.000	1.000	-	30	30	30	25	59
Honduras	-	-	2.000	5.000	5.000	5.000	-	-	128	56	61	66
Hungary	*54	*99	*79	*89	-	-	370	172	165	90	-	-
Iceland	1.000	1.000	1.000	1.000			652	622	623	592	-	-
India	2,710	2,989	2,859	3,085	-	-	60	62	62	53	-	-
Indonesia	33	42	46	52	58	66	221	237	227	232	220	229
Iran (Islamic Republic of)	172	48	48	45	-	-	69	236	257	251	-	-
Ireland	13	12	-	-	-	-	15	17	-	-	-	-
Israel	47	57	57	56	87	-	38	33	39	41	32	-
Italy	*304	*301	*331	*338	*350	-	116	121	108	101	96	-
Japan	3,606	3,492	3,333	3,330	3,030	3,077	34	36	36	36	36	36
Jordan	-	15	14	16	31	14	-	23	20	27	25	41
Korea, Republic of	728	743	760	812	756	830	45	42	39	36	39	40
Kyrgyzstan	4.000	4.000	5.000	5.000	6.000	-	1,384	1,403	2,082	1,963	1,364	-
Latvia	1.000	1.000	3.000	3.000	1.000	1.000	292	235	59	29	67	71
Lithuania	-	-	-	*6.000	*3.000	-	-	-	-	2.500	11	-
Luxembourg	*7.000	*6.000	*6.000	*6.000	-	-	133	148	144	141	-	-
Malaysia	28	37	36	40	42	-	171	168	178	183	174	-
Mexico	25	25	25	24	23	23	722	710	618	554	531	502
Mongolia	-	-	-	-	3.000	-	-	-	-	-	146	-
Mozambique	-	-	-	-	1.000	-	-	-	-	-	36	-
Myanmar	21	20	18	15	10	-	668	677	704	839	168	-
Netherlands Antilles	-	-	-	2.000	-	-	-	-	-	-	-	-
New Zealand	*84	*94	*96	*99	-	-	40	32	35	37	-	-
Norway	41	41	35	30	35	-	255	250	273	294	257	-
Pakistan	-	14	-	-	-	-	-	38	-	-	-	-
Panama	3.000	3.000	-	-	-	-	-	-	-	-	-	-
Paraguay	-	3.000	-	-	-	-	-	71	-	-	-	-
Peru	113	123	124	-	-	-	92	74	70	-	-	-

Continued.

Depending on the table, * means *Enterprises, Engaged,* or *Factor Values*; ± means *Basis Unspecified*; § means *shown in millions*. For additional notes and sources, see Appendix I.

Establishments and Number Engaged

- Continued -

Country	Establishments or Enterprises (number)						Employees per Establishment					
	1990	1991	1992	1993	1994	1995	1990	1991	1992	1993	1994	1995
Philippines	88	97	26	30	-	-	33	30	123	113	-	-
Poland	*23	*21	*22	*20	-	-	1,304	1,524	1,045	1,100	-	-
Portugal	*316	*308	*346	*358	*368	-	16	18	16	16	15	-
Romania	*23	-	-	-	197	-	-	-	-	-	-	-
Russian Federation	-	-	-	*416	*531	*832	-	-	-	795	733	266
Singapore	21	21	20	22	21	-	39	39	39	32	38	-
Slovakia	-	6.000	6.000	9.000	8.000	-	-	1,546	1,391	885	808	-
Slovenia	28	*17	*24	*65	*63	*114	-	-	-	-	-	-
South Africa	-	116	-	-	-	-	-	172	-	-	-	-
Spain	468	463	465	-	-	-	46	45	42	-	-	-
Sri Lanka	2.000	3.000	3.000	2.000	-	-	129	86	117	56	-	-
Suriname	-	-	12	-	-	-	-	-	-	-	-	-
Sweden	84	88	80	78	81	-	114	103	104	97	96	-
Thailand	148	123	-	-	-	-	19	99	-	-	-	-
Tunisia	-	-	-	105	113	112	-	-	-	62	57	55
Turkey	81	83	124	114	106	-	248	188	120	121	115	-
Ukraine	25	44	55	26	29	28	-	-	909	1,808	862	857
United Kingdom	1,093	1,080	1,089	1,695	1,365	-	54	51	46	31	36	-
United Republic of Tanzania	4.000	3.000	-	-	-	-	212	232	-	-	-	-
USSR (Former)	*161.	-	-	-	-	-	2,727	-	-	-	-	-
United States of America	-	-	3,987	-	-	-	-	-	58	-	-	-
Venezuela	97	108	84	65	81	68	156	138	156	195	149	184
Yugoslavia	36	48	37	105	104	121	-	423	519	170	170	143
Zambia	2.000	-	-	-	2.000	-	69	-	-	-	50	-
Zimbabwe	7.000	6.000	6.000	5.000	6.000	-	200	250	233	260	200	-

Employment and Compensation of Employees

Country	Number Employed/Engaged (000)						Wage/Salary per Employee ($)					
	1990	1991	1992	1993	1994	1995	1990	1991	1992	1993	1994	1995
Albania	-	-	-	-	2.158	2.185	-	-	-	-	489	1,075
Argentina	6.104	-	-	8.040	7.944	-	8,302	-	-	18,844	-	-
Armenia	*4.521	*4.292	-	-	-	-	-	-	-	-	-	-
Australia	28	*28	*26	-	-	-	28,404	30,024	30,656	-	-	-
Austria	9.600	9.000	8.000	6.984	6.746	6.700	27,750	29,491	33,067	33,470	34,911	-
Azerbaijan	8.358	8.164	8.670	7.899	7.419	-	2,258	2,493	-	-	-	-
Bahrain	-	-	3.084	-	-	-	-	-	26,355	-	-	-
Bangladesh	0.073	0.050	0.459	-	-	-	-	547	895	-	-	-
Belgium	12	11	10	-	-	-	-	-	-	-	-	-
Bolivia	0.983	0.979	0.871	1.082	1.099	1.140	2,575	2,927	3,394	3,036	2,986	2,746
Bosnia & Herzegovina	5.838	5.912	-	-	-	-	5,792	5,654	-	-	-	-
Bulgaria	12	11	11	11	11	11	2,402	868	1,605	2,226	1,824	2,263
Cameroon	0.403	0.316	0.413	0.381	-	-	16,679	17,914	16,896	-	-	-
Canada	44	41	40	39	38	38	37,028	-	-	37,743	34,937	35,846
Chile	8.507	8.608	7.884	8.225	7.710	7.801	10,130	11,657	14,290	14,306	16,382	19,415
China	*860	*880	*920	*920	*1,010	*1,010	-	-	-	-	-	-
China (Hong Kong SAR)	*1.700	*1.600	*1.400	*1.600	*1.300	-	11,176	13,994	15,318	17,775	19,907	-
China (Taiwan Province)	19	20	21	23	24	25	11,373	12,439	14,389	13,534	14,036	14,870
Colombia	2.200	2.270	2.497	2.456	2.447	2.468	2,693	2,701	3,326	3,807	4,544	5,278
Costa Rica	0.315	0.311	0.316	0.340	-	-	3,632	3,102	3,588	3,872	-	-
Croatia	5.580	4.980	4.040	2.650	2.520	2.520	5,280	3,733	1,339	1,456	1,887	3,017
Czechoslovakia (Former)	24	21	-	-	-	-	2,437	1,777	-	-	-	-
Czech Republic	14	12	8.000	9.000	-	-	2,348	1,724	2,256	2,695	-	-
Denmark	1.855	1.808	1.745	1.557	1.625	1.652	28,744	28,623	31,141	29,501	31,236	35,941
Ecuador	0.302	0.358	0.388	0.399	0.424	0.428	4,593	5,698	5,400	5,677	4,317	6,109
Egypt	*21	*20	*20	*19	*20	*20	2,740	1,811	1,714	2,034	2,150	2,331
El Salvador	-	-	*0.162	-	*0.020	-	-	-	-	-	3,537	-
Finland	4.300	4.200	4.200	4.200	4.200	-	32,822	31,605	30,680	25,034	28,612	-

Continued.

Depending on the table, * means *Enterprises*, *Engaged*, or *Factor Values*; ± means *Basis Unspecified*; § means *shown in millions*. For additional notes and sources, see Appendix I.

515

Employment and Compensation of Employees

- Continued -

Country	Number Employed/Engaged (000)						Wage/Salary per Employee ($)					
	1990	1991	1992	1993	1994	1995	1990	1991	1992	1993	1994	1995
FYR Macedonia	2.016	2.215	1.818	1.817	1.087	0.844	4,821	5,019	2,649	2,774	3,130	4,085
France	48	47	46	43	42	41	-	-	-	-	-	-
Germany	-	*133	*116	*104	*96	-	-	29,348	35,459	34,497	37,151	-
Germany (Western Part)	*105	*111	*105	-	-	-	31,565	33,012	37,180	-	-	-
Ghana	-	-	-	*2.483	*2.553	*2.625	-	-	-	6,492	4,849	4,948
Greece	*5.703	*5.456	*5.470	*5.501	*5.435	*5.423	22,422	22,285	23,852	22,374	25,250	29,329
Guatemala	-	*0.030	*0.030	*0.030	*0.025	*0.059	-	504	490	461	570	6,817
Honduras	*0.209	*0.198	*0.256	*0.280	*0.307	*0.332	1,692	2,380	2,285	2,512	1,540	1,443
Hungary	20	17	13	8.000	7.090	8.220	3,285	3,475	4,179	3,887	3,770	3,282
Iceland	0.652	0.622	0.623	0.592	0.534	-	38,888	-	38,406	31,978	-	-
India	*162	*186	*177	*163	*185	*190	1,450	1,280	1,233	1,108	1,146	1,174
Indonesia	7.282	9.944	10	12	13	15	1,174	1,309	1,543	1,478	1,950	2,601
Iran (Islamic Republic of)	12	11	12	11	-	-	5,095	6,938	6,731	7,869	-	-
Ireland	0.200	0.200	0.185	0.190	0.180	0.190	17,413	22,617	24,618	21,651	22,906	25,112
Israel	1.800	1.900	2.200	2.300	2.800	2.583	19,013	19,398	17,005	16,438	17,910	19,694
Italy	35	36	36	34	34	-	38,352	-	-	33,285	34,493	-
Japan	*123	*124	*121	*120	*109	*110	32,118	36,219	-	-	-	-
Jordan	0.299	0.350	0.284	0.425	0.761	0.574	3,808	4,166	3,982	4,808	3,564	3,631
Korea, Republic of	33	32	30	30	29	33	10,551	12,037	13,200	13,606	15,489	18,924
Kyrgyzstan	*5.536	*5.611	*10	*9.815	*8.187	-	738	572	-	-	-	-
Latvia	0.292	0.235	0.178	0.088	0.067	0.071	-	-	-	-	-	-
Lithuania	-	-	0.022	0.015	0.032	-	-	-	-	-	471	-
Luxembourg	0.931	0.885	0.865	0.847	0.865	0.865	28,478	29,913	33,657	34,433	38,153	-
Malaysia	4.800	6.200	6.400	7.300	7.300	8.500	3,820	3,941	4,711	5,183	5,862	6,480
Mexico	18	18	15	13	12	12	5,300	6,710	7,761	9,185	9,538	5,426
Mongolia	-	-	-	-	*0.438	-	-	-	-	-	-	-
Mozambique	-	-	-	-	0.036	-	-	-	-	-	-	-
Myanmar	14	14	13	13	1.682	-	1,238	1,308	1,841	2,524	6,166	-
New Zealand	3.334	3.000	3.326	3.636	-	-	-	-	-	-	-	-
Norway	10	10	9.568	8.813	9.006	8.700	35,958	36,879	-	39,363	-	-
Pakistan	0.488	0.533	0.557	-	-	-	1,520	1,470	1,514	-	-	-
Paraguay	-	0.212	-	-	-	-	-	-	-	-	-	-
Peru	*10	*9.072	*8.622	-	-	-	5,135	6,894	5,887	-	-	-
Philippines	2.900	2.900	3.200	3.400	-	-	2,440	2,949	3,062	3,825	-	-
Poland	30	32	23	22	24	27	1,807	2,536	3,133	3,324	2,867	2,960
Portugal	*5.164	*5.596	*5.597	*5.640	*5.643	*5.652	-	-	-	-	-	-
Russian Federation	-	-	-	331	389	221	-	-	-	1,636	2,645	3,730
Singapore	*0.815	*0.824	*0.788	*0.696	*0.807	*0.752	14,266	16,139	19,931	20,360	21,984	23,971
Slovakia	-	*9.278	*8.346	*7.968	*6.466	-	-	1,813	2,234	2,570	2,910	-
South Africa	21	20	19	16	16	18	9,515	9,868	11,608	13,121	13,360	16,711
Spain	22	21	20	19	19	19	23,291	24,886	27,681	24,868	24,179	27,552
Sri Lanka	*0.258	*0.258	*0.351	*0.111	-	-	1,258	1,508	1,273	1,039	-	-
Sweden	9.600	*9.090	*8.298	7.552	7.814	7.778	22,755	26,222	27,827	22,623	24,881	27,743
Thailand	2.813	12	-	-	-	-	1,317	5,679	-	-	-	-
Tunisia	-	-	-	6.555	6.490	6.212	-	-	-	-	-	-
Turkey	20	16	15	14	12	12	7,407	10,991	10,470	11,710	6,886	9,554
Ukraine	-	-	50	47	25	24	-	-	3,487	545	591	864
United Kingdom	59	55	50	52	49	49	24,243	25,390	26,466	22,840	24,096	25,551
United Republic of Tanzania	0.846	0.695	-	-	-	-	582	670	-	-	-	-
USSR (Former)	439	-	-	-	-	-	2,820	-	-	-	-	-
United States of America	247	236	230	232	241	255	29,919	30,466	31,591	32,052	32,672	33,227
Uruguay	*0.245	*0.256	*0.444	*0.414	*0.339	*0.320	3,381	3,098	6,380	6,833	6,710	7,479
Venezuela	15	15	13	13	12	12	5,888	7,814	6,642	8,428	4,561	6,130
Yugoslavia	-	20	19	18	18	17	-	-	-	-	1,703	1,195
Zambia	0.139	-	-	-	0.100	-	613	-	-	-	1,267	-
Zimbabwe	1.400	1.500	1.400	1.300	1.200	1.148	3,676	3,890	2,342	2,104	2,311	2,516

Depending on the table, * means *Enterprises, Engaged,* or *Factor Values*; ± means *Basis Unspecified*; § means *shown in millions*. For additional notes and sources, see Appendix I.

Female Workers: Total and Percent of Employment

Country	Female Workers (000)						Female Workers as % of Total Employment					
	1990	1991	1992	1993	1994	1995	1990	1991	1992	1993	1994	1995
Afghanistan	0.010	0.008	-	-	-	-	-	-	-	-	-	-
Albania	-	-	-	-	0.694	0.676	-	-	-	-	32.16	30.94
Argentina	-	-	-	-	0.475	-	-	-	-	-	5.98	-
Australia	2.098	-	-	-	-	-	7.49	-	-	-	-	-
Austria	1.700	2.000	1.000	1.326	1.254	-	17.71	22.22	12.50	18.99	18.59	-
Bahrain	-	-	0.075	-	-	-	-	-	2.43	-	-	-
Bosnia & Herzegovina	0.863	0.798	-	-	-	-	14.78	13.50	-	-	-	-
Bulgaria	-	-	-	3.200	3.100	3.000	-	-	-	30.19	29.25	27.27
Canada	4.000	4.000	-	-	-	-	9.09	9.76	-	-	-	-
Chile	0.192	0.232	0.191	0.248	0.263	-	2.26	2.70	2.42	3.02	3.41	-
China (Taiwan Province)	4.354	4.687	4.904	5.195	5.449	5.636	22.49	23.66	23.08	22.79	22.39	22.70
Colombia	-	0.295	-	0.369	0.358	-	-	13.00	-	15.02	14.63	-
Croatia	1.000	0.830	0.670	0.360	0.420	0.430	17.92	16.67	16.58	13.58	16.67	17.06
Czechoslovakia (Former)	4.100	3.900	-	-	-	-	17.08	18.57	-	-	-	-
Czech Republic	4.000	4.000	3.000	3.000	-	-	28.57	33.33	37.50	33.33	-	-
Denmark	0.437	0.389	0.350	-	-	-	23.56	21.52	20.06	-	-	-
Egypt	-	0.328	0.685	0.614	-	-	-	1.66	3.34	3.25	-	-
El Salvador	-	-	-	-	0.003	-	-	-	-	-	15.00	-
FYR Macedonia	0.698	0.618	0.507	0.501	0.523	0.508	34.62	27.90	27.89	27.57	48.11	60.19
Germany (Western Part)	16	-	-	-	-	-	15.24	-	-	-	-	-
Ghana	-	-	-	0.104	-	-	-	-	-	4.19	-	-
Hungary	6.000	-	2.000	2.000	-	-	30.00	-	15.38	25.00	-	-
Indonesia	-	-	-	2.609	1.258	1.283	-	-	21.66	9.84	8.49	
Iran (Islamic Republic of)	-	0.247	0.288	0.246	-	-	-	2.18	2.34	2.18	-	-
Italy	-	3.764	3.964	3.746	3.791	-	-	10.37	11.07	10.97	11.22	-
Jordan	0.009	0.012	0.006	0.008	0.026	0.014	3.01	3.43	2.11	1.88	3.42	2.44
Korea, Republic of	4.000	4.407	4.034	3.953	4.048	4.354	12.31	13.97	13.47	13.34	13.80	13.22
Malaysia	0.800	1.000	1.200	1.300	1.400	-	16.67	16.13	18.75	17.81	19.18	-
Mongolia	-	-	-	-	0.089	-	-	-	-	-	20.32	-
Morocco	-	-	-	-	-	0.015	-	-	-	-	-	-
Myanmar	1.758	1.680	1.775	1.859	0.293	-	12.54	12.40	14.00	14.78	17.42	-
New Zealand	0.456	0.351	0.425	0.432	-	-	13.68	11.70	12.78	11.88	-	-
Panama	0.005	-	-	-	-	-	-	-	-	-	-	-
Philippines	-	-	0.475	0.530	-	-	-	-	14.84	15.59	-	-
Portugal	0.264	0.346	0.290	0.297	0.421	-	5.11	6.18	5.18	5.27	7.46	-
Sri Lanka	0.012	0.016	0.043	0.018	-	-	4.65	6.20	12.25	16.22	-	-
Sweden	1.800	-	-	-	-	-	18.75	-	-	-	-	-
Thailand	0.230	1.372	-	-	-	-	8.18	11.32	-	-	-	-
Turkey	0.681	-	-	-	-	-	3.39	-	-	-	-	-
United Kingdom	5.762	-	7.000	-	-	-	9.77	-	14.00	-	-	-
United Republic of Tanzania	0.047	0.038	-	-	-	-	5.56	5.47	-	-	-	-
United States of America	-	-	174	176	185	200	-	-	75.65	75.86	76.76	78.43
Zambia	-	-	-	-	0.003	-	-	-	-	-	3.00	-

Output and Output per Employee

Country	Output ($ bil.)						Output per Employee ($)					
	1990	1991	1992	1993	1994	1995	1990	1991	1992	1993	1994	1995
Albania	-	-	-	-	*0.025	*0.042	-	-	-	-	11,376	19,081
Argentina	±0.642	-	-	0.948	-	-	105,241	-	-	117,928	-	-
Armenia	0.002	0.005	0.030	0.267	0.004	-	522	1,056	-	-	-	-
Australia	*9.462	*9.416	*8.072	-	-	-	337,946	336,301	310,464	-	-	-
Austria	2.564	2.264	2.327	1.837	1.916	2.288	267,040	251,542	290,916	263,079	283,961	341,532
Azerbaijan	±0.163	±0.221	±0.019	±0.035	±0.009	-	19,505	27,066	2,167	4,407	1,262	-
Bahrain	-	-	±0.623	-	-	-	-	-	201,968	-	-	-
Bangladesh	0.000	0.000	0.006	-	-	-	4,161	4,919	12,753	-	-	-
Belgium	4.807	3.385	3.438	2.745	3.272	4.042	400,612	294,361	333,763	-	-	-
Bolivia	0.081	0.085	0.084	0.087	0.093	0.100	82,410	86,401	96,799	80,322	84,365	87,906

Continued.

Depending on the table, * means *Enterprises, Engaged,* or *Factor Values*; ± means *Basis Unspecified*; § means *shown in millions.* For additional notes and sources, see Appendix I.

517

Output and Output per Employee

- Continued -

Country	Output ($ bil.)						Output per Employee ($)					
	1990	1991	1992	1993	1994	1995	1990	1991	1992	1993	1994	1995
Bosnia & Herzegovina	±0.369	±0.365	-	-	-	-	63,292	61,809	-	-	-	-
Bulgaria.	±0.384	±0.218	±0.229	±0.217	±0.298	±0.407	31,221	19,473	21,641	20,441	28,086	36,970
Canada	8.305	7.890	7.355	7.061	8.026	8.229	188,745	192,448	183,875	181,063	211,205	217,130
Chile.	4.986	4.527	5.036	4.140	4.534	5.451	586,090	525,848	638,776	503,340	588,075	698,785
China	±11	±11	±13	±17	±14	±16	12,385	12,243	13,977	18,382	13,812	16,269
China (Hong Kong SAR) . .	0.336	0.398	0.612	0.653	0.497		197,770	248,835	437,120	408,339	382,605	
China (Taiwan Province) . . .	±3.091	±3.265	±3.395	±4.087	±4.593	±4.955	159,657	164,849	159,791	179,270	188,730	199,553
Colombia	0.129	0.121	0.109	0.122	0.157	0.184	58,825	53,449	43,767	49,490	64,267	74,574
Costa Rica	0.005	0.005	0.005	0.006	0.006	0.007	14,535	16,806	16,645	17,185		
Cote d'Ivoire	±0.007	±0.004	±0.004	±0.004	-	-	-	-	-	-	-	-
Croatia	±0.280	±0.240	±0.100	-	-	-	50,181	48,284	24,715	-	-	-
Czechoslovakia (Former) . . .	±1.047	±0.384	-	-	-	-	43,617	18,285	-	-	-	-
Denmark	*0.183	*0.175	*0.190	*0.154	*0.177	*0.209	98,775	96,766	109,058	98,686	108,623	126,642
Ecuador	0.024	0.031	0.031	0.033	0.038	0.043	80,216	87,660	78,785	83,562	89,927	100,940
Egypt	*0.679	*0.416	*0.442	*0.415	*0.466	*0.534	32,978	21,106	21,545	21,975	23,900	26,688
El Salvador.					±0.001						41,486	
Finland	±1.952	±1.637	±1.751	±1.487	±1.822		454,048	389,760	416,829	354,040	433,891	
FYR Macedonia	±0.043	±0.055	±0.036	±0.025	±0.014	±0.012	21,532	24,937	19,884	14,015	12,990	13,844
France	±15	±13	±15	±12	±13	±17	313,666	284,072	318,603	278,732	318,804	403,108
Germany	-	20	20	17	19		-	151,634	176,723	163,434	195,656	
Germany (Western Part) . . .	20	19	19	-	-	-	194,348	171,916	183,310	-	-	-
Ghana	-	-	-	0.171	0.139	0.146	-	-	-	68,843	54,540	55,659
Greece	*1.192	*1.125	*1.154	*1.092	*1.222	*1.419	209,074	206,122	211,023	198,528	224,918	261,746
Guatemala.	-	0.002	0.002	0.001	0.002	0.003	-	51,025	51,367	45,693	70,837	52,881
Honduras	0.002	0.002	0.002	0.002	0.002	0.002	8,250	8,358	6,760	6,945	6,744	6,322
Hungary	±1.138	±0.787	±0.509	±0.376	±0.315	±0.403	56,878	46,281	39,120	46,951	44,364	48,968
Iceland	0.166	0.140	±0.139	±0.125			255,192	225,616	223,100	210,420		
India	*3.686	*3.720	*3.717	*3.072	*3.648	*3.898	22,797	19,963	21,052	18,825	19,722	20,503
Indonesia	0.663	0.643	0.845	0.610	0.946	1.138	91,020	64,657	80,741	50,649	73,956	75,268
Iran (Islamic Republic of) . . .	1.003	0.856	1.044	0.768	-		85,004	75,622	84,715	68,000		
Ireland	*0.046	*0.041	*0.041	*0.037	*0.037	*0.044	232,172	204,362	218,982	195,078	204,982	228,995
Israel	0.174	0.173	0.184	0.190	0.278	0.263	96,717	90,987	83,734	82,652	99,276	102,008
Italy	±8.187	±7.558	±8.364	±6.724	±7.846	-	232,219	208,193	233,635	196,912	232,313	
Japan	±39	±40	±38	±38	±40	±48	315,399	323,934	317,658	320,069	364,513	438,791
Jordan	0.020	0.025	0.033	0.028	0.039	0.036	66,905	71,420	114,696	66,609	50,874	62,738
Korea, Republic of	4.636	5.025	4.802	4.865	6.090	9.386	142,639	159,350	160,347	164,234	207,597	285,066
Kyrgyzstan	0.127	0.092	0.186	0.023	0.035		22,946	16,474	17,846	2,312	4,280	
Latvia	0.000	0.000	0.001	0.001	0.001	0.003	265	616	3,913	6,577	16,530	36,209
Lithuania	-	-	0.000	0.000	0.000		-	-	590	813	2,042	
Luxembourg	*0.193	*0.179	*0.189	*0.201	*0.230	-	207,282	202,773	218,305	237,206	266,260	
Malaysia	*0.460	*0.512	*0.588	*0.604	*0.751	*0.917	95,860	82,584	91,883	82,801	102,832	107,895
Mexico	2.224	1.853	1.741	1.584	1.774	1.872	123,164	104,415	112,646	119,144	145,342	161,990
Mozambique		±0.001	-	±0.002	±0.002		-	-	-	-	44,496	
Myanmar	0.037	0.054	0.050	0.052	0.103	-	2,625	3,976	3,919	4,114	61,295	
Norway	±3.604	±3.231	±3.001	±2.598	±3.284	±3.612	345,374	315,600	313,615	294,768	364,596	415,131
Pakistan	0.005	0.006	0.007				11,047	11,445	11,886			
Peru	1.747	1.318	1.266				167,461	145,263	146,811			
Philippines	0.552	0.455	0.460	0.587	-		190,193	156,985	143,866	172,566		
Poland	±2.335	±2.039	±1.846	±1.740	±2.144	±2.957	77,832	63,732	80,256	79,088	87,875	109,209
Portugal	0.369	0.336	0.330	0.266	0.310	0.373	71,388	60,056	58,983	47,235	54,974	66,031
Romania	±0.990	±0.886	±0.464	±0.390	-		-	-	-	-	-	
Russian Federation	-	22	23	6.473	7.697	11	-	-	19,567	19,788	47,869	
Singapore	*0.119	*0.115	*0.128	*0.123	*0.140	*0.143	146,169	139,293	162,301	176,450	173,786	189,623
Slovakia	-	±0.285	±0.255	±0.220	±0.206	-	-	30,700	30,535	27,552	31,867	
Slovenia				0.129	0.130	0.424						
South Africa	±1.929	±1.761	±1.661	±1.547	±1.690	±2.310	91,840	88,074	87,418	96,673	105,621	131,694
Spain	*4.844	*4.432	*4.341	±3.180	±3.674	±4.948	223,833	213,521	221,617	166,204	198,125	265,211
Sri Lanka	0.009	0.006	0.007	0.005			33,958	23,205	18,857	44,859		
Sweden	*2.240	*2.576	*2.102	*1.577	*1.932	*2.159	233,359	283,395	253,362	208,791	247,261	277,618
Thailand	0.059	1.641	-	-	-		21,148	135,373	-	-	-	

Continued.

Depending on the table, * means *Enterprises*, *Engaged*, or *Factor Values*; ± means *Basis Unspecified*; § means *shown in millions*. For additional notes and sources, see Appendix I.

Output and Output per Employee
- Continued -

Country	Output ($ bil.)						Output per Employee ($)					
	1990	1991	1992	1993	1994	1995	1990	1991	1992	1993	1994	1995
Turkey	1.557	1.129	1.203	1.314	1.075	1.497	77,521	72,216	81,140	94,964	88,391	123,538
Ukraine	*1.633	*1.451	*2.212	*0.470	*0.370	*0.493	-	-	44,241	9,991	14,790	20,526
United Kingdom	*11	*8.876	8.898	8.447	9.202	9.686	179,661	161,384	177,951	162,451	187,799	199,383
United Republic of Tanzania	*0.025	*0.046	-	-	-	-	29,390	66,191	-	-	-	-
USSR (Former)	±19	-	-	-	-	-	43,850	-	-	-	-	-
United States of America	*59	*53	*54	*53	*61	*73	237,004	223,983	235,796	228,565	251,241	284,510
Uruguay	0.007	0.006	0.018	0.018	0.015	0.016	26,850	24,264	40,995	44,102	45,194	50,374
Venezuela	1.619	1.393	1.261	1.387	1.394	-	107,207	93,522	96,231	109,408	115,277	-
Yugoslavia	±0.115	±0.131	±0.250	-	±0.461	±0.278	-	6,456	13,037	-	26,036	16,074
Zambia	0.002	-	-	-	0.001	-	14,165	-	-	-	9,303	-
Zimbabwe	*0.036	*0.037	*0.035	*0.025	*0.026	*0.028	25,969	24,893	24,678	18,897	21,268	24,422

Value Added and Value Added per Employee

Country	Value Added ($ bil.)						Value Added per Employee ($)					
	1990	1991	1992	1993	1994	1995	1990	1991	1992	1993	1994	1995
Argentina	±0.305	-	-	0.238	-	-	50,022	-	-	29,605	-	-
Armenia	0.001	0.001	-	-	-	-	151	333	-	-	-	-
Australia	*3.791	*3.867	*3.653	-	-	-	135,379	138,099	140,508	-	-	-
Austria	0.434	0.431	0.424	0.332	0.373	0.440	45,249	47,876	53,030	47,490	55,274	65,713
Bangladesh	0.000	0.000	0.001	-	-	-	1,030	1,093	2,293	-	-	-
Belgium	*1.140	*0.803	*0.823	*0.657	*0.787	*0.972	95,016	69,839	79,941	-	-	-
Bolivia	0.010	0.012	0.015	0.026	0.024	0.026	10,540	12,516	17,221	24,123	22,112	23,039
Bosnia & Herzegovina	±0.139	±0.137	-	-	-	-	23,746	23,187	-	-	-	-
Bulgaria	-	±0.073	±0.094	±0.113	±0.122	±0.201	-	6,474	8,824	10,689	11,494	18,316
Cameroon	0.015	0.012	0.009	-	-	-	36,739	37,051	22,256	-	-	-
Canada	3.222	2.994	2.747	2.690	3.303	3.398	73,238	73,020	68,669	68,967	86,910	89,655
Chile	1.717	1.522	1.736	1.217	1.304	1.562	201,843	176,828	220,157	147,980	169,099	200,219
China	±2.050	±2.015	±2.502	±4.527	±3.057	±3.617	2,384	2,290	2,719	4,921	3,027	3,581
China (Hong Kong SAR)	0.040	0.057	0.051	0.068	0.054	-	23,485	35,870	36,450	42,579	41,207	-
China (Taiwan Province)	0.672	0.724	0.730	0.779	0.875	0.998	34,732	36,566	34,374	34,178	35,969	40,215
Colombia	0.056	0.052	0.042	0.044	0.055	0.064	25,340	23,027	16,674	18,020	22,453	26,013
Costa Rica	0.001	0.001	0.001	0.001	0.001	0.002	3,460	4,068	4,042	4,067	-	-
Cote d'Ivoire	0.007	0.004	0.004	0.004	-	-	-	-	-	-	-	-
Croatia	±0.106	±0.088	±0.042	-	-	-	19,004	17,620	10,285	-	-	-
Czechoslovakia (Former)	±0.236	-	-	-	-	-	9,842	-	-	-	-	-
Denmark	*0.073	*0.068	*0.074	*0.061	*0.069	*0.081	39,109	37,703	42,283	39,357	42,296	48,917
Ecuador	0.002	0.006	0.002	0.012	0.008	0.009	8,199	18,049	6,116	29,041	18,152	21,988
Egypt	*0.176	*0.072	*0.077	*0.082	*0.092	*0.105	8,524	3,673	3,765	4,351	4,722	5,261
El Salvador	-	-	-	-	0.000	-	-	-	-	-	6,280	-
Finland	±0.363	±0.300	±0.315	±0.300	±0.375	-	84,412	71,417	74,893	71,316	89,385	-
FYR Macedonia	±0.016	±0.022	±0.020	±0.011	±0.006	±0.005	8,073	9,991	10,807	6,250	5,792	5,363
France	±4.534	±3.956	±4.557	±3.990	±4.549	±5.586	93,862	83,822	99,939	92,571	108,318	135,580
Germany (Western Part)	7.733	7.576	7.912	6.684	-	-	73,641	68,210	75,000	-	-	-
Ghana	-	-	-	0.050	0.043	0.045	-	-	-	20,090	16,932	17,280
Greece	*0.347	*0.284	*0.286	*0.270	*0.301	*0.349	60,845	52,010	52,342	49,077	55,388	64,326
Honduras	*0.001	*0.001	*0.001	*0.001	*0.001	*0.001	3,452	3,463	3,163	3,472	3,101	2,831
Hungary	±0.201	±0.113	±0.066	±0.070	±0.059	±0.074	10,049	6,636	5,110	8,781	8,320	8,972
Iceland	0.036	0.018	±0.024	±0.017	-	-	55,283	28,619	39,260	29,100	-	-
India	*0.654	*0.691	*0.626	*0.500	*0.596	*0.638	4,042	3,706	3,545	3,064	3,222	3,356
Indonesia	*0.188	*0.176	*0.333	*0.187	*0.339	*0.388	25,821	17,679	31,795	15,558	26,550	25,687
Iran (Islamic Republic of)	0.436	0.324	0.579	0.294	-	-	36,912	28,661	46,940	26,047	-	-
Ireland	*0.010	*0.005	*0.005	*0.005	*0.005	*0.006	49,751	26,656	28,987	25,453	26,901	29,414
Israel	0.061	0.061	0.070	0.064	0.094	0.093	33,617	32,099	31,978	27,807	33,685	35,823
Italy	±1.788	±1.695	±1.852	±1.516	±1.826	-	50,710	46,692	51,729	44,388	54,070	-
Japan	±12	±13	±13	±14	±15	±17	97,366	102,909	106,430	117,281	133,293	152,417
Jordan	0.009	0.007	0.010	0.012	0.018	0.009	29,582	20,910	36,583	27,757	23,531	15,479
Korea, Republic of	1.201	1.475	1.376	1.450	1.788	2.745	36,949	46,769	45,940	48,964	60,944	83,365

Continued.

Depending on the table, * means *Enterprises, Engaged,* or *Factor Values*; ± means *Basis Unspecified*; § means *shown in millions*. For additional notes and sources, see Appendix I.

Value Added and Value Added per Employee
- Continued -

Country	Value Added ($ bil.)						Value Added per Employee ($)					
	1990	1991	1992	1993	1994	1995	1990	1991	1992	1993	1994	1995
Latvia	-	-	-	*0.000	*0.000	*0.001	-	-	-	4,616	4,430	8,974
Luxembourg	*0.062	*0.059	*0.061	*0.063	*0.072	-	66,533	66,907	70,982	74,462	83,155	-
Malaysia	*0.063	*0.071	*0.119	*0.118	*0.161	*0.195	13,109	11,448	18,524	16,157	22,085	22,983
Mexico	0.706	0.599	0.667	0.649	0.680	0.719	39,116	33,735	43,166	48,780	55,672	62,187
Myanmar	-	-	-	±0.060	±0.081		-	-	-	4,778	48,260	
New Zealand	0.139	-	-	-	-	-	41,723	-				
Norway	±0.826	±0.680	±0.660	±0.590	±0.743	±0.818	79,157	66,377	68,931	66,890	82,488	93,978
Pakistan	0.001	0.001	0.001	-	-	-	2,387	2,188	2,337	-	-	-
Peru	0.930	0.597	0.464	-	-	-	89,099	65,785	53,827	-	-	-
Philippines	0.117	0.032	0.076	0.262			40,283	10,993	23,678	77,066	-	
Poland	±0.951	±0.217	±0.194	±0.166	±0.195	±0.261	31,688	6,778	8,417	7,565	8,001	9,648
Portugal	0.081	0.084	0.094	0.079	0.091	0.110	15,636	15,000	16,722	13,984	16,166	19,413
Romania	±0.022	±0.130	±0.024	±0.031	-		-	-	-	-	-	
Russian Federation				±3.600	±4.308	±5.413	-	-	-	10,882	11,075	24,458
Singapore	*0.041	*0.039	*0.046	*0.043	*0.052	*0.053	49,794	47,096	58,147	61,473	64,418	70,242
Slovakia	-	-	-	±0.049	±0.040	-	-	-	-	6,200	6,255	-
Slovenia	-	-	-	0.026	0.014	0.092	-	-	-	-	-	-
South Africa	*0.641	*0.875	*0.850	*0.792	*0.864	*1.180	30,534	43,765	44,751	49,476	54,020	67,288
Spain	*1.275	*1.044	*1.088	±0.718	±0.832	±1.260	58,932	50,286	55,566	37,521	44,858	67,534
Sri Lanka	0.003	0.002	0.003	0.003	-	-	10,836	6,531	8,093	25,794	-	-
Sweden	*0.640	*0.462	*0.525	*0.372	*0.451	*0.504	66,699	50,794	63,223	49,201	57,711	64,747
Thailand	0.011	0.305	-	-	-	-	3,793	25,164	-	-	-	-
Turkey	0.580	0.422	0.429	0.429	0.353	0.490	28,901	26,964	28,957	31,025	29,002	40,476
United Kingdom	*2.786	*2.457	*2.583	*2.631	*3.012	*3.174	47,215	44,666	51,661	50,589	61,475	65,333
United Republic of Tanzania	*0.003	*0.007	-	-	-	-	2,975	9,795	-	-	-	-
United States of America	*18	*16	*18	*17	*22	*24	70,891	65,847	76,222	75,091	89,971	95,271
Uruguay	0.003	0.003	0.010	0.011	0.008	0.009	12,897	10,538	21,961	25,367	24,288	27,072
Venezuela	0.788	0.420	0.326	0.439	0.668	2.193	52,164	28,166	24,895	34,658	55,257	175,517
Yugoslavia	±0.031	±0.042	±0.104	-	±0.143	±0.096	-	2,050	5,403	-	8,099	5,563
Zambia	0.001	-	-	-	0.000	-	8,052	-	-	-	2,809	-
Zimbabwe	*0.013	*0.014	*0.016	*0.009	*0.009	*0.010	9,104	9,296	11,189	6,929	7,280	8,333

Capital Investment

Country	Gross Fixed Capital Formation ($ mil.)						Per Establishment ($ mil)					
	1990	1991	1992	1993	1994	1995	1990	1991	1992	1993	1994	1995
Argentina	-	-	-	29	-	-	-	-	-	0.063	-	-
Australia	416	-	-	-	-	-	1.823	-	-	-	-	-
Austria	118	148	173	102	87	-	2.068	2.599	3.198	1.992	1.774	-
Bahrain	-	-	19	-	-	-	-	-	4.794	-	-	-
Bangladesh	0.006	-	0.180	-	-	-	0.002	-	0.016	-	-	-
Belgium	128	136	114	81	62	125	-	-	-	-	-	-
Bolivia	0.558	0.529	4.495	1.487	0.487	-	0.093	0.088	0.749	0.248	0.097	-
Bosnia & Herzegovina	12	-	-	-	-	-	1.783	-	-	-	-	-
Bulgaria	11	4.047	182	4.103	9.952	12	1.248	0.450	17	0.108	0.226	0.342
Canada	1,862	-	-	-	-	-	6.672	-	-	-	-	-
Chile	46	60	67	107	119	-	2.003	2.617	3.046	4.282	5.662	-
China (Hong Kong SAR)	1.284	8.236	6.976	16	10	-	0.008	0.084	0.078	0.156	0.113	-
Colombia	7.741	2.704	4.901	2.057	-3.608	-	0.228	0.080	0.136	0.054	-0.088	-
Croatia	9.340	0.074	0.205	1.328	-	-	0.623	0.004	0.006	0.025	-	-
Czechoslovakia (Former)	63	-	-	-	-	-	2.861	-	-	-	-	-
Denmark	5.817	14	-	-	-	-	0.075	0.178	-	-	-	-
Ecuador	-	7.137	5.769	-1.721	1.883	14	-	1.784	1.154	-0.344	0.377	2.258
Egypt	94	96	26	47	-	-	2.419	3.856	0.875	2.139	-	-
El Salvador	-	-	-	-	0.001	0.001	-	-	-	-	0.001	-
Finland	56	78	38	54	41	-	2.538	3.404	1.791	2.847	2.160	-
FYR Macedonia	0.629	0.129	0.073	0.665	0.120	0.397	0.210	0.022	0.006	0.039	0.005	0.017
France	1,685	1,521	1,695	1,854	2,023	1,835	-	-	-	-	-	-

Continued.

Depending on the table, * means *Enterprises, Engaged,* or *Factor Values;* ± means *Basis Unspecified;* § means *shown in millions.* For additional notes and sources, see Appendix I.

Capital Investment

- Continued -

Country	Gross Fixed Capital Formation ($ mil.)						Per Establishment ($ mil)					
	1990	1991	1992	1993	1994	1995	1990	1991	1992	1993	1994	1995
Germany (Western Part) . . .	1,005	1,018	1,196	1,119	-	-	2.171	2.104	2.451	-	-	-
Greece	51	42	67	-	-	-	1.178	0.983	1.488	-	-	-
Hungary	23	8.791	11	7.745	-	-	0.417	0.089	0.145	0.087	-	-
India	198	310	444	411	-	-	0.073	0.104	0.155	0.133	-	-
Indonesia	30	78	38	34	47	26	0.921	1.868	0.835	0.645	0.806	0.400
Iran (Islamic Republic of) . . .	25	198	128	47	-	-	0.145	4.125	2.673	1.040	-	-
Ireland	1.161	0.808	-	-	-	-	0.089	0.067	-	-	-	-
Israel	7.936	13	12	25	19	-	0.169	0.223	0.207	0.454	0.221	-
Italy	598	567	391	282	315	-	1.966	1.883	1.180	0.835	0.899	-
Japan	2,065	2,658	3,237	2,977	1,878	-	0.573	0.761	0.971	0.894	0.620	-
Jordan	-	0.023	-	0.059	3.991	-	-	0.002	-	0.004	0.129	-
Korea, Republic of	338	535	900	1,284	571	859	0.464	0.721	1.185	1.582	0.755	1.035
Latvia	-0.000	-0.001	0.000	-	0.007	-	-0.000	-0.001	0.000	-	0.007	-
Luxembourg	38	20	17	-	-	-	5.408	3.290	2.768	-	-	-
Malaysia	38	81	40	44	103	-	1.347	2.201	1.123	1.107	2.441	-
Mexico	33	39	-	-	-	-	1.318	1.571	-	-	-	-
Myanmar	9.875	9.930	13	6.009	13	-	0.470	0.496	0.747	0.401	1.339	-
Netherlands	-	-	-	55	-	-	-	-	-	-	-	-
Norway	140	137	122	56	80	-	3.413	3.352	3.495	1.870	2.295	-
Pakistan	-	0.220	-	-	-	-	-	0.016	-	-	-	-
Peru	29	16	3.613	-	-	-	0.257	0.130	0.029	-	-	-
Philippines	5.389	22	14	17	-	-	0.061	0.228	0.523	0.561	-	-
Poland	76	92	109	76	-	-	3.302	4.372	4.977	3.802	-	-
Portugal	24	-16.957	35	28	20	-	0.075	-0.055	0.101	0.078	0.054	-
Romania	129	60	-	6.668	26	-	5.613	-	-	-	0.134	-
Singapore	7.253	5.132	5.505	14	4.818	-	0.345	0.244	0.275	0.647	0.229	-
Slovenia	20	7.762	9.386	12	5.520	12	0.701	0.457	0.391	0.178	0.088	0.109
Spain	320	428	269	118	128	109	0.684	0.924	0.577	-	-	-
Sri Lanka	0.075	0.791	0.185	0.012	-	-	0.037	0.264	0.062	0.006	-	-
Thailand	14	8.724	-	-	-	-	0.094	0.071	-	-	-	-
Turkey	78	45	40	38	35	-	0.961	0.540	0.323	0.330	0.326	-
Ukraine	71	11	26	3.940	5.402	4.078	2.833	0.246	0.481	0.152	0.186	0.146
United Kingdom	354	297	322	-	213	-	0.323	0.275	0.295	-	0.156	-
United Republic of Tanzania . .	1.743	1.214	-	-	-	-	0.436	0.405	-	-	-	-
United States of America . . .	1,930	1,760	1,897	1,677	2,260	2,139	-	-	0.476	-	-	-
Uruguay	0.135	0.132	0.333	0.221	-	-	-	-	-	-	-	-
Venezuela	319	154	127	280	5.131	874	3.292	1.430	1.510	4.306	0.063	13
Yugoslavia	2.554	1.315	1.108	-	6.193	2.443	0.071	0.027	0.030	-	0.060	0.020
Zambia	0.005	-	-	-	-	-	0.003	-	-	-	-	-

Indexes of Industrial Production

Country	Index of Industrial Production (1990=100)						Index of Employment (1990=100)					
	1990	1991	1992	1993	1994	1995	1990	1991	1992	1993	1994	1995
Albania	100	-	-	-	-	-	-	-	-	-	-	-
Algeria	100	165	195	215	196	222	-	-	-	-	-	-
Angola	100	100	100	100	-	-	-	-	-	-	-	-
Argentina	100	102	-	-	-	-	100	-	-	132	130	-
Armenia	100	-	-	-	-	-	100	95	-	-	-	-
Australia	100	101	101	107	-	-	100	100	93	-	-	-
Austria	100	96	96	87	97	100	100	94	83	73	70	70
Azerbaijan	-	-	-	-	-	-	100	98	104	95	89	-
Bahrain	100	101	137	-	-	-	-	-	-	-	-	-
Bangladesh	100	-	-	-	-	-	100	68	629	-	-	-
Barbados	100	100	100	100	100	100	-	-	-	-	-	-
Belgium	100	98	89	95	103	106	100	96	86	-	-	-
Belize	100	100	100	100	100	100	-	-	-	-	-	-
Bhutan	100	-	-	-	-	-	-	-	-	-	-	-

Continued.

Indexes of Industrial Production
- Continued -

Country	Index of Industrial Production (1990=100)						Index of Employment (1990=100)					
	1990	1991	1992	1993	1994	1995	1990	1991	1992	1993	1994	1995
Bolivia	100	118	119	130	134	103	100	100	89	110	112	116
Bosnia & Herzegovina	-	-	-	-	-	-	100	101	-	-	-	-
Botswana	100	100	100	-	-	-	-	-	-	-	-	-
Brazil	100	94	94	101	111	109	-	-	-	-	-	-
Bulgaria	100	73	70	82	92	88	100	91	86	86	86	89
Burundi	100	100	100	100	-	-	-	-	-	-	-	-
Cambodia	100	100	100	-	-	-	-	-	-	-	-	-
Cameroon	100	97	-	-	-	-	100	78	102	95	-	-
Canada	100	109	110	120	120	121	100	93	91	89	86	86
Central African Republic	100	100	100	100	100	-	-	-	-	-	-	-
Chile	100	97	104	100	96	101	100	101	93	97	91	92
China	100						100	102	107	107	117	117
China (Hong Kong SAR)	100	98	94	85	84	83	100	94	82	94	76	-
China (Taiwan Province)	100	109	121	134	143	144	100	102	110	118	126	128
Colombia	100	99	108	106	110	115	100	103	114	112	111	112
Congo	100	100	100	100	-	-	-	-	-	-	-	-
Costa Rica	100	-	-	-	-	-	100	99	100	108	-	-
Croatia	100	-	-	-	-	-	100	89	72	47	45	45
Cuba	100	94	-	-	-	-	-	-	-	-	-	-
Cyprus	100	100	100	100	100	100	-	-	-	-	-	-
Czechoslovakia (Former)	100	61	48	-	-	-	100	88	-	-	-	-
Czech Republic	100	-	-	-	-	-	100	86	57	64	-	-
Dem. Rep. of the Congo	100	69	38	-	-	-	-	-	-	-	-	-
Denmark	100	97	96	82	87	89	100	97	94	84	88	89
Dominican Republic	100	97	-	-	-	-	-	-	-	-	-	-
Ecuador	100	-	-	-	-	-	100	119	128	132	140	142
Egypt	100	112	91	90	97	103	100	96	100	92	95	97
El Salvador	100	100	-	-	-	-	-	-	-	-	-	-
Ethiopia and Eritrea	100	100	100	-	-	-	-	-	-	-	-	-
Fiji	100	100	100	100	100	100	-	-	-	-	-	-
Finland	100	87	91	97	106	116	100	98	98	98	98	-
FYR Macedonia	100	-	-	-	-	-	100	110	90	90	54	42
France	100	98	101	94	102	104	100	98	94	89	87	85
Gabon	100	-	-	-	-	-	-	-	-	-	-	-
Gambia	100	-	-	-	-	-	-	-	-	-	-	-
Germany (Eastern Part)	100	-	-	-	-	-	-	-	-	-	-	-
Germany (Western Part)	100	98	91	87	88	94	100	106	100	-	-	-
Ghana	100	-	-	-	-	-	-	-	-	-	-	-
Greece	100	105	106	101	112	114	100	96	96	96	95	95
Guatemala	100	100	-	-	-	-	-	-	-	-	-	-
Guyana	100	100	100	100	100	100	-	-	-	-	-	-
Haiti	100	100	100	100	100	100	-	-	-	-	-	-
Honduras	100	118	113	122	148	275	100	95	122	134	147	159
Hungary	100	67	53	47	39	49	100	85	65	40	35	41
Iceland	100	103	103	-	-	-	100	95	96	91	82	-
India	100	110	104	121	134	137	100	115	109	101	114	118
Indonesia	100	-	-	-	-	-	100	137	144	165	176	208
Iran (Islamic Republic of)	100	113	134	-	-	-	100	96	104	96	-	-
Ireland	100	89	81	85	81	87	100	100	93	95	90	95
Israel	100	111	116	121	139	166	100	106	122	128	156	143
Italy	100	107	108	103	-	-	100	103	102	97	96	-
Japan	100	104	99	98	100	103	100	101	98	98	89	89
Jordan	100	-	-	-	-	-	100	117	95	142	255	192
Korea, Republic of	100	111	116	129	140	154	100	97	92	91	90	101
Kuwait	100	100	100	100	-	-	-	-	-	-	-	-
Kyrgyzstan	100	-	-	-	-	-	100	101	188	177	148	-
Latvia	100	-	-	-	-	-	100	80	61	30	23	24
Libyan Arab Jamahiriya	100	-	-	-	-	-	-	-	-	-	-	-
Lithuania	100	-	-	-	-	-	-	-	-	-	-	-
Luxembourg	100	96	101	106	115	128	100	95	93	91	93	-

Continued.

Depending on the table, * means *Enterprises*, *Engaged*, or *Factor Values*; ± means *Basis Unspecified*; § means *shown in millions*. For additional notes and sources, see Appendix I.

Indexes of Industrial Production

- Continued -

Country	Index of Industrial Production (1990=100)						Index of Employment (1990=100)					
	1990	1991	1992	1993	1994	1995	1990	1991	1992	1993	1994	1995
Macau	100	100	100	-	-	-	-	-	-	-	-	-
Madagascar	100	100	100	100	-	-	-	-	-	-	-	-
Malawi	100	100	100	100	100	100	-	-	-	-	-	-
Malaysia	100	99	115	120	137	155	100	129	133	152	152	177
Mali	100	100	100	100	100	100	-	-	-	-	-	-
Malta	100	100	100	100	-	-	-	-	-	-	-	-
Mexico	100	97	97	104	108	117	100	98	86	74	68	64
Mongolia	100	-	-	-	-	-	-	-	-	-	-	-
Morocco	100	106	105	102	106	126	-	-	-	-	-	-
Mozambique	100	-	-	-	-	-	-	-	-	-	-	-
Myanmar	100	173	131	-	-	-	100	97	90	90	12	-
Namibia	100	97	98	-	-	-	-	-	-	-	-	-
Nepal	100	100	100	-	-	-	-	-	-	-	-	-
New Zealand	100	99	93	-	-	-	100	90	100	109	-	-
Nicaragua	100	100	100	-	-	-	-	-	-	-	-	-
Norway	100	101	99	101	110	106	100	98	92	84	86	83
Pakistan	100	-	-	-	-	-	100	109	114	-	-	-
Panama	100	146	169	196	170	202	-	-	-	-	-	-
Papua New Guinea	100	100	100	-	-	-	-	-	-	-	-	-
Paraguay	100	103	106	127	79	79	-	-	-	-	-	-
Peru	100	119	119	123	136	148	100	87	83	-	-	-
Philippines	100	97	156	-	-	-	100	100	110	117	-	-
Poland	100	76	74	75	87	101	100	107	77	73	81	90
Portugal	100	89	96	90	95	100	100	108	108	109	109	109
Puerto Rico	100	-	-	-	-	-	-	-	-	-	-	-
Republic of Moldova	100	-	-	-	-	-	-	-	-	-	-	-
Romania	100	74	52	55	57	67	-	-	-	-	-	-
Russian Federation	100	-	-	-	-	-	-	-	-	-	-	-
Saudi Arabia	100	-	-	-	-	-	-	-	-	-	-	-
Senegal	100	100	100	100	100	100	-	-	-	-	-	-
Singapore	100	104	116	111	123	112	100	101	97	85	99	92
Slovakia	100	-	-	-	-	-	-	-	-	-	-	-
Slovenia	100	88	81	79	82	89	-	-	-	-	-	-
Somalia	100	-	-	-	-	-	-	-	-	-	-	-
South Africa	100	97	91	89	98	124	100	95	90	76	76	84
Spain	100	101	96	94	103	110	100	96	91	88	86	86
Sri Lanka	100	95	64	72	94	53	100	100	136	43	-	-
Sudan	100	-	-	-	-	-	-	-	-	-	-	-
Suriname	100	98	103	-	-	-	-	-	-	-	-	-
Swaziland	100	100	100	-	-	-	-	-	-	-	-	-
Sweden	100	92	96	106	125	126	100	95	86	79	81	81
Switzerland	100	90	90	83	83	86	-	-	-	-	-	-
Syrian Arab Republic	100	98	109	112	82	74	-	-	-	-	-	-
Thailand	100	71	69	-	-	-	100	431	-	-	-	-
Tonga	100	-	-	-	-	-	-	-	-	-	-	-
Trinidad and Tobago	100	100	100	100	100	100	-	-	-	-	-	-
Tunisia	100	103	99	97	97	96	-	-	-	-	-	-
Turkey	100	82	81	91	77	93	100	78	74	69	61	60
Uganda	100	139	177	241	362	455	-	-	-	-	-	-
Ukraine	100	-	-	-	-	-	-	-	-	-	-	-
United Arab Emirates	100	-	-	-	-	-	-	-	-	-	-	-
United Kingdom	100	91	88	85	85	84	100	93	85	88	83	82
United Republic of Tanzania	100	102	112	129	106	-	100	82	-	-	-	-
USSR (Former)	100	-	-	-	-	-	100	-	-	-	-	-
United States of America	100	97	99	102	111	113	100	96	93	94	98	103
Uruguay	100	83	95	80	-	-	100	104	181	169	138	131
Venezuela	100	101	86	-	-	-	100	99	87	84	80	83
Yugoslavia	100	-	-	-	-	-	-	-	-	-	-	-

Continued.

Indexes of Industrial Production

- Continued -

Country	Index of Industrial Production (1990=100)						Index of Employment (1990=100)					
	1990	1991	1992	1993	1994	1995	1990	1991	1992	1993	1994	1995
Yugoslavia (Former)	100	80	-	-	-	-	-	-	-	-	-	-
Zambia	100	88	103	94	83	71	100	-	-	-	72	-
Zimbabwe	100	102	90	74	83	79	100	107	100	93	86	82

Representative Companies in Sector

Name	Address	Telephone	Fax	Employment	Y
Philip Services Corp.	100 King St. W, Hamilton, Ontario L8N 4J6, Canada	905-521-1600	905-521-5522	12,000	98
Alcan Aluminium Ltd.	1188 Sherbrooke St. W, Montreal, Quebec H3A 3G1, Canada	514-848-8000	514-848-8115	11,000	98
Alcan Smelters & Chemicals Ltd.	1188 Sherbrooke St. W, Montreal, Quebec H3A 3G2, Canada	514-848-8000	514-848-8115	10,000	98
Olin Corp.	PO Box 4500, Norwalk, Connecticut 06856-4500, U.S.A.	203-750-3000	203-750-3595	10,000	97
Kaiser Aluminum Corp.	PO Box 572887, Houston, Texas 77257, U.S.A.	713-267-3777		9,567	97
Raychem Corp.	300 Constitution Dr., Menlo Park, California 94025-1164, U.S.A.	415-361-3333	415-361-4913	8,650	97
Inco Ltd.	145 King St. W Ste. 1500, Toronto, Ontario M5H 4B7, Canada	416-361-7511	416-361-7781	8,600	98
Tambang Timah PT	Jl. Jend Sudirman Pangkal Pinang No. 51, Bangka 12730, Indonesia	71722355	717 22323	8,300	97
British Alcan Aluminium PLC	Chalfont Park, Gerrards Cross SL9 0QB, United Kingdom	753 887373	753 889667	7,718	93
Commercial Metals Co.	7800 Stemmons Frwy., Dallas, Texas 75247-4227, U.S.A.	214-689-4300	214-689-5586	7,150	97
CBA Cia. Brasileira de Aluminio	Praca Ramos de Azevedo, 3rd Fl., Sao Paulo 01037-912, Brazil	11 2225144	11 2224162	7,000	97
Hindustan Cables Ltd.	9 Elgin Rd., Calcutta 700020, India	332471808	332471657	6,790	97
Intermet Corp.	5445 Corporate Dr., 200, Troy, Michigan 48098-2683, U.S.A.	248-952-2500	248-952-2501	6,500	97
NKT Holding AS	NKT Alle 1, Brondby DK-2605, Denmark	43482000	43961820	6,425	93
Cookson America Inc.	1 Cookson Place, Providence, Rhode Island 02903, U.S.A.	401-521-1000	401-521-7258	6,000	97
Elkem AS	PO Box 4282 Torshov, Oslo N-0401, Norway	22 450100	22 450155	6,000	93
ZSNP Aluminium Works AS	Ziar nad Hronom 965 63, Slovakia	857 2201	857 2240	5,880	93
Zelezorudne Bane SP	Stara Cesta 3, Spisska Nova Ves 052 54, Slovakia	965 23009	965 21370	5,800	93
Cominco Ltd.	200 Burrard St. Ste. 500, Vancouver, British Columbia V6C 3L7, Canada	604-682-0611	604-685-3091	5,743	98
ACX Technologies Inc.	16000 Table Mountain Pkwy., Golden, Colorado 80403, U.S.A.	303-271-7000	303-271-7003	5,600	97
Noranda Aluminum Inc.	1000 Corporate Centre Dr., 300, Franklin, Tennessee 37067, U.S.A.	615-771-5700	615-377-4301	5,600	97
Heywood Williams Group PLC	Edgerton Rd., Huddersfield HD3 3AR, United Kingdom	484 435477	484 547511	5,588	93
Amphenol Corp.	PO Box 5030, Wallingford, Connecticut 06492-7530, U.S.A.	203-265-8900	203-265-8516	5,459	95
West Company Inc.	PO Box 645, Exton, Pennsylvania 19341-0645, U.S.A.	610-594-2900	610-648-2424	5,210	97
Essex Group Inc.	PO Box 1601, Fort Wayne, Indiana 46801-1601, U.S.A.	219-461-4000	219-461-4150	5,100	97
Southern Peru Copper Corp.	180 Maiden Ln., New York, New York 10038, U.S.A.	212-510-2000	212-510-1887	5,035	95
Siecor Corp.	489 Siecor Park, Hickory, North Carolina 28601, U.S.A.	704-327-5000	704-327-5993	5,000	97
Oneida Ltd.	5-9 Pembroke Rd., Ruislip HA4 8NQ, United Kingdom	895 639452		4,982	91
Newman Tonks Group PLC	Birmingham Business Park, Birmingham B37 7YX, United Kingdom	21 7177777	21 7177776	4,749	93
Agrium Inc.	426-10333 Southport Rd. SW, Calgary, Alberta T2W 3X6, Canada	403-258-4600	403-258-4692	4,500	98
Belden Inc.	7701 Forsyth Blvd., 800, St. Louis, Missouri 63105, U.S.A.	314-854-8000	314-983-5294	4,200	96
General Cable Industries Inc.	4 Tesseneer Dr., Highland Heights, Kentucky 41076, U.S.A.	606-572-8000	606-572-8440	4,000	97
Taihan Electric Wire Co., Ltd.	Kuro-Gu, Seoul, Republic of Korea	2 3169114	2 7572942	3,936	97
General Cable Corp.	4 Tesseneer Dr., Highland Heights, Kentucky 41076, U.S.A.	606-572-8000	606-572-8458	3,900	96
Quanex Corp.	1900 W. Loop S., 1500, Houston, Texas 77027, U.S.A.	713-961-4600	713-877-5333	3,771	97
Wyman-Gordon Co.	PO Box 8001, North Grafton, Massachusetts 01536-8001, U.S.A.	508-839-4441	508-839-7500	3,650	97
Mostostal, Zabrze	Wolnosci 191, Zabrze PL-41-800, Poland	32 713221	32 713151	3,600	93
Alcan Aluminum Corp.	6060 Parkland Blvd., Cleveland, Ohio 44124-4185, U.S.A.	440-423-6600	440-423-6665	3,500	97
BICC Cables Corp.	1 Crosfield Ave., West Nyack, New York 10994, U.S.A.	914-353-4000	914-353-4032	3,500	95
Wolverine Tube Inc.	1525 Perimeter Pkwy., Ste. 210, Huntsville, Alabama 35806, U.S.A.	205-353-1310	205-351-2312	3,377	96
Andrew Corp.	10500 West 153rd St., Orland Park, Illinois 60462, U.S.A.	708-349-3300	708-873-8954	3,213	97
Hulett Aluminum Profiles Pty. Ltd.	PO Box 74, Pietermaritzburg 3200, Republic of South Africa	331 956911	331 946335	3,077	93
Alcan Aluminio do Brasil SA	Av. Paulista 1106, 14/15th Fl., Sao Paulo 01310-000, Brazil	11 2520722	11 2520123	3,000	97
Cable Design Technologies Corp.	661 Andersen Dr., Foster Plz. 7, Pittsburgh, Pennsylvania 15220, U.S.A.	412-937-2300	412-937-9690	2,900	97
Varlen Corp.	PO Box 3089, Naperville, Illinois 60566-7089, U.S.A.	630-420-0400	630-420-7123	2,850	97
Indonesia Asahan Aluminium PT	Jl. Jend Sudirman Kav 61-62, Jakarta 12190, Indonesia	212520185	21 5201278	2,700	97
Szopienice	ul. Lwowska 23, Katowice PL-40-390, Poland	32 591240	32 593038	2,700	93
Inco Europe Ltd.	50 Victoria St., London SW1H 0XB, United Kingdom	71 931773		2,621	94
Divpac	Mobeni Box 12650, Durban 4000, Republic of South Africa	31 4690641	31 420705	2,600	93

Continued

Representative Companies in Sector
- Continued -

Name	Address	Tele-phone	Fax	Employ-ment	Y
McKechnie UK Ltd.	Aldridge, Walsall WS9 8DS, United Kingdom			2,567	93
Handy and Harman	250 Park Ave., New York, New York 10177, U.S.A.	212-661-2400		2,562	97
Carol Cable Company Inc.	249 Roosevelt Ave., Pawtucket, Rhode Island 02860, U.S.A.	401-728-7000	401-723-2670	2,500	
Johnson Matthey Inc.	456 Devon Park Dr., Wayne, Pennsylvania 19087, U.S.A.	610-341-8300	610-341-8259	2,500	96
Interlake Corp.	550 Warrenville Rd., Lisle, Illinois 60532-4387, U.S.A.	630-852-8800	630-719-7277	2,491	97
Mueller Industries Inc.	2959 N. Rock Rd., Wichita, Kansas 67226-1191, U.S.A.	316-636-6300	316-682-9650	2,350	97
CommScope Inc.	PO Box 1729, Hickory, North Carolina 28602, U.S.A.	704-324-2200		2,300	97
Taiwan Semiconductor Manufacturing Co., Ltd.	Science Based Industrial Park, Hsinchu, Taiwan	35 780221	35 782875	2,300	93
KCM SA	Assenovgradsko Shosse, Plovdiv BG-4009, Bulgaria	32 23496	32 269044	2,287	93
Titanium Metals Corp.	1999 Broadway, 4300, Denver, Colorado 80202, U.S.A.	303-296-5600	303-296-5640	2,250	96
Atomic Energy Corp. of South Africa Ltd.	PO Box 582, Pretoria 0001, Republic of South Africa	12 3163270	12 3163222	2,200	97
Brush Wellman Inc.	17876 St. Clair Ave., Cleveland, Ohio 44110, U.S.A.	216-486-4200	216-383-4091	2,160	97
Speciality Packaging (UK) PLC	Rock Valley Rd., Mansfield NG18 2EZ, United Kingdom	623 22651	623 653481	2,134	93
Kety	Ul. Kosciuszki 111, Kety PL-32-650, Poland	381 52251	381 53094	2,115	93
Albras Aluminio Brasileiro SA	Rua do Mercado 11, 21/21st Fl., Rio de Janeiro 20010-120, Brazil	21 2920052	21 2929259	2,100	96
ACS Industries Inc. (Woonsocket, Rhode Island)	160 Hamlet Ave., Woonsocket, Rhode Island 02895, U.S.A.	401-769-4700	401-762-9135	2,000	97
Century Aluminum Co.	2511 Garden Rd., Monterey, California 93940, U.S.A.	408-642-9300	408-642-9399	2,000	97
Engelhard Corp. Catalysts and Chemicals	101 Wood Ave., Iselin, New Jersey 08830, U.S.A.	732-205-6000	732-205-5915	2,000	97
Commonwealth Aluminum Corp.	1200 Meidinger Twr., Louisville, Kentucky 40202, U.S.A.	502-589-8100		1,997	96
Hopkinsons Group PLC	Trafford Park, Manchester M17 1HP, United Kingdom	61 8728291		1,989	94
Easco Inc.	706 S. State St., Girard, Ohio 44420, U.S.A.	330-545-4311		1,983	96
Aberdare Cables Pty. Ltd.	PO Box 1679, Edenvale 1610, Republic of South Africa	11 6094020	11 6092432	1,980	94
International Aluminum Corp.	767 Monterey Pass Rd., Monterey Park, California 91754, U.S.A.	213-264-1670		1,900	95
Myers Industries Inc.	1293 S. Main St., Akron, Ohio 44301, U.S.A.	330-253-5592	330-253-6568	1,882	97
Norsk Hydro UK Ltd.	69 London Rd., Twickenham TW1 1EE, United Kingdom	81 8911366	81 8921686	1,862	93
Noranda Mining and Exploration Inc.	PO Box 3000 Sta. Main, Bathurst, New Brunswick E2A 3Z8, Canada	506-546-6671	506-547-6076	1,800	98
Curtiss-Wright Corp.	1200 Wall St. W., Lyndhurst, New Jersey 07071-0635, U.S.A.	201-896-8400	201-438-5680	1,700	96
Fabryka Kabli, Ozarow	ul. Poznanska 55, Ozarow PL-05-850, Poland	2 6286431	2 6286433	1,700	93
Superior Telecom Inc.	1790 Broadway, New York, New York 10019, U.S.A.	212-757-3333		1,678	96
Aluminium de Grece SAIC	1 Sekeri Str., Athens GR-10671, Greece	1 3693602	3693615	1,630	97
Ertl Company Inc.	PO Box 500, Dyersville, Iowa 52040-0500, U.S.A.	319-875-2000	319-875-8263	1,600	97
Polmo Praszka SA	ul. Kaliszka 72, Praszka PL-46-320, Poland	83 4444		1,501	93
Allvac	PO Box 5030, Monroe, North Carolina 28111-5030, U.S.A.	704-289-4511	704-289-4018	1,500	97
Walsin Lihwa Electric Corp.	Min Sheng E. Rd., Sec. 3, 12th Fl., 1127, Taipei 106, Taiwan	2 7192211	2 7193382	1,499	91
Norandal USA Inc.	PO Box 2087, Brentwood, Tennessee 37024-2087, U.S.A.	615-371-1250	615-371-1251	1,400	97
Shaw Industries Ltd.	25 Bethridge Rd., Toronto, Ontario M9W 1M7, Canada	416-743-7111	416-743-7199	1,400	98
V.A.W. of America Inc.	PO Box 667, Ellenville, New York 12428, U.S.A.	914-647-7510	914-647-7535	1,400	96
Meridian Technologies Inc.	1700-2 St. Clair Ave. W, Toronto, Ontario M4V 1L5, Canada	416-922-2050	416-922-4282	1,345	98
Csepel Muvek Femmu	PO Box 49, Budapest H-1751, Hungary	1 2773152	1 2766260	1,300	93
Namsun Aluminium Co. Ltd.	Tong-Gu, Taegu, Republic of Korea	53 9806313	53 9832001	1,300	97
Aavid Thermal Technologies Inc.	1 Eagle Sq., Ste. 509, Concord, New Hampshire 03301, U.S.A.	603-224-1117		1,279	96
Woodhead Industries Inc.	2150 E. Lake Cook Rd., 400, Buffalo Grove, Illinois 60089, U.S.A.	847-465-8300	847-272-8126	1,259	97
Sociedad Espanola de Carburos Metalicos, SA	Aragon, 300, Barcelona E-08009, Spain	93 2902600	93 2902611	1,250	97
Cerro Metal Products Co.	PO Box 388, Bellefonte, Pennsylvania 16823, U.S.A.	814-355-6217	814-355-6227	1,200	97
Intalco Aluminum Corp.	PO Box 937, Ferndale, Washington 98248, U.S.A.	360-384-7061	360-384-6185	1,200	97
Lexington Precision Corp.	767 3rd Ave., New York, New York 10017, U.S.A.	212-319-4657	212-949-1588	1,200	96
Voksel Electric PT	Jakarta Barat, Jakarta 11120, Indonesia	216592196	21 6297866	1,134	97
INCO Engineered Products Ltd.	Melbourne, Derby DE7 1FE, United Kingdom	332 864900		1,112	94
Deloro Stellite Inc.	101 S. Hanley Rd., 300, St. Louis, Missouri 63105, U.S.A.	314-862-2666		1,100	95
Wah Chang	PO Box 460, Albany, Oregon 97321, U.S.A.	541-926-4211	541-967-6994	1,100	95

Continued

Representative Companies in Sector
- Continued -

Name	Address	Tele-phone	Fax	Employ-ment	Y
Lindberg Corp.	6133 N. River Rd., 700, Rosemont, Illinois 60018, U.S.A.	708-823-2021	708-823-0795	1,071	96
Aluminio Espanol, SA	Jose Abascal, 2-4, 30 Y 40, Madrid E-28003, Spain	91 4484100	91 4487657	1,033	97
Gunung Gahapi Sakti PT	Jalan Medan Belawan Km 10, Medan 20112, Indonesia	61651380	61 513091	1,020	97
Fansteel Inc.	1 Tantalum Pl., North Chicago, Illinois 60064, U.S.A.	847-689-4900	847-539-8140	1,000	97
Okonite Company Inc.	PO Box 340, Ramsey, New Jersey 07446, U.S.A.	201-825-0300	201-825-9026	1,000	94
Ormet Primary Aluminum Corp.	PO Box 164, Hannibal, Ohio 43931, U.S.A.	614-483-1341	614-483-2745	1,000	96
Padaeng Industry Co., Ltd.	Klongtoey, Bangkok 10110, Thailand	2611110	2611109	1,000	94
Stern Metals Inc.	PO Box 2018, Attleboro, Massachusetts 02703, U.S.A.	508-222-7400	508-222-2162	1,000	94
Tandy Wire and Cable	3500 McCart Ave., Fort Worth, Texas 76110, U.S.A.	817-924-5789	817-923-3345	1,000	97
Triangle Wire and Cable Inc.	10 Lincoln Center Blvd., Lincoln, Rhode Island 02865, U.S.A.	401-729-5400	401-729-5450	1,000	95
Metal Improvement Company Inc.	10 Forest Ave., Paramus, New Jersey 07652-5214, U.S.A.	201-843-7800	201-843-3460	980	95
RMI Titanium Co.	PO Box 269, Niles, Ohio 44446-0269, U.S.A.	330-544-7700	330-544-7796	953	97
Aluar-Aluminio Argentino SAIC	Luis Pasteur 4600, Victoria 1644, Argentina	1 7258000	1 7258091	952	96
ATC International	PO Box 663, Brits 0250, Republic of South Africa	1211 502130	1211 502072	950	95
Stahl Specialty Co.	PO Box 6, Kingsville, Missouri 64061, U.S.A.	816-597-3322	816-597-3485	950	97
Tredegar Aluminum	PO Box 428, Newnan, Georgia 30264, U.S.A.	404-253-2020	404-254-7716	950	94
C.R. Smith Glaziers Dunfermline Ltd.	Gardeners St., Dunfermline KY12 0RM, United Kingdom	383 732181	383 739095	948	92
Madeco SA	San Miguel, Santiago, Chile	2 5201616	2 5201100	900	96
Reycan L.P.	290 Rue Saint-Laurent, Cap-De-La-Madeleine, Quebec G8T 6G7, Canada	819-373-6363	819-373-6425	900	98
Glynwed Tubes & Fittings Ltd.	Bessemer Rd., Welwyn Garden City AL7 1HH, United Kingdom	707 326333		882	93
Tremont Corp.	1999 Broadway, Ste. 4300, Denver, Colorado 80202, U.S.A.	303-296-5652	303-296-5650	880	95
CasTech Aluminum Group Inc.	2630 El Presidio St., Long Beach, California 90810, U.S.A.	310-886-8300		874	94
Homeshield Fabricated Div.	525 Dunham Rd., 60, St. Charles, Illinois 60174-1490, U.S.A.	708-513-4100		874	94
Consolidated Metco Inc.	PO Box 83201, Portland, Oregon 97283-0201, U.S.A.	503-286-5741	503-240-5443	850	97
Ciena Corp.	920 Elkridge Landings Rd., Linthicum Heights, Maryland 21090, U.S.A.	410-865-8500	410-865-8600	841	97
Lawson Mardon Star Ltd.	50 Portland Pl., London W1N 3DG, United Kingdom			841	93
Unihold Group Ltd.	PO Box 59404, Kengray 2100, Republic of South Africa	11 6157017	11 6158149	830	95
MNP Corp.	PO Box 189002, Utica, Michigan 48318-9002, U.S.A.	810-254-1320	810-726-5663	825	97
AB Rexroth Mecman	Varuvagen 7, Stockholm S-125 81, Sweden	8 7279200	8 998476	800	93
Hartzell Manufacturing Inc.	PO Box 64529, St. Paul, Minnesota 55164-0529, U.S.A.	612-646-9456	612-643-2300	800	97
LaBarge Inc.	PO. Box 14499, St. Louis, Missouri 63178-4499, U.S.A.	314-997-0800		800	97
Metalplast	Ul. Lukowska 7/9, Oborniki Wielkopolskie PL-64-600, Poland	668 68510	61 527796	800	93
ITT Automotive Precision Die Castings	3000 University Dr., Auburn Hills, Michigan 48326, U.S.A.	810-340-3000	810-340-3719	750	94
Pimalco Corp.	6833 W. Willis Rd., Chandler, Arizona 85226, U.S.A.	520-796-1098	520-796-0369	750	97
AFC Cable Systems Inc.	50 Kennedy Plaza, Ste. 1250, Providence, Rhode Island 02903-2360, U.S.A.	401-453-2000	401-453-2009	749	96
Ferriere Nord, SpA	Fr. Rivoli, Osoppo I-33010, Italy	432 981826	432 981812	735	97
Kelsey Industries PLC	Wood Lane End, Hemel Hempstead HP2 4RQ, United Kingdom	442 233233	442 69554	732	94
Alucon Manufacturing Co., Ltd.	Samrong Nua Muang, Samut Prakan 10270, Thailand	3980147		727	94
Alcoa Manufacturing Great Britain Ltd.	PO Box 68, Swansea SA1 1XH, United Kingdom	792 873301	792 879723	708	93
Alcatel Canada Wire	140 Allstate Parkway, Markham, Ontario L3R 0Z7, Canada	905-944-4300	905-944-4333	700	98
Amcan Casting Ltd.	10 Hillyard St., Hamilton, Ontario L8L 7X3, Canada	905-527-9178	905-527-6821	700	98
Anodizing Inc.	7933 N.E. 21st Ave., Portland, Oregon 97211, U.S.A.	503-285-0404	503-286-0356	700	94
Budidharma Jakarta, PT	Jln Jend Gatot Subroto Kav 22, Jakarta 12930, Indonesia	21514756	21 4401147	700	97
Chase Brass Industries Inc.	State Rte. 15, Montpelier, Ohio 43543, U.S.A.	419-485-3193	419-485-8150	700	97
Derlan Manufacturing Inc.	145 King St. E Ste. 500, Toronto, Ontario M5C 2Y7, Canada	416-364-5852	416-362-5334	700	98
Industrias Metalurgicas Unidas SA	Calle 50 53-107 Copacabana, Medellin, Colombia	4 2740222	4 2745704	700	97
Times Fiber Communications Inc.	PO Box 384, Wallingford, Connecticut 06492, U.S.A.	203-265-8500	203-265-8422	700	95
Wabash Alloys	PO Box 466, Wabash, Indiana 46992, U.S.A.	219-563-7461	219-563-5997	700	94
Blue Circle Cement Pty. Ltd.	PO Box 1320, Parklands 2121, Republic of South Africa	11 7882160	11 4471725	694	95
USKO Ltd.	PO Box 48, Vereeniging 1930, Republic of South Africa	16 4508200	16 233406	689	97
CCX Inc.	1901 Roxborough Rd., 205, Charlotte, North Carolina 28211, U.S.A.	704-365-0560	704-365-8195	675	97
Bangkok Cable Co., Ltd.	Pathumwan, Bangkok 10330, Thailand	2 2544550	2 2536028	660	94

Product Tables
Copper, Blister and Other Unrefined

Unit of Measure: Metric tons.

	1990		1991		1992		1993		1994		1995		1996	
	Value	%	Value	%	Value	%	Value	%	Value	%	Value	%	Value	%
Total Production	8,977,042	100.0	9,432,355	100.0	9,698,329	100.0	9,842,529	100.0	9,961,798	100.0	10,117,082	100.0	10,957,049	100.0
Regions														
Africa	944,790	10.5	760,000	8.1	729,800	7.5	594,600	6.0	501,481	5.0	466,999	4.6	481,500	4.4
America, North	1,859,100	20.7	1,827,800	19.4	2,010,476	20.7	2,146,322	21.8	2,175,800	21.8	2,208,080	21.8	2,240,890	20.5
America, South	1,493,433	16.6	1,534,500	16.3	1,555,700	16.0	1,609,400	16.4	1,613,680	16.2	1,821,260	18.0	2,104,740	19.2
Asia	3,293,524	36.7	3,610,919	38.3	3,709,868	38.3	3,759,486	38.2	3,764,854	37.8	3,855,331	38.1	4,051,576	37.0
Europe	1,186,262	13.2	1,510,616	16.0	1,493,525	15.4	1,511,681	15.4	1,623,983	16.3	1,623,453	16.0	1,907,343	17.4
Oceania	199,933	2.2	188,520	2.0	198,960	2.1	221,040	2.2	282,000	2.8	141,960	1.4	171,000	1.6
Leading Producers														
Japan	1,351,401	15.1	1,417,774	15.0	1,492,350	15.4	1,526,679	15.5	1,434,552	14.4	1,534,318	15.2	1,568,045	14.3
United States of America	1,158,500	12.9	1,120,000	11.9	1,180,000	12.2	1,270,000	12.9	1,310,000	13.2	1,250,000	12.4	1,300,000	11.9
Chile	1,110,033	12.4	1,076,400	11.4	1,095,000	11.3	1,173,600	11.9	1,148,280	11.5	1,322,760	13.1	1,589,640	14.5
Canada	525,200	5.9	532,800	5.6	552,436	5.7	576,322	5.9	560,040	5.6	618,240	6.1	612,690	5.6
Zambia	355,400	4.0	300,500	3.2	380,200	3.9	337,800	3.4	265,200	2.7	237,700	2.3	262,000	2.4

Copper, Refined, Unwrought (Total Production)

Unit of Measure: Metric tons.

	1990		1991		1992		1993		1994		1995		1996	
	Value	%	Value	%	Value	%	Value	%	Value	%	Value	%	Value	%
Total Production	10,782,266	100.0	11,512,168	100.0	12,216,234	100.0	12,337,165	100.0	12,072,613	100.0	12,603,397	100.0	13,388,591	100.0
Regions														
Africa	721,200	6.7	651,400	5.7	627,100	5.1	603,200	4.9	543,500	4.5	496,700	3.9	482,300	3.6
America, North	2,690,300	25.0	2,728,100	23.7	2,870,300	23.5	3,009,399	24.4	2,957,028	24.5	3,060,157	24.3	3,145,509	23.5
America, South	1,374,341	12.7	1,398,343	12.1	1,651,350	13.5	1,516,002	12.3	1,503,033	12.4	1,735,757	13.8	2,032,000	15.2
Asia	3,395,109	31.5	3,640,009	31.6	3,892,677	31.9	3,918,009	31.8	3,872,150	32.1	4,313,247	34.2	4,460,906	33.3
Europe	2,356,463	21.9	2,824,116	24.5	2,839,207	23.2	2,954,555	23.9	2,841,702	23.5	2,723,936	21.6	2,949,875	22.0
Oceania	244,853	2.3	270,200	2.3	335,600	2.7	336,000	2.7	355,200	2.9	273,600	2.2	318,000	2.4
Leading Producers														
United States of America	2,017,400	18.7	2,000,000	17.4	2,140,000	17.5	2,250,000	18.2	2,230,000	18.5	2,280,000	18.1	2,340,000	17.5
Japan	1,007,976	9.3	1,076,283	9.3	1,160,859	9.5	1,188,776	9.6	1,119,168	9.3	1,187,959	9.4	1,251,373	9.3
Chile	990,841	9.2	1,012,800	8.8	1,242,300	10.2	1,093,200	8.9	1,080,000	8.9	1,288,800	10.2	1,518,000	11.3
China	561,500	5.2	560,000	4.9	659,000	5.4	733,000	5.9	736,100	6.1	1,079,704	8.6	1,120,500	8.4
Germany	-		521,500	4.5	581,469	4.8	632,079	5.1	591,859	4.9	483,887	3.8	535,748	4.0

Copper, Primary, Refined

Unit of Measure: Metric tons.

	1990		1991		1992		1993		1994		1995		1996	
	Value	%	Value	%	Value	%	Value	%	Value	%	Value	%	Value	%
Total Production	9,166,315	100.0	9,636,371	100.0	10,188,183	100.0	10,189,746	100.0	10,190,377	100.0	10,414,947	100.0	11,060,603	100.0
Regions														
Africa	712,800	7.8	643,200	6.7	618,870	6.1	594,990	5.8	537,450	5.3	490,700	4.7	476,300	4.3
America, North	2,170,897	23.7	2,228,343	23.1	2,322,392	22.8	2,416,858	23.7	2,440,100	23.9	2,514,816	24.1	2,593,100	23.4
America, South	1,324,741	14.5	1,398,315	14.5	1,651,415	16.2	1,516,017	14.9	1,503,049	14.7	1,735,773	16.7	2,032,016	18.4
Asia	3,081,787	33.6	3,291,889	34.2	3,383,879	33.2	3,391,538	33.3	3,370,750	33.1	3,458,483	33.2	3,541,344	32.0
Europe	1,658,190	18.1	1,839,424	19.1	1,908,027	18.7	1,958,343	19.2	2,007,828	19.7	1,965,575	18.9	2,123,842	19.2
Oceania	217,900	2.4	235,200	2.4	303,600	3.0	312,000	3.1	331,200	3.3	249,600	2.4	294,000	2.7
Leading Producers														
United States of America	1,576,600	17.2	1,580,000	16.4	1,710,000	16.8	1,790,000	17.6	1,840,000	18.1	1,930,000	18.5	2,010,000	18.2
Japan	947,687	10.3	1,005,410	10.4	1,087,019	10.7	1,126,321	11.1	1,064,727	10.4	1,112,777	10.7	1,181,070	10.7
Chile	990,841	10.8	1,012,800	10.5	1,242,300	12.2	1,093,200	10.7	1,080,000	10.6	1,288,800	12.4	1,518,000	13.7
Canada	468,401	5.1	511,243	5.3	501,892	4.9	521,158	5.1	495,500	4.9	481,716	4.6	475,900	4.3
Belgium	*456,492	5.0	*456,492	4.7	*456,492	4.5	*456,492	4.5	*456,492	4.5	*456,492	4.4	*456,492	4.1

Product Tables

Copper, Secondary, Refined

Unit of Measure: Metric tons.

	1990 Value	%	1991 Value	%	1992 Value	%	1993 Value	%	1994 Value	%	1995 Value	%	1996 Value	%
Total Production	1,248,006	100.0	1,596,852	100.0	1,645,475	100.0	1,683,843	100.0	1,524,732	100.0	1,659,441	100.0	1,716,040	100.0
Regions														
Africa	8,400	0.7	8,200	0.5	8,200	0.5	8,200	0.5	6,000	0.4	6,000	0.4	6,000	0.3
America, North	519,363	41.6	497,802	31.2	550,860	33.5	592,532	35.2	518,900	34.0	547,300	33.0	555,400	32.4
America, South	49,600	4.0	50,407	3.2	50,971	3.1	51,534	3.1	52,097	3.4	52,661	3.2	53,224	3.1
Asia	66,889	5.4	124,587	7.8	130,465	7.9	84,045	5.0	76,182	5.0	96,149	5.8	127,863	7.5
Europe	576,755	46.2	880,856	55.2	872,979	53.1	923,532	54.8	847,553	55.6	933,332	56.2	949,553	55.3
Oceania	27,000	2.2	35,000	2.2	32,000	1.9	24,000	1.4	24,000	1.6	24,000	1.4	24,000	1.4
Leading Producers														
United States of America	440,800	35.3	418,000	26.2	433,000	26.3	460,000	27.3	392,000	25.7	352,000	21.2	333,000	19.4
Germany	-		318,300	19.9	339,316	20.6	361,670	21.5	338,979	22.2	369,100	22.2	352,400	20.5
Japan	60,289	4.8	70,873	4.4	73,840	4.5	62,455	3.7	54,441	3.6	75,182	4.5	70,303	4.1
Italy	83,000	6.7	83,400	5.2	76,000	4.6	90,300	5.4	84,000	5.5	98,000	5.9	85,800	5.0
Mexico	31,164	2.5	53,045	3.3	80,452	4.9	92,100	5.5	94,900	6.2	104,400	6.3	139,100	8.1

Copper-Base Alloys

Unit of Measure: Metric tons.

	1990 Value	%	1991 Value	%	1992 Value	%	1993 Value	%	1994 Value	%	1995 Value	%	1996 Value	%
Total Production	2,355,158	100.0	2,479,174	100.0	2,480,650	100.0	2,445,387	100.0	2,159,460	100.0	2,902,612	100.0	2,693,065	100.0
Regions														
America, North	703,825	29.9	704,318	28.4	774,978	31.2	767,940	31.4	878,457	40.7	828,100	28.5	858,980	31.9
America, South	633	0.0	599	0.0	566	0.0	215	0.0	259	0.0	295	0.0	263	0.0
Asia	1,355,543	57.6	1,414,725	57.1	1,387,304	55.9	1,352,028	55.3	902,714	41.8	1,695,107	58.4	1,479,677	54.9
Europe	295,157	12.5	359,531	14.5	317,802	12.8	325,204	13.3	378,030	17.5	379,109	13.1	354,146	13.2
Leading Producers														
China	*1,233,680	52.4	*1,233,680	49.8	*1,233,680	49.7	*1,233,680	50.4	778,196	36.0	1,571,881	54.2	1,350,964	50.2
United States of America	655,000	27.8	655,000	26.4	724,000	29.2	740,300	30.3	846,400	39.2	796,400	27.4	815,400	30.3
United Kingdom	99,600	4.2	113,850	4.6	128,100	5.2	130,383	5.3	158,401	7.3	146,700	5.1	126,700	4.7
Japan	89,152	3.8	91,955	3.7	82,620	3.3	82,034	3.4	*90,441	4.2	*92,225	3.2	*94,009	3.5
Italy	80,300	3.4	75,700	3.1	76,400	3.1	81,300	3.3	89,100	4.1	106,500	3.7	106,100	3.9

Copper Bars, Rods, Angles, Shapes, Sections

Unit of Measure: Metric tons.

	1990 Value	%	1991 Value	%	1992 Value	%	1993 Value	%	1994 Value	%	1995 Value	%	1996 Value	%
Total Production	920,857	100.0	1,119,205	100.0	1,160,823	100.0	1,234,743	100.0	1,332,678	100.0	1,413,380	100.0	1,285,265	100.0
Regions														
Africa	5,294	0.6	4,256	0.4	4,390	0.4	3,753	0.3	3,502	0.3	3,421	0.2	3,449	0.3
America, North	60,598	6.6	57,094	5.1	59,531	5.1	73,884	6.0	100,538	7.5	105,091	7.4	98,145	7.6
America, South	32,942	3.6	37,234	3.3	41,527	3.6	45,979	3.7	50,256	3.8	54,635	3.9	53,994	4.2
Asia	208,729	22.7	248,327	22.2	273,716	23.6	292,619	23.7	289,718	21.7	308,982	21.9	299,044	23.3
Europe	553,047	60.1	752,826	67.3	760,785	65.5	759,441	61.5	876,911	65.8	897,402	63.5	783,612	61.0
Oceania	60,247	6.5	19,467	1.7	20,874	1.8	59,067	4.8	11,754	0.9	43,848	3.1	47,020	3.7
Leading Producers														
Germany	-		221,414	19.8	218,443	18.8	187,917	15.2	223,340	16.8	225,327	15.9	182,462	14.2
France	105,223	11.4	108,736	9.7	108,785	9.4	100,174	8.1	153,725	11.5	124,951	8.8	108,794	8.5
United Kingdom	88,453	9.6	85,768	7.7	82,641	7.1	85,379	6.9	87,442	6.6	93,555	6.6	79,308	6.2
United States of America	59,900	6.5	56,300	5.0	58,500	5.0	70,800	5.7	95,400	7.2	97,900	6.9	88,900	6.9
Turkey	62,073	6.7	69,407	6.2	100,093	8.6	112,524	9.1	92,178	6.9	99,735	7.1	90,220	7.0

Product Tables
Copper Wire

Unit of Measure: Metric tons.

	1990 Value	%	1991 Value	%	1992 Value	%	1993 Value	%	1994 Value	%	1995 Value	%	1996 Value	%
Total Production	4,797,757	100.0	4,652,781	100.0	4,914,047	100.0	4,785,978	100.0	5,275,244	100.0	5,224,271	100.0	5,438,759	100.0
Regions														
Africa	89,108	1.9	88,185	1.9	85,405	1.7	86,236	1.8	83,053	1.6	87,547	1.7	76,301	1.4
America, North	1,552,759	32.4	1,490,109	32.0	1,554,750	31.6	1,670,678	34.9	1,812,919	34.4	1,789,961	34.3	1,996,539	36.7
America, South	195,090	4.1	190,389	4.1	209,814	4.3	122,977	2.6	226,044	4.3	129,485	2.5	142,287	2.6
Asia	384,075	8.0	453,653	9.8	407,173	8.3	391,322	8.2	516,557	9.8	592,374	11.3	572,047	10.5
Europe	2,554,066	53.2	2,410,378	51.8	2,643,210	53.8	2,505,348	52.3	2,616,753	49.6	2,614,108	50.0	2,644,463	48.6
Oceania	22,659	0.5	20,067	0.4	13,695	0.3	9,416	0.2	19,918	0.4	10,796	0.2	7,121	0.1
Leading Producers														
United States of America	1,500,900	31.3	1,450,600	31.2	1,514,000	30.8	1,644,768	34.4	1,783,143	33.8	1,761,869	33.7	1,966,692	36.2
Germany	595,715	12.4	582,916	12.5	585,188	11.9	512,663	10.7	518,483	9.8	569,019	10.9	619,555	11.4
France	494,460	10.3	475,971	10.2	615,522	12.5	460,838	9.6	491,475	9.3	367,148	7.0	359,065	6.6
Italy	294,000	6.1	275,000	5.9	287,000	5.8	302,000	6.3	308,200	5.8	344,300	6.6	334,000	6.1
United Kingdom	216,741	4.5	193,068	4.1	250,018	5.1	256,660	5.4	305,777	5.8	325,262	6.2	320,632	5.9

Copper Plates, Sheets, Strip, Foil

Unit of Measure: Metric tons.

	1990 Value	%	1991 Value	%	1992 Value	%	1993 Value	%	1994 Value	%	1995 Value	%	1996 Value	%
Total Production	1,153,175	100.0	1,520,828	100.0	1,485,728	100.0	1,475,578	100.0	1,634,604	100.0	1,879,586	100.0	1,847,129	100.0
Regions														
Africa	2,275	0.2	4,281	0.3	1,058	0.1	9,441	0.6	2,034	0.1	5,092	0.3	4,452	0.2
America, North	134,632	11.7	121,484	8.0	128,846	8.7	141,281	9.6	185,080	11.3	184,048	9.8	193,941	10.5
America, South	18,470	1.6	19,837	1.3	21,204	1.4	22,597	1.5	23,868	1.5	25,172	1.3	25,616	1.4
Asia	365,336	31.7	397,320	26.1	361,903	24.4	366,373	24.8	396,719	24.3	428,390	22.8	438,395	23.7
Europe	632,462	54.8	977,906	64.3	972,717	65.5	935,886	63.4	1,026,903	62.8	1,236,884	65.8	1,184,725	64.1
Leading Producers														
Germany	-		325,919	21.4	339,989	22.9	336,447	22.8	391,180	23.9	563,951	30.0	539,684	29.2
Japan	213,045	18.5	225,681	14.8	204,760	13.8	214,139	14.5	230,873	14.1	249,823	13.3	242,096	13.1
United States of America	130,200	11.3	116,100	7.6	124,300	8.4	132,900	9.0	160,400	9.8	167,900	8.9	170,800	9.2
Korea, Republic of	103,560	9.0	107,307	7.1	100,402	6.8	91,312	6.2	102,008	6.2	110,090	5.9	*122,753	6.6
Italy	79,100	6.9	87,800	5.8	90,300	6.1	85,600	5.8	87,100	5.3	89,300	4.8	83,200	4.5

Copper Tubes and Pipes

Unit of Measure: Metric tons.

	1990 Value	%	1991 Value	%	1992 Value	%	1993 Value	%	1994 Value	%	1995 Value	%	1996 Value	%
Total Production	1,215,936	100.0	1,542,491	100.0	1,577,407	100.0	1,571,810	100.0	1,719,147	100.0	1,821,074	100.0	1,867,905	100.0
Regions														
Africa	638	0.1	49	0.0	402	0.0	466	0.0	485	0.0	504	0.0	524	0.0
America, North	413,894	34.0	439,834	28.5	478,151	30.3	502,085	31.9	546,066	31.8	569,375	31.3	654,545	35.0
America, South	16,904	1.4	18,127	1.2	19,350	1.2	20,575	1.3	21,815	1.3	23,024	1.3	25,517	1.4
Asia	262,401	21.6	312,458	20.3	285,989	18.1	274,351	17.5	323,291	18.8	369,873	20.3	361,373	19.3
Europe	522,099	42.9	772,023	50.1	793,514	50.3	774,333	49.3	827,490	48.1	858,298	47.1	825,947	44.2
Leading Producers														
United States of America	346,000	28.5	364,200	23.6	390,100	24.7	410,100	26.1	449,300	26.1	473,100	26.0	542,100	29.0
Germany	-		235,690	15.3	248,288	15.7	237,468	15.1	255,751	14.9	263,199	14.5	240,411	12.9
Japan	212,940	17.5	239,039	15.5	218,827	13.9	202,240	12.9	234,563	13.6	254,456	14.0	262,022	14.0
United Kingdom	66,836	5.5	67,733	4.4	63,609	4.0	66,996	4.3	76,004	4.4	78,062	4.3	77,585	4.2
France	68,050	5.6	75,168	4.9	74,297	4.7	66,063	4.2	76,931	4.5	75,900	4.2	74,700	4.0

Product Tables
Nickel, Unwrought

Unit of Measure: Metric tons.

	1990 Value	%	1991 Value	%	1992 Value	%	1993 Value	%	1994 Value	%	1995 Value	%	1996 Value	%
Total Production	821,433	100.0	831,916	100.0	860,693	100.0	861,541	100.0	901,911	100.0	955,272	100.0	1,011,612	100.0
Regions														
Africa	39,642	4.8	38,175	4.6	37,715	4.4	41,789	4.9	43,618	4.8	40,663	4.3	43,393	4.3
America, North	198,624	24.2	186,488	22.4	193,088	22.4	176,889	20.5	160,789	17.8	182,290	19.1	199,193	19.7
America, South	31,625	3.8	34,000	4.1	34,900	4.1	35,400	4.1	37,300	4.1	41,400	4.3	39,300	3.9
Asia	341,136	41.5	358,630	43.1	369,705	43.0	380,520	44.2	395,852	43.9	421,765	44.2	439,292	43.4
Europe	133,106	16.2	130,823	15.7	128,415	14.9	124,210	14.4	147,223	16.3	139,812	14.6	163,023	16.1
Oceania	77,300	9.4	83,800	10.1	96,870	11.3	102,733	11.9	117,129	13.0	129,343	13.5	127,412	12.6
Leading Producers														
Canada	126,800	15.4	120,300	14.5	132,900	15.4	123,100	14.3	105,100	11.7	121,500	12.7	126,693	12.5
Japan	22,275	2.7	23,658	2.8	22,038	2.6	23,108	2.7	25,311	2.8	26,824	2.8	26,564	2.6
Australia	45,000	5.5	49,400	5.9	57,500	6.7	55,000	6.4	67,000	7.4	77,000	8.1	74,000	7.3
Norway	57,811	7.0	58,729	7.1	55,686	6.5	56,818	6.6	67,955	7.5	53,237	5.6	61,582	6.1
New Caledonia	32,300	3.9	34,400	4.1	39,370	4.6	47,733	5.5	50,129	5.6	52,343	5.5	53,412	5.3

Alumina, Calcined Equivalent

Unit of Measure: Metric tons.

	1990 Value	%	1991 Value	%	1992 Value	%	1993 Value	%	1994 Value	%	1995 Value	%	1996 Value	%
Total Production	38,843,111	100.0	40,172,111	100.0	40,096,484	100.0	41,076,211	100.0	40,880,329	100.0	41,666,794	100.0	42,861,293	100.0
Regions														
America, North	9,186,000	23.6	9,376,000	23.3	9,211,000	23.0	9,479,000	23.1	9,254,000	22.6	8,624,000	20.7	8,900,000	20.8
America, South	4,479,000	11.5	4,548,000	11.3	4,715,000	11.8	4,860,000	11.8	4,666,000	11.4	5,113,000	12.3	5,213,000	12.2
Asia	7,502,111	19.3	7,853,111	19.5	7,886,111	19.7	7,994,111	19.5	7,922,611	19.4	8,598,111	20.6	9,009,111	21.0
Europe	5,803,000	14.9	6,041,000	15.0	5,840,073	14.6	5,866,600	14.3	5,636,718	13.8	5,825,683	14.0	5,821,182	13.6
Oceania	11,873,000	30.6	12,354,000	30.8	12,444,300	31.0	12,876,500	31.3	13,401,000	32.8	13,506,000	32.4	13,918,000	32.5
Leading Producers														
Australia	11,231,000	28.9	11,703,000	29.1	11,783,000	29.4	12,221,000	29.8	12,761,000	31.2	12,940,000	31.1	13,318,000	31.1
United States of America	5,230,000	13.5	5,230,000	13.0	5,190,000	12.9	5,290,000	12.9	4,860,000	11.9	4,530,000	10.9	4,700,000	11.0
Jamaica	2,869,000	7.4	3,015,000	7.5	2,917,000	7.3	3,009,000	7.3	3,224,000	7.9	3,030,000	7.3	3,200,000	7.5
Suriname	1,531,000	3.9	1,510,000	3.8	1,574,000	3.9	1,507,000	3.7	1,498,000	3.7	1,589,000	3.8	1,643,000	3.8
Brazil	1,655,000	4.3	1,743,000	4.3	1,833,000	4.6	1,853,000	4.5	1,868,000	4.6	1,883,000	4.5	1,870,000	4.4

Aluminum, Unwrought (Total Production)

Unit of Measure: Metric tons.

	1990 Value	%	1991 Value	%	1992 Value	%	1993 Value	%	1994 Value	%	1995 Value	%	1996 Value	%
Total Production	23,836,733	100.0	25,140,431	100.0	25,957,366	100.0	26,472,844	100.0	26,943,964	100.0	27,587,904	100.0	27,459,808	100.0
Regions														
Africa	572,608	2.4	570,637	2.3	574,796	2.2	575,731	2.2	543,003	2.0	552,664	2.0	989,223	3.6
America, North	8,193,154	34.4	8,397,300	33.4	8,936,700	34.4	9,129,268	34.5	8,888,883	33.0	8,995,669	32.6	9,456,092	34.4
America, South	1,805,098	7.6	2,041,473	8.1	2,007,357	7.7	2,066,757	7.8	2,093,006	7.8	2,184,998	7.9	2,228,744	8.1
Asia	6,353,767	26.7	6,746,465	26.8	7,157,900	27.6	7,470,920	28.2	8,052,587	29.9	8,614,844	31.2	8,738,506	31.8
Europe	5,379,606	22.6	5,856,756	23.3	5,797,013	22.3	5,604,669	21.2	5,650,185	21.0	5,618,433	20.4	5,615,906	20.5
Oceania	1,532,500	6.4	1,527,800	6.1	1,483,600	5.7	1,625,500	6.1	1,716,300	6.4	1,621,296	5.9	431,337	1.6
Leading Producers														
United States of America	6,441,000	27.0	6,411,000	25.5	6,802,000	26.2	6,635,000	25.1	6,389,000	23.7	6,565,000	23.8	6,867,000	25.0
Canada	1,635,054	6.9	1,889,300	7.5	2,057,800	7.9	2,398,868	9.1	2,349,683	8.7	2,268,992	8.2	2,384,210	8.7
Japan	1,140,628	4.8	1,148,467	4.6	1,112,219	4.3	1,044,160	3.9	1,215,339	4.5	1,227,269	4.4	1,237,794	4.5
Australia	1,268,000	5.3	1,264,600	5.0	1,234,000	4.8	1,340,800	5.1	1,439,000	5.3	1,340,000	4.9	138,670	0.5
Brazil	995,600	4.2	1,206,000	4.8	1,260,400	4.9	1,248,800	4.7	1,275,645	4.7	1,304,800	4.7	1,343,000	4.9

Commodity data are provided by the United Nations Statistical Division. The symbol * means that data are estimated. For additional notes, see Appendix I.

531

Product Tables
Aluminum, Unwrought, Primary

Unit of Measure: Metric tons.

	1990 Value	%	1991 Value	%	1992 Value	%	1993 Value	%	1994 Value	%	1995 Value	%	1996 Value	%
Total Production	18,689,785	100.0	19,645,071	100.0	19,437,519	100.0	20,054,667	100.0	19,985,828	100.0	20,494,594	100.0	21,557,664	100.0
Regions														
Africa	572,581	3.1	570,610	2.9	574,769	3.0	575,704	2.9	542,976	2.7	552,637	2.7	989,126	4.6
America, North	5,672,220	30.3	5,985,363	30.5	6,031,171	31.0	6,028,984	30.1	5,582,561	27.9	5,580,059	27.2	5,929,492	27.5
America, South	1,723,666	9.2	1,945,947	9.9	1,886,160	9.7	1,940,312	9.7	1,970,167	9.9	2,026,009	9.9	2,045,561	9.5
Asia	4,539,490	24.3	4,811,388	24.5	5,192,892	26.7	5,542,633	27.6	5,887,852	29.5	6,387,277	31.2	6,434,578	29.8
Europe	4,687,013	25.1	4,838,262	24.6	4,315,627	22.2	4,383,634	21.9	4,349,172	21.8	4,390,316	21.4	4,544,640	21.1
Oceania	1,494,815	8.0	1,493,500	7.6	1,436,900	7.4	1,583,400	7.9	1,653,100	8.3	1,558,296	7.6	1,614,267	7.5
Leading Producers														
United States of America	4,048,000	21.7	4,121,000	21.0	4,042,000	20.8	3,695,000	18.4	3,299,000	16.5	3,375,000	16.5	3,577,000	16.6
Canada	1,567,395	8.4	1,821,600	9.3	1,971,800	10.1	2,308,868	11.5	2,254,683	11.3	2,171,992	10.6	2,283,210	10.6
Australia	1,235,105	6.6	1,235,000	6.3	1,194,000	6.1	1,306,000	6.5	1,384,000	6.9	1,285,000	6.3	1,329,600	6.2
China	854,300	4.6	900,000	4.6	1,096,400	5.6	1,255,400	6.3	1,498,310	7.5	1,869,678	9.1	1,896,200	8.8
Brazil	930,585	5.0	1,139,000	5.8	1,193,300	6.1	1,172,000	5.8	1,184,645	5.9	1,188,100	5.8	1,197,400	5.6

Aluminum, Unwrought, Secondary

Unit of Measure: Metric tons.

	1990 Value	%	1991 Value	%	1992 Value	%	1993 Value	%	1994 Value	%	1995 Value	%	1996 Value	%
Total Production	5,712,851	100.0	5,696,620	100.0	6,426,600	100.0	6,537,743	100.0	7,109,744	100.0	7,425,570	100.0	7,620,826	100.0
Regions														
Africa	27	0.0	27	0.0	27	0.0	27	0.0	27	0.0	27	0.0	27	0.0
America, North	2,520,959	44.1	2,411,900	42.3	2,905,500	45.2	3,099,900	47.4	3,310,300	46.6	3,415,600	46.0	3,526,600	46.3
America, South	81,000	1.4	94,500	1.7	120,900	1.9	126,000	1.9	137,300	1.9	154,200	2.1	182,800	2.4
Asia	1,847,262	32.3	1,935,037	34.0	1,964,979	30.6	1,928,281	29.5	2,164,744	30.4	2,227,567	30.0	2,294,928	30.1
Europe	1,225,903	21.5	1,220,855	21.4	1,388,494	21.6	1,341,435	20.5	1,434,173	20.2	1,582,477	21.3	1,551,371	20.4
Oceania	37,700	0.7	34,300	0.6	46,700	0.7	42,100	0.6	63,200	0.9	45,700	0.6	65,100	0.9
Leading Producers														
United States of America	2,393,000	41.9	2,290,000	40.2	2,760,000	42.9	2,940,000	45.0	3,090,000	43.5	3,190,000	43.0	3,290,000	43.2
Japan	1,090,112	19.1	1,096,415	19.2	1,073,730	16.7	1,005,639	15.4	1,174,587	16.5	1,180,824	15.9	1,191,484	15.6
Italy	349,600	6.1	348,000	6.1	353,100	5.5	346,100	5.3	375,500	5.3	412,300	5.6	376,600	4.9
France	208,284	3.6	217,164	3.8	222,408	3.5	202,831	3.1	227,421	3.2	253,600	3.4	236,800	3.1
Austria	87,158	1.5	116,900	2.1	45,400	0.7	43,300	0.7	52,500	0.7	93,500	1.3	97,500	1.3

Aluminum Bars, Rods, Angles, Shapes, Sections

Unit of Measure: Metric tons.

	1990 Value	%	1991 Value	%	1992 Value	%	1993 Value	%	1994 Value	%	1995 Value	%	1996 Value	%
Total Production	4,427,802	100.0	4,700,329	100.0	4,742,594	100.0	5,122,099	100.0	5,265,923	100.0	5,523,810	100.0	5,367,344	100.0
Regions														
Africa	1,157	0.0	1,153	0.0	1,155	0.0	1,278	0.0	1,291	0.0	1,305	0.0	1,242	0.0
America, North	1,390,101	31.4	1,277,529	27.2	1,370,872	28.9	1,578,977	30.8	1,464,507	27.8	1,817,165	32.9	1,644,313	30.6
America, South	11,826	0.3	12,490	0.3	13,154	0.3	13,760	0.3	14,901	0.3	13,446	0.2	15,609	0.3
Asia	1,323,435	29.9	1,356,886	28.9	1,299,942	27.4	1,281,096	25.0	1,377,564	26.2	1,390,756	25.2	1,451,580	27.0
Europe	1,595,740	36.0	1,959,798	41.7	1,962,247	41.4	2,143,540	41.8	2,319,193	44.0	2,209,347	40.0	2,169,987	40.4
Oceania	105,543	2.4	92,473	2.0	95,224	2.0	103,448	2.0	88,467	1.7	91,791	1.7	84,614	1.6
Leading Producers														
United States of America	1,267,101	28.6	1,146,796	24.4	1,234,767	26.0	1,415,900	27.6	1,304,155	24.8	1,655,900	30.0	1,439,000	26.8
Japan	1,113,126	25.1	1,127,896	24.0	1,067,527	22.5	1,039,558	20.3	1,109,643	21.1	1,108,120	20.1	1,166,278	21.7
Germany	-		369,280	7.9	367,037	7.7	344,722	6.7	382,705	7.3	415,981	7.5	403,088	7.5
Italy	301,800	6.8	338,800	7.2	349,300	7.4	335,400	6.5	370,800	7.0	367,200	6.6	337,500	6.3
France	159,063	3.6	152,539	3.2	146,692	3.1	144,157	2.8	161,340	3.1	152,838	2.8	154,257	2.9

Commodity data are provided by the United Nations Statistical Division. The symbol * means that data are estimated. For additional notes, see Appendix I.

Product Tables
Aluminum Wire

Unit of Measure: Metric tons.

	1990 Value	%	1991 Value	%	1992 Value	%	1993 Value	%	1994 Value	%	1995 Value	%	1996 Value	%
Total Production	852,951	100.0	877,719	100.0	854,726	100.0	802,351	100.0	784,869	100.0	919,582	100.0	899,275	100.0
Regions														
Africa	32,564	3.8	29,835	3.4	31,626	3.7	31,221	3.9	31,659	4.0	32,499	3.5	33,996	3.8
America, North	264,550	31.0	266,356	30.3	264,157	30.9	272,188	33.9	284,690	36.3	349,125	38.0	360,176	40.1
Asia	196,574	23.0	208,746	23.8	175,108	20.5	158,881	19.8	156,761	20.0	193,288	21.0	178,347	19.8
Europe	349,491	41.0	364,662	41.5	377,366	44.2	335,243	41.8	308,593	39.3	343,155	37.3	326,757	36.3
Oceania	9,772	1.1	8,120	0.9	6,469	0.8	4,818	0.6	3,166	0.4	1,515	0.2	-	
Leading Producers														
United States of America	246,772	28.9	255,447	29.1	254,617	29.8	264,794	33.0	278,815	35.5	342,000	37.2	351,000	39.0
France	102,984	12.1	105,444	12.0	107,244	12.5	73,848	9.2	70,740	9.0	108,469	11.8	99,524	11.1
India	154,955	18.2	150,758	17.2	120,014	14.0	109,464	13.6	103,491	13.2	141,353	15.4	128,094	14.2
Japan	23,485	2.8	26,899	3.1	26,285	3.1	26,029	3.2	24,154	3.1	24,650	2.7	23,125	2.6
Belgium	38,803	4.5	39,979	4.6	48,435	5.7	45,270	5.6	36,030	4.6	45,823	5.0	33,942	3.8

Aluminum Plates, Sheets, Strip, Foil

Unit of Measure: Metric tons.

	1990 Value	%	1991 Value	%	1992 Value	%	1993 Value	%	1994 Value	%	1995 Value	%	1996 Value	%
Total Production	7,931,357	100.0	8,954,158	100.0	9,249,828	100.0	9,271,517	100.0	10,560,768	100.0	11,157,382	100.0	10,912,668	100.0
Regions														
Africa	51,108	0.6	49,698	0.6	48,146	0.5	38,247	0.4	42,253	0.4	67,824	0.6	61,909	0.6
America, North	3,836,950	48.4	3,819,741	42.7	4,123,360	44.6	4,085,876	44.1	4,954,594	46.9	4,578,566	41.0	4,394,013	40.3
America, South	5,767	0.1	6,173	0.1	6,578	0.1	5,992	0.1	5,857	0.1	6,267	0.1	6,909	0.1
Asia	1,524,929	19.2	1,736,300	19.4	1,571,545	17.0	1,625,595	17.5	1,738,680	16.5	1,860,984	16.7	1,929,014	17.7
Europe	2,305,656	29.1	3,117,119	34.8	3,256,892	35.2	3,254,319	35.1	3,539,716	33.5	4,359,512	39.1	4,270,033	39.1
Oceania	206,947	2.6	225,127	2.5	243,308	2.6	261,488	2.8	279,668	2.6	284,229	2.5	250,790	2.3
Leading Producers														
United States of America	3,798,590	47.9	3,783,108	42.2	4,097,054	44.3	4,034,047	43.5	4,909,980	46.5	4,533,600	40.6	4,350,000	39.9
Germany	-		895,033	10.0	951,071	10.3	888,310	9.6	1,018,493	9.6	1,828,808	16.4	1,766,199	16.2
Japan	1,197,761	15.1	1,264,647	14.1	1,208,564	13.1	1,167,826	12.6	1,301,532	12.3	1,362,438	12.2	1,381,855	12.7
United Kingdom	318,173	4.0	269,420	3.0	328,893	3.6	309,326	3.3	346,441	3.3	359,199	3.2	295,020	2.7
Italy	255,600	3.2	259,500	2.9	294,800	3.2	300,500	3.2	330,100	3.1	357,400	3.2	346,400	3.2

Aluminum Tubes and Pipes

Unit of Measure: Metric tons.

	1990 Value	%	1991 Value	%	1992 Value	%	1993 Value	%	1994 Value	%	1995 Value	%	1996 Value	%
Total Production	302,809	100.0	328,477	100.0	310,264	100.0	330,886	100.0	338,754	100.0	361,420	100.0	371,261	100.0
Regions														
Africa	1,993	0.7	1,825	0.6	1,994	0.6	1,895	0.6	1,796	0.5	1,697	0.5	1,576	0.4
America, North	134,527	44.4	134,681	41.0	121,109	39.0	131,224	39.7	137,349	40.5	171,612	47.5	183,855	49.5
America, South	109	0.0	103	0.0	96	0.0	77	0.0	73	0.0	34	0.0	-	
Asia	66,298	21.9	67,379	20.5	62,778	20.2	64,679	19.5	67,099	19.8	70,055	19.4	74,375	20.0
Europe	99,357	32.8	124,339	37.9	124,287	40.1	133,011	40.2	132,437	39.1	118,022	32.7	111,455	30.0
Oceania	526	0.2	150	0.0	-		-		-		-		-	
Leading Producers														
United States of America	132,224	43.7	132,340	40.3	118,799	38.3	129,923	39.3	136,228	40.2	171,000	47.3	183,000	49.3
Japan	58,015	19.2	60,767	18.5	56,462	18.2	54,443	16.5	56,028	16.5	56,890	15.7	60,719	16.4
Germany	-		30,167	9.2	30,983	10.0	26,614	8.0	27,840	8.2	24,060	6.7	20,279	5.5
Italy	17,000	5.6	20,000	6.1	20,000	6.4	20,000	6.0	20,000	5.9	18,900	5.2	17,200	4.6
Sweden	10,351	3.4	8,833	2.7	9,529	3.1	10,771	3.3	13,582	4.0	13,862	3.8	*12,759	3.4

Commodity data are provided by the United Nations Statistical Division. The symbol * means that data are estimated. For additional notes, see Appendix I.

533

Product Tables

Lead, Refined, Unwrought (Total Production)

Unit of Measure: Metric tons.

	1990		1991		1992		1993		1994		1995		1996	
	Value	%	Value	%	Value	%	Value	%	Value	%	Value	%	Value	%
Total Production	5,372,671	100.0	5,717,688	100.0	5,981,493	100.0	5,962,889	100.0	5,971,973	100.0	6,242,067	100.0	6,412,369	100.0
Regions														
Africa	145,829	2.7	142,867	2.5	147,500	2.5	137,800	2.3	119,640	2.0	134,552	2.2	136,945	2.1
America, North	1,683,558	31.3	1,618,043	28.3	1,673,980	28.0	1,667,051	28.0	1,713,166	28.7	1,866,988	29.9	1,937,970	30.2
America, South	186,100	3.5	180,223	3.2	192,853	3.2	204,347	3.4	148,396	2.5	153,283	2.5	160,231	2.5
Asia	1,490,809	27.7	1,561,888	27.3	1,708,533	28.6	1,746,636	29.3	1,782,365	29.8	1,947,480	31.2	2,040,560	31.8
Europe	1,652,576	30.8	1,998,467	35.0	2,011,227	33.6	1,964,055	32.9	1,968,406	33.0	1,914,364	30.7	1,931,863	30.1
Oceania	213,800	4.0	216,200	3.8	247,400	4.1	243,000	4.1	240,000	4.0	225,400	3.6	204,800	3.2
Leading Producers														
United States of America	1,326,600	24.7	1,231,000	21.5	1,220,000	20.4	1,230,000	20.6	1,280,000	21.4	1,390,000	22.3	1,430,000	22.3
United Kingdom	329,378	6.1	311,014	5.4	346,795	5.8	364,013	6.1	352,466	5.9	320,704	5.1	345,574	5.4
China	296,500	5.5	319,700	5.6	366,000	6.1	410,296	6.9	467,926	7.8	607,880	9.7	715,500	11.2
Germany	-		362,510	6.3	354,322	5.9	334,081	5.6	331,684	5.6	202,401	3.2	141,532	2.2
Japan	195,639	3.6	207,315	3.6	274,500	4.6	276,143	4.6	250,428	4.2	220,479	3.5	237,041	3.7

Lead, Secondary, Refined, Soft

Unit of Measure: Metric tons.

	1990		1991		1992		1993		1994		1995		1996	
	Value	%	Value	%	Value	%	Value	%	Value	%	Value	%	Value	%
Total Production	1,957,987	100.0	1,966,675	100.0	1,954,153	100.0	1,956,824	100.0	2,047,138	100.0	2,287,355	100.0	2,406,492	100.0
Regions														
Africa	4,700	0.2	6,600	0.3	7,800	0.4	6,800	0.3	7,000	0.3	10,100	0.4	8,100	0.3
America, North	558,115	28.5	528,122	26.9	557,128	28.5	513,314	26.2	625,022	30.5	687,879	30.1	770,167	32.0
America, South	72,300	3.7	71,000	3.6	68,267	3.5	77,027	3.9	36,053	1.8	36,600	1.6	46,700	1.9
Asia	388,949	19.9	436,015	22.2	433,141	22.2	442,178	22.6	447,781	21.9	456,078	19.9	462,248	19.2
Europe	916,923	46.8	905,537	46.0	872,017	44.6	898,704	45.9	911,282	44.5	1,076,299	47.1	1,095,677	45.5
Oceania	17,000	0.9	19,400	1.0	15,800	0.8	18,800	1.0	20,000	1.0	20,400	0.9	23,600	1.0
Leading Producers														
United States of America	461,489	23.6	422,000	21.5	453,000	23.2	444,000	22.7	527,000	25.7	584,000	25.5	652,000	27.1
United Kingdom	173,505	8.9	146,676	7.6	147,990	7.6	154,453	7.9	161,430	7.9	170,998	7.5	177,466	7.4
Germany	*176,575	9.0	203,600	10.4	171,100	8.8	169,700	8.7	161,900	7.9	*176,575	7.7	*176,575	7.3
Italy	102,200	5.2	96,500	4.9	84,300	4.3	92,900	4.7	95,100	4.6	135,000	5.9	143,900	6.0
Canada	96,465	4.9	105,946	5.4	103,936	5.3	69,107	3.5	97,800	4.8	103,641	4.5	117,914	4.9

Lead-Base Alloys

Unit of Measure: Metric tons.

	1990		1991		1992		1993		1994		1995		1996	
	Value	%	Value	%	Value	%	Value	%	Value	%	Value	%	Value	%
Total Production	146,175	100.0	153,567	100.0	168,242	100.0	153,609	100.0	164,379	100.0	204,047	100.0	192,872	100.0
Regions														
Africa	37,184	25.4	30,309	19.7	36,197	21.5	28,434	18.5	26,944	16.4	34,653	17.0	45,001	23.3
America, North	32,660	22.3	32,660	21.3	32,660	19.4	32,660	21.3	32,660	19.9	32,660	16.0	32,660	16.9
America, South	3,857	2.6	4,092	2.7	4,327	2.6	2,550	1.7	2,776	1.7	7,961	3.9	4,825	2.5
Asia	16,587	11.3	22,284	14.5	28,436	16.9	14,452	9.4	9,952	6.1	18,059	8.9	21,668	11.2
Europe	55,886	38.2	64,222	41.8	66,621	39.6	75,513	49.2	92,047	56.0	110,714	54.3	88,718	46.0
Leading Producers														
South Africa	37,184	25.4	30,309	19.7	36,197	21.5	28,434	18.5	26,944	16.4	34,653	17.0	45,001	23.3
United States of America	*32,660	22.3	*32,660	21.3	*32,660	19.4	*32,660	21.3	*32,660	19.9	*32,660	16.0	*32,660	16.9
Sweden	20,486	14.0	21,925	14.3	19,202	11.4	18,300	11.9	22,703	13.8	20,894	10.2	*20,193	10.5
Spain	12,792	8.8	19,563	12.7	24,769	14.7	36,252	23.6	48,355	29.4	67,907	33.3	46,509	24.1
Denmark	*11,177	7.6	*11,177	7.3	*11,177	6.6	*11,177	7.3	*11,177	6.8	*11,177	5.5	*11,177	5.8

Commodity data are provided by the United Nations Statistical Division. The symbol * means that data are estimated. For additional notes, see Appendix I.

Product Tables
Zinc, Unwrought (Total Production)

Unit of Measure: Metric tons.

	1990 Value	%	1991 Value	%	1992 Value	%	1993 Value	%	1994 Value	%	1995 Value	%	1996 Value	%
Total Production	6,927,053	100.0	7,616,090	100.0	7,897,468	100.0	8,244,541	100.0	8,286,246	100.0	8,360,613	100.0	8,517,786	100.0
Regions														
Africa	63,673	0.9	60,516	0.8	55,508	0.7	40,377	0.5	21,200	0.3	32,967	0.4	28,776	0.3
America, North	1,058,406	15.3	1,127,785	14.8	1,185,275	15.0	1,258,142	15.3	1,268,966	15.3	1,309,230	15.7	1,317,030	15.5
America, South	308,500	4.5	353,900	4.6	348,714	4.4	381,462	4.6	412,200	5.0	391,687	4.7	406,858	4.8
Asia	2,870,954	41.4	3,006,857	39.5	3,125,620	39.6	3,333,988	40.4	3,491,057	42.1	3,485,500	41.7	3,553,928	41.7
Europe	2,326,300	33.6	2,742,132	36.0	2,852,891	36.1	2,894,072	35.1	2,772,848	33.5	2,824,729	33.8	2,876,695	33.8
Oceania	299,220	4.3	324,900	4.3	329,460	4.2	336,500	4.1	319,975	3.9	316,500	3.8	334,500	3.9
Leading Producers														
Japan	687,461	9.9	730,829	9.6	729,454	9.2	695,687	8.4	665,502	8.0	663,562	7.9	599,053	7.0
Canada	591,786	8.5	660,552	8.7	671,702	8.5	661,920	8.0	692,880	8.4	720,360	8.6	716,467	8.4
China	551,800	8.0	612,100	8.0	718,900	9.1	891,351	10.8	1,077,612	13.0	1,076,715	12.9	1,163,800	13.7
Germany	-		345,712	4.5	383,117	4.9	380,948	4.6	350,878	4.2	345,031	4.1	352,340	4.1
United States of America	358,412	5.2	376,000	4.9	399,000	5.1	382,000	4.6	356,000	4.3	363,000	4.3	366,000	4.3

Zinc, Unwrought, Primary

Unit of Measure: Metric tons.

	1990 Value	%	1991 Value	%	1992 Value	%	1993 Value	%	1994 Value	%	1995 Value	%	1996 Value	%
Total Production	6,327,811	100.0	6,885,133	100.0	7,119,504	100.0	7,216,948	100.0	7,231,016	100.0	7,267,802	100.0	7,321,674	100.0
Regions														
Africa	90,021	1.4	90,004	1.3	84,996	1.2	72,465	1.0	57,903	0.8	69,755	1.0	65,564	0.9
America, North	962,694	15.2	1,004,785	14.6	1,058,275	14.9	1,116,142	15.5	1,129,966	15.6	1,178,230	16.2	1,177,030	16.1
America, South	301,213	4.8	345,552	5.0	338,914	4.8	371,462	5.1	404,518	5.6	382,702	5.3	397,835	5.4
Asia	2,580,336	40.8	2,676,201	38.9	2,718,117	38.2	2,771,964	38.4	2,803,543	38.8	2,822,123	38.8	2,831,479	38.7
Europe	2,098,827	33.2	2,448,191	35.6	2,594,242	36.4	2,552,914	35.4	2,520,086	34.9	2,502,992	34.4	2,519,766	34.4
Oceania	294,720	4.7	320,400	4.7	324,960	4.6	332,000	4.6	315,000	4.4	312,000	4.3	330,000	4.5
Leading Producers														
Japan	664,178	10.5	702,752	10.2	696,863	9.8	666,428	9.2	656,288	9.1	652,711	9.0	589,219	8.0
Canada	591,786	9.4	660,552	9.6	671,702	9.4	661,920	9.2	692,880	9.6	720,360	9.9	716,467	9.8
China	*444,524	7.0	*479,810	7.0	*515,095	7.2	*550,381	7.6	*585,667	8.1	*620,952	8.5	*656,238	9.0
Australia	294,720	4.7	320,400	4.7	324,960	4.6	332,000	4.6	315,000	4.4	312,000	4.3	330,000	4.5
Germany	-		299,484	4.3	328,596	4.6	331,183	4.6	300,536	4.2	237,600	3.3	226,800	3.1

Zinc, Unwrought, Secondary

Unit of Measure: Metric tons.

	1990 Value	%	1991 Value	%	1992 Value	%	1993 Value	%	1994 Value	%	1995 Value	%	1996 Value	%
Total Production	361,875	100.0	452,931	100.0	477,010	100.0	483,838	100.0	488,024	100.0	524,961	100.0	553,981	100.0
Regions														
America, North	95,708	26.4	122,457	27.0	127,623	26.8	141,000	29.1	139,000	28.5	131,000	25.0	140,000	25.3
America, South	7,300	2.0	8,300	1.8	9,800	2.1	10,000	2.1	8,000	1.6	9,000	1.7	9,000	1.6
Asia	143,977	39.8	155,668	34.4	163,061	34.2	164,092	33.9	146,657	30.1	151,984	29.0	154,528	27.9
Europe	110,390	30.5	162,005	35.8	172,026	36.1	164,245	33.9	189,392	38.8	228,477	43.5	245,952	44.4
Oceania	4,500	1.2	4,500	1.0	4,500	0.9	4,500	0.9	4,975	1.0	4,500	0.9	4,500	0.8
Leading Producers														
United States of America	95,708	26.4	122,457	27.0	127,623	26.8	141,000	29.1	139,000	28.5	131,000	25.0	140,000	25.3
Germany	-		46,228	10.2	54,521	11.4	49,765	10.3	59,342	12.2	107,431	20.5	125,540	22.7
FYR Macedonia	*38,200	10.6	*38,200	8.4	38,200	8.0	*38,200	7.9	*38,200	7.8	*38,200	7.3	*38,200	6.9
Japan	23,283	6.4	28,077	6.2	32,591	6.8	29,259	6.0	9,214	1.9	10,851	2.1	9,834	1.8
France	12,096	3.3	12,648	2.8	14,076	3.0	13,649	2.8	29,316	6.0	*20,339	3.9	*19,324	3.5

Commodity data are provided by the United Nations Statistical Division. The symbol * means that data are estimated. For additional notes, see Appendix I.

535

Product Tables
Zinc-Base Alloys

Unit of Measure: Metric tons.

	1990		1991		1992		1993		1994		1995		1996	
	Value	%	Value	%	Value	%	Value	%	Value	%	Value	%	Value	%
Total Production	333,025	100.0	446,417	100.0	435,995	100.0	434,159	100.0	460,898	100.0	481,587	100.0	481,022	100.0
Regions														
America, North	160,332	48.1	160,332	35.9	160,332	36.8	160,332	36.9	160,332	34.8	160,332	33.3	160,332	33.3
America, South	22,803	6.8	22,846	5.1	22,889	5.2	22,953	5.3	23,246	5.0	23,531	4.9	23,287	4.8
Asia	5,639	1.7	6,307	1.4	6,486	1.5	5,823	1.3	5,815	1.3	5,987	1.2	5,777	1.2
Europe	144,251	43.3	256,932	57.6	246,288	56.5	245,051	56.4	271,504	58.9	291,737	60.6	291,626	60.6
Leading Producers														
United States of America	*160,332	48.1	*160,332	35.9	*160,332	36.8	*160,332	36.9	*160,332	34.8	*160,332	33.3	*160,332	33.3
Germany	-		119,545	26.8	113,958	26.1	77,382	17.8	77,778	16.9	73,839	15.3	69,900	14.5
Italy	60,800	18.3	56,600	12.7	52,300	12.0	59,300	13.7	65,500	14.2	68,600	14.2	70,200	14.6
Spain	28,439	8.5	25,788	5.8	23,920	5.5	52,769	12.2	62,734	13.6	77,461	16.1	69,727	14.5
United Kingdom	*32,621	9.8	*32,981	7.4	*33,341	7.6	*33,701	7.8	*34,061	7.4	*34,420	7.1	*34,780	7.2

Zinc Plates, Sheets, Strip, Foil

Unit of Measure: Metric tons.

	1990		1991		1992		1993		1994		1995		1996	
	Value	%	Value	%	Value	%	Value	%	Value	%	Value	%	Value	%
Total Production	347,856	100.0	359,504	100.0	365,379	100.0	369,719	100.0	373,809	100.0	374,816	100.0	375,211	100.0
Regions														
America, South	50,315	14.5	51,721	14.4	53,127	14.5	54,532	14.7	55,938	15.0	56,788	15.2	58,079	15.5
Asia	47,888	13.8	48,269	13.4	48,721	13.3	47,531	12.9	47,299	12.7	48,944	13.1	48,598	13.0
Europe	249,653	71.8	259,514	72.2	263,531	72.1	267,656	72.4	270,573	72.4	269,084	71.8	268,533	71.6
Leading Producers														
France	93,708	26.9	94,920	26.4	89,532	24.5	92,376	25.0	92,100	24.6	*90,798	24.2	*90,415	24.1
Germany	*78,268	22.5	78,268	21.8	*78,268	21.4	*78,268	21.2	*78,268	20.9	*78,268	20.9	*78,268	20.9
Chile	*40,473	11.6	*40,473	11.3	*40,473	11.1	*40,473	10.9	*40,473	10.8	*40,473	10.8	*40,473	10.8
Japan	*26,882	7.7	*26,375	7.3	*25,868	7.1	*25,361	6.9	*24,855	6.6	*24,348	6.5	*23,841	6.4
Italy	-		9,000	2.5	10,800	3.0	13,600	3.7	14,900	4.0	16,800	4.5	16,100	4.3

Tin, Unwrought (Total Production)

Unit of Measure: Metric tons.

	1990		1991		1992		1993		1994		1995		1996	
	Value	%	Value	%	Value	%	Value	%	Value	%	Value	%	Value	%
Total Production	263,906	100.0	232,531	100.0	230,792	100.0	237,971	100.0	240,825	100.0	247,697	100.0	252,439	100.0
Regions														
Africa	3,266	1.2	3,083	1.3	2,525	1.1	2,282	1.0	1,268	0.5	1,813	0.7	1,495	0.6
America, North	13,077	5.0	13,248	5.7	16,490	7.1	13,840	5.8	12,676	5.3	12,580	5.1	11,982	4.7
America, South	48,478	18.4	40,807	17.5	40,591	17.6	41,897	17.6	36,439	15.1	37,709	15.2	36,733	14.6
Asia	166,632	63.1	154,435	66.4	157,334	68.2	161,701	68.0	175,049	72.7	181,273	73.2	190,873	75.6
Europe	31,870	12.1	20,334	8.7	13,252	5.7	17,742	7.5	14,948	6.2	13,567	5.5	10,576	4.2
Oceania	584	0.2	624	0.3	600	0.3	508	0.2	446	0.2	755	0.3	780	0.3
Leading Producers														
Malaysia	48,864	18.5	43,008	18.5	45,598	19.8	40,079	16.8	37,990	15.8	39,433	15.9	38,051	15.1
China	35,800	13.6	36,400	15.7	39,600	17.2	51,617	21.7	67,764	28.1	67,659	27.3	71,500	28.3
Indonesia	38,000	14.4	38,000	16.3	35,900	15.6	38,300	16.1	39,000	16.2	44,218	17.9	48,960	19.4
Brazil	35,111	13.3	26,000	11.2	27,300	11.8	27,200	11.4	20,700	8.6	19,800	8.0	19,800	7.8
Thailand	19,979	7.6	14,937	6.4	10,910	4.7	8,618	3.6	7,634	3.2	8,246	3.3	10,983	4.4

Commodity data are provided by the United Nations Statistical Division. The symbol * means that data are estimated. For additional notes, see Appendix I.

Product Tables
Tin, Unwrought, Primary

Unit of Measure: Metric tons.

	1990		1991		1992		1993		1994		1995		1996	
	Value	%	Value	%	Value	%	Value	%	Value	%	Value	%	Value	%
Total Production	235,922	100.0	206,257	100.0	217,151	100.0	221,074	100.0	226,556	100.0	234,972	100.0	242,436	100.0
Regions														
Africa	2,389	1.0	2,208	1.1	1,658	0.8	1,430	0.6	461	0.2	891	0.4	573	0.2
America, North	7,915	3.4	5,166	2.5	5,488	2.5	4,531	2.0	3,652	1.6	3,647	1.6	3,641	1.5
America, South	48,378	20.5	40,483	19.6	40,191	18.5	41,497	18.8	36,039	15.9	37,309	15.9	36,333	15.0
Asia	159,022	67.4	147,216	71.4	153,882	70.9	157,999	71.5	171,094	75.5	177,780	75.7	187,104	77.2
Europe	17,835	7.6	10,860	5.3	15,693	7.2	15,358	6.9	15,124	6.7	14,889	6.3	14,305	5.9
Oceania	384	0.2	324	0.2	240	0.1	258	0.1	186	0.1	455	0.2	480	0.2
Leading Producers														
Malaysia	48,864	20.7	43,008	20.9	45,598	21.0	40,079	18.1	37,990	16.8	39,433	16.8	38,051	15.7
China	35,800	15.2	36,400	17.6	39,600	18.2	51,617	23.3	67,764	29.9	67,659	28.8	71,500	29.5
Indonesia	38,000	16.1	38,000	18.4	35,900	16.5	38,300	17.3	39,000	17.2	44,218	18.8	48,960	20.2
Brazil	35,111	14.9	25,776	12.5	27,000	12.4	26,900	12.2	20,400	9.0	19,500	8.3	19,500	8.0
Thailand	15,512	6.6	11,255	5.5	10,908	5.0	8,616	3.9	7,629	3.4	8,243	3.5	10,981	4.5

Tin, Unwrought, Secondary

Unit of Measure: Metric tons.

	1990		1991		1992		1993		1994		1995		1996	
	Value	%	Value	%	Value	%	Value	%	Value	%	Value	%	Value	%
Total Production	20,956	100.0	21,529	100.0	21,471	100.0	19,908	100.0	18,714	100.0	17,387	100.0	16,363	100.0
Regions														
Africa	70	0.3	70	0.3	60	0.3	45	0.2	58	0.3	40	0.2	21	0.1
America, North	7,339	35.0	10,619	49.3	13,900	64.7	12,200	61.3	11,908	63.6	11,810	67.9	11,212	68.5
America, South	349	1.7	350	1.6	350	1.6	350	1.8	350	1.9	350	2.0	350	2.1
Asia	395	1.9	386	1.8	277	1.3	277	1.4	180	1.0	178	1.0	294	1.8
Europe	12,603	60.1	9,804	45.5	6,524	30.4	6,786	34.1	5,958	31.8	4,709	27.1	4,186	25.6
Oceania	200	1.0	300	1.4	360	1.7	250	1.3	260	1.4	300	1.7	300	1.8
Leading Producers														
United States of America	7,139	34.1	10,420	48.4	13,700	63.8	12,000	60.3	11,700	62.5	11,600	66.7	11,000	67.2
United Kingdom	5,900	28.2	3,575	16.6	100	0.5	100	0.5	100	0.5	100	0.6	100	0.6
France	2,652	12.7	2,916	13.5	3,180	14.8	3,444	17.3	2,676	14.3	*1,395	8.0	*1,062	6.5
Belgium	*2,300	11.0	*2,300	10.7	*2,300	10.7	*2,300	11.6	*2,300	12.3	*2,300	13.2	2,300	14.1
Spain	200	1.0	200	0.9	200	0.9	200	1.0	200	1.1	100	0.6	50	0.3

Cadmium, Unwrought

Unit of Measure: Metric tons.

	1990		1991		1992		1993		1994		1995		1996	
	Value	%	Value	%	Value	%	Value	%	Value	%	Value	%	Value	%
Total Production	19,087	100.0	20,517	100.0	20,920	100.0	19,315	100.0	20,156	100.0	20,273	100.0	20,857	100.0
Regions														
Africa	261	1.4	210	1.0	173	0.8	90	0.5	79	0.4	74	0.4	72	0.3
America, North	3,821	20.0	3,631	17.7	3,679	17.6	3,066	15.9	3,181	15.8	3,732	18.4	4,117	19.7
America, South	520	2.7	387	1.9	636	3.0	706	3.7	842	4.2	845	4.2	845	4.1
Asia	7,277	38.1	8,790	42.8	8,565	40.9	8,186	42.4	8,273	41.0	7,990	39.4	8,497	40.7
Europe	6,570	34.4	6,419	31.3	6,866	32.8	6,316	32.7	6,871	34.1	6,795	33.5	6,644	31.9
Oceania	638	3.3	1,080	5.3	1,001	4.8	951	4.9	910	4.5	838	4.1	682	3.3
Leading Producers														
Japan	2,451	12.8	2,889	14.1	2,986	14.3	2,832	14.7	2,629	13.0	2,652	13.1	*2,949	14.1
Canada	1,431	7.5	1,787	8.7	1,963	9.4	1,888	9.8	2,129	10.6	2,349	11.6	2,540	12.2
Belgium	1,960	10.3	1,810	8.8	1,549	7.4	1,573	8.1	1,556	7.7	1,710	8.4	1,580	7.6
United States of America	1,678	8.8	1,680	8.2	1,620	7.7	1,090	5.6	1,010	5.0	1,270	6.3	1,530	7.3
Germany	-		-		961	4.6	1,056	5.5	1,145	5.7	1,150	5.7	1,150	5.5

Product Tables
Magnesium, Unwrought (Total Production)

Unit of Measure: Metric tons.

	1990		1991		1992		1993		1994		1995		1996	
	Value	%	Value	%	Value	%	Value	%	Value	%	Value	%	Value	%
Total Production	454,041	100.0	439,637	100.0	478,836	100.0	463,998	100.0	487,571	100.0	588,611	100.0	546,523	100.0
Regions														
America, North	220,867	48.6	216,343	49.2	220,492	46.0	214,000	46.1	218,900	44.9	255,100	43.3	258,000	47.2
America, South	8,700	1.9	7,800	1.8	7,200	1.5	11,300	2.4	11,300	2.3	11,300	1.9	10,600	1.9
Asia	147,344	32.5	144,679	32.9	180,526	37.7	174,885	37.7	195,455	40.1	258,644	43.9	211,118	38.6
Europe	77,130	17.0	70,814	16.1	70,618	14.7	63,813	13.8	61,916	12.7	63,567	10.8	66,805	12.2
Leading Producers														
United States of America	194,141	42.8	181,831	41.4	193,992	40.5	191,000	41.2	190,000	39.0	207,000	35.2	204,000	37.3
Norway	48,222	10.6	44,618	10.1	32,727	6.8	27,177	5.9	27,600	5.7	28,000	4.8	30,000	5.5
Russian Federation	-		-		40,000	8.4	30,000	6.5	35,400	7.3	37,500	6.4	35,000	6.4
Japan	36,151	8.0	28,718	6.5	20,097	4.2	20,686	4.5	22,421	4.6	11,767	2.0	8,175	1.5
Canada	26,726	5.9	34,512	7.9	26,500	5.5	23,000	5.0	28,900	5.9	48,100	8.2	54,000	9.9

Magnesium, Unwrought, Primary

Unit of Measure: Metric tons.

	1990		1991		1992		1993		1994		1995		1996	
	Value	%	Value	%	Value	%	Value	%	Value	%	Value	%	Value	%
Total Production	344,522	100.0	348,000	100.0	330,545	100.0	322,871	100.0	344,887	100.0	437,065	100.0	373,973	100.0
Regions														
America, North	166,059	48.2	165,800	47.6	163,447	49.4	158,052	49.0	156,081	45.3	172,111	39.4	165,141	44.2
America, South	9,700	2.8	9,700	2.8	9,700	2.9	9,700	3.0	9,700	2.8	9,700	2.2	9,700	2.6
Asia	113,244	32.9	116,348	33.4	115,495	34.9	120,267	37.2	128,671	37.3	205,238	47.0	149,533	40.0
Europe	55,519	16.1	56,153	16.1	41,904	12.7	34,852	10.8	50,434	14.6	50,016	11.4	49,599	13.3
Leading Producers														
United States of America	139,333	40.4	131,288	37.7	136,947	41.4	132,000	40.9	128,000	37.1	142,000	32.5	133,000	35.6
Norway	33,147	9.6	36,388	10.5	25,445	7.7	19,748	6.1	*32,127	9.3	*32,094	7.3	*32,060	8.6
Canada	26,726	7.8	34,512	9.9	26,500	8.0	*26,052	8.1	*28,081	8.1	*30,111	6.9	*32,141	8.6
China	5,900	1.7	8,600	2.5	10,500	3.2	13,233	4.1	24,009	7.0	93,593	21.4	*36,078	9.6
France	14,640	4.2	14,051	4.0	13,453	4.1	10,983	3.4	*14,903	4.3	*15,236	3.5	*15,569	4.2

Commodity data are provided by the United Nations Statistical Division. The symbol * means that data are estimated. For additional notes, see Appendix I.

ISIC 3810
METAL PRODUCTS

ISIC 3810—Metal Products—includes a diverse grouping of fabricated metal products, including razor blades, locksmiths' ware, general hardware, tanks and vats, boilers, metal gas cylinders, cables; also, nails, screws, nuts, bolts, and rivets, etc.; large metal containers; and non-electrical central-heating apparatus (radiators, etc.).

The products of this industry are classified, in the U.S. system, under SIC group 34, Fabricated Metal Products.

Summary Statistics

		1990 Value	N	1991 Value	N	1992 Value	N	1993 Value	N	1994 Value	N	1995 Value	N
Establishments or enterprises	(number)	225,358	86	243,117	98	281,854	92	233,061	91	220,985	73	136,618	35
Number employed	(000)	10,071	95	10,149	99	8,995	93	9,435	95	9,226	86	8,017	65
Total Wages	($ mil.)	131,139	91	136,954	92	147,415	86	133,069	86	140,069	78	119,352	58
Wage/salary per employee	($)	8,594	89	9,126	90	14,594	85	11,133	85	9,122	77	11,133	58
Female workers	(000)	557	35	315	33	1,227	30	1,238	33	1,245	27	1,200	16
as % of total employment*	(%)	17	35	16	33	18	30	18	33	20	27	23	16
Output	($ bil.)	622	95	634	94	717	89	667	89	703	81	656	63
per employee	($ 000)	44	91	46	92	1,107	87	225	86	48	78	59	61
per establishment	($ 000)	2,395	82	2,342	88	40,676	81	5,405	78	2,059	62	1,531	31
Value added	($ bil.)	277	84	288	85	307	79	306	80	284	71	302	59
per employee	($ 000)	19	81	20	82	22	77	20	76	21	68	26	56
per establishment	($ 000)	999	73	959	79	1,088	71	754	67	862	52	640	27
Capital investment	($ mil.)	22,374	70	21,723	65	27,317	60	23,472	54	19,511	43	11,918	22
per employee	($ 000)	4	66	3	61	3	58	2	51	3	41	3	21
per establishment	($ 000)	239	66	165	61	180	57	145	48	99	39	76	19

Data presented above are drawn from the detailed tables that follow. Columns headed 'N' show the number of countries that provided valid data for inclusion. Values are not strictly comparable one year to the next or one row to the next because the number of countries that report varies from period to period and row to row. However, a general indication of magnitudes can be gleaned from reviewing this summary—especially in earlier years in which more reports are available. For detailed explanations, see Appendix I. *The average for those countries reporting both total and female employment.

Establishments and Number Engaged

Country	Establishments or Enterprises (number)						Employees per Establishment					
	1990	1991	1992	1993	1994	1995	1990	1991	1992	1993	1994	1995
Albania	-	-	-	23	21	20	-	-	-	73	109	55
Angola	-	-	16	16	-	-	-	-	34	18	-	-
Argentina	-	-	-	13,488	15,187	-	-	-	-	3.750	3.624	-
Armenia	59	57	56	53	52	-	164	149	-	-	-	-
Australia	4,828	6,628	6,528	-	-	-	22	15	14	-	-	-
Austria	896	869	888	878	883	-	65	66	62	61	63	-
Azerbaijan	172	160	151	151	140	-	53	53	46	45	34	-
Bahrain	-	-	131	-	-	-	-	-	12	-	-	-
Bangladesh	823	978	952	-	-	-	22	23	20	-	-	-
Barbados	11	11	11	9.000	21	-	50	54	46	38	34	-
Belgium	3,651	3,714	3,785	3,439	3,438	-	-	-	-	-	-	-
Belize	28	30	32	-	-	-	-	6.000	5.250	-	-	-
Benin	-	-	59	59	-	-	-	-	-	-	-	-
Bolivia	89	101	109	122	117	132	12	12	11	11	10	9.811
Bosnia & Herzegovina	108	107	-	-	4.000	-	332	320	-	-	259	-
Bulgaria	*150	*159	*180	*1,265	*2,048	*2,546	548	415	284	35	27	16
Burundi	*8.000	*9.000	-	-	-	-	44	38	-	-	-	-
Canada	3,913	3,653	3,467	3,287	3,186	-	35	34	31	32	34	-
Cape Verde	16	16	23	29	24	-	-	-	-	-	-	-
Central African Republic	-	*3.000	*3.000	*3.000	*3.000	-	-	-	-	-	-	-
Chile	143	147	150	160	179	-	149	138	143	142	138	-
China	*33,179	*32,832	*32,238	*28,427	*29,311	*30,728	55	57	57	68	67	63
China (Hong Kong SAR)	6,729	5,855	5,618	4,612	4,262	-	7.609	7.925	7.583	7.069	6.335	-
Colombia	604	580	628	622	600	-	47	45	48	53	54	-
Costa Rica	363	358	353	369	-	-	13	12	13	14	-	-
Croatia	619	889	1,279	1,776	2,221	2,491	85	43	24	16	12	10
Cyprus	900	905	904	906	1,038	1,119	2.997	2.993	3.021	3.079	2.922	2.733
Czechoslovakia (Former)	*81	*99	-	-	-	-	2,025	1,081	-	-	-	-
Czech Republic	-	*75	*186	*277	-	-	-	827	446	307	-	-
Denmark	2,173	2,201	2,197	-	-	-	21	20	20	-	-	-
Ecuador	133	148	150	152	145	131	50	53	52	51	55	43
Egypt	446	419	409	392	-	-	111	116	129	119	-	-
El Salvador	-	19	25	39	34	52	-	-	54	35	40	39
Equatorial Guinea	3.000	-	-	-	-	-	5.000	-	-	-	-	-
Ethiopia	13	13	13	14	35	40	117	111	118	112	51	51
Fiji	70	59	62	-	-	-	11	13	12	-	-	-
Finland	740	831	721	698	672	611	43	36	34	33	35	34
FYR Macedonia	*83	*158	*310	*332	*461	*612	131	72	33	30	24	14
Gabon	-	19	16	12	13	12	-	38	34	44	51	48
Gambia	-	-	-	7.000	-	-	-	-	-	8.429	-	-
Germany	-	6,897	7,061	7,119	6,990	-	-	114	-	99	96	-
Germany (Western Part)	6,061	6,271	6,371	-	-	-	109	109	107	-	-	-
Ghana	-	-	-	41	-	-	-	-	-	87	-	-
Greece	535	528	544	-	-	-	38	34	33	-	-	-
Grenada	-	-	13	9.000	-	-	-	-	11	10	-	-
Guatemala	-	25	24	24	24	24	-	90	105	98	98	87
Honduras	-	-	43	57	57	57	-	-	65	54	59	72
Hungary	*558	*1,244	*1,636	*1,969	-	-	90	35	20	21	-	-
Iceland	496	488	504	511	-	-	4.327	4.016	3.685	3.405	-	-
India	6,964	6,890	7,038	7,496	-	-	33	34	34	32	-	-
Indonesia	632	647	690	723	797	958	127	147	170	163	163	154
Iran (Islamic Republic of)	853	372	434	400	-	-	40	83	80	81	-	-
Iraq	-	19	20	-	-	-	-	332	400	-	-	-
Ireland	597	570	-	-	-	-	21	21	-	-	-	-
Israel	1,422	1,632	1,721	1,894	1,908	-	27	26	25	23	24	-
Italy	*2,876	*2,930	*4,879	*4,948	*5,084	-	59	59	49	47	46	-
Japan	53,216	52,369	50,676	51,542	48,450	49,830	17	18	18	17	18	18
Jordan	1,789	1,851	2,245	2,299	2,722	3,066	1.978	2.245	3.041	2.364	2.511	2.432
Kenya	237	321	327	330	377	377	50	37	37	37	33	35
Korea, Republic of	5,873	6,274	6,813	9,174	9,741	10,694	30	28	24	20	20	19

Continued.

Depending on the table, * means *Enterprises, Engaged,* or *Factor Values;* ± means *Basis Unspecified;* § means *shown in millions.* For additional notes and sources, see Appendix I.

Establishments and Number Engaged

- Continued -

Country	Establishments or Enterprises (number)						Employees per Establishment					
	1990	1991	1992	1993	1994	1995	1990	1991	1992	1993	1994	1995
Kuwait	665	619	608	606	600	-	8.684	7.779	9.038	9.625	9.793	-
Kyrgyzstan	643	616	388	389	354	-	59	51	51	35	31	-
Latvia	96	118	126	156	68	160	161	93	65	30	60	26
Lesotho	74	-	-	-	-	-	15	-	-	-	-	-
Liechtenstein	-	58	-	-	-	-	-	22	-	-	-	-
Lithuania	-	-	-	*69	*101	-	-	-	-	145	105	-
Luxembourg	*137	*137	*130	*140	-	-	33	35	37	33	-	-
Macau	282	265	254	243	225	206	3.156	2.962	2.925	2.872	3.400	2.549
Malawi	*9.000	*9.000	*9.000	*9.000	*9.000	-	156	144	122	111	122	-
Malaysia	504	560	578	652	642	-	65	69	76	72	76	-
Mauritius	39	31	35	33	35	33	29	33	34	36	37	44
Mexico	218	212	203	195	188	183	237	243	240	212	216	199
Mongolia	18	19	25	19	28	26	42	38	27	58	51	40
Morocco	377	417	422	457	440	456	44	53	60	59	53	53
Mozambique	17	17	19	19	20	-	146	141	128	118	110	-
Myanmar	4.000	4.000	4.000	4.000	-	-	288	281	319	318	-	-
Namibia	-	-	-	-	32	-	-	-	-	-	45	-
Nepal	136	198	-	229	247	-	21	20	-	19	21	-
Netherlands	881	969	1,051	1,060	1,063	-	73	71	66	62	61	-
Netherlands Antilles	-	-	-	60	-	-	-	-	-	22	-	-
New Zealand	*2,316	*2,362	*2,389	*2,509	-	-	7.359	6.762	6.719	6.918	-	-
Norway	836	772	511	389	367	-	25	26	35	33	34	-
Pakistan	-	211	-	-	-	-	-	56	-	-	-	-
Panama	55	56	-	-	-	-	20	21	-	-	-	-
Paraguay	-	11	-	-	-	-	-	49	-	-	-	-
Peru	1,600	1,612	1,652	-	-	-	9.586	8.672	8.402	-	-	-
Philippines	7,404	8,170	633	600	-	-	4.943	4.945	45	51	-	-
Poland	*823	*522	*505	*489	-	-	238	333	317	297	-	-
Portugal	*11,597	*11,485	*11,742	*11,473	*11,521	-	6.871	7.657	7.123	7.005	6.742	-
Puerto Rico	182	194	191	194	179	200	19	17	15	15	16	15
Republic of Moldova	25	31	30	20	19	20	1,274	918	916	858	218	247
Romania	-	709	1,371	1,507	2,484	-	-	261	132	102	55	-
Russian Federation	-	-	-	*5,834	*4,172	*6,927	-	-	-	72	84	46
Saint Lucia	-	2.000	3.000	4.000	4.000	4.000	-	-	-	-	-	-
Saint Vincent & the Grenadines .	-	-	-	1.000	1.000	1.000	-	-	-	-	-	61
Senegal	18	10	11	11	12	-	45	58	53	50	49	-
Sierra Leone	-	-	-	34	-	-	-	-	-	25	-	-
Singapore	479	518	528	545	561	-	59	59	57	60	60	-
Slovakia	-	65	65	121	218	-	-	392	347	194	180	-
Slovenia	503	*2,851	*3,612	*5,632	*6,338	*4,813	-	-	-	-	-	-
South Africa	-	3,955	-	-	-	-	-	33	-	-	-	-
Spain	16,630	16,729	16,689	-	-	-	8.747	8.658	8.433	-	-	-
Sri Lanka	114	83	57	80	-	-	36	51	49	44	-	-
Swaziland	27	17	17	15	-	-	21	25	24	33	-	-
Sweden	1,667	1,641	1,450	1,348	1,399	-	44	40	40	37	38	-
Thailand	329	890	-	-	-	-	162	73	-	-	-	-
Tonga	-	11	10	11	9.000	-	-	3.909	4.300	4.909	4.333	-
Trinidad and Tobago	195	195	181	183	177	-	7.318	8.692	8.315	7.383	7.994	-
Tunisia	-	-	-	497	507	502	-	-	-	18	19	22
Turkey	389	366	828	812	767	-	104	100	51	54	53	-
Ukraine	15,805	16,949	16,533	14,672	14,610	14,405	35	32	28	29	24	22
United Kingdom	14,981	14,366	14,264	26,170	27,729	-	22	22	21	14	14	-
United Republic of Tanzania . .	85	83	-	-	-	-	55	47	-	-	-	-
United States of America . . .	-	-	47,758	-	-	-	-	-	26	-	-	-
Venezuela	1,162	1,194	1,209	984	1,007	902	30	29	30	32	28	31
Yugoslavia	658	1,201	2,903	4,272	4,352	4,280	-	84	34	22	21	20
Zambia	67	-	-	-	77	-	85	-	-	-	56	-
Zimbabwe	171	177	171	161	178	-	92	80	85	82	85	-

Depending on the table, * means *Enterprises*, *Engaged*, or *Factor Values*; ± means *Basis Unspecified*; § means *shown in millions*. For additional notes and sources, see Appendix I.

541

Employment and Compensation of Employees

Country	Number Employed/Engaged (000)						Wage/Salary per Employee ($)					
	1990	1991	1992	1993	1994	1995	1990	1991	1992	1993	1994	1995
Albania	-	-	-	1.685	2.297	1.109	-	-	-	-	833	866
Angola	-	*0.954	*0.541	*0.281	-	-	-	-	-	-	-	-
Argentina	76	-	-	51	55	-	6,848	-	-	10,930	-	-
Armenia	*9.686	*8.509	-	-	-	-	-	-	-	-	-	-
Australia	106	*98	*89	-	-	-	19,157	20,909	19,952	-	-	-
Austria	58	57	55	54	55	54	24,670	25,736	29,959	28,634	29,320	36,422
Azerbaijan	9.052	8.480	7.013	6.780	4.766	-	2,259	2,127	-	438	-	-
Bahrain	-	-	1.604	-	-	-	-	-	4,288	-	-	-
Bangladesh	18	22	19	-	-	-	471	489	452	-	-	-
Barbados	0.550	0.590	0.505	0.338	0.707	0.599	10,702	10,271	10,988	11,982	10,401	11,701
Belize	-	0.180	0.168	-	-	-	-	3,817	5,301	-	-	-
Bermuda	0.035	0.036	0.034	0.034	0.033	0.024	-	-	-	-	-	-
Bolivia	1.053	1.212	1.209	1.282	1.201	1.295	1,268	1,352	1,470	1,592	1,750	1,741
Bosnia & Herzegovina	36	34	-	-	1.035	-	2,621	2,199	-	-	-	-
Bulgaria	82	66	51	45	56	40	1,864	612	996	1,178	961	1,263
Burundi	0.355	0.340	-	-	-	-	3,948	4,051	-	-	-	-
Canada	135	123	109	106	109	111	25,540	26,731	26,201	24,541	24,078	24,487
Central African Republic	0.031	-	-	-	-	-	5,806	-	-	-	-	-
Chile	21	20	21	23	25	26	4,512	4,909	5,906	6,336	6,924	8,076
China	*1,830	*1,870	*1,840	*1,920	*1,960	*1,930	-	-	-	-	-	-
China (Hong Kong SAR)	*51	*46	*43	*33	*27	-	7,532	8,642	9,732	11,222	12,034	-
China (Taiwan Province)	240	243	251	258	270	265	9,520	10,349	12,184	12,202	13,082	13,643
Colombia	28	26	30	33	33	33	2,098	2,111	2,176	2,453	3,258	3,884
Costa Rica	4.539	4.315	4.637	5.211	-	-	2,615	2,334	2,654	3,014	-	-
Croatia	53	38	31	29	27	25	3,387	3,667	1,487	1,731	2,171	3,570
Cyprus	2.697	2.709	2.731	2.790	3.033	3.058	9,089	9,837	11,026	10,394	10,661	14,189
Czechoslovakia (Former)	164	107	-	-	-	-	2,116	1,512	-	-	-	-
Czech Republic	67	62	83	85	-	-	2,104	1,494	1,987	2,409	-	-
Denmark	46	45	44	43	46	47	29,035	29,282	32,985	30,528	35,229	-
Ecuador	6.688	7.773	7.787	7.718	7.935	5.653	2,760	3,043	3,285	3,792	2,277	3,304
Egypt	*50	*49	*53	*47	*47	*47	1,921	2,011	1,549	1,512	1,529	1,558
El Salvador	-	-	*1.361	*1.352	*1.353	*2.032	-	-	-	2,113	2,992	2,794
Equatorial Guinea	0.015	-	-	-	-	-	1,587	-	-	-	-	-
Ethiopia	1.522	1.439	1.536	1.566	1.798	2.051	2,576	2,698	1,802	1,184	1,059	914
Fiji	0.756	0.779	0.758	0.909	0.917	-	3,347	3,589	3,656	3,239	3,534	-
Finland	32	30	25	23	23	21	28,090	26,661	24,758	20,137	23,270	30,340
FYR Macedonia	11	11	10	9.886	11	8.762	4,512	5,168	2,035	1,998	2,220	3,655
France	354	351	335	310	299	303	37,891	38,142	-	-	-	-
Gabon	-	0.714	0.548	0.523	0.665	0.573	-	23,125	18,187	22,918	11,647	16,055
Gambia	-	-	-	0.059	-	-	-	-	-	-	-	-
Germany	-	*787	-	*706	*673	-	-	-	-	30,937	32,724	-
Germany (Western Part)	*660	*684	*681	-	-	-	29,000	29,917	33,393	-	-	-
Ghana	-	-	-	*3.553	*3.653	*3.756	-	-	-	1,072	801	817
Greece	*20	*18	*18	*18	*17	*18	11,366	11,550	12,423	11,857	11,903	13,672
Grenada	-	-	*0.137	*0.093	-	-	-	-	-	-	-	-
Guatemala	-	*2.240	*2.525	*2.360	*2.351	*2.086	-	-	-	-	-	-
Honduras	*2.854	*2.956	*2.798	*3.066	*3.360	*4.089	2,559	2,192	2,642	2,721	2,652	2,187
Hungary	50	44	32	42	48	48	2,439	2,591	3,064	3,150	3,266	3,373
Iceland	2.146	1.960	1.857	1.740	1.730	-	27,784	32,744	30,077	23,816	-	-
India	*233	*231	*240	*237	*247	*263	1,235	1,033	1,086	1,011	1,153	1,342
Indonesia	80	95	117	118	130	147	822	1,136	1,432	1,169	1,240	1,849
Iran (Islamic Republic of)	34	31	35	33	-	-	3,945	4,753	5,137	4,892	-	-
Iraq	-	6.302	8.000	-	-	-	-	8,874	29,465	-	-	-
Ireland	13	12	11	11	12	12	18,242	18,195	19,713	17,176	18,033	19,864
Israel	38	43	43	44	47	48	21,746	21,763	22,960	20,499	21,185	24,764
Italy	169	172	238	233	234	-	32,469	34,600	35,982	29,142	29,586	-
Japan	*889	*917	*897	*900	*882	*882	26,632	30,802	34,039	38,989	-	-
Jordan	3.538	4.156	6.826	5.434	6.836	7.456	2,229	1,967	1,948	1,872	1,965	2,267
Kenya	12	12	12	12	12	13	1,885	1,712	1,601	977	1,138	1,471
Korea, Republic of	175	173	165	185	199	207	9,444	11,126	11,781	12,836	14,342	16,980

Continued.

Depending on the table, * means *Enterprises, Engaged,* or *Factor Values;* ± means *Basis Unspecified;* § means *shown in millions.* For additional notes and sources, see Appendix I.

Employment and Compensation of Employees

- Continued -

Country	Number Employed/Engaged (000)						Wage/Salary per Employee ($)					
	1990	1991	1992	1993	1994	1995	1990	1991	1992	1993	1994	1995
Kuwait	5.775	4.815	5.495	5.833	5.876	5.871	4,660	4,210	6,827	6,828	7,107	9,350
Kyrgyzstan	*38	*31	*20	*14	*11	-	-	-	-	-	-	-
Latvia	15	11	8.135	4.746	4.047	4.086	-	-	-	-	-	-
Lesotho	*1.143	-	-	-	-	-	-	-	-	-	-	-
Liechtenstein	-	*1.254	-	-	-	-	-	-	-	-	-	-
Lithuania	-	-	13	10	11	-	-	-	-	478	834	-
Luxembourg	4.506	4.752	4.781	4.676	5.077	-	26,006	27,509	30,590	29,485	31,152	-
Macau	0.890	0.785	0.743	0.698	0.765	0.525	4,550	5,260	5,784	6,424	6,777	6,992
Malawi	1.400	1.300	1.100	1.000	1.100	-	1,016	1,098	1,135	1,363	905	-
Malaysia	33	39	44	47	49	51	3,251	3,308	4,118	4,342	4,695	5,228
Mauritius	*1.137	1.021	1.192	1.188	1.303	1.452	2,834	3,119	4,059	5,099	4,423	4,574
Mexico	52	52	49	41	41	36	4,629	5,613	6,847	7,837	7,937	4,853
Mongolia	*0.758	*0.727	*0.685	*1.093	*1.419	*1.046	5,371	6,646	1,938	-	-	679
Morocco	17	22	25	27	23	24	5,178	4,349	4,299	4,171	4,620	5,080
Mozambique	2.476	2.404	2.433	2.245	2.204	-	804	619	512	464	462	-
Myanmar	1.153	1.122	1.274	1.271	1.268	-	1,305	1,337	1,476	1,705	1,806	-
Namibia	-	-	-	-	1.449	-	-	-	-	-	8,221	-
Nepal	2.841	4.054	-	4.433	5.180	-	514	512	-	435	416	-
Netherlands	64	69	70	65	65	64	26,019	26,249	29,422	28,466	29,314	34,882
Netherlands Antilles	-	-	-	1.314	-	-	-	-	-	18,449	-	-
New Zealand	17	16	16	17	-	-	17,619	16,458	15,823	19,656	-	-
Norway	21	20	18	13	13	13	27,554	27,868	36,152	28,414	30,186	34,514
Pakistan	11	12	12	-	-	-	1,329	1,277	1,313	-	-	-
Panama	*1.080	*1.154	*1.223	*1.266	*1.355	*1.407	4,957	5,489	5,560	5,587	5,667	5,730
Paraguay	-	0.542	-	-	-	-	-	-	-	-	-	-
Peru	*15	*14	*14	-	-	-	2,175	3,295	3,315	-	-	-
Philippines	37	40	28	31	-	-	1,437	1,474	2,022	1,862	-	-
Poland	196	174	160	145	153	163	1,172	1,787	2,116	2,371	2,750	3,576
Portugal	*80	*88	*84	*80	*78	*78	-	-	-	-	-	-
Puerto Rico	3.390	3.300	2.910	2.840	2.920	2.960	22,153	17,848	19,381	19,648	20,445	21,723
Republic of Moldova	32	28	27	17	4.138	4.941	-	3,336	-	-	459	512
Romania	189	185	181	153	136	-	1,739	1,081	690	837	912	-
Russian Federation	-	-	-	421	351	321	-	-	-	615	1,007	1,055
Saint Vincent & the Grenadines	-	-	-	-	-	0.061	-	-	-	-	-	-
Senegal	*0.817	*0.577	*0.585	*0.545	*0.583	-	5,228	6,997	6,833	7,335	4,282	-
Sierra Leone	-	-	-	*0.861	-	-	-	-	-	404	-	-
Singapore	*28	*31	*30	*33	*33	*38	10,927	12,364	14,282	15,096	17,276	19,681
Slovakia	-	*25	*23	*24	*39	-	-	1,628	1,994	2,250	2,269	-
South Africa	137	129	121	120	122	123	7,851	7,605	8,409	8,390	8,456	9,563
Spain	145	145	141	198	201	212	17,611	19,106	20,957	17,077	17,019	19,378
Sri Lanka	*4.153	*4.207	*2.801	*3.531	-	-	559	805	731	833	-	-
Swaziland	0.563	0.429	0.401	0.493	-	-	4,383	5,942	7,030	5,694	-	-
Sweden	73	*65	*58	51	54	54	23,215	25,448	27,768	21,780	23,364	28,446
Thailand	53	65	-	-	-	-	2,224	2,632	-	-	-	-
Tonga	0.090	0.043	0.043	0.054	0.039	-	2,585	2,081	2,244	2,248	2,350	-
Trinidad and Tobago	1.427	1.695	1.505	1.351	1.415	-	5,808	4,261	5,061	6,335	3,721	-
Tunisia	-	-	-	8.951	9.429	11	-	-	-	-	-	-
Turkey	40	36	43	44	41	42	5,599	7,309	6,314	6,763	4,374	5,579
Ukraine	561	544	469	427	354	320	-	2,866	2,353	426	508	581
United Kingdom	337	314	297	363	381	381	21,725	23,375	24,598	21,757	22,131	23,623
United Republic of Tanzania	4.679	3.923	-	-	-	-	-	440	-	-	-	-
USSR (Former)	602	-	-	-	-	-	2,214	-	-	-	-	-
United States of America	1,297	1,225	1,231	1,247	1,287	1,338	24,580	25,380	26,896	27,273	27,876	28,443
Uruguay	*7.407	*7.492	*7.125	*6.862	*7.621	*10	3,034	4,423	4,711	5,833	6,130	6,937
Venezuela	34	35	36	31	28	28	3,198	3,577	4,096	3,876	3,327	4,606
Yugoslavia	-	101	98	95	91	88	-	-	-	-	1,611	732
Zambia	5.718	-	-	-	4.275	-	2,179	-	-	-	1,660	-
Zimbabwe	16	14	15	13	15	15	3,545	3,145	2,525	2,507	2,416	2,702

Female Workers: Total and Percent of Employment

Country	Female Workers (000)						Female Workers as % of Total Employment					
	1990	1991	1992	1993	1994	1995	1990	1991	1992	1993	1994	1995
Albania	-	-	-	0.366	0.770	0.448	-	-	-	21.72	33.52	40.40
Angola	-	-	0.059	0.029	-	-	-	-	10.91	10.32	-	-
Argentina	-	-	-	-	4.385	-	-	-	-	-	7.97	-
Australia	18	-	-	-	-	-	16.89	-	-	-	-	-
Austria	12	12	12	11	11	-	21.07	21.05	21.82	19.85	19.87	-
Bahrain	-	-	0.011	-	-	-	-	-	0.69	-	-	-
Bangladesh	0.124	0.196	0.111	-	-	-	0.70	0.88	0.57	-	-	-
Bermuda	0.006	0.007	0.008	0.008	0.007	0.003	17.14	19.44	23.53	23.53	21.21	12.50
Bosnia & Herzegovina	8.320	7.071	-	-	-	-	23.20	20.65	-	-	-	-
Bulgaria	-	-	-	21	25	17	-	-	-	45.88	45.78	43.39
Canada	22	18	-	-	-	-	16.30	14.63	-	-	-	-
Chile	1.269	1.279	1.502	1.572	1.801	-	5.95	6.33	7.01	6.93	7.28	-
China (Taiwan Province)	75	78	83	86	89	88	31.30	31.95	33.18	33.25	33.04	33.00
Colombia	-	4.729	-	6.021	6.047	-	-	17.96	-	18.41	18.54	-
Croatia	12	9.790	7.660	7.140	6.380	5.830	23.27	25.59	24.99	24.71	23.60	23.25
Cyprus	0.379	0.425	0.443	0.414	0.468	-	14.05	15.69	16.22	14.84	15.43	-
Czechoslovakia (Former)	49	32	-	-	-	-	30.06	30.09	-	-	-	-
Czech Republic	30	27	27	28	-	-	44.78	43.55	32.53	32.94	-	-
Denmark	9.881	9.771	9.609	-	-	-	21.70	21.95	21.83	-	-	-
Egypt	-	2.737	2.774	2.540	-	-	-	5.63	5.27	5.46	-	-
El Salvador	-	-	-	0.122	0.133	0.200	-	-	-	9.02	9.83	9.84
Ethiopia	-	0.220	0.232	0.261	0.276	0.312	-	15.29	15.10	16.67	15.35	15.21
FYR Macedonia	1.878	1.667	1.462	1.480	1.248	0.846	17.29	14.69	14.28	14.97	11.39	9.66
Germany (Western Part)	145	-	-	-	-	-	21.96	-	-	-	-	-
Ghana	-	-	-	0.123	-	-	-	-	-	3.46	-	-
Hungary	17	-	12	10	-	-	34.00	-	37.50	23.81	-	-
Indonesia	-	-	-	28	31	33	-	-	-	23.88	23.79	22.11
Iran (Islamic Republic of)	-	0.731	0.748	0.701	-	-	-	2.36	2.16	2.16	-	-
Iraq	-	0.429	-	-	-	-	-	6.81	-	-	-	-
Ireland	1.700	1.500	-	-	-	-	13.49	12.71	-	-	-	-
Italy	-	30	39	38	40	-	-	17.34	16.50	16.26	17.03	-
Jordan	0.081	0.052	0.139	0.087	0.100	0.121	2.29	1.25	2.04	1.60	1.46	1.62
Kenya	0.661	0.683	0.692	0.706	-	-	5.58	5.75	5.78	5.79	-	-
Korea, Republic of	35	38	36	40	42	43	19.79	21.87	22.12	21.51	21.10	20.57
Macau	0.036	0.034	0.043	0.044	0.037	0.031	4.04	4.33	5.79	6.30	4.84	5.90
Malaysia	10	11	13	14	13	-	30.49	29.38	30.16	29.34	26.48	-
Mongolia	-	-	-	-	0.402	-	-	-	-	-	28.33	-
Morocco	-	-	-	-	-	1.815	-	-	-	-	-	7.48
Myanmar	0.132	0.132	0.136	0.216	0.200	-	11.45	11.76	10.68	16.99	15.77	-
Nepal	0.049	0.121	-	0.051	0.121	-	1.72	2.98	-	1.15	2.34	-
New Zealand	2.993	2.936	2.821	2.879	-	-	17.56	18.38	17.58	16.59	-	-
Panama	0.097	-	-	-	-	-	8.98	-	-	-	-	-
Philippines	-	-	3.807	5.025	-	-	-	-	13.40	16.31	-	-
Portugal	4.054	4.138	4.318	4.266	4.010	-	5.09	4.71	5.16	5.31	5.16	-
Puerto Rico	0.600	0.600	0.630	0.590	0.690	0.640	17.70	18.18	21.65	20.77	23.63	21.62
Republic of Moldova	-	-	-	-	1.033	1.381	-	-	-	-	24.96	27.95
Senegal	0.032	-	-	-	-	-	3.92	-	-	-	-	-
Sri Lanka	0.734	0.466	0.364	0.297	-	-	17.67	11.08	13.00	8.41	-	-
Sweden	16	-	-	-	-	-	21.40	-	-	-	-	-
Thailand	18	19	-	-	-	-	33.47	28.99	-	-	-	-
Turkey	3.634	-	-	-	-	-	9.00	-	-	-	-	-
United Kingdom	61	-	60	-	-	-	18.15	-	20.20	-	-	-
United Republic of Tanzania	0.541	0.477	-	-	-	-	11.56	12.16	-	-	-	-
United States of America	-	-	906	928	965	1,008	-	-	73.60	74.42	74.98	75.34
Zambia	-	-	-	-	0.148	-	-	-	-	-	3.46	-

Depending on the table, * means *Enterprises, Engaged,* or *Factor Values*; ± means *Basis Unspecified*; § means *shown in millions.* For additional notes and sources, see Appendix I.

Output and Output per Employee

Country	Output ($ bil.)						Output per Employee ($)					
	1990	1991	1992	1993	1994	1995	1990	1991	1992	1993	1994	1995
Albania	-	-	-	*0.003	*0.004	*0.006	-	-	-	1,628	1,758	5,632
Angola	-	-	±50	±4.333	-	-	-	-	§92,05	§15,42	-	-
Argentina	±2.900	-	-	3.563	-	-	38,025	-	-	70,442	-	-
Armenia	0.006	0.005	0.016	0.489	0.004	-	604	614	-	-	-	-
Australia	*9.840	*9.616	*8.496	-	-	-	92,829	98,121	95,464	-	-	-
Austria	5.915	5.926	6.557	6.037	6.555	8.013	102,158	103,963	119,212	112,599	118,715	147,720
Azerbaijan	±0.209	±0.188	±0.011	±0.043	±0.011	-	23,061	22,189	1,526	6,272	2,400	-
Bahrain	-	-	±0.041	-	-	-	-	-	25,712	-	-	-
Bangladesh	0.088	0.112	0.085	-	-	-	4,993	5,017	4,418	-	-	-
Barbados	0.038	0.038	0.027	0.027	0.040	0.039	68,209	64,355	53,578	79,001	56,200	64,531
Belize	0.004	0.005	0.006	-	-	-	-	25,114	35,098	-	-	-
Benin	±0.012	±0.016	±0.017	±0.014	-	-	-	-	-	-	-	-
Bolivia	0.022	0.027	0.027	0.026	0.027	0.024	21,005	22,388	22,724	19,935	22,849	18,335
Bosnia & Herzegovina	±0.836	±0.772	-	-	-	-	23,304	22,537	-	-	-	-
Bulgaria	±1.015	±0.321	±0.345	±0.270	±0.340	±0.332	12,343	4,861	6,751	6,011	6,107	8,268
Burundi	±0.019	±0.020	-	-	-	-	54,601	59,130	-	-	-	-
Canada	15	13	12	11	12	12	108,242	105,804	107,022	104,789	108,968	111,126
Cape Verde	±0.001											
Central African Republic	*0.002	-	-	-	-	-	52,132	-	-	-	-	-
Chile	0.865	0.882	1.063	1.122	1.238	1.515	40,585	43,633	49,588	49,492	50,015	58,294
China	±11	±12	±15	±23	±20	±20	5,970	6,180	7,888	11,769	10,111	10,241
China (Hong Kong SAR)	2.294	2.581	2.515	2.312	2.052	-	44,798	55,615	59,045	70,929	75,997	-
China (Taiwan Province)	±8.748	±9.448	±11	±14	±13	±15	36,406	38,856	41,809	55,110	49,694	56,017
Colombia	0.695	0.618	0.667	0.732	0.951	1.142	24,381	23,491	21,925	22,389	29,168	34,851
Costa Rica	0.085	0.086	0.107	0.124	0.132	0.142	18,726	19,966	23,059	23,766	-	-
Croatia	±0.903	±0.742	±0.389	-	-	-	17,138	19,391	12,682	-	-	-
Cyprus	0.159	0.172	0.171	0.151	0.159	0.209	58,821	63,457	62,642	54,244	52,442	68,383
Czechoslovakia (Former)	±1.824	±0.709	-	-	-	-	11,122	6,623	-	-	-	-
Denmark	*3.640	*3.724	*4.171	*3.795	*4.667	*5.639	79,918	83,651	94,737	87,744	101,702	120,498
Ecuador	0.149	0.186	0.210	0.238	0.245	0.209	22,225	23,884	26,937	30,856	30,859	36,935
Egypt	*0.635	*0.414	*0.802	*0.576	*0.606	*0.653	12,782	8,518	15,239	12,379	12,963	13,979
El Salvador	-	±0.010	-	±0.020	±0.016	±0.030	-	-	-	14,796	11,851	14,735
Equatorial Guinea	±0.000	-	-	-	-	-	7,346	-	-	-	-	-
Ethiopia	0.017	0.013	0.009	0.008	0.015	0.014	10,890	9,270	5,754	5,325	8,353	6,856
Fiji	0.026	0.028	0.030	0.030	0.033	-	33,731	36,026	39,713	33,441	36,493	-
Finland	±3.834	±3.176	±2.652	±2.324	±2.748	±3.289	120,195	106,223	107,792	101,950	117,443	156,838
FYR Macedonia	±0.173	±0.215	±0.113	±0.084	±0.093	±0.102	15,883	18,979	11,015	8,457	8,524	11,671
France	±42	±41	±43	±36	±40	±48	117,621	116,233	127,136	116,323	134,477	158,393
Gabon	-	*0.072	*0.047	*0.056	*0.037	*0.045	-	100,478	85,605	107,210	55,394	78,654
Germany	-	-	-	81	85	-	-	-	-	115,219	125,550	-
Germany (Western Part)	75	80	88	-	-	-	113,506	116,531	129,421	-	-	-
Ghana	-	-	-	0.051	0.042	0.044	-	-	-	14,472	11,465	11,701
Greece	*1.238	*1.158	*1.323	*1.226	*1.225	*1.412	60,908	63,969	73,199	69,788	70,020	80,362
Guatemala	-	0.072	0.072	0.056	0.085	0.070	-	31,934	28,641	23,611	35,952	33,318
Honduras	0.068	0.060	0.072	0.088	0.089	0.090	23,889	20,433	25,766	28,802	26,387	22,056
Hungary	±0.742	±0.728	±0.698	±0.777	±0.896	±0.923	14,840	16,543	21,825	18,507	18,743	19,096
Iceland	0.174	0.185	±0.175	±0.144	-	-	80,992	94,228	93,996	82,605	-	-
India	*3.345	*2.906	*2.784	*3.125	*3.786	*4.780	14,350	12,582	11,604	13,168	15,320	18,171
Indonesia	1.367	1.477	2.079	2.139	2.468	3.129	17,003	15,494	17,750	18,163	18,947	21,244
Iran (Islamic Republic of)	0.936	0.793	0.872	0.781	-	-	27,620	25,587	25,193	24,016	-	-
Iraq	-	*0.163	*0.396	-	-	-	-	25,889	49,558	-	-	-
Ireland	*1.098	*0.993	*1.047	*0.913	*1.005	*1.149	87,104	84,116	91,387	79,677	83,569	92,028
Israel	2.507	2.745	2.957	2.928	3.232	3.713	65,122	64,587	67,981	66,845	69,209	78,119
Italy	±22	±23	±34	±27	±29	-	131,650	134,788	140,770	113,680	122,895	-
Japan	±138	±164	±169	±181	±192	±212	155,627	178,306	188,293	200,629	217,894	240,234
Jordan	0.059	0.068	0.132	0.099	0.123	0.146	16,742	16,426	19,286	18,229	17,956	19,573
Kenya	*0.369	*0.404	*0.329	*0.203	*0.223	*0.299	31,154	33,943	27,472	16,675	17,924	22,731
Korea, Republic of	12	14	13	15	19	25	69,460	79,509	80,640	81,393	97,483	118,330
Kuwait	0.099	0.166	0.219	0.237	0.229	0.317	17,111	34,487	39,764	40,574	38,907	53,953
Kyrgyzstan	0.182	0.144	0.069	0.012	0.010	-	4,828	4,614	3,488	846	929	-

Continued.

Depending on the table, * means *Enterprises, Engaged,* or *Factor Values*; ± means *Basis Unspecified*; § means *shown in millions*. For additional notes and sources, see Appendix I.

545

Output and Output per Employee
- Continued -

Country	Output ($ bil.)						Output per Employee ($)					
	1990	1991	1992	1993	1994	1995	1990	1991	1992	1993	1994	1995
Latvia	0.002	0.003	0.014	0.013	0.022	0.040	115	256	1,685	2,637	5,467	9,741
Lithuania	-	-	0.019	0.020	0.034	-	-	-	1,540	1,968	3,187	-
Luxembourg	*0.555	*0.558	*0.595	*0.519	*0.596	-	123,169	117,433	124,443	110,919	117,467	-
Macau	0.019	0.019	0.020	0.019	0.021	0.017	20,919	24,355	26,615	27,509	26,886	33,315
Malawi	*0.027	*0.041	*0.031	*0.029	*0.018	-	19,579	31,446	28,383	29,118	16,680	-
Malaysia	*1.191	*1.484	*1.984	*2.137	*2.447	*2.878	36,317	38,258	44,979	45,766	49,836	56,057
Mauritius	0.029	0.027	0.025	0.045	0.033	0.041	25,238	26,876	20,829	37,895	25,314	28,246
Mexico	2.138	2.354	2.553	2.302	2.340	1.813	41,427	45,698	52,477	55,622	57,606	49,771
Mongolia	0.029	0.026	0.007	0.002	0.003	0.004	37,952	35,081	10,801	1,704	2,198	3,411
Morocco	±0.647	±0.736	±0.877	±0.741	±0.789	±0.956	38,609	33,241	34,728	27,630	33,847	39,392
Mozambique	-	±0.007	-	±0.006	±0.004	-	-	2,784	-	2,539	1,906	-
Myanmar	0.004	0.004	0.004	0.004	0.007	-	3,189	3,217	3,370	3,510	5,280	-
Namibia	-	-	-	-	0.058	-	-	-	-	-	40,291	-
Nepal	0.023	0.037	-	0.035	0.036	-	8,234	9,091	-	7,913	6,952	-
Netherlands	8.162	8.598	9.292	7.792	8.054	9.803	126,709	124,147	133,099	119,030	123,469	152,878
New Zealand	1.468	1.269	1.251	1.439	-	-	86,134	79,426	77,944	82,893	-	-
Norway	±2.102	±1.997	±1.950	±1.290	±1.380	±1.703	99,002	99,686	108,318	101,531	110,360	126,236
Pakistan	0.132	0.140	0.152	-	-	-	12,174	11,868	12,315	-	-	-
Panama	0.053	0.065	0.072	0.077	0.086	0.092	49,421	56,176	59,064	60,495	63,637	65,607
Peru	0.422	0.434	0.486	-	-	-	27,483	31,038	34,984	-	-	-
Philippines	0.454	0.513	0.495	0.528	-	-	12,414	12,695	17,437	17,139	-	-
Poland	±1.912	±2.159	±1.895	±2.017	±2.470	±3.402	9,753	12,409	11,847	13,907	16,108	20,936
Portugal	2.552	3.021	3.326	2.796	2.864	3.308	32,030	34,351	39,763	34,786	36,870	42,151
Puerto Rico	±0.415	±0.408	±0.403	±0.377	±0.403	±0.406	122,460	123,667	138,660	132,887	137,911	137,061
Republic of Moldova	0.000	0.423	0.015	0.035	0.023	0.023	0.020	14,871	546	2,017	5,565	4,738
Romania	±2.305	±1.393	±0.910	±0.877	-	-	12,188	7,533	5,024	5,716	-	-
Russian Federation	-	-	-	1.246	1.530	2.199	-	-	-	2,961	4,357	6,845
Senegal	0.038	-	-	-	-	-	46,998	-	-	-	-	-
Sierra Leone	-	-	-	0.003	-	-	-	-	-	3,719	-	-
Singapore	*2.098	*2.390	*2.678	*3.026	*3.666	*4.721	73,785	78,126	88,677	92,811	109,597	124,495
Slovakia	-	±0.277	±0.274	±0.253	±0.455	-	-	10,894	12,121	10,754	11,571	-
Slovenia	-	-	-	0.580	0.686	1.154	-	-	-	-	-	-
South Africa	±4.255	±4.336	±4.494	±4.059	±4.071	±4.634	31,058	33,612	37,144	33,824	33,370	37,782
Spain	*13	*14	*14	±15	±16	±21	92,775	96,237	100,530	73,435	78,069	97,560
Sri Lanka	0.026	0.034	0.024	0.037	-	-	6,257	8,175	8,466	10,495	-	-
Swaziland	*0.017	*0.018	*0.018	*0.018	-	-	29,947	42,235	45,746	37,191	-	-
Sweden	*8.402	*8.434	*8.019	*5.324	*6.543	*8.279	115,250	128,860	137,952	105,364	121,872	152,830
Thailand	3.232	2.207	-	-	-	-	60,560	34,154	-	-	-	-
Tonga	0.001	0.001	0.001	0.001	0.000	-	11,164	21,496	20,388	10,423	12,605	-
Trinidad and Tobago	±0.035	±0.043	±0.044	±0.039	±0.035	-	24,750	25,408	29,224	28,882	24,798	-
Turkey	2.037	2.127	2.712	2.954	2.074	2.724	50,427	58,379	63,714	67,385	51,084	65,636
Ukraine	*6.820	*6.376	*6.294	*1.354	*1.328	*1.352	12,156	11,721	13,420	3,172	3,753	4,224
United Kingdom	*31	*29	28	27	31	33	92,645	93,089	95,151	75,096	82,269	87,626
United Republic of Tanzania	*0.028	*0.034	-	-	-	-	5,967	8,571	-	-	-	-
USSR (Former)	±6.427	-	-	-	-	-	10,676	-	-	-	-	-
United States of America	*145	*140	*147	*155	*168	*181	111,827	114,057	119,629	124,153	130,477	135,342
Uruguay	0.154	0.197	0.186	0.207	0.241	0.370	20,810	26,333	26,109	30,170	31,645	35,742
Venezuela	0.948	1.106	1.250	1.109	0.894	1.221	27,551	31,786	34,526	35,466	31,406	43,681
Yugoslavia	±0.234	±0.236	±0.403	-	±0.864	±0.492	-	2,341	4,129	-	9,522	5,626
Zambia	0.073	-	-	-	0.037	-	12,763	-	-	-	8,540	-
Zimbabwe	*0.291	*0.279	*0.224	*0.189	*0.197	*0.217	18,434	19,619	15,460	14,315	13,074	14,726

Depending on the table, * means *Enterprises, Engaged,* or *Factor Values;* ± means *Basis Unspecified;* § means *shown in millions.* For additional notes and sources, see Appendix I.

Value Added and Value Added per Employee

Country	Value Added ($ bil.)						Value Added per Employee ($)					
	1990	1991	1992	1993	1994	1995	1990	1991	1992	1993	1994	1995
Argentina	±1.611	-	-	1.375	-	-	21,123	-	-	27,174	-	-
Australia	*4.215	*4.196	*4.322	-	-	-	39,763	42,820	48,562	-	-	-
Austria	2.534	2.480	2.773	2.527	2.829	3.467	43,758	43,510	50,426	47,129	51,233	63,908
Bangladesh	0.022	0.030	0.022	-	-	-	1,249	1,352	1,153	-	-	-
Barbados	0.010	0.013	0.010	0.011	0.018	0.015	18,674	22,186	18,823	31,312	25,037	25,622
Belize	0.001	0.001	0.001	-	-	-	-	7,339	7,238	-	-	-
Bolivia	0.006	0.010	0.010	0.009	0.010	0.009	6,070	7,937	8,023	6,681	8,547	6,782
Bosnia & Herzegovina	±0.402	±0.435	-	-	-	-	11,199	12,692	-	-	-	-
Bulgaria	-	±0.124	±0.135	±0.113	-	-	-	1,882	2,641	2,527	-	-
Burundi	0.002	0.002	-	-	-	-	6,888	7,193	-	-	-	-
Canada	6.454	5.813	5.179	4.876	5.250	5.446	47,804	47,260	47,515	45,996	48,169	49,043
Central African Republic	*0.000	-	-	-	-	-	11,730	-	-	-	-	-
Chile	0.366	0.339	0.437	0.487	0.523	0.639	17,191	16,753	20,412	21,500	21,120	24,602
China	±2.946	±2.993	±3.660	±6.914	±5.106	±4.597	1,610	1,601	1,989	3,601	2,605	2,382
China (Hong Kong SAR)	0.717	0.762	0.793	0.706	0.632	-	14,003	16,423	18,611	21,663	23,401	-
China (Taiwan Province)	3.167	3.707	4.527	4.649	4.868	5.016	13,179	15,243	18,023	18,022	18,027	18,909
Colombia	0.279	0.267	0.290	0.301	0.429	0.515	9,780	10,123	9,520	9,220	13,142	15,718
Costa Rica	0.020	0.020	0.025	0.029	0.030	0.032	4,321	4,602	5,418	5,564	-	-
Cote d'Ivoire	0.749	0.759	0.737	0.675	-	-	-	-	-	-	-	-
Croatia	±0.411	±0.363	±0.192	-	-	-	7,791	9,484	6,280			
Cyprus	0.055	0.060	0.060	0.056	0.058	0.071	20,358	22,223	22,070	19,972	19,134	23,323
Czechoslovakia (Former)	±0.602	-	-	-	-	-	3,669	-	-	-	-	-
Denmark	*1.837	*1.909	*2.138	*1.946	*2.395	*2.894	40,342	42,881	48,576	44,986	52,182	61,840
Ecuador	0.044	0.054	0.070	0.086	0.073	0.052	6,560	6,928	8,949	11,199	9,217	9,225
Egypt	*0.144	*0.131	*0.306	*0.236	*0.249	*0.269	2,906	2,703	5,824	5,076	5,323	5,752
El Salvador	-	0.006	-	0.010	0.012	0.028	-	-	-	7,073	9,075	14,004
Ethiopia	0.008	0.006	0.004	0.004	0.006	0.006	4,980	4,417	2,810	2,557	3,533	3,083
Fiji	0.005	0.006	0.007	0.007	0.007	-	7,244	7,451	8,766	7,162	7,815	-
Finland	±1.759	±1.450	±1.181	±0.971	±1.148	±1.119	55,154	48,488	47,997	42,567	49,055	53,373
FYR Macedonia	±0.090	±0.107	±0.063	±0.047	±0.052	±0.051	8,247	9,401	6,115	4,758	4,732	5,812
France	±20	±20	±21	±18	±20	±24	56,706	57,627	64,038	59,583	67,213	78,558
Gabon	-	*0.025	*0.016	*0.021	*0.014	*0.015	-	35,447	30,038	39,212	20,338	26,320
Germany (Western Part)	39	42	46	41	-	-	59,341	61,774	67,254	-	-	-
Ghana	-	-	-	0.021	0.018	0.019	-	-	-	5,857	4,937	5,038
Greece	*0.449	*0.422	*0.502	*0.465	*0.465	*0.536	22,089	23,295	27,779	26,493	26,586	30,520
Honduras	*0.021	*0.016	*0.016	*0.018	*0.018	*0.021	7,279	5,286	5,858	5,993	5,398	5,081
Hungary	±0.275	±0.230	±0.202	±0.262	±0.295	±0.300	5,507	5,229	6,304	6,231	6,176	6,212
Iceland	0.085	0.087	±0.072	±0.059	-	-	39,395	44,530	38,536	33,969	-	-
India	*0.614	*0.586	*0.520	*0.614	*0.738	*0.926	2,634	2,539	2,166	2,588	2,989	3,519
Indonesia	*0.402	*0.479	*0.744	*0.833	*0.963	*1.150	5,001	5,023	6,354	7,075	7,390	7,806
Iran (Islamic Republic of)	0.357	0.363	0.398	0.383	-	-	10,521	11,722	11,515	11,783	-	-
Iraq	-	*0.098	*0.287	-	-	-	-	15,490	35,852	-	-	-
Ireland	*0.469	*0.453	*0.478	*0.417	*0.458	*0.524	37,248	38,416	41,738	36,381	38,110	41,925
Israel	1.228	1.288	1.209	1.213	1.343	1.529	31,897	30,301	27,793	27,695	28,752	32,173
Italy	±8.014	±8.655	±13	±9.970	±10	-	47,398	50,442	52,580	42,736	44,654	-
Japan	±63	±75	±79	±85	±93	±103	70,759	81,786	87,795	94,794	105,325	116,693
Jordan	0.023	0.022	0.038	0.034	0.045	0.048	6,562	5,233	5,509	6,279	6,646	6,399
Kenya	*0.064	*0.062	*0.056	*0.035	*0.045	*0.051	5,405	5,219	4,717	2,854	3,579	3,867
Korea, Republic of	5.145	5.737	5.974	7.096	9.105	11	29,347	33,214	36,240	38,275	45,807	51,150
Kuwait	0.054	0.060	0.099	0.110	0.105	0.145	9,419	12,519	17,965	18,928	17,883	24,732
Latvia	-	-	-	*0.007	*0.013	*0.020	-	-	-	1,542	3,237	4,824
Luxembourg	*0.210	*0.219	*0.229	*0.203	*0.233	-	46,513	46,065	47,935	43,344	45,811	-
Macau	0.008	0.008	0.008	0.008	0.009	0.007	9,034	10,660	10,838	11,463	11,498	12,731
Malawi	*0.012	*0.018	*0.004	*0.003	*0.003	-	8,271	13,528	3,204	3,044	2,425	-
Malaysia	*0.316	*0.414	*0.597	*0.696	*0.764	*0.899	9,637	10,673	13,539	14,894	15,567	17,515
Mauritius	0.014	0.012	0.012	0.016	0.016	0.017	12,397	11,958	9,714	13,307	12,533	11,972
Mexico	0.945	1.053	1.099	1.000	0.996	0.771	18,311	20,454	22,594	24,148	24,526	21,174
Mongolia	0.016	0.008	0.003	0.000	0.002	0.001	21,367	10,504	3,821	384	1,239	1,367
Morocco	±0.166	±0.194	±0.225	±0.201	±0.198	±0.220	9,922	8,755	8,917	7,480	8,498	9,065
Myanmar	±0.002	±0.001	±0.002	±0.002	±0.002	-	1,513	936	1,246	1,443	1,716	-

Continued.

Value Added and Value Added per Employee
- Continued -

Country	Value Added ($ bil.)						Value Added per Employee ($)					
	1990	1991	1992	1993	1994	1995	1990	1991	1992	1993	1994	1995
Namibia	-	-	-	-	0.025	-	-	-	-	-	17,101	-
Nepal	0.008	0.011	-	0.014	0.011	-	2,793	2,622	-	3,225	2,121	-
Netherlands	2.904	3.158	3.404	2.953	3.042	3.632	45,072	45,594	48,759	45,113	46,631	56,646
New Zealand	0.480	0.468	0.476	0.587	-	-	28,162	29,291	29,635	33,799	-	-
Norway	±0.784	±0.779	±0.760	±0.531	±0.557	±0.681	36,912	38,854	42,211	41,773	44,570	50,465
Pakistan	0.042	0.040	0.045	-	-	-	3,827	3,412	3,641	-	-	-
Panama	0.017	0.019	0.021	0.022	0.024	0.025	15,820	16,873	17,152	17,268	17,580	17,808
Peru	0.180	0.166	0.163	-	-	-	11,714	11,843	11,731	-	-	-
Philippines	0.156	0.181	0.165	0.177	-	-	4,258	4,489	5,809	5,732	-	-
Poland	±1.081	±0.926	±0.864	±0.792	±0.956	±1.305	5,513	5,323	5,402	5,465	6,236	8,033
Portugal	0.826	1.018	1.138	0.979	1.029	1.188	10,360	11,573	13,605	12,183	13,246	15,137
Puerto Rico	±0.116	±0.114	±0.114	±0.112	±0.131	±0.139	34,159	34,667	39,107	39,331	44,795	46,824
Romania	±0.869	±0.473	±0.280	±0.309	-	-	4,597	2,556	1,544	2,017	-	-
Russian Federation	-	-	-	±0.752	±0.913	±1.030	-	-	-	1,787	2,600	3,208
Saint Lucia	-	±0.002	±0.002	±0.002	±0.002	±0.002	-	-	-	-	-	-
Senegal	0.011	0.012	0.013	0.011	0.011	-	13,982	21,238	22,055	20,943	18,728	-
Sierra Leone	-	-	-	0.002	-	-	-	-	-	2,197	-	-
Singapore	*0.730	*0.870	*1.011	*1.136	*1.331	*1.714	25,675	28,448	33,481	34,827	39,774	45,198
Slovakia	-	-	-	±0.163	±0.183	-	-	-	-	6,942	4,665	-
Slovenia	-	-	-	0.198	0.230	0.427	-	-	-	-	-	-
South Africa	*1.696	*1.532	*1.611	*1.454	*1.457	*1.655	12,379	11,878	13,312	12,117	11,939	13,492
Spain	*5.437	*5.805	*5.915	±5.824	±6.137	±7.649	37,381	40,082	42,030	29,437	30,578	36,024
Sri Lanka	0.010	0.013	0.009	0.017	-	-	2,362	2,975	3,188	4,762	-	-
Swaziland	*0.005	*0.006	*0.006	*0.005	-	-	8,744	15,145	14,193	10,775	-	-
Sweden	*4.448	*3.153	*3.096	*2.118	*2.688	*3.347	61,020	48,178	53,262	41,922	50,058	61,792
Thailand	0.605	1.086	-	-	-	-	11,333	16,805	-	-	-	-
Trinidad and Tobago	±0.014	±0.015	±0.014	±0.015	±0.012	-	9,477	8,710	9,464	11,204	8,206	-
Turkey	0.904	0.966	1.167	1.282	0.913	1.200	22,367	26,518	27,425	29,248	22,494	28,908
United Kingdom	*15	*14	*14	*15	*16	*17	44,563	45,358	46,626	39,957	42,027	44,806
United Republic of Tanzania	*0.003	*0.005	-	-	-	-	736	1,399	-	-	-	-
United States of America	*70	*68	*73	*78	*85	*90	54,248	55,257	59,574	62,199	65,703	67,466
Uruguay	0.073	0.105	0.100	0.107	0.125	0.192	9,830	14,036	14,015	15,661	16,433	18,568
Venezuela	0.336	0.378	0.453	0.414	0.351	1.032	9,775	10,863	12,514	13,235	12,317	36,946
Yugoslavia	±0.109	±0.119	±0.225	-	±0.417	±0.240	-	1,185	2,303	-	4,603	2,739
Zambia	0.042	-	-	-	0.016	-	7,393	-	-	-	3,854	-
Zimbabwe	*0.135	*0.146	*0.106	*0.080	*0.084	*0.092	8,571	10,268	7,331	6,073	5,545	6,238

Capital Investment

Country	Gross Fixed Capital Formation ($ mil.)						Per Establishment ($ mil)					
	1990	1991	1992	1993	1994	1995	1990	1991	1992	1993	1994	1995
Argentina	-	-	-	121	-	-	-	-	-	0.009	-	-
Australia	404	-	-	-	-	-	0.084	-	-	-	-	-
Austria	368	394	406	363	392	-	0.411	0.453	0.457	0.413	0.444	-
Bahrain	-	-	1.731	-	-	-	-	-	0.013	-	-	-
Bangladesh	0.810	1.448	0.693	-	-	-	0.001	0.001	0.001	-	-	-
Barbados	1.075	1.338	0.011	1.047	2.515	-	0.098	0.122	0.001	0.116	0.120	-
Bolivia	0.708	0.477	0.555	2.435	1.223	-	0.008	0.005	0.005	0.020	0.010	-
Bosnia & Herzegovina	7.986	-	-	-	-	-	0.074	-	-	-	-	-
Bulgaria	20	14	274	71	8.947	14	0.135	0.088	1.520	0.056	0.004	0.005
Cameroon	-	-	-	-	1.077	-	-	-	-	-	-	-
Canada	281	-	-	-	-	-	0.072	-	-	-	-	-
Cape Verde	0.002	0.002	-	-	-	-	0.000	0.000	-	-	-	-
Chile	45	32	51	63	61	-	0.317	0.221	0.342	0.397	0.339	-
China (Hong Kong SAR)	88	84	56	86	68	-	0.013	0.014	0.010	0.019	0.016	-
Colombia	23	23	28	20	16	-	0.038	0.040	0.045	0.032	0.026	-
Cote d'Ivoire	514	503	529	459	-	-	-	-	-	-	-	-
Croatia	13	20	4.491	4.701	5.793	2.151	0.021	0.022	0.004	0.003	0.003	0.001

Continued.

Depending on the table, * means *Enterprises, Engaged,* or *Factor Values;* ± means *Basis Unspecified;* § means *shown in millions.* For additional notes and sources, see Appendix I.

Capital Investment

- Continued -

Country	Gross Fixed Capital Formation ($ mil.)						Per Establishment ($ mil)					
	1990	1991	1992	1993	1994	1995	1990	1991	1992	1993	1994	1995
Cyprus	7.963	8.635	8.042	5.728	5.125	8.943	0.009	0.010	0.009	0.006	0.005	0.008
Czechoslovakia (Former)	138	-	-	-	-	-	1.699	-	-	-	-	-
Denmark	204	167	-	-	-	-	0.094	0.076	-	-	-	-
Ecuador	20	17	33	22	24	32	0.152	0.113	0.219	0.144	0.167	0.247
Egypt	28	16	22	13	-	-	0.062	0.039	0.053	0.033	-	-
El Salvador	-	0.743	-	0.890	1.575	1.765	-	0.039	-	0.023	0.046	0.034
Ethiopia	4.206	1.489	0.769	0.373	0.639	1.302	0.324	0.115	0.059	0.027	0.018	0.033
Fiji	0.791	1.465	1.354	-	-	-	0.011	0.025	0.022	-	-	-
Finland	236	158	71	56	74	94	0.319	0.190	0.099	0.080	0.110	0.153
FYR Macedonia	2.243	4.633	0.470	0.106	0.264	0.984	0.027	0.029	0.002	0.000	0.001	0.002
France	3,769	3,702	3,121	2,092	2,182	2,696	-	-	-	-	-	-
Gabon	-	6.696	1.398	0.968	0.892	1.254	-	0.352	0.087	0.081	0.069	0.105
Germany (Western Part)	-	-	4,174	3,665	-	-	-	-	0.655	-	-	-
Greece	91	82	103	-	-	-	0.171	0.156	0.190	-	-	-
Hungary	28	35	156	48	-	-	0.051	0.028	0.095	0.025	-	-
India	195	124	199	232	-	-	0.028	0.018	0.028	0.031	-	-
Indonesia	78	252	195	221	146	191	0.123	0.389	0.283	0.306	0.183	0.200
Iran (Islamic Republic of)	51	65	70	184	-	-	0.060	0.174	0.162	0.461	-	-
Ireland	53	48	-	-	-	-	0.088	0.085	-	-	-	-
Israel	103	183	151	175	191	-	0.072	0.112	0.088	0.092	0.100	-
Italy	1,221	1,219	1,736	1,253	1,438	-	0.425	0.416	0.356	0.253	0.283	-
Japan	4,848	6,889	7,272	6,763	5,665	-	0.091	0.132	0.144	0.131	0.117	-
Jordan	0.515	3.405	4.451	1.997	4.004	29	0.000	0.002	0.002	0.001	0.001	0.009
Korea, Republic of	1,286	1,115	1,286	1,373	1,810	2,011	0.219	0.178	0.189	0.150	0.186	0.188
Kuwait	4.277	41	4.229	6.884	7.513	-	0.006	0.066	0.007	0.011	0.013	-
Latvia	0.013	0.022	1.105	0.257	1.441	5.515	0.000	0.000	0.009	0.002	0.021	0.034
Luxembourg	25	29	22	-	-	-	0.180	0.214	0.170	-	-	-
Macau	0.466	0.495	0.747	0.415	1.067	0.191	0.002	0.002	0.003	0.002	0.005	0.001
Malawi	1.026	0.464	0.611	1.136	0.710	-	0.114	0.052	0.068	0.126	0.079	-
Malaysia	136	170	204	135	196	-	0.270	0.303	0.353	0.207	0.305	-
Mexico	61	101	-	-	-	-	0.281	0.476	-	-	-	-
Mongolia	53	27	7.371	1.299	3.284	-	2.935	1.437	0.295	0.068	0.117	-
Morocco	37	36	38	34	35	46	0.099	0.085	0.090	0.074	0.080	0.101
Myanmar	6.537	6.747	7.385	7.241	-	-	1.634	1.687	1.846	1.810	-	-
Nepal	9.057	-	-	-	-	-	0.067	-	-	-	-	-
Netherlands	546	481	503	382	-	-	0.620	0.496	0.478	0.360	-	-
New Zealand	43	-	-	-	-	-	0.019	-	-	-	-	-
Norway	80	73	86	35	32	-	0.096	0.095	0.168	0.090	0.088	-
Pakistan	-	5.526	-	-	-	-	-	0.026	-	-	-	-
Panama	1.074	2.379	-	-	-	-	0.020	0.042	-	-	-	-
Peru	5.974	8.379	13	-	-	-	0.004	0.005	0.008	-	-	-
Philippines	7.898	25	21	28	-	-	0.001	0.003	0.033	0.046	-	-
Poland	107	101	123	120	-	-	0.130	0.193	0.243	0.245	-	-
Portugal	202	206	254	142	219	-	0.017	0.018	0.022	0.012	0.019	-
Romania	-	-	22	37	49	-	-	-	0.016	0.025	0.020	-
Senegal	1.719	-	-	-	-	-	0.095	-	-	-	-	-
Singapore	145	161	190	323	307	-	0.303	0.311	0.361	0.593	0.547	-
Slovenia	36	29	25	34	26	42	0.071	0.010	0.007	0.006	0.004	0.009
Spain	487	460	449	725	624	768	0.029	0.028	0.027	-	-	-
Sri Lanka	1.129	2.360	0.895	2.941	-	-	0.010	0.028	0.016	0.037	-	-
Swaziland	0.098	0.103	0.039	0.137	-	-	0.004	0.006	0.002	0.009	-	-
Thailand	875	213	-	-	-	-	2.658	0.239	-	-	-	-
Trinidad and Tobago	-	-	1.812	1.925	1.148	-	-	-	0.010	0.011	0.006	-
Turkey	81	83	564	196	175	-	0.208	0.228	0.681	0.241	0.229	-
Ukraine	245	155	284	56	38	17	0.015	0.009	0.017	0.004	0.003	0.001
United Kingdom	1,184	848	723	-	1,093	-	0.079	0.059	0.051	-	0.039	-
United Republic of Tanzania	23	19	-	-	-	-	0.275	0.224	-	-	-	-
United States of America	3,780	3,200	3,740	3,831	4,551	5,658	-	-	0.078	-	-	-
Uruguay	3.502	3.576	5.439	7.763	-	-	-	-	-	-	-	-
Venezuela	33	56	55	66	35	289	0.028	0.047	0.046	0.067	0.035	0.321

Continued.

Depending on the table, * means *Enterprises, Engaged,* or *Factor Values;* ± means *Basis Unspecified;* § means *shown in millions.* For additional notes and sources, see Appendix I.

549

Capital Investment
- Continued -

Country	Gross Fixed Capital Formation ($ mil.)						Per Establishment ($ mil)					
	1990	1991	1992	1993	1994	1995	1990	1991	1992	1993	1994	1995
Yugoslavia	5.671	5.107	13	-	15	8.372	0.009	0.004	0.005	-	0.004	0.002
Zambia	15	-	-	-	-	-	0.220	-	-	-	-	-

Indexes of Industrial Production

Country	Index of Industrial Production (1990=100)						Index of Employment (1990=100)					
	1990	1991	1992	1993	1994	1995	1990	1991	1992	1993	1994	1995
Albania	100	-	-	-	-	-	-	-	-	-	-	-
Algeria	100	97	93	89	79	73	-	-	-	-	-	-
Angola	100	-	-	-	-	-	-	-	-	-	-	-
Argentina	100	130	-	-	-	-	100	-	-	66	72	-
Armenia	100	-	-	-	-	-	100	88	-	-	-	-
Australia	100	94	91	97	-	-	100	92	84	-	-	-
Austria	100	103	105	100	109	115	100	98	95	93	95	94
Azerbaijan	-	-	-	-	-	-	100	94	77	75	53	-
Bahrain	100	-	-	-	-	-	-	-	-	-	-	-
Bangladesh	100	-	-	-	-	-	100	126	109	-	-	-
Barbados	100	-	-	-	-	-	100	107	92	61	129	109
Belgium	100	101	97	90	92	99	-	-	-	-	-	-
Belize	100	73	53	39	29	21	-	-	-	-	-	-
Benin	100	-	-	-	-	-	-	-	-	-	-	-
Bermuda	-	-	-	-	-	-	100	103	97	97	94	69
Bolivia	100	104	111	116	136	132	100	115	115	122	114	123
Bosnia & Herzegovina	-	-	-	-	-	-	100	95	-	-	3	-
Brazil	100	94	94	101	111	109	-	-	-	-	-	-
Bulgaria	100	81	64	48	48	50	100	80	62	55	68	49
Burkina-Faso	100	-	-	-	-	-	-	-	-	-	-	-
Burundi	100	-	-	-	-	-	100	96	-	-	-	-
Cameroon	100	-	-	-	-	-	-	-	-	-	-	-
Canada	100	87	82	83	92	94	100	91	81	79	81	82
Central African Republic	100	-	-	-	-	-	100	-	-	-	-	-
Chile	100	100	124	139	141	151	100	95	101	106	116	122
China	100	-	-	-	-	-	100	102	101	105	107	105
China (Hong Kong SAR)	100	98	94	86	85	81	100	91	83	64	53	-
China (Taiwan Province)	100	113	118	113	120	125	100	101	105	107	112	110
Colombia	100	90	100	114	125	134	100	92	107	115	114	115
Costa Rica	100	95	108	123	127	116	100	95	102	115	-	-
Cote d'Ivoire	100	87	87	89	95	121	-	-	-	-	-	-
Croatia	100	-	-	-	-	-	100	73	58	55	51	48
Cuba	100	-	-	-	-	-	-	-	-	-	-	-
Cyprus	100	102	98	107	118	118	100	100	101	103	112	113
Czechoslovakia (Former)	100	66	59	-	-	-	100	65	-	-	-	-
Czech Republic	100	-	-	-	-	-	100	93	124	127	-	-
Dem. Rep. of the Congo	100	-	-	-	-	-	-	-	-	-	-	-
Denmark	100	105	107	103	117	122	100	98	97	95	101	103
Dominican Republic	100	-	-	-	-	-	-	-	-	-	-	-
Ecuador	100	114	133	147	-	-	100	116	116	115	119	85
Egypt	100	98	98	98	99	99	100	98	106	94	94	94
El Salvador	100	103	-	-	-	-	-	-	-	-	-	-
Equatorial Guinea	-	-	-	-	-	-	100	-	-	-	-	-
Ethiopia and Eritrea	100	-	-	-	-	-	-	-	-	-	-	-
Ethiopia	-	-	-	-	-	-	100	95	101	103	118	135
Fiji	100	-	-	-	-	-	100	103	100	120	121	-
Finland	100	86	80	82	94	108	100	94	77	71	73	66
FYR Macedonia	100	-	-	-	-	-	100	104	94	91	101	81
France	100	96	93	85	90	91	100	99	94	88	84	86
Gabon	100	-	-	-	-	-	-	-	-	-	-	-
Gambia	100	-	-	-	-	-	-	-	-	-	-	-

Continued.

Depending on the table, * means *Enterprises, Engaged,* or *Factor Values;* ± means *Basis Unspecified;* § means *shown in millions.* For additional notes and sources, see Appendix I.

Indexes of Industrial Production

- Continued -

Country	Index of Industrial Production (1990=100)						Index of Employment (1990=100)					
	1990	1991	1992	1993	1994	1995	1990	1991	1992	1993	1994	1995
Germany (Eastern Part)	100	-	-	-	-	-	-	-	-	-	-	-
Germany (Western Part)	100	105	107	101	107	110	100	104	103	-	-	-
Ghana	100	-	-	-	-	-	-	-	-	-	-	-
Greece	100	95	105	98	97	98	100	89	89	86	86	86
Guatemala	100	-	-	-	-	-	-	-	-	-	-	-
Guyana	100	100	100	100	100	100	-	-	-	-	-	-
Haiti	100	-	-	-	-	-	-	-	-	-	-	-
Honduras	100	107	113	141	167	193	100	104	98	107	118	143
Hungary	100	88	69	77	88	89	100	88	64	84	96	97
Iceland	-	-	-	-	-	-	100	91	87	81	81	-
India	100	93	87	85	96	116	100	99	103	102	106	113
Indonesia	100	85	94	90	101	115	100	119	146	146	162	183
Iran (Islamic Republic of) . . .	100	136	144	142	144	132	100	91	102	96	-	-
Iraq	100	-	-	-	-	-	100	-	-	-	-	-
Ireland	100	95	92	92	97	101	100	94	91	91	95	99
Israel	100	111	117	122	134	150	100	110	113	114	121	123
Italy	100	94	93	89	-	-	100	101	141	138	138	-
Jamaica	100	-	-	-	-	-	-	-	-	-	-	-
Japan	100	100	95	92	93	94	100	103	101	101	99	99
Jordan	100	-	-	-	-	-	100	117	193	154	193	211
Kenya	100	131	119	119	133	134	100	100	101	103	105	111
Korea, Republic of	100	108	103	102	118	134	100	99	94	106	113	118
Kuwait	100	-	-	-	-	-	100	83	95	101	102	102
Kyrgyzstan	100	-	-	-	-	-	100	82	52	37	29	-
Lao P.D.R.	100	-	-	-	-	-	-	-	-	-	-	-
Latvia	100	-	-	-	-	-	100	71	53	31	26	27
Lebanon	100	-	-	-	-	-	-	-	-	-	-	-
Lesotho	-	-	-	-	-	-	100	-	-	-	-	-
Liberia	100	-	-	-	-	-	-	-	-	-	-	-
Libyan Arab Jamahiriya . . .	100	-	-	-	-	-	-	-	-	-	-	-
Lithuania	100	-	-	-	-	-	-	-	-	-	-	-
Luxembourg	100	101	105	102	111	122	100	105	106	104	113	-
Macau	100	-	-	-	-	-	100	88	83	78	86	59
Madagascar	100	103	42	-	-	-	-	-	-	-	-	-
Malawi	-	-	-	-	-	-	100	93	79	71	79	-
Malaysia	100	119	170	268	314	342	100	118	134	142	150	157
Mali	100	101	106	93	106	125	-	-	-	-	-	-
Malta	100	95	87	82	-	-	-	-	-	-	-	-
Mauritius	100	-	-	-	-	-	100	90	105	104	115	128
Mexico	100	111	110	110	118	101	100	100	94	80	79	71
Mongolia	100	77	101	34	50	102	100	96	90	144	187	138
Morocco	100	107	122	113	117	123	100	132	151	160	139	145
Mozambique	100	-	-	-	-	-	100	97	98	91	89	-
Myanmar	100	108	95	-	-	-	100	97	110	110	110	-
Namibia	100	-	-	-	-	-	-	-	-	-	-	-
Nepal	100	-	-	-	-	-	100	143	-	156	182	-
Netherlands	100	102	101	93	94	99	100	108	108	102	101	100
New Zealand	100	-	-	-	-	-	100	94	94	102	-	-
Nicaragua	100	-	-	-	-	-	-	-	-	-	-	-
Niger	100	-	-	-	-	-	-	-	-	-	-	-
Nigeria	100	-	-	-	-	-	-	-	-	-	-	-
Norway	100	98	100	104	110	119	100	94	85	60	59	64
Pakistan	100	-	-	-	-	-	100	108	113	-	-	-
Panama	100	116	127	134	149	158	100	107	113	117	125	130
Papua New Guinea	100	-	-	-	-	-	-	-	-	-	-	-
Paraguay	100	109	103	85	49	49	-	-	-	-	-	-
Peru	100	108	99	133	158	220	100	91	90	-	-	-
Philippines	100	108	117	127	130	141	100	110	78	84	-	-
Poland	100	86	92	98	113	131	100	89	82	74	78	83
Portugal	100	100	95	89	94	95	100	110	105	101	97	98

Continued.

Indexes of Industrial Production

- Continued -

Country	Index of Industrial Production (1990=100)						Index of Employment (1990=100)					
	1990	1991	1992	1993	1994	1995	1990	1991	1992	1993	1994	1995
Puerto Rico	100	-	-	-	-	-	100	97	86	84	86	87
Republic of Moldova	100	-	-	-	-	-	100	89	86	54	13	16
Romania	100	87	62	56	48	58	100	98	96	81	72	-
Russian Federation	100	-	-	-	-	-	-	-	-	-	-	-
Saudi Arabia	100	-	-	-	-	-	-	-	-	-	-	-
Senegal.	100	103	90	98	89	96	100	71	72	67	71	-
Sierra Leone	100	-	-	-	-	-	-	-	-	-	-	-
Singapore	100	101	107	111	124	143	100	108	106	115	118	133
Slovakia	100	-	-	-	-	-	-	-	-	-	-	-
Slovenia	100	94	80	76	76	70	-	-	-	-	-	-
Somalia.	100	-	-	-	-	-	-	-	-	-	-	-
South Africa	100	96	97	92	93	98	100	94	88	88	89	90
Spain	100	95	92	81	87	96	100	100	97	136	138	146
Sri Lanka	100	94	101	124	154	213	100	101	67	85	-	-
Sudan	100	-	-	-	-	-	-	-	-	-	-	-
Suriname	100	-	-	-	-	-	-	-	-	-	-	-
Swaziland	100	-	-	-	-	-	100	76	71	88	-	-
Sweden.	100	86	75	71	85	97	100	90	80	69	74	74
Switzerland	100	105	104	100	102	104	-	-	-	-	-	-
Syrian Arab Republic	100	124	179	206	277	297	-	-	-	-	-	-
Thailand	100	106	-	-	-	-	100	121	-	-	-	-
Tonga	100	-	-	-	-	-	100	48	48	60	43	-
Trinidad and Tobago	100	126	156	138	179	199	100	119	105	95	99	-
Tunisia	100	99	87	85	83	81	-	-	-	-	-	-
Turkey	100	103	93	101	79	90	100	90	105	109	100	103
Uganda.	100	74	109	320	379	401	-	-	-	-	-	-
Ukraine	100	-	-	-	-	-	100	97	84	76	63	57
United Arab Emirates	100	-	-	-	-	-	-	-	-	-	-	-
United Kingdom	100	91	86	83	86	86	100	93	88	108	113	113
United Republic of Tanzania . .	100	89	100	31	21	-	100	84	-	-	-	-
USSR (Former)	100	-	-	-	-	-	100	-	-	-	-	-
United States of America . . .	100	95	100	104	111	115	100	94	95	96	99	103
Uruguay	100	91	89	79	89	126	100	101	96	93	103	140
Venezuela	100	-	-	-	-	-	100	101	105	91	83	81
Yugoslavia	100	-	-	-	-	-	-	-	-	-	-	-
Yugoslavia (Former).	100	45	-	-	-	-	-	-	-	-	-	-
Zambia	100	97	90	75	64	68	100	-	-	-	75	-
Zimbabwe	100	102	90	74	83	79	100	90	92	84	96	93

Depending on the table, * means *Enterprises, Engaged,* or *Factor Values;* ± means *Basis Unspecified;* § means *shown in millions.* For additional notes and sources, see Appendix I.

Representative Companies in Sector

Name	Address	Telephone	Fax	Employment	Y
Budd Co.	PO Box 2601, Troy, Michigan 48007-2601, U.S.A.	313-643-3500		10,000	96
Crane Co.	100 1st Stamford Pl., Stamford, Connecticut 06902, U.S.A.	203-363-7300	203-363-7295	10,000	96
Olin Corp.	PO Box 4500, Norwalk, Connecticut 06856-4500, U.S.A.	203-750-3000	203-750-3595	10,000	97
Pentair Inc.	1500 County Rd. B2 W., Ste. 400, St. Paul, Minnesota 55113-3105, U.S.A.	612-636-7920		10,000	97
General Signal Corp.	PO Box 10010, Stamford, Connecticut 06904-2010, U.S.A.	203-329-4100		9,900	97
Grupo Industrial Saltillo SA de CV	25280 Saltillo, Coahuila, Mexico	84 155620	84 153313	9,575	92
GenCorp Inc.	175 Ghent Rd., Fairlawn, Ohio 44333-3300, U.S.A.	330-869-4200	330-869-4211	9,460	97
Sequa Corp.	200 Park Ave., New York, New York 10166, U.S.A.	212-986-5500	212-370-1969	9,350	97
AMSTED Industries Inc.	205 N. Michigan Ave., Chicago, Illinois 60601, U.S.A.	312-645-1700	312-819-8523	9,000	95
Godrej & Boyce Manufacturing Co. Ltd.	Godrej Bhavan Homji St., Bombay 400 001, India	22 5171166	22 5172688	8,810	94
Tower Automotive Inc.	4508 IDS Ctr., Minneapolis, Minnesota 55402, U.S.A.	612-342-2310		8,750	97
Inco Ltd.	145 King St. W Ste. 1500, Toronto, Ontario M5H 4B7, Canada	416-361-7511	416-361-7781	8,600	98
Cooper Cameron Corp.	515 Post Oak Blvd., 1200, Houston, Texas 77027, U.S.A.	713-513-3300	713-513-3320	8,500	96
A.O. Smith Corp.	PO Box 23972, Milwaukee, Wisconsin 53223-3972, U.S.A.	414-359-4000	414-359-4248	8,400	97
General Electric Co. Aircraft Electronics and Defense Sys. D	600 Main St., Johnson City, New York 13790-1888, U.S.A.	607-770-3901		8,200	
MTD Products Inc.	PO Box 360900, Cleveland, Ohio 44136, U.S.A.	330-225-2600		8,000	95
Ball Corp.	PO Box 2407, Muncie, Indiana 47307-0407, U.S.A.	765-747-6100		7,900	97
Modine Manufacturing Co.	1500 DeKoven Ave., Racine, Wisconsin 53403-2552, U.S.A.	414-636-1200	414-636-1424	7,900	96
FKI Industries Inc.	425 Post Rd., Fairfield, Connecticut 06430, U.S.A.	203-255-7100	203-255-7101	7,800	95
Iwka AG	Gartenstr. 71, Karlsruhe D-76135, Germany	721 1430	721 143243	7,774	94
Short Brothers PLC	Airport Rd., Belfast BT3 9DZ, United Kingdom	232 458444	232 732974	7,551	94
SPX Corp.	PO Box 3301, Muskegon, Michigan 49443-3301, U.S.A.	616-724-5000	616-724-5720	7,125	96
T and N Industries Inc.	777 E. Eisenhower Pkwy., 600, Ann Arbor, Michigan 48108-3258, U.S.A.	313-663-6749	313-663-0629	7,100	95
HON Industries Inc.	PO Box 1109, Muscatine, Iowa 52761-7109, U.S.A.	319-264-7085	319-264-7217	6,900	97
Alliant Techsystems Inc.	600 2nd St., N.E., Hopkins, Minnesota 55343-8384, U.S.A.	612-931-6000	612-931-5423	6,800	97
Chicago Bridge and Iron Co.	1501 N. Division St., Plainfield, Illinois 60544-8984, U.S.A.	815-439-6000	815-439-6010	6,800	96
Sigma-Aldrich Corp.	3050 Spruce St., St. Louis, Missouri 63103, U.S.A.	314-771-5765	314-534-2674	6,666	97
Apogee Enterprises Inc.	7900 Xerxes Ave. S., Minneapolis, Minnesota 55431-1159, U.S.A.	612-835-1874	612-835-3196	6,553	96
Doornfontein Gold Mining Co. Ltd.	75 Fox St., Johannesburg 2001, Republic of South Africa			6,530	92
CCL Industries Inc.	800-105 Gordon Baker Rd., Willowdale, Ontario M2H 3P8, Canada	416-756-8500	416-756-8555	6,500	98
Scaw Metals	PO Box 61721, Marshalltown 2107, Republic of South Africa	11 9021001	11 9027702	6,400	95
Johnson Matthey PLC	Trafalgar Sq., London SW1Y 5BQ, United Kingdom	71 2698000	71 2698133	6,287	94
Elkem AS	PO Box 4282 Torshov, Oslo N-0401, Norway	22 450100	22 450155	6,000	93
H.B. Fuller Co.	PO Box 64683, St. Paul, Minnesota 55164-0683, U.S.A.	612-645-3401		6,000	97
Vectura Group Inc.	PO Box 52189, New Orleans, Louisiana 70152-2189, U.S.A.	504-529-8600		6,000	
Royal Ordnance PLC	Euxton, Chorley PR7 6AD, United Kingdom	257 265511		5,944	93
Cooper Oil Tool Div.	PO Box 1212, Houston, Texas 77251-1212, U.S.A.	713-939-2211	713-939-2620	5,900	97
ZSNP Aluminium Works AS	Ziar nad Hronom 965 63, Slovakia	857 2201	857 2240	5,880	93
Insilco Corp.	425 Metro Pl. N., 5th Fl., Dublin, Ohio 43017, U.S.A.	614-792-0468	614-791-3197	5,764	96
Blount International Inc.	PO Box 949, Montgomery, Alabama 36101-0949, U.S.A.	334-244-4000		5,700	97
Perindustrian Tentara Nasional Id Angkatan Darat PT	Jl. Jenderal Gatot Subroto Kiaracondong, Bandung, Indonesia	22312073	22 301222	5,500	97
Trencor Services Pty. Ltd.	PO Box 65081, Benmore 2010, Republic of South Africa	11 8834119	11 8834481	5,475	96
Jannock Ltd.	5205-40 King St. W, Toronto, Ontario M5H 3Y2, Canada	416-364-8586	416-364-9342	5,400	98
Weirton Steel Corp.	400 Three Springs Dr., Weirton, West Virginia 26062-4989, U.S.A.	304-797-2000	304-797-2887	5,373	97
Zurn Industries Inc.	1 Zurn Pl., Erie, Pennsylvania 16514-2000, U.S.A.	814-452-2111	814-459-3535	5,100	96
Hexcel Corp.	5794 W. Las Positas Blvd., Pleasanton, California 94588-8781, U.S.A.	510-847-9500	510-734-8276	5,013	96
Raymond Woollen Mills Ltd.	Bellard Estate, Bombay 400 038, India	22 2618321	22 2622010	5,010	94
Donnelly Corp.	414 E. 40th St., Holland, Michigan 49423-5368, U.S.A.	616-786-7000	616-786-6034	5,000	97
Griffon Corp.	100 Jericho Quadrangle, Jericho, New York 11753, U.S.A.	516-938-5544	516-938-5644	5,000	97
SPS Technologies Inc.	101 Greenwood Ave., 470, Jenkintown, Pennsylvania 19046, U.S.A.	215-517-2000	215-860-3034	4,966	97

Continued

Representative Companies in Sector
- Continued -

Name	Address	Telephone	Fax	Employment	Y
Envirodyne Industries Inc.	701 Harger Rd., 190, Oak Brook, Illinois 60521, U.S.A.	630-571-8800		4,900	96
Dexter Corp.	1 Elm St., Windsor Locks, Connecticut 06096, U.S.A.	860-292-7675	860-292-7673	4,800	97
Standex International Corp.	6 Manor Pkwy., Salem, New Hampshire 03079, U.S.A.	603-893-9701	603-893-7324	4,800	97
Bridon PLC	Carr Hill, Doncaster DN4 8JX, United Kingdom	302 344010	302 382263	4,764	93
Watts Industries Inc.	815 Chestnut St., North Andover, Massachusetts 01845-6098, U.S.A.	978-688-1811	978-688-5841	4,650	97
Great American Management Inc.	2 N. Riverside Plz., Chicago, Illinois 60606, U.S.A.	312-648-5656	312-454-9946	4,600	95
Shaw Group Inc.	11100 Mead Rd., 2nd Fl., Baton Rouge, Louisiana 70816, U.S.A.	504-296-1195		4,600	97
Greif Bros. Corp.	425 Winter Rd., Delaware, Ohio 43015, U.S.A.	740-549-6000		4,500	97
Linamar Corp.	301 Massey Rd., Guelph, Ontario N1K 1B2, Canada	519-836-7550	519-824-8479	4,500	98
Butler Manufacturing Co.	PO Box 419917, Kansas City, Missouri 64141-0917, U.S.A.	816-968-3000		4,350	96
Shaft Sinkers Pty. Ltd.	PO Box 783501, Sandton 2146, Republic of South Africa	11 4445600	11 4444410	4,324	97
Israel Military Industries Ltd.	64 Bailik Blvd., Ramat Hasharon 47205, Israel	3 5485617	3 5485729	4,300	97
Vermont American Corp.	2300 National City Twr., 101 S., Louisville, Kentucky 40202, U.S.A.	502-625-2000	502-625-2122	4,300	97
Keystone International Inc.	PO Box 40010, Houston, Texas 77040, U.S.A.	713-466-1176		4,250	96
Premdor Inc.	1600 Britannia Rd. E, Mississauga, Ontario L4W 1J2, Canada	905-670-6500	905-670-6520	4,250	98
Applied Power Inc.	PO Box 325, Milwaukee, Wisconsin 53201-0325, U.S.A.	414-781-6600	414-783-3790	4,235	97
AptarGroup Inc.	475 W. Terra Cotta Ave., E, Crystal Lake, Illinois 60014, U.S.A.	815-477-0424		4,100	97
Falcon Building Products Inc.	2 N. Riverside Plz., 1100, Chicago, Illinois 60606, U.S.A.	312-906-9700		4,100	96
U.S. Can Corp.	900 Commerce Dr., Oak Brook, Illinois 60521, U.S.A.	630-571-2500	630-573-0715	4,065	96
Amcast Industrial Corp.	PO Box 98, Dayton, Ohio 45401-0098, U.S.A.	513-298-7000		4,040	97
Hickory Springs Manufacturing Co.	PO Box 128, Hickory, North Carolina 28603, U.S.A.	704-328-2201	704-328-5501	4,000	97
KEC International Ltd.	Shastri Marg Kunda, Bombay 400 070, India	22 5115180		4,000	
Rafael Armament Development Authority	PO Box 2250, Haifa 31021, Israel	4 8794777	4 8794653	4,000	97
UOP	25 E. Algonquin Rd., Des Plaines, Illinois 60017-5017, U.S.A.	847-391-2000	708-391-2253	4,000	95
CSS Industries Inc.	1845 Walnut St., 800, Philadelphia, Pennsylvania 19103, U.S.A.	215-569-9900		3,924	97
Barnes Group Inc.	123 Main St., Bristol, Connecticut 06011-0489, U.S.A.	860-583-7070	860-589-3507	3,900	97
Alpine Group Inc.	1790 Broadway, 1500, New York, New York 10019-1412, U.S.A.	212-757-3333	212-757-3423	3,809	97
Commercial Intertech Corp.	PO Box 239, Youngstown, Ohio 44505, U.S.A.	216-746-8011	216-746-1148	3,805	97
Quanex Corp.	1900 W. Loop S., 1500, Houston, Texas 77027, U.S.A.	713-961-4600	713-877-5333	3,771	97
Valmont Industries Inc.	PO Box 358, Valley, Nebraska 68064, U.S.A.	402-359-2201	402-343-0668	3,751	97
Eljer Industries Inc.	17120 Dallas Pkwy., Dallas, Texas 75248, U.S.A.	214-407-2600		3,700	95
Wyman-Gordon Co.	PO Box 8001, North Grafton, Massachusetts 01536-8001, U.S.A.	508-839-4441	508-839-7500	3,650	97
C. Walker & Sons Ltd.	Guide, Blackburn BB1 2LJ, United Kingdom	254 55161	254 59820	3,590	90
Redpath Dorman Long Ltd.	PO Box 27, Darlington DL1 4DE, United Kingdom	325 381188		3,505	93
Alcan Aluminum Corp.	6060 Parkland Blvd., Cleveland, Ohio 44124-4185, U.S.A.	440-423-6600	440-423-6665	3,500	97
Skyline Corp.	PO Box 743, Elkhart, Indiana 46515-0743, U.S.A.	219-294-6521		3,500	97
TriMas Corp.	315 E. Eisenhower Pkwy., Ann Arbor, Michigan 48108, U.S.A.	313-747-7025	313-747-6565	3,500	95
Watts Regulator Co.	815 Chestnut St., North Andover, Massachusetts 01845, U.S.A.	978-688-1811	978-975-8350	3,500	97
Columbus McKinnon Corp.	140 Audubon Pkwy., Amherst, New York 14228-1197, U.S.A.	716-689-5400	716-689-5644	3,479	96
Lukens Inc.	50 S. 1st Ave., Coatesville, Pennsylvania 19320-0911, U.S.A.	610-383-2000	610-383-3324	3,450	96
Emco Ltd.	620 Richmond St., London, Ontario N6A 5J9, Canada	519-645-3900	519-645-3939	3,300	98
American Tool Companies Inc.	301 S. 13th St., 600, Lincoln, Nebraska 68508, U.S.A.	402-435-3300	402-435-3619	3,200	97
Moen Inc.	25300 Al Moen Dr., North Olmsted, Ohio 44070-8022, U.S.A.	440-323-3341		3,200	97
NIBCO Inc.	1516 Middlebury St., Elkhart, Indiana 46515, U.S.A.	219-295-3000	219-295-3307	3,200	95
Caribe General Electric	1590 Ponce de Leon Ave., Guaynabo 00926, Puerto Rico	787725-2966	787725-4735	3,150	92
Atlantic, Gulf and Pacific Co. of Manila, Inc.	351 Sn Gil J. Puyat Ave. Extension, Makati, Philippines	2 878071	2 8172684	3,130	91
Seco Tools AB	Bjornbacksvagen, Fagersta S-737 82, Sweden	223 40000	223 11860	3,093	93
Hulett Aluminum Profiles Pty. Ltd.	PO Box 74, Pietermaritzburg 3200, Republic of South Africa	331 956911	331 946335	3,077	93
Coopcam UK Ltd.	Houston Industrial Estate, Livingston EH54 5BZ, United Kingdom	506 31122		3,031	93

Continued

For sources and notes, see Appendix I.

Representative Companies in Sector
- Continued -

Name	Address	Tele-phone	Fax	Employ-ment	Y
Aarque Steel Corp.	PO Box 310, Jamestown, New York 14702-0310, U.S.A.	716-664-6014	716-664-5057	3,000	95
Alcoa Closure Systems International	2485 Directors Row, A, Indianapolis, Indiana 46241, U.S.A.	317-241-2595	317-390-5137	3,000	95
Aluma Systems USA Inc.	1111 N. Loop W., 700, Houston, Texas 77008, U.S.A.	713-802-9055		3,000	96
American Cast Iron Pipe Co.	PO Box 2727, Birmingham, Alabama 35202, U.S.A.	205-325-7701	205-325-8014	3,000	97
American Steel Foundries	10 S. Riverside Plz., 10th Fl., Chicago, Illinois 60606, U.S.A.	312-258-8000	312-258-5466	3,000	97
Clopay Corp.	312 Walnut St., 1600, Cincinnati, Ohio 45202-4036, U.S.A.	513-381-4800	513-762-3965	3,000	97
Collins and Aikman Plastics Inc.	201 W. Big Beaver Rd., 1010, Troy, Michigan 48084, U.S.A.	810-524-9650	810-524-9655	3,000	96
Deutsch Co.	2444 Wilshire Blvd 600, Santa Monica, California 90403, U.S.A.	909-849-7822	909-453-6467	3,000	
Femmunkas	Frangepan u. 7, Budapest H-1139, Hungary	1 1490570	1 1401340	3,000	93
Gould Electronics Inc.	35129 Curtis Blvd., Eastlake, Ohio 44095, U.S.A.	216-953-5000	216-953-5050	3,000	95
Northern Engraving Corp.	PO Box 377, Sparta, Wisconsin 54656, U.S.A.	608-269-6911	608-269-6735	3,000	95
Sandvik Inc.	1702 Nevins Rd., Fair Lawn, New Jersey 07410, U.S.A.	201-794-5000	201-794-5011	3,000	97
Steeledale Reinforcing & Engineering Industries Ltd.	Tulisa Park, Johannesburg 2197, Republic of South Africa	11 6132331	11 6131394	2,950	93
Varlen Corp.	PO Box 3089, Naperville, Illinois 60566-7089, U.S.A.	630-420-0400	630-420-7123	2,850	97
Jordan Industries Inc.	1751 Lake Cook Rd., 550, Deerfield, Illinois 60015, U.S.A.	847-945-5591	847-945-9645	2,800	95
Kuhlman Corp.	3 Skidaway Village Sq., Ste. 201, Savannah, Georgia 31411, U.S.A.	912-598-7809		2,782	96
Ductile Steel Processors Ltd.	131 Pentonville Rd., London N1 9NE, United Kingdom	21 7422366		2,770	90
Ameron International Corp.	245 S. Los Robles Ave., Pasadena, California 91101-2894, U.S.A.	626-683-4000	626-683-4060	2,761	97
L.S. Starrett Co.	121 Crescent St., Athol, Massachusetts 01331-1915, U.S.A.	508-249-3551	508-249-8495	2,740	97
EDA SA	Fabryczna 16, Poniatowa PL-24-320, Poland	4017	4017	2,700	93
Indomobil Suzuki International, PT	Wisma Indomobil 8th Flr., Jakarta 13330, Indonesia	218564530	21 4893048	2,700	97
McDermott Indonesia PT	Jl. H.R. Rasuna Said Kav C8-9, Jakarta 12940, Indonesia	215208611	21 5208607	2,700	97
Inco Europe Ltd.	50 Victoria St., London SW1H 0XB, United Kingdom	71 931773		2,621	94
Veitscher Magnesitwerke AG	Postfach 143, Vienna A-1011, Austria	222 515130	222 5134315	2,621	91
Divpac	Mobeni Box 12650, Durban 4000, Republic of South Africa	31 4690641	31 420705	2,600	93
Valhi Inc.	5430 Lyndon B. Johnson Fwy., 17, Dallas, Texas 75240-2697, U.S.A.	972-233-1700	972-385-0586	2,600	97
Drotovna, AS Hlohovec	Hlohovec 920 28, Slovakia	804 22241	804 22742	2,550	93
Primex Technologies Inc.	10101 9th St. N., St. Petersburg, Florida 33716-3807, U.S.A.	813-578-8100		2,550	97
Autosan SA	Lipinskiego ul. 109, Sanok PL-38-500, Poland	137 50126	137 50430	2,542	93
Aeroquip Industrial Americas Group	1695 Indian Wood Cir., Maumee, Ohio 43537, U.S.A.	419-891-7600		2,500	94
Bic Corp.	500 Bic Dr., Milford, Connecticut 06460, U.S.A.	203-783-2000	203-783-2086	2,500	93
ITW Finishing Systems	3939 W. 56th St., Indianapolis, Indiana 46254, U.S.A.	317-298-5000		2,500	
JSJ Corp.	700 Robbins Rd., Grand Haven, Michigan 49417, U.S.A.	616-842-6350	616-847-3112	2,500	95
Stanley-Bostitch Inc.	Briggs Dr., East Greenwich, Rhode Island 02818, U.S.A.	401-884-2500	401-885-3122	2,500	97
Yale Security Inc.	PO Box 25288, Charlotte, North Carolina 28229-8010, U.S.A.	704-283-2101	704-289-2875	2,500	97
Boiler Factory, Rafako	ul. Lakowa 33, Raciborz PL-47-400, Poland	335 2171	335 3427	2,450	93
Fibreboard Corp.	2200 Ross Ave., Ste. 3600, Dallas, Texas 75201-6731, U.S.A.	214-954-9500	214-954-9512	2,400	96
Hilti Inc.	5400 S. 122nd East Ave., Tulsa, Oklahoma 74146, U.S.A.	918-252-6000	800-879-7000	2,400	97
Huck International Inc.	3724 E. Columbia St., Tucson, Arizona 85714-3415, U.S.A.	520-747-9898	520-748-2142	2,400	97
Societe Tunisienne de Siderurgie Elfouladh	BP 23 & 24, Menzel Bourguiba 7050, Tunisia	2 60522	2 61533	2,350	93
Park Electrochemical Corp.	5 Dakota Dr., New Hyde Park, New York 11042, U.S.A.	516-354-4100	516-354-4128	2,340	97
Mueller Co.	PO Box 671, Decatur, Illinois 62525, U.S.A.	217-423-4471	217-425-7382	2,300	97
Drew Industries Inc.	200 Mamaroneck Ave., White Plains, New York 10601, U.S.A.	914-428-9098	914-428-4581	2,258	97
ASC Inc.	1 Sunroof Ctr., Southgate, Michigan 48195, U.S.A.	313-285-4911	313-246-2735	2,230	97
Atomic Energy Corp. of South Africa Ltd.	PO Box 582, Pretoria 0001, Republic of South Africa	12 3163270	12 3163222	2,200	97
Crane Valves Nuclear Operation	860 Remington Blvd., Bolingbrook, Illinois 60440, U.S.A.	630-226-4900		2,200	94
Elco Textron Inc.	PO Box 7009, Rockford, Illinois 61125-7009, U.S.A.	815-397-5151	815-395-8270	2,180	97
Kety	Ul. Kosciuszki 111, Kety PL-32-650, Poland	381 52251	381 53094	2,115	93
Boma Bisma Indra, PT	Jl. Jend. Sudirman Kav. 50, Surabaya 60246, Indonesia	31583836	61 514241	2,114	97
American Safety Razor Co.	PO Box 500, Staunton, Virginia 24402-0500, U.S.A.	540-248-8000	540-248-0522	2,100	97
Finomszerelvenygyar, Eger	PO Box 2, Eger H-3301, Hungary	36 11911	36 11112	2,100	93
ITEQ Inc.	2727 Allen Pkwy., Ste. 760, Houston, Texas 77019, U.S.A.	713-285-2700	713-520-8228	2,100	97
UNR Industries Inc.	332 S. Michigan Ave., Chicago, Illinois 60604-4385, U.S.A.	312-341-1234		2,100	93

Product Tables
Boilers, Steam Generating

Unit of Measure: Number.

	1990		1991		1992		1993		1994		1995		1996	
	Value	%	Value	%	Value	%	Value	%	Value	%	Value	%	Value	%
Total Production	106,787	100.0	105,881	100.0	110,563	100.0	97,706	100.0	99,796	100.0	95,060	100.0	100,125	100.0
Regions														
Africa	1,608	1.5	1,322	1.2	1,270	1.1	1,255	1.3	1,235	1.2	1,265	1.3	1,264	1.3
America, South	659	0.6	560	0.5	515	0.5	526	0.5	455	0.5	622	0.7	476	0.5
Asia	45,944	43.0	44,005	41.6	42,352	38.3	40,419	41.4	39,771	39.9	43,081	45.3	42,445	42.4
Europe	57,814	54.1	59,231	55.9	65,664	59.4	54,743	56.0	57,572	57.7	49,329	51.9	55,177	55.1
Oceania	763	0.7	763	0.7	763	0.7	763	0.8	763	0.8	763	0.8	763	0.8
Leading Producers														
Austria	*41,379	38.7	*41,379	39.1	47,851	43.3	39,484	40.4	43,106	43.2	35,075	36.9	*41,379	41.3
Japan	21,494	20.1	21,492	20.3	18,479	16.7	18,097	18.5	18,097	18.1	18,081	19.0	17,618	17.6
Korea, Republic of	*5,588	5.2	*5,588	5.3	*5,588	5.1	*5,588	5.7	*5,588	5.6	*5,588	5.9	*5,588	5.6
Ukraine	8,196	7.7	8,701	8.2	7,016	6.3	4,465	4.6	1,723	1.7	1,043	1.1	403	0.4
Turkey	6,804	6.4	3,678	3.5	3,096	2.8	2,329	2.4	1,744	1.7	4,377	4.6	3,591	3.6

Structures (Except Prefabs) and Parts of Structures, of Iron and Steel

Unit of Measure: 1,000 Metric tons.

	1990		1991		1992		1993		1994		1995		1996	
	Value	%	Value	%	Value	%	Value	%	Value	%	Value	%	Value	%
Total Production	3,286,491	100.0	3,286,057	100.0	3,285,842	100.0	3,281,976	100.0	3,288,684	100.0	3,525,607	100.0	3,044,143	100.0
Regions														
America, North	83	0.0	98	0.0	95	0.0	232	0.0	242	0.0	163	0.0	180	0.0
America, South	47	0.0	30	0.0	33	0.0	49	0.0	43	0.0	45	0.0	43	0.0
Asia	8,137	0.2	7,909	0.2	7,668	0.2	3,646	0.1	10,372	0.3	6,711	0.2	6,566	0.2
Europe	3,278,224	99.7	3,278,020	99.8	3,278,047	99.8	3,278,049	99.9	3,278,026	99.7	3,518,689	99.8	3,037,354	99.8
Leading Producers														
Germany	*3,276,910	99.7	*3,276,910	99.7	*3,276,910	99.7	*3,276,910	99.8	*3,276,910	99.6	3,517,573	99.8	3,036,247	99.7
Indonesia	*3,688	0.1	*3,688	0.1	*3,688	0.1	44	0.0	7,331	0.2	*3,688	0.1	*3,688	0.1
Japan	1,938	0.1	1,938	0.1	2,027	0.1	1,812	0.1	1,662	0.1	1,698	0.0	1,567	0.1
Russian Federation	1,838	0.1	1,581	0.0	1,222	0.0	1,030	0.0	590	0.0	507	0.0	464	0.0
Korea, Republic of	*637	0.0	*666	0.0	*695	0.0	*724	0.0	*753	0.0	*782	0.0	*811	0.0

Cans, Metal (Capacity Not Exceeding 300 Liters)

Unit of Measure: Thousands of units.

	1990		1991		1992		1993		1994		1995		1996	
	Value	%	Value	%	Value	%	Value	%	Value	%	Value	%	Value	%
Total Production	15,938,453	100.0	17,495,381	100.0	18,066,767	100.0	18,674,039	100.0	19,210,510	100.0	20,783,399	100.0	20,956,540	100.0
Regions														
Africa	298,819	1.9	298,795	1.7	316,050	1.7	296,750	1.6	294,422	1.5	179,631	0.9	318,840	1.5
America, North	2,351	0.0	2,248	0.0	2,644	0.0	5,000	0.0	5,000	0.0	5,000	0.0	6,000	0.0
America, South	325,537	2.0	369,007	2.1	429,198	2.4	569,650	3.1	423,361	2.2	185,024	0.9	430,015	2.1
Asia	4,872,308	30.6	6,304,575	36.0	6,649,321	36.8	6,692,990	35.8	7,738,036	40.3	9,463,872	45.5	9,327,291	44.5
Europe	10,439,439	65.5	10,520,755	60.1	10,669,554	59.1	11,109,649	59.5	10,749,690	56.0	10,949,871	52.7	10,874,394	51.9
Leading Producers														
United Kingdom	*6,520,292	40.9	*6,520,292	37.3	*6,520,292	36.1	6,741,421	36.1	6,299,164	32.8	*6,520,292	31.4	*6,520,292	31.1
Korea, Republic of	2,935,385	18.4	4,003,623	22.9	3,988,466	22.1	4,057,575	21.7	4,976,814	25.9	5,407,400	26.0	*5,454,378	26.0
Belgium	*1,984,000	12.4	*1,984,000	11.3	*1,984,000	11.0	*1,984,000	10.6	1,966,000	10.2	1,984,000	9.5	2,002,000	9.6
Greece	1,262,439	7.9	1,376,960	7.9	1,547,107	8.6	1,790,184	9.6	1,773,727	9.2	1,797,201	8.6	*1,723,535	8.2
Malaysia	802,512	5.0	1,024,980	5.9	1,221,000	6.8	1,182,213	6.3	1,318,617	6.9	1,590,660	7.7	1,364,169	6.5

Product Tables
Casks, Drums Etc. (Capacity Over 300 Liters)

Unit of Measure: Thousands of units.

	1990 Value	%	1991 Value	%	1992 Value	%	1993 Value	%	1994 Value	%	1995 Value	%	1996 Value	%
Total Production	104,590	100.0	104,952	100.0	105,249	100.0	106,434	100.0	105,769	100.0	106,965	100.0	105,701	100.0
Regions														
America, North	29	0.0	38	0.0	30	0.0	19	0.0	35	0.0	25	0.0	21	0.0
Asia	28,208	27.0	28,759	27.4	28,191	26.8	28,700	27.0	30,237	28.6	30,301	28.3	29,015	27.4
Europe	76,353	73.0	76,155	72.6	77,028	73.2	77,715	73.0	75,497	71.4	76,639	71.6	76,666	72.5
Leading Producers														
United Kingdom	*57,521	55.0	*57,521	54.8	*57,521	54.7	58,602	55.1	56,440	53.4	*57,521	53.8	*57,521	54.4
Indonesia	*18,068	17.3	*18,068	17.2	*18,068	17.2	18,068	17.0	*18,068	17.1	*18,068	16.9	*18,068	17.1
Austria	*17,901	17.1	*17,901	17.1	*17,901	17.0	*17,901	16.8	*17,901	16.9	17,901	16.7	*17,901	16.9
Malaysia	4,253	4.1	4,879	4.6	4,226	4.0	4,585	4.3	6,069	5.7	6,019	5.6	4,613	4.4
Hong Kong	*3,104	3.0	*3,104	3.0	*3,104	2.9	*3,104	2.9	*3,104	2.9	*3,104	2.9	*3,104	2.9

Compressed Gas Cylinders, Made of Metal

Unit of Measure: Number.

	1990 Value	%	1991 Value	%	1992 Value	%	1993 Value	%	1994 Value	%	1995 Value	%	1996 Value	%
Total Production	11,200,857	100.0	11,727,770	100.0	12,163,358	100.0	12,027,467	100.0	12,342,921	100.0	13,020,672	100.0	12,912,509	100.0
Regions														
Africa	127,361	1.1	126,866	1.1	126,370	1.0	125,875	1.0	125,379	1.0	124,883	1.0	124,388	1.0
America, North	1,274,092	11.4	1,277,234	10.9	1,280,376	10.5	1,283,517	10.7	1,286,659	10.4	1,289,801	9.9	1,292,942	10.0
America, South	852,064	7.6	884,417	7.5	1,166,576	9.6	1,548,974	12.9	1,909,658	15.5	2,370,489	18.2	2,083,216	16.1
Asia	6,476,803	57.8	6,677,346	56.9	6,650,047	54.7	6,771,286	56.3	6,889,675	55.8	7,114,353	54.6	7,335,347	56.8
Europe	1,997,230	17.8	2,288,599	19.5	2,466,683	20.3	1,824,509	15.2	1,658,242	13.4	1,647,839	12.7	1,603,309	12.4
Oceania	473,307	4.2	473,307	4.0	473,307	3.9	473,307	3.9	473,307	3.8	473,307	3.6	473,307	3.7
Leading Producers														
Iran, Islamic Republic of	*2,533,093	22.6	*2,670,552	22.8	*2,808,012	23.1	*2,945,471	24.5	*3,082,931	25.0	*3,220,390	24.7	*3,357,850	26.0
Turkey	*1,911,045	17.1	*1,911,045	16.3	*1,911,045	15.7	*1,911,045	15.9	*1,911,045	15.5	*1,911,045	14.7	*1,911,045	14.8
Mexico	*1,220,184	10.9	*1,220,184	10.4	*1,220,184	10.0	*1,220,184	10.1	*1,220,184	9.9	*1,220,184	9.4	1,220,184	9.4
Spain	1,009,770	8.0	1,318,940	11.2	1,557,980	12.8	*1,027,177	8.5	*985,527	8.0	*943,878	7.2	*902,228	7.0
Thailand	1,296,491	11.6	1,372,366	11.7	*1,297,824	10.7	*1,398,358	11.6	*1,498,893	12.1	*1,599,427	12.3	*1,699,961	13.2

Cables

Unit of Measure: Metric tons.

	1990 Value	%	1991 Value	%	1992 Value	%	1993 Value	%	1994 Value	%	1995 Value	%	1996 Value	%
Total Production	2,056,511	100.0	1,839,167	100.0	2,039,419	100.0	1,905,085	100.0	1,865,744	100.0	1,995,814	100.0	2,103,075	100.0
Regions														
Africa	86,025	4.2	68,040	3.7	73,397	3.6	76,326	4.0	75,731	4.1	75,866	3.8	75,892	3.6
America, North	262,914	12.8	236,013	12.8	227,537	11.2	239,717	12.6	290,084	15.5	297,020	14.9	393,096	18.7
America, South	5,885	0.3	5,973	0.3	6,061	0.3	15,133	0.8	8,694	0.5	7,170	0.4	9,250	0.4
Asia	425,310	20.7	341,023	18.5	472,882	23.2	412,836	21.7	343,520	18.4	365,543	18.3	350,353	16.7
Europe	1,268,425	61.7	1,178,221	64.1	1,251,055	61.3	1,155,905	60.7	1,144,583	61.3	1,249,119	62.6	1,274,484	60.6
Oceania	7,951	0.4	9,897	0.5	8,488	0.4	5,168	0.3	3,132	0.2	1,096	0.1	-	
Leading Producers														
Mexico	259,975	12.6	232,967	12.7	224,384	11.0	236,458	12.4	286,718	15.4	293,548	14.7	389,517	18.5
United Kingdom	*181,975	8.8	*181,975	9.9	*181,975	8.9	181,975	9.6	*181,975	9.8	*181,975	9.1	*181,975	8.7
Germany	-		-		152,069	7.5	136,926	7.2	146,133	7.8	219,361	11.0	215,727	10.3
Spain	202,547	9.8	189,436	10.3	179,734	8.8	127,368	6.7	131,054	7.0	174,485	8.7	182,331	8.7
Korea, Republic of	130,246	6.3	120,116	6.5	126,356	6.2	123,870	6.5	123,151	6.6	139,081	7.0	*138,891	6.6

Commodity data are provided by the United Nations Statistical Division. The symbol * means that data are estimated. For additional notes, see Appendix I.

557

Product Tables
Nails, Screws, Nuts, Bolts, Rivets Etc.

Unit of Measure: Metric tons.

	1990 Value	%	1991 Value	%	1992 Value	%	1993 Value	%	1994 Value	%	1995 Value	%	1996 Value	%
Total Production	4,549,553	100.0	5,082,962	100.0	4,884,726	100.0	4,424,001	100.0	4,776,778	100.0	5,153,901	100.0	5,081,480	100.0
Regions														
Africa	128,934	2.8	113,957	2.2	104,743	2.1	102,773	2.3	104,060	2.2	102,079	2.0	108,071	2.1
America, North	212,536	4.7	212,653	4.2	215,641	4.4	242,985	5.5	269,787	5.6	260,060	5.0	273,343	5.4
America, South	224,232	4.9	228,038	4.5	237,794	4.9	260,596	5.9	308,188	6.5	304,607	5.9	305,471	6.0
Asia	1,900,739	41.8	1,964,331	38.6	1,853,146	37.9	1,764,038	39.9	2,009,369	42.1	2,187,688	42.4	2,154,243	42.4
Europe	2,072,924	45.6	2,555,088	50.3	2,465,798	50.5	2,047,296	46.3	2,080,354	43.6	2,295,739	44.5	2,237,917	44.0
Oceania	10,188	0.2	8,896	0.2	7,604	0.2	6,312	0.1	5,020	0.1	3,728	0.1	2,436	0.0
Leading Producers														
Japan	1,014,868	22.3	1,004,552	19.8	936,943	19.2	860,950	19.5	832,541	17.4	857,693	16.6	868,357	17.1
Germany	-		603,070	11.9	608,689	12.5	509,138	11.5	522,892	10.9	642,924	12.5	598,268	11.8
India	360,224	7.9	354,405	7.0	351,181	7.2	358,095	8.1	*472,839	9.9	*504,479	9.8	*536,120	10.6
Korea, Republic of	277,118	6.1	362,762	7.1	335,621	6.9	367,097	8.3	396,440	8.3	420,947	8.2	*445,872	8.8
Brazil	185,753	4.1	190,333	3.7	192,179	3.9	217,830	4.9	262,686	5.5	252,360	4.9	256,695	5.1

Containers, One Cubic Meter and Over

Unit of Measure: Number.

	1990 Value	%	1991 Value	%	1992 Value	%	1993 Value	%	1994 Value	%	1995 Value	%	1996 Value	%
Total Production	10,880,871	100.0	10,992,951	100.0	11,057,031	100.0	10,938,371	100.0	9,218,631	100.0	9,022,833	100.0	9,422,447	100.0
Regions														
Africa	9,590,000	88.1	9,708,000	88.3	9,662,000	87.4	9,859,000	90.1	8,270,000	89.7	8,173,000	90.6	8,162,000	86.6
America, South	33,000	0.3	33,000	0.3	21,000	0.2	61,000	0.6	16,000	0.2	34,000	0.4	33,000	0.4
Asia	788,300	7.2	635,100	5.8	655,900	5.9	584,533	5.3	565,333	6.1	318,333	3.5	636,200	6.8
Europe	469,571	4.3	616,851	5.6	718,131	6.5	433,838	4.0	367,297	4.0	497,499	5.5	591,247	6.3
Leading Producers														
South Africa	9,590,000	88.1	9,708,000	88.3	9,662,000	87.4	9,859,000	90.1	8,270,000	89.7	8,173,000	90.6	8,162,000	86.6
Korea, Republic of	334,000	3.1	331,000	3.0	354,000	3.2	163,000	1.5	111,000	1.2	114,000	1.3	*252,800	2.7
Hong Kong	159,800	1.5	193,600	1.8	227,400	2.1	261,200	2.4	295,000	3.2	43,000	0.5	*225,067	2.4
Slovakia	*154,000	1.4	*154,000	1.4	*154,000	1.4	*154,000	1.4	77,000	0.8	119,000	1.3	266,000	2.8
Belarus	285,000	2.6	101,000	0.9	65,000	0.6	*150,333	1.4	*150,333	1.6	*150,333	1.7	*150,333	1.6

Central-Heating Apparatus, Non-Electric (Boilers, Radiators Etc.)

Unit of Measure: Number.

	1990 Value	%	1991 Value	%	1992 Value	%	1993 Value	%	1994 Value	%	1995 Value	%	1996 Value	%
Total Production	11,893,733	100.0	11,754,733	100.0	11,552,144	100.0	12,546,979	100.0	12,543,962	100.0	11,993,345	100.0	16,301,825	100.0
Regions														
America, North	2,646,400	22.3	2,563,900	21.8	2,481,400	21.5	2,398,900	19.1	2,316,400	18.5	2,233,900	18.6	2,151,400	13.2
Asia	591,667	5.0	613,667	5.2	610,078	5.3	598,667	4.8	595,667	4.7	594,667	5.0	595,667	3.7
Europe	8,655,667	72.8	8,577,167	73.0	8,460,667	73.2	9,549,412	76.1	9,631,895	76.8	9,164,778	76.4	13,554,759	83.1
Leading Producers														
Spain	3,632,000	30.5	2,677,000	22.8	3,728,000	32.3	*4,272,321	34.1	*4,637,226	37.0	*5,002,131	41.7	*5,367,036	32.9
United States of America	*2,646,400	22.3	*2,563,900	21.8	*2,481,400	21.5	*2,398,900	19.1	*2,316,400	18.5	*2,233,900	18.6	*2,151,400	13.2
Hungary	1,588,000	13.4	1,360,500	11.6	1,133,000	9.8	1,729,000	13.8	1,113,000	8.9	1,079,000	9.0	*1,424,724	8.7
Croatia	1,560,000	13.1	1,514,000	12.9	712,000	6.2	524,000	4.2	682,000	5.4	130,000	1.1	3,588,000	22.0
Germany	-		1,081,000	9.2	952,000	8.2	780,000	6.2	875,000	7.0	787,459	6.6	811,034	5.0

ISIC 3820
MACHINERY NEC

ISIC 3820—Machinery nec—includes industrial and other machinery not classified as "electrical machinery". This is a wide range of machines, ranging from agricultural machinery (ploughs, seeders) to household machines (refrigerators, washing machines). The vast majority of industrial production machines (machine tools, looms, etc.) are included. Also included are office machines. These, however, because of their importance in the current computer age, are also broken out as a separate industry in the next chapter.

The products of this industry are categorized under SIC 35—Industrial Machinery and Equipment in the U.S. system of classification. Household laundry and refrigeration equipment, however, is classified under Electronic and Other Electrical Equipment in the U.S. as SIC 363—Household Appliances.

Summary Statistics

		1990 Value	N	1991 Value	N	1992 Value	N	1993 Value	N	1994 Value	N	1995 Value	N
Establishments or enterprises	(number)	206,510	73	218,796	82	289,231	78	185,609	76	170,271	62	128,515	29
Number employed	(000)	28,457	83	22,418	86	21,297	84	20,190	84	19,236	75	16,809	58
Total Wages	($ mil.)	243,125	79	278,566	82	293,862	78	223,401	76	232,690	68	196,317	54
Wage/salary per employee	($)	10,147	77	10,988	79	22,488	76	15,479	75	11,633	67	13,173	54
Female workers	(000)	706	33	414	32	1,564	29	1,499	32	1,501	22	1,495	13
as % of total employment*	(%)	15	33	15	32	16	29	16	32	19	22	21	13
Output	($ bil.)	1,230	80	1,334	81	1,386	78	1,205	76	1,304	71	1,262	59
per employee	($ 000)	59	78	61	79	1,537	76	248	74	72	68	95	57
per establishment	($ 000)	5,199	70	5,176	75	45,528	69	9,333	66	4,630	54	4,224	28
Value added	($ bil.)	525	73	525	71	544	68	529	68	493	63	526	56
per employee	($ 000)	27	70	28	68	29	66	27	65	30	61	35	54
per establishment	($ 000)	2,170	63	2,000	64	2,103	59	1,530	56	1,743	46	1,278	25
Capital investment	($ mil.)	46,460	62	45,460	57	43,328	53	36,635	48	33,224	39	18,317	20
per employee	($ 000)	3	60	3	55	2	51	2	47	3	38	3	20
per establishment	($ 000)	328	58	200	53	176	49	143	43	139	35	100	17

Data presented above are drawn from the detailed tables that follow. Columns headed 'N' show the number of countries that provided valid data for inclusion. Values are not strictly comparable one year to the next or one row to the next because the number of countries that report varies from period to period and row to row. However, a general indication of magnitudes can be gleaned from reviewing this summary—especially in earlier years in which more reports are available. For detailed explanations, see Appendix I. *The average for those countries reporting both total and female employment.

Establishments and Number Engaged

Country	Establishments or Enterprises (number)						Employees per Establishment					
	1990	1991	1992	1993	1994	1995	1990	1991	1992	1993	1994	1995
Albania	-	-	-	13	19	19	-	-	-	117	89	70
Angola	-	-	4.000	4.000	-	-	-	-	25	24	-	-
Argentina	-	-	-	7,048	7,601	-	-	-	-	7.679	7.372	-
Armenia	60	60	59	73	85	-	1,243	1,112	-	-	-	-
Australia	2,499	3,623	3,633	-	-	-	28	18	17	-	-	-
Austria	833	856	665	668	649	-	88	86	93	87	89	-
Azerbaijan	767	750	791	1,016	585	-	82	78	57	42	61	-
Bangladesh	123	114	146	-	-	-	32	51	45	-	-	-
Belgium	882	898	891	1,045	1,086	-	-	-	-	-	-	-
Bolivia	27	28	28	36	38	40	9.630	11	10	11	9.921	9.550
Bosnia & Herzegovina	41	44	-	-	1.000	-	443	403	-	-	838	-
Brazil	2,293	-	2,083	1,878	-	-	151	-	143	145	-	-
Canada	4,694	4,345	4,157	4,000	3,910	-	32	32	31	32	36	-
Chile	73	72	71	79	73	-	170	177	175	150	173	-
China	*48,179	*48,189	*47,767	*44,310	*45,188	*48,332	205	200	202	163	163	147
China (Hong Kong SAR)	6,384	5,638	5,486	4,582	4,538	-	8.741	8.443	7.765	7.988	7.470	-
Colombia	339	331	369	365	349	-	47	49	52	54	58	-
Costa Rica	146	157	157	163	-	-	22	17	19	21	-	-
Croatia	356	463	523	642	760	843	123	65	50	37	28	22
Cyprus	280	280	280	280	324	348	4.832	4.739	4.864	4.775	3.664	3.221
Czechoslovakia (Former)	*244	*294	-	-	-	-	2,049	1,429	-	-	-	-
Czech Republic	-	*213	*338	*405	-	-	-	1,493	633	469	-	-
Denmark	4,573	4,680	4,644	-	-	-	18	16	17	-	-	-
Ecuador	18	24	24	24	23	36	39	34	30	28	28	59
Egypt	153	168	160	153	-	-	259	248	236	254	-	-
El Salvador	-	5.000	5.000	4.000	11	24	-	-	17	22	64	40
Ethiopia	-	1.000	1.000	1.000	1.000	1.000	-	190	177	185	178	175
Finland	648	733	707	657	665	818	81	65	61	62	63	64
FYR Macedonia	*20	*38	*62	*62	*77	*90	81	95	54	50	35	17
Gambia	-	-	-	1.000	-	-	-	-	-	9.000	-	-
Germany	-	7,512	7,393	7,223	6,990	-	-	202	175	157	147	-
Germany (Western Part)	6,281	6,454	6,476	-	-	-	187	182	174	-	-	-
Ghana	-	-	-	11	-	-	-	-	-	59	-	-
Greece	404	400	397	-	-	-	22	20	18	-	-	-
Grenada	-	-	1.000	1.000	-	-	-	-	4.000	4.000	-	-
Guatemala	-	6.000	6.000	6.000	6.000	6.000	-	104	98	102	116	155
Honduras	-	-	34	41	41	26	-	-	18	16	18	46
Hungary	*1,436	*3,311	*2,110	*2,417	-	-	95	38	44	24	-	-
India	8,067	8,168	8,554	8,977	-	-	59	59	59	54	-	-
Indonesia	208	221	248	266	269	322	147	150	141	136	138	136
Iran (Islamic Republic of)	890	198	217	212	-	-	66	131	114	113	-	-
Iraq	-	11	8.000	-	-	-	-	593	750	-	-	-
Ireland	334	317	-	-	-	-	48	49	-	-	-	-
Israel	372	338	359	426	467	-	26	27	26	25	26	-
Italy	*4,517	*4,607	*4,627	*4,686	*4,717	-	84	81	85	81	81	-
Japan	50,159	49,746	47,761	47,308	44,225	46,065	28	29	29	28	31	29
Jordan	59	61	65	63	142	156	13	19	25	26	16	15
Kenya	47	58	62	61	67	67	32	25	24	24	22	22
Korea, Republic of	6,808	8,344	8,925	10,901	11,330	12,203	31	31	28	25	26	24
Kuwait	22	23	23	24	24	-	213	119	179	188	196	-
Kyrgyzstan	23	22	24	30	28	-	1,878	1,859	1,429	929	636	-
Latvia	1,052	1,009	273	256	120	198	45	42	98	67	105	61
Lithuania	-	-	-	*107	*134	-	-	-	-	211	248	-
Luxembourg	*37	*36	*35	*36	-	-	76	74	74	68	-	-
Macau	56	57	24	31	24	23	2.464	2.860	3.583	3.194	2.292	3.478
Malaysia	402	438	433	506	512	-	66	78	82	89	91	-
Malta	23	25	27	30	31	-	28	27	22	19	18	-
Mauritius	5.000	10	11	11	13	13	93	71	61	57	54	55
Mexico	175	169	162	153	145	142	190	186	188	152	160	160
Mongolia	-	-	-	-	8.000	-	-	-	-	-	28	-

Continued.

Depending on the table, * means *Enterprises*, *Engaged*, or *Factor Values*; ± means *Basis Unspecified*; § means *shown in millions*. For additional notes and sources, see Appendix I.

Establishments and Number Engaged

- Continued -

Country	Establishments or Enterprises (number)						Employees per Establishment					
	1990	1991	1992	1993	1994	1995	1990	1991	1992	1993	1994	1995
Morocco	247	261	268	290	296	296	30	31	29	25	27	25
Mozambique	13	13	10	10	10	-	97	93	83	80	75	-
Netherlands	917	972	1,005	1,018	1,052	-	86	83	80	75	72	-
Netherlands Antilles	-	-	-	8.000	-	-	-	-	-	14	-	-
New Zealand	*2,757	*2,837	*2,774	*2,932	-	-	4.274	3.962	4.128	4.617	-	-
Norway	516	504	332	300	323	-	64	67	105	63	59	-
Pakistan	-	259	-	-	-	-	-	95	-	-	-	-
Panama	8.000	5.000	-	-	-	-	9.875	9.600	-	-	-	-
Paraguay	-	2.000	-	-	-	-	-	5.000	-	-	-	-
Peru	693	711	720	-	-	-	15	13	11	-	-	-
Philippines	1,959	2,168	540	488	-	-	12	11	38	40	-	-
Poland	*719	*743	*716	*683	-	-	579	483	416	392	-	-
Portugal	*2,407	*2,373	*2,515	*2,795	*2,945	-	17	17	15	14	13	-
Puerto Rico	60	59	56	51	55	47	57	48	48	48	46	45
Republic of Moldova	49	50	57	-	-	-	1,488	1,369	1,042	-	-	-
Romania		334	502	675	967	-		1,744	884	580	374	-
Russian Federation	-	-	-	*6,904	*9,707	*14,960	-	-	-	448	270	156
Senegal	1.000	1.000	1.000	1.000	1.000	-	429	561	245	216	116	-
Singapore	459	466	518	518	551	-	147	143	136	133	134	-
Slovakia	-	144	144	161	182	-	-	825	667	472	375	-
Slovenia	244	*210	*267	*314	*350	*1,134	-	-	-	-	-	-
South Africa	-	2,569	-	-	-	-	-	29	-	-	-	-
Spain	14,314	14,445	13,974	-	-	-	9.772	9.514	9.201	-	-	-
Sri Lanka	26	30	31	44	-	-	89	47	62	82	-	-
Sweden	1,272	1,423	1,316	1,177	1,184	-	78	73	72	72	68	-
Thailand	200	419	-	-	-	-	330	141	-	-	-	-
Trinidad and Tobago	1.000	13	14	18	15	-	8.000	16	16	15	20	-
Tunisia	-	-	-	755	755	781	-	-	-	7.931	8.426	7.271
Turkey	326	299	770	716	714	-	157	151	65	70	67	-
Ukraine	641	1,028	1,199	654	637	635	1,094	698	523	945	851	767
United Kingdom	23,168	22,307	22,747	13,073	14,494	-	24	23	22	33	32	-
United Republic of Tanzania	31	33	-	-	-	-	64	55	-	-	-	-
United States of America	-	-	76,759	-	-	-	-	-	25	-	-	-
Venezuela	278	314	308	253	247	278	51	58	56	64	57	53
Yugoslavia	144	214	339	439	454	612	-	342	182	133	126	91
Zambia	9.000	-	-	-	18	-	124	-	-	-	-	-
Zimbabwe	42	45	43	40	38	-	83	67	58	58	63	-

Employment and Compensation of Employees

Country	Number Employed/Engaged (000)						Wage/Salary per Employee ($)					
	1990	1991	1992	1993	1994	1995	1990	1991	1992	1993	1994	1995
Albania	-	-	-	1.517	1.691	1.328	-	-	-	-	710	941
Angola	-	*0.258	*0.100	*0.096	-	-	-	-	-	-	-	-
Argentina	43	-	-	54	56	-	7,137	-	-	13,244	-	-
Armenia	*75	*67	-	-	-	-	-	-	-	-	-	-
Australia	70	*67	*62	-	-	-	21,830	24,246	23,482	-	-	-
Austria	73	74	62	58	57	59	28,307	30,039	33,275	33,055	34,352	-
Azerbaijan	63	58	45	43	36	-	1,964	1,882	-	-	-	-
Bangladesh	3.994	5.822	6.594	-	-	-	690	1,014	942	-	-	-
Bermuda	0.011	0.009	0.012	0.008	0.006	-	-	-	-	-	-	-
Bolivia	0.260	0.298	0.288	0.383	0.377	0.382	-	-	-	-	-	-
Bosnia & Herzegovina	18	18	-	-	0.838	-	2,941	3,055	-	-	-	-
Brazil	347	-	298	272	-	-	8,947	-	8,126	9,725	-	-
Canada	148	138	128	129	140	151	26,777	28,316	27,851	27,268	26,676	27,057
Chile	12	13	12	12	13	13	7,262	7,965	9,185	9,796	10,038	12,200
China	*9,890	*9,630	*9,640	*7,240	*7,360	*7,090	-	-	-	-	-	-
China (Hong Kong SAR)	*56	*48	*43	*37	*34	-	8,600	9,829	11,660	12,722	13,455	-

Continued.

Depending on the table, * means *Enterprises, Engaged,* or *Factor Values;* ± means *Basis Unspecified;* § means *shown in millions.* For additional notes and sources, see Appendix I.

561

Employment and Compensation of Employees

- Continued -

Country	Number Employed/Engaged (000)						Wage/Salary per Employee ($)					
	1990	1991	1992	1993	1994	1995	1990	1991	1992	1993	1994	1995
China (Taiwan Province)	135	138	145	151	152	149	10,306	11,095	13,164	13,158	13,977	15,059
Colombia	16	16	19	20	20	20	1,834	1,942	2,143	2,458	3,287	3,687
Costa Rica	3.242	2.678	2.987	3.459	-	-	3,047	2,843	3,011	3,306	-	-
Croatia	44	30	26	24	21	18	4,495	4,162	1,464	1,590	2,175	3,146
Cyprus	1.353	1.327	1.362	1.337	1.187	1.121	9,547	10,236	11,537	11,084	11,964	15,616
Czechoslovakia (Former)	500	420	-	-	-	-	2,289	1,620				
Czech Republic	349	318	214	190	-	-	2,340	1,644	1,925	2,359	-	-
Denmark	81	77	77	74	79	83	30,189	30,826	34,153	32,226	35,717	-
Ecuador	0.710	0.810	0.722	0.678	0.647	2.113	2,018	2,488	2,937	3,275	2,044	1,961
Egypt	*40	*42	*38	*39	*40	*43	2,184	1,348	1,569	1,733	1,863	2,142
El Salvador	-	-	*0.084	*0.088	*0.699	*0.966	-	-	-	2,884	4,774	3,132
Ethiopia	-	0.190	0.177	0.185	0.178	0.175	-	2,039	1,635	1,052	1,125	1,034
Finland	52	48	43	41	42	52	31,051	29,109	27,557	22,388	25,813	33,811
FYR Macedonia	1.614	3.605	3.375	3.092	2.728	1.570	4,117	5,842	2,533	2,462	2,317	4,760
France	440	436	418	394	378	382	38,001	39,009	-	-	-	-
Gambia	-	-	-	0.009								
Germany	-	*1,514	*1,297	*1,134	*1,027	-	-	28,910	35,671	35,696	38,519	
Germany (Western Part)	*1,174	*1,174	*1,127	-	-	-	33,550	34,429	38,621	-	-	-
Ghana	-	-	-	*0.646	*0.664	*0.683	-	-	-	818	611	623
Greece	*8.759	*7.902	*7.265	*6.856	*6.898	*7.538	10,251	10,681	11,250	10,508	10,693	13,469
Grenada	-	-	*0.004	*0.004	-	-	-	-	-	-	-	-
Guatemala	-	*0.625	*0.589	*0.615	*0.694	*0.928	-	-	-	-	-	-
Honduras	*0.705	*0.649	*0.615	*0.674	*0.737	*1.203	1,694	1,746	4,166	2,667	2,303	1,504
Hungary	136	125	92	59	65	69	2,545	2,665	3,142	3,416	3,976	4,446
India	*478	*482	*505	*484	*495	*534	1,819	1,552	1,528	1,434	1,576	1,848
Indonesia	30	33	35	36	37	44	1,017	1,223	1,338	1,522	1,581	2,228
Iran (Islamic Republic of)	59	26	25	24	-	-	4,523	5,417	5,087	5,273	-	-
Iraq	-	6.521	6.000	-	-	-	-	10,180	26,987	-	-	-
Ireland	16	16	16	17	17	19	22,216	23,435	27,332	24,225	26,634	34,911
Israel	9.600	9.000	9.300	10	12	11	21,286	21,402	23,000	22,479	23,879	26,600
Italy	380	375	391	380	380	-	37,789	-	-	33,748	34,560	-
Japan	*1,406	*1,431	*1,403	*1,343	*1,353	*1,353	31,124	34,886	37,751	-	-	-
Jordan	0.792	1.171	1.599	1.625	2.316	2.348	2,326	2,426	2,322	2,097	2,604	2,956
Kenya	1.489	1.477	1.479	1.492	1.506	1.502	1,458	1,322	1,260	774	1,108	1,450
Korea, Republic of	209	259	249	272	291	298	10,287	11,840	12,926	13,611	14,941	17,844
Kuwait	4.679	2.736	4.120	4.504	4.698	4.543	3,168	4,973	6,243	7,073	8,487	9,833
Kyrgyzstan	*43	*41	*34	*28	*18	-	509	412	-	-	-	-
Latvia	47	42	27	17	13	12	-	-	-	-	-	-
Lithuania	-	-	30	23	33	-	-	-	-	-	944	-
Luxembourg	2.795	2.654	2.585	2.458	2.467	-	34,785	36,688		-	-	-
Macau	0.138	0.163	0.086	0.099	0.055	0.080	4,562	4,731	6,142	8,835	5,525	6,963
Malawi	0.619	-	-	-	-	-	-	-	-	-	-	-
Malaysia	27	34	36	45	47	51	3,571	3,967	4,608	4,617	4,962	5,635
Malta	0.645	0.681	0.582	0.555	0.546	-	9,718	9,324	11,817	10,226	11,013	-
Mauritius	*0.465	0.709	0.671	0.629	0.702	0.710	3,661	2,251	3,616	3,829	4,497	4,463
Mexico	33	31	30	23	23	23	5,673	6,823	8,166	9,612	9,632	5,859
Mongolia	-	-	-	-	*0.226	-	-	-	-	-	-	-
Morocco	7.334	7.983	7.672	7.394	7.987	7.374	3,871	3,827	4,427	4,480	4,843	5,161
Mozambique	1.262	1.207	0.832	0.801	0.754	-	923	786	427	-	488	-
Netherlands	79	81	81	77	75	74	28,052	28,602	32,322	31,336	33,695	-
Netherlands Antilles	-	-	-	0.115	-	-	-	-	-	11,314	-	-
New Zealand	12	11	11	14	-	-	18,087	17,568	16,446	-	-	-
Norway	33	34	35	19	19	18	34,430	35,320	-	31,962	36,308	-
Pakistan	22	25	26	-	-	-	1,933	2,073	2,132	-	-	-
Panama	*0.079	*0.048	*0.040	*0.040	*0.043	*0.046	3,899	4,167	4,211	4,271	4,336	4,346
Paraguay	-	0.010	-	-	-	-	-	-	-	-	-	-
Peru	*11	*9.307	*7.915	-	-	-	2,851	3,677	3,725	-	-	-
Philippines	24	25	21	20	-	-	1,577	1,799	2,388	1,712	-	-
Poland	416	359	298	268	273	280	1,308	1,868	2,351	2,474	2,936	4,088
Portugal	*41	*41	*38	*39	*39	*38	-	-	-	-	-	-

Continued.

Depending on the table, * means *Enterprises*, *Engaged*, or *Factor Values*; ± means *Basis Unspecified*; § means *shown in millions*. For additional notes and sources, see Appendix I.

Employment and Compensation of Employees
- Continued -

Country	Number Employed/Engaged (000)						Wage/Salary per Employee ($)					
	1990	1991	1992	1993	1994	1995	1990	1991	1992	1993	1994	1995
Puerto Rico	3.410	2.850	2.660	2.460	2.540	2.130	25,308	-	-	-	38,346	34,272
Republic of Moldova	73	68	59	-	-	-	-	3,198	-	-	-	-
Romania	603	583	444	391	362	-	1,806	1,124	759	931	968	-
Russian Federation	-	-	-	3,090	2,622	2,331	-	-	-	575	937	998
Senegal	*0.429	*0.561	*0.245	*0.216	*0.116	-	3,253	1,997	4,719	5,134	2,609	-
Singapore	*67	*66	*71	*69	*74	*83	10,526	12,021	13,830	15,094	16,316	17,930
Slovakia	-	*119	*96	*76	*68	-	-	1,540	1,754	1,956	2,242	-
South Africa	82	75	71	67	67	70	9,908	10,985	12,149	11,387	11,803	13,791
Spain	140	137	129	141	143	146	20,572	21,962	24,121	22,190	22,467	25,035
Sri Lanka	*2.308	*1.423	*1.930	*3.592	-	-	779	900	865	1,047	-	-
Sweden	100	*103	*95	85	80	77	25,440	29,072	31,945	25,538	26,262	34,365
Thailand	66	59	-	-	-	-	3,646	3,844	-	-	-	-
Trinidad and Tobago	0.008	0.208	0.230	0.268	0.295	-	2,671	6,791	4,914	3,975	4,005	-
Tunisia	-	-	-	5.988	6.362	5.679	-	-	-	1,910	1,790	2,259
Turkey	51	45	50	50	48	49	6,817	9,340	8,716	9,101	5,446	7,149
Ukraine	701	718	627	618	542	487	-	2,815	2,129	-	448	484
United Kingdom	557	523	501	434	461	466	24,554	26,819	28,417	24,889	25,738	27,725
United Republic of Tanzania	1.969	1.810	-	-	-	-	-	-	-	-	-	-
USSR (Former)	6,745	-	-	-	-	-	2,212	-	-	-	-	-
United States of America	2,066	1,954	1,922	1,927	1,999	2,096	29,976	30,885	32,738	33,261	34,233	34,811
Uruguay	*2.501	*2.833	*2.588	*2.807	*2.782	*2.833	2,798	3,674	3,999	5,018	5,469	5,335
Venezuela	14	18	17	16	14	15	3,519	3,806	4,299	3,785	3,761	4,188
Yugoslavia	-	73	62	58	57	56	-	-	-	-	1,603	759
Zambia	1.116	-	-	-	-	-	2,385	-	-	-	-	-
Zimbabwe	3.500	3.000	2.500	2.300	2.400	2.416	4,027	3,131	2,685	2,398	2,142	2,296

Female Workers: Total and Percent of Employment

Country	Female Workers (000)						Female Workers as % of Total Employment					
	1990	1991	1992	1993	1994	1995	1990	1991	1992	1993	1994	1995
Albania	-	-	-	0.337	0.609	0.419	-	-	-	22.21	36.01	31.55
Angola	-	-	0.006	0.005	-	-	-	-	6.00	5.21	-	-
Argentina	-	-	-	-	4.329	-	-	-	-	-	7.73	-
Australia	13	-	-	-	-	-	18.57	-	-	-	-	-
Austria	9.300	9.000	8.000	7.476	7.376	-	12.74	12.16	12.90	12.85	12.84	-
Bangladesh	0.079	0.081	0.068	-	-	-	1.98	1.39	1.03	-	-	-
Bermuda	0.001	0.002	0.001	0.001	-	-	9.09	22.22	8.33	12.50	-	-
Bosnia & Herzegovina	2.536	2.339	-	-	-	-	13.95	13.20	-	-	-	-
Canada	23	22	-	-	-	-	15.54	15.94	-	-	-	-
Chile	0.692	0.573	0.576	0.408	0.600	-	5.59	4.50	4.62	3.45	4.75	-
China (Taiwan Province)	45	46	48	51	53	52	33.20	33.38	33.21	33.71	34.57	34.60
Colombia	-	2.283	-	2.969	3.008	-	-	14.09	-	14.99	14.98	-
Croatia	6.670	4.450	3.770	3.520	3.140	2.660	15.23	14.80	14.48	14.85	14.88	14.57
Cyprus	0.181	0.163	0.201	0.181	0.169	-	13.38	12.28	14.76	13.54	14.24	-
Czechoslovakia (Former)	76	67	-	-	-	-	15.10	15.93	-	-	-	-
Czech Republic	111	99	56	50	-	-	31.81	31.13	26.17	26.32	-	-
Denmark	17	16	16	-	-	-	20.78	20.75	21.02	-	-	-
Egypt	-	1.870	1.977	2.061	-	-	-	4.50	5.23	5.30	-	-
El Salvador	-	-	-	0.010	0.092	0.078	-	-	-	11.36	13.16	8.07
Ethiopia	-	0.043	0.043	0.047	0.044	0.042	-	22.63	24.29	25.41	24.72	24.00
FYR Macedonia	0.272	0.330	0.377	0.395	0.330	0.195	16.85	9.15	11.17	12.77	12.10	12.42
Gambia	-	-	-	0.001	-	-	-	-	-	11.11	-	-
Germany (Western Part)	196	-	-	-	-	-	16.70	-	-	-	-	-
Ghana	-	-	-	0.033	-	-	-	-	-	5.11	-	-
Hungary	31	-	17	11	-	-	22.79	-	18.48	18.64	-	-
Indonesia	-	-	-	4.494	4.266	4.817	-	-	-	12.43	11.49	11.01
Iran (Islamic Republic of)	-	0.625	0.707	0.664	-	-	-	2.40	2.86	2.77	-	-
Iraq	-	0.643	-	-	-	-	-	9.86	-	-	-	-

Continued.

Depending on the table, * means *Enterprises, Engaged,* or *Factor Values;* ± means *Basis Unspecified;* § means *shown in millions.* For additional notes and sources, see Appendix I.

563

Female Workers: Total and Percent of Employment
- Continued -

Country	Female Workers (000)						Female Workers as % of Total Employment					
	1990	1991	1992	1993	1994	1995	1990	1991	1992	1993	1994	1995
Ireland	4.500	4.500	-	-	-	-	28.30	28.85	-	-	-	-
Italy	-	53	53	54	56	-	-	14.23	13.49	14.30	14.81	-
Jordan	0.034	0.041	0.048	0.045	0.108	0.137	4.29	3.50	3.00	2.77	4.66	5.83
Kenya	0.033	0.028	0.029	0.029	-	-	2.22	1.90	1.96	1.94	-	-
Korea, Republic of	33	49	43	50	53	55	15.97	18.92	17.39	18.44	18.35	18.32
Macau	0.008	0.019	0.016	0.013	0.011	0.007	5.80	11.66	18.60	13.13	20.00	8.75
Malaysia	6.700	11	12	16	15	-	25.19	33.72	34.17	35.84	31.33	-
Malta	0.018	0.020	0.031	0.029	0.031	-	2.79	2.94	5.33	5.23	5.68	-
Mongolia	-	-	-	-	0.035	-	-	-	-	-	15.49	-
Morocco	-	-	-	-	-	0.635	-	-	-	-	-	8.61
New Zealand	1.504	1.491	1.522	1.682	-	-	12.76	13.27	13.29	12.42	-	-
Panama	0.008	-	-	-	-	-	10.13	-	-	-	-	-
Philippines	-	-	3.181	4.388	-	-	-	-	15.37	22.39	-	-
Portugal	2.613	2.808	2.748	2.575	2.119	-	6.45	6.86	7.20	6.65	5.49	-
Puerto Rico	1.320	1.030	0.910	0.780	0.800	0.600	38.71	36.14	34.21	31.71	31.50	28.17
Sri Lanka	0.089	0.091	0.155	1.058	-	-	3.86	6.39	8.03	29.45	-	-
Sweden	18	-	-	-	-	-	17.65	-	-	-	-	-
Thailand	11	18	-	-	-	-	16.02	29.89	-	-	-	-
Turkey	2.947	-	-	-	-	-	5.74	-	-	-	-	-
United Kingdom	94	-	84	-	-	-	16.88	-	16.77	-	-	-
United Republic of Tanzania	0.220	0.206	-	-	-	-	11.17	11.38	-	-	-	-
United States of America	-	-	1,210	1,233	1,297	1,379	-	-	62.96	63.99	64.88	65.79

Output and Output per Employee

Country	Output ($ bil.)						Output per Employee ($)					
	1990	1991	1992	1993	1994	1995	1990	1991	1992	1993	1994	1995
Albania	-	-	-	*0.002	*0.005	*0.004	-	-	-	1,098	2,721	2,697
Angola	-	-	±11	±1.322	-	-	-	-	§111,9	§13,77	-	-
Argentina	±1.876	-	-	4.492	-	-	44,134	-	-	83,009	-	-
Armenia	0.025	0.029	0.124	2.871	0.015	-	335	435	-	-	-	-
Australia	*6.633	*6.986	*6.307	-	-	-	94,754	104,262	101,731	-	-	-
Austria	9.123	9.563	8.849	7.878	8.855	11	124,979	129,223	142,719	135,442	154,102	192,360
Azerbaijan	±0.849	±0.676	±0.027	±0.112	±0.023	-	13,466	11,612	608	2,602	651	-
Bangladesh	0.019	0.023	0.024	-	-	-	4,651	4,008	3,640	-	-	-
Bolivia	0.002	0.002	0.003	0.003	0.003	0.004	7,959	7,446	10,501	8,176	8,946	9,208
Bosnia & Herzegovina	±0.316	±0.207	-	-	-	-	17,357	11,664	-	-	-	-
Brazil	*19	-	*13	*14	-	-	53,470	-	43,418	53,109	-	-
Canada	17	16	15	15	19	20	113,732	113,784	117,314	119,395	132,229	135,284
Chile	0.449	0.494	0.596	0.641	0.679	0.861	36,285	38,752	47,816	54,140	53,701	65,497
China	±35	±37	±48	±60	±49	±49	3,539	3,892	5,026	8,306	6,595	6,962
China (Hong Kong SAR)	3.820	4.278	4.201	3.792	3.514	-	68,454	89,871	98,626	103,605	103,644	-
China (Taiwan Province)	±6.920	±7.460	±8.478	±9.973	±11	±11	51,193	54,037	58,531	65,877	69,470	75,578
Colombia	0.355	0.344	0.360	0.448	0.550	0.620	22,448	21,254	18,821	22,622	27,377	30,804
Costa Rica	0.039	0.036	0.050	0.065	0.067	0.065	11,930	13,595	16,679	18,705	-	-
Croatia	±1.110	±0.792	±0.441	-	-	-	25,347	26,349	16,919	-	-	-
Cyprus	0.057	0.061	0.061	0.056	0.057	0.063	42,471	45,853	44,851	42,183	48,105	56,398
Czechoslovakia (Former)	±6.312	±2.531	-	-	-	-	12,624	6,025	-	-	-	-
Denmark	*5.728	*5.661	*6.474	*5.975	*7.012	*8.944	70,986	73,463	84,433	80,233	88,964	108,292
Ecuador	0.010	0.011	0.010	0.010	0.010	0.056	13,575	14,076	14,373	14,269	15,291	26,722
Egypt	*0.582	*0.452	*0.385	*0.465	*0.519	*0.642	14,703	10,861	10,174	11,954	12,921	14,988
El Salvador	-	±0.001	-	±0.002	±0.024	±0.020	-	-	-	27,566	33,788	20,711
Ethiopia	-	0.001	0.000	0.000	0.001	0.001	-	3,733	2,794	1,168	2,834	3,743
Finland	±7.251	±5.381	±5.074	±4.502	±5.749	±10	138,374	112,810	118,273	111,166	137,527	193,417
FYR Macedonia	±0.029	±0.069	±0.042	±0.032	±0.023	±0.024	17,994	19,212	12,500	10,456	8,316	15,421
France	±62	±61	±63	±54	±56	±67	141,131	139,557	149,426	136,122	148,653	174,949
Germany	-	155	161	139	145	-	-	102,340	123,809	122,442	141,089	-
Germany (Western Part)	144	147	153	-	-	-	122,417	124,932	136,030	-	-	-

Continued.

Depending on the table, * means *Enterprises, Engaged,* or *Factor Values*; ± means *Basis Unspecified*; § means *shown in millions*. For additional notes and sources, see Appendix I.

Output and Output per Employee

- Continued -

Country	Output ($ bil.)						Output per Employee ($)					
	1990	1991	1992	1993	1994	1995	1990	1991	1992	1993	1994	1995
Ghana	-	-	-	0.005	0.004	0.004	-	-	-	8,065	6,392	6,520
Greece	*0.425	*0.390	*0.407	*0.359	*0.367	*0.502	48,480	49,325	56,075	52,372	53,208	66,655
Guatemala	-	0.014	0.014	0.019	0.026	0.054	-	21,825	24,114	31,671	37,024	58,395
Honduras	0.008	0.007	0.009	0.014	0.011	0.014	12,000	10,229	14,908	20,212	15,576	11,612
Hungary	±1.819	±1.822	±1.307	±1.242	±1.524	±1.762	13,378	14,580	14,208	21,056	23,486	25,535
India	*9.061	*8.046	*8.303	*7.198	*8.183	*10	18,949	16,704	16,442	14,861	16,533	19,645
Indonesia	0.504	0.731	0.646	0.818	1.019	1.345	16,536	22,084	18,534	22,624	27,464	30,738
Iran (Islamic Republic of)	1.563	0.672	0.658	0.601	-	-	26,630	25,856	26,599	25,053	-	-
Iraq	-	*0.119	*0.297	-	-	-	-	18,197	49,518	-	-	-
Ireland	*4.627	*4.198	*5.283	*4.783	*5.552	*8.482	290,987	269,086	323,639	289,580	325,187	447,966
Israel	0.669	0.639	0.775	0.867	1.037	1.087	69,644	71,032	83,386	82,548	85,743	96,339
Italy	±59	±57	±67	±54	±61	-	155,802	152,477	170,302	143,326	161,165	
Japan	±300	±345	±336	±345	±389	±454	213,587	241,023	239,304	256,693	287,779	335,580
Jordan	0.022	0.031	0.042	0.041	0.061	0.069	27,812	26,119	26,387	24,996	26,175	29,180
Kenya	*0.021	*0.020	*0.019	*0.011	*0.015	*0.052	14,067	13,291	12,592	7,627	9,951	34,694
Korea, Republic of	16	24	23	27	33	43	77,575	92,995	92,519	98,707	113,869	145,868
Kuwait	0.031	0.037	0.062	0.081	0.089	0.105	6,670	13,350	15,042	18,088	18,951	23,202
Kyrgyzstan	0.464	0.353	0.211	0.026	0.025	-	10,748	8,629	6,140	945	1,400	-
Latvia	0.005	0.010	0.053	0.063	0.076	0.100	102	236	1,976	3,644	6,040	8,270
Lithuania	-	-	0.051	0.041	0.138	-	-	-	1,715	1,831	4,150	-
Luxembourg	*0.355	*0.312	*0.305	*0.288	*0.308	-	127,094	117,534	117,883	117,028	124,653	-
Macau	0.002	0.005	0.002	0.002	0.001	0.002	15,606	30,278	21,558	21,427	22,536	19,017
Malaysia	*1.089	*1.725	*2.147	*2.686	*3.291	*4.159	40,955	50,587	60,126	59,422	70,618	81,052
Malta	0.016	0.021	0.025	0.026	0.028	-	24,708	31,560	43,155	46,366	51,064	-
Mauritius	0.007	0.007	0.008	0.007	0.007	0.009	14,426	10,048	11,865	11,071	10,628	12,530
Mexico	1.588	1.711	1.837	1.828	2.362	1.984	47,819	54,423	60,475	78,475	101,636	87,435
Morocco	±0.189	±0.184	±0.178	±0.166	±0.209	±0.209	25,709	23,019	23,174	22,412	26,175	28,361
Mozambique	-	±0.002	-	±0.001	±0.001	-	-	1,249	-	816	810	-
Netherlands	9.506	9.952	11	9.705	10	13	120,903	123,451	133,781	126,844	138,376	170,070
New Zealand	0.949	-	-	-	-	-	80,554	-	-	-	-	-
Norway	±6.932	±6.928	±8.023	±2.271	±2.829	±3.036	210,349	204,296	230,874	120,249	147,967	167,755
Pakistan	0.348	0.368	0.399	-	-	-	15,505	14,935	15,498	-	-	-
Panama	0.001	0.001	0.000	0.000	0.000	0.001	11,937	12,042	9,700	10,062	11,550	12,577
Peru	0.239	0.297	0.248	-	-	-	22,659	31,915	31,280	-	-	-
Philippines	0.200	0.236	0.268	0.387	-	-	8,285	9,497	12,958	19,724	-	-
Poland	±4.283	±3.887	±3.001	±3.070	±3.740	±5.373	10,296	10,827	10,072	11,445	13,693	19,164
Portugal	1.555	1.852	1.828	1.684	1.718	2.048	38,370	45,219	47,887	43,472	44,542	53,729
Puerto Rico	±2.057	±1.695	±1.416	±1.292	±1.735	±2.343	603,153	594,912	532,519	525,285	682,913	§1,100
Republic of Moldova	0.000	1.268	0.086	-	-	-	0.033	18,524	1,450	-	-	-
Russian Federation	-	-	-	7.676	11	13	-	-	-	2,484	4,015	5,784
Senegal	0.004	-	-	-	-	-	10,146	-	-	-	-	-
Singapore	*8.649	*9.448	*11	*15	*20	*25	128,466	142,115	159,103	222,860	269,870	300,330
Slovakia	-	±1.202	±0.828	±0.579	±0.564	-	-	10,117	8,613	7,623	8,254	-
Slovenia	-	-	-	0.882	0.971	1.243	-	-	-	-	-	-
South Africa	±3.737	±3.587	±3.600	±3.428	±3.252	±3.917	45,579	47,823	50,708	51,166	48,537	56,267
Spain	*14	*15	*14	±13	±15	±19	101,285	108,550	112,437	92,267	105,939	127,712
Sri Lanka	0.021	0.012	0.017	0.042	-	-	8,944	8,542	8,575	11,653	-	-
Sweden	*12	*15	*15	*8.919	*10	*14	125,228	149,577	159,861	105,493	130,559	176,856
Thailand	3.580	10	-	-	-	-	54,186	172,662	-	-	-	-
Trinidad and Tobago	±0.000	±0.005	±0.005	±0.005	±0.005	-	24,644	25,718	23,360	19,455	16,992	-
Tunisia	0.090	0.095	0.101	0.098	0.103	0.115	-	-	-	16,356	16,199	20,279
Turkey	3.323	3.615	4.014	4.634	3.223	4.380	64,723	79,999	80,591	92,243	67,154	88,645
Ukraine	*13	*12	*9.118	*2.008	*1.703	*1.596	18,926	16,318	14,542	3,250	3,142	3,277
United Kingdom	*65	*61	62	53	60	65	117,492	117,444	124,673	121,587	129,371	140,183
United Republic of Tanzania	*0.008	*0.006	-	-	-	-	3,989	3,547	-	-	-	-
USSR (Former)	±72	-	-	-	-	-	10,667	-	-	-	-	-
United States of America	*280	*267	*284	*304	*343	*379	135,760	136,807	147,983	157,763	171,724	180,981
Uruguay	0.040	0.047	0.046	0.060	0.065	0.064	16,076	16,694	17,639	21,417	23,316	22,733
Venezuela	0.427	0.646	0.652	0.581	0.490	0.545	29,842	35,310	37,482	35,804	34,552	37,079
Yugoslavia	±0.171	±0.162	±0.269	-	±0.424	±0.220	-	2,212	4,372	-	7,435	3,936

Continued.

Depending on the table, * means *Enterprises, Engaged,* or *Factor Values;* ± means *Basis Unspecified;* § means *shown in millions.* For additional notes and sources, see Appendix I.

Output and Output per Employee

- Continued -

Country	Output ($ bil.)						Output per Employee ($)					
	1990	1991	1992	1993	1994	1995	1990	1991	1992	1993	1994	1995
Zambia	0.015	-	-	-	-	-	13,705	-	-	-	-	-
Zimbabwe	*0.079	*0.059	*0.041	*0.030	*0.029	*0.032	22,526	19,545	16,332	13,166	12,014	13,111

Value Added and Value Added per Employee

Country	Value Added ($ bil.)						Value Added per Employee ($)					
	1990	1991	1992	1993	1994	1995	1990	1991	1992	1993	1994	1995
Argentina	±0.835	-	-	1.727	-	-	19,646	48,503	51,768	31,908	-	-
Australia	*3.070	*3.250	*3.210	-	-	-	43,850	48,503	51,768	-	-	-
Austria	3.292	3.430	3.100	2.820	3.288	4.201	45,098	46,355	49,999	48,486	57,215	71,046
Bangladesh	0.007	0.006	0.007	-	-	-	1,785	1,112	1,094	-	-	-
Bolivia	0.001	0.001	0.001	0.001	0.002	0.002	3,291	3,139	5,100	3,686	4,420	4,550
Bosnia & Herzegovina	±0.172	±0.200	-	-	-	-	9,466	11,282	-	-	-	-
Brazil	*12	-	*9.103	*9.866	-	-	35,281	-	30,566	36,236	-	-
Canada	7.576	6.817	6.536	7.077	7.835	8.598	51,191	49,397	51,062	54,860	55,967	57,013
Chile	0.168	0.171	0.220	0.214	0.272	0.344	13,606	13,415	17,643	18,047	21,485	26,177
China	±10	±11	±14	±18	±14	±13	1,023	1,102	1,422	2,514	1,851	1,890
China (Hong Kong SAR) . . .	1.078	1.072	1.121	0.941	0.936	-	19,320	22,530	26,326	25,706	27,608	-
China (Taiwan Province) . . .	2.379	2.689	3.459	3.475	3.556	3.281	17,598	19,479	23,877	22,952	23,403	22,000
Colombia	0.124	0.147	0.154	0.185	0.244	0.276	7,822	9,063	8,040	9,355	12,163	13,695
Costa Rica	0.013	0.013	0.017	0.021	0.022	0.021	4,053	4,689	5,714	6,168	-	-
Cote d'Ivoire	0.555	0.482	0.355	0.431	-	-	-	-	-	-	-	-
Croatia	±0.517	±0.409	±0.239	-	-	-	11,803	13,612	9,169	-	-	-
Cyprus	0.024	0.026	0.028	0.026	0.026	0.029	17,645	19,587	20,390	19,090	21,775	25,560
Czechoslovakia (Former) . .	±2.597	-	-	-	-	-	5,193	-	-	-	-	-
Denmark	*3.050	*3.066	*3.504	*3.235	*3.791	*4.831	37,800	39,785	45,695	43,441	48,097	58,499
Ecuador	0.003	0.004	0.004	0.003	0.004	0.016	3,997	4,488	5,283	4,857	6,098	7,530
Egypt	*0.142	*0.107	*0.093	*0.113	*0.125	*0.154	3,587	2,560	2,452	2,896	3,116	3,587
El Salvador	-	0.001	-	0.002	0.012	0.005	-	-	-	21,485	17,087	4,829
Ethiopia	-	0.001	0.000	-	0.000	0.000	-	3,049	1,675	-	1,755	1,778
Finland	±3.355	±2.333	±2.139	±1.810	±2.089	±3.281	64,019	48,901	49,853	44,690	49,981	63,014
FYR Macedonia	±0.011	±0.045	±0.033	±0.026	±0.014	±0.015	6,969	12,372	9,754	8,318	5,057	9,252
France	±25	±24	±24	±20	±20	±23	56,436	54,579	57,201	50,581	53,211	60,842
Germany (Western Part) . .	83	85	89	74	-	-	70,315	72,459	79,207	-	-	-
Ghana	-	-	-	0.002	0.001	0.001	-	-	-	2,326	1,961	2,000
Greece	*0.178	*0.170	*0.174	*0.154	*0.157	*0.216	20,339	21,557	24,013	22,439	22,812	28,632
Honduras	*0.004	*0.003	*0.004	*0.003	*0.004	*0.004	5,005	4,920	5,930	4,948	4,964	3,463
Hungary	±0.757	±0.642	±0.571	±0.482	±0.577	±0.657	5,563	5,136	6,212	8,167	8,898	9,517
India	*2.011	*1.861	*1.931	*1.686	*1.911	*2.437	4,207	3,864	3,823	3,481	3,861	4,565
Indonesia	*0.171	*0.289	*0.264	*0.244	*0.366	*0.414	5,598	5,683	8,719	6,759	9,871	9,456
Iran (Islamic Republic of) . .	0.764	0.317	0.336	0.322	-	-	13,008	12,176	13,577	13,440	-	-
Iraq	-	*0.049	*0.159	-	-	-	-	7,587	26,474	-	-	-
Ireland	*2.235	*1.943	*2.452	*2.222	*2.583	*3.961	140,586	124,539	150,211	134,503	151,296	209,185
Israel	0.279	0.279	0.290	0.323	0.390	0.407	29,036	31,055	31,177	30,758	32,222	36,031
Italy	±20	±20	±23	±18	±20	-	53,566	54,480	58,267	48,389	51,701	-
Japan	±127	±145	±141	±142	±159	±181	90,016	101,655	100,777	105,918	117,695	133,621
Jordan	0.009	0.013	0.015	0.012	0.019	0.020	11,902	10,677	9,398	7,608	8,408	8,654
Kenya	*0.005	*0.005	*0.004	*0.003	*0.004	*0.005	3,517	3,446	2,938	1,849	2,843	3,107
Korea, Republic of	7.004	10	9.695	11	14	18	33,482	38,702	38,994	40,964	46,823	59,020
Kuwait	0.019	0.026	0.045	0.059	0.062	0.076	3,966	9,544	10,935	13,089	13,266	16,641
Latvia	-	-	-	*0.039	*0.039	*0.055	-	-	-	2,257	3,106	4,571
Luxembourg	*0.158	*0.142	*0.151	*0.138	*0.148	-	56,379	53,537	58,322	56,080	59,987	-
Macau	0.001	0.002	0.001	0.001	0.001	0.001	9,445	13,741	6,980	13,062	9,884	9,958
Malaysia	*0.348	*0.533	*0.612	*0.800	*0.898	*1.131	13,101	15,639	17,132	17,704	19,273	22,050
Malta	0.008	0.009	0.012	0.010	0.011	-	13,151	13,152	20,505	17,678	20,578	-
Mauritius	0.004	0.004	0.000	0.001	0.003	0.003	7,770	5,803	192	1,946	4,545	4,504
Mexico	0.746	0.765	0.808	0.722	0.874	0.732	22,455	24,342	26,601	30,993	37,616	32,237
Morocco	-	-	-	±0.061	±0.076	±0.068	-	-	-	8,276	9,550	9,194

Continued.

Depending on the table, * means *Enterprises*, *Engaged*, or *Factor Values*; ± means *Basis Unspecified*; § means *shown in millions*. For additional notes and sources, see Appendix I.

Value Added and Value Added per Employee

- Continued -

Country	Value Added ($ bil.)						Value Added per Employee ($)					
	1990	1991	1992	1993	1994	1995	1990	1991	1992	1993	1994	1995
Netherlands	3.552	3.636	4.116	3.660	3.882	4.626	45,172	45,109	51,058	47,838	51,467	62,184
New Zealand	0.340	-	-	-	-	-	28,827	-	-	-	-	-
Norway	±1.590	±1.708	±1.810	±0.813	±0.951	±1.032	48,261	50,367	52,086	43,076	49,757	57,035
Pakistan	0.102	0.118	0.131	-	-	-	4,532	4,779	5,100	-	-	-
Panama	0.001	0.000	0.000	0.000	0.000	0.000	7,215	8,771	8,577	8,725	9,015	9,157
Peru	0.116	0.117	0.103	-	-	-	11,002	12,551	13,072	-	-	-
Philippines	0.084	0.100	0.126	0.110	-	-	3,457	4,013	6,109	5,597	-	-
Poland	±2.604	±1.979	±1.499	±1.328	±1.599	±2.280	6,260	5,514	5,030	4,954	5,854	8,132
Portugal	0.528	0.643	0.605	0.569	0.593	0.707	13,041	15,691	15,855	14,686	15,382	18,549
Puerto Rico	±0.668	±0.553	±0.503	±0.456	±0.572	±0.512	195,777	193,965	189,135	185,569	225,276	240,469
Russian Federation	-	-	-	±4.883	±6.932	±7.592	-	-	-	1,580	2,643	3,257
Senegal	0.002	0.002	0.002	0.001	0.001	-	4,212	2,907	7,016	6,180	6,118	-
Singapore	*2.737	*2.956	*3.538	*4.537	*5.417	*6.783	40,649	44,462	50,069	66,047	73,395	81,341
Slovakia	-	-	-	±0.261	±0.235	-	-	-	-	3,441	3,439	-
Slovenia	-	-	-	0.255	0.290	0.343	-	-	-	-	-	-
South Africa	*1.432	*1.464	*1.582	*1.503	*1.425	*1.711	17,459	19,517	22,277	22,439	21,273	24,573
Spain	*5.745	*6.184	*5.929	±5.492	±5.885	±6.724	41,070	44,999	46,113	38,828	41,292	46,134
Sri Lanka	0.009	0.006	0.009	0.014	-	-	3,720	3,883	4,896	4,020	-	-
Sweden	*6.226	*5.232	*5.246	*4.001	*4.454	*5.615	62,444	50,598	55,113	47,318	55,458	73,093
Thailand	1.929	6.828	-	-	-	-	29,203	115,756	-	-	-	-
Trinidad and Tobago	±0.000	±0.002	±0.002	±0.002	±0.002	-	6,247	11,056	10,407	8,647	8,010	-
Tunisia	0.013	0.013	0.014	0.015	0.017	0.019	-	-	2,462	2,617	3,354	
Turkey	1.423	1.388	1.703	1.823	1.485	2.022	27,709	30,720	34,196	36,275	30,934	40,921
United Kingdom	*30	*28	*28	*21	*25	*27	53,988	54,363	56,474	49,515	53,775	58,061
United Republic of Tanzania	*0.002	*0.002	-	-	-	-	1,221	1,034	-	-	-	-
United States of America	*145	*137	*147	*154	*177	*188	70,213	70,072	76,229	80,166	88,508	89,551
Uruguay	0.022	0.028	0.026	0.035	0.037	0.037	8,829	9,847	10,115	12,355	13,450	13,114
Venezuela	0.180	0.233	0.270	0.214	0.217	0.433	12,575	12,732	15,521	13,167	15,318	29,442
Yugoslavia	±0.086	±0.083	±0.173	-	±0.231	±0.121	-	1,129	2,814	-	4,049	2,162
Zambia	0.008	-	-	-	-	-	7,283	-	-	-	-	-
Zimbabwe	*0.043	*0.026	*0.022	*0.016	*0.014	*0.016	12,337	8,586	8,904	6,905	5,971	6,497

Capital Investment

Country	Gross Fixed Capital Formation ($ mil.)						Per Establishment ($ mil)					
	1990	1991	1992	1993	1994	1995	1990	1991	1992	1993	1994	1995
Argentina	-	-	-	147	-	-	-	-	-	0.021	-	-
Australia	208	-	-	-	-	-	0.083	-	-	-	-	-
Austria	412	416	373	301	366	-	0.495	0.485	0.561	0.450	0.564	-
Bangladesh	0.437	0.683	5.597	-	-	-	0.004	0.006	0.038	-	-	-
Bolivia	0.026	0.074	0.126	0.025	0.135	-	0.001	0.003	0.005	0.001	0.004	-
Bosnia & Herzegovina	15	-	-	-	-	-	0.364	-	-	-	-	-
Brazil	1,537	-	578	346	-	-	0.670	-	0.278	0.184	-	-
Cameroon	-	-	-	-	0.578	-	-	-	-	-	-	-
Canada	430	-	-	-	-	-	0.092	-	-	-	-	-
Chile	31	18	23	30	14	-	0.419	0.253	0.323	0.375	0.188	-
China (Hong Kong SAR)	162	161	187	97	113	-	0.025	0.029	0.034	0.021	0.025	-
Colombia	12	8.132	34	6.121	11	-	0.034	0.025	0.093	0.017	0.030	-
Croatia	34	23	6.430	5.436	10	2.936	0.095	0.049	0.012	0.008	0.013	0.003
Cyprus	3.963	2.676	2.880	2.465	1.707	2.488	0.014	0.010	0.010	0.009	0.005	0.007
Czechoslovakia (Former)	525	-	-	-	-	-	2.153	-	-	-	-	-
Denmark	306	258	-	-	-	-	0.067	0.055	-	-	-	-
Ecuador	1.500	6.582	1.969	1.068	0.789	6.975	0.083	0.274	0.082	0.045	0.034	0.194
Egypt	42	11	4.074	16	-	-	0.272	0.068	0.025	0.103	-	-
El Salvador	-	0.371	-	0.804	1.177	0.643	-	0.074	-	0.201	0.107	0.027
Ethiopia	-	0.157	0.085	0.006	0.014	0.005	-	0.157	0.085	0.006	0.014	0.005
Finland	374	226	148	40	139	322	0.577	0.309	0.209	0.061	0.208	0.394
FYR Macedonia	0.415	1.995	0.344	0.148	0.134	0.137	0.021	0.053	0.006	0.002	0.002	0.002

Continued.

Depending on the table, * means Enterprises, Engaged, or Factor Values; ± means Basis Unspecified; § means shown in millions. For additional notes and sources, see Appendix I.

567

Capital Investment

- Continued -

Country	Gross Fixed Capital Formation ($ mil.)						Per Establishment ($ mil)					
	1990	1991	1992	1993	1994	1995	1990	1991	1992	1993	1994	1995
France	2,766	2,476	2,265	1,946	2,042	2,803	-	-	-	-	-	-
Germany (Western Part)	7,730	7,752	6,942	4,920	-	-	1.231	1.201	1.072	-	-	-
Greece	15	15	11	-	-	-	0.038	0.038	0.027	-	-	-
Hungary	40	93	49	67	-	-	0.028	0.028	0.023	0.028	-	-
India	552	477	662	406	-	-	0.068	0.058	0.077	0.045	-	-
Indonesia	74	84	49	56	46	96	0.355	0.378	0.197	0.209	0.170	0.299
Iran (Islamic Republic of)	70	67	52	35	-	-	0.079	0.338	0.240	0.164	-	-
Ireland	111	154	-	-	-	-	0.332	0.485	-	-	-	-
Israel	15	21	20	35	40	-	0.041	0.062	0.054	0.083	0.085	-
Italy	2,573	2,355	2,252	1,752	2,035	-	0.570	0.511	0.487	0.374	0.431	-
Japan	12,107	16,205	14,615	12,356	12,259	-	0.241	0.326	0.306	0.261	0.277	-
Jordan	0.203	0.197	0.888	0.042	1.323	1.887	0.003	0.003	0.014	0.001	0.009	0.012
Korea, Republic of	1,372	2,161	2,325	3,137	2,767	3,557	0.202	0.259	0.261	0.288	0.244	0.292
Kuwait	0.110	11	5.454	8.533	5.121	-	0.005	0.474	0.237	0.356	0.213	-
Latvia	0.084	0.154	0.760	0.826	0.705	2.493	0.000	0.000	0.003	0.003	0.006	0.013
Luxembourg	14	11	11	-	-	-	0.374	0.293	0.328	-	-	-
Macau	0.388	1.781	0.038	0.006	0.031	0.015	0.007	0.031	0.002	0.000	0.001	0.001
Malaysia	218	168	163	246	231	-	0.543	0.384	0.376	0.486	0.451	-
Malta	1.202	1.111	0.780	0.217	1.111	-	0.052	0.044	0.029	0.007	0.036	-
Mexico	37	40	-	-	-	-	0.211	0.236	-	-	-	-
Morocco	7.522	10	12	7.635	9.236	12	0.030	0.039	0.044	0.026	0.031	0.040
Netherlands	517	466	471	401	-	-	0.564	0.480	0.469	0.393		
New Zealand	21	-	-	-	-	-	0.008	-	-	-	-	-
Norway	106	132	162	59	59	-	0.205	0.261	0.487	0.196	0.183	-
Pakistan	-	9.167	-	-	-	-	-	0.035	-	-	-	-
Panama	0.018	0.054	-	-	-	-	0.002	0.011	-	-	-	-
Peru	9.960	5.647	4.138	-	-	-	0.014	0.008	0.006	-	-	-
Philippines	16	26	16	20	-	-	0.008	0.012	0.029	0.040	-	-
Poland	183	192	130	118	-	-	0.254	0.259	0.182	0.173	-	-
Portugal	77	99	96	69	97	-	0.032	0.042	0.038	0.025	0.033	-
Romania	-	-	56	52	49	-	-	-	0.112	0.078	0.051	-
Senegal	0.051	-	-	-	-	-	0.051	-	-	-	-	-
Singapore	332	346	561	498	562	-	0.724	0.742	1.083	0.961	1.020	-
Spain	500	417	397	585	525	546	0.035	0.029	0.028	-	-	-
Sri Lanka	0.349	0.405	1.714	3.387	-	-	0.013	0.014	0.055	0.077	-	-
Swaziland	-	-	1.269	-	-	-	-	-	-	-	-	-
Thailand	1,083	256	-	-	-	-	5.416	0.611	-	-	-	-
Trinidad and Tobago	-	-	0.282	0.411	0.709	-	-	-	0.020	0.023	0.047	-
Tunisia	26	10	8.028	7.273	6.920	7.401	-	-	0.010	0.009	0.009	0.009
Turkey	105	144	148	207	111	-	0.322	0.483	0.192	0.289	0.155	-
Ukraine	528	239	90	11	20	12	0.823	0.232	0.075	0.016	0.032	0.020
United Kingdom	2,198	1,834	1,557	-	1,776	-	0.095	0.082	0.068	-	0.123	-
United Republic of Tanzania	4.311	4.896	-	-	-	-	0.139	0.148	-	-	-	-
United States of America	8,930	8,010	8,791	8,628	9,905	10,836	-	-	0.115	-	-	-
Uruguay	0.980	1.486	-0.671	0.649	-	-	-	-	-	-	-	-
Venezuela	16	25	31	12	13	105	0.058	0.080	0.100	0.048	0.054	0.378
Yugoslavia	6.376	3.096	6.716	-	3.662	1.962	0.044	0.014	0.020	-	0.008	0.003
Zambia	2.990	-	-	-	-	-	0.332	-	-	-	-	-

Indexes of Industrial Production

Country	Index of Industrial Production (1990=100)						Index of Employment (1990=100)					
	1990	1991	1992	1993	1994	1995	1990	1991	1992	1993	1994	1995
Albania	100	-	-	-	-	-	-	-	-	-	-	-
Algeria	100	95	84	75	55	46	-	-	-	-	-	-
Angola	100	-	-	-	-	-	-	-	-	-	-	-
Argentina	100	112	-	-	-	-	100	-	-	127	132	-
Armenia	100	-	-	-	-	-	100	89	-	-	-	-

Continued.

Indexes of Industrial Production

- Continued -

Country	Index of Industrial Production (1990=100)						Index of Employment (1990=100)					
	1990	1991	1992	1993	1994	1995	1990	1991	1992	1993	1994	1995
Australia	100	98	116	123	-	-	100	96	89	-	-	-
Austria	100	100	95	93	92	102	100	101	85	80	79	81
Azerbaijan	-	-	-	-	-	-	100	92	71	68	57	-
Bahrain	100	-	-	-	-	-	-	-	-	-	-	-
Bangladesh	100	104	107	112	61	33	100	146	165	-	-	-
Barbados	100	-	-	-	-	-	-	-	-	-	-	-
Belgium	100	95	87	81	86	91	-	-	-	-	-	-
Belize	100	100	100	100	100	100	-	-	-	-	-	-
Bermuda	-	-	-	-	-	-	100	82	109	73	55	-
Bolivia	100	104	111	116	136	132	100	115	111	147	145	147
Bosnia & Herzegovina	-	-	-	-	-	-	100	97	-	-	5	-
Brazil	100	90	81	95	115	110	100	-	86	78	-	-
Burkina-Faso	100	-	-	-	-	-	-	-	-	-	-	-
Burundi	100	100	100	100	-	-	-	-	-	-	-	-
Cameroon	100	-	-	-	-	-	-	-	-	-	-	-
Canada	100	92	93	105	137	164	100	93	86	87	95	102
Central African Republic	100	100	100	100	100	-	-	-	-	-	-	-
Chile	100	96	143	186	200	222	100	103	101	96	102	106
China	100	-	-	-	-	-	100	97	97	73	74	72
China (Hong Kong SAR)	100	98	94	85	84	83	100	85	76	66	61	-
China (Taiwan Province)	100	111	117	121	126	129	100	102	107	112	112	110
Colombia	100	98	116	130	142	143	100	103	121	125	127	127
Costa Rica	100	86	111	116	127	130	100	83	92	107	-	-
Cote d'Ivoire	100	52	124	100	129	-	-	-	-	-	-	-
Croatia	100	-	-	-	-	-	100	69	59	54	48	42
Cyprus	100	102	119	110	111	129	100	98	101	99	88	83
Czechoslovakia (Former)	100	69	48	-	-	-	100	84	-	-	-	-
Czech Republic	100	-	-	-	-	-	100	91	61	54	-	-
Dem. Rep. of the Congo	100	-	-	-	-	-	-	-	-	-	-	-
Denmark	100	99	103	97	109	120	100	96	95	92	98	102
Dominican Republic	100	-	-	-	-	-	-	-	-	-	-	-
Ecuador	100	117	137	152	-	-	100	114	102	95	91	298
Egypt	100	90	80	83	89	102	100	105	95	98	101	108
El Salvador	100	104	-	-	-	-	-	-	-	-	-	-
Ethiopia and Eritrea	100	100	100	-	-	-	-	-	-	-	-	-
Fiji	100	-	-	-	-	-	-	-	-	-	-	-
Finland	100	75	73	79	89	108	100	91	82	77	80	99
FYR Macedonia	100	-	-	-	-	-	100	223	209	192	169	97
France	100	93	87	78	83	88	100	99	95	90	86	87
Gabon	100	-	-	-	-	-	-	-	-	-	-	-
Gambia	100	-	-	-	-	-	-	-	-	-	-	-
Germany (Eastern Part)	100	-	-	-	-	-	-	-	-	-	-	-
Germany (Western Part)	100	99	90	79	81	84	100	100	96	-	-	-
Ghana	100	-	-	-	-	-	-	-	-	-	-	-
Greece	100	89	89	79	80	96	100	90	83	78	79	86
Guatemala	100	-	-	-	-	-	-	-	-	-	-	-
Guyana	100	100	100	100	100	100	-	-	-	-	-	-
Haiti	100	-	-	-	-	-	-	-	-	-	-	-
Honduras	100	107	113	141	167	193	100	92	87	96	105	171
Hungary	100	91	73	79	96	109	100	92	68	43	48	51
India	100	98	99	100	106	130	100	101	106	101	104	112
Indonesia	100	98	-	-	-	-	100	109	114	119	122	143
Iran (Islamic Republic of)	100	150	157	154	133	121	100	44	42	41	-	-
Iraq	100	-	-	-	-	-	-	-	-	-	-	-
Ireland	100	90	104	108	120	167	100	98	103	104	107	119
Israel	100	100	114	129	141	143	100	94	97	109	126	118
Italy	100	90	84	84	-	-	100	99	103	100	100	-
Jamaica	100	-	-	-	-	-	-	-	-	-	-	-
Japan	100	101	87	77	77	81	100	102	100	96	96	96
Jordan	100	-	-	-	-	-	100	148	202	205	292	296

Continued.

Depending on the table, * means *Enterprises, Engaged*, or *Factor Values*; ± means *Basis Unspecified*; § means *shown in millions*. For additional notes and sources, see Appendix I.

Indexes of Industrial Production

- Continued -

Country	Index of Industrial Production (1990=100)						Index of Employment (1990=100)					
	1990	1991	1992	1993	1994	1995	1990	1991	1992	1993	1994	1995
Kenya	100	76	72	71	76	59	100	99	99	100	101	101
Korea, Republic of	100	111	108	114	134	157	100	124	119	130	139	142
Kuwait	100	-	-	-	-	-	100	58	88	96	100	97
Kyrgyzstan	100	-	-	-	-	-	100	95	79	65	41	-
Latvia	100	-	-	-	-	-	100	89	56	36	26	25
Lebanon	100	-	-	-	-	-	-	-	-	-	-	-
Liberia	100	-	-	-	-	-	-	-	-	-	-	-
Libyan Arab Jamahiriya	100	-	-	-	-	-	-	-	-	-	-	-
Lithuania	100	-	-	-	-	-	-	-	-	-	-	-
Luxembourg	100	104	89	85	86	90	100	95	92	88	88	-
Macau	100	-	-	-	-	-	100	118	62	72	40	58
Madagascar	100	100	100	100	-	-	-	-	-	-	-	-
Malawi	100	-	-	-	-	-	100	-	-	-	-	-
Malaysia	100	165	220	206	276	323	100	128	134	170	175	193
Mali	100	101	106	93	106	125	-	-	-	-	-	-
Malta	100	137	158	195	-	-	100	106	90	86	85	-
Mauritius	100	-	-	-	-	-	100	152	144	135	151	153
Mexico	100	111	110	110	118	101	100	95	91	70	70	68
Mongolia	100	77	101	34	50	102	-	-	-	-	-	-
Morocco	100	107	122	113	117	123	100	109	105	101	109	101
Mozambique	100	-	-	-	-	-	100	96	66	63	60	-
Myanmar	100	108	95	-	-	-	-	-	-	-	-	-
Namibia	100	-	-	-	-	-	-	-	-	-	-	-
Nepal	100	100	100	-	-	-	-	-	-	-	-	-
Netherlands	100	101	98	97	102	107	100	103	103	97	96	95
New Zealand	100	-	-	-	-	-	100	95	97	115	-	-
Nicaragua	100	-	-	-	-	-	-	-	-	-	-	-
Niger	100	-	-	-	-	-	-	-	-	-	-	-
Nigeria	100	-	-	-	-	-	-	-	-	-	-	-
Norway	100	97	108	113	117	110	100	103	105	57	58	55
Pakistan	100	91	95	-	-	-	100	110	115	-	-	-
Panama	100	47	31	32	39	45	100	61	51	51	54	58
Papua New Guinea	100	-	-	-	-	-	-	-	-	-	-	-
Paraguay	100	110	125	99	77	77	-	-	-	-	-	-
Peru	100	114	98	80	110	144	100	88	75	-	-	-
Philippines	100	119	130	149	163	202	100	102	86	81	-	-
Poland	100	76	68	75	86	104	100	86	72	64	66	67
Portugal	100	89	80	72	70	73	100	101	94	96	95	94
Puerto Rico	100	-	-	-	-	-	100	84	78	72	74	62
Republic of Moldova	100	-	-	-	-	-	100	94	81	-	-	-
Romania	100	72	54	55	51	63	100	97	74	65	60	-
Russian Federation	100	-	-	-	-	-	-	-	-	-	-	-
Saudi Arabia	100	-	-	-	-	-	-	-	-	-	-	-
Senegal	100	118	88	84	74	113	100	131	57	50	27	-
Singapore	100	113	111	105	127	143	100	99	105	102	110	124
Slovakia	100	-	-	-	-	-	-	-	-	-	-	-
Slovenia	100	79	64	61	72	76	-	-	-	-	-	-
Somalia	100	100	100	-	-	-	-	-	-	-	-	-
South Africa	100	95	91	91	87	97	100	91	87	82	82	85
Spain	100	96	85	82	91	107	100	98	92	101	102	104
Sri Lanka	100	94	101	124	154	213	100	62	84	156	-	-
Sudan	100	-	-	-	-	-	-	-	-	-	-	-
Suriname	100	-	-	-	-	-	-	-	-	-	-	-
Sweden	100	87	86	76	89	104	100	104	95	85	81	77
Switzerland	100	105	104	100	102	104	-	-	-	-	-	-
Syrian Arab Republic	100	119	173	-	-	-	-	-	-	-	-	-
Thailand	100	92	-	-	-	-	100	89	-	-	-	-
Tonga	100	-	-	-	-	-	-	-	-	-	-	-
Trinidad and Tobago	100	102	100	96	89	125	100	2,600	2,875	3,350	3,688	-
Tunisia	100	121	118	99	100	89	-	-	-	-	-	-

Continued.

Indexes of Industrial Production

- Continued -

Country	Index of Industrial Production (1990=100)						Index of Employment (1990=100)					
	1990	1991	1992	1993	1994	1995	1990	1991	1992	1993	1994	1995
Turkey	100	111	110	142	112	132	100	88	97	98	93	96
Uganda	100	74	109	320	379	401	-	-	-	-	-	-
Ukraine	100	-	-	-	-	-	100	102	89	88	77	69
United Arab Emirates	100	-	-	-	-	-	-	-	-	-	-	-
United Kingdom	100	93	93	95	106	109	100	94	90	78	83	84
United Republic of Tanzania	100	118	100	123	149	-	100	92	-	-	-	-
USSR (Former)	100	-	-	-	-	-	100	-	-	-	-	-
United States of America	100	98	105	105	132	149	100	95	93	93	97	101
Uruguay	100	84	84	69	72	66	100	113	103	112	111	113
Venezuela	100	-	-	-	-	-	100	128	122	114	99	103
Yugoslavia	100	-	-	-	-	-	-	-	-	-	-	-
Yugoslavia (Former)	100	62	-	-	-	-	-	-	-	-	-	-
Zambia	100	97	90	75	64	68	100	-	-	-	-	-
Zimbabwe	100	102	90	74	83	79	100	86	71	66	69	69

Depending on the table, * means *Enterprises, Engaged,* or *Factor Values*; ± means *Basis Unspecified*; § means *shown in millions*. For additional notes and sources, see Appendix I.

571

Representative Companies in Sector

Name	Address	Tele-phone	Fax	Employ-ment	Y
Crane Co.	100 1st Stamford Pl., Stamford, Connecticut 06902, U.S.A.	203-363-7300	203-363-7295	10,000	96
Pentair Inc.	1500 County Rd. B2 W., Ste. 400, St. Paul, Minnesota 55113-3105, U.S.A.	612-636-7920		10,000	97
PMI Food Equipment Group	701 S. Ridge Ave., Troy, Ohio 45374, U.S.A.	937-332-3000	937-332-2142	10,000	97
General Signal Corp.	PO Box 10010, Stamford, Connecticut 06904-2010, U.S.A.	203-329-4100		9,900	97
Siemens Energy and Automation Inc.	3333 Old Milton Pkwy., Alpharetta, Georgia 30202, U.S.A.	770-751-2000	770-751-2001	9,500	97
Hiltiag	Schaan FL-9494, Liechtenstein	75 2111		9,400	89
Sequa Corp.	200 Park Ave., New York, New York 10166, U.S.A.	212-986-5500	212-370-1969	9,350	97
Bemis Company Inc.	222 S. 9th St., 2300, Minneapolis, Minnesota 55402-4099, U.S.A.	612-376-3000		9,275	97
Nortek Inc.	50 Kennedy Plz., Providence, Rhode Island 02903-2360, U.S.A.	401-751-1600	401-751-4610	9,262	97
Bumar-Labedy	Mechanikow St. 9, Gliwice PL-44-109, Poland	32 345111	32 346966	9,014	93
GE Appliances	Appliance Park, Louisville, Kentucky 40225, U.S.A.	502-452-4311	502-452-3389	9,000	96
Mannesmann Capital Corp.	450 Park Ave., 24th Fl., New York, New York 10022, U.S.A.	212-826-0040	212-826-0074	9,000	97
Godrej & Boyce Manufacturing Co. Ltd.	Godrej Bhavan Homji St., Bombay 400 001, India	22 5171166	22 5172688	8,810	94
Hussmann Corp.	12999 St. Charles Rock Rd., Bridgeton, Missouri 63044, U.S.A.	314-291-2000	314-298-6484	8,700	97
British Shipbuilders	136 Sandyford Rd., Newcastle-upon-Tyne NE2 1QE, United Kingdom	91 6326772	91 6326772	8,642	87
Inco Ltd.	145 King St. W Ste. 1500, Toronto, Ontario M5H 4B7, Canada	416-361-7511	416-361-7781	8,600	98
Cooper Cameron Corp.	515 Post Oak Blvd., 1200, Houston, Texas 77027, U.S.A.	713-513-3300	713-513-3320	8,500	96
Pall Corp.	2200 Northern Blvd., East Hills, New York 11548, U.S.A.	516-484-5400	516-484-3649	8,500	97
Hyundai Precision & Industry Co., Ltd.	Chongno-Gu, Seoul, Republic of Korea	2 7412211	2 7414244	8,482	97
Coltec Industries Inc.	2550 W. Tyvola Rd., Charlotte, North Carolina 28217-4543, U.S.A.	704-423-7000		8,153	96
Lennox International Inc.	PO Box 799900, Dallas, Texas 75379-9900, U.S.A.	972-497-5000	972-980-6599	8,000	95
MTD Products Inc.	PO Box 360900, Cleveland, Ohio 44136, U.S.A.	330-225-2600		8,000	95
Modine Manufacturing Co.	1500 DeKoven Ave., Racine, Wisconsin 53403-2552, U.S.A.	414-636-1200	414-636-1424	7,900	96
AGCO Corp.	4830 River Green Pkwy., Duluth, Georgia 30096-2568, U.S.A.	770-813-9200	770-246-6158	7,800	96
FKI Industries Inc.	425 Post Rd., Fairfield, Connecticut 06430, U.S.A.	203-255-7100	203-255-7101	7,800	95
Iwka AG	Gartenstr. 71, Karlsruhe D-76135, Germany	721 1430	721 143243	7,774	94
Stant Corp.	425 Commerce Dr., Richmond, Indiana 47374-2646, U.S.A.	317-962-6655	317-962-6866	7,600	96
Briggs and Stratton Corp.	PO Box 702, Milwaukee, Wisconsin 53201-0702, U.S.A.	414-259-5333	414-259-5313	7,560	97
Beloit Corp.	500 Lake Cook Dr., Deerfield, Illinois 60015, U.S.A.	847-236-6700	847-236-6701	7,500	97
Kennametal Inc.	PO Box 231, Latrobe, Pennsylvania 15650, U.S.A.	412-539-5000		7,500	97
Sunbeam Corp.	PO Box 9218, Delray Beach, Florida 33445, U.S.A.	561-243-2100		7,500	97
Outboard Marine Corp.	100 Sea Horse Dr., Waukegan, Illinois 60085, U.S.A.	847-689-6200	847-689-6082	7,442	97
Arcom SA	20 Kogalniceanu Blvd., Bucharest R-70607, Romania	1 159063	1 120129	7,344	93
SPX Corp.	PO Box 3301, Muskegon, Michigan 49443-3301, U.S.A.	616-724-5000	616-724-5720	7,125	96
Dresser-Rand Co.	1 Baron Steuben Pl., Corning, New York 14830, U.S.A.	607-937-6400	607-937-6406	7,100	94
Eagle Industries Inc.	2 N. Riverside Plz., 1100, Chicago, Illinois 60606, U.S.A.	312-906-8700	312-906-8372	7,000	94
NACCO Industries Inc.	5875 Landerbrook Dr., Mayfield Heights, Ohio 44124-4017, U.S.A.	216-449-9600		7,000	97
Carlisle Companies Inc.	250 S. Clinton St., 201, Syracuse, New York 13202-1258, U.S.A.	315-477-9108	315-474-2008	6,900	97
AMETEK Inc.	Station Sq., Paoli, Pennsylvania 19301, U.S.A.	610-647-2121	610-296-3412	6,700	97
Weatherford Enterra Inc.	PO Box 27608, Houston, Texas 77056-7608, U.S.A.	713-439-9400		6,578	96
Detroit Diesel Corp.	13400 W. Outer Dr., Detroit, Michigan 48239-4001, U.S.A.	313-592-5000		6,500	97
Gomel Casting Enterprise	ul. Mogilevskaya 16, Gomel 246010, Belarus	564733	564733	6,500	93
Oktobar	14.Oktobra 2, Krusevac YU-37000, Serbia	37 21502	37 29686	6,500	92
Thyssen Maschinenbau GmbH Werk Witten-Annen Ruhrpumpen	Postfach 6320, Witten 58453, Germany	2302 66101	2302 661	6,438	91
Whitman Corp.	3501 Algonquin Rd., Rolling Meadows, Illinois 60008, U.S.A.	847-818-5000		6,381	97
Donaldson Company Inc.	PO Box 1299, Minneapolis, Minnesota 55440, U.S.A.	612-887-3131		6,230	97
Federal Signal Corp.	1415 W. 22nd St., Oak Brook, Illinois 60521-9945, U.S.A.	708-954-2000		6,207	97
Nissan Motor Iberica, SA	Panama, 7, Barcelona E-08034, Spain	93 2908080	93 2907091	6,165	97

Continued

For sources and notes, see Appendix I.

Representative Companies in Sector
- Continued -

Name	Address	Tele-phone	Fax	Employ-ment	Y
Terex Corp.	500 Post Rd. E., 320, Westport, Connecticut 06880, U.S.A.	203-222-7170	203-222-7976	6,071	95
Armco Inc.	1 Oxford Centre, 301 Grant St., Pittsburgh, Pennsylvania 15219-1415, U.S.A.	412-255-9800	800-231-1054	6,000	97
Embraco Empresa Brasileira de Compresores	Rua Rui Barbosa 1020, Joinville 89219-901, Brazil	47 4412310	47 4412500	6,000	96
SKF USA Inc.	1100 1st Ave., King of Prussia, Pennsylvania 19406, U.S.A.	610-962-4300	610-265-1457	6,000	97
Solar Turbines Inc.	PO Box 85376, San Diego, California 92186-5376, U.S.A.	619-544-5000	619-544-5669	6,000	97
Smith International Inc.	PO Box 60068, Houston, Texas 77205-0068, U.S.A.	713-443-3370	713-233-5996	5,975	96
Lincoln Electric Co.	22801 St. Clair Ave., Cleveland, Ohio 44117-1199, U.S.A.	216-481-8100	216-486-1751	5,971	97
Cooper Oil Tool Div.	PO Box 1212, Houston, Texas 77251-1212, U.S.A.	713-939-2211	713-939-2620	5,900	97
Thiokol Corp.	15 W South Temple, Ste. 1600, Salt Lake City, Utah 84101, U.S.A.	801-629-2270		5,900	97
Schott Glaswerke/Schott Gruppe	Hattenbergstrassse 10, Mainz 55122, Germany	6131 660	6131 662000	5,899	93
Zelezorudne Bane SP	Stara Cesta 3, Spisska Nova Ves 052 54, Slovakia	965 23009	965 21370	5,800	93
Krones AG Hermann Kronseder Maschinenfabrik	Postfach 12 30, Neutraubling D-93073, Germany	9401 700	9401 702488	5,776	93
Blount International Inc.	PO Box 949, Montgomery, Alabama 36101-0949, U.S.A.	334-244-4000		5,700	97
Crompton and Knowles Corp.	1 Station Pl., Metro Ctr., Stamford, Connecticut 06902, U.S.A.	203-353-5400	203-353-5424	5,665	96
Amana Appliance	PO Box 8901, Amana, Iowa 52204-0001, U.S.A.	319-622-5511	319-622-2180	5,500	94
Camco International Inc.	PO Box 14484, Houston, Texas 77221, U.S.A.	713-747-4000	713-747-6751	5,500	97
Fabryka Maszyn, Glinik SA	ul. Michalusa 1, Gorlice PL-38-320, Poland	28200	28463	5,500	93
Mercury Marine	PO Box 1939, Fond du Lac, Wisconsin 54936-1939, U.S.A.	920-929-5000	414-929-5060	5,500	97
Perindustrian Tentara Nasional Id Angkatan Darat PT	Jl. Jenderal Gatot Subroto Kiaracondong, Bandung, Indonesia	22312073	22 301222	5,500	97
ZVL Kysucke Nove Mesto	ul. Kukucinova, Kysucke Nove Mesto 024 11, Slovakia	826 2620	826 2818	5,490	93
Clorox Co.	1221 Broadway, Oakland, California 94612, U.S.A.	510-271-7291	510-832-1463	5,300	97
Grove Worldwide	PO Box 21, Shady Grove, Pennsylvania 17256, U.S.A.	717-597-8121	717-597-4062	5,300	97
Joy Mining Machinery	177 Thorn Hill Rd., Warrendale, Pennsylvania 15086, U.S.A.	724-779-4500	724-779-4509	5,300	97
Goulds Pumps Inc.	300 WillowBrook Office Park, Fairport, New York 14450-4285, U.S.A.	716-387-6600	315-568-2418	5,250	96
Zurn Industries Inc.	1 Zurn Pl., Erie, Pennsylvania 16514-2000, U.S.A.	814-452-2111	814-459-3535	5,100	96
Howden Group PLC	Old Govan Rd., Renfrew PA4 8XJ, United Kingdom	41 8852245	41 4294244	5,028	94
Walbro Corp.	6242 Garfield St., Cass City, Michigan 48726-1325, U.S.A.	517-872-2131	517-872-2301	5,028	97
AGRA Inc.	335 8 Ave. SW Ste. 1900, Calgary, Alberta T2P 1C9, Canada	403-263-9606	403-263-9676	5,000	98
Baker Hughes INTEQ	PO Box 670968, Houston, Texas 77267-0968, U.S.A.	713-625-4200	713-625-5200	5,000	96
Crown Equipment Corp.	40 S. Washington St., New Bremen, Ohio 45869, U.S.A.	419-629-2311		5,000	95
Greenfield Industries Inc. (Augusta, Georgia)	2743 Perimeter Pkwy., Bldg. 100,, Augusta, Georgia 30909, U.S.A.	706-863-7708	706-860-8559	5,000	96
General Binding Corp.	1 GBC Plaza, Northbrook, Illinois 60062-4195, U.S.A.	847-272-3700	847-272-1389	4,800	97
Lam Research Corp.	4650 Cushing Pky., Fremont, California 94538, U.S.A.	510-659-0200	510-490-5026	4,800	97
Standex International Corp.	6 Manor Pkwy., Salem, New Hampshire 03079, U.S.A.	603-893-9701	603-893-7324	4,800	97
Thermo King Corp.	314 W. 90th St., Minneapolis, Minnesota 55420, U.S.A.	612-887-2200	612-887-2615	4,800	97
Bosal Afrika Pty. Ltd.	PO Box 1719, Pretoria 0001, Republic of South Africa	12 3333258	12 3333053	4,700	94
Watts Industries Inc.	815 Chestnut St., North Andover, Massachusetts 01845-6098, U.S.A.	978-688-1811	978-688-5841	4,650	97
Great American Management Inc.	2 N. Riverside Plz., Chicago, Illinois 60606, U.S.A.	312-648-5656	312-454-9946	4,600	95
New Holland North America Inc.	PO Box 1895, New Holland, Pennsylvania 17557-0903, U.S.A.	717-355-1121	717-355-1826	4,600	97
Winstone Ltd.	PO Box 395, Auckland, New Zealand	09 778877		4,600	87
Tuboscope Inc.	PO Box 808, Houston, Texas 77051, U.S.A.	713-799-5100	713-799-5151	4,598	97
Stewart and Stevenson Services Inc.	PO Box 1637, Houston, Texas 77251-1637, U.S.A.	713-868-7700		4,511	95
Copeland Corp.	1675 Campbell Rd., Sidney, Ohio 45365-0669, U.S.A.	937-498-3011	937-497-2713	4,500	95
Linamar Corp.	301 Massey Rd., Guelph, Ontario N1K 1B2, Canada	519-836-7550	519-824-8479	4,500	98
Premier Industrial Corp.	PO Box 94884, Cleveland, Ohio 44101-4884, U.S.A.	216-391-8300		4,500	95
Signode Corp.	3610 W. Lake Ave., Glenview, Illinois 60025, U.S.A.	708-724-6100	708-657-5192	4,500	95
Sulzer Inc.	200 Park Ave., New York, New York 10166, U.S.A.	212-949-0999	212-370-1138	4,500	97
Clark Equipment Co.	100 N. Michigan St., South Bend, Indiana 46634, U.S.A.	219-239-0100		4,410	94
Paterson Zochonis PLC	60 Whitworth St., Manchester M1 6LU, United Kingdom	61 2367111	61 2286719	4,377	94
Butler Manufacturing Co.	PO Box 419917, Kansas City, Missouri 64141-0917, U.S.A.	816-968-3000		4,350	96
Energomontaz-Polnoc SA	Przemyslowa St. 30, Warsaw PL-00-450, Poland	22 211041	22 296324	4,317	93
Asea Brown Boveri AS	Petersmindevej 1, Odense DK-5000, Denmark	66147080	66142580	4,300	93

Continued

Representative Companies in Sector
- Continued -

Name	Address	Tele-phone	Fax	Employ-ment	Y
Ingersoll International Inc.	707 Fulton Ave., Rockford, Illinois 61103, U.S.A.	815-987-6000	815-987-6725	4,300	97
Vermont American Corp.	2300 National City Twr., 101 S., Louisville, Kentucky 40202, U.S.A.	502-625-2000	502-625-2122	4,300	97
Applied Power Inc.	PO Box 325, Milwaukee, Wisconsin 53201-0325, U.S.A.	414-781-6600	414-783-3790	4,235	97
Asea Brown Boveri Ltda.	Vila Campesina, Osasco 06020-902, Brazil	11 7049111	11 7049993	4,100	98
AEG Corp.	180 Mount Airy Rd., PO Box 609, Basking Ridge, New Jersey 07920, U.S.A.	908-204-8900	908-204-8999	4,000	93
Energy Ventures Inc.	5 Post Oak Park, 1760, Houston, Texas 77027-3415, U.S.A.	713-297-8400	713-963-9785	4,000	96
Holstein & Kappert AG	Juchostrasse 20, Postfach 105026, Dortmund 44143, Germany	231 51850	23 585377	4,000	91
Nordson Corp.	28601 Clemens Rd., Westlake, Ohio 44145-1119, U.S.A.	440-892-1580	440-892-9507	4,000	97
Toro Co.	8111 Lyndale Ave. S., Bloomington, Minnesota 55420-1196, U.S.A.	612-888-8801		3,911	97
General Cable Corp.	4 Tesseneer Dr., Highland Heights, Kentucky 41076, U.S.A.	606-572-8000	606-572-8458	3,900	96
Imo Industries Inc.	1009 Lenox Dr., Lawrenceville, New Jersey 08648, U.S.A.	609-896-7600	609-896-7688	3,900	96
Mining Machinery Factory, Pioma SA	ul. R Dmowskiego 38, Piotrkow Trybunalski 97-300, Poland	470490	445082	3,880	93
Baldor Electric Co.	PO Box 2400, Fort Smith, Arkansas 72902, U.S.A.	501-646-4711	501-648-5752	3,843	97
Commercial Intertech Corp.	PO Box 239, Youngstown, Ohio 44505, U.S.A.	216-746-8011	216-746-1148	3,805	97
General Motors Corp. Electro-Motive Div.	9301 W. 55th St., La Grange, Illinois 60525, U.S.A.	708-387-6000	708-387-6501	3,800	97
IDEX Corp.	630 Dundee Rd., Northbrook, Illinois 60062, U.S.A.	847-498-7070	847-498-3940	3,800	97
ITT Flygt AB	Box 1309, Solna S-171 25, Sweden	471 17000	8 6276900	3,757	93
Valmont Industries Inc.	PO Box 358, Valley, Nebraska 68064, U.S.A.	402-359-2201	402-343-0668	3,751	97
Scotsman Industries Inc.	775 Corporate Woods Pkwy., Vernon Hills, Illinois 60061, U.S.A.	708-215-4500		3,750	97
Eljer Industries Inc.	17120 Dallas Pkwy., Dallas, Texas 75248, U.S.A.	214-407-2600		3,700	95
Thermadyne Holdings Corp.	101 S. Hanley Rd., Ste. 300, St. Louis, Missouri 63105, U.S.A.	314-721-5573	314-721-4822	3,563	97
Goodyear SA	Ave. T. Gordon Smith, Colmar-Berg L-7750, Luxembourg	81991	81992496	3,550	95
Caterpillar Belgium SA	Ave. des Etats-Unis 1, Gosselies B-6041, Belgium	71 252111	71 252527	3,500	95
Cubic Corp.	PO Box 85587, San Diego, California 92186-5587, U.S.A.	619-277-6780	619-277-1878	3,500	97
Hyster Co. North American Industrial Truck Div.	PO Box 847, Danville, Illinois 61834-0847, U.S.A.	217-443-7000	217-443-7494	3,500	97
McQuay International	PO Box 1551, Minneapolis, Minnesota 55440, U.S.A.	612-553-5330		3,500	97
TriMas Corp.	315 E. Eisenhower Pkwy., Ann Arbor, Michigan 48108, U.S.A.	313-747-7025	313-747-6565	3,500	95
Columbus McKinnon Corp.	140 Audubon Pkwy., Amherst, New York 14228-1197, U.S.A.	716-689-5400	716-689-5644	3,479	96
Zardoya Otis, SA	Plaza Del Liceo, 3, Madrid E-28043, Spain	91 3435100	91 3435279	3,456	97
Nesco Inc.	6140 Parkland Blvd., Mayfield Heights, Ohio 44124, U.S.A.	216-461-6000		3,400	95
Scitex Corporation Ltd.	PO Box 330, Herzliyya 46733, Israel	9 9597222	9 9502922	3,400	97
Esterline Technologies Corp.	10800 Northeast 8th St., Bellevue, Washington 98004, U.S.A.	425-453-9400	425-453-2916	3,360	97
Giddings and Lewis Inc.	PO Box 590, Fond du Lac, Wisconsin 54936-0590, U.S.A.	414-921-9400	414-929-4522	3,315	96
Duracraft Corp.	250 Turnpike Rd., Southborough, Massachusetts 01772, U.S.A.	508-490-7080		3,300	95
Thomas Industries Inc.	PO Box 35120, Louisville, Kentucky 40232, U.S.A.	502-893-4600		3,300	97
Moog Inc.	Plant 24, East Aurora, New York 14052-0018, U.S.A.	716-652-2000	716-687-4457	3,229	97
American Tool Companies Inc.	301 S. 13th St., 600, Lincoln, Nebraska 68508, U.S.A.	402-435-3300	402-435-3619	3,200	97
Hyosung Industries Co., Ltd.	Mapo-gu, Seoul 121-020, Republic of Korea	2 7076000	2 7140707	3,200	97
Bumar-Warynski SA	Kolejowa ul. 57, Warsaw PL-01-210, Poland	22 321164		3,154	93
National-Oilwell Inc.	5555 San Felipe, Houston, Texas 77056, U.S.A.	713-960-5100		3,138	97
Coopcam UK Ltd.	Houston Industrial Estate, Livingston EH54 5BZ, United Kingdom	506 31122		3,031	93
Silicon Valley Group Inc.	101 Metro Dr., Ste. 400, San Jose, California 95110, U.S.A.	408-441-6700	408-467-5867	3,009	97
EFACEC, Empresa Fabril de Maquinas Electricas, SA	Apartado 18, Sao Mamede de Infesta P-4466, Portugal	95 12015	95 18940	3,007	91
Agrikon	Kulso-Szegedi ut. 136, Kecskemet H-6000, Hungary	76 21340	76 22563	3,000	93
Alcoa Closure Systems International	2485 Directors Row, A, Indianapolis, Indiana 46241, U.S.A.	317-241-2595	317-390-5137	3,000	95
Bristol Compressors Inc.	15185 Industrial Park Rd., Bristol, Virginia 24202, U.S.A.	703-466-4121	703-645-2423	3,000	94
Manitowoc Company Inc.	PO Box 66, Manitowoc, Wisconsin 54221-0066, U.S.A.	920-684-4410	920-683-6277	3,000	97
McQuay International. Commercial Products Group	PO Box 1551, Minneapolis, Minnesota 55440, U.S.A.	612-553-5330	612-553-5177	3,000	95
Murry Inc.	219 Franklin Rd., Brentwood, Tennessee 37027, U.S.A.	615-373-6500	615-373-6554	3,000	97

Product Tables
Steam Turbines

Unit of Measure: Number.

	1990		1991		1992		1993		1994		1995		1996	
	Value	%	Value	%	Value	%	Value	%	Value	%	Value	%	Value	%
Total Production	1,861	100.0	1,948	100.0	2,310	100.0	2,111	100.0	2,029	100.0	2,074	100.0	2,110	100.0
Regions														
America, North	30	1.6	29	1.5	28	1.2	27	1.3	25	1.3	24	1.2	23	1.1
America, South	127	6.8	127	6.5	86	3.7	111	5.3	131	6.5	181	8.7	127	6.0
Asia	940	50.5	1,051	53.9	1,003	43.4	868	41.1	830	40.9	841	40.6	890	42.2
Europe	764	41.0	741	38.0	1,193	51.7	1,105	52.4	1,043	51.4	1,027	49.5	1,070	50.7
Leading Producers														
Japan	406	21.8	526	27.0	481	20.8	358	17.0	339	16.7	381	18.4	460	21.8
Germany	-		-		452	19.6	344	16.3	293	14.4	269	13.0	291	13.8
Colombia	*127	6.8	*127	6.5	86	3.7	111	5.3	131	6.5	181	8.7	*127	6.0
United Kingdom	127	6.8	133	6.8	139	6.0	144	6.8	150	7.4	*155	7.5	*160	7.6
Russian Federation	90	4.8	90	4.6	95	4.1	92	4.4	81	4.0	59	2.8	37	1.8

Engines, Diesel (Excluding Engines for Vehicles)

Unit of Measure: Number.

	1990		1991		1992		1993		1994		1995		1996	
	Value	%	Value	%	Value	%	Value	%	Value	%	Value	%	Value	%
Total Production	3,737,164	100.0	3,768,142	100.0	3,689,474	100.0	3,686,921	100.0	3,861,644	100.0	4,211,779	100.0	4,160,730	100.0
Regions														
Africa	9,289	0.2	9,271	0.2	9,253	0.3	9,235	0.3	9,217	0.2	9,199	0.2	9,182	0.2
America, North	206,106	5.5	184,245	4.9	178,215	4.8	196,895	5.3	220,276	5.7	245,762	5.8	242,472	5.8
America, South	56,253	1.5	56,253	1.5	56,227	1.5	56,229	1.5	56,283	1.5	56,273	1.3	56,253	1.4
Asia	3,001,698	80.3	2,841,625	75.4	2,803,591	76.0	2,834,614	76.9	2,959,103	76.6	3,344,594	79.4	3,322,829	79.9
Europe	463,818	12.4	676,748	18.0	642,188	17.4	589,948	16.0	616,765	16.0	555,951	13.2	529,994	12.7
Leading Producers														
India	1,741,154	46.6	1,677,178	44.5	1,682,273	45.6	1,674,407	45.4	1,765,856	45.7	1,988,373	47.2	1,990,572	47.8
Japan	1,165,266	31.2	1,056,628	28.0	1,031,287	28.0	1,035,724	28.1	1,057,982	27.4	1,226,173	29.1	1,217,144	29.3
United States of America	206,106	5.5	184,245	4.9	178,215	4.8	196,895	5.3	220,276	5.7	245,762	5.8	242,472	5.8
Germany	-		228,132	6.1	213,371	5.8	194,114	5.3	222,375	5.8	173,353	4.1	158,800	3.8
United Kingdom	100,635	2.7	86,791	2.3	72,947	2.0	59,102	1.6	53,423	1.4	*33,746	0.8	*20,135	0.5

Engines, Internal Combustion (Excluding Engines for Vehicles)

Unit of Measure: Number.

	1990		1991		1992		1993		1994		1995		1996	
	Value	%	Value	%	Value	%	Value	%	Value	%	Value	%	Value	%
Total Production	23,650,033	100.0	22,923,502	100.0	24,955,266	100.0	27,368,188	100.0	30,012,930	100.0	29,604,683	100.0	29,973,436	100.0
Regions														
Africa	1,155	0.0	774	0.0	393	0.0	12	0.0	-		-		-	
America, North	16,326,000	69.0	16,220,000	70.8	18,216,000	73.0	20,538,000	75.0	23,287,000	77.6	22,287,000	75.3	22,621,000	75.5
Asia	6,317,750	26.7	5,866,750	25.6	5,972,750	23.9	6,032,000	22.0	5,946,000	19.8	6,526,000	22.0	6,642,000	22.2
Europe	766,628	3.2	597,478	2.6	527,623	2.1	559,676	2.0	541,430	1.8	553,183	1.9	471,936	1.6
Oceania	238,500	1.0	238,500	1.0	238,500	1.0	238,500	0.9	238,500	0.8	238,500	0.8	238,500	0.8
Leading Producers														
United States of America	16,326,000	69.0	16,220,000	70.8	18,216,000	73.0	20,538,000	75.0	23,287,000	77.6	22,287,000	75.3	22,621,000	75.5
Japan	6,285,000	26.6	5,803,000	25.3	5,923,000	23.7	5,990,000	21.9	5,899,000	19.7	6,479,000	21.9	6,603,000	22.0
Poland	209,000	0.9	100,000	0.4	105,000	0.4	111,000	0.4	122,000	0.4	114,000	0.4	121,000	0.4
Sweden	*80,676	0.3	*36,190	0.2	-		-		-		-		-	
Australia	*238,500	1.0	*238,500	1.0	*238,500	1.0	*238,500	0.9	*238,500	0.8	*238,500	0.8	*238,500	0.8

Product Tables
Gas Turbines

Unit of Measure: Number.

	1990 Value	%	1991 Value	%	1992 Value	%	1993 Value	%	1994 Value	%	1995 Value	%	1996 Value	%
Total Production	679	100.0	696	100.0	650	100.0	503	100.0	477	100.0	487	100.0	478	100.0
Regions														
America, North	33	4.9	35	5.0	37	5.7	39	7.7	41	8.5	43	8.7	44	9.3
Asia	299	44.0	329	47.3	260	40.0	119	23.6	90	18.9	105	21.6	76	15.9
Europe	347	51.1	332	47.7	353	54.3	346	68.7	347	72.6	339	69.7	357	74.8
Leading Producers														
Russian Federation	263	38.7	228	32.8	184	28.3	74	14.7	42	8.8	56	11.5	29	6.1
United Kingdom	*125	18.3	*125	17.9	*125	19.2	*125	24.7	*125	26.1	*125	25.6	*125	26.1
Spain	122	18.0	107	15.4	127	19.6	*119	23.6	*119	24.9	*119	24.4	*119	24.9
Germany	*78	11.6	*78	11.3	*78	12.1	*78	15.6	*78	16.5	70	14.4	87	18.2
Japan	36	5.3	101	14.5	76	11.7	45	8.9	48	10.1	49	10.1	47	9.8

Hydraulic Turbines

Unit of Measure: Number.

	1990 Value	%	1991 Value	%	1992 Value	%	1993 Value	%	1994 Value	%	1995 Value	%	1996 Value	%
Total Production	1,348	100.0	551	100.0	601	100.0	624	100.0	546	100.0	708	100.0	605	100.0
Regions														
America, North	6	0.5	6	1.1	6	0.9	5	0.8	5	0.9	5	0.7	4	0.7
America, South	73	5.5	73	13.3	73	12.2	73	11.8	3	0.5	144	20.4	73	12.2
Asia	38	2.8	34	6.2	32	5.4	35	5.6	34	6.2	41	5.8	33	5.5
Europe	1,230	91.3	437	79.4	490	81.5	510	81.8	504	92.3	518	73.2	493	81.6
Leading Producers														
Finland	714	53.0	4	0.7	6	1.0	6	1.0	2	0.4	5	0.7	5	0.8
Denmark	*268	19.9	*268	48.6	*268	44.6	*268	43.0	*268	49.1	*268	37.9	*268	44.3
Colombia	*73	5.5	*73	13.3	*73	12.2	*73	11.8	3	0.5	144	20.4	*73	12.2
Romania	41	3.0	16	2.9	9	1.5	8	1.3	3	0.5	3	0.4	1	0.2
Czech Republic	*44	3.2	*44	7.9	*44	7.3	51	8.2	39	7.1	51	7.2	34	5.6

Cultivators, Scarifiers, Weeders, Hoes Etc.

Unit of Measure: Number.

	1990 Value	%	1991 Value	%	1992 Value	%	1993 Value	%	1994 Value	%	1995 Value	%	1996 Value	%
Total Production	1,145,234	100.0	1,137,763	100.0	1,191,821	100.0	1,034,817	100.0	977,428	100.0	941,530	100.0	935,238	100.0
Regions														
Africa	295,783	25.8	301,867	26.5	306,422	25.7	310,977	30.1	315,533	32.3	320,088	34.0	324,643	34.7
America, North	32,534	2.8	24,348	2.1	20,613	1.7	17,711	1.7	18,271	1.9	6,677	0.7	2,864	0.3
America, South	65	0.0	65	0.0	65	0.0	65	0.0	65	0.0	65	0.0	65	0.0
Asia	609,752	53.2	607,727	53.4	594,189	49.9	497,325	48.1	436,144	44.6	422,561	44.9	428,829	45.9
Europe	207,101	18.1	203,757	17.9	270,532	22.7	208,739	20.2	207,415	21.2	192,139	20.4	178,838	19.1
Leading Producers														
Congo	294,404	25.7	*300,488	26.4	*305,043	25.6	*309,598	29.9	*314,154	32.1	*318,709	33.9	*323,264	34.6
Japan	269,027	23.5	270,714	23.8	245,675	20.6	225,564	21.8	212,539	21.7	205,758	21.9	214,702	23.0
France	23,671	2.1	15,541	1.4	96,068	8.1	18,300	1.8	*18,323	1.9	*9,124	1.0	-	
Poland	32,730	2.9	16,272	1.4	16,216	1.4	15,725	1.5	19,727	2.0	16,957	1.8	19,379	2.1
Russian Federation	101,380	8.9	71,821	6.3	49,126	4.1	38,211	3.7	3,691	0.4	1,985	0.2	2,882	0.3

Product Tables

Harrows, Rotary, Animal or Tractor-Operated

Unit of Measure: Number.

	1990 Value	%	1991 Value	%	1992 Value	%	1993 Value	%	1994 Value	%	1995 Value	%	1996 Value	%
Total Production	600,352	100.0	439,358	100.0	396,873	100.0	397,646	100.0	324,882	100.0	232,157	100.0	211,563	100.0
Regions														
America, North	91,874	15.3	72,489	16.5	77,477	19.5	97,627	24.6	123,703	38.1	57,988	25.0	50,241	23.7
America, South	34,796	5.8	34,796	7.9	34,796	8.8	34,796	8.8	34,796	10.7	34,796	15.0	34,796	16.4
Asia	195,453	32.6	73,091	16.6	72,573	18.3	48,082	12.1	47,171	14.5	42,700	18.4	38,789	18.3
Europe	278,229	46.3	258,982	58.9	212,027	53.4	217,141	54.6	119,212	36.7	96,673	41.6	87,737	41.5
Leading Producers														
United States of America	88,840	14.8	69,390	15.8	74,313	18.7	94,399	23.7	120,410	37.1	*54,630	23.5	*46,818	22.1
Ukraine	158,037	26.3	153,527	34.9	98,367	24.8	114,081	28.7	19,632	6.0	6,716	2.9	2,497	1.2
Brazil	*34,796	5.8	*34,796	7.9	*34,796	8.8	*34,796	8.8	*34,796	10.7	*34,796	15.0	*34,796	16.4
Azerbaijan	132,630	22.1	15,136	3.4	24,300	6.1	4,110	1.0	5,477	1.7	1,395	0.6	-	
Belgium	19,186	3.2	13,131	3.0	*7,076	1.8	*1,022	0.3	-		-		-	

Ploughs, Animal or Tractor-Operated

Unit of Measure: Number.

	1990 Value	%	1991 Value	%	1992 Value	%	1993 Value	%	1994 Value	%	1995 Value	%	1996 Value	%
Total Production	810,018	100.0	784,478	100.0	718,211	100.0	639,052	100.0	610,850	100.0	572,991	100.0	542,857	100.0
Regions														
Africa	9,594	1.2	8,014	1.0	7,862	1.1	8,441	1.3	7,976	1.3	8,007	1.4	7,941	1.5
America, North	6,010	0.7	5,525	0.7	5,853	0.8	5,494	0.9	7,116	1.2	4,771	0.8	6,145	1.1
America, South	148	0.0	157	0.0	166	0.0	129	0.0	172	0.0	200	0.0	71	0.0
Asia	510,158	63.0	523,765	66.8	495,269	69.0	438,370	68.6	444,030	72.7	426,362	74.4	409,070	75.4
Europe	283,788	35.0	246,845	31.5	209,037	29.1	186,618	29.2	151,557	24.8	133,651	23.3	119,630	22.0
Oceania	320	0.0	172	0.0	24	0.0	-		-		-		-	
Leading Producers														
Korea, Republic of	*216,904	26.8	*216,904	27.6	*216,904	30.2	*216,904	33.9	*216,904	35.5	*216,904	37.9	*216,904	40.0
Belgium	52,199	6.4	51,282	6.5	49,061	6.8	*48,974	7.7	*47,998	7.9	*47,021	8.2	*46,045	8.5
Poland	31,794	3.9	14,694	1.9	13,485	1.9	14,158	2.2	19,696	3.2	19,092	3.3	15,128	2.8
France	23,560	2.9	18,178	2.3	15,071	2.1	13,240	2.1	5,875	1.0	-		-	
Russian Federation	85,705	10.6	81,684	10.4	68,658	9.6	20,794	3.3	13,024	2.1	3,981	0.7	1,600	0.3

Seeders, Planters and Transplanters

Unit of Measure: Number.

	1990 Value	%	1991 Value	%	1992 Value	%	1993 Value	%	1994 Value	%	1995 Value	%	1996 Value	%
Total Production	1,337,537	100.0	1,340,419	100.0	1,263,924	100.0	1,190,034	100.0	1,169,854	100.0	1,174,204	100.0	1,140,755	100.0
Regions														
Africa	1,628	0.1	1,532	0.1	1,436	0.1	1,598	0.1	1,446	0.1	1,806	0.2	1,627	0.1
America, North	432,191	32.3	491,202	36.6	482,754	38.2	432,938	36.4	439,060	37.5	445,173	37.9	451,152	39.5
America, South	388,078	29.0	388,075	29.0	388,072	30.7	388,069	32.6	388,067	33.2	388,064	33.0	388,063	34.0
Asia	303,264	22.7	283,061	21.1	249,963	19.8	235,133	19.8	219,196	18.7	210,753	17.9	186,934	16.4
Europe	210,072	15.7	174,780	13.0	140,136	11.1	130,942	11.0	120,939	10.3	127,469	10.9	112,248	9.8
Oceania	2,303	0.2	1,769	0.1	1,561	0.1	1,354	0.1	1,146	0.1	939	0.1	731	0.1
Leading Producers														
United States of America	428,629	32.0	488,220	36.4	480,097	38.0	*430,675	36.2	*436,730	37.3	*442,785	37.7	*448,841	39.3
Brazil	*388,063	29.0	*388,063	29.0	*388,063	30.7	*388,063	32.6	*388,063	33.2	*388,063	33.0	*388,063	34.0
Japan	91,141	6.8	87,019	6.5	80,540	6.4	84,980	7.1	85,837	7.3	86,713	7.4	70,614	6.2
Belgium	*41,160	3.1	*41,160	3.1	*41,160	3.3	*41,160	3.5	*41,160	3.5	*41,160	3.5	*41,160	3.6
France	38,227	2.9	30,743	2.3	28,823	2.3	22,550	1.9	19,859	1.7	*22,926	2.0	*20,281	1.8

Commodity data are provided by the United Nations Statistical Division. The symbol * means that data are estimated. For additional notes, see Appendix I.

577

Product Tables
Combine Harvester-Threshers

Unit of Measure: Number.

	1990 Value	%	1991 Value	%	1992 Value	%	1993 Value	%	1994 Value	%	1995 Value	%	1996 Value	%
Total Production	288,537	100.0	265,830	100.0	243,781	100.0	220,110	100.0	188,375	100.0	190,504	100.0	206,235	100.0
Regions														
Africa	567	0.2	510	0.2	530	0.2	291	0.1	176	0.1	332	0.2	305	0.1
America, North	15,648	5.4	12,617	4.7	10,304	4.2	6,586	3.0	5,430	2.9	4,275	2.2	3,120	1.5
America, South	4,694	1.6	3,351	1.3	4,216	1.7	5,729	2.6	7,807	4.1	5,415	2.8	4,814	2.3
Asia	220,946	76.6	218,207	82.1	196,632	80.7	175,721	79.8	143,184	76.0	147,656	77.5	164,438	79.7
Europe	45,986	15.9	30,447	11.5	31,401	12.9	31,086	14.1	31,080	16.5	32,129	16.9	32,861	15.9
Oceania	697	0.2	697	0.3	697	0.3	697	0.3	697	0.4	697	0.4	697	0.3
Leading Producers														
Japan	68,993	23.9	72,913	27.4	65,673	26.9	65,192	29.6	61,242	32.5	66,767	35.0	63,371	30.7
Russian Federation	65,736	22.8	55,356	20.8	42,165	17.3	32,989	15.0	12,063	6.4	6,241	3.3	2,515	1.2
United States of America	14,629	5.1	11,555	4.3	9,198	3.8	*5,436	2.5	*4,237	2.2	*3,039	1.6	*1,840	0.9
China	5,173	1.8	10,638	4.0	15,370	6.3	9,569	4.3	7,310	3.9	17,393	9.1	45,700	22.2
Belgium	7,007	2.4	6,388	2.4	5,769	2.4	*6,075	2.8	*5,913	3.1	*5,751	3.0	*5,589	2.7

Mowers, Animal or Tractor-Operated and Self-Propelled

Unit of Measure: Number.

	1990 Value	%	1991 Value	%	1992 Value	%	1993 Value	%	1994 Value	%	1995 Value	%	1996 Value	%
Total Production	1,450,440	100.0	1,366,637	100.0	1,436,450	100.0	1,403,615	100.0	1,426,768	100.0	1,495,691	100.0	1,660,477	100.0
Regions														
Africa	3,007	0.2	3,000	0.2	1,243	0.1	1,003	0.1	636	0.0	1,193	0.1	1,025	0.1
America, North	20,679	1.4	22,541	1.6	16,227	1.1	18,640	1.3	24,362	1.7	15,578	1.0	14,703	0.9
America, South	6	0.0	6	0.0	6	0.0	6	0.0	6	0.0	6	0.0	6	0.0
Asia	103,840	7.2	112,612	8.2	102,644	7.1	100,632	7.2	94,189	6.6	90,097	6.0	80,261	4.8
Europe	965,410	66.6	902,425	66.0	942,842	65.6	910,597	64.9	935,590	65.6	1,017,583	68.0	1,194,001	71.9
Oceania	357,497	24.6	326,053	23.9	373,489	26.0	372,737	26.6	371,985	26.1	371,233	24.8	370,480	22.3
Leading Producers														
Austria	*350,486	24.2	*376,049	27.5	*401,612	28.0	*427,175	30.4	*452,739	31.7	*478,302	32.0	*503,865	30.3
Australia	247,000	17.0	213,000	15.6	*257,880	18.0	*254,572	18.1	*251,264	17.6	*247,956	16.6	*244,648	14.7
Sweden	285,989	19.7	205,651	15.0	206,250	14.4	165,685	11.8	189,251	13.3	258,155	17.3	*319,575	19.2
New Zealand	*110,497	7.6	*113,053	8.3	*115,609	8.0	*118,165	8.4	*120,721	8.5	*123,277	8.2	*125,832	7.6
Denmark	120,233	8.3	138,657	10.1	157,082	10.9	144,196	10.3	141,267	9.9	90,634	6.1	*168,562	10.2

Rakes, Animal or Tractor-Operated and Self-Propelled

Unit of Measure: Number.

	1990 Value	%	1991 Value	%	1992 Value	%	1993 Value	%	1994 Value	%	1995 Value	%	1996 Value	%
Total Production	239,941	100.0	221,670	100.0	268,353	100.0	230,939	100.0	222,847	100.0	209,404	100.0	187,262	100.0
Regions														
America, North	16,670	6.9	11,348	5.1	10,779	4.0	10,857	4.7	14,533	6.5	9,894	4.7	12,599	6.7
Asia	88,314	36.8	85,144	38.4	139,362	51.9	99,569	43.1	89,398	40.1	84,914	40.6	61,706	33.0
Europe	134,555	56.1	124,776	56.3	117,810	43.9	120,111	52.0	118,515	53.2	114,195	54.5	112,555	60.1
Oceania	402	0.2	402	0.2	402	0.1	402	0.2	402	0.2	402	0.2	402	0.2
Leading Producers														
Kazakhstan	106	0.0	148	0.1	74,000	27.6	29,000	12.6	24,000	10.8	21,000	10.0	1,000	0.5
Belgium	*21,102	8.8	*21,102	9.5	*21,102	7.9	*21,102	9.1	*21,102	9.5	*21,102	10.1	*21,102	11.3
Germany	*20,582	8.6	*20,582	9.3	17,740	6.6	17,725	7.7	26,280	11.8	*20,582	9.8	*20,582	11.0
United Kingdom	*18,424	7.7	*18,424	8.3	*18,424	6.9	*18,424	8.0	18,424	8.3	*18,424	8.8	*18,424	9.8
France	17,556	7.3	11,852	5.3	9,881	3.7	14,130	6.1	7,565	3.4	*10,018	4.8	*8,952	4.8

Commodity data are provided by the United Nations Statistical Division. The symbol * means that data are estimated. For additional notes, see Appendix I.

Product Tables
Threshing Machines

Unit of Measure: Number.

	1990		1991		1992		1993		1994		1995		1996	
	Value	%	Value	%	Value	%	Value	%	Value	%	Value	%	Value	%
Total Production	84,123	100.0	80,110	100.0	68,080	100.0	61,903	100.0	64,405	100.0	63,680	100.0	62,762	100.0
Regions														
America, North	97	0.1	97	0.1	97	0.1	97	0.2	97	0.2	97	0.2	97	0.2
America, South	660	0.8	604	0.8	675	1.0	724	1.2	805	1.2	147	0.2	61	0.1
Asia	33,315	39.6	31,849	39.8	17,996	26.4	15,958	25.8	15,848	24.6	16,626	26.1	16,865	26.9
Europe	50,051	59.5	47,560	59.4	49,312	72.4	45,124	72.9	47,655	74.0	46,810	73.5	45,739	72.9
Leading Producers														
Japan	22,634	26.9	20,337	25.4	12,656	18.6	11,663	18.8	11,422	17.7	12,422	19.5	11,593	18.5
Korea, Republic of	5,506	6.5	5,423	6.8	*2,269	3.3	*435	0.7	-		-		-	
Poland	809	1.0	646	0.8	6	0.0	1	0.0	-		-			
Spain	3,468	4.1	1,145	1.4	3,543	5.2	773	1.2	491	0.8	1,064	1.7	-	
Turkey	1,691	2.0	2,700	3.4	1,078	1.6	1,703	2.8	2,298	3.6	1,537	2.4	2,602	4.1

Milking Machines

Unit of Measure: Number.

	1990		1991		1992		1993		1994		1995		1996	
	Value	%	Value	%	Value	%	Value	%	Value	%	Value	%	Value	%
Total Production	196,250	100.0	186,260	100.0	164,412	100.0	160,295	100.0	137,830	100.0	137,941	100.0	129,434	100.0
Regions														
America, North	9,124	4.6	9,124	4.9	9,124	5.5	9,124	5.7	9,124	6.6	9,124	6.6	9,124	7.0
America, South	1	0.0	-		-		-		-		-		-	
Asia	107,014	54.5	95,627	51.3	96,150	58.5	97,542	60.9	82,169	59.6	81,880	59.4	81,831	63.2
Europe	80,111	40.8	81,509	43.8	59,138	36.0	53,629	33.5	46,537	33.8	46,938	34.0	38,480	29.7
Leading Producers														
Belgium	*32,620	16.6	*31,316	16.8	*30,012	18.3	*28,708	17.9	*27,405	19.9	*26,101	18.9	*24,797	19.2
Latvia	21,796	11.1	34,546	18.5	13,446	8.2	13,121	8.2	7,024	5.1	7,090	5.1	4,125	3.2
Poland	5,802	3.0	1,092	0.6	911	0.6	730	0.5	261	0.2	709	0.5	456	0.4
France	5,622	2.9	3,352	1.8	2,657	1.6	-		-		-		-	
Russian Federation	30,742	15.7	17,692	9.5	13,716	8.3	16,241	10.1	1,006	0.7	528	0.4	496	0.4

Garden Tractors

Unit of Measure: Number.

	1990		1991		1992		1993		1994		1995		1996	
	Value	%	Value	%	Value	%	Value	%	Value	%	Value	%	Value	%
Total Production	1,569,921	100.0	1,840,488	100.0	1,873,066	100.0	1,411,215	100.0	1,796,333	100.0	2,537,503	100.0	2,556,070	100.0
Regions														
America, North	107,647	6.9	121,096	6.6	134,546	7.2	147,995	10.5	161,445	9.0	161,099	6.3	138,763	5.4
America, South	2,528	0.2	1,974	0.1	1,878	0.1	1,487	0.1	1,532	0.1	1,603	0.1	968	0.0
Asia	1,372,634	87.4	1,645,131	89.4	1,656,906	88.5	1,202,642	85.2	1,574,097	87.6	2,274,247	89.6	2,313,676	90.5
Europe	87,113	5.5	72,287	3.9	79,737	4.3	59,090	4.2	59,259	3.3	100,554	4.0	102,663	4.0
Leading Producers														
China	1,101,400	70.2	1,347,800	73.2	1,390,700	74.2	961,400	68.1	1,355,400	75.5	2,062,997	81.3	2,096,600	82.0
Japan	269,027	17.1	270,714	14.7	245,675	13.1	225,564	16.0	212,539	11.8	205,758	8.1	214,702	8.4
United States of America	107,647	6.9	121,096	6.6	134,546	7.2	147,995	10.5	161,445	9.0	161,099	6.3	138,763	5.4
Germany	-		-		17,435	0.9	13,212	0.9	11,009	0.6	52,502	2.1	53,433	2.1
Croatia	21,242	1.4	18,005	1.0	5,641	0.3	1,947	0.1	1,861	0.1	1,774	0.1	1,120	0.0

Commodity data are provided by the United Nations Statistical Division. The symbol * means that data are estimated. For additional notes, see Appendix I.

579

Product Tables

Tractors of 10 HP and Over, Except Industrial and Road Tractors

Unit of Measure: Number.

	1990 Value	%	1991 Value	%	1992 Value	%	1993 Value	%	1994 Value	%	1995 Value	%	1996 Value	%
Total Production	3,249,272	100.0	3,111,602	100.0	3,074,977	100.0	2,927,486	100.0	2,870,817	100.0	3,044,318	100.0	3,136,399	100.0
Regions														
Africa	6,133	0.2	5,197	0.2	5,516	0.2	5,933	0.2	4,677	0.2	4,084	0.1	4,598	0.1
America, North	1,134,432	34.9	1,136,372	36.5	1,212,768	39.4	1,205,348	41.2	1,273,561	44.4	1,439,973	47.3	1,520,790	48.5
America, South	66,342	2.0	67,773	2.2	71,656	2.3	73,999	2.5	77,939	2.7	80,023	2.6	83,673	2.7
Asia	1,280,065	39.4	1,249,253	40.1	1,165,236	37.9	1,058,834	36.2	973,236	33.9	959,032	31.5	967,459	30.8
Europe	762,300	23.5	653,008	21.0	619,801	20.2	583,372	19.9	541,403	18.9	561,205	18.4	559,878	17.9
Leading Producers														
United States of America	1,125,197	34.6	1,128,613	36.3	1,202,870	39.1	1,199,986	41.0	1,265,983	44.1	*1,432,666	47.1	*1,510,722	48.2
Japan	174,529	5.4	161,842	5.2	156,182	5.1	155,497	5.3	167,686	5.8	164,685	5.4	164,008	5.2
Belgium	108,885	3.4	108,008	3.5	103,208	3.4	*107,534	3.7	*107,653	3.7	*107,773	3.5	*107,892	3.4
Russian Federation	213,812	6.6	178,178	5.7	136,598	4.4	89,087	3.0	28,695	1.0	21,169	0.7	13,964	0.4
France	108,635	3.3	76,293	2.5	93,955	3.1	82,540	2.8	75,032	2.6	*83,461	2.7	*82,525	2.6

Fertilizer Distributors, Animal, Hand or Tractor-Operated

Unit of Measure: Number.

	1990 Value	%	1991 Value	%	1992 Value	%	1993 Value	%	1994 Value	%	1995 Value	%	1996 Value	%
Total Production	495,251	100.0	502,021	100.0	498,675	100.0	437,503	100.0	517,445	100.0	463,014	100.0	598,078	100.0
Regions														
Africa	282	0.1	1,045	0.2	663	0.1	694	0.2	724	0.1	755	0.2	785	0.1
America, North	227,027	45.8	212,686	42.4	253,087	50.8	203,108	46.4	263,931	51.0	218,332	47.2	329,335	55.1
Asia	42,475	8.6	44,645	8.9	19,994	4.0	14,611	3.3	13,335	2.6	11,846	2.6	10,865	1.8
Europe	221,135	44.7	240,008	47.8	220,982	44.3	214,830	49.1	234,883	45.4	227,198	49.1	251,897	42.1
Oceania	4,332	0.9	3,637	0.7	3,949	0.8	4,260	1.0	4,572	0.9	4,884	1.1	5,195	0.9
Leading Producers														
Mexico	203,686	41.1	192,828	38.4	232,711	46.7	180,078	41.2	238,943	46.2	195,383	42.2	305,829	51.1
Belgium	46,431	9.4	45,686	9.1	44,941	9.0	*44,196	10.1	*43,451	8.4	*42,705	9.2	*41,960	7.0
United Kingdom	45,195	9.1	52,499	10.5	59,802	12.0	67,106	15.3	94,669	18.3	*81,640	17.6	*88,277	14.8
Germany	-		24,496	4.9	17,756	3.6	14,021	3.2	18,297	3.5	32,215	7.0	53,660	9.0
Netherlands	*18,146	3.7	*15,174	3.0	*12,202	2.4	*9,230	2.1	*6,258	1.2	*3,286	0.7	*314	0.1

Drilling and Boring Machines

Unit of Measure: Number.

	1990 Value	%	1991 Value	%	1992 Value	%	1993 Value	%	1994 Value	%	1995 Value	%	1996 Value	%
Total Production	165,720	100.0	158,441	100.0	155,879	100.0	148,740	100.0	140,067	100.0	146,342	100.0	134,960	100.0
Regions														
Africa	412	0.2	210	0.1	122	0.1	30	0.0	170	0.1	163	0.1	157	0.1
America, North	10,966	6.6	9,790	6.2	8,397	5.4	7,669	5.2	8,823	6.3	10,465	7.2	7,927	5.9
Asia	67,192	40.5	62,602	39.5	47,986	30.8	35,719	24.0	37,589	26.8	45,899	31.4	41,007	30.4
Europe	87,150	52.6	85,839	54.2	99,374	63.8	105,322	70.8	93,486	66.7	89,814	61.4	85,869	63.6
Leading Producers														
Belgium	*37,211	22.5	*37,211	23.5	*37,211	23.9	*37,211	25.0	*37,211	26.6	*37,211	25.4	*37,211	27.6
Japan	40,171	24.2	33,929	21.4	22,973	14.7	14,496	9.7	11,936	8.5	14,678	10.0	16,414	12.2
Germany	-		-		15,085	9.7	21,222	14.3	15,955	11.4	13,049	8.9	11,730	8.7
United States of America	8,828	5.3	7,603	4.8	7,542	4.8	7,182	4.8	8,823	6.3	10,465	7.2	7,927	5.9
Russian Federation	16,192	9.8	16,020	10.1	12,835	8.2	10,607	7.1	5,291	3.8	5,021	3.4	3,088	2.3

Product Tables

Forging, Stamping and Die-Stamping Machines

Unit of Measure: Number.

	1990 Value	%	1991 Value	%	1992 Value	%	1993 Value	%	1994 Value	%	1995 Value	%	1996 Value	%
Total Production	49,094	100.0	45,105	100.0	34,311	100.0	23,698	100.0	15,010	100.0	13,370	100.0	11,472	100.0
Regions														
Asia	31,172	63.5	27,319	60.6	19,235	56.1	9,509	40.1	4,786	31.9	4,035	30.2	2,815	24.5
Europe	17,923	36.5	17,786	39.4	15,076	43.9	14,189	59.9	10,224	68.1	9,335	69.8	8,657	75.5
Leading Producers														
Russian Federation	27,302	55.6	23,936	53.1	16,532	48.2	7,451	31.4	3,114	20.7	2,184	16.3	1,237	10.8
Ukraine	10,858	22.1	10,813	24.0	7,854	22.9	6,078	25.6	2,520	16.8	1,379	10.3	614	5.4
United Kingdom	*1,386	2.8	*1,386	3.1	*1,386	4.0	*1,386	5.8	*1,386	9.2	*1,386	10.4	*1,386	12.1
Finland	770	1.6	667	1.5	*961	2.8	*1,059	4.5	*1,158	7.7	*1,256	9.4	*1,355	11.8
Kazakhstan	1,173	2.4	1,165	2.6	757	2.2	730	3.1	434	2.9	269	2.0	127	1.1

Grinding and Sharpening Machines

Unit of Measure: Number.

	1990 Value	%	1991 Value	%	1992 Value	%	1993 Value	%	1994 Value	%	1995 Value	%	1996 Value	%
Total Production	199,192	100.0	180,377	100.0	210,916	100.0	263,627	100.0	265,164	100.0	253,915	100.0	294,033	100.0
Regions														
Africa	14	0.0	14	0.0	14	0.0	14	0.0	15	0.0	15	0.0	15	0.0
America, North	58,276	29.3	49,187	27.3	54,585	25.9	56,068	21.3	61,100	23.0	66,439	26.2	65,737	22.4
Asia	34,963	17.6	33,949	18.8	25,616	12.1	22,392	8.5	21,021	7.9	21,608	8.5	22,609	7.7
Europe	105,940	53.2	97,228	53.9	130,701	62.0	185,153	70.2	183,029	69.0	165,854	65.3	205,671	69.9
Leading Producers														
United States of America	58,276	29.3	49,187	27.3	54,585	25.9	56,068	21.3	61,100	23.0	66,439	26.2	65,737	22.4
Germany	-		-		26,804	12.7	21,715	8.2	20,469	7.7	20,285	8.0	16,968	5.8
Spain	751	0.4	1,088	0.6	1,063	0.5	70,303	26.7	70,322	26.5	55,940	22.0	96,598	32.9
China	*13,361	6.7	*13,361	7.4	*13,361	6.3	*13,361	5.1	*13,361	5.0	*13,361	5.3	*13,361	4.5
Poland	13,048	6.6	5,879	3.3	11,065	5.2	5,187	2.0	7,471	2.8	4,002	1.6	6,042	2.1

Machining Centers

Unit of Measure: Number.

	1990 Value	%	1991 Value	%	1992 Value	%	1993 Value	%	1994 Value	%	1995 Value	%	1996 Value	%
Total Production	30,012	100.0	28,199	100.0	27,388	100.0	24,907	100.0	27,666	100.0	35,381	100.0	38,186	100.0
Regions														
America, North	2,327	7.8	2,316	8.2	2,676	9.8	3,917	15.7	5,362	19.4	7,305	20.6	8,442	22.1
Asia	21,547	71.8	20,030	71.0	15,188	55.5	13,640	54.8	14,414	52.1	18,456	52.2	20,416	53.5
Europe	6,138	20.5	5,853	20.8	9,524	34.8	7,350	29.5	7,890	28.5	9,621	27.2	9,328	24.4
Leading Producers														
Japan	15,820	52.7	13,766	48.8	8,548	31.2	6,420	25.8	6,461	23.4	9,665	27.3	11,324	29.7
United States of America	2,327	7.8	2,316	8.2	2,676	9.8	3,917	15.7	5,362	19.4	7,305	20.6	8,442	22.1
Germany	-		-		3,530	12.9	1,514	6.1	2,089	7.6	3,227	9.1	2,779	7.3
United Kingdom	1,778	5.9	1,665	5.9	1,552	5.7	1,439	5.8	1,326	4.8	*1,963	5.5	*2,093	5.5
Korea, Republic of	705	2.3	791	2.8	716	2.6	846	3.4	1,128	4.1	1,515	4.3	*1,366	3.6

Commodity data are provided by the United Nations Statistical Division. The symbol * means that data are estimated. For additional notes, see Appendix I.

581

Product Tables
Lathes

Unit of Measure: Number.

	1990 Value	%	1991 Value	%	1992 Value	%	1993 Value	%	1994 Value	%	1995 Value	%	1996 Value	%
Total Production	159,012	100.0	142,479	100.0	124,238	100.0	109,696	100.0	113,439	100.0	124,590	100.0	120,584	100.0
Regions														
Africa	270	0.2	273	0.2	310	0.2	194	0.2	118	0.1	236	0.2	245	0.2
America, North	3,247	2.0	2,658	1.9	2,409	1.9	3,042	2.8	3,662	3.2	4,643	3.7	4,190	3.5
America, South	86	0.1	92	0.1	97	0.1	103	0.1	155	0.1	112	0.1	78	0.1
Asia	109,039	68.6	98,871	69.4	79,756	64.2	74,975	68.3	78,029	68.8	85,114	68.3	83,751	69.5
Europe	46,369	29.2	40,586	28.5	41,666	33.5	31,383	28.6	31,476	27.7	34,485	27.7	32,320	26.8
Leading Producers														
China	*48,253	30.3	*48,253	33.9	*48,253	38.8	*48,253	44.0	*48,253	42.5	*48,253	38.7	*48,253	40.0
Japan	32,659	20.5	26,216	18.4	16,155	13.0	12,343	11.3	14,961	13.2	20,339	16.3	21,443	17.8
Korea, Republic of	10,597	6.7	11,324	7.9	6,643	5.3	6,931	6.3	10,265	9.0	12,606	10.1	*11,464	9.5
Russian Federation	14,747	9.3	9,850	6.9	7,079	5.7	6,506	5.9	3,807	3.4	3,269	2.6	2,095	1.7
Germany	-		-		7,689	6.2	4,755	4.3	5,322	4.7	8,232	6.6	6,375	5.3

Milling Machines (Cutters)

Unit of Measure: Number.

	1990 Value	%	1991 Value	%	1992 Value	%	1993 Value	%	1994 Value	%	1995 Value	%	1996 Value	%
Total Production	76,227	100.0	67,990	100.0	61,308	100.0	53,695	100.0	53,605	100.0	53,117	100.0	51,536	100.0
Regions														
Africa	267	0.4	150	0.2	103	0.2	81	0.2	124	0.2	135	0.3	138	0.3
America, North	4,787	6.3	2,772	4.1	2,581	4.2	3,386	6.3	4,087	7.6	4,747	8.9	4,102	8.0
America, South	142	0.2	142	0.2	142	0.2	142	0.3	142	0.3	142	0.3	142	0.3
Asia	28,566	37.5	27,291	40.1	21,566	35.2	19,086	35.5	18,689	34.9	18,211	34.3	18,140	35.2
Europe	42,465	55.7	37,635	55.4	36,917	60.2	31,000	57.7	30,563	57.0	29,883	56.3	29,015	56.3
Leading Producers														
China	*10,278	13.5	*10,278	15.1	*10,278	16.8	*10,278	19.1	*10,278	19.2	*10,278	19.3	*10,278	19.9
Japan	8,492	11.1	7,584	11.2	3,913	6.4	2,007	3.7	1,791	3.3	1,832	3.4	2,198	4.3
United States of America	4,787	6.3	2,772	4.1	2,581	4.2	3,386	6.3	4,087	7.6	4,747	8.9	4,102	8.0
Germany	*5,997	7.9	*5,997	8.8	*5,997	9.8	6,680	12.4	6,327	11.8	5,770	10.9	5,213	10.1
Spain	7,062	9.3	4,169	6.1	4,553	7.4	788	1.5	1,458	2.7	1,600	3.0	1,142	2.2

Other Metal-Cutting Machine-Tools

Unit of Measure: Number.

	1990 Value	%	1991 Value	%	1992 Value	%	1993 Value	%	1994 Value	%	1995 Value	%	1996 Value	%
Total Production	305,072	100.0	322,522	100.0	316,677	100.0	296,597	100.0	301,816	100.0	310,342	100.0	297,457	100.0
Regions														
Africa	44	0.0	200	0.1	100	0.0	65	0.0	112	0.0	91	0.0	71	0.0
America, North	14,942	4.9	12,796	4.0	11,838	3.7	13,347	4.5	14,803	4.9	18,696	6.0	17,013	5.7
America, South	153	0.1	153	0.0	153	0.0	153	0.1	153	0.1	153	0.0	153	0.1
Asia	221,839	72.7	245,506	76.1	251,447	79.4	243,305	82.0	247,691	82.1	252,323	81.3	255,842	86.0
Europe	68,093	22.3	63,867	19.8	53,138	16.8	39,727	13.4	39,057	12.9	39,078	12.6	24,378	8.2
Leading Producers														
China	134,500	44.1	163,900	50.8	*187,493	59.2	*193,299	65.2	*199,105	66.0	*204,912	66.0	*210,718	70.8
Japan	72,163	23.7	65,519	20.3	49,347	15.6	41,905	14.1	41,401	13.7	38,675	12.5	38,736	13.0
United States of America	14,942	4.9	12,796	4.0	11,838	3.7	13,347	4.5	14,803	4.9	18,696	6.0	17,013	5.7
Germany	*19,178	6.3	*19,178	5.9	*19,178	6.1	18,428	6.2	21,321	7.1	22,719	7.3	14,246	4.8
Austria	9,532	3.1	6,686	2.1	3,941	1.2	3,726	1.3	3,450	1.1	3,745	1.2	-	

Product Tables

Planing, Shaping and Slotting Machines

Unit of Measure: Number.

	1990		1991		1992		1993		1994		1995		1996	
	Value	%	Value	%	Value	%	Value	%	Value	%	Value	%	Value	%
Total Production	7,618	100.0	7,767	100.0	7,470	100.0	7,651	100.0	7,327	100.0	7,270	100.0	7,787	100.0
Regions														
Africa	12	0.2	15	0.2	3	0.0	-		-		-		-	
America, South	387	5.1	369	4.7	351	4.7	364	4.8	374	5.1	375	5.2	344	4.4
Asia	546	7.2	510	6.6	436	5.8	235	3.1	126	1.7	114	1.6	120	1.5
Europe	6,673	87.6	6,873	88.5	6,681	89.4	7,052	92.2	6,827	93.2	6,781	93.3	7,323	94.0
Leading Producers														
Croatia	*4,319	56.7	*4,481	57.7	*4,643	62.2	*4,805	62.8	*4,967	67.8	*5,128	70.5	*5,290	67.9
Slovakia	*1,038	13.6	*1,038	13.4	*1,038	13.9	*1,038	13.6	*1,038	14.2	789	10.9	1,287	16.5
Spain	173	2.3	225	2.9	209	2.8	*10	0.1	-		-		-	
Romania	398	5.2	200	2.6	1	0.0	147	1.9	10	0.1	33	0.5	21	0.3
Brazil	*340	4.5	*340	4.4	*340	4.6	*340	4.4	*340	4.6	*340	4.7	*340	4.4

Metal-Working Presses

Unit of Measure: Number.

	1990		1991		1992		1993		1994		1995		1996	
	Value	%	Value	%	Value	%	Value	%	Value	%	Value	%	Value	%
Total Production	74,868	100.0	90,665	100.0	122,831	100.0	79,210	100.0	95,369	100.0	94,004	100.0	94,158	100.0
Regions														
America, North	7,293	9.7	5,920	6.5	5,830	4.7	6,244	7.9	10,955	11.5	5,053	5.4	11,031	11.7
America, South	14,401	19.2	16,958	18.7	18,186	14.8	15,129	19.1	24,497	25.7	28,202	30.0	25,759	27.4
Asia	25,271	33.8	22,766	25.1	15,957	13.0	12,028	15.2	11,867	12.4	12,984	13.8	12,967	13.8
Europe	27,903	37.3	45,021	49.7	82,858	67.5	45,809	57.8	48,050	50.4	47,765	50.8	44,400	47.2
Leading Producers														
Germany	-		17,726	19.6	55,997	45.6	18,939	23.9	23,284	24.4	16,399	17.4	*15,859	16.8
Japan	22,571	30.1	19,173	21.1	12,458	10.1	9,516	12.0	9,531	10.0	10,512	11.2	10,061	10.7
Colombia	13,219	17.7	15,105	16.7	16,992	13.8	13,366	16.9	22,661	23.8	26,344	28.0	*24,102	25.6
United States of America	7,285	9.7	5,912	6.5	5,822	4.7	6,236	7.9	10,947	11.5	5,045	5.4	11,023	11.7
Spain	1,600	2.1	2,152	2.4	1,687	1.4	2,436	3.1	723	0.8	5,452	5.8	*2,435	2.6

Other Metal-Forming Machine-Tools

Unit of Measure: Number.

	1990		1991		1992		1993		1994		1995		1996	
	Value	%	Value	%	Value	%	Value	%	Value	%	Value	%	Value	%
Total Production	146,494	100.0	143,685	100.0	174,324	100.0	167,027	100.0	173,142	100.0	163,163	100.0	167,342	100.0
Regions														
America, North	44,459	30.3	35,600	24.8	39,185	22.5	45,408	27.2	55,060	31.8	36,241	22.2	42,622	25.5
America, South	4,547	3.1	4,547	3.2	4,547	2.6	4,547	2.7	4,547	2.6	4,547	2.8	4,547	2.7
Asia	2,662	1.8	2,400	1.7	1,472	0.8	783	0.5	124	0.1	318	0.2	369	0.2
Europe	94,827	64.7	101,138	70.4	129,120	74.1	116,289	69.6	113,411	65.5	122,057	74.8	119,804	71.6
Leading Producers														
United States of America	44,459	30.3	35,600	24.8	39,185	22.5	45,408	27.2	55,060	31.8	36,241	22.2	42,622	25.5
Germany	-		-		29,967	17.2	23,981	14.4	23,071	13.3	35,316	21.6	22,480	13.4
Spain	26,302	18.0	31,851	22.2	32,631	18.7	26,979	16.2	21,327	12.3	19,159	11.7	*31,416	18.8
Colombia	*4,547	3.1	*4,547	3.2	*4,547	2.6	*4,547	2.7	*4,547	2.6	*4,547	2.8	*4,547	2.7
Czech Republic	*4,180	2.9	*4,180	2.9	*4,180	2.4	*4,180	2.5	4,807	2.8	4,392	2.7	3,341	2.0

Commodity data are provided by the United Nations Statistical Division. The symbol * means that data are estimated. For additional notes, see Appendix I.

583

Product Tables
Rolling Mills for Rolling Metals

Unit of Measure: Number.

	1990		1991		1992		1993		1994		1995		1996	
	Value	%	Value	%	Value	%	Value	%	Value	%	Value	%	Value	%
Total Production	980	100.0	2,743	100.0	970	100.0	1,524	100.0	1,731	100.0	1,187	100.0	1,843	100.0
Regions														
America, South	1	0.1	1	0.0	1	0.1	1	0.1	1	0.1	1	0.1	1	0.1
Asia	842	85.9	2,633	96.0	888	91.5	1,454	95.4	1,657	95.7	1,089	91.7	1,590	86.3
Europe	137	14.0	108	4.0	81	8.4	69	4.5	73	4.2	97	8.2	252	13.7
Leading Producers														
Yugoslavia	-		1,922	70.1	250	25.8	787	51.6	1,022	59.0	455	38.3	942	51.1
Slovenia	*386	39.4	*386	14.1	*386	39.8	*386	25.3	*386	22.3	*386	32.5	386	20.9
Japan	435	44.4	312	11.4	245	25.3	255	16.7	243	14.0	234	19.7	260	14.1
Hungary	38	3.9	33	1.2	17	1.8	27	1.8	37	2.1	46	3.9	56	3.0
Spain	52	5.3	43	1.6	35	3.6	27	1.7	18	1.0	16	1.3	167	9.1

Machine-Tools for Working Wood

Unit of Measure: Number.

	1990		1991		1992		1993		1994		1995		1996	
	Value	%	Value	%	Value	%	Value	%	Value	%	Value	%	Value	%
Total Production	692,903	100.0	686,491	100.0	632,413	100.0	674,693	100.0	571,777	100.0	471,647	100.0	522,133	100.0
Regions														
America, North	11,589	1.7	11,589	1.7	11,589	1.8	177	0.0	9,023	1.6	17,090	3.6	20,067	3.8
America, South	62,565	9.0	70,314	10.2	78,063	12.3	70,025	10.4	106,732	18.7	7,751	1.6	82,550	15.8
Asia	246,659	35.6	262,113	38.2	214,769	34.0	169,928	25.2	161,590	28.3	151,737	32.2	125,267	24.0
Europe	372,090	53.7	342,474	49.9	327,991	51.9	434,562	64.4	294,432	51.5	295,069	62.6	294,248	56.4
Leading Producers														
Germany	*188,591	27.2	*188,591	27.5	*188,591	29.8	*188,591	28.0	*188,591	33.0	*188,591	40.0	188,591	36.1
Japan	140,709	20.3	148,551	21.6	101,776	16.1	76,372	11.3	70,051	12.3	63,592	13.5	51,807	9.9
Ecuador	61,528	8.9	69,264	10.1	77,000	12.2	68,831	10.2	105,296	18.4	6,662	1.4	*81,314	15.6
Austria	23,490	3.4	23,918	3.5	23,044	3.6	14,988	2.2	16,341	2.9	15,460	3.3	*17,608	3.4
Poland	39,349	5.7	26,385	3.8	14,728	2.3	13,594	2.0	15,267	2.7	14,954	3.2	10,877	2.1

Electro-Mechanical Hand Tools

Unit of Measure: Number.

	1990		1991		1992		1993		1994		1995		1996	
	Value	%	Value	%	Value	%	Value	%	Value	%	Value	%	Value	%
Total Production	43,801,376	100.0	55,898,376	100.0	55,581,501	100.0	52,674,501	100.0	53,651,501	100.0	50,498,560	100.0	48,530,725	100.0
Regions														
America, North	18,087,000	41.3	18,087,000	32.4	18,087,000	32.5	18,087,000	34.3	18,087,000	33.7	18,087,000	35.8	18,087,000	37.3
America, South	55,875	0.1	55,875	0.1	81,000	0.1	68,500	0.1	56,000	0.1	18,000	0.0	55,875	0.1
Asia	13,431,000	30.7	14,392,000	25.7	13,882,000	25.0	11,767,000	22.3	12,300,000	22.9	11,489,000	22.8	10,599,855	21.8
Europe	12,034,501	27.5	23,170,501	41.5	23,338,501	42.0	22,559,001	42.8	23,015,501	42.9	20,711,560	41.0	19,594,995	40.4
Oceania	193,000	0.4	193,000	0.3	193,000	0.3	193,000	0.4	193,000	0.4	193,000	0.4	193,000	0.4
Leading Producers														
United States of America	*18,087,000	41.3	*18,087,000	32.4	*18,087,000	32.5	*18,087,000	34.3	*18,087,000	33.7	*18,087,000	35.8	*18,087,000	37.3
Japan	12,759,000	29.1	13,697,000	24.5	13,141,000	23.6	10,950,000	20.8	11,302,000	21.1	10,533,000	20.9	10,098,000	20.8
Germany	-		11,577,000	20.7	11,651,000	21.0	10,568,000	20.1	10,806,000	20.1	8,447,059	16.7	7,159,994	14.8
Spain	1,038,000	2.4	633,000	1.1	741,000	1.3	810,000	1.5	795,000	1.5	613,000	1.2	569,000	1.2
Slovenia	*484,000	1.1	*484,000	0.9	*484,000	0.9	561,000	1.1	664,000	1.2	590,000	1.2	121,000	0.2

Product Tables

Spinning Machines

Unit of Measure: Number of spindles.

	1990		1991		1992		1993		1994		1995		1996	
	Value	%	Value	%	Value	%	Value	%	Value	%	Value	%	Value	%
Total Production	254,816	100.0	482,717	100.0	513,464	100.0	270,027	100.0	224,413	100.0	211,057	100.0	189,752	100.0
Regions														
Asia	17,740	7.0	388,382	80.5	326,693	63.6	93,905	34.8	54,565	24.3	38,304	18.1	27,115	14.3
Europe	237,076	93.0	94,335	19.5	186,771	36.4	176,122	65.2	169,848	75.7	172,754	81.9	162,637	85.7
Leading Producers														
Russian Federation	-		368,000	76.2	311,000	60.6	81,000	30.0	42,000	18.7	25,000	11.8	15,000	7.9
Poland	37,740	14.8	7,412	1.5	5,204	1.0	724	0.3	640	0.3	*9,471	4.5	*5,896	3.1
France	7,816	3.1	8,820	1.8	-		-		-		-		-	
Czech Republic	*18,264	7.2	*18,264	3.8	*18,264	3.6	18,264	6.8	*18,264	8.1	*18,264	8.7	*18,264	9.6
Japan	10,963	4.3	13,248	2.7	7,960	1.6	4,593	1.7	2,834	1.3	3,415	1.6	2,040	1.1

Looms

Unit of Measure: Number.

	1990		1991		1992		1993		1994		1995		1996	
	Value	%	Value	%	Value	%	Value	%	Value	%	Value	%	Value	%
Total Production	91,819	100.0	84,162	100.0	73,218	100.0	56,575	100.0	55,688	100.0	49,719	100.0	49,478	100.0
Regions														
Asia	77,619	84.5	74,728	88.8	62,641	85.6	45,743	80.9	44,885	80.6	39,579	79.6	40,306	81.5
Europe	14,200	15.5	9,433	11.2	10,577	14.4	10,831	19.1	10,803	19.4	10,140	20.4	9,172	18.5
Leading Producers														
Japan	23,458	25.5	25,856	30.7	26,799	36.6	19,558	34.6	20,167	36.2	17,245	34.7	14,451	29.2
Korea, Republic of	17,778	19.4	13,750	16.3	6,782	9.3	4,462	7.9	6,023	10.8	3,870	7.8	*8,838	17.9
Russian Federation	18,341	20.0	17,608	20.9	11,887	16.2	5,377	9.5	1,278	2.3	1,890	3.8	685	1.4
France	3,221	3.5	875	1.0	*2,877	3.9	*3,004	5.3	*3,132	5.6	*3,259	6.6	*3,387	6.8
Germany	*1,656	1.8	*1,656	2.0	*1,656	2.3	1,426	2.5	1,886	3.4	*1,656	3.3	*1,656	3.3

Printing Presses

Unit of Measure: Number.

	1990		1991		1992		1993		1994		1995		1996	
	Value	%	Value	%	Value	%	Value	%	Value	%	Value	%	Value	%
Total Production	82,305	100.0	95,096	100.0	83,956	100.0	82,136	100.0	82,717	100.0	82,544	100.0	81,661	100.0
Regions														
America, North	2,400	2.9	2,400	2.5	2,400	2.9	2,400	2.9	2,400	2.9	2,400	2.9	2,400	2.9
Asia	11,906	14.5	24,905	26.2	13,811	16.5	14,050	17.1	9,537	11.5	8,177	9.9	12,972	15.9
Europe	67,998	82.6	67,791	71.3	67,744	80.7	65,686	80.0	70,779	85.6	71,968	87.2	66,288	81.2
Leading Producers														
United Kingdom	*49,028	59.6	*49,028	51.6	*49,028	58.4	46,596	56.7	51,461	62.2	*49,028	59.4	*49,028	60.0
Germany	*18,369	22.3	*18,369	19.3	*18,369	21.9	*18,369	22.4	*18,369	22.2	21,161	25.6	15,577	19.1
Japan	7,751	9.4	6,686	7.0	5,196	6.2	4,295	5.2	3,823	4.6	4,145	5.0	4,683	5.7
Hong Kong	1,816	2.2	15,961	16.8	5,876	7.0	7,343	8.9	3,696	4.5	2,111	2.6	*6,202	7.6
United States of America	*2,400	2.9	*2,400	2.5	*2,400	2.9	*2,400	2.9	*2,400	2.9	*2,400	2.9	*2,400	2.9

Product Tables

Bulldozers

Unit of Measure: Number.

	1990 Value	%	1991 Value	%	1992 Value	%	1993 Value	%	1994 Value	%	1995 Value	%	1996 Value	%
Total Production	109,363	100.0	96,995	100.0	78,655	100.0	72,627	100.0	65,258	100.0	64,681	100.0	64,498	100.0
Regions														
Africa	22	0.0	22	0.0	22	0.0	22	0.0	22	0.0	22	0.0	22	0.0
America, North	8,781	8.0	7,193	7.4	5,846	7.4	7,299	10.1	7,032	10.8	6,764	10.5	6,497	10.1
America, South	1,569	1.4	1,569	1.6	1,569	2.0	1,569	2.2	1,569	2.4	1,569	2.4	1,569	2.4
Asia	91,187	83.4	82,665	85.2	69,619	88.5	62,669	86.3	56,501	86.6	56,210	86.9	56,308	87.3
Europe	7,701	7.0	5,469	5.6	1,548	2.0	1,042	1.4	134	0.2	115	0.2	102	0.2
Oceania	103	0.1	77	0.1	51	0.1	25	0.0	-		-		-	
Leading Producers														
Japan	17,639	16.1	12,666	13.1	9,743	12.4	8,940	12.3	10,954	16.8	10,240	15.8	10,643	16.5
United States of America	8,781	8.0	7,193	7.4	5,846	7.4	*7,299	10.1	*7,032	10.8	*6,764	10.5	*6,497	10.1
Russian Federation	14,131	12.9	12,431	12.8	12,226	15.5	6,498	8.9	2,161	3.3	2,404	3.7	2,669	4.1
Kazakhstan	13,328	12.2	10,288	10.6	3,494	4.4	4,234	5.8	695	1.1	521	0.8	247	0.4
Ukraine	7,507	6.9	5,375	5.5	1,444	1.8	932	1.3	32	0.0	12	0.0	4	0.0

Excavating Machines

Unit of Measure: Number.

	1990 Value	%	1991 Value	%	1992 Value	%	1993 Value	%	1994 Value	%	1995 Value	%	1996 Value	%
Total Production	282,408	100.0	273,625	100.0	228,102	100.0	216,416	100.0	225,928	100.0	232,451	100.0	222,750	100.0
Regions														
America, North	4,503	1.6	3,122	1.1	3,496	1.5	4,002	1.8	9,005	4.0	9,889	4.3	10,081	4.5
America, South	920	0.3	668	0.2	574	0.3	761	0.4	873	0.4	671	0.3	438	0.2
Asia	222,724	78.9	218,730	79.9	176,776	77.5	164,313	75.9	171,975	76.1	175,000	75.3	171,715	77.1
Europe	54,261	19.2	51,106	18.7	47,257	20.7	47,341	21.9	44,076	19.5	46,892	20.2	40,516	18.2
Leading Producers														
Japan	147,370	52.2	143,808	52.6	111,704	49.0	102,067	47.2	113,103	50.1	113,596	48.9	114,847	51.6
Germany	*20,239	7.2	*20,239	7.4	*20,239	8.9	*20,239	9.4	*20,239	9.0	23,065	9.9	17,414	7.8
United Kingdom	14,034	5.0	14,713	5.4	15,392	6.7	16,071	7.4	*17,086	7.6	*17,855	7.7	*18,625	8.4
Russian Federation	23,121	8.2	21,112	7.7	15,378	6.7	12,642	5.8	6,510	2.9	5,234	2.3	3,504	1.6
Korea, Republic of	10,667	3.8	12,441	4.5	9,362	4.1	9,961	4.6	13,532	6.0	17,456	7.5	*15,350	6.9

Graders and Levelers

Unit of Measure: Number.

	1990 Value	%	1991 Value	%	1992 Value	%	1993 Value	%	1994 Value	%	1995 Value	%	1996 Value	%
Total Production	100,869	100.0	103,432	100.0	86,731	100.0	90,318	100.0	95,355	100.0	91,078	100.0	99,026	100.0
Regions														
America, North	4,136	4.1	3,289	3.2	2,894	3.3	3,185	3.5	3,475	3.6	3,766	4.1	3,762	3.8
America, South	1,481	1.5	1,202	1.2	1,271	1.5	1,099	1.2	1,869	2.0	1,282	1.4	1,254	1.3
Asia	91,864	91.1	95,648	92.5	79,408	91.6	83,098	92.0	86,922	91.2	83,174	91.3	91,213	92.1
Europe	1,693	1.7	1,598	1.5	1,462	1.7	1,241	1.4	1,394	1.5	1,161	1.3	1,103	1.1
Oceania	1,695	1.7	1,695	1.6	1,695	2.0	1,695	1.9	1,695	1.8	1,695	1.9	1,695	1.7
Leading Producers														
Japan	83,479	82.8	88,510	85.6	73,378	84.6	77,540	85.9	83,567	87.6	80,838	88.8	89,040	89.9
United States of America	4,136	4.1	3,289	3.2	2,894	3.3	3,185	3.5	3,475	3.6	3,766	4.1	3,762	3.8
Russian Federation	4,841	4.8	4,135	4.0	3,420	3.9	3,415	3.8	1,468	1.5	1,187	1.3	1,386	1.4
Australia	*1,695	1.7	*1,695	1.6	*1,695	2.0	*1,695	1.9	*1,695	1.8	*1,695	1.9	*1,695	1.7
Brazil	1,481	1.5	1,202	1.2	1,271	1.5	1,099	1.2	1,869	2.0	1,282	1.4	1,254	1.3

Product Tables
Scrapers

Unit of Measure: Number.

	1990		1991		1992		1993		1994		1995		1996	
	Value	%	Value	%	Value	%	Value	%	Value	%	Value	%	Value	%
Total Production	13,987	100.0	12,324	100.0	6,661	100.0	5,908	100.0	5,717	100.0	5,358	100.0	4,497	100.0
Regions														
America, North	2,024	14.5	1,416	11.5	1,279	19.2	1,598	27.0	1,916	33.5	2,324	43.4	2,255	50.1
America, South	175	1.3	83	0.7	84	1.3	41	0.7	129	2.3	95	1.8	41	0.9
Asia	8,827	63.1	8,087	65.6	4,829	72.5	3,766	63.7	3,569	62.4	2,862	53.4	2,128	47.3
Europe	2,961	21.2	2,738	22.2	468	7.0	504	8.5	102	1.8	77	1.4	73	1.6
Leading Producers														
United States of America	1,885	13.5	1,260	10.2	1,106	16.6	1,407	23.8	1,709	29.9	2,100	39.2	2,013	44.8
Ukraine	2,824	20.2	2,645	21.5	382	5.7	423	7.2	20	0.3	-		-	
Russian Federation	1,360	9.7	1,348	10.9	192	2.9	19	0.3	*730	12.8	*730	13.6	*730	16.2
Belarus	1,847	13.2	1,831	14.9	442	6.6	264	4.5	69	1.2	74	1.4	52	1.2
Romania	40	0.3	-		-		-		-		-		-	

Concrete Mixers for Use at Construction Sites

Unit of Measure: Number.

	1990		1991		1992		1993		1994		1995		1996	
	Value	%	Value	%	Value	%	Value	%	Value	%	Value	%	Value	%
Total Production	407,807	100.0	416,815	100.0	406,891	100.0	407,871	100.0	409,658	100.0	408,563	100.0	395,667	100.0
Regions														
Africa	3,689	0.9	6,094	1.5	5,268	1.3	2,156	0.5	738	0.2	3,020	0.7	2,894	0.7
America, North	40,826	10.0	43,121	10.3	45,415	11.2	47,710	11.7	50,005	12.2	52,300	12.8	54,594	13.8
America, South	11,398	2.8	11,398	2.7	11,398	2.8	11,398	2.8	11,398	2.8	11,398	2.8	11,398	2.9
Asia	32,721	8.0	48,067	11.5	40,999	10.1	23,363	5.7	19,956	4.9	18,104	4.4	23,751	6.0
Europe	319,174	78.3	308,135	73.9	303,810	74.7	323,244	79.3	327,561	80.0	323,741	79.2	303,029	76.6
Leading Producers														
Germany	*103,263	25.3	*103,263	24.8	*103,263	25.4	106,941	26.2	116,080	28.3	104,860	25.7	85,172	21.5
United States of America	*40,826	10.0	*43,121	10.3	*45,415	11.2	*47,710	11.7	*50,005	12.2	*52,300	12.8	*54,594	13.8
Czech Republic	*20,755	5.1	*20,755	5.0	*20,755	5.1	20,755	5.1	*20,755	5.1	*20,755	5.1	*20,755	5.2
United Kingdom	*13,663	3.4	*13,663	3.3	*13,663	3.4	*13,663	3.3	*13,663	3.3	*13,663	3.3	*13,663	3.5
Spain	13,199	3.2	11,549	2.8	12,512	3.1	26,243	6.4	22,217	5.4	41,191	10.1	40,694	10.3

Commodity data are provided by the United Nations Statistical Division. The symbol * means that data are estimated. For additional notes, see Appendix I.

587

ISIC 3825
OFFICE, COMPUTING MACHINERY

ISIC 3825—Office, Computing Machinery—includes typewriters, calculating machines, computers and peripherals, and scales. This is a subindustry of ISIC 3820—Machinery nec—and is featured so that the important category of computing equipment, reported separately by some countries, can be more visible.

The products of this industry are classified under SIC 357—Computer and Office Equipment and SIC 3596—Scales and Balances, excluding Laboratory in the U.S. SIC system.

Summary Statistics

		1990 Value	N	1991 Value	N	1992 Value	N	1993 Value	N	1994 Value	N	1995 Value	N
Establishments or enterprises	(number)	8,620	34	9,280	38	12,298	39	7,438	33	4,497	23	5,231	7
Number employed	(000)	938	36	1,043	39	967	39	894	35	563	24	427	11
Total Wages	($ mil.)	26,832	32	32,288	35	32,356	36	25,720	31	15,206	23	11,716	11
Wage/salary per employee	($)	15,573	32	16,128	35	15,376	36	10,066	31	11,926	23	10,532	11
Female workers	(000)	64	12	31	10	129	9	116	10	111	6	110	3
as % of total employment*	(%)	34	12	33	10	42	9	44	10	40	6	54	3
Output	($ bil.)	205	34	232	37	236	36	213	31	129	22	103	10
per employee	($ 000)	106	31	114	34	112	33	100	29	126	22	139	10
per establishment	($ 000)	16,306	31	18,058	34	18,315	34	20,506	28	30,777	21	19,235	6
Value added	($ bil.)	88	33	90	33	87	30	80	28	43	18	40	8
per employee	($ 000)	43	30	47	30	43	27	35	24	39	18	42	8
per establishment	($ 000)	6,317	30	6,692	30	7,081	28	5,819	24	8,266	17	1,133	4
Capital investment	($ mil.)	5,811	15	6,855	19	8,801	22	6,910	19	2,857	12	2,539	4
per employee	($ 000)	5	14	4	18	6	19	4	17	5	12	5	4
per establishment	($ 000)	1,167	14	962	18	935	21	759	17	975	11	172	3

Data presented above are drawn from the detailed tables that follow. Columns headed 'N' show the number of countries that provided valid data for inclusion. Values are not strictly comparable one year to the next or one row to the next because the number of countries that report varies from period to period and row to row. However, a general indication of magnitudes can be gleaned from reviewing this summary—especially in earlier years in which more reports are available. For detailed explanations, see Appendix I. *The average for those countries reporting both total and female employment.

Establishments and Number Engaged

Country	Establishments or Enterprises (number)						Employees per Establishment					
	1990	1991	1992	1993	1994	1995	1990	1991	1992	1993	1994	1995
Australia	363	579	572	-	-	-	50	29	30	-	-	-
Bosnia & Herzegovina	1.000	1.000	-	-	-	-	2,334	2,075	-	-	-	-
Canada	225	189	190	182	173	-	71	85	89	82	92	-
China (Hong Kong SAR)	233	198	177	158	171	-	101	94	80	85	75	-
Colombia	11	10	12	13	12	-	27	29	29	31	35	-
Costa Rica	12	11	11	11	-	-	109	58	72	84	-	-
Denmark	175	169	190	-	-	-	17	16	15	-	-	-
Egypt	1.000	3.000	1.000	2.000	-	-	100	167	100	300	-	-
Finland	21	24	27	25	25	-	238	121	130	136	152	-
Germany	-	205	197	190	203	-	-	535	415	307	256	-
Germany (Western Part)	187	187	182	-	-	-	445	441	408	-	-	-
Ghana	-	-	-	1.000	-	-	-	-	-	25	-	-
Greece	6.000	7.000	7.000	-	-	-	14	17	16	-	-	-
Guatemala	-	1.000	1.000	1.000	1.000	1.000	-	66	69	69	69	69
Hungary	-	-	*119	*140	-	-	-	-	-	-	-	-
India	257	260	262	284	-	-	106	116	133	69	-	-
Indonesia	5.000	5.000	6.000	6.000	4.000	4.000	50	41	154	217	235	238
Iran (Islamic Republic of)	-	6.000	9.000	10	-	-	-	39	49	40	-	-
Italy	*101	*111	-	-	-	-	245	211	-	-	-	-
Japan	4,858	4,947	4,692	4,545	-	-	60	59	62	61	-	-
Korea, Republic of	335	426	444	558	606	666	56	57	46	48	47	46
Kuwait	-	-	-	1.000	-	-	-	-	-	27	-	-
Kyrgyzstan	2.000	2.000	2.000	3.000	3.000	-	2,290	2,228	1,534	617	296	-
Macau	1.000	1.000	1.000	1.000	-	-	-	-	-	-	-	-
Malaysia	13	22	24	28	28	-	262	355	388	471	414	-
Malta	-	-	-	-	5.000	-	-	-	-	-	12	-
Mexico	22	16	14	13	13	13	188	243	237	234	209	200
Mozambique	-	-	1.000	1.000	1.000	-	-	-	41	32	28	-
New Zealand	*127	*157	*147	*172	-	-	4.244	3.427	2.857	2.756	-	-
Norway	16	14	8.000	-	-	-	127	68	128	-	-	-
Peru	40	42	44	-	-	-	8.800	13	8.614	-	-	-
Philippines	118	164	37	21	-	-	15	13	62	176	-	-
Portugal	*107	*102	*88	*125	*173	-	10	10	9.432	8.384	6.757	-
Russian Federation	-	-	-	*780	*2,925	*4,529	-	-	-	140	32	22
Singapore	54	53	54	51	51	-	799	800	828	849	900	-
Slovakia	-	6.000	6.000	4.000	6.000	-	-	543	350	433	228	-
South Africa	-	48	-	-	-	-	-	-	-	-	-	-
Spain	53	50	41	-	-	-	85	94	88	-	-	-
Sri Lanka	1.000	-	1.000	2.000	-	-	88	-	44	499	-	-
Sweden	61	70	70	57	54	-	133	113	119	127	97	-
Thailand	4.000	19	-	-	-	-	28	68	-	-	-	-
Trinidad and Tobago	1.000	1.000	1.000	1.000	1.000	-	8.000	7.000	7.000	7.000	5.000	-
Turkey	6.000	6.000	18	19	22	-	63	49	28	27	31	-
Ukraine	-	-	80	24	17	17	-	-	-	1,458	1,706	1,353
United Kingdom	1,190	1,155	1,354	-	-	-	46	55	47	-	-	-
United States of America	-	-	3,198	-	-	-	-	-	80	-	-	-
Venezuela	13	13	10	9.000	1.000	1.000	15	31	30	36	22	24
Zambia	-	-	-	-	2.000	-	-	-	-	-	-	-

Employment and Compensation of Employees

Country	Number Employed/Engaged (000)						Wage/Salary per Employee ($)					
	1990	1991	1992	1993	1994	1995	1990	1991	1992	1993	1994	1995
Australia	18	*17	*17	-	-	-	21,385	27,819	25,260	-	-	-
Azerbaijan	0.974	0.784	0.541	0.309	0.238	-	2,769	2,448	-	-	-	-
Bosnia & Herzegovina	2.334	2.075	-	-	-	-	-	-	-	-	-	-
Canada	16	16	17	15	16	15	28,818	30,822	28,957	31,574	29,200	30,494
China (Hong Kong SAR)	*23	*19	*14	*13	*13	-	9,030	10,349	13,328	13,603	14,911	-

Continued.

Depending on the table, * means *Enterprises, Engaged,* or *Factor Values*; ± means *Basis Unspecified*; § means *shown in millions*. For additional notes and sources, see Appendix I.

Employment and Compensation of Employees

- Continued -

Country	Number Employed/Engaged (000)						Wage/Salary per Employee ($)					
	1990	1991	1992	1993	1994	1995	1990	1991	1992	1993	1994	1995
Colombia	0.294	0.288	0.347	0.399	0.415	0.268	1,835	2,101	2,547	2,867	3,529	3,368
Costa Rica	1.311	0.637	0.787	0.922	-	-	3,204	3,281	2,823	3,053	-	-
Denmark	2.894	2.722	2.807	-	-	-	-	-	-	-	-	-
Egypt	*0.100	*0.500	*0.100	*0.600	-	-	16,500	4,562	-	2,620	-	-
Finland	5.000	2.900	3.500	3.400	3.800	-	35,026	30,970	28,665	21,661	24,696	-
Germany	-	*110	*82	*58	*52	-	-	33,878	-	-	-	-
Germany (Western Part)	*83	*82	*74	-	-	-	-	-	-	-	-	-
Ghana	-	-	-	*0.025	-	-	-	-	-	-	-	-
Greece	*0.086	*0.116	*0.114	-	-	-	10,857	12,439	12,563	-	-	-
Guatemala	-	*0.066	*0.069	*0.069	*0.069	*0.069	-	-	-	-	-	-
India	*27	*30	*35	*19	-	-	2,137	1,717	2,007	1,429	-	-
Indonesia	0.250	0.207	0.924	1.303	0.942	0.952	1,048	1,092	1,563	1,033	1,036	1,650
Iran (Islamic Republic of)	-	0.233	0.443	0.397	-	-	-	4,395	3,601	2,836	-	-
Italy	25	23	-	-	-	-	-	-	-	-	-	-
Japan	*290	*293	*292	*278	-	-	26,912	30,606	33,557	39,627	-	-
Korea, Republic of	19	24	20	27	29	31	8,348	9,310	10,797	12,013	13,703	15,941
Kuwait	-	-	-	0.027	-	0.009	-	-	-	10,302	-	11,171
Kyrgyzstan	*4.580	*4.455	*3.068	*1.850	*0.888	-	893	683	-	-	-	-
Lithuania	-	-	6.372	4.403	-	-	-	-	-	-	-	-
Malaysia	3.400	7.800	9.300	13	12	-	2,142	2,867	3,183	2,952	3,715	-
Malta	-	-	-	-	0.059	-	-	-	-	-	12,600	-
Mexico	4.135	3.885	3.312	3.045	2.713	2.606	7,460	8,915	11,432	13,061	14,092	7,968
Mozambique	-	-	0.041	0.032	0.028	-	-	-	533	653	716	-
New Zealand	0.539	0.538	0.420	0.474	-	-	-	-	-	-	-	-
Norway	2.030	0.953	1.025	-	-	-	-	39,170	-	-	-	-
Peru	*0.352	*0.539	*0.379	-	-	-	2,880	2,994	2,350	-	-	-
Philippines	1.800	2.100	2.300	3.700	-	-	1,531	1,716	2,045	1,549	-	-
Portugal	*1.104	*1.048	*0.830	*1.048	*1.169	-	-	-	-	-	-	-
Russian Federation	-	-	-	109	94	98	-	-	-	-	646	818
Singapore	*43	*42	*45	*43	*46	-	9,148	10,246	11,742	12,968	13,880	-
Slovakia	-	*3.260	*2.099	*1.734	*1.368	-	-	1,321	1,635	1,743	2,509	-
Spain	4.491	4.699	3.604	-	-	-	26,354	27,855	32,820	-	-	-
Sri Lanka	*0.088	-	*0.044	*0.997	-	-	1,135	-	461	727	-	-
Sweden	8.100	*7.903	*8.320	7.247	5.247	-	27,741	32,181	34,953	28,623	29,169	-
Thailand	0.112	1.283	-	-	-	-	1,047	3,186	-	-	-	-
Trinidad and Tobago	0.008	0.007	0.007	0.007	0.005	-	2,671	10,185	10,185	8,009	13,502	-
Turkey	0.380	0.297	0.499	0.511	0.681	-	7,062	9,685	5,540	6,235	2,777	-
Ukraine	-	-	-	35	29	23	-	-	-	-	-	-
United Kingdom	55	64	63	-	-	-	29,383	33,186	35,252	-	-	-
United States of America	294	272	256	249	256	256	35,442	37,647	-	-	-	-
Uruguay	*0.325	*0.348	*0.319	*0.266	-	-	-	-	-	-	-	-
Venezuela	0.200	0.400	0.300	0.322	0.022	0.024	5,117	4,180	5,655	5,231	2,755	2,356

Female Workers: Total and Percent of Employment

Country	Female Workers (000)						Female Workers as % of Total Employment					
	1990	1991	1992	1993	1994	1995	1990	1991	1992	1993	1994	1995
Australia	6.244	-	-	-	-	-	34.69	-	-	-	-	-
Canada	5.000	5.000	-	-	-	-	31.25	31.25	-	-	-	-
Colombia	-	0.037	-	0.073	0.071	-	-	12.85	-	18.30	17.11	-
Denmark	0.833	0.772	0.767	-	-	-	28.78	28.36	27.32	-	-	-
Egypt	-	0.126	-	0.136	-	-	-	25.20	-	22.67	-	-
Germany (Western Part)	25	-	-	-	-	-	30.05	-	-	-	-	-
Indonesia	-	-	-	0.771	0.679	0.762	-	-	-	59.17	72.08	80.04
Iran (Islamic Republic of)	-	0.009	0.017	0.017	-	-	-	3.86	3.84	4.28	-	-
Italy	-	5.814	-	-	-	-	-	24.83	-	-	-	-
Korea, Republic of	8.300	13	9.144	12	12	13	44.62	52.40	45.08	44.21	41.08	43.11
Malaysia	2.500	5.900	6.800	9.600	7.400	-	73.53	75.64	73.12	72.73	63.79	-

Continued.

Depending on the table, * means Enterprises, Engaged, or Factor Values; ± means Basis Unspecified; § means shown in millions. For additional notes and sources, see Appendix I.

591

Female Workers: Total and Percent of Employment
- Continued -

Country	Female Workers (000)						Female Workers as % of Total Employment					
	1990	1991	1992	1993	1994	1995	1990	1991	1992	1993	1994	1995
Malta	-	-	-	-	0.007	-	-	-	-	-	11.86	-
New Zealand	0.129	0.141	0.093	0.118	-	-	23.93	26.21	22.14	24.89	-	-
Philippines	-	-	1.429	2.533	-	-	-	-	62.13	68.46	-	-
Sri Lanka	0.002	-	0.034	0.903	-	-	2.27	-	77.27	90.57	-	-
Sweden	2.600	-	-	-	-	-	32.10	-	-	-	-	-
Thailand	0.076	0.578	-	-	-	-	67.86	45.05	-	-	-	-
Turkey	0.058	-	-	-	-	-	15.26	-	-	-	-	-
United Kingdom	13	-	19	-	-	-	24.02	-	30.16	-	-	-
United States of America	-	-	92	90	91	96	-	-	35.94	36.14	35.55	37.50

Output and Output per Employee

Country	Output ($ bil.)						Output per Employee ($)					
	1990	1991	1992	1993	1994	1995	1990	1991	1992	1993	1994	1995
Australia	*1.987	*2.298	*2.214	-	-	-	110,417	135,154	130,234	-	-	-
Canada	3.197	3.308	3.367	3.186	4.489	3.853	199,799	206,751	198,074	212,387	280,554	256,553
China (Hong Kong SAR)	2.598	2.873	2.773	2.502	2.200	-	110,565	153,659	195,275	186,693	171,870	-
Colombia	0.004	0.004	0.006	0.006	0.008	0.005	13,544	15,396	16,867	16,141	18,510	19,648
Denmark	*0.237	*0.235	-	-	-	-	82,017	86,215	-	-	-	-
Egypt	*0.004	*0.012	*0.000	*0.012	-	-	43,000	24,368	899	20,414	-	-
Finland	±0.745	±0.523	±0.855	±0.867	±1.023	-	148,902	180,259	244,293	255,074	269,279	-
Germany	-	16	16	13	16	-	-	144,703	190,283	231,341	305,854	-
Germany (Western Part)	12	16	15	-	-	-	140,348	188,672	206,164	-	-	-
Greece	*0.004	*0.008	*0.008	-	-	-	50,763	72,694	68,843	-	-	-
Guatemala	-	0.000	0.000	0.000	0.000	0.000	-	2,492	2,363	2,405	4,826	6,375
India	*0.699	*0.857	*0.912	*0.512	-	-	25,542	28,319	26,123	26,285	-	-
Indonesia	0.003	0.002	0.006	0.044	0.010	0.041	13,299	11,583	6,088	33,746	10,854	42,661
Iran (Islamic Republic of)	-	0.004	0.009	0.008	-	-	-	16,775	20,675	19,810	-	-
Italy	±7.034	±5.854	-	-	-	-	284,452	249,984	-	-	-	-
Japan	±84	±96	±97	±103	-	-	288,837	329,085	331,541	371,358	-	-
Korea, Republic of	1.730	1.876	1.830	3.590	4.339	6.430	93,016	77,618	90,213	134,332	151,594	209,512
Kuwait	-	-	-	0.001	-	0.000	-	-	-	20,235	-	29,044
Kyrgyzstan	0.048	0.034	0.007	0.001	0.000	-	10,466	7,669	2,171	721	261	-
Lithuania	-	-	0.002	0.003	-	-	-	-	268	721	-	-
Macau	0.000	0.000	0.000	0.000	-	-	-	-	-	-	-	-
Malaysia	*0.118	*0.352	*0.522	*0.797	*0.914	-	34,763	45,183	56,085	60,374	78,796	-
Malta	-	-	-	-	0.002	-	-	-	-	-	41,207	-
Mexico	0.635	0.700	0.649	0.821	1.197	1.207	153,557	180,239	195,937	269,619	441,350	463,149
Norway	±0.378	±0.187	±0.184	-	-	-	186,194	196,497	179,370	-	-	-
Peru	0.012	0.036	0.007	-	-	-	33,118	66,313	17,227	-	-	-
Philippines	0.009	0.017	0.032	0.229	-	-	5,027	8,093	13,804	61,949	-	-
Portugal	0.044	0.065	0.043	0.053	0.078	-	40,146	62,048	51,932	51,045	66,568	-
Romania	±0.218	±0.081	±0.036	±0.063	-	-	-	-	-	-	-	-
Russian Federation	-	-	-	0.277	0.479	0.554	-	-	-	2,532	5,099	5,654
Singapore	*6.784	*7.293	*8.873	*13	*17	-	157,254	172,114	198,459	300,290	369,915	-
Slovakia	-	±0.016	±0.015	±0.013	±0.012	-	-	4,963	7,215	7,253	8,577	-
Spain	*1.049	*1.524	*1.316	-	-	-	233,586	324,247	365,174	-	-	-
Sri Lanka	0.001	-	0.000	0.015	-	-	7,091	-	1,761	15,245	-	-
Sweden	*1.478	*1.597	*1.722	*0.657	*0.611	-	182,505	202,117	206,980	90,692	116,407	-
Thailand	0.001	0.077	-	-	-	-	6,980	59,991	-	-	-	-
Trinidad and Tobago	±0.000	±0.000	±0.000	±0.000	±0.000	-	24,644	68,397	68,397	53,394	54,009	-
Turkey	0.028	0.024	0.042	0.060	0.036	-	73,643	80,709	83,690	117,403	52,273	-
Ukraine	*1.269	*0.630	*0.248	*0.062	*0.060	*0.037	-	-	-	1,772	2,058	1,625
United Kingdom	*15	*15	15	-	-	-	269,805	229,259	244,966	-	-	-
United States of America	*65	*59	*67	*70	*80	*91	220,238	218,566	263,234	280,598	313,586	355,047
Venezuela	0.006	0.008	0.007	0.015	0.000	-	27,825	19,185	24,814	47,494	4,285	-

    Depending on the table, * means *Enterprises, Engaged,* or *Factor Values*; ± means *Basis Unspecified*; § means *shown in millions*. For additional notes and sources, see Appendix I.

Value Added and Value Added per Employee

Country	Value Added ($ bil.)						Value Added per Employee ($)					
	1990	1991	1992	1993	1994	1995	1990	1991	1992	1993	1994	1995
Australia	*0.909	*1.060	-	-	-	-	50,477	62,330	-	-	-	-
Canada	1.106	0.951	0.860	1.000	0.989	1.233	69,099	59,461	50,613	66,662	61,786	82,063
China (Hong Kong SAR)	0.601	0.575	0.575	0.442	0.459	-	25,560	30,725	40,513	32,964	35,856	-
Colombia	0.002	0.002	0.003	0.003	0.004	0.003	6,772	8,079	8,847	7,592	9,523	9,919
Cote d'Ivoire	0.555	0.482	0.355	0.431	-	-	-	-	-	-	-	-
Denmark	*0.151	*0.152	-	-	-	-	52,091	55,888	-	-	-	-
Egypt	*0.002	*0.004	*0.000	*0.008	-	-	18,000	7,923	300	12,605	-	-
Finland	±0.390	±0.150	±0.255	±0.202	±0.181	-	78,086	51,588	72,841	59,315	47,709	-
Germany (Western Part)	9.433	12	12	8.366	-	-	113,384	149,870	157,569	-	-	-
Greece	*0.002	*0.005	*0.004	-	-	-	24,942	39,398	38,287	-	-	-
India	*0.162	*0.216	*0.259	*0.112	-	-	5,916	7,144	7,408	5,754	-	-
Indonesia	*0.001	*0.001	*0.002	*0.005	*0.004	*0.006	5,179	4,518	2,069	3,696	4,617	5,824
Iran (Islamic Republic of)	-	0.002	0.003	0.004	-	-	-	8,387	6,892	11,320	-	-
Italy	±1.496	±1.459	-	-	-	-	60,489	62,307	-	-	-	-
Japan	±31	±35	±34	±34	-	-	108,314	119,939	117,139	122,050	-	-
Korea, Republic of	0.719	0.744	0.681	1.172	1.493	1.965	38,665	30,784	33,551	43,849	52,158	64,031
Kuwait	-	-	-	0.001	-	0.000	-	-	-	19,254	-	23,458
Macau	0.000	0.000	0.000	0.000	-	-	-	-	-	-	-	-
Malaysia	*0.026	*0.087	*0.092	*0.145	*0.130	-	7,579	11,207	9,941	10,995	11,241	-
Malta	-	-	-	-	0.002	-	-	-	-	-	28,114	-
Mexico	0.245	0.245	0.214	0.240	0.330	-	59,208	62,978	64,675	78,929	121,703	-
Norway	±0.126	±0.059	±0.062	-	-	-	62,091	61,668	60,815	-	-	-
Peru	0.006	0.008	0.002	-	-	-	16,151	15,117	5,059	-	-	-
Philippines	0.005	0.009	0.012	0.042	-	-	2,994	4,142	5,130	11,434	-	-
Portugal	0.015	0.018	0.016	0.016	0.019	-	13,903	17,429	19,259	15,375	16,311	-
Romania	±0.125	±0.030	±0.007	±0.013	-	-	-	-	-	-	-	-
Russian Federation	-	-	-	±0.201	±0.310	±0.299	-	-	-	1,832	3,305	3,053
Singapore	*2.038	*2.136	*2.630	*3.593	*4.261	-	47,229	50,398	58,813	82,977	92,833	-
Slovakia	-	-	-	±0.005	±0.006	-	-	-	-	2,811	4,722	-
Spain	*0.391	*0.559	*0.314	-	-	-	87,125	118,985	87,168	-	-	-
Sri Lanka	0.000	-	-0.000	0.003	-	-	5,106	-	§-1,62	2,989	-	-
Sweden	*0.525	*0.479	*0.508	*0.338	*0.284	-	64,867	60,560	61,101	46,618	54,211	-
Thailand	0.000	0.061	-	-	-	-	3,141	47,773	-	-	-	-
Trinidad and Tobago	±0.000	±0.000	±0.000	±0.000	±0.000	-	6,247	22,632	22,632	18,688	16,878	-
Turkey	0.013	0.010	0.014	0.020	0.008	-	35,308	33,898	28,577	39,194	11,159	-
United Kingdom	*5.696	*5.287	*4.783	-	-	-	103,571	82,605	75,916	-	-	-
United States of America	*32	*28	*30	*29	*35	*36	107,551	101,985	115,602	118,213	134,867	141,625
Venezuela	0.004	0.005	0.006	0.011	0.000	0.000	19,936	12,144	19,110	33,953	3,061	5,419

Capital Investment

Country	Gross Fixed Capital Formation ($ mil.)						Per Establishment ($ mil)					
	1990	1991	1992	1993	1994	1995	1990	1991	1992	1993	1994	1995
China (Hong Kong SAR)	-	-	70	55	97	-	-	-	0.396	0.349	0.568	-
Colombia	-	0.085	0.249	0.102	0.703	-	-	0.009	0.021	0.008	0.059	-
Denmark	8.887	8.755	-	-	-	-	0.051	0.052	-	-	-	-
Finland	13	9.718	13	16	20	-	0.613	0.405	0.471	0.634	0.819	-
Germany (Western Part)	1,493	1,607	1,350	-	-	-	7.986	8.591	7.420	-	-	-
Greece	0.057	0.159	0.582	-	-	-	0.009	0.023	0.083	-	-	-
Hungary	-	-	0.924	30	-	-	-	-	0.008	0.212	-	-
India	37	36	51	32	-	-	0.144	0.139	0.195	0.114	-	-
Indonesia	-	-	-	2.875	-	0.250	-	-	-	0.479	-	0.062
Iran (Islamic Republic of)	-	0.321	1.172	1.382	-	-	-	0.053	0.130	0.138	-	-
Italy	201	191	-	-	-	-	1.992	1.721	-	-	-	-
Japan	3,778	4,328	4,232	3,813	-	-	0.778	0.875	0.902	0.839	-	-
Korea, Republic of	-	-	92	312	250	403	-	-	0.208	0.559	0.412	0.605
Kuwait	-	-	-	0.023	-	-	-	-	-	0.023	-	-
Malaysia	-	-	-	40	40	-	-	-	-	1.415	1.443	-

Continued.

Capital Investment

- Continued -

Country	Gross Fixed Capital Formation ($ mil.)						Per Establishment ($ mil)					
	1990	1991	1992	1993	1994	1995	1990	1991	1992	1993	1994	1995
Malta	-	-	-	-	0.156	-	-	-	-	-	0.031	-
Mexico	12	9.181	-	-	-	-	0.534	0.574	-	-	-	-
Netherlands	40	37	28	79	-	-	-	-	-	-	-	-
Norway	-	-	2.220	-	-	-	-	-	0.277	-	-	-
Peru	0.030	0.127	0.027	-	-	-	0.001	0.003	0.001	-	-	-
Philippines	0.617	1.892	11	13	-	-	0.005	0.012	0.309	0.599	-	-
Portugal	1.333	1.073	1.215	0.808	-5.109	-	0.012	0.011	0.014	0.006	-0.030	-
Singapore	188	197	354	316	372	-	3.475	3.719	6.556	6.189	7.297	-
Spain	38	36	18	-	-	-	0.722	0.728	0.445	-	-	-
Sri Lanka	-	-	1.115	2.630	-	-	-	-	1.115	1.315	-	-
Thailand	0.051	0.635	-	-	-	-	0.013	0.033	-	-	-	-
Trinidad and Tobago	-	-	-	-	0.084	-	-	-	-	-	0.084	-
Turkey	-	0.240	1.746	1.183	0.338	-	-	0.040	0.097	0.062	0.015	-
Ukraine	-	-	4.654	-0.972	0.488	-2.591	-	-	0.058	-0.040	0.029	-0.152
United Kingdom	-	390	272	-	-	-	-	0.338	0.201	-	-	-
United States of America	-	-	2,294	2,197	2,080	2,139	-	-	0.717	-	-	-

Indexes of Industrial Production

Country	Index of Industrial Production (1990=100)						Index of Employment (1990=100)					
	1990	1991	1992	1993	1994	1995	1990	1991	1992	1993	1994	1995
Australia	-	-	-	-	-	-	100	94	94	-	-	-
Azerbaijan	-	-	-	-	-	-	100	80	56	32	24	-
Bosnia & Herzegovina	-	-	-	-	-	-	100	89	-	-	-	-
Canada	-	-	-	-	-	-	100	100	106	94	100	94
China (Hong Kong SAR)	-	-	-	-	-	-	100	80	60	57	54	-
Colombia	-	-	-	-	-	-	100	98	118	136	141	91
Costa Rica	-	-	-	-	-	-	100	49	60	70	-	-
Denmark	-	-	-	-	-	-	100	94	97	-	-	-
Egypt	-	-	-	-	-	-	100	500	100	600	-	-
Finland	-	-	-	-	-	-	100	58	70	68	76	-
Germany (Western Part)	-	-	-	-	-	-	100	99	89	-	-	-
Greece	-	-	-	-	-	-	100	135	133	-	-	-
India	-	-	-	-	-	-	100	111	128	71	-	-
Indonesia	-	-	-	-	-	-	100	83	370	521	377	381
Italy	-	-	-	-	-	-	100	95	-	-	-	-
Japan	-	-	-	-	-	-	100	101	101	96	-	-
Korea, Republic of	-	-	-	-	-	-	100	130	109	144	154	165
Kyrgyzstan	-	-	-	-	-	-	100	97	67	40	19	-
Malaysia	-	-	-	-	-	-	100	229	274	388	341	-
Mexico	-	-	-	-	-	-	100	94	80	74	66	63
New Zealand	-	-	-	-	-	-	100	100	78	88	-	-
Norway	-	-	-	-	-	-	100	47	50	-	-	-
Peru	-	-	-	-	-	-	100	153	108	-	-	-
Philippines	-	-	-	-	-	-	100	117	128	206	-	-
Portugal	-	-	-	-	-	-	100	95	75	95	106	-
Singapore	-	-	-	-	-	-	100	98	104	100	106	-
Spain	-	-	-	-	-	-	100	105	80	-	-	-
Sri Lanka	-	-	-	-	-	-	100	-	50	1,133	-	-
Sweden	-	-	-	-	-	-	100	98	103	89	65	-
Thailand	-	-	-	-	-	-	100	1,146	-	-	-	-
Trinidad and Tobago	-	-	-	-	-	-	100	88	88	88	63	-
Turkey	-	-	-	-	-	-	100	78	131	134	179	-
United Kingdom	-	-	-	-	-	-	100	116	115	-	-	-
United States of America	-	-	-	-	-	-	100	93	87	85	87	87
Uruguay	-	-	-	-	-	-	100	107	98	82	-	-
Venezuela	-	-	-	-	-	-	100	200	150	161	11	12

Depending on the table, * means *Enterprises*, *Engaged*, or *Factor Values*; ± means *Basis Unspecified*; § means *shown in millions*. For additional notes and sources, see Appendix I.

Representative Companies in Sector

Name	Address	Tele-phone	Fax	Employ-ment	Y
Moore Corp. Ltd.	PO Box 78 Sta. 1st Can Place, Toronto, Ontario M5X 1G5, Canada	416-364-2600	416-364-1667	20,000	98
Amdahl Corp.	PO Box 3470, Sunnyvale, California 94088-3470, U.S.A.	408-746-6000	408-738-1051	9,900	96
Gateway 2000 Inc.	610 Gateway Dr., North Sioux City, South Dakota 57049-2000, U.S.A.	605-232-2000	605-232-2023	9,700	96
Imation Corp.	1 Imation Place, Oakdale, Minnesota 55128, U.S.A.	612-704-4000		9,400	97
Alcatel Standard Electrica, SA	Ramirez de Prado, 5, Madrid E-28045, Spain	91 5272121	91 5286122	9,348	97
Applied Magnetics Corp.	75 Robin Hill Rd., Goleta, California 93117, U.S.A.	805-683-5353	805-967-8227	8,500	97
Intergraph Corp.	Huntsville, Alabama 35894-0001, U.S.A.	205-730-2000	205-730-9441	8,400	96
Tektronix Inc.	PO Box 1000, Wilsonville, Oregon 97070-1000, U.S.A.	503-627-7111		8,392	97
Storage Technology Corp.	2270 South 88th St., Louisville, Colorado 80028-4309, U.S.A.	303-673-5151	303-673-8876	8,300	97
Tandem Computers Inc.	19333 Vallco Pkwy., Cupertino, California 95014-2599, U.S.A.	408-285-6000	408-285-4545	7,938	96
Hutchinson Technology Inc.	40 W. Highland Park, Hutchinson, Minnesota 55350-9784, U.S.A.	612-587-3797		7,181	97
Diebold Inc.	PO Box 3077, North Canton, Ohio 44720-8077, U.S.A.	330-489-4000		6,714	97
EMC Corp.	171 South St., Hopkinton, Massachusetts 01748-9103, U.S.A.	508-435-1000	508-497-6915	6,700	97
Unova Inc.	360 N. Crescent Dr., Beverly Hills, California 90210, U.S.A.	310-888-2500	310-888-2848	6,650	97
AST Research Inc.	PO Box 57005, Irvine, California 92619-7005, U.S.A.	714-727-4141	714-727-8845	6,595	95
JTS Corp.	166 Baypointe Pkwy., San Jose, California 95134, U.S.A.	408-468-1800		6,500	97
Standard Register Co.	600 Albany St., Dayton, Ohio 45401, U.S.A.	937-443-1000		6,400	97
Bay Networks Inc.	4401 Great America Pkwy., Santa Clara, California 95052-8185, U.S.A.	408-988-2400	408-988-5525	5,960	97
John H. Harland Co.	PO Box 105250, Atlanta, Georgia 30348, U.S.A.	404-981-9460		5,599	96
Data General Corp.	4400 Computer Dr., Westborough, Massachusetts 01580, U.S.A.	508-898-5000		5,100	97
Acer Inc.	156 Minsheng E. Rd., Sec.3, Taipei, Taiwan	2 5455288	2 5455308	5,000	91
GTECH Holdings Corp.	55 Technology Way, West Greenwich, Rhode Island 02817, U.S.A.	401-392-1000	401-392-1234	4,700	96
Micron Electronics Inc.	900 E. Karcher Rd., Nampa, Idaho 83687-3045, U.S.A.	208-898-3434	208-465-3424	4,620	97
Maxtor Corp.	510 Cottonwood Dr., Milpitas, California 95035, U.S.A.	408-432-1700	408-432-4510	4,500	97
Cherry Corp.	3600 Sunset Ave., Waukegan, Illinois 60087, U.S.A.	708-662-9200		4,367	96
Komag Inc.	1704 Automation Pkwy., San Jose, California 95131, U.S.A.	408-576-2000		4,101	96
Fujitsu America Inc.	3055 Orchard Dr., San Jose, California 95134, U.S.A.	408-432-1300	408-456-7064	4,000	97
Simplex Time Recorder Co.	1 Simplex Plz., Gardner, Massachusetts 01441, U.S.A.	978-632-2500		4,000	97
LSI Logic Corp.	1551 McCarthy Blvd., Milpitas, California 95035, U.S.A.	408-433-8000	408-434-6457	3,870	96
Capetronic USA (HK) Inc. Computer Products Div.	42001 Christy St., Fremont, California 94538, U.S.A.		510-623-1222	3,800	
BancTec Technologies Inc.	4851 Lyndon B. Johnson Pkwy., Dallas, Texas 75244, U.S.A.	972-341-4000	972-450-7986	3,700	96
IOMEGA Corp.	1821 W. Iomega Way, Roy, Utah 84067, U.S.A.	801-778-1000	801-778-3190	3,700	97
Schlumberger PLC	1 Kingsway, 1st Fl., London WC2B 6XH, United Kingdom	71 2407301		3,594	93
VeriFone Inc.	4988 Great America Pkwy., Santa Clara, California 95054, U.S.A.	408-496-0444	408-919-5509	3,550	97
Siemens Canada Ltd.	2185 Derry Rd. W, Mississauga, Ontario L5N 7A6, Canada	905-819-8000	905-819-5777	3,500	98
Telefongyar	PO Box 16, Budapest, Hungary	1 2526949	1 2529161	3,500	93
ACCO Europe PLC	Gate House Rd., Aylesbury HP19 3DY, United Kingdom	296 397444	296 895399	3,484	91
Scitex Corporation Ltd.	PO Box 330, Herzliyya 46733, Israel	9 9597222	9 9502922	3,400	97
Symbol Technologies Inc.	One Symbol Plaza, Holtsville, New York 11742, U.S.A.	516-738-2400		3,200	97
Intermec Technologies Corp.	6001 36th Ave., W., Everett, Washington 98203, U.S.A.	425-348-2600	425-355-9551	3,100	97
DataCard Corp.	PO Box 9355, Minneapolis, Minnesota 55343, U.S.A.	612-933-1223	612-933-7971	3,000	97
Tokheim Corp.	PO Box 360, Fort Wayne, Indiana 46801-0360, U.S.A.	219-470-4600	219-484-1110	2,903	97
Sequent Computer Systems Inc.	15450 S.W. Koll Pkwy., Beaverton, Oregon 97006-6063, U.S.A.	503-626-5700		2,818	97
Adaptec Inc.	691 S. Milpitas Blvd., Milpitas, California 95035, U.S.A.	408-945-8600	408-262-2533	2,794	96
National Computer Systems Inc.	11000 Prairie Lakes Dr., Minneapolis, Minnesota 55344, U.S.A.	612-829-3000	612-829-3186	2,700	96
Tesla, Stropkov AS	Hviedoslavova 37/46, Stropkov 091 01, Slovakia	938 3422	938 3466	2,600	93
NEC do Brasil SA	Villa Mariana, Sao Paulo 04109-900, Brazil	11 2389600	11 2515787	2,496	96
Key Tronic Corp.	PO Box 14687, Spokane, Washington 99214-0687, U.S.A.	509-928-8000	509-927-5248	2,434	97
AT&T Paradyne	8545 126th Ave., Largo, Florida 34649, U.S.A.	813-530-2000		2,400	
Monroe Systems for Business Inc.	1000 The American Rd., Morris Plains, New Jersey 07950-	201-993-2000	201-993-2482	2,400	95

Continued

Representative Companies in Sector
- Continued -

Name	Address	Tele-phone	Fax	Employ-ment	Y
	2497, U.S.A.				
Nashua Corp.	PO Box 2002, Nashua, New Hampshire 03061-2002, U.S.A.	603-880-2323	603-880-5671	2,398	96
Elta Electronics Industries Ltd.	PO Box 303, Ashdod 77268, Israel	8 8572410	8 8564568	2,382	97
Kunnan Enterprise, Ltd.	Tan Tzu Hsiang, Taichung, Taiwan	45 320183	45 333370	2,360	92
Freios Varga SA	Via Anhanguera s/n Km.147, Limera 13486-915, Brazil	194 401360	194 401444	2,356	98
Smith Corona Corp.	65 Locust Ave., New Canaan, Connecticut 06840, U.S.A.	203-972-1471		2,300	95
BancTec Inc.	4435 Spring Valley Rd., Dallas, Texas 75244, U.S.A.	214-450-7700	214-450-7867	2,274	94
Cray Research Inc.	655A Lone Oak Dr., Eagan, Minnesota 55121, U.S.A.	612-452-6650	612-683-3099	2,200	96
Welch Allyn Inc.	4341 State Street Rd., Skaneateles Falls, New York 13153, U.S.A.	315-685-4100	315-685-3361	2,100	97
StreamLogic Corp.	8450 Central Ave., Newark, California 94560, U.S.A.	510-608-4000	510-608-4010	2,069	95
Pitney Bowes Holdings Ltd.	Elizabeth Way, Harlow CM19 5BD, United Kingdom	279 416771	279 441723	2,054	93
Madge Networks Inc.	2310 N. 1st St., San Jose, California 95131-1011, U.S.A.	408-955-0700	408-955-0970	2,000	97
PAXAR Corp.	105 Corporate Park Dr., White Plains, New York 10604, U.S.A.	914-697-6800	914-697-6893	1,923	95
Bell Industries Inc.	11812 San Vicente Blvd., Los Angeles, California 90049-5069, U.S.A.	310-826-2355	310-447-3265	1,900	97
Gerber Scientific Inc.	83 Gerber Rd., W, South Windsor, Connecticut 06074, U.S.A.	860-644-1551	860-643-7039	1,900	97
Osicom Technologies Inc.	2800 28th St., Ste. 100, Santa Monica, California 90405, U.S.A.	310-581-4030		1,849	96
Wyse Technology Inc.	3471 North 1st St., San Jose, California 95134, U.S.A.	408-473-1200	408-922-5729	1,802	97
Chung Hsin Electric & Machinery Manufacturing Co. Ltd.	Kwei Shan Shiang, Taoyuan, Taiwan	3 3284170	3 3284155	1,800	93
Trigem Computer Inc.	Ansan-Shi, Kyonggi, Republic of Korea	345 4919528	345 4920467	1,800	97
Control Data Systems Inc.	4201 Lexington Ave., N, St. Paul, Minnesota 55126-6198, U.S.A.	612-415-2999	612-415-2000	1,700	96
Kyocera International Inc.	8611 Balboa Ave., San Diego, California 92123, U.S.A.	619-576-2600	619-569-9412	1,700	96
National Semiconductor UK Ltd.	Larkfield Industrial Estate, Greenock PA16 0EQ, United Kingdom	475 33733	475 38515	1,691	94
Exabyte Corp.	1685 38th St., Boulder, Colorado 80301, U.S.A.	303-442-4333	303-417-7170	1,648	96
GENICOM Corp.	14800 Conference Center Dr., 40, Chantilly, Virginia 22021-3806, U.S.A.	703-802-9200		1,638	97
AT&T Global Information Solutions Scotland Ltd.	Kingsway West, Dundee DD1 9QA, United Kingdom	382 611511		1,615	93
Hewlett-Packard (Canada) Ltd.	5150 Spectrum Way, Mississauga, Ontario L4W 5G2, Canada	905-206-4725	905-206-4739	1,600	98
Telxon Corp.	PO Box 5582, Akron, Ohio 44334-0582, U.S.A.	330-664-1000	330-664-2099	1,600	96
Toco Holdings Ltd.	PO Box 79496, Senderwood 2145, Republic of South Africa	11 8231957	11 8232423	1,600	91
Network Systems Corp.	7600 Boone Ave. N., Minneapolis, Minnesota 55428, U.S.A.	612-424-4888	612-424-1900	1,599	94
Computer Products Inc.	7900 Glades Rd., Ste. 500, Boca Raton, Florida 33434, U.S.A.	407-451-1000	407-451-1050	1,557	97
Tah Hsin Industrial Corp.	Taichung Industrial Park, Taichung, Taiwan	4 2522111	4 2525307	1,550	90
Micros Systems Inc.	12000 Baltimore Ave., Beltsville, Maryland 20705-1291, U.S.A.	301-210-6000	301-210-8021	1,534	97
Unisys Brasil Ltda.	Rua Teixeira de Freitas 31, 14thFl., Rio de Janeiro 20021-350, Brazil	21 2171133	21 2219435	1,523	96
Canberra Industries Inc.	800 Research Pkwy., Meriden, Connecticut 06450, U.S.A.	203-238-2351	203-235-1347	1,500	
Dynatech Corp.	3 New England Executive Park, Burlington, Massachusetts 01803-5087, U.S.A.	617-272-6100	617-272-2304	1,500	95
Korea Xerox Co., Ltd.	Chung-Gu, Seoul, Republic of Korea	2 3108500	2 3108555	1,500	97
Twinlock PLC	36 Croydon Rd., Beckenham BR3 4BH, United Kingdom	81 6504818	81 6503093	1,500	90
FileNet Corp.	3565 Harbor Blvd., Costa Mesa, California 92626-1429, U.S.A.	714-966-3400	714-966-3490	1,443	97
RICOH Corp.	5 Dedrick Pl., West Caldwell, New Jersey 07006, U.S.A.	201-882-2000	201-882-5840	1,377	94
Kronos Inc.	400 5th Ave., Waltham, Massachusetts 02154, U.S.A.	781-890-3232	781-890-8768	1,341	97
Kalamazoo PLC	Northfield, Birmingham B31 2RW, United Kingdom	21 4112345	21 4757566	1,322	93
Mannai Trading Co. Ltd.	PO Box 76, Doha, Qatar	412555	411982	1,300	
Pitney Bowes of Canada Ltd.	2200 Yonge St. Ste. 100, Toronto, Ontario M4S 3E1, Canada	416-489-2211	416-484-3972	1,300	98
British Olivetti Ltd.	154 Upper Richmond Rd., London SW15 2UR, United Kingdom	81 7856666	81 8743014	1,268	87
Digital Communications Associates Inc.	1000 Alderman Dr., Alpharetta, Georgia 30201-4199, U.S.A.	770-442-4000	770-442-4346	1,256	
Comverse Technology Inc.	170 Crossways Park Dr., Woodbury, New York 11797, U.S.A.	516-677-7200	516-677-7355	1,243	97
International Business Machines Korea Corp.	Yongdungpo-Gu, Seoul, Republic of Korea	2 7816114	2 7842910	1,211	97

Continued

Representative Companies in Sector
- Continued -

Name	Address	Telephone	Fax	Employ-ment	Y
Aydin Corp.	PO Box 349, Horsham, Pennsylvania 19044, U.S.A.	215-657-7510	215-657-3830	1,200	96
Hunt Corp.	2005 Market St., 1 Commerce Sq., Philadelphia, Pennsylvania 19103-7085, U.S.A.	215-732-7700		1,200	97
IBM Argentina	Ingeniero Butty 275, 2nd Fl., Buenos Aires 1300, Argentina	1 3130014	1 3197373	1,200	96
IBM Argentina SA	Ingeniero Butty 275, 2nd Fl., Buenos Aires 1300, Argentina	1 3130014	1 3197373	1,200	96
Titan Corp. (San Diego, California)	3033 Science Park Rd., San Diego, California 92121, U.S.A.	619-552-9500	619-552-9645	1,200	97
Unidata	Parklands, Johannesburg 2121, Republic of South Africa	11 4414111	11 4414601	1,200	93
Plasti-Line Inc.	PO Box 59043, Knoxville, Tennessee 37950-9043, U.S.A.	423-938-1511		1,160	95
DRS Technologies Inc.	5 Sylvan Way, Parsippany, New Jersey 07054, U.S.A.	201-898-1500	201-898-4730	1,107	96
SyQuest Technology Inc.	47071 Bayside Pkwy., Fremont, California 94538-6517, U.S.A.	510-226-4000		1,107	97
CalComp Technology Inc.	2411 W. La Palma Ave., Anaheim, California 92801-9808, U.S.A.	714-821-2000		1,100	97
Computalog Ltd.	530 8 Ave. SW Ste. 2000, Calgary, Alberta T2P 3S8, Canada	403-265-6060	403-218-2424	1,100	98
Hayes Microcomputer Products Inc.	PO Box 105203, Atlanta, Georgia 30348, U.S.A.	770-840-9200	770-441-1213	1,100	96
LeFebure Corp.	PO Box 2028, Cedar Rapids, Iowa 52406, U.S.A.	319-369-5000	319-366-7608	1,100	97
Xerox do Brasil Ltda.	Av. Rodriguez Alves 261/275, 5th Fl., Rio de Janeiro 20220-360, Brazil	271 1212	271 1646	1,100	98
OKI Europe Ltd.	Balfour Rd., Hounslow TW3 1HY, United Kingdom	81 5779000	81 5727444	1,022	94
CMC Industries Inc.	PO Box 831, Corinth, Mississippi 38834, U.S.A.	601-287-3771		1,014	96
StorMedia Inc.	390 Reed St., Santa Clara, California 95050-3118, U.S.A.	408-327-8000	408-727-4928	1,010	
Corel Corp.	100-1600 Carling Ave., Ottawa, Ontario K1Z 8R7, Canada	613-728-8200	613-761-9176	1,000	98
ElectroCom Automation Inc.	PO Box 95080, Arlington, Texas 76005-1080, U.S.A.	817-640-5690	817-695-5599	1,000	97
Shape Inc.	PO Box 366, Biddeford, Maine 04005, U.S.A.	207-282-6155	207-283-9130	1,000	95
FORE Systems Inc.	1000 Fore Dr., Warrendale, Pennsylvania 15086, U.S.A.	724-742-4444	724-742-7700	977	96
Stanford Telecommunications Inc.	PO Box 3733, Sunnyvale, California 94088, U.S.A.	408-745-0818		967	96
Autotote Corp.	100 Bellevue Rd., Newark, Delaware 19713, U.S.A.	302-737-4300		935	96
Axiohm Transaction Solutions Inc.	15070 Avenue of Science, San Diego, California 92128, U.S.A.	619-451-3485	619-451-3573	925	96
Dialogic Corp.	1515 Rte. 10, Parsippany, New Jersey 07054, U.S.A.	201-993-3000		905	96
Paradyne Corp.	PO Box 2826, Largo, Florida 33779-2826, U.S.A.	813-530-2000	813-530-8216	900	97
Innovex Inc.	530 11th Ave. S., Hopkins, Minnesota 55343-9904, U.S.A.	612-938-4155	612-938-7718	893	97
Printronix Inc.	PO Box 19559, Irvine, California 92623-9559, U.S.A.	714-863-1900	714-660-8682	885	96
PAR Technology Corp.	8383 Seneca Tpk., New Hartford, New York 13413-4991, U.S.A.	315-738-0600	315-738-0411	880	97
Intel Puerto Rico Inc.	Rd. 183, South Ind. Park, Las Piedras 00771, Puerto Rico	787720-8080		875	92
Diamond Multimedia Systems Inc.	2880 Junction Ave., San Jose, California 95134-1922, U.S.A.	408-325-7000	408-325-7070	864	97
Interstate Electronics Corp.	1001 E. Ball Rd., Anaheim, California 92803, U.S.A.	714-758-0500	714-758-4148	850	97
McRae Industries Inc.	402 N. Main St., Mount Gilead, North Carolina 27306, U.S.A.	910-439-6147	919-439-9596	850	96
Riva Group PLC	Westhoughton, Bolton BL5 3XW, United Kingdom	942 811441	942 810269	834	93
Weidmuller Ltd.	Power Station Rd., Sheerness ME12 3AB, United Kingdom	795 580999	795 661115	811	
Brandt Inc. (Watertown, Wisconsin)	705 S. 12th St., Watertown, Wisconsin 53094, U.S.A.	414-261-1780		800	
Ergon Inc.	PO Drawer 1639, Jackson, Mississippi 39215-1639, U.S.A.	601-948-3472		800	
Korea Computer Inc.	Mapo-gu, Seoul, Republic of Korea	2 7176531	2 7174967	800	97
Lexmark International Group Inc.	740 New Circle Rd., NW, Lexington, Kentucky 40511-1876, U.S.A.	606-232-2700		800	97
Evans and Sutherland Computer Corp.	600 Komas Dr., Salt Lake City, Utah 84108, U.S.A.	801-582-5847	801-588-4517	784	97
Twinhead International Corp.	235 Baochiao Rd., 2nd Fl., Lane 235, Hsin Tien, Taiwan	2 9179036	2 9172675	780	93
General Scanning Inc.	500 Arsenal St., Watertown, Massachusetts 02172, U.S.A.	617-924-1010	617-926-0708	779	96
Acer Sertek Inc.	11-15 Fl., 135 Chien Kuo N. Rd., Sec. 2, Taipei, Taiwan	2 5006363	2 5012521	778	93
Xircom Inc.	2300 Corporate Center Dr., Thousand Oaks, California 91320-1420, U.S.A.	805-376-9300	805-376-9311	770	97
Lotte Canon Co., Ltd.	Kangnam-Gu, Seoul, Republic of Korea	2 34500700	2 5585002	765	97
Zebra Technologies Corp.	333 Corporate Woods Pkwy., Vernon Hills, Illinois 60061-3109, U.S.A.	847-634-6700	847-634-1830	745	97
VTEL Corp.	108 Wild Basin Rd., Austin, Texas 78746, U.S.A.	512-314-2700	512-314-2792	734	97
Datapoint Corp.	8400 Datapoint Dr., San Antonio, Texas 78229-8500, U.S.A.	210-593-7000		705	96
QMS Inc.	PO Box 2153 Dept. 3297, Birmingham, Alabama 35287-3297, U.S.A.	334-433-6300		705	97
Escalade Inc.	817 Maxwell Ave., Evansville, Indiana 47717, U.S.A.	812-467-1200		700	96

Product Tables
Typewriters

Unit of Measure: Number.

	1990 Value	%	1991 Value	%	1992 Value	%	1993 Value	%	1994 Value	%	1995 Value	%	1996 Value	%
Total Production	8,463,107	100.0	8,476,750	100.0	7,837,943	100.0	7,124,666	100.0	6,643,025	100.0	6,067,536	100.0	6,454,415	100.0
Regions														
America, North	1,991,000	23.5	2,077,000	24.5	2,106,000	26.9	2,134,000	30.0	2,395,000	36.1	2,307,000	38.0	2,378,000	36.8
America, South	461,000	5.4	480,000	5.7	408,000	5.2	362,000	5.1	323,000	4.9	255,000	4.2	221,000	3.4
Asia	3,188,000	37.7	2,914,000	34.4	2,579,000	32.9	2,144,000	30.1	1,395,000	21.0	999,867	16.5	1,390,852	21.5
Europe	2,823,107	33.4	3,005,750	35.5	2,744,943	35.0	2,484,666	34.9	2,530,025	38.1	2,505,669	41.3	2,464,564	38.2
Leading Producers														
Japan	1,942,000	22.9	1,549,000	18.3	1,064,000	13.6	802,000	11.3	436,000	6.6	209,000	3.4	70,000	1.1
United States of America	*1,479,000	17.5	*1,479,000	17.4	*1,479,000	18.9	*1,479,000	20.8	*1,479,000	22.3	*1,479,000	24.4	*1,479,000	22.9
Germany	*590,750	7.0	858,000	10.1	623,000	7.9	388,000	5.4	494,000	7.4	*590,750	9.7	*590,750	9.2
Korea, Republic of	918,000	10.8	994,000	11.7	1,005,000	12.8	893,000	12.5	654,000	9.8	494,000	8.1	*1,074,210	16.6
Mexico	512,000	6.0	598,000	7.1	627,000	8.0	655,000	9.2	916,000	13.8	828,000	13.6	899,000	13.9

Calculating Machines

Unit of Measure: Thousands of units.

	1990 Value	%	1991 Value	%	1992 Value	%	1993 Value	%	1994 Value	%	1995 Value	%	1996 Value	%
Total Production	137,873	100.0	136,594	100.0	120,124	100.0	103,962	100.0	79,170	100.0	76,650	100.0	70,115	100.0
Regions														
America, North	5,476	4.0	5,431	4.0	5,386	4.5	5,383	5.2	5,381	6.8	5,381	7.0	5,381	7.7
America, South	1,938	1.4	1,463	1.1	739	0.6	1,047	1.0	761	1.0	900	1.2	632	0.9
Asia	129,303	93.8	128,631	94.2	112,887	94.0	96,232	92.6	72,230	91.2	69,256	90.4	62,951	89.8
Europe	1,152	0.8	1,065	0.8	1,107	0.9	1,296	1.2	794	1.0	1,109	1.4	1,146	1.6
Oceania	4	0.0	4	0.0	4	0.0	4	0.0	4	0.0	4	0.0	4	0.0
Leading Producers														
Japan	67,479	48.9	69,371	50.8	55,800	46.5	41,576	40.0	20,170	25.5	5,565	7.3	3,249	4.6
China	*34,759	25.2	*38,814	28.4	*42,868	35.7	*46,923	45.1	*50,977	64.4	*55,032	71.8	*59,086	84.3
Hong Kong	25,060	18.2	18,985	13.9	12,910	10.7	6,835	6.6	760	1.0	8,267	10.8	-	
United States of America	*5,381	3.9	*5,381	3.9	*5,381	4.5	*5,381	5.2	*5,381	6.8	*5,381	7.0	*5,381	7.7
Korea, Republic of	1,958	1.4	1,426	1.0	1,304	1.1	894	0.9	316	0.4	386	0.5	*616	0.9

Input or Output Units

Unit of Measure: Number

	1990 Value	%	1991 Value	%	1992 Value	%	1993 Value	%	1994 Value	%	1995 Value	%	1996 Value	%
Total Production	40,393,634	100.0	44,166,134	100.0	43,266,634	100.0	42,685,088	100.0	46,612,406	100.0	51,645,405	100.0	55,925,546	100.0
Regions														
America, North	6,881,000	17.0	7,543,500	17.1	8,206,000	19.0	8,868,500	20.8	9,531,000	20.4	10,193,500	19.7	10,856,000	19.4
Asia	18,168,330	45.0	20,443,330	46.3	18,971,330	43.8	18,345,330	43.0	20,079,330	43.1	22,840,330	44.2	26,463,330	47.3
Europe	15,344,304	38.0	16,179,304	36.6	16,089,304	37.2	15,471,258	36.2	17,002,076	36.5	18,611,575	36.0	18,606,217	33.3
Leading Producers														
Japan	16,224,000	40.2	18,499,000	41.9	17,027,000	39.4	16,401,000	38.4	18,135,000	38.9	20,896,000	40.5	24,519,000	43.8
United Kingdom	*7,595,758	18.8	*7,595,758	17.2	*7,595,758	17.6	*7,595,758	17.8	7,595,758	16.3	*7,595,758	14.7	*7,595,758	13.6
United States of America	*6,881,000	17.0	*7,543,500	17.1	*8,206,000	19.0	*8,868,500	20.8	*9,531,000	20.4	*10,193,500	19.7	*10,856,000	19.4
Germany	*3,256,546	8.1	*3,256,546	7.4	*3,256,546	7.5	1,934,000	4.5	2,879,000	6.2	4,036,681	7.8	4,176,504	7.5
Hong Kong	*1,873,330	4.6	*1,873,330	4.2	*1,873,330	4.3	*1,873,330	4.4	*1,873,330	4.0	*1,873,330	3.6	*1,873,330	3.3

Commodity data are provided by the United Nations Statistical Division. The symbol * means that data are estimated. For additional notes, see Appendix I.

Product Tables
Central and Peripheral Storage Units

Unit of Measure: Number.

	1990 Value	%	1991 Value	%	1992 Value	%	1993 Value	%	1994 Value	%	1995 Value	%	1996 Value	%
Total Production	13,407,984	100.0	13,410,413	100.0	13,416,127	100.0	13,426,025	100.0	13,364,778	100.0	13,462,829	100.0	13,424,520	100.0
Regions														
America, North	12,537,000	93.5	12,537,000	93.5	12,537,000	93.4	12,537,000	93.4	12,537,000	93.8	12,537,000	93.1	12,537,000	93.4
Asia	6,000	0.0	6,000	0.0	6,000	0.0	6,000	0.0	6,000	0.0	6,000	0.0	6,000	0.0
Europe	864,984	6.5	867,413	6.5	873,127	6.5	883,025	6.6	821,778	6.1	919,829	6.8	881,520	6.6
Leading Producers														
United States of America	*12,537,000	93.5	*12,537,000	93.5	*12,537,000	93.4	12,537,000	93.4	*12,537,000	93.8	*12,537,000	93.1	*12,537,000	93.4
United Kingdom	*413,718	3.1	*413,718	3.1	*413,718	3.1	409,501	3.1	417,935	3.1	*413,718	3.1	*413,718	3.1
Finland	*165,000	1.2	*165,000	1.2	*165,000	1.2	*165,000	1.2	*165,000	1.2	*165,000	1.2	*165,000	1.2
Germany	*141,311	1.1	*141,311	1.1	*141,311	1.1	157,000	1.2	84,000	0.6	182,933	1.4	*141,311	1.1
Spain	22,000	0.2	25,000	0.2	31,000	0.2	*29,429	0.2	*32,750	0.2	*36,071	0.3	*39,393	0.3

Scales, Industrial

Unit of Measure: Number.

	1990 Value	%	1991 Value	%	1992 Value	%	1993 Value	%	1994 Value	%	1995 Value	%	1996 Value	%
Total Production	650,361	100.0	868,591	100.0	851,502	100.0	839,550	100.0	842,849	100.0	925,773	100.0	922,845	100.0
Regions														
America, South	152,346	23.4	169,863	19.6	183,532	21.6	174,027	20.7	149,936	17.8	151,242	16.3	190,204	20.6
Asia	43,086	6.6	36,011	4.1	34,849	4.1	34,628	4.1	35,281	4.2	33,435	3.6	63,801	6.9
Europe	454,928	70.0	662,716	76.3	633,121	74.4	630,895	75.1	657,632	78.0	741,096	80.1	668,840	72.5
Leading Producers														
Germany	-		203,107	23.4	183,234	21.5	185,104	22.0	203,345	24.1	292,725	31.6	217,169	23.5
United Kingdom	*190,434	29.3	*190,434	21.9	*190,434	22.4	190,434	22.7	*190,434	22.6	*190,434	20.6	*190,434	20.6
Colombia	117,786	18.1	132,267	15.2	146,748	17.2	138,857	16.5	116,380	13.8	119,299	12.9	*159,875	17.3
Brazil	34,560	5.3	37,596	4.3	*36,784	4.3	*35,170	4.2	*33,556	4.0	*31,943	3.5	*30,329	3.3
Portugal	9,017	1.4	13,583	1.6	-		-		-		-		-	

Scales, Other Than Industrial (Retail, Commercial and Household Scales)

Unit of Measure: Number.

	1990 Value	%	1991 Value	%	1992 Value	%	1993 Value	%	1994 Value	%	1995 Value	%	1996 Value	%
Total Production	8,863,333	100.0	11,410,833	100.0	12,451,167	100.0	12,494,785	100.0	12,479,200	100.0	11,937,661	100.0	12,404,673	100.0
Regions														
America, South	16,000	0.2	17,000	0.1	18,000	0.1	16,000	0.1	20,038	0.2	20,637	0.2	21,236	0.2
Asia	840,667	9.5	941,667	8.3	903,000	7.3	715,000	5.7	669,000	5.4	635,667	5.3	787,804	6.4
Europe	8,006,667	90.3	10,452,167	91.6	11,530,167	92.6	11,763,785	94.1	11,790,162	94.5	11,281,357	94.5	11,595,633	93.5
Leading Producers														
France	2,103,000	23.7	4,539,000	39.8	5,325,000	42.8	5,830,000	46.7	5,731,000	45.9	*5,303,022	44.4	*5,587,358	45.0
Germany	*3,088,500	34.8	2,906,000	25.5	3,162,000	25.4	3,094,000	24.8	3,192,000	25.6	*3,088,500	25.9	*3,088,500	24.9
Korea, Republic of	539,000	6.1	644,000	5.6	614,000	4.9	416,000	3.3	347,000	2.8	350,000	2.9	*509,709	4.1
Slovenia	*244,000	2.8	*244,000	2.1	*244,000	2.0	245,000	2.0	266,000	2.1	236,000	2.0	229,000	1.8
Spain	25,000	0.3	219,000	1.9	255,000	2.0	52,000	0.4	78,000	0.6	101,000	0.8	*131,265	1.1

Commodity data are provided by the United Nations Statistical Division. The symbol * means that data are estimated. For additional notes, see Appendix I.

599

ISIC 3830
ELECTRICAL MACHINERY

ISIC 3830—Electrical Machinery—includes both industrial and consumer products driven by electricity or producing electricity. At one end of the spectrum are generators and electrical motors. At the other end are consumer products such as shavers, irons, and lamps. A subset of this industry, ISIC 3832—Radio, Television, etc., is reported on separately, as well, in the next chapter. Product tables for that subindustry, however, are included in this chapter.

This industry corresponds to SIC 36—Electronic & Other Electric Equipment in the U.S. classification scheme.

Summary Statistics

		1990		1991		1992		1993		1994		1995	
		Value	N	Value	N	Value	N	Value	N	Value	N	Value	N
Establishments or enterprises	(number)	102,225	79	114,554	90	139,095	87	113,437	81	103,402	70	79,853	35
Number employed	(000)	17,763	91	14,983	96	14,643	91	13,870	88	13,257	81	12,150	65
Total Wages	($ mil.)	210,355	88	245,077	88	264,342	84	203,757	80	203,148	72	176,165	59
Wage/salary per employee	($)	9,335	85	10,142	85	23,970	82	14,313	79	10,986	71	13,150	59
Female workers	(000)	1,583	35	997	33	1,998	30	1,956	33	1,906	25	1,607	15
as % of total employment*	(%)	34	35	33	33	34	30	36	33	37	25	42	15
Output	($ bil.)	1,161	90	1,295	89	1,391	86	1,253	81	1,276	75	1,321	64
per employee	($ 000)	60	87	64	87	647	84	201	80	78	73	92	63
per establishment	($ 000)	8,792	77	8,108	82	64,889	76	20,579	70	8,616	58	6,249	32
Value added	($ bil.)	498	83	508	82	539	77	551	75	484	68	571	62
per employee	($ 000)	26	79	25	78	29	74	28	72	30	66	37	60
per establishment	($ 000)	3,288	70	2,889	74	3,435	67	2,735	62	3,025	51	2,213	30
Capital investment	($ mil.)	57,910	68	59,104	61	56,219	58	44,934	53	48,965	43	32,888	24
per employee	($ 000)	3	65	4	58	4	56	3	52	4	43	4	24
per establishment	($ 000)	578	63	481	56	469	54	377	48	418	40	283	21

Data presented above are drawn from the detailed tables that follow. Columns headed 'N' show the number of countries that provided valid data for inclusion. Values are not strictly comparable one year to the next or one row to the next because the number of countries that report varies from period to period and row to row. However, a general indication of magnitudes can be gleaned from reviewing this summary—especially in earlier years in which more reports are available. For detailed explanations, see Appendix I. *The average for those countries reporting both total and female employment.

Establishments and Number Engaged

Country	Establishments or Enterprises (number)						Employees per Establishment					
	1990	1991	1992	1993	1994	1995	1990	1991	1992	1993	1994	1995
Albania	-	-	-	7.000	13	8.000	-	-	-	128	100	71
Angola	-	-	7.000	7.000	-	-	-	-	89	78	-	-
Argentina	-	-	-	3,596	3,918	-	-	-	-	9.211	8.435	-
Armenia	31	34	35	38	37	-	1,090	932	-	-	-	-
Australia	952	1,260	1,298	-	-	-	57	40	35	-	-	-
Austria	496	489	496	495	491	-	167	166	161	150	150	-
Azerbaijan	27	26	27	28	27	-	940	919	817	739	745	-
Bahrain	-	-	5.000	-	-	-	-	-	113	-	-	-
Bangladesh	396	401	400	-	-	-	43	45	33	-	-	-
Belgium	936	936	971	791	675	-	-	-	-	-	-	-
Belize	7.000	7.000	6.000	-	-	-	-	2.714	3.167	-	-	-
Bolivia	19	18	19	21	19	26	17	15	23	24	29	21
Bosnia & Herzegovina	37	42	-	-	1.000	-	495	367	-	-	13	-
Brazil	1,520	-	1,368	1,241	-	-	194	-	158	161	-	-
Bulgaria	*177	*186	*187	*2,943	*4,425	*5,138	820	614	467	22	12	9.887
Cameroon	*8.000	*5.000	*5.000	*4.000	*2.000	*1.000	86	94	145	155	206	266
Canada	1,378	1,281	1,223	1,176	1,142	-	89	89	90	88	88	-
Cape Verde	-	-	1.000	1.000	1.000	-	-	-	-	-	-	-
Central African Republic	-	*1.000	*1.000	-	-	-	-	-	-	-	-	-
Chile	26	27	29	29	29	-	151	137	151	149	148	-
China	*16,933	*22,847	*23,258	*23,446	*24,632	*27,668	193	156	157	161	161	150
China (Hong Kong SAR)	1,393	1,177	1,091	940	842	-	49	46	46	46	45	-
Colombia	214	199	226	214	205	-	85	88	86	91	97	-
Costa Rica	72	92	93	112	-	-	55	56	55	62	-	-
Croatia	534	806	1,138	1,529	1,848	1,991	68	36	23	16	12	10
Cyprus	95	92	92	91	107	115	7.305	7.011	6.891	6.747	5.187	4.496
Czechoslovakia (Former)	*106	*128	-	-	-	-	1,783	1,164	-	-	-	-
Czech Republic	-	*85	*128	*159	-	-	-	1,165	648	465	-	-
Denmark	1,350	1,396	1,342	-	-	-	23	21	19	-	-	-
Ecuador	52	57	60	53	51	49	80	78	69	67	70	73
Egypt	81	107	105	109	-	-	383	308	312	293	-	-
El Salvador	-	12	9.000	8.000	14	7.000	-	-	63	75	52	26
Ethiopia	1.000	1.000	1.000	1.000	2.000	2.000	99	97	100	100	54	56
Finland	236	250	257	266	264	325	121	104	95	91	107	115
FYR Macedonia	*75	*165	*360	*384	*527	*653	163	70	34	28	18	14
Gabon	-	1.000	1.000	1.000	1.000	1.000	-	33	32	32	31	17
Germany	-	4,472	4,462	4,486	4,407	-	-	279	247	226	214	-
Germany (Western Part)	3,854	3,955	4,002	-	-	-	268	263	253	-	-	-
Ghana	-	-	-	18	-	-	-	-	-	65	-	-
Greece	297	291	289	-	-	-	47	45	44	-	-	-
Grenada	-	-	2.000	2.000	2.000	-	-	-	44	14	16	-
Guatemala	-	9.000	9.000	9.000	9.000	9.000	-	73	72	83	87	84
Honduras	-	-	31	20	20	20	-	-	20	33	36	42
Hungary	*571	*1,190	*900	*1,056	-	-	210	80	78	56	-	-
India	4,856	4,835	5,120	5,110	-	-	80	81	79	79	-	-
Indonesia	266	293	327	349	407	459	225	251	263	307	354	358
Iran (Islamic Republic of)	119	232	252	262	-	-	145	211	202	188	-	-
Iraq	-	8.000	8.000	-	-	-	-	1,290	1,250	-	-	-
Ireland	289	297	-	-	-	-	76	75	-	-	-	-
Israel	554	597	633	648	743	-	79	75	72	75	70	-
Italy	*1,734	*1,772	*2,194	*2,190	*2,164	-	154	148	116	107	105	-
Japan	34,800	35,658	33,865	32,829	26,509	26,530	52	52	54	53	56	56
Jordan	13	14	9.000	12	37	34	51	38	73	63	40	41
Kenya	28	39	40	39	47	47	85	66	69	72	60	68
Korea, Republic of	6,566	6,545	6,800	7,899	8,202	8,816	68	62	58	50	50	48
Kuwait	22	20	36	36	36	-	25	24	27	32	38	-
Kyrgyzstan	15	15	15	17	20	-	1,399	1,303	1,165	968	526	-
Latvia	20	47	81	107	53	89	2,778	1,085	515	253	340	149
Liechtenstein	-	25	-	-	-	-	-	74	-	-	-	-
Lithuania	-	-	-	*36	*51	-	-	-	-	1,218	561	-

Continued.

Establishments and Number Engaged

- Continued -

Country	Establishments or Enterprises (number)						Employees per Establishment					
	1990	1991	1992	1993	1994	1995	1990	1991	1992	1993	1994	1995
Macau	19	10	38	31	27	21	47	53	17	21	33	64
Malaysia	422	492	530	594	609	-	506	525	532	560	608	-
Malta	43	49	48	59	54	-	56	47	54	48	56	-
Mauritius	23	20	17	18	17	18	40	48	51	45	45	49
Mexico	174	165	162	153	147	145	545	558	559	506	513	528
Mongolia	1.000	1.000	2.000	-	8.000	1.000	320	194	82	-	38	220
Morocco	132	117	118	113	111	108	85	99	86	93	90	99
Mozambique	9.000	9.000	7.000	7.000	7.000	-	135	122	136	111	80	-
Nepal	21	35	-	26	42	-	38	32	-	72	34	-
Netherlands	338	368	393	359	321	-	332	279	254	240	266	-
Netherlands Antilles	-	-	-	7.000	-	-	-	-	-	5.714	-	-
New Zealand	*607	*615	*585	*602	-	-	17	15	16	18	-	-
Norway	281	268	194	168	173	-	58	58	69	73	71	-
Pakistan	-	220	-	-	-	-	-	85	-	-	-	-
Panama	8.000	7.000	-	-	-	-	14	16	-	-	-	-
Paraguay	-	3.000	-	-	-	-	-	27	-	-	-	-
Peru	516	540	557	-	-	-	16	13	13	-	-	-
Philippines	286	338	296	265	-	-	266	266	345	372	-	-
Poland	*355	*363	*339	*318	-	-	665	515	425	406	-	-
Portugal	*1,173	*1,170	*1,104	*1,227	*1,312	-	35	38	41	37	35	-
Puerto Rico	122	125	113	108	96	90	142	131	138	139	178	223
Republic of Moldova	15	15	15	-	-	-	1,109	869	776	-	-	-
Romania	-	172	259	333	1,103	-	-	813	475	343	93	-
Russian Federation	-	-	-	*1,764	*1,883	*2,808	-	-	-	365	274	168
Saint Lucia	-	6.000	6.000	8.000	7.000	7.000	-	-	-	-	-	-
Saint Vincent & the Grenadines	-	-	-	1.000	1.000	1.000	-	-	-	-	-	-
Senegal	3.000	2.000	3.000	1.000	1.000	-	109	102	73	251	173	-
Singapore	323	334	346	352	355	-	315	305	283	267	276	-
Slovakia	-	63	63	66	77	-	-	820	663	525	399	-
Slovenia	616	*994	*1,258	*1,182	*1,459	*1,723	66	37	26	25	19	16
South Africa	-	1,201	-	-	-	-	-	77	-	-	-	-
Spain	2,107	2,007	1,957	-	-	-	52	54	53	-	-	-
Sri Lanka	22	25	19	36	-	-	74	96	82	91	-	-
Suriname	-	-	3.000	-	-	-	-	-	-	-	-	-
Sweden	437	529	493	457	414	-	126	118	119	111	131	-
Thailand	132	276	-	-	-	-	643	267	-	-	-	-
Trinidad and Tobago	47	34	26	25	23	-	33	43	54	48	46	-
Tunisia	-	-	-	199	227	225	-	-	-	33	38	47
Turkey	252	238	429	402	359	-	183	183	107	110	111	-
Ukraine	151	323	487	187	191	196	1,483	721	478	1,160	1,000	857
United Kingdom	10,029	9,786	10,172	9,048	9,892	-	51	48	43	34	34	-
United Republic of Tanzania	9.000	8.000	-	-	-	-	175	156	-	-	-	-
USSR (Former)	*1,607	-	-	-	-	-	2,400	-	-	-	-	-
United States of America	-	-	23,803	-	-	-	-	-	63	-	-	-
Venezuela	239	246	237	204	164	190	69	77	80	81	73	63
Yugoslavia	478	875	1,642	2,253	2,278	2,332	-	72	35	25	23	22
Zambia	12	-	-	-	10	-	113	-	-	-	-	-
Zimbabwe	60	63	60	49	54	-	107	100	103	112	102	-

Employment and Compensation of Employees

Country	Number Employed/Engaged (000)						Wage/Salary per Employee ($)					
	1990	1991	1992	1993	1994	1995	1990	1991	1992	1993	1994	1995
Albania	-	-	-	0.894	1.294	0.571	-	-	-	-	-	1,227
Angola	-	*0.788	*0.620	*0.546	-	-	-	-	-	-	-	-
Argentina	29	-	-	33	33	-	6,976	-	-	16,637	-	-
Armenia	*34	*32	-	-	-	-	-	-	-	-	-	-
Australia	54	*51	*45	-	-	-	20,275	21,846	21,258	-	-	-

Continued.

Depending on the table, * means *Enterprises*, *Engaged*, or *Factor Values*; ± means *Basis Unspecified*; § means *shown in millions*. For additional notes and sources, see Appendix I.

603

Employment and Compensation of Employees

- Continued -

Country	Number Employed/Engaged (000)						Wage/Salary per Employee ($)					
	1990	1991	1992	1993	1994	1995	1990	1991	1992	1993	1994	1995
Austria	83	81	80	74	74	74	27,924	30,100	33,075	34,493	35,803	-
Azerbaijan	25	24	22	21	20	-	1,956	2,059	-	-	-	-
Bahrain	-	-	0.564	-	-	-	-	-	6,644	-	-	-
Bangladesh	17	18	13	-	-	-	759	700	710	-	-	-
Belize	-	0.019	0.019	-	-	-	-	3,605	4,132	-	-	-
Bermuda	0.015	0.015	0.019	0.008	0.017	0.007	-	-	-	-	-	-
Bolivia	0.319	0.271	0.439	0.498	0.545	0.553	1,293	1,188	1,299	1,512	1,409	1,360
Bosnia & Herzegovina	18	15	-	-	0.013	-	2,932	2,508	-	-	-	-
Brazil	295	-	216	200	-	-	8,826	-	9,088	11,088	-	-
Bulgaria	145	114	87	66	54	51	1,749	575	867	1,161	997	1,216
Cambodia	*2.552	*2.204	-	-	-	-	-	-	-	-	-	-
Cameroon	0.690	0.469	0.725	0.620	0.412	0.266	8,836	9,871	9,406	8,669	5,762	7,999
Canada	123	114	110	104	101	100	28,032	29,929	30,032	28,881	28,472	32,817
Central African Republic	0.006	-	-	-	-	-	6,122	-	-	-	-	-
Chile	3.930	3.695	4.386	4.310	4.292	4.242	6,650	7,962	8,500	8,719	10,228	12,358
China	*3,260	*3,560	*3,640	*3,770	*3,960	*4,160	-	-	-	-	-	-
China (Hong Kong SAR)	*68	*55	*50	*44	*38	-	9,018	10,586	12,085	13,578	15,686	-
China (Taiwan Province)	459	454	461	463	478	494	9,553	10,649	12,440	12,994	14,237	14,924
Colombia	18	17	19	20	20	19	2,550	2,567	2,733	3,120	4,080	4,478
Costa Rica	3.988	5.173	5.093	6.972	-	-	3,389	3,160	4,480	3,953	-	-
Croatia	36	29	26	24	22	21	4,747	4,943	1,708	1,942	2,614	4,357
Cyprus	0.694	0.645	0.634	0.614	0.555	0.517	9,191	9,517	11,195	11,168	12,332	14,979
Czechoslovakia (Former)	189	149	-	-	-	-	2,057	1,439	-	-	-	-
Czech Republic	125	99	83	74	-	-	2,068	1,460	1,850	2,298	-	-
Denmark	32	29	25	25	26	26	31,177	31,681	34,014	32,077	36,427	-
Ecuador	4.170	4.419	4.154	3.539	3.576	3.570	2,824	3,164	3,468	4,242	2,186	2,150
Egypt	*31	*33	*33	*32	*32	*32	2,555	2,274	2,070	2,163	2,271	2,417
El Salvador	-	-	*0.564	*0.603	*0.726	*0.180	-	-	-	5,014	9,178	4,007
Ethiopia	0.099	0.097	0.100	0.100	0.109	0.112	2,049	2,381	1,863	1,256	982	982
Finland	28	26	24	24	28	37	28,472	28,053	26,720	21,898	25,442	32,715
FYR Macedonia	12	12	12	11	9.550	9.300	5,403	5,491	2,177	2,231	2,838	3,410
France	474	472	456	431	421	425	36,399	36,222	-	-	-	-
Gabon	-	0.033	0.032	0.032	0.031	0.017	-	17,294	17,591	16,334	12,376	16,852
Germany	-	*1,249	*1,104	*1,014	*942	-	-	28,818	35,479	35,793	38,102	-
Germany (Western Part)	*1,033	*1,040	*1,011	-	-	-	31,603	32,789	37,211	-	-	-
Ghana	-	-	-	*1.166	*1.199	*1.233	-	-	-	1,038	775	791
Greece	*14	*13	*13	*12	*12	*12	12,236	13,120	14,148	15,548	15,383	18,179
Grenada	-	-	*0.089	*0.027	0.033	-	-	-	-	-	-	-
Guatemala	-	*0.655	*0.650	*0.745	*0.780	*0.753	-	-	-	-	-	-
Honduras	*0.807	*0.942	*0.608	*0.662	*0.726	*0.832	1,727	1,185	2,461	2,046	2,078	3,863
Hungary	120	95	70	59	65	67	2,368	2,708	3,276	3,491	4,462	4,983
India	*389	*392	*403	*404	*422	*437	1,920	1,667	1,644	1,506	1,870	2,201
Indonesia	60	73	86	107	144	164	927	998	1,128	1,227	1,333	2,401
Iran (Islamic Republic of)	17	49	51	49	-	-	4,073	5,263	5,260	4,866	-	-
Iraq	-	10	10	-	-	-	-	8,029	17,961	-	-	-
Ireland	22	22	23	24	26	29	19,338	20,143	22,808	20,660	23,652	29,671
Israel	44	45	46	49	52	52	31,477	33,124	35,558	33,293	34,782	39,691
Italy	266	263	255	235	228	-	37,217	39,009	-	33,065	33,742	-
Japan	*1,824	*1,871	*1,817	*1,736	*1,489	*1,474	25,457	28,801	31,926	37,468	-	-
Jordan	0.665	0.526	0.656	0.755	1.491	1.381	2,777	3,180	2,464	2,548	2,855	3,581
Kenya	2.369	2.591	2.771	2.799	2.834	3.218	2,709	2,490	2,321	1,431	1,944	2,668
Korea, Republic of	449	404	392	395	409	426	9,137	10,813	11,354	11,954	13,122	16,395
Kuwait	0.557	0.473	0.966	1.150	1.353	1.180	6,425	3,677	9,140	9,099	9,100	12,299
Kyrgyzstan	*21	*20	*17	*16	*11	-	464	443	-	-	-	-
Latvia	56	51	42	27	18	13	-	-	-	-	-	-
Liechtenstein	-	*1.860	-	-	-	-	-	-	-	-	-	-
Lithuania	-	-	50	44	29	-	-	-	-	489	852	-
Macau	0.888	0.529	0.657	0.660	0.904	1.341	4,589	4,395	4,623	5,043	6,344	6,308
Malaysia	214	258	282	333	370	431	2,823	3,149	3,809	3,960	4,307	4,769
Malta	2.421	2.285	2.569	2.819	2.998	-	10,840	11,687	12,395	11,739	12,199	-

Continued.

Depending on the table, * means *Enterprises, Engaged,* or *Factor Values*; ± means *Basis Unspecified*; § means *shown in millions*. For additional notes and sources, see Appendix I.

Employment and Compensation of Employees

- Continued -

Country	Number Employed/Engaged (000)						Wage/Salary per Employee ($)					
	1990	1991	1992	1993	1994	1995	1990	1991	1992	1993	1994	1995
Mauritius	*0.914	0.964	0.871	0.805	0.770	0.883	2,495	3,102	3,570	4,019	5,199	4,651
Mexico	95	92	91	77	75	76	4,140	4,923	5,508	6,119	5,861	4,002
Mongolia	*0.320	*0.194	*0.164	-	*0.307	*0.220	837	2,437	960	-	-	-
Morocco	11	12	10	10	9.956	11	5,295	5,111	6,165	5,473	5,981	6,636
Mozambique	1.214	1.094	0.955	0.780	0.562	-	2,886	1,446	641	740	589	-
Nepal	0.801	1.114	-	1.861	1.444	-	588	618	-	545	501	-
Netherlands	112	103	100	86	85	85	30,239	31,651	34,315	33,863	36,688	-
Netherlands Antilles	-	-	-	0.040	-	-	-	-	-	11,425	-	-
New Zealand	11	9.353	9.623	11	-	-	15,678	17,892	16,719	-	-	-
Norway	16	16	13	12	12	12	31,948	33,237	-	33,608	35,276	-
Pakistan	19	19	20	-	-	-	2,074	2,101	2,161	-	-	-
Panama	*0.110	*0.114	*0.121	*0.132	*0.138	*0.137	5,855	5,904	6,221	6,837	7,179	7,131
Paraguay	-	0.081	-	-	-	-	-	-	-	-	-	-
Peru	*8.180	*7.251	*7.281	-	-	-	4,454	4,533	4,540	-	-	-
Philippines	76	90	102	99	-	-	2,423	2,505	2,889	2,904	-	-
Poland	236	187	144	129	128	125	1,220	1,789	2,763	2,601	3,021	4,242
Portugal	*41	*44	*45	*45	*46	*50	-	-	-	-	-	-
Puerto Rico	17	16	16	15	17	20	26,185	20,275	21,046	23,136	22,702	22,104
Republic of Moldova	17	13	12	-	-	-	-	3,405	-	-	-	-
Romania	171	140	123	114	102	-	1,774	1,115	687	846	921	-
Russian Federation	-	-	-	644	516	471	-	-	-	524	818	797
Senegal	*0.326	*0.205	*0.219	*0.251	*0.173	-	5,453	9,683	10,471	6,782	5,362	-
Singapore	*102	*102	*98	*94	*98	*100	9,836	11,293	13,270	14,722	16,939	19,355
Slovakia	-	*52	*42	*35	*31	-	-	1,428	1,741	1,944	2,245	-
Slovenia	41	37	33	30	27	28	9,441	7,092	7,054	6,697	8,311	9,716
South Africa	86	92	96	90	111	115	8,885	9,581	10,249	9,976	11,100	12,282
Spain	110	108	104	107	104	102	22,195	23,616	25,694	23,257	22,603	25,341
Sri Lanka	*1.632	*2.409	*1.557	*3.276	-	-	749	1,038	1,051	852	-	-
Sweden	55	*63	*58	51	54	49	24,703	28,220	31,185	24,729	26,564	39,146
Switzerland	*132	*128	*119	*115	*115	*113	-	-	-	-	-	-
Thailand	85	74	-	-	-	-	3,247	4,288	-	-	-	-
Tonga	0.003	0.005	-	-	-	-	3,383	-	-	-	-	-
Trinidad and Tobago	1.555	1.474	1.401	1.196	1.065	-	5,406	6,209	6,454	4,750	5,055	-
Tunisia	-	-	-	6.475	8.558	11	-	-	-	8,684	6,689	6,035
Turkey	46	44	46	44	40	43	7,581	10,159	10,187	11,429	6,502	8,514
Ukraine	224	233	233	217	191	168	3,001	2,855	2,092	-	405	452
United Kingdom	508	466	436	308	332	341	22,258	24,224	25,375	21,835	23,040	25,253
United Republic of Tanzania	1.571	1.248	-	-	-	-	-	589	-	-	-	-
USSR (Former)	3,857	-	-	-	-	-	2,190	-	-	-	-	-
United States of America	1,541	1,470	1,493	1,491	1,520	1,583	28,397	29,517	31,138	31,948	32,820	33,770
Uruguay	*4.748	*4.107	*4.051	*3.709	*3.718	*3.699	3,586	4,363	5,361	5,923	6,388	6,383
Venezuela	17	19	19	16	12	12	4,311	4,595	5,059	5,191	4,755	5,640
Yugoslavia	-	63	57	56	52	51	-	-	-	-	1,485	742
Zambia	1.350	-	-	-	-	-	3,683	-	-	-	-	-
Zimbabwe	6.400	6.300	6.200	5.500	5.500	5.297	4,770	4,217	2,888	2,719	2,503	2,790

Female Workers: Total and Percent of Employment

Country	Female Workers (000)						Female Workers as % of Total Employment					
	1990	1991	1992	1993	1994	1995	1990	1991	1992	1993	1994	1995
Albania	-	-	-	0.309	0.602	0.271	-	-	-	34.56	46.52	47.46
Angola	-	-	0.200	0.189	-	-	-	-	32.26	34.62	-	-
Argentina	-	-	-	-	6.648	-	-	-	-	-	20.12	-
Australia	17	-	-	-	-	-	30.80	-	-	-	-	-
Austria	30	28	26	24	23	-	35.66	34.57	32.50	31.91	31.49	-
Bahrain	-	-	0.012	-	-	-	-	-	2.13	-	-	-
Bangladesh	0.491	0.385	0.350	-	-	-	2.90	2.12	2.69	-	-	-
Bermuda	0.005	0.005	0.004	0.002	0.004	0.002	33.33	33.33	21.05	25.00	23.53	28.57

Continued.

Depending on the table, * means *Enterprises*, *Engaged*, or *Factor Values*; ± means *Basis Unspecified*; § means *shown in millions*. For additional notes and sources, see Appendix I.

605

Female Workers: Total and Percent of Employment

- Continued -

Country	Female Workers (000)						Female Workers as % of Total Employment					
	1990	1991	1992	1993	1994	1995	1990	1991	1992	1993	1994	1995
Bosnia & Herzegovina	7.449	6.846	-	-	-	-	40.63	44.47	-	-	-	-
Bulgaria	-	-	-	36	27	25	-	-	-	54.53	50.46	49.41
Canada	45	40	-	-	-	-	36.59	35.09	-	-	-	-
Chile	0.474	0.452	0.592	0.570	0.562	-	12.06	12.23	13.50	13.23	13.09	-
China (Taiwan Province)	250	245	245	245	251	260	54.58	53.94	53.07	52.82	52.41	52.69
Colombia	-	5.340	-	6.451	7.498	-	-	30.62	-	33.05	37.87	-
Croatia	14	11	10	9.390	8.490	8.250	38.97	39.53	39.53	39.12	39.32	40.03
Cyprus	0.239	0.224	0.227	0.231	0.193	-	34.44	34.73	35.80	37.62	34.77	-
Czechoslovakia (Former)	61	50	-	-	-	-	32.12	33.42	-	-	-	-
Czech Republic	60	49	37	33	-	-	48.00	49.49	44.58	44.59	-	-
Denmark	12	11	9.589	-	-	-	39.09	38.31	37.93	-	-	-
Egypt	-	5.749	5.849	5.509	-	-	-	17.42	17.83	17.27	-	-
El Salvador	-	-	-	0.249	0.136	0.039	-	-	-	41.29	18.73	21.67
Ethiopia	-	0.018	0.018	0.017	0.019	0.015	-	18.56	18.00	17.00	17.43	13.39
FYR Macedonia	3.332	3.229	2.572	2.653	2.626	2.500	27.20	27.87	21.29	24.47	27.50	26.88
Germany (Western Part)	389	-	-	-	-	-	37.67	-	-	-	-	-
Ghana	-	-	-	0.053	-	-	-	-	-	4.55	-	-
Hungary	54	-	31	27	-	-	45.00	-	44.29	45.76	-	-
Indonesia	-	-	-	56	79	86	-	-	-	51.90	54.78	52.49
Iran (Islamic Republic of)	-	6.087	6.236	6.210	-	-	-	12.46	12.27	12.64	-	-
Iraq	-	2.573	-	-	-	-	-	24.92	-	-	-	-
Ireland	12	11	-	-	-	-	52.49	51.35	-	-	-	-
Italy	-	79	66	58	56	-	-	30.03	25.80	24.84	24.68	-
Jordan	0.080	0.063	0.063	0.062	0.239	0.226	12.03	11.98	9.60	8.21	16.03	16.36
Kenya	0.166	0.173	0.188	0.190	-	-	7.01	6.68	6.78	6.79	-	-
Korea, Republic of	220	188	178	176	186	194	49.05	46.43	45.34	44.59	45.43	45.53
Macau	0.593	0.389	0.471	0.496	0.671	1.037	66.78	73.53	71.69	75.15	74.23	77.33
Malaysia	161	188	207	237	260	-	75.37	72.73	73.27	71.25	70.11	-
Malta	0.877	0.828	1.153	1.216	1.411	-	36.22	36.24	44.88	43.14	47.06	-
Mongolia	-	-	-	-	0.072	-	-	-	-	-	23.45	-
Morocco	-	-	-	-	-	4.807	-	-	-	-	-	45.10
Nepal	0.022	0.074	-	0.055	0.064	-	2.75	6.64	-	2.96	4.43	-
New Zealand	3.544	3.139	3.177	3.658	-	-	33.48	33.56	33.01	34.27	-	-
Panama	0.010	-	-	-	-	-	9.09	-	-	-	-	-
Philippines	-	-	71	66	-	-	-	-	-	69.73	66.53	-
Portugal	14	16	18	19	20	-	33.72	37.02	40.30	41.62	43.31	-
Puerto Rico	11	9.620	9.180	8.680	9.310	11	60.98	58.87	58.88	57.67	54.44	53.38
Senegal	0.004	-	-	-	-	-	1.23	-	-	-	-	-
Sri Lanka	0.620	0.433	0.297	0.797	-	-	37.99	17.97	19.08	24.33	-	-
Sweden	19	-	-	-	-	-	34.55	-	-	-	-	-
Thailand	34	34	-	-	-	-	40.56	46.45	-	-	-	-
Turkey	11	-	-	-	-	-	22.84	-	-	-	-	-
United Kingdom	152	-	139	-	-	-	29.93	-	31.88	-	-	-
United Republic of Tanzania	0.440	0.349	-	-	-	-	28.01	27.96	-	-	-	-
United States of America	-	-	930	933	966	1,014	-	-	62.29	62.58	63.55	64.06

Output and Output per Employee

Country	Output ($ bil.)						Output per Employee ($)					
	1990	1991	1992	1993	1994	1995	1990	1991	1992	1993	1994	1995
Albania	-	-	-	*0.003	*0.003	*0.003	-	-	-	2,959	2,188	5,485
Angola	-	-	±30	±5.751	-	-	-	-	§48,17	§10,53	-	-
Argentina	±2.285	-	-	3.649	-	-	79,513	-	-	110,163	-	-
Armenia	0.014	0.019	0.151	4.590	0.024	-	405	608	-	-	-	-
Australia	*5.226	*5.528	*4.949	-	-	-	96,774	108,389	109,984	-	-	-
Austria	11	11	11	11	12	14	127,109	136,022	142,486	144,403	159,945	194,346
Azerbaijan	±0.402	±0.354	±0.019	±0.056	±0.022	-	15,850	14,806	844	2,691	1,093	-
Bahrain	-	-	±0.103	-	-	-	-	-	183,143	-	-	-

Continued.

Depending on the table, * means *Enterprises, Engaged,* or *Factor Values*; ± means *Basis Unspecified*; § means *shown in millions*. For additional notes and sources, see Appendix I.

Output and Output per Employee

- Continued -

Country	Output ($ bil.)						Output per Employee ($)					
	1990	1991	1992	1993	1994	1995	1990	1991	1992	1993	1994	1995
Bangladesh	0.188	0.151	0.098	-	-	-	11,095	8,342	7,559	-	-	-
Belize	0.001	0.001	0.002	-	-	-	-	62,868	106,658	-	-	-
Bolivia	0.007	0.006	0.008	0.010	0.010	0.010	20,810	20,804	18,214	20,902	18,028	18,597
Bosnia & Herzegovina	±0.359	±0.289	-	-	-	-	19,597	18,789	-	-	-	-
Brazil	*19	-	*14	*19	-	-	62,825	-	65,262	95,625	-	-
Bulgaria	±1.854	±0.627	±0.490	±0.418	±0.372	±0.469	12,768	5,488	5,602	6,307	6,880	9,230
Cameroon	0.039	0.029	0.039	0.026	0.014	0.015	55,893	61,704	54,096	41,553	34,183	55,576
Canada	15	16	16	15	16	18	119,429	138,121	141,625	142,505	153,779	179,127
Central African Republic	*0.000	-	-	-	-	-	54,482	-	-	-	-	-
Chile	0.242	0.223	0.282	0.316	0.269	0.321	61,640	60,460	64,349	73,403	62,574	75,742
China	±29	±32	±39	±55	±50	±61	8,859	8,874	10,786	14,501	12,678	14,751
China (Hong Kong SAR)	4.836	4.283	4.549	4.297	4.574	-	70,910	78,307	91,165	98,324	119,430	-
China (Taiwan Province)	±29	±31	±34	±41	±45	±55	62,420	67,998	73,185	87,707	93,801	111,856
Colombia	0.645	0.592	0.599	0.609	0.743	0.769	35,444	33,926	30,844	31,208	37,508	41,107
Costa Rica	0.131	0.149	0.221	0.288	0.303	0.290	32,850	28,716	43,354	41,260	-	-
Croatia	±0.938	±0.936	±0.544	-	-	-	25,811	32,352	21,138	-	-	-
Cyprus	0.033	0.031	0.039	0.035	0.033	0.033	47,434	47,553	61,255	56,630	58,735	64,477
Czechoslovakia (Former)	±2.971	±0.920	-	-	-	-	15,720	6,172	-	-	-	-
Denmark	*2.622	*2.370	*2.710	*2.532	*2.935	*3.744	83,130	82,734	107,197	101,427	114,733	143,431
Ecuador	0.112	0.143	0.124	0.131	0.159	0.164	26,858	32,406	29,935	36,956	44,386	45,844
Egypt	*0.697	*0.534	*0.567	*0.679	*0.738	*0.813	22,498	16,188	17,276	21,295	22,885	25,058
El Salvador	-	±0.029	-	±0.047	±0.038	±0.003	-	-	-	78,649	51,829	17,081
Ethiopia	0.001	0.001	0.000	0.001	0.001	0.000	11,985	8,088	4,057	7,092	7,427	3,990
Finland	±3.838	±2.931	±3.099	±3.329	±5.002	±9.346	134,680	113,147	127,020	137,009	177,369	250,973
FYR Macedonia	±0.307	±0.350	±0.232	±0.174	±0.144	±0.175	25,031	30,218	19,181	16,010	15,098	18,831
France	±55	±56	±60	±55	±58	±69	116,330	118,722	131,784	127,060	138,788	161,479
Gabon	-	*0.007	*0.006	*0.004	*0.003	*0.003	-	222,673	173,198	115,549	90,871	200,812
Germany	-	131	144	131	137	-	-	104,984	130,441	129,429	145,860	-
Germany (Western Part)	124	127	139	-	-	-	120,196	121,796	137,821	-	-	-
Ghana	-	-	-	0.023	0.019	0.020	-	-	-	19,803	15,687	16,006
Greece	*1.211	*1.303	*1.436	*1.564	*1.553	*1.830	86,504	100,037	113,532	125,606	124,408	147,355
Guatemala	-	0.027	0.029	0.032	0.043	0.075	-	40,724	45,297	42,624	55,018	99,470
Honduras	0.022	0.018	0.022	0.024	0.022	0.029	26,966	18,758	35,438	36,763	30,104	34,737
Hungary	±1.911	±1.547	±1.302	±1.373	±2.017	±2.387	15,927	16,282	18,601	23,272	31,180	35,555
India	*8.896	*7.589	*8.104	*7.231	*9.761	*12	22,845	19,379	20,130	17,901	23,121	27,924
Indonesia	1.406	1.635	2.593	2.869	3.789	5.966	23,512	22,269	30,201	26,774	26,286	36,295
Iran (Islamic Republic of)	0.545	1.644	1.707	1.426	-	-	31,497	33,649	33,593	29,019	-	-
Iraq	-	*0.126	*0.456	-	-	-	-	12,252	45,595	-	-	-
Ireland	*3.164	*3.494	*4.304	*4.155	*5.474	*8.322	143,176	157,403	185,640	172,909	209,930	284,575
Israel	4.015	4.549	5.084	5.289	6.043	6.674	92,098	101,999	111,256	108,378	116,205	128,552
Italy	±39	±40	±39	±29	±31	-	147,505	150,718	152,981	123,192	136,111	-
Japan	±337	±392	±381	±409	±392	±456	184,834	209,635	209,831	235,885	263,025	309,295
Jordan	0.032	0.029	0.022	0.027	0.062	0.083	47,896	55,133	33,593	35,253	41,854	60,110
Kenya	*0.258	*0.279	*0.592	*0.367	*0.457	*0.135	109,048	107,754	213,719	130,958	161,283	42,054
Korea, Republic of	37	38	40	45	55	79	81,355	93,883	100,890	113,233	135,427	185,603
Kuwait	0.037	0.010	0.071	0.093	0.113	0.124	67,208	20,866	73,287	80,481	83,446	105,338
Kyrgyzstan	0.237	0.233	0.233	0.036	0.020	-	11,311	11,925	13,336	2,169	1,890	-
Latvia	0.009	0.015	0.064	0.067	0.073	0.073	167	304	1,543	2,489	4,072	5,466
Lithuania	-	-	0.164	0.210	0.117	-	-	-	3,262	4,796	4,102	-
Macau	0.037	0.028	0.026	0.037	0.051	0.057	41,385	53,639	39,384	55,925	56,837	42,733
Malaysia	*8.986	*12	*16	*20	*27	*35	42,068	47,542	56,846	60,154	71,941	80,848
Malta	0.536	0.650	0.868	0.747	0.951	-	221,240	284,368	337,747	265,124	317,193	-
Mauritius	0.020	0.027	0.027	0.027	0.030	0.034	22,017	27,697	31,205	34,153	39,359	38,235
Mexico	2.848	2.966	3.374	3.092	3.126	2.160	30,010	32,238	37,257	39,933	41,476	28,244
Mongolia	0.005	0.009	0.001	-	-	0.000	15,234	44,183	7,350	-	-	1,630
Morocco	±0.511	±0.471	±0.445	±0.328	±0.347	±0.374	45,454	40,550	43,730	31,268	34,881	35,124
Mozambique	-	±0.009	-	±0.003	±0.002	-	-	7,914	-	4,403	3,983	-
Nepal	0.011	0.016	-	0.024	0.015	-	13,858	14,192	-	12,858	10,066	-
Netherlands	17	16	17	12	13	16	147,135	154,742	169,657	139,501	153,993	187,386
New Zealand	0.845	-	-	-	-	-	79,801	-	-	-	-	-

Continued.

Output and Output per Employee

- Continued -

Country	Output ($ bil.)						Output per Employee ($)					
	1990	1991	1992	1993	1994	1995	1990	1991	1992	1993	1994	1995
Norway	±2.182	±2.048	±2.146	±1.637	±1.851	±2.287	134,128	131,355	159,742	133,480	149,994	191,024
Pakistan	0.509	0.521	0.566	-	-	-	26,419	27,922	28,974	-	-	-
Panama	0.007	0.010	0.011	0.015	0.017	0.017	67,455	84,009	94,510	113,345	123,924	122,566
Peru	0.529	0.355	0.411	-	-	-	64,664	48,971	56,425	-	-	-
Philippines	1.999	2.589	3.154	3.001	-	-	26,240	28,794	30,919	30,440	-	-
Poland	±2.694	±2.426	±1.944	±2.205	±2.558	±3.551	11,416	12,974	13,497	17,092	20,031	28,312
Portugal	2.629	2.861	3.190	2.910	3.119	4.206	63,899	64,722	71,179	64,406	67,961	83,495
Puerto Rico	±3.653	±3.322	±3.397	±3.843	±4.349	±4.965	211,137	203,286	217,890	255,322	254,351	247,025
Republic of Moldova	0.000	0.281	0.025	-	-	-	0.031	21,577	2,190	-	-	-
Romania	±2.996	±1.738	±0.779	±0.859	-	-	17,509	12,426	6,326	7,530	-	-
Russian Federation	-	-	-	2.331	2.517	3.566	-	-	-	3,618	4,877	7,571
Senegal	0.016	-	-	-	-	-	50,531	-	-	-	-	-
Singapore	*9.944	*11	*12	*13	*17	*20	97,705	108,063	127,071	142,780	175,876	201,692
Slovakia	-	±0.650	±0.576	±0.509	±0.496	-	-	12,593	13,792	14,689	16,134	-
Slovenia	1.464	1.168	1.091	0.976	1.106	1.390	35,969	31,829	32,908	32,599	40,260	50,392
South Africa	±3.365	±3.188	±3.048	±2.835	±3.078	±3.540	39,131	34,652	31,747	31,503	27,730	30,784
Spain	*15	*15	*14	±11	±12	±15	136,568	134,484	138,527	105,086	113,639	146,146
Sri Lanka	0.019	0.034	0.031	0.037	-	-	11,731	14,020	20,059	11,353	-	-
Sweden	*7.408	*9.364	*9.817	*6.779	*9.039	*13	134,697	149,449	167,930	133,149	166,336	265,326
Switzerland	21	20	20	19	21	28	159,453	158,399	169,105	162,063	179,952	243,318
Thailand	6.418	8.509	-	-	-	-	75,570	115,601	-	-	-	-
Tonga	0.000	-	-	-	-	-	42,678	-	-	-	-	-
Trinidad and Tobago	±0.056	±0.066	±0.063	±0.036	±0.041	-	36,188	44,926	45,062	30,516	38,415	-
Tunisia	0.395	0.447	0.540	0.525	0.577	0.681	-	-	-	81,013	67,366	63,726
Turkey	3.554	4.253	4.521	4.966	3.283	4.659	77,155	97,696	98,805	111,841	82,729	109,258
Ukraine	*3.758	*3.609	*3.332	*0.677	*0.608	*0.593	16,778	15,489	14,300	3,120	3,183	3,529
United Kingdom	*48	*45	46	32	38	43	94,710	97,634	105,557	105,008	115,463	125,630
United Republic of Tanzania	*0.014	*0.027	-	-	-	-	8,951	21,268	-	-	-	-
USSR (Former)	±44	-	-	-	-	-	11,473	-	-	-	-	-
United States of America	*204	*207	*228	*245	*269	*312	132,550	140,510	152,461	164,131	176,651	196,911
Uruguay	0.155	0.134	0.172	0.150	0.161	0.160	32,567	32,640	42,535	40,324	43,403	43,316
Venezuela	0.592	0.675	0.761	0.720	0.486	0.668	35,667	35,528	40,069	43,867	40,614	55,960
Yugoslavia	±0.185	±0.175	±0.327	-	±0.697	±0.416	-	2,782	5,721	-	13,512	8,190
Zambia	0.056	-	-	-	-	-	41,288	-	-	-	-	-
Zimbabwe	*0.151	*0.141	*0.116	*0.098	*0.110	*0.121	23,552	22,457	18,680	17,810	20,011	22,826

Value Added and Value Added per Employee

Country	Value Added ($ bil.)						Value Added per Employee ($)					
	1990	1991	1992	1993	1994	1995	1990	1991	1992	1993	1994	1995
Argentina	±1.025	-	-	1.122	-	-	35,675	-	-	33,884	-	-
Armenia	0.002	0.005	-	-	-	-	67	158	-	-	-	-
Australia	*2.466	*2.559	*2.572	-	-	-	45,674	50,169	57,155	-	-	-
Austria	3.926	3.973	4.292	4.121	4.369	5.335	47,306	49,052	53,651	55,631	59,150	71,919
Bangladesh	0.060	0.057	0.032	-	-	-	3,547	3,114	2,488	-	-	-
Belize	0.000	0.000	0.000	-	-	-	-	4,895	4,447	-	-	-
Bolivia	0.003	0.002	0.004	0.004	0.004	0.004	8,298	6,786	8,292	8,324	6,580	6,803
Bosnia & Herzegovina	±0.166	±0.134	-	-	-	-	9,047	8,673	-	-	-	-
Brazil	*12	-	*9.996	*14	-	-	41,404	-	46,219	67,827	-	-
Bulgaria	-	±0.316	±0.229	±0.190	±0.128	±0.181	-	2,764	2,625	2,865	2,362	3,563
Cameroon	0.009	0.009	0.010	0.005	0.003	0.004	13,036	18,600	14,075	8,476	7,655	13,579
Canada	7.465	7.035	6.842	6.604	7.052	8.083	60,690	61,711	62,201	63,501	69,820	81,153
Central African Republic	*0.000	-	-	-	-	-	12,243	-	-	-	-	-
Chile	0.125	0.115	0.151	0.183	0.155	0.185	31,805	31,188	34,498	42,413	36,002	43,588
China	±7.445	±7.938	±9.845	±15	±12	±15	2,284	2,230	2,705	4,077	3,123	3,566
China (Hong Kong SAR)	1.155	1.223	1.350	1.340	1.551	-	16,929	22,358	27,062	30,653	40,500	-
China (Taiwan Province)	8.500	9.356	10	11	12	14	18,537	20,618	22,524	23,920	24,914	27,815
Colombia	0.271	0.258	0.279	0.257	0.308	0.319	14,878	14,793	14,353	13,158	15,566	17,053

Continued.

Depending on the table, * means *Enterprises, Engaged,* or *Factor Values*; ± means *Basis Unspecified*; § means *shown in millions*. For additional notes and sources, see Appendix I.

Value Added and Value Added per Employee

- Continued -

Country	Value Added ($ bil.)						Value Added per Employee ($)					
	1990	1991	1992	1993	1994	1995	1990	1991	1992	1993	1994	1995
Costa Rica	0.033	0.038	0.055	0.069	0.071	0.068	8,287	7,269	10,815	9,861	-	-
Cote d'Ivoire	0.250	0.273	0.264	0.222	-	-	-	-	-	-	-	-
Croatia	±0.499	±0.518	±0.286	-	-	-	13,734	17,887	11,100	-	-	-
Cyprus	0.012	0.011	0.013	0.012	0.012	0.012	17,010	17,115	20,855	19,646	21,416	23,449
Czechoslovakia (Former) . . .	±0.894	-	-	-	-	-	4,731	-	-	-	-	-
Denmark	*1.319	*1.194	*1.362	*1.274	*1.468	*1.865	41,819	41,692	53,859	51,019	57,378	71,470
Ecuador	0.032	0.045	0.035	0.048	0.057	0.040	7,760	10,209	8,474	13,550	15,996	11,275
Egypt	*0.160	*0.148	*0.199	*0.191	*0.208	*0.228	5,177	4,498	6,054	5,993	6,434	7,039
El Salvador.	-	0.018	-	0.018	0.019	0.004				30,630	25,834	23,320
Ethiopia.	0.001	0.000	0.000	0.000	0.000	0.000	6,578	3,885	2,012	4,174	3,923	2,170
Finland	±1.832	±1.276	±1.433	±1.428	±1.724	±3.035	64,294	49,255	58,739	58,779	61,146	81,485
FYR Macedonia	±0.138	±0.137	±0.087	±0.067	±0.056	±0.065	11,250	11,855	7,203	6,207	5,880	6,964
France	±26	±26	±28	±26	±27	±31	54,369	54,654	61,104	59,739	63,840	73,920
Gabon	-	*0.003	*0.002	*0.002	*0.001	*0.001	-	78,306	65,053	49,001	28,586	49,142
Germany (Western Part) . . .	73	74	80	74	-	-	70,278	71,321	79,165	-	-	-
Ghana	-	-	-	0.009	0.008	0.008				7,961	6,709	6,846
Greece	*0.441	*0.499	*0.540	*0.589	*0.586	*0.690	31,543	38,271	42,667	47,336	46,906	55,610
Honduras	*0.007	*0.006	*0.005	*0.005	*0.005	*0.006	8,289	6,232	7,975	7,619	6,913	6,882
Hungary	±0.667	±0.466	±0.341	±0.414	±0.574	±0.661	5,556	4,903	4,876	7,023	8,864	9,840
India.	*2.009	*1.803	*1.897	*1.662	*2.232	*2.782	5,159	4,604	4,711	4,115	5,287	6,366
Indonesia	*0.403	*0.501	*0.944	*0.771	*1.108	*1.754	6,741	6,830	10,993	7,196	7,686	10,672
Iran (Islamic Republic of) . . .	0.350	0.778	0.780	0.668	-	-	20,236	15,924	15,354	13,606	-	-
Iraq	-	*0.041	*0.303	-	-	-		3,950	30,322	-	-	-
Ireland	*1.840	*2.021	*2.500	*2.420	*3.205	*4.897	83,249	91,043	107,833	100,720	122,909	167,474
Israel.	2.200	2.453	2.428	2.434	2.528	3.004	50,451	55,004	53,132	49,881	48,609	57,867
Italy	±15	±15	±16	±12	±12	-	56,307	57,597	62,130	50,043	51,015	-
Japan	±134	±156	±150	±163	±156	±185	73,401	83,117	82,482	93,958	104,461	125,696
Jordan	0.011	0.011	0.007	0.008	0.021	0.020	17,266	20,625	10,785	10,430	14,108	14,651
Kenya	*0.044	*0.047	*0.043	*0.027	*0.040	*0.052	18,420	17,959	15,682	9,609	14,101	16,072
Korea, Republic of	15	16	16	19	26	40	33,562	39,231	41,055	48,935	63,018	94,228
Kuwait	0.027	-0.013	0.027	0.034	0.040	0.046	49,322	§-27,4	27,689	29,456	29,805	38,642
Latvia	-	-	-	*0.036	*0.042	*0.032	-	-	-	1,325	2,358	2,387
Macau	0.007	0.004	0.006	0.006	0.010	0.015	8,024	8,459	9,475	8,883	11,374	10,839
Malaysia	*1.945	*2.641	*3.596	*4.426	*5.623	*7.324	9,107	10,216	12,748	13,295	15,186	17,012
Malta	0.061	0.055	0.067	0.066	0.088	-	25,125	24,002	25,932	23,496	29,387	-
Mauritius	0.005	0.007	0.007	0.007	0.009	0.009	5,690	7,237	7,820	8,172	11,129	9,863
Mexico	1.540	1.567	1.734	1.589	1.635	1.130	16,232	17,029	19,144	20,529	21,696	14,778
Mongolia	0.003	0.004	0.001	-	-	0.000	9,821	18,843	4,771	-	-	346
Morocco	±0.132	±0.127	±0.122	±0.113	±0.125	±0.138	11,701	10,924	11,985	10,802	12,529	12,942
Nepal	0.004	0.005	-	0.009	0.004	-	4,463	4,891	-	4,812	3,018	-
Netherlands	5.286	4.845	5.561	5.008	5.340	6.312	47,109	47,254	55,750	58,107	62,466	74,238
New Zealand	0.260	-	-	-	-	-	24,533	-	-	-	-	-
Norway	±0.751	±0.796	±0.759	±0.632	±0.670	±0.815	46,140	51,048	56,517	51,562	54,270	68,088
Pakistan	0.173	0.192	0.214	-	-	-	8,954	10,263	10,953	-	-	-
Panama.	0.003	0.003	0.004	0.005	0.006	0.006	26,500	28,684	32,837	40,207	44,327	43,794
Peru	0.235	0.142	0.129	-	-	-	28,784	19,623	17,673	-	-	-
Philippines.	0.775	0.985	1.011	1.043	-	-	10,171	10,953	9,915	10,575	-	-
Poland	±1.420	±0.972	±0.787	±0.819	±0.938	±1.289	6,018	5,198	5,468	6,352	7,345	10,282
Portugal	0.834	0.918	1.024	0.928	0.971	1.307	20,262	20,757	22,861	20,532	21,148	25,955
Puerto Rico	±1.409	±1.186	±1.289	±1.497	±1.419	±1.566	81,439	72,601	82,688	99,468	82,959	77,905
Romania	±1.208	±0.554	±0.245	±0.294	-	-	7,061	3,958	1,989	2,574	-	-
Russian Federation	-	-	-	±1.324	±1.412	±1.842	-	-	-	2,055	2,736	3,911
Saint Lucia.	-	±0.001	±0.001	±0.001	±0.001	±0.001	-	-	-	-	-	-
Senegal.	0.003	0.003	0.003	0.003	0.000	-	8,123	15,424	15,371	10,088	593	-
Singapore	*2.707	*3.076	*3.484	*3.913	*4.962	*5.798	26,597	30,195	35,607	41,703	50,581	57,916
Slovakia	-	-	-	±0.152	±0.151	-	-	-	-	4,375	4,925	-
Slovenia	0.546	0.363	0.414	0.345	0.397	0.465	13,405	9,888	12,478	11,526	14,441	16,847
South Africa	*0.970	*1.047	*1.188	*1.102	*1.521	*1.737	11,276	11,380	12,378	12,242	13,701	15,106
Spain	*5.978	*5.781	*5.567	±4.413	±4.534	±5.110	54,573	53,305	53,300	41,424	43,799	50,101
Sri Lanka	0.007	0.014	0.015	0.015	-	-	4,328	5,875	9,388	4,630	-	-

Continued.

Depending on the table, * means *Enterprises, Engaged,* or *Factor Values*; ± means *Basis Unspecified*; § means *shown in millions*. For additional notes and sources, see Appendix I.

Value Added and Value Added per Employee
- Continued -

Country	Value Added ($ bil.)						Value Added per Employee ($)					
	1990	1991	1992	1993	1994	1995	1990	1991	1992	1993	1994	1995
Sweden	*4.021	*3.007	*3.198	*2.568	*3.023	*4.140	73,108	47,994	54,706	50,443	55,634	84,139
Switzerland	8.476	7.909	7.856	7.343	8.087	11	64,070	61,839	66,019	63,627	70,631	94,894
Thailand	2.523	4.642	-	-	-	-	29,706	63,059	-	-	-	-
Trinidad and Tobago	±0.015	±0.024	±0.020	±0.014	±0.008	-	9,892	15,962	14,152	11,828	7,813	-
Tunisia	0.101	0.114	0.136	0.132	0.146	0.178	-	-	20,326	17,050	16,672	
Turkey	1.482	1.833	2.013	2.313	1.543	2.193	32,170	42,111	43,994	52,103	38,886	51,429
United Kingdom	*22	*21	*22	*15	*17	*19	44,010	46,037	49,348	47,502	52,593	57,136
United Republic of Tanzania	*0.004	*0.007	-	-	-	-	2,412	6,003	-	-	-	-
United States of America	*112	*112	*128	*135	*151	*180	72,940	76,027	85,423	90,856	99,659	113,757
Uruguay	0.069	0.067	0.087	0.082	0.089	0.088	14,622	16,230	21,586	22,212	23,938	23,906
Venezuela	0.245	0.248	0.296	0.245	0.188	0.538	14,779	13,027	15,565	14,919	15,717	45,050
Yugoslavia	±0.071	±0.078	±0.172	-	±0.319	±0.175	-	1,238	3,004	-	6,180	3,452
Zambia	0.030	-	-	-	-	-	22,023	-	-	-	-	-
Zimbabwe	*0.088	*0.088	*0.060	*0.050	*0.048	*0.053	13,700	13,901	9,733	9,138	8,707	9,917

Capital Investment

Country	Gross Fixed Capital Formation ($ mil.)						Per Establishment ($ mil)					
	1990	1991	1992	1993	1994	1995	1990	1991	1992	1993	1994	1995
Argentina	-	-	-	125	-	-	-	-	-	0.035	-	-
Australia	222	-	-	-	-	-	0.233	-	-	-	-	-
Austria	591	590	661	504	524	-	1.191	1.207	1.334	1.019	1.068	-
Bahrain	-	-	9.814	-	-	-	-	-	1.963	-	-	-
Bangladesh	3.688	1.312	1.284	-	-	-	0.009	0.003	0.003	-	-	-
Bolivia	0.070	0.113	0.772	0.203	1.036	-	0.004	0.006	0.041	0.010	0.055	-
Bosnia & Herzegovina	8.588	-	-	-	-	-	0.232	-	-	-	-	-
Brazil	1,098	-	496	368	-	-	0.722	-	0.362	0.297	-	-
Bulgaria	61	21	377	131	22	13	0.345	0.111	2.014	0.044	0.005	0.003
Cameroon	0.915	0.468	1.549	0.413	0.578	0.429	0.114	0.094	0.310	0.103	0.289	0.429
Canada	514	-	-	-	-	-	0.373	-	-	-	-	-
Chile	9.036	24	10	7.858	11	-	0.348	0.878	0.353	0.271	0.368	-
China (Hong Kong SAR)	252	202	307	207	323	-	0.181	0.171	0.281	0.220	0.384	-
Colombia	22	15	21	5.814	20	-	0.102	0.078	0.092	0.027	0.100	-
Cote d'Ivoire	242	237	249	215	-	-	-	-	-	-	-	-
Croatia	22	21	14	12	17	20	0.040	0.026	0.012	0.008	0.009	0.010
Cyprus	0.895	1.657	1.491	0.968	0.751	1.776	0.009	0.018	0.016	0.011	0.007	0.015
Czechoslovakia (Former)	236	-	-	-	-	-	2.223	-	-	-	-	-
Denmark	98	117	-	-	-	-	0.073	0.084	-	-	-	-
Ecuador	11	28	20	3.912	11	24	0.215	0.497	0.338	0.074	0.220	0.499
Egypt	16	23	9.195	39	-	-	0.194	0.216	0.088	0.354	-	-
El Salvador	-	3.713	-	3.325	0.901	0.163	-	0.309	-	0.416	0.064	0.023
Ethiopia	0.004	-	0.019	0.019	0.031	0.003	0.004	-	0.019	0.019	0.016	0.002
Finland	184	95	133	138	297	405	0.778	0.378	0.516	0.518	1.127	1.247
FYR Macedonia	8.513	12	0.981	2.780	1.998	2.050	0.114	0.070	0.003	0.007	0.004	0.003
France	4,492	4,010	3,614	2,409	2,585	2,890	-	-	-	-	-	-
Gabon	-	0.301	0.064	0.021	0.295	0.014	-	0.301	0.064	0.021	0.295	0.014
Germany (Western Part)	7,402	6,934	7,263	-	-	-	1.921	1.753	1.815	-	-	-
Greece	72	60	80	-	-	-	0.243	0.207	0.276	-	-	-
Hungary	67	71	108	92	-	-	0.117	0.059	0.120	0.088	-	-
India	510	353	404	476	-	-	0.105	0.073	0.079	0.093	-	-
Indonesia	124	205	268	172	211	315	0.467	0.700	0.820	0.494	0.517	0.686
Iran (Islamic Republic of)	24	119	140	96	-	-	0.199	0.511	0.557	0.366	-	-
Ireland	109	128	-	-	-	-	0.379	0.432	-	-	-	-
Israel	207	313	338	362	373	-	0.373	0.524	0.533	0.558	0.502	-
Italy	2,393	2,281	1,876	1,418	1,443	-	1.380	1.287	0.855	0.648	0.667	-
Japan	17,936	23,510	20,679	18,085	17,356	-	0.515	0.659	0.611	0.551	0.655	-
Jordan	-0.008	0.182	0.974	0.478	5.159	4.194	-0.001	0.013	0.108	0.040	0.139	0.123
Korea, Republic of	4,242	4,828	4,125	5,267	7,054	10,931	0.646	0.738	0.607	0.667	0.860	1.240

Continued.

Depending on the table, * means *Enterprises, Engaged,* or *Factor Values;* ± means *Basis Unspecified;* § means *shown in millions.* For additional notes and sources, see Appendix I.

Capital Investment

- Continued -

Country	Gross Fixed Capital Formation ($ mil.)						Per Establishment ($ mil)					
	1990	1991	1992	1993	1994	1995	1990	1991	1992	1993	1994	1995
Kuwait	1.603	9.673	26	0.460	3.567	-	0.073	0.484	0.714	0.013	0.099	-
Latvia	0.287	0.286	2.723	0.454	4.348	1.871	0.014	0.006	0.034	0.004	0.082	0.021
Macau	1.402	0.317	1.534	4.885	9.914	7.334	0.074	0.032	0.040	0.158	0.367	0.349
Malaysia	969	1,309	1,208	1,622	1,993	-	2.297	2.661	2.278	2.730	3.272	-
Malta	16	20	17	20	10	-	0.368	0.409	0.353	0.343	0.187	-
Mexico	76	139	-	-	-	-	0.438	0.843	-	-	-	-
Morocco	23	20	17	21	25	32	0.176	0.167	0.144	0.185	0.223	0.299
Nepal	2.554	-	-	-	-	-	0.122	-	-	-	-	-
Netherlands	958	631	814	636		-	2.835	1.714	2.072	1.773	-	-
New Zealand	21	-	-	-	-	-	0.035	-	-	-	-	-
Norway	79	84	76	52	52	-	0.280	0.314	0.391	0.309	0.299	-
Pakistan		23	-	-	-	-	-	0.106	-	-	-	-
Panama	0.749	1.971	-	-	-	-	0.094	0.282	-	-	-	-
Peru	8.226	5.090	4.190	-	-	-	0.016	0.009	0.008	-	-	-
Philippines	173	177	174	256	-	-	0.605	0.524	0.589	0.966	-	-
Poland	139	109	170	155	-	-	0.391	0.301	0.502	0.487	-	-
Portugal	94	252	143	209	95	-	0.080	0.216	0.129	0.170	0.073	-
Romania	-	-	15	118	23	-	-	-	0.058	0.353	0.021	-
Senegal	0.679						0.226					
Singapore	703	642	690	885	1,272	-	2.175	1.922	1.994	2.514	3.583	-
Slovenia	78	76	42	52	64	110	0.126	0.076	0.033	0.044	0.044	0.064
Spain	620	771	616	382	359	526	0.294	0.384	0.315	-	-	-
Sri Lanka	1.178	2.188	1.461	1.571	-	-	0.054	0.088	0.077	0.044	-	-
Thailand	1,070	548	-	-	-	-	8.109	1.986	-	-	-	-
Trinidad and Tobago	-	-	1.718	1.514	1.165	-	-	-	0.066	0.061	0.051	-
Tunisia	25	30	44	33	33	36	-	-	0.168	0.144	0.160
Turkey	225	208	178	215	165	-	0.893	0.873	0.415	0.534	0.461	-
Ukraine	144	79	45	2.625	8.837	13	0.951	0.246	0.093	0.014	0.046	0.065
United Kingdom	1,896	1,665	1,698	-	1,700	-	0.189	0.170	0.167	-	0.172	-
United Republic of Tanzania	6.285	10	-	-	-	-	0.698	1.304	-	-	-	-
United States of America	9,340	8,040	8,971	10,089	12,856	17,420	-	-	0.377	-	-	-
Uruguay	4.003	4.153	2.892	4.313	-	-	-	-	-	-	-	-
Venezuela	29	18	39	25	17	130	0.122	0.073	0.163	0.124	0.105	0.686
Yugoslavia	4.441	3.355	3.361	-	6.048	3.549	0.009	0.004	0.002	-	0.003	0.002
Zambia	7.032	-	-	-	-	-	0.586	-	-	-	-	-
Zimbabwe	13	22	8.951	6.721	7.558	-	0.225	0.346	0.149	0.137	0.140	-

Indexes of Industrial Production

Country	Index of Industrial Production (1990=100)						Index of Employment (1990=100)					
	1990	1991	1992	1993	1994	1995	1990	1991	1992	1993	1994	1995
Albania	100	-	-	-	-	-	-	-	-	-	-	-
Algeria	100	92	77	58	55	48	-	-	-	-	-	-
Angola	100	-	-	-	-	-	-	-	-	-	-	-
Argentina	100	98	-	-	-	-	100	-	-	115	115	-
Armenia	100	-	-	-	-	-	100	94	-	-	-	-
Australia	100	98	116	123	-	-	100	94	83	-	-	-
Austria	100	107	105	106	113	119	100	98	96	89	89	89
Azerbaijan	-	-	-	-	-	-	100	94	87	82	79	-
Bahrain	100	-	-	-	-	-	-	-	-	-	-	-
Bangladesh	100	76	77	79	88	101	100	107	77	-	-	-
Barbados	100	89	129	102	94	102	-	-	-	-	-	-
Belgium	100	95	87	81	86	91	-	-	-	-	-	-
Belize	100	101	101	90	86	100	-	-	-	-	-	-
Bermuda	-	-	-	-	-	-	100	100	127	53	113	47
Bolivia	100	104	111	116	136	132	100	85	138	156	171	173
Bosnia & Herzegovina	-	-	-	-	-	-	100	84	-	-	0	-
Brazil	100	94	82	93	111	127	100	-	73	68	-	-

Continued.

Depending on the table, * means *Enterprises, Engaged,* or *Factor Values*; ± means *Basis Unspecified*; § means *shown in millions.* For additional notes and sources, see Appendix I.

611

Indexes of Industrial Production
- Continued -

Country	Index of Industrial Production (1990=100)						Index of Employment (1990=100)					
	1990	1991	1992	1993	1994	1995	1990	1991	1992	1993	1994	1995
Bulgaria	100	64	41	39	39	39	100	79	60	46	37	35
Burkina-Faso	100	-	-	-	-	-	-	-	-	-	-	-
Burundi	100	100	100	100	-	-	-	-	-	-	-	-
Cambodia	100	-	-	-	-	-	100	86	-	-	-	-
Cameroon	100	-	-	-	-	-	100	68	105	90	60	39
Canada	100	93	97	98	108	122	100	93	89	85	82	81
Central African Republic	100	-	-	-	-	-	100	-	-	-	-	-
Chile	100	107	123	127	128	134	100	94	112	110	109	108
China	100	-	-	-	-	-	100	109	112	116	121	128
China (Hong Kong SAR)	100	103	105	107	111	121	100	80	73	64	56	-
China (Taiwan Province)	100	111	117	129	145	170	100	99	101	101	104	108
Colombia	100	91	106	114	120	111	100	96	107	107	109	103
Costa Rica	100	97	138	168	186	180	100	130	128	175	-	-
Cote d'Ivoire	100	61	53	24	12	21	-	-	-	-	-	-
Croatia	100	-	-	-	-	-	100	80	71	66	59	57
Cuba	100	-	-	-	-	-	-	-	-	-	-	-
Cyprus	100	105	133	147	134	146	100	93	91	88	80	74
Czechoslovakia (Former)	100	61	45	-	-	-	100	79	-	-	-	-
Czech Republic	100	-	-	-	-	-	100	79	66	59	-	-
Dem. Rep. of the Congo	100	-	-	-	-	-	-	-	-	-	-	-
Denmark	100	99	103	97	109	120	100	91	80	79	81	83
Dominican Republic	100	-	-	-	-	-	-	-	-	-	-	-
Ecuador	100	117	137	152	-	-	100	106	100	85	86	86
Egypt	100	106	97	91	95	97	100	106	106	103	104	105
El Salvador	100	105	-	-	-	-	-	-	-	-	-	-
Ethiopia and Eritrea	100	-	-	-	-	-	-	-	-	-	-	-
Ethiopia	-	-	-	-	-	-	100	98	101	101	110	113
Fiji	100	-	-	-	-	-	-	-	-	-	-	-
Finland	100	90	115	146	202	254	100	91	86	85	99	131
FYR Macedonia	100	-	-	-	-	-	100	95	99	89	78	76
France	100	98	95	92	93	98	100	100	96	91	89	90
Gabon	100	-	-	-	-	-	-	-	-	-	-	-
Gambia	100	-	-	-	-	-	-	-	-	-	-	-
Germany (Eastern Part)	100	-	-	-	-	-	-	-	-	-	-	-
Germany (Western Part)	100	105	104	97	104	104	100	101	98	-	-	-
Ghana	100	-	-	-	-	-	-	-	-	-	-	-
Greece	100	112	114	125	123	127	100	93	90	89	89	89
Guatemala	100	-	-	-	-	-	-	-	-	-	-	-
Guyana	100	100	100	100	100	100	-	-	-	-	-	-
Haiti	100	-	-	-	-	-	-	-	-	-	-	-
Honduras	100	122	143	201	136	157	100	117	75	82	90	103
Hungary	100	76	62	68	99	115	100	79	58	49	54	56
Iceland	100	-	-	-	-	-	-	-	-	-	-	-
India	100	93	100	85	107	128	100	101	103	104	108	112
Indonesia	100	69	70	73	80	86	100	123	144	179	241	275
Iran (Islamic Republic of)	100	160	181	175	142	139	100	282	294	284	-	-
Iraq	100	-	-	-	-	-	-	-	-	-	-	-
Ireland	100	107	121	134	169	234	100	100	105	109	118	132
Israel	100	106	114	124	132	144	100	102	105	112	119	119
Italy	100	99	95	93	-	-	100	99	96	88	85	-
Jamaica	100	-	-	-	-	-	-	-	-	-	-	-
Japan	100	107	97	96	104	116	100	103	100	95	82	81
Jordan	100	-	-	-	-	-	100	79	99	114	224	208
Kenya	100	134	130	116	117	131	100	109	117	118	120	136
Korea, Republic of	100	108	116	121	137	161	100	90	87	88	91	95
Kuwait	100	-	-	-	-	-	100	85	173	206	243	212
Kyrgyzstan	100	-	-	-	-	-	100	93	83	78	50	-
Lao P.D.R.	100	-	-	-	-	-	-	-	-	-	-	-
Latvia	100	-	-	-	-	-	100	92	75	49	32	24
Lebanon	100	-	-	-	-	-	-	-	-	-	-	-

Continued.

Indexes of Industrial Production

- Continued -

Country	Index of Industrial Production (1990=100)						Index of Employment (1990=100)					
	1990	1991	1992	1993	1994	1995	1990	1991	1992	1993	1994	1995
Liberia	100	-	-	-	-	-	-	-	-	-	-	-
Libyan Arab Jamahiriya	100	-	-	-	-	-	-	-	-	-	-	-
Lithuania	100	-	-	-	-	-	-	-	-	-	-	-
Luxembourg	100	83	83	82	91	108	-	-	-	-	-	-
Macau	100	-	-	-	-	-	100	60	74	74	102	151
Madagascar	100	-	-	-	-	-	-	-	-	-	-	-
Malawi	100	-	-	-	-	-	-	-	-	-	-	-
Malaysia	100	129	144	166	200	242	100	121	132	156	173	202
Mali	100	-	-	-	-	-	-	-	-	-	-	-
Malta	100	124	163	168	-	-	100	94	106	116	124	-
Mauritius	100	-	-	-	-	-	100	105	95	88	84	97
Mexico	100	111	110	110	118	101	100	97	95	82	79	81
Mongolia	100	77	101	34	50	102	100	61	51	-	96	69
Morocco	100	95	85	96	90	94	100	103	91	93	88	95
Mozambique	100	71	-	-	-	-	100	90	79	64	46	-
Myanmar	100	60	38	-	-	-	-	-	-	-	-	-
Namibia	100	-	-	-	-	-	-	-	-	-	-	-
Nepal	100	-	-	-	-	-	100	139	-	232	180	-
Netherlands	100	99	98	99	106	111	100	91	89	77	76	76
New Zealand	100	-	-	-	-	-	100	88	91	101	-	-
Nicaragua	100	-	-	-	-	-	-	-	-	-	-	-
Niger	100	-	-	-	-	-	-	-	-	-	-	-
Nigeria	100	-	-	-	-	-	-	-	-	-	-	-
Norway	100	94	96	102	109	118	100	96	83	75	76	74
Pakistan	100	95	73	-	-	-	100	97	101	-	-	-
Panama	100	127	149	194	219	213	100	104	110	120	125	125
Papua New Guinea	100	-	-	-	-	-	-	-	-	-	-	-
Paraguay	100	103	106	51	36	36	-	-	-	-	-	-
Peru	100	103	84	75	75	109	100	89	89	-	-	-
Philippines	100	117	129	144	170	202	100	118	134	129	-	-
Poland	100	79	78	87	95	111	100	79	61	55	54	53
Portugal	100	104	107	100	100	118	100	107	109	110	112	122
Puerto Rico	100	-	-	-	-	-	100	94	90	87	99	116
Republic of Moldova	100	-	-	-	-	-	100	78	70	-	-	-
Romania	100	75	50	53	81	103	100	82	72	67	60	-
Russian Federation	100	-	-	-	-	-	-	-	-	-	-	-
Saudi Arabia	100	-	-	-	-	-	-	-	-	-	-	-
Senegal	100	29	21	26	25	38	100	63	67	77	53	-
Singapore	100	109	109	113	125	131	100	100	96	92	96	98
Slovakia	100	-	-	-	-	-	-	-	-	-	-	-
Slovenia	100	84	68	70	86	100	100	90	81	74	68	68
Somalia	100	100	100	-	-	-	-	-	-	-	-	-
South Africa	100	94	87	85	93	99	100	107	112	105	129	134
Spain	100	101	92	94	106	121	100	99	95	97	94	93
Sri Lanka	100	94	101	124	154	213	100	148	95	201	-	-
Sudan	100	-	-	-	-	-	-	-	-	-	-	-
Suriname	100	-	-	-	-	-	-	-	-	-	-	-
Sweden	100	99	105	117	152	198	100	114	106	93	99	89
Switzerland	100	105	104	100	102	104	100	97	90	87	87	86
Syrian Arab Republic	100	148	198	-	-	-	-	-	-	-	-	-
Thailand	100	101	-	-	-	-	100	87	-	-	-	-
Tonga	100	-	-	-	-	-	100	167	-	-	-	-
Trinidad and Tobago	100	131	132	128	117	166	100	95	90	77	68	-
Tunisia	100	102	112	108	93	93	-	-	-	-	-	-
Turkey	100	120	118	122	104	128	100	94	99	96	86	93
Uganda	100	74	109	320	379	401	-	-	-	-	-	-
Ukraine	100	-	-	-	-	-	100	104	104	97	85	75
United Arab Emirates	100	-	-	-	-	-	-	-	-	-	-	-
United Kingdom	100	92	88	91	100	105	100	92	86	61	65	67
United Republic of Tanzania	100	128	135	134	112	-	100	79	-	-	-	-

Continued.

Depending on the table, * means *Enterprises, Engaged,* or *Factor Values*; ± means *Basis Unspecified*; § means *shown in millions*. For additional notes and sources, see Appendix I.

613

Indexes of Industrial Production

- Continued -

Country	Index of Industrial Production (1990=100)						Index of Employment (1990=100)					
	1990	1991	1992	1993	1994	1995	1990	1991	1992	1993	1994	1995
USSR (Former)	100	-	-	-	-	-	100	-	-	-	-	-
United States of America . . .	100	102	111	122	140	164	100	95	97	97	99	103
Uruguay	100	84	84	69	72	66	100	86	85	78	78	78
Venezuela	100	-	-	-	-	-	100	114	114	99	72	72
Yugoslavia	100	-	-	-	-	-	-	-	-	-	-	-
Yugoslavia (Former)	100	71	-	-	-	-	-	-	-	-	-	-
Zambia	100	97	90	75	64	68	100	-	-	-	-	-
Zimbabwe	100	102	90	74	83	79	100	98	97	86	86	83

   Depending on the table, * means *Enterprises, Engaged,* or *Factor Values*; ± means *Basis Unspecified*; § means *shown in millions.* For additional notes and sources, see Appendix I.

Representative Companies in Sector

Name	Address	Tele-phone	Fax	Employ-ment	Y
Northern Telecom Ltd.	8200 Dixie Rd. Ste. 100, Brampton, Ontario L6T 5P6, Canada	905-863-0000	905-863-8263	22,000	98
Crane Co.	100 1st Stamford Pl., Stamford, Connecticut 06902, U.S.A.	203-363-7300	203-363-7295	10,000	96
Zenith Electronics Corp.	1000 Milwaukee Ave., Glenview, Illinois 60025-2493, U.S.A.	847-391-7000	847-391-7253	10,000	97
General Signal Corp.	PO Box 10010, Stamford, Connecticut 06904-2010, U.S.A.	203-329-4100		9,900	97
Teleflex Inc.	630 W. Germantown Pike, Ste. 450, Plymouth Meeting, Pennsylvania 19462-1074, U.S.A.	610-834-6301	610-834-8228	9,700	96
Duracell Inc.	Berkshire Corporate Park, Bethel, Connecticut 06801, U.S.A.	203-796-4000	203-778-9016	9,600	96
Siemens Energy and Automation Inc.	3333 Old Milton Pkwy., Alpharetta, Georgia 30202, U.S.A.	770-751-2000	770-751-2001	9,500	97
Alcatel Standard Electrica, SA	Ramirez de Prado, 5, Madrid E-28045, Spain	91 5272121	91 5286122	9,348	97
Nortek Inc.	50 Kennedy Plz., Providence, Rhode Island 02903-2360, U.S.A.	401-751-1600	401-751-4610	9,262	97
GE Appliances	Appliance Park, Louisville, Kentucky 40225, U.S.A.	502-452-4311	502-452-3389	9,000	96
QUALCOMM Inc.	6455 Lusk Blvd., San Diego, California 92121-2779, U.S.A.	619-587-1121	619-658-2100	9,000	97
Raychem Corp.	300 Constitution Dr., Menlo Park, California 94025-1164, U.S.A.	415-361-3333	415-361-4913	8,650	97
Harris Corp. Harris Semiconductor	PO Box 883, Melbourne, Florida 32901, U.S.A.	407-724-7000	407-729-5691	8,500	97
A.O. Smith Corp.	PO Box 23972, Milwaukee, Wisconsin 53223-3972, U.S.A.	414-359-4000	414-359-4248	8,400	97
Harman International Industries Inc.	1101 Pennsylvania Ave. NW, Ste. 1010, Washington, District of Columbia 20004-2506, U.S.A.	202-393-1101		8,384	97
Pulse Engineering Inc.	PO Box 12235, San Diego, California 92112, U.S.A.	619-674-8100	619-674-8263	8,300	96
Tadiron Ltd.	29 Hamerkava St., Holon 58101, Israel	3 5573200	3 5573280	8,260	97
Northern Telecom Europe Ltd.	Stafferton Way, Maidenhead SL6 1AY, United Kingdom	628 812000	628 812810	8,172	93
Ceridian Corp.	8100 34th Ave. S, Minneapolis, Minnesota 55425-1604, U.S.A.	612-853-8100	612-853-7896	8,000	97
Leviton Manufacturing Company Inc.	59-25 Little Neck Pkwy., Little Neck, New York 11362, U.S.A.	718-229-4040	718-631-6439	8,000	95
Philips Consumer Electronics Co.	PO Box 14810, Knoxville, Tennessee 37914-1810, U.S.A.	615-521-4316		8,000	94
Valor Electronics Inc.	9715 Business Park Ave., San Diego, California 92131-1642, U.S.A.	619-537-2500	619-537-2727	8,000	97
Ball Corp.	PO Box 2407, Muncie, Indiana 47307-0407, U.S.A.	765-747-6100		7,900	97
Electronics Corp. of India Ltd.	ECIL (Post), Hyderabad 500762, India	842 623409	842 621802	7,900	93
Maquinas Agricolas Jacto SA	Rua Doctor Luiz Miranda 1650 Centro, Pompeia 01037-010, Brazil	11 2144411	11 2141781	7,800	97
Iwka AG	Gartenstr. 71, Karlsruhe D-76135, Germany	721 1430	721 143243	7,774	94
Analog Devices Inc.	PO Box 9106, Norwood, Massachusetts 02062-9106, U.S.A.	617-329-4700	800-262-5643	7,500	97
Eagle-Picher Industries Inc.	PO Box 779, Cincinnati, Ohio 45201, U.S.A.	513-721-7010	513-721-2341	7,500	95
Sunbeam Corp.	PO Box 9218, Delray Beach, Florida 33445, U.S.A.	561-243-2100		7,500	97
Anam Industrial Co., Ltd.	Kwangjin-gu, Seoul, Republic of Korea	2 4605114	2 4626354	7,450	97
Chicago Miniature Lamp Inc.	500 Chapman St., Canton, Massachusetts 02021, U.S.A.	781-828-2948	781-828-2012	7,408	97
Hubbell Inc.	584 Derby Milford Rd., Orange, Connecticut 06477-4024, U.S.A.	203-799-4100		7,405	94
Berg Electronics Corp.	101 S. Hanley Rd., Ste. 400, St. Louis, Missouri 63105, U.S.A.	314-726-1323		7,000	97
General Instrument Corp.	101 Tournament Dr., Horsham, Pennsylvania 19044, U.S.A.	215-323-1000	215-443-9454	7,000	97
MEMC Electronic Materials Inc.	501 Pearl Dr., St. Peters, Missouri 63376, U.S.A.	314-279-5500		7,000	96
NACCO Industries Inc.	5875 Landerbrook Dr., Mayfield Heights, Ohio 44124-4017, U.S.A.	216-449-9600		7,000	97
W.L. Gore and Associates Inc.	PO Box 9329, Newark, Delaware 19714, U.S.A.	302-738-4880	302-292-4159	7,000	97
Pittway Corp.	200 S. Wacker Dr., 700, Chicago, Illinois 60606-5802, U.S.A.	312-831-1070	312-831-0808	6,800	96
Diebold Inc.	PO Box 3077, North Canton, Ohio 44720-8077, U.S.A.	330-489-4000		6,714	97
AMETEK Inc.	Station Sq., Paoli, Pennsylvania 19301, U.S.A.	610-647-2121	610-296-3412	6,700	97
GTI Corp. (San Diego, California)	9715 Business Park Ave., San Diego, California 92131-1642, U.S.A.	619-537-2500		6,700	97
Cabletron Systems Inc.	PO Box 5005, Rochester, New Hampshire 03866-5005, U.S.A.	603-332-9400	603-337-2211	6,607	96
AlliedSignal Electronic and Avionics Systems	400 N. Rogers Rd., Olathe, Kansas 66062, U.S.A.	913-782-0400	913-791-1302	6,500	97
Oktobar	14.Oktobra 2, Krusevac YU-37000, Serbia	37 21502	37 29686	6,500	92
Sensormatic Electronics Corp.	PO Box 5037, Boca Raton, Florida 33431-0700, U.S.A.	561-989-7000	561-989-7017	6,500	97
Varian Associates Inc.	3050 Hansen Way, Palo Alto, California 94304-1000, U.S.A.	415-493-4000		6,500	97
NKT Holding AS	NKT Alle 1, Brondby DK-2605, Denmark	43482000	43961820	6,425	93
DSC Communications Corp.	1000 Coit Rd., Plano, Texas 75075-5813, U.S.A.	214-519-3000	214-519-2203	6,367	97
Federal Signal Corp.	1415 W. 22nd St., Oak Brook, Illinois 60521-9945, U.S.A.	708-954-2000		6,207	97

Continued

Representative Companies in Sector
- Continued -

Name	Address	Tele-phone	Fax	Employ-ment	Y
Hadco Corp.	12A Manor Pkwy., Salem, New Hampshire 03079, U.S.A.	603-898-8000	603-898-6227	6,142	97
Chunghwa Picture Tubes Ltd.	Yangmei Taoyuan, Tainan 326, Taiwan	3 4786121	3 4787072	6,000	92
Mitel Corp.	350 Legget Dr., Kanata, Ontario K2K 1X3, Canada	613-592-2122	613-592-4784	6,000	98
Rank America Inc.	5 Concourse Pkwy 2400, Atlanta, Georgia 30328, U.S.A.	404-392-9029	404-392-0585	6,000	
Sunbeam-Oster Household Products	PO Box 247, Laurel, Mississippi 39440, U.S.A.	601-425-7800	601-649-5680	6,000	94
Videoton Ipari Rt.	Berenyi U. 100, Szekesfehervar H-8000, Hungary	22 312730	22 314551	6,000	93
Lincoln Electric Co.	22801 St. Clair Ave., Cleveland, Ohio 44117-1199, U.S.A.	216-481-8100	216-486-1751	5,971	97
Bay Networks Inc.	4401 Great America Pkwy., Santa Clara, California 95052-8185, U.S.A.	408-988-2400	408-988-5525	5,960	97
ADC Telecommunications Inc.	PO Box 1101, Minneapolis, Minnesota 55440-1101, U.S.A.	612-938-8080	612-946-3292	5,924	97
Elco Holdings Ltd.	17 Haneviim St., Ramat Hasharon 47279, Israel	3 5121503	3 6834218	5,555	97
Flextronics International Ltd.	2241 Lundy Ave., San Jose, California 95131, U.S.A.	408-428-1300		5,518	96
ADFlex Solutions Inc.	2001 W. Chandler Blvd., Chandler, Arizona 85224, U.S.A.	602-963-4584		5,511	97
Amana Appliance	PO Box 8901, Amana, Iowa 52204-0001, U.S.A.	319-622-5511	319-622-2180	5,500	94
Nokia Telecommunications	Upseerinkatu 1, Espoo SF-02601, Finland	51151	51042460	5,500	94
Philips Components Discrete Products Div.	1440 Indiantown Rd., Jupiter, Florida 33458, U.S.A.	407-745-3300	407-881-3300	5,500	95
Singer Do Brasil Industria e Comercio Ltda.	Rua Hungria 574, 8th Fl., Sao Paulo 01455, Brazil	11 8147422	11 349770	5,470	98
Amphenol Corp.	PO Box 5030, Wallingford, Connecticut 06492-7530, U.S.A.	203-265-8900	203-265-8516	5,459	95
UCAR International Inc.	39 Old Ridgebury Rd., Danbury, Connecticut 06817-0001, U.S.A.	203-794-7000		5,380	97
Scientific-Atlanta Inc.	1 Technology Pkwy. S., Norcross, Georgia 30092-2967, U.S.A.	770-903-5000		5,343	97
Orion Electric Co., Ltd.	Kumi-Shi, Kyongbuk, Republic of Korea	546 4695000	546 4618779	5,074	97
Walbro Corp.	6242 Garfield St., Cass City, Michigan 48726-1325, U.S.A.	517-872-2131	517-872-2301	5,028	97
Bourns Inc.	1200 Columbia Ave., Riverside, California 92507, U.S.A.	909-781-5070	909-781-5378	5,000	97
Champion Spark Plug Company Div.	PO Box 910, Toledo, Ohio 43661, U.S.A.	419-535-2567	419-535-2332	5,000	93
Cooper Lighting	400 Busse Rd., Elk Grove Village, Illinois 60007, U.S.A.	847-956-8400	847-956-0043	5,000	97
Donnelly Corp.	414 E. 40th St., Holland, Michigan 49423-5368, U.S.A.	616-786-7000	616-786-6034	5,000	97
Griffon Corp.	100 Jericho Quadrangle, Jericho, New York 11753, U.S.A.	516-938-5544	516-938-5644	5,000	97
Hamilton Beach/Proctor-Silex Inc.	4421 Waterfront Dr., Glen Allen, Virginia 23060, U.S.A.	804-273-9777		5,000	97
Siecor Corp.	489 Siecor Park, Hickory, North Carolina 28601, U.S.A.	704-327-5000	704-327-5993	5,000	97
Teradyne Inc.	179 Lincoln St., Boston, Massachusetts 02111-2473, U.S.A.	617-482-2700	617-422-3440	5,000	97
Transelco Div.	1789 Transelco Dr., Penn Yan, New York 14527, U.S.A.	315-536-3357	315-536-8091	5,000	96
Itautec Philco	Distrito Industrial, Manaus 69075-000, Brazil	92 6151717	92 6151787	4,887	96
Atmel Corp.	2325 Orchard Pkwy., San Jose, California 95131, U.S.A.	408-441-0311	408-436-4200	4,589	97
Micro Switch Div.	11 W. Spring St., Freeport, Illinois 61032, U.S.A.	815-235-5500	815-235-6545	4,500	97
Ranco Inc.	8161 U.S. Rte. 42 N., Plain City, Ohio 43064, U.S.A.	614-873-9219		4,500	95
TDK U.S.A. Corp.	12 Harbor Park Dr., Port Washington, New York 11050, U.S.A.	516-625-0100	516-625-2923	4,500	97
Mecanismos Auxiliares Industriales, SA	Passeig de l'Estacio 14, Valls E-43800, Spain	977 617100	977 617320	4,430	97
International Rectifier Corp.	233 Kansas St., El Segundo, California 90245, U.S.A.	310-726-8000		4,385	97
Cherry Corp.	3600 Sunset Ave., Waukegan, Illinois 60087, U.S.A.	708-662-9200		4,367	96
Asea Brown Boveri AS	Petersmindevej 1, Odense DK-5000, Denmark	66147080	66142580	4,300	93
Applied Power Inc.	PO Box 325, Milwaukee, Wisconsin 53201-0325, U.S.A.	414-781-6600	414-783-3790	4,235	97
Coleman Company Inc.	1526 Cole Blvd., 300, Golden, Colorado 80401, U.S.A.	303-202-2400		4,200	96
Elektrosvit AS	Komarnanska cesta 3, Nove Zamky 940 37, Slovakia	817 22886	817 27866	4,200	93
Zelmer	Hoffmanowej Str. 19, Rzeszow PL-35-016, Poland	17 37431	17 36178	4,200	93
Komag Inc.	1704 Automation Pkwy., San Jose, California 95131, U.S.A.	408-576-2000		4,101	96
Tellabs Inc.	4951 Indiana Ave., Lisle, Illinois 60532, U.S.A.	630-378-8800		4,087	97
AEG Corp.	180 Mount Airy Rd., PO Box 609, Basking Ridge, New Jersey 07920, U.S.A.	908-204-8900	908-204-8999	4,000	93
Cooper Power Systems	PO Box 1640, Waukesha, Wisconsin 53187-1640, U.S.A.	414-524-3300	414-896-2381	4,000	97
Diora	Swidnicka ul. 38, Dzierzoniow PL-58-200, Poland	74 322200	74 318561	4,000	93
Fujitsu America Inc.	3055 Orchard Dr., San Jose, California 95134, U.S.A.	408-432-1300	408-456-7064	4,000	97
Simplex Time Recorder Co.	1 Simplex Plz., Gardner, Massachusetts 01441, U.S.A.	978-632-2500		4,000	97
M/A-COM Inc.	1011 Pawtucket Blvd., Lowell, Massachusetts 01853, U.S.A.	617-224-5600	617-224-5655	3,932	94
Augat Inc.	PO Box 448, Mansfield, Massachusetts 02048, U.S.A.	508-543-4300	508-543-7019	3,900	95
Imo Industries Inc.	1009 Lenox Dr., Lawrenceville, New Jersey 08648, U.S.A.	609-896-7600	609-896-7688	3,900	96

Continued

For sources and notes, see Appendix I.

Representative Companies in Sector
- Continued -

Name	Address	Tele-phone	Fax	Employ-ment	Y
CTS Corp.	905 West Blvd. N., Elkhart, Indiana 46514, U.S.A.	219-293-7511	219-293-6146	3,893	97
LSI Logic Corp.	1551 McCarthy Blvd., Milpitas, California 95035, U.S.A.	408-433-8000	408-434-6457	3,870	96
Baldor Electric Co.	PO Box 2400, Fort Smith, Arkansas 72902, U.S.A.	501-646-4711	501-648-5752	3,843	97
Integrated Device Technology Inc.	PO Box 58015, Santa Clara, California 95052-8015, U.S.A.	408-727-6116	408-492-8674	3,828	95
Alpine Group Inc.	1790 Broadway, 1500, New York, New York 10019-1412, U.S.A.	212-757-3333	212-757-3423	3,809	97
Valmont Industries Inc.	PO Box 358, Valley, Nebraska 68064, U.S.A.	402-359-2201	402-343-0668	3,751	97
Bose Corp.	The Mountain, Framingham, Massachusetts 01701-9168, U.S.A.	508-879-7330	508-872-6541	3,700	96
Jabil Circuit Inc.	10800 Roosevelt Blvd., St. Petersburg, Florida 33716, U.S.A.	813-577-9749	813-579-8529	3,661	97
Methode Electronics Inc.	7444 W. Wilson Ave., Chicago, Illinois 60656, U.S.A.	708-867-9600	708-867-9130	3,650	97
Energizer Power Systems Div.	Po Box 147114, Gainesville, Florida 32614-7114, U.S.A.	904-462-3911	904-462-6251	3,600	96
Jerrold Communications Div.	2200 Byberry Rd., Hatboro, Pennsylvania 19040, U.S.A.	215-674-4800		3,600	95
Eureka Co.	1201 E. Bell St., Bloomington, Illinois 61701, U.S.A.	309-828-2367	309-823-5203	3,500	94
Siemens Canada Ltd.	2185 Derry Rd. W, Mississauga, Ontario L5N 7A6, Canada	905-819-8000	905-819-5777	3,500	98
Telefongyar	PO Box 16, Budapest, Hungary	1 2526949	1 2529161	3,500	93
Motorola Israel Ltd.	3 Krementski St., Tel Aviv 67899, Israel	3 5658888	3 5624925	3,450	97
Siemens	Salto, Sao Paulo 05110, Brazil	11 7541444	11 7851838	3,419	96
ESCO Electronics Corp.	8888 Ladue Rd., 200, St. Louis, Missouri 63124-2090, U.S.A.	314-213-7200		3,400	97
Lucent Technologies Octel Messaging Div.	1001 Murphy Ranch Rd., Milpitas, California 95035-7912, U.S.A.	408-321-2000		3,400	97
Oak Industries Inc.	1000 Winter St., Waltham, Massachusetts 02154, U.S.A.	781-890-0400	781-890-8585	3,373	97
Giddings and Lewis Inc.	PO Box 590, Fond du Lac, Wisconsin 54936-0590, U.S.A.	414-921-9400	414-929-4522	3,315	96
SKC Ltd.	Chung-gu, Seoul, Republic of Korea	2 37085151	2 7529088	3,310	97
Duracraft Corp.	250 Turnpike Rd., Southborough, Massachusetts 01772, U.S.A.	508-490-7080		3,300	95
Thomas Industries Inc.	PO Box 35120, Louisville, Kentucky 40232, U.S.A.	502-893-4600		3,300	97
Hartono Istana Electronics PT	Jl. K. H. R. Asnawi No. 310, Kudus 59316, Indonesia	29123255	21 5303865	3,297	97
Automated Security Holdings PLC	25-26 Hampstead High St., Hemel Hempstead HP2 7TL, United Kingdom	71 4357161	71 7947645	3,230	93
Moog Inc.	Plant 24, East Aurora, New York 14052-0018, U.S.A.	716-652-2000	716-687-4457	3,229	97
Catalina Lighting Inc.	18191 N.W. 68th Ave., Miami, Florida 33015, U.S.A.	305-558-4777	305-558-3024	3,225	97
Andrew Corp.	10500 West 153rd St., Orland Park, Illinois 60462, U.S.A.	708-349-3300	708-873-8954	3,213	97
Hyosung Industries Co., Ltd.	Mapo-gu, Seoul 121-020, Republic of Korea	2 7076000	2 7140707	3,200	97
LG Information & Communications, Ltd.	Youngdungpo-Gu, Seoul 150-721, Republic of Korea	2 37771114	2 37772777	3,200	97
Philips Technologies	PO Box 868, Cheshire, Connecticut 06410, U.S.A.	203-271-6000	203-276-6100	3,200	94
Zwut	ul. Zupnicza 17, Warsaw PL-03-821, Poland	22 191069	22 199944	3,200	93
Caribe General Electric	1590 Ponce de Leon Ave., Guaynabo 00926, Puerto Rico	787725-2966	787725-4735	3,150	92
Bang & Olufsen AS	Peter Bangsvej 15, Struer DK-7600, Denmark	97851122	97855944	3,143	93
American Power Conversion Corp.	PO Box 278, West Kingston, Rhode Island 02892-9906, U.S.A.	401-789-5735	401-789-3710	3,110	97
E.C.I. Telecom Ltd.	30 Hasivim St., Petah Tikva 49130, Israel	3 9266555	3 9266711	3,094	97
Lucas Sei Wiring Systems Ltd.	Ystradgynlais, Swansea SA9 1FY, United Kingdom	639 842281	639 849853	3,032	93
EFACEC, Empresa Fabril de Maquinas Electricas, SA	Apartado 18, Sao Mamede de Infesta P-4466, Portugal	95 12015	95 18940	3,007	91
A.O. Smith Electrical Products Co.	531 N. 4th St., Tipp City, Ohio 45371, U.S.A.	513-667-2431		3,000	95
Avex Electronics Inc.	4807 Bradford Blvd., Huntsville, Alabama 35805, U.S.A.	205-722-6000	205-722-7428	3,000	97
Bakony Muvek	Pf. 78, Veszprem H-8201, Hungary	80 24022	80 27916	3,000	93
Coilcraft Inc.	1102 Silver Lake Rd., Cary, Illinois 60013, U.S.A.	847-639-2361	847-639-1469	3,000	97
Crouse-Hinds Div. als	PO Box 4999, Syracuse, New York 13221, U.S.A.	315-477-7000		3,000	95
Deutsch Co.	2444 Wilshire Blvd 600, Santa Monica, California 90403, U.S.A.	909-849-7822	909-453-6467	3,000	
East Penn Manufacturing Company Inc.	PO Box 147, Lyon Station, Pennsylvania 19536-0147, U.S.A.	610-682-6361	610-682-4781	3,000	97
General Semiconductor Inc.	10 Melville Park Rd., Melville, New York 11747, U.S.A.	516-847-3000		3,000	97
Gould Electronics Inc.	35129 Curtis Blvd., Eastlake, Ohio 44095, U.S.A.	216-953-5000	216-953-5050	3,000	95
Rauland North	2407 W. North Ave., Melrose Park, Illinois 60160, U.S.A.	708-345-4700		3,000	95
Robbins and Myers Inc.	1400 Kettering Tower, Dayton, Ohio 45423, U.S.A.	513-222-2610	513-225-3355	3,000	97
VLSI Technology Inc.	1109 McKay Dr., San Jose, California 95131, U.S.A.	408-434-3000	408-263-2511	2,948	96
Allen Group Inc.	25101 Chagrin Blvd., Beachwood, Ohio 44122-5619, U.S.A.	216-765-5800	216-765-0410	2,900	96

Product Tables
Ovens for Household Use (Electric and Non-Electric)

Unit of Measure: Number.

	1990 Value	%	1991 Value	%	1992 Value	%	1993 Value	%	1994 Value	%	1995 Value	%	1996 Value	%
Total Production	30,568,728	100.0	32,407,121	100.0	32,987,723	100.0	33,740,685	100.0	35,403,276	100.0	36,819,906	100.0	38,486,194	100.0
Regions														
America, North	9,948,000	32.5	9,404,000	29.0	10,259,000	31.1	10,129,000	30.0	10,043,000	28.4	10,925,451	29.7	11,102,044	28.8
America, South	49,714	0.2	52,571	0.2	55,429	0.2	68,619	0.2	63,810	0.2	63,000	0.2	68,257	0.2
Asia	7,901,333	25.8	8,909,333	27.5	8,814,333	26.7	9,808,333	29.1	11,804,333	33.3	11,568,069	31.4	12,682,066	33.0
Europe	12,669,681	41.4	14,041,216	43.3	13,858,961	42.0	13,734,733	40.7	13,492,133	38.1	14,263,386	38.7	14,633,827	38.0
Leading Producers														
United States of America	9,947,000	32.5	9,403,000	29.0	10,258,000	31.1	10,128,000	30.0	10,042,000	28.4	*10,924,451	29.7	*11,101,044	28.8
Korea, Republic of	6,061,000	19.8	7,174,000	22.1	7,172,000	21.7	8,279,000	24.5	10,209,000	28.8	10,487,000	28.5	*11,426,308	29.7
Germany	*3,979,000	13.0	4,638,000	14.3	4,061,000	12.3	3,577,000	10.6	3,640,000	10.3	*3,979,000	10.8	*3,979,000	10.3
Japan	324,000	1.1	297,000	0.9	294,000	0.9	231,000	0.7	227,000	0.6	217,000	0.6	231,000	0.6
France	1,818,000	5.9	2,576,000	7.9	2,764,000	8.4	2,797,000	8.3	2,853,000	8.1	*3,188,571	8.7	*3,392,643	8.8

Stoves, Ranges, Cookers (For Household, Hotel and Restaurant Use Etc.)

Unit of Measure: Number.

	1990 Value	%	1991 Value	%	1992 Value	%	1993 Value	%	1994 Value	%	1995 Value	%	1996 Value	%
Total Production	60,586,247	100.0	65,654,838	100.0	71,409,235	100.0	70,148,550	100.0	74,071,255	100.0	70,989,103	100.0	76,078,056	100.0
Regions														
Africa	212,179	0.4	226,893	0.3	242,607	0.3	225,321	0.3	247,397	0.3	228,432	0.3	128,451	0.2
America, North	11,586,520	19.1	11,881,113	18.1	12,955,707	18.1	13,905,800	19.8	14,532,433	19.6	14,383,617	20.3	14,903,444	19.6
America, South	2,048,115	3.4	2,188,665	3.3	2,365,329	3.3	2,101,643	3.0	1,946,071	2.6	1,689,973	2.4	2,227,612	2.9
Asia	27,518,850	45.4	28,511,817	43.4	28,774,783	40.3	29,369,500	41.9	29,797,133	40.2	29,710,098	41.9	30,195,012	39.7
Europe	18,949,333	31.3	22,679,601	34.5	26,844,059	37.6	24,357,990	34.7	27,393,860	37.0	24,856,558	35.0	28,537,048	37.5
Oceania	271,250	0.4	166,750	0.3	226,750	0.3	188,295	0.3	154,360	0.2	120,425	0.2	86,490	0.1
Leading Producers														
Japan	*21,180,333	35.0	*21,180,333	32.3	*21,180,333	29.7	*21,180,333	30.2	*21,180,333	28.6	*21,180,333	29.8	*21,180,333	27.8
United States of America	10,201,000	16.8	10,324,000	15.7	11,404,000	16.0	12,011,000	17.1	12,300,000	16.6	*11,808,451	16.6	*11,985,444	15.8
Spain	7,734,000	12.8	6,459,000	9.8	7,959,000	11.1	3,358,000	4.8	3,530,000	4.8	3,761,000	5.3	*5,924,343	7.8
Germany	-		3,473,000	5.3	3,396,000	4.8	3,348,000	4.8	3,433,000	4.6	4,058,500	5.7	4,684,000	6.2
France	3,521,000	5.8	3,380,000	5.1	3,578,000	5.0	3,761,000	5.4	3,827,000	5.2	*4,497,945	6.3	*4,644,347	6.1

Drying Machines for Household Use

Unit of Measure: Number.

	1990 Value	%	1991 Value	%	1992 Value	%	1993 Value	%	1994 Value	%	1995 Value	%	1996 Value	%
Total Production	9,607,939	100.0	8,918,657	100.0	9,624,687	100.0	9,439,771	100.0	9,251,359	100.0	8,875,333	100.0	8,962,670	100.0
Regions														
America, North	4,622,552	48.1	4,616,381	51.8	4,963,210	51.6	5,268,038	55.8	5,514,867	59.6	5,369,231	60.5	5,466,524	61.0
America, South	1,222	0.0	1,111	0.0	1,000	0.0	1,000	0.0	1,000	0.0	1,000	0.0	762	0.0
Asia	2,912,750	30.3	2,302,750	25.8	1,595,750	16.6	1,098,000	11.6	863,000	9.3	768,000	8.7	672,000	7.5
Europe	1,844,414	19.2	1,840,415	20.6	2,914,727	30.3	2,922,733	31.0	2,701,242	29.2	2,565,852	28.9	2,652,135	29.6
Oceania	227,000	2.4	158,000	1.8	150,000	1.6	150,000	1.6	171,250	1.9	171,250	1.9	171,250	1.9
Leading Producers														
United States of America	4,187,000	43.6	4,165,000	46.7	4,496,000	46.7	4,782,000	50.7	5,012,000	54.2	4,852,000	54.7	4,932,000	55.0
Germany	-		-		1,078,000	11.2	980,000	10.4	975,000	10.5	729,438	8.2	817,417	9.1
United Kingdom	*906,990	9.4	*906,990	10.2	*906,990	9.4	1,015,080	10.8	798,899	8.6	*906,990	10.2	*906,990	10.1
Russian Federation	2,228,000	23.2	1,604,000	18.0	954,000	9.9	506,000	5.4	376,000	4.1	266,000	3.0	231,000	2.6
Japan	632,000	6.6	639,000	7.2	596,000	6.2	556,000	5.9	444,000	4.8	454,000	5.1	390,000	4.4

Commodity data are provided by the United Nations Statistical Division. The symbol * means that data are estimated. For additional notes, see Appendix I.

Product Tables

Sewing Machines

Unit of Measure: Number.

	1990		1991		1992		1993		1994		1995		1996	
	Value	%	Value	%	Value	%	Value	%	Value	%	Value	%	Value	%
Total Production	17,395,946	100.0	17,007,289	100.0	17,115,696	100.0	16,666,593	100.0	15,587,575	100.0	18,394,655	100.0	13,387,853	100.0
Regions														
Africa	11,750	0.1	7,750	0.0	7,750	0.0	7,750	0.0	7,750	0.0	7,750	0.0	4,750	0.0
America, South	1,192,600	6.9	1,201,400	7.1	1,022,200	6.0	1,036,000	6.2	1,151,800	7.4	1,233,600	6.7	1,129,400	8.4
Asia	14,722,700	84.6	14,015,600	82.4	14,737,500	86.1	14,260,400	85.6	13,086,300	84.0	15,848,200	86.2	11,047,119	82.5
Europe	1,468,896	8.4	1,782,538	10.5	1,348,246	7.9	1,362,443	8.2	1,341,725	8.6	1,305,105	7.1	1,206,584	9.0
Leading Producers														
China	7,610,000	43.7	7,638,000	44.9	8,332,000	48.7	8,405,000	50.4	8,612,000	55.2	9,706,000	52.8	6,837,000	51.1
Japan	2,461,000	14.1	2,416,000	14.2	2,396,000	14.0	2,107,000	12.6	1,711,000	11.0	1,577,000	8.6	1,413,000	10.6
Brazil	1,128,000	6.5	1,134,000	6.7	952,000	5.6	963,000	5.8	1,076,000	6.9	1,155,000	6.3	1,048,000	7.8
Hong Kong	36,000	0.2	4,000	0.0	21,000	0.1	3,000	0.0	-		-		-	
Russian Federation	1,754,000	10.1	1,583,000	9.3	1,624,000	9.5	1,420,000	8.5	411,000	2.6	100,000	0.5	43,000	0.3

Air-Conditioning Machines

Unit of Measure: Number.

	1990		1991		1992		1993		1994		1995		1996	
	Value	%	Value	%	Value	%	Value	%	Value	%	Value	%	Value	%
Total Production	31,192,520	100.0	33,489,774	100.0	32,200,778	100.0	35,995,615	100.0	40,539,655	100.0	46,094,719	100.0	48,003,475	100.0
Regions														
Africa	134,750	0.4	90,000	0.3	74,000	0.2	74,000	0.2	47,000	0.1	46,143	0.1	20,202	0.0
America, North	8,524,230	27.3	8,967,663	26.8	9,637,096	29.9	10,253,279	28.5	10,869,462	26.8	11,485,645	24.9	12,101,828	25.2
America, South	598,000	1.9	388,000	1.2	324,000	1.0	493,000	1.4	487,000	1.2	635,000	1.4	587,000	1.2
Asia	20,320,253	65.1	22,383,003	66.8	20,501,754	63.7	23,517,255	65.3	27,593,339	68.1	32,000,756	69.4	32,881,107	68.5
Europe	1,615,288	5.2	1,661,108	5.0	1,663,928	5.2	1,658,081	4.6	1,542,854	3.8	1,927,175	4.2	2,413,338	5.0
Leading Producers														
Japan	18,077,000	58.0	20,047,000	59.9	18,272,000	56.7	21,615,000	60.0	25,550,000	63.0	29,259,000	63.5	30,350,000	63.2
United States of America	*8,035,933	25.8	*8,447,333	25.2	*8,858,733	27.5	*9,270,133	25.8	*9,681,533	23.9	*10,092,933	21.9	*10,504,333	21.9
Mexico	472,000	1.5	503,000	1.5	760,000	2.4	963,750	2.7	1,167,500	2.9	1,371,250	3.0	1,575,000	3.3
Korea, Republic of	655,000	2.1	836,000	2.5	826,000	2.6	658,000	1.8	850,000	2.1	1,487,000	3.2	*1,195,390	2.5
Iran, Islamic Republic of	*536,571	1.7	*559,429	1.7	*582,286	1.8	*605,143	1.7	*628,000	1.5	*650,857	1.4	*673,714	1.4

Industrial Refrigerators and Freezers

Unit of Measure: Number.

	1990		1991		1992		1993		1994		1995		1996	
	Value	%	Value	%	Value	%	Value	%	Value	%	Value	%	Value	%
Total Production	7,949,450	100.0	7,842,923	100.0	7,391,274	100.0	6,459,552	100.0	5,631,446	100.0	7,160,663	100.0	7,234,310	100.0
Regions														
Africa	109	0.0	109	0.0	109	0.0	109	0.0	109	0.0	109	0.0	109	0.0
America, North	82,246	1.0	90,339	1.2	84,115	1.1	183,491	2.8	214,851	3.8	166,477	2.3	194,963	2.7
America, South	78,014	1.0	78,062	1.0	78,612	1.1	81,561	1.3	83,001	1.5	83,785	1.2	82,888	1.1
Asia	2,379,059	29.9	2,621,921	33.4	2,344,046	31.7	2,219,930	34.4	2,547,534	45.2	2,361,875	33.0	2,359,695	32.6
Europe	5,351,075	67.3	4,989,196	63.6	4,816,745	65.2	3,902,464	60.4	2,709,603	48.1	4,467,720	62.4	4,511,609	62.4
Oceania	58,947	0.7	63,297	0.8	67,647	0.9	71,997	1.1	76,347	1.4	80,697	1.1	85,047	1.2
Leading Producers														
France	3,967,299	49.9	3,198,659	40.8	3,205,554	43.4	2,314,558	35.8	979,679	17.4	*2,749,312	38.4	*2,744,786	37.9
Japan	1,312,401	16.5	1,363,509	17.4	1,356,253	18.3	1,222,413	18.9	1,327,284	23.6	1,298,097	18.1	*1,352,479	18.7
Germany	-		494,648	6.3	399,142	5.4	393,290	6.1	390,726	6.9	307,355	4.3	269,638	3.7
United Kingdom	254,554	3.2	251,962	3.2	249,369	3.4	246,777	3.8	*244,185	4.3	*241,592	3.4	*239,000	3.3
Indonesia	16,800	0.2	218,922	2.8	124,719	1.7	171,810	2.7	469,369	8.3	317,762	4.4	*364,612	5.0

Commodity data are provided by the United Nations Statistical Division. The symbol * means that data are estimated. For additional notes, see Appendix I.

619

Product Tables

Pumps for Liquids, Except Liquid Elevators

Unit of Measure: Number.

	1990 Value	%	1991 Value	%	1992 Value	%	1993 Value	%	1994 Value	%	1995 Value	%	1996 Value	%
Total Production	41,845,358	100.0	57,771,976	100.0	58,979,080	100.0	59,547,322	100.0	60,508,903	100.0	80,529,216	100.0	80,605,027	100.0
Regions														
Africa	39,000	0.1	39,000	0.1	37,000	0.1	29,000	0.0	19,000	0.0	40,313	0.1	42,349	0.1
America, North	5,359,833	12.8	5,367,600	9.3	5,729,367	9.7	5,907,133	9.9	5,734,900	9.5	7,259,667	9.0	7,571,433	9.4
America, South	45,222	0.1	47,111	0.1	49,000	0.1	56,000	0.1	45,000	0.1	58,000	0.1	55,352	0.1
Asia	14,413,636	34.4	15,066,598	26.1	16,162,628	27.4	16,807,844	28.2	17,513,150	28.9	21,909,457	27.2	19,858,409	24.6
Europe	21,987,667	52.5	37,251,667	64.5	37,001,085	62.7	36,747,344	61.7	37,196,853	61.5	51,261,779	63.7	53,077,484	65.8
Leading Producers														
Germany	-		15,538,000	26.9	15,628,000	26.5	14,002,000	23.5	14,481,000	23.9	27,460,472	34.1	28,713,351	35.6
United States of America	5,186,000	12.4	5,123,000	8.9	5,494,000	9.3	5,767,000	9.7	5,610,000	9.3	7,163,000	8.9	7,446,000	9.2
China	3,664,000	8.8	3,992,000	6.9	5,450,000	9.2	6,632,000	11.1	6,453,000	10.7	10,606,000	13.2	8,875,000	11.0
France	3,848,000	9.2	3,898,000	6.7	3,611,000	6.1	4,256,000	7.1	4,306,000	7.1	*4,954,297	6.2	*5,221,136	6.5
Japan	3,468,000	8.3	3,688,000	6.4	3,407,000	5.8	3,173,000	5.3	3,615,000	6.0	3,566,000	4.4	3,798,000	4.7

Compressors

Unit of Measure: Number.

	1990 Value	%	1991 Value	%	1992 Value	%	1993 Value	%	1994 Value	%	1995 Value	%	1996 Value	%
Total Production	21,599,402	100.0	20,694,990	100.0	20,946,239	100.0	22,947,452	100.0	23,779,217	100.0	27,584,605	100.0	24,425,135	100.0
Regions														
Africa	1,440	0.0	1,031	0.0	897	0.0	412	0.0	23	0.0	-		-	
America, North	1,895,006	8.8	1,921,401	9.3	2,201,075	10.5	2,292,536	10.0	2,567,689	10.8	3,314,004	12.0	3,135,598	12.8
America, South	5,951	0.0	4,875	0.0	3,798	0.0	4,335	0.0	4,968	0.0	3,956	0.0	3,402	0.0
Asia	6,990,390	32.4	8,427,994	40.7	8,527,205	40.7	10,827,876	47.2	11,813,928	49.7	14,242,101	51.6	12,788,774	52.4
Europe	12,706,615	58.8	10,339,689	50.0	10,213,264	48.8	9,822,293	42.8	9,392,609	39.5	10,024,544	36.3	8,497,361	34.8
Leading Producers														
Germany	*8,179,814	37.9	*8,179,814	39.5	*8,179,814	39.1	*8,179,814	35.6	*8,179,814	34.4	8,841,405	32.1	7,518,223	30.8
Korea, Republic of	4,559,075	21.1	5,920,119	28.6	6,036,161	28.8	8,212,604	35.8	9,871,030	41.5	12,256,533	44.4	*10,733,803	43.9
United States of America	1,885,753	8.7	1,911,756	9.2	2,190,728	10.5	2,283,229	9.9	2,559,785	10.8	3,308,376	12.0	3,128,984	12.8
Lithuania	3,648,000	16.9	1,027,000	5.0	1,069,000	5.1	1,058,000	4.6	541,000	2.3	294,008	1.1	185,256	0.8
Turkey	849,611	3.9	925,969	4.5	1,139,297	5.4	1,191,787	5.2	1,114,293	4.7	1,527,463	5.5	1,605,157	6.6

Cranes

Unit of Measure: Number.

	1990 Value	%	1991 Value	%	1992 Value	%	1993 Value	%	1994 Value	%	1995 Value	%	1996 Value	%
Total Production	212,369	100.0	204,820	100.0	196,919	100.0	223,190	100.0	171,459	100.0	166,183	100.0	168,105	100.0
Regions														
Africa	385	0.2	392	0.2	373	0.2	329	0.1	141	0.1	316	0.2	307	0.2
America, North	6,089	2.9	7,426	3.6	8,029	4.1	8,720	3.9	10,628	6.2	8,371	5.0	8,692	5.2
America, South	147	0.1	137	0.1	127	0.1	194	0.1	338	0.2	215	0.1	198	0.1
Asia	41,631	19.6	41,328	20.2	40,597	20.6	53,466	24.0	18,831	11.0	16,406	9.9	18,004	10.7
Europe	164,117	77.3	155,538	75.9	147,793	75.1	160,481	71.9	141,521	82.5	140,874	84.8	140,905	83.8
Leading Producers														
Austria	*64,021	30.1	*64,021	31.3	*64,021	32.5	*64,021	28.7	*64,021	37.3	64,021	38.5	*64,021	38.1
Germany	*45,826	21.6	*45,826	22.4	*45,826	23.3	*45,826	20.5	*45,826	26.7	43,826	26.4	47,826	28.5
Russian Federation	20,952	9.9	19,161	9.4	16,168	8.2	14,092	6.3	6,246	3.6	4,438	2.7	4,433	2.6
Japan	18,333	8.6	16,506	8.1	13,192	6.7	10,235	4.6	8,370	4.9	9,256	5.6	11,346	6.7
Ukraine	15,613	7.4	12,594	6.1	9,081	4.6	22,072	9.9	2,946	1.7	1,358	0.8	465	0.3

Product Tables

Elevators, for Lifting Goods and Persons

Unit of Measure: Number.

	1990 Value	%	1991 Value	%	1992 Value	%	1993 Value	%	1994 Value	%	1995 Value	%	1996 Value	%
Total Production	6,494,793	100.0	6,498,364	100.0	6,490,765	100.0	6,476,203	100.0	6,390,580	100.0	6,557,099	100.0	6,492,494	100.0
Regions														
America, North	1,271	0.0	1,330	0.0	1,589	0.0	1,837	0.0	1,936	0.0	1,958	0.0	1,711	0.0
America, South	8,011	0.1	8,105	0.1	7,887	0.1	9,258	0.1	9,063	0.1	9,016	0.1	9,522	0.1
Asia	116,729	1.8	120,643	1.9	110,698	1.7	102,398	1.6	101,861	1.6	105,224	1.6	109,027	1.7
Europe	6,365,739	98.0	6,365,243	98.0	6,367,549	98.1	6,359,666	98.2	6,274,677	98.2	6,437,858	98.2	6,369,192	98.1
Oceania	3,043	0.0	3,043	0.0	3,043	0.0	3,043	0.0	3,043	0.0	3,043	0.0	3,043	0.0
Leading Producers														
Germany	*5,910,004	91.0	*5,910,004	90.9	*5,910,004	91.1	*5,910,004	91.3	*5,910,004	92.5	5,985,504	91.3	5,834,503	89.9
Czech Republic	*392,825	6.0	*392,825	6.0	*392,825	6.1	*392,825	6.1	307,176	4.8	393,602	6.0	477,698	7.4
Japan	34,653	0.5	37,887	0.6	33,109	0.5	30,738	0.5	32,774	0.5	35,703	0.5	41,468	0.6
Spain	19,911	0.3	19,364	0.3	22,743	0.4	18,507	0.3	17,114	0.3	18,860	0.3	*20,560	0.3
Austria	*12,652	0.2	*12,652	0.2	*12,652	0.2	*12,652	0.2	*12,652	0.2	12,652	0.2	*12,652	0.2

Fork-Lift Trucks

Unit of Measure: Number.

	1990 Value	%	1991 Value	%	1992 Value	%	1993 Value	%	1994 Value	%	1995 Value	%	1996 Value	%
Total Production	615,813	100.0	774,494	100.0	704,640	100.0	635,927	100.0	700,285	100.0	665,245	100.0	593,060	100.0
Regions														
Africa	1,594	0.3	1,224	0.2	1,280	0.2	1,335	0.2	1,484	0.2	1,271	0.2	1,227	0.2
America, North	33,486	5.4	30,114	3.9	26,743	3.8	23,371	3.7	20,000	2.9	16,629	2.5	13,257	2.2
Asia	197,750	32.1	194,986	25.2	166,143	23.6	135,515	21.3	141,036	20.1	159,442	24.0	165,072	27.8
Europe	381,932	62.0	547,118	70.6	509,422	72.3	474,653	74.6	536,712	76.6	486,851	73.2	412,452	69.5
Oceania	1,052	0.2	1,052	0.1	1,052	0.1	1,052	0.2	1,052	0.2	1,052	0.2	1,052	0.2
Leading Producers														
Germany	-		193,202	24.9	181,864	25.8	145,833	22.9	196,735	28.1	131,184	19.7	65,632	11.1
Japan	160,162	26.0	159,564	20.6	128,751	18.3	105,726	16.6	109,128	15.6	121,688	18.3	121,743	20.5
Bulgaria	58,075	9.4	27,669	3.6	12,316	1.7	6,403	1.0	5,711	0.8	11,640	1.7	3,903	0.7
United States of America	*33,486	5.4	*30,114	3.9	*26,743	3.8	*23,371	3.7	*20,000	2.9	*16,629	2.5	*13,257	2.2
Denmark	32,240	5.2	36,021	4.7	39,803	5.6	43,584	6.9	46,414	6.6	50,839	7.6	*54,151	9.1

Refrigerators for Household Use

Unit of Measure: Number.

	1990 Value	%	1991 Value	%	1992 Value	%	1993 Value	%	1994 Value	%	1995 Value	%	1996 Value	%
Total Production	59,624,705	100.0	65,952,886	100.0	65,619,600	100.0	68,754,819	100.0	73,335,960	100.0	75,658,330	100.0	75,312,000	100.0
Regions														
Africa	1,144,891	1.9	1,194,296	1.8	1,078,889	1.6	942,420	1.4	874,550	1.2	1,131,158	1.5	1,227,601	1.6
America, North	8,018,507	13.4	8,705,882	13.2	10,849,745	16.5	11,715,419	17.0	12,927,433	17.6	12,896,319	17.0	13,266,945	17.6
America, South	3,214,822	5.4	3,417,883	5.2	2,826,464	4.3	3,507,778	5.1	4,103,000	5.6	4,640,902	6.1	4,985,028	6.6
Asia	29,291,484	49.1	29,908,784	45.3	28,450,334	43.4	30,235,627	44.0	32,652,407	44.5	34,020,415	45.0	34,254,010	45.5
Europe	17,626,000	29.6	22,337,042	33.9	22,051,168	33.6	21,932,575	31.9	22,334,570	30.5	22,546,537	29.8	21,175,415	28.1
Oceania	329,000	0.6	389,000	0.6	363,000	0.6	421,000	0.6	444,000	0.6	423,000	0.6	403,000	0.5
Leading Producers														
United States of America	7,015,000	11.8	7,599,000	11.5	9,676,000	14.7	10,306,000	15.0	11,276,000	15.4	11,005,000	14.5	11,132,000	14.8
Japan	5,048,000	8.5	5,212,000	7.9	4,425,000	6.7	4,351,000	6.3	4,952,000	6.8	5,013,000	6.6	5,163,000	6.9
China	4,631,000	7.8	4,699,000	7.1	4,858,000	7.4	5,967,000	8.7	7,681,000	10.5	9,185,000	12.1	9,797,000	13.0
Italy	4,199,000	7.0	4,484,000	6.8	4,285,000	6.5	4,753,000	6.9	5,033,000	6.9	5,908,000	7.8	5,402,000	7.2
Germany	-		4,226,000	6.4	4,298,000	6.5	3,838,000	5.6	3,794,000	5.2	3,270,419	4.3	2,746,837	3.6

Product Tables

Ball, Roller or Needle Bearings

Unit of Measure: Thousands of units.

	1990		1991		1992		1993		1994		1995		1996	
	Value	%	Value	%	Value	%	Value	%	Value	%	Value	%	Value	%
Total Production	2,504,346	100.0	2,341,535	100.0	2,275,056	100.0	2,150,384	100.0	2,264,550	100.0	2,341,321	100.0	2,250,574	100.0
Regions														
America, North	1,154,000	46.1	1,097,000	46.8	1,277,000	56.1	1,381,000	64.2	1,639,000	72.4	1,614,000	68.9	1,631,000	72.5
America, South	2,393	0.1	2,538	0.1	2,682	0.1	2,047	0.1	1,977	0.1	9,511	0.4	4,555	0.2
Asia	1,041,000	41.6	967,000	41.3	784,000	34.5	563,000	26.2	453,000	20.0	556,000	23.7	473,000	21.0
Europe	306,953	12.3	274,997	11.7	211,374	9.3	204,337	9.5	170,573	7.5	161,810	6.9	142,019	6.3
Leading Producers														
United States of America	1,154,000	46.1	1,097,000	46.8	1,277,000	56.1	1,381,000	64.2	1,639,000	72.4	1,614,000	68.9	1,631,000	72.5
Russian Federation	784,000	31.3	748,000	31.9	582,000	25.6	409,000	19.0	294,000	13.0	304,000	13.0	239,000	10.6
Romania	100,696	4.0	91,701	3.9	70,000	3.1	79,000	3.7	88,000	3.9	91,000	3.9	97,000	4.3
India	109,000	4.4	87,000	3.7	94,000	4.1	97,000	4.5	134,000	5.9	231,000	9.9	215,000	9.6
Ukraine	151,124	6.0	143,401	6.1	114,000	5.0	89,000	4.1	46,000	2.0	31,000	1.3	*3,972	0.2

Washing Machines for Household Use

Unit of Measure: Number.

	1990		1991		1992		1993		1994		1995		1996	
	Value	%	Value	%	Value	%	Value	%	Value	%	Value	%	Value	%
Total Production	53,204,280	100.0	55,195,597	100.0	53,898,247	100.0	57,177,081	100.0	59,157,390	100.0	57,051,341	100.0	60,008,389	100.0
Regions														
Africa	288,000	0.5	289,000	0.5	242,000	0.4	252,000	0.4	264,000	0.4	254,000	0.4	241,000	0.4
America, North	7,415,343	13.9	7,450,857	13.5	7,654,371	14.2	7,945,136	13.9	8,404,900	14.2	8,046,664	14.1	8,432,429	14.1
America, South	1,124,424	2.1	1,642,545	3.0	1,960,000	3.6	2,407,000	4.2	2,675,000	4.5	2,634,000	4.6	3,045,400	5.1
Asia	28,388,953	53.4	29,571,348	53.6	27,941,799	51.8	30,038,137	52.5	30,325,425	51.3	28,526,933	50.0	30,192,412	50.3
Europe	15,644,560	29.4	15,946,846	28.9	15,805,077	29.3	16,206,808	28.3	17,174,065	29.0	17,279,743	30.3	17,831,148	29.7
Oceania	343,000	0.6	295,000	0.5	295,000	0.5	328,000	0.6	314,000	0.5	310,000	0.5	266,000	0.4
Leading Producers														
China	6,627,000	12.5	6,872,000	12.5	7,079,000	13.1	8,959,000	15.7	10,941,000	18.5	9,484,000	16.6	10,747,000	17.9
United States of America	6,428,000	12.1	6,404,000	11.6	6,566,000	12.2	6,739,000	11.8	7,081,000	12.0	6,605,000	11.6	6,873,000	11.5
Japan	5,576,000	10.5	5,587,000	10.1	5,225,000	9.7	5,163,000	9.0	5,042,000	8.5	4,876,000	8.5	5,006,000	8.3
Italy	4,372,000	8.2	5,044,000	9.1	5,140,000	9.5	5,693,000	10.0	6,251,000	10.6	6,996,000	12.3	7,135,000	11.9
Russian Federation	5,419,000	10.2	5,541,000	10.0	4,289,000	8.0	3,901,000	6.8	2,122,000	3.6	1,294,000	2.3	762,000	1.3

Generators for Hydraulic Turbines

Unit of Measure: Number.

	1990		1991		1992		1993		1994		1995		1996	
	Value	%	Value	%	Value	%	Value	%	Value	%	Value	%	Value	%
Total Production	88,132	100.0	96,985	100.0	82,844	100.0	86,802	100.0	80,206	100.0	85,382	100.0	83,187	100.0
Regions														
America, North	10	0.0	10	0.0	10	0.0	10	0.0	10	0.0	10	0.0	10	0.0
Asia	8,324	9.4	8,325	8.6	370	0.4	14,220	16.4	10,360	12.9	8,312	9.7	8,304	10.0
Europe	79,798	90.5	88,650	91.4	82,464	99.5	72,571	83.6	69,835	87.1	77,060	90.3	74,873	90.0
Leading Producers														
Germany	*30,231	34.3	36,548	37.7	34,803	42.0	26,202	30.2	23,371	29.1	*30,231	35.4	*30,231	36.3
Indonesia	*8,206	9.3	*8,206	8.5	253	0.3	14,106	16.3	10,259	12.8	*8,206	9.6	*8,206	9.9
Bulgaria	6,766	7.7	9,340	9.6	4,972	6.0	3,542	4.1	3,575	4.5	3,957	4.6	1,735	2.1
Spain	*1,684	1.9	*1,684	1.7	*1,684	2.0	*1,684	1.9	*1,684	2.1	*1,684	2.0	*1,684	2.0
Finland	220	0.2	183	0.2	109	0.1	247	0.3	281	0.4	328	0.4	332	0.4

Commodity data are provided by the United Nations Statistical Division. The symbol * means that data are estimated. For additional notes, see Appendix I.

Product Tables

Generators for Hydraulic Turbines - Kilowatts

Unit of Measure: Kilowatts.

	1990 Value	%	1991 Value	%	1992 Value	%	1993 Value	%	1994 Value	%	1995 Value	%	1996 Value	%
Total Production	4,289,760	100.0	3,731,684	100.0	5,588,225	100.0	7,800,317	100.0	5,584,409	100.0	4,409,441	100.0	2,614,127	100.0
Regions														
America, North	805,250	18.8	805,250	21.6	805,250	14.4	805,250	10.3	805,250	14.4	805,250	18.3	805,250	30.8
Asia	2,745,000	64.0	2,621,000	70.2	4,101,000	73.4	6,183,000	79.3	4,203,000	75.3	3,073,000	69.7	834,000	31.9
Europe	739,510	17.2	305,434	8.2	681,975	12.2	812,067	10.4	576,159	10.3	531,191	12.0	974,877	37.3
Leading Producers														
Japan	1,809,000	42.2	1,843,000	49.4	2,726,000	48.8	4,868,000	62.4	3,465,000	62.0	2,571,000	58.3	349,000	13.4
Russian Federation	936,000	21.8	778,000	20.8	1,375,000	24.6	1,315,000	16.9	738,000	13.2	502,000	11.4	485,000	18.6
United States of America	*805,250	18.8	*805,250	21.6	*805,250	14.4	*805,250	10.3	*805,250	14.4	*805,250	18.3	*805,250	30.8
Croatia	98,000	2.3	37,000	1.0	*301,200	5.4	*288,018	3.7	*274,836	4.9	*261,655	5.9	*248,473	9.5
Romania	144,000	3.4	164,000	4.4	164,000	2.9	75,000	1.0	47,000	0.8	*13,939	0.3	-	

Generators for Steam Turbines

Unit of Measure: Number.

	1990 Value	%	1991 Value	%	1992 Value	%	1993 Value	%	1994 Value	%	1995 Value	%	1996 Value	%
Total Production	265	100.0	252	100.0	253	100.0	258	100.0	250	100.0	255	100.0	264	100.0
Regions														
America, North	30	11.5	29	11.6	28	11.0	27	10.4	25	10.2	24	9.5	23	8.7
Asia	156	58.7	114	45.2	119	47.0	115	44.6	99	39.6	95	37.3	97	36.8
Europe	79	29.8	109	43.2	106	42.0	116	45.1	125	50.2	136	53.2	144	54.5
Leading Producers														
Japan	62	23.4	55	21.8	64	25.3	65	25.2	57	22.8	60	23.5	62	23.5
Russian Federation	72	27.2	49	19.4	55	21.7	50	19.4	42	16.8	35	13.7	35	13.3
Slovakia	31	11.7	48	19.0	*52	20.4	*58	22.5	*65	25.8	*71	27.8	*77	29.4
United States of America	*30	11.5	*29	11.6	*28	11.0	*27	10.4	*25	10.2	*24	9.5	*23	8.7

Generators for Steam Turbines - Kilowatts

Unit of Measure: 1,000 Kilowatts.

	1990 Value	%	1991 Value	%	1992 Value	%	1993 Value	%	1994 Value	%	1995 Value	%	1996 Value	%
Total Production	93,528	100.0	115,765	100.0	116,964	100.0	128,758	100.0	136,352	100.0	151,746	100.0	160,105	100.0
Regions														
Asia	28,900	30.9	32,707	28.3	25,932	22.2	25,931	20.1	22,397	16.4	26,292	17.3	23,400	14.6
Europe	64,629	69.1	83,058	71.7	91,032	77.8	102,827	79.9	113,955	83.6	125,455	82.7	136,705	85.4
Leading Producers														
Slovakia	58,261	62.3	78,350	67.7	*86,572	74.0	*98,150	76.2	*109,729	80.5	*121,307	79.9	*132,885	83.0
Japan	8,649	9.2	13,652	11.8	8,626	7.4	9,345	7.3	6,890	5.1	10,858	7.2	8,421	5.3
Russian Federation	7,082	7.6	5,899	5.1	4,163	3.6	3,455	2.7	2,389	1.8	2,328	1.5	1,886	1.2
United Kingdom	*3,350	3.6	*3,350	2.9	*3,350	2.9	*3,350	2.6	*3,350	2.5	*3,350	2.2	*3,350	2.1

Commodity data are provided by the United Nations Statistical Division. The symbol * means that data are estimated. For additional notes, see Appendix I.

623

Product Tables
Motors, Electric, Fractional Horsepower

Unit of Measure: Thousands of units.

	1990 Value	%	1991 Value	%	1992 Value	%	1993 Value	%	1994 Value	%	1995 Value	%	1996 Value	%
Total Production	457,095	100.0	531,994	100.0	583,515	100.0	662,607	100.0	631,689	100.0	550,165	100.0	549,681	100.0
Regions														
Africa	477	0.1	491	0.1	856	0.1	1,780	0.3	1,396	0.2	1,571	0.3	1,746	0.3
America, North	293,264	64.2	292,821	55.0	264,689	45.4	362,095	54.6	378,480	59.9	305,389	55.5	303,929	55.3
America, South	4,366	1.0	4,370	0.8	4,347	0.7	4,334	0.7	4,373	0.7	4,451	0.8	4,434	0.8
Asia	53,837	11.8	54,254	10.2	53,347	9.1	45,985	6.9	44,725	7.1	47,847	8.7	50,226	9.1
Europe	102,623	22.5	177,578	33.4	257,855	44.2	246,299	37.2	200,665	31.8	188,924	34.3	187,428	34.1
Oceania	2,528	0.6	2,480	0.5	2,421	0.4	2,113	0.3	2,048	0.3	1,984	0.4	1,919	0.3
Leading Producers														
United States of America	292,579	64.0	292,072	54.9	263,875	45.2	360,915	54.5	376,933	59.7	303,475	55.2	301,649	54.9
Germany	-		75,107	14.1	77,043	13.2	72,756	11.0	81,110	12.8	73,244	13.3	75,320	13.7
Georgia	-		-		78,356	13.4	63,956	9.7	9,260	1.5	4,609	0.8	4,023	0.7
Japan	27,828	6.1	28,037	5.3	26,603	4.6	21,875	3.3	22,348	3.5	22,390	4.1	22,701	4.1
Slovenia	*8,319	1.8	*8,319	1.6	*8,319	1.4	6,959	1.1	7,136	1.1	8,904	1.6	10,277	1.9

Motors, Electric, One Horsepower and Over

Unit of Measure: Number.

	1990 Value	%	1991 Value	%	1992 Value	%	1993 Value	%	1994 Value	%	1995 Value	%	1996 Value	%
Total Production	51,026,770	100.0	46,892,803	100.0	50,751,232	100.0	45,608,096	100.0	45,106,526	100.0	47,241,175	100.0	47,866,389	100.0
Regions														
Africa	764,147	1.5	830,156	1.8	896,166	1.8	962,175	2.1	1,028,185	2.3	1,094,194	2.3	1,160,204	2.4
America, North	7,163,000	14.0	6,825,000	14.6	6,655,000	13.1	8,201,000	18.0	9,604,000	21.3	9,008,703	19.1	9,361,235	19.6
America, South	756,000	1.5	756,000	1.6	756,000	1.5	756,000	1.7	756,000	1.7	756,000	1.6	756,000	1.6
Asia	12,854,995	25.2	14,028,636	29.9	11,122,076	21.9	9,473,164	20.8	10,259,651	22.7	14,109,404	29.9	14,091,958	29.4
Europe	29,046,850	56.9	23,995,700	51.2	30,849,145	60.8	25,727,379	56.4	22,954,779	50.9	21,753,429	46.0	21,962,015	45.9
Oceania	441,778	0.9	457,311	1.0	472,844	0.9	488,378	1.1	503,911	1.1	519,444	1.1	534,978	1.1
Leading Producers														
Poland	10,434,000	20.4	6,747,000	14.4	4,918,000	9.7	5,680,000	12.5	5,632,000	12.5	5,535,000	11.7	5,633,000	11.8
Japan	9,811,000	19.2	10,818,000	23.1	8,647,000	17.0	6,954,000	15.2	8,165,000	18.1	11,602,000	24.6	11,573,000	24.2
United States of America	7,026,000	13.8	6,662,000	14.2	6,483,000	12.8	8,041,000	17.6	9,466,000	21.0	*8,837,703	18.7	*9,221,235	19.3
Georgia	-		-		9,498,000	18.7	7,449,000	16.3	4,308,000	9.6	2,503,000	5.3	1,098,000	2.3
Germany	*3,909,250	7.7	4,349,000	9.3	4,229,000	8.3	3,408,000	7.5	3,651,000	8.1	*3,909,250	8.3	*3,909,250	8.2

Transformers, Less Than 5 KVA

Unit of Measure: Thousands of units.

	1990 Value	%	1991 Value	%	1992 Value	%	1993 Value	%	1994 Value	%	1995 Value	%	1996 Value	%
Total Production	124,360	100.0	131,955	100.0	130,850	100.0	133,698	100.0	152,928	100.0	158,475	100.0	172,076	100.0
Regions														
Africa	88	0.1	51	0.0	12	0.0	5	0.0	6	0.0	7	0.0	7	0.0
America, South	688	0.6	688	0.5	657	0.5	658	0.5	712	0.5	723	0.5	688	0.4
Asia	1,345	1.1	2,126	1.6	3,038	2.3	3,113	2.3	12,291	8.0	5,067	3.2	5,026	2.9
Europe	122,240	98.3	129,090	97.8	127,142	97.2	129,923	97.2	139,919	91.5	152,678	96.3	166,355	96.7
Leading Producers														
Germany	*49,883	40.1	55,304	41.9	50,894	38.9	44,840	33.5	48,493	31.7	*49,883	31.5	*49,883	29.0
Spain	9,531	7.7	11,063	8.4	10,112	7.7	*11,617	8.7	*12,516	8.2	*13,416	8.5	*14,315	8.3
Finland	4,309	3.5	3,322	2.5	3,052	2.3	3,403	2.5	8,465	5.5	23,100	14.6	34,566	20.1
Indonesia	645	0.5	1,262	1.0	45	0.0	29	0.0	10,345	6.8	*2,193	1.4	*1,655	1.0
Portugal	117	0.1	5	0.0	4,008	3.1	10,934	8.2	11,891	7.8	*7,055	4.5	*7,737	4.5

Commodity data are provided by the United Nations Statistical Division. The symbol * means that data are estimated. For additional notes, see Appendix I.

Product Tables

Transformers, 5 KVA and Over

Unit of Measure: Number.

	1990		1991		1992		1993		1994		1995		1996	
	Value	%	Value	%	Value	%	Value	%	Value	%	Value	%	Value	%
Total Production	4,157,963	100.0	4,215,301	100.0	4,052,550	100.0	4,127,873	100.0	4,268,059	100.0	4,628,341	100.0	4,549,254	100.0
Regions														
Africa	720	0.0	492	0.0	264	0.0	36	0.0	-		-		-	
America, North	20,785	0.5	21,309	0.5	14,651	0.4	27,059	0.7	39,467	0.9	51,875	1.1	64,283	1.4
America, South	25	0.0	24	0.0	26	0.0	26	0.0	25	0.0	25	0.0	25	0.0
Asia	1,142,949	27.5	1,208,009	28.7	1,057,680	26.1	1,119,649	27.1	1,233,070	28.9	1,543,930	33.4	1,416,462	31.1
Europe	2,904,902	69.9	2,888,770	68.5	2,875,115	70.9	2,868,173	69.5	2,874,452	67.3	2,903,349	62.7	2,931,206	64.4
Oceania	88,581	2.1	96,697	2.3	104,813	2.6	112,929	2.7	121,046	2.8	129,162	2.8	137,278	3.0
Leading Producers														
Japan	740,310	17.8	803,543	19.1	666,804	16.5	623,625	15.1	561,843	13.2	644,126	13.9	563,629	12.4
India	332,745	8.0	333,257	7.9	340,947	8.4	452,225	11.0	639,115	15.0	861,489	18.6	816,090	17.9
Greece	104,895	2.5	96,137	2.3	98,195	2.4	*110,694	2.7	*118,371	2.8	*126,048	2.7	*133,726	2.9
Australia	*88,581	2.1	*96,697	2.3	*104,813	2.6	*112,929	2.7	*121,046	2.8	*129,162	2.8	*137,278	3.0
Germany	*35,102	0.8	*35,102	0.8	*35,102	0.9	18,434	0.4	19,315	0.5	40,658	0.9	62,001	1.4

Meters, Electricity-Supply

Unit of Measure: Number.

	1990		1991		1992		1993		1994		1995		1996	
	Value	%	Value	%	Value	%	Value	%	Value	%	Value	%	Value	%
Total Production	53,066,570	100.0	56,095,072	100.0	58,298,269	100.0	62,113,933	100.0	71,067,774	100.0	79,669,001	100.0	79,637,105	100.0
Regions														
Africa	627,000	1.2	802,000	1.4	874,000	1.5	540,000	0.9	625,000	0.9	594,909	0.7	594,682	0.7
America, North	4,830,429	9.1	4,848,000	8.6	4,927,000	8.5	4,914,000	7.9	5,155,000	7.3	5,108,418	6.4	5,157,121	6.5
America, South	2,321,000	4.4	1,998,000	3.6	2,303,000	4.0	2,392,000	3.9	3,050,000	4.3	3,432,000	4.3	3,491,000	4.4
Asia	33,859,360	63.8	36,253,520	64.6	38,684,200	66.4	42,265,120	68.0	50,298,755	70.8	59,033,408	74.1	58,639,611	73.6
Europe	11,428,781	21.5	12,193,552	21.7	11,510,069	19.7	12,002,813	19.3	11,939,019	16.8	11,500,266	14.4	11,754,691	14.8
Leading Producers														
China	23,033,000	43.4	25,310,000	45.1	28,698,000	49.2	30,510,000	49.1	36,940,000	52.0	45,231,000	56.8	43,656,000	54.8
United States of America	4,830,429	9.1	4,848,000	8.6	4,927,000	8.5	4,914,000	7.9	5,155,000	7.3	*5,108,418	6.4	*5,157,121	6.5
India	2,868,000	5.4	3,295,000	5.9	2,760,000	4.7	5,270,000	8.5	5,490,000	7.7	5,680,000	7.1	7,180,000	9.0
Japan	3,178,000	6.0	3,159,000	5.6	2,716,000	4.7	2,500,000	4.0	2,930,000	4.1	3,301,000	4.1	3,203,000	4.0
Brazil	2,203,000	4.2	1,919,000	3.4	2,166,000	3.7	2,260,000	3.6	2,897,000	4.1	3,240,000	4.1	3,281,000	4.1

Electric Furnaces

Unit of Measure: Number.

	1990		1991		1992		1993		1994		1995		1996	
	Value	%	Value	%	Value	%	Value	%	Value	%	Value	%	Value	%
Total Production	558,919	100.0	660,714	100.0	612,784	100.0	570,540	100.0	531,547	100.0	1,148,476	100.0	1,232,460	100.0
Regions														
America, South	55,864	10.0	55,864	8.5	44,505	7.3	61,445	10.8	60,627	11.4	56,878	5.0	55,864	4.5
Asia	70,322	12.6	70,370	10.7	93,898	15.3	92,178	16.2	117,881	22.2	634,832	55.3	717,614	58.2
Europe	432,734	77.4	534,480	80.9	474,380	77.4	416,917	73.1	353,038	66.4	456,767	39.8	458,982	37.2
Leading Producers														
Turkey	-		-		24,135	3.9	22,610	4.0	47,212	8.9	563,372	49.1	645,202	52.4
Germany	*141,086	25.2	238,042	36.0	173,192	28.3	110,951	19.4	42,158	7.9	*141,086	12.3	*141,086	11.4
Colombia	*55,864	10.0	*55,864	8.5	44,505	7.3	61,445	10.8	60,627	11.4	56,878	5.0	*55,864	4.5
Hong Kong	*46,223	8.3	*46,223	7.0	*46,223	7.5	*46,223	8.1	*46,223	8.7	*46,223	4.0	*46,223	3.8
Slovakia	9,886	1.8	14,824	2.2	19,762	3.2	24,700	4.3	29,638	5.6	34,576	3.0	*36,883	3.0

ISIC 3832
RADIO, TELEVISION, ETC.

ISIC 3832—Radio, Television etc.—includes radio and television sets, telephones, transistors, sound recorders and reproducers, records, etc.—the consumer electronics category. Because of its world-wide importance, this subcategory of ISIC 3830—Electric Machinery, is provided as a separate chapter here. Product tables for this industry have been included in the previous chapter.

The ISIC corresponds to SIC 366—Communications Equipment in the U.S. classifications system.

Summary Statistics

		1990		1991		1992		1993		1994		1995	
		Value	N	Value	N	Value	N	Value	N	Value	N	Value	N
Establishments or enterprises	(number)	29,404	45	31,679	51	44,062	48	23,753	42	7,637	35	4,812	16
Number employed	(000)	3,714	48	4,242	52	4,045	48	3,319	43	2,175	34	1,738	20
Total Wages	($ mil.)	82,322	45	103,849	49	108,081	46	89,331	39	54,979	32	41,394	18
Wage/salary per employee	($)	10,829	45	11,928	49	12,592	46	9,143	39	10,035	32	9,509	18
Female workers	(000)	658	20	386	18	956	17	939	19	944	15	740	5
as % of total employment*	(%)	39	20	40	18	43	17	46	19	46	15	54	5
Output	($ bil.)	475	46	575	51	577	47	522	41	327	32	303	19
per employee	($ 000)	79	45	86	49	89	45	77	39	87	31	77	18
per establishment	($ 000)	13,479	43	13,562	48	12,715	45	12,500	38	16,225	30	5,083	13
Value added	($ bil.)	226	46	240	48	241	42	244	38	136	29	166	16
per employee	($ 000)	30	45	33	46	32	40	28	35	30	28	39	15
per establishment	($ 000)	5,484	43	5,304	45	4,483	41	4,096	35	5,014	27	1,847	10
Capital investment	($ mil.)	18,133	23	22,062	28	28,769	29	29,014	24	19,831	18	23,924	6
per employee	($ 000)	6	23	5	28	5	28	5	22	6	18	10	6
per establishment	($ 000)	1,541	23	975	28	737	29	903	22	1,307	17	734	5

Data presented above are drawn from the detailed tables that follow. Columns headed 'N' show the number of countries that provided valid data for inclusion. Values are not strictly comparable one year to the next or one row to the next because the number of countries that report varies from period to period and row to row. However, a general indication of magnitudes can be gleaned from reviewing this summary—especially in earlier years in which more reports are available. For detailed explanations, see Appendix I. *The average for those countries reporting both total and female employment.

Establishments and Number Engaged

Country	Establishments or Enterprises (number)						Employees per Establishment					
	1990	1991	1992	1993	1994	1995	1990	1991	1992	1993	1994	1995
Albania	-	-	-	1.000	1.000	1.000	-	-	-	180	135	88
Australia	51	74	79	-	-	-	59	27	25	-	-	-
Austria	63	63	58	59	53		311	286	293	269	302	
Bangladesh	35	23	24	-	-		72	95	83	-	-	
Bolivia	3.000	3.000	3.000	3.000	3.000	3.000	43	45	48	56	55	55
Bosnia & Herzegovina	16	17	-	-	-		253	209	-	-	-	
Chile	2.000	2.000	3.000	2.000	3.000	-	80	90	86	103	105	-
China (Hong Kong SAR)	625	535	484	367	300	-	69	63	65	80	83	-
Colombia	37	34	35	32	31	-	118	113	132	144	151	-
Costa Rica	19	23	24	23	-	-	21	44	59	96	-	-
Cyprus	7.000	7.000	7.000	6.000	6.000	6.000	2.857	4.000	4.571	2.333	2.000	-
Denmark	265	298	294	-	-		46	42	40	-	-	
Ecuador	13	13	13	14	13	10	84	72	65	56	53	88
Egypt	10	9.000	9.000	11			810	1,089	1,056	882	-	
El Salvador	-	-	-	2.000	3.000	1.000	-	-	-	37	28	75
Finland	66	79	74	83	85	-	174	139	142	139	171	-
FYR Macedonia	-	-	-	-	-	*110	-	-	-	-	-	1.800
Gabon	-	1.000	1.000	1.000	1.000	1.000	-	33	32	32	31	17
Germany	-	1,583	1,583	1,576	1,518	-	-	321	273	251	238	-
Germany (Western Part)	1,387	1,442	1,461	-	-		300	293	275	-	-	
Ghana	-	-	-	6.000	-		-	-	-	51	-	
Greece	46	45	43	-	-		60	71	69	-	-	
Guatemala	-	3.000	3.000	3.000	3.000	3.000	-	72	63	64	65	65
Honduras	-	-	5.000	1.000	1.000	1.000	-	-	33	180	197	224
Hungary	-	-	*432	*489			-	-	-	-		
India	1,173	1,193	1,194	1,207	-	-	98	99	104	99	-	-
Indonesia	80	97	115	118	159	183	338	361	375	485	510	495
Iran (Islamic Republic of)	-	19	24	24	-	-	-	565	481	464	-	-
Italy	*264	*277	-	-	-		258	233	-	-	-	
Japan	15,740	15,973	14,842	14,055			65	66	68	68		
Korea, Republic of	3,787	3,681	3,697	4,036	4,139	4,387	86	78	74	66	68	66
Macau	7.000	4.000	4.000	3.000	2.000	-	17	26	12	13	14	-
Malaysia	252	289	316	342	359	-	738	762	763	836	890	-
Malta	13	16	16	20	19	-	130	94	112	100	110	-
Mauritius	4.000	3.000	3.000	4.000	3.000	3.000	38	43	35	28	35	29
Mexico	70	65	64	60	58	56	612	612	623	602	615	627
Mongolia	1.000	1.000	1.000	-	1.000	1.000	320	194	38	-	8.000	-
Mozambique	-	-	-	-	1.000	·	-	-	-	-	40	
Myanmar	1.000	1.000	1.000	1.000	1.000	-	353	-	-	-	-	-
Nepal	3.000	6.000	-	5.000	4.000	-	34	37	-	43	55	-
New Zealand	*182	*187	*178	*196	-	-	13	9.845	10	10	-	-
Norway	92	87	62	-	-		69	70	83	-	-	
Pakistan	-	19	-	-	-		-	160	-	-	-	
Peru	164	170	165	-	-		13	7.494	5.206	-	-	
Philippines	102	146	134	118	-		530	423	525	572	-	
Portugal	*313	*327	*307	*356	*392		47	48	52	44	42	
Singapore	175	177	180	181	186	-	444	447	423	403	410	-
Slovakia	-	20	20	18	20	-	-	1,052	788	678	499	-
South Africa	-	203	-	-	-		-	-	-	-	-	
Spain	345	413	388	-	-		92	79	75	-	-	
Sri Lanka	5.000	4.000	4.000	7.000	-	-	73	150	90	123	-	-
Sweden	111	171	158	163	140		222	192	193	182	246	-
Thailand	39	86	-	-	-		914	411	-	-	-	
Trinidad and Tobago	14	14	10	8.000	7.000		16	18	22	27	25	
Tunisia	-	-	-	24	26	32	-	-	-	43	58	63
Turkey	51	53	105	101	79		355	338	191	182	193	
United Kingdom	3,719	3,676	3,727	-	-		65	60	55	-	-	
United States of America	-	-	13,675	-	-	-	-	-	64	-	-	-

Continued.

Depending on the table, * means *Enterprises, Engaged,* or *Factor Values*; ± means *Basis Unspecified*; § means *shown in millions.* For additional notes and sources, see Appendix I.

Establishments and Number Engaged

- Continued -

Country	Establishments or Enterprises (number)						Employees per Establishment					
	1990	1991	1992	1993	1994	1995	1990	1991	1992	1993	1994	1995
Venezuela	41	38	28	21	14	14	83	68	89	87	67	60
Zambia	2.000	-	-	-	1.000	-	20	-	-	-	-	-
Zimbabwe	9.000	9.000	9.000	6.000	5.000	-	156	156	133	133	160	-

Employment and Compensation of Employees

Country	Number Employed/Engaged (000)						Wage/Salary per Employee ($)					
	1990	1991	1992	1993	1994	1995	1990	1991	1992	1993	1994	1995
Albania	-	-	-	0.180	0.135	0.088	-	-	-	695	-	1,159
Australia	3.000	*2.000	*2.000	-	-	-	19,896	26,880	19,485	-	-	-
Austria	20	18	17	16	16	17	24,689	27,430	29,596	30,137	31,300	39,080
Bangladesh	2.518	2.181	1.987	-	-	-	898	764	711	-	-	-
Bolivia	0.128	0.134	0.143	0.169	0.164	0.166	1,248	1,351	1,190	1,401	1,560	1,502
Bosnia & Herzegovina	4.044	3.547	-	-	-	-	-	-	-	-	-	-
Chile.	0.159	0.179	0.257	0.207	0.315	0.201	8,870	8,591	6,171	7,721	11,106	11,979
China (Hong Kong SAR) . . .	*43	*34	*31	*29	*25	-	9,232	10,877	12,634	14,529	16,603	-
Colombia	4.379	3.844	4.605	4.599	4.682	4.124	1,979	1,897	2,015	2,379	2,938	3,256
Costa Rica	0.393	1.014	1.426	2.210	-	-	3,322	3,198	4,197	4,303	-	-
Cyprus	0.020	0.028	0.032	0.014	0.012	0.004	9,628	9,873	11,319	6,755	9,495	13,268
Denmark	12	12	12	-	-	-	31,183	32,525	35,384	-	-	-
Ecuador	1.094	0.933	0.842	0.779	0.689	0.881	1,638	1,932	2,213	2,320	1,511	1,620
Egypt	*8.100	*9.800	*9.500	*9.700	-	-	2,568	1,917	2,292	2,581	-	-
El Salvador.	-	-	-	*0.074	*0.083	*0.075	-	-	-	2,182	6,901	5,853
Finland	11	11	10	11	15	-	28,395	28,300	27,644	22,625	26,035	-
FYR Macedonia	-	-	-	-	-	0.198	-	-	-	-	-	5,051
Gabon	-	0.033	0.032	0.032	0.031	0.017	-	17,294	17,591	16,334	12,376	16,852
Germany	-	*508	*432	*395	*361	-	-	29,926	37,530	38,045	-	-
Germany (Western Part) . . .	*416	*422	*402	-	-	-	32,714	34,316	39,103	-	-	-
Ghana	-	-	-	*0.306	-	-	-	-	-	-	-	-
Greece	*2.776	*3.195	*2.977	-	-	-	12,763	14,016	14,494	-	-	-
Guatemala.	-	*0.216	*0.189	*0.192	*0.195	*0.194	-	-	-	-	-	-
Honduras	*0.284	*0.256	*0.165	*0.180	*0.197	*0.224	1,224	1,493	1,301	1,240	1,046	1,075
India.	*115	*118	*124	*119	-	-	1,743	1,466	1,483	1,440	-	-
Indonesia	27	35	43	57	81	91	883	813	1,083	1,166	1,305	2,496
Iran (Islamic Republic of) . .	-	11	12	11	-	-	-	5,359	5,319	4,648	-	-
Italy	68	64	-	-	-	-	39,750	-	-	-	-	-
Japan	*1,028	*1,049	*1,007	*956	-	-	24,495	28,066	31,262	37,617	-	-
Korea, Republic of	325	288	274	266	280	289	9,028	10,984	11,539	11,965	13,289	16,771
Lithuania	-	-	30	26	-	-	-	-	-	561	-	-
Macau	0.116	0.104	0.047	0.038	0.028	-	3,194	5,117	3,965	3,947	4,723	-
Malaysia	186	220	241	286	319	371	2,813	3,164	3,837	3,980	4,308	4,809
Malta	1.686	1.511	1.789	2.001	2.088	-	11,224	12,597	12,667	12,376	12,912	-
Mauritius	*0.152	0.128	0.104	0.113	0.106	0.086	2,410	2,943	4,557	3,259	4,622	-
Mexico	43	40	40	36	36	35	3,898	4,562	4,580	4,803	4,976	3,720
Mongolia	*0.320	*0.194	*0.038	-	*0.008	-	837	2,437	2,350	-	-	-
Mozambique	-	-	-	-	0.040	-	-	-	-	-	675	-
Myanmar	0.353	-	-	-	-	-	1,533	-	-	-	-	-
Nepal	0.101	0.221	-	0.214	0.221	-	753	608	-	468	547	-
New Zealand	2.329	1.841	1.783	2.015	-	-	-	-	-	-	-	-
Norway	6.364	6.079	5.144	-	-	-	33,637	37,402	-	-	-	-
Pakistan	-	3.034	-	-	-	-	-	2,291	-	-	-	-
Peru	*2.092	*1.274	*0.859	-	-	-	2,728	2,672	4,407	-	-	-
Philippines.	54	62	70	68	-	-	2,479	2,615	2,963	2,964	-	-
Portugal	*15	*16	*16	*16	*16	-	8,666	10,579	12,720	12,196	11,269	-
Singapore	*78	*79	*76	*73	*76	-	10,036	11,282	13,270	14,606	16,884	-
Slovakia	-	*21	*16	*12	*9.975	-	-	1,329	1,559	1,835	2,224	-
Spain	32	33	29	-	-	-	24,852	26,031	27,557	-	-	-
Sri Lanka	*0.364	*0.599	*0.361	*0.864	-	-	1,029	1,587	1,398	592	-	-

Continued.

Employment and Compensation of Employees

- Continued -

Country	Number Employed/Engaged (000)						Wage/Salary per Employee ($)					
	1990	1991	1992	1993	1994	1995	1990	1991	1992	1993	1994	1995
Sweden	25	*33	*31	30	34	-	26,304	29,052	31,654	25,090	27,785	-
Thailand	36	35	-	-	-	-	4,143	5,253	-	-	-	-
Trinidad and Tobago	0.228	0.250	0.217	0.213	0.177	-	4,232	5,078	5,115	4,387	4,100	-
Tunisia	-	-	-	1.023	1.497	2.018	-	-	-	-	-	-
Turkey	18	18	20	18	15	-	8,641	12,071	12,452	14,742	8,352	-
United Kingdom	240	220	204	-	-	-	23,490	25,479	26,701	-	-	-
United States of America	877	848	870	865	878	927	30,308	31,403	33,492	34,444	35,261	36,389
Uruguay	*0.418	*0.317	*0.310	*0.283	-	-	-	-	-	-	-	-
Venezuela	3.400	2.600	2.500	1.817	0.931	0.843	4,289	4,434	5,113	3,805	3,465	6,057
Zambia	0.040	-	-	-	-	-	6,255	-	-	-	-	-
Zimbabwe	1.400	1.400	1.200	0.800	0.800	-	3,414	2,771	2,503	2,260	2,301	-

Female Workers: Total and Percent of Employment

Country	Female Workers (000)						Female Workers as % of Total Employment					
	1990	1991	1992	1993	1994	1995	1990	1991	1992	1993	1994	1995
Albania	-	-	-	0.073	0.075	0.054	-	-	-	40.56	55.56	61.36
Australia	0.964	-	-	-	-	-	32.13	-	-	-	-	-
Austria	9.200	8.000	7.000	6.138	6.080	-	46.94	44.44	41.18	38.66	38.01	-
Bangladesh	0.390	0.256	0.261	-	-	-	15.49	11.74	13.14	-	-	-
Chile	0.026	0.042	0.082	0.054	0.055	-	16.35	23.46	31.91	26.09	17.46	-
Colombia	-	1.870	-	2.421	2.610	-	-	48.65	-	52.64	55.75	-
Cyprus	0.010	0.017	0.011	0.008	0.004	-	50.00	60.71	34.38	57.14	33.33	-
Denmark	5.380	5.300	4.905	-	-	-	43.94	42.72	41.31	-	-	-
Egypt	-	3.091	3.209	3.098	-	-	-	31.54	33.78	31.94	-	-
El Salvador	-	-	-	0.028	0.027	0.027	-	-	-	37.84	32.53	36.00
Germany (Western Part)	193	-	-	-	-	-	46.43	-	-	-	-	-
Indonesia	-	-	-	38	56	59	-	-	-	66.81	68.78	65.65
Iran (Islamic Republic of)	-	2.768	2.743	2.671	-	-	-	25.78	23.75	23.96	-	-
Italy	-	21	-	-	-	-	-	32.40	-	-	-	-
Korea, Republic of	180	142	132	128	136	142	55.43	49.50	48.11	47.96	48.68	49.31
Macau	0.058	0.065	0.031	0.023	0.018	-	50.00	62.50	65.96	60.53	64.29	-
Malaysia	145	167	183	211	230	-	78.21	75.70	75.94	73.87	71.95	-
Malta	0.620	0.538	0.842	0.941	1.032	-	36.77	35.61	47.07	47.03	49.43	-
Mongolia	-	-	-	-	0.003	-	-	-	-	-	37.50	-
Myanmar	0.078	-	-	-	-	-	22.10	-	-	-	-	-
Nepal	0.015	0.037	-	0.031	0.025	-	14.85	16.74	-	14.49	11.31	-
New Zealand	0.874	0.654	0.636	0.725	-	-	37.53	35.52	35.67	35.98	-	-
Philippines	-	-	54	51	-	-	-	-	77.28	75.12	-	-
Portugal	6.524	6.412	7.101	6.883	7.490	-	44.41	40.64	44.06	44.01	46.04	-
Sri Lanka	0.060	0.072	0.119	0.436	-	-	16.48	12.02	32.96	50.46	-	-
Sweden	9.400	-	-	-	-	-	38.21	-	-	-	-	-
Thailand	27	27	-	-	-	-	76.30	76.54	-	-	-	-
Turkey	5.377	-	-	-	-	-	29.71	-	-	-	-	-
United Kingdom	74	-	68	-	-	-	30.70	-	33.33	-	-	-
United States of America	-	-	492	488	505	538	-	-	56.55	56.42	57.52	58.04

Output and Output per Employee

Country	Output ($ bil.)						Output per Employee ($)					
	1990	1991	1992	1993	1994	1995	1990	1991	1992	1993	1994	1995
Albania	-	-	-	*0.002	*0.000	*0.002	-	-	-	8,764	27	17,435
Australia	*0.245	*0.276	*0.249	-	-	-	81,771	137,904	124,632	-	-	-
Austria	2.862	2.856	2.541	2.440	3.070	3.786	146,039	158,683	149,486	153,660	191,922	227,205
Bangladesh	0.050	0.030	0.020	-	-	-	19,864	13,869	10,272	-	-	-
Bolivia	0.002	0.002	0.001	0.002	0.002	0.003	13,295	15,069	7,984	10,208	13,243	15,246

Continued.

Depending on the table, * means *Enterprises, Engaged,* or *Factor Values;* ± means *Basis Unspecified;* § means *shown in millions.* For additional notes and sources, see Appendix I.

Output and Output per Employee

- Continued -

Country	Output ($ bil.)						Output per Employee ($)					
	1990	1991	1992	1993	1994	1995	1990	1991	1992	1993	1994	1995
Chile.	0.019	0.021	0.027	0.026	0.026	0.025	119,412	116,029	106,092	124,655	83,789	124,706
China (Hong Kong SAR) . . .	3.152	2.338	2.969	3.133	3.120	-	72,787	69,583	94,845	106,943	125,303	-
Colombia	0.127	0.108	0.105	0.120	0.136	0.133	29,099	27,990	22,828	26,064	28,998	32,131
Costa Rica	0.099	0.116	0.179	0.242	0.257	0.242	251,322	114,097	125,611	109,583		
Cyprus	0.000	0.001	0.001	0.000	0.000	0.000	21,335	20,750	21,042	18,684	29,841	34,277
Denmark	*1.285	*1.214	-	-	-	-	104,978	97,851	-	-	-	-
Ecuador	0.024	0.025	0.021	0.016	0.021	0.024	22,120	26,584	25,444	20,620	29,855	26,826
Egypt	*0.143	*0.110	*0.128	*0.146	-	-	17,648	11,186	13,485	15,067	-	-
El Salvador.	-	-	-	±0.008	±0.002	±0.002	-	-	-	108,926	19,842	20,898
Finland	±1.801	±1.259	±1.460	±1.913	±3.184	-	156,631	114,446	139,049	166,338	219,604	-
FYR Macedonia	-	-	-	-	-	±0.003	-	-	-	-	-	14,487
Gabon	-	*0.007	*0.006	*0.004	*0.003	*0.003	-	222,673	173,198	115,549	90,871	200,812
Germany	-	55	59	55	57	-	-	107,609	136,393	138,053	157,952	-
Germany (Western Part) . . .	51	54	58	-	-	-	121,495	126,749	143,996	-	-	-
Greece	*0.224	*0.377	*0.425	-	-	-	80,834	118,061	142,596	-	-	-
Guatemala	-	0.003	0.004	0.004	0.006	0.009	-	15,753	19,342	22,835	31,857	44,223
Honduras	0.005	0.004	0.002	0.002	0.002	0.006	16,129	14,161	11,804	10,064	8,888	26,238
India	*2.509	*2.340	*2.788	*2.284	-	-	21,857	19,841	22,469	19,116	-	-
Indonesia	0.618	0.612	1.307	1.411	2.076	3.145	22,894	17,459	30,347	24,668	25,582	34,729
Iran (Islamic Republic of) . . .	-	0.326	0.385	0.297	-	-	-	30,402	33,311	26,611	-	-
Italy	±10	±10	-	-	-	-	149,890	159,553	-	-	-	-
Japan	±196	±226	±213	±231	-	-	190,212	215,277	211,555	241,893	-	-
Korea, Republic of	27	28	29	33	42	61	82,069	97,249	104,772	125,632	151,619	212,700
Lithuania	-	-	0.087	0.111	-	-	-	-	2,934	4,255	-	-
Macau	0.004	0.005	0.000	0.000	0.000	-	38,040	51,650	5,548	6,151	10,402	-
Malaysia	*7.882	*11	*14	*18	*24	*31	42,400	49,145	58,891	62,355	74,747	84,153
Malta	0.497	0.582	0.802	0.684	0.882	-	294,592	385,317	448,191	341,824	422,452	-
Mauritius	0.007	0.008	0.009	0.007	0.007	-	48,823	59,248	84,458	58,920	68,128	-
Mexico	1.078	0.961	1.274	1.172	1.186	0.804	25,172	24,159	31,969	32,455	33,265	22,914
Mongolia	0.005	0.009	0.001	-	-	-	15,234	44,183	25,413	-	-	-
Myanmar	0.000	0.000	0.000	0.000	-	-	1,046	-	-	-	-	-
Nepal	0.002	0.005	-	0.003	0.003	-	15,845	20,405	-	11,819	12,274	-
Norway	±0.834	±0.834	±0.916	-	-	-	130,985	137,251	178,151	-	-	-
Pakistan	-	0.102	-	-	-	-	-	33,645	-	-	-	-
Peru	0.091	0.096	0.147	-	-	-	43,459	75,734	171,084	-	-	-
Philippines	1.354	1.778	2.161	2.054	-	-	25,034	28,777	30,734	30,429	-	-
Portugal	1.195	1.375	1.495	1.354	1.516	-	81,319	87,152	92,769	86,573	93,160	-
Romania	±0.758	±0.505	±0.191	±0.339	-	-	-	-	-	-	-	-
Singapore	*8.329	*9.135	*10	*11	*15	-	107,113	115,356	136,992	155,210	194,716	-
Slovakia	-	±0.188	±0.157	±0.138	±0.122	-	-	8,919	9,962	11,297	12,232	-
Spain	*5.730	*5.096	*4.417	-	-	-	180,966	155,312	152,345	-	-	-
Sri Lanka	0.005	0.012	0.007	0.004	-	-	14,400	19,587	18,047	4,221	-	-
Sweden	*3.915	*5.212	*5.590	*4.351	*6.362	-	159,127	158,550	183,075	146,733	184,891	-
Thailand	3.005	6.366	-	-	-	-	84,285	179,906	-	-	-	-
Trinidad and Tobago	±0.008	±0.013	±0.012	±0.010	±0.009	-	34,749	52,195	55,876	45,096	50,443	-
Turkey	1.797	2.260	2.433	2.633	1.499	-	99,262	126,323	121,105	143,193	98,302	-
United Kingdom	*25	*23	23	-	-	-	103,802	105,414	114,659	-	-	-
United States of America . . .	*117	*122	*138	*149	*165	*202	133,238	143,585	158,176	171,750	188,421	218,286
Venezuela	0.119	0.123	0.132	0.123	0.039	0.028	35,056	47,495	52,913	67,551	41,445	33,640
Zambia	0.003	-	-	-	-	-	77,672	-	-	-	-	-
Zimbabwe	*0.019	*0.019	*0.014	*0.011	*0.013	-	13,422	13,544	11,287	13,326	16,104	-

Depending on the table, * means *Enterprises, Engaged,* or *Factor Values*; ± means *Basis Unspecified*; § means *shown in millions*. For additional notes and sources, see Appendix I.

631

Value Added and Value Added per Employee

Country	Value Added ($ bil.)						Value Added per Employee ($)					
	1990	1991	1992	1993	1994	1995	1990	1991	1992	1993	1994	1995
Australia	*0.132	*0.134	-	-	-	-	44,010	67,004	-	-	-	-
Austria	0.872	0.792	0.794	0.783	0.910	1.145	44,469	43,998	46,683	49,309	56,918	68,724
Bangladesh	0.022	0.008	0.009				8,832	3,709	4,535			
Bolivia	0.001	0.000	0.001	0.001	0.001	0.001	4,629	3,689	4,434	3,718	5,135	5,912
Chile	0.010	0.010	0.015	0.019	0.021	0.015	61,057	55,783	59,893	93,613	67,303	75,823
China (Hong Kong SAR)	0.710	0.711	0.912	0.999	1.125	-	16,386	21,171	29,140	34,083	45,178	-
Colombia	0.048	0.050	0.048	0.053	0.057	0.055	10,912	13,137	10,427	11,493	12,174	13,343
Costa Rica	0.022	0.026	0.040	0.053	0.055	0.051	55,011	25,400	28,053	23,851	-	-
Cyprus	0.000	0.000	0.000	0.000	0.000	0.000	7,549	7,868	12,361	11,498	16,107	16,586
Denmark	*0.626	*0.611					51,097	49,272				
Ecuador	0.005	0.004	0.006	0.003	0.006	0.006	4,861	4,764	6,750	3,991	8,676	6,976
Egypt	*0.037	*0.042	*0.045	*0.035			4,617	4,324	4,745	3,648		
El Salvador	-	-	-	0.002	0.003	0.003	-	-	-	33,433	31,329	40,221
Finland	±0.876	±0.487	±0.602	±0.696	±1.017		76,155	44,308	57,321	60,480	70,121	
FYR Macedonia	-	-	-	-	-	±0.002	-	-	-	-	-	9,702
Gabon	-	*0.003	*0.002	*0.002	*0.001	*0.001	-	78,306	65,053	49,001	28,586	49,142
Germany (Western Part)	38	38	40	39	-	-	91,305	89,718	99,106	-	-	-
Greece	*0.097	*0.158	*0.174	-	-	-	35,094	49,585	58,572	-	-	-
Honduras	*0.002	*0.001	*0.001	*0.001	*0.001	*0.001	6,576	5,421	3,347	3,189	2,691	2,764
India	*0.506	*0.572	*0.619	*0.516			4,409	4,847	4,990	4,317		
Indonesia	*0.188	*0.214	*0.524	*0.353	*0.623	*0.768	6,964	6,094	12,166	6,169	7,670	8,485
Iran (Islamic Republic of)	-	0.176	0.177	0.160	-	-	-	16,384	15,334	14,313	-	-
Italy	±4.308	±4.203	-	-	-	-	63,193	65,166	-	-	-	-
Japan	±76	±87	±81	±89	-	-	73,715	82,782	80,118	92,647	-	-
Korea, Republic of	11	12	12	15	21	33	34,437	42,146	43,079	55,628	73,444	115,104
Macau	0.000	0.001	0.000	0.000	0.000	-	4,263	8,698	4,008	4,668	5,518	-
Malaysia	*1.637	*2.252	*3.074	*3.722	*4.782	*6.251	8,805	10,226	12,749	13,019	14,972	16,844
Malta	0.045	0.033	0.039	0.044	0.064	-	26,488	22,116	21,981	21,955	30,740	-
Mauritius	0.001	0.001	0.002	0.001	0.001	-	8,986	9,733	16,743	11,684	10,138	-
Mexico	0.694	0.649	0.764	0.727	0.753	-	16,208	16,313	19,184	20,149	21,138	-
Mongolia	0.003	0.004	0.001				9,821	18,843	16,509	-	-	-
Myanmar	±0.000	±0.000	±0.000	±0.000			375	-	-	-		
Nepal	0.001	0.001	-	0.001	0.001		8,428	5,830	-	6,041	5,157	
Norway	±0.317	±0.350	±0.328				49,803	57,575	63,833			
Pakistan	-	0.033	-	-	-	-	-	10,887	-	-	-	-
Peru	0.042	0.039	0.030	-	-	-	20,168	30,580	35,360	-	-	-
Philippines	0.532	0.707	0.691	0.684		-	9,840	11,436	9,834	10,133	-	-
Portugal	0.359	0.443	0.434	0.422	0.410		24,423	28,090	26,935	26,996	25,170	
Romania	±0.312	±0.168	±0.065	±0.129			-	-	-	-		
Singapore	*2.182	*2.468	*2.807	*3.219	*4.147		28,057	31,163	36,846	44,096	54,392	
Slovakia	-	-	-	±0.038	±0.041	-	-	-	-	3,116	4,098	-
Spain	*2.278	*2.098	*1.705	-	-	-	71,933	63,949	58,812	-	-	-
Sri Lanka	0.002	0.005	0.003	0.002	-	-	4,183	7,694	7,842	2,004	-	-
Sweden	*2.266	*1.594	*1.722	*1.533	*1.950	-	92,097	48,486	56,383	51,687	56,672	-
Thailand	1.549	3.661	-	-	-	-	43,445	103,462	-	-	-	-
Trinidad and Tobago	±0.003	±0.005	±0.003	±0.003	±0.002	-	14,034	18,909	15,791	14,564	11,157	-
Turkey	0.750	1.015	1.125	1.306	0.805	-	41,446	56,751	55,996	71,017	52,802	-
United Kingdom	*11	*11	*11	-	-	-	47,321	48,906	53,653	-	-	-
United States of America	*68	*69	*81	*86	*98	*125	77,457	81,014	92,956	99,326	112,164	134,672
Venezuela	0.047	0.039	0.045	0.038	0.014	0.018	13,721	14,825	17,819	20,651	15,276	20,835
Zambia	0.001	-	-	-	-	-	17,741	-	-	-	-	-
Zimbabwe	*0.010	*0.011	*0.006	*0.005	*0.002	-	7,003	7,647	4,940	5,755	2,776	-

Depending on the table, * means *Enterprises, Engaged,* or *Factor Values*; ± means *Basis Unspecified*; § means *shown in millions.* For additional notes and sources, see Appendix I.

Capital Investment

Country	Gross Fixed Capital Formation ($ mil.)						Per Establishment ($ mil)					
	1990	1991	1992	1993	1994	1995	1990	1991	1992	1993	1994	1995
Austria	187	194	243	144	217	-	2.969	3.076	4.188	2.435	4.103	-
Bangladesh	-	0.301	0.154	-	-	-	-	0.013	0.006	-	-	-
Bolivia	0.002	0.052	0.017	-	0.013	-	0.001	0.017	0.006	-	0.004	-
China (Hong Kong SAR)	-	-	229	215	257	-	-	-	0.473	0.585	0.857	-
Colombia	-	2.376	1.980	2.537	3.873	-	-	0.070	0.057	0.079	0.125	-
Cyprus	0.053	0.030	0.013	-	-	-	0.008	0.004	0.002	-	-	-
Denmark	63	39	-	-	-	-	0.239	0.130	-	-	-	-
Ecuador	-	1.731	1.664	0.787	0.995	6.076	-	0.133	0.128	0.056	0.077	0.608
El Salvador	-	-	-	0.055	0.075	0.070	-	-	-	0.027	0.025	0.070
Finland	71	59	68	90	215	-	1.079	0.753	0.924	1.084	2.530	-
Gabon	-	0.301	0.064	0.021	0.295	0.014	-	0.301	0.064	0.021	0.295	0.014
Germany (Western Part)	3,719	3,506	3,803	3,021	-	-	2.681	2.431	2.603	-	-	-
Greece	30	29	31	-	-	-	0.658	0.645	0.714	-	-	-
Hungary	-	-	24	30	-	-	-	-	0.056	0.061	-	-
India	166	140	165	204	-	-	0.142	0.117	0.139	0.169	-	-
Indonesia	65	90	168	102	117	145	0.807	0.930	1.465	0.865	0.733	0.790
Iran (Islamic Republic of)	-	27	31	14	-	-	-	1.418	1.288	0.586	-	-
Italy	566	522	-	-	-	-	2.144	1.886	-	-	-	-
Japan	11,686	14,862	12,112	10,764	-	-	0.742	0.930	0.816	0.766	-	-
Korea, Republic of	-	-	3,269	4,112	6,073	9,597	-	-	0.884	1.019	1.467	2.188
Macau	-	-	-	0.003	0.003	-	-	-	-	0.001	0.001	-
Malaysia	-	-	-	1,426	1,740	-	-	-	-	4.170	4.845	-
Malta	14	19	14	18	6.817	-	1.115	1.177	0.877	0.911	0.359	-
Mexico	23	44	-	-	-	-	0.334	0.678	-	-	-	-
Mongolia	3.143	-	-	-	-	-	3.143	-	-	-	-	-
Norway	-	-	31	-	-	-	-	-	0.508	-	-	-
Pakistan	-	13	-	-	-	-	-	0.709	-	-	-	-
Peru	0.974	1.773	0.726	-	-	-	0.006	0.010	0.004	-	-	-
Philippines	146	134	142	205	-	-	1.436	0.916	1.062	1.739	-	-
Portugal	31	69	51	103	-3.145	-	0.099	0.210	0.167	0.288	-0.008	-
Singapore	586	538	571	759	1,133	-	3.349	3.040	3.173	4.195	6.089	-
Spain	230	334	152	-	-	-	0.667	0.809	0.391	-	-	-
Sri Lanka	0.100	0.125	0.059	0.220	-	-	0.020	0.031	0.015	0.031	-	-
Thailand	480	364	-	-	-	-	12	4.234	-	-	-	-
Trinidad and Tobago	-	-	0.071	0.056	0.068	-	-	-	0.007	0.007	0.010	-
Turkey	63	126	64	77	56	-	1.240	2.374	0.608	0.761	0.711	-
United Kingdom	-	947	973	-	-	-	-	0.258	0.261	-	-	-
United States of America	-	-	6,621	7,726	10,015	14,176	-	-	0.484	-	-	-
Zambia	0.525	-	-	-	-	-	0.263	-	-	-	-	-

Indexes of Industrial Production

Country	Index of Industrial Production (1990=100)						Index of Employment (1990=100)					
	1990	1991	1992	1993	1994	1995	1990	1991	1992	1993	1994	1995
Australia	-	-	-	-	-	-	100	67	67	-	-	-
Austria	-	-	-	-	-	-	100	92	87	81	82	85
Bangladesh	-	-	-	-	-	-	100	87	79	-	-	-
Bolivia	-	-	-	-	-	-	100	105	112	132	128	130
Bosnia & Herzegovina	-	-	-	-	-	-	100	88	-	-	-	-
Chile	-	-	-	-	-	-	100	113	162	130	198	126
China (Hong Kong SAR)	-	-	-	-	-	-	100	78	72	68	58	-
Colombia	-	-	-	-	-	-	100	88	105	105	107	94
Costa Rica	-	-	-	-	-	-	100	258	363	562	-	-
Cyprus	-	-	-	-	-	-	100	140	160	70	60	20
Denmark	-	-	-	-	-	-	100	101	97	-	-	-
Ecuador	-	-	-	-	-	-	100	85	77	71	63	81
Egypt	-	-	-	-	-	-	100	121	117	120	-	-
Finland	-	-	-	-	-	-	100	96	91	100	126	-

Continued.

Indexes of Industrial Production

- Continued -

Country	Index of Industrial Production (1990=100)						Index of Employment (1990=100)					
	1990	1991	1992	1993	1994	1995	1990	1991	1992	1993	1994	1995
Germany (Western Part)	-	-	-	-	-	-	100	102	97	-	-	-
Greece	-	-	-	-	-	-	100	115	107	-	-	-
Honduras	-	-	-	-	-	-	100	90	58	63	69	79
India	-	-	-	-	-	-	100	103	108	104	-	-
Indonesia	-	-	-	-	-	-	100	130	160	212	301	335
Italy	-	±	-	-	-	-	100	95	-	-	-	-
Japan	-	-	-	-	-	-	100	102	98	93	-	-
Korea, Republic of	-	-	-	-	-	-	100	88	84	82	86	89
Macau	-	-	-	-	-	-	100	90	41	33	24	-
Malaysia	-	-	-	-	-	-	100	118	130	154	172	200
Malta	-	-	-	-	-	-	100	90	106	119	124	-
Mauritius	-	-	-	-	-	-	100	84	68	74	70	57
Mexico	-	-	-	-	-	-	100	93	93	84	83	82
Mongolia	-	-	-	-	-	-	100	61	12	-	3	-
Myanmar	-	-	-	-	-	-	100	-	-	-	-	-
Nepal	-	-	-	-	-	-	100	219	-	212	219	-
New Zealand	-	-	-	-	-	-	100	79	77	87	-	-
Norway	-	-	-	-	-	-	100	96	81	-	-	-
Peru	-	-	-	-	-	-	100	61	41	-	-	-
Philippines	-	-	-	-	-	-	100	114	130	125	-	-
Portugal	-	-	-	-	-	-	100	107	110	106	111	-
Singapore	-	-	-	-	-	-	100	102	98	94	98	-
Spain	-	-	-	-	-	-	100	104	92	-	-	-
Sri Lanka	-	-	-	-	-	-	100	165	99	237	-	-
Sweden	-	-	-	-	-	-	100	134	124	121	140	-
Thailand	-	-	-	-	-	-	100	99	-	-	-	-
Trinidad and Tobago	-	-	-	-	-	-	100	110	95	93	78	-
Turkey	-	-	-	-	-	-	100	99	111	102	84	-
United Kingdom	-	-	-	-	-	-	100	92	85	-	-	-
United States of America	-	±	-	-	-	-	100	97	99	99	100	106
Uruguay	-	-	-	-	-	-	100	76	74	68	-	-
Venezuela	-	-	-	-	-	-	100	76	74	53	27	25
Zambia	-	-	-	-	-	-	100	-	-	-	-	-
Zimbabwe	-	-	-	-	-	-	100	100	86	57	57	-

Depending on the table, * means *Enterprises, Engaged,* or *Factor Values*; ± means *Basis Unspecified*; § means *shown in millions*. For additional notes and sources, see Appendix I.

Product Tables

Television Receivers (Total)

Unit of Measure: Thousands of units.

	1990 Value	%	1991 Value	%	1992 Value	%	1993 Value	%	1994 Value	%	1995 Value	%	1996 Value	%
Total Production	141,143	100.0	138,487	100.0	136,936	100.0	140,161	100.0	144,014	100.0	146,402	100.0	146,667	100.0
Regions														
Africa	5,360	3.8	3,116	2.2	1,801	1.3	1,046	0.7	926	0.6	1,005	0.7	1,123	0.8
America, North	14,753	10.5	13,499	9.7	14,553	10.6	13,919	9.9	14,444	10.0	12,444	8.5	11,773	8.0
America, South	3,762	2.7	4,091	3.0	3,948	2.9	5,625	4.0	7,353	5.1	7,619	5.2	10,014	6.8
Asia	89,469	63.4	91,605	66.1	92,069	67.2	93,919	67.0	95,151	66.1	98,460	67.3	99,117	67.6
Europe	27,577	19.5	26,002	18.8	24,420	17.8	25,539	18.2	26,058	18.1	26,823	18.3	24,618	16.8
Oceania	222	0.2	175	0.1	144	0.1	113	0.1	82	0.1	51	0.0	22	0.0
Leading Producers														
China	26,847	19.0	26,914	19.4	28,678	20.9	30,330	21.6	32,833	22.8	34,962	23.9	35,418	24.1
Korea, Republic of	16,201	11.5	16,129	11.6	16,311	11.9	15,956	11.4	17,102	11.9	18,722	12.8	*20,123	13.7
Japan	15,132	10.7	15,640	11.3	14,253	10.6	12,840	9.2	11,192	7.8	9,022	6.2	7,568	5.2
United States of America	13,982	9.9	12,865	9.3	13,972	10.2	13,679	9.8	13,881	9.6	12,132	8.3	11,440	7.8
Kenya	4,186	3.0	*1,994	1.4	*709	0.5	-		-		-		-	

Radio Receivers (Total)

Unit of Measure: Thousands of units.

	1990 Value	%	1991 Value	%	1992 Value	%	1993 Value	%	1994 Value	%	1995 Value	%	1996 Value	%
Total Production	147,346	100.0	137,209	100.0	132,118	100.0	137,091	100.0	151,926	100.0	129,533	100.0	112,559	100.0
Regions														
Africa	1,477	1.0	1,558	1.1	1,504	1.1	1,328	1.0	1,345	0.9	1,393	1.1	1,392	1.2
America, North	3,186	2.2	3,666	2.7	4,156	3.1	2,859	2.1	1,523	1.0	944	0.7	366	0.3
America, South	5,167	3.5	5,350	3.9	5,497	4.2	4,117	3.0	4,686	3.1	4,751	3.7	2,964	2.6
Asia	121,315	82.3	112,369	81.9	103,549	78.4	111,418	81.3	125,922	82.9	107,262	82.8	92,682	82.3
Europe	15,801	10.7	13,830	10.1	16,939	12.8	16,859	12.3	17,904	11.8	14,598	11.3	14,535	12.9
Oceania	399	0.3	436	0.3	473	0.4	510	0.4	547	0.4	584	0.5	621	0.6
Leading Producers														
Hong Kong	8,182	5.6	6,145	4.5	6,508	4.9	11,018	8.0	3,390	2.2	1,698	1.3	-	
Malaysia	37,019	25.1	31,920	23.3	31,360	23.7	34,537	25.2	36,310	23.9	38,767	29.9	29,431	26.1
China	21,030	14.3	19,691	14.4	16,489	12.5	17,542	12.8	41,320	27.2	*21,295	16.4	*21,283	18.9
Singapore	*20,141	13.7	*20,807	15.2	*21,474	16.3	*22,140	16.2	*22,807	15.0	*23,474	18.1	*24,140	21.4
Japan	10,955	7.4	11,213	8.2	9,418	7.1	8,317	6.1	7,299	4.8	7,149	5.5	3,476	3.1

Telephones

Unit of Measure: Thousands of units.

	1990 Value	%	1991 Value	%	1992 Value	%	1993 Value	%	1994 Value	%	1995 Value	%	1996 Value	%
Total Production	124,890	100.0	127,490	100.0	128,993	100.0	136,611	100.0	163,407	100.0	201,223	100.0	188,102	100.0
Regions														
Africa	71	0.1	105	0.1	104	0.1	201	0.1	135	0.1	150	0.1	157	0.1
America, North	18,203	14.6	17,521	13.7	17,100	13.3	17,932	13.1	18,037	11.0	17,423	8.7	16,711	8.9
America, South	1,430	1.1	1,177	0.9	917	0.7	450	0.3	556	0.3	860	0.4	846	0.4
Asia	67,936	54.4	74,214	58.2	78,555	60.9	83,542	61.2	109,469	67.0	147,407	73.3	134,603	71.6
Europe	35,759	28.6	32,982	25.9	30,826	23.9	32,994	24.2	33,482	20.5	33,792	16.8	34,632	18.4
Oceania	1,491	1.2	1,491	1.2	1,491	1.2	1,491	1.1	1,728	1.1	1,592	0.8	1,153	0.6
Leading Producers														
China	8,800	7.0	14,820	11.6	19,820	15.4	26,636	19.5	57,229	35.0	99,563	49.5	79,608	42.3
United States of America	*17,001	13.6	*16,770	13.2	*16,539	12.8	*16,308	11.9	*16,077	9.8	*15,846	7.9	*15,615	8.3
Germany	*13,877	11.1	*13,877	10.9	*13,877	10.8	*13,877	10.2	*13,877	8.5	13,801	6.9	13,953	7.4
Hong Kong	11,268	9.0	9,225	7.2	7,182	5.6	5,139	3.8	3,096	1.9	1,053	0.5	*6,966	3.7
United Kingdom	*12,615	10.1	*12,615	9.9	*12,615	9.8	12,615	9.2	*12,615	7.7	*12,615	6.3	*12,615	6.7

Commodity data are provided by the United Nations Statistical Division. The symbol * means that data are estimated. For additional notes, see Appendix I.

635

Product Tables
Electronic Tubes

Unit of Measure: Thousands of units.

	1990		1991		1992		1993		1994		1995		1996	
	Value	%	Value	%	Value	%	Value	%	Value	%	Value	%	Value	%
Total Production	330,785	100.0	335,670	100.0	345,629	100.0	365,592	100.0	389,040	100.0	434,398	100.0	430,898	100.0
Regions														
America, North	23,666	7.2	22,137	6.6	20,609	6.0	19,080	5.2	17,552	4.5	16,023	3.7	14,495	3.4
America, South	2,497	0.8	2,497	0.7	2,497	0.7	2,497	0.7	2,497	0.6	2,497	0.6	2,497	0.6
Asia	295,758	89.4	303,951	90.6	314,296	90.9	334,906	91.6	359,592	92.4	403,608	92.9	404,281	93.8
Europe	8,864	2.7	7,084	2.1	8,227	2.4	9,108	2.5	9,400	2.4	12,270	2.8	9,625	2.2
Leading Producers														
Japan	*212,271	64.2	*225,857	67.3	*239,443	69.3	*253,029	69.2	*266,616	68.5	*280,202	64.5	*293,788	68.2
China	42,470	12.8	36,770	11.0	36,069	10.4	44,046	12.0	52,035	13.4	80,409	18.5	63,358	14.7
Korea, Republic of	41,017	12.4	40,717	12.1	38,355	11.1	37,778	10.3	40,863	10.5	42,938	9.9	*47,112	10.9
United States of America	*23,666	7.2	*22,137	6.6	*20,609	6.0	*19,080	5.2	*17,552	4.5	*16,023	3.7	*14,495	3.4
Poland	1,842	0.6	207	0.1	736	0.2	1,812	0.5	2,401	0.6	2,988	0.7	2,995	0.7

Sound Recorders

Unit of Measure: Thousands of units.

	1990		1991		1992		1993		1994		1995		1996	
	Value	%	Value	%	Value	%	Value	%	Value	%	Value	%	Value	%
Total Production	115,774	100.0	116,686	100.0	107,489	100.0	108,072	100.0	154,857	100.0	124,565	100.0	123,433	100.0
Regions														
Africa	167	0.1	185	0.2	202	0.2	220	0.2	238	0.2	255	0.2	273	0.2
America, North	916	0.8	663	0.6	712	0.7	740	0.7	776	0.5	923	0.7	989	0.8
Asia	104,728	90.5	106,423	91.2	97,431	90.6	98,532	91.2	145,900	94.2	115,434	92.7	114,216	92.5
Europe	9,962	8.6	9,415	8.1	9,143	8.5	8,579	7.9	7,943	5.1	7,954	6.4	7,956	6.4
Leading Producers														
Japan	43,670	37.7	48,232	41.3	39,402	36.7	36,862	34.1	32,049	20.7	30,515	24.5	24,636	20.0
China	30,235	26.1	28,737	24.6	32,318	30.1	36,429	33.7	83,956	54.2	*55,384	44.5	*59,629	48.3
Singapore	*16,335	14.1	*17,360	14.9	*18,385	17.1	*19,410	18.0	*20,435	13.2	*21,460	17.2	*22,485	18.2
Korea, Republic of	9,171	7.9	6,229	5.3	2,751	2.6	1,974	1.8	*6,981	4.5	*6,423	5.2	*5,864	4.8
Greece	*4,695	4.1	*4,695	4.0	*4,695	4.4	*4,695	4.3	*4,695	3.0	*4,695	3.8	*4,695	3.8

Sound Reproducers

Unit of Measure: Number.

	1990		1991		1992		1993		1994		1995		1996	
	Value	%	Value	%	Value	%	Value	%	Value	%	Value	%	Value	%
Total Production	18,905,889	100.0	26,154,082	100.0	22,791,379	100.0	18,090,634	100.0	20,746,460	100.0	21,599,782	100.0	19,449,180	100.0
Regions														
Africa	26,181	0.1	19,681	0.1	17,467	0.1	17,467	0.1	17,467	0.1	17,467	0.1	17,467	0.1
America, North	174,000	0.9	195,000	0.7	159,000	0.7	111,000	0.6	110,000	0.5	82,000	0.4	62,000	0.3
America, South	121,414	0.6	134,707	0.5	148,000	0.6	115,000	0.6	117,000	0.6	102,000	0.5	150,895	0.8
Asia	12,106,883	64.0	12,043,883	46.0	10,575,883	46.4	7,767,883	42.9	9,882,614	47.6	9,893,894	45.8	9,424,174	48.5
Europe	6,477,411	34.3	13,760,811	52.6	11,891,029	52.2	10,079,284	55.7	10,619,380	51.2	11,504,422	53.3	9,794,644	50.4
Leading Producers														
Japan	5,379,000	28.5	5,654,000	21.6	4,234,000	18.6	3,333,000	18.4	3,075,000	14.8	2,593,000	12.0	1,630,000	8.4
Germany	-		7,514,000	28.7	5,830,000	25.6	3,937,000	21.8	4,122,000	19.9	5,113,665	23.7	*3,350,732	17.2
Korea, Republic of	6,237,000	33.0	5,884,000	22.5	5,857,000	25.7	3,953,000	21.9	*6,316,731	30.4	*6,810,011	31.5	*7,303,291	37.6
United Kingdom	538,782	2.8	571,478	2.2	604,173	2.7	636,869	3.5	884,889	4.3	*769,851	3.6	*809,235	4.2

Product Tables
Vacuum Cleaners

Unit of Measure: Number.

	1990 Value	%	1991 Value	%	1992 Value	%	1993 Value	%	1994 Value	%	1995 Value	%	1996 Value	%
Total Production	45,702,866	100.0	56,493,769	100.0	55,867,392	100.0	64,950,963	100.0	52,361,748	100.0	52,612,460	100.0	55,183,557	100.0
Regions														
America, North	7,791,333	17.0	7,790,333	13.8	7,786,333	13.9	7,760,333	11.9	7,757,333	14.8	7,759,377	14.7	7,754,333	14.1
America, South	888,455	1.9	811,727	1.4	949,000	1.7	11,191,000	17.2	607,000	1.2	740,000	1.4	3,211,190	5.8
Asia	19,039,500	41.7	21,341,883	37.8	20,858,433	37.3	20,206,263	31.1	17,752,783	33.9	17,405,667	33.1	17,518,559	31.7
Europe	17,767,578	38.9	26,333,826	46.6	26,057,625	46.6	25,577,366	39.4	26,028,631	49.7	26,491,416	50.4	26,483,474	48.0
Oceania	216,000	0.5	216,000	0.4	216,000	0.4	216,000	0.3	216,000	0.4	216,000	0.4	216,000	0.4
Leading Producers														
United States of America	*7,754,333	17.0	*7,754,333	13.7	*7,754,333	13.9	*7,754,333	11.9	*7,754,333	14.8	*7,754,333	14.7	*7,754,333	14.1
Japan	6,851,000	15.0	6,981,000	12.4	6,465,000	11.6	6,331,000	9.7	6,355,000	12.1	6,595,000	12.5	6,708,000	12.2
Germany	-		7,155,000	12.7	6,546,000	11.7	5,793,000	8.9	5,982,761	11.4	6,172,521	11.7	5,983,613	10.8
Hong Kong	502,000	1.1	2,330,000	4.1	1,867,667	3.3	1,405,333	2.2	943,000	1.8	-		-	
Russian Federation	4,470,000	9.8	4,707,000	8.3	4,319,000	7.7	3,657,000	5.6	1,553,000	3.0	1,001,000	1.9	691,000	1.3

Shavers and Hair Clippers, Electric

Unit of Measure: Number.

	1990 Value	%	1991 Value	%	1992 Value	%	1993 Value	%	1994 Value	%	1995 Value	%	1996 Value	%
Total Production	32,209,750	100.0	31,763,750	100.0	37,764,500	100.0	34,430,500	100.0	32,386,250	100.0	32,292,253	100.0	30,410,739	100.0
Regions														
Asia	17,690,750	54.9	17,552,750	55.3	15,461,500	40.9	13,137,500	38.2	13,718,250	42.4	13,714,000	42.5	11,529,633	37.9
Europe	14,519,000	45.1	14,211,000	44.7	22,303,000	59.1	21,293,000	61.8	18,668,000	57.6	18,578,253	57.5	18,881,106	62.1
Leading Producers														
Japan	11,730,000	36.4	14,439,000	45.5	12,947,000	34.3	11,217,000	32.6	12,398,000	38.3	12,926,000	40.0	9,806,000	32.2
Germany	-		-		8,415,000	22.3	8,281,000	24.1	7,682,000	23.7	8,128,703	25.2	8,575,407	28.2
Ukraine	4,607,000	14.3	4,341,000	13.7	4,255,000	11.3	3,167,000	9.2	1,347,000	4.2	514,000	1.6	345,000	1.1
Hong Kong	5,332,000	16.6	2,485,000	7.8	1,885,750	5.0	1,286,500	3.7	687,250	2.1	88,000	0.3	*1,175,633	3.9
Russian Federation	*611,000	1.9	*611,000	1.9	*611,000	1.6	*611,000	1.8	*611,000	1.9	675,000	2.1	547,000	1.8

Heaters, Electric Space

Unit of Measure: Number.

	1990 Value	%	1991 Value	%	1992 Value	%	1993 Value	%	1994 Value	%	1995 Value	%	1996 Value	%
Total Production	27,453,061	100.0	27,220,460	100.0	25,957,574	100.0	33,326,109	100.0	29,243,742	100.0	25,185,753	100.0	25,821,537	100.0
Regions														
Africa	228,000	0.8	228,000	0.8	228,000	0.9	228,000	0.7	228,000	0.8	228,000	0.9	228,000	0.9
America, North	4,251,000	15.5	3,795,000	13.9	2,861,000	11.0	3,751,000	11.3	3,701,000	12.7	3,540,000	14.1	3,397,000	13.2
Asia	2,913,750	10.6	2,725,750	10.0	2,678,450	10.3	3,271,400	9.8	3,212,100	11.0	3,676,800	14.6	4,518,500	17.5
Europe	19,637,311	71.5	19,974,710	73.4	19,869,124	76.5	25,855,709	77.6	21,778,912	74.5	17,441,832	69.3	17,403,526	67.4
Oceania	423,000	1.5	497,000	1.8	321,000	1.2	220,000	0.7	323,731	1.1	299,121	1.2	274,511	1.1
Leading Producers														
United States of America	4,186,000	15.2	3,775,000	13.9	2,803,000	10.8	3,706,000	11.1	3,680,000	12.6	3,485,000	13.8	3,295,000	12.8
United Kingdom	*4,341,725	15.8	*4,341,725	16.0	*4,341,725	16.7	4,055,518	12.2	4,627,932	15.8	*4,341,725	17.2	*4,341,725	16.8
France	2,628,000	9.6	3,396,000	12.5	3,073,000	11.8	2,898,000	8.7	2,620,000	9.0	*2,751,473	10.9	*2,627,831	10.2
Germany	*3,364,500	12.3	3,865,000	14.2	3,350,000	12.9	3,011,000	9.0	3,232,000	11.1	*3,364,500	13.4	*3,364,500	13.0
Ukraine	1,275,000	4.6	1,154,000	4.2	1,709,000	6.6	9,340,000	28.0	4,574,000	15.6	258,000	1.0	180,000	0.7

Product Tables
Irons, Electric Smoothing

Unit of Measure: Number.

	1990		1991		1992		1993		1994		1995		1996	
	Value	%	Value	%	Value	%	Value	%	Value	%	Value	%	Value	%
Total Production	78,882,525	100.0	75,344,510	100.0	76,700,808	100.0	87,327,134	100.0	72,779,728	100.0	93,416,405	100.0	75,440,362	100.0
Regions														
Africa	23,600	0.0	13,000	0.0	13,000	0.0	13,000	0.0	13,000	0.0	13,000	0.0	13,000	0.0
America, North	7,200,000	9.1	4,018,000	5.3	3,941,000	5.1	11,782,818	13.5	8,716,329	12.0	11,954,839	12.8	12,089,350	16.0
America, South	5,546,380	7.0	4,134,333	5.5	3,750,000	4.9	5,190,000	5.9	5,055,000	6.9	5,626,000	6.0	4,869,371	6.5
Asia	31,895,917	40.4	32,249,683	42.8	34,594,450	45.1	37,516,714	43.0	28,176,169	38.7	44,156,267	47.3	27,211,467	36.1
Europe	33,613,128	42.6	34,325,993	45.6	33,798,858	44.1	32,221,101	36.9	30,215,731	41.5	31,062,800	33.3	30,653,674	40.6
Oceania	603,500	0.8	603,500	0.8	603,500	0.8	603,500	0.7	603,500	0.8	603,500	0.6	603,500	0.8
Leading Producers														
China	14,056,000	17.8	14,490,000	19.2	17,223,000	22.5	22,292,000	25.5	19,047,000	26.2	35,322,000	37.8	19,860,000	26.3
United States of America	6,560,000	8.3	3,598,000	4.8	3,276,000	4.3	*3,904,818	4.5	*3,473,329	4.8	*3,041,839	3.3	*2,610,350	3.5
Russian Federation	8,743,000	11.1	8,444,000	11.2	7,718,000	10.1	7,647,000	8.8	2,736,000	3.8	1,954,000	2.1	1,264,000	1.7
Germany	*4,740,500	6.0	*4,740,500	6.3	*4,740,500	6.2	5,163,000	5.9	4,318,000	5.9	*4,740,500	5.1	*4,740,500	6.3
France	4,308,000	5.5	6,350,000	8.4	6,768,000	8.8	6,421,000	7.4	6,865,000	9.4	*7,070,615	7.6	*7,429,231	9.8

Fuses, Electrical

Unit of Measure: Thousands of units.

	1990		1991		1992		1993		1994		1995		1996	
	Value	%	Value	%	Value	%	Value	%	Value	%	Value	%	Value	%
Total Production	1,468,715	100.0	1,447,731	100.0	1,443,306	100.0	1,224,641	100.0	1,376,718	100.0	1,720,346	100.0	1,518,511	100.0
Regions														
America, South	548	0.0	607	0.0	666	0.0	535	0.0	309	0.0	567	0.0	645	0.0
Asia	96,779	6.6	93,504	6.5	88,987	6.2	95,815	7.8	92,081	6.7	85,293	5.0	79,567	5.2
Europe	1,371,388	93.4	1,353,620	93.5	1,353,652	93.8	1,128,291	92.1	1,284,328	93.3	1,634,486	95.0	1,438,299	94.7
Leading Producers														
Germany	*735,679	50.1	*735,679	50.8	*735,679	51.0	524,482	42.8	650,358	47.2	980,805	57.0	787,071	51.8
Hungary	68,196	4.6	47,119	3.3	48,511	3.4	47,985	3.9	56,626	4.1	58,144	3.4	*56,007	3.7
Slovenia	*41,096	2.8	*41,096	2.8	*41,096	2.8	44,954	3.7	43,887	3.2	41,026	2.4	34,515	2.3
Iran, Islamic Republic of	*22,523	1.5	*22,523	1.6	*22,523	1.6	*22,523	1.8	*22,523	1.6	*22,523	1.3	*22,523	1.5
Turkey	21,608	1.5	16,984	1.2	12,448	0.9	14,716	1.2	11,348	0.8	6,720	0.4	6,804	0.4

Switches, Electric

Unit of Measure: Thousands of units.

	1990		1991		1992		1993		1994		1995		1996	
	Value	%	Value	%	Value	%	Value	%	Value	%	Value	%	Value	%
Total Production	4,681,548	100.0	5,256,235	100.0	5,342,626	100.0	5,471,206	100.0	5,931,016	100.0	6,789,038	100.0	7,041,381	100.0
Regions														
America, North	1,336,271	28.5	1,383,005	26.3	1,429,738	26.8	1,476,472	27.0	1,523,206	25.7	1,569,940	23.1	1,616,674	23.0
America, South	80,732	1.7	123,640	2.4	89,834	1.7	101,843	1.9	131,858	2.2	146,842	2.2	119,904	1.7
Asia	2,342,912	50.0	2,440,819	46.4	2,584,055	48.4	2,754,202	50.3	3,029,529	51.1	3,299,663	48.6	3,521,940	50.0
Europe	921,634	19.7	1,308,771	24.9	1,238,998	23.2	1,138,689	20.8	1,246,423	21.0	1,772,593	26.1	1,782,864	25.3
Leading Producers														
United States of America	*1,336,271	28.5	*1,383,005	26.3	*1,429,738	26.8	*1,476,472	27.0	*1,523,206	25.7	*1,569,940	23.1	*1,616,674	23.0
Japan	*1,335,111	28.5	*1,458,814	27.8	*1,582,516	29.6	*1,706,219	31.2	*1,829,921	30.9	*1,953,623	28.8	*2,077,326	29.5
Korea, Republic of	965,997	20.6	935,774	17.8	932,735	17.5	970,899	17.7	1,119,302	18.9	*1,250,133	18.4	*1,346,816	19.1
Germany	-		382,596	7.3	314,372	5.9	253,812	4.6	272,429	4.6	851,163	12.5	811,085	11.5
United Kingdom	*401,600	8.6	*401,600	7.6	*401,600	7.5	360,955	6.6	442,245	7.5	*401,600	5.9	*401,600	5.7

Product Tables
Wire and Cable, Insulated

Unit of Measure: Metric tons.

	1990 Value	%	1991 Value	%	1992 Value	%	1993 Value	%	1994 Value	%	1995 Value	%	1996 Value	%
Total Production	5,427,957	100.0	6,546,383	100.0	6,548,165	100.0	6,689,613	100.0	7,137,743	100.0	7,792,565	100.0	7,535,328	100.0
Regions														
Africa	97,111	1.8	111,944	1.7	105,031	1.6	99,651	1.5	108,293	1.5	112,521	1.4	121,338	1.6
America, North	1,386,467	25.5	1,330,285	20.3	1,382,231	21.1	1,634,095	24.4	1,786,897	25.0	2,269,996	29.1	2,371,174	31.5
America, South	115,214	2.1	129,909	2.0	125,718	1.9	127,125	1.9	126,681	1.8	133,874	1.7	148,226	2.0
Asia	1,390,415	25.6	1,371,458	20.9	1,280,071	19.5	1,303,453	19.5	1,551,100	21.7	1,782,375	22.9	1,440,792	19.1
Europe	2,411,188	44.4	3,557,551	54.3	3,604,470	55.0	3,467,276	51.8	3,499,713	49.0	3,419,229	43.9	3,380,389	44.9
Oceania	27,562	0.5	45,236	0.7	50,644	0.8	58,013	0.9	65,060	0.9	74,571	1.0	73,409	1.0
Leading Producers														
United States of America	1,357,238	25.0	1,301,447	19.9	1,351,940	20.6	1,573,266	23.5	1,720,951	24.1	2,218,907	28.5	2,295,223	30.5
Germany	-		1,208,143	18.5	1,206,971	18.4	1,054,884	15.8	1,076,559	15.1	931,045	11.9	864,928	11.5
Japan	*784,943	14.5	*784,943	12.0	*784,943	12.0	*784,943	11.7	*784,943	11.0	*784,943	10.1	*784,943	10.4
Spain	*233,276	4.3	*233,276	3.6	*233,276	3.6	186,022	2.8	229,964	3.2	258,311	3.3	258,809	3.4
Korea, Republic of	207,664	3.8	212,870	3.3	235,698	3.6	255,207	3.8	290,092	4.1	*278,995	3.6	*293,214	3.9

Batteries and Cells, Primary

Unit of Measure: Thousands of units.

	1990 Value	%	1991 Value	%	1992 Value	%	1993 Value	%	1994 Value	%	1995 Value	%	1996 Value	%
Total Production	38,988,226	100.0	43,775,336	100.0	39,872,094	100.0	39,537,411	100.0	40,780,181	100.0	48,767,015	100.0	53,979,166	100.0
Regions														
Africa	514,148	1.3	518,040	1.2	469,651	1.2	410,089	1.0	396,795	1.0	464,746	1.0	474,562	0.9
America, North	26,030,756	66.8	25,984,056	59.4	25,896,356	64.9	25,857,654	65.4	25,835,710	63.4	25,820,615	52.9	26,053,913	48.3
America, South	1,110,178	2.8	1,166,000	2.7	1,175,000	2.9	1,146,333	2.9	1,121,808	2.8	1,079,282	2.2	1,150,166	2.1
Asia	10,093,755	25.9	14,997,003	34.3	11,347,186	28.5	11,044,208	27.9	12,350,985	30.3	20,607,455	42.3	25,237,501	46.8
Europe	1,236,389	3.2	1,107,238	2.5	980,901	2.5	1,076,126	2.7	1,071,884	2.6	791,917	1.6	1,060,024	2.0
Oceania	3,000	0.0	3,000	0.0	3,000	0.0	3,000	0.0	3,000	0.0	3,000	0.0	3,000	0.0
Leading Producers														
United States of America	*25,379,000	65.1	*25,379,000	58.0	*25,379,000	63.7	*25,379,000	64.2	*25,379,000	62.2	*25,379,000	52.0	*25,379,000	47.0
Yugoslavia	-		2,721,000	6.2	1,394,500	3.5	68,000	0.2	1,350,000	3.3	8,110,000	16.6	12,497,000	23.2
Japan	4,113,000	10.5	4,283,000	9.8	4,385,000	11.0	4,532,000	11.5	4,479,000	11.0	4,716,000	9.7	4,523,000	8.4
Indonesia	1,348,000	3.5	1,407,000	3.2	1,842,000	4.6	1,920,000	4.9	2,104,000	5.2	2,718,000	5.6	*2,342,057	4.3
Hong Kong	1,938,000	5.0	3,713,000	8.5	882,000	2.2	1,279,000	3.2	1,155,000	2.8	1,295,000	2.7	*2,000,762	3.7

Accumulators, Electric, for Motor Vehicles

Unit of Measure: Thousands of units.

	1990 Value	%	1991 Value	%	1992 Value	%	1993 Value	%	1994 Value	%	1995 Value	%	1996 Value	%
Total Production	224,082	100.0	226,620	100.0	224,757	100.0	235,702	100.0	239,490	100.0	248,017	100.0	229,474	100.0
Regions														
Africa	4,031	1.8	4,166	1.8	4,260	1.9	4,337	1.8	4,388	1.8	4,680	1.9	4,593	2.0
America, North	69,185	30.9	69,791	30.8	69,829	31.1	73,748	31.3	76,768	32.1	73,161	29.5	74,718	32.6
America, South	5,047	2.3	5,582	2.5	4,894	2.2	6,083	2.6	6,741	2.8	7,541	3.0	8,049	3.5
Asia	60,954	27.2	60,483	26.7	59,100	26.3	65,059	27.6	64,081	26.8	63,079	25.4	65,042	28.3
Europe	82,494	36.8	84,273	37.2	84,394	37.5	84,242	35.7	85,326	35.6	97,415	39.3	74,978	32.7
Oceania	2,372	1.1	2,326	1.0	2,280	1.0	2,233	0.9	2,187	0.9	2,141	0.9	2,095	0.9
Leading Producers														
United States of America	*59,008	26.3	*59,008	26.0	*59,008	26.3	*59,008	25.0	*59,008	24.6	*59,008	23.8	*59,008	25.7
Germany	*53,340	23.8	*53,340	23.5	*53,340	23.7	*53,340	22.6	*53,340	22.3	65,283	26.3	41,397	18.0
Japan	29,387	13.1	28,121	12.4	26,692	11.9	23,868	10.1	23,814	9.9	24,512	9.9	24,811	10.8
France	11,316	5.0	12,665	5.6	11,664	5.2	11,280	4.8	11,870	5.0	*12,786	5.2	*13,195	5.8
Korea, Republic of	8,912	4.0	8,948	3.9	10,514	4.7	11,206	4.8	12,972	5.4	*13,216	5.3	*14,070	6.1

Product Tables

Lamps, Electric (Excluding Fluorescent Tubes)

Unit of Measure: Thousands of units.

	1990 Value	%	1991 Value	%	1992 Value	%	1993 Value	%	1994 Value	%	1995 Value	%	1996 Value	%
Total Production	15,184,183	100.0	17,179,038	100.0	19,939,937	100.0	16,380,704	100.0	15,638,057	100.0	16,329,467	100.0	17,226,477	100.0
Regions														
Africa	116,750	0.8	120,750	0.7	113,750	0.6	23,750	0.1	31,879	0.2	30,999	0.2	65,176	0.4
America, North	2,928,900	19.3	2,901,200	16.9	3,054,500	15.3	3,198,800	19.5	3,101,100	19.8	2,761,488	16.9	2,719,904	15.8
America, South	466,294	3.1	562,072	3.3	560,186	2.8	569,188	3.5	644,190	4.1	574,281	3.5	604,857	3.5
Asia	7,199,093	47.4	8,292,282	48.3	11,072,325	55.5	7,464,848	45.6	6,833,952	43.7	7,809,056	47.8	8,477,885	49.2
Europe	4,473,146	29.5	5,302,734	30.9	5,139,176	25.8	5,124,118	31.3	5,026,936	32.1	5,153,643	31.6	5,358,655	31.1
Leading Producers														
United States of America	2,751,000	18.1	2,724,000	15.9	2,866,000	14.4	2,981,000	18.2	2,930,000	18.7	*2,624,088	16.1	*2,556,204	14.8
Japan	1,868,000	12.3	1,964,000	11.4	1,904,000	9.5	1,878,000	11.5	1,835,000	11.7	1,999,000	12.2	2,200,000	12.8
Yugoslavia	50,000	0.3	1,045,000	6.1	3,963,000	19.9	105,000	0.6	546,000	3.5	1,360,000	8.3	1,499,000	8.7
Germany	-		999,000	5.8	957,000	4.8	896,000	5.5	896,000	5.7	943,888	5.8	1,015,507	5.9
Russian Federation	1,098,000	7.2	1,037,000	6.0	981,000	4.9	874,000	5.3	537,000	3.4	610,000	3.7	576,000	3.3

Tubes, Fluorescent

Unit of Measure: Thousands of units.

	1990 Value	%	1991 Value	%	1992 Value	%	1993 Value	%	1994 Value	%	1995 Value	%	1996 Value	%
Total Production	1,819,681	100.0	2,126,146	100.0	2,155,677	100.0	1,982,457	100.0	2,004,937	100.0	2,029,707	100.0	2,148,670	100.0
Regions														
America, North	502,852	27.6	524,925	24.7	569,048	26.4	608,317	30.7	614,333	30.6	620,356	30.6	640,502	29.8
America, South	59,164	3.3	55,978	2.6	63,317	2.9	65,693	3.3	73,833	3.7	72,113	3.6	76,802	3.6
Asia	732,145	40.2	851,460	40.0	803,596	37.3	716,133	36.1	708,413	35.3	734,571	36.2	841,281	39.2
Europe	525,521	28.9	693,783	32.6	719,717	33.4	592,315	29.9	608,359	30.3	602,667	29.7	590,086	27.5
Leading Producers														
United States of America	477,173	26.2	502,034	23.6	555,183	25.8	591,792	29.9	598,864	29.9	*608,371	30.0	*628,811	29.3
Japan	398,039	21.9	403,957	19.0	400,361	18.6	375,138	18.9	371,201	18.5	390,267	19.2	431,934	20.1
Germany	-		176,086	8.3	193,202	9.0	187,798	9.5	218,538	10.9	219,326	10.8	199,455	9.3
Korea, Republic of	81,952	4.5	76,658	3.6	98,837	4.6	119,148	6.0	115,401	5.8	*117,775	5.8	*125,164	5.8
United Kingdom	63,874	3.5	*69,138	3.3	*72,282	3.4	*75,425	3.8	*78,568	3.9	*81,712	4.0	*84,855	3.9

TRANSPORTATION EQUIPMENT

ISIC 3840—Transportation Equipment—includes all ships, aircraft, road vehicles, and rail equipment. Also included are bicycles; motorcycles; trailers and semi-trailers; wagons and vans; and children's vehicles such as perambulators. Two major subcategories—shipbuilding (ISIC 3841) and motor vehicles (ISIC 3843)—are also shown in separate chapters. All commodity and company tables for transportation equipment, however, are included in this chapter.

The U.S. classification system provides a 2-digit SIC grouping which covers the same ground. It is SIC 37—Transportation Equipment.

Summary Statistics

		1990		1991		1992		1993		1994		1995	
		Value	N	Value	N	Value	N	Value	N	Value	N	Value	N
Establishments or enterprises	(number)	61,757	81	65,804	90	81,415	89	68,936	83	60,686	66	49,389	31
Number employed	(000)	12,783	93	11,549	95	11,618	93	13,669	93	12,986	82	11,737	61
Total Wages	($ mil.)	224,255	89	219,296	89	235,705	87	208,980	84	210,352	72	174,070	55
Wage/salary per employee	($)	10,712	86	11,216	86	15,732	85	12,274	83	11,829	71	14,310	55
Female workers	(000)	472	34	253	32	1,418	31	1,368	33	1,345	25	1,311	14
as % of total employment*	(%)	14	34	15	32	14	31	16	33	16	25	20	14
Output	($ bil.)	1,417	93	1,407	92	1,575	90	1,538	86	1,601	75	1,520	59
per employee	($ 000)	64	88	66	89	395	87	86	84	77	73	102	58
per establishment	($ 000)	12,119	79	11,750	84	35,547	80	12,711	73	12,274	58	9,734	29
Value added	($ bil.)	489	83	492	81	529	77	535	74	475	65	502	56
per employee	($ 000)	25	78	25	78	28	75	26	72	28	64	36	55
per establishment	($ 000)	4,152	71	3,836	74	4,363	68	3,395	61	3,638	48	2,977	26
Capital investment	($ mil.)	52,669	69	68,794	63	69,471	59	57,604	53	48,815	42	30,404	23
per employee	($ 000)	3	66	4	61	4	57	4	51	4	42	4	23
per establishment	($ 000)	520	64	723	58	736	55	446	47	415	39	295	20

Data presented above are drawn from the detailed tables that follow. Columns headed 'N' show the number of countries that provided valid data for inclusion. Values are not strictly comparable one year to the next or one row to the next because the number of countries that report varies from period to period and row to row. However, a general indication of magnitudes can be gleaned from reviewing this summary—especially in earlier years in which more reports are available. For detailed explanations, see Appendix I. *The average for those countries reporting both total and female employment.

Establishments and Number Engaged

Country	Establishments or Enterprises (number)						Employees per Establishment					
	1990	1991	1992	1993	1994	1995	1990	1991	1992	1993	1994	1995
Albania	-	-	-	2.000	-	-	-	-	-	51	-	-
Angola	-	-	17	17	-	-	-	-	67	38	-	-
Argentina	-	-	-	3,524	3,828	-	-	-	-	20	19	-
Armenia	6.000	6.000	6.000	5.000	7.000	-	924	846	-	-	-	-
Australia	1,501	1,963	2,035	-	-	-	65	42	38	-	-	-
Austria	134	135	140	130	134	-	237	230	214	217	215	-
Azerbaijan	10	10	10	10	9.000	-	676	628	564	570	614	-
Bahrain	-	-	6.000	-	-	-	-	-	216	-	-	-
Bangladesh	184	184	177	-	-	-	64	61	71	-	-	-
Belgium	531	537	549	528	528	-	-	-	-	-	-	-
Belize	8.000	9.000	9.000	-	-	-	-	19	21	-	-	-
Benin	-	-	1.000	1.000	-	-	-	-	-	-	-	-
Bolivia	35	41	42	53	51	51	14	13	13	14	14	14
Bosnia & Herzegovina	36	38	-	-	-	-	557	525	-	-	-	-
Brazil	855	-	832	827	-	-	399	-	339	353	-	-
Bulgaria	*142	*122	*122	*2,142	*3,919	*4,539	585	555	471	24	13	11
Cameroon	*4.000	*7.000	*6.000	*5.000	*4.000	*1.000	65	63	39	42	42	11
Canada	1,469	1,346	1,273	1,224	1,210	-	139	140	151	153	160	-
Cape Verde	4.000	3.000	3.000	3.000	2.000	-	69	91	89	77	94	-
Central African Republic	-	*1.000	*1.000	-	-	-	-	-	-	-	-	-
Chile	44	46	50	44	45	-	194	173	241	265	254	-
China	*14,185	*14,527	*14,798	*15,439	*16,411	*19,445	-	158	162	219	210	190
China (Hong Kong SAR)	624	567	706	570	606	-	23	24	21	26	23	-
Colombia	252	249	305	283	253	-	75	72	73	83	85	-
Costa Rica	93	113	105	105	-	-	20	16	17	16	-	-
Croatia	141	177	246	323	381	431	265	197	114	81	64	55
Cyprus	76	76	76	75	93	101	5.645	5.395	4.921	4.547	4.656	4.505
Czechoslovakia (Former)	*74	*100	-	-	-	-	2,757	1,720	-	-	-	-
Czech Republic	-	*76	*141	*187	-	-	-	1,829	1,050	658	-	-
Denmark	846	852	838	-	-	-	31	30	30	-	-	-
Ecuador	37	43	43	48	58	59	58	63	63	70	68	50
Egypt	125	141	140	128	-	-	501	385	431	373	-	-
El Salvador	-	1.000	1.000	5.000	2.000	16	-	-	22	27	59	11
Ethiopia	2.000	2.000	2.000	2.000	3.000	3.000	208	204	185	183	129	135
Fiji	14	19	18	-	-	-	23	16	15	-	-	-
Finland	239	259	214	198	179	185	116	102	115	119	134	121
FYR Macedonia	*31	*72	*137	*154	*229	*338	228	144	71	58	37	22
Gambia	-	-	-	1.000	-	-	-	-	-	9.000	-	-
Germany	-	1,343	1,363	1,388	1,376	-	-	-	-	670	628	-
Germany (Western Part)	1,103	1,138	1,153	-	-	-	863	847	813	-	-	-
Ghana	-	-	-	7.000	-	-	-	-	-	77	-	-
Greece	215	210	212	-	-	-	105	106	103	-	-	-
Grenada	-	-	1.000	1.000	2.000	-	-	-	10	13	5.000	-
Guatemala	-	8.000	8.000	8.000	8.000	8.000	-	117	127	124	123	132
Honduras	-	-	9.000	11	11	11	-	-	31	28	30	39
Hungary	*416	*899	*254	*305	-	-	161	68	193	108	-	-
Iceland	59	70	70	65	-	-	11	8.557	7.800	8.277	-	-
India	4,890	5,213	5,552	5,966	-	-	132	129	129	119	-	-
Indonesia	436	441	490	513	535	577	198	224	198	195	215	223
Iran (Islamic Republic of)	82	111	110	105	-	-	388	321	379	305	-	-
Iraq	-	4.000	4.000	-	-	-	-	129	125	-	-	-
Ireland	120	120	-	-	-	-	63	67	-	-	-	-
Israel	192	216	232	236	215	-	94	86	82	77	74	-
Italy	*913	*932	*1,225	*1,180	*1,152	-	351	338	255	254	246	-
Japan	15,168	15,134	14,850	14,948	13,223	13,453	61	64	65	63	68	66
Jordan	17	20	21	22	35	35	5.765	6.800	8.476	34	31	31
Kenya	26	62	62	60	72	72	657	264	260	256	213	212
Korea, Republic of	2,819	3,134	3,182	3,475	3,686	4,090	86	78	77	76	78	75
Kuwait	21	16	14	14	15	-	97	162	187	198	67	-
Kyrgyzstan	4.000	4.000	4.000	4.000	4.000	-	1,267	1,169	1,094	994	543	-

Continued.

Depending on the table, * means *Enterprises, Engaged,* or *Factor Values*; ± means *Basis Unspecified*; § means *shown in millions.* For additional notes and sources, see Appendix I.

Establishments and Number Engaged

- Continued -

Country	Establishments or Enterprises (number)						Employees per Establishment					
	1990	1991	1992	1993	1994	1995	1990	1991	1992	1993	1994	1995
Latvia	26	53	66	53	43	58	953	424	274	282	315	227
Lithuania	-	-	-	*30	*39	-	-	-	-	368	329	-
Luxembourg	*15	*15	*16	*16	-	-	34	29	27	22	-	-
Macau	71	65	63	53	43	45	7.211	7.277	6.556	5.547	6.395	5.956
Malaysia	239	264	282	321	330	-	105	120	118	119	132	-
Malta	36	37	34	35	35	-	71	68	72	64	63	-
Mauritius	6.000	9.000	9.000	9.000	8.000	10	81	70	71	68	66	68
Mexico	155	152	149	145	140	137	738	775	739	633	591	566
Mongolia	-	-	-	-	2.000	-	-	-	-	-	45	-
Morocco	102	103	97	102	110	112	87	105	115	115	107	112
Mozambique	9.000	9.000	5.000	5.000	5.000	-	276	253	220	228	210	-
Netherlands	317	330	347	326	327	-	180	172	163	152	151	-
Netherlands Antilles	-	-	-	13	-	-	-	-	-	68	-	-
New Zealand	*933	*932	*947	*969	-	-	8.226	7.017	6.993	6.917	-	-
Norway	423	404	297	352	346	-	56	57	74	107	108	-
Pakistan	-	130	-	-	-	-	-	144	-	-	-	-
Panama	16	13	-	-	-	-	16	17	-	-	-	-
Paraguay	-	1.000	-	-	-	-	-	20	-	-	-	-
Peru	495	500	514	-	-	-	21	17	14	-	-	-
Philippines	598	653	266	257	-	-	38	35	93	93	-	-
Poland	*324	*335	*291	*391	-	-	852	701	704	506	-	-
Portugal	*749	*749	*772	*800	*831	-	53	46	43	41	44	-
Puerto Rico	17	19	19	21	18	20	56	47	47	49	55	33
Republic of Moldova	1.000	1.000	1.000	1.000	1.000	1.000	1,686	1,524	1,625	412	-	-
Romania	-	140	197	217	343	-	-	1,873	1,176	1,012	558	-
Russian Federation	-	-	-	*4,170	*2,850	*4,204	-	-	-	339	443	281
Senegal	16	11	10	8.000	10	-	71	108	90	47	155	-
Singapore	217	238	246	282	291	-	120	125	129	117	119	-
Slovakia	-	26	26	32	40	-	-	1,546	1,301	896	663	-
Slovenia	107	*69	*84	*97	*104	*117	209	293	217	157	143	108
South Africa	-	1,258	-	-	-	-	-	78	-	-	-	-
Spain	1,897	2,004	1,931	-	-	-	106	101	104	-	-	-
Sri Lanka	39	40	26	31	-	-	176	177	260	254	-	-
Suriname	-	-	2.000	-	-	-	-	-	38	-	-	-
Sweden	437	463	435	396	374	-	246	222	222	206	230	-
Thailand	337	697	-	-	-	-	130	112	-	-	-	-
Tonga	-	9.000	6.000	4.000	4.000	-	-	-	6.000	7.750	15	-
Trinidad and Tobago	25	44	43	43	38	-	23	19	20	17	7.237	-
Tunisia	-	-	-	514	543	567	-	-	-	13	14	15
Turkey	209	215	414	418	395	-	303	299	164	174	167	-
Ukraine	122	239	324	154	147	140	2,934	1,481	1,256	2,474	2,327	2,129
United Kingdom	4,254	4,233	4,333	3,704	4,356	-	125	117	108	113	93	-
United Republic of Tanzania	40	40	-	-	-	-	74	72	-	-	-	-
USSR (Former)	*903	-	-	-	-	-	3,210	-	-	-	-	-
United States of America	-	-	16,701	-	-	-	-	-	106	-	-	-
Venezuela	246	270	253	234	217	215	67	71	82	85	84	88
Yugoslavia	129	180	288	338	346	348	-	504	289	240	232	219
Zambia	8.000	-	-	-	11	-	120	-	-	-	51	-
Zimbabwe	56	57	56	54	53	-	126	131	111	109	108	-

Employment and Compensation of Employees

Country	Number Employed/Engaged (000)						Wage/Salary per Employee ($)					
	1990	1991	1992	1993	1994	1995	1990	1991	1992	1993	1994	1995
Albania	-	-	-	0.101	-	-	-	-	-	-	-	-
Angola	-	*2.075	*1.136	*0.650	-	-	-	-	-	-	-	-
Argentina	89	-	-	71	74	-	8,162	-	-	19,959	-	-
Armenia	*5.543	*5.077	-	-	-	-	-	-	-	-	-	-
Australia	98	*83	*77	-	-	-	22,070	25,786	22,947	-	-	-
Austria	32	31	30	28	29	29	25,614	28,799	32,402	31,584	32,957	39,868
Azerbaijan	6.756	6.277	5.638	5.701	5.526	-	2,091	2,371	-	483	-	-
Bahrain	-	-	1.296	-	-	-	-	-	12,430	-	-	-
Bangladesh	12	11	13	-	-	-	815	1,187	887	-	-	-
Belize	-	0.174	0.190	-	-	-	-	4,161	4,766	-	-	-
Bermuda	0.013	0.014	0.014	0.014	0.017	0.021	-	-	-	-	-	-
Bolivia	0.502	0.531	0.545	0.720	0.727	0.724	1,641	1,499	1,490	1,446	1,516	1,455
Bosnia & Herzegovina	20	20	-	-	-	-	3,426	3,176	-	-	-	-
Brazil	341	-	282	292	-	-	9,653	-	10,834	13,593	-	-
Bulgaria	83	68	58	52	52	49	2,042	741	1,225	1,647	1,249	1,492
Cambodia	*2.192	*1.894	-	-	-	-	-	-	-	-	-	-
Cameroon	0.260	0.441	0.232	0.209	0.168	0.011	8,448	10,345	15,812	10,054	6,636	6,557
Canada	204	188	192	187	193	198	31,835	33,813	33,050	33,186	33,093	34,779
Cape Verde	0.276	0.273	0.268	0.230	0.188	-	-	-	4,203	5,304	-	-
Central African Republic	0.100	-	-	-	-	-	5,877	-	-	-	-	-
Chile	8.521	7.936	12	12	11	11	4,199	4,885	6,003	5,593	7,097	7,835
China	-	*2,290	*2,400	*3,380	*3,450	*3,700	-	-	-	-	-	-
China (Hong Kong SAR)	*15	*14	*15	*15	*14	-	14,696	16,265	17,556	20,560	21,803	-
China (Taiwan Province)	133	132	138	142	144	142	12,449	14,150	16,231	16,309	16,900	17,183
Colombia	19	18	22	24	22	22	2,728	2,734	2,800	3,398	4,524	5,074
Costa Rica	1.832	1.766	1.792	1.720	-	-	3,382	2,627	3,980	4,241	-	-
Croatia	37	35	28	26	25	24	5,498	4,951	1,583	1,496	2,011	3,591
Cyprus	0.429	0.410	0.374	0.341	0.433	0.455	10,329	11,073	12,234	11,181	11,944	15,213
Czechoslovakia (Former)	204	172	-	-	-	-	2,239	1,550	-	-	-	-
Czech Republic	162	139	148	123	-	-	2,239	1,557	1,946	2,454	-	-
Denmark	26	26	25	24	29	31	30,123	31,290	35,644	33,372	36,937	-
Ecuador	2.155	2.701	2.727	3.367	3.926	2.938	3,002	3,665	4,100	4,848	2,212	2,756
Egypt	*63	*54	*60	*48	*48	*47	1,996	1,365	1,555	1,797	1,876	1,862
El Salvador	-	-	*0.022	*0.135	*0.118	*0.170	-	-	-	2,279	4,400	1,822
Ethiopia	0.416	0.409	0.370	0.365	0.386	0.404	3,621	3,950	2,936	1,919	1,837	1,897
Fiji	0.320	0.303	0.278	0.358	0.361	-	2,783	3,750	4,779	3,395	3,706	-
Finland	28	26	25	24	24	22	30,828	29,213	27,455	20,472	24,428	33,431
FYR Macedonia	7.078	10	9.717	8.980	8.460	7.316	4,822	6,528	2,034	2,286	2,714	3,572
France	536	529	521	506	489	487	35,229	35,444	39,858	-	-	-
Gambia	-	-	-	0.009	-	-	-	-	-	-	-	-
Germany	-	-	-	*930	*865	-	-	-	-	38,347	-	-
Germany (Western Part)	*952	*964	*938	-	-	-	35,359	36,849	-	-	-	-
Ghana	-	-	-	*0.539	*0.554	*0.570	-	-	-	-	-	-
Greece	*22	*22	*22	*18	*17	*16	15,901	16,934	17,652	21,231	22,719	28,297
Grenada	-	-	*0.010	*0.013	0.010	-	-	-	-	-	-	-
Guatemala	-	*0.933	*1.018	*0.990	*0.983	*1.054	-	-	-	-	-	-
Honduras	*0.434	*0.432	*0.279	*0.304	*0.334	*0.424	2,512	1,732	5,328	2,552	1,689	2,533
Hungary	67	61	49	33	36	45	2,430	2,585	3,054	3,856	3,837	4,183
Iceland	0.634	0.599	0.546	0.538	0.360	-	26,418	31,659	-	32,350	-	-
India	*644	*670	*718	*710	*770	*869	1,884	1,635	1,558	1,472	1,630	1,797
Indonesia	86	99	97	100	115	129	1,083	1,146	1,317	1,445	1,359	2,736
Iran (Islamic Republic of)	32	36	42	32	-	-	5,123	5,242	5,319	7,450	-	-
Iraq	-	0.517	0.500	-	-	-	-	9,068	19,196	-	-	-
Ireland	7.500	8.000	7.671	7.250	7,471	7.604	25,804	25,121	26,662	23,174	23,963	26,764
Israel	18	19	19	18	16	19	33,403	36,098	37,348	34,280	33,626	-
Italy	320	315	313	300	283	-	36,768	37,581	38,420	30,100	31,989	-
Japan	*929	*970	*959	*944	*904	*886	33,722	37,828	-	-	-	-
Jordan	0.098	0.136	0.178	0.741	1.085	1.081	2,551	2,591	2,181	2,498	2,907	2,833
Kenya	17	16	16	15	15	15	1,912	1,741	1,584	1,001	1,216	1,788
Korea, Republic of	243	244	245	263	289	307	13,572	14,685	16,447	17,750	20,272	23,175

Continued.

Depending on the table, * means *Enterprises, Engaged,* or *Factor Values*; ± means *Basis Unspecified*; § means *shown in millions*. For additional notes and sources, see Appendix I.

Employment and Compensation of Employees

- Continued -

Country	Number Employed/Engaged (000)						Wage/Salary per Employee ($)					
	1990	1991	1992	1993	1994	1995	1990	1991	1992	1993	1994	1995
Kuwait	2.030	2.592	2.614	2.768	1.009	2.194	3,167	2,085	13,762	12,835	12,452	17,899
Kyrgyzstan	*5.068	*4.677	*4.374	*3.977	*2.173	-	507	-	-	-	-	-
Latvia	25	22	18	15	14	13	-	-	-	-	-	-
Lithuania	-	-	13	11	13	-	-	-	-	522	1,004	-
Luxembourg	0.511	0.441	0.425	0.348	0.379	-	19,266	20,519	22,834	22,426	24,354	-
Macau	0.512	0.473	0.413	0.294	0.275	0.268	6,883	8,241	8,804	11,366	12,800	11,275
Malawi	0.587	-	-	-	-	-	-	-	-	-	-	-
Malaysia	25	32	33	38	43	51	3,712	4,128	5,025	5,144	5,356	6,084
Malta	2.573	2.527	2.446	2.256	2.202	-	10,425	11,678	11,683	10,940	12,790	-
Mauritius	*0.483	0.630	0.642	0.614	0.530	0.683	2,159	2,413	3,129	2,243	4,118	3,220
Mexico	114	118	110	92	83	78	5,555	6,670	8,567	9,907	10,867	6,096
Mongolia	-	-	-	-	*0.090	-	-	-	-	-	-	-
Morocco	8.846	11	11	12	12	12	6,350	5,532	5,714	5,438	6,015	6,459
Mozambique	2.487	2.276	1.102	1.138	1.049	-	711	414	536	563	834	-
Netherlands	57	57	57	49	49	50	28,526	29,206	32,698	30,877	32,362	37,442
Netherlands Antilles	-	-	-	0.888	-	-	-	-	-	17,537	-	-
New Zealand	7.675	6.540	6.622	6.703	-	-	29,170	29,218	25,922	28,152	-	-
Norway	24	23	22	38	37	40	30,190	31,238	39,861	33,278	34,168	39,697
Pakistan	20	19	20	-	-	-	2,359	2,477	2,547	-	-	-
Panama	*0.254	*0.225	*0.228	*0.262	*0.272	*0.270	6,764	6,764	6,885	7,273	7,436	7,469
Paraguay	-	0.020	-	-	-	-	-	-	-	-	-	-
Peru	*11	*8.601	*7.429	-	-	-	2,713	3,986	4,591	-	-	-
Philippines	23	23	25	24	-	-	2,171	2,110	2,856	2,841	-	-
Poland	276	235	205	198	219	223	1,325	1,873	2,508	2,687	3,115	3,759
Portugal	*39	*34	*34	*33	*36	*37	-	-	-	-	-	-
Puerto Rico	0.950	0.890	0.890	1.020	0.990	0.670	22,947	22,360	15,955	14,314	15,051	24,328
Republic of Moldova	1.686	1.524	1.625	0.412	-	-	-	4,636	-	-	-	-
Romania	346	262	232	220	192	-	1,892	1,208	833	973	1,064	-
Russian Federation	-	-	-	1,415	1,263	1,181	-	-	-	794	1,316	1,352
Senegal	*1.140	*1.183	*0.895	*0.375	*1.548	-	6,428	9,070	12,157	10,745	3,159	-
Singapore	*26	*30	*32	*33	*35	*35	13,306	14,212	17,479	18,545	20,254	22,646
Slovakia	-	*40	*34	*29	*27	-	-	1,514	1,806	1,980	2,377	-
Slovenia	22	20	18	15	15	13	9,510	6,936	6,107	6,454	7,665	9,228
South Africa	104	98	91	81	85	86	10,082	10,547	11,875	11,958	11,083	13,178
Spain	202	203	200	192	185	188	23,324	24,092	26,813	22,927	23,153	26,097
Sri Lanka	*6.848	*7.071	*6.771	*7.873	-	-	1,018	998	1,057	894	-	-
Suriname	*0.059	*0.072	*0.077	*0.091	-	-	17,092	17,896	14,551	19,700	-	-
Sweden	108	*103	*96	82	86	81	24,934	27,588	30,947	24,796	25,799	34,407
Thailand	44	78	-	0.031	0.058	-	3,468	3,817	-	3,706	1,554	-
Tonga	0.027	0.008	0.036	0.031	0.058	-	1,417	1,832	3,114	3,706	1,554	-
Trinidad and Tobago	0.583	0.822	0.865	0.712	0.275	-	5,126	5,067	5,759	4,383	3,437	-
Tunisia	-	-	-	6.553	7.377	8.316	-	-	-	5,405	4,936	4,932
Turkey	63	64	68	73	66	70	8,339	10,400	10,488	11,510	6,900	9,566
Ukraine	358	354	407	381	342	298	-	3,177	2,503	443	534	583
United Kingdom	531	496	466	417	405	405	26,022	27,901	29,281	25,857	27,179	28,759
United Republic of Tanzania	2.952	2.876	-	-	-	-	439	539	-	-	-	-
USSR (Former)	2,899	-	-	-	-	-	2,317	-	-	-	-	-
United States of America	1,914	1,760	1,777	1,725	1,707	1,690	35,084	36,449	37,799	38,963	-	-
Uruguay	*4.860	*4.611	*4.023	*3.546	*4.467	*2.913	4,684	6,286	6,205	7,184	7,949	7,635
Venezuela	17	19	21	20	18	19	4,444	4,846	5,854	5,616	5,749	5,675
Yugoslavia	-	91	83	81	80	76	-	-	-	-	1,301	556
Zambia	0.957	-	-	-	0.561	-	2,504	-	-	-	1,725	-
Zimbabwe	7.036	7.474	6.200	5.900	5.700	5.785	4,562	4,690	3,530	3,252	3,055	3,525

Depending on the table, * means *Enterprises*, *Engaged*, or *Factor Values*; ± means *Basis Unspecified*; § means *shown in millions*. For additional notes and sources, see Appendix I.

Female Workers: Total and Percent of Employment

Country	Female Workers (000)						Female Workers as % of Total Employment					
	1990	1991	1992	1993	1994	1995	1990	1991	1992	1993	1994	1995
Albania	-	-	-	0.012	-	-	-	-	-	11.88	-	-
Angola	-	-	0.157	0.076	-	-	-	-	13.82	11.69	-	-
Argentina	-	-	-	-	3.964	-	-	-	-	-	5.34	-
Australia	13	-	-	-	-	-	13.65	-	-	-	-	-
Austria	4.700	5.000	5.000	4.121	4.382	-	14.83	16.13	16.67	14.58	15.24	-
Bahrain	-	-	0.023	-	-	-	-	-	1.77	-	-	-
Bangladesh	0.034	0.099	0.045	-	-	-	0.29	0.89	0.36	-	-	-
Bermuda	0.001	0.001	0.001	0.001	0.002	0.002	7.69	7.14	7.14	7.14	11.76	9.52
Bosnia & Herzegovina	4.671	4.501	-	-	-	-	23.28	22.56	-	-	-	-
Bulgaria	-	-	-	16	16	15	-	-	-	30.23	30.53	29.61
Canada	31	27	-	-	-	-	15.20	14.36	-	-	-	-
Cape Verde	-	-	0.014	0.014	0.014	-	-	-	5.22	6.09	7.45	-
Chile	0.591	0.544	0.823	0.710	0.744	-	6.94	6.85	6.82	6.10	6.50	-
China (Taiwan Province)	29	29	32	34	34	35	21.63	22.28	23.15	23.77	23.95	24.39
Colombia	-	2.465	-	3.603	3.197	-	-	13.71	-	15.33	14.80	-
Croatia	6.420	6.030	4.780	4.070	3.780	3.670	17.17	17.30	17.11	15.62	15.40	15.52
Cyprus	0.082	0.081	0.064	0.057	0.074	-	19.11	19.76	17.11	16.72	17.09	-
Czechoslovakia (Former)	33	30	-	-	-	-	16.42	17.73	-	-	-	-
Czech Republic	50	44	49	41	-	-	30.86	31.65	33.11	33.33	-	-
Denmark	3.168	3.207	3.113	-	-	-	12.25	12.44	12.31	-	-	-
Egypt	-	1.913	2.172	1.716	-	-	-	3.52	3.60	3.59	-	-
El Salvador	-	-	-	0.017	0.007	0.013	-	-	-	12.59	5.93	7.65
Ethiopia	-	0.073	0.074	0.068	0.069	0.075	-	17.85	20.00	18.63	17.88	18.56
FYR Macedonia	1.568	1.856	1.367	1.337	1.073	0.688	22.15	17.86	14.07	14.89	12.68	9.40
Gambia	-	-	-	0.002	-	-	-	-	-	22.22	-	-
Germany (Western Part)	139	-	-	-	-	-	14.60	-	-	-	-	-
Hungary	17	-	10	9.000	-	-	25.37	-	20.41	27.27	-	-
Indonesia	-	-	-	12	14	16	-	-	-	11.95	12.14	12.54
Iran (Islamic Republic of)	-	0.848	1.062	0.874	-	-	-	2.38	2.54	2.73	-	-
Iraq	-	0.122	-	-	-	-	-	23.60	-	-	-	-
Ireland	0.600	0.700	-	-	-	-	8.00	8.75	-	-	-	-
Italy	-	39	22	21	22	-	-	12.35	7.13	7.12	7.71	-
Jordan	0.004	0.033	0.010	0.034	0.048	0.046	4.08	24.26	5.62	4.59	4.42	4.26
Kenya	0.404	0.337	0.328	0.314	-	-	2.37	2.06	2.03	2.05	-	-
Korea, Republic of	31	31	31	32	36	37	12.68	12.77	12.70	12.33	12.34	12.17
Macau	0.011	0.011	0.009	0.008	0.009	0.010	2.15	2.33	2.18	2.72	3.27	3.73
Malaysia	5.200	6.800	6.800	7.600	8.700	-	20.72	21.38	20.42	19.84	20.00	-
Malta	0.054	0.058	0.056	0.047	0.044	-	2.10	2.30	2.29	2.08	2.00	-
Mongolia	-	-	-	-	0.019	-	-	-	-	-	21.11	-
Morocco	-	-	-	-	-	2.679	-	-	-	-	-	21.45
New Zealand	1.360	1.119	1.178	1.105	-	-	17.72	17.11	17.79	16.49	-	-
Panama	0.021	-	-	-	-	-	8.27	-	-	-	-	-
Philippines	-	-	3.153	3.133	-	-	-	-	12.71	13.16	-	-
Portugal	4.691	2.677	2.692	2.620	3.601	-	11.89	7.79	8.03	8.00	9.96	-
Puerto Rico	0.500	0.470	0.430	0.530	0.510	0.290	52.63	52.81	48.31	51.96	51.52	43.28
Senegal	0.072	-	-	-	-	-	6.32	-	-	-	-	-
Sri Lanka	0.647	0.698	0.644	0.816	-	-	9.45	9.87	9.51	10.36	-	-
Sweden	21	-	-	-	-	-	19.61	-	-	-	-	-
Thailand	8.638	12	-	-	-	-	19.64	15.85	-	-	-	-
Turkey	4.293	-	-	-	-	-	6.78	-	-	-	-	-
United Kingdom	60	-	49	-	-	-	11.23	-	10.52	-	-	-
United Republic of Tanzania	0.255	0.252	-	-	-	-	8.64	8.76	-	-	-	-
United States of America	-	-	1,191	1,170	1,193	1,201	-	-	67.02	67.83	69.89	71.07
Zambia	-	-	-	-	0,027	-	-	-	-	-	4.81	-

Depending on the table, * means *Enterprises, Engaged*, or *Factor Values*; ± means *Basis Unspecified*; § means *shown in millions*. For additional notes and sources, see Appendix I.

Output and Output per Employee

Country	Output ($ bil.)						Output per Employee ($)					
	1990	1991	1992	1993	1994	1995	1990	1991	1992	1993	1994	1995
Albania	-	-	-	*0.000	-	-	-	-	-	485	-	-
Angola	-	-	±32	±0.777	-	-	-	-	§28,20	§1,195	-	-
Argentina	±4.550	-	-	9.006	-	-	51,024	-	-	127,509	-	-
Armenia	0.002	0.004	0.018	0.614	0.004	-	385	699	-	-	-	-
Australia	*12	*12	*9.906	-	-	-	126,090	138,589	128,648	-	-	-
Austria	4.745	5.410	6.114	5.339	5.819	7.131	149,699	174,524	203,810	188,843	202,335	248,344
Azerbaijan	±0.064	±0.059	±0.003	±0.007	±0.003	-	9,515	9,431	447	1,300	508	-
Bahrain	-	-	±0.063	-	-	-	-	-	48,439	-	-	-
Bangladesh	0.135	0.143	0.146	-	-	-	11,490	12,806	11,627	-	-	-
Belize	0.009	0.002	0.005	-	-	-	-	9,261	27,387	-	-	-
Bolivia	0.006	0.006	0.006	0.010	0.013	0.015	12,223	11,566	10,846	13,491	18,514	20,514
Bosnia & Herzegovina	±0.797	±0.925	-	-	-	-	39,703	46,371	-	-	-	-
Brazil	*24	-	*21	*30	-	-	69,331	-	75,778	101,752	-	-
Bulgaria	±1.367	±0.350	±0.335	±0.335	±0.316	±0.403	16,446	5,174	5,828	6,483	6,024	8,181
Cameroon	0.011	0.016	0.011	0.011	0.007	0.001	44,033	36,709	47,534	51,081	42,456	53,363
Canada	57	56	57	66	75	81	278,424	299,965	299,392	355,069	386,211	408,344
Cape Verde	±0.006	±0.004	±0.007	±0.004	±0.003	-	20,847	12,871	24,332	17,385	14,604	-
Central African Republic	*0.005	-	-	-	-	-	52,744	-	-	-	-	-
Chile	0.431	0.494	0.801	0.848	0.845	0.881	50,537	62,230	66,372	72,847	73,748	81,480
China	±15	±18	±28	±45	±37	±40	-	8,004	11,666	13,346	10,714	10,690
China (Hong Kong SAR)	0.539	0.572	0.704	0.762	0.766	-	37,184	42,077	46,645	52,187	54,742	-
China (Taiwan Province)	±12	±13	±15	±15	±16	±17	92,355	100,113	111,277	105,823	108,761	117,690
Colombia	1.167	1.016	0.975	1.402	1.682	1.895	61,407	56,504	44,049	59,640	77,894	87,604
Costa Rica	0.033	0.028	0.047	0.039	0.041	0.033	18,006	15,649	26,077	22,463	-	-
Cote d'Ivoire	±0.165	±0.160	±0.166	±0.124	-	-	-	-	-	-	-	-
Croatia	±0.955	±0.942	±0.604	-	-	-	25,532	27,029	21,613	-	-	-
Cyprus	0.019	0.019	0.021	0.017	0.021	0.026	44,973	46,236	56,203	49,110	48,901	56,150
Czechoslovakia (Former)	±3.984	±1.595	-	-	-	-	19,529	9,271	-	-	-	-
Denmark	*2.606	*2.888	*3.026	*2.664	*3.765	*4.679	100,719	112,042	119,679	111,017	128,986	152,244
Ecuador	0.145	0.193	0.215	0.312	0.405	0.339	67,171	71,559	78,825	92,638	103,078	115,520
Egypt	*0.633	*0.548	*0.599	*0.522	*0.555	*0.545	10,113	10,087	9,924	10,931	11,585	11,505
El Salvador	-	±0.002	-	±0.003	±0.003	±0.000	-	-	-	23,764	28,770	1,529
Ethiopia	0.014	0.013	0.010	0.014	0.024	0.052	34,330	32,801	26,441	39,036	63,205	128,756
Fiji	0.005	0.005	0.005	0.006	0.006	-	16,100	17,591	18,644	15,927	17,388	-
Finland	±3.864	±3.051	±2.989	±2.371	±2.865	±3.423	139,495	115,557	121,013	100,467	119,356	152,909
FYR Macedonia	±0.198	±0.223	±0.101	±0.088	±0.086	±0.076	28,041	21,451	10,436	9,815	10,165	10,428
France	±92	±88	±98	±84	±94	±109	171,337	166,017	188,204	165,957	192,170	224,475
Germany	-	-	-	155	168	-	-	-	-	166,141	194,065	-
Germany (Western Part)	163	178	188	-	-	-	171,368	185,049	200,038	-	-	-
Greece	*0.947	*1.057	*1.048	*1.046	*1.069	*1.260	42,111	47,537	48,082	57,825	61,877	77,069
Guatemala	-	0.014	0.017	0.016	0.027	0.020	-	15,198	16,749	16,307	27,242	18,622
Honduras	0.005	0.004	0.004	0.003	0.002	0.019	11,821	8,988	15,935	8,496	7,271	45,529
Hungary	±1.172	±1.218	±1.637	±1.758	±2.070	±3.192	17,498	19,961	33,412	53,271	57,783	71,466
Iceland	0.034	0.038	±0.044	±0.038	-	-	54,408	63,899	80,448	70,109	-	-
India	*9.725	*7.625	*8.589	*8.317	*10	*13	15,089	11,380	11,959	11,713	13,107	14,578
Indonesia	2.617	3.322	3.184	4.318	6.146	8.087	30,285	33,604	32,754	43,086	53,376	62,930
Iran (Islamic Republic of)	0.868	2.177	1.995	1.858			27,290	61,195	47,797	58,011		
Iraq	-	*0.023	*0.032	-	-	-	-	44,525	63,023	-	-	-
Ireland	*0.622	*0.566	*0.582	*0.483	*0.511	*0.575	82,875	70,719	75,859	66,686	68,436	75,621
Israel	1.403	1.526	1.495	1.371	1.316	1.711	77,494	82,489	78,677	75,745	82,279	90,543
Italy	±46	±43	±46	±31	±37	-	143,074	135,516	145,945	103,820	130,864	-
Japan	±321	±361	±388	±418	±424	±461	346,004	372,385	404,331	443,258	468,994	520,488
Jordan	0.002	0.003	0.003	0.036	0.062	0.042	23,851	19,878	18,407	48,421	56,898	38,677
Kenya	*0.353	*0.286	*0.498	*0.295	*0.590	*1.363	20,645	17,507	30,885	19,232	38,397	89,078
Korea, Republic of	28	31	34	38	48	59	115,167	125,887	140,268	145,008	164,432	192,211
Kuwait	0.012	0.009	0.067	0.079	0.048	0.092	5,761	3,460	25,749	28,700	47,542	42,092
Kyrgyzstan	0.122	0.144	0.144	0.017	0.005	-	24,119	30,756	32,979	4,314	2,333	-
Latvia	0.003	0.007	0.067	0.102	0.120	0.127	126	323	3,714	6,827	8,857	9,654
Lithuania	-	-	0.025	0.032	0.074	-	-	-	1,880	2,940	5,746	-
Luxembourg	*0.048	*0.047	*0.038	*0.032	*0.037	-	94,398	107,376	90,385	90,616	98,572	-

Continued.

Depending on the table, * means *Enterprises*, *Engaged*, or *Factor Values*; ± means *Basis Unspecified*; § means *shown in millions*. For additional notes and sources, see Appendix I.

Output and Output per Employee

- Continued -

Country	Output ($ bil.)						Output per Employee ($)					
	1990	1991	1992	1993	1994	1995	1990	1991	1992	1993	1994	1995
Macau	0.008	0.010	0.009	0.006	0.007	0.008	14,901	20,651	21,705	21,445	24,335	29,425
Malaysia	*1.701	*2.302	*2.446	*2.960	*3.715	*5.457	67,783	72,404	73,466	77,294	85,410	106,670
Malta	0.074	0.068	0.061	0.056	0.055	-	28,655	26,980	24,997	24,996	24,894	-
Mauritius	0.008	0.010	0.009	0.006	0.010	0.010	16,521	16,074	14,462	9,718	19,593	14,858
Mexico	11	15	17	15	15	13	99,230	123,124	153,599	168,383	177,594	172,759
Morocco	±0.513	±0.643	±0.636	±0.472	±0.517	±0.553	58,018	59,228	56,910	40,329	43,920	44,304
Mozambique	-	±0.007	-	±0.002	±0.003	-	-	3,042	-	1,814	2,496	-
Netherlands	9.798	9.677	11	7.965	8.630	11	171,472	170,688	189,800	161,058	175,227	214,729
New Zealand	1.260	0.921	0.968	1.062	-	-	164,208	140,775	146,185	158,423		-
Norway	±3.336	±3.536	±3.419	±4.903	±4.994	±6.171	140,088	152,227	155,331	129,838	133,885	156,062
Pakistan	0.597	0.610	0.663	-	-	-	29,803	32,704	33,935	-	-	-
Panama	0.009	0.006	0.007	0.008	0.009	0.009	34,846	28,822	29,431	31,642	32,507	32,642
Peru	0.426	0.380	0.325	-	-	-	40,368	44,180	43,784	-	-	-
Philippines	1.004	0.886	1.484	1.674	-	-	44,240	38,865	59,841	70,345	-	-
Poland	±3.463	±3.075	±3.092	±3.885	±5.022	±6.203	12,546	13,084	15,084	19,622	22,963	27,809
Portugal	2.448	2.282	2.576	2.023	2.305	2.825	62,042	66,420	76,805	61,790	63,738	75,617
Puerto Rico	±0.138	±0.160	±0.092	±0.098	±0.110	±0.115	145,126	179,551	103,258	95,784	111,313	171,791
Republic of Moldova	0.000	0.064	0.005	0.000	-	-	0.057	42,226	3,212	1,040	-	-
Romania	±2.523	±2.177	±1.400	±1.360	-	-	7,290	8,303	6,043	6,194		-
Russian Federation	-	-	-	8.940	11	12	-	-	-	6,316	8,463	9,938
Senegal	0.030	-	-	-	-	-	26,142	-	-	-	-	-
Singapore	*2.091	*2.323	*2.699	*2.841	*3.298	*3.716	80,012	78,135	85,151	86,462	95,229	107,167
Slovakia	-	±0.530	±0.504	±0.401	±0.495	-	-	13,173	14,898	14,008	18,683	-
Slovenia	1.168	1.068	0.993	0.908	1.065	1.355	52,125	52,750	54,386	59,522	71,418	107,241
South Africa	±6.764	±6.878	±6.852	±6.451	±6.692	±8.094	65,036	70,186	75,301	79,648	78,732	93,678
Spain	*35	*38	*41	±28	±32	±40	175,304	185,975	207,334	143,342	174,719	214,388
Sri Lanka	0.053	0.067	0.061	0.064	-	-	7,736	9,437	8,977	8,080	-	-
Suriname	*0.006	*0.006	*0.006	0.021	-	-	104,449	77,809	80,032	233,940		-
Sweden	*15	*18	*18	*11	*14	*19	142,475	175,143	191,485	133,759	161,790	228,070
Thailand	3.712	5.883	-	-	-	-	84,425	75,322	-	-	-	-
Tonga	0.000	0.000	0.000	0.000	0.001	-	6,390	55,455	9,815	8,717	9,259	-
Trinidad and Tobago	±0.021	±0.034	±0.045	±0.033	±0.011	-	35,940	41,174	52,421	45,748	40,077	-
Tunisia	0.218	0.234	0.253	0.261	0.290	0.327	-	-	-	39,819	39,277	39,269
Turkey	4.770	5.408	6.800	9.260	4.630	6.857	75,349	84,071	100,187	127,078	70,390	98,591
Ukraine	*4.211	*3.912	*6.145	*1.840	*1.532	*1.430	11,761	11,051	15,098	4,830	4,479	4,798
United Kingdom	*70	*65	72	60	69	74	131,867	130,595	153,801	144,944	171,081	182,221
United Republic of Tanzania	*0.054	*0.049	-	-	-	-	18,263	17,177	-	-	-	-
USSR (Former)	±34	-	-	-	-	-	11,685	-	-	-	-	-
United States of America	*385	*380	*417	*436	*477	*490	201,055	215,688	234,731	252,840	279,440	290,020
Uruguay	0.272	0.298	0.242	0.215	0.299	0.188	55,890	64,618	60,086	60,753	66,994	64,423
Venezuela	0.772	1.449	1.908	1.927	1.616	2.658	46,535	75,480	92,159	96,780	88,132	140,579
Yugoslavia	±0.252	±0.235	±0.223	-	±0.464	±0.249	-	2,586	2,686	-	5,788	3,278
Zambia	0.011	-	-	-	0.006	-	11,883	-	-	-	9,848	-
Zimbabwe	*0.225	*0.296	*0.258	*0.214	*0.178	*0.215	31,917	39,568	41,635	36,321	31,299	37,159

Value Added and Value Added per Employee

Country	Value Added ($ bil.)						Value Added per Employee ($)					
	1990	1991	1992	1993	1994	1995	1990	1991	1992	1993	1994	1995
Argentina	±2.140	-	-	2.431	-	-	24,005	-	-	34,418	-	-
Armenia	0.000	0.000	-	-	-	-	22	25	-	-	-	-
Australia	*5.379	*4.984	*5.189	-	-	-	55,168	60,049	67,393	-	-	-
Austria	1.652	1.814	2.014	1.669	1.858	2.264	52,110	58,507	67,122	59,032	64,593	78,839
Bangladesh	0.056	0.060	0.086	-	-	-	4,724	5,355	6,812	-	-	-
Belize	0.001	0.001	0.002	-	-	-	-	4,609	13,137	-	-	-
Bolivia	0.003	0.002	0.002	0.004	0.003	0.003	5,543	4,156	3,673	5,182	4,390	4,779
Bosnia & Herzegovina	±0.299	±0.347	-	-	-	-	14,880	17,380	-	-	-	-
Brazil	*12	-	*14	*17	-	-	36,296	-	47,913	59,507	-	-

Continued.

Depending on the table, * means *Enterprises*, *Engaged*, or *Factor Values*; ± means *Basis Unspecified*; § means *shown in millions*. For additional notes and sources, see Appendix I.

Value Added and Value Added per Employee

- Continued -

Country	Value Added ($ bil.)						Value Added per Employee ($)					
	1990	1991	1992	1993	1994	1995	1990	1991	1992	1993	1994	1995
Bulgaria.	-	±0.135	±0.131	±0.141	-	-	-	1,999	2,274	2,735	-	-
Cameroon	0.003	0.002	0.001	0.003	0.000	0.000	10,044	3,738	4,250	12,453	1,469	30,051
Canada	14	14	14	15	17	18	69,270	72,287	71,185	80,539	87,949	92,683
Central African Republic	*0.001	-	-	-	-	-	11,864		-	-	-	-
Chile.	0.153	0.144	0.248	0.249	0.214	0.223	17,924	18,164	20,554	21,370	18,697	20,662
China	±3.918	±4.601	±7.195	±12	±8.764	±9.641		2,009	2,998	3,584	2,540	2,606
China (Hong Kong SAR)	0.334	0.350	0.435	0.471	0.463	-	23,027	25,708	28,798	32,292	33,078	
China (Taiwan Province)	3.998	4.426	5.435	5.141	5.454	5.462	30,097	33,461	39,366	36,075	37,915	38,478
Colombia	0.332	0.340	0.294	0.341	0.435	0.490	17,500	18,937	13,261	14,520	20,157	22,630
Costa Rica	0.016	0.014	0.024	0.020	0.022	0.018	8,966	7,869	13,391	11,398		-
Cote d'Ivoire	0.151	0.145	0.151	0.113	-	-	-	-	-	-	-	-
Croatia	±0.339	±0.429	±0.342	-	-	-	9,069	12,320	12,239		-	-
Cyprus	0.009	0.009	0.009	0.008	0.010	0.012	20,535	21,825	22,828	22,540	23,503	26,454
Czechoslovakia (Former)	±0.903	-	-	-	-	-	4,427		-	-	-	-
Denmark	*1.128	*1.224	*1.284	*1.135	*1.570	*1.941	43,595	47,496	50,775	47,297	53,780	63,169
Ecuador	0.022	0.035	0.032	0.057	0.042	0.043	10,115	12,793	11,813	17,030	10,709	14,503
Egypt	*0.237	*0.227	*0.163	*0.154	*0.164	*0.161	3,792	4,174	2,706	3,223	3,422	3,397
El Salvador.	-	0.001	-	0.002	0.002	0.001	-	-	-	11,230	18,187	3,582
Ethiopia.	0.006	0.006	0.004	0.006	0.008	0.008	13,256	13,558	11,085	16,792	20,513	20,377
Fiji	0.001	0.002	0.001	0.002	0.002	-	4,518	5,467	5,392	4,675	5,104	-
Finland	±1.405	±1.163	±1.158	±0.926	±1.036	±1.181	50,734	44,061	46,863	39,255	43,178	52,765
FYR Macedonia	±0.103	±0.110	±0.067	±0.055	±0.054	±0.043	14,601	10,542	6,891	6,101	6,375	5,942
France	±29	±27	±29	±26	±29	±33	53,431	50,427	55,805	51,062	59,013	67,959
Germany (Western Part)	67	73	79	64	-	-	70,837	75,452	83,948	:		
Greece	*0.488	*0.555	*0.527	*0.539	*0.551	*0.649	21,711	24,972	24,180	29,804	31,892	39,723
Honduras	*0.002	*0.001	*0.002	*0.001	*0.001	*0.002	4,571	3,158	5,847	3,918	2,735	5,546
Hungary	±0.396	±0.290	±0.223	±0.342	±0.365	±0.485	5,911	4,759	4,557	10,368	10,181	10,854
Iceland	0.020	0.022	±0.025	±0.020	-	-	31,108	36,249	45,878	36,378	-	-
India.	*2.374	*1.905	*1.940	*1.890	*2.285	*2.853	3,684	2,843	2,701	2,662	2,966	3,285
Indonesia	*1.036	*1.004	*1.615	*2.144	*3.145	*3.451	11,995	10,157	16,614	21,397	27,318	26,858
Iran (Islamic Republic of)	0.531	1.149	0.690	0.640	-	-	16,712	32,300	16,530	19,992	-	-
Iraq	-	*0.012	*0.027	-	-	-	-	22,396	54,662		-	-
Ireland	*0.309	*0.259	*0.266	*0.220	*0.234	*0.264	41,194	32,411	34,641	30,306	31,256	34,765
Israel.	0.742	0.824	0.762	0.639	0.582	0.808	40,994	44,517	40,130	35,295	36,345	42,775
Italy	±15	±13	±14	±9.697	±11	-	45,399	42,841	43,556	32,311	38,733	
Japan	±96	±105	±110	±122	±126	±145	102,899	108,197	115,201	129,491	138,932	163,624
Jordan	0.001	0.001	0.001	0.010	0.023	0.012	7,115	5,215	4,709	13,026	20,857	10,890
Kenya	*0.039	*0.038	*0.032	*0.019	*0.026	*0.031	2,300	2,311	2,000	1,258	1,695	2,059
Korea, Republic of	10	12	14	14	18	23	42,166	51,035	55,980	54,444	63,126	75,714
Kuwait	0.002	0.006	0.051	0.052	0.024	0.061	1,132	2,212	19,632	18,628	23,297	27,885
Latvia	-	-	-	*0.038	*0.059	*0.064	-	-	-	2,574	4,342	4,850
Luxembourg	*0.019	*0.016	*0.014	*0.011	*0.013	-	36,717	36,522	31,909	31,396	34,181	-
Macau	0.004	0.005	0.004	0.004	0.004	0.004	7,096	10,508	10,817	12,409	12,739	13,510
Malaysia	*0.494	*0.656	*0.632	*0.764	*0.899	*1.308	19,678	20,628	18,979	19,951	20,675	25,562
Malta	0.027	0.023	0.015	0.007	0.013	-	10,404	9,071	6,100	3,225	5,804	-
Mauritius	0.004	0.005	0.005	0.002	0.004	0.004	7,912	7,961	7,576	3,267	8,394	6,545
Mexico	3.385	4.192	4.799	4.437	4.222	3.116	29,594	35,594	43,604	48,355	51,053	40,166
Morocco	±0.140	±0.158	±0.147	±0.150	±0.155	±0.172	15,787	14,585	13,179	12,797	13,173	13,780
Netherlands	2.464	2.430	2.695	2.086	2.203	2.641	43,125	42,856	47,548	42,189	44,730	52,706
New Zealand	0.322	0.334	0.290	0,275	-	-	42,005	50,998	43,799	41,058	-	-
Norway	±1.028	±1.002	±1.042	±1.713	±1.663	±2.049	43,158	43,125	47,349	45,369	44,595	51,821
Pakistan	0.132	0.121	0.135	-	-	-	6,566	6,471	6,905	-	-	-
Panama.	0.006	0.004	0.004	0.005	0.005	0.005	23,665	16,671	17,032	18,363	18,879	18,957
Peru	0.219	0.159	0.136	-	-	-	20,776	18,477	18,297	-	-	-
Philippines	0.258	0.223	0.390	0.464	-	-	11,360	9,771	15,725	19,488	-	-
Poland	±1.855	±1.200	±1.151	±1.374	±1.743	±2.138	6,722	5,107	5,615	6,939	7,971	9,583
Portugal	0.583	0.564	0.579	0.435	0.586	0.717	14,771	16,419	17,247	13,276	16,217	19,183
Puerto Rico	±0.068	±0.065	±0.045	±0.048	±0.053	±0.056	71,789	73,146	51,011	46,667	53,838	83,881
Romania	±0.704	±0.569	±0.365	±0.380	-	-	2,035	2,172	1,576	1,732	-	-
Russian Federation	-	-	-	±4.373	±5.142	±2.075	-	-	-	3,090	4,071	1,756

Continued.

Depending on the table, * means *Enterprises, Engaged,* or *Factor Values*; ± means *Basis Unspecified*; § means *shown in millions.* For additional notes and sources, see Appendix I.

649

Value Added and Value Added per Employee

- Continued -

Country	Value Added ($ bil.)						Value Added per Employee ($)					
	1990	1991	1992	1993	1994	1995	1990	1991	1992	1993	1994	1995
Senegal	0.017	0.013	0.013	0.008	0.007	-	15,330	10,988	14,513	20,803	4,693	-
Singapore	*0.890	*0.991	*1.220	*1.329	*1.492	*1.677	34,059	33,336	38,502	40,458	43,084	48,378
Slovakia	-	-	-	±0.120	±0.147					4,187	5,547	
Slovenia	0.287	0.158	0.152	0.175	0.187	0.227	12,827	7,792	8,311	11,454	12,563	17,945
South Africa	*1.704	*1.826	*2.104	*1.981	*2.054	*2.483	16,389	18,636	23,126	24,455	24,164	28,734
Spain	*10	*12	*13	±7.759	±8.371	±10	51,179	56,730	63,226	40,373	45,334	53,219
Sri Lanka	0.025	0.026	0.027	0.024	-	-	3,631	3,632	3,940	3,093	-	-
Sweden	*6.459	*4.501	*3.891	*3.748	*4.628	*5.982	60,027	43,722	40,381	45,842	53,717	73,467
Thailand	0.851	3.939					19,345	50,424	-	-	-	-
Trinidad and Tobago	±0.007	±0.012	±0.016	±0.007	±0.002		11,629	14,265	17,962	9,974	7,242	-
Tunisia	0.060	0.070	0.075	0.077	0.087	0.095	-	-	-	11,768	11,766	11,431
Turkey	1.743	2.126	2.594	3.637	1.903	2.822	27,531	33,052	38,219	49,908	28,930	40,570
United Kingdom	*29	*25	*26	*24	*26	*27	54,513	51,281	56,511	57,392	63,487	67,510
United Republic of Tanzania	*0.002	*0.015					611	5,388				
United States of America	*154	*158	*163	*175	*184	*184	80,475	89,920	91,965	101,278	108,080	109,114
Uruguay	0.129	0.149	0.132	0.104	0.144	0.090	26,491	32,296	32,931	29,266	32,287	31,043
Venezuela	0.198	0.480	0.615	0.723	0.641	2.411	11,927	24,977	29,708	36,303	34,943	127,494
Yugoslavia	±0.101	±0.097	±0.116	-	±0.250	±0.141	-	1,072	1,397	-	3,117	1,846
Zambia	0.007	-	-	-	0.002	-	7,519	-	-	-	3,185	-
Zimbabwe	*0.081	*0.138	*0.064	*0.073	*0.047	*0.056	11,463	18,444	10,246	12,394	8,296	9,754

Capital Investment

Country	Gross Fixed Capital Formation ($ mil.)						Per Establishment ($ mil)					
	1990	1991	1992	1993	1994	1995	1990	1991	1992	1993	1994	1995
Argentina	-	-	-	194	-	-	-	-	-	0.055	-	-
Australia	417	-	-	-	-	-	0.278	-	-	-	-	-
Austria	212	380	597	386	269	-	1.581	2.816	4.262	2.971	2.008	-
Bahrain	-	-	2.420	-	-	-	-	-	0.403	-	-	-
Bangladesh	0.822	5.083	0.616	-	-	-	0.004	0.028	0.003	-	-	-
Bolivia	0.406	0.137	0.025	0.292	0.037	-	0.012	0.003	0.001	0.006	0.001	-
Bosnia & Herzegovina	24	-	-	-	-	-	0.657	-	-	-	-	-
Brazil	1,171	-	545	522	-	-	1.370	-	0.655	0.632	-	-
Bulgaria	44	28	306	160	13	17	0.309	0.229	2.505	0.075	0.003	0.004
Cameroon	0.419	0.546	0.261	0.191	0.131	0.022	0.105	0.078	0.043	0.038	0.033	0.022
Canada	1,364	-	-	-	-	-	0.929	-	-	-	-	-
Cape Verde	0.004	0.004	-	-	-	-	0.001	0.001	-	-	-	-
Chile	9.275	6.870	15	20	12	-	0.211	0.149	0.304	0.452	0.278	-
China (Hong Kong SAR)	36	16	17	112	48	-	0.057	0.028	0.024	0.196	0.079	-
Colombia	17	26	38	15	8.940	-	0.068	0.106	0.125	0.052	0.035	-
Cote d'Ivoire	147	145	151	131	-	-	-	-	-	-	-	-
Croatia	37	130	25	4.214	1.729	1.774	0.260	0.733	0.100	0.013	0.005	0.004
Cyprus	0.407	0.622	1.611	0.763	0.580	0.865	0.005	0.008	0.021	0.010	0.006	0.009
Czechoslovakia (Former)	247	-	-	-	-	-	3.343	-	-	-	-	-
Denmark	85	101	-	-	-	-	0.101	0.118	-	-	-	-
Ecuador	11	17	33	-4.246	11	16	0.298	0.387	0.767	-0.088	0.193	0.268
Egypt	21	22	64	57	-	-	0.171	0.155	0.459	0.449	-	-
El Salvador	-	-	-	0.256	0.217	0.341	-	-	-	0.051	0.109	0.021
Ethiopia	0.169	0.207	0.185	1.654	0.207	0.665	0.084	0.103	0.092	0.827	0.069	0.222
Fiji	-0.163	0.281	0.257	-	-	-	-0.012	0.015	0.014	-	-	-
Finland	231	127	76	119	160	117	0.967	0.491	0.353	0.600	0.896	0.634
FYR Macedonia	1.785	5.851	0.767	1.581	1.912	1.205	0.058	0.081	0.006	0.010	0.008	0.004
France	5,889	5,979	6,271	4,390	5,191	5,463	-	-	-	-	-	-
Germany (Western Part)	-	10,378	9,545	7,683	-	-	-	9.119	8.278	-	-	-
Greece	222	168	122	-	-	-	1.032	0.800	0.576	-	-	-
Hungary	47	128	329	66	-	-	0.112	0.142	1.296	0.215	-	-
India	524	425	581	684	-	-	0.107	0.082	0.105	0.115	-	-
Indonesia	227	225	218	183	163	593	0.521	0.509	0.444	0.357	0.305	1.027

Continued.

Depending on the table, * means *Enterprises, Engaged,* or *Factor Values*; ± means *Basis Unspecified*; § means *shown in millions.* For additional notes and sources, see Appendix I.

Capital Investment

- Continued -

Country	Gross Fixed Capital Formation ($ mil.)						Per Establishment ($ mil)					
	1990	1991	1992	1993	1994	1995	1990	1991	1992	1993	1994	1995
Iran (Islamic Republic of) . . .	28	93	92	78	-	-	0.346	0.835	0.840	0.746	-	-
Ireland	27	132	-	-	-	-	0.228	1.100	-	-	-	-
Israel.	48	68	90	91	103	-	0.248	0.315	0.389	0.386	0.477	-
Italy	3,131	3,322	4,224	3,483	3,210	-	3.429	3.564	3.448	2.952	2.787	-
Japan	15,457	21,030	22,211	16,619	14,715	-	1.019	1.390	1.496	1.112	1.113	-
Jordan	0.021	0.047	0.022	0.003	1.093	1.029	0.001	0.002	0.001	0.000	0.031	0.029
Korea, Republic of	3,842	5,210	3,832	6,455	5,867	7,422	1.363	1.662	1.204	1.858	1.592	1.815
Kuwait	0.007	31	6.570	3.238	0.772	-	0.000	1.967	0.469	0.231	0.051	-
Latvia	0.104	0.177	2.660	6.864	4.978	7.211	0.004	0.003	0.040	0.130	0.116	0.124
Luxembourg	1.496	1.640	2.675	-	-	-	0.100	0.109	0.167	-	-	-
Macau	0.147	0.116	0.420	-0.039	0.018	0.357	0.002	0.002	0.007	-0.001	0.000	0.008
Malaysia	90	168	331	309	363	-	0.376	0.636	1.172	0.962	1.099	-
Malta	1.256	2.084	2.223	1.068	0.926	-	0.035	0.056	0.065	0.031	0.026	-
Mexico	236	425	-	-	-	-	1.521	2.795	-	-	-	-
Morocco	27	33	19	18	21	26	0.269	0.324	0.192	0.177	0.194	0.236
Netherlands	350	358	348	188	-	-	1.105	1.084	1.003	0.578	-	-
New Zealand	18	-	-	-	-	-	0.019	-	-	-	-	-
Norway	94	80	88	147	127	-	0.221	0.199	0.298	0.418	0.366	-
Pakistan	-	12	-	-	-	-	-	0.093	-	-	-	-
Panama.	0.196	0.105	-	-	-	-	0.012	0.008	-	-	-	-
Peru	5.667	4.293	33	-	-	-	0.011	0.009	0.064	-	-	-
Philippines.	63	42	47	110	-	-	0.106	0.064	0.178	0.429	-	-
Poland	550	712	474	153	-	-	1.697	2.125	1.628	0.392	-	-
Portugal	112	118	79	126	549	-	0.150	0.158	0.102	0.157	0.660	-
Romania	-	-	29	35	28	-	-	-	0.148	0.161	0.080	-
Senegal.	1.271	-	-	-	-	-	0.079	-	-	-	-	-
Singapore	156	140	187	193	322	-	0.720	0.589	0.760	0.683	1.106	-
Slovenia	51	40	107	55	25	33	0.481	0.583	1.276	0.563	0.242	0.285
Spain	1,220	2,053	3,165	1,924	1,651	1,828	0.643	1.024	1.639	-	-	-
Sri Lanka	0.649	1.152	1.396	10	-	-	0.017	0.029	0.054	0.325	-	-
Thailand	282	700	-	-	-	-	0.838	1.004	-	-	-	-
Trinidad and Tobago . . .	-	-	0.235	1.757	0.203	-	-	-	0.005	0.041	0.005	-
Tunisia	12	14	11	11	11	11	-	-	0.021	0.020	0.019	-
Turkey	233	440	357	477	482	-	1.113	2.047	0.863	1.140	1.219	-
Ukraine	355	99	76	12	15	24	2.910	0.415	0.234	0.075	0.102	0.173
United Kingdom	3,736	3,720	3,035	-	2,300	-	0.878	0.879	0.701	-	0.528	-
United Republic of Tanzania . .	6.321	11	-	-	-	-	0.158	0.272	-	-	-	-
United States of America . . .	11,490	11,360	11,601	12,313	13,068	14,624	-	-	0.695	-	-	-
Uruguay	6.308	8.686	4.990	5.253	-	-	-	-	-	-	-	-
Venezuela	25	27	41	45	47	208	0.103	0.098	0.161	0.190	0.215	0.968
Yugoslavia	7.033	8.756	18	-	18	7.349	0.055	0.049	0.063	-	0.053	0.021
Zambia	0.619	-	-	-	-	-	0.077	-	-	-	-	-
Zimbabwe	16	14	17	9.811	3.988	-	0.290	0.238	0.298	0.182	0.075	-

Indexes of Industrial Production

Country	Index of Industrial Production (1990=100)						Index of Employment (1990=100)					
	1990	1991	1992	1993	1994	1995	1990	1991	1992	1993	1994	1995
Albania	100	-	-	-	-	-	-	-	-	-	-	-
Algeria	100	89	86	71	46	89	-	-	-	-	-	-
Angola	100	-	-	-	-	-	-	-	-	-	-	-
Argentina	100	118	-	-	-	-	100	-	-	79	83	-
Armenia.	100	-	-	-	-	-	100	92	-	-	-	-
Australia	100	91	95	117	-	-	100	85	79	-	-	-
Austria	100	102	102	97	104	110	100	98	95	89	91	91
Azerbaijan	-	-	-	-	-	-	100	93	83	84	82	-
Bahrain	100	-	-	-	-	-	-	-	-	-	-	-
Bangladesh	100	84	85	73	55	65	100	95	107	-	-	-

Continued.

Depending on the table, * means *Enterprises*, *Engaged*, or *Factor Values*; ± means *Basis Unspecified*; § means *shown in millions*. For additional notes and sources, see Appendix I.

651

Indexes of Industrial Production

- Continued -

Country	Index of Industrial Production (1990=100)						Index of Employment (1990=100)					
	1990	1991	1992	1993	1994	1995	1990	1991	1992	1993	1994	1995
Barbados	100	-	-	-	-	-	-	-	-	-	-	-
Belgium	100	96	94	92	94	95	-	-	-	-	-	-
Belize	100	-	-	-	-	-	-	-	-	-	-	-
Bermuda	-	-	-	-	-	-	100	108	108	108	131	162
Bolivia	100	104	111	116	136	132	100	106	109	143	145	144
Bosnia & Herzegovina	-	-	-	-	-	-	100	99	-	-	-	-
Brazil	100	100	98	118	134	139	100	-	83	86	-	-
Bulgaria	100	64	41	39	39	39	100	81	69	62	63	59
Burkina-Faso	100	76	79	-	-	-	-	-	-	-	-	-
Burundi	100	100	100	100	-	-	-	-	-	-	-	-
Cambodia	100	-	-	-	-	-	100	86	-	-	-	-
Cameroon	100	-	-	-	-	-	100	170	89	80	65	4
Canada	100	90	91	99	108	115	100	92	94	92	95	97
Cape Verde	-	-	-	-	-	-	100	99	97	83	68	-
Central African Republic	100	-	-	-	-	-	100	-	-	-	-	-
Chile	100	107	125	120	115	105	100	93	142	137	134	127
China	100	-	-	-	-	-	-	-	-	-	-	-
China (Hong Kong SAR)	100	95	93	85	81	78	100	94	104	101	97	-
China (Taiwan Province)	100	111	121	120	123	127	100	100	104	107	108	107
Colombia	100	84	100	142	158	159	100	95	116	124	114	114
Congo	100	-	-	-	-	-	-	-	-	-	-	-
Costa Rica	100	84	112	101	88	84	100	96	98	94	-	-
Cote d'Ivoire	100	97	119	103	112	78	-	-	-	-	-	-
Croatia	100	-	-	-	-	-	100	93	75	70	66	63
Cyprus	100	123	122	141	124	107	100	96	87	79	101	106
Czechoslovakia (Former)	100	72	53	-	-	-	100	84	-	-	-	-
Czech Republic	100	-	-	-	-	-	100	86	91	76	-	-
Dem. Rep. of the Congo	100	-	-	-	-	-	-	-	-	-	-	-
Denmark	100	107	102	95	124	133	100	100	98	93	113	119
Dominican Republic	100	-	-	-	-	-	-	-	-	-	-	-
Ecuador	100	123	134	139	-	-	100	125	127	156	182	136
Egypt	100	121	85	99	101	92	100	87	96	76	77	76
El Salvador	100	105	-	-	-	-	-	-	-	-	-	-
Ethiopia and Eritrea	100	-	-	-	-	-	-	-	-	-	-	-
Ethiopia	-	-	-	-	-	-	100	98	89	88	93	97
Fiji	100	-	-	-	-	-	100	95	87	112	113	-
Finland	100	90	93	72	94	107	100	95	89	85	87	81
FYR Macedonia	100	-	-	-	-	-	100	147	137	127	120	103
France	100	101	99	89	97	100	100	99	97	94	91	91
Gabon	100	-	-	-	-	-	-	-	-	-	-	-
Gambia	100	-	-	-	-	-	-	-	-	-	-	-
Germany (Eastern Part)	100	-	-	-	-	-	-	-	-	-	-	-
Germany (Western Part)	100	102	102	85	92	92	100	101	99	-	-	-
Ghana	100	-	-	-	-	-	-	-	-	-	-	-
Greece	100	104	106	83	73	77	100	99	97	80	77	73
Guatemala	100	-	-	-	-	-	-	-	-	-	-	-
Guyana	100	100	100	100	100	100	-	-	-	-	-	-
Haiti	100	-	-	-	-	-	-	-	-	-	-	-
Honduras	100	122	143	201	136	157	100	100	64	70	77	98
Hungary	100	73	47	60	70	106	100	91	73	49	53	67
Iceland	100	-	-	-	-	-	100	94	86	85	57	-
India	100	97	102	106	120	144	100	104	111	110	120	135
Indonesia	100	165	205	257	195	204	100	114	113	116	133	149
Iran (Islamic Republic of)	100	265	277	275	204	314	100	112	131	101	-	-
Iraq	100	-	-	-	-	-	-	-	-	-	-	-
Ireland	100	90	85	81	82	84	100	107	102	97	100	101
Israel	100	105	113	105	97	97	100	102	105	100	88	104
Italy	100	96	91	75	-	-	100	98	98	94	88	-
Jamaica	100	-	-	-	-	-	-	-	-	-	-	-
Japan	100	101	99	92	88	87	100	104	103	102	97	95

Continued.

Depending on the table, * means *Enterprises, Engaged,* or *Factor Values;* ± means *Basis Unspecified;* § means *shown in millions.* For additional notes and sources, see Appendix I.

Indexes of Industrial Production

- Continued -

Country	Index of Industrial Production (1990=100)						Index of Employment (1990=100)					
	1990	1991	1992	1993	1994	1995	1990	1991	1992	1993	1994	1995
Jordan	100	-	-	-	-	-	100	139	182	756	1,107	1,103
Kenya	100	104	95	79	85	83	100	96	95	90	90	90
Korea, Republic of	100	118	147	136	148	171	100	100	101	108	119	126
Kuwait	100	-	-	-	-	-	100	128	129	136	50	108
Kyrgyzstan	100	-	-	-	-	-	100	92	86	78	43	-
Lao P.D.R.	100	-	-	-	-	-	-	-	-	-	-	-
Latvia	100	-	-	-	-	-	100	91	73	60	55	53
Lebanon	100	-	-	-	-	-	-	-	-	-	-	-
Libyan Arab Jamahiriya	100	-	-	-	-	-	-	-	-	-	-	-
Lithuania	100	-	-	-	-	-	-	-	-	-	-	-
Luxembourg	100	98	76	66	74	88	100	86	83	68	74	-
Macau	100	-	-	-	-	-	100	92	81	57	54	52
Madagascar	100	-	-	-	-	-	-	-	-	-	-	-
Malawi	100	-	-	-	-	-	100	-	-	-	-	-
Malaysia	100	117	110	114	136	185	100	127	133	153	173	204
Mali	100	101	106	93	106	125	-	-	-	-	-	-
Malta	100	88	81	80	-	-	100	98	95	88	86	-
Mauritius	100	-	-	-	-	-	100	130	133	127	110	141
Mexico	100	111	110	110	118	101	100	103	96	80	72	68
Mongolia	100	77	101	34	50	102	-	-	-	-	-	-
Morocco	100	103	85	78	81	79	100	123	126	132	133	141
Mozambique	100	-	-	-	-	-	100	92	44	46	42	-
Myanmar	100	120	80	-	-	-	-	-	-	-	-	-
Namibia	100	-	-	-	-	-	-	-	-	-	-	-
Nepal	100	100	100	-	-	-	-	-	-	-	-	-
Netherlands	100	101	101	84	89	96	100	99	99	87	86	88
New Zealand	100	113	74	-	-	-	100	85	86	87	-	-
Nicaragua	100	-	-	-	-	-	-	-	-	-	-	-
Nigeria	100	-	-	-	-	-	-	-	-	-	-	-
Norway	100	104	106	101	109	118	100	98	92	159	157	166
Pakistan	100	113	108	-	-	-	100	93	97	-	-	-
Panama	100	60	61	75	79	78	100	89	90	103	107	106
Papua New Guinea	100	-	-	-	-	-	-	-	-	-	-	-
Paraguay	100	101	108	89	60	60	-	-	-	-	-	-
Peru	100	112	83	68	118	108	100	81	70	-	-	-
Philippines	100	113	128	143	163	208	100	100	109	105	-	-
Poland	100	62	75	83	101	105	100	85	74	72	79	81
Portugal	100	97	93	78	69	74	100	87	85	83	92	95
Puerto Rico	100	-	-	-	-	-	100	94	94	107	104	71
Republic of Moldova	100	-	-	-	-	-	100	90	96	24	-	-
Romania	100	72	54	55	51	63	100	76	67	63	55	-
Russian Federation	100	-	-	-	-	-	-	-	-	-	-	-
Saudi Arabia	100	-	-	-	-	-	-	-	-	-	-	-
Senegal	100	81	91	92	80	129	100	104	79	33	136	-
Singapore	100	105	105	104	112	113	100	114	121	126	133	133
Slovakia	100	-	-	-	-	-	-	-	-	-	-	-
Slovenia	100	82	68	56	55	57	100	90	82	68	67	56
Somalia	100	-	-	-	-	-	-	-	-	-	-	-
South Africa	100	99	89	88	92	103	100	94	88	78	82	83
Spain	100	97	98	77	84	97	100	101	99	95	92	93
Sri Lanka	100	94	101	124	154	213	100	103	99	115	-	-
Sudan	100	-	-	-	-	-	-	-	-	-	-	-
Suriname	100	-	-	-	-	-	100	122	131	154	-	-
Swaziland	100	100	100	-	-	-	-	-	-	-	-	-
Sweden	100	101	100	96	129	155	100	96	90	76	80	76
Switzerland	100	105	104	100	102	104	-	-	-	-	-	-
Syrian Arab Republic	100	124	179	206	277	297	-	-	-	-	-	-
Thailand	100	106	116	-	-	-	100	178	-	-	-	-
Tonga	100	-	-	-	-	-	100	30	133	115	215	-
Trinidad and Tobago	100	163	134	94	81	31	100	141	148	122	47	-

Continued.

Indexes of Industrial Production

- Continued -

Country	Index of Industrial Production (1990=100)						Index of Employment (1990=100)					
	1990	1991	1992	1993	1994	1995	1990	1991	1992	1993	1994	1995
Tunisia	100	101	76	82	82	-	-	-	-	-	-	-
Turkey	100	109	148	183	109	140	100	102	107	115	104	110
Uganda	100	133	147	189	202	189	-	-	-	-	-	-
Ukraine	100	-	-	-	-	-	100	99	114	106	96	83
United Arab Emirates	100	-	-	-	-	-	-	-	-	-	-	-
United Kingdom	100	94	91	90	93	93	100	93	88	79	76	76
United Republic of Tanzania . .	100	75	28	8	19	-	100	97	-	-	-	-
USSR (Former)	100	-	-	-	-	-	100	-	-	-	-	-
United States of America . . .	100	95	98	102	105	103	100	92	93	90	89	88
Uruguay	100	110	99	99	133	77	100	95	83	73	92	60
Venezuela	100	-	-	-	-	-	100	116	125	120	110	114
Yugoslavia	100	-	-	-	-	-	-	-	-	-	-	-
Yugoslavia (Former)	100	63	-	-	-	-	-	-	-	-	-	-
Zambia	100	97	90	75	64	68	100	-	-	-	59	-
Zimbabwe	100	97	96	56	91	95	100	106	88	84	81	82

Depending on the table, * means *Enterprises, Engaged,* or *Factor Values*; ± means *Basis Unspecified*; § means *shown in millions*. For additional notes and sources, see Appendix I.

Representative Companies in Sector

Name	Address	Tele-phone	Fax	Employ-ment	Y
General Motors of Canada Ltd.	1908 Colonel Sam Dr., Oshawa, Ontario L1H 8P7, Canada	905-644-5000	905-644-6273	32,701	98
Boeing Canada Technology Ltd., Winnipeg Division	99 Murray Park Rd., Winnipeg, Manitoba R3J 3M6, Canada	204-888-2300	204-888-2951	14,000	98
Chrysler Canada Ltd.	2450 Chrysler Ctr, Windsor, Ontario N8W 3X7, Canada	519-973-2000	519-973-2858	14,000	98
BICC Cables Ltd.	Chester Business Park, Chester CH4 9PZ, United Kingdom	244 688400	244 688401	10,000	90
Budd Co.	PO Box 2601, Troy, Michigan 48007-2601, U.S.A.	313-643-3500		10,000	96
Lockheed Martin Aeronautical Systems	86 S. Cobb Dr., Marietta, Georgia 30063, U.S.A.	770-494-5432	770-494-7518	10,000	97
Varity Kelsey-Hayes	12025 Tech Center Dr., Livonia, Michigan 48150-8010, U.S.A.	313-266-2600	313-266-4853	10,000	97
Bosch do Brasil Ltda., Robert	Via Anhager, Km. 98 s/n, Campinas 13001, Brazil	192 441272	192 423466	9,700	97
Saturn Corp.	1420 Stephenson Hwy., Troy, Michigan 48007-7025, U.S.A.	248-524-5000		9,600	97
Grupo Industrial Saltillo SA de CV	25280 Saltillo, Coahuila, Mexico	84 155620	84 153313	9,575	92
GenCorp Inc.	175 Ghent Rd., Fairlawn, Ohio 44333-3300, U.S.A.	330-869-4200	330-869-4211	9,460	97
Sequa Corp.	200 Park Ave., New York, New York 10166, U.S.A.	212-986-5500	212-370-1969	9,350	97
Precision Castparts Corp.	4650 S.W. Macadam, 440, Portland, Oregon 97201, U.S.A.	503-417-4855	503-417-4817	9,280	96
Sundstrand Corp.	PO Box 7003, Rockford, Illinois 61125-7003, U.S.A.	815-226-6000	815-226-2699	9,200	95
Asia Motors Co., Inc.	Yongdungpo-Gu, Seoul, Republic of Korea	2 7888497	2 7888678	9,145	97
DAF NV	Postbus 90065, Eindhoven NL-5643 TW, Netherlands	40 149111	40 144325	9,000	94
Sikorsky Aircraft Corp.	6900 Main St., Stratford, Connecticut 06497, U.S.A.	203-386-4000	203-386-7300	9,000	95
Morrison Knudsen Corp.	PO Box 73, Boise, Idaho 83729, U.S.A.	208-386-5000	208-386-7186	8,900	97
British Shipbuilders	136 Sandyford Rd., Newcastle-upon-Tyne NE2 1QE, United Kingdom	91 6326772	91 6326772	8,642	87
Westland Group PLC	Wland Works, Yeovil BA20 2YB, United Kingdom	935 75222	935 702278	8,536	93
A.O. Smith Corp.	PO Box 23972, Milwaukee, Wisconsin 53223-3972, U.S.A.	414-359-4000	414-359-4248	8,400	97
Coltec Industries Inc.	2550 W. Tyvola Rd., Charlotte, North Carolina 28217-4543, U.S.A.	704-423-7000		8,153	96
Mando Machinery Corp.	Kunpo-Shi, Kyonggi, Republic of Korea	343 506114	343 596377	8,100	97
Samsung Aerospace Industries Ltd.	Chung-gu, Seoul, Republic of Korea	2 7518853	2 7518590	8,022	97
Huffy Corp.	PO Box 1204, Dayton, Ohio 45401, U.S.A.	513-866-6251	513-865-5470	8,020	96
Denso International America Inc.	24777 Denso Dr., Southfield, Michigan 48086-5133, U.S.A.	248-350-7500	248-350-7772	8,000	97
Lockheed Martin Astronautics Group	PO Box 179, Denver, Colorado 80201, U.S.A.	303-977-3000	303-977-4902	8,000	97
Modine Manufacturing Co.	1500 DeKoven Ave., Racine, Wisconsin 53403-2552, U.S.A.	414-636-1200	414-636-1424	7,900	96
Stant Corp.	425 Commerce Dr., Richmond, Indiana 47374-2646, U.S.A.	317-962-6655	317-962-6866	7,600	96
Bombardier Aerospace Europe Ltd.	35 Basinghall St., London EC2V 5DB, United Kingdom			7,551	94
Short Brothers PLC	Airport Rd., Belfast BT3 9DZ, United Kingdom	232 458444	232 732974	7,551	94
GOSA	Industrijska 70, Smederevska Palanka YU-11420, Serbia	26 31253	26 31472	7,500	92
UNC Inc.	175 Admiral Cochrane Dr., Annapolis, Maryland 21401-7394, U.S.A.	410-266-7333		7,449	96
Outboard Marine Corp.	100 Sea Horse Dr., Waukegan, Illinois 60085, U.S.A.	847-689-6200	847-689-6082	7,442	97
Transwerk	Lynn East, Pretoria 0039, Republic of South Africa	12 8425100	12 8425039	7,238	96
Engineering Equip. Inc.	Ortigas Industrial Estate, Quezon City, Philippines	2 6350851	2 7220644	7,200	92
SPX Corp.	PO Box 3301, Muskegon, Michigan 49443-3301, U.S.A.	616-724-5000	616-724-5720	7,125	96
T and N Industries Inc.	777 E. Eisenhower Pkwy., 600, Ann Arbor, Michigan 48108-3258, U.S.A.	313-663-6749	313-663-0629	7,100	95
COFAP Cia. Fabricadora de Pecas	Santo Andre, Sao Paulo 09110-901, Brazil	411 8211	11 4400409	7,000	96
Excel Industries Inc. (Elkhart, Indiana)	PO Box 118, Elkhart, Indiana 46515-3118, U.S.A.	219-264-2131	219-264-2136	6,800	97
Iochpe-Maxion SA	Santo Amaro, Sao Paulo 04726-170, Brazil	11 5482533	11 5147717	6,767	98
Fabryka Samochodow w Lublinie	ul. Melgiewska 7/9, Lublin PL-20-952, Poland	81 62200	81 62936	6,561	93
Meggitt PLC	Cowgrave, Wimborne BH21 4EL, United Kingdom	202 841141	202 842478	6,529	93
Harvard Industries Inc.	3 Werner Way, Lebanon, New Jersey 08833-2223, U.S.A.	908-437-4100		6,476	97
Lancaster Colony Corp.	37 W. Broad St., Columbus, Ohio 43215, U.S.A.	614-224-7141	614-469-8219	6,400	97
Boeing Co. Boeing Helicopters Div.	PO Box 16858, Philadelphia, Pennsylvania 19142, U.S.A.	610-591-2121	610-591-2701	6,288	
Iveco Magirus AG	Nikolaus-Otto-Str. 25-27, Ulm D-89070, Germany	731 4081	731 481340	6,281	93
Federal Signal Corp.	1415 W. 22nd St., Oak Brook, Illinois 60521-9945, U.S.A.	708-954-2000		6,207	97
Nissan Motor Iberica, SA	Panama, 7, Barcelona E-08034, Spain	93 2908080	93 2907091	6,165	97
Terex Corp.	500 Post Rd. E., 320, Westport, Connecticut 06880, U.S.A.	203-222-7170	203-222-7976	6,071	95
Armco Inc.	1 Oxford Centre, 301 Grant St., Pittsburgh, Pennsylvania 15219-1415, U.S.A.	412-255-9800	800-231-1054	6,000	97
Siebe Appliance Controls	2809 Emerywood Pkwy., Richmond, Virginia 23294-3743,	804-756-6500	804-756-6564	6,000	96

Continued

Representative Companies in Sector
- Continued -

Name	Address	Tele-phone	Fax	Employ-ment	Y
Thonburi Automobile Assembly Plant Co. Ltd.	U.S.A. 3 Rachadamnoen Rd., Bangkok 10200, Thailand	2228143		6,000	94
Thiokol Corp.	15 W South Temple, Ste. 1600, Salt Lake City, Utah 84101, U.S.A.	801-629-2270		5,900	97
Peugeot Talbot Motor Co. PLC	PO Box 227, Coventry CV3 1LT, United Kingdom	203 884000	203 884001	5,889	93
Roadmaster Industries Inc.	250 Spring St. N.W., Atlanta, Georgia 30303, U.S.A.	404-586-9000		5,800	94
Pirelli UK PLC	11 Berkeley St., London W1X 6BU, United Kingdom	71 2428881	71 4301096	5,588	93
Ciadea SA	Fray Justo Santa Maria de Oro 1744, Buenos Aires 1414, Argentina	1 7782000	1 7782158	5,521	96
Avondale Industries Inc.	PO Box 50280, New Orleans, Louisiana 70150-0280, U.S.A.	504-436-2121		5,500	97
Perindustrian Tentara Nasional Id Angkatan Darat PT	Jl. Jenderal Gatot Subroto Kiaracondong, Bandung, Indonesia	22312073	22 301222	5,500	97
Sundstrand Aerospace	PO Box 7002, Rockford, Illinois 61125-7002, U.S.A.	815-226-6000	815-226-7488	5,400	95
Peugeot Talbot Espana, SA	Ctra. de Villaverde, Km. 7,500, Madrid E-28041, Spain	91 3472000	91 3473082	5,205	97
Dura Automotive Systems Inc.	4508 IDS Ctr., Minneapolis, Minnesota 55402, U.S.A.	612-342-2311	612-332-2012	5,200	97
Gulfstream Aerospace Corp.	PO Box 2206, Savannah, Georgia 31402-2206, U.S.A.	912-965-3000	912-965-4171	5,200	97
Tan Chong Motor Holdings BHD	62-68 Jalan Ipoh, Kuala Lumpur 51200, Malaysia	3 4427644	3 4418373	5,128	94
MascoTech Inc.	21001 Van Born Rd., Taylor, Michigan 48180, U.S.A.	313-274-7405	313-274-8959	5,100	96
Mack Trucks Inc.	PO Box M, Allentown, Pennsylvania 18105-5000, U.S.A.	610-709-3011	610-709-3308	5,000	97
Samior Pty. Ltd.	PO Box 411, Pretoria 0181, Republic of South Africa	12 8422381	12 8422087	5,000	93
Ford Argentina SA	Av. Henry Ford Y Panamericana, General Pacheco 1617, Argentina	1 7569000	1 7569001	4,800	96
Harley-Davidson Inc.	3700 W. Juneau Ave., Milwaukee, Wisconsin 53208, U.S.A.	414-342-4680		4,800	95
Bosal Afrika Pty. Ltd.	PO Box 1719, Pretoria 0001, Republic of South Africa	12 3333258	12 3333053	4,700	94
African Lakes Corp. PLC	24 Gilbert St., London W1Y 2EQ, United Kingdom	71 4938561	71 4090755	4,683	93
Rohr Inc.	PO Box 878, Chula Vista, California 91912, U.S.A.	619-691-4111	619-691-4103	4,600	97
Stewart and Stevenson Services Inc.	PO Box 1637, Houston, Texas 77251-1637, U.S.A.	713-868-7700		4,511	95
Clark-Hurth Components	1293 Glenway Dr., Statesville, North Carolina 28625, U.S.A.	704-873-2811	704-878-5616	4,500	97
Delta Motor Corp. - Export Div.	PO Box 1137, Port Elizabeth 6000, Republic of South Africa	41 4032418	41 435932	4,500	95
Gdanska Stocznia Remontowa	Ostrowiu ul. Na 1, Gdansk PL-80-958, Poland	58 371300		4,500	93
Great Dane Trailers Inc.	PO Box 67, Savannah, Georgia 31402, U.S.A.	912-232-4471	912-944-2497	4,500	95
Linamar Corp.	301 Massey Rd., Guelph, Ontario N1K 1B2, Canada	519-836-7550	519-824-8479	4,500	98
Marcopolo SA - Carrocerias e Onibus	Rua Marcopolo 280, Caixas do Sul 95086-460, Brazil	54 2224422	54 2226700	4,500	98
Ranco Inc.	8161 U.S. Rte. 42 N., Plain City, Ohio 43064, U.S.A.	614-873-9219		4,500	95
Superior Industries International Inc.	7800 Woodley Ave., Van Nuys, California 91406, U.S.A.	818-781-4973	818-780-3500	4,500	97
Zaklady Starachowickie, Star Spolka Akcyjna	1 Maja ul. 12, Starachowice PL-27-200, Poland	8831	7038	4,430	93
Clark Equipment Co.	100 N. Michigan St., South Bend, Indiana 46634, U.S.A.	219-239-0100		4,410	94
Standard Motor Products Inc.	37-18 Northern Blvd., Long Island City, New York 11101, U.S.A.	718-392-0200	718-729-4549	4,400	97
Wabash National Corp.	PO Box 6129, Lafayette, Indiana 47903-6129, U.S.A.	765-448-1591	765-447-9405	4,320	97
Kaman Corp.	PO Box 1, Bloomfield, Connecticut 06002, U.S.A.	203-243-8311		4,318	97
Coachmen Industries Inc.	PO Box 3300, Elkhart, Indiana 46515, U.S.A.	219-262-0123		4,274	97
Aerojet General Corp.	PO Box 13222, Sacramento, California 95813-6000, U.S.A.	916-355-1000	916-351-8668	4,200	97
Samcor Pty. Ltd.	PO Box 411, Pretoria 0001, Republic of South Africa	12 8422381	12 8422087	4,200	96
Automotive Products PLC	Tachbrook Rd., Leamington Spa CV313ER, United Kingdom	926 470000		4,108	93
Westland Helicopters Ltd.	Westland Works, Yeovil BA20 2YB, United Kingdom	935 702585	935 702131	4,100	
Fabryka Lozysk Tocznych, Iskra SA	ul. Jagiellonska, Kielce PL-25-734, Poland	41 666111	41 54599	4,090	93
Sanyang Industry Co., Ltd.	Neihu District, Taipei, Taiwan	2 7912161	2 7912160	4,073	93
Mercedes-Benz of South Africa Pty. Ltd.	PO Box 1717, Pretoria 0001, Republic of South Africa	12 3091500	12 3252939	4,058	96
Amcast Industrial Corp.	PO Box 98, Dayton, Ohio 45401-0098, U.S.A.	513-298-7000		4,040	97
Federal Motor PT	P. O. Box 3009, Jakarta 14350, Indonesia	214301407	21 4300230	4,000	97
New Venture Gear Inc.	1650 Research Dr., 300, Troy, Michigan 48083, U.S.A.	313-680-4900	313-680-4924	4,000	95
Parker Bertea Aerospace Group	18321 Jamboree Rd., Irvine, California 92715, U.S.A.	714-833-3000	714-851-3311	4,000	97
Rafael Armament Development Authority	PO Box 2250, Haifa 31021, Israel	4 8794777	4 8794653	4,000	97
Barnes Group Inc.	123 Main St., Bristol, Connecticut 06011-0489, U.S.A.	860-583-7070	860-589-3507	3,900	97

Continued

Representative Companies in Sector
- Continued -

Name	Address	Telephone	Fax	Employ-ment	Y
CTS Corp.	905 West Blvd. N., Elkhart, Indiana 46514, U.S.A.	219-293-7511	219-293-6146	3,893	97
Kvaerner Holding U.K. Ltd.	1048 Govan Rd., Glasgow G51 4XP, United Kingdom	41 4452351	41 4454455	3,882	93
A.B. Electronic Products Group PLC	Abercynon, Mountain Ash CF45 4SF, United Kingdom	443 740331	443 741676	3,829	92
Alpine Group Inc.	1790 Broadway, 1500, New York, New York 10019-1412, U.S.A.	212-757-3333	212-757-3423	3,809	97
General Motors Corp. Electro-Motive Div.	9301 W. 55th St., La Grange, Illinois 60525, U.S.A.	708-387-6000	708-387-6501	3,800	97
Yulon Motor Co. Ltd.	2 Tun Hwa South Rd., 16th Fl. Sec. 2, Taipei, Taiwan	2 7551515	2 7082887	3,800	92
Kenworth Truck Co.	PO Box 1000, Kirkland, Washington 98083, U.S.A.	425-828-5000	425-828-5088	3,500	97
Moog Automotive Inc.	PO Box 7224, St. Louis, Missouri 63177, U.S.A.	314-385-3400	314-381-6476	3,500	97
Polaris Industries L.P.	1225 Hwy. 169 N., Minneapolis, Minnesota 55441, U.S.A.	612-542-0500	612-542-0599	3,500	97
Skyline Corp.	PO Box 743, Elkhart, Indiana 46515-0743, U.S.A.	219-294-6521		3,500	97
TriMas Corp.	315 E. Eisenhower Pkwy., Ann Arbor, Michigan 48108, U.S.A.	313-747-7025	313-747-6565	3,500	95
Western Star Trucks Inc.	2076 Enterprise Way, Kelowna, British Columbia V1Y 6H8, Canada	250-860-3319	250-860-1252	3,500	98
General Dynamics Land Systems Inc.	38500 Mound Rd., Sterling Heights, Michigan 48311, U.S.A.	810-825-4000	810-825-4013	3,400	97
Hanjin Heavy Industries Co., Ltd.	Yongdo-gu, Pusan, Republic of Korea	51 4103114	51 4103337	3,300	97
Johnstown America Industries Inc.	980 N. Michigan Ave., 1000, Chicago, Illinois 60611, U.S.A.	312-280-8844	312-280-4820	3,300	97
Moog Inc.	Plant 24, East Aurora, New York 14052-0018, U.S.A.	716-652-2000	716-687-4457	3,229	97
Philips Technologies	PO Box 868, Cheshire, Connecticut 06410, U.S.A.	203-271-6000	203-276-6100	3,200	94
B/E Aerospace Inc.	1400 Corporate Center Way, Wellington, Florida 33414-2105, U.S.A.	561-791-5000	561-791-7900	3,140	96
Hayes Lemmerz International Inc.	38481 Huron River Dr., Romulus, Michigan 48174, U.S.A.	313-941-2000		3,110	96
Orscheln Co.	PO Box 280, Moberly, Missouri 65270, U.S.A.	816-263-4377	816-263-9391	3,100	
Aftermarket Technology Corp.	33309 1st Way S., A-206, Federal Way, Washington 98003, U.S.A.	206-838-0346		3,075	95
American Steel Foundries	10 S. Riverside Plz., 10th Fl., Chicago, Illinois 60606, U.S.A.	312-258-8000	312-258-5466	3,000	97
Bakony Muvek	Pf. 78, Veszprem H-8201, Hungary	80 24022	80 27916	3,000	93
Dorbyl Automotive Products	PO Box 8061, Elandsfontein 1406, Republic of South Africa	11 8225912	11 8225921	3,000	96
Manitowoc Company Inc.	PO Box 66, Manitowoc, Wisconsin 54221-0066, U.S.A.	920-684-4410	920-683-6277	3,000	97
Murry Inc.	219 Franklin Rd., Brentwood, Tennessee 37027, U.S.A.	615-373-6500	615-373-6554	3,000	97
Orbital Sciences Corp.	21700 Atlantic Blvd., Dulles, Virginia 20166, U.S.A.	703-406-5000		3,000	96
Woodward Governor Co.	5001 N. 2nd St., Rockford, Illinois 61125-7001, U.S.A.	815-877-7441		3,000	97
Parish Light Vehicle Structure Div.	PO Box 13459, Reading, Pennsylvania 19612, U.S.A.	215-371-7000	215-371-7095	2,992	95
Steeledale Reinforcing & Engineering Industries Ltd.	Tulisa Park, Johannesburg 2197, Republic of South Africa	11 6132331	11 6131394	2,950	93
Thor Industries Inc.	PO Box 629, Jackson Center, Ohio 45334-0629, U.S.A.	937-596-6849		2,934	97
Scania	Av. Jose Odorizzi 151, Sao Bernado do Campo 09810-902, Brazil	11 7529333	11 4512659	2,924	96
Petrol Trgovina d.d.	Dunajska cesta - del 50, Ljubljana 1001, Slovenia	61 1714234	61 1714812	2,909	95
Allen Group Inc.	25101 Chagrin Blvd., Beachwood, Ohio 44122-5619, U.S.A.	216-765-5800	216-765-0410	2,900	96
Brunswick Corp. US Marine Div.	PO Box 9029, Everett, Washington 98206, U.S.A.	360-435-5571	360-403-4232	2,900	97
Clarcor Inc.	PO Box 7007, Rockford, Illinois 61125, U.S.A.	815-962-8867	815-962-0417	2,872	97
Varlen Corp.	PO Box 3089, Naperville, Illinois 60566-7089, U.S.A.	630-420-0400	630-420-7123	2,850	97
Winnebago Industries Inc.	PO Box 152, Forest City, Iowa 50436, U.S.A.	515-582-3535	515-582-6966	2,830	97
Stanadyne Automotive Corp.	92 Deerfield Rd., Windsor, Connecticut 06095, U.S.A.	860-525-0821	860-683-4582	2,800	95
Halter Marine Group Inc.	13085 Industrial Seaway Rd., Gulfport, Mississippi 39505, U.S.A.	601-896-0029		2,769	95
Oshkosh Truck Corp.	PO Box 2566, Oshkosh, Wisconsin 54903-2566, U.S.A.	920-235-9150		2,750	97
Tractors Malaysia Holdings BHD	Subang, Petaling Jaya 47500, Malaysia	3 7346688	3 7346623	2,708	94
Indomobil Suzuki International, PT	Wisma Indomobil 8th Flr., Jakarta 13330, Indonesia	218564530	21 4893048	2,700	97
Magellan Aerospace Corp.	3160 Derry Rd. E, Mississauga, Ontario L4T 1A9, Canada	905-677-1889	905-677-5658	2,700	98
Puritan-Bennett Corp.	9401 Indian Creek Pkwy., Bldg. 4, Overland Park, Kansas 66210, U.S.A.	913-661-0444	913-661-0234	2,700	94
Roadmaster Corp.	PO Box 344, Olney, Illinois 62450, U.S.A.	618-393-2991		2,700	93
Renold PLC	Wythenshawe, Manchester M22 5WL, United Kingdom	61 4375221	61 4377782	2,610	94
Wix Filtration Products Div.	PO Box 1967, Gastonia, North Carolina 28053, U.S.A.	704-864-6711	704-864-1843	2,585	95
Henlys Group PLC	53 Thoebald St., Borehamwood WD6 4RT, United Kingdom	81 9539953		2,559	93
Primex Technologies Inc.	10101 9th St. N., St. Petersburg, Florida 33716-3807, U.S.A.	813-578-8100		2,550	97

Product Tables
Sea-Going Bulk Carriers, Launched

Unit of Measure: Number.

	1990 Value	1990 %	1991 Value	1991 %	1992 Value	1992 %	1993 Value	1993 %	1994 Value	1994 %	1995 Value	1995 %	1996 Value	1996 %
Total Production	439	100.0	752	100.0	887	100.0	153	100.0	152	100.0	455	100.0	475	100.0
Regions														
Asia	343	78.1	656	87.1	783	88.2	53	34.5	62	40.5	349	76.6	364	76.8
Europe	96	21.9	97	12.9	104	11.8	100	65.5	91	59.5	106	23.4	110	23.2
Leading Producers														
Indonesia	334	76.1	647	86.0	774	87.3	44	28.8	53	34.8	*340	74.7	*356	75.0
Germany	*74	16.9	*74	9.8	*74	8.3	*74	48.5	*74	48.6	75	16.5	73	15.4
Romania	7	1.6	8	1.1	16	1.8	11	7.2	4	2.6	18	4.0	25	5.3
Korea, Republic of	*6	1.4	*6	0.8	*6	0.7	*6	3.9	*6	3.9	*6	1.3	*6	1.3

Tankers, Launched

Unit of Measure: Number.

	1990 Value	1990 %	1991 Value	1991 %	1992 Value	1992 %	1993 Value	1993 %	1994 Value	1994 %	1995 Value	1995 %	1996 Value	1996 %
Total Production	319	100.0	380	100.0	456	100.0	430	100.0	368	100.0	365	100.0	350	100.0
Regions														
Africa	2	0.6	2	0.5	2	0.4	2	0.5	2	0.5	2	0.5	2	0.6
America, North	6	1.8	6	1.5	6	1.3	6	1.3	6	1.5	6	1.6	6	1.6
America, South	5	1.4	6	1.6	9	2.0	6	1.3	6	1.7	3	0.8	6	1.7
Asia	224	70.2	266	69.9	328	72.0	318	73.9	270	73.3	260	71.2	228	65.2
Europe	66	20.6	83	21.9	94	20.6	82	19.0	67	18.2	78	21.2	91	26.0
Oceania	17	5.3	17	4.5	17	3.7	17	4.0	17	4.6	17	4.7	17	4.9
Leading Producers														
Japan	181	56.7	203	53.4	258	56.5	264	61.4	163	44.3	171	46.8	128	36.6
Korea, Republic of	8	2.5	23	6.1	37	8.1	28	6.5	62	16.8	43	11.8	50	14.3
Australia	*16	5.0	*16	4.2	16	3.5	*16	3.7	*16	4.3	*16	4.4	*16	4.6
Sweden	*15	4.7	*17	4.4	*18	3.9	*19	4.5	*21	5.7	*22	6.1	*24	6.8
Singapore	9	2.8	20	5.3	10	2.2	5	1.2	20	5.4	19	5.2	17	4.9

Other Sea-Going Merchant Vessels, Launched

Unit of Measure: Number.

	1990 Value	1990 %	1991 Value	1991 %	1992 Value	1992 %	1993 Value	1993 %	1994 Value	1994 %	1995 Value	1995 %	1996 Value	1996 %
Total Production	1,184	100.0	1,312	100.0	1,244	100.0	1,217	100.0	1,074	100.0	1,030	100.0	1,196	100.0
Regions														
Africa	10	0.9	6	0.5	7	0.6	8	0.7	5	0.5	5	0.5	6	0.5
America, North	27	2.3	32	2.4	37	3.0	57	4.7	9	0.8	4	0.4	7	0.6
America, South	26	2.2	40	3.0	50	4.0	50	4.1	29	2.7	44	4.3	44	3.6
Asia	635	53.6	640	48.8	594	47.7	630	51.8	627	58.3	650	63.1	806	67.4
Europe	462	39.0	571	43.6	538	43.2	436	35.8	383	35.6	305	29.6	315	26.4
Oceania	25	2.1	23	1.7	19	1.5	35	2.9	22	2.1	22	2.2	18	1.5
Leading Producers														
Japan	387	32.7	335	25.5	303	24.4	351	28.9	413	38.4	417	40.5	386	32.3
Korea, Republic of	92	7.8	94	7.2	73	5.9	60	4.9	35	3.3	100	9.7	120	10.0
Spain	92	7.8	62	4.7	49	3.9	27	2.2	21	2.0	9	0.9	12	1.0
Macau	*65	5.5	*65	4.9	*65	5.2	28	2.3	38	3.5	42	4.1	151	12.6
Netherlands	36	3.0	45	3.4	29	2.3	41	3.4	45	4.2	36	3.5	47	3.9

Commodity data are provided by the United Nations Statistical Division. The symbol * means that data are estimated. For additional notes, see Appendix I.

Product Tables
Locomotives, Electric

Unit of Measure: Number.

	1990		1991		1992		1993		1994		1995		1996	
	Value	%	Value	%	Value	%	Value	%	Value	%	Value	%	Value	%
Total Production	2,013	100.0	2,143	100.0	2,301	100.0	2,416	100.0	2,598	100.0	2,531	100.0	2,771	100.0
Regions														
Asia	1,335	66.3	1,421	66.3	1,296	56.3	1,544	63.9	1,758	67.7	1,831	72.4	1,996	72.0
Europe	678	33.7	722	33.7	1,005	43.7	872	36.1	840	32.3	700	27.6	775	28.0
Leading Producers														
Korea, Republic of	207	10.3	277	12.9	218	9.5	544	22.5	741	28.5	835	33.0	*956	34.5
India	289	14.4	303	14.1	302	13.1	294	12.2	300	11.5	278	11.0	312	11.3
China	165	8.2	173	8.1	200	8.7	*225	9.3	*242	9.3	*258	10.2	*274	9.9
Germany	-		-		237	10.3	222	9.2	158	6.1	15	0.6	83	3.0
Italy	28	1.4	92	4.3	100	4.3	20	0.8	27	1.0	60	2.4	50	1.8

Locomotives, Diesel

Unit of Measure: Number.

	1990		1991		1992		1993		1994		1995		1996	
	Value	%	Value	%	Value	%	Value	%	Value	%	Value	%	Value	%
Total Production	3,720	100.0	3,844	100.0	3,964	100.0	3,924	100.0	3,945	100.0	3,981	100.0	3,983	100.0
Regions														
America, North	800	21.5	800	20.8	800	20.2	800	20.4	800	20.3	800	20.1	800	20.1
America, South	64	1.7	64	1.7	64	1.6	64	1.6	64	1.6	64	1.6	64	1.6
Asia	2,045	55.0	2,106	54.8	2,140	54.0	2,154	54.9	2,189	55.5	2,270	57.0	2,287	57.4
Europe	811	21.8	874	22.7	960	24.2	906	23.1	892	22.6	847	21.3	832	20.9
Leading Producers														
United States of America	*800	21.5	*800	20.8	*800	20.2	*800	20.4	*800	20.3	*800	20.1	*800	20.1
China	466	12.5	521	13.6	563	14.2	*615	15.7	*656	16.6	*697	17.5	*738	18.5
India	*139	3.7	*141	3.7	*143	3.6	*144	3.7	*146	3.7	*148	3.7	*150	3.8
Germany	-		173	4.5	149	3.8	141	3.6	123	3.1	85	2.1	75	1.9
Romania	61	1.6	12	0.3	9	0.2	7	0.2	4	0.1	1	0.0	-	

Rail Motor Passenger Vehicles

Unit of Measure: Number.

	1990		1991		1992		1993		1994		1995		1996	
	Value	%	Value	%	Value	%	Value	%	Value	%	Value	%	Value	%
Total Production	6,101	100.0	5,802	100.0	6,296	100.0	6,376	100.0	5,454	100.0	5,158	100.0	5,941	100.0
Regions														
Asia	4,155	68.1	3,940	67.9	4,256	67.6	4,445	69.7	3,576	65.6	3,364	65.2	4,212	70.9
Europe	1,945	31.9	1,861	32.1	2,039	32.4	1,931	30.3	1,878	34.4	1,794	34.8	1,729	29.1
Leading Producers														
Japan	2,227	36.5	1,831	31.6	1,966	31.2	1,974	31.0	1,950	35.8	1,541	29.9	1,775	29.9
India	1,927	31.6	2,108	36.3	2,289	36.4	2,470	38.7	1,625	29.8	1,822	35.3	2,436	41.0
Germany	*1,162	19.0	*1,162	20.0	*1,162	18.5	*1,162	18.2	*1,162	21.3	1,162	22.5	*1,162	19.6
France	216	3.5	156	2.7	216	3.4	144	2.3	156	2.9	*161	3.1	*146	2.5
Sweden	45	0.7	38	0.7	61	1.0	53	0.8	55	1.0	21	0.4	*41	0.7

Commodity data are provided by the United Nations Statistical Division. The symbol * means that data are estimated. For additional notes, see Appendix I.

659

Product Tables
Goods Wagons and Vans

Unit of Measure: Number.

	1990 Value	%	1991 Value	%	1992 Value	%	1993 Value	%	1994 Value	%	1995 Value	%	1996 Value	%
Total Production	200,700	100.0	191,800	100.0	187,161	100.0	186,487	100.0	186,071	100.0	187,404	100.0	193,733	100.0
Regions														
Africa	732	0.4	578	0.3	780	0.4	694	0.4	498	0.3	416	0.2	381	0.2
America, North	535	0.3	570	0.3	604	0.3	639	0.3	508	0.3	501	0.3	495	0.3
America, South	1,426	0.7	1,465	0.8	1,504	0.8	1,543	0.8	1,581	0.8	1,620	0.9	1,659	0.9
Asia	145,175	72.3	140,672	73.3	139,415	74.5	131,657	70.6	123,262	66.2	127,659	68.1	133,814	69.1
Europe	30,647	15.3	33,651	17.5	30,623	16.4	32,537	17.4	34,868	18.7	31,803	17.0	30,055	15.5
Oceania	22,184	11.1	14,864	7.7	14,234	7.6	19,417	10.4	25,354	13.6	25,405	13.6	27,328	14.1
Leading Producers														
China	18,597	9.3	18,500	9.6	21,636	11.6	*24,661	13.2	*25,623	13.8	*26,585	14.2	*27,547	14.2
India	24,157	12.0	25,778	13.4	25,350	13.5	19,532	10.5	14,694	7.9	19,041	10.2	23,658	12.2
Russian Federation	25,092	12.5	22,388	11.7	16,895	9.0	12,309	6.6	7,785	4.2	7,072	3.8	7,417	3.8
Australia	22,184	11.1	14,864	7.7	14,234	7.6	19,417	10.4	25,354	13.6	25,405	13.6	27,328	14.1

Rail Passenger Carriages

Unit of Measure: Number.

	1990 Value	%	1991 Value	%	1992 Value	%	1993 Value	%	1994 Value	%	1995 Value	%	1996 Value	%
Total Production	16,801	100.0	16,475	100.0	17,659	100.0	16,924	100.0	14,881	100.0	16,055	100.0	16,249	100.0
Regions														
Africa	479	2.8	444	2.7	409	2.3	374	2.2	339	2.3	304	1.9	269	1.7
America, North	102	0.6	77	0.5	54	0.3	32	0.2	30	0.2	26	0.2	17	0.1
America, South	2	0.0	1	0.0	1	0.0	1	0.0	1	0.0	1	0.0	1	0.0
Asia	6,895	41.0	6,790	41.2	7,584	42.9	7,811	46.2	7,445	50.0	7,346	45.8	7,536	46.4
Europe	9,323	55.5	9,164	55.6	9,611	54.4	8,706	51.4	7,065	47.5	8,379	52.2	8,427	51.9
Leading Producers														
Germany	*3,349	19.9	*3,349	20.3	4,548	25.8	3,353	19.8	2,145	14.4	*3,349	20.9	*3,349	20.6
China	1,866	11.1	1,674	10.2	1,652	9.4	*2,021	11.9	*2,092	14.1	*2,163	13.5	*2,235	13.8
India	1,502	8.9	1,584	9.6	2,615	14.8	2,460	14.5	*2,217	14.9	*2,324	14.5	*2,431	15.0
Russian Federation	1,225	7.3	1,013	6.1	961	5.4	997	5.9	709	4.8	489	3.0	449	2.8
Algeria	*479	2.8	*444	2.7	*409	2.3	*374	2.2	*339	2.3	*304	1.9	*269	1.7

Engines, Internal Combustion (Compression-Ignition), for Vehicles (Diesel Engines)

Unit of Measure: Number.

	1990 Value	%	1991 Value	%	1992 Value	%	1993 Value	%	1994 Value	%	1995 Value	%	1996 Value	%
Total Production	3,528,339	100.0	4,285,701	100.0	4,523,310	100.0	4,465,811	100.0	5,031,259	100.0	4,912,407	100.0	4,868,198	100.0
Regions														
Africa	7,503	0.2	6,613	0.2	6,210	0.1	6,388	0.1	5,011	0.1	5,374	0.1	5,100	0.1
America, North	440,574	12.5	390,438	9.1	473,814	10.5	625,377	14.0	677,676	13.5	733,349	14.9	688,686	14.1
America, South	195,798	5.5	195,798	4.6	195,798	4.3	195,798	4.4	195,798	3.9	195,798	4.0	195,798	4.0
Asia	212,110	6.0	254,805	5.9	190,815	4.2	204,053	4.6	97,226	1.9	93,591	1.9	80,669	1.7
Europe	2,672,354	75.7	3,438,047	80.2	3,656,674	80.8	3,434,195	76.9	4,055,548	80.6	3,884,294	79.1	3,897,945	80.1
Leading Producers														
Germany	*1,015,596	28.8	*1,015,596	23.7	1,103,963	24.4	909,940	20.4	1,032,884	20.5	*1,015,596	20.7	*1,015,596	20.9
United States of America	432,945	12.3	383,110	8.9	469,317	10.4	607,921	13.6	661,155	13.1	732,148	14.9	686,707	14.1
United Kingdom	424,923	12.0	488,684	11.4	552,446	12.2	616,207	13.8	975,562	19.4	*767,600	15.6	*825,509	17.0
Austria	28,540	0.8	761,335	17.8	820,801	18.1	720,553	16.1	798,435	15.9	819,433	16.7	*831,540	17.1
Brazil	*195,798	5.5	*195,798	4.6	*195,798	4.3	*195,798	4.4	*195,798	3.9	*195,798	4.0	*195,798	4.0

Commodity data are provided by the United Nations Statistical Division. The symbol * means that data are estimated. For additional notes, see Appendix I.

Product Tables

Engines, Internal Combustion (Spark-Ignition), for Vehicles (Gasoline Fueled)

Unit of Measure: Number.

	1990		1991		1992		1993		1994		1995		1996	
	Value	%	Value	%	Value	%	Value	%	Value	%	Value	%	Value	%
Total Production	29,618,488	100.0	30,551,536	100.0	31,429,038	100.0	34,324,969	100.0	36,638,690	100.0	36,036,476	100.0	35,798,591	100.0
Regions														
America, North	17,410,086	58.8	18,242,714	59.7	19,008,343	60.5	20,486,971	59.7	21,891,600	59.7	22,828,229	63.3	23,929,857	66.8
America, South	6,470	0.0	5,235	0.0	4,000	0.0	12,000	0.0	9,000	0.0	10,000	0.0	5,965	0.0
Asia	2,386,667	8.1	2,422,667	7.9	2,553,667	8.1	2,727,000	7.9	3,048,000	8.3	3,257,000	9.0	3,257,781	9.1
Europe	9,401,838	31.7	9,458,742	31.0	9,432,100	30.0	10,659,319	31.1	11,241,661	30.7	9,484,069	26.3	8,139,060	22.7
Oceania	413,429	1.4	422,179	1.4	430,929	1.4	439,679	1.3	448,429	1.2	457,179	1.3	465,929	1.3
Leading Producers														
United States of America	*15,726,086	53.1	*16,709,714	54.7	*17,693,343	56.3	*18,676,971	54.4	*19,660,600	53.7	*20,644,229	57.3	*21,627,857	60.4
Germany	*3,182,140	10.7	*3,182,140	10.4	*3,182,140	10.1	4,331,000	12.6	4,449,000	12.1	2,799,186	7.8	1,149,372	3.2
Mexico	1,684,000	5.7	1,533,000	5.0	1,315,000	4.2	1,810,000	5.3	2,231,000	6.1	2,184,000	6.1	2,302,000	6.4
Korea, Republic of	1,349,000	4.6	1,495,000	4.9	1,701,000	5.4	1,904,000	5.5	2,216,000	6.0	2,399,000	6.7	*2,417,114	6.8
Indonesia	*783,000	2.6	783,000	2.6	*783,000	2.5	*783,000	2.3	*783,000	2.1	*783,000	2.2	*783,000	2.2

Bodies for Motor Vehicles

Unit of Measure: Thousands of units.

	1990		1991		1992		1993		1994		1995		1996	
	Value	%	Value	%	Value	%	Value	%	Value	%	Value	%	Value	%
Total Production	226,420	100.0	258,759	100.0	291,009	100.0	323,227	100.0	355,474	100.0	387,986	100.0	420,203	100.0
Regions														
America, North	14	0.0	18	0.0	21	0.0	165	0.1	150	0.0	93	0.0	114	0.0
America, South	225,636	99.7	257,871	99.7	290,107	99.7	322,343	99.7	354,578	99.7	386,813	99.7	419,049	99.7
Asia	8	0.0	12	0.0	8	0.0	7	0.0	6	0.0	8	0.0	9	0.0
Europe	741	0.3	838	0.3	853	0.3	692	0.2	720	0.2	1,052	0.3	1,011	0.2
Oceania	20	0.0	20	0.0	20	0.0	20	0.0	20	0.0	20	0.0	20	0.0
Leading Producers														
Argentina	*225,635	99.7	*257,871	99.7	*290,106	99.7	*322,342	99.7	*354,577	99.7	*386,813	99.7	*419,048	99.7
Netherlands	15	0.0	14	0.0	14	0.0	10	0.0	11	0.0	12	0.0	-	
United Kingdom	*246	0.1	*246	0.1	*246	0.1	244	0.1	248	0.1	*246	0.1	*246	0.1
Spain	184	0.1	205	0.1	178	0.1	79	0.0	102	0.0	107	0.0	*149	0.0
Germany	-		-		0	0.0	0	0.0	0	0.0	319	0.1	250	0.1

Passenger Cars, Assembled from Imported Parts

Unit of Measure: Number.

	1990		1991		1992		1993		1994		1995		1996	
	Value	%	Value	%	Value	%	Value	%	Value	%	Value	%	Value	%
Total Production	2,920,826	100.0	3,175,250	100.0	3,546,045	100.0	3,952,045	100.0	4,078,596	100.0	4,506,064	100.0	4,718,857	100.0
Regions														
Africa	306,812	10.5	299,193	9.4	305,145	8.6	328,975	8.3	311,325	7.6	374,261	8.3	374,587	7.9
America, North	2,000	0.1	4,000	0.1	3,000	0.1	2,000	0.1	2,000	0.0	-		-	
America, South	80,714	2.8	93,857	3.0	118,000	3.3	119,803	3.0	115,209	2.8	111,962	2.5	101,039	2.1
Asia	470,000	16.1	584,467	18.4	635,170	17.9	798,873	20.2	703,576	17.3	855,982	19.0	890,617	18.9
Europe	2,017,300	69.1	2,137,733	67.3	2,453,730	69.2	2,669,636	67.6	2,918,626	71.6	3,140,893	69.7	3,334,545	70.7
Oceania	44,000	1.5	56,000	1.8	31,000	0.9	32,758	0.8	27,861	0.7	22,965	0.5	18,069	0.4
Leading Producers														
Belgium	1,160,000	39.7	1,064,000	33.5	1,094,000	30.9	1,089,000	27.6	1,210,000	29.7	1,178,000	26.1	1,154,000	24.5
United Kingdom	726,300	24.9	958,733	30.2	1,191,167	33.6	1,423,600	36.0	1,545,078	37.9	*1,792,835	39.8	*2,003,975	42.5
South Africa	225,000	7.7	223,000	7.0	197,000	5.6	220,000	5.6	217,000	5.3	275,000	6.1	271,000	5.7
Turkey	166,000	5.7	195,000	6.1	265,000	7.5	343,000	8.7	208,000	5.1	222,000	4.9	196,000	4.2
Malaysia	131,000	4.5	152,000	4.8	137,000	3.9	145,000	3.7	173,000	4.2	241,000	5.3	313,000	6.6

Commodity data are provided by the United Nations Statistical Division. The symbol * means that data are estimated. For additional notes, see Appendix I.

661

Product Tables

Passenger Cars, Produced

Unit of Measure: Number.

	1990 Value	%	1991 Value	%	1992 Value	%	1993 Value	%	1994 Value	%	1995 Value	%	1996 Value	%
Total Production	36,119,139	100.0	39,569,939	100.0	39,907,863	100.0	37,949,340	100.0	39,656,914	100.0	39,475,328	100.0	40,216,710	100.0
Regions														
Africa	10,000	0.0	9,000	0.0	7,000	0.0	4,000	0.0	6,000	0.0	9,000	0.0	13,000	0.0
America, North	7,632,000	21.1	7,061,000	17.8	7,384,000	18.5	7,655,000	20.2	8,427,615	21.3	7,833,890	19.8	7,880,851	19.6
America, South	348,000	1.0	407,000	1.0	559,000	1.4	679,000	1.8	705,000	1.8	498,000	1.3	514,000	1.3
Asia	13,726,139	38.0	13,536,939	34.2	13,165,709	33.0	12,543,713	33.1	12,046,968	30.4	12,149,222	30.8	12,445,188	30.9
Europe	14,042,000	38.9	18,278,000	46.2	18,522,155	46.4	16,782,627	44.2	18,161,331	45.8	18,691,216	47.3	19,058,672	47.4
Oceania	361,000	1.0	278,000	0.7	270,000	0.7	285,000	0.8	310,000	0.8	294,000	0.7	305,000	0.8
Leading Producers														
Japan	9,948,000	27.5	9,753,000	24.6	9,379,000	23.5	8,494,000	22.4	7,801,000	19.7	7,611,000	19.3	7,864,000	19.6
United States of America	6,081,000	16.8	5,441,000	13.8	5,684,000	14.2	5,956,000	15.7	6,614,000	16.7	*6,202,989	15.7	*6,153,664	15.3
Germany	-		4,647,000	11.7	4,895,000	12.3	3,875,000	10.2	4,222,000	10.6	4,467,500	11.3	4,713,000	11.7
France	3,293,000	9.1	3,190,000	8.1	3,326,000	8.3	2,837,000	7.5	3,176,000	8.0	*3,217,165	8.1	*3,235,673	8.0
Spain	1,696,000	4.7	1,787,000	4.5	1,817,000	4.6	1,774,000	4.7	2,146,000	5.4	2,254,000	5.7	2,334,000	5.8

Buses and Motor Coaches, Produced

Unit of Measure: Number.

	1990 Value	%	1991 Value	%	1992 Value	%	1993 Value	%	1994 Value	%	1995 Value	%	1996 Value	%
Total Production	570,396	100.0	575,240	100.0	620,112	100.0	674,614	100.0	682,077	100.0	709,299	100.0	703,230	100.0
Regions														
Africa	1,493	0.3	1,128	0.2	760	0.1	701	0.1	652	0.1	740	0.1	680	0.1
America, North	52,594	9.2	58,542	10.2	63,299	10.2	54,583	8.1	54,499	8.0	55,809	7.9	58,267	8.3
America, South	16,694	2.9	25,294	4.4	28,807	4.6	23,469	3.5	23,835	3.5	26,933	3.8	23,013	3.3
Asia	410,827	72.0	409,154	71.1	453,343	73.1	531,594	78.8	545,047	79.9	562,248	79.3	565,211	80.4
Europe	88,137	15.5	80,518	14.0	73,346	11.8	63,758	9.5	57,583	8.4	63,154	8.9	55,691	7.9
Oceania	651	0.1	604	0.1	556	0.1	509	0.1	462	0.1	415	0.1	368	0.1
Leading Producers														
India	146,769	25.7	145,804	25.3	149,062	24.0	210,419	31.2	*199,638	29.3	*213,562	30.1	*227,485	32.3
Korea, Republic of	92,995	16.3	84,356	14.7	122,273	19.7	151,790	22.5	175,333	25.7	192,235	27.1	*178,570	25.4
Japan	40,185	7.0	44,449	7.7	52,005	8.4	48,074	7.1	49,112	7.2	47,266	6.7	53,126	7.6
Russian Federation	51,925	9.1	51,598	9.0	48,209	7.8	47,866	7.1	50,049	7.3	39,773	5.6	38,343	5.5
United States of America	*44,700	7.8	*46,891	8.2	*49,082	7.9	*51,273	7.6	*53,464	7.8	*55,654	7.8	*57,845	8.2

Trucks, Including Articulated Vehicles, Assembled from Imported Parts

Unit of Measure: Number.

	1990 Value	%	1991 Value	%	1992 Value	%	1993 Value	%	1994 Value	%	1995 Value	%	1996 Value	%
Total Production	697,454	100.0	721,455	100.0	695,810	100.0	777,023	100.0	801,881	100.0	827,053	100.0	891,918	100.0
Regions														
Africa	137,988	19.8	128,471	17.8	120,082	17.3	119,314	15.4	142,767	17.8	173,735	21.0	158,582	17.8
America, North	1,714	0.2	2,308	0.3	2,303	0.3	1,695	0.2	1,241	0.2	1,142	0.1	973	0.1
America, South	35,609	5.1	44,300	6.1	55,068	7.9	43,865	5.6	41,084	5.1	34,194	4.1	31,219	3.5
Asia	375,538	53.8	365,117	50.6	346,842	49.8	446,050	57.4	415,723	51.8	419,209	50.7	492,690	55.2
Europe	133,105	19.1	167,759	23.3	161,306	23.2	156,726	20.2	192,881	24.1	191,773	23.2	202,640	22.7
Oceania	13,500	1.9	13,500	1.9	10,210	1.5	9,372	1.2	8,186	1.0	6,999	0.8	5,813	0.7
Leading Producers														
Thailand	236,221	33.9	206,172	28.6	223,680	32.1	323,508	41.6	324,780	40.5	*306,489	37.1	*328,159	36.8
South Africa	107,922	15.5	97,178	13.5	93,599	13.5	96,772	12.5	118,221	14.7	147,792	17.9	131,183	14.7
Belgium	65,667	9.4	88,937	12.3	70,532	10.1	56,244	7.2	*80,861	10.1	*84,472	10.2	*88,084	9.9
Portugal	*41,925	6.0	*44,424	6.2	*46,923	6.7	*49,421	6.4	*51,920	6.5	*54,419	6.6	*56,917	6.4
Malaysia	73,733	10.6	78,926	10.9	34,711	5.0	34,711	4.5	42,618	5.3	55,961	6.8	78,571	8.8

Product Tables
Trucks, Including Articulated Vehicles, Produced

Unit of Measure: Number.

	1990 Value	%	1991 Value	%	1992 Value	%	1993 Value	%	1994 Value	%	1995 Value	%	1996 Value	%
Total Production	12,611,886	100.0	12,419,755	100.0	13,051,806	100.0	12,533,447	100.0	12,911,922	100.0	12,959,586	100.0	13,242,603	100.0
Regions														
Africa	1,371	0.0	1,127	0.0	1,529	0.0	1,208	0.0	1,379	0.0	1,241	0.0	738	0.0
America, North	4,709,055	37.3	4,400,404	35.4	5,289,169	40.5	5,632,763	44.9	5,976,859	46.3	6,202,607	47.9	6,621,948	50.0
America, South	68,466	0.5	71,683	0.6	69,024	0.5	98,681	0.8	128,159	1.0	115,239	0.9	95,692	0.7
Asia	5,877,569	46.6	5,880,989	47.4	5,592,146	42.8	5,238,476	41.8	5,069,968	39.3	4,799,155	37.0	4,767,229	36.0
Europe	1,929,467	15.3	2,047,886	16.5	2,085,388	16.0	1,546,861	12.3	1,721,896	13.3	1,829,232	14.1	1,746,436	13.2
Oceania	25,958	0.2	17,666	0.1	14,550	0.1	15,459	0.1	13,661	0.1	12,111	0.1	10,561	0.1
Leading Producers														
United States of America	3,720,000	29.5	3,372,000	27.2	4,118,578	31.6	*4,571,956	36.5	*4,775,414	37.0	*4,978,872	38.4	*5,182,331	39.1
Japan	3,486,618	27.6	3,433,790	27.6	3,053,477	23.4	2,674,941	21.3	2,689,340	20.8	2,519,319	19.4	2,417,370	18.3
Canada	789,932	6.3	789,600	6.4	901,000	6.9	838,000	6.7	*984,086	7.6	*1,013,663	7.8	*1,043,240	7.9
France	539,796	4.3	461,640	3.7	494,124	3.8	373,200	3.0	453,344	3.5	*483,316	3.7	*484,298	3.7
Russian Federation	665,201	5.3	615,868	5.0	582,963	4.5	466,925	3.7	185,018	1.4	142,483	1.1	134,130	1.0

Road Tractors for Tractor-Trailer Combinations, Produced

Unit of Measure: Number.

	1990 Value	%	1991 Value	%	1992 Value	%	1993 Value	%	1994 Value	%	1995 Value	%	1996 Value	%
Total Production	340,093	100.0	345,575	100.0	382,447	100.0	351,203	100.0	331,785	100.0	329,439	100.0	329,832	100.0
Regions														
America, North	158,908	46.7	170,206	49.3	175,604	45.9	179,180	51.0	186,410	56.2	187,831	57.0	195,254	59.2
America, South	4,868	1.4	3,099	0.9	4,298	1.1	3,830	1.1	4,642	1.4	2,551	0.8	5,681	1.7
Asia	112,680	33.1	109,635	31.7	111,180	29.1	92,967	26.5	57,238	17.3	46,380	14.1	38,351	11.6
Europe	63,637	18.7	62,635	18.1	91,366	23.9	75,226	21.4	83,495	25.2	92,677	28.1	90,546	27.5
Leading Producers														
United States of America	*155,015	45.6	*161,470	46.7	*167,924	43.9	*174,378	49.7	*180,832	54.5	*187,287	56.9	*193,741	58.7
Belarus	100,650	29.6	95,502	27.6	96,063	25.1	82,371	23.5	42,879	12.9	27,953	8.5	26,815	8.1
Germany	-		-		32,146	8.4	20,476	5.8	23,269	7.0	26,558	8.1	29,846	9.0
Japan	12,021	3.5	14,124	4.1	15,108	4.0	10,587	3.0	14,350	4.3	18,418	5.6	11,527	3.5
France	11,280	3.3	12,108	3.5	8,988	2.4	7,200	2.1	10,896	3.3	*10,933	3.3	*11,091	3.4

Trailers and Semi-Trailers

Unit of Measure: Number.

	1990 Value	%	1991 Value	%	1992 Value	%	1993 Value	%	1994 Value	%	1995 Value	%	1996 Value	%
Total Production	1,017,319	100.0	1,137,281	100.0	1,010,955	100.0	930,354	100.0	940,669	100.0	946,073	100.0	831,841	100.0
Regions														
Africa	40,203	4.0	36,914	3.2	36,910	3.7	32,533	3.5	28,252	3.0	28,195	3.0	26,018	3.1
America, North	153,279	15.1	125,698	11.1	168,409	16.7	188,674	20.3	237,266	25.2	281,689	29.8	205,045	24.6
America, South	2,564	0.3	2,647	0.2	3,271	0.3	4,632	0.5	4,287	0.5	2,842	0.3	4,188	0.5
Asia	274,548	27.0	263,823	23.2	173,883	17.2	125,501	13.5	85,914	9.1	112,360	11.9	72,663	8.7
Europe	532,241	52.3	696,411	61.2	619,387	61.3	572,610	61.5	581,242	61.8	519,974	55.0	523,926	63.0
Oceania	14,484	1.4	11,790	1.0	9,096	0.9	6,402	0.7	3,708	0.4	1,014	0.1	-	
Leading Producers														
United States of America	149,117	14.7	122,361	10.8	165,268	16.3	185,741	20.0	234,287	24.9	279,144	29.5	202,912	24.4
Germany	-		203,174	17.9	179,359	17.7	171,457	18.4	178,702	19.0	134,912	14.3	129,334	15.5
Russian Federation	150,228	14.8	134,256	11.8	75,574	7.5	41,303	4.4	11,223	1.2	11,508	1.2	7,659	0.9
Poland	39,775	3.9	29,228	2.6	25,845	2.6	17,175	1.8	23,785	2.5	17,156	1.8	30,513	3.7
United Kingdom	38,720	3.8	29,818	2.6	20,916	2.1	12,014	1.3	10,223	1.1	*22,082	2.3	*20,479	2.5

Commodity data are provided by the United Nations Statistical Division. The symbol * means that data are estimated. For additional notes, see Appendix I.

663

Product Tables

Motorcycles, Scooters Etc.

Unit of Measure: Number.

	1990 Value	%	1991 Value	%	1992 Value	%	1993 Value	%	1994 Value	%	1995 Value	%	1996 Value	%
Total Production	12,386,175	100.0	12,752,888	100.0	12,985,637	100.0	13,974,368	100.0	15,793,756	100.0	19,317,371	100.0	20,440,177	100.0
Regions														
Africa	105,508	0.9	87,533	0.7	71,773	0.6	69,243	0.5	72,787	0.5	69,664	0.4	64,781	0.3
America, North	139,524	1.1	135,810	1.1	139,095	1.1	137,381	1.0	136,667	0.9	139,702	0.7	140,488	0.7
America, South	244,667	2.0	285,333	2.2	282,000	2.2	293,000	2.1	338,000	2.1	384,000	2.0	423,933	2.1
Asia	9,678,167	78.1	10,301,533	80.8	10,705,245	82.4	12,088,852	86.5	13,752,791	87.1	17,188,730	89.0	18,246,594	89.3
Europe	2,218,310	17.9	1,942,678	15.2	1,787,523	13.8	1,385,892	9.9	1,493,511	9.5	1,535,274	7.9	1,564,382	7.7
Leading Producers														
Japan	2,807,000	22.7	3,029,000	23.8	3,197,000	24.6	3,023,000	21.6	2,725,000	17.3	2,753,000	14.3	2,584,000	12.6
China	979,000	7.9	1,341,000	10.5	2,052,000	15.8	3,556,000	25.4	5,354,000	33.9	8,254,000	42.7	9,168,000	44.9
India	1,889,000	15.3	1,607,000	12.6	1,500,000	11.6	1,759,000	12.6	2,189,000	13.9	2,588,000	13.4	2,918,000	14.3
Italy	721,000	5.8	622,000	4.9	571,000	4.4	558,000	4.0	749,000	4.7	837,000	4.3	896,000	4.4
Thailand	594,000	4.8	600,000	4.7	*601,345	4.6	*636,918	4.6	*672,491	4.3	*708,064	3.7	*743,636	3.6

Bicycles

Unit of Measure: Thousands of units.

	1990 Value	%	1991 Value	%	1992 Value	%	1993 Value	%	1994 Value	%	1995 Value	%	1996 Value	%
Total Production	93,238	100.0	102,629	100.0	107,159	100.0	108,311	100.0	111,767	100.0	113,362	100.0	103,637	100.0
Regions														
Africa	322	0.3	288	0.3	244	0.2	229	0.2	261	0.2	250	0.2	257	0.2
America, North	5,284	5.7	5,069	4.9	5,024	4.7	5,229	4.8	4,988	4.5	4,499	4.0	4,594	4.4
America, South	2,270	2.4	2,988	2.9	3,344	3.1	4,607	4.3	4,268	3.8	4,080	3.6	3,855	3.7
Asia	70,878	76.0	77,317	75.3	82,679	77.2	83,116	76.7	87,448	78.2	90,148	79.5	81,509	78.6
Europe	14,434	15.5	16,930	16.5	15,844	14.8	15,120	14.0	14,803	13.2	14,385	12.7	13,422	13.0
Oceania	50	0.1	37	0.0	24	0.0	11	0.0	-		-		-	
Leading Producers														
China	31,416	33.7	36,768	35.8	40,836	38.1	41,496	38.3	43,649	39.1	44,722	39.5	33,612	32.4
Thailand	*10,709	11.5	*12,227	11.9	*13,746	12.8	*15,264	14.1	*16,783	15.0	*18,301	16.1	*19,820	19.1
Japan	7,969	8.5	7,448	7.3	7,286	6.8	6,858	6.3	6,702	6.0	6,580	5.8	6,138	5.9
India	6,684	7.2	7,150	7.0	6,964	6.5	7,721	7.1	8,907	8.0	9,912	8.7	10,895	10.5
United States of America	*4,707	5.0	*4,530	4.4	*4,352	4.1	*4,175	3.9	*3,998	3.6	*3,820	3.4	*3,643	3.5

Commercial Passenger and Cargo Planes

Unit of Measure: Number.

	1990 Value	%	1991 Value	%	1992 Value	%	1993 Value	%	1994 Value	%	1995 Value	%	1996 Value	%
Total Production	2,836	100.0	2,643	100.0	2,270	100.0	2,061	100.0	2,150	100.0	1,905	100.0	2,011	100.0
Regions														
America, North	2,341	82.5	2,172	82.2	1,847	81.4	1,685	81.8	1,775	82.5	1,517	79.6	1,649	82.0
America, South	151	5.3	170	6.4	167	7.4	165	8.0	162	7.5	159	8.4	157	7.8
Asia	7	0.2	9	0.4	9	0.4	9	0.5	9	0.4	9	0.5	9	0.5
Europe	337	11.9	292	11.0	247	10.9	202	9.8	204	9.5	219	11.5	196	9.7
Leading Producers														
United States of America	2,341	82.5	2,172	82.2	1,847	81.4	1,685	81.8	1,775	82.5	1,517	79.6	1,649	82.0
Germany	*200	7.1	*200	7.6	*200	8.8	*200	9.7	*200	9.3	212	11.1	188	9.3
Brazil	151	5.3	*170	6.4	*167	7.4	*165	8.0	*162	7.5	*159	8.4	*157	7.8
Poland	136	4.8	91	3.4	46	2.0	1	0.0	3	0.1	6	0.3	7	0.3

Commodity data are provided by the United Nations Statistical Division. The symbol * means that data are estimated. For additional notes, see Appendix I.

Product Tables
Perambulators and Push-Chairs for Babies

Unit of Measure: Number.

	1990		1991		1992		1993		1994		1995		1996	
	Value	%	Value	%	Value	%	Value	%	Value	%	Value	%	Value	%
Total Production	16,710,948	100.0	16,951,803	100.0	15,526,157	100.0	14,126,469	100.0	13,498,146	100.0	13,673,608	100.0	13,980,417	100.0
Regions														
America, South	52,455	0.3	52,727	0.3	53,000	0.3	43,000	0.3	49,000	0.4	22,000	0.2	42,371	0.3
Asia	10,072,583	60.3	10,145,000	59.8	8,492,917	54.7	7,641,833	54.1	7,459,750	55.3	7,576,500	55.4	7,765,274	55.5
Europe	6,585,910	39.4	6,754,075	39.8	6,980,240	45.0	6,441,636	45.6	5,989,396	44.4	6,075,108	44.4	6,172,771	44.2
Leading Producers														
France	*2,122,295	12.7	*2,286,238	13.5	*2,450,181	15.8	*2,614,124	18.5	*2,778,067	20.6	*2,942,010	21.5	*3,105,952	22.2
Russian Federation	3,229,000	19.3	3,087,000	18.2	1,401,000	9.0	629,000	4.5	223,000	1.7	103,000	0.8	53,000	0.4
United Kingdom	*716,615	4.3	*716,615	4.2	*716,615	4.6	850,656	6.0	582,574	4.3	*716,615	5.2	*716,615	5.1
Ukraine	1,157,000	6.9	1,117,000	6.6	725,000	4.7	410,000	2.9	137,000	1.0	59,000	0.4	11,000	0.1
Spain	570,000	3.4	688,000	4.1	631,000	4.1	569,250	4.0	507,500	3.8	445,750	3.3	384,000	2.7

Commodity data are provided by the United Nations Statistical Division. The symbol * means that data are estimated. For additional notes, see Appendix I.

665

ISIC 3841
SHIPBUILDING, REPAIR

ISIC 3841—Shipbuilding, Repair—includes the construction and repair of all tankers, sea-going merchant vessels, floating structures, yachts, and boats. This is a subcategory of ISIC 3840—Transportation Equipment; for product and company tables, see that chapter.

The products of this industry are classified, in the U.S. SIC system, under SIC 373—Ship and Boat Building and Repair.

Summary Statistics

		1990		1991		1992		1993		1994		1995	
		Value	N	Value	N	Value	N	Value	N	Value	N	Value	N
Establishments or enterprises	(number)	9,361	49	9,765	53	13,991	53	5,924	43	2,867	37	1,150	13
Number employed	(000)	695	53	748	55	852	53	663	44	581	35	360	18
Total Wages	($ mil.)	17,180	51	18,931	52	19,619	52	13,089	41	8,984	31	6,927	16
Wage/salary per employee	($)	12,175	50	13,476	51	14,182	51	12,310	41	10,569	31	9,518	16
Female workers	(000)	18	23	10	20	134	20	127	19	122	11	116	4
as % of total employment*	(%)	7	23	8	20	10	20	15	19	12	11	25	4
Output	($ bil.)	72	51	80	53	84	51	68	42	42	31	34	16
per employee	($ 000)	56	49	57	51	56	49	52	40	48	29	45	14
per establishment	($ 000)	5,490	47	6,248	49	6,655	48	5,347	38	5,730	28	3,062	9
Value added	($ bil.)	30	49	31	50	33	43	29	37	16	27	15	14
per employee	($ 000)	25	47	24	48	25	41	26	34	22	25	27	12
per establishment	($ 000)	2,239	45	2,215	46	2,562	41	2,170	33	2,049	24	1,463	7
Capital investment	($ mil.)	1,213	26	1,423	29	2,791	32	3,460	26	2,439	17	2,610	7
per employee	($ 000)	2	26	2	29	3	31	3	25	4	17	7	7
per establishment	($ 000)	291	26	390	29	343	32	352	24	438	16	754	6

Data presented above are drawn from the detailed tables that follow. Columns headed 'N' show the number of countries that provided valid data for inclusion. Values are not strictly comparable one year to the next or one row to the next because the number of countries that report varies from period to period and row to row. However, a general indication of magnitudes can be gleaned from reviewing this summary—especially in earlier years in which more reports are available. For detailed explanations, see Appendix I. *The average for those countries reporting both total and female employment.

Establishments and Number Engaged

Country	Establishments or Enterprises (number)						Employees per Establishment					
	1990	1991	1992	1993	1994	1995	1990	1991	1992	1993	1994	1995
Australia	296	446	471	-	-	-	34	20	19	-	-	-
Austria	5.000	5.000	5.000	5.000	5.000	-	280	200	200	75	-	-
Azerbaijan	6.000	6.000	6.000	6.000	5.000	-	695	666	629	604	638	-
Bahrain	-	-	4.000	-	-	-	-	-	320	-	-	-
Bangladesh	25	20	45	-	-	-	114	230	68	-	-	-
Canada	373	306	287	250	240	-	43	42	49	48	46	-
Cape Verde	-	-	3.000	3.000	2.000	-	-	-	89	77	94	-
Chile	11	9.000	14	10	9.000	-	157	173	359	468	453	-
China (Hong Kong SAR)	454	393	421	408	486	-	18	19	19	19	14	-
Colombia	19	17	15	8.000	9.000	-	75	65	88	91	67	-
Costa Rica	17	15	15	15	-	-	15	16	15	15	-	-
Cyprus	24	24	24	23	38	43	7.083	6.000	5.708	4.565	3.342	2.465
Denmark	526	518	516	-	-	-	32	32	32	-	-	-
Ecuador	3.000	3.000	3.000	4.000	4.000	4.000	31	26	40	20	24	15
Egypt	17	21	21	22	-	-	876	595	619	591	-	-
El Salvador	-	-	-	1.000	-	1.000	-	-	-	7.000	-	11
Fiji	11	13	13	-	-	-	22	15	14	-	-	-
Finland	85	97	76	73	65	-	136	115	143	145	172	-
Germany	-	124	125	123	122	-	-	499	399	360	334	-
Germany (Western Part)	98	95	95	-	-	-	342	351	343	-	-	-
Greece	131	127	129	-	-	-	89	91	87	-	-	-
Grenada	-	-	-	-	1.000	-	-	-	-	-	-	-
Honduras	-	-	5.000	4.000	4.000	4.000	-	-	15	20	22	32
Hungary	-	-	*31	*39	-	-	-	-	-	-	-	-
Iceland	59	70	70	65	-	-	11	8.557	7.800	8.277	-	-
India	151	185	204	241	-	-	120	115	135	127	-	-
Indonesia	141	142	166	156	155	157	111	101	104	127	110	115
Iran (Islamic Republic of)	-	14	15	10	-	-	-	186	166	86	-	-
Italy	*142	*147	-	-	-	-	179	170	-	-	-	-
Japan	2,214	2,192	2,151	2,163	-	-	34	35	37	37	-	-
Jordan	-	-	-	-	3.000	4.000	-	-	-	-	6.333	4.000
Kenya	4.000	4.000	4.000	4.000	4.000	4.000	117	126	126	125	130	137
Korea, Republic of	475	601	547	622	677	803	112	96	99	95	100	97
Kuwait	8.000	7.000	6.000	7.000	8.000	-	191	349	400	363	92	-
Kyrgyzstan	1.000	1.000	1.000	1.000	1.000	-	91	86	65	43	33	-
Macau	70	64	62	51	41	-	7.286	7.391	6.661	5.745	6.683	-
Malaysia	65	66	76	83	84	-	54	82	89	89	90	-
Malta	24	24	22	21	23	-	99	96	102	98	87	-
Mauritius	2.000	2.000	2.000	2.000	2.000	2.000	58	48	48	50	48	48
Mexico	7.000	6.000	6.000	5.000	5.000	5.000	367	448	266	203	243	95
Mozambique	-	-	-	-	2.000	-	-	-	-	-	139	-
Netherlands	141	140	150	143	140	-	105	108	103	98	-	-
New Zealand	*558	*544	*525	*532	-	-	2.986	3.156	3.240	3.632	-	-
Norway	307	294	220	-	-	-	49	51	65	-	-	-
Pakistan	-	4.000	-	-	-	-	-	1.188	-	-	-	-
Panama	8.000	6.000	-	-	-	-	19	19	-	-	-	-
Peru	81	84	87	-	-	-	35	31	28	-	-	-
Philippines	107	104	49	46	-	-	53	50	106	85	-	-
Portugal	*257	*262	*303	*295	*267	-	45	45	38	37	28	-
Senegal	6.000	-	-	-	4.000	-	124	-	-	-	303	-
Singapore	163	179	186	221	223	-	112	118	115	98	107	-
Slovakia	-	1.000	1.000	2.000	2.000	-	-	-	-	-	-	-
South Africa	-	162	-	-	-	-	-	-	-	-	-	-
Spain	716	693	617	-	-	-	46	44	46	-	-	-
Sri Lanka	10	8.000	7.000	6.000	-	-	239	292	342	464	-	-
Sweden	64	70	63	60	54	-	83	88	94	82	88	-
Thailand	27	36	-	-	-	-	50	54	-	-	-	-
Trinidad and Tobago	9.000	10	10	11	11	-	7.778	8.000	7.900	7.182	7.091	-
Tunisia	-	-	-	58	60	52	-	-	-	24	25	28
Turkey	19	19	37	35	36	-	295	273	139	133	116	-

Continued.

Depending on the table, * means *Enterprises, Engaged,* or *Factor Values;* ± means *Basis Unspecified;* § means *shown in millions.* For additional notes and sources, see Appendix I.

Establishments and Number Engaged

- Continued -

Country	Establishments or Enterprises (number)						Employees per Establishment					
	1990	1991	1992	1993	1994	1995	1990	1991	1992	1993	1994	1995
Ukraine	-	-	105	73	60	54	-	-	1,295	1,767	1,967	1,796
United Kingdom	1,403	1,362	1,451	-	-	-	46	43	37	-	-	-
United Republic of Tanzania	3.000	3.000	-	-	-	-	15	20	-	-	-	-
United States of America	-	-	4,507	-	-	-	-	-	36	-	-	-
Venezuela	18	20	17	17	13	17	50	45	59	49	47	62
Zambia	-	-	-	-	2.000	-	-	-	-	-	-	-

Employment and Compensation of Employees

Country	Number Employed/Engaged (000)						Wage/Salary per Employee ($)					
	1990	1991	1992	1993	1994	1995	1990	1991	1992	1993	1994	1995
Australia	10	*9.000	*9.000	-	-	-	23,672	25,798	25,163	-	-	-
Austria	1.400	1.000	1.000	0.376	-	-	29,652	-	24,479	-	-	-
Azerbaijan	4.171	3.995	3.777	3.621	3.190	-	2,210	2,457	-	574	-	-
Bahrain	-	-	1.280	-	-	-	-	-	12,544	-	-	-
Bangladesh	2.854	4.602	3.043	-	-	-	949	1,259	1,190	-	-	-
Canada	16	13	14	12	11	12	28,175	30,549	30,198	30,747	27,893	28,903
Cape Verde	0.276	0.273	0.268	0.230	0.188	-	-	-	4,203	5,304	-	-
Chile	1.727	1.555	5.027	4.684	4.075	-	5,236	5,482	6,885	5,762	8,204	-
China	-	-	-	-	*93	-	-	-	-	-	-	-
China (Hong Kong SAR)	*8.200	*7.500	*8.200	*7.800	*7.000	-	13,213	14,326	14,983	18,032	18,762	-
Colombia	1.425	1.097	1.327	0.730	0.603	0.540	2,706	2,997	2,393	2,746	3,249	3,782
Costa Rica	0.251	0.239	0.228	0.218	-	-	2,684	2,395	2,812	2,947	-	-
Cyprus	0.170	0.144	0.137	0.105	0.127	0.106	11,263	12,329	13,236	12,801	13,874	17,065
Denmark	17	17	16	-	-	-	31,203	32,667	37,616	-	-	-
Ecuador	0.093	0.077	0.121	0.080	0.097	0.059	2,031	1,887	2,160	1,941	1,023	1,018
Egypt	*15	*13	*13	*13	-	-	1,963	1,402	1,717	1,889	-	-
El Salvador	-	-	-	*0.007	-	*0.011	-	-	-	1,697	-	2,004
Fiji	0.247	0.200	0.176	-	-	-	2,892	4,587	5,810	-	-	-
Finland	12	11	11	11	11	-	33,058	30,749	29,708	21,813	26,347	-
Germany	-	*62	*50	*44	*41	-	-	23,759	30,572	30,909	33,724	-
Germany (Western Part)	*34	*33	*33	-	-	-	32,856	33,884	37,806	-	-	-
Greece	*12	*12	*11	-	-	-	16,545	17,353	18,370	-	-	-
Honduras	*0.115	*0.115	*0.074	*0.081	*0.089	*0.129	2,823	1,766	7,821	900	732	609
Iceland	0.634	0.599	0.546	0.538	0.360	-	26,418	31,659	-	32,350	-	-
India	*18	*21	*28	*31	-	-	1,705	1,353	1,135	1,207	-	-
Indonesia	16	14	17	20	17	18	1,120	1,450	1,662	1,820	1,377	2,952
Iran (Islamic Republic of)	-	2.611	2.487	0.855	-	-	-	4,090	3,459	3,608	-	-
Italy	25	25	-	-	-	-	33,460	36,516	-	-	-	-
Japan	*76	*77	*79	*79	-	-	31,625	35,960	-	-	-	-
Jordan	-	-	-	-	0.019	0.016	-	-	-	-	2,033	1,963
Kenya	0.468	0.506	0.506	0.499	0.521	0.547	3,119	3,019	2,949	1,792	2,237	3,189
Korea, Republic of	53	58	54	59	68	78	16,650	16,562	20,757	21,901	23,567	26,390
Kuwait	1.524	2.440	2.402	2.541	0.736	1.957	2,737	1,951	13,925	12,996	13,655	18,181
Kyrgyzstan	*0.091	*0.086	*0.065	*0.043	*0.033	-	585	447	-	-	-	-
Lithuania	-	-	7.669	6.083	-	-	-	-	-	577	-	-
Macau	0.510	0.473	0.413	0.293	0.274	-	6,903	8,237	8,799	11,380	12,822	-
Malaysia	3.500	5.400	6.800	7.400	7.600	6.157	4,616	4,990	6,125	6,400	6,322	7,340
Malta	2.378	2.308	2.238	2.052	2.009	-	10,338	11,688	11,643	10,916	12,767	-
Mauritius	*0.115	0.095	0.097	0.100	0.097	0.096	3,570	4,696	5,953	1,133	6,027	-
Mexico	2.566	2.688	1.597	1.017	1.215	0.477	4,669	5,142	7,489	9,063	8,498	4,020
Mozambique	-	-	-	-	0.277	-	-	-	-	-	1,227	-
Netherlands	15	15	16	14	-	-	28,945	30,232	32,609	31,314	-	-
New Zealand	1.666	1.717	1.701	1.932	-	-	-	-	-	-	-	-
Norway	15	15	14	-	-	-	29,258	30,388	38,706	-	-	-
Pakistan	-	4.752	-	-	-	-	-	2,205	-	-	-	-
Panama	*0.150	*0.111	-	-	-	-	8,033	9,595	-	-	-	-
Peru	*2.846	*2.636	*2.415	-	-	-	1,500	3,761	3,574	-	-	-

Continued.

Depending on the table, * means *Enterprises*, *Engaged*, or *Factor Values*; ± means *Basis Unspecified*; § means *shown in millions*. For additional notes and sources, see Appendix I.

669

Employment and Compensation of Employees
- Continued -

Country	Number Employed/Engaged (000)						Wage/Salary per Employee ($)					
	1990	1991	1992	1993	1994	1995	1990	1991	1992	1993	1994	1995
Philippines	5.700	5.200	5.200	3.900	-	-	2,064	1,806	2,503	2,484	-	-
Portugal	*12	*12	*12	*11	*7.463	-	9,263	10,070	12,287	10,192	10,112	-
Senegal	*0.741	-	-	-	*1.213	-	9,130	-	-	-	2,887	-
Singapore	*18	*21	*21	*22	*24	-	12,692	13,235	14,898	16,307	16,859	-
Spain	33	30	28	-	-	-	21,008	22,461	24,554		-	-
Sri Lanka	*2.391	*2.336	*2.392	*2.784	-	-	1,420	1,654	1,674	1,169	-	-
Sweden	5.300	*6.158	*5.948	4.928	4.748	-	26,012	28,720	32,047	24,589	26,239	-
Thailand	1.359	1.939	-	-	-	-	1,913	2,247	-	-	-	-
Trinidad and Tobago	0.070	0.080	0.079	0.079	0.078	-	3,599	5,596	7,061	5,441	5,193	-
Tunisia	-	-	-	1.376	1.498	1.458	-	-	-	-	-	-
Turkey	5.614	5.183	5.152	4.641	4.167	-	8,467	11,793	11,664	13,103	9,191	-
Ukraine	-	-	136	129	118	97	-	-	2,302	441	602	672
United Kingdom	64	58	53	-	-	-	24,777	26,335	27,302		-	-
United Republic of Tanzania	0.045	0.059	-	-	-	-	-	-	-	-	-	-
United States of America	175	162	163	158	150	143	26,686	27,778	28,405	27,797	29,387	30,210
Uruguay	*1.210	*1.510	*1.104	*0.972	-	-	-	-	-	-	-	-
Venezuela	0.900	0.900	1.000	0.831	0.616	1.053	3,222	3,931	7,634	3,882	2,339	3,985

Female Workers: Total and Percent of Employment

Country	Female Workers (000)						Female Workers as % of Total Employment					
	1990	1991	1992	1993	1994	1995	1990	1991	1992	1993	1994	1995
Australia	0.878	-	-	-	-	-	8.78	-	-	-	-	-
Austria	0.100	0.100	0.100	0.031	-	-	7.14	10.00	10.00	8.24	-	-
Bahrain	-	-	0.023	-	-	-	-	-	1.80	-	-	-
Bangladesh	0.010	0.062	0.019	-		-	0.35	1.35	0.62	-	-	-
Canada	2.000	1.000	-	-	-	-	12.50	7.69	-	-	-	-
Cape Verde	-	-	0.014	0.014	0.014	-	-	-	5.22	6.09	7.45	-
Chile	0.073	0.070	0.283	0.269	0.239	-	4.23	4.50	5.63	5.74	5.87	-
Colombia	-	0.089	-	0.061	0.059	-	-	8.11	-	8.36	9.78	-
Cyprus	0.022	0.014	0.010	0.008	0.006	-	12.94	9.72	7.30	7.62	4.72	-
Denmark	1.614	1.591	1.541	-	-	-	9.45	9.60	9.43	-	-	-
Egypt	-	0.568	0.525	0.563	-	-	-	4.54	4.04	4.33	-	-
El Salvador	-	-	-	0.007	-	0.001	-	-	-	100.00	-	9.09
Germany (Western Part)	2.000	-	-	-	-	-	5.97	-	-	-	-	-
Indonesia	-	-	-	1.069	0.964	1.029	-	-	-	5.38	5.67	5.68
Iran (Islamic Republic of)	-	0.044	0.039	0.018	-	-	-	1.69	1.57	2.11	-	-
Italy	-	1.019	-	-	-	-	-	4.08	-	-	-	-
Kenya	0.023	0.031	0.035	0.035	-	-	4.91	6.13	6.92	7.01	-	-
Korea, Republic of	3.700	4.006	3.919	3.930	5.095	5.543	6.94	6.94	7.27	6.67	7.53	7.14
Macau	0.011	0.011	0.009	0.008	0.009	-	2.16	2.33	2.18	2.73	3.28	-
Malaysia	0.200	0.400	0.500	0.500	0.600	-	5.71	7.41	7.35	6.76	7.89	-
Malta	0.036	0.035	0.033	0.028	0.027	-	1.51	1.52	1.47	1.36	1.34	-
New Zealand	0.135	0.146	0.159	0.155	-	-	8.10	8.50	9.35	8.02	-	-
Panama	0.007	-	-	-	-	-	4.67	-	-	-	-	-
Philippines	-	-	0.333	0.270	-	-	-	-	6.40	6.92	-	-
Portugal	0.276	0.375	0.354	0.327	0.228	-	2.40	3.18	3.06	2.99	3.06	-
Senegal	0.042	-	-	-	-	-	5.67	-	-	-	-	-
Sri Lanka	0.584	0.573	0.586	0.743	-	-	24.42	24.53	24.50	26.69	-	-
Sweden	0.500	-	-	-	-	-	9.43	-	-	-	-	-
Thailand	0.111	0.309	-	-	-	-	8.17	15.94	-	-	-	-
Turkey	0.157	-	-	-	-	-	2.80	-	-	-	-	-
United Kingdom	5.398	-	4.000	-	-	-	8.43	-	7.55	-	-	-
United Republic of Tanzania	0.006	0.008	-	-	-	-	13.33	13.56	-	-	-	-
United States of America	-	-	122	119	115	109	-	-	74.85	75.32	76.67	76.22

Depending on the table, * means *Enterprises, Engaged*, or *Factor Values*; ± means *Basis Unspecified*; § means *shown in millions*. For additional notes and sources, see Appendix I.

Output and Output per Employee

Country	Output ($ bil.)						Output per Employee ($)					
	1990	1991	1992	1993	1994	1995	1990	1991	1992	1993	1994	1995
Australia	*1.735	*1.275	*1.270	-	-	-	173,516	141,627	141,095	-	-	-
Austria	0.130	0.117	0.065	0.080			92,537	117,420	64,974	212,867	-	-
Azerbaijan	±0.027	±0.022	±0.001	±0.004	±0.002	-	6,514	5,474	298	1,238	745	-
Bahrain	-	-	±0.062	-	-	-	-	-	48,778	-	-	-
Bangladesh	0.045	0.026	0.009	-	-	-	15,775	5,587	3,088	-	-	-
Canada	1.697	1.519	1.150	1.132	0.952	1.295	106,059	116,825	82,143	94,308	86,542	107,938
Cape Verde	-	-	±0.007	±0.004	±0.003	-	-	-	24,332	17,385	14,604	-
Chile	0.077	0.065	0.133	0.148	0.141		44,608	41,769	26,467	31,493	34,497	-
China (Hong Kong SAR)	0.285	0.293	0.352	0.397	0.394		34,739	39,084	42,947	50,913	56,341	-
Colombia	0.062	0.053	0.018	0.007	0.009	0.021	43,313	48,330	13,507	9,581	15,541	39,778
Costa Rica	0.004	0.003	0.003	0.003	0.002	0.003	17,136	11,896	12,945	13,051	-	-
Cyprus	0.005	0.006	0.007	0.005	0.006	0.005	31,561	40,587	53,479	50,034	49,903	50,716
Denmark	*1.820	*1.975				-	106,510	119,154				
Ecuador	0.001	0.001	0.001	0.001	0.001	0.001	10,700	7,771	6,821	15,385	6,641	8,790
Egypt	*0.147	*0.062	*0.134	*0.088			9,869	4,958	10,331	6,796		
El Salvador	-	-	-	±0.000		±0.000	-	-	-	11,056		8,099
Fiji	0.004	0.003	0.003	-	-	-	14,748	17,449	17,299	-	-	-
Finland	±1.733	±1.460	±1.690	±1.450	±1.782		149,374	130,396	155,063	136,762	159,102	-
France	±4.351	±4.238	±4.730	±4.118	±3.556	±4.131	-	-	-	-	-	-
Germany	-	5.564	5.923	5.205	5.570		-	89,841	118,853	117,597	136,886	
Germany (Western Part)	4.994	4.709	4.892	-	-		149,000	141,207	150,273	-	-	-
Greece	*0.372	*0.441	*0.326				32,054	38,217	29,002	-	-	-
Honduras	0.001	0.001	0.001	0.000	0.000	0.000	6,357	5,095	8,148	2,201	1,877	1,380
Iceland	0.034	0.038	±0.044	±0.038			54,408	63,899	80,448	70,109		
India	*0.171	*0.196	*0.397	*0.411			9,395	9,188	14,449	13,414		
Indonesia	0.125	0.151	0.390	0.344	0.187	0.367	8,014	10,545	22,686	17,329	11,014	20,219
Iran (Islamic Republic of)	-	0.023	0.023	0.009			-	8,982	9,207	10,512		
Italy	±2.567	±3.135	-	-			101,103	125,451	-	-		
Japan	±16	±18	±22	±26			209,650	238,415	276,052	333,189		
Jordan	-	-	-	-	0.000	0.000	-	-	-	-	10,090	6,513
Korea, Republic of	5.195	6.329	7.791	7.737	9.762	12	97,459	109,668	144,478	131,255	144,351	160,945
Kuwait	0.006	0.006	0.056	0.068	0.038	0.074	3,942	2,622	23,324	26,933	51,339	37,851
Kyrgyzstan	0.001	0.001	0.000	0.000	0.000		10,243	5,963	1,977	297	412	-
Lithuania			0.010	0.021					1,300	3,411		
Macau	0.008	0.010	0.009	0.006	0.007	-	14,886	20,566	21,602	21,297	24,184	-
Malaysia	*0.100	*0.293	*0.253	*0.314	*0.403	*0.333	28,646	54,193	37,201	42,481	52,987	54,125
Malta	0.060	0.051	0.050	0.046	0.044	-	25,332	22,183	22,372	22,531	22,011	-
Mauritius	0.003	0.004	0.003	0.000	0.003		26,327	45,126	30,604	3,570	33,006	-
Mexico	0.015	0.016	0.010	0.012	0.015	0.014	5,751	5,813	6,237	11,648	12,305	29,169
Netherlands	2.233	2.144	2.392	1.937	-	-	150,307	141,261	154,281	138,245	-	-
Norway	±2.303	±2.630	±2.441	-	-	-	154,636	174,555	170,190	-	-	-
Pakistan	-	0.033					-	7,013	-	-	-	-
Panama	0.006	0.004					39,460	34,009				
Peru	0.159	0.074	0.055				55,936	28,229	22,961			
Philippines	0.051	0.042	0.064	0.114			8,869	8,062	12,227	29,269		
Portugal	0.445	0.531	0.468	0.347	0.300	-	38,618	45,054	40,451	31,703	40,233	-
Romania	±0.437	±0.224	±0.217	±0.142			-	-	-	-		
Senegal	0.017	-	-	-			23,282	-	-	-		
Singapore	*1.433	*1.620	*1.651	*1.637	*1.989		78,770	77,024	77,227	75,432	83,729	-
Spain	*2.598	*2.663	*2.744	-			79,697	87,601	97,271	-	-	-
Sri Lanka	0.034	0.033	0.035	0.030	-		14,145	14,109	14,687	10,723	-	-
Sweden	*0.566	*0.980	*0.993	*0.464	*0.572		106,788	159,197	166,895	94,138	120,564	-
Thailand	0.010	0.023	-	-			7,046	11,642				
Trinidad and Tobago	±0.004	±0.005	±0.005	±0.007	±0.007		61,269	59,240	61,392	89,181	85,255	-
Turkey	0.139	0.154	0.160	0.160	0.116		24,787	29,738	31,096	34,406	27,898	-
Ukraine	-	-	*1.292	*0.361	*0.399	*0.349	-	-	9,498	2,795	3,385	3,595
United Kingdom	*4.405	*4.189	4.345	-			68,834	72,231	81,972	-	-	-
United Republic of Tanzania	*0.003	*0.000	-	-			61,520	155	-	-	-	-
United States of America	*16	*15	*15	*15	*15	*15	90,571	89,630	93,448	94,380	101,927	106,476
Venezuela	0.014	0.018	0.026	0.017	0.009	-	15,541	20,084	26,340	20,987	14,506	-

Depending on the table, * means *Enterprises*, *Engaged*, or *Factor Values*; ± means *Basis Unspecified*; § means *shown in millions*. For additional notes and sources, see Appendix I.

671

Value Added and Value Added per Employee

Country	Value Added ($ bil.)						Value Added per Employee ($)					
	1990	1991	1992	1993	1994	1995	1990	1991	1992	1993	1994	1995
Australia	*0.793	*0.673	-	-	-	-	79,297	74,795	-	-	-	-
Austria	0.046	0.041	0.024	0.041	-	-	32,982	41,196	24,206	107,919	-	-
Bangladesh	0.008	0.015	0.003	-	-	-	2,691	3,343	1,122	-	-	-
Canada	0.694	0.672	0.612	0.713	0.571	0.645	43,388	51,698	43,731	59,427	51,925	53,760
Chile	0.035	0.024	0.081	0.075	0.081	-	20,208	15,529	16,162	16,049	19,969	-
China (Hong Kong SAR)	0.155	0.152	0.178	0.202	0.185	-	18,865	20,280	21,663	25,871	26,396	-
Colombia	0.040	0.035	0.008	0.003	0.004	0.013	27,944	31,643	6,310	4,288	7,377	23,936
Costa Rica	0.002	0.001	0.001	0.001	0.001	0.001	7,752	5,475	5,977	5,872	-	-
Cyprus	0.003	0.003	0.003	0.003	0.004	0.003	18,638	22,048	23,131	25,256	29,317	26,182
Denmark	*0.790	*0.859	-	-	-	-	46,223	51,830	-	-	-	-
Ecuador	0.000	0.000	0.000	0.000	0.000	0.000	2,717	1,887	2,715	3,530	2,652	3,800
Egypt	*0.082	*0.014	*0.052	*0.034	-	-	5,500	1,107	4,004	2,621	-	-
El Salvador	-	-	-	0.000	-	-	-	-	-	3,922	-	-
Fiji	0.001	0.001	0.001	-	-	-	4,439	6,179	5,126	-	-	-
Finland	±0.605	±0.522	±0.613	±0.539	±0.538	-	52,143	46,608	56,200	50,817	48,066	-
France	±1.032	±0.952	±1.071	±0.896	±0.573	±0.601	-	-	-	-	-	-
Germany (Western Part)	1.789	1.975	2.141	2.056			53,368	59,218	65,763			
Greece	*0.223	*0.267	*0.203	-	-	-	19,219	23,139	18,018	-	-	-
Honduras	*0.000	*0.000	*0.000	*0.000	*0.000	*0.000	3,565	2,211	3,594	1,007	798	664
Iceland	0.020	0.022	±0.025	±0.020			31,108	36,249	45,878	36,378	-	-
India	*0.043	*0.042	*0.057	*0.077	-	-	2,369	1,957	2,090	2,521	-	-
Indonesia	*0.061	*0.067	*0.254	*0.251	*0.119	*0.273	3,908	4,701	14,731	12,616	7,005	15,072
Iran (Islamic Republic of)	-	0.016	0.015	0.007	-	-	-	5,988	6,138	7,884	-	-
Italy	±0.750	±0.758	-	-	-	-	29,516	30,322	-	-	-	-
Japan	±5.843	±6.681	±8.496	±10	-	-	76,881	86,766	107,542	132,274	-	-
Jordan	-	-	-	-	0.000	0.000	-	-	-	-	5,045	4,729
Korea, Republic of	1.999	3.054	4.003	3.461	4.737	5.561	37,499	52,910	74,220	58,710	70,045	71,606
Kuwait	-0.001	0.005	0.046	0.047	0.019	0.055	-752.795	1,978	19,323	18,598	25,619	28,041
Macau	0.004	0.005	0.004	0.004	0.003	-	7,103	10,490	10,793	12,375	12,706	-
Malaysia	*0.022	*0.118	*0.102	*0.150	*0.139	*0.133	6,158	21,817	14,934	20,269	18,316	21,682
Malta	0.021	0.010	0.008	0.001	0.007	-	8,945	4,246	3,483	717	3,591	-
Mauritius	0.002	0.004	0.002	0.000	0.002	-	21,647	37,459	21,595	1,473	22,559	-
Mexico	0.016	0.016	0.011	0.013	0.014	-	6,287	6,001	7,161	12,542	11,909	-
Netherlands	0.604	0.608	0.682	0.599			40,663	40,063	43,981	42,726	-	-
Norway	±0.626	±0.657	±0.678		-	-	42,011	43,616	47,276	-	-	-
Pakistan	-	0.013	-	-	-	-	-	2,791	-	-	-	-
Panama	0.005	0.002	-	-	-	-	31,140	22,315	-	-	-	-
Peru	0.093	0.038	0.024	-	-	-	32,575	14,311	10,070	-	-	-
Philippines	0.025	0.023	0.039	0.059	-	-	4,366	4,493	7,568	15,212	-	-
Portugal	0.194	0.225	0.162	0.145	0.117	-	16,865	19,043	14,048	13,294	15,610	-
Romania	±0.160	±0.068	±0.070	±0.046	-	-	-	-	-	-	-	-
Senegal	0.010				0.004	-	13,795	-	-	-	3,639	-
Singapore	*0.582	*0.657	*0.710	*0.724	*0.840	-	31,993	31,251	33,230	33,354	35,368	-
Spain	*1.143	*1.157	*1.238	-	-	-	35,066	38,048	43,899	-	-	-
Sri Lanka	0.017	0.017	0.018	0.012	-	-	7,224	7,259	7,391	4,421	-	-
Sweden	*0.296	*0.256	*0.261	*0.185	*0.219	-	55,785	41,528	43,806	37,590	46,066	-
Thailand	0.004	0.007	-	-	-	-	3,077	3,678	-	-	-	-
Trinidad and Tobago	±0.001	±0.001	±0.000	±0.002	±0.001	-	7,377	9,805	2,783	19,871	10,170	-
Turkey	0.074	0.096	0.119	0.122	0.085	-	13,247	18,592	23,131	26,344	20,384	-
United Kingdom	*2.482	*2.154	*2.260	-	-	-	38,783	37,138	42,636	-	-	-
United Republic of Tanzania	*0.003	*-0.000	-	-	-	-	61,064	-773.369	-	-	-	-
United States of America	*8.560	*7.960	*8.545	*8.023	*8.033	*8.159	48,914	49,136	52,423	50,778	53,553	57,056
Venezuela	0.006	0.008	0.013	0.006	0.006	0.024	6,515	8,742	13,397	7,314	9,084	22,839

　　Depending on the table, * means *Enterprises, Engaged,* or *Factor Values;* ± means *Basis Unspecified;* § means *shown in millions.* For additional notes and sources, see Appendix I.

Capital Investment

Country	Gross Fixed Capital Formation ($ mil.)						Per Establishment ($ mil)					
	1990	1991	1992	1993	1994	1995	1990	1991	1992	1993	1994	1995
Austria	4.749	3.340	1.274	0.602	-	-	0.950	0.668	0.255	0.120	-	-
Bahrain	-	-	2.418	-	-	-	-	-	0.604	-	-	-
Bangladesh	-	4.372	0.282	-	-	-	-	0.219	0.006	-	-	-
China (Hong Kong SAR)	-	-	8.268	101	37	-	-	-	0.020	0.247	0.077	-
Colombia	-	4.047	25	0.204	0.064	-	-	0.238	1.647	0.026	0.007	-
Cyprus	0.085	0.058	0.213	0.022	0.189	0.066	0.004	0.002	0.009	0.001	0.005	0.002
Denmark	57	72	-	-	-	-	0.109	0.139	-	-	-	-
Ecuador	-	0.156	0.357	0.334	-0.051	0.034	-	0.052	0.119	0.084	-0.013	0.009
El Salvador	-	-	-	0.001	-	0.010	-	-	-	0.001	-	0.010
Fiji	-0.192	0.129	0.090	-	-	-	-0.017	0.010	0.007	-	-	-
Finland	58	30	38	85	125	-	0.681	0.304	0.499	1.164	1.923	-
Germany (Western Part)	206	160	163	231	-	-	2.103	1.687	1.719	-	-	-
Greece	8.233	15	22	-	-	-	0.063	0.118	0.172	-	-	-
Hungary	-	-	0.025	-	-	-	-	-	0.001	-	-	-
India	5.313	7.475	9.607	6.526	-	-	0.035	0.040	0.047	0.027	-	-
Indonesia	24	19	81	67	12	309	0.173	0.134	0.487	0.430	0.075	1.966
Iran (Islamic Republic of)	-	6.467	1.247	0.113	-	-	-	0.462	0.083	0.011	-	-
Italy	157	96	-	-	-	-	1.105	0.653	-	-	-	-
Japan	428	631	916	872	-	-	0.193	0.288	0.426	0.403	-	-
Korea, Republic of	-	-	907	2,322	1,740	1,977	-	-	1.659	3.733	2.570	2.462
Kuwait	0.007	31	6.543	3.066	0.758	-	0.001	4.444	1.090	0.438	0.095	-
Macau	-	-	-	-0.039	0.018	-	-	-	-	-0.001	0.000	-
Malaysia	-	-	-	48	81	-	-	-	-	0.576	0.966	-
Malta	1.136	1.641	1.566	0.691	0.743	-	0.047	0.068	0.071	0.033	0.032	-
Mexico	0.734	-	-	-	-	-	0.105	-	-	-	-	-
Netherlands	68	70	85	46	-	-	0.483	0.500	0.569	0.324	-	-
Norway	-	-	61	-	-	-	-	-	0.277	-	-	-
Pakistan	-	0.407	-	-	-	-	-	0.102	-	-	-	-
Panama	0.054	0.023	-	-	-	-	0.007	0.004	-	-	-	-
Peru	1.385	0.848	0.568	-	-	-	0.017	0.010	0.007	-	-	-
Philippines	3.867	9.134	7.408	50	-	-	0.036	0.088	0.151	1.088	-	-
Portugal	27	21	7.785	-699	11	-	0.105	0.080	0.026	-2.369	0.041	-
Senegal	0.389	-	-	-	-	-	0.065	-	-	-	-	-
Singapore	93	76	75	67	175	-	0.571	0.426	0.405	0.302	0.784	-
Spain	56	71	79	-	-	-	0.078	0.103	0.128	-	-	-
Sri Lanka	0.624	1.018	1.338	8.430	-	-	0.062	0.127	0.191	1.405	-	-
Thailand	7.063	2.265	-	-	-	-	0.262	0.063	-	-	-	-
Trinidad and Tobago	-	-	0.047	1.439	0.169	-	-	-	0.005	0.131	0.015	-
Turkey	4.600	4.075	2.765	7.738	14	-	0.242	0.214	0.075	0.221	0.388	-
Ukraine	-	-	15	3.351	2.957	4.002	-	-	0.141	0.046	0.049	0.074
United Kingdom	-	86	80	-	-	-	-	0.063	0.055	-	-	-
United Republic of Tanzania	0.226	-	-	-	-	-	0.075	-	-	-	-	-
United States of America	-	-	192	238	240	320	-	-	0.043	-	-	-

Indexes of Industrial Production

Country	Index of Industrial Production (1990=100)						Index of Employment (1990=100)					
	1990	1991	1992	1993	1994	1995	1990	1991	1992	1993	1994	1995
Australia	-	-	-	-	-	-	100	90	90	-	-	-
Austria	-	-	-	-	-	-	100	71	71	27	-	-
Azerbaijan	-	-	-	-	-	-	100	96	91	87	76	-
Bangladesh	-	-	-	-	-	-	100	161	107	-	-	-
Canada	-	-	-	-	-	-	100	81	88	75	69	75
Cape Verde	-	-	-	-	-	-	100	99	97	83	68	-
Chile	-	-	-	-	-	-	100	90	291	271	236	-
China (Hong Kong SAR)	-	-	-	-	-	-	100	91	100	95	85	-
Colombia	-	-	-	-	-	-	100	77	93	51	42	38
Costa Rica	-	-	-	-	-	-	100	95	91	87	-	-

Continued.

Depending on the table, * means *Enterprises*, *Engaged*, or *Factor Values*; ± means *Basis Unspecified*; § means *shown in millions*. For additional notes and sources, see Appendix I.

673

Indexes of Industrial Production

- Continued -

Country	Index of Industrial Production (1990=100)						Index of Employment (1990=100)					
	1990	1991	1992	1993	1994	1995	1990	1991	1992	1993	1994	1995
Cyprus	-	-	-	-	-	-	100	85	81	62	75	62
Denmark	-	-	-	-	-	-	100	97	96	-	-	-
Ecuador	-	-	-	-	-	-	100	83	130	86	104	63
Egypt	-	-	-	-	-	-	100	84	87	87	-	-
Fiji	-	-	-	-	-	-	100	81	71	-	-	-
Finland	-	-	-	-	-	-	100	97	94	91	97	-
Germany (Western Part)	-	-	-	-	-	-	100	99	97	-	-	-
Greece	-	-	-	-	-	-	100	99	97	-	-	-
Honduras	-	-	-	-	-	-	100	100	64	70	77	112
Iceland	-	-	-	-	-	-	100	94	86	85	57	-
India	-	-	-	-	-	-	100	117	151	169	-	-
Indonesia	-	-	-	-	-	-	100	91	110	127	109	116
Italy	-	-	-	-	-	-	100	98	-	-	-	-
Japan	-	-	-	-	-	-	100	101	104	104	-	-
Kenya	-	-	-	-	-	-	100	108	108	107	111	117
Korea, Republic of	-	-	-	-	-	-	100	108	101	111	127	146
Kuwait	-	-	-	-	-	-	100	160	158	167	48	128
Kyrgyzstan	-	-	-	-	-	-	100	95	71	47	36	-
Macau	-	-	-	-	-	-	100	93	81	57	54	-
Malaysia	-	-	-	-	-	-	100	154	194	211	217	176
Malta	-	-	-	-	-	-	100	97	94	86	84	-
Mauritius	-	-	-	-	-	-	100	83	84	87	84	83
Mexico	-	-	-	-	-	-	100	105	62	40	47	19
Netherlands	-	-	-	-	-	-	100	102	104	94	-	-
New Zealand	-	-	-	-	-	-	100	103	102	116	-	-
Norway	-	-	-	-	-	-	100	101	96	-	-	-
Panama	-	-	-	-	-	-	100	74	-	-	-	-
Peru	-	-	-	-	-	-	100	93	85	-	-	-
Philippines	-	-	-	-	-	-	100	91	91	68	-	-
Portugal	-	-	-	-	-	-	100	102	100	95	65	-
Senegal	-	-	-	-	-	-	100	-	-	-	164	
Singapore	-	-	-	-	-	-	100	116	117	119	131	
Spain	-	-	-	-	-	-	100	93	87	-	-	
Sri Lanka	-	-	-	-	-	-	100	98	100	116	-	
Sweden	-	-	-	-	-	-	100	116	112	93	90	
Thailand	-	-	-	-	-	-	100	143	-	-	-	
Trinidad and Tobago	-	-	-	-	-	-	100	114	113	113	111	
Turkey	-	-	-	-	-	-	100	92	92	83	74	-
United Kingdom	-	-	-	-	-	-	100	91	83	-	-	
United Republic of Tanzania	-	-	-	-	-	-	100	131	-	-	-	
United States of America	-	-	-	-	-	-	100	93	93	90	86	82
Uruguay	-	-	-	-	-	-	100	125	91	80	-	-
Venezuela	-	-	-	-	-	-	100	100	111	92	68	117

    Depending on the table, * means *Enterprises, Engaged,* or *Factor Values;* ± means *Basis Unspecified;* § means *shown in millions.* For additional notes and sources, see Appendix I.

ISIC 3843
MOTOR VEHICLES

ISIC 3843—Motor Vehicles—includes the building of passenger cars, motor coaches and buses, trucks, road tractors and tractor trailers, and trailers and semi-trailers. This is a subcategory of ISIC 3840—Transportation Equipment; for product and company tables, see that chapter.

The products of this industry are classified, in the U.S. SIC system, under SIC 371—Motor Vehicles and Equipment.

Summary Statistics

		1990 Value	N	1991 Value	N	1992 Value	N	1993 Value	N	1994 Value	N	1995 Value	N
Establishments or enterprises	(number)	26,382	51	29,613	55	28,269	53	25,562	47	9,813	38	7,927	17
Number employed	(000)	3,409	54	4,373	56	4,109	54	3,671	49	2,326	37	1,450	22
Total Wages	($ mil.)	97,182	51	134,276	54	137,605	53	79,766	45	43,000	35	12,722	20
Wage/salary per employee	($)	9,534	50	10,284	53	10,710	52	8,189	45	9,264	35	8,780	20
Female workers	(000)	253	24	101	21	69	18	49	19	47	11	34	4
as % of total employment*	(%)	13	24	13	21	11	18	12	19	13	11	8	4
Output	($ bil.)	743	52	957	54	985	52	718	45	372	34	236	21
per employee	($ 000)	79	49	85	51	90	49	78	43	87	32	109	19
per establishment	($ 000)	16,524	48	20,714	51	22,054	47	16,793	41	20,272	32	13,544	14
Value added	($ bil.)	234	49	246	48	243	42	218	39	64	30	61	18
per employee	($ 000)	26	47	28	46	29	40	25	36	27	28	36	16
per establishment	($ 000)	5,175	46	5,786	45	6,121	39	4,222	35	4,846	28	2,494	11
Capital investment	($ mil.)	28,344	27	39,631	32	39,562	30	21,747	24	5,599	17	5,318	6
per employee	($ 000)	7	27	8	32	9	29	10	23	11	17	6	6
per establishment	($ 000)	5,019	27	4,579	32	4,773	30	5,521	23	6,111	17	487	6

Data presented above are drawn from the detailed tables that follow. Columns headed 'N' show the number of countries that provided valid data for inclusion. Values are not strictly comparable one year to the next or one row to the next because the number of countries that report varies from period to period and row to row. However, a general indication of magnitudes can be gleaned from reviewing this summary—especially in earlier years in which more reports are available. For detailed explanations, see Appendix I. *The average for those countries reporting both total and female employment.

Establishments and Number Engaged

Country	Establishments or Enterprises (number)						Employees per Establishment					
	1990	1991	1992	1993	1994	1995	1990	1991	1992	1993	1994	1995
Albania	-	-	-	1.000	-	-	-	-	-	69	-	-
Australia	995	1,226	1,254	-	-	-	68	46	41	-	-	-
Austria	94	94	96	87	90	-	251	250	240	253	252	-
Azerbaijan	1.000	1.000	1.000	1.000	1.000	-	1,527	1,322	1,072	1,118	1,175	-
Bahrain	-	-	2.000	-	-	-	-	-	8.000	-	-	-
Bangladesh	114	107	84	-	-	-	28	33	36	-	-	-
Bolivia	33	39	40	50	48	48	15	13	13	14	14	14
Bosnia & Herzegovina	26	28	-	-	-	-	618	571	-	-	-	-
Canada	854	800	758	748	747	-	156	154	168	174	181	-
Central African Republic	-	*1.000	*1.000	-	-	-	-	-	-	-	-	-
Chile	25	27	27	26	28	-	145	146	157	163	173	-
China (Hong Kong SAR)	121	138	230	127	85	-	9.091	7.246	6.957	10	13	-
Colombia	182	186	239	227	202	-	75	71	70	81	82	-
Costa Rica	54	75	68	66	-	-	9.222	6.093	6.926	8.924	-	-
Cyprus	52	52	52	52	55	58	4.981	5.115	4.558	4.538	5.564	6.017
Denmark	230	228	212	-	-	-	27	26	28	-	-	-
Ecuador	33	38	38	41	51	54	61	67	67	79	75	53
Egypt	95	113	115	101	-	-	433	311	362	289	-	-
El Salvador	-	-	-	3.000	-	14	-	-	-	7.000	-	11
Fiji	3.000	6.000	5.000	-	-	-	24	17	20	-	-	-
Finland	129	130	105	93	83	*29	69	61	65	70	78	-
FYR Macedonia	-	-	-	-	-	*29	-	-	-	-	-	133
Germany	-	944	968	996	991	-	-	908	835	735	691	-
Germany (Western Part)	795	828	846	-	-	-	1,003	974	927	-	-	-
Ghana	-	-	-	3.000	-	-	-	-	-	96	-	-
Greece	57	56	57	-	-	-	53	49	50	-	-	-
Grenada	-	-	-	-	1.000	-	-	-	-	-	10	-
Guatemala	-	5.000	5.000	5.000	5.000	5.000	-	115	115	114	116	132
Honduras	-	-	3.000	3.000	3.000	3.000	-	-	47	51	56	65
Hungary	-	-	*179	*208	-	-	-	-	-	-	-	-
India	2,832	3,051	3,379	3,471	-	-	106	102	101	97	-	-
Indonesia	196	204	220	235	241	259	194	252	200	186	223	236
Iran (Islamic Republic of)	-	75	73	72	-	-	-	379	477	365	-	-
Italy	*568	*582	-	-	-	-	363	346	-	-	-	-
Japan	11,184	11,201	10,997	11,098	-	-	71	74	74	72	-	-
Jordan	17	20	21	22	32	31	5.765	6.800	8.476	34	33	34
Kenya	17	54	54	52	61	61	211	65	64	67	54	54
Korea, Republic of	2,064	2,237	2,338	2,534	2,693	2,928	85	76	75	73	75	73
Kuwait	13	9.000	8.000	7.000	7.000	-	39	17	27	32	39	-
Kyrgyzstan	3.000	3.000	3.000	3.000	3.000	-	1,659	1,530	1,436	1,311	713	-
Malaysia	118	137	144	171	180	-	135	142	131	130	147	-
Malta	8.000	9.000	8.000	10	9.000	-	8.500	9.889	9.375	8.800	10	-
Mauritius	3.000	6.000	6.000	6.000	5.000	7.000	118	86	86	81	83	82
Mexico	123	123	122	121	118	116	847	880	856	723	666	645
Mozambique	-	-	2.000	2.000	1.000	-	-	-	111	103	112	-
Myanmar	2.000	2.000	2.000	2.000	2.000	-	1,458	1,385	1,303	1,264	1,113	-
New Zealand	*276	*288	*317	*320	-	-	19	14	13	12	-	-
Pakistan	-	71	-	-	-	-	-	115	-	-	-	-
Panama	8.000	7.000	-	-	-	-	13	16	-	-	-	-
Peru	361	362	374	-	-	-	19	15	12	-	-	-
Philippines	228	254	191	185	-	-	36	33	87	95	-	-
Portugal	*419	*407	*389	*418	*464	-	59	48	49	45	45	-
Russian Federation	-	-	-	*2,636	*2,267	*3,609	-	-	-	345	359	216
Senegal	9.000	-	-	-	4.000	-	40	-	-	-	73	-
Singapore	18	19	19	19	21	-	79	83	82	81	63	-
Slovakia	-	19	19	24	27	-	-	1,133	938	598	488	-
South Africa	-	974	-	-	-	-	-	-	-	-	-	-
Spain	1,014	1,081	1,080	-	-	-	144	135	133	-	-	-
Sri Lanka	21	22	16	19	-	-	23	45	62	60	-	-
Sweden	277	286	256	236	223	-	284	260	271	248	290	-

Continued.

Depending on the table, * means *Enterprises, Engaged,* or *Factor Values*; ± means *Basis Unspecified*; § means *shown in millions.* For additional notes and sources, see Appendix I.

Establishments and Number Engaged

- Continued -

Country	Establishments or Enterprises (number)						Employees per Establishment					
	1990	1991	1992	1993	1994	1995	1990	1991	1992	1993	1994	1995
Thailand	157	447	-	-	-	-	200	153	-	-	-	-
Trinidad and Tobago	16	-	-	-	-	-	32	-	-	-	-	-
Tunisia	-	-	-	422	450	476	-	-	-	11	12	13
Turkey	179	184	352	349	327	-	256	257	144	163	154	-
Ukraine	-	-	164	41	45	46	-	-	561	2,244	1,889	1,652
United Kingdom	2,067	2,050	2,066	-	-	-	129	120	113	-	-	-
United Republic of Tanzania	32	33	-	-	-	-	82	77	-	-	-	-
Venezuela	208	229	220	206	195	183	73	77	87	91	90	95
Zambia	7.000	-	-	-	7.000	-	116	-	-	-	-	-
Zimbabwe	44	45	44	43	41	-	121	126	116	112	120	-

Employment and Compensation of Employees

Country	Number Employed/Engaged (000)						Wage/Salary per Employee ($)					
	1990	1991	1992	1993	1994	1995	1990	1991	1992	1993	1994	1995
Albania	-	-	-	0.069	-	-	-	-	-	-	-	-
Australia	68	*57	*52	-	-	-	20,796	24,891	20,532	-	-	-
Austria	24	23	23	22	23	23	25,100	28,263	33,001	31,224	33,409	-
Azerbaijan	1.527	1.322	1.072	1.118	1.175	-	1,949	2,176	-	406	-	-
Bahrain	-	-	0.016	-	-	-	-	-	3,324	-	-	-
Bangladesh	3.198	3.486	3.040	-	-	-	986	1,019	980	-	-	-
Bolivia	0.490	0.516	0.528	0.680	0.686	0.693	1,629	1,484	1,469	1,397	1,477	1,424
Bosnia & Herzegovina	16	16	-	-	-	-	-	-	-	-	-	-
Canada	133	123	127	130	135	135	31,724	33,756	33,119	33,355	33,614	35,080
Chile	3.614	3.941	4.236	4.227	4.843	-	5,078	5,063	6,260	6,493	7,482	-
China (Hong Kong SAR)	*1.100	*1.000	*1.600	*1.300	*1.100	-	9,336	9,908	10,658	13,723	15,880	-
Colombia	14	13	17	18	17	18	2,777	2,735	2,683	3,282	4,474	5,092
Costa Rica	0.498	0.457	0.471	0.589	-	-	2,189	1,874	1,963	2,772	-	-
Cyprus	0.259	0.266	0.237	0.236	0.306	0.349	9,716	10,393	11,655	10,461	11,144	14,650
Denmark	6.107	6.041	6.026	-	-	-	28,072	28,236	31,040	-	-	-
Ecuador	2.004	2.556	2.538	3.237	3.814	2.864	3,090	3,774	4,254	4,957	2,246	2,797
Egypt	*41	*35	*42	*29	-	-	1,884	1,329	1,472	1,721	-	-
El Salvador	-	-	-	*0.021	-	*0.153	-	-	-	2,301	-	1,822
Fiji	0.073	0.103	0.102	-	-	-	2,414	2,125	3,001	-	-	-
Finland	8.900	7.900	6.800	6.500	6.500	-	27,476	26,619	25,030	16,841	20,331	-
FYR Macedonia	-	-	-	-	-	3.869	-	-	-	-	-	3,170
Germany	-	*857	*808	*732	*685	-	-	35,384	-	39,454	-	-
Germany (Western Part)	*797	*806	*784	-	-	-	35,492	37,074	-	-	-	-
Ghana	-	-	-	*0.287	-	-	-	-	-	-	-	-
Greece	*3.000	*2.762	*2.838	-	-	-	11,461	12,316	13,869	-	-	-
Grenada	-	-	-	-	0.010	-	-	-	-	-	-	-
Guatemala	-	*0.577	*0.574	*0.569	*0.582	*0.660	-	-	-	-	-	-
Honduras	*0.219	*0.219	*0.142	*0.154	*0.169	*0.194	3,076	2,389	6,244	4,563	2,953	5,131
India	*299	*311	*341	*335	-	-	2,087	1,811	1,688	1,626	-	-
Indonesia	38	51	44	44	54	61	1,056	1,064	1,240	1,362	1,400	2,775
Iran (Islamic Republic of)	-	28	35	26	-	-	-	5,518	5,428	8,013	-	-
Italy	206	201	-	-	-	-	36,409	36,366	-	-	-	-
Japan	*789	*829	*812	*800	-	-	33,990	38,147	-	-	-	-
Jordan	0.098	0.136	0.178	0.741	1.066	1.065	2,551	2,591	2,181	2,498	2,923	2,846
Kenya	3.591	3.496	3.472	3.487	3.265	3.275	2,541	2,242	2,103	1,265	1,767	2,422
Korea, Republic of	175	170	176	185	202	213	12,719	14,198	15,307	16,511	19,364	22,140
Kuwait	0.506	0.152	0.212	0.227	0.273	0.237	4,460	4,239	11,913	11,028	9,207	15,582
Kyrgyzstan	*4.977	*4.591	*4.309	*3.934	*2.140	-	506	-	-	-	-	-
Lithuania	-	-	3.108	2.836	-	-	-	-	447	409	-	-
Malawi	0.587	-	-	-	-	-	-	-	-	-	-	-
Malaysia	16	20	19	22	26	33	3,727	4,067	4,961	4,998	5,289	6,121
Malta	0.068	0.089	0.075	0.088	0.091	-	8,582	7,444	8,134	6,753	8,460	-
Mauritius	*0.353	0.518	0.519	0.488	0.416	0.573	1,715	1,963	2,518	2,357	3,574	-

Continued.

Depending on the table, * means *Enterprises, Engaged,* or *Factor Values;* ± means *Basis Unspecified;* § means *shown in millions.* For additional notes and sources, see Appendix I.

677

Employment and Compensation of Employees

- Continued -

Country	Number Employed/Engaged (000)						Wage/Salary per Employee ($)					
	1990	1991	1992	1993	1994	1995	1990	1991	1992	1993	1994	1995
Mexico	104	108	104	88	79	75	5,700	6,860	8,719	10,028	11,013	6,146
Mozambique	-	-	0.223	0.206	0.112	-	-	-	-	-	491	-
Myanmar	2.916	2.769	2.606	2.529	2.226	-	1,441	1,509	1,592	1,965	2,114	-
New Zealand	5.265	4.015	4.184	3.992	-	-	-	-	-	-	-	-
Pakistan	-	8.157	-	-	-	-	-	3,062	-	-	-	-
Panama	*0.104	*0.114					4,933	4,009	-	-	-	-
Peru	*7.033	*5.356	*4.505	-	-	-	3,226	4,136	5,092	-	-	-
Philippines	8.300	8.400	17	18	-	-	2,785	2,175	2,974	2,925	-	-
Portugal	*25	*19	*19	*19	*21	-	7,322	8,266	10,012	8,831	9,958	-
Russian Federation	-	-	-	910	814	778	-	-	-	847	1,414	1,555
Senegal	*0.360	-	-	-	*0.292		1,224	-	-	-	4,250	
Singapore	*1.421	*1.584	*1.553	*1.539	*1.324	-	12,045	12,795	19,642	14,641	23,811	-
Slovakia	-	*22	*18	*14	*13		-	1,545	1,804	1,998	2,419	
Spain	146	146	144	-	-	-	23,561	24,289	27,038	-	-	-
Sri Lanka	*0.475	*0.980	*0.989	*1.138	-		631	795	807	954	-	
Swaziland	-	0.270	0.270	0.300	-		-	5,240	5,383	4,700	-	
Sweden	79	*74	*70	59	65	-	24,427	27,164	30,388	24,433	25,510	-
Thailand	31	68	-	-	-		3,720	3,973	-	-	-	
Trinidad and Tobago	0.513	-	-	-	-		5,335	-	-	-	-	
Tunisia	-	-	-	4.547	5.265	6.046	-	-	-	-	-	-
Turkey	46	47	51	57	50	-	8,405	10,180	10,097	10,810	6,140	-
Ukraine	-	-	92	92	85	76	-	-	2,975	496	575	569
United Kingdom	267	245	233	-	-	-	25,589	27,502	29,133	-	-	-
United Republic of Tanzania	2.630	2.537	-	-			470	570	-	-		
Uruguay	*3.322	*2.751	*2.478	*2.044	-		-	-	-	-	-	
Venezuela	15	18	19	19	18	17	4,620	5,005	5,867	5,737	5,903	5,832
Zambia	0.811	-	-	-	-		2,502	-	-	-	-	
Zimbabwe	5.327	5.669	5.100	4.800	4.900	-	4,170	5,040	3,087	2,784	3,183	-

Female Workers: Total and Percent of Employment

Country	Female Workers (000)						Female Workers as % of Total Employment					
	1990	1991	1992	1993	1994	1995	1990	1991	1992	1993	1994	1995
Albania	-	-	-	0.007	-	-	-	-	-	10.14	-	-
Australia	11	-	-	-	-	-	15.88	-	-	-	-	-
Austria	3.700	4.000	4.000	3.441	3.727	-	15.68	17.02	17.39	15.64	16.43	-
Bangladesh	0.020	0.020	0.019	-	-	-	0.63	0.57	0.63	-	-	-
Canada	21	19	-	-	-	-	15.79	15.45	-	-	-	-
Chile	0.200	0.216	0.257	0.211	0.285	-	5.53	5.48	6.07	4.99	5.88	-
Colombia	-	1.779	-	2.600	2.339	-	-	13.42	-	14.20	14.10	-
Cyprus	0.060	0.067	0.054	0.049	0.068	-	23.17	25.19	22.78	20.76	22.22	-
Denmark	1.020	1.024	1.014	-	-	-	16.70	16.95	16.83	-	-	-
Egypt	-	0.984	1.366	0.890	-	-	-	2.80	3.28	3.05	-	-
El Salvador	-	-	-	0.005	-	0.012	-	-	-	23.81	-	7.84
Germany (Western Part)	121	-	-	-	-	-	15.18	-	-	-	-	-
Indonesia	-	-	-	3.358	3.656	4.529	-	-	-	7.67	6.82	7.41
Iran (Islamic Republic of)	-	0.754	0.964	0.784	-	-	-	2.65	2.77	2.98	-	-
Italy	-	29	-	-	-	-	-	14.41	-	-	-	-
Jordan	0.004	0.033	0.010	0.034	0.048	0.046	4.08	24.26	5.62	4.59	4.50	4.32
Kenya	0.109	0.089	0.082	0.082	-	-	3.04	2.55	2.36	2.35	-	-
Korea, Republic of	25	25	25	26	28	30	14.25	14.56	14.21	14.14	13.95	13.95
Malaysia	3.900	4.800	4.500	5.100	5.700	-	24.53	24.62	23.94	22.87	21.59	-
Malta	0.005	0.009	0.004	0.004	0.003	-	7.35	10.11	5.33	4.55	3.30	-
Myanmar	0.459	0.559	0.575	0.567	0.584	-	15.74	20.19	22.06	22.42	26.24	-
New Zealand	1.098	0.841	0.865	0.818	-	-	20.85	20.95	20.67	20.49	-	-
Panama	0.014	-	-	-	-	-	13.46	-	-	-	-	-
Philippines	-	-	1.835	2.561	-	-	-	-	11.05	14.55	-	-
Portugal	4.102	1.887	1.972	1.990	2.597	-	16.49	9.73	10.35	10.48	12.45	-

Continued.

Depending on the table, * means *Enterprises, Engaged,* or *Factor Values;* ± means *Basis Unspecified;* § means *shown in millions.* For additional notes and sources, see Appendix I.

Female Workers: Total and Percent of Employment
- Continued -

Country	Female Workers (000)						Female Workers as % of Total Employment					
	1990	1991	1992	1993	1994	1995	1990	1991	1992	1993	1994	1995
Senegal	0.029	-	-	-	-	-	8.06	-	-	-	-	-
Sri Lanka	0.012	0.058	0.058	0.068	-	-	2.53	5.92	5.86	5.98	-	-
Sweden	17	-	-	-	-	-	21.45	-	-	-	-	-
Thailand	8.517	10	-	-	-	-	27.12	15.25	-	-	-	-
Turkey	3.441	-	-	-	-	-	7.52	-	-	-	-	-
United Kingdom	31	-	26	-	-	-	11.62	-	11.16	-	-	-
United Republic of Tanzania . .	0.210	0.203	-	-	-	-	7.98	8.00	-	-	-	-

Output and Output per Employee

Country	Output ($ bil.)						Output per Employee ($)					
	1990	1991	1992	1993	1994	1995	1990	1991	1992	1993	1994	1995
Albania	-	-	-	*0.000	-	-	-	-	-	284	-	-
Australia	*9.171	*8.847	*7.330	-	-	-	134,869	155,209	140,964	-	-	-
Austria	3.910	4.564	5.278	4.507	5.068	6.336	165,690	194,212	229,490	204,887	223,437	270,929
Azerbaijan	±0.023	±0.019	±0.001	±0.002	±0.000	-	14,755	14,547	790	1,971	158	-
Bahrain	-	-	±0.000	-	-	-	-	-	21,277	-	-	-
Bangladesh	0.056	0.060	0.095	-	-	-	17,365	17,206	31,137	-	-	-
Bolivia	0.005	0.005	0.005	0.008	0.012	0.013	10,766	10,130	9,483	12,109	17,309	19,295
Canada	48	48	50	60	67	72	361,634	392,844	394,645	458,283	494,315	529,160
Chile	0.304	0.378	0.589	0.630	0.640	-	84,236	95,795	138,954	148,997	132,067	-
China (Hong Kong SAR) . . .	0.034	0.045	0.064	0.062	0.052	-	30,459	45,038	40,129	47,731	46,934	-
Colombia	0.997	0.843	0.819	1.224	1.445	1.714	72,964	63,601	48,906	66,819	87,125	95,524
Costa Rica	0.015	0.015	0.027	0.021	0.016	0.012	29,627	33,094	57,963	35,286	-	-
Cyprus	0.014	0.013	0.014	0.011	0.015	0.020	53,775	49,294	57,778	48,699	48,484	57,801
Denmark	*0.550	*0.606	-	-	-	-	90,062	100,341	-	-	-	-
Ecuador	0.143	0.192	0.213	0.310	0.403	0.339	71,312	74,982	83,968	95,803	105,773	118,244
Egypt	*0.384	*0.417	*0.397	*0.334	-	-	9,345	11,883	9,552	11,433	-	-
El Salvador	-	-	-	±0.000	-	±0.000	-	-	-	14,341	-	875
Fiji	0.002	0.002	0.002	-	-	-	20,673	17,865	20,965	-	-	-
Finland	±1.446	±1.017	±0.845	±0.529	±0.637	-	162,457	128,711	124,196	81,309	97,989	-
FYR Macedonia	-	-	-	-	-	±0.038	-	-	-	-	-	9,726
France	±63	±60	±67	±56	±67	±78	-	-	-	-	-	-
Germany	-	160	171	135	148	-	-	187,222	211,178	184,309	215,773	-
Germany (Western Part) . . .	145	159	169	-	-	-	181,668	197,606	215,102	-	-	-
Greece	*0.335	*0.356	*0.451	-	-	-	111,764	128,852	158,858	-	-	-
Guatemala	-	0.003	0.004	0.004	0.005	0.005	-	5,790	6,208	7,120	8,927	7,583
Honduras	0.002	0.002	0.002	0.002	0.002	0.019	11,082	8,462	17,231	15,614	13,381	98,590
India	*5.570	*4.111	*4.764	*4.618	-	-	18,597	13,214	13,982	13,774	-	-
Indonesia	1.704	1.855	1.193	1.930	3.017	3.796	44,843	36,039	27,097	44,074	56,252	62,129
Iran (Islamic Republic of) . . .	-	2.005	1.778	1.585	-	-	-	70,450	51,019	60,333	-	-
Italy	±34	±31	-	-	-	-	166,662	152,387	-	-	-	-
Japan	±293	±329	±350	±375	-	-	371,352	397,011	430,796	468,480	-	-
Jordan	0.002	0.003	0.003	0.036	0.062	0.042	23,851	19,878	18,407	48,421	57,732	39,160
Korea, Republic of	22	23	25	28	35	44	123,765	133,513	142,201	153,546	172,169	207,459
Kuwait	0.006	0.003	0.011	0.011	0.010	0.018	11,242	16,911	53,223	48,473	37,306	77,106
Kyrgyzstan	0.121	0.143	0.144	0.017	0.005	-	24,373	31,220	33,447	4,358	2,363	-
Lithuania	-	-	0.010	0.008	-	-	-	-	3,112	2,835	-	-
Malaysia	*1.274	*1.606	*1.659	*2.022	*2.642	*4.131	80,097	82,377	88,256	90,670	100,073	124,550
Malta	0.002	0.003	0.002	0.003	0.003	-	33,309	37,256	31,279	32,752	34,014	-
Mauritius	0.003	0.003	0.003	0.003	0.005	-	8,615	6,451	6,128	5,748	11,082	-
Mexico	11	14	17	15	15	13	107,234	132,536	161,033	174,956	185,380	178,259
Myanmar	0.037	0.029	0.028	0.043	0.046	-	12,527	10,388	10,868	16,928	20,601	-
Pakistan	-	0.480	-	-	-	-	-	58,892	-	-	-	-
Panama	0.003	0.003	-	-	-	-	28,192	23,772	-	-	-	-
Peru	0.249	0.281	0.236	-	-	-	35,444	52,514	52,470	-	-	-
Philippines	0.751	0.440	1.272	1.380	-	-	90,484	52,352	76,657	78,406	-	-
Portugal	1.891	1.636	1.971	1.573	1.681	-	76,024	84,345	103,471	82,851	80,573	-

Continued.

Output and Output per Employee
- Continued -

Country	Output ($ bil.)						Output per Employee ($)					
	1990	1991	1992	1993	1994	1995	1990	1991	1992	1993	1994	1995
Romania	±1.172	±1.255	±0.801	±0.783	-	-	-	-	-	-	-	-
Russian Federation	-	-	-	6.575	7.408	9.512	-	-	-	7,222	9,101	12,227
Senegal	0.011	-	-	-	-	-	31,730	-	-	-	-	-
Singapore	*0.100	*0.083	*0.136	*0.129	*0.138	-	70,242	52,093	87,868	83,680	104,033	-
Slovakia	-	±0.319	±0.283	±0.215	±0.258	-	-	14,802	15,900	14,968	19,583	-
South Africa	±4.597	±4.589	±4.483	-	-	-	-	-	-	-	-	-
Spain	*30	*32	*35	-	-	-	203,681	220,002	245,878	-	-	-
Sri Lanka	0.008	0.028	0.021	0.015	-	-	16,290	28,578	21,705	12,925	-	-
Swaziland	-	-	*0.006	*0.005	-	-	-	-	22,123	17,923	-	-
Sweden	*12	*14	*14	*8.626	*12	-	157,005	192,453	206,233	147,099	179,166	-
Thailand	2.771	5.432	-	-	-	-	88,226	79,412	-	-	-	-
Trinidad and Tobago	±0.017	-	-	-	-	-	32,484	-	-	-	-	-
Turkey	4.397	4.893	6.240	8.652	4.216	-	96,126	103,284	122,971	152,542	83,585	-
Ukraine	-	-	*2.458	*0.787	*0.565	*0.503	-	-	26,716	8,556	6,649	6,620
United Kingdom	*42	*37	42	-	-	-	158,240	152,090	178,392	-	-	-
United Republic of Tanzania	*0.050	*0.047	-	-	-	-	18,883	18,570	-	-	-	-
Venezuela	0.748	1.413	1.860	1.903	1.604	2.627	49,525	80,301	97,401	101,188	91,477	150,543
Zambia	0.006	-	-	-	-	-	7,992	-	-	-	-	-
Zimbabwe	*0.178	*0.258	*0.225	*0.186	*0.161	-	33,341	45,500	44,110	38,786	32,803	-

Value Added and Value Added per Employee

Country	Value Added ($ bil.)						Value Added per Employee ($)					
	1990	1991	1992	1993	1994	1995	1990	1991	1992	1993	1994	1995
Australia	*3.784	*3.533	-	-	-	-	55,653	61,974	-	-	-	-
Austria	1.323	1.476	1.688	1.325	1.557	1.928	56,065	62,798	73,401	60,244	68,663	82,441
Bangladesh	0.038	0.030	0.070	-	-	-	11,800	8,497	23,081	-	-	-
Bolivia	0.003	0.002	0.002	0.003	0.003	0.003	5,421	4,010	3,474	5,073	4,191	4,672
Canada	9.856	9.479	9.638	11	13	14	74,105	77,064	75,893	86,517	94,545	100,318
Chile	0.083	0.087	0.133	0.126	0.110	-	22,988	22,186	31,417	29,915	22,616	-
China (Hong Kong SAR)	0.017	0.015	0.028	0.025	0.028	-	15,521	14,670	17,521	19,391	25,173	-
Colombia	0.263	0.269	0.239	0.287	0.385	0.433	19,224	20,301	14,253	15,677	23,198	24,134
Costa Rica	0.006	0.007	0.012	0.009	0.007	0.005	12,882	14,634	25,714	15,256	-	-
Cyprus	0.006	0.006	0.005	0.005	0.006	0.009	21,780	21,704	22,654	21,331	21,091	26,537
Denmark	*0.241	*0.255	-	-	-	-	39,475	42,186	-	-	-	-
Ecuador	0.021	0.034	0.031	0.057	0.042	0.042	10,622	13,273	12,369	17,538	10,933	14,769
Egypt	*0.112	*0.196	*0.085	*0.111	-	-	2,736	5,587	2,032	3,813	-	-
El Salvador	-	-	-	0.000	-	0.001	-	-	-	7,827	-	3,583
Fiji	0.000	0.000	0.001	-	-	-	4,782	4,085	5,851	-	-	-
Finland	±0.491	±0.363	±0.331	±0.175	±0.277	-	55,151	45,919	48,720	26,906	42,677	-
FYR Macedonia	-	-	-	-	-	±0.020	-	-	-	-	-	5,060
France	±20	±18	±19	±17	±21	±23	-	-	-	-	-	-
Germany (Western Part)	59	64	69	55	-	-	74,467	79,244	88,181	-	-	-
Greece	*0.089	*0.093	*0.126	-	-	-	29,712	33,701	44,236	-	-	-
Honduras	*0.001	*0.001	*0.001	*0.001	*0.001	*0.002	5,140	3,825	7,762	7,205	4,985	11,681
India	*1.371	*1.132	*1.120	*1.074	-	-	4,579	3,637	3,287	3,204	-	-
Indonesia	*0.724	*0.380	*0.539	*0.969	*1.593	*1.465	19,062	7,387	12,238	22,126	29,701	23,976
Iran (Islamic Republic of)	-	1.096	0.601	0.448	-	-	-	38,521	17,254	17,061	-	-
Italy	±9.937	±8.904	-	-	-	-	48,262	44,265	-	-	-	-
Japan	±85	±93	±95	±104	-	-	107,327	111,691	117,046	129,935	-	-
Jordan	0.001	0.001	0.001	0.010	0.023	0.012	7,115	5,215	4,709	13,026	21,138	10,982
Korea, Republic of	7.781	8.561	9.125	10	12	17	44,541	50,254	51,826	55,097	60,240	77,945
Kuwait	0.003	0.001	0.005	0.004	0.005	0.006	6,809	5,953	23,134	18,963	17,037	26,598
Malaysia	*0.397	*0.447	*0.407	*0.464	*0.606	*0.956	24,961	22,929	21,624	20,816	22,963	28,828
Malta	0.001	0.002	0.001	0.001	0.002	-	11,644	20,315	15,304	16,986	17,588	-
Mauritius	0.001	0.001	0.002	0.001	0.002	-	2,859	1,418	3,256	3,042	4,511	-
Mexico	3.290	4.082	4.758	4.369	4.163	-	31,579	37,698	45,570	49,918	52,993	-
Myanmar	±0.005	±0.009	±0.009	±0.013	±0.010	-	1,674	3,188	3,306	4,995	4,586	-

Continued.

Depending on the table, * means *Enterprises, Engaged,* or *Factor Values;* ± means *Basis Unspecified;* § means *shown in millions.* For additional notes and sources, see Appendix I.

Value Added and Value Added per Employee

- Continued -

Country	Value Added ($ bil.)						Value Added per Employee ($)					
	1990	1991	1992	1993	1994	1995	1990	1991	1992	1993	1994	1995
Pakistan	-	0.095	-	-	-	-	-	11,634	-	-	-	-
Panama	0.001	0.001	-	-	-	-	12,885	11,175	-	-	-	-
Peru	0.120	0.113	0.099	-	-	-	17,067	21,177	21,966	-	-	-
Philippines	0.159	0.080	0.310	0.325	-	-	19,110	9,496	18,696	18,476	-	-
Portugal	0.350	0.297	0.369	0.256	0.317	-	14,072	15,316	19,370	13,503	15,195	-
Romania	±0.254	±0.295	±0.191	-	-	-	-	-	-	-	-	-
Russian Federation	-	-	-	±3.563	±3.524	±0.876	-	-	-	3,913	4,330	1,126
Senegal	0.007	-	-	-	0.003	-	19,569	-	-	-	8,759	-
Singapore	*0.033	*0.031	*0.048	*0.042	*0.055	-	23,546	19,286	30,915	27,584	41,299	-
Slovakia	-	-	-	±0.074	±0.066	-	-	-	-	5,157	5,028	-
Spain	*7.874	*9.052	*9.899	-	-	-	53,815	61,983	68,675	-	-	-
Sri Lanka	0.002	0.006	0.006	0.004	-	-	3,994	5,616	6,155	3,454	-	-
Swaziland	-	-	*0.001	*0.001	-	-	-	-	3,709	3,163	-	-
Sweden	*4.982	*3.167	*2.478	*2.812	*3.676	-	63,227	42,613	35,656	47,955	56,891	-
Thailand	0.687	3.790	-	-	-	-	21,867	55,402	-	-	-	-
Trinidad and Tobago	±0.006	-	-	-	-	-	12,209	-	-	-	-	-
Turkey	1.546	1.829	2.215	3.232	1.636	-	33,805	38,616	43,641	56,974	32,429	-
United Kingdom	*14	*12	*14	-	-	-	53,371	47,289	58,729	-	-	-
United Republic of Tanzania	*-0.001	*0.015	-	-	-	-	-491.221	5,865	-	-	-	-
Venezuela	0.189	0.467	0.596	0.715	0.634	2.381	12,502	26,512	31,196	38,005	36,172	136,489
Zambia	0.004	-	-	-	-	-	4,413	-	-	-	-	-
Zimbabwe	*0.068	*0.129	*0.048	*0.067	*0.044	-	12,737	22,700	9,469	13,921	8,907	-

Capital Investment

Country	Gross Fixed Capital Formation ($ mil.)						Per Establishment ($ mil)					
	1990	1991	1992	1993	1994	1995	1990	1991	1992	1993	1994	1995
Austria	177	346	552	354	233	-	1.881	3.678	5.754	4.065	2.589	-
Bahrain	-	-	0.003	-	-	-	-	-	0.001	-	-	-
Bangladesh	-	0.328	0.154	-	-	-	-	0.003	0.002	-	-	-
Bolivia	0.356	0.105	0.025	0.292	0.037	-	0.011	0.003	0.001	0.006	0.001	-
China (Hong Kong SAR)	-	-	1.421	1.939	-0.906	-	-	-	0.006	0.015	-0.011	-
Colombia	-	21	10	7.221	1.013	-	-	0.115	0.043	0.032	0.005	-
Cyprus	0.322	0.564	1.398	0.740	0.391	0.798	0.006	0.011	0.027	0.014	0.007	0.014
Denmark	23	25	-	-	-	-	0.100	0.111	-	-	-	-
Ecuador	-	16	32	-4.624	11	16	-	0.427	0.844	-0.113	0.218	0.292
El Salvador	-	-	-	-	-	0.329	-	-	-	-	-	0.023
Fiji	0.030	0.152	0.167	-	-	-	0.010	0.025	0.033	-	-	-
Finland	64	64	20	11	11	-	0.496	0.491	0.189	0.115	0.131	-
Germany (Western Part)	8,604	9,549	8,542	-	-	-	11	12	10	-	-	-
Greece	14	13	12	-	-	-	0.239	0.239	0.209	-	-	-
Hungary	-	-	328	62	-	-	-	-	1.831	0.296	-	-
India	361	313	431	514	-	-	0.128	0.103	0.127	0.148	-	-
Indonesia	135	170	70	77	100	201	0.687	0.832	0.318	0.328	0.417	0.775
Iran (Islamic Republic of)	-	77	87	73	-	-	-	1.021	1.185	1.007	-	-
Italy	2,366	2,370	-	-	-	-	4.166	4.072	-	-	-	-
Japan	14,566	19,768	20,569	15,117	-	-	1.302	1.765	1.870	1.362	-	-
Korea, Republic of	-	-	2,782	3,751	3,814	5,097	-	-	1.190	1.480	1.416	1.741
Kuwait	-	0.359	0.027	0.172	0.013	-	-	0.040	0.003	0.025	0.002	-
Malaysia	-	-	-	226	221	-	-	-	-	1.322	1.226	-
Malta	0.047	0.223	0.195	0.309	0.087	-	0.006	0.025	0.024	0.031	0.010	-
Mexico	232	423	-	-	-	-	1.890	3.441	-	-	-	-
Myanmar	219	221	228	225	187	-	109	111	114	113	94	-
Pakistan	-	9.672	-	-	-	-	-	0.136	-	-	-	-
Panama	0.142	0.082	-	-	-	-	0.018	0.012	-	-	-	-
Peru	4.250	3.185	32	-	-	-	0.012	0.009	0.086	-	-	-
Philippines	44	8.588	34	52	-	-	0.195	0.034	0.177	0.280	-	-
Portugal	79	90	64	821	516	-	0.189	0.222	0.165	1.964	1.113	-

Continued.

Capital Investment
- Continued -

Country	Gross Fixed Capital Formation ($ mil.)						Per Establishment ($ mil)					
	1990	1991	1992	1993	1994	1995	1990	1991	1992	1993	1994	1995
Senegal	0.882	-	-	-	-	-	0.098	-	-	-	-	-
Singapore	3.926	4.472	4.323	7.941	33	-	0.218	0.235	0.228	0.418	1.564	-
Spain	1,027	1,869	2,968	-	-	-	1.013	1.729	2.748	-	-	-
Sri Lanka	0.025	0.118	0.046	0.834	-	-	0.001	0.005	0.003	0.044	-	-
Swaziland	-	-	-	0.220	-	-	-	-	-	-	-	-
Thailand	203	687	-	-	-	-	1.294	1.536	-	-	-	-
Turkey	214	424	336	446	466	-	1.197	2.306	0.955	1.278	1.424	-
Ukraine	-	-	14	4.344	6.101	3.536	-	-	0.087	0.106	0.136	0.077
United Kingdom	-	3,147	2,445	-	-	-	-	1.535	1.184	-	-	-
United Republic of Tanzania	4.460	9.591	-	-	-	-	0.139	0.291	-	-	-	-
Zambia	0.510	-	-	-	-	-	0.073	-	-	-	-	-

Indexes of Industrial Production

Country	Index of Industrial Production (1990=100)						Index of Employment (1990=100)					
	1990	1991	1992	1993	1994	1995	1990	1991	1992	1993	1994	1995
Australia	-	-	-	-	-	-	100	84	76	-	-	-
Austria	-	-	-	-	-	-	100	100	97	93	96	99
Azerbaijan	-	-	-	-	-	-	100	87	70	73	77	-
Bangladesh	-	-	-	-	-	-	100	109	95	-	-	-
Bolivia	-	-	-	-	-	-	100	105	108	139	140	141
Bosnia & Herzegovina	-	-	-	-	-	-	100	100	-	-	-	-
Canada	-	-	-	-	-	-	100	92	95	98	102	102
Chile	-	-	-	-	-	-	100	109	117	117	134	-
China (Hong Kong SAR)	-	-	-	-	-	-	100	91	145	118	100	-
Colombia	-	-	-	-	-	-	100	97	123	134	121	131
Costa Rica	-	-	-	-	-	-	100	92	95	118	-	-
Cyprus	-	-	-	-	-	-	100	103	92	91	118	135
Denmark	-	-	-	-	-	-	100	99	99	-	-	-
Ecuador	-	-	-	-	-	-	100	128	127	162	190	143
Egypt	-	-	-	-	-	-	100	85	101	71	-	-
Fiji	-	-	-	-	-	-	100	141	140	-	-	-
Finland	-	-	-	-	-	-	100	89	76	73	73	-
Germany (Western Part)	-	-	-	-	-	-	100	101	98	-	-	-
Greece	-	-	-	-	-	-	100	92	95	-	-	-
Honduras	-	-	-	-	-	-	100	100	65	70	77	89
India	-	-	-	-	-	-	100	104	114	112	-	-
Indonesia	-	-	-	-	-	-	100	135	116	115	141	161
Italy	-	-	-	-	-	-	100	98	-	-	-	-
Japan	-	-	-	-	-	-	100	105	103	101	-	-
Jordan	-	-	-	-	-	-	100	139	182	756	1,088	1,087
Kenya	-	-	-	-	-	-	100	97	97	97	91	91
Korea, Republic of	-	-	-	-	-	-	100	98	101	106	116	122
Kuwait	-	-	-	-	-	-	100	30	42	45	54	47
Kyrgyzstan	-	-	-	-	-	-	100	92	87	79	43	-
Malawi	-	-	-	-	-	-	100	-	-	-	-	-
Malaysia	-	-	-	-	-	-	100	123	118	140	166	209
Malta	-	-	-	-	-	-	100	131	110	129	134	-
Mauritius	-	-	-	-	-	-	100	147	147	138	118	162
Mexico	-	-	-	-	-	-	100	104	100	84	75	72
Myanmar	-	-	-	-	-	-	100	95	89	87	76	-
New Zealand	-	-	-	-	-	-	100	76	79	76	-	-
Panama	-	-	-	-	-	-	100	110	-	-	-	-
Peru	-	-	-	-	-	-	100	76	64	-	-	-
Philippines	-	-	-	-	-	-	100	101	200	212	-	-
Portugal	-	-	-	-	-	-	100	78	77	76	84	-
Senegal	-	-	-	-	-	-	100	-	-	-	81	-
Singapore	-	-	-	-	-	-	100	111	109	108	93	-

Continued.

Depending on the table, * means *Enterprises, Engaged,* or *Factor Values;* ± means *Basis Unspecified;* § means *shown in millions.* For additional notes and sources, see Appendix I.

Indexes of Industrial Production

- Continued -

Country	Index of Industrial Production (1990=100)						Index of Employment (1990=100)					
	1990	1991	1992	1993	1994	1995	1990	1991	1992	1993	1994	1995
Spain	-	-	-	-	-	-	100	100	99	-	-	-
Sri Lanka	-	-	-	-	-	-	100	206	208	240	-	-
Sweden	-	-	-	-	-	-	100	94	88	74	82	-
Thailand	-	-	-	-	-	-	100	218	-	-	-	-
Trinidad and Tobago	-	-	-	-	-	-	100	-	-	-	-	-
Turkey	-	-	-	-	-	-	100	104	111	124	110	-
United Kingdom	-	-	-	-	-	-	100	92	87	-	-	-
United Republic of Tanzania . .	-	-	-	-	-	-	100	96	-	-	-	-
Uruguay	-	-	-	-	-	-	100	83	75	62	-	-
Venezuela	-	-	-	-	-	-	100	117	126	125	116	116
Zambia	-	-	-	-	-	-	100	-	-	-	-	-
Zimbabwe	-	-	-	-	-	-	100	106	96	90	92	-

Depending on the table, * means *Enterprises*, *Engaged*, or *Factor Values*; ± means *Basis Unspecified*; § means *shown in millions*. For additional notes and sources, see Appendix I.

683

ISIC 3850
PROFESSIONAL GOODS

ISIC 3850—Professional Goods—includes scientific and professional instruments, cameras, watches, clocks, and similar products, parts, and components.

The products of this industry are classified, in the U.S. SIC system, under the 2-digit classification 38—Instruments and Related Equipment.

Summary Statistics

		1990 Value	N	1991 Value	N	1992 Value	N	1993 Value	N	1994 Value	N	1995 Value	N
Establishments or enterprises	(number)	23,691	69	25,597	77	40,395	69	26,879	67	27,817	56	18,363	27
Number employed	(000)	3,545	78	3,078	82	3,039	77	3,332	76	3,234	68	2,843	57
Total Wages	($ mil.)	51,860	72	56,673	75	59,988	71	52,279	68	56,083	61	50,743	49
Wage/salary per employee	($)	9,998	71	10,284	73	11,647	70	10,527	67	11,479	60	13,586	49
Female workers	(000)	216	31	122	29	575	27	533	27	496	21	445	10
as % of total employment*	(%)	42	31	43	29	43	27	45	27	42	21	47	10
Output	($ bil.)	222	73	238	75	251	71	244	69	263	65	255	54
per employee	($ 000)	48	72	51	74	58	70	58	67	62	63	71	53
per establishment	($ 000)	5,437	63	4,697	68	4,569	62	6,246	56	5,378	49	4,579	25
Value added	($ bil.)	121	68	123	66	132	61	138	61	140	56	143	51
per employee	($ 000)	24	66	26	64	29	60	28	58	31	55	36	50
per establishment	($ 000)	2,167	58	2,129	59	1,995	52	2,153	47	2,437	40	2,085	22
Capital investment	($ mil.)	7,810	54	8,405	52	7,878	46	6,295	41	6,844	32	4,500	16
per employee	($ 000)	3	53	2	51	2	46	3	41	3	32	3	16
per establishment	($ 000)	285	52	259	50	168	45	189	38	234	30	140	14

Data presented above are drawn from the detailed tables that follow. Columns headed 'N' show the number of countries that provided valid data for inclusion. Values are not strictly comparable one year to the next or one row to the next because the number of countries that report varies from period to period and row to row. However, a general indication of magnitudes can be gleaned from reviewing this summary—especially in earlier years in which more reports are available. For detailed explanations, see Appendix I. *The average for those countries reporting both total and female employment.

Establishments and Number Engaged

Country	Establishments or Enterprises (number)						Employees per Establishment					
	1990	1991	1992	1993	1994	1995	1990	1991	1992	1993	1994	1995
Albania	-	-	-	1.000	2.000	2.000	-	-	-	72	103	58
Argentina	-	-	-	1,191	1,382	-	-	-	-	5.465	5.035	-
Armenia	20	20	26	32	28	-	1,709	1,507	-	-	-	-
Australia	588	957	1,113	-	-	-	22	13	12	-	-	-
Austria	81	78	76	75	73	-	84	90	92	79	77	-
Bahrain	-	-	2.000	-	-	-	-	-	14	-	-	-
Bangladesh	21	54	23	-	-	-	13	9.556	13	-	-	-
Belgium	484	476	486	606	635	-	-	-	-	-	-	-
Bolivia	15	17	18	18	18	18	10	9.941	11	11	11	11
Bosnia & Herzegovina	6.000	6.000	-	-	-	-	578	608	-	-	-	-
Canada	526	469	448	397	378	-	33	35	32	33	37	-
Cape Verde	1.000	1.000	1.000	1.000	1.000	-	-	-	-	-	-	-
Chile	7.000	7.000	6.000	6.000	8.000	-	88	93	108	108	104	-
China	*4,078	*4,095	*4,050	*4,975	*5,165	*5,637	154	161	178	173	174	153
China (Hong Kong SAR)	1,720	1,369	1,424	1,124	932	-	20	20	17	17	18	-
Colombia	73	72	79	75	63	-	53	56	54	63	58	-
Costa Rica	25	33	35	24	-	-	11	9.121	8.714	12	-	-
Croatia	51	73	90	102	123	133	30	19	16	13	9.268	9.474
Cyprus	3.000	3.000	3.000	3.000	10	13	2.000	2.333	2.333	3.333	4.700	4.462
Czechoslovakia (Former)	*21	*24	-	-	-	-	952	750	-	-	-	-
Czech Republic	-	*21	*53	*71	-	-	-	810	509	324	-	-
Denmark	423	446	454	-	-	-	36	32	30	-	-	-
Ecuador	11	11	11	10	11	5.000	47	48	45	41	34	27
Egypt	17	19	18	18	-	-	388	574	406	378	-	-
El Salvador	-	2.000	-	7.000	3.000	1.000	-	-	-	55	36	8.000
Finland	94	103	102	112	112	134	62	54	51	48	50	53
FYR Macedonia	*6.000	*27	*51	*54	*79	*98	37	8.926	6.765	6.481	5.684	4.204
Germany	-	1,373	1,407	1,421	1,414	-	-	139	114	102	96	-
Germany (Western Part)	1,363	1,309	1,321	-	-	-	106	111	108	-	-	-
Greece	20	19	19	-	-	-	38	42	40	-	-	-
Guatemala	-	3.000	3.000	3.000	3.000	3.000	-	74	81	78	80	81
Honduras	-	-	15	10	10	10	-	-	5.333	8.700	9.600	16
Hungary	*489	*895	*505	*579	-	-	90	39	50	28	-	-
India	739	861	796	940	-	-	67	58	70	62	-	-
Indonesia	53	56	66	61	66	73	63	68	115	103	145	201
Iran (Islamic Republic of)	41	28	34	31	-	-	78	134	154	171	-	-
Ireland	81	88	-	-	-	-	98	95	-	-	-	-
Israel	58	90	80	88	77	-	52	42	49	45	51	-
Italy	*411	*417	*802	*802	*789	-	83	82	71	75	76	-
Japan	6,045	5,959	5,529	5,387	7,051	7,064	36	37	37	34	36	35
Jordan	4.000	2.000	3.000	2.000	52	66	62	96	81	73	5.250	6.409
Kenya	1.000	1.000	1.000	1.000	1.000	6.000	181	214	223	244	266	49
Korea, Republic of	1,130	891	955	1,211	1,301	1,388	40	37	32	27	27	25
Kyrgyzstan	2.000	2.000	2.000	5.000	3.000	-	2,756	2,432	2,109	649	710	-
Latvia	4.000	24	37	35	23	52	551	88	30	20	36	13
Liechtenstein	-	2.000	-	-	-	-	-	2.500	-	-	-	-
Lithuania	-	-	-	*16	*35	-	-	-	-	546	260	-
Macau	2.000	2.000	2.000	2.000	-	-	315	329	292	197	-	-
Malaysia	26	35	37	34	37	-	562	534	500	585	605	-
Malta	10	13	12	11	10	-	131	101	112	100	136	-
Mauritius	13	19	21	20	20	18	105	87	70	75	75	99
Mexico	6.000	6.000	6.000	6.000	5.000	5.000	437	461	417	446	440	386
Mongolia	3.000	4.000	3.000	2.000	6.000	-	73	23	16	168	20	-
Morocco	29	28	28	29	35	34	41	43	32	29	30	28
Mozambique	1.000	1.000	1.000	1.000	1.000	-	72	84	91	88	86	-
Nepal	1.000	1.000	-	1.000	1.000	-	-	-	-	-	-	-
Netherlands	98	95	98	103	189	-	73	77	78	76	37	-
Netherlands Antilles	-	-	-	10	-	-	-	-	-	-	-	-
New Zealand	*86	*84	*84	*84	-	-	9.919	11	13	5.690	-	-
Norway	57	56	38	55	55	-	28	33	46	62	65	-

Continued.

Depending on the table, * means *Enterprises, Engaged,* or *Factor Values*; ± means *Basis Unspecified*; § means *shown in millions*. For additional notes and sources, see Appendix I.

Establishments and Number Engaged

- Continued -

Country	Establishments or Enterprises (number)						Employees per Establishment					
	1990	1991	1992	1993	1994	1995	1990	1991	1992	1993	1994	1995
Pakistan	-	59	-	-	-	-	-	67	-	-	-	-
Panama	11	5.000	-	-	-	-	11	18	-	-	-	-
Paraguay	-	4.000	-	-	-	-	-	20	-	-	-	-
Peru	101	104	106	-	-	-	15	11	11	-	-	-
Philippines	12	16	16	19	-	-	242	175	213	389	-	-
Poland	*105	*98	*86	*78	-	-	352	296	267	256	-	-
Portugal	*226	*306	*315	*336	*409	-	19	17	15	15	13	-
Puerto Rico	80	92	84	80	76	72	154	151	163	179	176	171
Romania	-	73	112	122	209	-	-	599	341	202	122	-
Russian Federation	-	-	-	*1,967	*1,879	*2,643	-	-	-	200	163	100
Singapore	57	61	60	58	56	-	145	141	147	143	159	-
Slovakia	-	14	14	19	19	-	-	1,005	866	536	480	-
Slovenia	35	-	-	-	-	*474	-	-	-	-	-	-
South Africa	-	288	-	-	-	-	-	24	-	-	-	-
Spain	622	616	612	-	-	-	14	15	14	-	-	-
Sri Lanka	9.000	7.000	2.000	7.000	-	-	32	39	24	46	-	-
Suriname	-	-	1.000	-	-	-	-	-	22	-	-	-
Sweden	178	186	177	179	184	-	92	96	85	84	93	-
Thailand	16	30	-	-	-	-	342	347	-	-	-	-
Tunisia	-	-	-	86	79	89	-	-	-	6.872	6.089	7.663
Turkey	38	44	75	85	89	-	118	106	68	64	55	-
Ukraine	111	238	249	110	120	121	1,982	908	900	1,427	1,150	967
United Kingdom	2,555	2,476	2,586	3,676	4,281	-	33	32	29	39	31	-
United Republic of Tanzania	4.000	4.000	-	-	-	-	29	29	-	-	-	-
USSR (Former)	*388	-	-	-	-	-	1,588	-	-	-	-	-
United States of America	-	-	15,244	-	-	-	-	-	56	-	-	-
Venezuela	53	58	51	43	44	47	53	53	59	61	52	58
Yugoslavia	33	60	120	152	154	157	-	75	36	27	25	24
Zambia	-	-	-	-	1.000	-	-	-	-	-	44	-
Zimbabwe	13	11	11	10	10	-	7.692	9.091	9.091	10	20	-

Employment and Compensation of Employees

Country	Number Employed/Engaged (000)						Wage/Salary per Employee ($)					
	1990	1991	1992	1993	1994	1995	1990	1991	1992	1993	1994	1995
Albania	-	-	-	0.072	0.206	0.116	-	-	-	-	645	1,361
Argentina	6.525	-	-	6.509	6.959	-	6,606	-	-	11,956		
Armenia	*34	*30	-	-	-	-	-	-	-	-	-	-
Australia	13	*12	*13	-	-	-	19,477	20,257	19,118	-	-	-
Austria	6.800	7.000	7.000	5.908	5.654	5.932	19,957	22,525	26,793	28,041	28,538	34,156
Bahrain	-	-	0.027	-	-	-	-	-	5,615	-	-	-
Bangladesh	0.267	0.516	0.289	-	-	-	444	-	444	-	-	-
Bermuda	0.007	0.008	0.008	0.004	0.003	0.003	-	-	-	-	-	-
Bolivia	0.151	0.169	0.190	0.197	0.198	0.200	2,102	2,115	2,144	1,942	1,781	1,716
Bosnia & Herzegovina	3.467	3.648	-	-	-	-	2,622	2,445	-	-	-	-
Cambodia	*0.528	*0.456	-	-	-	-	-	-	-	-	-	-
Canada	17	16	14	13	14	14	26,504	27,845	28,291	28,143	25,368	28,301
Chile	0.617	0.650	0.645	0.651	0.835	0.720	5,071	5,604	6,700	7,537	8,206	10,105
China	*630	*660	*720	*860	*900	*860	-	-	-	-	-	-
China (Hong Kong SAR)	*34	*27	*24	*20	*16	-	8,603	9,315	10,864	11,873	13,080	-
China (Taiwan Province)	40	41	39	34	33	33	8,989	9,430	11,228	11,392	12,225	12,584
Colombia	3.900	4.054	4.282	4.711	3.655	4.144	2,394	2,553	2,998	3,307	3,572	4,694
Costa Rica	0.267	0.301	0.305	0.284	-	-	2,365	2,253	3,019	2,614	-	-
Croatia	1.550	1.410	1.410	1.340	1.140	1.260	4,435	4,092	1,449	1,553	2,107	3,382
Cyprus	0.006	0.007	0.007	0.010	0.047	0.058	7,294	6,171	10,159	8,652	10,649	13,955
Czechoslovakia (Former)	20	18	-	-	-	-	2,228	1,432	-	-	-	-
Czech Republic	17	17	27	23	-	-	2,196	1,397	1,743	2,180	-	-
Denmark	15	14	14	14	15	15	32,785	33,802	37,370	35,984	-	-

Continued.

Depending on the table, * means *Enterprises, Engaged,* or *Factor Values*; ± means *Basis Unspecified*; § means *shown in millions*. For additional notes and sources, see Appendix I.

Employment and Compensation of Employees

- Continued -

Country	Number Employed/Engaged (000)						Wage/Salary per Employee ($)					
	1990	1991	1992	1993	1994	1995	1990	1991	1992	1993	1994	1995
Ecuador	0.521	0.530	0.498	0.407	0.372	0.136	4,505	5,326	5,195	3,384	1,166	1,672
Egypt	*6.600	*11	*7.300	*6.800	*6.682	*5.028	3,182	1,217	2,195	1,657	1,700	1,811
El Salvador	-	-	*0.040	*0.382	*0.107	*0.008			-	2,431	5,571	1,985
Finland	5.800	5.600	5.200	5.400	5.600	7.060	31,304	30,256	29,910	23,384	27,045	34,478
FYR Macedonia	0.224	0.241	0.345	0.350	0.449	0.412	3,866	5,177	2,261	2,597	2,939	5,046
France	68	68	64	60	58	58	37,537	37,294	-	-	-	-
Germany	-	*190	*161	*145	*135	-	-	24,533	33,135	31,141	33,163	-
Germany (Western Part)	*144	*145	*143	-	-	-	28,901	29,662	33,323	-	-	-
Greece	*0.766	*0.804	*0.759	*0.634	*0.605	*0.572	10,707	10,277	11,114	13,483	14,433	17,992
Guatemala	-	*0.221	*0.244	*0.233	*0.241	*0.243	-	703	1,027	989	839	1,230
Honduras	*0.148	*0.124	*0.080	*0.087	*0.096	*0.161	2,792	2,463	4,120	2,741	2,772	1,687
Hungary	44	35	25	16	16	16	2,739	2,974	3,442	3,682	3,676	4,104
India	*49	*50	*56	*59	*58	*60	1,549	1,257	1,340	1,372	1,393	1,574
Indonesia	3.319	3.797	7.619	6.278	9.555	15	692	675	897	1,122	1,356	1,478
Iran (Islamic Republic of)	3.200	3.754	5.225	5.310	-	-	3,796	4,266	4,030	4,033	-	-
Ireland	7.900	8.400	9.122	9.509	9.893	11	20,656	21,155	23,556	20,765	21,879	24,834
Israel	3.000	3.800	3.900	4.000	3.900	4.190	26,287	25,172	25,755	26,854	30,996	31,939
Italy	34	34	57	60	60	-	34,798	36,714	-	34,602	34,497	-
Japan	*216	*219	*202	*183	*256	*246	26,795	30,066	32,521	37,445	-	-
Jordan	0.246	0.192	0.242	0.145	0.273	0.423	2,173	2,195	2,759	3,234	1,368	2,022
Kenya	0.181	0.214	0.223	0.244	0.266	0.292	1,197	1,141	1,119	704	887	1,200
Korea, Republic of	45	33	31	33	35	35	8,500	9,965	10,638	11,504	12,453	15,408
Kyrgyzstan	*5.513	*4.864	*4.219	*3.246	*2.131	-	630	435	-	-	-	-
Latvia	2.203	2.101	1.094	0.712	0.827	0.683	-	-	-	-	-	-
Liechtenstein	-	*0.005	-	-	-	-	-	-	-	-	-	-
Lithuania	-	-	12	8.741	9.100	-	-	-	-	428	715	-
Macau	0.631	0.659	0.584	0.393	-	-	4,187	4,395	5,037	12,214	-	-
Malaysia	15	19	19	20	22	23	2,418	2,882	3,775	3,912	4,219	4,986
Malta	1.312	1.319	1.343	1.103	1.357	-	9,555	10,206	10,015	9,351	9,578	-
Mauritius	*1.365	1.647	1.462	1.508	1.504	1.780	3,169	2,462	2,514	2,506	3,051	2,686
Mexico	2.620	2.763	2.499	2.679	2.198	1.928	5,165	5,875	7,043	5,186	6,421	4,338
Mongolia	*0.219	*0.092	*0.047	*0.337	*0.121	*0.016	734	3,197	1,150	-	-	-
Morocco	1.183	1.198	0.891	0.853	1.046	0.967	2,872	2,876	3,418	3,278	3,532	3,996
Mozambique	0.072	0.084	0.091	0.088	0.086	-	2,048	1,021	725	557	582	-
Netherlands	7.203	7.321	7.654	7.842	6.971	6.835	25,389	24,912	30,016	28,699	31,606	38,091
Netherlands Antilles	-	-	-	0.008	-	-	-	-	-	14,386	-	-
New Zealand	0.853	0.890	1.051	0.478	-	-	13,998	9,108	7,168	-	-	-
Norway	1.585	1.873	1.761	3.409	3.559	3.832	35,478	37,389	-	37,008	-	-
Pakistan	3.654	3.970	4.152	-	-	-	1,275	1,138	1,171	-	-	-
Panama	*0.126	*0.089	*0.092	*0.091	*0.097	*0.098	6,071	6,719	7,062	6,929	7,524	7,673
Paraguay	-	0.078	-	-	-	-	-	-	-	-	-	-
Peru	*1.492	*1.181	*1.136	-	-	-	3,123	3,912	4,777	-	-	-
Philippines	2.900	2.800	3.400	7.400	-	-	2,284	2,664	3,286	1,921	-	-
Poland	37	29	23	20	19	17	1,200	1,894	2,257	2,624	3,269	5,480
Portugal	*4.227	*5.053	*4.845	*5.038	*5.295	*4.915						
Puerto Rico	12	14	14	14	13	12	26,753	17,448	19,306	20,126	24,063	24,089
Romania	50	44	38	25	25	-	1,743	1,103	680	833	901	-
Russian Federation	-	-	-	394	307	265	-	-	-	475	823	801
Singapore	*8.263	*8.577	*8.841	*8.298	*8.899	*9.000	10,570	12,393	13,814	15,035	17,170	18,814
Slovakia	-	*14	*12	*10	*9.129	-	-	1,345	1,611	1,984	2,229	-
South Africa	7.000	7.000	7.000	6.000	8.000	7.860	8,117	8,174	9,417	8,773	8,132	8,531
Spain	8.846	9.158	8.871	23	20	21	18,721	20,376	22,809	21,024	20,965	23,929
Sri Lanka	*0.284	*0.270	*0.048	*0.321	-	-	643	623	-	685	-	-
Suriname	*0.021	*0.021	*0.022	*0.022	-	-	8,003	10,671	10,186	15,279	-	-
Sweden	16	*18	*15	15	17	17	28,959	29,759	34,193	26,736	27,681	30,304
Switzerland	*39	*38	*36	*35	*37	*37	-	-	-	-	-	-
Thailand	5.472	10	-	-	-	-	1,504	2,531	-	-	-	-
Tunisia	-	-	-	0.591	0.481	0.682	-	-	-	-	-	-
Turkey	4.484	4.668	5.105	5.399	4.928	5.126	6,326	7,805	7,439	7,369	3,913	6,995
Ukraine	220	216	224	157	138	117	-	2,584	1,480	-	-	-

Continued.

    Depending on the table, * means *Enterprises, Engaged,* or *Factor Values;* ± means *Basis Unspecified;* § means *shown in millions.* For additional notes and sources, see Appendix I.

Employment and Compensation of Employees

- Continued -

Country	Number Employed/Engaged (000)						Wage/Salary per Employee ($)					
	1990	1991	1992	1993	1994	1995	1990	1991	1992	1993	1994	1995
United Kingdom	85	78	76	143	133	131	21,828	23,440	24,642	22,974	24,318	26,759
United Republic of Tanzania . .	0.115	0.117	-	-	-	-	446	-	-	-	-	-
USSR (Former)	616	-	-	-	-	-	2,280	-	-	-	-	-
United States of America . . .	902	855	853	825	776	757	33,204	35,170	36,346	36,777	38,246	39,124
Uruguay	*0.973	*0.989	*0.642	*0.652	*0.743	*0.715	3,898	5,566	4,590	5,438	5,535	5,992
Venezuela	2.800	3.100	3.000	2.604	2.286	2.717	4,592	4,735	7,420	6,076	6,460	6,283
Yugoslavia	-	4.500	4.300	4.100	3.900	3.800	-	-	-	-	2,131	849
Zambia	-	-	-	-	0.044	-	-	-	-	-	1,399	-
Zimbabwe	0.100	0.100	0.100	0.100	0.200	0.155	6,536	11,377	3,730	3,708	2,147	2,074

Female Workers: Total and Percent of Employment

Country	Female Workers (000)						Female Workers as % of Total Employment					
	1990	1991	1992	1993	1994	1995	1990	1991	1992	1993	1994	1995
Albania	-	-	-	0.022	0.070	0.079	-	-	-	30.56	33.98	68.10
Argentina	-	-	-	-	1.581	-	-	-	-	-	22.72	-
Australia	5.527	-	-	-	-	-	42.52	-	-	-	-	-
Austria	3.200	3.000	3.000	2.732	2.619	-	47.06	42.86	42.86	46.24	46.32	-
Bahrain	-	-	0.002	-	-	-	-	-	7.41	-	-	-
Bangladesh	0.011	0.013	0.010	-	-	-	4.12	2.52	3.46	-	-	-
Bermuda	0.003	0.004	0.004	0.002	0.001	0.001	42.86	50.00	50.00	50.00	33.33	33.33
Bosnia & Herzegovina . . .	2.880	2.436	-	-	-	-	83.07	66.78	-	-	-	-
Canada	7.000	6.000	-	-	-	-	40.46	36.81	-	-	-	-
Chile	0.132	0.145	0.147	0.157	0.129	-	21.39	22.31	22.79	24.12	15.45	-
China (Taiwan Province) . . .	22	23	22	20	20	19	56.80	56.86	56.99	58.77	59.10	58.80
Colombia	-	1.915	-	2.448	1.905	-	-	47.24	-	51.96	52.12	-
Croatia	0.720	0.660	0.600	0.560	0.410	0.490	46.45	46.81	42.55	41.79	35.96	38.89
Cyprus	0.003	0.005	0.006	0.007	0.032	-	50.00	71.43	85.71	70.00	68.09	-
Czechoslovakia (Former) . .	4.800	4.300	-	-	-	-	24.00	23.89	-	-	-	-
Czech Republic	8.000	7.000	13	10	-	-	47.06	41.18	48.15	43.48	-	-
Denmark	6.620	5.931	6.057	-	-	-	43.16	41.92	44.32	-	-	-
Egypt	-	1.566	0.440	1.870	-	-	-	14.37	6.03	27.50	-	-
El Salvador	-	-	-	0.242	0.036	-	-	-	-	63.35	33.64	-
Germany (Western Part) . . .	58	-	-	-	-	-	40.22	-	-	-	-	-
Hungary	18	-	9.000	7.000	-	-	40.91	-	36.00	43.75	-	-
Indonesia	-	-	-	3.036	5.228	8.908	-	-	-	48.36	54.71	60.72
Iran (Islamic Republic of) . .	-	0.412	1.055	1.071	-	-	-	10.97	20.19	20.17	-	-
Ireland	4.500	4.800	-	-	-	-	56.96	57.14	-	-	-	-
Italy	-	13	19	19	19	-	-	38.00	34.05	32.34	31.63	-
Jordan	0.085	0.077	0.078	0.051	0.067	0.096	34.55	40.10	32.23	35.17	24.54	22.70
Kenya	0.025	0.033	0.035	0.038	-	-	13.81	15.42	15.70	15.57	-	-
Korea, Republic of	18	15	13	14	14	14	40.72	43.94	42.53	42.51	39.09	38.36
Macau	0.360	0.402	0.354	0.209	-	-	57.05	61.00	60.62	53.18	-	-
Malaysia	11	13	13	14	16	-	72.60	71.12	69.73	72.36	69.64	-
Malta	0.928	0.920	0.948	0.735	0.938	-	70.73	69.75	70.59	66.64	69.12	-
Mongolia	-	-	-	-	0.045	-	-	-	-	-	37.19	-
Morocco	-	-	-	-	-	0.340	-	-	-	-	-	35.16
New Zealand	0.408	0.468	0.534	0.216	-	-	47.83	52.58	50.81	45.19	-	-
Panama	0.031	-	-	-	-	-	24.60	-	-	-	-	-
Philippines	-	-	2.515	5.787	-	-	-	-	-	73.97	78.20	-
Portugal	1.396	1.569	1.592	1.615	1.739	-	33.03	31.05	32.86	32.06	32.84	-
Puerto Rico	8.240	8.960	8.520	8.510	8.020	7.510	67.05	64.41	62.24	59.55	60.12	61.06
Sri Lanka	0.113	0.072	0.036	0.079	-	-	39.79	26.67	75.00	24.61	-	-
Sweden	5.000	-	-	-	-	-	30.67	-	-	-	-	-
Thailand	-	7.535	-	-	-	-	-	72.48	-	-	-	-
Turkey	1.195	-	-	-	-	-	26.65	-	-	-	-	-
United Kingdom	27	-	24	-	-	-	31.91	-	31.58	-	-	-
United Republic of Tanzania . .	0.031	0.032	-	-	-	-	26.96	27.35	-	-	-	-

Continued.

Depending on the table, * means *Enterprises, Engaged,* or *Factor Values;* ± means *Basis Unspecified;* § means *shown in millions.* For additional notes and sources, see Appendix I.

689

Female Workers: Total and Percent of Employment

- Continued -

Country	Female Workers (000)						Female Workers as % of Total Employment					
	1990	1991	1992	1993	1994	1995	1990	1991	1992	1993	1994	1995
United States of America . . .	-	-	435	419	405	395	-	-	51.00	50.79	52.19	52.18
Zambia	-	-	-	-	0.004	-	-	-	-	-	9.09	-

Output and Output per Employee

Country	Output ($ bil.)						Output per Employee ($)					
	1990	1991	1992	1993	1994	1995	1990	1991	1992	1993	1994	1995
Albania	-	-	-	*0.000	*0.001	*0.002	-	-	-	2,722	6,905	15,047
Argentina	±0.209	-	-	0.482	-	-	32,027	-	-	73,980	-	-
Armenia.	0.011	0.012	0.032	2.650	0.016	-	311	403	-	-	-	-
Australia	*0.980	*0.995	*1.031	-	-	-	75,361	82,911	79,299	-	-	-
Austria	0.455	0.530	0.654	0.558	0.533	0.687	66,985	75,760	93,405	94,497	94,224	115,795
Bahrain	-	-	±0.001	-	-	-	-	-	31,422	-	-	-
Bangladesh	0.001	0.001	0.001	-	-	-	4,323	2,171	3,553	-	-	-
Bolivia	0.003	0.003	0.004	0.003	0.003	0.003	17,189	15,425	18,676	14,431	15,447	17,408
Bosnia & Herzegovina . .	±0.155	±0.125	-	-	-	-	44,674	34,185	-	-	-	-
Canada	1.740	1.763	1.555	1.512	1.538	1.659	100,567	108,167	108,769	116,270	109,842	121,890
Chile.	0.018	0.022	0.027	0.031	0.043	0.046	29,571	33,248	41,246	47,766	51,210	63,404
China	±2.302	±2.568	±3.311	±6.347	±4.925	±5.097	3,654	3,891	4,599	7,381	5,472	5,927
China (Hong Kong SAR) . .	2.794	2.763	2.976	2.614	2.764	-	81,708	101,577	122,991	134,065	167,536	-
China (Taiwan Province) . .	±1.931	±2.118	±2.228	±1.872	±1.908	±1.934	48,860	52,170	57,114	55,158	57,100	58,830
Colombia	0.124	0.126	0.112	0.120	0.115	0.170	31,779	30,958	26,113	25,396	31,367	41,062
Croatia	±0.029	±0.052	±0.026	-	-	-	18,926	36,609	18,247	-	-	-
Cyprus	0.001	0.001	0.001	0.001	0.004	0.004	86,433	95,958	123,810	85,513	88,396	71,490
Czechoslovakia (Former) . .	±0.192	±0.077	-	-	-	-	9,610	4,297	-	-	-	-
Denmark	*1.042	*1.019	*1.132	*1.137	*1.424	*1.706	67,930	72,024	82,810	80,609	93,651	110,270
Ecuador	0.012	0.018	0.017	0.010	0.014	0.002	22,818	33,469	34,760	23,468	38,971	12,424
Egypt	*0.096	*0.060	*0.055	*0.035	*0.036	*0.030	14,553	5,471	7,509	5,186	5,409	5,992
El Salvador.	-	±0.001	-	±0.005	±0.003	±0.000	-	-	-	14,217	28,124	3,598
Finland	±0.655	±0.638	±0.606	±0.577	±0.702	±1.120	112,853	113,925	116,602	106,852	125,292	158,595
FYR Macedonia	±0.002	±0.005	±0.004	±0.004	±0.005	±0.006	10,532	22,608	12,204	11,380	10,827	13,477
France	±7.619	±7.463	±7.996	±7.270	±7.764	±9.002	112,208	109,428	124,348	120,357	134,088	153,875
Germany	-	15	17	15	15	-	-	80,481	105,750	101,926	110,070	-
Germany (Western Part) . .	15	15	17	-	-	-	100,582	103,032	115,423	-	-	-
Greece	*0.034	*0.036	*0.040	*0.038	*0.039	*0.045	44,342	45,310	52,288	59,472	63,664	79,363
Guatemala	-	0.009	0.010	0.012	0.012	0.007	-	39,493	41,501	50,646	47,830	27,469
Honduras	0.002	0.002	0.002	0.001	0.001	0.001	14,251	13,793	18,823	16,511	15,586	8,151
Hungary	±0.669	±0.624	±0.353	±0.326	±0.329	±0.382	15,210	17,815	14,103	20,362	20,549	23,393
India.	*0.703	*0.596	*0.630	*0.720	*0.721	*0.859	14,232	11,897	11,331	12,265	12,532	14,381
Indonesia	0.027	0.041	0.119	0.101	0.138	0.220	8,197	10,705	15,627	16,145	14,492	15,007
Iran (Islamic Republic of) . .	0.066	0.082	0.099	0.113	-	-	20,691	21,865	18,990	21,370	-	-
Ireland	*0.903	*0.969	*1.170	*1.075	*1.180	*1.514	114,344	115,317	128,243	113,005	119,236	135,129
Israel.	0.250	0.301	0.307	0.299	0.349	0.388	83,490	79,325	78,724	74,732	89,413	92,684
Italy	±3.998	±4.151	±7.553	±6.770	±6.901	-	116,560	121,751	132,859	113,177	115,820	-
Japan	±31	±35	±34	±34	±52	±58	142,991	160,433	166,319	185,212	203,816	236,400
Jordan	0.005	0.004	0.005	0.002	0.003	0.005	20,111	20,153	19,956	16,769	10,528	11,518
Kenya	*0.010	*0.009	*0.008	*0.006	*0.007	*0.006	53,040	40,770	36,188	24,025	25,487	21,308
Korea, Republic of	2.604	2.143	2.156	2.494	3.164	3.883	58,255	64,589	70,074	75,724	90,801	109,886
Kyrgyzstan.	0.075	0.037	0.017	0.003	0.004	-	13,598	7,591	4,000	948	1,695	-
Latvia	0.000	0.000	0.001	0.003	0.005	0.008	95	136	749	4,079	5,916	11,216
Lithuania	-	-	0.020	0.021	0.030	-	-	-	1,691	2,420	3,314	-
Macau	0.010	0.012	0.010	0.005	-	-	16,005	17,843	17,266	13,884	-	-
Malaysia	*0.335	*0.596	*0.601	*0.722	*0.980	*1.087	22,934	31,886	32,466	36,266	43,733	46,836
Malta	0.046	0.051	0.054	0.038	0.055	-	35,350	39,015	39,981	34,200	40,755	-
Mauritius	0.042	0.035	0.041	0.023	0.020	0.033	31,122	21,425	27,851	15,060	13,531	18,728
Mexico	0.182	*0.240	0.245	0.365	0.288	0.217	69,281	87,017	98,010	136,370	130,952	112,796
Mongolia	0.001	0.003	0.000	0.000	0.000	-	6,768	33,225	7,599	154	755	-
Morocco	±0.024	±0.029	±0.025	±0.024	±0.025	±0.025	20,512	23,967	28,131	27,736	23,893	26,035

Continued.

Depending on the table, * means *Enterprises, Engaged,* or *Factor Values*; ± means *Basis Unspecified*; § means *shown in millions*. For additional notes and sources, see Appendix I.

Output and Output per Employee

- Continued -

Country	Output ($ bil.)						Output per Employee ($)					
	1990	1991	1992	1993	1994	1995	1990	1991	1992	1993	1994	1995
Mozambique	-	±0.000	-	±0.000	±0.000	-	-	149	-	70	46	-
Netherlands	0.612	0.640	0.851	0.858	0.815	0.963	85,011	87,448	111,148	109,372	116,868	140,848
New Zealand	0.077	-	-	-	-	-	90,287	-	-	-	-	-
Norway	±0.200	±0.266	±0.253	±0.483	±0.598	±0.744	126,491	142,063	143,732	141,541	168,047	194,080
Pakistan	0.043	0.044	0.048	-	-	-	11,689	11,174	11,596	-	-	-
Panama	0.006	0.005	0.006	0.006	0.007	0.007	45,198	59,528	63,253	61,900	68,319	69,946
Peru	0.037	0.037	0.041	-	-	-	24,636	31,071	35,659	-	-	-
Philippines	0.016	0.023	0.032	0.089	-	-	5,603	8,162	9,442	12,000	-	-
Poland	±0.255	±0.235	±0.212	±0.233	±0.277	±0.415	6,885	8,099	9,225	11,667	14,557	24,422
Portugal	0.101	0.160	0.185	0.179	0.209	0.228	23,998	31,680	38,199	35,562	39,513	46,310
Puerto Rico	±1.801	±1.778	±2.033	±2.217	±2.250	±2.306	146,529	127,800	148,532	155,171	168,628	187,472
Russian Federation	-	-	-	0.954	1.115	0.906	-	-	-	2,423	3,630	3,420
Singapore	*0.432	*0.498	*0.581	*0.673	*0.746	*0.833	52,265	58,071	65,720	81,067	83,879	92,501
Slovakia	-	±0.119	±0.102	±0.110	±0.105	-	-	8,442	8,393	10,760	11,489	-
Slovenia	-	-	-	0.188	0.212	0.310	-	-	-	-	-	-
South Africa	±0.253	±0.288	±0.330	±0.314	±0.312	±0.325	36,165	41,181	47,135	52,361	38,966	41,403
Spain	*0.726	*0.793	*0.856	±1.923	±1.921	±2.271	82,101	86,540	96,535	84,358	94,018	107,725
Sri Lanka	0.001	0.001	0.000	0.001	-	-	3,076	2,230	618	3,597	-	-
Suriname	*0.001	*0.002	*0.002	0.003	-	-	26,677	80,032	76,394	152,788	-	-
Sweden	*1.909	*2.342	*2.340	*1.787	*2.211	*2.405	117,123	131,631	155,448	118,313	129,323	142,151
Switzerland	7.213	6.917	7.364	7.234	8.020	9.407	185,893	183,475	205,706	206,694	219,737	254,234
Thailand	0.088	0.224	-	-	-	-	16,078	21,532	-	-	-	-
Turkey	0.169	0.188	0.331	0.383	0.271	0.374	37,702	40,310	64,874	70,937	55,047	72,966
Ukraine	*3.173	*2.271	*1.019	*0.210	*0.189	*0.156	14,422	10,513	4,547	1,337	1,371	1,337
United Kingdom	*6.870	*6.952	7.451	13	13	14	80,819	89,131	98,033	87,539	98,792	108,921
United Republic of Tanzania	*0.000	*0.001	-	-	-	-	3,834	4,329	-	-	-	-
USSR (Former)	±6.125	-	-	-	-	-	9,943	-	-	-	-	-
United States of America	*116	*118	*124	*126	*128	*130	128,160	138,491	145,936	153,103	164,535	172,225
Uruguay	0.032	0.041	0.026	0.032	0.037	0.038	33,253	41,134	40,000	48,417	49,424	53,729
Venezuela	0.087	0.124	0.105	0.150	0.214	0.246	31,229	40,101	34,856	57,777	93,715	90,438
Yugoslavia	±0.010	±0.009	±0.017	-	±0.029	±0.017	-	2,064	3,847	-	7,531	4,505
Zambia	-	-	-	-	0.000	-	-	-	-	-	10,733	-
Zimbabwe	*0.004	*0.003	*0.003	*0.004	*0.004	*0.003	40,850	32,089	33,371	37,080	20,245	18,994

Value Added and Value Added per Employee

Country	Value Added ($ bil.)						Value Added per Employee ($)					
	1990	1991	1992	1993	1994	1995	1990	1991	1992	1993	1994	1995
Argentina	±0.112	-	-	0.216	-	-	17,190	-	-	33,197	-	-
Australia	*0.498	*0.491	*0.504	-	-	0.313	38,341	40,904	38,782	-	-	-
Austria	0.222	0.250	0.295	0.258	0.242	0.313	32,594	35,665	42,146	43,669	42,800	52,807
Bangladesh	0.000	0.000	0.000	-	-	-	1,170	741	1,421	-	-	-
Bolivia	0.001	0.001	0.002	0.001	0.001	0.001	8,583	7,099	10,165	5,597	5,931	6,717
Bosnia & Herzegovina	±0.101	±0.107	-	-	-	-	29,129	29,300	-	-	-	-
Canada	0.926	0.986	0.869	0.876	0.879	0.952	53,503	60,509	60,748	67,377	62,767	69,965
Chile	0.009	0.012	0.015	0.019*	0.025	0.027	13,837	18,375	23,193	29,827	29,939	37,439
China	±0.843	±0.930	±1.167	±2.122	±1.499	±1.467	1,338	1,409	1,621	2,467	1,665	1,706
China (Hong Kong SAR)	0.536	0.494	0.511	0.464	0.456	-	15,686	18,171	21,119	23,813	27,627	-
China (Taiwan Province)	0.595	0.662	0.740	0.687	0.700	0.660	15,062	16,309	18,977	20,237	20,940	20,085
Colombia	0.070	0.071	0.067	0.062	0.064	0.095	17,919	17,396	15,683	13,196	17,460	22,860
Cote d'Ivoire	0.029	0.025	0.019	0.039	-	-						
Croatia	±0.020	±0.030	±0.015	-	-	-	13,136	21,630	10,673	-	-	-
Cyprus	0.000	0.000	0.000	0.000	0.001	0.001	23,341	18,513	29,524	21,932	19,653	20,818
Czechoslovakia (Former)	±0.084	-	-	-	-	-	4,178	-	-	-	-	-
Denmark	*0.622	*0.613	*0.680	*0.683	*0.853	*1.021	40,571	43,316	49,770	48,398	56,095	66,013
Ecuador	0.003	0.007	0.005	0.002	0.005	0.001	5,565	13,214	9,409	5,750	12,726	5,448
Egypt	*0.037	*0.022	*0.009	*0.011	*0.012	*0.010	5,652	2,013	1,292	1,657	1,729	1,912
El Salvador	-	0.001	-	0.003	0.002	0.000	-	-	-	8,274	18,413	343

Continued.

Depending on the table, * means *Enterprises*, *Engaged*, or *Factor Values*; ± means *Basis Unspecified*; § means *shown in millions*. For additional notes and sources, see Appendix I.

691

Value Added and Value Added per Employee

- Continued -

Country	Value Added ($ bil.)						Value Added per Employee ($)					
	1990	1991	1992	1993	1994	1995	1990	1991	1992	1993	1994	1995
Finland	±0.344	±0.332	±0.325	±0.304	±0.332	±0.476	59,335	59,347	62,465	56,214	59,347	67,468
FYR Macedonia	±0.001	±0.003	±0.003	±0.002	±0.003	±0.003	6,587	13,375	7,762	6,780	5,826	7,409
France	±4.109	±3.982	±4.224	±3.850	±4.093	±4.676	60,520	58,394	65,689	63,741	70,684	79,938
Germany (Western Part)	8.011	8.427	9.422	8.191	-	-	55,545	57,944	65,842	-	-	-
Greece	*0.016	*0.017	*0.019	*0.018	*0.018	*0.022	21,290	21,338	24,682	28,196	30,183	37,626
Honduras	*0.001	*0.001	*0.001	*0.000	*0.000	*0.001	4,018	4,484	9,204	5,287	5,172	3,474
Hungary	±0.293	±0.202	±0.146	±0.143	±0.142	±0.160	6,650	5,762	5,832	8,933	8,864	9,780
India	*0.165	*0.165	*0.154	*0.182	*0.182	*0.216	3,341	3,294	2,766	3,099	3,163	3,608
Indonesia	*0.010	*0.012	*0.030	*0.019	*0.053	*0.073	2,988	3,234	3,994	2,989	5,590	4,971
Iran (Islamic Republic of)	0.024	0.037	0.060	0.056	-	-	7,610	9,891	11,394	10,579	-	-
Ireland	*0.611	*0.635	*0.770	*0.708	*0.778	*1.002	77,377	75,640	84,363	74,436	78,644	89,386
Israel	0.125	0.156	0.060	0.147	0.161	0.149	41,828	40,990	15,328	36,659	41,300	35,468
Italy	±1.761	±1.821	±3.558	±3.141	±3.178	-	51,345	53,405	62,593	52,512	53,333	-
Japan	±13	±14	±14	±14	±23	±26	59,249	65,624	68,951	78,527	88,971	106,013
Jordan	0.002	0.002	0.002	0.001	0.001	0.002	7,842	9,996	9,316	8,917	4,465	5,025
Kenya	*0.002	*0.001	-	-	-	*0.000	9,648	5,708	-	-	-	1,332
Korea, Republic of	1.144	0.951	1.009	1.093	1.424	1.807	25,590	28,657	32,792	33,180	40,864	51,152
Latvia	-	-	-	*0.002	*0.003	*0.005	-	-	-	2,567	3,570	7,817
Macau	0.004	0.004	0.004	0.002	-	-	5,699	6,285	6,278	4,869	-	-
Malaysia	*0.097	*0.163	*0.162	*0.185	*0.235	*0.277	6,642	8,706	8,764	9,298	10,484	11,938
Malta	0.019	0.024	0.023	0.017	0.026	-	14,487	18,177	17,241	15,420	18,867	-
Mauritius	0.013	0.011	0.008	0.009	0.008	0.010	9,528	6,676	5,366	5,757	5,227	5,454
Mexico	0.081	0.097	0.093	0.131	0.104	0.077	31,019	35,209	37,373	48,978	47,099	40,030
Mongolia	0.001	0.000	0.000	0.000	0.000	-	3,098	5,252	2,950	48	358	-
Morocco	±0.007	±0.008	±0.008	±0.007	±0.007	±0.007	5,846	6,903	8,939	8,195	6,752	7,629
Netherlands	0.308	0.305	0.390	0.397	0.378	0.447	42,772	41,642	50,967	50,601	54,208	65,331
New Zealand	0.024	-	-	-	-	-	27,996	-	-	-	-	-
Norway	±0.082	±0.103	±0.105	±0.183	±0.226	±0.280	51,604	55,014	59,773	53,796	63,500	73,115
Pakistan	0.011	0.011	0.012	-	-	-	3,068	2,769	2,955	-	-	-
Panama	0.003	0.003	0.003	0.003	0.004	0.004	25,889	34,753	37,066	36,228	40,221	41,224
Peru	0.021	0.016	0.015	-	-	-	14,055	13,868	13,572	-	-	-
Philippines	0.011	0.013	0.018	0.039	-	-	3,943	4,796	5,315	5,303	-	-
Poland	±0.173	±0.129	±0.130	±0.123	±0.144	±0.214	4,669	4,447	5,651	6,128	7,567	12,594
Portugal	0.036	0.067	0.080	0.077	0.090	0.099	8,404	13,348	16,594	15,261	17,062	20,074
Puerto Rico	±0.889	±0.842	±0.940	±1.064	±1.024	±1.115	72,360	60,510	68,692	74,437	76,769	90,626
Russian Federation	-	-	-	±0.675	±0.846	±0.568	-	-	-	1,716	2,753	2,145
Singapore	*0.200	*0.230	*0.288	*0.343	*0.381	*0.423	24,258	26,772	32,549	41,356	42,863	47,034
Slovakia	-	-	-	±0.053	±0.049	-	-	-	-	5,236	5,370	-
Slovenia	-	-	-	0.065	0.085	0.115	-	-	-	-	-	-
South Africa	*0.160	*0.184	*0.245	*0.235	*0.233	*0.244	22,914	26,333	35,063	39,084	29,164	31,055
Spain	*0.375	*0.412	*0.441	±0.862	±0.837	±0.947	42,413	44,961	49,680	37,795	40,978	44,921
Sri Lanka	0.000	0.000	0.000	0.000	-	-	1,406	1,041	346	1,354	-	-
Sweden	*1.166	*0.955	*0.965	*0.818	*0.978	*1.063	71,518	53,665	64,125	54,185	57,195	62,844
Switzerland	2.443	2.421	2.664	2.619	2.927	3.462	62,953	64,216	74,404	74,820	80,204	93,558
Thailand	0.058	0.126	-	-	-	-	10,650	12,133	-	-	-	-
Turkey	0.087	0.094	0.172	0.187	0.121	0.181	19,321	20,129	33,691	34,617	24,467	35,352
United Kingdom	*3.661	*3.947	*4.071	*6.652	*7.201	*7.831	43,067	50,601	53,561	46,515	54,140	59,681
United Republic of Tanzania	*0.000	*0.000	-	-	-	-	758	1,170	-	-	-	-
United States of America	*77	*77	*83	*85	*86	*86	84,834	90,222	97,081	103,615	110,707	113,618
Uruguay	0.019	0.025	0.014	0.019	0.023	0.024	19,674	25,436	21,375	29,714	30,348	33,000
Venezuela	0.037	0.064	0.057	0.044	0.112	0.153	13,250	20,769	18,852	16,951	48,796	56,343
Yugoslavia	±0.005	±0.006	±0.012	-	±0.017	±0.010	-	1,241	2,773	-	4,483	2,692
Zambia	-	-	-	-	0.000	-	-	-	-	-	3,097	-
Zimbabwe	*0.002	*0.002	*0.002	*0.002	*0.002	*0.002	24,510	15,461	19,041	19,930	10,491	9,843

     Depending on the table, * means *Enterprises, Engaged,* or *Factor Values*; ± means *Basis Unspecified*; § means *shown in millions.* For additional notes and sources, see Appendix I.

Capital Investment

Country	Gross Fixed Capital Formation ($ mil.)						Per Establishment ($ mil)					
	1990	1991	1992	1993	1994	1995	1990	1991	1992	1993	1994	1995
Argentina	-	-	-	10	-	-	-	-	-	0.008	-	-
Australia	49	-	-	-	-	-	0.084	-	-	-	-	-
Austria	25	31	26	40	41	-	0.306	0.400	0.344	0.530	0.556	
Bahrain	-	-	0.125	-	-	-	-	-	0.063	-	-	-
Bangladesh	-	-	0.026	-	-	-	-	-	0.001	-	-	-
Bolivia	0.061	0.076	-0.064	0.036	0.063	-	0.004	0.004	-0.004	0.002	0.003	
Bosnia & Herzegovina	0.268	-	-	-	-	-	0.045	-	-	-	-	-
Canada	71	-	-	-	-	-	0.135	-	-	-	-	-
Chile	0.351	0.378	1.081	1.591	0.509	-	0.050	0.054	0.180	0.265	0.064	
China (Hong Kong SAR)	63	79	63	46	69	-	0.036	0.058	0.044	0.040	0.074	
Colombia	17	1.899	1.699	2.393	0.365	-	0.229	0.026	0.022	0.032	0.006	
Croatia	0.730	0.312	0.165	0.976	1.990	0.008	0.014	0.004	0.002	0.010	0.016	0.000
Cyprus	-	0.019	0.016	0.318	0.092	0.323	-	0.006	0.005	0.106	0.009	0.025
Czechoslovakia (Former)	17	-	-	-	-	-	0.796	-	-	-	-	-
Denmark	33	47	-	-	-	-	0.077	0.105	-	-	-	-
Ecuador	-0.130	1.213	1.131	-0.220	-0.554	-0.167	-0.012	0.110	0.103	-0.022	-0.050	-0.033
Egypt	4.250	1.260	4.014	-4.508		-	0.250	0.066	0.223	-0.250	-	-
El Salvador	-	0.248	-	0.027	0.052	-0.016	-	0.124	-	0.004	0.017	-0.016
Finland	39	25	21	23	28	27	0.414	0.243	0.207	0.206	0.250	0.201
FYR Macedonia	0.050	0.280	-	-	-	0.095	0.008	0.010	-	-	-	0.001
Germany (Western Part)	758	792	757	-	-	-	0.556	0.605	0.573	-	-	-
Greece	1.640	1.158	1.280	-	-	-	0.082	0.061	0.067	-	-	-
Hungary	29	17	12	13	-	-	0.059	0.019	0.023	0.022	-	-
India	60	52	82	100	-	-	0.081	0.061	0.103	0.106	-	-
Indonesia	1.628	2.051	44	36	61	61	0.031	0.037	0.672	0.589	0.926	0.838
Iran (Islamic Republic of)	5.289	5.341	9.148	12		-	0.129	0.191	0.269	0.396	-	-
Ireland	67	72	-	-	-	-	0.821	0.819	-	-	-	-
Israel	12	21	19	22	17	-	0.205	0.229	0.234	0.253	0.220	
Italy	189	172	328	244	259	-	0.461	0.414	0.409	0.305	0.329	
Japan	1,485	1,982	1,563	1,205	1,565	-	0.246	0.333	0.283	0.224	0.222	
Jordan	0.494	0.113	0.010	0.027	0.009	0.417	0.123	0.057	0.003	0.014	0.000	0.006
Korea, Republic of	175	187	153	185	211	329	0.155	0.210	0.160	0.153	0.162	0.237
Latvia	0.006	-0.020	0.001	0.026	0.517	0.232	0.002	-0.001	0.000	0.001	0.022	0.004
Macau	0.156	0.080	0.059	-0.358	-	-	0.078	0.040	0.029	-0.179	-	-
Malaysia	89	118	33	35	55	-	3.427	3.366	0.881	1.028	1.493	-
Malta	1.454	1.217	0.802	2.584	3.016	-	0.145	0.094	0.067	0.235	0.302	
Mexico	2.378	0.645	-	-	-	-	0.396	0.108	-	-	-	-
Mongolia	0.036	0.032	0.009	0.001	0.000	-	0.012	0.008	0.003	0.000	0.000	
Morocco	2.305	2.527	1.054	1.936	4.346	1.991	0.079	0.090	0.038	0.067	0.124	0.059
Netherlands	39	39	39	43	-	-	0.398	0.411	0.400	0.413	-	-
New Zealand	1.791	-	-	-	-	-	0.021	-	-	-	-	-
Norway	8.627	6.170	4.990	16	25	-	0.151	0.110	0.131	0.290	0.448	
Pakistan	-	1.253	-	-	-	-	-	0.021	-	-	-	-
Panama	0.048	0.219	-	-	-	-	0.004	0.044	-	-	-	-
Peru	0.997	0.870	1.479	-	-	-	0.010	0.008	0.014	-	-	-
Philippines	1.234	0.764	1.921	8.407	-	-	0.103	0.048	0.120	0.442	-	-
Poland	15	14	25	14	-	-	0.140	0.145	0.294	0.173	-	-
Portugal	6.720	4.125	10	4.515	9.675	-	0.030	0.013	0.033	0.013	0.024	
Romania	-	-	1.844	6.311	1.917	-	-	-	0.016	0.052	0.009	
Singapore	64	39	46	37	73	-	1.117	0.639	0.765	0.636	1.307	
Spain	29	27	28	31	28	68	0.047	0.044	0.047	-	-	-
Sri Lanka	0.111	0.033	0.009	0.078	-	-	0.012	0.005	0.004	0.011	-	-
Thailand	24	87	-	-	-	-	1.470	2.915	-	-	-	-
Turkey	16	5.034	22	15	19	-	0.414	0.114	0.299	0.178	0.213	-
Ukraine	109	55	-0.451	2.965	0.922	-0.355	0.978	0.229	-0.002	0.027	0.008	-0.003
United Kingdom	257	265	269	-	541	-	0.101	0.107	0.104	-	0.126	
United Republic of Tanzania	0.410	0.329	-	-	-	-	0.103	0.082	-	-	-	-
United States of America	4,030	4,240	4,302	4,104	3,823	3,982	-	-	0.282	-	-	-

Continued.

Depending on the table, * means *Enterprises, Engaged,* or *Factor Values;* ± means *Basis Unspecified;* § means *shown in millions.* For additional notes and sources, see Appendix I.

693

Capital Investment

- Continued -

Country	Gross Fixed Capital Formation ($ mil.)						Per Establishment ($ mil)					
	1990	1991	1992	1993	1994	1995	1990	1991	1992	1993	1994	1995
Uruguay	1.175	2.438	1.272	2.837	-	-	-	-	-	-	-	-
Venezuela	9.168	3.379	2.428	35	5.293	30	0.173	0.058	0.048	0.808	0.120	0.642
Yugoslavia	0.254	0.360	0.445	-	1.080	0.230	0.008	0.006	0.004	-	0.007	0.001

Indexes of Industrial Production

Country	Index of Industrial Production (1990=100)						Index of Employment (1990=100)					
	1990	1991	1992	1993	1994	1995	1990	1991	1992	1993	1994	1995
Albania	100	-	-	-	-	-	-	-	-	-	-	-
Algeria	100	96	-	-			-	-	-	-	-	-
Angola	100	100	100	100			-	-	-	-	-	-
Argentina	100	138	-	-			100	-	-	100	107	
Armenia	100						100	88	-	-	-	
Australia	100	97	97	111	-		100	92	100	-	-	
Austria	100	-	-	-	-		100	103	103	87	83	87
Bahrain	100	-	-	-	-		-	-	-	-	-	
Bangladesh	100	-	-	-	-		100	193	108	-	-	
Barbados	100	100	100	100	100	100	-	-	-	-	-	
Belgium	100	105	105	95	107	127	-	-	-	-	-	-
Belize	100	100	100	100	100	100	-	-	-	-	-	-
Bermuda	-	-	-	-	-	-	100	114	114	57	43	43
Bolivia	100	104	111	116	136	132	100	112	126	130	131	132
Bosnia & Herzegovina	-	-	-	-	-		100	105	-	-	-	
Botswana	100	100	100				-	-	-	-	-	
Brazil	100						-	-	-	-	-	
Burkina-Faso	100	100	100				-	-	-	-	-	
Burundi	100	100	100	100			-	-	-	-	-	
Cambodia	100						100	86	-	-	-	
Cameroon	100	100	100	100	100	100	-	-	-	-	-	
Canada	100	105	106	107	115	122	100	94	83	75	81	79
Central African Republic	100	100	100	100	100		-	-	-	-	-	
Chile	100	132	-	-	-		100	105	105	106	135	117
China	100						100	105	114	137	143	137
China (Hong Kong SAR)	100	103	105	107	111	121	100	80	71	57	48	
China (Taiwan Province)	100	104	106	97	99	96	100	103	99	86	85	83
Colombia	100	106	88	85	89	118	100	104	110	121	94	106
Congo	100	100	100	100	-		-	-	-	-	-	
Costa Rica	100						100	113	114	106	-	-
Cote d'Ivoire	100	100	100	100	100	100	-	-	-	-	-	-
Croatia	100	-	-	-	-	-	100	91	91	86	74	81
Cyprus	100	103	118	116	131	140	100	117	117	167	783	967
Czechoslovakia (Former)	100	59	44	-	-		100	90	-	-	-	
Czech Republic	100						100	100	159	135	-	
Dem. Rep. of the Congo	100	100	100	-	-		-	-	-	-	-	
Denmark	100	95	96	102	118	122	100	92	89	92	99	101
Dominican Republic	100	-	-	-	-		-	-	-	-	-	
Ecuador	100	87	116	153	-		100	102	96	78	71	26
Egypt	100	283	271	255	251	194	100	165	111	103	101	76
El Salvador	100	100	-	-	-		-	-	-	-	-	-
Ethiopia and Eritrea	100	100	100	-	-		-	-	-	-	-	
Fiji	100	100	100	100	100	100	-	-	-	-	-	
Finland	100	88	87	97	103	113	100	97	90	93	97	122
FYR Macedonia	100						100	108	154	156	200	184
France	100	105	104	104	106	105	100	100	95	89	85	86
Gabon	100	-	-	-	-		-	-	-	-	-	
Gambia	100	-	-	-	-		-	-	-	-	-	
Germany (Eastern Part)	100	-	-	-	-		-	-	-	-	-	
Germany (Western Part)	100	104	102	93	93	95	100	101	99	-	-	-

Continued.

Depending on the table, * means *Enterprises, Engaged,* or *Factor Values*; ± means *Basis Unspecified*; § means *shown in millions*. For additional notes and sources, see Appendix I.

Indexes of Industrial Production

- Continued -

Country	Index of Industrial Production (1990=100)						Index of Employment (1990=100)					
	1990	1991	1992	1993	1994	1995	1990	1991	1992	1993	1994	1995
Ghana	100	-	-	-	-	-	-	-	-	-	-	-
Greece	100	116	109	82	67	65	100	105	99	83	79	75
Guatemala	100	-	-	-	-	-	-	-	-	-	-	-
Guyana	100	100	100	100	100	100	-	-	-	-	-	-
Haiti	100	100	100	100	100	100	-	-	-	-	-	-
Honduras	100	122	143	201	136	157	100	84	54	59	65	109
Hungary	100	82	40	50	50	57	100	80	57	36	36	37
Iceland	100	-	-	-	-	-	-	-	-	-	-	-
India	100	78	74	76	71	81	100	101	113	119	116	121
Indonesia	100	-	-	-	-	-	100	114	230	189	288	442
Iran (Islamic Republic of)	100	-	-	-	-	-	100	117	163	166	-	-
Iraq	100	-	-	-	-	-	-	-	-	-	-	-
Ireland	100	101	112	118	124	145	100	106	115	120	125	142
Israel	100	105	124	124	140	-	100	127	130	133	130	140
Italy	100	85	88	78	-	-	100	99	166	174	174	-
Japan	100	104	85	75	70	68	100	101	94	85	119	114
Jordan	100	-	-	-	-	-	100	78	98	59	111	172
Kenya	100	-	-	-	-	-	100	118	123	135	147	161
Korea, Republic of	100	106	106	118	136	153	100	74	69	74	78	79
Kuwait	100	-	-	-	-	-	-	-	-	-	-	-
Kyrgyzstan	100	-	-	-	-	-	100	88	77	59	39	-
Lao P.D.R.	100	-	-	-	-	-	-	-	-	-	-	-
Latvia	100	-	-	-	-	-	100	95	50	32	38	31
Lebanon	100	100	100	100	-	-	-	-	-	-	-	-
Libyan Arab Jamahiriya	100	100	100	100	-	-	-	-	-	-	-	-
Lithuania	100	-	-	-	-	-	-	-	-	-	-	-
Luxembourg	100	-	-	-	-	-	-	-	-	-	-	-
Macau	100	-	-	-	-	-	100	104	93	62	-	-
Madagascar	100	100	100	100	-	-	-	-	-	-	-	-
Malawi	100	100	100	100	100	100	-	-	-	-	-	-
Malaysia	100	-	-	-	-	-	100	128	127	136	153	159
Mali	100	100	100	100	100	100	-	-	-	-	-	-
Malta	100	113	119	98	-	-	100	101	102	84	103	-
Mauritius	100	-	-	-	-	-	100	121	107	110	110	130
Mexico	100	111	110	110	118	101	100	105	95	102	84	74
Mongolia	100	77	101	34	50	102	100	42	21	154	55	7
Morocco	100	-	-	-	-	-	100	101	75	72	88	82
Mozambique	100	-	-	-	-	-	100	117	126	122	119	-
Myanmar	100	100	100	100	100	100	-	-	-	-	-	-
Nepal	100	-	-	-	-	-	-	-	-	-	-	-
Netherlands	100	-	-	-	-	-	100	102	106	109	97	95
New Zealand	100	-	-	-	-	-	100	104	123	56	-	-
Nicaragua	100	-	-	-	-	-	-	-	-	-	-	-
Nigeria	100	-	-	-	-	-	-	-	-	-	-	-
Norway	100	105	107	113	123	134	100	118	111	215	225	242
Pakistan	100	-	-	-	-	-	100	109	114	-	-	-
Panama	100	101	109	105	122	125	100	71	73	72	77	78
Paraguay	100	19	19	15	9	9	-	-	-	-	-	-
Peru	100	110	121	124	183	185	100	79	76	-	-	-
Philippines	100	123	112	125	153	163	100	97	117	255	-	-
Poland	100	82	82	95	106	134	100	78	62	54	51	46
Portugal	100	-	-	-	-	-	100	120	115	119	125	116
Puerto Rico	100	-	-	-	-	-	100	113	111	116	109	100
Republic of Moldova	100	-	-	-	-	-	-	-	-	-	-	-
Romania	100	89	63	63	54	73	100	87	76	49	51	-
Russian Federation	100	-	-	-	-	-	-	-	-	-	-	-
Saudi Arabia	100	-	-	-	-	-	-	-	-	-	-	-
Senegal	100	100	100	100	100	100	-	-	-	-	-	-
Singapore	100	99	78	65	65	48	100	104	107	100	108	109
Slovakia	100	-	-	-	-	-	-	-	-	-	-	-

Continued.

Depending on the table, * means *Enterprises*, *Engaged*, or *Factor Values*; ± means *Basis Unspecified*; § means *shown in millions*. For additional notes and sources, see Appendix I.

695

Indexes of Industrial Production

- Continued -

Country	Index of Industrial Production (1990=100)						Index of Employment (1990=100)					
	1990	1991	1992	1993	1994	1995	1990	1991	1992	1993	1994	1995
Slovenia	100	101	95	97	106	125	-	-	-	-	-	-
Somalia	100	100	100	-	-	-	-	-	-	-	-	-
South Africa	100	107	119	119	119	115	100	100	100	86	114	112
Spain	100	104	106	93	100	102	100	104	100	258	231	238
Sri Lanka	100	94	101	124	154	213	100	95	17	113	-	-
Sudan	100	100	100	-	-	-	-	-	-	-	-	-
Suriname	100	-	-	-	-	-	100	100	105	105	-	-
Swaziland	100	100	100	-	-	-	-	-	-	-	-	-
Sweden	100	93	93	102	107	105	100	109	92	93	105	104
Switzerland	100	96	111	117	105	100	100	97	92	90	94	95
Syrian Arab Republic	100	100	100	100	100	100	-	-	-	-	-	-
Thailand	100	-	-	-	-	-	100	190	-	-	-	-
Tonga	100	-	-	-	-	-	-	-	-	-	-	-
Tunisia	100	-	-	-	-	-	-	-	-	-	-	-
Turkey	100	-	-	-	-	-	100	104	114	120	110	114
Uganda	100	100	100	100	100	100	-	-	-	-	-	-
Ukraine	100	-	-	-	-	-	100	98	102	71	63	53
United Arab Emirates	100	-	-	-	-	-	-	-	-	-	-	-
United Kingdom	100	96	94	98	94	96	100	92	89	168	156	154
United Republic of Tanzania	100	-	-	-	-	-	100	102	-	-	-	-
USSR (Former)	100	-	-	-	-	-	100	-	-	-	-	-
United States of America	100	101	101	102	101	102	100	95	95	91	86	84
Uruguay	100	104	106	104	117	113	100	102	66	67	76	73
Venezuela	100	-	-	-	-	-	100	111	107	93	82	97
Yugoslavia	100	-	-	-	-	-	-	-	-	-	-	-
Yugoslavia (Former)	100	73	-	-	-	-	-	-	-	-	-	-
Zambia	100	97	90	75	64	68	-	-	-	-	-	-
Zimbabwe	100	98	75	194	173	109	100	100	100	100	200	155

      Depending on the table, * means *Enterprises, Engaged,* or *Factor Values;* ± means *Basis Unspecified;* § means *shown in millions.* For additional notes and sources, see Appendix I.

Representative Companies in Sector

Name	Address	Tele-phone	Fax	Employ-ment	Y
Honeywell Inc. Space and Aviation Control	PO Box 21111, Phoenix, Arizona 85036, U.S.A.	602-436-2311	602-436-3000	10,000	97
Mallinckrodt Medical Inc.	PO Box 5840, St. Louis, Missouri 63134, U.S.A.	314-654-2000		10,000	97
Teleflex Inc.	630 W. Germantown Pike, Ste. 450, Plymouth Meeting, Pennsylvania 19462-1074, U.S.A.	610-834-6301	610-834-8228	9,700	96
C.R. Bard Inc.	730 Central Ave., Murray Hill, New Jersey 07974, U.S.A.	908-277-8000		9,550	97
Alcon Laboratories Inc.	6201 South Fwy., Fort Worth, Texas 76134-2099, U.S.A.	817-293-0450	817-551-4629	9,500	97
Siemens Energy and Automation Inc.	3333 Old Milton Pkwy., Alpharetta, Georgia 30202, U.S.A.	770-751-2000	770-751-2001	9,500	97
GenCorp Inc.	175 Ghent Rd., Fairlawn, Ohio 44333-3300, U.S.A.	330-869-4200	330-869-4211	9,460	97
Imation Corp.	1 Imation Place, Oakdale, Minnesota 55128, U.S.A.	612-704-4000		9,400	97
Thermo Instrument Systems Inc.	860 W. Airport Fwy., 301, Hurst, Texas 76054, U.S.A.	817-685-2171	817-685-2173	9,398	97
Sundstrand Corp.	PO Box 7003, Rockford, Illinois 61125-7003, U.S.A.	815-226-6000	815-226-2699	9,200	95
Tektronix Inc.	PO Box 1000, Wilsonville, Oregon 97070-1000, U.S.A.	503-627-7111		8,392	97
General Electric Co. Aircraft Electronics and Defense Sys. D	600 Main St., Johnson City, New York 13790-1888, U.S.A.	607-770-3901		8,200	
Ceridian Corp.	8100 34th Ave. S, Minneapolis, Minnesota 55425-1604, U.S.A.	612-853-8100	612-853-7896	8,000	97
Gambro Healthcare Inc.	1185 Oak St., Lakewood, Colorado 80215-4407, U.S.A.	303-232-6800		8,000	97
Mallinckrodt Inc.	PO Box 5840, St. Louis, Missouri 63134, U.S.A.	314-654-2000		8,000	97
J. Bibby & Sons PLC	16 Stratford Pl., London W1N 9AF, United Kingdom	71 6296243	71 4090556	7,976	93
Ball Corp.	PO Box 2407, Muncie, Indiana 47307-0407, U.S.A.	765-747-6100		7,900	97
Chemed Corp.	255 E. 5th St., Cincinnati, Ohio 45202-4726, U.S.A.	513-762-6900		7,886	96
Iwka AG	Gartenstr. 71, Karlsruhe D-76135, Germany	721 1430	721 143243	7,774	94
Eagle-Picher Industries Inc.	PO Box 779, Cincinnati, Ohio 45201, U.S.A.	513-721-7010	513-721-2341	7,500	95
Sola International Inc.	2420 Sand Hill Rd., 200, Menlo Park, California 94025, U.S.A.	650-324-6868		7,500	96
Hubbell Inc.	584 Derby Milford Rd., Orange, Connecticut 06477-4024, U.S.A.	203-799-4100		7,405	94
AGFA Gevaert NV	Septestraat 27, Mortsel B-2640, Belgium	3 4447110	3 4447102	7,300	94
W.L. Gore and Associates Inc.	PO Box 9329, Newark, Delaware 19714, U.S.A.	302-738-4880	302-292-4159	7,000	97
AMETEK Inc.	Station Sq., Paoli, Pennsylvania 19301, U.S.A.	610-647-2121	610-296-3412	6,700	97
AlliedSignal Electronic and Avionics Systems	400 N. Rogers Rd., Olathe, Kansas 66062, U.S.A.	913-782-0400	913-791-1302	6,500	97
Varian Associates Inc.	3050 Hansen Way, Palo Alto, California 94304-1000, U.S.A.	415-493-4000		6,500	97
Sybron International Corp.	411 E. Wisconsin Ave., Milwaukee, Wisconsin 53202, U.S.A.	414-274-6600	414-274-6561	6,300	97
Donaldson Company Inc.	PO Box 1299, Minneapolis, Minnesota 55440, U.S.A.	612-887-3131		6,230	97
Allergan Inc.	PO Box 19534, Irvine, California 92713-9534, U.S.A.	714-752-4500	714-955-6987	6,100	97
Guidant Corp.	PO Box 44906, Indianapolis, Indiana 46244-0906, U.S.A.	317-971-2000	317-971-2040	6,017	97
Baxter Healthcare Corp. of Puerto Rico	PO Box 2025, Carolina 00984, Puerto Rico	787750-7373		6,000	92
Siebe Appliance Controls	2809 Emerywood Pkwy., Richmond, Virginia 23294-3743, U.S.A.	804-756-6500	804-756-6564	6,000	96
Orion Corp.	Orionintie 1, Espoo SF-02101, Finland	4291	4293815	5,900	94
United States Surgical Corp.	150 Glover Ave., Norwalk, Connecticut 06856, U.S.A.	203-845-1000	203-845-4125	5,776	97
Stryker Corp.	PO Box 4085, Kalamazoo, Michigan 49003-4085, U.S.A.	616-385-2600	616-385-1062	5,691	97
Perkin-Elmer Corp.	761 Main Ave., Norwalk, Connecticut 06859-0001, U.S.A.	203-762-1000	203-762-6000	5,685	97
Elco Holdings Ltd.	17 Haneviim St., Ramat Hasharon 47279, Israel	3 5121503	3 6834218	5,555	97
Foxboro Co.	33 Commercial St., Foxboro, Massachusetts 02035, U.S.A.	508-543-8750	508-549-6750	5,500	97
Perindustrian Tentara Nasional Id Angkatan Darat PT	Jl. Jenderal Gatot Subroto Kiaracondong, Bandung, Indonesia	22312073	22 301222	5,500	97
DENTSPLY International Inc.	PO Box 872, York, Pennsylvania 17405-0872, U.S.A.	717-845-7511	717-854-2343	5,300	97
Nellcor Puritan Bennett Inc.	4280 Hacienda Dr., Pleasanton, California 94588, U.S.A.	510-463-4000	510-463-4450	5,000	97
Teradyne Inc.	179 Lincoln St., Boston, Massachusetts 02111-2473, U.S.A.	617-482-2700	617-422-3440	5,000	97
Zimmer Inc.	PO Box 708, Warsaw, Indiana 46581-0708, U.S.A.	219-267-6131		5,000	97
Bayer Corp. Diagnostics Div.	511 Benedict Ave., Tarrytown, New York 10591, U.S.A.	914-631-8000	914-524-2132	4,900	97
Ohmeda Inc.	110 Allen Rd., Liberty Corner, New Jersey 07938, U.S.A.	908-647-9200		4,800	96
Millipore Corp.	80 Ashby Rd., Bedford, Massachusetts 01730-2271, U.S.A.	781-275-9200	781-553-3110	4,754	97
Landis and Staefa Inc.	1000 Deerfield Pkwy., Buffalo Grove, Illinois 60089, U.S.A.	847-215-1000		4,500	97
Ranco Inc.	8161 U.S. Rte. 42 N., Plain City, Ohio 43064, U.S.A.	614-873-9219		4,500	95
Sulzer Inc.	200 Park Ave., New York, New York 10166, U.S.A.	212-949-0999	212-370-1138	4,500	97

Continued

Representative Companies in Sector
- Continued -

Name	Address	Tele-phone	Fax	Employ-ment	Y
Invacare Corp.	PO Box 4028, Elyria, Ohio 44036, U.S.A.	216-329-6000	216-366-9008	4,470	97
ABB Kent Holdings PLC	Lea Rd., Luton LU1 3AE, United Kingdom	582 31255	582 421115	4,327	87
Sunrise Medical Inc.	2382 Faraday Ave., Ste. 200, Carlsbad, California 92008-7220, U.S.A.	760-930-1500	760-930-1580	4,254	97
Mine Safety Appliances Co.	PO Box 426, Pittsburgh, Pennsylvania 15230, U.S.A.	412-967-3000	412-967-3460	4,200	97
International General Electric USA Ltd.	Hammersmith, London W6 8BX, United Kingdom	81 7419900	81 7419460	4,026	93
AEG Corp.	180 Mount Airy Rd., PO Box 609, Basking Ridge, New Jersey 07920, U.S.A.	908-204-8900	908-204-8999	4,000	93
Hill Rom Company Inc.	1069 State Rte. 46, Batesville, Indiana 47006-9167, U.S.A.	812-934-7777	812-934-8189	4,000	97
STERIS Corp.	5960 Heisley Rd., Mentor, Ohio 44060, U.S.A.	216-354-2600		4,000	96
Maxxim Medical Inc.	10300 49th St., N., Clearwater, Florida 33762-5000, U.S.A.	813-561-2100	813-561-2180	3,958	97
Imo Industries Inc.	1009 Lenox Dr., Lawrenceville, New Jersey 08648, U.S.A.	609-896-7600	609-896-7688	3,900	96
St. Jude Medical Inc.	1 Lillehei Plz., St. Paul, Minnesota 55117-9983, U.S.A.	612-483-2000	612-482-8318	3,620	97
Block Drug Company Inc.	257 Cornelison Ave., Jersey City, New Jersey 07302-9988, U.S.A.	201-434-3000	201-434-3032	3,600	95
KLA-Tencor Corp.	1 Technology Dr., Milpitas, California 95035, U.S.A.	408-434-4200	408-434-4266	3,600	97
Smith and Nephew Richards Inc.	1450 Brooks Rd., Memphis, Tennessee 38116, U.S.A.	901-396-2121	901-348-6151	3,600	96
Schlumberger PLC	1 Kingsway, 1st Fl., London WC2B 6XH, United Kingdom	71 2407301		3,594	93
Genzyme Corp.	1 Kendall Sq., Cambridge, Massachusetts 02139, U.S.A.	617-252-7500	617-374-7368	3,516	96
Boehringer Mannheim Corp.	PO Box 50457, Indianapolis, Indiana 46250, U.S.A.	317-845-2000	317-576-7129	3,500	97
Chiron Diagnostics Corp.	333 Coney St., East Walpole, Massachusetts 02032, U.S.A.	508-668-5000	508-660-4591	3,500	97
Cubic Corp.	PO Box 85587, San Diego, California 92186-5587, U.S.A.	619-277-6780	619-277-1878	3,500	97
Siemens Canada Ltd.	2185 Derry Rd. W, Mississauga, Ontario L5N 7A6, Canada	905-819-8000	905-819-5777	3,500	98
Spirax-Sarco Engineering PLC	Cirencester Rd., Cheltenham GL53 8ER, United Kingdom	242 521361	242 224163	3,489	94
ESCO Electronics Corp.	8888 Ladue Rd., 200, St. Louis, Missouri 63124-2090, U.S.A.	314-213-7200		3,400	97
Oak Industries Inc.	1000 Winter St., Waltham, Massachusetts 02154, U.S.A.	781-890-0400	781-890-8585	3,373	97
Esterline Technologies Corp.	10800 Northeast 8th St., Bellevue, Washington 98004, U.S.A.	425-453-9400	425-453-2916	3,360	97
Giddings and Lewis Inc.	PO Box 590, Fond du Lac, Wisconsin 54936-0590, U.S.A.	414-921-9400	414-929-4522	3,315	96
Marquette Medical Systems	8200 W. Tower Ave., Milwaukee, Wisconsin 53223, U.S.A.	414-355-5000	414-355-3790	3,240	97
Moog Inc.	Plant 24, East Aurora, New York 14052-0018, U.S.A.	716-652-2000	716-687-4457	3,229	97
Johnson & Johnson Management Ltd.	Roxborough Way, Maidenhead SL6 3UG, United Kingdom	628 822222	628 826818	3,097	94
Leica PLC	Clifton Rd., Cambridge CB3 8EL, United Kingdom			3,000	87
Orbital Sciences Corp.	21700 Atlantic Blvd., Dulles, Virginia 20166, U.S.A.	703-406-5000		3,000	96
Tillotson Corp.	59 Waters Ave., Everett, Massachusetts 02149, U.S.A.	617-387-9400	617-389-9639	3,000	95
Tokheim Corp.	PO Box 360, Fort Wayne, Indiana 46801-0360, U.S.A.	219-470-4600	219-484-1110	2,903	97
Allen Group Inc.	25101 Chagrin Blvd., Beachwood, Ohio 44122-5619, U.S.A.	216-765-5800	216-765-0410	2,900	96
Baxter Healthcare Corp. Baxter Diagnostics Inc.	PO Box 520672, Miami, Florida 33152, U.S.A.	305-592-2311		2,840	
NU-kote Holding Inc.	17950 Preston, Ste. 690, Dallas, Texas 75252, U.S.A.	972-250-2785		2,800	96
DePuy Inc.	700 Orthopaedic Dr., Warsaw, Indiana 46581-0988, U.S.A.	219-267-8143		2,780	95
L.S. Starrett Co.	121 Crescent St., Athol, Massachusetts 01331-1915, U.S.A.	508-249-3551	508-249-8495	2,740	97
Balzers Ltd.	Balzers FL-9496, Liechtenstein	75 3884111	75 3885407	2,723	92
Advanced Technology Laboratories Inc.	PO Box 3003, Bothell, Washington 98041-3003, U.S.A.	206-487-7000	206-487-7885	2,703	96
Puritan-Bennett Corp.	9401 Indian Creek Pkwy., Bldg. 4, Overland Park, Kansas 66210, U.S.A.	913-661-0444	913-661-0234	2,700	94
Tesla, Stropkov AS	Hviedoslavova 37/46, Stropkov 091 01, Slovakia	938 3422	938 3466	2,600	93
BMC Industries Inc.	2 Appletree Sq., 850, Minneapolis, Minnesota 55423, U.S.A.	612-851-6000	612-851-6050	2,597	97
Figgie International Inc.	4420 Sherwin Rd., Willoughby, Ohio 44094, U.S.A.	216-953-2700	216-951-1724	2,582	95
Biomet Inc.	PO Box 587, Warsaw, Indiana 46581-0587, U.S.A.	219-267-6639	219-267-8137	2,550	97
Fluke Corp.	PO Box 9090, Everett, Washington 98206-9090, U.S.A.	206-347-6100	206-356-5116	2,525	97
Coastcast Corp.	3025 E. Victoria St., Compton, California 90221-5616, U.S.A.	310-638-0595		2,517	95
Astronautics Corporation of America	PO Box 523, Milwaukee, Wisconsin 53201-0523, U.S.A.	414-447-8200	414-447-8231	2,500	96
Bio-Rad Laboratories Inc.	1000 Alfred Nobel Dr., Hercules, California 94547, U.S.A.	510-724-7000	510-741-5817	2,500	96
ITW Finishing Systems	3939 W. 56th St., Indianapolis, Indiana 46254, U.S.A.	317-298-5000		2,500	
Kodak Brasileira Com. e Ind. SA	Av. Maria Coelho de Aguiar 215, Sao Paulo 05804-900, Brazil	11 8486000	11 8445605	2,500	96
Leo Pharmaceutical Products	Industriparken 55, Ballerup DK-2750, Denmark	44923800	44941516	2,500	93
Starkey Labs Inc.	6700 Washington Ave. S, Eden Prairie, Minnesota 55344,	612-941-6401	612-828-9262	2,500	97

Continued

For sources and notes, see Appendix I.

Representative Companies in Sector
- Continued -

Name	Address	Tele-phone	Fax	Employ-ment	Y
	U.S.A.				
Talley Industries Inc.	2702 N. 44th St., 100A, Phoenix, Arizona 85008, U.S.A.	602-957-7711		2,494	96
Bristol-Myers Squibb Holdings Ltd.	Ickenham, Uxbridge UB10 8NS, United Kingdom	895 639911	895 636975	2,459	93
Tech-Sym Corp.	10500 Westoffice Dr., 200, Houston, Texas 77042-5391, U.S.A.	713-785-7790	713-780-3524	2,459	97
Bayer Inc.	77 Belfield Rd., Toronto, Ontario M9W 1G6, Canada	416-248-0771	416-248-1438	2,450	98
Photo-Me International PLC	Great Bookham, Leatherhead KT23 3EU, United Kingdom	372 453399	372 459064	2,427	94
ALARIS Medical Inc.	10221 Wateridge Circle, San Diego, California 92121-2733, U.S.A.	619-458-7000	619-458-7760	2,400	96
Elta Electronics Industries Ltd.	PO Box 303, Ashdod 77268, Israel	8 8572410	8 8564568	2,382	97
Pacific Scientific Co.	620 Newport Center Dr., 700, Newport Beach, California 92660, U.S.A.	714-720-1714	714-720-1083	2,375	97
Simpson Industries Inc.	47603 Halyard Dr., Plymouth, Michigan 48170-2429, U.S.A.	313-207-6200	313-207-6500	2,355	97
Litton Industries Inc. Guidance and Control Systems	5500 Canoga Ave., Woodland Hills, California 91367, U.S.A.	818-715-4040	818-712-7219	2,300	97
International Paper Holdings (UK) Ltd.	Rd. 3 Industrial Estate, Winsford CW7 3RJ, United Kingdom			2,270	94
Arrow International Inc.	PO Box 12888, Reading, Pennsylvania 19612, U.S.A.	610-378-0131	610-374-5360	2,264	97
Serv-Tech Inc.	PO Box 4334, Houston, Texas 77210, U.S.A.	713-644-9974	713-644-0731	2,243	95
Brown and Sharpe Manufacturing Co.	200 Frenchtown Rd., North Kingstown, Rhode Island 02852-2937, U.S.A.	401-886-2000	401-888-2762	2,223	97
Radiometer AS	Emdrupvej 72, Copenhagen DK-2400, Denmark	39696311	31678111	2,200	93
Raytheon E. Systems Inc. Falls Church Div.	7700 Arlington Blvd., Falls Church, Virginia 22046, U.S.A.	703-560-5000	703-280-4627	2,200	97
CONMED Corp.	310 Broad St., Utica, New York 13501, U.S.A.	315-797-8375	315-797-0321	2,161	97
Tecnol Medical Products Inc.	7201 Industrial Park Blvd., Fort Worth, Texas 76180, U.S.A.	817-581-6424		2,150	96
Coherent Inc.	PO Box 54980, Santa Clara, California 95056-0980, U.S.A.	408-764-4000		2,131	97
FII Group PLC	48 George St., London W1H 5PG, United Kingdom	71 9358463	71 4860705	2,107	93
American Safety Razor Co.	PO Box 500, Staunton, Virginia 24402-0500, U.S.A.	540-248-8000	540-248-0522	2,100	97
ELMOT	Westerplatte 29, Swidnica PL-58-100, Poland	52 5821	52 4224	2,100	93
Instituto Nacional de Seguro	Calles 9-9B, San Jose 1000, Costa Rica	2235800	2553427	2,100	
Welch Allyn Inc.	4341 State Street Rd., Skaneateles Falls, New York 13153, U.S.A.	315-685-4100	315-685-3361	2,100	97
Halma PLC	Rectory Way, Amersham HP7 0DE, United Kingdom	494 721111	494 728032	2,099	94
Thermo Optek Corp.	8 E. Forge Pkwy., Franklin, Massachusetts 02038-3157, U.S.A.	508-541-7111	508-541-0140	2,065	96
Elscint Ltd.	Science Industrial Center, Haifa 31004, Israel	4 8310310	4 8550962	2,026	97
3M Canada Co.	1840 Oxford St. E, London, Ontario N5V 3R6, Canada	519-451-2500	519-452-6262	2,000	98
Cordis Corp.	PO Box 025700, Miami Lakes, Florida 33102-5700, U.S.A.	305-824-2000	305-824-2080	2,000	97
Ganz Measuring Instruments Factory	Ullui ut 64, Budapest H-1191, Hungary	1 1470740	1 1271025	2,000	93
National Patent Development Corp.	9 West 57th St., New York, New York 10019, U.S.A.	212-826-8500	212-230-9545	2,000	97
Roper Industries Inc.	160 Ben Burton Rd., Bogart, Georgia 30622, U.S.A.	706-369-7170		2,000	97
MTS Systems Corp.	14000 Technology Dr., Eden Prairie, Minnesota 55344-2290, U.S.A.	612-937-4000	612-937-4095	1,981	97
OPTEK Technology Inc.	1215 W. Crosby Rd., Carrollton, Texas 75006, U.S.A.	214-323-2200	214-323-7009	1,971	97
Thai Kawasumi Co., Ltd.	Pathumwan, Bangkok 10330, Thailand	2548229	2548239	1,922	94
Gerber Scientific Inc.	83 Gerber Rd., W, South Windsor, Connecticut 06074, U.S.A.	860-644-1551	860-643-7039	1,900	97
W. Haking Industries Thailand Ltd.	Kwaeng Sa-Maedum, Bangkok 10150, Thailand	4150047	4153528	1,900	97
Life Sciences International PLC	51 Aldwych, London WC2B 4LS, United Kingdom	71 2403445		1,893	93
Waters Corp.	34 Maple St., Milford, Massachusetts 01757, U.S.A.	508-478-2000		1,880	95
Kollmorgen Corp.	1601 Trapelo Rd., Waltham, Massachusetts 02154, U.S.A.	617-890-5655	617-890-7150	1,863	97
Alvis PLC	215 Vauxhall Bridge Rd., London SW1V 1EN, United Kingdom	71 8218080	71 9317433	1,832	93
Kearfott Guidance and Navigation Corp.	150 Totowa Rd., Wayne, New Jersey 07474-0595, U.S.A.	201-785-6993	201-785-5900	1,800	96
SCIMED Life Systems Inc.	1 Scimed Pl., Maple Grove, Minnesota 55311-1566, U.S.A.	612-494-1700	612-550-7777	1,800	93
United Industrial Corp.	18 E. 48th St., New York, New York 10017, U.S.A.	212-752-8787	212-838-4629	1,800	97
Wesley-Jessen Co.	400 W. Superior St., Chicago, Illinois 60610, U.S.A.	312-751-6357		1,800	93
EL OP Electro Optics Industries Ltd.	Kiryat Weizman, Rehovot 76327, Israel	8 9386404	8 9386237	1,790	97
Acuson Corp.	PO Box 7393, Mountain View, California 94039-7393, U.S.A.	415-969-9112		1,777	97

Product Tables
Gas Meters

Unit of Measure: Number

	1990 Value	%	1991 Value	%	1992 Value	%	1993 Value	%	1994 Value	%	1995 Value	%	1996 Value	%
Total Production	15,954,632	100.0	17,780,105	100.0	19,442,355	100.0	19,175,663	100.0	15,776,597	100.0	16,520,242	100.0	17,060,601	100.0
Regions														
America, North	2,528,250	15.8	2,315,250	13.0	2,359,250	12.1	2,309,000	12.0	2,560,000	16.2	2,462,000	14.9	2,615,000	15.3
Asia	6,809,000	42.7	8,810,000	49.5	8,531,000	43.9	7,847,000	40.9	5,211,000	33.0	5,394,000	32.7	6,015,600	35.3
Europe	6,617,383	41.5	6,654,855	37.4	8,552,105	44.0	9,019,663	47.0	8,005,597	50.7	8,664,242	52.4	8,430,001	49.4
Leading Producers														
Japan	6,322,000	39.6	8,230,000	46.3	7,594,000	39.1	6,763,000	35.3	3,938,000	25.0	3,989,000	24.1	4,373,000	25.6
United Kingdom	*5,131,383	32.2	*5,131,383	28.9	*5,131,383	26.4	5,764,545	30.1	4,498,220	28.5	*5,131,383	31.1	*5,131,383	30.1
United States of America	2,355,000	14.8	2,142,000	12.0	2,186,000	11.2	2,159,000	11.3	2,401,000	15.2	2,374,000	14.4	2,319,000	13.6
Germany	-		-		1,881,000	9.7	1,719,000	9.0	1,956,000	12.4	1,967,224	11.9	1,719,724	10.1
Korea, Republic of	487,000	3.1	580,000	3.3	937,000	4.8	1,084,000	5.7	1,273,000	8.1	1,405,000	8.5	*1,642,600	9.6

Liquid Meters

Unit of Measure: Number

	1990 Value	%	1991 Value	%	1992 Value	%	1993 Value	%	1994 Value	%	1995 Value	%	1996 Value	%
Total Production	15,462,500	100.0	19,123,500	100.0	24,297,000	100.0	24,353,000	100.0	26,828,000	100.0	25,864,133	100.0	26,445,203	100.0
Regions														
America, North	5,067,000	32.8	4,701,000	24.6	4,876,000	20.1	5,015,000	20.6	5,378,000	20.0	4,612,000	17.8	4,747,000	18.0
America, South	172,500	1.1	172,500	0.9	221,000	0.9	229,000	0.9	122,000	0.5	118,000	0.5	172,500	0.7
Asia	4,701,000	30.4	4,936,000	25.8	5,485,000	22.6	4,858,000	19.9	5,154,000	19.2	4,740,000	18.3	5,087,923	19.2
Europe	5,108,000	33.0	8,866,000	46.4	13,233,000	54.5	13,735,000	56.4	15,624,000	58.2	15,810,133	61.1	15,819,780	59.8
Oceania	414,000	2.7	448,000	2.3	482,000	2.0	516,000	2.1	550,000	2.1	584,000	2.3	618,000	2.3
Leading Producers														
Germany	-		-		5,874,000	24.2	5,810,000	23.9	7,425,000	27.7	7,601,133	29.4	7,594,857	28.7
United States of America	4,784,000	30.9	4,418,000	23.1	4,593,000	18.9	4,821,000	19.8	5,080,000	18.9	4,327,000	16.7	4,392,000	16.6
Japan	3,296,000	21.3	3,302,000	17.3	3,452,000	14.2	3,347,000	13.7	3,644,000	13.6	3,531,000	13.7	3,327,000	12.6
United Kingdom	*1,019,000	6.6	*1,019,000	5.3	*1,019,000	4.2	*1,019,000	4.2	*1,019,000	3.8	*1,019,000	3.9	*1,019,000	3.9
Korea, Republic of	1,139,000	7.4	1,368,000	7.2	1,877,000	7.7	1,369,000	5.6	1,010,000	3.8	943,000	3.6	*1,494,923	5.7

Thermostats

Unit of Measure: Number

	1990 Value	%	1991 Value	%	1992 Value	%	1993 Value	%	1994 Value	%	1995 Value	%	1996 Value	%
Total Production	50,187,958	100.0	44,925,319	100.0	48,847,178	100.0	68,372,516	100.0	75,854,520	100.0	72,597,858	100.0	77,181,821	100.0
Regions														
America, North	43,347,000	86.4	38,296,000	85.2	43,053,000	88.1	44,105,000	64.5	45,518,000	60.0	37,408,000	51.5	40,456,000	52.4
America, South	131,125	0.3	131,125	0.3	117,000	0.2	128,500	0.2	140,000	0.2	139,000	0.2	131,125	0.2
Asia	3,000	0.0	3,000	0.0	3,000	0.0	3,000	0.0	3,000	0.0	3,000	0.0	3,000	0.0
Europe	6,706,833	13.4	6,495,194	14.5	5,674,178	11.6	24,136,016	35.3	30,193,520	39.8	35,047,858	48.3	36,591,696	47.4
Leading Producers														
United States of America	43,347,000	86.4	38,296,000	85.2	43,053,000	88.1	44,105,000	64.5	45,518,000	60.0	37,408,000	51.5	40,456,000	52.4
Spain	2,100,000	4.2	1,902,000	4.2	1,128,000	2.3	19,569,000	28.6	26,324,000	34.7	31,273,000	43.1	30,513,000	39.5
Czech Republic	*4,307,333	8.6	*4,307,333	9.6	*4,307,333	8.8	*4,307,333	6.3	3,568,000	4.7	3,530,000	4.9	5,824,000	7.5
United Kingdom	121,000	0.2	153,000	0.3	*166,200	0.3	*177,255	0.3	*188,309	0.2	*199,364	0.3	*210,418	0.3
Colombia	*131,125	0.3	*131,125	0.3	117,000	0.2	128,500	0.2	140,000	0.2	139,000	0.2	*131,125	0.2

Commodity data are provided by the United Nations Statistical Division. The symbol * means that data are estimated. For additional notes, see Appendix I.

Product Tables

Binoculars and Refracting Telescopes

Unit of Measure: Number.

	1990		1991		1992		1993		1994		1995		1996	
	Value	%	Value	%	Value	%	Value	%	Value	%	Value	%	Value	%
Total Production	4,899,249	100.0	5,077,249	100.0	4,461,249	100.0	4,521,250	100.0	4,114,250	100.0	3,615,730	100.0	3,827,674	100.0
Regions														
America, North	9,000	0.2	9,000	0.2	9,000	0.2	5,000	0.1	18,000	0.4	4,000	0.1	9,000	0.2
Asia	4,329,867	88.4	4,517,867	89.0	3,902,867	87.5	3,966,867	87.7	3,546,867	86.2	3,059,867	84.6	3,271,772	85.5
Europe	560,382	11.4	550,382	10.8	549,383	12.3	549,383	12.2	549,383	13.4	551,863	15.3	546,902	14.3
Leading Producers														
Japan	2,290,000	46.7	2,679,000	52.8	2,530,000	56.7	2,895,000	64.0	2,471,000	60.1	2,178,000	60.2	2,023,000	52.9
Hong Kong	1,023,000	20.9	638,000	12.6	*253,000	5.7	-		-		-		-	
Korea, Republic of	769,000	15.7	953,000	18.8	872,000	19.5	824,000	18.2	828,000	20.1	634,000	17.5	*1,000,905	26.1
Macau	*247,867	5.1	*247,867	4.9	*247,867	5.6	247,867	5.5	*247,867	6.0	*247,867	6.9	*247,867	6.5
Germany	*156,202	3.2	*156,202	3.1	*156,202	3.5	*156,202	3.5	*156,202	3.8	158,683	4.4	153,722	4.0

Optical and Analytical Instruments (Microprojectors, Microscopes Etc.)

Unit of Measure: Number.

	1990		1991		1992		1993		1994		1995		1996	
	Value	%	Value	%	Value	%	Value	%	Value	%	Value	%	Value	%
Total Production	5,261,909	100.0	6,021,273	100.0	6,400,636	100.0	6,911,288	100.0	18,590,127	100.0	11,318,401	100.0	12,167,243	100.0
Regions														
Asia	5,034,909	95.7	5,525,273	91.8	5,993,636	93.6	6,501,000	94.1	18,182,000	97.8	10,634,857	94.0	11,443,400	94.1
Europe	227,000	4.3	496,000	8.2	407,000	6.4	410,288	5.9	408,127	2.2	683,544	6.0	723,843	5.9
Leading Producers														
Indonesia	4,006,909	76.1	4,507,273	74.9	5,007,636	78.2	5,508,000	79.7	17,189,000	92.5	*9,631,857	85.1	*10,437,400	85.8
Hong Kong	*807,000	15.3	*807,000	13.4	*807,000	12.6	*807,000	11.7	*807,000	4.3	*807,000	7.1	*807,000	6.6
Germany	-		283,000	4.7	198,000	3.1	193,000	2.8	190,000	1.0	464,578	4.1	504,038	4.1
Japan	221,000	4.2	209,000	3.5	174,000	2.7	183,000	2.6	184,000	1.0	195,000	1.7	198,000	1.6
Hungary	25,000	0.5	11,000	0.2	7,000	0.1	*15,288	0.2	*16,127	0.1	*16,966	0.1	*17,805	0.1

Cameras, Photographic

Unit of Measure: Number.

	1990		1991		1992		1993		1994		1995		1996	
	Value	%	Value	%	Value	%	Value	%	Value	%	Value	%	Value	%
Total Production	54,004,031	100.0	58,637,364	100.0	55,321,130	100.0	66,897,216	100.0	74,282,372	100.0	78,942,027	100.0	89,028,003	100.0
Regions														
America, North	20,889,500	38.7	20,889,500	35.6	20,889,500	37.8	21,084,000	31.5	20,949,000	28.2	20,128,000	25.5	21,397,000	24.0
Asia	32,175,531	59.6	36,676,364	62.5	33,385,197	60.3	44,883,764	67.1	52,481,464	70.7	57,980,164	73.4	66,830,036	75.1
Europe	939,000	1.7	1,071,500	1.8	1,046,433	1.9	929,452	1.4	851,908	1.1	833,863	1.1	800,968	0.9
Leading Producers														
United States of America	*18,134,000	33.6	*18,134,000	30.9	*18,134,000	32.8	*18,134,000	27.1	*18,134,000	24.4	*18,134,000	23.0	*18,134,000	20.4
Japan	16,955,000	31.4	17,903,000	30.5	14,637,000	26.5	12,546,000	18.8	11,942,000	16.1	11,403,000	14.4	12,256,000	13.8
China	2,132,000	3.9	4,782,000	8.2	5,265,000	9.5	19,305,000	28.9	28,300,000	38.1	33,261,000	42.1	41,208,000	46.3
Hong Kong	5,339,000	9.9	5,999,000	10.2	5,814,000	10.5	6,377,000	9.5	6,079,000	8.2	7,216,000	9.1	*6,742,829	7.6
Mexico	*2,755,500	5.1	*2,755,500	4.7	*2,755,500	5.0	2,950,000	4.4	2,815,000	3.8	1,994,000	2.5	3,263,000	3.7

Commodity data are provided by the United Nations Statistical Division. The symbol * means that data are estimated. For additional notes, see Appendix I.

701

Product Tables
Frames and Mountings for Spectacles, Goggles or The Like

Unit of Measure: Number.

	1990		1991		1992		1993		1994		1995		1996	
	Value	%	Value	%	Value	%	Value	%	Value	%	Value	%	Value	%
Total Production	71,054,527	100.0	37,634,226	100.0	55,618,769	100.0	40,008,540	100.0	61,806,733	100.0	50,271,034	100.0	53,850,144	100.0
Regions														
America, North	199,250	0.3	199,250	0.5	199,250	0.4	355,000	0.9	242,000	0.4	176,000	0.4	24,000	0.0
America, South	5,000	0.0	4,000	0.0	5,000	0.0	29,000	0.1	31,000	0.1	31,000	0.1	35,000	0.1
Asia	55,830,333	78.6	23,235,889	61.7	35,473,289	63.8	19,075,022	47.7	42,320,756	68.5	31,247,822	62.2	35,447,701	65.8
Europe	15,019,944	21.1	14,195,087	37.7	19,941,230	35.9	20,549,518	51.4	19,212,977	31.1	18,816,211	37.4	18,343,443	34.1
Leading Producers														
Korea, Republic of	17,311,000	24.4	15,623,000	41.5	11,934,000	21.5	10,173,000	25.4	8,648,000	14.0	8,087,000	16.1	*9,364,500	17.4
Hong Kong	35,937,000	50.6	5,189,000	13.8	22,780,000	41.0	8,207,000	20.5	28,373,000	45.9	21,082,000	41.9	*24,043,868	44.6
Austria	3,636,000	5.1	3,542,000	9.4	3,884,000	7.0	3,660,000	9.1	4,012,000	6.5	3,631,000	7.2	*3,898,945	7.2
Germany	-		-		6,292,000	11.3	4,601,000	11.5	3,115,000	5.0	2,382,935	4.7	1,913,138	3.6
Hungary	1,584,000	2.2	1,324,000	3.5	1,192,000	2.1	1,005,250	2.5	818,500	1.3	631,750	1.3	445,000	0.8

Watches

Unit of Measure: Thousands of units.

	1990		1991		1992		1993		1994		1995		1996	
	Value	%	Value	%	Value	%	Value	%	Value	%	Value	%	Value	%
Total Production	808,155	100.0	845,417	100.0	824,929	100.0	892,394	100.0	1,145,585	100.0	1,172,470	100.0	1,152,865	100.0
Regions														
Africa	113	0.0	113	0.0	113	0.0	113	0.0	113	0.0	113	0.0	113	0.0
Asia	791,589	98.0	823,028	97.4	803,214	97.4	871,403	97.6	1,124,410	98.2	1,150,768	98.1	1,131,158	98.1
Europe	16,453	2.0	22,275	2.6	21,602	2.6	20,878	2.3	21,062	1.8	21,589	1.8	21,594	1.9
Leading Producers														
Japan	344,398	42.6	389,789	46.1	374,146	45.4	390,596	43.8	393,187	34.3	396,802	33.8	413,746	35.9
Hong Kong	180,746	22.4	169,783	20.1	158,820	19.3	147,856	16.6	133,651	11.7	140,714	12.0	*111,771	9.7
China	86,713	10.7	78,248	9.3	86,105	10.4	151,830	17.0	453,937	39.6	481,913	41.1	479,760	41.6
Russian Federation	60,121	7.4	61,553	7.3	57,842	7.0	60,093	6.7	25,879	2.3	17,800	1.5	7,563	0.7
Korea, Republic of	21,139	2.6	21,223	2.5	19,883	2.4	19,306	2.2	16,744	1.5	16,079	1.4	*23,914	2.1

Clocks, With Watch Movements

Unit of Measure: Thousands of units.

	1990		1991		1992		1993		1994		1995		1996	
	Value	%	Value	%	Value	%	Value	%	Value	%	Value	%	Value	%
Total Production	81,424	100.0	92,524	100.0	92,212	100.0	115,617	100.0	136,786	100.0	174,942	100.0	138,722	100.0
Regions														
Africa	84	0.1	68	0.1	69	0.1	60	0.1	64	0.0	60	0.0	45	0.0
Asia	36,782	45.2	42,942	46.4	47,104	51.1	74,491	64.4	95,370	69.7	131,039	74.9	95,069	68.5
Europe	44,558	54.7	49,514	53.5	45,039	48.8	41,066	35.5	41,352	30.2	43,843	25.1	43,608	31.4
Leading Producers														
China	30,557	37.5	36,601	39.6	40,222	43.6	69,252	59.9	90,544	66.2	126,265	72.2	89,600	64.6
Germany	*16,534	20.3	21,759	23.5	17,513	19.0	13,267	11.5	13,597	9.9	*16,534	9.5	*16,534	11.9
France	1,460	1.8	884	1.0	776	0.8	632	0.5	827	0.6	*565	0.3	*429	0.3
Indonesia	1,807	2.2	2,535	2.7	3,959	4.3	3,196	2.8	2,793	2.0	2,737	1.6	*3,522	2.5
Armenia	2,869	3.5	2,095	2.3	1,119	1.2	375	0.3	344	0.3	377	0.2	216	0.2

Product Tables

Clocks, Instrument-Panel & Similar Type

Unit of Measure: Number.

	1990 Value	%	1991 Value	%	1992 Value	%	1993 Value	%	1994 Value	%	1995 Value	%	1996 Value	%
Total Production	9,334,250	100.0	9,756,250	100.0	16,409,250	100.0	11,352,400	100.0	11,479,000	100.0	11,086,545	100.0	11,623,045	100.0
Regions														
Asia	1,797,250	19.3	2,222,250	22.8	1,985,250	12.1	1,617,000	14.2	1,871,000	16.3	1,625,000	14.7	1,546,000	13.3
Europe	7,537,000	80.7	7,534,000	77.2	14,424,000	87.9	9,735,400	85.8	9,608,000	83.7	9,461,545	85.3	10,077,045	86.7
Leading Producers														
Germany	-		-		6,891,000	42.0	2,202,000	19.4	2,075,000	18.1	1,928,945	17.4	2,544,845	21.9
India	*1,538,250	16.5	*1,538,250	15.8	*1,538,250	9.4	1,352,000	11.9	1,739,000	15.1	1,583,000	14.3	1,479,000	12.7
Belarus	254,000	2.7	286,000	2.9	190,000	1.2	186,000	1.6	30,000	0.3	14,000	0.1		
Yugoslavia	-		393,000	4.0	252,000	1.5	71,000	0.6	100,000	0.9	23,000	0.2	62,000	0.5
Slovenia	*5,000	0.1	*5,000	0.1	*5,000	0.0	8,000	0.1	2,000	0.0	*5,000	0.0	*5,000	0.0

Clocks, Other, Electric and Non-Electric

Unit of Measure: Thousands of units.

	1990 Value	%	1991 Value	%	1992 Value	%	1993 Value	%	1994 Value	%	1995 Value	%	1996 Value	%
Total Production	147,489	100.0	113,119	100.0	105,431	100.0	126,759	100.0	115,862	100.0	112,291	100.0	122,596	100.0
Regions														
Asia	144,124	97.7	109,808	97.1	102,299	97.0	123,514	97.4	112,971	97.5	109,256	97.3	119,564	97.5
Europe	3,341	2.3	3,287	2.9	3,108	2.9	3,221	2.5	2,867	2.5	3,011	2.7	3,008	2.5
Oceania	24	0.0	24	0.0	24	0.0	24	0.0	24	0.0	24	0.0	24	0.0
Leading Producers														
Japan	87,221	59.1	87,427	77.3	71,699	68.0	*93,909	74.1	*97,018	83.7	*100,127	89.2	*103,236	84.2
Hong Kong	44,475	30.2	7,848	6.9	16,775	15.9	13,885	11.0	3,230	2.8	2,751	2.4	*11,474	9.4
Belarus	9,736	6.6	11,984	10.6	11,408	10.8	13,647	10.8	11,010	9.5	4,847	4.3	2,662	2.2
United Kingdom	*2,218	1.5	*2,218	2.0	*2,218	2.1	2,372	1.9	2,065	1.8	*2,218	2.0	*2,218	1.8
Korea, Republic of	2,590	1.8	2,080	1.8	2,058	2.0	1,862	1.5	1,623	1.4	1,439	1.3	*2,116	1.7

ISIC 3900
INDUSTRIES NEC

ISIC 3900—Industries nec—includes products not classified in one of the other manufacturing ISICs. Major categories include musical instruments, writing devices, and zippers.

The products of this industry are classified, in the U.S. SIC system, under the 2-digit classification 39—Miscellaneous Manufacturing Industries.

Summary Statistics

		1990 Value	N	1991 Value	N	1992 Value	N	1993 Value	N	1994 Value	N	1995 Value	N
Establishments or enterprises	(number)	89,752	91	87,577	100	107,302	93	94,082	86	82,052	73	55,168	34
Number employed	(000)	5,464	99	4,324	101	4,290	96	4,208	93	4,369	87	3,872	64
Total Wages	($ mil.)	26,548	93	26,341	95	28,677	88	24,367	86	25,862	76	24,568	57
Wage/salary per employee	($)	7,732	91	7,989	92	8,935	87	8,377	84	7,888	75	9,607	57
Female workers	(000)	342	37	247	34	461	30	542	32	452	29	410	15
as % of total employment*	(%)	39	36	37	34	38	30	44	32	41	29	49	15
Output	($ bil.)	157	95	154	99	156	93	160	84	158	77	155	62
per employee	($ 000)	40	91	42	94	48	88	46	81	48	73	58	59
per establishment	($ 000)	2,337	83	2,659	91	1,884	83	17,055	72	2,664	61	1,587	30
Value added	($ bil.)	59	89	60	91	64	83	68	81	68	73	69	61
per employee	($ 000)	18	84	18	84	20	77	19	75	19	68	23	56
per establishment	($ 000)	878	76	716	83	703	73	910	67	996	56	562	28
Capital investment	($ mil.)	3,715	66	3,674	61	3,543	54	3,137	51	3,193	39	1,807	21
per employee	($ 000)	2	62	3	58	3	52	4	49	2	38	4	20
per establishment	($ 000)	139	62	100	58	136	52	170	47	79	37	105	19

Data presented above are drawn from the detailed tables that follow. Columns headed 'N' show the number of countries that provided valid data for inclusion. Values are not strictly comparable one year to the next or one row to the next because the number of countries that report varies from period to period and row to row. However, a general indication of magnitudes can be gleaned from reviewing this summary—especially in earlier years in which more reports are available. For detailed explanations, see Appendix I. *The average for those countries reporting both total and female employment.

Establishments and Number Engaged

Country	Establishments or Enterprises (number)						Employees per Establishment					
	1990	1991	1992	1993	1994	1995	1990	1991	1992	1993	1994	1995
Albania	-	-	-	-	4.000	3.000	-	-	-	-	60	17
Algeria	*1,248	*1,427	*1,313	-	*1,652	-	5.544	4.954	5.784	-	3.943	-
Argentina	-	-	-	1,878	2,121	-	-	-	-	4.075	3.813	-
Armenia	13	25	24	15	18	-	425	212	-	-	-	-
Australia	1,064	1,823	1,916	-	-	-	12	7.131	6.785	-	-	-
Austria	55	49	52	46	44	-	129	143	135	137	145	-
Azerbaijan	98	93	90	102	55	-	53	58	57	30	43	-
Bahamas	44	20	-	-	-	-	4.045	7.350	-	-	-	-
Bahrain	-	-	162	-	-	-	-	-	6.562	-	-	-
Bangladesh	306	292	338	-	-	-	21	24	25	-	-	-
Barbados	5.000	4.000	6.000	6.000	10		29	6.750	4.167	8.167	6.300	
Belgium	986	951	895	556	564	-	9.635	9.569	9.609			
Belize	12	12	13	-	-	-	-	17	19	-	-	-
Benin	-	-	1.000	1.000	-	-	-	-	-	-	-	-
Bolivia	25	25	24	27	27	27	21	21	22	23	50	50
Bosnia & Herzegovina	3.000	3.000	-	-	1.000	-	249	198	-	-	298	-
Botswana	56	69	75	96	102	168	82	77	65	46	44	29
Bulgaria	*556	*622	*439	*3,996	*2,904	*1,599	323	210	224	23	7.369	6.379
Burundi	*4.000	*4.000	-	-	-	-	11	13	-	-	-	-
Cambodia	572	1,097	-	-	-	-	2.288	2.008	-	-	-	-
Cameroon	*1.000	*4.000	*4.000	*5.000	*6.000	*2.000	234	91	66	76	49	95
Canada	2,848	2,520	2,334	2,157	2,047	-	17	18	17	18	20	-
Cape Verde	2.000	2.000	2.000	2.000	2.000		-	-	-	-	-	
Central African Republic	*5.000	*5.000	*5.000	*5.000	*6.000	*5.000	28	27	14	20	-	
Chile	15	18	15	11	12	-	104	94	104	102	103	-
China	*35,716	*29,237	*29,119	*21,865	*22,309	*22,429	72	82	81	93	113	96
China (Hong Kong SAR)	3,697	2,725	2,582	2,708	2,835	-	8.602	8.734	8.792	8.235	6.843	-
Colombia	182	175	195	182	168	-	48	46	49	54	53	-
Costa Rica	100	109	99	113	-	-	33	31	39	34	-	-
Croatia	46	67	99	130	179	208	45	30	17	12	7.933	6.202
Cyprus	397	395	392	381	393	399	2.733	2.630	2.515	2.360	2.310	2.178
Czechoslovakia (Former)	*40	*50	-	-	-	-	1,150	540	-	-	-	-
Czech Republic	-	*40	*100	*138	-	-	-	575	280	319	-	-
Denmark	1,323	1,467	1,459	-	-	-	8.469	7.604	7.164	-	-	-
Ecuador	22	32	32	31	32	33	38	42	41	46	43	38
Egypt	36	34	38	44	-	-	42	50	45	52	-	-
El Salvador	-	10	-	-	19	25	-	-	-	-	63	52
Fiji	49	44	42	-	-	-	5.735	5.591	6.857	-	-	-
Finland	86	93	91	87	80	91	42	38	36	38	40	38
FYR Macedonia	*14	*35	*77	*78	*107	*137	117	45	21	19	13	8.825
Gabon	-	1.000	1.000	-	-	-	-	-	-	-	-	-
Gambia	-	-	-	2.000	-	-	-	-	-	1.500	-	-
Germany	-	758	706	679	626		-	95	85	79	81	
Germany (Western Part)	605	617	609	-	-	-	89	88	86	-	-	-
Ghana	-	-	-	10	-	-	-	-	-	129	-	-
Greece	132	126	126	-	-	-	20	18	17	-	-	-
Grenada	-	-	1.000	1.000	2.000	-	-	-	-	-	4.500	-
Guatemala	-	9.000	9.000	9.000	9.000	9.000	-	37	42	43	48	53
Honduras	-	-	69	64	64	30	-	-	23	27	30	72
Hungary	*311	*792	*314	*353	-	-	116	38	73	17	-	-
Iceland	145	144	158	157	-	-	3.372	3.667	3.133	3.096	-	-
India	1,020	982	1,074	1,155	-	-	42	47	45	52	-	-
Indonesia	242	289	332	344	384	442	125	157	167	205	193	176
Iran (Islamic Republic of)	108	21	28	31	-	-	26	49	45	43	-	-
Iraq	-	-	1.000	-	-	-	-	-	33	-	-	-
Ireland	100	81	-	-	-	-	29	37	-	-	-	-
Israel	262	283	315	364	373	-	18	18	18	17	18	-
Italy	*758	*795	*781	*802	*788	-	48	46	47	46	47	-
Jamaica	24	24	-	-	-	-	76	24	-	-	-	-
Japan	13,735	13,176	12,555	12,848	13,795	14,011	16	16	18	16	16	15

Continued.

Depending on the table, * means *Enterprises, Engaged,* or *Factor Values;* ± means *Basis Unspecified;* § means *shown in millions.* For additional notes and sources, see Appendix I.

Establishments and Number Engaged

- Continued -

Country	Establishments or Enterprises (number)						Employees per Establishment					
	1990	1991	1992	1993	1994	1995	1990	1991	1992	1993	1994	1995
Jordan	43	42	38	38	86	91	4.860	4.119	2.632	2.974	8.302	4.286
Kenya	94	165	226	231	264	264	38	20	15	15	13	14
Korea, Republic of	2,757	2,637	2,463	2,864	2,784	2,838	33	29	26	21	20	19
Kuwait	191	178	190	190	190	-	6.728	5.461	6.337	5.532	6.216	-
Latvia	500	691	65	55	29	66	37	26	52	33	103	37
Lesotho	29	-	2.000	3.000	3.000	-	26	-	202	373	376	-
Liechtenstein	-	16	-	-	-	-	-	2.875	-	-	-	-
Lithuania	-	-	-	*27	*40	-	-	-	-	1,464	77	-
Macau	106	78	92	83	67	65	67	62	44	45	57	51
Malaysia	164	195	207	250	242	-	110	105	100	81	83	-
Malta	86	80	78	77	76	-	9.186	11	11	11	13	-
Mauritius	70	40	46	49	45	47	54	81	73	65	69	74
Mexico	36	36	34	34	34	33	177	179	196	186	196	165
Mongolia	48	50	58	110	22	36	24	32	16	15	58	134
Morocco	31	29	28	27	27	25	21	20	25	25	23	26
Mozambique	2.000	2.000	1.000	1.000	2.000	-	133	122	80	75	72	-
Myanmar	2.000	2.000	2.000	2.000	2.000	-	1,826	1,735	1,688	1,723	1,693	-
Namibia	-	-	-	-	8.000	-	-	-	-	-	15	-
Nepal	26	37	-	57	43	-	40	24	-	23	16	-
Netherlands	53	54	58	63	56	-	50	48	47	46	45	-
Netherlands Antilles	-	-	-	21	-	-	-	-	-	-	-	-
New Zealand	*880	*954	*1,003	*1,085	-	-	3.326	2.868	2.653	2.852	-	-
Norway	85	94	69	51	51	-	29	33	45	50	50	-
Oman	1.000	1.000	1.000	2.000	-	8.000	-	3.000	6.000	31	-	15
Pakistan	-	74	-	-	-	-	-	134	-	-	-	-
Paraguay	-	2.000	-	-	-	-	-	5.000	-	-	-	-
Peru	482	497	506	-	-	-	12	11	10	-	-	-
Philippines	1,822	1,874	348	319	-	-	14	15	80	82	-	-
Poland	*242	*192	*141	*126	-	-	256	219	170	159	-	-
Portugal	*2,109	*2,053	*1,966	*1,979	*1,988	-	7.114	7.635	6.849	7.146	7.266	-
Puerto Rico	58	51	51	43	47	45	47	45	37	43	42	42
Qatar	13	14	28	29	3.000	-	7.615	7.786	6.357	6.172	31	-
Republic of Moldova	28	4.000	7.000	-	-	-	385	2,911	2,282	-	-	-
Romania	-	1,152	1,415	3,901	1,417	-	-	7.986	5.371	1.923	5.152	-
Russian Federation	-	-	-	*17,997	*7,217	*10,598	-	-	-	21	33	25
Saint Lucia	-	5.000	5.000	5.000	5.000	5.000	-	-	-	-	-	-
Saudi Arabia	-	-	73	-	-	-	-	-	69	-	-	-
Senegal	2.000	1.000	-	1.000	2.000	-	8.500	7.000	-	2.000	1.500	-
Sierra Leone	-	-	-	3.000	-	-	-	-	-	10	-	-
Singapore	142	140	144	151	149	-	45	43	37	34	31	-
Slovakia	-	17	17	24	26	-	-	261	209	143	126	-
Slovenia	55	*1,353	*2,387	*2,317	*2,902	*768	-	-	-	-	-	-
South Africa	-	1,130	-	-	-	-	-	22	-	-	-	-
Spain	2,038	2,034	2,047	-	-	-	12	12	11	-	-	-
Sri Lanka	64	56	53	63	-	-	81	110	129	142	-	-
Suriname	-	-	6.000	-	-	-	-	-	16	-	-	-
Swaziland	4.000	5.000	5.000	4.000	-	-	5.500	3.400	6.000	3.500	-	-
Sweden	80	82	78	67	62	-	36	37	37	40	38	-
Syrian Arab Republic	-	581	593	717	-	-	-	-	-	-	-	-
Thailand	344	562	-	-	-	-	221	117	-	-	-	-
Tonga	2.000	-	-	-	-	-	3.500	-	-	-	-	-
Trinidad and Tobago	83	73	68	67	64	-	24	25	27	30	31	-
Tunisia	-	-	-	165	175	179	-	-	-	17	14	14
Turkey	60	55	106	106	112	-	78	78	49	53	50	-
United Kingdom	8,735	8,168	8,465	8,669	11,513	-	8.128	7.958	7.797	7.383	6.949	-
United Republic of Tanzania	20	15	-	-	-	-	51	58	-	-	-	-
USSR (Former)	*805	-	-	-	-	-	1,118	-	-	-	-	-
United States of America	-	-	24,583	-	-	-	-	-	14	-	-	-
Venezuela	170	175	169	148	153	127	35	36	37	39	32	40
Yugoslavia	68	108	222	338	356	355	-	99	45	28	26	26

Continued.

Depending on the table, * means *Enterprises*, *Engaged*, or *Factor Values*; ± means *Basis Unspecified*; § means *shown in millions*. For additional notes and sources, see Appendix I.

707

Establishments and Number Engaged

- Continued -

Country	Establishments or Enterprises (number)						Employees per Establishment					
	1990	1991	1992	1993	1994	1995	1990	1991	1992	1993	1994	1995
Zambia	8.000	-	-	-	8.000	-	42	-	-	-	30	-
Zimbabwe	46	47	46	44	44	-	39	36	39	66	59	-

Employment and Compensation of Employees

Country	Number Employed/Engaged (000)						Wage/Salary per Employee ($)					
	1990	1991	1992	1993	1994	1995	1990	1991	1992	1993	1994	1995
Albania	8.906	-	-	-	0.239	0.050	-	-	-	-	668	3,771
Algeria	6.919	7.069	7.594	7.035	6.514	6.868	5,824	2,826	2,346	2,763	2,316	2,144
Argentina	5.614	-	-	7.653	8.087	-	4,920	-	-	10,951	-	-
Armenia	*5.526	*5.291	-	-	-	-	-	-	-	-	-	-
Australia	13	*13	*13	-	-	-	16,520	17,201	15,837	-	-	-
Austria	7.100	7.000	7.000	6.314	6.395	6.310	22,607	22,133	25,831	26,714	26,546	32,425
Azerbaijan	5.166	5.354	5.134	3.024	2.350	-	2,004	2,167	-	-	-	-
Bahamas	0.178	0.147	-	-	-	-	12,247	13,109	-	-	-	-
Bahrain	-	-	1.063	-	-	-	-	-	-	3,853	-	-
Bangladesh	6.560	6.965	8.439	-	-	-	-	-	-	-	-	-
Barbados	0.146	0.027	0.025	0.049	0.063	0.053	5,240	7,019	9,160	6,224	7,675	7,742
Belgium	9.500	9.100	8.600	-	-	-	15,592	15,653	17,346	-	-	-
Belize	-	0.204	0.243	-	-	-	-	2,054	2,280	-	-	-
Bermuda	0.017	0.015	0.029	0.020	0.045	0.025	-	-	-	-	-	-
Bolivia	0.526	0.528	0.538	0.633	1.337	1.355	649	921	1,060	1,066	801	772
Bosnia & Herzegovina	0.748	0.593	-	-	0.298	-	2,539	2,856	-	-	-	-
Botswana	4.600	5.300	4.900	4.400	4.500	4.901	-	-	3,125	3,227	-	-
Bulgaria	180	131	98	93	21	10	-	-	-	855	754	1,199
Burundi	0.046	0.050	-	-	-	-	3,300	3,239	-	-	-	-
Cambodia	*1.309	*2.203	-	-	-	-	-	-	-	-	-	-
Cameroon	0.234	0.363	0.265	0.379	0.292	0.190	11,443	8,574	10,236	8,489	6,094	5,399
Canada	49	45	39	38	40	40	19,869	20,987	21,532	20,561	19,314	19,240
Central African Republic	0.139	0.136	0.070	0.101	-	-	5,549	7,715	8,150	6,888	-	-
Chile	1.553	1.684	1.554	1.120	1.232	1.237	3,254	3,554	4,204	6,044	5,498	6,343
China	*2.560	*2.400	*2.350	*2.030	*2.530	*2.160	-	-	-	-	-	-
China (Hong Kong SAR)	*32	*24	*23	*22	*19	-	8,958	10,408	11,274	11,750	12,826	-
China (Taiwan Province)	123	117	113	102	100	94	8,253	8,965	10,708	10,740	11,586	12,355
Colombia	8.700	8.124	9.596	9.892	8.965	8.841	2,108	1,945	2,073	2,467	3,543	3,882
Costa Rica	3.267	3.430	3.841	3.827	-	-	1,962	1,872	2,159	2,599		-
Croatia	2.050	2.000	1.670	1.550	1.420	1.290	4,268	4,380	1,578	1,919	2,518	3,751
Cyprus	1.085	1.039	0.986	0.899	0.908	0.869	8,025	8,623	9,085	9,129	9,528	12,205
Czechoslovakia (Former)	46	27	-	-	-	-	1,962	1,445	-	-	-	-
Czech Republic	24	23	28	44	-	-	2,136	1,504	1,832	2,182	-	-
Denmark	11	11	10	11	11	11	24,199	24,976	27,786	26,872	29,031	32,951
Ecuador	0.827	1.336	1.303	1.431	1.387	1.241	1,682	1,602	1,842	1,963	1,176	1,402
Egypt	*1.500	*1.700	*1.700	*2.300	*2.284	*2.037	2,200	989	1,269	1,328	1,338	1,246
El Salvador	-	-	*0.754	-	*1.193	*1.309	-	-	-	-	3,095	2,384
Fiji	0.281	0.246	0.288	0.322	0.325	-	2,059	2,771	2,770	2,318	2,528	-
Finland	3.600	3.500	3.300	3.300	3.200	3.495	24,785	24,205	22,527	17,607	20,006	26,279
FYR Macedonia	1.641	1.577	1.653	1.455	1.400	1.209	4,706	5,383	2,043	2,556	2,579	3,874
France	100	94	89	83	81	80	27,004	28,853	34,361	-	-	-
Gambia	-	-	-	0.003	-	-	-	-	-	-	-	-
Germany	-	*72	*60	*54	*51	-	-	20,109	25,125	25,363	27,080	-
Germany (Western Part)	*54	*55	*52	-	-	-	23,508	24,053	26,829	-	-	-
Ghana	-	-	-	*1.288	*1.324	*1.362	-	-	-	-	-	-
Greece	*2.574	*2.284	*2.191	*1.913	*1.826	*1.728	8,024	8,621	9,307	10,899	11,664	14,527
Grenada	-	-	-	-	0.009	-	-	-	-	-	-	-
Guatemala	-	*0.329	*0.374	*0.383	*0.433	*0.479	-	-	-	-	-	438
Honduras	*1.739	*1.796	*1.563	*1.712	*1.933	*2.160	1,992	1,733	2,303	1,382	1,075	1,123
Hungary	36	30	23	6.000	6.650	6.780	1,787	1,877	2,178	2,866	2,641	2,484
Iceland	0.489	0.528	0.495	0.486	0.551	-	36,444	35,393	-	37,964	-	-

Continued.

Depending on the table, * means *Enterprises*, *Engaged*, or *Factor Values*; ± means *Basis Unspecified*; § means *shown in millions*. For additional notes and sources, see Appendix I.

Employment and Compensation of Employees

- Continued -

Country	Number Employed/Engaged (000)						Wage/Salary per Employee ($)					
	1990	1991	1992	1993	1994	1995	1990	1991	1992	1993	1994	1995
India	*43	*47	*49	*60	*59	*60	1,076	942	946	867	882	861
Indonesia	30	46	55	71	74	78	529	551	545	651	752	1,082
Iran (Islamic Republic of)	2.800	1.019	1.255	1.345	-	-	3,283	4,263	3,988	3,385	-	-
Iraq	-	-	0.033	-	-	-	-	-	15,882	-	-	-
Ireland	2.900	3.000	3.019	2.660	2.962	3.243	15,955	15,886	17,354	15,092	15,609	18,082
Israel	4.800	5.100	5.700	6.200	6.900	6.657	14,673	13,765	14,197	13,051	14,632	16,042
Italy	36	37	36	37	37	-	38,591	27,515	31,623	25,552	25,951	-
Jamaica	1.823	0.583	0.582	0.551	0.605	-	1,139	2,321	1,628	-	-	-
Japan	*220	*217	*220	*210	*219	*214	22,697	26,067	28,927	33,359	36,857	-
Jordan	0.209	0.173	0.100	0.113	0.714	0.390	1,801	2,063	1,809	1,609	2,044	2,244
Kenya	3.592	3.325	3.423	3.398	3.469	3.579	1,477	1,296	1,289	843	902	1,322
Korea, Republic of	91	77	64	61	56	53	7,712	8,977	9,882	10,891	12,034	13,976
Kuwait	1.285	0.972	1.204	1.051	1.181	1.176	6,004	4,996	6,883	7,675	7,265	9,786
Latvia	18	18	3.401	1.832	2.984	2.420	-	-	-	-	-	-
Lesotho	*0.763	-	*0.404	*1.119	*1.127	-	-	-	677	1,231	1,417	-
Liechtenstein	-	*0.046	-	-	-	-	-	-	-	-	-	-
Lithuania	-	-	43	40	3.098	-	-	-	-	-	648	-
Macau	7.152	4.855	4.004	3.705	3.814	3.300	4,413	4,446	4,411	4,709	4,983	4,971
Malaysia	18	20	21	20	20	24	1,966	2,157	2,795	3,239	3,433	3,945
Malta	0.790	0.890	0.892	0.860	0.978	-	9,392	9,563	9,839	9,932	10,225	-
Mauritius	*3.751	3.221	3.336	3.209	3.105	3.486	2,047	2,559	2,394	2,889	2,937	3,014
Mexico	6.390	6.428	6.664	6.329	6.665	5.437	4,195	5,063	6,372	7,291	7,003	4,669
Mongolia	*1.175	*1.619	*0.919	*1.689	*1.285	*4.816	11,429	8,032	5,852	5,282	7,138	-
Morocco	0.638	0.571	0.698	0.663	0.634	0.651	3,803	4,425	4,698	4,055	4,627	4,677
Mozambique	0.265	0.243	0.080	0.075	0.144	-	544	453	-	-	-	-
Myanmar	3.652	3.469	3.375	3.446	3.386	-	1,447	1,490	1,598	1,930	2,131	-
Namibia	-	-	-	-	0.121	-	-	-	-	-	4,203	-
Nepal	1.045	0.879	-	1.308	0.704	-	442	502	-	-	452	-
Netherlands	2.624	2.582	2.754	2.910	2.521	2.472	23,022	24,650	27,256	27,938	30,232	36,432
Netherlands Antilles	-	-	-	0.017	-	-	-	-	-	7,756	-	-
New Zealand	2.927	2.736	2.661	3.094	-	-	13,870	12,910	12,537	14,155	-	-
Norway	2.497	3.140	3.124	2.547	2.569	2.527	26,039	23,678	29,449	20,090	20,738	23,467
Oman	-	*0.003	*0.006	*0.062	-	*0.116	-	3,463	4,329	5,320	-	7,792
Pakistan	7.390	9.928	10	-	-	-	1,085	1,015	1,044	-	-	-
Paraguay	-	0.010	-	-	-	-	-	4,165	-	-	-	-
Peru	*5.604	*5.417	*5.181	-	-	-	2,285	2,811	3,481	-	-	-
Philippines	25	28	28	26	-	-	1,421	1,417	1,697	1,714	-	-
Poland	62	42	24	20	22	26	1,247	1,965	2,159	2,131	2,156	2,541
Portugal	*15	*16	*13	*14	*14	*14	-	-	-	-	-	-
Puerto Rico	2.700	2.270	1.890	1.850	1.980	1.870	21,778	20,176	21,111	21,946	21,515	23,155
Qatar	0.099	0.109	0.178	0.179	0.092	-	8,936	5,542	3,017	4,253	7,041	-
Republic of Moldova	11	12	16	-	1.465	-	-	2,664	-	-	491	-
Romania	10	9.200	7.600	7.500	7.300	-	1,667	1,151	803	949	926	-
Russian Federation	-	-	-	385	236	260	-	-	-	489	893	938
Saudi Arabia	-	-	5.048	-	-	-	-	-	-	-	-	-
Senegal	*0.017	*0.007	-	*0.002	*0.003	-	4,321	2,532	-	1,766	1,801	-
Sierra Leone	-	-	-	*0.031	-	-	-	-	-	404	-	-
Singapore	*6.454	*5.962	*5.322	*5.059	*4.683	*3.600	8,375	10,013	11,601	12,301	15,455	16,934
Slovakia	-	*4.429	*3.560	*3.442	*3.274	-	-	1,310	1,541	1,747	1,916	-
South Africa	25	25	24	23	22	22	5,148	5,447	6,457	6,094	6,490	6,538
Spain	24	24	23	30	28	28	15,543	16,573	18,348	14,956	14,982	17,040
Sri Lanka	*5.174	*6.181	*6.832	*8.976	-	-	556	734	769	706	-	-
Suriname	*0.078	*0.067	*0.094	*0.055	-	-	1,436	2,508	1,192	-	-	-
Swaziland	0.022	0.017	0.030	0.014	-	-	756	5,784	2,410	4,158	-	-
Sweden	2.900	*3.020	*2.853	2.655	2.337	2.337	22,138	24,424	26,010	21,697	23,139	25,431
Thailand	76	66	-	-	-	-	943	2,186	-	-	-	-
Tonga	0.007	-	-	-	-	-	1,450	-	-	-	-	-
Trinidad and Tobago	2.002	1.804	1.807	2.016	2.004	-	5,400	6,822	7,852	4,607	4,910	-
Tunisia	-	-	-	2.784	2.519	2.429	-	-	-	3,955	4,440	5,375
Turkey	4.707	4.283	5.144	5.653	5.571	5.443	4,398	5,037	4,809	4,719	2,667	4,565

Continued.

Depending on the table, * means *Enterprises*, *Engaged*, or *Factor Values*; ± means *Basis Unspecified*; § means *shown in millions*. For additional notes and sources, see Appendix I.

709

Employment and Compensation of Employees

- Continued -

Country	Number Employed/Engaged (000)						Wage/Salary per Employee ($)					
	1990	1991	1992	1993	1994	1995	1990	1991	1992	1993	1994	1995
Ukraine	-	-	159	131	116	113	-	-	1,876	-	491	580
United Kingdom	71	65	66	64	80	76	17,379	18,380	19,060	16,165	18,415	19,210
United Republic of Tanzania	1.027	0.871	-	-	-	-	-	466	-	-	-	-
USSR (Former)	900	-	-	-	-	-	2,100	-	-	-	-	-
United States of America	369	347	351	365	368	383	19,919	21,009	22,781	23,068	23,418	24,063
Uruguay	*2.079	*2.078	*2.039	*1.257	*1.381	*1.343	2,281	2,755	2,830	3,522	3,707	3,970
Venezuela	6.000	6.300	6.300	5.800	4.920	5.083	3,077	3,171	3,865	4,062	3,147	4,173
Yugoslavia	-	11	10	9.500	9.300	9.100	-	-	-	-	1,754	842
Zambia	0.335	-	-	-	0.242	-	1,072	-	-	-	757	-
Zimbabwe	1.800	1.700	1.800	2.900	2.600	2.500	2,678	2,402	1,876	1,204	1,435	1,663

Female Workers: Total and Percent of Employment

Country	Female Workers (000)						Female Workers as % of Total Employment					
	1990	1991	1992	1993	1994	1995	1990	1991	1992	1993	1994	1995
Afghanistan	0.012	-	-	-	-	-	-	-	-	-	-	-
Albania	-	-	-	-	0.047	0.018	-	-	-	-	19.67	36.00
Algeria	0.684	0.916	0.708		-	-	9.89	12.96	9.32		-	-
Argentina	-	-	-	-	2.192	-	-	-	-	-	27.11	-
Australia	4.118	-	-	-	-	-	31.68	-	-	-	-	-
Austria	2.600	2.000	3.000	2.315	2.310	-	36.62	28.57	42.86	36.66	36.12	-
Bangladesh	1.353	1.387	1.574	-	-	-	20.63	19.91	18.65	-	-	-
Bermuda	0.006	0.004	0.013	0.010	0.022	0.008	35.29	26.67	44.83	50.00	48.89	32.00
Bosnia & Herzegovina	0.350	0.275	-	-	-	-	46.79	46.37	-	-	-	-
Botswana	-	-	-	-	-	2.484	-	-	-	-	-	50.68
Bulgaria	-	-	-	59	14	6.000	-	-	-	64.05	63.08	58.82
Canada	18	16	-	-	-	-	36.73	35.56	-	-	-	-
Chile	0.559	0.508	0.486	0.311	0.364	-	35.99	30.17	31.27	27.77	29.55	-
China (Taiwan Province)	69	67	64	57	54	50	56.47	57.11	57.07	55.73	53.92	53.02
Colombia	-	3.282	-	4.464	4.112	-	-	40.40	-	45.13	45.87	-
Croatia	1.140	1.160	0.960	0.910	0.630	0.900	55.61	58.00	57.49	58.71	44.37	69.77
Cyprus	0.367	0.380	0.433	0.259	0.261	-	33.82	36.57	43.91	28.81	28.74	-
Czechoslovakia (Former)	18	10	-	-	-	-	38.48	37.41	-	-	-	-
Czech Republic	14	12	20	21	-	-	58.33	52.17	71.43	47.73	-	-
Denmark	5.639	5.650	5.357	-	-	-	50.33	50.65	51.25	-	-	-
Egypt	-	0.417	0.227	0.330	-	-	-	24.53	13.35	14.35	-	-
El Salvador	-	-	-	-	0.634	0.532	-	-	-	-	53.14	40.64
FYR Macedonia	0.423	0.346	0.371	0.352	0.341	0.252	25.78	21.94	22.44	24.19	24.36	20.84
Gambia	-	-	-	0.003	-	-	-	-	-	100.00	-	-
Germany (Western Part)	27	-	-	-	-	-	50.42	-	-	-	-	-
Hungary	19	-	4.000	3.000	-	-	52.78	-	17.39	50.00	-	-
Indonesia	-	-	-	49	52	54	-	-	-	68.91	69.89	69.21
Iran (Islamic Republic of)	-	0.059	0.085	0.083	-	-	-	5.79	6.77	6.17	-	-
Ireland	1.500	1.600	-	-	-	-	51.72	53.33	-	-	-	-
Italy	-	20	16	16	16	-	-	54.13	44.06	44.42	42.24	-
Jordan	0.029	0.038	0.009	0.010	0.250	0.094	13.88	21.97	9.00	8.85	35.01	24.10
Kenya	0.950	0.818	0.849	0.843	-	-	26.45	24.60	24.80	24.81	-	-
Korea, Republic of	44	37	30	28	25	23	49.01	48.53	46.52	45.24	44.47	43.92
Kuwait	0.002	0.018	0.001	0.025	0.003	-	0.16	1.85	0.08	2.38	0.25	-
Lesotho	-	-	-	0.809	0.877	-	-	-	-	72.30	77.82	-
Macau	4.614	3.249	2.731	2.611	2.801	2.488	64.51	66.92	68.21	70.47	73.44	75.39
Malaysia	11	13	13	12	12	-	62.78	62.75	61.06	59.41	59.41	-
Malta	0.305	0.351	0.330	0.308	0.377	-	38.61	39.44	37.00	35.81	38.55	-
Mongolia	-	-	-	-	0.605	-	-	-	-	-	47.08	-
Morocco	-	-	-	-	-	0.264	-	-	-	-	-	40.55
Myanmar	1.013	0.944	0.974	1.005	0.857	-	27.74	27.21	28.86	29.16	25.31	-
Nepal	0.069	0.158	-	0.350	0.190	-	6.60	17.97	-	26.76	26.99	-
New Zealand	1.148	1.051	0.995	1.105	-	-	39.22	38.41	37.39	35.71	-	-

Continued.

Depending on the table, * means *Enterprises*, *Engaged*, or *Factor Values*; ± means *Basis Unspecified*; § means *shown in millions*. For additional notes and sources, see Appendix I.

Female Workers: Total and Percent of Employment
- Continued -

Country	Female Workers (000)						Female Workers as % of Total Employment					
	1990	1991	1992	1993	1994	1995	1990	1991	1992	1993	1994	1995
Philippines	-	-	17	16	-	-	-	-	60.54	61.93	-	-
Portugal	1.867	1.761	1.782	1.412	1.566	-	12.44	11.24	13.23	9.98	10.84	-
Puerto Rico	1.580	1.300	1.040	1.000	1.000	0.950	58.52	57.27	55.03	54.05	50.51	50.80
Republic of Moldova	-	-	-	-	0.432	-	-	-	-	-	29.49	-
Sri Lanka	3.520	4.314	4.747	6.811	-	-	68.03	69.79	69.48	75.88	-	-
Sweden	1.200	-	-	-	-	-	41.38	-	-	-	-	-
Thailand	53	40	-	-	-	-	69.28	60.36	-	-	-	-
Turkey	1.239	-	-	-	-	-	26.32	-	-	-	-	-
United Kingdom	32	-	26	-	-	-	44.64	-	39.39	-	-	-
United Republic of Tanzania	0.137	0.114	-	-	-	-	13.34	13.09	-	-	-	-
United States of America	-	-	244	255	260	269	-	-	69.52	69.86	70.65	70.23
Zambia	-	-	-	-	0.043	-	-	-	-	-	17.77	-

Output and Output per Employee

Country	Output ($ bil.)						Output per Employee ($)					
	1990	1991	1992	1993	1994	1995	1990	1991	1992	1993	1994	1995
Albania	-	-	-	-	*0.000	*0.001	-	-	-	-	588	25,429
Algeria	0.176	0.120	0.092	0.087	0.058	0.075	25,444	16,924	12,164	12,383	8,841	10,935
Argentina	±0.231	-	-	0.600	-	-	41,172	-	-	78,429	-	-
Armenia	0.013	0.038	0.185	16	0.076	-	2,326	7,105	-	-	-	-
Australia	*0.925	*0.987	*0.939	-	-	-	71,154	75,934	72,229	-	-	-
Austria	0.614	0.598	0.722	0.654	0.693	0.815	86,501	85,401	103,090	103,520	108,428	129,103
Azerbaijan	±0.190	±0.485	±0.007	±0.007	±0.003	-	36,688	90,641	1,319	2,327	1,065	-
Bahamas	0.014	0.014	-	-	-	-	78,652	95,238	-	-	-	-
Bahrain	-	-	±0.090	-	-	-	-	-	84,961	-	-	-
Bangladesh	0.063	0.056	0.054	-	-	-	9,644	8,023	6,376	-	-	-
Barbados	0.002	0.001	0.001	0.001	0.002	0.002	11,339	38,056	40,300	24,143	30,159	31,367
Belize	0.000	0.001	0.001	-	-	-	-	6,953	6,006	-	-	-
Bolivia	0.017	0.016	0.035	0.085	0.146	0.134	32,720	30,978	65,303	134,113	109,015	98,571
Bosnia & Herzegovina	±0.057	±0.059	-	-	-	-	75,598	99,439	-	-	-	-
Botswana	-	-	±0.087	±0.099	-	-	-	-	17,799	22,426	-	-
Bulgaria	±2.011	±0.470	±0.481	±0.520	±0.299	±0.305	11,182	3,595	4,886	5,600	13,985	29,876
Burundi	±0.001	±0.001	-	-	-	-	31,785	29,905	-	-	-	-
Cameroon	0.014	0.016	0.019	0.020	0.010	0.010	58,501	44,919	73,151	53,216	33,494	51,772
Canada	3.265	3.160	2.904	2.759	3.039	3.028	66,640	70,214	74,460	72,618	75,974	75,676
Central African Republic	*0.003	*0.003	*0.001	*0.002	-	-	19,025	19,887	17,433	18,112	-	-
Chile	0.026	0.030	0.034	0.033	0.035	0.040	16,483	17,547	21,954	29,388	28,227	32,538
China	±7.413	±8.008	±9.863	±14	±12	±13	2,896	3,337	4,197	6,749	4,830	5,935
China (Hong Kong SAR)	1.631	1.654	1.706	1.918	1.619	-	51,305	69,509	75,134	86,027	83,438	-
China (Taiwan Province)	±6.608	±6.703	±6.787	±5.174	±4.660	±4.605	53,749	57,148	60,324	50,487	46,833	48,774
Colombia	0.178	0.160	0.150	0.173	0.238	0.257	20,464	19,697	15,663	17,454	26,498	29,085
Costa Rica	0.010	0.009	0.012	0.011	0.014	0.009	3,078	2,497	3,050	2,772	-	-
Croatia	±0.047	±0.047	±0.022	-	-	-	22,714	23,464	12,889	-	-	-
Cyprus	0.066	0.070	0.062	0.056	0.061	0.069	60,400	67,807	62,533	62,280	67,258	79,267
Czechoslovakia (Former)	±0.601	±0.299	-	-	-	-	13,068	11,056	-	-	-	-
Denmark	*0.838	*0.903	*1.030	*1.044	*1.169	*1.299	74,789	80,942	98,529	95,740	105,029	119,702
Ecuador	0.010	0.012	0.013	0.015	0.016	0.017	12,534	9,236	9,650	10,345	11,603	13,698
Egypt	*0.040	*0.029	*0.026	*0.021	*0.022	*0.018	26,600	17,265	15,293	9,142	9,437	8,819
El Salvador	-	±0.006	-	-	±0.012	±0.012	-	-	-	-	10,408	9,430
Fiji	0.009	0.009	0.009	0.010	0.010	-	32,146	34,685	29,550	29,507	32,184	-
Finland	±0.324	±0.314	±0.297	±0.248	±0.287	±0.386	90,002	89,657	89,907	75,117	89,739	110,391
FYR Macedonia	±0.033	±0.034	±0.013	±0.016	±0.016	±0.015	19,808	21,379	7,839	10,717	11,409	12,102
France	±9.117	±8.643	±9.184	±8.555	±9.129	±11	90,992	92,340	103,307	102,953	112,701	132,679
Gabon	-	*0.000	*0.000	-	-	-	-	-	-	-	-	-
Germany	-	5.303	5.502	-	5.246	-	-	73,589	92,066	-	103,501	-
Germany (Western Part)	4.969	5.083	5.296	-	-	-	92,782	93,245	100,983	-	-	-
Greece	*0.119	*0.117	*0.118	*0.121	*0.124	*0.146	46,078	51,160	54,074	63,395	67,845	84,497

Continued.

Depending on the table, * means *Enterprises, Engaged,* or *Factor Values*; ± means *Basis Unspecified*; § means *shown in millions*. For additional notes and sources, see Appendix I.

711

Output and Output per Employee
- Continued -

Country	Output ($ bil.)						Output per Employee ($)					
	1990	1991	1992	1993	1994	1995	1990	1991	1992	1993	1994	1995
Guatemala	-	0.015	0.018	0.018	0.026	0.028	-	46,969	48,555	48,107	59,072	57,426
Honduras	0.015	0.013	0.014	0.012	0.011	0.013	8,812	7,474	8,725	7,115	5,617	5,791
Hungary	±0.277	±0.240	±0.100	±0.100	±0.116	±0.121	7,691	7,984	4,361	16,704	17,447	17,876
Iceland	0.091	0.089	±0.093	±0.082	-	-	186,454	168,768	187,922	169,307	-	-
India	*0.555	*0.611	*0.890	*1.083	*1.221	*1.214	12,930	13,130	18,246	18,189	20,700	20,383
Indonesia	0.180	0.234	0.407	0.526	0.546	0.641	5,935	5,138	7,327	7,458	7,355	8,247
Iran (Islamic Republic of)	0.061	0.018	0.046	0.055	-	-	21,925	17,260	36,490	40,930	-	-
Iraq	-	-	*0.009	-	-	-	-	-	263,081	-	-	-
Ireland	*0.251	*0.249	*0.274	*0.210	*0.242	*0.308	86,636	83,145	90,918	78,836	81,788	95,101
Israel	0.387	0.381	0.449	0.489	0.579	0.616	80,700	74,763	78,762	78,933	83,892	92,512
Italy	±5.694	±5.688	±6.315	±5.840	±6.358	-	157,930	154,353	173,219	157,451	171,200	-
Jamaica	0.009	0.007	0.004	-	-	-	5,103	11,425	7,691	-	-	-
Japan	±32	±37	±40	±43	±49	±52	146,074	168,719	181,639	204,779	221,632	244,972
Jordan	0.004	0.002	0.002	0.001	0.016	0.010	18,569	9,006	22,397	5,031	22,171	24,660
Kenya	*0.105	*0.111	*0.055	*0.034	*0.034	*0.019	29,157	33,237	16,140	9,945	9,669	5,433
Korea, Republic of	3.766	3.800	3.497	3.587	3.816	4.387	41,474	49,612	54,243	58,846	67,984	82,738
Kuwait	0.041	0.029	0.035	0.067	0.030	0.062	31,611	29,520	28,837	63,459	25,399	52,314
Latvia	0.004	0.009	0.008	0.011	0.015	0.017	191	496	2,299	6,235	5,001	7,109
Lithuania	-	-	0.123	0.113	0.011	-	-	-	2,871	2,846	3,661	-
Macau	0.224	0.155	0.120	0.119	0.133	0.095	31,337	31,900	29,970	32,119	34,950	28,856
Malaysia	*0.282	*0.444	*0.608	*0.589	*0.549	*0.855	15,651	21,764	29,253	29,167	27,160	36,345
Malta	0.046	0.062	0.081	0.085	0.087	-	58,008	69,572	90,363	98,463	88,768	-
Mauritius	0.049	0.053	0.054	0.061	0.061	0.070	12,999	16,346	16,206	19,001	19,736	20,111
Mexico	0.174	0.208	0.229	0.225	0.238	0.132	27,303	32,387	34,399	35,481	35,732	24,297
Mongolia	0.109	0.075	0.045	0.065	0.049	0.011	92,781	46,539	48,759	38,264	38,443	2,257
Morocco	±0.010	±0.010	±0.013	±0.012	±0.013	±0.012	16,355	17,298	18,793	17,680	20,224	18,707
Mozambique	-	±0.000	-	±0.000	±0.000	-	-	1,417	-	1,538	682	-
Myanmar	0.010	0.015	0.015	0.101	0.090	-	2,761	4,451	4,531	29,306	26,692	-
Namibia	-	-	-	-	0.001	-	-	-	-	-	7,031	-
Nepal	0.006	0.003	-	0.006	0.004	-	5,539	3,512	-	4,781	5,489	-
Netherlands	0.291	0.293	0.335	0.375	0.348	0.412	110,715	113,308	121,621	128,776	138,206	166,551
New Zealand	0.250	0.193	0.194	0.218	-	-	85,259	70,687	72,798	70,600	-	-
Norway	±0.200	±0.229	±0.217	±0.163	±0.174	±0.194	80,036	72,803	69,535	64,033	67,674	76,629
Oman	-	0.000	0.000	0.002	-	0.018	-	23,377	38,961	32,258	-	152,172
Pakistan	0.073	0.086	0.094	-	-	-	9,856	8,700	9,028	-	-	-
Paraguay	-	±0.000	-	-	-	-	-	2,984	-	-	-	-
Peru	0.135	0.176	0.227	-	-	-	24,140	32,532	43,739	-	-	-
Philippines	0.203	0.216	0.223	0.246	-	-	8,079	7,863	8,002	9,451	-	-
Poland	±0.424	±0.419	±0.192	±0.172	±0.209	±0.307	6,840	9,975	8,012	8,590	9,438	11,833
Portugal	0.453	0.535	0.566	0.570	0.597	0.689	30,226	34,132	42,011	40,306	41,354	50,572
Puerto Rico	±0.260	±0.270	±0.246	±0.249	±0.352	±0.362	96,243	119,119	130,000	134,541	177,879	193,316
Qatar	0.002	0.001	0.002	0.002	0.002	-	16,650	12,602	9,260	12,278	26,875	-
Republic of Moldova	0.000	0.301	0.044	-	-	-	0.046	25,888	2,739	-	-	-
Romania	±0.419	±0.268	±0.139	±0.097	-	-	41,490	29,170	18,330	12,999	-	-
Russian Federation	-	12	7.043	1.275	1.702	2.088	-	-	-	3,314	7,226	8,032
Senegal	0.000	-	-	-	-	-	9,939	-	-	-	-	-
Sierra Leone	-	-	-	0.000	-	-	-	-	-	3,723	-	-
Singapore	*0.471	*0.489	*0.478	*0.468	*0.570	*0.470	72,981	82,029	89,881	92,604	121,671	130,510
Slovakia	-	±0.036	±0.031	±0.032	±0.030	-	-	8,042	8,588	9,442	9,122	-
Slovenia	-	-	-	0.069	0.069	0.099	-	-	-	-	-	-
South Africa	±0.827	±0.695	±0.664	±0.702	±0.673	±0.670	33,069	27,813	27,671	30,511	30,606	31,083
Spain	*2.171	*2.261	*2.389	±2.239	±2.245	±2.604	91,954	96,003	103,921	75,614	79,789	92,182
Sri Lanka	0.084	0.096	0.118	0.140	-	-	16,272	15,491	17,344	15,596	-	-
Suriname	*0.001	*0.002	*0.003	0.004	-	-	14,365	33,446	35,759	81,487	-	-
Swaziland	*0.000	*0.000	*0.000	*0.000	-	-	5,273	14,962	8,048	19,151	-	-
Sweden	*0.264	*0.387	*0.365	*0.248	*0.263	*0.291	90,882	128,112	128,065	93,358	112,412	124,634
Syrian Arab Republic	*0.038	*0.034	*0.087	*0.079	*0.089	*0.098						
Thailand	1.431	2.040	-	-	-	-	18,833	31,087	-	-	-	-
Tonga	0.000	-	-	-	-	-	7,249	-	-	-	-	-
Trinidad and Tobago	±0.049	±0.055	±0.054	±0.047	±0.049	-	24,642	30,540	29,701	23,499	24,483	-

Continued.

Depending on the table, * means *Enterprises*, *Engaged*, or *Factor Values*; ± means *Basis Unspecified*; § means *shown in millions*. For additional notes and sources, see Appendix I.

Output and Output per Employee

- Continued -

Country	Output ($ bil.)						Output per Employee ($)					
	1990	1991	1992	1993	1994	1995	1990	1991	1992	1993	1994	1995
Tunisia	0.130	0.124	0.137	0.131	0.139	0.169	-	-	-	47,167	54,982	69,650
Turkey	0.158	0.158	0.189	0.208	0.167	0.215	33,473	36,826	36,660	36,814	29,967	39,511
United Kingdom	*5.761	*5.287	5.650	3.869	6.410	6.334	81,137	81,334	85,609	60,459	80,130	83,753
United Republic of Tanzania	*0.005	*0.003	-	-	-	-	4,802	3,348	-	-	-	-
USSR (Former)	±24						26,181					
United States of America	*35	*35	*37	*40	*41	*44	94,119	99,510	105,519	108,890	112,372	114,822
Uruguay	0.026	0.045	0.066	0.044	0.051	0.053	12,743	21,687	32,584	34,861	36,912	39,707
Venezuela	0.114	0.122	0.137	0.125	0.100	0.134	18,969	19,389	21,671	21,589	20,418	26,450
Yugoslavia	±0.015	±0.017	±0.027	-	±0.055	±0.024	-	1,622	2,685	-	5,940	2,638
Zambia	0.004	-	-	-	0.001	-	10,551	-	-	-	4,157	-
Zimbabwe	*0.025	*0.021	*0.017	*0.016	*0.016	*0.018	14,070	12,527	9,597	5,541	6,324	7,392

Value Added and Value Added per Employee

Country	Value Added ($ bil.)						Value Added per Employee ($)					
	1990	1991	1992	1993	1994	1995	1990	1991	1992	1993	1994	1995
Albania	±0.004	-	-	-	-	-	472	-	-	-	-	-
Algeria	0.079	0.052	0.041	0.039	0.026	0.034	11,358	7,367	5,458	5,595	3,989	4,941
Argentina	±0.097	-	-	0.233	-	-	17,204	-	-	30,381	-	-
Australia	*0.445	*0.475	*0.463	-	-	-	34,195	36,559	35,623	-	-	-
Austria	0.249	0.248	0.285	0.262	0.263	0.319	35,056	35,384	40,742	41,487	41,140	50,598
Bahamas	0.005	0.005	-	-	-	-	27,803	35,483	-	-	-	-
Bangladesh	0.018	0.012	0.011	-	-	-	2,712	1,769	1,345	-	-	-
Barbados	0.001	0.001	0.001	0.000	0.001	0.001	3,942	23,759	22,600	7,133	13,151	13,500
Belize	0.000	0.001	0.001	-	-	-	-	3,816	2,591	-	-	-
Bolivia	0.008	0.002	0.018	0.015	0.029	0.027	16,154	3,857	33,550	24,110	21,973	19,902
Bosnia & Herzegovina	±0.026	±0.027	-	-	-	-	34,137	45,254	-	-	-	-
Botswana	-	-	±0.024	±0.036	±0.037		-	-	4,972	8,085	8,211	
Bulgaria	-	±0.056	±0.081	±0.103	±0.055	±0.058	-	426	821	1,111	2,548	5,730
Burundi	0.001	0.001	-	-	-	-	16,019	16,958	-	-	-	-
Cameroon	0.006	0.007	0.011	0.009	0.005	0.004	25,303	18,837	41,772	23,948	16,667	20,361
Canada	1.706	1.632	1.547	1.465	1.626	1.620	34,807	36,271	39,670	38,553	40,641	40,485
Central African Republic	*0.001	*0.001	*0.000	*0.001	-	-	7,399	8,080	6,692	9,091	-	-
Chile	0.014	0.015	0.021	0.020	0.019	0.022	9,320	9,011	13,391	17,459	15,591	17,975
China	±2.125	±2.272	±2.673	±4.510	±3.287	±3.204	830	947	1,138	2,222	1,299	1,483
China (Hong Kong SAR)	0.432	0.389	0.434	0.467	0.442	-	13,596	16,339	19,128	20,950	22,790	-
China (Taiwan Province)	2.015	1.886	1.915	1.769	1.520	1.327	16,392	16,084	17,019	17,262	15,274	14,056
Colombia	0.084	0.079	0.082	0.094	0.140	0.152	9,662	9,706	8,529	9,510	15,634	17,177
Costa Rica	0.004	0.003	0.005	0.004	0.005	0.003	1,164	958	1,187	1,051	-	-
Cote d'Ivoire	0.037	0.032	0.042	0.046	-	-	-	-	-	-	-	-
Croatia	±0.028	±0.031	±0.012	-	-	-	13,495	15,583	7,410	-	-	-
Cyprus	0.019	0.019	0.018	0.016	0.018	0.018	17,798	18,430	18,208	17,898	19,311	20,961
Czechoslovakia (Former)	±0.192	-	-	-	-	-	4,178	-	-	-	-	-
Denmark	*0.489	*0.535	*0.610	*0.618	*0.692	*0.769	43,653	47,963	58,363	56,690	62,167	70,858
Ecuador	0.003	0.004	0.004	0.005	0.005	0.005	4,226	2,728	3,405	3,259	3,381	4,384
Egypt	*0.008	*0.009	*0.011	*0.004	*0.004	*0.003	5,467	5,278	6,378	1,754	1,814	1,672
El Salvador	-	0.004	-	-	0.007	0.008	-	-	-	-	5,766	6,340
Fiji	0.001	0.001	0.001	0.001	0.002	-	4,114	5,054	4,796	4,264	4,651	-
Finland	±0.168	±0.160	±0.156	±0.129	±0.151	±0.181	46,708	45,782	47,152	39,203	47,262	51,904
FYR Macedonia	±0.019	±0.021	±0.010	±0.011	±0.011	±0.009	11,382	13,220	6,110	7,892	7,755	7,684
France	±4.319	±3.992	±4.329	±4.253	±4.540	±5.153	43,102	42,654	48,695	51,174	56,054	64,169
Gabon	-	*0.000	*0.000	-	-	-	-	-	-	-	-	-
Germany (Western Part)	2.849	2.945	3.092	2.833	-	-	53,206	54,021	58,948	-	-	-
Greece	*0.047	*0.048	*0.052	*0.050	*0.051	*0.061	18,328	21,124	23,513	26,334	28,182	35,099
Haiti	-	±0.008	±0.003	±0.002	±0.002	±0.001	-	-	-	-	-	-
Honduras	*0.006	*0.006	*0.006	*0.004	*0.004	*0.004	3,555	3,176	3,784	2,448	1,930	1,848
Hungary	±0.101	±0.061	±0.038	±0.039	±0.041	±0.041	2,811	2,027	1,668	6,474	6,237	6,061
Iceland	0.027	0.028	±0.030	±0.029	-	-	55,928	53,056	61,138	60,195	-	-

Continued.

Depending on the table, * means *Enterprises, Engaged,* or *Factor Values*; ± means *Basis Unspecified*; § means *shown in millions*. For additional notes and sources, see Appendix I.

713

Value Added and Value Added per Employee

- Continued -

Country	Value Added ($ bil.)						Value Added per Employee ($)					
	1990	1991	1992	1993	1994	1995	1990	1991	1992	1993	1994	1995
India	*0.092	*0.131	*0.156	*0.344	*0.389	*0.386	2,135	2,813	3,197	5,779	6,587	6,488
Indonesia	*0.061	*0.086	*0.201	*0.210	*0.205	*0.232	2,018	1,896	3,630	2,973	2,765	2,984
Iran (Islamic Republic of)	0.030	0.008	0.014	0.035	-	-	10,600	7,671	10,947	25,895	-	-
Iraq	-	-	*0.002	-	-	-	-	-	52,909	-	-	-
Ireland	*0.132	*0.126	*0.139	*0.106	*0.123	*0.157	45,520	42,165	46,147	39,909	41,518	48,438
Israel	0.120	0.118	0.196	0.127	0.136	0.188	24,902	23,229	34,458	20,517	19,782	28,167
Italy	±1.890	±1.479	±1.760	±1.499	±1.538	-	52,435	40,136	48,274	40,418	41,415	-
Jamaica	0.006	0.004	0.003	-	-	-	3,459	7,481	5,422	-	-	-
Japan	±14	±16	±17	±19	±21	±22	62,410	73,276	79,460	88,472	94,353	101,745
Jordan	0.002	0.001	0.000	0.000	0.005	0.004	7,271	3,616	3,794	3,422	6,704	9,711
Kenya	*0.018	*0.016	-	-	-	*0.016	5,102	4,881	-	-	-	4,455
Korea, Republic of	1.769	1.792	1.696	1.758	1.903	2.251	19,482	23,399	26,314	28,838	33,901	42,453
Kuwait	0.017	0.012	0.019	0.020	0.017	0.026	12,883	12,762	15,914	19,121	14,719	22,483
Latvia	-	-	-	*0.006	*0.010	*0.009	-	-	-	3,544	3,285	3,644
Macau	0.068	0.045	0.032	0.036	0.035	0.025	9,510	9,331	7,911	9,633	9,233	7,475
Malaysia	*0.111	*0.153	*0.190	*0.178	*0.165	*0.261	6,182	7,504	9,153	8,789	8,161	11,085
Malta	0.021	0.027	0.022	0.025	0.030	-	26,019	30,807	24,618	29,438	31,151	-
Mauritius	0.016	0.017	0.015	0.018	0.017	0.020	4,226	5,397	4,596	5,458	5,631	5,742
Mexico	0.096	0.109	0.117	0.116	0.121	0.067	15,049	16,914	17,561	18,296	18,120	12,319
Mongolia	0.029	0.026	0.021	0.022	0.023	0.003	24,833	16,149	22,438	13,005	17,618	565
Morocco	±0.004	±0.004	±0.005	±0.004	±0.005	±0.004	6,086	7,040	7,048	6,164	7,198	6,655
Myanmar	±0.004	±0.005	±0.004	±0.043	±0.034	-	1,102	1,518	1,140	12,390	10,133	-
Namibia	-	-	-	-	0.000	-	-	-	-	-	3,957	-
Nepal	0.003	0.001	-	0.003	0.001	-	2,411	1,405	-	2,079	1,711	-
Netherlands	0.111	0.109	0.128	0.154	0.136	0.160	42,277	42,257	46,460	52,916	53,825	64,865
New Zealand	0.086	0.071	0.073	0.084	-	-	29,371	25,820	27,501	27,262	-	-
Norway	±0.089	±0.101	±0.105	±0.079	±0.081	±0.090	35,763	32,128	33,557	31,048	31,603	35,785
Pakistan	0.022	0.025	0.028	-	-	-	2,994	2,486	2,653	-	-	-
Paraguay	-	±0.000	-	-	-	-	-	1,996	-	-	-	-
Peru	0.066	0.057	0.096	-	-	-	11,704	10,534	18,589	-	-	-
Philippines	0.093	0.089	0.105	0.117	-	-	3,723	3,237	3,757	4,518	-	-
Poland	±0.258	±0.119	±0.091	±0.075	±0.088	±0.127	4,166	2,843	3,789	3,729	3,986	4,901
Portugal	0.141	0.159	0.162	0.158	0.172	0.196	9,412	10,155	12,029	11,179	11,909	14,415
Puerto Rico	±0.113	±0.105	±0.093	±0.091	±0.087	±0.100	41,704	46,344	49,101	49,459	43,788	53,209
Qatar	0.001	0.001	0.001	0.001	0.001	-	11,100	7,561	4,630	6,139	14,931	-
Romania	±0.227	±0.110	±0.040	±0.032	-	-	22,510	11,952	5,213	4,245	-	-
Russian Federation	-	-	-	±0.605	±0.822	±0.911	-	-	-	1,572	3,491	3,504
Saint Lucia	-	±0.000	±0.000	±0.000	±0.000	±0.000	-	-	-	-	-	-
Senegal	0.000	0.000	-	-0.000	0.000	-	6,482	2,532	-	§-7,06	1,201	-
Sierra Leone	-	-	-	0.000	-	-	-	-	-	2,194	-	-
Singapore	*0.114	*0.117	*0.113	*0.108	*0.111	*0.094	17,727	19,578	21,235	21,297	23,757	25,982
Slovakia	-	-	-	±0.013	±0.013	-	-	-	-	3,673	3,860	-
Slovenia	-	-	0.027	0.028	0.038	-	-	-	-	-	-	-
South Africa	*0.448	*0.406	*0.445	*0.476	*0.458	*0.455	17,918	16,253	18,554	20,690	20,800	21,140
Spain	*0.870	*0.873	*0.942	±0.864	±0.790	±0.913	36,859	37,069	40,963	29,184	28,077	32,336
Sri Lanka	0.010	0.035	0.041	0.026	-	-	2,021	5,661	5,983	2,923	-	-
Swaziland	*0.000	*0.000	*0.000	*0.000	-	-	896	3,116	1,696	5,734	-	-
Sweden	*0.157	*0.148	*0.141	*0.113	*0.111	*0.123	54,180	48,984	49,368	42,684	47,597	52,492
Syrian Arab Republic	*0.005	*0.004	*0.010	*0.013	*0.012	*0.013	-	-	-	-	-	-
Thailand	0.498	1.012	-	-	-	-	6,548	15,421	-	-	-	-
Trinidad and Tobago	±0.024	±0.022	±0.019	±0.017	±0.018	-	11,761	12,035	10,769	8,296	9,003	-
Tunisia	0.031	0.030	0.037	0.039	0.041	0.051	-	-	-	14,172	16,457	21,166
Turkey	0.084	0.075	0.097	0.093	0.081	0.104	17,754	17,573	18,783	16,426	14,532	19,032
United Kingdom	*2.786	*2.625	*2.869	*2.420	*3.430	*3.381	39,235	40,381	43,474	37,819	42,879	44,703
United Republic of Tanzania	*0.002	*0.001	-	-	-	-	1,488	1,409	-	-	-	-
United States of America	*19	*19	*20	*22	*23	*24	50,732	53,487	58,308	59,581	62,065	62,762
Uruguay	0.015	0.019	0.020	0.019	0.022	0.023	7,068	8,957	9,897	15,294	16,117	17,297
Venezuela	0.056	0.058	0.067	0.048	0.040	0.122	9,318	9,217	10,702	8,240	8,033	24,028

Continued.

Depending on the table, * means *Enterprises, Engaged,* or *Factor Values*; ± means *Basis Unspecified*; § means *shown in millions.* For additional notes and sources, see Appendix I.

Value Added and Value Added per Employee

- Continued -

Country	Value Added ($ bil.)						Value Added per Employee ($)					
	1990	1991	1992	1993	1994	1995	1990	1991	1992	1993	1994	1995
Yugoslavia	±0.007	±0.009	±0.019	-	±0.026	±0.013	-	825	1,848	-	2,782	1,376
Zambia	0.002	-	-	-	0.000	-	5,439	-	-	-	1,364	-
Zimbabwe	*0.013	*0.011	*0.009	*0.008	*0.010	*0.011	7,058	6,572	4,864	2,712	3,723	4,343

Capital Investment

Country	Gross Fixed Capital Formation ($ mil.)						Per Establishment ($ mil)					
	1990	1991	1992	1993	1994	1995	1990	1991	1992	1993	1994	1995
Argentina	-	-	-	20	-	-	-	-	-	0.011	-	-
Australia	34	-	-	-	-	-	0.032	-	-	-	-	-
Austria	21	21	55	32	29	-	0.374	0.425	1.062	0.686	0.659	-
Bahrain	-	-	0.372	-	-	-	-	-	0.002	-	-	-
Bangladesh	0.422	0.410	0.565	-	-	-	0.001	0.001	0.002	-	-	-
Barbados	0.194	0.065	0.036	0.041	0.073	-	0.039	0.016	0.006	0.007	0.007	-
Bolivia	0.040	0.071	0.271	0.248	0.498	-	0.002	0.003	0.011	0.009	0.018	-
Bosnia & Herzegovina	1.144	-	-	-	-	-	0.381	-	-	-	-	-
Botswana	-	-	23	6.975	-	-	-	-	-	0.307	0.073	-
Bulgaria.	41	8.066	84	30	7.290	0.819	0.074	0.013	0.192	0.008	0.003	0.001
Cameroon	0.301	0.202	6.245	7.406	0.146	0.345	0.301	0.051	1.561	1.481	0.024	0.172
Central African Republic . . .	0.088	0.046	-	0.226	-	-	0.018	0.009	-	0.045	-	-
Chile.	0.849	1.718	1.434	2.316	0.693	-	0.057	0.095	0.096	0.211	0.058	-
China (Hong Kong SAR) . . .	36	62	50	54	32	-	0.010	0.023	0.020	0.020	0.011	-
Colombia	8.512	5.467	7.171	5.790	5.322	-	0.047	0.031	0.037	0.032	0.032	-
Croatia	0.950	0.667	0.149	0.342	-	0.064	0.021	0.010	0.002	0.003	-	0.000
Cyprus	2.687	1.624	2.160	0.990	1.697	2.291	0.007	0.004	0.006	0.003	0.004	0.006
Czechoslovakia (Former) . . .	60	-	-	-	-	-	1.504	-	-	-	-	-
Denmark	53	70	-	-	-	-	0.040	0.048	-	-	-	-
Ecuador	1.202	2.873	1.856	1.890	2.291	3.526	0.055	0.090	0.058	0.061	0.072	0.107
Egypt	1.350	1.050	25	-1.038	-	-	0.038	0.031	0.645	-0.024	-	-
El Salvador.	-	0.248	-	-	2.708	3.397	-	0.025	-	-	0.143	0.136
Fiji	0.386	0.213	0.369	-	-	-	0.008	0.005	0.009	-	-	-
Finland	16	11	13	6.740	10	13	0.190	0.116	0.140	0.077	0.130	0.147
FYR Macedonia	0.815	36	1.171	0.343	0.116	0.526	0.058	1.038	0.015	0.004	0.001	0.004
Germany (Western Part) . . .	248	293	257	233	-	-	0.409	0.476	0.422	-	-	-
Greece	6.731	3.199	5.031	-	-	-	0.051	0.025	0.040	-	-	-
Hungary	12	8.309	6.153	7.310	-	-	0.037	0.010	0.020	0.021	-	-
India.	21	15	28	35	-	-	0.021	0.015	0.026	0.030	-	-
Indonesia	29	35	64	29	28	298	0.121	0.122	0.193	0.084	0.072	0.675
Iran (Islamic Republic of) . . .	5.627	2.085	4.492	2.560	-	-	0.052	0.099	0.160	0.083	-	-
Ireland	12	7.593	-	-	-	-	0.116	0.094	-	-	-	-
Israel.	10	20	17	21	24	-	0.040	0.070	0.056	0.057	0.065	-
Italy	168	173	197	150	166	-	0.221	0.218	0.252	0.187	0.211	-
Japan	746	1,076	1,169	962	1,301	-	0.054	0.082	0.093	0.075	0.094	-
Jordan	-	-	-	0.043	0.033	0.253	-	-	-	0.001	0.000	0.003
Korea, Republic of . . .	259	265	175	271	215	217	0.094	0.100	0.071	0.094	0.077	0.076
Kuwait	0.233	0.866	0.119	0.007	0.356	-	0.001	0.005	0.001	0.000	0.002	-
Latvia	0.152	0.050	1.146	0.605	0.436	0.383	0.000	0.000	0.018	0.011	0.015	0.006
Macau	4.140	2.624	2.770	3.401	3.772	-1.036	0.039	0.034	0.030	0.041	0.056	-0.016
Malaysia	22	23	20	25	33	-	0.133	0.119	0.097	0.101	0.135	-
Malta	2.634	6.978	6.028	4.377	11	-	0.031	0.087	0.077	0.057	0.141	-
Mexico	4.087	7.182	-	-	-	-	0.114	0.199	-	-	-	-
Mongolia	25	10	2.829	4.091	4.855	-	0.522	0.210	0.049	0.037	0.221	-
Morocco	0.121	0.345	0.703	0.430	0.869	0.234	0.004	0.012	0.025	0.016	0.032	0.009
Nepal	3.711	-	-	-	-	-	0.143	-	-	-	-	-
Netherlands	22	22	26	84	-	-	0.414	0.406	0.441	1.333	-	-
New Zealand	4.776	-	-	-	-	-	0.005	-	-	-	-	-
Norway	7.988	12	10	9.444	6.659	-	0.094	0.126	0.150	0.185	0.131	-
Oman	0.049	0.031	0.078	2.951	-	3.779	0.049	0.031	0.078	1.475	-	0.472

Continued.

Depending on the table, * means *Enterprises*, *Engaged*, or *Factor Values*; ± means *Basis Unspecified*; § means *shown in millions*. For additional notes and sources, see Appendix I.

Capital Investment
- Continued -

Country	Gross Fixed Capital Formation ($ mil.)						Per Establishment ($ mil)					
	1990	1991	1992	1993	1994	1995	1990	1991	1992	1993	1994	1995
Pakistan	-	3.113	-	-	-	-	-	0.042	-	-	-	-
Peru	20	1.826	7.473	-	-	-	0.041	0.004	0.015	-	-	-
Philippines	4.977	4.767	14	14	-	-	0.003	0.003	0.039	0.043	-	-
Poland	19	18	16	22	-	-	0.080	0.096	0.113	0.175	-	-
Portugal	24	22	22	30	45	-	0.011	0.011	0.011	0.015	0.023	-
Romania	74	121	-	2.614	4.315	-	-	0.105	-	0.001	0.003	-
Singapore	24	23	20	25	32	1.131	0.172	0.161	0.140	0.166	0.213	-
Slovenia	0.530	1.451	1.587	2.649	1.296	1.131	0.010	0.001	0.001	0.001	0.000	0.001
Spain	65	71	66	82	63	67	0.032	0.035	0.032	-	-	-
Sri Lanka	0.924	1.780	1.221	3.174	-	-	0.014	0.032	0.023	0.050	-	-
Swaziland	0.014	-	-	-	-	-	0.003	-	-	-	-	-
Thailand	670	291	-	-	-	-	1.948	0.518	-	-	-	-
Trinidad and Tobago	2.830	4.837	3.741	3.326	2.464	-	0.034	0.066	0.055	0.050	0.039	-
Tunisia	8.200	8.761	11	13	13	15	-	-	-	0.078	0.073	0.083
Turkey	8.817	5.034	7.712	81	14	-	0.147	0.092	0.073	0.768	0.129	-
United Kingdom	173	152	148	-	198	-	0.020	0.019	0.018	-	0.017	-
United Republic of Tanzania	0.523	0.461	-	-	-	-	0.026	0.031	-	-	-	-
United States of America	720	730	953	841	927	1,165	-	-	0.039	-	-	-
Uruguay	0.132	1.799	0.398	0.429	-	-	-	-	-	-	-	-
Venezuela	3.945	4.682	6.420	4.217	0.902	15	0.023	0.027	0.038	0.028	0.006	0.119
Yugoslavia	0.338	0.294	1.780	-	3.982	0.731	0.005	0.003	0.008	-	0.011	0.002
Zambia	0.028	-	-	-	-	-	0.004	-	-	-	-	-

Indexes of Industrial Production

Country	Index of Industrial Production (1990=100)						Index of Employment (1990=100)					
	1990	1991	1992	1993	1994	1995	1990	1991	1992	1993	1994	1995
Albania	100	-	-	-	-	-	100	-	-	-	3	1
Algeria	100	48	-	-	-	-	100	102	110	102	94	99
Angola	100	-	-	-	-	-	-	-	-	-	-	-
Argentina	100	138	-	-	-	-	100	-	-	136	144	-
Armenia	100	-	-	-	-	-	100	96	-	-	-	-
Australia	100	97	97	111	-	-	100	100	100	-	-	-
Austria	100	-	-	-	-	-	100	99	99	89	90	89
Azerbaijan	-	-	-	-	-	-	100	104	99	59	45	-
Bahamas	100	-	-	-	-	-	100	83	-	-	-	-
Bahrain	100	-	-	-	-	-	-	-	-	-	-	-
Bangladesh	100	95	104	126	142	168	100	106	129	-	-	-
Barbados	100	100	92	91	102	113	100	18	17	34	43	36
Belgium	100	105	105	95	107	127	100	96	91	-	-	-
Belize	100	-	-	-	-	-	-	-	-	-	-	-
Bermuda	-	-	-	-	-	-	100	88	171	118	265	147
Bolivia	100	104	111	116	136	132	100	100	102	120	254	258
Bosnia & Herzegovina	-	-	-	-	-	-	100	79	-	-	40	-
Botswana	100	-	-	-	-	-	100	115	107	96	98	107
Brazil	100	-	-	-	-	-	-	-	-	-	-	-
Bulgaria	100	96	85	88	63	49	100	73	55	52	12	6
Burkina-Faso	100	-	-	-	-	-	-	-	-	-	-	-
Burundi	100	-	-	-	-	-	100	109	-	-	-	-
Cambodia	100	-	-	-	-	-	100	168	-	-	-	-
Cameroon	100	-	-	-	-	-	100	155	113	162	125	81
Canada	100	97	92	90	99	97	100	92	80	78	82	82
Central African Republic	100	-	-	-	-	-	100	98	50	73	-	-
Chile	100	92	90	85	78	79	100	108	100	72	79	80
China	100	-	-	-	-	-	100	94	92	79	99	84
China (Hong Kong SAR)	100	95	93	85	81	78	100	75	71	70	61	-
China (Taiwan Province)	100	101	99	87	78	73	100	95	92	83	81	77
Colombia	100	103	106	120	121	117	100	93	110	114	103	102

Continued.

     Depending on the table, * means *Enterprises, Engaged,* or *Factor Values;* ± means *Basis Unspecified;* § means *shown in millions.* For additional notes and sources, see Appendix I.

Indexes of Industrial Production

- Continued -

Country	Index of Industrial Production (1990=100)						Index of Employment (1990=100)					
	1990	1991	1992	1993	1994	1995	1990	1991	1992	1993	1994	1995
Congo	100	100	100	100	-	-	-	-	-	-	-	-
Costa Rica	100	96	101	105	136	112	100	105	118	117	-	-
Cote d'Ivoire	100	117	160	117	167	180	-	-	-	-	-	-
Croatia	100	-	-	-	-	-	100	98	81	76	69	63
Cuba	100	-	-	-	-	-	-	-	-	-	-	-
Cyprus	100	103	118	116	131	140	100	96	91	83	84	80
Czechoslovakia (Former) . . .	100	76	64	-	-	-	100	59	-	-	-	-
Czech Republic	100	-	-	-	-	-	100	96	117	183	-	-
Dem. Rep. of the Congo . . .	100	-	-	-	-	-	-	-	-	-	-	-
Denmark	100	106	110	118	122	117	100	100	93	97	99	97
Dominican Republic	100	-	-	-	-	-	-	-	-	-	-	-
Ecuador	100	87	116	153	-	-	100	162	158	173	168	150
Egypt	100	283	271	255	251	194	100	113	113	153	152	136
El Salvador	100	100	-	-	-	-	-	-	-	-	-	-
Ethiopia and Eritrea	100	100	100	-	-	-	-	-	-	-	-	-
Fiji	100	-	-	-	-	-	100	88	102	115	116	-
Finland	100	94	94	102	106	107	100	97	92	92	89	97
FYR Macedonia	100	-	-	-	-	-	100	96	101	89	85	74
France	100	106	105	105	107	106	100	93	89	83	81	80
Gabon	100	-	-	-	-	-	-	-	-	-	-	-
Gambia	100	-	-	-	-	-	-	-	-	-	-	-
Germany (Eastern Part) . . .	100	-	-	-	-	-	-	-	-	-	-	-
Germany (Western Part) . . .	100	103	97	90	89	90	100	102	98	-	-	-
Ghana	100	-	-	-	-	-	-	-	-	-	-	-
Greece	100	137	57	35	32	37	100	89	85	74	71	67
Guatemala	100	-	-	-	-	-	-	-	-	-	-	-
Guyana	100	-	-	-	-	-	-	-	-	-	-	-
Haiti	100	-	-	-	-	-	-	-	-	-	-	-
Honduras	100	121	122	126	143	167	100	103	90	98	111	124
Hungary	100	85	61	68	78	80	100	83	64	17	18	19
Iceland	100	-	-	-	-	-	100	108	101	99	113	-
India	100	104	105	136	143	136	100	108	114	139	137	139
Indonesia	100	-	-	-	-	-	100	150	183	232	245	256
Iran (Islamic Republic of) . . .	100	-	-	-	-	-	100	36	45	48	-	-
Iraq	100	-	-	-	-	-	-	-	-	-	-	-
Ireland	100	96	97	85	94	109	100	103	104	92	102	112
Israel	100	105	124	124	140	-	100	106	119	129	144	139
Italy	100	85	89	78	-	-	100	102	101	103	103	-
Jamaica	100	-	-	-	-	-	100	32	32	30	33	-
Japan	100	93	98	89	74	67	100	99	100	95	100	97
Jordan	100	-	-	-	-	-	100	83	48	54	342	187
Kenya	100	118	118	118	118	126	100	93	95	95	97	100
Korea, Republic of	100	96	86	-	-	-	100	84	71	67	62	58
Kuwait	100	-	-	-	-	-	100	76	94	82	92	92
Kyrgyzstan	100	-	-	-	-	-	-	-	-	-	-	-
Lao P.D.R.	100	-	-	-	-	-	-	-	-	-	-	-
Latvia	100	-	-	-	-	-	100	96	18	10	16	13
Lebanon	100	-	-	-	-	-	-	-	-	-	-	-
Lesotho	100	-	-	-	-	-	100	-	53	147	148	-
Libyan Arab Jamahiriya . . .	100	-	-	-	-	-	-	-	-	-	-	-
Lithuania	100	-	-	-	-	-	-	-	-	-	-	-
Luxembourg	100	-	-	-	-	-	-	-	-	-	-	-
Macau	100	-	-	-	-	-	100	68	56	52	53	46
Madagascar	100	-	-	-	-	-	-	-	-	-	-	-
Malawi	100	108	103	92	89	93	-	-	-	-	-	-
Malaysia	100	-	-	-	-	-	100	113	116	112	112	131
Mali	100	100	100	100	100	100	-	-	-	-	-	-
Malta	100	165	206	214	-	-	100	113	113	109	124	-
Mauritius	100	-	-	-	-	-	100	86	89	86	83	93
Mexico	100	98	133	132	136	109	100	101	104	99	104	85

Continued.

Depending on the table, * means *Enterprises*, *Engaged*, or *Factor Values*; ± means *Basis Unspecified*; § means *shown in millions*. For additional notes and sources, see Appendix I.

717

Indexes of Industrial Production

- Continued -

Country	Index of Industrial Production (1990=100)						Index of Employment (1990=100)					
	1990	1991	1992	1993	1994	1995	1990	1991	1992	1993	1994	1995
Mongolia	100	-	-	-	-	-	100	138	78	144	109	410
Morocco	100	-	-	-	-	-	100	89	109	104	99	102
Mozambique	100	-	-	-	-	-	100	92	30	28	54	-
Myanmar	100	-	-	-	-	-	100	95	92	94	93	-
Namibia	100	-	-	-	-	-	-	-	-	-	-	-
Nepal	100	-	-	-	-	-	100	84	-	125	67	-
Netherlands	100	-	-	-	-	-	100	98	105	111	96	94
New Zealand	100	-	-	-	-	-	100	93	91	106	-	-
Nicaragua	100	-	-	-	-	-	-	-	-	-	-	-
Nigeria	100	-	-	-	-	-	-	-	-	-	-	-
Norway	100	105	105	116	124	121	100	126	125	102	103	101
Oman	100	-	-	-	-	-	-	-	-	-	-	-
Pakistan	100	-	-	-	-	-	100	134	141	-	-	-
Panama	100	97	103	124	126	-	-	-	-	-	-	-
Paraguay	100	103	105	81	66	66	-	-	-	-	-	-
Peru	100	110	121	124	183	185	100	97	92	-	-	-
Philippines	100	123	112	125	153	163	100	110	111	104	-	-
Poland	100	73	69	77	88	109	100	68	39	32	36	42
Portugal	100	-	-	-	-	-	100	104	90	94	96	91
Puerto Rico	100	-	-	-	-	-	100	84	70	69	73	69
Qatar	100	-	-	-	-	-	100	110	180	181	93	-
Republic of Moldova	100	-	-	-	-	-	100	108	148	-	14	-
Romania	100	-	-	-	-	-	100	91	75	74	72	-
Russian Federation	100	-	-	-	-	-	-	-	-	-	-	-
Saudi Arabia	100	-	-	-	-	-	-	-	-	-	-	-
Senegal	100	-	-	-	-	-	100	41	-	12	18	-
Sierra Leone	100	-	-	-	-	-	-	-	-	-	-	-
Singapore	100	99	78	65	65	48	100	92	82	78	73	56
Slovakia	100	-	-	-	-	-	-	-	-	-	-	-
Slovenia	100	77	84	97	92	91	-	-	-	-	-	-
Somalia	100	-	-	-	-	-	-	-	-	-	-	-
South Africa	100	87	82	91	88	81	100	100	96	92	88	86
Spain	100	113	-	-	-	-	100	100	97	125	119	120
Sri Lanka	100	-	-	-	-	-	100	119	132	173	-	-
Sudan	100	-	-	-	-	-	-	-	-	-	-	-
Suriname	100	-	-	-	-	-	100	86	121	71	-	-
Swaziland	100	-	-	-	-	-	100	77	136	64	-	-
Sweden	100	94	89	83	92	92	100	104	98	92	81	81
Switzerland	100	96	111	117	105	100	-	-	-	-	-	-
Syrian Arab Republic	100	169	438	-	-	-	-	-	-	-	-	-
Thailand	100	-	-	-	-	-	100	86	-	-	-	-
Togo	100	-	-	-	-	-	-	-	-	-	-	-
Tonga	100	-	-	-	-	-	100	-	-	-	-	-
Trinidad and Tobago	100	111	114	131	148	148	100	90	90	101	100	-
Tunisia	100	-	-	-	-	-	-	-	-	-	-	-
Turkey	100	-	-	-	-	-	100	91	109	120	118	116
Uganda	100	164	150	127	180	220	-	-	-	-	-	-
United Arab Emirates	100	-	-	-	-	-	-	-	-	-	-	-
United Kingdom	100	87	84	83	83	77	100	92	93	90	113	107
United Republic of Tanzania	100	98	8	15	22	-	100	85	-	-	-	-
USSR (Former)	100	-	-	-	-	-	100	-	-	-	-	-
United States of America	100	98	100	106	109	110	100	94	95	99	100	104
Uruguay	100	104	106	104	117	113	100	100	98	60	66	65
Venezuela	100	-	-	-	-	-	100	105	105	97	82	85
Yugoslavia	100	-	-	-	-	-	-	-	-	-	-	-
Yugoslavia (Former)	100	102	-	-	-	-	-	-	-	-	-	-
Zambia	100	97	90	75	64	68	100	-	-	-	72	-
Zimbabwe	100	98	75	194	173	109	100	94	100	161	144	139

Depending on the table, * means *Enterprises, Engaged,* or *Factor Values;* ± means *Basis Unspecified;* § means *shown in millions.* For additional notes and sources, see Appendix I.

Representative Companies in Sector

Name	Address	Tele-phone	Fax	Employ-ment	Y
Kimball International Inc.	1600 Royal St., Jasper, Indiana 47549-1001, U.S.A.	812-482-1600		8,949	97
Sapalux Ltd.	101 Oakley Rd., Luton LU4 9RJ, United Kingdom	582 491234		8,263	93
Tunzl PLC	110 Park St., London W1Y 3RB, United Kingdom	71 4954950	71 4954953	8,046	93
Huffy Corp.	PO Box 1204, Dayton, Ohio 45401, U.S.A.	513-866-6251	513-865-5470	8,020	96
Doornfontein Gold Mining Co. Ltd.	75 Fox St., Johannesburg 2001, Republic of South Africa			6,530	92
AlliedSignal Electronic and Avionics Systems	400 N. Rogers Rd., Olathe, Kansas 66062, U.S.A.	913-782-0400	913-791-1302	6,500	97
Jostens Inc.	5501 Norman Center Dr., Minneapolis, Minnesota 55437, U.S.A.	612-830-3300		6,300	96
Federal Signal Corp.	1415 W. 22nd St., Oak Brook, Illinois 60521-9945, U.S.A.	708-954-2000		6,207	97
Takata Inc.	2500 Takata Dr., Auburn Hills, Michigan 48326, U.S.A.	248-377-6130	248-373-5186	6,000	95
NV Sumatra Tobacco Trading Co.	Pematang, Medan 20232, Indonesia	61517139	622 23410	5,000	97
Perfekta Enterprises Ltd.	141 Connaught Rd. W., Hong Kong, Hong Kong	5475131	5592450	5,000	
Oneida Ltd.	5-9 Pembroke Rd., Ruislip HA4 8NQ, United Kingdom	895 639452		4,982	91
Acushnet Co.	PO Box 965, Fairhaven, Massachusetts 02719-0965, U.S.A.	508-979-2000		4,780	97
Wallace Computer Services Inc.	2275 E. Cabot Dr., Lisle, Illinois 60532-3630, U.S.A.	630-588-5000	630-588-5105	4,610	97
Oneida Ltd.	Kenwood Ave., Oneida, New York 13421-2899, U.S.A.	315-361-3000	315-829-3950	4,525	96
Zainco Traders	PO Box 413, Moradabad 244 001, India	26337	11 3316467	4,500	
Spalding and Evenflo Companies Inc.	PO Box 30101, Tampa, Florida 33630, U.S.A.	813-204-5200	813-204-5219	4,400	97
Kaman Corp.	PO Box 1, Bloomfield, Connecticut 06002, U.S.A.	203-243-8311		4,318	97
Gorenje Gospodinjski Aparati d.d.	Partizanska 12, Velenje 3320, Slovenia	63 853321	63 851745	4,290	95
AptarGroup Inc.	475 W. Terra Cotta Ave., E, Crystal Lake, Illinois 60014, U.S.A.	815-477-0424		4,100	97
K2 Inc.	PO Box 22252, Los Angeles, California 90040, U.S.A.	213-724-2800		3,800	97
Ford France SA	BP 32, Blanquefort F-33290, France	56 954000	56 954378	3,624	91
Stylo PLC	Apperley Bridge, Bradford BD10 0NW, United Kingdom	274 617761	274 616111	3,608	94
Block Drug Company Inc.	257 Cornelison Ave., Jersey City, New Jersey 07302-9988, U.S.A.	201-434-3000	201-434-3032	3,600	95
Mattel Indonesia PT	Jl. Jababeka V Kav 4-6, Bekasi, Indonesia	218934044	21 8934810	3,359	97
LG International Corp.	Youngdungpo-Gu, Seoul 150-606, Republic of Korea	2 37735031	2 7857762	3,330	97
Batesville Casket Company Inc.	1 Batesville Blvd., Batesville, Indiana 47006-9166, U.S.A.	812-934-7500	812-934-8302	3,300	97
CML Group Inc.	524 Main St., Acton, Massachusetts 01720, U.S.A.	978-264-4155		3,100	97
AB Wilh. Becker	Lovholmsgrand 12, Stockholm S-117 83, Sweden	8 7756400	8 7442181	3,050	93
Mannington Mills Inc.	PO Box 30, Salem, New Jersey 08079, U.S.A.	609-935-3000	609-339-5813	3,000	95
Revoz d.d.	Dunajska 22, Ljubljana 1001, Slovenia	61 1326203	61 1331178	2,876	95
Oneida Silversmiths Div.	Kenwood Ave., Oneida, New York 13421, U.S.A.	315-361-3000		2,850	96
NU-kote Holding Inc.	17950 Preston, Ste. 690, Dallas, Texas 75252, U.S.A.	972-250-2785		2,800	96
Roadmaster Corp.	PO Box 344, Olney, Illinois 62450, U.S.A.	618-393-2991		2,700	93
Wilson Sporting Goods Co.	8700 W. Bryn Mawr, Chicago, Illinois 60631, U.S.A.	773-714-6400	773-714-4565	2,700	97
International Game Technology Inc.	PO Box 10120, Reno, Nevada 89502, U.S.A.	702-448-7777	702-448-0777	2,600	97
Coastcast Corp.	3025 E. Victoria St., Compton, California 90221-5616, U.S.A.	310-638-0595		2,517	95
Bic Corp.	500 Bic Dr., Milford, Connecticut 06460, U.S.A.	203-783-2000	203-783-2086	2,500	93
Samyang Tongsang Co., Ltd.	Kangnam-gu, Seoul, Republic of Korea	2 34533963	2 34530756	2,500	97
Stanley-Bostitch Inc.	Briggs Dr., East Greenwich, Rhode Island 02818, U.S.A.	401-884-2500	401-885-3122	2,500	97
W.H. Brady Co.	PO Box 571, Milwaukee, Wisconsin 53201-0571, U.S.A.	414-358-6600		2,500	97
Parker Pen Holdings Ltd.	Parker House, New Haven BN9 0AU, United Kingdom	273 513233	273 513589	2,437	93
Kunnan Enterprise, Ltd.	Tan Tzu Hsiang, Taichung, Taiwan	45 320183	45 333370	2,360	92
Donit Pletilnica d.o.o.	Cesta Majde Silc 1, Sodrazica 1317, Slovenia	61 866211	61 866360	2,302	95
Berol Corp.	105 W. Park Dr., Brentwood, Tennessee 37027, U.S.A.	615-370-9700	615-371-1903	2,300	95
Little Tikes Co.	2180 Barlow Rd., Hudson, Ohio 44236, U.S.A.	330-650-3000	330-650-3349	2,200	
Town and Country Corp.	25 Union St., Chelsea, Massachusetts 02150, U.S.A.	617-884-8500		2,200	95
Jason Inc.	411 E. Wisconsin Ave., 2500, Milwaukee, Wisconsin 53202, U.S.A.	414-277-9300		2,172	97
Callaway Golf Co.	2285 Rutherford Rd., Carlsbad, California 92008-8815, U.S.A.	619-931-1771		2,152	96
Blyth Industries Inc. (Greenwich, Connecticut)	100 Field Point Rd, Greenwich, Connecticut 06830-6442, U.S.A.	203-661-1926		2,100	96
Halma PLC	Rectory Way, Amersham HP7 0DE, United Kingdom	494 721111	494 728032	2,099	94
First Alert Inc.	3901 Liberty Street Rd., Aurora, Illinois 60504-2495, U.S.A.	630-851-7330	708-851-1331	2,095	96
Pentland Group PLC	Finchley, London N3 2QL, United Kingdom	81 3462600	81 3462700	2,070	91

Continued

Representative Companies in Sector
- Continued -

Name	Address	Tele-phone	Fax	Employ-ment	Y
FHT Holdings Ltd.	Bridge St., Walsall WS1 1JZ, United Kingdom	922 723372	922 644495	2,058	93
Coats Crafts North America	PO Box 24998, Greenville, South Carolina 29616-2498, U.S.A.	803-234-0331		2,044	94
BRK Brands Inc.	3901 Liberty Street Rd., Aurora, Illinois 60504-8122, U.S.A.	630-851-7330	630-851-9015	2,040	95
Arthur J. Prescott International Ltd.	1-3 Pedder St., Central, Hong Kong, Hong Kong	5246135	8400195	2,025	
Spalding Sports Worldwide	PO Box 901, Chicopee, Massachusetts 01021-0901, U.S.A.	413-536-1200		2,000	97
Steinway Musical Instruments Inc.	600 Industrial Pkwy., Elkhart, Indiana 46516, U.S.A.	219-522-1675		1,959	96
Steklarna Boris Kidric d.d.	Ulica talcev 1nal, Rogaska Slatina 3250, Slovenia	63 814611	63 814744	1,912	95
Semo Co., Ltd.	Kangnam-gu, Seoul 135-080, Republic of Korea	2 34588114	2 5655656	1,891	96
Exhibitgroup Inc.	200 Geary, Roselle, Illinois 60172, U.S.A.	630-307-2400	630-351-7794	1,865	97
Stuart Entertainment Inc.	3211 Nebraska Ave., Council Bluffs, Iowa 51501, U.S.A.	712-323-1488	712-795-2508	1,840	96
Alliance Gaming Corp.	6601 Bermuda Rd., Las Vegas, Nevada 89119, U.S.A.	702-435-4200	702-454-0542	1,805	97
SZ ZJ Acroni d.o.o.	Cesta zelezarjev 8, Jesenice 4270, Slovenia	64 861441	64 861379	1,777	95
Dixon Ticonderoga Co.	PO Box 958413, Heathrow, Florida 32795-8413, U.S.A.	407-875-9000		1,770	97
Bell Sports Corp.	15170 N. Hayden Rd., Ste. 1, Scottsdale, Arizona 85260-2571, U.S.A.	602-951-0033	602-951-0511	1,700	97
Simon UK Ltd.	Unit A305, Brooklands Industrial Park, Weybridge KT13 0YU, United Kingdom	932 350500	61 4912472	1,667	93
Graphic Controls Corp.	PO Box 1271, Buffalo, New York 14240, U.S.A.	716-853-7500	800-347-2420	1,650	97
Binney and Smith Inc.	1100 Church Ln., Easton, Pennsylvania 18042, U.S.A.	215-253-6271	215-250-5768	1,600	94
Ertl Company Inc.	PO Box 500, Dyersville, Iowa 52040-0500, U.S.A.	319-875-2000	319-875-8263	1,600	97
Russ Berrie and Company Inc.	111 Bauer Dr., Oakland, New Jersey 07436, U.S.A.	201-337-9000	201-337-9634	1,580	96
Cinkarna - Metalursko Kemicna Industrija d.d.	Kidriceva 26, Celje 3001, Slovenia	63 33112	63 411178	1,554	95
SLM International Inc.	30 Rockfeller Plz., 4314, New York, New York 10112-4399, U.S.A.	212-332-1610		1,550	95
Tah Hsin Industrial Corp.	Taichung Industrial Park, Taichung, Taiwan	4 2522111	4 2525307	1,550	90
Baldwin Piano and Organ Co.	422 Wards Corner Rd., Loveland, Ohio 45140-8390, U.S.A.	513-576-4500	513-576-4679	1,530	96
Arden Corp.	26899 Northwestern Hwy., Southfield, Michigan 48034-8420, U.S.A.	248-355-1101	248-355-1230	1,500	97
Klimmex SA, Industria e Comercio	Rua Augusto Klimmek 325, Centro, Sao Bento do Sul 89290, Brazil	476 330022	476 331669	1,500	96
Matthews International Corp.	2 Northshore Ctr., Pittsburgh, Pennsylvania 15212-5851, U.S.A.	412-442-8200	412-442-8290	1,500	97
Sport Maska Inc.	2 Place Alexis Nihon Ste. 800, Westmount, Quebec H3Z 3C2, Canada	514-932-1118	514-932-6043	1,500	98
SRF Ltd.	9-10 Bahadur Shah Zafar Marg, New Delhi 110 002, India	11 3318155	11 3324052	1,500	94
Wellington Leisure Products Inc.	1140 Monticello Rd., Madison, Georgia 30650, U.S.A.	706-342-1916	706-342-0407	1,500	97
YKK (U.S.A.) Inc.	1251 Valley Brook Ave., Lyndhurst, New Jersey 07071, U.S.A.	201-935-4200	201-473-2581	1,500	95
Rawlings Sporting Goods Company Inc.	1859 Intertech Dr., Fenton, Missouri 63026, U.S.A.	314-349-3500	314-349-3588	1,420	97
Iskra Industrija Kondenzatorjev in Opreme d.d.	Vrtaca 1, Semic 8333, Slovenia	68 67709	68 67110	1,417	95
YKK Indonesia Zipper Co. Ltd. PT	Pusat, Jakarta, Indonesia	21331708	21 3106751	1,400	97
Livarna Maribor d.o.o.	Oresko nabrezje 9, Maribor 2001, Slovenia	62 212961	62 27155	1,377	95
Allied Partnership Group PLC	Piccadilly House 55 Piccadilly, York Y01 1PL, United Kingdom	904 646893	904 610972	1,370	90
TAM Gospodarska Vozila d.o.o.	Ptujska 184, Maribor 2001, Slovenia	62 413659	62 414055	1,368	95
Johnson Worldwide Associates Inc.	PO Box 901, Sturtevant, Wisconsin 53177, U.S.A.	414-884-1500	414-631-4426	1,366	97
Addis Ltd.	Ware Rd., Hertford SG13 7HL, United Kingdom	992 584221	992 553050	1,343	92
Day-Timers Inc.	1 Willow Ln., East Texas, Pennsylvania 18046, U.S.A.	610-398-1151	610-398-5520	1,300	97
York Group Inc.	9430 Old Katy Rd., Houston, Texas 77055, U.S.A.	713-984-5500	713-984-5569	1,300	96
Ekco Group Inc.	98 Spit Brook Rd., Ste. 102, Nashua, New Hampshire 03062-5738, U.S.A.	603-888-1212	603-888-1427	1,250	96
Swank Inc.	6 Hazel St., Attleboro, Massachusetts 02703, U.S.A.	508-222-3400	508-226-9598	1,250	97
Metal Ravne d.o.o.	Koroska 14, Ravne na Koroskem 2390, Slovenia	60221596	60223924	1,247	95
Javor Koncern d.o.o.	Kolodvorska 9A, Pivka 6257, Slovenia	67 51010	67 51674	1,236	95
Everbrite Inc.	PO Box 20020, Greenfield, Wisconsin 53220, U.S.A.	414-529-3500	414-529-7191	1,200	97
Hunt Corp.	2005 Market St., 1 Commerce Sq., Philadelphia, Pennsylvania 19103-7085, U.S.A.	215-732-7700		1,200	97
Karsten Manufacturing Corp.	PO Box 9990, Phoenix, Arizona 85068, U.S.A.	602-870-5000	602-687-4482	1,200	97

Continued

Representative Companies in Sector
- Continued -

Name	Address	Tele-phone	Fax	Employ-ment	Y
Rose Art Industries Inc.	6 Regent St., Livingston, New Jersey 07039, U.S.A.	973-535-1313		1,200	97
Titan Kamnik d.d.	Kovinarska 28, Kamnik 1240, Slovenia	61 816221	61 811284	1,200	95
Kovinoplastika Loz d.d.	C. 19.oktobra 57, Stari trg pri Lozu 1386, Slovenia	61 707422	61 707594	1,163	95
Plasti-Line Inc.	PO Box 59043, Knoxville, Tennessee 37950-9043, U.S.A.	423-938-1511		1,160	95
Talum d.o.o.	Tovarniska 10, Kidricevo SLO-62325, Slovenia	62 796110	62 796269	1,154	95
Polzela d.d.	Polzela 171, Polzela 3313, Slovenia	63 720111	63 720260	1,147	95
Meridian Sports Inc.	625 Madison Ave., New York, New York 10022, U.S.A.	212-527-4413		1,134	95
Biro Bic Ltd.	Park Royal, London NW10 7SG, United Kingdom	81 9654060	81 4531128	1,114	92
A.T. Cross Co.	One Albion Rd., Lincoln, Rhode Island 02865, U.S.A.	401-333-1200	401-334-2861	1,100	97
LeFebure Corp.	PO Box 2028, Cedar Rapids, Iowa 52406, U.S.A.	319-369-5000	319-366-7608	1,100	97
Scripto Tokai Corp.	PO Box 5555, Fontana, California 92334-5555, U.S.A.	909-360-2100	909-360-2130	1,100	
Zippo Manufacturing Co.	33 Barbour St., Bradford, Pennsylvania 16701, U.S.A.	814-368-2700	814-362-1350	1,100	97
Excalibur Group PLC	Handsworth, Birmingham B21 8LH, United Kingdom	21 5532228	21 5537892	1,089	94
CJC Holdings Inc.	PO Box 149056, Austin, Texas 78714-9056, U.S.A.	512-444-0571	512-444-7618	1,061	95
Stol d.d.	Ljubljanska 45, Kamnik 1240, Slovenia	61 812211	61 811979	1,058	95
Hamlin Inc.	612 E. Lake St., Lake Mills, Wisconsin 53551, U.S.A.	920-648-3000	920-648-3001	1,050	97
WMS Industries Inc.	3401 N. California Ave., Chicago, Illinois 60618, U.S.A.	773-961-1111	773-961-1099	1,047	97
Boosey & Hawkes PLC	Deansbrook Rd., Edgware HA8 9BB, United Kingdom	81 9527711	81 9511314	1,043	93
Gillette UK Ltd.	Great West Rd., Isleworth TW7 5NP, United Kingdom	81 5601234	81 8476165	1,043	93
IMPOL d.o.o.	Partizanska 38, Slovenska Bistrica 2310, Slovenia	62 817521	62 817219	1,041	95
Domel DOO	Otoki 21, Zelezniki 4228, Slovenia	64 66441	64 67150	1,027	95
Gibson Guitar Corp.	1818 Elm Hill Pike, Nashville, Tennessee 37210, U.S.A.	615-871-4500	615-889-5509	1,000	97
Selmer Company Inc.	PO Box 310, Elkhart, Indiana 46515, U.S.A.	219-522-1675	219-522-0334	1,000	97
Spoldzielnia RzemieslInicza Wielobranzowa	Rynek ul. 2, Lublin PL-20-111, Poland	81 28568	81 28568	1,000	93
Sport Maska Inc.	6375 Rue Picard, St.-Hyacinthe, Quebec J2S 1H3, Canada	514-773-5041	514-773-5312	1,000	98
Strathmore Paper Co.	39 S. Broad St., Westfield, Massachusetts 01085, U.S.A.	413-568-9111	413-568-4887	1,000	95
Zebco-MotorGuide Div.	PO Box 270, Tulsa, Oklahoma 74101-0270, U.S.A.	918-836-5581		1,000	97
Brushindo Cemerlang PT	Wisma Bisnis Indonesia 12th Fl., Jakarta, Indonesia	215307130	21 5307142	989	97
Harrow Industries Inc.	2627 E. Beltline Ave. S.E., Grand Rapids, Michigan 49546, U.S.A.	616-942-1440	616-942-2170	984	97
Kolektor d.o.o.	Vojkova 10, Idrija 5280, Slovenia	65 71411	65 72358	977	95
Avon Canada Inc.	5500 Rte Trans-Canada, Pointe-Claire, Quebec H9R 1B6, Canada	514-695-3371	514-630-5400	970	98
Dunlop Slazenger International Ltd.	Carr Gate, Wakefield WF1 1AA, United Kingdom	924 828222		965	93
Bell Sports Inc.	15170 N. Hayden Rd., 1, Scottsdale, Arizona 85260-2512, U.S.A.	602-951-0033	602-951-0511	960	97
Metalna d.d.	Zagrebska 20, Maribor 2001, Slovenia	62 412511	62 412160	960	95
Colombiana de Comercio Ltda.	Carrera 30 10-25, Bogota, Colombia	1 2776300	1 2017981	955	97
Axiohm Transaction Solutions Inc.	15070 Avenue of Science, San Diego, California 92128, U.S.A.	619-451-3485	619-451-3573	925	96
Steklarna Hrastnik d.d.	Cesta 1.maja 14, Hrastnik 1430, Slovenia	60143622	60144418	907	95
Ace Novelty Company Inc.	13434 N.E. 16th St., Bellevue, Washington 98005, U.S.A.	425-486-4660	425-486-4955	900	97
Bettanin Industrial SA	Rodovia Br 116 Km. 12, Parada 30, Esteio 93270-000, Brazil	51 4734222	51 4732508	900	97
Candle Corporation of America	999 E. Touhy Ave., 450, Des Plaines, Illinois 60018, U.S.A.	847-294-1100		900	97
Fisher-Price Inc.	636 Girard Ave. E, East Aurora, New York 14052-1824, U.S.A.	716-687-3000		900	97
Radece Papir d.d.	Njivice 7, Radece 1433, Slovenia	60181302	60185112	897	95
Jelovica d.d.	Kidriceva 58, Skofja Loka SLO-4220, Slovenia	64 6130	64 634261	889	95
Nestle Norge AS	PO Box 1, Sandvika N-1301, Norway	67 986400	67 986401	870	93
Rosenvasser David Ltd.	3 Hasanda St., Petah Tikva 49130, Israel	3 9243444	3 9242165	870	97
OroAmerica Inc.	443 N Varney St., Burbank, California 91502, U.S.A.	818-848-5555	818-841-4342	851	96
WorldCrisa Corp.	PO Box 5020, Wallingford, Connecticut 06492, U.S.A.	203-265-8000	203-265-8024	850	94
Video Lottery Technologies Inc.	115 Perimeter Center Pl., Atlanta, Georgia 30346, U.S.A.	770-481-1800		830	96
Mikohn Gaming Corp.	PO Box 98686, Las Vegas, Nevada 89119, U.S.A.	702-896-3890	702-263-1770	824	96
Daktronics Inc.	PO Box 5128, Brookings, South Dakota 57006-5128, U.S.A.	605-697-4000	605-697-4700	816	97
Brio AB/Brio Toy	Box 63, Osby S-283 00, Sweden	479 19000	479 14724	813	
GM Nameplate Inc.	2040 15th Ave. W, Seattle, Washington 98119, U.S.A.	206-284-2200	206-284-3705	800	97
Group Tomos - Promo d.o.o.	Smarska 4, Koper-Capodistria 6001, Slovenia	66 31111	66 33207	800	95
Mobile Mini Inc.	1834 West 3rd St., Tempe, Arizona 85281-2467, U.S.A.	602-894-6311		800	97
Pinkerton Group Inc.	6600 W. Broad St., Richmond, Virginia 23230, U.S.A.	804-287-3220	804-287-3208	800	97

Product Tables
Pianos

Unit of Measure: Number.

	1990		1991		1992		1993		1994		1995		1996	
	Value	%	Value	%	Value	%	Value	%	Value	%	Value	%	Value	%
Total Production	981,830	100.0	971,348	100.0	935,054	100.0	858,555	100.0	799,048	100.0	770,040	100.0	822,937	100.0
Regions														
America, North	181,100	18.4	181,100	18.6	181,100	19.4	181,100	21.1	181,100	22.7	181,100	23.5	181,100	22.0
Asia	704,596	71.8	696,057	71.7	668,471	71.5	600,475	69.9	547,213	68.5	515,796	67.0	572,901	69.6
Europe	96,135	9.8	94,191	9.7	85,483	9.1	76,980	9.0	70,735	8.9	73,144	9.5	68,936	8.4
Leading Producers														
Japan	264,878	27.0	231,197	23.8	213,312	22.8	177,180	20.6	173,411	21.7	167,831	21.8	*160,384	19.5
Korea, Republic of	232,650	23.7	259,507	26.7	254,109	27.2	228,193	26.6	211,891	26.5	212,463	27.6	*279,502	34.0
United States of America	*181,100	18.4	*181,100	18.6	*181,100	19.4	*181,100	21.1	*181,100	22.7	*181,100	23.5	*181,100	22.0
China	36,175	3.7	48,958	5.0	64,116	6.9	*66,229	7.7	*70,872	8.9	*75,515	9.8	*80,158	9.7
Russian Federation	56,953	5.8	49,707	5.1	40,908	4.4	27,351	3.2	11,002	1.4	6,833	0.9	2,080	0.3

Musical Instruments, String

Unit of Measure: Number.

	1990		1991		1992		1993		1994		1995		1996	
	Value	%	Value	%	Value	%	Value	%	Value	%	Value	%	Value	%
Total Production	5,039,629	100.0	4,384,753	100.0	4,138,887	100.0	3,730,384	100.0	3,142,359	100.0	2,931,542	100.0	3,086,430	100.0
Regions														
America, North	81,571	1.6	81,000	1.8	81,000	2.0	81,000	2.2	81,000	2.6	81,000	2.8	81,000	2.6
America, South	8,778	0.2	8,889	0.2	9,000	0.2	6,000	0.2	7,000	0.2	7,000	0.2	7,229	0.2
Asia	4,397,871	87.3	3,732,056	85.1	3,519,806	85.0	3,140,556	84.2	2,620,306	83.4	2,402,554	82.0	2,531,500	82.0
Europe	551,409	10.9	562,809	12.8	529,081	12.8	502,828	13.5	434,053	13.8	440,988	15.0	466,702	15.1
Leading Producers														
Korea, Republic of	1,204,000	23.9	1,017,000	23.2	1,066,000	25.8	1,062,000	28.5	942,000	30.0	973,000	33.2	*1,170,905	37.9
Japan	672,000	13.3	597,000	13.6	612,000	14.8	440,000	11.8	354,000	11.3	323,000	11.0	*344,545	11.2
Russian Federation	992,000	19.7	865,000	19.7	652,000	15.8	517,000	13.9	268,000	8.5	170,000	5.8	168,000	5.4
Spain	154,000	3.1	199,000	4.5	201,000	4.9	167,000	4.5	191,000	6.1	229,000	7.8	262,000	8.5
Ukraine	194,000	3.8	162,000	3.7	129,000	3.1	123,000	3.3	40,000	1.3	31,000	1.1	26,000	0.8

Organs

Unit of Measure: Number.

	1990		1991		1992		1993		1994		1995		1996	
	Value	%	Value	%	Value	%	Value	%	Value	%	Value	%	Value	%
Total Production	29,546	100.0	33,090	100.0	38,774	100.0	75,164	100.0	66,401	100.0	64,836	100.0	60,285	100.0
Regions														
America, North	12,812	43.4	12,812	38.7	12,812	33.0	12,812	17.0	12,812	19.3	12,812	19.8	12,812	21.3
Asia	16,445	55.7	19,722	59.6	25,462	65.7	61,923	82.4	53,166	80.1	51,571	79.5	47,050	78.0
Europe	290	1.0	555	1.7	499	1.3	429	0.6	423	0.6	453	0.7	422	0.7
Leading Producers														
United States of America	*12,812	43.4	*12,812	38.7	*12,812	33.0	*12,812	17.0	*12,812	19.3	*12,812	19.8	*12,812	21.3
Korea, Republic of	11,106	37.6	11,760	35.5	15,039	38.8	16,825	22.4	24,602	37.1	28,608	44.1	*22,677	37.6
Indonesia	4,131	14.0	6,755	20.4	9,216	23.8	43,667	58.1	27,265	41.1	22,071	34.0	*23,166	38.4
Slovenia	*627	2.1	*627	1.9	*627	1.6	851	1.1	719	1.1	312	0.5	*627	1.0

Commodity data are provided by the United Nations Statistical Division. The symbol * means that data are estimated. For additional notes, see Appendix I.

Product Tables

Musical Instruments, Wind

Unit of Measure: Number.

	1990 Value	%	1991 Value	%	1992 Value	%	1993 Value	%	1994 Value	%	1995 Value	%	1996 Value	%
Total Production	8,689,743	100.0	10,103,160	100.0	10,227,877	100.0	10,424,355	100.0	9,139,382	100.0	10,373,079	100.0	10,571,531	100.0
Regions														
America, North	5,138,000	59.1	5,138,000	50.9	5,138,000	50.2	5,138,000	49.3	5,138,000	56.2	5,138,000	49.5	5,138,000	48.6
America, South	14,500	0.2	14,500	0.1	2,000	0.0	2,000	0.0	2,000	0.0	52,000	0.5	14,500	0.1
Asia	1,321,550	15.2	1,550,967	15.4	1,765,683	17.3	2,028,400	19.5	618,950	6.8	1,567,038	15.1	1,691,332	16.0
Europe	2,215,693	25.5	3,399,693	33.6	3,322,194	32.5	3,255,955	31.2	3,380,432	37.0	3,616,041	34.9	3,727,699	35.3
Leading Producers														
United States of America	*5,138,000	59.1	*5,138,000	50.9	*5,138,000	50.2	*5,138,000	49.3	*5,138,000	56.2	*5,138,000	49.5	*5,138,000	48.6
Germany	-		1,188,000	11.8	1,119,500	10.9	1,051,000	10.1	1,164,000	12.7	1,409,347	13.6	1,531,006	14.5
Indonesia	752,500	8.7	1,001,667	9.9	1,250,833	12.2	1,500,000	14.4	90,000	1.0	*1,064,538	10.3	*1,155,158	10.9
Japan	267,000	3.1	240,000	2.4	247,000	2.4	249,000	2.4	259,000	2.8	236,000	2.3	*258,124	2.4
Kazakhstan	62,000	0.7	62,000	0.6	23,000	0.2	24,000	0.2	14,000	0.2	1,000	0.0	-	

Dolls

Unit of Measure: Thousands of units.

	1990 Value	%	1991 Value	%	1992 Value	%	1993 Value	%	1994 Value	%	1995 Value	%	1996 Value	%
Total Production	295,685	100.0	307,754	100.0	515,042	100.0	335,829	100.0	353,292	100.0	345,125	100.0	380,734	100.0
Regions														
Africa	248	0.1	248	0.1	248	0.0	248	0.1	248	0.1	248	0.1	248	0.1
America, North	79,425	26.9	79,429	25.8	79,433	15.4	79,784	23.8	79,805	22.6	79,442	23.0	78,745	20.7
America, South	1,092	0.4	1,178	0.4	1,264	0.2	670	0.2	505	0.1	733	0.2	869	0.2
Asia	198,535	67.1	207,596	67.5	413,293	80.2	239,594	71.3	260,051	73.6	252,023	73.0	290,674	76.3
Europe	16,385	5.5	19,303	6.3	20,803	4.0	15,534	4.6	12,683	3.6	12,679	3.7	10,198	2.7
Leading Producers														
Korea, Republic of	*157,942	53.4	*166,131	54.0	*174,320	33.8	*182,509	54.3	*190,699	54.0	*198,888	57.6	*207,077	54.4
United States of America	*77,100	26.1	*77,100	25.1	*77,100	15.0	*77,100	23.0	*77,100	21.8	*77,100	22.3	*77,100	20.3
Indonesia	16,560	5.6	15,094	4.9	222,701	43.2	40,951	12.2	61,782	17.5	49,194	14.3	*81,242	21.3
Russian Federation	21,491	7.3	24,125	7.8	14,154	2.7	14,401	4.3	6,721	1.9	3,574	1.0	2,082	0.5
Spain	5,619	1.9	6,146	2.0	5,424	1.1	6,953	2.1	8,437	2.4	8,474	2.5	6,702	1.8

Fountain Pens, Ball-Point Pens, Propelling Pencils Etc.

Unit of Measure: Thousands of units.

	1990 Value	%	1991 Value	%	1992 Value	%	1993 Value	%	1994 Value	%	1995 Value	%	1996 Value	%
Total Production	8,499,291	100.0	9,670,357	100.0	9,486,118	100.0	9,438,125	100.0	9,425,505	100.0	9,653,589	100.0	10,279,750	100.0
Regions														
Africa	4,214	0.0	3,606	0.0	3,050	0.0	4,846	0.1	653	0.0	7,004	0.1	2,374	0.0
America, North	2,136,922	25.1	2,197,120	22.7	2,259,581	23.8	2,247,131	23.8	2,258,268	24.0	2,332,150	24.2	2,674,968	26.0
America, South	89,005	1.0	91,029	0.9	93,054	1.0	72,700	0.8	113,253	1.2	114,587	1.2	97,624	0.9
Asia	4,478,438	52.7	4,650,375	48.1	4,459,846	47.0	4,633,786	49.1	4,502,588	47.8	4,699,959	48.7	4,927,067	47.9
Europe	1,732,070	20.4	2,669,584	27.6	2,611,947	27.5	2,421,021	25.7	2,492,102	26.4	2,441,247	25.3	2,519,075	24.5
Oceania	58,642	0.7	58,642	0.6	58,642	0.6	58,642	0.6	58,642	0.6	58,642	0.6	58,642	0.6
Leading Producers														
Japan	2,503,265	29.5	2,459,887	25.4	2,487,977	26.2	2,438,775	25.8	2,489,805	26.4	2,577,495	26.7	*2,749,214	26.7
United States of America	*1,832,000	21.6	*1,832,000	18.9	*1,832,000	19.3	*1,832,000	19.4	*1,832,000	19.4	*1,832,000	19.0	*1,832,000	17.8
Germany	-		986,934	10.2	922,128	9.7	898,738	9.5	958,283	10.2	1,008,240	10.4	1,056,219	10.3
China	716,680	8.4	*717,059	7.4	*750,664	7.9	*784,270	8.3	*817,875	8.7	*851,481	8.8	*885,086	8.6
Mexico	304,922	3.6	365,120	3.8	427,581	4.5	415,131	4.4	426,268	4.5	500,150	5.2	842,968	8.2

Commodity data are provided by the United Nations Statistical Division. The symbol * means that data are estimated. For additional notes, see Appendix I.

723

Product Tables
Pencils, Crayons Etc.

Unit of Measure: Thousands of units.

	1990 Value	%	1991 Value	%	1992 Value	%	1993 Value	%	1994 Value	%	1995 Value	%	1996 Value	%
Total Production	7,245,018	100.0	6,656,510	100.0	6,592,859	100.0	5,704,173	100.0	6,417,879	100.0	6,137,672	100.0	6,356,046	100.0
Regions														
America, North	2,275,590	31.4	2,325,630	34.9	2,375,670	36.0	2,425,460	42.5	2,528,500	39.4	2,473,540	40.3	2,575,580	40.5
America, South	49,868	0.7	42,267	0.6	34,667	0.5	35,667	0.6	36,667	0.6	34,667	0.6	23,667	0.4
Asia	4,150,785	57.3	3,522,785	52.9	3,451,785	52.4	2,613,543	45.8	3,178,826	49.5	2,966,109	48.3	3,085,373	48.5
Europe	768,776	10.6	765,828	11.5	730,738	11.1	629,504	11.0	673,886	10.5	663,356	10.8	671,426	10.6
Leading Producers														
United States of America	*1,679,840	23.2	*1,729,180	26.0	*1,778,520	27.0	*1,827,860	32.0	*1,877,200	29.2	*1,926,540	31.4	*1,975,880	31.1
Japan	952,000	13.1	806,000	12.1	808,000	12.3	765,000	13.4	661,000	10.3	709,000	11.6	*686,829	10.8
Indonesia	1,501,000	20.7	1,413,000	21.2	1,291,000	19.6	689,000	12.1	1,357,000	21.1	1,147,000	18.7	*1,443,552	22.7
Mexico	*578,250	8.0	*578,250	8.7	*578,250	8.8	578,000	10.1	631,000	9.8	526,000	8.6	578,000	9.1
China	446,000	6.2	472,000	7.1	534,000	8.1	*490,758	8.6	*506,041	7.9	*521,324	8.5	*536,607	8.4

Slide Fasteners (Zippers)

Unit of Measure: 1,000 Meters.

	1990 Value	%	1991 Value	%	1992 Value	%	1993 Value	%	1994 Value	%	1995 Value	%	1996 Value	%
Total Production	687,483	100.0	725,572	100.0	778,963	100.0	592,418	100.0	646,601	100.0	614,425	100.0	669,971	100.0
Regions														
Africa	3,057	0.4	3,057	0.4	3,057	0.4	3,057	0.5	3,057	0.5	3,057	0.5	3,057	0.5
America, South	18,480	2.7	15,703	2.2	16,159	2.1	5,966	1.0	20,671	3.2	21,784	3.5	15,419	2.3
Asia	410,886	59.8	362,770	50.0	426,720	54.8	274,887	46.4	332,245	51.4	311,517	50.7	371,244	55.4
Europe	248,169	36.1	338,504	46.7	328,841	42.2	305,676	51.6	289,148	44.7	277,940	45.2	280,250	41.8
Oceania	6,891	1.0	5,538	0.8	4,185	0.5	2,833	0.5	1,480	0.2	127	0.0	-	
Leading Producers														
Korea, Republic of	189,067	27.5	204,190	28.1	199,908	25.7	182,777	30.9	188,014	29.1	184,943	30.1	*247,860	37.0
Germany	-		98,280	13.5	80,271	10.3	70,281	11.9	68,499	10.6	69,150	11.3	78,585	11.7
Indonesia	123,072	17.9	98,821	13.6	137,702	17.7	8,390	1.4	*94,930	14.7	*98,326	16.0	*101,723	15.2
Spain	*64,783	9.4	*64,783	8.9	*64,783	8.3	*64,783	10.9	*64,783	10.0	*64,783	10.5	64,783	9.7
Russian Federation	65,745	9.6	27,148	3.7	55,998	7.2	50,667	8.6	22,640	3.5	8,767	1.4	4,322	0.6

Commodity data are provided by the United Nations Statistical Division. The symbol * means that data are estimated. For additional notes, see Appendix I.

Part II

COUNTRY PROFILES

World Averages

Population 5,926,467,000. Percent of world population: 100.000

	1990		1991		1992		1993		1994		1995	
	Value	%	Value	%	Value	%	Value	%	Value	%	Value	%
Establishments or enterprises (number)	2,304,282	100.000	2,383,469	100.000	2,808,669	100.000	2,055,572	100.000	1,789,529	100.000	1,226,867	100.000
Total employment (000)	209,215	100.000	186,729	100.000	182,482	100.000	182,111	100.000	159,152	100.000	142,223	100.000
Female employees (000)	13,319	100.000	8,994	100.000	21,509	100.000	22,306	100.000	21,100	100.000	19,107	100.000
Salaries and Wages ($ bil.)	2,154	100.000	2,377	100.000	2,554	100.000	2,040	100.000	1,951	100.000	1,658	100.000
Output ($ bil.)	14,312	100.000	15,369	100.000	16,439	100.000	14,506	100.000	13,789	100.000	13,337	100.000
Value added ($ bil.)	5,598	100.000	5,669	100.000	5,983	100.000	5,969	100.000	5,120	100.000	5,493	100.000
Capital investment ($ mil.)	592,894	100.000	623,756	100.000	639,121	100.000	543,567	100.000	443,960	100.000	312,283	100.000
Employees per establishment (number)	74	100.0	68	100.0	62	100.0	71	100.0	67	100.0	70	100.0
Output per establishment ($ mil.)	4	100.0	5	100.0	5	100.0	5	100.0	5	100.0	4	100.0
Value added per establishment ($ mil.)	2	100.0	2	100.0	2	100.0	2	100.0	2	100.0	2	100.0
Capital investment per estab. ($ mil.)	0.265	100.0	0.294	100.0	0.287	100.0	0.253	100.0	0.235	100.0	0.286	100.0
Payroll per employee ($)	13,681	100.0	18,099	100.0	20,077	100.0	16,238	100.0	17,560	100.0	17,476	100.0
Females as % of total employment (%)	32.0	100.0	32.1	100.0	46.3	100.0	49.9	100.0	52.4	100.0	58.4	100.0
Output per employee ($)	68,630	100.0	81,398	100.0	89,466	100.0	80,062	100.0	87,373	100.0	92,545	100.0
Value added per employee ($)	32,852	100.0	33,772	100.0	36,007	100.0	32,043	100.0	35,496	100.0	39,496	100.0
Capital investment per employee ($)	5,506	100.0	6,131	100.0	6,195	100.0	5,716	100.0	6,540	100.0	7,468	100.0

Columns headed % show percent of the world total or of the world ratio. Ratios closest to 100 are closest to the world average. Values higher than 100 are above, values lower than 100 are below world average. *nec* stands for *not elsewhere classified*. Ratios, where shown, are frequently based on a subset of total data; for example, payroll per employee is averaged from all cases where *both* payroll and employment data are present. Dividing *total* salaries/wages by *total* employment will not necessarily produce a matching value.

Albania

Top Ranked Employing Sectors	Top Ranked Output Sectors
3110 Food Products	3530 Petroleum Refineries
3690 Nonmetal Products nec	3720 Nonferrous metals
3720 Nonferrous metals	3110 Food Products
3710 Iron and Steel	3690 Nonmetal Products nec
3530 Petroleum Refineries	3710 Iron and Steel

Population 3,331,000. Percent of world population: 0.056

	1990		1991		1992		1993		1994		1995	
	Value	%	Value	%	Value	%	Value	%	Value	%	Value	%
Establishments or enterprises (number)	-	-	-	-	-	-	460	0.022	366	0.020	233	0.019
Total employment (000)	39	0.019	-	-	-	-	66	0.036	40	0.025	22	0.015
Female employees (000)	-	-	-	-	-	-	32	0.146	18	0.084	7	0.035
Salaries and Wages ($ bil.)	0.016	0.001	-	-	-	-	0.026	0.001	0.030	0.002	0.025	0.002
Output ($ bil.)	-	-	-	-	-	-	0.231	0.002	0.174	0.001	0.173	0.001
Value added ($ bil.)	0.024		-	-	-	-	-	-	-	-	-	-
Capital investment ($ mil.)	-	-	-	-	-	-	-	-	-	-	-	-
Employees per establishment (number)	-	-	-	-	-	-	144	204.5	110	163.4	96	136.8
Output per establishment ($ mil.)	-	-	-	-	-	-	0.502	10.7	0.483	9.6	1	33.4
Value added per establishment ($ mil.)	-	-	-	-	-	-	-	-	-	-	-	-
Capital investment per estab. ($ mil.)	-	-	-	-	-	-	-	-	-	-	-	-
Payroll per employee ($)	417	3.1	-	-	-	-	397	2.4	754	4.3	1,153	6.6
Females as % of total employment (%)	-	-	-	-	-	-	49.0	98.1	44.8	85.5	30.8	52.8
Output per employee ($)	-	-	-	-	-	-	3,478	4.3	4,391	5.0	10,501	11.3
Value added per employee ($)	602	1.8	-	-	-	-	-	-	-	-	-	-
Capital investment per employee ($)	-	-	-	-	-	-	-	-	-	-	-	-

Columns headed % show percent of the world total or of the world ratio. Ratios closest to 100 are closest to the world average. Values higher than 100 are above, values lower than 100 are below world average. *nec* stands for *not elsewhere classified*. Ratios, where shown, are frequently based on a subset of total data; for example, payroll per employee is averaged from all cases where *both* payroll and employment data are present. Dividing *total* salaries/wages by *total* employment will not necessarily produce a matching value.

Algeria

	Top Ranked Employing Sectors	Top Ranked Output Sectors
	3900 Industries nec	3900 Industries nec

Population 30,481,000. Percent of world population: 0.514

	1990 Value	%	1991 Value	%	1992 Value	%	1993 Value	%	1994 Value	%	1995 Value	%
Establishments or enterprises (number)	1,248	0.054	1,427	0.060	1,313	0.047	-	-	1,652	0.092	-	-
Total employment (000)	7	0.003	7	0.004	8	0.004	7	0.004	7	0.004	7	0.005
Female employees (000)	0.684	0.005	0.916	0.010	0.708	0.003	-	-	-	-	-	-
Salaries and Wages ($ bil.)	0.040	0.002	0.020	0.001	0.018	0.001	0.019	0.001	0.015	0.001	0.015	0.001
Output ($ bil.)	0.176	0.001	0.120	0.001	0.092	0.001	0.087	0.001	0.058	-	0.075	0.001
Value added ($ bil.)	0.079	0.001	0.052	0.001	0.041	0.001	0.039	0.001	0.026	0.001	0.034	0.001
Capital investment ($ mil.)	-		-		-		-		-		-	
Employees per establishment (number)	6	7.5	5	7.3	6	9.3	-	-	4	5.9	-	-
Output per establishment ($ mil.)	0.141	3.3	0.084	1.8	0.070	1.3	-	-	0.035	0.7	-	-
Value added per establishment ($ mil.)	0.063	3.8	0.036	2.2	0.032	1.5	-	-	0.016	0.9	-	-
Capital investment per estab. ($ mil.)	-		-		-		-		-		-	
Payroll per employee ($)	5,824	42.6	2,826	15.6	2,346	11.7	2,763	17.0	2,316	13.2	2,144	12.3
Females as % of total employment (%)	9.9	30.8	13.0	40.4	9.3	20.1	-	-	-	-	-	-
Output per employee ($)	25,444	37.1	16,924	20.8	12,164	13.6	12,383	15.5	8,841	10.1	10,935	11.8
Value added per employee ($)	11,358	34.6	7,367	21.8	5,458	15.2	5,595	17.5	3,989	11.2	4,941	12.5
Capital investment per employee ($)	-		-		-		-		-		-	

Columns headed % show percent of the world total or of the world ratio. Ratios closest to 100 are closest to the world average. Values higher than 100 are above, values lower than 100 are below world average. *nec* stands for *not elsewhere classified*. Ratios, where shown, are frequently based on a subset of total data; for example, payroll per employee is averaged from all cases where *both* payroll and employment data are present. Dividing *total* salaries/wages by *total* employment will not necessarily produce a matching value.

Angola

	Top Ranked Employing Sectors	Top Ranked Output Sectors
	3840 Transportation Equipment	3140 Tobacco
	3830 Electrical Machinery	3530 Petroleum Refineries
	3210 Textiles	3210 Textiles
	3140 Tobacco	3830 Electrical Machinery
	3810 Metal Products	3810 Metal Products

Population 10,865,000. Percent of world population: 0.183

	1990 Value	%	1991 Value	%	1992 Value	%	1993 Value	%	1994 Value	%	1995 Value	%
Establishments or enterprises (number)	-	-	-	-	83	0.003	83	0.004	-	-	-	-
Total employment (000)	-	-	11	0.006	7	0.004	3	0.002	-	-	-	-
Female employees (000)	-	-	-	-	1	0.007	0.627	0.003	-	-	-	-
Salaries and Wages ($ bil.)	-	-	-	-	3	0.130	0.781	0.038	-	-	-	-
Output ($ bil.)	-	-	-	-	409	2.500	98	0.675	-	-	-	-
Value added ($ bil.)	-	-	-	-	-		-		-	-	-	-
Capital investment ($ mil.)	-	-	-	-	-		-		-	-	-	-
Employees per establishment (number)	-	-	-	-	89	142.5	36	51.4	-	-	-	-
Output per establishment ($ mil.)	-	-	-	-	4,924	90,431.2	1,180	25,122.4	-	-	-	-
Value added per establishment ($ mil.)	-	-	-	-	-		-		-	-	-	-
Capital investment per estab. ($ mil.)	-	-	-	-	-		-		-	-	-	-
Payroll per employee ($)	-		-		452,389	2,253.3	259,893	1,600.5	-	-	-	-
Females as % of total employment (%)	-	-	-	-	20.0	43.2	20.9	41.8	-	-	-	-
Output per employee ($)	-		-		55,593,261	62,138.7	32,572,728	40,684.6	-	-	-	-
Value added per employee ($)	-	-	-	-	-		-		-	-	-	-
Capital investment per employee ($)	-	-	-	-	-		-		-	-	-	-

Columns headed % show percent of the world total or of the world ratio. Ratios closest to 100 are closest to the world average. Values higher than 100 are above, values lower than 100 are below world average. *nec* stands for *not elsewhere classified*. Ratios, where shown, are frequently based on a subset of total data; for example, payroll per employee is averaged from all cases where *both* payroll and employment data are present. Dividing *total* salaries/wages by *total* employment will not necessarily produce a matching value.

For additional notes and sources, see Appendix I.

Argentina

Top Ranked Employing Sectors	Top Ranked Output Sectors
3110 Food Products	3110 Food Products
3840 Transportation Equipment	3840 Transportation Equipment
3820 Machinery nec	3530 Petroleum Refineries
3810 Metal Products	3520 Chemical Products nec
3210 Textiles	3130 Beverages

Population 36,265,000. Percent of world population: 0.612

	1990 Value	%	1991 Value	%	1992 Value	%	1993 Value	%	1994 Value	%	1995 Value	%
Establishments or enterprises (number)	-	-	-	-	-	-	90,386	4.400	101,605	5.700	-	-
Total employment (000)	942	0.450	-	-	-	-	859	0.472	887	0.558	-	-
Female employees (000)	-	-	-	-	-	-	-	-	151	0.714	-	-
Salaries and Wages ($ bil.)	6	0.296	-	-	-	-	12	0.609	-	-	-	-
Output ($ bil.)	79	0.552	-	-	-	-	91	0.625	-	-	-	-
Value added ($ bil.)	31	0.557	-	-	-	-	30	0.496	-	-	-	-
Capital investment ($ mil.)	-	-	-	-	-	-	3,734	0.687	-	-	-	-
Employees per establishment (number)	-	-	-	-	-	-	10	13.5	9	13.0	-	-
Output per establishment ($ mil.)	-	-	-	-	-	-	1	21.3	-	-	-	-
Value added per establishment ($ mil.)	-	-	-	-	-	-	0.328	20.4	-	-	-	-
Capital investment per estab. ($ mil.)	-	-	-	-	-	-	0.041	16.3	-	-	-	-
Payroll per employee ($)	6,767	49.5	-	-	-	-	14,453	89.0	-	-	-	-
Females as % of total employment (%)	-	-	-	-	-	-	-	-	17.0	32.4	-	-
Output per employee ($)	83,878	122.2	-	-	-	-	105,488	131.8	-	-	-	-
Value added per employee ($)	33,080	100.7	-	-	-	-	34,484	107.6	-	-	-	-
Capital investment per employee ($)	-	-	-	-	-	-	4,347	76.1	-	-	-	-

Columns headed % show percent of the world total or of the world ratio. Ratios closest to 100 are closest to the world average. Values higher than 100 are above, values lower than 100 are below world average. *nec* stands for *not elsewhere classified*. Ratios, where shown, are frequently based on a subset of total data; for example, payroll per employee is averaged from all cases where *both* payroll and employment data are present. Dividing *total* salaries/wages by *total* employment will not necessarily produce a matching value.

Armenia

Top Ranked Employing Sectors	Top Ranked Output Sectors
3820 Machinery nec	3900 Industries nec
3830 Electrical Machinery	3110 Food Products
3850 Professional Goods	3830 Electrical Machinery
3210 Textiles	3850 Professional Goods
3110 Food Products	3820 Machinery nec

Population 3,422,000. Percent of world population: 0.058

	1990 Value	%	1991 Value	%	1992 Value	%	1993 Value	%	1994 Value	%	1995 Value	%
Establishments or enterprises (number)	675	0.029	704	0.030	666	0.024	700	0.034	714	0.040	-	-
Total employment (000)	354	0.169	324	0.173	86	0.047	76	0.042	52	0.032	66	0.047
Female employees (000)	-	-	-	-	-	-	-	-	-	-	-	-
Salaries and Wages ($ bil.)	0.026	0.001	0.034	0.001	0.027	0.001	1	0.054	-	-	0.007	-
Output ($ bil.)	0.164	0.001	0.303	0.002	1	0.007	49	0.335	0.260	0.002	-	-
Value added ($ bil.)	0.010	-	0.029	0.001	-	-	-	-	-	-	-	-
Capital investment ($ mil.)	-	-	-	-	-	-	-	-	-	-	-	-
Employees per establishment (number)	525	712.3	460	678.0	416	668.6	333	472.8	297	441.4	-	-
Output per establishment ($ mil.)	0.243	5.7	0.431	9.4	2	33.8	69	1,478.2	0.365	7.3	-	-
Value added per establishment ($ mil.)	0.026	1.6	0.072	4.2	-	-	-	-	-	-	-	-
Capital investment per estab. ($ mil.)	-	-	-	-	-	-	-	-	-	-	-	-
Payroll per employee ($)	75	0.5	104	0.6	318	1.6	14,613	90.0	-	-	105	0.6
Females as % of total employment (%)	-	-	-	-	-	-	-	-	-	-	-	-
Output per employee ($)	464	0.7	937	1.2	3,592	4.0	145,616	181.9	843	1.0	-	-
Value added per employee ($)	55	0.2	174	0.5	-	-	-	-	-	-	-	-
Capital investment per employee ($)	-	-	-	-	-	-	-	-	-	-	-	-

Columns headed % show percent of the world total or of the world ratio. Ratios closest to 100 are closest to the world average. Values higher than 100 are above, values lower than 100 are below world average. *nec* stands for *not elsewhere classified*. Ratios, where shown, are frequently based on a subset of total data; for example, payroll per employee is averaged from all cases where *both* payroll and employment data are present. Dividing *total* salaries/wages by *total* employment will not necessarily produce a matching value.

Australia

Top Ranked Employing Sectors	Top Ranked Output Sectors
3110 Food Products	3110 Food Products
3810 Metal Products	3840 Transportation Equipment
3420 Printing, Publishing	3810 Metal Products
3840 Transportation Equipment	3720 Nonferrous metals
3820 Machinery nec	3420 Printing, Publishing

Population 18,613,000. Percent of world population: 0.314

	1990		1991		1992		1993		1994		1995	
	Value	%	Value	%	Value	%	Value	%	Value	%	Value	%
Establishments or enterprises (number)	32,791	1.400	43,760	1.800	44,391	1.600	-	-	-	-	-	-
Total employment (000)	1,163	0.556	1,094	0.586	1,014	0.556	-	-	-	-	-	-
Female employees (000)	303	2.300	-	-	-	-	-	-	-	-	-	-
Salaries and Wages ($ bil.)	24	1.100	25	1.100	23	0.883	-	-	-	-	-	-
Output ($ bil.)	150	1.000	154	1.000	140	0.852	-	-	-	-	-	-
Value added ($ bil.)	63	1.100	65	1.100	55	0.921	-	-	-	-	-	-
Capital investment ($ mil.)	5,448	0.919	-	-	-	-	-	-	-	-	-	-
Employees per establishment (number)	35	48.1	25	36.8	23	36.7	-	-	-	-	-	-
Output per establishment ($ mil.)	5	108.2	4	76.9	3	58.0	-	-	-	-	-	-
Value added per establishment ($ mil.)	2	117.3	1	87.6	1	63.0	-	-	-	-	-	-
Capital investment per estab. ($ mil.)	0.179	67.5	-	-	-	-	-	-	-	-	-	-
Payroll per employee ($)	20,930	153.0	23,048	127.3	22,231	110.7	-	-	-	-	-	-
Females as % of total employment (%)	26.0	81.3	-	-	-	-	-	-	-	-	-	-
Output per employee ($)	129,183	188.2	141,110	173.4	138,163	154.4	-	-	-	-	-	-
Value added per employee ($)	54,213	165.0	59,167	175.2	60,885	169.1	-	-	-	-	-	-
Capital investment per employee ($)	5,358	97.3	-	-	-	-	-	-	-	-	-	-

Columns headed % show percent of the world total or of the world ratio. Ratios closest to 100 are closest to the world average. Values higher than 100 are above, values lower than 100 are below world average. nec stands for not elsewhere classified. Ratios, where shown, are frequently based on a subset of total data; for example, payroll per employee is averaged from all cases where both payroll and employment data are present. Dividing total salaries/wages by total employment will not necessarily produce a matching value.

Austria

Top Ranked Employing Sectors	Top Ranked Output Sectors
3830 Electrical Machinery	3830 Electrical Machinery
3820 Machinery nec	3110 Food Products
3810 Metal Products	3820 Machinery nec
3110 Food Products	3810 Metal Products
3320 Furniture, Fixtures	3840 Transportation Equipment

Population 8,134,000. Percent of world population: 0.137

	1990		1991		1992		1993		1994		1995	
	Value	%	Value	%	Value	%	Value	%	Value	%	Value	%
Establishments or enterprises (number)	9,470	0.411	9,243	0.388	8,909	0.317	8,688	0.423	8,476	0.474	-	-
Total employment (000)	715	0.342	695	0.372	667	0.366	624	0.343	617	0.388	604	0.424
Female employees (000)	208	1.600	201	2.200	189	0.880	172	0.769	167	0.792	-	-
Salaries and Wages ($ bil.)	19	0.861	19	0.810	21	0.810	19	0.946	20	1.000	24	1.400
Output ($ bil.)	104	0.727	106	0.689	111	0.674	100	0.693	109	0.787	128	0.959
Value added ($ bil.)	36	0.636	36	0.639	38	0.638	35	0.585	38	0.740	45	0.816
Capital investment ($ mil.)	7,331	1.200	7,229	1.200	7,950	1.200	6,143	1.100	5,911	1.300	-	-
Employees per establishment (number)	75	102.4	75	110.8	75	120.5	72	101.9	73	108.2	-	-
Output per establishment ($ mil.)	11	259.3	11	249.5	12	228.5	12	246.2	13	254.8	-	-
Value added per establishment ($ mil.)	4	229.4	4	232.0	4	203.2	4	249.5	4	253.8	-	-
Capital investment per estab. ($ mil.)	0.774	292.3	0.782	266.2	0.892	310.9	0.707	279.2	0.698	296.6	-	-
Payroll per employee ($)	25,944	189.6	27,706	153.1	30,985	154.3	30,936	190.5	32,225	183.5	39,093	223.7
Females as % of total employment (%)	29.2	91.0	29.0	90.2	28.4	61.3	27.5	55.1	27.1	51.6	-	-
Output per employee ($)	145,602	212.2	152,322	187.1	166,138	185.7	161,032	201.1	175,912	201.3	211,866	228.9
Value added per employee ($)	49,821	151.7	52,095	154.3	57,173	158.8	55,940	174.6	61,411	173.0	74,293	188.1
Capital investment per employee ($)	10,261	186.3	10,403	169.7	11,913	192.3	9,846	172.3	9,577	146.4	-	-

Columns headed % show percent of the world total or of the world ratio. Ratios closest to 100 are closest to the world average. Values higher than 100 are above, values lower than 100 are below world average. nec stands for not elsewhere classified. Ratios, where shown, are frequently based on a subset of total data; for example, payroll per employee is averaged from all cases where both payroll and employment data are present. Dividing total salaries/wages by total employment will not necessarily produce a matching value.

Azerbaijan

Top Ranked Employing Sectors	Top Ranked Output Sectors
3210 Textiles	3530 Petroleum Refineries
3110 Food Products	3110 Food Products
3820 Machinery nec	3210 Textiles
3830 Electrical Machinery	3510 Industrial Chemicals
3510 Industrial Chemicals	3820 Machinery nec

Population 7,856,000. Percent of world population: 0.133

	1990		1991		1992		1993		1994		1995	
	Value	%	Value	%	Value	%	Value	%	Value	%	Value	%
Establishments or enterprises (number)	3,740	0.162	3,614	0.152	5,066	0.180	6,070	0.295	4,037	0.226	-	-
Total employment (000)	405	0.194	390	0.209	358	.196	330	0.181	315	0.198	-	-
Female employees (000)	-	-	-	-	-	-	-	-	-	-	-	-
Salaries and Wages ($ bil.)	0.688	0.032	0.673	0.028	0.028	0.001	0.115	0.006	0.041	0.002	-	-
Output ($ bil.)	9	0.060	10	0.065	0.415	0.003	1	0.008	0.624	0.005	-	-
Value added ($ bil.)	-	-	-	-	-	-	-	-	-	-	-	-
Capital investment ($ mil.)	-	-	-	-	-	-	-	-	-	-	-	-
Employees per establishment (number)	104	141.7	104	153.4	68	109.4	52	74.0	75	111.3	-	-
Output per establishment ($ mil.)	2	53.8	3	59.9	0.082	1.5	0.197	4.2	0.155	3.1	-	-
Value added per establishment ($ mil.)	-	-	-	-	-	-	-	-	-	-	-	-
Capital investment per estab. ($ mil.)	-	-	-	-	-	-	-	-	-	-	-	-
Payroll per employee ($)	1,697	12.4	1,728	9.5	79	0.4	347	2.1	129	0.7	-	-
Females as % of total employment (%)	-	-	-	-	-	-	-	-	-	-	-	-
Output per employee ($)	21,803	31.8	26,398	32.4	1,203	1.3	3,775	4.7	2,063	2.4	-	-
Value added per employee ($)	-	-	-	-	-	-	-	-	-	-	-	-
Capital investment per employee ($)	-	-	-	-	-	-	-	-	-	-	-	-

Columns headed % show percent of the world total or of the world ratio. Ratios closest to 100 are closest to the world average. Values higher than 100 are above, values lower than 100 are below world average. *nec* stands for *not elsewhere classified*. Ratios, where shown, are frequently based on a subset of total data; for example, payroll per employee is averaged from all cases where *both* payroll and employment data are present. Dividing *total* salaries/wages by *total* employment will not necessarily produce a matching value.

Bahamas

Top Ranked Employing Sectors	Top Ranked Output Sectors
3130 Beverages	3130 Beverages
3420 Printing, Publishing	3520 Chemical Products nec
3520 Chemical Products nec	3420 Printing, Publishing
3110 Food Products	3690 Nonmetal Products nec
3220 Wearing Apparel	3900 Industries nec

Population 280,000. Percent of world population: 0.005

	1990		1991		1992		1993		1994		1995	
	Value	%	Value	%	Value	%	Value	%	Value	%	Value	%
Establishments or enterprises (number)	256	0.011	203	0.009	-	-	-	-	-	-	-	-
Total employment (000)	2	0.001	2	0.001	-	-	-	-	-	-	-	-
Female employees (000)	-	-	-	-	-	-	-	-	-	-	-	-
Salaries and Wages ($ bil.)	0.038	0.002	0.042	0.002	-	-	-	-	-	-	-	-
Output ($ bil.)	0.169	0.001	0.191	0.001	-	-	-	-	-	-	-	-
Value added ($ bil.)	0.075	0.001	0.096	0.002	-	-	-	-	-	-	-	-
Capital investment ($ mil.)	-	-	-	-	-	-	-	-	-	-	-	-
Employees per establishment (number)	9	11.9	11	15.6	-	-	-	-	-	-	-	-
Output per establishment ($ mil.)	0.660	15.6	0.941	20.5	-	-	-	-	-	-	-	-
Value added per establishment ($ mil.)	0.292	17.8	0.473	28.0	-	-	-	-	-	-	-	-
Capital investment per estab. ($ mil.)	-	-	-	-	-	-	-	-	-	-	-	-
Payroll per employee ($)	16,726	122.3	19,610	108.4	-	-	-	-	-	-	-	-
Females as % of total employment (%)	-	-	-	-	-	-	-	-	-	-	-	-
Output per employee ($)	75,346	109.8	89,003	109.3	-	-	-	-	-	-	-	-
Value added per employee ($)	33,364	101.6	44,769	132.6	-	-	-	-	-	-	-	-
Capital investment per employee ($)	-	-	-	-	-	-	-	-	-	-	-	-

Columns headed % show percent of the world total or of the world ratio. Ratios closest to 100 are closest to the world average. Values higher than 100 are above, values lower than 100 are below world average. *nec* stands for *not elsewhere classified*. Ratios, where shown, are frequently based on a subset of total data; for example, payroll per employee is averaged from all cases where *both* payroll and employment data are present. Dividing *total* salaries/wages by *total* employment will not necessarily produce a matching value.

Bahrain

Top Ranked Employing Sectors	Top Ranked Output Sectors
3220 Wearing Apparel	3720 Nonferrous metals
3720 Nonferrous metals	3530 Petroleum Refineries
3110 Food Products	3110 Food Products
3530 Petroleum Refineries	3830 Electrical Machinery
3320 Furniture, Fixtures	3690 Nonmetal Products nec

Population 616,000. Percent of world population: 0.010

	1990		1991		1992		1993		1994		1995	
	Value	%	Value	%	Value	%	Value	%	Value	%	Value	%
Establishments or enterprises (number)	-	-	-	-	2,307	0.082	-	-	-	-	-	-
Total employment (000)	-	-	-	-	30	0.017	-	-	-	-	-	-
Female employees (000)	-	-	-	-	3	0.015	-	-	-	-	-	-
Salaries and Wages ($ bil.)	-	-	-	-	0.300	0.012	-	-	-	-	-	-
Output ($ bil.)	-	-	-	-	2	0.011	-	-	-	-	-	-
Value added ($ bil.)	-	-	-	-	-		-	-	-	-	-	-
Capital investment ($ mil.)	-	-	-	-	117	0.018	-	-	-	-	-	-
Employees per establishment (number)	-	-	-	-	13	21.1	-	-	-	-	-	-
Output per establishment ($ mil.)	-	-	-	-	0.811	14.9	-	-	-	-	-	-
Value added per establishment ($ mil.)	-	-	-	-	-		-	-	-	-	-	-
Capital investment per estab. ($ mil.)	-	-	-	-	0.051	17.8	-	-	-	-	-	-
Payroll per employee ($)	-	-	-	-	9,893	49.3	-	-	-	-	-	-
Females as % of total employment (%)	-	-	-	-	11.1	24.0	-	-	-	-	-	-
Output per employee ($)	-	-	-	-	61,654	68.9	-	-	-	-	-	-
Value added per employee ($)	-	-	-	-	-		-	-	-	-	-	-
Capital investment per employee ($)	-	-	-	-	3,867	62.4	-	-	-	-	-	-

Columns headed % show percent of the world total or of the world ratio. Ratios closest to 100 are closest to the world average. Values higher than 100 are above, values lower than 100 are below world average. *nec* stands for *not elsewhere classified*. Ratios, where shown, are frequently based on a subset of total data; for example, payroll per employee is averaged from all cases where *both* payroll and employment data are present. Dividing *total* salaries/wages by *total* employment will not necessarily produce a matching value.

Bangladesh

Top Ranked Employing Sectors	Top Ranked Output Sectors
3210 Textiles	3210 Textiles
3220 Wearing Apparel	3110 Food Products
3110 Food Products	3220 Wearing Apparel
3520 Chemical Products nec	3520 Chemical Products nec
3140 Tobacco	3710 Iron and Steel

Population 127,567,000. Percent of world population: 2.153

	1990		1991		1992		1993		1994		1995	
	Value	%	Value	%	Value	%	Value	%	Value	%	Value	%
Establishments or enterprises (number)	34,882	1.500	37,205	1.600	38,023	1.400	-	-	-	-	-	-
Total employment (000)	1,558	0.745	1,675	0.897	1,743	0.955	-	-	-	-	-	-
Female employees (000)	155	1.200	186	2.100	188	0.872	-	-	-	-	-	-
Salaries and Wages ($ bil.)	1	0.051	1	0.044	1	0.041	-	-	-	-	-	-
Output ($ bil.)	7	0.050	7	0.048	8	0.046	-	-	-	-	-	-
Value added ($ bil.)	2	0.043	2	0.043	3	0.042	-	-	-	-	-	-
Capital investment ($ mil.)	275	0.046	231	0.037	253	0.040	-	-	-	-	-	-
Employees per establishment (number)	45	60.6	45	66.3	46	73.7	-	-	-	-	-	-
Output per establishment ($ mil.)	0.205	4.8	0.199	4.3	0.197	3.6	-	-	-	-	-	-
Value added per establishment ($ mil.)	0.069	4.2	0.065	3.9	0.067	3.2	-	-	-	-	-	-
Capital investment per estab. ($ mil.)	0.011	4.1	0.006	2.1	0.007	2.3	-	-	-	-	-	-
Payroll per employee ($)	708	5.2	629	3.5	599	3.0	-	-	-	-	-	-
Females as % of total employment (%)	10.0	31.3	11.2	34.8	10.8	23.3	-	-	-	-	-	-
Output per employee ($)	4,589	6.7	4,412	5.4	4,306	4.8	-	-	-	-	-	-
Value added per employee ($)	1,555	4.7	1,453	4.3	1,453	4.0	-	-	-	-	-	-
Capital investment per employee ($)	268	4.9	138	2.3	146	2.4	-	-	-	-	-	-

Columns headed % show percent of the world total or of the world ratio. Ratios closest to 100 are closest to the world average. Values higher than 100 are above, values lower than 100 are below world average. *nec* stands for *not elsewhere classified*. Ratios, where shown, are frequently based on a subset of total data; for example, payroll per employee is averaged from all cases where *both* payroll and employment data are present. Dividing *total* salaries/wages by *total* employment will not necessarily produce a matching value.

For additional notes and sources, see Appendix I.

Barbados

	Top Ranked Employing Sectors	Top Ranked Output Sectors
	3110 Food Products	3110 Food Products
	3220 Wearing Apparel	3810 Metal Products
	3810 Metal Products	3520 Chemical Products nec
	3520 Chemical Products nec	3220 Wearing Apparel
	3900 Industries nec	3900 Industries nec

Population 259,000. Percent of world population: 0.004

	1990		1991		1992		1993		1994		1995	
	Value	%	Value	%	Value	%	Value	%	Value	%	Value	%
Establishments or enterprises (number)	62	0.003	47	0.002	55	0.002	54	0.003	118	0.007	-	-
Total employment (000)	4	0.002	4	0.002	3	0.002	3	0.001	4	0.002	4	0.003
Female employees (000)	-	-	-	-	-	-	-	-	-	-	-	-
Salaries and Wages ($ bil.)	0.033	0.002	0.026	0.001	0.025	0.001	0.024	0.001	0.039	0.002	0.036	0.002
Output ($ bil.)	0.194	0.001	0.128	0.001	0.123	0.001	0.153	0.001	0.217	0.002	0.202	0.002
Value added ($ bil.)	0.050	0.001	0.046	0.001	0.048	0.001	0.053	0.001	0.083	0.002	0.075	0.001
Capital investment ($ mil.)	6	0.001	8	0.001	3	-	4	0.001	7	0.002	-	-
Employees per establishment (number)	58	78.6	78	115.7	57	91.9	47	66.5	32	47.2	-	-
Output per establishment ($ mil.)	3	73.8	3	59.2	2	41.2	3	60.5	2	36.6	-	-
Value added per establishment ($ mil.)	0.811	49.5	0.982	58.2	0.869	41.2	0.983	61.1	0.703	39.8	-	-
Capital investment per estab. ($ mil.)	0.095	35.8	0.162	55.1	0.049	17.1	0.082	32.5	0.067	28.6	-	-
Payroll per employee ($)	9,060	66.2	7,155	39.5	8,072	40.2	9,427	58.1	10,288	58.6	9,723	55.6
Females as % of total employment (%)	-	-	-	-	-	-	-	-	-	-	-	-
Output per employee ($)	53,920	78.6	34,610	42.5	39,221	43.8	60,565	75.6	57,893	66.3	54,406	58.8
Value added per employee ($)	13,995	42.6	12,513	37.1	15,204	42.2	20,966	65.4	22,106	62.3	20,224	51.2
Capital investment per employee ($)	1,635	29.7	2,064	33.7	860	13.9	1,757	30.7	2,121	32.4	-	-

Columns headed % show percent of the world total or of the world ratio. Ratios closest to 100 are closest to the world average. Values higher than 100 are above, values lower than 100 are below world average. *nec* stands for *not elsewhere classified*. Ratios, where shown, are frequently based on a subset of total data; for example, payroll per employee is averaged from all cases where *both* payroll and employment data are present. Dividing *total* salaries/wages by *total* employment will not necessarily produce a matching value.

Belgium

	Top Ranked Employing Sectors	Top Ranked Output Sectors
	3110 Food Products	3510 Industrial Chemicals
	3210 Textiles	3210 Textiles
	3710 Iron and Steel	3710 Iron and Steel
	3420 Printing, Publishing	3420 Printing, Publishing
	3220 Wearing Apparel	3320 Furniture, Fixtures

Population 10,175,000. Percent of world population: 0.172

	1990		1991		1992		1993		1994		1995	
	Value	%	Value	%	Value	%	Value	%	Value	%	Value	%
Establishments or enterprises (number)	21,913	0.951	21,932	0.920	21,745	0.774	20,748	1.000	20,423	1.100	-	-
Total employment (000)	395	0.189	388	0.208	378	0.207	-	-	-	-	-	-
Female employees (000)	-	-	-	-	-	-	-	-	-	-	-	-
Salaries and Wages ($ bil.)	12	0.546	12	0.498	13	0.501	11	0.557	12	0.607	-	-
Output ($ bil.)	58	0.406	55	0.355	59	0.358	52	0.361	59	0.427	72	0.543
Value added ($ bil.)	26	0.456	24	0.431	27	0.446	24	0.406	27	0.527	33	0.606
Capital investment ($ mil.)	6,883	1.200	7,151	1.100	7,217	1.100	5,425	0.998	5,965	1.300	7,347	2.400
Employees per establishment (number)	19	25.2	18	27.0	18	29.3	-	-	-	-	-	-
Output per establishment ($ mil.)	4	84.7	3	75.1	4	71.0	4	78.8	4	84.2	-	-
Value added per establishment ($ mil.)	0.983	60.0	0.975	57.8	1	51.3	1	64.4	1	65.5	-	-
Capital investment per estab. ($ mil.)	0.213	80.6	0.232	79.0	0.252	87.7	0.194	76.6	0.219	93.2	-	-
Payroll per employee ($)	19,223	140.5	20,018	110.6	22,340	111.3	-	-	-	-	-	-
Females as % of total employment (%)	-	-	-	-	-	-	-	-	-	-	-	-
Output per employee ($)	164,632	239.9	155,215	190.7	170,217	190.3	-	-	-	-	-	-
Value added per employee ($)	56,735	172.7	54,850	162.4	61,760	171.5	-	-	-	-	-	-
Capital investment per employee ($)	11,988	217.7	13,281	216.6	14,176	228.8	-	-	-	-	-	-

Columns headed % show percent of the world total or of the world ratio. Ratios closest to 100 are closest to the world average. Values higher than 100 are above, values lower than 100 are below world average. *nec* stands for *not elsewhere classified*. Ratios, where shown, are frequently based on a subset of total data; for example, payroll per employee is averaged from all cases where *both* payroll and employment data are present. Dividing *total* salaries/wages by *total* employment will not necessarily produce a matching value.

Belize

Top Ranked Employing Sectors	Top Ranked Output Sectors
3110 Food Products	3110 Food Products
3310 Wood Products	3130 Beverages
3320 Furniture, Fixtures	3310 Wood Products
3130 Beverages	3690 Nonmetal Products nec
3690 Nonmetal Products nec	3810 Metal Products

Population 230,000. Percent of world population: 0.004

	1990		1991		1992		1993		1994		1995	
	Value	%	Value	%	Value	%	Value	%	Value	%	Value	%
Establishments or enterprises (number)	314	0.014	322	0.014	330	0.012	-	-	-	-	-	-
Total employment (000)	-	-	5	0.003	6	0.003	-	-	-	-	-	-
Female employees (000)	-	-	-	-	-	-	-	-	-	-	-	-
Salaries and Wages ($ bil.)	0.017	0.001	0.021	0.001	0.024	0.001	-	-	-	-	-	-
Output ($ bil.)	0.170	0.001	0.178	0.001	0.197	0.001	-	-	-	-	-	-
Value added ($ bil.)	0.043	0.001	0.046	0.001	0.049	0.001	-	-	-	-	-	-
Capital investment ($ mil.)	-	-	-	-	-	-	-	-	-	-	-	-
Employees per establishment (number)	-	-	17	25.3	18	28.7	-	-	-	-	-	-
Output per establishment ($ mil.)	0.564	13.3	0.576	12.6	0.624	11.5	-	-	-	-	-	-
Value added per establishment ($ mil.)	0.143	8.7	0.149	8.8	0.154	7.3	-	-	-	-	-	-
Capital investment per estab. ($ mil.)	-	-	-	-	-	-	-	-	-	-	-	-
Payroll per employee ($)	-	-	4,033	22.3	4,227	21.1	-	-	-	-	-	-
Females as % of total employment (%)	-	-	-	-	-	-	-	-	-	-	-	-
Output per employee ($)	-	-	33,590	41.3	34,955	39.1	-	-	-	-	-	-
Value added per employee ($)	-	-	8,663	25.7	8,616	23.9	-	-	-	-	-	-
Capital investment per employee ($)	-	-	-	-	-	-	-	-	-	-	-	-

Columns headed % show percent of the world total or of the world ratio. Ratios closest to 100 are closest to the world average. Values higher than 100 are above, values lower than 100 are below world average. *nec* stands for *not elsewhere classified*. Ratios, where shown, are frequently based on a subset of total data; for example, payroll per employee is averaged from all cases where *both* payroll and employment data are present. Dividing *total* salaries/wages by *total* employment will not necessarily produce a matching value.

Benin

Top Ranked Employing Sectors	Top Ranked Output Sectors
	3110 Food Products
	3690 Nonmetal Products nec
	3210 Textiles
	3130 Beverages
	3220 Wearing Apparel

Population 6,101,000. Percent of world population: 0.103

	1990		1991		1992		1993		1994		1995	
	Value	%	Value	%	Value	%	Value	%	Value	%	Value	%
Establishments or enterprises (number)	-	-	-	-	398	0.014	398	0.019	-	-	-	-
Total employment (000)	-	-	-	-	-	-	-	-	-	-	-	-
Female employees (000)	-	-	-	-	-	-	-	-	-	-	-	-
Salaries and Wages ($ bil.)	-	-	-	-	-	-	-	-	-	-	-	-
Output ($ bil.)	0.512	0.004	0.538	0.004	0.621	0.004	0.525	0.004	-	-	-	-
Value added ($ bil.)	-	-	-	-	-	-	-	-	-	-	-	-
Capital investment ($ mil.)	-	-	-	-	-	-	-	-	-	-	-	-
Employees per establishment (number)	-	-	-	-	-	-	-	-	-	-	-	-
Output per establishment ($ mil.)	-	-	-	-	2	29.3	1	28.6	-	-	-	-
Value added per establishment ($ mil.)	-	-	-	-	-	-	-	-	-	-	-	-
Capital investment per estab. ($ mil.)	-	-	-	-	-	-	-	-	-	-	-	-
Payroll per employee ($)	-	-	-	-	-	-	-	-	-	-	-	-
Females as % of total employment (%)	-	-	-	-	-	-	-	-	-	-	-	-
Output per employee ($)	-	-	-	-	-	-	-	-	-	-	-	-
Value added per employee ($)	-	-	-	-	-	-	-	-	-	-	-	-
Capital investment per employee ($)	-	-	-	-	-	-	-	-	-	-	-	-

Columns headed % show percent of the world total or of the world ratio. Ratios closest to 100 are closest to the world average. Values higher than 100 are above, values lower than 100 are below world average. *nec* stands for *not elsewhere classified*. Ratios, where shown, are frequently based on a subset of total data; for example, payroll per employee is averaged from all cases where *both* payroll and employment data are present. Dividing *total* salaries/wages by *total* employment will not necessarily produce a matching value.

For additional notes and sources, see Appendix I.

Bermuda

Top Ranked Employing Sectors	Top Ranked Output Sectors
3420 Printing, Publishing	
3110 Food Products	
3130 Beverages	
3530 Petroleum Refineries	
3310 Wood Products	

Population 62,000. Percent of world population: 0.001

	1990		1991		1992		1993		1994		1995	
	Value	%	Value	%	Value	%	Value	%	Value	%	Value	%
Establishments or enterprises (number)	-	-	-	-	-	-	-	-	-	-	-	-
Total employment (000)	1	0.001	1	0.001	1	0.001	1	0.001	1	0.001	0.925	0.001
Female employees (000)	0.394	0.003	0.397	0.004	0.389	0.002	0.371	0.002	0.370	0.002	0.312	0.002
Salaries and Wages ($ bil.)	-	-	-	-	-	-	-	-	-	-	-	-
Output ($ bil.)	-	-	-	-	-	-	-	-	-	-	-	-
Value added ($ bil.)	-	-	-	-	-	-	-	-	-	-	-	-
Capital investment ($ mil.)	-	-	-	-	-	-	-	-	-	-	-	-
Employees per establishment (number)	-	-	-	-	-	-	-	-	-	-	-	-
Output per establishment ($ mil.)	-	-	-	-	-	-	-	-	-	-	-	-
Value added per establishment ($ mil.)	-	-	-	-	-	-	-	-	-	-	-	-
Capital investment per estab. ($ mil.)	-	-	-	-	-	-	-	-	-	-	-	-
Payroll per employee ($)	-		-		-		-		-		-	
Females as % of total employment (%)	35.6	111.1	36.4	113.4	36.2	78.1	35.7	71.4	33.9	64.6	33.7	57.8
Output per employee ($)	-	-	-	-	-	-	-	-	-	-	-	-
Value added per employee ($)	-	-	-	-	-	-	-	-	-	-	-	-
Capital investment per employee ($)	-	-	-	-	-	-	-	-	-	-	-	-

Columns headed % show percent of the world total or of the world ratio. Ratios closest to 100 are closest to the world average. Values higher than 100 are above, values lower than 100 are below world average. *nec* stands for *not elsewhere classified*. Ratios, where shown, are frequently based on a subset of total data; for example, payroll per employee is averaged from all cases where *both* payroll and employment data are present. Dividing *total* salaries/wages by *total* employment will not necessarily produce a matching value.

Bolivia

Top Ranked Employing Sectors	Top Ranked Output Sectors
3110 Food Products	3110 Food Products
3130 Beverages	3530 Petroleum Refineries
3210 Textiles	3130 Beverages
3690 Nonmetal Products nec	3900 Industries nec
3420 Printing, Publishing	3690 Nonmetal Products nec

Population 7,826,000. Percent of world population: 0.132

	1990		1991		1992		1993		1994		1995	
	Value	%	Value	%	Value	%	Value	%	Value	%	Value	%
Establishments or enterprises (number)	1,707	0.074	1,808	0.076	1,895	0.067	1,947	0.095	1,949	0.109	1,998	0.163
Total employment (000)	39	0.019	41	0.022	44	0.024	46	0.025	48	0.030	49	0.035
Female employees (000)	-	-	-	-	-	-	-	-	-	-	-	-
Salaries and Wages ($ bil.)	0.076	0.004	0.083	0.003	0.092	0.004	0.100	0.005	0.112	0.006	0.120	0.007
Output ($ bil.)	2	0.011	2	0.012	2	0.011	2	0.014	2	0.016	2	0.018
Value added ($ bil.)	0.721	0.013	0.896	0.016	0.854	0.014	0.821	0.014	0.948	0.019	0.984	0.018
Capital investment ($ mil.)	62	0.010	64	0.010	68	0.011	118	0.022	70	0.016	-	-
Employees per establishment (number)	23	31.3	23	33.5	23	37.7	24	33.8	25	36.5	25	35.1
Output per establishment ($ mil.)	0.945	22.3	1	22.8	0.994	18.3	1	21.5	1	23.1	1	28.4
Value added per establishment ($ mil.)	0.422	25.8	0.496	29.4	0.451	21.4	0.422	26.2	0.487	27.6	0.492	27.9
Capital investment per estab. ($ mil.)	0.036	13.7	0.036	12.1	0.036	12.6	0.061	23.9	0.036	15.2	-	-
Payroll per employee ($)	1,922	14.1	2,011	11.1	2,070	10.3	2,154	13.3	2,341	13.3	2,438	14.0
Females as % of total employment (%)	-	-	-	-	-	-	-	-	-	-	-	-
Output per employee ($)	40,968	59.7	46,064	56.6	42,461	47.5	42,438	53.0	47,405	54.3	47,760	51.6
Value added per employee ($)	18,303	55.7	21,834	64.6	19,239	53.4	17,708	55.3	19,813	55.8	20,013	50.7
Capital investment per employee ($)	1,566	28.4	1,567	25.6	1,539	24.8	2,550	44.6	1,454	22.2	-	-

Columns headed % show percent of the world total or of the world ratio. Ratios closest to 100 are closest to the world average. Values higher than 100 are above, values lower than 100 are below world average. *nec* stands for *not elsewhere classified*. Ratios, where shown, are frequently based on a subset of total data; for example, payroll per employee is averaged from all cases where *both* payroll and employment data are present. Dividing *total* salaries/wages by *total* employment will not necessarily produce a matching value.

For additional notes and sources, see Appendix I.

Bosnia & Herzegovina

Top Ranked Employing Sectors	Top Ranked Output Sectors
3840 Transportation Equipment	3110 Food Products
3720 Nonferrous metals	3840 Transportation Equipment
3230 Leather and Products	3810 Metal Products
3850 Professional Goods	3710 Iron and Steel
3240 Footwear	3310 Wood Products

Population 3,366,000. Percent of world population: 0.057

	1990		1991		1992		1993		1994		1995	
	Value	%	Value	%	Value	%	Value	%	Value	%	Value	%
Establishments or enterprises (number)	999	0.043	1,034	0.043	-	-	-	-	116	0.006	-	-
Total employment (000)	403	0.193	374	0.200	-	-	-	-	19	0.012	-	-
Female employees (000)	153	1.100	140	1.600	-	-	-	-	-	-	-	-
Salaries and Wages ($ bil.)	0.990	0.046	0.855	0.036	-	-	-	-	-	-	-	-
Output ($ bil.)	9	0.062	9	0.058	-	-	-	-	-	-	-	-
Value added ($ bil.)	4	0.064	4	0.071	-	-	-	-	-	-	-	-
Capital investment ($ mil.)	252	0.042	-	-	-	-	-	-	-	-	-	-
Employees per establishment (number)	403	547.2	362	533.3	-	-	-	-	164	243.2	-	-
Output per establishment ($ mil.)	10	226.3	9	197.2	-	-	-	-	-	-	-	-
Value added per establishment ($ mil.)	4	239.1	4	250.8	-	-	-	-	-	-	-	-
Capital investment per estab. ($ mil.)	0.274	103.3	-	-	-	-	-	-	-	-	-	-
Payroll per employee ($)	2,745	20.1	2,567	14.2	-	-	-	-	-	-	-	-
Females as % of total employment (%)	42.4	132.3	41.9	130.4	-	-	-	-	-	-	-	-
Output per employee ($)	24,446	35.6	25,809	31.7	-	-	-	-	-	-	-	-
Value added per employee ($)	9,993	30.4	12,070	35.7	-	-	-	-	-	-	-	-
Capital investment per employee ($)	698	12.7	-	-	-	-	-	-	-	-	-	-

Columns headed % show percent of the world total or of the world ratio. Ratios closest to 100 are closest to the world average. Values higher than 100 are above, values lower than 100 are below world average. *nec* stands for *not elsewhere classified*. Ratios, where shown, are frequently based on a subset of total data; for example, payroll per employee is averaged from all cases where *both* payroll and employment data are present. Dividing *total* salaries/wages by *total* employment will not necessarily produce a matching value.

Botswana

Top Ranked Employing Sectors	Top Ranked Output Sectors
3110 Food Products	3110 Food Products
3900 Industries nec	3130 Beverages
3130 Beverages	3900 Industries nec

Population 1,448,000. Percent of world population: 0.024

	1990		1991		1992		1993		1994		1995	
	Value	%	Value	%	Value	%	Value	%	Value	%	Value	%
Establishments or enterprises (number)	114	0.005	132	0.006	160	0.006	189	0.009	197	0.011	309	0.025
Total employment (000)	11	0.005	12	0.006	11	0.006	10	0.006	10	0.007	12	0.008
Female employees (000)	-	-	-	-	-	-	-	-	-	-	4	0.023
Salaries and Wages ($ bil.)	-	-	-	-	0.052	0.002	0.050	0.002	-	-	-	-
Output ($ bil.)	-	-	-	-	0.539	0.003	0.497	0.003	-	-	-	-
Value added ($ bil.)	-	-	-	-	0.124	0.002	0.120	0.002	0.122	0.002		
Capital investment ($ mil.)	-	-	-	-	48	0.008	30	0.005	-	-	-	-
Employees per establishment (number)	97	132.1	89	131.8	66	106.6	55	78.1	53	79.2	38	54.0
Output per establishment ($ mil.)	-	-	-	-	3	61.9	3	56.0	-	-	-	-
Value added per establishment ($ mil.)	-	-	-	-	0.776	36.8	0.635	39.4	0.620	35.2	-	-
Capital investment per estab. ($ mil.)	-	-	-	-	0.302	105.1	0.157	61.9	-	-	-	-
Payroll per employee ($)	-	-	-	-	4,865	24.2	4,833	29.8	-	-	-	-
Females as % of total employment (%)	-	-	-	-	-	-	-	-	-	-	37.4	64.0
Output per employee ($)	-	-	-	-	50,875	58.9	47,825	59.7	-	-	-	-
Value added per employee ($)	-	-	-	-	11,716	32.5	11,532	36.0	11,632	32.8	-	-
Capital investment per employee ($)	-	-	-	-	4,552	73.5	2,849	49.8	-	-	-	-

Columns headed % show percent of the world total or of the world ratio. Ratios closest to 100 are closest to the world average. Values higher than 100 are above, values lower than 100 are below world average. *nec* stands for *not elsewhere classified*. Ratios, where shown, are frequently based on a subset of total data; for example, payroll per employee is averaged from all cases where *both* payroll and employment data are present. Dividing *total* salaries/wages by *total* employment will not necessarily produce a matching value.

For additional notes and sources, see Appendix I.

Brazil

Top Ranked Employing Sectors	Top Ranked Output Sectors
3110 Food Products	3110 Food Products
3840 Transportation Equipment	3840 Transportation Equipment
3820 Machinery nec	3830 Electrical Machinery
3210 Textiles	3820 Machinery nec
3830 Electrical Machinery	3210 Textiles

Population 169,807,000. Percent of world population: 2.865

	1990		1991		1992		1993		1994		1995	
	Value	%	Value	%	Value	%	Value	%	Value	%	Value	%
Establishments or enterprises (number)	18,387	0.798	-	-	16,777	0.597	15,973	0.777	-	-	-	-
Total employment (000)	2,573	1.200	-	-	2,276	1.200	2,227	1.200	-	-	-	-
Female employees (000)	-		-		-		-		-		-	
Salaries and Wages ($ bil.)	18	0.829	-	-	15	0.601	18	0.902	-	-	-	-
Output ($ bil.)	148	1.000	-	-	125	0.762	152	1.000	-	-	-	-
Value added ($ bil.)	82	1.500	-	-	75	1.300	92	1.500	-	-	-	-
Capital investment ($ mil.)	9,092	1.500	-	-	4,739	0.741	5,877	1.100	-	-	-	-
Employees per establishment (number)	140	189.8	-	-	136	218.2	139	197.8	-	-	-	-
Output per establishment ($ mil.)	8	189.6	-	-	7	137.1	10	202.8	-	-	-	-
Value added per establishment ($ mil.)	4	272.6	-	-	4	212.3	6	358.7	-	-	-	-
Capital investment per estab. ($ mil.)	0.494	186.7	-	-	0.282	98.4	0.368	145.3	-	-	-	-
Payroll per employee ($)	6,943	50.7	-	-	6,740	33.6	8,267	50.9	-	-	-	-
Females as % of total employment (%)	-		-		-		-		-		-	
Output per employee ($)	57,385	83.6	-	-	55,040	61.5	68,325	85.3	-	-	-	-
Value added per employee ($)	31,919	97.2	-	-	32,975	91.6	41,423	129.3	-	-	-	-
Capital investment per employee ($)	3,534	64.2	-	-	2,082	33.6	2,639	46.2	-	-	-	-

Columns headed % show percent of the world total or of the world ratio. Ratios closest to 100 are closest to the world average. Values higher than 100 are above, values lower than 100 are below world average. *nec* stands for *not elsewhere classified*. Ratios, where shown, are frequently based on a subset of total data; for example, payroll per employee is averaged from all cases where *both* payroll and employment data are present. Dividing *total* salaries/wages by *total* employment will not necessarily produce a matching value.

Bulgaria

Top Ranked Employing Sectors	Top Ranked Output Sectors
3210 Textiles	3540 Petroleum, Coal Products
3110 Food Products	3110 Food Products
3830 Electrical Machinery	3710 Iron and Steel
3840 Transportation Equipment	3510 Industrial Chemicals
3220 Wearing Apparel	3130 Beverages

Population 8,240,000. Percent of world population: 0.139

	1990		1991		1992		1993		1994		1995	
	Value	%	Value	%	Value	%	Value	%	Value	%	Value	%
Establishments or enterprises (number)	2,090	0.091	2,212	0.093	2,231	0.079	23,119	1.100	30,846	1.700	32,378	2.600
Total employment (000)	1,119	0.535	899	0.481	754	0.413	660	0.363	612	0.385	563	0.396
Female employees (000)	-		-		-		363	1.600	329	1.600	287	1.500
Salaries and Wages ($ bil.)	2	0.081	0.498	0.021	0.705	0.028	0.879	0.043	0.688	0.035	0.822	0.050
Output ($ bil.)	21	0.148	8	0.051	8	0.046	7	0.050	7	0.054	10	0.072
Value added ($ bil.)	-	-	2	0.039	2	0.039	2	0.038	0.703	0.014	1	0.019
Capital investment ($ mil.)	700	0.118	248	0.040	4,664	0.730	1,264	0.232	533	0.120	452	0.145
Employees per establishment (number)	535	726.2	406	599.0	338	543.3	29	40.5	20	29.5	17	24.8
Output per establishment ($ mil.)	10	239.5	4	77.9	3	62.6	0.312	6.6	0.242	4.8	0.296	7.2
Value added per establishment ($ mil.)	-	-	0.883	52.3	0.919	43.6	0.087	5.4	0.058	3.3	0.086	4.9
Capital investment per estab. ($ mil.)	0.335	126.5	0.112	38.2	2	728.5	0.055	21.6	0.017	7.3	0.014	4.9
Payroll per employee ($)	1,863	13.6	649	3.6	1,076	5.4	1,332	8.2	1,124	6.4	1,461	8.4
Females as % of total employment (%)	-		-		-		54.9	110.0	53.8	102.5	51.1	87.4
Output per employee ($)	18,949	27.6	8,797	10.8	10,092	11.3	10,932	13.7	12,183	13.9	17,029	18.4
Value added per employee ($)	-	-	2,173	6.4	2,720	7.6	3,060	9.5	2,647	7.5	4,118	10.4
Capital investment per employee ($)	626	11.4	276	4.5	6,189	99.9	1,914	33.5	871	13.3	803	10.7

Columns headed % show percent of the world total or of the world ratio. Ratios closest to 100 are closest to the world average. Values higher than 100 are above, values lower than 100 are below world average. *nec* stands for *not elsewhere classified*. Ratios, where shown, are frequently based on a subset of total data; for example, payroll per employee is averaged from all cases where *both* payroll and employment data are present. Dividing *total* salaries/wages by *total* employment will not necessarily produce a matching value.

Burundi

Top Ranked Employing Sectors	Top Ranked Output Sectors
3210 Textiles	3110 Food Products
3110 Food Products	3130 Beverages
3130 Beverages	3810 Metal Products
3810 Metal Products	3210 Textiles
3520 Chemical Products nec	3510 Industrial Chemicals

Population 5,537,000. Percent of world population: 0.093

	1990		1991		1992		1993		1994		1995	
	Value	%	Value	%	Value	%	Value	%	Value	%	Value	%
Establishments or enterprises (number)	73	0.003	91	0.004	-	-	-	-	-	-	-	-
Total employment (000)	6	0.003	6	0.003	-	-	-	-	-	-	-	-
Female employees (000)	-	-	-	-	-	-	-	-	-	-	-	-
Salaries and Wages ($ bil.)	0.021	0.001	0.021	0.001	-	-	-	-	-	-	-	-
Output ($ bil.)	0.205	0.001	0.197	0.001	-	-	-	-	-	-	-	-
Value added ($ bil.)	0.104	0.002	0.113	0.002	-	-	-	-	-	-	-	-
Capital investment ($ mil.)	-	-	-	-	-	-	-	-	-	-	-	-
Employees per establishment (number)	84	114.5	70	102.5	-	-	-	-	-	-	-	-
Output per establishment ($ mil.)	3	66.3	2	47.2	-	-	-	-	-	-	-	-
Value added per establishment ($ mil.)	1	86.9	1	73.3	-	-	-	-	-	-	-	-
Capital investment per estab. ($ mil.)	-	-	-	-	-	-	-	-	-	-	-	-
Payroll per employee ($)	3,430	25.1	3,360	18.6	-	-	-	-	-	-	-	-
Females as % of total employment (%)	-	-	-	-	-	-	-	-	-	-	-	-
Output per employee ($)	33,284	48.5	31,126	38.2	-	-	-	-	-	-	-	-
Value added per employee ($)	16,867	51.3	17,781	52.6	-	-	-	-	-	-	-	-
Capital investment per employee ($)	-	-	-	-	-	-	-	-	-	-	-	-

Columns headed % show percent of the world total or of the world ratio. Ratios closest to 100 are closest to the world average. Values higher than 100 are above, values lower than 100 are below world average. *nec* stands for *not elsewhere classified*. Ratios, where shown, are frequently based on a subset of total data; for example, payroll per employee is averaged from all cases where *both* payroll and employment data are present. Dividing *total* salaries/wages by *total* employment will not necessarily produce a matching value.

Cambodia

Top Ranked Employing Sectors	Top Ranked Output Sectors
3210 Textiles	
3110 Food Products	
3220 Wearing Apparel	
3310 Wood Products	
3690 Nonmetal Products nec	

Population 11,340,000. Percent of world population: 0.191

	1990		1991		1992		1993		1994		1995	
	Value	%	Value	%	Value	%	Value	%	Value	%	Value	%
Establishments or enterprises (number)	1,147	0.050	1,580	0.066	-	-	-	-	-	-	-	-
Total employment (000)	78	0.037	72	0.038	-	-	-	-	-	-	-	-
Female employees (000)	-	-	-	-	-	-	-	-	-	-	-	-
Salaries and Wages ($ bil.)	-	-	-	-	-	-	-	-	-	-	-	-
Output ($ bil.)	-	-	-	-	-	-	-	-	-	-	-	-
Value added ($ bil.)	-	-	-	-	-	-	-	-	-	-	-	-
Capital investment ($ mil.)	-	-	-	-	-	-	-	-	-	-	-	-
Employees per establishment (number)	5	6.4	3	4.0	-	-	-	-	-	-	-	-
Output per establishment ($ mil.)	-	-	-	-	-	-	-	-	-	-	-	-
Value added per establishment ($ mil.)	-	-	-	-	-	-	-	-	-	-	-	-
Capital investment per estab. ($ mil.)	-	-	-	-	-	-	-	-	-	-	-	-
Payroll per employee ($)	-	-	-	-	-	-	-	-	-	-	-	-
Females as % of total employment (%)	-	-	-	-	-	-	-	-	-	-	-	-
Output per employee ($)	-	-	-	-	-	-	-	-	-	-	-	-
Value added per employee ($)	-	-	-	-	-	-	-	-	-	-	-	-
Capital investment per employee ($)	-	-	-	-	-	-	-	-	-	-	-	-

Columns headed % show percent of the world total or of the world ratio. Ratios closest to 100 are closest to the world average. Values higher than 100 are above, values lower than 100 are below world average. *nec* stands for *not elsewhere classified*. Ratios, where shown, are frequently based on a subset of total data; for example, payroll per employee is averaged from all cases where *both* payroll and employment data are present. Dividing *total* salaries/wages by *total* employment will not necessarily produce a matching value.

Cameroon

Top Ranked Employing Sectors	Top Ranked Output Sectors
3110 Food Products	3310 Wood Products
3310 Wood Products	3110 Food Products
3210 Textiles	3130 Beverages
3130 Beverages	3210 Textiles
3510 Industrial Chemicals	3140 Tobacco

Population 15,029,000. Percent of world population: 0.254

	1990		1991		1992		1993		1994		1995	
	Value	%	Value	%	Value	%	Value	%	Value	%	Value	%
Establishments or enterprises (number)	77	0.003	160	0.007	137	0.005	119	0.006	149	0.008	116	0.009
Total employment (000)	42	0.020	41	0.022	40	0.022	37	0.020	31	0.020	29	0.021
Female employees (000)	-	-	-	-	-	-	-	-	-	-	-	-
Salaries and Wages ($ bil.)	0.318	0.015	0.302	0.013	0.292	0.011	0.238	0.012	0.105	0.005	0.108	0.006
Output ($ bil.)	2	0.011	1	0.009	1	0.009	1	0.009	0.845	0.006	1	0.008
Value added ($ bil.)	0.630	0.011	0.653	0.012	0.571	0.010	0.329	0.006	0.304	0.006	0.350	0.006
Capital investment ($ mil.)	89	0.015	61	0.010	82	0.013	122	0.022	63	0.014	96	0.031
Employees per establishment (number)	337	456.6	198	292.0	221	354.9	237	336.7	210	311.6	254	362.1
Output per establishment ($ mil.)	13	317.6	8	170.5	10	174.6	9	193.0	6	112.8	9	223.5
Value added per establishment ($ mil.)	7	425.6	3	165.4	3	146.4	2	145.6	2	115.9	3	170.8
Capital investment per estab. ($ mil.)	1	435.6	0.384	130.6	0.599	208.7	1	405.4	0.409	173.9	0.824	288.2
Payroll per employee ($)	7,662	56.0	7,351	40.6	7,334	36.5	6,959	42.9	3,370	19.2	3,653	20.9
Females as % of total employment (%)	-	-	-	-	-	-	-	-	-	-	-	-
Output per employee ($)	40,219	58.6	38,706	47.6	41,964	46.9	38,665	48.3	27,036	30.9	36,380	39.3
Value added per employee ($)	15,184	46.2	15,904	47.1	14,350	39.9	9,821	30.6	9,740	27.4	11,881	30.1
Capital investment per employee ($)	3,428	62.3	1,936	31.6	2,714	43.8	4,325	75.7	1,949	29.8	3,247	43.5

Columns headed % show percent of the world total or of the world ratio. Ratios closest to 100 are closest to the world average. Values higher than 100 are above, values lower than 100 are below world average. *nec* stands for *not elsewhere classified*. Ratios, where shown, are frequently based on a subset of total data; for example, payroll per employee is averaged from all cases where *both* payroll and employment data are present. Dividing *total* salaries/wages by *total* employment will not necessarily produce a matching value.

Canada

Top Ranked Employing Sectors	Top Ranked Output Sectors
3840 Transportation Equipment	3840 Transportation Equipment
3110 Food Products	3110 Food Products
3820 Machinery nec	3820 Machinery nec
3420 Printing, Publishing	3410 Paper and Products
3810 Metal Products	3830 Electrical Machinery

Population 30,675,000. Percent of world population: 0.518

	1990		1991		1992		1993		1994		1995	
	Value	%	Value	%	Value	%	Value	%	Value	%	Value	%
Establishments or enterprises (number)	42,292	1.800	38,542	1.600	36,628	1.300	34,989	1.700	33,991	1.900	-	-
Total employment (000)	2,185	1.000	2,038	1.100	1,973	1.100	1,942	1.100	1,964	1.200	1,981	1.400
Female employees (000)	553	4.200	510	5.700	-	-	-	-	-	-	-	-
Salaries and Wages ($ bil.)	62	2.900	61	2.600	58	2.300	56	2.700	55	2.800	57	3.500
Output ($ bil.)	383	2.700	368	2.400	360	2.200	377	2.600	404	2.900	434	3.300
Value added ($ bil.)	139	2.500	129	2.300	126	2.100	128	2.100	141	2.800	148	2.700
Capital investment ($ mil.)	15,016	2.500	-	-	-	-	-	-	-	-	-	-
Employees per establishment (number)	52	70.1	53	77.9	54	86.6	56	78.7	58	85.8	-	-
Output per establishment ($ mil.)	9	213.5	10	207.9	10	180.4	11	229.2	12	236.2	-	-
Value added per establishment ($ mil.)	3	200.3	3	198.1	3	162.8	4	227.5	4	235.1	-	-
Capital investment per estab. ($ mil.)	0.406	153.2	-	-	-	-	-	-	-	-	-	-
Payroll per employee ($)	28,364	207.3	30,070	166.1	29,593	147.4	28,713	176.8	28,086	159.9	28,975	165.8
Females as % of total employment (%)	25.4	79.1	25.1	78.0	-	-	-	-	-	-	-	-
Output per employee ($)	175,085	255.1	180,428	221.7	182,326	203.8	193,967	242.3	205,605	235.3	218,908	236.5
Value added per employee ($)	63,540	193.4	63,222	187.2	63,703	176.9	65,998	206.0	71,725	202.1	74,654	189.0
Capital investment per employee ($)	8,284	150.4	-	-	-	-	-	-	-	-	-	-

Columns headed % show percent of the world total or of the world ratio. Ratios closest to 100 are closest to the world average. Values higher than 100 are above, values lower than 100 are below world average. *nec* stands for *not elsewhere classified*. Ratios, where shown, are frequently based on a subset of total data; for example, payroll per employee is averaged from all cases where *both* payroll and employment data are present. Dividing *total* salaries/wages by *total* employment will not necessarily produce a matching value.

Cape Verde

Top Ranked Employing Sectors	Top Ranked Output Sectors
3840 Transportation Equipment	3520 Chemical Products nec
3420 Printing, Publishing	3522 Drugs and Medicines
3240 Footwear	3140 Tobacco
3520 Chemical Products nec	3840 Transportation Equipment
3140 Tobacco	3420 Printing, Publishing

Population 400,000. Percent of world population: 0.007

	1990		1991		1992		1993		1994		1995	
	Value	%	Value	%	Value	%	Value	%	Value	%	Value	%
Establishments or enterprises (number)	309	0.013	307	0.013	577	0.021	649	0.032	690	0.039	-	
Total employment (000)	0.883	-	0.933	-	0.852	-	0.826	-	0.679	-	-	
Female employees (000)	0.036	-	0.033	-	0.060	-	0.061	-	0.063	-	-	
Salaries and Wages ($ bil.)	0.000	-	0.000	-	0.003	-	0.003	-	0.000	-	-	
Output ($ bil.)	0.016	-	0.011	-	0.034	-	0.027	-	0.025	-	-	
Value added ($ bil.)	-		-		-		-		-		-	
Capital investment ($ mil.)	0.025		0.011		-		-		-		-	
Employees per establishment (number)	32	43.9	26	39.0	30	48.9	28	39.1	25	37.3	-	
Output per establishment ($ mil.)	0.077	1.8	0.280	6.1	2	44.7	2	51.9	3	56.0	-	
Value added per establishment ($ mil.)	-		-		-		-		-		-	
Capital investment per estab. ($ mil.)	0.000	0.0	0.000	0.1	-		-		-		-	
Payroll per employee ($)	4,386	32.1	4,574	25.3	4,265	21.2	5,061	31.2	4,554	25.9	-	
Females as % of total employment (%)	56.3	175.5	47.1	146.8	9.9	21.4	11.4	22.7	13.8	26.3	-	
Output per employee ($)	22,699	33.1	19,235	23.6	48,137	53.8	43,981	54.9	47,576	54.5	-	
Value added per employee ($)	-		-		-		-		-		-	
Capital investment per employee ($)	14	0.3	13	0.2	-		-		-		-	

Columns headed % show percent of the world total or of the world ratio. Ratios closest to 100 are closest to the world average. Values higher than 100 are above, values lower than 100 are below world average. nec stands for not elsewhere classified. Ratios, where shown, are frequently based on a subset of total data; for example, payroll per employee is averaged from all cases where both payroll and employment data are present. Dividing total salaries/wages by total employment will not necessarily produce a matching value.

Central African Republic

Top Ranked Employing Sectors	Top Ranked Output Sectors
3310 Wood Products	3310 Wood Products
3110 Food Products	3110 Food Products
3140 Tobacco	3140 Tobacco
3130 Beverages	3130 Beverages
3520 Chemical Products nec	3520 Chemical Products nec

Population 3,376,000. Percent of world population: 0.057

	1990		1991		1992		1993		1994		1995	
	Value	%	Value	%	Value	%	Value	%	Value	%	Value	%
Establishments or enterprises (number)	5	-	56	0.002	54	0.002	52	0.003	48	0.003	37	0.003
Total employment (000)	4	0.002	0.136	-	0.070	-	0.101	-	-		-	
Female employees (000)	-		-		-		-		-		-	
Salaries and Wages ($ bil.)	0.018	0.001	0.001	-	0.001	-	0.001	-	-		-	
Output ($ bil.)	0.132	0.001	0.003	-	0.001	-	0.002	-	-		-	
Value added ($ bil.)	0.058	0.001	0.001	-	-0.000	-	0.001	-	-		-	
Capital investment ($ mil.)	-2.123	-	0.046	-	-		0.226	-	-		-	
Employees per establishment (number)	28	37.7	27	40.1	14	22.5	20	28.7	-		-	
Output per establishment ($ mil.)	0.529	12.5	0.541	11.8	0.244	4.5	0.366	7.8	-		-	
Value added per establishment ($ mil.)	0.206	12.6	0.220	13.0	-		0.100	6.2	-		-	
Capital investment per estab. ($ mil.)	0.018	6.7	0.009	3.1	-		0.045	17.8	-		-	
Payroll per employee ($)	4,626	33.8	7,715	42.6	8,150	40.6	6,888	42.4	-		-	
Females as % of total employment (%)	-		-		-		-		-		-	
Output per employee ($)	33,594	48.9	19,887	24.4	17,433	19.5	18,112	22.6	-		-	
Value added per employee ($)	14,741	44.9	8,080	23.9	6,692	18.6	9,091	28.4	-		-	
Capital investment per employee ($)	634	11.5	339	5.5	-		2,238	39.2	-		-	

Columns headed % show percent of the world total or of the world ratio. Ratios closest to 100 are closest to the world average. Values higher than 100 are above, values lower than 100 are below world average. nec stands for not elsewhere classified. Ratios, where shown, are frequently based on a subset of total data; for example, payroll per employee is averaged from all cases where both payroll and employment data are present. Dividing total salaries/wages by total employment will not necessarily produce a matching value.

Chile

Top Ranked Employing Sectors	Top Ranked Output Sectors
3110 Food Products	3110 Food Products
3810 Metal Products	3720 Nonferrous metals
3210 Textiles	3530 Petroleum Refineries
3310 Wood Products	3410 Paper and Products
3520 Chemical Products nec	3520 Chemical Products nec

Population 14,788,000. Percent of world population: 0.249

	1990		1991		1992		1993		1994		1995	
	Value	%	Value	%	Value	%	Value	%	Value	%	Value	%
Establishments or enterprises (number)	1,829	0.079	1,899	0.080	1,955	0.070	2,004	0.097	2,025	0.113	-	
Total employment (000)	336	0.161	347	0.186	369	0.202	373	0.205	375	0.236	372	0.262
Female employees (000)	66	0.498	71	0.793	75	0.347	73	0.329	76	0.358	-	
Salaries and Wages ($ bil.)	2	0.077	2	0.081	2	0.094	3	0.126	3	0.153	3	0.209
Output ($ bil.)	24	0.166	26	0.168	30	0.185	31	0.217	35	0.251	40	0.301
Value added ($ bil.)	10	0.178	11	0.192	13	0.218	13	0.223	15	0.293	18	0.320
Capital investment ($ mil.)	762	0.128	1,782	0.286	1,459	0.228	1,240	0.228	1,224	0.276	-	
Employees per establishment (number)	184	249.5	183	269.1	189	303.9	186	264.0	185	275.3	-	
Output per establishment ($ mil.)	13	305.9	14	297.0	16	285.7	16	334.1	17	340.1	-	
Value added per establishment ($ mil.)	5	332.8	6	340.2	7	316.5	7	413.2	7	420.8	-	
Capital investment per estab. ($ mil.)	0.460	173.7	1	352.3	0.824	287.2	0.677	267.4	0.664	282.2	-	
Payroll per employee ($)	4,945	36.1	5,585	30.9	6,490	32.3	6,912	42.6	7,940	45.2	9,321	53.3
Females as % of total employment (%)	19.7	61.5	20.6	64.0	20.2	43.7	19.7	39.4	20.1	38.4	-	
Output per employee ($)	70,454	102.7	74,649	91.7	82,337	92.0	84,311	105.3	92,253	105.6	107,982	116.7
Value added per employee ($)	29,651	90.3	31,449	93.1	35,299	98.0	35,742	111.5	40,012	112.7	47,216	119.5
Capital investment per employee ($)	2,554	46.4	5,797	94.5	4,467	72.1	3,718	65.0	3,647	55.8	-	

Columns headed % show percent of the world total or of the world ratio. Ratios closest to 100 are closest to the world average. Values higher than 100 are above, values lower than 100 are below world average. *nec* stands for *not elsewhere classified*. Ratios, where shown, are frequently based on a subset of total data; for example, payroll per employee is averaged from all cases where *both* payroll and employment data are present. Dividing *total* salaries/wages by *total* employment will not necessarily produce a matching value.

China

Top Ranked Employing Sectors	Top Ranked Output Sectors
3820 Machinery nec	3830 Electrical Machinery
3210 Textiles	3210 Textiles
3830 Electrical Machinery	3820 Machinery nec
3510 Industrial Chemicals	3110 Food Products
3840 Transportation Equipment	3710 Iron and Steel

Population 1,236,915,000. Percent of world population: 20.871

	1990		1991		1992		1993		1994		1995	
	Value	%	Value	%	Value	%	Value	%	Value	%	Value	%
Establishments or enterprises (number)	436,173	18.900	435,409	18.300	432,829	15.400	303,264	14.800	322,501	18.000	341,780	27.900
Total employment (000)	49,380	23.600	52,710	28.200	53,360	29.200	50,660	27.800	42,413	26.600	42,090	29.600
Female employees (000)	-		-		-		-		-		-	
Salaries and Wages ($ bil.)	-		-		-		-		-		-	
Output ($ bil.)	341	2.400	361	2.400	440	2.700	597	4.100	401	2.900	435	3.300
Value added ($ bil.)	88	1.600	92	1.600	113	1.900	182	3.100	108	2.100	107	1.900
Capital investment ($ mil.)	-		-		-		-		-		-	
Employees per establishment (number)	117	158.7	121	178.4	123	198.3	135	190.8	135	200.9	123	175.7
Output per establishment ($ mil.)	0.782	18.5	0.830	18.1	1	18.7	2	33.0	1	25.5	1	30.8
Value added per establishment ($ mil.)	0.202	12.3	0.211	12.5	0.261	12.4	0.469	29.1	0.344	19.5	0.313	17.7
Capital investment per estab. ($ mil.)	-		-		-		-		-		-	
Payroll per employee ($)	-		-		-		-		-		-	
Females as % of total employment (%)	-		-		-		-		-		-	
Output per employee ($)	6,602	9.6	6,855	8.4	8,241	9.2	11,788	14.7	9,481	10.9	10,331	11.2
Value added per employee ($)	1,706	5.2	1,746	5.2	2,115	5.9	3,597	11.2	2,542	7.2	2,542	6.4
Capital investment per employee ($)	-		-		-		-		-		-	

Columns headed % show percent of the world total or of the world ratio. Ratios closest to 100 are closest to the world average. Values higher than 100 are above, values lower than 100 are below world average. *nec* stands for *not elsewhere classified*. Ratios, where shown, are frequently based on a subset of total data; for example, payroll per employee is averaged from all cases where *both* payroll and employment data are present. Dividing *total* salaries/wages by *total* employment will not necessarily produce a matching value.

China (Hong Kong SAR)

Top Ranked Employing Sectors	Top Ranked Output Sectors
3220 Wearing Apparel	3220 Wearing Apparel
3210 Textiles	3210 Textiles
3420 Printing, Publishing	3830 Electrical Machinery
3830 Electrical Machinery	3820 Machinery nec
3820 Machinery nec	3420 Printing, Publishing

Population 6,707,000. Percent of world population: 0.113

	1990		1991		1992		1993		1994		1995	
	Value	%	Value	%	Value	%	Value	%	Value	%	Value	%
Establishments or enterprises (number)	56,267	2.400	47,530	2.000	45,352	1.600	37,534	1.800	35,043	2.000	-	-
Total employment (000)	900	0.430	766	0.410	699	0.383	598	0.328	514	0.323	-	-
Female employees (000)	-	-	-	-	-	-	-	-	-	-	-	-
Salaries and Wages ($ bil.)	8	0.370	8	0.326	8	0.310	8	0.368	7	0.367	-	-
Output ($ bil.)	52	0.362	52	0.336	53	0.325	49	0.340	47	0.340	-	-
Value added ($ bil.)	15	0.260	15	0.262	16	0.264	15	0.243	14	0.276	-	-
Capital investment ($ mil.)	1,548	0.261	1,469	0.236	2,103	0.329	1,794	0.330	1,368	0.308	-	-
Employees per establishment (number)	16	21.7	16	23.8	15	24.8	16	22.6	15	21.8	-	-
Output per establishment ($ mil.)	0.920	21.7	1	23.7	1	21.6	1	28.5	1	27.1	-	-
Value added per establishment ($ mil.)	0.259	15.8	0.313	18.5	0.349	16.6	0.394	24.5	0.411	23.3	-	-
Capital investment per estab. ($ mil.)	0.030	11.5	0.033	11.4	0.047	16.5	0.049	19.2	0.040	16.9	-	-
Payroll per employee ($)	8,854	64.7	10,116	55.9	11,365	56.6	12,726	78.4	14,154	80.6	-	-
Females as % of total employment (%)	-	-	-	-	-	-	-	-	-	-	-	-
Output per employee ($)	57,491	83.8	67,384	82.8	76,412	85.4	83,422	104.2	92,649	106.0	-	-
Value added per employee ($)	16,197	49.3	19,399	57.4	22,646	62.9	24,581	76.7	27,941	78.7	-	-
Capital investment per employee ($)	2,055	37.3	2,256	36.8	3,055	49.3	3,037	53.1	2,703	41.3	-	-

Columns headed % show percent of the world total or of the world ratio. Ratios closest to 100 are closest to the world average. Values higher than 100 are above, values lower than 100 are below world average. *nec* stands for *not elsewhere classified*. Ratios, where shown, are frequently based on a subset of total data; for example, payroll per employee is averaged from all cases where *both* payroll and employment data are present. Dividing *total* salaries/wages by *total* employment will not necessarily produce a matching value.

China (Taiwan Province)

Top Ranked Employing Sectors	Top Ranked Output Sectors
3830 Electrical Machinery	3830 Electrical Machinery
3810 Metal Products	3510 Industrial Chemicals
3210 Textiles	3110 Food Products
3820 Machinery nec	3840 Transportation Equipment
3840 Transportation Equipment	3710 Iron and Steel

Population 21,908,000. Percent of world population: 0.370

	1990		1991		1992		1993		1994		1995	
	Value	%	Value	%	Value	%	Value	%	Value	%	Value	%
Establishments or enterprises (number)	-	-	-	-	-	-	-	-	-	-	-	-
Total employment (000)	2,445	1.200	2,400	1.300	2,410	1.300	2,397	1.300	2,439	1.500	2,400	1.700
Female employees (000)	1,104	8.300	1,077	12.000	1,066	5.000	1,044	4.700	1,049	5.000	1,024	5.400
Salaries and Wages ($ bil.)	24	1.100	26	1.100	31	1.200	31	1.500	34	1.700	35	2.100
Output ($ bil.)	169	1.200	183	1.200	197	1.200	204	1.400	218	1.600	246	1.800
Value added ($ bil.)	53	0.953	60	1.100	67	1.100	68	1.100	70	1.400	73	1.300
Capital investment ($ mil.)	-	-	-	-	-	-	-	-	-	-	-	-
Employees per establishment (number)	-	-	-	-	-	-	-	-	-	-	-	-
Output per establishment ($ mil.)	-	-	-	-	-	-	-	-	-	-	-	-
Value added per establishment ($ mil.)	-	-	-	-	-	-	-	-	-	-	-	-
Capital investment per estab. ($ mil.)	-	-	-	-	-	-	-	-	-	-	-	-
Payroll per employee ($)	9,838	71.9	10,950	60.5	12,862	64.1	13,110	80.7	13,937	79.4	14,698	84.1
Females as % of total employment (%)	45.1	140.9	44.9	139.7	44.2	95.6	43.6	87.2	43.0	82.0	42.7	73.1
Output per employee ($)	69,237	100.9	76,145	93.5	81,558	91.2	85,030	106.2	89,496	102.4	102,471	110.7
Value added per employee ($)	21,817	66.4	24,925	73.8	27,880	77.4	28,305	88.3	28,662	80.7	30,509	77.2
Capital investment per employee ($)	-	-	-	-	-	-	-	-	-	-	-	-

Columns headed % show percent of the world total or of the world ratio. Ratios closest to 100 are closest to the world average. Values higher than 100 are above, values lower than 100 are below world average. *nec* stands for *not elsewhere classified*. Ratios, where shown, are frequently based on a subset of total data; for example, payroll per employee is averaged from all cases where *both* payroll and employment data are present. Dividing *total* salaries/wages by *total* employment will not necessarily produce a matching value.

For additional notes and sources, see Appendix I.

Colombia

Top Ranked Employing Sectors	Top Ranked Output Sectors
3110 Food Products	3110 Food Products
3220 Wearing Apparel	3520 Chemical Products nec
3210 Textiles	3130 Beverages
3520 Chemical Products nec	3210 Textiles
3810 Metal Products	3510 Industrial Chemicals

Population 38,581,000. Percent of world population: 0.651

	1990		1991		1992		1993		1994		1995	
	Value	%	Value	%	Value	%	Value	%	Value	%	Value	%
Establishments or enterprises (number)	8,279	0.359	8,047	0.338	8,781	0.313	8,441	0.411	8,235	0.460	-	-
Total employment (000)	573	0.274	575	0.308	677	0.371	681	0.374	679	0.426	675	0.474
Female employees (000)	-	-	176	2.000	-	-	231	1.000	231	1.100	-	-
Salaries and Wages ($ bil.)	1	0.065	1	0.061	2	0.072	2	0.105	3	0.139	3	0.187
Output ($ bil.)	25	0.174	25	0.159	25	0.151	27	0.188	34	0.250	40	0.297
Value added ($ bil.)	10	0.172	10	0.173	11	0.176	11	0.178	15	0.288	17	0.309
Capital investment ($ mil.)	1,054	0.178	918	0.147	1,351	0.211	905	0.167	515	0.116	-	-
Employees per establishment (number)	69	93.9	71	105.3	77	124.0	81	114.5	82	122.4	-	-
Output per establishment ($ mil.)	3	71.1	3	66.4	3	52.1	3	68.6	4	83.1	-	-
Value added per establishment ($ mil.)	1	71.1	1	72.1	1	56.9	1	78.4	2	101.5	-	-
Capital investment per estab. ($ mil.)	0.140	52.8	0.114	38.8	0.154	53.6	0.107	42.3	0.062	26.6	-	-
Payroll per employee ($)	2,440	17.8	2,503	13.8	2,719	13.5	3,135	19.3	3,999	22.8	4,584	26.2
Females as % of total employment (%)	-	-	30.7	95.6	-	-	33.8	67.8	34.1	64.9	-	-
Output per employee ($)	43,481	63.4	42,645	52.4	36,787	41.1	39,924	49.9	50,702	58.0	58,733	63.5
Value added per employee ($)	16,841	51.3	17,044	50.5	15,548	43.2	15,636	48.8	21,707	61.2	25,120	63.6
Capital investment per employee ($)	2,157	39.2	1,598	26.1	1,995	32.2	1,329	23.3	758	11.6	-	-

Columns headed % show percent of the world total or of the world ratio. Ratios closest to 100 are closest to the world average. Values higher than 100 are above, values lower than 100 are below world average. *nec* stands for *not elsewhere classified*. Ratios, where shown, are frequently based on a subset of total data; for example, payroll per employee is averaged from all cases where *both* payroll and employment data are present. Dividing *total* salaries/wages by *total* employment will not necessarily produce a matching value.

Costa Rica

Top Ranked Employing Sectors	Top Ranked Output Sectors
3110 Food Products	3110 Food Products
3220 Wearing Apparel	3130 Beverages
3210 Textiles	3520 Chemical Products nec
3830 Electrical Machinery	3510 Industrial Chemicals
3520 Chemical Products nec	3830 Electrical Machinery

Population 3,605,000. Percent of world population: 0.061

	1990		1991		1992		1993		1994		1995	
	Value	%	Value	%	Value	%	Value	%	Value	%	Value	%
Establishments or enterprises (number)	4,911	0.213	4,927	0.207	4,920	0.175	5,126	0.249	-	-	-	-
Total employment (000)	143	0.068	142	0.076	154	0.084	161	0.088	-	-	-	-
Female employees (000)	-	-	-	-	-	-	-	-	-	-	-	-
Salaries and Wages ($ bil.)	0.400	0.019	0.376	0.016	0.469	0.018	0.527	0.026	-	-	-	-
Output ($ bil.)	3	0.024	3	0.022	4	0.025	5	0.031	5	0.035	5	0.041
Value added ($ bil.)	1	0.018	1	0.018	1	0.021	1	0.022	1	0.028	2	0.028
Capital investment ($ mil.)	-	-	-	-	-	-	-	-	-	-	-	-
Employees per establishment (number)	29	39.4	29	42.5	31	50.2	31	44.6	-	-	-	-
Output per establishment ($ mil.)	0.696	16.4	0.702	15.3	0.832	15.3	0.896	19.1	-	-	-	-
Value added per establishment ($ mil.)	0.210	12.8	0.214	12.7	0.259	12.3	0.265	16.5	-	-	-	-
Capital investment per estab. ($ mil.)	-	-	-	-	-	-	-	-	-	-	-	-
Payroll per employee ($)	2,802	20.5	2,647	14.6	3,053	15.2	3,270	20.1	-	-	-	-
Females as % of total employment (%)	-	-	-	-	-	-	-	-	-	-	-	-
Output per employee ($)	23,960	34.9	24,207	29.7	26,534	29.7	28,467	35.6	-	-	-	-
Value added per employee ($)	7,222	22.0	7,399	21.9	8,254	22.9	8,429	26.3	-	-	-	-
Capital investment per employee ($)	-	-	-	-	-	-	-	-	-	-	-	-

Columns headed % show percent of the world total or of the world ratio. Ratios closest to 100 are closest to the world average. Values higher than 100 are above, values lower than 100 are below world average. *nec* stands for *not elsewhere classified*. Ratios, where shown, are frequently based on a subset of total data; for example, payroll per employee is averaged from all cases where *both* payroll and employment data are present. Dividing *total* salaries/wages by *total* employment will not necessarily produce a matching value.

Cote d'Ivoire

Top Ranked Employing Sectors	Top Ranked Output Sectors
	3690 Nonmetal Products nec
	3530 Petroleum Refineries
	3310 Wood Products
	3210 Textiles
	3840 Transportation Equipment

Population 15,446,000. Percent of world population: 0.261

	1990		1991		1992		1993		1994		1995	
	Value	%	Value	%	Value	%	Value	%	Value	%	Value	%
Establishments or enterprises (number)	-	-	-	-	-	-	-	-	-	-	-	-
Total employment (000)	-	-	-	-	-	-	-	-	-	-	-	-
Female employees (000)	-	-	-	-	-	-	-	-	-	-	-	-
Salaries and Wages ($ bil.)	-	-	-	-	-	-	-	-	-	-	-	-
Output ($ bil.)	3	0.021	3	0.017	3	0.017	3	0.019	-	-	-	-
Value added ($ bil.)	7	0.128	7	0.120	7	0.112	6	0.107	-	-	-	-
Capital investment ($ mil.)	915	0.154	897	0.144	941	0.147	816	0.150	-	-	-	-
Employees per establishment (number)	-	-	-	-	-	-	-	-	-	-	-	-
Output per establishment ($ mil.)	-	-	-	-	-	-	-	-	-	-	-	-
Value added per establishment ($ mil.)	-	-	-	-	-	-	-	-	-	-	-	-
Capital investment per estab. ($ mil.)	-	-	-	-	-	-	-	-	-	-	-	-
Payroll per employee ($)	-	-	-	-	-	-	-	-	-	-	-	-
Females as % of total employment (%)	-	-	-	-	-	-	-	-	-	-	-	-
Output per employee ($)	-	-	-	-	-	-	-	-	-	-	-	-
Value added per employee ($)	-	-	-	-	-	-	-	-	-	-	-	-
Capital investment per employee ($)	-	-	-	-	-	-	-	-	-	-	-	-

Columns headed % show percent of the world total or of the world ratio. Ratios closest to 100 are closest to the world average. Values higher than 100 are above, values lower than 100 are below world average. *nec* stands for *not elsewhere classified*. Ratios, where shown, are frequently based on a subset of total data; for example, payroll per employee is averaged from all cases where *both* payroll and employment data are present. Dividing *total* salaries/wages by *total* employment will not necessarily produce a matching value.

Croatia

Top Ranked Employing Sectors	Top Ranked Output Sectors
3110 Food Products	3110 Food Products
3220 Wearing Apparel	3840 Transportation Equipment
3210 Textiles	3510 Industrial Chemicals
3810 Metal Products	3520 Chemical Products nec
3840 Transportation Equipment	3830 Electrical Machinery

Population 4,672,000. Percent of world population: 0.079

	1990		1991		1992		1993		1994		1995	
	Value	%	Value	%	Value	%	Value	%	Value	%	Value	%
Establishments or enterprises (number)	4,900	0.213	6,507	0.273	8,875	0.316	11,918	0.580	14,839	0.829	16,472	1.300
Total employment (000)	546	0.261	449	0.240	389	0.213	371	0.204	354	0.222	338	0.238
Female employees (000)	227	1.700	187	2.100	163	0.757	157	0.702	151	0.715	145	0.760
Salaries and Wages ($ bil.)	2	0.113	2	0.088	0.611	0.024	0.644	0.032	0.840	0.043	1	0.078
Output ($ bil.)	17	0.120	16	0.101	9	0.055	-	-	-	-	-	-
Value added ($ bil.)	7	0.122	6	0.111	4	0.071	-	-	-	-	-	-
Capital investment ($ mil.)	394	0.066	457	0.073	287	0.045	222	0.041	251	0.057	265	0.085
Employees per establishment (number)	112	151.3	69	101.6	44	70.6	31	44.1	24	35.4	21	29.3
Output per establishment ($ mil.)	4	82.9	2	52.2	1	18.6	-	-	-	-	-	-
Value added per establishment ($ mil.)	1	85.2	0.963	57.1	0.476	22.6	-	-	-	-	-	-
Capital investment per estab. ($ mil.)	0.081	30.4	0.070	23.9	0.032	11.3	0.019	7.4	0.017	7.4	0.016	5.7
Payroll per employee ($)	4,449	32.5	4,641	25.6	1,569	7.8	1,737	10.7	2,374	13.5	3,812	21.8
Females as % of total employment (%)	41.5	129.4	41.6	129.6	41.8	90.4	42.3	84.6	42.7	81.4	42.9	73.5
Output per employee ($)	31,488	45.9	34,733	42.7	23,100	25.8	-	-	-	-	-	-
Value added per employee ($)	12,515	38.1	13,972	41.4	10,839	30.1	-	-	-	-	-	-
Capital investment per employee ($)	722	13.1	1,018	16.6	736	11.9	600	10.5	730	11.2	789	10.6

Columns headed % show percent of the world total or of the world ratio. Ratios closest to 100 are closest to the world average. Values higher than 100 are above, values lower than 100 are below world average. *nec* stands for *not elsewhere classified*. Ratios, where shown, are frequently based on a subset of total data; for example, payroll per employee is averaged from all cases where *both* payroll and employment data are present. Dividing *total* salaries/wages by *total* employment will not necessarily produce a matching value.

Cyprus

Top Ranked Employing Sectors	Top Ranked Output Sectors
3110 Food Products	3110 Food Products
3220 Wearing Apparel	3690 Nonmetal Products nec
3810 Metal Products	3220 Wearing Apparel
3690 Nonmetal Products nec	3810 Metal Products
3310 Wood Products	3130 Beverages

Population 749,000. Percent of world population: 0.013

	1990		1991		1992		1993		1994		1995	
	Value	%	Value	%	Value	%	Value	%	Value	%	Value	%
Establishments or enterprises (number)	7,330	0.318	7,295	0.306	7,279	0.259	7,185	0.350	7,679	0.429	7,987	0.651
Total employment (000)	44	0.021	44	0.024	44	0.024	42	0.023	41	0.026	41	0.029
Female employees (000)	23	0.170	22	0.247	23	0.105	20	0.088	19	0.088		
Salaries and Wages ($ bil.)	0.382	0.018	0.403	0.017	0.457	0.018	0.427	0.021	0.452	0.023	0.591	0.036
Output ($ bil.)	2	0.016	2	0.015	3	0.015	2	0.015	2	0.017	3	0.021
Value added ($ bil.)	0.814	0.015	0.845	0.015	0.951	0.016	0.867	0.015	0.930	0.018	1	0.019
Capital investment ($ mil.)	115	0.019	116	0.019	127	0.020	104	0.019	106	0.024	122	0.039
Employees per establishment (number)	6	8.2	6	8.9	6	9.8	6	8.3	5	7.9	5	7.3
Output per establishment ($ mil.)	0.308	7.3	0.322	7.0	0.348	6.4	0.311	6.6	0.310	6.2	0.353	8.5
Value added per establishment ($ mil.)	0.111	6.8	0.116	6.9	0.131	6.2	0.121	7.5	0.121	6.9	0.133	7.5
Capital investment per estab. ($ mil.)	0.016	5.9	0.016	5.4	0.017	6.1	0.015	5.7	0.014	5.9	0.015	5.4
Payroll per employee ($)	8,617	63.0	9,188	50.8	10,271	51.2	10,157	62.5	11,083	63.1	14,562	83.3
Females as % of total employment (%)	50.9	158.9	50.6	157.6	50.7	109.5	46.9	93.9	45.8	87.3		
Output per employee ($)	50,833	74.1	53,591	65.8	57,009	63.7	53,166	66.4	58,490	66.9	69,376	75.0
Value added per employee ($)	18,344	55.8	19,243	57.0	21,394	59.4	20,618	64.3	22,832	64.3	26,186	66.3
Capital investment per employee ($)	2,584	46.9	2,634	43.0	2,852	46.0	2,482	43.4	2,612	39.9	3,012	40.3

Columns headed % show percent of the world total or of the world ratio. Ratios closest to 100 are closest to the world average. Values higher than 100 are above, values lower than 100 are below world average. *nec* stands for *not elsewhere classified*. Ratios, where shown, are frequently based on a subset of total data; for example, payroll per employee is averaged from all cases where *both* payroll and employment data are present. Dividing *total* salaries/wages by *total* employment will not necessarily produce a matching value.

Czech Republic

Top Ranked Employing Sectors	Top Ranked Output Sectors
3820 Machinery nec	
3840 Transportation Equipment	
3710 Iron and Steel	
3210 Textiles	
3110 Food Products	

Population 10,286,000. Percent of world population: 0.174

	1990		1991		1992		1993		1994		1995	
	Value	%	Value	%	Value	%	Value	%	Value	%	Value	%
Establishments or enterprises (number)	-	-	1,177	0.049	2,257	0.080	2,934	0.143	-	-	-	-
Total employment (000)	1,577	0.754	1,391	0.745	1,287	0.705	1,202	0.660	-	-	-	-
Female employees (000)	674	5.100	594	6.600	542	2.500	506	2.300	-	-	-	-
Salaries and Wages ($ bil.)	4	0.163	2	0.092	2	0.098	3	0.139	-	-	-	-
Output ($ bil.)	-	-	-	-	-	-	-	-	-	-	-	-
Value added ($ bil.)	-	-	-	-	-	-	-	-	-	-	-	-
Capital investment ($ mil.)	-	-	-	-	-	-	-	-	-	-	-	-
Employees per establishment (number)	-	-	1,182	1,741.9	570	917.1	410	581.1	-	-	-	-
Output per establishment ($ mil.)	-	-	-	-	-	-	-	-	-	-	-	-
Value added per establishment ($ mil.)	-	-	-	-	-	-	-	-	-	-	-	-
Capital investment per estab. ($ mil.)	-	-	-	-	-	-	-	-	-	-	-	-
Payroll per employee ($)	2,223	16.2	1,580	8.7	1,939	9.7	2,353	14.5	-	-	-	-
Females as % of total employment (%)	42.7	133.4	42.7	133.0	42.1	91.0	42.1	84.3	-	-	-	-
Output per employee ($)	-	-	-	-	-	-	-	-	-	-	-	-
Value added per employee ($)	-	-	-	-	-	-	-	-	-	-	-	-
Capital investment per employee ($)	-	-	-	-	-	-	-	-	-	-	-	-

Columns headed % show percent of the world total or of the world ratio. Ratios closest to 100 are closest to the world average. Values higher than 100 are above, values lower than 100 are below world average. *nec* stands for *not elsewhere classified*. Ratios, where shown, are frequently based on a subset of total data; for example, payroll per employee is averaged from all cases where *both* payroll and employment data are present. Dividing *total* salaries/wages by *total* employment will not necessarily produce a matching value.

Denmark

Top Ranked Employing Sectors	Top Ranked Output Sectors
3110 Food Products	3110 Food Products
3820 Machinery nec	3820 Machinery nec
3420 Printing, Publishing	3810 Metal Products
3810 Metal Products	3840 Transportation Equipment
3840 Transportation Equipment	3520 Chemical Products nec

Population 5,334,000. Percent of world population: 0.090

	1990		1991		1992		1993		1994		1995	
	Value	%	Value	%	Value	%	Value	%	Value	%	Value	%
Establishments or enterprises (number)	27,359	1.200	28,161	1.200	27,689	0.986	-	-	-	-	-	-
Total employment (000)	576	0.275	562	0.301	547	0.300	472	0.259	498	0.313	507	0.356
Female employees (000)	193	1.400	187	2.100	182	0.846	-	-	-	-	-	-
Salaries and Wages ($ bil.)	17	0.778	17	0.696	18	0.704	14	0.699	17	0.862	20	1.200
Output ($ bil.)	61	0.425	60	0.394	59	0.357	54	0.373	64	0.464	77	0.574
Value added ($ bil.)	27	0.485	28	0.492	27	0.443	24	0.409	29	0.567	35	0.639
Capital investment ($ mil.)	2,861	0.482	3,149	0.505	-	-	-	-	-	-	-	-
Employees per establishment (number)	21	28.6	20	29.4	20	31.8	-	-	-	-	-	-
Output per establishment ($ mil.)	2	52.4	2	46.8	2	41.4	-	-	-	-	-	-
Value added per establishment ($ mil.)	0.993	60.6	0.990	58.6	1	48.4	-	-	-	-	-	-
Capital investment per estab. ($ mil.)	0.105	39.7	0.113	38.3	-	-	-	-	-	-	-	-
Payroll per employee ($)	29,077	212.5	29,445	162.7	32,852	163.6	30,197	186.0	33,759	192.3	40,052	229.2
Females as % of total employment (%)	33.4	104.3	33.3	103.7	33.3	71.9	-	-	-	-	-	-
Output per employee ($)	105,484	153.7	107,696	132.3	121,230	135.5	114,384	142.9	128,513	147.1	151,075	163.2
Value added per employee ($)	47,159	143.6	49,625	146.9	54,879	152.4	51,714	161.4	58,324	164.3	69,257	175.4
Capital investment per employee ($)	5,004	90.9	5,650	92.1	-	-	-	-	-	-	-	-

Columns headed % show percent of the world total or of the world ratio. Ratios closest to 100 are closest to the world average. Values higher than 100 are above, values lower than 100 are below world average. *nec* stands for *not elsewhere classified*. Ratios, where shown, are frequently based on a subset of total data; for example, payroll per employee is averaged from all cases where *both* payroll and employment data are present. Dividing *total* salaries/wages by *total* employment will not necessarily produce a matching value.

Ecuador

Top Ranked Employing Sectors	Top Ranked Output Sectors
3110 Food Products	3110 Food Products
3210 Textiles	3530 Petroleum Refineries
3130 Beverages	3410 Paper and Products
3520 Chemical Products nec	3840 Transportation Equipment
3810 Metal Products	3520 Chemical Products nec

Population 12,337,000. Percent of world population: 0.208

	1990		1991		1992		1993		1994		1995	
	Value	%	Value	%	Value	%	Value	%	Value	%	Value	%
Establishments or enterprises (number)	1,705	0.074	1,945	0.082	1,948	0.069	1,916	0.093	1,943	0.109	1,951	0.159
Total employment (000)	131	0.063	144	0.077	145	0.080	141	0.078	143	0.090	132	0.093
Female employees (000)	-	-	-	-	-	-	-	-	-	-	-	-
Salaries and Wages ($ bil.)	0.393	0.018	0.466	0.020	0.502	0.020	0.582	0.029	0.331	0.017	0.369	0.022
Output ($ bil.)	4	0.031	5	0.036	6	0.035	6	0.044	10	0.070	8	0.061
Value added ($ bil.)	1	0.024	2	0.029	2	0.031	2	0.036	4	0.081	3	0.049
Capital investment ($ mil.)	487	0.082	943	0.151	794	0.124	-302.642	-0.056	609	0.137	663	0.212
Employees per establishment (number)	76	102.5	73	107.5	74	118.3	72	102.2	72	106.9	66	94.6
Output per establishment ($ mil.)	2	53.7	2	54.2	3	46.0	3	59.9	4	82.6	4	85.1
Value added per establishment ($ mil.)	0.564	34.4	0.627	37.2	0.626	29.7	0.783	48.7	2	92.6	0.879	49.8
Capital investment per estab. ($ mil.)	0.255	96.2	0.373	126.8	0.392	136.7	0.178	70.2	0.256	109.0	0.222	77.8
Payroll per employee ($)	2,993	21.9	3,241	17.9	3,450	17.2	4,119	25.4	2,322	13.2	2,786	15.9
Females as % of total employment (%)	-	-	-	-	-	-	-	-	-	-	-	-
Output per employee ($)	34,044	49.6	38,169	46.9	39,360	44.0	45,034	56.2	67,391	77.1	61,989	67.0
Value added per employee ($)	10,186	31.0	11,563	34.2	12,915	35.9	15,221	47.5	29,232	82.4	20,170	51.1
Capital investment per employee ($)	4,369	79.3	6,555	106.9	5,461	88.1	-	-	4,275	65.4	5,013	67.1

Columns headed % show percent of the world total or of the world ratio. Ratios closest to 100 are closest to the world average. Values higher than 100 are above, values lower than 100 are below world average. *nec* stands for *not elsewhere classified*. Ratios, where shown, are frequently based on a subset of total data; for example, payroll per employee is averaged from all cases where *both* payroll and employment data are present. Dividing *total* salaries/wages by *total* employment will not necessarily produce a matching value.

For additional notes and sources, see Appendix I.

Egypt

Top Ranked Employing Sectors	Top Ranked Output Sectors
3210 Textiles	3110 Food Products
3110 Food Products	3530 Petroleum Refineries
3690 Nonmetal Products nec	3210 Textiles
3520 Chemical Products nec	3520 Chemical Products nec
3710 Iron and Steel	3710 Iron and Steel

Population 66,050,000. Percent of world population: 1.114

	1990		1991		1992		1993		1994		1995	
	Value	%	Value	%	Value	%	Value	%	Value	%	Value	%
Establishments or enterprises (number)	9,088	0.394	8,864	0.372	9,156	0.326	8,987	0.437	-	-	-	-
Total employment (000)	1,439	0.688	1,473	0.789	1,437	0.788	1,393	0.765	1,057	0.664	1,081	0.760
Female employees (000)	-		158	1.800	173	0.805	178	0.796				
Salaries and Wages ($ bil.)	3	0.133	2	0.083	2	0.080	2	0.106	2	0.089	2	0.115
Output ($ bil.)	26	0.178	18	0.117	20	0.122	22	0.149	19	0.141	22	0.166
Value added ($ bil.)	7	0.134	5	0.095	6	0.106	7	0.111	6	0.117	7	0.121
Capital investment ($ mil.)	1,650	0.278	1,380	0.221	1,232	0.193	1,040	0.191	-		-	
Employees per establishment (number)	158	214.8	166	244.9	157	252.5	155	219.8	-		-	
Output per establishment ($ mil.)	3	66.3	2	44.2	2	40.4	2	51.2	-		-	
Value added per establishment ($ mil.)	0.823	50.2	0.608	36.0	0.695	33.0	0.739	45.9	-		-	
Capital investment per estab. ($ mil.)	0.200	75.4	0.172	58.7	0.148	51.5	0.127	50.3	-		-	
Payroll per employee ($)	1,983	14.5	1,339	7.4	1,416	7.1	1,547	9.5	1,640	9.3	1,771	10.1
Females as % of total employment (%)	-		10.7	33.5	12.0	26.0	12.7	25.5				
Output per employee ($)	17,725	25.8	12,216	15.0	14,003	15.7	15,530	19.4	18,404	21.1	20,459	22.1
Value added per employee ($)	5,198	15.8	3,660	10.8	4,425	12.3	4,768	14.9	5,773	16.3	6,263	15.9
Capital investment per employee ($)	1,532	27.8	1,220	19.9	1,139	18.4	989	17.3				

Columns headed % show percent of the world total or of the world ratio. Ratios closest to 100 are closest to the world average. Values higher than 100 are above, values lower than 100 are below world average. *nec* stands for *not elsewhere classified*. Ratios, where shown, are frequently based on a subset of total data; for example, payroll per employee is averaged from all cases where *both* payroll and employment data are present. Dividing *total* salaries/wages by *total* employment will not necessarily produce a matching value.

El Salvador

Top Ranked Employing Sectors	Top Ranked Output Sectors
3220 Wearing Apparel	3110 Food Products
3210 Textiles	3130 Beverages
3110 Food Products	3210 Textiles
3520 Chemical Products nec	3530 Petroleum Refineries
3130 Beverages	3690 Nonmetal Products nec

Population 5,752,000. Percent of world population: 0.097

	1990		1991		1992		1993		1994		1995	
	Value	%	Value	%	Value	%	Value	%	Value	%	Value	%
Establishments or enterprises (number)	-		403	0.017	358	0.013	522	0.025	608	0.034	740	0.060
Total employment (000)	-		-		36	0.019	53	0.029	58	0.037	63	0.044
Female employees (000)	-		-		-		22	0.097	25	0.117	27	0.140
Salaries and Wages ($ bil.)	-		0.101	0.004	-		0.156	0.008	0.242	0.012	0.169	0.010
Output ($ bil.)	-		1	0.007	0.582	0.004	1	0.010	1	0.011	2	0.011
Value added ($ bil.)	-		0.516	0.009	-		0.586	0.010	0.653	0.013	0.785	0.014
Capital investment ($ mil.)	-		63	0.010	-		90	0.017	140	0.032	98	0.031
Employees per establishment (number)	-		-		97	156.7	101	143.8	96	142.1	85	121.3
Output per establishment ($ mil.)	-		3	67.1	2	40.6	3	56.8	3	64.5	2	49.8
Value added per establishment ($ mil.)	-		1	87.3	-		1	69.7	1	60.9	1	60.2
Capital investment per estab. ($ mil.)	-		0.169	57.4	-		0.174	68.8	0.231	98.1	0.139	48.5
Payroll per employee ($)	-		-		-		2,938	18.1	4,162	23.7	2,692	15.4
Females as % of total employment (%)	-		-		-		41.8	83.7	42.3	80.6	42.6	72.9
Output per employee ($)	-		-		20,686	23.1	26,310	32.9	33,494	38.3	24,217	26.2
Value added per employee ($)	-		-		11,070	34.5	11,070	34.5	11,217	31.6	12,486	31.6
Capital investment per employee ($)	-		-		-		1,709	29.9	2,409	36.8	1,642	22.0

Columns headed % show percent of the world total or of the world ratio. Ratios closest to 100 are closest to the world average. Values higher than 100 are above, values lower than 100 are below world average. *nec* stands for *not elsewhere classified*. Ratios, where shown, are frequently based on a subset of total data; for example, payroll per employee is averaged from all cases where *both* payroll and employment data are present. Dividing *total* salaries/wages by *total* employment will not necessarily produce a matching value.

Equatorial Guinea

Top Ranked Employing Sectors	Top Ranked Output Sectors
3320 Furniture, Fixtures	3320 Furniture, Fixtures
3110 Food Products	3110 Food Products
3690 Nonmetal Products nec	3520 Chemical Products nec
3520 Chemical Products nec	3690 Nonmetal Products nec
3810 Metal Products	3810 Metal Products

Population 454,000. Percent of world population: 0.008

	1990		1991		1992		1993		1994		1995	
	Value	%	Value	%	Value	%	Value	%	Value	%	Value	%
Establishments or enterprises (number)	59	0.003	-	-	-	-	-	-	-	-	-	-
Total employment (000)	0.395	-	-	-	-	-	-	-	-	-	-	-
Female employees (000)	-	-	-	-	-	-	-	-	-	-	-	-
Salaries and Wages ($ bil.)	0.001	-	-	-	-	-	-	-	-	-	-	-
Output ($ bil.)	0.002	-	-	-	-	-	-	-	-	-	-	-
Value added ($ bil.)	-	-	-	-	-	-	-	-	-	-	-	-
Capital investment ($ mil.)	-	-	-	-	-	-	-	-	-	-	-	-
Employees per establishment (number)	7	9.1	-	-	-	-	-	-	-	-	-	-
Output per establishment ($ mil.)	0.035	0.8	-	-	-	-	-	-	-	-	-	-
Value added per establishment ($ mil.)	-	-	-	-	-	-	-	-	-	-	-	-
Capital investment per estab. ($ mil.)	-	-	-	-	-	-	-	-	-	-	-	-
Payroll per employee ($)	1,740	12.7	-	-	-	-	-	-	-	-	-	-
Females as % of total employment (%)	-	-	-	-	-	-	-	-	-	-	-	-
Output per employee ($)	5,300	7.7	-	-	-	-	-	-	-	-	-	-
Value added per employee ($)	-	-	-	-	-	-	-	-	-	-	-	-
Capital investment per employee ($)	-	-	-	-	-	-	-	-	-	-	-	-

Columns headed % show percent of the world total or of the world ratio. Ratios closest to 100 are closest to the world average. Values higher than 100 are above, values lower than 100 are below world average. *nec* stands for *not elsewhere classified*. Ratios, where shown, are frequently based on a subset of total data; for example, payroll per employee is averaged from all cases where *both* payroll and employment data are present. Dividing *total* salaries/wages by *total* employment will not necessarily produce a matching value.

Ethiopia

Top Ranked Employing Sectors	Top Ranked Output Sectors
3210 Textiles	3110 Food Products
3110 Food Products	3130 Beverages
3130 Beverages	3210 Textiles
3690 Nonmetal Products nec	3230 Leather and Products
3220 Wearing Apparel	3840 Transportation Equipment

Population 58,390,000. Percent of world population: 0.985

	1990		1991		1992		1993		1994		1995	
	Value	%	Value	%	Value	%	Value	%	Value	%	Value	%
Establishments or enterprises (number)	313	0.014	288	0.012	283	0.010	289	0.014	499	0.028	501	0.041
Total employment (000)	82	0.039	84	0.045	83	0.045	82	0.045	88	0.055	90	0.063
Female employees (000)	-	-	27	0.302	27	0.126	27	0.121	28	0.134	29	0.149
Salaries and Wages ($ bil.)	0.131	0.006	0.129	0.005	0.100	0.004	0.068	0.003	0.073	0.004	0.069	0.004
Output ($ bil.)	1	0.008	1	0.007	0.629	0.004	0.535	0.004	0.734	0.005	0.801	0.006
Value added ($ bil.)	0.584	0.010	0.553	0.010	0.340	0.006	0.299	0.005	0.379	0.007	0.380	0.007
Capital investment ($ mil.)	66	0.011	44	0.007	26	0.004	61	0.011	38	0.009	32	0.010
Employees per establishment (number)	263	357.1	292	429.9	292	469.7	284	402.9	177	262.6	180	257.0
Output per establishment ($ mil.)	4	90.8	4	80.1	2	40.8	2	39.4	1	29.2	2	38.7
Value added per establishment ($ mil.)	2	113.9	2	113.8	1	56.9	1	64.4	0.759	43.1	0.758	43.0
Capital investment per estab. ($ mil.)	0.212	80.1	0.154	52.6	0.091	31.6	0.212	83.7	0.076	32.2	0.064	22.2
Payroll per employee ($)	1,596	11.7	1,532	8.5	1,206	6.0	823	5.1	827	4.7	768	4.4
Females as % of total employment (%)	-	-	32.4	100.8	32.7	70.7	32.9	66.0	32.2	61.3	31.6	54.1
Output per employee ($)	14,617	21.3	12,608	15.5	7,609	8.5	6,516	8.1	8,317	9.5	8,875	9.6
Value added per employee ($)	7,094	21.6	6,585	19.5	4,108	11.4	3,646	11.4	4,293	12.1	4,212	10.7
Capital investment per employee ($)	806	14.6	528	8.6	310	5.0	746	13.1	428	6.6	353	4.7

Columns headed % show percent of the world total or of the world ratio. Ratios closest to 100 are closest to the world average. Values higher than 100 are above, values lower than 100 are below world average. *nec* stands for *not elsewhere classified*. Ratios, where shown, are frequently based on a subset of total data; for example, payroll per employee is averaged from all cases where *both* payroll and employment data are present. Dividing *total* salaries/wages by *total* employment will not necessarily produce a matching value.

For additional notes and sources, see Appendix I.

Fiji

Top Ranked Employing Sectors	Top Ranked Output Sectors
3110 Food Products	3110 Food Products
3310 Wood Products	3310 Wood Products
3420 Printing, Publishing	3520 Chemical Products nec
3320 Furniture, Fixtures	3810 Metal Products
3810 Metal Products	3420 Printing, Publishing

Population 803,000. Percent of world population: 0.013

	1990 Value	%	1991 Value	%	1992 Value	%	1993 Value	%	1994 Value	%	1995 Value	%
Establishments or enterprises (number)	525	0.023	487	0.020	481	0.017	-	-	-	-	-	-
Total employment (000)	13	0.006	11	0.006	12	0.007	13	0.007	13	0.008	-	-
Female employees (000)	-	-	-	-	-	-	-	-	-	-	-	-
Salaries and Wages ($ bil.)	0.049	0.002	0.053	0.002	0.051	0.002	0.051	0.003	0.058	0.003	-	-
Output ($ bil.)	0.501	0.003	0.523	0.003	0.535	0.003	0.554	0.004	0.633	0.005	-	-
Value added ($ bil.)	0.107	0.002	0.112	0.002	0.112	0.002	0.115	0.002	0.131	0.003	-	-
Capital investment ($ mil.)	33	0.006	19	0.003	23	0.004	-	-	-	-	-	-
Employees per establishment (number)	24	32.3	22	31.9	25	41.0	-	-	-	-	-	-
Output per establishment ($ mil.)	0.954	22.5	1	23.4	1	20.4	-	-	-	-	-	-
Value added per establishment ($ mil.)	0.203	12.4	0.231	13.7	0.232	11.0	-	-	-	-	-	-
Capital investment per estab. ($ mil.)	0.063	23.8	0.039	13.4	0.049	17.0	-	-	-	-	-	-
Payroll per employee ($)	3,923	28.7	5,006	27.7	4,131	20.6	4,084	25.1	4,471	25.5	-	-
Females as % of total employment (%)	-	-	-	-	-	-	-	-	-	-	-	-
Output per employee ($)	40,059	58.4	49,536	60.9	43,636	48.8	44,311	55.3	48,859	55.9	-	-
Value added per employee ($)	8,523	25.9	10,643	31.5	9,105	25.3	9,227	28.8	10,091	28.4	-	-
Capital investment per employee ($)	2,643	48.0	1,811	29.5	1,914	30.9	-	-	-	-	-	-

Columns headed % show percent of the world total or of the world ratio. Ratios closest to 100 are closest to the world average. Values higher than 100 are above, values lower than 100 are below world average. *nec* stands for *not elsewhere classified*. Ratios, where shown, are frequently based on a subset of total data; for example, payroll per employee is averaged from all cases where *both* payroll and employment data are present. Dividing *total* salaries/wages by *total* employment will not necessarily produce a matching value.

Finland

Top Ranked Employing Sectors	Top Ranked Output Sectors
3820 Machinery nec	3410 Paper and Products
3410 Paper and Products	3820 Machinery nec
3110 Food Products	3110 Food Products
3830 Electrical Machinery	3830 Electrical Machinery
3420 Printing, Publishing	3310 Wood Products

Population 5,149,000. Percent of world population: 0.087

	1990 Value	%	1991 Value	%	1992 Value	%	1993 Value	%	1994 Value	%	1995 Value	%
Establishments or enterprises (number)	6,683	0.290	7,111	0.298	6,512	0.232	6,148	0.299	5,916	0.331	5,508	0.449
Total employment (000)	525	0.251	487	0.261	444	0.244	421	0.231	427	0.268	323	0.227
Female employees (000)	-	-	-	-	-	-	-	-	-	-	-	-
Salaries and Wages ($ bil.)	16	0.723	14	0.591	12	0.474	9	0.451	11	0.552	10	0.620
Output ($ bil.)	94	0.659	79	0.516	74	0.450	64	0.441	79	0.575	72	0.537
Value added ($ bil.)	34	0.607	27	0.471	26	0.431	23	0.381	28	0.539	24	0.432
Capital investment ($ mil.)	7,800	1.300	5,343	0.857	4,342	0.679	3,186	0.586	3,785	0.852	2,943	0.942
Employees per establishment (number)	78	106.5	68	101.0	68	109.8	68	97.0	72	107.1	61	87.2
Output per establishment ($ mil.)	14	332.9	11	242.7	11	208.5	10	221.6	13	266.4	14	327.8
Value added per establishment ($ mil.)	5	310.1	4	222.3	4	188.1	4	229.5	5	264.6	4	254.3
Capital investment per estab. ($ mil.)	1	440.6	0.751	255.7	0.667	232.3	0.518	204.6	0.640	271.9	0.556	194.7
Payroll per employee ($)	29,685	217.0	28,841	159.4	27,230	135.6	21,888	134.8	25,258	143.8	31,818	182.1
Females as % of total employment (%)	-	-	-	-	-	-	-	-	-	-	-	-
Output per employee ($)	179,667	261.8	162,652	199.8	166,392	186.0	152,204	190.1	185,843	212.7	221,673	239.5
Value added per employee ($)	64,726	197.0	54,786	162.2	58,081	161.3	54,037	168.6	64,685	182.2	73,508	186.1
Capital investment per employee ($)	14,868	270.0	10,968	178.9	9,770	157.7	7,577	132.6	8,872	135.7	9,111	122.0

Columns headed % show percent of the world total or of the world ratio. Ratios closest to 100 are closest to the world average. Values higher than 100 are above, values lower than 100 are below world average. *nec* stands for *not elsewhere classified*. Ratios, where shown, are frequently based on a subset of total data; for example, payroll per employee is averaged from all cases where *both* payroll and employment data are present. Dividing *total* salaries/wages by *total* employment will not necessarily produce a matching value.

Former Yugoslav Republic of Macedonia

Top Ranked Employing Sectors	Top Ranked Output Sectors
3220 Wearing Apparel	3110 Food Products
3110 Food Products	3140 Tobacco
3210 Textiles	3220 Wearing Apparel
3830 Electrical Machinery	3830 Electrical Machinery
3810 Metal Products	3510 Industrial Chemicals

Population 2,009,000. Percent of world population: 0.034

	1990		1991		1992		1993		1994		1995	
	Value	%	Value	%	Value	%	Value	%	Value	%	Value	%
Establishments or enterprises (number)	662	0.029	1,312	0.055	2,399	0.085	2,682	0.130	3,510	0.196	4,949	0.403
Total employment (000)	185	0.089	177	0.095	169	0.093	155	0.085	145	0.091	133	0.094
Female employees (000)	72	0.544	67	0.746	61	0.286	57	0.256	54	0.256	46	0.242
Salaries and Wages ($ bil.)	0.884	0.041	1	0.045	0.358	0.014	0.351	0.017	0.373	0.019	0.485	0.029
Output ($ bil.)	4	0.031	5	0.033	3	0.017	2	0.014	2	0.014	2	0.017
Value added ($ bil.)	2	0.033	2	0.039	1	0.021	0.992	0.017	0.769	0.015	1	0.019
Capital investment ($ mil.)	95	0.016	135	0.022	72	0.011	57	0.011	42	0.009	44	0.014
Employees per establishment (number)	280	380.2	135	199.1	71	113.5	58	81.7	41	61.5	27	38.5
Output per establishment ($ mil.)	7	157.1	4	85.2	1	22.0	0.783	16.7	0.538	10.7	0.455	11.0
Value added per establishment ($ mil.)	3	171.0	2	98.6	0.527	25.0	0.370	23.0	0.219	12.4	0.203	11.5
Capital investment per estab. ($ mil.)	0.144	54.3	0.103	35.1	0.031	10.9	0.022	8.7	0.012	5.3	0.009	3.3
Payroll per employee ($)	4,771	34.9	6,064	33.5	2,111	10.5	2,268	14.0	2,565	14.6	3,590	20.5
Females as % of total employment (%)	40.9	127.8	39.6	123.5	37.8	81.8	38.4	76.8	39.3	74.9	38.5	65.9
Output per employee ($)	23,784	34.7	28,969	35.6	16,943	18.9	13,585	17.0	12,991	14.9	16,851	18.2
Value added per employee ($)	10,017	30.5	12,328	36.5	7,470	20.7	6,420	20.0	5,291	14.9	7,510	19.0
Capital investment per employee ($)	508	9.2	764	12.5	436	7.0	378	6.6	300	4.6	354	4.7

Columns headed % show percent of the world total or of the world ratio. Ratios closest to 100 are closest to the world average. Values higher than 100 are above, values lower than 100 are below world average. *nec* stands for *not elsewhere classified*. Ratios, where shown, are frequently based on a subset of total data; for example, payroll per employee is averaged from all cases where *both* payroll and employment data are present. Dividing *total* salaries/wages by *total* employment will not necessarily produce a matching value.

France

Top Ranked Employing Sectors	Top Ranked Output Sectors
3840 Transportation Equipment	3110 Food Products
3110 Food Products	3840 Transportation Equipment
3830 Electrical Machinery	3830 Electrical Machinery
3820 Machinery nec	3820 Machinery nec
3810 Metal Products	3810 Metal Products

Population 58,805,000. Percent of world population: 0.992

	1990		1991		1992		1993		1994		1995	
	Value	%	Value	%	Value	%	Value	%	Value	%	Value	%
Establishments or enterprises (number)	-	-	-	-	-	-	-	-	-	-	-	-
Total employment (000)	4,351	2.100	4,283	2.300	4,144	2.300	3,951	2.200	3,841	2.400	3,839	2.700
Female employees (000)	-	-	-	-	-	-	-	-	-	-	-	-
Salaries and Wages ($ bil.)	166	7.700	165	7.000	177	6.900	-	-	-	-	-	-
Output ($ bil.)	762	5.300	741	4.800	795	4.800	708	4.900	765	5.500	891	6.700
Value added ($ bil.)	281	5.000	273	4.800	291	4.900	267	4.500	283	5.500	326	5.900
Capital investment ($ mil.)	29,720	5.000	28,207	4.500	26,387	4.100	19,378	3.600	21,405	4.800	25,415	8.100
Employees per establishment (number)	-	-	-	-	-	-	-	-	-	-	-	-
Output per establishment ($ mil.)	-	-	-	-	-	-	-	-	-	-	-	-
Value added per establishment ($ mil.)	-	-	-	-	-	-	-	-	-	-	-	-
Capital investment per estab. ($ mil.)	-	-	-	-	-	-	-	-	-	-	-	-
Payroll per employee ($)	34,529	252.4	34,984	193.3	38,646	192.5	-	-	-	-	-	-
Females as % of total employment (%)	-	-	-	-	-	-	-	-	-	-	-	-
Output per employee ($)	156,153	227.5	154,690	190.0	170,238	190.3	159,518	199.2	176,206	201.7	205,111	221.6
Value added per employee ($)	58,859	179.2	58,340	172.7	63,941	177.6	61,695	192.5	66,728	188.0	77,135	195.3
Capital investment per employee ($)	10,504	190.8	10,069	164.2	9,726	157.0	7,497	131.2	8,527	130.4	10,087	135.1

Columns headed % show percent of the world total or of the world ratio. Ratios closest to 100 are closest to the world average. Values higher than 100 are above, values lower than 100 are below world average. *nec* stands for *not elsewhere classified*. Ratios, where shown, are frequently based on a subset of total data; for example, payroll per employee is averaged from all cases where *both* payroll and employment data are present. Dividing *total* salaries/wages by *total* employment will not necessarily produce a matching value.

For additional notes and sources, see Appendix I.

Gabon

Top Ranked Employing Sectors	Top Ranked Output Sectors
3110 Food Products	3530 Petroleum Refineries
3310 Wood Products	3130 Beverages
3130 Beverages	3110 Food Products
3810 Metal Products	3810 Metal Products
3530 Petroleum Refineries	3310 Wood Products

Population 1,208,000. Percent of world population: 0.020

	1990		1991		1992		1993		1994		1995	
	Value	%	Value	%	Value	%	Value	%	Value	%	Value	%
Establishments or enterprises (number)	-	-	92	0.004	92	0.003	87	0.004	84	0.005	95	0.008
Total employment (000)	-	-	8	0.004	7	0.004	7	0.004	7	0.005	8	0.005
Female employees (000)	-	-	-	-	-	-	-	-	-	-	-	-
Salaries and Wages ($ bil.)	-	-	0.132	0.006	0.131	0.005	0.125	0.006	0.072	0.004	0.082	0.005
Output ($ bil.)	-	-	0.640	0.004	0.632	0.004	0.609	0.004	0.390	0.003	0.488	0.004
Value added ($ bil.)	-	-	0.230	0.004	0.228	0.004	0.232	0.004	0.128	0.003	0.145	0.003
Capital investment ($ mil.)	-	-	48	0.008	29	0.005	21	0.004	14	0.003	25	0.008
Employees per establishment (number)	-	-	92	135.5	81	130.2	85	120.5	87	128.5	80	113.6
Output per establishment ($ mil.)	-	-	7	151.5	7	126.1	7	149.1	5	92.3	5	124.4
Value added per establishment ($ mil.)	-	-	2	148.0	2	117.6	3	165.5	2	86.7	2	86.6
Capital investment per estab. ($ mil.)	-	-	0.530	180.4	0.322	112.1	0.237	93.5	0.169	71.9	0.266	92.9
Payroll per employee ($)	-	-	16,030	88.6	17,771	88.5	16,962	104.5	9,943	56.6	10,878	62.2
Females as % of total employment (%)	-	-	-	-	-	-	-	-	-	-	-	-
Output per employee ($)	-	-	77,994	95.8	85,736	95.8	82,414	102.9	53,674	61.4	64,538	69.7
Value added per employee ($)	-	-	28,029	83.0	30,935	85.9	31,358	97.9	17,670	49.8	19,210	48.6
Capital investment per employee ($)	-	-	5,888	96.0	3,973	64.1	2,787	48.8	1,956	29.9	3,335	44.7

Columns headed % show percent of the world total or of the world ratio. Ratios closest to 100 are closest to the world average. Values higher than 100 are above, values lower than 100 are below world average. *nec* stands for *not elsewhere classified*. Ratios, where shown, are frequently based on a subset of total data; for example, payroll per employee is averaged from all cases where *both* payroll and employment data are present. Dividing *total* salaries/wages by *total* employment will not necessarily produce a matching value.

Gambia

Top Ranked Employing Sectors	Top Ranked Output Sectors
3110 Food Products	
3420 Printing, Publishing	
3130 Beverages	
3220 Wearing Apparel	
3320 Furniture, Fixtures	

Population 1,292,000. Percent of world population: 0.022

	1990		1991		1992		1993		1994		1995	
	Value	%	Value	%	Value	%	Value	%	Value	%	Value	%
Establishments or enterprises (number)	-	-	-	-	-	-	148	0.007	-	-	-	-
Total employment (000)	-	-	-	-	-	-	3	0.001	-	-	-	-
Female employees (000)	-	-	-	-	-	-	0.655	0.003	-	-	-	-
Salaries and Wages ($ bil.)	-	-	-	-	-	-	-	-	-	-	-	-
Output ($ bil.)	-	-	-	-	-	-	-	-	-	-	-	-
Value added ($ bil.)	-	-	-	-	-	-	-	-	-	-	-	-
Capital investment ($ mil.)	-	-	-	-	-	-	-	-	-	-	-	-
Employees per establishment (number)	-	-	-	-	-	-	17	24.1	-	-	-	-
Output per establishment ($ mil.)	-	-	-	-	-	-	-	-	-	-	-	-
Value added per establishment ($ mil.)	-	-	-	-	-	-	-	-	-	-	-	-
Capital investment per estab. ($ mil.)	-	-	-	-	-	-	-	-	-	-	-	-
Payroll per employee ($)	-	-	-	-	-	-	-	-	-	-	-	-
Females as % of total employment (%)	-	-	-	-	-	-	28.1	56.3	-	-	-	-
Output per employee ($)	-	-	-	-	-	-	-	-	-	-	-	-
Value added per employee ($)	-	-	-	-	-	-	-	-	-	-	-	-
Capital investment per employee ($)	-	-	-	-	-	-	-	-	-	-	-	-

Columns headed % show percent of the world total or of the world ratio. Ratios closest to 100 are closest to the world average. Values higher than 100 are above, values lower than 100 are below world average. *nec* stands for *not elsewhere classified*. Ratios, where shown, are frequently based on a subset of total data; for example, payroll per employee is averaged from all cases where *both* payroll and employment data are present. Dividing *total* salaries/wages by *total* employment will not necessarily produce a matching value.

Germany

Top Ranked Employing Sectors	Top Ranked Output Sectors
3820 Machinery nec	3840 Transportation Equipment
3830 Electrical Machinery	3820 Machinery nec
3840 Transportation Equipment	3830 Electrical Machinery
3810 Metal Products	3110 Food Products
3110 Food Products	3810 Metal Products

Population 82,079,000. Percent of world population: 1.385

	1990		1991		1992		1993		1994		1995	
	Value	%	Value	%	Value	%	Value	%	Value	%	Value	%
Establishments or enterprises (number)	-	-	54,694	2.300	54,033	1.900	52,360	2.500	51,128	2.900	-	-
Total employment (000)	-	-	8,850	4.700	7,745	4.200	8,660	4.800	7,990	5.000	-	-
Female employees (000)	-	-	-	-	-	-	-	-	-	-	-	-
Salaries and Wages ($ bil.)	-	-	225	9.500	268	10.500	299	14.700	295	15.100	-	-
Output ($ bil.)	-	-	1,035	6.700	1,128	6.900	1,226	8.500	1,291	9.400	-	-
Value added ($ bil.)	-	-	-	-	-	-	-	-	-	-	-	-
Capital investment ($ mil.)	-	-	-	-	-	-	-	-	-	-	-	-
Employees per establishment (number)	-	-	169	249.0	170	273.1	165	234.6	158	234.8	-	-
Output per establishment ($ mil.)	-	-	23	496.0	27	492.1	25	531.2	27	534.4	-	-
Value added per establishment ($ mil.)	-	-	-	-	-	-	-	-	-	-	-	-
Capital investment per estab. ($ mil.)	-	-	-	-	-	-	-	-	-	-	-	-
Payroll per employee ($)	-	-	27,962	154.5	34,656	172.6	34,555	212.8	36,880	210.0	-	-
Females as % of total employment (%)	-	-	-	-	-	-	-	-	-	-	-	-
Output per employee ($)	-	-	128,415	157.8	155,024	173.3	149,861	187.2	170,127	194.7	-	-
Value added per employee ($)	-	-	-	-	-	-	-	-	-	-	-	-
Capital investment per employee ($)	-	-	-	-	-	-	-	-	-	-	-	-

Columns headed % show percent of the world total or of the world ratio. Ratios closest to 100 are closest to the world average. Values higher than 100 are above, values lower than 100 are below world average. *nec* stands for *not elsewhere classified*. Ratios, where shown, are frequently based on a subset of total data; for example, payroll per employee is averaged from all cases where *both* payroll and employment data are present. Dividing *total* salaries/wages by *total* employment will not necessarily produce a matching value.

Ghana

Top Ranked Employing Sectors	Top Ranked Output Sectors
3310 Wood Products	3720 Nonferrous metals
3210 Textiles	3310 Wood Products
3110 Food Products	3110 Food Products
3520 Chemical Products nec	3520 Chemical Products nec
3420 Printing, Publishing	3140 Tobacco

Population 18,497,000. Percent of world population: 0.312

	1990		1991		1992		1993		1994		1995	
	Value	%	Value	%	Value	%	Value	%	Value	%	Value	%
Establishments or enterprises (number)	-	-	-	-	-	-	559	0.027	-	-	-	-
Total employment (000)	-	-	-	-	-	-	90	0.050	81	0.051	83	0.058
Female employees (000)	-	-	-	-	-	-	6	0.027	-	-	-	-
Salaries and Wages ($ bil.)	-	-	-	-	-	-	0.100	0.005	0.077	0.004	0.080	0.005
Output ($ bil.)	-	-	-	-	-	-	1	0.008	0.947	0.007	0.994	0.007
Value added ($ bil.)	-	-	-	-	-	-	0.526	0.009	0.456	0.009	0.479	0.009
Capital investment ($ mil.)	-	-	-	-	-	-	-	-	-	-	-	-
Employees per establishment (number)	-	-	-	-	-	-	161	229.0	-	-	-	-
Output per establishment ($ mil.)	-	-	-	-	-	-	3	56.6	-	-	-	-
Value added per establishment ($ mil.)	-	-	-	-	-	-	1	74.8	-	-	-	-
Capital investment per estab. ($ mil.)	-	-	-	-	-	-	-	-	-	-	-	-
Payroll per employee ($)	-	-	-	-	-	-	1,388	8.5	1,037	5.9	1,058	6.1
Females as % of total employment (%)	-	-	-	-	-	-	8.5	17.0	-	-	-	-
Output per employee ($)	-	-	-	-	-	-	16,152	20.2	12,796	14.6	13,059	14.1
Value added per employee ($)	-	-	-	-	-	-	7,314	22.8	6,164	17.4	6,290	15.9
Capital investment per employee ($)	-	-	-	-	-	-	-	-	-	-	-	-

Columns headed % show percent of the world total or of the world ratio. Ratios closest to 100 are closest to the world average. Values higher than 100 are above, values lower than 100 are below world average. *nec* stands for *not elsewhere classified*. Ratios, where shown, are frequently based on a subset of total data; for example, payroll per employee is averaged from all cases where *both* payroll and employment data are present. Dividing *total* salaries/wages by *total* employment will not necessarily produce a matching value.

For additional notes and sources, see Appendix I.

Greece

Top Ranked Employing Sectors	Top Ranked Output Sectors
3110 Food Products	3110 Food Products
3210 Textiles	3530 Petroleum Refineries
3220 Wearing Apparel	3210 Textiles
3690 Nonmetal Products nec	3520 Chemical Products nec
3810 Metal Products	3130 Beverages

Population 10,662,000. Percent of world population: 0.180

	1990		1991		1992		1993		1994		1995	
	Value	%	Value	%	Value	%	Value	%	Value	%	Value	%
Establishments or enterprises (number)	9,013	0.391	8,899	0.373	8,980	0.320	-	-	-	-	-	-
Total employment (000)	409	0.195	384	0.206	370	0.203	311	0.171	307	0.193	304	0.214
Female employees (000)	-	-	-	-	-	-	-	-	-	-	-	-
Salaries and Wages ($ bil.)	5	0.225	5	0.200	5	0.193	4	0.199	4	0.212	5	0.291
Output ($ bil.)	34	0.239	33	0.218	34	0.210	29	0.202	30	0.221	36	0.266
Value added ($ bil.)	11	0.196	11	0.198	12	0.206	10	0.175	11	0.211	13	0.230
Capital investment ($ mil.)	1,793	0.302	1,576	0.253	1,725	0.270	-	-	-	-	-	-
Employees per establishment (number)	45	61.5	43	63.6	41	66.3	-	-	-	-	-	-
Output per establishment ($ mil.)	4	89.6	4	82.0	4	70.5	-	-	-	-	-	-
Value added per establishment ($ mil.)	1	74.4	1	74.6	1	65.0	-	-	-	-	-	-
Capital investment per estab. ($ mil.)	0.199	75.1	0.177	60.3	0.192	66.9	-	-	-	-	-	-
						-						
Payroll per employee ($)	11,850	86.6	12,346	68.2	13,336	66.4	13,026	80.2	13,493	76.8	15,839	90.6
Females as % of total employment (%)	-	-	-	-	-	-	-	-	-	-	-	-
Output per employee ($)	83,716	122.0	87,145	107.1	93,202	104.2	94,340	117.8	99,005	113.3	116,665	126.1
Value added per employee ($)	26,872	81.8	29,160	86.3	33,259	92.4	33,627	104.9	35,137	99.0	41,471	105.0
Capital investment per employee ($)	4,386	79.7	4,103	66.9	4,661	75.2	-	-	-	-	-	-

Columns headed % show percent of the world total or of the world ratio. Ratios closest to 100 are closest to the world average. Values higher than 100 are above, values lower than 100 are below world average. *nec* stands for *not elsewhere classified*. Ratios, where shown, are frequently based on a subset of total data; for example, payroll per employee is averaged from all cases where *both* payroll and employment data are present. Dividing *total* salaries/wages by *total* employment will not necessarily produce a matching value.

Grenada

Top Ranked Employing Sectors	Top Ranked Output Sectors
3130 Beverages	
3110 Food Products	
3220 Wearing Apparel	
3810 Metal Products	
3522 Drugs and Medicines	

Population 96,000. Percent of world population: 0.002

	1990		1991		1992		1993		1994		1995	
	Value	%	Value	%	Value	%	Value	%	Value	%	Value	%
Establishments or enterprises (number)	-	-	-	-	102	0.004	85	0.004	73	0.004	56	0.005
Total employment (000)	-	-	-	-	2	0.001	2	0.001	1	0.001	1	0.001
Female employees (000)	-	-	-	-	-	-	-	-	-	-	-	-
Salaries and Wages ($ bil.)	-	-	-	-	-	-	-	-	-	-	-	-
Output ($ bil.)	-	-	-	-	-	-	-	-	-	-	-	-
Value added ($ bil.)	-	-	-	-	-	-	-	-	-	-	-	-
Capital investment ($ mil.)	-	-	-	-	-	-	-	-	-	-	-	-
Employees per establishment (number)	-	-	-	-	16	25.5	18	25.9	19	28.2	21	30.4
Output per establishment ($ mil.)	-	-	-	-	-	-	-	-	-	-	-	-
Value added per establishment ($ mil.)	-	-	-	-	-	-	-	-	-	-	-	-
Capital investment per estab. ($ mil.)	-	-	-	-	-	-	-	-	-	-	-	-
Payroll per employee ($)	-	-	-	-	-	-	-	-	-	-	-	-
Females as % of total employment (%)	-	-	-	-	-	-	-	-	-	-	-	-
Output per employee ($)	-	-	-	-	-	-	-	-	-	-	-	-
Value added per employee ($)	-	-	-	-	-	-	-	-	-	-	-	-
Capital investment per employee ($)	-	-	-	-	-	-	-	-	-	-	-	-

Columns headed % show percent of the world total or of the world ratio. Ratios closest to 100 are closest to the world average. Values higher than 100 are above, values lower than 100 are below world average. *nec* stands for *not elsewhere classified*. Ratios, where shown, are frequently based on a subset of total data; for example, payroll per employee is averaged from all cases where *both* payroll and employment data are present. Dividing *total* salaries/wages by *total* employment will not necessarily produce a matching value.

Guatemala

	Top Ranked Employing Sectors	Top Ranked Output Sectors
	3110 Food Products	3110 Food Products
	3210 Textiles	3130 Beverages
	3130 Beverages	3520 Chemical Products nec
	3520 Chemical Products nec	3522 Drugs and Medicines
	3710 Iron and Steel	3690 Nonmetal Products nec

Population 12,008,000. Percent of world population: 0.203

	1990		1991		1992		1993		1994		1995	
	Value	%	Value	%	Value	%	Value	%	Value	%	Value	%
Establishments or enterprises (number)	-	-	432	0.018	426	0.015	427	0.021	425	0.024	426	0.035
Total employment (000)	-	-	69	0.037	72	0.039	73	0.040	74	0.047	79	0.056
Female employees (000)	-	-	-		-		-		-		-	
Salaries and Wages ($ bil.)	-	-	0.018	0.001	0.023	0.001	0.022	0.001	0.023	0.001	0.029	0.002
Output ($ bil.)	-	-	3	0.021	4	0.022	4	0.029	5	0.033	5	0.038
Value added ($ bil.)	-	-	-		-		-		-		-	
Capital investment ($ mil.)	-	-	-		-		-		-		-	
Employees per establishment (number)	-	-	159	234.1	169	271.6	171	242.0	175	259.7	186	264.8
Output per establishment ($ mil.)	-	-	8	166.8	8	152.6	10	208.8	11	212.5	12	290.4
Value added per establishment ($ mil.)	-	-	-		-		-		-		-	
Capital investment per estab. ($ mil.)	-	-	-		-		-		-		-	
Payroll per employee ($)	-	-	269	1.5	315	1.6	304	1.9	314	1.8	366	2.1
Females as % of total employment (%)	-	-	-		-		-		-		-	
Output per employee ($)	-	-	48,090	59.1	49,209	55.0	57,489	71.8	61,119	70.0	64,645	69.9
Value added per employee ($)	-	-	-		-		-		-		-	
Capital investment per employee ($)	-	-	-		-		-		-		-	

Columns headed % show percent of the world total or of the world ratio. Ratios closest to 100 are closest to the world average. Values higher than 100 are above, values lower than 100 are below world average. *nec* stands for *not elsewhere classified*. Ratios, where shown, are frequently based on a subset of total data; for example, payroll per employee is averaged from all cases where *both* payroll and employment data are present. Dividing *total* salaries/wages by *total* employment will not necessarily produce a matching value.

Honduras

	Top Ranked Employing Sectors	Top Ranked Output Sectors
	3220 Wearing Apparel	3110 Food Products
	3110 Food Products	3130 Beverages
	3310 Wood Products	3520 Chemical Products nec
	3210 Textiles	3220 Wearing Apparel
	3320 Furniture, Fixtures	3690 Nonmetal Products nec

Population 5,862,000. Percent of world population: 0.099

	1990		1991		1992		1993		1994		1995	
	Value	%	Value	%	Value	%	Value	%	Value	%	Value	%
Establishments or enterprises (number)	-	-	-	-	803	0.029	874	0.043	904	0.051	865	0.071
Total employment (000)	85	0.040	113	0.060	125	0.069	138	0.076	156	0.098	183	0.128
Female employees (000)	-	-	-		-		-		-		-	
Salaries and Wages ($ bil.)	0.186	0.009	0.190	0.008	0.242	0.009	0.244	0.012	0.228	0.012	0.287	0.017
Output ($ bil.)	2	0.012	2	0.011	2	0.012	2	0.013	2	0.014	2	0.017
Value added ($ bil.)	0.410	0.007	0.412	0.007	0.492	0.008	0.501	0.008	0.479	0.009	0.591	0.011
Capital investment ($ mil.)	-	-	-		-		-		-		-	
Employees per establishment (number)	-	-	-	-	156	250.5	157	223.2	173	256.7	211	301.8
Output per establishment ($ mil.)	-	-	-	-	2	45.1	2	46.2	2	42.8	3	62.0
Value added per establishment ($ mil.)	-	-	-	-	0.612	29.0	0.573	35.6	0.529	30.0	0.683	38.7
Capital investment per estab. ($ mil.)	-	-	-		-		-		-		-	
Payroll per employee ($)	2,194	16.0	1,679	9.3	1,933	9.6	1,773	10.9	1,457	8.3	1,649	9.4
Females as % of total employment (%)	-	-	-		-		-		-		-	
Output per employee ($)	20,277	29.5	15,220	18.7	15,779	17.6	13,804	17.2	12,464	14.3	12,104	13.1
Value added per employee ($)	4,852	14.8	3,651	10.8	3,930	10.9	3,641	11.4	3,063	8.6	3,230	8.2
Capital investment per employee ($)	-	-	-		-		-		-		-	

Columns headed % show percent of the world total or of the world ratio. Ratios closest to 100 are closest to the world average. Values higher than 100 are above, values lower than 100 are below world average. *nec* stands for *not elsewhere classified*. Ratios, where shown, are frequently based on a subset of total data; for example, payroll per employee is averaged from all cases where *both* payroll and employment data are present. Dividing *total* salaries/wages by *total* employment will not necessarily produce a matching value.

For additional notes and sources, see Appendix I.

Hungary

Top Ranked Employing Sectors	Top Ranked Output Sectors
3110 Food Products	3110 Food Products
3820 Machinery nec	3840 Transportation Equipment
3830 Electrical Machinery	3830 Electrical Machinery
3220 Wearing Apparel	3530 Petroleum Refineries
3810 Metal Products	3820 Machinery nec

Population 10,208,000. Percent of world population: 0.172

	1990 Value	%	1991 Value	%	1992 Value	%	1993 Value	%	1994 Value	%	1995 Value	%
Establishments or enterprises (number)	7,193	0.312	14,712	0.617	14,914	0.531	17,995	0.875	-	-	-	-
Total employment (000)	1,117	0.534	1,006	0.539	860	0.471	747	0.410	773	0.486	785	0.552
Female employees (000)	461	3.500	-	-	357	1.700	315	1.400	-	-	-	-
Salaries and Wages ($ bil.)	3	0.129	3	0.113	3	0.107	3	0.126	3	0.151	3	0.198
Output ($ bil.)	25	0.175	23	0.152	22	0.133	22	0.149	24	0.176	27	0.202
Value added ($ bil.)	7	0.125	6	0.110	6	0.104	7	0.110	7	0.140	8	0.140
Capital investment ($ mil.)	1,161	0.196	1,104	0.177	2,152	0.337	1,654	0.304	-	-	-	-
Employees per establishment (number)	155	210.7	68	100.8	62	100.2	45	63.4	-	-	-	-
Output per establishment ($ mil.)	3	82.3	2	34.6	2	29.1	1	27.5	-	-	-	-
Value added per establishment ($ mil.)	0.974	59.4	0.424	25.1	0.451	21.4	0.393	24.4	-	-	-	-
Capital investment per estab. ($ mil.)	0.161	60.9	0.075	25.5	0.144	50.3	0.092	36.4	-	-	-	-
Payroll per employee ($)	2,495	18.2	2,661	14.7	3,175	15.8	3,432	21.1	3,806	21.7	4,185	23.9
Females as % of total employment (%)	41.3	128.8	-	-	41.5	89.7	42.2	84.6	-	-	-	-
Output per employee ($)	22,454	32.7	23,210	28.5	25,468	28.5	28,850	36.0	31,319	35.8	34,378	37.1
Value added per employee ($)	6,270	19.1	6,207	18.4	7,242	20.1	8,786	27.4	9,245	26.0	9,816	24.9
Capital investment per employee ($)	1,039	18.9	1,097	17.9	1,946	31.4	1,816	31.8	-	-	-	-

Columns headed % show percent of the world total or of the world ratio. Ratios closest to 100 are closest to the world average. Values higher than 100 are above, values lower than 100 are below world average. *nec* stands for *not elsewhere classified*. Ratios, where shown, are frequently based on a subset of total data; for example, payroll per employee is averaged from all cases where *both* payroll and employment data are present. Dividing *total* salaries/wages by *total* employment will not necessarily produce a matching value.

Iceland

Top Ranked Employing Sectors	Top Ranked Output Sectors
3110 Food Products	3110 Food Products
3420 Printing, Publishing	3420 Printing, Publishing
3810 Metal Products	3810 Metal Products
3320 Furniture, Fixtures	3720 Nonferrous metals
3690 Nonmetal Products nec	3900 Industries nec

Population 271,000. Percent of world population: 0.005

	1990 Value	%	1991 Value	%	1992 Value	%	1993 Value	%	1994 Value	%	1995 Value	%
Establishments or enterprises (number)	2,560	0.111	2,563	0.108	2,614	0.093	2,656	0.129	-	-	-	-
Total employment (000)	23	0.011	23	0.012	21	0.012	21	0.011	21	0.013	-	-
Female employees (000)	-	-	-	-	-	-	-	-	-	-	-	-
Salaries and Wages ($ bil.)	0.609	0.028	0.674	0.028	0.671	0.026	0.554	0.027	-	-	-	-
Output ($ bil.)	3	0.019	3	0.018	3	0.018	2	0.017	-	-	-	-
Value added ($ bil.)	0.788	0.014	0.852	0.015	0.977	0.016	0.825	0.014	-	-	-	-
Capital investment ($ mil.)	-	-	-	-	-	-	-	-	-	-	-	-
Employees per establishment (number)	9	12.4	9	13.3	8	13.1	8	11.0	-	-	-	-
Output per establishment ($ mil.)	1	24.7	1	24.2	1	20.8	0.939	20.0	-	-	-	-
Value added per establishment ($ mil.)	0.308	18.8	0.333	19.7	0.374	17.7	0.311	19.3	-	-	-	-
Capital investment per estab. ($ mil.)	-	-	-	-	-	-	-	-	-	-	-	-
Payroll per employee ($)	25,942	189.6	29,115	160.9	31,451	156.7	26,862	165.4	-	-	-	-
Females as % of total employment (%)	-	-	-	-	-	-	-	-	-	-	-	-
Output per employee ($)	114,009	166.1	122,769	150.8	138,526	154.8	120,912	151.0	-	-	-	-
Value added per employee ($)	33,538	102.1	36,805	109.0	45,827	127.3	40,015	124.9	-	-	-	-
Capital investment per employee ($)	-	-	-	-	-	-	-	-	-	-	-	-

Columns headed % show percent of the world total or of the world ratio. Ratios closest to 100 are closest to the world average. Values higher than 100 are above, values lower than 100 are below world average. *nec* stands for *not elsewhere classified*. Ratios, where shown, are frequently based on a subset of total data; for example, payroll per employee is averaged from all cases where *both* payroll and employment data are present. Dividing *total* salaries/wages by *total* employment will not necessarily produce a matching value.

India

Top Ranked Employing Sectors	Top Ranked Output Sectors
3210 Textiles	3110 Food Products
3110 Food Products	3210 Textiles
3840 Transportation Equipment	3710 Iron and Steel
3140 Tobacco	3510 Industrial Chemicals
3820 Machinery nec	3840 Transportation Equipment

Population 984,004,000. Percent of world population: 16.604

	1990		1991		1992		1993		1994		1995	
	Value	%	Value	%	Value	%	Value	%	Value	%	Value	%
Establishments or enterprises (number)	126,630	5.500	129,115	5.400	138,430	4.900	141,331	6.900	-	-	-	-
Total employment (000)	9,477	4.500	9,527	5.100	10,140	5.600	10,111	5.600	8,078	5.100	8,390	5.900
Female employees (000)	-	-	-	-	-	-	-	-	-	-	-	-
Salaries and Wages ($ bil.)	13	0.605	11	0.462	12	0.463	11	0.532	9	0.473	11	0.641
Output ($ bil.)	176	1.200	149	0.969	160	0.976	155	1.100	140	1.000	163	1.200
Value added ($ bil.)	33	0.583	27	0.472	29	0.490	32	0.533	28	0.548	32	0.590
Capital investment ($ mil.)	12,515	2.100	11,064	1.800	12,893	2.000	13,567	2.500	-	-	-	-
Employees per establishment (number)	75	101.5	74	108.8	73	117.8	72	101.5	-	-	-	-
Output per establishment ($ mil.)	1	32.8	1	25.1	1	21.3	1	23.3	-	-	-	-
Value added per establishment ($ mil.)	0.258	15.7	0.207	12.3	0.212	10.1	0.225	14.0	-	-	-	-
Capital investment per estab. ($ mil.)	0.099	37.3	0.086	29.2	0.093	32.5	0.096	37.9	-	-	-	-
Payroll per employee ($)	1,375	10.1	1,152	6.4	1,167	5.8	1,073	6.6	1,144	6.5	1,268	7.3
Females as % of total employment (%)	-	-	-	-	-	-	-	-	-	-	-	-
Output per employee ($)	18,576	27.1	15,626	19.2	15,827	17.7	15,330	19.1	17,390	19.9	19,380	20.9
Value added per employee ($)	3,444	10.5	2,809	8.3	2,893	8.0	3,148	9.8	3,472	9.8	3,864	9.8
Capital investment per employee ($)	1,321	24.0	1,161	18.9	1,272	20.5	1,342	23.5	-	-	-	-

Columns headed % show percent of the world total or of the world ratio. Ratios closest to 100 are closest to the world average. Values higher than 100 are above, values lower than 100 are below world average. *nec* stands for *not elsewhere classified*. Ratios, where shown, are frequently based on a subset of total data; for example, payroll per employee is averaged from all cases where *both* payroll and employment data are present. Dividing *total* salaries/wages by *total* employment will not necessarily produce a matching value.

Indonesia

Top Ranked Employing Sectors	Top Ranked Output Sectors
3210 Textiles	3110 Food Products
3110 Food Products	3210 Textiles
3310 Wood Products	3840 Transportation Equipment
3220 Wearing Apparel	3310 Wood Products
3140 Tobacco	3830 Electrical Machinery

Population 212,942,000. Percent of world population: 3.593

	1990		1991		1992		1993		1994		1995	
	Value	%	Value	%	Value	%	Value	%	Value	%	Value	%
Establishments or enterprises (number)	18,694	0.811	18,680	0.784	20,139	0.717	20,639	1.000	21,551	1.200	24,361	2.000
Total employment (000)	3,147	1.500	3,560	1.900	3,942	2.200	4,241	2.300	4,547	2.900	4,967	3.500
Female employees (000)	-	-	-	-	-	-	1,978	8.900	2,144	10.200	2,270	11.900
Salaries and Wages ($ bil.)	2	0.100	3	0.112	4	0.139	4	0.197	4	0.227	8	0.454
Output ($ bil.)	47	0.332	55	0.359	66	0.404	79	0.542	91	0.661	110	0.821
Value added ($ bil.)	17	0.300	19	0.332	25	0.417	27	0.460	35	0.692	38	0.683
Capital investment ($ mil.)	6,451	1.100	8,725	1.400	5,940	0.929	4,961	0.913	6,577	1.500	7,306	2.300
Employees per establishment (number)	168	228.4	191	280.9	196	314.8	206	291.5	211	313.3	204	291.0
Output per establishment ($ mil.)	3	59.9	3	64.3	3	60.5	4	81.1	4	84.1	4	108.8
Value added per establishment ($ mil.)	0.899	54.9	1	59.6	1	58.8	1	82.7	2	93.3	2	87.2
Capital investment per estab. ($ mil.)	0.345	130.3	0.467	159.0	0.295	102.8	0.241	95.0	0.305	129.8	0.300	104.9
Payroll per employee ($)	686	5.0	745	4.1	897	4.5	948	5.8	974	5.5	1,516	8.7
Females as % of total employment (%)	-	-	-	-	-	-	46.6	93.4	47.1	89.9	45.7	78.3
Output per employee ($)	15,078	22.0	15,485	19.0	16,838	18.8	18,533	23.1	20,055	23.0	22,046	23.8
Value added per employee ($)	5,342	16.3	5,282	15.6	6,331	17.6	6,479	20.2	7,794	22.0	7,554	19.1
Capital investment per employee ($)	2,050	37.2	2,451	40.0	1,507	24.3	1,170	20.5	1,447	22.1	1,471	19.7

Columns headed % show percent of the world total or of the world ratio. Ratios closest to 100 are closest to the world average. Values higher than 100 are above, values lower than 100 are below world average. *nec* stands for *not elsewhere classified*. Ratios, where shown, are frequently based on a subset of total data; for example, payroll per employee is averaged from all cases where *both* payroll and employment data are present. Dividing *total* salaries/wages by *total* employment will not necessarily produce a matching value.

For additional notes and sources, see Appendix I.

Iran (Islamic Republic of)

Top Ranked Employing Sectors	Top Ranked Output Sectors
3210 Textiles	3110 Food Products
3110 Food Products	3710 Iron and Steel
3690 Nonmetal Products nec	3210 Textiles
3830 Electrical Machinery	3840 Transportation Equipment
3710 Iron and Steel	3830 Electrical Machinery

Population 68,960,000. Percent of world population: 1.164

	1990		1991		1992		1993		1994		1995	
	Value	%	Value	%	Value	%	Value	%	Value	%	Value	%
Establishments or enterprises (number)	10,518	0.456	6,301	0.264	6,782	0.241	6,576	0.320	-	-	-	-
Total employment (000)	653	0.312	793	0.424	820	0.449	783	0.430	-	-	-	-
Female employees (000)	-	-	47	0.519	49	0.226	46	0.208	-	-	-	-
Salaries and Wages ($ bil.)	3	0.125	4	0.163	4	0.167	4	0.197	-	-	-	-
Output ($ bil.)	18	0.128	24	0.159	25	0.153	24	0.165	-	-	-	-
Value added ($ bil.)	8	0.150	11	0.197	11	0.183	10	0.170	-	-	-	-
Capital investment ($ mil.)	798	0.135	1,566	0.251	1,923	0.301	1,672	0.308	-	-	-	-
Employees per establishment (number)	62	84.2	126	185.4	121	194.4	119	168.9	-	-	-	-
Output per establishment ($ mil.)	2	41.0	4	84.4	4	68.3	4	77.3	-	-	-	-
Value added per establishment ($ mil.)	0.800	48.8	2	105.1	2	76.4	2	95.8	-	-	-	-
Capital investment per estab. ($ mil.)	0.076	28.6	0.249	84.6	0.284	98.8	0.254	100.4	-	-	-	-
Payroll per employee ($)	4,134	30.2	4,897	27.1	5,207	25.9	5,130	31.6	-	-	-	-
Females as % of total employment (%)	-	-	5.9	18.3	5.9	12.8	5.9	11.8	-	-	-	-
Output per employee ($)	28,029	40.8	30,792	37.8	30,767	34.4	30,512	38.1	-	-	-	-
Value added per employee ($)	12,892	39.2	14,101	41.8	13,324	37.0	12,961	40.4	-	-	-	-
Capital investment per employee ($)	1,243	22.6	1,976	32.2	2,346	37.9	2,135	37.4	-	-	-	-

Columns headed % show percent of the world total or of the world ratio. Ratios closest to 100 are closest to the world average. Values higher than 100 are above, values lower than 100 are below world average. *nec* stands for *not elsewhere classified*. Ratios, where shown, are frequently based on a subset of total data; for example, payroll per employee is averaged from all cases where *both* payroll and employment data are present. Dividing *total* salaries/wages by *total* employment will not necessarily produce a matching value.

Iraq

Top Ranked Employing Sectors	Top Ranked Output Sectors
3690 Nonmetal Products nec	3690 Nonmetal Products nec
3210 Textiles	3530 Petroleum Refineries
3110 Food Products	3510 Industrial Chemicals
3830 Electrical Machinery	3210 Textiles
3530 Petroleum Refineries	3110 Food Products

Population 21,722,000. Percent of world population: 0.367

	1990		1991		1992		1993		1994		1995	
	Value	%	Value	%	Value	%	Value	%	Value	%	Value	%
Establishments or enterprises (number)	-	-	266	0.011	621	0.022	-	-	-	-	-	-
Total employment (000)	-	-	121	0.065	121	0.066	-	-	-	-	-	-
Female employees (000)	-	-	26	0.291	-	-	-	-	-	-	-	-
Salaries and Wages ($ bil.)	-	-	1	0.046	2	0.081	-	-	-	-	-	-
Output ($ bil.)	-	-	5	0.031	13	0.080	-	-	-	-	-	-
Value added ($ bil.)	-	-	2	0.040	6	0.102	-	-	-	-	-	-
Capital investment ($ mil.)	-	-	-	-	-	-	-	-	-	-	-	-
Employees per establishment (number)	-	-	267	392.8	193	309.9	-	-	-	-	-	-
Output per establishment ($ mil.)	-	-	10	220.8	21	386.6	-	-	-	-	-	-
Value added per establishment ($ mil.)	-	-	4	226.1	10	464.0	-	-	-	-	-	-
Capital investment per estab. ($ mil.)	-	-	-	-	-	-	-	-	-	-	-	-
Payroll per employee ($)	-	-	8,952	49.5	17,238	85.9	-	-	-	-	-	-
Females as % of total employment (%)	-	-	21.6	67.3	-	-	-	-	-	-	-	-
Output per employee ($)	-	-	38,776	47.6	108,734	121.5	-	-	-	-	-	-
Value added per employee ($)	-	-	18,589	55.0	50,464	140.2	-	-	-	-	-	-
Capital investment per employee ($)	-	-	-	-	-	-	-	-	-	-	-	-

Columns headed % show percent of the world total or of the world ratio. Ratios closest to 100 are closest to the world average. Values higher than 100 are above, values lower than 100 are below world average. *nec* stands for *not elsewhere classified*. Ratios, where shown, are frequently based on a subset of total data; for example, payroll per employee is averaged from all cases where *both* payroll and employment data are present. Dividing *total* salaries/wages by *total* employment will not necessarily produce a matching value.

For additional notes and sources, see Appendix I.

Ireland

Top Ranked Employing Sectors	Top Ranked Output Sectors
3110 Food Products	3110 Food Products
3830 Electrical Machinery	3820 Machinery nec
3820 Machinery nec	3830 Electrical Machinery
3810 Metal Products	3520 Chemical Products nec
3210 Textiles	3510 Industrial Chemicals

Population 3,619,000. Percent of world population: 0.061

	1990 Value	%	1991 Value	%	1992 Value	%	1993 Value	%	1994 Value	%	1995 Value	%
Establishments or enterprises (number)	4,374	0.190	4,158	0.174	-	-	-	-	-	-	-	-
Total employment (000)	193	0.092	194	0.104	195	0.107	195	0.107	199	0.125	205	0.144
Female employees (000)	62	0.469	62	0.690	-	-	-	-	-	-	-	-
Salaries and Wages ($ bil.)	4	0.188	4	0.176	5	0.190	4	0.214	5	0.253	6	0.371
Output ($ bil.)	33	0.232	33	0.215	39	0.240	36	0.249	42	0.305	54	0.408
Value added ($ bil.)	15	0.268	16	0.273	19	0.312	17	0.287	20	0.394	26	0.482
Capital investment ($ mil.)	1,143	0.193	1,264	0.203	-	-	-	-	-	-	-	-
Employees per establishment (number)	41	55.8	43	63.7	-	-	-	-	-	-	-	-
Output per establishment ($ mil.)	7	156.8	7	150.8	-	-	-	-	-	-	-	-
Value added per establishment ($ mil.)	3	174.5	3	181.7	-	-	-	-	-	-	-	-
Capital investment per estab. ($ mil.)	0.263	99.4	0.304	103.5	-	-	-	-	-	-	-	-
Payroll per employee ($)	20,956	153.2	21,523	118.9	24,832	123.7	22,436	138.2	24,884	141.7	30,022	171.8
Females as % of total employment (%)	32.3	100.8	32.0	99.7	-	-	-	-	-	-	-	-
Output per employee ($)	171,314	249.6	170,551	209.5	202,252	226.1	185,441	231.6	211,541	242.1	265,234	286.6
Value added per employee ($)	77,469	235.8	79,862	236.5	95,500	265.2	88,004	274.6	101,714	286.6	129,145	327.0
Capital investment per employee ($)	6,376	115.8	7,029	114.6	-	-	-	-	-	-	-	-

Columns headed % show percent of the world total or of the world ratio. Ratios closest to 100 are closest to the world average. Values higher than 100 are above, values lower than 100 are below world average. *nec* stands for *not elsewhere classified*. Ratios, where shown, are frequently based on a subset of total data; for example, payroll per employee is averaged from all cases where *both* payroll and employment data are present. Dividing *total* salaries/wages by *total* employment will not necessarily produce a matching value.

Israel

Top Ranked Employing Sectors	Top Ranked Output Sectors
3830 Electrical Machinery	3830 Electrical Machinery
3810 Metal Products	3110 Food Products
3110 Food Products	3810 Metal Products
3220 Wearing Apparel	3510 Industrial Chemicals
3420 Printing, Publishing	3840 Transportation Equipment

Population 5,644,000. Percent of world population: 0.095

	1990 Value	%	1991 Value	%	1992 Value	%	1993 Value	%	1994 Value	%	1995 Value	%
Establishments or enterprises (number)	8,441	0.366	9,276	0.389	9,373	0.334	10,047	0.489	10,330	0.577	-	-
Total employment (000)	281	0.134	295	0.158	303	0.166	318	0.174	326	0.205	332	0.234
Female employees (000)	-	-	-	-	-	-	-	-	-	-	-	-
Salaries and Wages ($ bil.)	6	0.270	6	0.262	7	0.265	7	0.327	7	0.373	8	0.504
Output ($ bil.)	23	0.160	25	0.161	27	0.164	27	0.187	29	0.213	34	0.251
Value added ($ bil.)	10	0.182	11	0.195	10	0.168	10	0.167	10	0.205	12	0.224
Capital investment ($ mil.)	1,082	0.182	1,451	0.233	1,591	0.249	1,967	0.362	2,263	0.510	-	-
Employees per establishment (number)	33	45.1	32	46.8	32	51.3	31	44.1	32	46.8	-	-
Output per establishment ($ mil.)	3	64.0	3	57.9	3	51.9	3	56.2	3	56.5	-	-
Value added per establishment ($ mil.)	1	69.0	1	66.3	1	50.3	0.976	60.6	1	57.6	-	-
Capital investment per estab. ($ mil.)	0.128	48.4	0.156	53.2	0.170	59.2	0.196	77.3	0.219	93.1	-	-
Payroll per employee ($)	20,746	151.6	21,113	116.7	22,291	111.0	20,981	129.2	22,325	127.1	25,155	143.9
Females as % of total employment (%)	-	-	-	-	-	-	-	-	-	-	-	-
Output per employee ($)	81,542	118.8	83,737	102.9	88,970	99.4	85,391	106.7	90,096	103.1	100,915	109.0
Value added per employee ($)	34,012	103.5	35,252	104.4	33,225	92.3	31,385	97.9	32,219	90.8	36,979	93.6
Capital investment per employee ($)	3,855	70.0	4,924	80.3	5,325	86.0	6,295	110.1	6,946	106.2	-	-

Columns headed % show percent of the world total or of the world ratio. Ratios closest to 100 are closest to the world average. Values higher than 100 are above, values lower than 100 are below world average. *nec* stands for *not elsewhere classified*. Ratios, where shown, are frequently based on a subset of total data; for example, payroll per employee is averaged from all cases where *both* payroll and employment data are present. Dividing *total* salaries/wages by *total* employment will not necessarily produce a matching value.

                                                                    For additional notes and sources, see Appendix I.

Italy

Top Ranked Employing Sectors	Top Ranked Output Sectors
3820 Machinery nec	3820 Machinery nec
3840 Transportation Equipment	3110 Food Products
3810 Metal Products	3840 Transportation Equipment
3830 Electrical Machinery	3210 Textiles
3210 Textiles	3830 Electrical Machinery

Population 56,783,000. Percent of world population: 0.958

	1990		1991		1992		1993		1994		1995	
	Value	%	Value	%	Value	%	Value	%	Value	%	Value	%
Establishments or enterprises (number)	32,476	1.400	32,789	1.400	38,675	1.400	39,079	1.900	38,941	2.200	-	-
Total employment (000)	3,181	1.500	3,164	1.700	2,957	1.600	2,836	1.600	2,798	1.800	-	-
Female employees (000)	-	-	818	9.100	758	3.500	729	3.300	718	3.400	-	-
Salaries and Wages ($ bil.)	113	5.300	117	4.900	114	4.500	88	4.300	90	4.600	-	-
Output ($ bil.)	557	3.900	552	3.600	528	3.200	417	2.900	462	3.400	-	-
Value added ($ bil.)	168	3.000	168	3.000	168	2.800	133	2.200	141	2.800	-	-
Capital investment ($ mil.)	32,904	5.500	32,034	5.100	27,330	4.300	19,721	3.600	20,535	4.600	-	-
Employees per establishment (number)	98	132.9	96	142.2	77	123.4	73	103.3	72	107.0	-	-
Output per establishment ($ mil.)	17	404.8	17	366.6	14	251.2	11	228.0	12	236.7	-	-
Value added per establishment ($ mil.)	5	316.0	5	303.2	4	206.3	3	211.9	4	206.4	-	-
Capital investment per estab. ($ mil.)	1	382.5	0.977	332.5	0.709	247.0	0.506	199.8	0.529	224.8	-	-
Payroll per employee ($)	35,550	259.8	36,976	204.3	38,664	192.6	31,153	191.8	32,015	182.3	-	-
Females as % of total employment (%)	-	-	25.9	80.5	25.6	55.3	25.7	51.5	25.7	49.0	-	-
Output per employee ($)	175,051	255.1	174,389	214.2	178,384	199.4	147,089	183.7	165,223	189.1	-	-
Value added per employee ($)	52,863	160.9	53,041	157.1	56,680	157.4	46,878	146.3	50,504	142.3	-	-
Capital investment per employee ($)	10,343	187.8	10,125	165.1	9,241	149.2	6,953	121.7	7,340	112.2	-	-

Columns headed % show percent of the world total or of the world ratio. Ratios closest to 100 are closest to the world average. Values higher than 100 are above, values lower than 100 are below world average. *nec* stands for *not elsewhere classified*. Ratios, where shown, are frequently based on a subset of total data; for example, payroll per employee is averaged from all cases where *both* payroll and employment data are present. Dividing *total* salaries/wages by *total* employment will not necessarily produce a matching value.

Jamaica

Top Ranked Employing Sectors	Top Ranked Output Sectors
3110 Food Products	3110 Food Products
3220 Wearing Apparel	3530 Petroleum Refineries
3130 Beverages	3130 Beverages
3520 Chemical Products nec	3140 Tobacco
3320 Furniture, Fixtures	3320 Furniture, Fixtures

Population 2,635,000. Percent of world population: 0.045

	1990		1991		1992		1993		1994		1995	
	Value	%	Value	%	Value	%	Value	%	Value	%	Value	%
Establishments or enterprises (number)	635	0.028	635	0.027	-	-	-	-	-	-	-	-
Total employment (000)	59	0.028	58	0.031	59	0.033	60	0.033	58	0.036	-	-
Female employees (000)	-	-	-	-	-	-	-	-	-	-	-	-
Salaries and Wages ($ bil.)	0.141	0.007	0.121	0.005	0.112	0.004	-	-	-	-	-	-
Output ($ bil.)	2	0.011	1	0.009	1	0.008	-	-	-	-	-	-
Value added ($ bil.)	0.531	0.009	0.433	0.008	0.367	0.006	-	-	-	-	-	-
Capital investment ($ mil.)	-	-	-	-	-	-	-	-	-	-	-	-
Employees per establishment (number)	90	122.6	88	129.3	-	-	-	-	-	-	-	-
Output per establishment ($ mil.)	5	114.7	4	88.9	-	-	-	-	-	-	-	-
Value added per establishment ($ mil.)	2	98.3	1	78.0	-	-	-	-	-	-	-	-
Capital investment per estab. ($ mil.)	-	-	-	-	-	-	-	-	-	-	-	-
Payroll per employee ($)	3,972	29.0	3,568	19.7	3,202	15.9	-	-	-	-	-	-
Females as % of total employment (%)	-	-	-	-	-	-	-	-	-	-	-	-
Output per employee ($)	34,269	49.9	30,604	37.6	27,556	30.8	-	-	-	-	-	-
Value added per employee ($)	11,032	33.6	9,726	28.8	8,672	24.1	-	-	-	-	-	-
Capital investment per employee ($)	-	-	-	-	-	-	-	-	-	-	-	-

Columns headed % show percent of the world total or of the world ratio. Ratios closest to 100 are closest to the world average. Values higher than 100 are above, values lower than 100 are below world average. *nec* stands for *not elsewhere classified*. Ratios, where shown, are frequently based on a subset of total data; for example, payroll per employee is averaged from all cases where *both* payroll and employment data are present. Dividing *total* salaries/wages by *total* employment will not necessarily produce a matching value.

Japan

Top Ranked Employing Sectors	Top Ranked Output Sectors
3830 Electrical Machinery	3840 Transportation Equipment
3820 Machinery nec	3830 Electrical Machinery
3110 Food Products	3820 Machinery nec
3840 Transportation Equipment	3110 Food Products
3810 Metal Products	3810 Metal Products

Population 125,932,000. Percent of world population: 2.125

	1990		1991		1992		1993		1994		1995	
	Value	%	Value	%	Value	%	Value	%	Value	%	Value	%
Establishments or enterprises (number)	491,532	21.300	485,358	20.400	467,281	16.600	463,833	22.600	382,133	21.400	387,012	31.500
Total employment (000)	13,988	6.700	14,225	7.600	13,957	7.600	13,580	7.500	10,408	6.500	10,305	7.200
Female employees (000)	-		-		-		-		-		-	
Salaries and Wages ($ bil.)	375	17.400	431	18.100	465	18.200	524	25.700	436	22.300	477	28.800
Output ($ bil.)	3,006	21.000	3,404	22.100	3,478	21.200	3,732	25.700	2,913	21.100	3,256	24.400
Value added ($ bil.)	1,174	21.000	1,332	23.500	1,369	22.900	1,486	24.900	1,209	23.600	1,362	24.800
Capital investment ($ mil.)	144,913	24.400	189,229	30.300	182,961	28.600	167,536	30.800	110,655	24.900	-	-
Employees per establishment (number)	28	38.6	29	43.2	30	48.0	29	41.5	27	40.4	27	38.0
Output per establishment ($ mil.)	6	144.4	7	152.8	7	136.7	8	171.3	8	151.6	8	203.7
Value added per establishment ($ mil.)	2	145.8	3	162.6	3	139.0	3	199.0	3	179.5	4	199.3
Capital investment per estab. ($ mil.)	0.295	111.3	0.390	132.7	0.392	136.4	0.361	142.6	0.290	123.1	-	-
Payroll per employee ($)	26,843	196.2	30,287	167.3	33,283	165.8	38,596	237.7	41,885	238.5	46,296	264.9
Females as % of total employment (%)	-		-		-		-		-		-	
Output per employee ($)	214,874	313.1	239,299	294.0	249,160	278.5	274,816	343.3	279,873	320.3	315,988	341.4
Value added per employee ($)	83,929	255.5	93,615	277.2	98,084	272.4	109,418	341.5	116,186	327.3	132,171	334.6
Capital investment per employee ($)	10,360	188.1	13,303	217.0	13,109	211.6	12,337	215.8	10,632	162.6	-	-

Columns headed % show percent of the world total or of the world ratio. Ratios closest to 100 are closest to the world average. Values higher than 100 are above, values lower than 100 are below world average. *nec* stands for *not elsewhere classified*. Ratios, where shown, are frequently based on a subset of total data; for example, payroll per employee is averaged from all cases where *both* payroll and employment data are present. Dividing *total* salaries/wages by *total* employment will not necessarily produce a matching value.

Jordan

Top Ranked Employing Sectors	Top Ranked Output Sectors
3110 Food Products	3110 Food Products
3690 Nonmetal Products nec	3530 Petroleum Refineries
3810 Metal Products	3510 Industrial Chemicals
3520 Chemical Products nec	3690 Nonmetal Products nec
3220 Wearing Apparel	3520 Chemical Products nec

Population 4,435,000. Percent of world population: 0.075

	1990		1991		1992		1993		1994		1995	
	Value	%	Value	%	Value	%	Value	%	Value	%	Value	%
Establishments or enterprises (number)	9,163	0.398	9,460	0.397	11,660	0.415	11,943	0.581	12,483	0.698	13,781	1.100
Total employment (000)	48	0.023	55	0.030	69	0.038	74	0.040	89	0.056	91	0.064
Female employees (000)	5	0.036	6	0.063	7	0.032	7	0.029	9	0.044	9	0.049
Salaries and Wages ($ bil.)	0.136	0.006	0.150	0.006	0.187	0.007	0.209	0.010	0.248	0.013	0.280	0.017
Output ($ bil.)	2	0.014	2	0.014	3	0.018	3	0.021	4	0.027	4	0.030
Value added ($ bil.)	0.644	0.012	0.669	0.012	0.803	0.013	0.887	0.015	1	0.022	1	0.020
Capital investment ($ mil.)	20	0.003	38	0.006	67	0.010	45	0.008	165	0.037	116	0.037
Employees per establishment (number)	5	7.0	6	8.6	6	9.5	6	8.7	7	10.5	7	9.4
Output per establishment ($ mil.)	0.209	4.9	0.226	4.9	0.250	4.6	0.249	5.3	0.298	5.9	0.291	7.1
Value added per establishment ($ mil.)	0.067	4.1	0.071	4.2	0.069	3.3	0.074	4.6	0.089	5.0	0.079	4.5
Capital investment per estab. ($ mil.)	0.002	0.8	0.004	1.4	0.006	2.2	0.004	1.5	0.013	5.7	0.008	3.0
Payroll per employee ($)	2,841	20.8	2,724	15.1	2,720	13.5	2,836	17.5	2,798	15.9	3,070	17.6
Females as % of total employment (%)	10.2	31.8	10.5	32.7	10.0	21.7	8.9	17.7	10.5	19.9	10.4	17.7
Output per employee ($)	42,162	61.4	38,795	47.7	42,441	47.4	40,473	50.6	42,021	48.1	44,027	47.6
Value added per employee ($)	13,404	40.8	12,130	35.9	11,682	32.4	12,067	37.7	12,508	35.2	11,973	30.3
Capital investment per employee ($)	454	8.3	747	12.2	1,070	17.3	672	11.7	2,031	31.1	1,307	17.5

Columns headed % show percent of the world total or of the world ratio. Ratios closest to 100 are closest to the world average. Values higher than 100 are above, values lower than 100 are below world average. *nec* stands for *not elsewhere classified*. Ratios, where shown, are frequently based on a subset of total data; for example, payroll per employee is averaged from all cases where *both* payroll and employment data are present. Dividing *total* salaries/wages by *total* employment will not necessarily produce a matching value.

For additional notes and sources, see Appendix I.

Kenya

Top Ranked Employing Sectors	Top Ranked Output Sectors
3110 Food Products	3110 Food Products
3210 Textiles	3840 Transportation Equipment
3840 Transportation Equipment	3520 Chemical Products nec
3810 Metal Products	3530 Petroleum Refineries
3520 Chemical Products nec	3810 Metal Products

Population 28,337,000. Percent of world population: 0.478

	1990		1991		1992		1993		1994		1995	
	Value	%	Value	%	Value	%	Value	%	Value	%	Value	%
Establishments or enterprises (number)	2,050	0.089	2,961	0.124	3,289	0.117	3,316	0.161	3,740	0.209	3,747	0.305
Total employment (000)	208	0.099	208	0.111	209	0.114	213	0.117	208	0.131	215	0.151
Female employees (000)	23	0.170	23	0.256	23	0.107	24	0.106	-		-	
Salaries and Wages ($ bil.)	0.336	0.016	0.279	0.012	0.266	0.010	0.171	0.008	0.213	0.011	0.303	0.018
Output ($ bil.)	8	0.055	8	0.050	9	0.052	6	0.038	7	0.049	9	0.065
Value added ($ bil.)	0.909	0.016	0.895	0.016	0.794	0.013	0.509	0.009	0.690	0.013	0.889	0.016
Capital investment ($ mil.)	-		-		-		-		-		-	
Employees per establishment (number)	98	132.3	67	99.4	61	98.2	62	87.5	56	82.6	57	81.9
Output per establishment ($ mil.)	4	91.2	3	56.9	3	47.3	2	35.4	2	36.2	2	55.7
Value added per establishment ($ mil.)	0.415	25.3	0.280	16.6	0.239	11.3	0.152	9.5	0.184	10.4	0.221	12.5
Capital investment per estab. ($ mil.)	-		-		-		-		-		-	
Payroll per employee ($)	1,619	11.8	1,397	7.7	1,322	6.6	834	5.1	1,023	5.8	1,409	8.1
Females as % of total employment (%)	10.9	34.0	11.1	34.5	11.1	23.9	11.1	22.3	-		-	
Output per employee ($)	43,165	62.9	42,169	51.8	46,134	51.6	29,524	36.9	35,550	40.7	43,497	47.0
Value added per employee ($)	4,982	15.2	4,881	14.5	4,385	12.2	2,762	8.6	3,403	9.6	4,167	10.5
Capital investment per employee ($)	-		-		-		-		-		-	

Columns headed % show percent of the world total or of the world ratio. Ratios closest to 100 are closest to the world average. Values higher than 100 are above, values lower than 100 are below world average. *nec* stands for *not elsewhere classified*. Ratios, where shown, are frequently based on a subset of total data; for example, payroll per employee is averaged from all cases where *both* payroll and employment data are present. Dividing *total* salaries/wages by *total* employment will not necessarily produce a matching value.

Korea, Republic of

Top Ranked Employing Sectors	Top Ranked Output Sectors
3830 Electrical Machinery	3830 Electrical Machinery
3840 Transportation Equipment	3840 Transportation Equipment
3820 Machinery nec	3820 Machinery nec
3210 Textiles	3710 Iron and Steel
3810 Metal Products	3110 Food Products

Population 46,417,000. Percent of world population: 0.783

	1990		1991		1992		1993		1994		1995	
	Value	%	Value	%	Value	%	Value	%	Value	%	Value	%
Establishments or enterprises (number)	81,297	3.500	84,915	3.600	87,557	3.100	103,066	5.000	106,142	5.900	111,523	9.100
Total employment (000)	3,896	1.900	3,702	2.000	3,587	2.000	3,658	2.000	3,742	2.400	3,773	2.700
Female employees (000)	1,539	11.600	1,393	15.500	1,302	6.100	1,280	5.700	1,280	6.100	1,264	6.600
Salaries and Wages ($ bil.)	37	1.700	41	1.700	43	1.700	48	2.300	55	2.800	66	4.000
Output ($ bil.)	337	2.400	370	2.400	386	2.400	425	2.900	500	3.600	645	4.800
Value added ($ bil.)	134	2.400	155	2.700	163	2.700	180	3.000	217	4.200	285	5.200
Capital investment ($ mil.)	30,332	5.100	42,144	6.800	46,617	7.300	50,816	9.300	56,621	12.800	79,978	25.600
Employees per establishment (number)	48	65.0	44	64.3	41	65.9	35	50.3	35	52.4	34	48.3
Output per establishment ($ mil.)	4	97.9	4	94.8	4	81.1	4	87.7	5	93.7	6	140.0
Value added per establishment ($ mil.)	2	100.5	2	108.4	2	88.6	2	108.4	2	116.0	3	144.8
Capital investment per estab. ($ mil.)	0.440	166.3	0.586	199.6	0.532	185.5	0.493	194.6	0.533	226.7	0.717	250.9
Payroll per employee ($)	9,553	69.8	11,115	61.4	12,065	60.1	13,045	80.3	14,629	83.3	17,538	100.4
Females as % of total employment (%)	39.5	123.3	37.6	117.2	36.3	78.4	35.0	70.1	34.2	65.2	33.5	57.4
Output per employee ($)	86,526	126.1	99,853	122.7	107,726	120.4	116,102	145.0	133,728	153.1	170,943	184.7
Value added per employee ($)	34,368	104.6	41,958	124.2	45,564	126.5	49,179	153.5	58,010	163.4	75,570	191.3
Capital investment per employee ($)	10,255	186.2	15,044	245.4	12,995	209.8	13,890	243.0	15,130	231.4	21,197	283.8

Columns headed % show percent of the world total or of the world ratio. Ratios closest to 100 are closest to the world average. Values higher than 100 are above, values lower than 100 are below world average. *nec* stands for *not elsewhere classified*. Ratios, where shown, are frequently based on a subset of total data; for example, payroll per employee is averaged from all cases where *both* payroll and employment data are present. Dividing *total* salaries/wages by *total* employment will not necessarily produce a matching value.

Kuwait

Top Ranked Employing Sectors	Top Ranked Output Sectors
3220 Wearing Apparel	3530 Petroleum Refineries
3110 Food Products	3110 Food Products
3810 Metal Products	3690 Nonmetal Products nec
3530 Petroleum Refineries	3810 Metal Products
3820 Machinery nec	3220 Wearing Apparel

Population 1,913,000. Percent of world population: 0.032

	1990		1991		1992		1993		1994		1995	
	Value	%	Value	%	Value	%	Value	%	Value	%	Value	%
Establishments or enterprises (number)	4,095	0.178	3,934	0.165	4,046	0.144	4,044	0.197	4,030	0.225	-	-
Total employment (000)	58	0.028	42	0.022	57	0.031	62	0.034	60	0.038	57	0.040
Female employees (000)	0.002		0.018		0.001		0.025		0.003		-	
Salaries and Wages ($ bil.)	0.415	0.019	0.298	0.013	0.624	0.024	0.669	0.033	0.662	0.034	0.853	0.051
Output ($ bil.)	6	0.039	1	0.009	4	0.026	5	0.036	7	0.047	7	0.054
Value added ($ bil.)	2	0.039	0.605	0.011	2	0.032	2	0.038	3	0.055	3	0.057
Capital investment ($ mil.)	135	0.023	401	0.064	250	0.039	279	0.051	268	0.060	-	
Employees per establishment (number)	14	19.3	11	15.6	14	22.7	15	21.7	15	22.0	-	
Output per establishment ($ mil.)	1	32.1	0.370	8.1	1	19.6	1	27.5	2	32.2	-	
Value added per establishment ($ mil.)	0.535	32.6	0.154	9.1	0.478	22.7	0.567	35.2	0.700	39.7	-	
Capital investment per estab. ($ mil.)	0.033	12.5	0.102	34.7	0.062	21.6	0.069	27.2	0.066	28.2	-	
Payroll per employee ($)	7,129	52.1	7,169	39.6	10,911	54.3	10,789	66.4	11,086	63.1	14,876	85.1
Females as % of total employment (%)	0.2	0.5	1.9	5.8	0.1	0.2	2.4	4.8	0.3	0.5	-	
Output per employee ($)	95,644	139.4	35,024	43.0	75,609	84.5	84,155	105.1	109,341	125.1	124,498	134.5
Value added per employee ($)	37,635	114.6	14,553	43.1	33,831	94.0	36,966	115.4	47,233	133.1	54,683	138.5
Capital investment per employee ($)	2,364	42.9	9,638	157.2	4,383	70.8	4,501	78.7	4,482	68.5	-	

Columns headed % show percent of the world total or of the world ratio. Ratios closest to 100 are closest to the world average. Values higher than 100 are above, values lower than 100 are below world average. *nec* stands for *not elsewhere classified*. Ratios, where shown, are frequently based on a subset of total data; for example, payroll per employee is averaged from all cases where *both* payroll and employment data are present. Dividing *total* salaries/wages by *total* employment will not necessarily produce a matching value.

Kyrgyzstan

Top Ranked Employing Sectors	Top Ranked Output Sectors
3210 Textiles	3210 Textiles
3110 Food Products	3110 Food Products
3820 Machinery nec	3720 Nonferrous metals
3810 Metal Products	3820 Machinery nec
3830 Electrical Machinery	3690 Nonmetal Products nec

Population 4,522,000. Percent of world population: 0.076

	1990		1991		1992		1993		1994		1995	
	Value	%	Value	%	Value	%	Value	%	Value	%	Value	%
Establishments or enterprises (number)	2,909	0.126	2,875	0.121	1,993	0.071	2,063	0.100	1,860	0.104	-	-
Total employment (000)	287	0.137	272	0.145	239	0.131	208	0.114	152	0.095	-	-
Female employees (000)	-	-	-	-	-	-	-	-	-	-	-	-
Salaries and Wages ($ bil.)	0.099	0.005	0.082	0.003	0.045	0.002	0.012	0.001	0.015	0.001		
Output ($ bil.)	5	0.036	6	0.039	4	0.022	0.600	0.004	0.486	0.004		
Value added ($ bil.)	-		-		-		-		-			
Capital investment ($ mil.)	-		-		-		-		-			
Employees per establishment (number)	99	134.0	94	139.3	120	193.1	101	143.4	82	122.4	-	
Output per establishment ($ mil.)	2	41.5	2	45.7	2	33.6	0.291	6.2	0.261	5.2	-	
Value added per establishment ($ mil.)	-		-		-		-		-		-	
Capital investment per estab. ($ mil.)	-		-		-		-		-		-	
Payroll per employee ($)	346	2.5	302	1.7	189	0.9	58	0.4	96	0.5	-	
Females as % of total employment (%)	-		-		-		-		-		-	
Output per employee ($)	17,802	25.9	22,177	27.2	15,220	17.0	2,877	3.6	3,195	3.7	-	
Value added per employee ($)	-		-		-		-		-		-	
Capital investment per employee ($)	-		-		-		-		-		-	

Columns headed % show percent of the world total or of the world ratio. Ratios closest to 100 are closest to the world average. Values higher than 100 are above, values lower than 100 are below world average. *nec* stands for *not elsewhere classified*. Ratios, where shown, are frequently based on a subset of total data; for example, payroll per employee is averaged from all cases where *both* payroll and employment data are present. Dividing *total* salaries/wages by *total* employment will not necessarily produce a matching value.

For additional notes and sources, see Appendix I.

Latvia

Top Ranked Employing Sectors	Top Ranked Output Sectors
3110 Food Products	3110 Food Products
3310 Wood Products	3310 Wood Products
3830 Electrical Machinery	3840 Transportation Equipment
3210 Textiles	3130 Beverages
3840 Transportation Equipment	3210 Textiles

Population 2,385,000. Percent of world population: 0.040

	1990		1991		1992		1993		1994		1995	
	Value	%	Value	%	Value	%	Value	%	Value	%	Value	%
Establishments or enterprises (number)	4,982	0.216	5,496	0.231	2,322	0.083	2,852	0.139	1,691	0.094	2,940	0.240
Total employment (000)	346	0.165	329	0.176	280	0.153	196	0.108	164	0.103	154	0.108
Female employees (000)	-	-	-	-	-	-	-	-	-	-	-	-
Salaries and Wages ($ bil.)	-	-	-	-	-	-	-	-	-	-	-	-
Output ($ bil.)	0.069	-	0.166	0.001	0.990	0.006	1	0.008	2	0.011	2	0.014
Value added ($ bil.)	-	-	-	-	-	-	0.498	0.008	0.705	0.014	0.828	0.015
Capital investment ($ mil.)	2	-	2	-	35	0.006	51	0.009	73	0.016	200	0.064
Employees per establishment (number)	69	94.1	60	88.2	121	193.8	69	97.5	97	143.9	52	74.7
Output per establishment ($ mil.)	0.014	0.3	0.030	0.7	0.426	7.8	0.416	8.9	0.899	17.9	0.630	15.2
Value added per establishment ($ mil.)	-	-	-	-	-	-	0.175	10.8	0.417	23.7	0.282	15.9
Capital investment per estab. ($ mil.)	0.000	0.1	0.000	0.1	0.015	5.3	0.018	7.1	0.043	18.3	0.068	23.8
Payroll per employee ($)	-	-	-	-	-	-	-	-	-	-	-	-
Females as % of total employment (%)	-	-	-	-	-	-	-	-	-	-	-	-
Output per employee ($)	200	0.3	504	0.6	3,537	4.0	6,060	7.6	9,274	10.6	12,035	13.0
Value added per employee ($)	-	-	-	-	-	-	2,542	7.9	4,305	12.1	5,383	13.6
Capital investment per employee ($)	5	0.1	5	0.1	126	2.0	270	4.7	445	6.8	1,303	17.5

Columns headed % show percent of the world total or of the world ratio. Ratios closest to 100 are closest to the world average. Values higher than 100 are above, values lower than 100 are below world average. *nec* stands for *not elsewhere classified*. Ratios, where shown, are frequently based on a subset of total data; for example, payroll per employee is averaged from all cases where *both* payroll and employment data are present. Dividing *total* salaries/wages by *total* employment will not necessarily produce a matching value.

Lesotho

Top Ranked Employing Sectors	Top Ranked Output Sectors
3110 Food Products	
3810 Metal Products	
3900 Industries nec	
3130 Beverages	
3690 Nonmetal Products nec	

Population 2,090,000. Percent of world population: 0.035

	1990		1991		1992		1993		1994		1995	
	Value	%	Value	%	Value	%	Value	%	Value	%	Value	%
Establishments or enterprises (number)	256	0.011	-	-	7	-	23	0.001	24	0.001	-	-
Total employment (000)	5	0.002	-	-	0.985	0.001	5	0.003	4	0.003	-	-
Female employees (000)	-	-	-	-	-	-	2	0.010	2	0.010	-	-
Salaries and Wages ($ bil.)	-	-	-	-	0.002	-	0.012	0.001	0.012	0.001	-	-
Output ($ bil.)	-	-	-	-	-	-	-	-	-	-	-	-
Value added ($ bil.)	-	-	-	-	-	-	-	-	-	-	-	-
Capital investment ($ mil.)	-	-	-	-	-	-	-	-	-	-	-	-
Employees per establishment (number)	19	25.6	-	-	124	199.7	203	287.9	171	253.9	-	-
Output per establishment ($ mil.)	-	-	-	-	-	-	-	-	-	-	-	-
Value added per establishment ($ mil.)	-	-	-	-	-	-	-	-	-	-	-	-
Capital investment per estab. ($ mil.)	-	-	-	-	-	-	-	-	-	-	-	-
Payroll per employee ($)	-	-	-	-	1,896	9.4	2,478	15.3	3,292	18.7	-	-
Females as % of total employment (%)	-	-	-	-	-	-	47.5	95.0	49.9	95.1	-	-
Output per employee ($)	-	-	-	-	-	-	-	-	-	-	-	-
Value added per employee ($)	-	-	-	-	-	-	-	-	-	-	-	-
Capital investment per employee ($)	-	-	-	-	-	-	-	-	-	-	-	-

Columns headed % show percent of the world total or of the world ratio. Ratios closest to 100 are closest to the world average. Values higher than 100 are above, values lower than 100 are below world average. *nec* stands for *not elsewhere classified*. Ratios, where shown, are frequently based on a subset of total data; for example, payroll per employee is averaged from all cases where *both* payroll and employment data are present. Dividing *total* salaries/wages by *total* employment will not necessarily produce a matching value.

Liechtenstein

Top Ranked Employing Sectors	Top Ranked Output Sectors
3830 Electrical Machinery	
3810 Metal Products	
3110 Food Products	
3420 Printing, Publishing	
3510 Industrial Chemicals	

Population 32,000. Percent of world population: 0.000

	1990		1991		1992		1993		1994		1995	
	Value	%	Value	%	Value	%	Value	%	Value	%	Value	%
Establishments or enterprises (number)	-	-	150	0.006	-	-	-	-	-	-	-	-
Total employment (000)	-	-	4	0.002	-	-	-	-	-	-	-	-
Female employees (000)	-	-	-	-	-	-	-	-	-	-	-	-
Salaries and Wages ($ bil.)	-	-	-	-	-	-	-	-	-	-	-	-
Output ($ bil.)	-	-	-	-	-	-	-	-	-	-	-	-
Value added ($ bil.)	-	-	-	-	-	-	-	-	-	-	-	-
Capital investment ($ mil.)	-	-	-	-	-	-	-	-	-	-	-	-
Employees per establishment (number)	-	-	28	40.9	-	-	-	-	-	-	-	-
Output per establishment ($ mil.)	-	-	-	-	-	-	-	-	-	-	-	-
Value added per establishment ($ mil.)	-	-	-	-	-	-	-	-	-	-	-	-
Capital investment per estab. ($ mil.)	-	-	-	-	-	-	-	-	-	-	-	-
Payroll per employee ($)	-	-	-	-	-	-	-	-	-	-	-	-
Females as % of total employment (%)	-	-	-	-	-	-	-	-	-	-	-	-
Output per employee ($)	-	-	-	-	-	-	-	-	-	-	-	-
Value added per employee ($)	-	-	-	-	-	-	-	-	-	-	-	-
Capital investment per employee ($)	-	-	-	-	-	-	-	-	-	-	-	-

Columns headed % show percent of the world total or of the world ratio. Ratios closest to 100 are closest to the world average. Values higher than 100 are above, values lower than 100 are below world average. *nec* stands for *not elsewhere classified*. Ratios, where shown, are frequently based on a subset of total data; for example, payroll per employee is averaged from all cases where *both* payroll and employment data are present. Dividing *total* salaries/wages by *total* employment will not necessarily produce a matching value.

Lithuania

Top Ranked Employing Sectors	Top Ranked Output Sectors
3110 Food Products	3110 Food Products
3210 Textiles	3210 Textiles
3820 Machinery nec	3130 Beverages
3830 Electrical Machinery	3820 Machinery nec
3220 Wearing Apparel	3830 Electrical Machinery

Population 3,600,000. Percent of world population: 0.061

	1990		1991		1992		1993		1994		1995	
	Value	%	Value	%	Value	%	Value	%	Value	%	Value	%
Establishments or enterprises (number)	-	-	-	-	-	-	1,333	0.065	1,682	0.094	-	-
Total employment (000)	-	-	-	-	484	0.265	423	0.232	313	0.197	-	-
Female employees (000)	-	-	-	-	-	-	-	-	-	-	-	-
Salaries and Wages ($ bil.)	-	-	-	-	0.181	0.007	0.205	0.010	0.327	0.017	-	-
Output ($ bil.)	-	-	-	-	2	0.012	2	0.014	2	0.015	-	-
Value added ($ bil.)	-	-	-	-	-	-	-	-	-	-	-	-
Capital investment ($ mil.)	-	-	-	-	-	-	-	-	-	-	-	-
Employees per establishment (number)	-	-	-	-	-	-	254	360.5	186	276.6	-	-
Output per establishment ($ mil.)	-	-	-	-	-	-	1	27.8	1	25.2	-	-
Value added per establishment ($ mil.)	-	-	-	-	-	-	-	-	-	-	-	-
Capital investment per estab. ($ mil.)	-	-	-	-	-	-	-	-	-	-	-	-
Payroll per employee ($)	-	-	-	-	375	1.9	484	3.0	1,043	5.9	-	-
Females as % of total employment (%)	-	-	-	-	-	-	-	-	-	-	-	-
Output per employee ($)	-	-	-	-	4,042	4.5	4,906	6.1	6,800	7.8	-	-
Value added per employee ($)	-	-	-	-	-	-	-	-	-	-	-	-
Capital investment per employee ($)	-	-	-	-	-	-	-	-	-	-	-	-

Columns headed % show percent of the world total or of the world ratio. Ratios closest to 100 are closest to the world average. Values higher than 100 are above, values lower than 100 are below world average. *nec* stands for *not elsewhere classified*. Ratios, where shown, are frequently based on a subset of total data; for example, payroll per employee is averaged from all cases where *both* payroll and employment data are present. Dividing *total* salaries/wages by *total* employment will not necessarily produce a matching value.

For additional notes and sources, see Appendix I.

Luxembourg

	Top Ranked Employing Sectors	Top Ranked Output Sectors
	3710 Iron and Steel	3710 Iron and Steel
	3810 Metal Products	3810 Metal Products
	3110 Food Products	3110 Food Products
	3820 Machinery nec	3520 Chemical Products nec
	3520 Chemical Products nec	3820 Machinery nec

Population 425,000. Percent of world population: 0.007

	1990		1991		1992		1993		1994		1995	
	Value	%	Value	%	Value	%	Value	%	Value	%	Value	%
Establishments or enterprises (number)	447	0.019	441	0.019	421	0.015	430	0.021	-	-	-	-
Total employment (000)	22	0.010	22	0.012	21	0.012	20	0.011	21	0.013	-	-
Female employees (000)	-	-	-	-	-	-	-	-	-	-	-	-
Salaries and Wages ($ bil.)	0.666	0.031	0.681	0.029	0.715	0.028	0.661	0.032	0.710	0.036	-	-
Output ($ bil.)	4	0.027	4	0.023	4	0.023	3	0.023	4	0.026	-	-
Value added ($ bil.)	1	0.021	1	0.018	1	0.019	0.943	0.016	1	0.020	-	-
Capital investment ($ mil.)	311	0.052	245	0.039	157	0.025	-	-	-	-	-	-
Employees per establishment (number)	49	66.3	50	73.2	51	82.0	47	67.1	-	-	-	-
Output per establishment ($ mil.)	9	201.0	8	176.3	9	162.4	8	165.8	-	-	-	-
Value added per establishment ($ mil.)	3	159.1	2	137.9	3	130.1	2	136.2	-	-	-	-
Capital investment per estab. ($ mil.)	0.696	262.6	0.556	189.3	0.372	129.8	-	-	-	-	-	-
Payroll per employee ($)	30,480	222.8	31,103	171.8	33,295	165.8	32,517	200.2	34,432	196.1	-	-
Females as % of total employment (%)	-	-	-	-	-	-	-	-	-	-	-	-
Output per employee ($)	174,256	253.9	162,916	200.1	173,454	193.9	164,678	205.7	174,471	199.7	-	-
Value added per employee ($)	53,378	162.5	46,869	138.8	53,781	149.4	46,361	144.7	49,834	140.4	-	-
Capital investment per employee ($)	14,236	258.5	11,199	182.6	7,303	117.9	-	-	-	-	-	-

Columns headed % show percent of the world total or of the world ratio. Ratios closest to 100 are closest to the world average. Values higher than 100 are above, values lower than 100 are below world average. *nec* stands for *not elsewhere classified*. Ratios, where shown, are frequently based on a subset of total data; for example, payroll per employee is averaged from all cases where *both* payroll and employment data are present. Dividing *total* salaries/wages by *total* employment will not necessarily produce a matching value.

Macau

	Top Ranked Employing Sectors	Top Ranked Output Sectors
	3220 Wearing Apparel	3220 Wearing Apparel
	3210 Textiles	3210 Textiles
	3900 Industries nec	3900 Industries nec
	3830 Electrical Machinery	3690 Nonmetal Products nec
	3420 Printing, Publishing	3830 Electrical Machinery

Population 429,000. Percent of world population: 0.007

	1990		1991		1992		1993		1994		1995	
	Value	%	Value	%	Value	%	Value	%	Value	%	Value	%
Establishments or enterprises (number)	2,390	0.104	2,185	0.092	2,054	0.073	1,913	0.093	1,674	0.094	1,436	0.117
Total employment (000)	66	0.032	62	0.033	55	0.030	50	0.027	47	0.029	43	0.030
Female employees (000)	42	0.312	40	0.449	37	0.170	33	0.150	32	0.153	31	0.160
Salaries and Wages ($ bil.)	0.274	0.013	0.277	0.012	0.269	0.011	0.269	0.013	0.267	0.014	0.253	0.015
Output ($ bil.)	2	0.012	2	0.011	2	0.011	2	0.012	2	0.012	2	0.013
Value added ($ bil.)	0.502	0.009	0.514	0.009	0.513	0.009	0.467	0.008	0.463	0.009	0.441	0.008
Capital investment ($ mil.)	39	0.007	53	0.009	28	0.004	76	0.014	50	0.011	35	0.011
Employees per establishment (number)	28	37.6	28	41.6	27	43.1	26	36.9	28	41.6	30	42.6
Output per establishment ($ mil.)	0.725	17.1	0.796	17.3	0.842	15.5	0.875	18.6	1	20.4	1	28.3
Value added per establishment ($ mil.)	0.210	12.8	0.235	13.9	0.250	11.9	0.244	15.2	0.277	15.7	0.307	17.4
Capital investment per estab. ($ mil.)	0.017	6.6	0.026	9.0	0.015	5.1	0.040	15.7	0.030	12.6	0.025	8.6
Payroll per employee ($)	4,137	30.2	4,500	24.9	4,892	24.4	5,407	33.3	5,701	32.5	5,901	33.8
Females as % of total employment (%)	62.7	195.8	65.5	204.0	66.7	144.2	67.4	135.0	69.1	131.8	71.4	122.3
Output per employee ($)	26,167	38.1	28,223	34.7	31,453	35.2	33,683	42.1	36,673	42.0	39,181	42.3
Value added per employee ($)	7,580	23.1	8,338	24.7	9,341	25.9	9,398	29.3	9,884	27.8	10,288	26.0
Capital investment per employee ($)	614	11.2	908	14.8	536	8.7	1,535	26.9	1,061	16.2	826	11.1

Columns headed % show percent of the world total or of the world ratio. Ratios closest to 100 are closest to the world average. Values higher than 100 are above, values lower than 100 are below world average. *nec* stands for *not elsewhere classified*. Ratios, where shown, are frequently based on a subset of total data; for example, payroll per employee is averaged from all cases where *both* payroll and employment data are present. Dividing *total* salaries/wages by *total* employment will not necessarily produce a matching value.

Malawi

Top Ranked Employing Sectors	Top Ranked Output Sectors
3110 Food Products	3110 Food Products
3210 Textiles	3130 Beverages
3140 Tobacco	3210 Textiles
3130 Beverages	3510 Industrial Chemicals
3310 Wood Products	3520 Chemical Products nec

Population 9,840,000. Percent of world population: 0.166

	1990		1991		1992		1993		1994		1995	
	Value	%	Value	%	Value	%	Value	%	Value	%	Value	%
Establishments or enterprises (number)	64	0.003	53	0.002	53	0.002	53	0.003	53	0.003	-	-
Total employment (000)	39	0.019	22	0.012	23	0.012	22	0.012	24	0.015	-	-
Female employees (000)	-		-		-		-		-			
Salaries and Wages ($ bil.)	0.032	0.001	0.038	0.002	0.039	0.002	0.034	0.002	0.022	0.001		
Output ($ bil.)	0.478	0.003	0.516	0.003	0.547	0.003	0.506	0.003	0.289	0.002		
Value added ($ bil.)	0.077	0.001	0.124	0.002	0.193	0.003	0.111	0.002	0.102	0.002		
Capital investment ($ mil.)	25	0.004	36	0.006	61	0.010	59	0.011	30	0.007		
Employees per establishment (number)	536	727.0	425	627.1	441	709.6	422	598.0	471	698.8	-	
Output per establishment ($ mil.)	9	221.3	10	220.3	11	197.1	10	211.2	6	112.6		
Value added per establishment ($ mil.)	2	92.3	2	143.9	4	179.3	2	135.3	2	112.9		
Capital investment per estab. ($ mil.)	0.482	182.0	0.708	240.9	1	417.9	1	458.5	0.596	253.4		
Payroll per employee ($)	1,125	8.2	1,772	9.8	1,726	8.6	1,600	9.9	936	5.3		
Females as % of total employment (%)	-		-		-		-		-			
Output per employee ($)	16,854	24.6	23,767	29.2	24,323	27.2	23,538	29.4	12,037	13.8		
Value added per employee ($)	2,719	8.3	5,709	16.9	8,564	23.8	5,166	16.1	4,231	11.9		
Capital investment per employee ($)	867	15.7	1,664	27.1	2,718	43.9	2,755	48.2	1,267	19.4		

Columns headed % show percent of the world total or of the world ratio. Ratios closest to 100 are closest to the world average. Values higher than 100 are above, values lower than 100 are below world average. *nec* stands for *not elsewhere classified*. Ratios, where shown, are frequently based on a subset of total data; for example, payroll per employee is averaged from all cases where *both* payroll and employment data are present. Dividing *total* salaries/wages by *total* employment will not necessarily produce a matching value.

Malaysia

Top Ranked Employing Sectors	Top Ranked Output Sectors
3830 Electrical Machinery	3830 Electrical Machinery
3310 Wood Products	3110 Food Products
3110 Food Products	3840 Transportation Equipment
3550 Rubber Products	3310 Wood Products
3220 Wearing Apparel	3820 Machinery nec

Population 20,933,000. Percent of world population: 0.353

	1990		1991		1992		1993		1994		1995	
	Value	%	Value	%	Value	%	Value	%	Value	%	Value	%
Establishments or enterprises (number)	7,400	0.321	8,058	0.338	8,273	0.295	9,225	0.449	9,273	0.518	-	-
Total employment (000)	1,073	0.513	1,257	0.673	1,340	0.734	1,512	0.830	1,618	1.000	1,758	1.200
Female employees (000)	588	4.400	684	7.600	731	3.400	809	3.600	845	4.000	-	-
Salaries and Wages ($ bil.)	3	0.149	4	0.170	5	0.202	6	0.299	7	0.360	9	0.516
Output ($ bil.)	47	0.329	60	0.389	73	0.443	86	0.591	107	0.777	133	0.997
Value added ($ bil.)	12	0.216	16	0.274	19	0.316	22	0.375	26	0.509	32	0.586
Capital investment ($ mil.)	3,319	0.560	4,703	0.754	5,594	0.875	8,742	1.600	8,978	2.000		
Employees per establishment (number)	145	196.6	156	229.9	162	260.5	164	232.5	175	259.2	-	-
Output per establishment ($ mil.)	6	150.3	7	161.7	9	161.6	9	197.9	12	229.7	-	
Value added per establishment ($ mil.)	2	99.6	2	114.3	2	108.3	2	150.9	3	159.4		
Capital investment per estab. ($ mil.)	0.493	186.2	0.644	219.3	0.750	261.3	0.948	374.1	0.968	411.5		
Payroll per employee ($)	2,991	21.9	3,220	17.8	3,844	19.1	4,039	24.9	4,341	24.7	4,861	27.8
Females as % of total employment (%)	54.8	171.0	54.4	169.5	54.5	117.8	53.5	107.1	52.2	99.6	-	
Output per employee ($)	43,930	64.0	47,590	58.5	54,328	60.7	56,702	70.8	66,211	75.8	75,635	81.7
Value added per employee ($)	11,260	34.3	12,364	36.6	14,095	39.1	14,820	46.3	16,101	45.4	18,298	46.3
Capital investment per employee ($)	3,996	72.6	4,873	79.5	5,470	88.3	5,781	101.1	5,548	84.8	-	

Columns headed % show percent of the world total or of the world ratio. Ratios closest to 100 are closest to the world average. Values higher than 100 are above, values lower than 100 are below world average. *nec* stands for *not elsewhere classified*. Ratios, where shown, are frequently based on a subset of total data; for example, payroll per employee is averaged from all cases where *both* payroll and employment data are present. Dividing *total* salaries/wages by *total* employment will not necessarily produce a matching value.

Malta

Top Ranked Employing Sectors	Top Ranked Output Sectors
3220 Wearing Apparel	3830 Electrical Machinery
3830 Electrical Machinery	3110 Food Products
3110 Food Products	3220 Wearing Apparel
3840 Transportation Equipment	3420 Printing, Publishing
3320 Furniture, Fixtures	3900 Industries nec

Population 380,000. Percent of world population: 0.006

	1990		1991		1992		1993		1994		1995	
	Value	%	Value	%	Value	%	Value	%	Value	%	Value	%
Establishments or enterprises (number)	1,505	0.065	1,532	0.064	1,557	0.055	1,570	0.076	1,549	0.087	-	-
Total employment (000)	29	0.014	29	0.015	29	0.016	29	0.016	29	0.018	-	-
Female employees (000)	11	0.080	10	0.113	10	0.048	10	0.046	10	0.049	-	-
Salaries and Wages ($ bil.)	0.284	0.013	0.298	0.013	0.319	0.013	0.288	0.014	0.314	0.016	-	-
Output ($ bil.)	2	0.015	2	0.015	3	0.017	3	0.017	3	0.022	-	-
Value added ($ bil.)	0.538	0.010	0.533	0.009	0.569	0.010	0.513	0.009	0.590	0.012	-	-
Capital investment ($ mil.)	66	0.011	108	0.017	87	0.014	83	0.015	88	0.020	-	-
Employees per establishment (number)	19	26.4	19	27.8	19	29.9	18	26.1	19	27.8	-	-
Output per establishment ($ mil.)	1	32.7	2	33.1	2	33.7	2	34.0	2	38.0	-	-
Value added per establishment ($ mil.)	0.358	21.8	0.348	20.6	0.366	17.3	0.327	20.3	0.381	21.6	-	-
Capital investment per estab. ($ mil.)	0.044	16.7	0.071	24.0	0.056	19.4	0.053	20.9	0.057	24.1	-	-
Payroll per employee ($)	9,699	70.9	10,326	57.1	11,041	55.0	9,952	61.3	10,811	61.6	-	-
Females as % of total employment (%)	36.5	113.9	35.4	110.4	35.8	77.3	35.3	70.7	35.7	68.1	-	-
Output per employee ($)	71,148	103.7	80,637	99.1	98,839	110.5	86,741	108.3	101,948	116.7	-	-
Value added per employee ($)	18,381	56.0	18,476	54.7	19,676	54.6	17,762	55.4	20,325	57.3	-	-
Capital investment per employee ($)	2,269	41.2	3,743	61.1	2,998	48.4	2,876	50.3	3,033	46.4	-	-

Columns headed % show percent of the world total or of the world ratio. Ratios closest to 100 are closest to the world average. Values higher than 100 are above, values lower than 100 are below world average. *nec* stands for *not elsewhere classified*. Ratios, where shown, are frequently based on a subset of total data; for example, payroll per employee is averaged from all cases where *both* payroll and employment data are present. Dividing *total* salaries/wages by *total* employment will not necessarily produce a matching value.

Mauritius

Top Ranked Employing Sectors	Top Ranked Output Sectors
3220 Wearing Apparel	3220 Wearing Apparel
3110 Food Products	3110 Food Products
3210 Textiles	3210 Textiles
3900 Industries nec	3130 Beverages
3130 Beverages	3690 Nonmetal Products nec

Population 1,168,000. Percent of world population: 0.020

	1990		1991		1992		1993		1994		1995	
	Value	%	Value	%	Value	%	Value	%	Value	%	Value	%
Establishments or enterprises (number)	1,079	0.047	997	0.042	1,017	0.036	1,016	0.049	989	0.055	967	0.079
Total employment (000)	115	0.055	118	0.063	119	0.065	115	0.063	112	0.071	111	0.078
Female employees (000)	-		-		-		-		-		-	
Salaries and Wages ($ bil.)	0.219	0.010	0.243	0.010	0.317	0.012	0.296	0.014	0.337	0.017	0.367	0.022
Output ($ bil.)	2	0.012	2	0.012	2	0.013	2	0.014	2	0.016	2	0.018
Value added ($ bil.)	0.531	0.009	0.562	0.010	0.686	0.011	0.660	0.011	0.731	0.014	0.803	0.015
Capital investment ($ mil.)	-		-		-		-		-		-	
Employees per establishment (number)	107	144.6	118	174.0	120	193.6	117	165.6	117	173.4	118	168.6
Output per establishment ($ mil.)	2	38.3	2	39.4	2	38.5	2	43.6	2	44.8	3	64.1
Value added per establishment ($ mil.)	0.492	30.0	0.563	33.4	0.695	33.0	0.669	41.6	0.740	41.9	0.874	49.5
Capital investment per estab. ($ mil.)	-		-		-		-		-		-	
Payroll per employee ($)	1,900	13.9	2,060	11.4	2,664	13.3	2,569	15.8	2,884	16.4	3,340	19.1
Females as % of total employment (%)	-		-		-		-		-		-	
Output per employee ($)	15,208	22.2	15,321	18.8	17,416	19.5	17,565	21.9	18,823	21.5	22,132	23.9
Value added per employee ($)	4,618	14.1	4,772	14.1	5,770	16.0	5,736	17.9	6,219	17.5	7,303	18.5
Capital investment per employee ($)	-		-		-		-		-		-	

Columns headed % show percent of the world total or of the world ratio. Ratios closest to 100 are closest to the world average. Values higher than 100 are above, values lower than 100 are below world average. *nec* stands for *not elsewhere classified*. Ratios, where shown, are frequently based on a subset of total data; for example, payroll per employee is averaged from all cases where *both* payroll and employment data are present. Dividing *total* salaries/wages by *total* employment will not necessarily produce a matching value.

Mexico

Top Ranked Employing Sectors	Top Ranked Output Sectors
3110 Food Products	3840 Transportation Equipment
3130 Beverages	3110 Food Products
3840 Transportation Equipment	3510 Industrial Chemicals
3830 Electrical Machinery	3710 Iron and Steel
3520 Chemical Products nec	3520 Chemical Products nec

Population 98,553,000. Percent of world population: 1.663

	1990		1991		1992		1993		1994		1995	
	Value	%	Value	%	Value	%	Value	%	Value	%	Value	%
Establishments or enterprises (number)	3,584	0.156	3,495	0.147	3,404	0.121	3,302	0.161	3,211	0.179	3,150	0.257
Total employment (000)	1,216	0.581	1,199	0.642	1,158	0.634	1,017	0.558	959	0.603	938	0.659
Female employees (000)	-	-	-	-	-	-	-	-	-	-	-	-
Salaries and Wages ($ bil.)	6	0.290	7	0.313	9	0.341	9	0.429	9	0.436	5	0.307
Output ($ bil.)	79	0.554	91	0.591	100	0.609	97	0.672	100	0.723	83	0.622
Value added ($ bil.)	31	0.552	35	0.612	38	0.642	38	0.631	39	0.762	23	0.412
Capital investment ($ mil.)	2,170	0.366	3,924	0.629	-	-	-	-	-	-	-	-
Employees per establishment (number)	339	460.1	343	505.7	340	546.9	308	436.8	299	443.6	298	424.8
Output per establishment ($ mil.)	22	522.2	26	565.8	29	540.4	30	628.5	31	617.0	26	637.7
Value added per establishment ($ mil.)	9	526.1	10	588.3	11	535.3	11	709.0	12	689.4	8	472.3
Capital investment per estab. ($ mil.)	0.605	228.6	1	382.8	-	-	-	-	-	-	-	-
Payroll per employee ($)	5,140	37.6	6,200	34.3	7,521	37.5	8,613	53.0	8,867	50.5	5,421	31.0
Females as % of total employment (%)	-	-	-	-	-	-	-	-	-	-	-	-
Output per employee ($)	65,214	95.0	75,693	93.0	86,539	96.7	95,856	119.7	103,884	118.9	88,496	95.6
Value added per employee ($)	25,418	77.4	28,944	85.7	33,176	92.1	37,067	115.7	40,689	114.6	30,289	76.7
Capital investment per employee ($)	1,785	32.4	3,280	53.5	-	-	-	-	-	-	-	-

Columns headed % show percent of the world total or of the world ratio. Ratios closest to 100 are closest to the world average. Values higher than 100 are above, values lower than 100 are below world average. *nec* stands for *not elsewhere classified*. Ratios, where shown, are frequently based on a subset of total data; for example, payroll per employee is averaged from all cases where *both* payroll and employment data are present. Dividing *total* salaries/wages by *total* employment will not necessarily produce a matching value.

Mongolia

Top Ranked Employing Sectors	Top Ranked Output Sectors
3110 Food Products	3110 Food Products
3220 Wearing Apparel	3210 Textiles
3310 Wood Products	3690 Nonmetal Products nec
3210 Textiles	3230 Leather and Products
3900 Industries nec	3900 Industries nec

Population 2,579,000. Percent of world population: 0.043

	1990		1991		1992		1993		1994		1995	
	Value	%	Value	%	Value	%	Value	%	Value	%	Value	%
Establishments or enterprises (number)	340	0.015	332	0.014	387	0.014	434	0.021	621	0.035	459	0.037
Total employment (000)	58	0.028	57	0.030	42	0.023	38	0.021	49	0.031	46	0.032
Female employees (000)	-	-	-	-	-	-	-	-	26	0.122	-	-
Salaries and Wages ($ bil.)	0.106	0.005	0.102	0.004	0.034	0.001	0.017	0.001	0.023	0.001	0.024	0.001
Output ($ bil.)	1	0.008	1	0.007	0.499	0.003	0.205	0.001	0.221	0.002	0.250	0.002
Value added ($ bil.)	0.376	0.007	0.383	0.007	0.214	0.004	0.068	0.001	0.087	0.002	0.101	0.002
Capital investment ($ mil.)	1,188	0.200	681	0.109	184	0.029	32	0.006	43	0.010	-	-
Employees per establishment (number)	171	232.6	172	253.2	109	174.9	87	123.5	80	118.2	101	144.5
Output per establishment ($ mil.)	3	79.8	3	68.8	1	23.7	0.472	10.0	0.372	7.4	0.545	13.2
Value added per establishment ($ mil.)	1	67.6	1	68.4	0.553	26.2	0.157	9.7	0.147	8.3	0.220	12.4
Capital investment per estab. ($ mil.)	4	1,327.2	2	715.4	0.486	169.5	0.075	29.5	0.076	32.3	-	-
Payroll per employee ($)	1,820	13.3	1,792	9.9	807	4.0	440	2.7	476	2.7	520	3.0
Females as % of total employment (%)	-	-	-	-	-	-	-	-	53.6	102.2	-	-
Output per employee ($)	19,724	28.7	18,424	22.6	11,871	13.3	5,427	6.8	4,585	5.2	5,420	5.9
Value added per employee ($)	6,462	19.7	6,738	20.0	5,093	14.1	1,805	5.6	1,813	5.1	2,184	5.5
Capital investment per employee ($)	20,593	374.0	12,220	199.3	4,470	72.2	866	15.2	917	14.0	-	-

Columns headed % show percent of the world total or of the world ratio. Ratios closest to 100 are closest to the world average. Values higher than 100 are above, values lower than 100 are below world average. *nec* stands for *not elsewhere classified*. Ratios, where shown, are frequently based on a subset of total data; for example, payroll per employee is averaged from all cases where *both* payroll and employment data are present. Dividing *total* salaries/wages by *total* employment will not necessarily produce a matching value.

For additional notes and sources, see Appendix I.

Morocco

Top Ranked Employing Sectors	Top Ranked Output Sectors
3220 Wearing Apparel	3110 Food Products
3110 Food Products	3210 Textiles
3210 Textiles	3810 Metal Products
3810 Metal Products	3220 Wearing Apparel
3840 Transportation Equipment	3840 Transportation Equipment

Population 29,114,000. Percent of world population: 0.491

	1990 Value	%	1991 Value	%	1992 Value	%	1993 Value	%	1994 Value	%	1995 Value	%
Establishments or enterprises (number)	3,326	0.144	4,710	0.198	4,753	0.169	4,836	0.235	4,075	0.228	4,082	0.333
Total employment (000)	202	0.096	303	0.162	345	0.189	354	0.194	317	0.199	320	0.225
Female employees (000)	-	-	-	-	-	-	-	-	-	-	177	0.928
Salaries and Wages ($ bil.)	0.647	0.030	0.890	0.037	0.967	0.038	0.921	0.045	1	0.053	1	0.070
Output ($ bil.)	5	0.036	9	0.059	10	0.059	9	0.064	8	0.059	9	0.066
Value added ($ bil.)	2	0.032	3	0.049	3	0.050	3	0.049	2	0.041	2	0.042
Capital investment ($ mil.)	397	0.067	550	0.088	545	0.085	620	0.114	434	0.098	480	0.154
Employees per establishment (number)	61	82.3	64	94.9	72	116.6	73	103.8	78	115.6	78	111.7
Output per establishment ($ mil.)	2	36.9	2	41.7	2	37.2	2	40.6	2	39.5	2	52.0
Value added per establishment ($ mil.)	0.369	22.5	0.626	37.1	0.661	31.3	0.603	37.4	0.517	29.3	0.568	32.2
Capital investment per estab. ($ mil.)	0.102	38.5	0.122	41.5	0.120	41.8	0.128	50.6	0.107	45.3	0.118	41.1
Payroll per employee ($)	3,205	23.4	3,319	18.3	3,153	15.7	2,895	17.8	3,268	18.6	3,652	20.9
Females as % of total employment (%)	-	-	-	-	-	-	-	-	-	-	55.5	95.0
Output per employee ($)	25,781	37.6	29,720	36.5	27,967	31.3	26,050	32.5	25,504	29.2	27,414	29.6
Value added per employee ($)	5,843	17.8	9,431	27.9	8,795	24.4	8,234	25.7	6,635	18.7	7,258	18.4
Capital investment per employee ($)	1,682	30.5	1,913	31.2	1,666	26.9	1,753	30.7	1,369	20.9	1,502	20.1

Columns headed % show percent of the world total or of the world ratio. Ratios closest to 100 are closest to the world average. Values higher than 100 are above, values lower than 100 are below world average. *nec* stands for *not elsewhere classified*. Ratios, where shown, are frequently based on a subset of total data; for example, payroll per employee is averaged from all cases where *both* payroll and employment data are present. Dividing *total* salaries/wages by *total* employment will not necessarily produce a matching value.

Mozambique

Top Ranked Employing Sectors	Top Ranked Output Sectors
3110 Food Products	3110 Food Products
3210 Textiles	3130 Beverages
3310 Wood Products	3690 Nonmetal Products nec
3810 Metal Products	3520 Chemical Products nec
3420 Printing, Publishing	3210 Textiles

Population 18,641,000. Percent of world population: 0.315

	1990 Value	%	1991 Value	%	1992 Value	%	1993 Value	%	1994 Value	%	1995 Value	%
Establishments or enterprises (number)	320	0.014	320	0.013	247	0.009	247	0.012	255	0.014	-	-
Total employment (000)	76	0.037	71	0.038	54	0.030	52	0.029	48	0.030	-	-
Female employees (000)	-	-	-	-	-	-	-	-	-	-	-	-
Salaries and Wages ($ bil.)	0.102	0.005	0.054	0.002	0.026	0.001	0.021	0.001	0.021	0.001	-	-
Output ($ bil.)	-	-	0.297	0.002	-	-	0.166	0.001	0.152	0.001	-	-
Value added ($ bil.)	-	-	-	-	-	-	-	-	-	-	-	-
Capital investment ($ mil.)	-	-	-	-	-	-	-	-	-	-	-	-
Employees per establishment (number)	239	323.8	221	325.7	219	352.8	212	300.8	187	277.7	-	-
Output per establishment ($ mil.)	-	-	0.896	19.5	-	-	0.696	14.8	0.627	12.5	-	-
Value added per establishment ($ mil.)	-	-	-	-	-	-	-	-	-	-	-	-
Capital investment per estab. ($ mil.)	-	-	-	-	-	-	-	-	-	-	-	-
Payroll per employee ($)	1,339	9.8	757	4.2	483	2.4	409	2.5	439	2.5	-	-
Females as % of total employment (%)	-	-	-	-	-	-	-	-	-	-	-	-
Output per employee ($)	-	-	4,053	5.0	-	-	3,239	4.0	3,294	3.8	-	-
Value added per employee ($)	-	-	-	-	-	-	-	-	-	-	-	-
Capital investment per employee ($)	-	-	-	-	-	-	-	-	-	-	-	-

Columns headed % show percent of the world total or of the world ratio. Ratios closest to 100 are closest to the world average. Values higher than 100 are above, values lower than 100 are below world average. *nec* stands for *not elsewhere classified*. Ratios, where shown, are frequently based on a subset of total data; for example, payroll per employee is averaged from all cases where *both* payroll and employment data are present. Dividing *total* salaries/wages by *total* employment will not necessarily produce a matching value.

Myanmar

Top Ranked Employing Sectors	Top Ranked Output Sectors
3310 Wood Products	3720 Nonferrous metals
3900 Industries nec	3710 Iron and Steel
3522 Drugs and Medicines	3900 Industries nec
3720 Nonferrous metals	3140 Tobacco
3320 Furniture, Fixtures	3310 Wood Products

Population 47,305,000. Percent of world population: 0.798

	1990		1991		1992		1993		1994		1995	
	Value	%	Value	%	Value	%	Value	%	Value	%	Value	%
Establishments or enterprises (number)	221	0.010	220	0.009	214	0.008	196	0.010	178	0.010	-	-
Total employment (000)	70	0.034	64	0.034	63	0.034	61	0.034	49	0.031	-	-
Female employees (000)	17	0.126	15	0.170	15	0.072	14	0.065	15	0.070	-	-
Salaries and Wages ($ bil.)	0.097	0.005	0.099	0.004	0.106	0.004	0.131	0.006	0.117	0.006	-	-
Output ($ bil.)	0.468	0.003	0.494	0.003	0.557	0.003	0.657	0.005	0.957	0.007	-	-
Value added ($ bil.)	0.342	0.006	0.343	0.006	0.375	0.006	0.496	0.008	0.627	0.012	-	-
Capital investment ($ mil.)	964	0.163	961	0.154	1,032	0.161	926	0.170	1,046	0.238	-	-
Employees per establishment (number)	353	478.7	326	480.3	329	529.5	353	501.1	276	409.4	-	-
Output per establishment ($ mil.)	2	54.7	2	53.5	3	52.5	4	79.0	5	106.7	-	-
Value added per establishment ($ mil.)	2	126.3	2	123.2	2	109.9	3	174.1	4	218.9	-	-
Capital investment per estab. ($ mil.)	5	1,905.9	5	1,720.8	6	2,066.5	5	2,162.9	6	2,569.2	-	-
Payroll per employee ($)	1,379	10.1	1,533	8.5	1,683	8.4	2,135	13.1	2,356	13.4	-	-
Females as % of total employment (%)	24.8	77.3	24.8	77.1	25.5	55.2	24.7	49.5	31.6	60.2	-	-
Output per employee ($)	6,615	9.6	7,635	9.4	8,806	9.8	10,679	13.3	19,180	22.0	-	-
Value added per employee ($)	6,103	18.6	6,808	20.2	7,535	20.9	8,082	25.2	12,889	36.3	-	-
Capital investment per employee ($)	15,436	280.3	16,906	275.7	17,925	289.3	16,589	290.2	23,383	357.5	-	-

Columns headed % show percent of the world total or of the world ratio. Ratios closest to 100 are closest to the world average. Values higher than 100 are above, values lower than 100 are below world average. *nec* stands for *not elsewhere classified*. Ratios, where shown, are frequently based on a subset of total data; for example, payroll per employee is averaged from all cases where *both* payroll and employment data are present. Dividing *total* salaries/wages by *total* employment will not necessarily produce a matching value.

Namibia

Top Ranked Employing Sectors	Top Ranked Output Sectors
3110 Food Products	3110 Food Products
3810 Metal Products	3130 Beverages
3690 Nonmetal Products nec	3810 Metal Products
3130 Beverages	3690 Nonmetal Products nec
3420 Printing, Publishing	3420 Printing, Publishing

Population 1,622,000. Percent of world population: 0.027

	1990		1991		1992		1993		1994		1995	
	Value	%	Value	%	Value	%	Value	%	Value	%	Value	%
Establishments or enterprises (number)	-	-	-	-	-	-	-	-	227	0.013	-	-
Total employment (000)	-	-	-	-	-	-	-	-	19	0.012	-	-
Female employees (000)	-	-	-	-	-	-	-	-	-	-	-	-
Salaries and Wages ($ bil.)	-	-	-	-	-	-	-	-	0.119	0.006	-	-
Output ($ bil.)	-	-	-	-	-	-	-	-	0.932	0.007	-	-
Value added ($ bil.)	-	-	-	-	-	-	-	-	0.359	0.007	-	-
Capital investment ($ mil.)	-	-	-	-	-	-	-	-	-	-	-	-
Employees per establishment (number)	-	-	-	-	-	-	-	-	85	126.0	-	-
Output per establishment ($ mil.)	-	-	-	-	-	-	-	-	4	81.6	-	-
Value added per establishment ($ mil.)	-	-	-	-	-	-	-	-	2	89.8	-	-
Capital investment per estab. ($ mil.)	-	-	-	-	-	-	-	-	-	-	-	-
Payroll per employee ($)	-	-	-	-	-	-	-	-	6,192	35.3	-	-
Females as % of total employment (%)	-	-	-	-	-	-	-	-	-	-	-	-
Output per employee ($)	-	-	-	-	-	-	-	-	48,375	55.4	-	-
Value added per employee ($)	-	-	-	-	-	-	-	-	18,651	52.5	-	-
Capital investment per employee ($)	-	-	-	-	-	-	-	-	-	-	-	-

Columns headed % show percent of the world total or of the world ratio. Ratios closest to 100 are closest to the world average. Values higher than 100 are above, values lower than 100 are below world average. *nec* stands for *not elsewhere classified*. Ratios, where shown, are frequently based on a subset of total data; for example, payroll per employee is averaged from all cases where *both* payroll and employment data are present. Dividing *total* salaries/wages by *total* employment will not necessarily produce a matching value.

For additional notes and sources, see Appendix I.

Nepal

Top Ranked Employing Sectors	Top Ranked Output Sectors
3210 Textiles	3210 Textiles
3690 Nonmetal Products nec	3110 Food Products
3220 Wearing Apparel	3220 Wearing Apparel
3110 Food Products	3690 Nonmetal Products nec
3520 Chemical Products nec	3140 Tobacco

Population 23,698,000. Percent of world population: 0.400

	1990 Value	1990 %	1991 Value	1991 %	1992 Value	1992 %	1993 Value	1993 %	1994 Value	1994 %	1995 Value	1995 %
Establishments or enterprises (number)	2,619	0.114	4,489	0.188	-	-	4,563	0.222	4,755	0.266	-	-
Total employment (000)	166	0.079	223	0.120	-	-	226	0.124	245	0.154	-	-
Female employees (000)	39	0.292	56	0.619	-	-	46	0.206	53	0.250	-	-
Salaries and Wages ($ bil.)	0.067	0.003	0.095	0.004	-	-	0.080	0.004	0.091	0.005	-	-
Output ($ bil.)	0.705	0.005	0.939	0.006	-	-	0.911	0.006	1	0.008	-	-
Value added ($ bil.)	0.291	0.005	0.389	0.007	-	-	0.354	0.006	0.413	0.008	-	-
Capital investment ($ mil.)	252	0.042	-	-	-	-	-	-	-	-	-	-
Employees per establishment (number)	63	85.9	50	73.4	-	-	50	70.3	51	76.4	-	-
Output per establishment ($ mil.)	0.270	6.4	0.209	4.6	-	-	0.200	4.3	0.222	4.4	-	-
Value added per establishment ($ mil.)	0.111	6.8	0.087	5.1	-	-	0.078	4.8	0.087	4.9	-	-
Capital investment per estab. ($ mil.)	0.106	39.9	-	-	-	-	-	-	-	-	-	-
Payroll per employee ($)	407	3.0	426	2.4	-		352	2.2	374	2.1	-	-
Females as % of total employment (%)	23.5	73.2	24.9	77.6	-	-	20.4	40.8	21.6	41.1	-	-
Output per employee ($)	4,256	6.2	4,205	5.2	-	-	4,028	5.0	4,319	4.9	-	-
Value added per employee ($)	1,758	5.4	1,740	5.2	-	-	1,564	4.9	1,689	4.8	-	-
Capital investment per employee ($)	1,613	29.3	-	-	-	-	-	-	-	-	-	-

Columns headed % show percent of the world total or of the world ratio. Ratios closest to 100 are closest to the world average. Values higher than 100 are above, values lower than 100 are below world average. *nec* stands for *not elsewhere classified*. Ratios, where shown, are frequently based on a subset of total data; for example, payroll per employee is averaged from all cases where *both* payroll and employment data are present. Dividing *total* salaries/wages by *total* employment will not necessarily produce a matching value.

Netherlands

Top Ranked Employing Sectors	Top Ranked Output Sectors
3110 Food Products	3110 Food Products
3830 Electrical Machinery	3510 Industrial Chemicals
3820 Machinery nec	3830 Electrical Machinery
3810 Metal Products	3530 Petroleum Refineries
3420 Printing, Publishing	3820 Machinery nec

Population 15,731,000. Percent of world population: 0.265

	1990 Value	1990 %	1991 Value	1991 %	1992 Value	1992 %	1993 Value	1993 %	1994 Value	1994 %	1995 Value	1995 %
Establishments or enterprises (number)	6,322	0.274	6,659	0.279	6,884	0.245	6,840	0.333	7,164	0.400	-	-
Total employment (000)	792	0.379	795	0.426	787	0.432	734	0.403	683	0.429	678	0.477
Female employees (000)	-		-		-		-		-		-	
Salaries and Wages ($ bil.)	23	1.100	24	0.994	26	1.000	24	1.200	24	1.200	28	1.700
Output ($ bil.)	156	1.100	154	1.000	164	0.996	144	0.995	148	1.100	175	1.300
Value added ($ bil.)	45	0.812	45	0.796	49	0.814	45	0.750	45	0.880	53	0.964
Capital investment ($ mil.)	8,607	1.500	8,189	1.300	8,359	1.300	7,066	1.300	-	-	-	-
Employees per establishment (number)	125	170.0	119	176.0	114	184.0	107	152.2	99	147.1	-	-
Output per establishment ($ mil.)	25	581.5	23	503.6	24	436.8	21	449.4	21	425.7	-	-
Value added per establishment ($ mil.)	7	439.0	7	401.5	7	335.7	7	406.5	7	371.0	-	-
Capital investment per estab. ($ mil.)	1	535.1	1	434.7	1	438.5	1	407.6	-	-	-	-
Payroll per employee ($)	29,044	212.3	29,715	164.2	33,103	164.9	32,457	199.9	34,688	197.5	40,757	233.2
Females as % of total employment (%)	-		-		-		-		-		-	
Output per employee ($)	196,522	286.3	193,625	237.9	207,922	232.4	196,801	245.8	216,099	247.3	257,307	278.0
Value added per employee ($)	57,393	174.7	56,766	168.1	61,846	171.8	61,015	190.4	66,019	186.0	78,023	197.5
Capital investment per employee ($)	11,196	203.3	10,608	173.0	10,930	176.4	9,572	167.5	-	-	-	-

Columns headed % show percent of the world total or of the world ratio. Ratios closest to 100 are closest to the world average. Values higher than 100 are above, values lower than 100 are below world average. *nec* stands for *not elsewhere classified*. Ratios, where shown, are frequently based on a subset of total data; for example, payroll per employee is averaged from all cases where *both* payroll and employment data are present. Dividing *total* salaries/wages by *total* employment will not necessarily produce a matching value.

Netherlands Antilles

	Top Ranked Employing Sectors	Top Ranked Output Sectors
	3810 Metal Products	
	3840 Transportation Equipment	
	3110 Food Products	
	3420 Printing, Publishing	
	3130 Beverages	

Population 206,000. Percent of world population: 0.004

	1990		1991		1992		1993		1994		1995	
	Value	%	Value	%	Value	%	Value	%	Value	%	Value	%
Establishments or enterprises (number)	-	-	-	-	-	-	403	0.020	-	-	-	-
Total employment (000)	-	-	-	-	-	-	5	0.003	-	-	-	-
Female employees (000)	-	-	-	-	-	-	-	-	-	-	-	-
Salaries and Wages ($ bil.)	-	-	-	-	-	-	0.071	0.003	-	-	-	-
Output ($ bil.)	-	-	-	-	-	-	-	-	-	-	-	-
Value added ($ bil.)	-	-	-	-	-	-	-	-	-	-	-	-
Capital investment ($ mil.)	-	-	-	-	-	-	-	-	-	-	-	-
Employees per establishment (number)	-	-	-	-	-	-	13	17.8	-	-	-	-
Output per establishment ($ mil.)	-	-	-	-	-	-	-	-	-	-	-	-
Value added per establishment ($ mil.)	-	-	-	-	-	-	-	-	-	-	-	-
Capital investment per estab. ($ mil.)	-	-	-	-	-	-	-	-	-	-	-	-
Payroll per employee ($)	-	-	-	-	-	-	14,094	86.8	-	-	-	-
Females as % of total employment (%)	-	-	-	-	-	-	-	-	-	-	-	-
Output per employee ($)	-	-	-	-	-	-	-	-	-	-	-	-
Value added per employee ($)	-	-	-	-	-	-	-	-	-	-	-	-
Capital investment per employee ($)	-	-	-	-	-	-	-	-	-	-	-	-

Columns headed % show percent of the world total or of the world ratio. Ratios closest to 100 are closest to the world average. Values higher than 100 are above, values lower than 100 are below world average. *nec* stands for *not elsewhere classified*. Ratios, where shown, are frequently based on a subset of total data; for example, payroll per employee is averaged from all cases where *both* payroll and employment data are present. Dividing *total* salaries/wages by *total* employment will not necessarily produce a matching value.

New Zealand

	Top Ranked Employing Sectors	Top Ranked Output Sectors
	3110 Food Products	3110 Food Products
	3810 Metal Products	3810 Metal Products
	3420 Printing, Publishing	3410 Paper and Products
	3310 Wood Products	3420 Printing, Publishing
	3820 Machinery nec	3840 Transportation Equipment

Population 3,625,000. Percent of world population: 0.061

	1990		1991		1992		1993		1994		1995	
	Value	%	Value	%	Value	%	Value	%	Value	%	Value	%
Establishments or enterprises (number)	18,749	0.814	19,258	0.808	19,446	0.692	20,630	1.000	-	-	-	-
Total employment (000)	233	0.112	220	0.118	228	0.125	230	0.126	-	-	-	-
Female employees (000)	66	0.493	61	0.684	63	0.294	66	0.295	-	-	-	-
Salaries and Wages ($ bil.)	2	0.106	2	0.081	2	0.070	2	0.107	-	-	-	-
Output ($ bil.)	20	0.142	12	0.080	12	0.073	13	0.091	-	-	-	-
Value added ($ bil.)	7	0.129	4	0.066	3	0.058	4	0.062	-	-	-	-
Capital investment ($ mil.)	614	0.104	-	-	-	-	-	-	-	-	-	-
Employees per establishment (number)	12	16.9	11	16.9	12	18.8	11	16.0	-	-	-	-
Output per establishment ($ mil.)	2	39.2	2	36.8	2	29.7	2	35.6	-	-	-	-
Value added per establishment ($ mil.)	0.430	26.2	0.513	30.4	0.462	21.9	0.467	29.0	-	-	-	-
Capital investment per estab. ($ mil.)	0.050	18.9	-	-	-	-	-	-	-	-	-	-
Payroll per employee ($)	20,133	147.2	19,299	106.6	17,777	88.5	19,890	122.5	-	-	-	-
Females as % of total employment (%)	28.8	89.7	28.5	88.7	28.2	61.0	28.6	57.2	-	-	-	-
Output per employee ($)	119,395	174.0	117,706	144.6	113,227	126.6	119,403	149.1	-	-	-	-
Value added per employee ($)	33,620	102.3	35,749	105.9	32,385	89.9	33,409	104.3	-	-	-	-
Capital investment per employee ($)	3,605	65.5	-	-	-	-	-	-	-	-	-	-

Columns headed % show percent of the world total or of the world ratio. Ratios closest to 100 are closest to the world average. Values higher than 100 are above, values lower than 100 are below world average. *nec* stands for *not elsewhere classified*. Ratios, where shown, are frequently based on a subset of total data; for example, payroll per employee is averaged from all cases where *both* payroll and employment data are present. Dividing *total* salaries/wages by *total* employment will not necessarily produce a matching value.

For additional notes and sources, see Appendix I.

Norway

Top Ranked Employing Sectors	Top Ranked Output Sectors
3110 Food Products	3110 Food Products
3840 Transportation Equipment	3840 Transportation Equipment
3420 Printing, Publishing	3720 Nonferrous metals
3820 Machinery nec	3420 Printing, Publishing
3810 Metal Products	3510 Industrial Chemicals

Population 4,420,000. Percent of world population: 0.075

	1990		1991		1992		1993		1994		1995	
	Value	%	Value	%	Value	%	Value	%	Value	%	Value	%
Establishments or enterprises (number)	7,074	0.307	6,855	0.288	4,459	0.159	3,877	0.189	3,993	0.223	-	-
Total employment (000)	312	0.149	304	0.163	277	0.152	230	0.126	235	0.148	237	0.167
Female employees (000)	-		-		-		-		-		-	
Salaries and Wages ($ bil.)	9	0.430	9	0.390	11	0.428	7	0.333	7	0.383	9	0.530
Output ($ bil.)	58	0.408	57	0.371	55	0.334	37	0.255	42	0.304	49	0.366
Value added ($ bil.)	16	0.286	16	0.275	14	0.239	11	0.177	12	0.229	14	0.251
Capital investment ($ mil.)	2,051	0.346	1,799	0.288	2,469	0.386	1,163	0.214	1,350	0.304	-	-
Employees per establishment (number)	44	59.7	44	65.5	62	99.9	59	84.0	59	87.4	-	-
Output per establishment ($ mil.)	8	195.1	8	181.0	12	228.2	10	205.5	11	209.0	-	-
Value added per establishment ($ mil.)	2	138.0	2	134.7	3	153.8	3	171.0	3	167.1	-	-
Capital investment per estab. ($ mil.)	0.317	119.7	0.287	97.6	0.558	194.6	0.303	119.8	0.339	143.9	-	-
Payroll per employee ($)	29,721	217.2	30,422	168.1	39,898	198.7	30,298	186.6	31,953	182.0	37,206	212.9
Females as % of total employment (%)	-		-		-		-		-		-	
Output per employee ($)	187,646	273.4	187,076	229.8	200,674	224.3	164,766	205.8	179,012	204.9	206,852	223.5
Value added per employee ($)	51,361	156.3	51,212	151.6	52,333	145.3	47,001	146.7	50,174	141.4	58,328	147.7
Capital investment per employee ($)	7,709	140.0	6,913	112.7	9,018	145.6	5,179	90.6	5,764	88.1	-	-

Columns headed % show percent of the world total or of the world ratio. Ratios closest to 100 are closest to the world average. Values higher than 100 are above, values lower than 100 are below world average. *nec* stands for *not elsewhere classified*. Ratios, where shown, are frequently based on a subset of total data; for example, payroll per employee is averaged from all cases where *both* payroll and employment data are present. Dividing *total* salaries/wages by *total* employment will not necessarily produce a matching value.

Oman

Top Ranked Employing Sectors	Top Ranked Output Sectors
3900 Industries nec	3900 Industries nec

Population 2,364,000. Percent of world population: 0.040

	1990		1991		1992		1993		1994		1995	
	Value	%	Value	%	Value	%	Value	%	Value	%	Value	%
Establishments or enterprises (number)	1	-	1	-	1	-	2	-	-	-	8	0.001
Total employment (000)	-	-	0.003	-	0.006	-	0.062	-	-	-	0.116	-
Female employees (000)	-	-	-	-	-	-	-	-	-	-	-	-
Salaries and Wages ($ bil.)	-	-	0.000	-	0.000	-	0.000	-	-	-	0.001	-
Output ($ bil.)	-	-	0.000	-	0.000	-	0.002	-	-	-	0.018	-
Value added ($ bil.)	-	-	-	-	-	-	-	-	-	-	-	-
Capital investment ($ mil.)	0.049	-	0.031	-	0.078	-	3	0.001	-	-	4	0.001
Employees per establishment (number)	-	-	3	4.4	6	9.7	31	44.0	-	-	15	20.7
Output per establishment ($ mil.)	-	-	0.070	1.5	0.234	4.3	1	21.3	-	-	2	53.4
Value added per establishment ($ mil.)	-	-	-	-	-	-	-	-	-	-	-	-
Capital investment per estab. ($ mil.)	0.049	18.6	0.031	10.6	0.078	27.2	1	582.4	-	-	0.472	165.2
Payroll per employee ($)	-	-	3,463	19.1	4,329	21.6	5,320	32.8	-	-	7,792	44.6
Females as % of total employment (%)	-	-	-		-		-		-		-	
Output per employee ($)	-	-	23,377	28.7	38,961	43.5	32,258	40.3	-	-	152,172	164.4
Value added per employee ($)	-	-	-		-		-		-		-	
Capital investment per employee ($)	-	-	10,390	169.4	12,987	209.6	47,591	832.6	-	-	32,579	436.3

Columns headed % show percent of the world total or of the world ratio. Ratios closest to 100 are closest to the world average. Values higher than 100 are above, values lower than 100 are below world average. *nec* stands for *not elsewhere classified*. Ratios, where shown, are frequently based on a subset of total data; for example, payroll per employee is averaged from all cases where *both* payroll and employment data are present. Dividing *total* salaries/wages by *total* employment will not necessarily produce a matching value.

Pakistan

Top Ranked Employing Sectors	Top Ranked Output Sectors
3210 Textiles	3210 Textiles
3110 Food Products	3110 Food Products
3710 Iron and Steel	3530 Petroleum Refineries
3520 Chemical Products nec	3520 Chemical Products nec
3820 Machinery nec	3710 Iron and Steel

Population 135,135,000. Percent of world population: 2.280

	1990		1991		1992		1993		1994		1995	
	Value	%	Value	%	Value	%	Value	%	Value	%	Value	%
Establishments or enterprises (number)	-	-	6,385	0.268	-	-	-	-	-	-	-	-
Total employment (000)	584	0.279	900	0.482	651	0.357	-	-	-	-	-	-
Female employees (000)	-	-	-	-	-	-	-	-	-	-	-	-
Salaries and Wages ($ bil.)	1	0.048	2	0.065	1	0.047	-	-	-	-	-	-
Output ($ bil.)	15	0.103	22	0.141	17	0.104	-	-	-	-	-	-
Value added ($ bil.)	4	0.080	6	0.113	5	0.087	-	-	-	-	-	-
Capital investment ($ mil.)	-	-	1,100	0.176	-	-	-	-	-	-	-	-
Employees per establishment (number)	-	-	141	207.7	-	-	-	-	-	-	-	-
Output per establishment ($ mil.)	-	-	3	74.1	-	-	-	-	-	-	-	-
Value added per establishment ($ mil.)	-	-	1.000	59.2	-	-	-	-	-	-	-	-
Capital investment per estab. ($ mil.)	-	-	0.172	58.7	-	-	-	-	-	-	-	-
Payroll per employee ($)	1,764	12.9	1,716	9.5	1,844	9.2	-	-	-	-	-	-
Females as % of total employment (%)	-	-	-	-	-	-	-	-	-	-	-	-
Output per employee ($)	25,150	36.6	24,130	29.6	26,265	29.4	-	-	-	-	-	-
Value added per employee ($)	7,657	23.3	7,093	21.0	7,999	22.2	-	-	-	-	-	-
Capital investment per employee ($)	-	-	1,223	19.9	-	-	-	-	-	-	-	-

Columns headed % show percent of the world total or of the world ratio. Ratios closest to 100 are closest to the world average. Values higher than 100 are above, values lower than 100 are below world average. *nec* stands for *not elsewhere classified*. Ratios, where shown, are frequently based on a subset of total data; for example, payroll per employee is averaged from all cases where *both* payroll and employment data are present. Dividing *total* salaries/wages by *total* employment will not necessarily produce a matching value.

Panama

Top Ranked Employing Sectors	Top Ranked Output Sectors
3110 Food Products	3110 Food Products
3220 Wearing Apparel	3130 Beverages
3420 Printing, Publishing	3690 Nonmetal Products nec
3690 Nonmetal Products nec	3410 Paper and Products
3130 Beverages	3220 Wearing Apparel

Population 2,736,000. Percent of world population: 0.046

	1990		1991		1992		1993		1994		1995	
	Value	%	Value	%	Value	%	Value	%	Value	%	Value	%
Establishments or enterprises (number)	969	0.042	910	0.038	-	-	-	-	-	-	-	-
Total employment (000)	36	0.017	37	0.020	39	0.021	40	0.022	42	0.026	43	0.030
Female employees (000)	10	0.077	-	-	-	-	-	-	-	-	-	-
Salaries and Wages ($ bil.)	0.187	0.009	0.201	0.008	0.217	0.008	0.234	0.011	0.257	0.013	0.269	0.016
Output ($ bil.)	1	0.010	2	0.010	2	0.010	2	0.012	2	0.014	2	0.015
Value added ($ bil.)	0.587	0.010	0.598	0.011	0.631	0.011	0.683	0.011	0.744	0.015	0.743	0.014
Capital investment ($ mil.)	34	0.006	58	0.009	-	-	-	-	-	-	-	-
Employees per establishment (number)	38	51.1	41	60.4	-	-	-	-	-	-	-	-
Output per establishment ($ mil.)	2	36.6	2	37.9	-	-	-	-	-	-	-	-
Value added per establishment ($ mil.)	0.610	37.2	0.662	39.2	-	-	-	-	-	-	-	-
Capital investment per estab. ($ mil.)	0.036	13.5	0.064	21.8	-	-	-	-	-	-	-	-
Payroll per employee ($)	5,163	37.7	5,418	29.9	5,590	27.8	5,794	35.7	6,135	34.9	6,301	36.1
Females as % of total employment (%)	28.4	88.6	-	-	-	-	-	-	-	-	-	-
Output per employee ($)	41,161	60.0	42,430	52.1	43,469	48.6	44,799	56.0	46,906	53.7	46,604	50.4
Value added per employee ($)	16,185	49.3	16,161	47.9	16,275	45.2	16,925	52.8	17,778	50.1	17,436	44.1
Capital investment per employee ($)	949	17.2	1,563	25.5	-	-	-	-	-	-	-	-

Columns headed % show percent of the world total or of the world ratio. Ratios closest to 100 are closest to the world average. Values higher than 100 are above, values lower than 100 are below world average. *nec* stands for *not elsewhere classified*. Ratios, where shown, are frequently based on a subset of total data; for example, payroll per employee is averaged from all cases where *both* payroll and employment data are present. Dividing *total* salaries/wages by *total* employment will not necessarily produce a matching value.

For additional notes and sources, see Appendix I.

Paraguay

Top Ranked Employing Sectors	Top Ranked Output Sectors
3110 Food Products	3900 Industries nec
3130 Beverages	
3690 Nonmetal Products nec	
3520 Chemical Products nec	
3210 Textiles	

Population 5,291,000. Percent of world population: 0.089

	1990		1991		1992		1993		1994		1995	
	Value	%	Value	%	Value	%	Value	%	Value	%	Value	%
Establishments or enterprises (number)	-	-	152	0.006	-	-	-	-	-	-	-	-
Total employment (000)	-	-	12	0.006	-	-	-	-	-	-	-	-
Female employees (000)	-	-	-	-	-	-	-	-	-	-	-	-
Salaries and Wages ($ bil.)	-	-	0.000	-	-	-	-	-	-	-	-	-
Output ($ bil.)	-	-	0.000	-	-	-	-	-	-	-	-	-
Value added ($ bil.)	-	-	0.000	-	-	-	-	-	-	-	-	-
Capital investment ($ mil.)	-	-	-	-	-	-	-	-	-	-	-	-
Employees per establishment (number)	-	-	78	114.3	-	-	-	-	-	-	-	-
Output per establishment ($ mil.)	-	-	0.015	0.3	-	-	-	-	-	-	-	-
Value added per establishment ($ mil.)	-	-	0.010	0.6	-	-	-	-	-	-	-	-
Capital investment per estab. ($ mil.)	-	-	-	-	-	-	-	-	-	-	-	-
Payroll per employee ($)	-	-	4,165	23.0	-	-	-	-	-	-	-	-
Females as % of total employment (%)	-	-	-	-	-	-	-	-	-	-	-	-
Output per employee ($)	-	-	2,984	3.7	-	-	-	-	-	-	-	-
Value added per employee ($)	-	-	1,996	5.9	-	-	-	-	-	-	-	-
Capital investment per employee ($)	-	-	-	-	-	-	-	-	-	-	-	-

Columns headed % show percent of the world total or of the world ratio. Ratios closest to 100 are closest to the world average. Values higher than 100 are above, values lower than 100 are below world average. nec stands for not elsewhere classified. Ratios, where shown, are frequently based on a subset of total data; for example, payroll per employee is averaged from all cases where both payroll and employment data are present. Dividing total salaries/wages by total employment will not necessarily produce a matching value.

Peru

Top Ranked Employing Sectors	Top Ranked Output Sectors
3110 Food Products	3110 Food Products
3210 Textiles	3530 Petroleum Refineries
3220 Wearing Apparel	3720 Nonferrous metals
3810 Metal Products	3210 Textiles
3520 Chemical Products nec	3130 Beverages

Population 26,111,000. Percent of world population: 0.441

	1990		1991		1992		1993		1994		1995	
	Value	%	Value	%	Value	%	Value	%	Value	%	Value	%
Establishments or enterprises (number)	18,288	0.794	18,852	0.791	19,406	0.691	-	-	-	-	-	-
Total employment (000)	342	0.164	303	0.162	281	0.154	-	-	-	-	-	-
Female employees (000)	-	-	-	-	-	-	-	-	-	-	-	-
Salaries and Wages ($ bil.)	1	0.058	1	0.057	1	0.048	-	-	-	-	-	-
Output ($ bil.)	16	0.115	17	0.110	17	0.101	-	-	-	-	-	-
Value added ($ bil.)	9	0.153	8	0.138	7	0.114	-	-	-	-	-	-
Capital investment ($ mil.)	488	0.082	361	0.058	448	0.070	-	-	-	-	-	-
Employees per establishment (number)	19	25.4	16	23.7	14	23.3	-	-	-	-	-	-
Output per establishment ($ mil.)	0.902	21.3	0.895	19.5	0.857	15.7	-	-	-	-	-	-
Value added per establishment ($ mil.)	0.467	28.5	0.416	24.7	0.350	16.6	-	-	-	-	-	-
Capital investment per estab. ($ mil.)	0.027	10.1	0.019	6.5	0.023	8.0	-	-	-	-	-	-
Payroll per employee ($)	3,655	26.7	4,453	24.6	4,381	21.8	-	-	-	-	-	-
Females as % of total employment (%)	-	-	-	-	-	-	-	-	-	-	-	-
Output per employee ($)	48,174	70.2	55,682	68.4	59,177	66.1	-	-	-	-	-	-
Value added per employee ($)	24,957	76.0	25,896	76.7	24,164	67.1	-	-	-	-	-	-
Capital investment per employee ($)	1,426	25.9	1,193	19.5	1,593	25.7	-	-	-	-	-	-

Columns headed % show percent of the world total or of the world ratio. Ratios closest to 100 are closest to the world average. Values higher than 100 are above, values lower than 100 are below world average. nec stands for not elsewhere classified. Ratios, where shown, are frequently based on a subset of total data; for example, payroll per employee is averaged from all cases where both payroll and employment data are present. Dividing total salaries/wages by total employment will not necessarily produce a matching value.

Philippines

	Top Ranked Employing Sectors	Top Ranked Output Sectors
	3110 Food Products	3110 Food Products
	3220 Wearing Apparel	3530 Petroleum Refineries
	3830 Electrical Machinery	3830 Electrical Machinery
	3210 Textiles	3520 Chemical Products nec
	3520 Chemical Products nec	3130 Beverages

Population 77,726,000. Percent of world population: 1.311

	1990		1991		1992		1993		1994		1995	
	Value	%	Value	%	Value	%	Value	%	Value	%	Value	%
Establishments or enterprises (number)	85,247	3.700	89,879	3.800	12,671	0.451	11,831	0.576	-	-	-	-
Total employment (000)	1,267	0.606	1,264	0.677	1,132	0.621	1,061	0.582	-	-	-	-
Female employees (000)	-	-	-	-	509	2.400	475	2.100	-	-	-	-
Salaries and Wages ($ bil.)	2	0.111	3	0.106	3	0.116	3	0.134	-	-	-	-
Output ($ bil.)	29	0.201	30	0.196	35	0.213	35	0.244	-	-	-	-
Value added ($ bil.)	10	0.187	11	0.191	13	0.213	13	0.222	-	-	-	-
Capital investment ($ mil.)	1,537	0.259	1,385	0.222	1,699	0.266	2,213	0.407	-	-	-	-
Employees per establishment (number)	15	20.2	14	20.7	89	143.8	90	127.2	-	-	-	-
Output per establishment ($ mil.)	0.337	8.0	0.336	7.3	3	50.8	3	63.7	-	-	-	-
Value added per establishment ($ mil.)	0.123	7.5	0.121	7.2	1	47.8	1	69.4	-	-	-	-
Capital investment per estab. ($ mil.)	0.018	6.8	0.015	5.2	0.134	46.7	0.187	73.8	-	-	-	-
Payroll per employee ($)	1,890	13.8	1,999	11.0	2,620	13.1	2,573	15.8	-	-	-	-
Females as % of total employment (%)	-	-	-	-	44.9	97.1	44.8	89.7	-	-	-	-
Output per employee ($)	22,663	33.0	23,878	29.3	30,977	34.6	33,361	41.7	-	-	-	-
Value added per employee ($)	8,259	25.1	8,584	25.4	11,274	31.3	12,465	38.9	-	-	-	-
Capital investment per employee ($)	1,213	22.0	1,095	17.9	1,500	24.2	2,086	36.5	-	-	-	-

Columns headed % show percent of the world total or of the world ratio. Ratios closest to 100 are closest to the world average. Values higher than 100 are above, values lower than 100 are below world average. *nec* stands for *not elsewhere classified*. Ratios, where shown, are frequently based on a subset of total data; for example, payroll per employee is averaged from all cases where *both* payroll and employment data are present. Dividing *total* salaries/wages by *total* employment will not necessarily produce a matching value.

Poland

	Top Ranked Employing Sectors	Top Ranked Output Sectors
	3110 Food Products	3110 Food Products
	3820 Machinery nec	3530 Petroleum Refineries
	3840 Transportation Equipment	3840 Transportation Equipment
	3210 Textiles	3710 Iron and Steel
	3810 Metal Products	3130 Beverages

Population 38,607,000. Percent of world population: 0.651

	1990		1991		1992		1993		1994		1995	
	Value	%	Value	%	Value	%	Value	%	Value	%	Value	%
Establishments or enterprises (number)	6,417	0.278	6,744	0.283	6,440	0.229	6,381	0.310	-	-	-	-
Total employment (000)	3,014	1.400	2,671	1.400	2,322	1.300	2,166	1.200	2,266	1.400	2,309	1.800
Female employees (000)	-	-	-	-	-	-	-	-	-	-	-	-
Salaries and Wages ($ bil.)	4	0.176	5	0.210	6	0.219	5	0.266	7	0.341	9	0.534
Output ($ bil.)	46	0.322	49	0.321	46	0.283	49	0.338	59	0.430	78	0.586
Value added ($ bil.)	23	0.411	20	0.349	20	0.330	18	0.308	22	0.428	29	0.521
Capital investment ($ mil.)	2,973	0.501	3,580	0.574	3,360	0.526	2,770	0.510	-	-	-	-
Employees per establishment (number)	470	637.2	396	583.7	361	579.9	339	481.5	-	-	-	-
Output per establishment ($ mil.)	7	169.6	7	159.2	7	132.5	8	163.4	-	-	-	-
Value added per establishment ($ mil.)	4	218.9	3	174.0	3	145.5	3	178.8	-	-	-	-
Capital investment per estab. ($ mil.)	0.463	174.9	0.531	180.7	0.522	181.8	0.434	171.4	-	-	-	-
Payroll per employee ($)	1,258	9.2	1,869	10.3	2,407	12.0	2,502	15.4	2,934	16.7	3,835	21.9
Females as % of total employment (%)	-	-	-	-	-	-	-	-	-	-	-	-
Output per employee ($)	15,293	22.3	18,456	22.7	20,018	22.4	22,617	28.2	26,167	29.9	33,829	36.6
Value added per employee ($)	7,637	23.2	7,417	22.0	8,507	23.6	8,481	26.5	9,681	27.3	12,390	31.4
Capital investment per employee ($)	986	17.9	1,340	21.9	1,447	23.4	1,279	22.4	-	-	-	-

Columns headed % show percent of the world total or of the world ratio. Ratios closest to 100 are closest to the world average. Values higher than 100 are above, values lower than 100 are below world average. *nec* stands for *not elsewhere classified*. Ratios, where shown, are frequently based on a subset of total data; for example, payroll per employee is averaged from all cases where *both* payroll and employment data are present. Dividing *total* salaries/wages by *total* employment will not necessarily produce a matching value.

Portugal

Top Ranked Employing Sectors	Top Ranked Output Sectors
3220 Wearing Apparel	3110 Food Products
3210 Textiles	3210 Textiles
3110 Food Products	3530 Petroleum Refineries
3810 Metal Products	3220 Wearing Apparel
3240 Footwear	3830 Electrical Machinery

Population 9,928,000. Percent of world population: 0.168

	1990 Value	%	1991 Value	%	1992 Value	%	1993 Value	%	1994 Value	%	1995 Value	%
Establishments or enterprises (number)	68,037	3.000	68,068	2.900	67,084	2.400	68,695	3.300	69,735	3.900	-	-
Total employment (000)	1,184	0.566	1,204	0.645	1,147	0.629	1,122	0.616	1,102	0.693	1,003	0.705
Female employees (000)	252	1.900	249	2.800	238	1.100	231	1.000	225	1.100	-	-
Salaries and Wages ($ bil.)	1	0.050	1	0.050	1	0.053	1	0.056	1	0.055	-	-
Output ($ bil.)	54	0.378	57	0.371	63	0.382	57	0.396	62	0.448	63	0.472
Value added ($ bil.)	16	0.286	17	0.296	19	0.310	18	0.305	20	0.386	20	0.371
Capital investment ($ mil.)	3,923	0.662	4,550	0.729	3,956	0.619	3,580	0.659	4,231	0.953	-	-
Employees per establishment (number)	17	23.6	18	26.1	17	27.5	16	23.2	16	23.5	-	-
Output per establishment ($ mil.)	0.796	18.8	0.838	18.2	0.937	17.2	0.837	17.8	0.885	17.6	-	-
Value added per establishment ($ mil.)	0.235	14.4	0.247	14.6	0.276	13.1	0.265	16.4	0.283	16.1	-	-
Capital investment per estab. ($ mil.)	0.058	21.8	0.067	22.8	0.059	20.6	0.052	20.6	0.061	25.8	-	-
Payroll per employee ($)	6,454	47.2	7,411	40.9	9,007	44.9	8,042	49.5	8,240	46.9	-	-
Females as % of total employment (%)	21.4	66.7	20.7	64.4	20.9	45.2	20.7	41.5	20.5	39.0	-	-
Output per employee ($)	45,738	66.6	47,353	58.2	54,760	61.2	51,236	64.0	55,990	64.1	62,783	67.8
Value added per employee ($)	13,522	41.2	13,941	41.3	16,145	44.8	16,201	50.6	17,918	50.5	20,317	51.4
Capital investment per employee ($)	3,314	60.2	3,780	61.6	3,448	55.7	3,191	55.8	3,838	58.7	-	-

Columns headed % show percent of the world total or of the world ratio. Ratios closest to 100 are closest to the world average. Values higher than 100 are above, values lower than 100 are below world average. *nec* stands for *not elsewhere classified*. Ratios, where shown, are frequently based on a subset of total data; for example, payroll per employee is averaged from all cases where *both* payroll and employment data are present. Dividing *total* salaries/wages by *total* employment will not necessarily produce a matching value.

Puerto Rico

Top Ranked Employing Sectors	Top Ranked Output Sectors
3220 Wearing Apparel	3520 Chemical Products nec
3830 Electrical Machinery	3522 Drugs and Medicines
3110 Food Products	3830 Electrical Machinery
3850 Professional Goods	3130 Beverages
3240 Footwear	3110 Food Products

Population 3,857,000. Percent of world population: 0.065

	1990 Value	%	1991 Value	%	1992 Value	%	1993 Value	%	1994 Value	%	1995 Value	%
Establishments or enterprises (number)	1,778	0.077	1,818	0.076	1,640	0.058	1,729	0.084	1,714	0.096	1,709	0.139
Total employment (000)	112	0.053	105	0.056	105	0.057	104	0.057	102	0.064	105	0.074
Female employees (000)	53	0.397	51	0.562	50	0.232	48	0.214	45	0.211	46	0.239
Salaries and Wages ($ bil.)	3	0.129	3	0.125	3	0.124	3	0.165	4	0.182	4	0.220
Output ($ bil.)	37	0.256	38	0.246	40	0.245	43	0.295	48	0.345	52	0.392
Value added ($ bil.)	11	0.203	12	0.206	13	0.218	14	0.242	15	0.290	16	0.292
Capital investment ($ mil.)	-		-		-		-		-		-	
Employees per establishment (number)	66	89.4	60	89.2	66	106.1	63	89.0	62	92.8	64	92.0
Output per establishment ($ mil.)	17	409.1	17	370.8	20	368.7	20	432.5	23	451.5	25	606.3
Value added per establishment ($ mil.)	5	302.4	4	258.0	5	236.6	5	338.8	6	318.9	6	341.8
Capital investment per estab. ($ mil.)	-		-		-		-		-		-	
Payroll per employee ($)	20,543	150.2	18,597	102.8	19,212	95.7	20,387	125.5	21,182	120.6	20,984	120.1
Females as % of total employment (%)	66.9	208.8	65.7	204.6	65.4	141.3	63.9	128.0	61.2	116.7	60.7	104.0
Output per employee ($)	165,655	241.4	168,617	207.2	172,784	193.1	184,955	231.0	207,641	237.6	217,887	235.4
Value added per employee ($)	62,861	191.3	61,479	182.0	67,947	188.7	73,961	230.8	77,444	218.2	79,949	202.4
Capital investment per employee ($)	-		-		-		-		-		-	

Columns headed % show percent of the world total or of the world ratio. Ratios closest to 100 are closest to the world average. Values higher than 100 are above, values lower than 100 are below world average. *nec* stands for *not elsewhere classified*. Ratios, where shown, are frequently based on a subset of total data; for example, payroll per employee is averaged from all cases where *both* payroll and employment data are present. Dividing *total* salaries/wages by *total* employment will not necessarily produce a matching value.

Qatar

Top Ranked Employing Sectors	Top Ranked Output Sectors
3110 Food Products	3530 Petroleum Refineries
3320 Furniture, Fixtures	3510 Industrial Chemicals
3510 Industrial Chemicals	3710 Iron and Steel
3420 Printing, Publishing	3110 Food Products
3710 Iron and Steel	3420 Printing, Publishing

Population 697,000. Percent of world population: 0.012

	1990		1991		1992		1993		1994		1995	
	Value	%	Value	%	Value	%	Value	%	Value	%	Value	%
Establishments or enterprises (number)	499	0.022	502	0.021	856	0.030	845	0.041	809	0.045	-	-
Total employment (000)	9	0.004	9	0.005	12	0.006	10	0.005	11	0.007	-	-
Female employees (000)	-	-	-	-	-	-	-	-	-	-	-	-
Salaries and Wages ($ bil.)	0.124	0.006	0.110	0.005	0.132	0.005	0.130	0.006	0.138	0.007	-	-
Output ($ bil.)	1	0.008	1	0.008	1	0.008	1	0.008	1	0.009	-	-
Value added ($ bil.)	0.877	0.016	0.842	0.015	0.791	0.013	0.685	0.011	0.675	0.013	-	-
Capital investment ($ mil.)	-	-	-	-	-	-	-	-	-	-	-	-
Employees per establishment (number)	18	24.8	19	27.3	14	22.2	12	17.0	14	20.3	-	-
Output per establishment ($ mil.)	2	55.4	2	51.9	1	27.3	1	29.1	1	29.8	-	-
Value added per establishment ($ mil.)	2	107.5	2	101.8	0.982	46.6	0.833	51.7	0.853	48.4	-	-
Capital investment per estab. ($ mil.)	-	-	-	-	-	-	-	-	-	-	-	-
Payroll per employee ($)	13,642	99.7	11,835	65.4	11,405	56.8	13,103	80.7	12,734	72.5	-	-
Females as % of total employment (%)	-	-	-	-	-	-	-	-	-	-	-	-
Output per employee ($)	128,338	187.0	128,422	157.8	107,361	120.0	113,999	142.4	109,687	125.5	-	-
Value added per employee ($)	96,344	293.3	91,001	269.5	69,289	192.4	69,589	217.2	62,500	176.1	-	-
Capital investment per employee ($)	-	-	-	-	-	-	-	-	-	-	-	-

Columns headed % show percent of the world total or of the world ratio. Ratios closest to 100 are closest to the world average. Values higher than 100 are above, values lower than 100 are below world average. *nec* stands for *not elsewhere classified*. Ratios, where shown, are frequently based on a subset of total data; for example, payroll per employee is averaged from all cases where *both* payroll and employment data are present. Dividing *total* salaries/wages by *total* employment will not necessarily produce a matching value.

Republic of Moldova

Top Ranked Employing Sectors	Top Ranked Output Sectors
3820 Machinery nec	3110 Food Products
3110 Food Products	3130 Beverages
3220 Wearing Apparel	3820 Machinery nec
3210 Textiles	3230 Leather and Products
3130 Beverages	3900 Industries nec

Population 4,458,000. Percent of world population: 0.075

	1990		1991		1992		1993		1994		1995	
	Value	%	Value	%	Value	%	Value	%	Value	%	Value	%
Establishments or enterprises (number)	532	0.023	466	0.020	474	0.017	333	0.016	345	0.019	354	0.029
Total employment (000)	373	0.178	329	0.176	320	0.176	155	0.085	129	0.081	120	0.084
Female employees (000)	-	-	-	-	-	-	-	-	78	0.370	70	0.365
Salaries and Wages ($ bil.)	0.000	-	1	0.045	-	-	0.047	0.002	0.058	0.003	0.064	0.004
Output ($ bil.)	0.000	-	13	0.082	0.978	0.006	0.652	0.004	0.659	0.005	0.766	0.006
Value added ($ bil.)	-	-	-	-	-	-	-	-	-	-	-	-
Capital investment ($ mil.)	-	-	-	-	-	-	-	-	-	-	-	-
Employees per establishment (number)	701	950.9	706	1,041.2	676	1,086.7	466	660.5	371	551.6	339	483.9
Output per establishment ($ mil.)	0.000	0.0	27	587.4	2	37.9	2	41.7	2	38.1	2	52.5
Value added per establishment ($ mil.)	-	-	-	-	-	-	-	-	-	-	-	-
Capital investment per estab. ($ mil.)	-	-	-	-	-	-	-	-	-	-	-	-
Payroll per employee ($)	0.006	-	3,261	18.0	-	-	355	2.2	446	2.5	538	3.1
Females as % of total employment (%)	-	-	-	-	-	-	-	-	60.5	115.4	58.4	100.0
Output per employee ($)	0.054	-	38,172	48.9	3,053	3.4	4,203	5.3	5,134	5.9	6,390	6.9
Value added per employee ($)	-	-	-	-	-	-	-	-	-	-	-	-
Capital investment per employee ($)	-	-	-	-	-	-	-	-	-	-	-	-

Columns headed % show percent of the world total or of the world ratio. Ratios closest to 100 are closest to the world average. Values higher than 100 are above, values lower than 100 are below world average. *nec* stands for *not elsewhere classified*. Ratios, where shown, are frequently based on a subset of total data; for example, payroll per employee is averaged from all cases where *both* payroll and employment data are present. Dividing *total* salaries/wages by *total* employment will not necessarily produce a matching value.

Romania

Top Ranked Employing Sectors	Top Ranked Output Sectors
3820 Machinery nec	3110 Food Products
3210 Textiles	3710 Iron and Steel
3220 Wearing Apparel	3530 Petroleum Refineries
3840 Transportation Equipment	3840 Transportation Equipment
3320 Furniture, Fixtures	3210 Textiles

Population 22,396,000. Percent of world population: 0.378

	1990		1991		1992		1993		1994		1995	
	Value	%	Value	%	Value	%	Value	%	Value	%	Value	%
Establishments or enterprises (number)	314	0.014	8,772	0.368	15,476	0.551	21,973	1.100	32,519	1.800	-	-
Total employment (000)	2,447	1.200	2,232	1.200	1,907	1.000	1,731	0.951	1,551	0.974	-	-
Female employees (000)	-	-	-	-	-	-	-	-	-	-	-	-
Salaries and Wages ($ bil.)	4	0.194	2	0.101	1	0.052	1	0.072	1	0.071	-	-
Output ($ bil.)	44	0.310	33	0.213	21	0.126	22	0.149	-	-	-	-
Value added ($ bil.)	12	0.223	9	0.153	6	0.093	6	0.108	-	-	-	-
Capital investment ($ mil.)	999	0.168	773	0.124	212	0.033	874	0.161	772	0.174	-	-
Employees per establishment (number)	1,898	2,574.4	369	544.6	185	297.1	113	160.9	77	115.0	-	-
Output per establishment ($ mil.)	28	653.2	3	59.5	0.956	17.6	0.725	15.4	-	-	-	-
Value added per establishment ($ mil.)	8	473.0	0.756	44.8	0.269	12.7	0.228	14.2	-	-	-	-
Capital investment per estab. ($ mil.)	1	507.1	0.181	61.7	0.037	13.0	0.032	12.7	0.024	10.1	-	-
Payroll per employee ($)	1,709	12.5	1,075	5.9	698	3.5	849	5.2	890	5.1	-	-
Females as % of total employment (%)	-	-	-	-	-	-	-	-	-	-	-	-
Output per employee ($)	11,173	16.3	8,684	10.7	5,758	6.4	6,537	8.2	-	-	-	-
Value added per employee ($)	3,570	10.9	2,489	7.4	1,630	4.5	1,996	6.2	-	-	-	-
Capital investment per employee ($)	796	14.5	778	12.7	141	2.3	271	4.7	226	3.5	-	-

Columns headed % show percent of the world total or of the world ratio. Ratios closest to 100 are closest to the world average. Values higher than 100 are above, values lower than 100 are below world average. *nec* stands for *not elsewhere classified*. Ratios, where shown, are frequently based on a subset of total data; for example, payroll per employee is averaged from all cases where *both* payroll and employment data are present. Dividing *total* salaries/wages by *total* employment will not necessarily produce a matching value.

Russian Federation

Top Ranked Employing Sectors	Top Ranked Output Sectors
3820 Machinery nec	3110 Food Products
3110 Food Products	3710 Iron and Steel
3840 Transportation Equipment	3820 Machinery nec
3690 Nonmetal Products nec	3840 Transportation Equipment
3710 Iron and Steel	3510 Industrial Chemicals

Population 146,861,000. Percent of world population: 2.478

	1990		1991		1992		1993		1994		1995	
	Value	%	Value	%	Value	%	Value	%	Value	%	Value	%
Establishments or enterprises (number)	-	-	-	-	-	-	90,610	4.400	83,002	4.600	120,087	9.800
Total employment (000)	-	-	-	-	-	-	16,818	9.200	14,915	9.400	13,909	9.800
Female employees (000)	-	-	-	-	-	-	-	-	-	-	-	-
Salaries and Wages ($ bil.)	-	-	-	-	-	-	12	0.593	17	0.876	17	1.000
Output ($ bil.)	-	-	322	2.100	240	1.500	106	0.732	132	0.960	176	1.300
Value added ($ bil.)	-	-	-	-	-	-	49	0.822	61	1.200	62	1.100
Capital investment ($ mil.)	-	-	-	-	-	-	-	-	-	-	-	-
Employees per establishment (number)	-	-	-	-	-	-	186	263.3	180	266.9	116	165.3
Output per establishment ($ mil.)	-	-	-	-	-	-	1	24.9	2	31.7	1	35.5
Value added per establishment ($ mil.)	-	-	-	-	-	-	0.542	33.6	0.738	41.9	0.520	29.5
Capital investment per estab. ($ mil.)	-	-	-	-	-	-	-	-	-	-	-	-
Payroll per employee ($)	-	-	-	-	-	-	719	4.4	1,146	6.5	1,238	7.1
Females as % of total employment (%)	-	-	-	-	-	-	-	-	-	-	-	-
Output per employee ($)	-	-	-	-	-	-	6,310	7.9	8,873	10.2	12,652	13.7
Value added per employee ($)	-	-	-	-	-	-	2,918	9.1	4,110	11.6	4,490	11.4
Capital investment per employee ($)	-	-	-	-	-	-	-	-	-	-	-	-

Columns headed % show percent of the world total or of the world ratio. Ratios closest to 100 are closest to the world average. Values higher than 100 are above, values lower than 100 are below world average. *nec* stands for *not elsewhere classified*. Ratios, where shown, are frequently based on a subset of total data; for example, payroll per employee is averaged from all cases where *both* payroll and employment data are present. Dividing *total* salaries/wages by *total* employment will not necessarily produce a matching value.

Saint Vincent & the Grenadines

Top Ranked Employing Sectors	Top Ranked Output Sectors
3220 Wearing Apparel	
3110 Food Products	
3130 Beverages	
3410 Paper and Products	
3810 Metal Products	

Population 120,000. Percent of world population: 0.002

	1990		1991		1992		1993		1994		1995	
	Value	%	Value	%	Value	%	Value	%	Value	%	Value	%
Establishments or enterprises (number)	-	-	-	-	-	-	30	0.001	32	0.002	40	0.003
Total employment (000)	-	-	-	-	-	-	0.850	-	0.852	0.001	0.940	0.001
Female employees (000)	-	-	-	-	-	-	0.248	0.001	0.249	0.001	-	-
Salaries and Wages ($ bil.)	-	-	-	-	-	-	0.003	-	0.003		-	-
Output ($ bil.)	-	-	-	-	-	-	-	-	-		-	-
Value added ($ bil.)	-	-	-	-	-	-	-	-	-		-	-
Capital investment ($ mil.)	-	-	-	-	-	-	-	-	-		-	-
Employees per establishment (number)	-	-	-	-	-	-	31	44.7	29	43.6	25	36.3
Output per establishment ($ mil.)	-	-	-	-	-	-	-	-	-	-	-	-
Value added per establishment ($ mil.)	-	-	-	-	-	-	-	-	-	-	-	-
Capital investment per estab. ($ mil.)	-	-	-	-	-	-	-	-	-	-	-	-
Payroll per employee ($)	-	-	-	-	-	-	3,873	23.9	4,037	23.0	-	-
Females as % of total employment (%)	-	-	-	-	-	-	29.2	58.4	29.2	55.7	-	-
Output per employee ($)	-	-	-	-	-	-	-	-	-	-	-	-
Value added per employee ($)	-	-	-	-	-	-	-	-	-	-	-	-
Capital investment per employee ($)	-	-	-	-	-	-	-	-	-	-	-	-

Columns headed % show percent of the world total or of the world ratio. Ratios closest to 100 are closest to the world average. Values higher than 100 are above, values lower than 100 are below world average. *nec* stands for *not elsewhere classified*. Ratios, where shown, are frequently based on a subset of total data; for example, payroll per employee is averaged from all cases where *both* payroll and employment data are present. Dividing *total* salaries/wages by *total* employment will not necessarily produce a matching value.

Senegal

Top Ranked Employing Sectors	Top Ranked Output Sectors
3110 Food Products	3110 Food Products
3210 Textiles	3530 Petroleum Refineries
3840 Transportation Equipment	3510 Industrial Chemicals
3510 Industrial Chemicals	3210 Textiles
3520 Chemical Products nec	3520 Chemical Products nec

Population 9,723,000. Percent of world population: 0.164

	1990		1991		1992		1993		1994		1995	
	Value	%	Value	%	Value	%	Value	%	Value	%	Value	%
Establishments or enterprises (number)	248	0.011	164	0.007	176	0.006	155	0.008	197	0.011	-	-
Total employment (000)	39	0.019	30	0.016	27	0.015	25	0.014	33	0.021	-	-
Female employees (000)	3	0.025									-	-
Salaries and Wages ($ bil.)	0.202	0.009	0.149	0.006	0.154	0.006	0.132	0.006	0.105	0.005	-	-
Output ($ bil.)	2	0.013	-	-	-	-	-	-	-	-	-	-
Value added ($ bil.)	0.467	0.008	0.360	0.006	0.390	0.007	0.341	0.006	0.385	0.008	-	-
Capital investment ($ mil.)	272	0.046	-	-	-	-	-	-	-	-	-	-
Employees per establishment (number)	158	214.8	184	271.4	153	246.2	161	228.1	169	250.8	-	-
Output per establishment ($ mil.)	7	174.6									-	-
Value added per establishment ($ mil.)	2	115.0	2	130.0	2	110.0	2	141.1	2	110.8	-	-
Capital investment per estab. ($ mil.)	1	418.1	-	-	-	-	-	-	-	-	-	-
Payroll per employee ($)	5,148	37.6	4,918	27.2	6,101	30.4	5,699	35.1	3,148	17.9	-	-
Females as % of total employment (%)	8.6	26.8	-	-	-	-	-	-	-	-	-	-
Output per employee ($)	46,698	68.0									-	-
Value added per employee ($)	11,902	36.2	11,913	35.3	15,463	42.9	14,661	45.8	11,565	32.6	-	-
Capital investment per employee ($)	6,940	126.0	-	-	-	-	-	-	-	-	-	-

Columns headed % show percent of the world total or of the world ratio. Ratios closest to 100 are closest to the world average. Values higher than 100 are above, values lower than 100 are below world average. *nec* stands for *not elsewhere classified*. Ratios, where shown, are frequently based on a subset of total data; for example, payroll per employee is averaged from all cases where *both* payroll and employment data are present. Dividing *total* salaries/wages by *total* employment will not necessarily produce a matching value.

For additional notes and sources, see Appendix I.

Sierra Leone

Top Ranked Employing Sectors	Top Ranked Output Sectors
3110 Food Products	3110 Food Products
3130 Beverages	3130 Beverages
3420 Printing, Publishing	3140 Tobacco
3320 Furniture, Fixtures	3420 Printing, Publishing
3810 Metal Products	3810 Metal Products

Population 5,080,000. Percent of world population: 0.086

	1990		1991		1992		1993		1994		1995	
	Value	%	Value	%	Value	%	Value	%	Value	%	Value	%
Establishments or enterprises (number)	-	-	-	-	-	-	255	0.012	-	-	-	-
Total employment (000)	-	-	-	-	-	-	13	0.007	-	-	-	-
Female employees (000)	-	-	-	-	-	-	-	-	-	-	-	-
Salaries and Wages ($ bil.)	-	-	-	-	-	-	0.007	-	-	-	-	-
Output ($ bil.)	-	-	-	-	-	-	0.148	0.001	-	-	-	-
Value added ($ bil.)	-	-	-	-	-	-	0.070	0.001	-	-	-	-
Capital investment ($ mil.)	-	-	-	-	-	-	-	-	-	-	-	-
Employees per establishment (number)	-	-	-	-	-	-	52	73.2	-	-	-	-
Output per establishment ($ mil.)	-	-	-	-	-	-	0.731	15.6	-	-	-	-
Value added per establishment ($ mil.)	-	-	-	-	-	-	0.347	21.6	-	-	-	-
Capital investment per estab. ($ mil.)	-	-	-	-	-	-	-	-	-	-	-	-
Payroll per employee ($)	-	-	-	-	-	-	627	3.9	-	-	-	-
Females as % of total employment (%)	-	-	-	-	-	-	-	-	-	-	-	-
Output per employee ($)	-	-	-	-	-	-	12,870	16.1	-	-	-	-
Value added per employee ($)	-	-	-	-	-	-	6,117	19.1	-	-	-	-
Capital investment per employee ($)	-	-	-	-	-	-	-	-	-	-	-	-

Columns headed % show percent of the world total or of the world ratio. Ratios closest to 100 are closest to the world average. Values higher than 100 are above, values lower than 100 are below world average. *nec* stands for *not elsewhere classified*. Ratios, where shown, are frequently based on a subset of total data; for example, payroll per employee is averaged from all cases where *both* payroll and employment data are present. Dividing *total* salaries/wages by *total* employment will not necessarily produce a matching value.

Singapore

Top Ranked Employing Sectors	Top Ranked Output Sectors
3830 Electrical Machinery	3820 Machinery nec
3820 Machinery nec	3830 Electrical Machinery
3810 Metal Products	3810 Metal Products
3840 Transportation Equipment	3840 Transportation Equipment
3420 Printing, Publishing	3510 Industrial Chemicals

Population 3,490,000. Percent of world population: 0.059

	1990		1991		1992		1993		1994		1995	
	Value	%	Value	%	Value	%	Value	%	Value	%	Value	%
Establishments or enterprises (number)	4,165	0.181	4,260	0.179	4,397	0.157	4,505	0.219	4,530	0.253	-	-
Total employment (000)	494	0.236	504	0.270	503	0.276	496	0.272	514	0.323	375	0.264
Female employees (000)	-	-	-	-	-	-	-	-	-	-	-	-
Salaries and Wages ($ bil.)	5	0.240	6	0.252	7	0.273	8	0.368	9	0.449	7	0.441
Output ($ bil.)	51	0.357	56	0.367	64	0.388	75	0.517	94	0.685	70	0.523
Value added ($ bil.)	17	0.295	19	0.328	21	0.356	25	0.416	30	0.579	23	0.415
Capital investment ($ mil.)	2,980	0.503	2,978	0.477	3,412	0.534	4,316	0.794	5,234	1.200	-	-
Employees per establishment (number)	118	160.8	118	174.2	114	184.0	110	156.1	113	168.4	-	-
Output per establishment ($ mil.)	12	289.3	13	288.4	14	266.3	17	354.4	21	414.4	-	-
Value added per establishment ($ mil.)	4	242.0	4	258.8	5	229.9	6	342.4	7	371.3	-	-
Capital investment per estab. ($ mil.)	0.716	270.2	0.699	237.9	0.776	270.5	0.958	378.2	1	491.1	-	-
Payroll per employee ($)	10,484	76.6	11,913	65.8	13,847	69.0	15,147	93.3	17,056	97.1	19,510	111.6
Females as % of total employment (%)	-	-	-	-	-	-	-	-	-	-	-	-
Output per employee ($)	103,412	150.7	112,000	137.6	126,711	141.6	151,309	189.0	183,855	210.4	185,909	200.9
Value added per employee ($)	33,461	101.9	36,962	109.4	42,346	117.6	50,103	156.4	57,733	162.6	60,794	153.9
Capital investment per employee ($)	6,038	109.7	5,914	96.5	6,782	109.5	8,707	152.3	10,192	155.8	-	-

Columns headed % show percent of the world total or of the world ratio. Ratios closest to 100 are closest to the world average. Values higher than 100 are above, values lower than 100 are below world average. *nec* stands for *not elsewhere classified*. Ratios, where shown, are frequently based on a subset of total data; for example, payroll per employee is averaged from all cases where *both* payroll and employment data are present. Dividing *total* salaries/wages by *total* employment will not necessarily produce a matching value.

Slovakia

Top Ranked Employing Sectors	Top Ranked Output Sectors
3820 Machinery nec	3110 Food Products
3110 Food Products	3710 Iron and Steel
3810 Metal Products	3530 Petroleum Refineries
3830 Electrical Machinery	3510 Industrial Chemicals
3210 Textiles	3820 Machinery nec

Population 5,393,000. Percent of world population: 0.091

	1990		1991		1992		1993		1994		1995	
	Value	%	Value	%	Value	%	Value	%	Value	%	Value	%
Establishments or enterprises (number)	-	-	1,121	0.047	1,071	0.038	1,323	0.064	1,572	0.088	-	-
Total employment (000)	-	-	707	0.379	613	0.336	547	0.300	514	0.323	-	-
Female employees (000)	-	-	-	-	-	-	-	-	-	-	-	-
Salaries and Wages ($ bil.)	-	-	1	0.046	1	0.044	1	0.055	1	0.061	-	-
Output ($ bil.)	-	-	13	0.082	12	0.071	10	0.069	11	0.078	-	-
Value added ($ bil.)	-	-	-	-	-	-	3	0.049	3	0.062	-	-
Capital investment ($ mil.)	-	-	-	-	-	-	-	-	-	-	-	-
Employees per establishment (number)	-	-	632	931.0	573	921.7	414	587.7	347	515.8	-	-
Output per establishment ($ mil.)	-	-	11	246.3	11	201.3	8	160.9	7	143.5	-	-
Value added per establishment ($ mil.)	-	-	-	-	-	-	2	144.7	2	121.9	-	-
Capital investment per estab. ($ mil.)	-	-	-	-	-	-	-	-	-	-	-	-
Payroll per employee ($)	-	-	1,531	8.5	1,835	9.1	2,045	12.6	2,332	13.3	-	-
Females as % of total employment (%)	-	-	-	-	-	-	-	-	-	-	-	-
Output per employee ($)	-	-	17,898	22.0	19,127	21.4	18,238	22.8	20,783	23.8	-	-
Value added per employee ($)	-	-	-	-	-	-	5,520	17.2	6,186	17.4	-	-
Capital investment per employee ($)	-	-	-	-	-	-	-	-	-	-	-	-

Columns headed % show percent of the world total or of the world ratio. Ratios closest to 100 are closest to the world average. Values higher than 100 are above, values lower than 100 are below world average. *nec* stands for *not elsewhere classified*. Ratios, where shown, are frequently based on a subset of total data; for example, payroll per employee is averaged from all cases where *both* payroll and employment data are present. Dividing *total* salaries/wages by *total* employment will not necessarily produce a matching value.

Slovenia

Top Ranked Employing Sectors	Top Ranked Output Sectors
3830 Electrical Machinery	3710 Iron and Steel
3840 Transportation Equipment	3830 Electrical Machinery
3310 Wood Products	3510 Industrial Chemicals
	3840 Transportation Equipment
	3110 Food Products

Population 1,972,000. Percent of world population: 0.033

	1990		1991		1992		1993		1994		1995	
	Value	%	Value	%	Value	%	Value	%	Value	%	Value	%
Establishments or enterprises (number)	3,770	0.164	11,976	0.502	17,387	0.619	19,740	0.960	23,480	1.300	19,861	1.600
Total employment (000)	83	0.040	73	0.039	66	0.036	59	0.032	55	0.035	52	0.037
Female employees (000)	-	-	-	-	-	-	-	-	-	-	-	-
Salaries and Wages ($ bil.)	0.749	0.035	0.497	0.021	0.427	0.017	0.375	0.018	0.425	0.022	0.484	0.029
Output ($ bil.)	3	0.023	3	0.018	3	0.015	12	0.084	14	0.099	18	0.132
Value added ($ bil.)	1	0.018	0.668	0.012	0.693	0.012	4	0.072	4	0.084	6	0.102
Capital investment ($ mil.)	437	0.074	436	0.070	366	0.057	386	0.071	378	0.085	665	0.213
Employees per establishment (number)	81	109.6	22	32.2	15	24.9	14	19.9	11	15.6	15	22.0
Output per establishment ($ mil.)	3	77.5	0.847	18.4	0.595	10.9	0.618	13.1	0.589	11.7	0.904	21.9
Value added per establishment ($ mil.)	1	61.1	0.201	11.9	0.162	7.7	0.218	13.5	0.185	10.5	0.289	18.4
Capital investment per estab. ($ mil.)	0.158	59.7	0.043	14.5	0.028	9.6	0.024	9.4	0.019	8.3	0.043	15.1
Payroll per employee ($)	9,056	66.2	6,850	37.8	6,444	32.1	6,352	39.1	7,728	44.0	9,277	53.1
Females as % of total employment (%)	-	-	-	-	-	-	-	-	-	-	-	-
Output per employee ($)	40,644	59.2	38,792	47.7	38,401	42.9	39,659	49.5	48,669	55.7	64,236	69.4
Value added per employee ($)	12,405	37.8	9,204	27.3	10,466	29.1	10,991	34.3	13,296	37.5	16,347	41.4
Capital investment per employee ($)	1,679	30.5	1,637	26.7	2,403	38.8	1,852	32.4	1,698	26.0	2,849	38.2

Columns headed % show percent of the world total or of the world ratio. Ratios closest to 100 are closest to the world average. Values higher than 100 are above, values lower than 100 are below world average. *nec* stands for *not elsewhere classified*. Ratios, where shown, are frequently based on a subset of total data; for example, payroll per employee is averaged from all cases where *both* payroll and employment data are present. Dividing *total* salaries/wages by *total* employment will not necessarily produce a matching value.

For additional notes and sources, see Appendix I.

South Africa

Top Ranked Employing Sectors	Top Ranked Output Sectors
3110 Food Products	3110 Food Products
3220 Wearing Apparel	3840 Transportation Equipment
3810 Metal Products	3710 Iron and Steel
3830 Electrical Machinery	3520 Chemical Products nec
3840 Transportation Equipment	3810 Metal Products

Population 42,835,000. Percent of world population: 0.723

	1990		1991		1992		1993		1994		1995	
	Value	%	Value	%	Value	%	Value	%	Value	%	Value	%
Establishments or enterprises (number)	-	-	25,524	1.100	-	-	-	-	-	-	-	-
Total employment (000)	1,525	0.729	1,484	0.795	1,441	0.790	1,208	0.663	1,215	0.763	1,244	0.875
Female employees (000)	-	-	-	-	-	-	-	-	-	-	-	-
Salaries and Wages ($ bil.)	12	0.545	12	0.513	13	0.520	10	0.509	11	0.546	12	0.745
Output ($ bil.)	80	0.561	81	0.525	82	0.496	66	0.457	67	0.489	78	0.583
Value added ($ bil.)	23	0.414	24	0.425	26	0.432	24	0.409	25	0.493	29	0.533
Capital investment ($ mil.)	-		-		-		-		-		-	
Employees per establishment (number)	-	-	63	92.6	-	-	-	-	-	-	-	-
Output per establishment ($ mil.)	-	-	3	69.9	-	-	-	-	-	-	-	-
Value added per establishment ($ mil.)	-	-	1	60.5	-	-	-	-	-	-	-	-
Capital investment per estab. ($ mil.)	-		-		-		-		-		-	
Payroll per employee ($)	7,701	56.3	8,214	45.4	9,219	45.9	8,594	52.9	8,768	49.9	9,924	56.8
Females as % of total employment (%)	-	-	-	-	-	-	-	-	-	-	-	-
Output per employee ($)	45,072	65.7	46,547	57.2	48,458	54.2	43,213	54.0	43,643	49.9	49,443	53.4
Value added per employee ($)	15,193	46.2	16,245	48.1	17,938	49.8	16,214	50.6	16,723	47.1	19,058	48.3
Capital investment per employee ($)	-		-		-		-		-		-	

Columns headed % show percent of the world total or of the world ratio. Ratios closest to 100 are closest to the world average. Values higher than 100 are above, values lower than 100 are below world average. *nec* stands for *not elsewhere classified*. Ratios, where shown, are frequently based on a subset of total data; for example, payroll per employee is averaged from all cases where *both* payroll and employment data are present. Dividing *total* salaries/wages by *total* employment will not necessarily produce a matching value.

Spain

Top Ranked Employing Sectors	Top Ranked Output Sectors
3110 Food Products	3110 Food Products
3810 Metal Products	3840 Transportation Equipment
3840 Transportation Equipment	3810 Metal Products
3820 Machinery nec	3820 Machinery nec
3690 Nonmetal Products nec	3520 Chemical Products nec

Population 39,134,000. Percent of world population: 0.660

	1990		1991		1992		1993		1994		1995	
	Value	%	Value	%	Value	%	Value	%	Value	%	Value	%
Establishments or enterprises (number)	151,752	6.600	147,985	6.200	143,449	5.100	-	-	-	-	-	-
Total employment (000)	2,269	1.100	2,251	1.200	2,160	1.200	2,109	1.200	2,069	1.300	2,072	1.500
Female employees (000)	-	-	-	-	-	-	-	-	-	-	-	-
Salaries and Wages ($ bil.)	44	2.000	46	2.000	49	1.900	40	2.000	39	2.000	44	2.700
Output ($ bil.)	323	2.300	335	2.200	345	2.100	240	1.700	260	1.900	318	2.400
Value added ($ bil.)	108	1.900	113	2.000	116	1.900	77	1.300	80	1.600	94	1.700
Capital investment ($ mil.)	13,324	2.200	16,116	2.600	16,360	2.600	10,376	1.900	9,322	2.100	12,018	3.800
Employees per establishment (number)	15	20.3	15	22.4	15	24.2	-	-	-	-	-	-
Output per establishment ($ mil.)	2	50.3	2	49.3	2	44.2	-	-	-	-	-	-
Value added per establishment ($ mil.)	0.709	43.3	0.762	45.2	0.810	38.4	-	-	-	-	-	-
Capital investment per estab. ($ mil.)	0.088	33.1	0.109	37.1	0.114	39.7	-	-	-	-	-	-
Payroll per employee ($)	19,357	141.5	20,594	113.8	22,855	113.8	19,070	117.4	18,978	108.1	21,309	121.9
Females as % of total employment (%)	-	-	-	-	-	-	-	-	-	-	-	-
Output per employee ($)	142,551	207.7	148,626	182.6	159,676	178.5	113,708	142.0	125,439	143.6	153,434	165.8
Value added per employee ($)	47,447	144.4	50,109	148.4	53,779	149.4	36,366	113.5	38,616	108.8	45,445	115.1
Capital investment per employee ($)	5,872	106.6	7,160	116.8	7,573	122.2	4,921	86.1	4,505	68.9	5,802	77.7

Columns headed % show percent of the world total or of the world ratio. Ratios closest to 100 are closest to the world average. Values higher than 100 are above, values lower than 100 are below world average. *nec* stands for *not elsewhere classified*. Ratios, where shown, are frequently based on a subset of total data; for example, payroll per employee is averaged from all cases where *both* payroll and employment data are present. Dividing *total* salaries/wages by *total* employment will not necessarily produce a matching value.

Sri Lanka

Top Ranked Employing Sectors	Top Ranked Output Sectors
3220 Wearing Apparel	3110 Food Products
3110 Food Products	3220 Wearing Apparel
3210 Textiles	3210 Textiles
3550 Rubber Products	3140 Tobacco
3900 Industries nec	3130 Beverages

Population 18,934,000. Percent of world population: 0.319

	1990		1991		1992		1993		1994		1995	
	Value	%	Value	%	Value	%	Value	%	Value	%	Value	%
Establishments or enterprises (number)	3,249	0.141	2,750	0.115	2,513	0.089	2,788	0.136	-	-	-	-
Total employment (000)	321	0.154	320	0.171	318	0.174	364	0.200	-	-	-	-
Female employees (000)	174	1.300	177	2.000	173	0.803	208	0.935	-	-	-	-
Salaries and Wages ($ bil.)	0.191	0.009	0.236	0.010	0.243	0.010	0.296	0.014	-	-	-	-
Output ($ bil.)	2	0.016	3	0.016	3	0.017	3	0.022	-	-	-	-
Value added ($ bil.)	1	0.020	1	0.020	1	0.021	1	0.023	-	-	-	-
Capital investment ($ mil.)	251	0.042	134	0.022	153	0.024	186	0.034	-	-	-	-
Employees per establishment (number)	99	134.2	116	171.6	127	203.7	131	185.2	-	-	-	-
Output per establishment ($ mil.)	0.715	16.9	0.917	20.0	1	20.0	1	23.8	-	-	-	-
Value added per establishment ($ mil.)	0.340	20.7	0.422	25.0	0.492	23.4	0.491	30.5	-	-	-	-
Capital investment per estab. ($ mil.)	0.077	29.1	0.049	16.6	0.061	21.2	0.067	26.4	-	-	-	-
Payroll per employee ($)	594	4.3	738	4.1	765	3.8	812	5.0	-	-	-	-
Females as % of total employment (%)	54.1	168.8	55.2	172.0	54.3	117.3	57.3	114.7	-	-	-	-
Output per employee ($)	7,226	10.5	7,879	9.7	8,596	9.6	8,579	10.7	-	-	-	-
Value added per employee ($)	3,436	10.5	3,621	10.7	3,887	10.8	3,762	11.7	-	-	-	-
Capital investment per employee ($)	780	14.2	419	6.8	481	7.8	511	8.9	-	-	-	-

Columns headed % show percent of the world total or of the world ratio. Ratios closest to 100 are closest to the world average. Values higher than 100 are above, values lower than 100 are below world average. *nec* stands for *not elsewhere classified*. Ratios, where shown, are frequently based on a subset of total data; for example, payroll per employee is averaged from all cases where *both* payroll and employment data are present. Dividing *total* salaries/wages by *total* employment will not necessarily produce a matching value.

Suriname

Top Ranked Employing Sectors	Top Ranked Output Sectors
3110 Food Products	3110 Food Products
3310 Wood Products	3130 Beverages
3130 Beverages	3140 Tobacco
3520 Chemical Products nec	3520 Chemical Products nec
3220 Wearing Apparel	3310 Wood Products

Population 428,000. Percent of world population: 0.007

	1990		1991		1992		1993		1994		1995	
	Value	%	Value	%	Value	%	Value	%	Value	%	Value	%
Establishments or enterprises (number)	-	-	-	-	278	0.010	-	-	-	-	-	-
Total employment (000)	6	0.003	6	0.003	6	0.003	6	0.003	-	-	-	-
Female employees (000)	-	-	-	-	-	-	-	-	-	-	-	-
Salaries and Wages ($ bil.)	0.064	0.003	0.073	0.003	0.096	0.004	0.129	0.006	-	-	-	-
Output ($ bil.)	0.402	0.003	0.499	0.003	0.663	0.004	1	0.010	-	-	-	-
Value added ($ bil.)	-	-	-	-	-	-	-	-	-	-	-	-
Capital investment ($ mil.)	-	-	-	-	-	-	-	-	-	-	-	-
Employees per establishment (number)	-	-	-	-	24	38.1	-	-	-	-	-	-
Output per establishment ($ mil.)	-	-	-	-	3	46.5	-	-	-	-	-	-
Value added per establishment ($ mil.)	-	-	-	-	-	-	-	-	-	-	-	-
Capital investment per estab. ($ mil.)	-	-	-	-	-	-	-	-	-	-	-	-
Payroll per employee ($)	10,810	79.0	12,033	66.5	15,416	76.8	23,100	142.3	-	-	-	-
Females as % of total employment (%)	-	-	-	-	-	-	-	-	-	-	-	-
Output per employee ($)	67,657	98.6	82,006	100.7	106,894	119.5	262,192	327.5	-	-	-	-
Value added per employee ($)	-	-	-	-	-	-	-	-	-	-	-	-
Capital investment per employee ($)	-	-	-	-	-	-	-	-	-	-	-	-

Columns headed % show percent of the world total or of the world ratio. Ratios closest to 100 are closest to the world average. Values higher than 100 are above, values lower than 100 are below world average. *nec* stands for *not elsewhere classified*. Ratios, where shown, are frequently based on a subset of total data; for example, payroll per employee is averaged from all cases where *both* payroll and employment data are present. Dividing *total* salaries/wages by *total* employment will not necessarily produce a matching value.

Swaziland

Top Ranked Employing Sectors	Top Ranked Output Sectors
3110 Food Products	3110 Food Products
3410 Paper and Products	3130 Beverages
3220 Wearing Apparel	3410 Paper and Products
3130 Beverages	3220 Wearing Apparel
3810 Metal Products	3810 Metal Products

Population 966,000. Percent of world population: 0.016

	1990 Value	%	1991 Value	%	1992 Value	%	1993 Value	%	1994 Value	%	1995 Value	%
Establishments or enterprises (number)	43	0.002	93	0.004	93	0.003	85	0.004	-	-	-	-
Total employment (000)	11	0.005	11	0.006	11	0.006	12	0.006	-	-	-	-
Female employees (000)	-	-	-	-	-	-	-	-	-	-	-	-
Salaries and Wages ($ bil.)	0.025	0.001	0.061	0.003	0.066	0.003	0.059	0.003	-	-	-	-
Output ($ bil.)	0.411	0.003	0.308	0.002	0.322	0.002	0.282	0.002	-	-	-	-
Value added ($ bil.)	0.179	0.003	0.106	0.002	0.132	0.002	0.079	0.001	-	-	-	-
Capital investment ($ mil.)	15	0.003	18	0.003	17	0.003	25	0.005	-	-	-	-
Employees per establishment (number)	250	338.7	118	173.7	122	196.0	133	189.3	-	-	-	-
Output per establishment ($ mil.)	10	225.5	3	72.9	3	63.1	3	70.1	-	-	-	-
Value added per establishment ($ mil.)	4	254.6	1	68.0	1	67.6	0.926	57.5	-	-	-	-
Capital investment per estab. ($ mil.)	0.349	131.9	0.208	70.8	0.184	64.2	0.320	126.2	-	-	-	-
Payroll per employee ($)	2,345	17.1	5,497	30.4	5,790	28.8	5,164	31.8	-	-	-	-
Females as % of total employment (%)	-	-	-	-	-	-	-	-	-	-	-	-
Output per employee ($)	38,266	55.8	28,393	34.9	28,040	31.3	24,508	30.6	-	-	-	-
Value added per employee ($)	16,711	50.9	9,740	28.8	11,504	31.9	6,838	21.3	-	-	-	-
Capital investment per employee ($)	1,399	25.4	1,661	27.1	1,411	22.8	2,190	38.3	-	-	-	-

Columns headed % show percent of the world total or of the world ratio. Ratios closest to 100 are closest to the world average. Values higher than 100 are above, values lower than 100 are below world average. *nec* stands for *not elsewhere classified*. Ratios, where shown, are frequently based on a subset of total data; for example, payroll per employee is averaged from all cases where *both* payroll and employment data are present. Dividing *total* salaries/wages by *total* employment will not necessarily produce a matching value.

Sweden

Top Ranked Employing Sectors	Top Ranked Output Sectors
3840 Transportation Equipment	3840 Transportation Equipment
3820 Machinery nec	3820 Machinery nec
3810 Metal Products	3830 Electrical Machinery
3110 Food Products	3110 Food Products
3830 Electrical Machinery	3410 Paper and Products

Population 8,887,000. Percent of world population: 0.150

	1990 Value	%	1991 Value	%	1992 Value	%	1993 Value	%	1994 Value	%	1995 Value	%
Establishments or enterprises (number)	9,758	0.423	10,337	0.434	9,628	0.343	9,011	0.438	8,974	0.501	-	-
Total employment (000)	899	0.430	910	0.487	839	0.460	742	0.407	764	0.480	582	0.409
Female employees (000)	233	1.800	-	-	-	-	-	-	-	-	-	-
Salaries and Wages ($ bil.)	23	1.000	25	1.100	26	1.000	18	0.884	20	1.000	18	1.100
Output ($ bil.)	148	1.000	169	1.100	170	1.000	113	0.781	137	0.992	123	0.923
Value added ($ bil.)	66	1.200	49	0.857	48	0.802	40	0.663	47	0.927	42	0.762
Capital investment ($ mil.)	-	-	-	-	-	-	-	-	-	-	-	-
Employees per establishment (number)	92	125.0	88	129.7	87	140.2	82	116.7	85	126.4	-	-
Output per establishment ($ mil.)	15	358.2	16	356.3	18	323.9	13	267.8	15	303.2	-	-
Value added per establishment ($ mil.)	7	411.8	5	278.4	5	236.4	4	272.9	5	300.0	-	-
Capital investment per estab. ($ mil.)	-	-	-	-	-	-	-	-	-	-	-	-
Payroll per employee ($)	25,092	183.4	27,542	152.2	30,589	152.4	24,312	149.7	25,578	145.7	30,507	174.6
Females as % of total employment (%)	26.0	81.0	-	-	-	-	-	-	-	-	-	-
Output per employee ($)	164,719	240.0	185,829	228.3	202,292	226.1	152,810	190.9	179,170	205.1	211,457	228.5
Value added per employee ($)	73,246	223.0	53,403	158.1	57,153	158.7	53,388	166.6	62,138	175.1	71,834	181.9
Capital investment per employee ($)	-	-	-	-	-	-	-	-	-	-	-	-

Columns headed % show percent of the world total or of the world ratio. Ratios closest to 100 are closest to the world average. Values higher than 100 are above, values lower than 100 are below world average. *nec* stands for *not elsewhere classified*. Ratios, where shown, are frequently based on a subset of total data; for example, payroll per employee is averaged from all cases where *both* payroll and employment data are present. Dividing *total* salaries/wages by *total* employment will not necessarily produce a matching value.

Switzerland

Top Ranked Employing Sectors	Top Ranked Output Sectors
3830 Electrical Machinery	3830 Electrical Machinery
3420 Printing, Publishing	3420 Printing, Publishing
3850 Professional Goods	3850 Professional Goods
3210 Textiles	3410 Paper and Products
3410 Paper and Products	3210 Textiles

Population 7,260,000. Percent of world population: 0.123

	1990		1991		1992		1993		1994		1995	
	Value	%	Value	%	Value	%	Value	%	Value	%	Value	%
Establishments or enterprises (number)	-	-	-	-	-	-	-	-	-	-	-	-
Total employment (000)	292	0.139	283	0.152	262	0.144	252	0.139	252	0.159	250	0.176
Female employees (000)	-	-	-	-	-	-	-	-	-	-	-	-
Salaries and Wages ($ bil.)	-	-	-	-	-	-	-	-	-	-	-	-
Output ($ bil.)	43	0.299	41	0.265	41	0.248	38	0.260	43	0.313	55	0.409
Value added ($ bil.)	17	0.309	16	0.289	17	0.277	15	0.257	18	0.342	22	0.403
Capital investment ($ mil.)	-	-	-	-	-	-	-	-	-	-	-	-
Employees per establishment (number)	-	-	-	-	-	-	-	-	-	-	-	-
Output per establishment ($ mil.)	-	-	-	-	-	-	-	-	-	-	-	-
Value added per establishment ($ mil.)	-	-	-	-	-	-	-	-	-	-	-	-
Capital investment per estab. ($ mil.)	-	-	-	-	-	-	-	-	-	-	-	-
Payroll per employee ($)	-	-	-	-	-	-	-	-	-	-	-	-
Females as % of total employment (%)	-	-	-	-	-	-	-	-	-	-	-	-
Output per employee ($)	146,874	214.0	144,185	177.1	155,458	173.8	149,214	186.4	170,807	195.5	217,943	235.5
Value added per employee ($)	59,309	180.5	57,977	171.7	63,114	175.3	60,765	189.6	69,475	195.7	88,463	224.0
Capital investment per employee ($)	-	-	-	-	-	-	-	-	-	-	-	-

Columns headed % show percent of the world total or of the world ratio. Ratios closest to 100 are closest to the world average. Values higher than 100 are above, values lower than 100 are below world average. *nec* stands for *not elsewhere classified*. Ratios, where shown, are frequently based on a subset of total data; for example, payroll per employee is averaged from all cases where *both* payroll and employment data are present. Dividing *total* salaries/wages by *total* employment will not necessarily produce a matching value.

Syrian Arab Republic

Top Ranked Employing Sectors	Top Ranked Output Sectors
	3900 Industries nec

Population 16,673,000. Percent of world population: 0.281

	1990		1991		1992		1993		1994		1995	
	Value	%	Value	%	Value	%	Value	%	Value	%	Value	%
Establishments or enterprises (number)	-	-	581	0.024	593	0.021	717	0.035	-	-	-	-
Total employment (000)	-	-	-	-	-	-	-	-	-	-	-	-
Female employees (000)	-	-	-	-	-	-	-	-	-	-	-	-
Salaries and Wages ($ bil.)	-	-	-	-	-	-	-	-	-	-	-	-
Output ($ bil.)	0.038	-	0.034	-	0.087	0.001	0.079	0.001	0.089	0.001	0.098	0.001
Value added ($ bil.)	0.005	-	0.004	-	0.010	-	0.013	-	0.012	-	0.013	-
Capital investment ($ mil.)	-	-	-	-	-	-	-	-	-	-	-	-
Employees per establishment (number)	-	-	-	-	-	-	-	-	-	-	-	-
Output per establishment ($ mil.)	-	-	0.059	1.3	0.147	2.7	0.110	2.3	-	-	-	-
Value added per establishment ($ mil.)	-	-	0.006	0.4	0.017	0.8	0.018	1.1	-	-	-	-
Capital investment per estab. ($ mil.)	-	-	-	-	-	-	-	-	-	-	-	-
Payroll per employee ($)	-	-	-	-	-	-	-	-	-	-	-	-
Females as % of total employment (%)	-	-	-	-	-	-	-	-	-	-	-	-
Output per employee ($)	-	-	-	-	-	-	-	-	-	-	-	-
Value added per employee ($)	-	-	-	-	-	-	-	-	-	-	-	-
Capital investment per employee ($)	-	-	-	-	-	-	-	-	-	-	-	-

Columns headed % show percent of the world total or of the world ratio. Ratios closest to 100 are closest to the world average. Values higher than 100 are above, values lower than 100 are below world average. *nec* stands for *not elsewhere classified*. Ratios, where shown, are frequently based on a subset of total data; for example, payroll per employee is averaged from all cases where *both* payroll and employment data are present. Dividing *total* salaries/wages by *total* employment will not necessarily produce a matching value.

For additional notes and sources, see Appendix I.

Thailand

Top Ranked Employing Sectors	Top Ranked Output Sectors
3210 Textiles	3420 Printing, Publishing
3220 Wearing Apparel	3820 Machinery nec
3110 Food Products	3110 Food Products
3840 Transportation Equipment	3830 Electrical Machinery
3830 Electrical Machinery	3220 Wearing Apparel

Population 60,037,000. Percent of world population: 1.013

	1990		1991		1992		1993		1994		1995	
	Value	%	Value	%	Value	%	Value	%	Value	%	Value	%
Establishments or enterprises (number)	12,157	0.528	15,847	0.665	-	-	-	-	-	-	-	-
Total employment (000)	2,126	1.000	1,972	1.100	-		-		-		-	
Female employees (000)	1,142	8.600	1,065	11.800	-		-		-		-	
Salaries and Wages ($ bil.)	5	0.243	6	0.242	-		-		-		-	
Output ($ bil.)	112	0.784	125	0.810	-		-		-		-	
Value added ($ bil.)	37	0.661	76	1.300	-		-		-		-	
Capital investment ($ mil.)	20,956	3.500	9,186	1.500	-		-		-		-	
Employees per establishment (number)	175	237.3	124	183.4	-		-		-		-	
Output per establishment ($ mil.)	9	218.0	8	171.2	-		-		-		-	
Value added per establishment ($ mil.)	3	185.8	5	284.0	-		-		-		-	
Capital investment per estab. ($ mil.)	2	651.1	0.580	197.3	-		-		-		-	
Payroll per employee ($)	2,461	18.0	2,914	16.1	-		-		-		-	
Females as % of total employment (%)	53.8	168.0	54.0	168.2	-	-	-		-	-	-	
Output per employee ($)	52,798	76.9	63,153	77.6	-		-		-		-	
Value added per employee ($)	17,407	53.0	38,522	114.1	-		-		-		-	
Capital investment per employee ($)	9,857	179.0	4,658	76.0	-	-	-		-	-	-	-

Columns headed % show percent of the world total or of the world ratio. Ratios closest to 100 are closest to the world average. Values higher than 100 are above, values lower than 100 are below world average. *nec* stands for *not elsewhere classified*. Ratios, where shown, are frequently based on a subset of total data; for example, payroll per employee is averaged from all cases where *both* payroll and employment data are present. Dividing *total* salaries/wages by *total* employment will not necessarily produce a matching value.

Tonga

Top Ranked Employing Sectors	Top Ranked Output Sectors
3110 Food Products	3110 Food Products
3420 Printing, Publishing	3520 Chemical Products nec
3230 Leather and Products	3230 Leather and Products
3320 Furniture, Fixtures	3420 Printing, Publishing
3840 Transportation Equipment	3840 Transportation Equipment

Population 108,000. Percent of world population: 0.002

	1990		1991		1992		1993		1994		1995	
	Value	%	Value	%	Value	%	Value	%	Value	%	Value	%
Establishments or enterprises (number)	5	-	46	0.002	48	0.002	45	0.002	35	0.002	-	-
Total employment (000)	1	0.001	0.852	-	0.494	-	0.379	-	0.319	-	-	
Female employees (000)	-	-	-		-		-		-		-	
Salaries and Wages ($ bil.)	0.002	-	0.002	-	0.001	-	0.001	-	0.001		-	
Output ($ bil.)	0.012	-	0.012	-	0.007	-	0.007	-	0.005		-	
Value added ($ bil.)	-	-	-		-		-		-		-	
Capital investment ($ mil.)	-	-	-		-		-		-		-	
Employees per establishment (number)	4	5.7	9	13.2	10	16.6	8	11.9	9	13.5	-	-
Output per establishment ($ mil.)	0.220	5.2	0.116	2.5	0.154	2.8	0.145	3.1	0.156	3.1	-	-
Value added per establishment ($ mil.)	-	-	-		-		-		-		-	
Capital investment per estab. ($ mil.)	-	-	-		-		-		-		-	
Payroll per employee ($)	1,842	13.5	2,017	11.1	1,910	9.5	2,246	13.8	2,959	16.8	-	-
Females as % of total employment (%)	-	-	-		-		-		-		-	
Output per employee ($)	11,090	16.2	13,721	16.9	14,917	16.7	17,199	21.5	17,073	19.5	-	-
Value added per employee ($)	-	-	-		-		-		-		-	
Capital investment per employee ($)	-	-	-	-	-	-	-	-	-	-	-	-

Columns headed % show percent of the world total or of the world ratio. Ratios closest to 100 are closest to the world average. Values higher than 100 are above, values lower than 100 are below world average. *nec* stands for *not elsewhere classified*. Ratios, where shown, are frequently based on a subset of total data; for example, payroll per employee is averaged from all cases where *both* payroll and employment data are present. Dividing *total* salaries/wages by *total* employment will not necessarily produce a matching value.

For additional notes and sources, see Appendix I.

Trinidad and Tobago

Top Ranked Employing Sectors	Top Ranked Output Sectors
3110 Food Products	3510 Industrial Chemicals
3220 Wearing Apparel	3110 Food Products
3130 Beverages	3710 Iron and Steel
3420 Printing, Publishing	3130 Beverages
3900 Industries nec	3690 Nonmetal Products nec

Population 1,117,000. Percent of world population: 0.019

	1990		1991		1992		1993		1994		1995	
	Value	%	Value	%	Value	%	Value	%	Value	%	Value	%
Establishments or enterprises (number)	1,447	0.063	1,734	0.073	1,647	0.059	1,644	0.080	1,801	0.101	-	-
Total employment (000)	36	0.017	39	0.021	39	0.021	40	0.022	41	0.026	-	-
Female employees (000)	-		-		-		-		-		-	
Salaries and Wages ($ bil.)	0.226	0.010	0.258	0.011	0.272	0.011	0.210	0.010	0.198	0.010	-	-
Output ($ bil.)	2	0.012	2	0.014	2	0.013	2	0.014	2	0.018	-	-
Value added ($ bil.)	0.552	0.010	0.787	0.014	0.666	0.011	0.663	0.011	0.915	0.018	-	-
Capital investment ($ mil.)	65	0.011	141	0.023	183	0.029	118	0.022	521	0.117	-	-
Employees per establishment (number)	26	35.1	22	33.0	23	37.7	25	34.8	23	33.5	-	-
Output per establishment ($ mil.)	1	28.1	1	27.6	1	23.6	1	25.9	1	27.1	-	-
Value added per establishment ($ mil.)	0.516	31.5	0.454	26.9	0.404	19.2	0.403	25.0	0.508	28.8	-	-
Capital investment per estab. ($ mil.)	0.155	58.5	0.203	69.0	0.132	46.0	0.072	28.6	0.341	144.8	-	-
Payroll per employee ($)	6,173	45.1	6,636	36.7	7,048	35.1	5,219	32.1	4,882	27.8	-	-
Females as % of total employment (%)	-		-		-		-		-		-	
Output per employee ($)	46,100	67.2	56,541	69.5	54,835	61.3	49,644	62.0	60,570	69.3	-	-
Value added per employee ($)	16,030	48.8	20,260	60.0	17,240	47.9	16,434	51.3	22,537	63.5	-	-
Capital investment per employee ($)	2,844	51.7	5,966	97.3	4,929	79.6	2,935	51.3	13,650	208.7	-	-

Columns headed % show percent of the world total or of the world ratio. Ratios closest to 100 are closest to the world average. Values higher than 100 are above, values lower than 100 are below world average. nec stands for not elsewhere classified. Ratios, where shown, are frequently based on a subset of total data; for example, payroll per employee is averaged from all cases where both payroll and employment data are present. Dividing total salaries/wages by total employment will not necessarily produce a matching value.

Tunisia

Top Ranked Employing Sectors	Top Ranked Output Sectors
3220 Wearing Apparel	3110 Food Products
3110 Food Products	3220 Wearing Apparel
3210 Textiles	3510 Industrial Chemicals
3690 Nonmetal Products nec	3210 Textiles
3810 Metal Products	3690 Nonmetal Products nec

Population 9,380,000. Percent of world population: 0.158

	1990		1991		1992		1993		1994		1995	
	Value	%	Value	%	Value	%	Value	%	Value	%	Value	%
Establishments or enterprises (number)	-		-		-		11,724	0.570	11,574	0.647	11,812	0.963
Total employment (000)	-		-		-		272	0.149	283	0.178	298	0.209
Female employees (000)	-		-		-		-		-		-	
Salaries and Wages ($ bil.)	0.839	0.039	0.884	0.037	1	0.042	1	0.050	1	0.053	1	0.079
Output ($ bil.)	8	0.056	9	0.055	10	0.060	9	0.063	10	0.075	12	0.090
Value added ($ bil.)	2	0.036	2	0.039	3	0.042	2	0.041	3	0.056	3	0.061
Capital investment ($ mil.)	473	0.080	522	0.084	546	0.086	524	0.096	553	0.124	573	0.184
Employees per establishment (number)	-		-		-		23	32.9	24	36.3	25	35.9
Output per establishment ($ mil.)	-		-		-		1	24.5	1	26.8	2	37.8
Value added per establishment ($ mil.)	-		-		-		0.312	19.4	0.374	21.2	0.438	24.8
Capital investment per estab. ($ mil.)	-		-		-		0.066	26.2	0.073	30.9	0.075	26.1
Payroll per employee ($)	-		-		-		4,950	30.5	4,848	27.6	5,773	33.0
Females as % of total employment (%)	-		-		-		-		-		-	
Output per employee ($)	-		-		-		43,736	54.6	48,081	55.0	52,871	57.1
Value added per employee ($)	-		-		-		11,873	37.1	13,326	37.5	14,862	37.6
Capital investment per employee ($)	-		-		-		2,524	44.2	2,586	39.5	2,530	33.9

Columns headed % show percent of the world total or of the world ratio. Ratios closest to 100 are closest to the world average. Values higher than 100 are above, values lower than 100 are below world average. nec stands for not elsewhere classified. Ratios, where shown, are frequently based on a subset of total data; for example, payroll per employee is averaged from all cases where both payroll and employment data are present. Dividing total salaries/wages by total employment will not necessarily produce a matching value.

For additional notes and sources, see Appendix I.

Turkey

Top Ranked Employing Sectors	Top Ranked Output Sectors
3210 Textiles	3210 Textiles
3110 Food Products	3110 Food Products
3220 Wearing Apparel	3530 Petroleum Refineries
3840 Transportation Equipment	3710 Iron and Steel
3710 Iron and Steel	3840 Transportation Equipment

Population 64,567,000. Percent of world population: 1.090

	1990		1991		1992		1993		1994		1995	
	Value	%	Value	%	Value	%	Value	%	Value	%	Value	%
Establishments or enterprises (number)	6,354	0.276	6,161	0.258	12,728	0.453	12,001	0.584	11,479	0.641	-	-
Total employment (000)	1,248	0.597	1,148	0.615	1,245	0.682	1,232	0.677	1,176	0.739	965	0.679
Female employees (000)	207	1.600	-		-		-		-		-	
Salaries and Wages ($ bil.)	8	0.377	10	0.427	10	0.410	11	0.541	6	0.328	7	0.395
Output ($ bil.)	90	0.630	93	0.604	107	0.652	121	0.832	92	0.670	98	0.736
Value added ($ bil.)	36	0.641	39	0.685	45	0.752	51	0.861	39	0.761	41	0.747
Capital investment ($ mil.)	5,531	0.933	4,506	0.722	5,334	0.835	5,677	1.000	4,606	1.000	-	
Employees per establishment (number)	196	266.6	186	274.7	98	157.3	103	145.6	102	152.2	-	
Output per establishment ($ mil.)	14	335.1	15	328.0	8	154.6	10	214.1	8	160.1	-	
Value added per establishment ($ mil.)	6	344.7	6	373.5	4	167.7	4	265.9	3	192.6	-	
Capital investment per estab. ($ mil.)	0.871	329.0	0.731	248.9	0.419	146.0	0.473	186.7	0.401	170.5	-	
Payroll per employee ($)	6,500	47.5	8,845	48.9	8,410	41.9	8,951	55.1	5,446	31.0	6,789	38.8
Females as % of total employment (%)	18.0	58.0	-		-		-		-		-	
Output per employee ($)	72,253	105.3	80,766	99.2	86,079	96.2	97,942	122.3	78,557	89.9	101,677	109.9
Value added per employee ($)	28,746	87.5	33,825	100.2	36,135	100.4	41,702	130.1	33,129	93.3	42,503	107.6
Capital investment per employee ($)	4,432	80.5	3,924	64.0	4,286	69.2	4,607	80.6	3,915	59.9	-	

Columns headed % show percent of the world total or of the world ratio. Ratios closest to 100 are closest to the world average. Values higher than 100 are above, values lower than 100 are below world average. *nec* stands for *not elsewhere classified*. Ratios, where shown, are frequently based on a subset of total data; for example, payroll per employee is averaged from all cases where *both* payroll and employment data are present. Dividing *total* salaries/wages by *total* employment will not necessarily produce a matching value.

Ukraine

Top Ranked Employing Sectors	Top Ranked Output Sectors
3110 Food Products	3110 Food Products
3820 Machinery nec	3710 Iron and Steel
3810 Metal Products	3510 Industrial Chemicals
3840 Transportation Equipment	3820 Machinery nec
3710 Iron and Steel	3690 Nonmetal Products nec

Population 50,125,000. Percent of world population: 0.846

	1990		1991		1992		1993		1994		1995	
	Value	%	Value	%	Value	%	Value	%	Value	%	Value	%
Establishments or enterprises (number)	63,976	2.800	71,586	3.000	73,815	2.600	63,643	3.100	65,952	3.700	68,690	5.600
Total employment (000)	4,950	2.400	4,848	2.600	5,220	2.900	4,854	2.700	4,342	2.700	4,001	2.800
Female employees (000)	-		-		-		-		-		-	
Salaries and Wages ($ bil.)	9	0.407	14	0.595	13	0.506	2	0.099	2	0.111	2	0.141
Output ($ bil.)	122	0.855	127	0.827	140	0.853	31	0.217	29	0.214	28	0.211
Value added ($ bil.)	-		-		-		-		-		-	
Capital investment ($ mil.)	3,093	0.522	1,991	0.319	2,397	0.375	477	0.088	624	0.141	372	0.119
Employees per establishment (number)	78	105.9	68	99.9	69	110.4	74	105.3	64	95.2	56	79.8
Output per establishment ($ mil.)	2	44.7	2	38.5	2	34.9	0.495	10.5	0.447	8.9	0.411	9.9
Value added per establishment ($ mil.)	-		-		-		-		-		-	
Capital investment per estab. ($ mil.)	0.048	18.3	0.028	9.5	0.032	11.3	0.008	3.0	0.009	4.0	0.005	1.9
Payroll per employee ($)	2,816	20.6	2,916	16.1	2,477	12.3	417	2.6	500	2.8	583	3.3
Females as % of total employment (%)	-		-		-		-		-		-	
Output per employee ($)	24,053	35.0	25,794	31.7	27,665	30.9	6,668	8.3	6,979	8.0	7,344	7.9
Value added per employee ($)	-		-		-		-		-		-	
Capital investment per employee ($)	605	11.0	408	6.7	473	7.6	101	1.8	148	2.3	97	1.3

Columns headed % show percent of the world total or of the world ratio. Ratios closest to 100 are closest to the world average. Values higher than 100 are above, values lower than 100 are below world average. *nec* stands for *not elsewhere classified*. Ratios, where shown, are frequently based on a subset of total data; for example, payroll per employee is averaged from all cases where *both* payroll and employment data are present. Dividing *total* salaries/wages by *total* employment will not necessarily produce a matching value.

United Kingdom

Top Ranked Employing Sectors	Top Ranked Output Sectors
3110 Food Products	3110 Food Products
3820 Machinery nec	3840 Transportation Equipment
3840 Transportation Equipment	3820 Machinery nec
3810 Metal Products	3830 Electrical Machinery
3830 Electrical Machinery	3520 Chemical Products nec

Population 58,970,000. Percent of world population: 0.995

	1990		1991		1992		1993		1994		1995	
	Value	%	Value	%	Value	%	Value	%	Value	%	Value	%
Establishments or enterprises (number)	155,303	6.700	147,210	6.200	151,721	5.400	143,534	7.000	153,574	8.600	-	-
Total employment (000)	5,762	2.800	5,315	2.800	5,172	2.800	3,986	2.200	4,052	2.500	4,051	2.800
Female employees (000)	1,537	11.500	-	-	1,461	6.800						
Salaries and Wages ($ bil.)	131	6.100	132	5.500	135	5.300	87	4.300	94	4.800	101	6.100
Output ($ bil.)	716	5.000	682	4.400	742	4.500	469	3.200	523	3.800	568	4.300
Value added ($ bil.)	314	5.600	297	5.200	312	5.200	206	3.500	229	4.500	249	4.500
Capital investment ($ mil.)	27,102	4.600	34,476	5.500	31,297	4.900	-	-	20,210	4.600	-	-
Employees per establishment (number)	37	50.3	36	53.2	34	54.8	28	39.4	26	39.2	-	-
Output per establishment ($ mil.)	5	110.2	5	101.0	5	89.8	3	69.5	3	67.7	-	-
Value added per establishment ($ mil.)	2	123.6	2	119.6	2	97.5	1	89.2	1	84.6	-	-
Capital investment per estab. ($ mil.)	0.190	71.6	0.234	79.7	0.206	71.9	-	-	0.128	54.6	-	-
Payroll per employee ($)	22,822	166.8	24,796	137.0	26,168	130.3	21,853	134.6	23,104	131.6	25,017	143.1
Females as % of total employment (%)	27.1	84.7	-	-	28.2	61.0						
Output per employee ($)	126,413	184.2	128,358	157.7	143,509	160.4	117,607	146.9	129,092	147.7	140,278	151.6
Value added per employee ($)	54,569	166.1	55,908	165.5	60,293	167.4	51,696	161.3	56,563	159.4	61,529	155.8
Capital investment per employee ($)	5,649	102.6	6,487	105.8	6,051	97.7	-	-	4,867	74.4	-	-

Columns headed % show percent of the world total or of the world ratio. Ratios closest to 100 are closest to the world average. Values higher than 100 are above, values lower than 100 are below world average. nec stands for not elsewhere classified. Ratios, where shown, are frequently based on a subset of total data; for example, payroll per employee is averaged from all cases where both payroll and employment data are present. Dividing total salaries/wages by total employment will not necessarily produce a matching value.

United Republic of Tanzania

Top Ranked Employing Sectors	Top Ranked Output Sectors
3110 Food Products	3110 Food Products
3210 Textiles	3210 Textiles
3510 Industrial Chemicals	3130 Beverages
3130 Beverages	3690 Nonmetal Products nec
3410 Paper and Products	3410 Paper and Products

Population 30,609,000. Percent of world population: 0.517

	1990		1991		1992		1993		1994		1995	
	Value	%	Value	%	Value	%	Value	%	Value	%	Value	%
Establishments or enterprises (number)	990	0.043	994	0.042	-	-	-	-	-	-	-	-
Total employment (000)	159	0.076	165	0.088	-	-	-	-	-	-	-	-
Female employees (000)	35	0.264	36	0.399	-	-	-	-	-	-	-	-
Salaries and Wages ($ bil.)	0.047	0.002	0.057	0.002								
Output ($ bil.)	0.946	0.007	1	0.007								
Value added ($ bil.)	0.131	0.002	0.172	0.003								
Capital investment ($ mil.)	749	0.126	1,155	0.185								
Employees per establishment (number)	161	217.8	166	244.9	-	-	-	-	-	-	-	-
Output per establishment ($ mil.)	0.956	22.6	1	25.1	-	-	-	-	-	-	-	-
Value added per establishment ($ mil.)	0.132	8.1	0.173	10.3	-	-	-	-	-	-	-	-
Capital investment per estab. ($ mil.)	0.756	285.5	1	396.7	-	-	-	-	-	-	-	-
Payroll per employee ($)	297	2.2	346	1.9	-	-	-	-	-	-	-	-
Females as % of total employment (%)	22.1	68.9	21.7	67.6	-	-	-	-	-	-	-	-
Output per employee ($)	5,954	8.7	6,924	8.5	-	-	-	-	-	-	-	-
Value added per employee ($)	824	2.5	1,042	3.1	-	-	-	-	-	-	-	-
Capital investment per employee ($)	4,710	85.5	6,995	114.1	-	-	-	-	-	-	-	-

Columns headed % show percent of the world total or of the world ratio. Ratios closest to 100 are closest to the world average. Values higher than 100 are above, values lower than 100 are below world average. nec stands for not elsewhere classified. Ratios, where shown, are frequently based on a subset of total data; for example, payroll per employee is averaged from all cases where both payroll and employment data are present. Dividing total salaries/wages by total employment will not necessarily produce a matching value.

For additional notes and sources, see Appendix I.

United States of America

Top Ranked Employing Sectors	Top Ranked Output Sectors
3820 Machinery nec	3840 Transportation Equipment
3840 Transportation Equipment	3110 Food Products
3830 Electrical Machinery	3820 Machinery nec
3420 Printing, Publishing	3830 Electrical Machinery
3110 Food Products	3510 Industrial Chemicals

Population 270,312,000. Percent of world population: 4.561

	1990 Value	%	1991 Value	%	1992 Value	%	1993 Value	%	1994 Value	%	1995 Value	%
Establishments or enterprises (number)	-	-	-	-	518,321	18.500	-	-	-	-	-	-
Total employment (000)	19,936	9.500	19,080	10.200	19,244	10.500	19,218	10.600	19,334	12.100	19,728	13.900
Female employees (000)	-	-	-	-	13,033	60.600	13,095	58.700	13,316	63.100	13,671	71.600
Salaries and Wages ($ bil.)	548	25.400	543	22.800	575	22.500	585	28.700	607	31.100	632	38.100
Output ($ bil.)	3,347	23.400	3,299	21.500	3,513	21.400	3,650	25.200	3,911	28.400	4,249	31.900
Value added ($ bil.)	1,577	28.200	1,559	27.500	1,691	28.300	1,760	29.500	1,901	37.100	2,064	37.600
Capital investment ($ mil.)	101,580	17.100	98,540	15.800	132,133	20.700	130,996	24.100	143,388	32.300	166,180	53.200
Employees per establishment (number)	-	-	-	-	37	59.7	-	-	-	-	-	-
Output per establishment ($ mil.)	-	-	-	-	7	124.5	-	-	-	-	-	-
Value added per establishment ($ mil.)	-	-	-	-	3	154.8	-	-	-	-	-	-
Capital investment per estab. ($ mil.)	-	-	-	-	0.255	88.8	-	-	-	-	-	-
Payroll per employee ($)	27,473	200.8	28,443	157.2	29,859	148.7	30,465	187.6	31,388	178.8	32,047	183.4
Females as % of total employment (%)	-	-	-	-	67.7	146.3	68.1	136.5	68.9	131.3	69.3	118.7
Output per employee ($)	167,911	244.7	172,911	212.4	182,567	204.1	189,903	237.2	202,277	231.5	215,381	232.7
Value added per employee ($)	79,097	240.8	81,734	242.0	87,889	244.1	91,602	285.9	98,325	277.0	104,622	264.9
Capital investment per employee ($)	5,804	105.4	5,893	96.1	6,866	110.8	6,816	119.3	7,416	113.4	8,424	112.8

Columns headed % show percent of the world total or of the world ratio. Ratios closest to 100 are closest to the world average. Values higher than 100 are above, values lower than 100 are below world average. nec stands for not elsewhere classified. Ratios, where shown, are frequently based on a subset of total data; for example, payroll per employee is averaged from all cases where both payroll and employment data are present. Dividing total salaries/wages by total employment will not necessarily produce a matching value.

Uruguay

Top Ranked Employing Sectors	Top Ranked Output Sectors
3110 Food Products	3110 Food Products
3210 Textiles	3520 Chemical Products nec
3220 Wearing Apparel	3530 Petroleum Refineries
3810 Metal Products	3130 Beverages
3420 Printing, Publishing	3210 Textiles

Population 3,285,000. Percent of world population: 0.055

	1990 Value	%	1991 Value	%	1992 Value	%	1993 Value	%	1994 Value	%	1995 Value	%
Establishments or enterprises (number)	-	-	-	-	-	-	-	-	-	-	-	-
Total employment (000)	192	0.092	183	0.098	167	0.092	151	0.083	138	0.086	135	0.095
Female employees (000)	-	-	-	-	-	-	-	-	-	-	-	-
Salaries and Wages ($ bil.)	0.540	0.025	0.636	0.027	0.678	0.027	0.747	0.037	0.820	0.042	0.850	0.051
Output ($ bil.)	5	0.036	6	0.042	7	0.040	6	0.044	7	0.053	7	0.056
Value added ($ bil.)	2	0.043	3	0.052	3	0.050	3	0.050	3	0.066	3	0.063
Capital investment ($ mil.)	132	0.022	173	0.028	180	0.028	228	0.042	-	-	-	-
Employees per establishment (number)	-	-	-	-	-	-	-	-	-	-	-	-
Output per establishment ($ mil.)	-	-	-	-	-	-	-	-	-	-	-	-
Value added per establishment ($ mil.)	-	-	-	-	-	-	-	-	-	-	-	-
Capital investment per estab. ($ mil.)	-	-	-	-	-	-	-	-	-	-	-	-
Payroll per employee ($)	3,261	23.8	4,029	22.3	4,649	23.2	5,588	34.4	6,011	34.2	6,331	36.2
Females as % of total employment (%)	-	-	-	-	-	-	-	-	-	-	-	-
Output per employee ($)	31,428	45.8	40,196	49.4	44,481	49.7	47,194	58.9	52,876	60.5	55,040	59.5
Value added per employee ($)	14,177	43.2	18,441	54.6	20,160	56.0	22,047	68.8	24,627	69.4	25,393	64.3
Capital investment per employee ($)	788	14.3	1,082	17.7	1,222	19.7	1,695	29.7	-	-	-	-

Columns headed % show percent of the world total or of the world ratio. Ratios closest to 100 are closest to the world average. Values higher than 100 are above, values lower than 100 are below world average. nec stands for not elsewhere classified. Ratios, where shown, are frequently based on a subset of total data; for example, payroll per employee is averaged from all cases where both payroll and employment data are present. Dividing total salaries/wages by total employment will not necessarily produce a matching value.

For additional notes and sources, see Appendix I.

Venezuela

Top Ranked Employing Sectors	Top Ranked Output Sectors
3110 Food Products	3110 Food Products
3810 Metal Products	3840 Transportation Equipment
3520 Chemical Products nec	3510 Industrial Chemicals
3710 Iron and Steel	3530 Petroleum Refineries
3220 Wearing Apparel	3710 Iron and Steel

Population 22,803,000. Percent of world population: 0.385

	1990 Value	%	1991 Value	%	1992 Value	%	1993 Value	%	1994 Value	%	1995 Value	%
Establishments or enterprises (number)	10,606	0.460	11,186	0.469	10,986	0.391	9,532	0.464	9,435	0.527	9,386	0.765
Total employment (000)	531	0.254	567	0.304	565	0.310	523	0.287	502	0.315	530	0.373
Female employees (000)	-		-		-		-		-		-	
Salaries and Wages ($ bil.)	2	0.107	3	0.118	3	0.114	3	0.134	2	0.114	3	0.170
Output ($ bil.)	27	0.191	30	0.198	32	0.197	32	0.220	28	0.202	34	0.257
Value added ($ bil.)	13	0.240	13	0.231	15	0.247	13	0.220	12	0.232	32	0.590
Capital investment ($ mil.)	1,576	0.266	1,683	0.270	1,990	0.311	2,580	0.475	2,381	0.536	6,685	2.100
Employees per establishment (number)	50	67.9	51	74.8	51	82.7	55	77.8	53	79.0	56	80.6
Output per establishment ($ mil.)	3	60.8	3	59.4	3	54.2	3	71.4	3	58.7	4	95.4
Value added per establishment ($ mil.)	1	77.3	1	69.4	1	63.8	1	85.6	1	71.5	3	195.7
Capital investment per estab. ($ mil.)	0.158	59.6	0.160	54.3	0.192	66.8	0.288	113.5	0.268	113.8	0.755	264.3
Payroll per employee ($)	4,326	31.6	4,961	27.4	5,154	25.7	5,228	32.2	4,453	25.4	5,333	30.5
Females as % of total employment (%)	-		-		-		-		-		-	
Output per employee ($)	51,416	74.9	53,727	66.0	57,393	64.2	61,151	76.4	55,572	63.6	70,989	76.7
Value added per employee ($)	25,291	77.0	23,092	68.4	26,160	72.7	25,111	78.4	23,723	66.8	61,195	154.9
Capital investment per employee ($)	3,393	61.6	3,381	55.1	4,023	64.9	5,630	98.5	5,412	82.8	14,249	190.8

Columns headed % show percent of the world total or of the world ratio. Ratios closest to 100 are closest to the world average. Values higher than 100 are above, values lower than 100 are below world average. *nec* stands for *not elsewhere classified*. Ratios, where shown, are frequently based on a subset of total data; for example, payroll per employee is averaged from all cases where *both* payroll and employment data are present. Dividing *total* salaries/wages by *total* employment will not necessarily produce a matching value.

Yugoslavia

Top Ranked Employing Sectors	Top Ranked Output Sectors
3110 Food Products	3110 Food Products
3810 Metal Products	3520 Chemical Products nec
3840 Transportation Equipment	3810 Metal Products
3210 Textiles	3830 Electrical Machinery
3220 Wearing Apparel	3130 Beverages

Population 10,526,000. Percent of world population: 0.178

	1990 Value	%	1991 Value	%	1992 Value	%	1993 Value	%	1994 Value	%	1995 Value	%
Establishments or enterprises (number)	4,596	0.199	7,820	0.328	16,656	0.593	22,839	1.100	23,537	1.300	24,903	2.000
Total employment (000)	-	-	938	0.502	887	0.486	853	0.469	825	0.518	808	0.568
Female employees (000)	-		-		-		-		-		-	
Salaries and Wages ($ bil.)	-		-		-		-		1	0.071	0.635	0.038
Output ($ bil.)	3	0.020	3	0.020	6	0.035	-		11	0.079	7	0.049
Value added ($ bil.)	1	0.021	1	0.023	3	0.051	-		5	0.097	3	0.051
Capital investment ($ mil.)	85	0.014	80	0.013	148	0.023	-		305	0.069	170	0.054
Employees per establishment (number)	-		120	176.8	53	85.7	37	53.0	35	52.0	32	46.3
Output per establishment ($ mil.)	0.625	14.8	0.388	8.5	0.345	6.3	-	-	0.460	9.1	0.262	6.3
Value added per establishment ($ mil.)	0.250	15.3	0.170	10.1	0.185	8.8	-	-	0.212	12.0	0.113	6.4
Capital investment per estab. ($ mil.)	0.018	7.0	0.010	3.5	0.009	3.1	-	-	0.013	5.5	0.007	2.4
Payroll per employee ($)	-		-		-		-		1,675	9.5	786	4.5
Females as % of total employment (%)	-		-		-		-		-		-	
Output per employee ($)	-	-	3,236	4.0	6,469	7.2	-	-	13,130	15.0	8,061	8.7
Value added per employee ($)	-	-	1,416	4.2	3,471	9.6	-	-	6,046	17.0	3,471	8.8
Capital investment per employee ($)	-	-	86	1.4	167	2.7	-	-	369	5.6	210	2.8

Columns headed % show percent of the world total or of the world ratio. Ratios closest to 100 are closest to the world average. Values higher than 100 are above, values lower than 100 are below world average. *nec* stands for *not elsewhere classified*. Ratios, where shown, are frequently based on a subset of total data; for example, payroll per employee is averaged from all cases where *both* payroll and employment data are present. Dividing *total* salaries/wages by *total* employment will not necessarily produce a matching value.

For additional notes and sources, see Appendix I.

Zambia

Top Ranked Employing Sectors	Top Ranked Output Sectors
3110 Food Products	3110 Food Products
3210 Textiles	3130 Beverages
3810 Metal Products	3520 Chemical Products nec
3310 Wood Products	3210 Textiles
3130 Beverages	3830 Electrical Machinery

Population 9,461,000. Percent of world population: 0.160

	1990		1991		1992		1993		1994		1995	
	Value	%	Value	%	Value	%	Value	%	Value	%	Value	%
Establishments or enterprises (number)	433	0.019	-	-	-	-	-	-	632	0.035	-	-
Total employment (000)	67	0.032	-	-	-	-	-	-	53	0.033	-	-
Female employees (000)	-	-	-	-	-	-	-	-	4	0.021	-	-
Salaries and Wages ($ bil.)	0.113	0.005	-	-	-	-	-	-	0.103	0.005	-	-
Output ($ bil.)	1	0.008	-	-	-	-	-	-	0.745	0.005	-	-
Value added ($ bil.)	0.534	0.010	-	-	-	-	-	-	0.310	0.006	-	-
Capital investment ($ mil.)	176	0.030	-	-	-	-	-	-	-	-	-	-
Employees per establishment (number)	154	208.8	-	-	-	-	-	-	91	134.9	-	-
Output per establishment ($ mil.)	3	61.3	-	-	-	-	-	-	1	25.5	-	-
Value added per establishment ($ mil.)	1	75.3	-	-	-	-	-	-	0.532	30.2	-	-
Capital investment per estab. ($ mil.)	0.407	153.6	-	-	-	-	-	-	-	-	-	-
Payroll per employee ($)	1,692	12.4	-	-	-	-	-	-	1,944	11.1	-	-
Females as % of total employment (%)	-	-	-	-	-	-	-	-	8.2	15.7	-	-
Output per employee ($)	16,862	24.6	-	-	-	-	-	-	14,119	16.2	-	-
Value added per employee ($)	8,017	24.4	-	-	-	-	-	-	5,876	16.6	-	-
Capital investment per employee ($)	2,643	48.0	-	-	-	-	-	-	-	-	-	-

Columns headed % show percent of the world total or of the world ratio. Ratios closest to 100 are closest to the world average. Values higher than 100 are above, values lower than 100 are below world average. *nec* stands for *not elsewhere classified*. Ratios, where shown, are frequently based on a subset of total data; for example, payroll per employee is averaged from all cases where *both* payroll and employment data are present. Dividing *total* salaries/wages by *total* employment will not necessarily produce a matching value.

Zimbabwe

Top Ranked Employing Sectors	Top Ranked Output Sectors
3110 Food Products	3110 Food Products
3210 Textiles	3710 Iron and Steel
3810 Metal Products	3210 Textiles
3220 Wearing Apparel	3130 Beverages
3710 Iron and Steel	3810 Metal Products

Population 11,044,000. Percent of world population: 0.186

	1990		1991		1992		1993		1994		1995	
	Value	%	Value	%	Value	%	Value	%	Value	%	Value	%
Establishments or enterprises (number)	1,061	0.046	1,077	0.045	1,057	0.038	1,003	0.049	1,030	0.058	-	-
Total employment (000)	188	0.090	192	0.103	224	0.123	176	0.096	170	0.107	157	0.110
Female employees (000)	-	-	-	-	-	-	-	-	-	-	-	-
Salaries and Wages ($ bil.)	0.733	0.034	0.681	0.029	0.515	0.020	0.431	0.021	0.444	0.023	0.466	0.028
Output ($ bil.)	5	0.033	5	0.031	4	0.026	4	0.025	4	0.027	4	0.029
Value added ($ bil.)	2	0.039	2	0.042	2	0.033	2	0.027	1	0.029	2	0.029
Capital investment ($ mil.)	313	0.053	527	0.085	262	0.041	218	0.040	215	0.048	-	-
Employees per establishment (number)	177	240.4	178	263.0	212	340.9	175	248.3	165	245.5	-	-
Output per establishment ($ mil.)	4	104.1	4	96.2	4	74.4	4	76.3	4	70.9	-	-
Value added per establishment ($ mil.)	2	126.1	2	130.4	2	88.6	2	101.0	1	82.3	-	-
Capital investment per estab. ($ mil.)	0.481	181.4	0.794	270.4	0.403	140.5	0.354	139.8	0.341	144.9	-	-
Payroll per employee ($)	3,901	28.5	3,545	19.6	2,297	11.4	2,452	15.1	2,605	14.8	2,976	17.0
Females as % of total employment (%)	-	-	-	-	-	-	-	-	-	-	-	-
Output per employee ($)	24,890	36.3	24,762	30.4	19,112	21.4	20,466	25.6	21,562	24.7	24,996	27.0
Value added per employee ($)	11,658	35.5	12,341	36.5	8,814	24.5	9,287	29.0	8,775	24.7	10,206	25.8
Capital investment per employee ($)	2,337	42.4	3,776	61.6	1,546	25.0	1,748	30.6	1,787	27.3	-	-

Columns headed % show percent of the world total or of the world ratio. Ratios closest to 100 are closest to the world average. Values higher than 100 are above, values lower than 100 are below world average. *nec* stands for *not elsewhere classified*. Ratios, where shown, are frequently based on a subset of total data; for example, payroll per employee is averaged from all cases where *both* payroll and employment data are present. Dividing *total* salaries/wages by *total* employment will not necessarily produce a matching value.

For additional notes and sources, see Appendix I.

Appendix I

Source Notes

General Source Note

The statistical materials presented in the third edition of *Manufacturing Worldwide* (*MW*) are drawn from data collected by the United Nations and made available to The Gale Group for this publication by special arrangement.

The U.N. data came to the author on diskette in two databases; one held data coded by country, by ISIC, and by special accounts; the second held data by commodity and by country. ISIC statistics were the United Nations Industrial Development Organization (UNIDO) Industrial Statistics Database, 1998. Both the 3-digit and 4-digit databases were used. As in the last edition, data in dollar-denominated formats were provided by UNIDO.

Company data were extracted from Gale's *World Business Directory*. Company listings were taken from Gale's computerized database in March 1999. Entries were sorted by ISIC and by employment and further extracted to obtain the top companies in each sector (up to 180).

What follows is a more detailed discussion of methods used under the topics of ISIC Extraction, Accounts Used, Explanation of Accounts, Dollar Conversions, Commodity Data Processing, and Country Profiles (Part II).

ISIC Extraction

Data were extracted for 37 ISIC categories. Of these, 28 were major 3-digit ISIC groups and 9 were more detailed subdivisions of selected 3-digit sectors. A clear listing of these industries is available by looking at the Table of Contents.

The ISICs were selected, first, because they correspond to those on which the U.N. regularly reports in its *Industrial Statistics Yearbook* series. This, in effect, facilitates comparisons between that publication and this one. Second, the industries presented provide an excellent overview of

all manufacturing activities and most countries report data in the selected ISIC formats.

The general rule for extracting data was the following:

If a country reported data in the standard format for an ISIC, its data were included. If it did not, its data were excluded.

Another limitation was by year. Only data for 1990 through 1995 were extracted. Countries in the database that had data for earlier years only were excluded.

Accounts Used

Data reported by UNIDO are arranged under 14 different "accounts" or categories. All of these were used. Some were combined before presentation. The accounts are the following:

{ Establishments	- either one or the other
{ Enterprises	
{ Persons Engaged	- one of the two
{ Employees	
Wages/Salaries of Employees	
{ Output, Factor Values	- one of the three
{ Output, Producer Prices	
{ Output (undefined)	
{ Value Added, Factor Values	- one of the three
{ Value Added, Producer Prices	
{ Value Added (undefined)	
Gross Fixed Capital Formation	
Female Employees	
Index of Industrial Production	

Listing of Countries Reviewed

Afghanistan	Congo	Iceland	Myanmar	Solomon Islands
Albania	Cook Islands	India	Namibia	Somalia
Algeria	Costa Rica	Indonesia	Nepal	South Africa
Angola	Cote d'Ivoire	Iran (Islamic Republicof)	Netherlands	Spain
Argentina	Croatia	Iraq	Netherlands Antilles	Sri Lanka
Armenia	Cuba	Ireland	New Zealand	Sudan
Australia	Cyprus	Israel	Nicaragua	Suriname
Austria	Czechoslovakia (Former)	Italy	Niger	Swaziland
Azerbaijan	Czech Republic	Jamaica	Nigeria	Sweden
Bahamas	Dem. Rep. of the Congo	Japan	Norway	Switzerland
Bahrain	Denmark	Jordan	Oman	Syrian Arab Republic
Bangladesh	Dominican Republic	Kenya	Pakistan	Thailand
Barbados	Ecuador	Korea, Republic of	Panama	Togo
Belgium	Egypt	Kuwait	Papua New Guinea	Tonga
Belize	El Salvador	Kyrgyzstan	Paraguay	Trinidad and Tobago
Benin	Equatorial Guinea	Lao P.D.R.	Peru	Tunisia
Bermuda	Ethiopia and Eritrea	Latvia	Philippines	Turkey
Bhutan	Ethiopia	Lebanon	Poland	Uganda
Bolivia	Fiji	Lesotho	Portugal	Ukraine
Bosnia & Herzegovina	Finland	Liberia	Puerto Rico	United Arab Emirates
Botswana	Former Yug. R. of	Libyan Arab Jamahiriya	Qatar	United Kingdom
Brazil	Macedonia	Liechtenstein	Republic of Moldova	United Republic of
Bulgaria	France	Lithuania	Romania	Tanzania
Burkina-Faso	Gabon	Luxembourg	Russian Federation	USSR (Former)
Burundi	Gambia	Macau	Rwanda	United States of America
Cambodia	Germany	Madagascar	Saint Lucia	Uruguay
Cameroon	Germany (Eastern Part)	Malawi	Saint Vincent & the	Venezuela
Canada	Germany (Western Part)	Malaysia	Grenadines	Yemen
Cape Verde	Ghana	Mali	Samoa	Yemen (Northern Part)
Central African Republic	Greece	Malta	Saudi Arabia	Yemen (SouthernPart)
Chad	Grenada	Mauritania	Senegal	Yugoslavia
Chile	Guatemala	Mauritius	Seychelles	Yugoslavia (Former)
China	Guyana	Mexico	Sierra Leone	Zambia
China (Hong Kong SAR)	Haiti	Mongolia	Singapore	Zimbabwe
China (Taiwan Province)	Honduras	Morocco	Slovakia	
Colombia	Hungary	Mozambique	Slovenia	

Items marked with the "{" symbol were combined, i.e., one of the items was used, with appropriate footnoting in the tables. For example, some countries report activity in establishments, others use counts of enterprises. We used either establishments or enterprises and footnoted the values to show which was used.

We began with a listing of 175 countries, shown in the tabulation above. The list includes current entities as well as former countries. Because of an absence of data in some countries, this list was reduced to 167 countries for which data were available at some level, for some ISICs during the period 1990 through 1995. The list includes three separate incarnations of Germany, the former Czechoslovakia, the former USSR, and the former Yugoslavia, and Macedonia; hence the actual count of countries covered is 161. In this edition, data are presented on 23 more countries than were available in the 2nd edition.

Counties for which data were not available are Cook Islands, Mauritania, Rwanda, Samoa, Seychelles, Solo-

mon Islands, and Yemen divided into a northern and a southern part.

Explanation of Accounts

Establishments and Enterprises. *MW* shows either the number of establishments or the number of enterprises engaged in an economic activity. Establishments are used by default. When data are reported as counts of enterprises, values are marked with an asterisk (*). Establishments, typically, are single locations (a factory, an operation) where a single activity is pursued. Enterprises, typically, are legal entities (e.g., a corporation, a partnership, a state-owned industry); they may have one or more locations where activity is pursued.

When a country's data are in establishments, respective ratios are based on establishment counts. If the data are in the form of enterprises, ratios referring to establishments are, in actuality, based on enterprise counts.

Persons Employed and Engaged. Data for a country are available in the form of persons employed, persons engaged, or both. When both categories were available, persons employed was used as the basis of total employment and for calculating ratios related to total employment such as Wages/Salaries of Employees. When only the number of persons "engaged" was available, it was used.

The category of *Persons Employed* typically includes paid employees only and excludes working proprietors, active business partners, and unpaid family members. *Persons Engaged* includes employees, working proprietors, active business partners, and unpaid family members. Therefore it is the more inclusive term.

In *MW*, employment is used by default; if persons engaged is used, the values are marked with an asterisk (*).

Employees and Production Workers. The term *Employee* includes all workers engaged in an industry, including production workers. In the first edition, data on "operatives" (i.e., production workers) were reported as a separate category. These data were not available for this edition; however, data on females in the workforce were reported, although not uniformly.

Female Employment. This category, a first in this edition, reports somewhat less than uniformly the number of female employees as well as total employment. Ratios showing females as a percent of total employment are reported, obviously, only when both total and female employment were available for an ISIC.

Wages and Salaries. All wage/salary data are annual figures per worker. Average pay, the world over, ranges from a high in the $30,000+ range to a low of a few hundred dollars a year. Wage rates are also relative to cost of living. In some of the former communist countries, for example, the costs of housing, education, recreation, and many other items were controlled or held artificially low; hence low rates of pay sufficed to provide workers a living.

Output. Output, when shown, is shown in current dollars. Output refers to the production of the industry in a country. Changes in inventories of raw materials, work-in-progress, and of finished goods are added to (or deducted from) actual shipments to obtain the final value.

Two variants to the value of output are reported. Valuation in *factor values* means that data exclude all indirect taxes charged on production and include all subsidies received for the production activity; *producers' price valuation* means that all indirect taxes are included and all subsidies excluded. Data elements using factor values are footnoted. When neither of these methods of valuation is used, the basis of the value is unspecified. Such values are footnoted accordingly.

Value Added. Value added, when shown, is shown in dollars.

Value added is the measure of value added to materials coming into an industry by the industry's efforts (labor, management, capital, innovation). An establishment purchases raw, semifinished, and finished raw materials; fuels or power; services; etc. It sells a product. The difference between the cost of its purchases and the sales price it receives is value added, the establishment's contribution to Gross Domestic Product.

Value added, like output, may be reported in factor values or in producers' prices (see above). If so, data elements are footnoted accordingly.

Capital Formation. One account is provided on capital investment expenditures: Gross Fixed Capital Formation; it includes all capital expenditures on equipment and buildings.

Index of Industrial Production. Again in this edition, indices of industrial production are provided, with a base year of 1990. This means that industrial production in 1990 was designated as 100. Indexes were calculated to show how other years compare with 1990. Values higher than 100 are greater than 1990 and those below 100 are lower than 1990. An index of employment was produced to provide an indirect method of viewing how employment and production correlate.

Dollar Conversions

Data elements reported in local currency values are shown in current dollars using the conversions provided by UNIDO. The phrase "current dollars" means that conversions are made without removing the effects of inflation.

Commodity Data Processing

Data showing details of production were obtained from the U.N. in order to show production statistics at a level below the ISIC grouping. These data are referred to as "commodities" because they are published as "commodity production statistics" by the U.N.

What The Tables Contain. The *MW* Product Tables show production, from 1990 through 1996, of specific items. Tables are arranged so that they fall under the appropriate ISIC category. Total and regional production is shown in physical quantities (rather than in currency values). Also shown are the top five producer nations and their results, which may be actual or estimated.

How Estimates Are Developed. The data obtained by the U.N. from member countries and other sources are frequently incomplete. In its own publications, the U.N. uses various methods of filling gaps in data series. Similar methods were used by the author in preparing the Product Tables for *MW*. The method used is explained below.

All available values from the year 1981 through the present were obtained; please note that 1981 data were used in estimates even though only data for 1990 and later years were published.

Using the available data, 1981 through 1996, the following method was applied:

1. If data were available for every year, the data were used as received.

2. If data were available for a single year only, all years were assigned the data for that single year.

3. If data were available for less than five years but more than one, the average of the data was assigned to all missing years.

4. If data were available for five or more years but for fewer than all of the years required, estimates for the missing years were obtained using a linear regression and the least squares method.

5. If gaps in data series were detected, gaps were filled using extrapolation.

The method just described is very similar to the method used by the U.N. with exceptions being (1) that the base year used for *MW* was 1981 whereas the U.N. may be using another base year in its published commodity production statistics and (2) in *MW* gaps are filled using extrapolation.

Country Profiles

A total of 133 country profiles are provided in Part II of *MW* down from 141 in the last edition. The former communist countries were excluded this time. Some countries also had insufficient information for inclusion. A table showing world averages begins the presentation.

Country profiles were developed by (1) adding all data for the categories shown (establishment counts, employment, salaries/wages, etc.) to develop appropriate totals; (2) leading employment sectors were selected and ranked based on largest employment by ISIC; (3) leading output sectors were selected and ranked based on ISICs with the largest total output; finally, (3) ratios were calculated.

Country Profile Ratios. Ratios were developed by cumulating data for ratio pairs, e.g., salaries/wages and employment. If both members of a pair were present for an ISIC, the values were added to a temporary subtotal, e.g., employment to the employment subtotal and salaries/wages to a temporary salaries/wages subtotal. After all of the ISICs were polled, the ratio was then calculated using the subtotals. This means that only legitimate "pairs" of values are reflected in the average ratios shown. It is important to keep this in mind because using absolute totals as shown in the profile (total salaries/wages, total employment) to calculate a ratio does not always provide legitimate value. In many cases, the database will have employment for an ISIC but will not have salaries/wages for that ISIC or vice versa; hence using raw, cumulated totals may tend to over- or understate the ratio shown.

World Averages. The same method described above was used to calculate world averages which are shown as the first table in the Country Profiles (Part II).

World Averages are reflected throughout the rest of the profiles in columns labelled with the "%" symbol. These columns show the country's and item's percent of world totals. By comparing these numbers to the percent of world population, also shown, the user can determine how a country is faring, relative to its population share of the world, in a given category.

Population Ratios. Population counts used in world and country population figures were drawn from the Statistical Abstract of the United States, 1998, and are for the year 1998.

Alphabetical Index

This Alphabetical Index holds references to all subject matter covered, countries mentioned in tables, and corporations shown in the corporate tables. Entries for countries typically show the major industries on which data are reported. All numerical references are to page numbers.

3M Canada Co, pp. 233, 699
A. Schulman Inc, p. 428
Aalborg Portland AS, p. 480
Aaron Rents Inc, p. 213
Aarque Steel Corp, pp. 500, 555
Aavid Thermal Technologies Inc, p. 526
A.B. Electronic Products Group PLC, p. 657
AB Rexroth Mecman, p. 527
AB Wilh. Becker, pp. 341, 719
ABB Kent Holdings PLC, p. 698
Abbas Ali Hazeem & Sons Co, p. 460
Aberdare Cables Pty. Ltd, p. 526
Aberdeen Service Co. North Sea Ltd, p. 371
Abrasives, pp. 465, 484
ABT Building Products Corp, pp. 194, 428
Acadia Polymers Inc, p. 408
Accessories Associates Inc, p. 173
ACCO Europe PLC, pp. 213, 595
ACCO-Rexel Ltd, pp. 214, 263
ACCO USA Inc, pp. 213, 262
Accumulators, p. 639
Ace Novelty Company Inc, p. 721
Acer Inc, p. 595
Acer Sertek Inc, p. 597
Acetaldehyde, p. 288
Acetates, p. 289
Acetic acid, p. 290
Acetone, p. 289
Acetylene, p. 282
Acetylsalicylic acid, p. 291
Ache Laboratorios Farmaceuticos SA, p. 342
Achema, p. 280
Acheson Colloids Co, p. 389
Acindar Industria Arg. de Aceros SA, p. 499
Acme Boot Company Inc, p. 173
Acme Metals Inc, p. 500
Acominas Aco Minas Gerais SA, p. 499
Acos Villares SA, p. 499
Acrylic polymers, p. 307
Acrylonitrile, p. 292
ACS Industries Inc, p. 99
ACS Industries Inc., p. 526
Action Performance Companies Inc, p. 460

Activated carbon, p. 299
Acushnet Co, pp. 172, 719
Acuson Corp, p. 699
ACX Technologies Inc, pp. 231, 443, 525
Adaptec Inc, p. 595
ADC Telecommunications Inc, p. 616
Addis Ltd, pp. 429, 720
Addison Wesley Longman, p. 262
Ade Textile Industries PT, p. 98
Adel Rootstein Display Mannequins, p. 461
ADFlex Solutions Inc, p. 616
Adolph Coors Co, p. 58
Advanced Drainage Systems Inc, p. 429
Advanced Lighting Technologies, p. 460
Advanced Technology Laboratories, p. 698
Adwest Group PLC, p. 428
AECI Industrial Chemicals, p. 280
AEG Corp, pp. 574, 616, 698
AEP Industries, Inc, p. 427
Aerojet General Corp, p. 656
Aeroquip Corp, pp. 406, 427
Aeroquip Industrial Americas Group, pp. 406, 555
AFC Cable Systems Inc, p. 527
AFG Industries Ltd, p. 460
Afghanistan
— beverages, p. 48
— chemical products nec, p. 332
— drugs and medicines, p. 352
— food products, p. 8
— industrial chemicals, p. 269
— industries nec, p. 710
— nonferrous metals, p. 517
— nonmetal products nec, p. 470
— paper and products, p. 222
— printing, publishing, p. 252
— rubber products, p. 397
— spinning, weaving, etc., p. 114
— synthetic resins, etc., p. 322
— textiles, p. 88
— wood products, p. 184
Africa Glass Holdings Pty. Ltd, p. 460
African Lakes Corp. PLC, p. 656
African Timber & Plywood Ghana, p. 194

Aftermarket Technology Corp, p. 657
A.G. Barr PLC, p. 59
AGCO Corp, p. 572
AGFA Gevaert NV, p. 697
Agio Tobacco Co. Ltd, p. 79
Agip Espana, SA, p. 390
AGRA Inc, p. 573
Agricultural chemicals, p. 265
Agricultural machinery, p. 559
Agrifoods International Cooperative, p. 20
Agrikon, p. 574
Agrium Inc, pp. 279, 525
Agro Fellesslakteri, p. 156
Agroexport d.d, p. 156
Agrovale Agro Industrias Do Vale Do, p. 19
Aguila Enterprises Co., Ltd, p. 172
A.H. Belo Corp, p. 263
Ahlstrom Riihimaen Lasi Oy, p. 460
Ainsworth Lumber Co. Ltd, p. 195
Air-conditioning machines, p. 619
Air Products PLC, p. 280
Aircraft, p. 641
Airsprung Group PLC, p. 214
Ajover, p. 408
AK Steel Holding Corp, p. 499
Akro Corp, p. 407
Akzo Nobel Chemicals Inc, pp. 279, 342
Akzo Nobel Coatings Inc, p. 342
Akzo Nobel Inc, p. 341
A.L. Gebhardt, p. 156
Aladdin Industries Inc, pp. 214, 428
Aladdin Mills Inc, p. 98
ALARIS Medical Inc, p. 699
Albania
— basic chemicals, excl fertilizers, pp. 312-315
— beverages, pp. 44, 46, 48-49, 54
— chemical products nec, pp. 328, 330, 332-333, 338
— electrical machinery, pp. 602-603, 605-606, 611
— food products, pp. 4, 6, 8-9, 14
— footwear, pp. 160-161, 163-164, 169
— furniture, fixtures, pp. 200, 202, 204-205, 210
— glass and products, pp. 448-449, 451-452, 456
— industrial chemicals, pp. 266-267, 269-270, 275
— industries nec, pp. 706, 708, 710-711, 713, 716
— iron and steel, pp. 488-489, 491-492, 496
— leather and products, pp. 144-145, 147-148, 153
— machinery nec, pp. 560-561, 563-564, 568
— metal products, pp. 540, 542, 544-545, 550
— motor vehicles, pp. 676-679
— nonferrous metals, pp. 514-515, 517, 521
— nonmetal products nec, pp. 466, 468, 470-472, 475
— paper and products, pp. 218, 220, 222-223, 225, 228
— petroleum refineries, pp. 360-363, 367
— petroleum, coal products, p. 386
— plastic products nec, pp. 414, 416-418, 423
— pottery, china, etc., pp. 432, 440

Albania continued:
— printing, publishing, pp. 248, 250, 252-253, 255, 258
— professional goods, pp. 686-687, 689-690, 694
— profile, p. 727
— pulp, paper, etc., pp. 240-243
— radio, television, etc., pp. 628-630
— rubber products, pp. 394-395, 397-398, 403
— spinning, weaving, etc., pp. 112-115
— textiles, pp. 84, 86, 88-89, 94
— tobacco, pp. 66, 68, 70-71, 75
— transportation equipment, pp. 642, 644, 646-647, 651
— wearing apparel, pp. 122, 124, 126-127, 131
— wood products, pp. 180, 182, 184-185, 190
Albany International Corp, p. 97
Albemarle Corp, p. 279
Albras Aluminio Brasileiro SA, p. 526
Albright & Wilson Ltd, p. 278
Alcan Aluminio do Brasil SA, p. 525
Alcan Aluminium Ltd, pp. 278, 525
Alcan Aluminum Corp, pp. 525, 554
Alcan Smelters & Chemicals Ltd, p. 525
Alcatel Canada Wire, p. 527
Alcatel Standard Electrica, SA, pp. 595, 615
Alcoa Closure Systems International, pp. 555, 574
Alcoa Manufacturing Great Britain, p. 527
Alcon Laboratories Inc, pp. 341, 697
Alen Mak Perfumery & Cosmetics Ad, p. 343
Alenco Div, p. 429
Alenco Holdings Ltd, pp. 407, 429
Alexander and Baldwin Inc, p. 20
Alexandra Workwear PLC, p. 137
Alfold Porcelan Edenygyar Rt, p. 443
Algeria
— beverages, p. 54
— chemical products nec, p. 338
— electrical machinery, p. 611
— food products, p. 14
— footwear, p. 169
— furniture, fixtures, p. 210
— glass and products, p. 456
— industrial chemicals, p. 275
— industries nec, pp. 706, 708, 710-711, 713, 716
— iron and steel, p. 496
— leather and products, p. 153
— machinery nec, p. 568
— metal products, p. 550
— nonferrous metals, p. 521
— nonmetal products nec, p. 475
— paper and products, p. 228
— petroleum refineries, p. 367
— petroleum, coal products, p. 386
— plastic products nec, p. 423
— pottery, china, etc., p. 440
— printing, publishing, p. 258
— professional goods, p. 694
— profile, p. 728

Algeria continued:
— rubber products, p. 403
— textiles, p. 94
— tobacco, p. 75
— transportation equipment, p. 651
— wearing apparel, p. 131
— wood products, p. 190
Algoma Steel Inc, p. 499
Alhos Export-Import, p. 137
Align-Rite International Inc, p. 461
Alimentos Kern de Guatemala SA, p. 60
Alkaloida Chemical Co, p. 342
Alkyd resins, p. 305
Allana Cold Storage Ltd, p. 156
Allegheny Teledyne Inc, p. 500
Allen-Edmonds Shoe Corp, p. 173
Allen Group Inc, pp. 617, 657, 698
Allergan Inc, p. 697
Alliance Gaming Corp, p. 720
Alliance Tire Co. Ltd, p. 408
Alliant Techsystems Inc, p. 553
Allied Colloids Group PLC, p. 279
Allied Colloids Ltd, p. 280
Allied Domecq Spirits & Wine, p. 60
Allied Partnership Group PLC, p. 720
Allied Textile Cos. PLC, p. 98
Allied Tube and Conduit Co, p. 501
AlliedSignal Electronic and Avionics, pp. 615, 697, 719
Allsteel Inc, p. 214
Alltrista Corp, pp. 407, 459
Allvac, p. 526
Alony Haim Combine Ltd, p. 481
Alpargatas SAIC, p. 99
Alpha Therapeutic Corp, p. 342
Alpharma Inc, p. 342
Alpina d.d, p. 172
Alpina Productos Alimenticios SA, p. 20
Alpine Group Inc, pp. 554, 617, 657
Aluar-Aluminio Argentino SAIC, p. 527
Alucon Manufacturing Co., Ltd, p. 527
Aluma Systems USA Inc, p. 555
Alumina, p. 531
Aluminio Espanol, SA, p. 527
Aluminium de Grece SAIC, p. 526
Aluminum, p. 513
Aluminum shapes, p. 532
Aluminum oxide, p. 296
Aluminum plates, sheets, strip, foil, p. 533
Aluminum sulfate, p. 297
Aluminum tubes and pipes, p. 533
Aluminum wire, p. 533
Alvis PLC, p. 699
ALZ NV, p. 501
Alza Corp, p. 343
Amalgamated Metal Corp. PLC, p. 500
Amana Appliance, pp. 573, 616

Amcan Casting Ltd, p. 527
Amcast Industrial Corp, pp. 554, 656
Amdahl Corp, p. 595
Amerada Hess Corp, p. 370
American Bank Note Co, p. 263
American Banknote Corp, p. 263
American Biltrite Inc, pp. 232, 406, 479
American Business Information Inc, p. 264
American Business Products Inc, pp. 232, 263
American Cast Iron Pipe Co, pp. 500, 555
American Consumer Products Inc, p. 407
American Drew, p. 215
American Fuel Cell and Coated, p. 408
American Home Products Holdings, p. 343
American Homestar Corp, p. 194
American Israeli Paper Mills Ltd, p. 233
American Marketing Industries Inc, p. 136
American Media Inc., p. 263
American Olean Tile Co, p. 479
American Power Conversion Corp, p. 617
American Safety Razor Co, pp. 555, 699
American Seating Co, p. 215
American Steel Foundries, pp. 500, 555, 657
American Tool Companies Inc, pp. 554, 574
American Tourister, p. 174
American Trading and Production, p. 173
American Woodmark Corp, p. 194
Ameriscribe Corp, p. 263
Ameriwood Industries International, p. 215
Ameron International Corp, pp. 479, 555
Ames Industries Inc, p. 408
AMETEK Inc, pp. 572, 615, 697
Amgen Inc, p. 341
Amino plastics, p. 305
Amity Leather Products Co, p. 172
Ammonia, p. 295
Amoco Fabrics and Fibers Co, p. 97
Ampco-Pittsburgh Corp, p. 501
Amphenol Corp, pp. 525, 616
Amrep Inc, p. 389
AMSTED Industries Inc, pp. 499, 553
Analog Devices Inc, p. 615
Anam Industrial Co., Ltd, p. 615
Anchor Hocking Corp, p. 459
Anchor Hocking Specialty Glass Co, p. 460
Ancock Ingram International, p. 342
Andersen Corp, pp. 193, 427, 459
Anderson Millwork Ltd, p. 461
Andoria SA, p. 500
Andres Wines Ltd, p. 60
Andrew Corp, pp. 525, 617
Andrikian Trading Co. Ltd, p. 80
Aneka Regalindo PT, p. 174
Angelica Uniform Group, p. 136
Anglian Group PLC, pp. 194, 459

Angola
— beverages, p. 54
— chemical products nec, p. 338
— electrical machinery, pp. 602-603, 605-606, 611
— food products, p. 14
— footwear, p. 169
— furniture, fixtures, p. 210
— glass and products, p. 456
— industrial chemicals, p. 275
— industries nec, p. 716
— iron and steel, p. 496
— leather and products, p. 153
— machinery nec, pp. 560-561, 563-564, 568
— metal products, pp. 540, 542, 544-545, 550
— nonferrous metals, p. 521
— nonmetal products nec, p. 475
— paper and products, pp. 218, 220, 222-223, 228
— petroleum refineries, pp. 360-363, 367
— petroleum, coal products, p. 386
— plastic products nec, p. 423
— printing, publishing, pp. 248, 250, 252-253, 258
— professional goods, p. 694
— profile, p. 728
— rubber products, p. 403
— textiles, pp. 84, 86, 88-89, 94
— tobacco, pp. 66, 68, 70-71, 75
— transportation equipment, pp. 642, 644, 646-647, 651
— wearing apparel, pp. 122, 124, 126-127, 131
— wood products, p. 190
Aniline, p. 292
Animal feeds, p. 3
Anodizing Inc, p. 527
Antarctica da Amazonia SA Industria, p. 59
A.O. Smith Corp, pp. 553, 615, 655
A.O. Smith Electrical Products Co, p. 617
A.P. Green Industries Inc, p. 480
Apco Industries Co. Ltd, p. 390
Apogee Enterprises Inc, pp. 459, 553
Apotex Inc, p. 342
Apparel, p. 121
Apparel Group, p. 137
Appliances, p. 559
Applied Magnetics Corp, p. 595
Applied Power Inc, pp. 554, 574, 616
AptarGroup Inc, pp. 554, 719
Aquascutum Group PLC, p. 137
Aranda Tertice Mills Pty. Ltd, p. 137
ARCO Chemical Co, p. 279
ARCO Products Co, p. 370
Arcom SA, p. 572
Arden Corp, pp. 137, 407, 720
Ares Serono SA, p. 341
Argentina
— beverages, pp. 44, 46, 48-49, 51, 53-54
— chemical products nec, pp. 328, 330, 332-333, 335-336, 338

Argentina continued:
— electrical machinery, pp. 602-603, 605-606, 608, 610-611
— food products, pp. 4, 6, 8-9, 11, 13-14
— footwear, pp. 160-161, 163-164, 166-167, 169
— furniture, fixtures, pp. 200, 202, 204-205, 207-208, 210
— glass and products, pp. 448-449, 451-452, 454-456
— industrial chemicals, pp. 266-267, 269-270, 272, 274-275
— industries nec, pp. 706, 708, 710-711, 713, 715-716
— iron and steel, pp. 488-489, 491-493, 495-496
— leather and products, pp. 144-145, 147-148, 150-151, 153
— machinery nec, pp. 560-561, 563-564, 566-568
— metal products, pp. 540, 542, 544-545, 547-548, 550
— nonferrous metals, pp. 514-515, 517, 519-521
— nonmetal products nec, pp. 466, 468, 470-472, 474-475
— paper and products, pp. 218, 220, 222-223, 225-226, 228
— petroleum refineries, pp. 360-364, 366-367
— petroleum, coal products, pp. 380-386
— plastic products nec, pp. 414, 416-418, 420, 422-423
— pottery, china, etc., pp. 432-433, 435-438, 440
— printing, publishing, pp. 248, 250, 252-253, 255, 257-258
— professional goods, pp. 686-687, 689-691, 693-694
— profile, p. 729
— rubber products, pp. 394-395, 397-398, 400, 402-403
— textiles, pp. 84, 86, 88-89, 91-92, 94
— tobacco, pp. 66, 68, 70-72, 74-75
— transportation equipment, pp. 642, 644, 646-648, 650-65
— wearing apparel, pp. 122, 124, 126-128, 130-131
— wood products, pp. 180, 182, 184-185, 187-188, 190
Argha Karya Prima Industry PT, p. 406
Argus Press Ltd, p. 263
Aris Industries Inc, p. 135
Aris Isotoner Inc, p. 174
Arisco Produtos Alimenticios Ltda, p. 19
Arjo Wiggings Belgium SA, p. 233
Armco Inc, pp. 499, 573, 655
Armenia
— beverages, pp. 44, 46, 49, 51, 54
— chemical products nec, pp. 328, 330, 333, 335, 338
— electrical machinery, pp. 602-603, 606, 608, 611
— food products, pp. 4, 6, 9, 11, 14
— footwear, pp. 160-161, 164, 169
— furniture, fixtures, pp. 200, 202, 205, 210
— glass and products, pp. 448-449, 452, 456
— industrial chemicals, pp. 266-267, 270, 272, 275
— industries nec, pp. 706, 708, 711, 716
— iron and steel, pp. 488-489, 492-493, 496
— leather and products, pp. 144-145, 148, 153
— machinery nec, pp. 560-561, 564, 568
— metal products, pp. 540, 542, 545, 550
— nonferrous metals, pp. 514-515, 517, 519, 521
— nonmetal products nec, pp. 466, 468, 471, 475
— paper and products, pp. 218, 220, 223, 225, 228
— petroleum, coal products, pp. 380-381, 383, 386
— plastic products nec, pp. 414, 416, 418, 420, 423
— pottery, china, etc., pp. 432-433, 436, 440
— printing, publishing, pp. 248, 250, 253, 255, 258

Armenia continued:
— professional goods, pp. 686-687, 690, 694
— profile, p. 729
— rubber products, pp. 394-395, 398, 400, 403
— textiles, pp. 84, 86, 89, 91, 94
— tobacco, pp. 66, 68, 71-72, 75
— transportation equipment, pp. 642, 644, 647-648, 651
— wearing apparel, pp. 122, 124, 127-128, 131
— wood products, pp. 180, 182, 185, 187, 190
Armitage Shanks Ltd, p. 443
Armitage Shanks South Africa Pty, p. 443
Armstrong World Industries Ltd, p. 429
Arno Hentschel GmbH, p. 390
Arrow Industries Inc, p. 233
Arrow International Inc, p. 699
Art, p. 461
Arthur J. Prescott International Ltd, p. 720
Artificial resins and plastic materials, p. 304
Asahimas Flat Glass Co. Ltd. PT, p. 459
Asbestos-cement articles, p. 483
Asbestos yarn, p. 465
ASC Inc, p. 555
Asea Brown Boveri AS, pp. 19, 573, 616
Asea Brown Boveri Ltda, p. 574
Ashland Chemical Co, p. 278
Ashland Petroleum Co, p. 370
Ashley Furniture Industries Inc, p. 213
Ashworth Inc, p. 408
Asia Motors Co., Inc, p. 655
Asia Pacific Breweries Ltd, p. 60
Asian Design Manufacturing Corp, p. 215
Asland, SA, p. 481
Asphalt, p. 359
Asphalt products, p. 359
Associated Furniture Co. Ltd, pp. 97, 193
Associated Furniture Companies, p. 213
Associated Newspapers PLC, p. 263
Associated Pan Malaysia Cement, p. 480
AST Research Inc, p. 595
Astronautics Corporation of America, p. 698
A.T. Cross Co, p. 721
AT&T Global Information Solutions, p. 596
AT&T Paradyne, p. 595
ATC Communications Group Inc, p. 263
ATC International, p. 527
Atchison Casting Corp, p. 500
ATH Albarus Transmissoes, p. 407
Athens Papermill Co. SA, p. 233
Atlantic Packaging Products Ltd, p. 233
Atlantic Steel Industries Inc, p. 501
Atlantic Veneer do Brasil SA, p. 193
Atlantic, Gulf and Pacific Co. of, p. 554
Atlantis Plastics Inc, p. 428
Atmel Corp, p. 616
Atomic Energy Corp. of South Africa, pp. 280, 500, 526, 555

Augat Inc, p. 616
Aurora Bahana PT, p. 173
Austin Reed Group PLC, p. 137
Australia
— basic chemicals, excl fertilizers, pp. 312-316, 318
— beverages, pp. 44, 46, 48-49, 51, 53-54
— chemical products nec, pp. 328, 330, 332-333, 335-336, 338
— drugs and medicines, pp. 350-354, 356
— electrical machinery, pp. 602-603, 605-606, 608, 610-611
— food products, pp. 4, 6, 8-9, 11, 13-14
— footwear, pp. 160-161, 163-164, 166-167, 169
— furniture, fixtures, pp. 200, 202, 204-205, 207-208, 210
— glass and products, pp. 448-449, 451-452, 454-456
— industrial chemicals, pp. 266-267, 269-270, 272, 274-275
— industries nec, pp. 706, 708, 710-711, 713, 715-716
— iron and steel, pp. 488-489, 491-493, 495-496
— leather and products, pp. 144-145, 147-148, 150-151, 153
— machinery nec, pp. 560-561, 563-564, 566-567, 569
— metal products, pp. 540, 542, 544-545, 547-548, 550
— motor vehicles, pp. 676-680, 682
— nonferrous metals, pp. 514-515, 517, 519-521
— nonmetal products nec, pp. 466, 468, 470-472, 474-475
— office, computing machinery, pp. 590-594
— paper and products, pp. 218, 220, 222-223, 225-226, 228
— petroleum refineries, pp. 360-364, 366-367
— petroleum, coal products, pp. 380-386
— plastic products nec, pp. 414, 416-418, 420, 422-423
— pottery, china, etc., pp. 432-433, 435-438, 440
— printing, publishing, pp. 248, 250, 252-253, 255, 257-258
— professional goods, pp. 686-687, 689-691, 693-694
— profile, p. 730
— pulp, paper, etc., pp. 240-245
— radio, television, etc., pp. 628-630, 632-633
— rubber products, pp. 394-395, 397-398, 400, 402-403
— shipbuilding, repair, pp. 668-673
— spinning, weaving, etc., pp. 112-116, 118
— synthetic resins, etc., pp. 320-323, 325
— textiles, pp. 84, 86, 88-89, 91-92, 94
— tobacco, pp. 66, 68, 70-72, 74-75
— transportation equipment, pp. 642, 644, 646-648, 650-651
— wearing apparel, pp. 122, 124, 126-128, 130-131
— wood products, pp. 180, 182, 184-185, 187-188, 190
Austria
— basic chemicals, excl fertilizers, pp. 312-318
— beverages, pp. 44, 46, 48-49, 51, 53-54
— chemical products nec, pp. 328, 330, 332-333, 335-336, 338
— drugs and medicines, pp. 350-356
— electrical machinery, pp. 602, 604-606, 608, 610-611
— food products, pp. 4, 6, 8-9, 11, 13-14
— footwear, pp. 160-161, 163-164, 166-167, 169
— furniture, fixtures, pp. 200, 202, 204-205, 207-208, 210
— glass and products, pp. 448-449, 451-452, 454-456
— industrial chemicals, pp. 266-267, 269-270, 272, 274-275
— industries nec, pp. 706, 708, 710-711, 713, 715-716

Austria continued:
- iron and steel, pp. 488-489, 491-493, 495-496
- leather and products, pp. 144-145, 147-148, 150-151, 153
- machinery nec, pp. 560-561, 563-564, 566-567, 569
- metal products, pp. 540, 542, 544-545, 547-548, 550
- motor vehicles, pp. 676-682
- nonferrous metals, pp. 514-515, 517, 519-521
- nonmetal products nec, pp. 466, 468, 470-472, 474-475
- paper and products, pp. 218, 220, 222-223, 225-226, 228
- petroleum refineries, pp. 360-364, 366-367
- petroleum, coal products, pp. 380-386
- plastic products nec, pp. 414, 416-418, 420, 422-423
- pottery, china, etc., pp. 432-433, 435-438, 440
- printing, publishing, pp. 248, 250, 252-253, 255, 257-258
- professional goods, pp. 686-687, 689-691, 693-694
- profile, p. 730
- pulp, paper, etc., pp. 240-245
- radio, television, etc., pp. 628-630, 632-633
- rubber products, pp. 394-395, 397-398, 400, 402-403
- shipbuilding, repair, pp. 668-673
- textiles, pp. 84, 86, 88-89, 91-92, 94
- tobacco, pp. 66, 68, 70-72, 74-75
- transportation equipment, pp. 642, 644, 646-648, 650-651
- wearing apparel, pp. 122, 124, 126-128, 130-131
- wood products, pp. 180, 182, 184-185, 187-188, 190

Authentic Fitness Corp, p. 137
Autobar Group Ltd, p. 428
Autoglass International UK Ltd, pp. 193, 459
Automated Security Holdings PLC, p. 617
Automobiles, p. 675
Automotive Plastic Technologies Inc, p. 429
Automotive Products PLC, p. 656
Autosan SA, p. 555
Autotote Corp, p. 597
Avex Electronics Inc, p. 617
Aviation fuel, p. 359
Aviation gasoline, p. 373
Avis Industrial Corp, p. 501
Avon Canada Inc, p. 721
Avon Cosmetics Ltd, p. 342
Avon Polymer Products Ltd, p. 429
Avon Rubber PLC, p. 406
Avondale Inc, p. 97
Avondale Industries Inc, p. 656
A.W. Chesterton Co, p. 407
Axiohm Transaction Solutions Inc, pp. 597, 721
Aydin Corp, p. 597
Azerbaijan
- basic chemicals, excl fertilizers, pp. 312-313, 315, 318
- beverages, pp. 44, 46, 49, 54
- chemical products nec, pp. 328, 330, 333, 338
- drugs and medicines, pp. 350-351, 353, 356
- electrical machinery, pp. 602, 604, 606, 611
- food products, pp. 4, 6, 9, 15
- footwear, pp. 160-161, 164, 169
- furniture, fixtures, pp. 200, 202, 205, 210

Azerbaijan continued:
- glass and products, pp. 448-449, 452, 456
- industrial chemicals, pp. 266-267, 270, 275
- industries nec, pp. 706, 708, 711, 716
- iron and steel, pp. 488-489, 492, 496
- leather and products, pp. 144-145, 148, 153
- machinery nec, pp. 560-561, 564, 569
- metal products, pp. 540, 542, 545, 550
- motor vehicles, pp. 676-677, 679, 682
- nonferrous metals, pp. 514-515, 517, 521
- nonmetal products nec, pp. 466, 468, 471, 475
- office, computing machinery, pp. 590, 594
- paper and products, pp. 218, 220, 223, 228
- petroleum refineries, pp. 360-361, 363, 367
- petroleum, coal products, pp. 380-381, 383, 386
- plastic products nec, pp. 414, 416, 418, 423
- pottery, china, etc., pp. 432-433, 436, 440
- printing, publishing, pp. 248, 250, 253, 258
- profile, p. 731
- rubber products, pp. 394-395, 398, 403
- shipbuilding, repair, pp. 668-669, 671, 673
- spinning, weaving, etc., pp. 112-113, 115, 118
- synthetic resins, etc., pp. 321, 325
- textiles, pp. 84, 86, 89, 94
- tobacco, pp. 66, 68, 71, 75
- transportation equipment, pp. 642, 644, 647, 651
- wood products, pp. 180, 182, 185, 190

B A Blacktop Ltd, p. 390
B/E Aerospace Inc, p. 657
Babolna Mezogazdasagi Termelo, p. 18
Bacon, p. 22
Badak Natural Gas Liquefaction Co, p. 370
Badan Tekstil Nasional PT, pp. 99, 136
Bahamas
- beverages, pp. 44, 46, 49, 51, 54
- chemical products nec, pp. 328, 330, 333, 335, 338
- food products, pp. 4, 6, 9, 11, 15
- furniture, fixtures, pp. 200, 202, 205, 207, 210
- industries nec, pp. 706, 708, 711, 713, 716
- iron and steel, pp. 488-489, 492-493, 496
- nonmetal products nec, pp. 466, 468, 471-472, 475
- printing, publishing, pp. 248, 250, 253, 255, 258
- profile, p. 731
- textiles, pp. 84, 86, 89, 91, 94
- wearing apparel, pp. 122, 124, 127-128, 132
Bahrain
- basic chemicals, excl fertilizers, pp. 312-315, 317
- beverages, pp. 44, 46, 48-49, 53-54
- chemical products nec, pp. 328, 330, 333, 336, 338
- drugs and medicines, pp. 350-351, 353, 355
- electrical machinery, pp. 602, 604-606, 610-611
- food products, pp. 4, 6, 8-9, 13, 15
- footwear, pp. 160-161, 163-164, 167, 169
- furniture, fixtures, pp. 200, 202, 204-205, 208, 210
- glass and products, p. 456
- industrial chemicals, pp. 266-267, 269-270, 274-275

Bahrain continued:
—industries nec, pp. 706, 708, 711, 715-716
—iron and steel, pp. 488-489, 491-492, 495-496
—leather and products, pp. 144-145, 148, 153
—machinery nec, p. 569
—metal products, pp. 540, 542, 544-545, 548, 550
—motor vehicles, pp. 676-677, 679, 681
—nonferrous metals, pp. 514-515, 517, 520-521
—nonmetal products nec, pp. 466, 468, 470-471, 474-475
—paper and products, pp. 218, 220, 222-223, 226, 228
—petroleum refineries, pp. 360-363, 366-367
—petroleum, coal products, p. 386
—plastic products nec, pp. 414, 416-417, 419, 422-423
—pottery, china, etc., pp. 432-433, 436, 438, 440
—printing, publishing, pp. 248, 250, 252-253, 257-258
—professional goods, pp. 686-687, 689-690, 693-694
—profile, p. 732
—rubber products, p. 403
—shipbuilding, repair, pp. 668-671, 673
—spinning, weaving, etc., pp. 112-113, 115
—synthetic resins, etc., pp. 320-322
—textiles, pp. 84, 86, 89, 94
—tobacco, p. 75
—transportation equipment, pp. 642, 644, 646-647, 650-651
—wearing apparel, pp. 122, 124, 126-127, 130, 132
—wood products, pp. 180, 182, 184-185, 190
Baik San Co., Ltd, p. 156
Bailey Corp, p. 428
Bairdwear Racke Ltd, p. 136
Bajaj Hindustan Ltd, pp. 18, 278, 479
Baked goods, p. 3
Baker Furniture Co, p. 214
Baker Hughes INTEQ, p. 573
Baker Petrolite, p. 343
Bakers, p. 20
Bakony Muvek, pp. 617, 657
Bakrie & Brothers, PT, p. 500
Bakrie Sumatra Plantation PT, p. 406
Balances, p. 589
Baldor Electric Co, pp. 574, 617
Baldwin Piano and Organ Co, p. 720
Ball bearings, p. 622
Ball Corp, pp. 427, 553, 615, 697
Ballet Makers Inc, p. 173
Baltek Corp, p. 195
Balzers Ltd, p. 698
BancTec Inc, p. 596
BancTec Technologies Inc, p. 595
Bandag Inc, p. 406
Bang & Olufsen AS, p. 617
Bangkok Cable Co., Ltd, p. 527
Bangkok Produce Merchandising, p. 18
Bangkok Rubber Co., Ltd, p. 172
Bangkok Steel Industry Co., Ltd, p. 501
Bangkok Weaving Mills Ltd, p. 99

Bangladesh
—basic chemicals, excl fertilizers, pp. 312-318
—beverages, pp. 44, 46, 48-49, 51, 53-54
—chemical products nec, pp. 328, 330, 332-333, 335-336, 338
—drugs and medicines, pp. 350-356
—electrical machinery, pp. 602, 604-605, 607-608, 610-611
—food products, pp. 4, 6, 8-9, 11, 13, 15
—footwear, pp. 160-161, 163-164, 166-167, 169
—furniture, fixtures, pp. 200, 202, 204-205, 207-208, 210
—glass and products, pp. 448-449, 451-452, 454-456
—industrial chemicals, pp. 266, 268-270, 272, 274-275
—industries nec, pp. 706, 708, 710-711, 713, 715-716
—iron and steel, pp. 488-489, 491-493, 495-496
—leather and products, pp. 144-145, 147-148, 150-151, 153
—machinery nec, pp. 560-561, 563-564, 566-567, 569
—metal products, pp. 540, 542, 544-545, 547-548, 550
—motor vehicles, pp. 676-682
—nonferrous metals, pp. 514-515, 517, 519-521
—nonmetal products nec, pp. 466, 468, 470-472, 474-475
—paper and products, pp. 218, 220, 222-223, 225-226, 228
—petroleum refineries, pp. 360-364, 366-367
—petroleum, coal products, pp. 380-386
—plastic products nec, pp. 414, 416, 418-420, 422-423
—pottery, china, etc., pp. 432-433, 435-438, 440
—printing, publishing, pp. 248, 250, 252-253, 255, 257-258
—professional goods, pp. 686-687, 689-691, 693-694
—profile, p. 732
—pulp, paper, etc., pp. 240-245
—radio, television, etc., pp. 628-630, 632-633
—rubber products, pp. 394-395, 397-398, 400, 402-403
—shipbuilding, repair, pp. 668-673
—spinning, weaving, etc., pp. 112-118
—synthetic resins, etc., pp. 320-323, 325
—textiles, pp. 84, 86, 88-89, 91-92, 94
—tobacco, pp. 66, 68, 70-72, 74, 76
—transportation equipment, pp. 642, 644, 646-648, 650-651
—wearing apparel, pp. 122, 124, 126-128, 130, 132
—wood products, pp. 180, 182, 184-185, 187-188, 190
Bank Bros & Son Ltd, p. 157
Banks Lumber Company Inc, p. 195
Banner Pharmacaps Inc, p. 343
Banta Corp, pp. 231, 262
Bantam Doubleday Dell Publishing, p. 264
Barbados
—beverages, p. 54
—chemical products nec, pp. 328, 330, 333, 335-336, 338
—electrical machinery, p. 611
—food products, pp. 4, 6, 9, 11, 13, 15
—footwear, p. 169
—furniture, fixtures, p. 210
—glass and products, p. 456
—industrial chemicals, pp. 274-275
—industries nec, pp. 706, 708, 711, 713, 715-716
—iron and steel, p. 496
—leather and products, p. 153

Barbados continued:
— machinery nec, p. 569
— metal products, pp. 540, 542, 545, 547-548, 550
— nonferrous metals, p. 521
— nonmetal products nec, p. 475
— paper and products, p. 228
— petroleum refineries, p. 367
— petroleum, coal products, p. 386
— pottery, china, etc., p. 440
— printing, publishing, p. 258
— professional goods, p. 694
— profile, p. 733
— textiles, p. 94
— tobacco, p. 76
— transportation equipment, p. 652
— wearing apparel, pp. 122, 124, 127-128, 130, 132
— wood products, p. 190
Barbour Corporation Inc, p. 156
Barnes Group Inc, pp. 554, 656
Barry Manufacturing Company Inc, p. 174
Barton Inc, p. 60
Bassett Furniture Industries Inc, p. 213
Bastos du Canada Ltee, p. 80
BAT Benelux SA NV, p. 79
B.A.T. Espana, SA, p. 79
BAT UK & Export Ltd, p. 79
Bata Chile SA, p. 406
Bata Industries Ltd, p. 172
Bata Malaysia BHD, p. 172
Bata SA Marocaine, p. 173
Batesville Casket Company Inc, p. 719
Battenfeld-American Inc, p. 390
Battenfeld Grease and Oil Corp, p. 390
Battenfeld Grease (Canada) Ltd, p. 371
Batteries and cells, p. 639
Batts Inc, pp. 195, 429
Bavaria BV, p. 60
Baxter Healthcare Corp. Baxter, pp. 342, 698
Baxter Healthcare Corp. of Puerto, pp. 341, 427, 697
Bay Networks Inc, pp. 595, 616
Bayer Corp. Diagnostics Div, pp. 341, 697
Bayer do Brasil SA, p. 341
Bayer Inc, pp. 280, 699
Bayer PLC, p. 343
Bayly Corp, p. 137
B.B. Walker Co, p. 174
BBC Enterprises Ltd, p. 264
Beacon Sweets & Chocolates Pty, p. 20
Beales Hunter PLC, p. 137
Beams, p. 179
Beatson Clark PLC, p. 460
Bed linen, pp. 83, 106, 111
Bedsprings, p. 199
Beef and veal, p. 21
Beer, pp. 43, 62
Bekaert Corp, p. 501

Bekaert Textiles NV, p. 99
Belapatfalvi Cement Es Meszipari Rt, p. 481
Belden Inc, p. 525
Belgin Madeni Yaglar AS, p. 390
Belgium
— beverages, pp. 44, 46, 49, 51, 53-54
— chemical products nec, pp. 333, 335, 338
— drugs and medicines, pp. 354-355
— electrical machinery, pp. 602, 611
— food products, pp. 4, 6, 11, 13, 15
— footwear, pp. 160-161, 167, 169
— furniture, fixtures, pp. 202, 205, 207-208, 210
— glass and products, pp. 448-449, 456
— industrial chemicals, pp. 270, 272, 274-275
— industries nec, pp. 706, 708, 716
— iron and steel, pp. 489, 492-493, 495-496
— leather and products, pp. 144-145, 151, 153
— machinery nec, pp. 560, 569
— metal products, pp. 540, 550
— nonferrous metals, pp. 515, 517, 519-521
— nonmetal products nec, pp. 468, 475
— paper and products, pp. 218, 220, 223, 225-226, 228
— petroleum refineries, pp. 360, 367
— petroleum, coal products, pp. 380, 386
— plastic products nec, pp. 414, 416, 423
— pottery, china, etc., pp. 433, 440
— printing, publishing, pp. 248, 250, 253, 255, 257-258
— professional goods, pp. 686, 694
— profile, p. 733
— pulp, paper, etc., pp. 244-245
— rubber products, pp. 394-395, 398, 400, 402-403
— textiles, pp. 84, 86, 89, 91-92, 94
— tobacco, pp. 66, 68, 71-72, 74, 76
— transportation equipment, pp. 642, 652
— wearing apparel, pp. 122, 124, 127, 129-130, 132
— wood products, pp. 182, 185, 187-188, 190
Belize
— beverages, pp. 44, 46, 49, 51, 54
— chemical products nec, p. 328
— electrical machinery, pp. 602, 604, 607-608, 611
— food products, pp. 4, 6, 9, 11, 15
— furniture, fixtures, pp. 200, 202, 205, 207, 210
— glass and products, p. 448
— industrial chemicals, p. 266
— industries nec, pp. 706, 708, 711, 713, 716
— iron and steel, p. 496
— machinery nec, p. 569
— metal products, pp. 540, 542, 545, 547, 550
— nonferrous metals, p. 521
— nonmetal products nec, pp. 466, 468, 471-472, 475
— paper and products, pp. 218, 220, 223, 225, 228
— petroleum refineries, p. 367
— petroleum, coal products, p. 386
— plastic products nec, p. 414
— pottery, china, etc., p. 440
— printing, publishing, pp. 248, 250, 253, 255, 258

Belize continued:
— professional goods, p. 694
— profile, p. 734
— rubber products, p. 394
— tobacco, pp. 66, 68, 71-72, 76
— transportation equipment, pp. 642, 644, 647-648, 652
— wood products, pp. 180, 182, 185, 187, 190
Bell Group UK Holdings Ltd, p. 264
Bell Industries Inc, p. 596
Bell Sports Corp, p. 720
Bell Sports Inc, p. 721
Belmont Homes Inc, p. 194
Beloit Corp, p. 572
Belopal, p. 459
Belting, p. 393
Bemis Company Inc, pp. 231, 427, 572
Bemrose Corp. PLC, p. 264
Bender SA, Curtume, p. 156
Benin
— basic chemicals, excl fertilizers, p. 312
— beverages, pp. 44, 49, 54
— chemical products nec, pp. 328, 333, 338
— drugs and medicines, p. 350
— food products, pp. 4, 9, 15
— footwear, pp. 164, 169
— furniture, fixtures, pp. 205, 210
— industrial chemicals, pp. 266, 270, 275
— industries nec, p. 706
— leather and products, pp. 148, 153
— metal products, pp. 540, 545, 550
— nonmetal products nec, pp. 466, 471, 475
— paper and products, pp. 218, 223, 228
— plastic products nec, pp. 419, 423
— printing, publishing, pp. 248, 253, 258
— profile, p. 734
— spinning, weaving, etc., p. 112
— synthetic resins, etc., p. 320
— textiles, pp. 84, 89, 94
— tobacco, pp. 66, 71, 76
— transportation equipment, p. 642
— wearing apparel, pp. 127, 132
— wood products, pp. 180, 185, 190
Benjamin Moore and Co, p. 342
Bennett Industries Inc, p. 173
Bentonite Co, p. 481
Benzene, p. 282
Berg Electronics Corp, p. 615
Beringer Wine Estates, p. 60
Berkline Corp, p. 213
Berlex Laboratories Inc, p. 343
Berli Jucker Co. Ltd, pp. 233, 343
Berlina PT, p. 428
Berlitz International Inc, p. 262
Bermuda
— beverages, pp. 46, 48, 54
— electrical machinery, pp. 604-605, 611

Bermuda continued:
— food products, pp. 6, 8, 15
— footwear, pp. 161, 163, 169
— furniture, fixtures, pp. 202, 204, 210
— glass and products, pp. 449, 451, 456
— industrial chemicals, pp. 268-269, 275
— industries nec, pp. 708, 710, 716
— machinery nec, pp. 561, 563, 569
— metal products, pp. 542, 544, 550
— nonmetal products nec, pp. 468, 470, 476
— petroleum refineries, pp. 361-362, 367
— pottery, china, etc., pp. 433, 435, 440
— printing, publishing, pp. 250, 252, 258
— professional goods, pp. 687, 689, 694
— profile, p. 735
— transportation equipment, pp. 644, 646, 652
— wearing apparel, pp. 124, 126, 132
— wood products, pp. 182, 184, 190
Bernstein Group PLC, p. 214
Berol Corp, p. 719
Bertrand Construction L'Orignal Inc, p. 390
Bespak PLC, p. 429
Besse Forest Products Group, p. 195
Best Chairs Inc, p. 215
Best Foods Div, p. 19
Best Manufacturing Co, p. 137
Bestform Foundations Inc, p. 136
Betagro Co. Ltd, p. 19
Bettanin Industrial SA, p. 721
Betterroads Asphalt Corp, p. 389
Beverage America Inc, p. 59
Beverages, p. 43
BHP Hawaii Inc, p. 370
Bhutan
— beverages, p. 54
— chemical products nec, p. 338
— food products, p. 15
— furniture, fixtures, p. 210
— glass and products, p. 456
— industrial chemicals, p. 275
— iron and steel, p. 496
— nonferrous metals, p. 521
— nonmetal products nec, p. 476
— paper and products, p. 228
— petroleum refineries, p. 367
— petroleum, coal products, p. 386
— plastic products nec, p. 423
— pottery, china, etc., p. 440
— printing, publishing, p. 259
— rubber products, p. 403
— tobacco, p. 76
— wood products, p. 190
Bibb Co, pp. 97, 135
Biber Paper Converting Ltd, p. 264
Bic Corp, pp. 555, 719
Bical Ltda, p. 173

Alphabetical Index

BICC Cables Corp, p. 525
BICC Cables Ltd, p. 655
Bicycles, pp. 641, 664
Bicycles tires, p. 393
Big Flower Holdings Inc, p. 262
Bil Mar Foods Inc, p. 20
Billets (iron), p. 487
Binhdinh Exported Rattan Enterprise, p. 195
Binney and Smith Inc, p. 720
Binoculars and refracting telescopes, p. 701
Bio-Rad Laboratories Inc, pp. 342, 698
Biological products, p. 349
Biomet Inc, p. 698
Bird Corp., p. 390
Birmid Components Ltd, p. 501
Birmingham Steel Corp, p. 501
Biro Bic Ltd, p. 721
Biscuits, p. 38
Bitum Petrochemicals Industries Ltd, p. 390
Bitumen, p. 376
Biwater Industries Ltd, p. 501
Blandin Paper Co, p. 195
Blankbooks, p. 247
Blankets, pp. 83, 106, 111
Blended liquors, p. 43
Blessings Corp, p. 429
Block Drug Company Inc, pp. 341, 698, 719
Blockboard, p. 197
Blocks (ceramic), p. 465
Blooms (iron), p. 487
Blount International Inc, pp. 553, 573
Blouses, p. 139
Blown glass, p. 447
Blue Circle Canada Inc, p. 480
Blue Circle Cement, p. 479
Blue Circle Cement Pty. Ltd, pp. 481, 527
Blyth Industries Inc., p. 719
BMC Industries Inc, p. 698
Boardman Bros. Pty. Ltd, p. 371
Boards, p. 179
Boats, p. 667
Bodies for motor vehicles, p. 661
Bodilsen Holding AS, p. 215
Body Shop International PLC, p. 342
Boehringer Mannheim Corp, p. 698
Boeing Canada Technology Ltd., p. 655
Boeing Co. Boeing Helicopters Div, p. 655
Boiler Factory, Rafako, p. 555
Boilers, pp. 539, 556
Bolivia
— basic chemicals, excl fertilizers, pp. 312-313, 315-318
— beverages, pp. 44, 46, 49, 51, 53-54
— chemical products nec, pp. 328, 330, 333, 335-336, 338
— drugs and medicines, pp. 350-351, 353-356
— electrical machinery, pp. 602, 604, 607-608, 610-611
— food products, pp. 4, 6, 9, 11, 13, 15

Bolivia continued:
— footwear, pp. 160-161, 164, 166, 168-169
— furniture, fixtures, pp. 200, 202, 205, 207-208, 210
— glass and products, pp. 448-449, 452, 454-456
— industrial chemicals, pp. 266, 268, 270, 272, 274-275
— industries nec, pp. 706, 708, 711, 713, 715-716
— iron and steel, pp. 488-489, 492-493, 495-496
— leather and products, pp. 144-145, 148, 150-151, 153
— machinery nec, pp. 560-561, 564, 566-567, 569
— metal products, pp. 540, 542, 545, 547-548, 550
— motor vehicles, pp. 676-677, 679-682
— nonferrous metals, pp. 514-515, 517, 519-520, 522
— nonmetal products nec, pp. 466, 468, 471-472, 474, 476
— paper and products, pp. 218, 220, 223, 225-226, 228
— petroleum refineries, pp. 360-361, 363-364, 366-367
— plastic products nec, pp. 414, 416, 419-420, 422-423
— pottery, china, etc., pp. 432-433, 436-438, 440
— printing, publishing, pp. 248, 250, 253, 255, 257, 259
— professional goods, pp. 686-687, 690-691, 693-694
— profile, p. 735
— pulp, paper, etc., pp. 240-241, 243-245
— radio, television, etc., pp. 628-630, 632-633
— rubber products, pp. 394-395, 398, 400, 402-403
— spinning, weaving, etc., pp. 112-113, 115-118
— synthetic resins, etc., pp. 320-325
— textiles, pp. 84, 86, 89, 91-92, 94
— tobacco, pp. 66, 68, 71-72, 74, 76
— transportation equipment, pp. 642, 644, 647-648, 650, 65
— wearing apparel, pp. 122, 124, 127, 129-130, 132
— wood products, pp. 180, 182, 185, 187-188, 190
Bolts, p. 539
Boma Bisma Indra, PT, p. 555
Bombardier Aerospace Europe Ltd, p. 655
Bonar Inc, pp. 233, 429
Bonnita Pty. Ltd, pp. 19, 58
Boohung Co., Ltd, p. 137
Book publishing, p. 247
Boon Rawd Brewery Co., Ltd, p. 59
Boosey & Hawkes PLC, pp. 264, 721
Boowoon Mulsan Co., Ltd, p. 174
Borden Decorative Products Ltd, p. 233
Borden UK Ltd, p. 428
Border Sheepskins Ltd, p. 157
Borregaard Industries Ltd, p. 233
Borsodchem Rt, pp. 231, 279
Bosal Afrika Pty. Ltd, pp. 573, 656
Bosal Benelux NV, p. 215
Bosch do Brasil Ltda., Robert, p. 655
Bose Corp, p. 617
Boskor Timber Processors, p. 195
Bosnia & Herzegovina
— basic chemicals, excl fertilizers, pp. 312-313, 318
— beverages, pp. 44, 46, 48-49, 51, 53-54
— chemical products nec, pp. 328, 330, 332-333, 335-336, 338
— drugs and medicines, pp. 350-351, 356

Bosnia & Herzegovina continued:
— electrical machinery, pp. 602, 604, 606-608, 610-611
— food products, pp. 4, 6, 8-9, 11, 13, 15
— footwear, pp. 160-161, 163-164, 166, 168-169
— furniture, fixtures, pp. 200, 202, 204-205, 207-208, 210
— glass and products, pp. 448-449, 451-452, 454-456
— industrial chemicals, pp. 266, 268-270, 272, 274-275
— industries nec, pp. 706, 708, 710-711, 713, 715-716
— iron and steel, pp. 488-489, 491-493, 495-496
— leather and products, pp. 144, 146-148, 150-151, 153
— machinery nec, pp. 560-561, 563-564, 566-567, 569
— metal products, pp. 540, 542, 544-545, 547-548, 550
— motor vehicles, pp. 676-677, 682
— nonferrous metals, pp. 514-515, 517-520, 522
— nonmetal products nec, pp. 466, 468, 470-472, 474, 476
— office, computing machinery, pp. 590, 594
— paper and products, pp. 218, 220, 222-223, 225-226, 228
— petroleum refineries, pp. 360-363, 365-367
— petroleum, coal products, pp. 380-386
— plastic products nec, pp. 414, 416, 418-420, 422-423
— pottery, china, etc., pp. 432-433, 435-438, 440
— printing, publishing, pp. 248, 250, 252-253, 255, 257, 259
— professional goods, pp. 686-687, 689-691, 693-694
— profile, p. 736
— pulp, paper, etc., pp. 240-241, 245
— radio, television, etc., pp. 628-629, 633
— rubber products, pp. 394, 396-398, 400, 402-403
— spinning, weaving, etc., pp. 112-113, 118
— synthetic resins, etc., pp. 320-321, 325
— textiles, pp. 84, 86, 88-89, 91-92, 94
— tobacco, pp. 66, 68, 70-71, 73-74, 76
— transportation equipment, pp. 642, 644, 646-648, 650, 652
— wearing apparel, pp. 122, 124, 126-127, 129-130, 132
— wood products, pp. 180, 182, 184-185, 187-188, 190
Bostinco BV, p. 215
Botanicals, p. 349
Botswana
— beverages, pp. 44, 46, 48-49, 51, 53-54
— food products, pp. 4, 6, 8-9, 11, 13, 15
— glass and products, p. 456
— industries nec, pp. 706, 708, 710-711, 713, 715-716
— iron and steel, p. 496
— nonferrous metals, p. 522
— nonmetal products nec, p. 476
— petroleum refineries, p. 367
— petroleum, coal products, p. 386
— pottery, china, etc., p. 440
— professional goods, p. 694
— profile, p. 736
— tobacco, p. 76
Bottles, p. 413
Boulton & Paul Manufacturing Ltd, p. 194
Bourns Inc, p. 616
Bowater Inc, p. 231
Bowater Packaging Ltd, p. 231
Bowater Windows Ltd, p. 428

Bowes Publishers Ltd, p. 263
Bowne and Company Inc, p. 263
Boxes, p. 217
BP Belgium NV, p. 408
BP Chemicals Ltd, p. 278
BP Oil Espana, SA, p. 371
BP Turkey Refining Ltd, p. 372
BPB Paper & Packaging Ltd, p. 232
BPP Holdings PLC, p. 264
Brake Bros. Foodservice Ltd, p. 20
Brandt Inc., p. 597
Brandy, p. 43
Brasperola Industria e Comercio SA, p. 98
Brasway PLC, p. 408
Brazil
— beverages, pp. 44, 46, 49, 51, 53-54
— chemical products nec, p. 338
— drugs and medicines, pp. 350-351, 353-356
— electrical machinery, pp. 602, 604, 607-608, 610-611
— food products, pp. 4, 6, 9, 11, 13, 15
— footwear, p. 169
— furniture, fixtures, pp. 200, 202, 205, 207-208, 210
— glass and products, p. 456
— industrial chemicals, p. 275
— industries nec, p. 716
— iron and steel, p. 496
— leather and products, p. 153
— machinery nec, pp. 560-561, 564, 566-567, 569
— metal products, p. 550
— nonferrous metals, p. 522
— nonmetal products nec, p. 476
— paper and products, pp. 218, 220, 223, 225-226, 228
— petroleum refineries, p. 367
— petroleum, coal products, p. 386
— plastic products nec, pp. 414, 416, 419-420, 422-423
— pottery, china, etc., p. 440
— printing, publishing, pp. 248, 250, 253, 255, 257, 259
— professional goods, p. 694
— profile, p. 737
— rubber products, pp. 394, 396, 398, 400, 402-403
— textiles, pp. 84, 86, 89, 91-92, 94
— tobacco, pp. 66, 68, 71, 73-74, 76
— transportation equipment, pp. 642, 644, 647-648, 650, 652
— wearing apparel, p. 132
— wood products, pp. 180, 182, 185, 187-188, 190
Bread, pp. 3, 37
Bricks, p. 465
Bridge of Weir Leather Co. Ltd, p. 156
Bridgestone/Firestone Canada Inc, p. 406
Bridgestone/Firestone do Brasil, p. 406
Bridon PLC, p. 554
Briggs Amasco Ltd, p. 480
Briggs and Stratton Corp, p. 572
Briggs Industries Inc, p. 443
Brintons Ltd, p. 99
Brio AB/Brio Toy, p. 721

Bristol Compressors Inc, p. 574
Bristol Evening Post PLC, p. 264
Bristol-Myers Squibb Holdings Ltd, pp. 342, 699
Britannia Products Ltd, p. 264
British Alcan Aluminium PLC, p. 525
British Bakeries Ltd, p. 18
British Bata Shoe Co. Ltd, p. 173
British Olivetti Ltd, p. 596
British Shipbuilders, pp. 572, 655
British Tissues Ltd, p. 194
Britoil PLC, p. 372
BRK Brands Inc, p. 720
Broadleaved sawn wood, p. 179
Brodart Co, p. 214
Brooke Bond Foods Ltd, p. 20
Brooke Group Ltd, pp. 79, 264
Brookwood Furniture Company Inc, p. 215
Brown and Sharpe Manufacturing, p. 699
Brown-coal briquettes, p. 391
Brown-Forman Corp, pp. 58, 172, 443
Brown Shoe Co. Of Canada, Ltd, p. 173
Broyhill Furniture Industries Inc, p. 213
Brunswick Corp. US Marine Div, p. 657
Brush Wellman Inc, p. 526
Brushindo Cemerlang PT, p. 721
BSC Footwear Supplies Ltd, p. 173
BSN Vidrio Espana, SA, p. 460
Buchanan Forest Products Ltd, p. 194
Bucina, AS, p. 194
Buckeye Steel Castings Co, p. 501
Budapesti Harisnyagyar, p. 98
Budd Co, pp. 427, 553, 655
Budgens PLC, p. 19
Budidharma Jakarta, PT, p. 527
Buil Leather Co., Ltd, p. 156
Building bricks, p. 482
Bukoza AS, pp. 193, 279
Bulgaria
— beverages, pp. 44, 46, 48-49, 51, 53-54
— chemical products nec, pp. 328, 330, 332-333, 335-336, 338
— electrical machinery, pp. 602, 604, 606-608, 610, 612
— food products, pp. 4, 6, 8-9, 11, 13, 15
— footwear, pp. 160, 162-164, 166, 168-169
— furniture, fixtures, pp. 200, 202, 204-205, 207-208, 210
— glass and products, pp. 448-449, 451-452, 454-456
— industrial chemicals, pp. 266, 268-270, 272, 274-275
— industries nec, pp. 706, 708, 710-711, 713, 715-716
— iron and steel, pp. 488-489, 491-493, 495-496
— leather and products, pp. 144, 146-148, 150-151, 153
— metal products, pp. 540, 542, 544-545, 547-548, 550
— nonferrous metals, pp. 514-515, 517-520, 522
— nonmetal products nec, pp. 466, 468, 470-472, 474, 476
— paper and products, pp. 218, 220, 222-223, 225-226, 228
— petroleum refineries, pp. 365, 367
— petroleum, coal products, pp. 380-386

Bulgaria continued:
— plastic products nec, pp. 414, 416, 418-420, 422-423
— pottery, china, etc., pp. 432-433, 435-438, 440
— printing, publishing, pp. 248, 250, 252-253, 255, 257, 259
— profile, p. 737
— rubber products, pp. 394, 396-398, 400, 402-403
— textiles, pp. 84, 86, 88-89, 91-92, 94
— tobacco, pp. 66, 68, 70-71, 73-74, 76
— transportation equipment, pp. 642, 644, 646-647, 649-650, 652
— wearing apparel, pp. 122, 124, 126-127, 129-130, 132
— wood products, pp. 180, 182, 184-185, 187-188, 190
Bulldozers, p. 586
Bum Jin Co., Ltd, p. 157
Bumar-Labedy, pp. 499, 572
Bumar-Warynski SA, p. 574
Burberrys Ltd, p. 136
Burkina-Faso
— beverages, p. 54
— chemical products nec, p. 338
— electrical machinery, p. 612
— food products, p. 15
— footwear, p. 169
— furniture, fixtures, p. 210
— glass and products, p. 456
— industrial chemicals, p. 275
— industries nec, p. 716
— leather and products, p. 153
— machinery nec, p. 569
— metal products, p. 550
— nonmetal products nec, p. 476
— paper and products, p. 228
— petroleum refineries, p. 367
— petroleum, coal products, p. 386
— plastic products nec, p. 423
— pottery, china, etc., p. 440
— printing, publishing, p. 259
— professional goods, p. 694
— rubber products, p. 403
— textiles, p. 94
— tobacco, p. 76
— transportation equipment, p. 652
— wearing apparel, p. 132
— wood products, p. 190
Burtons Gold Medal Biscuits Ltd, p. 20
Burtonwood Brewery PLC, p. 60
Burundi
— beverages, pp. 44, 46, 49, 51, 54
— chemical products nec, pp. 328, 330, 333, 335, 338
— electrical machinery, p. 612
— food products, pp. 4, 6, 9, 11, 15
— footwear, pp. 160, 162, 164, 166, 169
— glass and products, p. 456
— industrial chemicals, pp. 266, 268, 270, 272, 275
— industries nec, pp. 706, 708, 711, 713, 716
— iron and steel, p. 496

Burundi continued:
— leather and products, pp. 144, 146, 148, 150, 153
— machinery nec, p. 569
— metal products, pp. 540, 542, 545, 547, 550
— nonferrous metals, p. 522
— nonmetal products nec, pp. 466, 468, 471-472, 476
— paper and products, pp. 218, 220, 223, 225, 228
— petroleum refineries, p. 367
— petroleum, coal products, p. 386
— plastic products nec, pp. 414, 416, 419-420, 423
— pottery, china, etc., p. 440
— printing, publishing, pp. 248, 250, 253, 255, 259
— professional goods, p. 694
— profile, p. 738
— rubber products, p. 403
— textiles, pp. 84, 86, 89, 91, 94
— tobacco, pp. 66, 68, 71, 73, 76
— transportation equipment, p. 652
— wearing apparel, pp. 122, 124, 127, 129, 132
Busana Rama Textile & Garment PT, p. 136
Buses, p. 675
Buses and motor coaches, p. 662
Bush Boake Allen Inc, pp. 59, 342
Bush Industries Inc, p. 213
Butler Manufacturing Co, pp. 193, 554, 573
Butter, pp. 3, 26
Butyl alcohol, p. 286
Butylenes, p. 282
Buxton Co, p. 174
BXL Plastics Ltd, p. 429
Byucksan Corp, p. 480
C and J Clark America Inc, p. 172
C. Walker & Sons Ltd, p. 554
Cable Design Technologies Corp, p. 525
Cable Design Technologies Inc, p. 408
Cables, pp. 539, 557
Cabletron Systems Inc, p. 615
Cabot Corp, p. 279
Cadbury Ltd, p. 18
Cadbury Schweppes Australia Ltd, p. 18
Cadillac Rubber and Plastics Inc, p. 408
Cadmium, p. 537
Cadmus Communications Corp, p. 263
Cagle's Inc, p. 20
Calcados Daiby Ltda, p. 172
Calcium carbide, p. 299
CalComp Technology Inc, p. 597
Calculating machines, pp. 589, 598
California Products Corp, p. 389
Callaway Golf Co, p. 719
CalMat Co, pp. 389, 480
Calzado Olympic SA, pp. 173, 408
Calzados Azaleia, p. 172
Cambodia
— beverages, pp. 46, 54
— chemical products nec, pp. 330, 338

Cambodia continued:
— electrical machinery, pp. 604, 612
— food products, pp. 6, 15
— footwear, pp. 162, 169
— furniture, fixtures, pp. 202, 210
— glass and products, pp. 448-449, 456
— industrial chemicals, pp. 268, 275
— industries nec, pp. 706, 708, 716
— iron and steel, p. 496
— leather and products, p. 153
— nonferrous metals, p. 522
— nonmetal products nec, pp. 468, 476
— paper and products, pp. 218, 220, 228
— petroleum refineries, pp. 361, 367
— petroleum, coal products, p. 386
— plastic products nec, pp. 416, 424
— pottery, china, etc., pp. 433, 440
— printing, publishing, pp. 248, 250, 259
— professional goods, pp. 687, 694
— profile, p. 738
— rubber products, pp. 394, 396, 403
— textiles, pp. 86, 94
— tobacco, pp. 68, 76
— transportation equipment, pp. 644, 652
— wearing apparel, pp. 124, 132
— wood products, pp. 182, 190
Cambrex Corp, p. 280
Camco International Inc, p. 573
Cameras, pp. 685, 701
Cameroon
— beverages, pp. 44, 46, 49, 51, 53-54
— chemical products nec, pp. 330, 335, 338
— electrical machinery, pp. 602, 604, 607-608, 610, 612
— food products, pp. 4, 6, 9, 11, 13, 15
— footwear, p. 169
— furniture, fixtures, pp. 200, 202, 205, 207-208, 210
— glass and products, pp. 449, 454, 456
— industrial chemicals, pp. 268, 272, 275
— industries nec, pp. 706, 708, 711, 713, 715-716
— iron and steel, pp. 489, 493, 496
— leather and products, p. 153
— machinery nec, pp. 567, 569
— metal products, pp. 548, 550
— nonferrous metals, pp. 515, 519, 522
— nonmetal products nec, pp. 468, 473, 476
— paper and products, pp. 218, 220, 223, 225-226, 228
— pottery, china, etc., pp. 433, 437, 440
— printing, publishing, pp. 248, 250, 253, 255, 257, 259
— professional goods, p. 694
— profile, p. 739
— textiles, pp. 84, 86, 89, 91-92, 94
— tobacco, pp. 66, 68, 71, 73-74, 76
— transportation equipment, pp. 642, 644, 647, 649-650, 652
— wearing apparel, pp. 122, 124, 127, 129-130, 132
— wood products, pp. 180, 182, 185, 187-188, 190
Camoplast Inc, pp. 408, 429

Campbell Soup Co. Ltd, p. 59
CamVec Corp, p. 428
Canada
— basic chemicals, excl fertilizers, pp. 312-316, 318
— beverages, pp. 44, 46, 48-49, 51, 53-54
— chemical products nec, pp. 328, 330, 332-333, 335-336, 338
— drugs and medicines, pp. 350-354, 356
— electrical machinery, pp. 602, 604, 606-608, 610, 612
— food products, pp. 4, 6, 8-9, 11, 13, 15
— footwear, pp. 160, 162-164, 166, 168-169
— furniture, fixtures, pp. 200, 202, 204-205, 207-208, 210
— glass and products, pp. 448, 450-452, 454-455, 457
— industrial chemicals, pp. 266, 268-270, 272, 274-275
— industries nec, pp. 706, 708, 710-711, 713, 716
— iron and steel, pp. 488-489, 491-493, 495-496
— leather and products, pp. 144, 146-148, 150-151, 153
— machinery nec, pp. 560-561, 563-564, 566-567, 569
— metal products, pp. 540, 542, 544-545, 547-548, 550
— motor vehicles, pp. 676-680, 682
— nonferrous metals, pp. 514-515, 517-520, 522
— nonmetal products nec, pp. 466, 468, 470-471, 473-474, 476
— office, computing machinery, pp. 590-594
— paper and products, pp. 218, 220, 222-223, 225-226, 228
— petroleum refineries, pp. 360-363, 365-367
— petroleum, coal products, pp. 380-381, 383-386
— plastic products nec, pp. 414, 416, 418-420, 422, 424
— pottery, china, etc., pp. 432-433, 436-438, 440
— printing, publishing, pp. 248, 250, 252-253, 255, 257, 259
— professional goods, pp. 686-687, 689-691, 693-694
— profile, p. 739
— pulp, paper, etc., pp. 240-245
— rubber products, pp. 394, 396-398, 400, 402-403
— shipbuilding, repair, pp. 668-673
— spinning, weaving, etc., pp. 112-116, 118
— synthetic resins, etc., pp. 320-323, 325
— textiles, pp. 84, 86, 88-89, 91-92, 94
— tobacco, pp. 66, 68, 70-71, 73, 76
— transportation equipment, pp. 642, 644, 646-647, 649-650, 652
— wearing apparel, pp. 122, 124, 126-127, 129-130, 132
— wood products, pp. 180, 182, 184-185, 187-188, 190
Canada Safeway Ltd, p. 18
Canadian Forest Products Ltd, pp. 193, 231
Canadian Occidental Petroleum Ltd, p. 370
Canandaigua Wine Company Inc, p. 58
Canberra Industries Inc, p. 596
Candle Corporation of America, p. 721
Canfor Corp, pp. 193, 231
Canron Inc, p. 429
Canroof Corp. Inc, p. 390
Cans, p. 556
Cantol Ltd, p. 390
Cape Building Products Ltd, p. 460
Cape Cobra Pty. Ltd, p. 157

Cape Verde
— beverages, pp. 44, 49, 53
— chemical products nec, pp. 328, 330, 333, 336, 338
— drugs and medicines, pp. 350-353, 356
— electrical machinery, p. 602
— food products, p. 4
— footwear, pp. 160, 162, 164, 168-169
— furniture, fixtures, p. 200
— glass and products, pp. 448, 450, 452
— industrial chemicals, p. 266
— industries nec, p. 706
— leather and products, p. 144
— metal products, pp. 540, 545, 548
— nonferrous metals, p. 514
— nonmetal products nec, p. 466
— pottery, china, etc., pp. 432-433, 436, 439
— printing, publishing, pp. 248, 250, 253, 257, 259
— professional goods, p. 686
— profile, p. 740
— shipbuilding, repair, pp. 668-671, 673
— spinning, weaving, etc., p. 112
— textiles, p. 84
— tobacco, pp. 66, 68, 70-71, 74, 76
— transportation equipment, pp. 642, 644, 646-647, 650, 652
— wearing apparel, pp. 122, 127, 130
— wood products, pp. 180, 185, 188
Capetronic USA (HK) Inc. Computer, p. 595
Capital Mercury Shirt Corp, p. 136
Caradon Bathrooms Ltd, p. 443
Caradon Everest Ltd, p. 460
Caraustar Industries Inc, p. 231
Carbon bisulfide, p. 294
Carbon black, pp. 327, 346
Carbon tetrachloride, p. 284
Carborundum Co, p. 479
Cargill Agricola SA, p. 19
Cargill Ltd, pp. 19, 279
Carhartt Inc, p. 136
Caribe General Electric, pp. 554, 617
Carless Refining & Marketing Ltd, p. 371
Carlisle Companies Inc, pp. 406, 427, 572
Carlsberg-Tetley UK Ltd, p. 60
Carlton Cards Ltd, p. 263
Carol Cable Company Inc, p. 526
Carolina Leaf Tobacco Company, p. 79
Carolina Mills Inc, pp. 98, 213
Carolina Shoe Co, p. 173
Carpenter Co, p. 427
Carpenter Steel Div, p. 500
Carpenter Technology Corp, p. 499
Carpet mills, p. 83
Carpets, pp. 83, 109, 111
Carriere Bernier Ltee, p. 390
Carron Phoenix Ltd, p. 443
Carter-Wallace Inc, pp. 279, 341
Carters Drinks Group Ltd, p. 60

Carters Gold Medal Soft Drinks, Ltd, p. 60
Carton de Colombia SA, p. 233
Cashway Building Centres Ltd, p. 194
Casks, p. 557
CasTech Aluminum Group Inc, p. 527
Castings, p. 487
Castle Cement Ltd, p. 480
Catalina, p. 137
Catalina Lighting Inc, p. 617
Catecu SA, p. 407
Caterpillar Belgium SA, p. 574
Caterpillar Paving Products Inc, p. 481
Caustic soda, p. 296
Cavalier Homes Inc, p. 194
Cawsl Corp, p. 499
CBA Cia. Brasileira de Aluminio, p. 525
CBR Brewing Company Inc, p. 59
CCL Industries Inc, pp. 231, 341, 553
CCU Compania Cervecerias Unidas, p. 58
CCX Inc, pp. 460, 527
Cellulosic continuous filaments, p. 309
Cellulosic fibers, p. 83
Cellulosic staple and tow, p. 304
Celulosa Argentina SA, p. 233
Cement, p. 483
Cementos de Honduras SA, p. 480
Cementos del Caribe SA, p. 481
Cementos del Nare SA, p. 481
Cementos Progreso SA, p. 480
Cementownia Ozarow SA, p. 480
Cements, p. 465
Cenex/Land O'Lakes Ag Services, p. 279
Centex Construction Products Inc, p. 480
Centex Corp, p. 479
Central African Republic
— basic chemicals, excl fertilizers, p. 312
— beverages, pp. 44, 46, 49, 51, 55
— chemical products nec, pp. 328, 330, 333, 335, 338
— electrical machinery, pp. 602, 604, 607-608, 612
— food products, pp. 4, 6, 9, 11, 15
— furniture, fixtures, pp. 200, 202, 205, 207, 210
— glass and products, p. 457
— industrial chemicals, pp. 266, 268, 271-272, 275
— industries nec, pp. 706, 708, 711, 713, 715-716
— iron and steel, p. 496
— machinery nec, p. 569
— metal products, pp. 540, 542, 545, 547, 550
— motor vehicles, p. 676
— nonferrous metals, p. 522
— nonmetal products nec, pp. 466, 476
— paper and products, p. 228
— petroleum refineries, p. 367
— petroleum, coal products, p. 386
— plastic products nec, pp. 414, 424
— pottery, china, etc., pp. 432, 440
— printing, publishing, pp. 248, 250, 253, 255, 259

Central African Republic continued:
— professional goods, p. 694
— profile, p. 740
— rubber products, pp. 394, 403
— textiles, pp. 84, 91-92
— tobacco, pp. 66, 68, 71, 73, 76
— transportation equipment, pp. 642, 644, 647, 649, 652
— wearing apparel, p. 122
— wood products, pp. 180, 182, 185, 187, 190
Central and peripheral storage units, p. 599
Central Bottling Co. Ltd, p. 59
Central-heating apparatus, p. 539
— non-electric, p. 558
Central Newspapers Inc, p. 262
Central Pre-Mix Concrete Co, p. 481
Central Sprinkler Corp, p. 428
Centrilift, p. 501
CentroTextil, pp. 136, 172
Century Aluminum Co, p. 526
Century Furniture Industries, p. 214
Century Lubricants Co, p. 389
Ceradyne Inc, p. 443
Cerart International Co., Ltd, p. 444
Cereal breakfast food, p. 37
Cereals, p. 3
Ceridian Corp, pp. 615, 697
Cerro Metal Products Co, p. 526
Cerveceria Hondurena SA, p. 58
Cerveceria y Malteria Paysandu SA, p. 60
Cerveceria Y Malteria Quilmes SA ic, p. 58
CF and I Steel Corp, p. 500
C.F. Sauer Company Inc, p. 60
CFC Holdings Corp, p. 58
Chad
— beverages, pp. 44, 55
— chemical products nec, p. 328
— drugs and medicines, p. 350
— food products, pp. 4, 15
— glass and products, p. 457
— nonmetal products nec, p. 476
— paper and products, p. 228
— pottery, china, etc., p. 440
— printing, publishing, pp. 248, 259
— textiles, p. 84
— tobacco, pp. 66, 76
Champion Laboratories Inc, p. 370
Champion Papel Celulose Ltda, p. 280
Champion Products Inc., p. 135
Champion Spark Plug Company Div, p. 616
Chaparral Steel Co, p. 501
Chardon Rubber Co, p. 408
Charles Wells Ltd, p. 59
Chase Brass Industries Inc, p. 527
Cheese, pp. 3, 27
Cheil Glass Ind. Co., Ltd, p. 461
Cheler Corp, p. 264

Chemed Corp, p. 697
Chemicals, pp. 265, 311
Chemlite Holdings Pty. Ltd, p. 371
Chemlon AS, p. 278
Cheng Loong Co. Ltd, p. 232
Cherry Corp, pp. 595, 616
Chesapeake Corp, pp. 213, 231
Chesterfield Manufacturing Corp, p. 99
Chevron Canada Ltd, p. 371
Chevron Chemical Co, p. 279
Chevron U.S.A. Products Company, p. 370
Chewing tobacco, p. 65
Chic By H.I.S. Inc, p. 135
Chicago Bridge and Iron Co, p. 553
Chicago Miniature Lamp Inc, p. 615
Chicago Raw Hide Manufacturing, p. 406
Chickens, p. 3
Chief Industries Inc, p. 195
Child Craft Industries Inc, p. 215
Children's clothing, p. 121
Children's vehicles, p. 641
Chile
—basic chemicals, excl fertilizers, pp. 312-316, 318
—beverages, pp. 44, 46, 48-49, 51, 53, 55
—chemical products nec, pp. 328, 330, 332-333, 335-336,
   338
—drugs and medicines, pp. 350-354, 356
—electrical machinery, pp. 602, 604, 606-608, 610, 612
—food products, pp. 4, 6, 8-9, 11, 13, 15
—footwear, pp. 160, 162-164, 166, 168-169
—furniture, fixtures, pp. 200, 202, 204-205, 207-208, 210
—glass and products, pp. 448, 450-452, 454-455, 457
—industrial chemicals, pp. 266, 268-269, 271-272, 274-275
—industries nec, pp. 706, 708, 710-711, 713, 715-716
—iron and steel, pp. 488-489, 491-493, 495-496
—leather and products, pp. 144, 146-148, 150-151, 153
—machinery nec, pp. 560-561, 563-564, 566-567, 569
—metal products, pp. 540, 542, 544-545, 547-548, 550
—motor vehicles, pp. 676-680, 682
—nonferrous metals, pp. 514-515, 517-520, 522
—nonmetal products nec, pp. 466, 468, 470-471, 473-474,
   476
—paper and products, pp. 218, 220, 222-223, 225-226, 228
—petroleum refineries, pp. 360-363, 365-367
—petroleum, coal products, pp. 380-386
—plastic products nec, pp. 414, 416, 418-420, 422, 424
—pottery, china, etc., pp. 432-433, 435-437, 439-440
—printing, publishing, pp. 248, 250, 252-253, 255, 257, 259
—professional goods, pp. 686-687, 689-691, 693-694
—profile, p. 741
—pulp, paper, etc., pp. 240-245
—radio, television, etc., pp. 628-633
—rubber products, pp. 394, 396-398, 400, 402-403
—shipbuilding, repair, pp. 668-673
—spinning, weaving, etc., pp. 112-116, 118
—synthetic resins, etc., pp. 320-323, 325

Chile continued:
—textiles, pp. 84, 86, 88-89, 91-92, 94
—tobacco, pp. 66, 68, 70-71, 73-74, 76
—transportation equipment, pp. 642, 644, 646-647, 649-650,
   652
—wearing apparel, pp. 122, 124, 126-127, 129-130, 132
—wood products, pp. 180, 182, 184-185, 187-188, 190
China
—basic chemicals, excl fertilizers, p. 312
—beverages, pp. 44, 46, 49, 51, 55
—chemical products nec, pp. 328, 330, 333, 335, 338
—electrical machinery, pp. 602, 604, 607-608, 612
—food products, pp. 4, 6, 9, 11, 15
—furniture, fixtures, pp. 200, 202, 205, 207, 210
—glass and products, pp. 448, 450, 452, 454, 457
—industrial chemicals, pp. 266, 268, 271-272, 275
—industries nec, pp. 706, 708, 711, 713, 716
—iron and steel, pp. 488-489, 492-493, 496
—leather and products, pp. 144, 146, 148, 150, 153
—machinery nec, pp. 560-561, 564, 566, 569
—metal products, pp. 540, 542, 545, 547, 550
—nonferrous metals, pp. 514-515, 518-519, 522
—nonmetal products nec, pp. 466, 468, 471, 473, 476
—paper and products, pp. 218, 220, 223, 225, 228
—petroleum refineries, pp. 360-361, 363, 365, 367
—petroleum, coal products, pp. 380-381, 383-384, 386
—plastic products nec, pp. 414, 416, 419-420, 424
—pottery, china, etc., pp. 432-433, 436-437, 440
—printing, publishing, pp. 248, 250, 253, 255, 259
—professional goods, pp. 686-687, 690-691, 694
—profile, p. 741
—pulp, paper, etc., p. 240
—rubber products, pp. 394, 396, 398, 400, 403
—shipbuilding, repair, p. 669
—textiles, pp. 84, 86, 89, 91, 94
—tobacco, pp. 66, 68, 71, 73, 76
—transportation equipment, pp. 642, 644, 647, 649, 652
—wood products, pp. 180, 182, 185, 187, 190
China (Hong Kong SAR)
—basic chemicals, excl fertilizers, pp. 312-313, 315-318
—beverages, pp. 44, 46, 49, 51, 53, 55
—chemical products nec, pp. 328, 330, 333, 335-336, 338
—drugs and medicines, pp. 350-351, 353-356
—electrical machinery, pp. 602, 604, 607-608, 610, 612
—food products, pp. 4, 6, 9, 11, 13, 15
—footwear, pp. 160, 162, 164, 166, 168-169
—furniture, fixtures, pp. 200, 202, 205, 207-208, 210
—glass and products, pp. 448, 450, 452, 454-455, 457
—industrial chemicals, pp. 266, 268, 271-272, 274-275
—industries nec, pp. 706, 708, 711, 713, 715-716
—iron and steel, pp. 488-489, 492-493, 495-496
—leather and products, pp. 144, 146, 148, 150-151, 153
—machinery nec, pp. 560-561, 564, 566-567, 569
—metal products, pp. 540, 542, 545, 547-548, 550
—motor vehicles, pp. 676-677, 679-682
—nonferrous metals, pp. 514-515, 518-520, 522

China (Hong Kong SAR) continued:
—nonmetal products nec, pp. 466, 468, 471, 473-474, 476
—office, computing machinery, pp. 590, 592-594
—paper and products, pp. 218, 220, 223, 225-226, 228
—petroleum refineries, pp. 360, 367
—petroleum, coal products, pp. 380-381, 383-386
—plastic products nec, pp. 414, 416, 419-420, 422, 424
—pottery, china, etc., pp. 432-433, 436-437, 439-440
—printing, publishing, pp. 248, 250, 253, 255, 257, 259
—professional goods, pp. 686-687, 690-691, 693-694
—profile, p. 742
—pulp, paper, etc., pp. 240-241, 243-245
—radio, television, etc., pp. 628-629, 631-633
—rubber products, pp. 394, 396, 398, 400, 402-403
—shipbuilding, repair, pp. 668-669, 671-673
—spinning, weaving, etc., pp. 112-113, 115-118
—synthetic resins, etc., pp. 320-325
—textiles, pp. 84, 86, 89, 91-92, 94
—tobacco, pp. 66, 68, 71, 73-74, 76
—transportation equipment, pp. 642, 644, 647, 649-650, 652
—wearing apparel, pp. 122, 124, 127, 129-130, 132
—wood products, pp. 180, 182, 185, 187-188, 190
China (Taiwan Province)
—beverages, pp. 46, 48-49, 51, 55
—chemical products nec, pp. 330, 332-333, 335, 338
—electrical machinery, pp. 604, 606-608, 612
—food products, pp. 6, 8-9, 11, 15
—footwear, pp. 162-164, 166, 169
—furniture, fixtures, pp. 202, 204-205, 207, 210
—glass and products, pp. 450-452, 454, 457
—industrial chemicals, pp. 268-269, 271-272, 275
—industries nec, pp. 708, 710-711, 713, 716
—iron and steel, pp. 489, 491-492, 494, 496
—leather and products, pp. 146-148, 150, 153
—machinery nec, pp. 562-564, 566, 569
—metal products, pp. 542, 544-545, 547, 550
—nonferrous metals, pp. 515, 517-519, 522
—nonmetal products nec, pp. 468, 470-471, 473, 476
—paper and products, pp. 220, 222-223, 225, 228
—petroleum refineries, pp. 361-363, 365, 367
—petroleum, coal products, pp. 381-384, 386
—plastic products nec, pp. 416, 418-420, 424
—pottery, china, etc., pp. 433, 435-437, 440
—printing, publishing, pp. 250, 252-253, 255, 259
—professional goods, pp. 687, 689-691, 694
—profile, p. 742
—rubber products, pp. 396-398, 400, 403
—textiles, pp. 86, 88-89, 91, 94
—tobacco, pp. 68, 70-71, 73, 76
—transportation equipment, pp. 644, 646-647, 649, 652
—wearing apparel, pp. 124, 126-127, 129, 132
—wood products, pp. 182, 184-185, 187, 190
China (products), p. 431
Chiquita Processed Food, LLC, p. 20
Chiron Diagnostics Corp, p. 698
Chlorine, p. 292

Cho Kwang Leather Co., Ltd, p. 156
Chocolate and chocolate products, p. 41
Chocolates Garoto SA, p. 20
Chomerics Inc, p. 408
Chonbang Co., Ltd, p. 98
Chong Kun Dang Corp, p. 343
Choongwae Pharma Corp, p. 343
Chris-Craft Industries Inc, p. 429
Chromcraft Revington Inc, p. 213
Chrome, p. 513
Chrysler Canada Ltd, p. 655
Chun Yuan Steel Industry Co., Ltd, p. 501
Chung Hsin Electric & Machinery, p. 596
Chunghwa Picture Tubes Ltd, p. 616
Chunmi Lubricants Co., Ltd, p. 371
Church & Co. PLC, p. 172
Ciadea SA, p. 656
Ciba-Geigy PLC, p. 341
Cie. Franco-Tunisienne des Petroles, p. 371
Ciena Corp, p. 527
Cigarette paper, p. 237
Cigarettes, pp. 65, 81
Cigars, pp. 65, 81
Cincinnati Milacron UK Ltd, p. 174
Cinkarna - Metalursko Kemicna, p. 720
Cipla Ltd, pp. 280, 342
Cisper Da Amazonia SA, p. 461
Citation Corp, p. 499
CITGO Asphalt Refining Co, p. 371
CITGO Petroleum Corp, p. 370
Citrosuco Paulista SA, p. 20
CJC Holdings Inc, p. 721
Clarcor Inc, pp. 232, 479, 657
Clarin AGEAS.A, p. 262
Clark Equipment Co, pp. 573, 656
Clark-Hurth Components, p. 656
Clarson Enterprises, Inc, p. 215
Clayton Homes Inc, p. 193
Cleansers, p. 327
Cleo Inc, p. 233
Cleveland Chair Co, p. 214
C.L.O. Glass Ltd, p. 461
Clocks, pp. 685, 702-703
Clopay Corp, pp. 193, 427, 555
Clorox Co, pp. 18-19, 58, 279, 341, 573
Clothing, p. 121
Cluett, Peabody and Company Inc, p. 135
CMC Industries Inc, p. 597
CML Group Inc, p. 719
Coachmen Industries Inc, pp. 193, 656
Coal briquettes, p. 379
Coal products, p. 379
Coalite Products Ltd, p. 370
Coastal Lumber Co, p. 194
Coastal Refining and Marketing Inc, p. 371
Coastcast Corp, pp. 698, 719

Coats, pp. 121, 140
Coats Crafts North America, pp. 99, 720
Coca-Cola & Schweppes Beverages, p. 58
Coca-Cola Beverages Ltd, p. 58
Coca-Cola Bottlers Ulster Ltd, p. 60
Coca-Cola Bottling Company, p. 58
Coca Cola de Argentina SA, p. 58
Coca Cola-Femsa, p. 58
Cocoa, pp. 3, 40
Codan Gummi AS, p. 408
COFAP Cia. Fabricadora de Pecas, p. 655
Coffee, pp. 3, 41
Coherent Inc, p. 699
Coilcraft Inc, p. 617
Coke, pp. 379, 391
Coking gases, p. 379
Cold Spring Granite Co, p. 480
Cold Storage (Malaysia) BHD, pp. 59, 501
Coleman Company Inc, pp. 135, 427, 616
Collins and Aikman Plastics Inc, p. 555
Colombia
— basic chemicals, excl fertilizers, pp. 312-318
— beverages, pp. 44, 46, 48-49, 51, 53, 55
— chemical products nec, pp. 328, 330, 332-333, 335-336,
　　338
— drugs and medicines, pp. 350-356
— electrical machinery, pp. 602, 604, 606-608, 610, 612
— food products, pp. 4, 6, 8-9, 11, 13, 15
— footwear, pp. 160, 162-164, 166, 168-169
— furniture, fixtures, pp. 200, 202, 204-205, 207-208, 210
— glass and products, pp. 448, 450-452, 454-455, 457
— industrial chemicals, pp. 266, 268-269, 271-272, 274-275
— industries nec, pp. 706, 708, 710-711, 713, 715-716
— iron and steel, pp. 488-489, 491-492, 494-496
— leather and products, pp. 144, 146-148, 150-151, 153
— machinery nec, pp. 560, 562-564, 566-567, 569
— metal products, pp. 540, 542, 544-545, 547-548, 550
— motor vehicles, pp. 676-682
— nonferrous metals, pp. 514-515, 517-520, 522
— nonmetal products nec, pp. 466, 468, 470-471, 473-474,
　　476
— office, computing machinery, pp. 590-594
— paper and products, pp. 218, 220, 222-223, 225-226, 228
— petroleum refineries, pp. 360-363, 365-367
— petroleum, coal products, pp. 380-386
— plastic products nec, pp. 414, 416, 418-420, 422, 424
— pottery, china, etc., pp. 432-433, 435-437, 439-440
— printing, publishing, pp. 248, 250, 252-253, 255, 257, 259
— professional goods, pp. 686-687, 689-691, 693-694
— profile, p. 743
— pulp, paper, etc., pp. 240-245
— radio, television, etc., pp. 628-633
— rubber products, pp. 394, 396-398, 400, 402-403
— shipbuilding, repair, pp. 668-673
— spinning, weaving, etc., pp. 112-118
— synthetic resins, etc., pp. 320-325

Colombia continued:
— textiles, pp. 84, 86, 88-89, 91-92, 94
— tobacco, pp. 66, 68, 70-71, 73-74, 76
— transportation equipment, pp. 642, 644, 646-647, 649-650,
　　652
— wearing apparel, pp. 122, 124, 126-127, 129-130, 132
— wood products, pp. 180, 182, 184-185, 187-188, 190
Colombiana de Comercio Ltda, p. 721
Colony Glass, p. 461
Coltec Industries Inc, pp. 572, 655
Columbus McKinnon Corp, pp. 499, 554, 574
Comar Inc, p. 429
Combine harvester-threshers, p. 578
Cominco Ltd, pp. 278, 499, 525
Commerce Clearing House Inc, p. 262
Commercial Intertech Corp, pp. 554, 574
Commercial Metals Co, p. 525
Commercial passenger and cargo planes, p. 664
Commodore Corp, p. 195
Commonwealth Aluminum Corp, p. 526
Commonwealth Plywood Co. Ltd, p. 194
CommScope Inc, p. 526
Communications Quebecor Inc, p. 262
Companhia Antarctica Paulista, p. 58
Compania Siderurgica Huachipato, p. 499
Companias CIC SA, p. 214
Complexe Textile de Bujumbura, p. 99
Compressed gas cylinders, p. 557
Compressors, p. 620
Computalog Ltd, p. 597
Computer peripherals, p. 589
Computer Products Inc, p. 596
Computers, pp. 559, 589
Comverse Technology Inc, p. 596
Conaprole Cooperativa Nacional de, p. 20
Conchatoro Vina Concha y Toro SA, p. 60
Concrete, pp. 485-486
Concrete blocks and bricks, p. 484
Concrete mixers for use at construction sites, p. 587
Concrete pipe, p. 485
Conde Nast Publications Ltd, p. 264
Condere Corp, p. 408
Condiments, p. 3
Cone Mills Corp, p. 97
Conestoga Wood Specialties Inc, p. 195
Confecciones Colombia SA, p. 137
Confortluxe, p. 215
Congo
— beverages, p. 55
— food products, p. 15
— footwear, p. 169
— industries nec, p. 717
— iron and steel, p. 496
— nonferrous metals, p. 522
— paper and products, p. 228
— printing, publishing, p. 259

Congo continued:
— professional goods, p. 694
— tobacco, p. 76
— transportation equipment, p. 652
Congoleum Corp, p. 429
Coniferous sawn wood, p. 179
CONMED Corp, p. 699
Conoco UK Ltd, p. 370
Consani Engineering Pty. Ltd, p. 501
Consol Glass, p. 459
Consolidated Cigar Corp, p. 79
Consolidated Fiber Glass Products, p. 389
Consolidated Furniture Corp, p. 213
Consolidated Graphics Inc, p. 264
Consolidated Metco Inc, p. 527
Consolidated Papers Inc, p. 231
Consoltex Inc, p. 97
Consumers' Co-Operative Refineries, p. 371
Consumers Packaging Inc, p. 459
Containers, p. 558
Contessa-Skogar, pp. 173, 408
Contico International Inc, p. 428
Continental Can Company Inc, pp. 232, 427
Continental China Pty. Ltd, p. 443
Continental General Tire Inc, p. 406
Continental Materials Corp, p. 481
Continental UK Group Holdings Ltd, p. 406
Control Data Systems Inc, p. 596
Convenience Foods Ltd, p. 18
Converse Inc, pp. 136, 172
Converted paper products, p. 217
Cookson America Inc, pp. 479, 525
Coopcam UK Ltd, pp. 554, 574
Cooper Cameron Corp, pp. 553, 572
Cooper Companies Inc, p. 343
Cooper Lighting, p. 616
Cooper Oil Tool Div, pp. 553, 573
Cooper Power Systems, p. 616
Cooper Tire and Rubber Co, p. 406
Coordinated Apparel Inc, p. 136
Coors Ceramics Co, pp. 443, 480
Copeland Corp, p. 573
Copiers, p. 559
Copper, pp. 513, 528-529
Copper bars
— rods, angles, shapes, sections, p. 529
Copper-base alloys, p. 529
Copper plates
— sheets, strip, foil, p. 530
Copper sulfate, p. 297
Copper tubes and pipes, p. 530
Copper wire, p. 530
Copperweld Corp, p. 501
Corah PLC, p. 137
Cordage, pp. 83, 110-111
Cordis Corp, p. 699

Corel Corp, p. 597
Corhart Refractories Corp, p. 481
Corn Products International Inc, p. 19
Corning Consumers Products Co, p. 460
Corning Ltd, p. 460
Corobrik Pty. Ltd, p. 479
Coronet Industries Inc, p. 99
Cosco Inc, p. 214
Cosipa Cia. Siderurgica Paulista, p. 499
Costa Rica
— basic chemicals, excl fertilizers, pp. 312-313, 315-316, 318
— beverages, pp. 44, 46, 49, 51, 55
— chemical products nec, pp. 328, 330, 333, 335, 338
— drugs and medicines, pp. 350-351, 353-354, 356
— electrical machinery, pp. 602, 604, 607, 609, 612
— food products, pp. 4, 6, 9, 11, 15
— footwear, pp. 160, 162, 164, 166, 169
— furniture, fixtures, pp. 200, 202, 205, 207, 210
— glass and products, pp. 448, 450, 452, 454, 457
— industrial chemicals, pp. 266, 268, 271-272, 275
— industries nec, pp. 706, 708, 711, 713, 717
— iron and steel, pp. 488-489, 496
— leather and products, pp. 144, 146, 148, 150, 153
— machinery nec, pp. 560, 562, 564, 566, 569
— metal products, pp. 540, 542, 545, 547, 550
— motor vehicles, pp. 676-677, 679-680, 682
— nonferrous metals, pp. 514-515, 518-519, 522
— nonmetal products nec, pp. 466, 468, 471, 473, 476
— office, computing machinery, pp. 590-591, 594
— paper and products, pp. 218, 220, 223, 225, 228
— petroleum refineries, pp. 360-361, 363, 365, 367
— petroleum, coal products, pp. 380-381, 383-384, 386
— plastic products nec, pp. 414, 416, 419-420, 424
— pottery, china, etc., pp. 432-433, 436-437, 440
— printing, publishing, pp. 248, 250, 253, 255, 259
— professional goods, pp. 686-687, 694
— profile, p. 743
— pulp, paper, etc., pp. 240-241, 245
— radio, television, etc., pp. 628-629, 631-633
— rubber products, pp. 394, 396, 398, 400, 403
— shipbuilding, repair, pp. 668-669, 671-673
— spinning, weaving, etc., pp. 112-113, 115-116, 118
— synthetic resins, etc., pp. 320-323, 325
— textiles, pp. 84, 86, 89, 91, 94
— tobacco, pp. 66, 68, 71, 73, 76
— transportation equipment, pp. 642, 644, 647, 649, 652
— wearing apparel, pp. 122, 124, 127, 129, 132
— wood products, pp. 180, 182, 185, 187, 190
Cote d'Ivoire
— beverages, pp. 49, 51, 55
— chemical products nec, pp. 333, 335, 338
— electrical machinery, pp. 609-610, 612
— food products, p. 15
— footwear, pp. 164, 166
— furniture, fixtures, p. 210

Cote d'Ivoire continued:
— glass and products, pp. 452, 454, 457
— industrial chemicals, pp. 272, 275
— industries nec, pp. 713, 717
— leather and products, p. 150
— machinery nec, pp. 566, 569
— metal products, pp. 547-548, 550
— nonferrous metals, pp. 518-519
— nonmetal products nec, pp. 471, 473, 476
— office, computing machinery, p. 593
— petroleum refineries, pp. 363, 365
— petroleum, coal products, pp. 383-384
— plastic products nec, p. 424
— pottery, china, etc., p. 440
— printing, publishing, p. 259
— professional goods, pp. 691, 694
— profile, p. 744
— rubber products, pp. 398, 400, 403
— spinning, weaving, etc., pp. 115-116
— textiles, pp. 89, 91, 94
— tobacco, p. 76
— transportation equipment, pp. 647, 649-650, 652
— wearing apparel, p. 132
— wood products, pp. 185, 187-188, 190
Cotonifico Guilherme Giorgi SA, p. 99
Cott Corp, p. 59
Cotton, p. 83
Cotton linters, p. 100
Cotton Textile Mills, Ltd, p. 99
Cotton woven fabrics, p. 103
Cotton yarn, pp. 101-102
Courage Ltd, p. 58
Courier Corp, p. 264
Courtaulds Chemicals Holdings Ltd, p. 280
Courtaulds Films Holdings Ltd, p. 429
Courtaulds Packaging Ltd, p. 429
Cowtown Boot Company Inc, p. 174
C.R. Bard Inc, p. 697
C.R. Smith Glaziers Dunfermline Ltd, p. 527
Craftco Industries Inc, p. 444
Crain-Drummond Inc, p. 264
Crane Co, pp. 553, 572, 615
Crane Ltd, p. 501
Crane Valves Nuclear Operation, p. 555
Cranes, p. 620
Cray Research Inc, p. 596
Cream, p. 3
Creation Windows Inc, p. 460
Crestbrook Forest Industries Ltd, p. 195
Cristaleria Peldar S.A, p. 459
Croatia
— beverages, pp. 44, 46, 48-49, 51, 53, 55
— chemical products nec, pp. 328, 330, 332-333, 335-336, 338
— electrical machinery, pp. 602, 604, 606-607, 609-610, 612
— food products, pp. 4, 6, 8-9, 11, 13, 15

Croatia continued:
— footwear, pp. 160, 162-163, 165-166, 168-169
— furniture, fixtures, pp. 200, 202, 204-205, 207-208, 210
— glass and products, pp. 448, 450-452, 454-455, 457
— industrial chemicals, pp. 266, 268, 270-272, 274-275
— industries nec, pp. 706, 708, 710-711, 713, 715, 717
— iron and steel, pp. 488-489, 491-492, 494-496
— leather and products, pp. 144, 146-148, 150, 152-153
— machinery nec, pp. 560, 562-564, 566-567, 569
— metal products, pp. 540, 542, 544-545, 547-548, 550
— nonferrous metals, pp. 514-515, 517-520, 522
— nonmetal products nec, pp. 466, 468, 470-471, 473-474, 476
— paper and products, pp. 218, 220, 222-223, 225, 227-22
— petroleum refineries, pp. 360-361, 363, 365-367
— petroleum, coal products, pp. 380-386
— plastic products nec, pp. 414, 416, 418-419, 421-422, 42
— pottery, china, etc., pp. 432, 434-437, 439-440
— printing, publishing, pp. 248, 250, 252-253, 255, 257, 25
— professional goods, pp. 686-687, 689-691, 693-694
— profile, p. 744
— rubber products, pp. 394, 396-398, 400, 402-403
— textiles, pp. 84, 86, 88-89, 91-92, 94
— tobacco, pp. 66, 68, 70-71, 73-74, 76
— transportation equipment, pp. 642, 644, 646-647, 649-65 652
— wearing apparel, pp. 122, 124, 126-127, 129-130, 132
— wood products, pp. 180, 182, 184-185, 187-188, 190
Crompton and Knowles Corp, pp. 18, 278, 341, 573
Cross Oil Refining and Marketing, pp. 371, 390
Crotty Corp, p. 408
Crouse-Hinds Div. als, p. 617
Crown Berger Ltd, p. 280
Crown Central Petroleum Corp, p. 370
Crown Corp. Ltd, pp. 18, 58
Crown Crafts Inc, pp. 99, 136
Crown Equipment Corp, p. 573
Crown Footwear Pty. Ltd, p. 173
Crown Packaging Ltd, p. 233
Crown Vantage Inc, p. 232
Crucibles, p. 465
Crude steel
— for castings, p. 504
— ingots, p. 505
Crystal Clear Industries Inc, p. 460
CSC Industries Inc, p. 501
CSC Ltd, p. 501
Csepel Muvek Femmu, p. 526
CSS Industries Inc, pp. 232, 262, 554
CTS Corp, pp. 617, 657
Cuba
— beverages, p. 55
— electrical machinery, p. 612
— food products, p. 15
— footwear, p. 169
— furniture, fixtures, p. 210

Cuba continued:
— glass and products, p. 457
— industries nec, p. 717
— iron and steel, p. 496
— leather and products, p. 153
— metal products, p. 550
— nonferrous metals, p. 522
— nonmetal products nec, p. 476
— paper and products, p. 228
— pottery, china, etc., p. 440
— printing, publishing, p. 259
— textiles, p. 94
— tobacco, p. 76
— wearing apparel, p. 132
— wood products, p. 190
Cubic Corp, pp. 232, 574, 698
Cudahy Tanning Company Inc, p. 156
Culinar Inc., Culinar Canada Division, p. 20
Culp Inc, p. 98
Cultivators, p. 576
Curtice-Burns Foods Inc, p. 19
Curtidos San Luis SA, p. 174
Curtiembre Becas Sca, p. 156
Curtis 1000 Inc, pp. 233, 264
Curtiss-Wright Corp, p. 526
Curtitagui Curtiembres de Itagui, p. 156
Curtume Aimore SA, p. 156
Curtume Alianca SA, p. 156
Curtume Scuck SA, p. 174
Cut stock, p. 143
Cut stone, p. 465
CV Samson, p. 444
Cyclops Industries Inc, p. 499
Cydsa SA y Subsidiarias, pp. 97, 278
Cygne Designs Inc, p. 135
Cyprus
— basic chemicals, excl fertilizers, pp. 312-318
— beverages, pp. 44, 46, 48-49, 51, 53, 55
— chemical products nec, pp. 328, 330, 332-333, 335-336, 338
— drugs and medicines, pp. 350-356
— electrical machinery, pp. 602, 604, 606-607, 609-610, 612
— food products, pp. 4, 6, 8-9, 11, 13, 15
— footwear, pp. 160, 162-163, 165-166, 168-169
— furniture, fixtures, pp. 200, 202, 204-205, 207-208, 210
— glass and products, pp. 448, 450-452, 454-455, 457
— industrial chemicals, pp. 266, 268, 270-272, 274-275
— industries nec, pp. 706, 708, 710-711, 713, 715, 717
— iron and steel, p. 496
— leather and products, pp. 144, 146-148, 150, 152-153
— machinery nec, pp. 560, 562-564, 566-567, 569
— metal products, pp. 540, 542, 544-545, 547, 549-550
— motor vehicles, pp. 676-682
— nonferrous metals, p. 522
— nonmetal products nec, pp. 466, 468, 470-471, 473-474, 476

Cyprus continued:
— paper and products, pp. 218, 220, 222-223, 225, 227-228
— petroleum refineries, pp. 360-361, 363, 365-367
— petroleum, coal products, p. 386
— plastic products nec, pp. 414, 416, 418-419, 421-422, 424
— pottery, china, etc., pp. 432, 434-437, 439-440
— printing, publishing, pp. 248, 250, 252-253, 255, 257, 259
— professional goods, pp. 686-687, 689-691, 693-694
— profile, p. 745
— radio, television, etc., pp. 628-633
— rubber products, pp. 394, 396-398, 400, 402-403
— shipbuilding, repair, pp. 668-674
— spinning, weaving, etc., pp. 112-118
— textiles, pp. 84, 86, 88-89, 91, 93-94
— tobacco, pp. 66, 68, 70-71, 73-74, 76
— transportation equipment, pp. 642, 644, 646-647, 649-650, 652
— wearing apparel, pp. 122, 124, 126-127, 129-130, 132
— wood products, pp. 180, 182, 184-185, 187-188, 190
Cytec Industries Inc, p. 278
Czech Republic
— beverages, pp. 44, 46, 48, 55
— chemical products nec, pp. 328, 330, 332, 338
— electrical machinery, pp. 602, 604, 606, 612
— food products, pp. 4, 6, 8, 15
— footwear, pp. 160, 162-163, 169
— furniture, fixtures, pp. 200, 202, 204, 210
— glass and products, pp. 448, 450-451, 457
— industrial chemicals, pp. 266, 268, 270, 276
— industries nec, pp. 706, 708, 710, 717
— iron and steel, pp. 488, 490-491, 497
— leather and products, pp. 144, 146-147, 153
— machinery nec, pp. 560, 562-563, 569
— metal products, pp. 540, 542, 544, 550
— nonferrous metals, pp. 514-515, 517, 522
— nonmetal products nec, pp. 466, 468, 470, 476
— paper and products, pp. 218, 220, 222, 228
— petroleum refineries, pp. 360-361, 363, 367
— petroleum, coal products, pp. 380-382, 386
— plastic products nec, pp. 414, 416, 418, 424
— pottery, china, etc., pp. 432, 434-435, 440
— printing, publishing, pp. 248, 250, 252, 259
— professional goods, pp. 686-687, 689, 694
— profile, p. 745
— rubber products, pp. 394, 396-397, 403
— textiles, pp. 84, 86, 88, 94
— tobacco, pp. 66, 68, 70, 76
— transportation equipment, pp. 642, 644, 646, 652
— wearing apparel, pp. 122, 124, 126, 132
— wood products, pp. 180, 182, 184, 190
Czechoslovakia (Former)
— beverages, pp. 44, 46, 48-49, 51, 53, 55
— chemical products nec, pp. 328, 330, 332-333, 335, 337-338
— electrical machinery, pp. 602, 604, 606-607, 609-610, 612
— food products, pp. 4, 6, 8-9, 11, 13, 15

Czechoslovakia (Former) continued:
—footwear, pp. 160, 162-163, 165-166, 168-169
—furniture, fixtures, pp. 200, 202, 204-205, 207-208, 210
—glass and products, pp. 448, 450-452, 454-455, 457
—industrial chemicals, pp. 266, 268, 270-272, 274, 276
—industries nec, pp. 706, 708, 710-711, 713, 715, 717
—iron and steel, pp. 488-489, 491-492, 494-495, 497
—leather and products, pp. 144, 146-148, 150, 152-153
—machinery nec, pp. 560, 562-564, 566-567, 569
—metal products, pp. 540, 542, 544-545, 547, 549-550
—nonferrous metals, pp. 514-515, 517-520, 522
—nonmetal products nec, pp. 466, 468, 470-471, 473-474, 476
—paper and products, pp. 218, 220, 222-223, 225, 227-228
—petroleum refineries, pp. 360-361, 363, 365-367
—petroleum, coal products, pp. 380-386
—plastic products nec, pp. 414, 416, 418-419, 421-422, 424
—pottery, china, etc., pp. 432, 434-437, 439-440
—printing, publishing, pp. 248, 250, 252-253, 255, 257, 259
—professional goods, pp. 686-687, 689-691, 693-694
—rubber products, pp. 394, 396-397, 399-400, 402-403
—textiles, pp. 84, 86, 88-89, 91, 93-94
—tobacco, pp. 66, 68, 70-71, 73-74, 76
—transportation equipment, pp. 642, 644, 646-647, 649-650, 652
—wearing apparel, pp. 122, 124, 126-127, 129-130, 132
—wood products, pp. 180, 182, 184-185, 187, 189-190
D. Crupi & Sons Ltd, p. 389
Da-E Trading Co., Ltd, p. 157
Dae Shin Trading Co., Ltd, p. 173
Dae Sung Corp, p. 174
Dae Won Kang-Up Co., Ltd, p. 500
Dae Woo Leather Industrial Co., Ltd, p. 157
Daelim Trading Co., Ltd, p. 501
Daeseong Co., Ltd, p. 461
DAF NV, p. 655
Dai Nippon Printing Indonesia PT, p. 264
Dainong Corp, p. 97
Dairy products, p. 3
Daishowa Inc, p. 233
Daks Simpson Group PLC, p. 136
Daks-Simpson Ltd, p. 136
Daktronics Inc, p. 721
Dal-Tile International Inc, p. 479
Dalgety Spillers Foods Ltd, p. 19
Dan River Inc, p. 97
Dana Corp. Plumley Div, p. 407
Danapak AMBA, pp. 233, 428
Daniel Montgomery & Son Ltd, p. 443
Danish Crown Amba, p. 18
Danish Steel Works Ltd, p. 501
Dannimac Ltd, p. 137
Darby Group PLC, p. 460
Dart Container Corp, p. 427
Data Business Forms Ltd, p. 264
Data Documents Inc, p. 264

Data General Corp, p. 595
DataCard Corp, pp. 427, 595
Datapoint Corp, p. 597
David S. Smith Packaging Ltd, p. 233
David Whitehead Textiles Ltd, pp. 97, 135
Day Runner Inc, p. 264
Day-Timers Inc, p. 720
D.C. Thomson & Co., Ltd, p. 263
De la Rue PLC, p. 262
Dead Sea Works Ltd, p. 280
Decorel Inc, p. 195
Del Laboratories Inc, p. 343
Delami PT, p. 135
Deloro Stellite Inc, p. 526
Delta Galil Industries Ltd, pp. 97, 135
Delta Motor Corp. - Export Div, p. 656
Delta Tanning Corp, p. 157
Delta Woodside Industries Inc, pp. 97, 135
Democratic Republic of the Congo
—beverages, p. 55
—chemical products nec, p. 338
—electrical machinery, p. 612
—food products, p. 15
—footwear, p. 169
—furniture, fixtures, p. 210
—glass and products, p. 457
—industrial chemicals, p. 276
—industries nec, p. 717
—iron and steel, p. 497
—leather and products, p. 153
—machinery nec, p. 569
—metal products, p. 550
—nonferrous metals, p. 522
—nonmetal products nec, p. 476
—paper and products, p. 228
—petroleum refineries, p. 367
—petroleum, coal products, p. 386
—plastic products nec, p. 424
—pottery, china, etc., p. 440
—printing, publishing, p. 259
—professional goods, p. 694
—rubber products, p. 403
—textiles, p. 94
—tobacco, p. 76
—transportation equipment, p. 652
—wearing apparel, p. 132
—wood products, p. 190
DeMenno-Kerdoon, p. 371
Denmark
—basic chemicals, excl fertilizers, pp. 312-318
—beverages, pp. 44, 46, 48-49, 51, 53, 55
—chemical products nec, pp. 328, 330, 332-333, 335, 337-338
—drugs and medicines, pp. 350-356
—electrical machinery, pp. 602, 604, 606-607, 609-610, 61-
—food products, pp. 4, 6, 8-9, 12-13, 15

Denmark continued:
—footwear, pp. 160, 162-163, 165-166, 168-169
—furniture, fixtures, pp. 200, 202, 204-205, 207, 209-210
—glass and products, pp. 448, 450-452, 454-455, 457
—industrial chemicals, pp. 266, 268, 270-272, 274, 276
—industries nec, pp. 706, 708, 710-711, 713, 715, 717
—iron and steel, pp. 488, 490-492, 494-495, 497
—leather and products, pp. 144, 146-148, 150, 152-153
—machinery nec, pp. 560, 562-564, 566-567, 569
—metal products, pp. 540, 542, 544-545, 547, 549-550
—motor vehicles, pp. 676-682
—nonferrous metals, pp. 514-515, 517-520, 522
—nonmetal products nec, pp. 466, 468, 470-471, 473-474, 476
—office, computing machinery, pp. 590-594
—paper and products, pp. 218, 220, 222-223, 225, 227-228
—petroleum refineries, pp. 360-361, 363, 365-367
—petroleum, coal products, pp. 380-386
—plastic products nec, pp. 414, 416, 418-419, 421-422, 424
—pottery, china, etc., pp. 432, 434-437, 439-440
—printing, publishing, pp. 248, 250, 252-253, 255, 257, 259
—professional goods, pp. 686-687, 689-691, 693-694
—profile, p. 746
—pulp, paper, etc., pp. 240-245
—radio, television, etc., pp. 628-633
—rubber products, pp. 394, 396-397, 399-400, 402-403
—shipbuilding, repair, pp. 668-674
—spinning, weaving, etc., pp. 112-116, 118
—synthetic resins, etc., pp. 320-325
—textiles, pp. 84, 86, 88-89, 91, 93-94
—tobacco, pp. 66, 68, 70-71, 73-74, 76
—transportation equipment, pp. 642, 644, 646-647, 649-650, 652
—wearing apparel, pp. 122, 124, 126-127, 129-130, 132
—wood products, pp. 180, 182, 184-185, 187, 189-190
Denso International America Inc, p. 655
DENTSPLY International Inc, p. 697
Department 56 Inc, pp. 443, 461
DePuy Inc, p. 698
Derlan Manufacturing Inc, p. 527
Destilaria Alta Mogiana Ltda, p. 280
Detergents, p. 327
Detroit Diesel Corp, p. 572
Deutsch Co, pp. 555, 617
Devon Group Inc, p. 263
Dewhirst Ladieswear Ltd, p. 136
Dewhirst Ltd, p. 136
Dexter Corp, pp. 97, 341, 554
Diagnostic Products Corp, p. 343
Dial Corp, p. 342
Dialogic Corp, p. 597
Diamond Multimedia Systems Inc, p. 597
Diamond Rug and Carpet Mills Inc, p. 98
Dichloromethane, p. 285
Diebold Inc, pp. 595, 615
Diethelm Singapore Pte. Ltd, p. 214

Diethyl ether, p. 287
Digital Communications Associates, p. 596
Dimol Manufacturing Pty. Ltd, p. 390
DIMON Inc, p. 79
DIMON International Inc, p. 79
Diora, p. 616
Dispatch Printing Co, p. 264
Distillate fuel oils, pp. 359, 375
Distilled alcoholic beverages, pp. 43, 61
DiverseyLever, p. 341
Divpac, pp. 525, 555
Dixie Group Inc, p. 97
Dixie Toga SA, p. 233
Dixon Ticonderoga Co, pp. 389, 480, 720
DMIB BHD, p. 407
Dofasco Inc, p. 499
Dole Thailand Ltd, p. 20
Dolls, p. 723
Doman Industries Ltd, pp. 193, 232
Domel DOO, p. 721
Dominican Republic
—beverages, p. 55
—chemical products nec, p. 338
—electrical machinery, p. 612
—food products, p. 15
—footwear, p. 169
—furniture, fixtures, p. 210
—glass and products, p. 457
—industrial chemicals, p. 276
—industries nec, p. 717
—iron and steel, p. 497
—leather and products, p. 153
—machinery nec, p. 569
—metal products, p. 550
—nonferrous metals, p. 522
—nonmetal products nec, p. 476
—paper and products, p. 228
—petroleum refineries, p. 367
—petroleum, coal products, p. 386
—plastic products nec, p. 424
—pottery, china, etc., p. 440
—printing, publishing, p. 259
—professional goods, p. 694
—rubber products, p. 403
—textiles, p. 94
—tobacco, p. 76
—transportation equipment, p. 652
—wearing apparel, p. 132
—wood products, p. 190
Don & Low Holdings Ltd, p. 408
Donaldson Company Inc, pp. 572, 697
Dong-A Pharmaceutical Co., Ltd, p. 342
Dong Ah Tire Ind. Co., Ltd, p. 408
Dong Sung Co., Ltd, p. 156
Dongil Rubber Belt Co., Ltd, p. 407
Dongkuk Steel Mill Co., Ltd, p. 501

Dongsuh Furniture Co., Ltd, p. 215
Dongwon Industries Co., Ltd, p. 19
Donit Pletilnica d.o.o, p. 719
Donna Karan International Inc, pp. 137, 172
Donnelly Corp, pp. 459, 553, 616
Donnkenny Apparel Inc, p. 137
Donnkenny Inc, p. 137
Donohue Inc, pp. 193, 231
Doornfontein Gold Mining Co. Ltd, pp. 553, 719
Dooyang Corp, p. 157
Dorbyl Automotive Products, p. 657
Dorel Industries Inc, p. 213
Dorogi Szenbanyak RT, p. 280
Dott Industries Inc, p. 460
Dow Corning Corp, p. 278
Dow Quimica SA, p. 280
Dravo Corp, p. 481
Drawing mills, p. 513
Dresser-Rand Co, p. 572
Dresses, pp. 121, 140
Drevoindustria AS, p. 194
Drew Industries Inc, p. 555
Drexel Heritage Furnishings Inc, p. 213
Dreyer's Grand Ice Cream Inc, p. 19
DRG Plastics Ltd, p. 429
Drilling and boring machines, p. 580
Drotovna, AS Hlohovec, pp. 500, 555
DRS Technologies Inc, p. 597
Drugs, pp. 327, 349
Drying machines for household use, p. 618
DSC Communications Corp, p. 615
DSM Engineering Plastic Products, p. 429
Du Pont Engineering Products SA, p. 98
Dubek Ltd, p. 79
Ductile Steel Processors Ltd, p. 555
Dulux Pty. Ltd. Republic of South, p. 343
Dunaferr Fejleszto es Karbantarto, p. 501
Duncan Enterprises, p. 443
Dunlop Ltd, pp. 406, 427
Dunlop Slazenger International Ltd, p. 721
Dunlop Tire Corp, p. 406
Dunn Edwards Corp, p. 343
Dura Automotive Systems Inc, p. 656
Duracell Inc, p. 615
Duracraft Corp, pp. 574, 617
Duropack Holding AG-Gruppe, p. 233
Durospand Co., Ltd, p. 195
Duslo SP, p. 279
Dusun Durian Plantations Ltd, p. 279
Dyersburg Corp, p. 97
Dyestuffs, p. 299
Dynatech Corp, p. 596
E and J Gallo Winery, p. 58
E/M Corp, p. 389
Eagle Industries Inc, pp. 427, 443, 572
Eagle Ottawa Leather Co, p. 156

Eagle-Picher Industries Inc, pp. 479, 615, 697
Eagon Industrial Co., Ltd, p. 195
Earthenware, p. 431
Easco Inc, p. 526
East Asiatic Co. Malaysia BHD, p. 342
East Penn Manufacturing Company, p. 617
Eastern Asia Woods Industrial Corp, p. 214
Eastern Counties Newspapers, p. 263
Eastham Refinery Ltd, p. 371
E.B. Eddy Forest Products Ltd, pp. 193, 232
ECC International, pp. 280, 480
Ecco Indonesia PT, pp. 156, 172
Eccolet Sko AS, p. 172
E.C.I. Telecom Ltd, p. 617
Eclipse Blinds PLC, p. 98
Ecuador
— basic chemicals, excl fertilizers, pp. 312-313, 315-318
— beverages, pp. 44, 46, 49, 51, 53, 55
— chemical products nec, pp. 328, 330, 333, 335, 337-338
— drugs and medicines, pp. 350-351, 353-356
— electrical machinery, pp. 602, 604, 607, 609-610, 612
— food products, pp. 4, 6, 10, 12-13, 15
— footwear, pp. 160, 162, 165-166, 168-169
— furniture, fixtures, pp. 200, 202, 205, 207, 209-210
— glass and products, pp. 448, 450, 452, 454-455, 457
— industrial chemicals, pp. 266, 268, 271-272, 274, 276
— industries nec, pp. 706, 708, 711, 713, 715, 717
— iron and steel, pp. 488, 490, 492, 494-495, 497
— leather and products, pp. 144, 146, 149-150, 152-153
— machinery nec, pp. 560, 562, 564, 566-567, 569
— metal products, pp. 540, 542, 545, 547, 549-550
— motor vehicles, pp. 676-677, 679-682
— nonferrous metals, pp. 514-515, 518-520, 522
— nonmetal products nec, pp. 466, 468, 471, 473-474, 476
— paper and products, pp. 218, 220, 223, 225, 227-228
— petroleum refineries, pp. 361, 363, 365-367
— petroleum, coal products, pp. 380-381, 383-386
— plastic products nec, pp. 414, 416, 419, 421-422, 424
— pottery, china, etc., pp. 432, 434, 436-437, 439-440
— printing, publishing, pp. 248, 250, 254-255, 257, 259
— professional goods, pp. 686, 688, 690-691, 693-694
— profile, p. 746
— pulp, paper, etc., pp. 240-241, 243-245
— radio, television, etc., pp. 628-629, 631-633
— rubber products, pp. 394, 396, 399-400, 402-403
— shipbuilding, repair, pp. 668-669, 671-674
— spinning, weaving, etc., pp. 112-113, 115-118
— synthetic resins, etc., pp. 320-325
— textiles, pp. 84, 86, 89, 91, 93-94
— tobacco, pp. 68, 71, 73-74, 76
— transportation equipment, pp. 642, 644, 647, 649-650, 6
— wearing apparel, pp. 122, 124, 127, 129-130, 132
— wood products, pp. 180, 182, 185, 187, 189-190
EDA SA, p. 555
Edward C. Levy Co, pp. 389, 480
EFACEC, Empresa Fabril de, pp. 574, 617

Efes Pilsen Beer and Malt Group, p. 60

Eggs, p. 3

Egypt
— basic chemicals, excl fertilizers, pp. 312-316, 318
— beverages, pp. 44, 46, 48-49, 51, 53, 55
— chemical products nec, pp. 328, 330, 332-333, 335, 337-338
— drugs and medicines, pp. 350-354, 356
— electrical machinery, pp. 602, 604, 606-607, 609-610, 612
— food products, pp. 4, 6, 8, 10, 12-13, 15
— footwear, pp. 160, 162-163, 165-166, 168-169
— furniture, fixtures, pp. 200, 202, 204-205, 207, 209-210
— glass and products, pp. 448, 450-452, 454-455, 457
— industrial chemicals, pp. 266, 268, 270-272, 274, 276
— industries nec, pp. 706, 708, 710-711, 713, 715, 717
— iron and steel, pp. 488, 490-492, 494-495, 497
— leather and products, pp. 144, 146-147, 149-150, 152-153
— machinery nec, pp. 560, 562-564, 566-567, 569
— metal products, pp. 540, 542, 544-545, 547, 549-550
— motor vehicles, pp. 676-680, 682
— nonferrous metals, pp. 514-515, 517-520, 522
— nonmetal products nec, pp. 466, 468, 470-471, 473-474, 476
— office, computing machinery, pp. 590-594
— paper and products, pp. 218, 220, 222-223, 225, 227-228
— petroleum refineries, pp. 360-361, 363, 365-367
— petroleum, coal products, pp. 380-386
— plastic products nec, pp. 414, 416, 418-419, 421-422, 424
— pottery, china, etc., pp. 432, 434-437, 439-440
— printing, publishing, pp. 248, 250, 252, 254-255, 257, 259
— professional goods, pp. 686, 688-691, 693-694
— profile, p. 747
— pulp, paper, etc., pp. 240-245
— radio, television, etc., pp. 628-633
— rubber products, pp. 394, 396-397, 399-400, 402-403
— shipbuilding, repair, pp. 668-672, 674
— spinning, weaving, etc., pp. 112-116, 118
— synthetic resins, etc., pp. 320-323, 325
— textiles, pp. 84, 86, 88-89, 91, 93-94
— tobacco, pp. 66, 68, 70-71, 73-74, 76
— transportation equipment, pp. 642, 644, 646-647, 649-650, 652
— wearing apparel, pp. 122, 124, 126-127, 129-130, 132
— wood products, pp. 180, 182, 184-185, 187, 189-190

Eichsfelder Zementwerke, p. 480

Ekco Group Inc, pp. 429, 720

EL OP Electro Optics Industries Ltd, p. 699

El Salvador
— basic chemicals, excl fertilizers, pp. 312-317
— beverages, pp. 44, 46, 48-49, 51, 53, 55
— chemical products nec, pp. 328, 330, 332-333, 335, 337-338
— drugs and medicines, pp. 350-355
— electrical machinery, pp. 602, 604, 606-607, 609-610, 612
— food products, pp. 4, 6, 8, 10, 12-13, 15
— footwear, pp. 160, 162-163, 165-166, 168-169

El Salvador continued:
— furniture, fixtures, pp. 200, 202, 204-205, 207, 209-210
— glass and products, pp. 448, 450-452, 454-455, 457
— industrial chemicals, pp. 266, 268, 270-272, 274, 276
— industries nec, pp. 706, 708, 710-711, 713, 715, 717
— iron and steel, pp. 488, 490-492, 494-495, 497
— leather and products, pp. 144, 146-147, 149-150, 152-153
— machinery nec, pp. 560, 562-564, 566-567, 569
— metal products, pp. 540, 542, 544-545, 547, 549-550
— motor vehicles, pp. 676-681
— nonferrous metals, pp. 514-515, 517-520, 522
— nonmetal products nec, pp. 466, 468, 470-471, 473-474, 476
— paper and products, pp. 218, 220, 222-223, 225, 227-228
— petroleum refineries, pp. 360-361, 363, 365-367
— petroleum, coal products, pp. 380-386
— plastic products nec, pp. 414, 416, 418-419, 421-422, 424
— pottery, china, etc., pp. 432, 434-437, 439-440
— printing, publishing, pp. 248, 250, 252, 254, 256-257, 259
— professional goods, pp. 686, 688-691, 693-694
— profile, p. 747
— pulp, paper, etc., pp. 240-245
— radio, television, etc., pp. 628-633
— rubber products, pp. 394, 396-397, 399-400, 402-403
— shipbuilding, repair, pp. 668-673
— spinning, weaving, etc., pp. 112-117
— synthetic resins, etc., pp. 320-324
— textiles, pp. 84, 86, 88-89, 91, 93-94
— tobacco, pp. 66, 68, 70-71, 73-74, 76
— transportation equipment, pp. 642, 644, 646-647, 649-650, 652
— wearing apparel, pp. 122, 124, 126-127, 129-130, 132
— wood products, pp. 180, 182, 184-185, 187, 189-190

Elana JSC, p. 278

Elcanto Shoe Co., Ltd, p. 173

Elco Holdings Ltd, pp. 616, 697

Elco Textron Inc, pp. 428, 555

Elcor Corp, p. 389

Eldridge Pope & Co. PLC, p. 59

Electric furnaces, p. 625

Electrical insulators, p. 431

Electrical machinery, p. 601

Electro-mechanical hand tools, p. 584

ElectroCom Automation Inc, p. 597

Electronic tubes, p. 636

Electronics Corp. of India Ltd, p. 615

Elektro Isola AS, p. 444

Elektrosvit AS, pp. 279, 616

Elevators, p. 621

Elexmetal, p. 429

Elf Aquitaine Inc, p. 278

Elf Lubricants North America, p. 390

Elf Oil Co. Ltd, p. 370

Elite Industries Ltd, p. 19

Elizabeth Arden Co, p. 343

Eljer Industries Inc, pp. 554, 574

Elkem AS, pp. 499, 525, 553
Elmo-Calf AB, p. 156
ELMOT, p. 699
Elscint Ltd, p. 699
Elta Electronics Industries Ltd, pp. 596, 699
Embotelladora Central SA, p. 60
Embotelladora Tica, SA, p. 60
Embraco Empresa Brasileira de, p. 573
EMC Corp, p. 595
Emco Ltd, pp. 389, 427, 554
Enerchem International Inc, p. 371
Energizer Power Systems Div, p. 617
Energomontaz-Polnoc SA, pp. 499, 573
Energy Ventures Inc, p. 574
Engelhard Corp, p. 278
Engelhard Corp. Catalysts and, pp. 280, 526
Engelhard Corp. Pigments and, p. 480
Engineering Equip. Inc, pp. 499, 655
Engines, pp. 575, 660-661
Engraph Inc, p. 233
Ennis Business Forms Inc, p. 264
Enterprise Oil Norge Ltd, p. 371
Envases Industriales Hondurenos, p. 156
Envirodyne Industries Inc, pp. 231, 427, 554
Epwin Group PLC, p. 429
Equatorial Guinea
— beverages, pp. 44, 46, 49, 55
— chemical products nec, pp. 328, 330, 333, 338
— food products, pp. 4, 6, 10, 15
— furniture, fixtures, pp. 200, 202, 205, 210
— metal products, pp. 540, 542, 545, 550
— nonmetal products nec, pp. 466, 468, 471, 476
— printing, publishing, pp. 248, 250, 254, 259
— profile, p. 748
— wearing apparel, pp. 122, 124, 127, 132
ERA Group PLC, p. 215
Eregli Iron and Steel Works Co, pp. 278, 479, 499
Ergon Inc, pp. 370, 597
Ergonbedrijven, p. 99
Erna Djuliawati PT, p. 193
Erskine House Group PLC, p. 194
Ertl Company Inc, pp. 526, 720
Escalade Inc, p. 597
ESCO Electronics Corp, pp. 617, 698
Esquire Co., Ltd, p. 172
ESSEF Corp, p. 428
Essex Group Inc, p. 525
Esso Malaysia BHD, p. 389
Esso Petroleum Co. Ltd, p. 370
Esso Singapore Private Ltd, p. 370
Estee Lauder Inc, p. 341
Esterline Technologies Corp, pp. 574, 698
Ethan Allen Inc, p. 213
Ethan Allen Interiors Inc, p. 213
Ethanediol, p. 286

Ethiopia
— beverages, pp. 44, 46, 48-49, 51, 53, 55
— chemical products nec, pp. 328, 330, 332-333, 335, 337
  338
— electrical machinery, pp. 602, 604, 606-607, 609-610, 61
— food products, pp. 4, 6, 8, 10, 12-13, 15
— footwear, pp. 160, 162, 164-166, 168-169
— furniture, fixtures, pp. 200, 202, 204-205, 207, 209-210
— glass and products, pp. 448, 450-452, 454-455, 457
— industrial chemicals, pp. 266, 268, 270-272, 274, 276
— iron and steel, pp. 488, 490-492, 494-495, 497
— leather and products, pp. 144, 146-147, 149-150, 152-15
— machinery nec, pp. 560, 562-564, 566-567
— metal products, pp. 540, 542, 544-545, 547, 549-550
— nonmetal products nec, pp. 466, 468, 470-471, 473-474
  476
— paper and products, pp. 218, 220, 222-223, 225, 227-22
— plastic products nec, pp. 414, 416, 418-419, 421-422, 4
— printing, publishing, pp. 248, 250, 252, 254, 256-257, 25
— profile, p. 748
— rubber products, pp. 394, 396, 398-400, 402-403
— textiles, pp. 84, 86, 88-89, 91, 93-94
— tobacco, pp. 66, 68, 70-71, 73-74, 76
— transportation equipment, pp. 642, 644, 646-647, 649-6
  652
— wearing apparel, pp. 122, 124, 126-127, 129-130, 132
— wood products, pp. 180, 182, 184-185, 187, 189-190
Ethiopia and Eritrea
— beverages, p. 55
— chemical products nec, p. 338
— electrical machinery, p. 612
— food products, p. 15
— footwear, p. 169
— furniture, fixtures, p. 210
— glass and products, p. 457
— industrial chemicals, p. 276
— industries nec, p. 717
— iron and steel, p. 497
— leather and products, p. 153
— machinery nec, p. 569
— metal products, p. 550
— nonferrous metals, p. 522
— nonmetal products nec, p. 476
— paper and products, p. 228
— petroleum refineries, p. 367
— petroleum, coal products, p. 387
— plastic products nec, p. 424
— pottery, china, etc., p. 440
— printing, publishing, p. 259
— professional goods, p. 694
— rubber products, p. 403
— textiles, p. 94
— tobacco, p. 76
— transportation equipment, p. 652
— wearing apparel, p. 132
— wood products, p. 190

Ethyl alcohol, pp. 43, 61
Ethyl Corp, p. 370
Ethylene, p. 283
Ethylene oxide, p. 288
Ethylene-vinyl acetate copolymers, p. 306
Etienne Aigner Inc, pp. 156, 174
Etz Lavud Ltd, p. 194
Eureka Co, p. 617
Eurofloor SA, p. 408
European Touch, p. 157
Evanite Fiber Corp, p. 460
Evans and Sutherland Computer, p. 597
Everbrite Inc, p. 720
E.W. Scripps Co, p. 262
Exabyte Corp, p. 596
Excalibur Group PLC, p. 721
Excavating machines, p. 586
Excel Industries Inc., p. 655
Exhibitgroup Inc, p. 720
Explosives, pp. 327, 347
Express Dairy Ltd, p. 19
Express Foods Group International, p. 18
Express Gifts Ltd, p. 264
Express Newspapers PLC, p. 263
Fab. De Licores y Alcoholes De, p. 60
Fab. Industries Inc, p. 99
Fabric mills, p. 83
Fabric weaving, p. 111
Fabrica de Calzado Ecco, SA, p. 174
Fabricated metal products, p. 539
Fabrics, pp. 83, 111
Fabryka Dywanow Kowary SA, p. 99
Fabryka Kabli, Ozarow, pp. 99, 501, 526
Fabryka Lozysk Tocznych, Iskra SA, p. 656
Fabryka Maszyn, Glinik SA, p. 573
Fabryka Samochodow w Lublinie, p. 655
Fabryka Urzadzen Gornictwa, p. 500
Falcon Building Products Inc, p. 554
Falcon Products Inc, p. 214
Falconbridge Ltd, p. 279
Falk Corp, p. 501
Fansteel Inc, p. 527
Farah Inc, p. 135
Farbglashutte Lauscha GmbH, p. 461
Farinaceous preparations, p. 38
Farmer Brothers Co, p. 59
Fasteners, p. 705
Fasty, p. 98
Fats, p. 3
Fayette Tubular Products Inc, p. 407
Federal Motor PT, p. 656
Federal Signal Corp, pp. 572, 615, 655, 719
Federated Co-operatives Ltd, pp. 194, 370
Federico Meiners Limitada SA, p. 156
Fedgas Pty. Ltd, p. 371
Fel-Pro Inc, p. 406

Femmunkas, p. 555
Fenix SA, p. 136
Fenton Art Glass Co, p. 460
Ferriere Nord, SpA, p. 527
Ferro-chromium, p. 503
Ferro Corp, pp. 278, 427, 479
Ferro-nickel, p. 504
Ferro-silicon, p. 504
Ferrous metals, p. 487
Ferrum SA, p. 195
Fertilizer distributors, p. 580
Fertilizers, p. 265
Feuer Leather Corp, p. 156
FHT Holdings Ltd, p. 720
Fiberboard, pp. 217, 239
— compressed, p. 238
Fibers, p. 83
Fibreboard Corp, pp. 428, 479, 555
Fidelity Tire Manufacturing Co, p. 408
Field Group PLC, p. 233
Figgie International Inc, p. 698
FII Group PLC, pp. 172, 699
Fiji
— beverages, p. 55
— chemical products nec, pp. 328, 330, 333, 335, 337-338
— electrical machinery, p. 612
— food products, pp. 4, 6, 10, 12-13, 15
— furniture, fixtures, pp. 200, 202, 205, 207, 209-210
— glass and products, p. 457
— industrial chemicals, p. 276
— industries nec, pp. 706, 708, 711, 713, 715, 717
— machinery nec, p. 569
— metal products, pp. 540, 542, 545, 547, 549-550
— motor vehicles, pp. 676-677, 679-682
— nonferrous metals, p. 522
— nonmetal products nec, p. 476
— paper and products, pp. 218, 220, 223, 225, 227-228
— petroleum refineries, p. 367
— petroleum, coal products, p. 387
— plastic products nec, pp. 414, 416, 419, 421-422, 424
— pottery, china, etc., p. 440
— printing, publishing, pp. 248, 250, 254, 256-257, 259
— professional goods, p. 694
— profile, p. 749
— rubber products, pp. 394, 396, 399-400, 402-403
— shipbuilding, repair, pp. 668-669, 671-674
— tobacco, p. 76
— transportation equipment, pp. 642, 644, 647, 649-650, 652
— wood products, pp. 180, 182, 185, 187, 189-190
FileNet Corp, p. 596
Film and sheet, p. 413
FINA Inc, p. 370
Findlay Industries Inc, p. 135
Finland
— basic chemicals, excl fertilizers, pp. 312-313, 315-318
— beverages, pp. 44, 46, 49, 51, 53, 55

Finland continued:
— chemical products nec, pp. 328, 330, 333, 335, 337-338
— drugs and medicines, pp. 350-351, 353-356
— electrical machinery, pp. 602, 604, 607, 609-610, 612
— food products, pp. 4, 6, 10, 12-13, 15
— footwear, pp. 160, 162, 165-166, 168-169
— furniture, fixtures, pp. 200, 202, 205, 207, 209-210
— glass and products, pp. 448, 450, 452, 454-455, 457
— industrial chemicals, pp. 266, 268, 271-272, 274, 276
— industries nec, pp. 706, 708, 711, 713, 715, 717
— iron and steel, pp. 488, 490, 492, 494-495, 497
— leather and products, pp. 144, 146, 149-150, 152-153
— machinery nec, pp. 560, 562, 564, 566-567, 569
— metal products, pp. 540, 542, 545, 547, 549-550
— motor vehicles, pp. 676-677, 679-682
— nonferrous metals, pp. 514-515, 518-520, 522
— nonmetal products nec, pp. 466, 468, 471, 473-474, 476
— office, computing machinery, pp. 590-594
— paper and products, pp. 218, 220, 223, 225, 227-228
— petroleum refineries, pp. 360-361, 363, 365-367
— petroleum, coal products, pp. 380-381, 383-385, 387
— plastic products nec, pp. 414, 416, 419, 421-422, 424
— pottery, china, etc., pp. 432, 434, 436-437, 439-440
— printing, publishing, pp. 248, 250, 254, 256-257, 259
— professional goods, pp. 686, 688, 690, 692-694
— profile, p. 749
— pulp, paper, etc., pp. 240-241, 243-245
— radio, television, etc., pp. 628-629, 631-633
— rubber products, pp. 394, 396, 399-400, 402, 404
— shipbuilding, repair, pp. 668-669, 671-674
— spinning, weaving, etc., pp. 112-113, 115-118
— synthetic resins, etc., pp. 320-325
— textiles, pp. 84, 86, 89, 91, 93-94
— tobacco, pp. 66, 68, 71, 73-74, 76
— transportation equipment, pp. 642, 644, 647, 649-650, 652
— wearing apparel, pp. 122, 124, 127, 129-130, 132
— wood products, pp. 180, 182, 185, 187, 189-190
Finomszerelvenygyar, Eger, p. 555
Fiona Footwear Ltd, p. 173
Firestone De La Argentina SA IC, p. 407
First Alert Inc, p. 719
First Brands Corp, pp. 231, 389, 479
Fish
— frozen, p. 30
— salted, dried or smoked, p. 30
— tinned, p. 31
Fish and fish products, p. 3
Fisher-Price Inc, pp. 214, 721
Fitch Holdings PLC, p. 137
Fixtures, p. 199
Fjeldhammer Brug AS, pp. 389, 429
FKI Industries Inc, pp. 553, 572
Flambeau Corp, p. 429
Flat glass, p. 447
Flax yarn, p. 102
Fletcher Challenge Canada Ltd, p. 232

Flexsteel Industries Inc, p. 213
Flextronics International Ltd, p. 616
Floating structures, p. 667
Floor covering, p. 110
Flor-Quim, Inc, p. 390
Florida Rock Industries Inc, p. 479
Florsheim Shoe Co, p. 172
Flour, p. 3
— cereal, other than wheat, p. 36
— wheat, p. 36
Flower Indonesia PT, p. 172
Flowers Industries Inc, p. 18
Flue-Cured Tobacco Cooperative, p. 80
Fluke Corp, p. 698
Fluoroware Inc, p. 429
Fly Bridges Enterprise Co., Ltd.--, p. 461
F.M. Thorpe Manufacturing Co, p. 215
FMC Corp. (UK) Ltd, p. 501
Foam products, p. 413
Foamex International Inc, p. 278
Foamex L.P, p. 278
Fonicia Kereskedelmi Kft, p. 174
Food products, p. 3
Foodarama Supermarkets Inc, p. 19
Foodbrands America Inc, p. 20
Footwear, p. 159
— house, p. 176
— leather, children's, p. 175
— leather, men's, p. 175
— leather, women's, p. 176
— other, p. 176
— total production, excluding rubber footwear, p. 175
Ford Argentina SA, p. 656
Ford France SA, pp. 193, 279, 719
Ford Motor Co. Glass Div, p. 459
FORE Systems Inc, p. 597
Forest Laboratories Inc, p. 343
Forging machines, p. 581
Forgings, p. 487
Fork-lift trucks, p. 621
Form-O-Uth, p. 135
Formic acid, p. 290
Formosa Plastics Corporation U.S.A, p. 279
Formosa Taffeta Co., Ltd, p. 97
Forstmann and Company Inc, p. 98
Forsythe Lubrication Association Ltd, p. 390
Fort Howard Corp, p. 231
Forus SA, p. 173
Fossil Inc, p. 174
Foundries, p. 487
Fountain pens
— ball-point pens, propelling pencils etc., p. 723
Fownes Brothers and Company Inc, p. 174
Fox Valley Corp, p. 233
Foxboro Co, p. 697
Frame Textile Corp. Ltd, pp. 97, 135, 278

Frames for spectacles, p. 702

France
— beverages, pp. 46, 49, 51, 55
— chemical products nec, pp. 330, 333, 335, 337-338
— drugs and medicines, pp. 353-354
— electrical machinery, pp. 604, 607, 609-610, 612
— food products, pp. 6, 10, 12, 15
— footwear, pp. 162, 165-166, 169
— furniture, fixtures, pp. 202, 205, 207, 210
— glass and products, pp. 450, 452, 454-455, 457
— industrial chemicals, pp. 268, 271, 273-274, 276
— industries nec, pp. 708, 711, 713, 717
— iron and steel, pp. 490, 492, 494-495, 497
— leather and products, pp. 146, 149-150, 153
— machinery nec, pp. 562, 564, 566, 568-569
— metal products, pp. 542, 545, 547, 549-550
— motor vehicles, pp. 679-680
— nonferrous metals, pp. 516, 518-520, 522
— nonmetal products nec, pp. 468, 471, 473-474, 476
— paper and products, pp. 220, 223, 225, 227-228
— petroleum refineries, pp. 361, 364-365, 367
— petroleum, coal products, p. 387
— plastic products nec, pp. 416, 419, 421, 424
— pottery, china, etc., p. 440
— printing, publishing, pp. 250, 254, 256-257, 259
— professional goods, pp. 688, 690, 692, 694
— profile, p. 750
— rubber products, pp. 396, 399-400, 404
— shipbuilding, repair, pp. 671-672
— textiles, pp. 86, 89, 91, 94
— tobacco, pp. 68, 71, 73, 76
— transportation equipment, pp. 644, 647, 649-650, 652
— wearing apparel, pp. 124, 127, 129, 132
— wood products, pp. 182, 185, 187, 190

Frank Industrial Corp. Ltd, p. 156
Franklin Corp., p. 214
Franklin Covey Co, p. 262
Fraser Papers Inc, p. 232
Frederick Cooper PLC, p. 501
Freios Varga SA, p. 596
Fresh meat, p. 3
Friesland Frico Domo Cooperatie BA, p. 18
Frigobras Cia. Brasileira de, p. 18
Fruit, p. 39
Fruit & vegetable juices, p. 28
Fruits, p. 3
— dried, p. 27
— frozen, p. 29
— tinned or bottled, p. 29
Fuchs (UK) PLC, p. 371
Fuh Ching Leather Co., Ltd, p. 157
Fujitsu America Inc, pp. 595, 616
Fuller, Smith & Turner PLC, p. 59
Furnel International Ltd, p. 193
Furniture, p. 199
Furnsteel Pty. Ltd, p. 214

Furon Co, p. 427
Fuses, p. 638
F.W. Farnsworth Ltd, p. 18
G. A. Boulet Inc, p. 174
G and K Services Inc, p. 135
G. Heileman Brewing Company Inc, p. 59

Gabon
— basic chemicals, excl fertilizers, pp. 312-313, 315-317
— beverages, pp. 44, 46, 49, 51, 53, 55
— chemical products nec, pp. 328, 330, 333, 335, 337-338
— electrical machinery, pp. 602, 604, 607, 609-610, 612
— food products, pp. 4, 6, 10, 12-13, 15
— footwear, p. 169
— furniture, fixtures, pp. 200, 202, 205, 207, 209-210
— glass and products, p. 457
— industrial chemicals, pp. 266, 268, 271, 273-274, 276
— industries nec, pp. 706, 711, 713, 717
— iron and steel, p. 497
— leather and products, p. 153
— machinery nec, p. 569
— metal products, pp. 540, 542, 545, 547, 549-550
— nonferrous metals, p. 522
— nonmetal products nec, pp. 466, 468, 471, 473-474, 476
— paper and products, pp. 218, 220, 223, 225, 227, 229
— petroleum refineries, pp. 360-361, 364-367
— petroleum, coal products, p. 387
— plastic products nec, p. 424
— pottery, china, etc., p. 440
— printing, publishing, pp. 248, 250, 254, 256-257, 259
— professional goods, p. 694
— profile, p. 751
— radio, television, etc., pp. 628-629, 631-633
— rubber products, p. 404
— synthetic resins, etc., pp. 320-324
— textiles, p. 94
— tobacco, pp. 66, 68, 71, 73-74, 76
— transportation equipment, p. 652
— wearing apparel, pp. 122, 124, 127, 129-130, 132
— wood products, pp. 180, 182, 185, 187, 189-190

GAF Corp, p. 479
Gajah Tunggal, Pt, p. 406
Galey and Lord Inc, p. 97
Gali Industries Ltd, p. 173

Gambia
— beverages, pp. 44, 46, 48, 55
— chemical products nec, p. 338
— electrical machinery, p. 612
— food products, pp. 4, 6, 8, 15
— footwear, p. 169
— furniture, fixtures, pp. 200, 202, 210
— glass and products, p. 457
— industrial chemicals, p. 276
— industries nec, pp. 706, 708, 710, 717
— iron and steel, p. 497
— leather and products, pp. 144, 146-147, 153
— machinery nec, pp. 560, 562-563, 569

Gambia continued:
— metal products, pp. 540, 542, 550
— nonferrous metals, p. 522
— nonmetal products nec, pp. 466, 468, 470, 476
— paper and products, p. 229
— petroleum refineries, p. 367
— petroleum, coal products, pp. 380-382, 387
— plastic products nec, pp. 414, 416, 418, 424
— pottery, china, etc., p. 440
— printing, publishing, pp. 248, 250, 252, 259
— professional goods, p. 694
— profile, p. 751
— rubber products, pp. 394, 396, 398, 404
— spinning, weaving, etc., pp. 112-114
— textiles, pp. 84, 86, 88, 94
— tobacco, p. 76
— transportation equipment, pp. 642, 644, 646, 652
— wearing apparel, pp. 122, 124, 126, 132
— wood products, pp. 180, 182, 184, 190
Gambro Healthcare Inc, p. 697
Gamrat Plastic Works, p. 428
Ganz Measuring Instruments Factory, p. 699
Garan Inc, p. 136
Garden State Tanning Inc, p. 156
Garden tractors, p. 579
Garland Company Inc, p. 389
Garnar Booth PLC, p. 156
Garware Nylons Ltd, pp. 97, 279
Gas cylinders, p. 539
Gas meters, p. 700
Gas produced by cokeries, p. 392
Gas turbines, p. 576
Gaseosas Lux SA, p. 60
Gaskets, p. 393
Gasoline, p. 359
Gates UK Ltd, pp. 173, 407
Gateway 2000 Inc, p. 595
Gatic SACFI Y A, p. 406
Gator Industries Inc, p. 173
Gatorade, p. 58
Gawih Jaya PT, p. 79
Gaylord Container Corp, p. 232
GB Glass Ltd, p. 459
G.D. Searle and Co, p. 341
Gdanska Stocznia Remontowa, p. 656
GE Appliances, pp. 572, 615
GE Superabrasives, p. 480
Gemco-Ware Inc, p. 460
Gemtron Corp, p. 459
Gencor Industries Inc, p. 389
GenCorp Inc, pp. 427, 553, 655, 697
Genentech Inc, p. 341
General Binding Corp, p. 573
General Cable Corp, pp. 525, 574
General Cable Industries Inc, p. 525
General Cigar Holdings Inc, p. 79

General Directorate of Fine Textiles, p. 98
General Dynamics Land Systems, p. 657
General Electric Co. Aircraft, pp. 553, 697
General Foods Ltd, p. 19
General hardware, p. 539
General Instrument Corp, p. 615
General Motors Corp. Electro-Motive, pp. 574, 657
General Motors of Canada Ltd, p. 655
General Nutrition Companies Inc, p. 341
General Scanning Inc, p. 597
General Semiconductor Inc, p. 617
General Signal Corp, pp. 553, 572, 615
Generators, p. 601
Generators for hydraulic turbines, p. 622
Generators for hydraulic turbines - kilowatts, p. 623
Generators for steam turbines, p. 623
Generators for steam turbines - kilowatts, p. 623
Genesco Inc, pp. 135, 172
Genesee Corp, p. 60
Geneva Steel Co, p. 500
GENICOM Corp, p. 596
Genzyme Corp, pp. 341, 698
Geon Co, p. 280
George Koch Sons Inc, p. 479
Georgia Boot Inc, p. 173
Georgia Marble Co, p. 481
Gerber Plumbing Fixtures Corp, p. 443
Gerber Products Co, pp. 18, 135, 427
Gerber Scientific Inc, pp. 596, 699
Germany
— beverages, pp. 44, 46, 49
— chemical products nec, pp. 328, 330, 333
— drugs and medicines, pp. 350-351, 353
— electrical machinery, pp. 602, 604, 607
— food products, pp. 4, 6, 10
— footwear, pp. 160, 162, 165
— furniture, fixtures, pp. 200, 202, 205
— glass and products, pp. 448, 450, 452
— industrial chemicals, pp. 266, 268
— industries nec, pp. 706, 708, 711
— iron and steel, pp. 488, 490, 492
— leather and products, pp. 144, 146, 149
— machinery nec, pp. 560, 562, 564
— metal products, pp. 540, 542, 545
— motor vehicles, pp. 676-677, 679
— nonferrous metals, pp. 514, 516, 518
— nonmetal products nec, pp. 466, 468, 471
— office, computing machinery, pp. 590-592
— paper and products, pp. 218, 220, 223
— plastic products nec, pp. 414, 416, 419
— pottery, china, etc., pp. 432, 434, 436
— printing, publishing, pp. 248, 250, 254
— professional goods, pp. 686, 688, 690
— profile, p. 752
— pulp, paper, etc., pp. 240-241, 243
— radio, television, etc., pp. 628-629, 631

Germany continued:
— rubber products, pp. 394, 396, 399
— shipbuilding, repair, pp. 668-669, 671
— spinning, weaving, etc., pp. 112-113, 115
— textiles, pp. 84, 86, 89
— tobacco, pp. 66, 68, 71
— transportation equipment, pp. 642, 644, 647
— wearing apparel, pp. 122, 124, 127
— wood products, pp. 180, 182, 185

Germany (Eastern Part)
— beverages, p. 55
— chemical products nec, p. 338
— electrical machinery, p. 612
— food products, p. 15
— footwear, p. 169
— furniture, fixtures, p. 210
— glass and products, p. 457
— industrial chemicals, p. 276
— industries nec, p. 717
— iron and steel, p. 497
— leather and products, p. 153
— machinery nec, p. 569
— metal products, p. 551
— nonferrous metals, p. 522
— nonmetal products nec, p. 476
— paper and products, p. 229
— petroleum refineries, p. 367
— petroleum, coal products, p. 387
— plastic products nec, p. 424
— pottery, china, etc., p. 440
— printing, publishing, p. 259
— professional goods, p. 694
— rubber products, p. 404
— textiles, p. 94
— tobacco, p. 76
— transportation equipment, p. 652
— wearing apparel, p. 132
— wood products, p. 190

Germany (Western Part)
— beverages, pp. 44, 46, 48-49, 51, 53, 55
— chemical products nec, pp. 328, 330, 332-333, 335, 337-338
— drugs and medicines, pp. 350-356
— electrical machinery, pp. 602, 604, 606-607, 609-610, 612
— food products, pp. 4, 6, 8, 10, 12-13, 15
— footwear, pp. 160, 162, 164-166, 168, 170
— furniture, fixtures, pp. 200, 202, 204-205, 207, 209-210
— glass and products, pp. 448, 450-452, 454-455, 457
— industrial chemicals, pp. 266, 268, 270, 273, 276
— industries nec, pp. 706, 708, 710-711, 713, 715, 717
— iron and steel, pp. 488, 490-492, 494-495, 497
— leather and products, pp. 144, 146, 148-150, 152-153
— machinery nec, pp. 560, 562-564, 566, 568-569
— metal products, pp. 540, 542, 544-545, 547, 549, 551
— motor vehicles, pp. 676-682
— nonferrous metals, pp. 514, 516-519, 521-522

Germany (Western Part) continued:
— nonmetal products nec, pp. 466, 468, 470-471, 473-474, 476
— office, computing machinery, pp. 590-594
— paper and products, pp. 218, 220, 222-223, 225, 227, 229
— petroleum refineries, p. 363
— plastic products nec, pp. 414, 416, 418-419, 421-422, 424
— pottery, china, etc., pp. 432, 434-437, 439-440
— printing, publishing, pp. 248, 250, 252, 254, 256-257, 259
— professional goods, pp. 686, 688-690, 692-694
— pulp, paper, etc., pp. 240-246
— radio, television, etc., pp. 628-634
— rubber products, pp. 394, 396, 398-400, 402, 404
— shipbuilding, repair, pp. 668-674
— spinning, weaving, etc., pp. 112-118
— synthetic resins, etc., p. 323
— textiles, pp. 84, 86, 88-89, 91, 93-94
— tobacco, pp. 66, 68, 70-71, 73-74, 76
— transportation equipment, pp. 642, 644, 646-647, 649-650, 652
— wearing apparel, pp. 122, 124, 126-127, 129-130, 132
— wood products, pp. 180, 182, 184-185, 187, 189-190

Gessy Lever Ltda. Industrial, p. 341
GFPT Co., Ltd, p. 19
G.H. Bass and Co, p. 172
G.H. Bass Caribbean, Inc, p. 173

Ghana
— basic chemicals, excl fertilizers, pp. 312-313
— beverages, pp. 44, 46, 48-49, 51, 55
— chemical products nec, pp. 328, 330, 332-333, 335, 338
— drugs and medicines, pp. 350-351
— electrical machinery, pp. 602, 604, 606-607, 609, 612
— food products, pp. 4, 6, 8, 10, 12, 15
— footwear, pp. 160, 162, 170
— furniture, fixtures, pp. 200, 202, 204-205, 207, 210
— glass and products, pp. 448, 450, 457
— industrial chemicals, pp. 266, 268, 270-271, 273, 276
— industries nec, pp. 706, 708, 717
— iron and steel, pp. 488, 490-492, 494, 497
— leather and products, pp. 144, 146, 153
— machinery nec, pp. 560, 562-563, 565-566, 569
— metal products, pp. 540, 542, 544-545, 547, 551
— motor vehicles, pp. 676-677
— nonferrous metals, pp. 514, 516-519, 522
— nonmetal products nec, pp. 466, 468, 476
— office, computing machinery, pp. 590-591
— paper and products, pp. 218, 220, 222-223, 225, 229
— petroleum refineries, pp. 360-361, 367
— petroleum, coal products, pp. 380-381, 387
— plastic products nec, pp. 414, 416, 418-419, 421, 424
— pottery, china, etc., pp. 432, 434, 440
— printing, publishing, pp. 248, 250, 252, 254, 256, 259
— professional goods, p. 695
— profile, p. 752
— pulp, paper, etc., pp. 240-241
— radio, television, etc., pp. 628-629

Ghana continued:
— rubber products, pp. 394, 396, 398-400, 404
— spinning, weaving, etc., pp. 112-113
— synthetic resins, etc., pp. 320-321
— textiles, pp. 84, 86, 88-89, 91, 94
— tobacco, pp. 66, 68, 70-71, 73, 76
— transportation equipment, pp. 642, 644, 652
— wearing apparel, pp. 122, 124, 132
— wood products, pp. 180, 182, 184-185, 187, 190
Ghana Rubber Products Ltd, p. 174
Giant Industries Inc, p. 370
Gibaut Hnos Manufacturas De, p. 156
Gibson Greetings Inc, pp. 231, 262
Gibson Guitar Corp, p. 721
Giddings and Lewis Inc, pp. 574, 617, 698
Giesserei und Glasformenbau, p. 461
Gillette do Brasil Ltda, p. 343
Gillette UK Ltd, p. 721
Gilman Paper Co, p. 232
Giroflex SA, p. 215
Glass, pp. 462-463
— bottles, p. 463
— fibers, p. 462
— products, p. 447
— wool, p. 447
— yarn, p. 447
Glassware, p. 447
Glaxo Operations UK Ltd, p. 341
Glaxo Pharmaceuticals UK Ltd, p. 342
Glaxo Wellcome Inc, pp. 341, 343
Glaxochem Ltd, p. 342
Gleason Corp., p. 408
Glen Raven Mills Inc, p. 98
Glenmore Distilleries Co, p. 60
Glycerine, p. 287
Glynwed Steels Ltd, p. 500
Glynwed Tubes & Fittings Ltd, p. 527
GM Nameplate Inc, p. 721
Godrej & Boyce Manufacturing Co, pp. 213, 553, 572
Goetze Corporation of America, p. 407
Gokaldas Exports, p. 136
Gold, p. 513
Gold Eagle Co, p. 390
Goldcorp Inc, p. 481
Golden Books Family Entertainment, p. 264
Golden Building Material Co., Ltd, p. 444
Golden Poultry Company Inc, p. 18
Gomel Casting Enterprise, p. 572
Gomma Gomma Holdings Pty. Ltd, p. 214
Gomme Ltd, p. 215
Good Humor Ice Cream, p. 20
Good Tables Inc, p. 215
Goodfit Manufacturing Corp, p. 174
Goods wagons and vans, p. 660
Goody Products Inc, pp. 407, 429
Goodyear do Brasil Produtos de, p. 406

Goodyear Indonesia PT, p. 407
Goodyear SA, pp. 406, 499, 574
Goodyear Thailand Ltd, p. 408
Gorenje Gospodinjski Aparati d.d, p. 719
GOSA, p. 655
Goshen Rubber Company Inc, p. 407
Gould Electronics Inc, pp. 555, 617
Goulds Pumps Inc, p. 573
G.R. Holdings PLC, p. 156
Graboplast Textil-Es Muborgyarto, p. 429
Graders and levelers, p. 586
Grandoe Corp, p. 156
Granite Construction Inc, pp. 389, 479
Graphic Controls Corp, pp. 407, 720
Gravure printing, p. 247
Gray Beverage Inc, p. 60
Gray Communications Systems Inc, p. 264
Graymont Ltd, p. 481
Greases, p. 359
Great American Industries Inc, p. 408
Great American Management Inc, pp. 443, 554, 573
Great Dane Trailers Inc, p. 656
Great Lakes Carbon Corp, p. 389
Great Lakes Chemical Corp, p. 278
Greb International, Kodiak Division, p. 174
Grede Foundries Inc, p. 499
Greece
— basic chemicals, excl fertilizers, pp. 312-313, 315-318
— beverages, pp. 44, 46, 50-51, 53, 55
— chemical products nec, pp. 328, 330, 333, 335, 337-338
— drugs and medicines, pp. 350-351, 353-356
— electrical machinery, pp. 602, 604, 607, 609-610, 612
— food products, pp. 4, 7, 10, 12-13, 15
— footwear, pp. 160, 162, 165-166, 168, 170
— furniture, fixtures, pp. 200, 202, 205, 207, 209-210
— glass and products, pp. 448, 450, 453-455, 457
— industrial chemicals, pp. 266, 268, 271, 273-274, 276
— industries nec, pp. 706, 708, 711, 713, 715, 717
— iron and steel, pp. 488, 490, 492, 494-495, 497
— leather and products, pp. 144, 146, 149-150, 152-153
— machinery nec, pp. 560, 562, 565-566, 568-569
— metal products, pp. 540, 542, 545, 547, 549, 551
— motor vehicles, pp. 676-677, 679-682
— nonferrous metals, pp. 514, 516, 518-519, 521-522
— nonmetal products nec, pp. 466, 468, 471, 473-474, 476
— office, computing machinery, pp. 590-594
— paper and products, pp. 218, 220, 223, 225, 227, 229
— petroleum refineries, pp. 360-361, 364-367
— petroleum, coal products, pp. 380-381, 383-385, 387
— plastic products nec, pp. 414, 416, 419, 421-422, 424
— pottery, china, etc., pp. 432, 434, 436-437, 439-440
— printing, publishing, pp. 248, 250, 254, 256-257, 259
— professional goods, pp. 686, 688, 690, 692-693, 695
— profile, p. 753
— pulp, paper, etc., pp. 240-241, 243-246
— radio, television, etc., pp. 628-629, 631-634

Greece continued:
— rubber products, pp. 394, 396, 399-400, 402, 404
— shipbuilding, repair, pp. 668-669, 671-674
— spinning, weaving, etc., pp. 112-113, 115-118
— synthetic resins, etc., pp. 320-325
— textiles, pp. 84, 86, 89, 91, 93-94
— tobacco, pp. 66, 68, 71, 73-74, 76
— transportation equipment, pp. 642, 644, 647, 649-650, 652
— wearing apparel, pp. 122, 124, 127, 129, 131-132
— wood products, pp. 180, 182, 185, 187, 189-190
Green Bay Packaging Inc, p. 232
Greenalls Midlands Ltd, p. 58
Greene King PLC, p. 59
Greene, Tweed and Co, p. 408
Greenfield Industries Inc., p. 573
Greggs PLC, p. 18
Greif Board Corp, p. 231
Greif Bros. Corp, pp. 193, 554
Grenada
— beverages, pp. 44, 46
— chemical products nec, pp. 328, 330
— drugs and medicines, pp. 350-351
— electrical machinery, pp. 602, 604
— food products, pp. 4, 7
— furniture, fixtures, pp. 200, 202
— glass and products, pp. 448, 450
— industrial chemicals, pp. 266, 268
— industries nec, pp. 706, 708
— iron and steel, pp. 488, 490
— leather and products, pp. 144, 146
— machinery nec, pp. 560, 562
— metal products, pp. 540, 542
— motor vehicles, pp. 676-677
— nonmetal products nec, pp. 466, 468
— paper and products, pp. 218, 220
— printing, publishing, pp. 248, 251
— profile, p. 753
— pulp, paper, etc., pp. 240-241
— rubber products, pp. 394, 396
— shipbuilding, repair, p. 668
— synthetic resins, etc., pp. 320-321
— textiles, pp. 84, 86
— tobacco, pp. 66, 68
— transportation equipment, pp. 642, 644
— wearing apparel, pp. 122, 124
— wood products, pp. 180, 182
Griffon Corp, pp. 193, 231, 553, 616
Grinding and sharpening machines, p. 581
Grinding wheels, p. 465
Group Tomos - Promo d.o.o, p. 721
Grove Worldwide, p. 573
Grupo Billi, p. 342
Grupo Canguro SA, p. 156
Grupo Embotelladoras Unidas SA de, p. 58
Grupo Empresarial Maya SA de CV, p. 480
Grupo Industrial Saltillo SA de CV, pp. 443, 479, 553, 655

Grupo Zillo Lorenzetti, p. 407
G.T.C. Transcontinental Group Ltd, p. 262
GTECH Holdings Corp, p. 595
GTI Corp., p. 615
Guardian Media Group PLC, p. 262
Guatemala
— basic chemicals, excl fertilizers, pp. 312-313, 315
— beverages, pp. 44, 46, 50, 55
— chemical products nec, pp. 328, 330, 333, 338
— drugs and medicines, pp. 350-351, 353
— electrical machinery, pp. 602, 604, 607, 612
— food products, pp. 4, 7, 10, 15
— footwear, pp. 160, 162, 165, 170
— furniture, fixtures, pp. 200, 202, 205, 210
— glass and products, pp. 448, 450, 453, 457
— industrial chemicals, pp. 266, 268, 271, 276
— industries nec, pp. 706, 708, 712, 717
— iron and steel, pp. 488, 490, 492, 497
— leather and products, pp. 144, 146, 149, 153
— machinery nec, pp. 560, 562, 565, 569
— metal products, pp. 540, 542, 545, 551
— motor vehicles, pp. 676-677, 679
— nonferrous metals, pp. 514, 516, 518, 522
— nonmetal products nec, pp. 466, 468, 471, 476
— office, computing machinery, pp. 590-592
— paper and products, pp. 218, 220, 223, 229
— petroleum refineries, pp. 360-361, 364, 368
— petroleum, coal products, pp. 380-381, 383, 387
— plastic products nec, pp. 414, 416, 419, 424
— pottery, china, etc., pp. 432, 434, 436, 440
— printing, publishing, pp. 248, 251, 254, 259
— professional goods, pp. 686, 688, 690, 695
— profile, p. 754
— pulp, paper, etc., pp. 240-241, 243
— radio, television, etc., pp. 628-629, 631
— rubber products, pp. 394, 396, 399, 404
— spinning, weaving, etc., pp. 112-113, 115
— synthetic resins, etc., pp. 320-321
— textiles, pp. 84, 86, 89, 94
— tobacco, pp. 66, 68, 71, 76
— transportation equipment, pp. 642, 644, 647, 652
— wearing apparel, pp. 122, 124, 127, 132
— wood products, pp. 180, 182, 185, 190
Guess ? Inc, p. 136
Guidant Corp, p. 697
Guilford Mills Inc, p. 97
Guilford Mills Inc. Apparel Home, p. 98
Guilford of Maine Inc, p. 99
Guinness Anchor BHD, p. 59
Guinness Brewing Worldwide Ltd, p. 59
Gul Ahmed Textile Mills Ltd, p. 97
Gulf Glass Manufacturing Co. KSC, p. 460
Gulf Oil Refining Ltd, p. 371
Gulf States Asphalt Company Inc, p. 390
Gulf States Steel Incorporated of, p. 500
Gulfstream Aerospace Corp, p. 656

Gumarne Barum, AS, p. 406
Gunlocke Co, p. 215
Gunung Gahapi Sakti PT, p. 527
Guyana
— beverages, p. 55
— chemical products nec, p. 338
— electrical machinery, p. 612
— food products, p. 15
— footwear, p. 170
— furniture, fixtures, p. 210
— glass and products, p. 457
— industrial chemicals, p. 276
— industries nec, p. 717
— iron and steel, p. 497
— leather and products, p. 153
— machinery nec, p. 569
— metal products, p. 551
— nonferrous metals, p. 522
— nonmetal products nec, p. 476
— paper and products, p. 229
— petroleum refineries, p. 368
— petroleum, coal products, p. 387
— plastic products nec, p. 424
— pottery, china, etc., p. 440
— printing, publishing, p. 259
— professional goods, p. 695
— rubber products, p. 404
— textiles, p. 94
— tobacco, p. 76
— transportation equipment, p. 652
— wearing apparel, p. 132
— wood products, p. 190
G.W. Padley Holdings Ltd, p. 19
Gyproc AB, p. 481
H & R Johnson Tiles Ltd, p. 480
H L Blachford Ltd, p. 390
Hackman Tabletop Oy Ab, pp. 443, 460
Hadco Corp, p. 616
Hadinata Brothers & Co. PT, p. 214
Hadtex Indosyntec PT, p. 97
Hae Yang Knitting Factory Ltd, p. 80
Haggar Clothing Co, p. 135
Haggar Corp, p. 135
Haitai Beverage Co., Ltd, p. 58
Haiti
— beverages, pp. 51, 55
— chemical products nec, p. 338
— electrical machinery, p. 612
— food products, pp. 12, 15
— footwear, p. 170
— furniture, fixtures, p. 210
— glass and products, p. 457
— industrial chemicals, p. 276
— industries nec, pp. 713, 717
— iron and steel, pp. 494, 497
— leather and products, p. 153

Haiti continued:
— machinery nec, p. 569
— metal products, p. 551
— nonferrous metals, p. 522
— nonmetal products nec, pp. 473, 476
— paper and products, p. 229
— petroleum refineries, p. 368
— petroleum, coal products, p. 387
— plastic products nec, p. 424
— pottery, china, etc., p. 440
— printing, publishing, p. 259
— professional goods, p. 695
— rubber products, p. 404
— textiles, p. 94
— tobacco, pp. 73, 76
— transportation equipment, p. 652
— wearing apparel, p. 132
— wood products, p. 190
Hajdusagi Borgyar, p. 174
Hal Industries Inc, p. 390
Halford Hide & Leather Co. Ltd, p. 157
Halkis Cement Co. SA, p. 480
Hall China Co, p. 443
Halla Cement Manufacturing Corp, p. 480
Halma PLC, pp. 699, 719
Halter Marine Group Inc, p. 657
Hamilton Beach/Proctor-Silex Inc, p. 616
Hamlin Inc, p. 721
Hampshire Group Ltd, pp. 98, 136
Hampson Industries PLC, p. 213
Hampton Industries Inc, p. 137
Han Kuk Fiber Glass Co., Ltd, p. 459
Hanbee Industries Inc, p. 173
Handy and Harman, p. 526
Hanes Hosiery Inc, p. 97
Hanil Synthetic Fiber Co., Ltd, p. 98
Hanjin Heavy Industries Co., Ltd, p. 657
Hanjoo Corp, p. 137
Hankook Chinaware Co., Ltd, p. 460
Hankook Cosmetics Co., Ltd, p. 343
Hankook Shell Oil Co., Ltd, p. 371
Hankook Tire Manufacturing Co., p. 406
Hankuk Electric Glass Co. Ltd, p. 459
Hankuk Glass Industries Inc, p. 459
Hansol Paper Co. Ltd, p. 233
Harald Halberg Tobaksfabrikker AS, p. 80
Harcourt Brace and Co, p. 262
Harcros Chemicals UK Ltd, p. 343
Hard-coal briquettes, p. 391
Hard-Rock Paving Co. Ltd, p. 390
Harley-Davidson Inc, p. 656
Harman International Industries Inc, p. 615
Harper Collins Publishers Ltd, p. 263
HarperCollins Publishers Inc, p. 263
Harris Corp. Harris Semiconductor, p. 615
Harrow Industries Inc, pp. 195, 721

Harrows, p. 577

Hart Holding Company Inc, p. 99

Harte-Hanks Communications Inc, p. 262

Hartmarx Corp, p. 135

Hartono Istana Electronics PT, p. 617

Hartung Agalite Glass Co, p. 461

Hartzell Manufacturing Inc, p. 527

Harvard Industries Inc, p. 655

Havatampa Inc, p. 79

Hayes Lemmerz International Inc, p. 657

Hayes Microcomputer Products Inc, p. 597

H.B. Fuller Automotive Products Inc, p. 389

H.B. Fuller Co, pp. 341, 553

Headlam Group PLC, p. 174

Heaters, p. 637

Heavy distillates, p. 359

Heavy leather, p. 158

Helmerich and Payne Inc, p. 279

Hemisphere Oil Co. Inc, pp. 371, 390

Henkel Corp, p. 279

Henkel Corp. Chemicals Group, p. 343

Henlys Group PLC, p. 657

Henredon Furniture Industries Inc, pp. 214-215

Henry Barrett Group PLC, p. 501

Hepworth Building Products Ltd, p. 480

Her Majesty's Stationery Office, p. 263

Heracles General Cement Co. SA, p. 480

Herbst GmbH, p. 444

Hercules Inc, p. 278

Herend Porcelain Factory, p. 443

Herman Miller Inc, p. 213

Heto Holten AS, p. 371

Heublein do Brasil, p. 60

Hevea-Cameroun, p. 406

Hewlett-Packard (Canada) Ltd, p. 596

Hexcel Corp, pp. 97, 278, 553

Heywood Williams Group PLC, pp. 427, 459, 525

H.H. Brown Shoe Company Inc, p. 172

HI-TEC Sports PLC, p. 174

Hiag AG, p. 194

Hiang Seng Fibre Container Co. Ltd, p. 233

Hickory Chair Co, p. 215

Hickory Springs Manufacturing Co, pp. 406, 554

Hickory White Co, p. 215

Hickson International PLC, pp. 279, 427

Hides, pp. 3, 24, 143

Highlands & Lowlands BHD, p. 18

Hill Rom Company Inc, pp. 213, 698

Hill's Pet Nutrition Inc, p. 20

Hilti Inc, p. 555

Hiltiag, p. 572

Hindustan Cables Ltd, pp. 499, 525

Hindustan Lever Ltd, pp. 18, 135

Hindustan Spinning & Weaving Mills, p. 97

Hiram Walker & Sons Ltd, p. 60

Hiram Walker Group Ltd, p. 278

Hispano-Suiza International PLC, p. 215

Hitchiner Manufacturing Company, p. 499

HL & H Mining Timber, p. 193

HLTH Timber Processors Pty. Ltd, pp. 194, 213

Hoechst UK Ltd, pp. 280, 342, 406

Hoescht do Brasil Quimica e, p. 280

Hoffman Engineering Co, p. 213

Hollinee Corp, p. 460

Hollis PLC, pp. 193, 427

Holly Corp, p. 370

Holstein & Kappert AG, pp. 58, 574

Home Improvement Holdings Ltd, p. 213

Homeshield Fabricated Div, p. 527

HON Industries Inc, pp. 213, 479, 553

Honduras

— basic chemicals, excl fertilizers, pp. 312-313, 315-316, 318

— beverages, pp. 44, 46, 50-51, 55

— chemical products nec, pp. 328, 330, 333, 335, 338

— drugs and medicines, pp. 350-351, 353-354, 356

— electrical machinery, pp. 602, 604, 607, 609, 612

— food products, pp. 4, 7, 10, 12, 15

— footwear, pp. 160, 162, 165-166, 170

— furniture, fixtures, pp. 200, 202, 205, 207, 211

— glass and products, p. 457

— industrial chemicals, pp. 266, 268, 271, 273, 276

— industries nec, pp. 706, 708, 712-713, 717

— iron and steel, pp. 488, 490, 492, 494, 497

— leather and products, pp. 144, 146, 149-150, 153

— machinery nec, pp. 560, 562, 565-566, 569

— metal products, pp. 540, 542, 545, 547, 551

— motor vehicles, pp. 676-677, 679-680, 682

— nonferrous metals, pp. 514, 516, 518-519, 522

— nonmetal products nec, pp. 466, 468, 471, 473, 476

— paper and products, pp. 218, 220, 223, 225, 229

— petroleum refineries, pp. 360-361, 364-365, 368

— petroleum, coal products, p. 387

— plastic products nec, pp. 414, 416, 419, 421, 424

— pottery, china, etc., pp. 432, 434, 436-437, 440

— printing, publishing, pp. 248, 251, 254, 256, 259

— professional goods, pp. 686, 688, 690, 692, 695

— profile, p. 754

— radio, television, etc., pp. 628-629, 631-632, 634

— rubber products, pp. 394, 396, 399-400, 404

— shipbuilding, repair, pp. 668-669, 671-672, 674

— spinning, weaving, etc., pp. 112-113, 115-116, 118

— textiles, pp. 84, 86, 89, 91, 94

— tobacco, pp. 66, 68, 71, 73, 76

— transportation equipment, pp. 642, 644, 647, 649, 652

— wearing apparel, pp. 122, 124, 127, 129, 132

— wood products, pp. 180, 182, 185, 187, 190

Honeywell Inc. Space and Aviation, p. 697

Hooker Furniture Corp, p. 214

Hoop and strip

— cold-reduced, p. 509

— hot-rolled, p. 510

Hoops Ltd, p. 18
Hopkinsons Group PLC, pp. 500, 526
Horace Small Apparel Co, p. 137
Horizon Industries Inc., p. 98
Horton Homes Inc, p. 194
Hose, p. 393
Houghton International Inc, p. 389
Houghton Mifflin Co, p. 263
House of Blend AB, p. 79
House of Borkum Riff AB, p. 79
Household and sanitary paper, p. 236
Household appliances, p. 559
Household furniture, p. 199
Household ware of other ceramic materials, p. 445
Household ware of porcelain or china, p. 445
Houston Chronicle Publishing Co, p. 263
Howden Group PLC, p. 573
H.P. Bulmer Ltd, p. 59
HS Chemical Co., Ltd, p. 407
HS Corp, p. 173
HTH Kokkener AS, p. 215
Hubbell Inc, pp. 615, 697
Huber Tricot GmbH, p. 137
Huck International Inc, p. 555
Huffy Corp, pp. 655, 719
Hulera Centroamericana SA, p. 174
Hulett Aluminum Profiles Pty. Ltd, pp. 525, 554
Huls America Inc, p. 429
Hungary
— basic chemicals, excl fertilizers, pp. 312, 317
— beverages, pp. 44, 46, 48, 50-51, 53, 55
— chemical products nec, pp. 328, 330, 332-333, 335, 337, 339
— drugs and medicines, pp. 350, 355
— electrical machinery, pp. 602, 604, 606-607, 609-610, 612
— food products, pp. 4, 7-8, 10, 12-13, 15
— footwear, pp. 160, 162, 164-166, 168, 170
— furniture, fixtures, pp. 200, 202, 204-205, 207, 209, 211
— glass and products, pp. 448, 450-451, 453-455, 457
— industrial chemicals, pp. 266, 268, 270-271, 273-274, 276
— industries nec, pp. 706, 708, 710, 712-713, 715, 717
— iron and steel, pp. 488, 490-492, 494-495, 497
— leather and products, pp. 144, 146, 148-150, 152-153
— machinery nec, pp. 560, 562-563, 565-566, 568-569
— metal products, pp. 540, 542, 544-545, 547, 549, 551
— motor vehicles, pp. 676, 681
— nonferrous metals, pp. 514, 516-519, 521-522
— nonmetal products nec, pp. 466, 468, 470-471, 473-474, 476
— office, computing machinery, pp. 590, 593
— paper and products, pp. 218, 220, 222-223, 225, 227, 229
— petroleum refineries, pp. 360-361, 363-366, 368
— petroleum, coal products, pp. 380-381, 383-385, 387
— plastic products nec, pp. 414, 416, 418-419, 421-422, 424
— pottery, china, etc., pp. 432, 434-437, 439-440
— printing, publishing, pp. 248, 251-252, 254, 256-257, 259

Hungary continued:
— professional goods, pp. 686, 688-690, 692-693, 695
— profile, p. 755
— pulp, paper, etc., pp. 240, 245
— radio, television, etc., pp. 628, 633
— rubber products, pp. 394, 396, 398-400, 402, 404
— shipbuilding, repair, pp. 668, 673
— spinning, weaving, etc., pp. 112, 117
— synthetic resins, etc., pp. 320, 324
— textiles, pp. 84, 86, 88-89, 91, 93-94
— tobacco, pp. 66, 68, 70-71, 73, 75-76
— transportation equipment, pp. 642, 644, 646-647, 649-6 652
— wearing apparel, pp. 122, 124, 126-127, 129, 131-132
— wood products, pp. 180, 182, 184-185, 187, 189-190
Hunt Corp, pp. 597, 720
Hunt Corp., p. 214
Hunt-Wesson Inc, p. 18
Hunter PLC, p. 193
Huntsman Corp, pp. 278, 427
Huntway Partners L.P., p. 371
Hussmann Corp, p. 572
Huta Bobrek, p. 500
Huta-Buczek SA, p. 500
Huta Szkla, Jaroslaw SA, p. 459
Huta-Zabrze SA, p. 500
Huta Zawiercie SA, p. 499
Hutchinson Technology Inc, p. 595
Hwa Fong Rubber Industrial Co. Ltd, p. 406
Hwa Shin Instrument Co., Ltd, p. 444
Hyde Athletic Industries Inc, p. 174
Hydrated alumina, p. 296
Hydraulic turbines, p. 576
Hydrochloric acid, p. 293
Hydrogen peroxide, p. 298
Hygena Ltd, p. 214
Hyosung Industries Co., Ltd, pp. 574, 617
Hyster Co. North American Industrial, p. 574
Hyundai Pipe Co., Ltd, p. 501
Hyundai Precision & Industry Co., p. 572
IBM Argentina, p. 597
IBM Argentina SA, p. 597
Ibstock PLC, pp. 193, 232, 443, 479
Ice-cream, pp. 3, 27
Iceland
— beverages, pp. 44, 46, 50, 52, 55
— chemical products nec, pp. 328, 330, 333, 335, 339
— electrical machinery, p. 612
— food products, pp. 4, 7, 10, 12, 15
— footwear, pp. 160, 162, 165-166, 170
— furniture, fixtures, pp. 200, 202, 205, 207, 211
— glass and products, pp. 448, 450, 453-454, 457
— industrial chemicals, pp. 266, 268, 271, 273, 276
— industries nec, pp. 706, 708, 712-713, 717
— iron and steel, pp. 488, 490, 492, 494, 497
— leather and products, pp. 144, 146, 149-150, 153

Iceland continued:
— metal products, pp. 540, 542, 545, 547, 551
— nonferrous metals, pp. 514, 516, 518-519, 522
— nonmetal products nec, pp. 466, 468, 471, 473, 476
— paper and products, pp. 218, 220, 223, 225, 229
— petroleum refineries, p. 368
— petroleum, coal products, pp. 380-381, 383-384, 387
— plastic products nec, pp. 414, 416, 419, 421, 424
— pottery, china, etc., pp. 432, 434, 436-437, 441
— printing, publishing, pp. 248, 251, 254, 256, 259
— professional goods, p. 695
— profile, p. 755
— pulp, paper, etc., pp. 240-241, 243-244, 246
— rubber products, p. 404
— shipbuilding, repair, pp. 668-669, 671-672, 674
— spinning, weaving, etc., pp. 112-113, 115-116, 118
— textiles, pp. 84, 86, 89, 91, 94
— tobacco, pp. 66, 68, 76
— transportation equipment, pp. 642, 644, 647, 649, 652
— wearing apparel, pp. 122, 124, 127, 129, 132
— wood products, pp. 180, 182, 185, 187, 191
Icicle Seafoods Inc, p. 20
Icollantas Industria Colombiana de, p. 407
Ideal-Standard Ltd, p. 443
IDEX Corp, p. 574
IDEXX Laboratories Inc, p. 342
Ifo Sanitar AB, p. 460
Iglas (Persero) PT, pp. 460, 481
Igloo Products Corp, p. 428
Ikeda Hoover Ltd, p. 215
Il Yang Pharm Co., Ltd, p. 343
Ilshin Industrial Co., Ltd, p. 174
Ilshin Spinning Co., Ltd, p. 99
I.M. Lockhat, p. 137
Imacc Corp, p. 428
Imasco Ltd, p. 80
Imation Corp, pp. 595, 697
IMC Global Inc, p. 278
Imo Industries Inc, pp. 574, 616, 698
Impex Inc. 2000 C.A, p. 390
IMPOL d.o.o, p. 721
Imprimeries Transcontinental Inc, p. 262
INCO Engineered Products Ltd, p. 526
Inco Europe Ltd, pp. 280, 525, 555
Inco Ltd, pp. 278, 525, 553, 572
Inco United States Inc, p. 500
IND Coope Burton Brewery Ltd, p. 59
India
— basic chemicals, excl fertilizers, pp. 312-313, 315-318
— beverages, pp. 44, 46, 50, 52-53, 55
— chemical products nec, pp. 328, 330, 333, 335, 337, 339
— drugs and medicines, pp. 350-351, 353-356
— electrical machinery, pp. 602, 604, 607, 609-610, 612
— food products, pp. 4, 7, 10, 12-13, 16
— footwear, pp. 160, 162, 165-166, 168, 170
— furniture, fixtures, pp. 200, 202, 205, 207, 209, 211

India continued:
— glass and products, pp. 448, 450, 453-455, 457
— industrial chemicals, pp. 266, 268, 271, 273-274, 276
— industries nec, pp. 706, 709, 712, 714-715, 717
— iron and steel, pp. 488, 490, 492, 494-495, 497
— leather and products, pp. 144, 146, 149-150, 152-153
— machinery nec, pp. 560, 562, 565-566, 568-569
— metal products, pp. 540, 542, 545, 547, 549, 551
— motor vehicles, pp. 676-677, 679-682
— nonferrous metals, pp. 514, 516, 518-519, 521-522
— nonmetal products nec, pp. 466, 468, 471, 473-474, 476
— office, computing machinery, pp. 590-594
— paper and products, pp. 218, 220, 223, 225, 227, 229
— petroleum refineries, pp. 360-361, 364-366, 368
— petroleum, coal products, pp. 380-381, 383-385, 387
— plastic products nec, pp. 414, 416, 419, 421-422, 424
— pottery, china, etc., pp. 432, 434, 436-437, 439, 441
— printing, publishing, pp. 248, 251, 254, 256-257, 259
— professional goods, pp. 686, 688, 690, 692-693, 695
— profile, p. 756
— pulp, paper, etc., pp. 240-241, 243-246
— radio, television, etc., pp. 628-629, 631-634
— rubber products, pp. 394, 396, 399-400, 402, 404
— shipbuilding, repair, pp. 668-669, 671-674
— spinning, weaving, etc., pp. 112-113, 115-118
— synthetic resins, etc., pp. 320-325
— textiles, pp. 84, 86, 89, 91, 93-94
— tobacco, pp. 66, 68, 71, 73, 75-76
— transportation equipment, pp. 642, 644, 647, 649-650, 652
— wearing apparel, pp. 122, 124, 127, 129, 131-132
— wood products, pp. 180, 182, 185, 187, 189, 191
Indian Head Div, p. 195
Indian Head Industries Inc, p. 408
Indocement Tunggal, pp. 19, 97
Indocement Tunggal Prakarsa PT, p. 479
Indomobil Suzuki International, PT, pp. 427, 555, 657
Indonesia
— basic chemicals, excl fertilizers, pp. 312-318
— beverages, pp. 44, 46, 48, 50, 52-53, 55
— chemical products nec, pp. 328, 330, 332-333, 335, 337, 339
— drugs and medicines, pp. 350-356
— electrical machinery, pp. 602, 604, 606-607, 609-610, 612
— food products, pp. 4, 7-8, 10, 12-13, 16
— footwear, pp. 160, 162, 164-166, 168, 170
— furniture, fixtures, pp. 200, 202, 204-205, 207, 209, 211
— glass and products, pp. 448, 450-451, 453-455, 457
— industrial chemicals, pp. 266, 268, 270-271, 273-274, 276
— industries nec, pp. 706, 709-710, 712, 714-715, 717
— iron and steel, pp. 488, 490-492, 494-495, 497
— leather and products, pp. 144, 146, 148-150, 152-153
— machinery nec, pp. 560, 562-563, 565-566, 568-569
— metal products, pp. 540, 542, 544-545, 547, 549, 551
— motor vehicles, pp. 676-682
— nonferrous metals, pp. 514, 516-519, 521-522
— nonmetal products nec, pp. 466, 468, 470-471, 473-474,

Indonesia continued:
476
— office, computing machinery, pp. 590-594
— paper and products, pp. 218, 220, 222-223, 225, 227, 229
— petroleum refineries, pp. 360-361, 363-366, 368
— petroleum, coal products, pp. 380-385, 387
— plastic products nec, pp. 414, 416, 418-419, 421-422, 424
— pottery, china, etc., pp. 432, 434-437, 439, 441
— printing, publishing, pp. 248, 251-252, 254, 256-257, 259
— professional goods, pp. 686, 688-690, 692-693, 695
— profile, p. 756
— pulp, paper, etc., pp. 240-246
— radio, television, etc., pp. 628-634
— rubber products, pp. 394, 396, 398-400, 402, 404
— shipbuilding, repair, pp. 668-674
— spinning, weaving, etc., pp. 112-118
— synthetic resins, etc., pp. 320-325
— textiles, pp. 84, 86, 88-89, 91, 93-94
— tobacco, pp. 66, 68, 70-71, 73, 75-76
— transportation equipment, pp. 642, 644, 646-647, 649-650, 652
— wearing apparel, pp. 122, 124, 126-127, 129, 131-132
— wood products, pp. 180, 182, 184-185, 187, 189, 191
Indonesia Asahan Aluminium PT, p. 525
Indorama Synthetics, PT, p. 97
INDRESCO Inc, p. 479
Induroman Industrias Roman SA, p. 60
Induscuer S.C.A, p. 156
Industex Eastern Cape Pty. Ltd, pp. 99, 136, 406, 428
Industri Keramik Angsa Daya PT, p. 481
Industria Centroamericana de Vidrio, p. 460
Industria de Ceramica Costarricense, p. 443
Industrial Acoustics Company Inc, p. 481
Industrial chemicals, p. 265
Industrial de Gaseosas SA, p. 59
Industrial refrigerators and freezers, p. 619
Industrias Alimenticias Noel SA, p. 19
Industrias Metalurgicas Unidas SA, p. 527
Industrija usnja d.d, p. 156
Industry Modernization & Mining, p. 195
Ingersoll International Inc, p. 574
Ingots for tubes, p. 505
Inini d.d, p. 157
Inland Cement Ltd, p. 481
Inner tubes, pp. 393, 409
Innovex Inc, p. 597
Inorganics, p. 265
Inoue Rubber Indonesia PT, p. 407
Input or output units, p. 598
Insecticides, p. 303
Insilco Corp, pp. 262, 341, 553
Instituto Nacional de Seguro, pp. 342, 699
Instrochem SP, p. 279
Instrument Specialties Company Inc, p. 408
Instruments, p. 685
Insulating materials, p. 465

Insulators, p. 431
Intalco Aluminum Corp, p. 526
Integrated Device Technology Inc, p. 617
Intel Puerto Rico Inc, p. 597
Interbake Foods Inc, p. 20
Intercraft Co, pp. 194, 428
Interface Inc., pp. 97, 278
Intergraph Corp, p. 595
Interlake Corp, pp. 213, 428, 526
Interlake Material Handling Div, p. 214
Intermagnetics General Corp, p. 443
Intermec Technologies Corp, pp. 232, 595
Intermet Corp, pp. 499, 525
Intermet Foundries Inc, p. 499
Intermoda, p. 136
International Aluminum Corp, p. 526
International Business Machines, p. 596
International Distillers & Vintners Ltd, p. 60
International Flavors and Fragrances, p. 279
International Forest Products Ltd, p. 193
International Game Technology Inc, p. 719
International General Electric USA, p. 698
International Group Inc, p. 371
International MultiFoods Corp, p. 18
International Paper Holdings (UK), pp. 194, 699
International Rectifier Corp, p. 616
International Specialty Products Inc, p. 280
International Thomson Publishing, p. 262
Interstate Electronics Corp, p. 597
Intertape Polymer Group Inc, pp. 99, 233
Inti Indorayon Utama PT, pp. 232, 280
Invacare Corp, pp. 213, 698
Invergordon Distillers Holdings PLC, p. 60
Iochpe-Maxion SA, p. 655
IOMEGA Corp, p. 595
Ionics Inc, p. 59
IPC Inc, p. 232
IPC Magazines Ltd. Overseas, p. 263
IPEX Inc, p. 428
IPL Ltd, p. 263
Iran (Islamic Republic of)
— basic chemicals, excl fertilizers, pp. 312-317
— beverages, pp. 44, 46, 48, 50, 52-53, 55
— chemical products nec, pp. 328, 330, 332-333, 335, 337, 339
— drugs and medicines, pp. 350-355
— electrical machinery, pp. 602, 604, 606-607, 609-610, 612
— food products, pp. 4, 7-8, 10, 12-13, 16
— footwear, pp. 160, 162, 164-166, 168, 170
— furniture, fixtures, pp. 200, 202, 204-205, 207, 209, 211
— glass and products, pp. 448, 450-451, 453-455, 457
— industrial chemicals, pp. 266, 268, 270-271, 273-274, 276
— industries nec, pp. 706, 709-710, 712, 714-715, 717
— iron and steel, pp. 488, 490-492, 494-495, 497
— leather and products, pp. 144, 146, 148-150, 152-153
— machinery nec, pp. 560, 562-563, 565-566, 568-569

Iran (Islamic Republic of) continued:
— metal products, pp. 540, 542, 544-545, 547, 549, 551
— motor vehicles, pp. 676-681
— nonferrous metals, pp. 514, 516-519, 521-522
— nonmetal products nec, pp. 466, 468, 470-471, 473-474, 476
— office, computing machinery, pp. 590-593
— paper and products, pp. 218, 220, 222-223, 225, 227, 229
— petroleum refineries, pp. 360, 362-366, 368
— petroleum, coal products, pp. 380-385, 387
— plastic products nec, pp. 414, 416, 418-419, 421-422, 424
— pottery, china, etc., pp. 432, 434-436, 438-439, 441
— printing, publishing, pp. 248, 251-252, 254, 256-257, 259
— professional goods, pp. 686, 688-690, 692-693, 695
— profile, p. 757
— pulp, paper, etc., pp. 240-245
— radio, television, etc., pp. 628-633
— rubber products, pp. 394, 396, 398-400, 402, 404
— shipbuilding, repair, pp. 668-673
— spinning, weaving, etc., pp. 112-117
— synthetic resins, etc., pp. 320-324
— textiles, pp. 84, 86, 88-89, 91, 93, 95
— tobacco, pp. 66, 68, 70-71, 73, 75-76
— transportation equipment, pp. 642, 644, 646-647, 649, 651-652
— wearing apparel, pp. 122, 124, 126-127, 129, 131-132
— wood products, pp. 180, 182, 184-185, 187, 189, 191

Iraq
— beverages, pp. 44, 46, 48, 50, 52, 55
— chemical products nec, pp. 328, 330, 332-333, 335, 339
— electrical machinery, pp. 602, 604, 606-607, 609, 612
— food products, pp. 4, 7-8, 10, 12, 16
— footwear, pp. 160, 162, 164-166, 170
— furniture, fixtures, pp. 202, 204-205, 207, 211
— glass and products, pp. 448, 450-451, 453-454, 457
— industrial chemicals, pp. 266, 268, 270-271, 273, 276
— industries nec, pp. 706, 709, 712, 714, 717
— leather and products, pp. 144, 146, 149-150, 153
— machinery nec, pp. 560, 562-563, 565-566, 569
— metal products, pp. 540, 542, 544-545, 547, 551
— nonmetal products nec, pp. 466, 468, 470-471, 473, 476
— paper and products, pp. 218, 220, 222-223, 225, 229
— petroleum refineries, pp. 360, 362-365, 368
— petroleum, coal products, pp. 380-384, 387
— plastic products nec, pp. 414, 416, 418-419, 421, 424
— pottery, china, etc., pp. 432, 434-436, 438, 441
— printing, publishing, pp. 248, 251-252, 254, 256, 259
— professional goods, p. 695
— profile, p. 757
— rubber products, pp. 394, 396, 398-399, 401, 404
— textiles, pp. 84, 86, 88-89, 91, 95
— tobacco, pp. 66, 68, 70-71, 73, 76
— transportation equipment, pp. 642, 644, 646-647, 649, 652
— wearing apparel, pp. 122, 124, 126-127, 129, 132
— wood products, pp. 182, 184-185, 187, 191

Ireland
— beverages, pp. 44, 46, 48, 50, 52-53, 55
— chemical products nec, pp. 330, 332-333, 335, 339
— electrical machinery, pp. 602, 604, 606-607, 609-610, 612
— food products, pp. 4, 7-8, 10, 12, 14, 16
— footwear, pp. 160, 162, 164-166, 168, 170
— furniture, fixtures, pp. 200, 202, 204-205, 207, 209, 211
— glass and products, pp. 448, 450-451, 453-455, 457
— industrial chemicals, pp. 268, 270-271, 273, 276
— industries nec, pp. 706, 709-710, 712, 714-715, 717
— iron and steel, pp. 488, 490-492, 494-495, 497
— leather and products, pp. 144, 146, 148-150, 152, 154
— machinery nec, pp. 560, 562, 564-566, 568-569
— metal products, pp. 540, 542, 544-545, 547, 549, 551
— nonferrous metals, pp. 514, 516, 518-519, 521-522
— nonmetal products nec, pp. 466, 468, 470-471, 473-474, 476
— paper and products, pp. 218, 220, 222-223, 225, 227, 229
— plastic products nec, pp. 414, 416, 418-419, 421-422, 424
— pottery, china, etc., pp. 432, 434-436, 438-439, 441
— printing, publishing, pp. 248, 251-252, 254, 256-257, 259
— professional goods, pp. 686, 688-690, 692-693, 695
— profile, p. 758
— rubber products, pp. 394, 396, 398-399, 401-402, 404
— textiles, pp. 84, 86, 88-89, 91, 93, 95
— tobacco, pp. 66, 68, 70-71, 73, 75, 77
— transportation equipment, pp. 642, 644, 646-647, 649, 651-652
— wearing apparel, pp. 122, 124, 126-127, 129, 131-132
— wood products, pp. 180, 182, 184-185, 187, 189, 191

Iris Manufacturier de Bas Inc, p. 99
Iron, p. 487
Irons, p. 601
— electric smoothing, p. 638
Irvin & Johnson Ltd, pp. 18, 58
Irving Tanning Co, p. 156
Iskra Industrija Kondenzatorjev in, p. 720
Isofoam Insulating Materials Plants, p. 444
Ispat Indo PT, p. 501

Israel
— beverages, pp. 44, 46, 50, 52-53, 55
— chemical products nec, pp. 335, 339
— electrical machinery, pp. 602, 604, 607, 609-610, 612
— food products, pp. 4, 7, 10, 12, 14, 16
— footwear, pp. 160, 162, 165, 167-168, 170
— furniture, fixtures, pp. 200, 202, 205, 207, 209, 211
— glass and products, pp. 448, 450, 453-455, 457
— industrial chemicals, pp. 266, 268, 271, 273-274, 276
— industries nec, pp. 706, 709, 712, 714-715, 717
— iron and steel, pp. 488, 490, 492, 494-495, 497
— leather and products, pp. 144, 146, 149-150, 152, 154
— machinery nec, pp. 560, 562, 565-566, 568-569
— metal products, pp. 540, 542, 545, 547, 549, 551
— nonferrous metals, pp. 514, 516, 518-519, 521-522
— nonmetal products nec, pp. 466, 468, 471, 473-474, 476
— paper and products, pp. 218, 220, 223, 225, 227, 229

Israel continued:
— petroleum refineries, pp. 365, 368
— petroleum, coal products, pp. 384, 387
— plastic products nec, pp. 414, 416, 419, 421-422, 424
— pottery, china, etc., pp. 432, 434, 436, 438-439, 441
— printing, publishing, pp. 248, 251, 254, 256-257, 259
— professional goods, pp. 686, 688, 690, 692-693, 695
— profile, p. 758
— rubber products, pp. 394, 396, 399, 401-402, 404
— textiles, pp. 84, 86, 89, 91, 93, 95
— tobacco, pp. 66, 68, 71, 73, 75, 77
— transportation equipment, pp. 642, 644, 647, 649, 651-652
— wearing apparel, pp. 122, 124, 127, 129, 131-132
— wood products, pp. 180, 182, 185, 187, 189, 191
Israel Chemicals Ltd, p. 278
Israel Military Industries Ltd, p. 554
Italy
— basic chemicals, excl fertilizers, pp. 312-318
— beverages, pp. 44, 46, 48, 50, 52-53, 55
— chemical products nec, pp. 328, 330, 332-333, 335, 337, 339
— electrical machinery, pp. 602, 604, 606-607, 609-610, 612
— food products, pp. 4, 7-8, 10, 12, 14, 16
— footwear, pp. 160, 162, 164-165, 167-168, 170
— furniture, fixtures, pp. 200, 202, 204-205, 207, 209, 211
— glass and products, pp. 448, 450-451, 453-455, 457
— industrial chemicals, pp. 266, 268, 270-271, 273-274, 276
— industries nec, pp. 706, 709-710, 712, 714-715, 717
— iron and steel, pp. 488, 490-492, 494-495, 497
— leather and products, pp. 144, 146, 148-150, 152, 154
— machinery nec, pp. 560, 562, 564-566, 568-569
— metal products, pp. 540, 542, 544-545, 547, 549, 551
— motor vehicles, pp. 676-682
— nonferrous metals, pp. 514, 516-519, 521-522
— nonmetal products nec, pp. 466, 468, 470-471, 473-474, 476
— office, computing machinery, pp. 590-594
— paper and products, pp. 218, 220, 222-223, 225, 227, 229
— petroleum refineries, pp. 360, 362-366, 368
— petroleum, coal products, pp. 380-385, 387
— plastic products nec, pp. 414, 416, 418-419, 421-422, 424
— pottery, china, etc., pp. 432, 434-436, 438-439, 441
— printing, publishing, pp. 248, 251-252, 254, 256-257, 259
— professional goods, pp. 686, 688-690, 692-693, 695
— profile, p. 759
— pulp, paper, etc., pp. 240-246
— radio, television, etc., pp. 628-634
— rubber products, pp. 394, 396, 398-399, 401-402, 404
— shipbuilding, repair, pp. 668-674
— synthetic resins, etc., pp. 320-325
— textiles, pp. 84, 86, 88-89, 91, 93, 95
— tobacco, pp. 66, 68, 70-71, 73, 75, 77
— transportation equipment, pp. 642, 644, 646-647, 649, 651-652
— wearing apparel, pp. 122, 124, 126-127, 129, 131-132
— wood products, pp. 180, 182, 184-185, 187, 189, 191

Itautec Philco, p. 616
ITEQ Inc, p. 555
Ithaca Industries Inc, pp. 97, 135
ITT Automotive Precision Die, p. 527
ITT Flygt AB, p. 574
ITT Higbie Baylock Manufacturing, p. 406
ITW Finishing Systems, pp. 555, 698
ITW Ltd, p. 428
IVAX Corp, p. 341
Iveco Magirus AG, p. 655
Iwka AG, pp. 553, 572, 615, 697
J & J Fashions Ltd, p. 136
J and J Snack Foods Corp, p. 59
J and L Specialty Steel Inc, p. 501
J. Bibby & Sons PLC, pp. 18, 231, 459, 697
Ja/Mont UK Ltd, p. 233
Jabil Circuit Inc, p. 617
Jabur Pneus SA, p. 408
Jackets, p. 138
Jaclyn Inc, p. 174
Jacob's Bakery Ltd, p. 20
Jaeger Tailoring Ltd, p. 137
Jafora-Tabori Ltd, p. 60
Jamaica
— beverages, pp. 44, 46, 50, 52, 55
— chemical products nec, pp. 328, 330, 339
— electrical machinery, p. 612
— food products, pp. 5, 7, 10, 12, 16
— footwear, pp. 165, 167, 170
— furniture, fixtures, pp. 200, 202, 205, 207, 211
— glass and products, p. 457
— industrial chemicals, pp. 266, 268, 276
— industries nec, pp. 706, 709, 712, 714, 717
— leather and products, pp. 149-150, 154
— machinery nec, p. 569
— metal products, p. 551
— nonmetal products nec, pp. 468, 476
— paper and products, pp. 218, 220, 229
— petroleum refineries, pp. 360, 364-365, 368
— petroleum, coal products, pp. 380, 387
— plastic products nec, pp. 414, 416, 424
— pottery, china, etc., p. 441
— printing, publishing, pp. 249, 251, 259
— profile, p. 759
— rubber products, pp. 394, 396, 404
— textiles, pp. 84, 86, 95
— tobacco, pp. 66, 68, 71, 73, 77
— transportation equipment, p. 652
— wearing apparel, pp. 122, 124, 132
— wood products, pp. 180, 182, 185, 187, 191
James Seddon UK Ltd, p. 136
James Walker Group Ltd, p. 428
Jams, p. 28
Jannock Ltd, pp. 427, 479, 553
Jantzen Inc, p. 137

Japan
—basic chemicals, excl fertilizers, pp. 312-313, 315-318
—beverages, pp. 44, 47, 50, 52-53, 55
—chemical products nec, pp. 328, 330, 333, 335, 337, 339
—drugs and medicines, pp. 350-351, 353-356
—electrical machinery, pp. 602, 604, 607, 609-610, 612
—food products, pp. 5, 7, 10, 12, 14, 16
—footwear, pp. 160, 162, 165, 167-168, 170
—furniture, fixtures, pp. 200, 202, 205, 207, 209, 211
—glass and products, pp. 448, 450, 453-455, 457
—industrial chemicals, pp. 266, 268, 271, 273-274, 276
—industries nec, pp. 706, 709, 712, 714-715, 717
—iron and steel, pp. 488, 490, 492, 494-495, 497
—leather and products, pp. 144, 146, 149-150, 152, 154
—machinery nec, pp. 560, 562, 565-566, 568-569
—metal products, pp. 540, 542, 545, 547, 549, 551
—motor vehicles, pp. 676-677, 679-682
—nonferrous metals, pp. 514, 516, 518-519, 521-522
—nonmetal products nec, pp. 466, 468, 471, 473-474, 476
—office, computing machinery, pp. 590-594
—paper and products, pp. 218, 220, 223, 225, 227, 229
—petroleum refineries, pp. 360, 362, 364-366, 368
—petroleum, coal products, pp. 380-381, 383-385, 387
—plastic products nec, pp. 414, 416, 419, 421-422, 424
—pottery, china, etc., pp. 432, 434, 436, 438-439, 441
—printing, publishing, pp. 249, 251, 254, 256-257, 259
—professional goods, pp. 686, 688, 690, 692-693, 695
—profile, p. 760
—pulp, paper, etc., pp. 240-241, 243-246
—radio, television, etc., pp. 628-629, 631-634
—rubber products, pp. 394, 396, 399, 401-402, 404
—shipbuilding, repair, pp. 668-669, 671-674
—spinning, weaving, etc., pp. 112-113, 115-118
—synthetic resins, etc., pp. 320-325
—textiles, pp. 84, 86, 89, 91, 93, 95
—tobacco, pp. 66, 68, 71, 73, 75, 77
—transportation equipment, pp. 642, 644, 647, 649, 651-652
—wearing apparel, pp. 122, 124, 127, 129, 131-132
—wood products, pp. 180, 182, 185, 187, 189, 191
Japfa Comfeed Indonesia PT, p. 18
Jarrold & Sons Ltd, p. 264
Jason Inc, pp. 194, 719
Jasper Rubber Products Inc, p. 408
Javor Koncern d.o.o, p. 720
JDI Group Inc, p. 213
J.E. Baker Co, p. 481
Jel Sert Co, p. 60
Jelovica d.d, p. 721
Jerrold Communications Div, p. 617
Jet fuel, pp. 359, 373
Jeyes Group PLC, p. 343
Jia Hsing Enterprise Co. Ltd, p. 156
Jim Beam Brands Co, p. 59
Jinyork Enterprise Co., Ltd, p. 157
Jiyajeerao Cotton Mills Ltd, pp. 97, 278
J.M. Smucker Co, p. 59

JMK International Inc, p. 408
Joan Fabrics Corp, p. 99
John Carr Group PLC, p. 194
John Crane Inc, p. 406
John Fyfe Ltd, p. 481
John H. Harland Co, pp. 262, 595
John Middleton Inc, p. 80
John Tams Group PLC, pp. 443, 460
John Wiley and Sons Inc, p. 263
Johns Manville Corp, pp. 231, 479
Johnson & Johnson Management, p. 698
Johnson Controls UK Ltd, p. 195
Johnson Matthey Inc, p. 526
Johnson Matthey PLC, p. 553
Johnson Worldwide Associates Inc, pp. 137, 720
Johnston Industries Inc, p. 98
Johnstown America Industries Inc, p. 657
Joists, p. 179
Jones Apparel Group Inc, p. 136
Jones Stroud Holdings PLC, p. 407
Joongwon Co., Ltd, p. 174
Jordan
—basic chemicals, excl fertilizers, pp. 312-318
—beverages, pp. 45, 47-48, 50, 52-53, 55
—chemical products nec, pp. 328, 330, 332-333, 335, 337, 339
—drugs and medicines, pp. 350-353, 355-356
—electrical machinery, pp. 602, 604, 606-607, 609-610, 612
—food products, pp. 5, 7-8, 10, 12, 14, 16
—footwear, pp. 160, 162, 164-165, 167-168, 170
—furniture, fixtures, pp. 200, 203-205, 207, 209, 211
—glass and products, pp. 448, 450-451, 453-455, 457
—industrial chemicals, pp. 266, 268, 270-271, 273-274, 276
—industries nec, pp. 707, 709-710, 712, 714-715, 717
—iron and steel, pp. 488, 490-492, 494-495, 497
—leather and products, pp. 144, 146, 148-150, 152, 154
—machinery nec, pp. 560, 562, 564-566, 568-569
—metal products, pp. 540, 542, 544-545, 547, 549, 551
—motor vehicles, pp. 676-680, 682
—nonferrous metals, pp. 514, 516-519, 521-522
—nonmetal products nec, pp. 466, 468, 470-471, 473, 475, 477
—paper and products, pp. 218, 220, 222-223, 225, 227, 229
—petroleum refineries, pp. 360, 362-366, 368
—petroleum, coal products, p. 387
—plastic products nec, pp. 414, 416, 418-419, 421-422, 424
—pottery, china, etc., pp. 432, 434-436, 438-439, 441
—printing, publishing, pp. 249, 251-252, 254, 256-257, 260
—professional goods, pp. 686, 688-690, 692-693, 695
—profile, p. 760
—pulp, paper, etc., pp. 240-246
—rubber products, pp. 394, 396, 398-399, 401-402, 404
—shipbuilding, repair, pp. 668-669, 671-672
—spinning, weaving, etc., pp. 112-118
—synthetic resins, etc., pp. 320-325
—textiles, pp. 84, 86, 88-89, 91, 93, 95

Jordan continued:
— tobacco, pp. 66, 68, 70-71, 73, 75, 77
— transportation equipment, pp. 642, 644, 646-647, 649, 651, 653
— wearing apparel, pp. 122, 124, 126-127, 129, 131, 133
— wood products, pp. 180, 182, 184-185, 187, 189, 191
Jordan Industries Inc, pp. 263, 555
Jos. A. Bank Clothiers Inc, p. 137
Joseph Holt PLC, p. 59
Josip Kras, p. 19
Jostens Inc, pp. 135, 262, 719
Jostens Inc. School Products Group, p. 262
Journal Communications Inc, p. 262
Journal Register Co, p. 262
Joy Mining Machinery, p. 573
J.P. Stevens and Company Inc, p. 97
JSJ Corp, pp. 428, 555
JTS Corp, p. 595
Juan Minetti SA, p. 481
Jumbo Enterprise Co.Ltd, p. 461
Jumping Jacks Div, p. 174
Junckers Industrier AS, pp. 194, 214, 233, 343
Justin Boot Co, p. 172
Justin Industries Inc, pp. 172, 262, 479
Jute, p. 83
Jute fabrics, p. 105
Jute yarn, pp. 83, 103
K-Products Inc, p. 137
K2 Inc, pp. 135, 719
Kabool Ltd, p. 99
Kaiser Aluminum Corp, p. 525
Kalamazoo PLC, p. 596
Kalbe Farma PT, p. 459
Kaman Corp, pp. 656, 719
Kangar Consolidated Industries PT, p. 460
Karim Jute Mills Ltd, p. 98
Karsten Manufacturing Corp, p. 720
Katepal Oy, p. 390
KCM SA, pp. 280, 526
Kearfott Guidance and Navigation, p. 699
KEC International Ltd, p. 554
Kedaung Industrial Ltd, pp. 232, 459
Keller Industries Inc, p. 214
Kellogg Co. of Great Britain, Ltd, p. 20
Kelly-Springfield Tire Co, p. 406
Kelsey Industries PLC, p. 527
Kenda Rubber Industrial Co. Ltd, p. 407
Kennametal Inc, p. 572
Kenney Manufacturing Co, p. 215
Kenworth Truck Co, p. 657
Kenya
— basic chemicals, excl fertilizers, pp. 312-314, 318
— beverages, pp. 47-48, 50, 52, 55
— chemical products nec, pp. 328, 330, 332-333, 335, 339
— drugs and medicines, pp. 350-352, 356
— electrical machinery, pp. 602, 604, 606-607, 609, 612

Kenya continued:
— food products, pp. 5, 7-8, 10, 12, 16
— footwear, pp. 160, 162, 164-165, 167, 170
— furniture, fixtures, pp. 200, 203-205, 207, 211
— glass and products, pp. 448, 450, 452-454, 457
— industrial chemicals, pp. 266, 268, 270-271, 273, 276
— industries nec, pp. 707, 709-710, 712, 714, 717
— leather and products, pp. 144, 146, 148-149, 151, 154
— machinery nec, pp. 560, 562, 564-566, 570
— metal products, pp. 540, 542, 544-545, 547, 551
— motor vehicles, pp. 676-678, 682
— nonmetal products nec, pp. 466, 468, 470-471, 473, 47
— paper and products, pp. 218, 220, 222-223, 225, 229
— petroleum refineries, pp. 360, 362-365, 368
— plastic products nec, pp. 414, 416, 418-419, 421, 424
— pottery, china, etc., pp. 432, 434-436, 438, 441
— printing, publishing, pp. 249, 251-252, 254, 256, 260
— professional goods, pp. 686, 688-690, 692, 695
— profile, p. 761
— pulp, paper, etc., pp. 240-242, 246
— rubber products, pp. 394, 396, 398-399, 401, 404
— shipbuilding, repair, pp. 668-670, 674
— spinning, weaving, etc., pp. 112-114, 118
— textiles, pp. 84, 86, 88-89, 91, 95
— tobacco, pp. 68, 70-71, 73, 77
— transportation equipment, pp. 642, 644, 646-647, 649,
— wearing apparel, pp. 122, 124, 126-127, 129, 133
— wood products, pp. 180, 182, 184-185, 187, 191
Kenyon UK, p. 370
Kerosene, pp. 359, 374
Kerr-McGee Corp, p. 279
Kertas Leces (Persero) PT, p. 233
Kessler Industries Inc., p. 215
Kety, pp. 526, 555
Keumkang Ltd, p. 459
Kewaunee Scientific Corp, p. 215
Key Tronic Corp, p. 595
Keystone Consolidated Industries, p. 500
Keystone International Inc, p. 554
Kia Steel Co., Ltd, p. 500
Kieleckie Zaklady Wyrobow, p. 233
Kimball International Inc, pp. 193, 213, 719
Kimberly-Clark Inc, p. 232
Kimberly-Clark Ltd, p. 232
Kimble Glass Inc, p. 459
Kinetic Concepts Inc, p. 213
Kingsley Enamels Ltd, p. 444
Kingston-Warren Corp, p. 407
Kingsway Group PLC, p. 214
Kirkhill Rubber Co, p. 408
Kirsch, p. 213
Kitan Consolidated Ltd, p. 98
Kitchenware, p. 431
Kiwi Holdings Ltd, pp. 280, 342
KLA-Tencor Corp, p. 698
Klaussner Furniture Industries Inc, p. 213

Klein Tools Inc, pp. 173, 389
Kleinert's Inc, p. 137
Klimmex SA, Industria e Comercio, p. 720
Knape and Vogt Manufacturing Co, p. 214
Knitted fabrics, pp. 83, 107
Knitted outer garments
—other, p. 109
Knitted sports shirts, p. 108
Knitted sweaters, p. 108
Knitted undergarments, p. 108
Knitting mills, p. 83
Knoll Inc, pp. 98, 213
Knoll North America Corp, p. 215
Knott's Berry Farm Foods Inc, p. 20
Koch Refining Co, p. 370
Kodak Brasileira Com. e Ind. SA, p. 698
Kodeco Batu Licin Plywood PT, p. 194
Kokomo Sanitary Pottery Div, p. 443
Kolektor d.o.o, p. 721
Kollmorgen Corp, p. 699
Kolon Industries Inc, p. 97
Komag Inc, pp. 595, 616
Konus Konum d.o.o, p. 174
Korea Computer Inc, p. 597
Korea Iron & Steel Co., Ltd, p. 501
Korea Kumho Petrochemical Co., p. 408
Korea Xerox Co., Ltd, p. 596
Korea, Republic of
—basic chemicals, excl fertilizers, pp. 312-318
—beverages, pp. 45, 47-48, 50, 52-53, 55
—chemical products nec, pp. 328, 330, 332-333, 335, 337,
   339
—drugs and medicines, pp. 350-353, 355-356
—electrical machinery, pp. 602, 604, 606-607, 609-610, 612
—food products, pp. 5, 7-8, 10, 12, 14, 16
—footwear, pp. 160, 162, 164-165, 167-168, 170
—furniture, fixtures, pp. 200, 203-205, 207, 209, 211
—glass and products, pp. 448, 450, 452-455, 457
—industrial chemicals, pp. 266, 268, 270-271, 273-274, 276
—industries nec, pp. 707, 709-710, 712, 714-715, 717
—iron and steel, pp. 488, 490-492, 494-495, 497
—leather and products, pp. 144, 146, 148-149, 151-152, 154
—machinery nec, pp. 560, 562, 564-566, 568, 570
—metal products, pp. 540, 542, 544-545, 547, 549, 551
—motor vehicles, pp. 676-682
—nonferrous metals, pp. 514, 516-519, 521-522
—nonmetal products nec, pp. 466, 469-471, 473, 475, 477
—office, computing machinery, pp. 590-594
—paper and products, pp. 218, 220, 222-223, 225, 227, 229
—petroleum refineries, pp. 360, 362-366, 368
—petroleum, coal products, pp. 380-385, 387
—plastic products nec, pp. 414, 416, 418-419, 421-422, 424
—pottery, china, etc., pp. 432, 434-436, 438-439, 441
—printing, publishing, pp. 249, 251-252, 254, 256-257, 260
—professional goods, pp. 686, 688-690, 692-693, 695
—profile, p. 761

Korea, Republic of continued:
—pulp, paper, etc., pp. 240-246
—radio, television, etc., pp. 628-634
—rubber products, pp. 394, 396, 398-399, 401-402, 404
—shipbuilding, repair, pp. 668-674
—spinning, weaving, etc., pp. 112-118
—synthetic resins, etc., pp. 320-325
—textiles, pp. 84, 86, 88-89, 91, 93, 95
—tobacco, pp. 66, 68, 70-71, 73, 75, 77
—transportation equipment, pp. 642, 644, 646-647, 649,
   651, 653
—wearing apparel, pp. 122, 124, 126-127, 129, 131, 133
—wood products, pp. 180, 183-185, 187, 189, 191
Koryo Co., Ltd, p. 173
Kovinoplastika Loz d.d, p. 721
KPT Industries, Ltd, p. 481
K.R. Edwards Leaf Tobacco, p. 79
Kraft Canada Inc, p. 19
Kraft paper, pp. 217, 239
Krause's Furniture Inc, p. 214
Krka, Tovarna Zdravil d.d, p. 342
Kroehler-Dunmore Furniture, p. 215
Krones AG Hermann Kronseder, p. 573
Kronos Inc, p. 596
Krueger International Inc, p. 213
Kruger Inc, p. 231
Kuan Show International Corp, p. 173
Kuhlman Corp, p. 555
Kukje Corp, p. 172
Kumbong Container PT, p. 500
Kumkang Shoe Manufacturing Co., p. 172
Kunnan Enterprise, Ltd, pp. 172, 596, 719
Kuwait
—basic chemicals, excl fertilizers, pp. 312-313, 315-318
—beverages, pp. 45, 47, 50, 52-53, 55
—chemical products nec, pp. 328, 331, 333, 335, 337, 339
—electrical machinery, pp. 602, 604, 607, 609, 611-612
—food products, pp. 5, 7, 10, 12, 14, 16
—footwear, pp. 160, 162, 165, 167-168, 170
—furniture, fixtures, pp. 200, 203, 205, 207, 209, 211
—glass and products, pp. 448, 450, 453-455, 457
—industrial chemicals, pp. 266, 268, 271, 273-274, 276
—industries nec, pp. 707, 709-710, 712, 714-715, 717
—iron and steel, pp. 488, 490, 492, 494-495, 497
—leather and products, pp. 144, 146, 149, 151-152, 154
—machinery nec, pp. 560, 562, 565-566, 568, 570
—metal products, pp. 541, 543, 545, 547, 549, 551
—motor vehicles, pp. 676-677, 679-682
—nonferrous metals, p. 522
—nonmetal products nec, pp. 466, 469, 471, 473, 475, 477
—office, computing machinery, pp. 590-593
—paper and products, pp. 218, 221, 223, 225, 227, 229
—petroleum refineries, pp. 360, 362, 364-366, 368
—petroleum, coal products, pp. 380-381, 383-385, 387
—plastic products nec, pp. 414, 416, 419, 421-422, 424
—pottery, china, etc., p. 441

Kuwait continued:
— printing, publishing, pp. 249, 251, 254, 256-257, 260
— professional goods, p. 695
— profile, p. 762
— rubber products, pp. 394, 396, 399, 401-402, 404
— shipbuilding, repair, pp. 668-669, 671-674
— textiles, pp. 85-86, 89, 91, 93, 95
— tobacco, p. 77
— transportation equipment, pp. 642, 645, 647, 649, 651, 653
— wearing apparel, pp. 122, 125, 127, 129, 131, 133
— wood products, pp. 180, 183, 185, 187, 189, 191
Kuwait National Bottling Co, p. 60
Kvaerner Holding U.K. Ltd, p. 657
K.W.V, p. 60
Kyocera International Inc, p. 596
Kyrgyzstan
— beverages, pp. 45, 47, 50, 55
— chemical products nec, pp. 328, 331, 333, 339
— drugs and medicines, pp. 350-351, 353, 356
— electrical machinery, pp. 602, 604, 607, 612
— food products, pp. 5, 7, 10, 16
— footwear, pp. 160, 162, 165, 170
— furniture, fixtures, pp. 200, 203, 206, 211
— glass and products, pp. 448, 450, 453, 457
— industrial chemicals, pp. 266, 268, 271, 276
— industries nec, p. 717
— iron and steel, p. 497
— leather and products, pp. 144, 146, 149, 154
— machinery nec, pp. 560, 562, 565, 570
— metal products, pp. 541, 543, 545, 551
— motor vehicles, pp. 676-677, 679, 682
— nonferrous metals, pp. 514, 516, 518, 522
— nonmetal products nec, pp. 466, 469, 471, 477
— office, computing machinery, pp. 590-592, 594
— paper and products, pp. 219, 221, 223, 229
— petroleum refineries, p. 368
— petroleum, coal products, pp. 380-381, 383, 387
— plastic products nec, pp. 414, 416, 419, 424
— pottery, china, etc., p. 441
— printing, publishing, pp. 249, 251, 254, 260
— professional goods, pp. 686, 688, 690, 695
— profile, p. 762
— rubber products, pp. 394, 396, 399, 404
— shipbuilding, repair, pp. 668-669, 671, 674
— spinning, weaving, etc., pp. 112-113, 115, 118
— textiles, pp. 85, 87, 89, 95
— tobacco, pp. 66, 69, 71, 77
— transportation equipment, pp. 642, 645, 647, 653
— wearing apparel, pp. 122, 125, 127, 133
— wood products, pp. 180, 183, 186, 191
Kyung Nam Wool Textile Ind. Co., p. 99
Kyungbang Ltd, p. 99
La Bilbaina, SA, p. 156
L.A. Gear Inc, p. 173
LAB Chrysotile Inc, p. 480

LaBarge Inc, p. 527
Labod d.d, p. 136
Laboratory ceramics, p. 431
Lackawanna Leather, p. 156
Laclede Steel Co, p. 501
LaCrosse Footwear Inc, pp. 172, 407
LADD Furniture Inc, p. 213
Ladrillera Santafe, p. 481
Laemthong Corp. Ltd, p. 18
Lafarge Canada Inc, pp. 389, 479
Lafarge Corp, p. 479
Laitram Corp, p. 429
Lam Research Corp, p. 573
Lamb-Weston Inc, p. 19
Lambert Howarth Group PLC, p. 172
Lamino Oy, p. 481
Lampart Vegyipari Gepgyar, p. 460
Lamps, p. 601
— electric, p. 640
Lancaster Colony Corp, pp. 18, 459, 655
Lancaster Leaf Tobacco Company of, p. 79
Lance Inc, p. 19
Lancer Industries Inc, pp. 99, 233
Landesverlag Holding GmbH-, p. 264
Landis and Staefa Inc, p. 697
Langeberg Foods International, p. 18
Lao P.D.R
— beverages, p. 55
— chemical products nec, p. 339
— electrical machinery, p. 612
— food products, p. 16
— footwear, p. 170
— furniture, fixtures, p. 211
— glass and products, p. 457
— industrial chemicals, p. 276
— industries nec, p. 717
— leather and products, p. 154
— metal products, p. 551
— nonmetal products nec, p. 477
— paper and products, p. 229
— plastic products nec, p. 424
— printing, publishing, p. 260
— professional goods, p. 695
— rubber products, p. 404
— textiles, p. 95
— tobacco, p. 77
— transportation equipment, p. 653
— wearing apparel, p. 133
— wood products, p. 191
Laporte PLC, p. 278
Lapp Insulator Co, p. 443
Lard, p. 24
Lathes, p. 582
Latvia
— beverages, pp. 45, 47, 50, 52-53, 55
— chemical products nec, pp. 329, 331, 334-335, 337, 339

Latvia continued:
— electrical machinery, pp. 602, 604, 607, 609, 611-612
— food products, pp. 5, 7, 10, 12, 14, 16
— footwear, pp. 160, 162, 165, 167-168, 170
— furniture, fixtures, pp. 201, 203, 206-207, 209, 211
— glass and products, pp. 448, 450, 453-454, 456-457
— industrial chemicals, pp. 266, 268, 271, 273-274, 276
— industries nec, pp. 707, 709, 712, 714-715, 717
— iron and steel, pp. 488, 490, 492, 494-495, 497
— leather and products, pp. 144, 146, 149, 151-152, 154
— machinery nec, pp. 560, 562, 565-566, 568, 570
— metal products, pp. 541, 543, 546-547, 549, 551
— nonferrous metals, pp. 514, 516, 518, 520-522
— nonmetal products nec, pp. 466, 469, 471, 473, 475, 477
— paper and products, pp. 219, 221, 223, 225, 227, 229
— petroleum refineries, pp. 360, 362, 364-366, 368
— plastic products nec, pp. 414, 416, 419, 421-422, 424
— pottery, china, etc., pp. 432, 434, 436, 438-439, 441
— printing, publishing, pp. 249, 251, 254, 256, 258, 260
— professional goods, pp. 686, 688, 690, 692-693, 695
— profile, p. 763
— rubber products, pp. 394, 396, 399, 401-402, 404
— textiles, pp. 85, 87, 89, 91, 93, 95
— tobacco, pp. 66, 69, 71, 73, 75, 77
— transportation equipment, pp. 643, 645, 647, 649, 651, 653
— wearing apparel, pp. 123, 125, 127, 129, 131, 133
— wood products, pp. 180, 183, 186-187, 189, 191
Laura Ashley Holdings PLC, p. 97
Laura Ashley Ltd, p. 136
Lawson Mardon Group UK Ltd, pp. 232, 428
Lawson Mardon Star Ltd, p. 527
Lawson Mardon Wheaton Inc, pp. 427, 459
L.B.L. Inc, p. 460
Lead, pp. 513, 534
Lead-base alloys, p. 534
Lead oxides, p. 295
Leather, p. 143
— finishing, p. 143
Leather Jacket Land, p. 157
Lebanon
— beverages, p. 55
— electrical machinery, p. 612
— food products, p. 16
— furniture, fixtures, p. 211
— glass and products, p. 457
— industries nec, p. 717
— leather and products, p. 154
— machinery nec, p. 570
— metal products, p. 551
— nonmetal products nec, p. 477
— paper and products, p. 229
— petroleum refineries, p. 368
— petroleum, coal products, p. 387
— plastic products nec, p. 424
— pottery, china, etc., p. 441

Lebanon continued:
— printing, publishing, p. 260
— professional goods, p. 695
— rubber products, p. 404
— textiles, p. 95
— tobacco, p. 77
— transportation equipment, p. 653
— wood products, p. 191
Lee Apparel Company Inc, p. 135
Lee Enterprises Inc, p. 262
Leegin Creative Leather Products, p. 173
LeFebure Corp, pp. 597, 721
Lehigh Portland Cement Co, p. 479
Lehigh Safety Shoe Co, p. 174
Leica PLC, p. 698
Len, p. 98
Lennox International Inc, p. 572
Lenox Inc, pp. 443, 459
Lentex, pp. 137, 428
Leo Pharmaceutical Products, pp. 20, 342, 698
Leprino Foods Co, p. 20
Leslie Fay Companies Inc, p. 135
Leslie's Poolmart Inc, p. 428
Lesotho
— beverages, pp. 45, 47-48, 55
— chemical products nec, pp. 329, 331-332, 339
— food products, pp. 5, 7, 9, 16
— furniture, fixtures, pp. 201, 203-204, 211
— industries nec, pp. 707, 709-710, 717
— iron and steel, pp. 488, 490-491, 497
— metal products, pp. 541, 543, 551
— nonmetal products nec, pp. 467, 469-470, 477
— printing, publishing, pp. 249, 251-252, 260
— profile, p. 763
Letterpress printing, p. 247
Lever Brothers Ltd, p. 342
Levi Strauss & Co. (Canada) Inc, p. 136
Levi Strauss UK Ltd, p. 137
Leviton Manufacturing Company Inc, p. 615
Lexington Precision Corp, pp. 407, 526
Lexmark International Group Inc, p. 597
LG Information & Communications, p. 617
LG International Corp, p. 719
Libbey Glass Inc, p. 459
Libbey Inc, pp. 443, 459
Liberia
— beverages, p. 55
— chemical products nec, p. 339
— electrical machinery, p. 613
— food products, p. 16
— footwear, p. 170
— furniture, fixtures, p. 211
— glass and products, p. 457
— industrial chemicals, p. 276
— machinery nec, p. 570
— metal products, p. 551

Liberia continued:
— nonmetal products nec, p. 477
— paper and products, p. 229
— plastic products nec, p. 424
— printing, publishing, p. 260
— tobacco, p. 77
— wood products, p. 191
Liberty Footwear Co, p. 173
Liberty Homes Inc, p. 195
Libyan Arab Jamahiriya
— beverages, p. 55
— chemical products nec, p. 339
— electrical machinery, p. 613
— food products, p. 16
— footwear, p. 170
— furniture, fixtures, p. 211
— glass and products, p. 457
— industrial chemicals, p. 276
— industries nec, p. 717
— iron and steel, p. 497
— leather and products, p. 154
— machinery nec, p. 570
— metal products, p. 551
— nonferrous metals, p. 522
— nonmetal products nec, p. 477
— paper and products, p. 229
— petroleum refineries, p. 368
— petroleum, coal products, p. 387
— plastic products nec, p. 424
— pottery, china, etc., p. 441
— printing, publishing, p. 260
— professional goods, p. 695
— rubber products, p. 404
— textiles, p. 95
— tobacco, p. 77
— transportation equipment, p. 653
— wearing apparel, p. 133
— wood products, p. 191
Liechtenstein
— chemical products nec, pp. 329, 331
— electrical machinery, pp. 602, 604
— food products, pp. 5, 7
— industrial chemicals, pp. 266, 268
— industries nec, pp. 707, 709
— metal products, pp. 541, 543
— paper and products, pp. 219, 221
— printing, publishing, pp. 249, 251
— professional goods, pp. 686, 688
— profile, p. 764
Life Sciences International PLC, p. 699
Life Technologies Inc, p. 343
Light leather, p. 158
Lilly Industries Inc, p. 342
Lilly Industries Ltd, p. 342
Lin Pac Containers International Ltd, p. 428
Lin Pac Mouldings Ltd, p. 428

Lin Pac Plastics International Ltd, p. 428
Linamar Corp, pp. 554, 573, 656
Lincoln Electric Co, pp. 573, 616
Lindberg Corp, p. 527
Linen, p. 83
Linen fabrics, p. 104
Lionheart PLC, p. 460
Liquefied petroleum gas, pp. 359, 377
Liquid meters, p. 700
Lisca d.d, p. 137
Lister & Co. PLC, p. 408
Lithuania
— basic chemicals, excl fertilizers, pp. 313, 315
— beverages, pp. 45, 47, 50, 56
— chemical products nec, pp. 329, 331, 334, 339
— drugs and medicines, pp. 351, 353
— electrical machinery, pp. 602, 604, 607, 613
— food products, pp. 5, 7, 10, 16
— footwear, pp. 160, 162, 165, 170
— furniture, fixtures, pp. 201, 203, 206, 211
— glass and products, pp. 448, 450, 453, 457
— industrial chemicals, pp. 266, 268, 271, 276
— industries nec, pp. 707, 709, 712, 717
— iron and steel, pp. 488, 490, 492, 497
— leather and products, pp. 144, 146, 149, 154
— machinery nec, pp. 560, 562, 565, 570
— metal products, pp. 541, 543, 546, 551
— motor vehicles, pp. 677, 679
— nonferrous metals, pp. 514, 516, 518, 522
— nonmetal products nec, pp. 467, 469, 471, 477
— office, computing machinery, pp. 591-592
— paper and products, pp. 219, 221, 224, 229
— petroleum refineries, p. 368
— petroleum, coal products, pp. 380-381, 383, 387
— plastic products nec, pp. 414, 416, 419, 424
— pottery, china, etc., pp. 432, 434, 436, 441
— printing, publishing, pp. 249, 251, 254, 260
— professional goods, pp. 686, 688, 690, 695
— profile, p. 764
— pulp, paper, etc., pp. 241, 243
— radio, television, etc., pp. 629, 631
— rubber products, pp. 394, 396, 399, 404
— shipbuilding, repair, pp. 669, 671
— spinning, weaving, etc., pp. 113, 115
— synthetic resins, etc., pp. 321-322
— textiles, pp. 85, 87, 89, 95
— tobacco, pp. 66, 69, 71, 77
— transportation equipment, pp. 643, 645, 647, 653
— wearing apparel, pp. 123, 125, 127, 133
— wood products, pp. 181, 183, 186, 191
Little Tikes Co, pp. 213, 719
Litton Industries Inc. Guidance and, p. 699
Litton Poly-Scientific, p. 460
Liu Chiao Industrial Co. Ltd, p. 173
Livarna Maribor d.o.o, p. 720
Liz Claiborne Inc, p. 135

LK-Products Oy, p. 443
Lo Chin Seng Co. Ltd, p. 156
Loceria Colombiana SA, p. 459
Lockheed Martin Aeronautical, p. 655
Lockheed Martin Astronautics Group, p. 655
Locksmiths' ware, p. 539
Locomotives, p. 659
Londero Mosaik ApS, p. 444
London Asiatic Rubber & Produce, pp. 20, 279
Lone Star Industries Inc, p. 480
Lone Star Steel Co, p. 500
Lone Star Technologies Inc, p. 500
Longaberger Co, pp. 193, 443
Longman Group Ltd, p. 264
Longview Fibre Co, p. 232
Looms, pp. 559, 585
Lord Corp, p. 343
Lord Corp. Chemical Products Div, p. 389
L'Oreal UK Ltd, p. 343
Lotte Canon Co., Ltd, p. 597
Lotte Chilsung Beverage Co. Ltd, p. 58
Lotte Confectionery Co., Ltd, p. 18
Lotus Ltd, p. 173
Louisiana Land and Exploration Co, p. 370
Low & Bonar PLC, p. 427
Lozier Corp, p. 214
LPG, p. 359
LPS Laboratories Inc, p. 390
LRC Products Ltd, p. 407
L.S. Starrett Co, pp. 555, 698
LSB Industries Inc, p. 407
LSI Industries Inc, p. 264
LSI Logic Corp, pp. 595, 617
Lubelskie Zaklady Naprawy, p. 501
Lubricating oils, pp. 359, 375
Lubricating Specialties Co, p. 390
Lubrizol Canada Ltd, p. 390
Lubrizol Corp, pp. 279, 389
Lucas Sei Wiring Systems Ltd, p. 617
Lucent Technologies Octel, p. 617
Luckytex Thailand Co., Ltd, p. 98
Ludlow Corp, p. 429
Lukens Inc, pp. 499, 554
Lukens Steel Co, p. 500
Lunday-Thagard Co, p. 390
Luxembourg
— chemical products nec, pp. 329, 331, 334-335, 337, 339
— electrical machinery, p. 613
— food products, pp. 5, 7, 10, 12, 14, 16
— furniture, fixtures, p. 211
— industries nec, p. 717
— iron and steel, pp. 488, 490, 493-495, 497
— machinery nec, pp. 560, 562, 565-566, 568, 570
— metal products, pp. 541, 543, 546-547, 549, 551
— nonferrous metals, pp. 514, 516, 518, 520-522
— petroleum refineries, p. 368

Luxembourg continued:
— professional goods, p. 695
— profile, p. 765
— rubber products, p. 404
— transportation equipment, pp. 643, 645, 647, 649, 651, 653
— wood products, p. 191
Luxguard I SA, p. 460
Lykes Meat Group Inc, p. 20
Lyondell Petrochemical Co, p. 370
M/A-COM Inc, p. 616
M.A. Hanna Co, p. 278
MAAX Inc, p. 443
Macaroni, p. 3
Macaroni and noodle products, p. 37
Macau
— beverages, pp. 45, 47-48, 50, 52-53, 56
— chemical products nec, pp. 329, 331-332, 334-335, 337, 339
— drugs and medicines, pp. 350-353, 355-356
— electrical machinery, pp. 603-604, 606-607, 609, 611, 613
— food products, pp. 5, 7, 9-10, 12, 14, 16
— footwear, pp. 160, 162, 164-165, 167-168, 170
— furniture, fixtures, pp. 201, 203-204, 206-207, 209, 211
— glass and products, pp. 448, 450, 452-454, 456-457
— industrial chemicals, p. 276
— industries nec, pp. 707, 709-710, 712, 714-715, 717
— iron and steel, p. 497
— leather and products, pp. 144, 146, 148-149, 151-152, 154
— machinery nec, pp. 560, 562, 564-566, 568, 570
— metal products, pp. 541, 543-544, 546-547, 549, 551
— nonferrous metals, p. 523
— nonmetal products nec, pp. 467, 469-471, 473, 475, 477
— office, computing machinery, pp. 590, 592-593
— paper and products, pp. 219, 221-222, 224-225, 227, 229
— petroleum refineries, p. 368
— petroleum, coal products, p. 387
— plastic products nec, pp. 414, 417-419, 421-422, 424
— pottery, china, etc., pp. 432, 434-436, 438-439, 441
— printing, publishing, pp. 249, 251, 253-254, 256, 258, 260
— professional goods, pp. 686, 688-690, 692-693, 695
— profile, p. 765
— pulp, paper, etc., pp. 240-244, 246
— radio, television, etc., pp. 628-634
— rubber products, pp. 394, 396, 398-399, 401-402, 404
— shipbuilding, repair, pp. 668-674
— spinning, weaving, etc., pp. 112-118
— textiles, pp. 85, 87-88, 90-91, 93, 95
— tobacco, pp. 67, 69-70, 72-73, 75, 77
— transportation equipment, pp. 643, 645-646, 648-649, 651, 653
— wearing apparel, pp. 123, 125-127, 129, 131, 133
— wood products, pp. 181, 183-184, 186-187, 189, 191
Macedonia (FYR)
— basic chemicals, excl fertilizers, pp. 312-313, 315-316
— beverages, pp. 44, 46, 48-49, 51, 53, 55

Macedonia (FYR) continued:
— chemical products nec, pp. 328, 330, 332-333, 335, 337-338
— drugs and medicines, pp. 353-354
— electrical machinery, pp. 602, 604, 606-607, 609-610, 612
— food products, pp. 4, 6, 8, 10, 12-13, 15
— footwear, pp. 160, 162, 164-166, 168-169
— furniture, fixtures, pp. 200, 202, 204-205, 207, 209-210
— glass and products, pp. 448, 450, 452, 454-455, 457
— industrial chemicals, pp. 266, 268, 270-272, 274, 276
— industries nec, pp. 706, 708, 710-711, 713, 715, 717
— iron and steel, pp. 488, 490-492, 494-495, 497
— leather and products, pp. 144, 146-147, 149-150, 152-153
— machinery nec, pp. 560, 562-564, 566-567, 569
— metal products, pp. 540, 542, 544-545, 547, 549-550
— motor vehicles, pp. 676-677, 679-680
— nonferrous metals, pp. 514, 516-520, 522
— nonmetal products nec, pp. 466, 468, 471, 473-474, 476
— paper and products, pp. 218, 220, 222-223, 225, 227-228
— petroleum refineries, pp. 360-361, 363-367
— petroleum, coal products, p. 387
— plastic products nec, pp. 414, 416, 419, 421-422, 424
— pottery, china, etc., pp. 432, 434, 436-437, 439-440
— printing, publishing, pp. 248, 250, 252, 254, 256-257, 259
— professional goods, pp. 686, 688, 690, 692-694
— profile, p. 750
— pulp, paper, etc., pp. 240-241, 243-244
— radio, television, etc., pp. 628-629, 631-632
— rubber products, pp. 394, 396, 399-400, 402, 404
— synthetic resins, etc., pp. 320-323
— textiles, pp. 84, 86, 88-89, 91, 93-94
— tobacco, pp. 66, 68, 70-71, 73-74, 76
— transportation equipment, pp. 642, 644, 646-647, 649-650, 652
— wearing apparel, pp. 122, 124, 126-127, 129-130, 132
— wood products, pp. 180, 182, 184-185, 187, 189-190
MacFarlane Group Clansman PLC, p. 233
Machine tools, p. 559
— for wood, p. 584
Machining centers, p. 581
Mack Printing Company Inc, p. 264
Mack Trucks Inc, p. 656
MacMillan Bloedel Inc, pp. 194, 232
MacMillan Ltd, p. 263
Macmillan Publishing Co, p. 262
Macpell Industries Ltd, p. 137
Madagascar
— beverages, p. 56
— chemical products nec, p. 339
— electrical machinery, p. 613
— food products, p. 16
— footwear, p. 170
— glass and products, p. 458
— industrial chemicals, p. 276
— industries nec, p. 717
— iron and steel, p. 497

Madagascar continued:
— leather and products, p. 154
— machinery nec, p. 570
— metal products, p. 551
— nonferrous metals, p. 523
— nonmetal products nec, p. 477
— paper and products, p. 229
— petroleum refineries, p. 368
— petroleum, coal products, p. 387
— plastic products nec, p. 424
— pottery, china, etc., p. 441
— printing, publishing, p. 260
— professional goods, p. 695
— rubber products, p. 404
— textiles, p. 95
— tobacco, p. 77
— transportation equipment, p. 653
— wearing apparel, p. 133
Madeco SA, p. 527
Madge Networks Inc, p. 596
Madix Inc, p. 214
Mafco Consolidated Group Inc, p. 79
Magazine publishing, p. 247
Magellan Aerospace Corp, p. 657
Magla Products Inc, p. 408
Magnesium, p. 538
Magnet Ltd, p. 193
Magyar Viscosa Rt, pp. 99, 428
Maidenform Inc, p. 135
Makhteshim Chemical Works Ltd, p. 280
Malawi
— basic chemicals, excl fertilizers, pp. 313, 318
— beverages, pp. 45, 47, 50, 52-53, 56
— chemical products nec, pp. 329, 331, 334-335, 337, 339
— drugs and medicines, pp. 350-351, 357
— electrical machinery, p. 613
— food products, pp. 5, 7, 10, 12, 14, 16
— footwear, pp. 160, 162, 170
— furniture, fixtures, pp. 203, 211
— glass and products, p. 458
— industrial chemicals, pp. 266, 268, 271, 273-274, 276
— industries nec, p. 717
— iron and steel, p. 497
— leather and products, pp. 144, 146, 154
— machinery nec, pp. 562, 570
— metal products, pp. 541, 543, 546-547, 549, 551
— motor vehicles, pp. 677, 682
— nonferrous metals, p. 523
— nonmetal products nec, pp. 467, 469, 471, 473, 475, 47
— paper and products, p. 229
— petroleum refineries, p. 368
— petroleum, coal products, p. 387
— plastic products nec, pp. 417, 425
— pottery, china, etc., p. 441
— printing, publishing, p. 260
— professional goods, p. 695

Malawi continued:
— profile, p. 766
— rubber products, pp. 396, 404
— spinning, weaving, etc., pp. 112-113, 118
— textiles, pp. 85, 87, 90-91, 93, 95
— tobacco, pp. 67, 69, 72-73, 75, 77
— transportation equipment, pp. 645, 653
— wearing apparel, pp. 123, 125, 133
— wood products, pp. 183, 191
Malayawata Steel BHD, p. 501
Malaysia
— basic chemicals, excl fertilizers, pp. 312-318
— beverages, pp. 45, 47-48, 50, 52-53, 56
— chemical products nec, pp. 329, 331-332, 334-335, 337, 339
— drugs and medicines, pp. 350-353, 355-357
— electrical machinery, pp. 603-604, 606-607, 609, 611, 613
— food products, pp. 5, 7, 9-10, 12, 14, 16
— footwear, pp. 160, 162, 164-165, 167-168, 170
— furniture, fixtures, pp. 201, 203-204, 206-207, 209, 211
— glass and products, pp. 448, 450, 452-454, 456, 458
— industrial chemicals, pp. 267-268, 270-271, 273-274, 276
— industries nec, pp. 707, 709-710, 712, 714-715, 717
— iron and steel, pp. 488, 490-491, 493-495, 497
— leather and products, pp. 144, 146, 148-149, 151-152, 154
— machinery nec, pp. 560, 562, 564-566, 568, 570
— metal products, pp. 541, 543-544, 546-547, 549, 551
— motor vehicles, pp. 676-682
— nonferrous metals, pp. 514, 516-518, 520-521, 523
— nonmetal products nec, pp. 467, 469-470, 472-473, 475, 477
— office, computing machinery, pp. 590-594
— paper and products, pp. 219, 221-222, 224-225, 227, 229
— petroleum refineries, pp. 360, 362-366, 368
— petroleum, coal products, pp. 380-385, 387
— plastic products nec, pp. 415, 417-419, 421-422, 425
— pottery, china, etc., pp. 432, 434-436, 438-439, 441
— printing, publishing, pp. 249, 251, 253-254, 256, 258, 260
— professional goods, pp. 686, 688-690, 692-693, 695
— profile, p. 766
— pulp, paper, etc., pp. 240-246
— radio, television, etc., pp. 628-634
— rubber products, pp. 394, 396, 398-399, 401-402, 404
— shipbuilding, repair, pp. 668-674
— spinning, weaving, etc., pp. 112-118
— synthetic resins, etc., pp. 320-325
— textiles, pp. 85, 87-88, 90-91, 93, 95
— tobacco, pp. 67, 69-70, 72-73, 75, 77
— transportation equipment, pp. 643, 645-646, 648-649, 651, 653
— wearing apparel, pp. 123, 125-126, 128-129, 131, 133
— wood products, pp. 181, 183-184, 186-187, 189, 191
Malaysian Mosaics BHD, p. 480
Malaysian Tobacco Co. Ltd, p. 79
Maleic anhydride, p. 291
Malette Inc, p. 195

Mali
— beverages, p. 56
— electrical machinery, p. 613
— food products, p. 16
— industries nec, p. 717
— iron and steel, p. 497
— machinery nec, p. 570
— metal products, p. 551
— nonferrous metals, p. 523
— nonmetal products nec, p. 477
— professional goods, p. 695
— tobacco, p. 77
— transportation equipment, p. 653
Mallinckrodt Inc, pp. 278, 341, 697
Mallinckrodt Medical Inc, p. 697
Malt, pp. 43, 62
Malta
— beverages, pp. 45, 47-48, 50, 52-53, 56
— chemical products nec, p. 339
— electrical machinery, pp. 603-604, 606-607, 609, 611, 613
— food products, pp. 5, 7, 9-10, 12, 14, 16
— footwear, pp. 160, 162, 164-165, 167-168, 170
— furniture, fixtures, pp. 201, 203-204, 206-207, 209, 211
— glass and products, pp. 448, 450, 452-454, 456, 458
— industrial chemicals, pp. 267-268, 270-271, 273-274, 276
— industries nec, pp. 707, 709-710, 712, 714-715, 717
— iron and steel, p. 497
— leather and products, pp. 144, 146, 148-149, 151-152, 154
— machinery nec, pp. 560, 562, 564-566, 568, 570
— metal products, p. 551
— motor vehicles, pp. 676-682
— nonferrous metals, p. 523
— nonmetal products nec, pp. 467, 469-470, 472-473, 475, 477
— office, computing machinery, pp. 590-594
— paper and products, pp. 219, 221-222, 224-225, 227, 229
— petroleum refineries, p. 368
— petroleum, coal products, p. 387
— plastic products nec, pp. 415, 417-419, 421, 423, 425
— pottery, china, etc., pp. 432, 434-436, 438-439, 441
— printing, publishing, pp. 249, 251, 253-254, 256, 258, 260
— professional goods, pp. 686, 688-690, 692-693, 695
— profile, p. 767
— radio, television, etc., pp. 628-634
— rubber products, pp. 395-396, 398-399, 401-402, 404
— shipbuilding, repair, pp. 668-674
— spinning, weaving, etc., pp. 112-118
— textiles, pp. 85, 87-88, 90-91, 93, 95
— tobacco, pp. 67, 69-70, 72-73, 75, 77
— transportation equipment, pp. 643, 645-646, 648-649, 651, 653
— wearing apparel, pp. 123, 125-126, 128-129, 131, 133
— wood products, pp. 181, 183-184, 186-187, 189, 191
Manchester Tobacco Co. Ltd, p. 79
Mando Machinery Corp, p. 655
Maneklal Harilal Mills Ltd, p. 98

Manganese, p. 513
Manitowoc Company Inc, pp. 574, 657
Mannai Trading Co. Ltd, p. 596
Mannesmann Capital Corp, p. 572
Mannington Mills Inc, pp. 98, 193, 479, 719
Manor Bakeries Ltd, p. 20
Mansfield Brewery PLC, p. 58
Mansfield Plumbing Products Inc, p. 443
Manufacturas Quintero, p. 156
Manufacture de Tabacs de l'Ouest, p. 79
Manufacture de Tabacs Heintz van, p. 79
MAPCO Inc, pp. 278, 370
Maple Leaf Foods Inc, p. 18
Maquinas Agricolas Jacto SA, p. 615
Marathon Oil Co. Texas Refining Div, p. 371
Marcopolo SA - Carrocerias, p. 656
Margarine, pp. 3, 31
Markische Faser Aktiengesellschaft, p. 99
Marley PLC, p. 479
Marmoran Pty. Ltd, p. 342
Marquette Medical Systems, p. 698
Mars GB Ltd, p. 20
Marshall Food Group Ltd, p. 19
Marston, Thompson and Evershed, p. 58
Martin Industries Inc, p. 214
Martin International Holdings PLC, p. 135
Martin Marietta Materials Inc, p. 479
Martin Mills Inc, p. 99
Marvel Entertainment Group Inc, p. 263
Mary Kay Inc, p. 341
Mas Mail, p. 174
MascoTech Inc, pp. 193, 656
Masland Corp, pp. 98, 136
Mason Shoe Manufacturing Inc, p. 173
Masonite Corp, p. 193
Massalin Particulares SA, p. 79
Mastellone Hnos. SA, p. 20
Master Chemical Corp, p. 390
Mastercraft Fabrics L.L.C, p. 98
Mastics, p. 345
Material Sciences Corp, p. 429
Matheson & MacMillan Ltd, p. 390
Mattel Indonesia PT, p. 719
Matthews International Corp, pp. 264, 720
Mattress supports, p. 216
Mattresses, pp. 199, 216
Mauritius
— basic chemicals, excl fertilizers, pp. 312-313, 315-316, 318
— beverages, pp. 45, 47, 50, 52, 56
— chemical products nec, pp. 329, 331, 334-335, 339
— drugs and medicines, pp. 350, 352-353, 355, 357
— electrical machinery, pp. 603, 605, 607, 609, 613
— food products, pp. 5, 7, 10, 12, 16
— footwear, pp. 160, 162, 165, 167, 170
— furniture, fixtures, pp. 201, 203, 206-207, 211

Mauritius continued:
— glass and products, pp. 448, 450, 453-454, 458
— industrial chemicals, pp. 267-268, 271, 273, 276
— industries nec, pp. 707, 709, 712, 714, 717
— leather and products, pp. 145-146, 149, 151, 154
— machinery nec, pp. 560, 562, 565-566, 570
— metal products, pp. 541, 543, 546-547, 551
— motor vehicles, pp. 676-677, 679-680, 682
— nonmetal products nec, pp. 467, 469, 472-473, 477
— paper and products, pp. 219, 221, 224-225, 229
— petroleum refineries, p. 368
— petroleum, coal products, p. 387
— plastic products nec, pp. 415, 417, 419, 421, 425
— pottery, china, etc., pp. 432, 434, 436, 438, 441
— printing, publishing, pp. 249, 251, 254, 256, 260
— professional goods, pp. 686, 688, 690, 692, 695
— profile, p. 767
— radio, television, etc., pp. 628-629, 631-632, 634
— rubber products, pp. 395-396, 399, 401, 404
— shipbuilding, repair, pp. 668-669, 671-672, 674
— spinning, weaving, etc., pp. 112, 115-116
— textiles, pp. 85, 87, 90-91, 95
— tobacco, pp. 67, 69, 72-73, 77
— transportation equipment, pp. 643, 645, 648-649, 653
— wearing apparel, pp. 123, 125, 128-129, 133
— wood products, pp. 181, 183, 186-187, 191
Maverick Tube Corp, p. 501
Maxtor Corp, p. 595
Maxxim Medical Inc, pp. 135, 698
May & Baker Exports Ltd, p. 342
McCain Foods Ltd, p. 20
McClatchy Newspapers Inc, p. 262
McCormick and Company Inc, pp. 18, 58, 427
McCormick International Div, p. 60
McDermott Indonesia PT, p. 555
McGuire-Nicholas Co., Inc, p. 173
McKechnie UK Ltd, p. 526
McLouth Steel, p. 501
McQuay International, p. 574
McQuay International. Commercial, p. 574
McRae Industries Inc, pp. 173, 597
MD Foods Amba, pp. 18, 58
MD Foods PLC, p. 19
Mead Fine Paper Div, p. 233
Meal, p. 3
Meal and groats of all cereals, p. 36
Meals, frozen, p. 23
Meat by-products, p. 3
Meat products, p. 3
Meat, tinned, p. 24
Mecanismos Auxiliares Industriales, p. 616
Medeva PLC, p. 342
Media General Inc, pp. 231, 262
Mediakreasi Lokanusa Industry, p. 461
Medicinals, p. 349
Medicines, pp. 327, 349

Medusa Corp, p. 480
Meggitt PLC, p. 655
Meiwa Indonesia, PT, p. 407
Melba-Wain England Ltd, p. 444
Melham Holdings Ltd, p. 264
Melton Medes Ltd, pp. 98, 232, 406
MEM Company Inc, p. 174
MEMC Electronic Materials Inc, p. 615
Menasha Corp, pp. 231, 427
Men's and boys' jackets, p. 121
Mercedes-Benz of South Africa Pty, p. 656
Merck Sharp & Dohme Ltd, p. 343
Mercur SA, p. 408
Mercury Marine, p. 573
Meredith Corp, p. 263
Meridian Inc, p. 214
Meridian Sports Inc, p. 721
Meridian Technologies Inc, p. 526
Merrill Corp, p. 263
Metal containers, p. 539
Metal Improvement Company Inc, p. 527
Metal products, p. 539
Metal Ravne d.o.o, p. 720
Metal-working presses, p. 583
Metallurg Inc, p. 480
Metallurgical furnace liners, p. 465
Metalna d.d, p. 721
Metalplast, p. 527
Metalrax Group PLC, p. 428
Metals, ferrous, p. 487
Meters, electricity-supply, p. 625
Methanol, pp. 285, 289
Methode Electronics Inc, p. 617
Metroland Printing, Publishing &, p. 264
Mexico
—basic chemicals, excl fertilizers, pp. 312-313, 315-318
—beverages, pp. 45, 47, 50, 52-53, 56
—chemical products nec, pp. 329, 331, 334-335, 337, 339
—drugs and medicines, pp. 350, 352-353, 355-357
—electrical machinery, pp. 603, 605, 607, 609, 611, 613
—food products, pp. 5, 7, 10, 12, 14, 16
—footwear, pp. 160, 162, 165, 167-168, 170
—furniture, fixtures, pp. 201, 203, 206-207, 209, 211
—glass and products, pp. 448, 450, 453-454, 456, 458
—industrial chemicals, pp. 267-268, 271, 273-274, 276
—industries nec, pp. 707, 709, 712, 714-715, 717
—iron and steel, pp. 488, 490, 493-495, 497
—leather and products, p. 154
—machinery nec, pp. 560, 562, 565-566, 568, 570
—metal products, pp. 541, 543, 546-547, 549, 551
—motor vehicles, pp. 676, 678-682
—nonferrous metals, pp. 514, 516, 518, 520-521, 523
—nonmetal products nec, pp. 467, 469, 472-473, 475, 477
—office, computing machinery, pp. 590-594
—paper and products, pp. 219, 221, 224-225, 227, 229
—petroleum refineries, p. 368

Mexico continued:
—petroleum, coal products, pp. 380-381, 383-385, 387
—plastic products nec, pp. 415, 417, 419, 421, 423, 425
—pottery, china, etc., pp. 432, 434, 436, 438-439, 441
—printing, publishing, pp. 249, 251, 254, 256, 258, 260
—professional goods, pp. 686, 688, 690, 692-693, 695
—profile, p. 768
—pulp, paper, etc., pp. 240-241, 243-246
—radio, television, etc., pp. 628-629, 631-634
—rubber products, pp. 395-396, 399, 401-402, 404
—shipbuilding, repair, pp. 668-669, 671-674
—synthetic resins, etc., pp. 320-325
—textiles, pp. 85, 87, 90-91, 93, 95
—tobacco, pp. 67, 69, 72-73, 75, 77
—transportation equipment, pp. 643, 645, 648-649, 651, 653
—wearing apparel, pp. 123, 125, 128-129, 131, 133
—wood products, pp. 181, 183, 186-187, 189, 191
MFI Furniture Group PLC, p. 213
MGN Ltd, p. 264
Mi-Chang Oil Industries Co., Ltd, p. 371
Miami Herald Publishing Company, p. 263
Michael Foods Inc, p. 20
Michelin North America (Canada), p. 406
Michelin Tyre PLC, p. 406
Micro Switch Div, p. 616
Micron Electronics Inc, p. 595
Micros Systems Inc, p. 596
Midland News Association Ltd, p. 264
Midland Newspapers Ltd, p. 264
Mikasa Inc, pp. 443, 459
Mikohn Gaming Corp, p. 721
Milbank Manufacturing Co, p. 408
Milk, p. 3
Milk and cream
—condensed, p. 25
—dried, p. 26
Milking machines, p. 579
Millennium Chemicals Inc, p. 279
Millennium Inorganic Chemicals Inc, p. 280
Millennium Petrochemicals Inc, p. 280
Miller Paving Ltd, p. 389
Milling machines, p. 582
Millipore Corp, p. 697
Mine Safety Appliances Co, p. 698
Mineral waters, pp. 43, 62
Minerals Technologies Inc, p. 280
Miniature Precision Components Inc, p. 429
Mining Machinery Factory, Pioma SA, p. 574
Minnesota Mining & Manufacturing, pp. 341, 481
Minnesota Rubber, pp. 407, 429
Mirror Group Newspapers PLC, p. 263
Mississippi Chemical Corp, p. 280
Mitel Corp, p. 616
Mitie Group PLC, p. 370
MNP Corp, p. 527
Mobil Chemical Co. Plastics Div, p. 231

Mobil Chemical Company Inc, pp. 232, 279, 370

Mobil Holdings Ltd, p. 370

Mobil Oil Co. Ltd, p. 370

Mobil Oil do Brasil Industria e, p. 390

Mobile Mini Inc, p. 721

Modine Manufacturing Co, pp. 553, 572, 655

Modus, p. 137

Moen Inc, p. 554

Molinos Rio De La Plata SA, p. 19

Molson Breweries, p. 58

Molson Breweries, Ontario Division, p. 59

Monarch Cement Co, p. 481

Monarch Tile Inc, p. 481

Mongolia

— beverages, pp. 45, 47-48, 50, 52-53, 56

— chemical products nec, pp. 329, 331-332, 334, 336-337, 339

— electrical machinery, pp. 603, 605-607, 609, 613

— food products, pp. 5, 7, 9-10, 12, 14, 16

— footwear, pp. 160, 162, 164-165, 167-168, 170

— furniture, fixtures, pp. 201, 203-204, 206-207, 209, 211

— glass and products, pp. 448, 450, 452-454, 456

— industrial chemicals, pp. 267, 269-271, 273

— industries nec, pp. 707, 709-710, 712, 714-715, 718

— iron and steel, pp. 488, 490-491, 493-494, 497

— leather and products, pp. 145-146, 148-149, 151-152, 154

— machinery nec, pp. 560, 562, 564, 570

— metal products, pp. 541, 543-544, 546-547, 549, 551

— nonferrous metals, pp. 514, 516-517, 523

— nonmetal products nec, pp. 467, 469-470, 472-473, 475, 477

— paper and products, pp. 219, 221-222, 224-225

— petroleum, coal products, pp. 380, 382-384

— plastic products nec, p. 425

— pottery, china, etc., pp. 432, 434-436, 438-439, 441

— printing, publishing, pp. 249, 251, 253-254, 256, 258, 260

— professional goods, pp. 686, 688-690, 692-693, 695

— profile, p. 768

— radio, television, etc., pp. 628-634

— rubber products, p. 404

— spinning, weaving, etc., pp. 112-113, 115-118

— textiles, pp. 85, 87-88, 90-91, 93, 95

— transportation equipment, pp. 643, 645-646, 653

— wearing apparel, pp. 123, 125-126, 128-129, 131, 133

— wood products, pp. 181, 183-184, 186-187, 189, 191

Monroe Systems for Business Inc, p. 595

Monsanto Co. Agricultural Sector, p. 279

Monsey Bakor Inc, p. 390

Monsey Products Co, p. 389

Moody's Investors Service, p. 264

Moog Automotive Inc, p. 657

Moog Inc, pp. 574, 617, 657, 698

Moore and Munger Marketing Inc, p. 371

Moore Business Forms and Systems, p. 262

Moore Corp. Ltd, pp. 262, 595

Moran Holdings PLC, p. 18

Morgan-Grampian PLC, p. 264

Morgan Products Ltd, p. 194

Morland & Co. PLC, p. 59

Morocco

— beverages, pp. 45, 47-48, 50, 52-53, 56

— chemical products nec, pp. 332, 336, 339

— electrical machinery, pp. 603, 605-607, 609, 611, 613

— food products, pp. 5, 7, 9-10, 12, 14, 16

— footwear, pp. 164, 170

— furniture, fixtures, pp. 204, 209, 211

— glass and products, pp. 452, 458

— industrial chemicals, pp. 270, 273, 276

— industries nec, pp. 707, 709-710, 712, 714-715, 718

— iron and steel, pp. 491, 497

— leather and products, pp. 148, 154

— machinery nec, pp. 561-562, 564-566, 568, 570

— metal products, pp. 541, 543-544, 546-547, 549, 551

— nonferrous metals, pp. 517, 523

— nonmetal products nec, pp. 470, 477

— paper and products, pp. 219, 221-222, 224-225, 227, 22

— petroleum refineries, p. 368

— petroleum, coal products, p. 387

— plastic products nec, pp. 415, 417-419, 421, 423, 425

— pottery, china, etc., pp. 435, 441

— printing, publishing, pp. 249, 251, 253-254, 256, 258, 26

— professional goods, pp. 686, 688-690, 692-693, 695

— profile, p. 769

— rubber products, pp. 395-396, 398-399, 401-402, 404

— textiles, pp. 85, 87-88, 90-91, 93, 95

— tobacco, pp. 67, 69-70, 72-73, 75, 77

— transportation equipment, pp. 643, 645-646, 648-649, 6 653

— wearing apparel, pp. 123, 125-126, 128-129, 131, 133

— wood products, pp. 184, 189, 191

Morris Cohen Underwear Ltd, p. 137

Morrison Construction Ltd, p. 480

Morrison Knudsen Corp, p. 655

Mosby Inc, p. 264

Mosinee Paper Corp, p. 233

Mossop Leather, p. 156

Mostaren Brezno AS, p. 500

Mostostal Krakow SA, p. 500

Mostostal, Zabrze, pp. 499, 525

Motor coaches, p. 675

Motor gasoline, pp. 359, 373

Motor Oil Hellas Corinth Refineries, p. 370

Motor vehicles, p. 675

Motorcycle tires, p. 393

Motorcycles, pp. 641, 664

Motorola Israel Ltd, p. 617

Motors, p. 601

— electric, fractional horsepower, p. 624

— electric, one horsepower and over, p. 624

Mount Vernon Mills Inc, p. 97

Mowers, p. 578

Mozambique
— beverages, pp. 45, 47, 50, 56
— chemical products nec, pp. 329, 331, 334, 339
— drugs and medicines, pp. 350, 352
— electrical machinery, pp. 603, 605, 607, 613
— food products, pp. 5, 7, 10, 16
— footwear, pp. 161-162, 165, 170
— furniture, fixtures, pp. 201, 203, 206, 211
— glass and products, pp. 449-450, 453, 458
— industrial chemicals, pp. 267, 269, 271, 276
— industries nec, pp. 707, 709, 712, 718
— iron and steel, pp. 488, 490, 493, 497
— leather and products, pp. 145-146, 149, 154
— machinery nec, pp. 561-562, 565, 570
— metal products, pp. 541, 543, 546, 551
— motor vehicles, pp. 676, 678
— nonferrous metals, pp. 514, 516, 518, 523
— nonmetal products nec, pp. 467, 469, 472, 477
— office, computing machinery, pp. 590-591
— paper and products, pp. 219, 221, 224, 229
— petroleum refineries, pp. 360, 362, 364, 368
— petroleum, coal products, pp. 383, 387
— plastic products nec, pp. 415, 417, 419, 425
— pottery, china, etc., pp. 432, 434, 436, 441
— printing, publishing, pp. 249, 251, 254, 260
— professional goods, pp. 686, 688, 691, 695
— profile, p. 769
— pulp, paper, etc., pp. 240-241
— radio, television, etc., pp. 628-629
— rubber products, pp. 395-396, 399, 404
— shipbuilding, repair, pp. 668-669
— spinning, weaving, etc., pp. 112-113
— textiles, pp. 85, 87, 90, 95
— tobacco, pp. 67, 69, 72, 77
— transportation equipment, pp. 643, 645, 648, 653
— wearing apparel, pp. 123, 125, 128, 133
— wood products, pp. 181, 183, 186, 191
MTD Products Inc, pp. 553, 572
MTS Systems Corp, p. 699
Muebles y Almacenamiento Tecnico, p. 215
Mueller Co, p. 555
Mueller Industries Inc, pp. 428, 500, 526
Mufflers, p. 465
Mulgrave Machine Works Ltd, p. 461
Mulia Glass PT, p. 459
Multinutrient fertilizers, pp. 302-303
Municipal Ready Mix Ltd, p. 390
Munro and Co, p. 172
Mura European Fashion Design d.d, p. 136
Murry Inc, pp. 574, 657
Musical instruments, p. 705
— string, p. 722
— wind, p. 723
Mutton and lamb, p. 21
Myanmar
— basic chemicals, excl fertilizers, pp. 312-318

Myanmar continued:
— beverages, p. 56
— chemical products nec, pp. 332, 339
— drugs and medicines, pp. 350, 352-353, 355-357
— electrical machinery, p. 613
— food products, p. 16
— footwear, pp. 161-162, 164-165, 167-168, 170
— furniture, fixtures, pp. 201, 203-204, 206-207, 209, 211
— glass and products, pp. 449-450, 452-454, 456, 458
— industrial chemicals, p. 276
— industries nec, pp. 707, 709-710, 712, 714, 718
— iron and steel, pp. 488, 490-491, 493-495, 497
— leather and products, pp. 145-146, 148-149, 151-152, 154
— machinery nec, p. 570
— metal products, pp. 541, 543-544, 546-547, 549, 551
— motor vehicles, pp. 676, 678-682
— nonferrous metals, pp. 514, 516-518, 520-521, 523
— nonmetal products nec, pp. 467, 477
— paper and products, p. 229
— petroleum refineries, p. 368
— petroleum, coal products, p. 387
— plastic products nec, pp. 415, 418-419, 421, 423, 425
— pottery, china, etc., p. 441
— printing, publishing, pp. 249, 251, 253-254, 258, 260
— professional goods, p. 695
— profile, p. 770
— pulp, paper, etc., pp. 240-241, 243-246
— radio, television, etc., pp. 628-632, 634
— rubber products, pp. 395-396, 398-399, 401-402, 404
— spinning, weaving, etc., pp. 112, 114-118
— textiles, p. 95
— tobacco, pp. 67, 69-70, 72-73, 75, 77
— transportation equipment, p. 653
— wearing apparel, p. 133
— wood products, pp. 181, 183-184, 186, 188-189, 191
Myers Industries Inc, p. 526
Mylan Laboratories Inc, p. 343
Mylan Pharmaceuticals Inc, p. 343
Nabisco Biscuit Co, p. 18
NACCO Industries Inc, pp. 572, 615
Nafta Lendava d.o.o, p. 370
Nails, pp. 539, 558
Nalco Chemical Co, p. 278
Nam Chung Co., Ltd, p. 156
Namdinh Silk Weaving Co, p. 99
Namibia
— beverages, pp. 45, 47, 50, 52, 56
— electrical machinery, p. 613
— food products, pp. 5, 7, 10, 12, 16
— furniture, fixtures, pp. 201, 203, 206, 208, 211
— industrial chemicals, p. 276
— industries nec, pp. 707, 709, 712, 714, 718
— machinery nec, p. 570
— metal products, pp. 541, 543, 546, 548, 551
— nonferrous metals, p. 523
— nonmetal products nec, pp. 467, 469, 472-473

Namibia continued:
— paper and products, pp. 219, 221, 224, 226
— printing, publishing, pp. 249, 251, 254, 256, 260
— profile, p. 770
— tobacco, p. 77
— transportation equipment, p. 653
— wood products, pp. 181, 183, 186, 188, 191
Namsun Aluminium Co. Ltd, p. 526
Nanjing Ferro Alloy Plant, p. 500
Naphta, p. 359
Naphthalene, p. 283
Naphthas, p. 374
Nashua Corp, pp. 232, 596
National Agricultural and Food Corp, p. 19
National Automobiles, p. 390
National Beef Packing Company, p. 19
National Beverage Corp, p. 59
National Brands Ltd.--Becketts, p. 59
National Computer Systems Inc, p. 595
National Education Corp, p. 263
National Electrical Coil, p. 460
National Glass Ltd, p. 461
National Gypsum Co, p. 479
National-Oilwell Inc, p. 574
National Patent Development Corp, p. 699
National Refractories and Minerals, p. 481
National Semiconductor UK Ltd, p. 596
National Spinning Company Inc, p. 99
National-Standard Co, p. 501
National Starch and Chemical Co, pp. 18, 278
National Steel Corp, p. 499
Natura, p. 342
Navajo Refining Co, p. 371
NBTY Inc, p. 343
NCT Leather Ltd, p. 156
NEC do Brasil SA, p. 595
Negev Ceramics Ltd, p. 443
Nellcor Puritan Bennett Inc, p. 697
Nelson House Forest Industries, p. 372
Nepal
— basic chemicals, excl fertilizers, p. 312
— beverages, pp. 45, 47-48, 50, 52-53, 56
— chemical products nec, pp. 329, 331-332, 334, 336-337,
   339
— drugs and medicines, pp. 350, 352-355, 357
— electrical machinery, pp. 603, 605-607, 609, 611, 613
— food products, pp. 5, 7, 9-10, 12, 14, 16
— footwear, pp. 161-162, 164-165, 167-168, 170
— furniture, fixtures, pp. 201, 203-204, 206, 208-209, 211
— glass and products, pp. 449-450, 452-453, 458
— industrial chemicals, pp. 267, 276
— industries nec, pp. 707, 709-710, 712, 714-715, 718
— iron and steel, pp. 488, 490-491, 493-495, 497
— leather and products, pp. 145-146, 148-149, 151-152, 154
— machinery nec, p. 570
— metal products, pp. 541, 543-544, 546, 548-549, 551

Nepal continued:
— nonferrous metals, p. 523
— nonmetal products nec, pp. 467, 469-470, 472-473, 475,
   477
— paper and products, pp. 219, 221-222, 224, 226-227, 22
— petroleum refineries, p. 368
— petroleum, coal products, p. 387
— plastic products nec, pp. 415, 417-419, 421, 423, 425
— pottery, china, etc., p. 441
— printing, publishing, pp. 249, 251, 253-254, 256, 258, 26
— professional goods, pp. 686, 695
— profile, p. 771
— radio, television, etc., pp. 628-632, 634
— rubber products, pp. 395-396, 398-399, 401-402, 404
— spinning, weaving, etc., pp. 112, 114-116, 118
— textiles, pp. 85, 87-88, 90, 92-93, 95
— tobacco, pp. 67, 69-70, 72-73, 75, 77
— transportation equipment, p. 653
— wearing apparel, pp. 123, 125-126, 128-129, 131, 133
— wood products, pp. 181, 183-184, 186, 188-189, 191
Neptun, p. 172
Nesco Inc, p. 574
Nesher Israel Cement Enterprises, p. 481
Neste Alfa Oy, p. 390
Nestle Canada Inc, p. 19
Nestle Norge AS, p. 721
Netherlands
— beverages, pp. 45, 47, 50, 52, 54, 56
— chemical products nec, pp. 329, 331, 334, 336-337, 339
— drugs and medicines, pp. 350, 352, 354-357
— electrical machinery, pp. 603, 605, 607, 609, 611, 613
— food products, pp. 5, 7, 10, 12, 14, 16
— footwear, pp. 161-162, 165, 167-168, 170
— furniture, fixtures, pp. 201, 203, 206, 208-209, 211
— glass and products, pp. 449-450, 453-454, 456, 458
— industrial chemicals, pp. 267, 269, 271, 273-274, 276
— industries nec, pp. 707, 709, 712, 714-715, 718
— iron and steel, p. 495
— leather and products, pp. 145-146, 149, 151, 154
— machinery nec, pp. 561-562, 565, 567-568, 570
— metal products, pp. 541, 543, 546, 548-549, 551
— nonferrous metals, p. 521
— nonmetal products nec, pp. 467, 469, 472-473, 475, 477
— office, computing machinery, p. 594
— paper and products, pp. 219, 221, 224, 226-227, 229
— petroleum refineries, pp. 360, 362, 364-365, 368
— petroleum, coal products, pp. 380, 382-383, 385, 387
— plastic products nec, pp. 415, 417, 419, 421, 423, 425
— pottery, china, etc., pp. 432, 434, 436, 438-439, 441
— printing, publishing, pp. 249, 251, 254, 256, 258, 260
— professional goods, pp. 686, 688, 691-693, 695
— profile, p. 771
— pulp, paper, etc., pp. 240-241, 243-246
— rubber products, pp. 395-396, 399, 401-402, 404
— shipbuilding, repair, pp. 668-669, 671-674
— spinning, weaving, etc., pp. 112, 114-116, 118

Netherlands continued:
—textiles, pp. 85, 87, 90, 92-93, 95
—tobacco, pp. 67, 69, 72-73, 75, 77
—transportation equipment, pp. 643, 645, 648-649, 651, 653
—wearing apparel, pp. 123, 125, 128-129, 133
—wood products, pp. 181, 183, 186, 188-189, 191
Netherlands Antilles
—beverages, pp. 45, 47
—chemical products nec, pp. 329, 331
—electrical machinery, pp. 603, 605
—food products, pp. 5, 7
—footwear, pp. 161-162
—furniture, fixtures, pp. 201, 203
—industrial chemicals, pp. 267, 269
—industries nec, pp. 707, 709
—machinery nec, pp. 561-562
—metal products, pp. 541, 543
—nonferrous metals, p. 514
—nonmetal products nec, pp. 467, 469
—paper and products, pp. 219, 221
—petroleum refineries, p. 368
—plastic products nec, pp. 415, 417
—printing, publishing, pp. 249, 251
—professional goods, pp. 686, 688
—profile, p. 772
—textiles, pp. 85, 87
—tobacco, pp. 67, 69
—transportation equipment, pp. 643, 645
—wearing apparel, pp. 123, 125
—wood products, pp. 181, 183
Network Systems Corp, p. 596
New England Business Service Inc, p. 263
New Holland North America Inc, p. 573
New Oji Paper Co., Ltd, p. 231
New South Africa Garment, p. 135
New Venture Gear Inc, p. 656
New Zealana Steel Ltd, p. 501
New Zealand
—basic chemicals, excl fertilizers, pp. 312-318
—beverages, pp. 45, 47-48, 52, 56
—chemical products nec, pp. 329, 331-332, 334, 336-337, 339
—drugs and medicines, pp. 350, 352-353, 357
—electrical machinery, pp. 603, 605-607, 609, 611, 613
—food products, pp. 5, 7, 9-10, 12, 14, 16
—footwear, pp. 161-162, 164, 167, 170
—furniture, fixtures, pp. 201, 203-204, 208, 211
—glass and products, pp. 449-450, 452, 458
—industrial chemicals, pp. 267, 269-271, 273-274, 276
—industries nec, pp. 707, 709-710, 712, 714-715, 718
—iron and steel, pp. 488, 490-491, 494, 497
—leather and products, pp. 145-146, 148, 151, 154
—machinery nec, pp. 561-562, 564-565, 567-568, 570
—metal products, pp. 541, 543-544, 546, 548-549, 551
—motor vehicles, pp. 676, 678, 682
—nonferrous metals, pp. 514, 516-517, 520, 523

New Zealand continued:
—nonmetal products nec, pp. 467, 469-470, 477
—office, computing machinery, pp. 590-592, 594
—paper and products, pp. 219, 221-222, 224, 226-227, 229
—petroleum refineries, pp. 360, 362-366, 368
—petroleum, coal products, pp. 380, 382-383, 385, 387
—plastic products nec, pp. 415, 417-418, 420-421, 423, 425
—pottery, china, etc., pp. 432, 434-435, 441
—printing, publishing, pp. 249, 251, 253-254, 256, 258, 260
—professional goods, pp. 686, 688-689, 691-693, 695
—profile, p. 772
—pulp, paper, etc., pp. 240-241, 243-246
—radio, television, etc., pp. 628-630, 634
—rubber products, pp. 395, 397-399, 401-402, 404
—shipbuilding, repair, pp. 668-670, 674
—spinning, weaving, etc., pp. 112, 114, 118
—synthetic resins, etc., pp. 320-325
—textiles, pp. 85, 87-88, 90, 92-93, 95
—tobacco, pp. 67, 69-70, 73, 77
—transportation equipment, pp. 643, 645-646, 648-649, 651, 653
—wearing apparel, pp. 123, 125-126, 129, 133
—wood products, pp. 181, 183-184, 188, 191
Newcor Inc, p. 429
Newman Tonks Group PLC, pp. 427, 525
News International PLC, p. 262
Newspaper publishing, p. 247
Newsprint, pp. 217, 236, 239
NIBCO Inc, p. 554
Nicaragua
—beverages, p. 56
—chemical products nec, p. 339
—electrical machinery, p. 613
—food products, p. 16
—footwear, p. 170
—furniture, fixtures, p. 211
—glass and products, p. 458
—industrial chemicals, p. 276
—industries nec, p. 718
—iron and steel, p. 497
—leather and products, p. 154
—machinery nec, p. 570
—metal products, p. 551
—nonferrous metals, p. 523
—nonmetal products nec, p. 477
—paper and products, p. 229
—petroleum refineries, p. 368
—petroleum, coal products, p. 387
—plastic products nec, p. 425
—pottery, china, etc., p. 441
—printing, publishing, p. 260
—professional goods, p. 695
—rubber products, p. 404
—textiles, p. 95
—tobacco, p. 77
—transportation equipment, p. 653

Nicaragua continued:
—wearing apparel, p. 133
—wood products, p. 191
Nickel, pp. 513, 531
Niger
—beverages, p. 56
—chemical products nec, p. 339
—electrical machinery, p. 613
—food products, p. 16
—leather and products, p. 154
—machinery nec, p. 570
—metal products, p. 551
—nonmetal products nec, p. 477
—paper and products, p. 229
—pottery, china, etc., p. 439
—printing, publishing, p. 260
—textiles, pp. 93, 95
—wearing apparel, p. 133
—wood products, p. 191
Nigeria
—beverages, p. 56
—chemical products nec, p. 339
—electrical machinery, p. 613
—food products, p. 16
—footwear, p. 170
—furniture, fixtures, p. 211
—glass and products, p. 458
—industrial chemicals, p. 277
—industries nec, p. 718
—iron and steel, p. 497
—leather and products, p. 154
—machinery nec, p. 570
—metal products, p. 551
—nonmetal products nec, p. 477
—paper and products, p. 229
—petroleum, coal products, p. 387
—plastic products nec, p. 425
—pottery, china, etc., p. 441
—printing, publishing, p. 260
—professional goods, p. 695
—rubber products, p. 404
—textiles, p. 95
—tobacco, p. 77
—transportation equipment, p. 653
—wearing apparel, p. 133
—wood products, p. 191
Nissan Motor Iberica, SA, pp. 572, 655
Nitric acid, p. 293
Nitrogenous fertilizers, p. 300
NKT Holding AS, pp. 443, 499, 525, 615
NL Industries Inc, p. 279
NMC Group PLC, p. 429
Nobel Cigars AS, p. 79
Nobel Industries Sweden (UK) Ltd, p. 342
Nokia Telecommunications, p. 616
Noma Inc., Danbel Industries, p. 461

Non-cellulosic continuous fibers, p. 308
Non-cellulosic staple and tow, p. 304
Nonferrous metals, p. 513
Nong Shim Co., Ltd, p. 18
Noodles, p. 3
Norampac Inc, p. 232
Noranda Aluminum Inc, p. 525
Noranda Mining and Exploration Inc, pp. 280, 526
Norandal USA Inc, p. 526
Norco Windows Inc, p. 193
Norcross Footwear Inc, p. 173
Nordson Corp, p. 574
Norsk Hydro UK Ltd, pp. 428, 526
Nortek Inc, pp. 572, 615
North American Biologicals Inc, p. 342
Northern Engraving Corp, p. 555
Northern Foods Grocery Group Ltd, p. 19
Northern Telecom Europe Ltd, p. 615
Northern Telecom Ltd, p. 615
Northern Upholstery Group Ltd, p. 215
Northern Upholstery Ltd, p. 215
Northwest Hardwoods Div, p. 195
Northwestern Steel and Wire Co, p. 500
Northwood Inc, pp. 194, 232
Norton Chemical Process Products, p. 389
Norton Performance Plastics Corp, p. 428
Norton SA Industria e Comercio, p. 480
Norwalk Furniture Corp, p. 214
Norway
—basic chemicals, excl fertilizers, pp. 312-313, 315-318
—beverages, pp. 45, 47, 50, 52, 56
—chemical products nec, pp. 329, 331, 334, 336-337, 339
—drugs and medicines, pp. 350, 352, 354-357
—electrical machinery, pp. 603, 605, 608-609, 611, 613
—food products, pp. 5, 7, 10, 12, 14, 16
—footwear, pp. 161-162, 165, 167-168, 170
—furniture, fixtures, pp. 201, 203, 206, 208-209, 211
—glass and products, pp. 449-450, 453-454, 456, 458
—industrial chemicals, pp. 267, 269, 271, 273-274, 277
—industries nec, pp. 707, 709, 712, 714-715, 718
—iron and steel, pp. 488, 490, 493-495, 497
—leather and products, pp. 145-146, 149, 151-152, 154
—machinery nec, pp. 561-562, 565, 567-568, 570
—metal products, pp. 541, 543, 546, 548-549, 551
—nonferrous metals, pp. 514, 516, 518, 520-521, 523
—nonmetal products nec, pp. 467, 469, 472-473, 475, 477
—office, computing machinery, pp. 590-594
—paper and products, pp. 219, 221, 224, 226-227, 229
—petroleum refineries, pp. 360, 362, 364-366, 368
—petroleum, coal products, pp. 380, 382-383, 385, 387
—plastic products nec, pp. 415, 417, 420-421, 423, 425
—pottery, china, etc., pp. 432, 434, 436, 438-439, 441
—printing, publishing, pp. 249, 251, 254, 256, 258, 260
—professional goods, pp. 686, 688, 691-693, 695
—profile, p. 773
—pulp, paper, etc., pp. 240-241, 243-246

Norway continued:
— radio, television, etc., pp. 628-629, 631-634
— rubber products, pp. 395, 397, 399, 401-402, 404
— shipbuilding, repair, pp. 668-669, 671-674
— spinning, weaving, etc., pp. 112, 114-115, 117-118
— synthetic resins, etc., pp. 320-321, 323, 325
— textiles, pp. 85, 87, 90, 92-93, 95
— tobacco, pp. 67, 69, 72-73, 77
— transportation equipment, pp. 643, 645, 648-649, 651, 653
— wearing apparel, pp. 123, 125, 128-129, 131, 133
— wood products, pp. 181, 183, 186, 188-189, 191
Nova Chemicals Ltd, pp. 280, 370
Novo Nordisk AS, pp. 278, 341
Novopharm Ltd, p. 343
Nozzles, p. 465
NRI Industries Inc, p. 408
NS Group Inc, p. 500
NU-kote Holding Inc, pp. 698, 719
Nucor Corp, p. 499
Nunn Bush Shoe Co, p. 173
Nusantara Plywood PT, p. 193
NutraSweet Kelco Co, p. 280
Nutrifood Indonesia, PT, p. 60
Nuts, p. 539
NV Sumatra Tobacco Trading Co, pp. 79, 719
Nylex (Malaysia) SDN BHD, p. 407
O Globo Empresa Jornalistica, p. 263
Oak Industries Inc, pp. 617, 698
Oakwood Homes Corp, p. 193
Occidental Chemical Corp, p. 278
ODL Inc, p. 460
OEMV - MAPETROL d.o.o, p. 371
Office furniture, p. 199
Office machines, pp. 559, 589
Office National, p. 59
Offset printing, p. 247
Oglebay Norton Co, p. 480
Ohmeda Inc, pp. 341, 697
Oil
— cotton-seed, crude, p. 33
— cotton-seed, refined, p. 33
— groundnut, crude, p. 34
— groundnut, refined, p. 34
— olive, crude, p. 34
— olive, refined, p. 35
— soya bean, crude, p. 32
— soya bean, refined, p. 33
Oil-Dri Corporation of America, p. 481
Oil Refineries Ltd, p. 370
Oils, pp. 3, 32, 35
OKI Europe Ltd, p. 597
Okonite Company Inc, p. 527
Oktobar, pp. 572, 615
Olbernhauer Glas, Pech & Kunte, p. 460
Old Bond Street Holding Co. Ltd, p. 343
Olin Corp, pp. 278, 341, 525, 553

Oliver Rubber Co, p. 408
Oman
— industries nec, pp. 707, 709, 712, 715, 718
— profile, p. 773
Oneida Ltd, pp. 443, 459, 525, 719
Oneida Silversmiths Div, p. 719
Oneita Industries Inc, pp. 98, 136
Operadora de Factorias SA, p. 460
OPTEK Technology Inc, p. 699
Optica Industrial, SA, p. 461
Optical and analytical instruments, p. 701
Optical glass, p. 447
Orbital Sciences Corp, pp. 657, 698
Oregon Steel Mills Inc, p. 500
Organs, pp. 705, 722
Oriental Brewery Co., Ltd, p. 58
Oriental Precision & Engineering, p. 499
Orion Corp, pp. 341, 697
Orion Electric Co., Ltd, p. 616
Orlik Tobacco Co. AS, p. 79
Ormet Primary Aluminum Corp, p. 527
OroAmerica Inc, pp. 79, 721
Orscheln Co, p. 657
Ortel Corp, p. 460
Orthopedic shoes, p. 159
Oscar Mayer Foods Corp, p. 20
Osem Investments Ltd, pp. 20, 58
OSF Inc, p. 214
OshKosh B'Gosh Inc, pp. 135, 172
Oshkosh Truck Corp, p. 657
Osicom Technologies Inc, p. 596
Osothsapha Teck Heng Yoo Co. Ltd, p. 342
O'Sullivan Corp, p. 429
O'Sullivan Industries Inc, p. 213
Outboard Marine Corp, pp. 572, 655
Outer garments, p. 83
Ovens for household use, p. 618
Overcoats, pp. 121, 138
Owens-Brockway Div, p. 427
Owens-Corning Canada Inc, pp. 459, 480
Owens Corning Fiberglas Ltd, p. 460
Oxford Industries Inc, p. 135
Oy Hartwall Ab, p. 58
Oy P.C. Rettig Ab, p. 79
Ozeta Odevne Zavody, Akciova, p. 135
Ozeta Trencin, pp. 18, 97
P. Baillargeon Ltee, p. 390
Pabrik Kertas Tjiwi Kimia PT, pp. 231, 278
Pabst Brewing Co, p. 59
Pacific Coast Building Products Inc, pp. 194, 479
Pacific Industries Inc, p. 460
Pacific Scientific Co, p. 699
Packaging papers and board, pp. 217, 239
Packaging Resources Inc, p. 429
Packing containers of paper or paperboard, p. 238
Packings, p. 393

Padaeng Industry Co., Ltd, p. 527
Paints, pp. 327, 344
Pakistan
— basic chemicals, excl fertilizers, pp. 312-313, 315-317
— beverages, pp. 45, 47, 50, 52, 54, 56
— chemical products nec, pp. 329, 331, 334, 336-337, 339
— drugs and medicines, pp. 350, 352, 354-356
— electrical machinery, pp. 603, 605, 608-609, 611, 613
— food products, pp. 5, 7, 10, 12, 14, 16
— footwear, pp. 161-162, 165, 167-168, 170
— furniture, fixtures, pp. 201, 203, 206, 208-209, 211
— glass and products, pp. 449-450, 453-454, 456, 458
— industrial chemicals, pp. 267, 269, 271, 273-274, 277
— industries nec, pp. 707, 709, 712, 714, 716, 718
— iron and steel, pp. 488, 490, 493-495, 497
— leather and products, pp. 145-146, 149, 151-152, 154
— machinery nec, pp. 561-562, 565, 567-568, 570
— metal products, pp. 541, 543, 546, 548-549, 551
— motor vehicles, pp. 676, 678-679, 681
— nonferrous metals, pp. 514, 516, 518, 520-521, 523
— nonmetal products nec, pp. 467, 469, 472-473, 475, 477
— paper and products, pp. 219, 221, 224, 226-227, 229
— petroleum refineries, pp. 360, 362, 364-366, 368
— petroleum, coal products, pp. 380, 382-383, 385, 387
— plastic products nec, pp. 415, 417, 420-421, 423, 425
— pottery, china, etc., pp. 433-434, 436, 438-439, 441
— printing, publishing, pp. 249, 251, 254, 256, 258, 260
— professional goods, pp. 687-688, 691-693, 695
— profile, p. 774
— pulp, paper, etc., pp. 240-241, 243-245
— radio, television, etc., pp. 628-629, 631-633
— rubber products, pp. 395, 397, 399, 401-402, 404
— shipbuilding, repair, pp. 668-669, 671-673
— spinning, weaving, etc., pp. 112, 114-115, 117
— synthetic resins, etc., pp. 320-321, 323-324
— textiles, pp. 85, 87, 90, 92-93, 95
— tobacco, pp. 67, 69, 72-73, 75, 77
— transportation equipment, pp. 643, 645, 648-649, 651, 653
— wearing apparel, pp. 123, 125, 128-129, 131, 133
— wood products, pp. 181, 183, 186, 188-189, 191
Pall Corp, p. 572
Pall Europe Ltd, p. 233
Palliser Furniture Ltd, p. 213
Palm Harbor Homes Inc, p. 193
Paloma Sladkogorska Tovarna, p. 232
Panama
— basic chemicals, excl fertilizers, pp. 312-318
— beverages, pp. 45, 47-48, 50, 52, 54, 56
— chemical products nec, pp. 329, 331-332, 334, 336-337, 339
— drugs and medicines, pp. 350, 352-357
— electrical machinery, pp. 603, 605-606, 608-609, 611, 613
— food products, pp. 5, 7, 9-10, 12, 14, 16
— footwear, pp. 161, 163-165, 167-168, 170
— furniture, fixtures, pp. 201, 203-204, 206, 208-209, 211
— glass and products, pp. 449-450, 452-454, 456, 458

Panama continued:
— industrial chemicals, pp. 267, 269-271, 273-274, 277
— industries nec, p. 718
— iron and steel, pp. 489, 491, 497
— leather and products, pp. 145, 147-149, 151-152, 154
— machinery nec, pp. 561-562, 564-565, 567-568, 570
— metal products, pp. 541, 543-544, 546, 548-549, 551
— motor vehicles, pp. 676, 678-679, 681-682
— nonferrous metals, pp. 514, 517, 523
— nonmetal products nec, pp. 467, 469-470, 472-473, 475, 477
— paper and products, pp. 219, 221-222, 224, 226-227, 22
— petroleum refineries, p. 368
— petroleum, coal products, p. 387
— plastic products nec, pp. 415, 417-418, 420-421, 423, 4
— pottery, china, etc., pp. 433-436, 438, 441
— printing, publishing, pp. 249, 251, 253-254, 256, 258, 26
— professional goods, pp. 687-689, 691-693, 695
— profile, p. 774
— rubber products, pp. 395, 397-399, 401-402, 404
— shipbuilding, repair, pp. 668-674
— spinning, weaving, etc., p. 114
— textiles, pp. 85, 87-88, 90, 92-93, 95
— tobacco, pp. 67, 69-70, 72-73, 75, 77
— transportation equipment, pp. 643, 645-646, 648-649, 6
653
— wearing apparel, pp. 123, 125-126, 128-129, 131, 133
— wood products, pp. 181, 183-184, 186, 188-189, 191
Panca Wana Indonesia PT, p. 195
Pang Rim Co., Ltd, p. 98
Paper, pp. 217, 239
Paper Industries Corp. of the, pp. 193, 231
Paperboard, pp. 217, 239
Paperboard containers, p. 217
Papua New Guinea
— beverages, p. 56
— chemical products nec, p. 339
— electrical machinery, p. 613
— food products, p. 16
— footwear, p. 170
— furniture, fixtures, p. 211
— glass and products, p. 458
— industrial chemicals, p. 277
— iron and steel, p. 497
— machinery nec, p. 570
— metal products, p. 551
— nonferrous metals, p. 523
— nonmetal products nec, p. 477
— paper and products, p. 229
— petroleum refineries, p. 368
— petroleum, coal products, p. 387
— plastic products nec, p. 425
— pottery, china, etc., p. 441
— printing, publishing, p. 260
— rubber products, p. 404
— textiles, p. 95

Papua New Guinea continued:
—tobacco, p. 77
—transportation equipment, p. 653
—wood products, p. 191
PAR Technology Corp, p. 597
Para Press SA, p. 461
Paradyne Corp, p. 597
Paraffin wax, p. 376
Paragon Trade Brands Inc, p. 233
Paraguay
—beverages, pp. 45, 47, 56
—chemical products nec, pp. 329, 331, 339
—electrical machinery, pp. 603, 605, 613
—food products, pp. 5, 7, 16
—footwear, pp. 161, 163, 170
—furniture, fixtures, pp. 201, 203, 211
—glass and products, pp. 449-450, 458
—industrial chemicals, pp. 267, 269, 277
—industries nec, pp. 707, 709, 712, 714, 718
—iron and steel, pp. 489-490, 498
—leather and products, pp. 145, 147, 154
—machinery nec, pp. 561-562, 570
—metal products, pp. 541, 543, 551
—nonferrous metals, pp. 514, 516, 523
—nonmetal products nec, pp. 467, 469, 477
—paper and products, pp. 219, 221, 229
—petroleum refineries, pp. 360, 362, 368
—petroleum, coal products, pp. 380, 382, 387
—plastic products nec, pp. 415, 417, 425
—pottery, china, etc., pp. 433-434, 441
—printing, publishing, pp. 249, 251, 260
—professional goods, pp. 687-688, 695
—profile, p. 775
—rubber products, pp. 395, 397, 404
—textiles, pp. 85, 87, 95
—tobacco, pp. 67, 69, 77
—transportation equipment, pp. 643, 645, 653
—wearing apparel, pp. 123, 125, 133
—wood products, p. 191
Parish Light Vehicle Structure Div, p. 657
Park Electrochemical Corp, pp. 232, 555
Park-Ohio Industries Inc, pp. 280, 428, 479
Parke Davis & Co. Ltd, p. 343
Parker Bertea Aerospace Group, p. 656
Parker Pen Holdings Ltd, p. 719
Parkland Industries Ltd, p. 371
Parkland Refining Ltd, p. 371
Parmalat Dairy & Bakery Inc, p. 19
Particle board, pp. 179, 197
Passenger cars, p. 675
—assembled from imported parts, p. 661
—produced, p. 662
Pasteur Merieux Connaught, p. 343
Pastries, p. 3
Pastry, p. 38
Pataling Rubber Estates Ltd, pp. 20, 279

Paterson Zochonis PLC, pp. 19, 341, 573
Patrick Industries Inc, p. 194
Pauls PLC, pp. 19, 279
Pavages Maska Inc, p. 390
Paving mixtures, p. 359
PAXAR Corp, pp. 99, 263, 596
Paz Lubricants and Chemicals Ltd, p. 390
PBH UK Ltd, p. 231
Pegeg & Co, p. 173
Peko Trzic d.d, p. 172
Pellimport, SA, p. 157
Pencils, pp. 705, 724
Penguin Putnam Inc, p. 263
Penn Engineering and, p. 429
Pens, p. 705
Pentair Inc, pp. 231, 553, 572
Pentland Group PLC, pp. 136, 172, 719
Pepsi-Cola Canada Beverages, p. 58
Pepsi-Cola Northwest, p. 59
Pepsico - IVI SA, p. 60
Pepsico Holdings Ltd, p. 19
Perambulators, p. 641
Perambulators and push-chairs for babies, p. 665
Perawang Lumber PT, p. 194
Perfekta Enterprises Ltd, pp. 427, 719
Perindustrian Tentara Nasional Id, pp. 553, 573, 656, 697
Perkin-Elmer Corp, p. 697
Perlmooser Zementwerke AG-, p. 480
Permafreeze, p. 444
Perrier Group of America Inc, p. 58
Perrigo Co, p. 341
Peru
—basic chemicals, excl fertilizers, pp. 312, 314-318
—beverages, pp. 45, 47, 50, 52, 54, 56
—chemical products nec, pp. 329, 331, 334, 336-337, 339
—drugs and medicines, pp. 350, 352, 354-357
—electrical machinery, pp. 603, 605, 608-609, 611, 613
—food products, pp. 5, 7, 10, 12, 14, 16
—footwear, pp. 161, 163, 165, 167-168, 170
—furniture, fixtures, pp. 201, 203, 206, 208-209, 211
—glass and products, pp. 449-450, 453-454, 456, 458
—industrial chemicals, pp. 267, 269, 271, 273-274, 277
—industries nec, pp. 707, 709, 712, 714, 716, 718
—iron and steel, pp. 489-490, 493-495, 498
—leather and products, pp. 145, 147, 149, 151-152, 154
—machinery nec, pp. 561-562, 565, 567-568, 570
—metal products, pp. 541, 543, 546, 548-549, 551
—motor vehicles, pp. 676, 678-679, 681-682
—nonferrous metals, pp. 514, 516, 518, 520-521, 523
—nonmetal products nec, pp. 467, 469, 472-473, 475, 477
—office, computing machinery, pp. 590-594
—paper and products, pp. 219, 221, 224, 226-227, 229
—petroleum refineries, pp. 360, 362, 364-366, 368
—petroleum, coal products, pp. 380, 382-383, 385-387
—plastic products nec, pp. 415, 417, 420-421, 423, 425
—pottery, china, etc., pp. 433-434, 436, 438-439, 441

Peru continued:
— printing, publishing, pp. 249, 251, 254, 256, 258, 260
— professional goods, pp. 687-688, 691-693, 695
— profile, p. 775
— pulp, paper, etc., pp. 240-241, 243-246
— radio, television, etc., pp. 628-629, 631-634
— rubber products, pp. 395, 397, 399, 401-402, 404
— shipbuilding, repair, pp. 668-669, 671-674
— spinning, weaving, etc., pp. 112, 114, 116-118
— synthetic resins, etc., pp. 320-321, 323-325
— textiles, pp. 85, 87, 90, 92-93, 95
— tobacco, pp. 67, 69, 72-73, 75, 77
— transportation equipment, pp. 643, 645, 648-649, 651, 653
— wearing apparel, pp. 123, 125, 128-129, 131, 133
— wood products, pp. 181, 183, 186, 188-189, 191
Peter Black Holdings PLC, pp. 172, 213, 342
Peter Black Keighley Ltd, p. 173
Petindo Jaya Sakti PT, p. 408
Petro-Canada, Petro-Canada, p. 389
Petroecuador, p. 370
Petrol Trgovina d.d, p. 657
Petroleos del Norte, SA, p. 370
Petroleum coke, pp. 359, 376
Petroleum Corp. of N.Z, p. 370
Petroleum refineries, p. 359
Peugeot Talbot Espana, SA, p. 656
Peugeot Talbot Motor Co. PLC, p. 656
P.H. Glatfelter Co, p. 232
Pharma Plast International AS, pp. 407, 429
Pharmaceutical preparations, p. 349
Pharr Yarns Inc, p. 98
Phenol, p. 287
Phenolic and cresylic plastics, p. 305
Philip Morris Espana, SA, p. 79
Philip Morris Ltd, p. 79
Philip Services Corp, p. 525
Philippines
— basic chemicals, excl fertilizers, pp. 312, 314-318
— beverages, pp. 45, 47-48, 50, 52, 54, 56
— chemical products nec, pp. 329, 331-332, 334, 336-337, 339
— drugs and medicines, pp. 350, 352-357
— electrical machinery, pp. 603, 605-606, 608-609, 611, 613
— food products, pp. 5, 7, 9-10, 12, 14, 16
— footwear, pp. 161, 163-165, 167-168, 170
— furniture, fixtures, pp. 201, 203-204, 206, 208-209, 211
— glass and products, pp. 449-450, 452-454, 456, 458
— industrial chemicals, pp. 267, 269-271, 273-274, 277
— industries nec, pp. 707, 709, 711-712, 714, 716, 718
— iron and steel, pp. 489-491, 493-495, 498
— leather and products, pp. 145, 147-149, 151-152, 154
— machinery nec, pp. 561-562, 564-565, 567-568, 570
— metal products, pp. 541, 543-544, 546, 548-549, 551
— motor vehicles, pp. 676, 678-679, 681-682
— nonferrous metals, pp. 515-518, 520-521, 523
— nonmetal products nec, pp. 467, 469-470, 472-473, 475,

Philippines continued:
    477
— office, computing machinery, pp. 590-594
— paper and products, pp. 219, 221-222, 224, 226-227, 22
— petroleum refineries, pp. 360, 362-366, 368
— petroleum, coal products, pp. 380, 382-383, 385-387
— plastic products nec, pp. 415, 417-418, 420-421, 423, 4:
— pottery, china, etc., pp. 433-436, 438-439, 441
— printing, publishing, pp. 249, 251, 253-254, 256, 258, 26
— professional goods, pp. 687-689, 691-693, 695
— profile, p. 776
— pulp, paper, etc., pp. 240-246
— radio, television, etc., pp. 628-634
— rubber products, pp. 395, 397-399, 401-402, 404
— shipbuilding, repair, pp. 668, 670-674
— spinning, weaving, etc., pp. 112, 114-118
— synthetic resins, etc., pp. 320-325
— textiles, pp. 85, 87-88, 90, 92-93, 95
— tobacco, pp. 67, 69-70, 72-73, 75, 77
— transportation equipment, pp. 643, 645-646, 648-649, 6
    653
— wearing apparel, pp. 123, 125-126, 128-129, 131, 133
— wood products, pp. 181, 183-184, 186, 188-189, 191
Philips Components Discrete, p. 616
Philips Consumer Electronics Co, p. 615
Philips Technologies, pp. 617, 657
Phillips 66 Div, p. 370
Phonecia America-Israel (Flat Glass), p. 460
Phosphatic fertilizers, pp. 300-301
Phosphoric acid, p. 294
Photo-Me International PLC, p. 699
Photographic film, p. 327
Photographic paper, p. 327
Photronics Inc, p. 459
Phthalic anhydride, p. 290
Pianos, pp. 705, 722
Picvue Electronics, Ltd, p. 459
Pieles Costarricenses SA, p. 156
Pig iron, p. 487
— foundry, p. 502
— steel-making, p. 503
Pikolin, SA, p. 214
Pilkington Distribution Services Ltd, p. 459
Pilkington Glass Ltd, p. 459
Pilkington Insulation Ltd, p. 480
Pilliod Furniture Inc, p. 214
Piloimpregna, p. 195
Pimalco Corp, p. 527
Pinkerton Group Inc, pp. 79, 721
Pioneer Hi-Bred International Inc, p. 278
Pipe, p. 487
Pirelli Armstrong Tire Corp, p. 407
Pirelli Neumaticos, SA, p. 407
Pirelli UK PLC, p. 656
Pirelli UK Tyres Ltd, pp. 172, 406
Pitney Bowes Holdings Ltd, p. 596

Pitney Bowes of Canada Ltd, p. 596
Pittway Corp, pp. 262, 615
Piyavat Rubber Industry Co., Ltd, p. 172
Placas do Parana SA, p. 195
Planika, p. 172
Planing machines, p. 583
Planing mills, p. 179
Planks, p. 179
Plascon Paints Natal Pty. Ltd, p. 342
Plasti-Line Inc, pp. 597, 721
Plastic footwear, p. 412
Plastic pipe, p. 413
Plastic plate, p. 413
Plastic products, p. 413
Plastic Specialties and, pp. 280, 406
Plastics, pp. 265, 319
Plastika Nitra, p. 428
Plates
— 3 to 4.75 mm, p. 507
— over 4.75 mm, p. 507
— glass, p. 447
Platinum, p. 513
Players (sound), p. 627
Playtex Products Inc, pp. 233, 407, 428
Playthe, p. 137
PLM Redfearn Ltd, p. 459
Ploughs, pp. 559, 577
Plum Creek Timber Company L.P, p. 194
Pluma Inc, p. 136
Plysu PLC, p. 428
Plywood, pp. 179, 197
PMC Inc., pp. 279, 427
PMI Food Equipment Group, p. 572
Poland
— beverages, pp. 45, 47, 50, 52, 54, 56
— chemical products nec, pp. 329, 331, 334, 336-337, 339
— electrical machinery, pp. 603, 605, 608-609, 611, 613
— food products, pp. 5, 7, 10, 12, 14, 16
— footwear, pp. 161, 163, 165, 167-168, 170
— furniture, fixtures, pp. 201, 203, 206, 208-209, 211
— glass and products, pp. 449, 451, 453-454, 456, 458
— industrial chemicals, pp. 267, 269, 271, 273-274, 277
— industries nec, pp. 707, 709, 712, 714, 716, 718
— iron and steel, pp. 489-490, 493-495, 498
— leather and products, pp. 145, 147, 149, 151-152, 154
— machinery nec, pp. 561-562, 565, 567-568, 570
— metal products, pp. 541, 543, 546, 548-549, 551
— nonferrous metals, pp. 515-516, 518, 520-521, 523
— nonmetal products nec, pp. 467, 469, 472-473, 475, 477
— paper and products, pp. 219, 221, 224, 226-227, 229
— petroleum refineries, pp. 360, 362, 364-366, 368
— petroleum, coal products, pp. 380, 382-383, 385-387
— plastic products nec, pp. 415, 417, 420-421, 423, 425
— pottery, china, etc., pp. 433-434, 437-439, 441
— printing, publishing, pp. 249, 251, 254, 256, 258, 260
— professional goods, pp. 687-688, 691-693, 695

Poland continued:
— profile, p. 776
— rubber products, pp. 395, 397, 399, 401-402, 404
— textiles, pp. 85, 87, 90, 92-93, 95
— tobacco, pp. 67, 69, 72, 74-75, 77
— transportation equipment, pp. 643, 645, 648-649, 651, 653
— wearing apparel, pp. 123, 125, 128-129, 131, 133
— wood products, pp. 181, 183, 186, 188-189, 191
Polaris Industries L.P, p. 657
Polfa, p. 342
Polgat Ltd, pp. 98, 136
Polishes and creams, p. 346
Polmo Praszka SA, pp. 407, 526
Poludniowe Zaklady Przemyslu, p. 172
Poly Unggul PT, p. 408
Polyacetals, p. 308
Polyamides, p. 307
Polychrome Corp, p. 263
Polyethylene, p. 306
Polymer Group Inc, p. 99
Polypropylene, p. 306
Polysindo Eka Perkasa PT, p. 99
Polystyrene, p. 307
Polyvinyl chloride, p. 308
Polzela d.d, p. 721
Ponderosa Industrial SA de CV, p. 195
Poongsan Corp, p. 499
Pope and Talbot Inc, pp. 194, 232
Porcelain products, p. 431
Porcelanas Lladro, SA, p. 443
Porcelanosa, SA, p. 481
Pork, fresh, p. 21
Portsmouth & Sunderland, p. 263
Portugal
— basic chemicals, excl fertilizers, pp. 312, 314-318
— beverages, pp. 45, 47-48, 50, 52, 54, 56
— chemical products nec, pp. 329, 331-332, 334, 336-337, 339
— drugs and medicines, pp. 350, 352-357
— electrical machinery, pp. 603, 605-606, 608-609, 611, 613
— food products, pp. 5, 7, 9-10, 12, 14, 16
— footwear, pp. 161, 163-165, 167-168, 170
— furniture, fixtures, pp. 201, 203-204, 206, 208-209, 211
— glass and products, pp. 449, 451-454, 456, 458
— industrial chemicals, pp. 267, 269-271, 273-274, 277
— industries nec, pp. 707, 709, 711-712, 714, 716, 718
— iron and steel, pp. 489-491, 493-494, 496, 498
— leather and products, pp. 145, 147-149, 151-152, 154
— machinery nec, pp. 561-562, 564-565, 567-568, 570
— metal products, pp. 541, 543-544, 546, 548-549, 551
— motor vehicles, pp. 676, 678-679, 681-682
— nonferrous metals, pp. 515-518, 520-521, 523
— nonmetal products nec, pp. 467, 469-470, 472-473, 475, 477
— office, computing machinery, pp. 590-594
— paper and products, pp. 219, 221-222, 224, 226-227, 229

Alphabetical Index

Portugal continued:
— petroleum refineries, pp. 360, 362-366, 368
— petroleum, coal products, pp. 380, 382-383, 385-387
— plastic products nec, pp. 415, 417-418, 420-421, 423, 425
— pottery, china, etc., pp. 433-435, 437-439, 441
— printing, publishing, pp. 249, 251, 253-254, 256, 258, 260
— professional goods, pp. 687-689, 691-693, 695
— profile, p. 777
— pulp, paper, etc., pp. 240-246
— radio, television, etc., pp. 628-634
— rubber products, pp. 395, 397-399, 401-402, 404
— shipbuilding, repair, pp. 668, 670-674
— spinning, weaving, etc., pp. 112, 114-118
— synthetic resins, etc., pp. 320-325
— textiles, pp. 85, 87-88, 90, 92-93, 95
— tobacco, pp. 67, 69-70, 72, 74-75, 77
— transportation equipment, pp. 643, 645-646, 648-649, 651, 653
— wearing apparel, pp. 123, 125-126, 128-129, 131, 133
— wood products, pp. 181, 183-184, 186, 188-189, 191
Potassic fertilizers, pp. 301-302
Potlatch Corp, pp. 193, 231
Potlatch Corp. Western Wood, p. 194
Pottery, p. 431
Pou Chen Corp, pp. 406, 427
Poultry, p. 3
— dressed, fresh, p. 22
Powerine Oil Co, p. 371
Prakash Tubes Ltd, p. 500
Precidio Inc, p. 444
Precious metals, p. 513
Precision Castparts Corp, pp. 499, 655
Precoria Portland Cement Co. Ltd, p. 479
Preglejka AS, p. 195
Premdor Inc, pp. 193, 554
Premier Brands Ltd, pp. 19, 58
Premier Industrial Corp, p. 573
Premier Spring Industries Pty. Ltd, p. 215
Premium Beverage Packers Inc, p. 60
Prepared animal feeds, p. 42
Prepared leaf, p. 65
Prepared meats, p. 3
President Baking Company Inc, p. 20
President Enterprises Corp, p. 58
Press Corp. of South Africa Ltd, pp. 231, 262
Pressed glass, p. 447
Pretty Products Inc, p. 408
Pride Companies L.P, p. 371
PRIDE Industries Inc, p. 194
Prime Tanning Company Inc, p. 156
PRIMEDIA Inc, p. 262
Primex Technologies Inc, pp. 555, 657
Prince Corp., p. 135
Pringle of Scotland Ltd, p. 136
Printer's ink, p. 327
Printers' ink, p. 346

Printing, p. 247
Printing and writing paper, pp. 217, 239
Printing presses, p. 585
Printpack Inc, p. 232
Printpak Ltd, p. 233
Printronix Inc, p. 597
Prochnik Co, p. 136
Procter & Gamble Inc, pp. 20, 232, 342
Productora Tabacalera de Colombia, p. 80
Professional goods, p. 685
Professional instruments, p. 685
Promon Eletronica, p. 233
Propylene, p. 283
Propylene glycol, p. 288
Protela Ltda, p. 99
Przyjazn, pp. 370, 499
PTT Exploration & Production Co., p. 371
Publishing, p. 247
Puerto Rican Cement Co., Inc, p. 481
Puerto Rican Cement Company Inc, p. 480
Puerto Rico
— beverages, pp. 45, 47, 50, 52, 56
— chemical products nec, pp. 334, 336, 339
— drugs and medicines, pp. 350, 354
— electrical machinery, pp. 603, 605-606, 608-609, 613
— food products, pp. 5, 7, 10, 12, 16
— footwear, pp. 161, 163, 170
— furniture, fixtures, pp. 201, 203-204, 211
— glass and products, pp. 449, 451, 458
— industrial chemicals, pp. 271, 273, 277
— industries nec, pp. 707, 709, 711-712, 714, 718
— iron and steel, p. 498
— leather and products, pp. 145, 147, 154
— machinery nec, pp. 561, 563-565, 567, 570
— metal products, pp. 541, 543-544, 546, 548, 552
— nonferrous metals, p. 523
— nonmetal products nec, pp. 467, 469, 472, 477
— paper and products, pp. 219, 221-222, 224, 226, 229
— petroleum refineries, pp. 364, 368
— petroleum, coal products, pp. 384, 387
— plastic products nec, pp. 415, 417, 425
— pottery, china, etc., pp. 433-435, 441
— printing, publishing, pp. 249, 251, 253-254, 256, 260
— professional goods, pp. 687-689, 691-692, 695
— profile, p. 777
— rubber products, pp. 395, 397, 405
— textiles, pp. 85, 87-88, 90, 92, 95
— tobacco, pp. 67, 69-70, 72, 74, 77
— transportation equipment, pp. 643, 645-646, 648-649, 6
— wearing apparel, pp. 123, 125-126, 128-129, 133
— wood products, pp. 181, 183, 191
Pulaski Furniture Corp, p. 214
Pulau Sambu PT, p. 20
Pulitzer Publishing Co, p. 263
Pulp, pp. 217, 239
Pulp mills, pp. 217, 239

Pulp of fibers other than wood, p. 234
Pulse Engineering Inc, p. 615
Pumps, p. 620
PureTec Corp, p. 428
Puritan-Bennett Corp, pp. 280, 657, 698
Pyramid Handbags Inc, p. 173
Qatar
—beverages, pp. 45, 47, 50, 52, 56
—chemical products nec, pp. 329, 331, 334, 336, 339
—food products, pp. 5, 7, 10, 12, 16
—furniture, fixtures, pp. 201, 203, 206, 208, 211
—glass and products, p. 449
—industrial chemicals, pp. 267, 269, 271, 273, 277
—industries nec, pp. 707, 709, 712, 714, 718
—iron and steel, pp. 489-490, 493-494, 498
—leather and products, pp. 145, 147, 149, 151, 154
—paper and products, pp. 219, 221, 224, 226, 229
—petroleum refineries, pp. 360, 362, 364-365, 368
—petroleum, coal products, pp. 380, 382, 384-385, 387
—plastic products nec, pp. 415, 417, 420-421, 425
—printing, publishing, pp. 249, 251, 254, 256, 260
—profile, p. 778
—rubber products, pp. 395, 397, 399, 401, 405
—textiles, pp. 85, 87, 90, 92, 95
—wood products, pp. 181, 183, 186, 188, 191
QIT-Fer & Titane Inc, p. 500
QMS Inc, p. 597
Quaker Chemical Corp, p. 389
Quaker Fabric Corp, p. 99
Quaker Oats Co. of Canada Ltd, p. 60
Quaker State Corp, p. 370
Quaker State Inc, pp. 371, 390
QUALCOMM Inc, p. 615
Quanex Corp, pp. 499, 525, 554
Quartz, p. 459
Quebecor Printing Inc, pp. 231, 262
Queen Carpet Corp, p. 97
Quicklime, p. 483
Quintie Confectionery Ltd, p. 20
R. Griggs & Co. Ltd, p. 174
RA Multiproduct, pp. 194, 500
Radece Papir d.d, p. 721
Radiators, p. 539
Radio receivers, p. 635
Radiometer AS, p. 699
Radios, pp. 601, 627
Rafael Armament Development, pp. 554, 656
Rafters, p. 179
Rahman Hydraulic Tin BHD, p. 408
Rail equipment, p. 641
Rail motor passenger vehicles, p. 659
Rail passenger carriages, p. 660
Railway sleepers, p. 179
Railway track material, p. 510
Raincoats, p. 121
—men's and boys', p. 138

Raincoats continued:
—women's and girls', p. 140
Rakes, p. 578
Ralston Purina Canada Inc, p. 481
Ranco Inc, pp. 616, 656, 697
Rank America Inc, p. 616
Rauch Industries Inc, p. 460
Rauland Div, p. 617
Raven Industries Inc, p. 428
Raw sugar, p. 39
Rawlings Sporting Goods Company, pp. 137, 720
Raychem Corp, pp. 427, 525, 615
Raychem Ltd, p. 429
Raymond Woollen Mills Ltd, pp. 97, 479, 553
Rayonier Inc, pp. 193, 232
Raytech Corp, p. 480
Raytheon E. Systems Inc. Falls, p. 699
Razor blades, p. 539
Reader's Digest Association Inc, p. 262
Readicut International PLC, pp. 97, 279, 406
Readson Ltd, p. 98
Recalit SA, p. 389
Reckitt and Colman Inc, p. 342
Reckitt & Colman Products Ltd, p. 342
Recorders, p. 627
Records (sound), p. 627
Red Comercial del Calzado, SA, p. 174
Red Wing Shoe Company Inc, p. 172
Redland Aggregates Ltd, p. 480
Redland Roof Tiles Ltd, p. 481
Redland Stone Products Co, p. 389
Redpath Dorman Long Ltd, p. 554
Reed Business Publishing Ltd, p. 264
Reed International Books Ltd, p. 263
Refinadora Costarricense de, p. 370
Refined sugar, p. 39
Refractory cements, p. 465
Refractory ceramic products, p. 465
Refrigerators, p. 559
Refrigerators for household use, p. 621
Reichhold Chemicals Inc, p. 279
Reilly Industries Inc, p. 389
Reliable Manufacturing Works, Inc, pp. 215, 481
Relyon Group PLC, p. 214
Renfro Corp, p. 98
Renold PLC, p. 657
Repair of ships, p. 667
Repsol Derivados, SA, p. 371
Repsol Quimica SA, p. 389
Repsol (UK) Ltd, p. 371
Republic Engineered Steels Inc, p. 499
Republic Group Inc, p. 481
Republic of Moldova
—beverages, pp. 45, 47-48, 50, 56
—chemical products nec, pp. 329, 331-332, 334, 339
—electrical machinery, pp. 603, 605, 608, 613

Republic of Moldova continued:
—food products, pp. 5, 7, 9-10, 16
—footwear, pp. 161, 163-165, 170
—furniture, fixtures, pp. 201, 203-204, 206, 211
—glass and products, pp. 449, 451-453, 458
—industrial chemicals, pp. 267, 269-271, 277
—industries nec, pp. 707, 709, 711-712, 718
—iron and steel, pp. 489-491, 493, 498
—leather and products, pp. 145, 147-149, 154
—machinery nec, pp. 561, 563, 565, 570
—metal products, pp. 541, 543-544, 546, 552
—nonferrous metals, p. 523
—nonmetal products nec, pp. 467, 469-470, 472, 477
—paper and products, pp. 219, 221-222, 224, 229
—petroleum refineries, p. 368
—petroleum, coal products, pp. 380, 382, 384, 387
—plastic products nec, pp. 415, 417-418, 420, 425
—pottery, china, etc., p. 441
—printing, publishing, pp. 249, 251, 253-254, 260
—professional goods, p. 695
—profile, p. 778
—rubber products, p. 405
—textiles, pp. 85, 87-88, 90, 95
—tobacco, pp. 67, 69-70, 72, 77
—transportation equipment, pp. 643, 645, 648, 653
—wearing apparel, pp. 123, 125-126, 128, 133
—wood products, pp. 181, 183-184, 186, 191
Republic Tobacco Co, p. 79
Residual fuel oils, p. 375
Residuals, p. 359
Resins, p. 319
Retorts, p. 465
Reunion Industries Inc, p. 429
Revoz d.d, p. 719
Rex Trueform Clothing Co. Ltd, p. 136
Rexene Corp, p. 370
Reycan L.P, p. 527
Reynolds and Reynolds Co, p. 262
R.G. Barry Corp, pp. 98, 172
R.G. Carter Holdings Ltd, pp. 195, 480
RHC/Spacemaster Corp, p. 214
Rhodia SA, p. 280
Rhone-Poulenc Agrochimie, p. 278
Rhone-Poulenc Rorer Holdings Ltd, p. 341
Rich Products Corp, p. 18
Richards PLC, pp. 99, 136
RICOH Corp, p. 596
Rider Enterprise Co. Ltd, p. 174
Riser Foods Inc, p. 18
Riva Group PLC, p. 597
River Island Clothing Co. Ltd, p. 135
River Oaks Furniture Inc, p. 214
Riverside Manufacturing Co, p. 137
Riverwood International Corp, p. 231
Rivets, p. 539
Riviana Foods Inc, p. 20

R.J. Reynolds Tobacco (UK) Ltd, p. 80
RJR-Macdonald Inc, p. 79
RMI Titanium Co, p. 527
Road tractors, pp. 663, 675
Road vehicles, p. 641
Roadmaster Corp, pp. 657, 719
Roadmaster Industries Inc, p. 656
Roanoke Electric Steel Corp, p. 501
Robbins and Myers Inc, p. 617
Robert McBride Ltd, p. 342
Robert Mondavi Corp, p. 60
Robinson Manufacturing Company, p. 137
Rock-Tenn Co, p. 231
Rocky Shoes and Boots Inc, p. 173
Rods of glass, p. 447
Rods of steel, p. 487
Rogers Corp, p. 408
Rohr Inc, p. 656
Rolling mills, p. 513
Rolling mills for rolling metals, p. 584
Romania
—beverages, pp. 45, 50, 52, 54, 56
—chemical products nec, pp. 329, 334, 336-337, 339
—electrical machinery, pp. 603, 605, 608-609, 611, 613
—food products, pp. 5, 10, 12, 14, 16
—footwear, pp. 161, 165, 167-168, 170
—furniture, fixtures, pp. 201, 203, 206, 208-209, 211
—glass and products, pp. 449, 453-454, 456, 458
—industrial chemicals, pp. 267, 271, 273, 275, 277
—industries nec, pp. 707, 709, 712, 714, 716, 718
—iron and steel, pp. 489, 493-494, 496, 498
—leather and products, pp. 145, 149, 151-152, 154
—machinery nec, pp. 561, 563, 568, 570
—metal products, pp. 541, 543, 546, 548-549, 552
—motor vehicles, pp. 680-681
—nonferrous metals, pp. 515, 518, 520-521, 523
—nonmetal products nec, pp. 467, 472-473, 475, 477
—office, computing machinery, pp. 592-593
—paper and products, pp. 219, 221, 224, 226-227, 230
—petroleum refineries, pp. 360, 362, 364-366, 368
—petroleum, coal products, pp. 380, 382, 384-387
—plastic products nec, pp. 415, 420-421, 423, 425
—pottery, china, etc., pp. 433, 437-439, 441
—printing, publishing, pp. 249, 251, 254, 256, 258, 260
—professional goods, pp. 687-688, 693, 695
—profile, p. 779
—radio, television, etc., pp. 631-632
—rubber products, pp. 395, 399, 401-402, 405
—shipbuilding, repair, pp. 671-672
—textiles, pp. 85, 87, 90, 92-93, 95
—tobacco, pp. 67, 69, 72, 74-75, 77
—transportation equipment, pp. 643, 645, 648-649, 651, ●
—wearing apparel, pp. 123, 125, 128, 130-131, 133
—wood products, pp. 181, 183, 186, 188-189, 191
Romatex Ltd, p. 97
Roper Industries Inc, p. 699

Rose Art Industries Inc, pp. 195, 721
Rosenvasser David Ltd, p. 721
Ross Catherall Group PLC, p. 444
Ross Young's Holdings Ltd, p. 18
Rotem Amfert Negev Ltd, p. 280
Rothmans International PLC, p. 79
Rothmans of Pall Mall Indonesia PT, p. 79
Rothmans of Pall Mall Malaysia BHD, p. 79
Rothmans, Benson & Hedges Inc, p. 79
Rouge Steel Co, p. 500
Roussel Laboratories Ltd, p. 343
Rowe Furniture Corp, p. 214
Royal Beech-Nut Pty. Ltd, p. 59
Royal China and Porcelain, p. 444
Royal Doulton (UK) Ltd, pp. 443, 459
Royal Greenland Export, p. 19
Royal Group Technologies Ltd, p. 427
Royal Ordnance PLC, p. 553
RPC Group PLC, p. 429
RPM Inc., p. 341
Rubatex Corp, p. 408
Rubber
— hardened, p. 411
— reclaimed, pp. 393, 410
— synthetic, p. 303
— transmission, conveyor, elevator belts etc., p. 412
— unhardened vulcanized piping and tubing, p. 411
— unhardened vulcanized plates, sheets etc., p. 411
Rubber belting, p. 393
Rubber footwear, pp. 393, 412
Rubber products, p. 393
Rudne Bane SP, p. 500
Ruentex Industrial Co. Ltd, p. 98
Rugby Group PLC, pp. 459, 479, 499
Russ Berrie and Company Inc, pp. 194, 263, 720
Russian Federation
— basic chemicals, excl fertilizers, pp. 312, 314-316
— beverages, pp. 45, 47, 50, 52, 56
— chemical products nec, pp. 329, 331, 334, 336, 339
— drugs and medicines, pp. 350, 352, 354-355
— electrical machinery, pp. 603, 605, 608-609, 613
— food products, pp. 5, 7, 10, 12, 16
— footwear, pp. 161, 163, 165, 167, 170
— furniture, fixtures, pp. 201, 203, 206, 208, 211
— glass and products, pp. 449, 451, 453, 455, 458
— industrial chemicals, pp. 267, 269, 272-273, 277
— industries nec, pp. 707, 709, 712, 714, 718
— iron and steel, pp. 489-490, 493-494, 498
— leather and products, pp. 145, 147, 149, 151, 154
— machinery nec, pp. 561, 563, 565, 567, 570
— metal products, pp. 541, 543, 546, 548, 552
— motor vehicles, pp. 676, 678, 680-681
— nonferrous metals, pp. 515-516, 518, 520, 523
— nonmetal products nec, pp. 467, 469, 472-473, 477
— office, computing machinery, pp. 590-593
— paper and products, pp. 219, 221, 224, 226, 230

Russian Federation continued:
— petroleum refineries, pp. 360, 362, 364-365, 368
— petroleum, coal products, pp. 380, 382, 384-385, 387
— plastic products nec, pp. 415, 417, 420-421, 425
— pottery, china, etc., pp. 433-434, 437-438, 441
— printing, publishing, pp. 249, 251, 254, 256, 260
— professional goods, pp. 687-688, 691-692, 695
— profile, p. 779
— pulp, paper, etc., pp. 240-241, 243-244
— rubber products, pp. 395, 397, 399, 401, 405
— spinning, weaving, etc., pp. 112, 114, 116-117
— synthetic resins, etc., pp. 320-321, 323
— textiles, pp. 85, 87, 90, 92, 95
— tobacco, pp. 67, 69, 72, 74, 77
— transportation equipment, pp. 643, 645, 648-649, 653
— wearing apparel, pp. 123, 125, 128, 130, 133
— wood products, pp. 181, 183, 186, 188, 191
Rustom Mills & Industries Ltd, p. 99
RWE-DEA AG fur Mineraloel und, pp. 278, 370, 389
S. Jerome & Sons Holdings PLC, p. 408
Saez Merino, SA, p. 137
Safelite Glass Corp, p. 459
Safeskin Corp, p. 406
Safety-Kleen Canada Inc, p. 389
Saigon Leather Products and, p. 174
St. Austell Brewery Co. Ltd, p. 60
St. Glass Oblosuveggyarto Es, p. 459
St. Joe Paper Co, p. 233
St. John Knits Inc, pp. 136, 341
St. Joseph Printing Ltd, p. 264
St. Jude Medical Inc, p. 698
St. Lawrence Cement Inc, p. 479
Saint Lucia
— basic chemicals, excl fertilizers, pp. 312, 316
— beverages, pp. 45, 52
— chemical products nec, pp. 329, 336
— electrical machinery, pp. 603, 609
— food products, pp. 5, 12
— furniture, fixtures, pp. 201, 208
— industrial chemicals, pp. 267, 273
— industries nec, pp. 707, 714
— metal products, pp. 541, 548
— nonmetal products nec, p. 467
— paper and products, pp. 219, 226
— plastic products nec, pp. 415, 421
— pottery, china, etc., pp. 433, 438
— printing, publishing, pp. 249, 256
— pulp, paper, etc., pp. 240, 244
— rubber products, pp. 395, 401
— synthetic resins, etc., p. 320
— textiles, pp. 85, 92
— tobacco, pp. 67, 74
— wearing apparel, pp. 123, 130
Saint Vincent & the Grenadines
— beverages, pp. 45, 47-48
— chemical products nec, pp. 329, 331

Saint Vincent & the Grenadines continued:
— electrical machinery, p. 603
— food products, pp. 5, 7, 9
— industrial chemicals, pp. 267, 269
— metal products, pp. 541, 543
— paper and products, pp. 219, 221-222
— plastic products nec, p. 415
— profile, p. 780
— pulp, paper, etc., pp. 240-242
— tobacco, pp. 67, 69-70
— wearing apparel, pp. 123, 125-126
Salant Corp, p. 135
Sam Hyun Co., Ltd, p. 443
Sam Woo Co., Ltd, p. 156
Sam Yang Co., Ltd, p. 97
Sam Yang Foods Co., Ltd, p. 20
Sam Yung Chemical Co., Ltd, p. 174
Samator Gas Industry Group PT, p. 280
Sambo Chinaware Co., Ltd, p. 461
Samcor Pty. Ltd, p. 656
Samior Pty. Ltd, p. 656
Samsonite Corp, p. 172
Samsung Aerospace Industries Ltd, p. 655
Samsung Corning Co., Ltd, p. 459
Samyang Tongsang Co., Ltd, p. 719
San East UK PLC, p. 136
San Miguel Brewery Ltd, p. 60
Sandals, p. 159
Sandals and similar light footwear, p. 177
Sanderson Farms Inc, p. 18
Sandomierz Glass Works, p. 459
Sandoz Holdings Great Britain Ltd, p. 342
Sandpaper, p. 465
Sandvik Inc, pp. 500, 555
Sang-A Tech Corp, p. 174
Sango Ceramic Indonesia PT, pp. 233, 443
Sanitary ceramic fittings, p. 445
Sanitary fixtures, p. 431
Sans Fibres Pty. Ltd, p. 98
Santo Arden Co., Ltd, p. 444
Sanyang Industry Co., Ltd, p. 656
Sapalux Ltd, p. 719
Saskatchewan Wheat Pool, pp. 19, 60, 263, 279
Sasko Pty. Ltd, p. 18
Saturn Corp, p. 655
Satya Raya Indah Wood-Based, p. 194
Sauder Woodworking Co, p. 213
Saudi Arabia
— beverages, p. 56
— chemical products nec, p. 339
— electrical machinery, p. 613
— food products, p. 16
— footwear, p. 170
— furniture, fixtures, p. 211
— glass and products, p. 458
— industrial chemicals, p. 277

Saudi Arabia continued:
— industries nec, pp. 707, 709, 718
— iron and steel, p. 498
— leather and products, p. 154
— machinery nec, p. 570
— metal products, p. 552
— nonferrous metals, p. 523
— nonmetal products nec, p. 477
— paper and products, p. 230
— petroleum refineries, p. 368
— petroleum, coal products, p. 388
— plastic products nec, p. 425
— pottery, china, etc., p. 441
— printing, publishing, p. 260
— professional goods, p. 695
— rubber products, p. 405
— textiles, p. 95
— tobacco, p. 77
— transportation equipment, p. 653
— wearing apparel, p. 133
— wood products, p. 191
Sausages, p. 23
Sawmills, p. 179
Sawnwood
— broadleaved, p. 196
— coniferous, p. 196
SBW Ltd, p. 98
SCA Packaging Ltd, p. 232
Scales, pp. 559, 589, 599
Scancem International Ans, p. 481
Scandinavian Mobility Export, p. 215
Scania, p. 657
Scapa Group PLC, pp. 97, 499
Scaw Metals, p. 553
Schein Pharmaceutical Inc, p. 342
Schering Agrochemicals Ltd, p. 280
Schering Holdings Ltd, p. 280
Schering Manati Inc, p. 343
Schering-Plough Pharmaceuticals, p. 341
Schlumberger PLC, pp. 595, 698
Schnadig Corp, p. 215
Schnitzer Steel Industries Inc, p. 501
Scholastic Corp, p. 262
Scholl PLC, pp. 172, 343
Schott Glass Technologies Inc, p. 460
Schott Glaswerke/Schott Gruppe, pp. 459, 573
Schult Homes Corp, p. 194
Schultz Sav-O Stores Inc, p. 59
Schwan's Sales Enterprises Inc, p. 19
Schweitzer-Mauduit International Inc, p. 232
Schweppes, SA, p. 59
Scientific-Atlanta Inc, p. 616
Scientific Games Holdings Corp, p. 264
Scientific instruments, p. 685
SCIMED Life Systems Inc, p. 699
Scitex Corporation Ltd, pp. 574, 595

Scotsman Industries Inc, p. 574
Scott Ltd, p. 233
Scott Paper Ltd, p. 233
Scottish Heritable Trust PLC, p. 98
Scottish Tanning Industries Ltd, p. 156
Scotts Co, p. 280
Scrapers, p. 587
Screws, p. 539
Scripto Tokai Corp, p. 721
Se An Industrial Co., Ltd, p. 371
Se Won Co., Ltd, p. 172
Sea-going bulk carriers, p. 658
Sea-going vessels, p. 667
Seagram Distillers PLC, p. 58
Sealed Air Corp, p. 427
Sealright Company Inc, pp. 233, 428
Seamless Tubes Ltd, p. 501
Sebago Inc, p. 173
Seco Tools AB, p. 554
Seddon Group Ltd, p. 195
Seed oils, p. 3
Seeders, pp. 559, 577
Seil Leather Co, p. 157
Selmer Company Inc, p. 721
Semen Cibinong PT, p. 480
Semen Gresik (Persero) PT, p. 480
Semen Padang (Persero) PT, p. 479
Semen Tonasa (Persero) PT, p. 480
Semi-chemical pulp, pp. 217, 239
Semi-trailers, pp. 641, 675
Semis for tubes, p. 505
Semivitreous products, p. 431
Semo Co., Ltd, p. 720
Seneca Foods Corp, p. 20
Senegal
— basic chemicals, excl fertilizers, pp. 312, 314-318
— beverages, pp. 45, 47-48, 50, 52, 54, 56
— chemical products nec, pp. 329, 331-332, 334, 336-337, 339
— drugs and medicines, pp. 350, 352-357
— electrical machinery, pp. 603, 605-606, 608-609, 611, 613
— food products, pp. 5, 7, 9, 11-12, 14, 16
— furniture, fixtures, pp. 201, 203-204, 206, 208-209, 211
— glass and products, p. 458
— industrial chemicals, pp. 267, 269-270, 272-273, 275, 277
— industries nec, pp. 707, 709, 712, 714, 718
— iron and steel, p. 498
— machinery nec, pp. 561, 563, 565, 567-568, 570
— metal products, pp. 541, 543-544, 546, 548-549, 552
— motor vehicles, pp. 676, 678-682
— nonferrous metals, p. 523
— nonmetal products nec, pp. 467, 469-470, 472-473, 475, 477
— paper and products, pp. 219, 221-222, 224, 226-227, 230
— petroleum refineries, pp. 360, 362-366, 368
— petroleum, coal products, pp. 380, 382, 385

Senegal continued:
— plastic products nec, pp. 415, 417-418, 420-421, 423, 425
— pottery, china, etc., p. 441
— printing, publishing, pp. 249, 251, 253-254, 256, 258, 260
— professional goods, p. 695
— profile, p. 780
— rubber products, pp. 395, 397, 401
— shipbuilding, repair, pp. 668, 670-674
— spinning, weaving, etc., pp. 112, 114-118
— synthetic resins, etc., pp. 320-325
— textiles, pp. 85, 87-88, 90, 92-93, 95
— tobacco, pp. 67, 69-70, 72, 74-75, 77
— transportation equipment, pp. 643, 645-646, 648, 650-651, 653
— wearing apparel, pp. 123, 125, 128, 130-131, 133
— wood products, pp. 181, 183-184, 186, 188-189, 191
Sensormatic Electronics Corp, p. 615
Sentrachem Ltd, p. 278
Sepatu Bata PT, p. 172
Sequa Corp, pp. 553, 572, 655
Sequent Computer Systems Inc, p. 595
Serinco Djaja Marmer Industries PT, p. 480
Serm Suk Co. Ltd, p. 58
Serv-Tech Inc, p. 699
Seshin Commercial Co., Ltd, p. 501
Sewing machines, p. 619
Seyang Ind. Co., Ltd, p. 389
Shaft Sinkers Pty. Ltd, pp. 499, 554
Shape Inc, pp. 429, 597
Shaped stone, p. 465
Shasta Beverages Inc, p. 60
Shavers, p. 601
Shavers and hair clippers, p. 637
Shaw Group Inc, p. 554
Shaw Industries Ltd, p. 526
Shaw UK Holdings Ltd, p. 173
Sheet rubber, p. 393
Sheets
— electrical, p. 508
— galvanized, p. 509
— glass, p. 447
Sheets under 3 mm
— cold-rolled, uncoated, p. 508
— hot-rolled, p. 508
Sheffield Insulations Group PLC, p. 481
Sheffren's Hides & Skins Ltd, p. 157
Shelby Williams Industries Inc, p. 214
Sheldahl Inc, p. 429
Shell Canada Ltd, p. 370
Shell Espana, SA, p. 389
Shell International Petroleum Co, p. 370
Shelman Swiss-Hellenic Wood, p. 195
Shelter Components Corp, p. 480
Shepherd Neame Ltd, p. 60
Sherman International Corp, p. 481
Sherrill Furniture Co, p. 214

Alphabetical Index

Sherston Earl Vineyards, p. 444
Sherwin-Williams Diversified Brands, p. 342
Sherwood Group PLC, pp. 98, 135
Shin Kwang Co., Ltd, p. 156
Shin Won Co, p. 137
Shin Woo Co., Ltd, p. 136
Shinkong Synthetic Fibers Corp, pp. 98, 280, 428
Ships, p. 641
Shirts, pp. 121, 141
Shoe soles, p. 143
Shoes, p. 159
Shorewood Packaging Corp, pp. 232, 263
Short Brothers PLC, pp. 553, 655
Shorts, p. 121
Shree Ram Mills Ltd, p. 99
Shuford Industries Inc, pp. 99, 233
Siam Agro Industry, p. 18
Siam Cement Co., Ltd, p. 479
Siam Fibre-Cement Co., Ltd, p. 479
Siam Iron and Steel Co. Ltd, p. 481
Siam Tyre Co. Ltd, p. 407
Siebe Appliance Controls, pp. 655, 697
Siebe North Inc, p. 407
Siecor Corp, pp. 525, 616
Siemens, p. 617
Siemens Canada Ltd, pp. 595, 617, 698
Siemens Energy and Automation, pp. 443, 572, 615, 697
Sierra Leone
—basic chemicals, excl fertilizers, pp. 312, 314
—beverages, pp. 45, 47, 50, 52, 56
—chemical products nec, pp. 329, 331
—drugs and medicines, pp. 350, 352
—food products, pp. 5, 7, 11-12, 16
—footwear, pp. 161, 163
—furniture, fixtures, pp. 201, 203, 206, 208, 211
—glass and products, pp. 449, 451
—industrial chemicals, pp. 267, 269
—industries nec, pp. 707, 709, 712, 714, 718
—leather and products, pp. 145, 147
—metal products, pp. 541, 543, 546, 548, 552
—nonmetal products nec, pp. 467, 469
—paper and products, pp. 219, 221, 224, 226, 230
—petroleum refineries, pp. 360, 362
—plastic products nec, pp. 415, 417
—pottery, china, etc., pp. 433-434
—printing, publishing, pp. 249, 251, 254, 256, 260
—profile, p. 781
—pulp, paper, etc., pp. 240-241
—spinning, weaving, etc., pp. 112, 114
—textiles, pp. 85, 87, 90, 92, 95
—tobacco, pp. 67, 69, 72, 74, 77
—wearing apparel, pp. 123, 125, 128, 130, 133
—wood products, pp. 181, 183, 186, 188, 191
Sigma-Aldrich Corp, pp. 278, 341, 553
Signal P Co, p. 427
Signode Corp, p. 573

Sikorsky Aircraft Corp, p. 655
Silentnight Holdings PLC, p. 213
Silicon Valley Group Inc, p. 574
Silk, p. 83
Silk fabrics, p. 104
Silver, p. 513
Simard-Beaudry Inc, p. 389
Simon UK Ltd, p. 720
Simplex Time Recorder Co, pp. 595, 616
Simpson Industries Inc, p. 699
Simpson Manufacturing Company, p. 195
Simpson Paper Co, p. 231
Simpson Timber Co, p. 194
Simula Inc, p. 215
Singapore
—beverages, pp. 45, 47, 50, 52, 54, 56
—chemical products nec, pp. 329, 331, 334, 336-337, 339
—drugs and medicines, pp. 350, 352, 354-357
—electrical machinery, pp. 603, 605, 608-609, 611, 613
—food products, pp. 5, 7, 11-12, 14, 16
—footwear, pp. 161, 163, 165, 167-168, 170
—furniture, fixtures, pp. 201, 203, 206, 208-209, 211
—industrial chemicals, pp. 267, 269, 272-273, 275, 277
—industries nec, pp. 707, 709, 712, 714, 716, 718
—iron and steel, pp. 489-490, 493-494, 496, 498
—leather and products, pp. 145, 147, 149, 151-152, 154
—machinery nec, pp. 561, 563, 565, 567-568, 570
—metal products, pp. 541, 543, 546, 548-549, 552
—motor vehicles, pp. 676, 678, 680-682
—nonferrous metals, pp. 515-516, 518, 520-521, 523
—nonmetal products nec, pp. 467, 469, 472, 474-475, 477
—office, computing machinery, pp. 590-594
—paper and products, pp. 219, 221, 224, 226-227, 230
—plastic products nec, pp. 415, 417, 420-421, 423, 425
—printing, publishing, pp. 249, 251, 255-256, 258, 260
—professional goods, pp. 687-688, 691-693, 695
—profile, p. 781
—radio, television, etc., pp. 628-629, 631-634
—rubber products, pp. 395, 397, 399, 401-402, 405
—shipbuilding, repair, pp. 668, 670-674
—spinning, weaving, etc., pp. 112, 114, 116-118
—synthetic resins, etc., pp. 320-321, 323-325
—textiles, pp. 85, 87, 90, 92-93, 95
—tobacco, pp. 67, 69, 72, 74-75, 77
—transportation equipment, pp. 643, 645, 648, 650-651, 65
—wearing apparel, pp. 123, 125, 128, 130-131, 133
—wood products, pp. 181, 183, 186, 188-189, 191
Singer Do Brasil Industria e, p. 616
Singer Furniture Co, p. 214
SKC Ltd, p. 617
Skeena Cellulose Inc, p. 195
SKF USA Inc, p. 573
Skins, p. 143
—undressed, p. 25
Skirts, pp. 121, 141
Skogaholms Brod, AB, p. 20

Skyline Corp, pp. 193, 554, 657
Slacks, p. 121
Sleaford Trading Co. Ltd, p. 371
Slide fasteners, p. 724
Slippers, p. 159
SLM International Inc, pp. 137, 720
Slovakfarma AS, pp. 342, 459
Slovakia
—basic chemicals, excl fertilizers, pp. 312, 314-316
—beverages, pp. 45, 47, 50, 52, 56
—chemical products nec, pp. 329, 331, 334, 336, 339
—drugs and medicines, pp. 350, 352, 354-355
—electrical machinery, pp. 603, 605, 608-609, 613
—food products, pp. 5, 7, 11-12, 16
—footwear, pp. 161, 163, 165, 167, 170
—furniture, fixtures, pp. 201, 203, 206, 211
—glass and products, pp. 449, 451, 453, 455, 458
—industrial chemicals, pp. 267, 269, 272-273, 277
—industries nec, pp. 707, 709, 712, 714, 718
—iron and steel, pp. 489-490, 493-494, 498
—leather and products, pp. 145, 147, 149, 151, 154
—machinery nec, pp. 561, 563, 565, 567, 570
—metal products, pp. 541, 543, 546, 548, 552
—motor vehicles, pp. 676, 678, 680-681
—nonferrous metals, pp. 515-516, 518, 520, 523
—nonmetal products nec, pp. 467, 469, 472, 474, 477
—office, computing machinery, pp. 590-593
—paper and products, pp. 219, 221, 224, 226, 230
—petroleum refineries, pp. 361-362, 364-365, 368
—petroleum, coal products, pp. 380, 388
—plastic products nec, pp. 415, 417, 420-421, 425
—pottery, china, etc., pp. 433-434, 437-438, 441
—printing, publishing, pp. 249, 251, 255-256, 260
—professional goods, pp. 687-688, 691-692, 695
—profile, p. 782
—pulp, paper, etc., pp. 240, 242-244
—radio, television, etc., pp. 628-629, 631-632
—rubber products, pp. 395, 397, 399, 401, 405
—shipbuilding, repair, p. 668
—spinning, weaving, etc., pp. 112, 114, 116-117
—synthetic resins, etc., pp. 320-321, 323-324
—textiles, pp. 85, 87, 90, 92, 95
—tobacco, p. 67
—transportation equipment, pp. 643, 645, 648, 650, 653
—wearing apparel, pp. 123, 125, 128, 130, 133
—wood products, pp. 181, 183, 186, 188, 191
Slovenia
—beverages, pp. 45, 50, 52, 54, 56
—chemical products nec, pp. 329, 334, 336-337, 339
—electrical machinery, pp. 603, 605, 608-609, 611, 613
—food products, pp. 5, 11, 13-14, 17
—footwear, pp. 161, 165, 167-168, 171
—furniture, fixtures, pp. 201, 206, 208-209, 212
—glass and products, pp. 449, 458
—industrial chemicals, pp. 267, 272-273, 275, 277
—industries nec, pp. 707, 712, 714, 716, 718

Slovenia continued:
—iron and steel, pp. 489, 493-494, 496, 498
—leather and products, pp. 145, 149, 151-152, 154
—machinery nec, pp. 561, 565, 567, 570
—metal products, pp. 541, 546, 548-549, 552
—nonferrous metals, pp. 515, 518, 520-521, 523
—nonmetal products nec, pp. 467, 475, 477
—paper and products, pp. 219, 224, 226-227, 230
—petroleum refineries, pp. 361, 368
—petroleum, coal products, pp. 380, 388
—plastic products nec, pp. 415, 420-421, 423, 425
—pottery, china, etc., pp. 433, 439, 441
—printing, publishing, pp. 249, 255-256, 258, 260
—professional goods, pp. 687, 691-692, 696
—profile, p. 782
—rubber products, pp. 395, 399, 401-402, 405
—textiles, pp. 85, 90, 92, 95
—tobacco, pp. 67, 72, 74-75, 77
—transportation equipment, pp. 643, 645, 648, 650-651, 653
—wearing apparel, pp. 123, 128, 130, 133
—wood products, pp. 181, 183, 186, 188-189, 191
Smead Manufacturing Co, p. 232
SMED International Inc, p. 214
Smith & Nephew Ltd, p. 343
Smith and Nephew Richards Inc, p. 698
Smith Corona Corp, p. 596
Smith International Inc, p. 573
Smoking tobacco, p. 65
Smurfit Carton de Colombia SA, p. 232
Smurfit Espana, SA, p. 233
Snuff, p. 65
Soap, pp. 327, 345
Sociedad Anonima Damm, p. 60
Sociedad Anonima el Aguila, p. 59
Sociedad Espanola de Carburos, pp. 343, 526
Societat Catalana de Petrolis, SA, p. 389
Societe Anonyme des Brasseries du, p. 58
Societe Congolaise des Brasseries, p. 60
Societe de Limonaderies et, p. 59
Societe Generale des Industries, p. 97
Societe Industrielle des Tabacs du, p. 79
Societe Nationale de Boissons, p. 59
Societe Nationale Pour l'Industrie, p. 343
Societe Tunisienne de Siderurgie, pp. 500, 555
Societe Zaire Shell, p. 370
Socks, pp. 83, 107, 111
Soda ash, p. 297
Soda Sanayii AS, pp. 280, 480
Soda wood pulp, pp. 217, 239
Sodaso, p. 278
Sodium silicates, p. 298
Sodium sulfates, p. 298
Soft drinks, pp. 43, 63
Sola International Inc, p. 697
Solar Turbines Inc, p. 573
Solo Cup Co, p. 232

Somalia
— beverages, p. 56
— chemical products nec, p. 339
— electrical machinery, p. 613
— food products, p. 17
— footwear, p. 171
— furniture, fixtures, p. 212
— glass and products, p. 458
— industrial chemicals, p. 277
— industries nec, p. 718
— iron and steel, p. 498
— leather and products, p. 154
— machinery nec, p. 570
— metal products, p. 552
— nonferrous metals, p. 523
— nonmetal products nec, p. 477
— paper and products, p. 230
— petroleum refineries, p. 368
— petroleum, coal products, p. 388
— plastic products nec, p. 425
— printing, publishing, p. 260
— professional goods, p. 696
— rubber products, p. 405
— textiles, p. 95
— tobacco, p. 77
— transportation equipment, p. 653
— wearing apparel, p. 133
Sound recorders, p. 636
Sound reproducers, p. 636
South Africa
— basic chemicals, excl fertilizers, pp. 312, 315
— beverages, pp. 45, 47, 50, 52, 56
— chemical products nec, pp. 329, 331, 334, 336, 340
— drugs and medicines, pp. 351, 354
— electrical machinery, pp. 603, 605, 608-609, 613
— food products, pp. 5, 7, 11, 13, 17
— footwear, pp. 161, 163, 165, 167, 171
— furniture, fixtures, pp. 201, 203, 206, 208, 212
— glass and products, pp. 449, 451, 453, 455, 458
— industrial chemicals, pp. 267, 269, 272-273, 277
— industries nec, pp. 707, 709, 712, 714, 718
— iron and steel, pp. 489-490, 493-494, 498
— leather and products, pp. 145, 147, 149, 151, 154
— machinery nec, pp. 561, 563, 565, 567, 570
— metal products, pp. 541, 543, 546, 548, 552
— motor vehicles, pp. 676, 680
— nonferrous metals, pp. 515-516, 518, 520, 523
— nonmetal products nec, pp. 467, 469, 472, 474, 477
— office, computing machinery, p. 590
— paper and products, pp. 219, 221, 224, 226, 230
— petroleum refineries, pp. 361-362, 364-365, 368
— petroleum, coal products, pp. 380, 382, 384-385, 388
— plastic products nec, pp. 415, 417, 420-421, 425
— pottery, china, etc., pp. 433-434, 437-438, 441
— printing, publishing, pp. 249, 251, 255-256, 260
— professional goods, pp. 687-688, 691-692, 696

South Africa continued:
— profile, p. 783
— pulp, paper, etc., pp. 240, 243
— radio, television, etc., p. 628
— rubber products, pp. 395, 397, 399, 401, 405
— shipbuilding, repair, p. 668
— spinning, weaving, etc., pp. 112, 116
— synthetic resins, etc., pp. 320, 323
— textiles, pp. 85, 87, 90, 92, 95
— tobacco, pp. 67, 69, 72, 74, 77
— transportation equipment, pp. 643, 645, 648, 650, 653
— wearing apparel, pp. 123, 125, 128, 130, 133
— wood products, pp. 181, 183, 186, 188, 191
South African Breweries, p. 58
Southdown Inc, p. 479
Southeast Atlantic Corp, p. 60
Southeast Publishing Ventures Inc, p. 262
Southern Energy Homes Inc, p. 194
Southern Newspapers PLC, p. 263
Southern Peru Copper Corp, p. 525
Southwestern Petroleum Corp, p. 390
Southwestern Portland Cement Co, p. 481
Spain
— basic chemicals, excl fertilizers, pp. 312, 314-318
— beverages, pp. 45, 47, 50, 52, 54, 56
— chemical products nec, pp. 329, 331, 334, 336-337, 340
— drugs and medicines, pp. 351-352, 354-357
— electrical machinery, pp. 603, 605, 608-609, 611, 613
— food products, pp. 5, 7, 11, 13-14, 17
— footwear, pp. 161, 163, 165, 167-168, 171
— furniture, fixtures, pp. 201, 203, 206, 208-209, 212
— glass and products, pp. 449, 451, 453, 455-456, 458
— industrial chemicals, pp. 267, 269, 272-273, 275, 277
— industries nec, pp. 707, 709, 712, 714, 716, 718
— iron and steel, pp. 489-490, 493-494, 496, 498
— leather and products, pp. 145, 147, 149, 151-152, 154
— machinery nec, pp. 561, 563, 565, 567-568, 570
— metal products, pp. 541, 543, 546, 548-549, 552
— motor vehicles, pp. 676, 678, 680-683
— nonferrous metals, pp. 515-516, 518, 520-521, 523
— nonmetal products nec, pp. 467, 469, 472, 474-475, 47
— office, computing machinery, pp. 590-594
— paper and products, pp. 219, 221, 224, 226-227, 230
— petroleum refineries, pp. 361-362, 364-366, 368
— petroleum, coal products, pp. 381-382, 384-386, 388
— plastic products nec, pp. 415, 417, 420-421, 423, 425
— pottery, china, etc., pp. 433-434, 437-439, 441
— printing, publishing, pp. 249, 251, 255-256, 258, 260
— professional goods, pp. 687-688, 691-693, 696
— profile, p. 783
— pulp, paper, etc., pp. 240, 242-246
— radio, television, etc., pp. 628-629, 631-634
— rubber products, pp. 395, 397, 399, 401-402, 405
— shipbuilding, repair, pp. 668, 670-674
— spinning, weaving, etc., pp. 112, 114, 116-118
— synthetic resins, etc., pp. 320-321, 323-325

Spain continued:
—textiles, pp. 85, 87, 90, 92-93, 95
—tobacco, pp. 67, 69, 72, 74-75, 77
—transportation equipment, pp. 643, 645, 648, 650-651, 653
—wearing apparel, pp. 123, 125, 128, 130-131, 133
—wood products, pp. 181, 183, 186, 188-189, 192
Spalding and Evenflo Companies, pp. 213, 427, 719
Spalding Sports Worldwide, p. 720
Spanjaard Ltd, p. 390
Spartech Corp, p. 428
Speciality Packaging (UK) PLC, pp. 428, 526
SpecTran Corp, p. 460
Spencers Inc, pp. 136, 233
Spiegeleisen and ferro-manganese, p. 502
Spinning, p. 111
Spinning machines, p. 585
Spirax-Sarco Engineering PLC, p. 698
Spoldzielnia Rzemieslnicza, p. 721
Spolem PSS, p. 59
Sport Maska Inc, pp. 720-721
Sports shirts, pp. 83, 111
Sports shoes, p. 159
Springs Window Fashions Div, pp. 194, 213
Spruce Falls Inc, p. 195
SPS Technologies Inc, p. 553
SPX Corp, pp. 193, 553, 572, 655
SR Kent UK Ltd, p. 135
SRF Ltd, pp. 172, 720
Sri Lanka
—basic chemicals, excl fertilizers, pp. 312, 314-315, 317-318
—beverages, pp. 45, 47, 49-50, 52, 54, 56
—chemical products nec, pp. 329, 331-332, 334, 336-337, 340
—drugs and medicines, pp. 351-357
—electrical machinery, pp. 603, 605-606, 608-609, 611, 613
—food products, pp. 5, 7, 9, 11, 13-14, 17
—footwear, pp. 161, 163-165, 167-168, 171
—furniture, fixtures, pp. 201, 203-204, 206, 208-209, 212
—glass and products, pp. 449, 451-453, 455-456, 458
—industrial chemicals, pp. 267, 269-270, 272-273, 275, 277
—industries nec, pp. 707, 709, 711-712, 714, 716, 718
—iron and steel, pp. 489-491, 493-494, 496, 498
—leather and products, pp. 145, 147-149, 151-152, 154
—machinery nec, pp. 561, 563-565, 567-568, 570
—metal products, pp. 541, 543-544, 546, 548-549, 552
—motor vehicles, pp. 676, 678-683
—nonferrous metals, pp. 515-518, 520-521, 523
—nonmetal products nec, pp. 467, 469-470, 472, 474-475, 477
—office, computing machinery, pp. 590-594
—paper and products, pp. 219, 221-222, 224, 226-227, 230
—plastic products nec, pp. 415, 417-418, 420-421, 423, 425
—pottery, china, etc., pp. 433-435, 437-439, 441
—printing, publishing, pp. 249, 251, 253, 255-256, 258, 260
—professional goods, pp. 687-689, 691-693, 696

Sri Lanka continued:
—profile, p. 784
—pulp, paper, etc., pp. 240, 242-246
—radio, television, etc., pp. 628-634
—rubber products, pp. 395, 397-399, 401-402, 405
—shipbuilding, repair, pp. 668, 670-674
—spinning, weaving, etc., pp. 113-118
—synthetic resins, etc., pp. 320-325
—textiles, pp. 85, 87-88, 90, 92-93, 95
—tobacco, pp. 67, 69-70, 72, 74-75, 77
—transportation equipment, pp. 643, 645-646, 648, 650-651, 653
—wearing apparel, pp. 123, 125-126, 128, 130-131, 133
—wood products, pp. 181, 183-184, 186, 188-189, 192
Staffordshire Tableware, Ltd, pp. 443, 459
Stahl Specialty Co, p. 527
Stanadyne Automotive Corp, p. 657
Standard Commercial Corp, p. 99
Standard Commercial Tobacco Co, pp. 79-80
Standard Industries Ltd, pp. 97, 278
Standard Motor Products Inc, p. 656
Standard Products (Canada) Ltd, p. 407
Standard Register Co, pp. 231, 262, 595
Standard Textile Company Inc, p. 137
Standex International Corp, pp. 262, 554, 573
Stanford Telecommunications Inc, p. 597
Stanley-Bostitch Inc, pp. 555, 719
Stanley Furniture, p. 213
Stanley Furniture Company Inc, p. 213
Stant Corp, pp. 572, 655
Stanton y Compania SA, p. 407
Starcan Corp, p. 408
Starkey Labs Inc, p. 698
Starsax Ltda., Calcados, p. 173
State Meat and Fish Institution, p. 18
Statyba Stock Co, p. 480
Steam turbines, p. 575
Stearns Manufacturing Co, p. 408
Steel, p. 487
Steel castings in the rough state, p. 511
Steel forgings, p. 512
Steelcase Canada Ltd, p. 215
Steeledale Reinforcing &, pp. 193, 479, 500, 555, 657
Steinway Musical Instruments Inc, p. 720
Steklarna Boris Kidric d.d, p. 720
Steklarna Hrastnik d.d, p. 721
Stepan Co, p. 343
STERIS Corp, p. 698
Sterling Chemicals Inc, p. 280
Stern Metals Inc, p. 527
Steven Madden Ltd, p. 174
Stewart and Stevenson Services Inc, pp. 573, 656
Stiefel Laboratories Inc, p. 343
Stimson Lumber Co, p. 194
Stirling Group PLC, p. 136
Stockings, pp. 83, 111

Stoddard Sekers International PLC, p. 97
Stokely USA Inc, p. 20
Stol, p. 721
Stone, cut and shaped, p. 465
Stone-Consolidated Corp, pp. 193, 231
Stoneware, p. 431
Stora Papyrus AB, p. 231
Storage Technology Corp, p. 595
StorMedia Inc, p. 597
Stoves, ranges, cookers, p. 618
Strathmore Paper Co, p. 721
StreamLogic Corp, p. 596
Stride Rite Corp, p. 172
Stroehmann Bakeries L.C, p. 20
Stroh Brewery Co, p. 58
Structures, p. 556
Stryker Corp, p. 697
Stuart Entertainment Inc, p. 720
Sturm, Ruger and Company Inc, p. 500
Stylo PLC, pp. 172, 719
Styrene, p. 281
Sucocitrico Cutrale SA, p. 19
Sudan
—beverages, p. 56
—chemical products nec, p. 340
—electrical machinery, p. 613
—food products, p. 17
—footwear, p. 171
—glass and products, p. 458
—industrial chemicals, p. 277
—industries nec, p. 718
—iron and steel, p. 498
—leather and products, p. 154
—machinery nec, p. 570
—metal products, p. 552
—nonferrous metals, p. 523
—nonmetal products nec, p. 477
—paper and products, p. 230
—petroleum refineries, p. 369
—petroleum, coal products, p. 388
—plastic products nec, p. 425
—pottery, china, etc., p. 441
—printing, publishing, p. 260
—professional goods, p. 696
—rubber products, p. 405
—textiles, p. 95
—tobacco, p. 77
—transportation equipment, p. 653
—wearing apparel, p. 133
Sugar, p. 3
Sugar confectionery, p. 40
Suh Il Ceramics Co., Ltd, p. 461
Suits, p. 121
—men's and boys', p. 139
—women's and girls', p. 141
Sukwang Corp, p. 136

Sulfur, p. 281
Sulfuric acid, p. 293
Sulphate, pp. 217, 239
Sulphite pulp, pp. 217, 239
Sulzer Inc, pp. 573, 697
Sun Media Corp, p. 263
Sunbeam Corp, pp. 572, 615
Sunbeam-Oster Household Products, p. 616
Sundstrand Aerospace, p. 656
Sundstrand Corp, pp. 655, 697
Sung San Corp, p. 157
Sunkyong Industries Ltd, p. 97
Sunnen Products Co, p. 481
Sunrise Medical Inc, pp. 213, 698
Suntory International Corp, p. 58
Suparma PT, p. 233
Super Sagless Inc, p. 215
Superior Graphite Co, p. 389
Superior Industries International Inc, p. 656
Superior Surgical Manufacturing, p. 136
Superior Telecom Inc, p. 526
Superior Tobacco Co. NV, p. 80
Superphosphates, p. 301
Supertiendas y Droguerias Olimpica, p. 135
Suprachem, p. 371
Supreme Slipper Manufacturing, p. 174
Suriname
—beverages, pp. 45, 47, 50, 56
—chemical products nec, pp. 329, 331, 334, 340
—electrical machinery, pp. 603, 613
—food products, pp. 5, 7, 11, 17
—footwear, pp. 161, 163, 166, 171
—furniture, fixtures, pp. 201, 203, 206, 212
—industries nec, pp. 707, 709, 712, 718
—leather and products, p. 145
—machinery nec, p. 570
—metal products, p. 552
—nonferrous metals, pp. 515, 523
—paper and products, pp. 219, 221, 224, 230
—plastic products nec, pp. 415, 417, 420, 425
—printing, publishing, pp. 249, 251, 255, 260
—professional goods, pp. 687-688, 691, 696
—profile, p. 784
—rubber products, pp. 395, 397, 399, 405
—tobacco, pp. 67, 69, 72, 77
—transportation equipment, pp. 643, 645, 648, 653
—wearing apparel, pp. 123, 125, 128, 133
—wood products, pp. 181, 183, 186, 192
Suzano, p. 279
Swan Mills Ltd, p. 99
Swank Inc, pp. 173, 343, 720
Swaziland
—basic chemicals, excl fertilizers, pp. 312, 314-315, 317
—beverages, pp. 45, 47, 50, 52, 54, 56
—food products, pp. 5, 8, 11, 13-14, 17
—footwear, pp. 161, 163, 166-168, 171

Swaziland continued:
—furniture, fixtures, pp. 201, 203, 206, 208-209
—glass and products, pp. 449, 451, 453, 455-456
—industrial chemicals, pp. 267, 269, 272-273, 275
—industries nec, pp. 707, 709, 712, 714, 716, 718
—iron and steel, p. 498
—leather and products, p. 145
—machinery nec, p. 568
—metal products, pp. 541, 543, 546, 548-549, 552
—motor vehicles, pp. 678, 680-682
—nonferrous metals, p. 523
—nonmetal products nec, pp. 467, 469, 472, 474-475
—paper and products, pp. 219, 221, 224, 226-227, 230
—plastic products nec, pp. 415, 417, 420-421, 423
—pottery, china, etc., p. 441
—printing, publishing, pp. 249, 251, 255-256, 258, 260
—professional goods, p. 696
—profile, p. 785
—pulp, paper, etc., pp. 240, 242-245
—rubber products, p. 405
—spinning, weaving, etc., pp. 113-114, 116-117
—tobacco, p. 77
—transportation equipment, p. 653
—wearing apparel, pp. 123, 125, 128, 130-131
Sweaters, pp. 83, 111
Swedal International Inc, p. 461
Sweden
—basic chemicals, excl fertilizers, pp. 312, 314, 316-317
—beverages, pp. 45, 47, 49-50, 52, 56
—chemical products nec, pp. 329, 331-332, 334, 336, 340
—drugs and medicines, pp. 351-355, 357
—electrical machinery, pp. 603, 605-606, 608, 610, 613
—food products, pp. 5, 8-9, 11, 13, 17
—footwear, pp. 161, 163-164, 166-167, 171
—furniture, fixtures, pp. 201, 203-204, 206, 208, 212
—glass and products, pp. 449, 451-453, 455, 458
—industrial chemicals, pp. 267, 269-270, 272-273, 277
—industries nec, pp. 707, 709, 711-712, 714, 718
—iron and steel, pp. 489-491, 493-494, 498
—leather and products, pp. 145, 147-149, 151, 154
—machinery nec, pp. 561, 563-565, 567, 570
—metal products, pp. 541, 543-544, 546, 548, 552
—motor vehicles, pp. 676, 678-681, 683
—nonferrous metals, pp. 515-518, 520, 523
—nonmetal products nec, pp. 467, 469-470, 472, 474, 477
—office, computing machinery, pp. 590-594
—paper and products, pp. 219, 221-222, 224, 226, 230
—petroleum refineries, pp. 361-365, 369
—petroleum, coal products, pp. 381-385, 388
—plastic products nec, pp. 415, 417-418, 420, 422, 425
—pottery, china, etc., pp. 433-435, 437-438, 441
—printing, publishing, pp. 249, 251, 253, 255-256, 260
—professional goods, pp. 687-689, 691-692, 696
—profile, p. 785
—pulp, paper, etc., pp. 240, 242-244, 246
—radio, television, etc., pp. 628, 630-632, 634

Sweden continued:
—rubber products, pp. 395, 397-398, 400-401, 405
—shipbuilding, repair, pp. 668, 670-672, 674
—spinning, weaving, etc., pp. 113-118
—synthetic resins, etc., pp. 320-325
—textiles, pp. 85, 87-88, 90, 92, 95
—tobacco, pp. 67, 69-70, 72, 74, 77
—transportation equipment, pp. 643, 645-646, 648, 650, 653
—wearing apparel, pp. 123, 125-126, 128, 130, 133
—wood products, pp. 181, 183-184, 186, 188, 192
Sweet Seventeen Shoes CV, p. 174
Swewi Svendborg AS, p. 156
Swift Armour SA Industria e, p. 18
Swire Pacific Holdings Inc, p. 59
Swisher International Group Inc, p. 79
Switches, p. 638
Switzerland
—beverages, p. 56
—chemical products nec, p. 340
—electrical machinery, pp. 605, 608, 610, 613
—food products, p. 17
—footwear, p. 171
—furniture, fixtures, p. 212
—glass and products, p. 458
—industrial chemicals, p. 277
—industries nec, p. 718
—iron and steel, p. 498
—leather and products, p. 154
—machinery nec, p. 570
—metal products, p. 552
—nonferrous metals, p. 523
—nonmetal products nec, p. 477
—paper and products, pp. 221, 224, 226, 230
—petroleum refineries, p. 369
—petroleum, coal products, p. 388
—plastic products nec, p. 425
—pottery, china, etc., p. 441
—printing, publishing, pp. 252, 255-256, 260
—professional goods, pp. 688, 691-692, 696
—profile, p. 786
—rubber products, p. 405
—textiles, pp. 87, 90, 92, 96
—tobacco, p. 77
—transportation equipment, p. 653
—wearing apparel, p. 133
—wood products, p. 192
SWP Group PLC, p. 461
Sybron International Corp, p. 697
Symbol Technologies Inc, p. 595
Synthetic fiber yarn, p. 83
Synthetic Industries Inc, p. 98
Synthetic resins, pp. 265, 319
SyQuest Technology Inc, p. 597
Syracuse China Co, p. 443
Syratech Corp, p. 215

Syrian Arab Republic
— beverages, p. 56
— chemical products nec, p. 340
— electrical machinery, p. 613
— food products, p. 17
— footwear, p. 171
— furniture, fixtures, p. 212
— glass and products, p. 458
— industrial chemicals, p. 277
— industries nec, pp. 707, 712, 714, 718
— iron and steel, p. 498
— leather and products, p. 154
— machinery nec, p. 570
— metal products, p. 552
— nonferrous metals, p. 523
— nonmetal products nec, p. 477
— paper and products, p. 230
— petroleum refineries, p. 369
— petroleum, coal products, p. 388
— plastic products nec, p. 425
— pottery, china, etc., p. 441
— printing, publishing, p. 260
— professional goods, p. 696
— profile, p. 786
— rubber products, p. 405
— textiles, p. 96
— tobacco, p. 78
— transportation equipment, p. 653
— wearing apparel, p. 133
— wood products, p. 192
Syringes, p. 447
Systems Bio-Industries Inc, p. 60
SZ ZJ Acroni d.o.o, p. 720
Szopienice, p. 525
T and N Industries Inc, pp. 553, 655
Tabacalera Andina SA, p. 79
Tabacalera Centroamericana SA, p. 79
Tabacalera Costarricense SA, p. 79
Tabacalera Nacional SA, p. 79
Tabacalera, SA, p. 79
Tableware, p. 431
Tadiron Ltd, p. 615
TAE HWA Indonesia PT, p. 172
Taejon Leather Industrial Co., Ltd, p. 156
Tah Hsin Industrial Corp, pp. 99, 136, 214, 428, 596, 720
Taihan Electric Wire Co., Ltd, p. 525
Taihan Textile Co., Ltd, p. 99
Taiwan Fairmost International Corp, p. 444
Taiwan Glass Industry Corp, p. 459
Taiwan Semiconductor, p. 526
Taiyang Trading Co, p. 174
Takata Inc, pp. 135, 719
Takovo, p. 19
Talley Industries Inc, p. 699
Talum d.o.o, p. 721
TAM Gospodarska Vozila d.o.o, p. 720

Tambang Timah PT, pp. 479, 525
Tambour Paint and Chemicals Ltd, p. 343
Tambrands Inc, p. 232
Tan Chong Motor Holdings BHD, p. 656
Tandem Computers Inc, p. 595
Tandy Brands Accessories Inc, p. 173
Tandy Wire and Cable, p. 527
Tankers, p. 667
— launched, p. 658
Tanks, p. 539
Tanneries Modernes de la Manouba, p. 156
Tanning
— leather, p. 143
Tanzania Saruji Corp, p. 479
Tapei Hisz, p. 215
Tar, p. 379
Tarconord AS, p. 371
Taris Pamuk Tarim Satis Koop, pp. 99, 342
Tars, p. 392
Tartaric acid, p. 291
Taulell, SA, p. 481
Taunton Cider PLC, p. 60
Taunton Cider Trading Ltd, p. 60
Taurus Hungarian Rubber Works, p. 406
Tayun Products Ind. Taiwan Inc, p. 156
TDK U.S.A. Corp, p. 616
Tech-Sym Corp, p. 699
Tecnol Medical Products Inc, p. 699
Tecnoquimicas SA, p. 459
Teepak Inc, p. 427
Teijin Indonesia Fiber Corp. PT, p. 99
Tejicondor SA Tejidos El Condor, p. 342
Teka Industrial, SA, p. 443
Teka Tecelagem Kuehnrich SA, p. 97
Tektronix Inc, pp. 595, 697
Teleflex Inc, pp. 615, 697
Telefongyar, pp. 595, 617
Telegraph PLC, p. 264
Telephones, pp. 601, 627, 635
Television receivers, p. 635
Television sets, p. 627
Tellabs Inc, p. 616
Telvision sets, p. 601
Telxon Corp, p. 596
Tembec Inc, p. 231
Tempa Glass Industries Ltd, p. 461
Tempo Beer Industries Ltd, p. 59
Temtex Industries Inc, p. 481
Tender Tootsies Group Ltd, p. 173
Teneria el Progreso SA, p. 157
Teneria Moderna Franco Espanola, pp. 156, 174
Teneria Pirro Antonia Gomez, Ltda, p. 157
Teneria Primenca, SA, p. 156
Teradyne Inc, pp. 616, 697
Terex Corp, pp. 573, 655
Terra Footwear Ltd, p. 174

Terra Industries Inc, p. 279

Tesla, Stropkov AS, pp. 595, 698

Tesoro Alaska Petroleum Co, p. 371

Tesoro Petroleum Corp, p. 370

Teva Pharmeceutical Industries Ltd, p. 341

Texas Boot Co, p. 173

Texas Industries Inc, pp. 479, 499

Texas Olefins Co, p. 371

Texas Petrochemicals Corp, p. 371

Texas Refinery Corp, p. 389

Textile Manufacturing Co. Djaya PT, p. 97

Textiles, p. 83

Thai Bridgestone Co. Ltd, p. 407

Thai Durable Textile Co. Ltd, p. 98

Thai Garment Export Co. Ltd, p. 135

Thai Glass Industries Ltd, p. 459

Thai Kawasumi Co., Ltd, p. 699

Thai Oil Refinery, p. 370

Thai Pineapple Co., Ltd, p. 20

Thai Pure Drinks Ltd, p. 58

Thai Wacoal Co. Ltd, p. 136

Thaibinh Export Leather Products, p. 174

Thailand
— basic chemicals, excl fertilizers, pp. 313-314, 316-318
— beverages, pp. 45, 47, 49-50, 52, 54, 56
— chemical products nec, pp. 329, 331-332, 334, 336-337, 340
— drugs and medicines, pp. 351-357
— electrical machinery, pp. 603, 605-606, 608, 610-611, 613
— food products, pp. 5, 8-9, 11, 13-14, 17
— footwear, pp. 161, 163-164, 166-168, 171
— furniture, fixtures, pp. 201, 203-204, 206, 208-209, 212
— glass and products, pp. 449, 451-453, 455-456, 458
— industrial chemicals, pp. 267, 269-270, 272-273, 275, 277
— industries nec, pp. 707, 709, 711-712, 714, 716, 718
— iron and steel, pp. 489-491, 493-494, 496, 498
— leather and products, pp. 145, 147-149, 151-152, 154
— machinery nec, pp. 561, 563-565, 567-568, 570
— metal products, pp. 541, 543-544, 546, 548-549, 552
— motor vehicles, pp. 677-683
— nonferrous metals, pp. 515-518, 520-521, 523
— nonmetal products nec, pp. 467, 469-470, 472, 474-475, 477
— office, computing machinery, pp. 590-594
— paper and products, pp. 219, 221-222, 224, 226-227, 230
— petroleum refineries, pp. 361-366, 369
— petroleum, coal products, pp. 381-386, 388
— plastic products nec, pp. 415, 417-418, 420, 422-423, 425
— pottery, china, etc., pp. 433-435, 437-439, 442
— printing, publishing, pp. 249, 252-253, 255, 257-258, 261
— professional goods, pp. 687-689, 691-693, 696
— profile, p. 787
— pulp, paper, etc., pp. 240, 242-246
— radio, television, etc., pp. 628, 630-634
— rubber products, pp. 395, 397-398, 400-402, 405
— shipbuilding, repair, pp. 668, 670-674

Thailand continued:
— spinning, weaving, etc., pp. 113-118
— synthetic resins, etc., pp. 320-325
— textiles, pp. 85, 87-88, 90, 92-93, 96
— tobacco, pp. 67, 69-70, 72, 74-75, 78
— transportation equipment, pp. 643, 645-646, 648, 650-651, 653
— wearing apparel, pp. 123, 125-126, 128, 130-131, 133
— wood products, pp. 181, 183-184, 186, 188-189, 192

Thanomwongse Service Co. Ltd, pp. 371, 389

Therma-Tru Corp, p. 195

Thermadyne Holdings Corp, p. 574

Thermo Instrument Systems Inc, p. 697

Thermo King Corp, p. 573

Thermo Optek Corp, p. 699

Thermostats, p. 700

Thiokol Corp, pp. 573, 656

Thomas slag, p. 502

Thomas Industries Inc, pp. 574, 617

Thomas Legget & Sons Ltd, p. 157

Thomas Nelson Inc, p. 264

Thomas Roberts Westminster Ltd, p. 481

Thomaston Mills Inc, p. 98

Thomasville Furniture Industries Inc, p. 213

Thomson Corp., The, p. 263

Thomson Crown Wood Products Inc, p. 215

Thonburi Automobile Assembly, p. 656

Thor Industries Inc, p. 657

Thorn Apple Valley Inc, p. 19

Thorntons PLC, p. 20

Threshing machines, p. 579

Thyssen Maschinenbau GmbH Werk, p. 572

Tiles, p. 465
— floor and wall, p. 482
— roofing, made of clay, p. 482

Tillotson Corp, pp. 406, 698

Tillotson Healthcare Corp, p. 407

Timber Products Co, p. 195

Timber Products Co. Medford, p. 195

Timberland Co, pp. 135, 172

TimberWest Forest Ltd, p. 195

Time/System International AS, p. 156

Times Fiber Communications Inc, p. 527

Tin, pp. 513, 536-537

Tinplate, p. 509

Tires for agricultural and other off-the-road vehicles, p. 409

Tires for bicycles and motorcycles, p. 410

Tires for road motor vehicles
— excluding bicycles and motorcycles, p. 410

Titan Cement Co. SA, p. 480

Titan Corp., p. 597

Titan Kamnik d.d, p. 721

Titan PO Factory of Metal Products, p. 501

Titan SA, p. 156

Titanium Metals Corp, p. 526

Titanium oxides, p. 295
Tiz, Zemun, p. 136
TJ International Inc, p. 193
Tobacco, p. 65
— manufactured, pp. 65, 82
— prepared leaf, p. 81
Tobacco Processors Inc, p. 79
Tobacco stemming and redrying, p. 65
Tobacna Tovarna d.o.o, p. 79
Toco Holdings Ltd, pp. 343, 596
Todd & Duncan Ltd, p. 99
Togo
— footwear, p. 171
— furniture, fixtures, p. 212
— industries nec, p. 718
— leather and products, p. 154
— paper and products, p. 230
— printing, publishing, p. 261
— textiles, p. 96
— wearing apparel, p. 133
— wood products, p. 192
Toilet goods, p. 327
Tokheim Corp, pp. 595, 698
Tolko Industries Ltd, p. 195
Tolmex, SA de CV, p. 479
Tolo d.d, p. 174
Toluene, p. 284
Tom James Co, p. 137
Tommy Hilfiger Corp, p. 137
Tong Kook Spinning Co., Ltd, p. 98
Tong Yang Cement Corp, p. 480
Tong Yang Confectionery Corp, p. 19
Tong Yang Moolsan Co., Ltd, p. 501
Tong Yang Nylon Co., Ltd, p. 97
Tong Yang Polyester Co., Ltd, p. 99
Tonga
— beverages, pp. 47, 50, 56
— chemical products nec, pp. 329, 331, 334, 340
— electrical machinery, pp. 605, 608, 613
— food products, pp. 8, 11, 17
— footwear, pp. 163, 166, 171
— furniture, fixtures, pp. 203, 206, 212
— industrial chemicals, p. 277
— industries nec, pp. 707, 709, 712, 718
— iron and steel, p. 498
— leather and products, pp. 147, 149, 154
— machinery nec, p. 570
— metal products, pp. 541, 543, 546, 552
— nonferrous metals, p. 523
— paper and products, pp. 221, 224, 230
— petroleum refineries, p. 369
— petroleum, coal products, p. 388
— plastic products nec, p. 425
— printing, publishing, pp. 249, 252, 255, 261
— professional goods, p. 696
— profile, p. 787

Tonga continued:
— rubber products, p. 405
— textiles, pp. 85, 87, 90, 96
— tobacco, p. 78
— transportation equipment, pp. 643, 645, 648, 653
— wearing apparel, pp. 123, 125, 128, 133
— wood products, pp. 181, 183, 186, 192
Tony Lama Co, p. 173
Topps Company Inc, p. 264
Toro Co, p. 574
Torrebiarte Sucs Miguel, p. 172
Towelling, pp. 83, 106, 111
Tower Automotive Inc, p. 553
Town and Country Corp, p. 719
Town and Country Shoes Inc, p. 173
Tractor trailers, p. 675
Tractors, p. 580
Tractors Malaysia Holdings BHD, p. 657
Trailers, pp. 641, 675
Trailers and semi-trailers, p. 663
Transelco Div, pp. 278, 479, 616
Transformers
— 5 kva and over, p. 625
— less than 5 kva, p. 624
Transistors, p. 627
Transwerk, p. 655
Treasure Craft Co, p. 443
Tredegar Aluminum, p. 527
Tredegar Industries Inc, p. 427
Trek Plastics CC, p. 444
Tremont Corp, p. 527
Trencor Services Pty. Ltd, p. 553
Triangle Pacific Corp, pp. 193, 213
Triangle Wire and Cable Inc, p. 527
Triarc Companies Inc, p. 59
Trichloroethylene, p. 285
Trigem Computer Inc, p. 596
TriMas Corp, pp. 406, 554, 574, 657
Trina Inc, p. 174
Trinidad and Tobago
— basic chemicals, excl fertilizers, pp. 313-314, 316-317
— beverages, pp. 45, 47, 50, 52, 54, 56
— chemical products nec, pp. 329, 331, 334, 336-337, 340
— drugs and medicines, pp. 351-352, 354-357
— electrical machinery, pp. 603, 605, 608, 610-611, 613
— food products, pp. 5, 8, 11, 13-14, 17
— footwear, pp. 161, 163, 166-168, 171
— furniture, fixtures, pp. 201, 203, 206, 208-209, 212
— glass and products, pp. 449, 451, 453, 455-456, 458
— industrial chemicals, pp. 267, 269, 272-273, 275, 277
— industries nec, pp. 707, 709, 712, 714, 716, 718
— iron and steel, pp. 489-490, 493-494, 496, 498
— leather and products, pp. 145, 147, 149, 151-152, 155
— machinery nec, pp. 561, 563, 565, 567-568, 570
— metal products, pp. 541, 543, 546, 548-549, 552
— motor vehicles, pp. 677-678, 680-681, 683

Trinidad and Tobago continued:
— nonferrous metals, p. 523
— nonmetal products nec, pp. 467, 469, 472, 474-475, 477
— office, computing machinery, pp. 590-594
— paper and products, pp. 219, 221, 224, 226-227, 230
— petroleum refineries, p. 369
— petroleum, coal products, p. 388
— plastic products nec, pp. 415, 417, 420, 422-423, 425
— pottery, china, etc., pp. 433-434, 437-439, 442
— printing, publishing, pp. 249, 252, 255, 257-258, 261
— profile, p. 788
— radio, television, etc., pp. 628, 630-634
— rubber products, pp. 395, 397, 400-401, 403, 405
— shipbuilding, repair, pp. 668, 670-674
— textiles, pp. 85, 87, 90, 92-93, 96
— tobacco, pp. 67, 69, 72, 74-75, 78
— transportation equipment, pp. 643, 645, 648, 650-651, 653
— wearing apparel, pp. 123, 125, 128, 130-131, 133
— wood products, pp. 181, 183, 186, 188-189, 192
Tropicana Dole Beverages North, pp. 19, 459
Trousers, p. 121
— men's and boys', p. 139
Trucks, pp. 641, 675
— including articulated vehicles, assembled from
    imported parts, p. 662
— including articulated vehicles, produced, p. 663
Truval Manufacturers CC, p. 80
Tubes
— fluorescent, p. 640
— glass, p. 447
— seamless, p. 511
— steel, p. 487
— welded, p. 511
Tubos e Conexoes Tigre Ltda, p. 406
Tuboscope Inc, p. 573
Tudor Corp. Ltd, p. 157
Tudor PLC, p. 461
Tulip International AS, p. 19
Tultex Corp, pp. 97, 135
Tunisia
— basic chemicals, excl fertilizers, pp. 313-314, 316-317
— beverages, pp. 45, 47, 50, 52, 54, 56
— chemical products nec, pp. 329, 331, 334, 336-337, 340
— drugs and medicines, pp. 351-352, 354-356
— electrical machinery, pp. 603, 605, 608, 610-611, 613
— food products, pp. 5, 8, 11, 13-14, 17
— footwear, pp. 161, 163, 171
— furniture, fixtures, pp. 201, 203, 212
— glass and products, pp. 449, 451, 453, 455-456, 458
— industrial chemicals, pp. 267, 269, 272-273, 275, 277
— industries nec, pp. 707, 709, 713-714, 716, 718
— iron and steel, pp. 489-490, 493-494, 496, 498
— leather and products, pp. 145, 147, 155
— machinery nec, pp. 561, 563, 565, 567-568, 570
— metal products, pp. 541, 543, 552
— motor vehicles, pp. 677-678

Tunisia continued:
— nonferrous metals, pp. 515-516, 523
— nonmetal products nec, pp. 467, 469, 472, 474-475, 477
— paper and products, pp. 219, 221, 230
— petroleum refineries, pp. 361-362, 369
— petroleum, coal products, pp. 381-382, 388
— plastic products nec, pp. 415, 417, 420, 422-423, 425
— pottery, china, etc., pp. 433-434, 437-439, 442
— printing, publishing, pp. 249, 252, 261
— professional goods, pp. 687-688, 696
— profile, p. 788
— pulp, paper, etc., pp. 240, 242
— radio, television, etc., pp. 628, 630
— rubber products, pp. 395, 397, 400-401, 403, 405
— shipbuilding, repair, pp. 668, 670
— spinning, weaving, etc., pp. 113-114
— synthetic resins, etc., pp. 320-321
— textiles, pp. 85, 87, 90, 92-93, 96
— tobacco, pp. 67, 69, 72, 74-75, 78
— transportation equipment, pp. 643, 645, 648, 650-651, 654
— wearing apparel, pp. 123, 125, 128, 130-131, 133
— wood products, pp. 181, 183, 192
Tunzl PLC, pp. 231, 427, 719
Tupperware Corp, p. 427
Turkey
— basic chemicals, excl fertilizers, pp. 313-314, 316-318
— beverages, pp. 45, 47, 49, 51-52, 54, 56
— chemical products nec, pp. 329, 331-332, 334, 336-337,
    340
— drugs and medicines, pp. 351-357
— electrical machinery, pp. 603, 605-606, 608, 610-611, 613
— food products, pp. 5, 8-9, 11, 13-14, 17
— footwear, pp. 161, 163-164, 166-168, 171
— furniture, fixtures, pp. 201, 203-204, 206, 208-209, 212
— glass and products, pp. 449, 451-453, 455-456, 458
— industrial chemicals, pp. 267, 269-270, 272-273, 275, 277
— industries nec, pp. 707, 709, 711, 713-714, 716, 718
— iron and steel, pp. 489-491, 493-494, 496, 498
— leather and products, pp. 145, 147-148, 150-152, 155
— machinery nec, pp. 561, 563-565, 567-568, 571
— metal products, pp. 541, 543-544, 546, 548-549, 552
— motor vehicles, pp. 677-683
— nonferrous metals, pp. 515-517, 519-521, 523
— nonmetal products nec, pp. 467, 469-470, 472, 474-475,
    477
— office, computing machinery, pp. 590-594
— paper and products, pp. 219, 221-222, 224, 226-227, 230
— petroleum refineries, pp. 361-366, 369
— petroleum, coal products, pp. 381-382, 384-386, 388
— plastic products nec, pp. 415, 417-418, 420, 422-423, 425
— pottery, china, etc., pp. 433-435, 437-439, 442
— printing, publishing, pp. 249, 252, 255, 257-258, 261
— professional goods, pp. 687-689, 691-693, 696
— profile, p. 789
— pulp, paper, etc., pp. 240, 242-246
— radio, television, etc., pp. 628, 630-634

Turkey continued:
— rubber products, pp. 395, 397-398, 400-401, 403, 405
— shipbuilding, repair, pp. 668, 670-674
— spinning, weaving, etc., pp. 113-118
— synthetic resins, etc., pp. 320-325
— textiles, pp. 85, 87-88, 90, 92-93, 96
— tobacco, pp. 67, 69-70, 72, 74-75, 78
— transportation equipment, pp. 643, 645-646, 648, 650-651, 654
— wearing apparel, pp. 123, 125, 128, 130-131, 133
— wood products, pp. 181, 183-184, 186, 188-189, 192
Turkeys, p. 3
Turtle Wax Inc, p. 342
Tuscarora Inc, p. 428
TVK Rt, pp. 18, 278, 427
Twine, pp. 83, 111
Twinhead International Corp, p. 597
Twinlock PLC, p. 596
Tye-Sil Corp. Ltd, p. 215
Tyler Corp, p. 501
Typewriters, pp. 559, 589, 598
UCAR International Inc, p. 616
Uganda
— beverages, p. 56
— chemical products nec, p. 340
— electrical machinery, p. 613
— food products, p. 17
— footwear, p. 171
— furniture, fixtures, p. 212
— glass and products, p. 458
— industrial chemicals, p. 277
— industries nec, p. 718
— iron and steel, p. 498
— leather and products, p. 155
— machinery nec, p. 571
— metal products, p. 552
— nonferrous metals, p. 523
— nonmetal products nec, p. 478
— paper and products, p. 230
— petroleum refineries, p. 369
— petroleum, coal products, p. 388
— plastic products nec, p. 425
— pottery, china, etc., p. 442
— printing, publishing, p. 261
— professional goods, p. 696
— rubber products, p. 405
— textiles, p. 96
— tobacco, p. 78
— transportation equipment, p. 654
— wearing apparel, p. 133
— wood products, p. 192
Ukraine
— basic chemicals, excl fertilizers, p. 314
— beverages, pp. 45, 47, 51, 54, 56
— chemical products nec, pp. 329, 331, 334, 337, 340
— drugs and medicines, pp. 351-352, 354, 356

Ukraine continued:
— electrical machinery, pp. 603, 605, 608, 611, 613
— food products, pp. 5, 8, 11, 14, 17
— footwear, pp. 161, 163, 166, 168, 171
— furniture, fixtures, pp. 201, 203, 206, 209, 212
— glass and products, pp. 449, 451, 453, 456, 458
— industrial chemicals, pp. 267, 269, 272, 275, 277
— industries nec, p. 710
— iron and steel, pp. 489-490, 493, 496, 498
— leather and products, pp. 145, 147, 150, 152, 155
— machinery nec, pp. 561, 563, 565, 568, 571
— metal products, pp. 541, 543, 546, 549, 552
— motor vehicles, pp. 677-678, 680, 682
— nonferrous metals, pp. 515-516, 519, 521, 523
— nonmetal products nec, pp. 467, 469, 472, 475, 478
— office, computing machinery, pp. 590-592, 594
— paper and products, pp. 219, 221, 224, 227, 230
— petroleum refineries, pp. 361-362, 364, 366, 369
— petroleum, coal products, pp. 381-382, 384, 386, 388
— plastic products nec, pp. 415, 417, 420, 423, 425
— pottery, china, etc., pp. 433, 435, 437, 439, 442
— printing, publishing, pp. 249, 252, 255, 258, 261
— professional goods, pp. 687-688, 691, 693, 696
— profile, p. 789
— pulp, paper, etc., pp. 240, 242-243, 245
— rubber products, pp. 395, 397, 400, 403, 405
— shipbuilding, repair, pp. 669-671, 673
— spinning, weaving, etc., pp. 113-114, 116, 118
— synthetic resins, etc., pp. 320-321, 323-324
— textiles, pp. 85, 87, 90, 93, 96
— tobacco, pp. 67, 69, 72, 75, 78
— transportation equipment, pp. 643, 645, 648, 651, 654
— wearing apparel, pp. 123, 125, 128, 131, 134
— wood products, pp. 181, 183, 186, 189, 192
Ultramar PLC, p. 370
Unaka Company Inc, p. 214
UNC Inc, p. 655
Undergarments, pp. 83, 111
Underwear, p. 121
— men's and boys', p. 142
— women's and girls', p. 142
Unicord Public Co. Ltd, p. 18
Unidata, p. 597
Unifi Inc, p. 97
Unifi Inc. Yadkinville Div, p. 99
Unifirst Corp, p. 135
Uniflex, Inc, p. 174
Unihold Group Ltd, p. 527
Unijax Div, p. 343
Unilever Canada Ltd, pp. 19, 58, 279, 341
Unilever De Argentina SA, p. 341
UniMark Group Inc, p. 19
Union Camp Corp, pp. 59, 233, 280
Union Industries Corp, Ltd, p. 173
Union Spinning Mills Pty. Ltd, p. 98
Union Steel Manufacturing Co., Ltd, p. 501

Alphabetical Index

Union Textiles Industries Corp. Ltd, p. 97
Uniroyal Chemical Company Inc, p. 279
Uniroyal Chemical Corp, p. 280
Uniroyal Engelbert Tyres Ltd, p. 407
Uniroyal Technology Corp, p. 429
Unisys Brasil Ltda, p. 596
United Arab Emirates
— beverages, p. 56
— chemical products nec, p. 340
— electrical machinery, p. 613
— food products, p. 17
— footwear, p. 171
— furniture, fixtures, p. 212
— glass and products, p. 458
— industrial chemicals, p. 277
— industries nec, p. 718
— iron and steel, p. 498
— leather and products, p. 155
— machinery nec, p. 571
— metal products, p. 552
— nonferrous metals, p. 523
— nonmetal products nec, p. 478
— paper and products, p. 230
— petroleum refineries, p. 369
— petroleum, coal products, p. 388
— plastic products nec, p. 425
— pottery, china, etc., p. 442
— printing, publishing, p. 261
— professional goods, p. 696
— rubber products, p. 405
— textiles, p. 96
— tobacco, p. 78
— transportation equipment, p. 654
— wearing apparel, p. 134
— wood products, p. 192
United Distillers Blending and, p. 59
United Distillers PLC, pp. 19, 58
United Glass Co. WLL, p. 461
United Glass Ltd, p. 459
United Industrial Corp, p. 699
United Kingdom
— basic chemicals, excl fertilizers, pp. 313-314, 316-318
— beverages, pp. 45, 47, 49, 51-52, 54, 56
— chemical products nec, pp. 329, 331-332, 334, 336-337, 340
— drugs and medicines, pp. 351-357
— electrical machinery, pp. 603, 605-606, 608, 610-611, 613
— food products, pp. 5, 8-9, 11, 13-14, 17
— footwear, pp. 161, 163-164, 166-168, 171
— furniture, fixtures, pp. 201, 203-204, 206, 208-209, 212
— glass and products, pp. 449, 451-453, 455-456, 458
— industrial chemicals, pp. 267, 269-270, 272-273, 275, 277
— industries nec, pp. 707, 710-711, 713-714, 716, 718
— iron and steel, pp. 489, 491-494, 496, 498
— leather and products, pp. 145, 147-148, 150-152, 155
— machinery nec, pp. 561, 563-565, 567-568, 571

United Kingdom continued:
— metal products, pp. 541, 543-544, 546, 548-549, 552
— motor vehicles, pp. 677-683
— nonferrous metals, pp. 515-517, 519-521, 523
— nonmetal products nec, pp. 467, 469-470, 472, 474-475, 478
— office, computing machinery, pp. 590-594
— paper and products, pp. 219, 221-222, 224, 226-227, 230
— petroleum refineries, pp. 361-366, 369
— petroleum, coal products, pp. 381-386, 388
— plastic products nec, pp. 415, 417-418, 420, 422-423, 425
— pottery, china, etc., pp. 433, 435, 437-439, 442
— printing, publishing, pp. 249, 252-253, 255, 257-258, 261
— professional goods, pp. 687, 689, 691-693, 696
— profile, p. 790
— pulp, paper, etc., pp. 241-246
— radio, television, etc., pp. 628, 630-634
— rubber products, pp. 395, 397-398, 400-401, 403, 405
— shipbuilding, repair, pp. 669-674
— spinning, weaving, etc., pp. 113-118
— synthetic resins, etc., pp. 320-325
— textiles, pp. 85, 87-88, 90, 92-93, 96
— tobacco, pp. 67, 69-70, 72, 74-75, 78
— transportation equipment, pp. 643, 645-646, 648, 650-651, 654
— wearing apparel, pp. 123, 125-126, 128, 130-131, 134
— wood products, pp. 181, 183-184, 186, 188-189, 192
United Malt and Grain Distillers Ltd, p. 59
United Methodist Publishing House, p. 264
United Refining Co, p. 370
United Republic of Tanzania
— basic chemicals, excl fertilizers, pp. 313-318
— beverages, pp. 45, 47, 49, 51-52, 54, 56
— chemical products nec, pp. 329, 331-332, 334, 336-337, 340
— drugs and medicines, pp. 351-357
— electrical machinery, pp. 603, 605-606, 608, 610-611, 613
— food products, pp. 5, 8-9, 11, 13-14, 17
— footwear, pp. 161, 163-164, 166-168, 171
— furniture, fixtures, pp. 201, 203-204, 206, 208-209, 212
— glass and products, pp. 449, 451-453, 455-456, 458
— industrial chemicals, pp. 267, 269-270, 272-273, 275, 277
— industries nec, pp. 707, 710-711, 713-714, 716, 718
— iron and steel, pp. 489, 491-494, 496, 498
— leather and products, pp. 145, 147-148, 150-152, 155
— machinery nec, pp. 561, 563-565, 567-568, 571
— metal products, pp. 541, 543-544, 546, 548-549, 552
— motor vehicles, pp. 677-683
— nonferrous metals, pp. 515-517, 519-521, 523
— nonmetal products nec, pp. 467, 469-470, 472, 474-475, 478
— paper and products, pp. 219, 221-222, 224, 226-227, 230
— petroleum refineries, pp. 361-366, 369
— petroleum, coal products, p. 388
— plastic products nec, pp. 415, 417-418, 420, 422-423, 425
— pottery, china, etc., pp. 433, 435, 437-439, 442

United Republic of Tanzania continued:
— printing, publishing, pp. 249, 252-253, 255, 257-258, 261
— professional goods, pp. 687, 689, 691-693, 696
— profile, p. 790
— pulp, paper, etc., pp. 241-246
— rubber products, pp. 395, 397-398, 400-401, 403, 405
— shipbuilding, repair, pp. 669-674
— spinning, weaving, etc., pp. 113-118
— textiles, pp. 85, 87-88, 90, 92-93, 96
— tobacco, pp. 67, 69-70, 72, 74-75, 78
— transportation equipment, pp. 643, 645-646, 648, 650-651, 654
— wearing apparel, pp. 123, 125-126, 128, 130-131, 134
— wood products, pp. 181, 183-184, 186, 188-189, 192
U.S. Can Corp, p. 554
U.S. Furniture Industries Inc, p. 215
United States Gypsum Co, p. 479
U.S. Intec Inc, p. 389
United States of America
— basic chemicals, excl fertilizers, pp. 313-318
— beverages, pp. 45, 47, 49, 51-52, 54, 57
— chemical products nec, pp. 329, 331-332, 334, 336-337, 340
— drugs and medicines, pp. 351-357
— electrical machinery, pp. 603, 605-606, 608, 610-611, 614
— food products, pp. 6, 8-9, 11, 13-14, 17
— footwear, pp. 161, 163-164, 166-168, 171
— furniture, fixtures, pp. 201, 203-204, 206, 208-209, 212
— glass and products, pp. 449, 451-453, 455-456, 458
— industrial chemicals, pp. 267, 269-270, 272-273, 275, 277
— industries nec, pp. 707, 710-711, 713-714, 716, 718
— iron and steel, pp. 489, 491-494, 496, 498
— leather and products, pp. 145, 147-148, 150-152, 155
— machinery nec, pp. 561, 563-565, 567-568, 571
— metal products, pp. 541, 543-544, 546, 548-549, 552
— nonferrous metals, pp. 515-517, 519-521, 523
— nonmetal products nec, pp. 467, 469-470, 472, 474-475, 478
— office, computing machinery, pp. 590-594
— paper and products, pp. 219, 221-222, 224, 226-227, 230
— petroleum refineries, pp. 361-366, 369
— petroleum, coal products, pp. 381-386, 388
— plastic products nec, pp. 415, 417-418, 420, 422-423, 425
— pottery, china, etc., pp. 433, 435, 437-439, 442
— printing, publishing, pp. 249, 252-253, 255, 257-258, 261
— professional goods, pp. 687, 689-693, 696
— profile, p. 791
— pulp, paper, etc., pp. 241-246
— radio, television, etc., pp. 628, 630-634
— rubber products, pp. 395, 397-398, 400-401, 403, 405
— shipbuilding, repair, pp. 669-674
— spinning, weaving, etc., pp. 113-119
— synthetic resins, etc., pp. 320-325
— textiles, pp. 85, 87-88, 90, 92-93, 96
— tobacco, pp. 67, 69-70, 72, 74-75, 78
— transportation equipment, pp. 643, 645-646, 648, 650-651,

United States of America continued:
654
— wearing apparel, pp. 123, 125-126, 128, 130-131, 134
— wood products, pp. 181, 183-184, 186, 188-189, 192
United States Surgical Corp, p. 697
United Uniform Services PLC, p. 137
Unitog Co, p. 135
Universal Foods Corp, pp. 19, 341
Universal Forest Products, Inc, p. 194
Universal Furniture Industries Inc, p. 213
Uniwax SA, p. 371
Unocal Corp, p. 370
Unova Inc, p. 595
UNR Industries Inc, pp. 500, 555
UOP, pp. 279, 554
Uruguay
— basic chemicals, excl fertilizers, pp. 314, 318
— beverages, pp. 47, 51-52, 54, 57
— chemical products nec, pp. 331, 334, 336-337, 340
— drugs and medicines, pp. 352, 357
— electrical machinery, pp. 605, 608, 610-611, 614
— food products, pp. 8, 11, 13-14, 17
— footwear, pp. 163, 166-167, 169, 171
— furniture, fixtures, pp. 203, 206, 208-209, 212
— glass and products, pp. 451, 453, 455-456, 458
— industrial chemicals, pp. 269, 272-273, 275, 277
— industries nec, pp. 710, 713-714, 716, 718
— iron and steel, pp. 491, 493, 495-496, 498
— leather and products, pp. 147, 150-152, 155
— machinery nec, pp. 563, 565, 567-568, 571
— metal products, pp. 543, 546, 548-549, 552
— motor vehicles, pp. 678, 683
— nonferrous metals, pp. 516, 519-521, 523
— nonmetal products nec, pp. 469, 472, 474-475, 478
— office, computing machinery, pp. 591, 594
— paper and products, pp. 221, 224, 226-227, 230
— petroleum refineries, pp. 362, 364, 366-367, 369
— petroleum, coal products, pp. 382, 384-386, 388
— plastic products nec, pp. 417, 420, 422-423, 425
— pottery, china, etc., pp. 435, 437-439, 442
— printing, publishing, pp. 252, 255, 257-258, 261
— professional goods, pp. 689, 691-692, 694, 696
— profile, p. 791
— pulp, paper, etc., pp. 242, 246
— radio, television, etc., pp. 630, 634
— rubber products, pp. 397, 400-401, 403, 405
— shipbuilding, repair, pp. 670, 674
— spinning, weaving, etc., pp. 114, 119
— synthetic resins, etc., pp. 321, 325
— textiles, pp. 87, 90, 92-93, 96
— tobacco, pp. 69, 72, 74-75, 78
— transportation equipment, pp. 645, 648, 650-651, 654
— wearing apparel, pp. 125, 128, 130-131, 134
— wood products, pp. 183, 186, 188-189, 192
USKO Ltd, p. 527

USSR (Former)
—beverages, pp. 45, 47, 51, 57
—chemical products nec, pp. 329, 331, 334, 340
—electrical machinery, pp. 603, 605, 608, 614
—food products, pp. 5, 8, 11, 17
—footwear, pp. 161, 163, 166, 171
—furniture, fixtures, pp. 201, 203, 206, 212
—glass and products, pp. 449, 451, 453, 458
—industrial chemicals, pp. 267, 269, 272, 277
—industries nec, pp. 707, 710, 713, 718
—iron and steel, pp. 489, 491, 493, 498
—leather and products, pp. 145, 147, 150, 155
—machinery nec, pp. 563, 565, 571
—metal products, pp. 543, 546, 552
—nonferrous metals, pp. 515-516, 519, 523
—nonmetal products nec, pp. 467, 469, 472, 478
—paper and products, pp. 219, 221, 224, 230
—petroleum refineries, pp. 361-362, 364, 369
—petroleum, coal products, pp. 381-382, 384, 388
—plastic products nec, pp. 415, 417, 420, 425
—pottery, china, etc., pp. 433, 435, 437, 442
—printing, publishing, pp. 249, 252, 255, 261
—professional goods, pp. 687, 689, 691, 696
—rubber products, pp. 395, 397, 400, 405
—textiles, pp. 85, 87, 90, 96
—tobacco, pp. 67, 69, 72, 78
—transportation equipment, pp. 643, 645, 648, 654
—wearing apparel, pp. 123, 125, 128, 134
—wood products, pp. 181, 183, 186, 192
UST Inc, pp. 58, 79
V. Schierholzesche, p. 444
Vacuum cleaners, p. 637
Valassis Communications Co, p. 264
Valero Energy Corp, p. 370
Valhi Inc, pp. 20, 194, 555
Valmont Industries Inc, pp. 554, 574, 617
Valor Electronics Inc, p. 615
Valspar Corp, p. 341
Valvoline Co, p. 389
Vanguard Travellers Bag Inc, p. 157
Vanities Unlimited, p. 135
Vanity Fair Mills Inc, p. 135
Vans, p. 641
Vans Inc, p. 407
Varian Associates Inc, pp. 615, 697
Varity Kelsey-Hayes, p. 655
Varlen Corp, pp. 427, 525, 555, 657
Varsity Spirit Corp, p. 136
Vats, p. 539
Vaughan-Bassett Furniture Co, p. 215
V.A.W. of America Inc, p. 526
Vectura Group Inc, p. 553
Vegetable oils, p. 3
Vegetables, p. 3
—frozen, p. 29
—tinned or bottled, p. 30

Veitscher Magnesitwerke AG, pp. 500, 555
Vencemos Mara C.A, p. 479
Veneer sheets, pp. 179, 196
Venezuela
—basic chemicals, excl fertilizers, pp. 313-314, 316-318
—beverages, pp. 45, 47, 51-52, 54, 57
—chemical products nec, pp. 329, 331, 334, 336-337, 340
—drugs and medicines, pp. 351-352, 354-355, 357
—electrical machinery, pp. 603, 605, 608, 610-611, 614
—food products, pp. 6, 8, 11, 13-14, 17
—footwear, pp. 161, 163, 166-167, 169, 171
—furniture, fixtures, pp. 201, 203, 206, 208-209, 212
—glass and products, pp. 449, 451, 453, 455-456, 458
—industrial chemicals, pp. 267, 269, 272-273, 275, 277
—industries nec, pp. 707, 710, 713-714, 716, 718
—iron and steel, pp. 489, 491, 493, 495-496, 498
—leather and products, pp. 145, 147, 150-152, 155
—machinery nec, pp. 561, 563, 565, 567-568, 571
—metal products, pp. 541, 543, 546, 548-549, 552
—motor vehicles, pp. 677-678, 680-681, 683
—nonferrous metals, pp. 515-516, 519-521, 523
—nonmetal products nec, pp. 467, 469, 472, 474-475, 478
—office, computing machinery, pp. 590-594
—paper and products, pp. 219, 221, 224, 226-227, 230
—petroleum refineries, pp. 361-362, 364, 366-367, 369
—petroleum, coal products, pp. 381-382, 384-386, 388
—plastic products nec, pp. 415, 417, 420, 422-423, 425
—pottery, china, etc., pp. 433, 435, 437-439, 442
—printing, publishing, pp. 250, 252, 255, 257-258, 261
—professional goods, pp. 687, 689, 691-692, 694, 696
—profile, p. 792
—pulp, paper, etc., pp. 241-244, 246
—radio, television, etc., pp. 629-632, 634
—rubber products, pp. 395, 397, 400-401, 403, 405
—shipbuilding, repair, pp. 669-672, 674
—spinning, weaving, etc., pp. 113-114, 116-117, 119
—synthetic resins, etc., pp. 320-321, 323-325
—textiles, pp. 85, 87, 90, 92-93, 96
—tobacco, pp. 67, 69, 72, 74-75, 78
—transportation equipment, pp. 643, 645, 648, 650-651, 654
—wearing apparel, pp. 123, 125, 128, 130-131, 134
—wood products, pp. 181, 183, 186, 188-189, 192
Ventra Group Inc, p. 429
VeriFone Inc, p. 595
Vermont American Corp, pp. 554, 574
Vernay Laboratories Inc, p. 408
Vesper Corp, p. 407
Vesuvius Crucible Company Inc, p. 481
Vesuvius U.S.A. Corp, pp. 480, 501
Vetements Peerless Inc, p. 136
VF Corp. United Kingdom Ltd, p. 137
Vicasa, SA, p. 459
Victaulic PLC, p. 429
Video Lottery Technologies Inc, p. 721
Videoton Ipari Rt, p. 616
Vidrala SA, p. 460

Vidriera Centroamericana, SA, p. 460
Vidrieria Leonesa, SA, p. 460
Vidrieria Rovira, SA, p. 460
Vidrieria y Cristaleria de Lamiaco SA, p. 460
Villeroy & Boch SARL, p. 443
Vincke, SA, p. 408
Vincor International Inc, p. 59
Vinegar, pp. 3, 41
Vinyl chloride, p. 286
Virco Manufacturing Corp, p. 213
Viscosity Oil Co, p. 390
Vitreous plumbing fixtures, p. 431
Viva Corp, p. 157
Vivian Regina Sales Ltd, p. 461
VLSI Technology Inc, p. 617
Voksel Electric PT, pp. 501, 526
VTEL Corp, p. 597
Vulcan Materials Co, pp. 278, 389
Vulkan d.d, p. 407
W. Haking Industries Thailand Ltd, p. 699
W.A. Adams Company Inc, p. 80
Wabash Alloys, p. 527
Wabash National Corp, p. 656
Wagon Storage Products Ltd, p. 214
Wagons, p. 641
Wah Chang, p. 526
Wainoco Oil Corp, p. 371
Walbro Corp, pp. 573, 616
Walker & Homer Group PLC, p. 214
Wallace Computer Services Inc, pp. 262, 719
Walsin Lihwa Electric Corp, p. 526
Walsworth Publishing Company Inc, p. 264
Walter Industries Inc, p. 499
Waltons Stationery Co.--Export Div, p. 263
Wampler Foods Inc, p. 20
Wampler-Longacre Chicken Inc, p. 19
Wardle Storeys PLC, pp. 137, 428
Warner-Lambert Canada Inc, p. 343
Warner-Lambert UK Ltd, p. 342
Warner's, p. 137
Washing machines, p. 559
Washing machines for household use, p. 622
Washing powder and detergents, p. 345
Washington Post Co, p. 262
Watches, pp. 685, 702
Waterford Wedgwood Holdings PLC, p. 479
Waters Corp, p. 699
Watts Industries Inc, pp. 554, 573
Watts Regulator Co, p. 554
Watts, Blake, Bearne & Co. PLC, p. 480
Waupaca Foundry Inc, p. 500
Wausau Paper Mills Co, p. 233
Wave-Cover International Co., Ltd, p. 174
Wavin BV, p. 427
Wayne-Dalton Corp, p. 195
WCI Steel Inc, p. 500

WD-40 Company, p. 390
Wearing apparel, p. 121
Wearwel International, PT, p. 137
Weather Shield Manufacturing Inc, p. 459
Weatherford Enterra Inc, p. 572
Weaving, p. 111
Webb Furniture Enterprises Inc, p. 215
Weider Nutrition International Inc, p. 60
Weidmuller Ltd, p. 597
Weinbrenner Shoe Company Inc, p. 174
Weirton Steel Corp, pp. 499, 553
Welch Allyn Inc, pp. 596, 699
Weldwood of Canada Ltd, pp. 193, 232
Wellborn Cabinet Inc, p. 195
Wellington Leisure Products Inc, p. 720
Wellman Inc, pp. 98, 279
Werner Ladder Co, pp. 194, 428
Wesley-Jessen Co, pp. 343, 699
West Company Inc, pp. 427, 525
West Fraser Timber Co. Ltd, pp. 193, 231
West Lumber Company Inc, p. 194
Western Industries Inc, p. 429
Western Star Trucks Inc, p. 657
Western Veneer & Lumber Co. Ltd, p. 194
Westland Group PLC, p. 655
Westland Helicopters Ltd, p. 656
Westminster Press Ltd, p. 263
Westroc Inc, p. 481
Westvaco Corp. Kraft Div, p. 233
WEYCO Group Inc, p. 173
Weyerhaeuser Canada Ltd, pp. 193, 231
Weyerhaeuser Paper Co, p. 231
Weyerhaeuser Saskatchewan Ltd., p. 195
W.H. Brady Co, pp. 232, 719
Wheaton Plastic Products Div, p. 428
Wheels, tires and axles, p. 512
White spirit, pp. 359, 374
Whitman Corp, pp. 58, 572
Whole Space Plastic Manufacturing, p. 215
WHX Corp, p. 499
Whyte & MacKay Distillers Ltd, p. 60
Wihuri Oy, p. 428
William Collins Sons & Co., Ltd, p. 263
William Cook PLC, p. 500
William E. Coutts Co. Ltd, pp. 233, 264
William Grant & Sons Ltd, p. 60
William H. Kaufman Inc, pp. 172, 214, 407
William Muir Bond 9 Ltd, p. 60
Williamson-Dickie Manufacturing Co, p. 135
Wilson Sporting Goods Co, p. 719
Wimpey Hobbs Ltd, p. 480
Wind instruments, p. 705
Wine, pp. 43, 61
Winnebago Industries Inc, p. 657
Winner Co. Ltd, p. 98
Winner Garment Manufacturing, p. 136

WinsLoew Furniture Inc, p. 215
Winstone Ltd, p. 573
Wire, pp. 487, 510
Wire and cable, p. 639
Wire rods, p. 506
Witco Corp, pp. 278, 389, 406, 427
Wix Filtration Products Div, p. 657
W.J. & W. Lang Ltd, p. 157
W.L. Gore and Associates Inc, pp. 615, 697
WLR Foods Inc, p. 18
Wm. Wrigley Jr. Co, p. 18
WMS Industries Inc, p. 721
Wolverine Tube Inc, p. 525
Wolverine World Wide Inc, p. 172
Women's and girls' blouses, p. 121
Women's stockings, p. 107
Woo Jung Corp, p. 99
Wood products, p. 179
Wood pulp, pp. 217, 239
— dissolving grades, p. 234
— mechanical, p. 234
— semi-chemical, p. 235
— soda and sulfate, p. 235
— sulfite, p. 235
Woodbridge Foam Corp, p. 215
Woodhead Industries Inc, pp. 407, 526
Woodward Governor Co, p. 657
Wool, p. 83
Wool yarn, pp. 100-101
Woollen woven fabrics, p. 104
World Carpets Inc, p. 98
WorldCrisa Corp, pp. 443, 721
Worthington Steel Co, p. 501
Woven fabrics of cellulosic fibers, p. 105
Woven fabrics of non-cellulosic fibers, p. 105
W.R. Meadows Inc, p. 389
Wrapping and packaging paper and paperboard, p. 237
Writing devices, p. 705
WTD Industries Inc, p. 195
Wyman-Gordon Co, pp. 525, 554
Wynn's International Inc, p. 389
Wyse Technology Inc, p. 596
Xerox do Brasil Ltda, p. 597
Xircom Inc, p. 597
Xylenes, p. 284
Yachts, p. 667
Yale Security Inc, p. 555
Yale-South Haven Inc, p. 407
Yarn and sewing thread of man-made fibers, p. 102
Yarn mills, p. 83
Yarn of other vegetable textile fibers, p. 103
Yarns, p. 83
Yattendon Investment Trust PLC, p. 263
Yemen
— beverages, p. 57
— food products, p. 17

Yemen continued:
— footwear, p. 171
— glass and products, p. 458
— leather and products, p. 155
— nonmetal products nec, p. 478
— paper and products, p. 230
— petroleum refineries, p. 369
— plastic products nec, p. 425
— printing, publishing, p. 261
— rubber products, p. 405
— textiles, p. 96
— tobacco, p. 78
— wearing apparel, p. 134
Yeo Hiap Seng Ltd, p. 60
Yi Wai Industrial Co., Ltd, p. 444
YIT-Industry, p. 501
YKK Indonesia Zipper Co. Ltd. PT, p. 720
YKK (U.S.A.) Inc, p. 720
Yoo Yang Moolsan Co., Ltd, p. 157
York Group Inc, p. 720
York Timbers Ltd, pp. 195, 214
Young & Co.'s Brewery PLC, p. 59
Ytong Industries Ltd, p. 481
Yu Jin Ind. Co., Ltd, p. 156
Yugoslavia
— beverages, pp. 45, 48, 51, 53-54, 57
— chemical products nec, pp. 329, 331, 334, 336-337, 340
— electrical machinery, pp. 603, 605, 608, 610-611, 614
— food products, pp. 6, 8, 11, 13-14, 17
— footwear, pp. 161, 163, 166-167, 169, 171
— furniture, fixtures, pp. 201, 203, 206, 208-209, 212
— glass and products, pp. 449, 451, 453, 455-456, 458
— industrial chemicals, pp. 267, 269, 272, 274-275, 277
— industries nec, pp. 707, 710, 713, 715-716, 718
— iron and steel, pp. 489, 491, 493, 495-496, 498
— leather and products, pp. 145, 147, 150-152, 155
— machinery nec, pp. 561, 563, 565, 567-568, 571
— metal products, pp. 541, 543, 546, 548, 550, 552
— nonferrous metals, pp. 515-516, 519-521, 523
— nonmetal products nec, pp. 467, 469, 472, 474-475, 478
— paper and products, pp. 219, 221, 224, 226, 228, 230
— petroleum refineries, pp. 361-362, 364, 366-367, 369
— petroleum, coal products, pp. 381-382, 384-386, 388
— plastic products nec, pp. 415, 417, 420, 422-423, 425
— pottery, china, etc., pp. 433, 435, 437-439, 442
— printing, publishing, pp. 250, 252, 255, 257-258, 261
— professional goods, pp. 687, 689, 691-692, 694, 696
— profile, p. 792
— rubber products, pp. 395, 397, 400-401, 403, 405
— textiles, pp. 85, 87, 90, 92-93, 96
— tobacco, pp. 67, 69, 72, 74-75, 78
— transportation equipment, pp. 643, 645, 648, 650-651, 654
— wearing apparel, pp. 123, 125, 128, 130-131, 134
— wood products, pp. 181, 183, 186, 188-189, 192
Yugoslavia (Former)
— beverages, p. 57

Alphabetical Index

Yugoslavia (Former) continued:
— chemical products nec, p. 340
— electrical machinery, p. 614
— food products, p. 17
— footwear, p. 171
— furniture, fixtures, p. 212
— glass and products, p. 458
— industrial chemicals, p. 277
— industries nec, p. 718
— iron and steel, p. 498
— leather and products, p. 155
— machinery nec, p. 571
— metal products, p. 552
— nonferrous metals, p. 524
— nonmetal products nec, p. 478
— paper and products, p. 230
— petroleum refineries, p. 369
— petroleum, coal products, p. 388
— plastic products nec, p. 426
— pottery, china, etc., p. 442
— printing, publishing, p. 261
— professional goods, p. 696
— rubber products, p. 405
— textiles, p. 96
— tobacco, p. 78
— transportation equipment, p. 654
— wearing apparel, p. 134
— wood products, p. 192
Yuhan-Kimberly Ltd, p. 233
Yule Catto Group Ltd, p. 195
Yulinda Duta Fashion, PT, p. 137
Yulon Motor Co. Ltd, p. 657
Yung Ha Industrial Co., Ltd, p. 156
Za-Ko RT, p. 136
ZA Pulawy SA, p. 279
Zainco Traders, p. 719
Zakaria Shahid Industries, p. 156
Zaklady Azotowe W Tarnowie-, pp. 278, 341
Zaklady Azotowe, Wloclawek SA, p. 279
Zaklady Celulozowo-Papiernicze, p. 232
Zaklady Chemiczne, Organika-, p. 280
Zaklady Przemyslu Bawelnianego, pp. 98, 279
Zaklady Przemyslu Tytoniowego, p. 79
Zaklady Starachowickie, Star Spolka, pp. 499, 656
Zaklady Wlokien Chemicznych, p. 279
Zala Butorgyar RT, p. 214
Zambia
— basic chemicals, excl fertilizers, pp. 313-318
— beverages, pp. 45, 48-49, 51, 53, 57
— chemical products nec, pp. 329, 331-332, 334, 336-337, 340
— drugs and medicines, pp. 351-357
— electrical machinery, pp. 603, 605, 608, 610-611, 614
— food products, pp. 6, 8-9, 11, 13-14, 17
— footwear, pp. 161, 163-164, 166-167, 171
— furniture, fixtures, pp. 201, 203-204, 206, 208-209, 212

Zambia continued:
— glass and products, pp. 449, 458
— industrial chemicals, pp. 267, 269-270, 272, 274-275, 27
— industries nec, pp. 708, 710-711, 713, 715-716, 718
— iron and steel, pp. 489, 491-493, 495-496, 498
— leather and products, pp. 145, 147-148, 150-151, 155
— machinery nec, pp. 561, 563, 566-568, 571
— metal products, pp. 541, 543-544, 546, 548, 550, 552
— motor vehicles, pp. 677-678, 680-683
— nonferrous metals, pp. 515-517, 519-521, 524
— nonmetal products nec, pp. 467, 469-470, 472, 474-475, 478
— office, computing machinery, p. 590
— paper and products, pp. 219, 221-222, 224, 226, 228, 23
— petroleum refineries, pp. 361, 369
— petroleum, coal products, pp. 381, 388
— plastic products nec, pp. 415, 417-418, 420, 422-423, 42
— pottery, china, etc., pp. 433, 438, 442
— printing, publishing, pp. 250, 252-253, 255, 257-258, 261
— professional goods, pp. 687, 689-692, 696
— profile, p. 793
— pulp, paper, etc., p. 241
— radio, television, etc., pp. 629-634
— rubber products, pp. 395, 397-398, 400-401, 403, 405
— shipbuilding, repair, p. 669
— spinning, weaving, etc., pp. 113-119
— synthetic resins, etc., p. 320
— textiles, pp. 85, 87-88, 90, 92-93, 96
— tobacco, pp. 67, 69-70, 72, 74, 78
— transportation equipment, pp. 643, 645-646, 648, 650-65 654
— wearing apparel, pp. 123, 125-126, 128, 130-131, 134
— wood products, pp. 181, 183-184, 186, 188-189, 192
Zardoya Otis, SA, p. 574
Zebco-MotorGuide Div, p. 721
Zebra Technologies Corp, p. 597
Zelezorudne Bane SP, pp. 499, 525, 573
Zelmer, p. 616
Zenco Sales Inc, p. 172
Zenith Electronics Corp, p. 615
Zero Enclosures Div, p. 174
Ziegler Chemical and Mineral Corp, p. 371
Zimbabwe
— beverages, pp. 45, 48, 51, 53-54, 57
— chemical products nec, pp. 329, 331, 334, 336-337, 340
— electrical machinery, pp. 603, 605, 608, 610-611, 614
— food products, pp. 6, 8, 11, 13-14, 17
— footwear, pp. 161, 163, 166-167, 169, 171
— furniture, fixtures, pp. 201, 203, 206, 208-209, 212
— glass and products, pp. 449, 451, 453, 455, 458
— industrial chemicals, p. 277
— industries nec, pp. 708, 710, 713, 715, 718
— iron and steel, pp. 489, 491, 493, 495, 498
— leather and products, pp. 145, 147, 150-151, 155
— machinery nec, pp. 561, 563, 566-567, 571
— metal products, pp. 541, 543, 546, 548, 552

Zimbabwe continued:
— motor vehicles, pp. 677-678, 680-681, 683
— nonferrous metals, pp. 515-516, 519-520, 524
— nonmetal products nec, pp. 467, 469, 472, 474, 478
— paper and products, pp. 219, 221, 224, 226, 228, 230
— petroleum refineries, p. 369
— petroleum, coal products, p. 388
— plastic products nec, pp. 415, 417, 420, 422-423, 426
— pottery, china, etc., pp. 433, 435, 437-438, 442
— printing, publishing, pp. 250, 252, 255, 257-258, 261
— professional goods, pp. 687, 689, 691-692, 696
— profile, p. 793
— radio, television, etc., pp. 629-632, 634
— rubber products, pp. 395, 397, 400-401, 403, 405
— textiles, pp. 85, 87, 90, 92-93, 96
— tobacco, pp. 67, 69, 72, 74-75, 78
— transportation equipment, pp. 643, 645, 648, 650-651, 654
— wearing apparel, pp. 123, 125, 128, 130-131, 134
— wood products, pp. 181, 183, 186, 188-189, 192
Zimco Industries Pty. Ltd, pp. 280, 342, 406
Zimmer Inc, p. 697
Zinc, pp. 513, 535
— alloys, p. 536
— oxide, p. 294
— plates, sheets, strip, foil, p. 536
Zippers, p. 705
Zippo Manufacturing Co, p. 721
Zortech International Ltd, p. 443
ZPL Zyrardow, p. 98
ZSNP Aluminium Works AS, pp. 278, 525, 553
Zsolnay Porcelangyar, p. 443
Zurn Industries Inc, pp. 553, 573